2017 IEEE 44th Photovoltaic Specialist Conference (PVSC 2017)

Washington, DC, USA
25-30 June 2017

Pages 706-1420

IEEE Catalog Number: CFP17PSC-POD
ISBN: 978-1-5090-5606-4

**Copyright © 2017 by the Institute of Electrical and Electronics Engineers, Inc.
All Rights Reserved**

Copyright and Reprint Permissions: Abstracting is permitted with credit to the source. Libraries are permitted to photocopy beyond the limit of U.S. copyright law for private use of patrons those articles in this volume that carry a code at the bottom of the first page, provided the per-copy fee indicated in the code is paid through Copyright Clearance Center, 222 Rosewood Drive, Danvers, MA 01923.

For other copying, reprint or republication permission, write to IEEE Copyrights Manager, IEEE Service Center, 445 Hoes Lane, Piscataway, NJ 08854. All rights reserved.

*** *This is a print representation of what appears in the IEEE Digital Library. Some format issues inherent in the e-media version may also appear in this print version.*

IEEE Catalog Number: CFP17PSC-POD
ISBN (Print-On-Demand):   978-1-5090-5606-4
ISBN (Online):   978-1-5090-5605-7
ISSN: 0160-8371

Additional Copies of This Publication Are Available From:

Curran Associates, Inc
57 Morehouse Lane
Red Hook, NY 12571 USA
Phone: (845) 758-0400
Fax: (845) 758-2633
E-mail: curran@proceedings.com
Web: www.proceedings.com

2017 IEEE 44th Photovoltaic Specialist Conference (PVSC 2017)

Washington, DC, USA
25-30 June 2017

Pages 706-1420

IEEE Catalog Number: CFP17PSC-POD
ISBN: 978-1-5090-5606-4

TABLE OF CONTENTS

OPEN CIRCUIT VOLTAGE CALCULATION USING TEMPERATURE AND IRRADIANCE 1
Andrew Melvin

EFFECT OF CL-DOPING IN ZNTEO ON PHOTOLUMINESCENCE AND PHOTOVOLTAIC
PROPERTIES OF ZNTEO-BASED INTERMEDIATE BAND SOLAR CELLS 3
T. Tanaka ; S. Tsutsumi ; Y. Okano ; K. Matsuo ; K. Saito ; Q. Guo ; M. Nishio ; T. Tayagaki ; K. M. Yu ; W. Walukiewicz

TOWARD LEAD HALIDE PEROVSKITE-BASED INTERMEDIATE BAND ABSORBERS 6
Matthew D. Sampson ; Ji-Sang Park ; Richard D. Schaller ; Maria K. Y. Chan ; Alex B. F. Martinson

TYPE-II QUANTUM DOTS FOR APPLICATION TO PHOTON RATCHET INTERMEDIATE
BAND SOLAR CELLS 10
Ryo Tamaki ; Yasushi Shoji ; Yoshitaka Okada

AN INVESTIGATION OF THE ROLE OF RECOMBINATION PROCESSES IN THE
OPERATION OF INAS/GAASL-XSBX QUANTUM DOT SOLAR CELLS 14
Y. Cheng ; A. J. Meleco ; A. J. Roeth ; V. R. Whiteside ; M. C. Debnath ; M. B. Santos ; T. D. Mishima ; S. Hatch ; H.Y. Liu ; I. R. Sellers

TEMPERATURE AND VOLTAGE-BIAS DEPENDENT TWO-STEP PHOTON ABSORPTION IN
INAS/GAASL AL0.3GAAS QUANTUM DOT IN A WELL SOLAR CELLS 18
Yushuai Dai ; Brittany L. Smith ; Michael A. Slocum ; Zachary S. Bittner ; Hyun Kum ; Julia D'Rozario ; Seth M. Hubbard

INCREASING CURRENT GENERATION BY PHOTON UP-CONVERSION IN A SINGLE-
JUNCTION SOLAR CELL WITH A HETERO-INTERFACE 23
Shigeo Asahi ; Kazuki Kusaki ; Toshiyuki Kaizu ; Takashi Kita

RTP-ASSISTED EX-SITU ANALYSIS OF (AG,CU)(IN,GA)SE2 FORMATION USING
SELENIZATION 26
Sina Soltanmohammad ; William N. Shafarman

ROLE OF EV+0.98 EV TRAP IN LIGHT SOAKING-INDUCED SHORT CIRCUIT CURRENT
INSTABILITY IN CIGS SOLAR CELLS 30
P. K. Paul ; T. Jarmar ; L. Stolt ; A. Rockett ; A. R. Arehart

STUDY OF DEFECT PROPERTIES IN CUGASE2 THIN-FILM SOLAR-CELLS USING
ADMITTANCE SPECTROSCOPY 33
Muhammad Monirul Islam ; Shogo Ishizuka ; Hajime Shibata ; Shigeru Niki ; Katsuhiro Akimoto ; Takeaki Sakurai

TRANSMISSIVE SPECTRUM-SPLITTING CONCENTRATOR PHOTOVOLTAIC CELLS AND
MODULES 37
Yaping Ji ; Qi Xu ; Brian Riggs ; John Robertson ; Kazi Islam ; Vince Romanin ; Dimitri D. Krut ; Jim H. Ermer ; Matthew D. Escarra

ALGAINP/GAAS TANDEM SOLAR CELLS FOR POWER CONVERSION AT 400 C AND 1000X
CONCENTRATION 42
Myles A. Steiner ; Emmett E. Perl ; John Simon ; Daniel J. Friedman ; Nikhil Jain ; Paul Sharps ; Claiborne Mcpheeters ; Minjoo L. Lee

GALNASP SOLAR CELLS GROWN BY HYDRIDE VAPOR PHASE EPITAXY FOR ONE-SUN &
LOW-CONCENTRATION III-V/SI PHOTOVOLTAICS 46
Nikhil Jain ; John Simon ; Kevin L. Schulte ; David R. Diercks ; Corinne E. Packard ; David Young ; Aaron J. Ptak

PHOTO-ELECTROCHEMICAL HYDROGEN GENERATION FROM INVERTED
METAMORPHIC MULTIJUNCTION III-VS 47
Todd G. Deutsch ; James L. Young ; Myles A. Steiner ; Henning Döscher ; Ryan M. France ; John A. Turner

ADVANCED SILICON THIN FILMS FOR HIGH-EFFICIENCY SILICON HETEROJUNCTION-
BASED SOLAR CELLS 50
A. Descoeudres ; C. Allebe ; N. Badel ; L. Barraud ; J. Champliaud ; G. Christmann ; L. Curvat ; F. Debrot ; A. Faes ; J. Geissbiihler ; J. Horzel ; A. Lachowicz ; J. Levrat ; S. Martin De Nicolas ; S. Nicolay ; B. Paviet-Salomon ; L.-L. Senaud ; A. Tomasi ; C. Ballif ; M. Despeisse

MOOXAND WOXBASED HOLE-SELECTIVE CONTACTS FOR WAFER-BASED SI SOLAR
CELLS 55
Stephanie Essig ; Julie Dréon ; Jérémie Werner ; Philipp Löper ; Stefaan De Wolf ; Mathieu Boccard ; Christophe Ballif

METAL NANOPARTICLE HOLE CONTACTS FOR SILICON SOLAR CELLS 59
James Bullock ; Zhaoran Xu ; Mark Hettick ; Yimao Wan ; Ali Javey

NEAR-FIELD TRANSPORT IMAGING APPLICATION OF PHOTOVOLTAIC MATERIALS 62
Chuanxiao Xiao ; Chun-Sheng Jiang ; John Moseley ; John Simon ; Kevin Schulte ; Aaron J. Ptak ; Steve Johnston ; Brian Gorman ; Mowafak Al-Jassim ; Nancy M. Haegel ; Helio Moutinho

APPLICATIONS OF DMD-BASED INHOMOGENEOUS ILLUMINATION PHOTOLUMINESCENCE IMAGING FOR SILICON WAFERS AND SOLAR CELLS 66
Yan Zhu ; Mattias Klaus Juhl ; Ziv Hameiri ; Thorsten Trupke

NUMERICAL MODEL TO EXTRACT MATERIALS PROPERTIES MAP FROM SPECTRALLY RESOLVED LUMINESCENCE IMAGES 70
Nicolas Paul ; Vincent Le Guen ; Daniel Ory ; Laurent Lombez

NON-DESTRUCTIVE CONTACT RESISTIVITY MEASUREMENTS ON SOLAR CELLS USING THE CIRCULAR TRANSMISSION LINE METHOD 74
Geoffrey Gregory ; Andrew M. Gabor ; Andrew Anselmo ; Rob Janoch ; Zhihao Yang ; Kristopher O. Davis

RADIATION RESISTANCE OF LOW COST HIGH EFFICIENCY TRIPLE JUNCTION SOLAR CELLS 76
Roberta Campesato ; Erminio Greco ; Giuseppe Gabetta ; Mariacristina Casale ; Gabriele Gori ; M. Sankaran ; Suresh E. Puthanveettil ; B. R. Uma ; M. Ravindra ; Sheeja Krishnan

AMORPHOUS SILICON CARBIDE REAR-SIDE PASSIVATION AND REFLECTOR LAYER STACKS FOR MULTI-JUNCTION SPACE SOLAR CELLS BASED ON GERMANIUM SUBSTRATES 83
Stefan Janz ; Charlotte Weiss ; Christian Mohr ; Rufi Kurstjens ; Bruno Boizot ; Bianca Fuhrmann ; Victor Khorenko

HOT CARRIER TRANSPORTATION DYNAMICS IN INAS/GAAS QUANTUM DOT SOLAR CELL 85
Tomah Sogabe ; Kohdai Nii ; Katsuyoshi Sakamoto ; Koichi Yarnaquchi ; Yoshitaka Okada

INTEGRATION OF CRACK-TOLERANT COMPOSITE GRIDLINES ON TRIPLE JUNCTION PHOTOVOLTAIC CELLS 88
Omar K. Abudayyeh ; Geoffrey K. Bradshaw ; Steven Whipple ; David M. Wilt ; Sang M. Han

SUBCELL LIGHT CURRENT- VOLTAGE CHARACTERIZATION OF IRRADIATED MULTIJUNCTION SOLAR CELL 93
Don Walker ; John Nocerino ; Yao Yue ; Colin J. Mann ; Simon H. Liu

ANALYTICAL METHOD FOR PREDICTING SPACECRAFT POWER GENERATION ON PARTIALLY SHADED SOLAR PANELS 96
Gordon Wu ; Bao Hoang

EVALUATING THE EMISSIVITY OF PSEUDOMORPHIC GLASS (PMG) 102
Ryan D. Beauchemin ; David M. Wilt ; Paul E. Hausgen

CHARACTERIZING THE IMPACT OF SOLAR SPECTRAL IRRADIANCE ON PV MODULE OUTPUT 107
M. Schweiger ; W. Herrmann

USE OF MEASURED AEROSOL OPTICAL DEPTH AND PRECIPITABLE WATER TO MODEL CLEAR SKY IRRADIANCE 110
Mark M. Mikofski ; Clifford W. Hansen ; William F. Holmgren ; Gregory M. Kimbal

RECENT ADVANCEMENTS IN THE NUMERICAL SIMULATION OF SURFACE IRRADIANCE FOR SOLAR ENERGY APPLICATIONS 116
Yu Xie ; Manajit Sengupta ; Chris Deline

OPTIMAL IRRADIANCE SENSOR PLACEMENT FOR PHOTOVOLTAIC SYSTEMS USING MUTUAL INFORMATION BASED GREEDY ALGORITHM IN GAUSSIAN PROCESS 120
Lian Lian Jiang ; R. Srivatsan ; Douglas L. Maskell

EVALUATING DIFFERENT UPSCALING APPROACHES TO DERIVE THE ACTUAL POWER OF DISTRIBUTED PV SYSTEMS 126
Sven Killinger ; Björn Müller ; Bernhard Wille-Haussmann ; Russell Mckenna

ADVANCES IN LONG-TERM SOLAR ENERGY PREDICTION AND PROJECT RISK ASSESSMENT METHODOLOGY 132
Alemu Tadesse ; Adam Kankiewicz ; Alex Kubiniec ; Richard Perez ; John Dise ; Thomas Hoff

DECOUPLING THIN FILM CDTE GROWTH FROM PACKAGING: TOWARD RECORD SPECIFIC POWER IN LOW COST POLYCRYSTALLINE PV 138
D. Clayton-Warwick ; M.D. Kempe ; M. S. Dabney ; T. M. Barnes ; C. A. Wolden ; M. O. Reese

JUNCTION ACTIVATION OF CDTE/CDS SOLAR CELL USING MGCL2 142
G. Angeles-Ordóñez ; E. Regalado-Pérez ; M.G. Reyes-Banda ; N. R. Mathews ; X. Mathew

VARIATION OF CU CONTENT OF SPRAYED CU(IN, GA)(S,SE)2SOLAR CELLS BASED ON A THIOL-AMINE SOLVENT MIXTURE 146
Panagiota Arnou ; Sona Ulicná ; Alexander Eeles ; Mustafa Togay ; Lewis D. Wright ; Andrei V. Malkov ; John M. Walls ; Jake W. Bowers

CUINSE2 ABSORBER LAYER GROWN UNDER COPPER EXCESS WITH A COPPER POOR SURFACE FORMED BY A KF POST DEPOSITION TREATMENT 151

Finn Babbe ; Hossam Elanzeery ; Michele Melchiorre ; Susanne Siebentritt

CU2ZNSNSE4SOLAR CELLS ONTO POLYIMIDE SUBSTRATES FABRICATED AT LOW TEMPERATURE 155

Ignacio Becerril-Romero ; Simón Lopez-Marino ; Moisés Espíndola-Rodríguez ; Laura Acebo ; Markus Neuschitzer ; Yudania Sánchez ; Edgardo Saucedo ; Paul Pistor

AN OPTIMIZED PHOTOLITHOGRAPHY RECIPE FOR CU(IN1-X,GAX)(SY,SE1-Y)2(CIGSSE) SOLAR CELLS 160

Xia Hao ; Shenghao Wang ; Katsuhiro Akimoto ; Takuya Kato ; Hiroki Sugimoto ; Takeaki Sakurai

EFFECTS OF CDCL2PASSIVATION ON THIN CDTE ABSORBERS FABRICATED BY CLOSE-SPACE SUBLIMATION 164

Anna Wojtowicz ; Alexandra M. Huss ; Jennifer A. Drayton ; James R. Sites

CDS1-XSEXWINDOW LAYER FOR CDTE PREPARED BY THE EXCHANGE OF S WITH SE IN CDS FILMS 170

Geethika K. Liyanage ; Adam B. Phillips ; Zhaoning Song ; Suneth C. Watthage ; Ramez H. Ahanzhamejhad ; Michael J. Heben

EFFECT OF ILLUMINATION ON THERMAL CDCL2TREATMENT OF CDTE 175

Sudhajit Misra ; Carina E. Hahn ; Vasilios Palekis ; Christos Ferekides ; Michael A. Scarpulla

CHALLENGES IN THE INDUSTRIAL PRODUCTION OF CZTS MONOGRAIN SOLAR CELLS 178

Gerhard Peharz ; Valentin Satzinger ; Sandra Pötz ; Gernot Oreski ; Theodoros Dimopoulos ; Stefan Edinger ; Wolfeanz Hackl ; Hannes Starkl ; Parichehr Esfandiari ; Peter Krabb ; Stefan Gahr ; Lukas Plessing ; Dieter Meissner

UNDERSTANDING INSTABILITIES AND DEGRADATION DUE TO MOISTURE INGRESS IN CU(IN, GA)SE2SOLAR CELLS 182

Grace Rajan ; Shankar Karki ; Isaac Butt ; Krishna Aryal ; Tyler J. Grassman ; Angus Rockett ; Sylvain Marsillac

CONTROL OF MOSE2 FORMATION IN HYDRAZINE-FREE SOLUTION-PROCESSED CIS/CIGS THIN FILM SOLAR CELLS 186

Sona Ulicná ; Panagiota Arnou ; Alexander Eeles ; Mustafa Togay ; Lewis D. Wright ; Ali Abbas ; Andrei V. Malkov ; John M. Walls ; Jake W. Bowers

GROWTH AND PROPERTIES OF EPITAXIAL CU(IN, GA)SE2THIN FILMS DEPOSITED BY THE THREE-STAGE PROCESS FOR SOLAR CELLS 192

Takeru Yamagami ; Yuta Ando ; Ishwor Khatri ; Mutsumi Sugiyama ; Tokio Nakada

IMPROVEMENT OF CIS SOLAR CELLS WITH KF POSTDEPOSITION FOLLOWING A SIMPLE TWO-STEP SELENIZATION PROCESS 195

Yang Zhang ; Robert E. Bartolo ; Sang Jik Kwon ; Mario Dagenais

THE TWINS STRUCTURE, ELECTRICAL PROPERTIES AND CELL PERFORMANCE OF MAGNETRON SPUTTERING DEPOSITED CHLORINE DOPED CDTE 198

Ziyao Zhu ; Fu-Kuo Chiang ; Zhongming Du ; Yufeng Zhang ; Xiangxin Liu

INVESTIGATION AND MITIGATION OF SHUNTS FOR HIGHER EFFICIENCY EPITAXIAL GASB/GASB AND GASB/GAAS SOLAR CELLS 202

George T. Nelson ; Bor-Chau Juang ; Steve Johnston ; Michael A. Slocum ; Zachary S. Bittner ; Ramesh B. Lagumavarapu ; Diana Huffaker ; Seth M. Hubbard

DEVELOPMENT OF GASB SOLAR CELLS ON GAAS BY MOVPE VIA INTERFACE MISFIT TECHNIQUE 206

Michael A. Slocum ; Alessandro Giussani ; Emily Kessler ; Phil Ahrenkiel ; George T. Nelson ; Seth M. Hubbard

FABRICATION OF INGAASP SOLAR CELLS FOR CONCENTRATOR APPLICATIONS 210

Mitchell F. Bennett ; Matthew P. Lumb ; Kenneth J. Schmieder ; Brent Fisher ; Eric A. Armour ; Robert J. Walters

DETAILED CHARACTERIZATION FOR TCAD SIMULATIONS OF GAAS0.76P0.24/SI1-YGEY/SI SINGLE JUNCTION SOLAR CELLS 213

Sabina Abdul Hadi ; Timothy Milakovich ; Eugene A. Fitzgerald ; Ammar Nayfeh

COMPARATIVE STUDY OF >2 EV LATTICE-MATCHED AND METAMORPHIC (AL)GAINP MATERIALS AND SOLAR CELLS GROWN BY MOCVD 215

Daniel J. Chmielewski ; Christine Jackson ; Jacob Boyer ; Daniel Lepkowski ; John A. Carlin ; Aaron R. Arehart ; Tyler J. Grassman ; Steven A. Ringel

PERFORMANCE OF GASB PHOTOVOLTAICS WITH GRAPHENE COATING 219

Benjamin P. Conlon ; Daniel J. Herrera ; Shaimaa A. Abdallah ; Jonathan O. Okafor ; Luke F. Lester

HIGH EFFICIENCY SINGLE-JUNCTION INGAP PHOTOVOLTAIC DEVICES UNDER LOW INTENSITY LIGHT ILLUMINATION 222

Yushuai Dai ; Hyun Kum ; Michael A. Slocum ; George T. Nelson ; Seth M. Hubbard

RADIATION RESISTANT OF UPRIGHT METAMORPHIC GAINP/GAINAS/GE TRIPLE JUNCTION SOLAR CELLS FOR SPACE USE..226

Liang Fang ; Abuduwayiti Aierken ; Zhen Pan ; Qiming Zhang ; Zhanhang Li ; Heini Maliya ; Wei Gao ; Hui Gao ; Ronghua Wan ; Bao Zhang ; He Wang ; Qi Guo

HIGH EFFICIENCY GLASS WAVEGUIDING SOLAR CONCENTRATOR..229

Chehao Hu ; Yusuf Dogan ; Matthew Morrison ; A. Nanda ; D. Ma ; R. Atkins ; C. K. Madsen

GAINASP/GAINAS TANDEM SOLAR CELL WITH 32.6% ONE-SUN EFFICIENCY..232

Nikhil Jain ; Kevin L. Schulte ; John F. Geisz ; Ryan M. France ; Myles A. Steiner

EVALUATION OF TANDEM EFFICIENCIES: DILUTE NITRIDE P-I-N (BULK OR MQWS) IN CONJUNCTION WITH PRACTICAL SI SOLAR CELLS..236

Khim Kharel ; Alexandre Freundlich

GALLIUM PHOSPHIDE NANOSTRUCTURE ON SILICON BY SILICA NANOSPHERES LITHOGRAPHY AND METAL ASSISTED CHEMICAL ETCHING..240

Sangpyeong Kim ; Chaomin Zhang ; Som Dahal ; Stuart Bowden ; Christiana B. Honsberg

EFFICIENCY ENHANCEMENT OF INGAP/INGAAS/GE SOLAR CELLS WITH GRADUALLY DOPED P-N JUNCTION ACTIVE LAYERS..244

Youngjo Kim ; Sang Hyun Jung ; Chang Zoo Kim ; Kangho Kim ; Hyun-Beom Shin ; Kyung Ho Park ; Won-Kyu Park ; Jaejin Lee ; Ho Kwan Kang

ANALYSIS OF INGAP OXIDE GROWTH RATE AT HIGH TEMPERATURES AND AMBIENT CONDITIONS FOR TERRESTRIAL PHOTOVOLTAIC APPLICATIONS..247

Nicole A. Kotulak ; Matthew P. Lumb ; Raymond Hoheisel ; Erin Cleveland ; Mitchell Bennett ; Phillip P. Jenkins ; Robert J. Walters

GRAIN BOUNDARIES IN THIN-FILM POLYCRYSTALLINE GAAS SOLAR CELLS: A SIMULATION STUDY..251

Khushboo Kumari ; Sushobhan Avasthi

TIME-RESOLVED PL MEASUREMENTS IN THE GROWTH OF HIGH VOLTAGE (AL)GAINP/GAAS SOLAR CELLS..255

Xinyi Li ; Wei Zhang ; Hongbo Lu

LOW-RESISTANCE AND HIGHLY-TRANSPARENT GASB-BASED TUNNEL JUNCTIONS..259

Matthew P. Lumb ; Shawn Mack ; Maria Gonzalez ; Kenneth J. Schmieder ; Mitchell F. Bennett ; Chaffra A. Affouda ; James E. Moore ; Robert J. Walters

MODULATED PHOTOCURRENT MEASUREMENTS IN DOUBLE JUNCTION SOLAR CELLS..263

Nicolás Márquez Peraca ; Behrang H. Hamadani

EFFECT OF ATMOSPHERIC ABSORPTION BANDS ON THE OPTIMAL DESIGN OF MULTIJUNCTION SOLAR CELLS..268

William E. Mcmahon ; Daniel J. Friedman ; John F. Geisz

EFFECTS OF CONTACT CONFIGURATION AND PERIMETER RECOMBINATION ON OPTIMAL CELL SIZE FOR HIGH CONCENTRATION PHOTOVOLTAICS..272

James E. Moore ; Matthew P. Lumb ; Kenneth J. Schmieder ; Robert J. Walters ; Brent Fisher ; Matt Meitl ; Scott Burroughs

NUMERICAL SIMULATION OF DEFECTS IN III-V PV CELLS: THE EFFECT OF VOLTAGE BIAS AND DOPING CONCENTRATION..276

Vasiliki Paraskeva ; Constantinos Lazarou ; Andreas Livera ; Venizelos Venizelou ; Maria Hadjipanayi ; George E. Georghiou

IMPROVEMENT OF OPEN-CIRCUIT VOLTAGE IN METAMORPHIC GASB CELLS GROWN ON GAAS SUBSTRATES BY USING AN INTERFACIAL MISFIT ARRAY AND AN ALSB BLOCKING LAYER..281

E. J. Renteria ; S. J. Addamane ; D. M. Shima ; A. Mansoori ; A. L. Soudachanh ; G. Balakrishnan

ENERGY YIELD EVALUATION FOR FIELD OPERATION OF SOLAR CELLS IN SINGAPORE: GAAS/GAAS TANDEM VS. GAAS SINGLE-JUNCTION SOLAR CELLS..284

Maung Thway ; Zekun Ren ; Kevin Nay Yaung ; Haohui Liu ; Zhe Liu ; Samuel Raj ; Soo Jin Chua ; Armin G. Aberle ; Tonio Buonassisi ; Ian Marius Peters ; Fen Lin

SIMULATION OF THE PERFORMANCES OF MULTIJUNCTION SOLAR CELLS WITH IMPROVED VOLTAGE BY TRANSFER AND SCATTERING MATRIX METHODS..290

Gianluca Timò ; Lucio Andreani

OPTIMIZED DESIGN OF BACK-CONTACT THIN-FILM GAAS SOLAR CELLS..294

Jia-Ling Tsai ; Chung-Yu Hong ; Tien-Chien Zhan ; Yuh-Renn Wu ; Albert Lin ; Peichen Yu

DESIGN CONSIDERATIONS ON GAINNAS SOLAR CELLS WITH BACK SURFACE REFLECTORS..297

Antti Tukiainen ; Arto Aho ; Timo Aho ; Ville Polojärvi ; Mircea Guina

QUANTITATIVE ELECTROLUMINESCENCE ANALYSIS OF TRIPLE JUNCTION SOLAR CELLS TO DETERMINE SUBCELL VOLTAGE-TEMPERATURE COEFFICIENTS..301

Kevin Tyler ; Geoffrey K. Bradshaw ; Sam Wilt ; David M. Wilt ; Richard R. King

PROGRESS TOWARDS DOUBLE-JUNCTION INGAN SOLAR CELL 305

Ehsan Vadiee ; Evan A. Clinton ; Heather Mcfavilen ; Alec M. Fischer ; Yi Fang ; Joshua J. Williams ; Christiana B. Honsberg ; William A. Doolittle ; Stephen M. Goodnick

A PHYSICS-BASED SIMULATION TOOL FOR LEAKAGE CURRENTS IN C-SI PV MODULES 309

John M. Waddle ; Saroj Dahal ; Marco Nardone

BROADBAND TA2O5 MOTH-EYE ANTIREFLECTION COATINGS FOR TANDEM SOLAR CELLS ON SI 315

Bo Yuan ; Brian Thibeault ; David Payne ; James Mutitu ; Ivan Perez-Wurfl ; Kevin Dobson ; Brianna Conrad ; Allen Barnett ; Robert L. Opila

CARRIER TRANSPORT IN POLYCRYSTALLINE SILICON AT HIGH OPTICAL INJECTION: TRANSIENT PHOTOCONDUCTANCE VS. NUMERICAL MODELING 319

Uchechi Anyanwu ; Christian Harris ; Andrey Semichaevsky

IMPROVING SILICON SURFACE PASSIVATION WITH A SILICON OXIDE LAYER GROWN VIA OZONATED DEIONIZED WATER 322

Sara Bakhshi ; Ngwe Zin ; Kristopher O. Davis ; Marshall Wilson ; Ismail Kashkoush ; Winston V. Schoenfeld

DEPOSITION OF SIOC BY PLASMA-FREE ULTRA-LOW-TEMPERATURE ALD (ULT-ALD) AND ITS PASSIVATION ON P-TYPE SILICON 326

Meixi Chen ; Naoto Noda ; Raphael Rochat ; Abhishek Iyer ; James H. Hack ; Changhee Ko ; Christian Dussarrat ; Robert L. Opila

A METHOD FOR QUANTITATIVELY INVESTIGATING THE REAR-SIDE PASSIVATION PERFORMANCE OF PERC CELLS 329

Tsung-Cheng Chen ; Yung-Sheng Lin ; Chen-Hao Ku ; Ting-Wei Kuo ; Cheng-Shun Hu ; Ching-Chang Wen

FIELD-EFFECT PASSIVATION BY NEGATIVE CHARGE ON BORON EMITTER AND BORON-DOPED SURFACES BY A NOVEL LOW-COST PLASMA CHARGE INJECTION 333

Eunhwan Cho ; Young-Woo Ok ; James Hwang ; Aditi Jain ; Vijay D. Upadhyaya ; John Keith Tate ; Ajeet Rohatgi

INDUSTRY RELEVANT RIE TEXTURING FOR MC-SI DIAMOND WIRE OR DIRECT WAFER® PRODUCT: OPTIMIZED REFLECTIVITY, UNIFORMITY, AND THROUGHPUT 337

Jose Luis Cruz-Campa ; Ray Fraser ; Rob Steeman ; John Linton

SHORT-CIRCUIT CURRENT-DENSITY ENHANCEMENT OF SILICON SOLAR CELLS USING PLASMONICS ANTIREFLECTIVE COATING AND LUMINESCENT DOWNSHIFTING 343

Sheng-Kai Feng ; Wen-Jeng Ho ; Guan-Yi Li ; Jheng-Jie Liu ; Hao-Yu Yang ; Ta-Wei Chuang

EXTREMELY LOW REFLECTIVITY NANOPOROUS BLACK SILICON SURFACE BY COPPER CATALYZED ETCHING FOR EFFICIENT SOLAR CELLS 346

K A S M Ehteshamul Haque ; Wenqi Duan ; Fatima Toor

IMPACT OF FRONT SIDE PYRAMID SIZE ON THE LIGHT TRAPPING PERFORMANCE OF WAFER BASED SILICON SOLAR CELLS AND MODULES 352

Oliver Höhn ; Nico Tucher ; Benedikt Bläsi

A STUDY OF BLISTER CONTROL OF AL2O3 THIN FILM DEPOSITED BY PLASMA-ASSISTED ATOMIC LAYER DEPOSITION AFTER FIRING PROCESS 356

Min Gu Kang ; Jeong In Lee ; Hee-Eun Song ; Myeong Sangjeong ; Kyung Taekjeong ; Hyo Sikchang

PYPVCELL: AN OPEN-SOURCE SOLAR CELL MODELING LIBRARY IN PYTHON 359

Kan-Hua Lee ; Kenji Araki ; Omar Elleuch ; Nobuaki Kojima ; Masafumi Yamaguchi

IMPROVEMENT IN SURFACE PASSIVATION OF C-SI USING GRADIENT-LAYERED A-SI:H FILM FOR HIGH EFFICIENCY SILICON HETEROJUNCTION SOLAR CELLS 363

Soonil Lee ; Leo Mathew ; Rajesh Rao ; Jae Hyun Kim ; Sanjay K. Banerjee ; Edward T. Yu

PHOTOVOLTAIC PERFORMANCE ENHANCEMENT OF TEXTURED SILICON SOLAR CELLS USING LUMINESCENT DOWN-SHIFTING METHYLAMMONIUM LEAD TRIBROMIDE PEROVSKITE NANOPHOSPHORS 367

Guan-Yi Li ; Wen-Jeng Ho ; Sheng-Kai Feng ; Hao-Yu Yang ; Ta-Wei Chuang ; Bang-Jin You ; Zong-Xian Lin ; Zong-Liang Tseng ; Lung-Chien Chen

SINX THIN FILMS WITH APPROPRIATE ANTIREFLECTION AND SHIFT-CONVERSION PROPERTIES FOR SILICON SOLAR CELLS 370

E. Men-Pérez ; J. Salazar ; A. Dutt ; J. Santoyo-Salazar ; G. Santana

NUMERICAL SIMULATION OF CRYSTALLINE SILICON SOLAR CELLS WITH FULL AREA METAL OXIDE REAR CONTACTS 373

James E. Moore ; Woojun Yoon ; Phillip P. Jenkins ; Robert J. Walters

INTERDIGITATED BACK CONTACT SILICON SOLAR CELL WITH PEROVSKITE LAYER FOR FRONT SURFACE PASSIVATION AND ULTRAVIOLET RADIATION STABILITY 377

Rahul Pandey ; Shivam Gupta ; Trijul Khatri ; Rishu Chaujar

POTENTIAL OF A-SI:H/C-SI HETEROJUNCTION SOLAR CELLS WITH VERY THIN WAFERS 381
Hitoshi Sai ; Hiroshi Umishio ; Takuya Matsui ; Shota Nunomura ; Tomoyuki Kawatsu ; Hidetaka Takato ; Koji Matsubara

MANIPULATING FIXED CHARGES IN ZRO2 BY DOPING FOR PASSIVATION AND ANTIREFLECTION ON WAFER-SI SOLAR CELLS 385
Woo Jung Shin ; Laidong Wang ; Wen-Hsi Huang ; Meng Tao

LOW TEMPERATURE ANTIREFLECTION COATING FOR SILICON SOLAR CELLS 389
O. S. Shinde ; Ej Schneller ; N. Dhere ; S. V. Ghaisas

RELATIONSHIP BETWEEN POWER LOSS AND VOLTAGE APPLIED TO SOLAR CELLS IN PID-AFFECTED SOLAR MODULES 392
Fumei Wang ; Baosong Duan ; Wenshuang He ; He Wang ; Hong Yang ; Chengfeng Su ; Bojie Su ; Xue Zhang ; Yunxue Cao ; Hui Zhao

A NEW LOW-COST AND LOW-TEMPERATURE CHEMICAL PASSIVATION PROCESS FOR LARGE AREA INDUSTRIAL SINGLE CRYSTALLINE SILICON WAFERS 396
Tarun S. Yadav ; K. Sandeep ; Ashok K. Sharma ; B. Spandana ; K.L. Narasimhan ; B.M. Arora ; Anil Kottantharayil ; Prabir K. Basu

EVALUATION OF ALD PASSIVATION LAYERS FOR INDUSTRIAL PERC PROCESS 399
Chang Youn Yoo ; Keunkee Hong ; Jisun Kim ; Eunjoo Lee ; Dong Seop Kim

QUANTITATIVE ANALYSIS OF ELECTROLUMINESCENCE AND INFRARED THERMAL IMAGES FOR AGED MONOCRYSTALLINE SILICON PHOTOVOLTAIC MODULES 402
Irene Berardone ; Juan Lopez Garcia ; Marco Paggi

GAP PASSIVATION STRUCTURE FOR SCALABLE N-TYPE INTERDIGITATED ALL BACK CONTACT SILICON HETERO-JUNCTION SOLAR CELL 408
Lei Zhang ; Ujjwal Das ; Steven Hegedus

PROPOSAL OF THE BANDGAP DESIGN USING THE SUN HEIGHT OF THE CULMINATION ON THE WINTER SOLSTICE 412
Kenji Araki ; Kan-Hua Lee ; Masafumi Yamaguchi

PHOTOEXCITED CARRIERS, PHONONS, AND THEIR SCATTERING MEASURED IN SEMICONDUCTOR JUNCTIONS BY TRANSIENT EXTREME ULTRAVIOLET SPECTROSCOPY 417
Scott K. Cushing ; Brett M. Marsh ; Mihai E. Vaida ; Lucas M. Carneiro ; Ilana J. Porter ; Angela Lee ; Stephen R. Leone

ON THE USE OF VOLTAGE MEASUREMENTS FOR DETERMINING CARRIER LIFETIME AT HIGH ILLUMINATION INTENSITY 420
Robert Dumbrell ; Mattias K. Juhl ; Thorsten Trupke ; Ziv Hameiri

HIGH RESOLUTION 3D CHEMICAL CHARACTERISATION OF A CADMIUM TELLURIDE SOLAR CELL BY DYNAMIC SIMS 424
Thomas Fiducia ; Kexue Li ; Chris Grovenor ; Kurt Barth ; Walajabad Sampath ; Michael Walls

HARSH OUTDOOR EVALUATION SETUP AND FIRST POWER PRODUCTION RESULTS FOR SI MINI-MODULES COVERED BY EU3+-BASED DOWN CONVERTERS 429
Benjamín González-Díaz ; Carlos Montes ; Joaquín Sanchiz ; Luis Ocaña ; Carlos Quinto ; Cecilio Hernández-Rodríguez ; Mari Paz Friend ; Manuel Cendagorta-Galarza ; David Cañadillas ; Ricardo Guerrero-Lemus

STUDY OF MICRO-STRUCTURAL PROPERTIES OF ZNO AND TIO2THIN FILM GROWN BY SPRAY PYROLYSIS 433
G. Gordillo ; J.M. Correa ; A.A. Ramirez ; E. A. Ramírez

NONLINEAR RESPONSE OF SILICON SOLAR CELLS 437
Behrang H. Hamadani ; Andrew Shore ; Howard W. Yoon ; Mark Campanelli

EXTENDED LINEAR INTERPOLATION/EXTRAPOLATION PROCEDURE FOR ACCURATE AND VERSATILE TRANSLATION OF THE I-V CURVES OF PV CELLS AND MODULES 441
Y. Hishikawa ; H. Ohshima ; M. Higa ; K. Yamagoe ; T. Takenouchi ; T. Doi

SEVERITY TEST WITH UNEVEN LOAD DUE TO WIND ACTION ON PHOTOVOLTAIC MODULE 445
Shu-Tsung Hsu

STANDARDIZED DURABILITY TEST FOR ORGANIC PHOTOVOLTAIC AND DYE SENSITIZED SOLAR CELL 448
Shu-Tsung Hsu ; Yean-San Long ; Teng-Chun Wu

SPATIAL THICKNESS UNIFORMITY AND STRUCTURAL EVALUATION OF RF SPUTTERED ZNO THIN FILMS FOR SOLAR CELL 451
Babar Hussain ; Taj M. Khan

LOCAL MEASUREMENTS OF SURFACE CAPACITANCE BY ELECTROSTATIC FORCE MICROSCOPY ON CU(IN, GA)SE2MATERIALS 455
Tomoaki Ishii ; Takashi Minemoto ; Takuji Takahashi

A COMPARISON OF SI-BASED CAMERAS FOR IMAGING LUMINESCENCE FROM PHOTOVOLTAIC MATERIALS AND DEVICES 459

Steve Johnston

BLISTERING OF AL2O3/A-SINX:H STACKS: ANALYSIS OF THE SUBMERGED PART OF THE ICEBERG BY COLORED PICOSECOND ACOUSTIC MICROSCOPY 464

Fabien Lebreton ; Arnaud Devos ; Etienne Drahi ; Patricia De Coux ; François Silva ; Sergej Filonovich ; Pere Roca I Cabarrocas

SELF-REFERENCE PROCEDURE TO REDUCE UNCERTAINTY IN MODULE CALIBRATION 467

D.H. Levi ; C.R. Osterwald ; S. Rummel ; L. Ottoson ; A. Anderberg

UNCERTAINTY EVALUATION OF PRIMARY REFERENCE PHOTOVOLTAIC CELL CALIBRATION UNDER OUTDOOR CONDITION IN TIBET 472

Haitao Liu ; Shiyu Sang ; Guomin Zhou ; Yonghui Zhai

REQUIREMENT OF ARTIFICIAL LIGHTING SIMULATOR FOR EVALUATION EMERGING PV PERFORMANCE RATING UNDER INDOOR ENVIRONMENT 476

Yean-San Long ; Shu-Tsung Hsu ; Teng-Chun Wu

NON-CONTACT VOLTAGE MEASUREMENT OF SOLAR CELL WITH ELECTROSTATIC VOLTMETER 480

Sakutaro Miyajima ; Kensuke Nishioka ; Yoshihiro Hishikawa

NREL'S CELL AND MODULE PERFORMANCE GROUP'S ASYMPTOTIC PMAX PROTOCOL FOR PEROVSKITE DEVICES 483

Tom Moriarty ; Dean Levi

OUTDOOR OPERATING TEMPERATURE MODELING OF PHOTOVOLTAIC MODULES INCLUDING TRANSIENT EFFECT 487

Soo-Young Oh ; Min-Soo Kim ; Won-Shup So ; Woo Kyoung Kim ; Jae Hak Jung ; Chinho Park ; Benazzouz Aboubakr ; Ikken Badr ; Naimi Zakaria ; Benlarabi Ahmed

PRIMARY REFERENCE CELL CALIBRATIONS WITH REDUCED MEASUREMENT UNCERTAINTY 490

C.R. Osterwald ; L. Ottoson ; R. Williams ; C. Mack ; T. Moriarty ; K.A. Emery ; D.H. Levi

IMPLEMENTATION OF NOVEL PIN CONNECTION AND TEST ROUTINE FOR IMPROVED ACCURACY IN I-V MEASUREMENTS 496

Samuel Raj ; Johnson Kai Chi Wong ; Mohan Krishan Bhan ; Evan Palmer ; Jian Wei Ho ; Sumukh Ramprasad ; Wang Junci ; Thomas Mueller ; Armin G. Aberle

A NEW METHOD TO QUANTIFY CONTACT RESISTANCE USING LOCALIZED-ILLUMINATION PHOTOLUMINESCENCE TECHNIQUE IN A SOLAR CELL 499

Amit Singh Rajput ; Samuel Raj ; Johnson Wong ; Armin G. Aberle

IMPROVEMENT OF THE PROPERTIES OF CZTS THIN FILMS PREPARED BY SPRAY PYROLYSIS USING DMSO IN ACETONE AS SOLVENT 503

E. A. Ramírez ; A. Ramírez ; G. Gordillo

ASSESSMENT OF CARRIER LIFETIMES AND SURFACE RECOMBINATION VELOCITY THROUGH SPECTRAL MEASUREMENTS 508

John Roller ; Behrang H. Hamadani

IMPACT OF SPACE RADIATION ENVIRONMENT ON CONCENTRATOR PHOTOVOLTAIC SYSTEMS 512

Pilar Espinet-Gonzalez ; Tatiana Vinogradova ; Michael D. Kelzenberg ; Alexander Messer ; Emily C. Warmann ; Chris Peterson ; Nina Vaidya ; Ali Naqavi ; Jing-Shun Huang ; Samuel P. Loke ; Don Walker ; Colin J. Mann ; Sergio Pellegrino ; Harry A. Atwater

EXTRACTING THE FIXED CHARGE DENSITY IN HFOX FILMS GROWN ON HIGHLY-DOPED P-SI SAMPLES 517

Alexander To ; Jie Cur ; Bram Hoex

NEAR-UNITY ULTRA-WIDEBAND THERMAL INFRARED EMISSION FOR SPACE SOLAR POWER RADIATIVE COOLING 521

Ali Naqavi ; Samuel P. Loke ; Michael D. Kelzenberg ; Emily C. Warmann ; Pilar Espinet-González ; Nina Vaidya ; Jing-Shun Huang ; Tatiana A. Roy ; Alexander J. Messer ; Tatiana G. Vinogradova ; Ali Hajimiri ; Sergio Pellegrino ; Harry A. Atwater

LINE-FOCUS AND POINT-FOCUS SPACE PHOTOVOLTAIC CONCENTRATORS USING ROBUST FRESNEL LENSES, 4-JUNCTION CELLS, & GRAPHENE RADIATORS 525

Mark O'Neill ; A.J. Mcdanal ; Michael Piszczor ; Matt Myers ; Paul Sharps ; Claiborne Mcpheeters ; Jeff Steinfedt

SIMULATION OF LIGHT TRAPPING STRUCTURES FOR ENHANCING RADIATION HARDNESS IN SPACE SOLAR CELLS 531

Nizami Z. Vagidov ; Kyle H. Montgomery ; Geoffrey K. Bradshaw ; David M. Wilt

AN ALTERNATIVE METHOD FOR SOLAR CELL INTEGRATION 537

Jessica Buckner ; Tracy Davis ; Eric Muskovin ; Bernard Carpenter

NIEL DOSE ANALYSIS ON TRIPLE JUNCTION CELLS 30% EFFICIENT AND RELATED SINGLE JUNCTIONS 541

Roberta Campesato ; Erminio Greco ; Mariacristina Casale ; Massimo Gervasi ; P.G. Rancoita ; Davide Rozza ; Mauro Tacconi ; Enos Gombia ; Aldo Kingma ; Carsten Baur

THIN AND FLEXIBLE TRIPLE JUNCTION CELLS 30% EFFICIENT: QUALIFICATION RESULTS AND FUTURE SPACE APPLICATIONS 545

Roberta Campesato ; Mariacristina Casale ; Giuseppe Gabetta ; Emilio Fernandez Lisbona ; Laurent D'Abrigeon

PRINTED ASSEMBLIES OF MICROSCALE TRIPLE-JUNCTION (3J) INVERTED METAMORPHIC (IMM) GAINP/GAAS/INGAAS SOLAR CELLS 549

Boju Gai ; John Geisz ; Daniel Friedman ; Jongseung Yoon

COMPARATIVE STUDY ON NONRADIATIVE RECOMBINATION CENTERS IN PROTON IRRADIATED INAS/GAAS QUANTUM DOT STRUCTURE BY TWO WAVELENGTH EXCITED PHOTOLUMINESCENCE 552

M. D. Haque ; N. Kamata ; S-I. Sato ; S. M. Hubbard

DESIGN AND PROTOTYPING EFFORTS FOR THE SPACE SOLAR POWER INITIATIVE 558

Michael D. Kelzenberg ; Pilar Espinct-Gonzalez ; Nina Vaidya ; Tatiana A. Roy ; Emily C. Warmann ; Ali Naqavi ; Samuel P. Loke ; Jing-Shun Huang ; Tatiana G. Vinogradova ; Alexander J. Messer ; Christophe Leclerc ; Eleftherios E. Gdoutos ; Fabien Royer ; Ali Hajimiri ; Sergio Pellegrino ; Harry A. Atwater

DEFECT CHARACTERIZATION OF III-V QUANTUM STRUCTURE SOLAR CELLS USING PHOTO-INDUCED CURRENT TRANSIENT SPECTROSCOPY 562

Shin-Ichiro Sato ; Takeyoshi Sugaya ; Tetsuya Nakamura ; Takeshi Ohshima

EFFECT OF LUMINESCENCE COUPLING BETWEEN INGAP AND GAAS SUBCELLS TO EXTERNAL QUANTUM EFFICIENCY IN TRIPLE-JUNCTION SOLAR CELLS 567

Mitsunobu Suga ; Mitsuru Imaizumi ; Tetsuya Nakamur ; Takeshi Ohshima

LIGHTWEIGHT CARBON FIBER MIRRORS FOR SOLAR CONCENTRATOR APPLICATIONS 572

Nina Vaidya ; Michael D. Kelzenberg ; Pilar Espinet-Gonzalez ; Tatiana G. Vinogradova ; Jing-Shun Huang ; Christophe Leclerc ; Ali Naqavi ; Emily C. Warmann ; Sergio Pellegrino ; Harry A. Atwater

GAAS SOLAR CELLS ON V-GROOVED SILICON VIA SELECTIVE AREA GROWTH 578

Michelle Vaisman ; Nikhil Jain ; Qiang Li ; Kei May Lau ; Adele C. Tamboli ; Emily L. Warren

HIGH TEMPERATURE ANNEALING OF IN1-XGAXN MQW SOLAR CELLS 582

Joshua J. Williams ; Heather Mcfavilen ; Steven Young ; Christiana B. Honsberg ; Stephen M. Goodnick

SOLAR PROBE PLUS ARRAY RELIABILITY 585

Anton Yanchilin ; Edward Gaddy

PHOTOVOLTAIC TEMPERATURE ESTIMATION MODEL FOR RAPID IRRADIANCE CHANGE CONDITIONS IN TROPICAL REGIONS USING HEURISTIC ALGORITHMS 589

R. Srivatsan ; Lian L. Jiang ; Douglas L. Maskell

ACCURACY OF CDTE PV ENERGY PREDICTIONS USING SPECTRAL CORRECTIONS 595

Mitchell Lee ; Kendra Passow ; Paul Wolffersdorff

PLANTPREDICT: SOLAR PERFORMANCE MODELING MADE SIMPLE 600

Kendra Passow ; Lauren Ngan ; Geoffrey Rich ; Mitch Lee ; Stephen Kaplan

INTEGRABILITY COMPARISON BETWEEN BIPV AND BAPV IN TROPICAL CONDITIONS: A BANGALORE CASE-STUDY 604

Gayathri Aaditya ; Roshan R Rao ; Monto Mani

A NEW PHOTOVOLTAIC SYSTEM TOPOLOGY THROUGH LOAD MANAGEMENT 608

Joseph A. Azzolini ; Meng Tao

FIRST STEP FOR POWER GENERATION AMOUNT ESTIMATION OF SOLAR MATCHING SYSTEM 613

Kazuya Hosokawa ; Toshiaki Yachi ; Yoichi Hirata ; Yasuyuki Watanabe

IRRADIANCE AND TEMPERATURE DISTRIBUTIONS AT HIGH LATITUDES: DESIGN IMPLICATIONS FOR PHOTOVOLTAIC SYSTEMS 619

Anne Gerdimenes ; Josefine Sclj

STEP-BY-STEP EVALUATION OF PHOTOVOLTAIC MODULE PERFORMANCE RELATED TO OUTDOOR PARAMETERS: EVALUATION OF THE UNCERTAINTY 626

Anne Migan Dubois ; Jordi Badosa ; Fausto Calderón-Obaldía ; Olivier Atlan ; Vincent Bourdin ; Marko Pavlov ; Dae Young Kim ; Yvan Bonnassieux

PERFORMANCE COMPARISONS OF A PV SYSTEM BY MONITORING SOLAR IRRADIANCE WITH DIFFERENT PYRANOMETERS 632

Yasuhiro Matsumoto ; J. Antonio Urbano ; Ramón Peña ; María De La Luz Olvera ; Nun Pitalúa ; Miguel A. Luna ; René Asomoza

FINANCIAL ANALYSIS OF A GRID-CONNECTED PHOTOVOLTAIC SYSTEM IN SOUTH FLORIDA 638

Hadis Moradi ; Amir Abtahi ; Ali Zilouchian

STUDY OF PHOTOVOLTAIC SYSTEMS MONITORING METHODS..................................643
  E. Ortega ; G. Aranguren ; M.J. Sáenz ; R. Gutiérrez ; J.C. Jimeno
GLOBAL DESIGN ASPECTS OF PERSISTENT AND AUTONOMOUS PV POWERED SYSTEMS 648
  I. M. Peters ; S. Watson ; N. Sahraei ; T. Buonassisi
HOW TO CHOOSE THE BEST EMPIRICAL MODEL FOR OPTIMUM ENERGY YIELD
PREDICTIONS 652
  Steve Ransome ; Juergen Sutterlueti
MODELING AND ANALYSIS OF PHOTOVOLTAIC ELECTROCHEMICAL SYSTEM USING
MODULE-LEVEL POWER ELECTRONICS 658
  Gowri M. Sriramagiri ; Nuha Ahmed ; Kevin D. Dobson ; Steven S. Hegedus
BETAVOLTAIC GENERATION FUNCTION IN SILICON..663
  A.V. Sachenko ; I.O. Sokolovskyi ; M. Evstigneev
MULTI-OBJECTIVE OPTIMIZATION FOR COLOR-TUNABILITY AND TRANSPARENCY IN
COLLOIDAL QUANTUM DOT SOLAR CELLS..667
  *Ebuka S. Arinze ; Botong Qiu ; Nathan Palmquist ; Yan Cheng ; Yida Lin ; Gabrielle Nyirjesy ; Gary Qian ;
  Susanna M. Thon*
CUBIC PHASE INXGA1-XN/GAN QUANTUM WELLS FOR THEIR APPLICATION TO
TANDEM SOLAR CELLS..670
  *C. A. Hernández-Gutiérrez ; Y. L. Casallas-Moreno ; Dagoberto Cardona ; Yu. Kudriavtsev ; A. Morales-Acevedo
  ; G. Santana-Rodríguez ; M. López-López*
MODELING OF P-I-N GAASPN/GAP MQWS SOLAR CELL: TOWARDS LATTICE MATCHED
III-V/SI TANDEM..673
  Khim Kharel ; Alexandre Freundlich
INP QUANTUM DOT INTERMEDIATE BAND SOLAR CELL GROWN VIA MOCVD 677
  Hyun Kum ; Yushuai Dai ; Michael Slocum ; Zachary Bittner ; Seth Hubbard
MODIFIED LIMITING EFFICIENCY FOR MULTIPLE EXCITON GENERATION SOLAR
CELLS 681
  Jongwon Lee ; Christiana B. Honsberg
A SIMPLE MONTE CARLO MODEL OF A HOT CARRIER CELL...685
  Tor Oskar Saetre
OPTIMIZATION OF SEMICONDUCTOR QUANTUM DOTS FOR LUMINESCENT SOLAR
CONCENTRATORS: MINIMIZING REABSORPTION LOSSES 690
  *Anatoli I. Shkrebtii ; Anatoliy V. Sachenko ; Igor O. Sokolovskyi ; Vitaliy P. Kostylyov ; Mykola R. Kulish ; Denis
  V. Khomcnko ; Mykhaylo A. Evstigneev*
DEVELOPMENT OF ABSORBER AND ENERGY SELECTIVE CONTACTS FOR HOT
CARRIER SOLAR CELLS...696
  *Santosh Shrestha ; Simon Chung ; Yuanxun Liao ; Wenkai Cao ; Neeti Gupta ; Yi Zhang ; Xiaoming Wen ; Gavin
  Conibeer*
GAASBI DEVICES FOR THERMAL ENERGY CONVERSION 701
  Margaret Stevens ; Abigail Licht ; Nicole Pfiester ; Emily Carlson ; Kevin Grossklaus ; Thomas E. Vandervelde
ANALYTIC JV-CHARACTERISTICS OF IDEAL IMPURITY PV-CELLS 706
  Rune Strandberg
PHOTOLUMINESCENCE PROPERTIES OF IN-PLANE ULTRAHIGH-DENSITY INAS
QUANTUM DOTS ON GAASSB/GAAS(001) FOR SOLAR CELL APPLICATIONS......................712
  Ryo Sugiyama ; Naoki Akimoto ; Tomah Sogabe ; Koichi Yamaguchi
CARRIER SELECTIVE BACK CONTACT (CSBC) SOLAR CELL USING TRANSITION METAL
OXIDES..716
  Astha Tyagi ; Kunal Ghosh ; Anil Kottantharayil ; Saurabh Lodha
ANALYSIS OF OPEN-CIRCUIT VOLTAGE AND CONVERSION EFFICIENCY IN QUANTUM-
DOT SOLAR CELLS VIA DETAILED-BALANCE-LIMIT THEORY......................................721
  Lin Zhu ; Hidefumi Akiyama ; Yoshihiko Kanemitsu
ZINC SELENIDE SURFACE PASSIVATION LAYER FOR SINGLE-CRYSTALLINE CZTSE
SOLAR CELLS 726
  Michael A. Lloyd ; Douglas Bishop ; Brian E. Mccandless ; Robert Birkmirc
USE OF SINGLE WALL CARBON NANOTUBE FILMS DOPED WITH TRIETHYLOXONIUM
HEXACHLORANTIMONATE AS A TRANSPARENT BACK CONTACT FOR CDTE SOLAR
CELLS 730
  *Fadhil K. Alfadhili ; Jacob M. Gibbs ; Geethika K. Liyanage ; Patrick W. Krantz ; Suneth C. Watthage ; Zhaoning
  Song ; Adam B. Phillips ; Michael J. Heben*
GRAIN AND GRAIN BOUNDARY GEOMETRICAL SHAPE CONSIDERATIONS ON SODIUM
AND POTASSIUM DIFFUSION THROUGH MOLYBDENUM FILMS 735
  Orlando Ayala ; Chinedum Akwari ; Tasnuva Ashrafee ; Shankar Karki ; Grace Rajan ; Sylvain Marsillac

SOLUTION-PROCESSED NICKEL-ALLOYED IRON PYRITE THIN FILM AS HOLE TRANSPORT LAYER IN CADMIUM TELLURIDE SOLAR CELLS 738

Ebin Bastola ; Khagendra P. Bhandari ; Randy J. Ellingson

USE OF CDS:O AND CDSE AS WINDOW LAYERS FOR CDTE PHOTOVOLTAICS 742

Tom Baines ; Guillaume. Zoppi ; Ken Durose ; Jonathan D. Major

APPLICATIONS OF HYBRID ORGANIC-INORGANIC METAL HALIDE PEROVSKITE THIN FILM AS A HOLE TRANSPORT LAYER IN CDTE THIN FILM SOLAR CELLS 748

Khagendra P. Bhandari ; Suneth C. Watthage ; Zhaoning Song ; Adam Phillips ; Michael J. Heben ; Randy J. Ellingson

MAGNESIUM-DOPED ZINC OXIDE AS A HIGH RESISTANCE TRANSPARENT LAYER FOR THIN FILM CDS/CDTE SOLAR CELLS 752

Francesco Bittau ; Elisa Artegiani ; Ali Abbas ; Daniele Menossi ; Alessandro Romeo ; Jake W. Bowers ; John M. Walls

INVESTIGATION OF ZNL-XMGXO:A1 FILM BY RATIO FREQUENCY MAGNETRON CO-SPUTTERING AS TRANSPARENT CONDUCTIVE OXIDE LAYER 757

Jakapan Chantana ; Yuya Ishino ; Takashi Minemoto

A NEW TCO/WINDOW-BUFFER FRONT STACK FOR CDTE SOLAR CELLS AND ITS IMPLEMENTATION 761

Alan E. Delahoy ; Xuehai Tan ; Akash Saraf ; Payal Patra ; Surya Manda ; Yunfei Chen ; Krishnakumar Velappan ; Bastian Siepchen ; Shou Peng ; Ken K. Chin

SYNTHESIS OF HIGH-QUALITY AZO POLYCRYSTALLINE FILMS VIA TARGET BIAS RADIO FREQUENCY MAGNETRON SPUTTERING 767

Zhongming Du ; Yufeng Zhang ; Xiangxin Liu

CLOSE-SPACE SUBLIMATED CDTE SOLAR CELLS WITH CO-SPUTTERED CDSXSE1-XALLOY WINDOW LAYERS 771

Corey R. Grice ; Maxwell Junda ; Alexander Archer ; Jian Li ; Yanfa Yan

EFFECTS OF GRAPHENE OXIDE BARRIER ON CU2ZNSNSXSE4-XTHIN FILM SOLAR CELLS 777

Woo-Lim Jeong ; Jung-Hong Min ; In-Young Kim ; Hae-Sun Kim ; Jin-Hyeok Kim ; Dong-Seon Lee

13% CDS/CDTE SOLAR CELL USING A NANOCOMPOSITE (CUS)X(ZNS)1-X THIN FILM HOLE TRANSPORT LAYER 781

Kamala Khanal Subedi ; Khagendra P. Bhandari ; Ebin Bastola ; Randy J. Ellingson

MOLYBDENUM OXIDE AND MOLYBDENUM NITRIDE BACK CONTACTS FOR THIN-FILM CDTE SOLAR CELLS 785

Anna Kindvall ; Jason Kephart ; Walajabad Sampath

INVESTIGATION AND OPTIMIZATION OF CD-FREE BUFFER LAYERS IN2S3 AND ZN(O, S) FOR CU2ZNSN(S, SE)4-BASED SOLAR CELLS 791

Willi Kogler ; Thomas Schnabel ; Andreas Bauer ; Stefanie Spiering ; Erik Ahlswede ; Michael Powalla

REAR CONTACT PASSIVATION FOR HIGH BANDGAP CU(IN, GA)SE2 SOLAR CELLS WITH VARYING ABSORBER THICKNESS AND FLAT GA PROFILE 796

Dorothea Ledinek ; Pedro Salome ; Carl Hägglund ; Marika Edoff

LASER ANNEALED BACK CONTACTS FOR CDTE SOLAR CELLS 802

Vasilios Palekis ; Shamara Collins ; Imran Khan ; Vamsi Evani ; Sudhajit Misra ; Michael A. Scarpulla ; Mark Lonergan ; Don Morel ; Chris Ferekides

ENHANCED ANTI-REFLECTIVE COATING FOR THIN FILM SOLAR CELLS 807

Grace Rajan ; Shankar Karki ; Robert W. Collins ; Sylvain Marsillac

INFLUENCE OF AGS LAYER INSERTION AT ABSORBER/ITO INTERFACE ON STRUCTURAL AND PHOTOVOLTAIC PROPERTIES OF ULTRATHIN CU(IN,GA)SE2 SOLAR CELLS 810

Muhammad Saifullah ; Jihye Gwak ; Kihwan Kim ; Joo Hyung Park ; Junsik Cho ; Jae Ho Yun

NOVEL, FACILE BACK SURFACE TREATMENT FOR CDTE SOLAR CELLS 815

Suneth C. Watthage ; Geethika K. Liyanage ; Zhaoning Song ; Fadhil K. Alfadhili ; Rabee B. Alkhayat ; Khagendra P. Bhandari ; Randy J. Ellingson ; Adam B. Phillips ; Michael J. Heben

OPTIMIZING CDS BUFFER LAYER FOR CIGS BASED THIN FILM SOLAR CELL 820

Weijie Zhang ; Korhan Demirkan ; Geordie Zapalac ; David Spaulding ; Jochen Titus ; Neil Mackie

INVESTIGATION OF INP DEFECT CHARACTERISTICS GROWN USING NOVEL TF-VLS TECHNIQUE 823

Abhinav Chikhalkar ; Alec Fischer ; Mark Hettick ; Ali Javey ; Richard R. King

INVESTIGATION OF FAST GROWTH GAAS-BASED SOLAR CELL ON REUSABLE SUBSTRATE BY METALORGANIC CHEMICAL VAPOR DEPOSITION 827

Chaomin Zhang ; Abhinav Chikhalkar ; Ehsan Vadiee ; Richard King ; Christiana Honsberg ; Eric Armour ; Yeongho Kim

DEVELOPMENT OF ALUMINUM EPILAYERS AS BUFFERS FOR GAINAS 831
Phil Ahrenkiel ; Nathan Smaglik ; Nikhil Pokharel ; Alessandro Giussani ; Michael A. Slocum ; Seth M. Hubbard

LASER CRYSTALLIZATION OF AMORPHOUS GERMANIUM ON TITANIUM NITRIDE-
COATED STEEL FOR LOW-COST GAAS SOLAR-CELLS 837
Saloni Chaurasia ; Srinivasan Raghavan ; Sushobhan Avasthi

HIGH QUALITY EPITAXIAL GERMANIUM ON SI (100) FOR LOW -COST III–V SOLAR-
CELLS 841
Saloni Chaurasia ; Srinivasan Raghavan ; Sushobhan Avasthi

CRYSTALLINITY CONTROL IN LOW-TEMPERATURE GROWTH OF POLY-CRYSTALLINE
GE BY ION BEAM DEPOSITION 845
S. I. Maximenko ; N. A. Mahadik ; P. P. Jenkins ; R. J. Walters ; A. Giussani ; E. L. Mcclure ; S. M. Hubbard ; C. Bailey

HIGH EFFICIENCY GAINP/GAAS DOUBLE JUNCTION SOLAR CELL ON SI SUBSTRATE
ASSISTED BY THE ELECTRON BEAM TREATMENT 849
Hyo Jin Kim ; Yong Whan Kim

ANALYSIS OF DEPOSITED RESIDUES AND ITS CLEANING PROCESS ON GAAS
SUBSTRATE AFTER EPITAXIAL LIFT-OFF 854
Tatsuya Nakata ; Kentaroh Watanabe ; Hassanet Sodabanlu ; Daiki Kimura ; Naoya Miyashita ; Yoshitaka Okada ; Yoshiaki Nakano ; Masakazu Sugiyama

ULTRATHIN SILICON-AN-INSULATOR (SOI) WAFER FOR COMPLIANT SUBSTRATE 858
Shinyoung Noh ; Anita Ho-Baillie ; Stephen Bremner ; Martin A. Green ; Xiaojing Hao

CHARACTERIZATION OF GAAS SOLAR CELLS GROWN BY HYDRIDE VAPOR PHASE
EPITAXY IN HORIZONTAL REACTOR 861
Ryuji Oshima ; Kikuo Makita ; Takeyoshi Sugaya ; Akinori Ubukata

FLEXIBLE GAAS SINGLE-JUNCTION SOLAR CELLS BASED ON SINGLE-CRYSTAL-LIKE
THIN-FILM MATERIALS DIRECTLY GROWN ON METAL TAPES 866
Sara Pouladi ; Monika Rathi ; Mojtaba Asadirad ; Pavel Dutta ; Seung Kyu Oh ; Devendra Khatiwada ; Shahab Shervin ; Yao Yao ; Venkat Selvamanickam ; Jae-Hyun Ryou

REDUCED DEFECT DENSITY IN SINGLE-CRYSTALLINE-LIKE GAAS THIN FILM ON
FLEXIBLE METAL SUBSTRATES BY USING SUPERLATTICE STRUCTURES 869
M. Rathi ; P. Dutta ; D. Khatiwada ; Y. Yao ; Y. Gao ; Y. Li ; S. Sun ; S. Pouladi ; S. Reed ; A. Khadimallah ; J. Ryou ; V. Selvamanickam ; N. Zheng ; P. Ahrenkiel

ECONOMIC ANALYSIS OF TRANSFER PRINTED III–V VIRTUAL SUBSTRATES 873
Kenneth J. Schmieder ; Matthew P. Lumb ; Michael K. Yakes ; Shawn Mack ; Mitchell F. Bennett ; Sergey I. Maximenko ; Laura B. Ruppalt ; Michael A. Meeker ; Chase T. Ellis ; Matthew Meitl ; Joseph G. Tischler ; Robert J. Walters

THIN FILMS OF ZINC-DOPED GAAS BY RF MAGNETRON SPUTTERING FOR USE IN
PHOTOVOLTAIC CELLS 876
Kirby Simon ; Kyle Cepeda ; Nishit Shetty ; Elijah Thimsen

SELF ALIGNED ALUMINUM SELECTIVE EMITTER FOR N-TYPE SI CELLS 881
San Theigi ; Robert C. Reedy ; Vincenzo Lasalvia ; Paul Stradins ; Benjamin G. Lee

HOW TO REALIZE SOLAR CELLS WITH LASER STRUCTURED PLATED NI-CU-CONTACTS
WITH EXCELLENT ADHESION AND HIGH FILL-FACTORS WITHOUT PARASITIC
PLATING 884
A. Büchler ; S. Kluska ; J. Bartsch ; B. Grübel ; A.A. Brand ; S. Gutscher ; M. Glatthaar

EXPLOITING THE POTENTIALS OF THE FRONT SURFACE FIELD (FSF) INDUSTRIAL
SILICON SOLAR CELL 888
Ahrar Ahmed Chowdhury ; Yu -Chen Hsu ; Veysel Unsur ; Abasifreke Ebong

PHOTOVOLTAIC PERFORMANCE OF SILICON SOLAR CELLS ENHANCED BY
PLASMONIC SILVER NANOPARTICLES OF VARIOUS DIMENSIONS DEPOSITING
THROUGH ANODIC ALUMINUM OXIDE TEMPLATE 893
Ta-Wei Chuang ; Wen-Jeng Ho ; Sheng-Kai Feng ; Jheng-Jie Liu ; Guan-Yi Li ; Hao-Yu Yang ; Yun-Chie Yang ; Cho-Chun Chiang ; Yao-Hui Chen

MITIGATION OF POTENTIAL-INDUCED DEGRADATION 896
Orry Faur ; Maria Faur

ELECTRODEPOSITION OF SI-LAYER THROUGH REDUCTION OF DIATOMACEOUS
EARTH FOR THE APPLICATION OF SOLAR-CELLS 900
Muhammad Monirul Islam ; Imane Abdellaoui ; Takeaki Sakurai ; Saad Hamzaoui ; Katsuhiro Akimoto

EFFECT OF SI CONTENT IN A1 PASTE ON LOCAL A1 REAR CONTACTS IN PERC CELL 904
Supawan Joonwichien ; Katsuhiko Shirasawa ; Satoshi Utsunomiya ; Hidetaka Takato

NEW SILVER PASTE METALLIZATION APPROACH ON P+ DIFFUSION ZONES OF SILICON
SOLAR CELLS 907
Yunjun Li ; Mohshi Yang ; Igor Pavlovsky ; Guoping Zeng

INFLUENCES OF ANNEALING AND DEFECT LIMITATION ON P-TYPE SILICON SOLAR CELL 911
Yu-Hsuan Lin ; Sung-Yu Chen ; Kuen-Yi Wu ; Chien-Hsun Chen ; Chen-Hsun Du ; Chun-Ming Yeh

REDUCED TEMPERATURE SILVER PASTE WITH LOW CONTACT RESISTANCE FOR ADVANCED SOLAR CELL APPLICATIONS 914
Ryan Mayberry ; Daniel Holzmann ; Gerd Schulz ; Lindsey Karpowich ; Mark Naylor ; Matthias Hoerteis

BSF ISLANDS FOR REDUCED RECOMBINATION IN IBC CELLS 917
Agnes A. Mewe ; Nicolas Guillevin ; Ilkay Cesar ; Antonius R. Burgers

THERMAL STABILITY OF HYDROGENATED BORON EMITTERS 921
Khaja H. Mohammed ; Larry C. Cousar ; Philip A. Mcmeans ; Garrett Z. Evans ; Douglas A. Hutchings ; Hameed A. Naseem ; Sergiu C. Pop

LIGHT INDUCED PLATING OF SILICON SOLAR CELLS USING BORIC ACID-FREE NICKEL CHEMISTRY 925
Krystal Munoz ; Lynne Michaelson ; Joseph Karas ; Tom Tyson ; James Rand ; Stuart Bowden

BAKING TEMPERATURE DEPENDENCE OF CU PASTE ON A1-BSF CELL PROPERTIES 931
Tomohiro Saito ; Tetsuya Fukuda ; Hoang Tri Hai ; Yuji Kurimoto ; Daisuke Ando ; Yuji Sutou ; Katsuhiko Shirasawa ; Junichi Koike

THE SILVER CONTACT AND FORMATION MECHANISM OF THE BORON EMITTER AND THE CURRENT FLOW MECHANISM OF THE SOLAR CELL ELECTRODE 935
Seunghyun Shin ; Soohyun Bae ; Sungeun Park ; Yoonmook Kang ; Hae-Seok Lee ; Donghwan Kim

LASER ANNEALING TO ENHANCE PERFORMANCE OF ALL-LASER-BASED SILICON BACK CONTACT SOLAR CELLS 937
Zeming Sun ; Mool C. Gupta

LARGE AREA N-TYPE SELECTIVE EMITTER CELLS USING LASER DOPING THROUGH BORON DOPED SCREEN PRINTED PASTE 940
Ajay D Upadhyaya ; Vijaykumar D Upadhyaya ; Brian Rounsaville ; Keeya Madani ; Ajeet Rohatgi ; Toru Hanada

METALLIZED BORON-DOPED BLACK SILICON EMITTERS FOR FRONT CONTACT SOLAR CELLS 944
Guillaume Von Gastrow ; Hele Savin ; Eric Calle ; Pablo Ortega ; Ramón Alcubilla ; Andreana Daniil ; Elias Z. Stutz ; Anna Fontcuberta I Morral ; Sebastian Husein ; Tara Nietzold ; Mariana Bertoni

CONTACT RESISTANCE MEASUREMENT FOR THERMALLY DIFFUSED POINT CONTACT BY LOCALIZED DIELECTRIC BREAKDOWN SOLAR CELLS 948
Qilin Ye ; Ned J. Western ; Anqi Liao ; Stephen P. Bremner

LOW TEMPERATURE REAR SURFACE METALLIZATION OF MULTI-CRYSTALLINE SILICON SOLAR CELLS FOR IMPROVED BULK LIFETIME 953
N. J. Western ; S. P. Bremner

INVESTIGATION OF HIGH PERFORMANCE PEROVSKITE-BASED SOLAR CELLS GROWN BY HYBRID CHEMICAL VAPOR DEPOSITION TECHNIQUE 958
Huseyin Cem Gokkaya ; Shen Qian ; Zhiwei Ren ; Annie Ng ; Charles Surya

ENHANCED PEROVSKITE SOLAR CELL PERFORMANCE USING FULL SPACE DEVICE OPTIMIZATION 963
Ahmer A.B. Baloch ; Shahzada P. Aly ; Mohammad I. Hossain ; Raka Jovanovic ; Nouar Tabet ; Fahhad H. Alharbi

MEASURING OPTICAL ABSORPTION IN ORGANIC PHOTOVOLTAICS USING MONOCHROMATED ELECTRON ENERGY-LOSS SPECTROSCOPY 966
Jessica A. Alexander ; Frank J. Scheltens ; David W. Mccomb ; Lawrence F. Drummy ; Michael F. Durstock ; James B. Gilchrist ; Sandrine Hentz

ADVANCED DEPOSITION OF PHOTO-CATALYTIC TIO2 FILM BY ATMOSPHERIC SPPS FOR DYE SENSITIZED SOLAR CELLS 970
Ifeanacho Anyadiegwu ; Dickson Kindole ; Geoffrey Kibiegon Ronoh ; Yoshimasa Noda ; Yasutaka Ando

CH3NH3PBI3-XBRXPEROVSKITE SOLAR CELLS VIA SPRAY ASSISTED TWO-STEP DEPOSITION: INFLUENCE OF BROMIDE ON THE DEVICE PERFORMANCE 976
Gaoda Chai ; Shiqiang Luo ; Shizhen Wang ; Hang Zhou

MODULATED STRUCTURE TO MAXIMIZE THE OPEN-CIRCUIT VOLTAGE WITH MODERATE BAND-GAP OF SMALL MOLECULE ORGANIC SOLAR CELLS-DFT APPROACH 980
Saravanan Chinnusamy ; Amita Munshi ; Sukanya Santhosh Kumar ; W. S. Sampath ; Milind S. Dangate

PEROVSKITE GRAIN SIZE MODULATION BY ANNEALING IN METHYL-AMINE ENVIRONMENT 986
Arun Singh Chouhan ; Naga Prathibha Jasti ; Srinivasan Raghavan ; Sushobhan Avasthi ; Shreyash Hadke

FE2O3AS AN ELECTRON TRANSPORT MATERIAL FOR ORGANO-METAL HALIDE PEROVSKITE SOLAR CELLS 989
Dallas Fisher ; Pravakar P. Rajbhandari ; Tara P. Dhakal

OPTICAL EVALUATION OF PEROVSKITE FILMS IN AND FOR SOLAR CELL DEVICE
STRUCTURES 993
  Kiran Ghimire ; Dewei Zhao ; Changlei Wang ; Yanfa Yan ; Nikolas J. Printraza
HYBRID ORGANIC-INORGANIC SOLAR CELLS WITH A BENZOQUINONE PASSIVATING
LAYER 999
  James Hack ; Abhishek Iyer ; Meixi Chen ; Nicole Kotulak ; Akirt Sridharan ; Robert Opila
PRECISE 1-V CURVE MEASUREMENT PROCEDURE FOR PEROVSKITE SOLAR CELLS:
APPLICATION TO VARIOUS TYPES OF DEVICES 1003
  Y. Hishikawa ; M. Yoshita ; H. Shimura ; A. Sasaki ; T. Ueda
ENHANCING THE CRYSTALLINE OF PLANAR-STRUCTURE CH3NH3PBI3PEROVSKITE
SOLAR CELLS VIA SANDWICH EVAPORATION TECHNIQUE 1006
  Po-Tsun Kuo ; Shang-Pang Lin ; Cheng-Shian Lin ; Ching-Fuh Lin
TOWARD HIGH PERFORMANCE ORGANIC-SILICON HYBRID SOLAR CELLS 1009
  Yi Lai ; Hong-Jhang Syu ; Ching-Fuh Lin
NICKEL OXIDE THIN FILMS BY RADIO FREQUENCY SPUTTER FOR INVERTED
PEROVSKITE SOLAR CELLS 1012
  Hyeonseok Lee ; Yu-Ting Huang ; Shien-Ping Feng
ANOMALOUS EFFICIENCY SCALING WITH DARK CURRENT IN PEROVSKITE SOLAR
CELLS 1015
  Vikas Nandal ; Pradeep R. Nair
NUMERICAL SIMULATION AND PERFORMANCE OPTIMIZATION OF PEROVSKITE
SOLAR CELL 1018
  Sai Naga Raghuram Nanduri ; Mahbube K. Siddiki ; Ghulam M. Chaudhry ; Yahya Z. Alharthi
PERFORMANCE PREDICTION FOR LARGE AREA PEROVSKITE SOLAR CELLS 1022
  Yojak Raote ; Hitarth Choubisa ; Pradeep R. Nair
PHOTOCONVERSION EFFICIENCY MODELING IN PEROVSKITE SOLAR CELLS 1025
  A.V. Sachenko ; V.P. Kostylyov ; A.V. Bobyl ; V.M. Vlasiuk ; I.O. Sokolovskyi ; E.I. Terukov ; M. Evstigneev
INFLUENCE OF MONO- AND DI-VALENT METAL ADDITIVES ON MORPHOLOGY AND
CHARGE CARRIER DYNAMICS OF CH3NH3PBI3PEROVSKITE 1030
  *Niraj Shrestha ; Suneth C. Watthage ; Zhaoning Song ; Paul J. Roland ; Adam B. Phillips ; Michael J. Heben ;
  Randall J. Ellingson*
EFFECT OF DUAL CATHODE BUFFER LAYER ON TERNARY ORGANIC SOLAR CELL 1034
  Ashish Singh ; T. Bhim Raju ; Anamika Dey ; Ritesh Kant Gupta ; Parameswar K. Iyer
COPPER PLATED TOP ELECTRODE FOR AN INVERTED ORGANIC PHOTOVOLTAIC 1037
  Malia Steward ; Zhan Shi ; Kyoung- Tae Kim ; Seungkeun Choi
INTERFACE BAND GAP AND CHARGE TRAPPING IN BULK HETEROJUNCTION SOLAR
CELLS 1040
  Marian Tzolov ; Maxwell Mcintyre
FABRICATION OF EFFICIENT CH3NH3PBI3 SOLAR CELLS IN AMBIENT AIR 1044
  Feng Wang ; Ye Zhongbiao ; Hojjatollah Sarvari ; Somin Park ; Kenneth Graham ; Yuetao Zhao ; Zhi David Chen
HIGH EFFICIENCY PEROVSKITE SOLAR CELLS BY A MODIFIED LOW-TEMPERATURE
SOLUTION PROCESS INTER-DIFFUSION METHOD 1048
  Yangyi Yao ; Wei-Lun Hsu ; Mario Dagenais
INTERFACIAL MODIFICATION OF SOL-GEL ZNO/AZO BILAYER AS HIGHLY EFFICIENT
ELECTRON TRANSPORT LAYER FOR PEROVSKITE SOLAR CELLS 1051
  Shang-Hsuan Wu ; Ming-Yi Lin ; Sheng-Hao Chang ; Wei-Chen Tu ; Chi-Wei Chu ; Via-Chung Chang
THE POTENTIAL OF BIFACIAL PHOTOVOLTAICS: A GLOBAL PERSPECTIVE 1055
  Xingshu Sun ; Mohammad R. Khan ; Amir Hanna ; Muhammad M. Hussain ; Muhammad A. Alam
PERFORMANCE ASSESSMENT OF STAND ALONE BIFACIAL SOLAR PANEL UNDER REAL
TIME CONDITIONS 1058
  Ahmer A.B. Baloch ; Maher Armoush ; Basel Hindi ; Abdelkader Bousselham ; Nouar Tabet
OPERATION AND PERFORMANCE ASSESSMENT OF GRID-CONNECTED PV SYSTEMS IN
OPERATION IN MAUI, HAWAII 1061
  Severine Busquet ; Jonathan Kobayashi ; Richard E. Rocheleau
A NOVEL MULTILEVEL SOLAR PANEL SYSTEM: IMPLEMENTATION AND
VERIFICATION 1067
  Tanmoy Debnath ; Syed N. Imtiaz ; Syed F. Nawaz ; Abdullah Al Mahmud ; Mosaddequr Rahman
PREDICTING POWER LOSS DUE TO MODULE MISMATCH IN UTILITY-SCALE
PHOTOVOLTAIC SYSTEMS 1071
  Stephen Kaplan ; Kendra Passow

APPLICATION OF SHAPED REFLECTORS TO INCREASE THE ENERGY HARVEST OF BIFACIAL PV SYSTEMS - ANALYZED WITH A MINIATURIZED TEST ARRAY 1077

Hartmut Nussbaumer ; Markus Klenk ; Nico Keller ; Dominic Heller ; Remo Kaslin ; Thomas Baumann ; Franz Baumgartner

TOWARDS NEW MODULE AND SYSTEM CONCEPTS FOR LINEAR SHADING RESPONSE 1081

Kostas Sinapis ; Tom T.H. Rooijakkers ; Lenneke H. Slooff ; Lars A.G. Okel ; Mark J. Jansen ; Anna J. Carr

PARTIAL SHADING ABATEMENT THROUGH CASCADED H-BRIDGE TOPOLOGY 1086

Steven Tidwell ; Joseph Latham ; Michael Mcintyre

DATA ANALYSIS FOR EFFECTIVE MONITORING OF PARTIALLY SHADED PHOTOVOLTAIC SYSTEMS 1090

Odysseas Tsafarakis ; Kostas Sinapis ; Wilfried G.J.H.M. Van Sark

BIFACIAL PHOTOVOLTAIC MODULE ENERGY YIELD CALCULATION AND ANALYSIS 1094

Christopher E. Valdivia ; Chu Tu Li ; Annie Russell ; Joan E. Haysom ; Rui Li ; David Lekx ; Mohsen M. Sepeher ; Dan Henes ; Karin Hinzer ; Henry P. Schriemer

DESIGN AND DEVELOPMENT OF A SOLAR PHOTOVOLTAIC MODULE DETECTION CONTROL SYSTEM BASED ON PLC 1100

Yiwang Wang ; Jili Zhang ; Kanglin Liu ; Houjun Tang ; Hui Pan ; Yan Lin ; Peter Yang ; Rui Wang

DETECTING CALIBRATION DRIFT AT GROUND TRUTH STATIONS A DEMONSTRATION OF SATELLITE IRRADIANCE MODELS' ACCURACY 1104

Richard Perez ; James Schlemmer ; Adam Kankiewicz ; John Dise ; Alemu Tadese ; Thomas Hoff

PERFORMANCE OF SOLAR RESOURCE MONITORING STATIONS IN HOT CLIMATE REGIONS 1110

Yahya Z. Alharthi ; Mahbube K. Siddiki ; Ghulam M. Chaudhry ; Saad Muaddi ; Ahmed Alahmed

FIRST RESULTS OF A LOW COST ALL-SKY IMAGER FOR CLOUD TRACKING AND INTRA-HOUR IRRADIANCE FORECASTING SERVING A PV-BASED SMART GRID IN LA GRACIOSA ISLAND 1116

David Cañadillas ; Walter Richardson ; Benjamín Gonzalez-Díaz ; Les E. Shephard ; Ricardo Guerrero Lemus

STATISTICAL ANALYSIS OF PV INSOLATION DATA 1122

Abdulmunim Guwaeder ; Rama Ramakumar

A COMPARISON OF PV POWER FORECASTS USING PVLIB-PYTHON 1127

William F. Holmgren ; Antonio T. Lorenzo ; Clifford Hansen

COMPARING THE TYPICAL GHI YEAR VS TYPICAL POWER YEAR 1132

Alex Kubiniec ; Adam Kankiewicz ; Alemu Tadesse

THE HOLY GRAIL OF RESOURCE ASSESSMENT: LOW COST GROUND-BASED MEASUREMENTS WITH GOOD ACCURACY 1134

Bill Marion ; Benjamin Smith

GLOBAL COMPARISON OF THE IMPACT OF TEMPERATURE AND PRECIPITABLE WATER ON CDTE AND SILICON SOLAR CELLS 1140

I. M. Peters ; L. Haohui ; T. Reindl ; T. Buonassisi

ESTIMATION OF MEAN MONTHLY GLOBAL SOLAR RADIATION USING MODEL BASED ON SUNSHINE HOURS FOR COLOMBIA 1143

Diego J. Rodríguez ; Johan Hernández ; Adolfo Jaramillo

IMPLEMENTATION OF SOLAR DIFFUSE CIE MODEL IN RAY TRACING PROGRAM FOR IRRADIANCE CALCULATIONS 1147

Liliana Ruiz Diaz ; Pierre-Alexandre Blanche ; Robert A. Norwood

INVESTIGATION OF CITY-LEVEL SITE-PAIR CORRELATIONS OF SOLAR VARIABILITY USING EMPIRICAL SATELLITE DATA 1151

Rhythm Singh ; Rangan Banerje

ULTRA-SHORT-TERM PHOTOVOLTAIC GENERATION FORECASTING MODEL BASED ON WEATHER CLUSTERING AND MARKOV CHAIN 1158

Jin Tan ; Changhong Deng

DAILY SOLAR IRRADIANCE PROFILE CHARACTERIZATION AND RAMP RATE ANALYSIS AT DIFFERENT TIME RESOLUTIONS 1163

Spyros Theocharides ; Venizelos Venizelou ; George Makrides ; George E. Georghiou

COMPARISON AND ANALYSIS OF INSTRUMENTS MEASURING PLANE OF ARRAY IRRADIANCE FOR ONE-AXIS TRACKING PV SYSTEMS 1169

Frank Vignola ; Chun-Yu Chiu ; Josh Peterson ; Michael Dooraghi ; Manajit Sengupta

A SKY IMAGE ANALYSIS SYSTEM FOR SUB-MINUTE PV PREDICTION 1175

Rodrigo Verschac ; Li Li ; Shohei Nobuhara ; Takekazu Kato

LARGE AREA NANOSTRUCTURE INTEGRATION FOR BROAD-SPECTRUM, OMNIDIRECTIONAL ANTIREFLECTION IMPROVEMENTS ON POLYMER PACKAGED, MECHANICALLY FLEXIBLE, EPITAXIAL LIFT-OFF III-V SOLAR CELLS ... 1181

Gabriel Cossio ; Jihwan Lee ; Gautham Ragunathan ; Andre Wibowo ; Sudersena Rao Tatavarti ; Kimberly Sablon ; Edward T. Yu

DEVELOPMENT OF BACK SURFACE TEXTURE FOR LIGHT MANAGEMENT IN EPITAXIAL LIFT OFF (ELO) QUANTUM DOT SOLAR CELLS ... 1184

Brittany L. Smith ; George T. Nelson ; Yushuai Dai ; Michael A. Slocum ; Andre Wibowo ; Rao Tatavarti ; Seth M. Hubbard

ENABLING HIGH-EFFICIENCY INAS/GAAS QUANTUM DOT SOLAR CELLS BY EPITAXIAL LIFT-OFF AND LIGHT MANAGEMENT ... 1189

F. Cappelluti ; A. P. Cédola ; A. Khalili ; Farid Elsehrawy ; G. Bauhuis ; P. Mulder ; J. Schermer ; G. Bissels ; T. Aho ; T. Niemi ; M. Guina ; D. Kim ; J. Wu ; H. Liu

CHARACTERIZATION OF ARSENIC DOPED CDTE LAYERS AND SOLAR CELLS ... 1193

Sachit Grover ; Xiaoping Li ; Wei Zhang ; Ming Yu ; Gang Xiong ; Markus Gloeckler ; Roger Malik

ENHANCING P-TYPE DOPING IN POLYCRYSTALLINE CDTE FILMS ... 1196

Brian Mccandless ; Wayne Buchanan ; Gowri Sriramagiri ; Christopher Thompson ; Joel Duenow ; David Albin ; Soren Jensen ; John Moseley ; M. Al-Jassim ; Wyatt K. Metzger

SPECTRAL AND CONCENTRATION SENSITIVITY OF MULTIJUNCTION SOLAR CELLS AT HIGH TEMPERATURE ... 1201

Daniel J. Friedman ; Myles A. Steiner ; Emmett E. Perl ; John Simon

ON THE USE OF TRANSPARENT CONDUCTIVE OXIDES IN HIGH CONCENTRATOR III-V MULTIJUNCTION SOLAR CELLS ... 1204

Ignacio Rey-Stolle ; Yeonbae Lee ; Iván Garcia ; Luis Cifuentes ; Kin Man Yu ; Carlos Algora ; Wladek Walukiewicz

COMPONENT INTEGRATION EFFECTS IN 4-JUNCTION SOLAR CELLS WITH DILUTE NITRIDE 1EV SUBCELL ... 1210

I. García ; M. Ochoa ; I. Lombardero ; L. Cifuentes ; P. Caño ; M. Hinojosa ; I. Rey-Stolle ; C. Algora ; A. D. Johnson ; J. I. Davies ; K.H. Tan ; W.K. Loke ; S. Wicaksono ; S. F. Yoon

BISMUTH SURFACTANT-MEDIATED GROWTH OF GANASSB(BI) SOLAR CELLS ... 1215

Aymeric Maros ; Chaomin Zhang ; Jongwon Lee ; Hongfeng Wang ; Stephen Bremner ; Nikolai Faleev ; Christiana B. Honsberg ; Richard. R. King

AMORPHOUS SILICON CARBIDE FOR SILICON SURFACE PASSIVATION IN CARRIER-SELECTIVE CONTACT DEVICES ... 1220

Mathieu Boccard ; Christophe Ballif ; Zachary C. Holman

SURFACE PASSIVATION OF BORON DIFFUSED JUNCTIONS BY BOROSILICATE GLASS AND IN SITU GROWN SILICON DIOXIDE INTERFACE LAYER ... 1222

Valentin D. Mihailetchi ; Haifeng Chu ; Jan Lossen ; Radovan Kopecek

IMPROVED LIGHT INCOUPLING IN PLANAR SOLAR CELLS VIA IMPROVED TEXTURE MORPHOLOGY OF PDMS SCATTERING LAYER ... 1228

Salman Manzoor ; Zhengshan J. Yu ; Asad Ali ; Waqar Ali ; Zachary C. Holman

DAMAGE-FREE LASER ABLATION FOR EMITTER PATTERNING OF SILICON HETEROJUNCTION INTERDIGITATED BACK-CONTACT SOLAR CELLS ... 1233

Menglei Xu ; Twan Bearda ; Miha Filipic ; Hariharsudan Sivaramakrishnan Radhakrishnan ; Maarten Debucquoy ; Ivan Gordon ; Jozef Szlufcik ; Jef Poortmans

BENEFITS OF A THERMAL DRIFT DURING ATOMIC LAYER DEPOSITION OF AL2O3FOR C-SI PASSIVATION ... 1237

Fabien Lebreton ; Andy Zauner ; Pavel Bulkin ; Francois Silva ; Sergej Filonovich ; Pere Roca I Cabarrocas

GROWTH DIFFERENCE OF AMORPHOUS SILICON BETWEEN PLASMA ENHANCED AND CATALYTIC CVD BASED ON SILICON HETEROJUNCTION SOLAR CELLS ... 1241

Liping Zhang ; Renfang Chen ; Zhuopeng Wu ; Chenguang Sun ; Fanying Meng ; Zhengxin Liu

DEVELOPING AN UNDERSTANDING-BASED SELECTION OF HYBRID-PEROVSKITE COMPOUNDS AND THE CU-IN HYBRID-PEROVSKITE (CIHP) FAMILY ... 1245

Alex Zunger ; G. Dalpian ; Qihang Liu ; L.B Abdalla ; L.L. Kazmerski

EFFECTS OF ELECTRON AND PROTON RADIATION ON PEROVSKITE SOLAR CELLS FOR SPACE SOLAR POWER APPLICATION ... 1248

Jing-Shun Huang ; Michael D. Kelzenberg ; Pilar Espinet-González ; Colin Mann ; Don Walker ; Ali Naqavi ; Nina Vaidya ; Emily Warmann ; Harry A. Atwater

TOWARDS PEROVSKITE SILICON TANDEM SOLAR CELLS WITH OPTIMIZED OPTICAL PROPERTIES ... 1253

Jan Christoph Goldschmidt ; Alexander J. Bett ; Patricia S.C. Schulze ; Nico Tucher ; Martin Bivour ; Markus Kohlstädt ; Seunghun Lee ; Simone Mastroianni ; Laura Mundt ; Markus Mundus ; Paul Ndione ; Karl Wienands ; Kristina Winkler ; Uli Würfel ; Martin Hermle ; Stefan W. Glunz

FIRST-PRINCIPLES DENSITY FUNCTIONAL THEORY CALCULATION OF METAL-SUBSTITUTED LEAD HALIDE PEROVSKITE 1256
Ji-Sang Park ; Matthew D. Sampson ; Alex B.F. Martinson ; Maria K.Y. Chan

ESTIMATING THE EFFECTS OF MODULE AREA ON THIN-FILM PHOTOVOLTAIC SYSTEM COSTS 1259
Kelsey A. W. Horowitz ; Ran Fu ; Xingshu Sun ; Tim Silverman ; Michael Woodhouse ; Muhammad A. Alam

COST ANALYSIS OF TANDEM MODULES 1264
Sarah E. Sofia ; Jonathan Mailoal ; Dirk Weiss ; Tonio Buonassisi ; Ian Marius Peters

CAUSE OF CURRENT-COLLECTION FAILURE OBSERVED INISC-REDUCTION PHASE OF PV CELLS AND MODULES EXPOSED TO ACETIC ACID 1268
Tadanori Tanahashi ; Norihiko Sakamoto ; Hajime Shibata ; Atsushi Masuda

COMPARISON OF PV MODULE PERFORMANCE BEFORE AND AFTER 11, 20, AND 25.5 YEARS OF FIELD EXPOSURE 1271
Jacob Rada ; Charles Chamberlin ; Peter Lehman ; Arne Jacobson

MARRYING QUALITY ASSURANCE WITH DESIGN ENGINEERING – A WINNING PARTNERSHIP! BUT, A CULTURAL DIVIDE? 1275
Sarah Kurtz ; Govind Ramu ; Robert Cornell ; Sumanth Lokanath ; Edward Hsi ; Tony Sample ; Masaaki Yamamichi ; George Kelly ; Ted Spooner ; Jonathan Previtali ; John Wohlgemuth

UPDATED EVALUATION OF SHOCK HAZARDS TO FIREFIGHTERS WORKING IN PROXIMITY OF PV SYSTEMS 1280
Olga Lavrova ; Jimmy E. Quiroz ; Jack Flicker ; Renee Gooding

GROWTH AND OPTIMIZATION OF GAINP/INP NANOWIRE TUNNEL DIODE 1286
Xulu Zeng ; Gaute Otnes ; Magnus Heurlin ; Magnus T Borgström

CATHODOLUMINESCENCE MAPPING FOR THE DETERMINATION OF N-TYPE DOPING IN SINGLE GAAS NANOWIRES 1289
Hung-Ling Chen ; Chalermchai Himwas ; Andrea Scaccabarozzi ; Pierre Rale ; Fabrice Oehler ; Aristide Lemaître ; Laurent Lombez ; Jean-François Guillemoles ; Maria Tchemycheva ; Jean-Christophe Harmand ; Andrea Cattoni ; Stéphane Collin

OPTICAL OPTIMIZATION OF PASSIVATED GAAS NANOWIRE SOLAR CELLS 1294
Kyle W. Robertson ; Ray R. Lapierre ; Jacob J. Krich

HIGH EFFICIENCY GAN NANOWIRE/SI PHOTOCATHODE FOR PHOTOELECTROCHEMICAL WATER SPLITTING 1299
Srinivas Vanka ; Sheng Chu ; Yichen Wang ; Ishiang Shih ; Hong Guo ; Zetian Mi

ANALYTIC DESCRIPTION OF THE IMPACT OF GRAIN BOUNDARIES ON VOC 1303
Paul Haney ; Benoit Gaury

ROLE OF TELLURIUM BUFFER LAYER ON CDTE SOLAR CELLS' ABSORBER/BACK-CONTACT INTERFACE 1308
Tao Song ; James R. Sites

SIMULTANEOUS EXAMINATION OF GRAIN-BOUNDARY POTENTIAL, RECOMBINATION, AND PHOTOCURRENT IN CDTE SOLAR CELLS USING DIVERSE NANOMETER-SCALE IMAGING 1312
C.S. Jiang ; H.R. Moutinho ; J. Moseley ; A. Kanevce ; J.N. Duenow ; E. Colegrove ; C. Xiao ; W.K. Metzger ; M.M. Al-Jassim

NANOPARTICLE/METAL REAR REFLECTORS FOR LOW- AND HIGH-TEMPERATURE SILICON SOLAR CELLS 1317
Syeda Qudsia ; Farah Qazi ; Mehwish Azher Javed ; Mathieu Boccard ; Zhengshan J. Yu ; Peter Firth ; Jonathan Bryan ; Zachary C. Holman

ABSORPTION IN EACH LAYER OF A SILICON HETEROJUNCTION SOLAR CELL 1322
Keith R. Mcintosh ; Malcolm D. Abbott ; Benjamin A. Sudbury ; Salman Manzoor ; Zhengshan J. Yu ; Mehdi Leilaeioun ; Jiatiwei Shi ; Zachary C. Holman

INVESTIGATIONS ON PLASMONIC COLOR TUNING COATING ON C-SI SOLAR CELLS 1329
Gerhard Peharz ; Wolfgang Waldhauser ; Christine Prietl ; Bettina Großschädl ; Martin C. Schubert ; Bernhard Michl

INVESTIGATION OF INTERFACE AND BULK LOCALIZED STATES IN A-SI:H SOLAR CELLS 1333
Adrien Bidiville ; Takuya Matsui ; Hitoshi Sai ; Koji Matsubara

EXPERIMENTAL AND THEORETICAL STUDY OF THE INFRARED EMISSIVITY OF CRYSTALLINE SILICON SOLAR CELLS 1339
Alberto Riverola ; Alexander Mellor ; Diego Alonso Alvarez ; Lourdes Ferre Llin ; Ilaria Guarracino ; Christos N. Markides ; Douglas Paul ; Daniel Chemisana ; Ned Ekins-Daukes

HIGH PERFORMANCE MOLECULAR DONORS FOR ORGANIC SOLAR CELLS, MATERIALS DESIGN AND DEVICE OPTIMIZATION 1342
Paul Geraghty ; Haotian Wang ; Calvin Lee ; Jegadesan Subbiah ; David Jones

ADVANCED OPTICAL MODELLING OF MICRO-TEXTURED SOLUTION-PROCESSED
SOLAR CELLS WITH CONSIDERATION OF SMALL-AREA EFFECTS............1346
*Benjamin Lipovšek ; Marko Jošt ; Andrej Campa ; Fei Gu ; Christoph J. Brabec ; Karen Forberich ; Janez Krc ;
Marko Tonic*

IDENTIFICATION OF DEGRADATION PATHWAYS OF ORGANIC SOLAR CELLS USING
INFRARED SPECTROSCOPY 1350
S. Shah ; R Biswas ; T. Koschny ; V L Dalal

A DEVICE-INDEPENDENT SCREENING TECHNIQUE FOR RAPIDLY IDENTIFYING NEXT
GENERATION OPV MATERIALS.........1354
Bryon W. Larson ; Andrew J. Ferguson ; Bertrand J. Tremolet De Villers ; Ross E. Larsen

NOVEL ANTHANTHRONE AND ANTHANTHRENE CO-POLYMERS AS P-TYPE
CONJUGATED SEMICONDUCTORS FOR ORGANIC PHOTOVOLTAICS.........1360
*Suru Vivian John ; Patrick Denk ; Christoph Ulbricht ; Herwig Heilbrunner ; Jean-Benoit Giguère ; Antoine
Lafleur-Lambert ; Jean-Francois Morin ; Emmanuel Iwuoha ; Daniel Ayuk Mbi Egbe*

REDUCING UV INDUCED DEGRADATION LOSSES OF SOLAR MODULES WITH C-SI
SOLAR CELLS FEATURING DIELECTRIC PASSIVATION LAYERS.........1366
*Robert Witteck ; Henning Schulte-Huxel ; Boris Veith-Wolf ; Malte Ruben Vogt ; Fabian Kiefer ; Marc Kontges ;
Robby Peibst ; Rolf Brendel*

LARGE-AREA JUNCTION DAMAGE IN POTENTIAL-INDUCED DEGRADATION OF C-SI
SOLAR MODULES 1371
*Chuanxiao Xiao ; Chun-Sheng Jiang ; Steve Johnston ; Steve P. Harvey ; Peter Hacke ; Brian Gorman ; Mowafak
Al-Jassim*

SEARCH FOR MICROSTRUCTURAL DEFECTS AS NUCLEI FOR PID-SHUNTS IN SILICON
SOLAR CELLS 1376
Volker Naumann ; Otwin Breitenstein ; Jan Bauer ; Christian Hagendorf

INVESTIGATING PID SHUNTING IN POLYCRYSTALLINE SILICON MODULES VIA MULTI-
SCALE, MULTI-TECHNIQUE CHARACTERIZATION 1381
*Steven P. Harvey ; John Moseley ; Adam Stokes ; Andrew Norman ; Brian Gorman ; Peter Hacke ; Steve Johnston
; Mowafak Al-Jassim*

POTENTIAL-INDUCED DEGRADATION OF A SI NITRIDE/CRYSTALLINE SI INTERFACE
OBSERVED THROUGH MINORITY CARRIER LIFETIME MEASUREMENT.........1385
Naoyuki Nishikawa ; Seira Yamaguchi ; Keisuke Ohdaira

FIELD INSPECTION OF PV MODULES: QUANTIFICATION OF EVA BROWNING LEVEL
USING AN IMAGE PROCESSING TOOL.........1389
Sushanth Gudla ; Govindasamy Tamizhmani

PREVENTING POTENTIAL-INDUCED DEGRADATION IN CRYSTALLINE SILICON PV
MODULES: RELATIONSHIP BETWEEN DEGRADATION AND BILL OF MATERIAL 1395
Alessandro Virtuani ; Eleonora Annigoni ; Christophe Ballif

IDENTIFYING REVERSE-BIAS BREAKDOWN SITES IN CUINXGA(1-X)SE2 1400
*Steve Johnston ; Elizabeth Palmiotti ; Andreas Gerber ; Harvey Guthrey ; Lorelle Mansfield ; Timothy J.
Silverman ; Mowafak Al-Jassim ; Angus Rockett*

HIMAWARI-8 ENABLED REAL-TIME DISTRIBUTED PV SIMULATIONS FOR
DISTRIBUTION NETWORKS 1405
Nicholas A. Engerer ; Jamie M. Bright ; Sven Killinger

REDUCED MEASUREMENT UNCERTAINTY IN PV MODULE BATCH TESTING 1411
Blagovest Mihaylov ; Bengt Jaeckel ; Juergen Arp ; Ralph Gottschalg

CLOUD MOTION IDENTIFICATION ALGORITHMS BASED ON ALL-SKY IMAGES TO
SUPPORT SOLAR IRRADIANCE FORECAST.........1415
Lydie Magnone ; Fabrizio Sossan ; Enrica Scolari ; Mario Paolone

AUTOMATIC DETECTION OF INACTIVE SOLAR CELL CRACKS IN
ELECTROLUMINESCENCE IMAGES 1421
Sergiu Spataru ; Peter Hacke ; Dezso Sera

APPLYING SPATIAL DOWNSCALING AND SMART PERSISTENCE TO PROVIDE AN
IMPROVED SOLAR FORECAST TO REDUCE COMMERCIAL DEMAND CHARGES.........1427
Alex Kubiniec ; Ted Belanger ; Adam Kankiewicz ; Skip Dise ; Nate Glasgow ; Alemu Tadesse

THERMAL CHARACTERISTICS OF PID-AFFECTED MONOCRYSTALLINE SILICON SOLAR
MODULES UNDER ILLUMINATED AND DARK CONDITIONS.........1430
*Pan Zhao ; Shuwen Guo ; He Wang ; Hong Yang ; Dengyuan Song ; Shiyu Sang ; Bojie Su ; Xue Zhang ; Yunxue
Cao ; Hui Zhao*

TARGETED EVALUATION OF UTILITY-SCALE AND DISTRIBUTED SOLAR FORECASTING 1435
Matthew Lave ; Robert J. Broderick ; Laurie Burnham

RECORD EFFICIENCIES FOR SELENIUM PHOTOVOLTAICS AND APPLICATION TO INDOOR SOLAR CELLS 1441

Douglas M. Bishop ; Teodor Todorov ; Yun Seog Lee ; Oki Gunawan ; Richard Haight

CLOSE-SPACED SUBLIMATION FOR SB2SE3SOLAR CELLS 1445

Laurie J. Phillips ; Peter Yates ; Oliver S. Hutter ; Tom Baines ; Leon Bowen ; Ken Durose ; Jonathan D. Major

FABRICATION OF COPPER ARSENIC SULFIDE THIN FILMS FROM NANOPARTICLES FOR APPLICATION IN SOLAR CELLS 1449

Scott A. Mcclary ; Joseph Andler ; Carol A. Handwerker ; Rakesh Agrawal

ORIENTATION CONTROLLED GE THIN FILMS ON GLASS BY AL-INDUCED CRYSTALLIZATION 1452

Kaveh Shervin ; Khim Kharel ; Alexandre Freundlich

IN-LINE POTASSIUM FLUORIDE TREATMENT OF CIGS ABSORBERS DEPOSITED ON FLEXIBLE SUBSTRATES IN A PRODUCTION-SCALE PROCESS TOOL 1455

Ryan Kaczynski ; Jinwoo Lee ; Jane Van Alsburg ; Baosheng Sang ; Urs Schoop ; Jeffrey Britt

LIGHT-SOAK AND DARK-HEAT INDUCED CHANGES IN CU(IN, GA)SE2 SOLAR CELLS: A MACROSCOPIC TO MICROSCOPIC STUDY 1459

Rouin Farshchi ; Benjamin Hickey ; Dmitry Poplavskyy

A NEW MODEL TO DETERMINE INSTALLED SYSTEM COST AND LCOE FOR ARPA-E'S MOSAIC MICRO-CONCENTRATOR PV PROGRAM 1463

Ran Fu ; Kelsey A.W. Horowitz ; Daniel W. Cunningham ; James Zahler

FIXED-TILT 660 × CONCENTRATING PHOTOVOLTAIC SYSTEM WITH 30% EFFICIENCY 1469

Alex J. Grede ; Jared S. Price ; Baomin Wang ; Michael V. Lipski ; Brent Fisher ; Kyu-Tae Lee ; Junwen He ; Gregory S. Brulo ; Xiaokun Ma ; Scott Burroughs ; Christopher D. Rahn ; Ralph G. Nuzzo ; John A. Rogers ; Noel C. Giebink

WAFER INTEGRATED MICRO-SCALE CONCENTRATING PHOTOVOLTAICS 1473

Tian Gu ; Duanhui Li ; Lan Li ; Bradley Jared ; Gordon Keeler ; Bill Miller ; William Sweatt ; Scott Paap ; Michael Saavedra ; Ujjwal Das ; Steve Hegedus ; Anna Tanke-Pedretti ; Juejun Hu

TOWARD STATIONARY CONCENTRATOR PHOTOVOLTAIC PANELS 1476

Peter Kozodoy ; Christopher Gladden ; Michael Pavilonis ; Tobias Wheeler ; Christopher Rhodes ; Chadwick Casper ; Kevin Schneider

CPV TECHNOLOGIES NOT RELYING ON PERFECTION OF TRACKERS 1479

Kenji Araki ; Yasuyuki Ota ; Kan-Hua Lee ; Kensuke Nishioka ; Masafumi Yamaguchi

THE GETTERING EFFECT OF DIELECTRIC FILMS FOR SILICON SOLAR CELLS 1485

A. Y. Liu ; C. Sun ; V. P. Markevich ; A. R. Peaker ; J. D. Murphy ; D. Macdonald

TABULA RASA: OXYGEN PRECIPITATE DISSOLUTION THOUGH RAPID HIGH TEMPERATURE PROCESSING IN SILICON 1491

Erin E. Looney ; Hannu S. Laine ; Mallory A. Jensen ; Amanda Youssef ; Vincenzo Lasalvia ; Paul Stradins ; Tonio Buonassisi

TOWARD EFFECTIVE GETTERING IN BORON-IMPLANTED SILICON SOLAR CELLS 1494

Hannu S. Laine ; Ville Vähänissi ; Zhengjun Liu ; Ernesto Magaña ; Ashley E. Morishige ; Jan Krügener ; Kristian Salo ; Hele Savin ; Barry Lai ; David P. Fenning

IMPACT OF THE INITIAL GROWTH INTERFACE ON THE GRAIN STRUCTURE IN HPMC-SI INGOT 1498

Giri Wahyu Alam ; Etienne Pihan ; Benoit Marie ; Nathalie Mangelinck-Noël

EFFECT OF CARBON CONCENTRATION AND GROWTH CONDITIONS ON OXYGEN PRECIPITATION BEHAVIOR IN N-TYPE CZ-SI 1504

Takuto Kojima ; Ryota Suzuki ; Kosuke Kinoshita ; Kyotaro Nakamura ; Atsushi Ogura ; Yoshio Ohshita ; Isao Masada ; Shoji Tachibana

NANO-IMAGING OF PERFORMANCE IN PHOTOVOLTAICS 1508

Elizabeth M. Tennyson ; Marina S. Leite

IMPLICATIONS OF CONDUCTIVE GRAIN BOUNDARIES IN CHLORINE-TREATED CDTE SOLAR CELLS 1511

Mohit Tuteja ; Vasilios Palekis ; Allen Hall ; Scott Maclaren ; Chris S. Ferekides ; Angus A. Rockett

IMAGING THE MULTI-TEMPORAL PHOTO-CARRIER DYNAMICS AT THE NANOMETER SCALE IN ORGANIC AND INORGANIC SOLAR CELLS 1516

Pablo A. Fernández Garrillo ; Lukasz Borowik ; Florent Caffy ; Renaud Demadrille ; Benjamin Grévin

NANOSCALE TOMOGRAPHIC CHARGE TRANSPORT IN POLYCRYSTALLINE CHALCOGENIDE ABSORBERS: CDTE VERSUS CIGS 1522

Justin L. Luria ; Andrew Moore ; Sun Yu ; Mark Aindow ; Bryan D. Huey

IMPROVING THE PV MODULE SINGLE-DIODE MODEL ACCURACY WITH TEMPERATURE DEPENDENCE OF THE SERIES RESISTANCE 1526

Kyumin Lee

CELL-TO-MODULE (CTM) ANALYSIS FOR PHOTOVOLTAIC MODULES WITH SHINGLED SOLAR CELLS ... 1531

Max Mittag ; Tobias Zech ; Martin Wiese ; David Blasi ; Matthieu Ebert ; Harry Wirth

A PRACTICAL IRRADIANCE MODEL FOR BIFACIAL PV MODULES ... 1537

Bill Marion ; Sara Macalpine ; Chris Deline ; Amir Asgharzadeh ; Fatima Toor ; Daniel Riley ; Joshua Stein ; Clifford Hansen

A DETAILED MODEL OF REAR-SIDE IRRADIANCE FOR BIFACIAL PV MODULES ... 1543

Clifford W. Hansen ; Renee Gooding ; Nathan Guay ; Daniel M. Riley ; Johnson Kallickal ; Donald Ellibee ; Amir Asgharzadeh ; Bill Marion ; Fatima Toor ; Joshua S. Stein

VIEW FACTOR MODEL AND VALIDATION FOR BIFACIAL PV AND DIFFUSE SHADE ON SINGLE-AXIS TRACKERS ... 1549

Marc Abou Anoma ; David Jacob ; Ben C. Bourne ; Jonathan A. Scholl ; Daniel M. Riley ; Clifford W. Hansen

A FAST QUASI-STATIC TIME SERIES (QSTS) SIMULATION METHOD FOR PV IMPACT STUDIES USING VOLTAGE SENSITIVITIES OF CONTROLLABLE ELEMENTS ... 1555

Xiaochen Zhangl ; Santiago Grijalva ; Matthew J. Reno ; Jeremiah Deboever ; Robert J. Broderick

FAST DETERMINATION OF DISTRIBUTION-CONNECTED PV IMPACTS USING A VARIABLE-TIME-STEP QUASI-STATIC TIME-SERIES APPROACH ... 1561

Barry Mather

SCALABILITY OF THE VECTOR QUANTIZATION APPROACH FOR FAST QSTS SIMULATION ... 1567

Jeremiah Deboever ; Santiago Grijalva ; Matthew J. Reno ; Xiaochen Zhang ; Robert J. Broderick

MACHINE LEARNING FOR RAPID QSTS SIMULATIONS USING NEURAL NETWORKS ... 1573

Matthew J. Reno ; Robert J. Broderick ; Logan Blakely

ALGORITHMIC ASPECTS OF A COMMERCIAL-GRADE DISTRIBUTION SYSTEM LOAD FLOW ENGINE ... 1579

Francis Therrien ; Marc Belletête ; Jean-Sébastien Lacroix ; Matthew J. Reno

RESONANT AND NON-RESONANT DIELECTRIC COATINGS FOR HIGH EFFICIENCY SOLAR CELLS ... 1585

Dongheon Ha ; Chen Gong ; Marina S. Leite ; Jeremy N. Munday

ENHANCED LIGHT TRAPPING IN THIN SILICON SOLAR CELLS USING EFFECTIVELY TRANSPARENT CONTACTS (ETCS) ... 1589

Rebecca Saive ; André Augusto ; Stuart G. Bowden ; Harry A. Atwater

ENHANCED POWER CONVERSION EFFICIENCY IN SINGLE NANOWIRE DEVICES THROUGH SYMMETRY BREAKING DESIGN ... 1594

Jian Zhou ; Yonggang Wu ; Zihuan Xia ; Xuefei Qin ; Zongyi Zhang

CDSE(TE)/CDS/CDSE RODS VS. CDTE/CDS/CDSE SPHERES: MORPHOLOGY-DEPENDENT CARRIER DYNAMICS FOR PHOTON UPCONVERSION ... 1598

Eric Y. Chen ; Zhuohui Li ; Christopher C. Milleville ; Kyle R. Lennon ; Matthew F. Doty

DRIFT-DIFFUSION INGAN/GAN SOLAR CELL SIMULATOR WITH OPTICAL MANAGEMENT ... 1603

Y. Fang ; D. Guo ; A. Fischer ; E. Vadiee ; C. Zhang ; J. Williams ; S. M. Goodnick ; D. Vasileska

PERFORMANCE ENHANCEMENT OF A GAAS SOLAR CELL WITH COLLOIDAL QUANTUM DOTS EMBEDDED IN TRENCHES ... 1606

Chia-Jhe Shu ; Yu-Ming Huang ; Shun-Chieh Hsu ; Jinn-Kong Shu ; Jia-Lin Tsai ; Pei-Chen Yu ; Yung-Jr Hung ; Chien-Chung Lin

ENHANCED PHOTORESPONSE OF INN DEVICES USING INDIUM-TIN OXIDE NANORODS ... 1610

Lung-Hsing Hsu ; Yuh-Jen Cheng ; Peichen Yu ; Hao-Chung Kuo ; Chien-Chung Lin

PLASMONIC SILVER STRUCTURES FOR IMPROVED PEROVSKITE PHOTOVOLTAIC PERFORMANCE ... 1614

Arul Varman Kesavan ; Arun D Rao ; Praveen C Ramamurthy

QUANTUM CUTTING LUMINESCENT PMMA FILMS CONTAINING CE3+ - YB3+ CODOPED YAG PHOSPHOR FOR SI CONCENTRATOR SOLAR CELLS ... 1619

Lu Li ; Chaogang Lou ; Huihui Cao

NUMERICAL EVALUATION ON THE NANO-ROD ARRAY ON A N-SIDE-UP THIN-FILM GAAS SOLAR CELLS ... 1623

Po-Ching Wu ; Yan-Zhang Lin ; Shun-Chieh Hsu ; Chia-Jhe Hsu ; Chien-Chung Lin

DOWN SHIFTED CONVERSION FOR ENHANCED HIT SOLAR CELL EFFICIENCY ... 1627

Albert S. Lin ; Parag Parashar ; Wei-Ming Huang ; Yi-Wen Huang ; Ding-Rung Jian ; Ming-Hsuan Kao ; Shi-Wei Chen ; Chang-Hong Shen ; Jia-Min Shieh ; Tzu-Yu Chen ; Chien-Chung Lin ; Hao-Chung Kuo

THE PLANAR THERMOPHOTOVOLTAIC SELECTIVE NEARLY-PERFECT ABSORBERS/EMITTERS ... 1631

Parag Parashar ; Ding-Rung Jian ; Weiming Huang ; Vi-Wen Huang ; Albert Lin

HYBRID PEDOT:PSS SILICON SOLAR CELLS WITH PENCIL ROD STRUCTURES 1635
Ruei-Ying Wu ; Liang-Chian You ; Hsin-Fei Meng ; Chun-Chi Chen ; Peichen Yu

PL STUDY OF PHOSPHORUS-DOPED CDTE EVT FILMS 1638
Shamara Collins ; Imran Khan ; Vamsi Evani ; Chih An Hsu ; Vasilios Palekis ; Don Morel ; Chris Ferekides

CHARACTERIZATION OF SINGLE-SOURCE DEPOSITED CLOSE-SPACE SUBLIMATION CDTEXSE1-XTHIN FILM SOLAR CELLS 1643
Corey R. Grice ; Jian Li ; Yanfa Yan

THE INFLUENCE OF THE CU-RICH/CU-POOR SEQUENCE ON THE PROPERTIES OF CU(IN, GA)SE2 FILMS DEPOSITED BY IN-LINE CO-EVAPORATION PROCESS 1648
He Wang ; Fang Fang Liu ; Yi Tong Yang ; Li You Yao ; Peng Gao ; Zhi Bin Xiao ; Qiang Sun

DETERMINATION AND MODELING OF INJECTION DEPENDENT SERIES RESISTANCE IN CIGS SOLAR CELLS 1651
Vito Huhn ; Bart E. Pieters ; Andreas Gerber ; Yael Augarten ; Uwe Rau

LARGE GRAIN GROWTH IN CU2ZNSNS4 THIN FILMS IN THE ABSENCE OF NA USING RAPID THERMAL ANNEALING 1656
J. L. Johnson ; A. Bhatia ; J. G. Bolke ; M. A. Scarpulla

CU2ZNSNS4THIN FILMS SYNTHESIZED BY COSPUTTERING AND RAPID THERMAL ANNEALING: EFFECTS OF COMPOSITION AND TEMPERATURE 1661
J.L. Johnson ; W.M. Hlaing Oo ; M. Karmarkar ; M.A. Scarpulla

EARTH-ABUNDANT CZTSSE THIN FILM SOLAR CELLS ON FLEXIBLE STAINLESS STEEL FOIL SUBSTRATES 1665
Hae-Sun Kim ; Woo-Lim Jeong ; Dong-Seon Lee

COMPARISON OF MGCL2AND CDCL2ACTIVATION TREATMENT FOR CDTE SOLAR CELLS: RECRYSTALLIZATION AND DEFECTS 1669
Daniele Menossi ; Elisa Artegiani ; Ivan Rimmaudo ; Alessia Le Donne ; Simona Binetti ; Juan Luis Pena ; Fabio Piccinelli ; Alessandro Romeo

CHARACTERIZATION OF CDTE PHOTOVOLTAIC DEVICES PASSIVATED USING HYDROGEN PLASMA 1674
Amit Munshi ; Piotr Kaminski ; Ali Abbas ; Shiva Tarun Chenna ; Sreeram Chandralal ; John Walls ; Walajabad Sampath

GROUP-V DOPING IMPACT ON CD-RICH CDTE SINGLE CRYSTALS GROWN BY TRAVELING-HEATER METHOD 1679
Akira Nagaoka ; Kenji Yoshino ; Yoshitaro Nose ; Darius Kuciauskas ; Michael A. Scarpulla

BAND-GAP ENGINEERING IN CU2ZNSN(S,SE)4SOLAR CELLS BY POST-SULPHURIZATION OF SELENIZED ABSORBER LAYERS 1682
Markus Neuwirth ; Elisabeth Seydel ; Heinz Kalt ; Michael Hetterich

IMPACT OF GA/III PROFILE ON VOLTAGE-DEPENDENT COLLECTION LOSSES IN CIGS SOLAR CELLS 1686
Dmitry Poplavskyy ; Jeff Bailey ; Rouin Farshchi ; David Spaulding

CL DIFFUSION IN CDTE SOLAR CELLS ACTIVATED BY GASEOUS CHCLF2ATMOSPHERE 1691
I. Rimmaudo ; R. Mis Fernandez ; V. Rejon ; A. Abbas ; F. Lisco ; J.M. Walls ; J.L. Peña

STABILITY OF CD1-XZNXTE ALLOYS UNDER CDTE PROCESSING CONDITIONS 1697
Yegor Samoilenko ; Colin A. Wolden

CIGSE ABSORBER PREPARATION: AN ALTERNATIVE TO H2SE 1701
O.S. Shinde ; E.J. Schenller ; S.R. Jadkar ; S.V Ghaisas ; N. Dhere

CHARGE CONTROLLED SEQUENTIAL ELECTRODEPOSITION FOR SYNTHESIS OF CU2ZNSNS4ON MO-COATED GLASS SUBSTRATE 1704
Ashish K. Singh ; Rajiv Dubey ; Manoj Neergat ; Kavaipatti R. Balasubramaniam

EFFECT OF DEPOSITED PRESSURE ON THE CDTE THIN FILMS BY CLOSED SPACE SUBLIMATION METHOD 1707
Yufeng Zhang ; Zhongming Du ; Xiangxin Liu

ANALYZING THE COST REDUCTION POTENTIAL OF III-V/SI HYBRID CONCENTRATOR PHOTOVOLTAIC SYSTEMS 1711
Kan-Hua Lee ; Kenji Araki ; Masafumi Yamaguchi

GENERALIZED NUMERICAL DESIGN OF AXIALLY-ASYMMETRICAL AND GRID-ARRANGED STATIC CPV ARRAY FOR MAXIMIZING ANNUAL ENERGY GENERATION 1714
Kenji Araki ; Kan-Hua Lee ; Masafumi Yamaguchi

SPECTRAL TRANSMITTANCE ANALYSIS OF LIQUIDS FOR HIGH CONCENTRATION III-V PHOTOVOLTAIC IMMERSION COOLING APPLICATIONS 1719
Xinyue Han ; Yongjie Guo

OPTICAL DESIGN FOR 2-TERMINAL III-V/SI SMAC MODULE 1724
Masaaki Baba ; Kikuo Makita ; Hidenori Mizuno ; Hidetaka Takato ; Takeyoshi Sugaya ; Noboru Yamada

DESIGN OF OPTICAL ELEMENTS FOR LOW PROFILE CPV PANEL WITH SUN TRACKING FOR ROOFTOP INSTALLATION 1728

Xinbing Liu ; Zhou Lu ; Riccardo Leto ; Carlton Brule ; Nanu Brates

MICRO CHIPLET PRINTER DEVELOPMENT FOR MOSAIC PROGRAM 1733

P.Y. Maeda ; Y. D. Wang ; S. Raychaudhuri ; J. Kalb ; D. K. Biegelsen ; R. Lujan ; Q. Wang ; Y. Wang ; J. Bert ; B. Rupp ; I. Matei ; L. Crawford ; A. Plochowietz ; E.M. Chow ; J.P. Lu ; V. Gupta

MICRO-OPTICAL TANDEM LUMINESCENT SOLAR CONCENTRATOR 1737

David R. Needell ; Zach Nett ; Ognjen Ilic ; Colton R. Bukowsky ; Junwen He ; Lu Xu ; Ralph G. Nuzzo ; Benjamin G. Lee ; John F. Geisz ; A. Paul Alivisatos ; Harry A. Atwater

INCREASE IN MAXIMUM POWER OF A-SI, C-SI AND GAAS.76P.24 SOLAR CELLS UNDER LOW CONCENTRATION 1741

Hiba Riaz ; Sabina Abdul Hadi ; Ammar Nayfeh

DESIGN AND EVALUATION OF PARTIAL CONCENTRATION III-V/SI MODULE WITH ENHANCED DIFFUSE SUNLIGHT TRANSMISSION 1743

Daisuke Sato ; Noboru Yamada ; Kan-Hua Lee ; Kenji Araki ; Masafumi Yamaguchi

CONTAMINATION CONTROL CHALLENGES ON SHJ SOLAR CELL PROCESSING 1747

G. Condorelli ; P. Rotoli ; A. Canino ; A. Battaglia ; W. Favre ; A. -S. Ozanne ; A. Moustafa ; A. Danel ; D. Muñoz ; P. -J. Ribeyron ; C. Gerardi

>23% SILICON HETEROJUNCTION SOLAR CELLS IN MEYER BURGER'S DEMO LINE: RESULTS OF PILOT PRODUCTION ON MASS PRODUCTION TOOLS 1752

J. Zhao ; M. König ; A. Wissen ; V. Breus ; D. Deckerl ; M. Fritzsche ; M. Schorch ; H. J. Nonnenmacher ; M. Leonhardt ; T. Große ; J. Hausmann ; A. Waltmger ; D. Landgraf ; S. Burkhardt ; H. Mehlich ; E. Vetter ; F. Schitthelm ; Y. Yao ; T. Söderström ; A. Richter ; D. Habermann ; S. Leu

EXPERIMENTAL AND SIMULATION STUDIES ON TIO2/SILICON HETEROJUNCTION DIODES 1755

Swasti Bhatia ; Neha Raorane ; Nimisha Sreekumar ; Pradeep R. Nair ; Aldrin Antony

A STUDY ON BLISTER FORMATION AND ELECTRICAL PROPERTIES UNDER VARIOUS ANNEALING CONDITION FOR TUNNELING OXIDE PASSIVATION LAYER 1758

Sungjin Choi ; Ka-Hyun Kim ; Min Gu Kang ; Jeong In Lee ; Donghwan Kim ; Hee-Eun Song

PROCESSING APPROACHES AND CHALLENGES OF INTERDIGITATED BACK CONTACT SI SOLAR CELLS 1761

Ujjwal Das ; Lei Zhang ; Steven Hegedus

FABRICATION OF CUI/A-SI:H/C-SI STRUCTURE FOR APPLICATION TO HOLE-SELECTIVE CONTACTS OF HETEROJUNCTION SI SOLAR CELLS 1765

Kazuhiro Gotoh ; Min Cui ; Nguyen Cong Thanh ; Koichi Koyama ; Isao Takahashi ; Yasuyoshi Kurokawa ; Hideki Matsumura ; Noritaka Usami

CHARACTERISTICS OF THIN CRYSTALLINE SILICON SOLAR CELLS WITH RIB STRUCTURE 1769

Yukimi Ichikawa ; Shuhei Yoshiba ; Masakazu Hirai ; Makoto Konagai

MEASUREMENT OF TIO2/P-SI SELECTIVE CONTACT PERFORMANCE USING A HETEROJUNCTION BIPOLAR TRANSISTOR WITH A SELECTIVE CONTACT EMITTER 1773

Janam Jhaveri ; Alexander Berg ; Sigurd Wagner ; James C. Sturm

EFFECT OF GROWTH AND POST-OXIDATION ANNEALING TEMPERATURE OF THERMALLY GROWN TUNNELING SIOX, ON THE IIMPLIED VOCOF PASSIVATED CONTACTS FOR C-SI BASED SOLAR CELLS 1777

Abhijit S. Kale ; William Nemeth ; Matthew Page ; Sumit Agarwal ; Paul Stradins

PARTIALLY CONTACTED SURFACES WITH CONTACT SIZE IN THE 1 µM RANGE FOR C-SI PERC SOLAR CELLS 1781

R. Khoury ; I. Martín ; G. López ; C. Jin ; J.M. López-González ; L. Zeyu ; P. Bulkin ; E.V. Johnson ; R. Alcubilla

ENTRANCE OF LOW COST FABRICATION OF BACK-CONTACT HETEROJUNCTION SOLAR CELLS BY USING PLASMA ION IMPLANTATION 1787

Koichi Koyama ; Keisuke Ohdaira ; Hideki Matsumura

TLM MEASUREMENTS VARYING THE INTRINSIC A-SI:H LAYER THICKNESS IN SILICON HETEROJUNCTION SOLAR CELLS 1790

Mehdi Leilaeioun ; William Weigand ; Pradyumna Muralidharan ; Mathieu Boccard ; Dragica Vasileska ; Stephen Goodnick ; Zachary Holman

SOLAR CELLS APPLICATION OF P-TYPE POLY-SI THIN FILM BY ALUMINUM INDUCED CRYSTALLIZATION 1794

Shota Masuda ; Kazuhiro Gotoh ; Isao Takahashi ; Kyotaro Nakamura ; Yoshio Ohshita ; Noritaka Usami

A SELF - CONSISTENTLY COUPLED DRIFT DIFFUSION AND MONTE CARLO SIMULATOR TO MODEL SILICON HETEROJUNCTION SOLAR CELLS 1797

Pradyumna Muralidharan ; Stuart Bowden ; Stephen M. Goodnick ; Dragica Vasileska

DOPANT PATTERNING BY PECVD AND MECHANICAL MASKING FOR PASSIVATED TUNNELING CONTACT IBC CELL ARCHITECTURES 1801
William Nemeth ; Vincenzo Lasalvia ; Benjamin G. Lee ; Abhijit Kale ; Paul Stradins

ALD ALUMINUM OXIDE AS A HOLE SELECTIVE TUNNELING CONTACT FOR CRYSTALLINE SILICON SOLAR CELLS 1804
Kortan Ögütman ; Kristopher O. Davis ; Winston V. Schoenfeld ; Michael Haslinger ; Sofie Robert ; Emanuele Cornagliotti ; Joachim John

SCREEN PRINTED, LARGE AREA BIFACIAL N-PERT CELLS WITH TUNNEL OXIDE PASSIVATED BACK CONTACT 1807
Young-Woo Ok ; Ajay D Upadhyaya ; Brian Rounsaville ; Ying-Yuan Huang ; Vijaykumar D Upadhyaya ; Ajeet Rohatgi

CORRELATION BETWEEN ELECTROLUMINESCENCE AND PHOTOCONVERSION EFFICIENCY IN A-SI:H/C-SI HETEROJUNCTION SOLAR CELLS 1811
A.V. Sachenko ; A.V. Bobyl ; V.N. Verbitskiy ; V.M. Vlasyuk ; D.M. Zhigunov ; V.P. Kostylyov ; I.O. Sokolovskyi ; E.I. Terukov ; P.A. Forsh ; M. Evstigneev

AN ISOTOPE STUDY OF HYDROGEN PASSIVATION OF POLY-SI/SIOXPASSIVATED CONTACTS FOR SI SOLAR CELLS 1817
Manuel Schnabel ; William Nemeth ; Bas W.H. Van De Loo ; Bart Macco ; Wilhelmus M.M. Kessels ; Paul Stradins ; David L. Young

ALLEVIATING HYDROGEN PLASMA DAMAGE TO AMORPHOUS/CRYSTALLINE SILICON INTERFACE PASSIVATION 1820
Jianwei Shi ; Zachary C. Holman

LARGE-AREA N-TYPE TOPCON CELLS WITH SCREEN-PRINTED CONTACT ON SELECTIVE BORON EMITTER FORMED BY WET CHEMICAL ETCH-BACK 1824
Yuguo Tao ; Felix Book ; Barbara Terheiden ; Viiaykumar Upadhvaya ; Keeya Madani ; Brian Rounsaville ; Eunhwan Cho ; Ajeet Rohatgi

HYDROGEN PLASMA POST-DEPOSITION TREATMENT FOR PASSIVATION OF A-SI/C-SI INTERFACE FOR HETEROJUNCTION SOLAR CELL BY CORRELATING OPTICAL EMISSION SPECTROSCOPY AND MINORITY CARRIER LIFETIME 1828
Anishkumar Soman ; Ugochukwu Nsofor ; Lei Zhang ; Ujjwal Das ; Tingyi Gu ; Steve Hegedus

MEASURING DIODE RESISTIVITY OF PASSIVATED CONTACTS 1832
San Theingi ; William Nemeth ; David L. Young ; Paul Stradins ; Benjamin G. Lee

ULTRA-THIN CRYSTALLINE SILICON SOLAR CELLS WITH NICKEL OXIDE INTERLAYER AS HOLE-SELECTIVE CONTACT 1835
Muyu Xue ; Raisul Islam ; Junyan Chen ; Zheng Lyu ; Yusi Chen ; Daniel Dewitt ; Albert Pleus ; Christian Tae ; Ching-Ying Lu ; Kai Zang ; Jieyang Jia ; Yijie Huo ; Ted Kamins ; Krishna Saraswat ; James Harris

CRYSTALLINE SI SOLAR CELLS WITH PASSIVATING, CARRIER-SELECTIVE NICKEL OXIDE CONTACTS 1838
Woojun Yoon ; James Moore ; David Scheiman ; Eunhwan Cho ; Young-Woo Ok ; Nicole Kotulak ; Phillip P. Jenkins ; Ajeet Rohatgi ; Robert J. Walters

GAP/SI HETEROJUNCTION SOLAR CELLS GROWN BY MOLECULAR BEAM EPITAXY 1841
Chaomin Zhang ; Ehsan Vadiee ; Richard R. King ; Christiana B. Honsberg

SPIN COATED NICKEL OXIDE AND VANADIUM OXIDE LAYERS ON SILICON FOR A CARRIER SELECTIVE CONTACT SOLAR CELL 1845
Jing Zhao ; Fa-Jun Ma, Jae-Yun ; Anita Ho-Baillie ; Stephen Bremner

QUANTIFICATION OF PV MODULE DISCOLORATION USING VISUAL IMAGE ANALYSIS 1850
Shashwata Chattopadhyay ; Chetan Singh Solanki ; Anil Kottantharayil ; K.L. Narasimhan ; Juzer Vasi ; Sai Tatapudi ; Govindasamy Tamizhmani

TEMPERATURE AND POWER STUDY OF ADHERED AND RACKED DOUBLE GLASS PHOTOVOLTAIC MODULES 1855
Volker Beutner ; Rubina Singh ; Cameron Stark

FIELD INSPECTION OF PV MODULES: QUANTITATIVE DETERMINATION OF PERFORMANCE LOSS DUE TO CELL CRACKS USING EL IMAGES 1858
Carlos A. Rodríguez Castañeda ; Shashwata Chattopadhyay ; Jaewon Oh ; Sai Tatapudi ; Govindasamy Tamizhmani ; Hailin Hu

SCALE UP DESIGNS FOR HAND-HELD LIGHT-WEIGHT TPV DC POWER SUPPLY 1863
L. M. Fraas ; J. E. Avery ; L. Minkin ; Hui She ; L. Ferguson

HIGH EFFICIENCY ANTI-REFLECTIVE COATING FOR PV MODULE GLASS 1869
Brennen M. Freiburger ; Corey S. Thompson ; Robert A. Fleming ; Douglas Hutchings ; Sergiu C. Pop

INVESTIGATION OF EFFICIENCY FOR PID-AFFECTED SOLAR MODULE AT NONSTANDARD TEST CONDITIONS 1873
Shuwen Guo ; Pan Zhao ; Weijing Huang ; Jipeng Chang ; He Wang ; Hong Yang ; Chengfeng Su ; Bojie Su ; Xue Zhang ; Yunxue Cao ; Hui Zhao

THERMAL UNIFORMITY MAPPING OF PV MODULES AND PLANTS..............................1877
Ashwini Pavgi ; Jaewon Oh ; Joseph Kuitche ; Sai Tatapudi ; Govindasamy Tamizhmani

CLIMATE-SPECIFIC THERMAL MODEL COEFFICIENTS FOR C-SI AND THIN-FILM PV MODULES..............................1883
Ashwini Pavgi ; Joseph Kuitche ; Jaewon Oh ; Govindasamy Tamizhmani

EFFECT OF THE THERMOPHYSICAL PROPERTIES OF A PHASE CHANGE MATERIAL ON THE ELECTRICAL OUTPUT OF A CONCENTRATED PHOTOVOLTAIC SYSTEM..............................1888
Jawad Sarwar ; Ahmed E. Abbas ; Konstantinos E. Kakosimos

PASSIVE COOLING OF PHOTOVOLTAICS WITH DESICCANTS..............................1893
Lin J. Simpson ; Jason Woods ; Nicolas Valderrama ; Alex Hill ; Nina Vincent ; Timothy Silverman

MODIFIED MAXIMUM POWER EXTRACTION TECHNIQUE FOR RAPIDLY CHANGING NUI AND DYNAMIC LOADS..............................1898
U Aswani ; S.P. Duttagupta ; T.I. Eldho ; B.V. Rao

REAL-TIME MONITORING OF PHOTO VOLTAIC RELIABILITY ONLY USING MAXIMUM POWER POINT - THE SUNS-VMP METHOD..............................1904
Xingshu Sun ; Haejun Chung ; Raghu Vamsi Krishna Chavali ; Peter Bermel ; Muhammad Ashraful Alam

PHOTOVOLTAIC MODULE DURABILITY AND RELIABILITY: ANALYSIS OF A 23-YEAR-OLD ARRAY OPERATING IN QUEBEC, CANADA..............................1908
Christopher Baldus-Jeursen ; Alexandre Côté ; Naveen Goswamy ; Tanya Deer ; Yves Poissant

ARE E-W TRACKERS A BETTER OPTION FOR FUTURE INVESTMENTS IN PV SECTOR-A DETAILED TECHNO-COMMERCIAL STUDY..............................1912
Rakesh Bohra ; Ramesh Rame Gowda ; Mani R. Krishnan

EXPERIMENTAL EVALUATION OF THE PERFORMANCE OF CRYSTALLINE SI PV MODULE DEGRADATION AFTER 15-YEARS OF FIELD EXPOSURE..............................1917
Denio A. Cassini ; Antonia Sônia A. C. Diniz ; Marcelo Machado Viana ; Michele C. C. De Oliveira ; F. C. Lins Vanessa De ; Roberto Zilles ; Lawrence L. Kazmerski

FIELD INVESTIGATIONS OF POTENTIAL-INDUCED DEGRADATION (PID) FOR CRYSTALLINE SILICON PV PANELS IN DIFFERENT CLIMATES..............................1922
Yifeng Chen ; Peter Hacke ; Yong Sheng Khoo ; Kaitlyn Vansant ; Zigang Wang ; Wei Luo ; Jing Chai ; Chris Deline ; Yan Wang ; Armin G. Aberle ; Pietro P. Altermatt ; Zhiqiang Feng ; Sarah Kurtz ; Pierre J. Verlinden

DETERMINING THE POWER RATE OF CHANGE OF 353 PLANT INVERTERS TIME-SERIES DATA ACROSS MULTIPLE CLIMATE ZONES, USING A MONTH-BY-MONTH DATA SCIENCE ANALYSIS..............................1927
Alan J. Curran ; Yang Hu ; Rojiar Haddadian ; Jennifer L. Braid ; David Meakin ; Timothy J. Peshek ; Roger H. French

PHOTOVOLTAIC ARRAY DIFFERENTIAL BACKSIDE EXPOSURE CONDITIONS: BACKSHEET DEGRADATION AND SITE DESIGN..............................1933
Andrew Fairbrother ; Julien Avenet ; Yadong Lyu ; Matthew Boyd ; Scott Julien ; Kai-Tak Wan ; Liang Ji ; Kenneth Boyce ; Sebastien Merzlic ; Amy Lefebvre ; Greg O'Brien ; Yu Wang ; Laura Bruckman ; Roger French ; Michael Kempe ; Brian Dougherty ; Xiaohong Gu

STUDY ON RANDOM FAILURE OF CRYSTALLINE SILICON SOLAR MODULES IN THE FIELD..............................1937
Xuefang Jiang ; Fumei Wang ; Ao Wang ; Hong Yang ; He Wang ; Jie Ding ; Junjun Zhang ; Jingsheng Huang

POTENTIAL INDUCED DEGRADATION (PID) POWER LOSS CORRELATION TO LEAKAGE AND REVERSE BIAS CURRENTS..............................1941
Michalis Florides ; Georgios Konstantinou ; Venizelos Venizelou ; George Makrides ; George E. Georghiou

PERFORMANCE STUDY OF VARIOUS PV MODULE TECHNOLOGIES IN DESERT CONDITIONS..............................1946
Jim J John ; Ammar Elnosh ; Anwar Almheiri ; Wadhah Alzahmi ; Marco Stefancich ; Pedro Banda

HIGH-SPEED MEASUREMENTS OF GENERATED POWER AND ITS RELATIONSHIP TO WEATHER OBSERVATIONS AT YOSHINOGARI MEGA SOLAR POWER PLANT..............................1950
Makoto Kasu ; Shigeomi Hara ; Takumi Uematsu

IMPACT OF MISSING DATA ON THE ESTIMATION OF PHOTOVOLTAIC SYSTEM DEGRADATION RATE..............................1954
Andreas Livera ; Alexander Phinikarides ; George Makrides ; George E. Georghiou

FIELD DEGRADATION AND FAILURES OF AGED CRYSTALLINE SILICON PV MODULES IN MEXICO..............................1959
D. Martínez Escobar ; P. A. Sánchez-Pérez ; Rocío De La Luz Santos Magdaleno ; José Ortega Cruz ; Sai Tatapudi ; Aarón Sánchez Juárez ; Govindasamy Tamizhmani

RAPID SHUTDOWN WITH PANEL LEVEL ELECTRONICS-A SUITABLE SAFETY MEASURE?..............................1965
Adam Cordova ; Christopher Merz ; Gerd Bettenwort ; Markus Hopf ; Hannes Knopf ; Joachim Laschinski

INVESTIGATING A NEW OPERATING POINT FOR PV PANELS SEEKING MAXIMUM LIFE SPAN......1968

Bechara Nehme ; Nacer K. M'sirdi ; Tilda Akiki

POWER GENERATION EVALUATION OF LARGE-SCALE PHOTOVOLTAIC SYSTEMS LOCATED ON INCLINED PLANE......1973

Naotaka Oka ; Yasuhito Takahashi ; Koji Fujiwara ; Kazuyuki Hidaka ; Hiroshi Morita

INVESTIGATING THE IMPACT OF SOLAR CELLS PARTIAL SHADING ON PHOTOVOLTAIC MODULES BY THERMOGRAPHY......1979

David Pera ; José A. Silva ; Sara Costa ; João M. Serra

ANNUAL DEGRADATION RATE AND ITS LINEARITY ANALYSIS USING METERED KWH DATA......1984

Christopher Raupp ; Govindasamy Tamizhmani

ELECTRICAL PERFORMANCE ANALYSIS OF A 27 KW GRID-CONNECTED PV SYSTEM WITH SOILING AND SHADING IN MORELOS MEXICO......1990

P. A. Sánchez-Pérez ; D. Martínez Escobar ; E. O. Ángel Ruiz ; R. Santos Magdaleno ; José Ortega Cruz ; A. Sánchez Juárez

MODIFIED STC CORRECTION PROCEDURE FOR ASSESSING PV MODULE DEGRADATION IN FIELD SURVEYS......1995

Hemant K. Singh ; R. Dubey ; S. Zachariah ; K. L. Narasimhan ; B. M. Arora ; A. Kottantharayil ; J. Vasi

DEGRADATION MODELS OF PHOTOVOLTAIC MODULE BACKSHEETS EXPOSED TO DIVERSE REAL WORLD CONDITION......2000

Yu Wang ; Sebastien Merzlic ; Andrew Fairbrother ; Scott Julien ; Lucas Fridman ; Camille Loyer ; Amy L. Lefebvre ; Gregory O'Brien ; Xiaohong Gu ; Liang Ji ; Ken Boyce ; Michael Kempe ; Kai-Tak Wan ; Roger H. French ; Laura S. Bruckman

ADDRESSING HOTSPOTS IN THE PRODUCT ENVIRONMENTAL FOOTPRINT OF CDTE PHOTOVOLTAICS......2005

Parikhit Sinha ; Andreas Wade

PHOTOVOLTAIC SMART HOME SYSTEM - DUBAI CASE STUDY......2011

Ammar Natsheh ; Marwa Aljaziri ; Maitha Moosa ; Gharibah Essa ; Hassa Moosa

DIRECT DRIVE PHOTOVOLTAIC MILK CHILLING EXPERIENCE IN KENYA......2014

Robert Foster ; Brian Jensen ; Brian Dugdill ; Wendy Hadley ; Bruce Knight ; Abudul Faraj ; Johnson Kyalo Mwove

COST OPTIMIZATION OF DECOMMISSIONING AND RECYCLING CDTE PV POWER PLANTS......2019

V. Fthenakis ; Z. Zhang ; J. -K Choi

CHALLENGES FOR DECISION MAKERS WHEN FEED-IN TARIFFS OR NET METERING SCHEMES CHANGE TO INCENTIVES DEPENDENT ON A HIGH SHARE OF SELF-CONSUMED ELECTRICITY......2025

Mattias Gustafsson

PROCEDURES TO MAKE PROJECTS ABOUT RENEWABLE ENERGY GENERATION CONNECTED TO THE GRID IN COLOMBIA......2031

J. A. Hernandez ; C. A. Arredondo ; D. J. Rodriguez

A CRITICAL ANALYSIS ON THE THIN CRYSTALLINE SILICON PV MODULE OF THE LIGHTWEIGHT PV SYSTEM......2035

Meixi Chen ; Abhishek Iyer ; Cheng-Hao Shih ; Lado Kurdgelashvili ; Robert Opila

PHOTOVOLTAIC MODULE MANUFACTURING COSTS, AVERAGE PRICES AND INDUSTRY BALANCE 2006–2016......2039

Paula Mints ; Zhengshan J Yu

SOLAR CELL AND WIND ENERGY REPLACEMENT OF POWER PLANTS GLOBALLY......2042

Larry Partain ; Shirley Hansen ; Dirk Bennett ; Richard Hansen ; Allan Newlands ; Lewis Fraas

ANALYSIS OF LIGHT ENVIRONMENT UNDER SOLAR PANELS AND CROP LAYOUT......2048

Deng Wang ; Yaojie Sun ; Yandan Lin ; Yuan Gao

INTERFACE EFFECTS OF ALKALI TREATMENT ON CU-RICH THIN FILM SOLAR CELLS......2054

Hossam Elanzeery ; Finn Babbe ; Anastasiya Zelenina ; Michele Melchiorre ; Susanne Siebentritt

INCREASEDVOCAND FF IN ZNO1-XSX-BUFFERED CUIN1-XGAXSE2SOLAR CELLS BY CADMIUM PARTIAL ELECTROLYTE TREATMENT......2058

Andreas Bauer ; Dimitrios Hariskos ; Wiltraud Wischmann

PASSIVATING AND CARRIER-SELECTIVE CONTACTS - BASIC REQUIREMENTS AND IMPLEMENTATION......2064

S.W. Glunz ; M. Bivour ; C. Messmer ; F. Feldmann ; R. Müller ; C. Reichel ; A. Richter ; F. Schindler ; J. Benick ; M. Hermle

FIRST-PRINCIPLES MODELING OF ALKALI METAL POST DEPOSITION TREATMENT EFFECTS IN CIGS SOLAR CELLS..2070
Maria Fedina ; Hannu-Pekka Komsa ; Ville Havu ; Martti J. Puska

EXPLORING SILICON CARBIDE- AND SILICON OXIDE-BASED LAYER STACKS FOR PASSIVATING CONTACTS TO SILICON SOLAR CELLS..2073
P. Löper ; G. Nogay ; P. Wyss ; M. Hyvl ; P. Procel ; J. Stuckelberger ; A. Ingenito ; I. Mack ; Q. Jeangros ; M. Ledinsky ; A. Fejfar ; C. Allebé ; J. Horzel ; M. Despeisse ; F. Crupi ; F.-J. Haug ; C. Ballif

EFFICIENT ELECTRON CONTACTS FORN-TYPE SILICON SOLAR CELLS USING MAGNESIUM METAL, OXIDE, AND FLUORIDE..2076
Yimao Wan ; Chris Samundsett ; James Bullock ; Di Yan ; Thomas Allen ; Jun Peng ; Jie Cui ; Mark Hettick ; Ali Javey ; Andres Cuevas

GRADED (ALZGA1-Z)XIN1-XP WINDOW-EMITTER STRUCTURES FOR IMPROVED SHORT-WAVELENGTH RESPONSE..2079
Jacob T. Boyer ; Daniel L. Lepkowski ; Daniel J. Chmielewski ; Steven A. Ringel ; Tyler J. Grassman

INTEGRATION OF QUANTUM DOTS AND QUANTUM WELLS INTO INGAAS METAMORPHIC SUBCELL FOR RADIATION HARD 3-J ELO IMM PHOTOVOLTAICS......................2084
Zachary S. Bittner ; Hyun Kum ; Michael A. Slocum ; George T. Nelson ; Rao Tatavarti ; Andre Wibowo ; Seth M. Hubbard

PROTON IRRADIATION OF 3J SOLAR CELLS AT LOW TEMPERATURE..2087
Seonyong Park ; Jacques C. Bourgoin ; Olivier Cavani ; Sandrine Picard ; Jérôme Bourcois ; Victor Khorenko ; Carsten Baur ; Bruno Boizot

ULTRA-THIN GAAS SOLAR CELLS: RADIATION TOLERANCE AND SPACE APPLICATIONS...............2091
Louise C. Hirstl ; Michael K. Yakes ; Jeffery. H. Warner ; Mitchell F. Bennett ; Kenneth J. Schmieder ; Stephanie Tomasulo ; Erin Cleveland ; Sergey Maximenko ; James Moore ; Robert J. Walters ; Phillip P. Jenkins

LARGE AREA MULTIJUNCTION III-V SPACE SOLAR CELLS OVER 31% EFFICIENCY......................2094
X.Q. Liu ; C. Fetzer ; P. Chiu ; M. Haddad ; X. Zhang ; R. Cravens ; D. Law ; J. Ermer ; J. Krogen ; S. Sharma ; J. Hanley

ADVANCED-ARCHITECTURE HIGH-EFFICIENCY SOLAR CELLS FOR LOW IRRADIANCE LOW TEMPERATURE (LILT) APPLICATIONS..2099
Andreea Boca ; Jonathan Grandidier ; Claiborne Mcpheeters ; Paul Sharps ; Philip Chiu ; Xing-Quan Liu ; James Ermer

ULTRA-LIGHTWEIGHT PV MODULE DESIGN FOR BUILDING INTEGRATED PHOTOVOLTAICS..2104
Ana C. Martins ; Valentin Chapuis ; Alessandro Virtuani ; Christophe Ballif

DESIGN IT WITH LSCS; AN EXPLORATION OF APPLICATIONS FOR LUMINESCENT SOLAR CONCENTRATOR PV TECHNOLOGIES..2109
Wouter Eggink ; Angèle Reinders

INVESTIGATING PV-BATTERY 3-TERMINAL INTEGRATION CONCEPT AS A SELF-SUSTAINING POWER SOLUTION..2114
Solomon N. Agbo ; Oleksandr Astakhov ; Uwe Rau ; Tsvetelina Merdzhanova

PERFORMANCE ASSESSMENT OF A BIPV ROOFING TILE IN OUTDOOR TESTING 2118
Cristina S. Polo Lopez ; Pierluigi Bonomo ; Francesco Frontini ; Vasco Medici ; Lorenzo Nespoli

LIFE CYCLE ASSESSMENT OF TRANSPARENT ORGANIC PHOTOVOLTAIC FOR WINDOW APPLICATIONS 2124
Annick Anctil ; Eunsang Lee ; Jack Stephan ; Anjali Munasinghe ; Christopher Traverse ; Richard R. Lunt

A REDUCED ORDER MODEL FOR A TOV STUDY IN A SOLAR PV PROJECT................................2128
Ahmad Abdullah ; Billy Yancey

CYBER SECURITY ASSESSMENT OF DISTRIBUTED ENERGY RESOURCES 2135
Cedric Carter ; Ifeoma Onunkwo ; Patricia Cordeiro ; Jay Johnson

EVALUATION OF FAST-FREQUENCY SUPPORT FUNCTIONS IN HIGH PENETRATION ISOLATED POWER SYSTEMS 2141
Mohamed Elkhatib ; Jason Neely ; Jay Johnson

LOSS OF UTILITY DETECTION CAPABILITIES FOR TODAY'S UTILITY INTERCONNECTED PHOTOVOLTAIC INVERTERS 2147
Sigifredo Gonzalez ; Gregory Kern ; Michael Ropp

PARAMETRIC PV GRID-SUPPORT FUNCTION CHARACTERIZATION FOR SIMULATION ENVIRONMENTS 2153
Javier Hernandez-Alvidrez ; Jay Johnson

COST ANALYSIS AND COST REDUCTION OPPORTUNITIES OF RESIDENTIAL PV SYSTEM IN THE JAPAN 2159
Izumi Kaizuka ; Haruki Yamaya ; Takashi Ohigashi ; Risa Kurihara ; Osamu Ikki

SUPPLY AND DEMAND CONSTRAINTS ON FUTURE PV POWER IN THE USA 2163
Paul A. Basore ; Wesley J. Cole

RESIDENTIAL PHOTOVOLTAIC ELECTRICITY GENERATION IN THE EUROPEAN UNION 2017-OPPORTUNITIES AND CHALLENGES 2167
Arnulf Jäger-Waldau ; Thomas Huld ; Sandor Szabo

INVESTIGATING NANOSCALE DETERMINANTS OF CHARGE COLLECTION IN QUASI-2D PEROVSKITE SOLAR CELLS 2170
Yanqi Luo ; Xueying Li ; Bat-El Cohen ; Barry Lai ; Lioz Etgar ; David P Penning

RECENT DEVELOPMENTS OF SOLAR PHOTOVOLTAIC SYSTEMS IN INDIA 2172
Saravanan Vasudevan ; Arumugam Murugesan

OPERANDO X-RAY DIFFRACTION FOR CHARACTERIZATION OF PHOTOVOLTAIC MATERIALS 2176
Laura T Schelhasl ; Jeffrey A. Christians ; Joseph J. Berry ; Michael F. T Oney ; Christopher J. Tassone ; Joseph M. Luther ; Kevin H. Stone

X-RAY BEAM INDUCED VOLTAGE: A NOVEL TECHNIQUE FOR ELECTRICAL NANOCHARACTERIZATION OF SOLAR CELLS 2179
Michael E. Stuckelberger ; Tara Nietzold ; Bradley M. West ; Barry Lai ; Jörg M. Maser ; Volker Rose ; Mariana I. Bertoni

ELECTRO-LUMINESCENT REFRIGERATION ENABLED BY HIGHLY EFFICIENT PHOTOVOLTAICS 2185
T. Patrick Xiao ; Kaifeng Chen ; Parthiban Santhanam ; Shanhui Fan ; Eli Yablonovitch

MULTIPLE QUANTUM WELLS AS SLOWED HOT CARRIER COOLING ABSORBERS IN HOT CARRIER CELLS 2186
Gavin Conibeer ; Yi Zhang ; Simon Chung ; Yuaxun Liao ; Stephen Bremner ; Santosh Shrestha

QUANTITATIVE OPTOELECTRONIC MEASUREMENTS OF CARRIER THERMODYNAMICS PROPERTIES IN QUANTUM WELL HOT CARRIER SOLAR CELL 2192
Dac-Trung Nguven ; Laurent Lombez ; François Gibelli ; Soline Boyer-Richard ; Alain Le Corre ; Olivier Durand ; Jean-François Guillemoles

ABSORPTION ENHANCEMENT IN INGAASP/INGAP QUANTUM WELL SOLAR CELLS 2195
Islam E.H. Sayed ; Nikhil Jain ; Myles A. Steiner ; John F. Geisz ; Salah M. Bedair

CARRIER COLLECTION MODEL AND DESIGN RULE FOR QUANTUM WELL SOLAR CELLS 2201
Kasidit Toprasertpong ; Boram Kim ; Yoshiaki Nakano ; Masakazu Sugiyama

INFLUENCE OF CONDUCTION BAND OFFSETS AT WINDOW/BUFFER AND BUFFER/ABSORBER INTERFACES ON THE ROLL-OVER OF J-V CURVES OF CIGS SOLAR CELLS 2205
Giovanna Sozzi ; Simone Di Napoli ; Roberto Menozzi ; Florian Werner ; Susanne Siebentritt ; Philip Jackson ; Wolfram Witte

OVERVIEW OF SURFACE PASSIVATION SCHEMES FOR THIN FILM SOLAR CELLS 2209
Ratan Kotipalli ; Bart Vermang

TOWARDS 10% STATE-OF-THE-ART PURE SULFIDE CU2ZNSNS4 SOLAR CELL BY MODIFYING THE INTERFACE CHEMISTRY 2213
Kaiwen Sun ; Jialiang Huang ; Steve Johnston ; Chang Yan ; Fangyang Liu ; Xiaojing Hao ; Martin Green

BAND GAP CHANGES OF THE CDS BUFFER INDUCED BY POST-ANNEALING OF CU2ZNSN(S,SE)4SOLAR CELLS 2216
Mario Lang ; Nicolas Schäfer ; Christian Huber ; Thomas Schnabe ; Heinz Kalt ; Michael Hetterich

22.61 % EFFICIENT FULLY SCREEN PRINTED PERC SOLAR CELL 2220
Weiwei Deng ; Feng Ye ; Ruimin Liu ; Yunpeng Li ; Haiyan Chen ; Zhen Xiong ; Yang Yang ; Yifeng Chen ; Yongqian Wang ; Pietro P. Altermatt ; Zhiqiang Feng ; Pierre J. Verlinden

HOW TO ACHIEVE 23% EFFICIENT LARGE-AREA CU PLATED N-PERT CELLS? 2227
Monica Aleman ; Angel Uruena ; Emanuele Cornagliotti ; Patrick Choulat ; Joachim John ; Richard Russell ; Sukvhinder Singh ; Loic Tous ; Wen-Cheng Sun ; Filip Duerinckx ; Jozef Szlufcik

MICROSTRUCTURE AND RECOMBINATION ACTIVITY OF GRAIN BOUNDARIES FROM FRONT AND REAR SIDE DURING A LID-CYCLE OF MC-PERC SOLAR CELLS 2232
Tabea Luka ; Marko Turek ; Stephan Großer ; Christian Hagendorf

THERMODYNAMIC EFFICIENCY LIMIT OF BIFACIAL SOLAR CELLS FOR VARIOUS SPECTRAL ALBEDOS 2236
Thomas C.R. Russell ; Rebecca Saive ; Harry A. Atwater

PROCESS-INDUCED DEGRADATION RESISTANT N-CZ WAFERS THROUGH TABULA RASA DEFECT ENGINEERING 2242
Vincenzo Lasalvia ; William Nemeth ; Matthew Page ; Wooseok Nam ; Youngsik Han ; Sungsun Baik ; Amanda Youssef ; Tonio Buonassisi ; Paul Stradins

DETECTION OF A SHIFTING BROMINE CONCENTRATION IN HYBRID PEROVSKITES BY X-RAY FLUORESCENCE MICROSCOPY 2245
Yanqi Luo ; Parisa Khoram ; Sarah Brittman ; Barry Lai ; Erik C. Garnett ; David P. Fenning

INFLUENCE OF GRAIN SIZE AND INTERFACES ON PHOTO-STABILITY OF PEROVSKITE SOLAR CELLS 2247

Istiaque Hossain ; Liang Zhang ; Ranjith Kottokkaran ; Mohamed El-Henawey ; Pranav Joshi ; Max Noack ; Vikram Dalal

COLD THOUGHTS ON PEROVSKITE FEVER 2251

Tao Xu ; Jue Gong

LBIC ANALYSIS OF PEROVSKITE BASED SOLAR CELLS STABILITY 2255

Carmen M. Ruiz ; Javier Ramos ; Richard Garuz ; Damien Barakel ; Jean Reusser ; Judikaël Le Rouzo

ASSESSING JOB GROWTH AND SUSTAINABILITY IN THE US PV INDUSTRY 2258

Brion Bob

ENSURING THE RELIABILITY OF PHOTOVOLTAIC POWER SYSTEMS USING INTERNATIONAL STANDARDS AND THE IECRE CONFORMITY ASSESSMENT SYSTEM 2263

George Kelly ; Adrian Häring ; Ted Spooner ; Greg Ball ; Sarah Kurtz ; Matthias Heinze ; Masaaki Yamamichi ; Govind Ramu

A FRAMEWORK TO CALCULATE UNCERTAINTIES FOR LIFETIME ENERGY YIELD PREDICTIONS OF PV SYSTEMS 2267

Bjorn Muller ; Peter Bostock ; Boris Farnung ; Christian Reise

INTEGRATED PV-RECYCLING-MORE EFFICIENT, MORE EFFECTIVE 2272

Wolfram Palitzsch ; Ulrich Loser

ANALYSIS OF GAINP SOLAR CELLS GROWN BY HYDRIDE VAPOR PHASE EPITAXY 2275

Kevin L. Schulte ; John Simon ; David L. Young ; Aaron J. Ptak

INVESTIGATION OF ADHESION FORCES BETWEEN DUST PARTICLES AND SOLAR GLASS 2280

H.R. Moutinho ; C.-S. Jiang ; B. To ; C. Perkins ; M. Muller ; M.M. Al-Jassim ; L. Simpson

ANTI-REFLECTIVE AND ANTI-SOILING PROPERTIES OF A KLEANBOOST™, A SUPERHYDROPHOBIC NANO-TEXTURED COATING FOR SOLAR GLASS 2285

Illya Nayshevsky ; Qianfeng Xu ; Gil Barahman ; Alan Lyons

MULTILAYER-GROWN ULTRATHIN NANOSTRUCTURED GAAS SOLAR CELLS 2291

Boju Gai ; Yukun Sun ; Minjoo Lee ; Jongseung Yoon

LABORATORY STUDIES OF PARTICLE CEMENTATION AND PV MODULE SOILING 2294

Craig L. Perkins ; Matthew Muller ; Lin Simpson

VIRTUAL SUBSTRATES FOR LOW-COST HIGH EFFICIENCY III-V PHOTOVOLTAICS 2298

Sean J. Babcock ; Marlene L. Lichty ; Shankar Karki ; Grace Rajan ; Sylvain Marsillac ; Elisabeth L. Mcclure ; Seth M. Hubbard ; Christopher G. Bailey

SEASONAL TRENDS OF SOILING ON PHOTOVOLTAIC SYSTEMS 2301

Leonardo Micheli ; Daniel Ruth ; Matthew Muller

INTERRELATIONSHIPS AMONG NON-UNIFORM SOILING DISTRIBUTIONS AND PV MODULE PERFORMANCE PARAMETERS, CLIMATE CONDITIONS, AND SOILING PARTICLE AND MODULE SURFACE PROPERTIES 2307

Lawrence L. Kazmerski ; Antonia Sonia A.C. Diniz ; Daniel Sena Braga ; Cristiana Brasil Maia ; Marcelo Machado Viana ; Suellen C. Costa ; Pedro P. Brito ; Cláudio Dias Campos ; Sergio De Morais Hanriot ; Leila R. De Oliveira Cruz

PV MODULE DURABILITY -CONNECTING FIELD RESULTS, ACCELERATED TESTING, AND MATERIALS 2312

T. John Trout ; W. Gambogi ; T. Felder ; K. R. Choudhury ; L. Garreau-Iles ; Y. Heta ; K. Stika

FEMTOSECOND VS NANOSECOND: AN ANALYSIS ON THE LASER ABLATION PROPERTIES OF DIELECTRIC LAYERS FOR SOLAR CELLS 2318

Jaffar Moideen Yacob Ali ; Vinodh Shanmugam ; Carlos D. Rodríguez-Gallegos ; Bianca Lim ; Armin Aberle ; Thomas Mueller

GROWTH OF MOS2 THIN FILMS WITH MICRODOME TEXTURE AS OMNIDIRECTIONAL LIGHT TRAP FOR SOLAR CELL APPLICATIONS 2324

Hussain M. Abouelkhair ; Nina A. Orlovskaya ; Robert E. Peale

STUDY OF SPATIAL DISTRIBUTION OF ELECTRICAL, OPTICAL AND STRUCTURAL PROPERTIES OF MAGNETRON SPUTTERED AZO THIN FILMS 2330

Mohit Agarwal ; Rajiv O Dusane

MULTIBAND FORMATION IN CR DOPED CUGAS2 THIN FILMS SYNTHESIZED BY CHEMICAL SPRAY PYROLYSIS 2334

Nazmul Ahsan ; Sivaperuman Kalainatharr ; Naoya Miyashita ; Takuya Hoshii ; Yoshitaka Okada

EFFECTS OF ANNEALING AND SUBSTRATE TEMPERATURE FOR SN-S THIN FILMS 2338

Yoji Akaki ; Kazuya Iwasaki ; Shigeyuki Nakamura ; Hideaki Araki

MOLYBDENUM OXIDE THIN FILMS FOR HETEROJUNCTION SOLAR CELLS 2342

A. Dominguez ; Ateet Dutt ; O. De Melo ; G. Santana

DUAL ION BEAM SPUTTERED TCO THIN FILMS: SPUTTER-INSTIGATED PLASMONIC
FEATURES FOR ULTRATHIN PHOTOVOLTAICS..2345
Vivek Garg ; Brajendra S. Sengar ; Vishnu Awasthi ; Shailendra Kumar ; Shaibal Mukherjee

COMBINATORIAL STUDY OF SN-TI-W-O TRANSPARENT CONDUCTING OXIDE THIN
FILMS FOR PHOTOVOLTAIC APPLICATIONS...2349
Michael N. Gona ; Patrick J. M. Isherwood ; Jake W. Bowers ; John M. Walls

BANDGAP AND ELECTRON AFFINITY OPTIMIZATION OF ZINC OXIDE FOR N-ZNO/P-SI
SINGLE HETEROJUNCTION SOLAR CELL..2355
Babar Hussain ; Aasma Aslam

MODELING AND OPTIMIZING THE EFFICIENCY OF A ZNO/ZNTE SOLAR CELL USING
SCAPS SOFTWARE...2358
Amal Kabalan ; Sam Roy ; Benjamin Chen

TERNARY PHOSPHIDE SEMICONDUCTOR INMG/ZN3P2SOLAR CELLS...2361
Ryoji Katsube ; Kenji Kazumi ; Yoshitaro Nose

NUMERICAL MODELING OF WSE2SOLAR CELLS...2364
H. Kyureghian ; M. Hilfiker ; E. Ediger ; V. Medic ; N.J. Ianno

BIAXIAL-TEXTURED TITANIUM NITRIDE THIN FILMS ON LOW-COST, FLEXIBLE METAL
SUBSTRATE AS A CONDUCTIVE BUFFER LAYER FOR THIN FILM SOLAR CELLS...................2368
*Yongkuan Li ; Yao Yao ; Ying Gao ; Sicong Sun ; Pavel Dutta ; Monika Rathi ; Jae-Hyun Ryou ; Venkat
Selvamanickam*

SNS BY IONIZED JET DEPOSITION FOR PHOTOVOLTAIC APPLICATIONS 2372
*Daniele Menossi ; Simone Di Mare ; Ivan Rimmaudo ; Elisa Artegiani ; Giampiero Tedeschi ; Juan Luis Pena ;
Fabio Piccinelli ; Andrei Salavei ; Alessandro Romeo*

EFFECT OF VALENCE BAND SPLITTING ON THE ABSORPTION SPECTRA OF
MONOLAYER MOS2 IN PRESENCE OF SULPHUR VACANCIES 2376
Himani Mishra ; Sitangshu Bhattacharya

THE STUDY OF SOME MATERIALS AS BUFFER LAYER IN COPPER ANTIMONY SULPHIDE
(CUSBS2) SOLAR CELL USING SCAPS 1-D...2381
Muteeu Olopade ; Adeyinka Adewoyin ; Michael Chendo ; Adewumi Bolaji

INFLUENCE OF HETERO-INTERFACES ON PHOTOVOLTAIC PERFORMANCE IN SOLAR
CELLS BASED ON ZNSNP2BULK CRYSTAL...2385
Shigeru Nakatsuka ; Shunsuke Akari ; Jakapan Chantana ; Takashi Minemoto ; Yoshitaro Nose

JUNCTION BY DIFFUSION OF ELEMENTAL SODIUM ALONE INTO BRIDGMAN CU(IN, GA)
SE2 2388
S. Park ; C. H. Champness ; S. Vanka ; Z. Mi ; I. Shih

OXYGEN SUBSTITUTION AND SULFUR VACANCIES IN NABIS2: A PB-FREE CANDIDATE
FOR SOLUTION PROCESSABLE SOLAR CELLS 2392
*Robert J Patterson ; Hongze Xia ; Long Hu ; Zhilong Zhang ; Lin Yuan ; Jianfeng Yang ; Weijian Chen ; Zihan
Chen ; Yijun Gao ; Yicong Hu ; Binesh Puthen Veettil ; John A. Stride ; Gavin Conibeer ; Shujuan Huang*

EFFECT OF ANNEALING ON PERFORMANCE OF SOLAR CELLS WITH NEW OXIDE
ABSORBER MN2V2O7..2395
Pramod Ravindra ; Eashwer Athresh ; Rajeev Ranjan ; Srinivasan Raghavan ; Sushobhan Avasthi

ELECTRO-OPTICAL PROPERTIES OF ZN2MO3O8THIN-FILMS: A NOVEL LOW-BANDGAP
SOLAR ABSORBER..2399
Pramod Ravindra ; Eashwer Athresh ; Rajeev Ranjan ; Srinivasan Raghavan ; Sushobhan Avasthi

LOW TEMPERATURE SOLUTION PROCESS FOR RANDOM HIGH ASPECT RATIO SILVER
NANOWIRE AS PROMISING TRANSPARENT CONDUCTIVE LAYER..2403
Arastoo Teymouri ; Supriya Pillai ; Zi Ouyang ; Xiaojing Hao ; Martin Green

OXYGEN INCORPORATION INTO SI NANOCRYSTAL/SIC MULTILAYERS....................................2407
Charlotte Weiss ; Andreas Reichert ; Johannes Hofmann ; Stefan Janz

DESIGN OF CASCADED HETEROSTRUCTURED P-I-I-N CDS/CDSE LOW COST SOLAR
CELL 2411
M. Zinaddinov ; S. Mil'shtein

FAST C-V METHOD TO MITIGATE EFFECTS OF DEEP LEVELS IN CIGS DOPING
PROFILES...2414
P. K. Paull ; J. Bailey ; G. Zapalac ; A. R. Arehart

CRYSTAL GROWTH PHENOMENA IN POLYCRYSTALLINE (CU)ZNTE/CDTE/CDS VIA
MOLECULAR DYNAMICS..2419
Rodolfo Aguirre ; Jose J. Chavez ; Xiao W. Zhou ; David Zubia

USING HIGH-RESOLUTION ANOMALOUS-SCATTERING X-RAY DIFFRACTION TO
OBSERVE OFF-STOICHIOMETRIC CU2ZNSNS4CRYSTAL STRUCTURES....................................2423
Christopher J. Bosson ; Max T. Birch ; Douglas P. Halliday ; Chiu C. Tang ; Peter D. Hatton

SIMULATION OF ZNMGO AS THE WINDOW LAYER FORCDTESOLAR CELLS 2427
Yunfei Chen ; Shou Peng ; Xin Cao ; Alan E. Delahoy ; Ken K. Chin

MODELING EFFECT OF DEFECTS ON EFFICIENCY OF NANOWIRE CDS-CDTE SOLAR
CELLS 2432
Hongmei Dang ; Esther Ososanya ; Nian Zhang ; Xiaohui Wang ; Hojjatollah Sarvari ; Vijay P. Singlr

ANALYTICAL DESCRIPTION OF CHARGED GRAIN BOUNDARY RECOMBINATION IN
POLYCRYSTALLINE THIN FILM SOLAR CELLS 2438
Benoit Gaury ; Paul M. Haney

IMAGING THE EFFECT OF CDSE WINDOW LAYERS IN CDTE PHOTOVOLTAICS 2443
John M. Howard ; Elizabeth M. Tennyson ; William B. Gunnarsson ; Naba R. Paudel ; Yanfa Yan ; Marina S.
Leite

INVESTIGATION OF TRAPS DENSITY AND POSITION IN ALKALI TREATED CU(IN,GA)SE2
THIN FILMS AND SOLAR CELLS 2446
Shankar Karki ; Pran K. Paul ; Grace Rajan ; Chinedum Akwari ; Angus Rockett ; Steven A Ringel ; Aaron R.
Arehart ; Sylvain Marsillac

THE EFFECT OF DEPOSITION STOICHIOMETRY AND POST-DEPOSITION TREATMENTS
ON DEEP DEFECTS IN CDTE 2449
Imran S. Khan ; Vamsi Evani ; Shamara Collins ; Chih An Hsu ; Vasilis Palekis ; Chris Ferekides

TESTING THE LIMITS OF MECHANICALLY-SCRIBED CIGS MICROCELLS 2453
Ombline Lafont ; Nicolas Vandamme ; Leia Ruffini ; Jia Yu ; Philip Jackson ; Jose Alvarez ; Daniel Lincot

PHOTOLUMINESCENCE IMAGING ANALYSIS OF DOPING IN THIN FILM CDS AND
CDS/CDTE DEVICES 2457
C. Potamialis ; F. Lisco ; B. Maniscalco ; M. Togay ; A. Abbas ; M. Biiss ; J.W. Bowers ; J.M. Waiis ; I.
Rimmaudo ; R. Mis Fernandez ; V. Rejon ; J.L. Peña

APPLICATION OF MAPPING SPECTROSCOPIC ELLIPSOMETRY FOR CDSE/CDTE SOLAR
CELLS: OPTIMIZATION OF LOW-TEMPERATURE PROCESSED DEVICES WITH ALL-
SPUTTERED SEMICONDUCTORS 2462
Mohammed A. Razooqi ; Adam B. Phillips ; Geethika K. Liyanage ; Fadhil K. Al-Fadhili ; Maxwell M. Junda ;
Nikolas J. Podraza ; Michael J. Heben ; Robert W. Collins ; Prakash Koirala

ASSESSING THE VALIDITY AND ACCURACY OF EFFECTIVE ELECTRONIC MATERIALS:
CAN 1D SIMULATIONS PREDICT POLYCRYSTALLINE DEVICE PERFORMANCE? 2467
Yubo Sun ; Allison Perna ; Sudhajit Misra ; Vasilios Palekis ; Chris Ferekides ; Jeffrey Aguiar ; Peter Bermel ;
Michael A. Scarpulla

CHARACTERIZING RECOMBINATION IN CDTE-BASED SOLAR CELLS BY THE
TEMPERATURE AND EXCITATION DEPENDENCE OF OPEN-CIRCUIT VOLTAGE AND
PHOTOLUMINESCENCE 2473
Craig H. Swartz ; Sanjoy Paul ; Corey R. Grice ; Yanfa Yan ; Lorelle Mansfield ; Sachit Grover ; Gang Xiong ;
Jian V. Li

EXPERIMENTAL EVIDENCE FOR CDS-RELATED TRANSPORT BARRIER IN THIN FILM
SOLAR CELLS AND ITS IMPACT ON ADMITTANCE SPECTROSCOPY 2478
Florian Werner ; Anastasiya Zelenina ; Susanne Siebentritt

TRANSPARENT CONDUCTIVE ADHESIVES FOR TANDEM SOLAR CELLS 2482
Talysa R. Klein ; Benjamin G. Lee ; Manuel Schnabel ; Emily L. Warren ; Pauls Stradins ; Adele C. Tamboli ;
Maikel F.A.M. Van Hest

MODELING THREE-TERMINAL III- V LSI TANDEM SOLAR CELLS 2488
Emily L. Warren ; Michael G. Deceglie ; Paul Stradins ; Adele C. Tamboli

WAFER BONDING APPROACHES FOR III-V ON SI MULTI-JUNCTION SOLAR CELLS 2492
Laura Vauche ; Elias Veinberg-Vidal ; Clément Weick ; Christophe Morales ; Vincent Larrey ; Christophe
Lecouvey ; Mickaël Martin ; Jérémy Da Fonseca ; Christophe Jany ; Thibaut Desrues ; Céline Brughera ;
Philippe Voarino ; Thierry Salvetat ; Frank Fournel ; Mathieu Baudrit ; Cécilia Dupré

DESIGN ARITHMETIC OF THE LATERAL III-V / SI HYBRID MODULE 2498
Kenji Araki ; Kyotaro Nakamura ; Kan-Hua Lee ; Takefumi Kamioka ; Yu-Cian Wang ; Nobuaki Kojima ; Yoshio
Ohshita ; Masafumi Yamaguchi

GAASP NANOWIRE SOLAR CELL DEVELOPMENT TOWARDS NANOWIRE/SI TANDEM
APPLICATIONS 2502
Enrique Barrigon ; Yang Chen ; Gaute Otnes ; Vilgaile Dagyte ; Nicklas Anttu ; Lars Samuelson ; Magnus
Borgström

DEMONSTRATION OF GAINP2/SI VOLTAGE MATCHED TANDEM SOLAR CELLS 2506
David C. Bobela ; Kenneth J. Schmieder ; Matthew P. Lumb ; James E. Moore ; Robert J Walters ; Eric A. Armour
; Leo Matthew ; Rajesh Rao ; Angelo Mascarenhas ; Kirstin Alberi

WAFER BONDED III–V ON SILICON MULTI -JUNCTION CELL WITH EFFICIENCY
BEYOND 31% 2511
Romain Cariou ; Jan Benick ; Paul Beutel ; Nico Tucher ; Martin Graf ; David Lackner ; Martin Hermle ; Stefan
W. Glunz ; Andreas W. Bett ; Frank Dimroth

INTEGRATION OF THIN AL FILMS ON INO.18GAO.82AS METAMORPHIC GRADE STRUCTURES FOR LOW-COST III- V PHOTOVOLTAICS..................................2514
Alessandro Giussani ; Michael A. Slocum ; Seth M. Hubbard ; Nathan Smaglik ; Nikhil Pokharel ; S. Phillip Ahrenkiel

TEMPERATURE DEPENDENT CHARACTERISTICS OF GAINP/GAAS/GAINNASSB SOLAR CELL UNDER SIMULATED AM0 SPECTRA..................................2520
Riku Isoaho ; Arto Aho ; Antti Tukiainen ; Mircea Guina

EFFICIENCY OF GAAS P/SI TWO-JUNCTION SOLAR CELLS WITH MULTI-QUANTUM WELLS: A REALISTIC MODELING WITH CARRIER COLLECTION EFFICIENCY..................................2524
Boram Kim ; Kasidit Toprasertpong ; Oliver Supplie ; Agnieszka Paszuk ; Thomas Hannappel ; Yoshiaki Nakano ; Masakazu Sugiyama

INVERSE METAMORPHIC III-V/EPI-SIGE TANDEM SOLAR CELL PERFORMANCE ASSESSED BY OPTICAL AND ELECTRICAL MODELING..................................2528
Raphaël Lachaurne ; Martin Foldyna ; Gwénaëlle Hamon ; Nicolas Vaissiére ; Jean Decobert ; Romain Cariou ; Pere Roca I Cabarrocas ; José Alvarez ; Jean-Paul Kleider

TOWARDS MONOLITHICALLY INTEGRATED GAAS ON SI TANDEM SOLAR CELL..................................2532
Zhen Liu ; Zekun Ren ; Haohui Liu ; Tonio Buonassisi ; Ian Marius Peters

ZNSIP2 THIN FILM GROWTH FOR SI-BASED TANDEM PHOTOVOLTAICS..................................2536
Aaron D. Martinez ; Elisa M. Miller ; Andrew G. Norman ; Paul Stradins ; Eric S. Toberer ; Adele C. Tamboli

IN SITU CONTROL OVER THE SUBLATTICE ORIENTATION OF GAP/SI(100): AS VIRTUAL SUBSTRATES FOR TANDEM ABSORBERS..................................2538
Aznieszka Paszuk ; Oliver Supplie ; Sebastian Brückner ; Matthias M. May ; Anja Dobrich ; Andreas Nägelein ; Boram Kim ; Yoshiaki Nakano ; Masakazu Sugiyama ; Peter Kleinschmidt ; Thomas Hannappel ; Thomas Hannappel

III-V/SI TANDEM CELL TO MODULE INTERCONNECTION - COMPARISON BETWEEN DIFFERENT OPERATION MODES..................................2543
Henning Schulte-Huxel ; Emily L. Warren ; Manuel Schnabel ; Paul Stradins ; Daniel Friedman ; Adele C. Tamboli

INGAP/GAAS/ITO/SI HYBRID TRIPLE-JUNCTION CELLS WITH GAAS/ITO BONDING INTERFACES..................................2548
Naoteru Shigekawa ; Tomoya Hara ; Tomoki Ogawa ; Jianbo Liang ; Takefumi Kamioka ; Kenji Araki ; Masafumi Yamaguchi

MEASUREMENTS OF POTENTIALS AT TAP CONTACTS AND ESTIMATION OF RESISTANCE ACROSS BONDING INTERFACES IN INGAP/GAAS/SI HYBRID TRIPLE-JUNCTION CELLS..................................2551
Naoteru Shigekawa ; Jianbo Liang

OPTIMIZATION OF A GAASP TOP CELL FOR IMPLEMENTATION IN A III-V/SI TANDEM STRUCTURE..................................2554
Amber C. Silvaggio ; Daniel L. Lepkowski ; Daniel J. Chmielewski ; Jacob T. Boyer ; Steven A. Ringel ; Tyler J. Grassman

THEORETICAL DESIGN OF PEROVSKITE/CDTE FOUR-TERMINAL TANDEM SOLAR CELLS..................................2558
Tao Tang ; Huan Zhang ; Xingzhi Du ; Yiming Lnr ; Hang Zhou

WAFER-BONDED ALGAAS///SI DUAL-JUNCTION SOLAR CELLS..................................2562
Elias Veinberg-Vidal ; Laura Vauche ; Clément Weick ; Jérémy Da Fonseca ; Christophe Jany ; Christophe Morales ; Christophe Lecouvey ; Thibaut Desrues ; Philippe Voarino ; Frank Fournel ; Anne Kaminski-Cachopo ; Alejandro Datas ; Pablo Garcia-Linares ; Mathieu Baudrit ; Pierre Mur ; Cécilia Dupré

ENHANCEMENT OF SI PHOTOVOLTAIC MODULE BY INTRODUCING III-V/SI HYBRID CONFIGURATIONS AND COST EVALUATIONS UNDER VARIOUS COST RATIOS OF III-V/SI PHOTOVOLTAICS..................................2566
Yu-Cian Wang ; Kenii Araki ; Kyotaro Nakamura ; Kan-Hua Lee ; Takefumi Kamioka ; Nobuaki Kojima ; Yoshio Ohshita ; Masafumi Yamaguchi

NUMERICAL SIMULATION OF P-TYPE FRONT JUNCTION PERL SILICON CELL FOR III-V LSI TANDEM DEVICES..................................2569
Chuqi Yi ; Fa-Jun Ma ; Anita Ho-Baillie ; Stephen Bremner

EPITAXIAL GAP LAYERS GROWN ON SI SUBSTRATES USING MIGRATION ENHANCED AND MOLECULAR BEAM EPITAXY..................................2573
Chaomin Zhang ; Allison Boley ; Nikolai Faleev ; David J. Smith ; Christiana B. Honsberg

INVESTIGATION OF CARRIER-INDUCED DEFECT BEHAVIOR IN P-TYPE MULTICRYSTALLINE SILICON..................................2576
Catherine E. Chan ; Tsun H. Fung ; David N.R. Payne ; Daniel Chen ; Malcolm D. Abbott ; Alison M. Ciesla ; Ran Chen ; Brett J. Hallam ; Stuart R. Wenham

MAGNETRON SPUTTERED HYDROGENATED SILICON THIN FILMS: ASSESSMENT FOR APPLICATION IN PHOTOVOLTAICS 2582

Dipendra Adhikari ; Maxwell M. Junda ; Sylvain X. Marsillac ; Robert W. Collins ; Nikolas J. Podraza

HIGH QUALITY AND THIN SILICON WAFER FOR NEXT GENERATION SOLAR CELLS 2588

Yoshio Ohshita ; Takuto Kojima ; Ryota Suzuki ; Kosuke Kinoshita ; Tomoyuki Kawatsu ; Kyotaro Nakamura ; Atsushi Ogura

FIRST DEMONSTRATION OF RADIAL JUNCTION SILICON NANOWIRE SOLAR MINI-MODULES PREPARED BY PECVD AND LASER SCRIBING 2593

Mutaz Al-Ghzaiwat ; Martin Foldyna ; Takashi Fuyuki ; Wanghua Chen ; Erik V. Johnson ; Jacques Meot ; Pere Roca I Cabarrocas

IMPACT OF INDUCED DEFECTS ON DEVICE PERFORMANCE IN SILICON HETEROJUNCTION SOLAR CELLS 2596

Pradeep Balaji ; André Augusto ; Stuart G. Bowden

LASER HYDROGENATION ON HEAVILY DISLOCATED CAST-MONO SILICON CELLS 2600

Alison M. Ciesla ; Catherine E. Chan ; Sisi Wang ; Malcolm D. Abbott ; Cheemun Chong ; Stuart R. Wenham

PERFORMANCE OPTIMIZATION OF SEMI-TRANSPARENT THIN-FILM AMORPHOUS SILICON SOLAR CELLS 2605

Yuan Gao ; Fai Tong Si ; Olindo Isabella ; Rudi Santbergen ; Guangtao Yang ; Jianfei Dong ; Guoqi Zhang ; Miro Zeman

LOW TEMPERATURE SPALLING OF SILICON: A CRACK PROPAGATION STUDY 2610

Pablo Guimera Coll ; Tine Uberg Nærland ; Nathan Stoddard ; Michael Stuckelberger ; Mariana Bertoni

NEW FINDINGS OF THERMAL EFFECT ON PM-SI:H SOLAR CELLS OPTOELECTRONIC PROPERTIES 2614

L. Hamui ; L. A. Górnez-González ; G. Santana

STUDY OF PV MODULE DEGRADATION RATE PREDICTION THROUGH CORRELATION OF FIELD-AGED AND ACCELERATED-AGED MODULE DEGRADATION DATA 2618

Babak T. Hamzavy ; William J. Grieco ; Brian J. Fields ; Cara S. Libby ; William B. Hobbs ; Olga Lavrova ; C. Birk Jones

ADVANCED ANALYSIS OF MULTI WIRE WAFERING PROCESSES 2622

Ringo Koepgel ; Samuel Brinnig ; Felix Kaule ; Hartmut Schwabe ; Stephan Schoenfelder

CONSIDERATION ON OPEN-CIRCUIT VOLTAGE OF SI HETEROJUNCTION SOLAR CELLS UNDER LOW CONCENTRATION CONDITION 2627

Makoto Konagai

CHARACTERIZATION OF MICROCRYSTALLINE SILICON THIN FILM SOLAR CELLS PREPARED BY HIGH WORKING PRESSURE PLASMA-ENHANCED CHEMICAL VAPOR DEPOSITION 2631

Jung-Dae Kwon ; Dong-Ho Kim ; Ji-Hoon Lee ; Myungkwan Song ; Myunghun Shin

ATOMIC-LAYER-DEPOSITEDV2O5-XFILMS AS A HIGHLY-EFFICIENT P-TYPE LAYER FOR THIN FILM A-SI SOLAR CELLS 2634

Ji-Hoon Lee ; Myungkwan Song ; Dong-Ho Kim ; Jung-Dae Kwon

A NOVEL DEFECT PASSIVATION METHOD FOR MULTICRYSTALLINE SI WAFER BY H2S REACTION 2637

Hsiang-Yu Liu ; Ujjwal K. Das ; Robert W. Birkmire

CARRIER TRANSPORTATION AT NOVEL SILVER PASTE CONTACT 2642

Takefumi Kamioka ; Satoshi Kamevama ; Kazuo Muramatsu ; Aki Tanaka ; Naotaka Iwata ; Kyotaro Nakamura ; Atsushi Ogura ; Yoshio Ohshita

INFLUENCE OF DEPOSITION PARAMETERS ON SILICON THIN FILMS DEPOSITED BY MAGNETRON SPUTTERING 2646

Grace Rajan ; Tejaswini Miryala ; Shankar Karki ; Robert W. Collins ; Nikolas Podraza ; Sylvain Marsillac

MINORITY CARRIER LIFETIME VARIATIONS IN MULTICRYSTALLINE SILICON WAFERS WITH TEMPERATURE AND INGOT POSITION 2651

Sissel Tind Søndergaard ; Jan Ove Odden ; Rune Strandberg

CUO NANOWIRES-BASED RADIAL HETERO-JUNCTION THIN FILM SILICON SOLAR CELLS WITH A HIGH OPEN-CIRCUIT VOLTAGE 2656

Xiaolin Sun ; Jiawen Lu ; Fan Yang ; Linwei Yu ; Jun Xu ; Ling Xu ; Kunji Chen

THE EFFECT OF CHEMICAL COMPOSITION ON POROUS ETCHING FOR EPI AND LIFT-OFF WAFER PROCESS 2660

Teng-Yu Wang ; Peng-Wei Chen ; Han-Wen Liu

ELECTRICAL AND OPTICAL PERFORMANCE OF SILICON SOLAR CELLS USING PLASMONICS INDIUM NANOPARTICLES LAYER EMBEDDED IN SIO2ANTIREFLECTIVE COATING 2664

Hao-Yu Yang ; Wen-Jeng Ho ; Sheng-Kai Feng ; Jheng-Jie Liu ; Ta-Wei Chuang ; Guan-Yi Li ; Yun-Chie Yang ; Cho-Chun Chiang ; Yao- Hui Chen

ELECTROLUMINESCENCE ANALYSIS FOR SEPARATION OF SERIES RESISTANCE FROM RECOMBINATION EFFECTS IN SILICON SOLAR CELLS WITH INTERDIGITATED BACK CONTACT DESIGN 2667
Nuha Ahmed ; Lei Zhang ; Ujjwal Das ; Steven Hegedus

INDOOR MEASUREMENT OF ANGLE RESOLVED LIGHT ABSORPTION BY BLACK SILICON 2672
Mekbib W. Amdemeskel ; Beniamino Iandolo ; Rasmus S. Davidsen ; Ole Hansen ; Gisele A. Dos Reis Benatto ; Nicholas Riedel ; Peter B. Poulsen ; Sune Thorsteinsson ; Anders Thorseth ; Carsten Dam-Hansen

IMPACT OF NON- FLAT PHOTOGENERATION AND CARRIER PROFILES ON THE LUMINESCENT EMISSION AND DETECTION OF SILICON SOLAR CELLS 2677
Nekane Azkona ; Federico Recart ; Pedro Rodríguez ; Vanesa Fano ; Aloña Otaegi ; Juan Carlos Jimeno

DEVELOPMENT OF OUTDOOR LUMINESCENCE IMAGING FOR DRONE-BASED PV ARRAY INSPECTION 2682
Gisele A. Dos Reis Benatto ; Nicholas Riedel ; Sune Thorsteinsson ; Peter B. Poulsen ; Anders Thorseth ; Carsten Dam-Hansen ; Claire Mantel ; Soren Forchhammer ; Kenn H. B. Frederiksen ; Jan Vedde ; Michael Petersen ; Henrik Voss ; Michael Messerschmidt ; Harsh Parikh ; Sergiu Spataru ; Dezso Sera

CLIMBING DRUM PEEL (CDP) TEST METHOD FOR CHARACTERIZING ADHESION IN FLEXIBLE PV MODULES 2688
Venkata Bheemreddy ; Kedar Hardikar

ACCURACY OF SOLAR SIMULATOR SPECTRAL DETERMINATION USING BAND-PASS FILTERING METHOD 2692
Weston Dobson ; Harrison Wilterdink ; Cassidy Sainsbury ; Adrienne Blum ; Justin Dinger ; Ronald A. Sinton ; Karsten Bothe ; David Hinken ; Martin Wolf

CORRELATION OF I-V CURVE PARAMETERS WITH MODULE-LEVEL ELECTROLUMINESCENT IMAGE DATA OVER 3000 HOURS DAMP-HEAT EXPOSURE 2697
Justin S. Fada ; Andrew J. Loach ; Alan J. Curran ; Jennifer L. Braid ; Shuying Yang ; Timothy J. Peshek ; Roger H. French

A NOVEL METHOD TO INVESTIGATE STOICHIOMETRY AND PERFORMANCE OF BURIED PASSIVATED CONTACTS UTILIZING TIME-OF-FLIGHT SIMS 2702
Steven P. Harvey ; William Nemeth ; Jeff Aguiar ; Craig Perkins ; Pauls Stradins

A COMPARISON BETWEEN QUASI-STEADY STATE AND TRANSIENT PHOTOCONDUCTANCE LIFETIMES IN SILICON INGOTS: SIMULATIONS AND MEASUREMENTS 2707
Mohsen Goodarzi ; Ronald Sinton ; Daniel Chung ; Bernhard Mitchell ; Thorsten Trupke ; Daniel Macdonald

NEW DEVELOPMENT IN GLOW DISCHARGE OPTICAL EMISSION SPECTROMETRY FOR THE CHARACTERIZATION AND THE THICKNESS MEASUREMENT OF LAYERS FOR PHOTOVOLTAIC APPLICATIONS 2711
Philippe Hunault ; Matthieu Chausseau ; Patrick Chaporr ; Sofia Gaiaschi ; Anais Loubar ; Muriel Bouttcmy ; Arnaud Etcheberry

DEEP LEVEL TRANSIENT SPECTROSCOPY MEASUREMENTS OF SILICON HETEROJUNCTION CELLS 2716
Sanchit Khatavkar ; C. V. Kannan ; Vijay Kumar ; P. R. Nair ; B. M. Arora

CHARACTERIZATION OF MODULES AND ARRAYS WITH SUNS VOC 2719
Alex Killam ; Stuart Bowden

A STUDY OF PERFORMANCE CHARACTERIZATION WITH REAR LIGHT SOURCE IN CONVENTIONAL BIFACIAL SOLAR CELLS 2723
Soo Min Kim ; Sang Hoon Jung ; Rae-Won Choi ; Yong Bae Kim ; Min Gu Kang ; Hee-Eun Sonp ; Gyu-Seok Choi

ELECTRICAL CHARACTERIZATION OF THE CARRIER TRANSPORT PROPERTIES IN ACU(IN,GA)SE2SOLAR CELL 2728
Roberto Lopez ; Sanjoy Paull ; Ingrid Repins ; Jian V. Li

SYSTEMATIC THERMALPHOTOVOLTAIC SOLAR CELL OPTIMIZATION 2732
Zheng Lyu ; Muyu Xue ; Junyan Chen ; Jieyang Jia ; Shanhui Fan ; James Harris

CHARACTERIZATION OF TELLURIUM AS A BACK CONTACT FOR CDTE SOLAR CELLS 2736
C.E. Moffett ; W.S. Sampath

ON THE DIFFERENT EXPLANATIONS OF THE RECOMBINATION CURRENTS WITH HIGH IDEALITY FACTOR IN SILICON SOLAR CELLS 2740
A. Otaegi ; V. Fano ; N. Azkona ; J. R. Gutiérrez ; J. C. Jimeno

IDENTIFICATION OF SHUNTS IN A MONOLITHIC MULTIJUNCTION GAAS/GAAS DEVICE BY SPECTROMETRIC CHARACTERIZATION 2744
Felipe Oviedo ; Liu Zhe ; Zekun Ren ; Kevin Nay Yaung ; Maung Thway ; Liu Haohui ; Tonio Buonassisi ; Ian Marius Peters

A SIMULATION STUDY ON RADIATIVE RECOMBINATION ANALYSIS IN CIGS SOLAR CELL 2749

Sanjoy Paul ; Roberto Lopez ; Md Dalim Mia ; Craig H. Swartz ; Jian V. Li

SIMULATION AND SPECTROSCOPY OF CARRIER RELAXATION IN GASB AND GAAS 2755

A.C. Scofield ; A.I. Hudson ; B.L. Liang ; B.C. Juang ; D.L. Huffaker ; S.M. Hubbard ; W.T. Lotshaw

COMPUTATIONAL DESIGN OF DOPANTS IN CDTE GRAIN BOUNDARIES FOR EFFICIENT PHOTOVOLTAICS 2759

Fatih G. Sen ; Tadas Paulauskas ; Ce Sun ; Moon Kim ; Robert F. Klie ; Maria K.Y. Chan

ANALYSES OF PHOTOVOLTAIC POWER PLANT PERFORMANCE ESTIMATES BASED ON DETAILED LABORATORY MODULE CHARACTERIZATIONS AND TYPICAL REAL-WORLD INPUT DATA SOURCES 2762

Rajeev Singh ; John L.R. Watts ; Kellen Gillispie

CRITICAL EVALUATION OF THE FOUNDATIONS OF SOLAR SIMULATOR STANDARDS 2765

Ronald A. Sinton ; Harrison Wilterdink ; Justin Dinger ; Adrienne L. Blum ; Weston Dobson ; Cassidy Sainsbury

IMPACT OF INFRARED OPTICAL PROPERTIES ON CRYSTALLINE SI AND THIN FILM CDTE SOLAR CELLS 2771

Indra Subedi ; Timothy J Silverman ; Michael Deceglie ; Nikolas J. Podraza

THE IMPACT OF IMPURITIES ON THE RELATIVE EFFICIENCIES OF SOLAR CELLS FROM DIFFERENT SILICON FEEDSTOCKS 2776

Muhammad Tayyib ; Aleksandr Dobroliubov ; Zekija Ramic ; Muhammad Nadeem Akarm ; Jan Ove Odden

ACCURACY EVALUATION OF ABSOLUTE ELECTROLUMINESCENCE-EFFICIENCY MEASUREMENTS OF SOLAR CELLS USING A SENSITIVITY-CALIBRATED-PHOTODETECTOR CONTACT METHOD 2781

Masahiro Yoshita ; Yoshihiro Hishikawa ; Yoshihiko Kanemitsu ; Hidefumi Akiyama

NANOMETER-SCALE CARRIER IMAGING OF POTENTIAL-INDUCED DEGRADATION IN C-SI SOLAR CELLS 2785

C.-S. Jiang ; C. Xiao ; H.R. Moutinho ; S. Johnston ; M.M. Al-Jassim ; X. Yang ; Y. Chen ; J. Ye

NREL EFFORTS TO ADDRESS SOILING ON PV MODULES 2789

Lin J. Simpson ; Matthew Muller ; Michael Deceglie ; Helio Moutinho ; Craig Perkins ; C. S. Jiang ; David C. Miller ; Leonardo Micheli ; Govindasamy Tamizhmani ; Sai Ravi Vasista Tatapudi ; Mowafak Al-Jassim

MODELING POTENTIAL-INDUCED DEGRADATION (PID) OF FIELD-EXPOSED CRYSTALLINE SILICON SOLAR PV MODULES: FOCUS ON A REGENERATION TERM 2794

Eleonora Annigoni ; Alessandro Virtuani ; Fanny Sculati-Meillaud ; Christophe Ballif

SOILING LOSS ON PV MODULES AT TWO LOCATIONS IN INDIA STUDIED USING A WATER BASED ARTIFICIAL SOILING METHOD 2799

Sonali Bhaduri ; Sachin Zachariah ; Lawrence L. Kazmcrski ; Balasubramaniam Kavaipatti ; Anil Kottantharayil

QUANTIFYING YEAR-TO-YEAR VARIATIONS IN SOLAR PANEL SOILING FROM PV ENERGY-PRODUCTION DATA 2804

Michael G. Deceglie ; Leonardo Micheli ; Matthew Muller

ACCURATELY MEASURING PV SOILING LOSSES WITH SOILING STATION EMPLOYING PV MODULE POWER MEASUREMENTS 2808

Michael Gostein ; Bill Stueve ; Mandy Chan

PERFORMANCE OF MONOCRYSTALLINE SILICON SOLAR CELL- INFLUENCE OF DUST ON ULTRA-VIOLET AND VISIBLE REGION DURING EARLY STAGE OF DEPOSITION 2811

Hemaprabha Elangovan ; Upasna Ranjan ; A K Jagdish ; Praveen C. Ramamurthy ; Kamanio Chattopadhyay

A COMPREHENSIVE STUDY OF LIGHT SOAKING EFFECT IN CDTE SOLAR CELLS 2816

D. Guo ; A. Moore ; D. Krasikov ; I. Sankin ; D. Vasileska

CORRECTION FOR METASTABILITY IN THE QUANTIFICATION OF PID IN THIN-FILM MODULE TESTING 2819

Peter Hacke ; Sergiu Spataru ; Steve Johnston

A FINE MODEL OF POWER DEGRADATION FOR CRYSTALLINE SILICON SOLAR MODULES 2823

Wenshuang Hea ; Baosong Duan ; Fumei Wang ; Ao Wang ; Jipeng Chang ; He Wang ; Hong Yang ; Jie Ding ; Junjun Zhang ; Jingsheng Huang

TEST METHODS FOR HYDROPHOBIC COATINGS ON SOLAR COVER GLASS 2827

Kenan Isbilir ; Biancamaria Maniscalco ; Ralph Gottschalg ; John Michael Walls

IMPACT OF DEGRADATION RATES ON SOLAR PV FINANCING FOR PROJECTS LOCATED IN THE UNITED STATES 2833

Rounak A. Kharait ; Phil Stiles ; Jarrett Carriere ; Larry Mcclung

ANALYSIS OF WIND DIRECTION AND SPEED MEASUREMENTS IN ARID REGION - A SITE EVALUATION USING DATA WITH LOW TEMPORAL RESOLUTION 2836

Elisabeth Klimm ; Felix Guischard ; Karl-Anders Weiss

FORECASTING ENVIRONMENTAL DEGRADATION POWER LOSS IN SOLAR PANELS WITH A PREDICTIVE CRACK OPENING TEST 2839
Jason L. Lincoln ; Andrew M. Gabor ; Eric J. Schneller ; Hubert Seigneur ; Joseph Walters ; Rob Janoch ; Andrew Anselmo ; Victor Huayamave ; Winston Schoenfeld

FLUORESCENCE IMAGING ON THE CROSS-SECTION OF PHOTOVOLTAIC LAMINATES AGED UNDER DIFFERENT UV INTENSITIES 2844
Yadong Lyu ; Jae Hyun Kim ; Xiaohong Gu

STATISTICAL ANALYSIS OF DEGRADATION DATA FOR C-SI MODULES OBSERVED IN INDIA IN 2016 2849
Chiranjibi Mahapatra ; Rajiv Dubey ; Shashwata Chattopadhyay ; Sachin Zachariah ; Sanjeev Sabnis

PROCESS INDUCED DEFLECTION AND STRESS ON ENCAPSULATED SOLAR CELLS 2854
Xiaodong Meng ; Michael Stuckelberger ; Peter Hacke ; Mariana Bertoni

A UNIFIED GLOBAL INVESTIGATION ON THE SPECTRAL EFFECTS OF SOILING LOSSES OF PV GLASS SUBSTRATES: PRELIMINARY RESULTS 2858
Leonardo Micheli ; Eduardo F. Fernández ; Greg P. Smestad ; Hameed Alrashidi ; Nabin Sarmah ; Nazmi Sellami ; Ibrahim A. I. Hassan ; Amal Kasry ; Gustavo Nofuentes ; Neeru Sood ; Bala Pesala ; S. Senthilarasu ; Florencia Almonacid ; K.S. Reddy ; Matthew Muller ; Tapas K. Mallick

REFERENCE: PROCEEDINGS OF THE IEEE PVSC CONF., 2017 THE DEVELOPMENT OF A DC BREAKDOWN VOLTAGE TEST FOR PHOTOVOLTAIC INSULATING MATERIALS 2864
David C. Miller ; Bernt Ake-Sultan ; Axel Borne ; Rene Eugen ; Bradley L. Givot ; Jiirgen Jung ; Steven W. Macmaster ; Byron K. Mcdanold ; Ulf H. Nilsson ; Nancy H. Phillips ; Ian A. Tappan ; Nick S. Bosco

FIELD-EVALUATION OF ELECTRODYNAMIC SCREENS FOR MAINTAINING HIGH OPTICAL EFFICIENCY OPERATION OF SOLAR COLLECTORS 2870
Cristian Morales ; Annie Bernard ; Ryan Eriksen ; Julius Yellowhair ; Sean Garner ; Ricci La Centra ; Alecia Griffin ; Alexis Lloyd ; Yujie Gao ; Ramakrishnan Lakshmanan ; Mark Horenstein ; Malay Mazumder

EFFECT OF REVERSE BIAS VOLTAGES ON SMALL SCALE GRIDDED CIGS SOLAR CELLS 2875
Soheyl Mortazavi ; Klaas Bakker ; Jome Carolus ; Michael Daenen ; Gabriela De Amorim Soares ; Henk Steijvers ; Arthur Weeber ; Mirjam Theelen

A METHOD TO EXTRACT SOILING LOSS DATA FROM SOILING STATIONS WITH IMPERFECT CLEANING SCHEDULES 2881
Matthew Muller ; Leonardo Micheli ; Alfredo A. Martinez-Morales

ANALYTICAL (S)TEM STUDIES OF DEFECTS ASSOCIATED WITH PID IN STRESSED SI PV MODULES 2887
Andrew Norman ; Adam Stokes ; John Moseley ; Steven Harvey ; Steve Johnston ; Harvey Guthrey ; Mowafak Al-Jassim

DESIGN, DEVELOPMENT, AND EVALUATION OF ELECTRODYNAMIC SCREENS FOR SELF-CLEANING SOLAR PANELS AND CONCENTRATING MIRRORS 2891
Annie Bernard ; Cristian Morales ; Ryan S. Eriksen ; Alecia C. Griffin ; Yujie Gao ; Ramakrishnan Lakshmanan ; Ricci La Centra ; Arash Sayyah ; Julius E. Yellowhair ; Sean M. Garner ; N Mark Horenstein ; Malay K. Mazumder

EVALUATING SOLAR CELL FRACTURE AS A FUNCTION OF MODULE MECHANICAL LOADING CONDITIONS 2897
Eric J. Schneller ; Andrew M. Gabor ; Jason Lincoln ; Rob Janoch ; Andrew Anselmo ; Joseph Walters ; Hubert Seigneur

COMPUTATIONAL STUDY OF THE EFFECT OF PHOTOVOLTAIC (PV) MODULE PARAMETERS ON STRESS DEVELOPMENT IN SILICON UNDER STATIC LOADING 2902
Saurabh Sethia ; Karan Shishir Yadav ; Sudharm Rathore ; Abhishek Shubhrant ; Aparna Singh

A SIMPLE METHOD FOR MEASURING SOLAR RADIATION INTENSITY BY IMAGE ANALYSES 2906
Akiko Takahashi ; Akinori Moriki ; Nobuyuki Yamada ; Jun Imai ; Shigeyuki Funabiki

DEGRADATION OF SOLDER BONDS IN FIELD AGED PV MODULES: CORRELATION WITH SERIES RESISTANCE INCREASE 2912
Abhishiktha Tummala ; Jaewon Oh ; Sai Tatapudi ; Govindasamy Tamizhmani

PERFORMANCE OF LIGHT AND DARK CURRENT-VOLTAGE CHARACTERISTICS FOR PID-AFFECTED MONOCRYSTALLINE SILICON SOLAR MODULES 2918
He Wang ; Pan Zhao ; Shuwen Guo ; Hong Yang ; Weijing Huang ; Shiyu Sang ; Bojie Su ; Xue Zhang ; Yunxue Cao ; Hui Zhao

SOILING RATES OF PV MODULES VS. THERMOPILE PYRANOMETERS 2923
Martin Waters ; Tejas Tirumalai ; Michael Gostein ; Bill Stueve

GRID INTEGRATION OF BUILDING SYSTEMS AND 1 MW PHOTOVOLTAIC ARRAY USING VOLTTRON 2926
David Raker ; Andrew Sellers ; Roshan Kini ; Michael Green ; Thomas Stuart ; Randall Ellingson ; Raghav Khanna ; Michael Heben

INTERCONNECTION STUDY OF DISTRIBUTED PV SYSTEMS BY INTERFACING MATLAB WITH OPENDSS AND GIS 2931
Joseph A. Ahamioje ; Hariharan Krishnaswami

NOVEL MPPT ALGORITHM FOR ACTIVE POWER CONTROL OF MULTI-LEVEL DUAL-ACTIVE BRIDGE PV CONVERTER IMPLEMENTED IN NI MYRIO 2936
Shilpa Marti ; Hariharan Krishnaswami

MODELING A GRID-CONNECTED PV/BATTERY MICROGRID SYSTEM WITH MPPT CONTROLLER 2941
Genesis Alvarez ; Hadis Moradi ; Mathew Smith ; Ali Zilouchian

>94.5%REDUCTION IN GRID-BUY ELECTRICITY AND ELIMINATION OF AM & PM ENERGY PEAKS/SPIKES BY OPTIMIZING ENERGY USAGE AND INTEGRATION OF CUSTOMER SELF-SUPPLY ROOFTOP SOLAR PV WITH ELECTRICAL & THERMAL (HOT & COLD) STORAGE BATTERIES: A CASE STUDY FOR RESIDENTIAL HAWAII 2947
John Borland ; Jay Moore ; Corpuz Poncho ; Takahiro Tanaka ; Harumi Mcclure

A SINGLE-STAGECˈUK-BASED TRANSFORMERLESS INVERTER FOR 1-Φ GRID-CONNECTED PV SYSTEMS 2952
Phani Kumar Chamarthi ; Amit Kumar Gupta ; Madhuwanti S. Joshi ; Vivek Agarwal

A STATE SPACE AVERAGE MODEL FOR DYNAMIC MICROGRID BASED SPACE STATION SIMULATIONS 2957
Rachid Darbali-Zamora ; Eduardo I. Ortiz-Rivera

BUCK CONVERTER AND SEPIC BASED ELECTRONIC POWER SUPPLY DESIGN WITH MPPT AND VOLTAGE REGULATION FOR SMALL SATELLITE APPLICATIONS 2963
Rachid Darbali-Zamora ; Nicolás Cobo-Yepes ; John E. Salazar-Duque ; Eduardo I. Ortiz-Rivera ; Amilcar A. Rincon-Charris

VIRTUAL POWER PLANT FEEDBACK CONTROL DESIGN FOR FAST AND RELIABLE ENERGY MARKET AND CONTINGENCY RESERVE DISPATCH 2969
Mohamed Elkhatib ; Jay Johnson ; David Schoenwald

INTELLIGENT SAMPLING OF PERIODS FOR REDUCED COMPUTATIONAL TIME OF TIME SERIES ANALYSIS OF PV IMPACTS ON THE DISTRIBUTION SYSTEM 2975
Jason Galtieri ; Matthew J. Reno

A PWM SCHEME TO REALISE TWO TIMES EFFECTIVE SWITCHING FREQUENCY WITH CONSTANT COMMON MODE VOLTAGE AND REACTIVE POWER CAPABILITY IN 1- Φ GRID-TIED TRANSFORMERLESS H6 PV INVERTER 2981
Amit Kumar Gupta ; Madhuwanti S. Joshi ; Vivek Agarwal

A SOLAR PV RETROFIT SOLUTION FOR RESIDENTIAL BATTERY INVERTERS 2986
Amit Kumar Gupta ; Vaibhav Pawar ; Madhuwanti S. Joshi ; Vivek Agarwal ; Deepak Chandran

COST BENEFIT AND ALTERNATIVES ANALYSIS OF DISTRIBUTION SYSTEMS WITH ENERGY STORAGE SYSTEMS 2991
Tom Harris ; Adarsh Nagarajan ; Murali Baggu ; Tom Bialek

EVALUATION OF PV HOSTING CAPACITIES OF DISTRIBUTION GRIDS WITH UTILIZATION OF SOLAR-ROOF-POTENTIAL-ANALYSES 2996
Gerd Heilscher ; Falko Ebe ; Basem Idlbi ; Jeromie Morris ; Florian Meier

EXPERIMENTAL DISTRIBUTION CIRCUIT VOLTAGE REGULATION USING DER POWER FACTOR, VOLT-VAR, AND EXTREMUM SEEKING CONTROL METHODS 3002
Jay Johnson ; Sigifredo Gonzalez ; Daniel B. Arnold

DYNAMIC SETPOINT CONTROL OF ELECTRIC HOT WATER HEATER TANKS FOR INCREASED INTEGRATION OF SOLAR PHOTOVOLTAIC SYSTEMS 3008
C. Birk Jones ; Monte Lunacek ; Matthew Lave ; Jay Johnson ; Robert Broderick

SPATIAL ANALYSIS OF RESIDENTIAL COMBINED PHOTOVOLTAIC AND BATTERY POTENTIAL: CASE STUDY UTRECHT, THE NETHERLANDS 3014
Geert Litjens ; Bala Bhavya Kausika ; Ernst Worrell ; Wilfried Van Sark

POWER BALANCE REQUIREMENTS FOR SUSTAINED ISLANDING OF INVERTER BASED DISTRIBUTED GENERATION 3020
Gregory A. Kern ; Michael Ropp ; Sigifredo Gonzalez

FULL-SCALE DEMONSTRATION OF DISTRIBUTION SYSTEM PARAMETER ESTIMATION TO IMPROVE LOW-VOLTAGE CIRCUIT MODELS 3025
Matthew Lave ; Matthew J. Reno ; Robert J. Broderick ; Jouni Peppanen

CREATION AND VALUE OF SYNTHETIC HIGH-FREQUENCY SOLAR INPUTS FOR DISTRIBUTION SYSTEM QSTS SIMULATIONS 3031
Matthew Lave ; Matthew J. Reno ; Robert J. Broderick

A DIRECT MAXIMUM POWER POINT SEARCH USING CURRENT-VOLTAGE BASED
POWER-LAW RELATION FOR PHOTOVOLTAIC SYSTEM UNDER UNIFORM IRRADIANCE 3038
   Hitesh K. Mehta ; Ashish K. Panchal

PASSIVITY BASED CONTROLLER FOR PHOTOVOLTAIC MODULES USING ZETA
CONVERTER 3044
   Daniel A. Merced Cirino ; Rachid Darbali Zamora ; Eduardo I. Ortiz Rivera

SIC SWITCH BASED SINGLE-STAGE BUCK-BOOST TRANSFORMERLESS MINI INVERTER
WITH LOW LEAKAGE CURRENT AND NEGLIGIBLE DC INJECTION 3050
   Soumya Ranjan Mohapatra ; Amit Kumar Gupta ; Madhuwanti S. Joshi ; Vivek Agarwal

OPEN SOURCE TOOLS FOR HIGH PERFORMANCE QUASI-STATIC-TIME-SERIES
SIMULATION USING PARALLEL PROCESSING 3055
   Davis Montenegro ; Roger C. Dugan ; Matthew J. Reno

MAXIMUM POWER POINT TRACKING OF PV MODULE BASED ON NEW EXPLICIT I-V
RELATION 3061
   Tejeswar Nukala ; A. K. Panchal

AN AUTOCORRELATION-BASED COPULA MODEL FOR PRODUCING REALISTIC CLEAR-
SKY INDEX AND PHOTOVOLTAIC POWER GENERATION TIME-SERIES 3067
   Joakim Munkhammar ; Joakim Widén

DYNAMIC RESPONSE OF MAXIMUM POWER POINT TRACKING USING PERTURB AND
OBSERVE ALGORITHM WITH MOMENTUM TERM 3073
   Gautam A. Raiker

A FRAMEWORK FOR COMPARING THE ECONOMIC PERFORMANCE AND ASSOCIATED
EMISSIONS OF GRID-CONNECTED BATTERY STORAGE SYSTEMS IN EXISTING
BUILDING STOCK: A NYISO CASE STUDY 3077
   Julian Do Nascimento Ricardo ; Vasilis Fthenakis

IMPROVING ANY ARBITRARY MPPT HILL CLIMBER WITH ANN ESTIMATIONS 3083
   Jesse Roberts ; Indranil Bhattacharya

INCREASING SOLAR PHOTOVOLTAIC PENETRATION USING THERMAL ENERGY
STORAGE 3088
   Alexander F. Routhier ; Christiana Honsberg

MODEL PREDICTIVE CONTROL OF GRID CONNECTED MODULAR MULTILEVEL
CONVERTER FOR INTEGRATION OF PHOTOVOLTAIC POWER SYSTEMS 3092
   Amir Shahirinia ; Amin Hajizadeh

MAXIMIZATION OF SELF-SUFFICIENCY WITH GRID CONSTRAINTS: PV GENERATORS,
WIND TURBINES AND STORAGE TO FEED TERTIARY SECTOR USERS 3096
   Filippo Spertino ; Jawad Ahmad ; Alessandro Ciocia ; Paolo Di Leo ; Francesco Giordano

SWITCHES CONTROLLING TO IMPLEMENT ADAPTIVE MULTILEVEL INVERTER ON PV
SYSTEM 3102
   Hadi Suhana ; Ngapuli I Sinisuka ; Muhammad Nurdin ; Yvon Besanger ; Vincent Debusschere

DEMAND RESPONSE FOR THE PROMOTION OF PHOTOVOLTAIC PENETRATION 3107
   Venizelos Venizelou ; Spyros Theocharides ; George Makrides ; Venizelos Efthymiou ; George E. Georghiou

GRIDDLER AI: NEW PARADIGM IN LUMINESCENCE IMAGE ANALYSIS USING
AUTOMATED FINITE ELEMENT METHODS 3113
   Johnson Wong ; Percis Teena ; Daniel Inns

INTERACTION OF O2IDIMERS WITH GA IN SI AND IMPLICATIONS FOR A
COMPREHENSIVE MODEL OF LIGHT- INDUCED DEGRADATION 3119
   Yu Jin ; Scott T. Dunham

NUMERICAL SIMULATION OF EBIC FOR ANALYSIS OF EXTENDED DEFECTS 3123
   Marco Nardone ; John Moseley ; Saroj Dahal ; Anuja V. Parikh ; John M. Waddle

COLLOIDAL QUANTUM DOT SOLAR CELL ELECTRICAL PARAMETER IMAGING USING
CAMERA-BASED HIGH-FREQUENCY HETERODYNE LOCK-IN CARRIEROGRAPHY 3129
   Lilei Hu ; Mengxia Liu ; Andreas Mandelis ; Qiming Sun ; Alexander Melnikov ; Edward H. Sargent

A NEW PERSPECTIVE ON POTENTIAL-INDUCED DEGRADATION OF THE SHUNTING
TYPE BY MICRO RAMAN-SPECTROSCOPY AND MICRO LIGHT-BEAM-INDUCED
CURRENT 3135
   A. Büchler ; H. Nagel ; M. Breitwieser ; S. Kluska ; F. D. Heinz ; M. C. Schubert ; M. Glatthaar ; S. Glunz

NANOSCALE DETECTION OF DEEP LEVELS IN CIGS USING ELECTRON ENERGY LOSS
SPECTROSCOPY 3139
   Julia I. Deitz ; Pran K. Paul ; Shankar Karki ; Sylvain Marsillac ; Aaron R. Arehart ; Tyler J. Grassman ; David W. Mccomb

MEASUREMENT OF CARRIER DYNAMICS IN PHOTOVOLTAIC CZTSE BY TIME-RESOLVED TERAHERTZ SPECTROSCOPY 3143
Siming Li ; Michael A. Lloyd ; Andrew A. Golembeski ; Brian E. Mccndless ; Jason B. Baxter

DECOUPLING GRAIN-BOUNDARY, GRAIN-INTERIOR, AND SURFACE RECOMBINATION WITH CATHODOLUMINESCENCE 3147
John Moseley ; Pierre Rale ; Stéphane Collin ; Ana Kanevce ; Eric Colegrove ; Joel Duenow ; Soren Jensen ; Wyatt K. Metzger ; Mowafak M. Al-Jassim

HIGH RESOLUTION THZ SCANNING FOR OPTIMIZATION OF DIELECTRIC LAYER OPENING PROCESS ON DOPED SI SURFACES 3150
P. Spinelli ; F.J.K. Danzl ; D. Deligiannls ; N. Guillevin ; A.R. Burgers ; S. Sawallich ; M. Nage ; I. Cesar

DEGRADATION ASSESSMENT OF FIELDED CIGS PHOTOVOLTAIC ARRAYS 3155
Bruce H. King ; Joshua S. Stein ; Daniel Riley ; C. Birk Jones ; Charles D. Robinson

APPLICATION OF IEC 61724 STANDARDS TO ANALYZE PV SYSTEM PERFORMANCE IN DIFFERENT CLIMATES 3161
Katherine A. Klise ; Joshua S. Stein ; Joseph Cunningham

EFFECTS OF URBAN ENVIRONMENT ON SOLAR PV PERFORMANCE 3167
Panagiotis Moraitis ; Bala Bhavya Kausika ; Wilfried G.J.H.M. Van Sark

IRRADIANCE MEASUREMENT CONSIDERATIONS FOR SYSTEM PERFORMANCE ASSESSMENT WHEN MANAGING FLEETS OF PHOTOVOLTAIC ASSETS ACROSS ASIA 3172
André M. Nobre ; Shravan Karthik ; Chenxi Liu ; Rohit Jaswal ; Rupesh Baker ; Raghav Malhotra ; Alan Khor

MACHINE LEARNING IN PV FAULT DETECTION, DIAGNOSTICS AND PROGNOSTICS: A REVIEW 3178
Sandy Rodrigues ; Helena Geirinhas Ramos ; F. Morgado-Dias

OUTDOOR FIELD PERFORMANCE FROM BIFACIAL PHOTOVOLTAIC MODULES AND SYSTEMS 3184
Joshua S. Stein ; Daniel Riley ; Matthew Lave ; Clifford Hansen ; Chris Deline ; Fatima Toor

DEFINING THRESHOLD VALUES OF ENCAPSULANT AND BACKSHEET ADHESION FOR PV MODULE RELIABILITY 3190
Nick Bosco ; Joshua Eafanti ; Sarah Kurtz ; Jared Tracy ; Reinhold Dauskardt

CHARACTERIZATIONS OF AGED GLASS/ETHYLENE VINYL ACETATE/GLASS USING FLUORESCENCE SPECTROSCOPY AND INSTRUMENTED INDENTATION 3195
Jae Hyun Kim ; Yadong Lyu ; David C. Miller ; Xiaohong Gu

ENCAPSULANT ADHESION TO SURFACE METALLIZATION ON PHOTOVOLTAIC CELLS 3200
Jared Tracy ; Nick Bosco ; Reinhold Dauskardt

IMPACT OF UV LIGHT INTENSITY ON PHOTODEGRADATION OF PV BACKSHEETS 3204
Xiaohong Gu ; Li-Chieh Yu ; Yadong Lyu ; Jae Hyun Kim ; Andrew Fairbrother ; Tinh Nguyen

SURVEY OF MECHANICAL DURABILITY OF PV BACKSHEETS 3208
Michael D. Kempe ; David C. Miller ; Allen Zielnik ; Daniel Montiel-Chicharro ; Jiang Zhu ; Ralph Gottschalg

SOLAR VARIABILITY REDUCTION USING OFF-MAXIMUM POWER POINT TRACKING AND BATTERY STORAGE 3214
Jason Galtieri ; Philip T. Krein

INTEGRATION OF ELECTROCHEMICAL CAPACITORS ON SILICON PHOTOVOLTAIC MODULES FOR RAPID-RESPONSE POWER BUFFERING 3220
Yu Jiang ; Xuanyi Shi ; Derwin Lau ; Da-Wei Wang ; Zi Ouyang ; Alison Lennon

DESIGN & EVALUATION OF A HYBRID SWITCHED CAPACITOR CIRCUIT WITH WIDE-BANDGAP DEVICES FOR COMPACT MVDC PV POWER CONVERSION 3224
J. Stewart ; J. Delhotal ; J. Richards ; J. Neely ; L. Rashkin ; J. D. Flicker ; R. Kaplar ; S. Gonzalez ; J. Lehr

SOLAR ENERGY FOR CLEAN AND AFFORDABLE WATER DESALINATION 3230
V. M. Fthenakis ; Adam A. Atia

GLOBAL RESIDENTIAL AIR-CONDITIONING SECTOR AS A DRIVER FOR PHOTOVOLTAIC INDUSTRY GROWTH DURING THE 21ST CENTURY 3236
Hannu S. Laine ; Jyri Salpakari ; Marius Peters ; Erin E. Looney ; Ashley E. Morishige ; Hele Savin ; Gregory Wilson ; Tonio Buonassisi

MEASURES TO REMOVE ECONOMIC NON-MARKET FAILURE AND INSTITUTIONAL BARRIERS THAT RESTRICT PHOTOVOLTAICS SELF-CONSUMPTION AND NET-METERING IN SPAIN 3240
Enrique Rosalcs-Ascnsio ; Juan A. Méndez ; Benjamín Gonzálcz-Díaz ; Ricardo Guerrero Lemus

COST COMPETITIVE CONCENTRATOR PHOTOVOLTAICS FOR SOLAR THERMAL APPLICATIONS 3245
Brian C. Riggs ; Richard E. Biedenham ; Chris Dougher ; Yaping Vera Ji ; Qi Xu ; Vince Romanin ; Daniel S. Codd ; James M. Zahler ; Matthew D. Escarra

PREDICTING THE EFFICIENCY OF THE SILICON BOTTOM CELL IN A TWO-TERMINAL TANDEM SOLAR CELL 3250
Zhengshan J. Yu ; Zachary C. Holman

MECHANICALLY STACKED 4-TERMINAL III-V/SI TANDEM SOLAR CELLS 3254
Stephanie Essig ; Christophe Allebe ; John F. Geisz ; Myles A. Steiner ; Loris Barraud ; Antoine Descoeudres ; J. Scott Ward ; Manuel Schnabel ; David L. Young ; Matthieu Despeisse ; Christophe Ballif ; Adele Tamboli

PEROVSKITE/SILICON TANDEM SOLAR CELLS: CHALLENGES TOWARDS HIGH-EFFICIENCY IN 4-TERMINAL AND MONOLITHIC DEVICES 3256
Jérémie Werner ; Florent Sahli ; Brett Kamino ; Davide Sacchetto ; Matthias Bräuninger ; Arnaud Walter ; Christophe Ballif ; Matthieu Despeisse ; Sylvain Nicolay ; Bjoern Niesen ; Raphaël Monnard ; Stefaan De Wolf ; Soo-Jin Moon ; Loris Barraud ; Bertrand Paviet-Salomon ; Jonas Geissbuehler ; Christophe Allebé

THE OUTCOME OF REPLACING SN COMPLETELY BY GE IN KESTERITE CU2ZNSNSE4SOLAR CELLS 3260
S. Sahayaraj ; G. Brammertz ; B. Vermang ; T. Schnabel ; E. Ahlswede ; Z. Huang ; S. Ranjbar ; M. Meuris ; J. Vleugels ; J. Poortmans

TRANSITION METAL OXIDES NANO-LAYERS AS EFFICIENT BACK ELECTRON REFLECTORS FOR CU2ZNSNSE4SOLAR CELLS 3265
Sergio Giraldo ; Moisés Espíndola-Rodríguez ; Florian Oliva ; Víctor Izquierdo-Roca ; Alejandro Pérez-Rodríguez ; Edgardo Saucedo

MIXED SULFUR AND SELENIUM ANNEALING STUDY OF COMPOUND-SPUTTERED BILAYER CU2ZNSNS4/ CU2ZNSNSE4PRECURSORS 3269
N. Ross ; S. Grini ; L. Vines ; C. Platzer-Björkman

REVEALING THE ROLE OF MN INCORPORATION IN CU2ZNSN(S, SE)4PHOTOVOLTAIC ABSORBER LAYER 3275
Stener Lie ; Joel M. R. Tan ; Wenjie Li ; Shin Woei Leow ; Oki Gunawan ; Doug Bishop ; Lydia H. Wong

NON-VACUUM SINGLE STEP SYNTHESIS OF LARGE-GRAIN SIZE CZTS PHOTO ABSORBER FOR THIN FILM SOLAR CELLS BY FLUX ASSISTED CHEMICAL SPRAY 3279
Ratheesh R. Thankalekshmi ; Navjot Kaur Sidhu ; A.C. Rastogi

RAMAN SCATTERING ASSESSMENT OF POINT DEFECTS IN KESTERITE SEMICONDUCTORS: UV RESONANT RAMAN CHARACTERIZATION FOR ADVANCED PHOTOVOLTAICS 3285
Florian Oliva ; Laia Arqués Farré ; Sergio Giraldo ; Mirjana Dimitrievska ; Paul Pistor ; Alejandro Martínez-Pérez ; Lorenzo Calvo-Barrio ; Edgardo Saucedo ; Alejandro Pérez-Rodríguez ; Victor Izquierdo-Roca

ASSESSING THE DEFECT RESPONSIBLE FOR LETID: TEMPERATURE- AND INJECTION-DEPENDENT LIFETIME SPECTROSCOPY 3290
Mallory A. Jensen ; Yan Zhu ; Erin E. Looney ; Ashley E. Morishige ; Carlos Vargas ; Ziv Hameiri ; Tonio Buonassisi

MICROSCOPIC DISTRIBUTION OF LUMINESCENCE FROM DISLOCATION CLUSTERS IN MULTICRYSTALLINE SILICON WAFERS 3295
H. T. Nguyen ; M. A. Jensen ; L. Li ; C. Samundsett ; H. C. Sio ; B. Lai ; T. Buonassisi ; D. Macdonald

DO GRAIN BOUNDARIES MATTER? ELECTRICAL AND ELEMENTAL IDENTIFICATION AT GRAIN BOUNDARIES IN LETID-AFFECTED P-TYPE MULTICRYSTALLINE SILICON 3300
Mallory A. Jensen ; Ashley E. Morishige ; Sagnik Chakraborty ; Romika Sharma ; Hang Cheong Sio ; Chang Sun ; Barry Lai ; Volker Rose ; Amanda Youssef ; Erin E. Looney ; Sarah Wieghold ; Jeremy Poindexter ; Juan-Pablo Correa-Baena ; Daniel Macdonald ; Joel B. Li ; Tonio Buonassisi

PERC SOLAR CELL PERFORMANCE PREDICTIONS FROM MULTICRYSTALLINE SILICON INGOT METROLOGY DATA 3304
Bernhard Mitchell ; Daniel Chung ; Qiuxiang He ; Hua Zhang ; Zhen Xiong ; Pietro P. Altermatt ; Peter Geelan-Small ; Thorsten Trupke

PHOTOLUMINESCENCE-IMAGING-BASED EVALUATION OF NON-UNIFORM CDTE DEGRADATION 3305
Steve Johnston ; David Albin ; Peter Hacke ; Steven P. Harvey ; Helio Moutinho ; Mowafak Al-Jassim ; Wyatt K. Metzger

MACHINE LEARNING AND CORRELATIVE MICROSCOPY: HOW 'BIG DATA' TECHNIQUES CAN BENEFIT THIN FILM SOLAR CELL CHARACTERIZATION 3309
Bradley M. West ; Michael Stuckelberger ; Tara Nietzold ; Barry Lai ; Jörg Maser ; Mariana I. Bertoni

METAL INDUCED CONTACT RECOMBINATION MEASURED BY QUASI-STEADY-STATE PHOTOLUMINESCENCE 3315
Robert Dumbrell ; Mattias K. Juhl ; Mengjie Li ; Thorsten Trupke ; Ziv Hameiri

USING TIME-OF-FLIGHT SIMS TO INVESTIGATE GROUP V DOPANT DISTRIBUTION IN CDTE 3319
Steven P. Harvey ; Eric Colegrove ; Brian Mccandless ; David Albin ; Mowafak Al-Jassim ; Wyatt K. Metzger

QUANTITATIVE ANALYSIS OF ACTIVE DOPANT DISTRIBUTION AND ESTIMATION OF
EFFECTIVE DIFFUSIVITY IN PHOSPHORUS- IMPLANTED EMITTER OF SI SOLAR CELL
USING SCANNING NONLINEAR DIELECTRIC MICROSCOPY 3323
    Kotaro Hirose ; Katsuto Tanahashi ; Hidetaka Takato ; Yasuo Cho

SIMULATION OF DRIVE-LEVEL CAPACITANCE PROFILING TO INTERPRET
MEASUREMENTS ON CU(IN, GA)SE2SCHOTTKY DEVICES 3327
    Geordie Zapalac ; Jeff Bailey

ANALYSIS OF THE IMPACT OF INSTALLATION PARAMETERS AND SYSTEM SIZE ON
BIFACIAL GAIN AND ENERGY YIELD OF PV SYSTEMS 3333
    Amir Asgharzadeh ; Tomas Lubenow ; Joseph Sink ; Bill Marion ; Chris Deline ; Clifford Hansen ; Joshua Stein ;
    Fatima Toor

DEPENDENCE OF STRING POWER ON ITS HEIGHT IN THE ARRAY IN YOSHINOGARI
MEGA SOLAR POWER PLANT 3339
    Shigeomi Hara ; Makoto Kasu ; Yasuki Masutomi

A BOTTOM-UP ENERGY SIMULATION FRAMEWORK TO ACCURATELY COMPARE PV
MODULE TOPOLOGIES UNDER NON-UNIFORM AND DYNAMIC OPERATING
CONDITIONS 3343
    Patrizio Manganiello ; Maro Baka ; Hans Goverde ; Tom Borgers ; Jonathan Govaerts ; Arvid Van Der Heide ;
    Eszter Voroshazi ; Francky Catthoor

A PERFORMANCE MODEL FOR BIFACIAL PV MODULES 3348
    Daniel Riley ; Clifford Hansen ; Joshua Stein ; Matthew Lave ; Johnson Kallickal ; Bill Marion ; Fatima Toor

ACCURATE MODELING OF PARTIALLY SHADED PV ARRAYS 3354
    Bennet Meyers ; Mark Mikofski

EVALUATION OF UNCERTAINTY IN PV PROJECT DESIGN: DEFINITION OF SCENARIOS
AND IMPACT ON ENERGY YIELD PREDICTIONS 3360
    Giorgio Belluardo ; Magnus Herz ; Ulrike Jahn ; Mauricio Richter ; David Moser

MONOCRYSTALLINE 1.7 EV MGCDTE DOUBLE-HETEROSTRUCTURE SOLAR CELL WITH
11.2% EFFICIENCY 3366
    Calli M. Campbell ; Xin-Hao Zhao ; Yuan Zhao ; Mathieu Boccard ; Cheng- Ying Tsai ; Jacob J. Becker ; Zachary
    Holman ; Yong-Hang Zhang

MBE GROWTH OF 1.7EV AL0.2GA0.8AS AND 1.42EV GAAS SOLAR CELLS ON SI USING
DISLOCATIONS FILTERS: AN ALTERNATIVE PATHWAY TOWARD III-V/ SI SOLAR CELLS
ARCHITECTURES 3370
    Arthur Onno ; Mingchu Tang ; Mu Wang ; Yurii Maidaniuk ; Mourad Benamara ; Yuriy I. Mazur ; Gregory J.
    Salamo ; Lars Oberbeck ; Jiang Wu ; Huiyun Liu

III- V/SI TANDEM CELLS UTILIZING INTERDIGITATED BACK CONTACT SI CELLS AND
VARYING TERMINAL CONFIGURATIONS 3371
    Manuel Schnabel ; Michael Rienacker ; Agnes Merkle ; Talysa R. Klein ; Nikhil Jain ; Stephanie Essig ; Henning
    Schulte-Huxel ; Emily Warren ; Maikel F.A.M. Van Hest ; John Geisz ; Jan Schmidt ; Rolf Brendel ; Robby Peibst
    ; Paul Stradins ; Adele Tamboli

TOWARDS HIGH-EFFICIENCY GAASP/SI TANDEM CELLS 3376
    S. Fan ; M. Vaisman ; K. Nay Yaung ; E. Perl ; D. Martín-Martín ; M. Leilaeioun ; Z. C. Holman ; M. L. Lee

CHARACTERIZATION OF HETEROEPITAXIAL GAAS FILMS GROWN ON SI USING
SELECTIVE AREA NUCLEATION 3381
    Emily L. Warren ; Emily A. Makoutz ; Michelle Vaisman ; Benjamin F. Bachman ; William E. Mcmahon ; Jeramy
    D. Zimmerman ; Adele C. Tamboli

EFFICIENT PHOTON UPCONVERSION IN SEMICONDUCTOR NANOSTRUCTURES:
CONSTRAINTS AND OPPORTUNITIES 3384
    Matthew F. Doty ; Eric Y. Chen ; Jing Zhang ; Diane G. Sellers ; Zhuohui Li ; Christopher C. Milleville ; Kyle
    Lennon ; Joshua M. O. Zide

ENHANCED ULTRA-THIN A-GE:H SOLAR CELLS BY PLASMONIC NANOPARTICLES
EMBEDDED IN THE OPTICAL RESONANT CAVITY 3388
    Brendan Brady ; Volker Steenhoff ; Benedikt Nickel ; Martin Vehse ; Alexander G. Brolo

NATIVE-METAL-OXIDE-COATED PLASMONIC ELECTRODE METASURFACES FOR
NANOPHOTONIC LIGHT TRAPPING AND EFFICIENT CHARGE COLLECTION 3393
    Deirdre M. O'Carroll ; Christopher E. Petoukhoff ; Zhongkai Cheng ; Zeqing Shen ; Catrice M. Carter

IN-GA PRECURSOR ISLANDS FOR CU(IN, GA)SE2MICRO-CONCENTRATOR SOLAR CELLS 3396
    Katharina Eylers ; Franziska Ringleb ; Berit Heidmann ; Sergiu Levcenco ; Thomas Unold ; Hagen W. Klemm ;
    Gina Peschel ; Alexander Fuhrich ; Thomas Teubner ; Thomas Schmidt ; Martina Schmid ; Torta Boeck

ADVANCES IN SILICON SURFACE TEXTURIZATION BY METAL ASSISTED CHEMICAL
ETCHING FOR PHOTOVOLTAIC APPLICATIONS 3402
    Sylvain Le Gall ; Raphaël Lachaume ; Encarnacion Torralba ; Mathieu Halbwax ; Vincent Magnin ; Taha El
    Assimi ; Marin Fouchier ; Joseph Harari ; Jean-Pierre Vilcot ; Christine Cachet-Vivier ; Stéphane Bastide

SINGLE CRYSTALLINE SUBSTRATES FOR III- V GROWTH VIA EXFOLIATION OF BULK SINGLE CRYSTALS..3406
Celeste L. Melamed ; Brenden R. Ortiz ; Aaron D. Martinez ; William E. Mcmahon ; Adele C. Tamboli ; Andrew G. Norman ; Eric S. Toberer

CUZNS HOLE CONTACTS ON MONOCRYSTALLINE CDTE SOLAR CELLS................................3410
Jacob J. Becker ; Xiaojie Xu ; Rachel Woods-Robinson ; Calli M. Campbell ; Maxwell Lassise ; Joel Ager ; Yong-Hang Zhang

THE EFFECT OF THE CDCL2 HEAT TREATMENT ON CDSEXTE1-X SOLAR CELLS......................3413
Chih An Hsu ; Vasilios Palekis ; Imran Khan ; Shamara Collins ; Don Morel ; Chris Ferekides

EFFECTS OF CDCL2TREATMENT ON THE LOCAL ELECTRONIC PROPERTIES OF POLYCRYSTALLINE CDTE MEASURED WITH PHOTOEMISSION ELECTRON MICROSCOPY..3417
Morgann Berg ; Jason M. Kephart ; Walajabad S. Sampath ; Taisuke Ohta ; Calvin Chan

POINT DEFECTS IN CDTE BULK SINGLE CRYSTALS GROWN IN CD-RICH CONDITIONS....................3422
Tursun Ablekim ; Santosh K. Swain ; Teresa M. Barnes ; Kelvin G. Lynn

OPTICAL PROPERTIES OFCDSE1-XSXANDCDSE1-YTEYALLOYS AND THEIR APPLICATION FOR CDTE PHOTOVOLTAICS..3426
Maxwell M. Junda ; Corey R. Grice ; Prakash Koirala ; Robert W. Collins ; Yanfa Yan ; Nikolas J. Podraza

BLISTERING OF MAGNETRON SPUTTERED THIN FILM CDTE DEVICES..................................3430
P.M. Kaminski ; S. Yilmaz ; A. Abbas ; F. Bittau ; J.W. Bowers ; R.C. Greenhalgh ; J.M. Walls

ENERGY YIELD IN HOT & SUNNY CLIMATES: IMPACT OF SILICON SOLAR CELL ARCHITECTURE AND CELL INTERCONNECTION..3435
Jan Haschke ; Johannes P. Seif ; Yannick Riesen ; Andrea Tomasi ; Jean Cattin ; Loïc Tous ; Patrick Choulat ; Monica Aleman ; Emanuele Comagliotti ; Angel Uruena ; Richard Russell ; Filip Duerinckx ; Jonathan Champliaud ; Jacques Levrat ; Amir A. Abdallah ; Brahim Aïssa ; Nouar Tabet ; Nicolas Wyrsch ; Matthieu Despeisse ; Jozef Szlufcik ; Stefaan De Wolf ; Christophe Ballif

NOVEL REAR SIDE METALLIZATION ROUTE FOR SI SOLAR CELLS USING A TRANSPARENT CONDUCTING ADHESIVE..3439
Manuel Schnabel ; Talysa R. Klein ; Benjamin G. Lee ; William Nemeth ; Vincenzo Lasalvia ; Maikel F.A.M. Van Hest ; Paul Stradins

MULTILAYER FOIL METALLIZATION FOR ALL BACK CONTACT CELLS..................................3442
David H. Levy ; David E. Carlson

ELECTROLUMINESCENCE EXCITATION SPECTROSCOPY: A NOVEL APPROACH TO NON-CONTACT QUANTUM EFFICIENCY MEASUREMENTS..3448
Kristopher O. Davis ; Greg S. Horner ; Joshua B. Gallon ; Leonid A. Vasilyev ; Kyle B. Lu ; Antonius B. Dirriwachter ; Terry B. Rigdon ; Eric J. Schneller ; Kortan Ogutman ; Richard K. Ahrenkiel

ILLUMINATED OUTDOOR LUMINESCENCE IMAGING OF PHOTOVOLTAIC MODULES......................3452
Timothy J Silverman ; Michael G. Deceglie ; Kaitlyn Vansant ; Steve Johnston ; Ingrid Repins

ELECTROLUMINESCENT IMAGE PROCESSING AND CELL DEGRADATION TYPE CLASSIFICATION VIA COMPUTER VISION AND STATISTICAL LEARNING METHODOLOGIES..3456
Justin S. Fada ; Mohammad A. Hossain ; Jennifer L. Braid ; Shuying Yang ; Timothy J Peshek ; Roger H. French

TOWARDS DEVELOPING A STANDARD FOR TESTING BIFACIAL PV MODULES: SINGLE-SIDE VERSUS DOUBLE-SIDE ILLUMINATION METHOD I-V MEASUREMENTS UNDER DIFFERENT IRRADIANCE AND TEMPERATURE..3462
Stefan Roest ; Witek Nawara ; Bas B. Van Aken ; Elias Garcia Goma

ELECTRICAL TRANSPORT PROPERTIES FROM LONG WAVELENGTH ELLIPSOMETRY......................3468
Prakash Uprety ; Maxwell M. Junda ; Indra Subedi ; Michael A. Slocum ; David V. Forbes ; Seth M. Rubbard ; Nikolas J. Podraza

IN SITU RAMAN MONITORING OF KESTERITE CU2ZNSNS4 PHASE FORMATION FROM SULFURIZATION OF SOL-GEL OXIDE PRECURSORS..3473
Osama Awadallah ; Joseph Hernandez ; Andriy Durygin ; Zhe Cheng

PERFORMANCE OF FIELD-AGED PV MODULES IN INDIA: RESULTS FROM 2016 ALL INDIA SURVEY OF PV MODULE RELIABILITY..3478
Rajiv Dubey ; Sachin Zachariah ; Shashwata Chattopadhyay ; Vivek Kuthanazhi ; Sugguna Rambabu ; Sonali Bhaduri ; Hemant K. Singh ; Archana Sinha ; Birinchi Bora ; Rajesh Kumar ; O. S. Sastry ; Chetan S. Solanki ; Anil Kottantharayil ; Brij M. Arora ; K. L. Narasimhan ; Juzer Vasi

INFERRING THE PERFORMANCE RATIO OF PV SYSTEMS DISTRIBUTED IN AN REGION: A REAL-CASE STUDY IN SOUTH TYROL..3482
Marco Pierro ; Giorgio Belluardo ; Philip Ingenhoven ; Cristina Cornaro ; David Moser

QUANTIFY PHOTOVOLTAIC MODULE DEGRADATION USING THE LOSS FACTOR MODEL PARAMETERS 3488

C. Birk Jones ; Bruce H. King ; Joshua S. Stein ; Justin S. Fada ; Alan J. Curran ; Roger H. French ; Erdmut Schnabel ; Michael Koehl ; Olga Lavrova

SIMULATING PV SYSTEM PERFORMANCE WITH COMPONENT RELIABILITY DISTRIBUTIONS 3494

Geoffrey T. Klisel ; Janine M. Freeman ; Olga Lavrova

LIFETIME AND DEGRADATION OF PRE-DAMAGED PV-MODULES – FIELD STUDY AND LAB TESTING 3500

Claudia Buerhop ; Sven Wirsching ; Simon Gehre ; Tobias Pickel ; Thilo Winkler ; Andreas Bemrrr ; Julia Merghcim ; Christian Camus ; Jens Hauch ; Christoph J. Brabec

IMM TRIPLE-JUNCTION SOLAR CELLS AND MODULES OPTIMIZED FOR SPACE AND TERRESTRIAL CONDITIONS 3506

Tatsuya Takamoto ; Hiroyuki Juso ; Kohsuke Ueda ; Hidetoshi Washio ; Hiroshi Yamaguchi ; Mitsuru Imaizumi ; Taishi Sumita ; Tetsuya Nakamura

VERY HIGH SPECIFIC POWER ELO SOLAR CELLS (>3 KW/KG) FOR UAV, SPACE, AND PORTABLE POWER APPLICATIONS 3511

D. Cardwell ; A. Kirk ; C. Stender ; A. Wibowo ; F. Tuminello ; M. Drees ; R. Chan ; M. Osowski ; N. Pan

ENHANCED ENDURANCE OF A UNMANNED AERIAL VEHICLES USING HIGH EFFICIENCY SI AND III-V SOLAR CELLS 3514

David Scheiman ; Raymond Hoheisel ; Daniel J Edwards ; Andrew Paulsen ; Justin Lorentzen ; Steve Carruthers ; Sam Carter ; Matthew Kelly ; Phillip Jenkins ; Robert Walters

HIGH PERFORMANCE, LIGHTWEIGHT GAAS SOLAR CELLS FOR AEROSPACE AND MOBILE APPLICATIONS 3520

Aarohi Vijh ; Lori Washington ; Robert C. Parenti

THROUGH-EPITAXIAL-VIA BACK-CONTACT MULTI-JUNCTION SOLAR CELLS FABRICATED USING EPITAXIAL LIFT-OFF 3524

Rekha Reddy ; Marilyn L. Nowakowski ; David Rowell ; Christopher L. Stender ; Christopher Youtsey

DESIGN OF INGAP/GAAS/LNGAAS MULTI-JUNCTION CELLS WITH REDUCED LAYER THICKNESSES USING LIGHT-TRAPPING REAR TEXTURE 3528

Lin Zhu ; Anurag Reddy ; Kentaroh Watanabe ; Masakazu Sugiyama ; Yoshiaki Nakano ; Hidefumi Akiyama

Author Index

Analytic JV-Characteristics of Ideal Impurity PV-Cells

Rune Strandberg

University of Agder, Department of Engineering Sciences, P.O. Box 509, NO-4898 Grimstad, Norway

Abstract—In this article the mathematical modeling of idealized impurity photovoltaic cells is greatly simplified through the derivation of analytic JV-characteristics. The resulting expressions are also facilitating the intuitive understanding of such photovoltaic devices. The new model is used to investigate the sensitivity of impurity photovoltaic cells to the absorption band width, the impurity related absorptivity and the external radiative efficiency. It is found that impurities with narrow absorption bands or low absorptivity can greatly reduce the efficiency of the device, even if they are fully radiative and have energy levels situated at optimal positions in the band gap. It is also found that increasing the number of impurity species improves the efficiency of devices with low impurity absorptivity or narrow absorption bands. A clear shortcoming of the model is its inability to handle overlapping absorption coefficients.

I. INTRODUCTION

In an impurity photovoltaic cell (IPVC) [1], impurity levels are used to excite electrons from the valence band to the conduction band in two-step processes similar to that taking place via the intermediate band in an intermediate band solar cell (IBSC). With one impurity level, the efficiency limit of an ideal IPVC is thus similar to that of an ideal IBSC [2]. The main difference between IPVCs and IBSCs is that the former is based on excitation via impurity states localized around individual impurity atoms, whereas the latter is based on excitation via de-localized states extending throughout the entire IB-material [3]. The difference allows IPVCs to consist of a host material with a mixture of different impurities, randomly distributed, that absorbs different parts of the solar spectrum. One striking example of optically active impurities is found in silicon carbide, which is transparent when pure, but turns brown or black by the presence of iron impurities. According to basic physics, absorbing transitions must also be emitting. Emission from impurities in silicon has indeed been observed, as well as utilized to identify particular impurity species [4].

Impurities are well known to allow electron-hole pairs to recombine non-radiatively via the cascade [5] or multi-phonon [6] mechanisms, which both goes under the umbrella Shockley-Read-Hall (SRH) recombination. It is thus possible that the recombination via the impurity states may have both radiative and non-radiative components. A model for SRH-recombination via optically active impurities was developed by Beaucarne and Green in 2003 [7].

Turning to device modeling, existing analysis of IPVCs with several different impurity levels have mostly been based on detailed balance models where a set of non-linear equations are solved numerically to obtain the IV-characteristic, as in Ref. [2].

In the present work, an analytic IV-characteristic for IPVCs is derived. The model can handle an arbitrary number of impurity levels with non-overlapping absorption coefficients. Since the concentration of impurities may be to small to allow unit absorptivity, the model takes into account the possibility to adjust the absorptivity. Non-radiative recombination is taken into account by incorporating the external radiative efficiency in the model. The development of this new model is based on a similar model for intermediate band solar cells, which was recently published [8]. The model for intermediate band solar cells, and the main assumptions for its validity, is therefore briefly described in the remainder of this introduction.

The link between detailed balance modeling [9] and the ideal diode equation [10] is the approximation

$$\left[e^{(E-qV_{ab})/kT} - 1 \right]^{-1} \approx e^{-(E-qV_{ab})/kT} \qquad (1)$$

which allows the current associated with radiative recombination between bands or states a and b to be expressed by

$$J_{ab} = J_{0,ab} e^{qV_{ab}/kT}. \qquad (2)$$

In the equations above, q is the elementary charge, k is Boltzmann's constant, T is the cell temperature and V_{ab} is the voltage associated with the splitting of the quasi-Fermi levels of the involved bands or levels. The approximation is valid whenever $E-qV_{ab}$, where E is the energy of the emitted photons, is larger than a few kT. The pre-exponential factor $J_{0,ab}$ is given by

$$J_{0,ab} = \frac{2\pi q}{h^3 c^2} \int_{E_l}^{E_h} E^2 \, e^{-E/kT} \, dE, \qquad (3)$$

where E_h and E_l are the upper and lower limits of the absorbing energy interval which allows transitions between the two bands or levels in question.

In Ref. [8], it was shown that whenever the approximation in Eq. (1) is valid, the JV-characteristic of an ideal intermediate band solar cell with non-overlapping absorption coefficients is well described by

$$J = J_G - J_{0,cv} e^{qV/kT} - \sqrt{J_{0,i}^2 e^{qV/kT} + \tfrac{1}{4}\Delta J_G^2}, \qquad (4)$$

where V is the cell voltage. The parameter J_G is a generation current density given by

$$J_G = J_{G,cv} + \tfrac{1}{2}\big(J_{G,ci} + J_{G,iv}\big), \qquad (5)$$

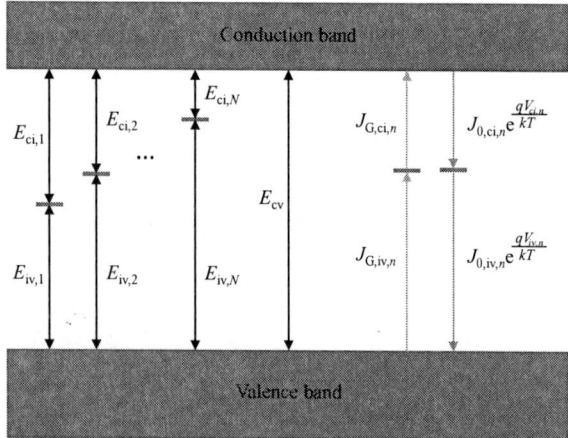

Fig. 1. Sketch where the left part shows the energy gaps involved in the modeling of a device with N impurities. The right part shows the generation and recombination currents transporting electrons to or from impurity type n.

where $J_{G,cv}$ is the generation current density across the entire band gap of the IB-material, and $J_{G,ci}$ and $J_{G,iv}$ are the generation currents across the respective sub-band gaps. $J_{0,i}$ is a pre-exponential constant determining the magnitude of the recombination current density taking place via the intermediate band. It is given by

$$J_{0,i} = \sqrt{J_{0,ci} J_{0,iv}}, \qquad (6)$$

where $J_{0,ci}$ and $J_{0,iv}$ are calculated by Eq. (3) with integration limits corresponding to the absorbing energy intervals for transitions across the respective energy gaps. The last term in (4), $\Delta J_{G,i}$, takes into account the penalty imposed on the cell current when there is a mismatch between the generation rates across the sub-band gaps. ΔJ_G is given by

$$\Delta J_G = |J_{G,ci} - J_{G,iv}|. \qquad (7)$$

The similarities between IBSCs and IPVCs give rise to obvious similarities in the JV-characteristics of the two types of devices. In the following, the letters c, i and v are used to denote various types of transitions. While c and v always denote the conduction and valence bands, as above, the letter i will be used to designate transitions involving impurity states or an intermediate band depending on the device type. The energy gaps to be taken into account in an IPVC are sketched Fig. 1.

II. THE MODEL

With non-overlapping absorption coefficients, each impurity level can be treated individually. As can be seen from Fig. 1, each of them has to obey the particle balance [2]

$$J_{G,iv,n} - J_{0,iv,n} e^{qV_{iv,n}/kT} = J_{G,ci,n} - J_{0,ci,n} e^{qV_{ci,n}/kT}. \qquad (8)$$

The indexing parameter n is used to identify the different impurity types. Applying the constraint $V = V_{ci,n} + V_{iv,n}$

allows eliminating $V_{vi,n}$ and establishing a relation between $V_{ci,n}$ and V through

$$e^{qV_{ci,n}/kT} = \frac{1}{2J_{0,ci,}} \left[J_{G,ci,n} - J_{G,iv,n} \right.$$
$$\left. - \sqrt{4 J_{0,ci,n} J_{0,iv,n} e^{qV/kT} + \Delta J_{G,n}^2} \right]. \qquad (9)$$

This result is valid for any impurity level, as long as the approximation in (1) holds. The starting point for the JV-characteristic of an IPVC with N types of different impurities is given by the particle balance for the conduction band which reads

$$J = J_{G,cv} - J_{0,cv} e^{qV/kT}$$
$$+ \sum_{n=1}^{N} \left[J_{G,ci,n} - J_{0,ci,n} e^{qV_{ci,n}/kT]} \right]. \qquad (10)$$

Using Eq. (9) to substitute for $\exp(qV_{ci,n}/kT)$ then gives

$$J = J_{G,cv} - J_{0,cv} e^{qV/kT}$$
$$+ \sum_{n=1}^{N} \left[J_{G,i,n} - \sqrt{J_{0,i,n}^2 e^{qV/kT} + \tfrac{1}{4}\Delta J_{G,n}^2} \right] \qquad (11)$$

with $J_{G,i,n} = \tfrac{1}{2}(J_{G,iv,n} + J_{G,ci,n})$. $J_{0,i,n}$ is calculated with Eqs. (3) and (6) adapted to each impurity type.

As was shown in Ref. [8], for the intermediate band solar cell, the impact of carrier relaxation in intermediate states [11] can be taken into account by adjusting the integration limits when calculating the value of $J_{0,i,n}$.

A. Absorptivity

To achieve full absorption of the incoming light, there has to be a sufficient number of impurity atoms present in the cell. If the concentration is low or the cell thin, the absorptivity for transitions via the impurity states will be less than 1. A reduced absorptivity can be incorporated by multiplying the terms in the sum in Eq. (10) with an absorptivity A_n associated with the impurities of type n [12]. Assuming equal absorptivity for excitations of electrons to and from the same impurity, the JV-characteristic with the absorptivity taken into account becomes

$$J = A_{cv} \left[J_{G,cv} - J_{0,cv} e^{qV/kT} \right]$$
$$+ \sum_{n=1}^{N} A_n \left[J_{G,i,n} - \sqrt{J_{0,i,n}^2 e^{qV/kT} + \tfrac{1}{4}\Delta J_{G,n}^2} \right], (12)$$

where a separate absorptivity A_{cv} is introduced for full band gap transitions. Although overlapping absorption coefficients in general results in complex modeling [13], the symmetry of a mid-gap impurity level allows this to be modeled as a special case. If such an impurity absorbs photons across E_{ci} and E_{iv} with the same probability, then $V_{ci} = V_{iv}$ and the recombination rates, as well as the generation rates, across the two sub-band gaps are equal. A symmetrical mid-gap impurity can thus be taken into account by adding the term

$$\tfrac{1}{2} A_{\mathrm{m}} \left[J_{\mathrm{G,m}} - J_{0,\mathrm{m}} e^{qV/2kT} \right],$$

to (12). The subscript m is used to indicate a mid-gap level. $J_{\mathrm{G,m}}$ equals the flux of incoming photons with energies allowing them to be absorbed in transitions involving the impurity level. $J_{0,\mathrm{m}}$ is given by (3). The reason for the factor of 1/2 is the fact that the two sub-band gaps have to share the number of absorbed and emitted photons.

B. External radiative efficiency

The external radiative efficiency (ERE) is a parameter that describes the ratio of the net radiative recombination rate to the total net recombination rate [14]. The total net recombination rate is calculated by dividing the net radiative recombination rate by the ERE. In the following, the ERE is treated as independent on the cell voltage, although in reality it will generally vary with the voltage [14]. Values of the ERE are assigned to the recombination routes involving the impurities. Optically inactive impurities may be modeled by assigning a value smaller than 1 for the ERE associated with recombination across the full band gap. They can also be taken into account by including separate terms for these in the sum over impurity types, where the generation rates are set to zero. Dividing the recombination current densities in Eqs. (8) and (10) by the respective EREs yields the JV-charactristic

$$
\begin{aligned}
J = & J_{\mathrm{G,cv}} - \frac{J_{0,\mathrm{cv}}}{\mathrm{ERE}_{\mathrm{cv}}} e^{qV/kT} \\
& + \sum_{n=1}^{N} \left[J_{\mathrm{G,i},n} - \sqrt{\frac{J_{0,\mathrm{i},n}^2}{\mathrm{ERE}_{\mathrm{i},n}^2} e^{qV/kT} + \tfrac{1}{4}\Delta J_{\mathrm{G},n}^2} \right],
\end{aligned} \tag{13}
$$

where, as above, the subscript identifies the various transition routes. $\mathrm{ERE}_{\mathrm{i},n}^2$ equals the product of the ERE for the recombination processes involving each impurity, that is $\mathrm{ERE}_{\mathrm{i},n}^2 = \mathrm{ERE}_{\mathrm{ci},n}\mathrm{ERE}_{\mathrm{iv},n}$.

Finally, the ERE and the absorptivity can both be included in a combined JV-characteristic which reads

$$
\begin{aligned}
J = & A_{\mathrm{cv}} \left[J_{\mathrm{G,cv}} - \frac{J_{0,\mathrm{cv}}}{\mathrm{ERE}_{\mathrm{cv}}} e^{qV/kT} \right] \\
& + \sum_{n=1}^{N} A_n \left[J_{\mathrm{G,i},n} - \sqrt{\frac{J_{0,\mathrm{i},n}^2}{\mathrm{ERE}_{\mathrm{i},n}^2} e^{qV/kT} + \tfrac{1}{4}\Delta J_{\mathrm{G},n}^2} \right]
\end{aligned} \tag{14}
$$

A mid-gap impurity with non-unity absorptivity and symmetric ERE can be taken into account by adding the term

$$\tfrac{1}{2} A_{\mathrm{m}} \left[J_{\mathrm{G,m}} - \frac{J_{0,\mathrm{m}}}{\mathrm{ERE}_{\mathrm{m}}} e^{qV/2kT} \right]$$

to the JV-characteristic.

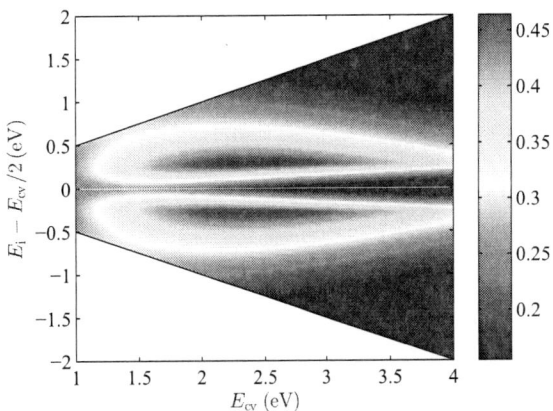

Fig. 2. The efficiency of an IPVC with an infinite absorption band width calculated for the unconcentrated 6000 K black body spectrum with the cell temperature set to 300 K. The vertical axis shows the position of the impurity level relative to the middle of the band gap. The color scale shows whether the IPVC has an efficiency higher than (green) or lower than (red) an ideal single junction cell. Values have not been calculated for mid-gap impurities.

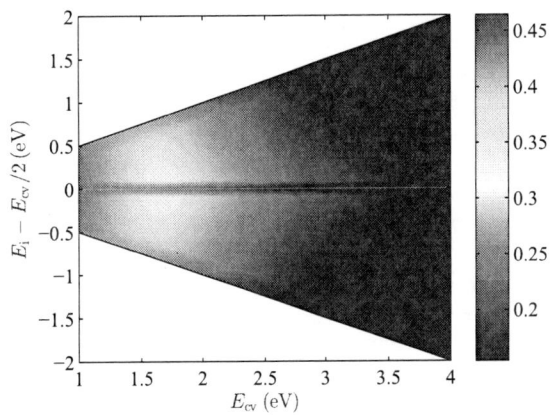

Fig. 3. The efficiency of an IPVC with $\Delta E = 0.2\,\mathrm{eV}$ calculated with all other parameters set as in Fig. 2.

III. Numerical Results and Discussion

In this section, the model derived above will be used to investigate the impact of the absorption band width, the absorptivity and the ERE. All calculations are performed using the unconcentrated 6000 K black body spectrum and a device temperature of 300 K. The limiting efficiency of 63.2 % for IPVCs with one impurity level is then reproduced. The absorption band width is controlled by allowing photons to be absorbed and emitted only in restricted energy intervals. The intervals go from the absorption threshold of the transition in question, to the smaller value of either the absorption threshold plus the absorption band witdh, ΔE, or the absorption threshold of a larger energy gap. Above the full band gap, all photons are assumed to be absorbed.

Figures 2 and 3 show how the absorption band width

impacts the efficiency of IPVCs operating at the radiative limit. Efficiency maps are shown for $\Delta E = \infty$ and $\Delta E = 0.2\,\text{eV}$, respectively. It is clear that reducing the absorption band width to 0.2 eV is devastating for the performance of the IPVS as it ends up with efficiencies below the limiting efficiency of single band gap cells. An impurity level will only be beneficial if its presence increases the short circuit current without a proportional reduction of the voltage. Due to the exponential fall-off of the integrand in Eq. (3), a broadening of the absorption band beyond a couple of kT will not have a major impact on $J_{0,n}$. We can say that the impurity reaches its recombinative potential for rather narrow absorption band widths. A further broadening will, however, allow more photons to be excited via the impurity state and be beneficial for the cell efficiency. Figs. 2 and 3 clearly shows that an impurity level may only be beneficial if it is associated with a sufficiently wide absorption band.

With reduced absorptivity, both generation and recombination via the impurity states is reduced. The exponential dependence of the recombination on the voltage means that the impurity may still act as an important recombination center and possibly degrade the cell voltage. In Fig. 4, efficiencies are shown with $\Delta E = \infty$ while the absorptivity for transitions via the impurity set to 0.1. It becomes clear that the reduced absorption has turned the impurities into net recombination centers which are degrading the cell efficiency to values below that of an ideal single band gap cell.

In Fig. 5 it is shown how the efficiency of a cell with a total band gap of 2.0 eV and an impurity level 1.5 eV above the valence band edge vary with the absorptivity of the impurity. Curves are plotted for three different values of the ERE, where the EREs for full band gap transitions and impurity transitions are, for convenience, assumed to be equal. Note that the curves have valleys where the impurity acts as a net recombination center. As the absorptivity approaches 1, the positive contribution to the current becomes sufficiently large to increase the efficiency to higher values than that of a single band gap cell with the selected ERE. The plot highlights the importance of having fully absorbing impurities in order to reach high efficiency with IPVCs.

A simple plot showing how the efficiency vary with the ERE of the recombination via the impurities and across the full band gap is shown in Fig. 6. The efficiency drops immediately when the impurity related ERE is reduced. As the ERE of the full band gap is reduced, the IPVC maintains a high efficiency until the recombination across the full band gap starts to become more important. The plot is calculated with a band gap of 2.0 eV and an impurity level 1.3 eV above the valence band edge.

Fig. 7 shows the efficiency calculated with two impurity levels. ΔE is set to infinity and the absorptivity set to 1. One impurity is placed in the middle of the band gap while the position of the other is varied according to the vertical axis of the plot. It is easily seen that efficiency improvements can be achieved for a larger range of impurity positions than with a single impurity.

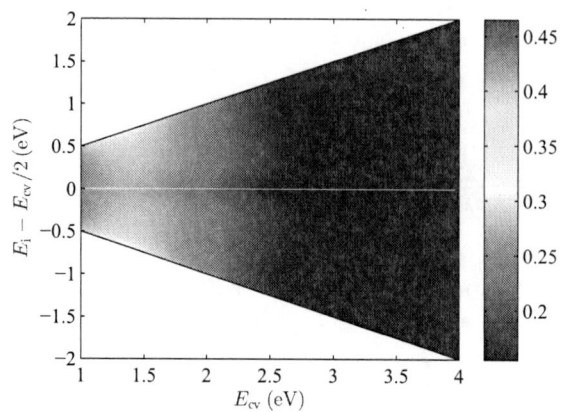

Fig. 4. The efficiency of an IPVC with the absorptivity for impurity related transitions set to 0.1. All other parameters are set as in Fig. 2.

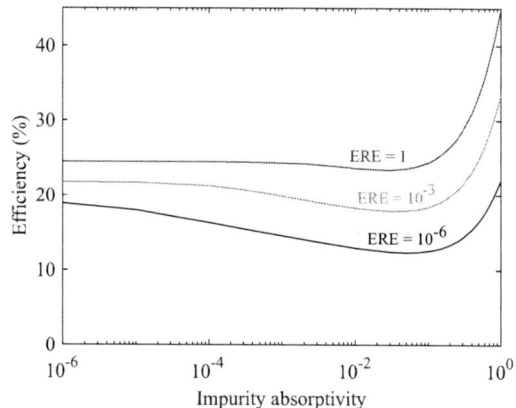

Fig. 5. The efficiency, plotted as a function of the impurity absorptivity, of IPVCs with a band gap of 2.0 eV and an impurity level at 1.5 eV above the valence band edge. The absorption band width is set to infinity. Curves are shown for three different values of the ERE where $ERE_i = ERE_{cv}$.

When a second impurity is added, the short circuit current of the device will typically increase. On the other hand, the exponential dependency of the recombination rates on the cell voltage limits the voltage loss when adding the second impurity. Therefore, one should expect a multi-level impurity cell to be less sensitive to narrow absorption bands or low absorptivity. This is indeed showing in the numerical results. Fig. 8 shows the efficiency calculated for a device with two impurities with an absorptivity of 0.1. By comparing Figs. 7 and Fig. 8 to Figs. 2 and 3, it becomes clear that an IPVC with two impurity levels is visibly more resistant to reduced absorptivity than its counterpart with only a single impurity level. The same is true for narrow absorption bands, although it is not shown here. Be aware that the results for impurities near the band edges are likely to violate the approximation in (1), which will overestimate the efficiency.

The non-overlap criterion constraining the absorption coef-

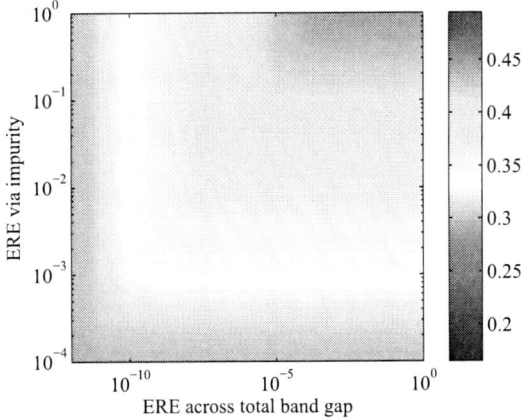

Fig. 6. The efficiency of an IPVC with a band gap of 2.0 eV and an impurity level 1.3 eV from the valence band edge. The impurity is fully absorbing and has a full absorption band width. The plot shows how the efficiency changes with the EREs of the full band gap as well as the impurity related energy gaps.

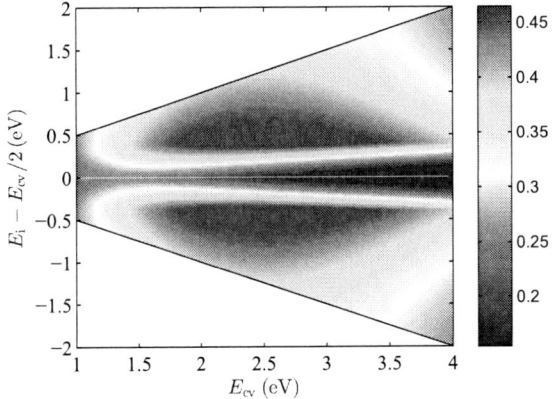

Fig. 7. The efficiency of an IPVC with two fully absorbing impurity levels of which one is placed in the middle of the full band gap. All other parameters are set as in Fig. 2.

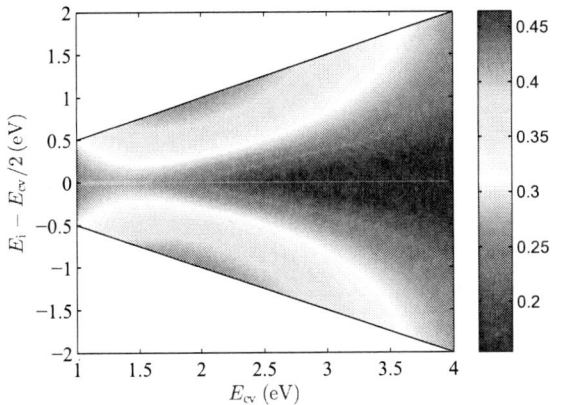

Fig. 8. A plot similar to that in Fig. 7, with the absorptivity for all transitions that involve an impurity level set to 0.1.

ficients represents a major weakness of the presented model. It is highly unlikely that impurities are found that fulfill this requirement. This is particularly true if the difference in energy between different impurity levels is small. Overlapping absorption coefficients are known to induce some extra loss mechanisms [13], but it can also improve the current matching for generation via intermediate energy levels and increase the efficiency for some combinations of band gaps in intermediate band solar cells [15]. Overlapping absorption coefficients may also improve the total absorption of photons, which will be beneficial if it is hard to achieve unity absorptivity by only one impurity type.

One issue which is of particular importance in multilevel systems is the need to partially fill the various levels. If some intermediate levels are close to completely full or completely empty, the photogeneration via them will be weak. With only one level, partial filling can be accomplished for example by doping [16]. With several levels, one will either have to establish a gradient in the electric potential across the structure [17] or rely on photofilling [18] to achieve the partial filling. With non-overlapping absorption coefficients the latter requires either concentrated light or a large current mismatch. It is yet unknown how the ability to photofill states will be affected by overlapping absorption coefficients.

The complexity of including an overlap between the absorption coefficients prevents a clear understanding of all aspects of general IPVCs at the time of writing.

IV. CONCLUDING SUMMARY

A model is developed which allows the JV-characteristic of IPVCs to be described by analytic expressions. The model may enhance the intuitive understanding of, and simplifies the modeling of, such photovoltaic devices. From numerical examples it is clear that:

- Even purely radiative impurities may reduce the efficiency of IPVCs to values below that of single band gap cells if the impurity related absorptivity is low.

- Impurities with a narrow absorption band may reduce the efficiency of IPVCs to values below that of single band gap cells. This is true even if the impurity level has an optimal position in the band gap.

- Increasing the number of impurity species makes the IPVC less sensitive to reduced absorptivity or narrow absorption bands.

REFERENCES

[1] M. Wolf, "Limitations and possibilities for improvement of photovoltaic solar energy converters: Part I: Considerations for earth's surface operation," *Proceedings of the IRE*, vol. 48, no. 7, pp. 1246–1263, 1960.

[2] A. S. Brown and M. A. Green, "Impurity photovoltaic effect: Fundamental energy conversion efficiency limits," *Journal of Applied Physics*, vol. 92, no. 3, pp. 1329–1336, aug 2002.

[3] A. Martí, L. Cuadra, N. López, and A. Luque, "Intermediate band solar cells: Comparison with shockley-read-hall recombination," *Semiconductors*, vol. 38, no. 8, pp. 946–949, aug 2004.

[4] E. Olsen and A. Flø, "Spectral and spatially resolved imaging of photoluminescence in multicrystalline silicon wafers," *Applied Physics Letters*, vol. 99, no. 1, p. 011903, jul 2011.

[5] M. Lax, "Cascade Capture of Electrons in Solids," *Physical Review*, vol. 119, no. 5, pp. 1502–1523, sep 1960.

[6] C. H. Henry and D. V. Lang, "Nonradiative capture and recombination by multiphonon emission in GaAs and GaP," *Physical Review B*, vol. 15, no. 2, pp. 989–1016, jan 1977.

[7] G. Beaucarne and M. A. Green, "A modified Shockley-Read-Hall theory including radiative transitions," *Solid-State Electronics*, vol. 47, no. 4, pp. 685–689, 2003.

[8] R. Strandberg, "Analytic JV-Characteristics of Ideal Intermediate Band Solar Cells and Solar Cells With Up and Downconverters," *IEEE Transactions on Electron Devices*, vol. 64, no. 5, pp. 2275–2282, may 2017.

[9] A. De Vos and H. Pauwels, "On the thermodynamic limit of photovoltaic energy conversion," *Applied Physics*, vol. 25, no. 2, pp. 119–125, 1981.

[10] W. Shockley, "The Theory of p-n Junctions in Semiconductors and p-n Junction Transistors," *Bell System Technical Journal*, vol. 28, no. 3, pp. 435–489, jul 1949.

[11] A. S. Brown and M. A. Green, "Impurity photovoltaic effect with defect relaxation: Implications for low band gap semiconductors such as silicon," *Journal of Applied Physics*, vol. 96, no. 5, pp. 2603–2609, 2004.

[12] M. A. Green, "Multiple band and impurity photovoltaic solar cells: General theory and comparison to tandem cells," *Progress in Photovoltaics: Research and Applications*, vol. 9, no. 2, pp. 137–144, mar 2001.

[13] L. Cuadra, A. Marti, and A. Luque, "Influence of the Overlap Between the Absorption Coefficients on the Efficiency of the Intermediate Band Solar Cell," *IEEE Transactions on Electron Devices*, vol. 51, no. 6, pp. 1002–1007, jun 2004.

[14] M. A. Green, "Radiative efficiency of state-of-the-art photovoltaic cells," *Progress in Photovoltaics: Research and Applications*, vol. 20, no. 4, pp. 472–476, jun 2012.

[15] R. Strandberg, "Theoretical efficiency of intermediate band solar cells with overlapping absorption coefficients for various combinations of band gaps," in *Conference Record of the IEEE Photovoltaic Specialists Conference*, 2012, pp. 97–100.

[16] A. Luque and A. Martí, "On the Partial Filling of the Intermediate Band in IB Solar Cells," *IEEE Transactions on Electron Devices*, vol. 57, no. 6, pp. 1201–1207, jun 2010.

[17] A. Lin and J. Phillips, "Resolving spectral overlap issue of intermediate band solar cells using non-uniform sub-bandgap state filling," *Progress in Photovoltaics: Research and Applications*, vol. 22, no. 10, pp. 1062–1069, oct 2014.

[18] R. Strandberg and T. Reenaas, "Photofilling of intermediate bands," *Journal of Applied Physics*, vol. 105, no. May, p. 124512, 2009.

Photoluminescence Properties of In-Plane Ultrahigh-Density InAs Quantum Dots on GaAsSb/GaAs(001) for Solar Cell Applications

Ryo Sugiyama[1], Naoki Akimoto[1], Tomah Sogabe[2] and Koichi Yamaguchi[1]

[1]Department of Engineering Science, The Univsersity of Electro-Communications, 1-5-1 Chofugaoka, Chofu, Tokyo 182-8585, Japan

[2] Info-Powered Energy System Research Center, The Univsersity of Electro-Communications, 1-5-1 Chofugaoka, Chofu, Tokyo 182-8585, Japan

Abstract — In-plane ultrahigh-density InAs quantum dots (QDs) were grown on GaAsSb/GaAs(001) for intermediated-band solar cell applications. Photoluminescence (PL) spectra under 785-nm-light excitation were originated from the type-II transition between the QD electron ground states (GS) and the GaAsSb valence band and did not changed at different measurement positions. It can be explained by a GS filling effect due to tunneling between QDs. For 860-nm-light excitation, PL spectra were obtained at a shorter wavelength region and depended on the measurement position. These PL spectra were composed into some crossed transitions, which were attributed to localized excitons between QD electron excited states and GaAsSb valence band.

Index Terms — quantum dot, InAs, ultrahigh density, GaAsSb, photoluminescence.

I. INTRODUCTION

Development of intermediate-band quantum-dot solar cells (IB-QDSCs) with high power conversion efficiency (PCE) over 50 % is the challenging theme [1]. We calculated the QD density dependence of IB-QDSCs using InAs QDs on GaAsSb/GaAs, based on a detailed balance model [2]. As a result, a total QD density of 3×10^{13} cm^{-2} was required for achieving at least 40% PCE under concentrated sunlight [2]. Therefore, in order to reduce the stacking number of QD layers, it is necessary to develop a fabrication technique of in-plane ultrahigh-density QDs. Recently, we demonstrated in-plane ultrahigh-density InAs QDs with $0.5–1.0 \times 10^{12}$ cm^{-2} on GaAs(001) substrates by Sb-mediated growth using a GaAsSb buffer layer [3] and an InAsSb wetting layer (WL) [4]. This Sb-mediated growth is a promising method for the fabrication of in-plane ultrahigh-density InAs QDs. However, little is known about their fundamental properties. Recently, it was found that abnormal photoluminescence (PL) properties of ultrahigh-density InAs QD layers was caused by in-plane tunnel coupling between QDs [4]. Understanding of carrier dynamics in laterally coupled QDs is important for their IBSC applications.

In this conference, we present the fabrication of in-plane ultrahigh-density InAs QDs on GaAsSb/GaAs(001) and their PL properties. The PL properties reveal fundamental processes of optical absorption and carrier transfer in laterally coupled QD layers having quasi-miniband.

II. EXPERIMENTAL PROCEDURE

The QD samples in this study were grown on the GaAs(001) substrates by solid-source MBE. In order to increase in-plane QD density, the InAs QDs were grown on a 3-nm-thick GaAsSb buffer layer. During the GaAsSb growth, a flux ratio of $Sb_4/(Sb_4+As_4)$ was kept at 5 %. After the InAs growth, the GaAs capping layer was grown at 470 °C. Fig. 1 shows a schematic diagram of QD sample (a) and an atomic force microscopy (AFM) image of in-plane ultrahigh-density InAs QDs on GaAsSb/GaAs (b). The InAs QD density was about 5×10^{11} cm^{-2} cm^{-2}. PL properties were measured by using a low-temperature micro-PL system. The diameter of an excitation laser beam was about 50 μm.

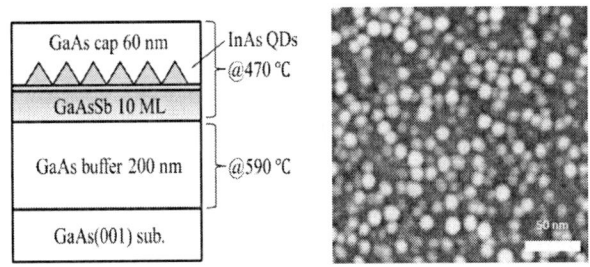

Fig. 1. Schematic diagram of sample structure (a) and an AFM image (b) of ultrahigh-density InAs QDs on GaAsSb/GaAs.

III. RESULTS AND DISCUSSION

First, PL properties of in-plane ultrahigh-density InAs QDs on GaAsSb/GaAs were measured at 15 K under 785-nm-light

978-1-5090-5606-4/17 $31.00 © 2017 IEEE

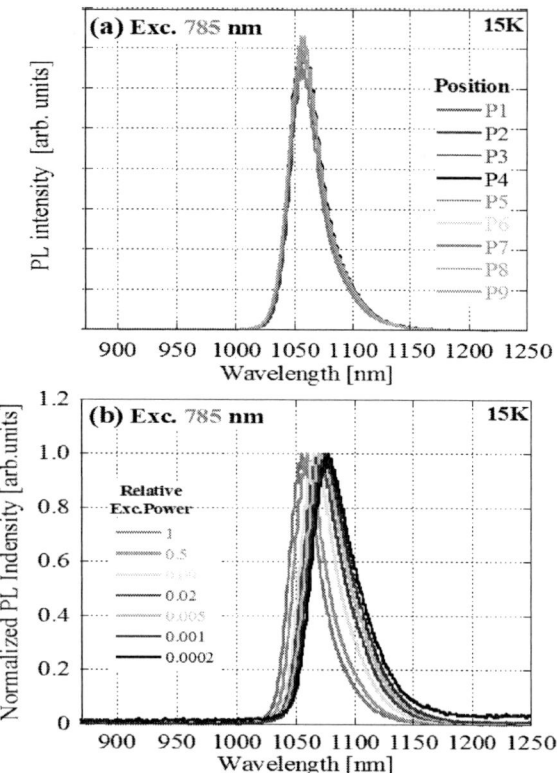

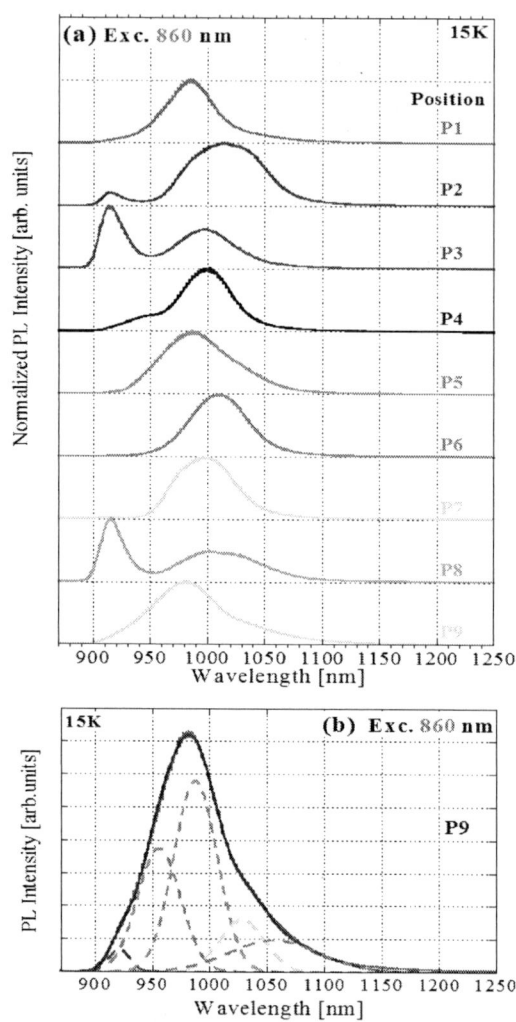

Fig. 2. Normalized PL spectra (15 K) of ultrahigh-density InAs QDs on GaAsSb/GaAs. PL spectra were measured at 9 different points (P1-P9) on the surface (a) and were measured at P1 as a function of excitation power intensity (b). The excitation light wavelength was 785 nm.

Fig. 3. Normalized PL spectra (15 K) of ultrahigh-density InAs QDs on GaAsSb/GaAs. PL spectra were measured at 9 different points (P1-P9) on the surface (a). A typical PL spectrum at P9 was decomposed into 5 peaks; 1060 nm, 1022 nm, 985 nm, 952 nm, and 920 nm (b). The excitation light wavelength was 860 nm.

excitation. Figure 2(a) shows PL spectra at 9 different positions (P1~P9, 3×3) on the sample surface. The spacing distance between each measurement point was 300 μm. All PL spectra indicated almost same intensity, peak wavelength (1060 nm) and FWHM (32 meV). From time-resolved PL spectra, it was found that the PL decay time of 7 ns was obtained at peak wavelength. It suggests a type-II transition between electrons in QD ground states and holes in GaAsSb valence band.

Figure 2(b) shows normalized PL spectra at position of P1 under 785-nm-light excitation as a function of excitation power. As the excitation power increased, the PL peak shifted continuously toward high energy side. As reported previously [4], since a separation distance between neighboring QDs is very narrow for ultrahigh-density QDs, the electrons can transfer to the lower ground state by tunneling. Therefore, the continuous blue shift of PL spectra can be explained by a filling effect of electrons at QD ground states with an inhomogeneous broadening. As the excitation power increases, electrons fill into larger QDs with lower ground states in the early stage and then are gradually incorporated into smaller QDs with higher ground states. As a result, the continuous

blue shift of PL spectrum occurred (Fig. 2(b)), and PL spectra obtained at different positions showed almost same spectra (Fig. 2(a)). It suggests the formation of quasi-miniband through in-plane tunnel coupling of QDs.

Next, PL properties of in-plane ultrahigh-density InAs QDs were measured at 15 K under 860-nm-light excitation, which is below GaAs barrier excitation. That is, the 860-nm-light excited carriers at the wetting layer (WL) and GaAsSb buffer layer. Figure 3(a) shows normalized PL spectra at 9 different positions (P1~P9, 3×3) on the sample surface. The PL spectrum indicated short wavelength peaks, compared with that obtained by 785-nm-light excitation. Here, it should be noted that PL spectra strongly depended on the measurement position.

Figure 3(b) shows a PL spectrum obtained at P9, which was decomposed into 5 peaks. All PL spectra of Fig. 3(a) could be well fitted by these 5 peaks; 1060 nm, 1022 nm, 985 nm, 952 nm, and 920 nm. As mentioned before, the 1060 nm peak was originated from a type-II transition between QD electron ground states (GS) and GaAsSb valence band. The long PL decay time of 6.5-8.0 ns were also obtained at a region from 950 nm to 1060 nm. In addition, the energy separation between neighboring peaks was an almost similar value of approximately 45 meV. Therefore, it is considered that 1022 nm, 980 nm, and 952 nm peaks are respectively based on type-II transitions between QD first, second and third excited states (ES1, ES2, ES3) and GaAsSb valence band.

Figure 4 shows the measurement position dependences of PL peak intensities of GS, ES1, ES2 and ES3 transitions. The PL peak originating from the excited states strongly depended on the measurement position. However, the PL intensity of GS transition was almost kept constant for all positions. This result is similar to Fig. 2(a). Therefore, electrons relaxed into QD GS are de-localized and can transfer to lower GS of neighboring QDs by tunneling. However, electrons at ES1, ES2 and ES3 of InAs QDs and holes in the GaAsSb buffer layer make localized excitons. Thereby, such crossed transitions between QD electron states and GaAsSb valence band were observed at high energy side [5]. Since these bound excitons are localized between the QD and the GaAsSb, these crossed transitions depend on the measurement positions (Fig. 3(a) and Fig. 4).

Figure 5 shows normalized PL spectra (15 K) of in-plane ultrahigh-density InAs QDs as a function of 860-nm excitation power. As the excitation power increased, the PL peak shifted slightly toward a short wavelength region. However, a blue

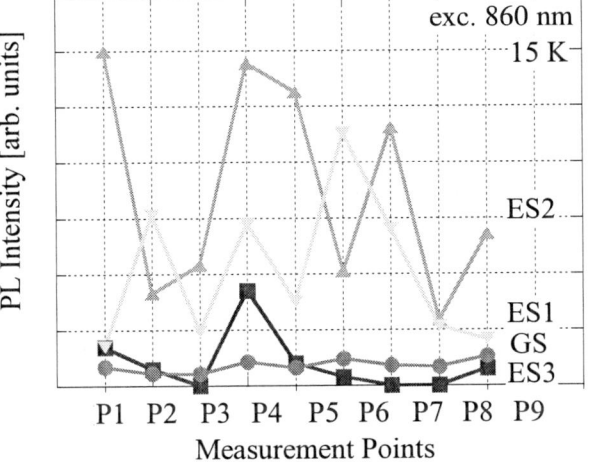

Fig. 4. PL spectra (15 K) of ultrahigh-density InAs QDs on GaAsSb/GaAs. PL spectra were measured at 9 different positions under 860-nm-light excitation. Measurement position dependences of PL peak intensities of GS (1060 nm), ES1 (1022 nm), ES2 (985 nm) and ES3 (952 nm) transitions.

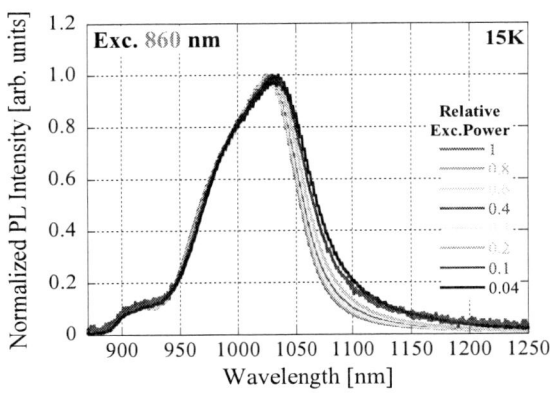

Fig. 5. Normalized PL spectra (15 K) of ultrahigh-density InAs QDs on GaAsSb/GaAs as a function of excitation power intensity. The excitation light wavelength was 860 nm.

shift significantly occurred only at a long wavelength region, not at a short wavelength region. As mentioned before, the PL spectra obtained by 860-nm-light excitation consist of GS, ES1, ES2 and ES3 transitions. Therefore, a blue shift of a long wavelength region is caused by a blue shift of GS transition. That is, a filling effect of GS due to tunneling induced the blue shift. On the contrary, the crossed transitions based on ES1, ES2 and ES3 did not shift, despite of high excitation power. It suggests that these excitons are localized between the QD and the GaAsSb.

IV. CONCLUSION

In-plane ultrahigh-density InAs QDs were grown on GaAsSb/GaAs(001) by MBE. The QD density was 5×10^{11} cm^{-2}. PL spectra were measured under 785-nm and 860-nm light excitations. For 785-nm-light excitation, PL spectra were originated from a type-II transition between QD electron GS and GaAsSb valence band and did not depend on the measurement position. It can be explained by a GS filling effect due to tunneling between the QDs. For 860-nm-light excitation, PL spectra were observed at a shorter wavelength region and depended on measurement position. These PL spectra were originated from some crossed transitions, which were attributed to localized bound excitons composed of electrons in the QD excited states and holes in the GaAsSb valence band. These results are useful to understand two-step photo-excitation through the quasi-miniband in IB-QDSCs using in-plane ultrahigh-density QDs.

ACKNOWLEDGEMENT

This work was supported by the Incorporated Administrative Agency New Energy and Industrial Technology Development Organization (NEDO) under the Ministry of Economy, Trade and Industry (METI), Japan.

REFERENCES

[1] A. Luque and A. Marti, "Increasing the Efficiency of Ideal Solar Cells by Photon Induced Transitions at Intermediate Levels", *Phys. Rev. Lett.*, **78**, 5014 (1997).

[2] K. Sakamoto, Y. Kondo, K. Uchida and K. Yamaguchi, "Quantum-Dot Density Dependence of Power Conversion Efficiency of Intermediate-Band Solar Cells", *J. Appl. Phys.* **112**, 124515 (2012).

[3] E. Saputra, J. Ohta, N. Kakuda and K. Yamaguchi, "Self-Formation of In-Plane Ultrahigh-Density InAs Quantum Dots on GaAsSb/GaAs(001)", *Appl. Phys. Express* **5**, 125502 (2012).

[4] K. Sameshima, T. Sano and K. Yamaguchi, "Self-Formation of Ultrahigh-Density (10^{12} cm^{-2}) InAs Quantum Dots on InAsSb/GaAs(001) and Their Photoluminescence Properties", *Appl. Phys. Express* **9**, 075501 (2016).

[5] Yu. I. Mazur, B. L. Liang, Zh. M. Wang, G. G. Tarasov, D. Guzun and G. J. Salamo, "Development of Continuum States in Photoluminescence of Self-Assembled InGaAs/GaAs Quantum Dots, *J. Appl. Phys.* **101**, 014301 (2007).

Carrier Selective Back Contact (CSBC) Solar Cell using Transition Metal Oxides

Astha Tyagi[1], Kunal Ghosh[2], Anil Kottantharayil[1], and Saurabh Lodha[*1]

[1]Indian Institute of Technology Bombay, Mumbai, Maharashtra, 400 076, India

[2]Indian Institute of Technology Mandi, Mandi, Himachal Pradesh, 175 001, India

[*]Email : saurabh.lodha@gmail.com

Abstract— **A solar cell with transition metal oxides as back contact carrier selective layers has been conceptualized and analyzed in this paper. We term such a solar cell as a carrier selective back contact solar cell (CSBC-SC). A CSBC-SC has both contacts on the rear and thus circumvents the trade-off between surface passivation, light trapping, and charge extraction, necessary for the front surface development of a high-performance CSC solar cell with contacts on both sides (front-rear CSC-SC). Further, it also eliminates optical losses due to metal shading and parasitic absorption occurring in low bandgap thin film layers. Therefore, an idealized CSBC-SC provides superior performance as compared to a front-rear CSC-SC, while simultaneously reducing the complexity of design rules. In this paper, the analysis of a CSBC-SC has been performed through an integrated multi-scale electro-optical numerical model developed in the commercial simulator Sentaurus. Results show that a non-ideal CSBC-SC outperforms a front-rear CSC-SC for rear surface recombination velocities (SRV) below 100 cm/s and bulk lifetime values above 100 microseconds. However, a CSBC-SC shows significant sensitivity to front SRV and the values need to be below 10 cm/s to outperform a front-rear CSC-SC.**

Index Terms — **back contact solar cell, carrier selective, CSBC, heterojunction solar cell, transition metal oxide.**

I. INTRODUCTION

Photovoltaic market is largely dominated by c-Si solar cells that depend primarily on *p-n* junction technology. Cost intensive process due to high-temperature processing and inability to obtain thermodynamically optimum cells [1] have paved the path for other alternatives. One such alternative is the carrier selective contact (CSC) solar cell that shows potential to provide high efficiency [2][3] without increasing the thermal budget. However, CSC solar cells as developed have contacts on both sides. Apart from the metal shading loss, the CSC solar cell has a unique limitation its front surface development requires an optimization between light trapping, passivation and charge extraction. This significantly constrains the design space, with a severe restriction on the choice of materials and the device parameters. As a result fabricated CSC solar cells show sub-optimal performance when compared to their thermodynamic potential. This shortcoming can be overcome by developing a solar cell with both carrier selective contacts on the rear. We term such a solar cell as a carrier selective back contact solar cell (CSBC-SC). In a

CSBC-SC, the front surface can be optimized for light trapping and passivation independently from the rear surface which is used for charge extraction. This opens up the design space while simultaneously reducing front surface metal shading and parasitic absorption losses that are present in a CSC solar cell.

In this paper, we have modeled and analyzed the performance of a CSBC-SC with transition metal oxides (TMO) as contact layers, through an integrated multi-scale electro-optical numerical model developed in the commercial simulator Sentaurus. The numerical model is described in Section II. In section III, we describe the structure of a CSBC-SC and compare the performance of an optimized CSBC-SC with an analogous CSC solar cell. In section IV, we analyze the performance of a CSBC-SC by varying key device parameters. Conclusions are presented in section V.

II. DESCRIPTION OF THE NUMERICAL MODEL

A summary of device parameters used for modeling of the CSBC device in Sentaurus TCAD [4] is given in Table I. The

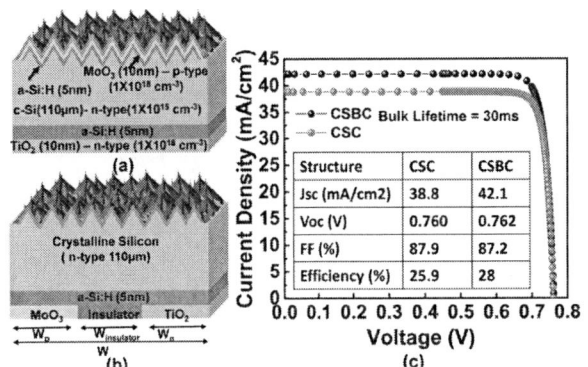

Fig. 1. Schematic and comparison of the proposed CSBC structure with CSC structure, a) Conventional Carrier Selective Contact (CSC) solar cell with a-Si passivation, b) proposed Carrier Selective Back Contact (CSBC) solar cell with a-Si passivation, and c) Comparison of both the structures for maximum achievable efficiency. The interface layers are assumed to be ideal, contacts are considered to be ohmic (series resistance effect due to front contact is not included) and effects of parasitic absorption and metal shading are taken into account.

TABLE I
SUMMARY OF DEVICE AND MATERIAL PARAMETERS USED IN THIS WORK

Material	Bulk (Silicon)	Passivation layer (a-Si)	TiO$_2$	MoO$_3$
Thickness	110µm	5nm [11]	10nm	10nm [11]
Doping	$1X10^{15}cm^{-3}$ (n type)	Intrinsic	$1X10^{18}cm^{-3}$ (n type)	$1X10^{18}cm^{-3}$ (p type)
Lifetime	30ms [3]	1µs	1ns	1ns
Electron Affinity	4.05eV	3.9eV [12]	4.0eV [13]	2.3eV [14]
Band gap @ 0K	1.16964eV	1.72eV [12]	3.2eV [13] [15]	3.0eV [14]
p-contact (MoO$_3$) width (W$_p$)/ n-contact (TiO$_2$) width (W$_n$)		50µm		
Insulator width (W$_{insulator}$)		50µm (fixed)		
Optical Confinement				
Si$_3$N$_4$		65nm [16]		
Back reflectivity		1		

TMO layers are doped to $1X10^{18}$ cm^{-3} for enhanced band bending at the c-Si interface, which subsequently assists in current flow across the heterointerface. The charge carrier transportation mechanism through a-Si layer is modeled through Trap-Assisted Tunneling (TAT) mechanism and the trap cross sections for enhanced emission probability are modeled using a Poole-Frenkel model [5]. For silicon, default mobility models [6][7] and recombination models [8] with doping dependence are used with a bulk lifetime of 30 ms. Schenk band gap narrowing [9] is incorporated to effectively model the impact of excess charge carriers generated in the device.

III. CSBC STRUCTURE AND COMPARISON WITH CSC SOLAR CELL

The structure discussed here consists of silicon as the absorbing material and thin TiO$_2$ and MoO$_3$ layers are used as electron and hole selection layers respectively. In addition, a thin passivation layer of hydrogenated amorphous silicon (a-Si:H) is present between the silicon and the TMO. The schematic of the proposed structure is shown in Fig. 1(b) and it is compared with a CSC structure (reference) shown in Fig. 1(a). To account for metal shading effect, it is assumed that 95% of incident light enters the reference structure. The surface passivation and metal contacts are assumed to be ideal to clearly understand the impact of each parameter on the performance of the device.

IV. RESULTS AND DISCUSSION

The comparison of CSBC structure with CSC shows a clear improvement of over 2% in efficiency, shown in Fig. 1(c), due to reduced metal shading losses and parasitic absorption only.

When lifetime variations are considered, as shown in Fig. 2, the CSBC structure provides better performance than CSC even at 100µs. Moreover with an increase in W (total device width as shown in Fig. 1(b)), for low bulk lifetime wafers

performance is worse due to less diffusion length of charge carriers and more contact spacing.

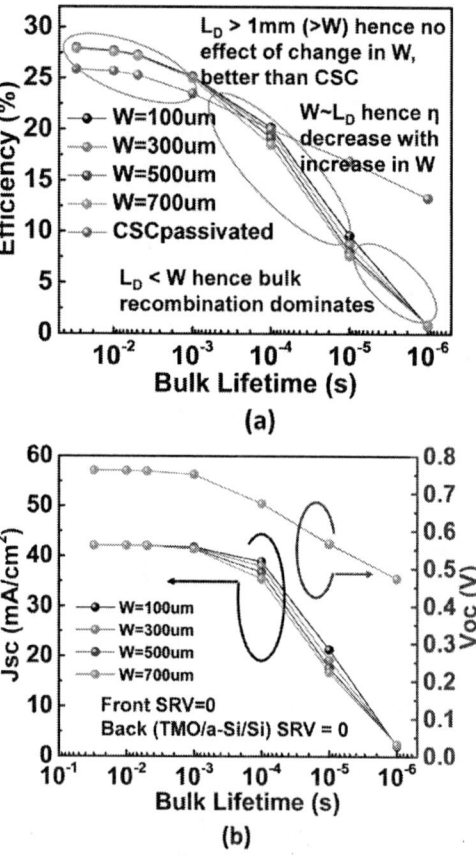

Fig. 2: Effect of variation of bulk lifetime on the performance of proposed structure, a) The efficiency of CSBC reduces for a lower bulk lifetime with an increase in pitch size due to higher diffusion length (L$_D$) requirement for the transportation of charge carriers. b) Variation of J$_{sc}$ drives the trends observed in the efficiency. The value of open circuit voltage remains constant with variation in W since the value of junction recombination current doesn't change, although as a whole the degradation of V$_{oc}$ is observed with a reduction in lifetime.

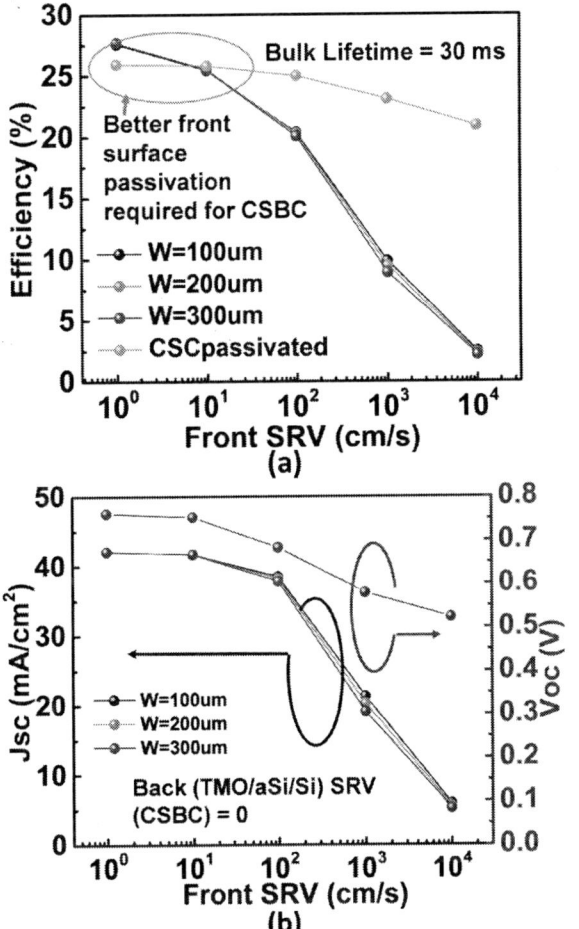

(a)

(b)

Fig. 3: Effect of variation of front surface recombination velocity on the performance of proposed CSBC structure, a) efficiency and, b) J_{sc} and V_{oc}. Values of both, J_{sc} and efficiency, drop significantly with increasing front SRV values. Since most of the carriers are generated near the top surface, there is a high probability of recombination if the front SRV is high. Hence increase in SRV causes reduction in V_{oc} also. There is no variation with respect to W because current collection junctions are at the rear end of the structure.

When diffusion length of charge carriers is large enough (for high bulk lifetime) compared to the total width of the device considered, there is no effect of variations in W for the simulated values. As the bulk lifetime of the wafer decreases, the diffusion length of charge carriers decreases and hence variation in W starts affecting the performance of CSBC structure. Even though $W_{insulator}$ is considered to be constant ($50\mu m$), the effective distance (path length) need to be travelled by charge carriers to the desired contact increases with W. In this case, devices with lower W values have better charge collection and hence better performance. For higher W values, diffusion length of charge carriers is not enough for collection and results in higher recombination and hence lower efficiency.

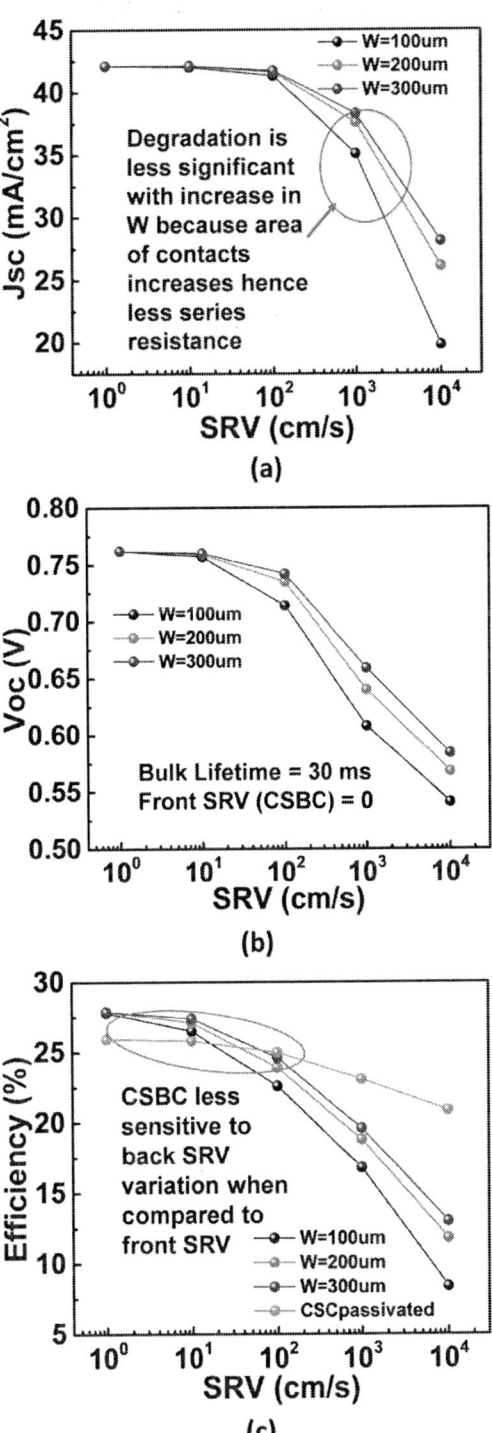

(a)

(b)

(c)

Fig. 4: Effect of variation of back surface recombination velocity on the performance of proposed structure, a) J_{sc}, b) V_{oc} and, c) efficiency. The values of J_{sc}, V_{oc} and efficiency drops with increasing back SRV but the drop is less significant than for front SRV. Performance degradation is less with an increase in W, explained by an increase in the area of the current collecting junction and hence reduction in series resistance.

978-1-5090-5606-4/17 $31.00 © 2017 IEEE

TABLE II

CSBC AND CSC – A COMPARISON

No parasitic absorption	J_{sc} improves in CSBC by around 1.4 mA/cm² over CSC
No metal shading	J_{sc} improves in CSBC by around 2 mA/cm² over CSC
Module design	Coplanar interconnection makes design simple in CSBC
Series resistance	Low R_s in CSC than CSC due to no front contact hence large contact area
Front surface passivation	Better front surface passivation required but lack of tradeoff between electrical and optical properties allows optimization for better surface passivation

For lower bulk lifetime (< 100μs), CSC outperforms CSBC structure. This is due to the fact that in CSBC structure, both type of charge carriers are collected at the rear contact hence J_{sc} values are more sensitive to bulk lifetime when compared to CSC structure.

Since charge carriers need to travel across the device to be collected in back contact solar cell, the CSBC structure is very sensitive to front surface recombination velocity (SRV) degradation. As shown in Fig. 3, front SRV should be maintained below 10 cm/s for better performance of the device when compared to the reference cell. There is no front contact required for collection of charge carriers in CSBC structure hence it is easier to optimize the front surface for better passivation and achieve lower SRV values. On the other hand, in case of both side contact solar cell, there is tradeoff between contact resistance and recombination current for efficient collection of charge carriers. Achieving better passivation quality in CSC structure may result in increased contact resistance affecting the fraction of charge carriers collected at the contact [10]. If charge carrier collection is favored SRV values can increase due to poorer passivation leading to increased recombination current at the contact.

The effect of variation on back (contact) SRV is less significant in the case of CSBC structure as shown in Fig. 4 when compared to front SRV values. For SRV values less than 100 cm/s back contact CSBC structure outperforms the CSC front and back contact design. With increase in W, the effect of an increase in SRV becomes less significant due to increase in the area of charge collecting junction and hence reduction in series resistance.

V. CONCLUSION

The proposed CSBC solar cell retains the benefits of CSC solar cell and also overcomes the shortcomings of parasitic absorption and metal shading loss. CSBC shows better performance for realistic lifetime values (> 100 μs) and back contact SRV (< 100 cm/s). In spite of high performance sensitivity to front SRV variation in CSBC structure, there is the benefit of optimal design of the front surface only for optical properties due to the absence of an electrical contact.

ACKNOWLEDGEMENT

The authors would like to thank Department of Science and Technology (DST), Government of India for providing funding for this project (DST/TM/SER|/D2Z(C)). We would also like to acknowledge Center of Excellence in Nanoelectronics (CEN) and National Center for Photovoltaic Research and Education (NCPRE), IIT Bombay for providing required facilities for carrying out this work.

REFERENCES

[1] A. Richter, S. W. Glunz, A. Richter, M. Hermle, and S. W. Glunz, "Crystalline Silicon Solar Cells Reassessment of the Limiting Efficiency for Crystalline Silicon Solar Cells," *IEEE J. Photovoltaics*, vol. 3, no. 4, pp. 1184–1191, 2013.

[2] R. Islam and K. C. Saraswat, "Metal/insulator/semiconductor carrier selective contacts for photovoltaic cells," *2014 IEEE 40th Photovolt. Spec. Conf. PVSC 2014*, pp. 285–289, 2014.

[3] H. Imran, T. M. Abdolkader, and N. Z. Butt, "Carrier-Selective NiO/Si and TiO 2/Si Contacts for Silicon Heterojunction Solar Cells," *IEEE Trans. Electron Devices*, vol. 63, no. 9, pp. 3584--3590, 2016.

[4] S. S. Workbench and J. Version, "Synopsys." Inc., Mountain View, CA, USA, 2014.

[5] L. Colalongo, M. Valdinoci, P. Baccarani, P. Migliorato, G. Tallarida, and C. Reita, "Numerical analysis of poly-TFT under off conditions," *Solid. State. Electron.*, vol. 41, no. 4, pp. 627--633, 1997.

[6] G. Masetti, M. Severi, and S. Solmi, "Modeling of Carrier Mobility Against Carrier Concentration in Arsenic-, Phosphorus-, and Boron-Doped Silicon," *IEEE Trans. Electron Devices*, vol. 30, no. 7, pp. 764–769, 1983.

[7] E. Conwell and V. F. Weisskopf, "Theory of impurity scattering in semiconductors," *Phys. Rev.*, vol. 77, no. 3, pp. 388–390, 1950.

[8] W. Shockley and W. T. Read, "Statistics of the Recombination of Holes and Electrons," *Phys. Rev.*, vol. 87, no. 46, pp. 835–842, 1952.

[9] A. Schenk, "Finite-temperature full random-phase approximation model of band gap narrowing for silicon device simulation," *J. Appl. Phys.*, vol. 84, no. 7, pp. 3684–

3695, 1998.

[10] J. Bullock, D. Yan, A. Cuevas, B. Demaurex, A. Hessler-Wyser, and S. De Wolf, "Passivated contacts to n+ and p+ silicon based on amorphous silicon and thin dielectrics," *2014 IEEE 40th Photovolt. Spec. Conf.*, pp. 3442–3447, 2014.

[11] C. Battaglia *et al.*, "Silicon heterojunction solar cell with passivated hole selective MoO x contact," *Appl. Phys. Lett.*, vol. 104, no. 11, p. 113902, 2014.

[12] K. Ghosh, "Heterojunction and Nanostructured Photovoltaic Device: Theory and Experiment," Diss. Arizona State University, 2011.

[13] X. Yang, Q. Bi, H. Ali, K. Davis, W. V. Schoenfeld, and K. Weber, "High-Performance TiO2-Based Electron-Selective Contacts for Crystalline Silicon Solar Cells," *Adv. Mater.*, pp. 5891–5897, 2016.

[14] S. Fonash, *Solar cell device physics*. Elsevier, 2010.

[15] S. Avasthi, W. E. McClain, G. Man, A. Kahn, J. Schwartz, and J. C. Sturm, "Hole-blocking titanium-oxide/silicon heterojunction and its application to photovoltaics," *Appl. Phys. Lett.*, vol. 102, no. 20, 2013.

[16] J. Geissbühler *et al.*, "22.5% Efficient Silicon Heterojunction Solar Cell With Molybdenum Oxide Hole Collector," *Appl. Phys. Lett.*, vol. 107, no. 8, 2015.

978-1-5090-5606-4/17 $31.00 © 2017 IEEE

Analysis of open-circuit voltage and conversion efficiency in quantum-dot solar cells via detailed-balance-limit theory

Lin Zhu[1,2], Hidefumi Akiyama[1,2] and Yoshihiko Kanemitsu[3]

1) Institute for Solid State Physics, University of Tokyo, Kashiwanoha, Kashiwa, Chiba 277-8581, Japan. 2) AIST-UTokyo Advanced Operando-Measurement Technology Open Innovation Laboratory (OPERANDO-OIL), 5-1-5 Kashiwanoha, Kashiwa, Chiba, 277-8589, Japan.3) Institute for Chemical Research, Kyoto University, Uji, Kyoto 611-0011, Japan.

Abstract — **Via detailed-balance analysis, we systematically studied conversion efficiency limit, short-circuit current, and open-circuit voltage (V_{oc}) of quantum dot (QD) solar cells for various absorptivity and bonding energy of QD below host-material absorption band. Our results quantitatively revealed that drops of V_{oc} and conversion efficiency in QD solar cells were caused by the intrinsic consequence of mid-gap absorption band, and that current gain in intermediate-band QD solar cells via multi-step infrared-light absorption must be large enough to compensate at least the intrinsic drop of V_{oc} to improve final conversion efficiency.**

Index Terms — **quantum dots, solar cells, below-bandgap absorption, intermediate band.**

I. INTRODUCTION

Quantum dot (QD) solar cells, inducing quantum confined energy levels within the bandgap of host materials can utilize sub-band photons to generate higher photon-current, have been widely researched to implement the concept of intermediate band (IB) solar cells predicted as detailed-balance efficiency limit of 63.1% under full concentration [1], in order to exceed the efficiency of conventional single- or multi-junction photovoltaic devices [1-12]. However, it has been known that significantly drops in open circuit voltage (V_{oc}) of QD solar cells always occur in experiments, while J_{sc} were improved only moderately [7,8], which lead to the lower practical conversion efficiency (η_{sc}) comparing with that of the bulk-host-material solar cells without QDs [10-12].

Primary reasons on the drops of V_{oc} and η_{sc} are interpreted that the strain accumulation due to the growth of QDs embedded in host material would induce more defects and/or dislocations. Such extrinsic degradations in crystal lattice may cause more non-radiative recombination losses, and degrade V_{oc} and η_{sc}. However, intrinsic drops of V_{oc} and η_{sc} due to mid-gap absorption band of QDs may co-exist. Therefore, it is important to analyze quantitatively such inherent drops of V_{oc} and η_{sc} in QD solar cells on the basis of detailed-balance-limit theory.

In this work, focusing on the intrinsic effects of the mid-gap absorption band induced by QD materials, we quantitatively and systematically study the drops in voltage and conversion efficiency in QD solar cells via detailed-balance-limit model [13]. Our results should be used to explain the empirical characteristics of QD solar cells, and evaluate the potentials and limitations of QD cells with varied designs.

II. ABSORPTION MODEL AND ANALYSIS METHODS

Figure 1 (a) shows the absorption spectrum of QD solar cells with well-known sub-band absorption tail modeled as a two-step function in this work. The bandgaps of host material and binding energy of QDs are defined as E_g and E_b, respectively. All cells are assumed to have perfect rear mirrors, absorptivity spectrum $a(E)$ is expressed by $a=1-Exp(-2\alpha L)$, with absorption coefficient (α) and cell thickness (L). The

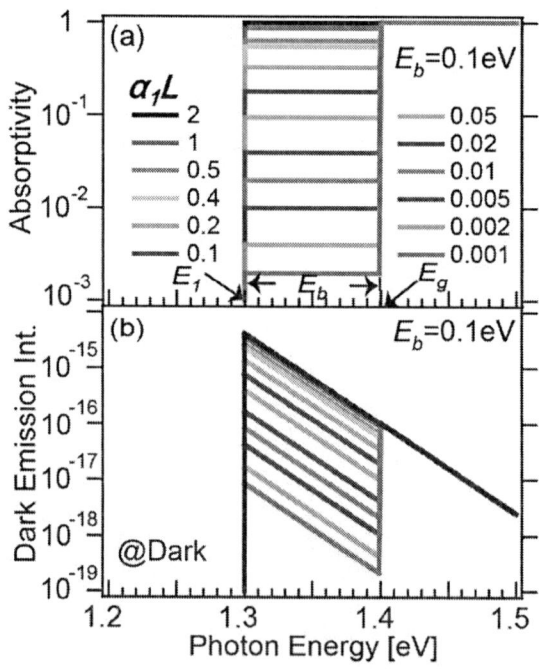

Fig. 1 (a) Absorptivity spectrum and (b) dark emission spectrum for QD solar cells with host material E_g=1.4 eV, binding energy of E_b =0.1eV and various QD absorptivity a_1.

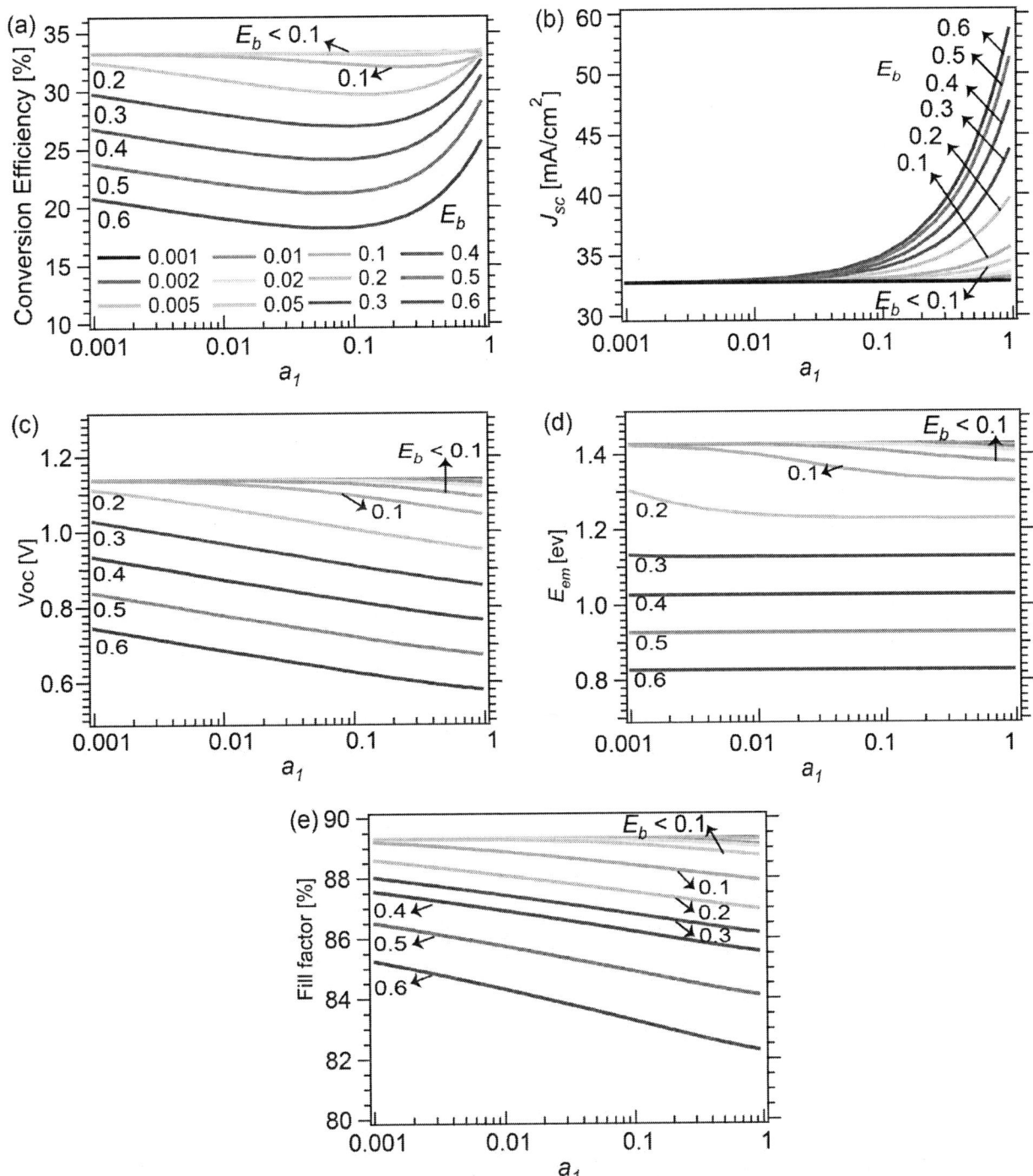

Fig. 2 (a) Conversion efficiency, (b) J_{sc}, (c) V_{oc} (d) effective bandgap and (e) fill factor of QD cells with E_g=1.4eV for varied a_1 and E_b.

product αL is fixed as greater than 5 and a is approximately equal to 1 above E_g. On the other hand, absorptivity a_1 or the product $\alpha_1 L$ between E_g and $E_1 = E_g - E_b$ are varied as parameters.

In realistic QDs the generated carriers promptly thermalize and spend most of their life in thermal equilibrium, thus the assumption of carriers in thermal equilibrium under optical pumping or current injection is approximately justified. We assume infinite carrier mobility, and assume that carriers in

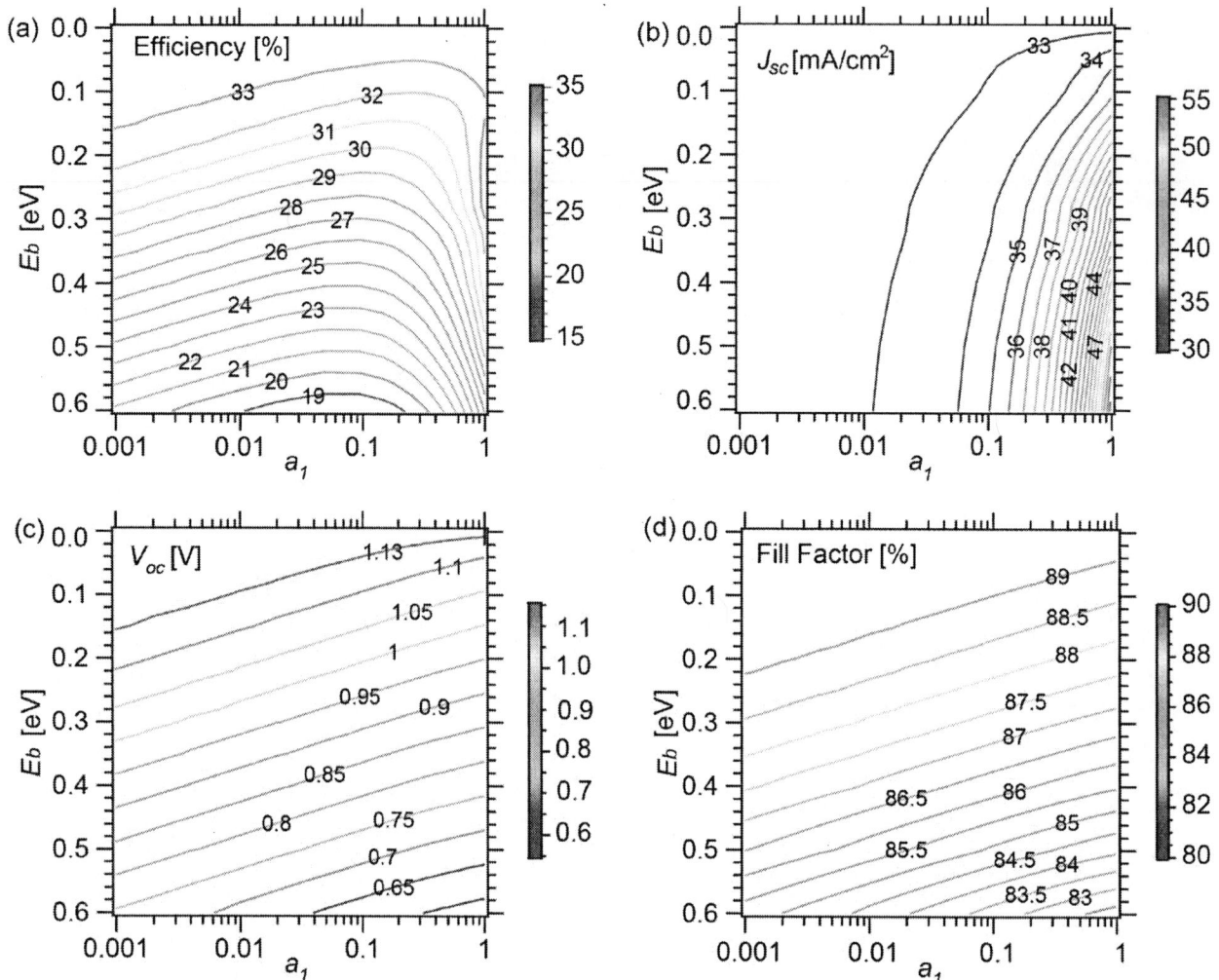

Fig. 3 Contour of conversion efficiency, J_{sc}, V_{oc} and fill factor of QD solar cells with E_g=1.4eV for varied a_1 and E_b.

QD levels and host band have the same quasi-Fermi levels. We neglect non-radiative recombination loss, which are not realistic, but justified to evaluate the ideal intrinsic upper limit of conversion efficiency.

Current-voltage (J-V) characteristics of QD solar cells are given by the carrier balance equation for single-junction solar cells in detailed-balance-limit model [13-15], only needing to change the function of absorptivity. Finally, we calculated the detailed-balance-limit conversion efficiency η_{sc} as a function of a_1 between 0 and 1. Considering a typical example of In$_x$Ga$_{1-x}$As QDs embedded in GaAs host material, we assume E_g=1.4 eV and E_b between 0.001 eV and 0.6 eV.

III. RESULTS AND DISCUSSION

Figure 1 (b) shows dark emission spectra in a log scale for various a_1 with E_b=0.1 eV. It clearly exhibits two peaks at E_g and E_1 emitting from host material and QDs respectively,

whose intensities change with the values of a_1. For large a_1 the QD emission is dominant, while the host material emission becomes dominant for very small a_1.

Figure 2 shows the detailed-balance-limit conversion efficiency η_{sc} and the corresponding J_{sc} and V_{oc} of QD solar cells, whose product is almost proportional to η_{sc}. All of η_{sc}, J_{sc} and V_{oc} for varied E_b respectively converge to 33%, 32.7 mA/cm^2 and 1.14 V at a_1=0 served as the values of a bulk solar cells without QDs, while at a_1=1 they are nothing but that of a bulk solar cell with bandgap of E_1. As expected, η_{sc} stays at 33% for various values of a_1, if binding energy of the QD (E_b) is smaller than 0.1 eV. However, for deep E_b above 0.1 eV, as a_1 increases, η_{sc} drops linearly from 33% in semi-logarithmic coordinate, then increases at the region of a_1>0.1. The origins of this behavior of η_{sc} can be clarify by those of J_{sc} and V_{oc}. The increase in J_{sc} is proportional to a_1, where the slopes increase as E_b increases. Thus, J_{sc} remains almost the same for a_1<0.01 in semi-logarithmic coordinate, and begins to increase for a_l>0.1. On the other hand, the drop of V_{oc} against a_l shows a linear-logarithm relation from 1.14 V to the values at a_l=1, which decreases as E_b increase. In Fig. 2 (d) for large E_b>0.1 eV, emission energy (E_{em}) defined as the center-of-mass energy in emission spectra immediately drops to E_l as a_l increases from 0, while for shallow E_b <0.1 eV drops only slightly and stays close to the host bandgap E_g. These are consistent with that the position of the dominant emission peak for large E_b>kT switches from E_g to E_l, as a_l increase shown in Fig. 1 (b). Therefore, E_{em} can be interpreted as effective band-gap energy to determine V_{oc} in the radiative limit, which can be used to explain the features of V_{oc} versus a_l and E_b in Fig. 2 (c).

Figure 3 exhibits the contour of η_{sc}, J_{sc} and V_{oc} of QD solar cells for varied a_l and E_b. The uppermost region close to E_b =0 are the values for bulk GaAs solar cells, served as references. According to the experimental reports on EQE [3,4,10,11], the absorptive tail of the most QD solar cells with GaAs as host materials would locate in the region of E_b between 0.1 eV and 0.3 eV, and a_l between 0.001 and 0.1. In such region, J_{sc} are almost unchanged, while obvious drops in V_{oc} already occur with the maximum drop of around 0.2V, which causes η_{sc} are within 5% absolute lower than that of bulk solar cells. Note that the above results exclude the extrinsic degradation effects, such as non-radiative recombination loss and oblique-band potential, which would further reduce the realistic η_{sc} and V_{oc}.

We need to remark that the present theoretical study only included single-photon absorption process, and current gain via multi-photon or multi-step photo-absorption processes in IB solar cells [1-5, 16] can make additional contribution to the present J_{sc}. However, current gain in IB QD solar cells via multi-step infrared-light absorption must be large enough to compensate at least the intrinsic drop of V_{oc} to gain the final conversion efficiency.

IV. CONCLUSIONS

This work quantitatively clarified that the changes in drops of V_{oc} and conversion efficiency caused by introduction of the various QDs density with different current gain separately stem from intrinsic consequence of the additional sub-band absorption band, when compared with host-material bulk solar cells. It consistently explains the empirical characteristics of QD solar cells and provides more realistic and more reasonable efficiency target for evaluation on QD solar cells.

ACKNOWLEDGEMENTS

This work was partly supported by JST-CREST, AIST-UT Operand OIL, JSPS KAKENHI (No. 26610081, 26390075), the Photon Frontier Network Program of MEXT, and JST-SENTAN. L.Z. thanks China Scholarship Council for financial support for her study in Japan.

REFERENCES

[1]. Luque, A., and Martí, A., "Increasing the efficiency of ideal solar cells by photon induced transitions at intermediate levels." *Physical Review Letters*, vol. 78, no. 26, pp. 5014, 1997.

[2]. Martí, A., Antolín, E., Stanley, C. R., Farmer, C. D., López, N., Díaz, P., Cánovas, E., Linares, P. G., and Luque, A., "Production of photocurrent due to intermediate-to-conduction-band transitions: a demonstration of a key operating principle of the intermediate-band solar cell." *Physical Review Letters*, vol. 97, no. 24, pp. 247701, 2006.

[3]. Luque, A., Martı, A., Stanley, C., López, N., Cuadra, L., Zhou, D., Pearson, JL and McKee, A., "General equivalent circuit for intermediate band devices: Potentials, currents and electroluminescence." *Journal of Applied Physics*, vol. 96, no. 1, pp. 903-909, 2004.

[4]. Martí, A., Antolín, E., Cánovas, E., López, N., Linares, P. G., Luque, A., C. R. Stanley and Farmer, C. D.. "Elements of the design and analysis of quantum-dot intermediate band solar cells." *Thin solid films*, vol. 516, no. 20, pp. 6716-6722, 2008.

[5]. Luque, A., Martí A. and Stanley C., "Understanding intermediate - band solar cells." *Nature Photonics* vol. 6, no. 3, pp. 146-152, 2012.

[6]. Martí, A., Lopez, N., Antolin, E., Canovas, E., Luque, A., Stanley, C. R., Farmer, C. D. and Diaz, P., "Emitter degradation in quantum dot intermediate band solar cells." *Applied Physics Letters*, vol. 90, no. 23, pp. 233510, 2007.

[7]. Hubbard, S. M., Cress, C. D., Bailey, C. G., Raffaelle, R. P., Bailey, S. G., and Wilt, D. M., "Effect of strain compensation on quantum dot enhanced GaAs solar cells." *Applied Physics Letters*, vol. 92, no.12. pp.123512, 2008.

[8]. Okada, Y., Oshima, R., and Takata, A. "Characteristics of InAs/GaNAs strain-compensated quantum dot solar cell." *Journal of Applied Physics*, vol. 106, no.2, pp. 4306, 2009.

[9]. Li, T., and Mario D., "Below-bandgap absorption in InAs/GaAs self-assembled quantum dot solar cells." *Progress in Photovoltaics: Research and Applications* vol. 23, no. 8, pp. 997-1002, 2015.

[10]. Guimard, D., Morihara, R., Bordel, D., Tanabe, K., Wakayama, Y., Nishioka, M., and Arakawa, Y., "Fabrication of InAs/GaAs quantum dot solar cells with enhanced photocurrent and without degradation of open circuit voltage." *Applied Physics Letters*, vol. 96, no. 20, pp. 3507, 2010.

[11]. Bailey, C. G., Forbes, D. V., Raffaelle, R. P., and Hubbard, S. M., "Near 1 V open circuit voltage InAs/GaAs quantum dot solar cells." *Applied Physics Letters*, vol.98, no. 16, pp. 163105.

[12]. Bailey, C. G., Forbes, D. V., Polly, S. J., Bittner, Z. S., Dai, Y., Mackos, C., Raffaelle. R. P. and Hubbard, S. M.,

"Open-circuit voltage improvement of InAs/GaAs quantum-dot solar cells using reduced InAs coverage." *IEEE Journal of Photovoltaics*, vol.2, no. 3, pp. 269-275, 2012.

[13]. Shockley, W. and Queisser H. J., "Detailed balance limit of efficiency of p-n junction solar cells." *Journal of applied physics*, vol.32, no.3, pp. 510-519,1961.

[14]. Zhu, L., Mochizuki, T., Yoshita, M., Chen, S., Kim, C., Akiyama, H., and Kanemitsu, Y., "Conversion efficiency limits and bandgap designs for multi-junction solar cells with internal radiative efficiencies below unity." *Optics Express*, vol. 24, no.10, pp. A740-A751, 2016.

[15]. Zhu, L., Kim, C., Yoshita, M., Chen, S., Sato, S., Mochizuki, T., Akiyama. H and Kanemitsu, Y., "Impact of sub-cell internal luminescence yields on energy conversion efficiencies of tandem solar cells: A design principle." *Appl. Phys. Lett*, vol.104, no.3, pp. 031118, 2014.

[16]. Nozawa. T.. and Arakawa Y.. "Detailed balance limit of the efficiencv of multilevel intermediate band solar cells." *Applied Physics Letters*, vol. 98, no.17, pp. 171108, 2011.

Zinc Selenide Surface Passivation Layer for Single-Crystalline CZTSe Solar Cells

Michael A. Lloyd[1,2], Douglas Bishop[3], Brian E. McCandless[2] and Robert Birkmire[1,2]

[1]Materials Science & Engineering, University of Delaware, Newark, DE, 19711, United States
[2]Institute of Energy Conversion, University of Delaware, Newark, DE, 19716, United States
[3]IBM T.J. Watson Research Center, Yorktown Heights, NY, 10598, United States

Abstract — **The use of thin ZnSe layers for interface passivation on the CZTSe/CdS junction in solar cell devices is presented. Single-crystalline CZTSe material of mm size scales was grown *via* a solid-state ingot growth mechanism. Thermally evaporated ZnSe produced an increase in integrated photoluminescence intensity of up to 450% when deposited onto a polished crystal surface. Device characterization showed V_{OC} increases as high as 150% with the implementation of a 10 nm ZnSe layer. Device performance was characterized as a function of ZnSe layer thickness. External quantum efficiency curves indicate that the presence of this high bandgap layer does not cause significant changes to the absorption spectrum. Temperature dependent *J-V* characteristics indicate that the ZnSe layer leads to an improved interface region.**

I. INTRODUCTION

$Cu_2ZnSn(S,Se)_4$-based (CZTSSe) solar cell performance has been limited to 12.6% efficiency with V_{OC} deficits on the order of 400 mV. Wang et al.[1] identified several limiting factors to the champion efficiency device through material and device characterization. High recombination rates at the CZTSSe/CdS interface were identified as one of the critical problems to overcome. Furthermore, two separate studies regarding interface properties within the CZTSSe/CdS system concluded a device-limiting band alignment. The first of these studies measured a conduction band spike, the later measuring a cliff-type alignment [2][3]. These conflicting results paired with the observed device limitations call for a better understanding of the CZTSSe surface and CdS interface properties.

There have been several studies published which emphasize the importance of a Zn-rich surface termination on device performance. Repins et al. correlated the level of Zn-termination with improved V_{OC} of pure selenide co-evaporated films[4]. Furthermore, Zn-rich surfaces were correlated with an increase in activation energy for recombination extracted from V_{OC} vs T plots, which is indicative of reduced interface recombination[5]. Buffière et al. determined that KCN treatments, a common treatment in kesterite cell fabrication, preferentially etches Cu and Sn related secondary phases[6]. This treatment resulted in improved device performance, when used in moderation.

The investigations in this work explore the effect of Zn-termination through the controlled deposition of a ZnSe passivation layer at the CZTSe/CdS interface of devices. As many convoluting effects exist in polycrystalline thin-film materials, pure-selenide, single-crystalline CZTSe will be used in these studies.

II. SAMPLE PREPARATION

A. Single-crystal growth and processing

Pure selenide crystals were fabricated via a solid-state ingot growth method in quartz ampoules in a similar fashion as that presented in ref. [7]. High-purity (5 or 6N) elemental sources of Cu, Sn, Zn and Se were used to synthesize ZnSe and Cu_2SnSe_3 precursor materials. These materials were pulverized into a powder and loaded into a quartz ampoule sealed below 10^{-6} torr. Crystal growth was promoted by a 20-day long growth step at 770°C. The resulting boule consisted of polycrystalline material from which single-crystals of CZTSe were extracted.

The nature of crystal surfaces resulting from the above growth technique is fundamentally different from those of thin-film processes. An optimization of polishing and chemical etching steps was previously developed which resulted in a single crystal device with an efficiency of 8.6% [7]. After polishing, ZnSe was deposited via thermal evaporation of Zn and Se precursor material at pressures below 10^{-6} torr through a 6 mm^2 mask. The temperature of the single crystalline specimens were held at 300°C for the duration of the deposition. Passivation layer thicknesses were controlled by the deposition time. Crystal specimen and their surface treatments discussed herein are shown in Table 1.

Devices on single-crystalline CZTSe were fabricated by the method described in ref [7]. CdS was deposited onto bare and passivated CZTSe surfaces by chemical bath deposition. Sputtered ZnO and ITO layers define the active device area and act as the top contact during characterization methods. Crystals are physically and electrically adhered to a Mo-coated SLG substrate with a high work-function carbon paste. An

Table 1. XRF-verified composition of crystal specimen. The ZnSe passivation layer thickness ≈ 100 nm. Information on the champion single crystal were previously reported [7].

Specimen	Passivation	Cu/Zn+Sn	Zn/Sn	Se/metals
Crystal A	ZnSe	0.94	1.09	1.18
Crystal B	ZnSe	0.95	1.08	1.18
Crystal C	Bare	0.95	1.08	1.18
Champion [7]	Bare	0.87	1.22	1.19

anneal at 220°C in ambient air was performed on finished devices for 4 minutes.

The thickness of single-crystal devices (1 - 1.5 mm) enables the reprocessing of the same crystal with different ZnSe layer thicknesses. The Mo-coated glass substrate is removed from the device by dissolving the carbon-paste in an acetone-based solution. CdS and window layers are removed in a dilute bath of 10% hydrochloric acid in DI water. Approximately 5 microns of surface are then removed via mechanical polishing with a 0.05 μm alumina slurry. The surface is then reinstated with the processes described above.

B. Characterization

X-ray fluorescence spectroscopy (XRF) calibrated by inductively coupled plasma spectroscopy (ICP) was used to quantify crystal composition. Photoluminescence (PL) spectra was obtained on a τau Science PixEL module capable of large area PL mapping. An 808 nm laser was used as the excitation source. PL spectra were obtained in 100 nm steps on an InGaAs detector with a set of band-pass filters ranging from 1000 - 1600 nm. PL yield, and correspondingly radiative recombination, is utilized as a rough metric for relative surface quality.

Device J-V characteristics were measured using an OAI TriSOL solar simulator calibrated to 100 mW/cm^2 via a silicon standard cell. External quantum efficiency (EQE) of devices was measured using an Oriel monochromator from 300 to 1400 nm using an AC method with a lock-in amplifier. The resulting EQE curves were normalized to the measured J_{SC} of the devices. Temperature-dependent J-V measurements were performed from -120°C to 40°C. The quartz-halogen source lamp was calibrated to the known J_{SC} value of a silicon reference cell at 25°C.

III. EXPERIMENTAL RESULTS

Fig. 1a shows the peak energy position of the photoluminescence spectra from crystal A, which was shown to have a Cu/Zn+Sn ratio of 0.95 with a Zn/Sn ratio of 1.09 via XRF measurements. This composition falls slightly Cu-rich and Zn-poor of those typical in high-performing devices [8].

The PL images shown in Fig. 1 highlight the PL improvements from a ~ 100 nm thick, circular ZnSe passivation layer. The intensity of the PL peak for crystal A is ~ 5.5 x larger for the passivated surface region compared to the un-passivated surface (Fig. 1a). The peak of the photoluminescence curves for each crystal in Fig. 1 lies near

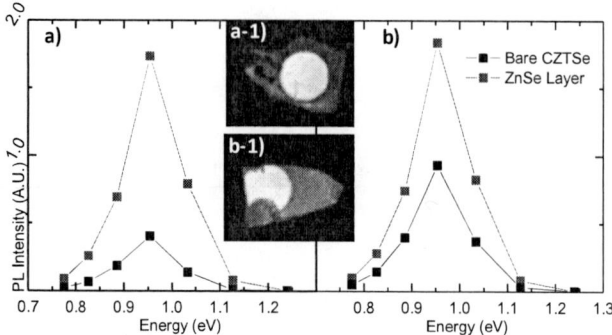

Fig. 1. Luminescence intensity as a function of energy. The blue curves were quantified in the circular regions in the insert whereas the black curve was measured on the bare surface of each crystal. Shown above is a), data for crystal A and b), for crystal B.

0.95 eV for both bare and ZnSe terminated surface regions, consistent with PL peak positions previously reported on bare CZTSe single crystals [7]. The similarity in peak position between surface terminations allows us to conclude that PL improvements are, at least in part, due to the improvement of the CZTSe surface, rather than absorption in the ZnSe (E_g = 2.7 eV). Crystal B showed a smaller PL improvement of 125% with a ZnSe layer, although the bare surface PL intensity was higher than that of crystal A. Both crystals have very similar compositions, therefore it is likely the baseline PL discrepancy is due to subtle differences during surface preparation. Both crystals exhibited similar absolute peak intensities upon ZnSe deposition. To further evaluate the effect of ZnSe, full devices were fabricated on single crystals with a ZnSe passivation layer.

Fig. 2 shows the J-V curves for two different crystals where crystal C represents the control device, without a passivation layer. On crystal A, devices were prepared with two different ZnSe thicknesses of 50 nm (red) and 10 nm (blue) using the reprocessing procedure discussed in Section II. Since composition differences may convolute device results, crystals A & C were chosen from the same growth process and were shown via XRF to have similar elemental ratios.

The baseline device from this set of samples shows shunted behavior with a V_{OC} of 150 mV. With a 50 nm ZnSe passivation layer, crystal A is characterized by a V_{OC} of 250 mV, a 67% increase from the baseline device. However, this improvement is accompanied by artifacts consistent with a large series resistance.

With a reduced ZnSe thickness of 10 nm, improvements in all parameters are observed. The V_{OC} increases to 372 mV, a 150% improvement from the control device. This voltage is comparable to the highest efficiency single-crystal device with a V_{OC} of 389 mV [7]. Device parameters for all the devices on crystal A are shown in Fig. 2. Devices on crystal B behave in a similar manner with ZnSe thickness as crystal A.

The following device characterization techniques were performed on a device fabricated on crystal B with a 10 nm

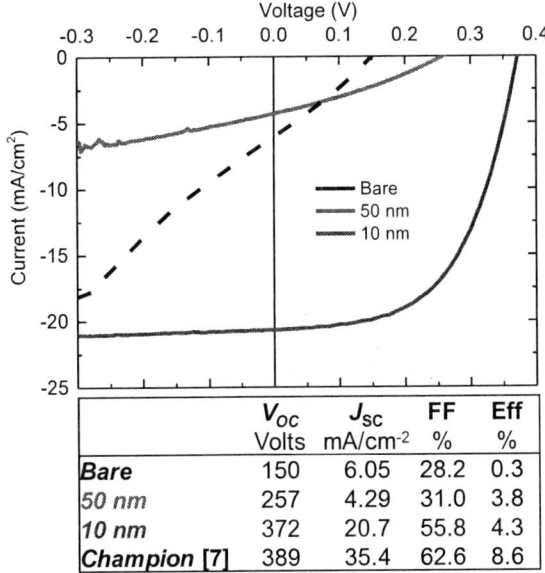

	V_{OC} Volts	J_{SC} mA/cm^{-2}	FF %	Eff %
Bare	150	6.05	28.2	0.3
50 nm	257	4.29	31.0	3.8
10 nm	372	20.7	55.8	4.3
Champion [7]	389	35.4	62.6	8.6

Fig. 2. Device performance for a baseline device (crystal C) without a passivation layer (black). Crystal A is represented by a red curve (50 nm) and blue curve (10 nm) to denote the ZnSe layer thickness in each reprocessing.

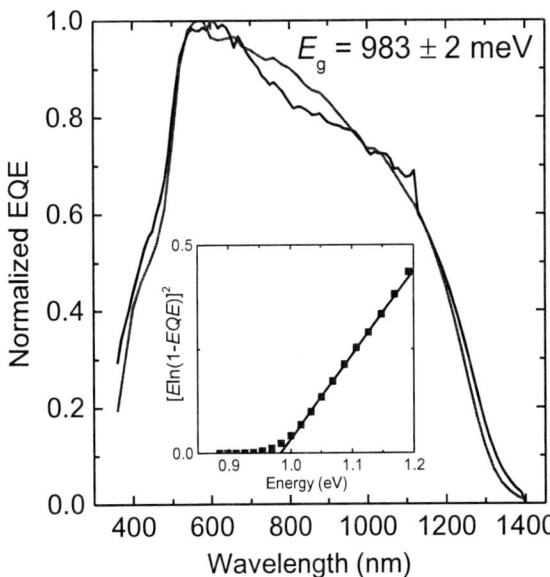

Fig. 3. Normalized EQE curves for the device with 10 nm ZnSe on crystal B (blue) compared to that of the champion single crystal device[7] (black). Inset: Bandgap extrapolation taking the *x*-intercept of a linear fit to the $[E\ln(1-EQE)]^2$ vs. E plot at the low energy tail of crystal B.

ZnSe layer. Fig. 3 compares the EQE of the device on crystal B to the previously reported 8.6% device in ref [7]. There does not appear to be a significant difference in shape between the two curves, indicating minimal absorption losses due to the ZnSe layer. A bandgap value of 980 ± 4meV was extracted from the low energy tail region of the EQE using the method described by Bag et al. [9]. This observation and the PL data indicate that the ZnSe contributes little to light absorption.

The temperature dependence of the *J-V* characteristics is shown in Fig. 4. At room temperature, the *J-V* curve is well behaved with a J_{SC} and *FF* of 31.5 mA/cm^2 and 57 % respectively. It is noted that the J_{SC} measured with this technique is higher than the values reported above because of a decrease in shading due to differences in contact prove size. With decreasing temperature, the shape of the *J-V* curve begins to change as forward bias transport across the junction becomes hindered at temperatures below 60°C. This is coupled with a drop in J_{SC} and *FF* reaching a 19% and 32% drop the lowest probed temperature (-120°C). Similar behavior was displayed in our champion single crystal device [7], although the effect is more dramatic in the case with a ZnSe layer. The dark *J-V* curves for this device do not show a change in behavior with temperature other than an expected shift to higher voltages.

The activation energy (E_A) obtained through the extrapolation of the $V_{OC}(T)$ to 0 K can provide insight into the dominant recombination behavior of a device [5]. Fig. 4 displays the behavior of the V_{OC} with temperature on the

device implementing a 10 nm ZnSe layer. The V_{OC} extrapolates to an E_A value of 979.9 mV, which is the E_g/q value as measured via EQE to within the error of our measurements. This provides evidence that the dominant recombination mechanisms occur in the bulk of the absorber rather than the junction interfaces.

IV. DISCUSSION AND PLANS

These results support previously reported observations that Zn-termination is beneficial to CZTSe device performance [4][6]. The single crystal device behavior further suggests an interplay between surface passivating properties and the current blocking nature of the insulating ZnSe. Although the presence of excess ZnSe at the CZTS/CdS interface is typically observed to be detrimental to device performance [10], the above results indicate that a controlled amount of the phase can be beneficial. Furthermore, the initial investigations appear to indicate that ZnSe passivation may loosen requirements for chemical/mechanical polishing on single-crystals. This is highlighted in Fig. 1b), which shows that, despite the lower non-passivated PL intensity for crystal A, the intensities of the passivated crystals are nearly identical.

The activation energy obtained from the $V_{OC}(T)$ curves matches the E_g/q obtained by EQE and is the highest measured to date on single-crystal devices fabricated by this collaboration. This indicates that the ZnSe layered device has

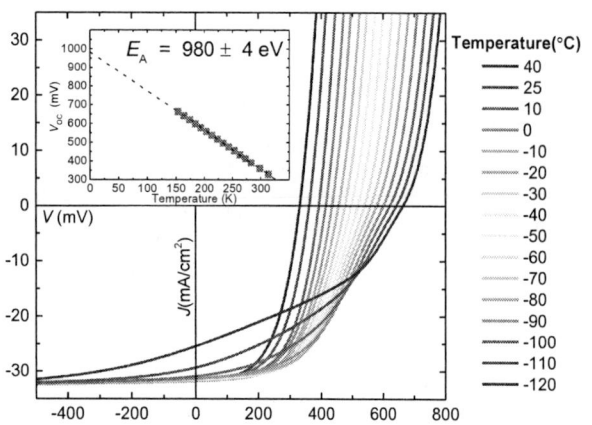

Fig. 4. *J-V* behavior of the 10 nm ZnSe device fabricated on crystal B. The device exhibits well-behaved curves until -60°C where a blocking of current in forward bias is evident. Inset: The V_{OC} as a function of temperature for this device with a linear fit to extrapolate an E_A of 980 ± 4 meV.

the lowest surface/bulk recombination contribution to date on this device structure and is mainly limited by the letter processes. The $V_{OC}(T)$ and PL results both provide evidence for the ability of ZnSe to passivate of the CZTSe surface. The PL intensity increase speaks towards an increase in radiative recombination, while the $V_{OC}(T)$ activation energy speaks to an improved junction interface quality. EQE curves further indicate that the ZnSe layer does not contribute to parasitic absorption in a significant way when compared to devices without this layer.

The blocking of current in forward bias at low temperature may be related to the ZnSe layer. The presence of similar behavior in our highest efficiency device, without the ZnSe layer, suggest that the new structure is not solely responsible for this behavior [7]. No current blocking is apparent at room temperature. Further investigation into layer thickness are required to provide insight into this effect.

In future investigations, we will explore the behavior of devices with varied ZnSe thicknesses between 1 nm and 10 nm utilizing the reprocessing technique discussed in Section II. Additionally, photoluminescence and device characterization for crystals covering a larger compositional space will be investigated. Relative *Cu* and *Zn* concentrations will be explored to provide insight into the efficacy of the ZnSe termination on different bulk environments.

V. ACKNOWLEDGEMENTS

The information, data, or work presented herein was funded in part by the U.S. Department of Energy, Energy Efficiency and Renewable Energy Program, under Award Number DE- EE0006334. This work was also supported in part by the National Science Foundation under Grant No. DMR-1508042.

REFERENCES

[1] W. Wang, M. T. Winkler, O. Gunawan, T. Gokmen, T. K. Todorov, Y. Zhu, and D. B. Mitzi, "Device Characteristics of CZTSSe Thin-Film Solar Cells with 12.6% Efficiency," *Adv. Energy Mater.*, pp. 1–5, Nov. 2013.

[2] R. Haight, A. R. Barkhouse, O. Gunawan, B. Shin, M. W. Copel, M. Hopstaken, and D. B. Mitzi, "Band alignment at the $Cu_2ZnSnS_xSe_{[1-x]4}$/CdS interface," *Appl. Phys. Lett.*, vol. 98, no. 25, p. 253502, 2011.

[3] A. Santoni, F. Biccari, C. Malerba, M. Valentini, R. Chierchia, and A. Mittiga, "Valence band offset at the CdS/Cu2ZnSnS4 interface probed by x-ray photoelectron spectroscopy," *J. Phys. D. Appl. Phys.*, vol. 46, no. 17, p. 175101, May 2013.

[4] I. L. Repins, J. V. Li, A. Kanevce, C. L. Perkins, K. X. Steirer, J. Pankow, G. Teeter, D. Kuciauskas, C. Beall, C. Dehart, J. Carapella, B. Bob, J.-S. Park, and S.-H. Wei, "Effects of deposition termination on Cu2ZnSnSe4 device characteristics," *Thin Solid Films*, pp. 2–5, Sep. 2014.

[5] D. Abou-Ras, T. Kirchartz, and U. Rau, Eds., *Advanced Characterization Techniques for Thin Film Solar Cells*. Weinheim, Germany: Wiley-VCH Verlag GmbH & Co. KGaA, 2011.

[6] M. Buffiere, G. Brammertz, S. Sahayaraj, M. Batuk, S. Khelifi, D. Mangin, A. A. El Mel, L. Arzel, J. Hadermann, M. Meuris, and J. Poortmans, "KCN Chemical Etch for Interface Engineering in Cu2ZnSnSe4 Solar Cells," *ACS Appl. Mater. Interfaces*, vol. 7, no. 27, pp. 14690–14698, 2015.

[7] M. A. Lloyd, D. Bishop, O. Gunawan, and B. McCandless, "Fabrication and performance limitations in single crystal Cu2ZnSnSe4 Solar Cells," in *2016 IEEE 43rd Photovoltaic Specialists Conference*, 2016, pp. 3636–3640.

[8] H. Katagiri, K. Jimbo, M. Tahara, H. Araki, and K. Oishi, "The influence of the composition ratio on CZTS-based thin film solar cells," *Mater. Res. Soc. Symp. Proc. Vol. 1165*, vol. 1165, pp. 1165-M04-1, 2009.

[9] S. Bag, O. Gunawan, T. Gokmen, Y. Zhu, T. K. Todorov, and D. B. Mitzi, "Low band gap liquid-processed CZTSe solar cell with 10.1% efficiency," *Energy Environ. Sci.*, vol. 5, no. 5, p. 7060, 2012.

[10] S. López-Marino, Y. Sánchez, M. Placidi, A. Fairbrother, M. Espindola-Rodríguez, X. Fontané, V. Izquierdo-Roca, J. López-García, L. Calvo-Barrio, A. Pérez-Rodríguez, and E. Saucedo, "ZnSe Etching of Zn-Rich Cu2ZnSnSe4: An Oxidation Route for Improved Solar-Cell Efficiency.," *Chemistry*, pp. 14814–14822, Oct. 2013.

Use of Single Wall Carbon Nanotube films doped with Triethyloxonium Hexachlorantimonate as a Transparent Back Contact for CdTe Solar Cells

Fadhil K. Alfadhili, Jacob M. Gibbs, Geethika K. Liyanage, Patrick W. Krantz, Suneth C. Watthage, Zhaoning Song, Adam B. Phillips, and Michael J. Heben

Wright Center for Photovoltaics Innovation and Commercialization, Department of Physics and Astronomy, University of Toledo, Toledo, OH, 43606, USA

Abstract — **Highly conductive transparent films were formed by doping single-wall carbon nanotubes (SWCNTs) networks with triethyloxonium hexachlorantimonate (OA). This doping can result in complete quenching of the SWCNT absorption features in the visible and infra-red (IR) range leaving an IR transparent film. These doped SWCNT films were investigated for possible use as a transparent back contact for CdTe-based devices. One particular goal is the development of tunnel junction for use in tandem cells. The transmittance of device stacks with and without an OA-doped SWCNT back contact was nearly identical. However, it was found that during the doping process, the OA interacted more strongly with the CdTe layer than the SWCNTs, resulting in reduced device performance. The average power conversion efficiency decreased from 11.8% for a standard Cu/Au back contacted device to 10.6% for the OA-doped SWCNT back contacted device.**

Index terms- CdTe, SWCNTs, transparent tunnel junction.

I. INTRODUCTION

Thin film tandem solar cells have attracted renewed attention recently because of the potential for an excellent combination of high power conversion efficiency (PCE) and easy manufacturing[1,2]. In their simplest form, tandem devices are a combination of two cells, a top cell consisting of a higher band gap material (e.g., ~1.7 eV), and a bottom cell with a lower energy band gap (e.g., ~1.1 eV). The top cell absorbs the high-energy photons and transmits infrared and near infrared light to the bottom cell[1,2]. A recent analysis of commercially available technologies indicates that CdTe-based devices (currently with a record efficiency of 22.1%[3]) could be an important wide band gap partner for high efficiency thin film tandem cells[4]. Although CdTe could be paired with $CuInSe_2$ or GaSb, there is also consideration of widening the band gap of CdTe through the incorporation of Mg or Zn[4]. While challenges remain for increasing the band gap of CdTe-based materials, progress has been made with recent studies achieving 9.3% efficiency for a CdMgTe solar cell with 1.6 eV band gap[5]. Even with such developments, however, a significant challenge for developing a high efficiency CdTe tandem solar cell, is finding an optically transparent tunnel junction that could enable monolithic integration[4]. Recent work with substrate configuration CdTe devices suggests that SWCNT films could act as a tunnel junction[6]. SWCNTs have also been used as a transparent layer for CdTe[7].

As a one dimensional material, SWCNTs possess sharp optical absorptions associated with the van Hove singularities in the density of states[8].The wavelengths at which these absorptions occur are largely determined by the diameter of the SWCNTs. Larger diameter SWCNTs, such as those produced by laser vaporization or arc discharge, have so-called S_{11} and S_{22} absorptions occurring at roughly 1800 nm and 1100 nm, respectively, while smaller diameter SWCNTs, such as those produced by the CoMoCat or HipCO processes, have S_{11} and S_{22} absorptions at roughly 1000 nm and 550 nm, respectively[9]. The absorption peaks in SWCNT films could be highly disadvantageous for tunnel junctions contacts since there will be a loss of photo-generated current in the bottom cell in the S_{11} and S_{22} wavelength regions.

It is possible to quench the SWCNTs absorption peaks through doping. For example, HNO_3 is a well-known p-type dopant that will completely quench the first absorption peak and partially quench the second absorption peak of large diameter tubes[10]. However, HNO_3 will also oxidize, etch, and, thus, destroy CdTe devices. In an effort to explore a potentially more gentle dopant, the present work explores the use of triethyloxonium hexachlorantimonate $((C_2H_5)_3O^+ SbCl_6^-;$ OA), which has been used to strongly dope SWCNTs[11] and completely quench the absorption features at all wavelengths shorter than 900 nm[12].Such doping would effectively remove all absorption features that would appear below the CdTe band gap, and would result in an IR transparent SWCNT film suitable for tandem devices. Here, we investigate OA doping of SWCNT films and the effect of doping on SWCNT back contacts to CdTe devices.

II. MATERIALS AND METHODS

CoMoCat SWCNT (SG65i, Southwest Nanotechnologies) films were prepared by spraying[13] sodium dodecylbenzenesulfonate (SDBS)-dispersed SWCNTs at 300 µl/min under nitrogen flowing at 7 std l min⁻¹ with an ultrasonic spray head (06-5108, Sonotek, 3W) onto heated (140 °C) soda lime glass and CdTe substrates. Twenty-five nanometers thick SWCNT films were prepared[14]. The

978-1-5090-5606-4/17 $31.00 © 2017 IEEE

SDBS was removed by soaking the films in water for 1 hour.

Doping of SWCNT films followed the method reported by Chandra et al[11]. The prepared SWCNT films were soaked in an OA/dichloroethane (DCE) solution (10 mg-ml^{-1}) for 30 to 300 s at 70 °C. Residual OA on the surface was removed by rinsing with acetone.

CdS and CdTe films were deposited on commercially available TEC 15M (Pilkington NA) substrates using a commercial vapor transport deposition process (Willard & Kelsey Solar Group). These samples were activated by applying a CdCl$_2$/methanol solution and heating at 387 °C for 30 min in a dry air flow. Samples were subsequently rinsed with methanol to remove the excess CdCl$_2$.

CdTe/SWCNT devices were completed with a 40 nm thick Au film, which was deposited by thermal evaporation. For comparison, standard CdTe devices were completed by evaporating 3 nm Cu/40 nm Au and annealing at 150 °C in air for 45 minutes. A solar cell area of 0.079 cm^2 was defined by laser scribing[15]. These devices were measured under simulated AM1.5 illumination to obtain the current density-voltage (J-V) characteristics. The sheet resistance of SWCNTs films were measured with a standard four-point probe system (Signatone-2400), and the optical absorbance of the sprayed SWCNT thin films were obtained using a Perkin-Elmer Lambda 1050 spectrophotometer. The CdTe films were characterized using scanning electron microscopy (SEM; Hitachi S-4800) and energy dispersive X-ray spectroscopy (EDS).

III. RESULTS AND DISCUSSION

Significant changes in the optical properties were observed after the SWCNT films were doped with OA (Fig. 1a). There was a significant decrease in the intensities of the absorption peaks at ~550 nm and ~1000 nm, corresponding to S$_{22}$ and S$_{11}$ transition energies for the SWCNTs, respectively. The quenching of the S$_{11}$ and S$_{22}$ features indicates that hole doping has occurred and that the Fermi level in the SWCNTs has been pushed deep into the valence band[11,12]. The complete quenching of the S$_{22}$ transitions demonstrated here indicates a higher degree of SWCNT doping than has been reported to date[11,12].

The doping also reduces the electrical resistance of the SWCNT film. As Fig. 1(b) shows, strong doping of the SWCNT film occurs in less than 30 s. The increase in the conductivity of the SWCNT film is consistent with an increase in the hole carrier density. The strong enhancement in optical transmission and conductivity of SWCNT films by OA doping, coupled with previous results[7,14], qualitatively suggests that SWCNTs films could act as a highly transparent back contact for the CdTe device.

To investigate this possibility, SWCNT films were deposited on CdTe device stacks by ultrasonic spraying (see

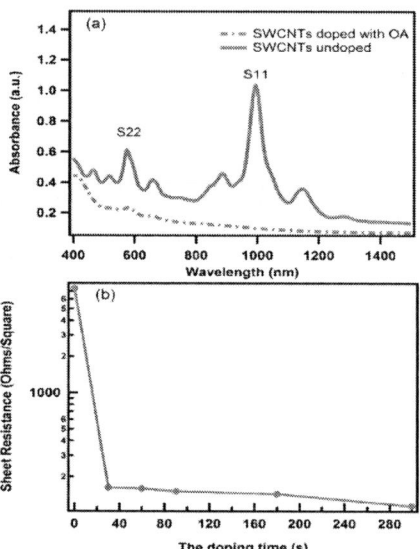

Fig. 1. (a) The absorbance spectra of a 100 nm thick CoMoCat SWCNT film on glass before and after doping with OA for 30 sec. (b) Variation of sheet resistance with doping time for SWCNTs films on glass.

Materials and Methods). Figure 2 shows the transmittance spectra of a bare CdTe device stack (TEC15M/ CdS (~150 nm)/ CdTe (~4 μm)) and the same device stack with undoped and OA-doped SWCNT films on it. The average transmittance of the bare device stack is 55% for the wavelengths range that is most likely to be important for tandem solar cells (850 to 1100 nm). However, when the 25 nm thick SWCNT film was added, the transmittance decreased across the same wavelength range was reduced to 40% with a minimum value of 36% corresponding to the S$_{11}$ absorption peak at ~1000 nm. As expected, the OA doping induced quenching of the S$_{11}$ peak such that the average transmittance below the CdTe band gap (850 to 1100 nm) increased to 53% .These results suggest that OA-doped SWCNT films could provide a transparent back contact with better performance than the current state-of-the-art transparent back contact[7].

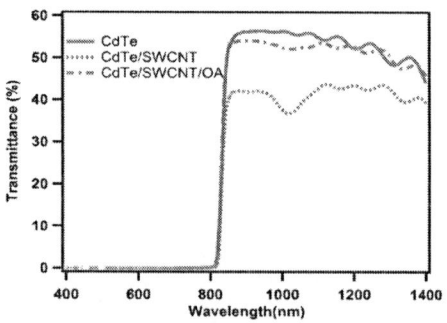

Fig. 2. The transmittance spectra of CdTe, CdTe/SWCNT, and CdTe/SWCNT/OA.

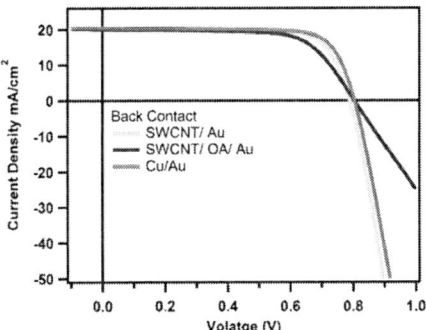

Fig. 3. J-V characteristic of the CdTe devices with different back contacts.

TABLE I
J-V CHARACTERISTICS OF CDTE DEVICES PREPARED WITH
DIFFERENT BACK CONTACTS

Back contact	V_{OC} (mV)	J_{sc} (mA/cm^2)	FF (%)	Eff (%)
Cu/Au	781± 3 804	20.3±0.3 20.4	74.7±1 76.5	11.8±0.4 12.5
SWCNT /Au	790± 2 789	20±0.31 20.5	73.8±0.7 74.2	11.6±0.2 12.0
SWCNT with OA for 5min/Au	805± 2 805	19.4±0.31 19.8	67.7±0.9 68.9	10.6±0.2 10.9

To determine the effect of OA doping on the device performance. CdTe devices with Cu/Au, SWCNT/Au, and OA-doped SWCNT/Au back contacts were fabricated. Fig 3 shows the J-V characteristics for the best device, and table I shows the PV parameters for devices prepared in the three device architectures. Note that more than 20 devices where measured and included in data set reported in table I, and that the SWCNT-OA doping (vie *infra*). As previously reported[14], the average efficiency of SWCNT back contact device matched the performance obtained from the standard Cu/Au devices, with an improvement in V_{OC} being offset by a decrease in FF. However, after exposing the SWCNT layer to the OA doping process, the efficiency was significantly reduced. Previous device modeling[14] suggested that reducing the Fermi level in the SWCNTs through doping would result in improved device performance by increasing the open circuit voltage (V_{OC}), and this was indeed observed. However, the FF was significantly impacted and the J-V curves displayed indications of high series resistance and, in

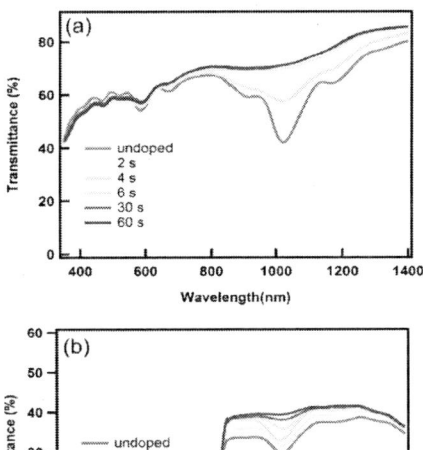

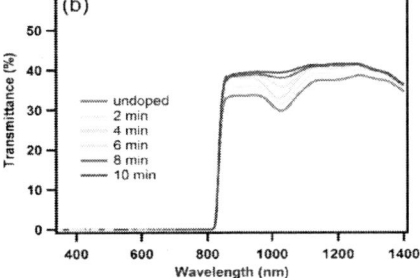

Fig 4. OA doping of SWCNT films at 70 °C as a function of time on (a) glass and (b) on CdTe.

some cases (not shown) a kink in the characteristic at V_{OC}. While the increase in the V_{OC} was expected based on our understanding of the behavior of the SWCNT back contact and doping, the increase in R_S (and accompanying reduction in FF) was not.

The data in Fig. 3 and Table I compares the performance of the devices with SWCNT films that were optimally doped. It is interesting to note that the kinetics of the doping of SWCNTs was very different depending on whether the SWCNT film was on CdTe or on glass. Figure 4 compares the time required to quench the S_{11} absorption transitions in SWCNT films prepared on glass (Fig. 4a) and CdTe (Fig. 4b). Complete doping of films on glass is achieved in a matter of seconds, while minutes are required for doping of films on CdTe.

To quantify the differences in SWCNT doping behavior on glass and CdTe substrates, we measured the activation energy for the process. Because the doping of SWCNTs on glass is rapid, it seems clear that doping occurs when the OA is in contact with the SWCNT. In this case, the rate-limiting step would be the diffusion of OA to the surface of the individual SWCNTs. As a result, a diffusion-limited model in the form of Eqn. 1 was used.

$$Ln\,[A_{S11}]_t = -Dt + Ln\,[A_{S11}]_0 \qquad (1)$$

Where D is the diffusion time (s^{-1}), $[A_{S11}]_t$ is the area under S11 curve at time t, $[A_{S11}]_0$ is the area under S11 curve at time 0, and t is time (s). To determine the diffusion time (D) for each temperature (27, 50, and 70 ℃) the area under the S11 curve was calculated for several doping times, and the results

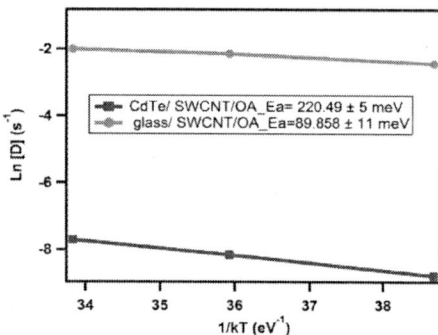

Fig. 5. Plot of ln (D) vs. 1/KT to evaluate Ea of OA doping for SWCNT on glass and CdTe.

were fit to equation (1). Figure 4 shows an example of the measured transmittance for doping completed at 70 °C.

The thermally activated diffusion process is expected to follow an Arrhenius behavior, following Eqn. 2.

$$D = Ae^{-\frac{Ea}{KT}} \qquad (2)$$

where E_a is the activation energy, T is the temperature, K is the Boltzmann constant, and A is a constant. Fitting the diffusion data to Eqn. 2 (Fig. 5) shows that the activation energy of OA doping of SWCNTs on CdTe is 2.45 times higher than that of doping SWCNT on glass. The significant difference in the activation energy indicates that the presence of CdTe is severely inhibiting SWCNT doping. This could be due to an interaction between the OA and CdTe, either repulsive or attractive, or due to an interaction between the SWCNTs and CdTe that prevents the OA from interacting with the SWCNTs.

To determine if the CdTe is interacting with the OA, bare CdTe device stacks underwent the OA doping process prior to device completion with Cu/Au and SWCNTs/Au. As shown in Fig. 6, the device performance severely degraded by the OA processing. In both cases, the J_{SC} and V_{OC} decreased relative to the untreated sample. In the case of the Cu/Au back contact devices, a kink in the J-V curve also appeared, indicating an increase in the barrier at the back contact[16]. These results

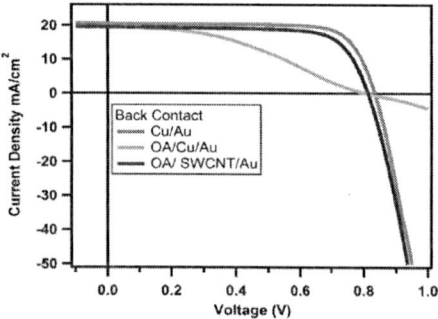

Fig. 6. J-V characteristic of the CdTe devices with different back contacts

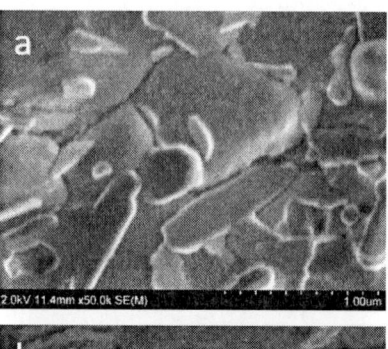

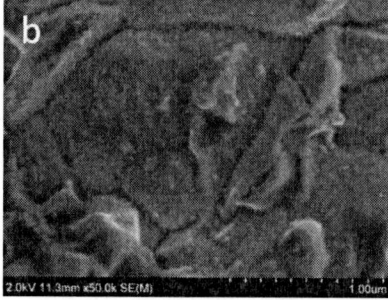

Fig. 7. SEM images of CdTe (a) before OA treated (b) after OA treated for 10 min.

indicate that there is an interaction between CdTe and OA, but do not suggest what type of interaction occurs.

To determine the type of interaction between CdTe and OA, the CdTe films were characterized using SEM and EDS. Fig 7a shows the SEM image of the pristine CdTe containing tightly packed grains with a size of ~1 µm. This morphology is commonly observed in the CdCl$_2$-treated CdTe films. After 10 minutes of OA treatment, though, the SEM image (Fig. 7b) looked very different. New features appear on the surface of the CdTe. EDS of the sample shows the presence of ~1.5 wt.% of Cl that is not present in the untreated sample. OA doping of the SWCNTs is thought to be through the interaction of the SbCl$_6^-$ ion[11]. Assuming the same mechanism for CdTe, the Cl detected in EDS is likely from this ion. The Sb was probably not detected because the sensitivity of EDS is not sufficiently high for such low loading levels. These results show that a Cl-containing chemical residue forms on the CdTe surface after exposure to the OA doping process.

The high activation energy for OA doping of SWCNTs on CdTe substrates, relative to on glass, is due to a preferential interaction of OA towards CdTe. This preferential interaction results in an OA film on the CdTe surface, which, in turn, reduces the device performance of SWCNT back contact devices exposed to an OA doping process. Evidently, the OA interaction with CdTe is stronger than the interaction with SWCNTs, suggesting that the CdTe surface getters OA species which, in turn, impacts the rate at which SWCNT doping is achieved. Consequently, CdTe devices where the SWCNT back contact is OA doped after deposition onto the CdTe will not result in high performance transparent back

978-1-5090-5606-4/17 $31.00 © 2017 IEEE

contacts. Current data indicates that a partially insulating film may be formed at the CdTe/SWCNT interface. Such a layer might increase the series resistance and impact the FF, as observed in Fig. 3. Future directions will include efforts to fabricate a high-performance device with a transparent SWCNT back contact by doping the SWCNTs prior to application to the CdTe.

IV. CONCLUSION

OA doped SWCNT films were investigated as a transparent back contact for CdTe solar cells. OA doping induces a complete quenching of S_{11} and S_{22} optical transitions resulting in increasing the transmittance of the SWCNT films and a decrease in the electrical resistance due to doping. The transmittance of the doped SWCNTs on CdTe was nearly identical to that of the bare CdTe sample after the SWCNTs were thoroughly doped. The OA doping process, however, adversely affected the device performance. It was found that the OA preferentially interacted with CdTe during the doping process and formed a reaction film on the CdTe surface. This film is believed to be the cause of the reduced device performance. Consequently, OA doping of the SWCNTs after application to the CdTe may not be a viable pathway towards preparing a transparent back contact or tunnel junction for CdTe solar cells.

V. ACKNOWLEDGEMENT

The work was supported by the National Science Foundation (Contract no. #ECCS-1665172), the Ohio Federal Research Network (Center for Materials and Manufacturing), the Air Force Research Laboratory (Contract no. # FA9453-11-C-0253), and the Wright Center Endowment for Photovoltaics Innovation and Commercialization. We gratefully acknowledge the donation of TEC15M samples from Pilkington NA and CdS/CdTe samples from Willard & Kelsey Solar Group.

REFERENCES

[1] L. Kranz, A. Abate, T. Feurer, F. Fu, E. Avancini, J. Löckinger, P. Reinhard, S. M. Zakeeruddin, M. Grätzel, S. Buecheler, and A. N. Tiwari, "High-efficiency polycrystalline thin film tandem solar cells," *The Journal of Physical Chemistry Letters,* vol. 6, no. 14, pp. 2676-2681, 2015.

[2] R. W. Crisp, G. F. Pach, J. M. Kurley, R. M. France, M. O. Reese, S. U. Nanayakkara, B. A. MacLeod, D. V. Talapin, M. C. Beard, and J. M. Luther, "Tandem solar cells from solution-processed CdTe and PbS quantum dots using a ZnTe–ZnO tunnel junction," *Nano Letters,* vol. 17, no. 2, pp. 1020-1027, 2017.

[3] http://www.nrel.gov/ncpv/images/efficiency_chart.jpg. (accessed June1,2017).

[4] J. P. Mailoa, M. Lee, M. Lee, T. Buonassisi, A. Panchula, and D. N. Weiss, "Energy-yield prediction for II-VI-based thin-film tandem solar cells," *Energy & Environmental Science,* vol. 9, no. 8, pp. 2644-2653, 2016.

[5] O. S. Martinez, D. Reyes-Coronado, and X. Mathew, "Cd1-xMgxTe thin films and top-cells for possible applications in tandem solar cells." pp. 74090V-74090V-7, 2009.

[6] R. R. Khanal, A. B. Phillips, Z. Song, Y. Xie, H. P. Mahabaduge, M. D. Dorogi, S. Zafar, G. T. Faykosh, and M. J. Heben, "Substrate configuration, bifacial CdTe solar cells grown directly on transparent single wall carbon nanotube back contacts," *Solar Energy Materials and Solar Cells,* vol. 157, pp. 35-41, 2016.

[7] T. M. Barnes, X. Wu, J. Zhou, A. Duda, J. v. d. Lagemaat, T. J. Coutts, C. L. Weeks, D. A. Britz, and P. Glatkowski, "Single-wall carbon nanotube networks as a transparent back contact in CdTe solar cells," *Applied Physics Letters,* vol. 90, no. 24, pp. 243503, 2007.

[8] R. Saito, M. Fujita, G. Dresselhaus, and M. S. Dresselhaus, "Electronic structure of chiral graphene tubules," *Applied Physics Letters,* vol. 60, no. 18, pp. 2204-2206, 1992.

[9] H. Kataura, Y. Kumazawa, Y. Maniwa, I. Umezu, S. Suzuki, Y. Ohtsuka, and Y. Achiba, "Optical properties of single-wall carbon nanotubes," *Synthetic Metals,* vol. 103, no. 1, pp. 2555-2558, 1999.

[10] H.-Z. Geng, K. K. Kim, K. P. So, Y. S. Lee, Y. Chang, and Y. H. Lee, "Effect of acid treatment on carbon nanotube-based flexible transparent conducting films," *Journal of the American Chemical Society,* vol. 129, no. 25, pp. 7758-7759, 2007.

[11] B. Chandra, A. Afzali, N. Khare, M. M. El-Ashry, and G. S.Tulevski,"Stable charge-transfer doping of transparent single-walled carbon nanotube films," *Chemistry of Materials,* vol. 22, no. 18, pp. 5179-5183, 2010.

[12] A. D. Avery, B. H. Zhou, J. Lee, E.-S. Lee, E. M. Miller, R. Ihly, D. Wesenberg, K. S. Mistry, S. L. Guillot, and B. L. Zink, "Tailored semiconducting carbon nanotube networks with enhanced thermoelectric properties," *Nature Energy,* vol. 1, pp. 16033, 2016.

[13] R. C. Tenent, T. M. Barnes, J. D. Bergeson, A. J. Ferguson, B. To, L. M. Gedvilas, M. J. Heben, and J. L. Blackburn, "Ultrasmooth, large-area, high-uniformity, conductive transparent single-walled-carbon-nanotube films for photovoltaics produced by ultrasonic spraying," *Advanced Materials,* vol. 21, no. 31, pp. 3210-3216, 2009.

[14] A. B. Phillips, R. R. Khanal, Z. Song, R. M. Zartman, J. L. DeWitt, J. M. Stone, P. J. Roland, V. V. Plotnikov, C. W. Carter, J. M. Stayancho, R. J. Ellingson, A. D. Compaan, and M. J. Heben, "Wiring-up carbon single wall nanotubes to polycrystalline inorganic semiconductor thin films: low-barrier, copper-free back contact to CdTe solar cells," *Nano Letters,* vol. 13, no. 11, pp. 5224-5232, 2013.

[15] A. B. Phillips, Z. Song, J. L. DeWitt, J. M. Stone, P. W. Krantz, J. M. Royston, R. M. Zeller, M. R. Mapes, P. J. Roland, M. D. Dorogi, S. Zafar, G. T. Faykosh, R. J. Ellingson, and M. J. Heben, "High speed, intermediate resolution, large area laser beam induced current imaging and laser scribing system for photovoltaic devices and modules," *Review of Scientific Instruments,* vol. 87, no. 9, pp. 093708, 2016.

[16] Alex Niemegeers, and Marc. Burgelman, "Effects of the Au/CdTe back contact on IV and CV characteristics of Au/CdTe/CdS/TCO solar cells,"*Journal of Applied Physics,* vol. 81, 1997.

Grain and Grain Boundary Geometrical Shape Considerations on Sodium and Potassium Diffusion Through Molybdenum Films

Orlando Ayala[1], Chinedum Akwari[1], Tasnuva Ashrafee[1], Shankar Karki[1], Grace Rajan[1], Sylvain Marsillac[1]

[1] Virginia Institute of Photovoltaics, Old Dominion University, Norfolk, VA 23529 USA

Abstract — Molybdenum (Mo) thin films were deposited on soda lime glass (SLG) using direct-current (DC) magnetron sputtering at different substrate temperatures to study the diffusion process of Sodium (Na) and Potassium (K). To quantify the concentration of both species, secondary ion mass spectrometry (SIMS) was used. In this paper, a grain diffusion model was developed that incorporates the effect of grain and grain boundary geometrical shape and size in the overall species concentration that SIMS data report. It was concluded that the ratio of grain boundary size and grain size is not constant along the Mo thin film and should be considered a key component when analyzing diffusion processes in thin films.

Index Terms — CIGS, molybdenum, thin film solar cells, sodium, potassium, diffusion.

I. INTRODUCTION

It has been found that higher efficiencies of Cu(In,Ga)Se$_2$ (CIGS) solar cells can be achieved by incorporating small amount of sodium (Na) or potassium (K) into the CIGS film to improve open-circuit voltage and fill factor [1,2]. A common substrate used for CIGS solar cells is a Molybdenum (Mo) coated soda-lime glass (SLG). The SLG acts as a source of alkali and the Mo thin film serves as transport gate for alkali diffusion from the SLG into the CIGS. It is therefore of interest to study the diffusivity of these alkali in Mo thin films and the diffusion mechanism that occur. To the best of our knowledge, very few studies have attempted to directly characterize the diffusion and provide diffusivity constants in Mo thin films [3] and none of them have considered directly the grain size effect on the SIMS intensity data, which might have biased the measured diffusion coefficient, as it will be shown later. Here, we attempt to investigate the effect of grain and grain boundary sizes on the SIMS data used to study diffusion driven concentration profiles of species.

II. EXPERIMENTAL DETAILS

Molybdenum thin films were fabricated onto SLG substrates using direct current magnetron sputtering, under 10 mTorr of argon, for 1.5 hours at 150W. The substrate temperature was kept constant during deposition, and was either room temperature (RT), 50°C, 100°C, 150°C, 200°C, or 250°C. The films were later characterized by scanning electron microscopy (SEM), X-ray diffraction (XRD), and secondary ion mass spectrometry (SIMS) to detect the effect of sodium and potassium.

III. THEORETICAL MODELING BACKGROUND

A. Relation between Species Concentration and SIMS Intensity Data

The output of the SIMS applied to a specimen is a signal intensity level that shows the presence of species. Such **intensity level (I) is proportional to the species mass (m)** the SIMS beam detects when sputtering a volume of $V_{beam} = L_{beam}^2 \cdot d$, where d is the depth the SIMS beam sputtered at a fix Δt and L_{beam} is its width (the beam cross sectional area is assumed to be a square). L_{beam} is related to the grains and grain boundaries size (L_g and L_b respectively) as follows: $L_{beam} = N_g\left(L_b + L_g\right)$, where N_g is the number of grains the SIMS beam is sputtering.

It can be proved that the beam intensity level is correlated to the average species concentration ($\bar{C}(y)$) in the sputtered volume from the definition of average concentration as follows:

$$\bar{C}(y) = \frac{\int_0^{L_{beam}} C(x,y)\,dx}{L_{beam}} = \frac{\int_0^{N_g(L_b+L_g)} C(x,y)\,dx}{N_g\left(L_b + L_g\right)}$$

or,

$$\bar{C}(y) = \frac{N_g\left[\int_0^{L_b} C_b(x,y)\,dx + \int_{L_b}^{L_b+L_g} C_g(x,y)\,dx\right]}{N_g\left(L_b + L_g\right)} \quad (1)$$

Assuming no variations in the direction perpendicular to the depth, species concentration is related to species mass as $dm_g = C_g(x,y) \cdot d \cdot L_{beam} \cdot dx$ for the mass in the grain. Using this and similar expression for the mass in the grain boundary, (1) can be written as:

$$\bar{C}(y) = \frac{N_g\left[\int_0^{L_b} dm_b + \int_{L_b}^{L_b+L_g} dm_g\right]}{d \cdot L_{beam} \cdot N_g\left(L_b + L_g\right)}$$

The numerator is directly proportional to the intensity level (I). Normalizing the right hand side with the concentration at y=0 and the left hand side with the intensity at y=0, it is finally proved that:

$$\frac{\bar{C}(y)}{\bar{C}_0} = \frac{I(y)}{I_0} \tag{2}$$

B. Relation between Average Species Concentration from SIMS and Average Species Concentration in the Grain and Grain Boundary

What the SIMS intensity level provides is the species average concentration over a number of grains. This overall concentration contains the information of both the species concentration in the grain and in the grain boundary. Introducing the definition of average species concentration in the grain ($\bar{C}_g(y)$) and in the grain boundary ($\bar{C}_b(y)$) into (1), we obtain:

$$\bar{C}(y) = \frac{L_g}{\left(L_b + L_g\right)} \bar{C}_g(y) + \frac{L_b}{\left(L_b + L_g\right)} \bar{C}_b(y) \tag{3}$$

It is well accepted that most of the species diffusion in Mo thin films occur through the grain boundaries [3]. If that is the case, (3) becomes:

$$\bar{C}(y) = \frac{L_b}{\left(L_b + L_g\right)} \bar{C}_b(y) \tag{4}$$

Therefore the average concentration provided by the SIMS intensity readings could mislead, as they show not only the average species concentration but also the effect of the grain boundary size to grain size ratio. This ratio could vary along the thin film thickness as shown in Figure 1. Thus, even if there is a typical concentration profile along the boundaries (Type C diffusion following Harrison [4]), the SIMS data could provide a profile that could be confused with a species transport process where there is diffusion through the grain as well (Type B diffusion following Harrison [4]) (Figure 1).

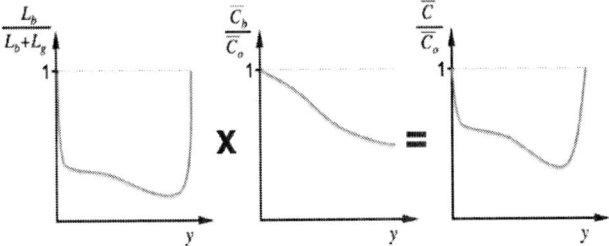

Fig. 1. Schematic of grain boundary size to total size ratio, typical concentration distribution along a straight grain boundary, and overall average concentration. This last one is what the SIMS data provides.

One last comment on the grain boundary size to grain size ratio is that it becomes unity at the interface between the thin film and the substrate and at the opposite end (at y=h) also known as accumulation layer [3]. This is because both the interfaces behave similarly to the grain boundaries in terms of the diffusion process.

IV. RESULTS AND DISCUSSION

The film thickness was obtained from cross-sectional SEMs (not shown) and the average grain size was estimated using Scherrer's equation [5]:

$$L_g = \frac{K\lambda}{\beta \cos(\theta)} \tag{5}$$

where K is the Scherrer constant and considered 0.90 for spherical particles, λ is the X-ray wavelength (0.154 nm for Cu$K\alpha$ radiation), β is the full width at half maximum (FWHM) of the peak obtained from XRD plots of the (110) peak (not shown), and θ is the Bragg angle. As far as the average grain boundary size (L_b), it was estimated to be 3 nm following Forest *et al.* [3]. Table 1 lists the size characteristics of the different Mo film samples. It also shows the grain boundary size to grain size ratios, which are in the same range as the porosity values obtained by Bommersbach *et al.* [6].

TABLE I
Film and grain sizes of Mo films for various temperatures.

T_{SS} (°C)	Film Thickness (nm)	Grain Size (L_g) (nm)	$\dfrac{L_b}{\left(L_b + L_g\right)}$	Average Potassium I/I_o
RT	550	5.0	0.375	0.503
50	450	7.8	0.278	0.288
100	500	8.5	0.261	0.248
150	490	10.4	0.224	0.180
200	530	15.0	0.167	0.158
250	500	29.7	0.092	0.161

Figure 2 shows the SIMS intensity data for the case of Na. The maximum intensity (I_{max}) is expected to be at the interface between SLG and Mo, and therefore was chosen to be the y=0 point, where $I_o=I_{max}$. The sputter time at I_{max} was found to correlate well with the film thickness, as expected. All the SIMS profiles, for both Na and K, were normalized and some of them are shown in Figure 3.

After analyzing the profiles, it was noted that the I/I_o profile for K correlated very closely to the grain boundary size to grain size ratio (see last two columns in Table I). It can be argued then that if the grain boundaries are saturated with K, the K concentration $\bar{C}_b(y)$ is constant thus the I/I_o profile is just showing the grain boundary size to grain size ratio profile. Figure 4 shows the likelihood of this hypothesis. Using the Potassium I/I_o profile as the grain boundary size to grain size ratio profile, the Na at grain boundary concentration profile can be obtained using (4). The result of this procedure is shown in Figure 5.

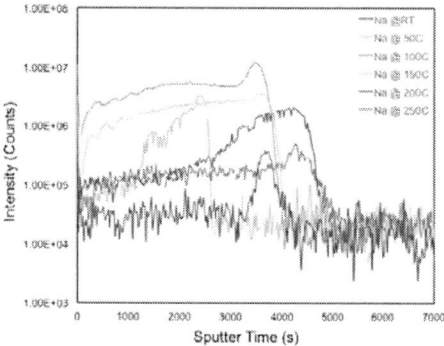

Fig. 2. SIMS depth profile of Sodium (Na) in the Molybdenum (Mo) thin films deposited at different substrate temperatures. Similar profiles were observed for the case of Potassium (K).

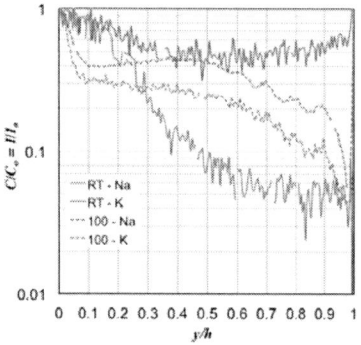

Fig. 3. Normalized SIMS depth profile of Na and K in Mo films with substrate temperatures of RT and 100°C.

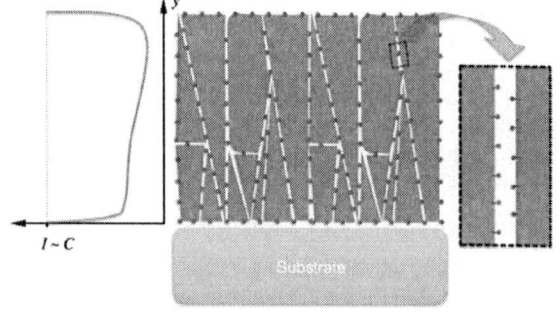

Fig. 4. Schematic of grain and grain boundary ensemble showing K saturated grain boundaries with corresponding SIMS depth profile.

It can be observed that for substrates temperatures of 100, 150, 200, and 250°C the Na concentration remains constant throughout the film thickness, but that it is not the case for RT and 50°C. These are therefore more appropriate profiles to use for determining diffusivity constants. It is important to mention that the profiles for higher temperatures show

constant values larger than unity due to the uncertainty selecting the right value for I_o to normalize the SIMS data. A better procedure to choose I_o will be developed in the future.

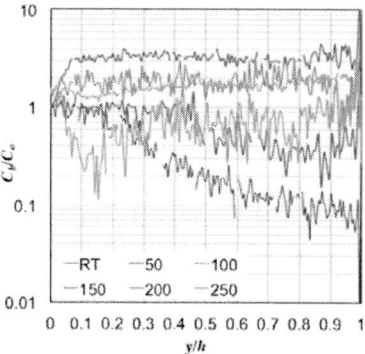

Fig. 5. Na concentration profiles along grain boundaries for different substrate temperatures.

V. CONCLUSION

In this paper, we looked at the grain boundary size to grain size ratio and demonstrated that it was extremely important to consider when studying species diffusion in solid thin films. Looking at temperature dependence of both K and Na diffusion, a clear difference was observed as a function of both species and temperature. This will be used in the next stage of our study to determine diffusivity constant for these alkalis as a function of temperature.

ACKNOWLEDGEMENT

This research was supported by the Department of Energy, under the Contract No. DE-EE0007141.

REFERENCES

[1] D. Rudmann, AF da Cunha, M Kaelin, Kurdesau F, Zogg H, Tiwari AN, and Bilger G. *Appl Phys Lett.* vol 84(7), pp: 1129–1131, 2004.

[2] A. Chirila, P. Reinhard, F. Pianezzi, P. Bloesch, A. Uhl, C. Fella, L. Kranz, D. Keller, C. Gretener, H. Hagendorfer, D. Jaeger, R. Erni, S. Nishiwaki, S. Buechelewr, and A. Tiwari, *Nature Materials*, DOI:10.1038/NMAT3789, 2013.

[3] R. V. Forest, E. Eser, B. E. McCandless, R. W. Birkmire, and J. G. Chen, *AIChE Journal,* vol. 60, No. 6, 2014.

[4] L. G. Harrison, "Influence of dislocations on diffusion kinetics in solids with particular reference to the alkali halides," *Trans. Faraday Soc.*, vol 57, pp: 1191-1199, 1961.

[5] N. Kasai, and M. Kakudo, "Springer series in chemical physics: preface," *Springer Series in Chemical Physics*, vol. 80, pp. 364-365, 2005.

[6] P. Bommersbach, L. Arzel, M. Tomassini, E. Gautron, C. Leyder, M. Urien, D. Dupuy, and N. Barreau, *Prog Photovoltaics Res Appl.*, vol 21(3), pp: 332–343, 2013.

Solution-Processed Nickel-Alloyed Iron Pyrite Thin Film as Hole Transport Layer in Cadmium Telluride Solar Cells

Ebin Bastola, Khagendra P. Bhandari, Randy J. Ellingson

Wright Center for Photovoltaics Innovation and Commercialization (PVIC), Department of Physics and Astronomy, University of Toledo, Toledo, Ohio, 43606, USA

Abstract — Here, we report the test of solution-processed nickel (Ni)-alloyed iron disulfide (FeS_2) pyrite nanocrystals (NCs) as back contact interface layer in cadmium telluride (CdTe) solar cells. The alloyed NCs were synthesized using the hot-injection colloidal method, and the thin films were fabricated using layer-by-layer drop-cast method as the back contacts to the CdTe solar cells. Based on this study, the average increase in device performance observed was 6% with $Ni_{0.05}Fe_{0.95}S_2$ and $Ni_{0.1}Fe_{0.9}S_2$ NCs compared to standard copper/gold (Cu/Au) back contacts while the device performance was sharply reduced with ≥ 20% Ni-alloyed iron pyrite NCs.

I. INTRODUCTION

Cadmium telluride (CdTe) solar cells often use copper/gold (Cu/Au) as the back contact materials in the laboratory scale.[1, 2] However, these Cu/Au back contacts do not facilitate the smooth transport of the photogenerated charge carriers. A potential barrier exists due to the low work function of Au (~ -5.0 eV)[3] compared to the deeply located (~ -5.7 eV)[4] valence band edge of the CdTe. Thus, a material with suitable band edge location is desired for the smooth extraction of holes and repel of electrons at the back contact of the CdS/CdTe devices. Several materials such as carbon nanotubes, copper thiocyanate (CuSCN), iron pyrite (FeS_2), zinc telluride (ZnTe), antimony telluride (Sb_2Te_3) and copper telluride (Cu_2Te), prepared by different methods including solution-processing, rf magnetron sputtering, chemical vapor transport have been tested as the back contacts in CdS/CdTe solar cells.[5-10] However, researchers are still working to find earth-abundant and efficient hole transport materials for CdTe solar cells.

Earth-abundant iron pyrite exhibits cubic crystal structure with an energy gap of ~0.95 eV and high absorption coefficient ($\geq 10^5$ cm^{-1}) in the visible and near IR spectral range. In addition, the valence band edge location (~5.3 eV), good conductivity, and high carrier density (~10^{19} cm^{-3}) make these materials promising to apply as hole transport layers (HTLs) in energy harvesting applications.[11] The iron dichalcogenide nanocrystals (NCs) can be synthesized using thermal injection method.[11, 12] Previously, colloidal iron pyrite NCs synthesized by thermal-injection route were deposited from solution to form an efficiency-improving HTL in CdTe/perovskite solar cells.[9, 13] Further, the researchers have been investigating the change in the optical and the electronic properties of the iron pyrite by alloying with other elements. For example, a theoretical study shows by replacing 10% sulfur atoms with oxygen (O) atoms in iron disulfide, the band gap energy can be increased to 1.2-1.3 eV, more suitable band gap for photovoltaic applications, from 0.95 eV.[14] Here, we seek to exploit the tolerance for alloying Ni with iron pyrite to study the properties and performance of $Ni_xFe_{1-x}S_2$ pyrite NCs as the interface layer with different material composition at the back contact interface layer in CdS/CdTe solar cells. The Ni-alloyed iron pyrite NCs were synthesized by a thermal injection synthesis method, and the thin film HTL at the CdTe back contact was fabricated via layer-by-layer drop-cast method. As compared to standard Cu/Au back contacts, the Ni-alloyed iron pyrite interface layer shows slightly enhanced device performance for $0 < x \leq 0.10$. Going to higher Ni-concentrations ($x \geq 0.20$), conversion efficiency diminishes substantially.

II. EXPERIMENTAL

The Ni-alloyed iron pyrite NCs were synthesized using a hot-injection colloidal method similar to that reported previously for $Co_xFe_{1-x}S_2$.[15] For the typical synthesis, 0.5 mmol of $FeBr_2$ and $NiBr_2$, 3 mmol trioctyalphosphine oxide (TOPO), and 10 mL oleylamine (OLA) were taken in a three-neck flask, placed under vacuum for ~15 mins to remove air and oxygen, and then purged with nitrogen in a standard Schlenk line system. Then, the metal precursor solution was refluxed at 170 °C for about 3 hrs under nitrogen flow with constant stirring. The sulfur (S) precursor was prepared by dissolving 3 mmol elemental S in 5 mL OLA by sonication, and it was swiftly injected to the three-neck flask at 170 °C. The temperature was immediately increased, and NCs were grown at 220 °C for about 2 hrs. The synthesized NCs were cleaned twice by a solvent/antisolvent method using chloroform/methanol and centrifugation, and the NCs were dry-stored for further characterization and device fabrication.

A 3 nm Cu layer was thermally deposited on the top of cadmium chloride ($CdCl_2$) treated CdS/CdTe devices, which were then annealed at 150 °C for about 30 mins. To complete the back contacts, the alloyed NCs with a concentration of ~8 mg/mL in chloroform were drop-casted onto the CdTe layer, and allowed to dry. Two layers were thus deposited, with each NC layer being treated by immersion for 1 min. in a 1 M hydrazine solution in methanol to remove the TOPO molecules and facilitate inter-NC coupling. Finally, 40 nm Au was deposited by thermal evaporation to complete the back contact. The performance of the CdS/CdTe solar cells with $Ni_xFe_{1-x}S_2$ NC film as the interface layer and Au as the back contact metal was tested by illuminating devices with simulated AM1.5G solar irradiance and measuring the current density vs. voltage. The CdS and CdTe films were deposited using a vapor transport deposition method.

III. RESULTS AND DISCUSSION

The synthesized $Ni_xFe_{1-x}S_2$ NCs have cubic crystal structure in pyrite phase. Based on the composition-dependent X-ray diffraction (XRD) pattern, as the Ni incorporates into the iron pyrite NCs the diffraction peaks shift towards smaller angles indicating the expected increase in the lattice constants of the alloyed NCs. Based on SEM imaging, a systematic decrease in the particle size was observed as the Ni-fraction in the NCs increased. Further, the majority charge carriers of the NC-based thin films can be controlled by material composition. The characteristics of these Ni-alloyed iron pyrite NCs including X-ray diffraction pattern, particle size distribution, optical properties, and electronic properties were similar to the previous report.[16]

For the control device, 3 nm Cu and 40 nm Au were deposited onto the $CdCl_2$-treated CdTe absorbing layer, and annealed at 150 °C for about 30 minutes to diffuse the Cu atoms into the CdTe. To study the effect of these alloyed NCs in the CdTe devices, $Ni_xFe_{1-x}S_2$ NC layers were fabricated after 3 nm Cu deposition and diffusion onto CdTe, and 40 nm Au was finally thermally evaporated on top of $Ni_xFe_{1-x}S_2$ to complete the back contact. The Ni concentration was varied from $x = 0$ to $x = 0.3$, and 5 different compositions were tested as the HTL. During the deposition of the NCs, a thickness of ~1.0 μm was preferred to avoid pinholes and achieve good coverage of the CdTe light-absorbing layer. After device fabrication, the current density vs. voltage characteristics and external quantum efficiency of the devices were measured.

The typical J-V characteristics of CdS/CdTe solar cells using different Ni-alloyed iron pyrite NCs as back contacts are shown in Fig. 1. In the case of Cu/Au back contacts, the best device has efficiency (η) of 11.6% with open-circuit voltage (V_{OC}) 826 mV, short-circuit current-density (J_{SC}) 19.12 mA cm^{-2}, and fill factor of 73.8%. The average parameters of twenty cells for Cu/Au back contacts is presented in Table I. When pure FeS_2 NC thin film was used as an interface layer, the best device efficiency was found to be 11.9 % with V_{OC} of 835 mV, J_{SC} of 20.4 mAcm^{-2}, and FF of 69.6%. For Cu/FeS₂-NC/Au back contact the average efficiency is increased by ~ 4% than Cu/Au back contacts (Table I). These results agree well with the previous report.[9] The device performance was enhanced mainly due to the increased V_{OC} and J_{SC}, though the FF was slightly reduced. Further, the external quantum

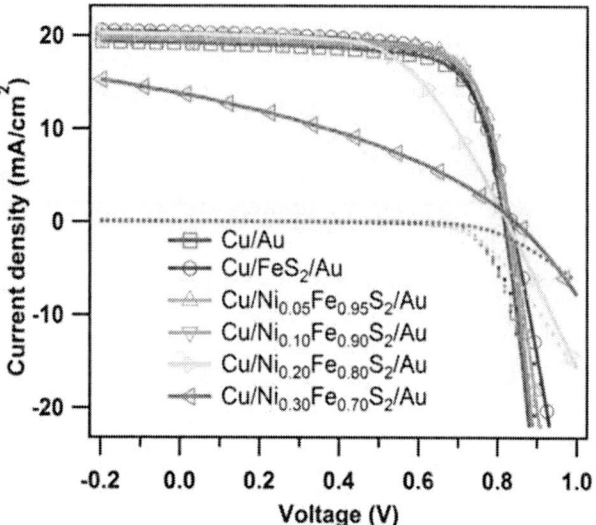

Fig. 1. Current-density and voltage (J-V) characteristics of CdTe solar cells with different $Ni_xFe_{1-x}S_2$ NCs as back contact interface layers

efficiency (EQE) was measured to verify the increase in J_{SC}. The EQE of the device with FeS_2 NCs as an interface layer is slightly higher than the control Cu/Au back contact in the range of 550 to 800 nm as shown in Fig. 2. The observed increase in device enhancement is attributed to the efficient transport of the photogenerated charge carriers towards the back contacts of the devices.[9, 13] Here, we want to test the effect of $Ni_xFe_{1-x}S_2$ in the performance of CdTe devices. When $Ni_{0.05}Fe_{0.95}S_2$ NC thin film is used, the average efficiency increased by ~ 6% to 12.2% whereas for $Ni_{0.1}Fe_{0.9}S_2$ NCs, the device performance was similar to the iron pyrite NCs as shown in the J-V characteristics (Fig. 1). The device efficiency with $Ni_{0.2}Fe_{0.8}S_2$ NCs as the interface material was found to be 4.9%, and with $Ni_{0.3}Fe_{0.7}S_2$ NCs, the device efficiency was

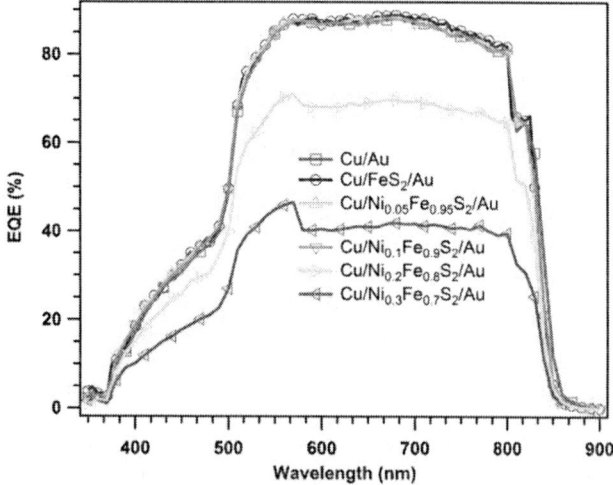

Fig. 2. External quantum efficiency (EQE) of CdTe solar cells with different $Ni_xFe_{1-x}S_2$ NCs as back contact interface layers.

TABLE I

SUMMARY OF THE *J-V* CHARACTERISTICS OF THE TWENTY CdS/CdTe CELLS WITH
$Ni_xFe_{1-x}S_2$ AS BACK CONTACT INTERFACE LAYER

Back Contacts		V_{oc} (mV)	J_{sc} (mA/cm^2)	Fill factor (*FF*%)	Efficiency (η%)
Cu/Au	Best	826	19.1	73.8	11.6
	Average	829 ± 2	18.9 ± 0.2	71.1 ± 2.3	11.2 ± 0.4
Cu/FeS$_2$/Au	Best	835	20.4	69.6	11.9
	Average	827 ± 2	19.8 ± 0.2	70.7 ± 0.7	11.6 ± 0.4
Cu/Ni$_{0.05}$Fe$_{0.95}$S$_2$/Au	Best	832	19.9	73.3	12.2
	Average	834 ± 3	19.6 ± 0.3	71.1 ± 1.8	11.8 ± 0.3
Cu/Ni$_{0.1}$Fe$_{0.9}$S$_2$/Au	Best	827	19.6	73.4	11.9
	Average	834 ± 2	19.1 ± 0.1	71.5 ± 0.9	11.5 ± 0.3
Cu/Ni$_{0.2}$Fe$_{0.8}$S$_2$/Au	Best	846	20.0	36.3	6.1
	Average	845 ± 2	17.7 ± 2.6	32.3 ± 2.6	4.9 ± 1.0
Cu/Ni$_{0.3}$Fe$_{0.7}$S$_2$/Au	Best	847	17.8	31.4	4.7
	Average	846± 1	10.5 ± 3.6	29.9 ± 1.3	2.7 ± 0.9

2.7%. The decrease in device performance was attributed to the decrease in J_{SC} and *FF* and the typical *J-V* characteristics are shown in Fig. 1. To verify the current-densities obtained from *J-V* measurements the EQEs were measured and presented in Fig. 2. Based on these EQE measurements, the current densities for $Ni_{0.2}Fe_{0.8}S_2$ and $Ni_{0.3}Fe_{0.7}S_2$ NCs interface layers were found to be 17.6 mA/cm^2 and 10.2 mA/cm^2 respectively which agree well with the values obtained from *J-V* measurement. This decrease in the device efficiency is also attributed to the strong decrease in the *FF* of the solar cells. The significant increase in the series resistance and the strong decrease in the shunt resistance of the devices for both $Ni_xFe_{1-x}S_2$ (*x* = 0.2, 0.3) alloyed NCs are the main cause of poor performance. Here, the behavior of $Ni_xFe_{1-x}S_2$ (*x* = 0.2, 0.3) as the back contact interface layer is different from other compositions (x ≤ 0.1). It is due to a change in conductivity and/or majority carrier type of the composite NCs. Based on a laboratory-constructed thermal-probe measurement, it shows n-type conductivity for $Ni_{0.2}Fe_{0.8}S_2$ (or higher) Ni-alloyed iron pyrite NCs, indicating electrons as the majority charge carriers. The increase in V_{OC} for $Ni_xFe_{1-x}S_2$ (*x* = 0.2, 0.3) NCs is different from our expectance for materials having n-type conductivity. The increase in series resistance indicates that materials are resistive than 5% or 10% Ni-alloyed NCs, and hence makes more difficult path for the smooth charge transport. During the fabrication of these back contact interface layer, no heat treatment was carried, and hence the intermixing of NCs and CdTe is minimum.

IV. Conclusion

Here, we have successfully demonstrated the application of solution-processed Ni-alloyed iron pyrite NC thin film as the back contact HTL for CdS/CdTe solar cells. The $Ni_xFe_{1-x}S_2$

NCs with low Ni concentration (≤10%) served as interface layer enhancing the device performance by up to 8% where as for higher concentration (≥20%), the device performance was sharply reduced. The increase in resistance, carrier concentration, and other electronic properties are responsible for the sharp decrease in the device performance. Further, we are currently working to understand the behavior of these alloyed NCs as back contact interface layer in CdTe devices.

V. Acknowledgement

This study was supported by the US National Science Foundation Sustainable Energy Pathways program under Grant CHE-1230246.

References

[1] E. Janik and R. Triboulet, "Ohmic contacts to p-type cadmium telluride and cadmium mercury telluride," *Journal of Physics D: Applied Physics,* vol. 16, p. 2333, 1983.

[2] C. R. Corwine, A. O. Pudov, M. Gloeckler, S. H. Demtsu, and J. R. Sites, "Copper inclusion and migration from the back contact in CdTe solar cells," *Solar Energy Materials and Solar Cells,* vol. 82, pp. 481-489, 5/30/ 2004.

[3] B. de Boer, A. Hadipour, M. M. Mandoc, T. van Woudenbergh, and P. W. M. Blom, "Tuning of Metal Work Functions with Self-Assembled Monolayers," *Advanced Materials,* vol. 17, pp. 621-625, 2005.

[4] J. L. Freeouf and J. M. Woodall, "Schottky barriers: An effective work function model," *Applied Physics Letters,* vol. 39, pp. 727-729, 1981.

[5] J. H. Yun, K. H. Kim, D. Y. Lee, and B. T. Ahn, "Back contact formation using Cu2Te as a Cu-doping source and as an electrode in CdTe solar cells," *Solar Energy Materials and Solar Cells,* vol. 75, pp. 203-210, 1// 2003.

[6] S. Hu, Z. Zhu, W. Li, L. Feng, L. Wu, J. Zhang, *et al.*, "Band diagrams and performance of CdTe solar cells with a Sb2Te3

back contact buffer layer," *AIP Advances,* vol. 1, p. 042152, 2011/12/01 2011.

[7] A. B. Phillips, R. R. Khanal, Z. Song, R. M. Zartman, J. L. DeWitt, J. M. Stone, *et al.*, "Wiring-up carbon single wall nanotubes to polycrystalline inorganic semiconductor thin films: low-barrier, copper-free back contact to CdTe solar cells," *Nano letters,* vol. 13, pp. 5224-5232, 2013.

[8] J. Li, D. R. Diercks, T. R. Ohno, C. W. Warren, M. C. Lonergan, J. D. Beach, *et al.*, "Controlled activation of ZnTe:Cu contacted CdTe solar cells using rapid thermal processing," *Solar Energy Materials and Solar Cells,* vol. 133, pp. 208-215, 2// 2015.

[9] K. P. Bhandari, P. Koirala, N. R. Paudel, R. R. Khanal, A. B. Phillips, Y. Yan, *et al.*, "Iron pyrite nanocrystal film serves as a copper-free back contact for polycrystalline CdTe thin film solar cells," *Solar Energy Materials and Solar Cells,* vol. 140, pp. 108-114, 9// 2015.

[10] N. R. Paudel and Y. Yan, "Application of copper thiocyanate for high open-circuit voltages of CdTe solar cells," *Progress in Photovoltaics: Research and Applications,* vol. 24, pp. 94-101, 2016.

[11] K. P. Bhandari, P. J. Roland, T. Kinner, Y. Cao, H. Choi, S. Jeong, *et al.*, "Analysis and characterization of iron pyrite nanocrystals and nanocrystalline thin films derived from bromide anion synthesis," *Journal of Materials Chemistry A,* vol. 3, pp. 6853-6861, 2015.

[12] E. Bastola, K. P. Bhandari, A. J. Matthews, N. Shrestha, and R. J. Ellingson, "Elemental anion thermal injection synthesis of nanocrystalline marcasite iron dichalcogenide FeSe2 and FeTe2," *RSC Advances,* vol. 6, pp. 69708-69714, 2016.

[13] A. J. Huckaba, P. Sanghyun, G. Grancini, E. Bastola, C. K. Taek, L. Younghui, *et al.*, "Exceedingly Cheap Perovskite Solar Cells Using Iron Pyrite Hole Transport Materials," *ChemistrySelect,* vol. 1, pp. 5316-5319, 2016.

[14] J. Hu, Y. Zhang, M. Law, and R. Wu, "Increasing the Band Gap of Iron Pyrite by Alloying with Oxygen," *Journal of the American Chemical Society,* vol. 134, pp. 13216-13219, 2012/08/15 2012.

[15] T. Kinner, K. P. Bhandari, E. Bastola, B. M. Monahan, N. O. Haugen, P. J. Roland, *et al.*, "Majority Carrier Type Control of Cobalt Iron Sulfide (CoxFe1–xS2) Pyrite Nanocrystals," *The Journal of Physical Chemistry C,* vol. 120, pp. 5706-5713, 2016/03/17 2016.

[16] E. Bastola, K. P. Bhandari, and R. J. Ellingson, "Application of composition controlled nickel-alloyed iron sulfide pyrite nanocrystal thin films as the hole transport layer in cadmium telluride solar cells," *Journal of Materials Chemistry C,* vol. 5, pp. 4996-5004, 2017.

Use of CdS:O and CdSe as Window layers for CdTe Photovoltaics

Tom Baines[1]*, Guillaume. Zoppi[2], Ken Durose[1] and Jonathan D. Major[1]

[1]Stephenson Institute for Renewable Energy and Department of Physics, University of Liverpool L69 7ZF, UK

[2]Department of Mathematics, Physics and Electrical Engineering, Northumbria University, Ellison Building, Newcastle upon Tyne NE1 8ST, UK

tbaines@liverpool.ac.uk

Abstract - CdTe devices comprising of a CdS:O/CdSe dual window layer structure have been fabricated and characterised. Devices produced using a wider band gap CdS:O show increased photocurrent at short wavelengths and device performance is increased from 10% to 12% in comparison to a CdS layer. Cells with the addition of a CdSe layer show evidence of a reduced band gap through the formation of $CdTe_{(1-x)}Se_x$ phases and increased collection at long wavelength. The overall performance however, is significantly reduced, due to the incomplete interdiffusion of the CdSe layer and Se diffusion into the CdS:O layer. Devices show significant loss at short wavelengths, through optical absorption by the 1.7 eV CdSe film and a reduced V_{oc} due to poor interfaces qualities.

Index Terms – Solar energy, Cadmium Compounds, Photovoltaic cells, Selenium.

I. Introduction

CdTe has established itself as the leading industrial thin film photovoltaic (PV) technology and a performance of over 22% has been achieved [1]. The performance has improved significantly in recent years from 16.7% to 22.1%, through the optimisation of the short circuit density (J_{sc}). Which has improved from 26.1 mA.cm^{-2} to 31.69 mA.cm^{-2} since 2012 [1-2].

The use of CdS as the window layer has long been one of the principal J_{sc} losses due to (i) a lower than optimal band gap of 2.5 eV and (ii) loss of carriers generated in this region due to a one sided depletion region [3-4].

However, removing the CdS window layer in order to improve collection in the short wavelength region of the spectrum without compromising other device parameters such as, open circuit voltage (V_{oc}) and fill factor (FF) has proven challenging.

One alternative developed by Wu *et al.* is the use of the nanostructured CdS grown in an oxygenated atmosphere to controllably increase the band gap from 2.5eV to > 3.5eV [5]. Integration of CdS:O into the CdTe device structure has led to an improvement in device J_{sc} and performance. Despite this device external quantum efficiency (EQE) responses still seem restricted in the blue region therefore, further opportunities need to be explored in order to capture this region.

Recently work has demonstrated that incorporation of CdSe into the device structure between the CdS and CdTe layers can lead to the formation of $CdTe_{1-x}Se_x$ phases which has a lower band gap than CdTe [6]. This reduction of the band gap near the CdSe/CdTe interface can increase performance through an enhanced EQE response and potentially increase carrier lifetime via band gap grading [6]. In order to retain a high V_{oc} it has been reported that a thin CdS layer is still required and the FF is significantly reduced for cells without a thick CdS layer due to the formation of unfavourable CdTe/TCO interfaces [4-6].

By combining a wider band gap CdS:O with CdSe layers one should expect no such performance losses and gives the opportunity to investigate the relationship between the CdS:O for different oxygen contents, with variations in the band gap grading through different CdSe thicknesses.

In this work we report development of a CdS:O/CdSe/CdTe device structure. The current density - voltage (JV) and EQE results will be presented. The influence of CdSe thickness on device performance will be investigated and secondary ion mass spectrometry (SIMS) used to examine the diffusion of Se throughout the device structure.

II. Experimental

2.1: Device fabrication:

CdTe solar cells were produced in the conventional superstrate configuration. All cells were deposited on NSG ltd soda lime **TEC**™ 15M glass. 120 nm of either CdS or CdS:O with varying oxygen concentration in the sputter ambient, were deposited via radio frequency (RF) sputtering at room temperature, using a power of 60W. Oxygen content was controlled via the argon/oxygen flow rates in a total chamber pressure of 5mTorr. CdSe layers were also deposited by RF sputtering at room temperature, at a chamber pressure of 5mTorr and power of 60W.

Close space sublimation (CSS) was used to deposit 5-6 µm of CdTe at a source and substrate temperature of 610°C and 510°C respectively. The CdTe growth was performed via a two stage process: 1) a high pressure growth at 30 Torr N_2 for 14 minutes and 2) a lower pressure growth at 1 Torr for 30 seconds. The growth at a higher pressure facilitates larger grain growth however, to overcome the possible formation of pinholes the deposition is completed by growth at a lower

TABLE I: DEMONSTRATES THE INFLUENCE OF INCREASING O_2 CONTENT INTO THE CdS LAYER ON DEVICE PERFORMANCE (AVERAGES $\pm$ STANDARD DEVIATION). EFFICIENCY (η), SHORT CIRCUIT DENSITY (J_{SC}), OPEN CIRCUIT VOLGTAGE (V_{OC}), FILL FACTOR (FF), SERIES RESISTANCE (R_S), SHUNT RESISTANCE (R_{SH}).

O_2 (%)	η (%)	J_{sc} (mA.cm^{-2})	V_{oc} (V)	FF (%)	R_s (Ω.cm^{-2})	R_{sh} (Ω.cm^{-2})
0	9.95	18.73	0.78	68.16	7.19	1357.57
	(7.73±0.16)	(16.84±0.14)	(0.73±0.04)	(61.53±0.29)	(9.13±0.39)	(1101.66±146.1)
3	10.45	22.48	0.75	61.22	11.47	708.72
	(9.85±0.33)	(22.29±0.32)	(0.74±0.01)	(59.22±1.33)	(11.63±0.55)	(748.1±89.71)
5	12.01	22.77	0.81	65.11	10.08	848.41
	(10.48±0.64)	(22.42±0.64)	(0.76±0.02)	(60.99±1.12)	(10.13±0.48)	(810.20±45.55)
7	1.69	10.31	0.5	32.91	57.41	116.46
	(0.72±0.16)	(8.36±0.53)	(0.43±0.05)	(19.81±2.17)	(586.41±135.57)	(132.44±15.68)

pressure to aid pinhole blocking [7].

All samples were then $MgCl_2$ treated at 430°C for 20 minutes in an air ambient [8]. Samples were etched with a nitric - phosphoric acid (NP) solution for 15 s following the Cl activation step, in order to removes contaminants and create a Te rich surface [8]. All cells were completed by thermally evaporating 0.25cm^2 gold contacts onto the back surface. No copper was intentionally used during device fabrication.

2.2: Characterisation:

JV characteristics were carried out using AM1.5 spectrum at 1000W/m^2 using a TS Space System solar simulator. EQE measurements were performed using a Bentham PVE 300 system. SIMS was performed using a Hiden Analytical gas ion and quadrupole detector. An O^{2+} ion gun was used to sputter the sample, using a beam energy of 5keV and beam current of 300nA and the depth profiles were then normalised. Transmission measurements were performed using a Shimadzu Solidspec 3700 spectrophotometer.

III. Results and Discussion

3.1: Effect of oxygen content in CdS:

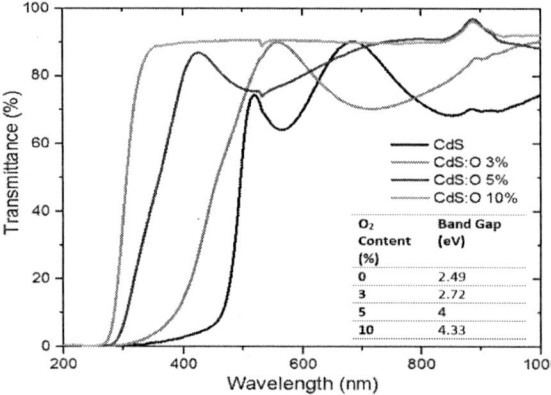

Figure 1: Shows how the transmission data changes as a function of O_2 incorporated into the CdS layer. Insert calculated band gaps from the tauc method.

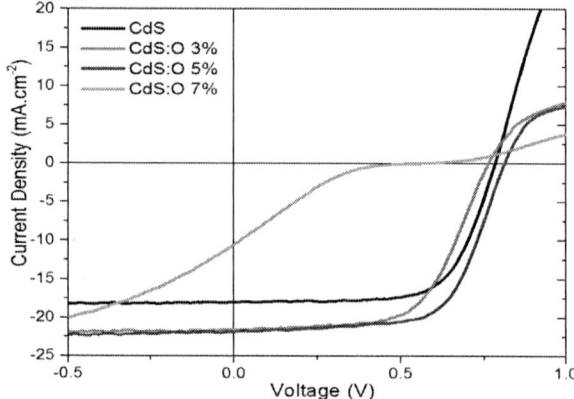

Figure 2: Current- Voltage (JV) characteristics for for CdTe devices using CdS, CdS:O 3%, CdS:O 5% and CdS:O 7% as the window layer.

Figure 1 shows how the transmission of CdS changes as a function of O_2 content and the extrapolated band gaps calculated from the tauc method. This shows that by increasing the O_2 content in the sputtering chamber the band gap of CdS can be increased from 2.5 eV to over 4 eV and also the band gap can be controlled by adding a specific amount of O_2 [5].

A series of cells were prepared with varying amounts of oxygen in the sputtering ambient during the CdS:O deposition, ranging from 0% to 7%. All other cell deposition and processing conditions were as described previously and identical. Table 1 gives average device performance parameters with JV highest efficiency curves shown in Figure 2.

These results show that incorporating only a small amount of oxygen into the CdS film, and the resultant band gap shift can have a dramatic impact on the J_{sc}. CdS:O with a 3% oxygen content improved the average J_{sc} from 16.84 mA.cm^{-2} to 22.48 mA.cm^{-2} which resulted in an average performance increase of over 2%. This confirms that by increasing the band gap of the window layer the current produced by the device can be maximised without having a detrimental effect on other device parameters.

Whilst the J_{sc} is not significantly improved by further increasing the oxygen content to 5% the V_{oc} has improved

978-1-5090-5606-4/17 $31.00 © 2017 IEEE

TABLE II: SHOWS THE INFLUENCE OF INCREASING THE THICKNESS OF CdSe ON DEVICE PERFORMANCE (AVERAGE).

CdSe Thickness (nm)	η (%)	J_{sc} (mA.cm^{-2})	V_{oc} (V)	FF (%)	R_s (Ω.cm^{-2})	R_{sh} (Ω.cm^{-2})
0	12.01 (10.48 ±0.64)	22.77 (22.42±0.32)	0.81 (0.76±0.02)	65.11 (60.99±1.12)	10.08 (10.13±0.48)	848.41 (810.20±45.55)
20	9.55 (8.54±0.52)	22.020 (20.99±0.49)	0.73 (0.70±0.01)	62.39 (57.49±1.72)	12.58 (10.45±0.92)	916.77 (821.40±47.98)
40	7.35 (7.05±0.16)	14.69 (14.63±0.12)	0.76 (0.76±0.01)	65.86 (63.21±1.34)	9.07 (11.36±0.90)	1464.53 (1320.73±109.75)
80	7.45 (6.35±0.44)	16.99 (15.56±0.51)	0.73 (0.71±0.01)	60.11 (57.19±1.56)	9.68 (12.27±1.15)	750.29 (969.84±106.66)
100	6.99 (6.70±0.12)	15.26 (15.14±0.22)	0.74 (0.73±0.01)	61.91 (60.89±1.07)	15.34 (12.62±1.41)	1312.83 (1077.52±116.25)

reaching a peak value of 0.81V and the series resistance (R_s) is also reduced compared to the 3% CdS:O film. This indicates that CdS:O with a 5% oxygen content may have better interfacial properties with CdTe.

Adding a higher, 7%, oxygen content into the film leads to a dramatic loss in performance with average efficiency decreasing to 0.72%, with all parameters being affected. The JV curves produced for devices with 7% CdS:O exhibit an uncharacteristic S – shape, suggesting a large barrier was formed at the CdS:O/CdTe interface which is limiting the extraction of generated carriers [9]. This could either be due to the formation of CdO phases owing to the high oxygen concentration or the potential formation of a spike in the conduction band due to the increased band gap.

3.2: Incorporation of CdSe:

Following initial optimisation CdS:O with 5% oxygen ambient was identified as the most suited to device fabrication. A series of cells with a 5% CdS:O layer were then fabricated with varying thickness of CdSe ranging from 0 to 100 nm, Table 2 shows the influence of CdSe on average and peak device performance. Figure 3a and b shows the JV curves and EQE spectra for these devices. From these results it is clear that incorporation of CdSe into the device structure has had a detrimental effect on device performance regardless of thickness.

Device J_{sc} is particularly affected falling from 22.42 mA.cm^{-2} for the cells with no CdSe to 15.14 mA.cm^{-2} when a 100 nm layer is incorporated. This can be explained by analysing the EQE spectra produced (Figure 3b), where it is clearly evident that the short wavelength response at < 650 nm is significantly reduced.

This would appear to be evidence of a remaining CdSe layer which hasn't been intermixing with the CdTe during either the CSS deposition or the MgCl$_2$ treatment. Ideally the CdSe layer should be completely absorbed by the CdTe to form a CdTe$_{(1-x)}$Se$_x$ intermixed region which has a lower band gap than CdTe due to the band bowing effect and thus enhance device photocurrent [6].

The EQE data does show enhanced absorption at long wavelength region (> 850 nm), indicating a CdTe$_{(1-}$

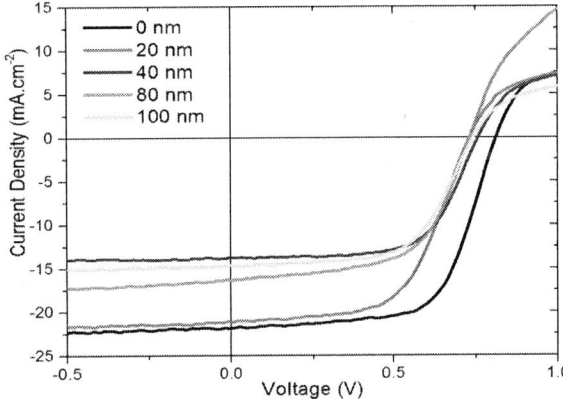

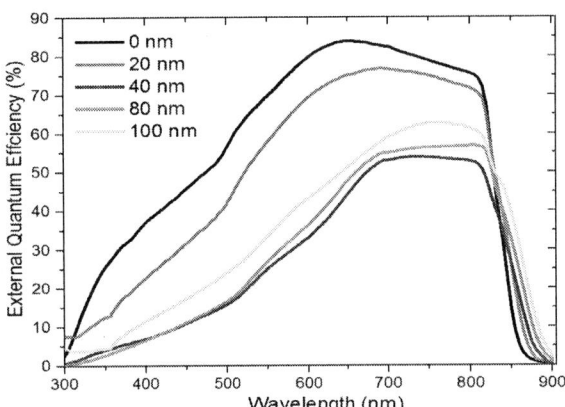

Figure 3: a) Illustrates the JV characteristics as a function of CdSe thickness, b) shows External Quantum Efficiency (EQE) response as a function of CdSe thickness ranging from 0 – 100 nm.

$_x)$Se$_x$intermixed region has indeed been formed but also indicate that the CdSe has not been completely consumed.

The decrease in band gap follows a trend with increasing the CdSe thickness, with 100 nm of CdSe demonstrating the most absorption at long wavelengths and therefore has the lowest band gap (1.39 eV) by estimation from the EQE absorption cut-off.

In addition to this, the inclusion of CdSe into the device structure has a detrimental effect on the V_{oc} of the device, this

TABLE III: SHOWS HOW THE PERFORMANCE OF CdTe/CdSe/CdS:O PV VARIES WITH INCREASING CHLORINE TREATMENT TIME FROM 20 TO 120 MINS.

Treatment Time (mins)	η (%)	J_{sc} (mA.cm^{-2})	V_{oc} (V)	FF (%)	R_s (Ω.cm^{-2})	R_{sh} (Ω.cm^{-2})
20	6.99	15.26	0.74	61.91	15.34	1312.83
	(6.59$\pm$0.11)	(15.14$\pm$0.15)	(0.72$\pm$0.01)	(60.84$\pm$0.68)	(13.40$\pm$1.09)	(1141.55$\pm$88.91)
60	6.68	14.87	0.71	63.30	10.36	1348.09
	(5.75$\pm$0.50)	(14.43$\pm$0.20)	(0.66$\pm$0.03)	(59.12$\pm$3.04)	(10.56$\pm$0.45)	(1290.25$\pm$113.4)
120	4.71	14.34	0.59	55.60	9.89	966.52
	(3.95$\pm$0.23)	(12.95$\pm$0.46)	(0.56$\pm$0.02)	(54.31$\pm$1.32)	(11.49$\pm$1.21)	(671.64$\pm$90.08)

could be due to a number of factors such as, poorer interface quality, a shift in the band alignment or introduction of defect states, all of which would limit V_{oc}.

3.3: Increase Chlorine Treatment Time:

It has often been suggested that the chlorine treatment step has a large influence on interdiffusion of CdS and CdTe layers. This however, seems to be less pronounced for high temperature deposition techniques such as CSS [10]. In order to investigate whether CdSe intermixing could be controlled via the chlorine treatment we varied the MgCl$_2$ activation step from 20 to 120 minutes for cells containing 100nm CdSe layer. The idea being to allow the CdSe to convert from the non-photoactive wuzite phase to the photoactive CdTe$_{(1-x)}$Se$_x$ zincblende structure [6].

Table 3 shows how the device parameters change with increased Cl treatment time using the CdTe/CdSe/CdS:O device structure with a 100 nm thick CdSe layer. Figure 4a and b show the JV curves and EQE produced for these cells.These results show that cells produced with a CdSe layer show a high level of robustness to the chlorine step with similar peak performance being achieved after a 60 minute treatment time.

It is only after a long treatment time of 120 minutes that the performance starts to deteriorate with a noticeable loss in both the V_{oc} and R_{sh}, indicating the cells have become over treated and shunting pathways have been created. The J_{sc} however, has largely been retained even though the cells have been over treated, peak J_{sc} drops from 15.26 mA.cm^{-2} to 14.34 mA.cm^{-2}. The JV and EQE data presented in Figure 4 indicate that not much has changed with respect to Se diffusion and junction position. Cells produced for treatment times of 20 and 60 mins show near identical curves and the cells treated for 120 minutes show a similar cut-offs at short and long wavelengths, with losses still primarily occurring at short wavelengths.

This suggests that the increased chlorine treatment time hasn't substantially increased the amount of Se diffusion into the CdTe and the non-photoactive CdSe is still present. An alternative explanation is that the Se could also have diffused into the CdS layer leading to the formation of CdS$_{(1-x)}$Se$_x$ phases which would be of a lower band gap than CdS:O and cause enhanced losses at short wavelengths.

The shortage of Se diffusion into the CdTe during the chlorine step could also be an indication that the Se has been utilised elsewhere primarily in the CdS:O layer. Alternatively the lack of Se diffusion during chloride treatment could be because CdTe deposited via CSS is large grained and typically doesn't recrystallize during the chlorine step, which in turn could limit the amount of Se diffusion. This effect has been observed for S diffusion in CSS grown CdTe PV, where very limited S diffusion occurred during the chlorine step [10].

Therefore, further optimisation of the CdTe deposition could be required to better control the Se diffusion as it is unable to be remedied by the post growth treatment.

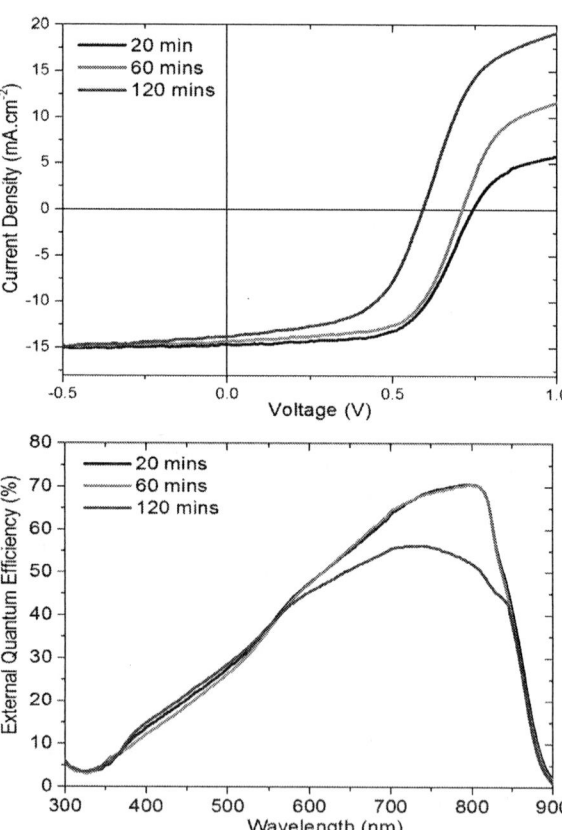

Figure 4: a) JV curves for cells with a chlorine treatment of 20, 60 and 120 mins with b) associated EQE response for these cells.

3.4: SIMS analysis of CdTe/CdSe/CdS:O devices:

Figure 5 shows the SIMS profile for S, Se and Te content as function of device thickness using a 100 nm CdSe layer and a 20 minute treatment time. From this data it is clearly evident that a large amount of Se is still present at the front surface. This confirms the suspicion that the CdSe layer hasn't been consumed by the CdTe, although it would appear there has been significant Se diffusion towards the back surface. The Se also appears to have diffused into the CdS:O layer, supporting the idea that $CdS_{(1-x)}Se_x$ phases may have been formed.

Both a remaining CdSe layer and the formation of $CdS_{(1-x)}Se_x$ phases would therefore, appear to indeed be responsible for the poor EQE response in these cells at short wavelength. Due to the depletion region lying entirely in the CdTe layer any absorption in these layers would be anticipated to cause enhanced losses [3]. Therefore, it would be appear that a combination of CdSe and CdS:O is detrimental to performance, which could be a fundamental limitation of this device structure.

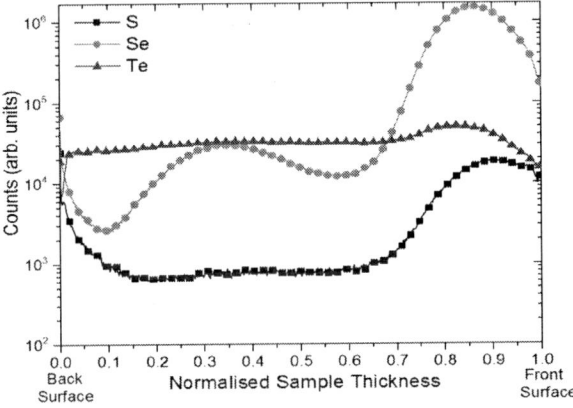

Figure 5: Shows the SIMS profile of the S, Se and Te content in CdTe/CdSe/CdS:O PV with 1 representing the front surface and 0 representing the back surface.

The SIMS data also shows that the nature of the Se diffusion into the CdTe layer is rather unexpected. One would anticipate a linear gradient towards the back contact, the hope being an efficient band gap grading had occurred. The data seems to suggest that while there is general gradient towards the back contact, there also seems to be an accumulation of Se in the middle of the CdTe layer. This is rather difficult to explain as there should be no interface at this point, the mid-CdTe layer, therefore accumulation of Se in this region is unexpected. This could be indicative of a number of things either, i) the device processing hasn't yet been perfected to achieve optimal Se diffusion, ii) there is some inherent limitation to the Se out-diffusion, iii) there is some discontinuity in the CdTe layer at this point or iv) the chloride treatment step may be serving to partially re-diffuse Se accumulated away from the back surface into the CdTe layer. Further investigation is required to understand the cause of this Se accumulation.

IV.   Conclusion

We have demonstrated that by integrating a CdS:O film with a 5% O_2 content can increases the optical band gap of CdS thus leading to a significant increase in device J_{sc} and performance.

However, the additional incorporation of CdSe into the device to generate band gap grading has so far proved a challenge with even a thin layer having a detrimental effect on device performance, in particular on device J_{sc}. The EQE and SIMS data presented suggests that a CdSe layer is still present in the device structure and contributing to a loss in photocurrent and voltage. The SIMS data also indicates that the Se is potentially diffusing into the CdS:O layer forming a $CdS_{(1-x)}Se_x$ phase.

Altering device processing was investigated by increasing the chloride treatment step from 20 mins to 120 mins to examine its influence on Se diffusion. The EQE data presented suggests this had very little effect on diffusion of the Se into the CdTe layer and as a result increasing the Cl step also had very little influence on device J_{sc}. In addition to this, the overall device performance was reduced due to enhanced resistive losses.

Further SIMS and TEM investigation is required in order to characterise fully what is happening to the Se layer during device processing in CdS:O based devices. In addition to this, further investigation of the CdTe growth needs to be undertaken in an attempt to understand how this can be used to control the Se diffusion. The additional incorporation of Cu into the devices structure will significantly improve device performance this will however, add another layer of complexity to the cell fabrication.

V.   References

[1]   M. A. Green, K. Emery, Y. Hishikawa, W. Warta, E. D. Dunlop, "Solar cell efficiency tables (version 49)," *Prog. Photovoltaics Res. Appl.*, vol. 25, pp. 3–13, 2017.

[2]   M. A. Green, K. Emery, Y. Hishikawa, W. Warta, E. D. Dunlop, "Solar cell efficiency table (version 39)," *Prog. Photovolt Res. Appl.*, vol. 20, pp. 12–20, 2012.

[3]   J. V. Li, A. F. Halverson, O. V. Sulima, S. Bansal, J. M. Burst, T. M. Barnes, T. A. Gessert, and D. H. Levi, "Theoretical analysis of effects of deep level, back contact, and absorber thickness on capacitance-voltage profiling of CdTe thin-film solar cells," *Sol. Energy Mater. Sol. Cells*, vol. 100, pp. 126–131, 2012.

[4]   N. R. Paudel and Y. Yan, "Enhancing the photo-currents of CdTe thin-film solar cells in both short and long wavelength regions," *Appl. Phys. Lett.*, vol. 105, no. 18, 2014.

[5]   X. Wu, Y. Yan, R. G. Dhere, Y. Zhang, J. Zhou, C. Perkins, and B. To, "Nanostructured CdS:O film: preparation, properties, and application," *Phys. Status Solidi*, vol. 1, no. 4, pp. 1062–1066, 2004.

[6]   J. D. Poplawsky, W. Guo, N. Paudel, A. Ng, K. More, D. Leonard, Y. Yan, "Structural and compositional dependence of the CdTexSe1−x alloy layer photoactivity in CdTe-based solar cells," *Nat. Commun.*, vol. 7, p. 12537, 2016.

[7]   J. D. Major, Y. Y. Proskuryakov, K. Durose, G. Zoppi, and I. Forbes, "Control of grain size in sublimation-grown CdTe,

and the improvement in performance of devices with systematically increased grain size," *Sol. Energy Mater. Sol. Cells*, vol. 94, no. 6, pp. 1107–1112, 2010.

[8]   J. D. Major, R. E. Treharne, L. J. Phillips, and K. Durose, "A low-cost non-toxic post-growth activation step for CdTe solar cells," *Nature*, vol. 511, no. 7509, pp. 334–337, 2014.

[9]   W. Tress and O. Inganäs, "Simple experimental test to distinguish extraction and injection barriers at the electrodes of (organic) solar cells with S-shaped current-voltage characteristics," *Sol. Energy Mater. Sol. Cells*, vol. 117, pp. 599–603, 2013.

[10]   A. A. Taylor, J. D. Major, G. Kartopu, D. Lamb, J. Duenow, R. G. Dhere, X. Maeder, S. J. C. Irvine, K. Durose, and B. G. Mendis, "A comparative study of microstructural stability and sulphur diffusion in CdS/CdTe photovoltaic devices," *Sol. Energy Mater. Sol. Cells*, vol. 141, pp. 341–349, 2015.

Applications of hybrid organic-inorganic metal halide perovskite thin film as a hole transport layer in CdTe thin film solar cells

Khagendra P. Bhandari, Suneth C. Watthage, Zhaoning Song, Adam Phillips, Michael J. Heben,

Randy J. Ellingson

Wright Center for Photovoltaic Innovation and Commercialization, Department of Physics and Astronomy, University of Toledo, 2801 W. Bancroft Street, Toledo, OH 43615

Abstract — We fabricate and characterize methylammonium lead halide perovskite film as a novel back contact to CdTe thin film solar cells. We prepare a ~500 nm perovskite film directly on a CdCl$_2$-activated and Cu-doped CdTe polycrystalline film, and complete the device with Au back contact. Results of this novel back contact are compared with the performance of a Cu/Au back contact. Our investigation shows that incorporation of a thin perovskite layer before the back contact metallization reduces the back contact barrier effect, improving both FF and V$_{OC}$ of the solar cells. These improvements are consistent with improved passivation and a lower or negligible valence band offset between CdTe and the perovskite, especially with the incorporation of the Br in perovskite. Our best device shows FF improvement by ~3% with the highest FF of 78%, and solar cell efficiency increases by 5% when MAPb(I$_{0.7}$Br$_{0.3}$)$_3$ is used as a hole transfer layer.

Index Terms – halide perovskite, hole-transport, cadmium telluride, interface, single-step, fill-factor, open circuit voltage.

I. INTRODUCTION

As the work functions of commonly available metals are smaller than the work function of CdTe, a Schottky junction is formed at a CdTe/metal interface.[1, 2] The diode opposing that of the main CdS/CdTe junction has a tendency to block the photo-generated holes reaching the back contact metal, severely degrading the device performance. Therefore, for CdTe solar cells, suitable hole transport layers have been a subject of interest to minimize or eliminate the back barrier effects. [3-6] In this perspective, we have developed hybrid organic-inorganic metal halide perovskite (methylammonium lead halide, CH$_3$NH$_3$PbX$_3$ or MAPbX$_3$, x = Cl, Br, I) as hole transport layers for CdS/CdTe solar cells.

The hybrid organic-inorganic metal halide perovskites, methylammonium lead iodide (CH$_3$NH$_3$PbI$_3$ or MAPbI$_3$) as a prototypical example, possess several outstanding optoelectronic properties that make them ideal for photovoltaic (PV) application. [7-11] As a result of the tremendous research efforts across the world, perovskite based solar cells are now comparable to other previously existing high efficient technologies at least in the laboratory scale.[12, 13] MAPbX$_3$ thin films are easily synthesized via solution processing, at low annealing temperature. In this paper, we have demonstrated the application of MAPb(I$_{1-x}$Br$_x$)$_3$ films as an interface or back contact layer to CdTe thin film solar cells. Detailed studies are ongoing but here, and we provide a description of our initial work along with our very promising initial results. The application of this material as a hole transport layer (HTL) – as the back contact to CdTe solar cells – is our first attempt and to the authors' knowledge this achievement is the first in scientific community.

II. EXPERIMENTAL METHODS

Cadmium sulfide (CdS), a window layer and cadmium telluride (CdTe), an absorber layer were fabricated by vapor transport deposition (VTD) mechanism. In the VTD process, saturated vapor of Cd and Te (S) species is transported to the substrate of interest where the polycrystalline CdTe (CdS) film forms.[14] The device structure for the CdS/CdTe solar cells used in this work is shown in Fig. 1, where CdS/CdTe layers were grown on a front transparent conducting electrode coated on sodalime glass substrate. About 3 nm Cu was deposited onto CdCl$_2$-treated CdTe and annealed in dry air at 150 $^\circ$C for 25 minutes. A thin layer of perovskite, MAPb(I$_{1-x}$Br$_x$)$_3$ of thickness ~500 nm was then deposited onto CdTe by spin coating. Finally, the device was completed by thermal evaporation of ~45 nm of Au on top of the MAPb(I$_{1-x}$Br$_x$)$_3$. For the standard Cu/Au back contact deposition, similar thicknesses of Cu and Au were sequentially deposited by thermal evaporation onto CdCl$_2$-treated CdTe, and Cu was diffused by annealing at 150 $^\circ$C for 40 minutes.

For both types of back contacts, the cell areas (0.8 cm^2) are defined by laser scribing. Current density-voltage measurements were performed in the dark and light at 1 sun AM1.5G illumination. External quantum efficiency (EQE) characterization was used to confirm J$_{SC}$ values.

A single-step deposition method was used to fabricate MAPb(I$_{1-x}$Br$_x$)$_3$ films on the CdTe layer. [7] Precursor solutions used for perovskite film fabrication were prepared as 0.8 M methylammonium iodide (MAI), and different concentrations of lead iodide (PbI$_2$) and lead bromide (PbBr$_2$) (the Pb concentration was set to 0.9 M) in a mixed solution of anhydrous dimethylformamide (DMF) and dimethyl sulfoxide (DMSO) (4:1, volume ratio). The precursor solution was spin coated onto the CdTe layer at 1000 rpm for 10 s and 4000 rpm for 30 s at room temperature (~20 $^\circ$C). During the second spinning step, 100 µL of chlorobenzene was dropped on the spinning sample 15 s prior to the end. Then the samples were annealed at 100 $^\circ$C for 30 min to form MAPb(I$_{1-x}$Br$_x$)$_3$ The value of *x* is varied from 0.1 to 0.7 and for the convenience to use in Figure and Table, MAPb(I$_{1-x}$Br$_x$)$_3$ is represented by PK, PK1, PK2, and PK3 for *x* = 0.1, 0.3, 0.5 and 0.7 respectively.

III. RESULTS AND DISCUSSION

Device structure of CdS/CTe solar cells with novel back contact is shown in Fig.1(a). Current density voltage (J-V) curves for the best devices using each of the two back contact designs are shown in Fig. 1 (b). In the perovskite back contact, J-V measurements were conducted in forward to reverse and reverse to forward bias voltage directions to test for hysteresis since changes have been frequently observed in perovskite thin film solar cells where perovskite is used as absorber layer.[15, 16] The J-V characteristics parameters for each type of contact are shown in boxes along with the J-V curves. Performance improvement of devices in novel back contact is compared with respect to the standard back contact in laboratory scale.

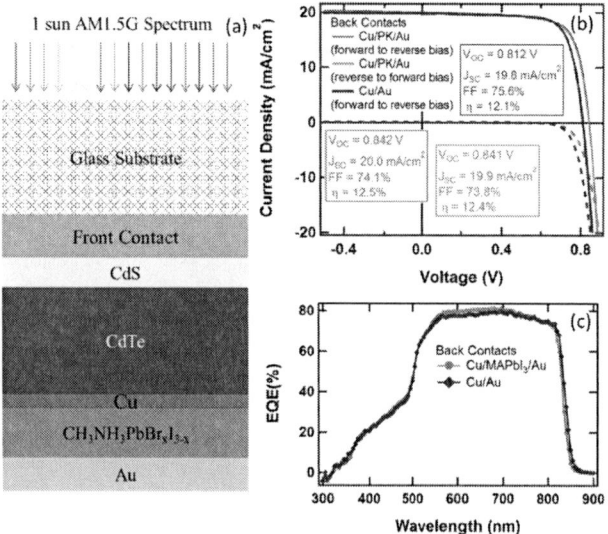

Fig. 1. (a) Device structure when MAPbI$_3$ is used as an interface layer, (b) current density-voltage characteristics at different back contacts; bias was applied in two directions, and (c) external quantum efficiency (EQE) for similar back contacts shown in (b).

From J-V curves and parameter values shown in Fig. 1, improvement in open circuit voltage (V$_{OC}$) can be clearly seen for novel back contact under identical condition. Short circuit current densities are identical in all back contacts, and slightly lower in comparison to normal CdTe solar cells due to slightly thicker CdS layer as is clear from the collection loss in the wavelength range from 400 nm to ~510 nm shown in external quantum efficiency data in Fig. 1(c). Preliminary and non-optimized novel back contacts already showed ~5% improvement in efficiency even though FF for the novel back contact type is slightly smaller than for the standard back contact.

Since the electron affinity of MAPbI$_3$ is ~4.1 eV [17] and the band gap is 1.55 eV [18], its valence band edge closely matches with the work function of Au. [3] This may create valence band offset between MAPbI$_3$ and CdTe and may generate a barrier to the photo-generated holes to reach the back contact metal. However, the band gap of MAPbI$_3$ can be tuned in the range of 1.55 to 2.3 eV by introducing Br to the formation of MAPb(I$_{1-x}$Br$_x$)$_3$ [19] When the band gap increases, top of the valence band shifts further from the vacuum level, reducing the valence band offset with the CdTe creating barrier-free interface. Optimization is still underway but our preliminary results at various values of x are shown in Fig. 2 and corresponding characteristic J-V parameters are shown in Table I. In Fig. 2, efficiency improvement by ~5% is observed when introducing a thin layer of MAPb(I$_{1-x}$Br$_x$)$_3$ with the values of x ranging from 0.1 to 0.7. As shown in Table I, we see ~3% increases in FF, with the introduction of thin layer of perovskite with respect to the standard back contact. This increase in FF is accompanied by a favorable reduction in the crossover between the illumination and dark current density-voltage as shown in Fig. 2. The crossover as seen in Cu/Au back contact is eliminated when a thin layer of MAPb(I$_{1-x}$Br$_x$)$_3$ is sandwiched between Cu and Au. Light-dependent forward current and voltage dependent carrier collection are responsible for factors causing crossover effect in thin film solar cells. [1, 20] CdTe thin films have very high densities of trap states (deep traps) within the bandgap of the material. Since the mobility of CdTe is comparatively smaller and minority carrier diffusion length is also comparatively shorter, propagation of carriers decreases and recombination current increases in the neutral region. At a certain forward bias, the diode current shows contribution of incident photons, diode ideality factor under illumination exceeds 2 and photocurrent of solar cells also depends on external bias voltage.[20]Due to the existence of back barrier in Cu/Au back contact and defect states present also on the CdTe/Au interface, recombination loss is more increased with the illumination in forward bias. On the other hand, when MAPb(I$_{1-x}$Br$_x$)$_3$ layer is included, recombination loss is minimized due to the decrease in contact barrier, and possibly to an improvement in passivation.

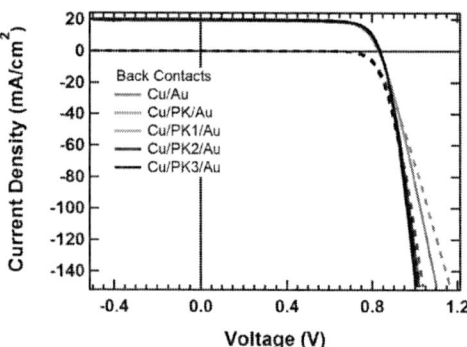

Fig. 2. Current density vs. voltage characteristics when scanning from forward to reverse bias voltage for CdS/CdTe/Cu/perovskite/Au device. In Fig. different concentration of bromine is incorporated in order to increase the bandgap of the MAPbI$_3$. The PK, PK1, PK2, and PK3 representations were made for MAPb(I$_{1-x}$Br$_x$)$_3$, when the value of x is 0.1, 0.3, 0.5 and 0.7 respectively.

TABLE I
PERFORMANCE PARAMETERS OF J-V CURVES SHOWN IN FIG. 2

Back Contacts	V_{OC} (V)	J_{SC} (mA/cm^2)	FF (%)	PCE (%)	R_S (Ω cm^2)	R_{Sh} (Ω cm^2)
Cu/Au	0.832	19.9	75.7	12.5	2.5	3210
Cu/PK/Au	0.838	19.6	77.2	12.7	3.2	2448
Cu/PK1/Au	0.836	20.2	77.5	13.1	3.2	3299
Cu/PK2/Au	0.839	19.3	77.8	12.6	3.4	3105
Cu/PK3/Au	0.841	20.0	77.1	13.0	3.4	3124

In the device structure of conventional perovskite solar cells, perovskite thin film interfaces with electron contact layer (TiO$_2$ or PCBM) coated either on transparent cathode (FTO) or on metal cathode depending on the n-i-p or p-i-n structures [7]. In analogy with this structure, we could also consider the CdTe film as an electron contact layer in our device structure. However, due to the bandgap and thickness of the CdTe (~1.5 eV, ~ 3 μm), we expect no measureable photogeneration of electron-hole pairs directly within the perovskite layer. A tellurium rich surface on CdTe is often employed to lower the back barrier effect and to increase the V_{OC} and FF in the devices.[21] In the process of the application of perovskite thin films on the CdTe surface, some co-authors discovered that non-toxic methylammonium lead iodide (MAI) can be used as an efficient etching material for CdTe surface to the formation of Te rich surface. That work is presented in another paper in this forum. Lastly, in the course of the studies described here, we have measured some notably high V_{OC} values (> 870 mV) in several devices on one of our test substrates. We are pursuing the ability to repeat and control the processing parameters that led to those exceptionally promising results.

IV. CONCLUSION

We have demonstrated that hybrid organic-inorganic metal halide perovskite, successful absorber layer in thin film solar cells, can be a promising hole-transport material in CdTe thin film solar cells. Perovskite MAPb(I$_{1-x}$Br$_x$)$_3$ thin films are fabricated by a solution-based approach at low temperature. Our initial results indicate that application of perovskite as an interface layer in CdTe solar cells enhances the performance of CdTe solar cells through a reduction in the Schottky barrier and introduction of electron-reflection property at the CdTe back contact. Our preliminary results showed ~5% increase in photo-conversion efficiency with the introduction of perovskite with respect to standard Cu/Au back contacts. Our experiment also opened a new avenue of research for perovskite solar cells that is CdTe thin film can be used as an electron transport layer in perovskite solar cells.

ACKNOWLEDGEMENT

The authors gratefully acknowledge financial support provided by the NSF's Sustainable Energy Pathways program (CHE-1230246) and the U.S. Air Force Research Laboratory, Space Vehicles Directorate (contract # FA9453-11-C-0253).

REFERENCES

[1] A. Niemegeers and M. Burgelman, "Effects of the Au/CdTe back contact on IV and CV characteristics of Au/CdTe/CdS/TCO solar cells," *Journal of Applied Physics,* vol. 81, pp. 2881-2886, 1997.

[2] S. Demtsu and J. Sites, "Effect of back-contact barrier on thin-film CdTe solar cells," *Thin Solid Films,* vol. 510, pp. 320-324, 2006.

[3] K. P. Bhandari, P. Koirala, N. R. Paudel, R. R. Khanal, A. B. Phillips, Y. Yan, *et al.*, "Iron pyrite nanocrystal film serves as a copper-free back contact for polycrystalline CdTe thin film solar cells," *Solar Energy Materials and Solar Cells,* vol. 140, pp. 108-114, 2015.

[4] A. B. Phillips, R. R. Khanal, Z. Song, R. M. Zartman, J. L. DeWitt, J. M. Stone, *et al.*, "Wiring-up carbon single wall nanotubes to polycrystalline inorganic semiconductor thin films: low-barrier, copper-free back contact to CdTe solar cells," *Nano letters,* vol. 13, pp. 5224-5232, 2013.

[5] K. P. Bhandari, X. Tan, P. Zereshki, F. K. Alfadhili, A. B. Phillips, P. Koirala, et al., "Thin film iron pyrite deposited by hybrid sputtering/co-evaporation as a hole transport layer for sputtered CdS/CdTe solar cells," *Solar Energy Materials and Solar Cells,* vol. 163, pp. 277-284, 4// 2017.

[6] E. Bastola, K. P. Bhandari, and R. J. Ellingson, "Application of composition controlled nickel-alloyed iron sulfide pyrite nanocrystal thin films as the hole transport layer in cadmium telluride solar cells," *Journal of Materials Chemistry C,* vol. 5, pp. 4996-5004, 2017.

[7] Z. Song, S. C. Watthage, A. B. Phillips, and M. J. Heben, "Pathways toward high-performance perovskite solar cells: review of recent advances in organo-metal halide perovskites for photovoltaic applications," *Journal of Photonics for Energy,* vol. 6, pp. 022001-022001, 2016.

[8] Y. Chen, H. Yi, X. Wu, R. Haroldson, Y. Gartstein, Y. Rodionov, et al., "Extended carrier lifetimes and diffusion in hybrid perovskites revealed by Hall effect and

photoconductivity measurements," *Nature Communications,* vol. 7, 2016.

[9] . De Wolf, J. Holovsky, S.-J. Moon, P. Löper, B. Niesen, M. Ledinsky, et al., "Organometallic halide perovskites: sharp optical absorption edge and its relation to photovoltaic performance," *The Journal of Physical Chemistry Letters,* vol. 5, pp. 1035-1039, 2014.

[10] G. E. Eperon, S. D. Stranks, C. Menelaou, M. B. Johnston, L. M. Herz, and H. J. Snaith, "Formamidinium lead trihalide: a broadly tunable perovskite for efficient planar heterojunction solar cells," *Energy & Environmental Science,* vol. 7, pp. 982-988, 2014.

[11] C. Wehrenfennig, G. E. Eperon, M. B. Johnston, H. J. Snaith, and L. M. Herz, "High Charge Carrier Mobilities and Lifetimes in Organolead Trihalide Perovskites," *Advanced Materials,* vol. 26, pp. 1584-1589, 2014.

[12] W. S. Yang, J. H. Noh, N. J. Jeon, Y. C. Kim, S. Ryu, J. Seo, et al., "High-performance photovoltaic perovskite layers fabricated through intramolecular exchange," *Science,* vol. 348, pp. 1234-1237, 2015.

[13] C.-H. Chiang, M. K. Nazeeruddin, M. Gratzel, and C.-G. Wu, "The synergistic effect of H2O and DMF towards stable and 20% efficiency inverted perovskite solar cells," *Energy & Environmental Science,* vol. 10, pp. 808-817, 2017.

[14] B. E. McCandless, R. W. Birkmire, and W. A. Buchanan, "Vapor transport deposition of cadmium telluride films," in *Photovoltaic Specialists Conference, 2002. Conference Record of the Twenty-Ninth IEEE,* 2002, pp. 547-550.

[15] E. L. Unger, E. T. Hoke, C. D. Bailie, W. H. Nguyen, A. R. Bowring, T. Heumuller, *et al.,* "Hysteresis and transient behavior in current-voltage measurements of hybrid-perovskite absorber solar cells," *Energy & Environmental Science,* vol. 7, pp. 3690-3698, 2014.

[16] H. J. Snaith, A. Abate, J. M. Ball, G. E. Eperon, T. Leijtens, N. K. Noel, et al., "Anomalous hysteresis in perovskite solar cells," *The journal of physical chemistry letters,* vol. 5, pp. 1511-1515, 2014.

[17] P. Schulz, E. Edri, S. Kirmayer, G. Hodes, D. Cahen, and A. Kahn, "Interface energetics in organo-metal halide perovskite-based photovoltaic cells," *Energy & Environmental Science,* vol. 7, pp. 1377-1381, 2014.

[18] G. Giorgi, J.-I. Fujisawa, H. Segawa, and K. Yamashita, "Small photocarrier effective masses featuring ambipolar transport in methylammonium lead iodide perovskite: a density functional analysis," *The journal of physical chemistry letters,* vol. 4, pp. 4213-4216, 2013.

[19] J. H. Noh, S. H. Im, J. H. Heo, T. N. Mandal, and S. I. Seok, "Chemical management for colorful, efficient, and stable inorganic–organic hybrid nanostructured solar cells," *Nano letters,* vol. 13, pp. 1764-1769, 2013.

[20] Z. Wang, Z. Cheng, A. E. Delahoy, and K. K. Chin, "A Study of light-sensitive ideality factor and voltage-dependent carrier collection of cdte solar cells in forward bias," *IEEE Journal of Photovoltaics,* vol. 3, pp. 843-851, 2013.

[21] C. Ferekides, U. Balasubramanian, R. Mamazza, V. Viswanathan, H. Zhao, and D. Morel, "CdTe thin film solar cells: device and technology issues," *Solar Energy,* vol. 77, pp. 823-830, 2004.

Magnesium-doped Zinc Oxide as a High Resistance Transparent Layer for thin film CdS/CdTe solar cells

Francesco Bittau[1], Elisa Artegiani[2], Ali Abbas[1], Daniele Menossi[2], Alessandro Romeo[2]
, Jake W. Bowers[1] and John M. Walls[1]

[1]Centre for Renewable Energy Systems Technology (CREST), Wolfson School of Mechanical, Electrical and Manufacturing Engineering, Loughborough University, Loughborough, Leicestershire, LE11 3TU, UK

[2]Laboratory for Applied Physics, Department of Computer Science, Univ. Verona, Strada Le Grazie 15, 37134, Verona, Italy

Abstract—**Magnesium-doped Zinc Oxide (MZO) was used as an alternative high resistance transparent layer for CdS/CdTe thin film solar cells. Thin films of MZO were deposited by RF magnetron sputtering and deposited on an Indium Tin Oxide contact (ITO). Thin film CdTe devices including a MZO high resistance transparent layer deposited at above 300 °C yielded a mean efficiency exceeding 10.5 %. This compares with an efficiency of 8.2 % without the MZO layer. The improvement in efficiency was due to a higher open circuit voltage and fill factor. Lowering the deposition temperature of MZO reduced the performance of the devices.**

Index Terms—**magnesium-doped zinc oxide, high resistance transparent layer, thin film, solar cells, CdTe.**

I. INTRODUCTION

Doping ZnO with MgO (E_G = 7.7 eV) leads to an energy bandgap increase ($E_G \approx 3.3$ eV [1]–[3]) through the formation of $Zn_{1-x}Mg_xO$ (MZO) [4]. The increase depends linearly upon the Mg content in the film, up to a Mg content of x = 0.46, at which point the band gap becomes $E_G = 4.2$ eV [5]. Experimental determination of the band alignment of MZO indicates that the larger band gap of MZO is almost exclusively due to an upshifting of the conduction band energy level [6], [7]. A similar behaviour occurs when ZnO is doped with Ca rather than Mg [8], [9]. In the range x = 0 to x = 0.46 $Zn_{1-x}Mg_xO$ maintains the typical hexagonal structure of ZnO. Above this doping level there is a gradual transition to the cubic structure of MgO [5]. There are several studies highlighting the importance of controlling the alignment between the band levels of absorber and window layers in chalcogenide solar cells [10]–[12]. These emphasize the importance of a slightly positive conduction band at the buffer/absorber interface to control the inversion of the absorber and recombination at the interface. The first applications of MZO were reported with copper indium gallium selenide (CIGS) thin film solar cells [13]–[17], where MZO was used as a replacement for the CdS buffer layer. The successful application of magnesium-doped zinc oxide to CdTe thin film solar cells has been reported recently [18]. Other semiconductors have been investigated for the tuneability of their energy band structure [19]–[23]. This study has focused on using MZO as an HRT layer with an emphasis on the effect of changing its conduction band

alignment with the adjacent semiconductors by varying the temperature.

II. EXPERIMENTAL DETAILS

Tin-doped indium oxide and magnesium-doped zinc oxide thin films were deposited by Radio-Frequency (RF) magnetron sputtering. Soda lime glass (SLG) was used as a superstrate. The glass was cleaned using a solution composed of 1/3 isopropanol, 1/3 acetone and 1/3 deionized water in an ultrasonic bath at 50 °C for 60 min. Thin films were deposited using an Orion 8 HV magnetron sputtering system (AJA international, USA) equipped with an AJA 600 series RF power supply. The 3" diameter ITO target contained 10 % SnO_2 and 90 % In_2O_3 Wt %. The 3" diameter MZO target contained 11 % MgO and 89 % ZnO Wt %. The glass superstrates were rotated at 10 rpm during deposition to enhance the uniformity of the films. The sputtering process was carried out at a constant power density of 3.5 W.cm^{-2} and at a pressure of 133.3 Pa using Ar as the working gas. Sputtering of MZO films was carried out in a 1 % O_2 in an Ar atmosphere. The temperature of the superstrate was kept constant at 450 °C for the deposition of ITO and ranged from 20 to 400 °C for MZO. The optical properties were investigated using a Cary Varian 5000 UV-VIS-NIR spectrophotometer. The composition of the films was measured using an X-ray photoelectron spectrometer (XPS) (Thermo Scientific K-alpha). Samples were processed into complete CdTe solar cells in the laboratories of University of Verona. ITO/MZO superstrates were coated with CdS and CdTe by thermal evaporation. The deposition process was carried out in a vacuum chamber at a pressure of 10^{-4} Pa with a Edwards XDS10 roughing pump and a Edwards ST-451 turbo-molecular pump. CdS was evaporated from a tungsten crucible at a deposition rate of 0.15 nm/sec. During deposition the substrate temperature was kept at 100 °C using halogen lamps. Before and after CdS deposition, the stack was annealed in vacuum at 450 °C for 30 minutes. CdTe was deposited from a special graphite Knudsen cell with an evaporation rate of 40 Å/sec. The deposition rate was controlled using an Intellemetrics IL-150 quartz controller. The CdTe activation treatment was performed using a CdCl$_2$

wet treatment. The solution was prepared by dissolving the $CdCl_2$ powder in methanol to form a saturated solution. The $CdCl_2$ powder was dried in a furnace at 0.1Pa before processing in solution. Typically, $\approx$250 μl is deposited in form of drops on the CdTe surface. The stack is then annealed in air at 380 °C for 30 minutes after a 15 minutes ramp from room temperature. Prior to back contact formation, the CdTe surface is treated with a solution of bromine (50 ul) and methanol (50 ml). This process removes residual $CdCl_2$ and forms a Te-rich layer. Subsequently, a 2 nm thick layer of Cu and a 50 nm thick layer of Au are deposited by thermal evaporation at room temperature in a vacuum of 10^{-3} Pa. The process is finished by annealing the structure for 20 min at 190 °C in air. Devices were characterized using current density-voltage (J-V) characteristics and cross-section images were obtained using transmission electron microscopy (TEM). Samples for TEM were prepared by focused ion beam milling using a dual beam FEI Nova 600 Nanolab. A standard in situ lift out method was used to prepare cross-sectional samples. An electron beam assisted platinum(e-Pt) over-layer was deposited onto the sample surface above the area to be analysed followed by an ion assisted layer to define the surface and homogenize the final thinning of the samples down to 100 nm. TEM analysis was carried out using a Tecnai F20 operating at 200 kV to investigate the detailed microstructure of the cell cross sections. Images were obtained using the bright field (BF) detector.

III. Characterization of Magnesium-doped Zinc Oxide films

Transmission plots of MZO films deposited at increasing temperatures are shown in Fig. 1(a). From the absorption edge in the UV region it is clear that the band gap of the films varies with the temperature of the substrate during deposition. The band gap of MZO films was estimated using the Tauc plot technique (Fig. 1(b)). The energy band gap of MZO films deposited at room temperature was estimated to be $E_G = 3.65$ eV, 0.35 eV higher than ZnO. This confirms that doping ZnO with Mg widens the optical band gap of the semiconductor. It was also observed that raising the temperature assists further increase in E_G. E_G increased from 3.65 eV at room temperature to 3.95 eV at 400 °C, as previously reported [24]. XPS analysis showed that increasing the temperature during MZO film deposition leads to an increased concentration of Mg atoms as shown in Fig. 2. The Mg concentration increases up to 400 °C while the Zn concentration decreases up to 300 °C. The oxygen concentration remains $\approx$41% at 20 °C, 100 °C and 400 °C and significantly increases at 200 and 300 °C. These results suggest that temperature enhances the inclusion of Mg ions in the MZO crystal structure. Evaporation of Zn or Mg during deposition can have an important role in this process. The vapourization temperature of Zn and Mg at a pressure of 1 mTorr (used during deposition) is $\approx$290 °C and $\approx$380 °C respectively [25]. At 300 °C and 400 °C, the Mg/Zn ratio increases significantly compared to films deposited at lower temperatures. Zn ions evaporate leaving free lattice

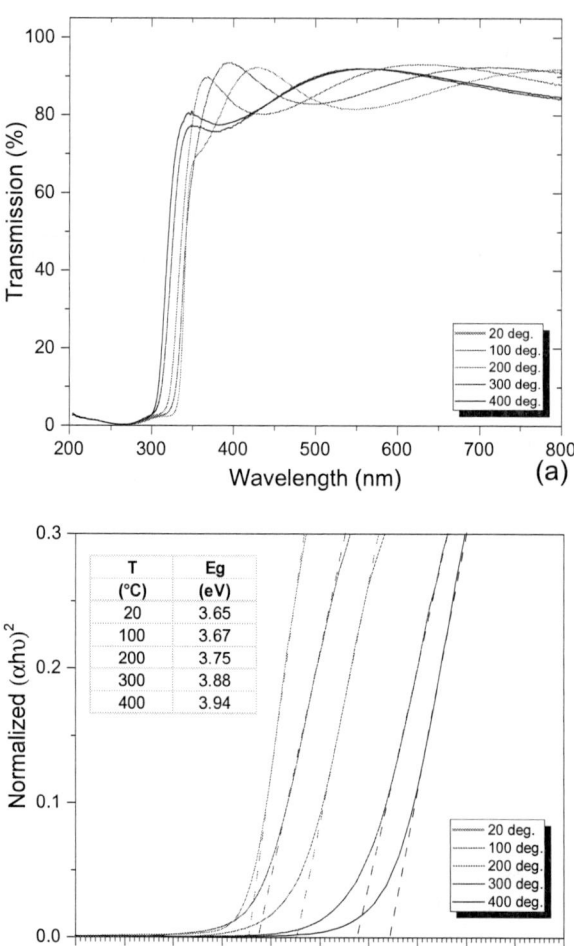

Fig. 1. The transmission spectra (a) and the Tauc plots (b) of MZO films deposited at temperature ranging between 20 °C and 400 °C showing respectively the shift of the absorption edge and the widening of the energy band gap in MZO films caused by the variation of the substrate temperature.

sites for Mg ions to occupy. Up to 200 °C the increasing Mg atomic concentration is not related to Zn evaporation but some other mechanism. Hwang et al have reported [24] that the increasing temperature causes a reduction of the Mg content in the films due to evaporation of Mg as measured using electron probe microanalysis. This is opposite to our findings. They also suggested that the increase in optical energy band gap occurs because raising temperature assists the replacement of Zn ions with Mg ions. Our work confirmed that temperature helps Mg ions replacing Zn ions. It was also confirmed that higher film deposition temperatures increase the MZO band gap. However the Mg concentration was found to grow by increasing temperature. The crystal structure of MZO has been investigated by XRD (Fig. 3). The (002) peak was observed, and the other peaks are associated with the underlying ITO film. The (002) peak was also observed for ZnO films [26] and is indexed from the crystallographic

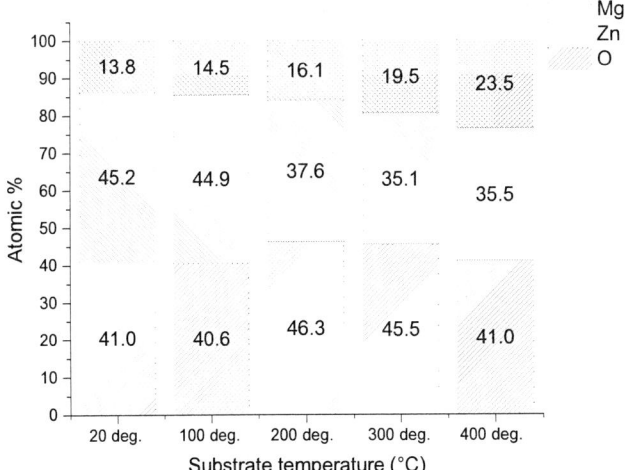

Fig. 2. The atomic percentage of Mg,ZnO and O in MZO films estimated by XPS analysis showing the variation of the atomic ratios in MZO films caused by the variation of the substrate temperature.

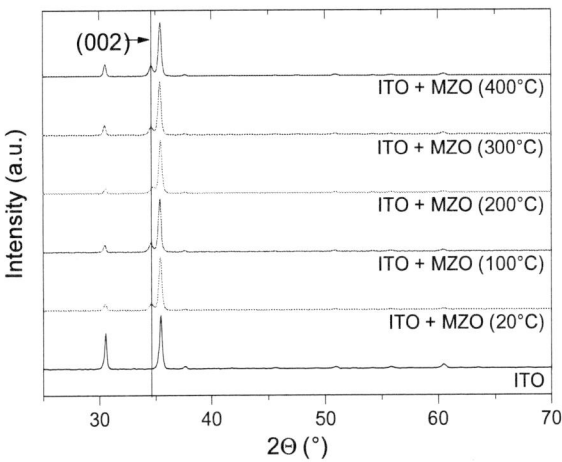

Fig. 3. The XRD patterns of MZO films deposited at different temperatures on ITO. The line highlights the position of the (002) peak, typical of the hexagonal structure of ZnO.

data of the ZnO hexagonal structure. Peaks (200) and (220) indexed from the crystallographic data of the cubic structure of MgO, are not visible [5] (ICDD 00-003-0752). ZnO has a band gap of 3.3 eV and a negative CBO with CdS of -0.3 eV. Depositing MZO at room temperature provides an MZO film with a band gap of 3.65 eV [7]. This corresponds to an almost flat conduction band alignment with CdS. Raising the deposition temperature of MZO increases the band gap to 3.95 eV at 400 °C. This corresponds to in a CBO with CdS of +0.35 eV.

A. Thermal Stability of MZO

Multiple annealing steps are repeated during the fabrication process of thin film CdTe solar cells. The band gap of MZO films is sensitive to the temperature of the substrate during

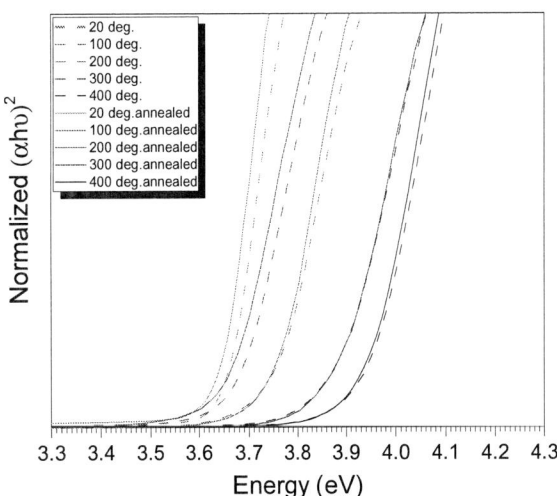

Fig. 4. Tauc plot of MZO films deposited at different temperatures (dashed lines) and of the same films annealed at 450 °C for 30 minutes(solid lines) showing the slight decrease of optical band gap that occurs after the MZO films are annealed at 450 °C for 30 minutes.

deposition. To test if the band gap of MZO can change during the fabrication process, a thermal annealing step was carried out at 450 °C for 30 minutes. This step simulates an equivalent annealing step used during the fabrication process. The band gaps of the films were extrapolated using the Tauc Plot technique (Fig. 4). The plot reveals that annealing the films at 450 °C slightly decreases the band gap of the films. The films deposited at lower temperatures result in a greater change in band gap. The change in all cases is small. The thin film deposited at room temperature changed the energy band gap from 3.65 eV before annealing to 3.63 eV post annealing. Similarly the film deposited at 100 °C decreased from 3.67 eV to 3.65 eV. These results suggest that the annealing steps carried out during the fabrication of CdS/CdTe devices do not significantly change the energy band gap of the MZO layers.

IV. CdS/CdTe Solar Cells with MZO HRTs

MZO films have been tested as HRT layers in CdS/CdTe solar cells with the structure ITO/MZO/CdS/CdTe/back contact. The performance of the devices was significantly affected by the deposition temperature of the MZO HRTs (Fig. 5). This suggests that the band gap of the MZO layer is playing a key role. The mean device efficiency increases from slightly below 5% achieved with a room temperature deposited MZO to $\approx$10.6 % at 400 °C. The higher efficiency is a consequence of improved Voc and FF. Voc and FF of 0.82 V and 66% respectively were achieved with a deposition temperature of 400 °C for the MZO layer. Temperature has a detrimental effect on Jsc at 100°C while a further increase of temperature gradually improves the current density. The current density reaches a maximum at 300 °C and 400 °C ($\approx$20 mA/cm^2). Devices not containing the MZO HRTs were fabricated and the results have been compared. Efficiency, Voc and FF all benefit from the addition of MZO deposited at high temperature. The highest current densities achieved with MZO are similar to those achieved without the layer. This indicates that the high transparency of the films deposited at high temperature does not greatly affect the device current density.

V. QUANTUM EFFICIENCY, TEM AND EDX ANALYSIS

The EQE of the highest performing device, with an MZO layer deposited at 400 °C is shown in Fig. 6. The absorption edge in the near infra-red (NIR) region lies at 850 nm, which corresponds to a typical band gap for the CdTe absorber of 1.45 eV [27]. Whilst the device performance improves with MZO deposition temperature with improved Voc and FF, the current density is limited by the high absorption in the blue region by the CdS. It is clear that to improve the device further, a reduction of thickness of the CdS layer is required. This would lead to higher photocurrent generation in the device. TEM images of the cross section of the sample show the layer by layer microstructure of the solar cell (Fig. 7). The thickness of the CdS layer has been estimated from the images to be in the range 250 nm - 300 nm. The CdS layer is sufficiently thick to absorb most of the radiation in the blue visible region. The CdTe layer is ≈ 7 μm in thickness. The CdTe grains develop across the full width of the layer.

VI. CONCLUSIONS

MZO films were deposited by magnetron sputtering at different substrate temperatures. It was confirmed that the addition of Mg to ZnO widens the band gap of the films. Increasing substrate temperature causes the band gap to widen further. These properties were utilized to produce MZO films with different Eg and to test them as HRT layers for CdTe thin film solar cells. The band gap of the MZO films ranged between 3.65 e V and 3.95 eV. A larger band gap was found to be beneficial for the Voc, FF and efficiency of devices. This suggests that the energy band alignment between HRT and CdS plays an important role in the performance of the solar cells. A thick CdS layer ($\approx$ 300nm) was used as a buffer layer leading to a low EQE response below 650 nm. A thinner CdS should significantly improve the current density and the efficiency of these devices. This study provides evidence that MZO is a promising material as a HRT layer for thin film CdTe solar cells and highlights the importance of energy band alignment between adjacent semiconductors.

ACKNOWLEDGMENT

The authors are grateful to the Engineering and Physical Science Research Council (EPSRC) (EP/J017361/1) for financial assistance under the EPSRC Supergen SuperSolar Hub.

REFERENCES

[1] R. Ondo-Ndong, G. Ferblantier, M. Al Kalfioui, A. Boyer, and A. Foucaran, "Properties of RF magnetron sputtered zinc oxide thin films," *Journal of Crystal Growth*, vol. 255, pp. 130–135, jul 2003.

[2] K. Ellmer and A. Klein, "ZnO and Its Applications," pp. 1–33, 2008.

[3] X. Zhang, H. Ma, Q. Wang, J. Ma, F. Zong, H. Xiao, F. Ji, and S. Hou, "Structural and optical properties of MgxZn1−xO thin films deposited by magnetron sputtering," *Physica B: Condensed Matter*, vol. 364, pp. 157–161, jul 2005.

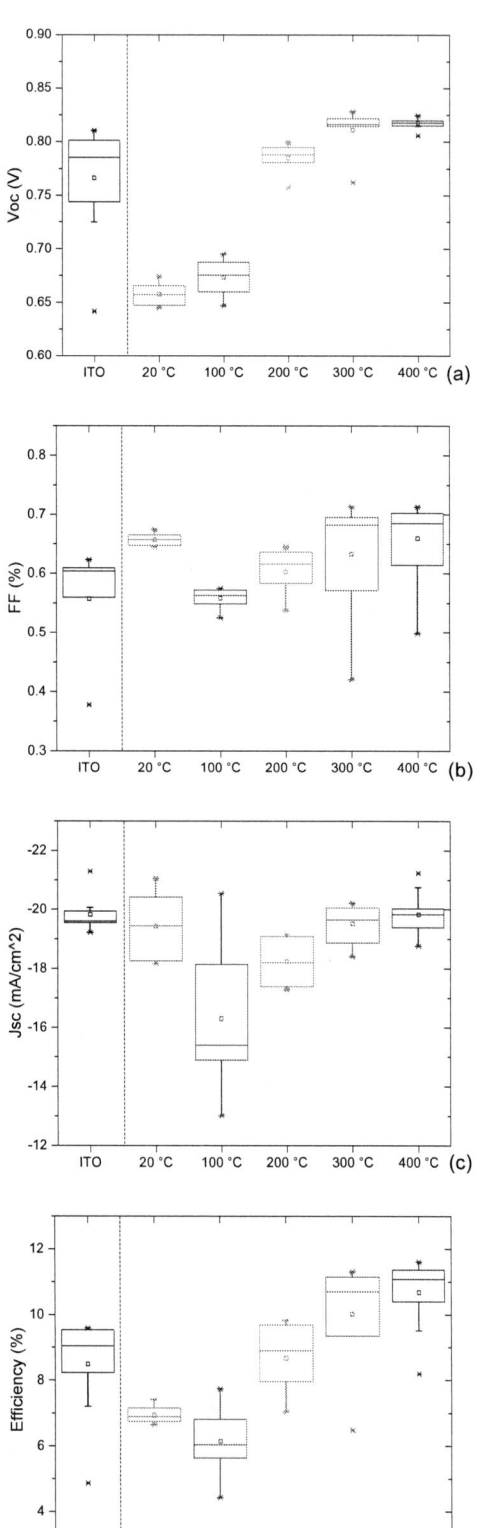

Fig. 5. Box Plots giving a statistical representation of Voc (a), FF (b), Jsc (c) and efficiency (d) of devices containing MZO deposited at increasing substrate temperature.

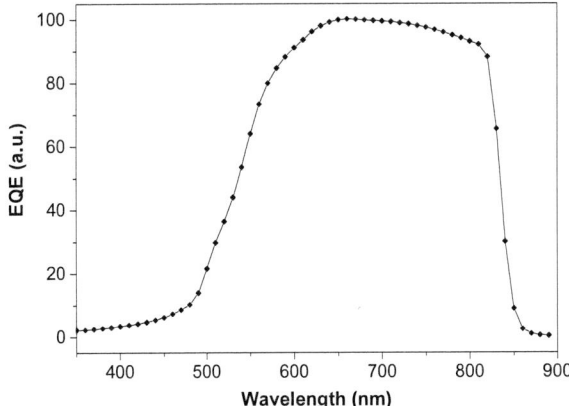

Fig. 6. The EQE spectra of a ITO/MZO/CdS/CdTe solar cell with MZO Eg = 3.94 eV showing the low spectral response of the best performing device below 600 nm is due to the thick CdS layer absorption losses.

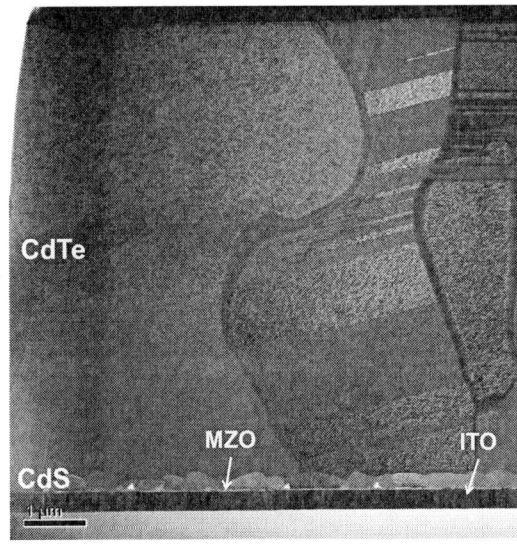

Fig. 7. A TEM cross-section of a full device. Starting from the top: the CdTe absorber, the CdS layer, the MZO HRT layer and the ITO layer.

[4] J.-H. Huang and C.-P. Liu, "The influence of magnesium and hydrogen introduction in sputtered zinc oxide thin films," *Thin Solid Films*, vol. 498, pp. 152–157, mar 2006.

[5] T. Minemoto, T. Negami, S. Nishiwaki, H. Takakura, and Y. Hamakawa, "Preparation of Zn1−xMgxO films by radio frequency magnetron sputtering," *Thin Solid Films*, vol. 372, pp. 173–176, sep 2000.

[6] A. Ohtomo, M. Kawasaki, I. Ohkubo, H. Koinuma, T. Yasuda, and Y. Segawa, "Structure and optical properties of ZnO/Mg0.2Zn0.8O superlattices," *Applied Physics Letters*, vol. 75, pp. 980–982, aug 1999.

[7] G. V. Rao, F. Säuberlich, and A. Klein, "Influence of Mg content on the band alignment at CdS/(Zn,Mg)O interfaces," *Applied Physics Letters*, vol. 87, p. 032101, jul 2005.

[8] L. Cao, J. Jiang, and L. Zhu, "Realization of band-gap engineering of ZnO thin films via Ca alloying," *Materials Letters*, vol. 100, pp. 201–203, jun 2013.

[9] K. P. Misra, K. C. Dubey, R. K. Shukla, and A. Srivastava, "Reduction in carrier concentration by calcium doping in ZnO thin films," in *2009 International Conference on Emerging Trends in Electronic and Photonic Devices & Systems*, pp. 495–496, IEEE, dec 2009.

[10] J. Sites and J. Pan, "Strategies to increase CdTe solar-cell voltage," *Thin Solid Films*, vol. 515, pp. 6099–6102, may 2007.

[11] M. Gloeckler and J. Sites, "Efficiency limitations for wide-band-gap chalcopyrite solar cells," *Thin Solid Films*, vol. 480-481, pp. 241–245, jun 2005.

[12] T. Song, A. Kanevce, and J. R. Sites, "Emitter/absorber interface of CdTe solar cells," *Journal of Applied Physics*, vol. 119, p. 233104, jun 2016.

[13] T. Minemoto, Y. Hashimoto, T. Satoh, T. Negami, H. Takakura, and Y. Hamakawa, "Cu(In,Ga)Se2 solar cells with controlled conduction band offset of window/Cu(In,Ga)Se2 layers," *Journal of Applied Physics*, vol. 89, pp. 8327–8330, jun 2001.

[14] T. Minemoto, Y. Hashimoto, W. Shamskolahi, T. Satoh, T. Negami, H. Takakura, and Y. Hamakawa, "Control of conduction band offset in wide-gap Cu(In,Ga)Se solar cells," *Solar Energy Materials and Solar Cells*, vol. 75, pp. 121–126, jan 2003.

[15] G.-R. Uhm, S. Y. Jang, Y. H. Jeon, H. K. Yoon, and H. Seo, "Optimized electronic structure of a Cu(In,Ga)Se2 solar cell with atomic layer deposited Zn(O,S) buffer layer for high power conversion efficiency," *RSC Advances*, vol. 4, no. 53, p. 28111, 2014.

[16] N. Naghavi, D. Abou-Ras, N. Allsop, N. Barreau, S. Bücheler, A. Ennaoui, C.-H. Fischer, C. Guillen, D. Hariskos, J. Herrero, R. Klenk, K. Kushiya, D. Lincot, R. Menner, T. Nakada, C. Platzer-Björkman, S. Spiering, A. Tiwari, and T. Törndahl, "Buffer layers and transparent conducting oxides for chalcopyrite Cu(In,Ga)(S,Se)2 based thin film photovoltaics: present status and current developments," *Progress in Photovoltaics: Research and Applications*, vol. 18, pp. 411–433, sep 2010.

[17] T. Törndahl, C. Platzer-Björkman, J. Kessler, and M. Edoff, "Atomic layer deposition of Zn1−xMgxO buffer layers for Cu(In,Ga)Se2 solar cells," *Progress in Photovoltaics: Research and Applications*, vol. 15, pp. 225–235, may 2007.

[18] J. Kephart, J. McCamy, Z. Ma, A. Ganjoo, F. Alamgir, and W. Sampath, "Band alignment of front contact layers for high-efficiency CdTe solar cells," *Solar Energy Materials and Solar Cells*, vol. 157, pp. 266–275, dec 2016.

[19] S. Sharbati and J. R. Sites, "Impact of the Band Offset for n-Zn(O,S)/p-Cu(In,Ga)Se Solar Cells," *IEEE Journal of Photovoltaics*, vol. 4, pp. 697–702, mar 2014.

[20] J. Perrenoud, S. Buecheler, L. Kranz, C. Fella, J. Skarp, and A. Tiwari, "Application of ZnO(1-x)S(x) as window layer in cadmium telluride solar cells," in *2010 35th IEEE Photovoltaic Specialists Conference*, vol. 2, pp. 000995–001000, IEEE, jun 2010.

[21] D. M. Meysing, C. A. Wolden, M. M. Griffith, H. Mahabaduge, J. Pankow, M. O. Reese, J. M. Burst, W. L. Rance, and T. M. Barnes, "Properties of reactively sputtered oxygenated cadmium sulfide (CdS:O) and their impact on CdTe solar cell performance," *Journal of Vacuum Science & Technology A: Vacuum, Surfaces, and Films*, vol. 33, p. 021203, mar 2015.

[22] D. E. Swanson, S. R. Hafner, W. S. Sampath, and J. D. Williams, "Development of plasma enhanced closed space sublimation for the deposition of CdS:O in CdTe solar cells," in *2013 IEEE 39th Photovoltaic Specialists Conference (PVSC)*, pp. 0434–0437, IEEE, jun 2013.

[23] J. M. Kephart, R. Geisthardt, and W. Sampath, "Sputtered, oxygenated CdS window layers for higher current in CdS/CdTe thin film solar cells," in *2012 38th IEEE Photovoltaic Specialists Conference*, pp. 000854–000858, IEEE, jun 2012.

[24] D.-K. Hwang, M.-C. Jeong, and J.-M. Myoung, "Effects of deposition temperature on the properties of Zn1−xMgxO thin films," *Applied Surface Science*, vol. 225, pp. 217–222, mar 2004.

[25] R. E. Honig, "Vapor pressure data for the solid and liquid elements," *RCA Review*, vol. 23, no. 4, pp. 567–586, 1962.

[26] F. Bittau, A. Abbas, K. Barth, J. Bowers, and J. Walls, "The effect of temperature on resistive ZnO layers and the performance of thin film CdTe solar cells," *Thin Solid Films*, nov 2016.

[27] B. E. McCandless and J. R. Sites, "Cadmium Telluride Solar Cells," in *Handbook of Photovoltaic Science and Engineering*, pp. 617–662, Chichester, UK: John Wiley & Sons, Ltd, jan 2005.

978-1-5090-5606-4/17 $31.00 © 2017 IEEE

Investigation of $Zn_{1-x}Mg_xO$:Al film by Ratio Frequency Magnetron Co-Sputtering as Transparent Conductive Oxide layer

Jakapan Chantana, Yuya Ishino, and Takashi Minemoto

Ritsumeikan University, Kusatsu, Shiga, 525-8577, Japan

Abstract — ZnO:Al is widely used as transparent conductive oxide (TCO) layer in $Cu(In,Ga)Se_2$ and $Cu(In,Ga)(S,Se)_2$-based solar cells. To widen the bandgap energy (E_g) of TCO layer and optimize the difference of conduction band minimum between TCO and the absorbers for the enhancement of solar cell performances, $Zn_{1-x}Mg_xO$:Al was systematically investigated as the alternative TCO layer. The $Zn_{1-x}Mg_xO$:Al films with different [Mg]/([Mg]+[Zn]) ratios were prepared by radio frequency (RF) magnetron co-sputtering from ZnO:Al and MgO targets. It is demonstrated that E_g of $Zn_{1-x}Mg_xO$:Al is increased with enhancing [Mg]/([Mg]+[Zn]), while the basic crystal structure of $Zn_{1-x}Mg_xO$:Al with [Mg]/([Mg]+[Zn]) in a range of 0-0.205 is ZnO. Moreover, $Zn_{1-x}Mg_xO$:Al films with small [Mg]/([Mg]+[Zn]) ratios of approximately 0.064-0.123 with low resistivity possesses the higher film quality and Hall mobility than those of ZnO:Al, while the carrier density is decreased with increasing [Mg]/([Mg]+[Zn]), which could be beneficial to lower the free carrier absorption. Consequently, $Zn_{1-x}Mg_xO$:Al films are promisingly utilized as the TCO layer.

I. INTRODUCTION

Thin-film solar cells are the promising solar cells for the reduction of production cost [1]. The polycrystalline chalcopyrite $Cu(In,Ga)Se_2$ (CIGSe) and $Cu(In,Ga)(S,Se)_2$ (CIGSSe)-based thin-film solar cells are the interesting thin-film solar cells as their high conversion efficiencies (η) above 20% with small area have been reported [2,3]. The standard structure of the thin-film solar cells is Glass/Mo/CIGSe or CIGSSe/CdS/ZnO/ZnO:Al. To increase the solar cell performances, conduction band offset (CBO) of the buffer/absorber interface in the solar cells has been developed to reduce the carrier recombination at the interface [4,5]. In addition to the CBO of the buffer/absorber interface, it was theoretically reported [6] that difference of conduction band minimum (E_C) between transparent conductive oxide (TCO) layer and absorber plays an important part in reducing the carrier recombination for enhancing the η, especially open-circuit voltage (V_{OC}) and fill factor (FF). ZnO:Al as TCO layer is widely used in the CIGSe and CIGSSe solar cells. However, ZnO:Al with electron affinity of about 4.3-4.6 eV [7-5] might not lead to the optimization of difference of E_C between TCO layer and the absorber because electron affinity of CIGSe is $4.35-0.421y-0.244y^2$ eV [4], where y denotes [Ga]/([Ga]+[In]) ratio of CIGSe. Moreover, ZnO:Al has a fixed E_C. Therefore, to optimize the difference of E_C between TCO and the absorber layers for high cell performances, the alternative material as the TCO layer with the ability to control E_C are

required. It was reported that the E_C of $Zn_{1-x}Mg_xO$ can be varied and move toward the vacuum level with increasing Mg content (x), namely E_g, where valence band maximum (E_V) is slightly changed when x is in a range of 0-0.17 [10]. Consequently, $Zn_{1-x}Mg_xO$:Al was considered as the alternative TCO layer in CIGSe and CIGSSe solar cells to experimentally optimize the difference of E_C between TCO layer and the absorber to boost cell performances.

In this work, $Zn_{1-x}Mg_xO$:Al films were prepared by co-sputtering method with different Mg contents, which is expressed as x = [Mg]/([Mg]+[Zn]). The Mg content of the films was adjusted by varying the sputtering power applied to ZnO:Al and MgO targets. The optical and electrical properties of $Zn_{1-x}Mg_xO$:Al were investigated for the application as the TCO layer.

II. EXPERIMENTAL METHODS

The $Zn_{1-x}Mg_xO$:Al films on soda-lime glass (SLG) substrates with the thickness in a range of about 800-1000 nm were deposited at room temperature by ratio frequency magnetron co-sputtering from ZnO:Al (Al_2O_3: 2 wt%-doped) (99.99%) and MgO (99.99%) targets. The target diameter and background pressure were 76.2 mm and 2.0×10^{-4} Pa, working pressure was 0.25 Pa, and Ar gas flow was fixed at 10 sccm. The shutters for ZnO:Al and MgO targets were opened simultaneously. The sample was water-cooled to maintain a substrate temperature at room temperature during the co-sputtering. The [Mg]/([Mg]+[Zn]) of $Zn_{1-x}Mg_xO$:Al films was controlled by varying the sputtering power applied to each target to control the sputtering rate. For the [Mg]/([Mg]+[Zn]) in a range of 0-0.205, the sputtering powers for MgO and ZnO:Al targets are adjusted as shown in Table I. The [Mg]/([Mg]+[Zn]) was measured by energy dispersive spectroscopy (EDS) operated at 4 kV. The electrical properties of the $Zn_{1-x}Mg_xO$:Al films were investigated by Hall-effect measurement system (ResiTest 8400, TOYO Corporation). The phase and crystallographic structure of the films were characterized by X-ray diffraction (XRD) operated at 45 kV and 40 mA using Cu Kα radiation (λ = 1.5405 Å). The optical transmittance and reflectance spectra of the films were recorded by UV-VIS-NIR spectrophotometer (UV-3600, Shimadzu). Bandgap energy (E_g) of $Zn_{1-x}Mg_xO$:Al films were derived from the plot of $(\alpha h\nu)^2$ as a function of photon energy (hν), where α is the absorption coefficient of their own films.

978-1-5090-5606-4/17 $31.00 © 2017 IEEE

TABLE I

Sputtering powers applied to ZnO:Al and MgO targets for
varying [Mg]/([Mg]+[Zn]) of $Zn_{1-x}Mg_xO$:Al films.

ZnO:Al (W)	MgO (W)	[Mg]/([Mg]+[Zn]) (x)
110	0	0
110	60	0.064
110	83	0.086
110	90	0.111
110	100	0.120
110	110	0.123
110	120	0.139
110	140	0.205

III. RESULTS AND DISCUSSION

The $Zn_{1-x}Mg_xO$:Al films on SLG substrates were grown with
varying [Mg]/([Mg]+[Zn]) from 0 (ZnO:Al) to 0.205 by
adjusting the powers applied to MgO and ZnO:Al targets as
shown in Table I. Figure 1 depicts (a) transmittance spectra of
$Zn_{1-x}Mg_xO$:Al films on SLG substrates and (b) their E_g as a
function of [Mg]/([Mg]+[Zn]) (0-0.205). It is demonstrated
that the absorption edge makes a blue shift as
[Mg]/([Mg]+[Zn]) increases from 0 to 0.205 in Fig 1(a). The α
of the $Zn_{1-x}Mg_xO$:Al films was calculated from their
transmittance and reflectance spectra, thereby leading to the
$(\alpha h\nu)^2$ plots as a function of $h\nu$ to estimate their optical E_g. The
derived optical E_g of $Zn_{1-x}Mg_xO$:Al films is consequently
enhanced from 3.54 eV for pure ZnO:Al (x=0) to 3.90 eV for
(x= 0.205) as shown in Fig. 1(b).

When the $h\nu$ is near or less than the optical E_g of the $Zn_{1-x}Mg_xO$:Al films, their absorption coefficient (α) shows the
exponential dependence on the $h\nu$ [11]:

$$\alpha = \alpha_0 \exp(\frac{h\nu}{E_u}) \qquad (1)$$

where α_0 is a constant. E_u is Urbach energy, correlating to the
width of the band tail, and can be utilized to observe the
impact of defects in the $Zn_{1-x}Mg_xO$:Al films. According to
Equation (1), the plot of $\ln \alpha$ as a function of $h\nu$ should give
the linear relation and E_u is obtained from the slope. The $\ln \alpha$
versus $h\nu$ for the $Zn_{1-x}Mg_xO$:Al films with different
[Mg]/([Mg]+[Zn]) ratios (0-0.205) was plotted to estimate the
E_u as depicted in Fig. 2. It is seen that E_us are significantly
decreased with small introduction of Mg content (x) in a range
of about 0.064-0.123, implying the improvement of films
quality. Moreover, the phase and crystallographic structure of
the films was next investigated by XRD

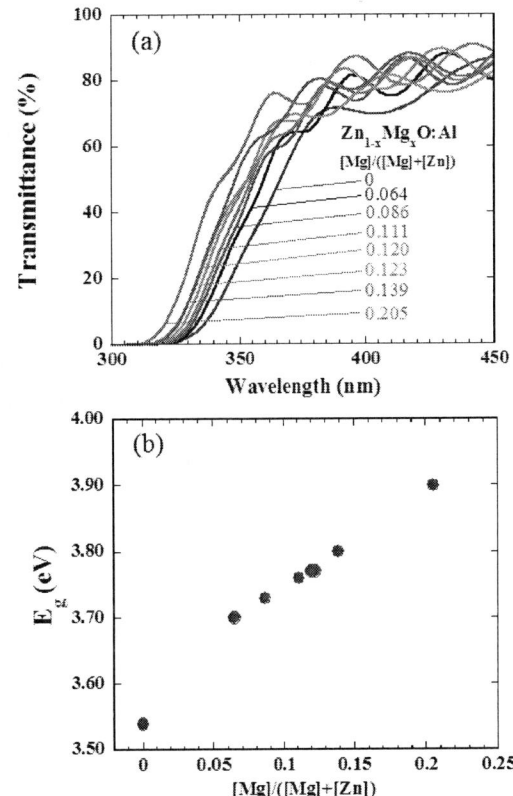

Fig. 1.　(a) Optical transmittance spectra of $Zn_{1-x}Mg_xO$:Al films on
SLG substrates with different [Mg]/([Mg]+[Zn]) ratios and (b) their
E_g as a function of the [Mg]/([Mg]+[Zn]) ratio.

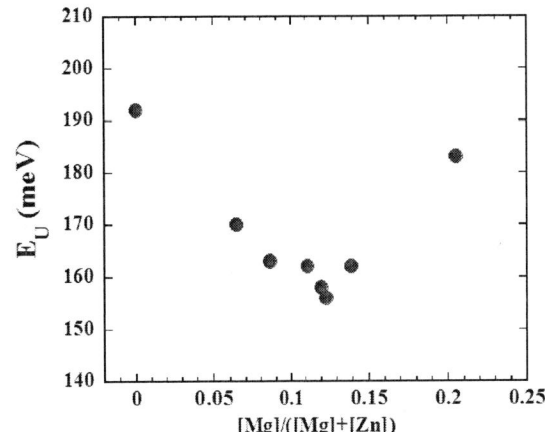

Fig. 2.　E_u of the the $Zn_{1-x}Mg_xO$:Al films as a function of
[Mg]/([Mg]+[Zn]) ratios.

Figure 3 illustrates XRD patterns of $Zn_{1-x}Mg_xO$:Al films
with different [Mg]/([Mg]+[Zn]) ratios. It is found that 002
peak indexed from the crystallographic data of hexagonal

structure of ZnO is observed in all $Zn_{1-x}Mg_xO$:Al films, where the peak position moves to higher 2θ with increasing [Mg]/([Mg]+[Zn]) from 0 to 0.205. In addition, full width at half maximum (FWHM) in Fig. 3 is decreased with small introduction of Mg content, implying the improvement of film quality, which is consistent with the result of E_u in Fig. 2.

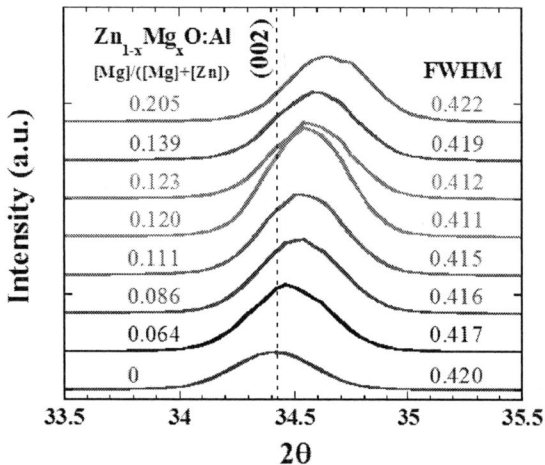

Fig. 3.   XRD patterns of $Zn_{1-x}Mg_xO$:Al films on SLG with different [Mg]/([Mg]+[Zn]) ratios.

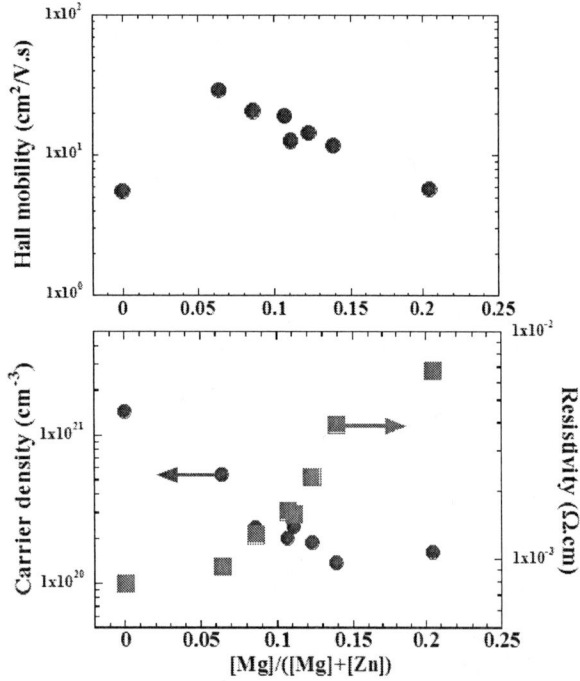

Fig. 4.   Hall mobility, carrier density, and resistivity of $Zn_{1-x}Mg_xO$:Al films on SLG as a function of [Mg]/([Mg]+[Zn]).

Next, the electrical properties of $Zn_{1-x}Mg_xO$:Al films were investigated. Figure 4 shows the electrical properties of $Zn_{1-x}Mg_xO$:Al films as a function of the [Mg]/([Mg]+[Zn]) ratio. In Fig. 4, Hall mobility is enhanced and takes a maximum value of 28.8 cm^2/V.s at [Mg]/([Mg]+[Zn]) of 0.064. The increase in Hall mobility of $Zn_{1-x}Mg_xO$:Al films with small Mg content in a range of about 0.064-0.123 is attributed to the improvement of the film quality as confirmed by the decreased E_u and FWHM in Figs. 2 and 3. On the other hand, the carrier density is decreased with increasing the [Mg]/([Mg]+[Zn]). As a result, the resistivity is increased.

According to the results, E_g of $Zn_{1-x}Mg_xO$:Al films from 3.54 to 3.90 eV is varied by [Mg]/([Mg]+[Zn]) from 0 to 0.205. Therefore, $Zn_{1-x}Mg_xO$:Al films with wider E_g compared with ZnO:Al can be used as the alternative TCO layer for the optimization of the E_C difference between TCO and the absorber layers for high cell performances as theoretically predicted [6].

IV. CONCLUSIONS

$Zn_{1-x}Mg_xO$:Al films on SLG substrates were prepared with different Mg contents. The E_g is increased with Mg content. Additionally, the small introduction of Mg content into $Zn_{1-x}Mg_xO$:Al films in a range of about 0.064-0.123 increases film quality and mobility but decreases carrier density, where the resistivity is still low for TCO application in the solar cell. It is notable that the decreased carrier density can suppress the free carrier absorption in the near infrared area in the TCO layer. Consequently, $Zn_{1-x}Mg_xO$:Al films are possibly used as the TCO layer.

V. ACKNOWLEDGEMENT

This work is partly supported by NEDO (Japan).

REFERENCES

[1]   Y. Hamakawa, *Thin-Film Solar Cells Next Generation Photovoltaics and Its Applications*. Springer: Heidelberg, 2004.
[2]   Kamada, T. Yagioka, S. Adachi, A. Handa, K. F. Tai, T. Kato, and H. Sugimoto, "New world record Cu(In,Ga)(Se,S)$_2$ thin film solar cell efficiency beyond 22%," in *43rd IEEE Photovoltaic Specialists Conference*, Oregon, June 2016, pp. 1287-1291.
[3]   P. Jackson, R. Wuerz, D. Hariskos, D. Hariskos, E. Lotter, W. Witte, and M. Powalla, "Effects of heavy alkali elements in Cu(In,Ga)Se$_2$ solar cells with efficiencies up to 22.6%," *Physica Status Solidi Rapid Research Letters* 2016, 10, 583-586. DOI: 10.1002/pssr.201600199
[4]   T. Minemoto, T. Matsui, H. Takakura, Y. Hamakawa, T. Negami, Y. Hashimoto, T. Uenoyama, and M. Kitagawa, "Theoretical analysis of the effect of conduction band offset of window/CIS layers on performance of CIS solar cells using device simulation," *Solar Energy Materials and Solar Cells* 2001, 67, 83-88.
[5]   T. Nakada, M. Hongo, and E. Hayashi, "Band offset of high efficiency CBD-ZnS/CIGS thin film solar cells," *Thin Solid Films* 2003, 431-432, 242-248.

[6] M. Murata, J. Chantana, N. Ashida, D. Hironiwa, and T. Minemoto, "Influence of conduction band minimum difference between transparent conductive oxide and absorber on photovoltaic performance of thin-film solar cell," *Japanese Journal of Applied Physics* 2015, 54, 032301.

[7] K. Jacobi, G. Zwicker, and A. Gutmann, "Work function, electron affinity and band bending of Zinc oxide surface," *Surface Science* 1984, 141, 109-125.

[8] J. A. Aranovich, D. Golmayo, A. L. Fahrenbruch, and R. H. Bube, "Photovoltaic properties of ZnO/CdTe heterojunctions prepared by spray pyrolysis," *Journal of Applied Physics* 1980, 51, 4260-4268.

[9] R. K. Swank, "Surface properties of II-VI compounds," *Physics Review* 1967, 153, 844-849.

[10]T. Minemoto, Y. Hashimoto, T. Satoh, T. Negami, H. Takakura, and Y. Hamakawa, "Cu(In,Ga)Se$_2$ solar cells with controlled conduction band offset of window/Cu(In,Ga)Se$_2$," *Journal of Applied Physics* 2001, 89, 8327-8330.

[11] V. Srikant, and D. R. Clarke, "Optical absorption edge of ZnO thin films: the effect of substrate," *Journal of Applied Physics* 1997, 81, 6357-6364.

A New TCO/window-buffer Front Stack for CdTe Solar Cells and its Implementation

Alan E. Delahoy[1], Xuehai Tan[1], Akash Saraf[1], Payal Patra[2], Surya Manda[1], Yunfei Chen[1], Krishnakumar Velappan[3], Bastian Siepchen[3], Shou Peng[4,5], and Ken K. Chin[1]

[1]CNBM New Energy Materials Research Center, Department of Physics,
New Jersey Institute of Technology, Newark, NJ, USA
[2]New Jersey Innovation Institute, Newark, NJ, USA
[3]CTF Solar GmbH, Dresden, Germany
[4]Bengbu Design and Research Institute for Glass Industry, Bengbu, China
[5]China Triumph International Engineering Co. Ltd., Shanghai, China

Abstract — A new TCO/window-buffer combination is presented for use in thin film CdTe solar cells. The TCO layer is cadmium tin oxide (CTO) and the window-buffer layer is tin magnesium oxide (TMO). Zinc magnesium oxide (ZMO), previously utilized by other groups and our group, was also prepared. The layers were produced by hollow cathode sputtering and were characterized by optical and Hall effect measurements. It was found that the carrier concentration of the window (buffer) layer at a fixed Mg concentration could be controlled over a wide range. CdTe solar cells were fabricated on glass/TCO/TMO and other bilayers without use of CdS. Quantum efficiency data confirmed the greatly improved blue response of the cells.

I. INTRODUCTION

The efficiency of solar cells based on thin film CdTe depends strongly on the nature and properties of the transparent conducting oxide (TCO) and buffer front stack layers as well as the doping and lifetime of the CdTe [1-4]. Proper engineering of the front stack offers several opportunities for improvement of J_{sc}, V_{oc}, and FF, namely: i) increased J_{sc} via reduction of TCO optical absorption and reduction of the blue loss stemming from buffer layer absorption; ii) increased FF via reduced TCO sheet resistance; and iii) increased V_{oc} via reduction of dark current and recombination at the absorber-buffer interface. The interface recombination is controlled by the size and sign of the absorber-buffer conduction band offset (CBO).

This work targets a high mobility, low absorption TCO and a wide gap buffer layer offering minimal blue loss and an optimized $\Delta E_c^{abs-buff}$ band offset. Modeling using SCAPS, both at NJIT and elsewhere (e.g. CSU), has shown that a type I heterojunction (straddling, with $\Delta E_c > 0$) is preferable to a type II heterojunction (staggered, with $\Delta E_c < 0$), with ΔE_c preferably lying in the range 0.1 – 0.3 eV. (In this paragraph, $\Delta E_c > 0$ means that the CBM rises in going from the absorber to the buffer.) If $\Delta E_c > 0.4$ eV the collection of the photocurrent would be hindered and temperature dependent.

We have explored various combinations of TCO and buffer, including cadmium tin oxide Cd_xSn_yO (CTO) as the TCO, and $Zn_{1-x}Mg_xO$ and $Sn_{1-x}Mg_xO$ as the buffer layer. The use of $Sn_{1-x}Mg_xO$ in this capacity has not previously been reported. The workhorse front end structure has, for several decades, been SnO_2:F/CdS. Record cells have been reported using the TCO/HRT/buffer combinations Cd_2SnO_4/$Zn_{1-x}Sn_xO$/CdS:O [1] and more recently SnO_2:F/$Zn_{1-x}Mg_xO$ which contains neither a separate HRT layer nor a CdS layer [5].

II. EXPERIMENTAL

In the work reported here, the metal oxides CTO, ZMO and TMO are deposited by hollow cathode sputtering (HCS) using a proprietary dual-cathode linear source operated in a reactive mode [6]. HCS is a versatile, scalable process for metal oxide deposition featuring low lattice damage in the deposited layers. Low cost metal targets are used and mixed metal oxides can be produced. A photo of the deposition system is shown in Fig. 1.

Fig. 1. Advanced hollow cathode sputtering system at NJIT.

Film thicknesses are determined using a Dektak IIA stylus profilometer. Optical properties are determined using a

Filmetrics F10-RT-UVX measurement system. Resistivity and mobility are determined using an Ecopia 21 HMS-3000 Hall system in a square van der Pauw configuration formed by etching. The magnetic field (0.57 T) was measured using an AlphaLab Hall magnetometer. Solar cell fabrication involves the steps of CdTe deposition on glass/TCO/ZMO by close-spaced sublimation ($T_s \leq 610$ °C), CdCl$_2$ treatment, NP etch, back contact application (including Cu$_x$O deposition and annealing), and scribing for cell definition. The use of O$_2$ during CSS [7] was discontinued for these experiments. Solar cells were characterized by J-V curves at AM1.5 100 mW/cm^2 obtained using a Newport xenon solar simulator and a Keithley 2401 sourcemeter, and by quantum efficiency measurements using a tungsten-halogen source with light dispersed by a grating monochromator and referenced to a calibrated silicon sensor from Gamma Scientific.

III. TCO LAYER PROPERTIES

To prepare cadmium tin oxide (CTO) the cathode was fitted with one pure Cd target and one pure Sn target. The principal variables studied were substrate temperature, target powers, and oxygen flow. The optical properties (spectral T, R, A and $T/(1-R)$) of a representative as-grown CTO layer are shown in Fig. 2. The properties are comparable to films produced in earlier work [8-10]. In contrast to some other TCOs, the free carrier absorption in CTO only becomes significant for $\lambda >$ 1000 nm and so does not reduce the J_{sc} of a CdTe-based solar cell (E_g^{CdTe} = 1.5 eV).

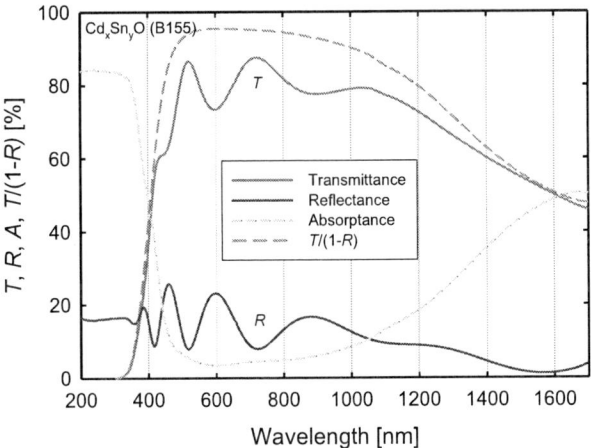

Fig. 2.   Specular spectral transmittance and reflectance of Cd$_x$Sn$_y$O produced by HCS.

We may extract the optical absorption coefficient $\alpha(\lambda)$ using the equation

$$\alpha(\lambda) = -\frac{1}{t}\ln\left(\frac{T(\lambda)}{1-R(\lambda)}\right) \qquad (1)$$

where t is the film thickness.   Since the fundamental absorption is both direct and allowed the band gap can be found from the usual $(\alpha E)^2$ versus E (Tauc) plot. The bandgap for CTO film B155 is E_g = 3.41 eV, with values typically lying between 3.25 – 3.41 eV.

Table I shows the electrical parameters of three of the more interesting CTO films. Film B265, with resistivity ρ = 3.78 x 10^{-4} Ω cm, had an excellent mobility μ = 46.5 cm^2/Vs, and a sheet resistance of 6.5 Ω/sq. We note that films with high carrier concentrations (e.g. B166, with n_e = 1.05 x 10^{21} cm^{-3}) tend to have lower mobilities because of increased ionized impurity scattering [6, 11].

TABLE I
ELECTRICAL PROPERTIES OF CD$_x$SN$_y$O, ZN$_{1-x}$MG$_x$O, SN$_{1-x}$MG$_x$O AND ZN$_x$SN$_y$MG$_z$O LAYERS PREPARED BY HOLLOW CATHODE SPUTTERING

Run #	Matl.	R_{sh} Ω/sq	Carrier conc. n_e (/cm^3)	Mobility μ (cm^2/Vs)	Resistivity ρ (Ω cm)
B265	CTO	6.5	3.55 x 10^{20}	46.5	3.78 x 10^{-4}
B160	CTO	7.9	4.08 x 10^{20}	41.9	3.65 x 10^{-4}
B166	CTO	4.3	1.05 x 10^{21}	30.5	1.95 x 10^{-4}
B184	ZMO	4.5k	1.05 x 10^{19}	8.64	6.89 x 10^{-2}
B189	ZMO	17M	5.52 x 10^{15}	1.94	5.83 x 10^{2}
B274	TMO	157M	1.84 x 10^{14}	6.0	5.63 x 10^{3}
B322	ZTMO	1.59M	1.16 x 10^{17}	5.2	1.04 x 10^{1}

Scrutiny of the deposition conditions for these and other CTO films suggests that the Cd/Sn ratio in the film plays a strong role in determining carrier concentration n_e, resistivity ρ, and also optical properties. A temperature-dependent Cd desorption process at the film surface appears to exist, and film mass measurements support this hypothesis. The Cd loss could be mitigated by increasing the oxygen partial pressure. We concluded that n_e declined with decreasing Cd/Sn ratio. Mamazza et al. co-sputtered CdO and SnO$_2$ at RT to produce amorphous CTO films that became polycrystalline after annealing above 580 °C [9]. In contrast, we have produced CTO films with $\mu >$ 40 cm^2/Vs in the as-deposited state. Meng et al. also produced CTO by RF magnetron sputtering but the films required annealing to attain a low resistivity [10].

Hall effect data confirms the n-type conductivity of all CTO films made to date. The nature of the donors was in the past commonly ascribed to point defects such as oxygen vacancies or interstitial Cd [12] but more recent work indicates that the oxygen vacancy is a very deep donor with transition energy at E_{CBM} − 0.66 eV while the Sn$_{Cd2}$ anti-site defect is a shallow donor at E_{CBM} − 0.05 eV with a low formation energy [13].

Figure 3 shows an uncalibrated depth profile of a CTO film on soda-lime glass. There would appear to be a slight gradient in the Cd/Sn ratio. (We plan to investigate this using optical emission spectroscopy conducted on the HC discharge.)

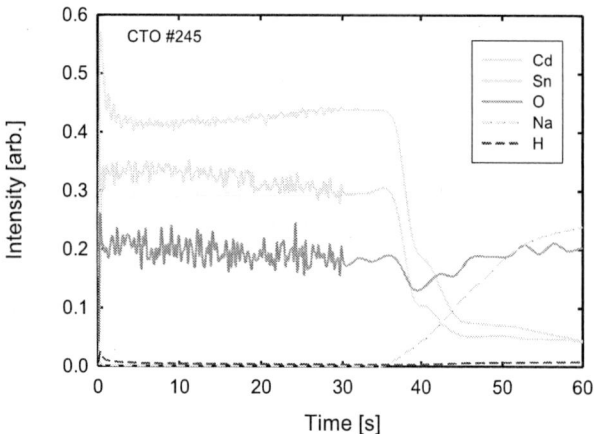

Fig. 3. Depth profile obtained using RF GD-OES of Cd_xSn_yO film produced by HCS.

IV. WINDOW LAYER PROPERTIES

To prepare $Zn_{1-x}Mg_xO$ films two $Zn_{0.85}Mg_{0.15}$ metal alloy targets were fitted to the cathode; $Sn_{1-x}Mg_xO$ films were prepared using two $Sn_{0.85}Mg_{0.15}$ targets; $Zn_xSn_yMg_zO$ films were prepared using one of each of the two target types.

The optical data for the ZMO film B189 is shown in Fig. 4. A sharp absorption edge is seen, blue shifted from that of ZnO.

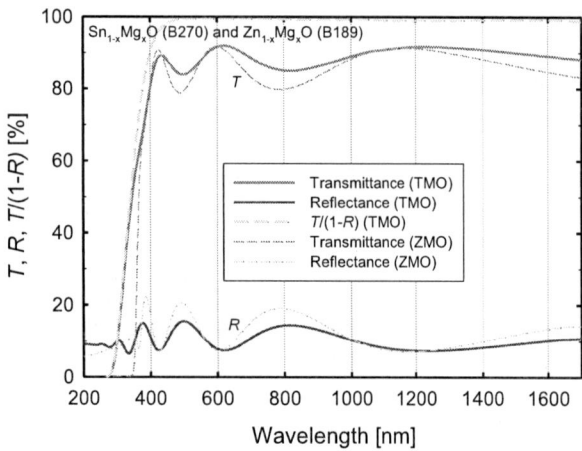

Fig. 4. Spectral transmittance and reflectance of $Zn_{1-x}Mg_xO$ produced by HCS.

Tauc plots for ZnO, ZMO film B240 (prepared using one Zn and one $Zn_{0.85}Mg_{0.15}$ target), and ZMO film 189 are shown in Fig. 5. The band gaps are 3.30, 3.38, and 3.505 eV, respectively. Also shown in Fig. 4 is data for TMO film B270. Both layers exhibit almost zero absorption for $\lambda > 400$ nm. The bandgap of the TMO film was found to be 4.24 eV, one of the largest values we have measured for this class of material.

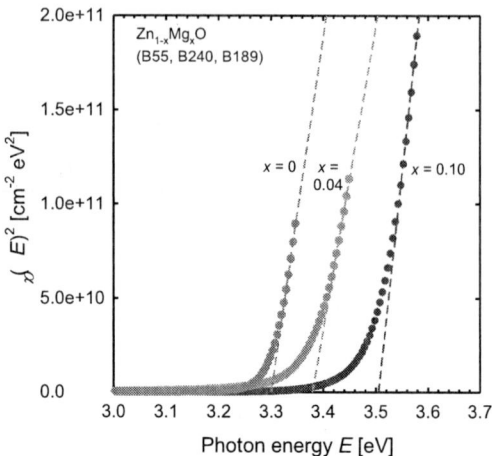

Fig. 5. Determination of band gap of $Zn_{1-x}Mg_xO$ from a plot of $(\alpha E)^2$ vs. photon energy E.

Regarding ZMO, the ionic radius of Mg^{2+} (0.57 Å) is close to that of Zn^{2+} (0.60 Å) [14] and allows it to substitute without change in the wurzite structure, at least for $x < 0.4$. The variation of bandgap of with x is found to be approximately linear and we deduce from a figure given in [15] that $E_g(x)$ is given by

$$E_g(x) = 3.296 + 2.19x \text{ eV} \qquad (2)$$

Based on this equation, the Mg contents in the two ZMO films B240 and B189 are $x = 0.04$ and 0.10, respectively. For larger values of x phase separation may occur with some MgO (rocksalt structure) occurring. We found that the ZMO films prepared using two $Zn_{0.85}Mg_{0.15}$ targets exhibited exceptionally bright photoluminescence even at room temperature [16].

The as-deposited ZMO films were n-type and, for a given Mg content, it was found that the carrier concentration could be varied over a wide range ($10^{15} - 10^{19}$ cm^{-3}) through control of the oxygen flow rate during deposition (see Fig. 6).

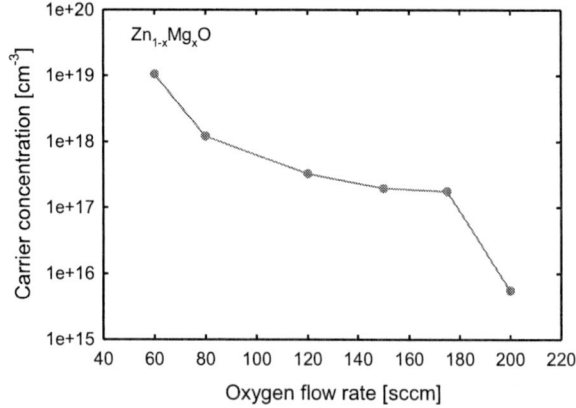

Fig. 6. Control of ZMO carrier density via flow rate of O_2 during deposition by HCS.

Detailed electrical properties for ZMO, TMO, and ZTMO films are given in Table I above.

V. DEVICE PROPERTIES

The superstrate devices being investigated in this program have the general structure: glass/TCO/window/CdTe/ohmic contact. With ZMO replacing CdS as the window layer, the band diagram of this type 1 (straddling) heterojunction (HJ) device becomes that shown in Fig. 7. The relevant material parameters needed to construct such a diagram are shown in Table II.

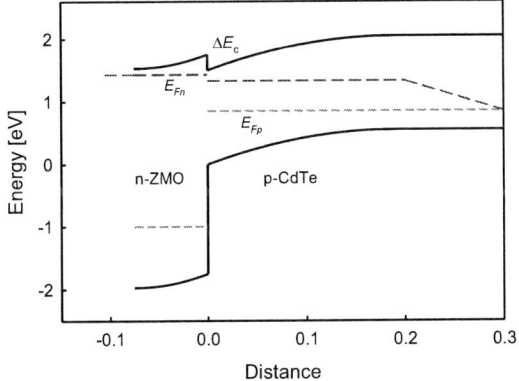

Fig. 7. Schematic band diagram for the ZMO/CdTe straddling heterojunction interface region showing a $\Delta E_c > 0$ band offset (several features not to scale).

TABLE II
FILM PARAMETERS (BANDGAP AND ELECTRON AFFINITY, IN EV) NEEDED TO CONSTRUCT THE BAND DIAGRAM

Property	FTO	CTO	CdS	ZnO	ZMO	CdTe
E_g (300 K)	≈ 4.2	3.3–3.7*	2.4	3.3	3.4–3.7	1.5
χ	4.8–5.3	4.2	4.5	4.4–4.5	4.0–4.2	4.3–4.4

From Anderson's rule, the CBO ΔE_c^{And} between the absorber and the window layers is defined as (and $= -\Delta E_c$ of page 1):

$$\Delta E_c^{And} = E_c^{abs} - E_c^{window} = \chi_{window} - \chi_{abs} \qquad (3)$$

By raising E_c of the $Zn_{1-x}Mg_xO$ window layer by introducing Mg, the magnitude of ΔE_c^{And} can be brought into the favorable range of $0.1 - 0.3$ eV (spike rather than cliff offset). The introduction of ZMO into thin-film solar cells was first studied in the context of CIGS [17-21] and later in CdTe [4,5].

Hollow cathode sputtering (HCS) was successfully used to prepare an entire front stack on glass suitable for CdTe solar cells that performed the roles of TCO, high resistance layer, window layer, and emitter. Using such stacks, viable CdTe

solar cells were fabricated. The J-V curve of such a cell prepared on glass coated with CTO and overcoated with a ZMO window layer is shown in Fig. 8. This preliminary experiment yielded a very good J_{sc} of 25.8 mA/cm². The reason for the poor V_{oc} is not known at present but could be due either to the use of CTO or ZMO, or to sub-optimal cell processing. It was decided after these early experiments that the number of uncertainties should be reduced and subsequent ZMO development was conducted on commercial FTO.

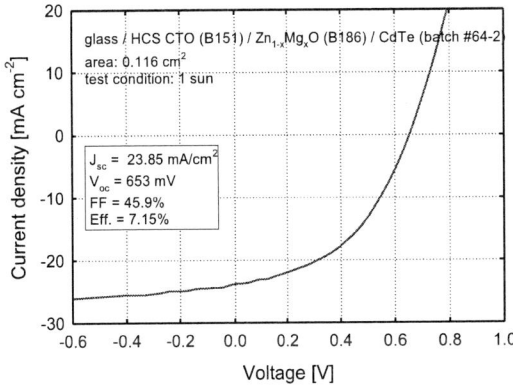

Fig. 8. Light J-V curve for CdS-free CdTe solar cell on HCS CTO as TCO and HCS ZMO as wide-gap window layer.

For the glass/FTO/ZMO/CdTe cells we observed that the linear dark $J-V$ curves maintained a conventional diode shape with low series resistance and high shunt resistance in the dark. Interestingly, the curves exhibited an increased turn-on voltage relative to that obtained using CdS, as shown in Fig. 9. The magnitude of the $J-V$ voltage shift was found to correlate with ZMO thickness, with O_2 flow also playing a role.

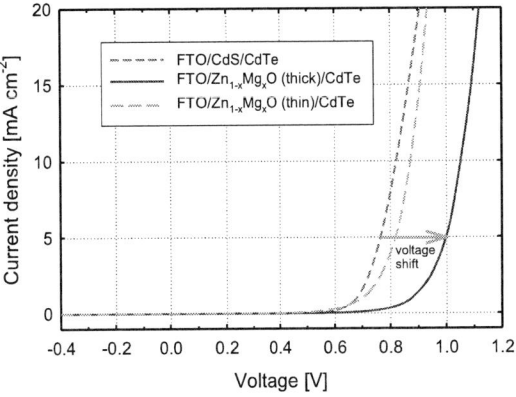

Fig. 9. Dark J-V characteristics for CdS/CdTe and two ZMO/CdTe solar cells on FTO revealing the substantial voltage shift obtainable with ZMO.

Regarding the light J-V characteristics for 1-sun illumination we have found that a large dark J-V voltage shift correlates with kinky light J-V curves showing photocurrent suppression and an inflection as V_{oc} is approached. This type of J-V curve is not uncommon in experimental devices. It has been observed in both CIGS and CdTe solar cells containing $Zn_{1-x}Mg_xO$ buffer layers with a Mg content greater than the optimal value [22, 23]. This is shown in Fig. 10 for a ZMO/CdTe cell on FTO that has an impressively high V_{oc} of 878 mV but only a 4.2% efficiency. Use of a higher CdTe deposition temperature and a thinner ZMO layer enabled the kink to be largely eliminated and a respectable efficiency of 10.5% to be achieved [16].

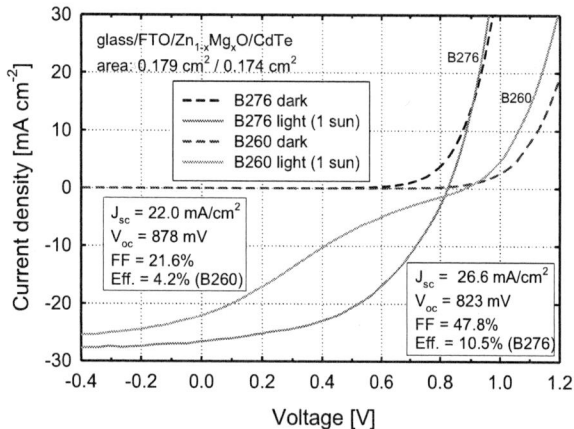

Fig. 10. Light and dark J-V curves for ZMO/CdTe solar cells with different ZMO thickness.

Since carriers have to surmount the ΔE_c barrier by thermionic emission (TE) it is clear that a sufficiently large ΔE_c offset will affect the transport properties of these HJ devices. For barriers > 0.4 eV, TE cannot support a J_{ph} of 25 mA cm^{-2} and so reduced collection will occur. If the dark current is modulated by a factor $\exp(-\Delta E_c/kT)$ this can account for the observed shift in the dark $J–V$ curve. We have conducted simulations of TCO/ZMO/CdTe devices using SCAPS and can qualitatively reproduce the photocurrent suppression if the doping of the ZMO is too low (Fig. 11).

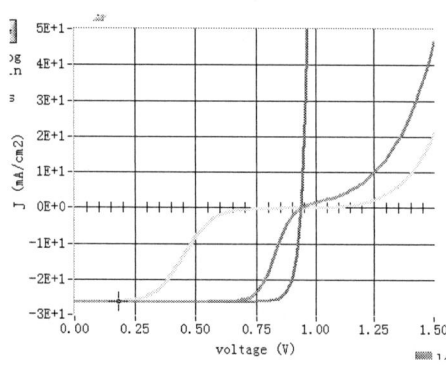

Fig. 11. SCAPS simulation of J-V curves of CdTe solar cells as a function of ZMO doping (ΔE_c = 0.4eV, TCO doping 10^{18} cm^{-3}). Green, light blue (overlapped), pink and red indicate ZMO doping of 0, 10^{14}, 10^{16} and 10^{18} cm^{-3}, respectively.

Using ZMO and TMO window layers in devices fully fabricated at NJIT with the structure glass/FTO/ZMO/CdTe, glass/CTO/ZMO/CdTe or glass/FTO/TMO/CdTe has enabled us to almost completely eliminate the blue loss previously caused by optical absorption in the dead CdS layer. The dramatic improvement in external quantum efficiency (EQE) for λ < 550 nm exhibited by devices using ZMO or TMO as a window layer is shown in Fig. 12. At 400 nm, for example, the EQE can be doubled. The short-circuit current density implied by integration of the measured EQE multiplied by the 100 mW cm^{-2} AM1.5 solar photon flux distribution is 25.8 mA cm^{-2} for a device with ZMO, and 25.4 mA cm^{-2} for TMO, compared to 20.4 mA cm^{-2} for a standard device with CdS.

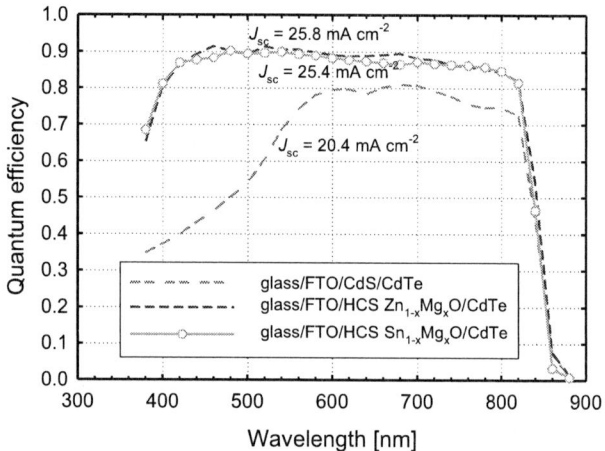

Fig. 12. External quantum efficiency versus wavelength for conventional CdTe cells on FTO/CdS and the new cells on FTO/ZMO or FTO/TMO.

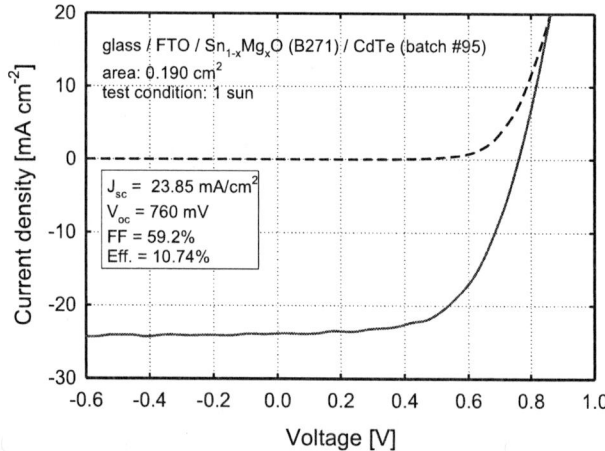

Fig. 13. Dark and light J-V curves for CdS-free CdTe solar cell with HCS $Sn_{1-x}Mg_xO$ (TMO) as a wide-gap window layer.

The use of TMO as the window layer in CdTe solar cells eliminated the photocurrent suppression often seen with ZMO.

Thus, the dark and light J-V curves for a TMO/CdTe cell are shown in Fig. 13. A good J_{sc} of 24 mA/cm^2 was achieved and the cell efficiency was 10.7%. The somewhat disappointing V_{oc} of 760 mV suggests that ΔE_c remains to be more favorably adjusted. While ZTMO has been prepared, results for cells using this type of window layer are not yet available.

VI. CONCLUSIONS

A custom-built hollow cathode sputtering (HCS) system was used to prepare high quality CTO, ZMO, TMO, and ZTMO layers for use in CdTe solar cells. CTO layers with electron mobilities up to 46.5 cm^2/Vs were demonstrated. Carrier concentrations in ZMO could be controlled over the range 10^{15} – 10^{19} /cm^3. High J_{sc}'s up to 26 mA/cm^2 demonstrate that ZMO and TMO provide a means to obtain competitive CdTe solar cells with improved blue response due to elimination of CdS. With ZMO, V_{oc}'s up to 878 mV were achieved. The flexibility of HCS to produce almost any type of metal oxide offers a route for comprehensive optimization of the entire front side (TCO/window layer) of thin film CdTe solar cells including control of ΔE_c at the absorber-buffer interface.

ACKNOWLEDGEMENTS

This work was generously supported by CTIEC under a Research Agreement with NJII. We are grateful to Dr. Philippe Hunault, Horiba Scientific, Edison, NJ for RF GD-OES measurements.

REFERENCES

[1] X. Wu, X., J. C. Keane, R. G. Dhere, C. DeHart, D. S. Albin, A. Duda, T. A. Gessert, S. Asher, D. H. Levi, and P. Sheldon, "16.5%-efficient CdS/CdTe polycrystalline thin-film solar cell," in *Proceedings of the 17th European photovoltaic solar energy conference*, vol. 995. London: James & James Ltd., 2001.

[2] J. N. Duenow, J. M. Burst, D. S. Albin, M. O. Reese, S. A. Jensen, S. W. Johnston, D. Kuciauskas, K. S. Swain, T Ablekim, K. G. Lynn, and A. L. Fahrenbruch, "Relationship of open-circuit voltage to CdTe hole concentration and lifetime," *IEEE Journal of Photovoltaics*, vol. 6, no. 6, pp. 1641-1644, 2016.

[3] J. M. Burst, J. N. Duenow, D. S. Albin, E. Colegrove, M. O. Reese, J. A. Aguiar, C-S. Jiang, M. K. Patel, M. M. Al-Jassim, D. Kuciauskas, and S. Swain, "CdTe solar cells with open-circuit voltage breaking the 1 V barrier," *Nature Energy*, vol. 1, p. 16015, 2016.

[4] T. Song, A. Kanevce, and J. R. Sites, "Emitter/absorber interface of CdTe solar cells," *Journal of Applied Physics*, vol. 119, no. 23, p. 233104, 2016.

[5] J. Sites, A. Munshi, J. Kephart, D. Swanson, and W. S. Sampath, "Progress and challenges with CdTe cell efficiency," in *43rd IEEE Photovoltaic Specialist Conference*, 2016, pp. 3632-3635.

[6] A. E. Delahoy, S. Falk, P. Patra, R. Korotkov, B. Siepchen, S. Peng, and K. K. Chin, "Metal oxides produced by hollow cathode sputtering, their utility in thin film photovoltaics, and

results for AZO," in *43rd IEEE Photovoltaic Specialist Conference*, 2016, pp. 1443-1448.

[7] V. Krishnakumar, B. Späth, C. Drost, C. Kraft, B. Siepchen, A. Delahoy, X. Tan, K. Chin, S. Peng, D. Hirsch, O. Zywitzki, T. Modes, and H. Morgner, "Close spaced sublimation deposition of CdTe layers with process gas oxygen for thin film solar cells," *Thin Solid Films*, 2016.

[8] X. Wu, W. P. Mulligan, and T. J. Coutts, "Recent developments in RF sputtered cadmium stannate films," *Thin Solid Films*, vol. 286, no. 1-2, pp. 274-276, 1996.

[9] R. Mamazza, D. L. Morel, and C. S. Ferekides, "Transparent conducting oxide thin films of Cd$_2$SnO$_4$ prepared by RF magnetron co-sputtering of the constituent binary oxides," *Thin Solid Films*, vol. 484, pp. 26-33, 2005.

[10] T. Meng, B. McCandless, W. Buchanan, E. Kimberly, and R. Birkmire, "Cadmium tin oxide thin films deposited by RF magnetron sputtering for photovoltaic applications,"*Journal of Alloys and Compounds*, vol. 556, pp. 39-44, 2013.

[11] A. E. Delahoy and S. Guo, Transparent Conductive Oxides for Photovoltaics, Chapter 17 in *Handbook of Photovoltaic Science and Engineering*, 2nd Edition, Edited by A. Luque and S. Hegedus. Chichester: Wiley, 2011, pp. 716-796.

[12] E. Leja, T. Stapiński, and K. Marszałek, "Electrical and optical properties of conducting n-type Cd$_2$SnO$_4$ thin films," *Thin Solid Films*, vol. 125, pp. 119-122, 1985.

[13] S. B. Zhang and S. H. Wei, "Self-doping of cadmium stannate in the inverse spinel structure," *Appl. Phys. Lett.*, vol. 80, pp. 1376-1378, 2002.

[14] R. D. Shannon, *Acta Cryst.*, vol. A32, pp. 751-767, 1976.

[15] M. Lorenz, E. M. Kaidashev, H. Von Wenckstern, V. Riede, C. Bundesmann, D. Spemann, G. Benndorf, H. Hochmuth, A. Rahm, H. C. Semmelhack, and M. Grundmann, "Optical and electrical properties of epitaxial (Mg,Cd)$_x$Zn$_{1-x}$O, ZnO, and ZnO:(Ga,Al) thin films on c-plane sapphire grown by pulsed laser deposition," *Solid-State Electronics*, vol. 47, pp. 2205-2209, 2003.

[16] A. E. Delahoy, S. Peng, P. Patra, S. Manda, A. Saraf, Y. Chen, X. Tan, and K. K. Chin, "Cadmium tin oxide and zinc magnesium oxide prepared by hollow cathode sputtering for CdTe photovoltaics," *MRS Advances*, 2017.

[17] T. Minemoto, Y. Hashimoto, W. Shams-Kolahi, T. Satoh, T. Negami, H. Takakura, and Y. Hamakawa, *Sol. Energy Mater. Sol. Cells*, vol. 75, pp. 121-126, 2003.

[18] A. Yamada, K. Matsubara, K. Sakurai, S. Ishizuka, H. Tampo, P. J. Fons, K. Iwata, and S. Niki, *Appl. Phys. Lett.*, vol. 85, pp. 5607-5609, 2004.

[19] J. V. Li, X. Li, Y. Yan, C-S. Jiang, W. K. Metzger, I. L. Repins, M. A. Contreras, and D. H. Levi, *J. Vac. Sci. Technol.*, vol. B 27, pp. 2384-2389, 2009.

[20] F. Erfurth, A. Grimm, J. Palm, T. P. Niesen, F. Reinert, L. Weinhardt, and E. Umbach, *Appl. Phys. Lett.*, vol. 98, p. 142107, 2011.

[21] C. S. Lee, S. Kim, Y. M. Shin, B. G. Park, B. T. Ahn, and H. Kwon, "Performance improvement in Cd-free Cu(In,Ga)Se$_2$ solar cells by modifying the electronic structure of the ZnMgO buffer layer," *RSC Adv.*, vol. 4, pp. 36784-36790, 2014.

[22] T. Törndahl, C. Platzer-Björkman, J. Kessler, and M. Edoff, "Atomic layer deposition of Zn$_{1-x}$Mg$_x$O buffer layers for Cu(In,Ga)Se$_2$ solar cells," *Prog. Photovolt: Res. Appl.*, vol. 15, pp. 225-235, 2007.

[23] J. M. Kephart, J. W. McCamy, Z. Ma, A. Ganjoo, F. M. Alamgir, and W. S. Sampath, "Band alignment of front contact layers for high-efficiency CdTe solar cells," *Sol. Energy Mater. Sol. Cells*, vol. 157, pp. 266-275, 2016.

Synthesis of High-Quality AZO Polycrystalline Films via Target Bias Radio Frequency Magnetron Sputtering

Zhongming Du [1,2], Yufeng Zhang [1,2], Xiangxin Liu [1,2] *

[1] The Key Laboratory of Solar Thermal and Photovoltaic System, Institute of Electrical Engineering, CAS, Beijing 100190, China;

[2] University of Chinese Academy of Sciences, Beijing 100049, China

Abstract: Ceramic ZnO:Al (AZO) target bias of radio frequency (r.f.) sputtering system declines with time. The target bias voltage considerably influenced the resistivity and deposition rate of AZO films. This work only involved the regulation of target bias of r.f. magnetron sputtering system via a direct current (d.c.) bias power. Deposition rate and electrical properties of AZO films were increased through modulation bias voltage of target.

Key words: AZO, electrical property, optical property, r.f. sputtering, target bias voltage.

I. INTRODUCTION

Low resistivity and high optical transmittance make AZO films relatively suitable as a candidate for transparent conductive oxide (TCO) layers in solar cells, such as CdTe and Cu(InGa)Se$_2$. For example, a CdTe solar cell with efficiency of 14% was obtained by employing AZO as front electrode in 2004 [1]. Among deposition methods of AZO, radio frequency (r.f.) magnetron sputtering ceramic target is widely used. The influence of substrate temperature [2], r.f. power [3,4], deposition pressure [4], oxygen partial pressure [5], target–substrate distance [6] and film thickness [7] on the properties of AZO films has been widely studied. However, the effect of target bias voltage is rarely studied. In this work, AZO polycrystalline thin films were deposited by r.f. magnetron sputtering. The target d.c. bias voltage was controlled by a d.c. electrical power. The effect of target bias on the AZO thin film properties was studied.

II. Main experimental

AZO thin films were deposited on Corning 7059 glass substrate through r.f. magnetron sputtering ZnO:Al (2 wt.% Al$_2$O$_3$) ceramic target. During the growth process, the target d.c. bias voltage was controlled and maintained via a d.c. power source at constant current mode. The d.c. power source was connected to the sputtering gun through an r.f. filter parallel with the r.f. This system only works when the d.c. source generates a bias larger than the self-bias of the target with r.f. alone.

III. Results and discussion

A. Electronic properties

Figure 1 shows the relationship between resistivity and target bias. Only one data point was recorded for 5.5 mm thick target (new target). When the r.f. power density of 3.94 W/cm^2 was used for the thick target, AZO film resistivity was low at approximately 5.62×10^{-4} Ω cm with self-bias of −112 V. However, film resistivity increased to 7.73×10^{-4} Ω cm when the target was thin at 2.5 mm (used target), and self-bias decreased to −86 V. When the bias was regulated to −112 V by the d.c. source, similar to the self-bias of the thick target, resistivity of the as-deposited film recovered to the same value. For the case of lower r.f. power density of 2.19 W/cm^2 on the thick target, the corresponding resistivity of AZO film was 9.82×10^{-4} Ω cm with the target self-bias of −72 V. This value also decreased with the increase of target bias. When sputtered with the same target bias voltage (between -86V and -112V in this work), the resistivity of AZO showed similar resistivity despite the different r.f. power densities (Figure 1). Thus, the resistivity of AZO was mainly affected by the target bias voltage.

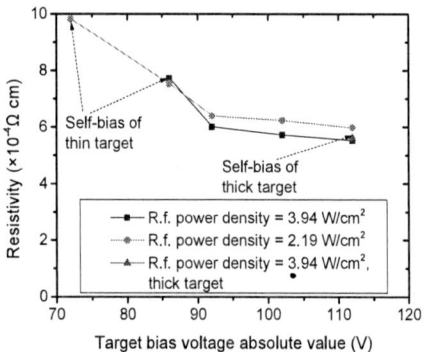

Fig. 1. AZO resistivity versus the absolute value of target bias voltage.

---

* Corresponding author: Tel.: +86-10-8254-7044; Fax: +86-10-8254-7041.
E-mail: shinelu@mail.iee.ac.cn (Xiangxin Liu)

Figure 2 demonstrates the variation in carrier concentration (a) and carrier mobility (b) of AZO films with target bias. The carrier concentration increased with target bias and peaked at ~4.2 × 10^{20} /cm^3 for the target bias of −112V, independent of target thickness and r.f. power density. Meanwhile, the carrier mobility also increased with target bias, peaked at 24 cm^2/(Vs) for around −90 V and slightly decreased thereafter. In general, the resistivity of AZO film was controlled by d.c. voltage applied to the ceramic target.

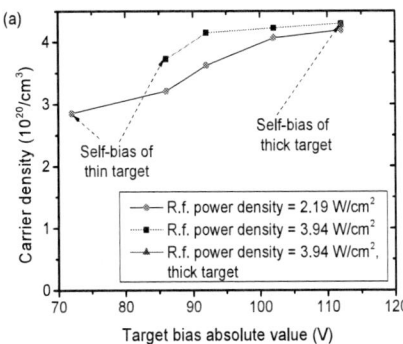

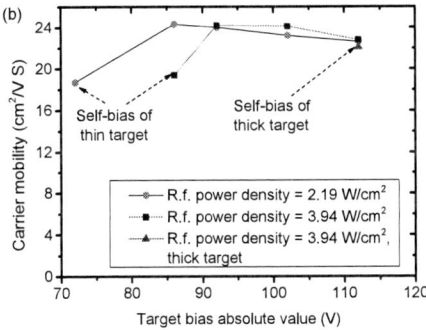

Fig. 2. Carrier concentration (a) and carrier mobility (b) of AZO films

Figure 3 shows the aluminum ratio and oxygen ratio of AZO films deposited via sputtering thin target. The Al ratio and O ratios of AZO film deposited by sputtering thick target were 0.94% and 40.42%, respectively. Al atom ratio difference was small and showed no increase with the absolute value of target bias. But, as described previously, the carrier concentration of AZO films increased with the absolute value of target bias. This difference can be ascribed to increase in effective Al substitution and Zn interstitial doping. When the target bias increased, more Al atoms could have enough kinetic energy to penetrate into the film. Consequently, effective Al doping in ZnO film will be achieved when target bias increased. Interstitial Zn is also a kind of shallow level donor in the ZnO films [8]. It can be formed by sputtering [9], and when target bias increased the Zn interstitial concentration might increase.

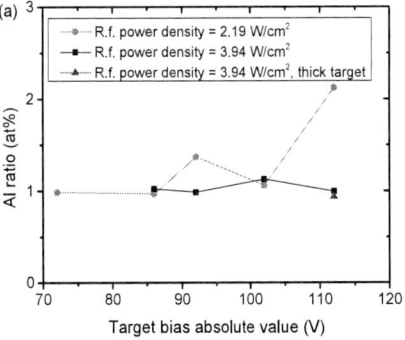

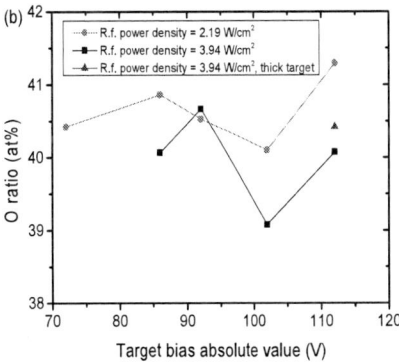

Fig. 3 The aluminum ratio and oxygen ratio of AZO films deposited via sputtering thin target.

B. Deposition rate

Figure 4 shows that the growth rate depends on the target bias voltage in r.f. sputtering. For r.f. power density of 2.19 W/cm^2, the growth rate increased from 10.54 nm/min to 25.14 nm/min when the target bias increased from −72 V (self-bias) to −112 V. This growth rate (25.14 nm/min) was equivalent to the one (24.81 nm/min) with r.f. power density of 3.94 W/cm^2 at self-bias. For r.f. power density of 3.94 W/cm^2, the growth rate increased from 24.81 nm/min to 46.33 nm/min when the target bias increased from −86 V (self-bias) to −112 V.

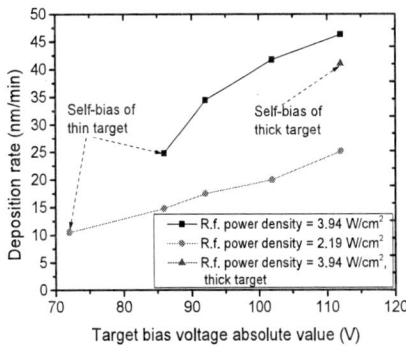

a)

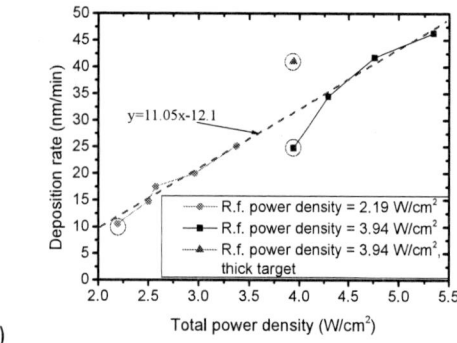

b)

Fig. 4. a) Relationship between AZO growth rate and target bias. b) Relationship between groth rate and total power density of r.f. and d.c.

Both r.f. power and d.c. power were consumed, and the deposition rate of AZO film when sputtered with a specific total power density (sum of r.f. and d.c.) could be predicted as a linear function. From this perspective, the growth rate could be maintained with relatively low r.f. power source. This strategy could reduce the requirements of costly high-power r.f. source in manufacturing line of sputtering AZO, therefore decreasing manufacturing cost.

C. Optical properties

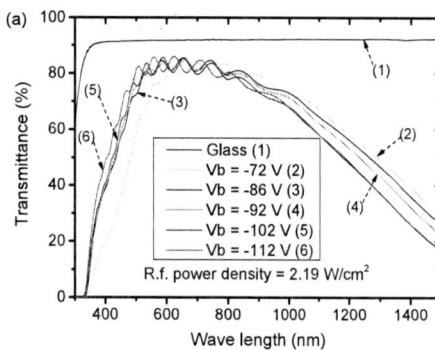

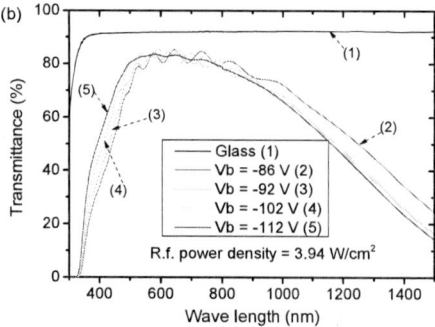

Fig. 5. Transmittance spectra of AZO films deposited with

different target bias voltages. (a) Transmittance of films deposited with r.f. power density of 2.19 W/cm^2, (b) transmittance of films deposited with r.f. power density of 3.94 W/cm^2.

Figure 5 shows the transmittance spectra of AZO films deposited on 7059 glass, in which 'V$_b$' means target bias voltage. These spectra were measured with 7059 glass substrate at a spectral range of 300–1500 nm. The optical transmittance of AZO films in shortwave range increased with target bias. This will be beneficial to solar cell applications in reducing optical loss in shortwave region.

As an example, $(hv\alpha)^2$ versus (hv) of AZO films deposited under power density of 2.19 W/cm^2 is plotted in Figure 6, in which intercepts of tangent lines show E$_g$ of AZO films.

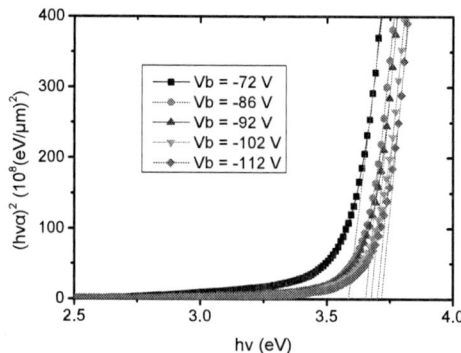

Fig. 6. Optical band gap of AZO films deposited under r.f. power density of 2.19 W/cm^2 with different target d.c. bias voltages.

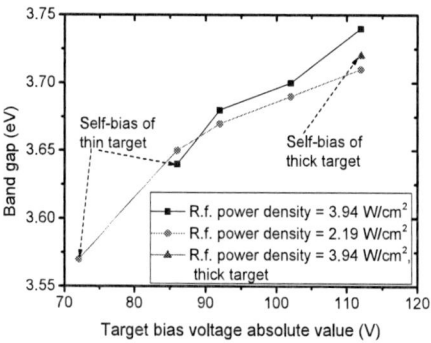

Fig. 7. Optical band gap of AZO versus absolute value of target bias voltage.

The optical band gaps of AZO films were calculated and depicted in Figure 7. For r.f. power density of 2.19 W/cm^2, E$_g$ of AZO films increased continuously from 3.57 eV to 3.71 eV when the target bias increased from −72 V to −112 V. For 3.94 W/cm^2, E$_g$ increased from 3.64 eV to 3.74 eV when the target bias increased from −86 V to −112 V.

However, at the same target bias, E_g difference among AZO films deposited through sputtering with r.f. power densities of 2.19 and 3.94 W/cm^2 is minimal. Therefore, target bias could be used to regulate the optical properties of r.f. sputtered AZO film.

D. Morphologies

Figure 8 shows the typical two-dimensional (2D) and three-dimensional (3D) AFM images of an AZO film. The surface morphologies of AZO films different slightly. Analysis software *NanoScope* Analysis demonstrated that the height difference in the measured surface (2 μm × 2 μm surface area) of AZO films was below 45 nm, the peak depth histogram of all films was around 20 nm and the diameter of stand-alone cylindrical little crystal grains was approximately 23 nm. Some films showed large grains clustered by stand-alone cylindrical crystal grains. In general, the RMS of AZO films were different slightly. Therefore, in our sputtering configuration, target bias voltage could not affect the morphologies of AZO films, and the roughness of AZO film surface could not be changed via regulating target bias voltage. In a mass production system, low morphology variation during regulation of deposition parameters may be beneficial to the uniformity and reproducibility of deposited films.

Fig. 8. 2D AFM image (left) and 3D (right) AFM image of AZO film sputtered at target bias voltage of −92 V and r.f power density of 2.19 W/cm^2.

IV. Conclusion

Growth rate, conductivity and optical bandgap of AZO films could be regulated by coupling a d.c. power with r.f. power to the sputtering target. This is particularly important for manufacturing, because these properties of conductive oxide films are always vary with the erosion depth of sputtering target, causing stability problems for film quality. Coupling d.c. bias with r.f. provides a simple and cost-effective method to maintain manufacturing stability.

References

[1] Akhlesh Gupta, and Alvin D. Compaan. "All-sputtered 14 % CdS/CdTe thin-film solar cell with ZnO:Al transparent conducting oxide," Applied Physics Letters, vol.85, pp. 684-686, 2004.
[2] Zhiyun Zhang, Chonggao Bao, Wenjing Yao, Shengqiang Ma, Lili Zhang, and Shuzeng Hou, "Influence of deposition temperature on the crystallinity of Al-doped ZnO thin films at glass substrates prepared by RF magnetron sputtering method," Superlattices and Microstructures, vol. 49, pp. 644–653, 2011.
[3] Y.M. Lu, W.S. Hwang, W.Y. Liu, and J.S. Yang, "Effect of RF power on optical and electrical properties of ZnO thin film by magnetron sputtering," Materials Chemistry and Physics, vol. 72, pp. 269–272, 2001.
[4] BoenHoung, Chi ShiungHsi, Bing Yi Hou, and Shen Li Fu, "Effect of process parameters on the growth and properties of impurity-doped zinc oxide transparent conducting thin films by RF magnetron sputtering," Vacuum, vol. 83, pp. 534–539, 2009.
[5] Takashi Tsuji, and Mitsuji Hirohashi, "Influence of oxygen partial pressure on transparency and conductivity of RF sputtered Al-doped ZnO thin films," Applied Surface Science, vol. 157, pp. 47–51, 2000.
[6] Luo Chen, and Xiaofang Bi. "Variations of microstructure, conductivity and transparency of Al-doped ZnO thin films prepared by radio frequency magnetron sputtering with target–substrate distances," Vacuum, vol. 82, pp. 1216–1219, 2008.
[7] Kang HyonRi, Yunbo Wang, Wen Li Zhou, Jun Xiong Gao, Xiao Jing Wang, and Jun Yu, "The structural properties of Al doped ZnO films depending on the thickness and their effect on the electrical properties," Applied Surface Science, vol. 258, pp. 1283–1289, 2011.
[8] J. R. Look, J. W. Hemsky, and J. R. Sizelove, "Residual native shallow donor in ZnO," Physical Review Letters, vol. 82, pp. 2552–2555, 1999.
[9] W Chamorro, D Horwat, P Pigeat, P Miska, S Migot, F Soldera, P Boulet, and F Mücklich, "Near-room temperature single-domain epitaxy of reactively sputtered ZnO films," Journal of Physics D: Applied Physics, vol. 46, pp. 235107, 2013.

Close-Space Sublimated CdTe Solar Cells with Co-Sputtered CdS$_x$Se$_{1-x}$ Alloy Window Layers

Corey R. Grice[1], Maxwell Junda[1], Alexander Archer[1], Jian Li[2], and Yanfa Yan[1]

[1]The University of Toledo, Toledo, Ohio, 43607, USA,

[2]Texas State University, San Marcos, Texas, 78666, USA

Abstract — **Thin film CdTe solar cells devices are fabricated using homogenous alloys of CdS-CdSe, deposited by a co-sputter process and of varying compositions, as the window layer materials. After close-space sublimation of the CdTe layer and subsequent device finishing processes, the cells are characterized in terms of electo-optical performance as well as structural and compositional properties.**

Index Terms — **CdTe solar cells, CdSe, CdS, window layer.**

I. INTRODUCTION

CdS has long been used as an n-type heterojunction partner for thin film CdTe solar cells[1-4] while also serving as a window layer that allows sufficient visible light to pass through and generate electron-hole pairs within the CdTe layer. Unfortunately, the transmission characteristics of pure CdS are not ideal for this purpose and a non-trivial amount of light is absorbed by this window layer, thus decreasing the maximum possible photocurrent generated within the CdTe layer. One interesting approach to increasing the device photocurrent has been to use CdSe as a window layer material [5]. CdSe has a lower bandgap than CdS and higher absorption coefficients in the visible wavelength range, which would initially suggest that it would serve as a poorer window layer substitute for CdS. However, CdSe is also known to have a high degree of interdiffusion with CdTe at high process temperatures, which allows for CdSe layers of sufficient thinness to be completely consumed by the CdTe during high temperature (over 600°C) depositions. The subsequent alloying of Se into the CdTe layer slightly decreases the bandgap of the CdTe, thus increasing the absorption of lower energy photons and enhancing the total photogenerated current in those devices. Additionally, the consumption of the CdSe layer via diffusion into the CdTe layer reduces the effective thickness of the window layer as well as the amount of light absorbed by this layer. The two effects combine to increase the photocurrent generated by the CdTe device.

Initial studies [5,6] on this type of device configuration have provided insight into the optimum thickness of CdSe to use for their construction, as well as supporting the use of an ultra-thin CdS window layer in conjunction with the CdSe layer, while subsequent studies [7,8] provided some additional understanding of the changes occurring within the CdTe layer

and, to a lesser extent, at the CdS/CdSe/CdTe interface. More recently, some groups have begun exploring Se incorporation by depositing CdTe$_x$Se$_{1-x}$ alloy layers directly onto the device window layers, followed by a pure CdTe layer afterwards, though this technique increases the complexity of the device fabrication process [9]. Another alternative approach to using bilayer windows of CdS/CdSe is to directly incorporate the Se into the CdS window layer. Some groups have already attempted this by depositing alternating ultra-thin layers of CdS and CdSe sequentially and then relying on the device processing temperatures to fuse them into an alloyed material [10], although there is uncertainty about the extent of the interdiffusion between layers of CdS and CdSe during CSS CdTe processing [11]. As such, these studies raise questions about the influence of CdS at the window layer with respect to Se diffusion into the CdTe layer.

This work is intended to provide a more definitive understanding of the influence that window layers of homogeneous CdS$_x$Se$_{1-x}$ alloys have upon the performance characteristics of CdTe solar cells. To begin, layers of CdS$_x$Se$_{1-x}$ alloys are prepared by co-sputtering from pure CdS and CdSe material sources, with x ranging from 0 to 1. These films are characterized in their as-deposited state in terms of optical and structural properties. CdTe solar cells are then prepared for each window layer (keeping the layer thickness constant at 100nm), with ~ 4 μm of CdTe deposited using a close-space sublimation (CSS) process followed by standard cell finishing techniques. The resulting devices are evaluated in terms of their performance characteristics using current-voltage (I-V) and external quantum efficiency (EQE) measurements. The device structure and composition is also evaluated using cross-sectional microscopy and energy dispersive x-ray spectroscopy (EDS). The results of this study provide insight into how the various window layer compositions affect the CdTe film near that interface as well as their effect upon overall device performance.

II. EXPERIMENTAL DETAILS

Commercially available soda-lime glass (Pilkington North America Inc.), both with pre-applied SnO$_2$/SnO$_2$:F layers

(TEC-12M) and without them, were used as substrates. These were treated using a detergent-based (Micro-90, International Products Corp.) cleaning procedure assisted by ultra-sonication to ensure chemically clean surfaces. CdS and CdSe were deposited using an RF magnetron sputtering system (AJA International) using between 10-60 W of RF power for each target (2 inch diameter) at 13.56 MHz. The deposition times ranged from 15-30 minutes for the samples deposited on bare soda-lime glass, while the films deposited on TEC-12M were typically done in 3-10 minutes. The ambient environment in the sputter chamber was maintained at a pressure of 10 mTorr with a constant gas flow rate of 23 sccm of argon with the substrates maintained at 270°C. Samples were allowed to cool to ambient temperature within the chamber before transferring to the CdTe deposition system.

Deposition of CdTe was performed using a custom built fused-silica tube furnace system [12], with a 3 mm separation distance between CdTe source material and the substrate. The furnace contained an environment of 0.5% O_2 by volume, balance He, which was maintained at a pressure of approximately 450 Torr during heating. Once the source material had reached a temperature of 660 °C, with the substrate at approximately 607 °C, the pressure within the furnace was decreased to 50 Torr for the duration of the deposition. After 80 seconds, the pressure was restored to the previous value to end the deposition process and the system was allowed to cool naturally to ambient temperature.

Sample surfaces were then treated with a saturated $CdCl_2$/methanol solution for activation annealing, which was performed in a similar tube furnace system at 390°C for 30 minutes, with ambient pressure dry air provided at approximately 5 scfh. After cooling to ambient temperature, the sample surfaces were rinsed with methanol to remove residual $CdCl_2$ and back contacts of Cu and Au metal bi-layer disks were thermally evaporated, with areas of 0.08 cm^2 defined using shadow masks and each sample having between 70 and 140 devices defined in this fashion. A subsequent annealing step at 200 °C for 20 minutes in a N_2 environment was performed to facilitate Cu diffuse at the device back contact region.

Optical characterization of the CdS_xSe_{1-x} films was accomplished using UV-vis transmission spectroscopy. Structural characteristics were measured using x-ray diffraction (XRD) and atomic force microscopy (AFM). Thicknesses were determined using spectroscopic ellipsometry (SE). Compositional measurements were obtained using EDS.

Device characterization was done by measuring I-V responses in near-dark and under AM 1.5G illumination. External quantum efficiencies were also measured at 0mV bias potentials. Additionally, transmission electron microscope (TEM) samples were prepared using Ga focused ion beam (FIB) milling. Cross sectional images were obtained from these samples as well as elemental line profiles using energy dispersive x-ray spectroscopy (EDS).

III. RESULTS AND DISCUSSIONS

A. CdS_xS_{1-x} Film Characterization

In order to evaluate the bulk properties of the co-sputtered CdS_xSe_{1-x} alloys, depositions in excess of 15 minutes were performed on bare glass substrates. These "thick" samples ranged between 300-600 nm in thickness, which is several times the typical window layer thickness of CdS/CdSe films used in most CdTe solar cells (50-100nm). This was done to ensure that the intrinsic properties of the material could be determined without being limited by the film thickness.

EDS measurements, of these films were conducted to determine the effective elemental composition of each film, with each CdS_xSe_{1-x} film subsequently described by the atomic basis x-parameter. The values of x (between 0 and 1) reported are the average of five measurements (~ 500 μm x 500 μm areas) at different locations across the film surface, the standard deviations of which were all below 0.7% at., thus confirming that films were highly uniform across the each sample.

The resulting transmission curves for the visible spectral range are shown in Figure 1, with the absorption edge clearly moving from short to long wavelengths (red-shifting) as the value of x decreases (becoming more Se-rich). This is expected since the bandgap of CdS is 2.4eV while CdSe is 1.7eV and the alloys are not known to experience sufficient "bowing effects" which would cause them to have bandgaps outside of this range. This has been observed for CdS_xSe_{1-x} produced by other deposition methods, such as chemical bath deposition and evaporation [13-15], although reports for sputtered alloys could not be found.

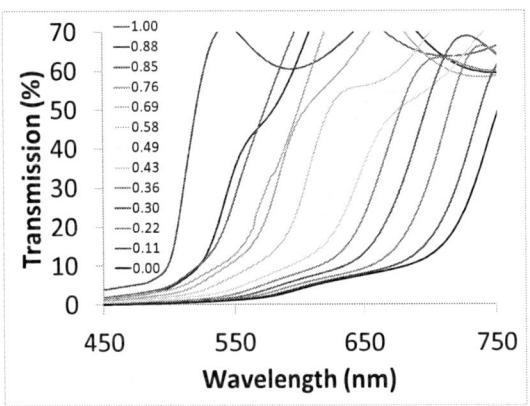

Fig. 1. Optical transmission measurements (not reflection-corrected) of "thick" CdS_xSe_{1-x} alloy films deposited on soda-lime glass with x values ranging from 0 to 1.

XRD measurements were taken from each "thick" film deposited on bare glass using a glancing incidence (2°) arrangement to minimize signal contributions from the substrate. The resulting diffraction patterns, shown in Figure

2, reveal that all of the samples show the same crystal structure (hexagonal P63mc, space group 186), with the lattice parameter increasing from a = 4.14 Å for pure CdS (x = 1) to a = 4.30Å for pure CdSe (x = 0) while the effective crystallite size decreased from 25nm to 15nm. Four example AFM surface images are shown in Figure 3 to demonstrate the surface morphology and roughness. From the images it can be seen that films consist of irregularly shaped grains, ranging in size from 20-250nm. This is roughly consistent with the crystallite size obtained from the XRD measurements if a relatively high defect density is present within the observed grains, which would likely have an adverse effect on the electrical properties of the materials. The film roughness for all samples ranged between 4-20nm, although the films with nearly equal proportions of S and Se had the highest roughness and also the largest grains.

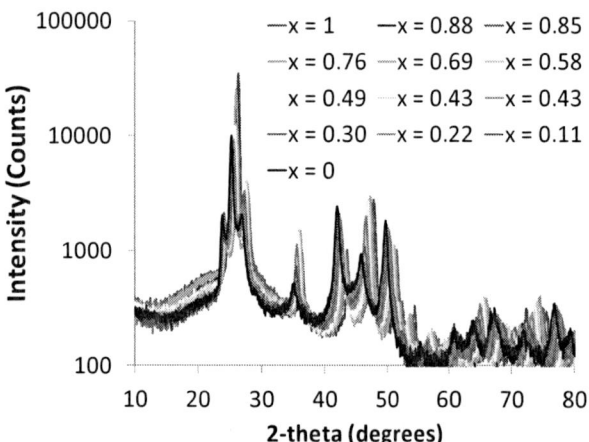

Fig. 2. X-ray diffraction patterns of "thick" CdS_xSe_{1-x} alloy films deposited on soda-lime glass with x values ranging from 0 to 1.

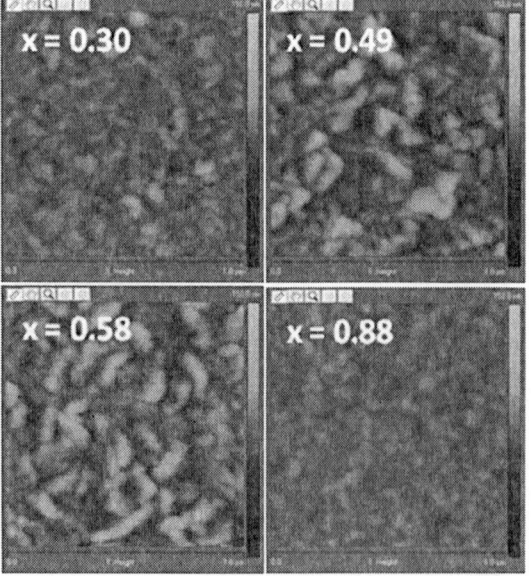

Fig. 3. Surface profile AFM images of select co-sputtered CdS_xSe_{1-x} films of compositions x = 0.30, 0.49, 0.58 and 0.88.

B. CdTe Device Characterization

Thin film CdTe solar cells fabricated using CdS_xSe_{1-x} window layers would intuitively be expected to exhibit a significant variation in their photocurrents as a function of the window layer composition due to the observed variance in the optical bandgap. Since the bandgap decreases monotonically with increasing Se content, the photocurrent would also be expected to decrease in a similar fashion. However, the increasing availability of Se within the window layer should permit enhanced Se diffusion into the CdTe layer, so there should also be a competing influence of increasing photocurrent with increased Se content due to reduced absorber layer bandgap ($CdTe_xSe_{1-x}$ alloying), increased window layer bandgap (becoming S-rich and Se-poor), and also from thickness reduction of the window layer (due to partial consumption by the CdTe absorber layer). This balance between increasing and decreasing photocurrent effects was observed from the short circuit current density (J_{sc}) values obtained from the I-V polarization curves of the devices and the integrated EQE curves, as shown in Figure 4. The individual spectral response curves from the EQE measurements are shown in Figure 5. This data clearly shows that the devices produced using the CdS_xSe_{1-x} alloy layers show decreased photocurrent compared to devices made with either pure CdS or CdSe window layers, both in terms of total photocurrent and at all wavelengths across the visible range (~ 400 – 800 nm), with the greatest reduction in photocurrent occurring at near-equal proportions of S and Se in the window layer. The decrease in photocurrent in the blue-green visible range (400 – 600 nm) suggests that a significant portion of those photons are being absorbed either in the window layer or within a region of S-alloyed CdTe (or $CdTe_xSe_{1-x}$) near the window layer and not producing photocurrent. An expanded view of the red and infrared portion of the EQE response curves, shown in Figure 6, do indicate that the degree of Se alloying into the CdTe absorber layer increases with increasing Se content of the window layer, as evidenced by the red-shift of the device absorption onset between ~ 840 nm and 920 nm. However, it is likely that the degree of Se diffusion from the window layers into the CdTe is insufficient to reduce the optical losses induced by the relatively low bandgaps of the CdS_xSe_{1-x} window layers.

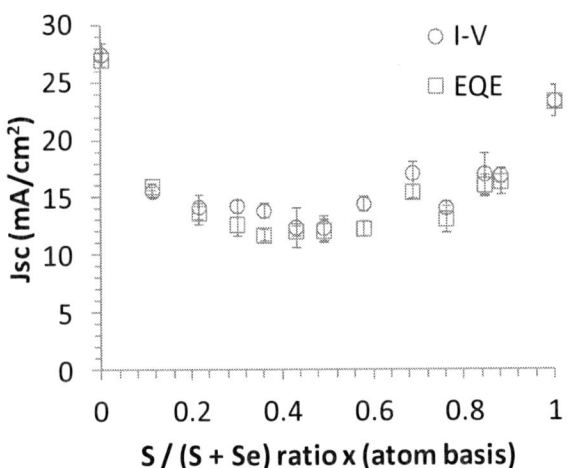

Fig. 4. Short-circuit current density (Jsc) values obtained from CdTe solar cell devices made using CdS_xSe_{1-x} window layers. Data points represent the average of the 50 best performing devices for each sample, with standard deviation error bars.

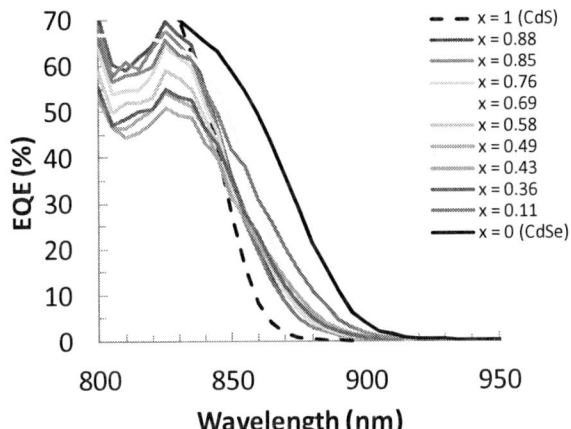

Fig. 6. Expanded view of the long-wavelength range of the external quantum efficiency response (at 0mV bias potential in near-dark) of CdTe devices made with various CdSxSe1-x window layers.

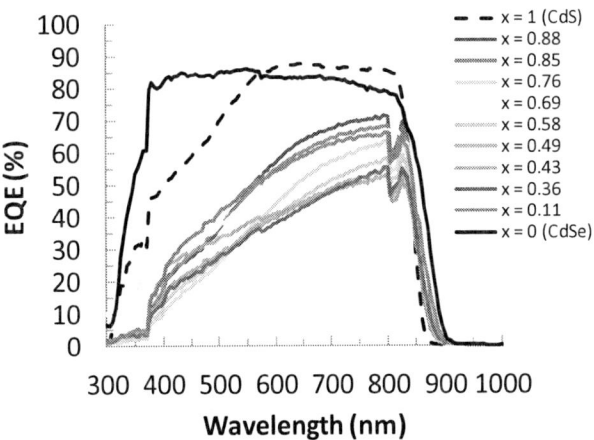

Fig. 5. External quantum efficiency response (at 0mV bias potential in near-dark) of CdTe devices made with various CdSxSe1-x window layers.

To gain further understanding of these photocurrent responses with respect to the evolution of the various CdS_xSe_{1-x} during device processing, cross-sectional TEM microscopy was performed for each window layer composition, coupled with EDS compositional line profiling to assess the relative distributions of S, Se and Te throughout the fabricated devices.

Figure 6 shows the TEM cross-sectional image and EDS compositional profile for the sample prepared with a pure CdS window layer (x = 1), while Figure 7 shows data for a sample prepared with pure a CdSe window layer (x = 0). Figure 6 shows a clearly discernible CdS layer in the TEM image, while the EDS profiles confirms that the sulfur appears to be strongly confined to the region near the front interface, although a small amount (< 5% at.) is detectable within the first ~ 300 nm of the CdTe layer. Figure 7 does not show any discernible CdSe layer in the TEM image, while the EDS data reveals that the Se content at the front contact of the device (at FTO interface) is approximately 7% (atom basis). The Se content proceeds to decline gradually into the bulk CdTe layer, ultimately becoming too low to measure at a depth of ~1.1μm from the FTO contact.

Cross sectional information from some of the devices made from the CdS_xSe_{1-x} alloy window layers are provided in Figure 8. The TEM images indicate that a distinct window layer remains near the FTO interface for many of the sulfur-rich alloy compositions (x > 0.5), but this layer becomes thinner and increasingly non-uniform in thickness as the sulfur content decreases. The EDS profiles reveal that the devices with x > 0.5 have very strong concentration of sulfur at the FTO interface, while those with x < 0.5 show relatively low sulfur concentrations immediately at the FTO interface compared to further (~ 100 nm) into the CdTe layer.

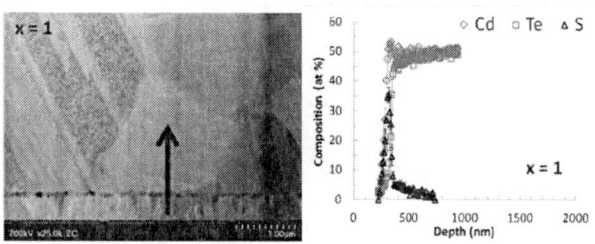

Fig. 6. TEM cross-sectional image (left) and EDS compositional profile (right) for CdTe device made with 100nm CdS window layer (x = 0). EDS data represent average of 3 to 5 separate line scans with standard deviation error bars. Arrow in cross-section image indicates approximate location and direction of EDS line scan.

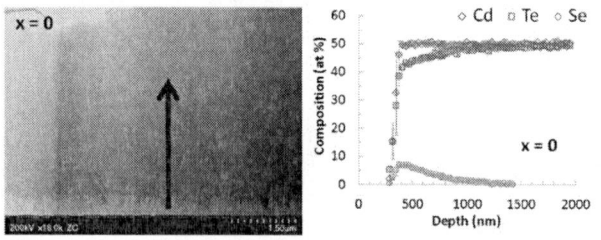

Fig. 7. TEM cross-sectional image (left) and EDS compositional profile (right) for CdTe device made with 100nm CdSe window layer (x = 1). EDS data represent average of 3 to 5 separate line scans with standard deviation error bars. Arrow in cross-section image indicates approximate location and direction of EDS line scan.

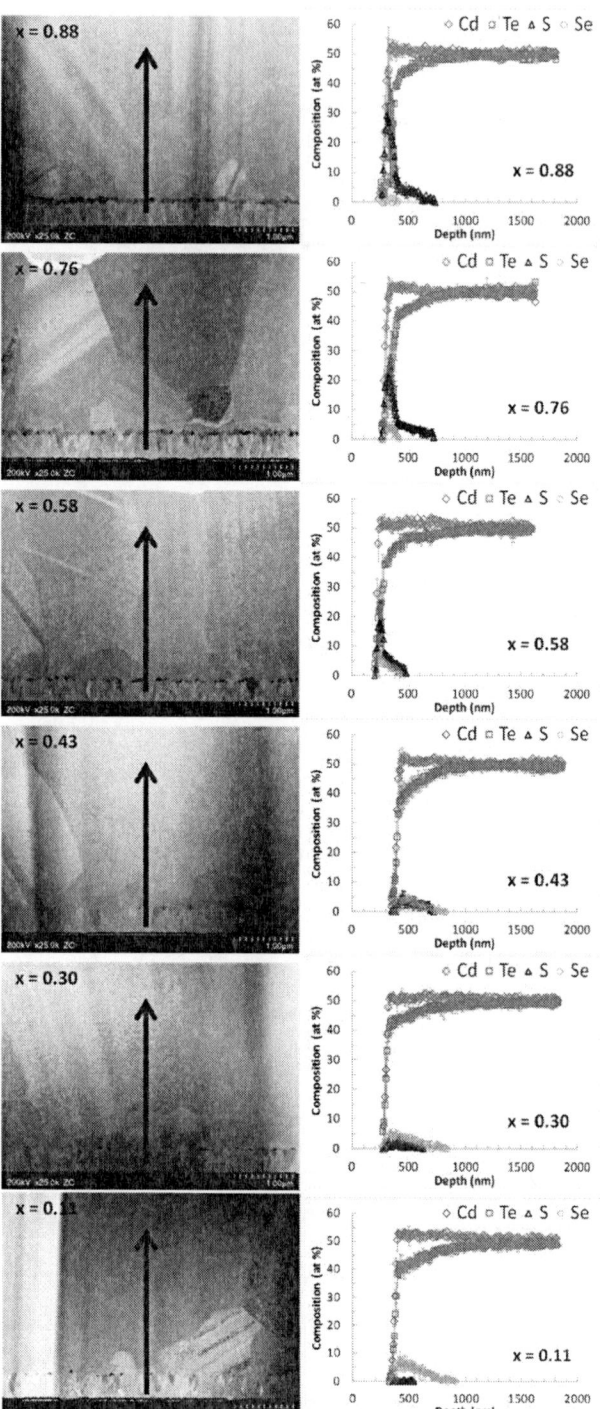

Fig. 8. TEM cross-sectional images (left) and EDS compositional profiles (right) for CdTe device made with 100nm window layers of various CdS_xSe_{1-x} alloys. EDS data represent average of 3 to 5 separate line scans with standard deviation error bars. Arrow in cross-section image indicates approximate location and direction of EDS line scan.

The Se profiles observed from these samples (both using pure and alloyed window layers) are rather different than what has been previously observed from devices made using a 17nm CdS/ 100nm CdSe bilayer window structure [8], where the residual Se content near the front interface was much higher (~20%, atom basis) as shown in Figure 9. Even though both the bilayer and x = 0 samples had a layer of 100nm pure CdSe as the exposed surface during the CdTe deposition, clearly the presence of CdS at the opposite side of the CdSe layer had a noticeable mitigating effect upon the degree of Se diffusion into the CdTe layer. This is also evident in the CdS_xSe_{1-x} alloy layer samples, where the presence of sulfur as even a minor component of the window layer significantly reduces the depth of the diffusion of Se into the bulk of the CdTe layer. These findings ultimately suggest that the deliberate incorporation of a small amount of CdS can be used to dramatically control the subsequent diffusion of Se into a CdTe absorber layer in thin film solar cells.

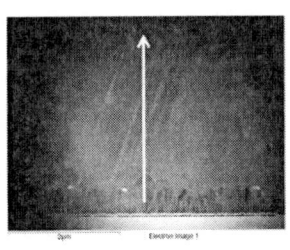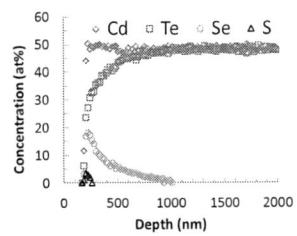

Fig. 9. TEM cross-sectional image (left) and EDS compositional profile (right) for CdTe device made with 17nm CdS / 100nm CdSe window bilayer (adapted from reference [8]). Arrow in cross-section image indicates approximate location and direction of EDS line scan.

IV. SUMMARY AND CONCLUSIONS

The results of this study provide additional insight into how the various window layer compositions containing both CdS and CdSe can affect the CdTe film, both the elemental composition near the front interface and into the bulk, as well as their effect upon overall device performance. The use of homogenous window layers of CdS_xSe_{1-x} alloys appear to have a net detrimental effect upon the photocurrent produced within the CdTe solar cells, most likely due to reduced bandgap of the residual window layers (resulting from the inhibited diffusion of Se from the alloy layer into the CdTe by the presence of S), thus reducing the amount of visible light reaching the CdTe layer. Also, the combined effects of interdiffusion of CdS, CdSe, and CdTe near the interface, can create a region of highly defective material that can increase carrier recombination within the device.

ACKNOWLEDGEMENTS

The authors would like to extend our acknowledgement to Dr. David Strickler from NSG, North America for supplying us FTO coated substrates. This work is partially supported by the U.S. Department of Energy F-PACE program and through Sunshot PVRD grant DE-EE-0007541.

REFERENCES

[1] J. Britt and C. Ferekides, "Thin film CdS/CdTe solar cell with 15.8% efficiency," *Applied Physics Letters*, vol. 62, pp. 2851-2852, 1993.

[2] X. Wu, "High-efficiency polycrystalline CdTe thin-film solar cells," *Solar Energy*, vol. 77, pp. 803-814, 2004.

[3] A. Bosio, A. Romeo, D. Menossi, S. Mazzamuto, and N. Romeo, "Review: The second-generation of CdTe and CuInGaSe(2) thin film PV modules," *Crystal Research and Technology*, vol. 46, iss. 8, pp. 857-864, 2011.

[4] N. R. Paudel, C. R. Grice, C. Xiao, and Y. Yan, "The effects of high temperature processing on the structural and optical properties of oxygenated CdS window layers in CdTe solar cells," *Journal of Applied Physics*, vol. 116, 044506, 2014.

[5] N. R. Paudel and Y. Yan, "Enhancing the photo-currents of CdTe thin-film solar cells in both short and long wavelength regions," *Applied Physics Letters*, vol. 105, iss. 18, p. 5, 2014.

[6] X. Y. Yang, Z. Bao, R. Luo, B. Liu, P. Tang, B. Li, J. Q. Zhang, W. Li, L. L. Wu and L. H. Feng, "Preparation and characterization of pulsed laser deposited CdS/CdSe bi-layer films for CdTe solar cell application," *Materials Science in Semiconductor Processing*, vol. 48, pp. 27-32, 2016.

[7] N. R. Paudel, J. D. Poplawsky, K. L. Moore, and Y. Yan, "Current Enhancement of CdTe-Based Solar Cells ," *IEEE Journal of Photovoltaics*, vol. , iss. 5, pp. 1492-1496, 2015.

[8] C. Grice, A. Archer, S. Basnet, N. R. Paudel, and Y. Yan, "Characterization of CdS/CdSe window layers in CdTe thin film solar cells," in 43rd IEEE Photovoltaic Specialist Conference, 2016, pp. 1459-1463.

[9] D. E. Swanson, J. R. Sites, and W. S. Sampath, "Co-sublimation of CdSexTe1-x layers for CdTe solar cells," *Solar Energy Materials and Solar Cells*, vol. 159, pp. 389-394, 2017.

[10] X. Y. Yang, B. Liu, B. Li, J. Q. Zhang, W. Li, L. L. Wu and L. H. Feng, "Preparation and characterization of pulsed laser deposited a novel CdS/CdSe composite window layer for CdTe thin film solar cell," *Applied Surface Science*, vol. 367, pp. 480-484, 2016.

[11] C.R. Grice, M. M. Junda, A. Archer, and Y. Yan, Manuscript in preparation.

[12] N. R. Paudel and Y. Yan, "Fabrication and characterization of high-efficiency CdTe-based thin-film solar cells on commercial SnO2:F-coated soda-lime glass substrates," *Thin Solid Films*, vol. 549, pp. 30-35, 2013.

[13] A. A. Al-Bassam and U. A. Elani, "Crystallographic Parameters and Band Gap of CdS x Se 1-x Mixed Crystals," *Journal of Energy and Power Engineering*, vol. 4, pp. 770-774, 2014.

[14] S. R. Deo, A. K. Singh, L. Deshmukh, N. P. Singh and M. P. Aleksandrova, "Studies on Structural, Morphological and Optical Properties of Chemically Deposited CdS1-xSex Thin Films," *Journal of Fluorescence*, vol. 26, iss. 2, pp. 459-469, 2016.

[15] K. Sivaramamoorthy, S. A. Bahadur, M. Kottaisamy and K. R. Murali, "Structural, optical and photoconductive properties of electron beam evaporated CdSxSe1-x films," *Crystal Research and Technology*, vol. 45, iss. 4, pp. 414-420, 2010.

Effects of graphene oxide barrier on $Cu_2ZnSnS_xSe_{4-x}$ thin film solar cells

Woo-Lim Jeong[1,2], Jung-Hong Min[1,2], In-Young Kim[1,2], Hae-Sun Kim[1,2], Jin-Hyeok Kim[3], and Dong-Seon Lee[1,2,*]

[1]School of Electrical Engineering and Computer Science, Gwangju Institute of Science and Technology, Gwangju 61005, South Korea

[2]Research Institute for Solar and Sustainable Energies, Gwangju 61005, South Korea

[3]Optoelectronics Convergence Research Center, Chonnam National University, 300 Youngbong-Dong, Puk-Gu, Gwangju 61186, South Korea

Abstract — We report the improved power conversion efficiency using graphene oxide (GO) barrier on $Cu_2ZnSnS_xSe_{4-x}$ (CZTSSe) solar cells. In order to verify the effect of the graphene oxide barrier, morphology was investigated for Mo substrate, reduced graphene oxide (rGO), GO, and thermally reduced GO (TrGO) by atomic force microscopy (AFM). The CZTSSe solar cell fabricated with the rGO substrate showed lower open-circuit voltage deficit than that of the reference cell resulting in the maximum conversion efficiency of 7.6%.

Index Terms — CZTSSe, graphene oxide, thin films, interfaces, solar cells.

I. INTRODUCTION

Among various photovoltaic devices, silicon solar cells and $Cu(In,Ga)Se_2$ (CIGS) solar cells are widely commercialized due to the high efficiency. Compared with the silicon and CIGS solar cells, $Cu_2ZnSnS_xSe_{4-x}$ (CZTSSe) solar cells have a relatively lower efficiency of 12.6% and there is still room for improvement.

Recently, interfacial optimization of the CZTSSe solar cells has been studied by many researchers as a way to resolve the open-circuit voltage deficit problem [1]. In addition, graphene oxide (GO) is actively discussed and studied as a contact material between CZTS and MoS layers due to the appropriate electrical properties [2]. In this study, we demonstrate the effects of the graphene oxide barrier using reduced graphene oxide (rGO), GO and thermally reduced graphene oxide (TrGO) between CZTSSe absorber layer and Mo back contact layer in order to improve electrical properties. From the CZTSSe solar cell with the rGO barrier, a maximum efficiency of 7.6% was obtained. The efficiency of CZTSSe solar cells with the rGO barrier is higher than that of the reference cell without the barrier.

II. EXPERIMENT

In the case of the reference cell, thin CZTSSe films and metal layers were deposited on molybdenum (Mo) coated soda lime glass (SLG) substrates. In the case of the GO cells, thin CZTSSe films and metal layers were deposited on GO/Mo/SLG substrates. The GO solution was prepared by mixing 30 mg of GO flake and 10 mL of deionized water. The rGO solution was prepared by chemical reduction of the GO solution. Then the GO and rGO solutions were spun one time and five times at 3000 rpm for 20 s, respectively. Additionally, TrGO sample was prepared by annealing of the GO coated Mo substrate at 600 °C for 300 s in a N_2 atmosphere.

The base pressure before Cu/Sn/Zn precursor deposition was less than 6.7×10^{-5} Pa. For the deposition of the Zn and Cu layers, DC magnetron sputtering was used. The Sn layers were deposited using a Sn metal-element target by RF magnetron sputtering. Under the 0.4 Pa pressure condition of Ar gas, CZT layers were deposited with 40 W (0.88 W/cm2) sputtering power. The precursors were sulfo-selenized in a graphite box with 0.5 g of Se powder and 0.02 g of SeS_2 powder. The sample was heated from room temperature to 300 °C over 1000 s and then maintained at 300 °C for 1500 s. Subsequently, it was heated from 300 °C to 550 °C over 1000 s and then maintained at 550 °C for 1100 s. These absorber films were KCN-etched (5 min in a 0.05 M aqueous solution) to remove Cu-Se binary phases, and then rinsed with deionized water. Right after the KCN-etching, a 70-nm CdS buffer layer was deposited by chemical bath deposition at 80 °C. Then 50 nm of intrinsic ZnO and 300 nm of aluminum-doped ZnO window-layer were deposited at room temperature and 150 °C, respectively. Lastly, 50 nm of Ni and 500 nm of Al top-grid were deposited through a shadow mask by electron beam evaporation.

978-1-5090-5606-4/17 $31.00 © 2017 IEEE

III. RESULTS AND DISCUSSION

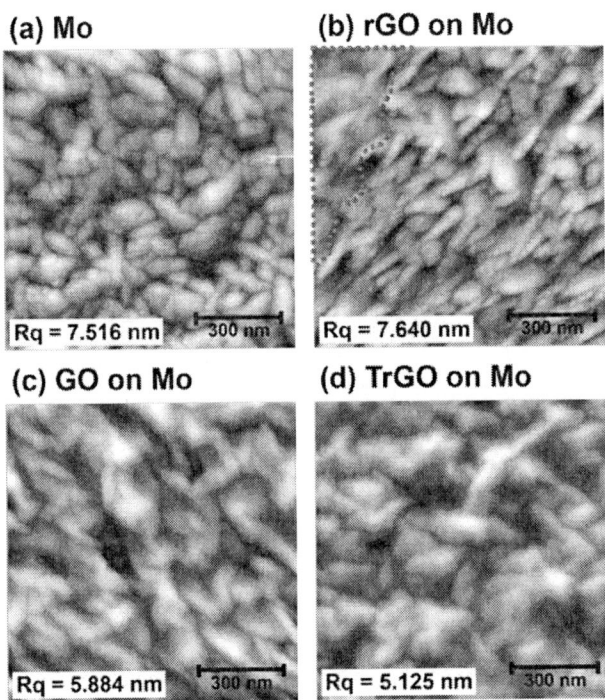

Fig. 1. AFM images and root mean square roughness of the (a) Mo substrate, (b) rGO on Mo substrate, (c) GO on Mo substrate, and (d) TrGO on Mo substrate.

Surface roughness and morphological characteristics of substrates were measured by atomic force microscopy (AFM, XE-100, Park systems), and their images and the root mean square roughness (Rq) values are shown in Fig. 1. The AFM image of the Mo substrate shows a rougher and agglomerated shape (Fig. 1 (a)). On the other hand, the AFM image of GO and TrGO on Mo substrate show a flatter surface due to the covering of the GO and TrGO (Fig. 1 (c) and (d)). In the case of the rGO, as seen in Fig. 1. (b), it is only partially covered by rGO (red dotted line). These shape of the surface may affect the electrical properties of the substrates.

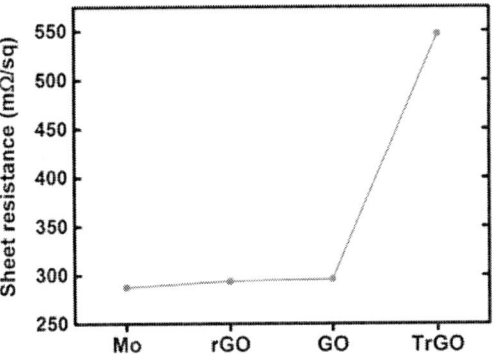

Fig. 2. Sheet resistance of the Mo, rGO, GO and TrGO substrates measured by the 4-point probe.

In order to verify the electrical properties of the substrates, sheet resistance of GO substrates was measured by 4-point probe at room temperature. Because of the conductive material molybdenum, Mo substrate showed the lowest sheet resistance as seen in Fig. 2. Following Mo substrate, rGO and GO substrate showed a slightly higher sheet resistance than that of the Mo substrate. However, TrGO showed the very high sheet resistance than that of other substrates. The high sheet resistance may be due to the high temperature (600 °C) of the reducing process of the TrGO.

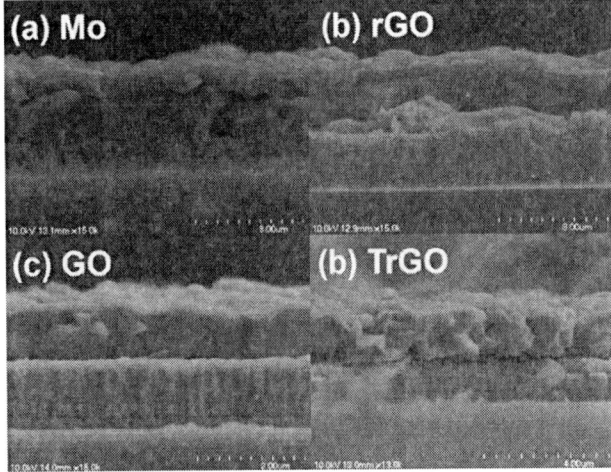

Fig. 3. Cross-sectional SEM images of the CZTSSe solar cells with (a) Mo substrate, (b) rGO substrate, (c) GO substrate, and (d) TrGO substrate.

The CZTSSe absorber layers were fabricated by sulfo-selenization of metallic precursors with the Mo and GO substrates. The same CdS buffer layer, iZO, AZO window layer and Ni/Al grid were deposited on the CZTSSe absorber layers. The cross-sectional SEM images of the completed cells

TABLE I
DEVICE CHARACTERISTICS OF THE CZTSSE SOLAR CELLS.

Cell Name	Efficiency (%)	V_{OC} (mV)	J_{SC} (mA/cm^2)	FF (%)	R_{sh} (Ωcm^2)	R_s (Ωcm^2)	E_g (eV)	V_{OC} deficit (V)
Reference cell	5.8	322.7	38.7	46.2	100	9.3	1.02	0.70
rGO barrier	7.6	366.8	39.3	52.5	250	8.1	1.01	0.64
GO barrier	6.7	359.6	40.3	46	1000	10.5	1.01	0.65
TrGO barrier	0.1	32.6	10.6	24.2	7	7.6	-	-

are shown in Fig. 3. It shows that the CZTSSe layer has a thickness of 1.5 µm. In the case of the rGO and GO, a lower density of voids has formed at the interface between CZTSSe and Mo as seen Fig. 3 (b) and (c). It may be due to the GO contact material between CZTSSe and Mo. But, SEM image of TrGO sample showed bad contact between CZTSSe and Mo. In addition, the CZTSSe absorber layer with TrGO substrate was easily peeled off during the experiment. The delamination of CZTSSe absorber layer may occurred due to the annealed TrGO substrate.

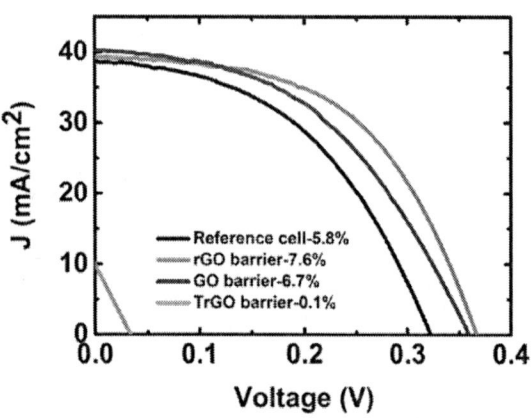

Fig. 4.   J-V characteristics of the CZTSSe solar cells.

J-V characteristics were obtained for the completed CZTSSe solar cells. Device characteristics of the CZTSSe solar cells are shown in Fig. 4 and Table I. The CZTSSe solar cell fabricated with the rGO substrate shows a lower series resistance (R_s) and a higher open-circuit voltage (V_{OC}) than that of the reference cell. On the other hand, the CZTSSe solar cell fabricated with the GO substrate shows a higher R_s and a higher V_{OC} than that of the reference cell. It may be due to the bad electrical property of the thick GO layer. In addition, the CZTSSe absorber layer deposited on the TrGO substrate was exfoliated because of the annealing effect. As a result, the reference cell, rGO, GO and TrGO cells show a power conversion efficiency of 5.8%, 7.6%, 6.7% and 0.1%, respectively.

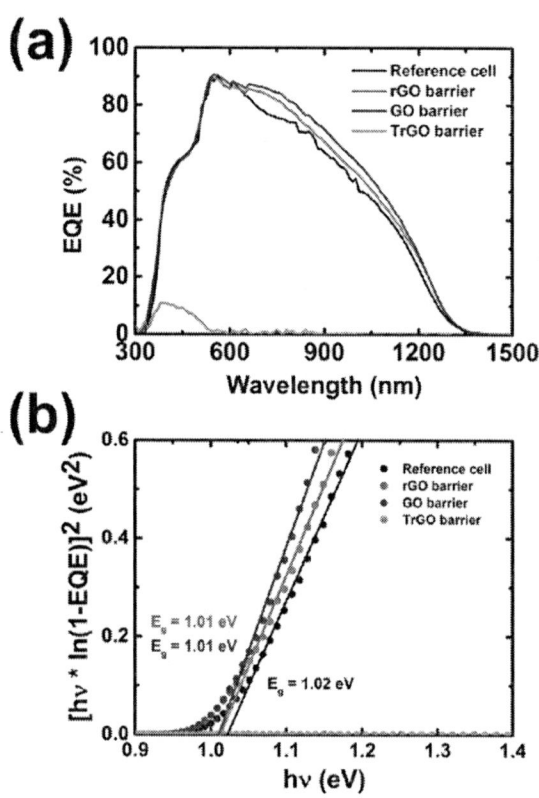

Fig. 5.   (a) External quantum efficiency and (b) plot to determine band gap energy of the CZTSSe solar cells fabricated with GO substrate.

To verify the external quantum efficiency (EQE) and band gap energy, we used incident photon to current efficiency measurement (IPCE, QEX7 system). From the Fig. 5 (a), the rGO and GO cells showed slightly higher EQE than that of the reference cell in longer wavelength region (600 nm to 1200 nm). But, TrGO cell showed very little EQE in short-wavelength region (300 nm to 600 nm) because of the remained CdS layer absorption. The estimation of the band gap of the CZTSSe solar cells are shown in Fig. 5 (b). All CZTSSe cells showed similar band gap energy 1.01 eV to 1.02

eV. From these values, V_{OC} deficit of the each cells were calculated as seen in Table I. The rGO and GO cells have lower V_{OC} deficit (0.64 and 0.65 V) than that of reference cell (0.7 V). It can be explained by formation of recombination centers [3]. Thus, the rGO and GO barriers may reduce the recombination centers in the vicinity of the CZTSSe absorber and Mo back contact.

IV. CONCLUSION

We have demonstrated the effects of the GO barrier on the CZTSSe solar cells. The rGO substrate exhibits a partially covered Mo surface by rGO. The CZTSSe solar cell fabricated with the rGO substrate shows a lower series resistance and a higher open-circuit voltage. In addition, the rGO cell has lower V_{OC} deficit than that of reference cell because of the reducing recombination centers. As a result, the solar cell fabricated with the rGO barrier shows the maximum conversion efficiency of 7.6%.

ACKNOWLEDGEMENT

This work was supported by the GIST Research Institute (GRI) grant funded by the GIST in 2017.

REFERENCES

[1] S. Giraldo, T. Thersleff, G. Larramona, M. Neuschitzer, P. Pistor, K. Leifer, A. Pérez-Rodríguez, C. Moisan, G. Dennler, E. Saucedo, "$Cu_2ZnSnSe_4$ solar cells with 10.6% efficiency through innovative absorber engineering with Ge superficial nanolayer", *Progress in Photovoltaics*, vol. 24, pp. 1359-1367, 2016.

[2] E. Ha, W. Liu, L. Wang, H.W. Man, L. Hu, S.C. Tsang, C.T. Chan, W.M. Kwok, L.Y. Lee, K.Y. Wong, "Cu_2ZnSnS_4/MoS_2-Reduced Graphene Oxide Heterostructure: Nanoscale Interfacial Contact and Enhanced Photocatalytic Hydrogen Generation", *Scientific Reports*, vol. 7, pp. 39411, 2017.

[3] J. Kim, S. Park, S. Ryu, J. Oh, B. Shin "Improving the open-circuit voltage of $Cu_2ZnSnSe_4$ thin film solar cells via interface passivation", *Progress in Photovoltaics*, vol. 25, pp. 308-317, 2017.

13% CdS/CdTe Solar Cell Using a Nanocomposite $(CuS)_x(ZnS)_{1-x}$ Thin Film Hole Transport Layer

Kamala Khanal Subedi, Khagendra P. Bhandari, Ebin Bastola, and Randy J. Ellingson

Wright Center for Photovoltaic Innovation and Commercialization, Department of Physics and Astronomy,

The University of Toledo, 2801 Bancroft Street, Toledo, OH 43606, USA

Abstract — **We report the properties of the earth-abundant p-type transparent conducting nanocomposite thin film $(CuS)_x$ $(ZnS)_{1-x}$, deposited by chemical bath deposition. Our films of 55 nm thickness show transparency > 65% over the visible region, and compact grains of size < 10 nm. We investigate the photovoltaic performance of CdTe solar cells using $(CuS)_x$ $(ZnS)_{1-x}$ as a back-contact interface layer. Our best device exhibits a *FF* value of 73 %, V_{OC} of 816 mV, and efficiency of 13.0 % when using a nanocomposite p-type transparent conductor based on Cu, Zn, and S as an interface layer before the back-contact metallization layer.**

Index Terms — **back-contact, chemical bath deposition, CdTe, nanocomposite**

I. INTRODUCTION

Vast carbon emissions associated principally with the combustion of fossil fuels has created an imbalance in Earth's radiative energy flux, resulting in evident warming of the atmosphere. The serious environmental consequences created by the use of fossils fuels have motivated continued pursuit of improvements in highly efficient, environmentally benign, and cost effective solar cells. Among the commercially available technologies, cadmium telluride (CdTe) remains the leader as measured by the shortest energy payback time and, equivalently, the largest energy return on energy invested. Thus, despite sales of CdTe PV modules comprising just ~5% of global sales, [1] CdTe remains an important commercial technology for PV generation.

The high work function of ~5.7 eV exhibited by polycrystalline CdTe for PV application presents a challenge for creating a low resistance and barrier-free back (hole selective) contact for use in the CdS/CdTe solar cell.[2-5] In analogy with their more prevalent n-type counterparts, p-type transparent conducting materials (TCMs) ideally exhibit high transparency to light and high electrical conductivity by hole transport.[5,6] Nanocomposite thin films consisting of $(CuS)_x(ZnS)_{1-x}$ have shown high optical transparency for the visible region, with a free hole concentration of $\sim 10^{21}$ cm^{-3} which supports high electrical conductivity.[7] Many efforts have been devoted to developing effective hole-selective contacts to CdTe polycrystalline solar cells, and this class of p-type nanocrystalline TCM offers several attractive properties for such an application, as has been demonstrated for the passivated c-Si HIT cell by Xu et al.[7]. P-type TCM's have been deposited by numerous methods such as chemical bath deposition [8], electrochemical deposition [9], and pulsed laser deposition [10]. Here, we use chemical bath deposition (CBD) to prepare semi-transparent $(CuS)_x(ZnS)_{1-x}$ for study as a hole transport layer at the CdTe back contact. CBD offers a straightforward and cost-effective method to deposit at an arbitrary thickness on almost any type of substrate on a large scale. [7,8,11]

II. EXPERIMENTAL DETAILS

Chemical bath deposition allows for a solution-based technique to prepare films on a glass substrate immersed in a dilute solution of precursor. First, the glass substrates were ultrasonically cleaned using Micro-90 and deionized water (DIW) for 30 mins, rinsed several times with DIW, cleaned in methanol for 15 minutes, and finally the substrates were dried using dry N_2 in air.

The precursor solution was prepared following the procedure reported previously.[7,8] For the $(CuS)_x(ZnS)_{1-x}$ film, three solutions were prepared: (1) with total concentration of 0.1 M of $CuSO_4$ and $Zn(CH_3COO)_2$ mixture with DIW for 50 ml solution; (2) 0.1 M concentration of thioactamide(C_2H_5NS) with DIW for 50 ml solution and, (3) 0.1 M concentration of edetate disodium salt dihydrate (Na_2EDTA) in DIW for 25 ml solution, for a total of 125 ml solution. Before starting the reaction, solution (1) and solution (3) were mixed thoroughly in an ultrasonic bath for 20 mins at a temperature 30°C, and subsequently combined and mixed with solution (2). The cleaned substrate was immediately placed vertically inside the solution and the beaker was covered by aluminum foil to prevent evaporation. The bath temperature was maintained at 80° C under the constant stirring. After 55 mins., the substrate coated with the nanocomposite film was removed from the solution, washed with DIW, and dried by N_2 in air at room temperature.

For the solar cell filmstack, CdTe and cadmium sulfide (CdS) layers were deposited by vapor transport deposition (VTD) onto TEC™ 15 (Pilkington, N.A.) glass substrates. The CdTe devices were treated with a saturated solution of $CdCl_2$ in methanol, annealed at 387 °C for 30 mins in dry air to advance grain growth, release interfacial strain, and facilitate sulfur and tellurium mixing at the CdS/CdTe interface. [4,12] Subsequently, the $CdCl_2$ treated devices were used instead of the glass substrates to deposit the $(CuS)_x(ZnS)_{1-x}$ nanocomposite thin films as interface, or hole-transport,

978-1-5090-5606-4/17 $31.00 © 2017 IEEE

layers. After the deposition, the devices were rinsed and dried as before. The back contact metallization was completed by depositing 3 nm Cu and 35 nm Au via thermal evaporation at room temperature, followed by 45 minutes annealing at 150°C in dry air to enable the diffusion of Cu as we normally do for the case of our common Cu/Au laboratory back contact to CdTe. Finally, the devices were scribed using a 532 nm laser to create an array of approximately 20 devices of active area 0.08 cm².

III. RESULTS AND DISCUSSION

Scanning electron microscopy (SEM) was used to study the morphology of the $(CuS)_x(ZnS)_{1-x}$ films for two compositions: $x = 0.64$, and $x = 0.80$. We observed that the surface morphology depends on the concentrations of copper and zinc within the film. From the SEM images of the two films (Fig. 1a and 1b), we can see that the grains are compact, and that the grain size for the $(CuS)_{0.64}(ZnS)_{0.36}$ is smaller (~10 nm) and more compact than the $(CuS)_{0.80}(ZnS)_{0.20}$ film (`~ 12 nm). Energy-dispersive X-ray spectroscopy (EDS) measurements, used to quantify film stoichiometry, shown in the Fig. 1a and 1b, were conducted at an operating voltage of 20 kV, operating current of 15 µA, and 15 mm working distance. The

films' composition ratio of the copper and zinc were measured using the EDS method, and found to be spatially uniform at 25K spatial resolution within measurement limits. The homogeneous nanocrystalline film was obtained through CBD via ion-by-ion process in which cluster by cluster growth is suppressed; this is essential for the formation of good-quality

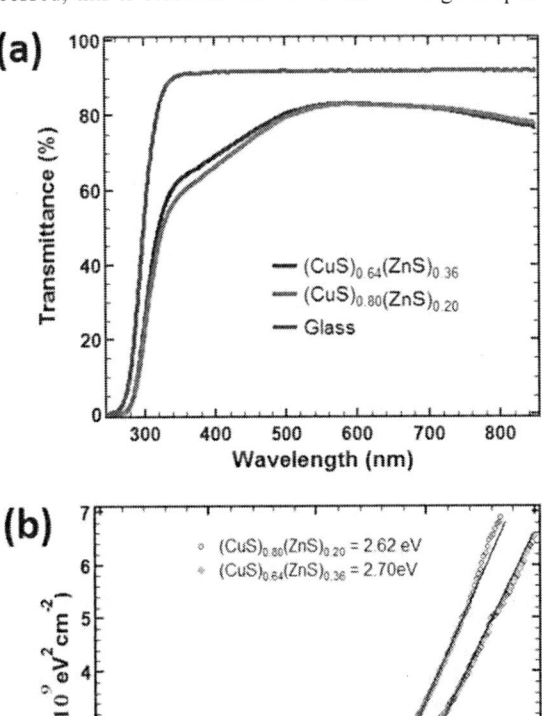

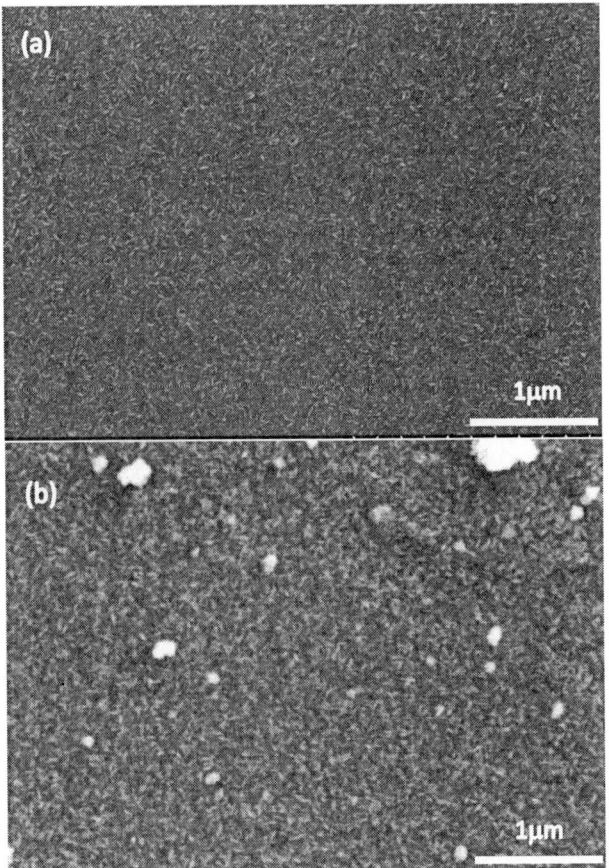

Fig. 1.   SEM images of (a) $(CuS)_{0.64}(ZnS)_{0.36}$ and (b) $(CuS)_{0.80}(ZnS)_{0.20}$ nanocomposite thin films.

Fig. 2.   a) Transmission spectra of $(CuS)_{0.64}(ZnS)_{0.36}$, $(CuS)_{0.80}(ZnS)_{0.20}$ and bare glass. (b) Band gap spectra ("Tauc plot") of $(CuS)_{0.64}(ZnS)_{0.36}$ and $(CuS)_{0.80}(ZnS)_{0.20}$.

films. [7]

We have used four-point probe to measure the sheet resistance of the nanocomposite (NC) thin film, and found a value of ~ 2,000 Ω/sq for a ~50 nm film thickness prepared on glass. Here the NC film exhibits amorphous structure whereas the Cu-free ZnS films shows cubic crystalline structure. With the addition of CuS, we observed the film to amorphize based on results of initial x-ray diffraction studies. [8] We have studied the transmittance spectra of the $(CuS)_{0.64}(ZnS)_{0.36}$ and $(CuS)_{0.80}(ZnS)_{0.20}$ nanocomposite thin films in the range of 250-850 nm for ~50 nm thick film. The maximum transmittance up to 83% for $(CuS)_{0.64}(ZnS)_{0.36}$ films and 82% for $(CuS)_{0.80}(ZnS)_{0.20}$ films were seen at 550 nm wavelength as

978-1-5090-5606-4/17 $31.00 © 2017 IEEE

shown in Fig. 2a. With increasing (ZnS) content, the transmittance does not increase rapidly. Instead the transparency appears to depend on the crystal grain size due to smaller grains which minimizes scattering loss, but aids transmittance.[13] Since the grain size of the $(CuS)_{0.64}(ZnS)_{0.36}$ is somewhat smaller than that of the $(CuS)_{0.80}(ZnS)_{0.20}$, the transmittance of the $(CuS)_{0.64}(ZnS)_{0.36}$ is greater than the transmittance of the $(CuS)_{0.80}(ZnS)_{0.20}$. The transmittance spectra of $(CuS)_{0.64}(ZnS)_{0.36}$, $(CuS)_{0.80}(ZnS)_{0.20}$, and glass are shown in the Fig 2(a). ZnS has a direct band gap with energy 3.7 eV and CuS has a direct band gap of energy 2.2 eV. With the combination of CuS with the ZnS the band gap of the NC is smaller than ZnS but it is greater than the CuS. We obtained the implied direct band gap values for $(CuS)_{0.64}(ZnS)_{0.36}$ and $(CuS)_{0.80}(ZnS)_{0.20}$ as 2.70 eV and 2.62 eV respectively as shown in Fig. 2(b). The thin films $(CuS)_x(ZnS)_{1-x}$ were applied as the back-contact layer of VTD CdTe/CdS devices. Fig. 3 shows the J-V curves of best cells

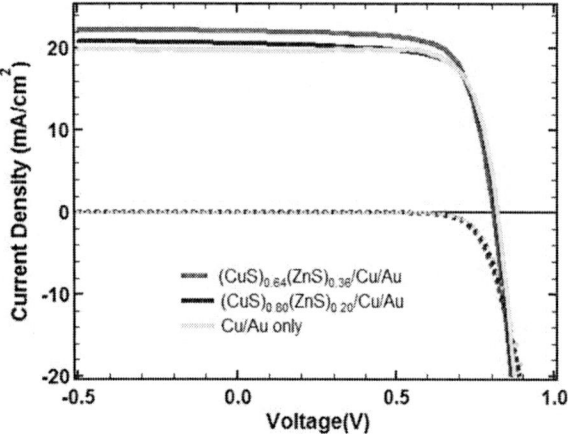

Fig. 3. J-V characteristics for CdTe solar cells using $(CuS)_{0.64}(ZnS)_{0.36}$, $(CuS)_{0.80}(ZnS)_{0.20}$. and Cu/Au as the back-contact layers. Dashed lines show dark J-V measurement results.

among 20 cells of each contact type.

In Fig.3, the solid curves represent measurements under simulated AM 1.5 illumination, and dashed curves represent measurements in the dark. Table I presents the open circuit voltage (V_{OC}), short-circuit current (J_{SC}), fill factor (FF), series resistance (R_s) and shunt resistance (R_{sh}) of the best cell and average of 20 cells of the device with two compositions of NC $(CuS)_{0.60}(ZnS)_{0.40}$, $(CuS)_{0.80}(ZnS)_{0.20}$ and Cu/Au as a back-contact layers. From the table below, with the NC as a back-contact layer the J_{sc} is improved significantly while V_{OC} and fill factor are not measurably improved. When we used the NC hole transport layer, the series resistance is slightly higher. We propose that an increase in the thickness of the NC layer may decrease the R_s, and we plan to study this thickness dependence to increase the V_{OC} and FF. [7]

Even though we have thermally evaporated Cu and Au, using Au only creates a significant Schottky barrier between the CdTe and Au/Cu interface because Cu content is very

small and mostly diffused into the body of the CdTe [14]. CdTe has a large work function and a deep valence band edge; normally available metals generate a barrier against the transfer of photo-generated holes at the back contact [15]. By

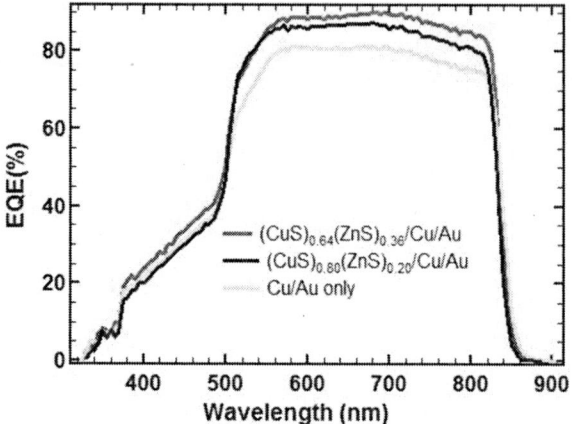

Fig. 4. External quantum efficiency (EQE) spectra of CdTe solar cells when using $(CuS)_x(ZnS)_{1-x}$ as the HTL.

using the NC thin film as the back-contact we can minimize the back-contact barrier height. However, given the relatively high bandgap energy of the NC films, the lack of improvement in V_{OC} and FF suggests that the HTL interfacial layer does not achieve an ideal combination of CdTe passivation, barrier height reduction for hole transport out to the metallization layer, and electron reflection as expected by a favorably large conduction band offset to reject the presence of, or transfer of, electrons near or through the back contact.

The external quantum efficiency (EQE) spectra as shown in the Fig. 4, in corporation of the NC thin film as a back-contact layer in the CdTe solar cell, show improvement in the current collection in the wave length 550-820 nm regions. EQE measures the collection probability of an electron- hole pair created by the photons at 0 bias and it measures over the different wavelength for the efficiency of the device. The EQE measurement verifies the Jsc values obtained from J-V measurement. In comparison with the two NC thin films, $(CuS)_{0.64}(ZnS)_{0.36}$ shows the better performance in current collection than $(CuS)_{0.80}(ZnS)_{0.20}$ and Cu/Au back contact only.

VI. CONCLUSION

The $(CuS)_x(ZnS)_{1-x}$ layer is easily prepared using a room-temperature, cost-effective, solution approach. We have achieved a high degree of control over the stoichiometry to prepare highly conductive $(CuS)_x(ZnS)_{1-x}$ NC thin films with high free hole concentration. These films perform very well as a low resistance back contact to polycrystalline CdTe solar cells. As demonstrated by the J-V curves, the $(CuS)_x(ZnS)_{1-x}$ hole transport layer can improve performance of the CdTe cells. So, by the development and application of p-type transparent conducting hole transport layer, PV technology can continue to evolve toward low-toxicity earth-abundant materials.

ACKNOWLEDGEMENT

The authors gratefully acknowledge financial support provided by the NSF's Sustainable Energy Pathways program (CHE-1230246) and the U.S. Air Force Research Laboratory, Space Vehicles Directorate (contract # FA9453-11-C-0253).

TABLE I
J-V PARAMETERS FOR THE BEST AND AVERAGE CELLS. DEVICE AREA = 0.080 CM^2.

Film composition	Backcontact of	V_{OC} (V)	J_{SC} (mA/cm^2)	FF (%)	PCE (%)	R_s (Ω cm^2)	R_{sh} (Ω cm^2)
$(CuS)_{0.64}(ZnS)_{0.36}$/Cu/Au	Best cell	0.805	22.1	73.3	13.0	2.8	2101
	Average	0.802	21.6	72.7	12.6	2.8	1781
$(CuS)_{0.80}(ZnS)_{0.20}$/Cu/Au	Best cell	0.816	20.6	73.0	12.3	2.4	1510
	Average	0.811	20.1	72.1	11.8	2.7	1333
Cu/Au only	Best cell	0.818	19.8	76.1	12.3	2.6	2878
	Average	0.816	19.7	75.6	12.1	2.5	2795

REFERENCES

[1] Jean, Joel, Patrick R. Brown, Robert L. Jaffe, Tonio Buonassisi, and Vladimir Bulović. "Pathways for solar photovoltaics." *Energy & Environmental Science* 8, no. 4 (2015): 1200-1219.

[2] R. K. Swank, "Surface properties of II-VI compounds," *Physical Review*, vol. 153, p. 844, 1967.

[3] J. Freeouf and J. Woodall, "Schottky barriers: An effective work function model," in *Electronic Structure of Metal-Semiconductor Contacts*, ed: Springer, 1990, pp. 154-156.

[4] K. P. Bhandari, P. Koirala, N. R. Paudel, R. R. Khanal, A. B. Phillips, Y. Yan, *et al.*, "Iron pyrite nanocrystal film serves as a copper-free back contact for polycrystalline CdTe thin film solar cells," *Solar Energy Materials and Solar Cells*, vol. 140, pp. 108-114, 2015.

[5] A. Banerjee and K. Chattopadhyay, "Recent developments in the emerging field of crystalline p-type transparent conducting oxide thin films," *Progress in Crystal Growth and Characterization of Materials*, vol. 50, pp. 52-105, 2005.

[6] R. Woods-Robinson, J. K. Cooper, X. Xu, L. T. Schelhas, V. L. Pool, A. Faghaninia, *et al.*, "P-Type Transparent Cu-Alloyed ZnS Deposited at Room Temperature," *Advanced Electronic Materials*, vol. 2, pp. 1500396-n/a, 2016.

[7] X. Xu, J. Bullock, L. T. Schelhas, E. Z. Stutz, J. J. Fonseca, M. Hettick, *et al.*, "Chemical bath deposition of p-type transparent, highly conducting $(CuS)_x:(ZnS)_{1-x}$ nanocomposite thin films and fabrication of Si heterojunction solar cells," *Nano letters*, vol. 16, pp. 1925-1932, 2016.

[8] D. E. Ortíz-Ramos, L. A. González, and R. Ramirez-Bon, "p-Type transparent Cu doped ZnS thin films by the chemical bath deposition method," *Materials Letters*, vol. 124, pp. 267-270, 2014.

[9] K. Yang and M. Ichimura, "Fabrication of transparent p-Type Cu_xZn_yS thin films by the electrochemical deposition method," *Japanese Journal of Applied Physics*, vol. 50, p. 040202, 2011.

[10] M. Dula, K. Yang, and M. Ichimura, "Photochemical deposition of a p-type transparent alloy semiconductor Cu_xZn_yS," *Semiconductor Science and Technology*, vol. 27, p. 125007, 2012.

[11] P. Nair, M. Nair, V. Garcıa, O. Arenas, Y. Pena, A. Castillo, *et al.*, "Semiconductor thin films by chemical bath deposition for solar energy related applications," *Solar Energy Materials and Solar Cells*, vol. 52, pp. 313-344, 1998.

[12] K. V. Krishna and V. Dutta, "Effect of in situ CdCl$_2$ treatment on spray deposited CdTe/ CdS heterostructure," *Journal of applied physics*, vol. 96, pp. 3962-3971, 2004.

[13] Y. Lu, X. Meng, G. Yi, and J. Jia, "In situ growth of CuS thin films on functionalized self-assembled monolayers using chemical bath deposition," *Journal of colloid and interface science*, vol. 356, pp. 726-733, 2011.

[14] Li, Jiaojiao, David R. Diercks, Timothy R. Ohno, Charles W. Warren, Mark C. Lonergan, Joseph D. Beach, and Colin A. Wolden. "Controlled activation of ZnTe: Cu contacted CdTe solar cells using rapid thermal processing." *Solar Energy Materials and Solar Cells* 133 (2015): 208-215.

[15] Demtsu, S. H., and J. R. Sites. "Effect of back-contact barrier on thin-film CdTe solar cells." *Thin Solid Films* 510, no. 1 (2006): 320-324.

Molybdenum Oxide and Molybdenum Nitride Back Contacts for Thin-Film CdTe Solar Cells

Anna Kindvall, Jason Kephart and Walajabad Sampath

Department of Mechanical Engineering, Colorado State University, Fort Collins, Colorado, 80526

Abstract — In thin-film CdTe solar cells, a high quality back contact is essential for high voltage and fill factor. Properties such as a high work function, stability to air and moisture, and an ohmic contact to the CdTe interface are important considerations. MoO_x and MoN_x thin films are promising alternative back contacts. Films were deposited with varying composition via DC reactive sputter deposition. Thin film properties were measured through characterization techniques such as four-point probe measurements, profilometry measurements, Hall measurements, XPS, and GAXRD. Devices were compared between current back contact types and alternatives such as molybdenum. Experimental results for MoO_x and MoN_x as back contact buffer layers are discussed in detail.

Index Terms – Thin film photovoltaics, back contacts, CdTe

I. INTRODUCTION

Cadmium telluride thin-film solar cells are inexpensive and research-scale devices have achieved efficiencies up to 22.1% [1]. For solar to continue to be a viable energy solution, research must continue to improve overall device efficiency while maintaining the ability to manufacture cells at low cost. P-type cadmium telluride (CdTe) is a highly resistive, high work function material [2]. The work function of p-CdTe is 5.7 eV which is higher than most metals [3]. To create a good back contact, tellurium (Te) has been used due to the high p-type conductivity of the material, however the work function of Te is lower than that of p-CdTe [2]. By placing a high work function, and relatively low resistance back contact behind the CdTe or in addition to the tellurium, the band alignment can be improved while creating an ohmic, uniform interface [3]. While the work function of molybdenum is only 4.36-4.95eV, it significantly increases in molybdenum nitride (MoN_x) and molybdenum oxide (MoO_x) [4]. The work function of MoN_x can range from 5.1-5.23 eV with low nitrogen contents [5]. MoO_3 has a high work function of 6.8 eV [6].

Currently at Colorado State University (CSU), standard solar cells are finished with several layers of back contacts. The first back contact layer is the tellurium coating and that is followed by a finishing coating of nickel paint. These back contacts assisted in developing a certified, 18.3% efficiency device [7]; however the nickel paint is not suitable for commercial modules. The paint is moisture sensitive and does not adhere well to the tellurium without added stresses or delamination [8]. Within weeks of creating a device, degradation can occur and cell efficiency drops.

Consequently, there is a need for a more robust back contact. Molybdenum oxides and nitrides provide good conductivity, have a high work function, and provide a robust film even with exposure to the environment. With current devices, the molybdenum oxide and or nitride films act as an intermediate layer between the CdTe or tellurium and back electrode.

II. METHODS

A. Device Fabrication

Device fabrication includes: soda-lime glass of thickness 3.2-mm, a transparent conductive oxide layer of fluorine-doped tin oxide, a window layer of 100 nm of magnesium zinc oxide ($Mg_xZn_{1-x}O$), 4.0 microns of CdTe, cadmium chloride ($CdCl_2$) (which diffuses between the grains of the CdTe), several monolayers of copper, and back contact layers. The first back contact layer consists of 20-50 nm layer of evaporated tellurium. In standard devices, after the tellurium, nickel-based paint is deposited and used as a baseline for testing as shown on the left of Fig. 1. For this research project, after the CdTe or tellurium, layers of MoO_x and MoN_x were deposited. The back electrode is sputter-deposited aluminum as displayed in the middle and right of Fig. 1.

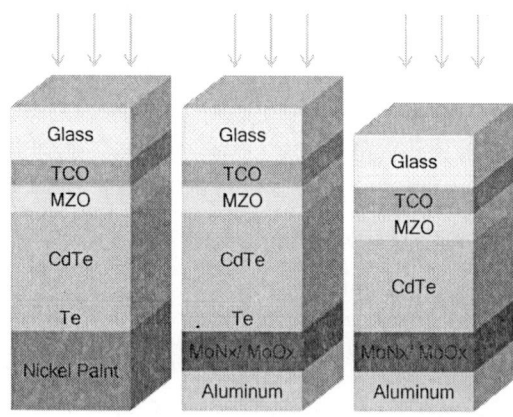

Fig. 1.   (Left) Baseline device layers at CSU. (Middle and right) The devices with MoO_x and MoN_x back contacts.

The molybdenum oxide, molybdenum nitride, and molybdenum thin film layers were deposited using a DC magnetron sputtering system with a two-inch Mo target. The back electrode of aluminum was deposited utilizing the same DC magnetron. The samples were placed 15 cm above the 5-cm magnetron. The chamber reached a base pressure of 10^{-6} Torr before beginning the deposition. Films were deposited at a power of 200 W and a pressure of 10 milliTorr. The composition of the films was controlled using two mass-flow controllers by varying the reactive gas flow into the chamber as a ratio of argon flow. Fig. 2 displays a schematic of the vacuum chamber and deposition equipment. Samples were held in a rotating eight-sample holder. Deposition times were adjusted based upon a calibrated rate to deposit the desired film thickness.

After depositing the molybdenum oxide, nitride, or pure molybdenum, the films were removed from the vacuum chamber. Samples were immediately stored in an argon-containing glovebox to minimize contamination prior to testing.

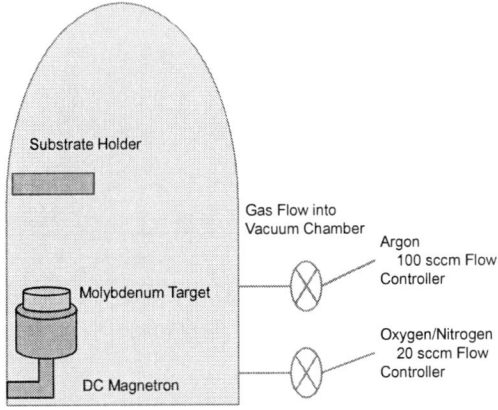

Fig. 2.   Vacuum chamber setup and deposition process.

B. Film Characterization

Several characterization methods have been used to analyze the properties of the reactively sputtered films, including: four-point probe measurements, profilometry measurements, Hall measurements, X-ray photoelectron spectroscopy (XPS), and X-ray diffraction (XRD). Using a four-point probe measurement system, film resistivity was measured. Thickness measurements were done using a profilometer and to calculate deposition rate. Hall measurements were done using an Ecopia HMS-3000 to determine carrier concentration and mobility. XPS measurements were taken using the PHI PE-5800 X-Ray Photoelectron Spectrometer to determine material properties. Glancing Angle X-ray diffraction (GAXRD) measurements were conducted using the Bruker D8 Discover Series I X-ray to analyze crystalline properties.

III. RESULTS AND DISCUSSION

Molybdenum oxide and molybdenum nitride thin films of varying composition were deposited by varying the reactive gas flow. Molybdenum films were deposited at a ratio of 10%, 20%, and 40% O_2/Ar and N_2/Ar contents. As a transition metal, Mo can have different oxidation states and different oxides and nitrides may have substantially different properties. For example MoO_2 is a metallic conductor, while MoO_3 is a mostly transparent insulator and is sensitive to water and air [9].

With increasing oxygen content, the MoO_x films become more transparent as seen in Fig. 3a and 3b. The film made without oxygen is a metallic gray, the 10% and 20% O_2/Ar samples are brown and slightly transparent, and the 40% O_2/Ar sample was almost completely transparent with a slight blue color, similar to evaporated MoO_3 films prepared by the authors.

To test whether the highest-oxygen film was predominately MoO_3 (water-soluble), the films were placed in a beaker of deionized water. The 40% O_2/Ar film changed after several minutes in the water. The blue film transformed to more of a light brown, yet still transparent film.

After 2 hours, the films were removed from the deionized water. The 40% O_2/Ar sample was completely transparent. The samples made with 10% and 20% O_2/Ar had dissolved partially and the films had a lighter color than originally deposited. The molybdenum film with no intentional oxygen had no change. All samples can be seen after treatment in Fig. 3c and 3d.

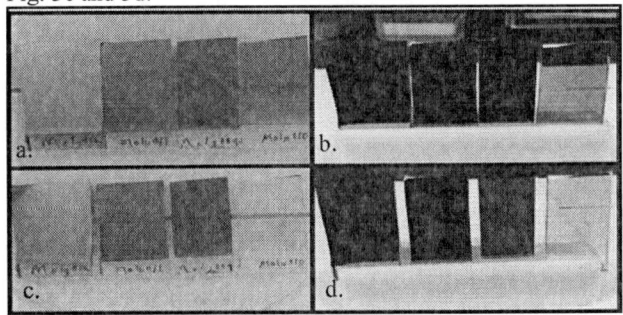

Fig. 3.   a. and b. (Top) As deposited, the four films containing different oxygen contents. (Increasing left to right from Mo to 40% O_2/Ar film) The films are displayed at two different angles.
c. and d. (Bottom) Samples after two hours of being placed in deionized water.

A. Resistivity and Profilometry Measurements

A four-point probe was used to measure the sheet resistance of each sample. In order to calculate the resistivity of the samples, the thickness of the sample was measured with a profilometer. For each sample, the resistivity increased with increasing oxygen content as seen in table 1.

978-1-5090-5606-4/17 $31.00 © 2017 IEEE

When depositing the molybdenum nitride films, all films had a metallic appearance. The nitrogen content appeared to increase resistivity of the film, but not as significantly as the molybdenum oxide films. The deposition rate remained fairly constant for the MoN_x films.

TABLE 1
SUMMARY OF RESISTIVITY AND DEPOSITION RATE

% O$_2$ or % N$_2$ in sputtered film	Average Resistivity Ω-cm	Deposition Rate nm/s
Mo	2.7×10^{-4}	0.26
10% O$_2$/Ar	4.8×10^{-2}	0.55
20% O$_2$/Ar	2.1	0.46
40% O$_2$/Ar	14.9	0.35
10% N$_2$/Ar	1.4×10^{-3}	0.30
20% N$_2$/Ar	1.7×10^{-2}	0.25
40% N$_2$/Ar	5.7×10^{-2}	0.30

B. Hall Measurements

For the Hall measurements, all samples were 100 nm thick. In order to obtain more accurate measurements, all samples were tested at the highest attainable current for each specific sample.

As displayed in table 2, the MoO_x films decreased in carrier concentration with increasing oxygen content. The mobility of the MoO_x films increased with increasing oxygen content. With the MoN_x films, the carrier concentration remained fairly constant with increasing nitrogen content.

TABLE 2
SUMMARY OF HALL MEASUREMENTS

Sample	Current	Bulk Concentration (cm^{-3})	Hall Mobility (cm^2/V-s)
Mo	3 mA	7.49×10^{22}	1.11×10^{-1}
10% O$_2$/Ar	1 mA	1.60×10^{22}	7.74×10^{-3}
20% O$_2$/Ar	100 μA	8.76×10^{20}	2.46×10^{-2}
40% O$_2$/Ar	10 nA	1.08×10^{14}	56.9
10% N$_2$/Ar	3 mA	2.44×10^{22}	1.50×10^{-1}
20% N$_2$/Ar	1 mA	3.40×10^{21}	5.93×10^{-2}
40% N$_2$/Ar	1mA	6.33×10^{22}	3.75×10^{-3}

C. X-ray Photoelectron Spectroscopy

Utilizing the molybdenum 3d peak in XPS, peak curves were used to identify different bonding environments due to oxygen or nitrogen incorporation. As seen in Fig. 4a with increasing oxygen contents, the peak positions move toward higher binding energy, corresponding MoO_x with higher x. Both the pure molybdenum and 40% O$_2$/Ar samples are predominately one pair of 3d peaks, while intermediate compositions have a combination of several intermediate binding energies.

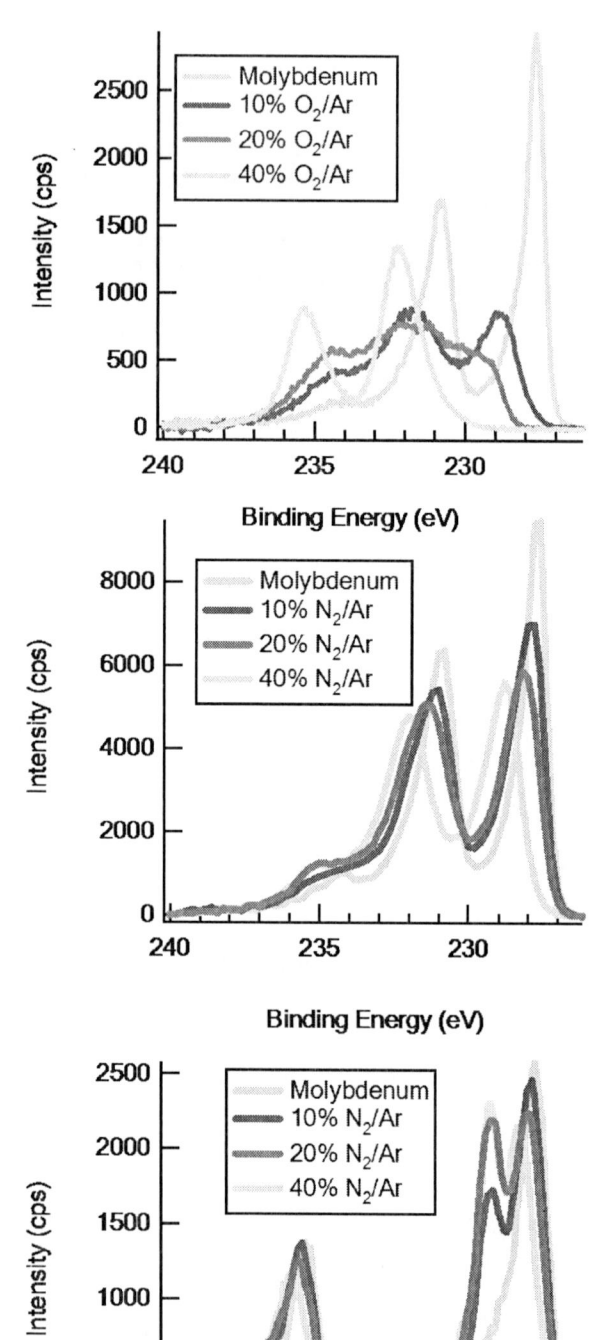

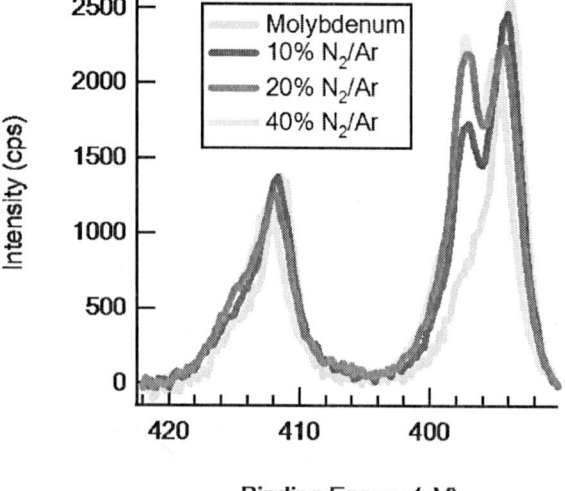

Fig. 4.   a. (Top) Molybdenum 3d peaks for MoO_x films. b. (Middle) Molybdenum 3d peaks for MoN_x. c. (Bottom) Molybdenum 3p peaks and nitrogen 1s peak for the MoN_x films.

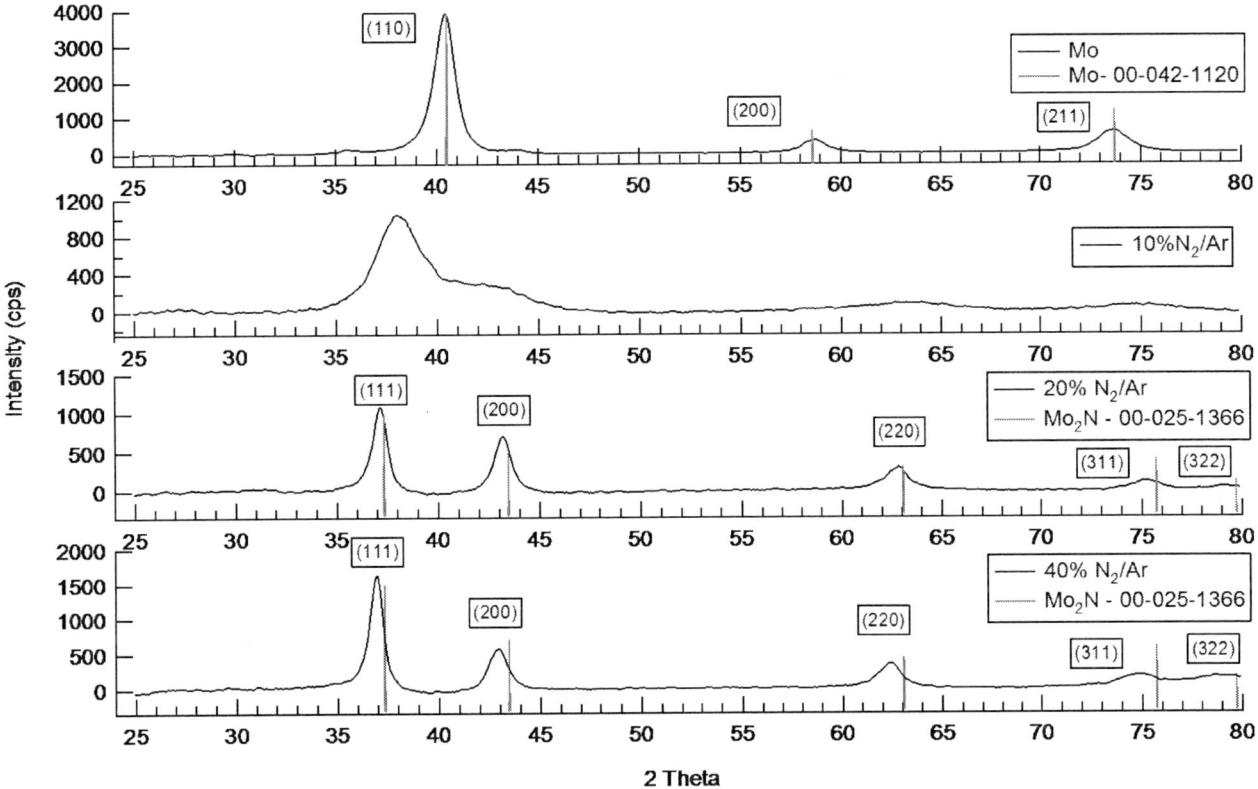

Fig. 5. GAXRD analysis of the Mo and MoN$_x$ films. With nitrogen, the MoN$_x$ films display crystalline properties and are compared to JCPDS XRD cards.

Analysis of the molybdenum 3d peaks was also conducted for nitride films. The 3d peaks moved toward higher binding energies with increased nitrogen as seen in Fig. 4b. The 3p peaks of the MoN$_x$ overlap the nitrogen peak. Metal nitrides tend to have a binding energy around 397eV [10]. With increasing nitrogen content in the film, there is an increased nitrogen peak as seen in Fig. 4c. The increased peak indicates that the nitrogen is incorporating into the film. With increasing nitrogen flow during deposition, the XPS nitrogen peak clearly increases relative to the molybdenum peak.

Expected binding energy values for the different molybdenum oxide and nitride states are found in Table 3.

TABLE 3
COMPARISON OF XPS BINDING ENERGIES

Element	Formula	Peak	Binding Energy (eV)	Ref.
Mo	MoO$_2$	3d3/2	232.3	[11]
Mo	MoO$_3$	3d3/2	235.8	[12]
Mo	Mo	3d3/2	231.0	[13]
Mo	Mo$_2$N	3d5/2	228.2	[14]
Mo	Mo$_2$N	3p3/2	394.3	[14]
Mo	Mo	3p3/2	394.0	[15]
N	Mo$_2$N	1s	397.55	[14]

D. Glancing Angle X-ray Diffraction

GAXRD measurements have been done on the molybdenum oxide and nitrogen containing films. For the molybdenum oxides, the reactively sputtered films appear to be amorphous-like, supporting the complex mixture of XPS binding energies.

The molybdenum nitride films displayed some crystallinity and are compared with the Mo$_2$N and Mo structures from the JCPDS cards which are displayed as the red lines in Fig. 5. Deposited film properties appear comparable to molybdenum nitride films from the JCPDS database.

E. Device Characterization

To characterize how the films behaved in a device structure, small area cells (0.6 cm^2) were created with the MoO$_x$ and MoN$_x$ buffer layers. The devices were coated in 100nm of molybdenum, molybdenum oxide, or molybdenum nitride. Devices were capped with an aluminum back electrode except for the baseline device which had nickel paint. Results are shown in Fig. 6 for open-circuit voltage and fill factor and J-V curves are displayed in Fig. 7 for the best devices.

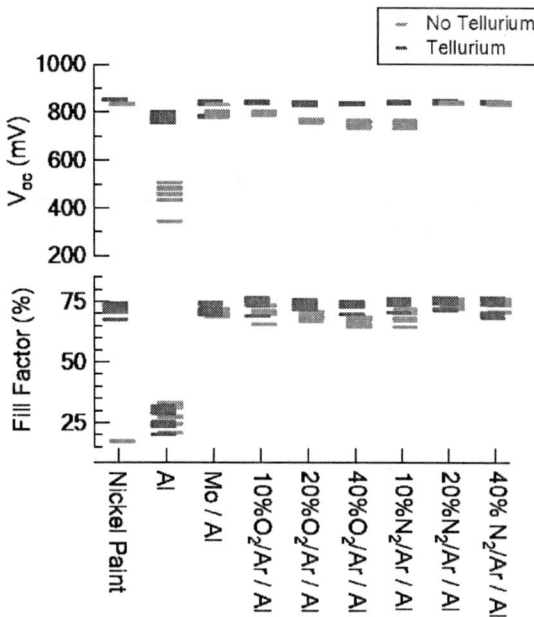

Fig. 6. Fill factor and V_{oc} for device back contacts with (blue) and without tellurium (red). Devices without Te did not perform as well as the devices with Te with the exception of the 20% and 40% $N_2/$ Ar devices.

To show the need for a buffer layer between the tellurium and aluminum, a device was created with only the aluminum back contact. The fill factor was very low and it performed poorly in comparison to other devices. Devices with any form of molybdenum, molybdenum oxide, or molybdenum nitride buffer layer performed well against the baseline device with tellurium. On the devices without tellurium, the 20% N_2/Ar and 40% N_2/Ar devices had higher V_{OC} and fill factor than the MoO_x devices.

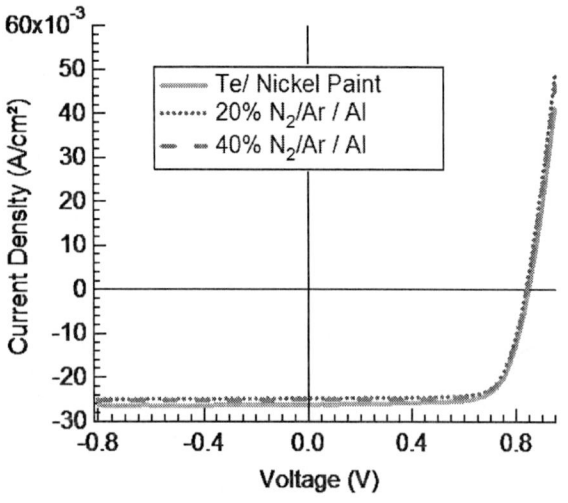

Fig. 7. J-V curves for the baseline device and MoN_x devices without tellurium.

The nickel paint sample with Te was a better device due only to a slight difference in short-circuit current. However, devices with the higher nitrogen contents in the MoN_x back contacts had promising results with and without tellurium. The samples were comparable in open-circuit voltage (V_{oc}) and fill factor. The 20% N_2/Ar and 40% N_2/Ar back contacts displayed consistently high values of short-circuit current (J_{sc}), V_{oc}, fill factor, and efficiency with and without tellurium.

After completing the device analysis of the different oxygen- and nitrogen-containing films, a second experiment was conducted using the 20% N_2/Ar molybdenum film, as it had displayed the most consistent, promising results from the first set of devices. With the same processing conditions as in the first tests of samples, half of the sample plate was coated in tellurium. Film thickness of the MoN_x was varied from 5 nm to 200 nm to gain a better understanding of how the film impacts devices and to find an ideal thickness. Shown in Fig. 8 are device parameters and Fig. 9 J-V curves for devices. The thicker the MoN_x layer, the better the devices performed. The results appear to converge to one point for devices with and without tellurium.

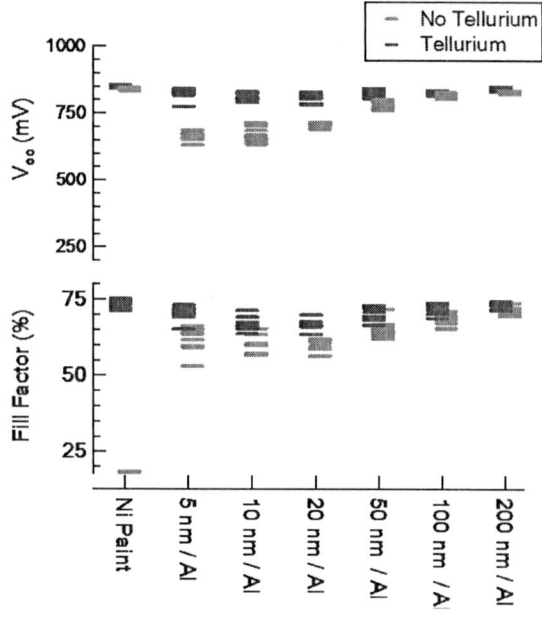

Fig. 8. Fill factor and V_{oc} for the thickness sweep of the 20% N_2/Ar molybdenum film. Devices with Te (blue) have fairly consistent performance, while the devices without Te (red) perform better as the MoN_x layer becomes thicker.

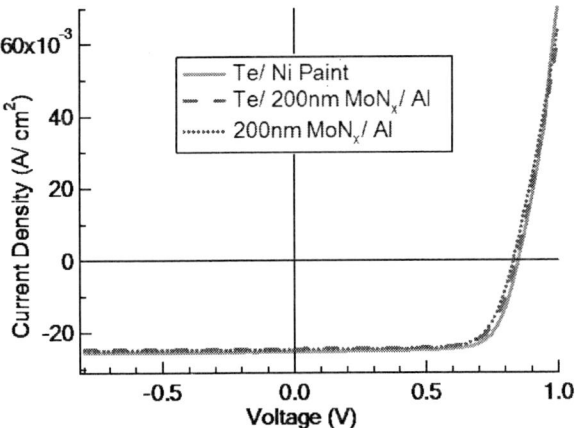

Fig. 9. Displays the 200 nm 20% N_2/Ar device curves with and without Te. The MoN_x devices are compared to the baseline device.

IV. CONCLUSIONS

Molybdenum oxide and nitride films were deposited by reactive DC sputter deposition and characterized with four-point probe measurements, profilometry measurements, Hall measurements, XPS, and GAXRD. The MoO_x films were found to be amorphous-like and tended to have increased resistivity with increased oxygen content. The MoN_x films displayed more crystallinity and were more conductive.

From the initial film characterization, full devices were made. From the MoO_x and MoN_x back contacts, the 20% N_2/Ar film displayed promising results. Molybdenum nitride with aluminum is a comparable back contact to current devices without the need for nickel paint. Thicker MoN_x films indicate that devices without any Te can achieve comparable efficiency to the best devices using the baseline back contact.

ACKNOWLEDGEMENTS

The authors acknowledge funding from the U.S Department of Energy Photovoltaics R&D: Small Innovative Projects in Solar program (DE-EE0007365). Anna Kindvall would like to thank John Williams, Daisy Williams, and Jennifer Drayton for their guidance throughout this project.

REFERENCES

[1] S. Krum and S. Haymore, "First Solar Achieves Yet Another Cell Conversion Efficiency World Record," *First Solar*, 2016. [Online]. Available: http://investor.firstsolar.com/releasedetail.cfm?ReleaseID=956479.

[2] W. Xia, H. Lin, H. N. Wu, C. W. Tang, I. Irfan, C. Wang, and Y. Gao, "Te/Cu bi-layer: A low-resistance back contact buffer for thin film CdS/CdTe solar cells," *Sol. Energy Mater. Sol. Cells*, vol. 128, pp. 411–420, 2014.

[3] H. Lin, Irfan, W. Xia, H. N. Wu, Y. Gao, and C. W. Tang, "MoO_x back contact for CdS/CdTe thin film solar cells: Preparation, device characteristics, and stability," *Sol. Energy Mater. Sol. Cells*, vol. 99, pp. 349–355, 2012.

[4] H. B. Michaelson, "Work functions of the elements," *J. Appl. Phys.*, 1950.

[5] S.-Y. Lin and Y.-S. Lai, "Effect of nitrogen on the physical properties and work Function of MoN_x cap layers on HfO_2 gate dielectrics," *ECS J. Solid State Sci. Technol.*, vol. 3, no. 12, pp. N161–N165, 2014.

[6] I. Irfan, A. James Turinske, Z. Bao, and Y. Gao, "Work function recovery of air exposed molybdenum oxide thin films," *Appl. Phys. Lett.*, vol. 101, no. 9, pp. 0–4, 2012.

[7] J. Sites, A. Munshi, J. Kephart, D. Swanson, and W. S. Sampath, "Progress and challenges with CdTe cell efficiency," *43rd IEEE Photovolt. Spec. Conf.*, pp. 3632–3635, 2016.

[8] J. a Drayton, D. D. Williams, R. M. Geisthardt, C. L. Cramer, J. D. Williams, and J. R. Sites, "Molybdenum oxide and molybdenum oxide-nitride back contacts for CdTe solar cells Molybdenum oxide and molybdenum oxide-nitride back contacts for CdTe solar cells," *J. Vac. Sci. Technol. A Vacuum, Surfaces, Film. vac*, vol. 33, no. 2015, p. 41201, 2015.

[9] C. Gretener, J. Perrenoud, L. Kranz, C. Baechler, S. Yoon, Y. E. Romanyuk, S. Buecheler, and A. N. Tiwari, "Development of MoOx thin films as back contact buffer for CdTe solar cells in substrate configuration," *Thin Solid Films*, vol. 535, pp. 193–197, 2013.

[10] "XPS Interpretation of Nitrogen", Xpssimplified.com, 2017, [Online]. Available: http://xpssimplified.com/elements/nitrogen.php. [Accessed: 1-Apr-2017].

[11] F. Jahan and B. Smith, "Investigation of solar selective and microstructural properties of molybdenum black immersion coatings on cobalt substrates", *Journal of Materials Science*, vol. 27, no. 3, pp. 625-636, 1992.

[12] M. Anwar, C. Hogarth and R. Bulpett, "An XPS study of amorphous MoO_3/SiO films deposited by co-evaporation", *Journal of Materials Science*, vol. 25, no. 3, pp. 1784-1788, 1990.

[13] B. Brox and I. Olefjord, "ESCA studies of MoO_2 and MoO_3", *Surface and Interface Analysis*, vol. 13, no. 1, pp. 3-6, 1988.

[14] R. Sanjinés, C. Wiemer, J. Almeida and F. Lévy, "Valence band photoemission study of the Ti-Mo-N system", *Thin Solid Films*, vol. 290-291, pp. 334-338, 1996.

[15] R. Nyholm and N. Martensson, "Core level binding energies for the elements Zr-Te (Z=40-52)", *Journal of Physics C: Solid State Physics*, vol. 13, no. 11, pp. L279-L284, 1980.

Investigation and optimization of Cd-free buffer layers In_2S_3 and $Zn(O,S)$ for $Cu_2ZnSn(S,Se)_4$-based solar cells

Willi Kogler, Thomas Schnabel, Andreas Bauer, Stefanie Spiering, Erik Ahlswede and Michael Powalla

Zentrum für Sonnenenergie- und Wasserstoff-Forschung (ZSW) Baden-Württemberg, Stuttgart, Baden-Württemberg, 70563, Germany

Abstract — **In this work we show that both of the investigated buffer materials, In_2S_3 and $Zn(O,S)$, were suitable for alternative Cd-free buffer layers for use in kesterite $Cu_2ZnSn(S,Se)_4$ solar cells. The In_2S_3 buffer was optimized by varying the layer thickness and applied heat treatment resulting in a maximum average efficiency of 4.6 %, compared to 6.1% for CdS. We also showed that the band alignment of absorber and the $Zn(O,S)$ buffer can be adjusted by variation of the sulphur content. The best average efficiency obtained from the $Zn(O,S)$ series was 4.6 %.**

Index Terms — **alternative buffer materials, green solution process, kesterite solar cells**

I. INTRODUCTION

$Cu_2ZnSn(S,Se)_4$ (CZTSSe) is a promising material for thin-film solar cells due to a high absorption coefficient ($<10^4$ cm^{-1}), the use of cheap and earth-abundant elements and a tunable band gap. However, up to date a maximum power conversion efficiency of 12.6 % [1] could be reached for CZTSSe, which lacks far behind the more advanced and already commercialized thin film technologies $Cu(In,Ga)Se_2$ (CIGS) or CdTe with record efficiencies of 22.6 % [2] and 22.1% [3], respectively. Among the reported limitations for kesterite solar cells [4], the high open-circuit voltage (V_{OC}) deficit is considered to be the bottle neck for high efficiencies, resulting from a high recombination at the buffer absorber heterojunction. Although the record efficiencies for CZTSSe as well as for CIGS were achieved using a CdS buffer, the devices suffer from parasitic absorption of this material system, due to its small band gap of 2.4 eV, and environmental issues due to the toxicity of Cd.

Alternative materials e.g. In_2S_3 and $Zn(O,S)$ were already successfully used in highly efficient CIGS devices [5]. These buffer materials were already adopted and investigated also for kesterite based solar cells, resulting in record efficiencies of 7.6 % and 6.5 % for In_2S_3 and $Zn(O,S)$, respectively [6], [7]. Different methods including chemical bath deposition (CBD), co-evaporation [8], chemical spray pyrolysis [9], sputtering [10] and atomic layer deposition (ALD) [11] were used to deposit the buffer layers. In this work In_2S_3 and $Zn(O,S)$ were deposited by ALD and sputtering on CZTSSe absorber for the first time, respectively. Therefore this paper contributes to a more complete comparison of different deposition techniques and a better understanding of these two

alternative material systems for use as buffer layers in kesterite based solar cells.

In terms of optimization a layer thickness variation between 3 nm and 27 nm and different annealing treatments were performed for In_2S_3 buffers. The $Zn(O,S)$ material system was optimized by variation of the concentration ratio $[S]/([S]+[O])$ to obtain an ideal band alignment to the CZTSSe absorber.

II. EXPERIMENTAL

A non-vacuum, solution-based approach is used to deposit the absorber layer. Tin (II) oxide (99.99 %, Sigma-Aldrich), zinc oxide (>99.0 %, Sigma-Aldrich) and copper (II) oxide (>99.0 %, Merck) were dissolved in an aqueous ammonium thioglycolate solution (~60 %, Aldrich Chemistry). The solution was then deposited onto molybdenum coated (550 nm) soda-lime glass (SLG) using doctor blading and annealed at 300°C in nitrogen atmosphere to evaporate the solvent and form the precursor layer. This procedure was repeated four times to obtain the desired precursor layer thickness. Subsequently the sample was placed in a graphite box in the presence of 400 mg selenium and annealed at 550°C for 15 min in nitrogen atmosphere inside an Annealsys AS-One 100 rapid thermal annealing furnace.

The respective buffer layers were deposited as follows: the CdS for the reference sample was deposited by chemical bath deposition (CBD) at 65°C. In_2S_3 was deposited by ALD inside an ASM microchemistry F-450 reactor at a deposition temperature of 210°C. Indium acetylacetonate ($In(CH_3COCHCOCH_3)_3$) and hydrogen sulfide (H_2S) were used as precursor materials and the layer thickness was controlled by the number of process cycles. $Zn(O,S)$ was sputtered from mixed $Zn(O,S)$ targets with respective $[S]/([S]+[O])$ concentration ratios of 0.2, 0.3, 0.4 and 0.5, inside an Ardenne lab sputter coater system CS730S at 10 μbar. A sputter power of 0.95 W/cm² was used. Finally, a thin layer intrinsic of ZnO (40 nm) and aluminum-doped ZnO (400 nm) was sputtered onto the buffer layer to form the front contact. The single solar cells on the sample were separated by mechanical scribing to obtain devices with a total area of 0.25 cm² each.

For solar cell characterization a Keithley 2400 source measuring unit under simulated AM 1.5 global solar irradiation with a WACOM 2-lamp sun simulator at 100 mW/cm² was used to obtain the current-voltage (J-V) curves.

978-1-5090-5606-4/17 $31.00 © 2017 IEEE

Temperature dependent *J-V* measurements were performed in the range of 7°C to 60°C. External quantum efficiency (EQE) measurements were performed using an Optosolar SR300 setup.

III. RESULTS

A. In₂S₃

In_2S_3 was selected as an alternative buffer material due to its higher (2.7 ± 0.1 eV [12]) and indirect band gap and hence superior absorption performance compared to the conventional CdS. This material system is well investigated for CIGS solar cells and a beneficial effect using annealing treatment was observed due to diffusion processes at the absorber/buffer interface [13]. Therefore the fabricated solar cells in this work were also annealed at different temperatures (180°C and 250°C) for 15 min at ambient conditions and compared to the as-prepared samples in terms of optimization. Fig. 1 shows the effect of the buffer thickness variation and annealing temperature on the solar cell efficiency. The samples with very thin buffer layers (3 nm and 5 nm) did not result in working devices, since no photocurrent could be detected. The remaining solar cells show an efficiency enhancement with increasing annealing temperature, which results from increasing short-circuit current (J_{SC}) and open-circuit voltage (V_{OC}). A beneficial effect of annealing is present for all layer thicknesses, but highest for 7 nm thick layers. Thicker layers show slightly higher values already without post deposition heat treatment, but cannot reach as high values as the sample containing a 7 nm In_2S_3 buffer layer.

This effect is commonly attributed to an intermixing of the absorber and buffer layer at the interface due to diffusion of copper from the absorber into the buffer and indium from the buffer into the absorber layer. Furthermore an efficiency decrease can be observed with increasing buffer thickness.

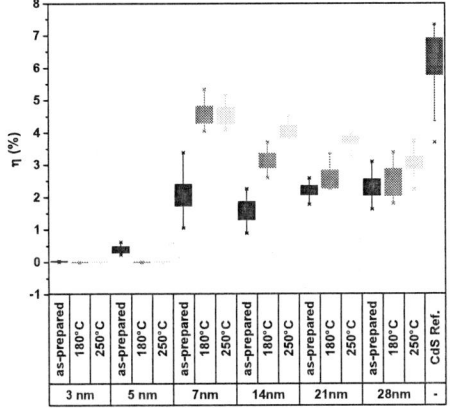

Fig. 1: Evolution of efficiency of the In_2S_3 samples: Shown are different In_2S_3 thicknesses and annealing temperatures in comparison with the CdS reference

The highest average power conversion efficiency of 4.6 % of this series was measured for the sample containing a thin (7 nm) In_2S_3 buffer layer annealed at 180°C. Although a comparably high V_{OC} could be obtained, the efficiency is still 1.5 % lower compared to the average value of the CdS reference sample due to a significantly lower J_{SC}. For a further investigation of the origin of the low J_{SC} a quantum efficiency measurement was performed on an average solar cell with 7 nm In_2S_3 buffer. The spectrum in Fig. 2 shows the comparison of the EQE of the best solar cell device containing an In_2S_3 buffer (7 nm, 180°C) and an average reference solar cell with CdS buffer. It can be seen that the device with the In_2S_3 buffer shows a superior performance in the UV region of the spectrum compared with the reference. In this region a higher current can be gained. However, a strong current loss is observed in the long-wavelength region of the spectrum compared to the reference. This indicates a worse charge carrier collection deeper inside the absorber layer.

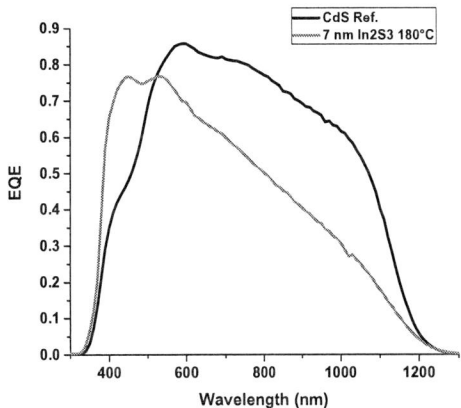

Fig. 2: EQE measurement of average solar cells with In_2S_3 and CdS buffer

However, despite a J_{SC} value lower than for the reference sample, a comparably high V_{OC} could be obtained. Therefore temperature dependent current-voltage curves were measured. Fig. 3 shows the respective measured data for the solar cells with In_2S_3 and CdS buffer, respectively. The measured data points were fitted and extrapolated to 0 K. The activation energy of the respective sample was determined from the intercept of the V_{OC} axis. Both investigated samples show a similar activation energy at about 1.02 eV. The deviation of activation energy from the band gap value is considered to be an indication for the dominating recombination mechanism in the device [14]. The band gap for the two compared solar cells

978-1-5090-5606-4/17 $31.00 © 2017 IEEE

was determined from the quantum efficiency spectrum and had a value of 1.07 eV. Since the activation energy deviates by just 50 meV from the band gap for both of the compared solar cells, it can be concluded that bulk recombination has a much higher contribution to the charge carrier recombination in the device than the interface recombination at the p-n heterojunction. The same activation energy value of both samples indicates a comparable band alignment, which is in accordance with their similar V_{OC} values.

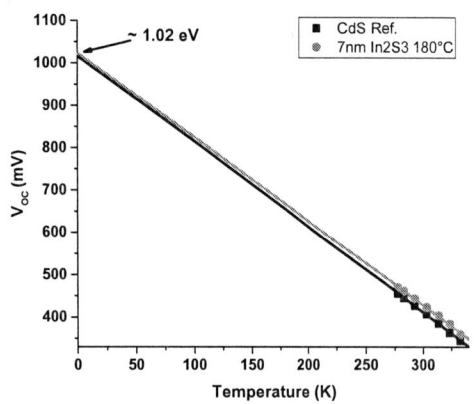

Fig. 3: Temperature dependent J-V measurement for In$_2$S$_3$ (red fit) containing solar cell and CdS reference (black fit)

B. Zn(O,S)

Zn(O,S) is well suitable for solar cell buffers due to its high band gap and adjustable band alignment to the absorber. With increasing sulfur content, the band gap can be adjusted between 3.3 eV for pure ZnO and 3.6 eV for pure ZnS, with a minimum band gap for mixed ratios, an effect referred to as band bowing [15]

Thereby parasitic absorption can be avoided in the UV region of the spectrum and the band alignment can be adjusted. Four different [S]/([S]+[O]) ratios of 0.2, 0.3, 0.4 and 0.5 were investigated in this work. Fig. 4 shows the comparison of power conversion efficiencies for the samples with different Zn(O,S) buffers and the CdS reference. Considering the solar cells with the alternative buffer material, it can be seen that the sample with a [S]/([S]+[O]) ratio of 30 % shows the best performance. This results from a comparably higher fill factor and J_{SC}. However, the average efficiency of the sample with the best Zn(O,S) composition is still 1.5 % lower than the average value of the CdS reference, which is due to a considerably lower V_{OC}, while a comparably high fill factor and slightly higher J_{SC} could be obtained.

The low V_{OC} most likely results from a non-ideal band alignment at the heterojunction. To get a better understanding of the dominating recombination mechanism a temperature dependent J-V measurement was performed and compared to the CdS reference, which is shown in Fig. 5.

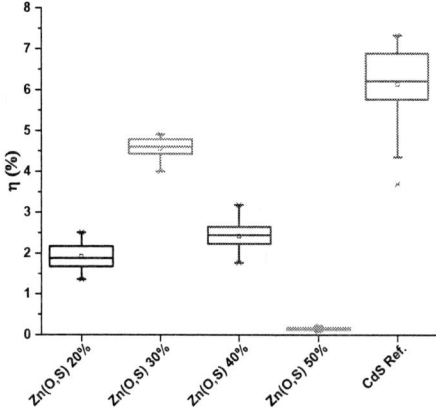

Fig. 4: Characteristic solar cell parameters for the samples with the different investigated Zn(O,S) buffer and the CdS reference

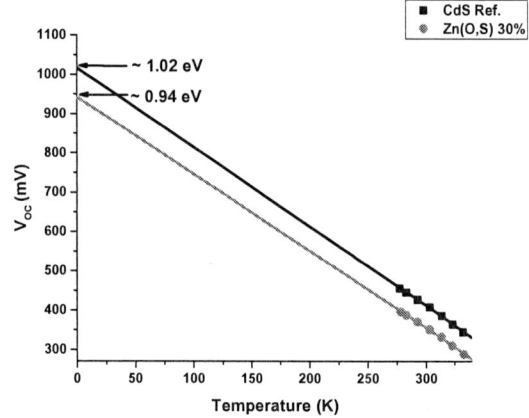

Fig. 5: Temperature dependent J-V measurement for Zn(O,S) (red fit) containing solar cell and CdS reference (black fit)

As can be seen in Fig. 5, the activation energy of the measured device with Zn(O,S) buffer is lower than the CdS reference and hence shows a bigger deviation from the band gap energy. This indicates a higher contribution of interface recombination at the heterojunction for the Zn(O,S) containing solar cell compared to the CdS reference, which is likely to be the reason for the low V_{OC}. Since the alternative buffer layer was deposited by sputtering, sputter damage cannot be excluded, which could be the reason for the poor quality of the buffer/absorber interface due to additionally introduced defects and therefore higher charge carrier recombination.

Finally, a quantum efficiency measurement was performed on a solar cell with a Zn(O,S) layer with a ration of [S]/([S]+[O]) = 0.3 and the CdS reference to investigate the origin of the high J_{SC} of the sample with alternative buffer material.

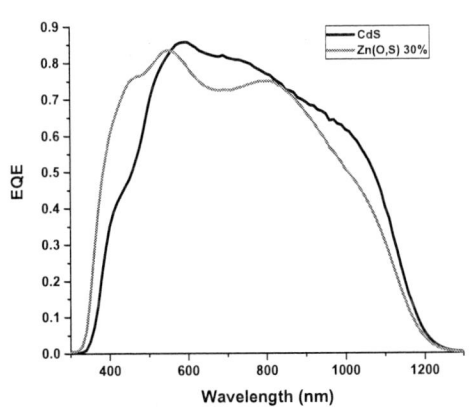

Fig. 6: EQE measurement of average solar cells with Zn(O,S) (S/(S+O) = 30 %) and CdS buffer

As can be seen from the EQE spectra in Fig. 6 and as expected, the current gain of the Zn(O,S) containing sample results from a superior performance in the UV region. The absorber band gap, determined from the linear region of the squared EQE spectrum was found to be 1.08 eV for both of the measured devices, as well as for the solar cell with In_2S_3 buffer. Therefore it can be concluded that the buffer layer has no influence on the absorber band gap.

IV. SUMMARY

In this work two alternative buffer materials, In_2S_3 and Zn(O,S), have been investigated and compared to a reference sample with CdS buffer. A variation of the layer thicknesses was performed for the samples with ALD deposited In_2S_3 buffer, resulting in better performance of thin buffer layers. Additionally, different annealing treatments were applied to the complete solar cells to initiate diffusion processes resulting in an intermixing at the heterojunction and therefore an enhanced interface quality with less charge carrier recombination. Thin layers of 7 nm and an annealing of the devices at 180°C for 15 min in air gave the best cell performance for the In_2S_3 series with a comparable V_{OC} to the CdS reference. Furthermore a similar band alignment as for the CdS reference could be shown by JV-T measurements. In agreement with the expectation, In_2S_3 showed a superior absorption behavior in the UV region of the spectrum than

CdS. ALD proved to be an interesting deposition technique for In_2S_3 buffers on CZTSSe. However, no higher J_{SC} could be obtained compared to the CdS reference, due to significant losses in the long-wave region of the spectrum. The origin of these losses is subject of ongoing investigations.

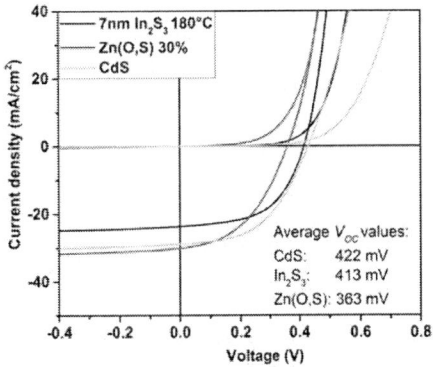

Fig. 7: Measured J-V curves of solar cells with different buffer materials, black: In_2S_3, red: Zn(O,S) and green: CdS

The Zn(O,S) buffer was deposited with different compositions ([S]/([S]+[O]) = 0.2, 0.3, 0.4 and 0.5) by sputtering from Zn(O,S) mixed targets. The sample with a [S]/([S]+[O]) – ratio of 0.3 was found to yield the best performance of this series due to a comparably higher fill factor and J_{SC}. In comparison to the CdS reference, the best Zn(O,S) sample showed a slightly higher J_{SC}, but lower V_{OC}. This might result from introduced defects at the buffer/absorber interface due to sputtering damage during the deposition process. The JV-T measurement also indicated a poor interface quality at the heterojunction. Hybrid material systems could be a good possibility to prevent the absorber layer from sputter damage. Ideally, hybrid buffer systems could combine the beneficial properties of the respective material, e.g. the comparably high V_{OC} of In_2S_3 and high J_{SC} of Zn(O,S).

ACKNOWLEDGEMENT

These activities were funded by the Federal Ministry of Education and Research (BMBF) under contract number 100251857 (FREE-INCA)

REFERENCES

[1] W. Wang, M. T. Winkler, O. Gunawan, T. Gokmen, T. K. Todoro, Y. Zhu and D. B. Mitzi, "Device Characteristics of CZTSSe Thin-Film Solar Cells with 12.6% Efficiency" *Adv. Energy Mater.*, 2014, 4, 1301465

[2] P. Jackson, R. Wuerz, D. Hariskos, E. Lotter, W. Witte, M. Powalla, "Effects of heavyalkali elements in Cu(In,Ga)Se2 solar cells with efficiencies up to 22.6%", *Phys. Status Solidi – Rapid Res. Lett.* 10 (2016) 583 – 586.

[3] Best Research-Cell Efficiencies, National Renewable Energy Laboratory (https://www.nrel.gov/pv/assets/images/efficiency-chart.png) (accessed April 2017)

[4] X. Liu, Y. Feng, H. Cui1, F. Liu, X. Hao, G. Conibeer, D. B. Mitzi and Martin Green, "The current status and future prospects of kesterite solar cells: a brief review", *Prog. Photovolt: Res. Appl.* (2016); 24:879–898

[5] T. Feurer, P. Reinhard, E. Avancini, B. Bissig, J. Löckinger, P. Fuchs, R. Carron, T. P. Weiss, J. Perrenoud, S. Stutterheim, S. Buecheler and A. N. Tiwari, " Progress in thin film CIGS photovoltaics – Research and development, manufacturing, and applications", *Prog. Photovolt: Res. Appl.*, 2016

[6] D. A. R. Barkhouse, R. Haight, N. Sakai, H. Hiroi, H. Sugimoto and D. B. Mitzi, "Cd-free buffer layer materials on $Cu_2ZnSn(S_xSe_{1-x})_4$: Band alignments with ZnO, ZnS, and In_2S_3" *Applied Physics Letters,* vol. 100, 193904, 2012

[7] M. Neuschitzer, K. Lienau, M. Guc, L. C. Barrio, S. Haass, J. M. Prieto, Y. Sanchez, M. Espindola-Rodriguez, Y. Romanyuk, A. Perez-Rodriguez, V. Izquierdo-Roca and E. Saucedo, "Towards high performance Cd-free CZTSe solar cells with a ZnS(O,OH) buffer layer: the influence of thiourea concentration on chemical bath deposition", *Journal of Physics D: Applied Physics* 49, 2016, 125602 (9pp)

[8] M. Buffière, N. Barreau, G. Brammertz, S. Sahayaraj, M. Meuris and J. Poortmans, "Development of co-evaporated In2S3 buffer layer for Cu2ZnSnSe4 thin film solar cells", IEEE, 4799

[9] D. B. Khadka, S.Y. Kim and J. H. Kim, "A Nonvacuum Approach for Fabrication of Cu2ZnSnSe4/In2S3 Thin Film Solar Cell and Optoelectronic Characterization", *The Journal of Physical Chemistry*, 2015, 119, 12226–12235

[10] P. Bras and J. Sterner, "Influence of H2S annealing and buffer layer on CZTS solar cells sputtered from a quaternary compound target", *Proceedings of the 40th IEEE PVSC.* IEEE, 2014

[11] H. K. Hong, I. Y. Kim, S. W. Shin, G. Y. Song, J. Y. Cho, M. G. Gang, J. C. Shin, J. H. Kim, J. Heo, "Atomic layer deposited zincoxysulfide n-type buffe rlayers fo rCu2ZnSn (S,Se)4 thin film solar cells", *Solar Energy Materials & Solar Cells* 155, 2016, 43–50

[12] S. Spiering, D. Hariskos, M. Powalla, N. Naghavi and D. Lincot, "CD-free Cu(In,Ga)Se2 thin-film solar modules with In2S3 buffer layer by ALCVD", *Thin Solid Films* 431 –432, 2003, 359–363

[13] M. Bär, N. Barreau, F. Couzinie-Devy, L. Weinhardt, R. G. Wilks, J. Kessle and C. Heske, "Impact of Annealing-Induced Intermixing on the Electronic Level Alignment at the In_2S_3/Cu(In,Ga)Se_2 Thin-Film Solar Cell Interface", *ACS Applied . Mateials &. Interfaces*, vol. 8, pp. 2120– 2124

[14] V. Nadenau, U. Rau, A. Jasenek, and H. W. Schock, "Electronic properties of CuGaSe2-based heterojunction solar cells. Part I. Transport analysis", *Journal of Applied Physics*, Vol. 87, No. 1, 2000

[15] M. Buffiere, S. Harel, C. Guillot-Deudon, l. Azrael, N. Barreau and J. Kessler, "Effect of the chemical composition of co-sputtered Zn(O,S) buffer layers on Cu(In,Ga)Se_2 solar cell performance", *Phyica. Status Solidi*, A 212, No. 2, pp. 282-290 2015

Rear contact passivation for high bandgap Cu(In,Ga)Se₂ solar cells with varying absorber thickness and flat Ga profile

Dorothea Ledinek[1], Pedro Salome[2], Carl Hägglund[1] and Marika Edoff[1]

[1]Uppsala Universitet, Department for Engineering Sciences, Solid State Electronics,
Uppsala, 75273, Sweden

[2]International Iberian Nanotechnology Laboratory, Braga, 4715-330, Portugal

Abstract — **In this study, six samples of Cu(In,Ga)Se₂ solar cells with a high band gap (1.31 eV) and a flat Ga profile ([Ga]/([Ga]+[In]) ≈ 0.60) were produced. For every nominal absorber thickness of 0.5, 1.0 and 1.5 μm, the Mo rear contact of one sample was passivated with a 27 nm thick Al₂O₃ passivation layer with point contact openings, while the other sample's rear contact remained unpassivated. For the passivated cells, mainly large gains in the short circuit current lead to an up to 21%, 13% and 14% (relative) higher power conversion efficiency compared to the unpassivated cells.**

Index Terms — **back contact, Cu(In,Ga)Se₂ CIGS, CIGSe, Schottky contact, passivation, rear contact, thin films**

I. INTRODUCTION

Recently, the power conversion efficiency record for Cu(In,Ga)Se₂ (CIGS) solar cells has increased significantly, mostly due to improvements in the absorber quality and the front contact. In addition, thinning down the absorber layer has been explored by different groups with the aim to save materials and to decrease production costs [1]–[6]. To curtail the increasing influence of recombination at the rear contact for thinner absorbers, Ga grading with a high Ga content at the rear contact has been successfully applied in state-of-the-art solar cell devices [2], [4]. If Ga replaces In in CuInSe₂, the bandgap increases mostly by elevating the conduction band edge. Thus, by having a large quantity of Ga at the back while keeping a low quantity at the front of the solar cell, a conduction band gradient is introduced and acts as a quasi-electrical field for electrons, lowering rear interface recombination. Another method to curb rear contact recombination is rear contact passivation [7], [8]. This passivation can be done by the introduction of an extra oxide layer (the passivation layer) between the absorber layer and the metal rear contact. Hole conduction through the electrically insulating passivation layer is achieved by point contact openings. The passivation layer works in two ways, namely through a field effect (favorable band bending of both the conduction and the valence band) and through chemical passivation due to a lower interface trap density[9], [10].

For solar cells with thin absorbers, light with long wavelengths is absorbed, and carriers are generated, close to the rear contact. If a strong Ga grading is used, the absorption of this light can be reduced and, in addition, the properties of the interface between the high Ga CIGS and the rear contact as well as the material properties of the high Ga CIGS attain importance. To model those cells and to distinguish recombination in the bulk, at the front interface and at the rear interface, respectively, requires good knowledge of the electrical properties of CIGS with a high Ga content and of the interface between high Ga CIGS and the rear contact. These models can then be used to optimize the Ga grading profiles for thin absorbers with and without rear surface passivation [11]. For high Ga absorbers modeling fails if Ga content is only taken into account by changing the position of the conduction band edge and thus neglecting the negative band-offset (cliff) in the conduction band at the CIGS-CdS interface, but doping, defect density [12], defect position within the band gap [13], minority carrier diffusion length [14] and grain boundary type-inversion [15] are held constant. Modelling then predicts too high open circuit voltages and fill factors and thus power conversion efficiencies.

In previous works our group studied the passivation of the rear contact interface for absorber layers without Ga grading and with a Ga content of about Ga/(Ga+In) ≈ 0.3 [7], [8], [16]. As even thin layers of Al₂O₃ hinder Na diffusion from the soda lime glass, a NaF precursor layer was applied on the Al₂O₃ passivation layer before evaporation of the CIGS absorber layer. In the present study, an ungraded absorber with much higher Ga content of Ga/(Ga+In) ≈ 0.60 was used and NaF was in-situ evaporated on top of the CIGS layer (post-deposition treatment). This study adds also to a previous work [17], in which a two dimensional matrix of solar cells with absorber layers with a Ga content [Ga]/([Ga]+[In]) of 0.15, 0.30, 0.45 and 0.60 and thickness of 0.5, 1.0 and 1.5 μm were studied, with no NaF added and without rear contact passivation. For the samples with a Ga content of [Ga]/([Ga]+[In]) ≈ 0.60 the EQE curves were nearly indistinguishable between CIGS thicknesses of 1.0 and 1.5 μm even for long wavelengths. In agreement with reference [14], a short collection length (minority carrier diffusion length plus space charge region) was proposed. Therefore, we were curious if a passivation effect would be observable in the solar cells with a 1.5 μm thick absorber layer in this study.

978-1-5090-5606-4/17 $31.00 © 2017 IEEE

II. Experimental

For this study a two dimensional matrix of six CIGS solar cell samples with a nominal Ga content Ga/(Ga+In)≈0.60 were produced. Three samples were passivated and three were non-passivated reference samples. The intended thickness of the CIGS absorber was 0.5 µm, 1.0 µm and 1.5 µm for the two cases, respectively.

For each solar cell sample the CIGS layer was co-evaporated onto a passivated or a non-passivated Mo rear contact on a soda-lime glass substrate with a size of 5x5 cm^2. Following the group's baseline procedure[18], a CdS buffer layer was deposited from a chemical bath on top of the CIGS, a ZnO/ZnO:Al window layer was sputtered and a front contact grid was evaporated.

The 27 nm thick Al_2O_3 passivation layer was deposited onto the Mo rear contact with atomic layer deposition (ALD) at 300°C using water and trimethylaluminium (TMA) as precursors and nitrogen as purge gas. A quadratic grid of openings, each with a length of 0.7 µm and a width of 0.4 µm and a pitch of 2 µm between the openings was carefully etched into the Al_2O_3 layer by reactive ion etching after defining the pattern using electron beam lithography as in reference[7].

For all samples the CIGS absorber layer was co-evaporated in a vacuum chamber at a base pressure of $3 \cdot 10^{-6}$ mbar. Constant evaporation rates were used for all elements using a quadrupole mass spectrometer rate control for the metal sources and temperature control for the Se source. Due to the constant evaporation rates, resulting films have no, or very small, elemental gradients. During the first 125 s after the shutter opened, the substrate temperature was 410°C and during the following 125 s the temperature was linearly raised to 530°C for all samples. To evaporate a 500 nm thick absorber layer the substrate temperature was kept constant until the end of the evaporation after 125 more seconds. This high temperature phase was prolonged to achieve 1.0 and 1.5 µm thick absorbers by multiplying the total evaporation time (i.e. all three stages combined) by a factor of 2 or 3 respectively. After the CIGS deposition, all substrates were allowed to cool down to 490°C and then kept at that temperature for 400 seconds during the post deposition treatment, where 0.5 mg NaF was evaporated. In each CIGS run one passivated and one unpassivated sample were deposited to reach the most similar process conditions possible. In the last step 32 solar cells were scribed on each sample.

For every CIGS evaporation process, the composition and thickness of four dedicated test samples with a glass-Mo-CIGS stack were determined by x-ray fluorescence (XRF), calibrated against a CIGS sample with known composition, a [Ga]/([Ga]+[In]) ratio of 0.40 and a thickness of 1.7 µm as measured by a profilometer. The thickness of the Al_2O_3 passivation layer and the complex refractive indices of Mo, the Al_2O_3 and the CIGS were determined by spectroscopic ellipsometry on different sample stacks with a Woollam VASE instrument. The absorption coefficient was then calculated and used to calculate the remaining light intensity reaching the back contact at different wavelengths. The reflectance at the unpassivated and the passivated back contact was also calculated with the transfer matrix method. The reflectance of the whole sample stacks was measured using a Perkin Elmer L900 spectrophotometer with an integrating sphere. The diameter and depth of the passivation layer contact openings were determined by atomic force microscopy (AFM) in tapping mode to control that the etching process was complete.

TABLE 1

SUMMARY OF SOLAR CELL PARAMETERS (ARITHMETIC AVERAGE ± STANDARD DEVIATION)

Absorber thickness in µm	Passivation	V_{OC} in V	J_{SC} in mA/cm^2	Fill factor in %	Efficiency in %	G_{dark} in mS/cm^2	R_{dark} in ohm·cm^2	G_{light} in mS/cm^2	R_{light} in ohm·cm^2
0.5	Yes	0.732±0.003	18.8±0.5	68±3	9.3±0.5	2.2±0.3	1.0±0.1	0.5±0.1	0.9±0.1
0.5	No	0.709±0.004	15.6±0.2	69±1	7.7±0.2	1.9±0.1	0.4±0.5	0.6±0.1	0.4±0.5
1.0	Yes	0.721±0.010	20.0±0.2	69±0	9.9±0.1	1.9±0.1	0.6±0.1	0.7±0.2	0.8±0.1
1.0	No	0.727±0.003	17.6±0.2	69±1	8.8±0.1	2.4±0.1	0.3±0.2	1.0±0.1	0.3±0.2
1.5	Yes	0.747±0.004	22.9±0.7	70±1	12.0±0.4	1.6±0.2	0.3±0.2	0.8±0.2	0.3±0.2
1.5	No	0.718±0.010	20.7±0.3	71±1	10.5±0.1	1.3±0.1	0.2±0.2	0.4±0.1	0.2±0.2

The current voltage (JV) curves were measured in the dark and under a halogen lamp with a four point probe on all cells. The light intensity was calibrated with a Si solar cell to obtain a photon flux corresponding to 1000 W/m^2 at AM 1.5. The temperature during the measurements was set to 25°C by a temperature stage cooled using a Peltier element. External quantum efficiencies (EQE) were measured on two to three representative cells under ambient light. For the calculation of the short circuit current J_{SC} and the efficiency of all cells on a sample, J_{SC} from the JV measurements was corrected for the spectral mismatch by the arithmetic average of the quotient of the J_{SC} from the JV measurements and the EQE measurements on that sample. Both the dark and the light JV curves of 8 good cells were fitted to the single diode model with a method based on reference [19] and the shunt conductance G, the series resistance R and their standard deviations were derived. The bandgap E_G was found by linear regression to the energy axis in a plot of the square of the EQE against energy and by the empirical formula E_G=1.010+0.626x-0.167x(1-x) eV [20].

III. RESULTS

The composition of the dummy CIGS samples used for the XRF measurements varied significantly due to the used deposition geometry while their thicknesses varied negligibly. The average CIGS layer thicknesses and standard deviations were 0.60 ± 0.01 µm, 0.85 ± 0.04 µm and 1.45 µm ± 0.09 µm, respectively. Cu/(In+Ga) varied from 0.83 to 0.89 and Ga/(Ga+In) from 0.55 to 0.61. As the dummy samples were placed furthest out in the deposition zone, these values denote maximum lateral composition and thickness differences.

According to the linear regression described in the experimental part, the bandgap is 1.31 ± 0.01 eV. The empirical formula [20] gives a band gap between 1.27 and 1.33 eV.

The arithmetic average and the standard deviation of the four standard solar cell parameters are given in table 1 for the eight best, representative cells for all samples, except for the thinnest, passivated sample. For this sample, the CIGS layer peeled from the Al_2O_3 layer and only 18 cells could be used for for the analysis. Therefore the average and the standard deviation were calculated only for the seven best, representative cells instead of eight. As the peeling continued over time and the current deteriorated, no EQE measurements are shown for this sample. The JV curves for the best solar cell of every sample can be seen in in fig. 1.

The short circuit current J_{SC} and the EQE increase with the thickness of the absorber layer. Even for the samples with a 1.5 µm absorber layer the J_{SC} is about 3 mA/cm^2 higher compared to the samples with a 1.0 µm absorber layer. The increasing EQE for the increasingly thicker samples (both passivated and un-passivated) can be explained by a long collection length (long minority carrier diffusion length and/or a large space charge region) in combination with increasing absorption when the thickness increases from 0.6 to 0.85 µm. Increasing the thickness further to 1.45 µm leads to a further increase of the current. The reflectance in the region around 800 nm is slightly higher for the 0.85 µm sample as compared to the 1.45 µm sample and the interference fringes indicate a smoother surface. The increase in EQE between sample with 0.85 and 1.45 µm thickness contrasts with the previously produced unpassivated samples[17], that were not post-deposition treated. For those samples the J_{SC} and EQE decreased even slightly for the sample with a 1.0 µm thick absorber layer compared to the one with a 1.5 µm absorber layer.

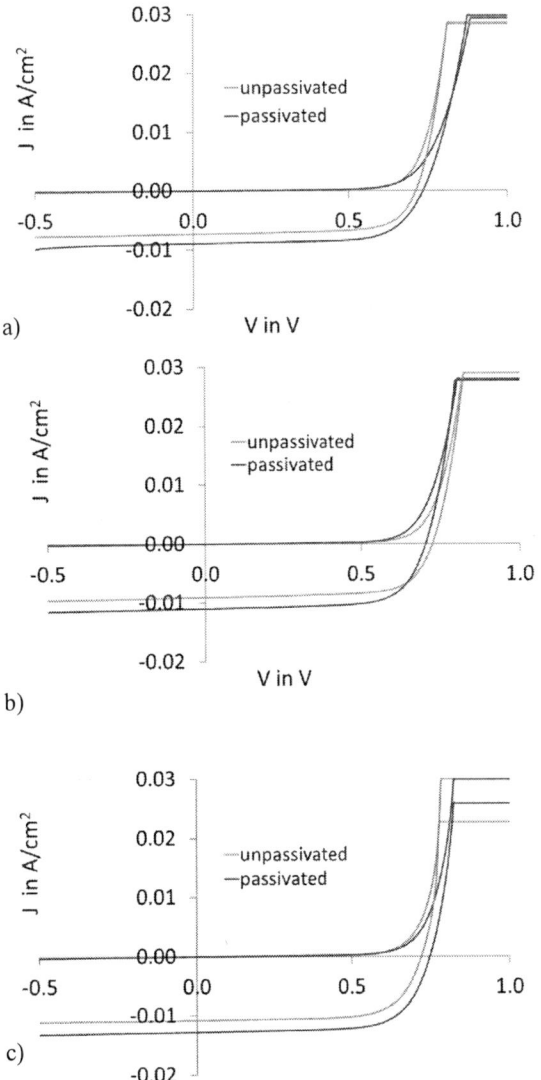

Fig. 1   Current-voltage curves of the best, representative cell on each sample for the samples with a a) 0.5 µm, b) 1.0 µm and c) 1.5 µm thick absorber layer.

The EQE between 530 and 900 nm wavelength (see fig. 2) is significantly higher for the passivated samples compared to the unpassivated ones for all investigated thicknesses. This difference does probably not only stem from optical effects: 1) As seen from fig. 2, the difference in reflectance at the whole sample stack of the passivated and unpassivated samples only explain part of the difference in EQE for the 1.0 µm sample and almost nothing for the 1.5 µm sample. 2) According to the absorption coefficient from the ellipsometry measurements, less than 5% of the incoming light with wavelengths below 800 nm is expected to reach the rear contact for the samples with 0.85 and 1.45 µm thick absorber layers. 3) More detailed transfer matrix calculations based on the spectroscopic ellipsometry analysis shows that the optical absorption in the

CIGS layer varies on the order of 0.1 mA/cm^2 with and without the 27 nm thick passivation layer, while the observed photocurrent difference is on the order of 1 mA/cm^2. Therefore, it appears that most of the additional photocurrent with the Al$_2$O$_3$ layer stems from a passivation effect. This is likely due to electron repulsion from the highly recombinative back contact, since Al$_2$O$_3$ is known to be negatively charged. In addition, a lower effective doping in the passivated samples as compared to the unpassivated may result from the blocking of Na diffusion, and if the post deposition treatment gives a lower effective Na concentration than Na in-diffusion from the glass, this would increase the width of the depletion region and thereby the collection length for the passivated samples.

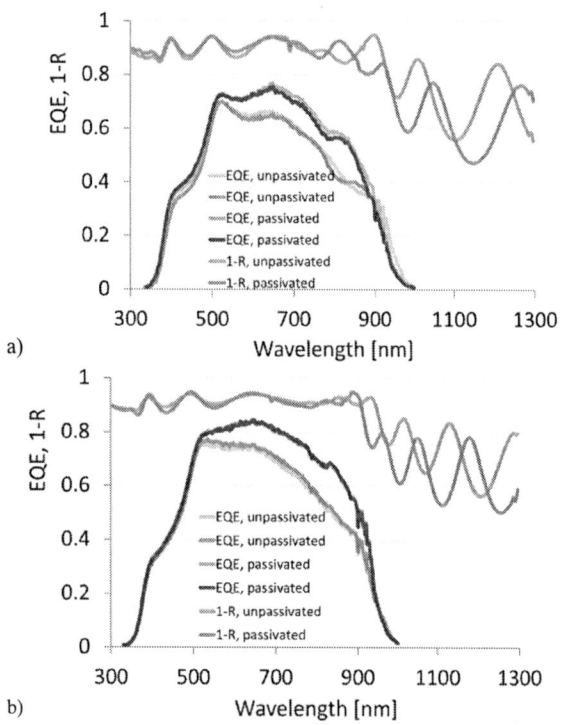

a)

b)

Fig. 2   EQE and 1-R (1-reflectance) of the complete solar cell stack for the samples with a a) 1.0 μm and b) 1.5 μm thick absorber layer.

V_{OC} is higher for the passivated samples than for the unpassivated samples for the samples with a 0.6 μm (+23 mV) and a 1.45 μm(+29 mV) thick absorber, but not for the sample with a 0.85 μm thick absorber. The FFs for all samples are about 70% and the differences are not significant, considering the standard deviation. The thin sample gained the most in efficiency by passivation, namely 21% (relative), the medium thick sample gained 13% (relative) and the thick one 14% (relative) compared to the respective reference samples. No roll-over in the current-voltage curves was observed at room temperature.

V_{OC} is governed by the recombination, where the most important recombination mechanism will dominate. If we assume that the interface recombination is similar for all devices, then differences in depletion region width, bulk recombination and back contact recombination will have to account for all the differences we see. In addition, a lower effective doping will have the effect of lowering the V_{OC} of the passivated samples, in contrast to what we observe. Without modeling, the task of distinguishing between all these effects is difficult, but we believe that back contact recombination may play a role for the voltage differences between the passivated and unpassivated devices also for the thickest sample. Thereby we need to assume that back contact recombination is not only a local effect at the back contact, but an effect of band bending, that extends further into the film. This is presently under further investigation.

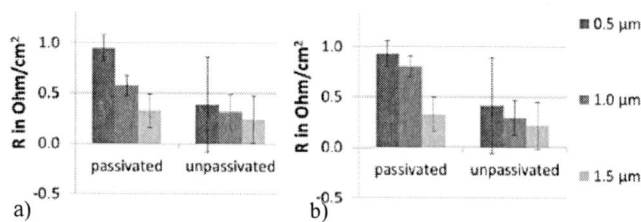

a)   b)

Fig. 3   a) R_{light} and b) R_{dark} and their standard deviations for the eight best, representative cells on every sample.

For the one-diode model the following trends can be seen: For the passivated samples G_{dark}, R_{light} and R_{dark} decrease with increasing absorber layer thickness, while there is no general trend for the unpassivated samples. For all absorber thicknesses R_{light} and R_{dark} (see Fig. 3) are larger for the passivated samples compared to the unpassivated references. The increase in R for thinner passivated absorbers can be explained by a smaller cross-sectional area for the lateral hole movement: the pitch between point contacts is 2 μm for all samples, but the height of the absorber layer varies between 0.6 and 1.45 μm (see fig. 2). This means that the lateral bulk resistance of the absorber layer contributes significantly to R, while the increase in contact resistance due to the smaller contact area is not dominant. All data are summarized in table 1.

IV. CONCLUSION

In this work, a comparison of rear surface passivated samples using Al$_2$O$_3$ with hole contact openings and unpassivated reference samples of CIGS solar cells with absorbers with a Ga content of [Ga]/([Ga]+[In]) ≈ 0.6 and a flat Ga profile has been made. Absorbers with three different thicknesses of 0.60, 0.85 and 1.45 μm, were compared. We observe a large improvement in short circuit current with the passivation when comparing passivated and unpassivated samples in combination with a slight improvement in voltage in the case of 0.6 and 1.45 μm thick absorbers. The thin sample gained the most in efficiency by passivation, namely 21% (relative), the medium thick sample gained 13% (relative) and the thick one 14% (relative) compared to the respective reference samples

The enhanced photocurrent with the passivation layer cannot be accounted for by optical effects. Rather, our data suggests it to be the result of improved current collection, which may occur through an increased depletion region width for the passivated samples and/or reduced back contact recombination. In order to explain both the low EQE response for the unpassivated samples in the IR region and the higher voltage for the passivated samples with 1.45 μm thick CIGS, back contact recombination is most probably not only a local effect at the back contact, but is the result of band bending that extends into the bulk of the CIGS film.

A higher series resistance for both passivated cells and for thinner absorber layers indicate that the additional series resistance is dominated by lateral hole transport in the CIGS to the point contacts in the passivation layer, and not by the contact resistance at the point contacts. Temperature dependent current voltage measurements on these samples can clarify this further and will be examined in subsequent work.

ACKNOWLEDGEMENTS

This work was funded by the Swedish Energy Agency and the Strategic Research program STandUP. We are grateful for input from our colleagues in the Thin Film Solar Cell group at the Ångström lab.

REFERENCES

[1] J. Malmström, O. Lundberg, and L. Stolt, "Potential for light trapping in Cu(In, Ga)Se$_2$ solar cells," in *3rd World Conference on Photovoltaic Energy Conversion May 11-18, 2003 Osaka, Japan*, 2003, pp. 344–347.

[2] O. Lundberg, M. Edoff, and L. Stolt, "The effect of Ga-grading in CIGS thin film solar cells," *Thin Solid Films*, vol. 480–481, pp. 520–525, Jun. 2005.

[3] Z. Jehl, F. Erfurth, N. Naghavi, L. Lombez, I. Gerard, M. Bouttemy, P. Tran-Van, a. Etcheberry, G. Voorwinden, B. Dimmler, W. Wischmann, M. Powalla, J. F. Guillemoles, and D. Lincot, "Thinning of CIGS solar cells: Part II: Cell characterizations," *Thin Solid Films*, vol. 519, no. 21, pp. 7212–7215, Aug. 2011.

[4] A. Han, Y. Zhang, W. Song, B. Li, W. Liu, and Y. Sun, "Structure, morphology and properties of thinned Cu(In, Ga)Se$_2$ films and solar cells," *Semicond. Sci. Technol.*, vol. 27, no. 3, p. 35022, Mar. 2012.

[5] J. Pettersson, T. Törndahl, C. Platzer-Björkman, A. Hultqvist, and M. Edoff, "Influence of absorber thickness on on Cu(In,Ga)Se2 solar cells with different buffer layers,"

IEEE J. photovoltaics, vol. 3, no. 4, p. 1376, 2013.

[6] E. Jarzembowski, M. Maiberg, F. Obereigner, K. Kaufmann, S. Krause, and R. Scheer, "Optical and electrical characterization of Cu(In,Ga)Se$_2$ thin film solar cells with varied absorber layer thickness," *Thin Solid Films*, vol. 576, pp. 75–80, Feb. 2015.

[7] B. Vermang, J. T. Watjen, C. Frisk, V. Fjallstrom, F. Rostvall, M. Edoff, P. Salome, J. Borme, N. Nicoara, and S. Sadewasser, "Introduction of Si PERC rear contacting design to boost efficiency of Cu(In,Ga)Se$_2$solar cells," *IEEE J. Photovoltaics*, vol. 4, no. 6, pp. 1644–1649, 2014.

[8] B. Vermang, V. Fjallström, X. Gao, and M. Edoff, "Improved rear surface passivation of Cu(In,Ga)Se$_2$ solar cells: A combination of an Al$_2$O$_3$ rear surface passivation layer and nanosized local rear point contacts," *IEEE J. Photovoltaics*, vol. 4, no. 1, pp. 486–492, 2014.

[9] W.-W. Hsu, J. Y. Chen, T.-H. Cheng, S. C. Lu, W.-S. Ho, Y.-Y. Chen, Y.-J. Chien, and C. W. Liu, "Surface passivation of Cu(In,Ga)Se$_2$ using atomic layer deposited Al$_2$O$_3$," *Appl. Phys. Lett.*, vol. 100, no. 2, p. 23508, 2012.

[10] R. Kotipalli, B. Vermang, J. Joel, R. Rajkumar, M. Edoff, and D. Flandre, "Investigating the electronic properties of Al$_2$O3/Cu(In,Ga)Se$_2$ interface," *AIP Adv.*, vol. 5, p. 107101, 2015.

[11] C. Frisk, C. Platzer-Björkman, J. Olsson, P. Szaniawski, J. T. Wätjen, V. Fjällström, P. Salome, and M. Edoff, "Optimizing Ga-profiles for highly efficient Cu(In,Ga)Se2 thin film solar cells in simple and complex defect models," *J. Phys. D. Appl. Phys.*, vol. 47, p. 485104, 2014.

[12] G. Hanna, A. Jasenek, U. Rau, and H. Schock, "Influence of the Ga-content on the bulk defect densities ofCu(In,Ga)Se$_2$," *Thin Solid Films*, vol. 387, pp. 71–73, 2001.

[13] J. T. Heath, J. D. Cohen, W. N. Shafarman, D. X. Liao, and A. A. Rockett, "Effect of Ga content on defect states in CuIn$_{1-x}$Ga$_x$Se$_2$ photovoltaic devices," *Appl. Phys. A Mater. Sci. Process.*, vol. 80, p. 4540, 2002.

[14] J. E. Phillips and W. N. Shafarman, "Analysis of Cu(In, Ga)Se$_2$ solar cells: Why performance decreases with increasing Ga content," *AIP Conf. Proc.*, vol. 120, pp. 120–125, 1999.

[15] W. Li, S. R. Cohen, K. Gartsman, and D. Cahen, "Ga Composition Dictates Macroscopic Photovoltaic and Nanoscopic Electrical Characteristics of Cu(In$_{1-x}$Ga$_x$)Se$_2$ Thin Films via Grain-Boundary-Type Inversion," *IEEE J. photovoltaics*, vol. 2, no. 2, pp. 191–195, 2012.

[16] B. Vermang, J. T. Wätjen, V. Fjällström, F. Rostvall, M. Edoff, R. Gunnarsson, I. Pilch, U. Helmersson, R. Kotipalli, F. Henry, and D. Flandre, "Highly reflective rear surface passivation design for ultra-thin Cu(In,Ga)Se$_2$ solar cells," *Thin Solid Films*, pp. 2–5, Oct. 2014.

[17] D. Ledinek, B. Vermang, and M. Edoff, "Thickness and Ga-content variations in co-evporated CIGS solar cells with flat Ga profile - an electrical characterization," in *29th*

European Photovoltaic Solar Energy Conference and Exhibition, 2014, no. 1, pp. 1832–1836.

[18] J. Lindahl, U. Zimmermann, and P. Szaniawski, "Inline Cu(In,Ga)Se$_2$ Co-evaporation for High-Efficiency Solar Cells and Modules," *ieeexplore.ieee.org*, vol. 3, no. 3, pp. 1–6, 2013.

[19] S. S. Hegedus and W. N. Shafarman, "Thin-film solar cells: device measurements and analysis," *Prog. Photovoltaics Res. Appl.*, vol. 12, no. 23, pp. 155–176, Mar. 2004.

[20] M. I. Alonso, M. Garriga, C. A. Durante Rincón, E. Hernández, and M. León, "Optical functions of chalcopyrite CuGa$_x$In$_{1-x}$Se$_2$ alloys," *Appl. Phys. A Mater. Sci. Process.*, vol. 74, no. 5, pp. 659–664, 2002.

Laser Annealed Back Contacts for CdTe Solar Cells

Vasilios Palekis[1], Shamara Collins[1], Imran Khan[1], Vamsi Evani[1], Sudhajit Misra[2], Michael A. Scarpulla[2], Mark Lonergan[3], Don Morel[1] and Chris Ferekides[1]

[1] Electrical Engineering, University of South Florida, Tampa, FL 33620, USA
[2] Materials Science and Engineering, University of Utah, Salt Lake City, UT 84112, USA
[3] Chemistry and Biochemistry, University of Oregon, Eugene, OR 97403, USA

Abstract — The CdTe solar cell back contact interface gets activated by means of thermal annealing. Depending on the back contact (BC) material the annealing time can vary between 20 - 60 minutes. In this study fast contact anneal times - <90 secs – are investigated using a 60 Watt dual diode 808nm laser. Two types of back contacts are used: (a) Cu-Graphite, and (b) MoN/Mo-Cu. Laser power density (LPD) and Cu concentration were varied to study their effect on device performance. Capacitance-voltage characteristics revealed a correlation between LPD and the net doping concentration in CdTe. Higher Cu concentration was found to be detrimental to the device performance, possibly due to excess Cu reaching the CdTe/CdS interface as indicated by secondary ion mass spectroscopy (SIMS) measurements. Solar cells exhibited open-circuit voltage (V_{OC}), short-circuit current (J_{SC}), and fill-factor (FF) similar to the baseline thermally anneal devices. To-date the LPD was optimized with best cell parameters being: V_{OC}= 830 mV, J_{SC}= 22 mA/cm^2, and FF= 70 %, for laser anneal time of 90 seconds.

I. INTRODUCTION

Cadmium telluride (CdTe)-based thin film solar cells continue to develop as a low-cost and high efficiency technology suitable for photovoltaic solar energy conversion. Recent developments have resulted in records for both cell (22.1 %) and module (18.6 %) efficiencies [1], [2].

The formation of stable, low resistance and non-rectifying contact to p-CdTe is one of the biggest challenges in the fabrication of high efficient solar cells due to the high work function of CdTe [3]. Use of lower work function metals result in the formation of Schottky barriers at the back contact. Low V_{OC} in CdTe solar cells has been attributed to barriers at the back contact and to the quality of the CdTe absorber [4]. Cu containing back contacts are often used as Cu increases the p-doping of CdTe near or at the back contact interface which causes the formation of ohmic contacts. However, during the back contact thermal anneal excess Cu can diffuse to the CdTe/CdS region that can be detrimental to device performance [5].

Lasers are appealing for use in semiconductor manufacturing as they are capable of high throughput and precise operations. The primary laser applications in PV today are scribing the cell interconnects and edge isolation. For the solar industry the use of laser processing can reduce the manufacturing costs through improved module efficiencies and higher throughput [6]. Compared to furnace and rapid thermal annealing, laser annealing provides a local source of heat as only the absorbing layer of the device is heated. Therefore higher temperatures can be reached almost instantaneously.

This work is a continuation of previous work where laser annealing was used for the CdCl$_2$ treatment of the devices, and was motivated by rapid thermal annealing results of the CdTe back contact, that suggested the potential for improved device stability [7], [8]. In this work near infra-red (NIR) laser annealing for times less than 90 secs resulted in similar device performance seen for the baseline thermal anneal process that requires 20 to 30 minutes.

II. EXPERIMENTAL

The CdTe cells discussed in this paper are of the superstrate configuration and were fabricated on corning EagleXG glass. The transparent contact (TC) is indium tin oxide (ITO) and was deposited by sputtering. Cadmium sulfide (CdS) and CdTe were then deposited by chemical bath deposition (CBD) and close-spaced sublimation (CSS) respectively; additional details on the fabrication of these films can be found in previously published work [9]. Cu/graphite paste and MoN/Mo/Cu were used as the back contacts. MoN, Mo, and Cu layers were deposited by RF sputtering at room temperature.

The BC laser annealing was carried out using a HighLight FAP DUO 60/810 60 watt dual diode laser (donated by Intevac, Inc.) operating at 808 nm. The 800 μm^2 square profile laser beam coming out of the fiber optic cable was expanded to a rectangular shaped beam with size 1 × 5 cm^2 using a combination of rectangular lenses. Laser power density and annealing time were varied from 1.5 to 6 W/cm^2 and 30 to 90 seconds respectively. For the MoN/Mo/Cu BC the Cu concentration was varied by depositing Cu to a thickness between 1 and 10 Å. Devices were laser annealed under a stationary beam. Although, precise direct measurements of the sample temperature during the laser anneal process were not very reliable during this work, it is estimated that within the LPD range and anneal times used the sample temperatures ranged from 180 to 400 °C.

CdTe solar cells were characterized using the current-voltage (J-V), spectral response (SR), capacitance-voltage (C-V), secondary ion mass spectroscopy (SIMS), and deep-level transient spectroscopy (DLTS). Devices were also thermally stressed to determine whether laser anneal offers any advantages for long term stability. J-V measurements were done using four-terminal connections with a Keithley 2410 source meter. An Oriel monochromator (model 74100) was used for the SR measurements. The intensity of the light source was calibrated using a standard silicon reference solar cell calibrated by the National Renewable Energy Laboratory. C-V measurements were performed using a HP 4194A impedance/gain-phase analyzer.

SIMS measurements were performed at the University of Oregon by professor Mark Lonergan's group. Details of the SIMS setup have been published in previous work [10].

III. RESULTS AND DISCUSSION

A. Cu-Graphite Contacts

Fig. 1 shows the effect of LPD on the V_{OC} and FF of devices with graphite back contact. There is an initial increase in device performance with increasing LPD; however at high LPDs a significant drop in the V_{OC}/FF is observed.

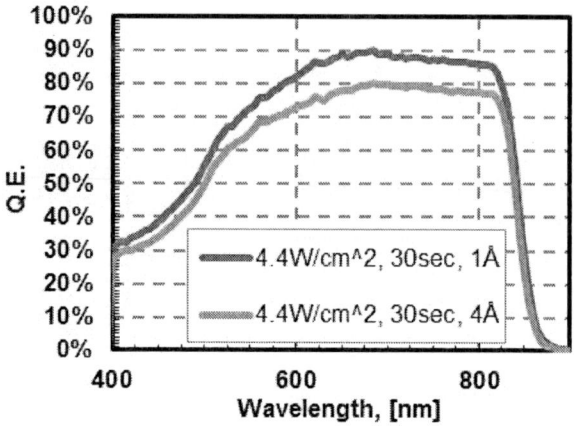

Fig. 2.   SR data for LA back contact for different amounts of Cu.

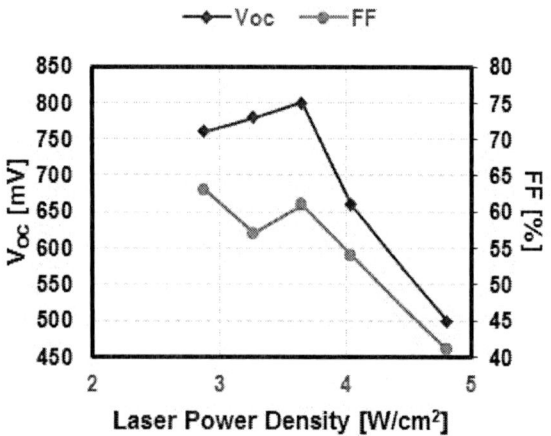

Fig. 1.   Effect of LPD on device performance.

Fig. 2 is a plot of the spectral response of laser annealed devices with different amounts of Cu (1 & 4 Å). The devices were annealed at the same LPD for 30 seconds.

This type of lower QE behavior at wavelengths between 500-800 nm for the device with higher Cu *concentration* has been observed in several laser annealed devices. Cu reaching the CdTe/CdS region increases interface recombination which can explain the no-wavelength dependence of carrier collection.

The carrier concentrations calculated from the slope of A^2/C^2-V characteristics for devices with 1 Å Cu and annealed at different LPD's are shown in Fig. 3. A correlation is observed between LPD and net doping. Increasing the laser power results in higher net doping concentration. This might provide a means of controlling the net doping of the device by adjusting the LPD.

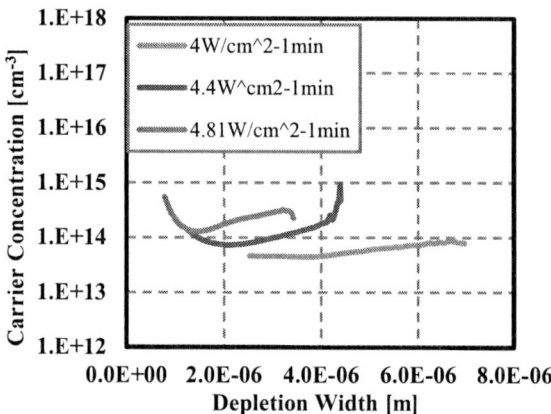

Fig. 3.   Net doping Vs LPD.

The above results suggest that LPD and amount of Cu are critical for optimum device performance, and suggest that laser-based fast annealing can be used to control the extent of Cu diffusion into the device which is critical for the performance and long-term stability of the solar cell.

B. MoN/Mo-Cu Contacts

MoN/Mo-Cu was used as the back contact for devices laser annealed for 30 seconds at different power densities. SR and J-V data are shown in Fig. 4. For low LPD's devices exhibit similar behavior. As the LPD increases to 4.62 W/cm² an initial decrease in Q.E. (independent of wavelength) is observed. This is similar to what was observed in Fig. 2 for *"Cu-graphite"* devices, and is indicative of increased interface recombination. At LPD's above 5 W/cm² the Q.E. continues to degrade, however, the additional degradation exhibits a wavelength dependence that suggests reduced minority carrier lifetime.

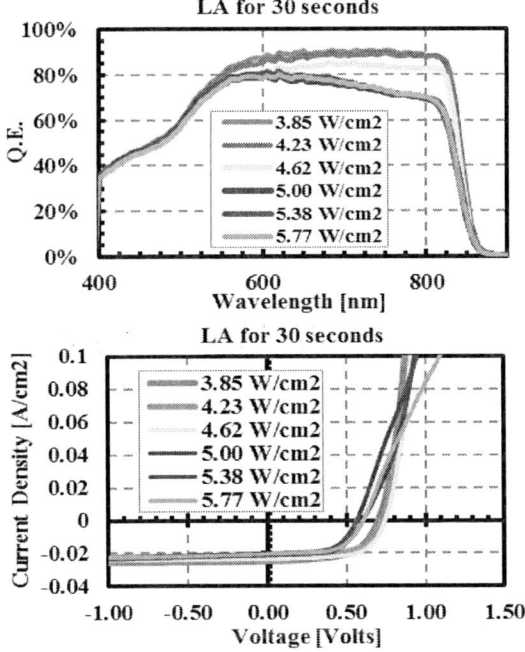

Fig. 4.   Laser annealed devices (30 seconds and 5Å Cu) using varying laser power density (LPD).

In order to better understand these device results, a set of devices were characterized using SIMS. The results are shown in Fig. 5; (these devices were annealed for longer times - 90 secs - than those in Fig. 4. Mo has many isotopes which causes interference with almost all visible Cd and Te isotopes. Thus for each sample, high counts of "Cd" in the Mo is seen, but this is believed to be MoO. Deeper into the device, the Mo counts drop off as expected, and the counts for Cd, Te, and Sn are more reliable. This particular interference is not an issue with the Cu profiles as Cu is lighter than Mo. The Cu SIMS profiles show that at higher LPD Cu diffusion reaches the CdTe/CdS interface. Cu accumulation at the junction interface can lead to increased interface recombination and therefore lower performance similar to what was observed in the *"Cu-Graphite"* devices shown in Figs. 1 and 2, and the device annealed @ LPD of 4.62 W/cm^2 in Fig. 4. Therefore, it appears that several competing mechanisms are at play during the laser anneal process: interface and bulk recombination. As the LPD (and therefore the sample temperature) is increased a significant amount of Cu can diffuse from the back contact and reach the CdTe/CdS interface leading to performance degradation due to interface recombination. At higher LPD, Cu related defects (such as Cu_i) begin to form resulting in lower lifetimes, and further performance degradation [11], [12].

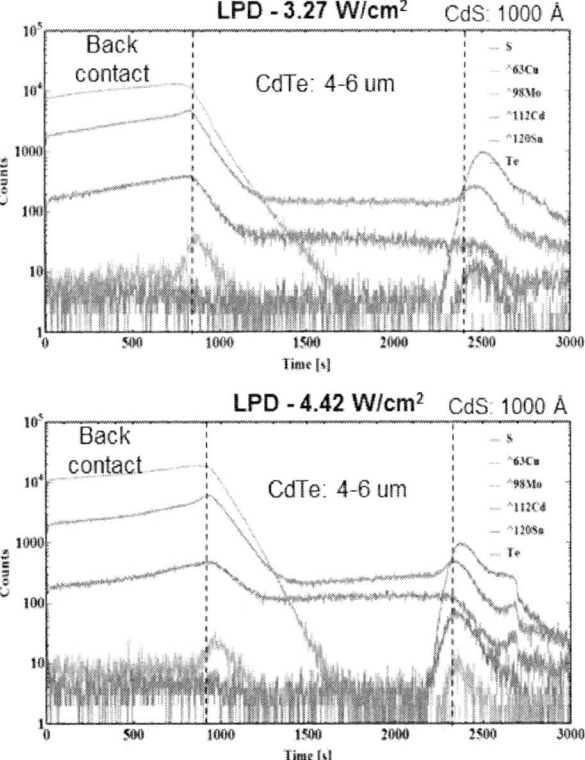

Fig. 5.   SIMS analysis for CdTe/CdS devices treated at 3.27 and 4.42 LPDs.

C. Thermal Stress Results

Cells with thermal anneal and laser-processed back contacts were thermally stressed in inert ambient under low vacuum at 80 and 100 °C; to-date cells have been stressed for 250 hours. Fig. 6 shows the device performance for a thermally and a laser treated at low LPD (3.08 W/cm^2) solar cells. The thermally stressed device is not altered by the stressed conditions used in this study. For the laser annealed device after an initial drop in V_{OC} and FF the performance improves as the solar cell undergoes thermal stress. The opposite trend was observed for solar cells treated at higher LPD. These observations are believed to be related to Cu diffusion and Cu-related defects. DLTS results (discussed below) suggest that the thermal stress impacts an important acceptor in CdTe cells – Cu_{Cd}.

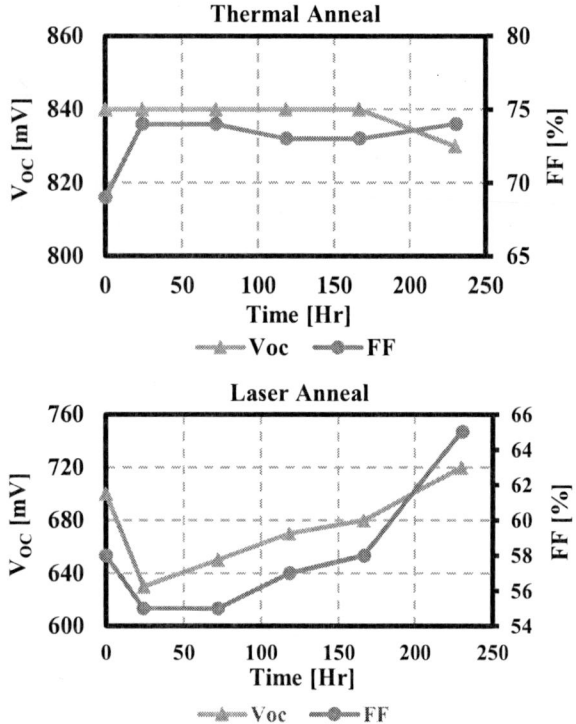

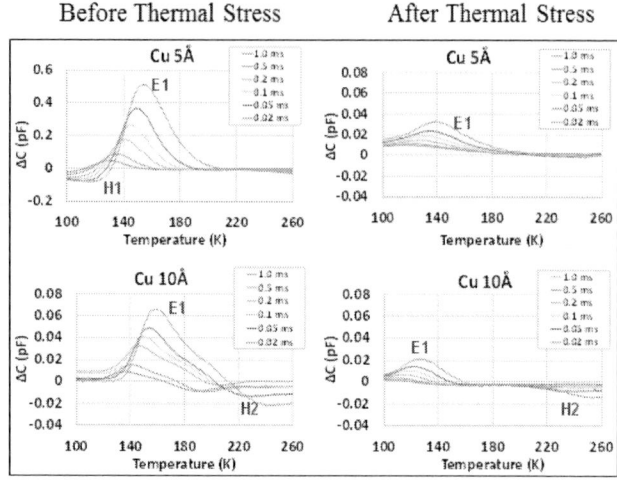

Fig. 7. DLTS spectra of LA CdTe devices with different amounts of Cu, before and after thermal stress.

Fig. 6. Thermal stability of thermal and laser annealed devices (5 Å Cu).

Fig. 7 shows the DLTS spectra for devices with 5 and 10 Å of Cu before and after thermal stress. Pre-thermal stress for the 5 Å Cu device shows a minority carrier trap E1 (E_C- 0.26 eV) possibly a Te_{Cd} or Cl_{Te} defect, and a shallow majority trap H1 most likely an acceptor defect Cu_{Cd}. Post-thermal stress shows the presence of only the minority carrier trap E1 [13]. The majority trap –related to p-doping- is eliminated; this is consistent with C-V measurements that indicated that the absorber was completely depleted. Higher Cu concentrations introduce a deep level majority trap H2, which is most likely related to Cu defects; further analysis and additional data are needed prior to assigning this trap to a specific defect. The behavior is similar to thermally annealed devices where the increase in Cu was found to result in the formation of deep majority traps [14].

IV. ACKNOWLEDGMENTS

This work was supported in full by the Department of Energy through the Bay Area Photovoltaic Consortium under award DE-EE0004946. The HighLight FAP DUO 60/810 Laser System was donated by Intevac Inc. S. Collins is supported by NSF GRFP (1144244).

V. CONCLUSIONS

It has been demonstrated that NIR laser annealing is a fast (<90 secs) and effective approach for the formation-activation of the CdTe back contact. Laser-annealed CdTe cells exhibited performance similar to baseline devices processed using conventional thermal annealing. High V_{OC} (830mV) and FF (70%) were achieved for 90 secs anneal time. The best cell fabricated to-date using laser anneal resulted in an efficiency of ~13%. Cu diffusion can be controlled by adjusting the LPD and anneal time. SIMS measurements for devices treated at high LPD show Cu reaching the CdS/CdTe interface. Non-optimum laser anneal can lead to the creation of Cu defects in the bulk and reduced lifetimes. Post-thermal stress of low Cu concentration cells revealed a minority carrier trap and an acceptor defect – most likely Cu_{Cd} – was eliminated. High Cu concentrations introduce a deep level majority trap, which is still present after thermal stress.

REFERENCES

[1] M. Gloeckler, *"Realization of the potential of CdTe thin-film PV"*, Proc. 43rd IEEE PVSC, 2016, pp. 1292-1292.

[2] M. A. Green, et al. "Solar cell efficiency tables (version 49): Solar cell efficiency tables (version 49)." *Progress in Photovoltaics* 25.NREL/JA--5J00-67687 (2016).

[3] A. Romeo, A. Terheggen, D. Abou-Ras, D. L. Batzner, F. J. Haug, M. Kalin, D. Rudmann, and A. N. Tiwari, *"Development of thin-film Cu(In,Ga)Se₂ and CdTe thin film solar cells"*, Progress in Photovoltaics 12, 93, 2004, pp. 93-111.

[4] M. Gloeckler, I. Sankin, and Z. Zhao, *"CdTe Solar cells at the threshold to 20% efficiency"*, IEEE Journal of Photovoltaics 3.4, 2013, pp. 1389-1393.

[5] T. A. Gessert, S. H. Wei, J. Ma, D. S. Albinm, R.G. Dhere, J. N. Duenow, D. Kuciauskas, A. Kanevce, T. M. Barnes, J. M. Burst, W. L. Rance, M. O. Reese, and H. R. Moutinho, *"Research strategies toward improving thin-film CdTe photovoltaic devices beyond 20% conversion efficiency"*, Solar Energy Materials and Solar cells, 119, 2013, pp. 149-155.

[6] H. Booth, *"Laser processing in industrial solar module manufacturing"*, J. of Laser Micro/Nanoengineering, Vol. 5, 2010, pp. 183-191.

[7] V. Palekis, S. Collins, M. Khan, V. Evani, S. Misra, M. A. Scarpulla, A. Abbas, J. Walls, and C. Ferekides, *"Near infrared laser CdCl₂ heat treatment for CdTe solar cells"*, Proc. 43rd IEEE PVSC, 2016, pp. 1498-1502.

[8] J. Li, J. D. Beach, and C. A. Wolden, *"Rapid thermal processing of ZnTe:Cu contacted CdTe solar cells"*, Proc. 40th IEEE PVSC, 2014, pp. 2360-2365.

[9] C.S. Ferekides, D. Marinskiy, V. Viswanathan, B. Tetali, V. Palekis, P. Selvaraj, and D.L. Morel, High efficiency CSS CdTe solar cells, *Thin Solid Films*, 361-362, 2000, pp. 520-526.

[10] J. Li, D. R. Diercks, T. R. Ohno, C. W. Warren, M. Lonergan, J. D. Beach, and C. A. Wolden, *"Controlled activation of ZnTe:Cu contacted CdTe solar cells using rapid thermal processing"*, Solar Energy Materials and Solar cells, 133, 2015, pp. 208-215.

[11] V. Komin, V. Viswanathan, B. Tetali, D. L. Morel, and C. S. Ferekides, *"Identification of defect levels in CdTe/CdS solar cells using deep level transient spectroscopy"*, Proc. 29th IEEE PVSC, 2002, pp. 736-739.

[12] T. A. Gessert, W. K. Metzger, P. Dippo, S. E. Asher, R. G. Dhere, and M. R. Young, *"Dependence of carrier lifetime on Cu-contacting temperature and ZnTe:Cu thickness in CdS/CdTe thin film solar cells"*, Thin Solid Films, 517, 2009, pp. 2370-2373.

[13] J. H. Yang, W. J. Yin, J. S. Park, J. Ma, and S. H. Wei, *"Review on first-principles study of defect properties of CdTe as a solar cell absorber"*, Semicond. Sci. Technol. 31, 2016.

[14] M. Khan, V. Evani, V. Palekis, S. Collins, D. Morel, and C. Ferekides, *"Study of defects in polycrystalline CdTe using DLTS"*, Proc. 43rd IEEE PVSC, 2016, pp. 2191-2194.

Enhanced Anti-reflective Coating for Thin Film Solar Cells

Grace Rajan,[1] Shankar Karki,[1] Robert W. Collins,[2] Sylvain Marsillac[1]

[1] Virginia Institute of Photovoltaic, Old Dominion University, Norfolk, VA, USA

[2] Department of Physics and Astronomy, The University of Toledo, Toledo, OH, USA

Abstract — In this paper, we describe a modified anti-reflective (AR) coating to enhance the efficiency of $Cu(In_{1-x}Ga_x)Se_2$ (CIGS) solar cells. The process is designed using in-situ real time spectroscopic ellipsometry and optical modeling allowing for the optimization of the properties of the AR coating. The optical model developed is based on a transfer matrix as well as accurate measurement of the dielectric function and thickness of each layer in the stack by spectroscopic ellipsometry. We demonstrate the influence of the substrate temperature on the properties of the AR coating, which can provide an increase in the short circuit current density by up to 10%.

Index Terms — CIGS, AR coating, ellipsometry, optical modeling, photovoltaic cells

I. INTRODUCTION

Several laboratories have been able to produce record devices over 22% efficiency for thin film solar cells based on $Cu(In,Ga)Se_2$ (CIGS) after major revisions of the deposition process[1]. It is important to continue to develop an effective light trapping mechanism to minimize the overall reflection losses. The power conversion efficiency of the device can be increased by applying an efficient anti-reflective (AR) coating. Magnesium fluoride (MgF_2) is the most widely used material for AR coatings in CIGS solar cells because it forms high quality films and has a low refractive index, n [2]. The thickness of the AR coating should be chosen such that destructive interference occurs between the light reflected from the CIGS cell interface and the AR coating surface, allowing reflections at the specific wavelength to be eliminated. This leads to the condition that the AR layer thickness should equal one quarter of the wavelength within the coating, or $d = \lambda/4 = \lambda_0/4n$, where λ_0 is the wavelength of the wave in vacuum [3]. In this paper, we describe an optical model to accurately predict the thickness of the AR layer that should be deposited at different substrate temperatures.

II. EXPERIMENTAL DETAILS

Molybdenum, chosen as back contact was deposited by DC magnetron sputtering on soda lime glass substrates. CIGS absorber layers were deposited by 3-stage process by thermal co-evaporation. The samples were dipped into a chemical bath to form a thin layer of CdS buffer layer after the CIGS deposition. High resistive ZnO layers along with ZnO:Al layers were deposited by RF sputtering to obtain a transparent window layer and Ni/Al/Ni grids were evaporated by e-beam evaporation for the electrical contacts. MgF_2 layers were deposited as the anti-reflective coating on the CIGS solar cells by e-beam evaporation at various substrate temperatures varying from room temperature to 100°C. The deposition was carried out at a low base pressure of 10^{-6} Torr and it was obtained by a turbo-molecular pump and the substrate holder was rotated at 15 rpm during the deposition to ensure uniform deposition. For the purpose of characterization, the layers were also deposited on well characterized silicon wafers with native oxide thickness of 2nm. Quartz crystal monitors were used to accurately monitor the deposition rate of the MgF_2 film during the deposition.

The optical properties of the layers were measured for different wavelengths during deposition by Real Time Spectroscopic Ellipsometry (RTSE). The RTSE measurements were carried out *in-situ* during film growth while spectroscopic ellipsometry measurements were carried out *ex-situ*. They both used a rotating compensator, multichannel instrument with an energy range of 0.75–6.5 eV at an angle of incidence of 65°.

III. OPTICAL MODELLING

A. Prediction of QE of CIGS device using SE and optical model.

A transfer matrix (TM) model was developed on a Matlab platform to calculate the optical interference and absorption in the multi-layer stack. The light waves were assumed to be planar and transversal such that the magnitudes of the electric field components in each layer were used to calculate the irradiance and absorption [2]. The relationship between the incident, reflected and transmitted amplitudes can be expressed as [6]:

$$\begin{bmatrix} I \\ R \end{bmatrix} = \begin{bmatrix} E_1^{f+} \\ E_1^{f-} \end{bmatrix} = I_1 \begin{bmatrix} E_1^{b+} \\ E_1^{b-} \end{bmatrix} = I_1 L_1 \begin{bmatrix} E_2^{f+} \\ E_2^{f-} \end{bmatrix} = I_1 L_1 I_2 \begin{bmatrix} E_2^{b+} \\ E_2^{b-} \end{bmatrix}$$

$$\begin{bmatrix} I \\ R \end{bmatrix} = I_1 L_1 I_2 \ldots \ldots I_n L_n I_{n+1} \begin{bmatrix} E_{n+1}^{b+} \\ E_{n+1}^{b-} \end{bmatrix} = \begin{bmatrix} S_{11} & S_{12} \\ S_{21} & S_{22} \end{bmatrix} \begin{bmatrix} T \\ 0 \end{bmatrix}$$

where E_n^{f+} is the incident forward amplitude on the nth interface, E_n^{b-} the incident backward amplitude, E_n^{f-} and E_n^{b+} the amplitudes that leaves the nth interface backward and forward, respectively, L is the layer matrix, I the interface matrix. The electric field amplitudes across the entire multilayer structure is calculated from the structure matrix. To ensure accuracy of the model, spectroscopic ellipsometry was employed to extract the optical properties of each layer within the multi-layer structure. The thickness and the optical properties of the component layers of the solar cells are extracted with SE and the maximum external quantum efficiency (QE) can be deduced by assuming perfectly specular reflection/transmission conditions [3]. By simulating the quantum efficiency variations, the short circuit current density (Jsc) in the solar cell can be thus deduced. The anti-reflection coating stack is also optimized by obtaining the reflectance spectra from the structure. The simulation of QE spectra is based on the assumption that all the photo generated carriers within the active layers are collected without any recombination [6]. Thus, a comparison between the simulated QE and experimental QE obtained from the measurement of a completed solar cell device can provide information on the electronic losses, as well as the spectral dependence of the losses. The optical model does not take into consideration the scattering of light at rough surfaces and interfaces, thus the modeled QE spectra can provide insight into the gain due to light trapping caused by scattering at rough surfaces and interfaces [6].

IV. RESULTS AND DISCUSSION

The process variables for the deposition of MgF_2 layer was varied and the growth conditions were studied using RTSE to identify and understand the optimum process conditions. The substrate temperature (T_{SS}) was changed from room temperature (RT) to 100°C to monitor the effect on the thin film. The refractive index of the film was extracted from the data obtained during RTSE. One can observe (Fig. 1) that the refractive index decreases from RT to 50°C, then increases for all wavelengths from 50°C to 100°C. The later can be a result of the faster molecular motions occurring at higher temperature [4]. At RT, the film is different and might be partially crystallized or remain amorphous; thus, the grain boundaries may not be well defined and result in a higher refractive index [5].

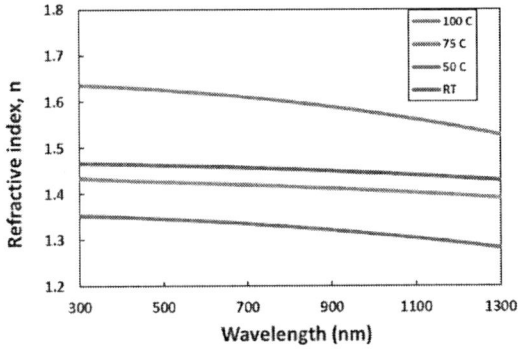

Fig. 1. Measured refractive index as a function of wavelength for MgF_2 films deposited for different T_{SS}, as extracted using RTSE [6].

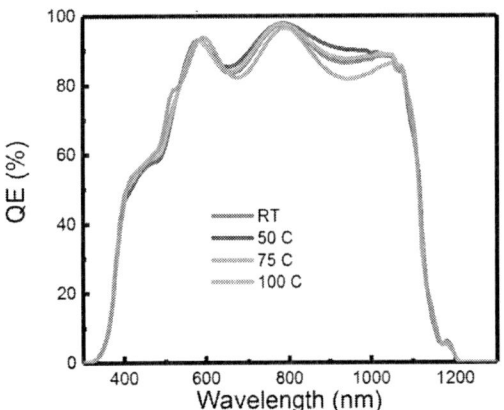

Fig. 2. Simulated QE spectra CIGS devices with varied AR layer deposited at different T_{SS}.

The TM optical modeling was used to predict the QE spectra (Fig. 2) and the maximum obtainable J_{SC} for CIGS solar cells with AR layer deposited at different T_{SS} (Fig. 3). The maximum value of J_{SC} can be observed for the cells with an AR layer deposited at 50°C with a thickness of 118 nm, for that specific structure. These results obtained using simulation results were used to guide the fabrication of AR coated CIGS devices with various T_{SS}. However, one has to keep in mind the effect of T_{SS} on the rest of the structure. Following this modeling, samples were loaded in the e-beam chamber after full characterization and a MgF_2 layer was deposited at different T_{SS}. The relative reflectance was measured in real time, but the deposition was stopped at the ideal thickness for these runs. After the run, the solar cells performance was measured. The average relative change (in percentage) in J-V parameters of the solar cell for AR coating deposited at different T_{SS} is reported in Table I. The experimental results are in good correlation with the modeling results. The results substantiate an enhanced performance of the devices with AR coating applied at 50°C. There is an increase in the J_{SC} by 10%, while AR coating deposited at RT only enhanced it by 5%. Even though higher J_{SC} values are reported for AR coating deposited at 75 °C and 100° C, the negative impact on Voc and FF reduces the overall efficiency of the cell, as the higher T_{SS} probably deteriorate the junction quality. It is interesting to notice here that the TM modeling for the 100°C deposition predicted a lower current than for the other temperature, while it is quite similar experimentally. Here again, the combination of predictive modeling with *in-situ* real time measurement proves to be very useful in obtaining the best results.

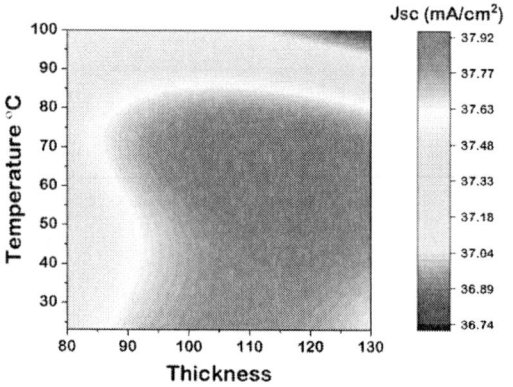

Fig.3. Simulation of the variation of J_{SC} for CIGS devices with varied thickness of MgF$_2$ layer deposited at different T_{SS}.

TABLE I

AVERAGE RELATIVE CHANGE (IN %) IN J-V PARAMETERS AT DIFFERENT Tss

Tss for AR coating	η (%)	J_{SC} (mA/cm^2)	Voc (V)	FF (%)
RT	+4	+4	0	0
50°C	+10	+9	0	+1
75°C	- 5	+15	- 2	- 15
100°C	- 11	+10	- 4	- 15

IV. CONCLUSION

The influence of substrate temperature on the properties of AR coating for CIGS applications was demonstrated via modeling and experimental results. The thickness of the AR layer was optimized using a transfer matrix model and RTSE. The experimental results confirmed the validity of the model and allowed for optimized devices to be fabricated. An optimized deposition of AR coating at room temperature provided an increase in the short circuit current density by around 5%, while the AR layer deposited at higher substrate temperature provided an increase of 10%. Future studies will involve characterizing these different AR layers deposited at different temperature to shine light on the morphological and optical changes.

ACKNOWLEDGEMENT

This research was supported in part by the DOE program under contract number DE-EE0007141.

DISCLAIMER

This report was prepared as an account of work sponsored by an agency of the United States Government. Neither the United States Government nor any agency thereof, nor any of their employees, makes any warranty, express or implied, or assumes any legal liability or responsibility for the accuracy, completeness, or usefulness of any information, apparatus, product, or process disclosed, or represents that its use would not infringe privately owned rights. Reference herein to any specific commercial product, process, or service by trade name, trademark, manufacturer, or otherwise does not necessarily constitute or imply its endorsement, recommendation, or favoring by the United States Government or any agency thereof. The views and opinions of authors expressed herein do not necessarily state or reflect those of the United States Government or any agency thereof.

REFERENCES

[1] Jackson P, Wuerz R, Hariskos D, Lotter E, Witte W, Powalla M,. "Effects of heavy alkali elements in Cu(In,Ga)Se$_2$ solar cells with efficiencies up to 22.6%", *Phys. Status Solidi RRL.* 2016; 1-4.
[2] Kaminski PM, Lisco F and Walls J M. Multilayer broadband anti-reflective coatings for more efficient thin film CdTe solar cells. *IEEE Journal of Photovoltaics* 2014; **4**: 452-456.
[3] Nubile P. Analytical design of antireflection coatings for silicon photovoltaic devices. *Thin Solid Films* 1999; **342**: 257-261.
[4] Yang H H, Park GC. A Study on the properties of MgF2 Anti-reflection film for Solar Cells. Transactions on Electrical and Electronic Materials. 2010; 11: 33-36.
[5] Pilvi T, Hatanpää T, Puukilaninen E, Arstila K, Bischoff M, Kaiser U, Kaiser N, Leskelä M, Ritala M,. Study of a novel ALD process for depositing MgF2 thin films. Journal of Materials Chemistry. 2007; 17: 5077-5083.
[6] G. Rajan et al. "Real-time Optimization of Anti-reflective Coatings for CIGS Solar Cells" 43rd IEEE PVSC, Portland, USA, June 2016.

Influence of AGS Layer Insertion at Absorber/ITO Interface on Structural and Photovoltaic Properties of Ultrathin Cu(In,Ga)Se$_2$ Solar Cells

Muhammad Saifullah[a,b], Jihye Gwak[a,b], Kihwan Kim[a], Joo Hyung Park[a], JunSik Cho[a,b], Jae Ho Yun[a,b]

[a]Korea Institute of Energy Research (KIER), Daejeon, 34129, Korea
[b]University of Science and Technology (UST), Daejeon, 34113, Korea

Abstract — **The insertion of a sulfurized-AgGa (AGS) layer at Cu(In,Ga)Se$_2$ (CIGSe)/ITO interface was found effective in improving the PV properties of ~300 nm thick CIGSe ($E_g \approx 1.5$ eV) absorber-based solar cells (deposition time (t_{dep}) = 90 min, substrate temperature (T_s) = 490 °C) [1]. This study explores the benefits of the AGS layer when T_s and t_{dep} for the CIGSe deposition are 560 °C and 15 min, respectively. The PV properties of a CIGSe solar cell modified with the AGS layer are relatively poor when T_s is high and t_{dep} is short.**

Index Terms — **Ultrathin, semitransparent, Cu(In,Ga)Se$_2$, sulfurized-AgGa, Cu(In,Ga)Se$_2$/ITO interface, GaO$_x$ interfacial layer materials.**

I. INTRODUCTION

The installation of stand-alone photovoltaic (PV) systems is becoming increasingly problematic with the increase in the population. This is becoming a more of a concern especially in countries with no free land. In such scenario, the dissemination of stand-alone PV systems can become cumbersome, so there is a need for designing and fabricating a new type of PV that could be utilized or integrated into various structures. Inducing the semitransparency in the solar cell is the one such way to exploit them in various technologies starting from the electronic gadgets, and signboard on road to building applications. This can certainly increase the PV installations due to their increased universal acceptance.

Cu(In,Ga)Se$_2$ (CIGSe) is an appealing PV material due to its direct and adjustable bandgap, high absorption coefficient, benign nature of grain boundaries, and high stability. The CIGSe thin film solar cell has attained the conversion efficiency of >22% which distinguishes it in the other PV technologies [2]. CIGSe has better prospects for semitransparent (ST) PV applications compared with other PV materials due to its high performance and stability. Punch-through type CIGSe ST solar cell is prepared by etching or laser scribing, which make their preparation process more intricate. In contrast, penetration-type CIGSe ST solar cell can be made easily by thinning the CIGSe layer to <500 nm [3] depending upon the absorber layer absorption coefficient which indirectly depends on their bandgap (E_g). The thickness of the absorber can be lessened by just shortening the deposition time. Several issues become prominent on ultra-thinning the photoactive layer. These challenges include high

bulk and back surface recombination and high shunt conductance which can seriously deteriorate the device performance. More details on these issues and their effect on the device efficiency can be found in ref [1, 4].

Previously, we proposed that the insertion of a sulfurized-AgGa (AGS) layer can considerably improve the efficiency of a CIGSe thin film solar cell, where the CIGSe layer thickness was <300 nm [1]. The performance enhancement was attributed to the increase in the grain size and shunt resistance and bandgap grading. GaO$_x$ usually forms at the CIGSe/transparent conducting oxide (TCO) interface when the substrate temperature (T_s) is high during the CIGSe film deposition [5]. In various studies, it is found that the formation of the spurious GaO$_x$ phase on the CIGSe/TCO interface limits the device performance [5, 6]. The formation of GaO$_x$ interfacial (IF) phase can be impeded by lowering the substrate temperature (T_s), therefore, T_s was fixed at 490 °C during our previous work [7]. Low T_s, besides suppressing the formation of GaO$_x$ interfacial phase, also keeps the resistivity of the back contact low [5]. The formation of GaO$_x$ IF phase can be further suppressed by the sulfurization of AgGa-precursor layer. During the sulfurization of AgGa-precursor layer, In$_2$O$_3$/SnO$_2$ in indium doped tin oxide (ITO) converts to In$_2$S$_3$/SnS$_2$ which can possibly hinder the formation of GaO$_x$ phase during the CIGSe deposition [6]. The combination of these two approaches is believed to offer an effective method preventing the formation of IF phases (Relevant results will be published soon). The T_s value of ~560 °C is advantageous for improving the absorber quality when it is being deposited on Mo back electrode. At elevated temperature, the advantage of improved absorber quality can be offset by the detrimental

TABLE I
DEPOSITION PARAMETERS OF THE REFERENCE (M0) AND THE MODIFIED ABSORBER (M1)

ID[a]	t_{AGS}[b] (nm)	t_{CIGSe}[c] (nm)	T_s[d] (°C)	t_{dep}[d] (min)
M0	0	300±20	560	17
M1	50±10	250±20	560	15

[a]Sample name. [b]Thickness of the AGS layer. [c]Thickness of the CIGSe layer. [d]The substrate temperature during the CIGSe layer deposition. [e]Active deposition time for growing the CIGSe film

effect of GaO_x IF phase if the CIGSe is being grown on ITO back contact. Simchi et al. reported that CIGSe solar cell ($E_g = 1.20$ eV) with the absorber thickness of only 300 nm exhibited conversion efficiency of 9.0% in backwall superstrate configuration when the absorber was deposited at $T_s = 550$ °C [6]. This intimates that at high T_s, the extent of increase in the ITO resistivity does not substantially degrade the device performance. The effect of AGS layer insertion on the CIGSe/ITO interface properties is crucial for the device. The aim of the present study is to investigate the effect of AGS layer on the CIGSe/ITO interfacial chemistry and CIGSe structure when the absorber is prepared at $T_s = 560$ °C. The effect of subsequent changes in the structure and interface on the corresponding device's PV parameters are also studied.

II. EXPERIMENTAL DETAILS

The CIGSe absorber ($E_g = 1.55$ eV) is deposited on commercial ITO-glass (sheet resistance ≈ 10 Ω $\square^{-1}$) employing a single stage co-evaporation process. The deposition time (t_{dep}) for growing a ~300 nm thick CIGSe thin film is ~17 min which is >5 times lower compared with t_{dep} in the previous study [1]. AGS layer is prepared by the sulfurization of AgGa-precursor layer (details can be seen in ref. [1]). Cu/(Ga+In) (CGI) and Ga/(Ga+In) (GGI) ratios in the CIGSe film are 0.72 and 0.77, respectively, whereas the Ag/Ga ratio in AGS layer is 0.60. Deposition parameters of two types of absorbers prepared in this study are summarized in Table 1.

Afterwards, device are made by the depositing CdS buffer layer (~60 nm), intrinsic ZnO (~50 nm), Al-doped zinc oxide (~350 nm), and Al grid (~1.0 μm). On a substrate (5 cm x 5 cm), twelve cells each with an area of 0.44 cm^2 are prepared by the mechanical scribing using a blade. The PV parameter are mainly given only for the best cells.

Structural characterizations are carried out using X-ray diffractometer (Rigaku, D/Max-2500, Japan) and Raman spectrometer. X-ray diffraction analysis is performed in grazing angle mode with grazing angle (α) of 6° using 10 mm wide X-ray (40 kV and 200mA). Δλ for each scan is 0.01°. Raman analysis is performed using a laser having the wavelength and power of 532 nm and 2 mW, respectively. Spot width during the Raman analysis is 0.7 μm. A Si film is taken as a reference whose Raman peak usually appears at 520.8 cm^{-1}. Raman peaks of the samples are adjusted with respect to the reference peak. Morphological analysis of CIGSe films is done with scanning electron microscopy (S-4700, Hitachi, Japan), whereas, corresponding atomic concentrations are measured with energy dispersive X-ray spectrometry (EDS). For composition analysis, CIGSe films are prepared on Mo back contact, in the similar run in which the film is deposited on ITO back contact. To see the CIGSe/ITO interface, transmission electron microscopy analysis is performed on transmission electron microscope (JEOL, JEM-ARM200F, USA). Line scans of the films are taken with EDS integrated with TEM. PV parameters of the

completed devices are measured on (K201-LAB 50, McScience, South Korea) under AM1.5G spectrum with an intensity of 100 mW/cm^2, whereas, their external quantum efficiencies (EQE) are taken on spectral response measurement system (K3100, McScience, South Korea).

III. RESULTS AND DISCUSSION

The structural analysis of the M0 and M1 were carried out by X-ray diffraction and Raman spectroscopy. The grazing incident X-ray diffraction (GIXRD) patterns of M0 and M1 on ITO back electrode with α of 6° are presented in the Fig. 1. In both M0 and M1, peaks at 27.43°, 45.59°, 53.85°, 66.17°, and 73.46° are because of the diffraction from (112), (220/204), (312/116), (400), and (332/316) planes of the CIGSe chalcopyrite phase (PDF #35-1101), respectively. The rest of diffraction peaks belong to In_2O_3 (PDF #06-0416), the source of which is the back contact. The peaks of the reference (M0) are slightly shifted to higher 2θ values compared with the peaks given in the PDF #35-1101. Fig. 1 shows that the (112) peak of the CIGSe film with the AGS layer (M1) is slightly shifted to lower angle compared with M0 which is the indication of Ag-alloying of the CIGSe absorber. Ag has a larger ionic radius compared with Cu, therefore, substitution of Cu with Ag results in the expansion of unit cell ascribable to the peak shift wo lower 2θ value. No impurity peak was detected in both of M0 and M1. Evidently shown in Fig. 1, both M0 and M1 have preferred growth along the (112) plane.

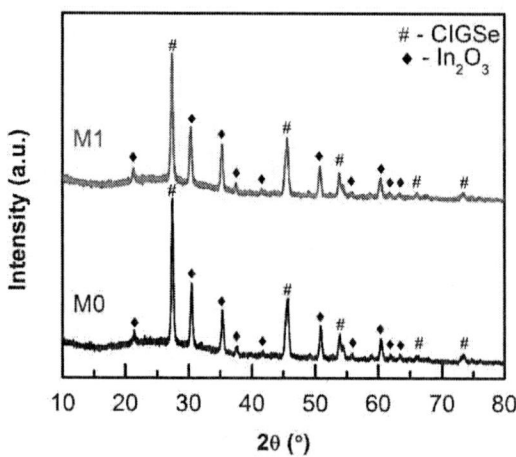

Fig. 1.   GIXRD patterns (α = 6°) of M0 and M1.

To detect the presence of order defect compound (ODC) or other impurity phases, Raman scattering analysis was performed. Raman spectra of M0 and M1 are shown in Fig. 2. A Raman peak at 179.12 cm^{-1} is due to the A_1 mode of CIGSe which emerges from the vibration of Se with respect to Cu, In, and Ga. The position of A_1 in M0 is in accordance with the previously reported value for the similar value of GGI, i.e. 0.77. The Raman peak related to the A_1 mode in M1 slightly shifted to lower energy. The force constant for Ag-Se bond is lower than the force constant of Cu-Se, which renders the A_1 mode in Ag-alloyed CIGSe to respond at slightly lower energy. Moreover, Fig. 2 also shows that the formation of ODC in M0 is minimal, the A_1 mode of ODC in M0 usually responds at 159.42 cm^{-1}. M1 showed slightly higher tendency to form ODC. The low value of Ag/Ga (0.60) in the AGS layer compared with CGI value (0.72) in CIGSe can consequence in the drop of (Ag+Cu)/(Ga+In) ratio which instigates the formation of ODC. The formation of ODC in a modified film was large when the t_{dep} for CIGSe film was prolonged (4-5 times higher than the t_{dep} used in this study). Diffusion of In from the ITO back contact to the CIGSe absorber was found to the reason for a large amount of ODC formation in the case of high t_{dep} value (details will be published separately). The t_{dep} of ~15 min in case of M1 is too short to lead In diffusion from the ITO to the CIGSe layer. The E_g values for M0 and M1 are also almost identical which also confirms that In in-diffusion is insignificant (see Fig. 3). A broad Raman band in the range of ~210-230 cm^{-1} emerges from E/B_2 and E/B_1 symmetry modes of a CIGSe absorber. Another peak at ~ 262.38 cm^{-1} represents the $Cu_{2-x}Se$ secondary phase [8] whose intensity is comparatively lower in M1 than in M0.

of M1 compared with M0, in comparison with modified sample that was prepared at low T_s and with longer t_{dep} [7]. Nevertheless, the grains of M1 are slightly larger and more uniform than M1.

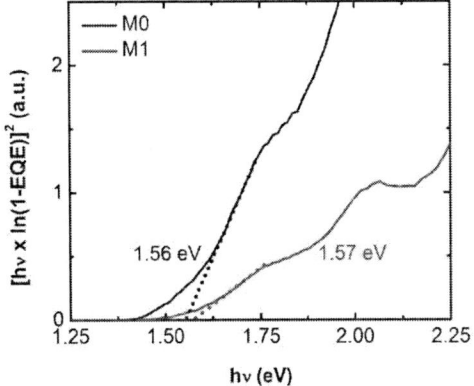

Fig. 3. E_g values of M0 and M1 calculated from EQE curves (E_g is calculated following the method is given in ref [7]).

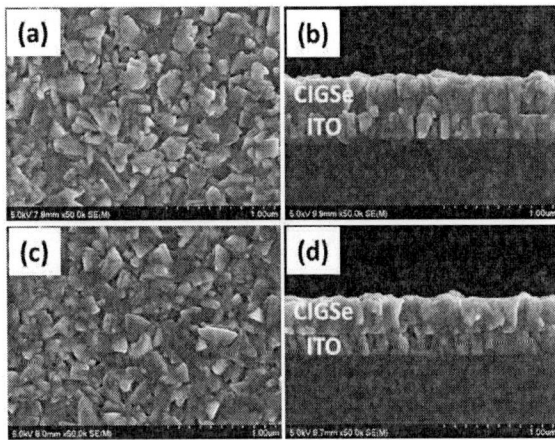

Fig. 4. SEM surface and cross-sectional images of M0 (a, b) and M1 (c, d).

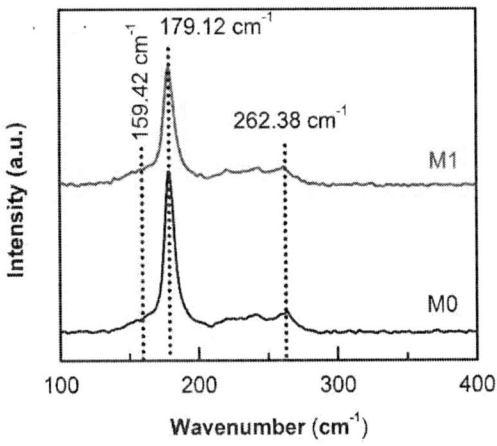

Fig. 2. Raman spectra of M0 and M1.

Morphology of the absorber layers and formation of a GaO_x IF phase at the CIGSe/ITO interface are assessed by scanning electron microscopy (SEM) and transmission electron microscopy (TEM), respectively. SEM and TEM images of M0 and M1 are shown in the Fig.4. It is apparent from the SEM surface and cross-sectional micrographs that at higher T_s there is not a substantial improvement in the grain morphology

TEM images in Fig. 5 (a, b) reveal that the GaO_x interfacial (IF) phase is more likely to form in M0 than in M1. The magnified images which zoom in the interface regions of both the samples are shown in Fig. 5 (b and d). The thickness of the GaO_x IF phase in M0 is ~35.8 nm (Fig. 5 (b)) which is significantly higher than the IF layer thickness in M1, i.e., ~4.3 nm (see Fig. 5 (d)). Another important feature of M1 is the formation of nanowire on its surface which can adversely affect the device fill factor (FF). The formation of such nanowire (Ag-rich) in an earlier study (T_s = 490 °C) [1] was only observed when the deposition time was ≥90 min and AGS layer was ≥90 nm. Line scans of M0 and M1 can be seen in the Fig. 6 which show that the width of a Ga segregation region is higher in M0 than in M1. This implies that the AGS layer effectively suppressed the formation of the GaO$_x$ IF phase. Hereafter, the effect of these structural and morphological changes on the PV of the corresponding cell will be described.

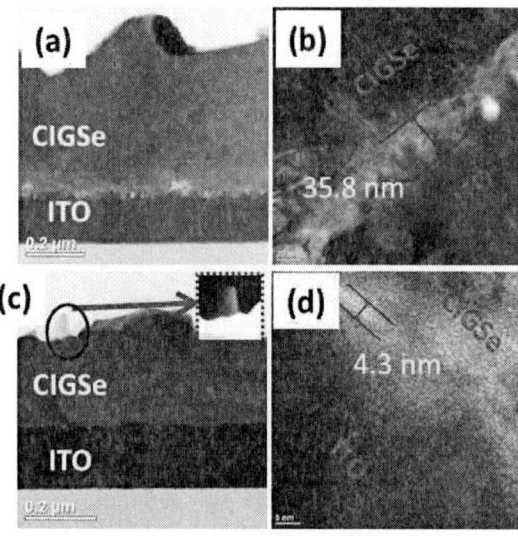

Fig. 5. TEM cross-sectional images (M0 (a) and M1 (c)) and high-resolution TEM scans of the CIGSe/ITO interface regions (M0 (b) and M1 (d)) (scale bars in images b and d are 20 nm and 5 nm, respectively). Inset in Fig. 5 (c) shows the nanowire on the surface of M1

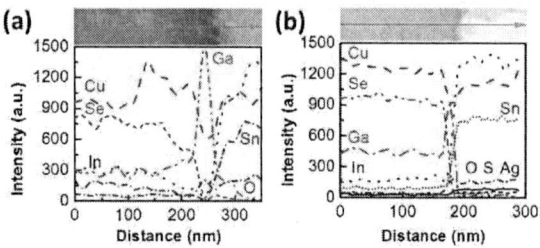

Fig. 6. Line scans of M0 (a) and M1 (b) taken through EDS setup embodied with TEM (images of interface region along which the line scanning is performed also given with the figures). Cu counts are incorrect due to the Cu grid used for TEM analysis.

Light-JV and EQE curves of the solar cells made from M0 (CIGSe without the AGS interlayer) and M1 (CIGSe with the AGS interlayer) are given in the Fig. 7. It is visible from the Fig. 7 (a) that the PV parameters of M1 (open circuit voltage (V_{oc}) = 0.76 V, short-circuit current density (J_{sc}) = 12.80 mA/cm^2, fill factor (FF) = 47.58%, and efficiency (η) = 4.62%) are ameliorated because of the introduction of the AGS layer. In contrast, the PV parameters of M0 are

significantly poorer than M1 (V_{oc} = 0.60 V, J_{sc} = 8.15 mA/cm^2, FF = 36.69%, and η = 1.80%). In particular, the lower V_{oc} of M0 than M1 is because of the high back surface recombination because of the absence of bandgap grading and in part, due to Cu$_{2-x}$Se secondary phase. Higher J_{sc} in M1 seems to be connected with reduced bulk defects by Ag-alloying [9] and favorable band grading near the rear interface accomplished with the Ga grading. EQE profiles in Fig. 7 (b) show that M1 exhibited improved spectral response over entire wavelength (350 ~ 900 nm) compared with the M0.

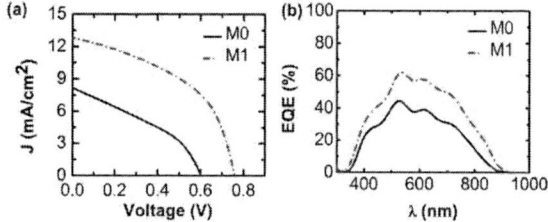

Fig. 7. Light-JV curves obtained under standard AM1.5 illumination (a) and the corresponding EQE profiles (b) for M0 and M1.

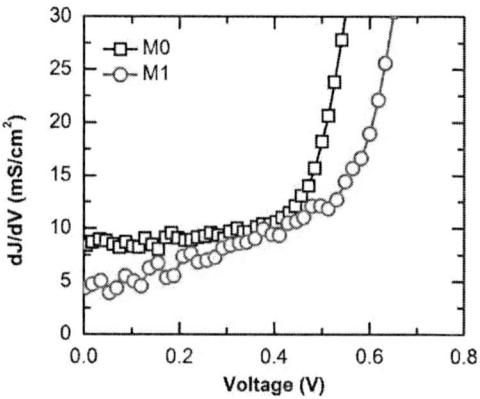

Fig. 8. dJ/dV versus voltage plot for calculating R_{sh} using the light-JV data (R_{sh} is calculated by the procedure given in ref. [10].

In an ultrathin CIGSe solar cell, shunt resistance (R_{sh}) is an important performance limiting factor. When the surface roughness becomes comparable with a thickness of the absorber, it results in the increase of shunt conductance because of the photoconductive buffer layer [11]. R_{sh} values of M0 and M1 are calculated to investigate their effect on the corresponding device FF (see Fig. 8). The R_{sh} value of M0 and M1 are 116.82 and 196.85 Ω cm^2, respectively. The R_{sh} value of M0 is lower than M1 due to mainly the Cu$_{2-x}$Se surface secondary phase and in part, high surface roughness. The R_{sh} value in M1 is more than 3 times lower than the R_{sh} of a similar cell that was prepared at T$_s$ = 490 °C and with longer

t_{dep} [1]. The R_{sh} in M1 seems to be ascribable to nanowire growth on the surface of the absorber (see Fig. 5 (c)) and $Cu_{2-x}Se$ phase. Raman analysis (Fig. 2) showed that M0 exhibited a greater amount of the $Cu_{2-x}Se$ secondary than M0, therefore, we believe that a greater number of shunting paths could be formed in M0 than in M1. However, both of M0 and M1, to a greater or lesser extent, exhibited limited device performance mainly due to low R_{sh}. This indicates that still there is a need to modify the deposition process in a way that could eliminate the formation of the $Cu_{2-x}Se$ phase. Post-deposition Se annealing is one such approach that can help $Cu_{2-x}Se$ phase to become a part of the absorber. Results in this study have shown that the modification of CIGSe absorber by AGS layer is beneficial for the cell performance but this modification can be of more utility if we increase the t_{dep} and decrease the T_s.

IV. CONCLUSIONS

Semitransparent solar cells can be integrated into the buildings, electronic gadgets, automobiles, yachts, etc. Reduction of the back surface recombination and GaO_x formation at CIGSe/ITO is imperative to improve the efficiency of ultrathin CIGSe solar cells. This study investigated the effected of AGS layer insertion at the CIGSe/ITO interface with the T_s and t_{dep} of 560 °C and 15 min, respectively. Raman analysis revealed the tendency of formation of ODC in M1 and $Cu_{2-x}Se$ both in M0 and M1. The interfacial GaO_x layer in M0 is thicker (~35.8 nm; no AGS layer) than in M1 (~4.3 nm; with AGS layer). The AGS layer was helpful in impeding the formation GaO_x interfacial phase even at a higher T_s of 560 °C. The efficiency of M1 (modified cell) is >2.5 times higher than M0 (reference). High V_{oc}, J_{sc}, and FF are responsible for high η in M1. The efficiency in semitransparent CIGSe solar cells modified with AGS layer insertion at the CIGSe/ITO interface can comparatively be enhanced more when T_s was 490 °C and t_{dep} was long.

ACKNOWLEDGEMENT

This work was conducted under the framework of the Research and Development Program of the Korea Institute of Energy Research (KIER) (B7-2421) and also partly supported by the Technology Development Program to Solve Climate Changes of the National Research Foundation (NRF) funded by the Ministry of Science, ICT & Future Planning (2016M1A2A2936753).

REFERENCES

[1] M. Saifullah, S. Ahn, J. Gwak, S. Ahn, K. Kim, J. Cho, J.H. Park, Y.J. Eo, A. Cho, J.-S. Yoo, J.H. Yun, Development of semitransparent CIGS thin-film solar cells modified with a sulfurized-AgGa layer for building applications, J. Mater. Chem. A, 4 (2016) 10542-10551.

[2] J. Xia, L. Chen, S. Yanagida, Application of polypyrrole as a counter electrode for a dye-sensitized solar cell, J. Mater. Chem., 21 (2011) 4644-4649.

[3] K. Kim, W.N. Shafarman, Alternative device structures for CIGS-based solar cells with semi-transparent absorbers, Nano Energy, 30 (2016) 488-493.

[4] M. Saifullah, J. Gwak, J.H. Yun, Comprehensive review on material requirements, present status, and future prospects for building-integrated semitransparent photovoltaics (BISTPV), J. Mater. Chem. A, 4 (2016) 8512-8540.

[5] T. Nakada, Y. Hirabayashi, T. Tokado, D. Ohmori, T. Mise, Novel device structure for $Cu(In,Ga)Se_2$ thin film solar cells using transparent conducting oxide back and front contacts, Sol. Energy, 77 (2004) 739-747.

[6] H. Simchi, J.K. Larsen, K. Kim, W.N. Shafarman, Improved performance of ultrathin $Cu(InGa)Se_2$ solar cells with a backwall superstrate configuration, IEEE Journal of Photovoltaics, 4 (2014) 1630-1635.

[7] M. Saifullah, J.H. Moon, S. Ahn, J. Gwak, S. Ahn, K. Kim, Y.J. Eo, J.H. Yun, Effect of Cu content on the photovoltaic properties of wide bandgap CIGS thin-film solar cells prepared by single-stage process, Current Applied Physics, 16 (2016) 1517-1522.

[8] D. Papadimitriou, N. Esser, C. Xue, Structural properties of chalcopyrite thin films studied by Raman spectroscopy, Phys. Status Solidi B, 242 (2005) 2633-2643.

[9] P.T. Erslev, J. Lee, G.M. Hanket, W.N. Shafarman, J.D. Cohen, The electronic structure of $Cu(In_{1-x}Ga_x)Se_2$ alloyed with silver, Thin Solid Films, 519 (2011) 7296-7299.

[10] S.S. Hegedus, W.N. Shafarman, Thin-film solar cells: device measurements and analysis, Prog. Photovoltaics, 12 (2004) 155-176.

[11] T. Negami, S. Nishiwaki, Y. Hashimoto, N. Kohara, Effect of absorber thickness on performance of $Cu(In,Ga)Se_2$ solar cells, in: 2nd World Conference on Photovoltaic Solar Energy Conversion, Piscataway USA, July, 1998, pp. 1181-1184.

Novel, Facile Back Surface Treatment for CdTe Solar Cells

Suneth C. Watthage, Geethika K. Liyanage, Zhaoning Song, Fadhil K. Alfadhili, Rabee B. Alkhayat, Khagendra P. Bhandari, Randy J. Ellingson, Adam B. Phillips, and Michael J. Heben

Wright Center for Photovoltaics Innovation and Commercialization, School for Solar and Advanced Renewable Energy, Department of Physics and Astronomy, University of Toledo, Toledo, OH, 43606, USA

Abstract — Formation of a Te-rich surface on the CdTe absorber is beneficial to the formation of a low barrier contact at the back junction in CdTe solar cells. Regardless of the CdTe fabrication method, surface etching processes are widely used to form Te-rich surfaces. Here, we show that methylammonium iodide (CH3NH3I, MAI) can simply and controllably react with the CdTe surface to produce elemental tellurium. Both X-ray diffraction and Raman spectroscopy confirmed formation of a Te layer. Devices constructed with the MAI-treated CdTe absorber exhibited higher photovoltaic parameters of open circuit voltage (V$_{oc}$) and fill factor (FF) relative to similar devices prepared without the MAI treatment. This indicates a reduction in the Schottky barrier at the back contact due to the MAI surface treatment. CdTe samples that were treated with 125 mM MAI solution and heated to 150 ^{0}C for 10 minutes showed the best power conversion efficiency (PCE) of 13.5% while without treating the standard device efficiency was 12.7%.

Index Terms — CdTe, surface treatment, Te-rich, Schottky barrier, back contact, methylammonium iodide.

I. INTRODUCTION

CdTe solar cell technology is one of the most well-established approaches for preparing high efficiency, low cost, and stable thin films photovoltaics (PV) [1]. With a direct band gap of 1.45 eV, CdTe is an ideal absorber material for single junction solar cells, and a power conversion efficiency (PCE) as high as 22.1% has recently been achieved [2].

Creating an efficient and stable back contact is necessary to achieve high efficiency CdTe solar cells that exhibit long-term stability [3]. Since CdTe has a high electron affinity, a high work function metal is required to form a low resistance ohmic contact; however, no metals have a work function high enough to form an ohmic contact with p-type CdTe. Therefore, metal/CdTe junctions typically result in a Schottky barrier. This increases the contact resistance and causes lower device performance [3, 4]. A common approach to overcome this problem is to use a thin intermediate degenerate semiconductor layer that increases the conductivity and forms a tunneling barrier between the CdTe and metal interface [3, 5]. Cu$_x$Te, the most often used degenerate semiconductor, is formed by depositing a thin Cu layer (~3 nm) on the CdTe surface and annealing the sample to induce intermixing [4]. To improve the Cu$_x$Te tunnel junction, a Te-rich surface on the CdTe layer is often employed. Wet-chemical etching processes, such as bromine-methanol (BrMeOH), which uses a diluted solution of bromine in methanol [3], nitric-phosphoric etch (NP etch), which uses a mixture of nitric and phosphoric acid in water [6, 7], and dry etching in Ar plasma [8] are the most commonly used surface etching techniques. Alternatively, deposition of a Te layer by either close space sublimation (CSS) [9] or chemical bath deposition (CBD) [10] may be used to form a Te-rich surface on the CdTe absorber.

Here, we introduce a wet-chemical process to form a Te layer on the CdTe surface using methylammonium iodide (CH3NH3I, MAI) solutions. Facile preparation, reduced toxicity, and high controllability are the main advantages of this process. MAI thin films, which were fabricated on the CdTe surface by spin-coating MAI/isopropanol (IPA) solutions, reacted with the CdTe surface to form a Te layer. The effect of MAI concentration and reaction time on the surface morphology, structure, and device performance were investigated. Current density-voltage (J-V) characteristics showed significant improvements in open circuit voltage (V$_{OC}$) and fill factor (FF) with the MAI treatment, indicating a reduced Schottky barrier height at the back contact of CdTe solar cells. Improved V$_{OC}$ and FF, resulted a higher PCE (13.5%) in the MAI-treated devices relative to standard devices prepared without any surface treatment (12.7%).

II. EXPERIMENTAL DETAILS

A. MAI Surface Treatment

CdS and CdTe were deposited onto fluorine-doped SnO$_2$ glass substrates using a commercial vapor transport deposition process. CdCl$_2$ was deposited on the CdTe film by dropper using a saturated methanolic solution. Samples were heated at 387 ^{0}C in dry air for 30 min to activate the device [4]. Excess CdCl$_2$ was removed by thorough rinsing with methanol. The CdTe surface was then covered with 500 µL of MAI in anhydrous IPA at different concentrations ranging from 67.5 mM to 500 mM. After letting the solution sit on the CdTe layer for 40 s, the sample was spun at 4000 rpm for 20 s to form a thin MAI film on the CdTe surface. The samples were then heated to 150 °C for 5 min, 10 min, or 15 min, followed by thorough rinsing using anhydrous IPA.

B. Film Characterization

X-ray diffraction (XRD) patterns of the films were recorded from 2θ = 20° to 50° with a 0.02° step size and a scanning speed of 0.5°/s using a Rigaku Ultima III X-ray diffractometer. Surface scanning electron microscope (SEM) images of the films were obtained using a field emission

978-1-5090-5606-4/17 $31.00 © 2017 IEEE

scanning electron microscope (Hitachi S-4800). Raman spectra of the films were obtained at room temperature from 50 to 500 cm^{-1} using a LabRam confocal scanning microspectrometer (Horiba-Jobin Yvon) equipped with a He-Ne laser with line excitation at 632 nm.

B. Solar Cell Preparation

After reacting the MAI thin films with the CdTe surface and cleaning the surface, 3 nm of Cu and 40 nm of Au were deposited by thermal evaporation without breaking the vacuum. The samples were then annealed at 150 °C for 45 min to form the Cu$_x$Te layer and back contact. Individual cells were scribed on a 3 mm x 3 mm grid [11]. The active area of the fabricated solar cells is 0.08 cm^2.

C. Solar Cell Characterization

J-V characteristics were measured using a Keithley 2440 digital source meter and a solar simulator (Newport model 91195A-1000) configured to simulate AM1.5 illumination. A NIST-traceable Si reference solar cell was used to calibrate the light intensity.

III. RESULTS AND DISCUSSION

A. XRD Spectra

XRD was used to investigate the effect of MAI treatment on the CdTe structure. The XRD spectra (Fig. 1) of the standard and treated CdTe samples all show three main diffraction peaks at 2θ of 23.8°, 39.3°, and 46.4°, corresponding to diffractions from the (111), (220), and (311) crystalline planes of cubic CdTe, respectively [12].

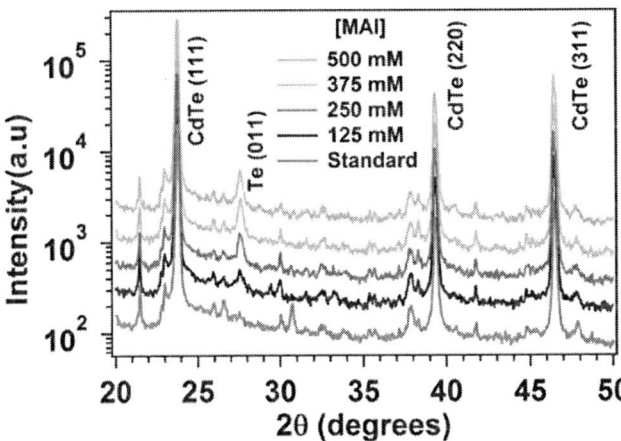

Fig. 1. X-ray diffraction patterns from MAI treated (using 125 mM, 250 mM, 375 mM, and 500 mM MAI solutions) and untreated CdTe/CdS samples.

After the MAI treatment, a new XRD peak at 27.5° appeared. This peak is likely due to hexagonal Te (PDF: 97-

009-6502) structure, suggesting that MAI treatment selectively removes Cd from the surface.

B. Variation of the Surface Morphology

To examine the impact of MAI treatment on the morphology of the CdTe surface, we obtained SEM images (Fig. 2) of the samples. Figure 2a shows the surface morphology of a CdCl$_2$ treated CdTe sample. The morphology of the CdTe surface was dramatically changed after MAI treatment. Figures 2b and 2c show the surface morphology of a CdTe sample after reacting with a MAI thin film formed by spin coating a 125 mM MAI solution on the CdTe surface. Here, small islands less than 100 nm in diameter were found on the CdTe surface. These islands evidently consist of Te. Reacting the CdTe surface with methylammonium bromide (MABr) (Fig. 2d) or methylammonium chloride (MACl) (Fig. 2e) thin films also showed the formation of these small islands. However, the size of the islands decreases in the order MAI > MABr > MACl.

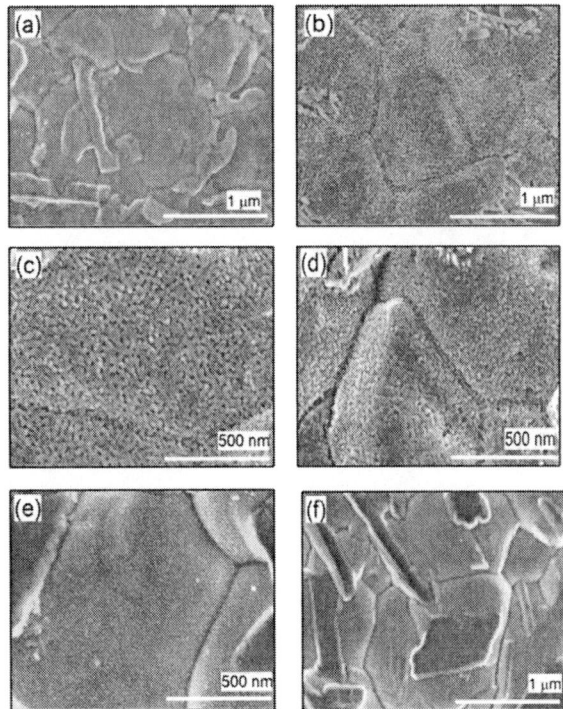

Fig. 2. SEM images of (a) CdCl$_2$ treated CdTe sample. Panels (b) and (c) show the surface morphology of a MAI treated sample using a 125 mM MAI solution and heated for 10 min, in low (b) and high (c) magnifications. Panels (d) and (e) show the surface morphology of (d) MABr and (e) MACl treated CdTe samples using 125 mM MABr and MACl solutions, respectively, and heated for 10 min. (f) Surface morphology of a CdTe sample, with the MAI rinsed prior to heating.

To determine at which step these Te islands were formed, MAI was deposited as above, but the sample was rinsed with

IPA prior to heating (henceforth referred to as the cleaned sample). Te islands were not observed on these cleaned samples (Fig. 2f) indicating heating is a necessary step for Cd removal.

C. Raman Spectroscopy Analysis

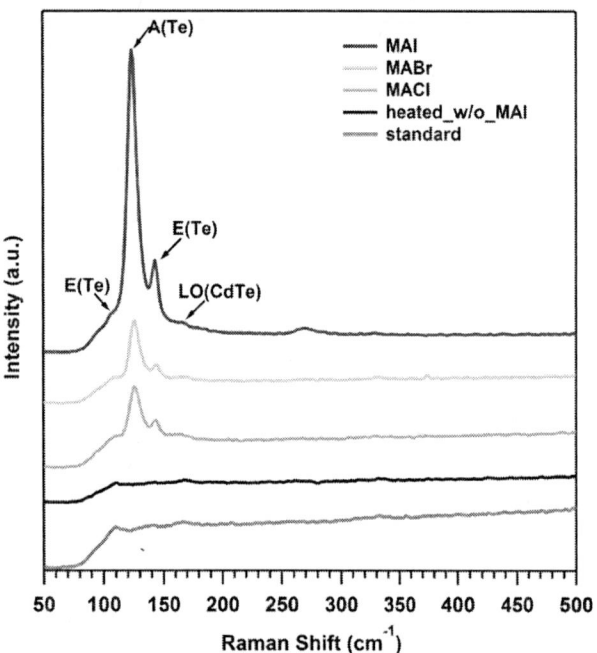

Fig. 3. Raman spectra from MAI-, MABr-, and MACl-treated samples using 125 mM MAI, MABr and MACl solutions, respectively, and heated for 10 min at 150 °C, a cleaned sample, and a standard.

To better understand the surface of the films, Raman spectroscopy analysis (Fig. 3) was performed on MAI-, MABr-, and MACl-treated samples using 125 mM MAI, MABr and MACl solutions, respectively and heated for 10 min at 150 °C, a cleaned sample, and a standard. Raman spectra of all the samples showed the presence of LO phonons of CdTe (LO(CdTe)) peak at 165 cm^{-1} and the E(Te) mode peak at 108 cm^{-1} [13, 14]. After the MAI, MABr, and MACl treatments, two new distinct peaks appeared at 123 cm^{-1} and 143 cm^{-1}. These peaks are likely due to the A(Te) and E(Te) modes [13, 14]. The calculated integrated peak intensity of the highest intense peak at 123 cm^{-1} was reduced in the order MAI (3245 a.u.) > MABr (746 a.u.) > MACl (666 a.u.), indicating a reduction of the Te volume in the order MAI, MABr, and MACl, respectively. This is consistent with the SEM observation of size reduction of the Te islands in the order MAI > MABr > MACl. Once again, no evidence of the Te islands were observed for the cleaned samples.

D. Formation of a Te-rich Surface

We recently showed that MAI, which is one of the main precursors used in the fabrication of lead halide perovskite solar cells, readily forms $(CH_3NH_3)_2CdI_4$ (MA_2CdI_4) perovskite with Cd^{2+} ions [15, 16]. In this study, MAI was mixed with $CdCl_2$ in an IPA solution, and MA_2CdI_4 perovskite spontaneously formed in solution at room temperature.

Here, the data from the cleaned samples indicate that the perovskites do not spontaneously form in solution. The Te island formation only occurs after the excess solution has been spun from the sample and heated. Therefore, the formation of a Te layer can be explained by the following reaction.

$$MAI(s) + CdTe(s) \xrightarrow{\Delta} MA_2CdI_4(s) + Te(s) \quad (1)$$

Because, MA_2CdI_4 perovskite is soluble in IPA[15, 16], the MA_2CdI_4 can be rinsed, leaving a Te layer behind on the CdTe film. This layer facilitates the efficient formation of a Cu_xTe layer during the Cu diffusion process. The MAI treatment appears to be self-limiting as significant changes in the surface morphology images were not observed with changes in the reaction time or MAI concentration. One of the reasons could be that the formed Te layer may form a capping layer for the CdTe and block the contact between the CdTe and the MAI film, hindering the formation of Te. The amount of MAI that react on the CdTe surface can also be a reason for this observation. The required MAI concentrations for a better contrast of the Te formation can be much higher than what we tested here.

Similar to MAI, MABr and MACl also form Te-rich surface by forming MA_2CdBr_4 and MA_2CdCl_4 perovskites, respectively. However, SEM and Raman spectroscopy analysis showed that the yield of the Te is reduced in the order MAI > MABr > MACl. Since all of the processing was similar for these reactions, the data indicates that the formation energy of MA_2CdX_4 (where X is I, Br, or Cl) perovskites reduces in the order Cl > Br > I.

E. Device Performances

Complete devices were fabricated to study the effect of MAI treatment on the device performances. Figure 4 shows the open circuit voltage (V_{oc}), short circuit current density (J_{sc}), Fill Factor (FF), and PCE data for devices reacted with MAI thin films fabricated with various MAI concentrations and reaction times. The MAI treatment mainly affected the V_{oc} and FF. All the treated samples showed higher average V_{oc} and FF than the standard samples. The V_{oc} and FF improvements with the Cu_xTe are consistent with the literature and attributed to a reduced Schottky barrier at the back contact [3, 5]. Samples treated with 125 mM MAI solution and heated for 10 min achieved the highest average V_{oc}, FF, and PCE of 824 ± 3 mV, 77.1 ± 0.6%, and 13.0 ± 0.2%, respectively. The average V_{oc}, FF, and PCE for devices made without MAI treatment were 771 ± 12 mV, 74.2 ± 1.8%, and 11.8 ± 0.4%, respectively. In addition to the improvements in the V_{oc}, FF, and PCE, the standard deviations of these photovoltaic parameters were also reduced with the MAI treatment. This is due to the precise control of the Cd removal process.

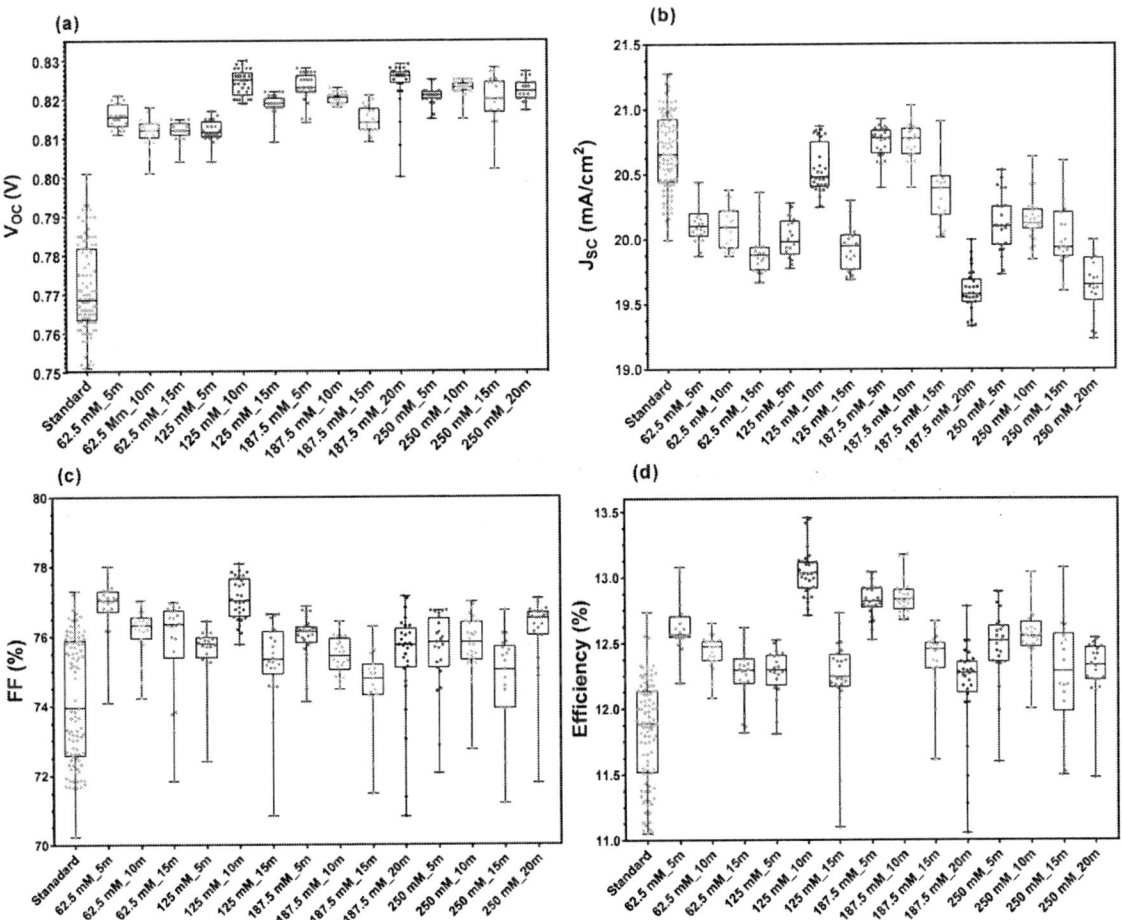

Fig. 4. Photovoltaic performances of (a) V_{oc}, (b) J_{sc}, (c) FF, and (d) PCE of MAI-treated (with various concentrations and reaction times) and untreated CdTe/CdS devices. Note that the average values presented in the figure were derived from measurements on more than 20 devices.

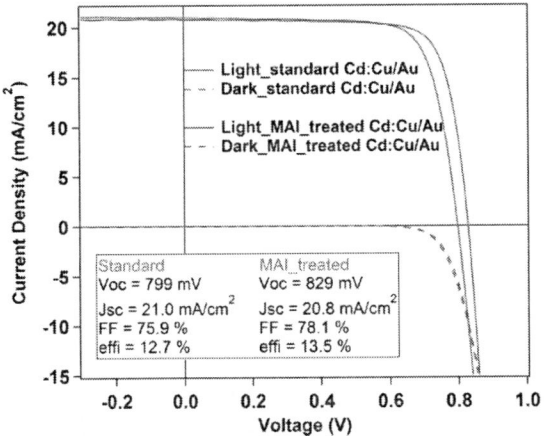

Fig. 5. J-V characteristics of champion devices from the MAI treated (using a 125 mM MAI solution and 10 min reaction time) and untreated standard CdTe/CdS samples.

The highest efficiency of 13.5% with V_{oc} of 829 mV, J_{sc} of 20.8 mA/cm^2, and a FF of 78.1% was also achieved using a 125 mM MAI solution with the 10 min reaction time. The best standard device prepared without treating the CdTe surface had a PCE of 12.7% with a V_{oc} of 799 mV, a J_{sc} of 21.0 mA/cm^2, and a FF of 75.9% (Fig. 5).

V. CONCLUSION

We have shown that methylammonium iodide can be simply and controllably reacted with CdTe surfaces to form a Te-rich surface. MAI extracts Cd from the CdTe surface by forming the $(CH_3NH_3)_2CdI_4$ perovskite. SEM and Raman spectra confirmed that thermal annealing is required for this reaction to occur. The MAI treatment reduces the Schottky barrier height at the back contact, resulting in improved V_{oc} and FF in CdTe solar cells. CdTe samples that were reacted with MAI thin films, formed by spin-coating a 125 mM MAI solution on the CdTe surface, obtained the highest device efficiency of 13.5%, while the champion device efficiency of the standard sample without a surface treatment was 12.7%.

978-1-5090-5606-4/17 $31.00 © 2017 IEEE

V. ACKNOWLEDGEMENT

This work was supported by the National Science Foundation (Contract no. #ECCS-1665172), the Ohio Federal Research Network (Center for Materials and Manufacturing), the Office of Naval Research (Contract No. N00014-17-I-2223), and the Wright Center Endowment for Photovoltaics Innovation and Commercialization. We thank Willard and Kelsey Solar Group for providing samples.

REFERENCES

[1] N. Strevel, L. Trippel, and M. Gloeckler, "Performance characterization and superior energy yield of first solar PV power plants in high-temperature conditions," *Photovoltaics international,* vol. 17, pp. 148-154, 2012.

[2] *NREL, "Solar Cell Efficiency Chart".* [http://www.nrel.gov/ncpv/images/efficiency_chart.j pg] accessed: Janurary, 2016.

[3] D. L. Bätzner, R. Wendt, A. Romeo, H. Zogg, and A. N. Tiwari, "A study of the back contacts on CdTe/CdS solar cells," *Thin Solid Films,* vol. 361–362, pp. 463-467, 2000.

[4] A. B. Phillips, R. R. Khanal, Z. Song, R. M. Zartman, J. L. DeWitt, J. M. Stone, P. J. Roland, V. V. Plotnikov, C. W. Carter, J. M. Stayancho, R. J. Ellingson, A. D. Compaan, and M. J. Heben, "Wiring-up Carbon Single Wall Nanotubes to Polycrystalline Inorganic Semiconductor Thin Films: Low-Barrier, Copper-Free Back Contact to CdTe Solar Cells," *Nano Letters,* vol. 13, pp. 5224-5232, 2013.

[5] J. Zhou, X. Wu, A. Duda, G. Teeter, and S. H. Demtsu, "The formation of different phases of CuxTe and their effects on CdTe/CdS solar cells," *Thin Solid Films,* vol. 515, pp. 7364-7369, 2007.

[6] S. Chun, S. Lee, Y. Jung, J. S. Bae, J. Kim, and D. Kim, "Wet chemical etched CdTe thin film solar cells," *Current Applied Physics,* vol. 13, pp. 211-216, 2013.

[7] J. Sarlund, M. Ritala, M. Leskelä, E. Siponmaa, and R. Zilliacus, "Characterization of etching procedure in preparation of CdTe solar cells," *Solar Energy Materials and Solar Cells,* vol. 44, pp. 177-190, 1996.

[8] C. S. Ferekides, U. Balasubramanian, R. Mamazza, V. Viswanathan, H. Zhao, and D. L. Morel, "CdTe thin film solar cells: device and technology issues," *Solar Energy,* vol. 77, pp. 823-830, 2004.

[9] N. Abbas Shah, A. Ali, Z. Ali, A. Maqsood, and A. K. S. Aqili, "Properties of Te-rich cadmium telluride thin films fabricated by closed space sublimation technique," *Journal of Crystal Growth,* vol. 284, pp. 477-485, 2005.

[10] R. Ochoa-Landín, O. Vigil-Galan, Y. V. Vorobiev, and R. Ramírez-Bon, "Chemically-deposited Te layers improving the parameters of back contacts for CdTe solar cells," *Solar Energy,* vol. 83, pp. 134-138, 2009.

[11] A. B. Phillips, Z. Song, J. L. DeWitt, J. M. Stone, P. W. Krantz, J. M. Royston, R. M. Zeller, M. R. Mapes, P. J. Roland, M. D. Dorogi, S. Zafar, G. T. Faykosh, R. J. Ellingson, and M. J. Heben, "High speed, intermediate resolution, large area laser beam induced current imaging and laser scribing system for photovoltaic devices and modules," *Review of Scientific Instruments,* vol. 87, p. 093708, 2016.

[12] E. R. Shaaban, N. Afify, and A. El-Taher, "Effect of film thickness on microstructure parameters and optical constants of CdTe thin films," *Journal of Alloys and Compounds,* vol. 482, pp. 400-404, 2009.

[13] A. Jarkov, S. Bereznev, O. Volobujeva, R. Traksmaa, A. Tverjanovich, A. Öpik, and E. Mellikov, "Photo-assisted electrodeposition of polypyrrole back contact to CdS/CdTe solar cell structures," *Thin Solid Films,* vol. 535, pp. 198-201, 2013.

[14] J.-W. Liu, F. Chen, M. Zhang, H. Qi, C.-L. Zhang, and S.-H. Yu, "Rapid Microwave-Assisted Synthesis of Uniform Ultralong Te Nanowires, Optical Property, and Chemical Stability," *Langmuir,* vol. 26, pp. 11372-11377, 2010.

[15] S. C. Watthage, Z. Song, N. Shrestha, A. B. Phillips, G. K. Liyanage, P. J. Roland, R. J. Ellingson, and M. J. Heben, "Enhanced Grain Size, Photoluminescence, and Photoconversion Efficiency with Cadmium Addition During the Two-Step Growth of $CH_3NH_3PbI_3$," *ACS Applied Materials & Interfaces,* 2016.

[16] S. C. Watthage, Z. Song, N. Shrestha, A. B. Phillips, G. K. Liyanage, P. J. Roland, R. J. Ellingson, and M. J. Heben, "Impact of Divalent Metal Additives on the Structural and Optoelectronic Properties of $CH_3NH_3PbI_3$ Perovskite Prepared by the Two-Step Solution Process," *MRS Advances,* pp. 1-6, 2017.

978-1-5090-5606-4/17 $31.00 © 2017 IEEE

Optimizing CdS Buffer Layer For CIGS Based Thin Film Solar Cell

Weijie Zhang, Korhan Demirkan, Geordie Zapalac, David Spaulding, Jochen Titus, Neil Mackie

MiaSolé Hi-Tech Corp, Santa Clara, CA, 95051, USA.

Abstract — **CdS buffer layer is one of the important factors that affect the conversion efficiency in CIGS-based solar cell. In our PVD CdS process optimization, different sputter powers in the early stage of CdS film growth were tested. Higher sputter power in the early stage of CdS growth improves cell efficiency and widens the process window. AES depth profile reveals a decrease of oxygen content at CdS/CIGS interface. Having fewer oxygen atoms at CdS/CIGS interface reduces the amount of deep acceptors in CIGS and results in improved cell efficiency.**

Index Terms — **PVD CdS, CIGS, sputter, Thin film.**

I. INTRODUCTION

Cu(In,Ga)Se$_2$ (CIGS) based solar cells are a promising thin film solar technology. As a leading thin film manufacturing company, Miasolé uses an all PVD CIGS technology based on a "roll-to-cell" platform where all the films are sputter-deposited sequentially onto a flexible stainless steel substrate in a multi-chamber vacuum system[1]. Miasolé has been able to increase flexible solar module efficiency at the manufacturing level from 13% to 16.5% over a 5 year period using this approach.

One factor related to the overall performance of a CIGS based solar cell is the choice of buffer layer to form the p-n heterojunction. Historically, the buffer layer has been a "wet" Chemical Bath Deposited (CBD) CdS layer, however this approach is not compatible with an all PVD technology and so a "dry" PVD CdS process has been developed at MiaSolé.

Previous work investigating the interfacial and chemical nature of the MiaSolé CIGS-PVD CdS interface by He et al [2] shows that epitaxial growth of PVD CdS on CIGS can be achieved using this "dry" PVD process. In addition, both Cubic and Hexagonal crystal phases have been shown to be present.

Further studies from He et al [3] also revealed significant elemental intermixing across the CIGS/PVD CdS heterojunction. It was found that Cu from CIGS migrates into the CdS crystal lattice with a concomitant migration of Cd into the near surface region of CIGS. These intermixing effects may have a critical role in the heterojunction properties resulting in a potential area for device and solar cell performance optimization.

In this paper we explore the relative changes in CdS deposition rate during the PVD CdS growth sequence and the impact on device properties with a view of intermixing at the interface and interfacial quality.

II. EXPERIMENTAL DETAILS

All the depositions for this study were carried out as a "roll-to-cell" manufacturing process in a custom multi-chamber PVD system, as shown in Figure 1.

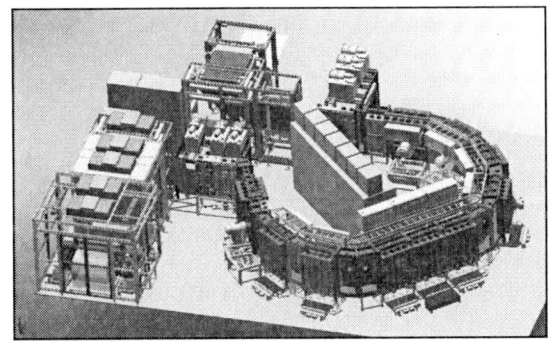

Fig1 Miasolé PVD Roll-to-Cell System

The MiaSolé "roll-to-cell" platform is a high vacuum system operating at 1-10 mTorr. The substrate is a 50μm thick stainless steel foil that continuously moves at fixed speed through the multi-chamber PVD system. A cross section of the full stack film illustrates the design and process steps in Figure 2. All layers are deposited sequentially with no vacuum break.

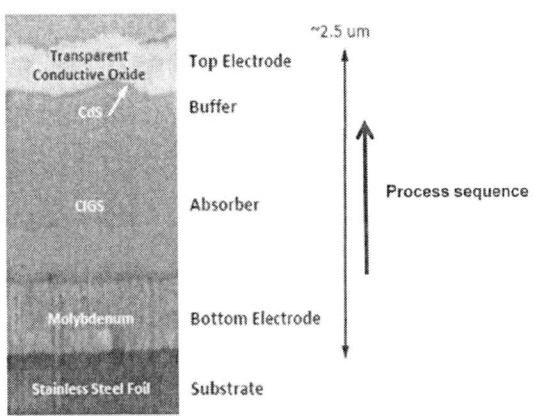

Fig2 Miasolé solar cell design and process sequence

A single vacuum chamber with multiple CdS sputter targets is used to deposit the entire CdS buffer layer. The resulting 30~90nm CdS films for this study were deposited in an Ar/O$_2$ environment. Total gas flow was controlled to between 200 and 500sccm during processing with optimum oxygen flow determined by analyzing the peak efficiency with a real-time in-line IV test. To maintain the same CdS thickness, total sputter power was kept constant. However, relative sputter power of the targets was adjusted in order to test the effect of higher initial-stage growth rate of the CdS film.

Film thickness was measured using an Oxford ED2000 x-ray fluorescence (XRF) system. The device performance for the resulting cells from the manufacturing line was measured using a custom flash test apparatus under 1 sun illumination. AES depth profile was obtained using PHI 660 Auger Electron Spectroscopy (AES) system with Zalar rotation.

III.   ANALYSIS OF RESULTS AND DISCUSSION

With constant gas flow and constant total sputter power, different relative sputter powers for the initial CdS growth stage were explored. Two test conditions were investigated: one with even power on all targets designated "flat" and one with twice the sputter power on the early stage CdS growth designated "graded". Figure 3 illustrates the individual target power for flat and grading conditions.

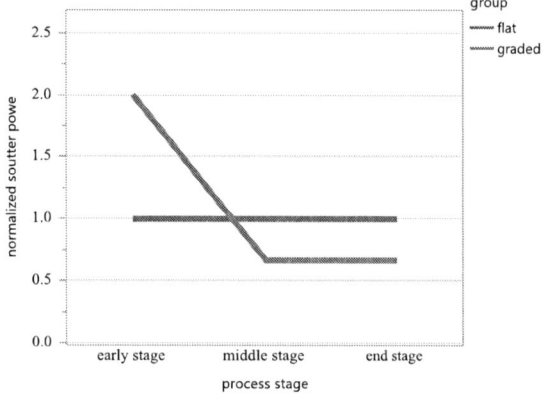

Fig.3 Sputter power for each target in flat and grading sputter power processes

Comparing the IV performance of "flat" power with "graded" in Figure 4 shows that "graded" had better solar cell performance due to higher Voc and FF without a change in Jsc. An absolute 0.9% improvement in cell efficiency was obtained utilizing "graded" power at the early stage of CdS growth.

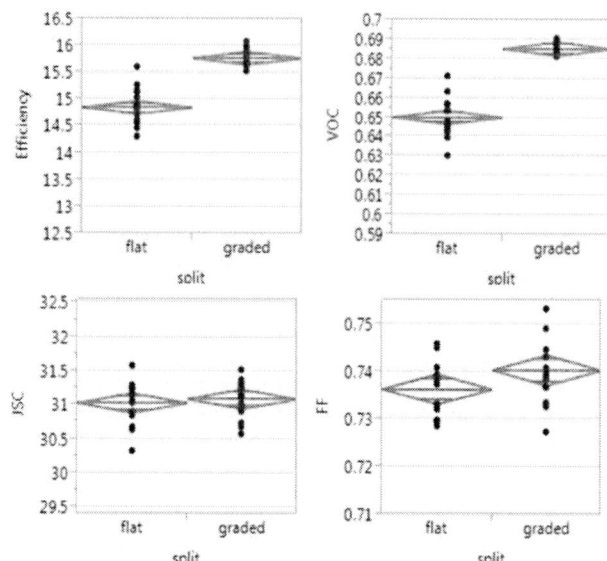

Fig.4 Cell efficiency comparisons for flat and graded sputter conditions

Optimum process window with respect to the oxygen partial pressure in the system was evaluated. At each sputter power setting, cell efficiency was continuously monitored at the output of the "roll-to-cell" system with oxygen flow varied from −20% below center point to 50% above. Data is illustrated in Figure 5. For "flat" power conditions we observed a small optimum where efficiency is reduced at higher and lower oxygen flow from the center point. With "graded" sputter power in the early stage growth of CdS we also observed reduced performance at oxygen flow below center point; however the process window is significantly larger with oxygen flows above the center point.

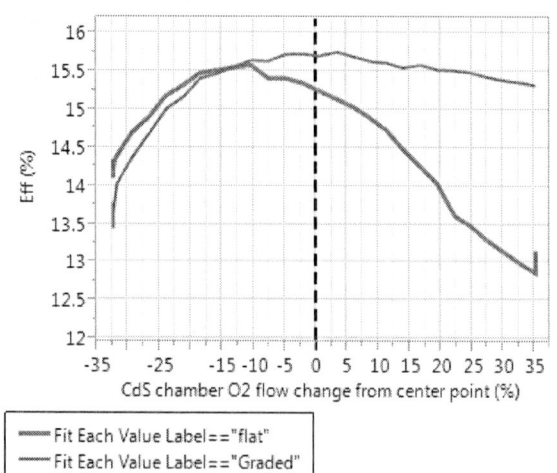

Fig.5 Cell efficiency vs. O2 flow

978-1-5090-5606-4/17 $31.00 © 2017 IEEE

AES depth profile was performed for both "flat" and "graded" power samples at the center point oxygen flow condition. Since the baseline process is deposited using a constant Ar/O2 process gas the oxygen content of the interface may be a critical factor in the observed performance difference.

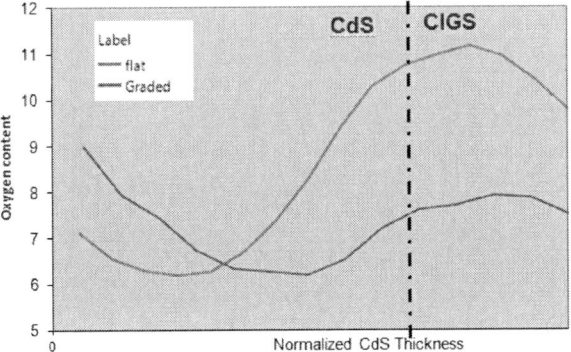

Fig.6 Oxygen content at CIGS/CdS interface through AES depth profile

A nearly 50% reduction in oxygen content at the CIGS layer was observed when "graded" sputter power was used compared with "flat" power. In the CdS layer nearest to the CIGS/CdS interface, oxygen content with the graded sputter power process is significantly less than that with flat power, as show in Figure 6.

These results show that increased sputter power at the early stage of CdS film growth resulted in less oxygen content in CIGS and at the CIGS/CdS interface. In the studies of Na-induced oxygenation of CIGS at CIGS/CdS interfaces, Kronik et al [4] and Su-Huai et al [5] determined that atomic oxygen at the surface of CIGS can substitute Se vacancy and convert to O_{Se}-. As the deep acceptor, Ose-, can promote recombination, hence this would decrease solar cell electrical performance.

The reduced oxygen content at the interface with the "graded" power configuration also matches the wider performance window with a higher oxygen flow due to reduced oxygen sensitivity at the CIGS/CdS interface.

IV. SUMMARY

Overall solar cell efficiency is improved using higher sputter power in the early stage of PVD CdS growth. We provide evidence to show that oxygen content in CIGS and CdS/CIGS interface was reduced with grading sputter power condition. One potential explanation is that lower atomic oxygen at CIGS/CdS interface will result in less deep acceptor O_{Se}- in CIGS which reduces carrier recombination and improves device efficiency. This leaning may also

result a modification of intermixing at this interface. Future work is planned in this area.

REFERENCE

[1] Neil Mackie, Robert Tas, John Corson, Atiye Bayman, "A Different Approach: CIGS using all PVD." 28th EU PVSEC, 2013

[2] X.Q.He, G. Brown, K. Demirkan, N. Mackie, V. Lordi, A./ Rockett, "Microstructure and chemical investigation of PVD-CdS/PVD-CuIn1-xGaxSe2 heterojunctions: a transmission electron microscopy study" IEEE Journal of Photovoltaic, 4, 1625 (2014)

[3] X .He, P. Ercius, J. Bailey, G.Zapalac, N. Mackie, A. Bayman, J.Varley, V. Lordi, and A. Rockett, "Cu rich domains and secondary phases in PVD-CdS/PVD-Cuin1-xGaxSe2 heterojunctions" in photovoltaic, specialist conference (PVSC), IEEE 42nd, 2015, pp1-3

[4] L.Kronik, U. Rau, J.F. Guillemoles, D. Braunnger, H.-W Schock and D. Caben, "Interface redox engineering of Cu(In, Ga)Se2-based solar cells: oxygen, sodium, and chemical ath effects." Thin Solid Films, vol.361-362, pp 353-359, 2000

[5] Su-Huai Wei, S. B. Zhang, Alex Zunger "Effects of Na on the electrical and structural properties of CuInSe2."Journal of Applied Physics, vol.85, pp7214-7218, 1999

Investigation of InP defect characteristics grown using novel TF-VLS technique

Abhinav Chikhalkar[1], Alec Fischer[1], Mark Hettick[2], Ali Javey[2] and Richard R. King[1]

[1]Arizona State University, Tempe, AZ, 85287, U. S. A.

[2]University of California, Berkeley, CA, 94720, U. S. A.

Abstract — The optoelectronic behavior of defects in InP grown using thin-film vapor-liquid-solid (TF-VLS) technique are characterized and analyzed in this paper. Cathodoluminescence spectroscopy is used to investigate the effect of Zn doping on the bulk defect states. The spatial distribution of electrically active defects is mapped using cathodoluminescence at the defect peak energy. CL mapping is also used to form a spatial distribution of defects corresponding to the deep level defect energy peak. Conductive AFM is used to analyze grain boundary conductivity. Light and dark surface potential measurements were carried out to better understand band bending at the grain boundaries. Understanding the recombination activity of defects in these low-cost III-V layers will help to build high efficiency single junction and multijunction solar cells from this material.

Index Terms — III-V semiconductor, Cathodoluminescence, conductive AFM, defects, TF-VLS.

I. INTRODUCTION

InP is a III-V semiconductor with a direct band gap of 1.34 eV, near the optimum value for a single-junction solar cell, and furthermore has relatively low unpassivated surface recombination velocity and high electron mobility. These properties make InP a very promising material for concentrator solar cell applications[1]. Efficiency as high as 22.1% has been reported for single crystal InP cells[2], but the higher capital cost of conventional III-V growth technology compared to silicon technology has inhibited scale up of III-V solar cells.

In addition to the properties above, a high absorption coefficient ($> 10^4$ cm^{-1}) makes InP thin-film an attractive option for solar cells[3]. A recently developed thin-film vapor-liquid-solid (TF-VLS) growth technique presents a cost effective and scalable way to synthesize high quality polycrystalline InP thin-film with grain sizes as large as several hundred microns[1,4-7]. A solar cell efficiency of 12.1% at 1-sun intensity (AM 1.5G) has been reported for a single junction cell using TF-VLS grown InP as absorber layer[8].

To improve the efficiency further and to understand the limitations of this technique, it is important to understand the role of grain boundaries and defects on the optoelectronic properties of the TF-VLS material. Photoluminescence (PL) imaging and electron beam induced current (EBIC) have been carried out previously to show grain structure and grain boundaries[1] at the micron scale.

In this study, the spatial and energy distribution of defects in non-intentionally doped n-InP and Zn doped p-InP is carried out using cathodoluminescence (CL). An effort to understand grain boundary conductivity is also carried out using conductive atomic force microscopy (cAFM). Future characterization techniques to develop our understanding about the nature of these defects are also suggested.

II. EXPERIMENTAL DETAILS

Fig 1 shows a schematic of the TF-VLS process for the growth of InP[1]. A 3 μm thick pure indium film is evaporated on molybdenum coated glass. A 40 nm SiO_2 layer is then deposited on the indium before phosphorization at high temperatures. The as-grown InP samples show n-type behavior. The n-type samples being characterized were obtained by skipping the Zn doping stage, and continuing etching of the SiO_2. The p-type samples were obtained by incorporating Zn in InP using vapor phase liquid source doping.

The effect of Zn doping on bulk defect characteristics is observed using cathodoluminescence (CL). The analysis is carried out at an accelerating voltage of 15 kV and a beam current of ~10 nA. Grain boundary conductivity is measured using conductive atomic force microscopy (cAFM). Surface potential measurements and band bending across grain boundaries were measured using Kelvin probe force microscopy (KPFM). Both cAFM and KPFM were carried out using a 20-nm Pt-Ir tip in a Bruker Multimode 8 AFM system.

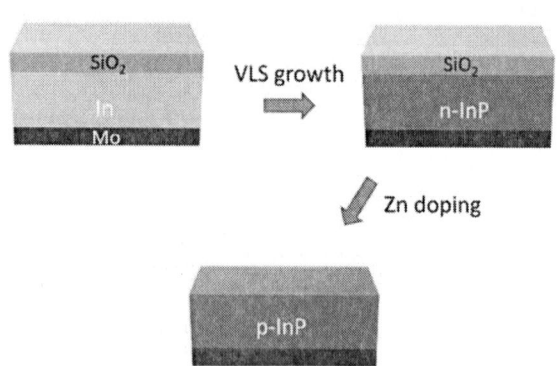

Fig 1. Schematic of TF-VLS process for the growth of Zn doped InP[1]

III. RESULTS AND DISCUSSION

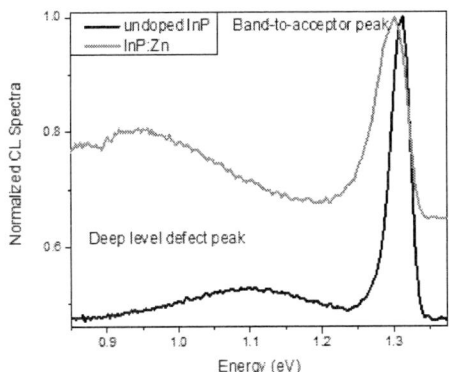

Fig. 2. CL spectra of undoped and doped InP, showing ~150 meV red shift of the BA peak with Zn doping

Cathodoluminescence (CL) spectra for Zn-doped InP (InP:Zn) and undoped InP are shown in Fig. 2. The deep level (DL) defect peak is visible even at room temperature in CL – something which could only be observed at low temperatures with photoluminescence spectroscopy. We also observe a large red shift of 150 meV in the deep level defect peak position of InP:Zn when compared to undoped InP, suggesting a key role of Zn in the bulk defect characteristics.

To understand the spatial distribution of the radiative recombination sites leading to the CL spectra of InP:Zn, CL imaging was carried out at 4.5K. At low temperature, we can differentiate the band-to-acceptor (BA) transition peak from the band-to-band transition peak (Fig 3b). The DL peak is also very pronounced at low temperatures (Fig 3a). Interestingly, within a grain, we observe that the regions that are relatively bright in the BA image are darker in the DL image, and vice versa. This could result from the deep level recombination pulling down the excess carrier concentration, thus reducing the luminescence from the BA transition. EDX mapping of Zn concentration could be used to understand the origin of spatial inhomogeneity of the DL peak.

The panchromatic CL image is shown in Fig 4. Line scan of intensity across the distinct grain boundary was carried out at room temperature to calculate the diffusion length of minority charge carriers. The value of ~3.4±2.1 μm obtained from this method is very close to the value previously extracted from EBIC measurements[8].

To understand the grain boundary characteristics further, scanning probe microscopy was used. In addition to the topology of the thin film, the electric potential was measured across grain boundaries using KPFM, and grain boundary conductivity was measured with cAFM. Fig. 5a. images the roughness of the film and Fig. 5(b) and 5(c) show the variation of surface conductivity with varying voltage. It is observed that the current through the grain core does change with voltage, following the expected semiconductor behavior of InP. The

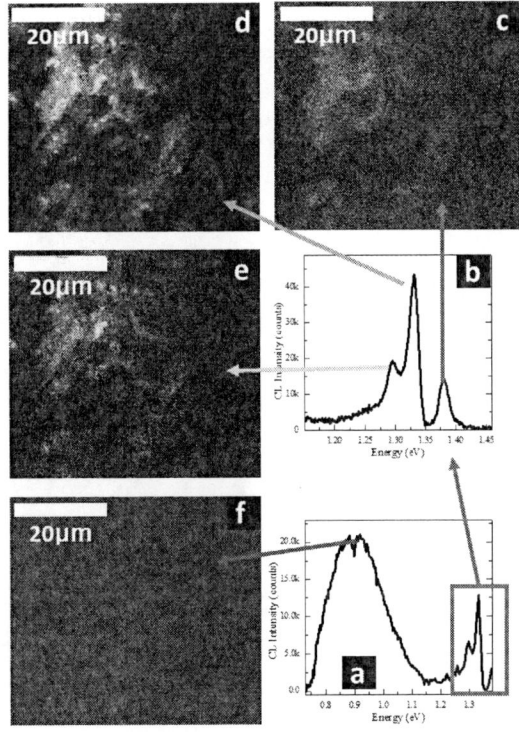

Fig 3. (a, b) CL spectra of InP:Zn at 4.5K, and CL images at (c) band-to-band peak, (d, e) band-acceptor peak with its phonon replica and (f) deep level defect peak

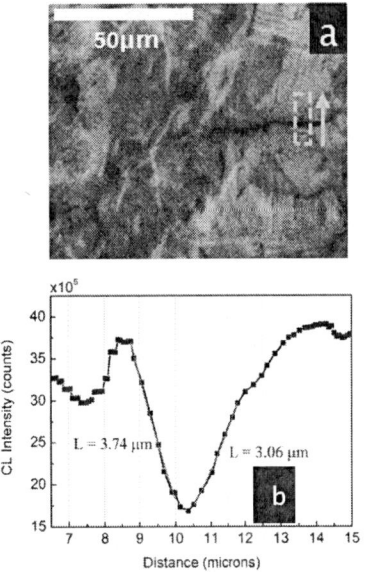

Fig 4. (a) Panchromatic CL image of Zn doped InP, and (b) CL line scan of change in luminescence intensity across a grain boundary

grain boundaries appear to be much more pronounced in current maps as compared to the depth profiles and show insulating behavior. This behavior may be attributable to the band bending near the grain boundaries which could create a local intrinsic region, thereby reducing the conductivity around grain boundaries drastically.

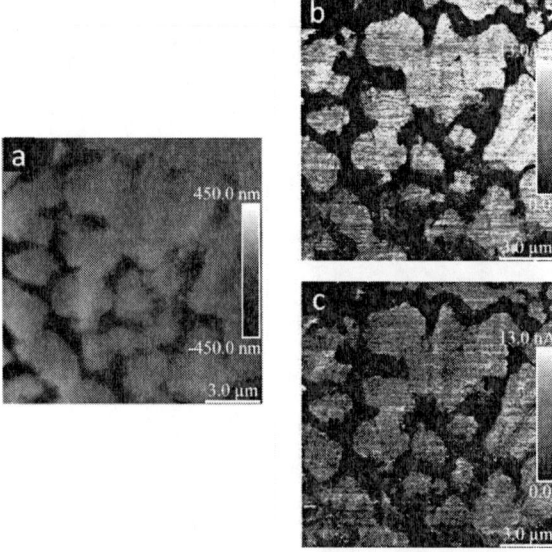

Fig. 5. (a) Depth profile of InP, (b) and (c) are current plots at 1.9V and 1.7V respectively

Surface potential measurements are plotted in Fig. 6(a) and 6(b) to quantify band bending at the grain boundaries in polycrystalline InP film before Zn doping, with and without white light illumination. We observe that the surface potential

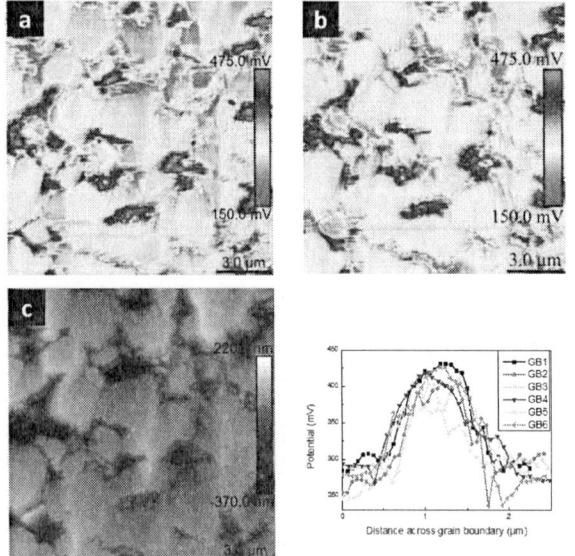

Fig. 6. Surface potential (a) without illumination and (b) with illumination. (c) surface morphology and (d) change in electric potential across the grain boundary

is approximately 275 meV. Using work function values of the Pt/Ir AFM tip (ϕ_m~5.4eV) and the surface potential value, the Fermi level at the surface is estimated to be ~5.675eV. We observe that the potential near grain boundaries increase to approximately 410 meV, suggesting that the bands bend by 135 meV at the grain boundaries as represented in Fig. 7.

Incidence of white light has shown to reduce both the potential over majority of the surface area and the average grain boundary potential. This behavior is expected since the material

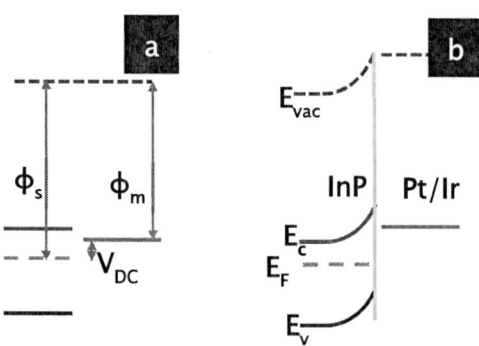

Fig. 7. (a) Band diagram outlining the measured surface potential, and (b) reduced surface potential because of band bending at the grain boundary

is lightly n-doped and addition of equal number of electrons and holes will move the Fermi level towards the valence band. We also observe that the change in potential in the grain core is much more pronounced than near the boundary, indicating that the trap state density tends to pin the Fermi level at the grain boundary.

IV. CONCLUSIONS

A preliminary investigation of optoelectronic properties of defects in a novel form of polycrystalline InP is presented. It is observed that Zn greatly affects the deep level defect states, causing a 150 meV red shift in its CL peak. Contrasting regions of complementary bright band-to-acceptor luminescence and dark deep level defect luminescence are observed, as well as regions with the opposite behavior of darker BA luminescence and bright deep level defect emission. This may be caused by suppressed excess carrier concentration in regions for which deep level recombination is especially active. The decrease in conductivity near grain boundaries as measured by conductive AFM may be caused by local formation of a depleted region near grain boundaries due to band bending. However, quantitative analysis of band bending at grain boundaries by Kelvin probe force microscopy indicate only a 135meV rise in conduction and valence bands at the grain boundaries, typically not enough to cause the dramatic contrast between conduction in the grain cores and insulating behavior at the grain

978-1-5090-5606-4/17 $31.00 © 2017 IEEE

boundaries. Thus, the preliminary defect characterization work presented here raises key questions to be resolved by future work. Engineering the grain boundary structure to increase the band bending could be used to shield photogenerated carriers from recombining readily, improving solar cell performance in this polycrystalline InP material.

V. ACKNOWLEDGEMENTS

We would like to thank LeRoy Eyring Center for Solid State Science at ASU for providing the conductive AFM technique, and QESST for funding this research.

REFERENCES

[1] H. Wang, C. M. Sutter-fella, P. Lobaccaro, M. Hettick, M. Zheng, D. Lien, D. W. Miller, C. W. Warren, E. T. Roe, M. C. Lonergan, H. L. Guthrey, N. M. Haegel, J. W. Ager, C. Carraro, R. Maboudian, J. He, and A. Javey, "Increased Optoelectronic Quality and Uniformity of Hydrogenated p-InP Thin Films," *Chem. Mater.*, vol. 28, pp. 4602–4607, 2016.

[2] M. S. Keavney, C. J.; Haven, V. E.; Vernon, "Emitter Structures in MOCVD InP solar cells," *Conf. Rec. Twenty First IEEE Photovolt. Spec. Conf.*, vol. 1, pp. 141–144, 1990.

[3] H. Burkhard, H. W. Dinges, E. Kuphal, H. Burkhard, H. W. Dinges, and E. Kuphal, "Optical properties of In(1-x)Ga(x)P(1-y)As(y) , InP , GaAs , and GaP determined by ellipsometry," *J. Appl. Phys.*, vol. 53, no. 1, pp. 655–662,

1982.

[4] R. Kapadia, Z. Yu, M. Hettick, J. Xu, M. S. Zheng, C. Y. Chen, A. D. Balan, D. C. Chrzan, and A. Javey, "Deterministic nucleation of InP on metal foils with the thin-film vapor-liquid-solid growth mode," *Chem. Mater.*, vol. 26, no. 3, pp. 1340–1344, 2014.

[5] R. Kapadia, Z. Yu, H. H. Wang, M. Zheng, C. Battaglia, M. Hettick, D. Kiriya, K. Takei, P. Lobaccaro, J. W. Beeman, J. W. Ager, R. Maboudian, D. C. Chrzan, and A. Javey, "A direct thin-film path towards low-cost large-area III-V photovoltaics," *Sci. Rep.*, vol. 3, no. 2275, pp. 1–7, 2013.

[6] M. Zheng, H. Wang, C. M. Sutter-fella, C. Battaglia, S. Aloni, X. Wang, J. Moore, J. W. Beeman, M. Hettick, M. Amani, W. Hsu, J. W. Ager, P. Bermel, M. Lundstrom, J. He, and A. Javey, "Thin-Film Solar Cells with InP Absorber Layers Directly Grown on Nonepitaxial Metal Substrates," *Adv. Energy Mater.*, vol. 5, no. 1501337, pp. 1–9, 2015.

[7] M. Zheng, K. Horowitz, M. Woodhouse, C. Battaglia, R. Kapadia, and A. Javey, "III-Vs at scale : a PV manufacturing cost analysis of the thin fi lm vapor − liquid − solid growth mode," *Prog. Photovoltaics Res. Appl.*, vol. 24, pp. 871–878, 2016.

[8] M. Zheng, H. P. Wang, C. M. Sutter-Fella, C. Battaglia, S. Aloni, X. Wang, J. Moore, J. W. Beeman, M. Hettick, M. Amani, W. T. Hsu, J. W. Ager, P. Bermel, M. Lundstrom, J. H. He, and A. Javey, "Thin-Film Solar Cells with InP Absorber Layers Directly Grown on Nonepitaxial Metal Substrates," *Adv. Energy Mater.*, vol. 5, no. 22, pp. 1–9, 2015.

Investigation of Fast Growth GaAs-based Solar Cell on Reusable Substrate by Metalorganic Chemical Vapor Deposition

Chaomin Zhang[1], Eric Armour[2], Yeongho Kim[3], Abhinav Chikhalkar[1], Ehsan Vadiee[1], Richard King[1], Christiana Honsberg[1]

[1] School of Electrical, Computer and Energy Engineering, Arizona State University, Tempe, AZ, 85287, USA

[2] Veeco Instruments Inc., MOCVD Operations, NJ, 08873, USA

[3] Division of Metrology for Future Technology, Korea Research Institute Standards and Science (KRISS), 267 Gajeong-ro, Yuseong-gu, Daejeon 305-600, KOREA

Abstract — **GaAs solar cells can be grown with a fast growth rate without sacrificing the device performance. This paper investigates the effect of fast growth rate with metalorganic chemical vapor deposition (MOCVD) on GaAs single-junction solar cell crystal quality and device performance. The minority carrier lifetime of the samples grown with a fast growth rate (~56 μm/hr) was investigated. The growth parameters including growth temperature, V/III, and doping densities were varied to study the lifetime degradation mechanisms in the GaAs solar cell. In addition, Bi, as a surfactant, was used during the fast growth to improve the surface roughness compared to the standard growth methods with the growth rate of ~14 μm/hr. An epitaxial lift-off process was also designed to investigate the GaAs epitaxial quality on reused substrate. The goal of this work was to achieve a high quality GaAs solar cell with a fast growth on a reused GaAs substrate.**

Index Terms — **high growth rate, GaAs solar cells, metalorganic chemical vapor deposition, minority carrier lifetime, semiconductor growth, TRPL**

I. INTRODUCTION

High efficiency solar cells produced with low fabrication costs are required to meet the increasing demand of renewable energies. Among all high efficiency solar cell technologies, multijunction solar cells offer the potential for achieving higher photovoltaic conversion efficiency by extending absorption wavelengths in the solar spectrum to higher values. It was demonstrated that the efficiency of the multijunction solar cells can reach up to 46% under high concentration (508 suns) [1].

Metalorganic chemical vapor deposition (MOCVD) is widely used as a cost-effective system to grow III-V compound semiconductors including solar cells. Thus, increasing the throughput of MOCVD by increasing its growth rate can further reduce the production costs. The extremely fast growth of GaAs-based solar cells was previously reported [3]-[5]. However, it was shown that the fast growth method can degrade the minority carrier lifetime and as a result, the

solar cell characteristics. In the present work, we showed that the solar cell performances of fast growth method can be preserved.

The cost analysis [2] of the lattice-matched (LM) multijunction solar cells shows that the epi-wafers (Ge or GaAs) used as substrates or bottom cells are the major cost drivers of today's concentrated photovoltaic (CPV) technology. Epitaxial lift-off (ELO) or substrate removal process can have a significant impact on the cost reduction of the solar cells. In this method, a sacrificial layer, or a release layer, is deposited on the substrate and the subsequent solar cell structure can be grown on top of this sacrificial layer. The release layer will be then etched off with a material-selective etchant. ELO process enables the III-V active layer to be bonded to a secondary substrate or another solar cell. This allows a significant cost reduction since high-cost III-V substrates can be reused. However, it is commonly observed that the conventional ELO process will result in significant substrate surface roughening and contamination by etch byproducts such as AlF_3 during ELO process. To reuse the substrates after ELO process for epitaxial growth, a polishing process is typically applied. Nevertheless, tens of microns of material from the substrate is consumed in order to regrow high quality epitaxial layers, which can eventually limit the potential substrate reusing number. Lee *et al.* [6] proposed that by depositing $In_{0.49}Ga_{0.51}P/GaAs/In_{0.49}Ga_{0.51}P$ lattice-matched protection layer system to preserve the GaAs substrate in its epi-ready condition before the release layer deposition. With both fast growth rate and reusable substrate, the manufacturing cost of GaAs-based solar cells can be significantly reduced.

In this work, the GaAs structures were grown with 56 μm/hr and varied growth conditions including different growth temperatures, V/III ratios, and applying Bi surfactant flux. Further, minority carrier lifetime and surface morphology are studied to investigate the effect of fast growth rate and ELO process on the GaAs solar cell performance. Finally, GaAs

978-1-5090-5606-4/17 $31.00 © 2017 IEEE

hetero-structures epixially grown on reused substrates with a high growth rate are explored.

II. EXPERIMENTAL DETAILS

The epi-structures were grown on p-type Zn-doped 5 degrees offcut (001) GaAs substrates using a Veeco K475i As/P MOCVD system. To investigate the minority carrier lifetime of the base layer, a lifetime test structure comprised of a 3000 nm p-type GaAs with a doping level of 5×10^{16} cm^{-3}, sandwiched between InGaP layers, were grown on the Zn-doped GaAs substrate at the two different growth rates of 14 μm/hr and 56 μm/hr, shown in Fig. 1(a). Moreover, a series of similar samples with the same structure were grown by changing the growth parameters such as arsine flux, Zn doping level, and growth temperature.

The TMBi precursor, 1.8×10^{-4} moles/minute and 9.0×10^{-5} moles/minute, which is about 1% and 0.5% of TMGa precursor (1.665×10^{-2} moles/min), was used as a surfactant during the fast growth of GaAs base layer to study the Bi surfactant impact on the GaAs fast growth samples.

The typical cell structure with the doping levels is shown in Fig. 1(b). A 250 nm GaAs buffer layer was deposited on the GaAs substrate, followed by In$_{0.49}$Ga$_{0.51}$P (100nm)/GaAs (100nm)/In$_{0.49}$Ga$_{0.51}$P (100nm) layers, to protect the substrate during the lift-off process. The lifetime structure was then grown on top of it. In order to reuse the GaAs substrate for the regrowth, smooth and clean GaAs surface should be provided. Concentrated HCl (37%) was used to etch off the InGaP layers and a mixed NH$_4$OH:H$_2$O$_2$:H$_2$O solution to etch the GaAs layer.

Time-resolved photoluminescence (TRPL) measurement is performed to evaluate the carrier lifetimes with the time-correlated single photon counting (TCSPC) system at room temperature. The lifetime structures are optically excited by a Ti: Sapphire laser with an excitation wavelength of 800 nm. The TRPL spectra are then detected at the GaAs band-edge (860 nm). Moreover, atomic force microscopy (AFM) was conducted by multimode scanning probe microscope (SPM) to study the surface morphology.

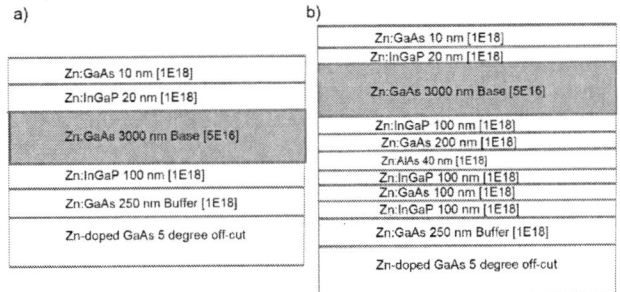

Fig. 1. (a) Schematic structure for TRPL measurement and (b) schematic device structure for lift-off process. (gray areas were grown with different growth rates (14 μm/hr and 56 μm/hr) and other epilayers were grown with the standard growth rate (14 μm/hr).

III. RESULTS AND DISCUSSIONS

A. Effect of different epitaxial growth conditions

The minority carrier lifetimes of grown samples, measured by TRPL system at room temperature, are listed in Table I. It is evident that the lifetimes of the samples grown with the high growth rate (56 μm/hr) are lower than that of samples grown with the standard growth rate (14 μm/hr). The higher growth temperature (680 °C) was chosen for samples with the fast growth rate to enhance the adatom migration and better surface mobility. In addition, it is shown that by increasing the arsine flux, the carrier lifetime reduces significantly. This low lifetime might be originated from inactive arsenic precipitants in the volume of the structure, resulted in higher the density of nonradiative recombination centers. However, there is no significant difference in the lifetime of samples with the V/III of 9 and 6. On the other hand, increasing the growth temperature to 700 °C does not show any major change in the lifetime. Nevertheless, the increase of the Zn doping level decreases the lifetime, which means higher Zn doping increases the Shockley-Read-Hall recombination centers.

The effect of different carrier lifetimes on open circuit (V_{OC}) and short circuit current (J_{SC}) are examined by using the PC1D simulation software. PC1D is a one-dimensional drift-diffusion solver and is commonly used for GaAs based solar cell simulation. From the PC1D simulation, the sample with the fast growth rate and the minority carrier lifetime of 115 μs has a V_{OC} of 1.031 V. The sample with the standard growth rate and the minority carrier lifetime of 165 μs has a V_{OC} of 1.035 V. Also, the J_{SC} of both samples with different minority carrier lifetimes are similar [5].

AFM images show that the surface of samples with the fast growth rate (Fig. 2(b)) is rougher than that of the samples with the standard growth rate (Fig. 2(a)). Bumps were observed on the fast growth samples, which may be originated from the lack of surface diffusion.

TABLE I

THE MINORITY CARRIER LIFETIME OF TRPL STRUCTURES
AT ROOM TEMPERATURE

TRPL samples		Lifetime (ns)
Standard growth (14 μm/hr, 640 °C)		165
Fast growth (56 μm/hr)	Baseline (680 °C, V/III=9)	115
	Increased arsine flux (680 °C, V/III=12)	87
	Decreased arsine flux (680 °C, V/III=6)	107
	Increased Zn doping level (2 times higher than the baseline)	91
	Increased growth temperature (20 °C higher than the baseline) (700 °C)	107

B. Effect of Bi surfactant

One of the common issues with the high growth rate method is that the adatom migration to the step edges can be hampered because the impinging sources can "trap" the surface species before they can get to the appropriate bonding sites. It is believed that the Bi atoms can help the migration process and make the surface smoother. Thus, Bi is used as a surfactant, to change the surface reconstruction of InGaP layer [7]. In order to confirm the effect of Bi surfactant on the surface kinetics of the samples grown at the high growth rate, surface roughness was measured using AFM. The surface of the samples with the fast growth rate and Bi surfactant are smoother (shown in Fig. 2(c) and Fig. 2(d)) than that of the samples with no Bi surfactant (Fig. 2(b)). This reveals that the Bi surfactant has a smoothing effect on the growth surface. The roughness of the sample with the fast growth rate and 0.5% Bi is even less than the sample with the standard growth rate. However, the minority carrier lifetimes for samples with 1% and 0.5% Bi surfactant are 110 ns and 107 ns, respectively, which are lower than the lifetime of standard samples but still comparable to the fast growth baseline sample.

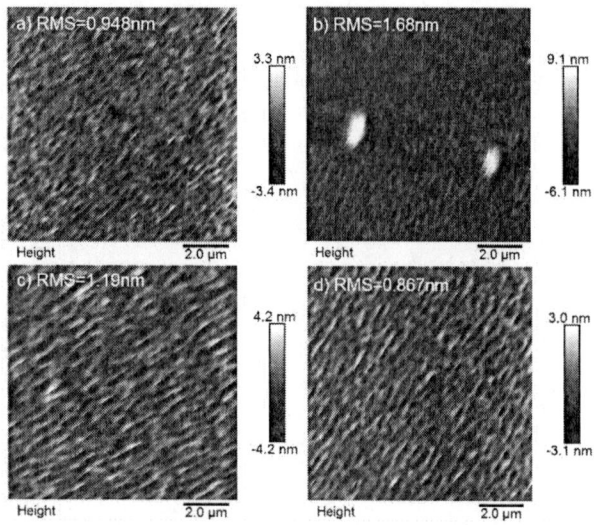

Fig. 2. 10×10 μm^2 atomic force microscopy (AFM) images of (a) the standard growth, (b) fast growth baseline, (c) fast growth with 1% Bi surfactant, and (d) fast growth with 0.5% Bi surfactant samples.

C. Lift-off on reusable substrate

The lift-off structure was grown, as shown in Fig. 1(b), and the minority lifetime was measured after the complete lift-off process. The minority carrier lifetime is 133 ns for the sample with the standard growth rate (14 μm/hr). This is lower than the lifetime sample with the structure shown Fig.1 (a). After etching off the epilayers (the layers on and including AlAs layer), we achieved 0.65 nm roughness. Furthermore, a 0.43 nm surface roughness was obtained after etching off the

"protection" layers (*i.e.* InGaP/GaAs/InGaP). A regrowth structure as Fig. 1(b) was conducted to two etched substrates at a standard growth rate (14 μm/hr) and a high growth rate (56 μm/hr), respectively. The surface roughness of standard regrowth sample is 0.526 nm, which is smooth and indicates high quality of epi-layers is achieved. 1.11 nm surface roughness was obtained from the fast regrowth sample, which is comparable to the initial growth (1.16 nm). Therefore, the results reveal that high quality GaAs epi-layers can be achieved at the fast growth rate on the reused GaAs substrates.

IV. CONCLUSION

The cost reduction of high efficiency concentrated solar cells by reducing the deposition time and reusing the substrates is promising for future multijunction solar cell applications. For GaAs-based solar cells, we have studied the minority carrier lifetime and the surface morphology of samples with different growth rates and different growth conditions. From the PC1D simulation, the device performances achieved by the fast growth rate method are comparable to the devices with the standard growth rates. To better understand the fast growth method, various growth conditions were conducted. Further inspection confirms that the Arsine flux has a significant impact on the defects formation. Also, the effect of growth temperature seems to be saturated at about 680 °C. We showed that Bi, acting as a surfactant, has improved the surface roughness, however, no significant lifetime improvement is observed. Substrate reuse by epitaxial lift-off process with the designed structure is also shown. The cost reduction of high-efficiency concentrated solar cells by reducing the deposition time and the substrate cost by utilizing these growth techniques is promising for future multijunction solar cell applications.

ACKNOWLEDGEMENTS

This material is based upon work primarily supported by the Engineering Research Center Program of the National Science Foundation and the Office of Energy Efficiency and Renewable Energy of the Department of Energy under NSF Cooperative Agreement No. EEC-1041895. We gratefully acknowledge the use of facilities within the LeRoy Eyring Center for Solid State Science at Arizona State University.

REFERENCES

[1] M. A. Green, K. Emery, Y. Hishikawa, W. Warta and E. D. Dunlop, Prog. Photovoltaics Res. Appl. 23 (1), 1-9 (2015).

[2] K. A. W. Horowitz, M. Woodhouse, H. Lee and G. P. Smestad, AIP Conf. Proc. 1679, 100001 (2015).

[3] C. Ebert, A. Parekh, Z. Pulwin, W. Zhang, D. Lee, and D. Byrnes, "Fast growth rate GaAs and InGaP for MOCVD grown triple junction solar cells," Photovoltaic Specialists Conference (PVSC), 2010 35th IEEE, 2010, pp. 002007-002011.

[4] K. J. Schmieder, E. A. Armour, M. P. Lumb, M. K. Yakes, Z. Pulwin, J. Frantz, and R. J. Walters, "Effect of Growth Temperature on GaAs Solar Cells at High MOCVD Growth Rates," IEEE J. Photovoltaics, vol. 7, no. 1, pp. 340–346, Jan. 2017.

[5] C. Zhang, Y. Kim, C. Ebert, N. N. Faleev, and C. B. Honsberg, "Influence of high growth rate on GaAs-based solar cells grown by metalorganic chemical vapor deposition," in 2015 IEEE 42nd Photovoltaic Specialist Conference (PVSC), 2015, pp. 1–5.

[6] K. Lee, J. D. Zimmerman, X. Xiao, K. Sun, and S. R. Forrest, "Reuse of GaAs substrates for epitaxial lift-off by employing protection layers," J. Appl. Phys., vol. 111, no. 3, p. 33527, Feb. 2012.

[7] S. W. Jun, R. T. Lee, C. M. Fetzer, J. K. Shurtleff, G. B. Stringfellow, C. J. Choi, and T.-Y. Seong, "Bi surfactant control of ordering and surface structure in GaInP grown by organometallic vapor phase epitaxy," J. Appl. Phys., 2000.

Development of Aluminum Epilayers as Buffers for GaInAs

Phil Ahrenkiel, Nathan Smaglik, and Nikhil Pokharel

South Dakota School of Mines & Technology, Rapid City, South Dakota, 57701 U.S.A.

Alessandro Giussani, Michael A. Slocum, and Seth M. Hubbard

Rochester Institute of Technology, Rochester, NY, 14623 U.S.A.

Abstract — **Aluminum could be used as a structural buffer and sacrificial layer that is lattice-matched on the (001) surface to GaInAs having a 45° in-plane rotation. Conventional MOCVD from trimethylaluminum results in nanocrystalline aluminum carbide. By using hydrogen-plasma assistance in the reactor inlet, we avoid carbide formation, allowing deposition of elemental aluminum. The microstructures of these films resulting from initial applications of this method are discussed.**

Index Terms — **aluminum, metamorphic, metalorganic chemical vapor deposition, plasma enhancement.**

I. INTRODUCTION

The earth-abundant element aluminum is widely appreciated by the microelectronics industry as an inexpensive and easily deposited electrical conductor. But as a group-III metal, Al also serves as a component of the III-V compound semiconductor system. Elemental Al can be lattice matched to $Ga_{0.82}In_{0.18}As$ (GaInAs) in the orientation GaInAs(001)[100]||Al(001)||[110] to serve as buffer layers or sacrificial layers for epitaxial lift-off. Thus, it is desirable to enable deposition of both elemental Al and Al-containing III-Vs in a single process.

Elemental Al deposited by physical vapor deposition typically consists of micro- or nano-scale crystalline grain sizes. But Al heteroepitaxy on cubic substrates remains an appealing possibility. Physical vapor deposition and molecular-beam epitaxy (MBE) have both been used to produce oriented epitaxial films. An ultra-fast evaporation method produced very thin, epitaxial Al films on (111) Si [1]. Sputtered Al on <111>-miscut (001) Si gave (110)-oriented, single-crystal films [2]. Epitaxial Al (001) films on (001) AlAs was shown to be possible by MBE if a thin layer of Ga or In is incorporated at the interface [3]. More recent work using MBE also produced (110)-oriented films on exact (001) Si, though the Al orientation changed upon post-growth annealing [4]. Single-crystal Al on (001) GaAs was used to study magnetic effects, but those films exhibited apparent roughness and pinholes [5]. The MBE deposition of epitaxial Al on a spinel substrate, and its subsequent use as a GaInAs template were studied for applications in photovoltaics [6].

Most industrial growth of III-V films for photovoltaics relies on metalorganic chemical vapor deposition (MOCVD). But both C and O incorporation during deposition of Al-containing III-Vs from conventional metalorganic precursors have remained challenges. Direct Al-C bonds in the most common

aluminum-alkyl precursors, particularly trimethylaluminum (TMAl), are not easily broken by pyrolysis [7]. Thus, a number of alternative Al precursors have been developed that more readily decompose under typical growth conditions [8]. Nonetheless, most of these precursors have other drawbacks [9], and it remains desirable to release Al from the organic components by other means. Note that a vast excess of group-V hydride (e.g., PH_3, AsH_3), compared to the group-III organometallics, is present during a conventional MOCVD process. Cracking of these hydride species at the growth surface results in the release of hydrogen radicals, which would not be present otherwise, even with an abundant flow of H_2 diluent.

In the chemical vapor deposition of group-IV materials (e.g., Si, Ge) the use of hydrogen-plasma enhancement is commonplace. But relatively little work has been reported on plasma enhancement for MOCVD of III-Vs, after demonstrations of its benefits by Huelsman et al. in the 1980s [10], [11]. Here, we first discuss the results of Al deposition from conventional MOCVD processes. We then demonstrate the use of hydrogen plasma for the deposition of Al from TMAl, as a potential step towards Al epitaxy by MOCVD.

II. EXPERIMENT

MOCVD was performed on a home-built system with a vertical quartz reactor. Typical growth conditions were 60 Torr with 720 sccm of purified H_2. Samples were loaded on POCO graphite susceptors, which were inductively heated at roughly 50 KHz to growth temperature T_g, measured by a thermocouple within the susceptor shaft. In addition to TMAl, trimethylgallium, trimethylindium, and AsH_3 were used as precursors. GaAs (001) substrates were obtained from AXT. Carbon-coated, 200-mesh Mo TEM grids were purchased from EMS. We obtained these grids with both standard carbon thickness (Mo:C) and extra-thick carbon (Mo:C-ET). For use in MOCVD, the grids were mounted on GaAs wafers with colloidal graphite at the edges.

A 13.56 MHz Seren radio-frequency generator rated at up to 100 W maximum output was connected through an RF-VII MN-500 manual impedance matching unit to terminals on RF antennas coupled to the reactor inlet. Impedance tuning was performed initially and throughout each growth run to achieve maximum forward power of nearly 100% into the load.

978-1-5090-5606-4/17 $31.00 © 2017 IEEE

Additional MOCVD was performed at RIT on an Aixtron system. Along with TMAl, that system is also equipped with a tritertiarybutylaluminum (TTBAl) source.

Transmission electron microscopy (TEM) was performed at 200 KV on a JEM-2100 LaB$_6$ with scanning TEM (STEM), as well as secondary electron (SEM) capability. Energy-dispersive X-ray spectrometry used an Oxford Inca system with a silicon drift detector. X-ray diffraction data were acquired with a Rigaku Ultima Plus power diffractometer.

III. RESULTS AND DISCUSSION

We discuss progress towards deposition of integration of Al buffer layers with III-Vs. Initial experiments using conventional MOCVD highlight the challenges of elemental Al deposition using standard deposition conditions. We then demonstrate the use of plasma enhancement to release Al from the TMAl precursor. Subsequent experiments are presented that highlight refinements of the method.

A. Conventional MOCVD using TMAl

We initially attempted growth of Al layers by conventional MOCVD. Deposition directly on TEM grids enabled quick turnaround for microstructural analysis. The resulting films were highly crystalline, containing thin platelets or whiskers inclined to the film surface. However, the lattice spacings in high-resolution images and diffraction patterns did not match fcc Al [Fig. 1]. EDX confirmed the presence of both Al and C in these films. XRD analysis of companion samples revealed peaks consistent with the rhombohedral Al$_4$C$_3$ compound.

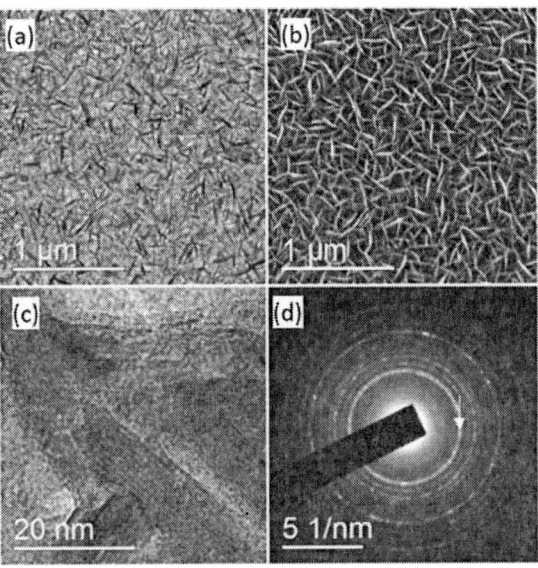

Fig. 1. Film synthesized by conventional MOCVD from TMAl on Mo:C TEM grids: (a) STEM bright-field; (b) SEM; (c) TEM lattice image; (d) electron-diffraction pattern consistent with Al$_4$C$_3$ (101/012 ring indicated by arrow.)

We also attempted incorporation of Al layers in III-V heterostructures by conventional MOCVD on GaAs substrates. A GaAs buffer layer was first grown under standard conditions. In some of the samples, a few alternating AlAs/GaAs layers were grown on top of the GaAs buffers. Next, TMAl in H$_2$ diluent was introduced for 20 min at various temperatures. In most cases, the structures were then capped with either GaAs or GaInAs.

Investigation of the resulting films in TEM plan-view showed that the III-V films at the tops of these structures were polycrystalline. Examination in cross section showed that, indeed, a layer was formed by the TMAl flow, but that the layer consisted of nanocrystalline domains intermixed with some amorphous regions [Fig. 2]. The lattice spacings of these domains again matched nanocrystalline Al$_4$C$_3$. Close inspection of the diffraction patterns from these samples show evidence of texturing in the orientation (001)[110]Al$_4$C$_3$||(001)[110]GaAs.

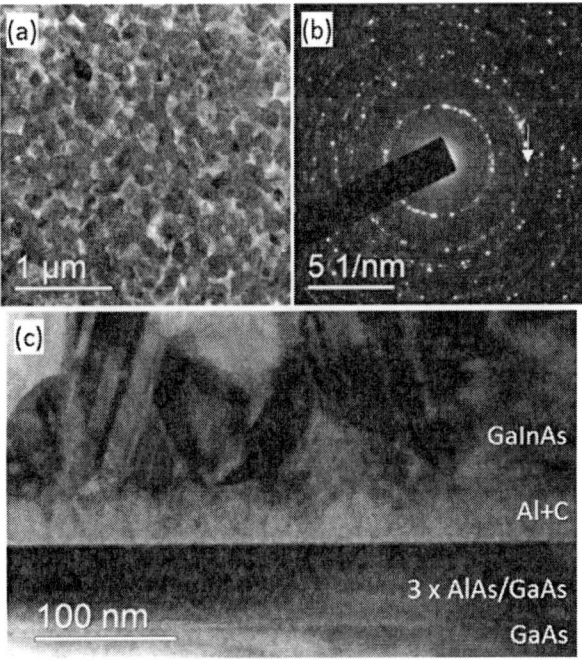

Fig. 2. TEM data showing polycrystalline structure of III-V films deposited on buffers layers synthesized by conventional MOCVD from TMAl: (a) plan-view bright-field image of GaAs grains; (b) plan-view electron diffraction pattern consistent with poly-GaAs (111 ring indicated by arrow); (c) cross-section showing GaInAs grains on an Al+C layer.

Additional growth runs performed at RIT with similar growth sequences, only using TTBAl as the Al precursor, also resulted in Al$_4$C$_3$. Other metalorganic precursors can be obtained from which it may be possible to grow elemental Al by conventional MOCVD [9]. Instead, we pursued the introduction of hydrogen plasma into the growth environment as a means to promote decomposition of TMAl.

B. PE-MOCVD using TMAl

We installed a ¾" quartz tube on the inlet of the MOCVD reactor, using Ultratorr compression fittings on each end. Several different setups have been studied to date. Three distinct configurations are addressed here: The first (C) used a brass, parallel-plate capacitor, rounded to conform to the outer contour of the quartz inlet tube. A second (L1) used a multi-turn copper coil wound coaxially around the reactor inlet. In a third configuration (L2), a coil was wound around the reactor neck. RF excitation at up to 100 W was applied to generate hydrogen plasma. The plasma can be maintained at pressures of up to about 10 Torr. Higher pressures required more RF power to sustain plasma. Other configurations, such as a transverse antenna geometry, have also been explored.

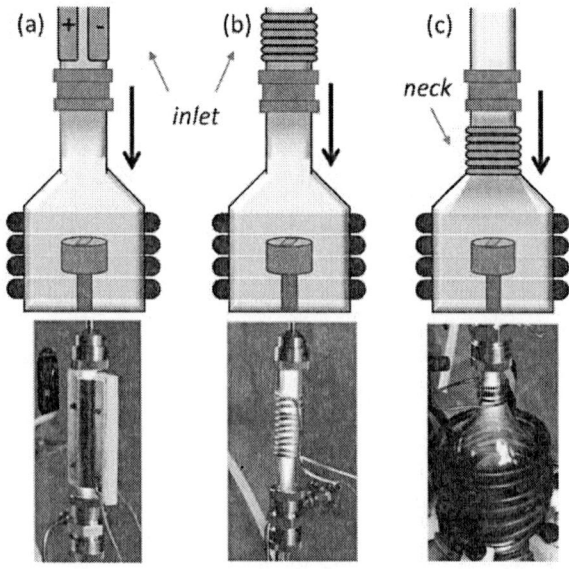

Fig. 3.   Three different configurations for PE-MOCVD. The top row shows diagrams of each. Photographs are shown below: (a) capacitive coupling to reactor inlet (C); (b) inductive coupling to reactor inlet (L1); (c) inductive coupling to reactor neck (L2). Inductive heating coils surround the reactor chamber.

In initial PE-MOCVD growth runs from TMAl, we used configuration C, with a relatively low RF power (30 W). Deposition was performed on Mo:C TEM grids at various T_g. At 600 °C and higher, the companion TEM grid showed a similar Al_4C_3 platelet structure to that observed without plasma. However, at 200 °C, the presence of larger clusters of elemental Al nanoparticles was observed [Fig. 4]. We encountered challenges in theses initial growth runs with substantial damage to the C support films during PE-MOCVD. Surviving clusters contained various shapes, including both rounded and faceted profiles, as well as straight needles. The crystal structure was confirmed by comparison of the electron-diffraction patterns with that from a common Al diffraction standard specimen. The spotty ring pattern indicates relatively large grain sizes. A thin oxide layer was evident on the surface of each Al nanoparticle.

Fig. 4.   Film synthesized by plasma-enhanced MOCVD from TMAl on Mo:C TEM grids: (a) STEM bright-field; (b) SEM; (c) TEM lattice image; (d) electron-diffraction pattern consistent with fcc Al (111 ring indicated by arrow).

EDX spectra provided further verification of the compositions of films on Mo:C grids. At 625 °C with no plasma, Al-C compounds were formed, with some O present. PE-MOCVD at 600 °C also resulted in Al-C, along with O, although a decrease in the C/Al atomic ratio was observed. At 200 °C with plasma, the C peak was substantially reduced, leaving only Al, along with an O peak [Fig. 5]. The presence of O is consistent with the observation by TEM of an amorphous layer on the surface of each Al particle. This formation of AlO_x on the particle surfaces is to be expected, as the MOCVD reactor is vented to air after each growth run, followed by exposure during transfer to the TEM.

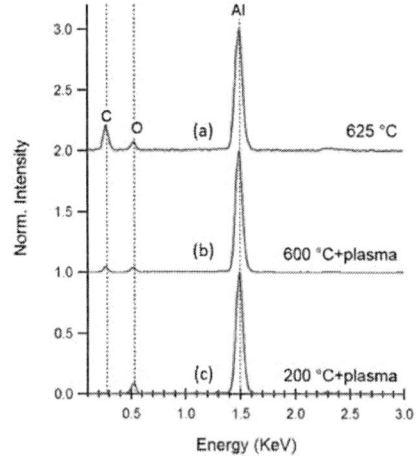

Fig. 5.   EDX spectra from films deposited by MOCVD from TMAl on Mo:C TEM grids: (a) 625 °C (no plasma); (b) 625 °C with plasma; (c) 200 °C with plasma.

In subsequent growth runs, we were able to increase the RF power to produce more film-like coverage of the TEM grids. The capacitive antenna was replaced with inductive-coil designs that allow better visibility of the reactor inlet and easier tuning with our currently available electronics. We then adopted the L1 configuration for a set of PE-MOCVD TMAl films on glass slides [Fig. 6]. At 600 °C, the deposition was brown, smooth and translucent. Powder XRD patterns showed a predominance of the rhombohedral Al_4C_3 phase. Very little deposition was evident at 500 °C. But at temperatures of 300-400 °C, the films were gray, dull, and rough. Powder XRD confirmed that these low-T_g films were composed of polycrystalline, elemental Al.

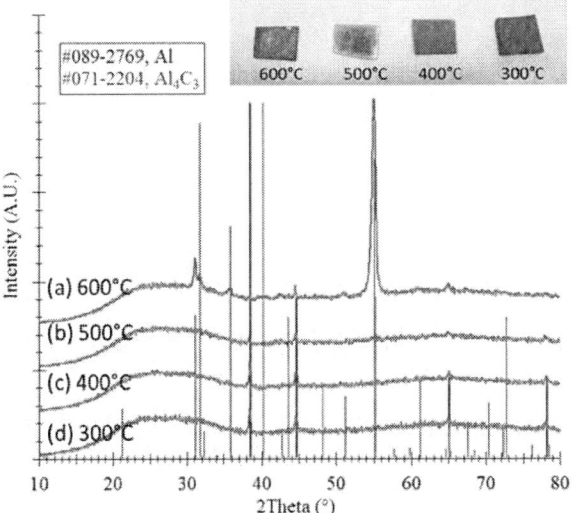

Fig. 6. Powder XRD patterns from PE-MOCVD films using TMAl on glass slides at various T_g. Standards are shown for comparison. The growth time was 60 min for the 600 °C sample and 30 min for the other three samples.

We used the PE-MOCVD method in configuration L1 to deposit Al on (001) GaAs miscut 2°-[101]. The Al layer was preceded by deposition of a GaAs buffer by conventional MOCVD in standard conditions. The resulting film appeared dull gray-white. TEM and SEM analysis showed large Al grains, with a surface roughness exceeding 100 nm [Fig. 7]. Powder XRD confirmed the presence of polycrystalline Al, with an enhancement of the 200 and 220 peaks. (Notice that the 220 peak is nearly coincident with the 400 peak of the GaAs substrate.)

Based on a set of measured XRD intensities I_i for various reflections corresponding to non-parallel reciprocal lattice vectors, and given their relative intensities r_i from the XRD powder database, the relative fraction of any particular orientation can be estimated by

$$f_i = \frac{I_i/r_i}{\sum_j I_j/r_j} \tag{1}$$

Using the measured 111, 200, 220, and 311 intensities, the Al film on GaAs is found to have 51% (100) and 42% (110) orientations.

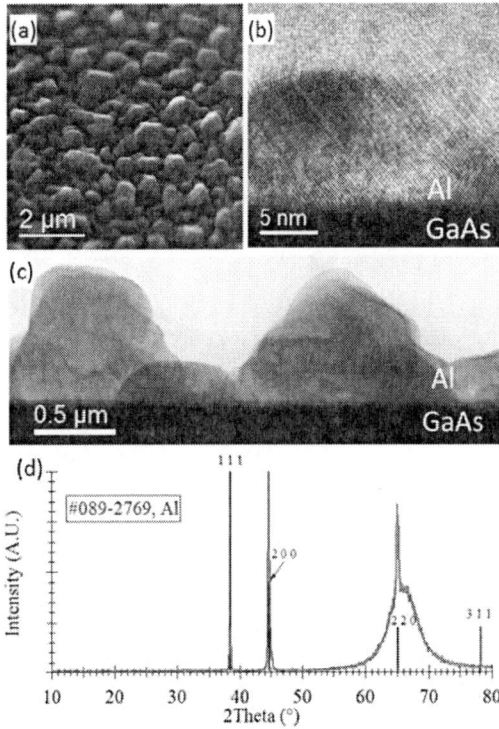

Fig. 7. PE-MOCVD Al film on (001) GaAs miscut 2°-[101]: (a) SEM image of surface; (b) <110> TEM cross-sectional lattice image showing interface; (c) Low-magnification TEM cross-sectional bright-field image; (d) XRD powder pattern and Al standard.

Detailed electron-diffraction analysis of the interface region was performed to determine the orientations present [Fig. 8].

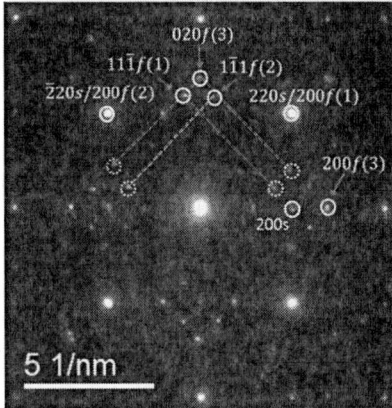

Fig. 8. Electron-diffraction pattern acquired on the [001] GaAs zone axis through the interface from PE-MOCVD Al on (001) GaAs miscut 2°-[101]. Substrate (s) and film (f) peaks are labeled. Film variants are indicated as 1, 2, and 3. Dashed lines indicate double-diffraction paths.

The diffraction pattern acquired in plan-view through the interface region contains additional spots that arise from double diffraction. The orientations of Al identified are indicated below. Two (011) orientations and a single (001) orientation can be identified:

1) (011)[100]Al‖(001)[110]GaAs
2) (011)[100]Al‖(001)[1$\bar{1}$0]GaAs
3) (001)[100]Al‖(001)[100]GaAs

Note that the targeted (001)[100]Al‖(001)[110]GaAs orientation was not observed.

C. PE-MOCVD refinement

In the L2 configuration, the substrate and susceptor are directly exposed to hydrogen plasma, which nearly fills the reactor. Initial attempts to deposit directly on standard Mo:C TEM grids with this method frequently resulted in loss of the C film, apparently due to exposure in this geometry of the growth surface to high-density plasma. Mo:C-ET grids were found to be more resilient. The resulting films were fairly dense, with grain sizes on the order of 50 nm, considerably smaller than those found in previous films.

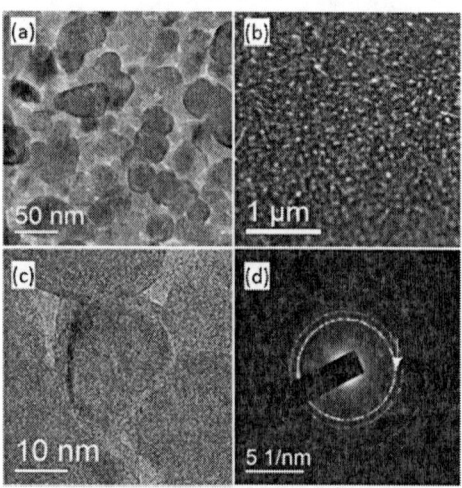

Fig. 9.   Film synthesized by plasma-enhanced MOCVD from TMAl on Mo:C-ET TEM grids: (a) STEM bright-field; (b) SEM; (c) TEM lattice image; (d) electron-diffraction pattern consistent with fcc Al (111 ring indicated by arrow).

We observed noticeable improvements and differences in PE-MOCVD Al films grown GaAs substrates using the L2 geometry. While plasma damage to the substrate must be considered, the plasma can also conceivably remove contaminants from the substrate prior to deposition. In these initial runs, in situ hydrogen plasma cleaning was attempted, but no GaAs buffer layers were grown. Films deposited on (001)-2°[101] GaAs substrates with T_g in the range 20-200°C had highly specular and reflective metallic finishes, clearly smoother than previous films. Dense, small grained films were

formed by growth at 20 °C. The roughness measured by AFM was approximately 3 nm.

At 100 °C, the grains appeared more columnar [Fig. 10]. The roughness in a 60-nm-thick film measured by AFM scans is 5 nm, with grain sizes on the order of 50 nm.

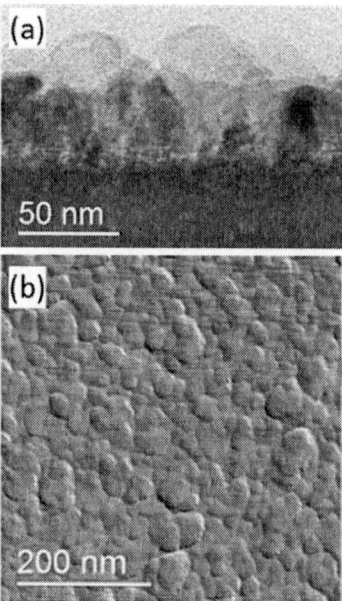

Fig. 10. Film synthesized by PE-MOCVD from TMAl in L2 configuration on (001)-2°[101] GaAs at 100°C: (a) TEM bright-field cross sectional image; (b) AFM scan showing topography corresponding to underlying grain structure.

Powder XRD and TEM diffraction analysis in both plan view and cross section [Fig. 11] showed (011) texturing of the Al film grown at 100°C in the L2 setup. From the TEM data, it is evident that two (011) variants are present:

1) (011)[100]Al‖(001)[110]GaAs
2) (011)[100]Al‖(001)[1$\bar{1}$0]GaAs

Note that Al (200) and GaAs (220) planes are nominally lattice-matched. However, the GaAs (001) face has four-fold projection symmetry, whereas the Al (110) face has only two-fold symmetry. Thus, while epitaxy in this orientation may occur along either GaAs in-plane <110> direction, any particular Al film grain is highly incommensurate along the orthogonal <110> direction. This structure has been reported elsewhere [2], [12], [13]. It was found that [111] miscut of the Si substrate allow the dominance of one (011) Al variant, resulting in single-crystal Al growth. Note that, the 2°-[101] miscut used here does not structurally differentiate between the two variants, though there are chemical disparities between the [111]A and B steps present on this surface. We anticipate that miscut towards either <111> direction may break the symmetry, possibly leading to preferential selection of a single (011) Al variant.

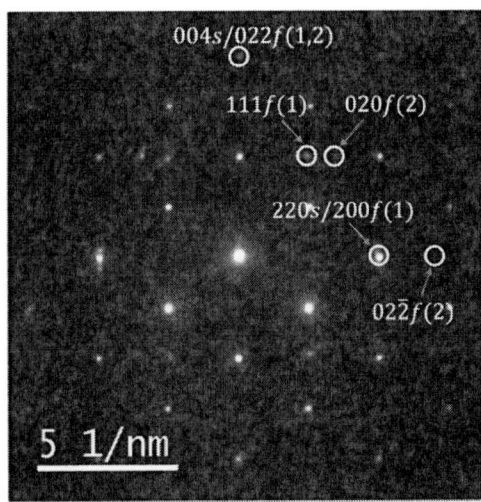

Fig. 11. Cross-sectional diffraction pattern from film and substrate on <110> GaAs zone axis of sample in Fig. 10: substrate (*s*) and film (*f*) peaks are labeled. Film variants are indicated as 1 and 2.

III. CONCLUSIONS

The presence of attendant radical species is essential to release fully the Al atom from its Al-C bonds in the TMAl molecule. In MOCVD of Al-containing III-Vs, these are provided through the cracking of hydride precursors [9]. In the absence of the companion hydride gas, the deposition of elemental Al from metalorganic precursors, using thermal decomposition alone, is challenging. However, the necessary radicals can be provided by the application of hydrogen plasma. The ability to introduce the plasma through a separate inlet could have advantages. We have shown that RF excitation with an antenna positioned close to the reactor or reactor inlet is sufficient to generate sufficient plasma density for Al deposition. This presents a possible method to incorporate Al into III-V semiconductor heterostructures using MOCVD with only minor modifications to an existing system.

At high growth temperatures, with the moderate plasma power levels we have studied, Al-C formation remains dominant. But at low temperature, Al can be released, even with relatively low plasma density, without carbide formation. It appears that, in this plasma power range, only partial cracking of the precursor is achieved by the plasma, leaving a significant fraction of the precursor to undergo thermal decomposition at the growth surface in the higher temperature range.

In the C and L1 antenna configurations, plasma is generated in the reactor inlet, releasing Al into the vapor. As it proceeds through the inlet stream, the Al is then subject to coalescence prior to condensing on the growth surface. Thus, the Al flux reaching the surface in this mode may largely consist of pre-nucleated Al nanoparticles, resulting in rough films with relatively large grains.

In the L2 configuration, plasma extends close to the growth surface. This allows maintaining the Al flux in vapor form, without nucleation, until penetration of the boundary layer near the growth surface. The change in orientation from (001) to (011) in this method indicates that rather small changes in the deposition environment can substantially influence the mode of crystal growth.

We continue to explore the potential of the PE-MOCVD method for Al epilayer deposition, while seeking to identify other applications of the method for photovoltaics.

ACKNOWLEDGEMENT

This work was supported by the U. S. Dept. of Energy through Award #DE-EE0007363.

REFERENCES

[1] I. Levine, A. Yoffe, A. Salomon, W. Li, Y. Feldman, and A. Vilan, "Epitaxial two dimensional aluminum films on silicon (111) by ultra-fast thermal deposition," *J. Appl. Phys.*, vol. 111, no. 12, p. 124320, 2012.

[2] S. Yokoyama and K. Okamoto, "Single-Crystal Growth of Al(110) on Vicinal Si(100) in Ultra-High-Vacuum Sputtering System," *Jpn. J. Appl. Phys.*, vol. 30, p. 3685, 1990.

[3] N. Maeda, M. Kawashima, and Y. Horikoshi, "Epitaxial growth of Al films on modified AIAs(001) surfaces," *J. Appl. Phys.*, vol. 74, no. 1, pp. 4461–4471, 1993.

[4] H. F. Liu, S. J. Chua, and N. Xiang, "Growth-temperature- and thermal-anneal-induced crystalline reorientation of aluminum on GaAs (100) grown by molecular beam epitaxy," *J. Appl. Phys.*, vol. 101, no. 5, p. 53510, 2007.

[5] S.-T. Lo, C. Chuang, S.-D. Lin, K. Y. Chen, C.-T. Liang, S.-W. Lin, J.-Y. Wu, and M.-R. Yeh, "Magnetotransport in an aluminum thin film on a GaAs substrate grown by molecular beam epitaxy," *Nanoscale Res. Lett.*, vol. 6, no. 1, p. 102, 2011.

[6] A. Ptak, T. Lin, A. Norman, and K. Alberi, "Methods of producing free-standing semiconductors using sacrificial buffer layers and recyclable substrates," 9,041,027 B2, 2015.

[7] N. Suzuki, C. Anayama, K. Masu, K. Tsubouchi, and N. Mikoshiba, "Pyrolysis and Photolysis of Trimethylaluminum," *Jpn. J. Appl. Phys.*, vol. 25, pp. 1236–2142, 1986.

[8] T. Kobayashi, A. Sekiguchi, N. Hosokawa, and T. Asakami, "Epitaxial Growth of Al on Si by Gas-Temperature-Controlled Chemical Vapor Deposition," *Jpn. J. Appl. Phys.*, vol. 27, p. L1775, 1988.

[9] G. B. Stringfellow, *Organometallic Vapor-Phase Epitaxy*, 2nd ed. San Diego: Academic Press, 1999.

[10] A. D. Huelsman, R. Reif, and C. G. Fonstad, "Plasma-enhanced metalorganic chemical vapor deposition of GaAs," *Appl. Phys. Lett.*, vol. 50, no. 4, pp. 206–208, 1987.

[11] A. D. Huelsman, L. Zien, and R. Reif, "Plasma-controlled deposition of GaAs and GaAsP by metalorganic chemical vapor deposition," *Appl. Phys. Lett.*, vol. 52, pp. 726–727, 1988.

[12] Y. Neo, T. Otoda, K. Sagae, H. Mimura, and K. Yokoo, "Epitaxial Growth of Aluminum on Silicon Substrates by Metalorganic Molecular Beam Epitaxy using Dimethyl-Ethylamine Alane," *Jpn. J. Appl. Phys.*, vol. 37, no. Part 1, No. 5A, pp. 2602–2605, 1998.

[13] G. Landgren, R. Ludeke, and C. Serrano, "Epitaxial Al Films on GaAs(100) Surfaces," *J. Cryst. Growth*, vol. 60, p. 393, 1982.

Laser Crystallization of Amorphous Germanium on Titanium Nitride-Coated Steel for Low-Cost GaAs Solar-Cells

Saloni Chaurasia, Srinivasan Raghavan, Sushobhan Avasthi

Indian Institute of Science, Bangalore, Karnataka, 560012, India

Abstract —High-efficiency III-V solar cells on low-cost substrates such as steel, are limited because a) the overlying III-V semiconductor stack gets contaminated with iron, b) it is not easy to deposit crystalline Ge thin-films on amorphous steel substrates. In this work, we report a TiN-based barrier layer that inhibits Fe diffusion into the overlying semiconductor. We also report a method to deposit crystalline germanium thin-film on steel on which high quality III-V layers can be deposited using laser annealing. PECVD amorphous germanium films were irradiated with a 532 nm laser with a fluence of $4 - 8 \times 10^8$ Jcm^{-2} which melted and re-crystallized the irradiated amorphous film into nano-crystalline germanium film. The re-crystallized films were characterized using SEM and Raman spectroscopy. Films irradiated at higher fluence and films with higher thicknesses, had more crystalline order. The work enables future development of lost-cost GaAs solar cells.

Index Terms — GaAs; Germanium; Laser Annealing; Solar -Cells.

I. INTRODUCTION

Thin film solar cells on flexible substrates form an attractive alternative to the conventional solar cells due to their lower cost and wide range of applications. Steel foil is an ideal substrate due to its stability at high temperatures and low cost. Integration of high quality GaAs thin film solar cells on steel substrates requires a) a diffusion barrier layer between steel and the overlying semiconductor stack to inhibit diffusion of iron which otherwise may get saturated with iron diffusing from steel into the device layers at high temperatures, deteriorating device performance [1] b) a high quality buffer layer of germanium for high efficiency solar cells on steel. Titanium nitride is proposed and characterized as candidate diffusion barrier for solar cells on steel in our work. Solar cells on titanium nitride coated steel are expected to be efficient.

High quality germanium on steel can be obtained by various method like crystallization of amorphous films of germanium at high temperatures. Amorphous Ge films can be crystallized rapid thermal [2], ion beam treatment [3], pulsed and CW laser annealing [4].

Therefore, in this work, we investigate crystallization of thin amorphous germanium films on titanium nitride laser annealing technique [5].

II. EXPERIMENTAL DETAILS

TiN was sputtered in an Ar and N_2 ambient at a substrate temperature of 300^0 C, on silicon and low carbon steel (100 micron) substrates.

Amorphous germanium films were deposited using an Oxford PECVD reactor at 350^0C, with 10 sccm of GeH_4, 100 sccm of N_2 and 100 sccm of H_2, and RF power density of 0.20 W/cm^2. Ge was deposited on bare silicon, TiN-coated silicon, and TiN-coated steel substrates. The silicon wafers were cleaned using standard RCA cleaning recipe, prior to Ge or TiN deposition.

For laser annealing, amorphous Ge films were exposed for 0.25s to a 532 nm CW laser focused to a spot size of 1 micron diameter, at 0.2mW to 26mW intensity, corresponding to fluence of 4 to 8x 10^8 Jcm^{-2}. The laser was scanned on an area of 40μm x 40μm. Laser annealing was done at ambient conditions. Recrystallized films were analyzed under Zeiss-Ultra 55 SEM and Horiba microPL with a Raman spectrometer attachment.

III. RESULTS AND ANALYSIS

A. *TiN as barrier layer*:

Fe diffuses very fast in semiconducting layer such as Ge or GaAs [6]. One way to reduce Fe contamination is to insert a diffusion barrier between semiconductor and steel (fig.1). To measure the diffusion coefficient of Fe in TiN, TiN-coated steel substrates were annealed from 600^0C – 900^0C in hydrogen ambient in a tube furnace. Next, the annealed films were characterized using secondary ion mass spectrometry (SIMS) (fig.2). The measured depth profile was fitted to an error-function. The extracted diffusion coefficients (Table 1) show that Fe diffusion coefficient in TiN are ~10^5 times lower than in GaAs [6]. The measured values compare favorably with previous reports [7].

978-1-5090-5606-4/17 $31.00 © 2017 IEEE

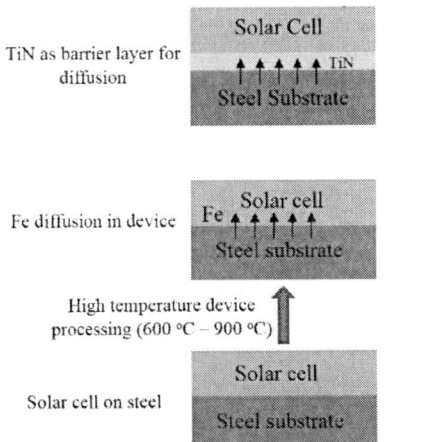

Fig. 1: Schematic representation of TiN layer a barrier for Fe diffusion to solar cell on steel

For device quality, the required thickness of the TiN layer, for a given thermal budget, can be calculated using error function profile (Fig.3). For typical processing conditions of GaAs solar cells (700 ^{0}C and 2 hrs) with C = 10^{12} / cm^3 and C_0 = 10^{22} / cm^3, just 1μm thick TiN is found to be sufficient. Based on the above analysis, for different thermal budgets, the required thickness of barrier layer can be obtained (fig.4).

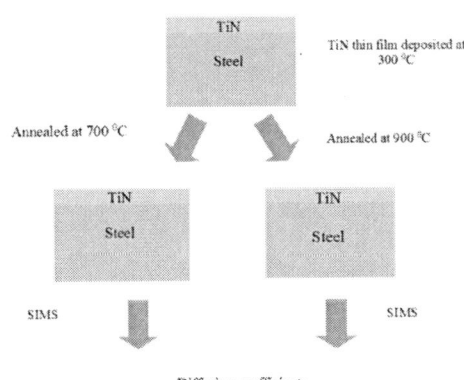

Fig. 2: Flow chart for SIMS characterization of TiN on Steel.

TABLE I: DIFFUSION COEFFICIENT OF FE IN TIN MEASURED BY SIMS

T (^{0}C)	300 ^{0}C	700 ^{0}C	900 ^{0}C
D_{Fe} in TiN (10^{-14} cm^2 s^{-1})	0.02 ± 0.001	0.05 ± 0.02	0.89 ± 0.05

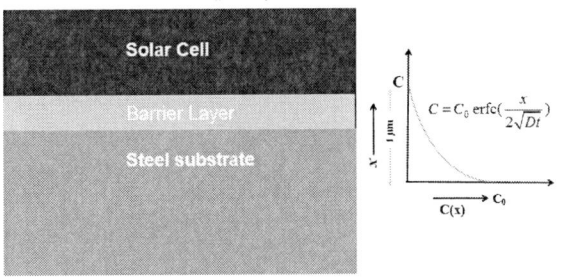

Fig:3 Schematic of TiN thickness calculation from erfc profile

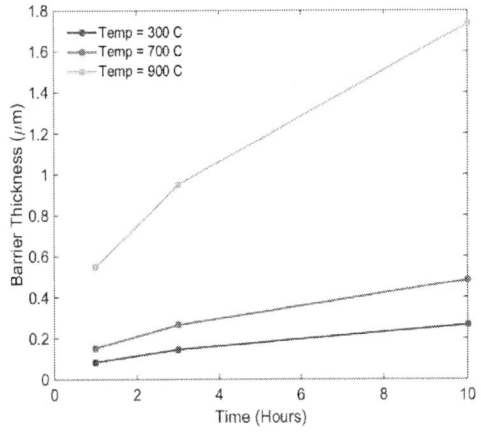

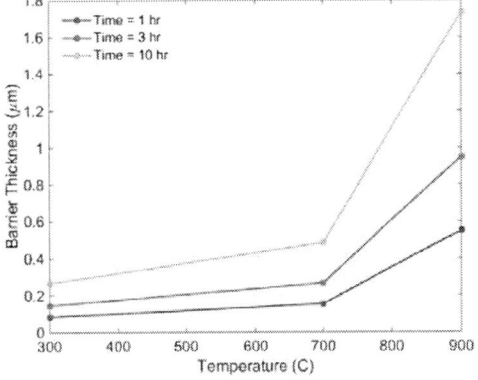

Fig.4 Thickness of the TiN barrier layer required for different thermal budgets using the measured value of diffusion coefficients

B. Crystallization of Germanium:

PECVD Ge films on both Si and TiN-coated/steel shows a broad Ge optical phonon peak at ~270 cm-1, indicating that the as-deposited films are amorphous.

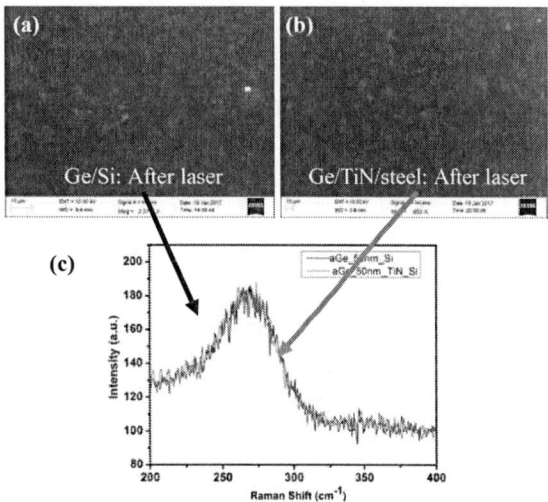

Fig.5 Thin a-Ge (50 nm) after irradiation at 8×10^8 J cm^{-2} on a) Si substrate, and b) TiN-coated/steel substrate. c) Raman spectrum of the two films.

Ge films were crystallized by a 532 nm laser. The crystallization is thickness dependent. Thin Ge film (50 nm) did not crystallize even at highest fluence of 8×10^8 Jcm^{-2}: no change was observed either in the morphology (Fig 5a) ,(Fig.5b) or in the Raman spectra (Fig. 5c). Ge is a very good conductor of heat with a thermal diffusivity of 0.36 cm^{-2} s^{-1}. During a 0.25 s exposure, heat will dissipate to a depth of 3 mm of Ge. For very thin-film this is a problem because the heat generated does not stay localized in the Ge layer long enough to cause re-crystallization. Thicker amorphous germanium (500 nm) film crystallize at laser fluence > 4×10^8 J cm^{-2}. SEM images of a-Ge-on-Si show nanostructures in the laser-exposed area (Fig.6a & Fig. 6b). Morphology of films exposed to higher fluence of 8×10^8 J cm^{-2} was very similar. Raman spectra also changes in the exposed portion of the films with a red shift in the Ge phonon peak from 270 cm-1 to 291 cm-1 and 294 cm-1, at fluence of 4×10^8 J cm^{-2} and 8×10^8 J cm^{-2}, respectively. The peak for crystalline Ge is expected at 300.6 cm-1, so the red shift is evidence of re-crystallization.

The peaks are not exactly at 300 cm-1 probably because of tensile stress due to the phase change. Films exposed to higher fluence (8×10^8 J cm^{-2}) show lower red-shift, suggesting

that the resulting films have lower tensile stress. The line width of the peaks also reduced from 12 cm-1 to 10 cm-1 at the low and high fluence, respectively. For comparison, linewidth of the Ge phonon line in Ge wafer is 3 cm-1. The broader line width in the recrystallized films may be due to factors like defects, variation in crystallite-sizes and phonon confinement in nanocrystals [8]. Asymmetry in peak shape at both the fluence is observed which can also attributed to nanocrystalline nature of the films.

a) Ge/Si: Lower Fluence

b) Ge/Si: Higher Fluence

Fig.6 SEM image of aGe(500nm)-on-Si a) irradiated at 4×10^8 J cm^{-2} and b) 8×10^8 J cm^{-2}. c) Raman spectra for the two

Laser annealing can also be used to re-crystallize a-Ge films deposited on TiN-coated steel. Films annealed at a fluence of 8×10^8 J cm^{-2} showed similar morphology to what was observed in Si substrates (Fig. 7a). Raman peaks were measured at 289 cm-1 with a linewidth of 13 cm-1 (Fig.7b), suggesting that re-crystallized Ge films on TiN-coated steel are more stressed and defective as compared to Ge films on Si.

a) Ge/TiN/Steel: High Fluence

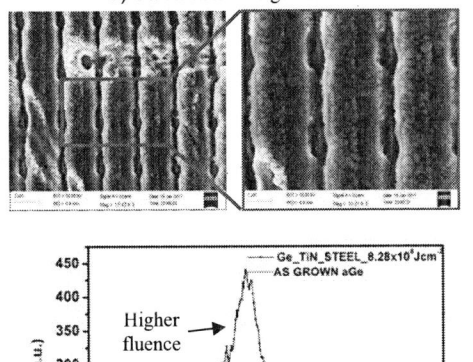

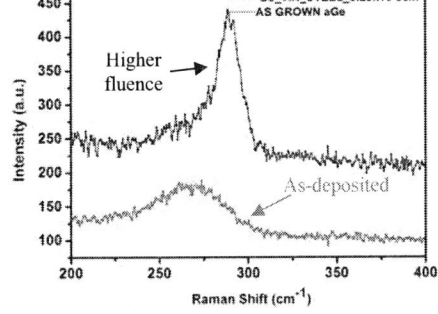

Fig.7 a) SEM image of aGe(500nm)/TiN/steel irradiated at 8×10^8 J cm^{-2}. b) Raman spectra of as grown and crystallized

IV. CONCLUSION

Thin-film III-V solar cells on low-cost substrates like silicon and steel suffer from two main problems: a) diffusion of Fe atoms in to the overlying semiconductor layers, and b) poor crystallinity of as-deposited Ge films. To prevent Fe contamination, Titanium nitride barrier layer has been demonstrated which reduces Fe diffusion by 5 orders of magnitude. The TiN layer is stable upto 900 ^{0}C on steel substrates. To get crystalline Ge films on steel substrates, a laser annealing process is demonstrated .Laser annealing is done by a rastering 532nm CW laser. Laser annealed films were crystallized. SEM and Raman spectroscopy show that film exposed to fluences between 4 & 8×10^8 J cm^{-2} are nano-crystalline. Higher fluence lead to better quality films with lower tensile stress and higher crystalline order. This work may pave way for efficient III-V solar cells on low-cost substrates such as silicon and steel.

REFERENCES

[1] J. B. Rem, M. C. V. de Leuw, J. Holleman, J. F. Verweij, "Furnace and rapid thermal crystallization of amorphous Ge$_x$Si$_{1-x}$ and Si for thin film transistors," *Thin solid films*, vol. 296, pp152-156, 1997.

[2] N. P. Stepina, A. V. Dvurechenskii, V. A. Armbrister, V. G. Kesler, P. L. Novikov, A. K. Gutakovskii, V. V. Kirienko, and Zh. V. Smagina, "Pulsed ion-beam induced nucleation and growth of Ge nanocrystals on SiO2,"*Applied Physics Letters,vol.* 90, p133120 ,2007.,

[3] Z. Said Bcar, YannLeroy, Frederic Antoni , E. Fogarassy," Modeling of CW laser diode irradiation of amorphous siliconfilms", *Applied Surface Science* ,vol.257,p.12 ,April 2011.

[4] R. Wuerz, A. Eicke, M. Frankenfeld, F. Kessler, M. Powalla, P. Rogin, O. Yazdani-Assl, "CIGS thin-film solar cells on steel substrates," *Thin Solid Films*, Vol.517, pp-2415-2418, 2009.

[5] Ziheng Liu, Xiaojing Hao, Jialiang Huang, Wei Li, Anita Ho-Baillie, Martin A. Green.," Diode laser annealing on Ge/Si (100) epitaxial films grown by magnetron sputtering," *Thin Solid Films,* April 2016.

[6] IA Prudaev, S. S. Khludkov," Diffusion and solubility of electrically active iron atoms in gallium arsenide," *Russian Physics Journal*, vol.51, p.11, 2008.

[7] J Garcia-Cespedes, J A lvarez-Garcia, X Zhang, J Hampshire andE Bertran, "Optimal deposition conditions of TiN barrier layers for the growth of vertically aligned carbon nanotubes onto metallic substrates,"*Appl. Phys.* Vol. 42,9pp,2009.

[8] Danilo Bersani, Pier Paolo Lottici Xing-Zhao Ding, "Phonon Confinement Effects in the Raman Scattering by TiO2 Nanocrystals, *Applied Physics Letters* vol.72(1),pp.73-75,1998

High Quality Epitaxial Germanium on Si (100) for low -cost III-V Solar-Cells

Saloni Chaurasia, Srinivasan Raghavan, Sushobhan Avasthi

Indian Institute of Science, Bangalore, Karnataka, 560012, INDIA

Abstract — Integration of high quality germanium on silicon is prerequisite for fabricating high efficiency low-cost III-V solar cells on silicon. In this work, we present a method of monolithic integration of epitaxial germanium on crystalline silicon wafers where a single-step anneal in hydrogen ambient is used to re-crystallize PECVD-deposited amorphous germanium into crystalline germanium. The re-crystallization progresses via liquid phase epitaxy by heating the germanium layer in inert ambient to just above its melting temperature of 937 ^{0}C. On cooling, the Ge layer gets oriented by the underlying Si (100) wafer, yielding an epitaxial Ge film on Si. SEM show that the films are continuous and void free. XRD pattern indicates highly oriented films. Symmetric Rocking curve measurements showed a thickness dependence with 1 μm thick film showing the lowest threading dislocation density (TDD) of the order of 10^8 cm^{-2}. AFM measurements indicate surface measurements ~10 nm. Quality of epitaxy depends on thickness of film.

Index Terms – Epitaxy; Germanium; Silicon; Solar -Cells

I. INTRODUCTION

The growth of epitaxial Germanium (Ge) on silicon is highly desirable due to emerging technologies such as Ge-on-Si infrared photo detectors [1], solar cells [2], opto-electronics [3], germanium-based laser [4] along with heterogeneous CMOS devices [5]. Furthermore, Ge is lattice matched to GaAs and AlAs [6]. So, germanium-on-silicon can be used as a platform for integrating III-V photovoltaic devices on to the lower cost silicon substrate [7]. Unfortunately, growing high-quality epitaxial Ge on Si is not easy. The 4% lattice mismatch between Si and Ge leads to a high density of dislocations, which degrades the device performance. Still several methods have been proposed to grow solid phase epitaxial Ge films on silicon, such as MBE [8] and also using vapor phase epitaxy such as two-step process using RPCVD [9] and three-step process using MOCVD [10]. Although the above-mentioned methods have their advantages and disadvantages, simpler and low-cost liquid phase epitaxy (LPE) growth technique is not explored enough. In this work, a relatively simple single-step anneal process is used to grow high quality epitaxial films of germanium on silicon (100) which proceeds via liquid phase epitaxy on Si (100). Effect of film thickness on quality of film was investigated. Films with 1 micron thickness had the lowest TDD. AFM measurements show a surface roughness of ~10 nm RMS. To understand how LPE process is different from VPE process, films were also deposited using LPCVD and compared with the films deposited using LPE.

II. EXPERIMENTAL DETAILS

A. Ge growth and annealing

1. Ge growth:

Amorphous germanium films of various thicknesses were deposited Si (100) wafers using OXFORD Plasma lab PECVD deposition system 350^{0}C, with GeH$_4$ 10 sccm, N$_2$ 100 sccm and H$_2$ 100 sccm, and RF power density of 0.20 W/cm^2. RCA cleaning followed by HF dip was done for Si wafers prior to deposition. RCA cleaning of silicon wafers was done to ensure removal of organic as well as metal contaminants from the substrate which might contaminate the deposited film and subsequently hinder the re-crystallization process. The depositions were preceded by a plasma cleaning of the PECVD chamber to avoid any contamination on the deposited film. Films were also deposited using LPCVD on RCA cleaned Si (100) wafers at a temperature of 450°C with GeH$_4$ 20 sccm at a pressure of 300 mtorr. The wafers were baked at a temperature of 850°C in hydrogen ambient for 10 minutes prior to deposition to remove any native oxide present on silicon surface.

2. Ge annealing:

The PECVD grown Ge layers were annealed above melting temperature of Ge (937^{0}C) in a clean tube furnace in hydrogen ambient devoid of chamber contamination above melting temperature. The films annealed in tube furnace were oriented in (100) following orientation of Si (100) beneath.

B. Ge Characterization

Both amorphous and re-crystallized films were observed under Zeiss ultra 55® Scanning Electron microscope. X-ray diffraction measurements were done using Rigaku Smartlab® X-ray diffractometer for films crystallized in tube furnace. High resolution rocking curve XRD measurements were done to estimate threading dislocation density on the basis of rocking curve FWHM. AFM measurements were done to find surface roughness of films. LPCVD deposited films were also characterized using SEM and XRD measurements.

III. RESULTS AND ANALYSIS

θ-2θ pattern for 300 nm thin film deposited using LPCVD process shows that the films are oriented in (100) direction (see fig. 1). However symmetric rocking curve for the same film shows a broad peak with FWHM of 1.5° indicating very high density of defects in the film (see fig. 2).

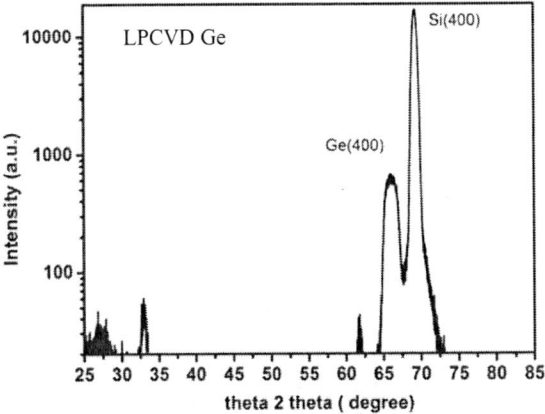

Fig. 1: XRD Theta 2 theta curve for 300 nm thin film deposited using LPCVD process showing orientation in (100) direction

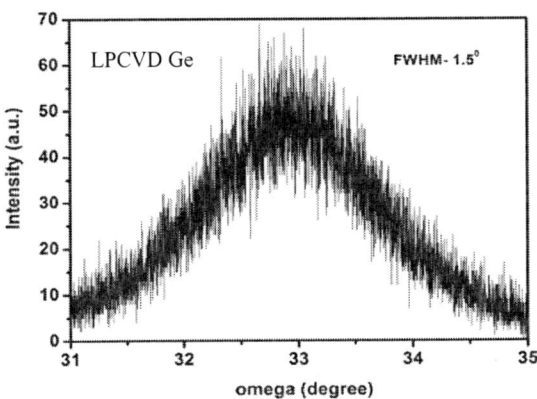

Fig. 2: XRD rocking curve for LPCVD deposited 300nm film indicating highly defective film

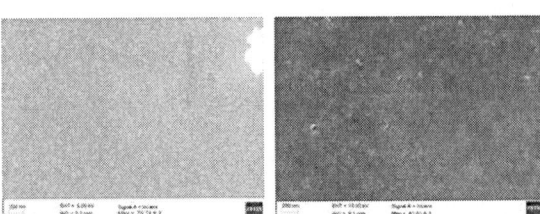

Fig 3.a) Amorphous Ge films deposited using PECVD b) hydrogen ambient furnace annealed void free films.

Fig.3 shows SEM image of amorphous Ge film deposited using PECVD before and after re-crystallization in tube furnace. As seen the amorphous films were converted into crystalline, smooth and void free films.

Roughness measurement is a requisite for device application and hence were measured for the films. The AFM scan was done on an area of 5x5 um for 1 micron films annealed in hydrogen ambient was done and found to be ~ 10 nm (see fig. 4).

Fig.4.AFM images of films annealed in hydrogen ambient.

XRD θ-2θ pattern shows that annealed Ge films are highly oriented in (100) direction following the orientation of underlying silicon. It is observed that the orientation of the deposited Ge film improves with its thickness. We believe that this can be due to more relaxation in the Ge film with increase in thickness, which is also marked by slight leftward shift in the Ge (400) peak for thicker films (see fig. 5).

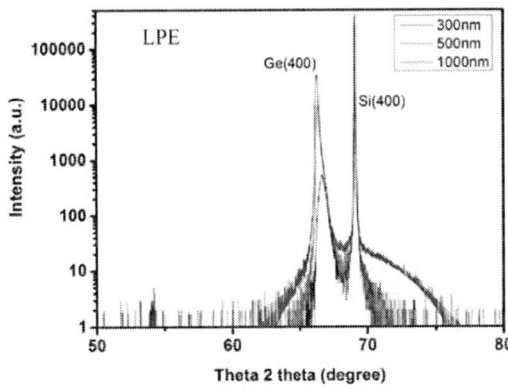

Fig.5 XRD Theta 2 theta curve for films annealed in nitrogen ambient showing orientation in (100) direction

The quality of epitaxial films in terms of FWHM were estimated using HRXRD rocking curve measurements. Rocking curve measurement was done in the vicinity of (400) Bragg peak (66^0) of germanium. It was found that with increase

978-1-5090-5606-4/17 $31.00 © 2017 IEEE

in film thickness, the rocking curve FWHM decreases resulting in decrease in TDD (fig.6).

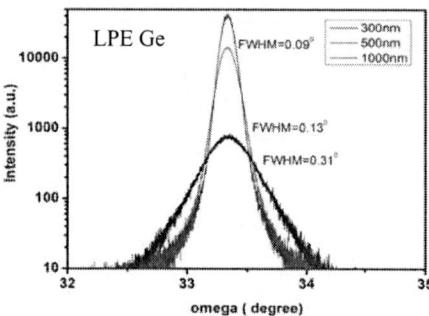

Fig.6: XRD rocking curve for films annealed in hydrogen ambient showing thickness dependence on TDD (FWHM)

The lowest TDD was found to be as low as $\sim 10^8$ cm^{-2} for 1 micron thick film and was calculated using [11]. It shows that thicker films have better quality as compared to thinner films. In-plane rocking curve measurements around Bragg's peak (221), (111), (110) also showed similar trend in FWHM being lowest ($\sim 0.1^0$) for 1 micron film. Symmetric rocking curve for LPCVD deposited films are also shown in fig.2. As can be observed from FWHM of 300 nm LPE film is as low as $\sim 0.31^0$ indicative of better quality and LPCVD film of same thickness having FWHM of 1.5^0 indicating highly defective film. Although θ-2θ scan shows orientation for both however in terms of quality, LPE films exceed LPCVD deposited films several orders of magnitude.

XRD phi scan was done to confirm in plane epitaxial relationship between Si substrate and germanium films crystallized using LPE. Fig.7 shows phi scan corresponding to (221) plane of germanium for hydrogen ambient annealed films. Phi scan for planes (110) and (111) also showed similar results. As evident from fig.7 that the film is epitaxial with cubic four-fold symmetry.

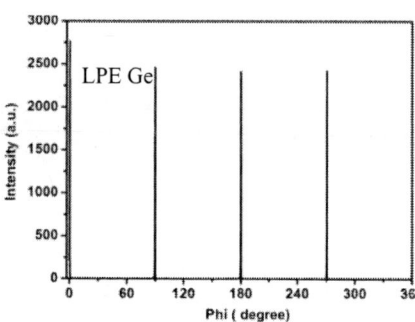

Fig.7: XRD phi scan for films annealed in hydrogen ambient showing epitaxy and cubic symmetry for Ge films.

IV. CONCLUSION

We have presented a method to grow epitaxial thin-film of germanium on silicon using a relatively simple one-step process. The films were crystallized by annealing at a temperature slightly higher than the melting point (937^0 C) of germanium in hydrogen ambient. The Si (100) wafer underneath oriented the Ge melt leading to high-quality Ge epitaxial films. SEM and XRD show that the resulting films were highly oriented and void free respectively. The rocking curve (Omega scan) confirmed that the1 micron films are highly-oriented with a FWHM of only ~ 0.09 degree. A comparison with LPCVD deposited films was done with films crystallized using LPE and LPE film was found be far superior in quality as indicated through rocking curve FWHM. Also, phi scan for LPE films confirms high quality epitaxial growth. The work provides a way to develop low-cost III-V solar cells on inexpensive silicon substrate by integrating high quality germanium on silicon.

REFERENCES

[1] Jurgen Michel, Jifeng Liu & Lionel C. Kimerling, "High performance Ge-on-Si photodetectors", *Nature Photonics* vol.4, pp.527 – 534, 2010.

[2] R. R. King, D. C. Law, K. M. Edmondson, C. M. Fetzer, G. S. Kinsey, H. Yoon, R. A. Sherif, and N. H. Karam, "40% efficient metamorphic GaInP/GaInAs/GeGaInP/GaInAs/Ge multijunction solar cells", *Applied Physics Letters* ,Vol 90, Issue 18, 2007.

[3] D. Marris-Morini, P. Chaisakul, M-S. Rouifed , J. Frigerio , G. Isella , D. Chrastina , L. Vivien, "Low energy consumption and high speed germanium-based optoelectronic devices," *Conference on Lasers & Electro-Optics Europe & International Quantum Electronics Conference CLEO EUROPE/IQEC*, Munich, pp. 1-1, 2013.

[4] Jifeng Liu, Xiaochen Sun, Rodolfo Camacho-Aguilera, Lionel C. Kimerling, and Jurgen Michel "Ge-on-Si laser operating at room temperature," *Opt. Lett.* Vol.35, pp.679-681,2010.

[5] M. V. Fischettia and S. E. Laux, "Band structure, deformation potentials, and carrier mobility in strained Si, Ge, and SiGe alloys," *J. Appl. Phys.* vol.80, 1996

[6] Mantu Kumar Hudaita, S.B. Krupanidhia, "Transmission electron microscopic study of GaAs/Ge heterostructures grown by low-pressure metal organic vapor phase epitaxy," *Materials Research Bulletin*, vol.35 ,pp.125–133, 2000.

[7] http://spectrum.ieee.org/semiconductors/materials/

[8] J. Liu, H. J. Kim, O. Hul'ko, and Y. H. Xie,S. Sahni, P. Bandaru, and E. Yablonovitch, "Ge films grown on Si substrates by molecular-beam epitaxy below 450°C," *J. Appl. Phys.* ,vol.96, 2004 .

[9] V.A. Shah, A. Dobbie, M. Myronov, D.R. Leadley, "High quality relaxed Ge layers grown directly on a Si(0 0 1) substrate" *Solid-State Electronics*,vol. 62pp.189–194, 2011.

[10] Kwang Hong Lee, Adam Jandl, Yew Heng Tan, Eugene A. Fitzgerald, and Chuan Seng Tan, "Growth and characterization of germanium epitaxial film on silicon (001) with germane precursor in metal organic chemical vapour deposition (MOCVD) chamber",*AIP Advances* vol.3, p.092123,2013.

[11] J.E. Ayers, "The measurement of threading dislocation densities in semiconductor crystals by X-ray diffraction," *vol.*135 pp.,71—77, 1994

Crystallinity Control in Low-Temperature Growth of Poly-Crystalline Ge by Ion Beam Deposition

S. I. Maximenko[1], N. A. Mahadik[1], A. Giussani[2], E. L. McClure[2], C. Bailey[3], S. M. Hubbard[2], P. P. Jenkins[1] and R. J. Walters[1]

[1]Naval Research Laboratory, Washington, DC, 20375, USA
[2]Rochester Institute of Technology, Rochester, NY, 14623, USA
[3]Old Dominion University, Norfolk, VA 23529, USA

Abstract — An ion beam sputtering process was used to grow polycrystalline Ge films at low temperatures (200-400 °C). The effect of transition metals (Sn, Al) and ion assisted beam was evaluated to enhance grain growth. It was found that the presence of the seed metals facilitates the grain growth even at low temperatures, below eutectic point. Modification of the microstructure was found impinging the growth of Ge films with a 100 eV ion assisted beam in the energy dose range 0-7 eV/atom. Ge layers deposited on Sn with ~4eV/atom doses had the largest grain size. X-ray diffraction (XRD) and Atomic Force Microscopy (AFM) were employed to characterize the film microstructure. The grown poly-Ge films were used to grow GaAs structures.

I. INTRODUCTION

Growth of Ge polycrystalline films on foreign substrates has attracted a great interest in recent years due to their potential applications in photovoltaic [1] and flexible electronics [2] as an approach to bypass native substrate size restrictions and reduce the overall cost of the fabricated devices. Unlike single crystal material, poly-Ge can be deposited as a thin film on a variety of amorphous or poly substrates like glass, high temperature plastic, or metal films.

Traditionally, Ge is the substrate of choice for GaAs epitaxy due to its close lattice match to GaAs (~0.07%). The highest efficiency solar cells are found in devices utilizing III-V materials with multijunction InGaP/InGaAs/GaAs/Ge (38%) or single junction GaAs devices (29%) [3]. While these results are reported for single crystal materials, it was shown that single junction GaAs solar cells grown on polycrystalline Ge exceed 18% efficiency [1]. Further increases in solar cell efficiency can be obtained by increasing the grain size of polycrystalline material and by improvements in grain surface passivation techniques [4].

The best quality Ge buffer films were achieved by growing on rocksalt type Y_2O_3/CaF_2 or MgO biaxially-textured structures which serve as transient nucleation layers for further Ge growth [5]. Unfortunately, Y_2O_3 is a non-conductive material. For a majority of applications, however, an electrically conductive seed layer is a necessary requirement.

Recently, research was focused on the growth of poly-Ge by recrystallization of amorphous Ge films through an annealing process [6]. This process produces larger grain sizes in comparison to the direct deposition of poly-Ge. Large grain size corresponds with a longer annealing time. Non-uniform film formation and prohibitively long recrystallization times (up to 10 hours) are still obstacles which need to be resolved. High temperature, ~ 600 °C, associated with the post-recrystallization process can be decreased significantly (to <250°C) using a metal induced crystallization (MIC) process [for example Ref. 7]. The decrease in the processing temperature is essential for substrates which cannot withstand this high-temperature process (glass or plastic substrates that are limited to ~500 °C). Several metals were reported to reduce crystallization temperatures with to a varying degree [8]. Some attractive candidates reported in the literature are presented in Table 1. Among them, Tin (Sn) has shown to have the lowest amorphous Ge crystallization (75-85 °C) and eutectic (232 °C) temperatures. Along with another low eutectic temperature metal, Al is being widely studied for MIC application.

Metal	Crystallization Temperature, °C	Eutectic Temperature, °C
Sn	75-85	232
Au	87-105	356
Zn	98-100	398
Al	121-141	424
Cu	190-197	640
Ni	194-209	762

Table 1: Crystallization temperatures of Ge for selected Ge-Metal combinations.

Direct deposition of poly-Ge, despite resulting in a small grain size film, is still an attractive approach because of the short process time and lower process complexity. However, a high process temperature is required to grow large size grains, or else deposited films have an amorphous or poor crystalline quality. In addition to using the metal catalyzed mechanism, the decrease in process temperature can be obtained by increasing the kinetic energy of the deposited atoms, analogous to a temperature increase [9]. Furthermore, it also reported an increase in the degree of grain preferential orientation as well as grain size of the polycrystalline films [10]. The increase in the energy of the sputtered atoms can be accomplished by using ion beam sputter (IBS) or assisted low energy energetic ions.

978-1-5090-5606-4/17 $31.00 © 2017 IEEE

The main intention of this work is to explore the effects of selected transition metals and ion assisted energy fluxes on the crystallinity in Ge films formed by low energy ion beam sputter techniques on amorphous type substrates in the temperature range of 200-400 °C. Growth of crystalline Ge and other semiconductor films by ion beam sputter (IBS) has been demonstrated in the past [11]. Among some advantages of IBS in contrast to other sputter techniques are the many options in the choice of substrate/film combinations, good adhesion and uniformity of deposited films, grain size and the degrees of freedom to manage growth, for example through energy of deposited particles.

II. EXPERIMENTAL DETAILS

To study Ge crystallization effects, films of Ge and seeding layers of selected transition metals with low crystallization temperature (Al, Sn) were deposited by ion beam sputtering using a Kaufman-type source with a 3-cm argon ion beam on substrates Si (100), thermally grown amorphous SiO_2 on Si (100), or 7059 corning glass substrates. The same Kaufman-type ion beam source was used for the ion assisted beam. Argon beam energy used for sputtering and assisted beam was set to 1000 and 100 eV accordingly with an operational pressure of 2E-4 Torr. The sputter rate for Ge was 1Å/s with film thicknesses of 200 nm. 40-nm-thick Tin and Aluminum seed layers were deposited just before Ge sputter at the same process temperature. Prior to deposition, the substrates were cleaned with low energy 100 eV ion assisted beam to remove physisorbed contaminants [12]. During the sputter process substrates were heated in the temperature range of 100-400 °C by a resistive type heater and the temperature was tracked by a thermocouple in close proximity to the sample coupon position. Afterward, the crystallinity of the sputtered Ge films were evaluated using XRD and AFM.

III. RESULTS

The effect of seed nucleation layers of Al and Sn on Ge film crystallinity as grown on amorphous SiO_2 deposited at 250 °C is presented in Fig. 1. AFM images show a pronounced difference in the surface morphology and grain size of sputtered films.

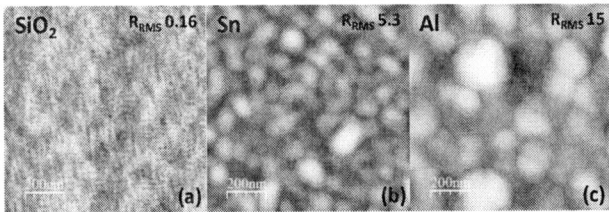

Fig. 1. AFM images of the surface morphology of 200 nm thick Ge films simultaneously deposited on bare Si(100)/SiO$_2$ and Si(100)/SiO$_2$ substrate with Sn and Al nucleation layer at 250 °C.

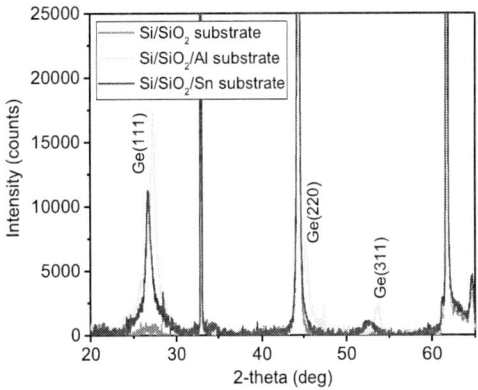

Fig. 2. XRD patterns of 200 nm thick Ge films simultaneously deposited on bare Si(100)/SiO$_2$ and Si(100)/SiO$_2$ substrate with Sn and Al nucleation layer at 250 °C.

Ge films deposited on samples with metal nucleation layers have a polycrystalline nature, evidence that the presence of a seed metal facilitates the grain growth. While Ge film simultaneously grown on a bare substrate has an amorphous structure. Corresponding XRD patterns captured from the samples agree with AFM observations as presented in Fig. 2. A strong signature of the poly-Ge peaks as indicated by the intensity of the (111), (220) and (311) is observed. This similar trend was observed sputtering Ge films on Si (100) or glass substrates, indicating there is a surface substrate independent mechanism of the growth at low process temperatures. The found grain formation in Sn-Ge and Al-Ge systems at temperatures above and below their eutectic points allows for making a conclusion about the presence of two different mechanisms of growth; liquid-solid and solid-solid correspondingly [13]. While Ge film deposited with Al nucleation layer has larger grain size, the film deposited onto Sn layer has more pronounced (111) preferential orientation from XRD data.

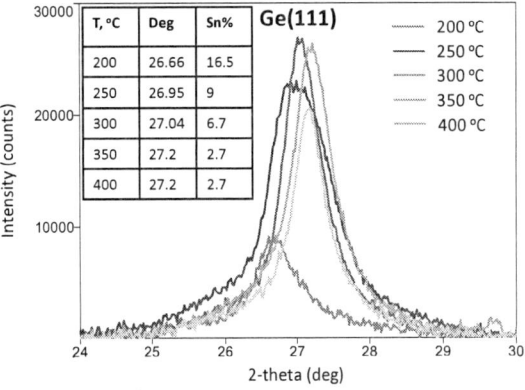

Fig. 3. Evolution of Ge(111) peak position as function of substrate temperature and estimated Ge-Sn alloy composition for 200 nm thick Ge films sputtered onto substrate with Sn nucleation layer.

To better understand the underlying growth mechanism, a set of Ge films was deposited on substrates covered with Sn nucleation layers at varying temperatures (200-400 °C). XRD data captured from samples for Ge (111) peak are presented in Fig. 3. A shift of the Ge (111) peak position deposited onto Sn layer is clearly seen in Fig. 2 and is related to the formation of Ge-Sn composition. Estimated from XRD data, the Ge-Sn alloy composition with calculated Sn content is summarized in Fig. 3. The Ge (111) peak intensity found from XRD results (Fig. 3) has straightforward correlation with the poly-Ge grain size estimated from AFM scans. Films grown in the temperature range of 300-350°C have the largest grain size (~250 nm).

To study the effect of energy addition on the deposition flux of Ge atoms, films were bombarded with Ar ions of various energy. Figure 4 presents the microstructure and grain size

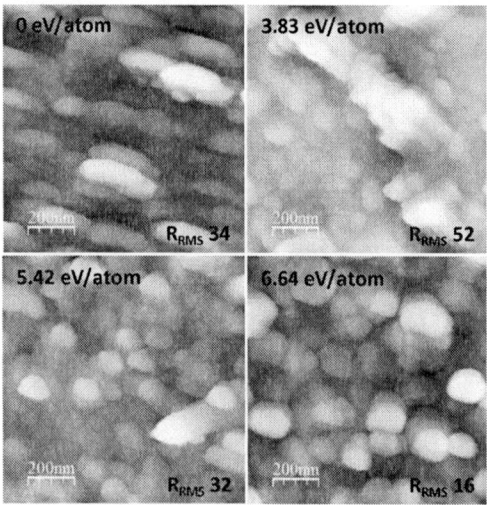

Fig. 4. AFM images of the surface morphology of Ge films deposited on substrate covered with ~40nm thick Sn nucleation layer at different energy doses.

changes for Ge films deposited onto Sn layers which were impinged with Ar ions at different energy doses at 350 °C. An increase in ion to atom ratio leads to gradual changes of the film structure. The largest grains were observed for Ge layers deposited at ~3.83 eV/atom doses, which correlates with earlier reports [14]. Larger energy doses lead to the creation of a large number of defects which results in a renucleation process and the formation of reduced size grains as can be seen in Fig. 4.

Further, to evaluate suitability of grown poly-Ge films for future GaAs epitaxy, photoluminescence (PL) test structures (Figure 5) were grown by MOCVD technique on virtual poly-Ge substrates with the mean grain size ~200 nm determined from AFM surface scans. The grain size found in grown GaAs films is ~4 times larger than grain size in poly-Ge substrates. Optical properties of the films were evaluated by room temperature PL spectroscopy excited by 532 nm wavelength laser. Strong peak PL intensity was observed from all samples

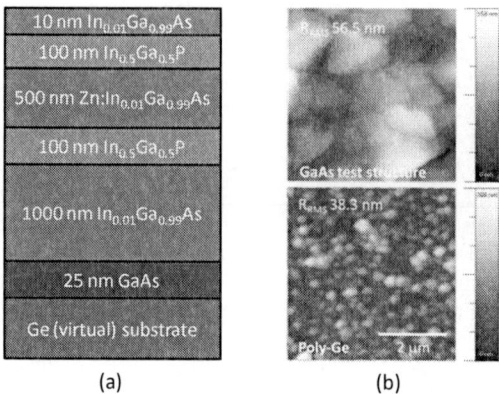

Fig. 5. (a) GaAs test structure. Growth Temperature 500 °C, active layer doped p with nominal [Zn] 1.1E18 cm^{-3}. (b) Surface morphology of virtual poly-Ge substrate and grown GaAs test structure.

evidencing a good crystallinity of the films. The PL peak position corresponding to near edge emission of GaAs was found at ~918 nm (1.35 eV). It is slightly red-shifted and has lower intensity in comparison with structures grown simultaneously on commercial polycrystalline preferentially (112)-oriented optical grade Ge wafers, where the PL peak appears at ~ 888 nm (1.4 eV). This is characteristic emission observed from $In_{0.01}Ga_{0.99}As$ samples doped with Zn. A similar red shift was reported by other researchers [15]. Typically, red shift can be attributed to band gap shrinkage, which has a straightforward correlation with the doping concentration of the material. Band gap shrinkage due to an increase in the doping is a well documented effect in III-V compound semiconductors [16]. It can be suggested that the dopant (Zn) incorporation rate is higher in small grain size films. In addition to the enhanced nonradioactive recombination on smaller grains it also degrades intensity and broadens the PL signal.

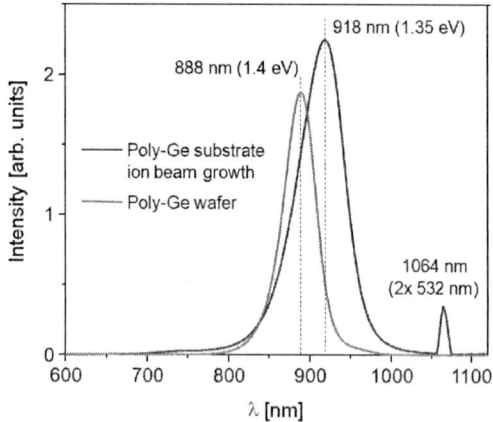

Fig. 6. PL spectroscopy of GaAs test structures grown on poly-Ge virtual substrate grown by ion beam sputter and on commercial bulk poly-Ge substrate. Neutral density filter (1.5 OD) was used to attenuate PL intensity from sample grown on commercial poly-Ge wafer.

IV. CONCLUSION

Ion beam deposition was evaluated for growth of polycrystalline Ge films at low process temperatures (<400 °C) on substrates with transition metals Al, Sn. It was found that the presence of seed metals facilitates the grain growth even at low temperatures, below the eutectic point. Modification of the microstructure was found to impinge growing Ge films with a 100 eV ion assisted beam. Ge layers deposited with ~4eV/atom doses had the largest grain size. GaAs films grown using MOCVD on virtual poly-Ge substrates showed an increased grain size (~4 times) and have strong red shifted near edge emission. Further improvement in the growth technology to increase the poly-Ge grain size, the ion beam deposited films, have the potential to be used as virtual substrates for III-V based solar cell structures.

ACKNOWLEDGEMENT

This work is sponsored by the Office of Naval Research (ONR)

REFERENCES

[1] R. Venkatasubramanian, B. C. O'Quinn, J. S. Hills, P. R. Sharps, M. L. Timmons, J. a. Hutchby, H. Field, R. Ahrenkiel, and B. Keyes, "18.2% (AM1.5) efficient GaAs solar cell on optical-grade\npolycrystalline Ge substrate," *Conf. Rec. Twenty Fifth IEEE Photovolt. Spec. Conf. - 1996*, pp. 31–36, 1996.

[2] M. Asadirad, Y. Gao, P. Dutta, S. Shervin, S. Sun, S. Ravipati, S.H. Kim, Y. Yao, K.H. Lee, A.P. Litvinchuk, V. Selvamanickam, and J.-H. Ryou, Adv. Electron. Mater. 2, (2016).

[3] M. A. Green, K. Emery, Y. Hishikawa, W. Warta, and E.D. Dunlop, Prog. Photovoltaics Res. Appl. 24, 905 (2016).

[4] E. L. McClure, Z. S. Bittner, M. A. Slocum, D. V. Forbes, and S. M. Hubbard, "Modeling the effects of using polycrystalline substrates for low cost III–V photovoltaics," in *2015 IEEE 42nd Photovoltaic Specialist Conference (PVSC)*, 2015, pp. 1–5.

[5] V. Selvamanickam, C. Jian, X. Xiong, G. Majkic, and E. Galtsyan, "Single-crystalline-like germanium thin films on glass substrates," in *2012 38th IEEE Photovoltaic Specialists Conference*, 2012, pp. 002592–002595.

[6] H. Abu-Safe, A. Hickerson, H. Zhong, H. Naseem, and S.-Q. Yu, "Selected area crystallization of amorphous Si and Ge thin films on glass substrates for solar cell and 3D-optoelectronic applications," in *2013 IEEE 39th Photovoltaic Specialists Conference (PVSC)*, 2013, pp. 1314–1317.

[7] T. Sadoh, M. Kurosawa, T. Hagihara, K. Toko, and M. Miyao, "Low-Temperature (~ 250°C) Cu-Induced Lateral Crystallization of Amorphous Ge on Insulator," *Electrochem. Solid-State Lett.*, vol. 14, no. 7, p. H274, Jul. 2011.

[8] F. Oki, Y. Ogawa, and Y. Fujiki, "Effect of Deposited Metals on the Crystallization Temperature of Amorphous Germanium Film," *Jpn. J. Appl. Phys.*, vol. 8, no. 8, pp. 1056–1056, Aug. 1969.

[9] F. A. Smidt, "Use of ion beam assisted deposition to modify the microstructure and properties of thin films," *Int. Mater. Rev.*, vol. 35, no. 1, pp. 61–128, Jan. 1990.

[10] A. Anders, "A structure zone diagram including plasma-based deposition and ion etching," *Thin Solid Films*, vol. 518, no. 15, pp. 4087–4090, May 2010.

[11] N. Mosleh, F. Meyer, C. Schwebel, C. Pellet, and M. Eizenberg, "Growth mode of Ge films on Si(100) substrate deposited by ion beam sputtering," *Thin Solid Films*, vol. 246, no. 1–2, pp. 30–34, 1994.

[12] H. Kaufman and J.M.E. Harper, J. Vac. Sci. Technol. A Vacuum, Surfaces, Film. **22**, 221 (2004).

[13] B. T. Richards, B. Gaskey, B. D. A. Levin, K. Whitham, D. Muller, and T. Hanrath, "Direct growth of germanium and silicon nanowires on metal films," *J. Mater. Chem. C*, vol. 2, no. 10, p. 1869, 2014.

[14] H. R. Kaufman and J. M. Harper, "Ion-assist applications of broad-beam ion sources," in *Optical Science and Technology, the SPIE 49th Annual Meeting*, 2004, pp. 50–68.

[15] P. Dutta, M. Rathi, N. Zheng, Y. Gao, Y. Yao, J. Martinez, P. Ahrenkiel, and V. Selvamanickam, "High mobility single-crystalline-like GaAs thin films on inexpensive flexible metal substrates by metal-organic chemical vapor deposition," *Appl. Phys. Lett.*, vol. 105, no. 9, pp. 92104–1, 2014.

[16] M. K. Hudait, "Zn incorporation and band gap shrinkage in p -type GaAs," *J. Appl. Phys.*, vol. 82, no. November, pp. 4931–4937, 1997.

High efficiency GaInP/GaAs double junction Solar cell on Si substrate assisted by the electron beam treatment

Hyo Jin Kim[1]*(hjk@kopti.re.kr) and Yong Whan Kim[2]

[1]Reasearch center for ICT and Photonics Energy, Korea Photonics Technology Institute, Gwangju ,

[2]Research and Development Team, Infovion Incorporation, Seoul, Korea

Abstract —We have studied the dependence of the growth of GaAs on Si Substrate on the flow rate of source for the GaAs seed layer. The GaAs buffer with the different growth condition of the inserting GaAs seed has been grown on Si (001) substrates by metalorganic chemical vapor deposition. Structural properties of GaAs buffer were investigated by scanning electron microscope (SEM) and high resolution X-ray diffraction (HRXRD). The misfit dislocation of GaAs seeds on Si substrates were decreased as the growth rate and V/III ratio of GaAs seed were increased and decreased, respectively. Also, we have investigated the improvement of GaAs buffer assisted by electron beam treatment. The peak position of GaAs buffer on Si was reached to GaAs Substrate assisted by e-beam treatment. Finally, the efficiency of GaInP/GaAs double junction solar cell on Si substrate was about 22.6 %.

Index Terms — Solar cell, III-V semiconductor, Si substrate, electron beam treatment, Thermal cycling annealing.

I. INTRODUCTION

Epitaxial growth of GaAs on Si substrate has attracted significant attention for many years due to the possibility of unique material combinations such as highly mismatched materials and polar-nonpolar materials [1]. The fabrication of GaAs solar cells on Si substrates has been reported since 1980 [2, 3]. Combination of GaAs active layer and Si substrate may provide high-efficiency, high radiation resistant, low-weight, low-cost, and large-area solar cells.

However, the large dislocation between Si and GaAs was occurred over 1×10^7 cm^{-3} [4] because of lattice mismatch (4 %), the difference in their thermal expansion coefficients, and the formation of antiphase domains (APDs) [4, 5] concerning the epitaxy of polar semiconductors on (001) non-polar semiconductor substrates. There are many techniques that can help to improve the quality of the GaAs/Si epilayers, such as in situ or postgrowth annealing [5], thermal cycle annealing [6], use of strained-layer-superlattice intermediate layers [7] use of impurity-doped layers [8], use of a sawtooth-paterned Si substrate [9] insertion of one monolayer of a group II element in place of Ga atoms at the GaAs/Si interface [10], and use of layered structure GaSe as a buffer layer [11].

Kawabe et al. Reported that when the Si (100) substrate has an inclined angle to the [011] direction, the APD disappears in the initial growth process, and also in the Si (001) substrate It was shown that single domain GaAs grows even if the conditions do not have a tilted angle. On the other hand, as a method to reduce lattice mismatch, a buffer layer such as Ge is grown between Si and GaAs [13], a method of growing InGaAs/GaAs or GaAsP/GaAs superlattice [7] The ELO method is proposed.

However, there are few reports on the growth of the GaAs buffer layer due to the growth conditions of the GaAs seed layer. There are also few reports on high-efficiency III-V double-junction solar cells on GaAs on Si substrates. In this study, we investigate the variation of the GaAs buffer layer with respect to the variation of the flow rate of the seed layer of GaAs. We also studied the characteristics of GaAs on Si by electron beam annealing. Finally, We have compared the characteristics of GaInP/GaAs double junction solar cell structure were grown on GaAs/Si substrate and GaAs substrate at the same time.

II. EXPERIMENTAL PROCEDURES

The GaAs epilayer was grown on (100) p+-Si substrates oriented 2°-off towards [011] by metalorganic chemical vapor deposition (MOCVD). Trimethylgallium (TMG) was used as a group III source and arsine (AsH$_3$) was used as a group V source. The substrates were cleaned in organic solvents using an ultrasonic bath and then in solutions of H$_2$SO$_4$: H$_2$O$_2$ (4 : 1 by volume) and HF : H$_2$O (1 : 1 by volume). As soon as the chemical process for the Si wafers was finished, the substrates were mounted onto a molybdenum susceptor. Before GaAs growth, a self-cleaning process was performed in H$_2$ atmosphere at 750~780 ℃, in the MOCVD reactor for 20 min in order to remove the native oxide on the surface of substrate.

The growth temperature was 700 ℃ except for the initial GaAs nucleation layer which was grown on Si substrate at 400 ℃, in the two-step growth technique. An initial 10 nm thick undoped GaAs nucleation layer and a 1.6 um thick n-GaAs buffer layer were grown. The detailed flow rate of TMG and AsH$_3$ of samples used in this study was represented in Table 1. Also, in order to remove the native oxide film on Si, preheating is performed at a temperature higher than 780 ℃. before growth. In this study, GaAs can be grown at a temperature of 750~780 ℃. only by the growth condition of the GaAs seed layer. Structural properties of GaAs buffer were investigated by scanning electron microscope (SEM) and double crystal X-ray diffraction (DCXRD).

978-1-5090-5606-4/17 $31.00 © 2017 IEEE

Commonly, the multi-junction solar cells are performed by the series connection of sub-cells and tunnel junctions. The ultra-high doped (>10^{19} cm^{-3}) pn-junction was required to grow the proper tunnel junction. Generally, an intrinsic doped p+-AlGaAs and a Te doped n+-GaAs are used for the high doped p- and n- region. It was noted that the low V/III ratio (≤ 10) growth was required to maximize the intrinsic doping for p-AlGaAs and minimize the optimum flow of Te precursors for the n-GaAs in order to reduce Te memory effect [2]. The growth temperature and V/III was 550℃ and 5 for tunnel junction and 650 ℃ and ~100 for other layers, respectively.

Table 1. The flow rate and growth time of samples used in this study

Sample NO.	TMGa (10^{-6}μmol/min)	AsH$_3$ (10^{-3}μmol/min)	Time (min.)
S-1	6.976	10.339	100
S-2	2.325	4.464	30
S-3	4.186	8.036	17
S-4	2.325	1.339	30
S-5	2.325	8.036	30

III. RESULTS AND DISCUSSION

GaAs on Si was initially grown at a low temperature of 10 nm thick GaAs nucleation layer and the temperature was raised to grow a 1.5 μm buffer layer. We have investigated the formation of GaAs buffer layer by varying the flow rate for the initial GaAs nucleation layer. Figure 1 represented the scanning electron microscopy (SEM) images of the shapes of the grown GaAs buffer layers when the flow and time of TMG and AsH$_3$ for the initial nucleation are selected as S-1, S-2 and S-3 conditions in Table 1.

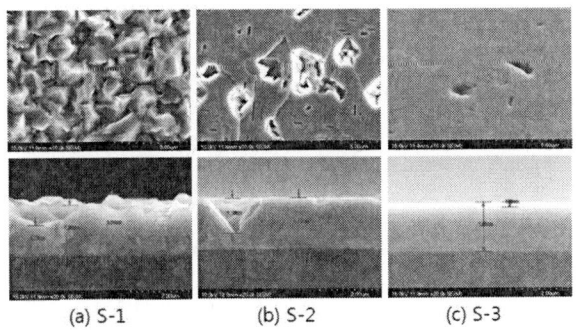

(a) S-1   (b) S-2   (c) S-3

Fig. 1. SEM image for surface morphology and cross-section of GaAs on Si with condition of (a)S-1, (b)S-2, and (c) S-3, respectively

The selected conditions gave a change in the growth rate until the GaAs nucleation layer was grown to a constant thickness. As shown in Fig. 1, the surface was improved when the growth rate was moderately high.

Figure 1 represented the SEM images of the shapes of the grown GaAs buffer layers when the flow and time of TMG and

AsH$_3$ for the initial nucleation are selected as S-4, S-2 and S-5 conditions in Table 1. The selected conditions gave a change in the V/III ratio until the GaAs nucleation layer was grown to a constant thickness. As shown in Fig. 2, the surface was improved dramatically when the V/III ratio of GaAs nucleation seed was moderately high.

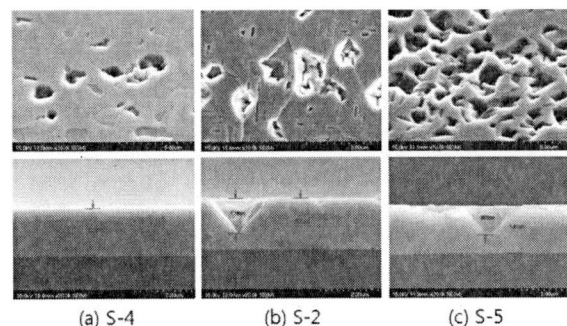

(a) S-4   (b) S-2   (c) S-5

Fig. 2. SEM image for surface morphology and cross-section of GaAs on Si with condition of (a)S-4, (b)S-2, and (c) S-5, respectively

Although the exact cause of this change is not known, our result has shown that the GaAs nucleation layer does not roam for chemical bonding on the Si substrate, and that the faster the GaAs nucleation layer is seated, the better the GaAs buffer layer grown thereon can be formed. The thermal cycling annealing (TCA) method was applied to grow the GaAs buffer layer as in other groups. Figure 3 shows the schematic diagram for the thermal cycling annealing (TCA) and the SEM image of GaAs buffer on Si substrate after TCA. Surprisingly, it was observed that the surface of GaAs on Si was slightly improved after TCA.

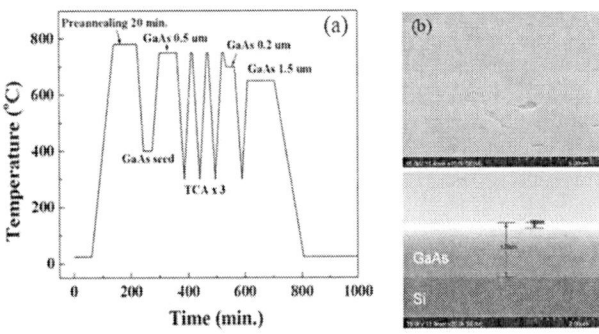

Fig. 3. (a) Schematic diagram of the growing procedures for GaAs on Si substrate with the thermal cycling annealing (TCA) and (b) SEM image of GaAs buffer on Si substrate after TCA

In addition, we have studied the method of improving the surface of GaAs on Si substrate by electron beam. There is a method of using an electron beam as a method capable of effectively heat-treating the surface with energy that does not damage the surface of the semiconductor. A self-made plasma electron beam source can irradiate a certain amount of electron beam to an area of about 50 x 50 mm^2. The surface characteristics of GaAs on Si were investigated by varying the acceleration voltage of the electron beam to 0.3 ~ 3 keV. Electron beam annealing (EBA) was performed in a separate chamber by taking samples out of MOCVD and operating pressure was 4.7 x 10^{-4} torr.

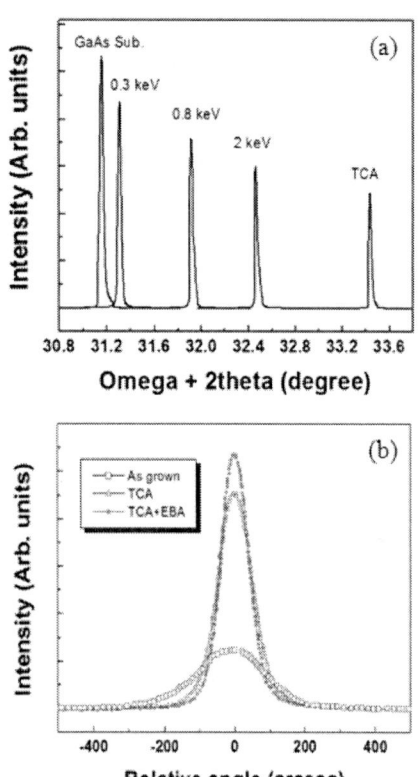

Fig. 4. (a) HRXRD data of GaAs on Si substrate treated with the EBA conditions and TCA, (b) FWHM of XRD peaks of GaAs on Si samples

Figure 4 (a) shows the peaks obtained by high-resolution X-ray diffraction (HRXRD) equipment on GaAs on Si samples investigated by varying the EBA acceleration voltage. In this case, the samples were subjected to TCA treatment in MOCVD. Figure 4 (b) shows the full-width half maximum (FWHM) of XRD peaks of GaAs on Si samples when grown without TCA (As-grown), grown with TCA treatment (TCA), and grown with TCA and EBA (TCA + EBA).

As shown in Fig. 1 (a), the XRD peak of the GaAs buffer layer grown on the Si substrate in MOCVD is slightly larger than the XRD peak of the GaAs substrate, and the lattice structure is slightly changed. It was surprisingly found that the peak of GaAs on Si shifted similar to the peak of GaAs substrate when energy was injected by EBA irradiation. It was confirmed that the structure, which was slightly reduced when GaAs having a larger lattice constant than that of Si was grown on the Si substrate, was moved to the original lattice structure

by the EBA rapid thermal annealing. More detailed studies need to be added later. As shown in Fig. 4 (b), the FWHM for the samples after as-grown, TCA and TCA + EBA were 228, 112 and 95 arcsec, respectively. Figure 5 shows the FWHM results for XRD of GaAs buffer layers grown on Si substrates in other group results and our result. Our results are represented as the star mark.

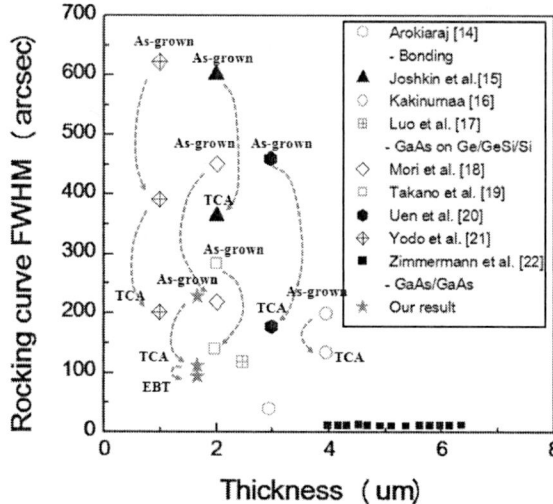

Fig. 5. FWHM of HRXRD peaks of GaAs buffer layers grown on Si substrates in other group results and our result

In order to verify the optimal conditions of GaAs on Si substrate we studied, the solar cell structure was grown on it.

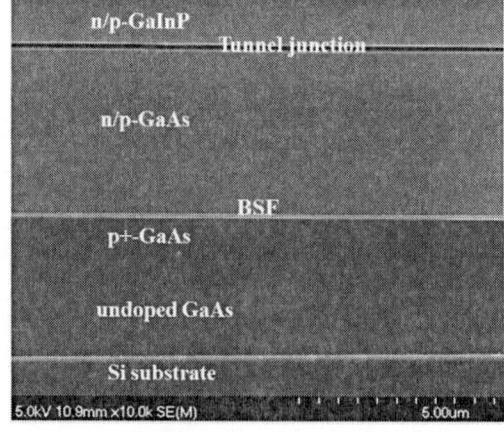

Fig. 6. SEM cross section images for GaInP/GaAs double junction solar cell grown on Si substrate

Figure 6 shows the SEM cross section image of GaInP/GaAs double junction solar cell structure grown on the optimal GaAs buffer layer on Si substrate obtained through TCA and EBA conditions. The buffer layer of the solar cell has an undoped GaAs layer between the p+-GaAs and the Si substrate. The reason is that the bottom electrode is connected to the p+-

GaAs layer to prevent current from flowing to the interface between Si and undoped GaAs.

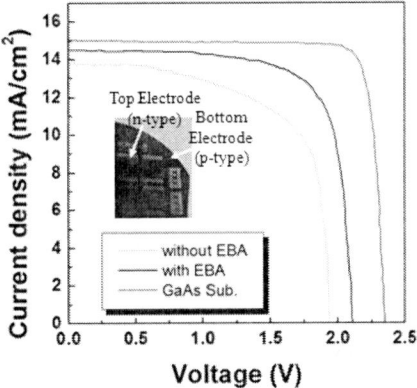

Fig. 7. Current density - voltage curve for GaInP/GaAs double junction solar cell grown on Si substrate

The solar cell structure was grown simultaneously on three substrate types which were GaAs grown on Si substrate without EBA, EBA treated GaAs grown on Si substrate, and GaAs substrate. Figure 7 shows the current-voltage curves plotted for solar cells fabricated with the same structure grown for these three substrates. As shown in the solar cell picture in Figure 7, both n-type and p-type electrodes were deposited on the front surface. The characteristics of the solar cells fabricated on the three substrates were represented in Table 2.

Table 2. Characteristics of GaInP/GaAs double junction solar cells

	Without EBA	With EBA	GaAs Sub.
Short circuit Current density (mA/cm^2)	13.76	14.41	15.3
Open circuit voltage (V)	1.94	2.10	2.34
Fill factor (%)	64.86	74.76	86.98
Efficiency (%)	17.31	22.62	31.12

The efficiency of solar cell grown on GaAs on Si without EBA was 55.6% of the efficiency of solar cell grown on GaAs substrate. On the other hand, the efficiency of solar cell grown on EBA treated GaAs on Si obtained 72.7% of the efficiency of solar cell grown on GaAs substrate.

IV. SUMMARY

We have studied the dependence of the formation of GaAs grown on Si Substrate on the flow rate of source for GaAs seed. As the V/III ratio and growth rate of GaAs nucleation seed were low and increased, the quality of GaAs buffers were improved. According to our result, it was confirmed that the better the GaAs buffer layer grown on thereon can be formed that the GaAs nucleation layer does not roam for chemical bonding on the Si substrate, and that the faster the GaAs nucleation layer is seated. Also, we have investigated the improvement of GaAs buffer assisted by electron beam treatment. The peak position of GaAs buffer on Si was reached to GaAs Substrate assisted by e-beam treatment. Finally, the efficiency of GaInP/GaAs double junction solar cell on Si substrate was about 22.6 %.

V. ACKNOWLEDGEMENT

This research was supported by Power standardization and certification support project in electrical power industry funded by the Ministry of Commerce, Industry and Energy, Korea.

REFERENCES

[1] V. I. Vdovin et al., "Antiphase boundaries in GaAs layers on Si and Ge", Journal of Crystal Growth 132, pp 477-482, 1993.

[2] H. J. Hovel and J. M. Woodall, "The effect of depletion region recombination currents on the efficiencies of Si and GaAs solar cells," in *10th IEEE Photovoltaic Specialist Conference*, p. 25, 1993.

[3] G. Wang et al., "A detailed study of H2 plasma passivation effects on GaAs/Si solar cell", Solar Energgy Materials & Solar cells 66, 599-605, 2001.

[4] H. Noge et al., " Antiphase domains in GaAs grown on (001)-oriented Si substrate by molecular beam epitaxy", J. Appl. Phys. 64, pp. 2246-2248, 1988.

[5] Y. ITOH, H. MORI, and M. Yamaguchi, "Anti-phase boundaries of GaAs on Si", Journal of Crystal Growth 103, 363-365, 1990.

[6] C. Choi, N. Otsuka, G. Munns, R. Houdre, H. Morkoc, S.L. Zhang, D. Levi and M.V. Klein., "Effect of in situ and ex situ annealing on dislocations in GaAs on Si substrates", Appl. Phys. Lett. 50, pp 992-994, 1987.

[7] M. Shimizu, M. Enatsu, M. Furukawa, T. Mizuki and T. Sakurai., GaAs on Si substrates", J. Cryst. Growth 93, pp. 475-480, 1988.

[8] N. Hayafuji, S. Ochi, M. Miyashita, M. Tsugami, T. Murotani and A. Kawagishi., "Effective of AlGaAs/GaAs superlattice in reducing dislocation in GaAs on Si", J. Cryst. Growth 93, pp.494-498, 1988.

[9] K. Nozawa and Y. Horikoshi., "Impurity Doping Effect on the Dislocation Density in GaAs on Si (100) Grown by Migration-Enhanced Epitaxy ", Jpn. J. Appl. Phys. 28, p. L1877-L1879, 1989.

[10] K. Ismail, F. Legoues, N.H. Karam, J. Carter and H.I. Smith., "High-quality GaAs on sawtooth-patterned Si substrates", Appl. Phys. Lett. 59, 1991, pp. 2418-2420.

[11] J.E. Palmer, T. Saitoh, T. Yodo and M. Tamura., "Growth and characterization of GaAs/GaSe/Si Heterostructures", Jpn. J. Appl. Phys. 32, 1993, p. L1126-L1129.

[12] Mitsuo Kawabe and Toshio Ueda, "Molecular Beam Epitaxy of Controlled Single Domain GaAs on Si(100)", Jpn. J. Appl. Phys. 25, 1986, pp. L285-L287.

[13] Y. D. Woo, T. W. Kang, T. W. Kim, "Improvement of the crystallinity of a GaAs epitaxial film grown on a Si substrate using a Si/SiGe/Ge buffer layer", Thin Solid Film 279, 166-168, 1996.

[14] Arokiaraj J et al. "High-quality GaAs on Si substrate by the epitaxial lift-off technique using SeS₂", Appl. Phys. Lett. 75, 3826, 1999.

[15] Joshkin V et al. " Biaxial compression in GaAs thin films grown on Si", J. Cryst. Growth 147, 13, 1995.

[16] Kakinuma H et al. " Reduction of threading dislocations in GaAs on Si by the use of intermediate GaAs buffer layers prepared under high V–III ratios", J. Cryst. Growth 205, 25, 1999.

[17] Luo G-L et al. " High-speed GaAs metal gate semiconductor field effect transistor structure grown on a composite Ge/Ge$_x$Si$_{1-x}$/Si substrate", J. Appl. Phys. 101, 084501, 2007.

[18] Mori H et al. "GaAs heteroepitaxy on an epitaxial Si surface with a low-temperature process ", Appl. Phys. Lett. 63 1963, 1993.

[19] Takano Y et al. " Reduction of threading dislocations by InGaAs interlayer in GaAs layers grown on Si substrates ", Appl. Phys. Lett. 73 2917, 1998.

[20] Uen W-Y, Ohori T, Nishinaga T, " Molecular beam epitaxy of gallium arsenide on 0.3°-misoriented epitaxial Si substrates" J. Cryst. Growth 156 133, 1995.

[21] Yodo T, Tamura M, " GaAs heteroepitaxial growth on Si substrates with thin Si interlayers in situ annealed at high temperatures ", J. Vac. Sci. Technol. B 13 1000, 1995.

[22] Zimmermann G et al. " In-situ formation of As-H functions by β-elimination of specific metalorganic arsenic compounds for the MOVPE of III/V semiconductors", J. Cryst. Growth 124 136, 1992.

Analysis of Deposited Residues and Its Cleaning Process on GaAs Substrate after Epitaxial Lift-Off

Tatsuya Nakata[1], Kentaroh Watanabe[2], Hassanet Sodabanlu[2], Daiki Kimura[1], Naoya Miyashita[2], Yoshitaka Okada[1,2], Yoshiaki Nakano[1,2], Masakazu Sugiyama[1,2]

[1]School of Engineering, The University of Tokyo, Bunkyo-ku, Tokyo, 113-8656, Japan
[2]Research Center for Advanced Science and Technology, Meguro-ku, Tokyo, 153-8904, Japan

Abstract — The selective wet etching of the AlAs release layer with a HF solution was investigated for epitaxial lift-off (ELO) process. By measuring the lateral etch depth, the etch rate was estimated. A stress by the deposited Au electrode on the thin device layer enhanced the etch rate.

For substrate reuse, the exposed substrate surface after the etching was observed by scanning electron microscopy (SEM). Two kinds of morphologic depositions were typically observed in this study. Streaky aligned depositions were found just after the ELO process. During long-term storage of the substrate, some parts of the depositions were gradually re-formed into micrometer-size crystals. These micro-crystals were revealed to be As_2O_3 upon the compositional characterization by energy dispersive X-ray spectrometry, and we found the annealing process removes the micro-crystals.

Index Terms — gallium arsenide, III-V semiconductor materials, thin-film solar cell, epitaxial lift-off, wet etching, substrate reuse.

I. INTRODUCTION

The III-V compound semiconductors are ideal for solar cell applications, since they can easily be combined into multi-junction solar cells yielding record efficiencies [1]. However, these high efficiency solar cells have not been used for terrestrial applications because of the high material cost. GaAs or Ge substrate is usually used for epitaxial crystal growth of III-V device layers. After the crystal growth, only the device layer is necessary for a solar cell, and hence by reusing the expensive substrates we can achieve cost reduction. The epitaxial lift-off (ELO) is known as a major substrate reuse method [2]. The ELO process is based on the selective etching of an AlAs release layer, which enables the lift-off of a thin film solar cell and the reuse of the substrate. By using a HF solution, the AlAs release layer is etched 10^6 times faster than both the GaAs device layer and the substrate [3]. Ideally, only the AlAs release layer is etched without affecting the device layer and the substrate.

However, indeed the ELO has some challenges to be improved such as a long process time and the high substrate surface roughness due to the deposition of byproducts. It is known that the As layer appears on the surface of the GaAs wafer after it is exposed to a HF solution, and micrometer-size crystals are re-formed after a certain period of time with the assist of light illumination during storage of the wafer in the atmosphere [4]. In this research, the GaAs substrate surfaces

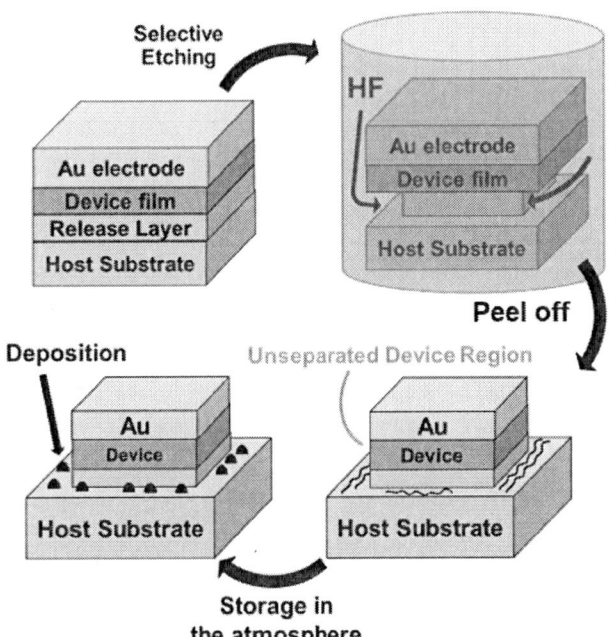

Fig. 1. Schematic illustration of the sample structure and selective etching process in a HF solution. After the process proceeded partly, the samples were taken out and the part of the lifted off device layer was removed to expose the surface of the host substrate. To compare the change of surface condition, the samples were stored for various periods of time after ELO process.

after the ELO process were characterized and the formation of deposit was confirmed. After micro-crystals were formed, the chemical composition of these crystals was identified. The cleaning of GaAs surface is very important for practical reuse of the substrate. Therefore, the application of annealing for the surface cleaning process was investigated.

II. EXPERIMENTAL DETAIL

All samples in this work were grown on the 2-inch p-type GaAs wafers with crystal orientation (100) using metal-organic vapor phase epitaxy (MOVPE). The schematic sample structure is depicted in Fig. 1. The p-i-n GaAs layers were grown on the 10-100 nm-thick AlAs release layer. Then,

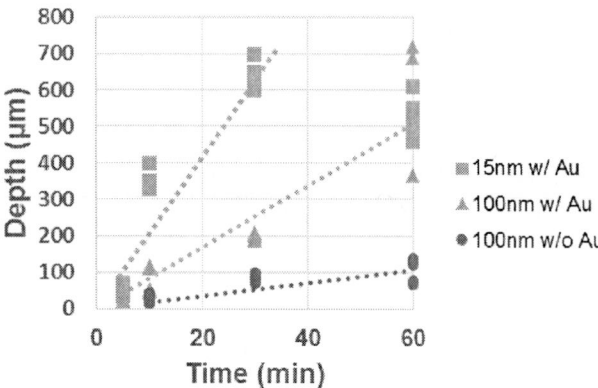

Fig. 2. Lateral etch depth of the release layers in the ELO as a function of the etch time. The effects of the release layer thickness (15 nm, 100 nm) and with or without Au metal electrode were compared.

approximately 700 nm-thick Au metal electrode was deposited on the GaAs device layer using a thermal evaporator. ELO processes were performed by putting the samples in a 9% HF aqueous solution. The etchant was used at room temperature and was not stirred during the experiment. After various periods of etching time, the etching process was stopped and the samples were rinsed by deionized water. After this selective etching process, the peripheral zone of the device layer was partly peeled off, and then we determined the etching depth which indicates the etch speed. The surface condition of the substrate was observed using scanning electron microscope (SEM).

After a long storage of the processed samples in the atmosphere, relatively large-size crystals were found on the part of the exposed GaAs surface. To compare the change of

surface condition, the samples were stored for various periods of time after ELO process. The effect of light illumination and substrate conduction type on the crystal formation process was also investigated. N-type (Si-doped, carrier density $=2 \times 10^{18} cm^{-3}$) and p-type (Zn-doped, carrier density $=1 \times 10^{19} cm^{-3}$) GaAs substrate were used for the comparison. To identify the chemical composition of these crystals, SEM-energy dispersive X-ray spectrometry (SEM-EDX) was carried out. To remove the crystals from the surface thermally, the effect of annealing process was investigated. A rapid thermal annealing was performed in the temperature range of 100 to 500 °C for various periods of time under nitrogen ambient.

III. RESULTS AND DISCUSSION

The lateral etch rate was estimated from the cross-sectional SEM images. When the samples without Au metal layer were put into a HF solution, the etch rate was ~110 μm/h. On the other hand, in the samples with the Au metal layer, the etch rate became approximately five times faster. Meanwhile, we occasionally found the device layer curling up after the partial ELO process. In the ELO process, some kind of stress is often applied to make the thin flexible device layer curl up [5, 6], which enables a HF solution to diffuse well into the etch front. In our case, it is thought that the deposited Au metal layer applied a stress to the thin device layer, which made a slit between the device layer and the substrate open, and this leads to the increased etch rate. In terms of the release layer thickness, the etch rate of 15 nm sample was two times faster than that of 100 nm sample. The etch depth of 15 nm sample was saturated around 600 μm. This might be due to the deposition at the etch slit, and this stagnated the etch process.

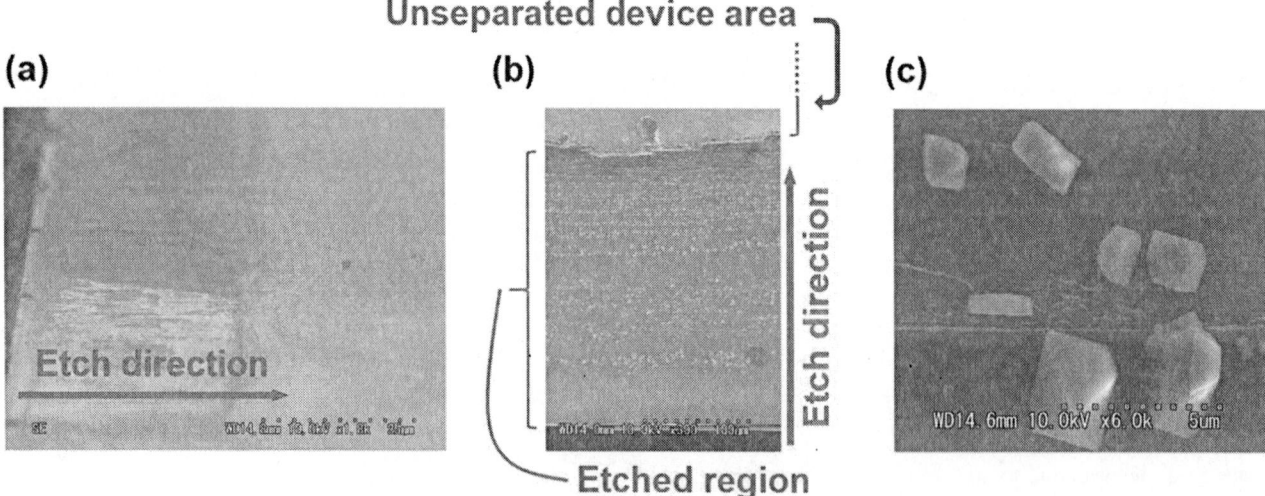

Fig. 3. SEM images of the exposed GaAs substrate surface after ELO process; (a) just after the partial ELO process, (b) after a long storage, (c) higher magnification of the micro-crystals.

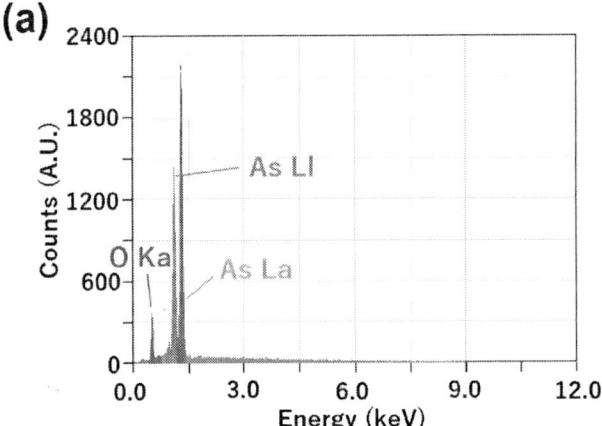

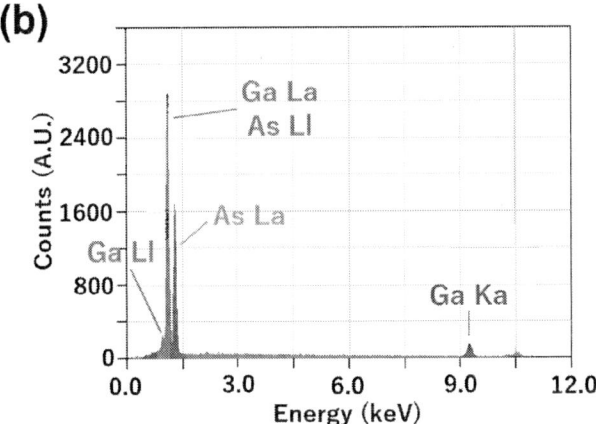

Fig. 4. Energy dispersive spectra of the deposited crystal (a) and the substrate surface not covered by the crystals (b).

After the partial ELO process and the exposure of the GaAs substrate surface, some samples were kept in the atmosphere under the illumination from the fluorescent light and the others were stored in the dark. From the SEM observation, specific streaky pattern was found just after the partial ELO process. The style of the deposition has a position dependence. The residue was deposited like a film around the edge, while like a ridge around the inner part (Fig. 3a). This result is likely related to how well the fresh HF solution and the reaction products can be exchanged through the narrow etch slit. The streaky pattern can be due to the depositions from the solution after the ELO reaction, and this might indicate the stagnation of the etchant exchange, where residues tend to be deposited.

As seen in Fig. 3b, many large-size crystals were observed on the surface after a long storage of the processed samples. To investigate the change of the surface condition, the samples were stored for various periods of time after the partial ELO process. SEM observation revealed that many small crystals appeared after a short period storage. After the longer period storage, larger crystals were also observed

simultaneously. The size and the density of the crystals seem to have some tendencies; around the position of the large crystals, less crystals can be observed. This might indicate that some small crystals were gradually re-formed up to a large crystal. In the same way, considering the similarity between the streaky pattern in the primary stage and the alignment of the micro-crystals after long storage, the micro-crystals should be relevant to the streaky pattern.

Then, to confirm the effect of the light illumination, another series of the processed samples stored in the dark were observed. It was found that the formation of the crystals was suppressed for these samples. This indicates that photo-generated carriers play some important roles and accelerate the crystal formation process [4, 7]. We also demonstrated similar experiment using a n-type GaAs substrate for comparison. Comparing to the case of p-type substrate, the number of the micro-crystals reduced drastically either for the samples with and without light illumination after long term storage. This result suggests that the formation of the crystals can be affected by the type of majority carrier of the surface GaAs layer.

Next, to identify the chemical composition of the deposited crystals, SEM-EDX analysis was carried out. Figure 4a shows the energy dispersive spectra of the micro-crystal. As shown in Fig. 4a, only As and O peaks were detected, while the peaks of Ga, Al (AlAs release layer) and F (hydrofluoric acid) were not observed. The relative peak intensities and sensitivity factors of elements suggest that the micro-crystal is most likely to be As_2O_3. To confirm this result, the energy dispersive spectra for the substrate surface not covered by the crystals was also measured (Fig. 4b). The results showed only Ga and As peaks, and the As composition was higher than stoichiometry of GaAs by approximately 10%. Therefore, this

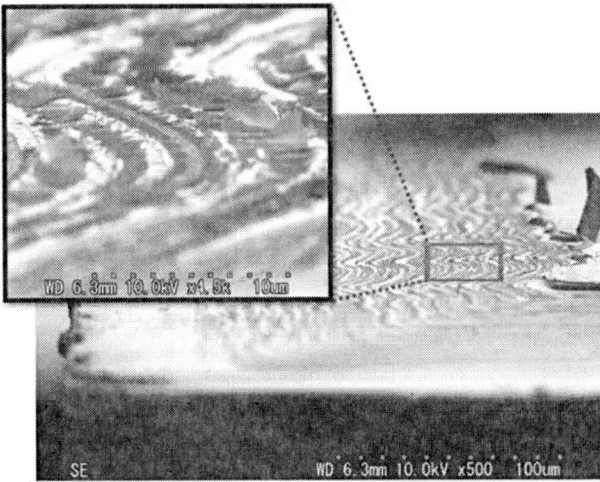

Fig. 5. SEM images of the GaAs substrate surface after annealing process (200 °C for 15 minutes).

result indicates that the pure GaAs surface or the thin elemental As layer on the GaAs substrate is present.

Finally, the effect of the annealing process on GaAs surface cleaning was investigated. By annealing at 200 °C for 15 minutes under nitrogen ambient, all micro-crystals were removed successfully. This result is consistent with the previous research [8]. However, as shown in Fig. 5, the streaky pattern of the deposition still remained on the surface. This indicates that a higher temperature may be required to remove the depositions completely. By annealing at 500 °C for 60 minutes, the amount of the depositions was reduced, but the removal was still incomplete.

IV. SUMMARY

In this work, deposition formation after the ELO and the removal process of the micro-crystals were studied. It was found that the Au metal layer on the device surface accelerated the ELO process through the application of stress and the resultant extension of the gap between the device layer and the substrate. After the ELO process, specific streaky pattern appeared on the surface, and the depositions seemed to re-form gradually into micrometer-size crystals during long term storage. Furthermore, formation of the crystals was accelerated by the light illumination. The micro-crystals were revealed to be As_2O_3 from the SEM-EDX analyses. We also found that the deposited crystals can be removed by the annealing process in nitrogen ambient, although complete removal of the streaky deposit would require longer annealing at higher temperature.

ACKNOWLEDGEMENT

A part of this study was supported by NEDO project: the research and development of ultra-high efficiency and low-cost III-V compound semiconductor solar cell modules. And we also acknowledge the Advanced Characterization Nanotechnology Platform of the University of Tokyo, supported by "Nanotechnology Platform" of the Ministry of Education, Culture, Sports, Science and Technology (MEXT), Japan.

REFERENCES

[1] M. A. Green, K. Emery, Y. Hishikawa, W. Warta, E. D. Dunlop, D. H. Levi, and A. W. Y. Ho-Baillie, "Solar cell efficiency tables (version 49)," *Progress in Photovoltaics*, vol. 25, pp. 3-13, 2017.

[2] M. Konagai, M. Sugimoto, and K. Takahashi, " High efficiency GaAs thin film solar cells by peeled film technology," *Journal of Crystal Growth*, vol. 45, pp. 277-280, 1978.

[3] M. M. A. J. Voncken, J. J. Schermer, G. J. Bauhis, P. Mulder, and P. K. Larsen, "Multiple release layer study of the intrinsic lateral etch rate of the epitaxial lift-off process," *Applied Physics A: Materials Science and Processing*, vol. 79, pp. 1801-1807, 2004.

[4] N.J. Smeenk, J. Engel, P. Mulder, G. J. Bauhis, G. M. M. W. Bissels, J. J. Schermer, E. Vlieg, and J. J. Kelly, "Arsenic formation on GaAs during etching in HF solutions: relevance for the epitaxial lift-off process," *ECS Journal of Solid State and Technology*, vol. 2, pp. 58-65, 2013.

[5] J. J. Schermer, P. Mulder, G. J. Bauhuis, M. M. A. J. Voncken, J. V. Deelen, E. Haverkamp, and P. K. Larsen, "Epitaxial Lift-Off for large area thin film III/V devices," *Physics Status Solidi A-Applications and Materials Science*, vol. 202, pp. 501-508, 2005.

[6] C. Cheng, K. Shiu, N. Li, S. Han, L. Shi and D. K. Sadana, "Epitaxial lift-off process for gallium arsenide substrate reuse and flexible electronics," *Nature Communications*, vol. 4, 1577, 2013.

[7] Y. A. Bioud, A. Boucherif, A. Balarouci, E. Paradis, D. Drouin, and R. Ares, "Chemical Composition of Nanoporous Layer Formed by Electrochemical Etching of p-Type GaAs," *Nanoscale Research Letter*, vol. 11, 2016.

[8] U. Resch, N. Esser, Y. S. Raptis, W. Richter, J. Wasserfall, A. Förster, and D. I. Westwood, "Arsenic passivation of MBE grown GaAs (100): structural and electronic properties of the decapped surfaces," *Surface Science*, vol. 269, pp. 797-803, 1992.

Ultrathin Silicon-on-Insulator (SOI) wafer for compliant substrate

Shinyoung Noh, Anita Ho-Baillie, Stephen Bremner, Martin A. Green, and Xiaojing Hao

University of New South Wales, Sydney, New South Wales, 2033, Australia

Abstract — **Compliant substrate is one of the most promising approaches for heteroepitaxy of the semiconductor devices. SOI wafer is the only commercial structure used for the compliant substrate researches. Yet, previous studies reported ineffectiveness of SOI as a compliant substrate for thickness regions above 10 nm. This was due to the interference of BOX layer. This research explored compliance of SOI thickness below 10 nm. 300 nm GaAs film was deposited by MBE on 4nm SOI and bulk Si wafer. Comparison showed reduced threading dislocation density of GaAs on ultrathin SOI and subsequently larger columnar grains.**

I. INTRODUCTION

Thin compliant substrate has been a promising approach for epitaxial growth of lattice-mismatches materials on a substrate. The concept of compliant substrate is on infinitely continuing pseudo-morphic layer which appears at the initial stage of epitaxial deposition methods such as molecular beam epitaxy(MBE). In a pseudo-morphic layer, deposited film is elastically stretched to accommodate lattice mismatch. Such structure is only maintained until a certain thickness which mostly ranges within a few nanometers. Once the deposited thickness reaches the critical thickness, misfit dislocations are formed and continued as threading dislocations. Lo first suggested that effective critical thickness(t_{eff}) of pseudomorphic growth can increase with substrate thickness(h_b) reduction and the pseudomorphic growth can continue infinitely if the substrate thickness reaches a thickness thinner than the critical thickness(t_c) [1]. Lo calculated their relations as equation (1)

$$\frac{1}{t_{eff}} \equiv \frac{1}{t_c} - \frac{1}{h_b} \qquad (1)$$

Extensive researches have been carried on in various types of surface patterns [2-4], twist-bonded wafer [5], and SOI wafer [6, 7]. Among the various types of experimental structures for compliant substrate, Silicon-on-Insulator (SOI) is the most industrial, yet, controversial structure as a compliant substrate [8, 9]. The controversy arises from the interference of Buried Oxide (BOX) underneath Si layer [8]. BOX reduces the effective critical thickness of deposited film and it was anticipated that SOI is unlikely to serve as compliant substrate. Rehder et al and David et al reported that ineffective compliance of SOI substrate for SiGe deposition [7, 8]. David et al tested compliance of SOI, by comparing films deposited with bulk Si, at the thinnest Si thickness, 12 nm [7].

Critical thicknesses for most of the device applications locate within 10 nm. t_c for 4% mismatched GaAs on Si wafer is 1.4 nm. Moreover, Ayers calculated that free-standing layer of 10 nm or below are expected to have relative compliance which should suffice for most of the device applications [9]. Therefore, compliance of SOI remains as a question and the compliance of SOI at the thickness region below 10 nm remains unexplored. We report a comparison of GaAs films deposited on 4 nm SOI and bulk Si wafer.

II. EXPERIMENTAL

Smartcut SOI wafers were purchased from Soitec France. Initial SOI wafer consisted of 10 Ω·cm B-doped 55nm Si device layer, 145 nm SiO2 layer (BOX) and 10 Ω·cm B-doped 730um Si handle wafer. After RCA clean, the wafer was thinned to 4 nm by thermal oxidation at 800˚C, HF dip, 5 h thermal oxidation at 900˚C, and repetition of HF dip and RCA cleaning until device Si layer thickness is measured to be 4 nm on ellipsometry. The thinning process is described in Fig 1. Reference bulk Si sample was prepared by removal of device Si layer and BOX using HNA for the consistency of Si profile. Two samples, 4 nm and bulk Si, were loaded into MBE after RCA clean and HF dip. In MBE chamber, Samples were deoxidized at 770˚C, and then 300 nm GaAs film was deposited after 10 MEE on Si surface.

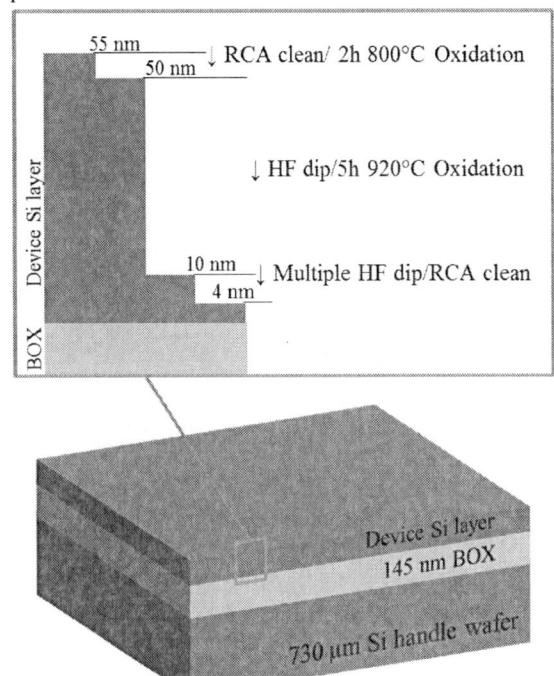

III. RESULTS AND DISCUSSION

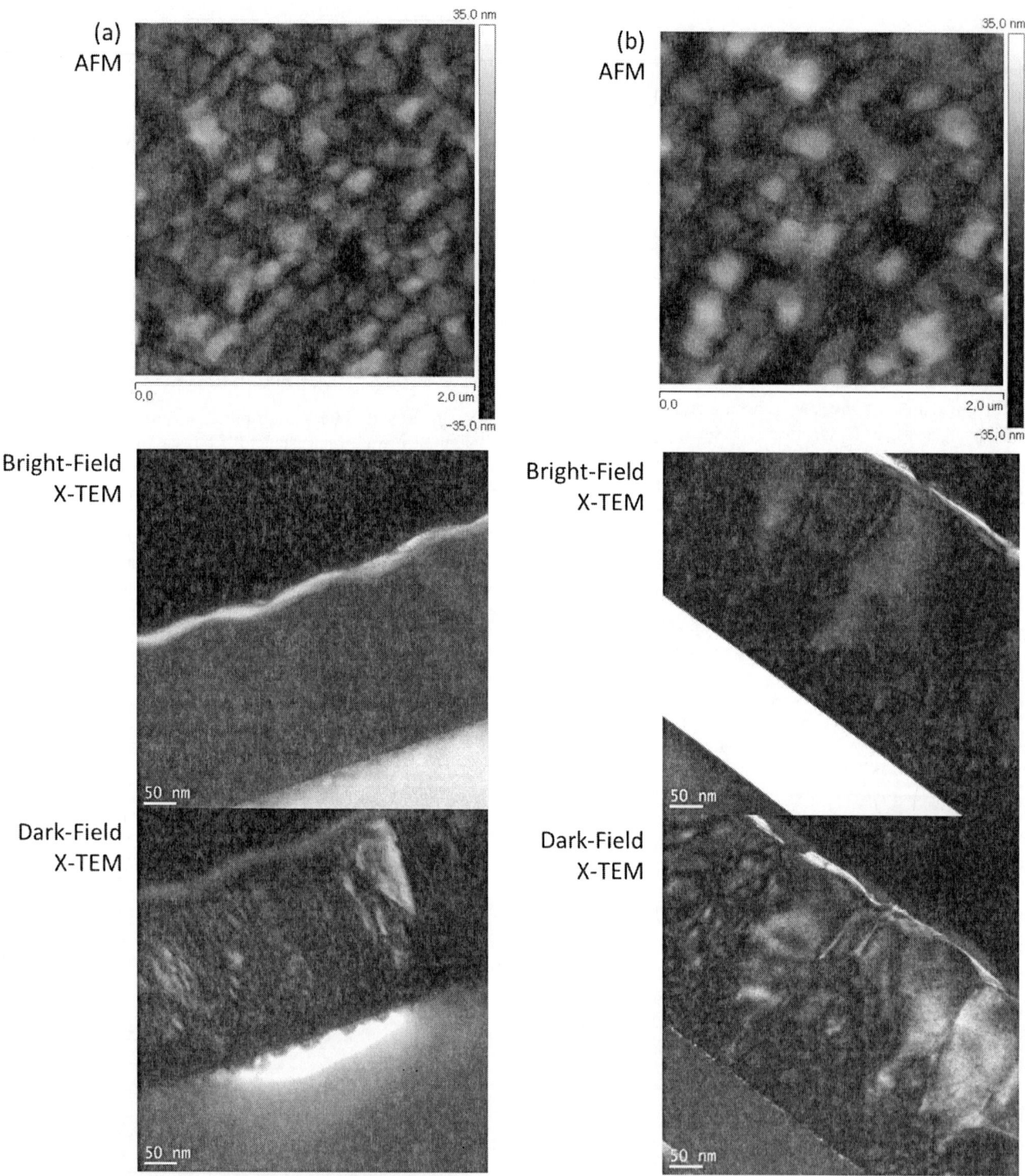

Fig. 2. AFM and TEM of deposited GaAs film on (a) bulk Si, (b) 4 nm SOI

Deposited GaAs film achieved higher quality. AFM on GaAs surface showed larger grains on 4 nm SOI. This can be seen in TEM images that larger grains are due to lower threading dislocation density on 4 nm SOI. Compliant substrate was not observed as high density of misfit dislocations were observed in both samples. Whether GaAs film on SOI had lower number of misfit dislocation formation is not clearly observed. However, it is evident that smaller number of defects reached the surface on 4 nm SOI which indicates that SOI provided compliance to the deposition.

IV. SUMMARY

We studied whether SOI at ultrathin thickness can provide compliance for heteroepitaxy of GaAs on Si. 300 nm GaAs was deposited on 4nm and Bulk Si wafer. It was observed on TEM and AFM that lesser number of misfit dislocations continued as threading dislocations and subsequently gained larger grains on 4 nm SOI.

REFERENCES

[1] Y. H. Lo, "New approach to grow pseudomorphic structures over the critical thickness", *Applied Physics Letters*, vol. 59, pp. 2311-2313, 1991.

[2] C. L. Chua, W. Y. Hsu, C. H. Lin, G. Christenson, and Y. H. Lo, "Overcoming the pseudomorphic critical thickness limit using compliant substrates", *Applied Physics Letters*, vol. 64, pp. 3640-3642, 1994.

[3] C. Carter-Coman, A. S. Brown, N. M. Jokerst, D. E. Dawson, R. Bicknell-Tassius, Z. C. Feng, K. C. Rajkumar, and G. Dagnall, "Strain Accommodation in Mismatched Layers by Molecular Beam Epitaxy: Introduction of a New Compliant Substrate Technology", *Joural of Electronic Materials*, vol. 25, pp. 1044-1048, 1996.

[4] A. M. Jones, J. L. Jewell, J. C. Mabon, E. E. Reuter, S. G. Bishop, S. D. Roh, and J. J. Coleman, "Long-wavelength InGaAs quantum wells grown without strain-induced warping on InGaAs compliant membranes above a GaAs substrate", *Applied Physics Letters*, vol. 74, pp. 1000-1002, 1999.

[5] F. E. Ejeckam, M. L. Seaford, Y. H. Lo, H. Q. Hou, and B. E. Hammons, "Dislocation-free InSb grown on GaAs compliant universal substrates", *Applied Physics Letters*, vol. 71, pp. 776-778, 1997.

[6] A. R. Powell, S. S. Iyer, and F. K. LeGoues, "New approach to the growth of low dislocation relaxed SiGe material", *Applied Physics Letters*, vol. 64, pp. 1856-1858, 1994.

[7] T. David, K. Liu, S. Fernandez, MI. Richard, A. Ronda, L. Favre, M. Abbarchi, A. Benkouider, JN. Aqua, M. Peters, P. Voorhees, O. Thomas, and I. Berbezier, "Remarkable Strength Characteristics of Defect-Free SiGe/Si Heterostructures Obtained by Ge Condensation", *Journal of Physical Chemistry C*, vol. 120, pp. 20333-20340, 2016.

[8] E. M. Rehder, C. K. Inoki, T. S. Kuan, and T. F. Kuech, "SiGe relaxation on silicon-on-insulator substrates: An experimental and modeling study", *Journal of Applied Physics*, vol. 94, pp. 7892-7903, 2003.

[9] J. E. Ayers, "Compliant Substrates for Heteroepitaxial Semiconductor Devices: Theory, Experiment, and Current Directions", *Journal of Electronic Materials*, vol. 37, pp. 1511-1523, 2008.

Characterization of GaAs solar cells grown by hydride vapor phase epitaxy in horizontal reactor

Ryuji Oshima[1], Kikuo Makita[1], Akinori Ubukata[2], and Takeyoshi Sugaya[1]

[1] Research Center for Photovoltaics, National Institute of Advanced Industrial Science and Technology (AIST), 1-1-1 Umezono, Tsukuba, Ibaraki 305-8568, Japan

[2] Taiyo Nippon Sanso Corporation, 10 Okubo, Tsukuba, Ibaraki 300-2611, Japan

Abstract — The use of highly efficient III-V multijunction devices is limited to space high-concentration terrestrial applications owing to their extremely high manufacturing cost. In this study, we characterized GaAs solar cells grown by hydride vapor-phase epitaxy (H-VPE) in a horizontal reactor, in order to develop a low-cost, high-throughput growth technique. Secondary ion mass spectroscopy showed that the doping transition zone during H-VPE growth was narrow at 40 nm. The characterization of the thickness distribution of 5-μm-thick GaAs films grown on 2-inch GaAs wafers showed that the distribution could be improved from 30.1% to 2.5% by modifying the gas flow pattern and by rotating the substrate. The GaAs solar cell grown with a InGaP window layer without using an antireflection coating exhibited short-circuit current density, open-circuit voltage, fill factor, and efficiency values of 16.2 mA/cm², 0.91 V, 85.93%, and 12.67%, respectively.

Index Terms - Gallium Arsenide, Hydride Vapor-Phase Epitaxy, Photovoltaics

I. INTRODUCTION

The efficiency of III–V multijunction solar cells has increased with the optimization of the device design [1,2]. However, to date, the use of highly efficient III-V multijunction devices has been limited to space and high-concentration terrestrial systems owing to their extremely high manufacturing cost when industry-standard metal-organic vapor-phase epitaxy (MOVPE) techniques are employed. The high cost of MOVPE is primarily owing to the use of expensive metal-organic sources for the group III precursors and the low throughput of the common MOVPE technique because of its low growth rate, which is of the order of several microns per hour. The long-term target of our group is to develop low-cost, high-throughput growth techniques for the implementation of large-scale terrestrial III-V/Si modules [3].

Hydride vapor phase epitaxy (H-VPE), in contrast to MOVPE, has a significantly higher growth rate of several hundred microns per hour. A higher growth rate can result in substantial cost reduction, as it results in a higher reactor throughput. In addition, H-VPE uses relatively cheaper group III metals, and the growth process typically involves a lower group V overpressure compared to that in the case of MOVPE [4,5]. H-VPE has been shown to be suitable for fabricating a variety of junction devices [6-8] as well as for producing bulk GaN templates [9]. Recently, Simon et al. developed vertical H-VPE systems for solar cell production and successfully fabricated highly efficient GaAs solar cells with a high growth

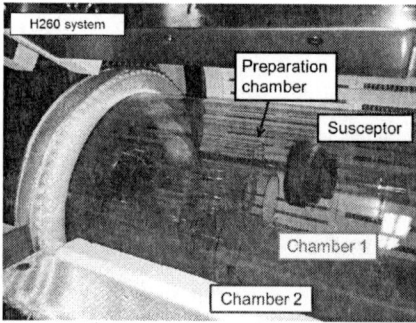

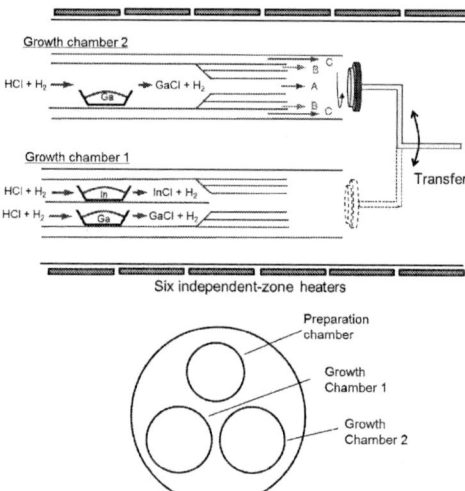

Fig. 1. (a) Photograph of custom-built quartz H-VPE reactor used in this work and (b) schematic of growth chambers in reactor.

rate (60 μm/h) [10-12]. In this study, we developed a horizontal H-VPE system for solar cell fabrication. Further, we evaluated the effects of the homogeneity of the gas flow pattern on the growth surface on the thickness and performance of the thus-fabricated GaAs solar cells.

II. EXPERIMENTS

All the samples were grown on just-oriented GaAs(001) substrates in a custom-built, horizontal hot-wall reactor (Taiyo Nippon Sanso, H260) at atmospheric pressure. The quartz reactor tube was designed to have three chambers, namely, two growth chambers and a preparation chamber (see Fig. 1(a)). The substrate was placed vertically in the growth chamber using a quartz mechanical transfer arm. The presence of the two growth chambers with different growth environments allowed for the continuous growth of multiple layers without interruption (see Fig. 1(b)). The source and substrate regions were heated using a six-zone clamshell resistance furnace. The temperature of the source region was set to 850 °C. Gaseous HCl in H_2 as the carrier was introduced through port A to react with the liquid Ga and In contained in the quartz boats, resulting in the formation of GaCl and InCl, which were used as the group III precursors. AsH_3 and PH_3, used as the group V precursors, and the dopants (DEZn and H_2S) were made to flow through port B while using H_2 as the carrier. Port C was used to form a curtain of H_2 gas at the highest flow rate (10 SLM), in order to prevent the contamination of the reactants present in the two growth chambers.

After the substrate had been loaded into the reactor, it was thermally cleaned in an AsH_3 flow in the preparation chamber. Then, multiple layers of epitaxial films were grown on it using the two growth chambers. The temperature of the substrate region was 710 °C for the GaAs samples. The growth rates for GaAs were 10 μm/h for an HCl(Ga) flow rate of 5.0 sccm and 30 μm/h for a flow rate of 20 sccm, where HCl(X) indicates the flow rate of HCl over the group III elemental metal. The V/III ratio was 5.0. For the growth of the samples with an InGaP layer, the temperature was lowered to 680 °C, because the incorporation efficiency of In is lower than that of Ga with respect to the growth of InGaP [10]. We could successfully grow lattice-matched InGaP films using an HCl(Ga) flow rate of 1.5 sccm and an HCl(In) flow rate of 20 sccm; the resulting growth rate was 12 μm/h.

To characterize the structures of the films grown by H-VPE, secondary ion mass spectroscopy (SIMS) was performed, in order to determine the impurity levels in the films. The thicknesses of the deposited films were determined by using an optical microscope to analyze the edges of the GaAs samples after they had been etched in a $HF/H_2O_2/H_2O$ solution [13].

The structures of the solar cells fabricated in this study are shown in Fig. 2. After the H-VPE growth process, a front AuGeNi/Au finger was formed and mesa isolation was applied

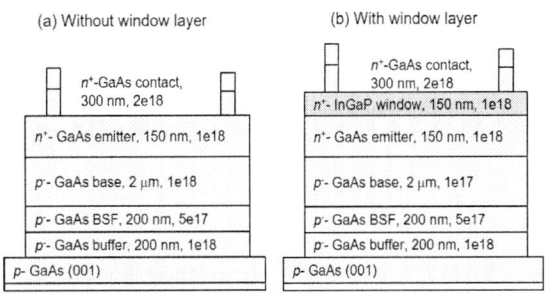

Fig. 2. Schematics of structures of GaAs solar cells fabricated (a) without and (b) with an InGaP window layer.

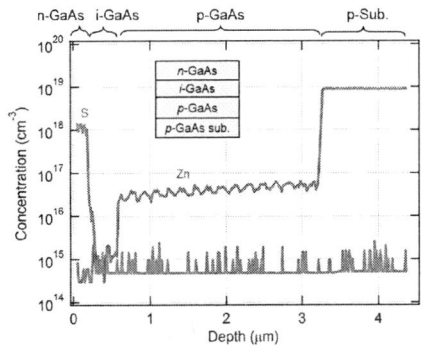

Fig. 3. SIMS measurements of *p-i-n* GaAs structure grown by H-VPE; the Zn and S doping profiles are shown.

using a standard photolithography system and an electron beam evaporator. Ti/Au was used as the back electrode. No antireflection coating (ARC) was applied. The cell size was 0.08 cm^2. To characterize the solar cells, external quantum efficiency (QE) measurements were performed under chopped monochromatic light at a constant photon flux of 1×10^{14} cm^{-2}. The current density–voltage (*J–V*) curves were measured using air mass 1.5 global (AM1.5G) illumination at a concentration of 1 sun.

III. RESULTS AND DISCUSSION

First, we analyzed the doping transition based on the doping profiles for the dopants Zn and S in a *p-i-n* GaAs homojunction structure as measured by SIMS; the results are shown in Fig. 3. The sample was fabricated using an HCl(Ga) flow rate of 5 sccm. The SIMS results did not provide any evidence of the interdiffusion of the dopants during the high-temperature H-VPE growth process. In addition, the doping transition zone of each dopant was found to be as narrow as 40 nm, which is similar to that reported in Ref. 11. In a separate experiment performed to measure the background carrier density from the electrochemical capacitance-voltage, the undoped GaAs film was determined to be an n-type one with a carrier density of 1.4×10^{16} cm^{-3} (not shown). This was in good agreement with the values corresponding to the density wherein Si atoms were incorporated in the GaAs films grown by H-VPE; this can be attributed to the reaction of the HCl gas with the quartz tube. Thus, these results indicated that the p-i-n junction was formed abruptly during the low-cost H-VPE growth process. However, small oscillation fringes observed in Zn dopant were due to the growth under after-mentioned inhomogeneous gas flow pattern with substrate rotation.

Figure 4 compares the thickness distributions of 5-µm-thick GaAs films grown on 2-inch wafers using the (a) conventional and (b) modified gas flow patterns. Figure 4 (c) shows that of the films grown under the modified gas flow pattern with substrate rotation at the rate of one rotation per minute. All the films were grown under nearly identical conditions and using an HCl(Ga) flow rate of 20 sccm. The thickness distribution ($\Delta t/t$) was found to be as high as 30.1% in the case of (a), which is much larger than that seen typically for films grown by MOVPE. In addition, the thickness was greater towards the lower end, suggesting that the reactant concentration was higher in the lower region of the growth chamber, owing to the gas flow being inhomogeneous because of the unoptimized reactor design and the specific gravity of the various gas species. As can be seen from Fig. 4(b), by modifying the shape of the nozzle of the growth chamber, $\Delta t/t$ could be reduced significantly to 3.2%, owing to the improved uniformity of the gas flow pattern. Furthermore, as can be seen from Fig. 4(c), it could be improved further, to 2.5%, by rotating the substrate during growth.

Figure 5 shows the distributions of the (a) short-circuit current density (J_{SC}), (b) open-circuit voltage (V_{OC}), (c) fill factor (FF), and (d) efficiency of the GaAs solar cells obtained from the J-V curves of the cells grown on 2-inch wafers. The left side shows the data for the cells grown without an InGaP window layer under conventional growth conditions (denoted as left cells) while the right side shows the data for the cells grown with a window layer using a modified nozzle and substrate rotation (denoted as right cells). The GaAs layers were fabricated at an HCl(Ga) flow rate of 20 sccm. No ARC was applied, in order to prevent the effect of the thickness distribution of the ARC from affecting the cell performance. Figure 5(a) shows that the distribution of J_{SC} for the right cells is more homogeneous than that for the left cells. Because J_{SC} is greater near the lower end in the case of the left cells, this result can be attributed primarily to the thickness distribution. Overall, the J_{SC} value of the right cells was greater than that of the left cells because of the insertion of the InGaP window layer. Thus, the window layer effectively reduced the recombination rate at the cell surface. Further, V_{OC} and FF show trends similar to that observed in the case of J_{SC}. In particular, a large difference in the V_{OC} values (of 30 mV) was

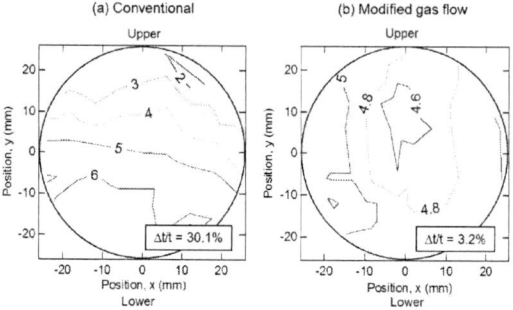

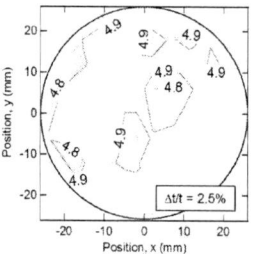

Fig. 4. Thickness distributions of GaAs films formed on 2-inch wafer under (a) conventional and (b) modified gas flow patterns. (c) Thickness distribution of film grown under modified gas flow pattern with substrate rotation.

observed in the case of the left cells. V_{OC} is generally determined using the following equation [14]:

$$V_{OC} = \frac{nkT}{q} \ln\left(\frac{J_{SC}}{J_0} + 1\right) \qquad (1)$$

where n, k, T, q, and J_0 are the diode factor, Boltzmann constant, measurement temperature, elementary charge, and reverse saturation current density, respectively. In a separate experiment performed to measure J_0 from the dark I-V characteristics, an J_0 of 5.0×10^{-11} mA/cm^2 was obtained in the upper position (open circle in Fig. 5(b)); this value is much larger than that (2.0×10^{-11} mA/cm^2) at the lower position (closed circle). Therefore, in addition to the thickness, the V/III ratio also showed variations during growth when unoptimized equipment was used, resulting in differences in the quality of the films grown on the wafers. Thus, improving the distribution of gas flow pattern was found to be crucial for increasing the performance of the H-VPE-grown III-V cells. The J_{SC}, V_{OC}, FF, and efficiency values of the best cell were 11.6 mA/cm^2, 0.90 V, 81.72%, and 8.53%, respectively, for the left cells, and 16.2 mA/cm^2, 0.91 V, 85.93%, and 12.67%, respectively, for the right cells. The relatively higher V_{OC} of 0.90 V was close to that of the GaAs solar cells grown by H-VPE by Simon et al. [10].

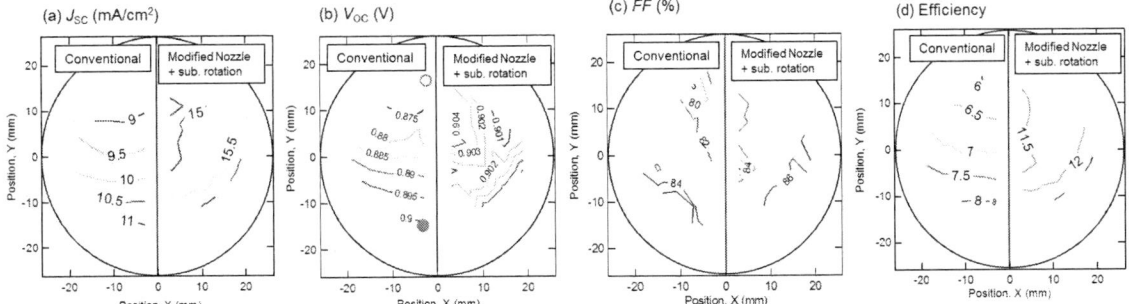

Fig. 5. Distributions of (a) J_{SC}, (b) V_{OC}, (c) FF, and (d) efficiency of GaAs solar cells obtained from J-V curves corresponding to cells grown on 2-inch wafers. Left side shows data for cells grown without InGaP window layer under conventional growth conditions while right side shows data for cells grown with window layer using modified nozzle and substrate rotation.

IV. CONCLUSIONS

We characterized GaAs films grown using a custom-built horizontal H-VPE system. The doping transitions zone for the dopants Zn and S were found to be as narrow as 40 nm, indicating that a *p-i-n* junction was formed abruptly during the low-cost H-VPE process. The thickness uniformity of the grown films was significantly affected by the design of the equipment used and could be improved from 30.1% to 2.5% by modifying the nozzle as well as by making the substrate rotate during the growth process. This improvement in the thickness distribution was found to be crucial for improving the performance of the H-VPE-grown GaAs cells. The cell efficiency was improved from 8.53% to 12.67% when a InGaP window layer was formed in the GaAs solar cells. This study is a major step toward the development of highly efficient III-V solar cells grown by the low-cost H-VPE technique.

REFERENCES

[1] S. P. Philipps, G. Peharz, R. Hoheisel, T. Homunf, N. M. Al-Abbadi, F. Dimroth, A.W. Bett, "Energy harvesting efficiency of III–V triple-junction concentrator solar cells under realistic spectral conditions" *Sol. Energ. Mat. Sol. Cells* , vol. 94, p. 869, 2010.

[2] D.C. Law, R.R. King, H. Yoon, A. Boca, C.M. Fetzer, S. Mesropian, T. Isshiki, M. Haddad, K.M. Edmondson, D. Bhusari, J. Yen, R.A, Sherif, H.A. Atwater, N.H. Karam, "Future technology pathways of terrestrial III-V multijunction solar cells for concentrator photovoltaic systems" *Sol. Energ. Mat. Sol. Cells*, vol. 94 p. 1314, 2010.

[3] K. Makita, H. Mizuno, R. Oshima, T. Tayagaki, M. Baba, N. Yamada, H. Takato, T. Sugaya, "Low concentration InGaP/GaAs/Si 3-junction solar cells with smart stack technology" *26th edition of the International Photovoltaic Science and Engineering Conference*, 1.2.3b, Singapore, 2016.

[4] R. Oshima, R. M. France, J. F. Geisz, A. G. Norman, M. A. Steiner, "Growth of lattice-matched GaInAsP grown on vicinal GaAs(001) substrates within the miscibility gap for solar cells" *J. Crystal Growth*, vol. 458, p. 1 2017.

[5] I. Garcia, I. Rey-Stolle, C. Algora, W. Stolz, K. Volz, "Influence of GaInP ordering on the electronic quality of concentrator solar cells" *J. Crystal Growth*, vol. 310, p. 5209, 2008.

[6] K. Makita, A. Gomyo, K. Taguchi, T. Suzuki, "Photoluminescence study of InGaAs grown on InP by hydride vapor phase epitaxy-Effect of O_2 injection and substrate orientation" *Appl. Phys. Lett.*, vol. 46 p. 1069, 1985.

[7] A. Usui, T. Matsumoto, M. Inai, "Room temperature cw operation of visible InGaAsP double heterostructure laser at 671 nm grown by hydride VPE" *Jpn. J. Appl. Phys.*, vol. 24 p. L163, 1985.

[8] T. Mizutani, M. Yoshida, A. Usui, H. Watanabe, T. Yuasa, I. Hayashi, "Vapor phase epitaxy of InGaAsP/InP DH structure by the dual-growth-chamber method" *Jpn. J. Appl. Phys.* vol. 19 p. L113, 1980.

[9] K. Koyama, H. Aida, S.-W. Kim, K. Ikejiri, T. Doi, T. Yamazaki, "Growth of thick GaN layers on laser-processed sapphire substrate by hydride vapor phase epitaxy" *J. Crystal Growth*, vol. 403, p. 38, 2014.

[10] J. Simon, K. L. Schulte, D. L. Young, N. M. Haegel, A. J. Ptak, "GaAs solar cells grown by hydride vapor-phase epitaxy and the development of GaInP cladding Layers" *IEEE J. Photovolt.*, vol. 6, p. 191, 2016.

[11] K. L. Schulte, W. L. Rance, R. C. Reedy, A. J. Ptak, D. L. Young, T. F. Keuch, "Controlled formation of GaAs pn junctions during hydride vapor phase epitaxy of GaAs" *J. Crystal Growth*, vol. 352, p. 253, 2012.

[12] J. Simon, K. L. Schulte, N. Jain, S. Johnson, M. Young, M. R. Young, D. L. Young, A. J. Ptak, "Upright and

Inverted Single-Junction GaAs Solar Cells Grown by Hydride Vapor Phase Epitaxy" *IEEE J. Photovolt.*, vol. 7, p. 157, 2017.

[13] A. R. Clawson, "Guide to references on III-V semiconductor chemical etching" *Mat. Sci. Eng. R*, vol. 31, p. 1, 2001.

[14] J. Nelson, "The Physics of Solar Cells" Imperial College Press, 2003.

Flexible GaAs Single-Junction Solar Cells Based on Single-Crystal-Like Thin-Film Materials Directly Grown on Metal Tapes

Sara Pouladi, Monika Rathi, Mojtaba Asadirad, Pavel Dutta, Seung Kyu Oh, Devendra Khatiwada, Shahab Shervin, Yao Yao, Venkat Selvamanickam, and Jae-Hyun Ryou

University of Houston, Houston, TX 77204-4006, USA

Abstract — **We demonstrate flexible single-junction III-V solar cells based on single-crystal-like GaAs thin films directly grown on a metal tape. The epitaxial (Al)GaAs thin-film structures are developed on flexible polycrystalline Hastelloy tapes using crystallinity-transformational buffer layers for photovoltaic devices with front illumination geometry. The GaAs solar cell shows promising photovoltaic characteristics under AM 1.5G 1-Sun illumination. An open-circuit voltage of 330mV, a short circuit current of 21mA/cm², and a fill factor of 73 result in a conversion efficiency of ~5%. Further improvement in semiconductor thin films quality, device design and fabrication process optimization would improve the device performance characteristics. This new technology has the potential for next-generation low-cost and high efficiency flexible photovoltaic devices.**

I. INTRODUCTION

Increasing demands for cost-effective and environmentally friendly energy sources are a compelling driving force to develop new energy harvesting and conversion technologies for further improved conversion efficiency while reducing manufacturing costs of the energy devices and systems. This is also the case for the most efficient III-V semiconductor-based solar cells (SCs) in the photovoltaics (PV). Unique optoelectronic material characteristics of the III-V compound semiconductors, including a wide range of bandgap energies and high absorption coefficients that enable the complete absorption of photons within a few micron thickness of the material and nearly full coverage of the entire terrestrial solar spectrum, have made them the best material for SCs with the highest efficiency. These materials have been traditionally used for the space PV applications where low weight and high efficiency are the main concerns and their high material and fabrication costs are not considered as a limitation. The state-of-the-art SC conversion efficiencies higher than 45% were reported from the III-V materials which rely on the epitaxial growth on expensive single-crystal wafers. Developing a technology that can provide high-quality epitaxial materials without employ expensive wafer substrates would significantly reduce the manufacturing cost of the SCs. We proposed a new PV technology that combines the efficiency advantages of high-quality epitaxial III-V materials with the cost-effective roll-to-roll processing of thin films on low-cost

easily scalable metal foils [1]. The similar concept was applied to demonstrate high-performance flexible transistors [2],[3]. Recently, the epitaxial growth of single-crystal-like *p*- and *n*-(Al)GaAs thin films on metal tapes by newly-designed roll-to-roll metalorganic chemical vapor deposition (MOCVD) system have been demonstrated [4],[5]. This technology has the potential to offer the conversion efficiencies close to 20% for single-junction (SJ) GaAs SCs, based on a 2-dimensional (2D) numerical simulation. Previously, we reported on the successful conversion of the new technology concept into the demonstration of working SC devices [6]. However, the conversion efficiencies were significantly lower than the theoretically estimated values. In this paper, we report on the improved PV characteristics of the III-V thin film single-junction solar cells which are grown directly on a flexible polycrystalline metal tape through transitional buffer layers from poly- to single-crystal-like thin film. The fabricated devices show the photon conversion efficiency of ~5% under AM1.5G illumination at 1 Sun.

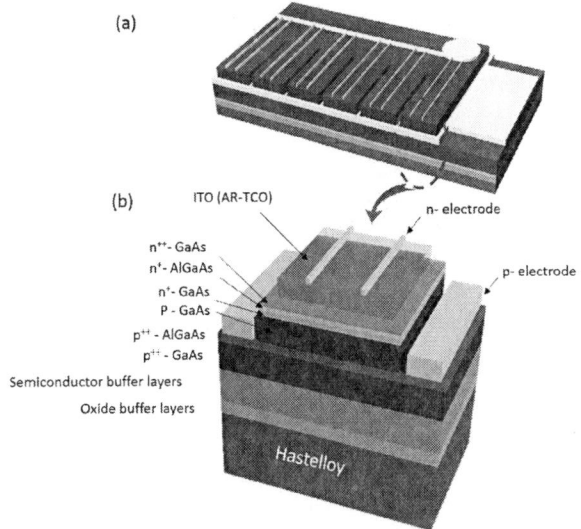

Fig. 1. (a) Schematic lateral device structure including interdigitated configuration of metal contacts. Lateral device geometry is used due to the presence of insulating buffer layers. (b) Schematic layer structure of a single-junction GaAs solar cell on a flexible substrate including oxide buffer and device layers.

II. DEVICE STRUCTURE MODELING AND DESIGN

Fig. 1 shows a schematic structure of the SJ GaAs SC device with a lateral geometry. Unlike the typical SCs on electrical conductive single-crystalline substrates with a vertical geometry, here a front contact for bottom electrode in addition to a top electrode is required due to the presence of resistive oxide buffer layers between metallic substrate and device structure. This lateral device geometry and different dynamics of charged carriers (electrons and holes) require a new epitaxial structure and device designs.

2D device simulation was used to define an optimized SC device structure based on evaluated threading dislocation density (TDD) and minority carrier lifetimes for electrons and holes of the GaAs on polycrystalline metal foil. The optimized layers' thickness and their doping concentrations based on the simulation results were used for the device structure growth, as shown in Table I.

TABLE I
DEVICE LAYER STRUCTURE

Layers	Thickness (nm)	Doping (cm^{-3})
n^{++}-GaAs	25	1×10^{19}
n^{+}-AlGaAs	35	1×10^{18}
n^{+}-GaAs	53	1×10^{18}
p-GaAs	850	1×10^{17}
p^{++}-AlGaAs	35	5×10^{18}
p^{++}-GaAs	1700	5×10^{18}

Sheet and contact resistances of contact layers plays a critical role in lateral device geometry due to longer and lateral travel distance of minority carriers before being collected by contact electrodes compared to vertical device geometry. linear grid spacing of metal electrodes for both bottom and front contact has to be determined considering the effect of contact shadowing and high sheet resistance of thin films.

III. DEVICE FABRICATION

Figure 2 shows the steps of device fabrication process. GaAs SC device structures were grown on flexible metal tapes (1.2 cm width). A series of buffer layers were employed to grow a biaxially textured Ge film on top of the polycrystalline substrate as a seed for GaAs epitaxial film growth. Fabrication of flexible SCs starts with mesa formation through a wet etch process of (Al)GaAs layers to define devices of 0.0016 to 0.5 cm^2. Ni/Zn/Au and Ni/Ge/Au electrodes for p-type and n-type ohmic contacts, respectively, were developed. For the device sizes of 0.1-0.5 cm^2, an additional step of ITO layer deposition was conducted before metallization of contacts to mitigate the effect of higher resistance. A thin layer of ~90 nm ITO was sputtered at room temperature. After lift-off process and photoresist removal, ITO annealed at ~400 °C for 20 min in

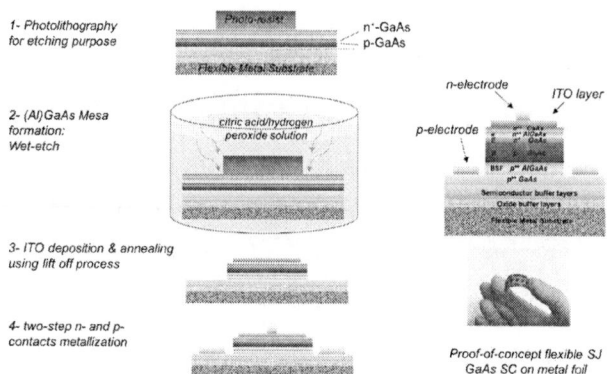

Fig. 2. Schematic illustration of device fabrication steps. The photograph of completed flexible SC device shown in this picture.

oven with oxygen to make a completely crystallized layer with transparency higher than 85%. This layer acts as both anti-reflection (AR) and transparent conductive electrode (TCE). In the presence of ITO layer, ohmic contact of Ti/Au was deposited on top of this TCE.

IV. RESULTS AND DISCUSSION

Fabricated flexible SJ GaAs thin-film SC devices were tested using a solar simulator. Figure 3 shows the J-V curve of the measured flexible SJ GaAs solar cell under AM1.5G illumination with a power density of 0.1 W/cm^2. The flexible GaAs SC shows an efficiency of ~5% with V_{oc} of ~0.33 V and J_{sc} of ~21 mA/cm^2. The values of contact resistance for n- and p-ohmic contacts were also calculated using TLM patterns and estimated to be 1.4×10^{-3} and 4×10^{-5} $\Omega \cdot$cm^2, respectively for devices without ITO TCE. Lower top contact resistance of 1.5×10^{-5} $\Omega \cdot$cm^2 was obtained for Ti/Au n-contact deposited on the ITO layer.

V_{oc} is lower than the simulated value for SCs on flexible substrate (~0.9 V) as shown in Fig. 4 and the value from comparable devices grown on GaAs wafers, which indicates

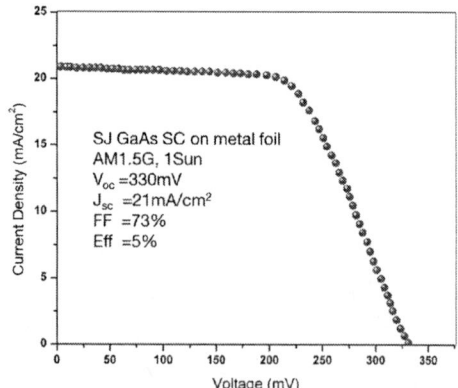

Fig. 3. J-V characteristics of the flexible SJ GaAs SC on a polycrystalline metal substrate.

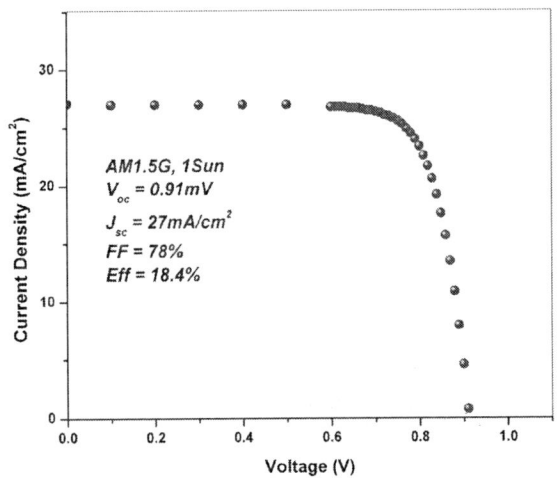

Fig. 4.   Simulated J-V characteristics of the SCs.

that the diffusion length (L_D) is shorter than required for efficient carrier collection. Lower V_{oc} could be related to bulk recombination of minority carriers as well as sidewall recombination (high surface recombination velocity particularly in lateral devices), which reduce the minority carrier lifetime and therefore diffusion length. Achieving higher material quality and applying sidewall passivation will improve the V_{oc}.

J_{sc} also needs to be further improved by optimizing doping level in the n$^+$-AlGaAs (window) and back contact layer (p^+-GaAs) and thickness of contact layers to reduce contact and sheet resistance. More control on mesa etch depth through chemical etching process can also improve J_{sc}, since the precise landing of p-electrode on relatively thick p^{++}-GaAs instead of thin p^{++}-AlGaAs is critical in reducing sheet resistance and increasing carrier injection. Electrode design also needs to be optimized more for better current collection particularly for these lateral devices.

The J_{sc} of devices with active areas larger than 0.1 cm^2 showed a drop from 21 to ~8-11 mA/cm^2. The series resistance can cause more drop in current collection compared to vertical devices. By applying an ITO layer below top contact electrode, improvement in the current collection of big devices has been observed (J_{sc} of ~15.5 mA/cm^2). Along with the optimizing the lateral devices, we are moving towards epitaxial back contact structure which will make developing vertical device geometry possible and would improve the SJ GaAs SC performance.

III. CONCLUSION

The proof-of-concept device demonstration showed conversion efficiency of ~5% under AM1.5G and 1sun condition. Here, V_{oc} of the device is lower than expected by

bulk and sidewall leakages originated from bulk material and un-passivated etched device areas which needs to be reduced to boost minority carriers' lifetime. The J_{sc} also needs to be improved by removing parasitic resistances in the device and applying more optimized fabrication process. Optimizing the SJ GaAs SC device structure growth and its fabrication process can lead to higher efficiencies and also extend it to a tandem III-V SC device for even higher efficiencies.

ACKNOWLEDGEMENT

This work is supported by funding from the U.S. Department of Energy (DOE) SunShot Initiative program (Award #: DEEE0006711), Advanced Superconductor Manufacturing Institute, and Texas Center for Superconductivity at the University of Houston.

The authors acknowledge the technical discussion with Dr. Ramesh Laghumavarapu at the University of California, Los Angeles.

REFERENCES

[1] V. Selvamanickam, S. Sambandam, A. Sundaram, S. Lee, A. Rar, X. Xiong, A. Alemu, C. Boney, and A. Freundlich, "Germanium films with strong in- plane and out-of-plane texture on flexible, randomly textured metal substrates," *J. Crystal Growth*, vol. 311, pp. 4553–4557, 2009.

[2] M. Asadirad, Y. Gao, P. Dutta, Y. Yao, S. Shervin, S. Sun, S. Ravipati, S.-H. Kim, K. H. Lee, V. Selvamanickam, and J.-H. Ryou., "High-performance flexible thin-film transistors based on single-crystal-like germanium on glass," *Adv. Electron. Mater.*, vol. 2, pp. 1600041-1–7, 2016.

[3] Y. Gao, M. Asadirad, Y. Yao, P. Dutta, E. Galstyan, S. Shervin, K. H. Lee, S. Pouladi, S. Sun, Y. Li, M. Rathi, J.-H. Ryou, and V. Selvamanickam, "High-performance flexible thin-film transistors based on single-crystal-like silicon epitaxially grown on metal tape by roll-to-roll continuous deposition process," *ACS Appl. Mater. Interfaces*, vol. 8, pp. 29565–29572, 2016.

[4] P. Dutta, M. Rathi, N. Zheng, Y. Gao, Y. Yao, J. Martinez, P. Ahrenkiel, and V. Selvamanickam, "High mobility single-crystalline-like GaAs thin films on inexpensive flexible metal substrates by metal-organic chemical vapor deposition," *Appl. Phys. Lett.* vol. 105, pp. 092104-1–5, 2014.

[5] M. Rathi, P. Dutta, N. Zheng, Y. Yao, Y Gao, S. Sun, A. Khadimallah, M. Thomas, M. Asadirad, P. Ahrenkiel, J.-H. Ryou, and V. Selvamanickam, "AlGaAs/GaAs DH and InGaP/GaAs DH grown by MOCVD on flexible metal substrates," *Proc. 43rd IEEE Photovoltaic Specialist Conference*, pp. 1926–1928, 2016.

[6] M. Asadirad, M. Rathi, S. Pouladi, Y. Yao, P. Dutta, S. Shervin, K. H. Lee, N. Zheng, P. Ahrenkiel, V. Selvamanickam, and J.-H. Ryou, "III-V thin film photovoltaic solar cells based on single-crystal-like GaAs grown on flexible metal tapes," *Proc. 43rd IEEE Photovoltaic Specialist Conference*, pp. 1954–1956, 2016.

978-1-5090-5606-4/17 $31.00 © 2017 IEEE

Reduced Defect Density in Single-Crystalline-like GaAs thin Film on Flexible Metal Substrates by using Superlattice Structures

M. Rathi[1], P. Dutta[1], D. Khatiwada[1], N. Zheng[2], Y. Yao[1], Y. Gao[1], S. Sun[1], Y. Li[1], S. Pouladi[1], P. Ahrenkiel[2], S. Reed[1], A. Khadimallah[1], J. Ryou[1] and V. Selvamanickam[1]

[1]Department of Mechanical Engineering, Advanced Superconductor Manufacturing Institute, Texas Center for Superconductivity, University of Houston, Houston, TX 77059, USA
[2]South Dakota School of Mines and Technology, Rapid City, SD 57701, USA

Abstract — High quality epitaxial and single-crystalline-like GaAs thin films grown by metal-organic chemical vapor deposition (MOCVD) on ion-beam textured buffers on flexible metal substrates have been reported earlier. Here we report significant reduction in defect density of the GaAs films by employing several techniques such as elevated temperature annealing/thermal cycling, low temperature GaAs growth and insertion of single $In_{1-x}Ga_xAs$, $In_{1-x}Ga_xAs/GaAs$ superlattice (SL) structures. The GaAs films were thoroughly investigated by transmission electron microscopy (TEM). Further, the flexible GaAs films were used to grow single-junction GaAs solar cells which exhibited improved device performances with efficiencies reaching 5 % at AM1.5. The results show promise for low-cost, flexible, lightweight and high-performance III-V multijunction solar cells which can be a potentially game-changing technology in the photovoltaic market.

Keywords: GaAs, Solar Cell, textured materials, MOCVD

I. INTRODUCTION

III-V photovoltaics by far have demonstrated highest efficiencies and are very popularly used in space application. But due to the overall cost (mainly from the expensive wafer substrates), its use in terrestrial application is limited. Several techniques to reduce the cost of III-V photovoltaics have been studied over the years, with lift-off technique to be the most popular one where the wafers are reused multiple times thereby reducing overall cost of the process. However, development of high quality III-V epitaxial layers directly on alternative substrates is desirable to achieve the maximum impact on cost-reduction. We are developing an innovative, non-mainstream approach to produce high-efficiency thin film III-V solar cells on low-cost, lightweight and flexible alternative substrates. Flexible substrate in addition allows large-area and high-volume roll-to-roll manufacturing.

Previously we have reported high quality, epitaxial, III-V thin films of GaAs, $Al_{1-x}Ga_xAs$, InP, $In_{1-x}Ga_xAs$, and $In_{1-x}Ga_xP$ grown by metal organic chemical vapor deposition (MOCVD) on flexible metal substrates using ion-beam textured epitaxial buffer templates [1-5]. The GaAs films grown exhibited single-crystalline-like structure, strong (00l) orientation, sharp in-and out-of-plane texture, low grain misorientation (< 2 degrees) and strong photoluminescence at room temperature. High carrier mobility and controllable p and n doping density were achieved in these flexible GaAs films [6]. Ion-beam assisted deposition

(IBAD) was used to achieve biaxially-textured epitaxial buffer layers on non-crystalline metal foils. However, the limitation was substantially higher defect densities in the GaAs layers which ranged from 1-5 $\times 10^9$ cm^{-2}. Therefore, it is critical to reduce the defect densities in order to make these films suitable for opto-electronic device applications.

This work describes the various defect mitigation techniques we have employed to reduce the defect densities in the flexible GaAs films. Our reference was the defect mitigation approaches which were previously employed in GaAs films grown on lattice-mismatched Si substrates [7]. Techniques for dislocation density reduction adopted in this work were classified into two methods 1) Insertion of strained buffer layers 2) Thermal processes. The flexibility and thermal expansion coefficient of the metal substrate and associated buffer layers added new unknowns and challenges to the defect-reduction processes. In this work, GaAs was grown on single-crystalline-like Ge films grown by sputtering on IBAD templates. [8] GaAs/Ge is a nearly lattice-matched system, but the underlying buffer layers consist of highly mismatched CeO_2/Ge interface where a high density of defects (misfit dislocations) originate and eventually propagate into the GaAs films. This leads to a substantially high defect density of $\sim 10^9$ cm^{-2}.

We investigated the effect of multiple defect reduction approaches and will present the effect of thermal annealing/cycling, low temperature GaAs growth as nucleation layer and insertion of single layer InGaAs and InGaAs/GaAs strained SL structures. Plan-view TEM imaging was used to estimate the defect densities. InGaAs/GaAs SLs with two-step GaAs growth and combined thermal annealing steps showed maximum reduction in defect density to the order OF $6\text{-}9 \times 10^7$ cm^{-2}. The optimum defect reduction technique was incorporated in the process and GaAs solar cell structures were fabricated. The flexible GaAs solar cells on metal substrates demonstrated efficiencies of about 5 % at AM 1.5.

II. EXPERIMENTAL DETAILS

GaAs and $In_{1-x}Ga_xAs$ thin films on flexible metal substrates were grown using MOCVD in a stationary reactor of a dual reactor MOCVD tool at University of Houston (UH). The tool has a traditional horizontal wafer-based quartz deposition chamber which accommodates 50 mm diameter wafers with

978-1-5090-5606-4/17 $31.00 © 2017 IEEE

load lock and glove box for sample transfer. Single-crystalline-like Ge templates were deposited in roll-to-roll RF sputtering system in meter lengths [8] and subsequently the tapes were cut into short pieces for III-V MOCVD process. Firstly, $In_{1-x}Ga_xAs$ films with varying "x" from 2 % to 14 % were optimized. Trimethyl Gallium (TMGa), Trimethyl Indium (TMIn), and Arsine (AsH_3) were used as Ga, In, and Arsine sources, respectively, with H_2 as the carrier gas purified by a palladium cell. The films were grown at temperature ranging from 600°C to 650°C, with pressure constant at 20 Torr. Scanning Electron Microscopy-Focused Ion Beam Imaging (SEM-FIB), photoluminescence spectroscopy and High Resolution X-ray Diffraction (HRXRD) data were used to optimize the morphology, thickness, and concentration of these films respectively. Once the individual elements of the SL structures were optimized, they were inserted in GaAs buffer for defect reduction. For all the samples, initially the Ge substrate was in-situ thermal annealed at 700-750°C for 10 mins to remove oxide layer, followed by lowering the temperature to 400°C to grow the first low temperature (LT) GaAs layer. This was followed by a standard temperature, second step of GaAs growth at 650°C, continued with a single strained $In_{0.11}Ga_{0.89}As$ layer. Further, another set of GaAs and $In_{0.07}Ga_{0.93}As$ was grown at 650°C followed by $In_{0.07}Ga_{0.93}As$/GaAs SL of 30 periods SL. This GaAs templates were then used to grow a complete 1J device structure consist of GaAs p-n junction, AlGaAs BSF, window and highly doped GaAs cap. Devices were fabricated on GaAs templates with and without the SL structure.

III. RESULTS AND DISCUSSION

Plan view TEM of selected GaAs thin films and the description of the samples is given in Figure 1 and table I respectively. As tabulated in Table I, S1-305 has100 nm single $In_xGa_{1-x}As$ layer (x=11%) sandwiched between GaAs, S1-311 has two sets of 20 periods $In_xGa_{1-x}As$/GaAs SL (x=11%) separated by 300 nm GaAs, S1-332 consists of 2 step GaAs growth with thermal cycling, single layer InGaAs, 40 periods of $In_xGa_{1-x}As$/GaAs SL. The defect density numbers are results of average defect density calculated from several images covering 40 to 50 um^2 area.

TABLE I
SUMMARY OF SAMPLES STUDIED WITH VARIOUS DEFECT
REDUCTION TECHNIQUES

Sample #	Defect reduction Scheme	LT-GaAs layer	Thermal cycling
S1-305	Single 100 nm InGaAs layer	No	No
S1-311	20 period SL/300 nm GaAs/20 period SL	No	Yes
S1-332	Single 100nm InGaAs/40 period SL	Yes	Yes

PV TEM images of the samples which underwent different defect reduction steps are shown in figure 2. The InGaAs/GaAs SL layers appeared to be highly effective to reduce the defects. A significant reduction of defect density, by six times was observed in S1-311. It was evident that the combination of thermal cycling, LT GaAs layer, single layer InGaAs and SL was effective in lowering the defect density. It has been reported and well documented that the strained SL and InGaAs layers can bend the threading dislocations parallel to the interface, preventing the dislocations from propagating to the surface [7]. The same phenomenon is expected to be responsible for defect reduction in this case too.

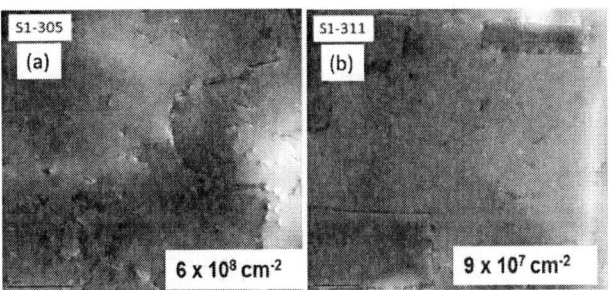

Fig. 1. PV-TEM of selected samples revealing defect densities

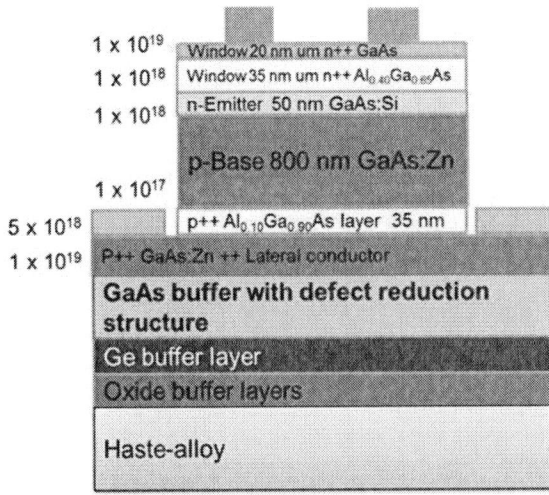

Fig. 2. Device structure grown on GaAs templates

A combination of defect reduction scheme, like S1-311, was employed to fabricate a single junction GaAs solar cell structure. It is important to note that all the defect mitigation steps were confined in the buffer GaAs. Figure 2 describes the structure of the solar cell showing the buffer layer and the architecture of the template. Due to the presence of oxide buffer layers, side-contacts were employed to collect the carriers. Figure 3 shows the TEM cross-section of a single-junction GaAs solar cell as described in Figure 2. In this structure, two sets of 20 period SL were incorporated as indicated by the circle in Figure 3 (a). Fig 3 (a) shows a significant defect reduction in the top GaAs cap layer compared to the bottom GaAs and Ge layers. Figure 4 (b) reveals the p-n junction, p++ layer and the location of few grain boundaries in active layer of solar cells. Fig 3 (c) shows the SAED pattern obtained from the SL region, as indicated by the circle in Fig 3 (a). The spotty pattern confirms the single-crystalline nature of the film. Moreover, absence of additional secondary spots indicates that the InGaAs/GaAs was strained and did not relax during growth.

The flexible GaAs solar cell prepared using the defect reduction scheme of S1-356 (as shown in Table I) was measured under AM1.5 intensity. Fig 4 shows the J-V curve of the device. The open-circuit voltage, short circuit current density and efficiency were 477 mV, 14.6 mA/cm^2 and 5.1% respectively. Devices without the defect mitigation processes showed significantly lower efficiencies < 2% [9]. The enhancement of the efficiency was attributed to the lower defect densities in the active layers due to defect mitigation steps employed in the GaAs film.

combinations to bring about a significant reduction in defect density. Bending of defects at the strained interfaces is considered the primary phenomenon for the substantial defect reduction in the GaAs layers. Efficiency of above 5% at 1 sun was achieved in flexible GaAs single-junction solar cells on metal substrates employing defect reduction processes.

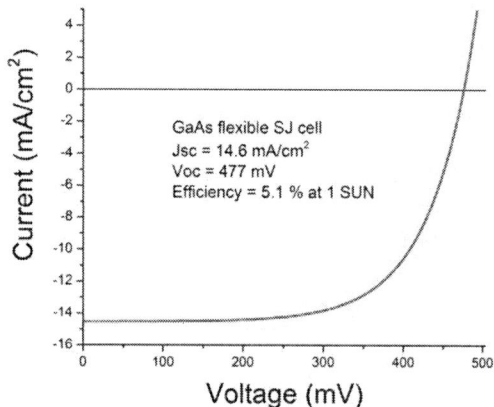

Fig 4. J-V curve of the 1J GaAs device grown on defect reduced GaAs templates

ACKNOWLEDGEMENTS

This work is supported by funding from the U.S. Department of Energy (DOE) SunShot Initiative program (Award #: DEEE0006711), Advanced Superconductor Manufacturing Institute, and Texas Center for Superconductivity at the University of Houston. The authors acknowledge the help received from Dr. Ramesh Laghumavarapu at the University of California, Los Angeles.

REFERENCES

[1] M. Rathi, et al., "AlGaAs/GaAs DH and InGaP/GaAs DH grown by MOCVD on flexible metal substrates," *Proc. 43rd IEEE Photovoltaic Specialist Conference*, pp.1926-1929,2016.

[2] M. Rathi, et al., " Thin Film III-V Photovoltaics using Single-Crystalline-Like, Flexible Substrates," *Proc. 42nd IEEE Photovoltaic Specialist Conference*, , 2015.

[3] P. Dutta, et al., " InP thin films with single-crystalline-like properties on flexible metal substrates photovoltaic application," *Proc. 43rd IEEE Photovoltaic Specialist Conference*, pp. 1892-1894, 2013.

[4] P. Dutta, et al., " Epitaxial thin film GaAs deposited by MOCVD on Low-Cost, Flexible Substrates for High Efficiency Photovoltaics," *Proc. 39th IEEE Photovoltaic Specialist Conference*, pp. 1926–1928, 2013.

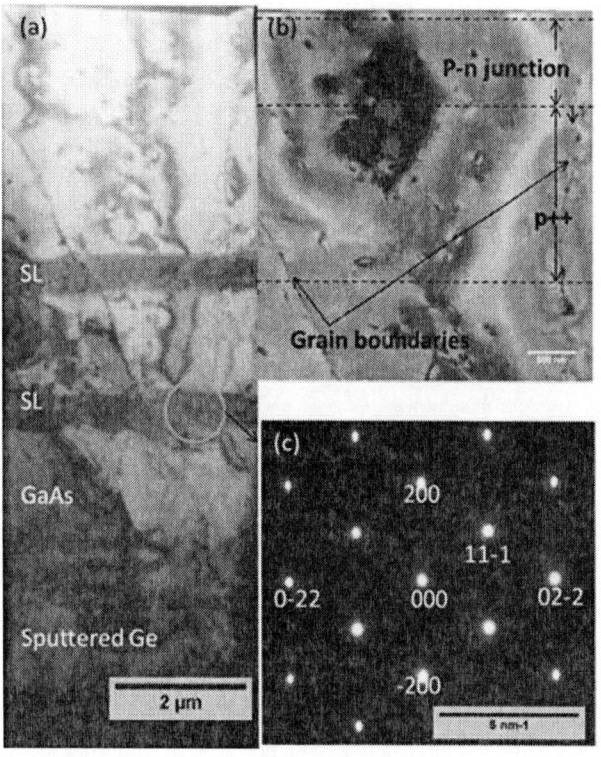

Fig. 3.(a) CS-TEM of 1J GaAs device grown on defect reduced GaAs templates (b) high magnification image of the top GaAs active layers (c) SAED pattern of the SL structure embedded in buffer

CONCLUSION

Defect mitigation techniques were employed to reduce defect density in GaAs thin films. Thermal annealing and cycling, low-temperature GaAs layer, single-layer strained InGaAs/GaAs SL structures were employed in certain

[5] P. Dutta, et al.," High mobility singlecrystalline-like GaAs thin films on inexpensive flexible metal substrates by metal-organic chemical vapor deposition", *Appl. Phys. Lett.* vol. 105, pp. 092104-1– 092104-5, 2014

[6] P. Dutta, et al., "High mobility single-crystalline-like GaAs thin films on inexpensive flexible metal substrates by metal-organic chemical vapor deposition," *Appl. Phys. Lett.* vol. 105, pp. 092104-1–5, 2014

[7] Hiroshi Okamoto, et al.," Dislocation Reduction in GaAs on Si by Thermal Cycles and InGaAs/GaAs Strained-Layer Superlattices ", Japanese Journal of Applied Physics, 26, 1987

[8] V. Selvamanickam, et. al," Germanium films with strong in plane and out-of-plane texture on flexible, randomly textured metal substrates ", *J. Crystal Growth*, vol. 311, pp. 4553-4557, 2009

[9] M. Asadirad, et al., "III-V thin film photovoltaic solar cells based on single-crystal-like GaAs grown on flexible metal tapes," *Proc. 43rd IEEE Photovoltaic Specialist Conference*, pp. 1954–1956, 2016.

Economic Analysis of Transfer Printed III-V Virtual Substrates

Kenneth J. Schmieder[*], Matthew P. Lumb[†*], Michael K. Yakes[*], Shawn Mack[*], Mitchell F. Bennett[‡*],
Sergey I. Maximenko[*], Laura B. Ruppalt[*], Michael A. Meeker[*], Chase T. Ellis[*], Matthew Meitl[§],
Joseph G. Tischler[*], Robert J. Walters[*]

[*]U.S. Naval Research Laboratory, Washington, DC 20375 USA
[†]George Washington University, Washington, DC 20052 USA
[‡]Sotera Defense Solutions, Annapolis Junction, MD 20701 USA
[§]Semprius Inc., Durham, NC 27713 USA

Abstract—We propose a novel methodology for III-V substrate cost reduction that does not rely on chemical-mechanical polishing (CMP) or low-throughput release layer etching. The details of this process are provided, and results of an economic model for GaAs and InP virtual substrates dictate 20x and 34x reductions to substrate cost, respectively. The impact of modeled assumptions are quantified in order to better understand the potential range of end-goal virtual substrate cost.

Index Terms—III-V semiconductor materials, photovoltaic cells, semiconductor growth

I. INTRODUCTION

Despite record efficiencies, cost remains an obstacle for III-V photovoltaics in terrestrial markets. While numerous factors have contributed, one inescapable component is the high cost of producing a III-V device. High-X concentrator photovoltaics (CPV) has enabled vastly improved utilization of cell material, but does so at the risk of increased design complexity. If III-V device costs could be significantly reduced, this could accelerate the commercialization of 1-sun or reduced-concentration III-V photovoltaics.

A recent analysis by J. S. Ward, *et al.* [1] has reported that 84% of the cost for a GaAs solar cell is associated with the substrate (assuming no substrate reuse), and only 7% and 9% associated with metal organic chemical vapor deposition (MOCVD) and device processing, respectively. It is therefore evident that reducing the cost of III-V photovoltaics necessitates a disruptive reduction to substrate cost.

There have been many promising efforts targeting the reduction of III-V substrate cost by reuse of the initial source substrate. These include epitaxial lift-off (ELO) [2], spalling [3], and ion bombardment [4] among others. The most mature of these is ELO, in which a selective etch layer is grown between the parent substrate and solar cell structure, enabling the use of a targeted chemical solution to etch-undercut the full solar cell stack, lifting it off from the parent substrate. The substrate, now separated from the solar cell stack, can be reused for subsequent growth runs with the potential to produce many solar cells from one parent substrate.

Due to the finite selectivity between common release layers and etch-stop layers, as well as the significant immersion time associated with ELO, the substrate surface may be roughened by the ELO process. Left untreated, this could lead to performance degradation in subsequently grown solar cells. In order to avoid this, chemical-mechanical polishing (CMP)

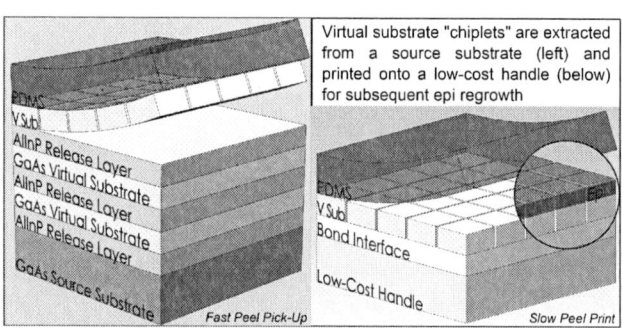

Fig. 1. A cartoon describing the TPVS concept. After MTP, epitaxial regrowth on each individual chiplet yields a PV device enabled by the atomically smooth, air-stable, and clean surfaces of this process.

is often executed to restore the wafer to a pristine, epi-ready state. This CMP process can be costly [1] and result in wafer breakage, limiting the number of times a substrate can be reused. If ELO can be executed with minimal use of CMP, the time associated with full-wafer release may still be a factor in limiting the cost savings of ELO substrate reuse technology.

Here, we propose an alternative technique for reducing substrate cost. The incumbent technologies focus on removal of epitaxial device layers in order to reuse the initial source substrate. Instead, we propose the use of an initial source substrate to produce many virtual substrates in one growth run, followed by the selective and sequential removal of each virtual substrate and bonding to separate low-cost handles. These virtual substrates will be epi-ready by leveraging a process, micro-transfer printing (MTP), that facilitates air-stable, atomically smooth, and clean surfaces. The concept will not require a CMP step and can be performed with a fast release process.

In this paper, the transfer printed virtual substrate (TPVS) concept is introduced; an estimation for the cost of GaAs and InP TPVS is detailed. The model is then extended to consider the impact that TPVS deployment can have on high-performance stacked multijunction photovoltaics.

II. TPVS METHODOLOGY

Figure 1 demonstrates the concept underlying TPVS. A source substrate is used to grow a periodic structure comprising release and virtual substrate layers. This stack should be made up of as many periods as possible in order to

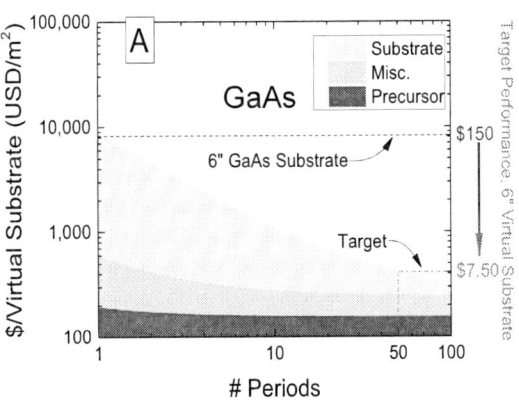 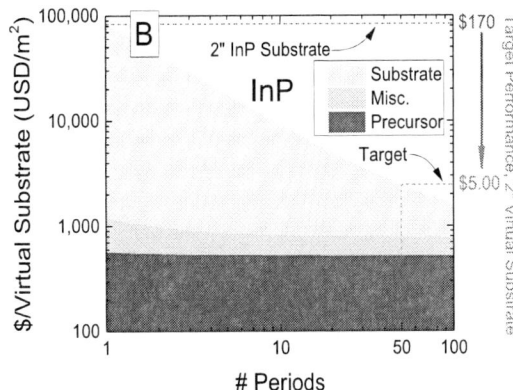

Fig. 2. Cost breakdown for a virtual substrate of A) GaAs and B) InP versus the number of periods successfully transferred from an initial source stack. Miscellaneous costs include MOCVD depreciation and labor.

maximize production of virtual substrates from a single source substrate. After growth, each virtual substrate layer will be sequentially removed by wet etching the underlying release layer in a manner consistent with MTP fabrication procedures. The details of MTP, a commercially demonstrated technique for the deterministic printing of microscale assemblies, can be found elsewhere in literature [5]. We apply the MTP technique for the parallel transfer of dense arrays of "chiplets" from source substrate to a low-cost handle. The transferred chiplets will comprise greater than 90% of the areal epitaxial layer grown on the source substrate. Once transferred to a handle, the TPVS structure will be epi-ready for subsequent regrowth. Following regrowth and fabrication, each chiplet will yield an individual device. In this way, one can produce as many virtual substrates from a single source substrate as can be successfully grown monolithically and transfer printed.

III. ECONOMIC ANALYSIS

Costs accounted for in this study include the initial source substrate as well as MOCVD precursors, depreciation, and labor. The cost of the handle, transfer printing capital cost & MTP labor are not accounted for in this study since accurate estimations cannot be made at this time. We intend to include these factors as we gain further insight. It is assumed that 6-inch GaAs and 2-inch InP substrates cost $150 and $170, respectively. MOCVD costs are computed for a commercially-available, high-throughput tool capable of running 7 6-inch wafers or 58 2-inch wafers simultaneously. Precursors accounted for in this study include trimethylgallium (TMGa), trimethylindium (TMIn), trimethylaluminum (TMAl), arsine (AsH$_3$), and phosphine (PH$_3$). It is assumed for the economic analysis that the periodic structures can be 1 μm thick, comprised of virtual substrate and release layers of 800 nm and 200 nm, respectively. This thickness is an important variable, and as such its impact on cost is considered later in the section.

Figure 2 provides the cost breakdown for GaAs and InP TPVS. Here, cost per unit area is provided along the primary y-axis, and the number of periods comprising the source stack on the x-axis. Miscellaneous cost includes MOCVD depreciation and labor. It is evident that cost reductions are driven by increasing the number of transferable TPVS periods in order to more effectively utilize each source substrate. If 50 periods can be successfully achieved, the TPVS cost will no longer be dominated by the substrate and instead becomes more balanced between substrate, precursors, and a combination of MOCVD depreciation and labor. This achievement would reduce GaAs TPVS cost to $7.50, a 20x reduction relative to the initial source substrate. The improvement for more expensive substrates, such as InP in Figure 2B, is even more dramatic. Here, a 2-inch substrate initially costing $170 could produce a 50 period TPVS stack for $5.00 per virtual substrate, a 34x cost reduction. This example also illustrates that even relatively expensive precursors, such as TMIn, are not a preventative barrier to low-cost virtual substrates produced via MOCVD growth.

It is found that, along with the number of successfully transferable periods, the thickness of each period has a notable impact on the overall cost. In Figure 3, GaAs TPVS is considered as a function of the thickness of each period. The ratio of virtual substrate thickness to release layer thickness is 4:1 in this analysis. The total source stack thickness is held constant at 50 μm, meaning that the total number of periods varies inversely with the thickness of each period. The cost increases linearly with period thickness and has a high impact on both the effective substrate cost as well as the precursor-related growth costs. If each period is required to be 5 μm thick, each 6-inch GaAs virtual substrate would cost approximately $37. This is believed to be a conservative estimate for the thickness of each period, as each period would resemble a structure proven compatible with MTP [6], [7]. Furthermore, recent MTP efforts from literature have demonstrated 650 nm thick GaAs-based transferred structures with 400 nm thick release layers[8], similar to our target period thickness. Of course, the relationship between virtual substrate thickness and compatibility with epitaxial regrowth is presently unknown,

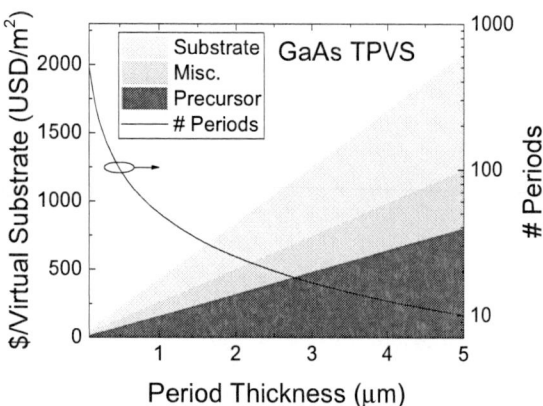

Fig. 3. Cost breakdown for GaAs TPVS as a function of individual period thickness. The total source stack thickness is held constant, assumed to be 50 μm.

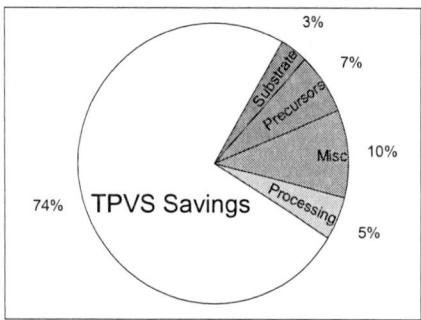

Fig. 4. Cost breakdown for stacked 6-junction solar cell from [10], demonstrating the dramatic reduction to device-level cost possible with the success of TPVS.

but expected to be better understood (and accounted for in cost models) in the near-term.

We finally consider the impact of successful TPVS deployment on next-generation stacked multijunction photovoltaics targeting 4+ junctions and record efficiencies [9]. We evaluate the design of a current matched 6-junction cell modeled to be capable of > 50% AM1.5D efficiency at 1000x discussed elsewhere [10], utilizing 3 GaAs-based junctions and 3 InP-based junctions. Here, the impact is expected to be particularly significant since the stacked multijunction architecture requires both GaAs and InP substrates to produce each 6-junction solar cell. The model accounts for TPVS cost (assumed to be $7.50 and $5.00 per 6-inch GaAs and 2-inch InP substrate, respectively), MOCVD depreciation, labor, and precursors, as well as device fabrication cost. The layer thicknesses and compositions are as designed for the 6-junction device modeled in [10]. The resulting cost breakdown is provided in Figure 4, indicating that the total multijunction device cost is reduced 74% by successful implementation of TPVS, as opposed to conventional substrates.

IV. CONCLUSIONS

TPVS has the potential for disruptive cost reduction of III-V photovoltaics, among other devices. The economic model has predicted virtual substrates that are 20 to 34x cheaper than conventional substrates. This has the potential for even further cost reductions by coupling it with alternative growth techniques having low source material costs, such as hydride vapor phase epitaxy (HVPE) [11]. The successful implementation of TPVS has been considered for next-generation stacked multijunction photovoltaics and indicates the potential for a 74% reduction to total device cost. Such a reduction can enable vastly greater design flexibility toward low-X CPV and 1-sun applications.

REFERENCES

[1] J. S. Ward, T. Remo, K. Horowitz, M. Woodhouse, B. Sopori, K. VanSant, and P. Basore *Prog. Photovolt: Res. Appl.*, vol. 24, pp. 1284–1292, 2016.

[2] J. Adams, V. Elarde, A. Hains, C. Stender, F. Tuminello, C. Youtsey, A. Wobowo, and M. Osowski, "Demonstration of multiple substrate reuses for inverted metamorphic solar cells," in *Proc. 37th IEEE Photovoltaic Specialists Conf.*, (Austin), June 2012.

[3] D. Shahrjerdi, S. W. Bedell, C. Bayram, C. C. Lubguban, K. Fogel, P. Lauro, J. A. Ott, M. Hopstaken, M. Gayness, and D. Sadana *Adv. Energy Mater.*, vol. 3, pp. 566–571, 2013.

[4] A. Tauzin, E. Lagoutte, T. Salvetat, J. Guelfucci, Y. Bogumilowicz, B. Imbert, F. Fournel, S. Reboh, F. P. Luce, C. Lecouvey, T. Chaira, V. Carron, H. Moriceau, J. Duvernay, T. Signamarcheix, C. Drazek, C. Charles-Alfred, B. Ghyselen, E. Guiot, T. Tibbits, P. Beutel, and F. Dimroth *AIP Conference Proceedings*, vol. 1679, p. 040009, 2015.

[5] M. A. Meitl, Z.-T. Zhu, V. Kumar, K. J. Lee, X. Feng, Y. Y. Huang, I. Adesida, R. G. Nuzzo, and J. A. Rogers *Nature Materials*, vol. 5, p. 33, 2006.

[6] J. Yoon, S. Jo, I. S. Chun, I. Jung, H.-S. Kim, M. Meitl, E. Menard, X. Li, J. J. Coleman, U. Paik, and J. A. Rogers *Nature*, vol. 465, pp. 329–334, 2010.

[7] X. Sheng, C. A. Bower, S. Bonafede, J. W. Wilson, B. Fisher, M. Meitl, H. Yuen, S. Wang, L. Shen, A. R. Banks, C. J. Corcoran, R. G. Nuzzo, S. Burroughs, and J. A. Rogers *Nature Materials*, vol. 13, pp. 593–598, 2014.

[8] B. Gai, Y. Sun, H. Lim, H. Chen, J. Faucher, M. L. Lee, and J. Yoon *ACS Nano*, 2017. in press, DOI:10.1021/acsnano.6b07605.

[9] F. Dimroth, M. Grave, P. Beutel, U. Fiedeler, C. Karcher, T. N. D. Tibbits, E. Oliva, G. Siefer, M. Schnachtner, A. Wekkeli, A. W. Bett, R. Krause, M. Piccin, N. Blanc, C. Drazek, E. Guiot, B. Ghyselen, T. Salvetat, A. Tauzin, T. Signamarcheix, A. Dobrich, T. Hannappel, and K. Schwarzburg *Prog. Photovolt: Res. Appl.*, vol. 22, pp. 277–282, 2014.

[10] M. P. Lumb, K. J. Schmieder, M. Gonzalez, S. Mack, M. K. Yakes, M. Meitl, S. Burroughs, C. Ebert, M. F. Bennett, D. V. Forbes, X. Sheng, J. A. Rogers, and R. J. Walters *AIP Conference Proceedings*, vol. 1679, p. 040007, 2016.

[11] J. Simon, K. L. Schulte, D. L. Young, N. M. Haegel, and A. J. Ptak *IEEE Journal of Photovoltaics*, vol. 6, pp. 191–195, 2016.

Thin Films of Zinc-Doped GaAs by RF Magnetron Sputtering for Use in Photovoltaic Cells

Kirby Simon, Kyle Cepeda, Nishit Shetty, and Elijah Thimsen

Washington University in St. Louis, St. Louis, Missouri, 63130, United States of America

Abstract — Radio frequency (RF) magnetron sputtering parameters that produce highly crystalline thin films of gallium arsenide (GaAs) were investigated. Literature results were reproduced to determine optimum conditions for sputtering highly crystalline films. At these desirable conditions, a noticeable increase in the Ga:As ratio was discovered, resulting from metallic gallium present on the film surface. A simple HCl treatment was developed to remove the excess gallium and return the film to the stoichiometric ratio of Ga:As. An *in-situ* zinc doping method was investigated and refined to sputter p-type GaAs films. Hall effect measurements confirmed that the Zn-doped films were p-type and the hole concentration increased with increasing Zn content.

Index Terms — crystalline, gallium arsenide, photovoltaic cells, p-type, sputter deposition.

I. INTRODUCTION

Thin-film gallium arsenide (GaAs) based devices have proven to have the highest conversion efficiencies among single-junction photovoltaic technologies.[1] However, the high costs associated with traditional production methods, namely metal-organic chemical vapor deposition (MOCVD) and epitaxial liftoff techniques, have hindered the widespread use of GaAs in commercial modules.[2] While some success has been achieved using close-space vapor transport as an alternative deposition process,[3]-[5] there are concerns about maintaining the integrity of the p/n junction at the very high temperatures required by that technique. Radio frequency (RF) magnetron sputtering could offer a lower cost alternative since it does not require expensive pyrophoric metal-organic precursors or toxic gases such as arsine. However, to date, little is known about depositing GaAs layers by sputtering for use in photovoltaic devices. This investigation of GaAs thin films for solar energy applications aims to better understand the process of sputtering and its impact on film growth as well as explore the potential of developing photovoltaic devices in a new, cost-effective, and less-hazardous way when compared to traditional methods.

Altering sputtering parameters results in varying degrees of film crystallinity and stability of the deposited GaAs. Previous reports have found that crystalline GaAs thin-films can be deposited by sputtering at high substrate temperatures, high RF powers, and low pressures.[6]-[10] We reproduced these results with the RF magnetron sputtering setup illustrated in Fig. 1, which utilizes s-guns. Highly crystalline GaAs thin films were deposited at high temperatures (>400 °C), high RF powers (>400 W), and low pressures (~1 mTorr). High deposition

rates, ranging from 1.2-4.2 µm/hr, were used throughout these experiments.

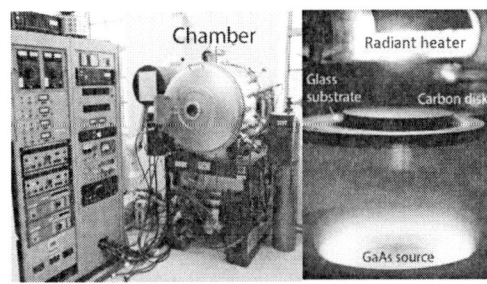

Fig. 1. Sputtering chamber and equipment rack (left). Sputtering setup inside chamber (right), in the process of depositing GaAs.

This work focuses on the deposition of highly crystalline, Zn-doped, p-type GaAs films by RF magnetron sputtering. Although there have been some reports on the *in-situ* doping of sputtered GaAs,[11] in this work we systematically investigated a simpler doping procedure to better understand the relationships between the sputtering parameters, specifically chamber pressure and substrate temperature, and the chemical composition of the resulting films. Various conditions were tested to determine how the sputtering parameters impacted the resulting Ga:As ratio and the incorporation of Zn as a p-type dopant. *In-situ* annealing was also investigated as a method to improve thin-film crystallinity. The films were characterized using various techniques, including x-ray fluorescence, x-ray diffraction, scanning electron microscopy, and Hall effect measurements.

II. EXPERIMENTAL DETAILS

The RF magnetron sputtering setup seen in Fig. 1 was used for the GaAs film deposition. A NESLAB HX-75 Series II Recirculating Chiller was installed to cool the sputtering chamber along with the Pfeiffer TMU 521 P turbomolecular pump to control pressure. The base pressure of the turbomolecular pump was 10^{-6} Torr. Unless otherwise stated, all depositions were carried out at 1.1 mTorr.

Single crystal sapphire substrates with (0001) orientation and VWR plain glass microscope slides were used as deposition substrates. Prior to deposition, the substrates were washed with acetone and isopropanol then blown dry before being placed on the substrate holder in the chamber. After pumping down to the base pressure of the Leybold TRIVAC D 65 BCS rough pump and purging the chamber with argon (5.0 UHP), a radiant heater was ramped up to establish the desired substrate temperature (Fig. 1). A thermocouple imbedded in the carbon disk resting on the substrates measured the substrate temperature. After an equilibration period, an argon plasma was ignited at a low RF power (40 W) with the shutter closed and a 10-minute pre-sputter commenced to clean the target (GaAs source). After the pre-sputtering period, the shutter was opened and the RF power was ramped up to the desired level of 400 W to initiate the deposition process. Post-deposition, the radiant heater was slowly ramped down to reduce thermal stress on the deposited films as they cooled.

For doped samples, 60 mg of metallic zinc flakes were placed on the target surface so that the plasma simultaneously sputtered Zn from the flakes while sputtering Ga and As from the target. Between experiments, the target was vacuumed and wiped clean with isopropanol to remove traces of zinc left behind from the previous experiment. This procedure was carried out for each experiment to control the mass of zinc that was incorporated into the GaAs film.

Various instruments were used for analysis of the sputtered films after deposition. X-ray fluorescence (XRF, Spectro Midex) was used to determine film thickness, gallium-to-arsenic ratio, and zinc dopant concentration. X-ray diffraction (XRD) was performed on the samples to characterize crystallinity. Crystallite size was determined from the full-width-half-max (FWHM) of the XRD peaks using the Scherrer equation (1). In the Scherrer equation, the calculated crystallite size (τ) is based on the x-ray wavelength (λ, $Cu_{k\alpha}$), the FWHM of the diffraction peak (β), the diffraction angle associated with the peak being analyzed (Θ), and a dimensionless shape factor equal to 0.9 (K).

$$\tau = K\lambda/\beta\cos\theta \qquad (1)$$

For electronic characterization of the deposited films, an Ecopia HMS-5000 Hall effect measurement system was utilized to determine carrier type, carrier mobility, and carrier concentration. The magnetic field strength was 0.52 T. Indium contacts were press-bonded to the four corners of the film in the van der Pauw electrode configuration. These measurements helped verify that the doping procedure was successful and describe the relationship between zinc-doping and various film characteristics.

The first set of experiments focused on understanding the relationship between substrate temperature and the crystallinity of the zinc-doped GaAs films. The second set of experiments consisted of varying chamber pressure while keeping all other deposition parameters constant to determine the impact pressure had on the deposition and doping process. The final set of experiments attempted an *in-situ* annealing procedure to further increase film crystallinity to increase carrier mobility.

This annealing procedure consisted of slowly ramping up the radiant heater after deposition to the desired annealing temperature (440-450 °C), holding at that temperature for a specified period of time, and slowly cooling down the chamber after annealing. The experimental results are reported below.

III. RESULTS

At the conditions that produce highly crystalline films, XRF results indicated that the Ga:As ratio deviated from stoichiometry. Low substrate temperatures near room temperature resulted in Ga:As of 1:1, while high temperature depositions near 450 °C had Ga-rich ratios as high as 1.3:1. Scanning electron microscopy (SEM) images indicated that the excess gallium was present as a metal on the film surface, as seen in Fig. 2.

Fig. 2.   SEM image of the surface of a sputtered film indicating the presence of metallic gallium.

An acid etch procedure was developed to selectively remove the metallic gallium from the surface and return the Ga:As ratio to stoichiometric. This etch consisted of dipping the films in concentrated (37%) HCl for 5 minutes. XRF measurements performed post-acid bath indicated that the etching procedure returned the ratio closer to 1:1 (Fig. 3).

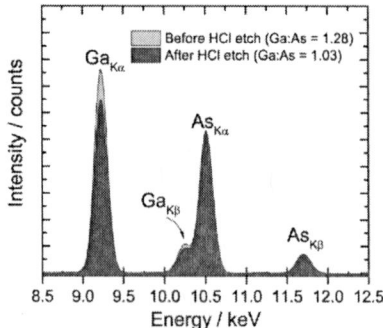

Fig. 3.   XRF results before and after acid etch indicating removal of metallic gallium from the film surface.

At low substrate temperatures, the GaAs films were nanocrystalline and the grain size increased significantly with increasing substrate temperature. Depositions were carried out keeping chamber pressure and RF power constant at 1.1 mTorr and 400 W respectively, while varying the substrate temperature. Substrate temperatures varied from room temperature up to 451 °C. Plots of the normalized XRD patterns are presented in Fig. 4. The plot shows the narrowing of the diffraction peaks as the substrate temperature increased.

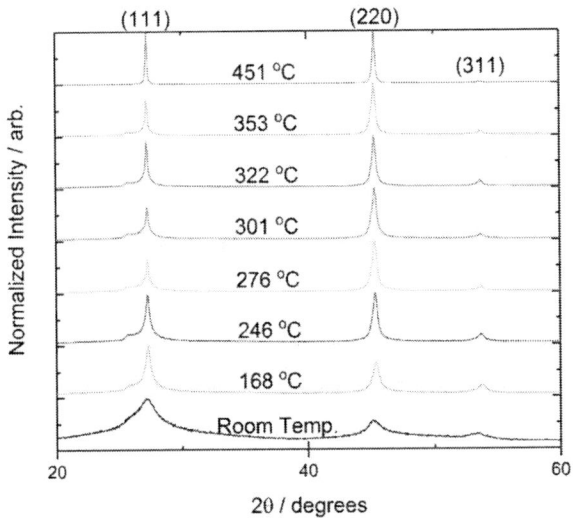

Fig. 4. Normalized XRD patterns of the films deposited at various temperatures onto microscope glass substrates.

The XRD patterns were used to calculate the FWHM of the various diffraction peaks. These values were then used in the Scherrer equation to find the associated crystallite sizes in the films as a function of substrate temperature (Fig. 5). A representative crystallite size for each film was determined from a linear fit of $K\lambda/\cos(\Theta)$ versus β, with the slope of the line corresponding to the crystallite size. The crystallite size increased significantly with increasing substrate temperature (Fig. 5). The room temperature point is plotted at 50 °C due to heating effects experienced from the deposition process.

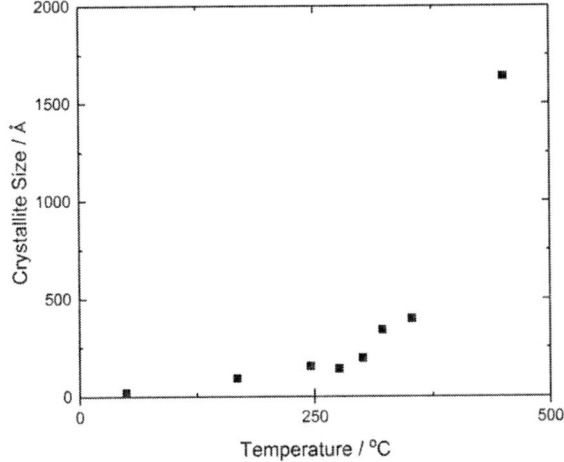

Fig. 5. Crystallite size of the films as a function of substrate temperature for a pressure of 1.1 mTorr and RF power of 400 W.

For a constant mass of flakes placed on the GaAs target, constant applied RF power, and constant chamber pressure, the amount of Zn incorporated into the films also depended on the substrate temperature. The concentration of zinc in the thin-films, obtained from XRF measurements, was determined to be inversely related to the substrate temperature, as seen in Fig. 6. At very high substrate temperatures of approximately 450 °C, the measured amount of Zn in the GaAs film was at or below the detection limit of the XRF measurement (3.2×10^{19} atoms/cm^3).

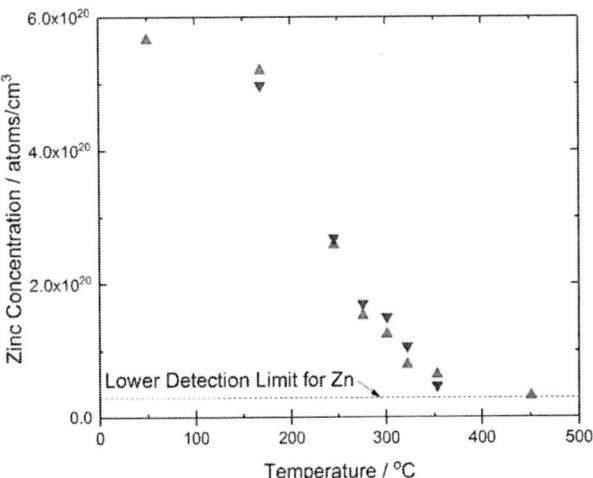

Fig. 6. Zinc concentration in sputtered GaAs thin-films as a function of substrate temperature. The amount of Zn flakes placed on the GaAs target was held constant at 60 mg, and the RF power and chamber pressure were held constant at 400 W and 1.1 mTorr, respectively.

Electrical resistivity measurements of the films deposited at various substrate temperatures are presented in Fig. 7. Each of these films had a different crystallite size and Zn concentration, since both parameters depended on substrate temperature (Figs. 5 and 6). Apart from an outlier, the resistivity decreased with decreasing substrate temperature. The decreasing grain size with decreasing substrate temperature (Fig. 5) would be expected to manifest as an increase in resistivity. The increase in Zn concentration with decreasing substrate temperature (Fig. 6) would be expected to decrease the resistivity, if Zn were acting as a p-type dopant. Since the resistivity decreases with decreasing substrate temperature, it appears that Zn acting as a p-type dopant is the more dominant effect compared to the effect of crystallite size. Thus, resistivity is plotted as a function of Zn concentration in Fig. 7.

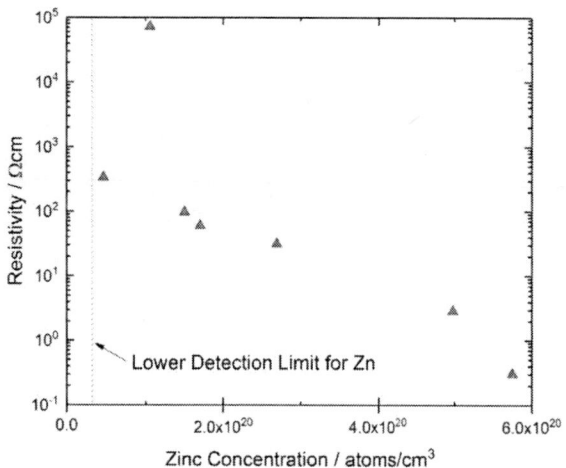

Fig. 7. Resistivity as a function of zinc concentration in films with the lower detection limit for zinc labeled.

Hall effect measurements performed on the films deposited throughout these experiments varied greatly from sample to sample. Many of the films tested had inconsistent Hall coefficient values, reducing the confidence in certain results. However, there were several films for which reliable Hall effect measurements were obtained. The sign of the Hall coefficient was positive for the Zn-doped samples, indicating p-type conduction. Fig. 8 presents the hole concentration plotted against the zinc concentration for several samples. The hole concentration increased with increasing Zn concentration, which is consistent with Zn acting as a p-type dopant. The hole concentration is within the desired range of 10^{16} to 10^{20} cm^{-3}.

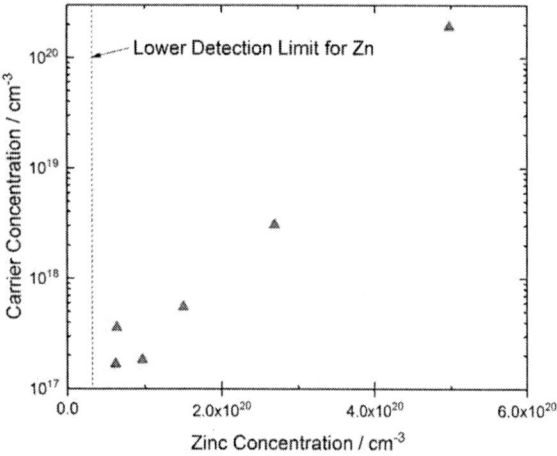

Fig. 8. Carrier concentration measured by the Hall effect compared to zinc concentration measured by XRF.

To improve the thin-film crystallinity, an *in-situ* annealing procedure was tested on the films post-deposition. After depositing GaAs layers at lower temperatures (353 °C) that allowed zinc to be incorporated into the films (Fig 6.), the

samples were put through a high-temperature anneal in the chamber at 440-450 °C for up to 3 hours in an attempt to increase the crystallite size. Unfortunately, the annealing procedure was ineffective, and the crystallite size did not change significantly. The crystallite size and diffraction pattern of the annealed films resembled more closely the films deposited at 353 °C than those deposited at 451 °C (Fig. 4). We are continuing to explore alternative strategies that allow for zinc to be incorporated into the film and for large crystallite sizes to be obtained.

IV. DISCUSSION

Results from the first set of experiments show that the crystallinity of the sputtered GaAs thin-films increases with increasing substrate temperature. The normalized XRD plot shows the diffraction peaks narrowing as substrate temperature increases, demonstrating an increasing degree of crystallinity (Fig. 4). The plot of crystallite size also confirms this trend of increasing crystallinity with substrate temperature (Fig. 5).

The trend in temperature also demonstrates the ability to controllably dope the GaAs films while sputtering. However, at high temperatures, in excess of 400 °C, the vapor pressure of zinc exceeds the chamber pressure at which the experiments were run (1.1 mTorr). In these instances, zinc was undetectable through XRF measurements, suggesting that the zinc evaporated from the film before it could be incorporated into the GaAs crystal lattice. To confidently dope the GaAs films, the chamber pressure should exceed the vapor pressure of Zn at a given substrate temperature. One might imagine that it would be possible to suppress Zn evaporation by increasing the chamber pressure while keeping all other parameters constant. Unfortunately, a number of parameters change with pressure. Films deposited at high pressures, from 13 up to 33 mTorr, experienced a loss of crystallinity, even when high substrate temperatures and RF powers were used. A decreasing deposition rate was also noted as the chamber pressure increased. Thus, it currently appears that lower pressures (1.1 mTorr) are more effective for depositing crystalline, p-type GaAs films by RF magnetron sputtering, and alternative means by which to incorporate Zn into the film and achieve a high degree of crystallinity must be developed.

Hall effect measurements of the films varied greatly in reliability from sample-to-sample, as indicated by the alternating sign of the measured Hall coefficient. This result is hypothesized to be related to the low carrier mobility values (less than 1 cm^2/Vs) reported for all of the films analyzed. While our experiments did confirm that the zinc doping procedure was successful in making the GaAs films p-type, higher reliability in future measurements is necessary to better understand the ability of these films to function in photovoltaic cells.

To improve carrier mobility in the films, the crystallite size must be further increased. We are currently exploring strategies to increase crystallite size while maintaining an appropriate p-type dopant concentration.

V. SIGNIFICANCE

By controlling substrate temperature, RF power, and chamber pressure, the deposition of zinc-doped GaAs by RF magnetron sputtering was determined to be a viable production method for highly crystalline, p-type thin-films. Further investigation and optimization of this process promises the idea that highly crystalline, highly conductive GaAs thin-films can be produced via RF magnetron sputtering. This notion presents the sputtering of GaAs as a low cost, low risk, viable alternative to current thin-film production methods. If this method can be fully studied and understood, production costs associated with GaAs thin-films for photovoltaic cells could potentially be reduced without sacrificing performance, making higher efficiency GaAs-based solar modules a more competitive option in the solar energy conversion industry.

ACKNOWLEDGEMENTS

The authors would like to thank Prof. Daniel Rode for useful discussions about the research, and thank the international center for energy, environment and sustainability (InCEES) for financial support.

REFERENCES

[1] M.A. Green, K. Emery, Y. Hishikawa, W. Warta, E. D. Dunlop, D. H. Levi, and A. W. Y. Ho-Baillie, "Solar cell efficiency tables (Version 49)," *Progress in Photovoltaics*, vol. 25, no. 1, 3-13, 2017.

[2] M. Woodhouse and A. Goodrich, *A Manufacturing Cost Analysis Relevant to Single- and Dual-Junction Photovoltaic Cells Fabricated with III-Vs and III-Vs Grown on Czochralski Silicon (Presentation).* 2014; p Medium: ED; Size: 92 pp.

[3] A. J Ritenour, R. C. Cramer, S. Levinrad, and S. W. Boettcher, "Efficient n-GaAs photoelectrodes grown by close-spaced vapor transport from solid source," *ACS Appl. Mater. Interfaces*, vol. 4, no. 1, 69-73, 2012.

[4] A. J. Ritenour, J.W. Boucher, R. DeLancey, A. L. Greenaway, S. Aloni, and S. W. Boettcher, "Doping and electronic properties of GaAs grown by close-spaced vapor transport from powder sources for scalable III-V photovoltaics," *Energy Environ. Sci.*, vol. 8, 278-285, 2015.

[5] J. W. Boucher, A. J. Ritenour, A. L. Greenaway, S. Aloni, and S. W. Boettcher, "Homojunction GaAs solar cells grown by close space vapor transport," in *40th IEEE Photovoltaic Specialist Conference (PVSC)*, 2014, pp. 0460-0464.

[6] B. El Hadadi, R. Lbibb, and A. Slaoui, "Influence of the deposition conditions on the crystallographic structure and the GaAs refraction index," *Vacuum*, vol. 83, no. 1, 1100-1105, 2009.

[7] H. Reuter, H. Schmitt, M. Böffgen, "Properties of sputtered amorphous and microcrystalline GaAs films," *Thin Solid Films*, vol. 254, 96-102, 1995.

[8] J.-L. Seguin, B. El Hadadi, H. Carchano, A. Fennouh, and K. Aguir, "High-temperature sputtered amorphous GaAs," *Journal of Non-Crystalline Solids*, vol. 183, 175-181, 1995.

[9] L.H. Ouyang, D. L. Rode, T. Zulkifli, B. Abraham-Shrauner, N. Lewis, and M. R. Freeman, "Hydrogenated amorphous and microcrystalline GaAs films prepared by radio-frequency magnetron sputtering," *Journal of Applied Physics*, vol. 91, no. 5, 3459, 2002.

[10] S. H. Baker, S. C. Bayliss, S. J. Gurman, N. Elgun, J. S. Bates, and E. A Davis, "The effect of varying substrate temperature on the structural and optical properties of sputtered GaAs films," *Journal of Physics: Condensed Matter*, vol. 5, 519-534, 1993.

[11] S. B. Hyder, "Thin film GaAs photocathodes deposited on single crystal sapphire by a modified rf sputtering technique," *Journal of Vacuum Science & Technology*, vol. 8, no. 1, 228-232, 1971.

Self Aligned Aluminum Selective Emitter for n-type Si Cells

San Theingi[1,2], Robert C. Reedy[1], Vincenzo LaSalvia[1], Paul Stradins[1], Benjamin G. Lee[1]

[1]National Renewable Energy Laboratory, Golden CO, 80401 USA

[2]Colorado School of Mines, Golden CO, 80401 USA

Abstract — We show a method for making a self-aligned selective emitter by annealing aluminum grid lines above the Si-Al eutectic temperature, and study the recombination properties using lifetime measurements. The dark saturation current density at the metal contact ($J_{0,metal}$) is determined as a function of minority carrier density using photoconductive decay. After annealing at 600 °C, $J_{0,metal}$ decreases from 1800 fA cm^{-2} to 500 fA cm^{-2} at an injection level of 1E15 cm^{-3}. This value of $J_{0,metal}$ would correspond to a very low emitter recombination contribution of ~10 - 25 fA cm^{-2} for a front metal grid with area coverage of 2 – 5%. In addition, $J_{0,metal}$ is found to be injection dependent. A low contact resistivity of ~0.1 mOhm cm^{-2} is obtained from transmission line measurement (TLM).

Index Terms — Aluminum, selective emitter, dark saturation current, silicon solar cells, surface passivation

I. INTRODUCTION

A selective emitter (SE), where the emitter is heavily doped locally underneath the metal contact grid and is lightly doped in the regions between the grids, provides a path to achieving high efficiency solar cells [1]. One of the efficiency losses in solar cells is through recombination at the metal contacts, which increases the dark saturation current of the device. With a SE, the field created by the heavily doped layer beneath the metal contact helps reflect the minority carriers away from it, thereby reducing the recombination at the metal/emitter interface and minimizing the dark saturation current due to the contact. In addition, the high surface doping concentration also reduces the contact resistance of the device [2].

Many approaches exist for forming a selective emitter, such as etch back, laser ablation, oxide masking and ion-implantation processes; however, these technologies add additional processing steps to cell fabrication and some of these processes also pose processing problems such as difficulty in aligning the grid to the heavily doped regions [1].

Here, we present a self-aligned technique for creating aluminum (Al) front grid n-type crystalline silicon boron-diffused SE solar cells. Fig. 1 shows a possible realization of such a device. Al is deposited at the openings for the front grid and then the sample is heated to above the Al-Si eutectic temperature (577 °C) to diffuse and form a p$^+$ layer underneath the metal. Aluminum, in this case, not only acts as a front grid for the solar cell, but also acts as a diffusion source for creating a heavily doped layer at the metal/emitter interface. Note that this differs from how Ag pastes are used to contact most (standard) Si solar cells.

In literature, alloyed Al paste/powder has been used in Al back surface field (BSF) and rear emitters solar cells, and has been proven to reduce the recombination surface velocity and produce moderately high efficiency solar cells [3-5]. In this paper, we study the recombination properties using symmetric lifetime test structures to investigate if the dark saturation current at the front contacts could be decreased by annealing Al at 600 °C to create a heavily doped layer underneath the contacts.

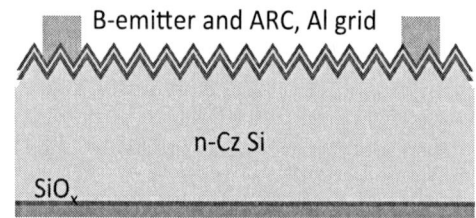

Fig. 1. Schematic illustration of Al front grid n-type B-diffused selective emitter solar cell.

II. EXPERIMENTAL

We investigate the recombination properties using a set of samples that contain test structures with different Al metal area coverage fraction (A_{metal}) ranging from 3% to 21%. Each test structure is a symmetric boron diffused, n-type Cz Si wafer with bulk resistivity of 3.5 Ω cm and emitter sheet resistance of ~110 Ω/sq. The test structure is passivated on both front and back using atomic layer deposited Al$_2$O$_3$. 25 μm diameter circular dot openings arranged in a square lattice are patterned on the front side using photolithography. Al is then deposited on the front side using e-beam evaporation to make metal contacts to the emitter through the circular openings. Thus, the front surface contains a mixture of silicon-metal interface and Al$_2$O$_3$ passivated surface whereas the backside is fully passivated with Al$_2$O$_3$. The top view and the cross-section of the test structure are shown in Fig 2.

The samples are then thermally annealed in a forming gas environment at 400 °C and 600 °C for 5 minutes and 1 minute respectively. Effective lifetime measurements are performed using a Sinton lifetime tester and the saturation current density of metal, $J_{0,metal}$, is extracted using the technique outlined in Deckers *et al.* [6]. Assuming that the injection level is uniform

978-1-5090-5606-4/17 $31.00 © 2017 IEEE

throughout the wafer, the effective lifetime (τ_{eff}) as a function of A_{metal} can be expressed as,

$$\frac{1}{\tau_{eff}} = \frac{1}{\tau_{bulk}} + 2J_{0,diel}\frac{N_D + \Delta p}{qn_i^2 W} + A_{metal}\left[J_{0,metal} - J_{0,diel}\right]\frac{N_D + \Delta p}{qn_i^2 W}$$

where τ_{bulk} is the carriers' lifetime in the bulk, $J_{0,diel}$ is saturation current density at the passivated surface, N_D is the doping density of the base wafer, Δp is the injection level, q is the elementary charge, n_i is the intrinsic carrier concentration, and W is the wafer thickness. Using the equation above, $J_{0,metal}$ can be extracted from the slope of the linear fit of inverse effective lifetime as a function of A_{metal}. An example of such a fit is shown in Fig 3.

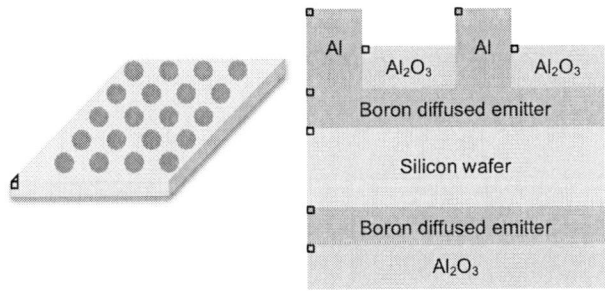

Fig. 2. Schematic representation of (left) top view and (right) cross section of the test structure.

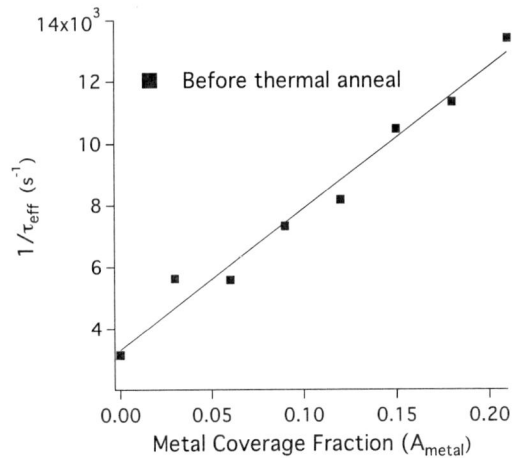

Fig. 3. Inverse effective lifetime as a function of metal coverage for test structure before thermal anneal measured at the injection level of 5×10^{15} cm^{-3}.

III. RESULTS & DISCUSSION

We extract $J_{0,metal}$ at different injection levels for different thermal anneals and the results are shown in Fig 4. Right after metallization and before any thermal treatment, the values of $J_{0,metal}$ are high – between 1600 fA cm^{-2} and 1800 fA cm^{-2}. This is to be expected because there is no passivation or highly doped layer underneath the metal to reduce the

recombination of the minority carriers in that region. The samples are then thermally annealed in a forming gas environment at 400 °C for 5 minutes. This decreases the $J_{0,metal}$ value to between 1400 fA cm^{-2} and 1600 fA cm^{-2}. It is possible that the thermal anneal improves the surface passivation since hydrogen is mobile at 400 °C and is able to move to passivate dangling bonds at interfaces. An additional forming gas anneal is performed at 600 °C for 1 minute. At temperatures above 577 °C, Al and Si form an Al-Si eutectic. Upon cooling below this temperature, Si regrows epitaxially from the bulk crystal, with some Al dopants remaining in the Si. Thus, 600 °C is chosen to study if Al can be used to create a p$^+$ doped layer underneath the metal to further reduce the recombination at the emitter-metal interface. The $J_{0,metal}$ values decrease significantly to ~500 fA cm^{-2} at the injection level of 1E15 cm^{-3} and ~1300 fA cm^{-2} for injection levels above 4E15 cm^{-3}. The area-weighted $J_{0e,metal}$ for a typical cell front metal grid, which has grid area coverage of 2 – 5%, would then be ~10 - 25 fA cm^{-2}. This very low value is comparable to the recombination current achieved for a selective emitter in a 25.1% cell from Fraunhofer ISE [7]. Specifically, their area-weighted recombination current for a B-doped selective emitter was 13 fA cm^{-2}.

After the 600 °C thermal anneal, we re-passivate the samples with Al$_2$O$_3$ to ensure that there is no passivation loss around the perimeter of the metal dots due to an undercut from the lithography process; any passivation loss can contribute an additional dark saturation current to the result. After re-passivation, we see a further decrease in $J_{0,metal}$ values to ~600 fA cm^{-2} at the injection level of 1E15 cm^{-3} and ~1100 fA cm^{-2} for injection levels above 4E15 cm^{-3}. This result is consistent with some passivation loss arising from our lithography processing, so we are currently addressing this issue with processing.

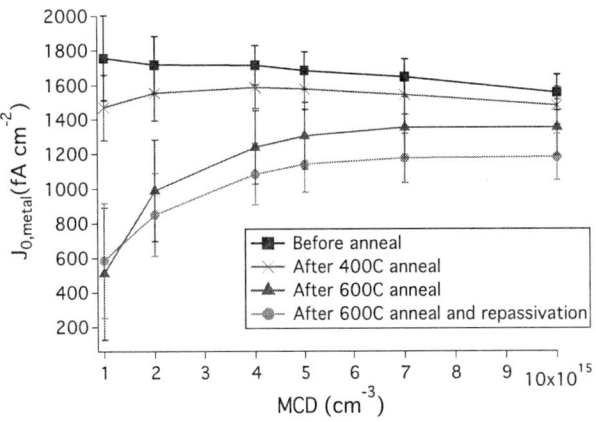

Fig. 4. Contact saturation current density as a function of minority carrier density before annealing (black squares), after annealing at 400 °C (green cross), after annealing at 600 °C (blue triangles), and after re-passivation with Al$_2$O$_3$ (red circles).

Another interesting behavior observed from this study is that the $J_{0,metal}$ value is dependent on the injection level; the value decreases with injection level for MCD lower that 4E15 and remains constant for MCD above 4E15. This can be understood from reasoning that field-effect passivation is more effective at lower MCD.

We determine the contact resistivity (ρ_c) using transmission line measurement (TLM) and a low ρ_c ~0.1 mOhm cm^{-2} is obtained. Thus we clearly see that a quite good low resistance contact is formed between the annealed Al and the B emitter.

IV.   CONCLUSION

We show that a self-aligned selective emitter for a B-diffused n-type Si cell can be formed by depositing Al and annealing the sample above the Si-Al eutectic temperature. We study the recombination properties using lifetime test structures and find that the dark saturation current from the contact decreased by more than half at the minority carrier density of 1E15 cm^{-3}. This allows a very low emitter recombination current of ~10 – 25 fA cm^{-2} for a cell with front grid metal area coverage of 2 – 5%. $J_{0,metal}$ is found to be injection dependent. In addition, TLM measurement gives a low ρ_c ~0.1 mOhm cm^{-2}. Since Al can act as both the front grid for the solar cell and create a local heavily doped region underneath the grid at the same time, this method gives a simpler route for producing high performance n-type Si cells.

ACKNOWLEDGMENT

Funding for this work was provided by the United Sates Department of Energy EERE contract DE-261 EE0006336 (FPACE-II) and under Contract No. DE-AC36-08GO28308 and SETP DE-EE00030301 (SuNLaMP) and under Contract No. DE-AC36-08GO28308.

The U.S. Government retains and the publisher, by accepting the article for publication, acknowledges that the U.S. Government retains a nonexclusive, paid-up, irrevocable, worldwide license to publish or reproduce the published form of this work, or allow others to do so, for U.S. Government purposes.

REFERENCES

[1] Rahman, Mohammad Ziaur. "Status of Selective Emitters for p-Type c-Si Solar Cells." *Optics and Photonics Journal*, no. 2, 2012, pp. 129–134.

[2] Jager, Ulrich et al. "Selective Emitter by Laser Doping from Phosphosilicate Glass." *EU PVSEC Proceedings*, 2009.

[3] Fellmeth, Tobias et al. "20.1% Efficient Silicon Solar Cell With Aluminum Back Surface Field." *IEEE Electron Device Letters*, vol. 32, no. 8, 2011, pp. 1101–1103.

[4] Bock, Robert et al. "The ALU$^+$ Concept: N-Type Silicon Solar Cells with Surface-Passivated Screen-Printed Aluminum-Alloyed Rear Emitter." *2009 34th IEEE Photovoltaic Specialists Conference (PVSC)*, 2009.

[5] Narasimha, Shreesh et al. "An Optimized Rapid Aluminum Back Surface Field Technique for Silicon Solar Cells." *IEEE Transactions on Electron Devices*, vol. 46, no. 7, 1999, pp. 1363–1370.

[6] Deckers, Jan et al. "Injection Dependent Emitter Saturation Current Density Measurement under Metallized Areas Using Photoconductance Decay." *28th European Photovoltaic Solar Energy Conference and Exhibition*, 2014.

[7] Glunz, Stefan et al. "The Irresistible Charm of a Simple Current Flow Pattern – 25% With a Solar Cell Featuring a Full-Area Back Contact." *31st European Photovoltaic Solar Energy Conference and Exhibition*, 2015.

How to realize solar cells with laser structured plated Ni-Cu-Contacts with Excellent Adhesion and High Fill-Factors without parasitic plating

A. Büchler, S. Kluska, J. Bartsch, B. Grübel, A.A. Brand, S. Gutscher, M. Glatthaar

Fraunhofer Institute for Solar Energy Systems, Heidenhofstraße 2, 79110 Freiburg, Germany

Abstract — **Investigating modifications of the silicon/metal interface during the manufacturing process and the influence on properties of laser-structured Ni-Cu-plated contacts enabled optimizing the metallization process. The micro pattern induced by ultrashort pulsed lasers ensures adhesion even without nickel silicide formation. Minimizing growth of laser-induced and native oxide at the surface of the laser patterned contact openings before plating enables high fill-factors without HF-treatment before metal deposition. Skipping the wet-chemical treatment eliminates parasitic plating on precursors with defective ARC. Compared to the state-of-the-art plating sequence, the approach presented features less process steps and enables improved solar cell aesthetics.**

I. INTRODUCTION

Plated nickel-copper contacts have the potential to outperform silver screen printing for silicon solar front side metallization [1, 2]. Replacing silver by copper reduces material cost. Nickel acts in this contact system as seed layer and diffusion barrier and should enable low contact resistivity even on lowly doped emitters. The manufacturing of plated nickel-copper contacts for mass production was intensively investigated in recent years [3, 4]. At Fraunhofer ISE [5] the state-of-the art standard process sequence to manufacture plated contact grids used to be as follows: Contact definition by local removal of the passivation layer (e.g. by laser ablation), removal of native oxide within the passivation layer opening (e.g. by HF-dip), immediate plating of the metal stack (e.g. by light-induced plating), silicide formation (e.g. thermally induced in a belt furnace). As reported in other publications and as observed in our lab, following non-optimized structuring, plating and silicide formation processes can lead to insufficient adhesion [6], to parasitic metal deposition [7] and to device-shunting due to nickel silicide spikes [8]. As mentioned in earlier publications, spiking can be avoided by reducing the silicidation temperature or by eluding the silicidation anneal [9]. While other authors stated the silicide formation was required to ensure *adhesion* [10], it is still questionable if this fact holds if ultrashort pulsed lasers are used for structuring [11]. In the following we will analyze if silicide formation increases the adhesion on silicon surfaces that feature an optimized microstructure due to laser interference, a common characteristic of ultrashort pulsed laser ablation. *Parasitic plating* (PP) (or ghost plating) names the unintended plating of metal within the solar cell area caused by pinholes in the passivation layer. PP impairs the aesthetics

and the cell performance by shading and surface recombination [12]. Native oxide can block the deposition at passivation layer impairments. But in order to reduce native oxides along the laser-ablated passivation layer opening solar cells are usually dipped into low concentrated HF before metal deposition. This oxide removal promotes excellent electrical properties and sufficient contact adhesion [13]. As negative side effect the HF treatment also increases parasitic plating at passivation layer defects, such as pinholes or process induced defects (e.g. by automatic or manual handling). Previous results suggest that a reduction of process induced oxide within the laser ablated area obviates the need of an oxide removal before deposition. Therefore, within the experiment presented we tried to minimize laser induced oxidation and to reduce the growth of native oxide by minimizing the time between laser ablation and metal deposition. In the following the process sequence without HF-pretreatment is called "Easy Plating", while "Reference plating" will refer to the process sequence with a wet-chemical treatment before deposition. The last part of the experiment aims to analyze weather the high fill-factors can be reached by the Easy Plating sequence.

II. EXPERIMENTAL

All results shown in this paper are generated using industrially fabricated Cz Al-BSF solar cell precursors. Precursors of several manufactures were used. The precursor front-side features a random pyramid texture and a silicon nitride passivation layer that is optimized for silver screen printing. To structure the metal deposition the passivation layer was locally ablated by a picosecond-pulsed UV laser system (Coherent Lumera Super Rapid, laser pulse duration<15 ps, 355 nm). The laser pulse energy was set to realize contact openings with an opening diameter of 20 μm. After laser ablation the solar cells that were metallized following the Easy Plating sequence were plated within 120 s while cells for the Reference plating were stored for one day and cleaned in 1% HF for 30 s immediately before plating. Using light-induced plating a metal stack of Ni (~1 μm), Cu (~10 μm) and Ag (~0.5 μm) was deposited in RENA inline tools. To analyze the effect of silicide formation certain cells were annealed in an inline belt furnace under forming gas atmosphere. Adhesion of busbars was determined by soldering interconnects and measuring the peel-off force at an angle of

90° to the busbar. The adhesion of fingers was determined qualitatively by optical evaluation. Microstructure was analyzed in a Zeiss Auriga SEM. The IV curve was measured with an inline flash tester. For the determination of the contact resistance a measurements using the transfer-length method (TLM) were conducted. The spatial distribution of local power losses was analyzed in a Dark Lock-In Thermography (DLIT) set-up.

III. RESULTS AND DISCUSSION

A. How to ensure adhesion of plated contacts

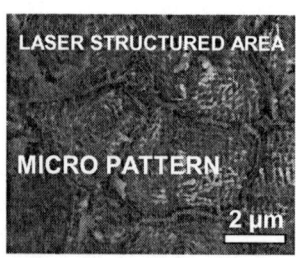 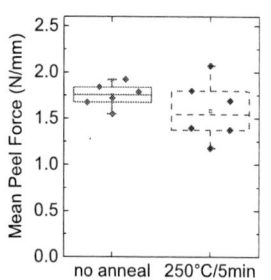

Fig. 1. *Left:* TopView SEM image: Passivation layer ablation with ultrashort pulsed lasers creates a rough microstructure within the laser contact opening. *Right:* Results of busbars adhesion test show high mean peel forces also for contacts without silicidation anneal. Surface micro structure promotes high mechanical adhesion

To ensure reliable adhesion for of the Ni-Cu-Ag contacts the adhesion force between silicon and nickel and between the three metal layers has to withstand the mechanical load due to stress induced by the thermal treatments within the metallization process, module fabrication and the temperature cycles once the cell is in usage. While fabrication both factors need to be considered: Reduction of induced mechanical stress and optimization of adhesion forces. There are several ways reported to promote adhesion force of plated contacts. The approaches can be sorted to approaches based on the formation of silicides along the nickel silicon interfaces [10] and approaches based on the formation of advantageous micro structures within the contact opening [9, 14]. If silicidation of the interface is needed to ensure adhesion, interface oxide has to be completely removed [15] and the formation of Kirkendall voids needs to be avoided [6]. In order to reduce process-induced mechanical stress within the metallization, plating parameters and temperature profiles of thermal treatments after metal deposition (e.g. the silicidation anneal) need to be adjusted [11]. Still it remained questionable if a suitable micro structure on the surface may overcome the need for silicide formation.

Experiments with a femto-second laser ablation proof the potential of laser-induced microstructures. As Fig. 1 shows the laser pulses induce an interference pattern into the silicon. The

sharp riffles ensured more than sufficient adhesion without post-plating anneal.

Experiments with a pico-second laser showed that adhesion results are exceeding 1 N/mm for different tested precursor materials after moderate post-plating anneal at 250°C/5min. However, we could not attribute the effect of the post-plating anneal to the formation of silicides, since no silicide are detectable in the SEM. If the temperature of the post-plating anneal exceeds 325°C we observed finger peel-off. As studies in the SEM proof, the fingers that peel off after inappropriate annealing, feature break out of silicon pyramids along the contact opening. Hence, break out is induced by high residual stress within the fingers due to the high temperature steps rather than by insufficient initial adhesion.

To ensure adhesion we conclude that usage of an ablation method that creates a microstructure on silicon and a post-plating anneal at moderate anneal (below 300°C) is recommended. An optimized micro structure on the silicon surface even allows high adhesion without post-plating anneal, as shown in experiments with a femto-second laser.

B. How to avoid parasitic plating

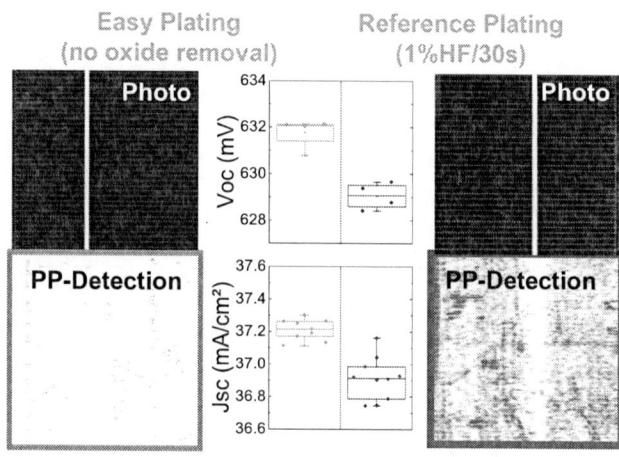

Fig. 2. Comparison of solar cells with high pinhole density of the passivation layer featuring Easy Plating (left) and Reference Plating contacts (right). *Sideways:* High resolution photographs of solar cell surface and graphics generated by particle analysis to visualize parasitic plating (PP). By Easy Plating the parasitic deposition is minimized. *Middle:* V_{oc} and J_{sc} values for solar cells with different plating procedures: Minimization of parasitic plating increases J_{sc} and V_{oc} due to less shading and less recombination.

Skipping the oxide removal before metal deposition minimizes parasitic plating (PP). Fig. 2 shows high resolution photographs of sections of two solar cells from the same precursor batch that were manufactured following the Easy Plating (left side) and the reference plating route (right side). The graphics below the photographs highlight the intensity of PP. As indicated in the diagrams in Fig. 2, the reduction of PP has – beside the aesthetical improvement – a slight positive effect on J_{sc} and V_{oc} due to less shading and less recombination, respectively. The precursors shown here have a high pinhole density in the passivation layer compared to usual industrial precursors.

C. How to realize good fill factors

In order to realize high fill factors the front-side metallization has to be optimized towards low contact resistivity and high shunt resistance. To lower the contact resistivity several aspects need to be considered before metal deposition: homogeneous, debris free ablation of the passivation layer and suppression of residual interface oxide before metal deposition. Further, the contact resistance can be improved by formation of nickel silicides along the nickel-silicon interface.

In the reference plating procedure the precursors are etched in 1% HF in a single side etching toll just before plating to remove the interface oxide within the contact opening. This enables low contact resistivity already without anneal and the effect of the post-plating anneal on the FF is negligible.

In the Easy Plating sequence there is no oxide removal step before plating but formation of interface oxides while processing is minimized. For Easy Plating contacts a moderate anneal of 200-300°C has a drastic effect on the contact resistivity. The contact resistance results of Easy Plating, compared to the reference plating sequence, vary for precursors from different manufactures. However it is a common feature that the contact resistance is in the range of 0,2-10 mΩcm² before the anneal and in the range of 0,1-1 mΩcm² after annealing. This effect is not fully understood yet, since we could not detect the formation of silicides. We rather assume the formation of conductive channels in the residual interface oxides.

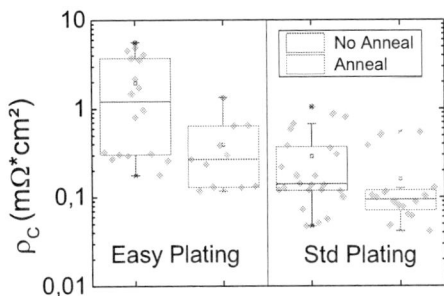

Fig. 3.   Specific contact resistances measured via TLM for contacts manufactured with Easy Plating and Reference plating.

For precursor material that is especially suited for Easy-Plating the measured FF was already higher compared to the reference plating route before post-plating anneal. Due to the anneal it improved further, resulting in a 0.5%abs. gain. The gain in FF can be attributed to higher shunt resistances of Easy Plating cells. The HF-dip in the reference plating route

dissolves native oxides also on the edge of the wafer, parasitic plating can lead to power leakage along the edges. Within the Easy Plating sequence native oxides passivated the wafer edges. This difference is detectable in DLIT images taken at +0.5 V, as shown in Figure 5.

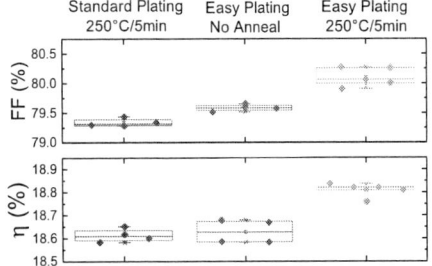

Fig. 4.   Fill factor (FF) and solar cell efficiency (η) results for different plating procedures: Reference Plating with silicidation anneal, Easy Plating without anneal and Easy Plating with silicidation anneal. For all cells interface oxide formation was minimized by lasering without pulse overlap and immediate plating after laser ablation.

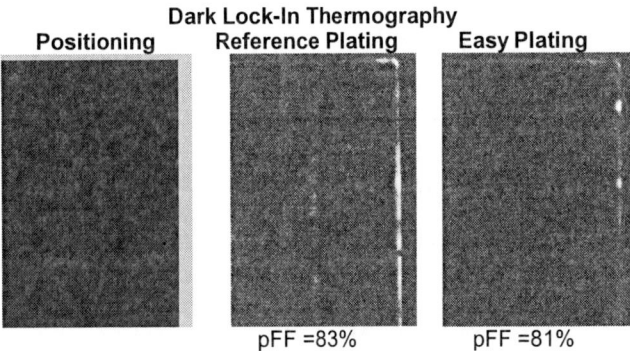

Fig. 5.   Distribution of local power losses at +0.5V in Dark Lock-In Thermography. The close-up shows the edge of a solar cell plated following the standard plating route and one plated following the Easy Plating route. Both solar cells are positioned as indicated by the scetch.

Realizing good fill-factors in the Easy Plating routes requires to suppress the formation of interface oxides and a moderate post-plating anneal. If the contact resistivity is optimized, the FF of the Easy Plating routes overcomes the reference plating due to advantages in the pFF.

V. CONCLUSION

We presented how laser structured plated contacts can be realized with excellent adhesion, no parasitic plating and high fill-factors. *Excellent adhesion* can be achieved by solely introducing a microstructure at the contact interface. We further explained that a moderate anneal ensures busbar and finger adhesion when pico-second pulsed lasers are used for structuring. Parasitic plating is eliminated by Easy Plating. In this process sequence the wet-chemical oxide removal before plating is skipped. Beside the perfect aesthetics the "Easy Plating" approach enables slight improvement of J_{sc} and V_{oc}. To realize *high fill-factors* of solar cells with "Easy Plating" contacts, interface oxides need to be minimized and a moderate post-plating anneal needs to be performed. By adjusting the laser ablation parameters and plating immediately after the contact structuring we were able to produce solar cells with fill-factors as high as with the reference plating route. The required post-plating anneal does not affect the busbar and finger adhesion. For material that is especially suited for Easy Plating, the FF results of the Easy Plating route are higher as the reference values. Since Easy Plating also reduces the power leakage due to parasitic plating along the wafer edge.

REFERENCES

[1] G. Beaucarne, G. Schubert, L. Tous, and J. Hoornstra, "Summary of the 6th Workshop on Metallization and Interconnection for Crystalline Silicon Solar Cells," *Energy Procedia*, vol. 98, pp. 2–11, 2016.

[2] S. Kluska *et al.,* "Electrical and Mechanical Properties of Plated Ni/Cu Contacts for Si Solar Cells," *Energy Procedia*, pp. 733–743, 2015.

[3] J. T. Horzel *et al.,* "Industrial Si Solar Cells With Cu-Based Plated Contacts," *IEEE J. Photovoltaics*, vol. 5, no. 6, pp. 1595–1600, 2015.

[4] A. Metz *et al.,* "Industrial high performance crystalline silicon solar cells and modules based on rear surface passivation technology," *Solar Energy Materials and Solar Cells*, vol. 120, pp. 417–425, 2014.

[5] G. Cimiotti, J. Bartsch, A. Kraft, A. Mondon, and M. Glatthaar, "Design Rules for Solar Cells with Plated Metallization," *Energy Procedia*, pp. 84–92, 2015.

[6] S. H. Lee, D. W. Lee, A. U. Rehman, J. W. Baik, and S. H. Lee, "Study of Annealing Temperature for Ni/Cu/Ag Plated Front Contact Single Crystalline Solar Cells,"

[7] M. S. Jeong *et al.,* "Use of antireflection layers to avoid ghost plating on Ni/Cu plated crystalline silicon solar cells," *Jpn. J. Appl. Phys.*, vol. 55, no. 3, p. 36502, 2016.

[8] M. C. Raval *et al.,* "Study of Nickel Silicide Formation and Associated Fill-Factor Loss Analysis for Silicon Solar Cells With Plated Ni-Cu Based Metallization," *IEEE J. Photovoltaics*, vol. 5, no. 6, pp. 1554–1562, 2015.

[9] S. Kluska *et al.,* "Micro characterization of laser structured solar cells with plated Ni–Ag contacts," *Solar Energy Materials and Solar Cells*, vol. 120, pp. 323–331, 2014.

[10] A. Mondon, M. Jawaid, J. Bartsch, M. Glatthaar, and S. Glunz, "Microstructure analysis of the interface situation and adhesion of thermally formed nickel silicide for plated nickel–copper contacts on silicon solar cells," *Solar Energy Materials and Solar Cells*, pp. 209–213, 2013.

[11] A. Büchler *et al.,* "Optimizing Adhesion of Laser Structured Plated Ni-Cu Contacts with Insights from Micro Characterization," *Energy Procedia*, pp. 913–918, 2016.

[12] M. Heinrich *et al.,* "Quantification of Front Side Metallization Area on Silicon Wafer Solar Cells for Background Plating Detection," *Energy Procedia*, pp. 717–724, 2015.

[13] A. Mondon *et al.,* "Nanoscale investigation of the interface situation of plated nickel and thermally formed nickel silicide for silicon solar cell metallization," *Applied Surface Science*, vol. 323, pp. 31–39, 2014.

[14] Alison Wenham, Chee Mun Chong, Sisi Wang, Ran Chen, Jingjia Ji, Zhengrong Shi, Ly Mai, Adeline, "Copper plated contacts for large-scale manufacturing," in *2016 IEEE 43rd Photovoltaic Specialists Conference (PVSC)*.

[15] A. Büchler *et al.,* "Interface oxides in femtosecond laser structured plated Ni-Cu-Ag contacts for silicon solar cells," *Solar Energy Materials and Solar Cells*, vol. 166, pp. 197–203, 2017.

Exploiting the Potentials of the Front Surface Field (FSF) Industrial Silicon Solar Cell

Ahrar Ahmed Chowdhury, Yu-Chen Hsu, Veysel Unsur and Abasifreke Ebong

Energy Production and Infrastructure Center, Department of Electrical and Computer Engineering, University of North Carolina at Charlotte, Charlotte, NC 28223-0001 USA

Abstract—This paper reports the PC2D simulation of an industrial size FSF solar cell with screen-printed Al alloyed back p-n junction. Best cell efficiency of 18.4% was modeled by PC2D confirming fill factor (FF) of 79.3%. Contour maps were generated to confirm the impact of FSF sheet resistance on the lateral voltage distribution. The maximum collection efficiency of ~97% was obtained at 1010 nm, which is close to the p-n junction and ~96% at 350 nm. The short circuit current was somewhat low because of lower emitter to the total cell area. Thus, this simple and cost-effective structure has great potential to reduce the cost of solar electricity.

Index Terms— Front Surface Field, n-type, PC2D, PC2D simulation, PhosTop, n-type solar cell

I. INTRODUCTION

The conventional high efficiency p type solar cell suffers from light induced degradation (LID) because of the boron-oxygen defects [1]. The reduction in the bulk lifetime decreases the efficiency. Additionally, performance degradation in p-type substrate can be associated with the metal impurities such as Fe and physical impurities (dislocations)[2] .The degradation is a strong function of Fe concentration in p type wafers [3]. Alternative to boron doping is gallium, which is held back by small segregation coefficient resulting in resistivity non-uniformity. The cost of gallium doped wafers is also a concern that is why the PV community is still making solar cells on boron-doped substrates.

The other alternative is the use of n-type wafers, which does not exhibit LID because of the absence of such defects that is plaguing the p-type substrate. To capitalize on the myriads advantages of n-type wafers there has been growing research interest in recent times in the PV community. According to the International Technology Roadmap for Photovoltaic (ITRPV 2016), it is predicted that n type wafers will dominate the market share after 2020 [4]. However, n-type production is still challenging compared to the p-type counterparts. This is due to additional processing steps and hence the increased production cost. To reduce cost of production, instead of adding steps, it should be subtracted. Therefore, research on various structures of n type solar cells are undertaken to maintain high performance without increasing cost of manufacturing.

Recently the n-type cell, tagged "PhosTop", which consists front surface field (FSF), n type base and rear alloyed Al as back

p-n junction was explored [5-14]. The premise of this structure is rooted in utilization of rear junction coupled with high minority carrier lifetime in n type wafer. There are two major processes involved in the formation of rear junction; (a) Al-Si alloy emitter on back side, and (b) Boron implant emitter. In case of Al-Si alloy emitter the rear side is screen printed with full Al back that facilitates the formation of emitter. This fabrication process requires only one high temperature phosphorous implant FSF formation. However, with the boron implant process typically two high temperature process sequence are involved to form (i) FSF and (ii) boron implant emitter. Thus, the additional process step can be eliminated by using screen printed Al and emitter formation during co-firing metal contact step. This advantage is critical to maintaining high yield in manufacturing.

In this work, FSF solar cell with Al-Si alloy rear junction emitter is presented. PC2D (Personal Computer Two Dimensional) [15] is used, for the first time, to model the electrical output parameters of the cell to ascertain the finger print of this structure. The two-dimensional simulation approach was adopted to elucidate the lateral conductivity of carrier transportation in the device.

II. CELL FABRICATION AND RESULTS

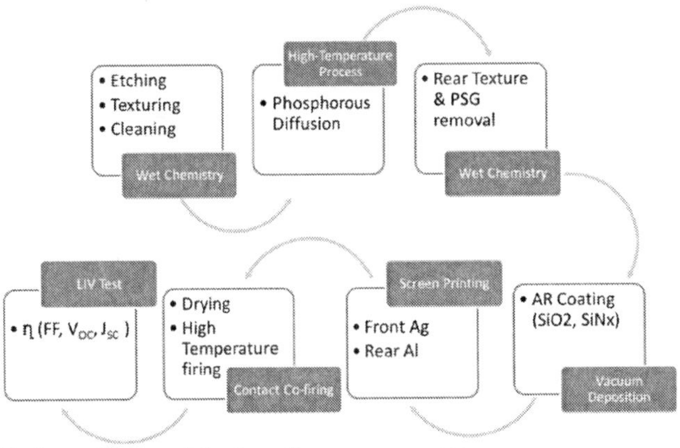

Fig.1.   Commercial solar cell process sequence

Cells were fabricated using the commercial process flow given in Fig. 1. A 10 Ω-cm base resistivity Czoklasky (CZ) n-type was used with FSF sheet resistance of 80 ohm/sq. The wafers had 180 μm thickness and 239 cm^2 surface area. The electrical output parameters of the FSF cells are presented in Table I. Fig. 2 shows the internal quantum efficiency (IQE) – orange, and the reflectance – blue, for the best cell.

TABLE I
ELECTRICAL OUTPUT PARAMETERS OF FSF CELLS MEASURED UNDER STANDARD TEST CONDITIONS (AM 1.5G, 100MW/CM^2, 25 C)

Cell ID	V_{OC} (mV)	J_{sc} (mA/cm^2)	FF (%)	η (%)
Best	638	36.4	79.3	18.42
Average (10 cells)	638	36.2	78.5	18.13

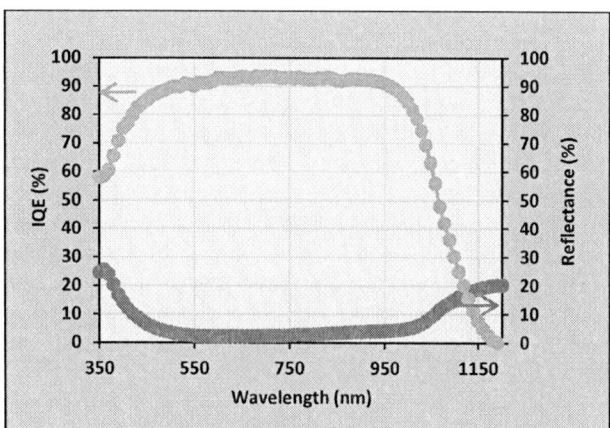

Fig. 2.   Internal quantum efficiency (IQE) – orange, and reflectance – blue, for the best FSF cell

III. PC2D SIMULATION OF FSF CELL

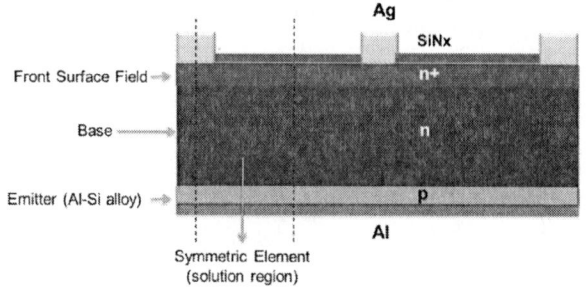

Fig.3.   Device schematic of FSF solar cell.

PC2D represents the simulation region of any width and height using a grid of 20 by 20 identical rectangular elements as described by Basore and Holmen [16]. PC2D simulation is based on a symmetrical unit that is compliant throughout the whole device area as shown in Fig. 3. In the simulation, metal contacts are assigned a J_{O1} value of 1000 fA/cm^2 and FSF region 57 fA/cm^2 and Al alloyed emitter 320 fA/cm^2. The p-n junction contributed a J_{O2} of 1 nA/cm^2. These recombination parameters modeled the experimental result of the best FSF cell as shown in Table II.

TABLE II
SIMULATION RESULTS FOR N TYPE FSF SOLAR CELL

Voc (mV)	Jsc (fA/cm$^{2)}$	FF	Eff (%)
638	36.3	79.1	18.4

Illumination of one sun (AM1.5G spectrum) was used in the simulation. Front surface is treated as 50% diffuse and 0% specular (i.e. 100% haze). Internal front surface reflectance is 5% for specular and 60% for diffuse. Metal contacted surfaces (front and rear) have 100% haze, with an internal specular reflectance of 80% and a diffuse reflectance of 70%. The cell is assumed to have an extrinsic series resistance (which encompasses gridlines, busbar and contact) of 0.15 Ωcm2 and the extrinsic shunt conductance of 0.2 mS/cm^2.

A. Front Surface Field

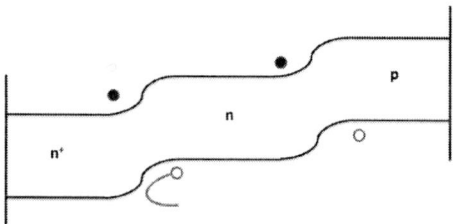

Fig. 4.   Energy band diagram of n$^+$-n-p solar cell (the sketch is not scaled)

The front surface field in the n-type cell creates a high-low junction, Fig. 4, which effectively repels the minority carrier holes to recombine in the front surface. This electric field created by FSF thus improves the front surface passivation. For a high base resistivity n-type material, the FSF contributes also to uniform lateral conductivity. Thus, the lower the surface doping concentration the more uniform lateral voltage drop as demonstrated in Figs. 5-6 for FSF sheet resistance values of 60 and 100 Ω/sq, respectively. Of the two different FSF sheet resistances, the 100 Ω/sq. shows a more uniform distributed lateral voltage than 60 Ω/sq. counterparts. This indicates that the for a high sheet resistance FSF, the n-type cells has better performance characteristic.

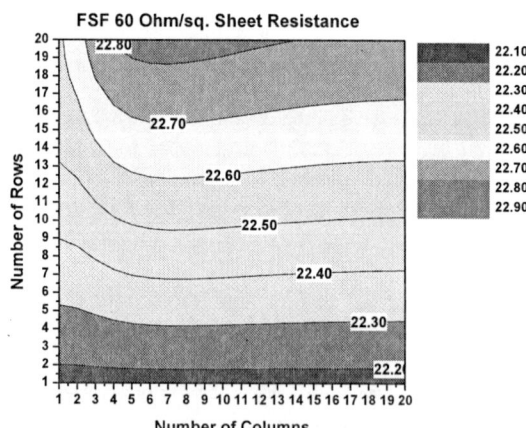

Fig.5.   Contour map for hole quasi-fermi potentials for 60 Ω/sq. FSF

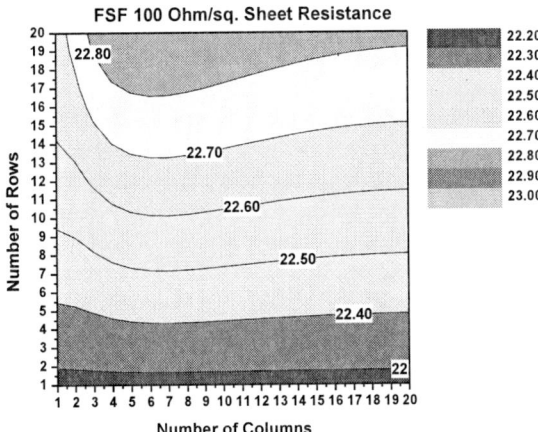

Fig.6.   Contour map for hole quasi-fermi potentials for 100 Ω/sq. FSF

B. Base Resistivity

For the n-type wafers the base resistance significantly impacts the cell performance. In the simulation, the contour plots were generated for different base resistivity to understand the injection-level dependence of the base at the maximum power point. Thus, unlike the p-type cells, for the n-type cells base introduces lateral conductivity that improves the overall cell performance. Two base resistivities; 1 Ω-cm and 5 Ω-cm-cm were considered as shown in Figs. 7-8. The contour plots represent quasi fermi potentials of minority carrier holes at maximum power points. The gradient of these potentials is essentially the flow of holes in the symmetric element of the device. It can be observed from the contour plots that, as the base resistivity increases from 1 Ω-cm to 5 Ω-cm, the fermi potentials decrease from 25 mV to 23mV. Thus, holes flow is increased in the 5 Ω-cm base resistivity.

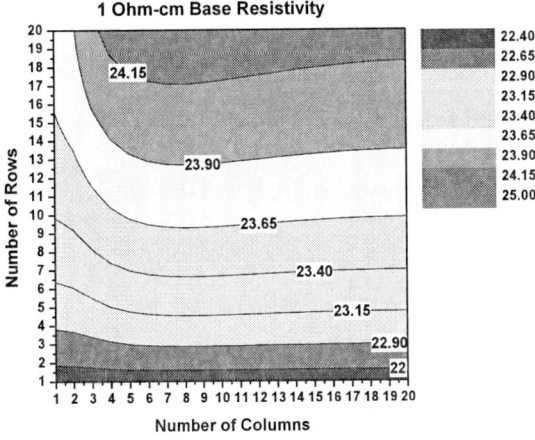

Fig.7.   Contour map of hole quasi-fermi potentials for 1Ω-cm base resistivity.

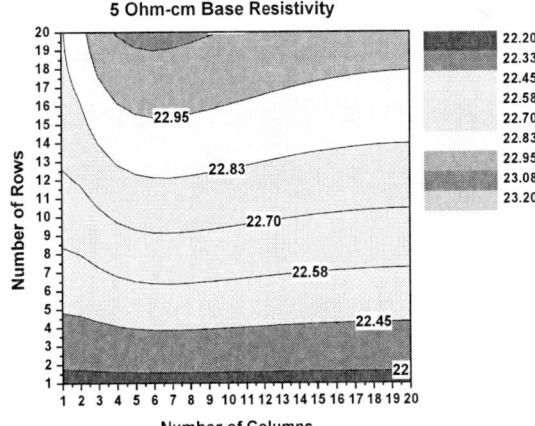

Fig.8.   Contour map of hole quasi-fermi potentials for 5Ω-cm base resistivity

C. Rear Junction collection efficiency

The Al-Si alloy at the rear surface creates the p-n junction, thus a back junction solar cell. Thus, high quality material is needed with high minority carrier lifetime for high collection efficiency. Simulation collection efficiency of the structure is as shown in Fig. 9. The maximum collection efficiency of ~97% is observed at 1010 nm, which is close to the p-n junction. But at 350 nm, a collection efficiency of ~ 96% is exhibited. Note that the minority carrier lifetime of ~300 μs was used in the simulation.

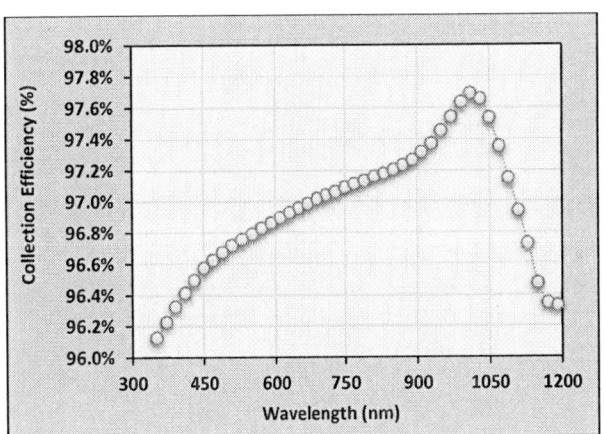

Fig.9.   Collection Efficiency of n+-n-p solar cell (the sketch is not to scaled)

IV. DISCUSSION

A. IV Characterization of solar cells

Table I summarizes the electrical output parameters of the fabricated FSF cell. The best efficiency of 18.4% was measured but average efficiency was 18.1%. The best FF of 79.3% compared to average of 78.5. The somewhat low FF is due to poor printing as was discovered after contact co-firing. The Sun_VOC measurement confirmed a pseudo FF of 83.1%, which indicates there is room for higher performance for this structure. Although the high base resistivity may dominant resistive loss as shown in Fig. 10, with precision printing, the gridline resistance can be decreased.

The relatively low J_{SC} value of ~36.1 mA/cm^2 can be attributed to the non-optimized peak front doping concentration and ARC thickness and refractive index. The smaller emitter (Fig. 11) than the wafer size due to 0.5 mm edge requirement around the back side to avoid shunting causes low J_{SC}. However, if the Al-Si alloy emitter is extended close to the wafer edge as demonstrated by use of implant boron [16]the J_{SC} will increase. Thus, by extending the emitter along with precise printing the efficiency will be raised to >21%. The V_{OC} of only ~638 mV was measured, which is ~2-7 mV lower than the p-type Al BSF counterpart [17].

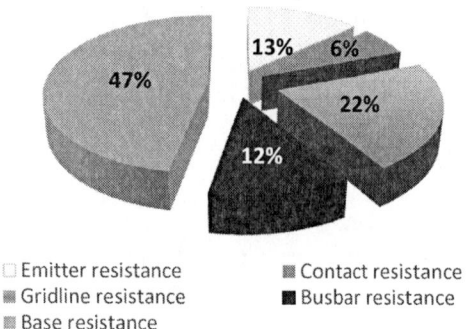

☐ Emitter resistance   ▨ Contact resistance
▨ Gridline resistance   ■ Busbar resistance
▨ Base resistance

Fig.10.   Series resistance components

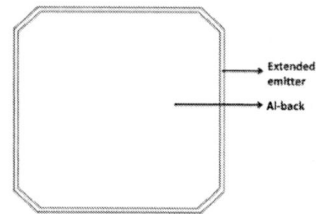

Fig. 11.   Extended back Al-alloyed emitter concept.

TABLE III
SUNS_V_{OC} MEASUREMENTS & IQE PC1D MATCHING

Parameter	Value
Lifetime (μs)	377
J_{01} (fA/cm^2)	4.9×10^{-13}
J_{02} (fA/cm^2)	4.8×10^{-9}
Pseudo FF	83.1
Pseudo Efficiency (%)	19.4
V_{OC} (mV)	642
n factor @ 0.1 sun	1.09
n factor @ 1 sun	1.01
FSRV (cm/s)	1×10^5
BSRV (cm/s)	500

V. CONCLUSIONS

An n-type FSF solar cell was fabricated and simulated with PC2D showing best efficiency of 18.4%. The impact of FSF, base resistivity and the collection efficiency was elucidated in the simulation results. The high sheet resistance FSF enables the uniform lateral voltage drop than the higher ones. The higher the base resistivity, the better the hole flow. Also, the collection efficiency increases from ~96%-97% from the front side at 350 nm to rear side at 1010 nm. J_{SC} of 36.4 mA/cm^2 was realized for the best cell due to use of non-optimum FSF peak surface concentration and less emitter due 0.5 mm back edge requirement for the Al-printing. This indicates if the emitter is extended to cover most of the back side, the FSF cells efficiency of >21% is achievable.

ACKNOWLEDGEMENTS

The authors would like to thank Paul Basore for assisting the discussion on the simulation model.

REFERENCES

[[1] K. Bothe, R. Sinton, and J. Schmidt, "Fundamental boron-oxygen-related carrier lifetime limit in mono- and multicrystalline silicon," *Progress in Photovoltaics: Research and Applications*, vol. 13, pp. 287-296, 2005.

[2] J. Lagowski, P. Edelman, A. M. Kontkiewicz, O. Milic, W. Henley, M. Dexter, *et al.*, "Iron detection in the part per quadrillion range in silicon using surface photovoltage and photodissociation of iron-boron pairs," *Applied Physics Letters*, vol. 63, pp. 3043-3045, 1993.

[3] H. H. A.A. Istratov, E.R. Weber, "Iron contamination in silicon technology," *Appl. Phys. A* pp. 489-534, 2004.

[4] (2016). Available: http://www.itrpv.net/Reports/

[5] C. Schmiga, H. Nagel, and J. Schmidt, "19% efficientn-type Czochralski silicon solar cells with screen-printed aluminium-alloyed rear emitter," *Progress in Photovoltaics: Research and Applications*, vol. 14, pp. 533-539, 2006.

[6] M. Rudiger, C. Schmiga, M. Rauer, M. Hermle, and S. W. Glunz, " Efficiency Potential of n-Type Silicon Solar Cells With Aluminum-Doped Rear p+ Emitter High efficiency bifacial back surface field solar cells." *IEEE Transactions on Electron Devices*, vol. 59, pp. 1295-1303, 2012.

[7] K. Meyer, C. Schmiga, R. Jesswein, M. Dupke, J. Lossen, H. J. Krokoszinski, *et al.*, "All screen-printed industrial n-type Czochralski silicon solar cells with aluminium rear emitter and selective front surface field," in *2010 35th IEEE Photovoltaic Specialists Conference*, 2010, pp. 003531-003535.

[8] J. M. P. Hacke, S. Yamanaka and D. L. Meier, "Efficiency Optimization of the n+/n/p+ 'PhosTop'Cell," *19th European Photovoltaic Solar Energy Conference, Paris, France.*, 2004.

[9] R. Bock, J. Schmidt, S. Mau, B. Hoex, and R. Brendel, "The ALU+ concept: n-type silicon solar cells with surface-passivated screenprinted aluminium-alloyed rear emitter," *IEEE Transactions on Electron Devices*, vol. 57, pp. 1966-1971, 2010.

[10] C. Schmiga, M. Rauer, M. Rüdiger, K. Meyer, J. Lossen, H.-J. Krokoszinski, M. Hermle, S.W. Glunz, "Aluminium-Doped p+ Silicon for Rear Emitters And Back Surface Fields: Results and Potentials of Industrial n- and p-Type Solar Cells," in *25th European Photovoltaic Solar Energy Conference and Exhibition / 5th World Conference on Photovoltaic Energy Conversion*, Valencia, Spain 2010.

[11] S. W. Glunz, J. Benick, D. Biro, M. Bivour, M. Hermle, D. Pysch, *et al.*, "n-type silicon - enabling efficiencies > 20% in industrial production," in *2010 35th IEEE Photovoltaic Specialists Conference*, 2010, pp. 000050-000056.

[12] F. Book, T. Wiedenmann, G. Schubert, H. Plagwitz, and G. Hahn, "Influence of the Front Surface Passivation Quality on Large Area n-Type Silicon Solar Cells with Al-Alloyed Rear Emitter," *Energy Procedia*, vol. 8, pp. 487-492, 2011.

[13] D. L. Meier, V. Chandrasekaran, H. P. Davis, A. M. Payne, X. Wang, V. Yelundur, *et al.*, "N-Type, Ion-Implanted Silicon Solar Cells and Modules," *IEEE Journal of Photovoltaics,* vol. 1, pp. 123-129, 2011.

[14] J. J. Valentin D. Mihailetchi, Alexander Edler, Radovan Kopecek, Rudolf Harney,Daniel Stichtenoth, Jan Lossen, Tim S. Böscke, Hans-Joachim Krokoszinski, "Screen Printed n-Type Silicon Solar Cells for Industrial Application," in *25th European Photovoltaic Solar Energy Conference and Exhibition*, 2010, pp. 1446 - 1448.

[15] P. A. Basore and K. Cabanas-Holmen, "PC2D: A Circular-Reference Spreadsheet Solar Cell Device Simulator," *IEEE Journal of Photovoltaics,* vol. 1, pp. 72-77, 2011.

[16] V. Chandrasekaran, H. P. Davis, A. M. Payne, V. Yelundur, A. Rohatgi, and D. L. Meier, "20% n-Type silicon solar cell fabricated by a simple process with an aluminum alloy rear junction and extended emitter," in *2016 IEEE 43rd Photovoltaic Specialists Conference (PVSC)*, 2016, pp. 2412-2416.

[17] A. Ebong, N. Chen, V. Unsur, A. Chowdhury, and B. Damiani, "Innovative Front Grid Design, Four-Streets and Five-Busbars (4S-5BB), for High Efficiency Industrial Al-BSF Silicon Solar Cell," *IEEE Electron Device Letters,* vol. 37, pp. 459-462, 2016.

Photovoltaic Performance of Silicon Solar Cells Enhanced by Plasmonic Silver Nanoparticles of Various Dimensions Depositing Through Anodic Aluminum Oxide Template

Ta-Wei Chuang, Wen-Jeng Ho*, Sheng-Kai Feng, Jheng-Jie Liu, Guan-Yi Li, Hao-Yu Yang, Yun-Chie Yang, Cho-Chun Chiang, and Yao-Hui Chen

Department of Electro-Optical Engineering, National Taipei University of Technology, Taipei 10608, Taiwan, R.O.C.

*: wjho@ntut.edu.tw

Abstract — The plasmonic scattering performances of periodic silver nanoparticles (Ag-NPs) deposited on silicon solar cells using anodic aluminum oxide (AAO) mask and e-beam evaporation are demonstrated. The dimension and profile of Ag-NPs are controlled by the pore diameter and configuration of AAO template. Optical reflectance, absorption, external quantum efficiency (EQE), and photovoltaic current-voltage measurements are used to reveal the contribution of plasmonic scattering induced by Ag-NPs of various dimensions. Larger dimension of Ag-NPs exhibited a high efficiency enhancement (from 10.85% to 12.50%) than that of smaller one (10.85% to 12.07%) because the improving in optical reflectance and EQE due to plasmonic scattering of NPs are appeared much high for the cell with large Ag-NPs beyond 650 nm wavelength.

Index Terms — Nanoparticles, Plasmonic scattering, Anodic aluminum oxide (AAO), Silicon solar cells

I. INTRODUCTION

Metallic particles have unique optical properties when the size of a metallic particle was reduced to the nano-scale because the optical properties can be greatly changed by the appearance of surface plasmons (SPs) [1, 2]. The localized surface plasmon resonance (LSPR) was discovered that increase coupling of light into silicon (Si) by strong optical far-field scattering and a strongly enhanced optical near-field around the particles [3]. Metallic particles can be obtained by a number of deposited methods. Typically, the metallic thin films deposited by thermal evaporation and subsequently annealed to obtain metallic particles which can obtain random sized and random distributed metallic particles. However, the periodic distribution and nano-scale metallic particles are needed some advanced techniques. To obtain nano-size controllable nanoparticles (NPs), the anodic aluminum oxide (AAO) templates have attracted considerable attention for being one of the most successful mask tooling [4].

In this study, the pore dimension and profile of the fabricated AAO template are examined by field emission scanning electron microscope (FE-SEM, LEO 1530). The optical and electrical properties of silver nanoparticles (Ag-NPs) of various dimensions using AAO template deposited on silicon solar cells are revealed by optical reflectance, optical absorption and external quantum efficiency (EQE) measurements. The photovoltaic performances of silicon solar cells enhanced by plasmonic Ag-NPs of various dimensions are confirmed by photovoltaic current-voltage (I-V) under one-sun AM 1.5G illumination.

II. EXPERIMENTS

The 5N purity aluminum (Al) foil was used as the starting material in the fabrication of the AAO template. The Al foil was cleaned by acetone and isopropyl alcohol, and then washed in deionized water. The cleaned Al foil was annealed under ambient atmospheric conditions at 300 °C for 3 hours. To minimize the impact of surface roughness, the Al foil was through chemical polishing to smooth the surface. Next, the polished Al foil was then anodized in a solution of 0.3 M oxalic acid ($H_2C_2O_4$) with an applied voltage of 60V, at 10 °C for 3 min. The AAO thin-film was formed on the Al template with self-ordering pores. The depth of the nanopores and their diameter can be controlled by anodization time and pore-widening time. The thickness of the AAO depends on the duration of the anodization process. Through control pore-widening time, respectively were 30 mins and 40 mins, to form two different diameter pores. The samples were then immersed in a solution of copper chloride ($CuCl_2$), hydrochloric acid (HCl), and DI water at a ratio of 1:1:20 by weight for 30-40 mins, in order to remove the Al substrate. The barrier layer of the thin AAO film was removed using chemical wet etching in a 6 wt% H_3PO_4 solution at 25 °C in order to obtain fully-opened pores at the bottom of the AAO layer. Finally, we characterized the morphologies and pore structures of the AAO templates by using the images of SEM and confirmed the pore distribution using image processing software of image-J. Figs. 1(a) and (b) show the top-view SEM images and the distribution profile of the fabricated AAO templates by pore-widening for 30-40 mins.

Plasmonic Si solar cells were begun by depositing a 20-nm-thick TiO_2 spacer layer on two bare Si solar cells (called samples A, B) using an e-beam evaporator. Sample-A was deposited Ag film to a thickness of 30 nm through the AAO template with larger dimension nanopores. On the other hand, Sample-B was deposited Ag film to a thickness of 30 nm through the AAO template with smaller dimension nanopores. After Ag films deposition and AAO template removed, the Sample-A and sample-B were annealed in an RTA chamber at 200 °C for 30 mins under ambient H_2 to transform the thin Ag films into Ag NPs. A schematic of the plasmonic Si solar cell with periodic distribution of Ag-NPs through AAO template is presented in Fig. 2. The optical reflectance of the bare cell and plasmonic cells was characterized using a UV/Vis/NIR spectrophotometer. The EQE responses of the bare solar cell and plasmonic Si solar cell were characterized from 350 to 1100 nm wavelength. The photovoltaic current density-voltage (J-V) characteristics of the bare and plasmonic Si solar cells were characterized under one-sun AM 1.5G solar simulation.

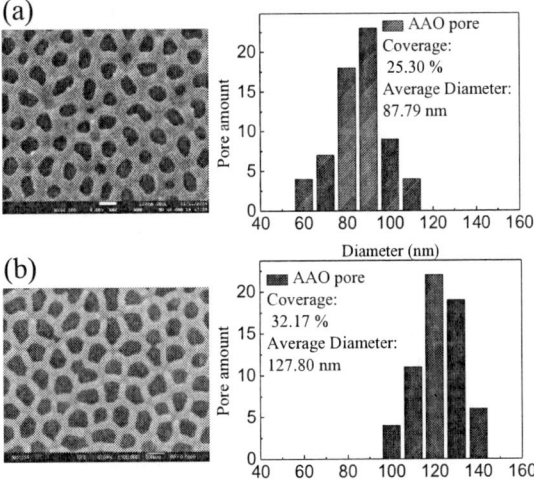

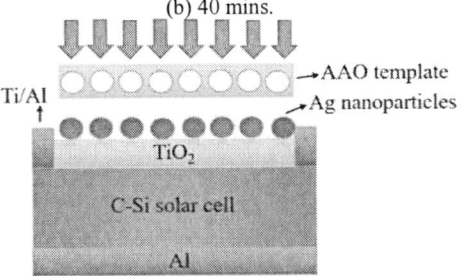

Fig. 1. The top-view SEM images and the distribution profile of the fabricated AAO templates by pore-widening for (a) 30 mins, (b) 40 mins.

Fig. 2. Schematic of plasmonic Si solar cell depositing Ag-NPs through AAO template

III. RESULTS AND DISCUSSIONS

The phenomenon of surface plasmon resonance (SPR) results in strong light scattering and absorption characteristics. Fig. 3 shows the absorption spectrum of samples with the following configurations: TiO_2/quartz-glass-substrate and Ag-NPs/TiO_2/quartz-glass-substrate. The plasmonics absorption range of Ag-NPs can be observed at 370-550 nm wavelengths

which well agreed to the surface plasmon resonance of Ag-NPs on the previous reported.

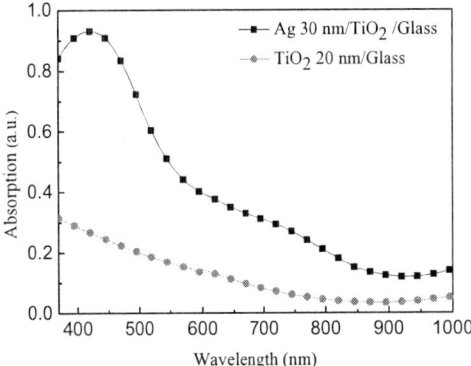

Fig. 3. Absorption spectrum.

Figure 4 shows the optical reflectance of a bare solar cell, a cell with a TiO_2 layer, and cells with different dimension Ag-NPs on a TiO_2 layer. The reflectance of the cell with a layer of TiO_2 was lower than that of the bare cell. The reflectance of the cell with the larger dimension Ag-NPs on a TiO_2 layer decreased to a level below that of the cell with only a TiO_2 layer beyond 475 nm wavelengths was due to the plasmonic forward scattering of incident photons induced by Ag-NPs. The reflectance of the cell with the larger and smaller Ag-NPs was higher than that of the cell with a TiO_2 layer in the wavelength range of 350-450 nm due to optical reflectivity of the Ag-NPs. Besides, the weighted average reflectance of the all evaluated cells is summarized in Table I. The lowest weighted average reflectance of 26.86% for the cell with larger Ag-NPs, particularly, plasmonic forward scattering of the larger Ag-NPs was higher than that others one in the wavelength range 500-1000 nm.

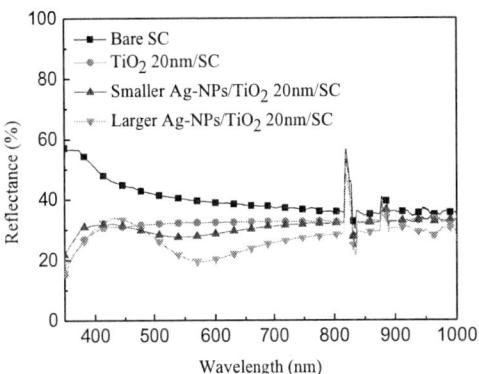

Fig. 4. Optical reflectance spectrum.

Figure 5 shows the EQE response of all solar cells tested in this work. The EQE values of the cell with Ag-NPs of larger dimension and of smaller dimension were lower than that of the cell with only a TiO_2 layer at wavelengths below 560 nm and 550 nm, respectively due to the optical reflection of the Ag-NPs. It should also be noted that the EQE values of the cell with Ag-NPs was lower than that of the bare solar cell at wavelengths of 350-550 nm due to the plasmonic resonance absorption of Ag-NPs. On the other hand, the EQE values increased when the wavelength of incident photons exceeded 550 nm, which also agrees with the results of optical reflectance, due to the plasmonic forward scattering induced by Ag-NPs. Furthermore, the EQE values of the cell with Ag-NPs of larger dimension were superior to that of the cell with smaller dimension. Therefore, larger periodic Ag-NPs as well as high forward plasmonic scattering for enhancing photovoltaic performance are expected.

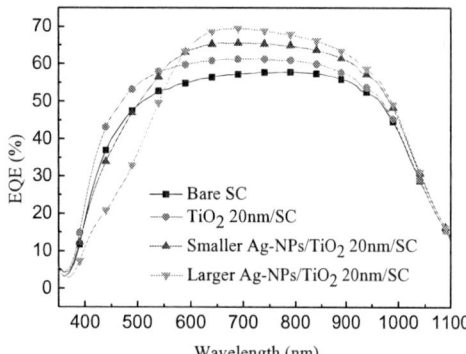

Fig. 5.   EQE response.

Figure 6 shows the photovoltaic J-V characteristics of all solar cells tested in this study under a one-sun AM 1.5G illumination and photovoltaic performance are summarized in Table II. The short circuit current (J_{sc}) of the solar cells were as follows: bare solar cell (25.99 mA/cm^2), cell containing a TiO$_2$ space layer (27.53 mA/cm^2), cell containing Ag-NPs of smaller dimension on a TiO$_2$ space layer (30.65 mA/cm^2), and cell containing Ag-NPs of larger dimension on a TiO$_2$ space layer (32.05 mA/cm^2). The short-circuit current density enhancement (ΔJ_{sc}) results demonstrated the photovoltaic performance of the plasmonic solar cell with Ag-NPs of larger dimension is better than smaller dimension one. This is based on the fact that J_{sc} is proportional to EQE and η is proportional to J_{sc}.

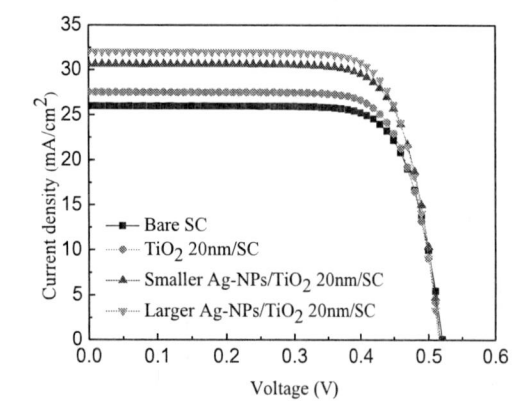

Fig. 6.   Photovoltaic J-V.

TABLE I
THE REFLECTANCE OF THE CELLS

380-1040 nm	Weighted average reflectance
Bare solar cell	40.78
TiO$_2$/Bare solar cell	33.78
Smaller Ag-NPs/TiO$_2$/Bare solar cell	30.97
Larger Ag-NPs/TiO$_2$/Bare solar cell	26.86

TABLE II
THE PHOTOVOLTAIC PERFORMANCES OF THE CELLS

	V_{oc} (mV)	J_{sc} (mA/cm^2)	F.F. (%)	η (%)	ΔJ_{sc} (%)	$\Delta \eta$ (%)
Bare SC	520.1	25.99	76.6	10.35	N/A	
TiO$_2$/Bare SC	517.4	27.53	76.2	10.85	5.93	4.83
Smaller Ag-NPs/TiO$_2$/Bare SC	517.3	30.65	76.2	12.07	17.9	16.6
Larger Ag-NPs/TiO$_2$/Bare SC	515.3	32.05	75.7	12.50	23.3	20.8

IV. CONCLUSION

Anodic aluminum oxide (AAO) templates with the pore dimension of 88-128 nm and coverage of 25-32% were fabricated. The periodic nano-scale Ag-NPs using AAO templates for film masking deposition were demonstrated. In this study, larger dimension Ag-NPs exhibited a superior plasmonic scattering than that of smaller dimension Ag-NPs one. The short-circuit current density of 32.05 mA/cm^2 of the cell with larger dimension of Ag-NPs were obtained, which it was higher than the cell with smaller dimension of Ag-NPs of 30.65 mA/cm^2.

ACKNOLOWLEDGEMENTS

The authors would like to thank the Ministry of Science and Technology (MOST) of the Republic of China for financial support under Grant MOST 103-2221-E-027-049-MY.

REFERENCES

[1] M. A. Garcia, "Surface plasmons in metallic nanoparticles: fundamentals and applications," *J. Phys. D: Appl. Phys.*, vol. 44, 283001, pp. 1-20, 2011.

[2] M. Schmid, P. Andare, P. Manley, "Plasmonic and photonic scattering and near fields of nanoparticles," *Nanoscale Res. Lett..,* vol. 9:50, pp. 1-11, 2015.

[3] E. Hutter, J. H. Fendler, "Exploitation of localized surface plasmon resonance," *Adv. Mater.*, vol. 16, pp. 1685-1706, 2004.

[4] J. Liang, H. Chik, A. Yin, J. Xu, "Two-dimensional lateral superlattices of nanostructures: Nonlithographic formation by anodic membrane template," J. Appl. Phys., vol. 91(4), pp. 2544-2546,2002.

Mitigation of Potential-Induced Degradation

Orry Faur, Maria Faur

Specmat Inc, North Olmsted, Ohio, USA

Abstract — since potential-induced degradation (PID) is driven by voltage, heat, and humidity, all PV systems can be susceptible to PID and the associated power output drop of 30 percent or more [1]. The failure mechanism depends on the module's voltage potential and leakage current to facilitate ion migration through the antireflection coating (ARC) from other parts of the module such as the glass, encapsulation, mount, or frame [2]. Sodium ions are believed to be the most likely culprits but debate is ongoing concerning other ions such as aluminum, magnesium, and calcium which are present in the soda-lime glass commonly used in panels [3]. There are some recognized methods for mitigation of potential-induced degradation (PID). Replacing soda-lime glass with quartz glass ensures that the solar glass is free of the suspected PID ions [2]. At the cell level, utilizing a less than ideal higher IR silicon nitride (SiN) in the ARC minimizes PID. Here we describe the Room Temperature Wet Chemistry Growth (RTWCG) technology, a novel technology for growing highly uniform amorphous SiOX layers into silicon substrates [4]. These RTWCG SiOX layers mitigate PID when used as the first dielectric layer in a stacked SiOX / SiN ARC. Unlike some other methods for mitigating PID which lower cell performances, the inclusion of a SiOX layer to the ARC passivates the c-Si surface and thus increases cell performances.

Index Terms — oxidation, passivation, photovoltaic cells, PID, potential-induced degradation, RTWCG, surface passivation.

I. INTRODUCTION

Potential-induced degradation (PID) is a phenomenon resulting in catastrophic power output losses of grid-connected photovoltaic systems. To build up the voltage output, solar panels are typically connected in series, creating a high electric potential difference between the solar cells and the module frame. This electric potential can drive ion mobility into the photovoltaic cell from the materials surrounding it (e.g. frame, glass, mount, surrounding environment). Since increases in maximum system voltage reduces system costs, it is inevitable that, absent PID mitigation, PID in future systems will only intensify [5].

Unavoidable conditions which promote PID in modern photovoltaic systems include environmental factors such as heat and humidity. Although certain materials are known to increase the likelihood of PID, market forces oblige manufacturers to use these inferior, yet commercially feasible, materials. It's known that soda-lime glass contains many of the prime suspects of PID (sodium, aluminum, magnesium, and calcium) [6], it is, nonetheless, the only commercially feasible glass available to manufacturers. And, even though a high system voltage is a major contributor to PID, installers prefer large arrays of series-connected modules to lower the costs from wire losses, and copper utilization.

One way PID can be mitigated at the cell level, is to utilize a silicon-rich silicon nitride ARC [7]. Unfortunately, although PID is mitigated, the higher index of refraction of the resulting ARC is optically inferior, with significant losses in Jsc. In one test, Amrani, et al observed that by increasing the index of refraction from 1.9 to 2.1, Jsc decreased from 24.8 mA/cm^2 to 21.1 mA/cm^2 [8]. This lowering in Jsc is problematic to the photovoltaic cell market which places a premium on cell efficiency. Higher efficiency reduces space requirements, encapsulation costs, and module fabrication costs.

The RTWCG technology is a wet chemical emersion process that results in the very rapid oxidation of silicon substrates to create a beneficial oxide (SiOX). Generally, the technology takes place in a single step lasting between about 30 and 80 seconds on a simple wet chemical bench (either inline or batch) [4].

The SEM micrograph (Fig. 1) shows an approximately 40 nanometers-thick RTWCG SiOX oxide layer which was grown in 60 seconds on a textured CZ single-crystalline silicon wafer. The SiOX layer is composed of silicon (Si), oxygen (O), and X can be either fluorine (F), carbon (C), nitrogen (N), or oxygen (O). The SiOX growth process can be precisely controlled to accurately create specific thicknesses at specific growth rates. The thickness range can be from 1 nm to 500 nm and the growth rate can range from as high as 10 nm per second to 10 nm per hour [4].

Fig. 1 The SiOX layer can be grown to a desired thickness irrespective of surface topography such as the pyramid structures found on textured CZ wafers.

The RTWCG SiOX process was created by Special Materials Research and Technology (Specmat) Inc. for the use in advanced cell designs including the PERC and PERT cell structures, as well as selective emitter schemes. This process relies on a chemical solution that oxidizes silicon surfaces, growing a highly uniform SiOX oxide layer into the wafer [4].

Since the silicon source is the substrate itself, this process changes the surface resistivity of the wafer. Depending on the diffusion profile of the wafer, the sheet rho delta for creating a 15 nm SiOX layer can range anywhere between 8 to 30 ohms per square [4].

II. EXPERIMENTAL

The degree to which the inclusion of a certain thickness of a bottom SiOX layer to a silicon nitride (SiN) antireflection coating ameliorates surface conditions, was determined by measuring the minority carrier lifetime of 180 sister CZ wafers. All minority carrier lifetime measurements were obtained using the quasi-steady-state photoconductance method (QSSPC) on a Sinton Instruments WCT-120.

The 180 wafers were textured, POCl diffused to 70 ohms per square, and the phosphosilicate glass was cleaned in a ten percent aqueous hydrofluoric solution bath. The cleaned wafers were randomized between two groups, the "stacked" group and the "baseline" group.

The stacked group, containing one hundred PSG-cleaned CZ wafers, was fully immersed into the RTWCG SiOX growth solution to grow a ten nanometers-thick SiOX layer into both faces. The growth process increased sheet resistivity by approximately nine ohms per square. A seventy nm-thick silicon nitride layer was deposited on both faces to form a SiN / SiOX / Si / SiOX / SiN structure. The remainder of the PSG-cleaned CZ wafers made up the baseline group on which a seventy-eight nm-thick silicon nitride layer was deposited on both faces, forming a SiN / Si / SiN structure. Both groups were then treated to a metal-firing thermal budget in a nine-zone firing furnace.

The two groups' reflectance curves were measured and averaged. Although the SiOX component lowers reflection in the blue region of the spectrum, as Fig. 2 shows, the reflectance minimums of the stacked and baseline groups were otherwise well matched. The averaged minimum reflectance of both the SiN-only group and the SiOX / SiN group were at 600 nm.

The optimal, ten nanometers-thick, SiOX thickness was determined experimentally by looking at the influence that the SiOX component thickness plays to the minority carrier lifetime of SiN / SiOX / Si / SiOX / SiN structures. Six groups of five sister CZ wafers were treated in a similar manner as the already previously described stacked group. The SiOX thickness tested were: 0 nm, 3 nm, 6 nm, 9 nm, 12 nm, and 15 nm. All the groups were thermally treated to standard metal firing conditions and the QSSPC minority carrier lifetime was measured using a WCT-120 lifetime tool.

Ten, sixty-cell, panels were built to determine the SiOX thickness of a SiOX / SiN stacked ARC, that will mitigate potential-induced degradation (PID) of CZ c-Si photovoltaic cells. Three panels utilized eight nanometers-thick SiOX, two

panels utilized ten nanometers-thick SiOX, two panels utilized fifteen nanometers-thick SiOX, and three panels utilized twenty nanometers-thick SiOX. The silicon nitride component of the SiOX / SiN stack had an index of refraction of 2.05.

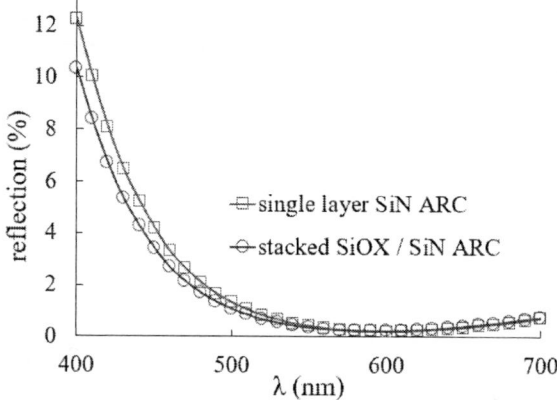

Fig. 2   A ten nanometers-thick SiOX component in a SiOX / SiN stacked ARC lowers the ARC's reflectance in the blue region of the spectrum.

Similarly, two panels were built to test the SiOX thickness necessary to mitigate PID for multicrystalline photovoltaic cells. One of the panels contained a five nanometers-thick SiOX component. The second panel contained a fifteen nanometers-thick SiOX component.

The PV modules were tested under bias of -1000 VDC for a total stress duration of 96 hours. The relative humidity was 85% at a temperature of 85°C. Panels were considered to fail if their power retention (initial power divided by final power) dropped below 95%.

In all experiments, inline wet chemical benches were used to create the SiOX layer. A Baker prototype, two-bath, wet bench was used in all lifetime studies. A commercially available, Rena wet bench was used for the potential-induced degradation tests. The process flow in Fig. 3 was employed to create the photovoltaic cells used in the PID panel tests.

step 1: texturing	step 4: PSG cleaning	step 7 screen printing metallization
step 2: phosphorus diffusion	step 5: RTWCG of SiOX layer	step 8 metallization firing
step 3: rear emitter etch	step 6 PECVD SiN deposition	

Fig. 3   All photovoltaic cells utilized in the PID testing were created using the same process flow.

III. RESULTS

Figures 4 and 5 show the degree to which the SiOX thickness, when employed as the bottom layer of a SiOX / SiNx stacked ARC, impacts PID for single crystalline and multicrystalline cells, respectively. In both trials, a specific SiOX thickness existed which when surpassed, the panels passed PID testing. For the single crystalline panels in Fig. 4 the SiOX thickness for PID mitigation was approximately twenty nanometers-thick, fifteen nanometers-thick for the multicrystalline panels in Fig. 5.

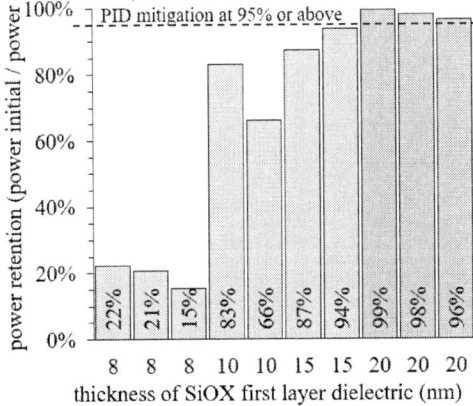

Fig. 4   The PID tests conducted on ten, sixty-cell, panels show that a 20 nm-thick SiOX layer mitigates PID for these CZ panels.

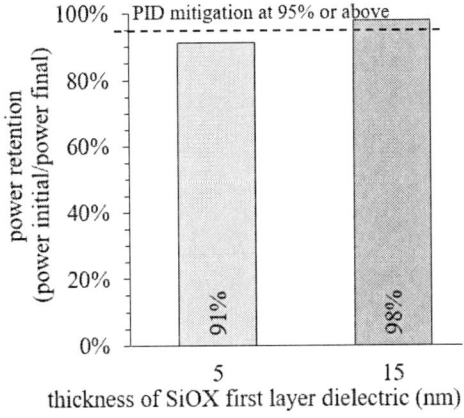

Fig. 5 –PID test on two, sixty-multicrystalline cell, panels show that a 15 nm-thick SiOX layer mitigated PID.

An approximately ten nanometers-thick SiOX layer provides the maximum amount of surface passivation, according to the lifetime data of Fig. 6. Thirty sister CZ wafers were split into six groups, each group containing five CZ wafers. The wafers in one of the groups did not contain a SiOX layer.

Using the ten-nanometer maximum passivation thickness of SiOX, the minority carrier lifetimes of a one hundred-wafer stacked group were compared to an eighty-wafer baseline group, and the results are shown in Fig. 7. The inclusion of the ten nanometers-thick SiOX component into the ARC accounted for an increase in minority carrier lifetime of 68 µs, from 93 µs to 161 µs.

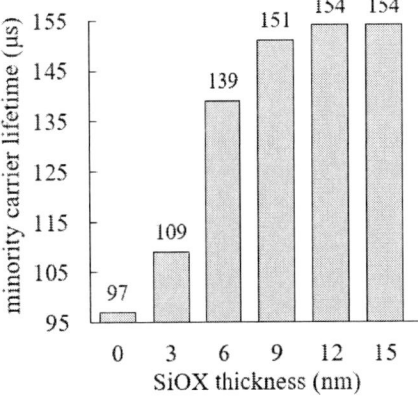

Fig. 6   An approximately ten nm-thick SiOX component layer of a SiOX / SiNx stacked ARC provides the maximum surface passivation.

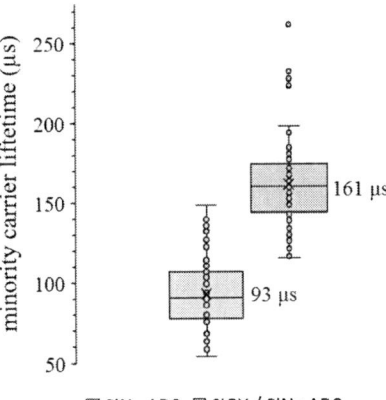

Fig. 7   The SiOX component of a SiOX / SiNx stacked ARC improved the minority carrier lifetime of CZ wafers, by nearly 70 µs.

IV. CONCLUSION

Amorphous SiOX thin films were grown into silicon wafers using the Room Temperature Wet Chemical Growth (RTWCG) process. Quasi-steady-state photoconductance (QSSPC) minority carrier lifetime measurements confirmed that it is possible to increase the minority carrier lifetime of silicon substrates by using the SiOX layer as the bottom layer of SiOX / SiN stacked antireflection coatings. Photovoltaic cell panels utilizing the SiOX / SiN stacked antireflection coating, wherein a minimal SiOX component thickness is reached, were shown to be free of potential-induced degradation.

All c-Si photovoltaic cell manufacturing facilities already make use of wet benches for either batch, or inline, phosphosilicate, or borosilicate, glass cleaning. The RTWCG SiOX process can utilize either type of chemical wet bench to clean wafers post-diffusion. But unlike standard post-diffusion cleaning solutions, the RTWCG solutions will concomitantly create a beneficial SiOX layer. As has been shown herein, the SiOX layer not only mitigates potential-induced degradation, but also increases the minority carrier lifetime of the silicon material. The RTWCG process can add a tremendous amount of, essentially capex-free, value to the post-diffusion cleaning step of any photovoltaics cell fab.

REFERENCES

[1]. S Hoffman, M. Koehl, "Effect of humidity and temperature on the potential-induced degradation," in Progress in photovoltaics, vol. 22, 2014, pp. 173-179.

[2]. Blieske, Stollwerck, "Glass and other encapsulation materials," Semiconductors and semimetals, vol. 89, 2013, pp. 200-215.

[3]. V. Naumann, et al, "Explanation of potential-induced degradation of the shunting type by Na decoration of stacking faults in Si solar cells," in Solar energy materials and solar cells, vol. 120, Part A, 2014, pp. 383-389.

[4]. Maria Faur, Horia Faur, Mircea Faur. "Compositions to facilitate room temperature growth of an oxide layer on a substrate." USA, PCTUS2011028190, June 03, 2014.

[5]. Wei Luo, et al, "Potential-induced degradation in photovoltaic modules: a critical review," Energy & environmental science, vol. 10, 2017, pp. 43-68.

[6]. M. Schütze, et al, "Laboratory study of potential induced degradation of silicon photovoltaic modules", 37th IEEE PVSC, Seattle, Washington, USA, 2011, pp. 821.

[7]. Wenguang Quan. "Solar cell and manufacture method thereof," USA, US20160284884 A1, Jun 8, 2016.

[8]. A. El Amrani, et al, "Determination of the suitable refractive index of solar cells silicon nitride," Superlattices and Microstructures, vol. 73, September 2014, pp 224–231.

Electrodeposition of Si-layer Through Reduction of Diatomaceous Earth for the Application of Solar-Cells

Muhammad Monirul Islam[1,2], Imane Abdellaoui[1], Takeaki Sakurai[1], Saad Hamzaoui[3], and Katsuhiro Akimoto[1]

[1]Division of Applied Physics, Faculty of Pure and Applied Sciences, University of Tsukuba, Tsukuba, Ibaraki 305-8573, Japan.

[2]Alliance for Research on North Africa (ARENA), Faculty of Pure and Applied Sciences, University of Tsukuba, Tsukuba, Ibaraki 305-8572, Japan.

[3]Laboratory of Electronic Microscopic and Materials Sciences, Faculty of Electrical Engineering, Oran University, Oran 31000, Algeria

Abstract — Silicon (Si) layer was formed on the silver (Ag) substrates using electrodeposition technique through electrochemical-reduction of the purified (99%) diatomaceous earth (SiO₂) at high temperature of 855° C using calcium chloride (CaCl₂) molten-salt. Electrodeposition was conducted using a three-electrode electrochemical -cell. Reduction of the SiO₂ was possible with applying negative potential of more than 0.7 volt on the Ag-electrode with respect to graphite reference-electrode. Raman spectroscopy and X-ray diffraction method confirm formation of the Si-layer on the Ag-substrate from SiO₂ using electrodeposition technique. Scanning electron microscopy indicates that deposited Si has morphology of rough surfaces.

I. INTRODUCTION

Silicon (Si) is the most abundant semiconductor-material found in form of silica (SiO_2) that contributes to the large-scale to macro and micro-electronics including photovoltaics. Thus, production cost of Si has significant impact on the application of solar electricity. For solar cells, Si with 99.9999% (6N), *i.e.*, solar-grade silicon (SOG-Si) is necessary. However, there remain several challenges in the current production of Si using carbothermic reduction which is associated with huge energy consumption and carbon-di-oxide (CO_2) emission [1]. We aim to develop a cost-effective technology for the large-scale production of Si, taking into account the environmental issue at the same time. In this paper, we have studied electrodeposition of the Si on the Ag-substrate obtained directly from the SiO_2 source-material. In particular, electrochemical-reduction of the SiO_2 using electrodeposition technique is very promising, because this is a low-cost method requiring less energy comparing to the current carbothermic process, and also environment-friendly [2,3]. According to thermodynamic properties of SiO_2, reduction occurs at high temperature of 850° C in $CaCl_2$ molten salts as of the Ellingham diagram [4] following the basic equation:

$$SiO_2 \text{ (Solid)} + 4e \rightarrow Si \text{ (reduced)} + 2O^{2-} \text{ (gas)} \quad (1)$$

Traditionally, to be used as starting material, silica (SiO_2) is obtained from silica-ores, sand *etc*. In this study, we

have used diatomaceous earth, an alternative source for high purity Si. Diatomaceous earth is a siliceous sedimentary rock mainly consists of fossilized diatoms, a kind of alage. It is chemically composed of mainly SiO_2 ranging from 80-95 %, along with Al_2O_3, iron-oxide *etc*. In this study, we have used diatomaceous earth (SiO_2) with 99% purity obtained after a wet chemical purification process. Thus, in this study, we have reported purification and reduction of diatomaceous earth to make Si-layer on the Ag- substrate. Various reduction potentials were applied during the electrodeposition technique to study the effect of the electrolysis potential on the quality of the deposited Si-layer.

II. EXPERIMENTAL

For purification of silica, the conventional process, an alkaline dissolution followed by acid precipitation of silica gel was adopted [5]. Experiment for the electrodeposition of Si has been carried out in a Al_2O_3- crucible placed inside a quartz electrochemical-cell. A vertical electrical furnace has been used to increase temperature of the electrochemical-cell up to 855^0 C required for electrochemical reduction of SiO_2 to occur. Electrochemical analysis was performed with a three-electrode configuration of the electrochemical-cell. Two separate graphite-rods were used as a counter-electrode (CE)/anode, and a reference-electrode (RE), respectively. An Ag-sheet has been used as a working-electrode (WE)/cathode as well as the substrate for the electrodeposition experiment. Chronoamperograms has been done at constant potential (E) applied between Ag-substrate (WE /cathode) and graphite-RE. Cyclic voltammetry (CV), and all the constant potential (chrono-amperometry, CA) electrolysis were carried out with an automatic polarization system (HSV-110, Hokuto Denko Co. Ltd., Japan).

32 gm of $CaCl_2$-electrolyte (anhydrous powder, 97%, Sigma-Aldrich, Japan), has been taken at the Al_2O_3-crucible, which was placed inside the quartz electrochemical-cell. 450 mg SiO_2 (diatomaceous earth) been taken in the same

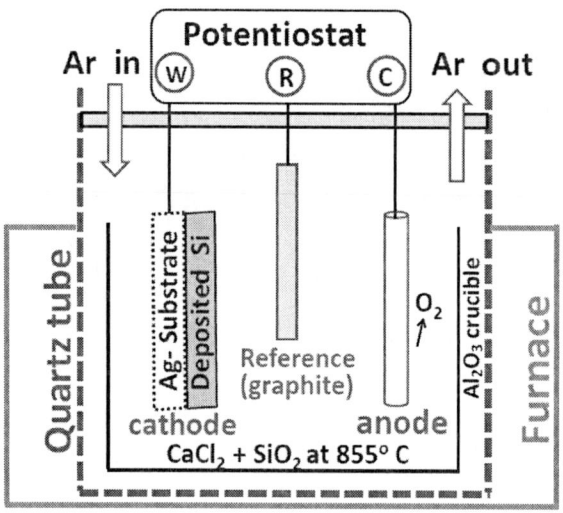

Fig. 1. Experimental setup for the electrodeposition system to conduct the electrochemical reduction of SiO₂ and formation of Si-layer.

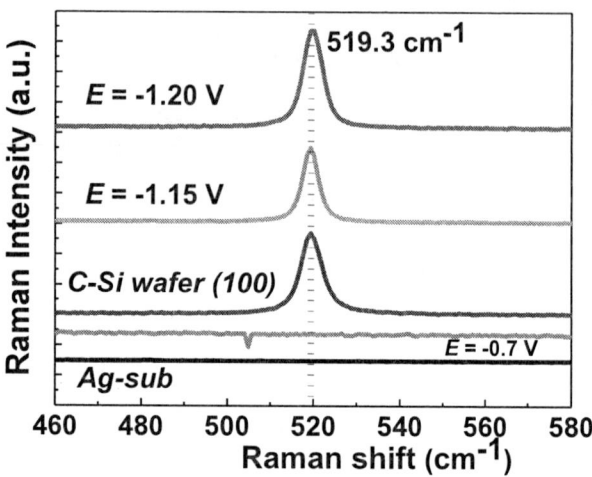

Fig. 2. Raman spectra of the Si-layer electrodeposited with various reduction potential applied on the Ag-substrate vs. graphite-RE.

Al₂O₃-crucible and mixed together with the CaCl₂-electrolyte, and the crucible was placed again inside the electrochemical-cell. Electrodeposition experiment was performed under Ar-atmosphere (flow rate ~ 150mL/min) at 855°C. After electrodeposition experiment, formed Si-layer on Ag-substrate has been washed ultrasonically using 0.2 mol% hydrochloric (HCl) acid for 10 minutes followed by washing with pure water.

Characterization of the electrodeposited Si-layer was performed using Raman spectroscopy, scanning electron microscope (SEM), and X-ray diffraction technique (XRD). XRD was taken in the range of 10° ~ 90° by philips X'pert diffractometer at θ-2θ mode with Cu-Kα (α =1.541837 Å) radiation operated at 45 kV and 40 mA. A 532-nm excitation line from a Nd:YAG laser source has been used for the Raman spectroscopy. All the characterization of the Si-layer has been done at room temperature (RT).

II. RESULTS AND DISCUSSIONS

In order to investigate the required potential which is necessary for the reduction of the SiO₂ (diatomaceous earth) in the molten CaCl₂ on the Ag-substrate, and also to understand the nucleation mechanism of Si on the Ag, we have performed CV. CV was performed with applying potential at a scan rate of 20 mV/s between Ag-substrate (WE) and graphite-RE at 855° C containing CaCl₂ and SiO₂ in the melt. It has been found from the CV curve (not shown) that reduction of the SiO₂ is possible from – 0.7 V to more negative potential vs. graphite-RE. Thus, we have performed electrodeposition using the Chronoamperograms (CA) technique that involves single

potential step i.e., applying a suitable reduction potential at Ag-WE vs. graphite-RE in forward direction, which produces a cathodic or reduction current only. We have selected several reduction potentials to investigate the effect of electrolysis potential on the quality of the formed Si-layer from SiO₂ in molten CaCl₂. The potential-window for the reduction was chosen in way so that it remains well below the reduction potential (above -1.4 V vs, graphite-RE, as found from CV curve) of Ca²⁺ i.e., potentials where it is possible to form Ca-Si related alloys.

Shown in Fig. 2 is the Raman spectra of the obtained Si-layer on the Ag-substrate electrodeposited with applying various reduction potential, such as E = -0.7, -1.15, and -1.20 V vs. graphite-RE. Raman spectra of an oriented Si-wafer and bare Ag-metal were also shown in the similar graph for comparison. Before deposition of Si (bare Ag-metal), no peak is detected in the region of interest of the Rama shift (black, bottom spectrum). In addition, no peak related to the formation of any Si-layer has been detected either for the sample deposited with E = -0.7 V (red, second spectrum from the bottom). While, a sharp and symmetric Raman-peak at ~ 519 cm⁻¹, similar to the peak-position of the Si-wafer, has been apparent for the sample electrodeposited with E = -1.15 and -1.2 V. For crystalline Si, a sharp peak, attributed to the Raman scattering by vibration of the Si-Si bonds in transverse optical (TO) vibrational mode has been reported to appear at 520 cm⁻¹ [6]. When the long-range order is lost, the TO peak becomes broad and shifts to lower wave numbers [7,8]. Thus, in case of amorphous silicon, the peak corresponding to TO phonon appears around 480 cm⁻¹. Nevertheless, it is clear from the Fig. 2 that for the deposited film is basically crystalline in nature.

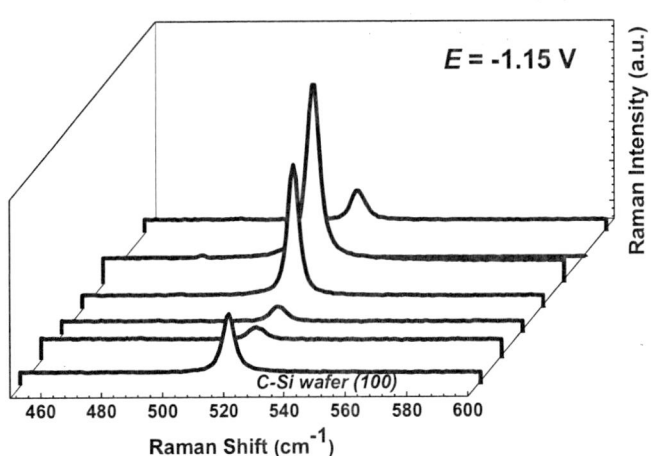

Fig. 3. Raman spectra at various point of the Si-layer on Ag-substrate electrodeposited with the reduction potential, E= -1.15 V vs graphite-RE.

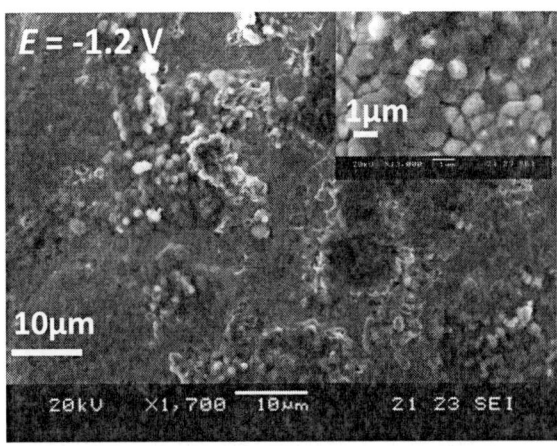

Fig. 4. SEM image of the surface of the Si-layer electrodeposited from diatomaceous earth (SiO_2) with the reduction potential of E= -1.2 V vs graphite-RE. Inset of the figure shows magnified version of the small crystallites region at 1 μm scale.

It has also been seen form the figure that in general, crystallinity of the reduced Si on the Ag-substrate increases with an increase in the reduction potential, as intensity of the Raman peak was found to increases with an increase in the reduction potential more negative with respect to the reference electrode. However, we could not find any clear correlation with broadening of the Raman peak with the reduction potential here in this study. Moreover, Raman spectra at various points of a similar sample (for example, sample electrodeposited with reduction potential, E = -1.15 V vs graphite-RE, Fig. 3) show that feature of the Raman phonon band line (Raman peak) is not similar across the sample associated with various points on the sample with peaks having lower intensity, asymmetrical broadening, and peak-shift to the lower wave number. It suggests that Si-layer electrodeposited for ~ 1 hr on the Ag-substrate is not uniform across the sample surface with the presence of small crystallites or nano-crystalline phase.

To investigate the morphology and structural properties of the deposited Si-layer, we have performed SEM. Shown in Fig. 4 is the surface-SEM images of the electrodeposited Si-layer on the Ag-substrate obtained with applying potential of E = -1.20 V vs. graphite-RE. SEM image reveals that distribution of deposited-Si on the surface is basically not homogeneous in nature. Clustering of small crystallites is evident with the grain size less than 1μm (inset of the figure). Moreover, morphology of the formed Si-layer exhibits rough surface. It seems short deposition period of 1 hr did not cover the surface of the sample homogeneously.

Finally, we have performed XRD for the deposited Si-layer as shown in Fig. 5. XRD also supports the formation of multi-crystalline Si-layer obtained through electrodeposition technique on the Ag-substrate. Presence of several strong peaks

on the XRD pattern which were related to Ag mainly originated from the thick Ag-substrate used in our study. Although we could not confirm from Raman spectra, relatively broad peak at XRD pattern around $2\theta \sim 22^O$ may suggests presence of amorphous like mixed phase in formed multi-crystalline Si in this study. It may also worth to mention that XRD in this study was severely affected by the tilting and roughness of the Ag-substrate which may hinder some Si-peaks and also may weaken intensity of some peaks.

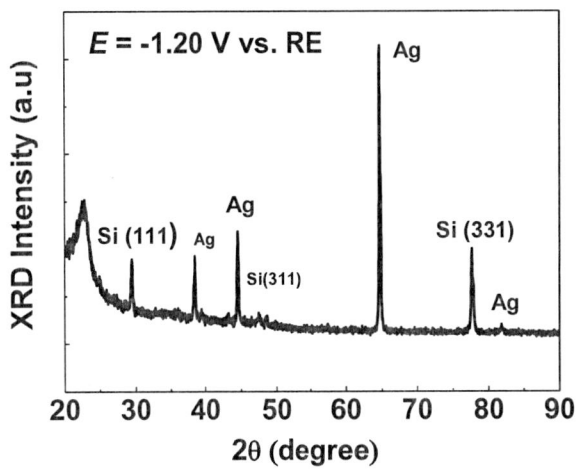

Fig. 5. XRD pattern of the Si-layer electrodeposited from diatomaceous earth (SiO_2) with the reduction potential of E= -1.2 V vs graphite-RE.

IV. CONCLUSION

In conclusion, we have successfully obtained Si-layer on the Ag-substrate using electrodeposition technique through electrochemical-reduction of the diatomaceous earth (SiO_2) in $CaCl_2$ melt at 855° C. A moderate negative potential above 0.7 volt or more negative with respect to the graphite-RE has been found suitable for the reduction of SiO_2. Multi-crystalline Si with very sharp peak at ~ 519 cm^{-1} at Raman spectroscopy was obtained. However, Raman measurement at various spot shows that peak intensity, broadening, and even peak positions are not always comparable which suggest the formation of non-homogeneous Si-layer on the relatively larger Ag-substrate with short time of electrodeposition. Surface image obtained by SEM reveals morphology with rough surface and formation of small crystallites with grain size less than 1 μm.

REFERENCES

[1] W. Zulehner, B. Elvers, S. Hawkin, W. Russey G. Schulz, *Ullmann's Encyclopedia of Industrial Chemistry*, VCH,Weinheim, 5th ed., Vol. A23, pp. 721–748, 1995. .

[2] S. K. Cho, F.-R. F. Fan, A. J. Bard, "Electrodeposition of Crystalline and Photoactive Silicon Directly from Silicon Dioxide Nanoparticles in Molten $CaCl_2$", *Angewandte Chemie International Edition*, vol. 51, pp.12740–12744, 2012.

[3] T. Nohira, K. Yasuda, and Y. Ito, "Pinpoint and bulk electrochemical reduction of insulating silicon dioxide to silicon", *Nature Materials*, vol. 2, pp. 397-401, 2003.

[4] Ellingham, H. J. T., "Transactions and Communications", *J. Soc. Chem. Ind. (London)*, Vol. 63, p125, 1944.

[5] Abdellaoui Imane, M. Monirul Islam, Saad Hamzaoui, Katsuhiro Akimoto, Takeaki Sakurai, "The Purification of Silica (diatomaceous earth) by Wet Chemical Process", *the 77th Japan Society of Applied Physics (JSAP) Autumn Meeting*, 15p-P13-4, Sep 13-16, 2016, Niigata, Japan.

[6] J. H. Parker Jr., D. W. Feldman, M. Ashkin, *Physical Review*, vol. 155, pp. 712-714, 1967.

[7] C. Smit, R. van Swaaij, H. Donker, A. Petit, W. M. M. Kessels, M. C. M. van de Sanden, Journal of Applied Physics, vol. 94, pp. 3582–3588, 2003.

[8] R. L. C. Vink, G.T. Barkema, W. F. van der Weg, *Physical Review B*, vol. 63, pp. 115210 (1-6), 2001.

Effect of Si content in Al paste on Local Al Rear Contacts in PERC Cell

Supawan Joonwichien, Katsuhiko Shirasawa, Satoshi Utsunomiya, and Hidetaka Takato

National Institute of Advanced Industrial Science and Technology (AIST), 2-2-9 Machiikedai, Koriyama, Fukushima 963-0298, Japan

Abstract — We investigated the effect of Si content in the Al paste on the characteristic of local Al contacts, and their impact on passivated emitter and rear cell (PERC) cell performance. We observed an increased open-circuit voltage (V_{oc}) for PERC cells using Al paste with Si. This can be attributed to a thicker and more uniform local aluminum–back surface field (Al–BSF), including a lower void density. By adding Si to the Al paste, the driving force for Si diffusion during alloying is believed to be reduced, causing the increased rejected Si to form a thicker Al–BSF during cooling. This behavior of the Si could prevent the formation of voids, leading to an improvement in the cell performance.

Index Terms — Al local contacts, Al paste, Back surface field, PERC solar cells, Void formation.

I. INTRODUCTION

Please follow these instructions carefully. Passivated emitter and rear cells (PERC) are currently being introduced into high-volume industrial production. An introduction of a dielectric layer above the local Al rear contacts increases the efficiency of the solar cells by minimizing the rear surface recombination at the passivated surface and the recombination of charge carriers at the contacted rear surface, respectively. One way to improve the local Al rear contacts is to modify the patterns of the laser contact opening (LCO) using laser ablation, thereby forming only local aluminum–back surface field (Al–BSF) regions during fire-through contact metallization. The fundamental mechanism of Al–BSF formation during firing can be understood as follows [1]–[3]: 1) During the ramp-up, Si begins to interdiffuse into the Al paste at ~300 °C, followed by Al melting at ~660 °C; 2) At the peak temperature, the maximum amount of Si is dissolved and distributed into the Al-paste matrix. The Al–Si melt at the interface then begins to penetrate into the Si bulk to form the p⁺–doped region (so-called local Al–BSF); 3) During the ramp-down, the Al–BSF is formed by both epitaxial recrystallization of Si from the Al–Si melt, which is heavily doped with Al atoms, and by incorporation of the Al atoms into the Si lattice. This local Al–BSF can reduce the recombination of charge carriers at the rear contact due to the built-in electric field.

In addition to an optimization of the LCO parameters and firing temperatures, other parameters regarding the Al paste, such as the size of the Al particles, the presence of the glass frit and the paste composition elements, should also be considered for improving local Al rear contacts. Rauer et al., [4] reported that by increasing the Si content in Al paste, the contact depth of point contacts is significantly decreased with an increase in thickness of the Al–p⁺ region. Netherthless, the mechanism remains unclear since an Al–BSF thickness relies on the contact geometry and several other process parameters.

In this work, we aim to study the effect of the presence of Si content in the Al paste on the formation of Al–BSF locally underneath the alloy (eutectic layer) and on the PERC cell performance. Si content in the Al paste offers better local contacts, resulting in a thicker Al–BSF below the rear contacts, which leads to an increase in V_{oc}. Si content in the Al paste is believed to suppress the driving force during alloying, thereby reducing a strong lateral Si diffusion into the paste. In addition, Si content in the paste could help to suppress void formation, demonstrating a lower density of voids after firing. This could be beneficial in improving the conversion efficiency of PERC cells.

II. EXPERIMENT

Use a two column format. All paragraphs of text, including the abstract, In this work, 156 mm × 156 mm PERC solar cells were fabricated on p-type Czochralski-Silicon (Cz-Si) wafers with a thickness of 200-μm and a resistivity of 2.0–2.2 Ω·cm. After standard texturing and cleaning, an n⁺–emitter with a sheet resistance of approximately 105 Ω/sq was formed using POCl₃ as a precursor gas. The rear side doped layer was subsequently removed. The rear surface is passivated with a 10 nm-thick atomic-layer-deposited-aluminum oxide (ALD-Al₂O₃) layer. Following this, a 190 nm-thick SiN$_x$ diffusion barrier was deposited using plasma-enhanced chemical vapor deposition (PECVD). An HF-dip was used to remove the oxide layer, prior to the deposition of an 80 nm PECVD-SiN$_x$ film on the front side above the n⁺–emitter. The samples then received a line pattern with LCO linewidth of 72 μm and a pitch of 1 mm. All types of PERC cells were subsequently annealed at 450 °C for 30 min to activate the field effect passivation [5, 6]. Prior to metallization, the samples were shortly dipped into diluted KOH and HF solutions, followed by rear side Al printing with two different Al PERC pastes (paste A and paste B). These two Al pastes were provided by the same supplier to avoid any differences caused by different manufacturing processes. Paste A has no Si content in the Al paste, and paste B has Si content. Finally, top side Ag printing was applied and these finished PERC cells were fired.

TABLE I
CURRENT–VOLTAGE (I–V) PARAMETERS OF PERC CELLS

Description	V_{oc} (mV)	J_{sc} (mA/cm^2)	FF (%)	η (%)
Sample 1_without Si	646.2	38.6	80.0	19.9
Sample 2_without Si	647.8	38.7	79.8	19.9
Sample 3_without Si	647.8	38.7	79.2	19.8
Sample 4_without Si	647.4	38.6	80.6	20.2
Sample 5_without Si	648.8	38.6	80.6	20.2
Sample 1_with Si	649.2	38.4	77.7	19.4
Sample 2_with Si	649.4	38.5	78.4	19.6
Sample 3_with Si	649.1	38.4	77.5	19.3
Sample 4_with Si	649.1	38.4	77.0	19.2
Sample 5_with Si	647.3	38.3	78.5	19.5

III. RESULTS AND DISCUSSION

Table 1 summarizes the current–voltage (I–V) parameters of the PERC cells under standard testing conditions. From these results, it is clear that the short-circuit current density (J_{sc}), fill factor (FF), and conversion efficiency (η) of PERC cells using paste A (no Si content) are higher than those using paste B (with Si content). Furthermore, PERC cells with paste B show an increase in open-circuit voltage V_{oc}.

Scanning electron microscopy (SEM) was used to analyze the local rear contact geometry and the thicknesses of the p$^+$–doped region after the firing-through process at 810 °C. Figure 1 shows cross-sectional SEM images of the local contacts, indicating the local contact geometry as well as the thickness of the local Al-BSF for different paste compositions (with and without Si content). Using paste A (no Si content) (Fig. 1a), the thickness of the local Al-BSF underneath the rear contacts ranges from 4–6 μm. Paste B (with Si content), however, offers a 6–8 μm increase in Al-BSF thickness, which would result in greater electron shielding. It is interesting to note that the contact depth (the eutectic layer below the Al paste formed during cooling) is significantly decreased with the use of paste B (Fig. 1b). This result is consistent with previous report [4] that contact depths after firing become shallower and Al–BSF thickness increase when Si is added to the Al paste. However, it remains unclear whether or not the contact depth would impact the final cell performance.

As has been well demonstrated in [1]–[3], during alloying at peak temperature, the maximum amount of Si from the contact sites is dissolved and distributed into the Al-paste matrix due to a strong Si concentration gradient within the Al-Si melt. At the same time, Al in the Al paste diffuses to the local contact sites to form the Al-BSF. During the ramp-down, excessive Si is rejected from the melt and diffuses back into the contact to recrystallize epitaxially on the Si surface. By adding Si to the Al paste, the driving force during alloying is suppressed, thereby reducing a strong lateral Si diffusion into the paste. It remains possible that additional Si in the Al paste would result in an increase in the amount of Si rejected from the Al-Si melt during cooling, leading to an accumulation of Si on the contact sites, as can be seen in the shallower contact sites. On the other hand, if the concentration of Si in the Al-Si melt is lower, this would result in a thinner Al–BSF or nonexistent Al–BSF, as well as void formation instead of eutectic layers [4].

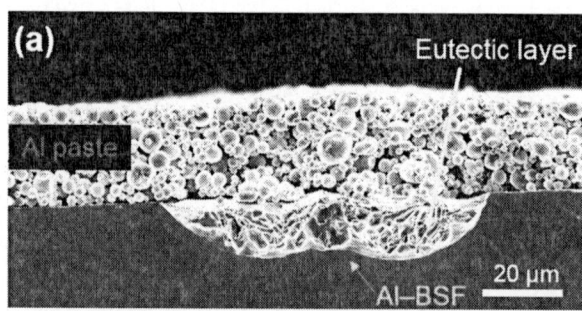

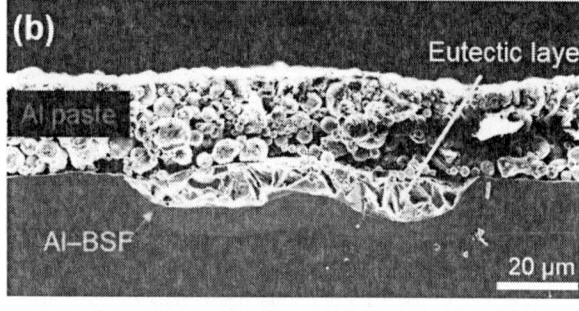

Fig. 1.   SEM cross-sectional images of local Al contacts using (a) Paste A (no Si content), and (b) Paste B (with Si content).

From the SEM analysis, voids were detected regardless of the paste composition. When paste A (no Si content) was used, we observed a high density of complete voids, and an Al–BSF thickness of ~3–4 μm. When paste B (with Si content) was used, voids were only rarely observed, and the voids had an

Al–BSF thickness comparable to that of deep filled contacts (~6–8 μm). In this case, although the Al-BSF of the voids had an impressive thickness, their presence would negatively impact cell performance. A complete void was formed when the void formed during cooling, before the eutectic Al–Si composition solidified [7].

The results observed in this study suggest that the thicker and more uniform Al–BSF produced in PERC cells using Al paste with Si content could be the one of the reasons for the higher V_{oc} indicated in Table 1. In our previous results (data not shown), the particle size of the Al paste has impacted the adhesion between the Al paste and the dielectric layers on the Si substrate, thereby affecting the field-effect passivation, indicated by V_{oc}. Therefore, in this work, Si content in the Al paste might have impacted the adhesion and conductivity between the Al layer and the passivation layers on the Si substrate. Furthermore, one might expect that a thicker Al–BSF should result in a higher FF. In our experiment, however, we found that PERC cells with a shallower Al–BSF resulted in an 80.6% higher FF. This observation might be explained by the different contact resistance of the finished cells caused by the presence of Si in the Al paste. Without considering the series resistance, the pseudo-FF was calculated using the parameters of the Suns–V_{oc} curves and the photocurrent values obtained. It was observed that the pseudo-FF of PERC cells with Al paste containing Si were higher than those using Al paste without Si. This issue can be solved by changing the Al layer thickness, which would mean that an optimum contact design would be required.

These results suggest that the composition of the Al paste has a direct impact on the local contact formation, and Si content in the Al paste is essential for improving local rear contacts, leading to a high efficiency for the PERC solar cells.

IV. CONCLUSION

We studied the effect of Si content in the Al paste on the formation of local contacts and on PERC cell performance. For this purpose, the rear side of PERC cells were metalized with two different Al pastes—one with Si content, and one with no Si content. We found that the Si content in the Al paste significantly impacted the I-V parameters of the PERC cells and the local contact formation. It was observed that PERC cells using Al paste with Si content showed an increase in V_{oc}, partly due to a thicker Al–BSF beneath the rear contacts. In this work, voids with an Al–BSF were observed, with Al–BSF thicknesses comparable to those of deep filled contacts. In addition, a significant decrease in the contact was observed, owing to an accumulation of the rejected excessive Si from the Al-Si melt during cooling. These results suggest that Si content in the Al paste is desirable for improving the formation of local rear contacts as well as for suppressing void formation, thereby enhancing the performance of PERC cells.

ACKNOWLEDGEMENT

The authors gratefully acknowledge the financial support of The New Energy and Industrial Technology Development Organization (NEDO) of Japan. The authors also wish to thank Mr. Yasuhiro Kida, and Mr. Masaaki Moriya, FREA-AIST for their technical support.

REFERENCES

[1] B. Sopori, V. Mehta, P. Rupnowski, H. Moutinho, A. Shaikh, C. Khadilkar, M. Bennett, and D. Carlson, "Studies on backside Al-contact formation in Si solar cells: fundamental mechanisms," in Proceedings of the Materials Research Society (MRS) Fall Meeting, Boston, Massachusetts, USA, 2008, p. 7-11.

[2] E. Urrejola, K. Peter, H. Plagwitz, and G. Schubert, "Al–Si alloy formation in narrow p-type Si contact areas for rear passivated solar cells," *Journal of Applied Physics*, vol. 107, 124516, 2010.

[3] C. Kranz, B. Wolpensinger, R. Brendel, and T. Dullweber, "Analysis of local aluminum rear contacts of bifacial PERC+ solar cells," *IEEE Journal of Photovoltaics*, vol. 6 (4), 830-836, 2016.

[4] M. Rauer, R. Woehl, K. Rühle, C. Schmiga, M. Hermle, M. Hörteis, and D. Biro, "Aluminum Alloying in Local Contact Areas on Dielectrically Passivated Rear Surfaces of Silicon Solar Cells," *IEEE Electron Device Letters*, vol. 32 (7), 916-918, 2011.

[5] J. Schmidt, A. Merkle, R. Brendel, B. Hoex, M.C.M. van de Sanden, and W.M.M. Kessels, "Surface passivation of high-efficiency silicon solar cells by atomic-layer-deposited Al2O3," *Progress in Photovoltaics: Research and Applications*, vol. 16, 461-466, 2008.

[6] B. Hoex, J.J.H. Gielis, M.C.M. van de Sanden, and W.M.M. Kessels, "On the c-Si surface passivation mechanism by the negative-charge-dielectric Al2O3," *Journal of Applied Physics*, vol. 104, 113703, 2008.

[7] T. Dullwerber and J. Schmidt, "Industrial silicon solar cells applying the passivated emitter and rear cell (PERC) concept–A review," *IEEE Journal of Photovoltaics*, vol. 6 (5), 1366-1381, 2016.

New Silver Paste Metallization Approach on p+ Diffusion Zones of Silicon Solar cells

Yunjun Li, Mohshi Yang, Igor Pavlovsky, and Guoping Zeng

Starsource Scientific LLC, Austin, Texas 78750, USA

Abstract — Silver paste containing no aluminum powder has been developed to form low resistance contact on p+ diffusion zones for silicon solar cells. Unlike conventional Ag/Al pastes for p+ silicon on solar cell, silver paste without metallic aluminum reduces the risk of "spiking" issues that may damage PN-junction during high temperature fire-through process. A low series resistance and high cell efficiency were demonstrated with this paste on n-PERT solar cells. This paste was found to be capable of forming low resistance contacts on both p+ and n+ silicon. This unique property may lead to a single print and co-firing process for IBC solar cell that can significantly reduce manufacturing costs.

Index Terms — silver paste, n-type solar cells, screen printing.

I. INTRODUCTION

High-efficiency n-type silicon solar cells have advantages over the current mainstream p-type silicon solar cells. In recent years, companies such as SunPower, Yingli, Panasonic, and LG, have been already producing n-type cells and modules. Among the n-type cell structures are PERT (Passivated Emitter Rear Totally diffused) [1], IBC (Interdigitated Back Contact) solar cells [2], and HIT (Heterojunction with Intrinsic Thin Layer) [3]. For n-type PERT and IBC solar cells [1], fire-through Ag/Al pastes can be used to form electric contact on p+ diffusion zones during cell manufacturing. Ag/Al silver pastes normally contain aluminum powder in formulations to reduce the contact resistance [4]. Including aluminum powders in paste formulations, however, creates aluminum spiking issue during high temperature fire-through process. Moreover, higher loading of aluminum powders in pastes leads to higher line resistance. In some cases, double-printing is required to reduce the line resistance and series resistance in solar cells.

To address those issues and further improve n-type cell performance, we developed a silver paste without including aluminum powders in our proprietary formulations. Specific contact resistivity as low as 3 mΩ.cm^2 has been achieved on p+ diffusion silicon layer on n-PERT solar cells after fire-through SiO$_2$/SiN$_x$ passivation layer. Due to no aluminum powders in the silver paste, low line resistivity below 4 $\mu\Omega$.cm could be obtained. This paste allows high aspect ratio contacts which can be printed 25 μm tall and 45μm wide. This new paste has the potential to improve the performance of both n-PERT and IBC solar cells.

II. EXPERIMENTAL

Silver paste is normally formulated with silver powders, glass frits, functional additives, organic vehicle, rheological modifiers, etc. The mixture is roll-milled to reduce particle size and adjust to proper viscosity for screen printing.

The developed paste was printed on n-PERT silicon wafers as described in Figure 1 to test the cell performance and to characterize the silver paste. Back side silver paste 2 is the silver paste that can form electric contact on n+ silicon, which is extensively used for front metallization of p-type silicon solar cells. The silicon wafers with diffusion layers and passivation layers were provided by a leading solar cell manufacturer and were fabricated in their production lines. We use silver paste Starsource XY-803 as the paste 2 for the testing. A reference silver paste, which is commercially available, is compared with our developed pastes in this study. All the pastes were printed using a DEK screen printer and fired with a belt furnace to make solar cells. The n-PERT cells are tested with Sinton FCT-350 solar simulator.

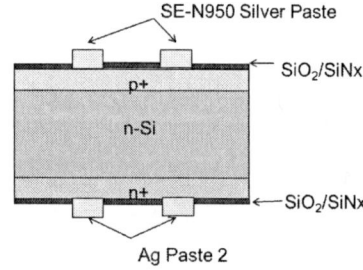

Fig. 1.   Aluminum-free paste SE-N950 is printed on the front of n-PERT solar cell.

SE-N900 is an aluminum-containing paste that was developed in our early development. This paste comprises of silver powders, aluminum powders, glass frit, functional additives, and organic vehicle. SE-N900 paste was compared with the reference paste in Table I below. The compared values for the reference paste are taken as 100 for the purpose of this test. SE-N950 is the paste that does not contain any aluminum in the formulation. It only includes silver powders, glass frit, functional additives, and organic vehicle. The best

978-1-5090-5606-4/17 $31.00 © 2017 IEEE

cell efficiency and lowest series resistance (Rs) were observed with paste SE-N950 as seen in Table I.

The paste SE-N900 was also tested and compared with the reference Ag/Al paste at a PERT cell production facility. In the production line, double printing is employed to reduce line resistance. Firstly, the reference Ag/Al paste or SE-N900 was printed as grid lines and the electric contact layer. After baking, the second pure silver paste was printed on the top to add the total thickness to over 20 μm. The testing results are listed in Table II. The cell efficiency of SE-N900 is lower than that of the reference paste due to a much lower Isc (short circuit current), resulting from a much wider line width than reference paste (72 μm vs. 52 μm). However, the Rs for SE-N900 is lower and Voc is higher than that of the reference paste.

TABLE I
INTERNAL CHARACTERIZATION RESULTS

Pastes	ΔVoc (V)	ΔIsc (A)	ΔPmax (W)	ΔFill factor	ΔRsh (ohm)	ΔRs (ohm)	ΔEfficiency (%)
Reference Paste	100	100	100	100	100	100	100
SE-n900	100.16	100.05	100.50	100.25	104.39	94.09	100.48
SE-n950	100.16	102.57	103.21	100.50	140.69	60.10	103.20

TABLE II
CUSTOMER TESTING COMPARISON

Customer Testing	ΔVoc (V)	ΔIsc (A)	ΔPmax (W)	ΔFill factor	ΔRsh (ohm)	ΔRs (ohm)	ΔEfficiency (%)
Production Baseline Paste	100	100	100	100	100	100	100
SE-N900	100.14	98.97	99.18	100.13	55.75	93.75	99.26

Although SE-N950 shows better results in Table I, customer testing results were not good because the front second silver printing paste did not stick on SE-N950 and resulted in many broken lines. This means that SE-N950 may not work for double printing for the time being. However, we have tried to modify SE-N950 paste to allow us to print high aspect ratio of grid lines. This may eliminate the second silver printing and reduce cell manufacturing cost.

A proprietary organic system was developed to modify SE-N950 silver paste that does not contain aluminum in formulations. Figure 3 shows the improved printed grid lines. With 34 μm opening screen, the grid lines can be printed 25 μm tall and 45μm wide. The surface of the printed grid lines is relatively smooth and continuous. A low line resistivity of SE-N950, 3.6 μΩ.cm, was measured. Therefore, with the relatively low line resistivity, it is feasible to print a single layer of grid lines with high aspect ratio on the front of n-PERT cells, and can still reach low Rs. We are working with our customers to approach this route. This will eliminate the

second silver printing on grid lines, which requires relatively good alignment for screen printers.

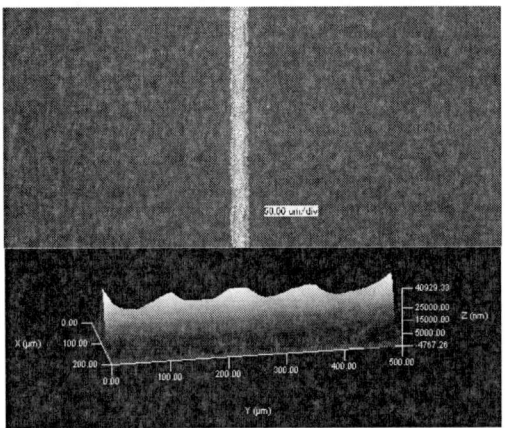

Fig. 2 Optical image of a printed grid line. Thickness and cross section morphology of grid are measured by Tencor surface profiler.

Since SE-N950 silver paste does not have aluminum, one interesting question to ask is that, whether the paste can make electric contact on n+ diffusion silicon or not. To answer this question, SE-N950 paste was printed on both sides of a PERT wafer to make a solar cell. After fire-through process, the cell was tested and 0.2% lower efficiency was obtained, comparing to a cell printed with a traditional back-side silver paste for n+ diffusion silicon. This indicates that SE-N950 silver paste can actually form electrical contacts both to n+ and p+ silicon.

Using a TLM (Transmission Line Method) technique and a four probe tester, the specific contact resistivity and line resistivity were tested with the n-type PERT solar cells in which both sides were printed with SE-N950 paste. The results are shown in Table III. Specific contact resistivity less than 3 mΩ.cm^2 was observed both on n+ and p+ silicon.

TABLE III
SUMARY OF SPECIFIC CONTACT RESISTIVITY

Ag Paste	ρ_c (mΩ. cm^2) on 80 Ω/Sq p+ silicon	ρ_c (mΩ. cm^2) on 70 Ω/Sq n+ silicon	Line Resistivity (μΩ. cm)
Starsource SE-N950	< 3	< 3	<4

The unique property of forming low resistance contact on both n+ and p+ silicon may allow the SE-N950 paste to be used for manufacturing IBC solar cells with a single print and a single firing. This simple process will reduce IBC manufacturing steps and help industry to commercialize the lower cost IBC solar cells.

III. DISCUSSION

It has been reported in literature that it is more difficult for a silver paste to make a good screen printed contact to Boron-doped emitter in n-type Si than to traditional n-doped emitters in p-type Si solar cell substrates. Aluminum powders are commonly considered as a necessary paste ingredient to form Ag/Al-alloyed contacts on p+ diffusion zones and create Al-doping in p+ silicon to reduce contact resistance [5]. Although the effects of aluminum addition to solar pastes have been extensively investigated [6], the mechanisms affecting the properties of the interface between Ag/Al paste and p+ doped emitter of n-type silicon solar cells are still not totally clear.

As reported in literature, specific contact resistivity ρ_c for pure Ag screen printed pastes on p+ Si is normally > 20 $m\Omega.cm^2$ [6]. Although addition of Al to a paste reduces ρ_c, it has been found that the unique glass frit chemistry and functional additives of SE-N950 silver paste play important roles in reducing the contact resistance without any addition of aluminum. The ρ_c of less than 3 $m\Omega.cm^2$ can be achieved with SE-N950 paste, which is even lower than that of the reference Ag/Al paste (4 $m\Omega.cm^2$) compared with the same testing tool. This indicates a different mechanism of formation of a contact in the interface between silver and p+ diffusion silicon. The mechanism of Ag/Al alloy contact formation on silicon as described in [6] is obviously not applicable to SE-N950 that does not contain aluminum. Moreover, SE-N950 paste can also form good contact on n+ silicon, as seen from Table III. This indicates that a possible similar contact formation mechanism may exist in both cases.

A few mechanisms have been proposed earlier to explain interfacial electron transport for fire-through silver pastes, such as (A) current transport via silver crystallites directly contacting silicon [8], and (B) electron tunneling via precipitated silver nanoparticles dispersed within an interfacial glass layer [7]. For Ag/Al paste, it is more likely that aluminum addition modifies glass frit properties that allows uniform and discontinuous interfacial glass layer and results in formation of Ag/Al alloy crystallites directly on silicon to form electrical contact. We believe that the unique glass formulation and functional additives in SE-N950 play a major role in creating of precipitated Ag crystallites directly on silicon to form the electrical contact. It is possible that a very thin SiO_2 layer is produced between Ag crystallites and silicon surface through oxidation reactions [7] between glass/or Ag_2O and underneath crystalline Si during high temperature fire-through process. The thin SiO_2 layer formed between Ag and Si could be as thin as 2 nm or below. This may produce a local metal-insulator-semiconductor (MIS) structure between Ag and silicon, resulting in an improved electron tunneling transport. Agrawal et al. recently reported that a lower barrier height was measured with a Ti/1nm TiO_2/n-Si MIS structure, indicating a 55% reduction compared to metal/n-Si control contact [9]. In our cases, the possibility of having the very thin oxide layer

between Ag and Si is significant with proper glass chemistry and the high density of precipitated Ag crystallites on silicon surface after high temperature fire-through process, as shown in Figure 3. This local MIS contact mechanism may explain why SE-N950 forms low-resistance contact both on n+ and p+ silicon.

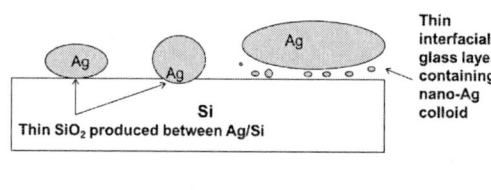

$$2Ag_2O + Si \longrightarrow SiO_2 + 4Ag$$
$$2PbO + Si \longrightarrow SiO_2 + 2Pb$$

Figure 3   Nano-meter thin SiO2 could be formed during firing through the chemical reactions.

IV. CONCLUSION

Silver paste containing no aluminum powder has been developed to form low resistance contact on p+ diffusion zones for silicon solar cells. This paste demonstrated a low series resistance and high cell efficiency for n-PERT solar cells. This paste was found to be capable of forming low resistance contacts on both p+ and n+ silicon. This unique property may lead to a single print and co-firing process for IBC solar cell that can significantly reduce manufacturing costs. A local MIS contact mechanism was proposed to explain how SE-N950 can form good contact on both n+ and p+ diffusion silicon.

REFERENCES

[1] Valentin D. Mihailetchi, Haifeng Chu, Giuseppe Galbiati, Corrado Comparroto, Andreas Halm, Radovan Kopecek, Energy Procedia , Vol. 77, 534-539 (2015).

[2] Smith et.al, "SunPower 's Maxeon Gen III solar cell: High efficient and energy yield", Proc. 39th IEEE PVSC, Tampa, 2013.

[3] Masuko K. Shigematsu, et. al, "Achievement of more than 25% conversion efficiency with crystalline silicon heterojunction solar cell", IEEE Journal of Photovoltaics, 4(6), 1435 (2014).

[4] Norihiko Takeda, US Patent 2012/0255605A, Oct. 11, 2012.

[5] S. Riegel, F. Mutter, T. Lauermann, B. Terheiden, G. Hahn, "Review onscreen printed metallization on p-type silicon", 3rd Workshop on Metallization for Crystalline Silicon Solar Cells, October 2011, Charleroi, Belgium.

[6] S. Fritzl, S. Riegel1, A. Herguth, M. König, M. Hörteis, G. Hahn, "Preservation of Si Surface Structure by Ag/Al Contact Spots –an Explanatory Model", 5thMetallization Workshop Konstanz, October , 2014.

[7] Jeremy D. Fields, Md. Imteyaz Ahmad, Vanessa L. Pool, Jiafan Yu, Douglas G. Van Campen, Philip A. Parilla , Michael F. Toney, and Maikel F.A.M. van Hest, "The formation

978-1-5090-5606-4/17 $31.00 © 2017 IEEE

mechanism for printed silver-contacts for silicon solar cells", Nature Communications, 2016.

[8] Vinodh Shanmugam, Jessen Cunnusamy, Ankit Khanna, Prabir Kanti Basu, Yi Zhang, Chilong Chen, Arno F. Stassen, Matthew B. Boreland, Thomas Mueller, Bram Hoex, and Armin G. Aberle, "Electrical and Microstructural Analysis of Contact Formation on Lightly Doped Phosphorus Emitters Using Thick-Film Ag Screen Printing Pastes", IEEE J. Photovoltaics, 4 (1), 168~174 (2014)

[9] Ashish Agrawal, Joyce Lin, et al., "Fermi level depinning and contact resistivity reduction using a reduced titania interlayer in n-silicon metal-insulator-semiconductor ohmic contacts", Appl. Phys. Lett., 104, 112101 (2014)

Influences of annealing and defect limitation on p-type silicon solar cell

Yu-Hsuan Lin, Sung-Yu Chen, Kuen-Yi Wu, Chien-Hsun Chen, Chen-Hsun Du, Chun-Ming Yeh

Green Energy and Environment Research Lab. (GEL), Industrial Technology Research Institute (ITRI),

Hsinchu, Taiwan, 31040, R.O.C.

* Corresponding author's e-mail: yuhsuanlin@itri.org.tw

Abstract — **In this paper, we identify the influences of different annealing conditions on the p-type Cz silicon wafer surfaces and solar cells. X-ray photoelectron spectroscopy (XPS) has been applied to surfaces of silicon wafers for evaluating the composition. In order to identify the properties of the defect, minority carrier lifetime measurement was performed. Annealing condition effected the silicon surface oxidation and the defect of silicon. The best cell efficiency of 20.07 % with an open circuit voltage (Voc) of 648 mV was achieved by annealed at an oxygen atmosphere.**

Index Terms —**passivation, lifetime, XPS, oxidation, silicon solar cells.**

I. INTRODUCTION

Screen-printed solar cell with a full-area Al-p+ back surface field (Al-BSF cells) made on Czochralski-grown silicon (Cz-Si) is the industrial standard in today's silicon solar cell production [1]. However, this cell type is limited by the relatively high recombination losses at the fully metalized rear side of the cell. A new type cell, passivated emitter and rear cell (PERC), will gradually replace Al-BSF cell due to the low recombination losses and high efficiencies. However, reduction in the rear surface recombination increases the sensitivity of other recombination channels for the solar cell, such as the bulk recombination [2]. Therefore, the bulk quality of the wafer plays a key role in an increase of a cell efficiency. The existence of boron-oxygen defects, thermal donors, other grown-in defects have been shown to have negative affect on the minority carrier lifetime for silicon [3]. The boron-oxygen center in silicon can be deactivated by illuminating and annealing. [4]

In this paper, we identified the influences of annealing gas and temperature on p-type Cz silicon wafer surfaces by XPS analysis and lifetime measurement. The solar cells applying different annealing conditions were evaluated.

II. EXPERIMENTAL DETAILS

Schematics of the P-type silicon wafers for lifetime measurement and solar cell are shown in Fig. 1. P-type silicon solar cells were fabricated on large area (156×156 mm²) Cz p-type Si wafers, with a thickness approximately 180 μm and

resistivity between 1.5 and 2 Ωcm. The process flows of annealed wafers and solar cell are shown in Fig. 2. Silicon wafers polished by NaOH solution were prepared for examining the influence of annealing and defect limitation on silicon wafer. Polished wafers were RCA cleaned and annealed with these four type conditions was shown in Table I . The wafer surface was studied by XPS after each one, and samples were identified as follows: Sample A annealed at 800 ℃ for 20 min under oxygen ambient and then annealed at 400 ℃ for 20 min under forming gas (FG) ambient. Sample B annealed by FG. Sample C without annealing as a reference wafer. Sample D only annealed by oxygen. All of these polished wafers were deposited SiNx onto both side surfaces by plasma-enhanced chemical vapor deposition (PECVD) for carrier lifetime measurement, shows as Fig. 1(a).

Table I
Summary of the main gas condition for silicon wafer

	A	B	C	D
O₂	○	×	×	○
FG	○	○	×	×

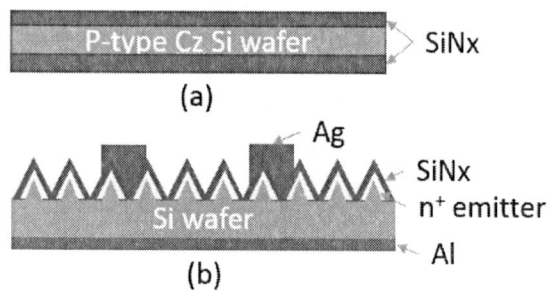

Fig. 1. Schematic of (a) sample for lifetime measurement (b) p-type silicon solar cell

The structure schematic of solar cell is shown in Fig. 1(b). For solar cell preparation, the manufacturing process can be roughly separated into the following stages. Firstly, the silicon wafer surface was textured with random pyramids in KOH /isopropanol solution. A wafer cleaning process was then applied before the phosphorous diffusion. Then, the emitter was formed by POCl$_3$ diffusion with a sheet resistance ~70 Ω/sq and subsequently the phosphsilicate glass (PSG) was removed by dipping hydrogen fluoride (HF). For comparison of the influence on emitter, these wafers were separated into four batch and treated by different annealing conditions. The SiNx was deposited on the front side of cells as an anti-reflection layer. Finally, the metallization was carried out by screen printing and co-firing process. The process flow is shown in Fig. 2.

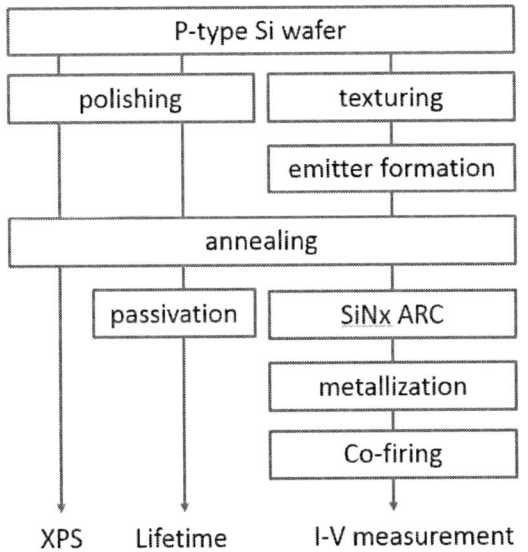

Fig. 2. Experimental flow chart in this work.

III. RESULTS AND DISCUSSION

Fig. 3(a) and (b) show the XPS spectra of the wafer surfaces with different annealing conditions. The Si 2p and O 1s core level spectra are extracted for different sample surfaces. The peak position at 99.4 eV corresponds to the p-type bulk silicon, and the peak position at 103.2 eV corresponds to silicon oxide [5-6]. In Fig. 3(a), the O 1s core level spectra of sample A and D have shifted to high binding energies of 103.6 and 103.4 eV, respectively. The Si 2p and O 1s signals at 103.5 eV and 532.7 eV, typical for SiO$_2$ [7]. Compare to all of these peaks, the intensity of peak of sample A and D near 103.5 eV are increasing. The ratio of Si/O was shown in Table II. After oxygen annealing, Si/O ratio is decreased. This indicates that the uppermost atomic monolayer of the silicon wafer has been oxidized to SiO$_2$ after annealing at an oxygen atmosphere.

The lifetime results were measured by a Sinton WCT120 system in QSS mode. Effective lifetime, τ_{eff} as a function of the injection density, measured on p-type CZ wafer passivated by SiNx after using four different annealing conditions. The

measurements are shown in the injection level ranging from 10^{13} to 10^{16} cm^{-3}. For comparison the effective lifetimes (τ_{eff}) of sample A, B, C and D are 52, 184, 120 and 54 μs which were determined at an injection level of 10^{15} cm^{-3}.

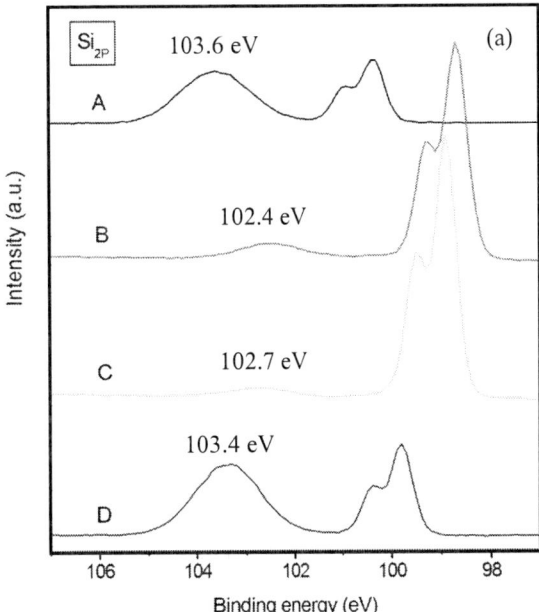

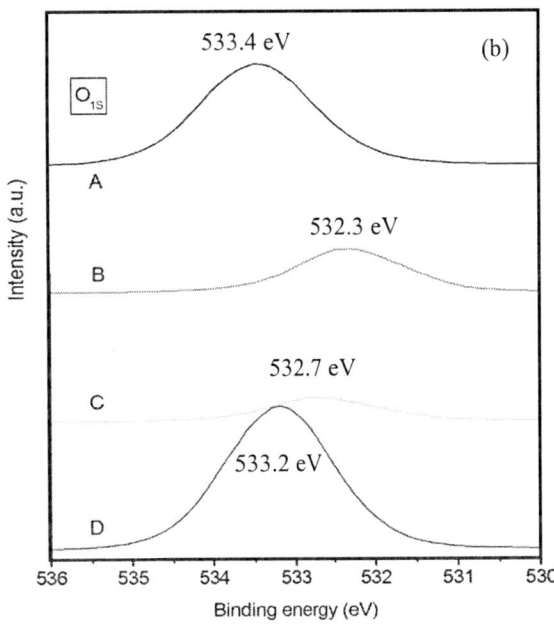

Fig. 3. XPS multi-region spectra of the different sample surfaces.

Table II

Summary of the surface atomic concentration and ratio of Si and O for different sample surfaces.

	A	B	C	D
%Si	38.3	63.7	75.6	41.9
%O	51.5	25.1	15.0	58.1
%C	10.3	11.2	9.4	-
Si/O	0.74	2.53	5.04	0.72

In order to analyze the influences of annealing and defect of silicon wafer surface on emitter, all samples were annealed with different conditions after emitter formation. Table III list the results of the processed solar cells. One-sun parameters measured under standard testing conditions of ~180 mm thick silicon solar cells with different annealing treatment. For oxygen annealing (cell A and D), the open-circuit voltage (Voc) and short-current density (Jsc) were increasing due to the SiO_2 passivation on silicon surface.The cell efficiency of $0.14\%_{abs}$ of the cell D versus the conventional cell C is mainly due to an increasing of Voc and Jsc.

Table III

Comparison of the I-V characteristics on different rear surface cells. All of these data are average value.

Cell	Jsc [mA/cm²]	Voc [mV]	F.F. [%]	Cell eff. [%]
A	38.07	649	80.4	19.86
B	38.00	643	81.2	19.83
C	37.95	644	81.2	19.86
D	38.14	648	80.9	20.00

III. CONCLUSIONS

In this paper, annealing and defect limitation on p-type silicon solar cell were reported. After annealing with different conditions, including gas and temperature, the silicon wafer surfaces were analyzed by XPS. Annealing condition effected the silicon surface oxidation and the defect of silicon. The best cell efficiency of 20.07 % and Voc of 648 mV were achieved by an oxygen annealing treatment.

ACKNOLEDGMENT

The financial support provided by Bureau of Energy, Ministry of Economic Affairs in Taiwan is gratefully acknowledged.

REFERENCES

[1] D. C. Walter, B. Lim and J. Schmidt, "Realistic efficiency potential of next-generation industrial Czochralski-grown silicon solar cells after deactivation of the boron-oxygen-related defect center," *Prog. Photovolt: Res. Appl.*, vol. 24, pp. 920-928, 2016.

[2] J. Schmidt, B. Lim, D. Walter, K. Bothe, S. Gatz, T. Dullweber, P. P. Altermatt," Impurity-related limitations of the next-generation industrial silicon solar cells," *IEEE Journal of Photovoltaics*, vol. 3, pp. 114-118, 2013.

[3] F. E. Rougieux, N. E. Grant, C. Barugkin, D. Macdonald and J. D. Murphy, " Influence of annealing and bulk hydrogenation on lifietime-limiting defects in nitrogen-doped floating zone silicon, " *IEEE Journal of Photovoltaics*, vol. 5, pp. 495-498, 2015.

[4] A. Herguth, G. Schubert, M. Kaes and G. Hahn, "Investigations on the long time behavior of the metastable boron-oxygen complex in crystalline silicon," *Prog. Photovoltaics: Res. Appl.*, vol. 16, pp. 135-140, 2008.

[5] J. Hedman, Y. Bear, A. Berndtson, M. Klasson, G. Leonhardt, R. Nilsson, C. Nordling, "Influence of doping on the electron spectrum of silicon", *J. Electron Spectrosc. Relat. Phenom.* vol. 1, pp. 101-104, 1972.

[6] Y. Chen, S. Zhong, M. Tan, W. Shen, " SiO_2 passivation layer grown by liquid phase deposition for silicon solar cell application, "Frontiers in Energy, vol. 11, pp. 52-59, 2017.

[7] J.F. Moulder, W.F. Stickle, P.E. Sobol, K.D. Bomben: *In Handbook of X-ray Photoelectron Spectroscopy*, ed. by J. Chastain, PHI, Eden Prairie, Minnesota, 1992.

Reduced Temperature Silver Paste with Low Contact Resistance for Advanced Solar Cell Applications

Ryan Mayberry, Daniel Holzmann, Gerd Schulz, Lindsey Karpowich, Mark Naylor, Matthias Hoerteis

Heraeus Precious Metals North America Conshohocken LLC, W. Conshohocken, PA, 19428, USA

Abstract — The electrical contact formation on a crystalline silicon (c-Si) solar cell is generally a well-known phenomenon in the PV industry. The glass component along with inorganic additives in the silver paste are responsible for the reactions with silicon nitride ARC, silicon and silver that may lead to a low contact resistance in a screen printed solar cell. However, the detailed chemistry that controls the thermodynamic and kinetic temperature dependence of these reactions remains widely unknown. Currently, the vast majority of industrial c-Si solar cells are fired at relatively high temperatures (760-810 °C). In this work, it will be demonstrated how an additive, hereby referred to as the activator, can be used in order to reduce the required temperature for forming a low resistance electrical contact by greater than 150°C. Both the formation and nature of silver crystallites and colloids formed on the silicon emitter region and in the glass layer are greatly affected by the activator additive; which is referred to as such due to its ability to abruptly alter the temperature of contact formation in a non-linear fashion. This will create an entirely new temperature range in which a good front grid contact can be formed. This has important implications on current and future cell processing technologies such as PERC, which shows many advantages versus traditional aluminum back surface field (Al-BSF) cells, easily shown by PERC's large gains in internal quantum efficiency (IQE) over a wide range of wavelengths [1,2].

I. INTRODUCTION

Screen-printed thick film metallization on crystalline silicon solar cells is the predominant processing technology in the PV industry today. Rapid advances in cell processing call for simultaneous advances in metallization paste as well as compatibility with limiting cell parameters, including final processing temperature. The effort in development of advanced cell types in recent years, specifically those in the PERC family, have been directed at reducing the effects of high temperature degradation which becomes prominent at firing temperatures above ~700 °C [3]. Without these advances, PERC technology would not be commercially viable due to the inability of current generation metallization pastes to form sufficient contact at such low temperatures. There is also a critical benefit of processing PERC cells at lower temperatures which stems from less diffusion of the SiO_2 passivation into the metallization paste; an inherent consequence of decreased thermal energy. Lessening this diffusion would lead to a higher purity passivation layer yielding higher performance. In addition to PERC cells, lower firing temperature paste may prove critical for the development and commercialization of passivated contact

solar cells - devices where a thin poly-Si layer is added to the backside of a (typically n-type) cell. This cell structure shows the benefit of decreased surface recombination as compared to thicker traditional dielectric stacks [4].

Overall, the development of low firing temperature pastes should result in reduced junction shunting and recombination while maintaining low contact resistance, leading to high fill factor (FF), high open circuit voltage (Voc), and high overall conversion efficiency (Eta). This would also allow for wafer development to focus on reduction of light induced degradation (LID), or more generally carrier induced degradation (CID), and other passivation methods which can result in far less surface and bulk recombination as well as decreasing recombination velocity in areas apart from the front grid metallization.

In this work, it will be demonstrated how additives can shift solar cell firing temperatures from 780°C to well below 700°C. This will lower the processing temperatures below that of the previously reported threshold for significant degradation effects on solar cells of multiple structures and types, minimizing additional processing steps that have been both suggested and already implemented to neutralize negative effects at higher temperatures [3],[5]-[6].

II. EXPERIMENTAL APPROACH

Typical thick film metallization pastes used on c-Si solar cells consist of 85-90% Ag powder, 5-10% organics, 1-5% glass frit and/or other inorganic additives. The silver pastes in this study were produced by thoroughly mixing each component, then homogenizing the paste through the use of a three roll mill. Once milled, the pastes were screen printed using a 3-busbar pattern onto P-type monocrystalline (Cz) silicon wafers with a sheet resistance of ~90 Ohm/sq. The printed wafers were dried at 150°C for 10 minutes in a box oven. After drying, the wafers were co-fired with aluminum paste (previously printed and dried on the rear side of the wafer) using an IR belt furnace at varying peak temperatures. Electrical performance was then analyzed through standard current-voltage (I-V) characterization methods.

In this study, a standard industrial front side Ag paste is used as a reference and as a base for activator additions. Two separate effects are investigated based on the additions. The first examines the concentration limits and effects of the activator additive while the second demonstrates the

978-1-5090-5606-4/17 $31.00 © 2017 IEEE

temperature dependency on contact formation of pastes with and without the activator additive. Performance will generally be represented by either conversion efficiency or FF. Fill factor is used in order to show the trend in contact behavior when comparing high and low temperatures to minimize the direct impact of factors such as the BSF formation and silver sintering, those of which can strongly impact overall efficiency. These factors can later be altered through optimization of paste components and selection of better passivated solar cells, both of which will lead to efficiencies comparable to or better than current standards without negating the effect seen by the activator addition.

A. Concentration dependence

Three increasing additive levels were investigated (+, ++, and +++), as shown in in Table 1. It has been previously demonstrated that without the addition of the activator, this paste system completely fails at temperatures below 750°C due to the inability to form sufficient contact at the emitter region. The effects of each level of the additive will be examined.

B. Temperature dependence

Using the optimum concentration from Section A, peak firing zone temperatures are varied over a range of 140°C in increments of 20°C from 660 to 800°C for the additive paste and from 740 to 800°C in increments of 20°C for the standard paste. The firing window, as it is generally referred to, is the temperature range over which the solar cell has consistent performance. The work shows how this window is shifted from moderate/high temperatures for current paste systems, to low temperatures with the use of our novel activator addition. This experiment also uses actual PERC cells in order to capture more accurate data at low temperatures.

III. RESULTS AND DISCUSSION

A. Concentration dependence

Three pastes with increasing amounts of the activator additive were tested. Each paste and its corresponding efficiency and FF can be seen in Table 1.

Table 1.
Summary of varying additive concentration fired at 680°C

	Paste A	Paste B	Paste C
Additive	+	++	+++
Efficiency (%)	8.6	18.2	18.4
Fill Factor (%)	<40	78.2	78.4

This result demonstrates that the activator additive possesses a threshold at which contact formation occurs, opposed to a linear effect on firing temperature. When under this threshold,

as seen in Paste A, there is a negligible effect resulting in a completely under-fired solar cell. In this scenario, there is insufficient formation of both Ag crystallites and colloids near the emitter/glass/silver interface. Above this threshold, the paste performs on a very similar level, as seen by the resemblance of Paste B to Paste C. This shows that contact formation at a given low firing temperature follows a step-wise behavior with a saturation limit at higher concentrations. While higher concentrations may show incremental gains in contact resistance, it will be more than negated by further damage to the emitter region of the solar cell. In this case, overall cell efficiency will demonstrate optimal performance at some moderate concentration.

B. Temperature dependence

Using Paste B from the previous experiment, a firing study was performed in order to determine the firing window of a current paste used for PERC application (standard) versus that of a similar paste with the activator additive (Paste B). As shown in both Figure 1 and Table 2, the results clearly demonstrate a large shift in firing window from an optimum temperature of 780°C to 680°C when the activator is added to

Table 2.
Summary of performance, additive vs no additive

	Standard	Paste B
Additive	No	Yes
Optimum Temp.	**780°C**	**680°C**
Fill Factor (%)	77.2	77.1
Rc (mOhm/cm^2)	2.2	2.1

the paste. This shift in optimal firing temperature can be fully explained by the ability of the activator additive to promote silver crystallite formation at far lower temperatures. This is

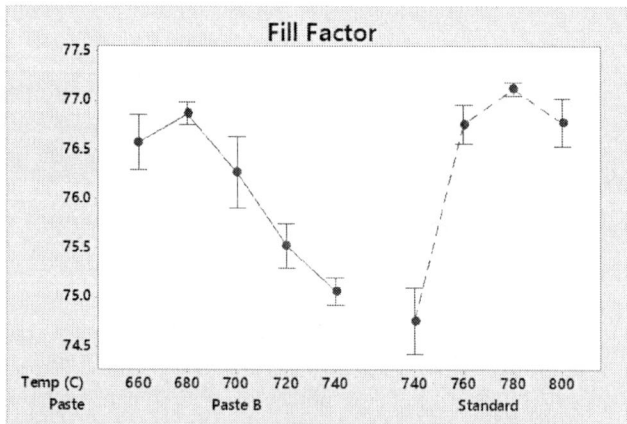

Figure 1. Fill Factor at increasing temperatures for Paste B (with activator) and Standard paste (no additive). The FF for Paste B is reasonable down to 660°C while the standard paste has a steep decline in FF due to insufficient contact at 740°C.

confirmed by scanning electron microscope images as well as direct measurement of contact resistance (Rc). It is also important to note that the activator additive does not extend the firing window of the paste but rather shifts it to lower temperatures.

This sheds further light on the mechanism in which the additive works, which is reducing the onset temperature of crystallite formation. Consequently, at higher temperatures the crystallite formation becomes overly aggressive and leads to a sharp decrease in shunt resistance, effectively causing an electrical short in the cell. This can be visualized in Figure 2. in which large Ag crystallites are formed and grown into the silicon surface. This will cause a disruption of the p-n junction and heavily damage the solar cell, causing a steep drop in performance. It would be interesting to study the possibility of combining this system with other inorganics such as further additives or glass to determine if it is possible to suppress the reaction at higher temperatures. If passivation materials can be found to protect the silicon and/or silicon nitride surface at high temperatures, the reaction could effectively be controlled to have a firing window ranging from the onset of contact formation all the way until the cell itself, not metallization paste, is the cause of decreased performance.

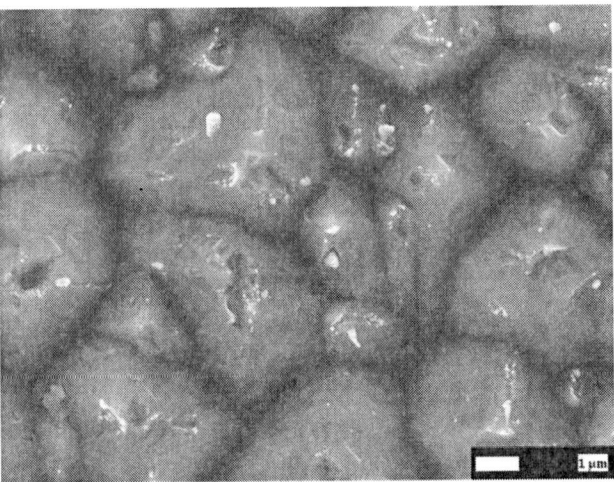

Figure 2. SEM image of an HF etched over-fired solar cell in which silver crystallites have penetrated deep into the emitter region of the silicon surface, commonly referred to as pitting.

IV. CONCLUSION

The primary findings of this work are that of the activator's ability to shift the firing temperature of a c-Si solar cell to far lower temperatures at which the metallization can maintain an optimal balance between low contact resistance and minimal cell damage. Performance losses in modern solar cells are largely due to damage of the emitter/surface/bulk (J01) and space charge region (J02) of the silicon cell both near and away from the metallization area. However, total losses due to factors apart from the metallization, compared to those caused by the metallization, are much greater. With a low firing temperature paste, these losses can be minimized through advanced wafer technologies which are inherently subject to less damage at lower temperatures. Therefore, the implications of this work are that wafer technologies can, and more importantly should, be developed irrespective of the need to reach processing temperatures in excess of 700°C. In addition to increased performance, this can also reduce the number of processing steps and overall energy needed, further reducing costs. Although this technology is in its early stages, a vast amount of further research can stem from these results. A brief list of these include: investigations on the redox behavior of the activator additive when added to the paste system, firing conditions altering belt speed, ramp rates and peak temperature to understand the kinetic versus thermodynamic causes of the lower firing temperature, and exploration of similar additives that may exhibit comparable effects. Maintaining low contact resistance at far lower temperatures then that currently used to process crystalline silicon solar cells could prove to be a vital step in the future progress of the photovoltaic industry.

REFERENCES

[1] S.W. Glunz, "High Efficiency Crystalline Silcon Solar Cells," *Advances in OptoElectronics*, vol. 2007, Article ID 97470, 2007.

[2] S. Gatz et al, "Analysis and Optimization of the Bulk and Rear Recombination of Screen-printed PERC Solar Cells," *Energy Procedia*, vol. 27, pp. 95-102, 2002.

[3] C. Chan et al, "Rapid Stabilization of High-Performance Multicrystalline type Silicon PERC Cells," *IEEE Journal of Photovoltaics*, vol. 6, issue 6, 2016.

[4] P. Stradins et al, "Passivated Tunneling Contacts to N-Type Wafer Silicon and Their Implementation into High Performance Solar Cells," *WCPEC-6: 6th World Conference on Photovoltaic Energy Conversion*, 2014.

[5] F. Kersten et al, "Degradation of Multicrystalline Silicon Solar Cells and Modules after Illumination at Elevated Temperatures," *Solar Energy Materials and Solar Cells*, vol. 142, pp. 83-86, 2015.

[6] J. Schmidt et al, "Advances in the Surface Passivation of Silicon Solar Cells," *Energy Procedia*, vol. 15, pp. 30-39, 2011.

BSF Islands For Reduced Recombination In IBC Cells

Agnes A. Mewe, Nicolas Guillevin, Ilkay Cesar and Antonius R. Burgers

ECN Solar Energy, Petten, NL-1755 LE, The Netherlands

Abstract — We present a back junction back contact cell with small isolated BSF "island" areas, surrounded by emitter area. This is an alternative for the more common IBC cell with interdigitated fingers. The optimal cell lay-out was determined by device simulations. We used discontinuous emitter and BSF contacts, which we connected through floating metal fingers. Additionally, we applied a lighter front floating emitter diffusion. The performance increase predicted by the simulations was experimentally confirmed. The investigated design leads to a strong recombination reduction in our industrial IBC Mercury cell, leading to a 0.6% absolute gain in efficiency.

Index Terms — device simulation, interdigitated back contact, photovoltaic cells, recombination, silicon.

I. INTRODUCTION

The IBC Mercury cell is ECN's 6 inch industrial Interdigitated Back Contact cell design with a front floating emitter (FFE) that mitigates electrical shading. Although efficiencies above 21% have been reached [1], the performance is currently limited by the open circuit voltage, due to high recombination. The recombination activity in the cell is dominated by the emitter contacts and the heavily doped BSF area. Therefore, reducing both the BSF area and the emitter contact fraction is a route to decrease the recombination in the cell and therefore enhance the cell performance. This was explored both by simulations and experimental work.

Depending on the contact width and the screen printing tolerances, a minimum width of the passivated BSF area is required, which is typically more than 300 μm. In a one-dimensional interdigitated finger design, the only option to reduce the BSF area fraction further is then to increase the emitter width, but this induces large transport losses. Therefore, we reduced the BSF length within the unit cell, and in this way we created "islands" of BSF surrounded by the rear side emitter.

Similar point-contact structures for the diffused areas have been studied before for IBC cells [2]-[3], but in these cases, the BSF islands were mainly created to study the electrical shading reduction benefits. In our case, electrical shading is not a major issue due to the collecting and transporting front floating emitter. Therefore, the BSF area reduction will mainly improve the passivation of the cell.

II. SIMULATIONS

We simulated the cell lay-out in Quokka by taking the smallest possible representative part of the cell, which involves the half-widths of both the BSF and the emitter for an IBC cell. This unit cell is indicated in the IBC cell cross-section in Fig. 1 by the dashed line.

The Quokka device simulations include the influence of the metal contact area, but not of the metal grid resistance. Also cell features beyond the unit cell, such as busbars and cell edges, are not taken into account.

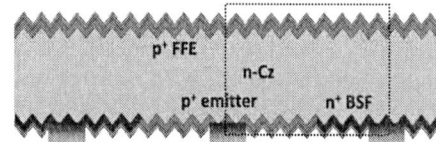

Fig. 1. Schematic cross-section of the IBC Mercury cell, with the standard unit cell indicated.

For the BSF island simulations, besides the width of the unit cell, also the unit length becomes a parameter, with the BSF length separately defined, as is shown in Fig. 2. This change leads to a 3D unit cell simulation.

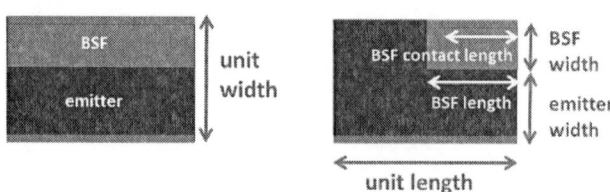

Fig. 2. Schematic view of the simulated rear side of the IBC Mercury cell, with the standard IBC unit cell (left) and BSF island lay-out (right).

A. Unit Cell Studies

In case of isolated BSF islands, the unit cell size determines the average travel distance for electrons to the BSF (i.e. transport losses), especially if the BSF fraction becomes very small. Additionally, a too small BSF contact area eventually becomes a limiting factor for effective transport.

To avoid these performance limitations, we decreased the unit cell size, which solved the BSF related transport limitations. The emitter contact length was kept equal to the unit cell length because of undesired current crowding effects. This caused the emitter contact to become very narrow, as the emitter contact faction was kept constant. However, in practice a certain minimum width is necessary due to screen-printing limitations. The solution to this will be discussed in the next section.

The lay-out of the optimized unit cell with BSF islands is sketched in Fig. 5 (left). The BSF area is reduced to 12.2% and the contact fractions of both emitter and BSF to 4% each. We calculate that a gain of 0.6% absolute in efficiency can be achieved compared to the reference case with 37% BSF fraction and 6% contact fractions.

B. Sensitivity Studies

As we use screen-printing for patterning the doped areas and for the metallization, there are two parameters that will become critical for a BSF islands design. Firstly, the size of the passivated BSF area needs to be large enough for good alignment of the BSF contact. Secondly, the emitter contact needs to have a certain minimum width because of metal paste printability. For both aspects, sensitivity studies were carried out to translate the optimal unit cell design from the simulations into a real unit cell that can be manufactured by screen printing.

The passivated BSF area fraction was varied between 8% and 60% and different shapes were used, to investigate how sensitive the area lay-out is to the performance. It appeared that the efficiency change is less than 0.05% absolute between 8 and 20% BSF area fraction, as shown in Fig. 3. The performance is mainly a trade-off between V_{oc} and FF. We concluded that we can use a BSF fraction of 20% as a safer option for metal contact alignment.

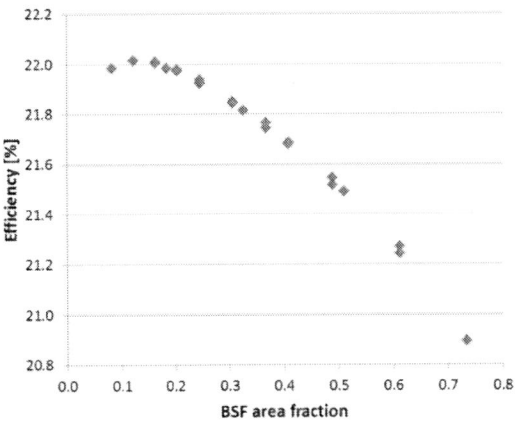

Fig. 3. Simulated cell efficiency as a function of BSF area fraction on the rear side of the IBC cell. A broad optimum between 8 and 20% BSF area fraction can be used.

The optimal emitter contact resulting from the simulations was quite narrow (less than 30 µm wide) due to the small unit cell width and small metal fraction. To make the contact design compatible with screen printing, a sensitivity study to wider and shorter emitter contact fingers was carried out, while keeping the contact coverage the same. The simulations showed that doubling the width and halving the length results in printable dimensions and with a minor efficiency loss of

0.05% absolute, as seen in Fig. 4. Therefore, this emitter contact design change was adopted.

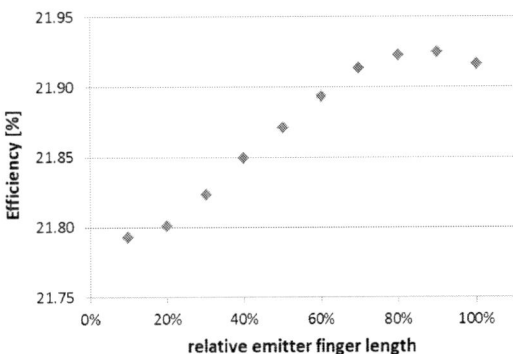

Fig. 4. Simulated cell efficiency as a function of emitter finger length relative to the unit cell length.

Besides the above modifications, we also changed the shape of the BSF area from square to circular. This is not expected to change the cell performance significantly, based on the sensitivity study of the BSF area size. The final two designs as used in the experiment are shown in Fig. 5 (right).

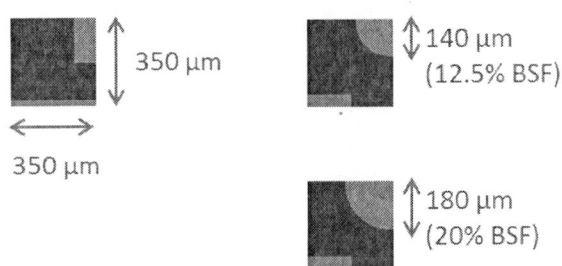

Fig. 5. Schematic view of the simulated 2D rear side of the IBC Mercury cell, after optimization of the unit cell size (left) and the two designs that followed from the sensitivity studies (right).

III. EXPERIMENTAL

The simulation results were used as input for the cell lay-out in the experiment. The experimental procedure and results are presented in the next sections.

A. Experimental Procedure

The two unit cell designs with different BSF area fractions, derived from the optimal simulation design and sensitivity studies, were applied to a full-size 6 inch IBC solar cell lay-out. As the metal contacts of both the emitter and BSF are interrupted, the contacts of each polarity are connected by screen printed metal fingers that do not etch through the SiN$_x$ passivation layer ("floating fingers").

As an additional parameter, we applied a light front floating emitter, which was combined with the smallest BSF islands, as this configuration is most sensitive to a further decrease of recombination sources. These cells are expected to yield the highest performance by an extra J_{sc} and V_{oc} increase, based on previously published passivation results of lightly doped emitters [4].

B. Experimental Results

The analysis of the cell results revealed that the cells with BSF islands were all shunted. We observed unexpected busbar paste spreading near the edges of the cell, short-circuiting to the fingers of opposite polarity.

To mitigate the influence of the shunts in the analysis, the edges were removed by laser scribe-and-break 10 mm from each edge, indicated with the red dashed lines in Fig. 6. This edge removal leads to edge recombination losses, especially due to the minority carrier transport of the front floating emitter. For fair comparison of the different cells, also some of the cells with the reference process were laser-cut.

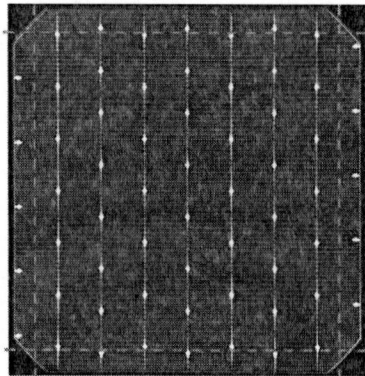

Fig. 6.   Position of the laser scribes to remove the cell edges.

The edge removal causes a 3% relative efficiency loss (exclusively due to J_{sc} loss, from 40.0 to 38.8 mA/cm²) in the reference cells. In Fig. 7, only the results of the reference cells after removal of the edges are presented.

As shown in Fig. 7, J_{sc} and V_{oc} both increase for the cells with BSF islands with respect to the reference. The average J_{sc} gain is 0.4 and 0.7 mA/cm² for medium and small islands, respectively, and the V_{oc} gain is 4 and 6 mV. This is in qualitative agreement with the simulation results (J_{sc} gain of 0.4 and 0.5 mA/cm², and V_{oc} gain of 7 and 9 mV).

The average FF loss of the cells with medium islands compared to the reference (0.4% absolute; not shown), is also well in line with the simulation results. For the cells with small BSF islands, we observe a higher FF loss than expected from the modeling. After further analysis of the cells, we can attribute most of the FF loss to pseudo-FF differences. The series resistance increase for the small BSF islands is relatively small. Most of cells with the small BSF islands also suffer from extremely low shunt resistance values, which results in even lower FF. Those cells are taken out of the analysis. The low shunt values are ascribed to misalignment

between the small BSF islands and the BSF metal contact, although this assumption needs to be verified by characterization.

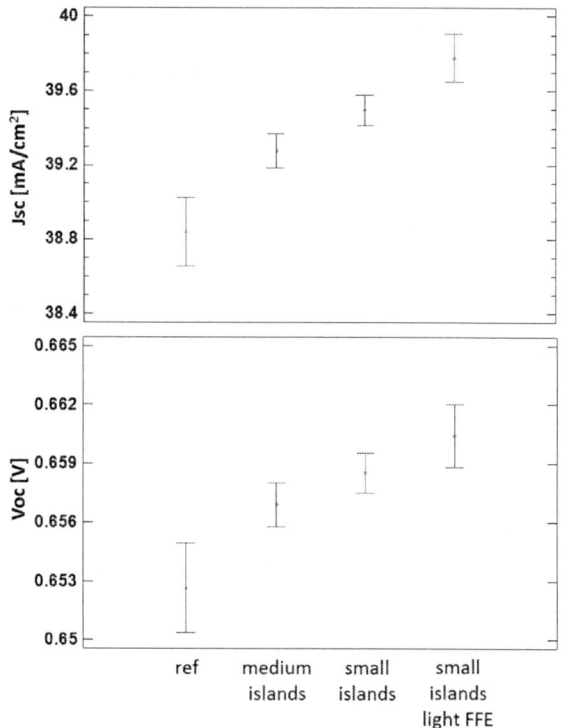

Fig. 7. J_{sc} and V_{oc} results of the cells after removal of the edges. Both J_{sc} and V_{oc} show a steady increase towards smaller BSF area fraction and lighter front side doping.

To complete the cell comparison, we compare the best cells of each group, as listed in Table 1. Although the reference cells show a lower performance level after the edge removal, it is clear that the use of small BSF areas and a small unit cell is beneficial, especially when it is combined with a lowly doped front floating emitter. The best cell of this type obtained a 0.6% absolute higher efficiency than the best reference cell.

TABLE I
SUMMARY OF BEST CELL RESULTS AFTER EDGE REMOVAL

	J_{sc} [mA/cm²]	V_{oc} [V]	FF [%]	Efficiency [%]
Reference	38.9	0.653	79.1	20.1
Medium islands	39.4	0.656	78.8	20.4
Small islands	39.7	0.657	77.7	20.3
Small islands and light FFE	39.9	0.663	77.9	20.6

IV. DISCUSSION

A. Device Simulations

The device simulations were of great assistance to define the optimal design parameters and sensitivity of the IBC cell. All resistance and recombination contributions could be separately listed in the simulation output, which was very helpful in identifying the limiting factors that needed further optimization.

The reduction of the highly recombinative BSF area fraction and of the metal contact fractions led to an increase in J_{sc} and V_{oc}. The observed trend in J_{sc} was not completely predicted by the simulations, so this needs further attention.

At the same time, a reduction of the unit cell size was necessary to keep the (majority) transport losses low. This is confirmed by the FF trend for medium islands, that follow the simulation and is almost as high as the FF of the reference.

B. Edge removal of cells

Some uncertainties arise in considering the impact of the edge removal to the cells with BSF islands compared to reference IBC cells. Although the properties of the cell edges are mostly similar for reference cells and BSF island cells, we cannot be sure that the impact is exactly the same, and therefore, we cannot be sure that the observed performance gain will be the same for full-size cells. For that reason, a final answer to this question can only be given for cells without edge cuts in a second experiment.

V. CONCLUSIONS

We presented an IBC cell with adjusted doping and contact pattern on the rear side, using small isolated BSF areas (BSF islands) surrounded by emitter, accompanied by a decrease of the unit cell size. Full-size 6 inch IBC cells were processed, and the cell edges were removed because of the shunts at these locations.

The BSF island design boosts the J_{sc} and V_{oc} of the cell, especially if it is combined with a lightly doped front floating emitter, resulting in 0.6% absolute increase in efficiency for the best cell, compared to the best reference cell.

The removal of the edges causes the lower efficiency level of the cell results. A repeat of the experiment on full-size cells will be performed to provide final conclusions on the performance of the BSF island design.

ACKNOWLEDGEMENTS

This work was supported by the Dutch Ministry of Economic Affairs within the TKI project IBCense.

REFERENCES

[1] P. Spinelli, P. Danzl, N. Guillevin, A. A Mewe, S. Sawallich, A. H. G. Vlooswijk, B. W. H. van de Loo, W. M. M. Kessels, M. Nagel and I. Cesar, "High-resolution sheet resistance mapping to unveil edge effects in industrial IBC solar cells", in *Energy Procedia* 92, 2016, p. 218.

[2] C. Reichel, F. Granek, M. Hermle and S. W. Glunz, "Back-contacted silicon solar cells with insulating thin films", at *npv workshop*, 2011

[3] J. Haschke, N. Mingirulli, and B. Rech, "Progress in point contacted rear silicon heterojunction solar cells", in *Energy Procedia* 27, 2012, p. 116

[4] A. A. Mewe, P. Spinelli, A. R. Burgers, A. H. G. Vlooswijk, N. Guillevin, E. J. Kossen and I. Cesar, "Emitter and contact optimization for high-efficiency IBC Mercury cells", in *Proc. 32nd EUPVSEC*, 2016, p. 760

Thermal Stability of Hydrogenated Boron Emitters

Khaja H. Mohammed[1], Larry C. Cousar[1], Sergiu C. Pop[3], Philip A. McMeans[1], Garrett Z. Evans[1], Hameed A. Naseem[2], & Douglas A. Hutchings[1]

1.   Picasolar, Inc., Fayetteville, AR, 72703, United States

2.   University of Arkansas Department of Electrical Engineering, Fayetteville, AR, 72701, United States

3.   Yingli Green Energy Americas, San Francisco, CA, 94108, United States

Abstract — **The Hydrogen Selective Emitter (HSE) is a self-aligned selective emitter designed for n-type silicon solar cells. In previous work the samples were processed by employing the Quinhydrone-methanol passivation applied to hydrogenated boron emitters and the photoconductance lifetime measurements were used to extract the emitter saturation current (J_{0e}). Prior evidence indicated that the conditions for hydrogenation could have an impact on the overall stability of the sample. In the present study we explore various conditions like: temperature, pressure, gas flow, etc. in order to understand the stability limitation of the samples where the hydrogenation process is applied as final step.**

Index terms — **atomic hydrogen, hydrogenation, photovoltaic cell, silicon, selective emitter.**

I. INTRODUCTION

Selective emitters are well understood but not widely adopted in industry. Traditionally the increased efficiency is offset by increased costs due to additional processing steps, alignment constraints and the use of extra materials. The increased efficiency is accomplished by improving the carrier selectivity in the emitter of the solar cell. Various methods have been explored to create selective emitters which include the etch-back selective emitter [1], a silicon ink emitter [2], and a laser doped selective emitter [3]. One of the major challenges with each of these approaches is the cost of implementation. The Hydrogen Selective Emitter (HSE) is a concept that incorporates all the benefits of a selective emitter while relying on a single, self-aligned, processing step. This is accomplished by utilizing the ability of atomic hydrogen to electrically deactivate the boron dopants [4].

The HSE process leverages the trap-limited diffusion of atomic hydrogen into boron-doped silicon [5] as the boron-hydrogen complexes form it electronically deactivates the dopant. Varying the time, temperature, pressure and gas flow rates can selectively control the amount and depth of boron deactivation. The emitter of high efficiency solar cells leverages a lower surface doping but deeper diffusion as shown in [6]. A similar or even lower surface doping can be achieved by deactivating dopants near the surface of a traditional emitter as shown in Fig. 1. Furthermore, this process can be accomplished in a single step post metallization where the gridlines block the hydrogen and result in a selective emitter structure.

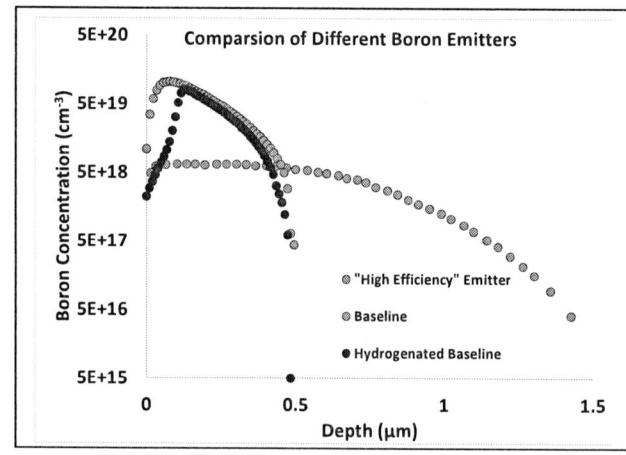

Fig. 1.   Comparison of different boron emitters

The goal of this work is to explore the thermal stability of boron-hydrogen complexes formed during the HSE process which may become unstable in the resulting device. Another motivation for looking into this phenomenon was to ensure that all important hydrogenation conditions are explored for a production setting that is stable above the thermal budget of the next processing step (encapsulation), which is typically done around 150°C.

II. EXPERIMENTAL DETAILS

A. Mechanisms of hydrogenation

The inactivation of boron using hydrogen has been reported to be followed by a trap-limited diffusion. A way to verify this theory is by performing experiments using boron-doped bulk wafers of varying doping concentrations. Maintaining the same hydrogenation conditions at each wafer should result in deactivation to varying depths. In theory it makes sense because when using an infinite source of atomic hydrogen at a

978-1-5090-5606-4/17 $31.00 © 2017 IEEE

concentration of 10^{18} (picked for illustrative purposes), boron concentrations below 10^{18} will rapidly deactivate and it will take much longer for concentrations over 10^{18}. The Spreading Resistance Profile (SRP) was recorded and the data is shown in Fig. 2. The results demonstrated that the depth/speed of hydrogenation is dependent on the concentration of boron. The atomic hydrogen rapidly diffuses through the lighter doped material in a consistent way with the trap-limited diffusion model. All profiles show the reduced doping characteristics at the surface with a return to the background/baseline doping at some depth into the silicon. The doping ranges used in this work correspond to the typical ranges of doping profiles used in n-type silicon solar cell fabrication.

The results are extremely compelling for the HSE concept. In an ideal situation we would have significant deactivation at the surface while quickly returning to the original doping profile. In order to achieve significant deactivation levels on lighter doped materials the hydrogen has to diffuse quite deep into the material and thus having a large impact on the sheet resistance. For heavily doped emitters, however, the sheer amount of boron available for deactivation means that we can primarily deactivate the surface without a significant change in the sheet resistance.

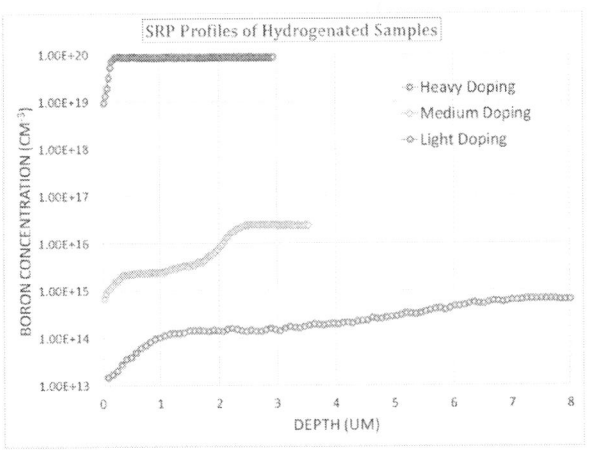

Fig. 2. SRP data from hydrogenation of uniformly doped planar wafers with varying boron concentration

B. Sample preparation

Industrial n-type solar cells were provided by Yingli Solar prior to nitride deposition in addition to fully completed solar cells. The doping profile of these cells was measured using ECV (Electrochemical Capacitance-Voltage) and the sheet resistance was recorded using a four-point probe.

The hydrogenation was performed in a custom-built chamber housed at Picasolar that utilizes hydrogen generated from deionized water. A custom-built plasma system is used to ionize molecular hydrogen into different ionic species of hydrogen. The system uses a resistive heater for in-situ substrate heating.

III. RESULTS AND DISCUSSION

A. Effect of Temperature

Experimental samples were processed under the same conditions except for the temperature which was modified from 130°C to 150°C. The hydrogenation process at 130°C achieved results right on target with post-hydrogenation sheet resistances of 100-120 Ω/square. In all cases, the samples showed no significant change in the sheet resistance after the first minutes at 150°C and 200°C, respectively. When moving above 200°C, at 250°C and 300°C they were slowly dehydrogenated as shown in Fig. 3.

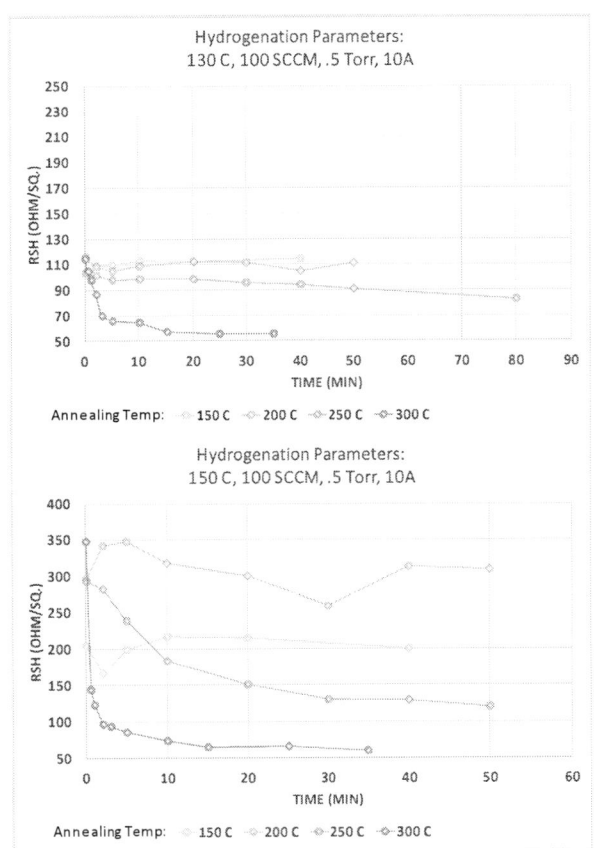

Fig. 3. Annealing study with the effect of temperature on hydrogenation

B. Effect of Pressure

It is well known that the processing pressure impacts the lifetime of atomic hydrogen in the chamber. Considering this concept a series of experiments at different pressure values and the sheet resistance was performed (Fig. 4). In each case, the devices showed no significant change in the sheet resistance after the first minutes at 150°C and 200°C, but they were slowly dehydrogenated at 250°C and quickly returning to the baseline doping at 300°C.

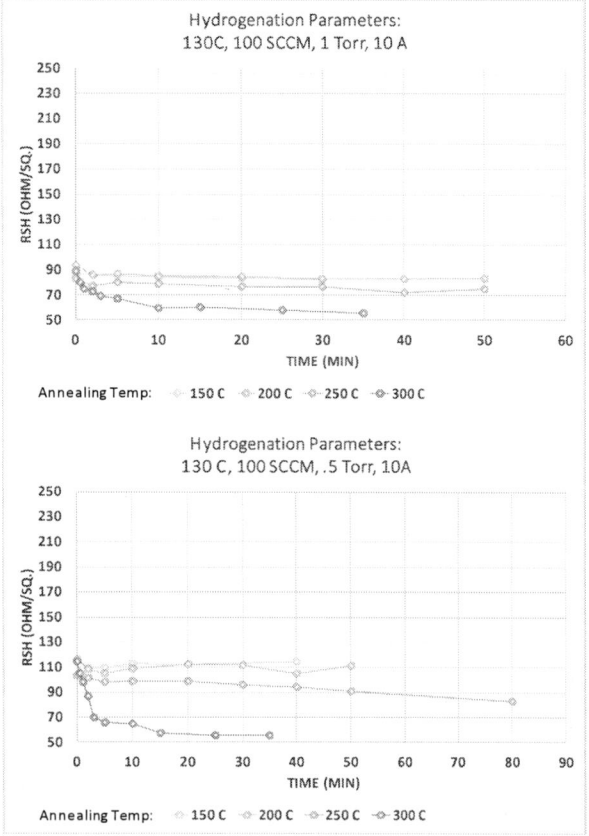

Fig. 4. Annealing study with the effect of pressure on hydrogenation

C. Effect of Gas Flow Rate

Upon exploring various gas flow parameters, it was revealed that an increased hydrogen flow rate should result in an increased atomic hydrogen concentration when assuming no other limiting factors. As a result, the increased flow rate turned into an increased sheet resistance but with a higher standard deviation (Fig. 5).

D. Hydrogenation on a solar cell

A custom built PIII process was incorporated in our chamber to hydrogenate through the standard industrial dielectrics. A full size n-type industrial solar cell from Yingli was diced into a $25cm^2$ solar cell. The cell was hydrogenated for just 15 seconds with a high voltage bias to accelerate the hydrogen ionic species towards the dielectric interface. TRIM simulations were used to determine the bias voltage. Table I shows the electrical characteristics of the solar cell before and after hydrogenation. Gains in Voc, Jsc and Efficiency were observed (Table I).

TABLE I
CELL PARAMETERS BEFORE AND AFTER HYDROGENATION

Sample	Voc (V)	Jsc (mA/cm²)	FF (%)	η (%)
Before H	0.628	38.34	80.02	22.49
After H	0.631	38.53	79.74	22.62
Δ	0.46%	0.48%	-0.36%	0.58%

In our previous work, we have already shown the impact of hydrogenation on reducing the J_{0e} using liquid passivation in conjunction with lifetime measurements [7]. The atomic hydrogen deactivates boron acceptors in emitter, reducing the probability of auger recombination, and thereby increasing Voc and Jsc.

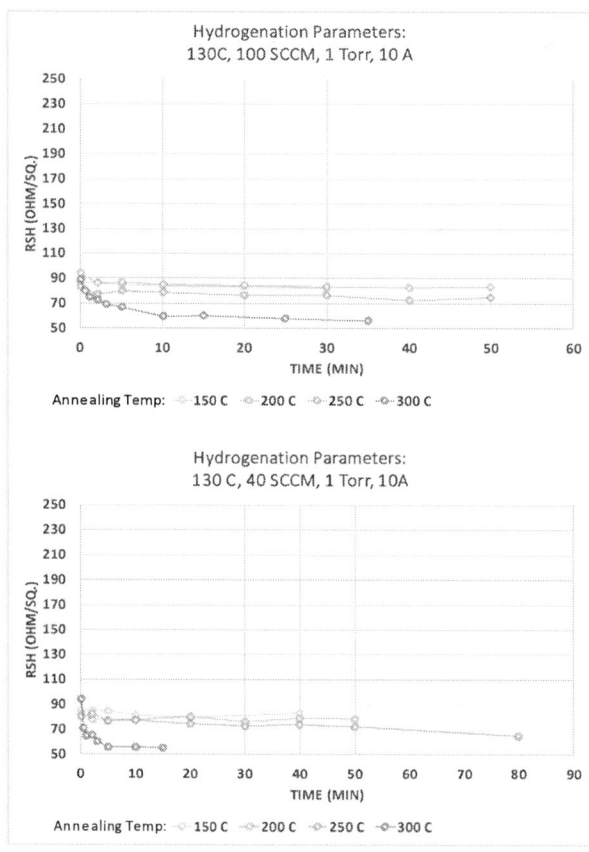

Fig. 5. Annealing study with the effect of flow rate on hydrogenation

IV. CONCLUSION

Early concerns of boron-hydrogen stability have shown to be manageable for a post-nitride HSE process. Not only we have succeeded in penetrating standard industry dielectrics but the process has been shown to be suitable to industry standard lamination temperature. We have demonstrated the ability to achieve target sheet resistance changes and will be using this knowledge to demonstrate further improvements in emitter saturation current density (J_{0e}) and solar cell efficiency.

REFERENCES

[1] A.Dastgheib-Shirazi, H. Haverkamp, B. Raabe, F. Book, and G. Hahn, "Selective emitter for industrial solar cell production: a wet chemical approach using a single side diffusion process." In *23rd European Photovoltaic Solar Energy Conference and Exhibition*, pp. 1197-1199, 2008.

[2] Antoniadis, Homer, F. Jiang, W. Shan, and Y. Liu. "All screen printed mass produced silicon ink selective emitter solar cells." In *Photovoltaic Specialists Conference (PVSC), 2010 35th IEEE*, pp. 001193-001196. IEEE, 2010.

[3] S. Adeline, J. Bovatsek, S. Wenham, B. Tjahjono, G. Xu, Y. Yao, B. Hallam et al. "18.5% laser-doped solar cell on CZ p-type silicon." In *Photovoltaic Specialists Conference (PVSC), 2010 35th IEEE*, pp. 000689-000694. IEEE, 2010.

[4] S.D. Shumate, M.G. Young, M.K. Hafeezuddin, D. Hutchings, L. Cousar, S.Q. Yu, and H. Naseem, "Progress on the hydrogen selective emitter for n-type solar cells," *Thirty-Ninth IEEE Photovoltaics Specialists Conference*, 2013.

[5] C.P.Herrero, M. Stutzmann, and A. Breitschwerdt, "Boron-hydrogen complexes in crystalline silicon," *Physical Review B*, vol. 43, no. 2, pp. 1555-1575, 1991.

[6] J. Shi, X. Li, D. Song, W. Yang, F. Li, F. Lang, S. C. Pop, W. Zhang, J. Wang, and B. Yu "Mass production and modeling of high efficiency n-PERT solar cells with ion implanted BSF/selective-BSF," *Solar Energy* vol. 142, 2017.

[7] M.G. Young, K. Schoelz, L. Cousar, H. Mohammed, S. Shumate, D. Hutchings, S. Pop, R. Schulze, J. Wang, H. Naseem, "Passivation studies with hydrogenated boron emitters," *Forty-Third IEEE Photovoltaics Specialists Conference*, 2016.

Light Induced Plating of Silicon Solar Cells Using Boric Acid-Free Nickel Chemistry

Krystal Munoz, Lynne Michaelson*, Joseph Karas[1], Tom Tyson, James Rand[2], Stuart Bowden[1]

Technic Inc, Cranston, RI, 02910, USA
[1]Ira A. Fulton Schools of Engineering, Arizona State University, Tempe, AZ, 85281, USA
[2]Core Energy Works, 304A Markus Ct, Newark, DE, 19713, USA
*contact and presenting author, lmichaelson@technic.com

Abstract — **A novel boric acid-free nickel plating chemistry has been developed to plate nickel onto silicon solar cells. This bath enables light induced plating (LIP) of nickel without the use of external rectification. The resulting deposit is low stress and has been shown to be an effective barrier to copper diffusion. Solar cells plated using this nickel bath demonstrate electrical, adhesion and reliability results similar to silver paste controls. In addition, these plated cells have lower contact resistance and higher metal conductivity than silver paste controls using similarly diffused wafers. The advantages of a boric acid-free nickel bath will be reviewed in detail.**

I. INTRODUCTION

Nickel-copper plated front contacts have been investigated by multiple research organizations and solar cell manufacturers over the last few years. [1-6] In addition to reduced cost, plated contacts may offer optical and electrical performance advantages. [1-7] Successful implementation of plated contacts into mainstream manufacturing requires both excellent performance of plated cells and commercial availability of plating chemistry and equipment that can be easily integrated into a solar cell manufacturing line. Metal plating is already established in high volume manufacturing in many industries, so the challenge is to adapt the chemistry and equipment to work with silicon solar cells. In addition, it is important to ensure that restricted chemicals are not included in the plating chemistries used. In 2010, boric acid, which is a component in commercially available nickel plating chemistries, was added to the candidate list of Substances of Very High Concern (SVHC) as part of REACH (European Union regulation concerning the Registration, Evaluation, Authorisation and restriction of Chemicals). Although boric acid chemistries can still be used, there are additional restrictions and requirements placed on the users.

This paper will discuss the formulation of a boric acid-free nickel plating bath that can be used to deposit nickel on silicon solar cells. Properties of the resulting nickel film will be discussed. In addition, performance of solar cells plated from this bath will be presented and compared to solar cells plated using boric acid containing nickel plating baths.

II. PLATING NICKEL ON SOLAR CELLS

Nickel plating has been around since the 1800s. In 1916, Professor Oliver P. Watts formulated an electrolyte that combined nickel sulfate, nickel chloride, and boric acid. [8] This formulation is the foundation of modern nickel electroplating. In 1950, Barrett reported commercial use of a nickel sulfamate bath in the US. [9] Nickel deposits from the sulfamate nickel plating bath are known to have a lower internal stress than those from a Watts bath; however, the sulfamate nickel bath costs more to make than the Watts nickel bath.[9] Both of these plating chemistries contain boric acid.

Watts and sulfamate nickel plating baths have been used to plate nickel on silicon solar cells. The nickel layer serves two purposes: 1) a source of nickel in order to form a nickel silicide (NiSi) ohmic contact and 2) a diffusion barrier for copper. In order to form this NiSi ohmic contact on a silicon solar cell, the ARC layer, typically silicon nitride (SiNx), must be removed in order to expose a clean silicon surface for nickel plating. Two of the more common approaches used for patterning the ARC layer are 1) masking & wet chemical etching and 2) laser ablation [10]. Table 1 lists two possible process flows that can be used to form plated contacts starting from the front side ARC patterning step. Once the ARC layer is removed, nickel can be deposited on the exposed silicon. The nickel silicide layer is formed by annealing, either after the deposition of the nickel layer, or after subsequent deposition of a copper layer over the nickel as illustrated in Table 1. For either approach, the nickel deposit must be uniform and low stress in order to function as a copper barrier and maintain good adhesion to the silicon surface. Many different factors can impact the quality of the nickel deposit such as the patterning process, cleaning steps prior to plating, equipment used, and plating chemistry.

Solar cells have been plated using direct electroplating and light induced plating (LIP) with and without rectification [5]. Direct electroplating requires that the areas to be plated, such as the bus bar and grid lines, are directly contacted to the rectifier which can be difficult to do with such fragile substrates. LIP takes advantage of electrons generated when

978-1-5090-5606-4/17 $31.00 © 2017 IEEE

the solar cell is illuminated so the electrical connection can be made to the backside of the solar cell loosening the requirements for the contact design and it can be performed with or without rectification. It is important to formulate the plating bath chemistry to be compatible with the method of deposition whether it is direct electroplating or LIP.

Table 1: Example ARC Patterning & Plating Flows

Laser Ablation	Mask & Wet Etch
Open ARC with Laser	Deposit Mask
Clean	Open ARC with Wet Etch
Nickel & Copper Plate	Clean
Anneal to form NiSi	Nickel Plate
Clean	Strip Resist
Tin Plate	**Anneal to form NiSi**
	Clean
	Nickel, Copper, & Tin Plate

III. BORIC ACID-FREE NICKEL PLATING BATH

A boric acid-free nickel plating bath was formulated in response to the restrictions being placed on boric acid as part of REACH. There are numerous advantages of this boric acid-free plating bath. In order to demonstrate these advantages, a lab scale study comparing this boric acid-free nickel chemistry with three boric acid containing nickel plating chemistries was performed. The pH, internal stress, and conductivity of the plating baths were monitored over a 5 week period. Table 2 lists the plating baths compared and summarizes the results. The boric acid plating baths tested fall into 3 different categories: 1) Watt's nickel (nickel sulfate / nickel chloride), 2) Nickel sulfamate, and 3) Technic proprietary formulation (Ni A) which is similar to the boric acid-free nickel bath but contains boric acid.

Table 2: Comparison of Nickel Plating Baths

	Watts Nickel	Nickel Sulfamate	Ni A w/ Boric Acid	Boric acid-free Ni A
Contains Boric Acid	Yes	Yes	Yes	No
pH target	4.0	4.0	4.0	2.0
Internal Stress over 5 weeks (MPa)	-20 (comp) to 20 (tensile)	-50 (comp) to 40 (tensile)	-15 (comp) to 15 (tensile)	-15 (comp) to 15 (tensile
LIP w/o rectificatio n	No	No	Yes	Yes
Technic Product Names	Watts Ni Semi bright	Technisol Ni 2420	N/A	Technisol Nickel D2428

As shown in Table 2, the operating pH of the boric acid-free nickel plating bath is 2.0 versus a pH of 4.0 for the boric acid containing plating baths. In addition, the operating pH of the boric acid-free plating bath is very stable over time. Figure 1 compares the pH monitored over 5 weeks for the four baths tested. The red arrow indicates when the pH was adjusted based on a drift of ± 0.10 from the operating pH. The boric acid-free Ni plating bath required only one adjustment over the 5 weeks; whereas all the other plating baths required 2 or more adjustments. In addition, the pH excursions for Ni A with boric acid and the Watts Ni bath were larger than any excursion for the boric acid-free nickel.

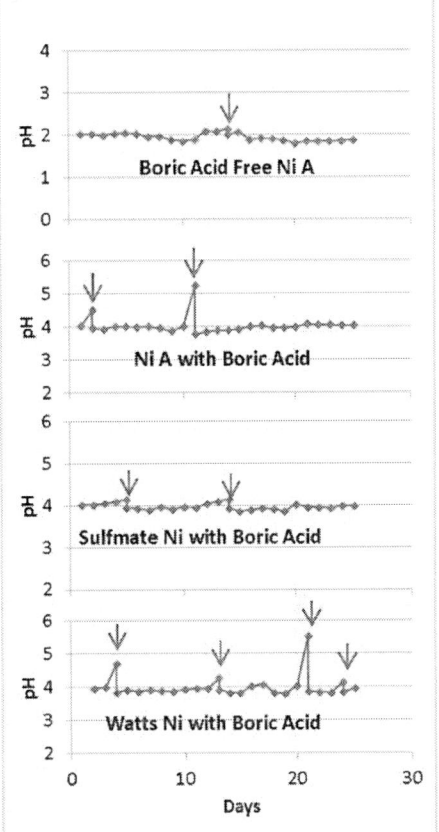

Fig. 1: pH monitoring over 5 weeks for nickel plating baths. Red arrow indicates pH adjustment when pH drifted ±0.10 from target.

In addition to the boric acid-free nickel bath having a stable pH over time, this bath has been formulated to plate nickel on silicon solar cells using LIP with or without rectification. Both the Watts and sulfamate plating baths are commonly used in other industries for nickel plating but are not specifically formulated for LIP plating on solar cells. Initially, the sulfamate nickel plating bath was used to plate nickel on a silicon solar cells using LIP without rectification; i.e. the backside of the solar cell is connected directly to the anode and the plating current is generated by illuminating the solar cell. This approach was not successful, with very little nickel plating observed unless rectification was used with the LIP; i.e. the backside of the solar cell is connected to the rectifier (DC power supply) and current is applied while illuminating the solar cell.

During the development of the boric acid-free nickel plating bath, the goal was to achieve a nickel deposit using LIP without rectification. A conductivity metric was identified to predict when a nickel plating solution is capable of LIP plating without rectification. This conductivity metric was used during the formulation of the boric acid-free nickel as a quick test to identify when the formulation was ready for LIP plating without rectification. The plating solutions in Table 2 were tested using the conductivity metric and the results are shown in Figure 2. The Ni A bath with and without boric acid has a conductivity that enables LIP plating without rectification; however, the Watts Ni and sulfamate formulations do not. In order to confirm these results, laser ablated solar cells were plated for 7 mins in plating equipment designed to immerse the front side of the solar cell in the plating solution and maintain a dry backside. The front side of the cell was illuminated and the dry backside was connected to the anode. Figure 3 compares the top down SEM images of grid lines plated using LIP without rectification in both the boric acid-free nickel bath and the sulfamate nickel bath. Figure 3a shows that nickel plating from the boric acid-free Ni bath did occur on the grid lines; i.e. the area in between the dotted white lines. Very little nickel plated from the sulfamate plating bath as shown in Figure 3b. In order to get a thicker layer, the time or light intensity would need to be increased. However, this demonstrates that the boric acid-free nickel bath is more conductive than the sulfamate nickel bath and is capable of plating nickel on the solar cell without the use of external rectification. Although LIP plating without rectification is possible with this plating chemistry, it may not be the preferred method of plating in high volume manufacturing. This, however, does demonstrate the unique properties of this boric acid-free nickel plating bath.

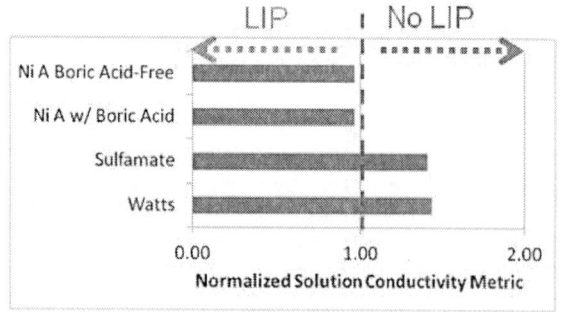

Fig. 2: Results from study used to predict if plating chemistry is capable of LIP without rectification.

In addition, the internal stress of the nickel deposited from each bath was measured using stress tabs from the Deposit Stress Analyzer System, Specialty Testing & Development Co. once a week over the 5 week period. As shown in Table 2, the Ni A baths with and without boric acid have the lowest max internal stress over the 5 week period. It is interesting to note that the max stress for the Watts nickel bath in this study was measured to be ~20 MPa which is much lower than the 185 MPa reported in literature. [11]

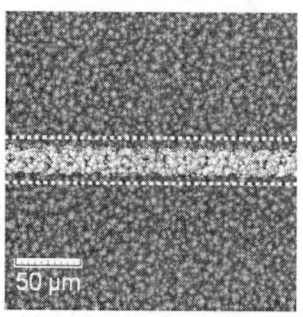 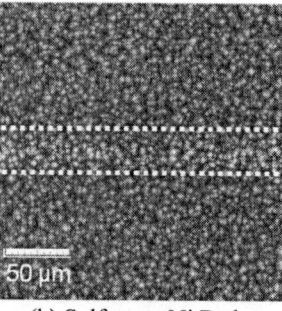

(a) Boric acid-free Ni Bath   (b) Sulfamate Ni Bath

Fig. 3: Top down SEM images of grid lines plated using LIP without rectification for two different nickel plating chemistries. (1000 X, dotted lines added to highlight location of grid line)

A 50 week bath study of the internal stress in the nickel deposited from the boric acid-free chemistry was also performed using the same method of measurement. Figure 4 shows the results of these measurements. The initial data point from the fresh bath shows a highly compressive value which requires further investigation. However, within one week of operation, the internal stress is measured to be 20 MPa compressive. The maximum tensile stress observed over this time period is 60 MPa. The green line in Figure 4 is the maximum expected internal stress of 55 MPa for nickel deposited from a sulfamate bath as reported in literature [11]. The majority of data points measured from the boric acid-free nickel fall below this line. For comparison, the red line in Figure 4 shows the maximum internal stress expected for a deposit from a Watts nickel bath from literature [11]. A nickel deposit with a low internal stress is expected to have better adhesion performance than a highly stressed film.

978-1-5090-5606-4/17 $31.00 © 2017 IEEE

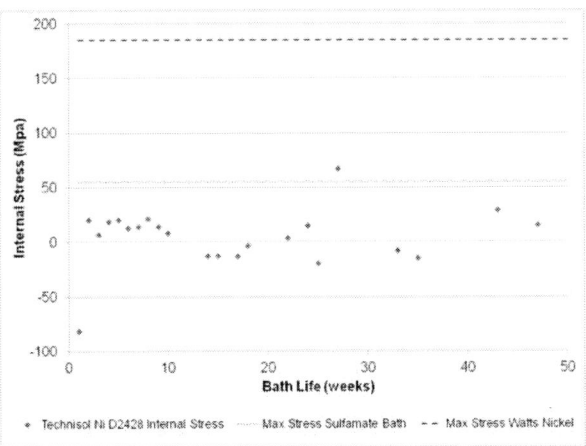

Fig. 4: Internal stress of nickel films deposited from a boric acid-free nickel plating bath over a 50 week bath life study.

IV. SOLAR CELL RESULTS

156 x 156 mm semi-fabricated (no metalized front grid) and fully processed silver paste control cells have been acquired by two different research organizations in order to test both process flows listed in Table 1. One set of cells, purchased from Energy Research Center of the Netherlands (ECN), were p-type mono-silicon textured cells with an 80 ohm/sq emitter, silicon nitride ARC layer, and fired Al/Ag backside. At Technic, these cells were patterned with a mask & wet etch process and plated and will be referred to as wet etch cells. The other set of cells, purchased from Fraunhofer ISE, were p-type Cz silicon, 1.5-4.0 ohm-cm, textured with a 90 ohm/sq emitter, silicon nitride ARC layer, and fired Al/Ag backside. A UV picoseconds laser ablation process was used by Fraunhofer ISE to pattern the front grid[12]. These cells were plated at Technic and will be referred to as laser ablated cells. Nickel plating was performed using the boric acid-free chemistry. Copper was plated over the nickel using a Technic low stress copper chemistry. For adhesion and reliability testing, a tin layer was deposited over the copper layer using a Technic tin chemistry specifically developed to be compatible with LIP.

Laser ablated, plated samples were IV tested using a Sinton FCT-400. 5 laser ablated, plated samples and 5 silver paste control samples from the same batch of cells were measured. Table 3 shows the results. The plated cells perform the same if not slightly better than the silver paste controls. The series resistance (R_s) is lower for the plated cells even though the plated metal lines on these cells are thinner; i.e. less volume of metal.

Table 3: IV Properties for Ag Paste Control vs. Laser ablated, plated cells

	% Eff	% FF	R_s (Ω-cm^2)	Grid W (um)	Grid Thk (um)
Ag Paste	18.96 ± 0.05	79.59 ± 0.19	0.56 ± 0.02	41.5 ± 9.0	20.2 ± 6.0
Technic Plate	19.08 ± 0.05	80.04 ± 0.21	0.46 ± 0.04	45.7 ± 5.9	13.8 ± 3.5

The contact and line resistivity was measured using the ContactSpot technique [13]. Figure 5 compares the effective contact resistivity of the silver paste control with cells plated from the boric acid-free Ni A bath and the sulfamate nickel bath. The plated cells have a median contact resistivity that is approximately half that of the silver paste controls. In addition, Table 4 shows that the plated lines are more than double the conductivity of the silver paste controls with a line resistivity very close to that of pure copper[13]. A larger sample size is needed to confirm if samples plated in the boric acid-free nickel are consistently lower in contact and line resistivity than samples plated in the sulfamate nickel bath. This study emphasizes that plated lines are more conductive and have a better contact resistance than silver paste controls.

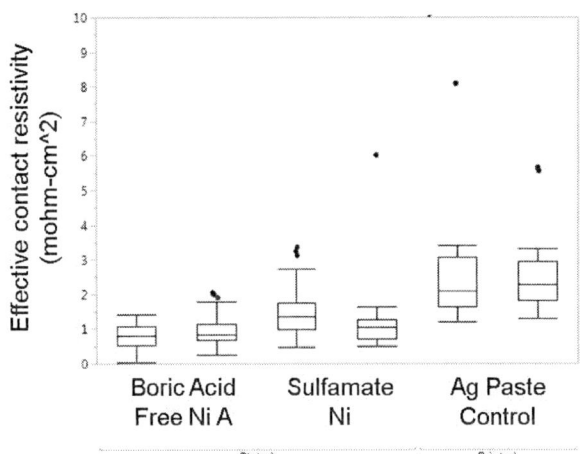

Fig. 5: Comparison of effective contact resistivity for plated contacts versus silver paste control.

Table 4: Comparison of line resistivity for plated cells versus silver paste control

	Line Resistivity (ohm-m)
Ag Paste Control	$5.0 \pm 0.4 \times 10^{-8}$
Plated from Ni A w/o boric acid	$1.9 \pm 0.1 \times 10^{-8}$
Plated from Sulfamate Ni	$2.2 \pm 0.1 \times 10^{-8}$
Pure Copper [13]	1.72×10^{-8}

The contact resistance and adhesion of plated cells is significantly impacted by the quality of the silicon / nickel interface. Figure 6 shows a SEM cross-section of a wet etch cell after nickel and copper plating imaged at Massachusetts Institute of Technology (MIT). This image shows that the interface between the silicon and the nickel is clean and uniform. The surface preparation of the silicon surface and the nickel plating process play a critical role in this resulting clean interface.

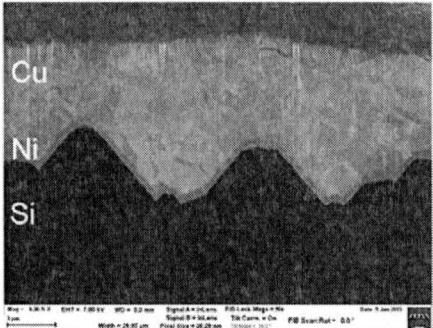

Fig. 6: SEM cross section of a Ni/Cu plated grid line.

Wet etch samples plated with Ni/Cu/Sn were peel tested at MIT [15] to determine the adhesion of the plated samples compared to silver paste controls. Ribbon similar to what is used in industry was first hand soldered onto the busbars of both the plated and silver paste cells. A 180° pull angle was used with a pulling speed of 200 mm/min using the ADMET eXpert 5600 Series Universal Testing System. This force tester gage has a load limit of 2.2 lbf (9.786 N) ~ 4.893 N/mm. The samples were secured to a vacuum plate for peel testing and the force required to pull the ribbon away from the cell was measured. The adhesion strength is the force per unit width of the ribbon. Figure 7 shows the maximum adhesion values for 10 plated samples compared to 5 silver paste control samples. The maximum adhesion values for the plated samples range from 3 to 4.5 N/mm. The silver paste control samples range from 2 to 3.5 N/m. The adhesion performance of the plated cells are as good as or better than the silver paste controls [15]. Previous work showed that the nickel film functioned as a barrier to copper diffusion based on dry heat annealing studies [16].

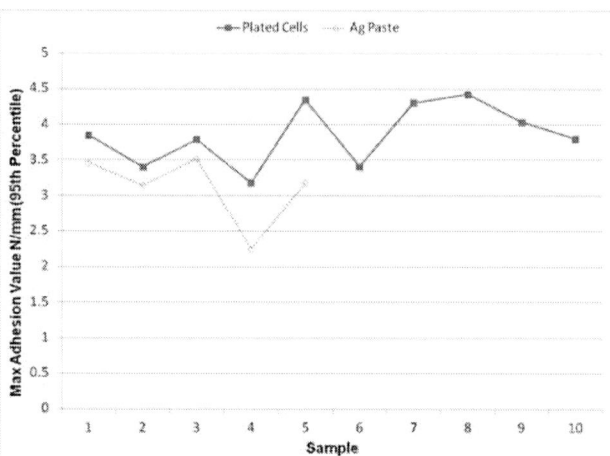

Fig. 7: Peel test results comparing the maximum adhesion value of plated cells versus silver paste cells

Initial reliability testing was performed on both wet etch and laser ablated samples. ASU fabricated single cell modules from the plated and silver paste control cells. Damp heat testing for 1000 hrs at 85°C/85% was performed at NREL and thermal cycling testing (-45°C to 85°C) was performed at Case Western Reserve University. This was a small sample set intended for an early look at reliability of the plated contacts. The plated cells used for this testing were similar in electrical performance to the silver paste controls. Table 5 shows the reliability results for laser ablated, plated cells compared to the silver paste controls. All plated cells perform similar to the Ag paste control and show less than 5% change in power post testing which is a passing result.

Table 5: Reliability results for laser ablated, plated cells in single cell modules.

Sample	Test	Δ Pmax (% rel)	Δ FF (% rel)	Δ Voc (% rel)	Δ Isc (% rel)	Δ Eff (% rel)
Control	1000 h DH	-0.7	-0.3	0.6	-1.0	-0.3
1	1000 h DH	-1.6	-0.8	0.5	-1.3	-0.8
2	1000 h DH	-0.9	-1.1	0.5	-0.3	-1.1
3	1000 h DH	-2.9	-1.9	0.3	-1.3	-1.9
4	1000 h DH	2.7	0.7	1.3	0.8	0.7
5	1000 h DH	1.5	-0.3	1.3	0.5	-0.3
Control	200 TC	-2.0	-1.1	0.0	-0.8	-2.2
1	200 TC	-2.1	-0.8	0.2	-1.3	-1.6
2	200 TC	-0.7	-0.4	0.0	-0.3	-0.5
3	200 TC	-3.2	-1.8	0.0	-1.3	-2.7
4	200 TC	-3.3	-4.1	0.2	0.8	-3.4

Table 6 shows the results for the wet etch cells. Unfortunately, there were not enough cells to have a silver paste control in the thermal cycling test. However, all the cells show passing results, with power changes less than 5%.

Further reliability tests are planned with a larger sample size and more silver paste controls.

Table 6: Reliability results for wet etch, plated cells in single cell modules

#	ID	Δ Pmax (% rel)	Δ FF (% rel)	Δ Voc (% rel)	Δ Isc (% rel)	Δ Eff (% rel)
Control	1000 hr DH	-0.4	0.3	0.6	-1.3	0.3
1	1000 hr DH	-2.8	-1.3	-0.3	-1.1	-1.3
2	1000 hr DH	-2.3	-1.1	0.2	-1.5	-1.1
3	1000 hr DH	-2.3	-1.4	0.0	-0.9	-1.4
4	1000 hr DH	0.0	1.0	0.8	-1.7	1.0
1	200 TC	-2.5	-1.1	-0.2	-1.1	-2.4
2	200 TC	-2.0	-0.4	0.0	-1.4	-1.9
3	200 TC	-1.6	-0.4	0.2	-1.2	-1.3
4	200 TC	-1.7	-0.3	0.2	-1.4	-1.2
5	200 TC	-3.2	-2.5	0.7	-1.2	-3.3
6	200 TC	-4.3	-2.2	-0.2	-1.8	-4.0

V. CONCLUSIONS

In conclusion, Technic has developed a boric acid-free nickel plating bath that can be used to plate nickel on silicon solar cells using LIP with or without rectification. To the best of our knowledge, there is no other boric acid-free nickel plating bath commercially available and specifically designed for LIP plating. This plating bath has a stable pH that requires little adjustment over the life of the bath. In addition, the nickel deposit produced is low stress which is important for good adhesion performance. The line and contact resistivity measured on laser ablated, plated cells is less than half that of the Ag paste controls. The cells used during this study were optimized for screen printing and not designed to take advantage of the low contact resistance and line resistivity of the plated contacts. Further studies will explore how to utilize plating to improve solar cell performance by modifying the emitter and front grid layout to bring out the true potential of plated contacts on silicon solar cells. This study also demonstrates that plated contacts have similar adhesion and reliability performance to silver paste contacts. By optimizing the plating chemistry, tooling, and solar cell design, it is expected that plated contacts can outperform silver paste contacts while saving money on materials costs.

ACKNOWLEDGEMENTS

A portion of this work is supported by the U.S. Department of Energy Office of Energy Efficiency and Renewable Energy under Award Number DE-EE0006814. Austin Akey, Mariela Castillo, and Tonio Buonassisi at MIT contributed SEM and Peel Testing results and discussion.

REFERENCES

[1] A. Letize et al, "Pilot scale production...electroplating process", Proc. of the 29th EUPVSEC 2014, pp. 1359-1362

[2] A. Mondon et al, "Plated Ni-Cu contacts...production", Proc. of the 29th EUPVSEC 2014, pp. 1286-1291

[3] R. Russell et al, "A simple copper...modules", Proc. of the 27th EUPVSEC 2012, pp. 538-543

[4] J. Hortzel et al, "Industrial silicon solar cells...contacts", IEEE Journal of Photovoltaics, Vol. 5, No. 6, November 2015

[5] A. Lennon et al, "Evolution of metal plating ...metallisation," *Prog. Photovoltaics Res. Appl.*, vol. 21, no. 7, pp. 1454–1468, 2013.

[6] L. Michaelson et al, "Improved contact ...process", Proc. of the 38th IEEE PVSC 2012, pp.

[7] G. Beaucarne, G. Schubert, and J. Hoornstra, "Summary of the 4th workshop ...cells," in *Energy Procedia*, 2013, vol. 43, pp. 2–11.

[8] G. A. Di Bari, "Electrodeposition of Nickel," in *Modern Electroplating*, 5th ed., M. Schlesinger and M. Paunovic, Eds. New York: Wiley, 2010, pp. 79–113.

[9] R.C. Barrett, "Nickel plating from the sulfamate bath", 41st Annual Convention of the American Electroplaters' Society, 1954

[10] M. Aleman et al., "Advances in ...ARC", Proc. of the 24th EPVSEC, 2009, pp. 1414-1

[11] G. Di Bari, "Nickel Plating", in *ASM Handbook, Vol. 5, Surface Engineering*, ASM International, Materials Park, OH 44073, 1994, p. 201

[12] S. Kluska et al, "Electrical and Mechanical ... Solar Cells", in *Energy Procedia*, 2015, vol. 77, pp.733-743

[13] S. Guo, G. Gregory, A. M. Gabor, W. V. Schoenfeld, and K. O. Davis, "Detailed investigation of TLM ...solar cells," *Sol. Energy*, vol. 151, pp. 163–172, 2017.

[14] http://hyperphysics.phy-astr.gsu.edu/hbase/Tables/rstiv.html

[15] J. Karas et al, "Addressing adhesion ....modules", Proc. of the 43rd IEEE PVSC 2016

[16] J. Karas et al, "Electrical characterization...contacts", Proc. of the 42nd IEEE PVSC 2015

Baking temperature dependence of Cu paste on Al-BSF cell properties

Tomohiro Saito[1], Tetsuya Fukuda[2], Hoang Tri Hai[1], Yuji Kurimoto[2], Daisuke Ando[1], Yuji Sutou[1], Katsuhiko Shirasawa[1], Junichi Koike[1,2]

1. Department of Materials Science, Tohoku University, Sendai 980-8579, Japan

2. Material Concept, Inc. Sendai 980-0845, Japan

Abstract — We investigated the baking temperature dependence of Al-BSF cell properties using Cu paste metallization and a metal oxide (MO) layer. The baking conditions of Cu paste were changed to find an optimum condition to achieve good adhesion, diffusion barrier property, and low contact resistance. Using 2 cm square cells, the conversion efficiency of 17.80% could be obtained, which was equivalent to the property of commercial Ag paste cells cut into the same size.

Index Terms — Cu paste, barrier layer, silicon, Al-BSF, photovoltaic cells.

I. INTRODUCTION

For cost reduction and further widespread use of silicon solar cells, replacing Ag with Cu has been awaited for years. Among various efforts to use Cu, electroplated (EP) Cu shows good cell performance [1, 2]. However, poor throughput and waste chemical treatment have hampered the use of the EP Cu technology in mass production level. Alternatively, Cu paste including low melting point alloys [3] or thermoset resin [4] has been proposed. However, solar cell properties and reliability have not reached the level of conventional Ag paste cells. One of the difficulties with Cu paste is easy diffusion of Cu into Si during baking of Cu paste at high temperature. The diffused Cu acts as carrier recombination centers, leading to the rapid degradation of cell performance. Therefore, in order to apply the screen-printed Cu paste for Si solar cells, a barrier layer is necessary to block Cu diffusion. In addition, the barrier layer should also have low contact resistivity and good adhesion between Cu and Si.

To solve these problems, we developed a multifunctional interface layer composed of a metal oxide (MO) that is to be placed between Si wafer and Cu wires [5]. The interface layers should satisfy excellent adherence, low contact resistivity and good diffusion barrier property. In this paper, we describe the relationship between baking condition of Cu paste and c-Si solar cell performance with an Al-BSF electrode on back surface using the multifunctional interface layer.

II. EXPERIMENTAL PROCEDURE

Mono-Si cells were purchased from a cell maker. The cells did not have bus and finger lines on its front surface, but had a SiN ARC layer and Al/Ag electrode. Full-size cells were cut into 2x2 cm² pieces to carry out experiments under different conditions. To obtain electrical contact between Cu and Si, the SiN layer was partially removed by photolithography and wet-chemical etching with buffered HF. The etched wafers were rinsed in ethanol using an ultrasonic bath, in 5%HF solution, and in de-ionized water, then blown dry with a blower. After that, an interface layer was deposited by RF magnetron sputtering, followed by screen printing of Cu paste. Finally, baking of the screen printed Cu paste was performed in a quartz tube furnace in two steps at various temperatures. Oxidation step was conducted in a mixed gas atmosphere of N_2+O_2, aiming at decomposition of organic substances and Cu oxide sintering. Reduction step followed to reduce the Cu oxide to metallic Cu in a forming gas ambient. Fig.1 shows a schematic drawing of an obtained cell.

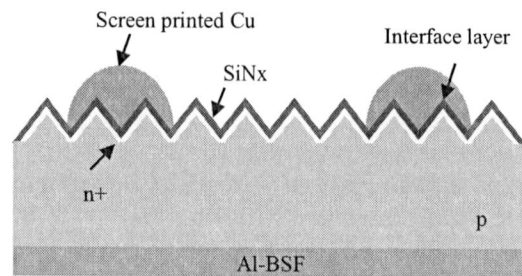

Fig.1 A schematic diagram of Al-BSF cells using a metal oxide (MO) interface layer and a Cu paste.

Baking behavior of Cu paste was investigated by measuring weight change of Cu paste with increasing temperature at a constant heating rate using thermogravimetric analysis (TGA). Electrical resistivity of baked Cu was measured with a four-point probe method. Adhesion strength of Cu on Si was examined by tape test. The test was performed three times for each samples in a given baking condition. Diffusion barrier property was examined by observing the formation of Cu silicide using SEM and TEM. Contact resistivity was measured with a TLM method. Solar cell performance was evaluated by measuring I-V curve using a solar simulator in a standard condition.

978-1-5090-5606-4/17 $31.00 © 2017 IEEE

III. RESULTS AND DISCUSSION

TGA results of Cu paste indicated that organic vehicle was removed around 375 °C. Since organic residue has a detrimental effect on paste sintering, it is very important to remove the organic substance from the paste. Therefore, the first baking step of oxidation should be carried out at higher temperature than 375 °C. Fig.2 shows electrical resistivity of oxidized and reduced Cu as a function of oxidation temperature. It is noted that TGA and XRD analysis indicated complete reduction to metallic Cu. Low resistivity of 3.5 to 5.5 μΩ.cm can be obtained, indicating that a good sintering property of the Cu paste.

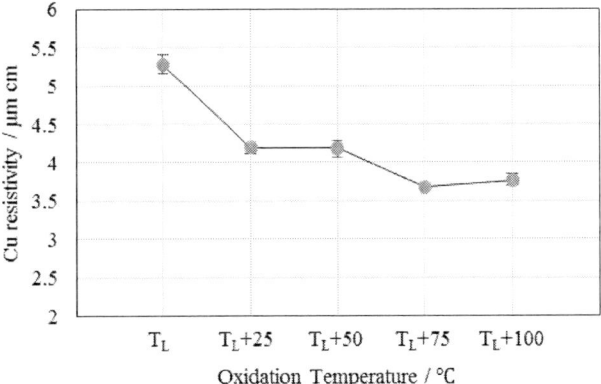

Fig.2 The resistivity of baked Cu paste as a function of oxidation temperature at a fixed reduction temperature of T_L.

Table1 shows the results of tape test in various baking conditions. In Table1, cross marks and triangle marks represent delamination by tape test on the first try, and on the second or third try, respectively, while circle marks represent no delamination by tape test after the third try. At the lowest reduction temperature denoted T_L, good adhesion was achieved. On the other hand, above T_L+75 °C, Cu paste was easily peeled off. Fig. 3 shows cross-sectional SEM images of the sample oxidized at T_L and at T_L+100 °C. At T_L, a continuous MO layer can be observed on textured Si surface, and a good bonding with sintered Cu particles. In contrast at T_L+100 °C, there is a gap between Cu and Si. It is considered that poor adhesion at higher reduction temperature is due to the lack of thermal stability of the MO layer resulting in partial evaporation or solid solution formation with Cu.

Fig.4 shows the reduction temperature dependence of Cu paste solar cell performance at a fixed oxidation temperature. Please note that the samples were without SiN films, which brings about open circuit voltage of approximately 600 mV because of the lack of surface passivation effect. Tape test results are overlaid on the data points. From T_L to T_L+50 °C in the reduction step, efficiency values remain nearly constant and adhesion is good. However, with increasing reduction temperature, conversion efficiency rapidly decreases and

Red. \ Oxi	T_L / °C	T_L+100 / °C
T_L / °C	O	O
T_L+25 / °C	△	O
T_L+50 / °C	△	O
T_L+75 / °C	×	×
T_L+100 / °C	×	×

Table1. The results of tape test in various baking conditions.

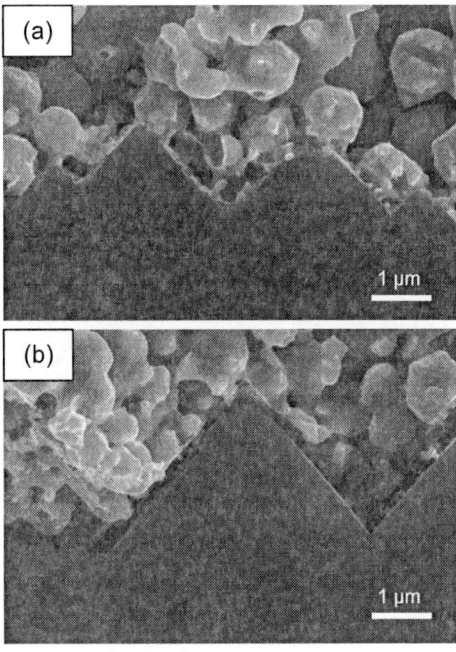

Fig.3 The cross-sectional SEM images of Cu/MO/Si cells annealed at reduction temperature of (a) T_L, (b) T_L+100 °C at a fixed oxidation temperature of T_L.

adhesion becomes very poor. The result suggests that contact resistivity would increase due to poor adhesion, leading to the increase of series resistance and the decrease of fill factor.

Fig.5 (a) shows the variation of conversion efficiency with increasing oxidation temperature from T_L to T_L+100 °C at a fixed reduction temperature. Similar to the reduction temperature dependence, efficiency value dramatically goes down at high oxidation temperature. As shown in Fig.5 (b), with increasing the oxidation temperature, series resistance increases, while fill factor decreases. TLM measurement indicated an increase of the contact resistivity of MO/Si interface. TEM observation also indicated the formation of an interfacial reaction layer composed of MSi_xO_y between MO and Si, as shown in Fig.6. In addition, the resistivity of the MO layer itself increased with increasing the oxidation temperature.

978-1-5090-5606-4/17 $31.00 © 2017 IEEE

Both the formation of the interfacial reaction layer and the higher resistivity of the MO layer appear to be a reason for the high contact resistivity of MO/Si, high Rs, low FF, and low efficiency.

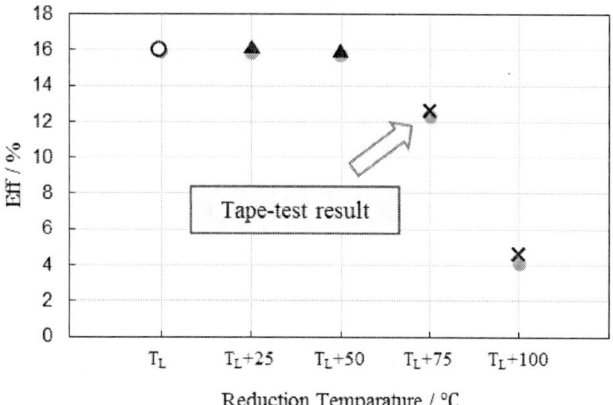

Fig.4 The reduction temperature dependence on Cu paste solar cell efficiency. Same data symbols are used as tape test results in Table 1.

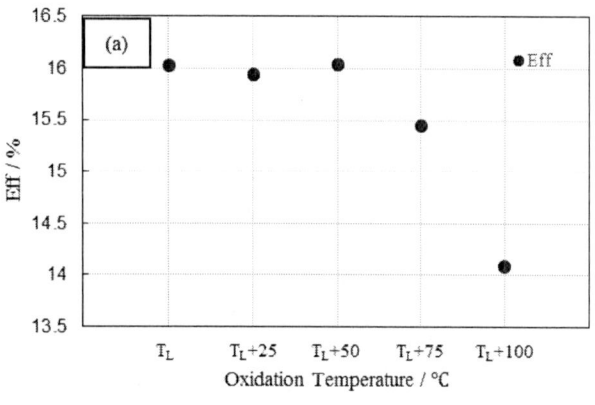

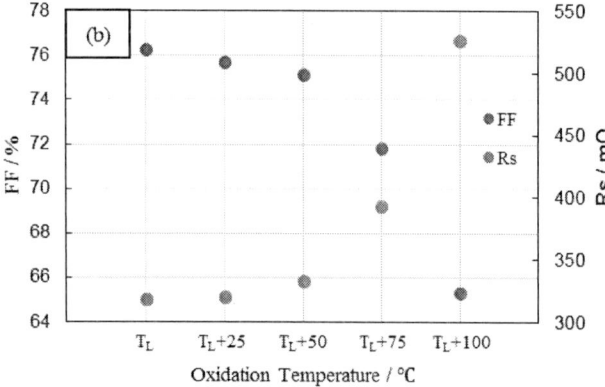

Fig.5 The results of (a) efficiency (b) fill factor and series resistance in Cu paste cell annealed at various oxidation temperatures.

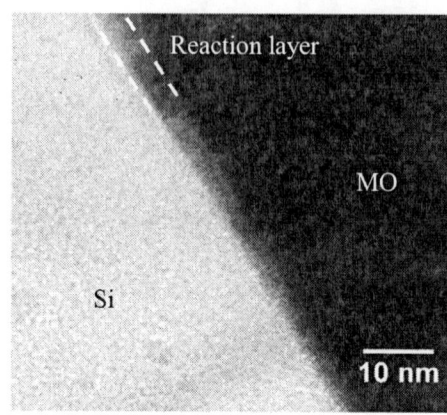

Fig.6 The cross-sectional TEM image of MO/Si interface region annealed at oxidation temperature of T_L+100 °C at a fixed reduction temperature of T_L.

Table 2 shows the results of best cell performance on 2 cm square chips with SiN. The high efficiency of 17.80 % can be obtained using Cu paste in Al-BSF cell under best oxidation and reduction conditions. For comparison, a commercial Ag paste cell was cut into 2x2 cm² chips and its cell parameters are listed in the table. The cell properties with Cu paste is found to be better than the Ag paste cells. Improvement in FF is strongly required. The contact resistance between MO/Si and the resistivity of a MO layer itself are controlling factors for series resistance and FF. As a follow-up work to further improve efficiency, it is necessary not only to reduce series resistance, but also to optimize the light transmittance properties of the MO layer with SiNx layer.

2cm sq. cell	E_{ff} [%]	FF [%]	Jsc [mA/cm²]	Voc [V]
Cu paste cell	17.80	77.25	36.35	0.634
Commercial Ag cell	17.54	78.33	35.41	0.632

Table2 Comparison of Al-BSF cell properties between Cu and Ag paste having a chip size of 2 cm.

IV. SUMMARY

We fabricated 2x2 cm² mono-Si Al-BSF solar cells using Cu paste together with a metal oxide (MO) interface layer. The influence of baking conditions on the Cu-paste cell properties was investigated. Two baking steps of oxidation and reduction were required to obtain well-sintered Cu. With increasing the baking temperature, solar cell properties were degraded because of poor adhesion and high contact resistivity at MO/Si interface. The best solar cell efficiency of 17.80% was obtained in the chip size of 2 cm square. The obtained results are superior to the efficiency of commercial Ag solar cell (17.54%) having the same size.

REFERENCES

[1] A. Dabirian, A. Lachowicz, J.-W. Schüttauf, B. Paviet-Salomon, M. Morales-Masis, A. Hessler-Wyser, M. Despeisse, and C. Ballif, "Metallization of Si heterojunction solar cells by nanosecond laser ablation and Ni-Cu plating," *Solar Energy Materials and Solar Cells*, vol.159, p.243 2017.

[2] Vikrant A. Chaudhari, and Chetan S. Solanki, "A novel two step metallization of Ni/Cu for low concentrator c-Si solar cells," *Solar Energy Materials and Solar Cells*, vol.94, p.2094 2010.

[3] M. Yoshida, H. Tokuhisa, U. Itoh, T. Kamata, I. Sumita, and S. Sekine, "Novel Low-Temperature-Sintering Type Cu-Alloy Pastes for Silicon Solar Cells," *Energy Procedia*, vol.21, p.66, 2012.

[4] K. Nakamura, T. Takahashi, and Y. Ohshita, "Novel Silver and Copper Pastes for n-Type Bi-Facial PERT Cell," in *31th EU PVSEC*, 2015.

[5] H. T. Hai, D. Ando, Y. Sutou, and J. Koike, "Replacement of Silver by Copper for Electrodes in c-Si Solar Cells," in *27th EU PVSEC*,2012

The silver contact and formation mechanism of the boron emitter and the current flow mechanism of the solar cell electrode

Seunghyun Shin[1], Soohyun Bae[1], Sungeun Park[1], Yoonmook Kang[2], Hae-seok Lee[2], and Donghwan Kim[1]

[1]Department of Materials Science and Engineering, Korea University, Anam-dong, Seoul, 02841, Republic of Korea

[2]KU KIST Green School, Graduate School of Energy and Environment, Korea University, Anam-dong, Seoul, 02841, Republic of Korea

Abstract — **As the efficiency of solar cells has increased, solar cells using n-type substrate have been used. There arises a problem that Silver(Ag) electrode formation is not performed well while boron is used for forming the emitter. This is connected to the series resistance(Rs) of the solar cell and directly causes reduction of Fill-factor(FF) and efficiency decrease. So, after making intentionally spikes by mixing Aluminum(Al), make an Ag/Al contact. However, the spike of Al is a factor that can cause the shunt problem. Therefore, in this study, we tried to make Ag contact with boron emitter, and studied about Ag crystallite formed in boron emitter through temperature, peak time, and current-injection. As the peak time increase, the phenomenon of receiving more heat and the increase of the amount of current increase the contact resistance and Ag crystallite to observe the contact resistance and current-path**

I. INTRODUCTION

N-type silicon wafers are advantageous for the design of high-efficiency solar cells because metal impurities are less detrimental to n-type silicon, which leads to a higher minority carrier lifetime. Additionally, n-type silicon solar cells do not undergo light-induced degradation, which reduces cell performance. However, contact problems arise for boron emitters that are applied to n-type silicon solar cells using the screen-printing technique. After firing with Ag paste, the contact resistance is higher than a phosphorus emitter. this problem can be resolved by adding aluminum to the Ag paste. However, this produces other problems, such as shunting behavior and a high line resistance. The Ag becomes less dense as the aluminum content increases. thus, pastes with little or no aluminum are beneficial. In this study, the reason why the Ag paste and metal paste forms a high contact resistance in the boron emitter is that the contact resistance is changed through the crystallite formation process, and the crystallite produced by the current-injection is observed. And actually where the current flows.

II. EXPERIMENTAL PROCEDURE

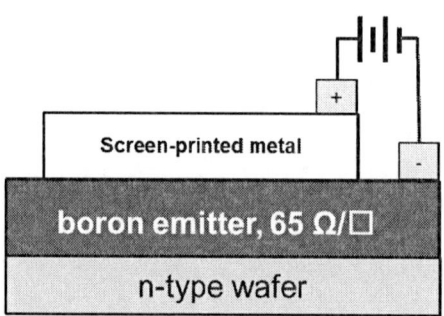

Fig. 1. Schematic of the sample structure

Pseudo-square, 4x4cm^2, n-type Czochralski-grown silicon wafers were used in this study. Their resistivities and thicknesses were 1 – 3 Ω•cm and 180um, respectively. The surfaces of the wafers were textured using a potassium hydroxide (KOH) solution. The textured wafers were cleaned with a hydrochloric acid/hydrogen peroxide mixture. Then, a boron emitter was formed from BBr3 and O2 in a tube furnace. The diffused samples were dipped in diluted hydrofluoric acid (HF, 5%) for 5 min to remove the borosilicate glass. The boron-rich layer was then etched for 30 s in an HF/nitric/acetic acid solution. The mixture was prepared from an aqueous solution. thus, the exact ratio of the mixture was 1:100:25:40 (HF/nitric/acetic acid/water). The edge of the boron emitter was etched using the acid mixture for 5 min. A conventional Ag paste and metal paste was used with the screen-printing method for front metallization, and an aluminum paste was used for the back metal, as shown in Fig. 1. All of the samples were dried for 1 min at 473 K with N2 gas. After drying, the Ag wires connected the screen-printed metal to the boron emitter and exposed the n-type wafer. First experiment is that the process was followed in a typical solar cell firing process. Then we observed the crystallite appearance of the surface with SEM. Second

978-1-5090-5606-4/17 $31.00 © 2017 IEEE

experimental is that current-injection to the fabricated sample to confirm that the contact resistance was lowered and shape of the metal crystallite confirmed. We varied the amount of injected current density from 0 to 5 A/cm^2 the peak wafer temperature was just below the Ag-Si eutectic temperature. We fabricated the samples three times and measured three times to determine the reproducibility and uncertainty.

III. RESULT & DISCUSSION

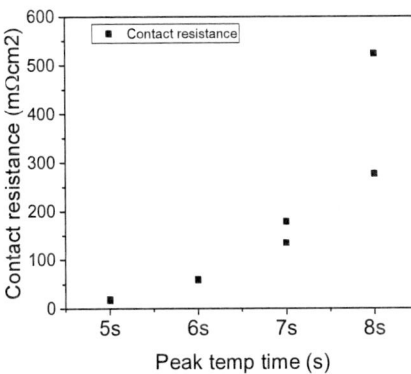

Fig. 2. Schematic of the sample structure

The process of forming the electrode through the general firing process starts when the peak temperature is reached. First, the peak temperature time increase to contact the boron emitter as shown Fig. 2. In case of phosphorous emitter, if peak temperature is high or time is long, over-firing occurs and cause the shunt but contact resistance is lowered. However, in case of boron emitter, contact resistance was observed to increase. The sample was observed with SEM

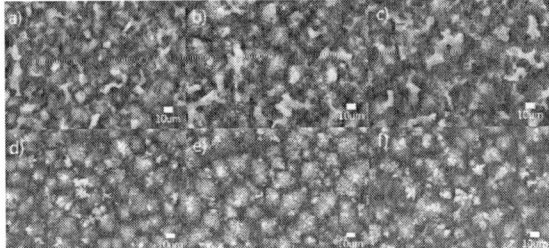

Fig. 3. SEM image to adjust peak temperature time. a) and d) is 3s, b) and e) is 5s, c) and f) is 6s

The SEM image of the sample with peak temperature time varying from 3 to 6 seconds. Peak temperature time 3s is a sample made following the most common firing process. In the SEM image, the glass that is completely transparent is covered with silver particles. 3s sample shows that the

amount of glass is small. The reason why the contact resistance increases is shown in b) and c). As time increases, more glass is generated and it seems to be due to increase of contact resistance.

In the experiment to decrease the contact resistance by increasing the firing process time, the amount of glass was increased and the contact resistance was increased. Thus, a current was applied to fix the process time and lower the contact resistance. The more current is applied when no current is applied, the more crystallite is produced and the lower the contact resistance. Therefore, the boron emitter contact shows better results when Ag is used, the shorter the process time and the higher the current.

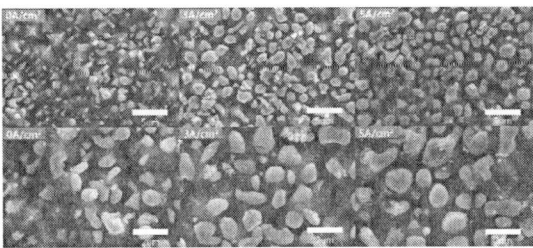

Fig. 4. SEM image to adjust current-flow using Cu paste.

The following experiment was carried out using the same drying and firing conditions as in the case of Cu paste and Ag paste. We confirmed by SEM whether current-injection actually has an effect on crystallite formation. As the current density increased, larger amounts of crystallite were formed and the size of crystallite was also increased.

IV. SUMMERY

In the case of silver paste without aluminum in boron emitter, it has a high contact resistance. As a method of increasing the firing process time, only the amount of glass is increased but the contact resistance tends to increase. However, it was confirmed that when the firing process time is shortened and the current is increased, the contact resistance is low. To reduce the contact resistance, a short process time is required to reduce the amount of glass and the current must be increased.

V. REFERENCES

[1] Lago, R. *et al.* Screen printing metallization of boron emitters. *Prog. Photovolt: Res. Appl.* 18 (2010).
[2] Korner, S. *et al.* Basic study on the influence of glass composition and aluminum content on the Ag/Al paste contact formation to boron emitter. *Energy Procedia* **67**, 20 ‐ 30 (2015).
[3] Kim , C. et al. Effects of Current-injection Firing with Ag Paste in a Boron Emitter, scientific reports, 6, 21553 (2016)

Laser Annealing to Enhance Performance of All-Laser-Based Silicon Back Contact Solar Cells

Zeming Sun and Mool C. Gupta

Charles L. Brown Department of Electrical and Computer Engineering, University of Virginia,
Charlottesville, VA, 22903, USA

Abstract — This study investigated the use of lasers for the fabrication of interdigitated back contact (IBC) solar cells where low-power laser annealing was used to improve the device performance. We have successfully demonstrated a novel concept of low-power laser annealing after the high-power laser doping which can effectively activate the dopants and increase the diffusion length as noticed by the decrease of sheet resistance. The low-power laser annealing decreased the series resistance and increased the fill factor to 77.2%, compared to 68.1% in non-annealed devices. The photovoltaic efficiency of 18.9% was achieved in laser annealed cells, compared to 16.5% in non-annealed cells.

Index — Laser annealing, laser processing, IBC silicon solar cells, dopant activation, dopant diffusion, defect removal.

I. INTRODUCTION

To achieve high-efficiency low-cost solar cell devices, a low-temperature, non-vacuum, photolithography-free fabrication process is desired. In doing this, high-power laser processing provides the noncontact, localized, and surface processing that can greatly simplify the fabrication steps and potentially replace high temperature processes. The laser processed silicon solar cells have been fabricated using laser doping [1-3], laser direct writing [4-5], laser microtexturing [6], and laser ablation [3].

Currently, the interdigitated back contact (IBC) solar cells have shown the efficiency of 24.4% [7]. However, the fabrication requires extremely complicated photolithography and etching procedures (16 steps in Ref. [7]) for patterning doping and contact regions. Also, multiple steps (7 steps in Ref. [7]) rely on high temperature furnaces and vacuum evaporators for making SiO$_2$ masks, diffusing dopants, and depositing metal contacts. Hence, laser application becomes attractive in fabricating IBC cells in order to achieve efficient, cost-effective, and high-throughput cells. The emerging laser doped IBC solar cells have demonstrated an efficiency of 23.2% [3].

However, a main challenge of laser processed solar cells lies in the ability to eliminate generation of induced crystal defects and formation of amorphous phases which will increase the carrier recombination and deteriorate the photovoltaic efficiency [8-10]. Furthermore, the dopant diffusion and activation still depend on high temperature furnace annealing conditions as described in literature [3,9-10].

To overcome these limitations, laser surface annealing becomes a very promising alternative technique and it can be integrated with the all-laser-based solar cell fabrication process.

In this work, we investigated lasers for solar cell fabrication processes such as laser selective doping, laser selective metal-transfer, laser diffusion of dopants, and laser annealing. We have avoided all the photolithography steps and nearly all high temperature processes. Furthermore, we investigated the laser annealing process parameters for their contribution to high efficiency photovoltaic devices.

II. ALL-LASER-BASED IBC SILICON SOLAR CELLS

Figure 1 presents the design of the all-laser-based IBC solar cell with the SiO$_2$ passivation (~15 nm) and SiN$_x$ ARC layer (~70 nm) on the microtextured front surface and the interdigitated aluminum emitter and phosphorus BSF doping on the rear side. The aluminum (Al) and titanium (Ti) metal contacts cover the p+ emitter and n+ BSF regions,

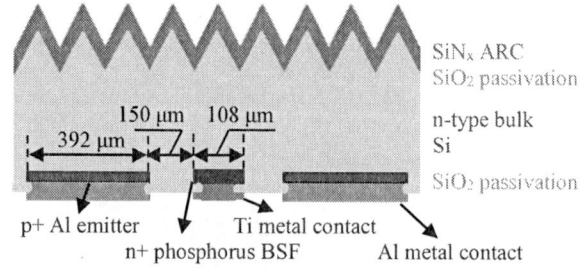

Fig. 1.Design of all-laser-based IBC solar cell.

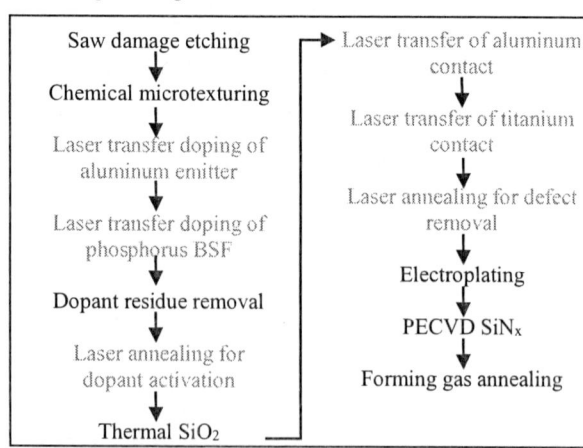

Fig. 2. Fabrication sequence of all-laser-based solar cells.

respectively. The cell has an active area of 1×1 cm^2.

Figure 2 shows the fabrication sequence. The starting wafers were n-type monocrystalline silicon (FZ) with orientation of <100>, thickness of 200 μm, and resistivity of ~1 Ω-cm. The native oxide was removed by diluted HF. The saw damage was etched using 25% KOH solution at 60 °C for 30 min. Afterwards, the wafers went through a standard cleaning procedure to remove any organic and ion contaminations [8,11]. The wafer's front side was then microtextured using 2% TMAH solution at 80 °C for 30 min.

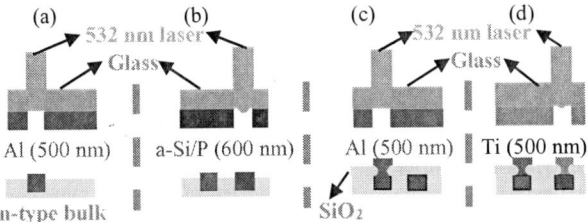

Fig. 3. Schematic of laser selective doping and metal transfer: (a) Aluminum emitter; (b) Phosphorus BSF; (c) Aluminum contact; (d) Titanium contact. (Dimensions not to scale)

The laser selective doping and metal-transfer processes are shown in Figure 3. The dopant films were laser transferred to the surface of bulk silicon after absorbing the laser energy. In order to drive the dopant atoms into the silicon, three laser shots were chosen with the scanning speed of 100 mm/s and the repetition rate of 20 kHz. To precisely pattern the doping regions, the samples were fixed in a holder and the movement was limited within ~30 μm. The transferred region was inspected by a high-resolution optical microscope. Following laser doping, the dopant residues were removed by diluted HF.

Afterwards, the laser annealing for dopant diffusion and activation was carried out. The activation effect of laser annealing was investigated through comparing the sheet resistance of annealed and non-annealed samples by four-point probe method. Next, the surface was passivated by the

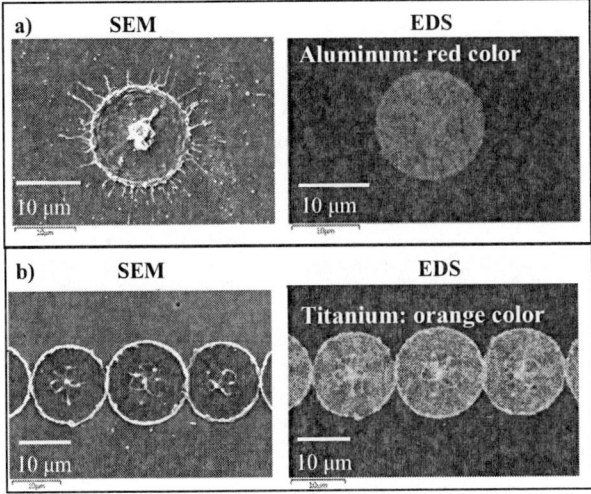

Fig. 4. SEM and EDS images for laser metal-transfer processes: (a) Aluminum contact; (b) Titanium contact.

thermally grown 15 nm SiO$_2$ layer.

The laser metal-transfer processes are shown in Figure 3. A high scanning speed of 300 mm/s together with a laser repetition rate of 20 kHz was chosen to generate less number of shots for avoiding a significant damage to the doped region. The transferred metal contacts are shown in Figure 4. Another laser annealing step was applied aiming to remove the defects.

To further improve light absorption, an anti-reflection coating (ARC) of 70 nm SiN$_x$ was deposited through PECVD.

To improve the metal contact quality and decrease the contact resistance, the electroplating and forming gas annealing were carried out.

III. RESULTS AND DISCUSSION

A. Sheet resistance after low-power laser annealing

Table I summarizes the sheet resistance of laser doped regions before and after laser annealing. The sheet resistance decreases significantly after laser annealing for both phosphorus and aluminum regions. For phosphorus, the sheet resistance decreases dramatically by around 4 times and the lowest sheet resistance is found to be 35.02 Ω/sq. Similarly, for aluminum, the lowest sheet resistance after laser annealing is found to be 141.48 Ω/sq.

The decrease of sheet resistance indicates the change of doping profile which will influence the device performance. It is highly likely that laser annealing leads to the dopant diffusion and activation. In order to validate this, we attempt to calculate the diffusion length of dopants during laser annealing.

According to literature [12-13], the diffusion coefficients D (at 1100 °C) for phosphorus and aluminum are 1.2×10^{-12} cm^2/s and 2.69×10^{-12} cm^2/s, respectively. There are 67 shots in one laser spot and thus the total annealing time τ in one laser spot will be 20 ms. According to the equation for diffusion length $L = \sqrt{D \times \tau}$, the estimated diffusion length is 1.5 nm and 2.3 nm for phosphorus and aluminum, respectively. These diffusion lengths are sufficient for dopant activation in the lattice, which can explain the observation of sheet resistance decrease.

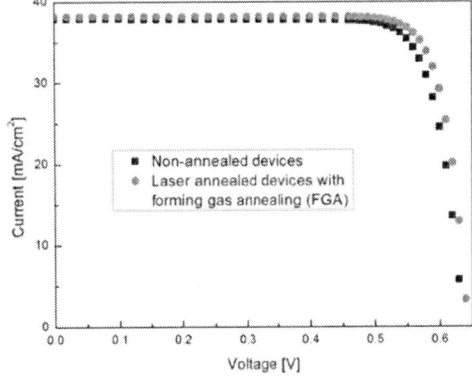

Fig. 5. I-V characteristics of laser annealed and non-annealed cells.

TABLE I
SUMMARY OF SHEET RESISTANCE AFTER LASER ANNEALING

Laser annealing for doped phosphorus			Laser annealing for doped aluminum		
Laser fluence for doping	Sheet resistance before laser annealing	Sheet resistance after laser annealing	Laser fluence for doping	Sheet resistance before laser annealing	Sheet resistance after laser annealing
1.50 J/cm^2	231.52 Ω/sq	45.76 Ω/sq	1.43 J/cm^2	317.03 Ω/sq	-
1.56 J/cm^2	200.53 Ω/sq	42.58 Ω/sq	1.50 J/cm^2	240.48 Ω/sq	156.17 Ω/sq
1.63 J/cm^2	140.01 Ω/sq	35.02 Ω/sq	1.56 J/cm^2	188.84 Ω/sq	141.48 Ω/sq
1.70 J/cm^2	123.06 Ω/sq	39.91 Ω/sq	1.63 J/cm^2	153.52 Ω/sq	-
1.76 J/cm^2	110.92 Ω/sq	86.95 Ω/sq	1.70 J/cm^2	210.92 Ω/sq	157.16 Ω/sq

B. I-V analysis of laser annealed all-laser-based solar cells

1-sun (100 mW/cm^2) light illumination I-V curves for laser annealed and non-annealed cells are shown in Figure 5. Table II summarizes the device performance under different annealing conditions. The fill factor (FF) increases from 68% to 73.4% and the series resistance drops from 2.47 Ω to 1.31 Ω after laser annealing.

The fill factor is further increased to 77.2% after forming gas annealing (FGA) under 400 °C for 4 min. This leads to the photovoltaic efficiency of 18.9% for all-laser-based back contact solar cells.

TABLE II
SUMMARY OF DEVICE PERFORMANCE UNDER DIFFERENT ANNEALING CONDITIONS

	Non-annealed devices	Laser annealed devices	Laser annealed devices with FGA
I_{sc} [mA/cm^2]	37.9	38.2	38.2
V_{oc} [V]	0.64	0.64	0.64
FF	68.0%	73.4%	77.2%
Efficiency	16.5%	17.9%	18.9%
R_s [Ω]	2.47	1.31	1.03
R_{sh} [Ω]	5k	8k	8k

IV. CONCLUSION

The low-power laser annealing concept is demonstrated to be effective in dopant activation and enhancing diffusion length for all-laser-based back-contact solar cells. After laser annealing, the sheet resistance decreases by 4 times, the fill factor increases significantly, and the series resistance drops by half. The photovoltaic efficiency of 18.9% is achieved in laser annealed cells, demonstrating an improvement of 2.4% in efficiency as compared to non-annealed cells. High-power laser processing with low-power laser annealing is viable to achieve high-efficiency low-cost solar cells.

V. ACKNOWLEDGEMENT

We acknowledge the support of NSF under the grant number CMMI-1436775, NASA Langley Professor Program, and NSF I/UCRC award.

REFERENCES

[1] S. J. Eisele, T. C. Röder, J. R. Köhler, and J. H. Werner, "18.9% efficient full area laser doped silicon solar cell," *Applied Physics Letters*, vol. 95, 133501, 2009.

[2] M. Dahlinger, B. Bazer-Bachi, T. C. Roder, J. R. Kohler, R. Zapf-Gottwick, and J. H. Werner, "Laser-doped back-contact solar cells," *IEEE Journal of Photovoltaics*, vol. 5, pp. 812-818, 2015.

[3] M. Dahlinger, K. Carstens, E. Hoffmann, J. R. Kohler, R. Zapf-Gottwick, and J. H. Werner, "23.2% laser processed back contact solar cell: fabrication, characterization and modeling," *Progress in Photovoltaics: Research and Applications*, vol. 25, pp. 192-200, 2017.

[4] E. Schneiderlochner, R. Preu, R. Ludemann, and S. W. Glunz, "Laser-fired rear contacts for crystalline silicon solar cells," *Progress in Photovoltaics: Research and Applications*, vol. 10, pp. 29-34, 2002.

[5] L. Wang, D. E. Charlson, and M. C. Gupta, "Silicon solar cells based on all-laser-transferred contacts," *Progress in Photovoltaics: Research and Applications*, vol. 23, pp. 61-68, 2015.

[6] V. V. Iyengar, B. K. Nayak, and M. C. Gupta, "Optical properties of silicon light trapping structures for photovoltaics," *Solar Energy Materials and Solar Cells*, vol. 94, pp. 2251-2257, 2010.

[7] E. Franklin, K. Fong, K. McIntosh, A. Fell, A. Blakers, T. Kho, D. Walter, D. Wang, N. Zin, M. Stocks, E. C. Wang, N. Grant, Y. Wan, Y. Yang, X. Zhang, Z. Feng, and P. J. Verlinden, "Design, fabrication and characterization of a 24.4% efficient interdigitated back contact solar cell," *Progress in Photovoltaics: Research and Applications*, vol. 24, pp. 411-427, 2016.

[8] Z. Sun and M. C. Gupta, "Laser annealing of silicon surface defects for photovoltaic applications," *Surface Science*, vol. 652, pp. 344-349, 2016.

[9] K. Ohmer, Y. Weng, J. R. Kohler, H. P. Strunk, and J. H. Werner, "Defect formation in silicon during laser doping," *IEEE Journal of Photovoltaics*, vol. 1, pp. 183-186, 2011.

[10] Z. Hameiri, T. Puzzer, L. Mai, A. B. Sproul, and S. R. Wenham, "Laser induced defects in laser doped solar cells," *Progress in Photovoltaics: Research and Applications*, vol. 19, pp. 391-405, 2010.

[11] Z. Sun and M. C. Gupta, "Laser induced defects in silicon solar cells and laser annealing," in *43th IEEE Photovoltaic Specialist Conference*, 2016.

[12] R. C. Miller and A. Savage, "Diffusion of aluminum in single crystal silicon," *Journal of Applied Physics*, vol. 27, pp. 1430-1432, 1956.

[13] S. Maekawa, "Diffusion of phosphorus into silicon," *Journal of the Physical Society of Japan*, vol. 17, pp. 1592-1597, 1962.

978-1-5090-5606-4/17 $31.00 © 2017 IEEE

Large Area N-type Selective Emitter cells using Laser Doping through Boron Doped Screen Printed Paste

Ajay D Upadhyaya [1], Vijaykumar D Upadhyaya[1], Brian Rounsaville[1], Keeya Madani[1], Toru Hanada[2], Ajeet Rohatgi [1]

[1]Georgia Institute of Technology, 777 Atlantic Drive, Atlanta GA 30332-0250, USA,
[2]Nanogram Corporation, Milpitas, CA 95035

Abstract — **In this paper, we report on laser doping though a screen printed boron paste to form selective emitter. We show that the selective emitter cell efficiency on n-PERT structure is higher by ~0.6% compared to the counterpart homogeneous emitter cell. In this study, we show that we can achieve great contact to a selective emitter with 180 ohm/sq lightly doped p^+ field and 50 ohm/sq p^{++} region formed by laser doping under the metal contact. This resulted in a total series resistance of ~0.6 ohm-cm^2 and a FF of >80%. We found that heavy doping under the contacts reduced the metal recombination induced degradation in Voc but the laser damage induced degradation prevented the expected gain in Voc. Work is in progress to anneal or eliminate the laser damage. Large area 239cm^2 n-PERT solar cells using this technology gave ~ 20.2% cell efficiency using conventional ion-implanted field region and screen-printed metallization.**

Index Terms- Selective Emitter, Laser Doping, Screen Printing, n-type cell.

I. INTRODUCTION

Most industrial solar cells in current production are limited to efficiencies of less than 21%, mainly because of high Auger recombination in the heavily doped boron (p^+) and phosphorus (n^+) regions, high metal-induced recombination at metal/Si contact interface, and high shading from screen-printed metal contact on the front.

A selective emitter/back surface field (BSF) is a very effective and manufacturable way to reduce J0 of the metallized and diffused regions simultaneously [1-5]. This is because heavy diffusion underneath the grid reduces the metal J0 and the light diffusion in between the grid lines reduces J0 of the field region. Our 2D device modeling showed that front and back selective doping can produce >40 mV higher Voc for n-PERT cells [6].

II. EXPERIMENTAL

To study the doping effect, N-type Cz (6") wafers were textured and cleaned using standard cleaning procedures followed by ion-implantation of boron on the front side. The wafers were then annealed at 1050C° using a standard tube furnace to yield a homogenous sheet resistance of ~180 ohm/sq. Following a short HF dip, DI rinse and dry, boron doped paste was screen printed over the entire front side of the wafer that had the homogeneous 180 ohm/sq. implanted emitter and dried in a belt dryer at ~200C° for ~10min. Using a 532nm wavelength laser, the fluence and speed conditions were varied

to form square patterns (Fig. 1) to obtain and measure different sheet resistance. To study the effect of heavily diffused regions and the side effects of laser damage on the Joe, first symmetric test samples were fabricated on textured high resistivity N-type Cz wafers with lightly boron diffused (ion-implanted 180 ohm/sq) emitters on both sides. Boron paste was then screen printed over the entire front and back of these wafers and dried. A laser pattern similar to the metal grid that would be used in a cell was created to form the laser-doped p^{++} regions. Laser fluence was varied to form different selective emitters on each quarter of the wafer as shown in Fig. 2. As shown in the Fig. 2, one quarter of the wafers was not laser doped to serve as reference and to measure the Joe of the 180-ohm/sq homogeneous emitter. After the laser doping, the excess boron

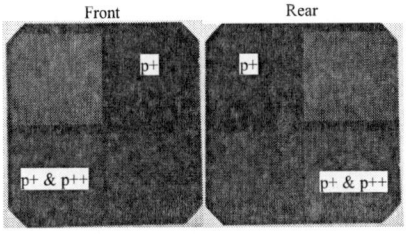

Fig. 2.   Laser doped patterns on symmetric structure to measure Joe

paste was removed by wet chemistry followed by an HF dip, DI rinse and dry. ALD Al$_2$O$_3$ was then deposited on both sides of the symmetric structure followed by a SiN cap. The implied Voc (iVoc), Joe, and lifetime were measured before and after the simulation firing for these test structures without screen-printed metal contacts.

Solar cells were fabricated using the n-PERT structure shown in Fig. 3 using the process flow described in Fig. 4. N-type Cz wafers with resistivity of 1-10 ohm-cm were textured on both

978-1-5090-5606-4/17 $31.00 © 2017 IEEE

sides followed by ion-implantation of boron on the front side. After a clean, the wafers were annealed at 1050C° to form the 180 ohm/sq homogeneous emitter. The rear side of the wafers was then planarized in KOH. After cleaning the wafers, the boron doping paste was screen printed on the entire front. Different laser conditions were used to drive in the dopant from the paste using the grid pattern. Reference cells were also

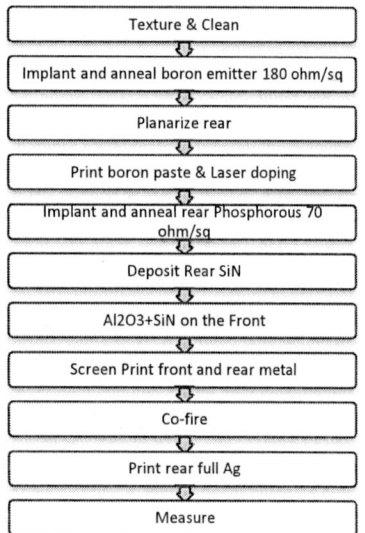

Fig. 3.   N-PERT structure with selective boron emitter.

fabricated on homogeneous 180ohm/sq emitter with no laser doping. After a short wet chemistry to remove the excess boron paste, the wafers received heavy phosphorous implantation on the rear. Wafers were then cleaned, annealed and oxidized in a tube furnace to form a homogeneous BSF with a sheet resistance of ~70 ohm/sq. The rear surface was then capped with SiN followed by a quick HF dip to remove the oxide on the front side.

Texture & Clean

Implant and anneal boron emitter 180 ohm/sq

Planarize rear

Print boron paste & Laser doping

Implant and anneal rear Phosphorous 70 ohm/sq

Deposit Rear SiN

Al2O3+SiN on the Front

Screen Print front and rear metal

Co-fire

Print rear full Ag

Measure

Fig. 4.   N-PERT Selective emitter process sequence.

The front side then received ~120Å of ALD Al_2O_3 and ~ 680 Å of SiN cap for passivation of boron emitter and AR coating. Metal grid was then carefully screen printed by aligning to the heavily boron doped p^{++} regions on the front side using a commercial metal paste. Point contacts were screen-printed on the rear of the cells followed by a co-firing of the front and rear contacts in a belt furnace. A low temperature Ag paste was

screen printed over the entire rear to connect the point contacts and dried at low temperature.

III. RESULTS AND ANALYSIS
(SHEET RESISTANCE AND JOE)

Figure 5 shows that as the laser fluence increases, the sheet resistance decreases and reaches ~5 ohm/sq for laser fluence of 5 J/cm^2.

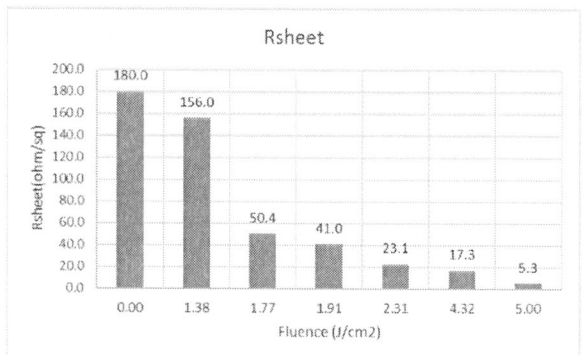

Fig. 5.   Sheet resistance vs laser fluence

The same laser doping conditions were applied to form selective emitter structures on symmetric lightly doped (180ohm/sq) boron emitter on high resistivity N-type Cz. These structures were passivated with an Al_2O_3/SiN stack and then fired in a belt furnace to simulate contact firing without metal.

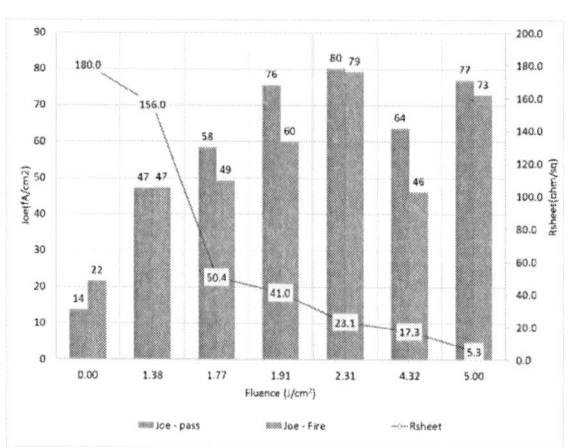

Fig. 6.   Joe & Sheet resistance vs laser fluence

Figure 6 shows that the Joe of the 180 ohm/sq field p^+ emitter, with no laser doping, was only ~14fA/cm^2. When p^{++} regions were formed by laser doping, Joe of the p^+-p^{++} selective emitters increased. Even with modest laser fluence of 1.38 J/cm^2, the Joe of the p^+-p^{++} selective emitter increased from 14 to ~47fA/cm^2 and remained high for all the other laser doping conditions.

The intended width of p^{++} regions was 150μm, however, as shown in Fig. 7, it increased up to 200μm which amounts to ~13% coverage. This could be due to the laser damage or heating effect beyond the intended 150µm width. Using the following simple equations and 13% area coverage, we can calculate the Joe of the heavily diffused p^{++} region to be ~ 215fA/cm^2 for the 1.38 J/cm^2 case with measured Joe of 47fA/cm^2 for the p^+-p^{++} region.

$$(_{p+}+_{p++}) = Joe_{p+} (1-f) + Joe_{p++} (f) = 47fA/cm^2$$
$$Joe_{p++} = [47\text{-}Joe_{p+} (1-f)] / f$$
$$Joe_{p++} = [47\text{-}(22x(1\text{-}0.13))]/0.13 = 215fA/cm^2$$

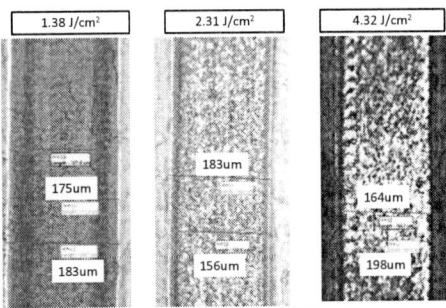

Fig. 7.   Microscope images of actual width of heavily doped area designed for 150µm design

Sugianto and Wenham et al. [7] showed that the laser damage can be recovered by annealing at ~350C° for about 5 min; however, in our case this treatment showed no improvement in Joe.

IV. N-PERT SELECTIVE EMITTER CELL RESULTS AND ANALYSIS

Table 1 shows the detailed LIV data of the n-PERT cells with and without selective emitter.

As shown in the Table I, the 180 ohm/sq homogeneous emitter cell suffers from high series resistance (1.2 ohm-cm^2) resulting in a low FF of 76.9%. However, both the selective emitter cells in Table 1 resulted in excellent FF of >80% due to much lower series resistance of ~0.6 ohm-cm^2. In addition, Table II shows that the Voc dropped 20mV due to high metal induced recombination in the case of the lightly doped homogeneous emitter cell compared to a drop of only about 8mV for the selective emitter cells.

TABLE I
DETAILED LIV DATA FOR N-PERT CELLS

Selective Emitter (rsheet) Field/Contact	Pre-metal iVoc (mV)	Cell Voc (mV)	Jsc (mA/cm²)	FF (%)	Eff (%)	RSeries (ohm-cm²)
Homogeneous Emitter						
No selective (180 ohm/sq)	0.674	0.654	38.6	76.9	19.4	1.20
Selective Emitter-1						
180/50	0.660	0.652	38.6	80.1	20.1	0.67
180/50	0.661	0.653	38.6	80.2	20.2	0.63
Selective Emitter-2						
180/150	0.660	0.652	37.8	80.5	19.9	0.61
180/150	0.664	0.652	38.2	80.4	20.0	0.65

TABLE II
VOC & LIFETIME ANALYSIS

Selective Emitter (rsheet) Field/Contact	Lifetime Fit (us) @5E15	Pre-metal iVoc (mV)	Cell Voc (mV)	Voc Change
No selective (180 ohm/sq)	~1242	0.674	0.654	-20mV
180/50	~597	0.660	0.652	-8mV
180/150		0.660	0.652	-8mV

Total Jo is the sum of the contributions from emitter (Joe), bulk (Job) and BSF (Job'). Job' of ~80fA/cm^2 for the phosphorous BSF was measured in a separate experiment. Job of 50fA/cm^2 and 100fA/cm^2 were calculated based on bulk lifetime of ~1200µs and ~600µs for homogeneous and selective emitter cells respectively as shown in Table 2 for a 5ohm-cm wafer. Next, we calculated the total Jo for the homogeneous and selective emitter cell structure before metallization as shown below.

a) Homogeneous emitter:
Joe+Job+Job' = 22+50+80 =152fA/cm^2 → 675mV
b) Selective emitter (without lifetime degradation Tau = ~1242µs):
Joe+Job+Job' = 47+50+80 =177fA/cm^2→ 671mV
c) Selective emitter (with lifetime degradation Tau = ~597µs):
Joe+Job+Job' = 47+100+80 =227fA/cm^2→ 664mV

Using the Voc equation below and assuming a Jsc of

$$V_{oc} = \frac{kT}{q} \ln\left(\frac{J_L}{J_{0e} + J_{0b}} + 1 \right).$$

39mA/cm^2, we should get an implied Voc of ~675mV for the homogeneous emitter cells and an implied Voc of 671mV for the selective emitter cells, if there was no laser induced lifetime degradation. However, in this experiment an implied Voc of 660-664mV was obtained for the selective emitter. Using a bulk lifetime of ~600µs and Job of 100fA/cm^2, we obtain a calculated total Jo of 227fA/cm^2 and an iVoc of ~664mV, which is in good aggrement with the measured values. Table 1 shows that even with the laser damage, the cell with selective emitter was ~0.6% higher than the homogeneous emitter cell, however we need to eliminate the laser damage and lifetime degradation to realize the full potential of this technology.

V. CONCLUSIONS

In this paper, we achieved excellent FF of 80.5% in selective emitter cells obtained by laser doping through a screen-printed boron paste. In comparison to the reference cell with 180 ohm/sq homogeneous emitter, the selective emitters gave ~0.6% higher efficiencies and ~12mV lower drop in Voc due to metallization. Analysis also showed that laser-induced damage limited the bulk lifetime and the cell Voc and efficiency. Efforts are underway to anneal this damage to reach the full potential of this technology.

ACKNOWLEDGEMENT

The authors would like to thank the DOE for funding this work in part under the PVRD project.

REFERENCES

[1] Hallam B, Urueñ A, Russell R, Aleman M, Abbott M, Dang C, Wenham S, Tous L, Poortmans J. Efficiency enhancement of i-PERC solar cell by implementation of a laser doped selective emitter. Sol Energ Mat Sol Cells 2015;134:89-98.

[2] G. Hahn, "Status of selective emitter technology," 2010.

[3] Tomizawa Y, Imamura T, Soeda M, Ikeda Y, Shiro T. Laser doping of boron-doped Si paste for high efficiency silicon solar cells. Jpn J Appl Phys 2015;54:08KD06.

[4] U. Jaeger, M. Okanovic, M. Hörteis, A. Grohe, and R. Preu, "Selective emitter by laser doping from phosphosilicate glass," in *Proceedings of the 24th European Photovoltaic Solar Energy Conference*, 2009, pp. 1740-1743.

[5] R. Low, A. Gupta, N. Bateman, D. Ramappa, P. Sullivan, W. Skinner, *et al.*, "High efficiency selective emitter enabled through patterned ion implantation," in *Photovoltaic Specialists Conference (PVSC), 2010 35th IEEE*, 2010, pp. 001440-001445.

[6] A. Tam, Y.-Y. Huang, Y.-W. Ok, A. Rohatgi, " Design of Selective Emitter Profiles for High Efficiency Solar Cells under Manufacturing Technology Costraints" in *Photovoltaic Specialists Conference (PVSC), 2016 43th IEEE*, 2016.

[7] A. Sugianto, Stuart Wenham et al., " 18.5% Laser-Doped solar cells on CZ P-type Silicon", in 35th IEEE PVSC , 06-2010.

Metallized Boron-Doped Black Silicon Emitters For Front Contact Solar Cells

Guillaume von Gastrow[1], Eric Calle[2], Pablo Ortega[2], Ramón Alcubilla[2], Andreana Daniil[3], Elias Z. Stutz[3], Anna Fontcuberta i Morral[3], Sebastian Husein[4], Tara Nietzold[4], Mariana Bertoni[4], and Hele Savin[1]

[1]Aalto University, Espoo, 02150, Finland
[2]Universitat Politècnica de Catalunya, Barcelona, 08034, Spain
[3]École Polytechnique Fédérale de Lausanne (EPFL), Lausanne, 1015, Switzerland
[4]Arizona State University, Tempe, Arizona, 85287, USA

Abstract — We study doping and metallization of black silicon (bSi) boron emitters formed by ion implantation or diffusion. We demonstrate that conformal metal layers can be deposited on bSi by electron beam evaporation. Raman spectroscopy shows that high boron concentrations ($4 \cdot 10^{19}$ cm^{-3}) are obtained in bSi by ion implantation, while maintaining emitter saturation current (J_{0e}) below 20 fA/cm^2 with Al$_2$O$_3$ passivation. In diffused bSi emitters, doping increases to twice the values of planar substrates, reaching values up to $7 \cdot 10^{20}$ cm^{-3}. Those doping values allow specific contact resistivities down to (0.3 ± 0.2) mΩ·cm^2 on boron-implanted bSi surfaces with nickel or aluminum contacts.

Index Terms — black silicon, contact resistance, emitter, doping

I. INTRODUCTION

Silicon solar cell surface nanostructures – also called black silicon, or bSi – provide a definite advantage in terms of reflectance reduction [1]. The efficiency of bSi front-side emitter solar cells remains however poor due to three main issues: 1) the high recombination due to large surface area [2], 2) the high Auger recombination caused by heavy diffusion of dopants [3], and 3) the lack of conformality of metal contacts due to the high aspect ratio of the structures [4]. While the surface recombination issue can be solved by ALD Al$_2$O$_3$ passivation [5, 6], the two others still remain recurrent issues with the standard emitter formation techniques. However, alternative methods exist and could allow the fabrication of performant bSi emitters without modification of the nanostructures. For instance, ion implantation is a promising doping technique for bSi, as it uses a fixed dopant dose and thus allows better doping control than diffusion in bSi structures, causing less recombination [7]. Electron beam evaporation could constitute a solution for contact formation, as it allows low deposition rates to produce conformal layers [8] and ensures limited surface damage [9]. Here we study whether ion implantation and electron beam evaporation can be combined to fabricate metallized bSi emitters with limited recombination activity and low contact resistance. In addition, doping is an important parameter to consider when optimizing emitter passivation and metallization, but it is typically difficult to measure in bSi structures. Successful silicon nanowire doping

measurements have been performed by CV measurements [10], although they remain complex due to the need for maintaining a contact with the nanowires. In this work, we utilize Raman spectroscopy, which has the advantage of being a contactless method. The first part of this paper focuses on doping mechanisms in bSi structures and the second one on contact performance.

II. EXPERIMENTAL DETAILS

The substrates were (100)-oriented magnetic Czochralski phosphorus-doped silicon wafers with a resistivity of 3.4 Ohm·cm and a thickness of 445 μm. Black silicon was produced by inductively-coupled reactive ion etching in a SF$_6$/O$_2$ plasma at a temperature of -120 °C. Boron implantation was performed at an energy of 10 keV and with a dose of $3 \cdot 10^{15}$ cm^{-2}. Then, a 20 min anneal was performed in N$_2$ ambient at temperatures of 950 °C and 1050 °C, followed by dry oxidation for 20 min at the same temperatures. Boron diffusions were performed at temperatures of 825 °C and 975 °C for 45 min with a solid source. The samples were then dipped in a 5 % HF solution for two minutes. The boron-rich layer was etched with low thermal oxidation (LTO) in pure O$_2$ atmosphere at 650 °C for 30 min, followed by another 5 % HF dip for 2 minutes. In addition, planar reference wafers were implanted and diffused using the same parameters.

Nickel contacts with thicknesses ranging from 100 nm to 400 nm were formed by e-beam evaporation at a rate of 0.1 nm/s and at an initial pressure of $6 \cdot 10^{-7}$ mbar. The rate was increased to 0.5 nm/s after a thickness of 200 nm was reached. Aluminum contacts of 1 μm were also deposited with e-beam evaporation at an initial rate of 0.1 nm/s and annealed in forming gas at 400 °C for 10 minutes. The nickel-coated samples were measured before and after post-deposition anneal. Annealing was performed at temperatures from 350 °C to 450 °C during 30 seconds to three minutes in N$_2$ ambient or in forming gas. Specific contact resistivity was measured with the transfer length method.

Raman spectroscopy measurements were performed at wavelengths of 405 nm and 532 nm, corresponding

978-1-5090-5606-4/17 $31.00 © 2017 IEEE

approximately to an absorption depth of 100 nm and 1 μm in planar silicon, respectively.

III. EMITTER DOPING

Excellent passivation results have been obtained in our previous study with atomic layer deposited Al_2O_3 on implanted bSi emitters annealed at 1000 °C and 1050 °C. An emitter saturation current (J_{0e}) of 20 fA/cm^2 was reported at both temperatures. Reasonably low J_{0e} of 30 fA/cm^2 and 110 fA/cm^2 were measured in emitters diffused at 825 °C and 975 °C, respectively [7]. Although this shows potential for integration in front-side-emitter solar cells, it is essential to determine whether such emitters also allow low-resistance contacts.

The corresponding sheet resistance values are summarized in Table I. In diffused emitters, doping increases with temperature, and consequently sheet resistance decreases. In implanted emitters however, the dopant dose is fixed, and sheet resistance remains stable once all dopants have been activated.

TABLE I
SHEET RESISTANCE OF PLANAR AND BLACK SILICON SAMPLES

Emitter formation	Sheet resistance (Ohm/sq)	
	planar	bSi
Diffusion, 825 °C	384 ± 1	260 ± 20
Diffusion, 975 °C	38 ± 2	21 ± 1
Implantation 950 °C	60 ± 1	200 ± 20
Implantation,1050 °C	56 ± 1	170 ± 5

Sheet resistance provides information on doping, which eventually indicates the potential for ohmic contact formation, but it remains unclear whether measurements on planar Si and on bSi are comparable. Additionally, doping can be measured by electrochemical capacitance voltage on planar samples (Fig. 1), but measurements on bSi become complex and unreliable.

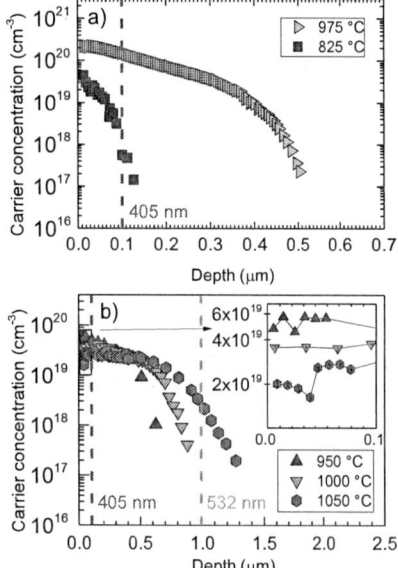

Fig. 1. Doping profiles measured by electrochemical capacitance voltage of a) boron-diffused emitters on planar samples at temperatures of 825 °C and 975 °C b) boron-implanted emitters on planar samples after annealing at 950 °C and 1050 °C. The dash lines indicate the absorption depth in silicon at wavelengths of 405 nm and 532 nm.

On the other hand, Raman spectroscopy is a contactless method that may help assessing the doping of bSi more accurately and allow comparison with planar samples. Examples of Raman spectroscopy measurements are shown in Fig. 2. The broadening and asymmetry of the Si TO-LO Raman peak can be related to the effective hole concentration via the Fano effect [11].

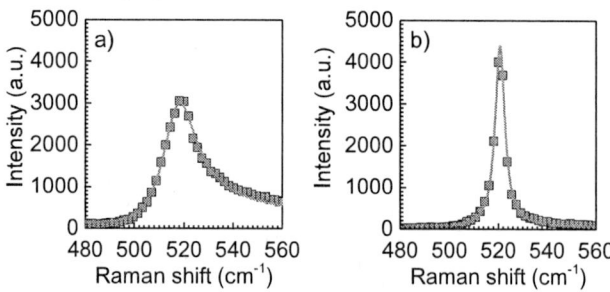

Fig. 2. Example of Raman spectroscopy measurements on a) a bSi sample diffused at 975 °C and b) a planar sample diffused at 975 °C. The Fano fits are indicated by red lines.

As an optical method, Raman spectroscopy can only assess the carrier concentration within the absorption length of the excitation wavelength λ_{exc} used. For λ_{exc} = 405 nm, this corresponds approximately to 100 nm. The results reported are thus effective carrier concentrations, which are convoluted along the whole corresponding absorption depth. In the case of bSi, one expects light to be confined at the surface. Raman spectroscopy provides information on the carrier concentration much closer to the surface than in the case of planar substrates and well below the absorption length of silicon. The effective hole concentrations obtained by Raman spectroscopy at different wavelengths are summarized in Fig. 3.

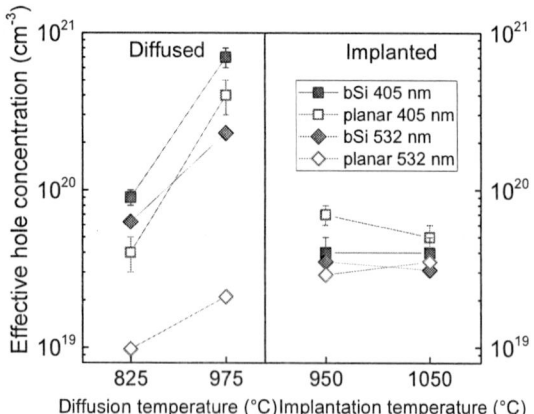

Fig. 3. Effective hole concentrations measured by Raman spectroscopy in diffused and in implanted emitters. Measurements were performed at wavelengths of 405 nm and 532 nm. The concentrations are effective values over the whole absorption depth, which corresponds to approximately 100 nm at a wavelength of 405 nm and 1 μm at a wavelength of 532 nm for planar silicon.

Relatively high hole concentrations – that reflect doping concentrations – are measured in all bSi emitters, which should thus allow performant electrical contacts [12]. The doping concentration measured in diffused bSi emitters is up to one order of magnitude higher than in the implanted emitters.

In diffused emitters, the comparison between the two excitation wavelengths indicates that doping at the surface (λ_{exc} = 405 nm) is higher than doping deeper in the bulk (λ_{exc} = 532 nm) for both bSi and planar substrates. At a wavelength of 532 nm, regions of low doping are also probed during the measurement, which is consistent with the low effective hole concentration obtained compared to the measurement performed at 405 nm. In addition, the diffused bSi samples seem to exhibit higher carrier concentrations than their planar reference at any given diffusion temperature.

Similarly in bSi implanted emitters, high hole concentrations over $3 \cdot 10^{19}$ cm^{-3} are detected. No clear variation in the hole density is observed between bSi samples and their planar reference or depending on implantation anneal temperature. This is consistent with the fixed nature of the implantation dose.

IV. METALLIZATION AND CONTACT RESISTANCE

The doping and hole densities reported in section III suggest that performant contacts could be obtained on bSi diffused and implanted emitters. The following section studies the conformality and specific contact resistivity of nickel (Ni) and aluminum (Al) contacts and bSi. Nickel has a potential for excellent adhesion and for low contact resistance on silicon, acts as a copper barrier layer, and serves as a seed layer for copper electroplating [13]. In addition, aluminum contacts were also fabricated.

Fig. 4 shows that electron-beam evaporation allows conformal coating of bSi; 250 nm of Ni deposited by electron beam evaporation are sufficient to fully cover the structures.

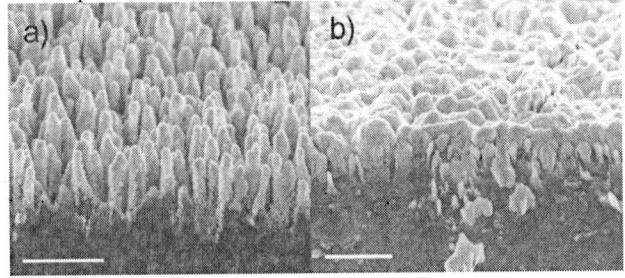

Fig. 4. SEM images of bSi structures metallized by electron gun evaporation with a) 100 nm of nickel and b) 250 nm of nickel. The scale bars represent 1 μm.

Nickel can provide a low contact resistance on silicon after formation of a silicide layer whose characteristics depend on the temperature [14]. The NiSi phase is thought to provide the lowest resistivity [15] and is usually obtained at very shorts anneals (from 30 seconds to 3 minutes) in a temperature range of 300 °C – 450 °C [16]. In this work however, post-deposition annealing did not seem to improve the contact resistivity of nickel. Fig. 5 displays the specific contact resistivity ρ_c in the different bSi emitters depending on the diffusion or implantation anneal temperature. The Ni samples were measured without post-deposition anneal, except for one set, which underwent annealing for 3 min at 400°C, as reported on the graph. It appears that excellent values down to 0.3 mΩ·cm^2 are obtained with Ni contacts on bSi without post-deposition annealing. It is possible that Ni annealing occured already during the evaporation process that caused severe heating of the samples for a long duration.

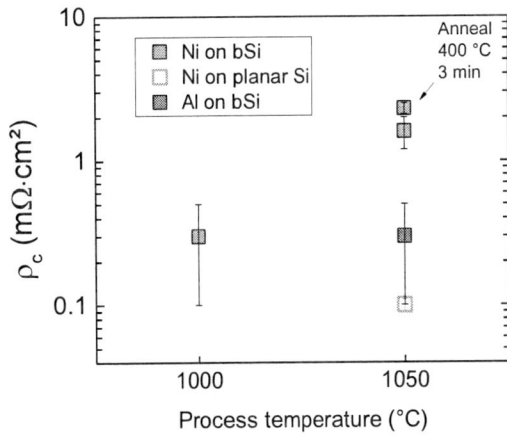

Fig. 5. Specific contact resistivity ρ_c measured in implanted bSi emitters at different implantation anneal temperatures. The results on Ni samples are reported without post-deposition annealing unless otherwise indicated. The Al sample was annealed at 400 °C for 10 min.

The implantation anneal temperature seems to affect the specific contact resistivity, possibly through surface doping. Although Raman measurements in Fig. 3 indicate a similar effective doping in the first 100 nm at the surface after implantation anneal at 1000 °C and 1050 °C, it is likely that the higher annealing temperature of 1050 °C reduces surface doping, thus causing high contact resistance. The value of 1 to 3 mΩ·cm^2 obtained with nickel after implantation anneal at 1050 °C can however be reduced to 0.3 mΩ·cm^2 with Al contacts. Measurements on planar reference samples implanted and annealed at 1050 °C indicate a very low specific contact resistivity compared to bSi samples. This may be explained by the high aspect ratio of bSi structures, causing shadowing and thus lower doping or no doping at all in some of the grooves between the spikes.

In diffused emitters, measurements were performed with a different set of structures that caued higher uncertainty. Thus, the specific contact resistivities measured ranged between 0,1 $m\Omega \cdot cm^2$ and 10 $m\Omega \cdot cm^2$.

V. Conclusions

We studied doping and contact formation bSi emitters. In bSi diffused emitters, the hole concentration near the surface was twice as high as in the planar references due to enhanced diffusion. In implanted emitters, an effective near-surface hole concentration of $(4 \pm 1) \cdot 10^{19}$ cm^{-3} was found both in planar and bSi samples. Consistent with these values, we reported excellent specific contact resistivities between 0.1 $m\Omega \cdot cm^2$ and 10 $m\Omega \cdot cm^2$ on black silicon emitters doped by boron implantation and by boron diffusion. Those emitters can also be efficiently passivated by ALD Al_2O_3 and display low emitter saturation current, as shown in our previous study. Consequently, both implanted and diffused black silicon emitters offer promising perspectives for solar cell applications.

Acknowledgements

This research was undertaken at the Micronova Nanofabrication Centre of Aalto University and in the cleanroom laboratory of the Micro and Nanotechnologies group at the Universitat Politècnica de Catalunya (UPC), Barcelona. This work was supported by the Spanish Ministry of Economy and Competitiveness MINECO (PCIN-2014-055) and Finnish TEKES (40329/14) agencies under Solar-Era.Net FP7 European Network. The work was also funded by the doctoral school of Aalto School of Electrical Engineering and the Project No. ENE2013-49984-EXP of the Spanish Ministry of Economy and Competitiveness (MINECO). AD, EZS and AFiM thank funding through the SNF Consolidator Grant 'Easeh'.

References

[1] P. B. Clapham and M. C. Hutley, "Reduction of Lens Reflexion by the "Moth Eye" Principle," Nature, vol. 244, pp. 281-282, 08/03, 1973.

[2] J. Oh, H. Yuan and H. M. Branz, "An 18.2%-efficient black-silicon solar cell achieved through control of carrier recombination in nanostructures," Nat Nano, vol. 7, pp. 743-748, print, 2012.

[3] S. Jeong, M. D. McGehee and Y. Cui, "All-back-contact ultra-thin silicon nanocone solar cells with 13.7% power conversion efficiency," Nat Commun, vol. 4, 12/16, 2013.

[4] S. Wang, Q. Li, K. Tao, R. Jia, S. Jiang, D. Wang and H. Dong, "Effective way to realize optimized carrier recombination and electrode contact for excellent electrical performance silicon nanostructure based solar cells," J. Mater. Sci. : Mater. Electron., vol. 27, pp. 4378-4384, 2016.

[5] M. Otto, M. Kroll, T. Käsebier, R. Salzer, A. Tünnermann and R. B. Wehrspohn. Extremely low surface recombination velocities in black silicon passivated by atomic layer deposition. Appl. Phys. Lett. 100(19), 2012.

[6] P. Repo, A. Haarahiltunen, L. Sainiemi, M. Yli-Koski, H. Talvitie, M. C. Schubert and H. Savin, "Effective Passivation of Black Silicon Surfaces by Atomic Layer Deposition," IEEE Journal of Photovoltaics, vol. 3, pp. 90-94, 2013.

[7] G. von Gastrow, P. Ortega, R. Alcubilla, S. Husein, T. Nietzold, M. Bertoni and H. Savin, "Recombination processes in passivated boron-implanted black silicon emitters," J. Appl. Phys., vol. 121, pp. 185706, 2017.

[8] J. Singh and D. Wolfe, "Review Nano and macro-structured component fabrication by electron beam-physical vapor deposition (EB-PVD)," J. Mater. Sci., vol. 40, pp. 1-26, 2005.

[9] A. Kuroyanagi, "Properties of aluminum-doped ZnO thin films grown by electron beam evaporation," Japanese Journal of Applied Physics, vol. 28, pp. 219, 1989.

[10] E. C. Garnett, Y. Tseng, D. R. Khanal, J. Wu, J. Bokor and P. Yang, "Dopant profiling and surface analysis of silicon nanowires using capacitance–voltage measurements," Nature Nanotechnology, vol. 4, pp. 311-314, 2009.

[11] B. G. Burke, J. Chan, K. A. Williams, Z. Wu, A. A. Puretzky and D. B. Geohegan, "Raman study of Fano interference in p‐type doped silicon," J. Raman Spectrosc., vol. 41, pp. 1759-1764, 2010.

[12] R. Müller, J. Benick, N. Bateman, J. Schön, C. Reichel, A. Richter, M. Hermle and S. W. Glunz, "Evaluation of implantation annealing for highly-doped selective boron emitters suitable for screen-printed contacts," Solar Energy Mater. Solar Cells, vol. 120, Part A, pp. 431-435, 1, 2014.

[13] A. Mondon, M. Jawaid, J. Bartsch, M. Glatthaar and S. Glunz, "Microstructure analysis of the interface situation and adhesion of thermally formed nickel silicide for plated nickel–copper contacts on silicon solar cells," Solar Energy Mater. Solar Cells, vol. 117, pp. 209-213, 2013.

[14] J. D. Lee, H. Y. Kwon and S. H. Lee, "Analysis of front metal contact for plated Ni/Cu silicon solar cell," Electronic Materials Letters, vol. 7, pp. 349-352, 2011.

[15] V. Teodorescu, L. Nistor, H. Bender, A. Steegen, A. Lauwers, K. Maex and J. Van Landuyt, "In situ transmission electron microscopy study of Ni silicide phases formed on (001) Si active lines," J. Appl. Phys., vol. 90, pp. 167-174, 2001.

[16] T. Ohguro, S. Nakamura, M. Koike, T. Morimoto, A. Nishiyama, Y. Ushiku, T. Yoshitomi, M. Ono, M. Saito and H. Iwai, "Analysis of resistance behavior in Ti-and Ni-salicided polysilicon films," IEEE Trans. Electron Devices, vol. 41, pp. 2305-2317, 1994.

978-1-5090-5606-4/17 $31.00 © 2017 IEEE

Contact Resistance Measurement for Thermally Diffused Point Contact by Localized Dielectric Breakdown Solar Cells

Qilin Ye, Ned J. Western, Anqi Liao and Stephen P. Bremner

School of Photovoltaic and Renewable Energy, Engineering, University of New South Wales, Sydney, NSW, 2052, Australia

Abstract — The Point contacting by Localized Dielectric Breakdown method utilizes dielectric breakdown above a laser doped region to form ohmic contact for a solar cell rear surface. A complete cell with the laser doping process replaced with a thermal diffusion has been fabricated giving 17.5% efficiency and 0.71 fill factor. Characterization indicates that the majority of the fill factor loss is due to unexpected high contact resistance at the rear surface. In this paper, we calculate the specific contact resistivity using two different methods giving comparable results. The potential cause for these high values is discussed along with possible solutions for further improving the cell fill factor. By modulating the doping profile, the impact of surface doping concentration on contact resistivity is demonstrated. Values for contact resistivity down to 0.3 mΩcm^2 have been achieved on control device with surface doping concentration of 6×10^{19} cm^{-3}. Low temperature annealing at 300 C is also able to improve contact resistivity to some extent.

Index Terms — Contact resistivity, dielectric breakdown, locally diffused, point contact, silicon solar cells.

I. INTRODUCTION

Point-contacting by Localized Dielectric Breakdown (PLDB) is a method to contact silicon by breaking down the dielectric above a laser doped region, being best suited to the rear surface of a PERL style cell structure [1], [2]. After applying a voltage to the gate metal, hard breakdown of the dielectric creates a conduction path between the gate metal and the laser doped silicon region. A large current density passing through the conduction path leads to local joule heating causing diffusion of the gate metal to the silicon resulting in ohmic contact [3].

In this work, the laser doped point contacts are replaced with thermal boron diffusions, avoiding the generation of laser induced defects and allowing improved control of the local doping profile. Measurements of completed cells with a standard screen printed front surface design demonstrate the successful application of the PLDB contacting method to thermally diffused point contacts. However, unusually high contact resistance at the rear surface limits the final fill factor. This is in contrast to earlier results which show values below 1 mΩ.cm^2 on specialized contact structures such as transmission line method (TLM) [1], [4].

In theory, a p-type surface doping concentration over 1×10^{18} cm^{-3} should be sufficient to achieve a contact resistivity of below 1 mΩ.cm^2 [5], [6]. Several authors have reported values between 1 and 100 $\mu\Omega$.cm^2 for various materials onto p-type silicon when the surface concentration is 10^{19} cm^{-3} [7]. Laser doping that utilizes the Al atoms from the AlO$_x$ passivation layer as a dopant source is one method to create contacts to p-type silicon. Yang et al. have reported values for contact resistivity as low as 10 $\mu\Omega$.cm^2 with a surface doping concentration of 1.9×10^{19} cm^{-3} on laser doped p$^+$ region [8]. Ernst et al. have also demonstrated laser doped contacts with slightly higher contact resistivity values below 100 $\mu\Omega$.cm^2 when the surface doping concentration is between 10 and 4×10^{19} cm^{-3} [9].

In this paper, we determine the value of the specific contact resistivity generated by PLDB using two different methods. The first method is to measure the lumped series resistance for the whole cell and subtract the known contributions to the series resistance from the front surface and wafer bulk. The second method is to measure the IV curve between a single point contact and the front metallization, where the resistance of this single contact is expected to dominate the total resistance. In this way the measurement is more sensitive to the contact resistance and allows independent verification of the calculated values. Control devices with higher surface boron surface doping concentration are fabricated to extract and compare their contact resistivity using the second method mentioned above. Annealing at several temperatures is performed to investigate the influence to contact resistivity.

II. METHOD

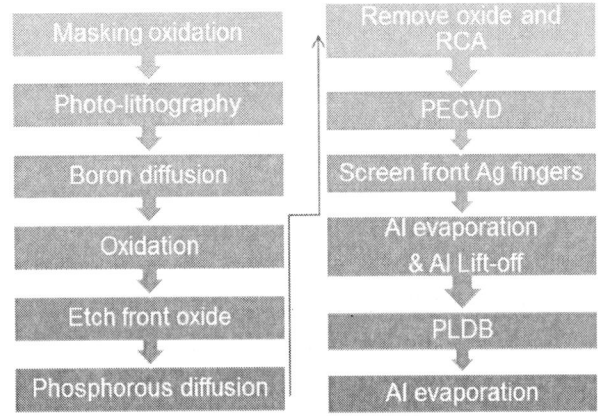

Fig. 1. Cell fabrication order with PLDB contacts at the rear surface

978-1-5090-5606-4/17 $31.00 © 2017 IEEE

Solar cells were fabricated from 2 Ω.cm p-type Czochralski silicon wafers, the complete process flow is given in Fig. 1. Photolithography was used to pattern an oxide layer as a mask for a thermal boron diffusion on the rear surface. The front phosphorus diffusion and screen printed Ag fingers are typical of a standard industrial design. A stack of 10 nm AlOx and 30 nm SiNx, both deposited by plasma enhanced chemical vapor deposition (PECVD), forms the rear passivation layer. The rear surface was metallized by thermally evaporated aluminium and photolithography with lift-off to create individual metal pads over each locally diffused region.

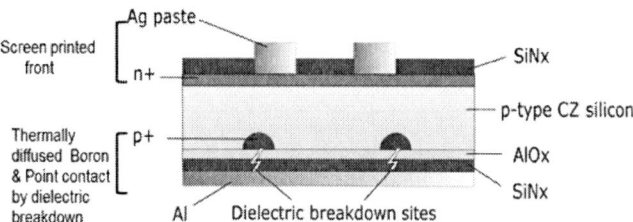

Fig. 2.   Schematic diagram of the final cell.

The PLDB method was applied to each of the metal pads to form ohmic contact between the rear metal and the local diffused region. At this stage, IV measurements were performed between the front metallization and a single point contact at the rear surface. A second aluminium layer was then evaporated on the entire rear surface to connect the individual metal contacts. Several measurements were then performed on the completed cells including light IV, dark IV and Suns V_{oc}. The finished cell structure is shown in Fig. 2 with the breakdown sites at the rear point contacts highlighted.

III. CELL RESULTS

The final cell results are given in Table I. High values for V_{oc} and J_{sc} demonstrate low recombination and low parasitic absorption at the rear surface respectively. A high total series resistance appears to be responsible for the low fill factor of 71% which reduces the final cell efficiency to 17.5%. The lumped series resistance for the cell was calculated by comparing the light IV and SunsV$_{oc}$ curves [10].

TABLE I
CELL RESULTS WITH PLDB REAR CONTACTS

Voc	Jsc	pFF	FF	η	R$_s$
634 mV	38.8 mA/cm^2	83%	71%	17.5%	1.86 Ωcm^2

The theoretical upper limit for the fill factor and the loss due to the measured series resistance can be estimated by the method outlined in [11]. This gives a maximum value of 72.5% demonstrating that the majority of the loss in fill factor is due to the series resistance. To investigate the cause for this series resistance in more detail we have calculate the contact resistivity at the rear surface using two different methods

IV. CONTACT RESISTIVITY ANALYSIS AND DISCUSSIONS

The current flow in the silicon bulk for both methods are shown in Fig. 3. Although the direction of current flow under cell illumination differs from that shown in the figure, the current path will be similar. The current flow in the dark for both devices is shown here for clarity.

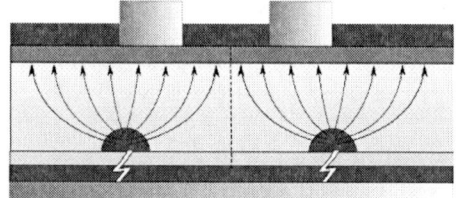

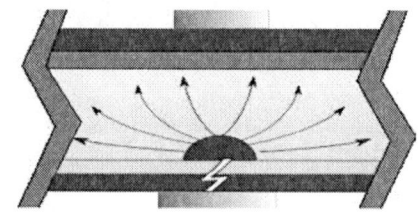

Fig. 3.   Current flow in the wafer bulk for Method 1 (top) and Method 2 (bottom)

A. Method 1: Series resistance of a complete cell

The contribution to the lumped cell series resistance from the front metallization and the phosphorus diffusion have been calculated in the standard way. The spreading resistance in the bulk of the silicon wafer around a single point contact can be approximated by the summation of two components in (1) as outlined in [12]

$$R_b = \rho_b W_b + S_c^2 \frac{\rho_b}{\pi d_c} \qquad (1)$$

Where R_b is the spreading resistance in the wafer bulk, ρ_b is the silicon resistivity, W_b is substrate thickness, d_c is the contact diameter, and S_c is the rear contact spacing. This method has been demonstrated to have adequate accuracy when the ratio of point contact pitch to wafer thickness is less than four. Accounting for the resistance at the front surface and silicon bulk allows estimation of the contact resistivity at the rear surface of 13.4 mΩ.cm^2. This is over an order of magnitude higher than previously reported values for series resistance.

B. Method 2: Single point contact

978-1-5090-5606-4/17 $31.00 © 2017 IEEE   949

The second method to measure the contact resistivity is to measure the IV curve of a single point contact before the second Al evaporation as outlined in the Method section. By limiting the current flow to a single rear point contact, the resistance of this contact dominates the total measured resistance as seen in Fig. 4, giving greater sensitivity to the measurement.

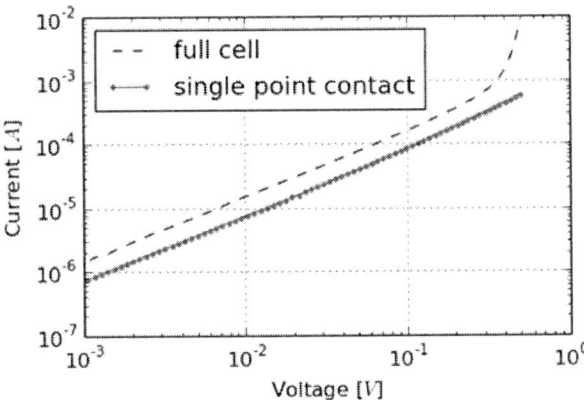

Fig. 4.   Current and voltage curves in the dark for a full cell and a single rear point contact. A lower current for the single point contact is observed over the full measurement range.

As the current flow in the bulk of the wafer differs compared to a complete cell, a new method for calculating the spreading resistance was required. An equation for spreading resistance was given by [13] and describes the spreading resistance around a single point contact.

$$R_b = \frac{\rho_b \tan^{-1}\left(\frac{2W_b}{d_c}\right)}{2\pi d_c} \tag{2}$$

Where ρ_b is the bulk resistivity, W_b is substrate thickness and d_c is the contact radius. The equation assumes a wafer of infinite size, since the point contact size is small compared to our wafer, this is a reasonable approximation in our case. The contact resistivity calculated using this method is 7.2 mΩ.cm^2. Comparison of several measurements on a single wafer show consistent values with a standard deviation of 6%. Although this is around half the magnitude of the value calculated in Method 1, it provides confirmation that the contact resistivity is indeed higher than expected. Table II shows a summary of both methods.

TABLE II
CONTACT RESISTIVITY VALUES FOR BOTH METHODS.

Method 1: complete cell	13.4 mΩcm^2
Method 2: single point contact	7.2 mΩcm^2

There are several potential causes for this larger than expected contact resistivity. The actual size of the dielectric breakdown site is not known so the full area of the diffused region is used for all calculations. The size of individual breakdown sites may be modulated by processing conditions such as dielectric type and composition and firing temperature for the front metallization. Method 2 outlined above will facilitate easier measurement of for the specific contact resistivity as entire cell devices do not need to be fabricated.

Previously reported measurements of PLDB devices were given for laser doped point contacts. The dopant profiles of these devices are inherently difficult to measure due to the small line widths and non-uniform lateral concentrations. The boron surface doping concentration for diffusions used in this work was measured by the electrochemical capacitance-voltage method, giving a value of 2×10^{19} cm^3.

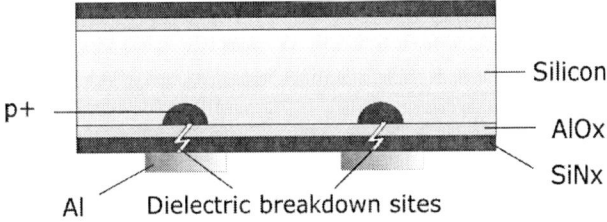

Fig. 5.   Schematic diagram of the control device used for contact resistivity measurements.

Additional contact resistivity measurements were undertaken on control samples without a phosphorous diffusion on the front surface. This simplified device shown in Fig. 5 allows a boron diffusion profile with a shallower diffusion depth and a higher surface doping concentration. In Fig. 6, the new boron doping profile has a surface doping concentration of 6×10^{19} cm^{-3} and a sheet resistivity of 50 $\Omega/\square$ as compared to the previous profile with a surface doping concentration of (2×10^{19} cm^{-3}) and 30 $\Omega/\square$.

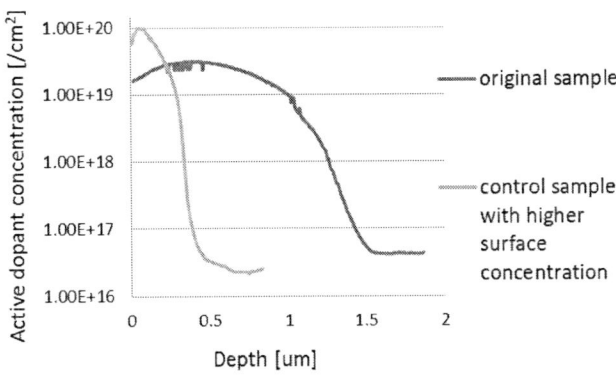

Fig. 6.   Electrochemical capacitance-voltage (ECV) profile of the boron diffused surface. The plotted data is recalculated using the sheet resistivity measured by four point probe.

After fabricating the rear surface with PLBD contacts, the entire front surface is then laser doped and evaporated with aluminum to create a large ohmic contact, replacing the emitter and front metallization in the measurements given above. IV measurements in the dark between the front metal and the rear single point contact have a resistance which is limited by the contact resistance of a single point contact and the current crowding in the silicon bulk to the point contact. This total resistance is then subtracted by the expected bulk resistance to determine the contact resistivity.

Table III summarized the contact resistivity calculated for two surface doping concentrations. The contact resistivity is 0.3 mΩ.cm² for the sample with surface doping concentration of 6×10^{19} cm⁻³, which is much lower compared to the value measure previously 7 mΩ.cm² on the sample with lower surface doping concentration of 2×10^{19} cm⁻³. This result shows that low contact resistivity (<1 mΩ.cm²) can be achieved by the doping profile with surface boron doping concentration of 6×10^{19} cm⁻³, which suggests that the cell series resistance contributed by the rear contact resistance is reduced if the doping profile is increased slightly.

TABLE III
CONTACT RESISTIVITY FOR TWO SURFACE DOPING
CONCENTRATIONS

Surface doping concentration(cm⁻³)	Specific contact resistivity (mΩcm²)
2×10^{19}	7.2
6×10^{19}	0.3

C. Effects of contact annealing

Annealing was performed to investigate the influence of thermal treatment on the PLDB contacts. Dark IV, light IV, SunsV$_{oc}$ and PLi measurement were taken before and after annealing at different temperature.

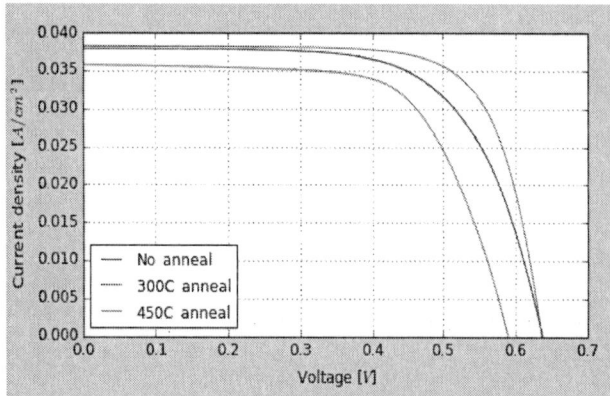

Fig. 7.   Light IV of a cell annealed for 30mins at 300C and 450C

The result of annealing on cell structures shows a significant improvement on the fill factor after 300 C anneal for 30 mins

as is seen in Fig. 7. The improvement in fill factor is due to a decrease in total series resistance. By comparing the light IV with Suns V$_{oc}$ measurement, the extracted series resistance decrease from 2.68 Ω.cm² to 1.10 Ω.cm². This shows that the contact resistivity at the rear surface decreased from 21.4 mΩ.cm² to 5.95 mΩ.cm². After a second anneal for 30 mins at 450 C, both short circuit current and open circuit voltage decrease. The open circuit voltage reduces from 640 mV to 590 mV. This decrease in V$_{oc}$ is thought to be due to a reaction between the rear dielectric and the evaporated aluminum layer, increasing the recombination at the rear surface.

A similar result is observed on other samples while the change in fill factor is not as significant. The fill factor slightly improved after 30 mins at 300 C, and slight decrease after 15mins at 400 C, suggesting that annealing should not exceed 400 C otherwise lifetime degradation occurs. 300 C anneal improves the series resistance from 1.86 Ωcm² to 1.25 Ω.cm², which reduce contact resistivity from 13.4 mΩ.cm² to 7.42 mΩ.cm² as is summarized in Table IV.

TABLE IV
SERIES RESISTANCE AND CORRESPONDENT CONTACT
RESISTIVITY BEFORE AND AFTER ANNEALING

Condition	Series resistance(Ωcm²)	Specific contact resistivity (mΩcm²)
Before anneal	1.86	13.4
After anneal	1.25	7.42

For metal-silicon contacts where the metal thermally evaporated onto bare silicon, annealing is typically performed in a reduction atmosphere to reduce interfacial oxide and improve the contact resistivity. The temperature is usually chosen to be below the Al-Si eutectic temperature of 580 C to avoid void formation and Al spiking. For laser doped contacts, it has also been reported that annealing at 250 - 300 C significantly reduces the contact resistivity and increase fill factor, but a slight V$_{oc}$ drop is observed as well [15].

V. CONCLUSIONS

A thermally diffused PLDB cell is fabricated from a p-type Cz silicon wafer. Efficiency of 17.5% and 0.71 fill factor is measured from the IV curve. The large fill factor loss is due to the series resistance result from the high rear contact resistivity. Two methods are used to calculate the contact resistivity to be 13.4 mΩ. cm² and 7.15 ± 0.45 mΩ. cm² respectively. The single point contact method is suggested to be used for future measurement since it is more convenient and accurate. The depletion of the surface boron doping concentration is believed to be the main reason for the high contact resistivity. Experiment with different surface doping concentration is conducted, the contact resistivity decrease from 7.15 to 0.3 mΩ.cm² if the surface doping concentration is 6×10^{19} cm^{-3}. Annealing at 300 C for 30 mins improves the contact resistivity from 13.4 to 7.42 mΩ.cm^2.

Experimenting on more surface doping concentration is required to determine the optimal doping profile and balance the contact resistivity with the contact recombination current for the future improvement of cell efficiency.

REFERENCES

[1] N. J. Western, A. Sung, S. R. Wenham, and S. P. Bremner, "Localized ohmic contact through a passivation dielectric for solar cell rear surface design," *Appl. Phys. Lett.*, vol. 102, no. 22, p. 222105, 2013.

[2] M. A. Green, "The Passivated Emitter and Rear Cell (PERC): From conception to mass production," *Sol. Energy Mater. Sol. Cells*, vol. 143, pp. 190–197, 2015.

[3] N. J. Western, I. Perez-Wurfl, S. R. Wenham, and S. P. Bremner, "Point-contacting by localised dielectric breakdown: Characterisation of a metallisation technique for the rear surface of a solar cell," *J. Appl. Phys.*, vol. 118, no. 4, p. 45711, 2015.

[4] N. J. Western and S. P. Bremner, "Contact Resistivity Measurements of Point-Contacting by Localised Dielectric Breakdown Structures by the Transmission Line Method," in *31st European Photovoltaic Solar Energy Conference and Exhibition*, 2015, pp. 629–632.

[5] D. K. Schroder and D. L. Meier, "Solar cell contact resistance - A review," *IEEE Trans. Electron Devices*, vol. ED-31, no. 5, pp. 637–647, 1984.

[6] C. Y. Chang, Y. K. Fang, and S. M. Sze, "Specific contact resistance of metal-semiconductor barriers," *Solid State Electron.*, vol. 14, no. 7, pp. 541–550, 1971.

[7] D. K. Schroder, *SEMICONDUCTOR MATERIAL AND DEVICE CHARACTERIZATION*, vol. 44, no. 4. Wiley-IEEE Press, 2006.

[8] X. Yang *et al.*, "Passivated contacts to laser doped p+ and n+ regions," *Sol. Energy Mater. Sol. Cells*, vol. 140, pp. 38–44, 2015.

[9] M. Ernst, D. Walter, A. Fell, B. Lim, and K. Weber, "Efficiency Potential of P-Type Al 2O 3/SiN x Passivated PERC Solar Cells With Locally Laser-Doped Rear Contacts," *IEEE J. Photovoltaics*, vol. 6, no. 3, pp. 624–631, 2016.

[10] D. Pysch, A. Mette, and S. W. Glunz, "A review and comparison of different methods to determine the series resistance of solar cells," *Sol. Energy Mater. Sol. Cells*, vol. 91, no. 18, pp. 1698–1706, 2007.

[11] M. A. Green, "Solar cell fill factors: General graph and empirical expressions," *Solid. State. Electron.*, vol. 24, no. 8, pp. 788–789, 1981.

[12] J. Zhao, A. Wang, and M. A. Green, "Series resistance caused by the localized rear contact in high efficiency silicon solar cells," *Sol. Energy Mater. Sol. Cells*, vol. 32, no. 1, pp. 89–94, 1994.

[13] R. H. Cox and H. Strack, "Ohmic contacts for GaAs devices," *Solid. State. Electron.*, vol. 10, no. 12, pp. 1213–1218, 1967.

[14] D. K. Schroder and D. L. Meier, "Solar cell contact resistance - a review," *Electron Devices, IEEE Trans.*, vol. 31, no. 5, pp. 637–647, 1984.

[15] M. Ernst, A. Fell, E. Franklin, and K. J. Weber, "Characterization of Recombination Properties and Contact Resistivity of Laser-Processed Localized Contacts From Doped Silicon Nanoparticle Ink and Spin-On Dopants," *IEEE J. Photovoltaics*, vol. 7, no. 2, pp. 1–8, 2017.

Low temperature rear surface metallization of multi-crystalline silicon solar cells for improved bulk lifetime

N. J. Western, S. P. Bremner

School of Photovoltaic and Renewable Energy Engineering,

University of New South Wales, Sydney, NSW, 2052, Australia

Abstract — **Multi-crystalline silicon PERC cells with Al-alloyed rear point contacts are limited in efficiency potential by light induced degradation, which is activated by high temperature firing of metallization pastes. Replacement of this rear surface design with a low temperature metallization technology will mitigate this degradation process and facilitate thermal processes that improve efficiency, such as extended gettering. Point-contacting by localized dielectric breakdown is a low temperature process to contact the rear surface of PERC style cell structures, here it has been applied to multicrystalline silicon for the first time. The implied open circuit voltage is preserved after contact formation successfully, demonstrating the separation of high temperature firing from metallization.**

I. INTRODUCTION

The efficiency of p-type multicrystalline silicon (mc-Si) solar cells is known to degrade after exposure to sunlight. The process is often called Light Induced Degradation (LID) or Light and elevated Temp Induced Degradation LeTID [1] and is particularly detrimental for PERC style cell architectures. Similar behavior is observed when carrier generation is created by forward bias in the dark and as such this process is more accurately referred to as Carrier Induced Degradation (CID) [2]. This decay in performance occurs on timescales longer than that observed in p-type crystalline silicon and it has been shown that this process is not due to B-O complexes or Fe contamination [1], [3]. While various causes for this degradation have been proposed, the root cause is as yet unknown and is currently still under investigation.

The susceptibility of mc-Si to this degradation process is influenced by a number of factors including the wafer supplier [1][4], the peak firing temperature [3]–[5], the cool down rate during firing [6] and the subsequent thermal history, with and without excess carrier injection [7]. Several authors have reported stabilization of this degradation by accelerating the degradation and regeneration process with high illumination intensities at elevated temperature [8], [9].

The magnitude of the effective lifetime degradation increases with increasing peak temperature and can result in a reduction of effective lifetime of over 70% [3]–[5]. The onset of this degradation occurs over 650°C [3], [5], which is well below that required to fire screen printed pastes at around 800°C [10]. Extended contact firing, or subsequent gettering after contact firing, is able to reduce the effects of lifetime

degradation. However, it has been shown that this treatment increases the contact resistivity of screen printed contacts [11].

This sensitivity of the degradation process to the thermal processing history highlights the complex nature of this defect/s. While novel thermal treatments are able to reduce the effects of lifetime degradation or accelerate regeneration, these processes may be incompatible with the temperature profile required for screen printed metallization pastes.

Another challenge with the fabrication of local Al-alloyed rear contacts in PERC style structures is the creation of voids under these contacts during firing [12]. These voids increase the contact resistance and enhance the local recombination. Potential solutions include careful control of the firing profile and aligned printing of small area Al paste over the contacts [10], [13]. While void formation can be prevented in some cases, it places restrictions on contact geometry and processing thermal budget, ultimately limiting the efficiency potential of mc-Si PERC cells [10].

While some type of high temperature firing step is usually required to improve the bulk lifetime of mc-Si [14], an ideal alternative metallization technology would be low temperature to allow independent optimization of this firing profile. This would facilitate modulation of the peak firing temperature, the cool down rate and any subsequent extended gettering or other thermal process without affecting the contact properties of the metallized regions.

An example of a potential replacement for screen printed Ag paste for front metallization is light induced Cu plating [15], [16] as it is low cost, low temperature and also benefits from using little or no precious metals. This work will utilize this front metallization technology while focusing on replacing the rear metallization of a mc-Si cell.

Point-contacting by Localized Dielectric Breakdown (PLDB) [17], [18] is a self-aligned method to form contact to a laser doped region. This technique applies a voltage between a metal contact and the wafer bulk to induce dielectric breakdown in the passivation layer leading to stable ohmic contact. It is performed at room temperature and thus has the benefits described above. In this work, results are given for mc-Si lifetime test structures fired at various peak temperatures and complete cells using low temperature contacting methods.

II. METHOD

Lifetime test structures shown in Figure 1 with no metallization were fabricated to assess changes in the effective bulk lifetime. These samples underwent the following standard treatments using commercially available tools: acidic texturing, RCA clean, POCl$_3$ diffusion (70 Ω/sq), rear etch to remove the diffusion, spin-on Filmtronics boron dopant source on the rear surface, laser doping at 15W and 0.5 m/s scanning speed, RCA cleaning, PECVD SiN$_x$ 75 nm on the front surface and PECVD AlO$_x$ / SiN$_x$ (5 nm/ 30 nm) stack on the rear surface.

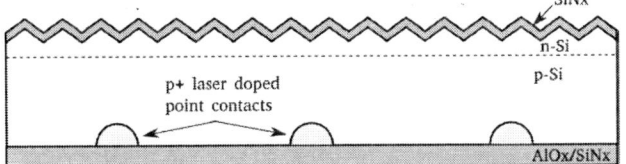

Figure 1 Lifetime test structure with no metallization. Laser doping was performed before deposition of the rear dielectric.

Three groups of sister wafers were used to compare the peak firing temperature reached in an industrial belt furnace. The wafers underwent either no high temperature process (Group A), a peak temperature of 500°C (Group B), or a peak temperature of 740°C (Group C). An accelerated degradation process was performed on a hotplate at 130°C with a 938 nm fibre-coupled laser at 25.9 kW/m^2 intensity for 10 s. This process is expected to be equivalent to several hundred hours at 75°C and 1 kW/m^2 as is commonly used to degrade mc-Si material [3]. These conditions are expected to degrade the effective lifetime to close to the minimum value expected for these wafers.

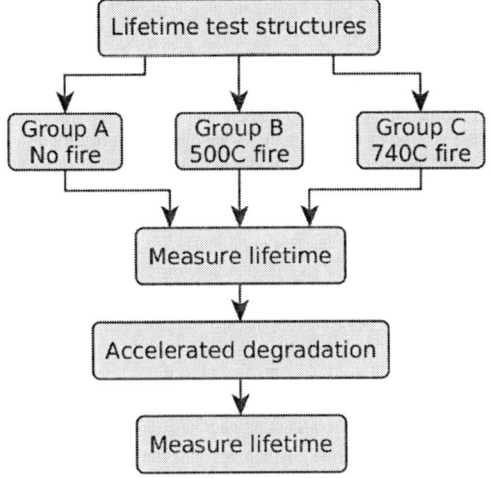

Figure 2 Process flow of lifetime test structures. Three groups of wafers received different thermal processing.

Measurements of effective lifetime were performed before and after degradation on a Sinton Labs WCT-120, quoted values were taken at an excess carrier concentration of 10^{15} cm^3.

Metallization on the rear surface was performed using the Point-contacting by Localized Dielectric Breakdown (PLDB) method by applying a voltage between the silicon bulk and an evaporated layer of aluminum on the rear surface. The dielectric breakdown in the rear passivation layer leads to ohmic contact between the aluminum and the laser doped regions while preserving the passivation quality of the surround dielectric. The full details of this technique can be viewed in refs [17]–[19]. Metallization of the front surface was created by laser ablation of the front dielectric and light induced plating of copper fingers [20].

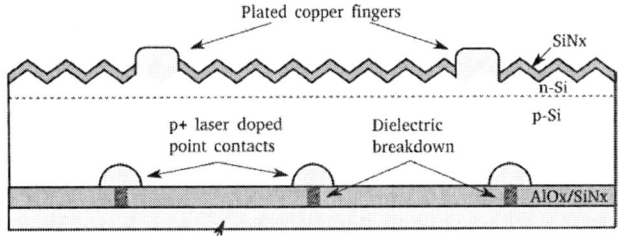

Figure 3 Completed cells with PLDB rear contacts and copper plated front fingers.

III. RESULTS

A. Lifetime degradation after firing

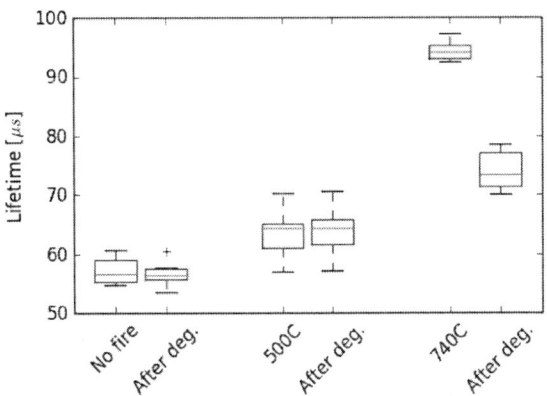

Figure 4 The effective lifetime before and after accelerated degradation. A decrease in lifetime in observed for high temperature firing. No change is seen for the other two groups of wafers.

The wafers in Group C that underwent the high temperature firing step show an expected decrease in effective lifetime of 22 % from 95 µs to 74 µs after accelerated degradation. Groups A and B show no significant change, as shown in Figure 4. The initial lifetime of Group B is 8 % higher than Group A showing benefit of the 500°C annealing step which appears to be due to increased surface passivation in this case.

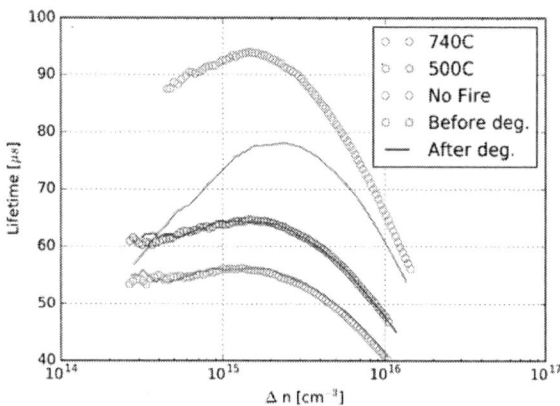

Figure 5. Injection level dependent lifetime for each wafer group, before and after degradation. A characteristic drop in low injection lifetime is observed only for wafers that received a high temperature firing step.

Figure 5 shows how the effective lifetime at low injection is reduced after high temperature firing. This is due to the activation of bulk defects in the grains of the mc-Si [5]. Lifetime curves are identical before and after degradation for the other two wafer groups. Limiting the peak firing temperature to below a threshold value, assumed to be around 650°C, prevents the onset of this degradation process, although this also limits the effective lifetime to lower initial values.

B. Front and rear metallization

To metallize the rear surface, aluminum is evaporated onto the rear surface before applying a voltage to induce dielectric breakdown. When applying the PLDB technique, this damage to the dielectric layer is limited to a small region above the laser doped contact so that the surrounding passivation layer is not affected. This is demonstrated in Figure 6 by photoluminescence imaging at 1 sun light intensity on BTi LIS-R2 imaging tool. Image registration and subtraction allows the difference in luminescence intensity to be calculated at each point on the cell before and after the PLDB contacting step.

A measurable and non-uniform decrease in photoluminescence intensity is observed after the process,

although this corresponds to average reduction in effective lifetime of only 4 %. It is not certain whether this reduction can be attributed to metal contact to the laser doped region or a slight reduction of the surrounding passivation quality, in any case it shows that voltage potential of the device is preserved during this metallization step. It is interesting to note that grain boundaries in the original image correspond to the grain boundary patterns in the subtracted image. This indicates that the difference in effective lifetime is smaller when the PLDB process is applied close to grain boundaries.

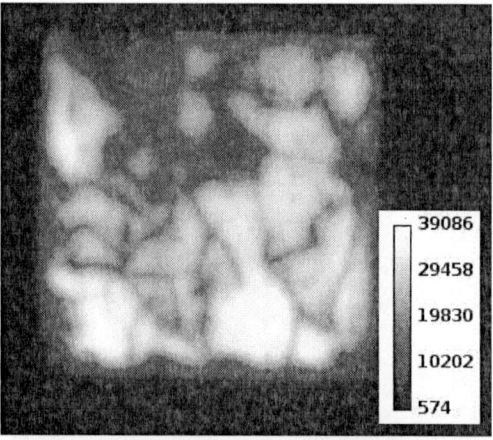

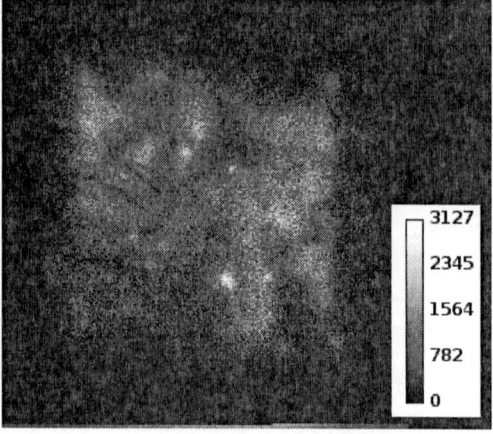

Figure 6 Photoluminescence image taken after the PLDB process (above). The difference in luminescence intensity before and after PLDB (below), note the difference in the intensity scale. Only a small decrease in effective lifetime is observed.

The sample shown in Figure 6 did not undergo a high temperature firing step. Similar results were found for wafers that had either a 500°C or 740°C firing step showing no significant decrease in luminescence intensity. This demonstrates that the rear metallization process is independent of the previous thermal history of the sample.

To metallize the front surface, a picosecond laser is used to ablate the dielectric and expose the silicon with a narrow line width of 15 um. Light induced plating is used to deposit copper onto the bare regions to create conductive metal fingers. Unfortunately, a larger than expected decrease in effective lifetime was seen after the laser ablation process reducing the implied open circuit voltage from over 640 mV to below 600 mV as evidenced in the photoluminescence image of Figure 7. This appears to be due to the composition of the SiN_x layer on the front surface which required a higher than usual laser fluence to ablate. Further difficulty in plating an adequate thickness of copper on the front fingers lead to a cell performance that was limited by series resistance. The efficiency of the final device was therefore limited in voltage and fill factor by the front metallization.

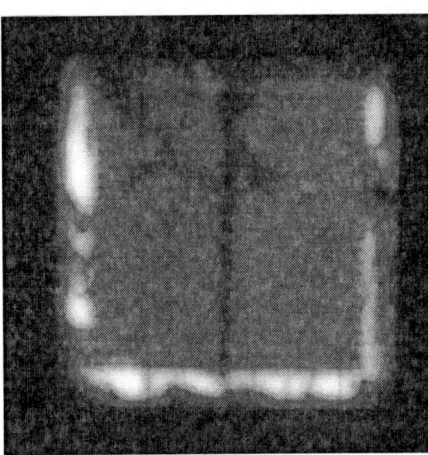

Figure 7 Photoluminescence image after laser ablation to the front surface. A large decrease in effective lifetime is due to laser damage to the emitter region.

III. CONCLUSION

Despite challenges in metallizing the front surface to make complete devices, the principle of low temperature processing for the rear surface of mc-Si solar cells has been demonstrated. Three groups of lifetime test structures were given either no firing step or a peak firing temperature of 500°C or 740°C. Only those wafers that received the high temperature process showed a decrease in effective lifetime during an accelerated degradation. Low temperature metallization of the rear surface of these wafers was performed using PLDB contacts which showed a minimal decrease in lifetime. Laser ablation and copper was used as a complementary low temperature process to metallize the front surface which unfortunately did not perform as intended. Resolution of these challenges is expected to facilitate the full potential of a metallization process that is independent of the thermal history. This will allow the inclusion of processes such as extended contact firing to reduce and perhaps eliminate the performance degradation that is observed in mc-Si wafers.

REFERENCES

[1] F. Kersten, P. Engelhart, H.-C. Ploigt, A. Stekolnikov, T. Lindner, F. Stenzel, M. Bartzsch, A. Szpeth, K. Petter, J. Heitmann, and J. W. Müller, "Degradation of multicrystalline silicon solar cells and modules after illumination at elevated temperature," *Sol. Energy Mater. Sol. Cells*, vol. 142, pp. 83–86, 2015.

[2] D. N. R. Payne, C. E. Chan, B. J. Hallam, B. Hoex, M. D. Abbott, S. R. Wenham, and D. M. Bagnall, "Rapid passivation of carrier-induced defects in p-type multi-crystalline silicon," *Sol. Energy Mater. Sol. Cells*, pp. 1–5, 2016.

[3] C. E. Chan, D. N. R. Payne, B. J. Hallam, M. D. Abbott, T. H. Fung, A. M. Wenham, B. S. Tjahjono, and S. R. Wenham, "Rapid Stabilization of High-Performance Multicrystalline P-type Silicon PERC Cells," *IEEE J. Photovoltaics*.

[4] K. Nakayashiki, J. Hofstetter, A. E. Morishige, T. T. A. Li, D. B. Needleman, M. A. Jensen, and T. Buonassisi, "Engineering Solutions and Root-Cause Analysis for Light-Induced Degradation in p-Type Multicrystalline Silicon PERC Modules," *IEEE J. Photovoltaics*, vol. 6, no. 4, pp. 860–868, 2016.

[5] D. Bredemeier, D. Walter, S. Herlufsen, and J. Schmidt, "Lifetime degradation and regeneration in multicrystalline silicon under illumination at elevated temperature," *AIP Adv.*, vol. 6, no. 3, 2016.

[6] R. Eberle, W. Kwapil, F. Schindler, M. C. Schubert, and S. W. Glunz, "Impact of the firing temperature profile on light induced degradation of multicrystalline silicon," *Phys. Status Solidi - Rapid Res. Lett.*, vol. 865, no. 12, pp. 861–865, 2016.

[7] C. Chan, T. H. Fung, M. Abbott, D. Payne, A. Wenham, B. Hallam, R. Chen, and S. Wenham, "Modulation of Carrier-Induced Defect Kinetics in Multi-Crystalline Silicon PERC Cells Through Dark Annealing," *Sol. RRL*, p. 1600028, 2017.

[8] D. N. R. Payne, C. E. Chan, B. J. Hallam, B. Hoex, M. D. Abbott, S. R. Wenham, and D. M. Bagnall, "Acceleration and mitigation of carrier-induced degradation in p-type multi-crystalline silicon," *Phys. Status Solidi - Rapid Res. Lett.*, vol. 10, no. 3, pp. 237–241, 2016.

[9] F. F. S. R. J. N. Karin Krauss Andreas A. Brand, "Fast regeneration processes to avoid light-induced degradation in multicrystalline silicon solar cells," *IEEE J. Photovoltaics*, vol. 6, no. 6, pp. 1427–1431, 2016.

[10] C. Kranz, U. Baumann, B. Wolpensinger, F. Lottspeich, M. Muller, P. Palinginis, R. Brendel, and T. Dullweber, "Void formation in screen-printed local aluminum contacts modeled by surface energy minimization," *Sol. Energy Mater. Sol. Cells*, vol. 158, pp. 11–18, 2016.

[11] A. Peral, A. Dastgheib-Shirazi, V. Fano, J. C. Jimeno, G. Hahn, and C. del Cañizo, "Impact of Extended Contact Cofiring on Multicrystalline Silicon Solar Cell Parameters," *IEEE J. Photovoltaics*, vol. 7, no. 1, pp. 91–96, 2017.

[12] S. Riegel, F. Mutter, T. Lauermann, B. Terheiden, and G. Hahn, "Review on screen printed metallization on p-type silicon," *Energy Procedia*, vol. 21, no. October 2011, pp. 14–23, 2011.

[13] J. Müller, K. Bothe, S. Gatz, H. Plagwitz, G. Schubert, and R. Brendel, "Contact formation and recombination at screen-printed local aluminum-alloyed silicon solar cell base contacts," *IEEE Trans. Electron Devices*, vol. 58, no. 10, pp. 3239–3245, 2011.

[14] J. Tan, A. Cuevas, D. MacDonald, T. Trupke, R. Bardos, and K. Roth, "On the electronic improvement of multi-crystalline silicon via gettering and hydrogenation," *Prog. Photovoltaics Res. Appl.*, vol. 16, no. 2, pp. 129–134, 2008.

[15] J. T. Horzel, Y. Shengzhao, N. Bay, M. Passig, D. Pysch, H. Kühnlein, H. Nussbaumer, and P. Verlinden, "Industrial Si Solar Cells with Cu-Based Plated Contacts," *IEEE J.*

Photovoltaics, vol. 5, no. 6, pp. 1595–1600, 2015.

[16] A. Lennon, Y. Yao, and S. Wenham, "Evolution of metal plating for silicon solar cell metallisation," *Prog. Photovoltaics Res. Appl.*, vol. 21, no. 7, pp. 1454–1468, 2013.

[17] N. J. Western, A. Sung, S. R. Wenham, and S. P. Bremner, "Localized ohmic contact through a passivation dielectric for solar cell rear surface design," *Appl. Phys. Lett.*, vol. 102, no. 22, p. 222105, 2013.

[18] N. J. Western, I. Perez-Wurfl, S. R. Wenham, and S. P. Bremner, "Point-contacting by localised dielectric breakdown: Characterisation of a metallisation technique for the rear surface of a solar cell," *J. Appl. Phys.*, vol. 118, no. 4, p. 45711, 2015.

[19] N. J. Western, I. Perez-Wurfl, S. R. Wenham, and S. P. Bremner, "Point-Contacting by Localized Dielectric Breakdown With Breakdown Fields Described by the Weibull Distribution," *Electron Devices, IEEE Trans.*, vol. 62, no. 6, pp. 1826–1830, 2015.

[20] X. Wang, P. C. Hsiao, W. Zhang, B. Johnston, A. Stokes, Q. Wei, A. Fell, S. Surve, Y. Shengzhao, P. Verlinden, and A. Lennon, "Untangling the Mysteries of Plated Metal Finger Adhesion: Understanding the Contributions from Plating Rate, Chemistry, Grid Geometry, and Sintering," *IEEE J. Photovoltaics*, vol. 6, no. 5, pp. 1167–1174, 2016.

Investigation of High Performance Perovskite-based Solar Cells Grown by Hybrid Chemical Vapor Deposition Technique

Huseyin Cem Gokkaya, Shen Qian, Zhiwei Ren, Annie Ng, and Charles Surya*

Department of Electronic and Information Engineering, The Hong Kong Polytechnic University, Hong Kong, China

*Correspondence author email: charles.surya@polyu.edu.hk

Abstract — **We report on a study of inverted-structure perovskite-based solar cells (PSCs). The perovskite layers were grown by hybrid chemical vapor deposition (HCVD) technique in which the PbI₂ layer was spin-coated onto the PTAA layer followed by the CVD-growth of the perovskite layer in a standard Si processing furnace. The enhancement in the device performance is attributed to the reduction in the bandgap states as indicated by low-frequency noise measurement performed on the perovskite layer over a wide range of temperatures. The results are corroborated by photothermal deflection spectroscopies and low-frequency noise measurement conducted on the perovskite films.**

I. Introduction

The development of cost-effective and highly efficient renewable energy sources is one of the important challenges for the 21st Century as continued dependence on fossil fuels will result in irreversible damages to the environment with devastating consequences [1,2]. It is projected by the International Energy Agency, nearly half of the net increase in electricity generation will come from renewables [3]. Hence, the development of high efficiency photovoltaic cells (PVCs) will be critical for meeting the future global energy demands [4]. Among the various PV materials, organic-inorganic perovskite thin films have drawn significant attention in recent years due to its tremendous enhancement in the power conversion efficiency (PCE) of the device [5,6].

Recent work by the author demonstrates significant reduction in the defect density by post-deposition annealing of the perovskite films in oxygen ambient [7]. The result corroborates with the theoretical analysis by Yin *et al* [8]. Our experimental data indicate passivation of material defects in the perovskite films prepared by standard solution technique due to thermal assisted annealing in oxygen. This had led to significant enhancement in the PCE of the device.

Oxygen annealing, it is believed, can be performed more effectively if done during the film growth process. To facilitate this, our group has developed a HCVD growth process for the deposition of CH₃NH₃PbI₃ (MAPI) thin films [9]. Here, PbI₂ is first spin-coated onto a glass/FTO/TiO₂ substrate. The substrate is then placed inside a standard silicon-processing furnace. A carrier gas consisting of N₂/O₂ mixture is used. The MAI powder is placed upstream to the substrate and is maintained at ~180°C to facilitate the sublimation of the material. Recent work shows that the optimal composition for the carrier gas is N₂:O₂ = 85%:15%. Also, it was found that the optimal substrate temperature for the HCVD growth of MAPI thin films is ~165°C. In this work, since the polymer (PTAA) is coated on the substrate first, the substrate temperature is maintained at 160°C during the HCVD process, which is a bit lower than the previous temperature used in order to prevent the material degradation. In the following paragraphs we present experimental results on the fabrication and characterization of PSCs with inverted structure. We compare the optoelectronic properties of the inverted PSCs fabricated both by standard solution technique and HCVD process.

II. Experiment

The structure and the energy band diagram of the device fabricated in this work are shown in the Fig. 1a and Fig. 1b below

a)

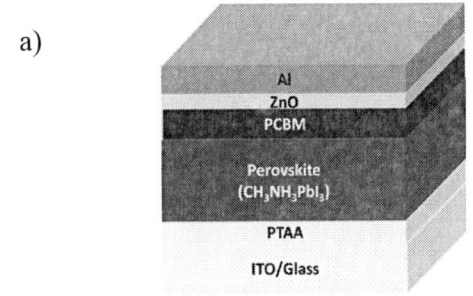

b)

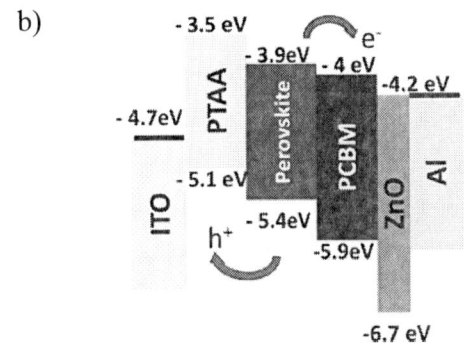

Fig. 1: (a) The structure and (b) the energy band diagram of the device.

A layer of PTAA, utilized as the HTL, is spin-coated onto an ITO-coated glass substrate followed by an annealing process at 110°C for 15 min inside the glovebox. The perovskite layer is deposited on the HTL by standard one-step solution technique, which will serve as the control device and the HCVD techniques. The ETL consists of a PCBM layer deposited by spin-coating process (20 mg/ml in chlorobenzene). A thin layer of ZnO, which serves as a hole-blocking layer is deposited by spin coating process of the ZnO nanoparticle ink (Sigma Aldrich Co.) at a spinning speed of 3500 rpm for 30 s and then annealed for 10 minutes at 80°C. A thickness of 100 nm aluminum was then thermally evaporated on the samples through a shadow mask. The device area is $0.1 cm^2$. The devices were encapsulated in an N_2 filled glove box before characterizations.

III. RESULT AND DISCUSSIONS

The plan view SEM images are shown in Fig. 2a and Fig. 2b for the comparison of the crystallinity of the perovskite layers grown by the two different processes.

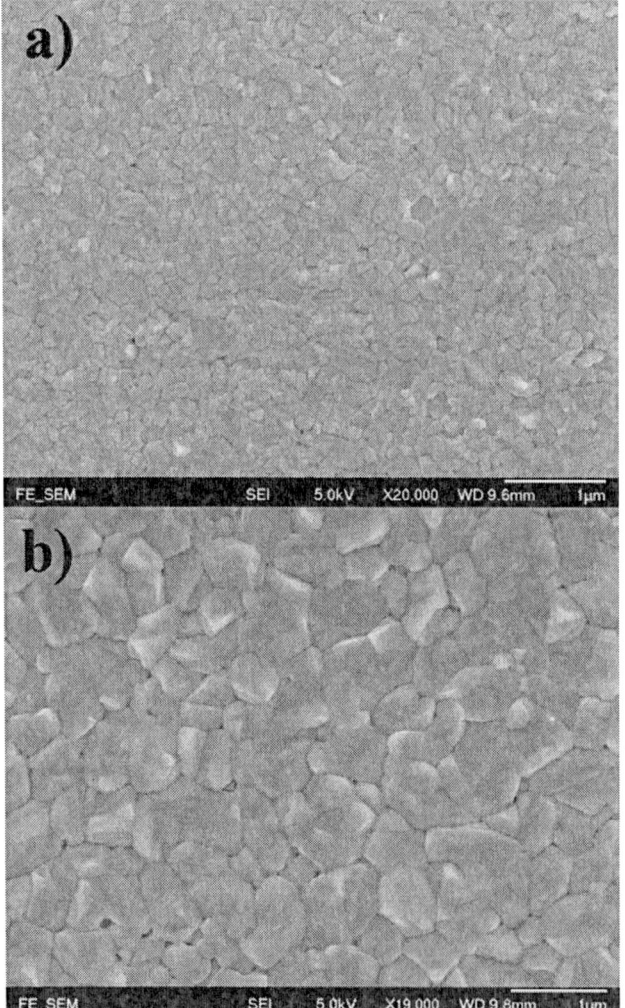

Fig. 2: The plan view SEM images for MAPI thin film grown by (a) one-step solution technique and (b) HCVD process.

The cross sectional SEM images of a complete device with the perovskite layer prepared by solution process and HCVD technique are shown in Fig. 3a and Fig. 3b respectively. Both of the plan view and the cross sectional SEM images clearly showed that the perovskite grown by the HCVD process exhibits larger grain size with substantial reduction in grain boundaries. It can be observed from the cross section of the device that the entire thickness of the perovskite layer grown by the HCVD is composed of a single crystal compared to the

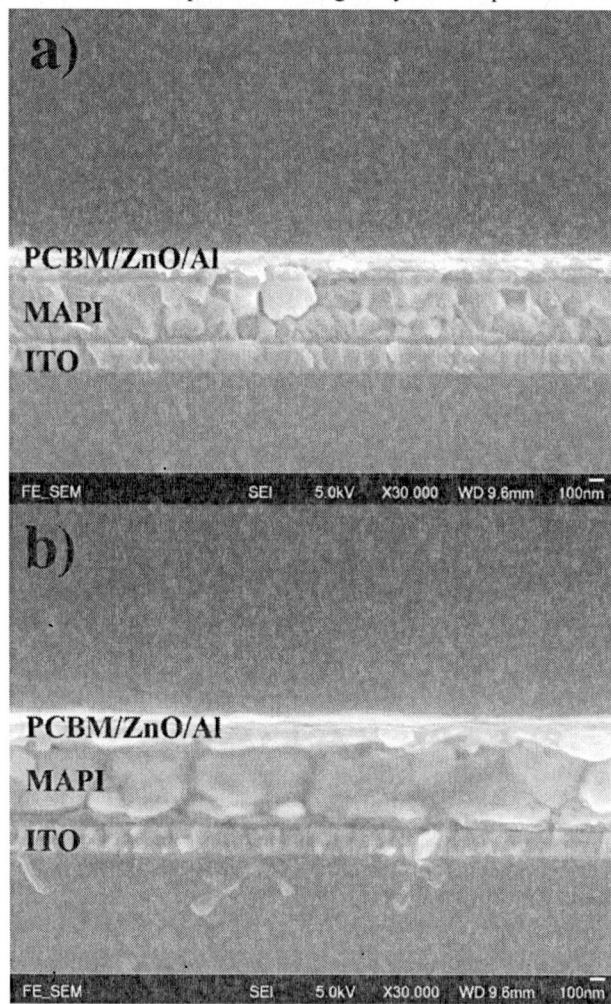

Fig. 3: Cross sectional SEM images for complete device (ITO/ PTAA/ MAPI/ PCBM/ ZnO/Al) with MAPI thin film grown by (a) one-step solution technique and (b) HCVD process.

solution processed perovskite film, which demonstrates much smaller crystallites. The results indicate significant enhancement in the crystallinity for the HCVD-grown perovskite layer compared to the one deposited by standard solution process

It is noted that the optimal growth temperature of this work is ~160°C, which is significantly higher than the standard reaction temperatures used in the solution process techniques, which are normally not higher than 100°C [10,11]. It is believed that a higher growth temperature is favorable to atomic motion during the crystallization process and thereby increases the grain size of the MAPI material. On the other hand, it is observed that using reaction temperatures >100°C, for solution processed perovskite, will lead to the decomposition of the MAPI due to the sublimation of MAI from the film [12]. In our case, the carrier gas is saturated with MAI, which suppresses the sublimation of the MAI from the film, and thus a higher growth temperature can be used without suffering any film degradation. It is believed that better crystallinity of the perovskite materials is desired for photovoltaic performance as carriers can be efficiently transported in a material with high order of crystallinity without trapping by the defects.

To investigate the effects of growth technique on the defect density in the devices we have conducted systematic low-frequency noise measurement on PSCs. Noise measurement is shown to be a highly sensitive and non-destructive technique in reflecting the defect density of the device. It was shown that the normalized low-frequency noise power spectral density (PSD) is given by

$$S_I(f) = 4\Delta I^2 \int_E \int_x \int_y \int_z N_T(E) \frac{\tau}{1+4\pi^2 f^2 \tau^2} dz dy dx dE \quad (1)$$

where ΔI is the fluctuation in the device current, I, arising from the fluctuation in the device current due to the capture of a single carrier by a localized state. For a thermally activated trapping and detrapping process, τ is given by

$$\tau = \tau_o \exp\left(\frac{E}{k_B T}\right) \quad (2)$$

where E is the activation energy and k_B is the Boltzmann constant. The Lorentzian is a sharply peaked function in Energy, simplifying Eq. 1 one obtains

$$N_T(E_p) = \frac{4Cf}{k_B T} S_I(f), \quad (3)$$

where $E_p = -k_B T \ln(\omega \tau_0)$. Thus, low-frequency noise measurement enables one to characterize the trap density directly from the devices. It is found that the PSD of the device critically depends on the parameters of the fabrication process. Experimental results of the low-frequency noise characterization performed on devices fabricated by standard one-step solution and HCVD techniques are shown in Fig. 4. It is observed that the noise level for perovskite sample prepared by the HCVD technique exhibits 3 orders lower in magnitude compared to the perovskite sample prepared by the conventional one-step solution process. It is a clear indication of the reduction in the trap density for perovskite films grown

by the HCVD technique, which has important implications on the performance of the device.

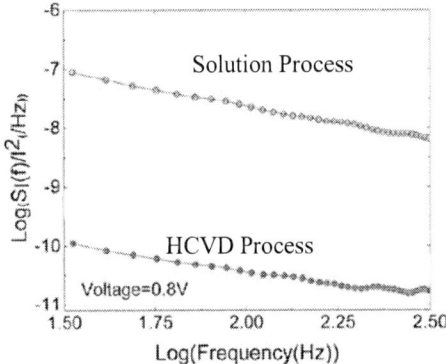

Fig. 4: Low-frequency noise power spectra measured from PSC fabricated by one-step solution technique (open circle) and HCVD technique (close circle).

This is because the defect states are widely known to be highly effective non-radiative recombination centers that have strong impact on the carrier lifetime. Figure 5a and 5b show the time-resolved photoluminescence (TRPL) data and photo-thermal deflection spectra (PDS) for the films grown by both one-step solution and HCVD techniques.

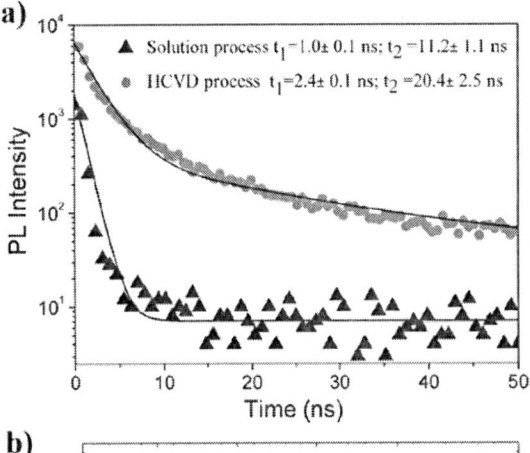

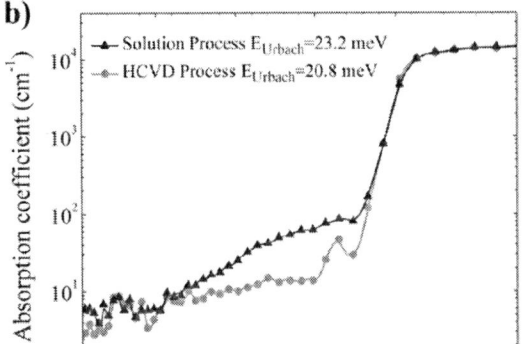

Fig. 5: (a) TRPL data for perovskite film grown by solution (triangle) and HCVD (circle) techniques; (b) PDS spectra for perovskite film grown by solution (triangle) and HCVD (circle) techniques.

It is believed that the result of TRPL can provide an insight into the trap density of the MAPI films prepared by different technique. The obtained TRPL data are well fitted to a bi-exponential decay function, in which two distinct lifetimes t_1 and t_2 can be determined from the data. A significant enhancement in the carrier lifetime from t_1=1.0 ns to 2.4 ns; t_2=11.2 ns to 20.4 ns is observed for the HCVD grown MAPI compared to the solution processed MAPI. The data suggest substantial reduction in the trap density and hence reduce the chance for carrier recombination at the defect sites for the samples fabricated by HCVD technique. The results of TRPL corroborate with the photo-thermal deflection spectra (PDS) obtained from the two types of samples deposited on quartz as shown in Fig. 5b. The PDS is an absorption characterization with high sensitivity down to the order of 10^{-4}. It can be used to detect tiny changes in the thermal state of the samples due to the nonradiative relaxation of photo-excited carriers. Therefore, it is commonly applied to characterize the energetic disorder as the exponential decay of the absorption below the bandgap with a characteristic energy named as Urbach energy. The shallow traps can be characterized by the magnitude of the Urbach energy, which reflects the steepness of the band tail states, located at the conduction band edge. A smaller Urbach energy stipulates a lower density of shallow traps and is, thus, a useful figure-of-merit for the film. It is found that the Urbach energy of the HCVD grown MAPI film is 20.8 meV, which is significantly lower than the 23.2 meV for the solution processed MAPI film. A smaller Urbach energy and lower sub-bandgap absorption obtained from PDS spectrum for the HCVD grown perovskite film clearly indicate lower defect density in this kind of material.

We have also performed the detailed characterization of the photovoltaic performance on both types of devices and the results are shown in Fig. 6 below. The PCE achieved for the solution-processed sample is 14.3 % while the PCE achieved for the HCVD grown sample is 15.4 %. The results clearly demonstrate the enhancement in the value of PCE as well as the reduction in the hysteresis for the HCVD-grown device. The *I-V* data are consistent with the experimental results on carrier lifetime and trap density measurement as demonstrated above.

III. CONCLUSION

HCVD process is successfully applied to inverted-structure PSCs, which can significantly improve the crystallinity of the perovskite film and, thereby, can reduce the defect density and low-frequency noise of the perovskite material compared to the perovskite films prepared by the conventional solution technique. HCVD-grown samples exhibit enhanced grain size, which improves the carrier transport properties leading to a higher short circuit current density (J_{SC}) by 2.4 mA/cm^2 (11.8 %) and a reduction in the

hysteresis. With device optimization our champion device reached 15.4% in PCE with 0.955 V in V_{OC}, 22.7 mA/cm^2 in J_{SC} and 71 % in FF.

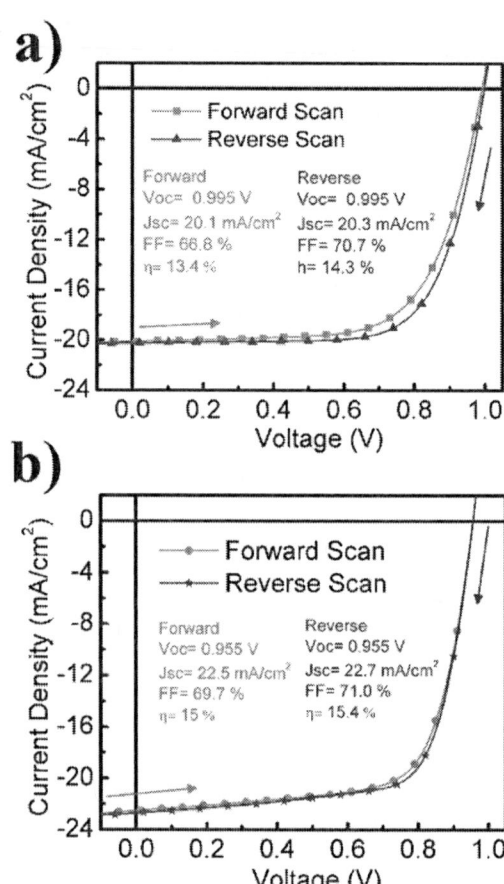

Fig. 6: Light *I-V* characteristics for the champion devices fabricated by (a.) one-step solution and (b.) HCVD processes.

IV. ACKNOWLEDGEMENTS

This work was supported by a GRF grant (Grant No. PolyU 152045/15E) and the Clarea Au Endowment Professorship. Special thanks are given to Prof. S. K. So and Mr. S. H. Cheung, Hong Kong Baptist University for PDS measurement.

V. REFERENCE

1. I. Dincer, "Renewable energy and sustainable development: a crucial review," *Renewable and Sustainable Energy Reviews*, vol. 4, pp. 157-175, 2000.
2. A. M. Omer, "Energy, environment and sustainable development," *Renewable and Sustainable Energy Reviews*, vol. 12, pp. 2265-2300, 2008.

3. International Energy Agency. 2014 Key World Energy Statistics,http://www.bp.com/content/dam/bp/en/corporate/pdf/bpstatistical-review-of-world-energy-2015-full-report.pdf, accessed: Jan, 2017.

4. H. J. Snaith, "Perovskites: The emergence of a new era for low-cost, high efficiency solar cells," *The Journal of Physical Chemistry Letters*, vol. 4, pp. 3623-3630, 2013.

5. M. Yang, T. Zhang, P. Schulz, Z. Li, G. Li, D. H. Kim, N. Guo, J. J. Berry, K. Zhu and Y. Zhao, "Facile fabrication of large-grain $CH_3NH_3PbI_{3-x}Br_x$ films for high-efficiency solar cells via CH_3NH_3Br-selective Ostwald ripening," *Nature Communications*, vol. 7, 12305, 2016.

6. H. Zhou, Q. Chen, G. Li, S. Luo, T. Song, H. Duan, Z. Hong, J. You, Y. Liu and Y. Yang, "Interface engineering of highly efficient perovskite solar cells," *Science*, vol. 345, pp. 542-546, 2014.

7. Z. Ren, A. Ng, Q. Shen, H. C. Gokkaya, J. Wang, L. Yang, W. Yiu, G. Bai, A. B. Djurišić, W. W. Leung, J. Hao, W. K. Chan and Charles Surya, "Thermal assisted oxygen annealing for high efficiency planar $CH_3NH_3PbI_3$ perovskite solar cells," *Scientific Reports*, vol. 4, 6752, 2014.

8. W. Yin, H. Chen, T. Shi, S. Wei, and Y. Yan, "Origin of high electronic quality in structurally disordered $CH_3NH_3PbI_3$ and the passivation effect of Cl and O at grain boundaries," *Advanced Electronic Materials*, vol. 1, 1500044, 2015.

9. A. Ng, Z. Ren, Q. Shen, S. H. Cheung, H. C. Gokkaya, S. K. So, A. B. Djurišić, Y. Wan, X.Wu, and C. Surya, "Crystal engineering for low defect density and high efficiency hybrid chemical vapor deposition grown perovskite solar cells," *ACS Applied Materials Interfaces*, vol. 8, pp. 32805-32814, 2016.

10. S. Sun, T. Salim, N. Mathews, M. Duchamp, C. Boothroyd, G. Xing, T. C. Sum and Y. M. Lam, "The origin of high efficiency in low-temperature solution-processable bilayer organometal halide hybrid solar cells," *Energy Environmental Science,* vol. 7, 399-407, 2014.

11. N. K. Noel, A. Abate, S. D. Stranks, E. S. Parrott, V. M. Burlakov, A. Goriely and H. J. Snaith, "Enhanced photoluminescence and solar cell performance via lewis base passivation of organic–inorganic lead halide perovskites, "*ACS Nano*, vol. 8, 9815, 2014.

12. B. Conings, J. Drijkoningen, N. Gauquelin, A. Babayigit, J. D'Haen, L. D'Oliesdaeger, A. Ethirajan, J. Verbeeck, J. Manca, E. Mosconi, F. D. Angelis and H. -G. Boyen, "Intrinsic thermal instability of methylammonium lead trihalide perovskite," *Advanced Energy Materials*, vol. 5, 1500477, 2015.

Enhanced Perovskite Solar Cell Performance Using Full Space Device Optimization

Ahmer A.B. Baloch[1], Shahzada P. Aly[1], Mohammad I. Hossain[2],
Raka Jovanovic[2], Nouar Tabet[1,2], and Fahhad H. Alharbi[1,2]

1. College of Science and Engineering, Hamad bin Khalifa University, Doha, Qatar
2. Qatar Environment & Energy Research Institute, Hamad bin Khalifa University, Doha, Qatar

Abstract — **A full-space material-independent optimization method to maximize the efficiency of hybrid perovskite solar cell is presented in this paper. The complete design space covers 23 parameters spanning the properties of electron and hole transport materials and contacts, and the thicknesses of all the device layers. The full space optimization identifies the parameters that can achieve the maximum efficiency. It is found that an efficiency of 26.63% can be achieved if the cell is without defects and under AM 1.5 spectrum; this is reduced to 23.42% if the defects are considered. These results can be used to guide the selection of new better matching materials to optimize the design of perovskite solar cell.**

Keywords— hybrid organic-inorganic perovskite, solar cell, device simulation, full-space optimization.

I. INTRODUCTION

Perovskite solar cells (PSC) have emerged as potential candidates for the next generation of solar cells due to a rise in their efficiency within a very short time span. However, in spite of significant advances, there are number of issues (such as the poor stability of perovskite and the inclusion of an expensive hole transport material (HTM) that needs to be resolved, before the commercialization of PSC [1-4]. It is believed that an alternative inorganic HTM to replace the expensive Spiro-OMeTAD shall allow better stability of the cell and a reduction in the production cost without compromising the efficiency. As a result, some less expensive and robust alternates of spiro-OMeTAD have been studied. Likewise, the electron transport material (ETM) has been evolving to increase the existing efficiencies. Currently the maximum power conversion efficiency (PCE) is 22.1% [5].

Figure 1 shows the typical device structure of the PSC along with energy band diagram. Parameters considered for full space optimization are shown for HTM, ETM, perovskite absorber and two contacts. More advanced PSC architecture can be employed for specific applications. The role of right material for HTM, ETM and light absorber layer is extremely vital for the proper operation of solar cell by allowing selective extraction of charges at particular contacts. Surprisingly, there are variable combinations of device structure for efficient PSCs. This diverse collection of architecture and materials illustrates how rich the field is; but on the other hand, is also a manifestation that the PSC devices are not optimized. Moreover, majority of the materials employed for ETM and HTM, and contacts are based –reasonably- on normally used

ones in solar cells field; for illustration, here is a non-inclusive record of those employed materials (Ref. [1,6] and the references within):

- ETM: TiO2, Al2O3, ZnO, ZrO2, and BCBM,
- Organic HTM: Spiro-OMETAD, P3HT, PTAA, PEDOT:PSS, and DPP-TT,
- Inorganic HTM: Cu2O, CuI, NiO, and CuSCN,
- Contact: Au, Ag, Al, FTO, and ITO.

Full space optimization parameters : Absorber, HTM, ETM and Contacts

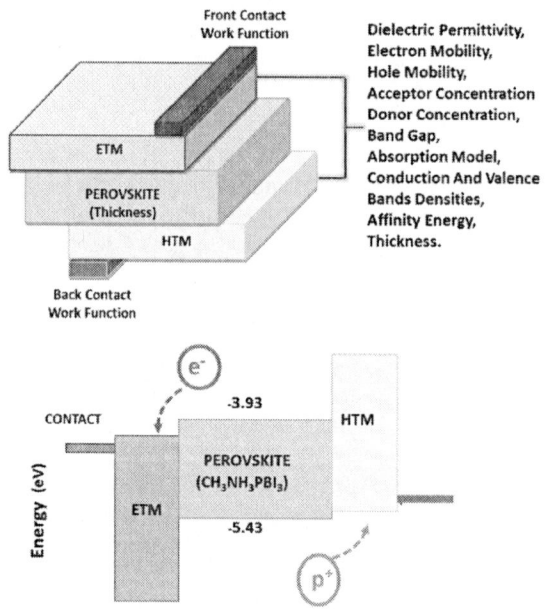

Fig. 1   The general device structure and the key parameters for full space optimization.

Therefore, it is justifiable to assume that there is possibility of better matching materials since most of them are commonly used materials in other solar cells and they are basically reused for PSCs along with the device concepts as well.

From the theoretical point of view, there are several studies examining the device structure of PSC [1,2,7-11], which can be classified in two categories. Category one involves simulation based on known-materials and encompasses the optimization of PSCs employing common ETMs, HTMs, and contacts. Thickness of the absorber layer and dopant concentrations are

978-1-5090-5606-4/17 $31.00 © 2017 IEEE   963

the primary optimization goals in these studies [1,2,7-10]. Category two, on the other hand, is directed towards particular device parameters such as band-offset [11] and contact work functions [9] with no preference for a specific material. Remarkably, majority of the numerical work shows that the efficiency of PSCs can surpass 23% with reasonable optimization. In this work, we conduct a full space material-independent optimization, where the essential device parameters are considered as shown in the next section.

II. OPTIMIZATION METHODOLOGY

The numerical analysis and the full space optimization have been carried out using the one-dimensional Solar Cell Capacitance Simulator (SCAPS) software interfaced with MATLAB. The simulation program solves the Poisson's equation, transport equation, and continuity equations numerically [12] while MATLAB handles the optimization. A primary step is to identify the type of solar cell and associated problem for optimization. For perovskite based solar cells, the issue of thickness is vital along with the selection of alternate HTM, ETM and contacts. Considering the need, the optimized space covers five layers of the device, namely: front contact, ETM, perovskite absorber, HTM, and back contact. The presented approach optimizes the device structure and identifies the properties of the optimal matching materials (except obviously for the perovskite absorbing layer) by selecting 23 optimum design parameters (listed below). For the absorbing $CH_3NH_3PbI_3$ properties, they are adopted from literature as reported in [1] and the references within except for the electron and hole mobilities where 2 cm2/V·s is used for both [7]. The key variable of the perovskite absorber is thickness, which is considered to optimize the performance based on its known absorption and intrinsic recombination properties. For the non-intrinsic recombination, two cases are independently considered; with and without defects. Here listed are the optimized parameters:

- Front and back contacts: work function,
- ETM & HTM layers: dielectric permittivity, electron mobility, hole mobility, acceptor and donor concentration, band gap, absorption model, conduction and valence bands densities, affinity energy, and thickness,
- Perovskite absorber: thickness,

All of these parameters are varied within very wide, yet physically reasonable, ranges to maximize the efficiency based on the following objective function

$$\max_{\boldsymbol{v}} \eta(\boldsymbol{v}) : \boldsymbol{v}_{\mathrm{LB}} \leq \boldsymbol{v} \leq \boldsymbol{v}_{UB} \qquad (1)$$

where η is the efficiency, $\boldsymbol{v} \in \mathrm{R}^N$ is a vector combining the used input parameters ($N = 23$ in this work), and $\boldsymbol{v}_{LB}$ and $\boldsymbol{v}_{UB}$ are respectively the lower and upper bounds of $\boldsymbol{v}$. It is bounded between the lower bound and the upper bound. The multivariate optimization is performed using the rich optimization toolbox in MATLAB. Local and global minima

solvers are employed to check the convergence on global solution for the optimized material data set and objective function of efficiency.

III. RESULTS AND DISCUSSION

The optimization using different local and global convergence algorithms have been carried out to ensure an optimum global data set and computational cost as shown in Fig. 2. Also, the comparison is used to analyze the optimization efficiency in term of minimum value and computational time (counted as number of function evaluations). Gradient based optimizer fmin performed better in terms of computational cost with 238 and 1384 function counts when compared with global optimizers such as genetic algorithm, particle swarms and pattern search at the same efficiency value the case of PSC with and without defects at 26.6% and 23.4 % efficiency. This is due to the smoothness of the drift-diffusion equations and convex nature of objective function.

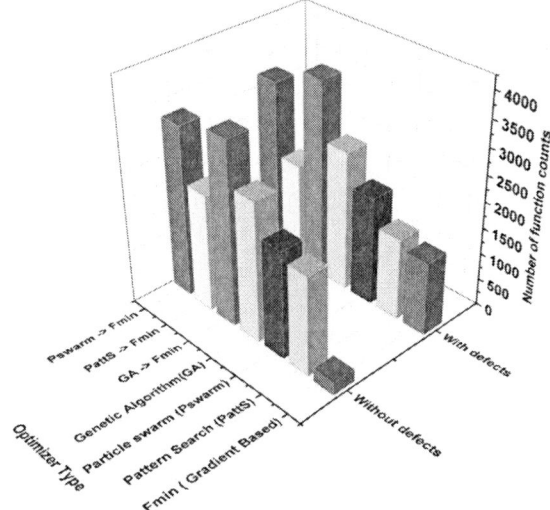

Fig. 2   Comparison of local and global minima optimizers using an efficiency of 26.6% for PSC without defects and 23.4% for PSC with defects.

Although, global optimizers yielded in a slow convergence due to its nature of initial guesses, they assisted in finding the optimal solution. This should be noted that hybrid optimizers such as pattern search to fmin and particle swarm to fmin showed less number of iterations due to the initial guesses already calculated by preceding optimizer.

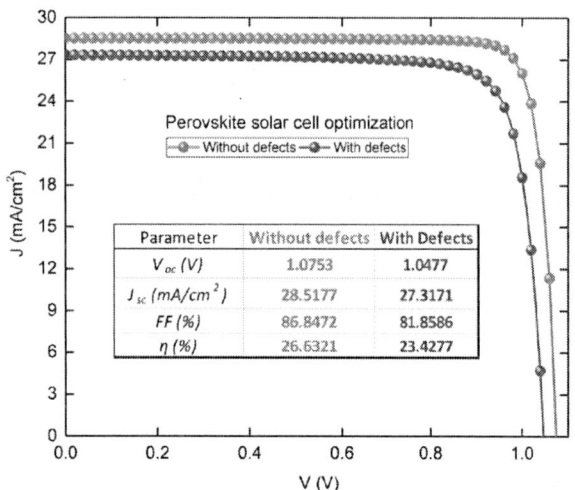

Fig. 3   IV curves for the optimum perovskite solar cell (with and without defects).

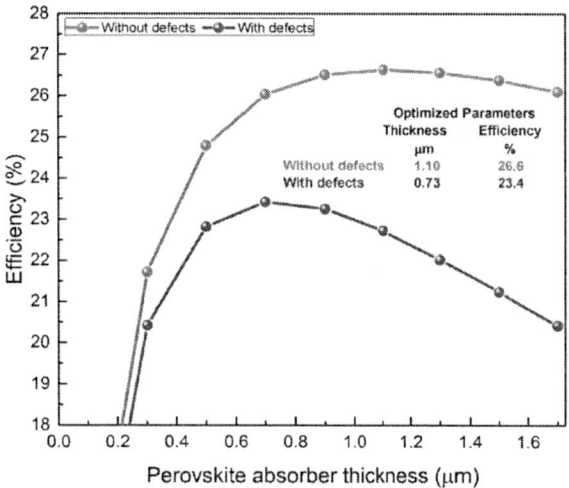

Fig. 4   Effect of perovskite thickness on the efficiency of PSC.

Optimal data set of materials for device layers were identified by full space optimization that can achieve the maximum conversion efficiency of 26.6% without defects and 23.4% with defects. The IV curves of the simulations are shown in Fig.3 with the fundamental solar cell parameters resulting from a device of 23 optimum physical properties of HTM, ETM, PSC and contacts. These results can be employed to pave a way for the selection of new better matching materials for ETM, HTM, and contacts.

Majority of the performance parameters found from full space optimization study were near to the experimentally reported values except for J_{sc}, which is around 23 mA/cm^2. There are various aspects that contributed to this. But, one of the most significant ones is the thickness of the absorber layer.

Fig. 4 displays that J_{sc} peaks around 1100 and 730 nm of the absorber layer thicknesses for the two studied cases because of the equilibrium between the absorption and recombination of the perovskites. Hence, the numerically reported thicknesses are larger than the thicknesses presently being fabricated which is around 450 nm.

IV. Conclusion

A coupled device simulation and full-space optimization technique is presented and it illustrates that efficiency of PSC can be enhanced to 26.6% by selecting optimum materials properties. This method can potentially guide the selection of new better matching materials for ETM, HTM, and contacts.

Acknowledgement

This work was made possible by the NPRP award [NPRP 6 - 931 - 2 – 382] from the Qatar National Research Fund (a member of The Qatar Foundation).

References

[1]   M.I. Hossain, F.H. Alharbi, N. Tabet, Copper oxide as inorganic hole transport material for lead halide perovskite based solar cells, Sol. Energy. 120 (2015) 370–380.

[2]   P. Yadav, K. Pandey, P. Bhatt, D. Raval, B. Tripathi, C. Kanth P., et al., Exploring the performance limiting parameters of perovskite solar cell through experimental analysis and device simulation, Sol. Energy. 122 (2015) 773–782.

[3]   F. El-Mellouhi, A. Marzouk, E.T. Bentria, S.N. Rashkeev, S. Kais, and F.H. Alharbi, Hydrogen Bonding and Stability of Hybrid Organic-Inorganic Perovskites, ChemSusChem 9 (2016) 2648-2655.

[4]   F. El-Mellouhi, E.T. Bentria, S.N. Rashkeev, S. Kais, and F.H. Alharbi, Enhancing Intrinsic Stability of Hybrid Perovskite Solar Cell by Strong, yet Balanced, Electronic Coupling, Scientific Reports 6 (2016) 30305.

[5]   Research Cell Efficiency Records, NREL.

[6]   J. Cui, H. Yuan, J. Li, X. Xu, Y. Shen, H. Lin, et al., Recent progress in efficient hybrid lead halide perovskite solar cells, Sci. Technol. Adv. Mater. 16 (2015) 036004.

[7]   K.R. Adhikari, S. Gurung, B.K. Bhattarai, B.M. Soucase, Comparative study on MAPbI3 based solar cells using different electron transporting materials, Phys. Status Solidi C 17 (2015) 13–17.

[8]   F. Liu, J. Zhu, J. Wei, Y. Li, M. Lv, S. Yang, et al., Numerical simulation: Toward the design of high-efficiency planar perovskite solar cells, Appl. Phys. Lett. 104 (2014) 253508.

[9]   T. Minemoto, M. Murata, Impact of work function of back contact of perovskite solar cells without hole transport material analyzed by device simulation, Curr. Appl. Phys. 14 (2014) 1428–1433.

[10] T. Minemoto, M. Murata, Device modeling of perovskite solar cells based on structural similarity with thin film inorganic semiconductor solar cells, J. Appl. Phys. 116 (2014).

[11] T. Minemoto, M. Murata, Theoretical analysis on effect of band offsets in perovskite solar cells, Sol. Energy Mater. Sol. Cells. 133 (2015) 8–14.

[12] M. Burgelman, P. Nollet, S. Degrave, Modelling polycrystalline semiconductor solar cells, Thin Solid Films. 361 (2000) 527–532.

Measuring Optical Absorption in Organic Photovoltaics Using Monochromated Electron Energy-Loss Spectroscopy

Jessica A. Alexander[1], Frank J. Scheltens[1], Lawrence F. Drummy[2], Michael F. Durstock[2], James B. Gilchrist[3], Sandrine Heutz[3], and David W. McComb[1]

[1]The Ohio State University, Columbus, OH, 43212, USA
[2]Air Force Research Laboratory, Wright-Patterson Air Force Base, OH, ZIP, USA
[3]Imperial College London, SW7 2AZ, UK

Abstract — The electronic properties of the materials utilized in organic photovoltaics (OPVs) play a key role in the performance of the device. Furthermore, the electronic structure at the interface between the donor and acceptor materials within the photoactive layer influences the ability of exciton charges to separate and, therefore, influences the current generation within the OPV device. Transmission electron microscopy can potentially provide information on the donor/acceptor interface but collecting spatially-resolved electronic information for organic materials has previously proven difficult, as these materials are extremely electron beam sensitive. In this contribution, we demonstrate how such electronic structure information, including the optical absorption spectrum, can be collected reliably and with nanometer scale spatial resolution for OPV materials using electron energy-loss spectroscopy (EELS) in a scanning transmission electron microscope (STEM).

I. INTRODUCTION

Due to their flexibility, low manufacturing costs, and lightweight, organic photovoltaics (OPVs) show promise for clean renewable electrical power [1]-[4]. However, the efficiencies of these devices are currently still too low for their implementation in consumer applications. Thus, it is critical to determine how to improve OPV device design such that the performance surpasses the record efficiency of 11.5% [5].

Improved understanding of the morphology and the local electronic structure at the interfaces between the acceptor and donor materials in the photoactive layer of the OPV device could give insight into the mechanisms of charge separation and current generation and thus lead to improved device efficiencies. Several studies have used transmission electron microscopy (TEM) to probe the morphology of the photoactive layer [6]-[10]. However, these studies have provided little insight into the local electronic structure of the donor-acceptor interfaces.

The local electronic structure of materials can be probed with nanometer spatial resolution using electron energy-loss spectroscopy (EELS) in a scanning transmission electron microscope (STEM). Specifically, the valence-loss region of the energy-loss spectrum can be used to probe transitions between the valence and conduction bands in the solid, and the spectrum can be processed to obtain the energy-dependent complex dielectric function, $\varepsilon(E)$. The real part, $\varepsilon_1(E)$, and imaginary part, $\varepsilon_2(E)$, of the dielectric function can be processed to obtain the optical absorption, $\alpha(E)$, spectrum and the energy of the peaks in the $\alpha(E)$ spectrum can be correlated

with specific single electron transitions from the valence band to the conduction band. Thus, STEM-EELS can be used to spatially map the optical absorption spectrum, and, therefore, probe the local electronic structure at the donor/acceptor interface within an OPV [11].

Collecting EELS data for organic materials is non-trivial. These materials are extremely beam-sensitive, exhibiting degradation when exposed to the electron beam. This is indicative of bonds breaking, meaning that the local chemistry of these materials has changed and is no longer representative of the pure material. Thus, before the donor/acceptor interface can be investigated with STEM-EELS, it is first necessary to determine how to collect reliable EELS data for materials used in OPVs.

II. ACQUISITION OF RELIABLE EELS DATA FOR OPV MATERIALS

In order to demonstrate that EELS data could be collected reliably for OPV materials, pure films were prepared of four materials commonly used in OPV devices – copper phthalocyanine (CuPc), fullerene (C_{60}), poly(3-hexylthiophene) (P3HT), and [6,6] phenyl C_{61} butyric acid methyl ester (PCBM) – as described in reference [12]. The optical properties of these films were investigated using variable-angle spectroscopic ellipsometry (VASE) [12]. This is a method in which a light beam (rather than an electron beam) is incident upon the sample, and, after interacting with the sample, undergoes phase shifts and amplitude changes, which can then be measured and analyzed [13]. From these data, it is possible to extract the energy-dependent refractive index and extinction coefficients, which can then be used to calculate $\varepsilon_1(E)$ and $\varepsilon_2(E)$ [13]. Furthermore, it is reasonable to assume that the samples undergo little to no damage during these measurements. Thus, any measurements made with VASE can be considered to be representative of the pure material. By comparing these standard VASE spectra to dielectric function obtained using EELS, it is possible to determine if the EELS measurements have been influenced by electron beam damage of the materials.

In order to minimize electron beam damage during measurements, a single spectrum STEM-EELS acquisition method was devised, which is described in detail in reference [12]. The complex dielectric functions were then extracted and compared to the data collected using VASE, and an example of this comparison is shown in Fig. 1 for C_{60} [12].

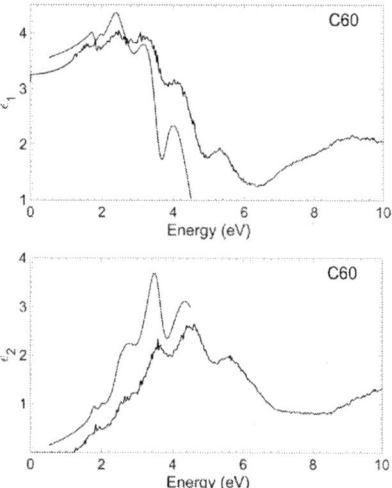

Fig. 1. Comparison of $\varepsilon_1(E)$ and $\varepsilon_2(E)$ data collected via VASE (blue) and EELS (black) for C_{60} [12].

If the sample had been damaged by the electron beam, it would be expected that peaks in $\varepsilon_1(E)$ and $\varepsilon_2(E)$ would either be missing or be of lower relative intensity when compared to the other peaks of the spectrum. However, for the $\varepsilon_1(E)$ and $\varepsilon_2(E)$ spectra shown in Fig. 1, the VASE and EELS data agree well in both the peak positions and their relative intensities [12]. There are no missing peaks in the EELS data and the small features in both spectra at approximately 2 eV that might have been dismissed as noise in the EELS data, are observed in the VASE data as well, indicating that the features are real signal [12]. Films of CuPc, P3HT, and PCBM were similarly compared and also agreed reasonably well, proving that the optimized EELS acquisition method can be trusted to collect reliable EELS data for these beam-sensitive materials [12].

From the EELS complex dielectric functions, the optical absorption spectra for all four of the materials were obtained (Fig. 2) [14]. This analysis illustrates the power of EELS, as it should be possible then to measure the optical absorption spectrum with high spatial resolution in OPV devices.

Interband transitions can be identified in the optical absorption spectrum of these materials by comparing the energies of the transitions to the energies of the peaks. This analysis is shown in Fig. 3 for C_{60}, in which peaks in the optical absorption spectrum have been correlated with known transitions from reference [15]. Thus, by spatially mapping the optical absorption spectrum at the interface of two materials, it will be possible to determine if there are any changes in the signal indicative of changes in the local electronic structure of the donor and acceptor materials.

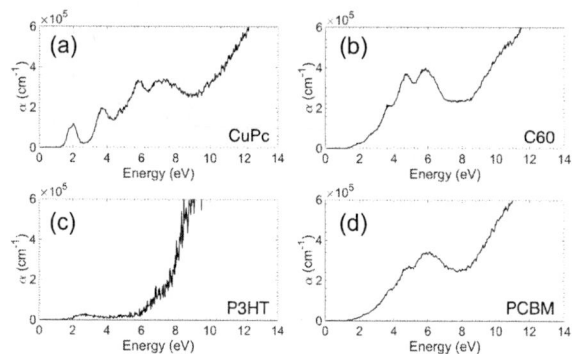

Fig. 2. Optical absorption spectra for (a) CuPc, (b) C_{60}, (c) P3HT, and (d) PCBM [14].

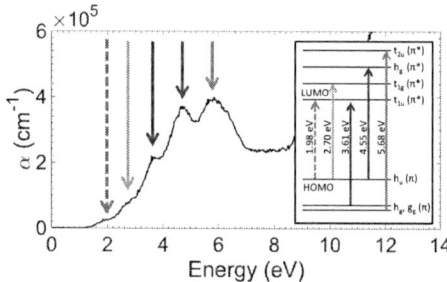

Fig. 3. Correlation of peaks the C_{60} optical absorption spectrum to known single electron transitions [15] from the valence band into the conduction band.

II. SPATIALLY-RESOLVED EELS MEASUREMENTS OF AN CUPC/C_{60} BILAYER STRUCTURE

Having demonstrated that EELS data could be collected for pure films of CuPc, C_{60}, P3HT, and PCBM, an organic bilayer structure consisting of a single interface between CuPc and C^-_{60} was analyzed to determine if reliable EELS data could also be collected with high spatial resolution [12]. First, a cross-section sample of the bilayer structure (shown in Fig. 4a) was prepared using a focused ion beam (FIB) method described in reference [16]. EELS data was then collected from the cross-section sample using the technique of spectrum imaging, in which EELS spectra are collected from specific areas of the sample. These spectra can then be summed together (to improve the signal-to-noise) and the resultant EELS spectra can be analyzed. This type of analysis is demonstrated in Fig. 4b in which spectra from areas of 448 nm in width and 4.8 nm in height have been summed together [12]. As the areas summed together move from the bulk C_{60} layer (at the top of the plot) through the interface and into the bulk CuPc layer (at the bottom of the plot), changes in the low-loss features (denoted by the arrows) and in the energy of the plasmon peak (the broad peak at 23-25 eV) are observed

[12]. This demonstrates that it is possible to collect EELS data with a spatial resolution of about 5 nm in OPV materials, which means that it is possible to probe the opto-electronic properties, including the optical absorption, of the donor/acceptor interface with similar spatial resolution [12].

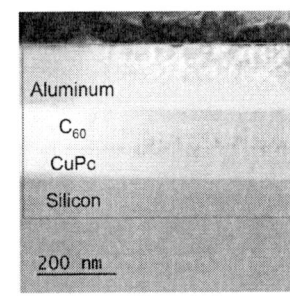

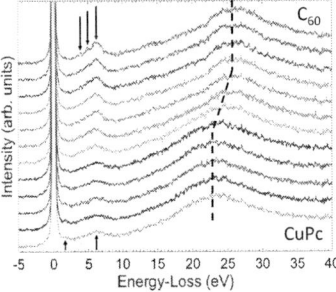

Fig. 4. (a) High-angle annular dark field scanning transmission electron microscopy (HAADF-STEM) image of the $CuPc/C_{60}$ bilayer structure in cross-section. The red box denotes the arrow from which spatially-resolved data was collected. (b) Low-loss EELS spectra moving from bulk C_{60} through the $C_{60}/CuPc$ interface and into the bulk CuPc layer in ~5 nm steps [12].

III. SUMMARY AND SIGNIFICANCE OF THE WORK

In summary, reliable EELS data were collected for various materials ($CuPc$, C_{60}, P3HT, and PCBM) used in OPV devices by utilizing an optimized acquisition method that minimized the beam damage incurred by the samples. From these data, the optical absorption spectra were extracted and correlated with known single electron transitions. Additionally, it was demonstrated that EELS data could be collected across the interface between two organic materials with high spatial resolution.

These preliminary results suggest that the goal of measuring the electronic structure at the donor/acceptor interface of an actual OPV device should be attainable. Ideally, we would like to use our techniques to collect spatially resolved EELS data at the donor/acceptor interface of multiple OPV devices, each with different power conversion efficiencies. Ultimately this would allow us to correlate the electronic structure of the donor/acceptor interface with the performance of the device to determine how to design future OPV devices with higher efficiencies.

REFERENCES

[1] Y. Liang, Z. Xu, J. Xia, S.T. Tsai, Y. Wu, G. Li, C. Ray, and L. Yu, "For the bright future-bulk heterojunction polymer solar cells with power conversion efficiency of 7.4%," *Advanced Materials*, vol. 22, no. 20, pp. E135-138, 2010.

[2] L. Hou, E. Wang, J. Bergqvist, B.V. Andersson, Z. Wang, C. Muller, M. Campoy-Quiles, M.R. Andersson, F. Zhang, and O. Inganas, "Lateral phase separation gradients in spin-coated thin films of high-performance polymer:fullerene photovoltaic blends," *Advanced Functional Materials*, vol. 21, no. 16, pp. 3169-3175, 2011.

[3] D. Angmo, J. Sweelssen, R. Andriessen, T. Galagan, and F.C. Krebs, "Inkjet printing of back electrodes for inverted polymer solar cells," *Advanced Energy Materials*, vol. 3, no. 9, 1230-1237, 2013.

[4] C. Koidis, S. Logothetidis, S. Kassavetis, C. Kapnopoulos, P.G. Karagiannidis, D. Georgiou, and A. Laskarakis, "Effect of process parameters on the morphology and nanostructure of roll-to-roll printed P3HT:PCBM thin films for organic photovoltaics," *Solar Energy Materials and Solar Cells*, vol. 112, pp. 36-46, 2013.

[5] NREL Best Research-Cell Efficiencies. (2017) At <http://www.nrel.gov/ncpv/images/efficicnecy_chart.jpg>

[6] D. Chen, A. Nakahara, D. Wei, D. Nordlund, and T.P. Russell, "P3HT/PCBM bulk heterojunction organic photovoltaics: correlating efficiency and morphology," *Nano Letters*, vol. 11, no. 2, pp. 798-801, 2007.

[7] M. Pfannmoller, H. Flugge, G. Benner, I. Wacker, C. Sommer, M. Hanselamnn, S. Schmale, H. Schmidt, F.A. Hamprecht, T. Rabe, W. Kowalsky, and R.R. Schroder, "Visualizing a homogenous blend in bulk heterojunction polymer solar cells by analytical electron microscopy," *Nano Letters*, vol. 11, no. 8, pp. 3099-3107, 2011.

[8] L.F. Drummy, R.J. Davis, D.L. Moore, M. Durstock, R.A. Vaia, and J.W.P. Hsu, "Molecular-scale and nanoscale morphology of P3HT:PCBM bulk heterojunctions: energy-filtered TEM and low dose HREM," *Chemistry of Materials*, vol. 23, no. 3, pp. 907-912, 2011.

[9] N. Rujisamphan, R.E. Murray, F. Deng, C. Ni, and S.I. Shah, "Study of the nanoscale morphology of polythiophene fibrils and a fullerene derivative," *ACS Applied Materials and Interfaces*, vol. 6, no. 15, pp. 11965-11972, 2014.

[10] S. van Bavel, E. Sourty, G. de With, K. Frolic, and J. Loos, "Relation between photoactive layer thickness, 3D morphology, and device performance in P3HT/PCBM bulk-heterojunctino solar cells," *Macromolecules*, vol. 42, no. 19, pp. 7396-7403, 2009.

[11] R.F. Egerton, *Electron Energy-Loss Spectroscopy in the Electron Microscope*. New York: Springer, 2003.

[12] J.A. Alexander, F.J. Scheltens, L.F. Drummy, M.F. Durstock, J.B. Gilchrist, S. Heutz, and D.W. McComb, "Measurement of optical properties in organic photovoltaic materials using monochromated electron energy-loss spectroscopy," *Journal of Materials Chemistry A*, vol. 4, pp. 13636-13645, 2016.

[13] H.G. Tompkins and W.A. McGahan, *Spectroscopic Ellipsometry and Reflectometry: A User's Guide*. New York: John Wiley & Sons, Inc., 1999.

[14] J.A. Alexander, F.J. Scheltens, L.F. Drummy, M.F. Durstock, F.S. Hage, Q.M. Ramasse, and D.W. McComb, "High-resolution monochromated electron energy-loss spectroscopy of organic photovoltaic materials," *Ultramicroscopy*, in press, 2017.

[15] M. Kelly, P. Etchegoin, D. Fuchs, W. Kratschmer, and K. Fostiropoulos, "Optical transitions of C_{60} films in the visible and

ultraviolet from spectroscopic ellipsometry," *Physical Review B*, vol. 46, no. 8, pp. 4963-4968, 1992.

[16] J.B. Gilchrist, T.H. Basey-Fisher, S.C. Chang, F. Scheltens, D.W. McComb, and S. Heutz, "Uncovering the buried interface in organic solar cells," *Advanced Functional Materials*, vol. 24, no. 41, pp. 6473-6483, 2014.

Advanced Deposition of Photo-catalytic TiO$_2$ Film by Atmospheric SPPS for Dye Sensitized Solar Cells

Ifeanacho Anyadiegwu[1], Dickson Kindole[1], Geoffrey Kibiegon Ronoh[1], Yoshimasa Noda[2], Yasutaka Ando[2]

[1]Graduate school of Engineering, Ashikaga Institute of Technology, Ashikaga, Tochigi, Japan, 326-0845

[2]Faculty of Engineering, Ashikaga Institute of Technology, Ashikaga, Tochigi, Japan, 326-0845

ABSTRACT

In this study, an environmentally friendly and cost effective technique, Atmospheric Solution Precursor Plasma Spray (ASPPS) was used to deposit photo-catalytic titanium oxide (TiO$_2$) film for the fabrication process of Dye-sensitized Solar Cell (DSSC) for low energy consuming devices in developing countries. Surface morphologies of the TiO$_2$ films were evaluated using optical micrographs. Film thickness and strength was measured using micro screw gauge and pencil scratch tester respectively. From these results it was confirmed that environmentally friendly semiconductor materials such as titanium dioxide, and natural dye, can be used for fabrication process of inexpensive photo-electronic solar cell using ASPPS technique.

Index Terms — Dye-sensitized solar cell, Photo-catalytic, Photovoltaic cells Titanium Oxide, Thermal spray,

I. INTRODUCTION

Recently, energy demand has been increasing around the world. Fossil fuel is still preferred owing to its relative availability, low price, and ease in their usage. However, since the fossil fuel reserves are rapidly decreasing, alternative energy resources which are environmentally friendly and low-cost are recommended to be explored and utilized[1-3]. Many people especially in developing countries live in rural areas far away from the grid. The installation and distribution costs are significantly higher for rural areas. Moreover, there is greater transmission line losses and poor reliability on electrical supply. In most of the rural areas and non-electrified sites, extension of utility grid lines experience series of complications such as high capital investment, high lead time, low load factor, poor voltage regulation and frequent power supply interruptions. As an alternative to meet energy demand, renewable energy technologies such as geothermal, hydropower, wind energy, bioenergy and solar energy has been much researched owing to their eco-friendliness and abundant availabilities even in rural areas. In addition, solar energy is an environmentally friendly technology used to harness the sun's energy into usable energy. Photovoltaic (PV) solar cell is one of several means of converting sunlight into useful energy. Since sunlight can last up to only six hours in many areas, solar charged battery systems are frequently used to provide power supply for complete 24hours a day. The electricity generated from PV solar cell can be used directly to power household utilities or can be stored in a battery for later use. PV solar cell

systems are generally much inexpensive means of electrical power generation especially to rural areas. As for semiconductor materials for making this solar cell, silicon and CdTe-based solar cells have been mass-produced and applied in many applications. However, they have many unavoidable disadvantages such as high production cost and their production process are associated with unfriendly environmental activities. Recently, as an alternative to the expensive silicon based solar cell, photo catalytic TiO$_2$ solar cell recognized as dye-sensitized solar cell (DSSC) has been among the most researched solar cells. This is owing to great promise of semiconductor TiO$_2$ material that has shown photo-stability, environmentally friendly, photo-catalytic activity and high solar energy conversion efficiency. Furthermore, due to its large band gap of 3.0-3.2eV, semiconductor TiO$_2$ has the ability to absorbs light even in the ultra-violet range. DSSCs have emerged as a promising option for harnessing sun's energy owing to low cost, production simplicity, environmental friendliness, than that of conventional silicon solar cells. The DSSCs use an organic dye extracted from plants to emulate the manner in which plants convert sunlight into useful energy. In DSSCs, TiO$_2$ films are fabricated with Platinum (Pt) and dye sensitizers to convert solar light into DC electricity. Dye-sensitizer plays a role of light harvesting and electrons transition. Various techniques have been practically used for deposition process of photo catalytic TiO$_2$ film such as sol-gel, MOCVD, Spray pyrolysis, and sputtering processes[3-5]. However, there are several problems associated with the processes such as low crystallinity, chemical homogeneity and low deposition rate. On the other hand, in this study, atmospheric solution precursor plasma spray (ASPPS) has been used to deposit photo catalytic TiO$_2$ film to overcome some of these problems. The authors utilized ASPPS techniques owing to its possibility of depositing unique TiO$_2$ films structure comprising of many required properties for its applications performance such as particle size, crystallinity and the morphology.

Atmospheric Solution precursor plasma spray (ASPPS) is a film deposition process using thermal plasma spray. The process has many advantages including high film deposition rate, simple equipment, etc. It has been widely used in various industrial applications due to excellent enhancement of surface properties. The solution-precursor plasma spray (SPPS) method is a new process for depositing thick ceramic coatings, where solution feedstock (liquid) is injected into plasma. This versatile method has several gains over the conventional plasma spray method, and it can be used to deposit nanostructured, porous coatings of a wide variety of oxide and non-oxide ceramics for a myriad of possible applications. However, to utilize ASPPS practically high equipment cost, limitation of sample size and high electrical power are the main challenges. In this study, in order to fabricate TiO_2 films with high strength and photocatalytic properties, deposition of amorphous and crystallized TiO_2 films was carried out by varying deposition distances (80 – 160 mm). Besides, in order to elevate thermal energy of plasma at low operating discharge power, N_2-dominant Ar/N_2 was used as the plasma working gas.

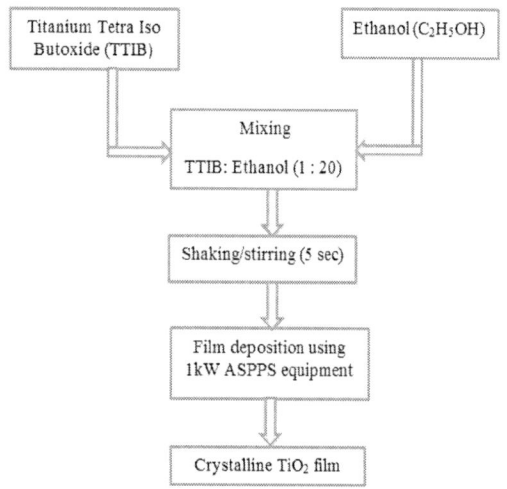

Fig.1.   Flow chart for preparation of feedstock and deposition of the crystalline TiO_2 film

II. EXPERIMENTAL PROCEDURES

A. Preparation of TiO_2 solution Precursor

Ethanol-diluted Titanium tetra iso butoxide (TTIB: $Ti(OC_4H_9)_4$) was used as feedstock for film deposition. All chemicals were of analytical grade and were used without further purification.

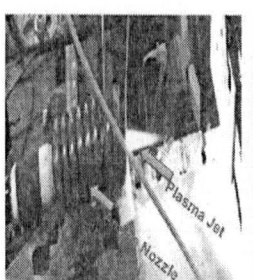
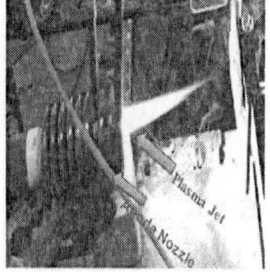

Fig.2. Ar/N_2 gas plasma jet at state and at combustion state

The precursor (TTIB) was added with the ethanol in the ratio of 1:20 for each deposition distance. For complete mixing of the feedstock, shaking process was performed continuously for approximately five seconds. Figure 1 shows the flow chart of preparation of feedstock and deposition process of crystalline TiO_2 thin films

B. Deposition of crystalline TiO_2 film

Photo-catalytic TiO_2 films were deposited on 15mm² and 1mm thick 304 stainless steel plate substrates. In order to improve quality of the films, in this study, a substrate with surface roughened grit-blasted was used.

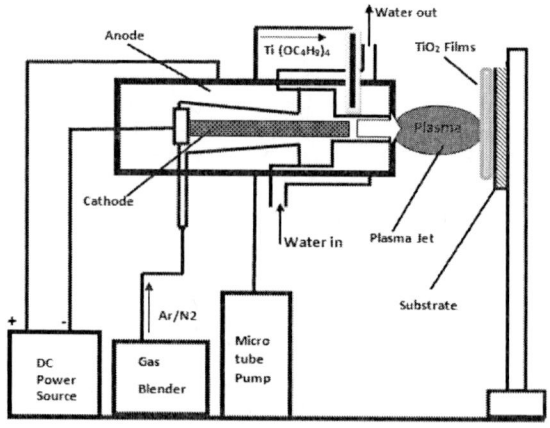

Fig.3.   Schematic Diagram of SPPS Film

Figure 2 and 3 respectively, shows the schematic diagram of 1kW-ASPPS equipment and the appearance of plasma jet before and after injection of feedstock. The feedstock was continuously injected by low cost precursor solution feed pump instead of conventional expensive mass flow controller. In addition, in order to maintain optimum system temperature, feedstock was fed without using carrier gas which cools down the plasma jet.

C. Characterization of crystalline TiO_2 films

The crystallinity and phase composition of each TiO_2 films were investigated using (Rigaku) X-ray diffraction with (CuKa, 40kV/100mA) radiation source in the rage of $20 – 90^0$. The optical properties and surface morphologies of crystallized TiO_2 films were evaluated using optical microscope. Moreover, methylene blue decomposition in ultraviolet light (Coper Electronic Co. PE-01) and wettability tests were carried out in order to confirm the photocatalytic properties of the deposited film. DSSC open circuit voltage and solar irradiance were measured using a digital multimeter (Custom M-04

12V A23) and a solar power meter (LA-1017), respectively. The film strength was evaluated using a pencil scratch tester (Coating tester kogyo, pencil scratch tester No. 850)

Table 1: Film Deposition Conditions

Plasma Working Gas	Ar/ N_2
Nitrogen gas flow rate	2.5 L/min
Argon gas flow rate	1.5 L/min
Operating discharge power	50A/ 20V
Deposition distance	80mm – 160mm
Deposition time	5 minutes
Feedstock materials	C_2H_5OH diluted TTIB
C_2H_5OH/ TTIB* Ratio	1 : 20
Feedstock feed rate	100ml/hr
Substrate	304 stainless steel plate / ITO glass

*TTIB: Titanium Tetra Iso Butoxide

III. EXPERIMENTAL RESULTS AND DISCUSSIONS

A. Influence of N_2-dominant plasma working gas on TiO₂ film crystallinity

Since the plasma spray process is relatively costly and complex, the N_2-dominant Ar/N_2 plasma gas was utilized to achieve high plasma thermal energy at a low discharge power. In order for argon plasma gas to attain a sufficiently high thermal energy for complete dissociation and ionization, it requires a high discharge power over 10 kW. Moreover, the N_2-dominant Ar/N_2 plasma gas requires approximately 1 kW to attain appropriate deposition thermal energy. The reason is that, a diatomic nitrogen plasma gas has a relatively higher energy at a given temperature than a monoatomic argon plasma gas.

Figure 2 shows the appearance of N_2-dominant Ar/N_2 plasma jets before and after feedstock (TTIB/ethanol) injection using vortex plasma generation nozzle. Even in the case of N_2 as the working gas, a plasma jet was not generated continuously at a N_2 flow rate of over 2.5 l/min, but was generated continuously at below 2.5 l/min. However, even under 2.5 l/min, extensive plasma jet fluctuation and intensive electrodes abrasion occurred. Also, when Ar at a flow rate of 1.5 l/min was added to the N_2 working gas, plasma jet was not generated continuously at over 2.5 l/min, plasma jet was generated without electrodes abrasion and intensive fluctuation of the plasma jet at below 2.5 l/min with enough thermal

energy to melt and evaporate precursor solution feedstock particles during flight as shown in figure 4. Although the thickness of the deposited film was deteriorating markedly with increasing deposition distance, over-100-μm-thick films could be deposited at each distance. It was noted that the microstructure and surface morphology of the film were synthesized by the accumulation of TiO_2 particles and these particles size increased at an optimum deposition distance and temperature. Therefore, in this study, the result indicates that the optimum thermal energy affects the crystal structure of the film, and TiO_2 particles were synthesized and grew during flight. In this study, wettability and methylene blue decoloration tests were carried out in order to confirm the photocatalytic properties of TiO_2 films. As for the wettability test, although partial deformation occurred during the deposition of the film at d=80 mm owing to deposition temperature, photocatalytic TiO_2 films were deposited at d=120 and 160 mm, and the surface contact angle were measured successfully. For the samples with films deposited at d=80 mm, the films showed hydrophilic properties with surface contact angles of over 24^0. On the other hand, the films deposited at d=120 and 160 mm showed hydrophobic properties with contact angles of over 70 and 90, respectively.

5 mm

Fig.4. Appearance of deposited film on the stainless steel substrate at d=120 mm

Figures 5, 6 show the schematic diagram of the equipment and the results of methylene blue decoloration test for the TiO_2 film. The titanium oxide film deposited substrate started to decolour methylene blue after only 6 hours of ultra violet irradiation and decoloured it almost perfectly after 24 hours of continuous operation. In the case of wettability, few drops of water was poured on the substrate before and after ultra violet radiation. A hydrophobic property of titanium oxide film was observed before UV irradiation whereas after UV irradiation, hydrophilic was observed.

It was observed that, the films deposited at d= 120 and 160 mm started to decrease methylene blue concentration after 2 hour under UV irradiation and decomposed almost completely (2.4% ppm) after 6 hour in the case of film deposited at d=120. On the other hand, it was noted that the absorption rate of methylene blue concentration was minimal (8.8% ppm) even after 6h under UV irradiation for the film deposited at d= 80 mm. Consequently, it was verified that practical photocatalytic TiO$_2$ films could be deposited by using this low-cost technique even at 1 kW using N$_2$-dominant Ar/N$_2$ as the plasma working gas.

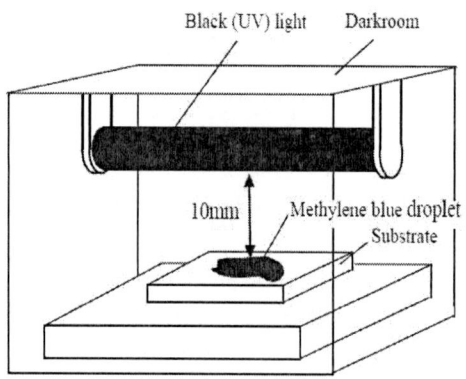

Fig.5. Illustration of methylene-blue decoloration test

a)   TiO$_2$ film

b)   Before UV irradiation

c)   After UV irradiation

Fig.6. Appearance of the sample in case of d=80mm during MB decoloration test
(T$_{UV}$: UV Irradiation time (6hrs), MB: Methylene-blue droplets)

As for the film strength measured by pencil scratch test, it was noted that the photocatalytic TiO$_2$ film deposited at d =120 mm had sufficiency strength to withstand 2H pencils scratch. On the other hand, films deposited at d = 80 and 160 mm showed weaker strength and can only with stand up to 1H and 1B respectively. From these results, it was proved that the degree of crystallinity, photocatalytic properties and strength of the film were increased and improved with increasing deposition distance.

B. Application of TiO$_2$ films to photovoltaic devices

In this study, in order to confirm the applicability of the photo-catalytic TiO$_2$ films to photovoltaic devices, the film was deposited on a 40 x 20 x 2^t mm conductive glass electrode positioned on water cooled substrate holder to protect from thermal damage. Figures 7(a) and 7(b) respectively shows the appearance of electrodes i.e. cathode electrode fabricated from photo-catalytic TiO$_2$ film and anode electrode fabricated by graphite.

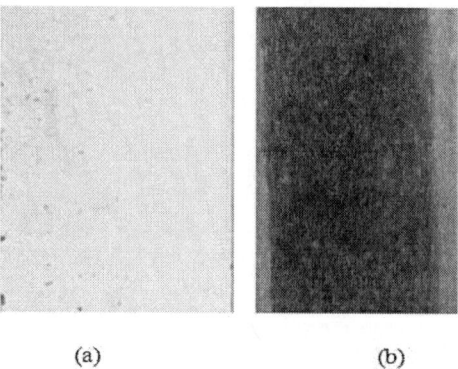

(a)   (b)

Fig. 7. Appearance of the photo-catalytic TiO$_2$ film and graphite electrodes
(a) Cathode electrode   (b) Anode electrode

Furthermore, in order to enhance sun-light harvesting capacity of the device, the dye extracted from hibiscus was infiltrated into the photo-catalytic TiO$_2$ films before assembling. Figure 8(a) and (b) respectively, shows the appearance and schematic diagram of photovoltaic device fabricated utilizing photocatalytic TiO$_2$ films deposited in this study. The photovoltaic device had an effective area of approximately 740 mm^2 and was able to generate 146.7mVoc at an average of 574W/m^2 of solar irradiance measured using radiation meter. Table 2 and Figure 9 respectively, shows appearance of photovoltaic device during testing and the comparison results readymade and developed device. From these results, the ASPPS technique was found to have high potential for high-rate TiO$_2$ film deposition with high crystallinity and photo-catalytic properties.

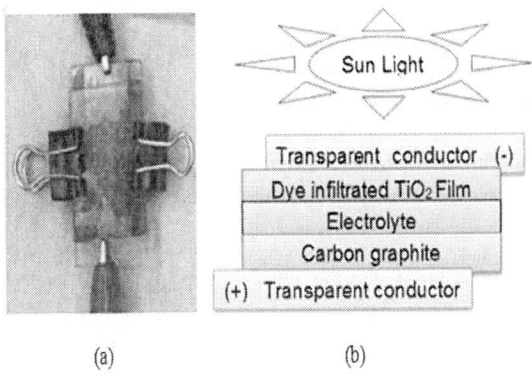

(a)   (b)

Fig.8. (a) Appearance and (b) Schematic diagram of photovoltaic device

Table 2: Results of photovoltaic test (in V)

Photovoltaic Device	Testing Conditions Without sunlight irradiation	Testing Conditions With sunlight irradiation
Developed device	0.00	0.14
Commercial device	0.00	0.15

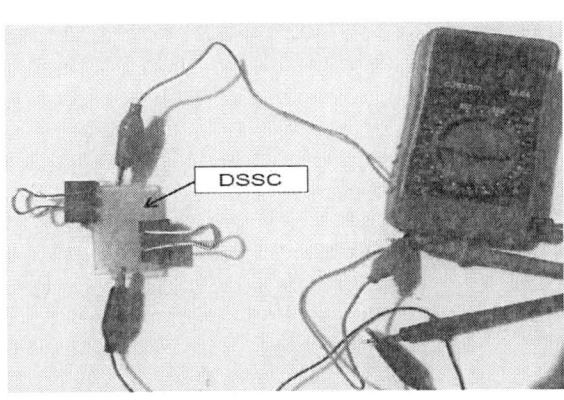

Fig.9. Appearance of photovoltaic device during testing

IV. CONCLUSION

A film with high photo-catalytic and crystallinity properties has been deposited by 1kW – ASPPS technique utilizing ethanol diluted titanium tetra-iso-butoxide as feedstock materials. Generally, in each deposition distance, film distribution with almost uniform thickness and porous structure were deposited even in the case of N_2-dominant Ar/N_2 utilization. In this study, although the film was deposited by a chemical reaction with heterogeneous nucleation, the structure of the film was varied from columnar to porous and the results of X-ray diffraction and Optical microscope clearly proved that, the degree of crystallinity and surface morphology can be controlled by varying deposition distances. Furthermore, photo-catalytic properties of the films were confirmed by wettability and methylene blue decoloration tests. The results proved that, the films had significant and enhanced photo-catalytic properties by showing hydrophobic and decolored methylene blue almost perfectly after 6hrs in a UV irradiation light. Therefore, in this study, it was proved that, the photo-catalytic properties of film can be controlled from hydrophilic to hydrophobic by varying deposition distance. It was also confirmed that the film deposited by this technique can be utilized for low-cost production of photovoltaic device that can generate up to 146.7mV$_{oc}$. The reasons why the open circuit voltage was not exactly the same as the commercial devices can be related to the low crystallinity of the film as well as the films defects formed during deposition.

ACKNOWLEDGEMENT

This study has been supported by JICA in Master's Degree and Internship Program of African Business Education Initiative for the Youth and Ashikaga Institute of Technology.

REFERENCES

[1] J.G. Bhadiyadra, M.V. Vaghani, *"A review on applicability of photocatalyst titanium dioxide for treatment of greywater"* Int. Journal of Engineering Research and Applications 5(Part-3,4) (2015) 102-105.

[2] Y. Ando, S. Tobe, H. Tahara, *"Photo-Catalytic TiO₂ Film Deposition by Atmospheric TPCVD"*, Vacuum, Vol.80, Nos.11-1, 2006, pp.1278-1283.

[3] Y. Ando, Y. Noda, A. Kobayashi, *"Al₂O₃ Film Deposition by Atmospheric Thermal Plasma Spray Using 1kW Class Electric Power Source"*, Frontier of Applied Plasma Technology, Vol.7 No.2, 2014, pp.75-80.

[4] A. Kobayashi, Y. Ando, S. Tobe, H. Tahara, " *Effect of poly-Ethyleneglycol Addition to feedstock material on deposited Titanium Oxide Film in Atmospheric Thermal plasma CVD"*

[5]T. Nakamura1, E. Matsubara, N. Sato, A. Muramatsu – *"Study on Fabrication of Titanium Oxide Films by Oxygen Pressure Controlled Pulsed Laser Deposition. Materials Transactions"*, Vol. 45, No. 7 (2004) pp. 2068 to 2072.

[6] A. Killinger, R. Gadow, G. Mauer, A. Guignard, R. Vaßen, D. Stover, *"Review of new developments in suspension and solution precursor thermal spray processes"* Journal of Thermal Spray Technology 20(4) (2011) 677-695.

[7] D. Chen, E.H. Jordan, M. Gell, X. Ma, *"Dense TiO₂ coating using the solution precursor plasma spray process"* J. Am. Ceram. Soc. 91(3) (2008) 865872.

[8] R.S. Lima, B.R. Marple, *"Thermal spray coatings engineered from nanostructured ceramic agglomerate powders for structural, thermal barrier and biomedical applications"* A review, Journal of Thermal Spray Technology 16(1) (2007) 40-63.

[9] L. Pawlowski, *"The science and engineering of thermal spray coatings"* 2nd Edition, England: John Wiley & Sons, Ltd, 2008.

[10] A. Fujishima, X. Zhang, D.A. Tryk, *"TiO_2 photocatalysis ans related surface phenomena"* Surface Science Reports 63 (2008) 515-582.

$CH_3NH_3PbI_{3-x}Br_x$ perovskite solar cells via spray assisted two-step deposition: influence of bromide on the device performance

Gaoda Chai, Shiqiang Luo, Shizhen Wang, Hang Zhou[*]

School of Electronic and Computer Engineering, Peking University ShenZhen Graduate School, Peking University, Shenzhen, 518055, China

Abstract — **In this study, we demonstrated full conversion of PbI_2 to perovskite via our spray assisted two-step deposition method and achieved solar cells with 13% efficiency. Moreover, the influence of adding MABr in the PbI_2 solution in the first step on the resulting solar cells is investigated for the first time. The perovskite solar cells prepared in this manner showed a significant increase in the power conversion efficiency after one-night storage, and a process of Br ions diffusion is proposed.**

I. INTRODUCTION

The availability of solar light and the maturity of semiconductor technology make photovoltaic a promising renewable energy technology. To develop inexpensive and high efficient solar cells, intensive research has been devoted to the organolead halide perovskite materials and remarkable milestones have been reported. For the highest power conversion efficiency (PCE), PCEs over 20% have been achieved by applying $MA_{1-y}FA_yPb(I_{1-x}Br_x)_3$ (MA short for CH_3NH_3 and FA short for NH_2CHNH_2) via the one-step precursor deposition method [1, 2] and the two-step deposition method [3]. A PCE of 21.1% was further achieved by optimizing the perovskite components as $Cs_{0.05}(MA_{0.17}\backslash FA_{0.83})_{0.95}Pb(I_{0.83}Br_{0.17})_3$ [4]. To obtain these PCEs, incorporating Br is necessary and this is because Br helps to stabilize the phase of $MA_{1-y}FA_yPb(I_{1-x}Br_x)_3$ [5]. Moreover , incorporating Br was reported to improve the stability of the perovskites against humidity [6].

Previously, incorporating Br was realized by one-step deposition of $MAPbI_{3-x}Br_x$ precursor solution [6], converting PbI_2 thin film in MAI+MABr mixed solution [7] or halide exchange of as-formed $MAPbI_3$ in MABr solution [8, 9]. However, while uniform $MAPbI_{3-x}Br_x$ could be easily formed by halide exchange of as-formed $MAPbI_3$ perovskite in MABr solution [8, 9], long dipping time was needed to convert PbI_2 in MAI+MABr solution to the $MAPbI_{3-x}Br_x$ perovskite [7]. However, the tendency of PCEs with Br concentration differed with the way of incorporating Br [6-9]. Thus, it is important to evaluate the PCEs of the Br containing perovskite solar cells by considering the difference of the deposition methods. In this study, we combine our research on MABr additives in PbI_2 with our two-step ultrasonic spray deposition technique. Our manuscript provides an alternative method for

incorporating Br and may help to explain the contradiction of the reported PCE trends.

II. EXPERIMENTAL PROCEDURES

Poly(3,4-ethylenedioxythiophene) poly(styrene-sulfonate) (PEDOT:PSS) (Heraeus Clevios P VP AI 4083)-isopropanol (1:3 by vol) solution was filtered through a 0.45 μm filter, acting as the hole transporting layer. 1 M PbI_2 with xM MABr (x =0, 0.1, 0.15, 0.2, 0.25, 0.3) (denoted as PbI_2+ xMABr) were dissolved in N,N-dimethylformamide (DMF). The mixed solutions of various molar weight ratios were stirred on a hot plate at 70 °C overnight. The methylammonium halide solution (10 mg/mL) in 2-propanol (IPA) were prepared at 60 °C for 1 h. A solution of [6,6]-phenyl C_{61}-butyric acid methyl ester (PCBM) (20 mg/ml) was prepared in chlorobenzene solution which was heated at 70 °C for 30 min, acting as the electron transporting layer. 2,9-dimethyl-4,7-diphenyl-1,10-phenanthroline (BCP) was dissolved in absolute ethyl alcohol with concentration of 0.5 mg/mL under stirring at 70 °C overnight.

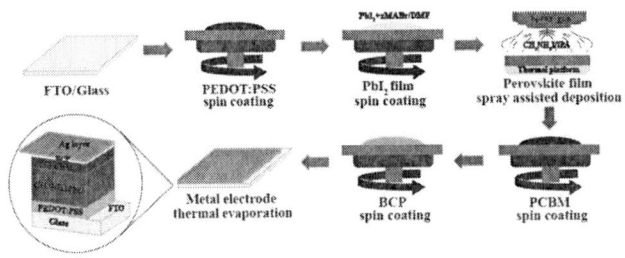

Fig 1.   Schematic of perovskite solar cells prepared by spray assisted two-step deposition method.

The preparation steps were given in Fig. 1. Pre-patterned fluorine doped tin oxide (FTO) glass substrates (15Ω/□) were cleaned though sonication in deionized water, acetone and alcohol for 15min each successively and dried with compressed nitrogen before use. The substrates were treated with UV light and ozone for 10 min to clean the surfaces and also to improve surface adhesion. The PEDOT:PSS solution was spin-coated onto FTO at 2000 rpm for 60 s, followed by annealing at 150 °C for 20 min in air to form a 30 nm thick layer. The substrates were transferred to a glovebox. For the first step, the $PbI_2 + x$MABr solution was spin coated on the PEDOT:PSS film at 6000 rpm for 30 s and then annealed at 100 °C for 60 s. For the second step, the MAI solution was ultrasonically spray-coated on the $PbI_2 + x$MABr film in ambient air and for 90 s. The spray system applied a pneumatic ultrasonic spraying system (Siansonic Z40S) equipped with a 36 kHz nozzle in nitrogen carrier. The distance from nozzle to substrate was 12 cm, high enough for better dispersion of MAI small droplets. During spraying, the substrates were kept at 75 °C and the infusion rate of MAI solution was kept as 150 μL/min. After that, all substrates were transferred to glovebox again for annealing at 100 °C for 60 min to convert to perovskite $MAPbI_{3-x}Br_x$ thin films (~300 nm). The perovskite thin films were then washed with IPA using the spinning drop method for 60 s. The PCBM layers were spin-coated on the perovskite layer at 1500 rpm for 45 s, followed by annealing at 100 °C for 30 min. BCP thin films were deposited on these substrates at 4000 rpm for 60 s. Finally, a 120 nm thick silver cathode was deposited through a patterned mask under a pressure <10^{-6} mbar giving an active area of 0.08 cm^2.

The phase identification of the films was achieved by X-ray diffractometer (Philips, X'Pert MPD) with Cu Kα radiation operating at 40 kV and 40 mA. The J-V characteristics of the perovskite solar cells were measured using a Keithley 2400 source meter under illumination of 100 mW/cm^2 (ABET TECHNOLOGIES, Sun 2000 Class A). The light intensity was calibrated with a standard silicon solar cell. The excitation and emission spectra of the perovskites were measured by Edinburgh FLS980 Spectrometer.

III. RESULTS AND DISCUSSION

To determine the phase, XRD patterns of the resulting perovskites are shown in Fig. 2. A perovskite phase in the lead salt can be observed when the MABr ratio reached 0.15 in Fig. 2A. In Fig. 2B, the diffraction peaks at 14.02°, 19.98°, 23.38°, 24.42°, 28.38°, 31.83°, 34.92°, 40.53°, and 43.08° correspond to the reflections from (110), (112), (211), (202), (220), (310), (312), (224), and (314) crystal planes of the tetragonal phase of $MAPbI_3$.[10] The peak at 12.6° is associated with the (100) crystal plane of PbI_2. The advantage of our spray assisted two-step deposition method is that PbI_2 was fully converted to $MAPbI_3$ [11] and this could be due to the optimized spraying time of MAI. However, strong peaks of PbI_2 were observed in Fig. 2B when the MABr ratio reached 0.25. As two phases were shown in the films spin-coated from MABr and PbI_2 mixed precursor solution (a newly formed $MAPbI_{3-x}Br_x$ phase and an unreacted PbI_2 phase) in Fig. 2A, we believe distinct bromine-rich and iodine-rich regions were formed after reaction of MAI solution with these two phases.

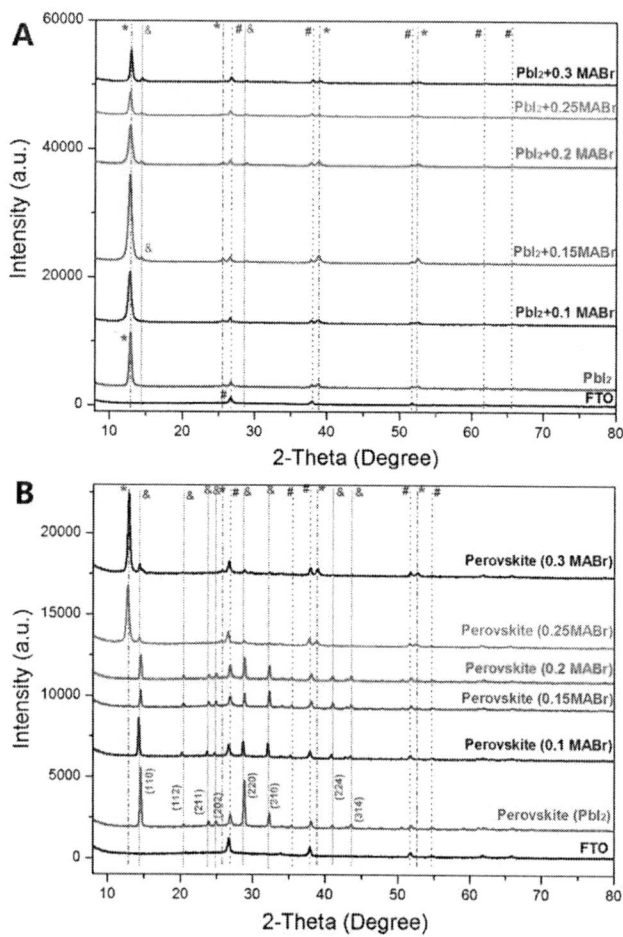

Fig. 2. X-ray diffraction (XRD) patterns of (A) $PbI_2 + x$MABr (x=0, 0.1, 0.15, 0.2 0.25 and 0.3) and (B) resulting perovskite are denoted as "Perovskite (xMABr)". Peaks from FTO substrates (JCPDS no.01-072-1147) are labelled with #, peaks from PbI_2 (JCPDS no.73-1754) are labelled with *, and peaks from the perovskite are labelled with &.

To further investigate the influence of adding MABr in PbI_2 on the solar cells performance, the PCEs of the resulting perovskite solar cells were shown in Fig. 3. In ref.[6], the $MAPbI_{3-x}Br_x$ was prepared by one-step deposition method, the PCEs initially increased and then decreased with Br addition. A similar trend of PCEs was also shown in ref.[9], where $MAPbI_{3-x}Br_x$ was prepared by halide exchange of $MAPbI_3$ in MABr solution. However, in ref. [7], the PCEs decreased at the beginning of incorporating Br by converting PbI_2 in MAI+MABr solution. Our results in Fig. 3A are more in accord with in ref. [7]. However, after one night storage in a sealed box with desiccant, the PCEs of the perovskite solar cell prepared from PbI_2+0.15MABr showed an obvious increase. We wonder if our results here can provide clues for these reports. Previously, several well-known factors which could enhance the PCE of perovskite, such as oxidation of hole transporting materials [12] and light soaking effect [13], have been reported. As the samples were kept in dark overnight, the light soaking effect from the previous tests should be negligible [13].

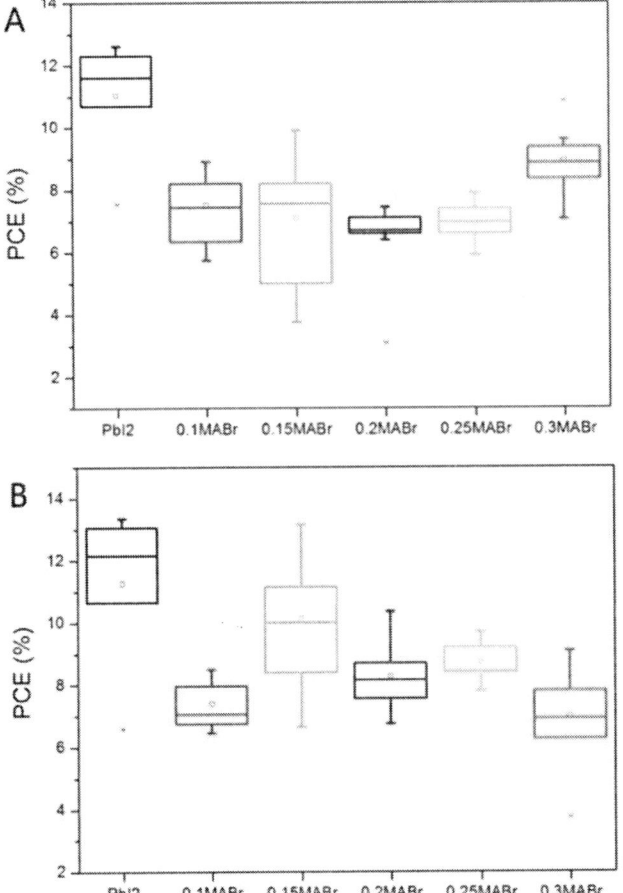

Fig. 3. Power conversion efficiency (PCE) of the perovskite solar cells based on "Perovskite (xMABr)" measured as prepared (A) and one day after (B).

The cyclic current-density/voltage (*J-V*) characteristics (forward and reverse scan directions) for the perovskite solar cells under under 100 mW cm^{-2} equivalent AM 1.5 illumination are shown in Fig. 4. While the PCE of the best non-MABr cell increased from 12.3% to 13.0% via forward scan (short-circuit to forward bias) and from 11.7% to 13.7% via backward scan (forward bias to short circuit) after one night storage, the PCE of the best 0.15MABr cell increased from 10.3% to 14.5% via forward scan and from 9.5% to 11.8% via backward scan. Thus, the average PCEs increased from 12.0% to 13.35% for the best non-MABr cell and from 9.9% to 13.15% for the best 0.15MABr cell after one night storage.

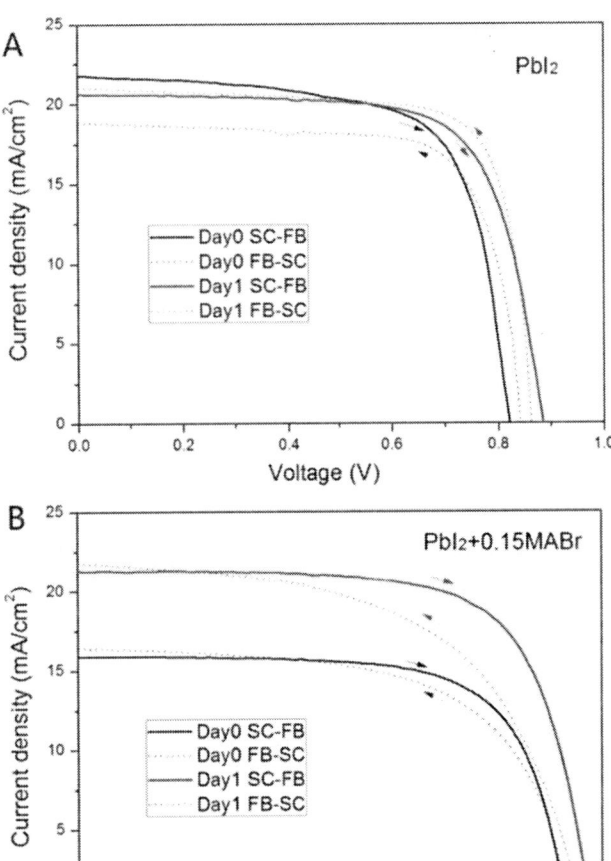

Fig. 4. Short-circuit to forward bias (SC-FB) and forward bias to short circuit (FB-SC) scanning of current density–voltage (J–V) curves of perovskite solar cells prepared from PbI$_2$ (A) and PbI$_2$+0.15MABr (B).

IV. CONCLUSION

In this study, incorporating Br was investigated by adding MABr into PbI$_2$ in the first step. Iodide-rich and bromide-rich regions were observed in the resulting perovskites. Full conversion of PbI$_2$ to perovskite was achieved by a spray-assisted two-step deposition method resulting in a cell with PCE of 13.35%, Jsc of 20.8 mA cm^{-2}, Voc of 0.87 V and FF of 73.55%. While the PCEs decreased after doping, the PCEs of the perovskite solar cell prepared from PbI$_2$+0.15MABr showed a significant increase after one night storage which is due to the diffusion of Br in the as-prepared perovskite thin films.

ACKNOWLEDGEMENT

Hang Zhou would like to acknowledge the Shenzhen Science and Technology Innovation Fund under Grant No. KQCX20140522143114399, and JCYJ20160229122349365.

REFERENCES

[1] T. Jesper Jacobsson, J.-P. Correa-Baena, M. Pazoki, M. Saliba, K. Schenk, M. Gratzel, *et al.*, "Exploration of the compositional space for mixed lead halogen perovskites for high efficiency solar cells," *Energy & Environmental Science,* vol. 9, pp. 1706-1724, 2016.

[2] M. Saliba, S. Orlandi, T. Matsui, S. Aghazada, M. Cavazzini, J.-P. Correa-Baena, *et al.*, "A molecularly engineered hole-transporting material for efficient perovskite solar cells," *Nature Energy,* vol. 1, p. 15017, 2016.

[3] W. S. Yang, J. H. Noh, N. J. Jeon, Y. C. Kim, S. Ryu, J. Seo, *et al.*, "High-performance photovoltaic perovskite layers fabricated through intramolecular exchange," *Science,* vol. 348, pp. 1234-1237, June 12, 2015 2015.

[4] M. Saliba, T. Matsui, J.-Y. Seo, K. Domanski, J.-P. Correa-Baena, M. K. Nazeeruddin, *et al.*, "Cesium-containing triple cation perovskite solar cells: improved stability, reproducibility and high efficiency," *Energy & Environmental Science,* vol. 9, pp. 1989-1997, 2016.

[5] N. J. Jeon, J. H. Noh, W. S. Yang, Y. C. Kim, S. Ryu, J. Seo, *et al.*, "Compositional engineering of perovskite materials for high-performance solar cells," *Nature,* vol. 517, pp. 476-480, 2015.

[6] J. H. Noh, S. H. Im, J. H. Heo, T. N. Mandal, and S. I. Seok, "Chemical Management for Colorful, Efficient, and Stable Inorganic-Organic Hybrid Nanostructured Solar Cells," *Nano Letters,* vol. 13, pp. 1764-1769, Apr 2013.

[7] S. A. Kulkarni, T. Baikie, P. P. Boix, N. Yantara, N. Mathews, and S. Mhaisalkar, "Band-gap tuning of lead halide perovskites using a sequential deposition process," *Journal of Materials Chemistry A,* vol. 2, pp. 9221-9225, 2014.

[8] W. Zhu, C. Bao, F. Li, X. Zhou, J. Yang, T. Yu, *et al.*, "An efficient planar-heterojunction solar cell based on wide-bandgap CH3NH3PbI2.1Br0.9 perovskite film for tandem cell application," *Chemical Communications,* vol. 52, pp. 304-307, 2015.

[9] W. Zhu, C. Bao, F. Li, T. Yu, H. Gao, Y. Yi, *et al.*, "A halide exchange engineering for CH3NH3PbI3−xBrx perovskite solar cells with high performance and stability," *Nano Energy,* vol. 19, pp. 17-26, 2016.

[10] T. Baikie, Y. N. Fang, J. M. Kadro, M. Schreyer, F. X. Wei, S. G. Mhaisalkar, *et al.*, "Synthesis and crystal chemistry of the hybrid perovskite (CH3NH3) PbI3 for solid-state sensitised solar cell applications," *Journal of Materials Chemistry A,* vol. 1, pp. 5628-5641, 2013.

[11] X. Zhonggao, C. Gaoda, Y. Wang, and H. Zhou, "Uniform perovskite photovoltaic thin films via ultrasonic spray assisted deposition method," in *Photovoltaic Specialist Conference (PVSC), 2015 IEEE 42nd,* 2015, pp. 1-4.

[12] Z. Ren, A. Ng, Q. Shen, H. C. Gokkaya, J. Wang, L. Yang, *et al.*, "Thermal Assisted Oxygen Annealing for High Efficiency Planar CH3NH3PbI3 Perovskite Solar Cells," *Sci. Rep.,* vol. 4, p. 6752, 2014.

[13] C. Zhao, B. Chen, X. Qiao, L. Luan, K. Lu, and B. Hu, "Revealing Underlying Processes Involved in Light Soaking Effects and Hysteresis Phenomena in Perovskite Solar Cells," *Advanced Energy Materials,* vol. 5, p. 1500279, 2015.

Modulated structure to maximize the open-circuit voltage with moderate band-gap of small molecule organic solar cells-DFT approach

Saravanan Chinnusamy[1], Amit Munshi[2], Sukanya Santhosh kumar[1], W. S. Sampath[2], Milind S. Dangate[1*]

[1]Department of Chemistry, Amrita School of Arts and Science, Amrita Vishwa Vidyapeetham,

Amrita University, Amritapuri, Kerala-690525, India

[2]Department of Mechanical Engineering, Colorado State University, Fort Collins, CO 80523, USA

ABSTRACT — Pre-synthesis of virtual screening of newly designed four small molecules was approached using Density Functional Theory method. 'Medium donor-vinylene-strong acceptor' concept was used to design four small molecules. Malononitrile is flanked on the donor moiety of benzo[1,2-b:6,5-b']dithiophene-4,5-dione(BDTD) which were blended with vinyl linked benzothiadiazole (BTD) acceptor moiety studied with strong electron withdrawing groups (EWG) such as F, CN, and NO_2 onto the benzothiadiazole unit. DFT and TD-DFT (Time Dependent Density Functional Theory) results of studied small molecules with 1.79eV – 2.06eV band-gap with simulated high open-circuit voltage of 1.38V – 1.49V and predicted power conversion efficiency of designed small molecules is shown using schraber diagram. Detailed investigation of this study is described in this manuscript.

Index Terms — Density functional theory, Organic solar cell, electron-hole distribution.

I. INTRODUCTION

Polymer-fullerene based bulk hetero-junction solar cells have demonstrated about 12% efficiency [1]. Enhancing open-circuit voltage of (BHJs) bulk hetero-junction solar cell is a task to be sort out in future to achieve higher power conversion efficiency (PCE). High open-circuit voltage of conjugated polymers and small molecules ≥1V very rarely reported with moderate power conversion efficiency (PCE).[2]–[4] Recently, small molecules have also demonstrated performance of about 11.3% efficiency with 1.68eV band-gap and open-circuit voltage (V_{OC}) of 0.95V as conjugated polymers.[5] The small molecules doesn't have any solubility issue in wider variety of organic solvents which is main advantage when compared to copolymer.[6] Organic

solar cells are not very stable under ambient operating conditions; however over 3.64% efficiency is demonstrated with considerable life-time of 5600 hours but this is significantly lower as compared to inorganic solar cells.[7] However, there is scope of achieving low to moderate band-gap materials with high open circuit voltage leading to achieve high efficient organic solar cells. Altering donor and acceptor units it can be ease to achieve high open-circuit voltage with conjugated polymers. *Thompson et al* reported an ideal donor conjugated polymer system should have 3.9eV of lowest unoccupied molecular orbital LUMO and 5.4eV of highest unoccupied molecular orbital HOMO; doing this we may be able to evade wasting of energies upon exciton dissociation.[8] High open-circuit voltage would result from lowered HOMO energy level of conjugated system. This would enhance the short-circuit current (Jsc) that may make easy intermolecular hole transport from adequate donor polymer to acceptor fullerene (PCBM).This may lead to improvement in fill-factor (FF).This may make it easy to balance hole and electron mobility between donor and acceptor polymer systems.[9] In inorganic materials such as silicon and CdTe open-circuit voltage approximately equal to the optical band–gap of the material(E_g).[10] Conversely, in an organic conjugated system open-circuit voltage is lower than optical band-gap making it difficult to dissociate an exciton at D-A interface. Narrow band-gap deep low-lying HOMO energy levels are required to enhance the short circuit current and open-circuit voltage. Electron withdrawing group substantially lower the band-gap and HOMO energy levels that also contributes to π ethylene linkage in between donor and acceptor units[11]. It is very important to choose adequate donor, acceptor and electron withdrawing groups to modulate and design the structures to achieve desired performance. Therefore, introducing π

978-1-5090-5606-4/17 $31.00 © 2017 IEEE

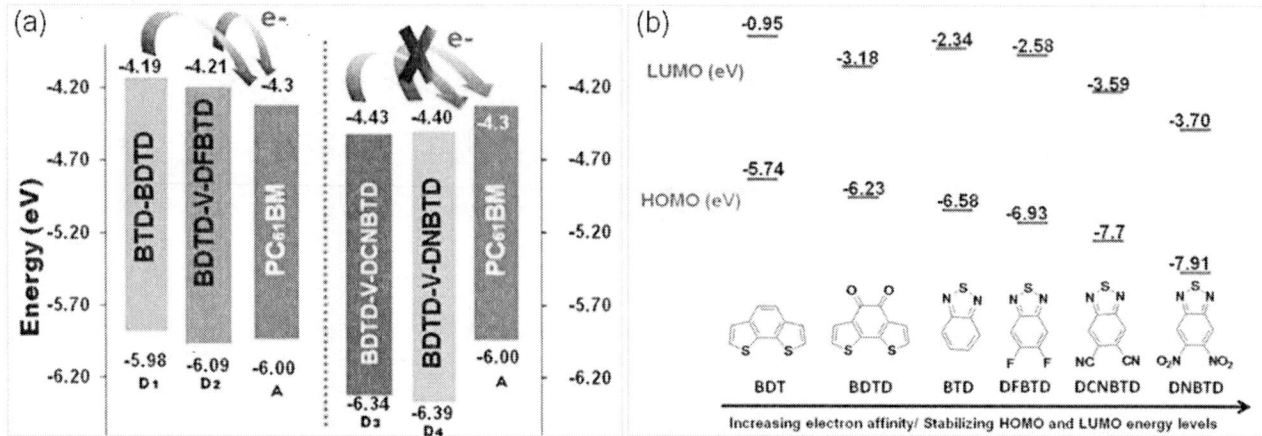

Figure 1. (a) Frontier molecular orbital energy levels of studied small molecules, (b) HOMO and LUMO energy levels of Donor (BDT and BDTD) and acceptor units (BTD-DNBTD).

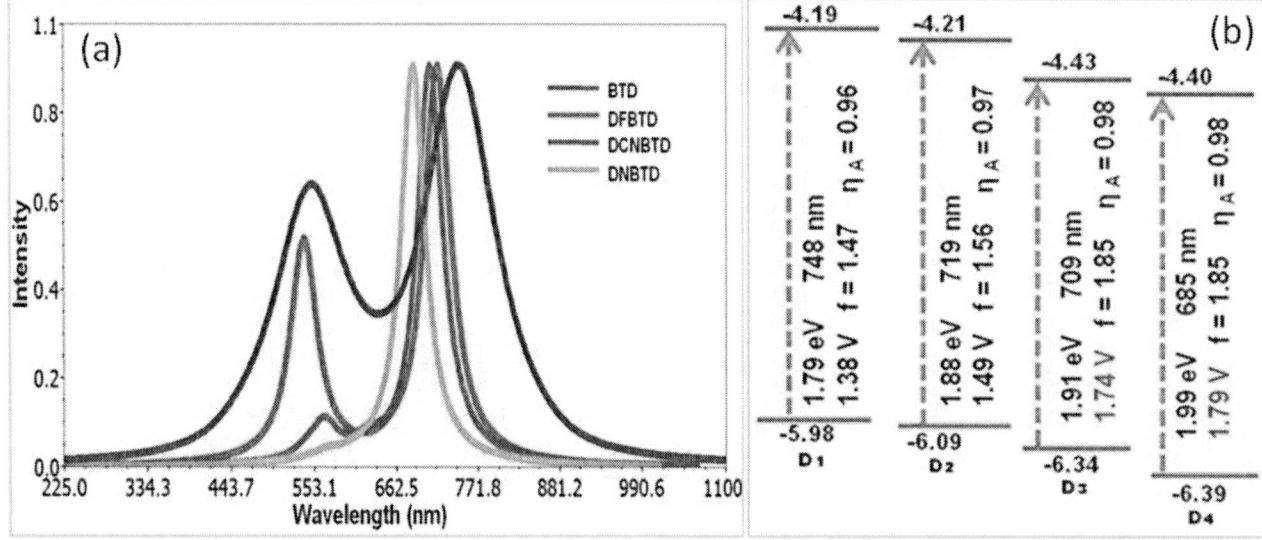

Figure 2. (a) Predicted UV-vis spectra of designed structures (b) Schematic energy levels of studied structures, simulated open-circuit voltage in V, Bang-gap (Eg) in eV, Light absorption co-efficient (f), λ_{max} in (nm) and light absorbtion efficiency (η_A).

extended ethylene linkage between donor and acceptor units, using benzothiadiazole as an acceptor, BDTD (Benzo[1,2-b:6,5-b']dithiophene-4,5-dione) as donor and electron withdrawing groups (EWG) such as F, CN and NO_2 should enhance the band-gap and open-circuit voltage of small molecules. Here, we have applied 'medium donor- π extended conjugated system-strong acceptor' concept to design structure and studied them theoretically with distinct electron withdrawing groups of designed structures. BTD-BDTD based small molecules may be promising candidates for potential materials in fabrication of high performance organic solar cells.

II. COMPUTATIONAL INVESTIGATIONS

All DFT and TD-DFT calculations have been carried out in Gaussian 9.0[12], Multiwfn 3.3.9 [13] and absorption spectra predicted using Gabedit 2.4.8[14] software packages. The ground state geometry of all small molecules were determined using full geometry optimization of its structural parameters with DFT/B3LYP/6-31G* basis set to minimize of molecules energy. Optimized ground state DFT calculations were performed at B3LYP/6-31G* basis set. All TD-DFT calculations were performed under vacuum condition with TD-DFT-PBE0/6-31G*level and without any structural symmetry constraints.[15]

III. RESULTS AND DISCUSSION

The optimized results of structures D_1, D_2 and D_3 show an excellent planar geometry which is preferable to increase the effective conjugation length (ECL) of the chain and to prevent bulk inter-molecular interaction between neighboring stacked π conjugated backbones[16]except D_4, because we assumed that two bulky nitro groups substituted onto benzothiadiazole ring introduce a steric effect that leads to non-planar geometry. Here, we have designed four small molecules based on EWG-D-V-A-V-D-EWG backbone. Strong electron withdrawing groups are flanked on ortho position of the sulfur atom of the thiophene unit which existing in the BDTD unit. Moreover, strong electron withdrawing atom F and groups such as CN, NO_2, and Malononitrile (MN) are important units in the conjugated system would tune both HOMO and LUMO energy levels. DFT study reveals that the frontier energy levels and band-gaps provide information of optical and electronic properties of molecules under study. Here, we have observed HOMO energy levels dramatically lowered from -5.98eV of D_1 to -6.39eV of D_4while LUMO lowered from -4.19eV of D_1 to -4.40eV of D_4. Furthermore, the HOMO and LUMO energy levels are lowered by 0.32eV and 0.25eV respectively. From Figure 1.b shows the electron accepting tendency of electron acceptor with and without EWG groups and importance of electron withdrawing atom and groups that tune the energy levels of conjugate systems. A separate unit of Dinitrobenzothiadiazole possesses -7.91eV and -3.7eV of HOMO and LUMO respectively. These energy levels lower from about 0.35eV to 1.33eV of HOMO and about 0.24eV to 1.36eV of LUMO. Figure 1.b clearly reveals that electron withdrawing atoms/groups are an important tool to tuning the HOMO and LUMO energy levels to some extent. However, the LUMO energy levels of donors should be lower than acceptor conjugated molecules because of the driving force of electron 0.3eV should be maintaining between acceptor and donor units in order to allow electron to jump to acceptor from donor moieties. In our study we have noticed that LUMO of donors D_1 and D_2 are -4.19eV and -4.21eV respectively which are lowered than corresponding acceptor unit of PCBM level of -4.3eV and their predicted driving forces are 0.11eV and 0.09eV respectively. These values are lower than historically believed. Energy gap of 0.3eV is needed for efficient charge separation. By contrast low driving force tends to increase the open-circuit voltage. Efficient charge separation and scant driving force affect the charge separation [17], [18] while external quantum efficiency is reduced by large driving force.[19]Impact of small driving force for charge separation studied by He Yan *et al* concluded that small driving force would reduce the columbic attraction between positive and negative charges which is required for efficient charge separation[19], [20]. Our DFT calculated results are in agreement with these results which are evident from figure 3.b of scharber predicted power conversion efficiency. These show D1 and D2 to have efficiency 9% and 8% (ideally). LUMO of donors D_3 and D_4were -4.43eV and -4.40eV and their driving forces were -0.13eV and -0.10eV respectively.

Former values are lower than PCBM and their calculated driving forces had negative value, thus they would strongly obstruct the charge separation at the donor-acceptor interface due to the scant driving force of <0eV. The open-circuit voltage in the photovoltaic cell is directly related to power conversion efficiency of conjugated system together with short circuit current (Jsc). This is directly dependent on the band-gap of the system. Moreover, the open-circuit voltage is directly related to the driving force of the material at the D-A interface[17]–[20]'[21]; thereby, these all factors decide the magnitude of fill-factor which decide power conversion efficiency of photovoltaic cell. Open-circuit voltage is expressed as.

$$\text{Voc} = \frac{1}{q}|E_{HOMO}^{DONOR}| - |E_{LUMO}^{ACCEPTOR}| - 0.3eV \qquad \rightarrow 1$$

Here, q is elementary charge; E_{HOMO}^{DONOR} is the energy level of donor; $E_{LUMO}^{ACCEPTOR}$ is energy level of acceptor and 0.3 eV is an empirical factor.[22] In theoretical method open circuit voltage can be determined by 1/q (q is the unit charge) times the energy level of highest occupied molecular orbital of the donor subtracted from 4.3eV of LUMO energy level of $PC_{60}BM$ with elementary charge of 0.3V which is an empirical factor. This is the energy difference between the LUMO of the donor and LUMO of acceptor. We have observed that Voc of designed small molecules D_1, D_2, D_3 and D_4 are 1.38eV, 1.49eV, 1.74eV, and 1.79V respectively. Yet open-circuit voltage results of D_3 and D_4 are neglected due to its insufficient driving force. Therefore, it would need acceptor bearing lower-lying LUMO energy levels than small molecule donors D_3 and D_4 in order to cause efficient charge separation. We have noticed that D1 small molecule's HOMO energy level only occurred at higher in energy than those of acceptor of PCBM, and it is energetically favorable for charge separation. By contrast D_2 HOMO energy level is slightly lower than PCBM (by 0.09eV) which is unfavorable for charge separation. The other two small molecules D_3 and D_4, HOMO energy levels were deep lower than PCBM that clearly reveal that it is strongly unfavorable for charge separation. The light absorption range and its intensity play vital role on the short-circuit current (Jsc) for photovoltaic devices. The simulated absorption spectra, oscillator strength (f) and light absorption of four newly designed small molecules are shown in figure 2b. The light absorption efficiency (η_A) is expressed as.

$$\eta_A = 1 - 10^{-f} \qquad \rightarrow 2$$

Here f is the (intensity) oscillator strength of representative absorption wavelength. Strong and broad light absorbing conjugated polymers lead to reduction in thickness of required light absorbing materials, optimizing voltage generation, fill-factor and improving free charge carriers. The electronic transition properties of newly designed small molecules were determined at TD-DFT-PBE0/6-31G* level and the simulated absorption spectra are shown in figure 3b. The predicted

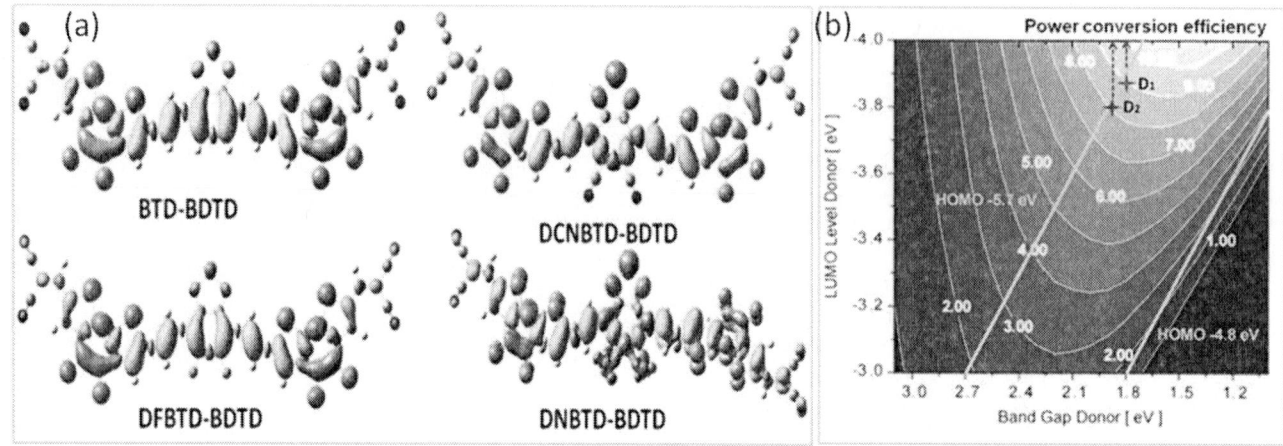

Figure 3.(a) Charge density difference plots of electronic transition for each studied small molecules (dark blue-electron, light blue-hole); (b) Scharber diagram to calculate the approximate the power conversion efficiencies of the studied small molecules

absorption maximum, oscillator strength and light absorption efficiency are depicted in figure 2b. All the scrutinized small molecules have better absorption spectrum and the maximum absorption for D_1, D_2, D_3 and D_4 are 748nm, 719nm, 709nm and 685 nm respectively. For studied small molecules all the systems are identical, and their light absorption efficiency was in the range of 0.96-0.98; therefore, these result revealed that these molecules could cultivate maximum light from the solar spectrum. All the investigated small molecule system were red shifted (>685nm); however, as we go from H to strong electron withdrawing group of NO_2 the absorption range had diminished by about 63nm. Because of the drastically lowered HOMO energy levels and no substantial lowering of LUMO caused the rise of the band gap by about 0.20eV of the system leading to reduce the absorption range (1240Eg).We have

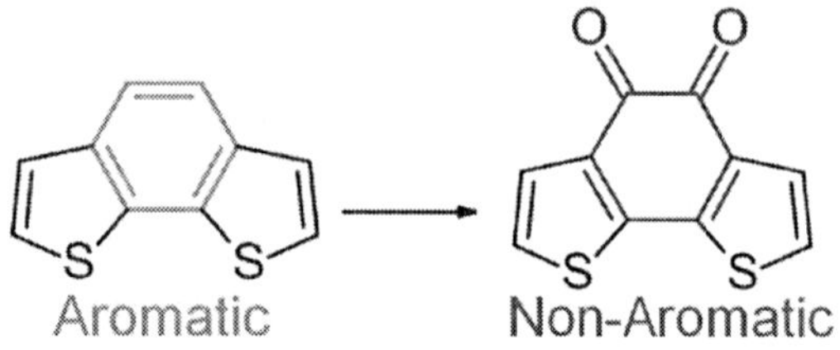

studied DFT calculations for separated units of donors and acceptors in order to get information about how do they behave while blending with different system. From figure 1b and BDTD units with deep lowered LUMO level than BDT; it can be inferred that phenyl group losses its aromatic character and it becomes non-aromatic when introducing two carbonyl groups onto BDT. These also have deep lower LUMO level of -3.18eV than BTD that has-2.34eV. After introduction of F atom onto the BTD, it produced LUMO of -2.58eVwhich is still was higher than BDTD. From figure 3a it can be inferred that the charge density difference plots of BTD-BDTD and DFBTD-BDTD show that the holes (light blue) accumulate on the BDT unit and vinylene linkage; whereas electrons (dark

blue) accumulate on the BDTD unit, especially on the carbonyl groups. This result clearly reveals that the electrons are transferred from BTD and DFBTD to the BDTD system on D1 and D2. Such above results drastically reverse on D3 and D4 that electrons were accumulated on the BDTD and holes were on the DFBTD and DNBTD unit. It clearly reveals that electron transferred from BDTD to DCNBTD and DNBTD units of D3 and D4 respectively. The question arises, how BDTD can behave as both acceptor as well as donor. We assume from the figure1b and figure 3a, LUMO levels of DCNBTD and DNBTD were -3.59eV and -3.7eV which were deep lowered than BDTD of -3.18eV, energy levels lowered about 0.41eV-0.52eV.This difference in LUMO level and lowered LUMO level of DCNBTD and DNBTD than BDTD have the tendency to pull an electron from the BDTD unit which demonstrated that strong electron withdrawing groups CN and NO_2 dramatically stabilized both LUMO and HOMO energy levels that lead to making them as strong electron acceptor than BTD. The band-gap increased by about 0.12eV-0.2eV. From figure 4 the schematic structure of BDT and BDTD shows that before introducing carbonyl group (C=O) onto the BTD, it would be as medium donor because of two electron rich thiophene rings fused with one weak donating benzene ring[23]. Also the LUMO level is higher than corresponding strong acceptor of BDT; whereas after introduced carbonyl group (C=O) onto the BDT its LUMO energy level is substantially lowered by the carbonyl group (C=O) as compared to its prelude. BTD led to phenyl ring lose its aromatic character and it's clearly demonstrated that BDTD could behave both electron acceptor and donor. Intuitively we assumed that the ability of donating and accepting electrons, perhaps due to the structural arrangement of BDTD, two fused thiophene can behave as either strong electron withdrawing unit when it blending with lower LUMO energy levels than itself and the carbonyl group tend to increase the electron accepting tendency when it is LUMO energy level is deep lowered than blending unit in the D-A conjugated system or maybe behave as ambipolar.

Figure 4. Schematic structure of BDT and BDTD

IV. CONCLUSION

DFT investigation is an important tool in engineering the energy levels of organic conjugated polymers that help design desired materials. In this work, we theoretically designed four donor small molecules based on 'Medium donor-vinylene-strong acceptor' backbone in order to enhance the open-circuit voltage and decrease the driving force between LUMO of the donor and acceptor. This would lead to improvement the power conversion efficiency of organic solar cells. Many parameters such as E_{HOMO}, E_{LUMO}, E_{gap}, absorption spectra, Voc, light absorption efficiency and charge density distribution in the excited state have been calculated by DFT and TD-DFT method with different hybrid functions. The calculated results demonstrate that small molecule D_1 shows better performance among four small molecules studied here with a band-gap of 1.79eV and Voc of 1.39 V with the predicted PCE 9% by schraber method. We consider that electron withdrawing group substituted small molecules are highly promising candidates because of the tuned HOMO and LUMO energy levels. Small molecule of D_1 unit BDTD has deep lowered LUMO than well-known strong acceptor of benzothiadiazole. Conceptually it must be an acceptor, yet it provided good outcome than others. These results reveal that benzo[1,2-b:6,5-b']dithiophene-4,5-dione(BDTD) may behave like ambipolar materials and its application may not be limited to organic electronics but also for inorganic solar cells to improve the charge mobility.

ACKNOWLEDGEMENT

This work was supported by the Department of science and technology (No-CS-0722013), Government of India.

REFERENCES

[1] J. Zhao et al., "Efficient organic solar cells processed from hydrocarbon solvents," Nature Energy, vol. 1, no. 2, p. 15027, 2016.

[2] C. L. Chochos et al., "Rational Design of High-Performance Wide-Bandgap (≈2 eV) Polymer Semiconductors as Electron Donors in Organic Photovoltaics Exhibiting High Open Circuit Voltages (≈1 V)," Macromolecular Rapid Communications, vol. 201600614. p. 1600614, 2016.

[3] M. Dolores Perez, C. Borek, S. R. Forrest, and M. E. Thompson, "Molecular and morphological influences on the open circuit voltages of organic photovoltaic devices," Journal of the American Chemical Society, vol. 131, no. 26, pp. 9281–9286, 2009.

[4] D. Dang, P. Zhou, linrui Duan, X. Bao, R. Yang, and W. Zhu, "An efficient method to achieve the balanced open circuit voltage and short circuit current density in polymers

solar cells," J. Mater. Chem. A, vol. 4, pp. 8291–8297, 2016.

[5] D. Deng et al., "Fluorination-enabled optimal morphology leads to over 11% efficiency for inverted small-molecule organic solar cells," Nature Communications, vol. 7, pp. 1–9, 2016.

[6] J. Min et al., "Solubility Based Identification of Green Solvents for Small Molecule Organic Solar Cells Solubility Based Identifi cation of Green Solvents for Small Molecule Organic Solar Cells," no. March, 2014.

[7] R. Cheacharoen et al., "Solar Energy Materials & Solar Cells Assessing the stability of high performance solution processed small molecule solar cells," Solar Energy Materials and Solar Cells, vol. 161, no. December 2016, pp. 368–376, 2017.

[8] B. C. Thompson and J. M. J. Fréchet, "Polymer-fullerene composite solar cells," Angewandte Chemie - International Edition, vol. 47, no. 1, pp. 58–77, 2008.

[9] C. Liu, K. Wang, X. Gong, and A. J. Heeger, "Low bandgap semiconducting polymers for polymeric photovoltaics," Chem. Soc. Rev., vol. 45, p. , 2015.

[10] H. Wang et al., "Multifunctional TiO2 nanowires-modified nanoparticles bilayer film for 3D dye-sensitized solar cells," Optoelectronics and Advanced Materials, Rapid Communications, vol. 4, no. 8, pp. 1166–1169, 2010.

[11] F. Jr, "DFT and TD-DFT study on the structural and optoelectronic characteristics of chemically modified donor-acceptor conjugated oligomers for organic polymer solar cells," no. August, 2016.

[12] M. J. Frisch et al., "Gaussian 09 software," Gaussian Inc., Wallingford, 2009.

[13] T. Lu and F. Chen, "Multiwfn: A multifunctional wavefunction analyzer," Journal of Computational Chemistry, vol. 33, no. 5, pp. 580–592, 2012.

[14] A. Allouche, "Gabedit—a graphical user interface for computational chemistry softwares," Journal of computational chemistry, vol. 32, no. 1, pp. 174–182, 2011.

[15] B. Hess, "P-LINCS : A Parallel Linear Constraint Solver for Molecular Simulation," pp. 116–122, 2008.

[16] G. Conboy et al., "SI: To bend or not to bend – are heteroatom interactions within conjugated molecules effective in dictating conformation and planarity?," Mater. Horiz., vol. 3, no. 4, pp. 333–339, 2016.

[17] D. Veldman, S. C. J. Meskers, and R. A. J. Janssen, "The energy of charge-transfer states in electron donor-acceptor blends: insight into the energy losses in organic solar cells," Advanced Functional Materials, vol. 19, no. 12, pp. 1939–1948, 2009.

[18] K. Vandewal et al., "Quantification of Quantum Efficiency and Energy Losses in Low Bandgap Polymer:Fullerene Solar Cells with High Open-Circuit Voltage," Advanced Functional Materials, vol. 22, no. 16, pp. 3480–3490, 2012.

[19] J. Liu et al., "Fast charge separation in a non-fullerene organic solar cell with a small driving force," Nature Energy, vol. 1, no. 9, p. 16089, 2016.

[20] Z. Ma et al., "Structure–property relationships of oligothiophene–isoindigo polymers for efficient bulk-heterojunction solar cells," Energy & Environmental Science, vol. 7, no. 1, pp. 361–369, 2014.

[21] A. Fradon, E. Cloutet, G. Hadziioannou, C. Brochon, and F. Castet, "Optical properties of donor–acceptor conjugated copolymers: A computational study," Chemical Physics

Letters, vol. 678, pp. 9–16, 2017.

[22] M. C. Scharber *et al.*, "Design rules for donors in bulk-heterojunction solar cells - Towards 10 % energy-conversion efficiency," *Advanced Materials*, vol. 18, no. 6, pp. 789–794, 2006.

[23] J. Huang and H. Huang, *Introduction to Organic Solar Cells*. 2014.

Perovskite Grain Size modulation by annealing in Methyl-Amine Environment

Arun Singh Chouhan[1], Naga Prathibha Jasti[1], Shreyash Hadke[2], Srinivasan Raghavan[1] and Sushobhan Avasthi[1]

[1] Centre for Nanoscience and Engineering, Indian Institute of Science, Bangalore, Karnataka, India, 560012

[2] Energy Research Institute, Nanyang Technological University, Singapore, 639798

Abstract — **Here we present a novel vapor-annealing technique to improve morphology of pure-iodide perovskite ($CH_3NH_3PbI_3$) thin-films and get pin-hole-free coverage with > 10μm grain size. X-Ray diffraction (XRD) measurement confirms that the annealed films are phase pure perovskite. The larger grain size leads to a significant increase in the effective recombination lifetime of charge carriers, with the best films showing lifetime of >50 μs. Solar cells fabricated with the large grain perovskite thin film shows significant improvement in short-circuit current density (J_{sc}) and open-circuit voltage (V_{oc}).**

I. INTRODUCTION

Methyl-ammonium lead halide perovskite ($CH_3NH_3PbX_3$, X: I, Cl, Br) have been widely studied as a solar absorber for high efficiency thin-film solar cells [1-2]. However, getting highly uniform and reproducible perovskite devices is still a challenge. Perovskite film morphology plays a vital role on device performance [3-4]. Unfortunately, morphology of typical spin-coated devices depends on so many parameters -- annealing conditions [5], solvent, precursor solution concentration [6], ambient [7], etc.; that it is hard to get consistency.

Here we present a novel vapor-annealing method to consistently grow perovskite thin films with larger grains. The method involves annealing spin-coated perovskite film in methyl-amine (MA) vapors in a custom-designed reactor. Under optimized conditions, the annealing process yields films with >10μm grains, and charge carrier recombination lifetime of >50 μs. Also, the method produces pin-hole free films that are suitable for large-area devices.

II. ANNEALING SETUP AND PROCEDURE

Thin films of $CH_3NH_3PbI_3$ were deposited using spin coating. As-deposited films had needle like structures with undefined grain-size and poor coverage as shown in Fig. 2(a). The morphology of the as-deposited films can be improved by the "MA-healing" process [8] where the perovskite films are exposed to MA vapors for short time (typically ~3s) at room temperature. The resulting films have much improved coverage, but the average grain size is just ~200nm (MA-RT) as shown in Fig. 2(b). One way to improve upon the MA-healing process is to increase the annealing temperature and

time of the MA exposure. The intuition is that high-temperature and longer time will allow more time for grain-growth, leading to better-quality films. Fig. 1(a) shows custom-designed glass reactor constructed for this process. The reactor connects to a source of MA vapor, which in this case is 40% MA solution in water. On the other end, reactor is connected to a pumping setup. For annealing, the as-deposited films are first loaded into the reactor. Next the reactor is pumped and then the valve connecting to the MA source is turned on. The whole setup is then heated to high-temperature somewhere between 80°C and 140°C for 60 mins in an oven. Finally, the reactor is allowed to cool down naturally to room temperature. The temperature cycle for reactor treatment is as shown in Fig. 1(b).

Perovskite thin-films show improvement in grain size and coverage with increase in annealing temperature as shown in Fig. 2(c-f) and best films are obtained at 140°C with an average grain size of 10 μm (MA-140°C), a 20x improvement over MA-RT films. X-Ray Diffraction (XRD) measurements confirm the annealed perovskite films are phase pure. In fact annealed thin-films have sharper and more pronounce peaks, showing improvement in crystalline quality as shown in Fig. 3. Peaks corresponding to impurities like PbI_2 are absent.

(a)

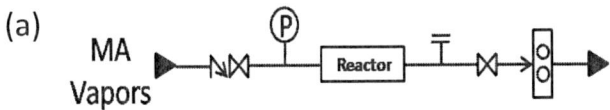

(b)

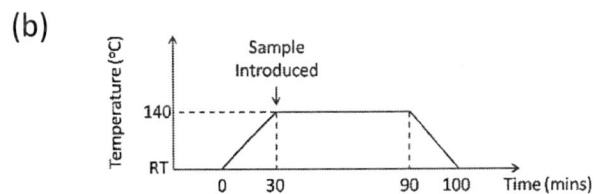

Fig. 1. (a) P & ID diagram of reactor (b) Temperature cycle for the process.

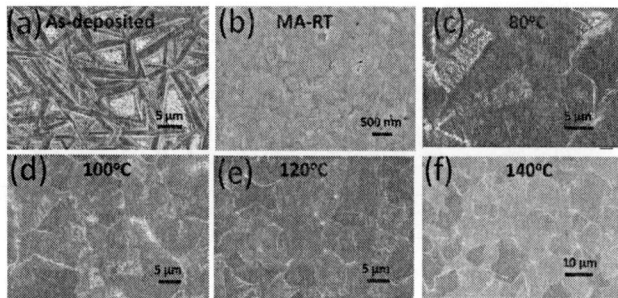

Fig. 2. SEM images (a) As-deposited (b) MA vapor healed (MA-RT). Reactor treated perovskite thin film at (c) 80°C (d) 100°C (e) 120°C (f) 140°C

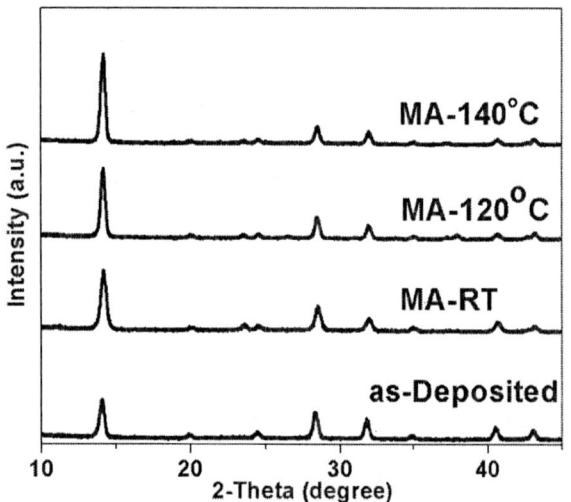

Fig. 3. XRD patterns of As-deposited, MA vapor healed and Reactor treated perovskite thin film at 120°C and 140°C

III. CARRIER LIFETIME MEASUREMENT USING MICROWAVE DETECTED PHOTOCONDUCTIVITY (MDP)

The improved micro-structure of the thin-films should translate into lower recombination losses. To confirm, carrier recombination lifetimes were measured using cavity-enhanced microwave detected photoconductivity (MDP) method. A commercial tool, MDP-Spot by Freiburg Instruments was used for this purpose [9]. Under AM 1.5 conditions, recombination processes in a defective material like perovskite is dominated by SRH-type recombination mechanism. The recombination can either occur in the bulk or at the grain-boundaries. In photoconductive decay an "effective" carrier lifetime (τ_{eff}) is measured which is the contribution from both the processes. MDP measurements were done at a laser excitation power of 180 mW. The probing spot was ~ 1 mm². Fig. 4 shows the normalized photoconductive decay of perovskite films with exponential characteristics. Compared to as-deposited and MA-RT films, perovskite film with large-grains have slower

photoconductive decay, confirming the improvement in carrier recombination lifetime. Quantitatively, the lifetime measured on MA-140°C thin film is 54 µs, which is 155% and 56% higher than the As-deposited and 'MA-healed' (MA–RT) films, respectively.

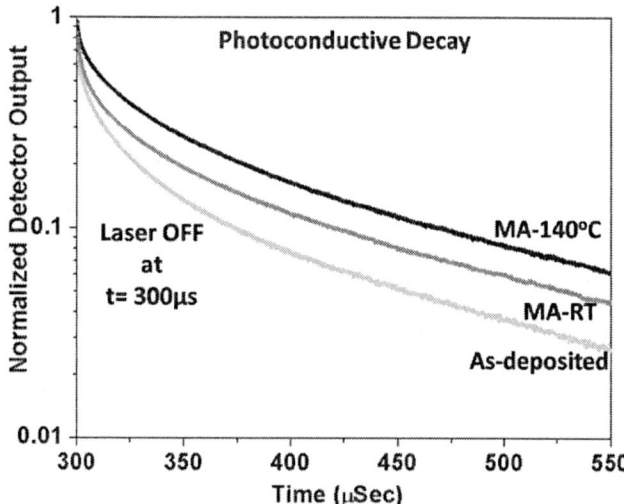

Fig. 4. Normalized photoconductive decay measured by the MDP decay measurement. The decay is modelled as an exponential, characterized by a recombination lifetime (τ_{eff}).

IV. ELECTRICAL MEASUREMENTS

Pure-iodide perovskite devices were fabricated to observe the effect of improved morphology on device performance. Devices were fabricated with perovskite thin films of 200 nm grains (MA-RT) and 10 µm grains (MA-140°C). Device architecture used was the typical n-i-p structure as shown in Fig. 5(a).

Fig. 5(b) shows typical J-V characteristics obtained under AM1.5 illumination. The annealed films with large grains showed an improvement in both J_{sc} and V_{oc} by 2.11 mA/cm² and 60mV respectively as mentioned in Table I. This improvement can be directly attributed to the increase in τ_{eff} of the perovskite films: from ~ 34 µs for films with 200 nm grains to ~54 µs for films with 10 µm grains.

TABLE I

PERFORMANCE DATA FOR MA-RT AND MA-140°C DEVICES

Device	V_{oc} (V)	J_{sc} (mA/cm²)	F.F. (%)	η (%)
MA-RT	0.89	18.14	53.6	8.6
MA-140°C	0.95	20.25	57.1	11.0

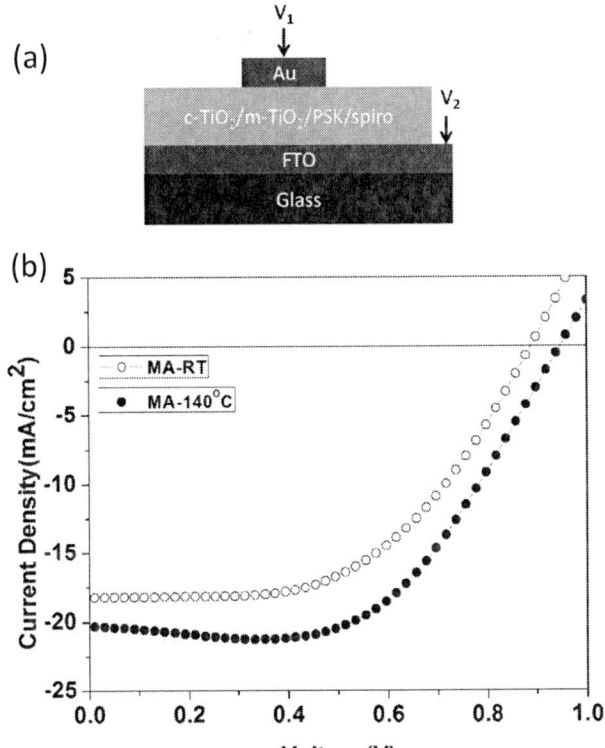

(a)

(b)

Fig. 5. (a) Device stack (b) J-V characteristics comparison between MA vapor healed (MA-RT) and Reactor treated perovskite thin film (MA-140°C) devices.

V. SUMMARY

A Methyl-amine vapor-annealing method is demonstrated that consistently yields perovskite thin-films with large grains up-to 10 μm. The 20x increase in grain size also improved the carrier recombination lifetime, with best films having effective lifetime of >50 μs, a 155% improvement over as-deposited thin films. Devices fabricated with perovskite films having large grains (MA-140°C) shows improved V_{oc} and J_{sc} as compared to small grains (MA-RT).

ACKNOWLEDGEMENT

This work is supported in part under the U.S.–India Partnership to Advance Clean Energy-Research (PACE-R) for the Solar Energy Research Institute for India and the United States (SERIIUS), funded jointly by the U.S. Department of Energy (Office of Science, Office of Basic Energy Sciences, and Energy Efficiency and Renewable Energy, Solar Energy Technology Program, under Subcontract DE-AC36-08GO28308 to the National Renewable Energy Laboratory, Golden, Colorado) and the Government of India, through the Department of Science and Technology under Subcontract IUSSTF/JCERDC-SERIIUS/2012 dated 22nd Nov. 2012. The work is also supported by Department of Science and Technology (DST), Government of India, under project reference no: SB/S3/EECE/0163/2014. Authors are also grateful to Praveen Ramamurthy, Rajeev Ranjan and Manish Jain, Indian Institute of Science (IISc) for helpful discussions.

REFERENCES

[1]   J.-P. Correa-Baena et al., "Unbroken Perovskite: Interplay of Morphology, Electro-optical Properties, and Ionic Movement," Adv. Mater., vol. 28, no. 25, pp. 5031–5037, Jul. 2016.

[2]   D. Bi et al., "Efficient luminescent solar cells based on tailored mixed-cation perovskites," Sci. Adv., vol. 2, no. 1, 2016.

[3]   Y. Zhao et al., "Organic–inorganic hybrid lead halide perovskites for optoelectronic and electronic applications," Chem. Soc. Rev., vol. 45, no. 3, pp. 655–689, 2016.

[4]   P. Fan et al., "High-performance perovskite CH3NH3PbI3 thin films for solar cells prepared by single-source physical vapour deposition," Sci. Rep., vol. 6, no. 1, p. 29910, Sep. 2016.

[5]   W. Nie et al., "High-efficiency solution-processed perovskite solar cells with millimeter-scale grains," Science (80-.)., vol. 347, no. 6221, 2015.

[6]   A. Sharenko and M. F. Toney, "Relationships between Lead Halide Perovskite Thin-Film Fabrication, Morphology, and Performance in Solar Cells," J. Am. Chem. Soc., vol. 138, no. 2, pp. 463–470, Jan. 2016.

[7]   W. Huang, J. S. Manser, P. V. Kamat, and S. Ptasinska, "Evolution of Chemical Composition, Morphology, and Photovoltaic Efficiency of CH3NH3PbI3 Perovskite under Ambient Conditions," Chem. Mater., vol. 28, no. 1, pp. 303–311, Jan. 2016.

[8]   Z. Zhou et al., "Methylamine-Gas-Induced Defect-Healing Behavior of CH3NH3PbI3 Thin Films for Perovskite Solar Cells," Angew. Chemie Int. Ed., vol. 54, no. 33, pp. 9705–9709, Aug. 2015.

[9]   R. Guo et al., "Influence of deep level defects on carrier lifetime in CdZnTe:In," J. Appl. Phys., vol. 117, no. 9, p. 94502, Mar. 2015.

Fe_2O_3 as an Electron Transport Material for Organo-Metal Halide Perovskite Solar Cells

Dallas Fisher, Pravakar P. Rajbhandari, and Tara P. Dhakal[1]

Department of Electrical and Computer Engineering, and Center for Autonomous Solar Power (CASP),
Binghamton University, Binghamton, NY 13902, USA

Abstract — **Sputtered iron (III) oxide (Fe_2O_3) was used as an electron transporting material for methylammonium lead iodide perovskite. The fill factor and efficiency of the solar cell improved as the layer thickness increased from 17 to 43 nm. The performance dropped when increasing the thickness to 67 nm due to parasitic absorption of the iron oxide. Open circuit voltage of the devices was 0.2-0.4 V below record efficiency cells, which is attributed to large work function of the iron oxide layer. However, a thin layer of PCBM between the Fe_2O_3 and the perovskite layers improved the open circuit voltage. Introduction of dopants in the iron oxide layer may further result in improved device performance.**

Index Terms — **Iron Oxide (Fe_2O_3), Electron Transport Layer (ETL), Interface, Perovskite.**

I. INTRODUCTION

The rapid growth of organo-metal halide perovskites has attracted considerable attention within the photovoltaic community. Efficiencies of perovskite devices has improved rapidly from 3.8% in 2009, to 21.0% as of 2016 [1], [2]. These devices are quickly approaching the theoretical maximum power conversion efficiency (PCE), but lack the required long-term stability for deployment. Instability of the device from water and humidity exposure is well documented [3], however the interfacial layers also play a critical role in the absorber stability. The most common transparent electron transport layer (ETL) for perovskite applications is titanium dioxide (TiO_2). It has been shown that that TiO_2 experiences photocatalytic activity under ultraviolet (UV) illumination causing degradation of the perovskite. Under continuous illumination titanium dioxide can oxidize two iodide (I^-) anions to produce elemental iodide (I_2) and 2 free electrons ($2e^-$) [3]. The electrons and iodine react with the perovskite lattice to produce volatile hydrogen iodide (HI) and methylamine (CH_3NH_2) gases [3]. Titanium dioxide is the most popular ETL owing to its transparent nature, however other ETLs such as ZnO have been explored [4], [5]. Iron (III) oxide on the other hand shows significantly lower photocatalytic activity compared to both zinc (II) oxide and TiO_2 [6].

II. EXPERIMENTAL DETAILS

Fluorine doped tin oxide (FTO) substrates were washed sequentially in soap solution, deionized water, acetone, and 2-propanol with sonication. Titanium dioxide was deposited by spraying 50 mM titanium diisopropoxide bis(acetylacetonate) in 2-propanol on the FTO substrate at 450°C. Iron (III) oxide (Fe_2O_3) was grown through rf sputtering of Fe_2O_3 target under argon atmosphere at 10^{-3} torr and 100 W power. The sputtering time was varied from 5, 7, 10, 13, and 20 minutes. The growth rate was about 3.4 nm/min. For each sputtering time a sample with and without phenyl-C61-butyric acid methyl ester (PCBM) was prepared. PCBM is known to passivate trap states near the TiO_2/perovskite interface [7] and is theorized to act similarly with Fe_2O_3. Methylammonium lead iodide ($MAPbI_3$) was synthesized through non-stoichiometric ratio of methylammonium iodide and lead (II) iodide in γ-butryolactone/dimethyl sulfoxide solvent (10% excess lead iodide). Toluene was cast onto the spinning substrate to improve the film morphology. The samples were annealed at 130 °C for 10 minutes to promote crystallization. Spiro-OMeTAD (N2, N2, N2′, N2′ , N7 , N7 , N7′ , N7′ -octakis(4-methoxyphenyl)-9, 9′-spirobi[9H-fluorene]-2, 2′ ,7 ,7′-tetramine) was dissolved as 55 mg/mL in chlorobenzene along with 14 μL of lithium stock solution (520 mg lithium bis(trifluoromethylsulfonyl)imide in acetonitrile) and 22 μL *tert*-butylpyridine. Samples were placed in dry air with desiccant overnight. 100 nm of gold was evaporated on the samples through a shadow mask at ~10^{-6} Torr. Samples were tested under a calibrated solar simulator with a light aperture mask to define the active area.

III. RESULTS AND DISCUSSION

The titanium dioxide (40 nm) control sample shows remarkable improvement in the voltage, current, and fill factor with PCBM (see table I). PCBM plays an integral role in the passivation of traps at the ETL/perovskite interface. This resulted in reduced interfacial recombination, and improved fill factor for the device with TiO_2 as the ETL layer. Similarly, with the iron oxide samples all but the 7-minute sample showed improvement in overall device performance with the PCBM interlayer (Table I, and Fig. 1). The improvement was mainly observed in the open circuit voltage (V_{oc}) of the devices.

For devices prepared without PCBM, Fe_2O_3 devices performed better than the TiO_2 devices. Although, Fe_2O_3 device efficiencies improved with PCBM, the increase was not as high as it was for TiO_2 devices. For TiO_2 devices, V_{oc}, short-circuit current density (J_{sc}), and fill factor (FF) all improved, but the increase in FF and the J_{sc} was significant.

---

[1] Corresponding author e-mail: tdhakal@binghamton.edu

TABLE I
BEST TiO$_2$ AND Fe$_2$O$_3$ CELL PERFORMANCES

ETL Materials		W/O PCBM	With PCBM	Device properties
TiO$_2$		0.98	0.99	V_{oc}
		9.80	15.99	J_{sc}
		0.43	0.62	FF
		4.11	9.76	η
		0.16	0.16	Area (cm^2)
Fe$_2$O$_3$	5 min	0.64	0.73	V_{oc}
		18.02	17.67	J_{sc}
		0.51	0.52	FF
		5.84	6.71	η
		0.09	0.09	Area (cm^2)
	7 min	0.72	0.74	V_{oc}
		19.71	17.33	J_{sc}
		0.66	0.52	FF
		9.34	6.66	η
		0.04	0.16	Area (cm^2)
	10 min	0.64	0.82	V_{oc}
		17.40	19.08	J_{sc}
		0.49	0.51	FF
		5.46	7.96	η
		0.09	0.16	Area (cm^2)
	13 min	0.69	0.84	V_{oc}
		18.65	17.34	J_{sc}
		0.60	0.54	FF
		7.71	7.90	η
		0.09	0.04	Area (cm^2)
	20 min	0.57	0.79	V_{oc}
		15.72	14.42	J_{sc}
		0.58	0.50	FF
		5.21	5.71	η
		0.16	0.16	Area (cm^2)

In most cases of Fe$_2$O$_3$ devices, the short circuit current density appeared to be lowered when PCBM was used. This was due to the absorption in the PCBM layer. In addition, as the Fe$_2$O$_3$ thickness increased, the devices (for both with and without PCBM cases) showed a marked decrease in current density due to light absorbance in the Fe$_2$O$_3$ layer.

Fig. 1 shows current-voltage (IV) characteristics of the Fe$_2$O$_3$ devices with and without PCBM interlayer. Except an outlier for 7 min Fe$_2$O$_3$ layer, the open circuit voltage clearly improves with PCBM layer. On the other hand, current density and fill factor decreased with PCBM layer. The slight decrease in fill factor could due to the loss in current density caused by the absorption of light in the PCBM layer.

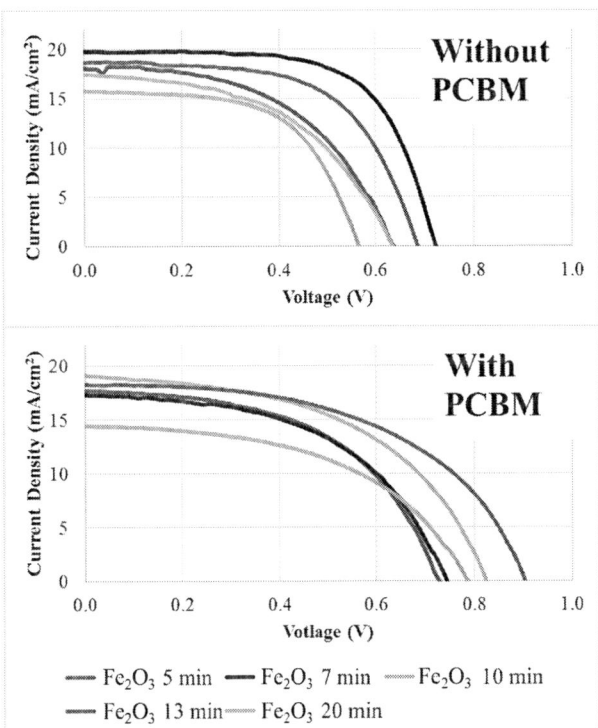

Fig. 1. Current-voltage characteristics of each best Fe$_2$O$_3$ cell with and without PCBM.

External quantum efficiency (EQE) measurements reveal that the 10, 13 and 20 minute Fe$_2$O$_3$ samples have reduced efficiency in the 300-500 nm region due to absorption of Fe$_2$O$_3$ (Fig. 2). Despite the parasitic absorption, device performance improves from the 5 to 13 minute samples due to increased voltage and current density. This points to reduced recombination at the Fe$_2$O$_3$/perovskite interface with increasing Fe$_2$O$_3$ thickness. The sputtered films of 5, 7, 10, 13, and 20 minutes were 17 nm, 23 nm, 33 nm, 43, and 67 nm respectively. Reported hole diffusion length in Fe$_2$O$_3$ are 2-4 nm, but up to 20 nm has been recorded [8], [9]. Interestingly, carriers are extracted efficiently from the perovskite even up to 67 nm. The main reduction in efficiency in the thickest film (20 nm) was due to parasitic light absorption.

One of the significant feature of the devices with PCBM is that the quantum efficiency improved for all cases at the higher wavelength side. This shows that PCBM worked well as a hole blocking layer and allowed for charge collection deeper into the perovskite layer.

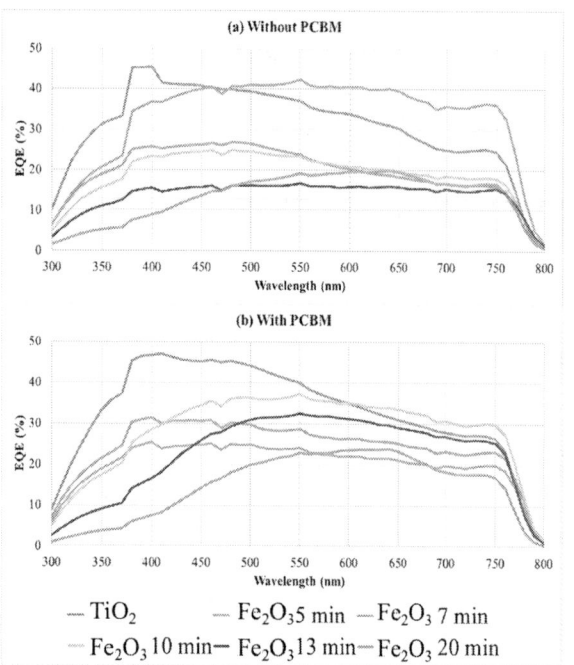

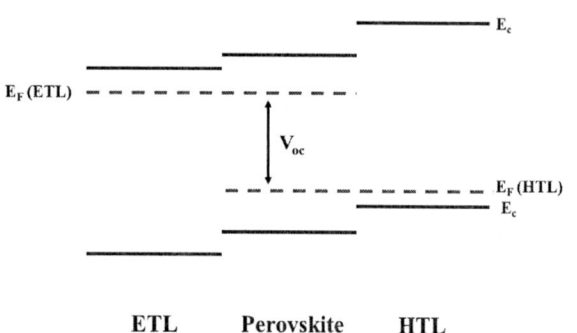

Fig. 2. Absolute external quantum efficiency (EQE). (a) without PCBM interlayer, and (b) with PCBM interlayer.

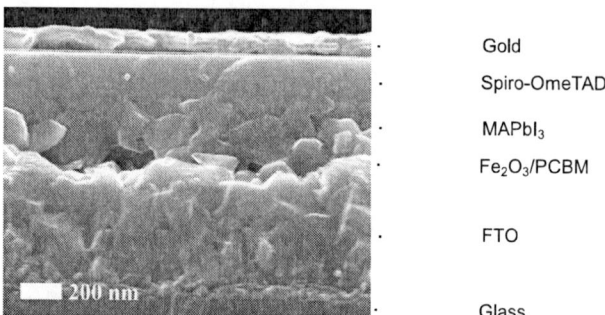

Fig. 3. Band diagram of ETL/perovskite/HTL. Dashed lines represent the Fermi level of ETL/HTL.

Current density of the iron oxide samples is excellent when accounting for parasitic absorption of Fe_2O_3. The open circuit voltage however is severely lacking for methylammonium lead iodide perovskite with Fe_2O_3 ETL. Record efficiency cells show a V_{oc} of 1.1-1.2 V, whereas the Fe_2O_3 cells display 0.6-0.8 V. In the organic n-i-p structure the V_{oc} is indicative of the Fermi levels (work functions) of the interfacial layers (Fig. 3). The hole transporting material (spiro-OMeTAD) is doped with lithium salt to push the Fermi level downward, thereby increasing the V_{oc} and device performance. Due to the low V_{oc} it can be assumed that the work function of the Fe_2O_3 is lower than that of the TiO_2. Doping the Fe_2O_3 could result in higher open circuit voltages and higher device performance. Additionally, a thin insulating layer such as aluminum oxide,

can reduce Columbic interactions at the interface resulting in improved V_{oc} without decrease in current density [10].

Fig. 4 shows a cross sectional image of a 13 min. device. The FTO displays large surface roughness which may play a role in the low fill factor in thin Fe_2O_3 layers due to current shunting. The perovskite layer exhibits grains >150 nm, and is compact (voids are from sample cleavage). The spiro-OMeTAD displays a continuous, pinhole-free, and smooth layer.

Fig. 4. Scanning electron microscope image of the 13 minute Fe_2O_3 sample with PCBM. Visible is gold, spiro-OMeTAD, perovskite, FTO, glass. Fe_2O_3 and PCBM are not visible.

IV. CONCLUSION

Iron (III) oxide (Fe_2O_3) has been demonstrated as an electron transporting material for use in organo-metal halide perovskites. When compared to titanium dioxide, iron oxide shows reduced photocatalytic activity, which may prevent oxidative damage at the ETL/perovskite interface. Iron oxide showed decreased V_{oc}, and fill factor, but high current density. Increasing the Fe_2O_3 thickness from 17 to 43 nm reduced interfacial recombination and improved the overall device performance. Further increasing the thickness to 67 nm resulted in lower current density attributed to parasitic absorption in the 300-500 nm region. PCBM was important to reducing traps at the interface and increase of the device open circuit voltage. Semi-transparent Fe_2O_3 has shown promise as more stable alternative to TiO_2 interfaces for perovskite solar cells. Further optimization must focus on interfacial engineering to improve conductivity of the ETL, and thereby increasing device V_{oc}. Doping of the Fe_2O_3 by co-sputtering or diffusion techniques may result in higher work function of the iron oxide, further increase in device V_{oc}, and ultimately an increase in device efficiency. In addition, the anomaly observed for 7-min device will be investigated in the future.

ACKNOWLEDGEMENT

Scanning Electron Microscope (SEM) images: Advanced Diagnostics Laboratory (ADL), S³IP, Binghamton University, Binghamton, NY

REFERENCES

[1] A. Kojima, K. Teshima, Y. Shirai, and T. Miyasaka, "Organometal Halide Perovskites as Visible-Light Sensitizers for Photovoltaic Cells," no. April, 2009.

[2] Dyesol, "EPFL Achieves 21% Efficiency for Perovskites," 2015.

[3] G. Niu, X. Guo, and L. Wang, "Review of Recent Progress in Chemical Stability of Perovskite Solar Cells," *J. Mater. Chem. A*, vol. 2, p. Advance, 2015.

[4] Z.-L. Tseng, C.-H. Chiang, and C.-G. Wu, "Surface Engineering of ZnO Thin Film for High Efficiency Planar Perovskite Solar Cells.," *Sci. Rep.*, vol. 5, p. 13211, 2015.

[5] W. Qiu, M. Buffière, G. Brammertz, U. W. Paetzold, L. Froyen, P. Heremans, and D. Cheyns, "High efficiency perovskite solar cells using a PCBM/ZnO double electron transport layer and a short air-aging step," *Org. Electron. physics, Mater. Appl.*, vol. 26, pp. 30–35, 2015.

[6] M. Chirita and I. Grozescu, "Fe2O3 – Nanoparticles, Physical Properties and Their Photochemical And Photoelectrochemical Applications," vol. 54, no. 68, pp. 1–8, 2009.

[7] A. H. Ip, L. N. Quan, M. M. Adachi, J. J. McDowell, J. Xu, D. H. Kim, and E. H. Sargent, "A two-step route to planar perovskite cells exhibiting reduced hysteresis," *Appl. Phys. Lett.*, vol. 106, no. 14, p. 143902, 2015.

[8] F. L. Souza, K. P. Lopes, E. Longo, and E. R. Leite, "The influence of the film thickness of nanostructured alpha-Fe2O3 on water photooxidation," *Phys Chem Chem Phys*, vol. 11, no. 8, pp. 1215–1219, 2009.

[9] H.-J. Ahn, M.-J. Kwak, J.-S. Lee, K.-Y. Yoon, and J.-H. Jang, "Nanoporous hematite structures to overcome short diffusion lengths in water splitting," *J. Mater. Chem. A*, vol. 2, no. OCTOBER 2014, pp. 19999–20003, 2014.

[10] Y. Zhong, A. Tada, S. Izawa, K. Hashimoto, and K. Tajima, "Enhancement of VOC without loss of JSC in organic solar cells by modification of donor/acceptor interfaces," *Adv. Energy Mater.*, vol. 4, no. 5, 2014.

Optical Evaluation of Perovskite Films in and for Solar Cell Device Structures

Kiran Ghimire, Dewei Zhao, Changlei Wang, Yanfa Yan, and Nikolas J. Podraza

Wright Center for Photovoltaics Innovation and Commercialization & Department of Physics and Astronomy, University of Toledo, Toledo, Ohio 43606, United States of America

Abstract — Optical characterization, monitoring, and simulations have been applied to study perovskite films and solar cells. Shifts in the absorption band edge to photon energies as low as ~1.2 eV are obtained from the complex dielectric function ($\varepsilon = \varepsilon_1 + i\varepsilon_2$) spectra of solution processed mixed tin (Sn) and lead (Pb) based $(FASnI_3)_{1-x}(MAPbI_3)_x$ films. Real time spectroscopic ellipsometry (RTSE) has monitored vapor deposition of methylammonium lead iodide ($MAPbI_3$) films and evaporation of underlying electron selective C_{60} in the solar cell structure. Spectra in ε serve as input for external quantum efficiency (EQE) simulations for comparison to experimental EQE.

Index Terms — low band gap perovskite, in situ monitoring, EQE simulation, spectroscopic ellipsometry.

I. INTRODUCTION

Organo-lead halide perovskite based materials have become a very important absorbing material in solar cells with efficiency of 22.1 % [1]. Lower band gap perovskites are of interest for incorporation as a bottom solar cell in tandem devices to better exploit more of the solar spectrum. Combining formamidinium tin iodide ($FASnI_3$) and methylammonium lead iodide ($MAPbI_3$) precursor solutions form high-quality $(FASnI_3)_{1-x}(MAPbI_3)_x$ which has already produced a 17.6% efficient solar cell collecting photons with energies as low as ~1.2 eV [2]. $MAPbI_3$ perovskites possess many desirable properties and initial single junction device performance is high, however understanding how to control material structure during deposition and the ultimate limits of device performance are still in progress.

We have studied the optical properties in terms of complex dielectric function ($\varepsilon = \varepsilon_1 + i\varepsilon_2$) spectra of $(FASnI_3)_{1-x}(MAPbI_3)_x$ perovskite films using spectroscopic ellipsometry to observe changes in absorption band edge and higher energy critical points energies (CPs) with composition [3]. The absorption band edge energy has been found to decrease to near ~1.2 eV up to x = 0.40 then increases to near ~1.4 eV as $FASnI_3$ is approached. CPs above the absorption band edge also vary substantially with composition.

Real time spectroscopic ellipsometry (RTSE) has also been applied in the solar cell device structure to study the growth evolution of vapor deposited $MAPbI_3$ and evaporation of an underlying electron selective layer (C_{60}) on commercially available SnO_2:F coated glass. The effective thicknesses of accumulated $MAPbI_3$ perovskite as well as lead iodide (PbI_2) and methylammonium iodide (MAI) components in this device

relevant structure are tracked as functions of time by analysis of RTSE data. This monitoring of $MAPbI_3$ growth in the solar cell structure builds from our previous RTSE studies of growth evolution and post-deposition degradation and decomposition [4]-[6]. Spectra in ε of C_{60}, $MAPbI_3$, Spiro-OMeTAD, and other components are used as input for simulations of external quantum efficiency (EQE) of perovskite solar cells. These measured optical properties, growth evolution of component materials, and EQE simulations of devices are expected to help better understand and optimize the fabrication and design of perovskite solar cells.

II. EXPERIMENTAL DETAILS

$(FASnI_3)_{1-x}(MAPbI_3)_x$ (x = 0.00, 0.20, 0.35, 0.40, 0.60, and 1.00) perovskite films are fabricated by solution processing by combining $FASnI_3$ and $MAPbI_3$ precursors as described in Ref. [7] on native oxide coated crystalline silicon (c-Si) wafers. Spectroscopic ellipsometry measurements are performed on $(FASnI_3)_{1-x}(MAPbI_3)_x$ films over the spectral range 1.00 to 5.89 eV at 60° and 70° angles of incidence as described in Ref. [3].

A vapor deposited $MAPbI_3$ film is grown on the top of thin layer C_{60}, deposited on commercially available SnO_2:F glass. In-situ RTSE measurements are performed during vapor deposition of a $MAPbI_3$ perovskite film as described in Ref. [5] and evaporation of the underlying C_{60} film every 2 sec at 70° angle of incidence over a spectral range from 0.74 to 5 eV. A thicker C_{60} film is also made on native oxide coated c-Si to obtain reference optical property spectra with spectroscopic ellipsometry measurement performed at 70° angle of incidence from 0.74 to 5.89 eV. Multichannel spectroscopic single rotating compensator ellipsometer instruments (M-2000, J. A. Woollam Co.) are used for all measurements.

EQE simulations are performed by accounting for coherent multiple reflections between film and interfacial layers in the complete solar cell structure using a scattering matrix formalism [8]. Input parameters include spectra in ε of each component layer in the device along with associated layer thicknesses. EQE of single junction $MAPbI_3$-based solar cells are measured (PV Measurements Inc. QE System).

978-1-5090-5606-4/17 $31.00 © 2017 IEEE

III. RESULTS AND DISCUSSION

A. Optical response of mixed $(FASnI_3)_{1-x}(MAPbI_3)_x$ perovskite films

A divided spectral range analysis procedure [3]-[6], [9], [10] is applied to extract the structural parameters including $(FASnI_3)_{1-x}(MAPbI_3)_x$ film thicknesses, surface roughness thicknesses, and void fractions in surface layers as assessed using Bruggeman effective medium approximation [11]. After structural parameters are identified, numerical inversion is used to extract ε over the full spectral range.

Fig. 1 shows ε_2 from 1.00 to 5.89 eV and the near absorption band edge absorption coefficient (α) obtained from ε for $(FASnI_3)_{1-x}(MAPbI_3)_x$ perovskite films. Spectra in $\alpha = 4\pi k/\lambda$ where $\varepsilon = N^2 = (n + ik)^2$ and λ is the wavelength of light. Fig. 2 shows the above absorption band edge CP energies obtained by fitting the first derivative of ε_2, $d\varepsilon_2/dE$ as described in [3].

The absorption band edge, defined as the photon energy at which $\alpha = 4000$ cm^{-1} here, decreases from its maximum for MAPbI$_3$ with increasing FASnI$_3$ proportion, reaching a minimum ~1.2 eV for $(FASnI_3)_{0.60}(MAPbI_3)_{0.40}$ as illustrated in Figs. 1 and 3. With further addition of FASnI$_3$ it increases to near ~ 1.4 eV. The absorption band edge energies at $\alpha =$

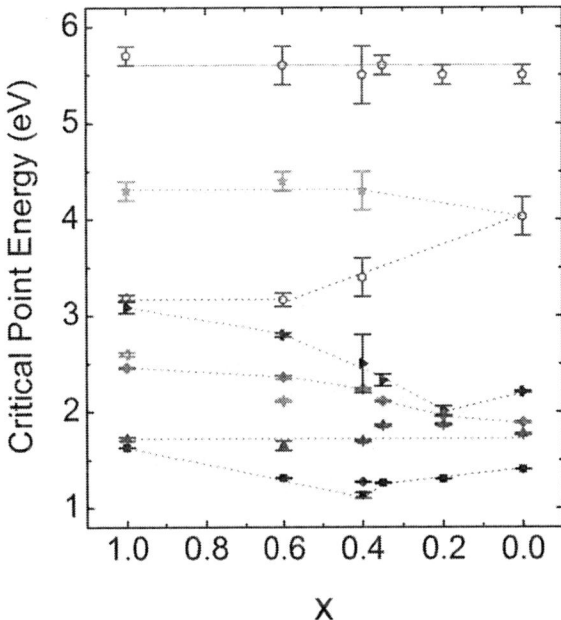

Fig. 2.   Critical point energies (CPs) obtained by fitting $d\varepsilon_2/dE$ of $(FASnI_3)_{1-x}(MAPbI_3)_x$ perovskite films. The dotted line represents the most likely red and blue shifts of the CPs.

4000 cm^{-1} and lowest CP energies obtained from fitting $d\varepsilon_2/dE$ lie close to each other and shift similarly with composition as shown in Fig. 3.

CPs found near 1.70-1.87 eV and 5.3-5.7 eV remain relatively stable for all compositions. The CP at 4.30 eV for MAPbI$_3$ decreases in magnitude with increasing FASnI$_3$ content for x = 0.40 and 0.60 with sensitivity lost for x = 0.35. Relative to CPs for MAPbI3, the CP at 2.46 eV red shifts, the CP at 3.09 eV red shifts until x = 0.20 followed by a blue shift for x = 0.00, and the CP at 3.19 eV blue shifts. The most likely red and blue shifting are represented by dotted lines in Fig. 3. The features at 2.60 eV for MAPbI$_3$ and 2.11 eV for x = 0.60 film are not observed in the other films. Overall, the maximum amplitude CP at 3.09 eV corresponds to a more

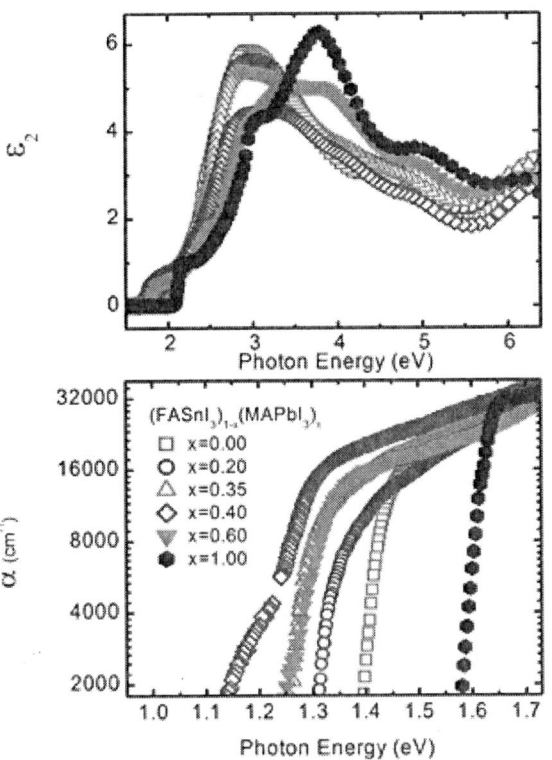

Fig. 1.   (Top panel) Imaginary part of complex dielectric function ($\varepsilon = \varepsilon_1 + i\varepsilon_2$) spectra (top panel) and (Bottom panel) absorption coefficient (α) of $(FASnI_3)_{1-x}(MAPbI_3)_x$ perovskite films.

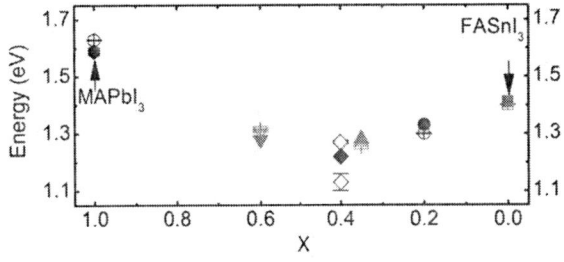

Fig. 3.   The absorption band edge energies at $\alpha = 4000$ cm^{-1} (solid symbols) and lowest CP energies obtained from fitting $d\varepsilon_2/dE$ (open symbols).

MAPbI₃-like material while maximum amplitude feature at 2.21 eV corresponds to material with more FASnI₃ characteristics.

B. RTSE monitoring of C₆₀ electron selective layer deposition

Fig. 4 shows the schematic of a MAPbI₃ perovskite film deposited on top of an underlying evaporated C₆₀ layer on commercially available SnO₂:F coated glass. C₆₀ is first thermally evaporated at 420°C on top of SnO₂:F coated glass with the effective thickness monitored by RTSE is applied to find the effective thickness during deposition as shown in Fig. 5. This effective thickness consists of the sum of all layer thicknesses weighted by the effective C₆₀ material within that layer as assessed by using a Bruggeman effective medium approximation. Reference spectra in N of C₆₀ shown in Fig. 6 is extracted from another thicker film on a specular c-Si wafer, and optical properties for SnO₂:F are determined from in-situ measurement prior to C₆₀ deposition. These reference optical properties are used in analysis of RTSE data so that only structural parameters are fit. The desired thickness ~60 Å is obtained over 7 minutes of deposition. Some deviation of effective thickness from a linear relationship is due highly rough surface of SnO₂:F and some difficulty in discerning small fractions of C₆₀ within the SnO₂:F roughness.

The direct optical band gap of C₆₀ is obtained from extrapolating $\alpha^2 = 0$ using a linear relationship as in Fig. 7. This result yields band gap of 2.35 ± 0.05 eV, which is in good agreement to values reported at 2.2 and 2.3 eV for single crystal [12] and thin film [13] C₆₀.

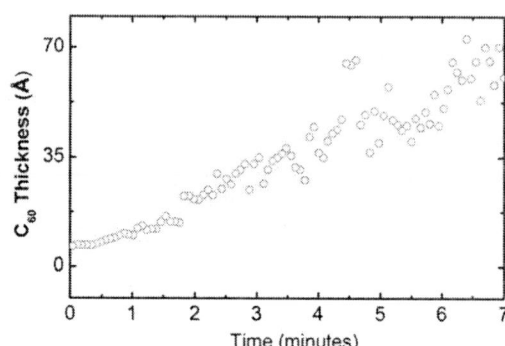

Fig. 5. Thickness evolution of thermally evaporated C₆₀ on commercially available SnO₂:F coated glass obtained from real time spectroscopic ellipsometry (RTSE).

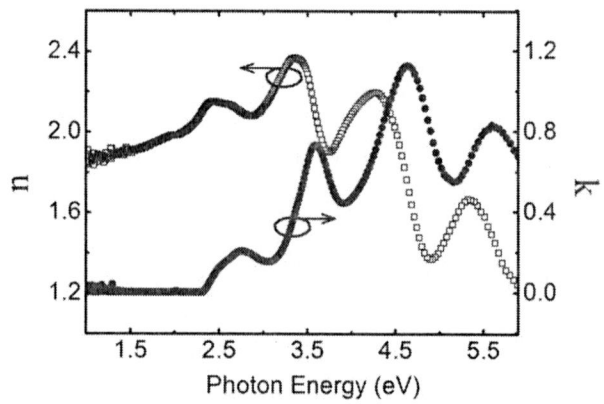

Fig. 6. Complex refractive index ($N = n + ik$) of a thermally evaporated C₆₀ thin film.

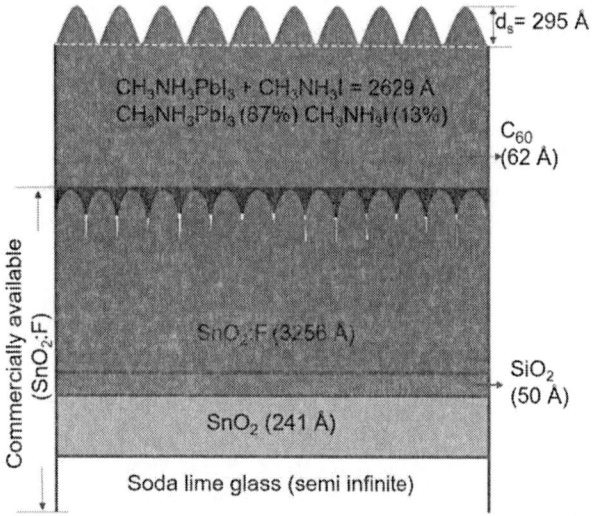

Fig. 4. Schematic of MAPbI₃ perovskite thin film deposited on top of an underlying C₆₀ layer on commercial SnO₂:F coated glass.

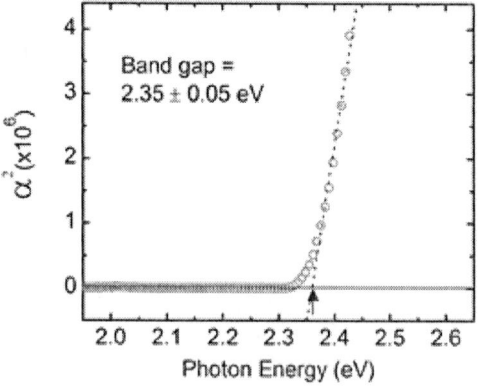

Fig. 7. Direct band gap for C₆₀ thin film.

C. RTSE monitoring of vapor deposited MAPbI₃ in the solar cell device configuration

MAPbI₃ is vapor deposited on top of C₆₀ by using PbI₂ and MAI source materials. Spectra in ε for the vapor deposited film is modeled using Bruggeman effective medium approximation of ε for vapor deposited MAPbI₃, evaporated MAI, and evaporated PbI₂ with variable material fractions [4]. The fluctuation of deposition parameters from its optimized conditions are likely to produce phase segregated or unreacted MAI and PbI₂ materials in addition to the desired MAPbI₃ [5].

Fig. 8. shows effective thicknesses of MAPbI₃ perovskite and PbI₂ and MAI components extracted from in-situ RTSE as functions of time. The growth of the MAPbI₃ in the early stage (0 to 6 minutes) is accompanied with excess unreacted PbI₂. After 6 mins, all unreacted PbI₂ is quickly used up and the unreacted or phase segregated MAI begins to accumulate throughout the remainder of the deposition. The final film is found to be 2460 ± 20 Å thick MAPbI₃ with 0.130 ± 0.006 material fraction residual MAI. Monitoring such behavior as functions of deposition parameters may be used to help achieve better control during solar cell fabrication.

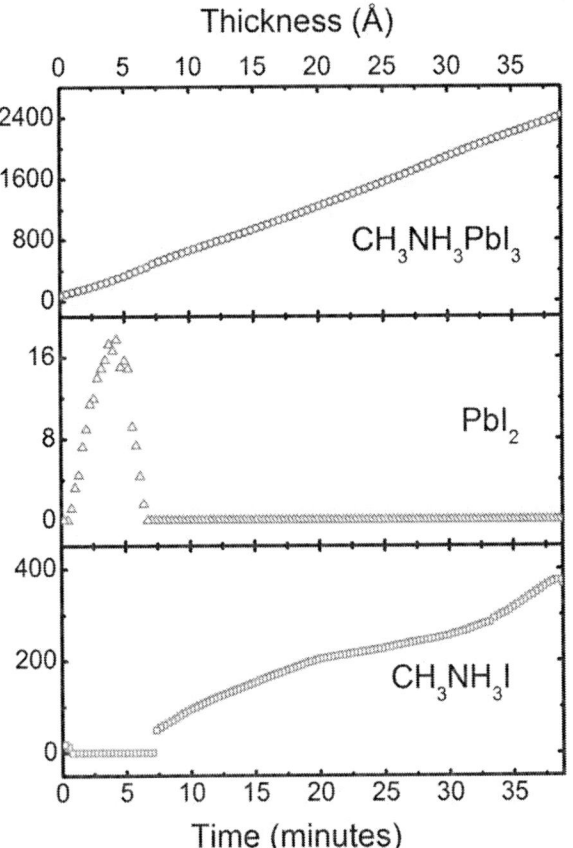

Fig. 8. Growth evolution in the form of effective material thicknesses obtained from RTSE during vapor deposition of MAPbI₃ using PbI₂ and CH₃NH₃I sources.

D. External quantum efficiency (EQE) simulation

The RTSE studies of C₆₀ and MAPbI₃ along with adaptation of previous successful modeling approaches for CdTe solar cells [8] have been used to help design a structural-optical model for EQE simulations of superstrate devices with perovskite absorbers. Fig 9. shows experimental and simulated EQE spectra for a solar cell with a MAPbI₃ absorber deposited on top of a C₆₀ electron selective layer on commercial SnO₂:F coated glass followed by hole selective Spiro-OMeTAD and back contact Au layers. Reference ε of each component in device configuration has been extracted separately from ellipsometric analysis and are used to calculate the structural parameters, i.e. the thickness of each component layers in a through-the-glass spectroscopic ellipsometry measurement. The structural model consists of light incident onto a 2.22 mm soda lime glass superstrate / 411 Å SnO₂ coating / 2565 Å SnO₂:F coating / a 695 Å 92.2% SnO₂:F + 7.8% void interface / 66 Å C₆₀ electron selective layer / 3973 Å perovskite absorber / a 626 Å 34.8% Spiro-OMeTAD (34.8%) + 65.2% perovskite interface / 1764 Å Spiro-OMeTAD / optically opaque Au back contact. Interfacial spectra in ε are represented using Bruggeman effective medium approximation. The thicknesses extracted other than MAPbI₃ layer are fixed and Fig. 9 shows the impact of small variations in the absorber layer thickness on EQE. The experimental EQE is best matched with simulated EQE of 4350 Å MAPbI₃. This thickness is within 10% of that obtained by ellipsometry on another similarly configured sample. The good agreement between simulated and experimental EQE here indicates that the model, as well as structural and optical parameters defining it, are realistic.

Building from the aforementioned model, Fig. 10 shows simulated EQE for devices incorporating 4350 Å thick (FASnI₃)₁₋ₓ(MAPbI₃)ₓ (x = 0.00, 0.20, 0.35, 0.40, 0.60, 1.00)

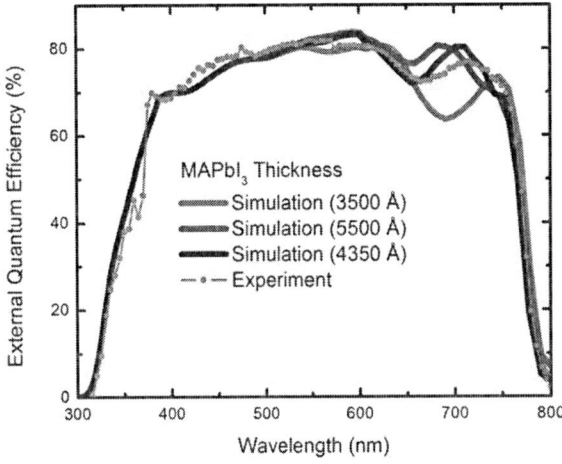

Fig. 9. Simulated external quantum efficiency (EQE) spectra of perovskite solar cells with different thickness MAPbI₃ layers compared to experimental EQE.

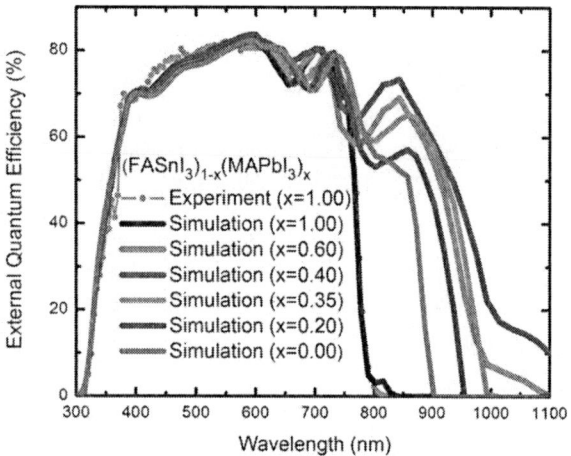

Fig. 10. Simulated EQE spectra of $(FASnI_3)_{1-x}(MAPbI_3)_x$ (x = 0.00, 0.20, 0.35, 0.40, 0.60, and 1.00) perovskite solar cells. Experimental EQE for a device incorporating MAPbI₃ is also shown.

perovskite absorber layers. The current generating layer in the simulation is assumed to be only the $(FASnI_3)_{1-x}(MAPbI_3)_x$ layer. Short circuit current densities (J_{sc}) are calculated from simulated EQE assuming all photo-generated electron-hole pairs are collected. J_{sc} values obtained are 23.08, 24.30, 25.73, 27.76, 25.85, and 19.27 mA/cm² for devices with x = 0.00, 0.20, 0.35, 0.40, 0.60, and 1.00 absorbers, respectively. The J_{sc} = 28.6 mA /cm² reported for a $(FASnI_3)_{0.60}(MAPbI_3)_{0.40}$ based solar cell [2] is close to our calculated value of 27.43 mA/cm². Other J_{sc} values reported here are also in accordance with their absorption edge energy trend. Application of these EQE models may help identify sources of any optical and electrical loss in these devices.

IV. CONCLUSION

Spectra in ε of $(FASnI_3)_{1-x}(MAPbI_3)_x$ (x = 0.00, 0.20, 0.35, 0.40, and 0.60) thin films have been obtained by spectroscopic ellipsometry to yield variations in absorption band edge and CP transitions with composition. A maximum amplitude CP at 3.09 eV indicates a more MAPbI₃-like material while a maximum amplitude CP at 2.21 eV corresponds to more FASnI₃ like material. The lowest energy CP and absorption band edge correspond to each other and show minimum values at x = 0.40. Growth evolution of vapor deposited MAPbI₃ perovskite film and electron selective layer C₆₀ on device structure has been performed using RTSE monitoring. Models successfully simulating EQE of perovskite solar cells have been developed. Simulations of solar cells with $(FASnI_3)_{1-x}(MAPbI_3)_x$ (x = 0.00, 0.20, 0.35, 0.40, 0.60, and 1.00) EQE have been performed and J_{sc} of those devices are calculated.

ACKNOWLEDGEMENT

This work was supported by University of Toledo start-up funds, the Ohio Department of Development (ODOD) Ohio Research Scholar Program (Northwest Ohio Innovators in Thin Film Photovoltaics, Grant No. TECH 09-025), the U.S. Department of Energy SunShot Initiative under the Next Generation Photovoltaics 3 program (DE-FOA-0000990), the National Science Foundation (CHE-1230246 and DMR-1534586), and the Office of Naval Research (Contract No. N00014-17-I-2223).

REFERENCES

[1] M. A. Green, K. Emery, Y. Hishikawa, W. Warta, E. D. Dunlop, D. H. Levi, A. Ho-Baillie, "Solar cell efficiency tables (Version 49)," *Progress in Photovoltaics: Research and Applications*, vol. 25, pp. 3-13, 2017.

[2] D. Zhao, Y. Yu, C. Wang, W. Liao, N. Shrestha, C.R. Grice, A.J. Cimaroli, L. Guan, R.J. Ellingson, K. Zhu, and X. Zhao, "Low-bandgap mixed tin–lead iodide perovskite absorbers with long carrier lifetimes for all-perovskite tandem solar cells," *Nature Energy*, vol. 2, p. 17018, 2017.

[3] K. Ghimire, D. Zhao, Y. Yan, and N. Podraza, " Optical response of mixed methylammonium lead iodide and formamidinium tin iodide perovskite thin films," submitted to *Applied Physics Letters*, 2017.

[4] K. Ghimire, D. Zhao, A. Cimaroli, W. Ke, Y. Yan, and N. Podraza, "Optical monitoring of CH₃NH₃PbI₃ thin films upon atmospheric exposure." *Journal of Physics D: Applied Physics*, 49(40), pp. 405102, 2016.

[5] K. Ghimire, A. Cimaroli, F. Hong, T. Shi, N. Podraza, and Y. Yan. "Spectroscopic ellipsometry studies of CH₃NH₃PbX₃ thin films and their growth evolution." in *42ⁿᵈ IEEE Photovoltaic Specialist Conference*, 2015, pp.1-5.

[6] K. Ghimire, D. Zhao, A. Cimaroli, W. Ke, M. Junda, Y. Yan, N. Podraza, "Optical properties and degradation monitoring of CH₃NH₃PbI₃." in *43ʳᵈ IEEE Photovoltaic Specialist Conference*, 2016, pp.89-94.

[7] W. Liao, D. Zhao, Y. Yu, N. Shrestha, K. Ghimire, C.R. Grice, C. Wang, Y. Xiao, A.J. Cimaroli, R.J. Ellingson, and N.J. Podraza, "Fabrication of efficient low-bandgap perovskite solar cells by combining formamidinium tin iodide with methylammonium lead iodide," *Journal of the American Chemical Society*, 138(38), pp. 12360-12363, 2016.

[8] P. Koirala, J. Li, H. P. Yoon, P. Aryal, S. Marsillac, A. A. Rockett, N. J. Podraza, R. W. Collins, "Through-the-glass spectroscopic ellipsometry for analysis of CdTe thin-film solar cells in the superstrate configuration", *Progress in Photovoltaics: Research and Applications*, vol. 24, pp. 1055-1067, 2016.

[9] L. Karki Gautam, H. Haneef, M. M. Junda, D. B. Saint John, and N. J. Podraza, "Approach for extracting complex dielectric function spectra in weakly-absorbing regions," *Thin Solid Films*, vol. 571, pp. 548-553, 2014.

[10] K. Ghimire, H. F. Haneef, R. W. Collins, and N. J. Podraza, "Optical properties of single crystal Gd₃Ga₅O₁₂ from the infrared to ultraviolet," *Physica status solidi (b)*, vol. 252, pp. 2191-2198, 2015.

[11] H. Fujiwara, J. Koh, P. I. Rovira, and R. W. Collins. "Assessment of effective-medium theories in the analysis of

nucleation and microscopic surface roughness evolution for semiconductor thin films." *Physical Review B*, vol. 61, pp. 10832, 2000.

[12] P. N. Saeta, B. I. Greene, A. R. Kortan, N. Kopylov, and F. A. Thiel. "Optical studies of single-crystal C_{60}." *Chemical physics letters* 190 (3-4), 184-186, 1992.

[13] R. W. Lof, M. A. Van Veenendaal, B. Koopmans, H. T. Jonkman, and G. A. Sawatzky. "Band gap, excitons, and Coulomb interaction in solid C_{60}." *Physical Review Letters* 68, (26), 3924, 1992.

Hybrid Organic-Inorganic Solar Cells with a Benzoquinone Passivating Layer

James Hack, Abhishek Iyer, Meixi Chen, Nicole Kotulak, Akirt Sridharan, Robert Opila

University of Delaware, Newark, DE, 19711, USA

Abstract — Devices were demonstrated making use of benzoquinone passivation of silicon surfaces in a PEDOT:PSS-Silicon hybrid device structure. It is shown that the insertion of the benzoquinone layer does not provide an impediment to current flow across the PEDOT:PSS – Silicon interface, and device efficiencies of 9.6% can be obtained. Devices using this passivation exhibited exceptionally high external quantum efficiencies in the low wavelength region, indicative of a high-quality front surface. Multiple rear contact schemes were evaluated, with Al giving the best results. Front-side texturing is shown to boost the short-circuit current of the devices by 20%, yielding significant gains to device fill factors.

Index Terms — benzoquinone, hybrid organic-inorganic, silicon passivation, PEDOT:PSS, solar cell characterization.

I. INTRODUCTION

The goal of any photovoltaic device is to separate photoexcited charge carriers and extract the energy of these excited electron states before they are able to recombine with lower energy hole states. Recombination of electron and hole can occur via one of several pathways (Radiative, Shockley-Read-Hall, and Auger) which result in emission of excess energy as heat or photon emission. Surfaces and interfaces of the crystal tend to be regions of high recombination due to the breaking of crystal symmetry and the higher concentration of impurities. Effective surface passivation is necessary to extend the lifetime of excited carriers by decreasing the availability of surface states with energies lying within the bandgap. Typically this is achieved either by saturating dangling bonds to reduce the number of intermediate energy states or introducing a surface dipole to induce band bending at the surface, repelling minority carriers.

A. Benzoquinone Passivation

Previous studies in the Opila group have focused on the use of quinhydrone derivatives, such as benzoquinone, for passivation of silicon <100> surfaces. Quinhydrone is a solid charge-transfer complex consisting of an equimolar mixture of benzoquinone (BQ) and hydroquinone. When dissolved in a solvent these two constituents form a two-electron redox couple. Quinhydrone was first shown as a passivant by Takato et al.[1] Studies by Chhabra et al.[2] showed the ability to achieve high minority carrier lifetimes in both p-type and n-type silicon by treating hydrogen-terminated silicon surfaces with solutions of quinhydrone in methanol. Subsequent studies provided evidence that benzoquinone was the more active constituent in the quinhydrone complex[3].

The precise mechanism by which the silicon <100> surface is passivated in benzoquinone and methanol solution is still under active study, with Kotulak suggesting a free radical bonding mechanism by which both quinone radicals and methanol radicals bond to the silicon surface[4]. XPS studies[5] have shown evidence of both quinones and methanol bonded to the surface appearing to form essentially a monolayer at the surface.

In recent years many groups have produced research on a hybrid organic-inorganic device structure using PEDOT:PSS deposited on silicon[6]–[9]. This structure uses silicon as the primary light absorber, utilizing PEDOT:PSS as a selective contact. Such a device structure enables the creation of a silicon-based junction through solution-based processing. In this device structure, the PEDOT:PSS layer is able to induce an inversion layer within the silicon. However further improvement to the passivation of the silicon at this interface should be able to improve the open circuit voltage and achieve stronger inversion.[10]

Based on this desire, we propose a device structure that uses a thin passivating layer to further promote carrier inversion and achieve high open circuit voltages, while using an organic semiconductor to serve as a carrier-selective transparent conductive layer. This device design is modeled to mirror the basic band structure used in the high-efficiency HIT solar cell architecture. Here, a thin benzoquinone layer can serve to passivate the silicon, while a thicker PEDOT:PSS layer facilitates transfer of holes.

II. DEVICE FABRICATION

The device structure (Fig. 1) is fabricated through simple solution-based processing. First, an n-type silicon wafer (1-5 ohm-cm) is cleaned in Piranha and HF acid solutions to produce a hydrogen terminated surface, following the protocol previously described by Kotulak[3]. Then the wafer is immersed for 2 hrs in a 0.1M solution of benzoquinone in methanol. After this surface passivation is complete, a layer of PEDOT:PSS (Clevios PH 1000) is spin-coated onto the substrate. The PEDOT:PSS solution includes 7% dimethyl sulfoxide (DMSO) to increase the conductivity (vide infra) and 0.25% Triton X-100 (from Sigma Aldrich) to act as a surfactant. Metal contacts are added via electron beam physical vapor deposition. For the front contact, a silver pattern is deposited through a shadow mask. The rear contact consists of aluminum covering the entire back surface.

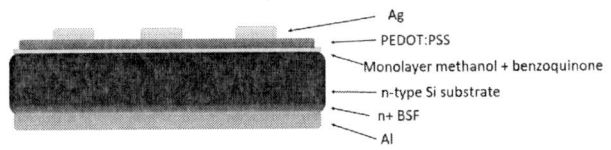

Fig. 1: Device structure of the PEDOT:PSS-BQ-Si solar cell. Surface texturing not depicted.

The conducting polymer used in this device, Poly(3,4-ethylenedioxy-thiophene):poly-(styrenesulfonate) (PEDOT:PSS) is a mixture of two ionic polymers, forming a macromolecular salt. These ionomers are typically distributed as a dispersion of gelled particles in water, which can be deposited by spin coating and subsequently dried. The particles consist of PEDOT-rich conductive cores, with PSS-rich outer shells. The addition of a solvent such as DMSO to the solution helps to decrease the thickness of the PSS-rich shells, facilitating conduction. Ensuring good uniformity of the PEDOT:PSS film coverage is crucial for a well operating device. This can be aided through the use of a surfactant, filtering of the solution, and careful control of the spin recipe.

A. Device Results

Different candidates were attempted for both the rear contacts, with Al yielding the lowest contact resistance. It is most desirable to create an ohmic contact to reduce resistive losses and avoid the creation of an additional junction at the contacts. Ni, Al and Ti/Pd/Ag stacks were each deposited via e-beam deposition as candidates for the back contacts. Although Ni and Ti/Pd/Ag are more commonly used for n-type contacts, Transmission Line Measurements yielded higher resistance values for these contacts and in some cases showed a non-linear voltage dependence.

To further address concerns of an inefficient rear contact limiting performance, devices were fabricated with a back-surface field (BSF). Wafers with a 100 Ohm/square sheet resistance BSF were created by POCl3 diffusion. This n+ doped region creates band bending which should reduce the rear side surface recombination, allowing us to focus our efforts on the front surface of the device.

Initial devices were seen to suffer from low short circuit currents. Improved light retention through surface texturing can boost current output of solar cells. This was achieved using wafers which had been textured in a KOH bath. At low concentrations (<2%), KOH etches silicon anisotropically in the <100> plane, creating a random pyramidal textured surface. Wafers with texturing and a BSF were provided by our collaborators at Arizona State University.

Our initial devices used a small device area to ensure good film uniformity, at 3.25mm x 3.25mm. These devices suffered losses due to edge effects. Since the minority carrier diffusion length in n-type wafers of these doping levels is about 0.5mm, almost half of the device area is within one diffusion length of a device edge. Increasing the device

dimensions to 1cmx1cm greatly decreased the energy losses due to recombination at the device edges.

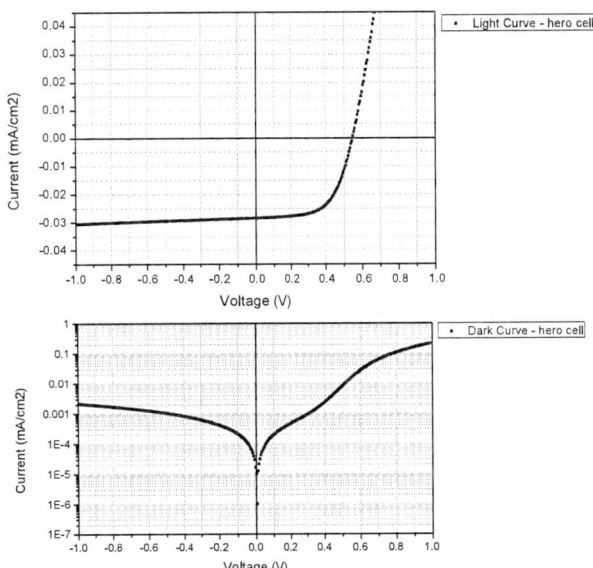

Fig. 2: IV curve of best device. (top) light curve, (bottom) dark curve graphed logarithmically

Initial devices showed power conversion efficiencies of around 3%, but processing improvements including steps detailed above have improved performance of the best device to 9.6% efficiency. This value is much more in line with the performance of Si-PEDOT:PSS devices from literature, which typically perform between about 8-13% efficient[8][11]. IV curves and relevant parameters of the hero device are detailed below. Results are summarized in Fig. 2 and Table 1. It is seen that the introduction of the benzoquinone passivating layer does not provide a barrier to charge transfer. It is expected that high quality passivation should enable higher open-circuit voltages, however this may require further optimization of the benzoquinone-PEDOT:PSS interface.

TABLE I
SUMMARY OF DEVICE VALUES

Jsc (mA/cm2)	Voc (V)	n (%)	FF (%)	Rs (Ohm cm2)	Rsh (Ohm cm2)	Area (cm2)
28.6	.545	9.6	61.6	1.769	425.336	1

From the external quantum efficiency data (Fig. 3) we can see that our devices perform quite well at short wavelengths, outperforming comparison diffused junction silicon cells. Since most short wavelength light is absorbed near the front of the cell this suggests a well-passivated front surface. The higher integrated short circuit current measured is most likely due to differences in finger shading between measurement techniques. Due to the relatively high contact resistance between PEDOT:PSS and Ag, our devices feature a relatively high shading ratio of 11%.

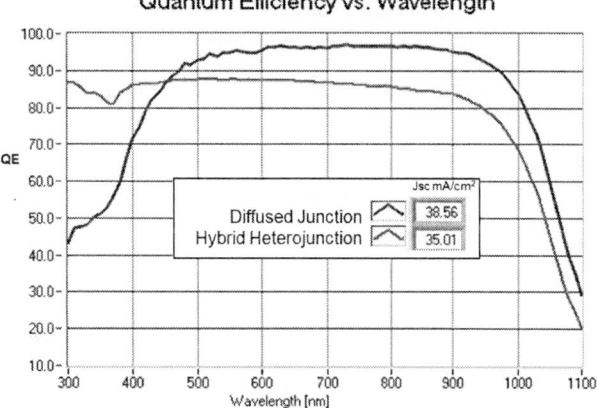

Fig. 3: External quantum efficiency measurements with integrated short circuit current of devices. Hybrid device on KOH textured wafer with BSF (blue), Conventional diffused junction silicon device for comparison (blue).

UV-Vis reflectance measurements (Fig. 4) show that that a large portion of the external quantum efficiency losses were due to reflection. The addition of wafer texturing was highly effective in reducing reflectance, which stays fairly uniform at about 8% across most usable wavelengths. This was able to boost total short circuit current output from the cells by 20%. The PEDOT:PSS layer has a dual purpose as an anti-reflection coating. It may be possible to further drive down reflection by adding in additional layers to gradate the refractive index n.

Reflectance of PEDOT:PSS coated wafers

Fig. 4: UV-Vis reflectance data for a textured and untextured wafer after BQ and PEDOT:PSS treatment

CONCLUSIONS

The viability of a hybrid organic-inorganic solar cell using benzoquinone-methanol as a surface passivant on silicon was demonstrated. Device efficiencies were improved to 9.6% by increasing the device dimensions and the inclusion of texturing and a back-surface field.

Future work will better characterize the interfaces occurring between Si, BQ and PEDOT:PSS. We aim to build a better understanding of energy levels and charge transport across these interfaces and to identify what materials characteristics are necessary to enable a passivated carrier-selective interface.

ACKNOWLEDGEMENT

Reflectance and quantum efficiency measurements taken at the Institute of Energy Conversion under the guidance of Steven Hegedus and Kevin Dobson. Wafers with texturing and BSF obtained from Solar Power Labs at Arizona State University under the supervision of Som Dahal, Rameshwari Ghimire and Stuart Bowden.

This material is based upon work primarily supported by the Engineering Research Center Program of the National Science Foundation and the Office of Energy Efficiency and Renewable Energy of the Department of Energy under NSF Cooperative Agreement No. EEC-1041895. Any opinions, findings and conclusions or recommendations expressed in this material are those of the author(s) and do not necessarily reflect those of the National Science Foundation or Department of Energy. Work at the ASU Solar Power Lab was supported by the Nanotechnology Collaborative Infrastructure Southwest (NCI-SW) with funding from NSF- ECCS-1542160.

REFERENCES

[1]   H. Takato, I. Sakata, and R. Shimokawa, "Quinhydrone/Methanol Treatment for the Measurement of Carrier Lifetime in Silicon Substrates," *Japan Soc. Appl. Phys.*, vol. 41, no. Pt. 2, No. 8A, pp. 870–872, 2002.

[2]   B. Chhabra, S. Bowden, R. L. Opila, and C. B. Honsberg, "High effective minority carrier lifetime on silicon substrates using quinhydrone-methanol passivation," *Appl. Phys. Lett.*, vol. 96, no. 2010, pp. 13–16, 2010.

[3]   N. Kotulak, M. Chen, N. Schreiber, and R. L. Opila, "Time and Light-Dependence of Electrical Passivation of c-Si Surfaces with Quinhydrone Constituent Solutions," no. Mcd, pp. 1529–1533, 2014.

[4]   N. A. Kotulak, M. Chen, N. Schreiber, K. Jones, and R. L. Opila, "Examining the free radical bonding mechanism of benzoquinone– and hydroquinone–methanol passivation of silicon surfaces," *Appl. Surf. Sci.*, vol. 354, pp. 469–474, 2015.

[5]   B. Chhabra, C. Weiland, R. L. Opila, and C. B. Honsberg, "Surface characterization of quinhydrone-methanol and

iodine-methanol passivated silicon substrates using X-ray photoelectron spectroscopy," *Phys. Status Solidi*, vol. 208, no. 1, pp. 86–90, 2011.

[6] J. P. Thomas and K. T. Leung, "Defect-minimized PEDOT:PSS/planar-si solar cell with very high efficiency," *Adv. Funct. Mater.*, vol. 24, pp. 4978–4985, 2014.

[7] K. A. Nagamatsu, S. Member, S. Avasthi, J. Jhaveri, and J. C. Sturm, "A 12 % Efficient Silicon / PEDOT : PSS Heterojunction Solar Cell Fabricated at <100 ∘ C," *IEEE J. Photovoltaics*, vol. 4, no. 1, pp. 260–264, 2014.

[8] S. Mahato, L. G. Gerling, C. Voz, R. Alcubilla, and J. Puigdollers, "PEDOT:PSS as an alternative hole selective contact for ITO-free hybrid crystalline silicon solar cell," *IEEE J. Photovoltaics*, vol. Accepted, pp. 1–6, 2016.

[9] S. Li, Z. Pei, F. Zhou, Y. Liu, H. Hu, S. Ji, and C. Ye, "Flexible Si/PEDOT:PSS hybrid solar cells," *Nano Res.*, vol. 8, no. 10, pp. 3141–3149, 2015.

[10] R. Liu, S. T. Lee, and B. Sun, "13.8% Efficiency Hybrid Si/Organic Heterojunction Solar Cells with MoO3 Film as Antireflection and Inversion Induced Layer," *Adv. Mater.*, vol. 26, no. 34, pp. 6007–6012, 2014.

[11] J. He, P. Gao, M. Liao, X. Yang, Z. Ying, S. Zhou, J. Ye, and Y. Cui, "Realization of 13.6% Efficiency on 20 μm Thick Si/Organic Hybrid Heterojunction Solar Cells via Advanced Nanotexturing and Surface Recombination Suppression," *ACS Nano*, vol. 9, no. 6, pp. 6522–6531, 2015.

Precise I-V Curve Measurement Procedure for Perovskite Solar Cells: Application to Various Types of Devices

Y. Hishikawa, M. Yoshita, H. Shimura, A. Sasaki and T. Ueda, Research Center for Photovoltaics, National Institute of Advanced Industrial Science and Technology (AIST), Japan

Abstract — **Precise performance measurement procedure for perovskite PV devices was applied to various types of cells. The study was focused on confirming the applicability and usefulness of the proposed measurement procedure by using devices prepared by different institutes. The procedure includes comparison of the I-V curves at various sweep time, and evaluation of P_{max} under a fixed voltage for precise characterization. The results indicated that the procedure is versatilely applicable to various types of perovskite solar cells, and precisely define the I-V curves and P_{max}.**

I. INTRODUCTION

Precise performance characterization is essential for developing and improving the performance of PV devices under the R&D stage as well as on the mass production stage. Recent development of novel PV devices has induced requirement for new consideration and update of the characterization technologies, based on their specific electrical and optical properties [1]. Among them, perovskite solar cells have recently attracted much attention, because of their high conversion efficiency raging 18% - 20% and potential low cost. They show very special features which affect the performance characterization, such as complex hysteresis of the current-voltage (I-V) curves and very slow temporal response. Those features show clear dependence on the device structure, which makes the performance characterization more complicated. Although procedures for precisely characterizing the performance have been investigated [2]-[6], their details remain to be confirmed by various device structures.

The present study confirms the validity of procedures which the authors have recently proposed [5] for precisely determining their performance, by using various kinds of perovskite devices made at different institutes.

II. MEASUREMENT PROCEDURE

I-V curves of perovskite solar cells measured under different sweep conditions are shown in Figs. 1(a) and 1(b). It has been recently proposed that the following procedure is appropriate for precisely characterizing the perovskite PV devices [5].

(1) I-V curves are measured at a wide range of sweep time, e.g., in the order of second to minute, for confirming the perspective of hysteresis. It may persist even at a long sweep time of 10 min.

(2) If the I-V curves of the forward and reverse sweeps agree within acceptable range (depending on the purpose of measurement) at a long-enough sweep time, they can be regarded as the stable performance. Although the criteria for the sweep time is not established at the present stage, the present results indicate that confirmation by a wide range of sweep time is recommended.

(3) When the hysteresis is persistent even for a long sweep time, the stable P_{max} can be determined by examining the temporal trend of the output current by fixing the bias voltage at near V_{mp} after forward and reverse bias histories.

III. RESULTS AND DISCUSSION

Application of the procedure (3) to two kinds of perovskite solar cells are shown in Figs. 2(a) and 2(b), which demonstrates that P_{max} can be defined for devices with various types of hysteresis.

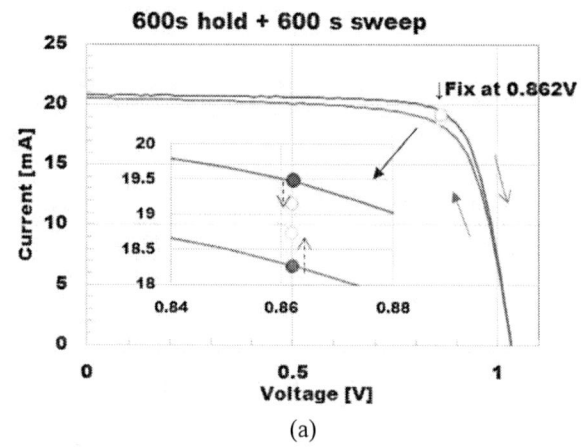

(a)

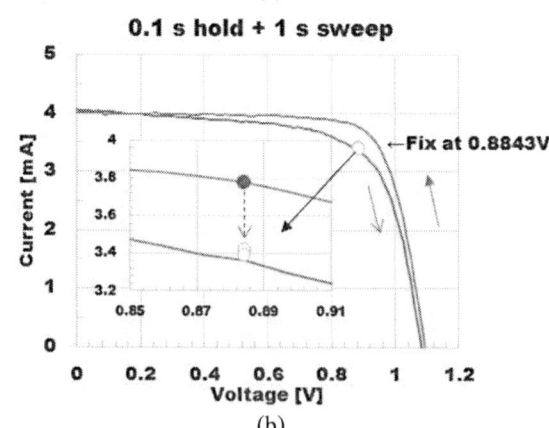

(b)

Fig. 1 Examples of the I-V curves of perovskite solar cells measured under different sweep conditions, which show (a) small and (b) moderate hysteresis effects. The stable output current at a fixed voltage near V_{mp} is shown by open circles.

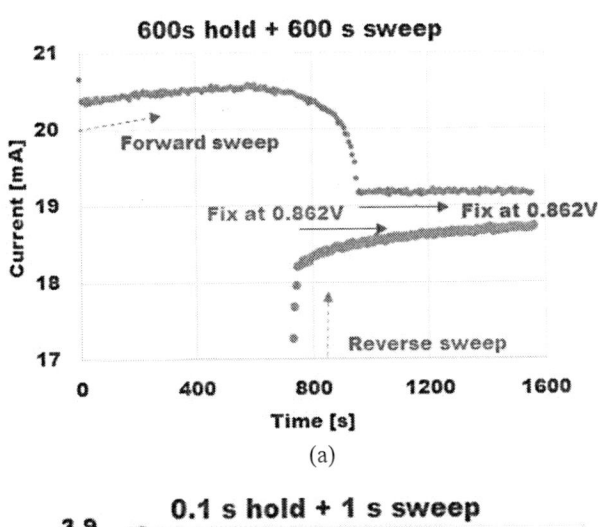

(a)

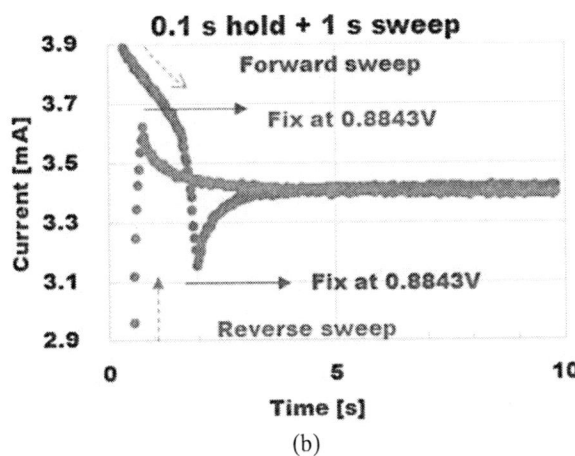

(b)

Fig. 2 Temporal variation of the output current of the same samples as in Figs. 1(a) and 1(b) at fixed voltages near V_{mp}.

Although the hysteresis effect is very dependent on the sweep conditions such as the sweep time etc. [5], the stable output current under fixed bias voltage is usually not very dependent on the bias voltage history, as exemplified in Figs. 3(a) and 3(b). This demonstrates the usefulness of measurement at a fixed voltage. It is noted that there are some samples such as shown in Figs. 1(a) and 2(a), which show slight difference of a few percent even at a fixed voltage [5]. This suggests that the hysteresis effect includes a component which is persistent for more than 10 minutes, and is dependent on the bias voltage history. This is an important information for investigating the precise performance of the devices.

It was also pointed out by the previous study [5] that the relative spectral response of perovskite solar cell is not sensitive to the measurement conditions such as the chopper frequency and bias light level, which is also shown in Fig. 4. Temporal responses of a perovskite solar cell to the chopped monochromatic light during the spectral response measurement were almost independent of the chopping frequency as seen in Fig. 5. It is a clear contrast to the dye-sensitized solar cells (DSCs) [7][8], which show slow responses both to the variation of bias voltage and irradiance level. The result suggests that its photo-response time is within about 1.1 ms, which is responsible for the insensibility of the measurement results to the chopper frequency.

In summary, the present study has confirmed the usefulness and applicability of the measurement procedure of perovskite solar cells proposed in the previous study [5]. The results indicated that the procedure is versatilely applicable to various types of perovskite solar cells, and is useful for precisely defining their I-V curves and P_{max}. The fast temporal response of perovskite solar cells to chopped monochromatic light is shown, which is consistent with the insensitivity of the measured spectral response to the chopper frequency.

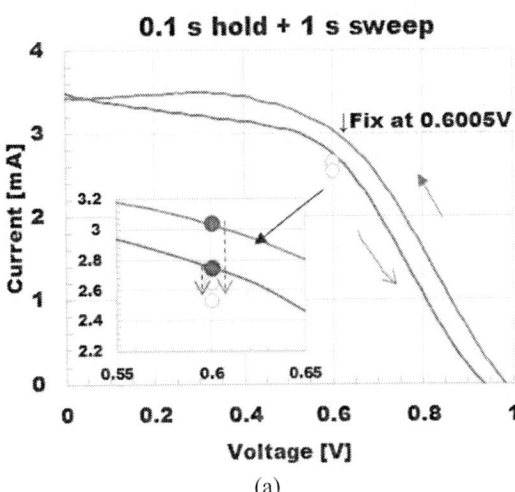

(a)

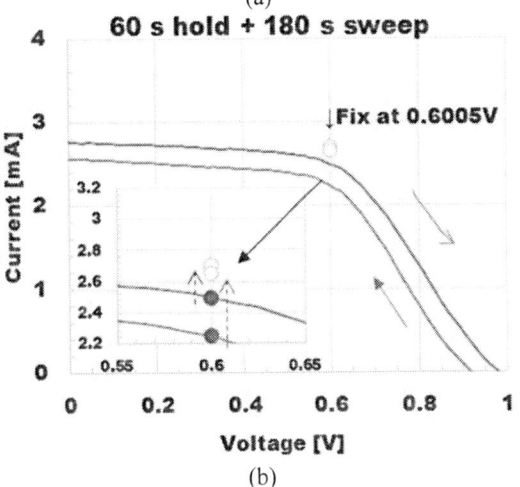

(b)

978-1-5090-5606-4/17 $31.00 © 2017 IEEE

Fig. 3 I-V curves of a perovskite solar cell at sweep times of (a) 1 s and (b) 180 s. The stable output current at a fixed voltage near V_{mp} is shown by open circles.

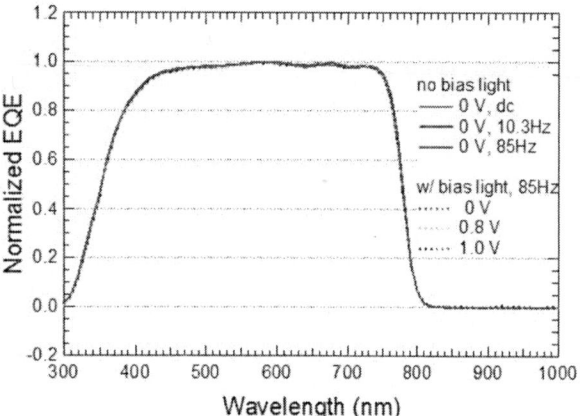

Fig. 4 Relative quantum efficiency spectra of a perovskite solar cell measured under various bias light and voltage levels.

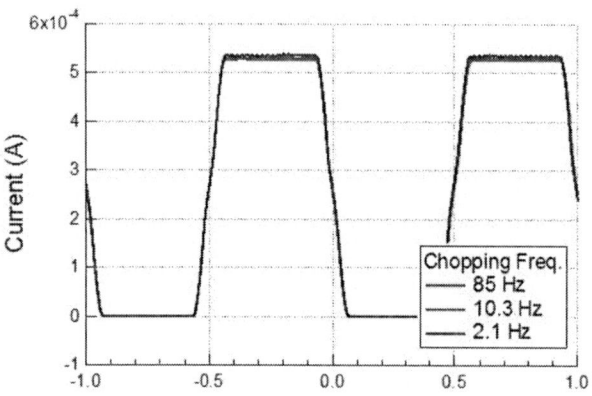

Fig. 5 Temporal responses of the same perovskite solar cell as in Fig. 4 under chopped monochromatic light (λ = 550 nm, 4×10^{15} photons /s·cm^2) under short circuit condition. No bias light was applied.

Acknowledgement: This study is supported in part by NEDO under METI.

REFERENCES

[1] Y. Hishikawa "Traceable Performance Characterization of State-of-the-Art PV Devices" Proceedings of the 27th EUPVSEC, Frankfurt, (2012), pp 2954-2960

[2] H. J. Snaith, A. Abate, J. M. Ball, G. E. Eperon, T. Leijtens, N. K. Noel, S. D. Stranks, J. T. Wang, K. Wojciechowski, and W. Zhang, "Anomalous Hysteresis in Perovskite Solar Cells", J. Phys. Chem. Lett. 2014, 5, 1511−1515.

[3] W. Tress, N. Marinova, T. Moehl, S. M. Zakeeruddin, M. K. Nazeeruddin, M. Gratzel, " Understanding the rate-dependent J–V hysteresis, slow time component, and aging in CH3NH3PbI3 perovskite solar cells: the role of a compensated electric field", Energy & Environmental Science (2015) [8] 995-1004.

[4] L. Cojocaru, S. Uchida, P. V. V. Jayaweera, S. Kaneko, J. Nakazaki, T. Kubo, H. Segawa, "Origin of the Hysteresis in I-V Curves for Planar Structure Perovskite Solar Cells Rationalized with a Surface Boundary Induced Capacitance Model", Chem Lett. 2015, 44(12), 1750-1752

[5] Y. Hishikawa, H. Shimura, T. Ueda, A. Sasaki, and Y. Ishii, "Precise Performance Characterization of Perovskite Solar Cells", Current Applied Physics 16 (2016) 896-904, DOI: 10.1016/j.cap.2016.05.002

[6] E. Zimmermann, K. K. Wong, M. Müller, H. Lu et al., "Characterization of perovskite solar cells: Towards a reliable measurement protocol", APL Materials 4 (2016), 091901

[7] Y. Hishikawa, M. Yanagida and N. Koide, "Performance Characterization of the Dye-Sensitized Solar Cells", Proceedings of the 31st IEEE PVSC, Lake Buena Vista (2005) 67-70

[8] G. Bardizza, D. Pavanello, H. Müllejans and T. Sample, "Spectral responsivity measurements of DSSC devices at low chopping frequency (1 Hz)", Prog. Photovol. Res. Appl. 24 (2016) 428-435

Enhancing the Crystalline of Planar-Structure $CH_3NH_3PbI_3$ Perovskite Solar Cells via Sandwich Evaporation Technique

Po-Tsun Kuo[1, a], Shang-Pang Lin[1, a], Cheng-shian Lin[1, a], Ching-Fuh Lin[1,2,3,4, b]

1. Graduate Institute of Photonics and Optoelectronics, National Taiwan University, No. 1, Sec. 4, Roosevelt Road, Taipei 10617, Taiwan.

2. Graduate Institute of Electronics Engineering, National Taiwan University, No. 1, Sec. 4, Roosevelt Road, Taipei 10617, Taiwan.

3. Department of Electrical Engineering, National Taiwan University, No 1, Sec 4, Roosevelt Road, Taipei 10617, Taiwan.

4. Innovative Photonics Advanced Research Center, National Taiwan University, Taipei, Taiwan, No 1, Sec 4, Roosevelt Road, Taipei 10617, Taiwan.

[a]Tel: +886-2-33663700#405, E-mail: r04941003@ntu.edu.tw

[b]Tel: +886-2-33663540, E-mail: lincf@ntu.edu.tw

Abstract—In this research, the sandwich evaporation technique (SET) process was exploited to fabricate perovskite solar cells. In the SET process, the use of the seed layer MAI and double-interdiffusion exhibit the great uniformity and coverage of the film in the ascendant. Also, the quality of the perovskite is further enhanced with the deposition environment controlling (DEC) method. The solvent-induced process effectively decelerates the reaction rate to achieve better film formation. The maximum grain size is beyond than 900nm and the highest efficiency is more than 14.5%.

I. INTRODUCTION

Due to significant progress in converting power efficiency recently [1,2], organometallic halide perovskite has become the center of attention for the next generation photovoltaics. In addition to being a low-cost material, perovskite also exhibits strong solar absorption, excellent carrier diffusion length, and easy fabrication. Various methods (e.g., solution process [3], thermal evaporation [4] and vapor-assisted solution (VSVP)process [5]) and light absorbers (e.g., $CH_3NH_3PbI_3$, $CH_3NH_3PbI_{3-x}Cl_x$ and CH_3NH_3PbBr) are built upon either planar-heterojunction structure or mesoporous structure. Furthermore, some methods, including doping engineering and solvent engineering, are confirmed to optimize the quality of the perovskite. In the solution process, the film formation usually suffers from the poor crystallization owing to fast reaction, the uniformity of coverage and limitation of mixing the precursor. In comparison, the traditional thermal evaporation is almost free from the problems mentioned in solution process. However, the cost issue for mass production prohibits commercialization. The VSVP process takes the advantages of the tradition solution process and the conventional thermal

evaporation. The film of PbI_2 is deposited by spin-coating. Then CH_3NH_3I (MAI) is evaporated to convert the PbI_2 into perovskite. The hybrid concept not only fixes the shortcomings in solution process to optimize the crystallization, but also avoids the high-cost issue in the traditional thermal evaporation. Nonetheless, now, the reports on VSVP process are frequently exploited to TiO_2-based system [6]. The formation of TiO_2 requires high temperature which is higher than 500°C. In addition, The TiO_2-based devices exhibit the undesired hysteresis phenomenon. In this research, a sandwich evaporation technique (SET) process was established to modify the primitive VSVP process. The concept of double-interdiffusion and seed layer promotes the crystallinity of the perovskite film significantly. For further enhancing the quality, the deposition environment controlling (DEC) method was applied to the SET process to decelerate the reaction speed. Eventually, the high power conversion efficiency (PCE) is boosted over than 14.5% and the maximum grain size was beyond 900nm.

II. EXPERIMENT

ITO-coated substrates were sequentially cleaned by acetone, isopropyl alcohol, methanol and then treated with plasma for 15 minute. All the solution-based coating and annealing steps are carried out in atmosphere. The PEDOT-4083, which were first filtered by 0.2um PVDF, was spin-coated at 3000rpm for 30sec on ITO substrate and annealed at 120°C for 15min. The seed layer MAI (15mg/ml dissolved in N,N-Dimethylmethanamide) was coated at 3000 rpm for 30 sec and annealed at 120°C for 10 min. PbI_2 was deposited latter by conventional thermal evaporation and annealed at 100°C for 30min. Vapor MAI was evaporated in the homemade SET chamber with the pressure of 0.5 Torr and the temperature of 130 °C. The dark brown perovskite layer about 340nm was formed. The substrates were annealed at 120°C for 30mins. Then $PC_{60}BM$ (20mg/ml dissolved in Chlorobenzene (CB)) was coated on the perovskite layer with 1300rpm for 60sec. At the end, the Bathocuproine (BCP) was evaporated as hole-blocking layer for 4.5 nm and silver was evaporated for 120nm as the back electrode. All of our devices were measured in the ambient atmosphere at room temperature. Current density-voltage (J-V) characteristics were measured with Keithley 2400 source under AM 1.5G at an intensity of $100Mw/cm^2$. X-ray diffraction(XRD) data was collected using an X-ray powder diffractometer PW3040 with Cu-Kα radiation. Scanning electron microscopy (SEM) images was analyzed on JEOL JSM-7600F microscope using 10 kV field emission.

III. Result and discussion

In the research, a homemade SET chamber was utilized to fabricate the perovskite solar cells in the SET process. The schematic structure of our devices were $ITO/PEDOT:PSS/MAPbI_3/PC_{60}BM/BCP/Ag$, as shown in Fig.1. The $MAI-PbI_2-MAI$ structure was constructed to improve the uniformity of the perovskite film. In Fig.2, the SEM images showed the morphology of perovskite fabricated by different methods. Due to the fast reaction rate happened in the solution-based process, the crystal was minor and irregular. The poor crystallization might lead to high recombination of electrons and holes in the perovskite layer. Moreover, there were plenty of excess MAI deposited on the surface which was harmful to extracting the electrons. In comparison, the quality of $MAPbI_3$ formed by the $MAI-PbI_2-MAI$ structure was improved significantly. Not only the uniformity was promoted, but also the crystal grew into a larger size. The better the crystalline was, the better efficiency could be achieved.

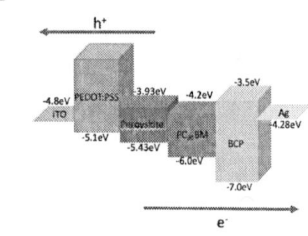

Fig.1.   The schematic structure of the devices.

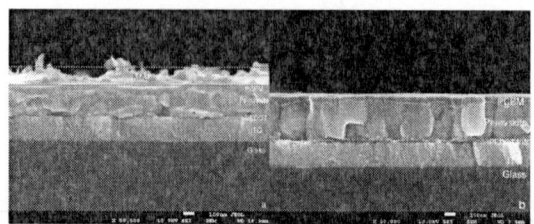

Fig.2.   The cross-section SEM images of perovskite: (a) produced by the solution process, (b) produced by the SET process.

To investigate the influence of annealing steps on the properties of the cells, the devices were annealed with different time in atmosphere. In Fig.3, the short-circuit current (J_{sc}) was slightly raised and open-circuit voltage (V_{oc}) and fill factor (FF) were promoted obviously. The reason was deduced that annealing for 30mins would fully convert PbI_2 into $MAPbI_3$. However, annealing more than 30mins would cause the decomposition of $MAPbI_3$, so a small peak of PbI_2 was observed at 2θ of 12.65° after annealing for 60mins and 90mins, respectively. The tendency revealed that annealing for proper time could optimize the crystallinity to enhance the performance of the devices.

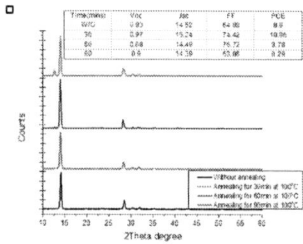

Fig.3.   The analysis of XRD measurement and the performance of the device.

To further optimize the crystalline, the DEC method was introduced in the original SET process. The solvents (e.g. Ether, Toluene, and H_2O) were applied at the beginning of the SET process. The main principle was similar to the solvent engineering [7], which is frequently applied in the traditional solution process. By treating with the additive solvents, the reacting rate of MAI and PbI_2 was reduced. As a result, the crystallization of the grain could be greatly promoted. In the case of the H_2O, some of the research [8] proved that with adequate humidity of environment condition, the $MAPbI_3$ can become more uniform to yield better efficiency of cells. However, this solvent-induced method is barely utilized in conventional thermal evaporation. The solvents were easily exhausted before reaching high vacuum level. On the contrary, the solvents could be exploited to SET process due to the fast-reaching to the working-pressure. TABLE I shows that the additive solvents could promote the short-circuit current (J_{sc}) from 14.99 mA/cm^2 to 19.71 mA/cm^2. To investigate the reason, the grain condition was measured. The Fig.4 showed the cross-section SEM images of the MAPbI₃ treated with w/o, Ether, CB, and H_2O respectively. The crystalline was truly improved by the additive solvents. The average grain size merged larger and with less boundaries in the vertical direction. Nonetheless, the uniformity varied with the solvents. The smoothness of MaPbI₃ with ether was the best of all and the worst one was applied with H_2O. Excitons could be extracted into free electrons and holes due to the less recombination in the perovskite caused by boundaries. Hence, the better crystal condition achieved by the additive solvent would give rise to better performance of the J_{sc} and efficiency. Also, the crystalline being optimized with the DEC method could surpass 900nm, as shown in Fig.5.

TABLE I. Each characteristic of the devices while applying DEC method with none/Ether/CB/H_2O.

Solvent	Voc (V)	Jsc (mA/cm²)	FF (%)	PCE (%)
W/O	0.98	14.99	77.95	11.40
Ether	0.95	19.71	77.94	14.53
CB	0.95	16.37	79.61	12.38
H_2O	0.95	17.28	75.43	12.46

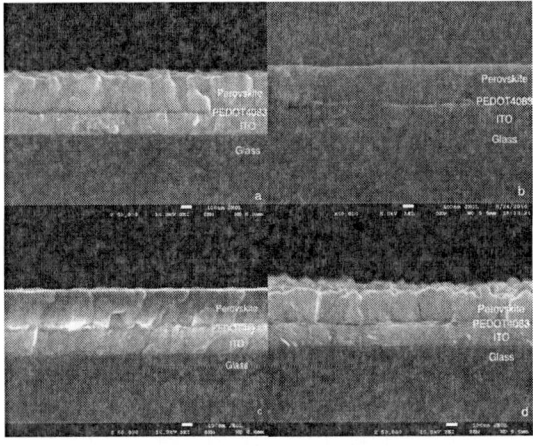

Fig.4. The cross-section SEM images of utilizing the DEC method with (a)W/O, (b)Ether, (c)CB, and (d) H_2O.

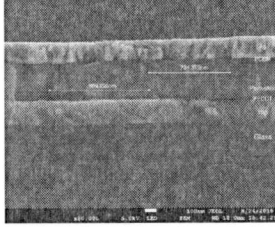

Fig.5. The cross-section SEM of DEC method with Ether.

IV. Conclusion

In conclusion, with a homemade SET chamber and SET process to manufacture the perovskite solar cells, the MAI-PbI₂-MAI structure effectively converts into MAPbI₃ with less boundaries. Further optimized by the DEC method with ether, the best cell achieves 0.95V of V_{oc}, 19.81mA/cm² of J_{sc}, 77.94% of FF, and yield 14.53% with grain size exceeding 900 nm. Moreover, The SET process gives the great potential for mass production in commercialization.

V. ACKNOWLEDGMENT

We gratefully acknowledge Ministry of Science and Technology of Taiwan and National Taiwan University for supporting the research with the contract numbers: MOST 105-3113-E-002-010, MOST 105-2119-M-002-009, MOST 103-2221-E-002 -132 -MY3, MOST 104-2221-E-002 -139 -MY3.

References

[1] Martin A. Green, Anita Ho-Baillie & Henry J. Snaith, "The emergence of perovskite solar cells," Nature Photonics 8, 506–514 (2014).

[2] S. Shahbazi, F. Tajabadi, H.-S. Shiu, R. Sedighi, E. Jokar, S. Gholipour, N. Taghavini, de S. Afshar and E. W.-G. Diau, "An easy method to modify PEDOT:PSS/perovskite interfaces for solar cells with efficiency exceeding 15%," RSC Adv.6, 65594, 2016.

[3] Dianyi Liu & Timothy L. Kelly, "Perovskite solar cells with a planar heterojunction structure prepared using room-temperature solution processing techniques," Nature Photonics 8, 133–138 2014.

[4] Mingzhen Liu1, Michael B. Johnston1 & Henry J. Snaith1, "Efficient planar heterojunction perovskite solar cells by vapour deposition," Nature 501, 395–398, Sep19, 2013.

[5] Qi Chen, Huanping Zhou, Ziruo Hong, Song Luo, Hsin-Sheng Duan, Hsin-Hua Wang, Yongsheng Liu, Gang Li, and Yang Yang, Department of Materials Science and Engineering and California NanoSystems Institute, University of California, Los Angeles, California 90095, United States, "Planar Heterojunction Perovskite Solar Cells via Vapor-Assisted Solution Process," J. Am. Chem. Soc., pp 622–625, 2014, 136 (2).

[6] Kyung Taek Cho, Sanghyun Paek, G. Grancini, Cristina Roldán Carmona, Peng Gao, Yong Hui Lee and Mohammad Khaja Nazeeruddin , "Highly efficient perovskite solar cells with a compositional engineered perovskite/hole transporting material interface," Energy Environ. Sci., 2017

[7] Nam Joong Jeon, Jun Hong Noh, Young Chan Kim, Woon Seok Yang, Seungchan Ryu & Sang Il Seok,"Solvent engineering for high-performance inorganic–organic hybrid perovskite solar cells," Nature Materials 13, 897–903 (2014)

[8] Qing Zhou, Zhiwen Jin, Hui Li & Jizheng Wang,"Enhancing performance and uniformity of CH3NH3PbI3−xClx perovskite solar cells by air-heated-oven assisted annealing under various humidities," cientific Reports 6, Article number: 21257 (2016).

Toward High Performance Organic-Silicon Hybrid Solar Cells

Yi Lai[1], Hong-Jhang Syu[1] and Ching-Fuh Lin[1,2,3,4]

[1]Graduate Institute of Photonics and Optoelectronics, National Taiwan University, Taipei, Taiwan (R.O.C.).
[2]Graduate Institute of Electronics Engineering, National Taiwan University, Taipei, Taiwan (R.O.C.).
[3]Department of Electrical Engineering, National Taiwan University, Taipei, Taiwan (R.O.C.).
[4]Innovative Photonics Advanced Research Center, National Taiwan University, Taipei, Taiwan (R.O.C.).
(Tel: +886-2-3366-3540. FAX: +886-2-2364-2603. E-mail address: *lincf@ntu.edu.tw*)

Abstract—**We propose methods of enhancing the efficiency of organic hybrid silicon solar cells. First, we use two-step metal-assisted method to fabricate silicon nanostructures to reduce the reflectivity. However, the organic material (PEDOT:PSS) has large absorption of light, so the silicon absorption is reduced. To solve this problem, PEDOT:PSS is diluted using IPA before it is deposited. Nonetheless, the adhesion of the thin PEDOT is bad, so it cannot be used together with ITO glass. Then, we use silver nanowires (AgNWs) to improve the contact between PEDOT:PSS and front electrode. Finally, the hybrid silicon solar cells have improved efficiency of 13.37%.**

I. INTRODUCTION

In recent years, crystalline silicon (c-Si) solar cells still dominate the PV market, on account of their nontoxicity, high efficiency, and supply of abundant material for the fabrication [1]. However, one drawback is that their manufacture involves expensive, high temperature dopant diffusion or ion implantation, and anti-reflection (AR) coating in vacuum. Hybrid photovoltaic devices incorporating organic and inorganic materials are thus receiving great attention because of their easy fabrication and low cost compared to conventional p–n junction Si solar cells. The polymer poly (3,4-ethylenedioxythiophene): polystyrenesulfonate (PEDOT:PSS) is a widely used organic material for hybrid devices, because it is transparent, highly conductive, and water-soluble in its metallic state. On the other hand, to improve device optical absorption and to replace conventional AR coating, nanostructures, like nanowires, nanorods, or nanocones, are fabricated on a Si surface to form Si-nanostructure/organic hybrid solar cells. Therefore, silicon nanostructures (SiNS) arrays can be applied to reduce reflection so that we can approach low reflection and high absorption based on light-trapping effect, which are absent for bulk silicon. Here, we choose the solution-processed and metal-assisted chemical etching (MacEtch) for SiNS arrays fabricatiodn [3-4]. MacEtch is low cost, and low energy consumption compared molecular beam epitaxy (MBE) and chemical vapor deposition (CVD), which need high vacuum and high temperature. Furthermore, because PEDOT:PSS is an absorption material [7,9], we used many solvents, like IPA, mix with PEDOT:PSS. It can assist us reduce the PEDOT absorption and thickness. However, it cause another problems, like the thin PEDOT:PSS has bad adhesion, so PEDOT:PSS can't be

used together with ITO. Thus, we use metal grid front electrode replace ITO glass. In this step, the PEDOT:PSS and front electrode doesn't contact as good as ITO glass. Finally, we took silver nanowires (AgNWs) to enhance our device conductivity and contact between PEDOT:PSS and front electrode. The schematic diagram of hybrid silicon solar cells is shown in Figure 1(a). The band diagram is shown in Figure 1(b).

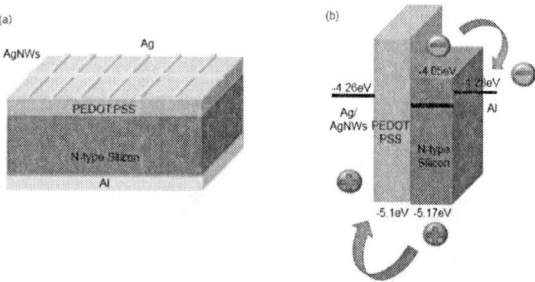

Figure 1. (a) Schematic diagram of device structure. (b) Band diagram of device structure.

II. EXPERIMENT

A. Fabrication of silicon nanostructures

In this experiment, 660-μm thickness N-type (100) mono-crystalline silicon with resistivity of 2-4 ohm-cm^2 was used. SiNS was fabricated by two-step metal-assisted chemical etching (MacEtch) [3,8]. First, the wafers were subsequently cleaned in acetone (ACE), de-ionized water (DI water), piranha solution (H_2SO_4:H_2O_2) at 120°C, isopropanol alcohol (IPA) and dried by nitrogen blow. In between, the wafer was rinsed with DI water. After cleaning, Si wafers were immersed in a MacEtch aqueous solution containing $AgNO_3$/HF for 5 seconds. This step is for silver deposition. Then, the silver-coated wafer was immersed in an aqueous solution with HF and H_2O_2 for 25 seconds. In this step, vertically aligned (100) Si nanostructures were formed. The as-prepared SiNS arrays were soaked in dilute HNO_3 to dissolve the silver dendrites. Afterwards, all samples were immersed in dilute HF to remove oxide. Besides, the chemical polishing etch (CPE) method was utilized to remove surface metal contamination and defect [8]. The CPE method was dipping 5 seconds in HNO_3 and HF mixture solution. Finally, the samples were soaked in dilute HF again to remove oxide and dried by nitrogen blow.

978-1-5090-5606-4/17 $31.00 © 2017 IEEE

B. Device fabrication

Rear metal Aluminum (Al) 120 nm were deposited successively on the back side of silicon wafer by thermal evaporation machine. Then, we select PEDOT:PSS to serve as our organic material. And mix IPA to dilute PEDOT:PSS. The PEDOT:PSS solution was spin-coat on SiNS and annealed under 120°C for 10 minutes in atmosphere. Then, spin casting AgNWs on device and annealed under 120°C for 10 minutes in atmosphere. Finally, Ag metal grid 150 nm was deposited on the front side by e-gun.

C. Characterization

For JV characteristic measurement, the device was illuminated under 1 sun AM 1.5 G 100mW/cm^2 by solar simulator SUN 2000, Abet Technologies, Inc, and measured by Keithley 2400 source meter. Scanning Electron Microscope (SEM) pictures of silicon nanostructures were observed by Hitachi S-4800 FE-SEM. Optical properties were measured by PerkinElmer Lambda 35 UV/V is system with an integrating sphere.

III. RESULTS AND DISCUSSION

The absorption cruve of PEDOT:PSS and dilute PEDOT:PSS are shown in Figure 2. We can see the dilute PEDOT:PSS has less absorption. And then we fabricated those two devices with different PEDOT:PSS. Its result shown in Table 1.

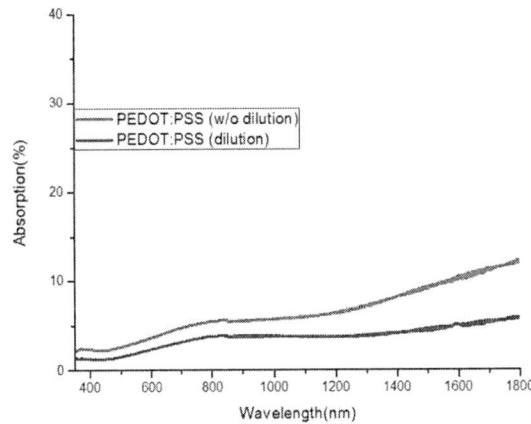

Figure 2. PEDOT:PSS (w/o dilution) and PEDOT:PSS (dilution) absorption of plane glass.

Coating Material	Photovoltaic parameters			
	$V_{oc}(V)$	$J_{sc}(mA/cm^2)$	$FF(\%)$	$PCE(\%)$
W/o dilution	0.4186	35.415	46.172	6.846
Dilution	0.4087	36.752	43.003	6.459

Table 1. Photovoltaic parameters of SiNS/PEDOT:PSS hybrid solar cell.

The result show the photocurrent which has obvious promotion, but the V_{oc} and fill factor (FF) doesn't better in this situation. We can see the SEM picture of w/o dilution and dilution in Figure 3(a) & (b). The dilute PEDOT will permeate into the nanostructures, it would cause many defects between the PEDOT and front electrode. For example, when hole generated by light incident to silicon, it must travel a long distance. It is in the Figure 3(c) & (d).

To solve the long distance which hole need to travel, we apply AgNWs to enhance the contact. Because AgNWs has small size and excellent conductivity, it would reduce the travel distance. It is shown in Figure 4.

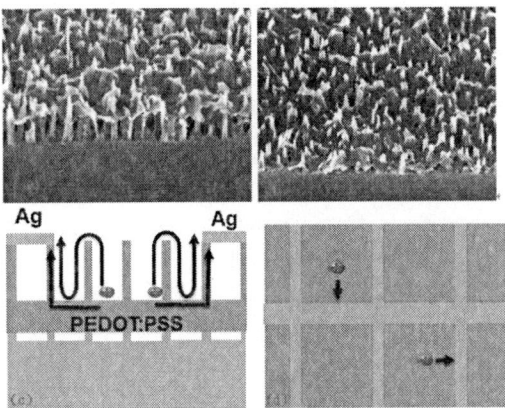

Figure 3. (a) SEM image of w/o dilute PEDOT coating on SiNS. (b) SEM image of dilute PEDOT coating on SiNS. (c) The cross section schematic of the hole travel distance. (d) The top view schematic of the hole travel distance.

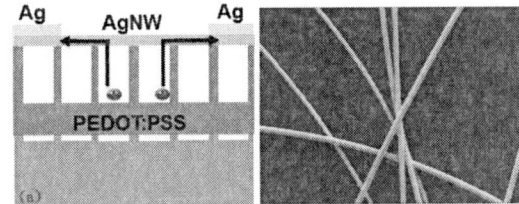

Figure 4. (a) The cross section schematic of casting AgNWs on device. (b) SEM image of casting AgNWs on SiNS.

Finally, we completed the device which has high performance. The external quantum efficiency (EQE) and JV curve shown in Figure 5 and Figure 6, respectively. Table 2 is the photovoltaic parameters of the device. After casting AgNWs on SiNS, the device performance can achieve power conversion efficiency of 13.37%, V_{oc} of 0.5592 V, J_{sc} of 35.79 mA/cm^2 and FF of 66.8%.

Coating Material	Photovoltaic parameters			
	$V_{oc}(V)$	$J_{sc}(mA/cm^2)$	$FF(\%)$	$PCE(\%)$
With AgNWs	0.5592	35.79	66.8	13.37

Table 2. Photovoltaic parameters of hybrid solar cell.

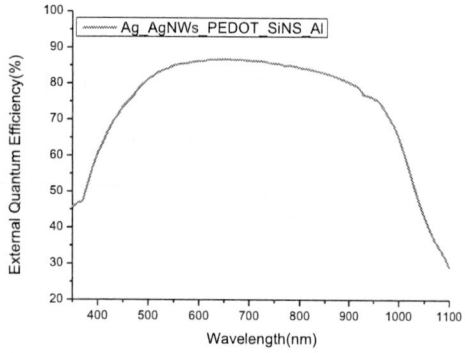

Figure 5. The EQE curve of hybrid solar cells.

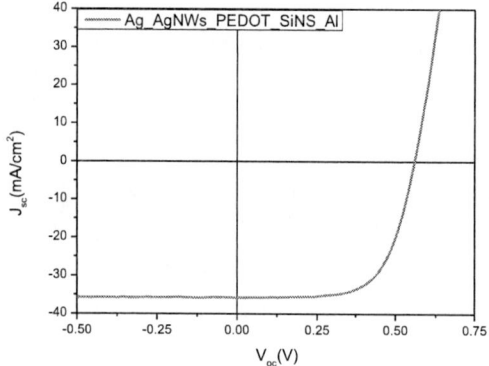

Figure 6. The JV curve of hybrid solar cells.

IV. CONCLUSION

We illustrate high performance of organic hybrid silicon solar cells, which can be easy process and low cost. Here, the SiNS can also give larger junction areas for carriers generation. Furthermore, the dilution can reduce the PEDOT:PSS absorption so that promote SiNS absorption. Besides, the device can be assist by AgNWs which improve the contact and reduce the defect in PEDOT and Ag Accordingly, organic hybrid silicon solar cells achieve PCE of 13.37%, V_{oc} of 0.5592 V, J_{sc} of 35.79 mA/cm^2, FF of 66.8%.

ACKNOWLEDGMENT

We gratefully acknowledge Ministry of Science and Technology of Taiwan and National Taiwan University for supporting the research with the contract numbers: MOST 103-2221-E-002 -132 -MY3, MOST 105-3113-E-002-010, MOST 105-2119-M-002-009, MOST 104-2221-E-002 - 139 -MY3.

REFERENCES

[1] S. W. Glunz, R. Preu and D. Biro, Crystalline silicon solar cells: State-of-the-art and future developments, ed. W. van Sark, Comprehensive Renewable Energy, Elsevier, Oxford, 2012, pp. 1–63.

[2] S. Wang, B. D. Weil, Y. Li, K. X. Wang, E. Garnett, S. Fan, and Y. Cui, "Large-Area Free-Standing Ultrathin SingleCrystal Silicon as Processable Materials," Nano Letters, 13, pp. 4393-4398, 2013.

[3] Z. Huang, N. Geyer, P. Werner, J. D. Boor, and U.Gosele, "Metal-Assisted Chemical Etching of Silicon: A Review," Adv. Mater. , 23, pp. 285-308, 2011.

[4] S. Thiyagu, C. Hsueh, C. Liu, H. Syu, T. Lin and C. Lin, "Hybrid organic–inorganic heterojunction solar cells with 12% efficiency by utilizing flexible film-silicon with a hierarchical surface," Nanoscale, 6, pp. 3361-3366, 2014.

[5] H. Syu, S. Shiu, Y. Hung, C. Hsueh, T. Lin, T. Subramani, S. Lee and C. Lin, Prog. Photovoltaics, 2013, 21, 1400–1410.

[6] Liu, Q., Ohki, T:, Liu, D., Sugawara, H., Ishikawa, R., Ueno, K., & Shirai, H. "Efficient organic/polycrystalline silicon hybrid solar cells." Nano Energy, 11, 260-266. (2015).

[7] Zhao, Z., Chen, X., Liu, Q., Wu, Q., Zhu, J., Dai, S., & Yang, S. "Efficiency enhancement of polymer solar cells via zwitterion doping in PEDOT: PSS hole transport layer." Organic Electronics, 27, 232-239. (2015).

[8] Subramani, T., Syu, H. J., Liu, C. T., Hsueh, C. C., Yang, S. T., & Lin, C. F. "Low-pressure-assisted coating method to improve interface between PEDOT: PSS and silicon nanotips for high-efficiency organic/inorganic hybrid solar cells via solution process." ACS applied materials & interfaces, 8(3), 2406-2415. (2016).

[9] Kuan-Ying Ho, Chi-Kang Li, Hong-Jhang Syu, Yi Lai, Ching-Fuh Lin and Yuh-Renn Wu*, "Analysis of the PEDOT:PSS/Si nanowire hybrid solar cell with a tail state model", J. Appl. Phys, 120, 215501, (2016).

Nickel Oxide Thin Films by Radio Frequency Sputter for Inverted Perovskite Solar Cells

Hyeonseok Lee, Yu-Ting Huang, and Shien-Ping Feng

Department of Mechanical Engineering, The University of Hong Kong, Hong Kong

Abstract — **The Fabrication of efficient hole transport layer (HTL) is one of essential components for high efficiency of inverted perovskite solar cells because the HTL plays an important role for hole transport, electron blocking, and light penetration for light energy conversion. In contrary to conventional solution process, we here fabricate the nickel oxide thin films by radio frequency sputter as HTL and investigate their optical and electrical properties. The conversion efficiency is reached to 12.39% with the control of deposition parameters.**

I. INTRODUCTION

Perovskite solar cells have become great interest for photovoltaic research owing to their striking progress on research only in few years. The light-to-electricity conversion efficiency has already been reached to more than 20% [1] because the perovskite have superior properties as an efficient active layer: a direct band gap of ~1.5 eV, a high absorption coefficient of $>10^4$ cm^{-1} at 550 nm, a low exciton binding energy of 20-30 meV, and long carrier diffusion length of 0.1-1µm, a simple fabrication process [2]. Widely adopted nip structure incorporating TiO$_2$ electron transport layer for the fabrication of perovskite solar cells have some disadvantages relating to instability in long term operation and severe hysteresis because of unbalanced carrier diffusion lengths. To solve these problems, inverted structure adopting different hole transport materials such as poly (3,4-ethylenedioxythiophene) poly (styrene sulfonate) (PEDOT:PSS) and nickel oxide (NiO) that lead pin structure have been suggested because the structure is favorable to achieve the balanced carrier diffusion length. However, the long term stability of perovskite solar cells with PEDOT:PSS is still questionable owing to its hygroscopic and acidic nature[2-4]. In this aspect, NiO would be better option for efficient and long term stable hole transport layers (HTL) owing to its inorganic nature, wide bandgap (3.6 eV), and favorable band alignment to perovskite layer [2, 5, 6].

For the deposition of the NiO layer, solution process through spin coating are predominant [2, 6]. Yet, the solution process is not favorable to obtain uniform and pinhole-free films. For this reason, physical vapor deposition such as sputtering would be one of solutions because deposition parameters are controllable in exact manner and conformal and compact films are easily formed via this technique. However, since Cui et al. [5] have reported their seminal work for sputtered NiO films for the inverted perovskite solar cells, there have no promising results that have been reported with

sputter and no breakthrough has been reported in conversion efficiency over 10%.

Here we demonstrate efficient inverted perovskite solar cells based on radio frequency (RF) sputtered NiO thin films. The conversion efficiency exhibits 12.39% with controlled deposition parameters. Optical and electrical properties are investigated.

II. EXPERIMENTAL DETAILS

NiO thin films were deposited by RF sputter from 80mm diameter NiO target. The base pressure was 5×10^{-4} Pa and the deposition pressure varied from 2 Pa to 6 Pa. The sputtering was carried out under argon atmosphere with 250 W power without any substrate heating on cleaned indium tin oxide (ITO) glasses. Prepared NiO thin films were subjected to annealing at 200 °C for 1h.

Perovskite layers were prepared, following the reference [6]. 2.3g of PbI$_2$, 0.8g of CH$_3$NH$_3$I were dissolved in the solvent that mixed with 1.5 mL of DMSO and 3.5 mL of DMF at 70 °C for 12h. Prepared solution were spread out by spin coating at 1000 rpm for 5s and 5000 rpm for 30s sequentially. During the second spin coating, 130 µl of chlorobenzene was dropped onto the samples. Resulted samples were dried at 100 °C for 30min.

Prepared [6,6]-phenyl C61-butyric acid methyl ester (PCBM) solution (20 mg/mL in chlorobenzene) was subsequently spin-coated onto NiO substrates at 1500 rpm for 30s. Bathocuproine (BCP) layers were deposited by spin coater at 4000 rpm for 40s from the BCP solution (0.5 mg/mL in anhydrous absolute ethanol). Finally, Ag metal contact was deposited by thermal evaporator.

III. RESULTS AND DISCUSSION

Photovoltaic conversion efficiency (PCE) of perovskite solar cells incorporating NiO thin films was evaluated under AM1.5 illumination as shown in Fig. 1 and Table 1.

The solar cells fabricated with NiO HTL prepared under 2 and 4 Pa of deposition pressure exhibit similar performance with small difference in FF. Among the samples, the perovskite solar cells with NiO thin films sputtered under 4 Pa deposition pressure shows remarkable performance. In case of the champion cell, J$_{sc}$ is 14.83 mA/cm^2 and PCE is 10.08%.

978-1-5090-5606-4/17 $31.00 © 2017 IEEE

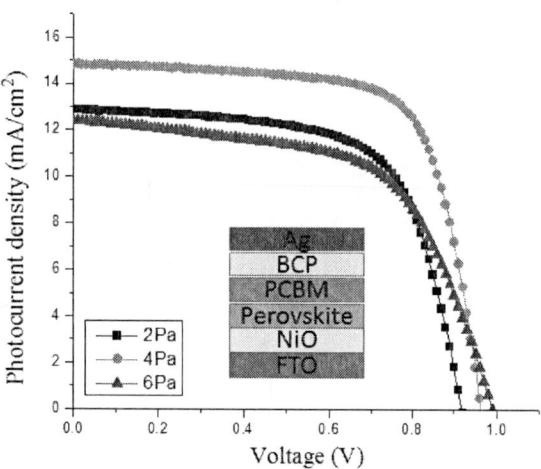

Fig. 1. J-V characteristics of inverted perovskite solar cells incorporating NiO thin films deposited under varied deposition pressure. The inset is the structure of the solar cells

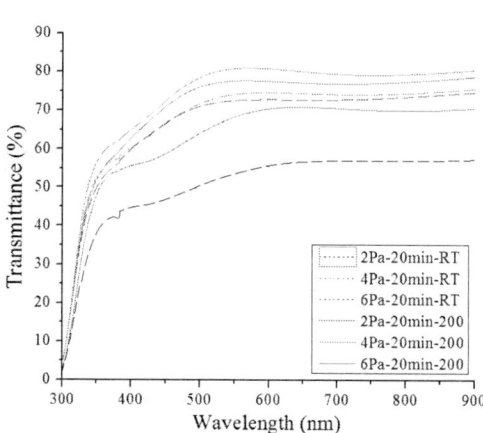

Fig. 2. Transmittance spectra of NiO thin films on ITO glasses sputtered with different deposition pressures. RT and 200 indicate the samples without and with annealing at 200 °C.

Deposition pressure (Pa)	Deposition time (min)	J_{sc} (mA/cm^2)	V_{oc} (V)	FF	PCE (%)
2	20	10.72	0.927	0.655	6.51
2	20	12.87	0.919	0.654	7.74*
4	20	13.12	0.948	0.656	8.15
4	20	14.83	0.964	0.706	10.08*
6	20	11.89	0.959	0.540	6.17
6	20	12.38	0.990	0.597	7.33*

Table 1. J-V characteristic parameters of the inverted perovskite solar cells. Asterisk(*) indicates the performance from the champion cell.

This outstanding performance of 4 Pa NiO sample is possibly resulted from the excellent optical transparency of the NiO thin films as shown in Fig. 2. 4 Pa NiO samples shows superior transparency through entire wavelength measured in comparison with other samples, even with and without annealing. This can contribute to improved photocurrent and PCE owing to larger availability of incident photon flux for efficient light absorption by perovskite.

Fig. 3 shows the work functions of NiO films measured by Kelvin probe. The work functions of the NiO samples are increased from 4.51 to 4.87 eV as the deposition pressure increased from 2 to 6 Pa for as-deposited samples. After annealing at 200 °C for 1h, all work functions of the NiO films are increased, indicating the enhancement of carrier concentration. Yet, considering the work function of ITO (4.6 eV), 2 Pa NiO films are more favorable in efficient collection of carriers at Ag/NiO interface because lower energy barrier is formed because of smaller difference in work functions.

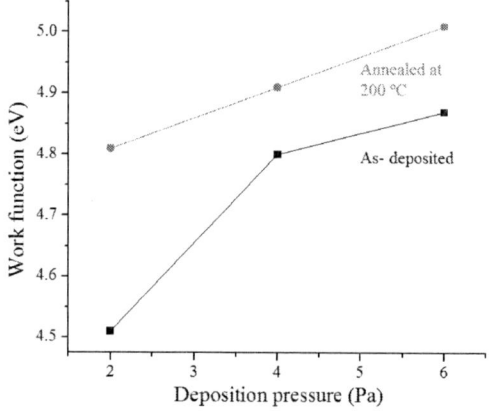

Fig. 3. Work function of NiO thin films by kelvin probe.

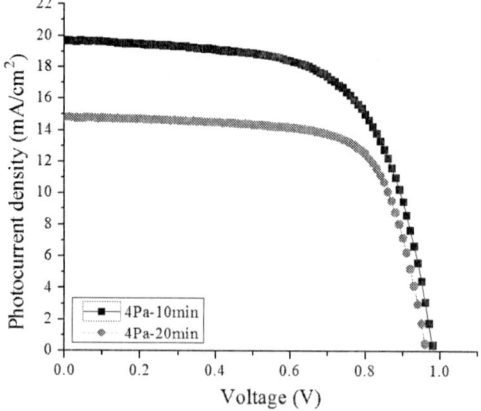

Fig. 4. Performance of inverted perovskite solar cells depending on deposition time of NiO.

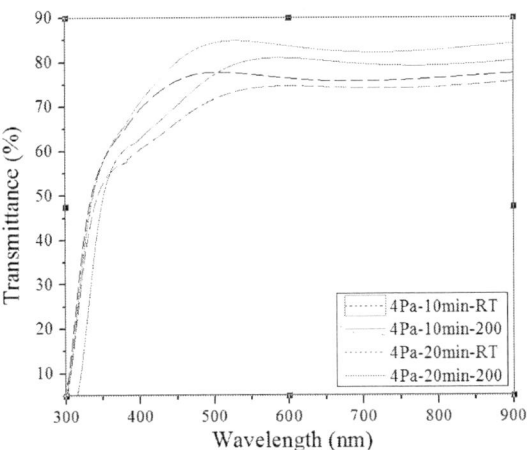

Fig. 5.   Transmittance spectra for deposition-time varied NiO thin films.

Therefore, it is speculated that this offset between transmittance and work function is the main reason for similar performance in 2 Pa and 6 Pa samples. In other words, 4 Pa possesses advantageous transmittance and work function for high efficiency.

In order to enhance the performance of solar cells further, the shorter time deposition (10 min) was implemented and inverted perovskite solar cells were fabricated with the NiO samples. All photovoltaic parameters are improved with NiO deposited for 10 min (4Pa-10min) in Fig. 4. The champion cell exhibits 19.69 mA/cm^2 and its PCE reached to 12.39 %. In the same manner, the improvement of performance can be explained by enhanced transmittance in 4Pa-10min-200 as shown in Fig. 5. In addition, decreased work function (4.91 eV → 4.87 eV) is also helpful in achieving higher performance of the solar cells with efficient carrier collection.

IV. CONCLUSIONS

Inverted perovskite solar cells incorporating RF sputtered NiO thin films as HTL are demonstrated. The performance of the solar cells were largely affected by transmittance of sputtered NiO thin films and work function of the NiO films was also influential factor for high efficiency of solar cells. With further control of deposition parameters, 19.69 mA/cm^2 of J_{sc} and 12.39% of PCE were achieved.

REFERENCES

[1]   W. S. Yang *et al.*, "High-performance photovoltaic perovskite layers fabricated through intramolecular exchange," *Science,* vol. 348, no. 6240, pp. 1234-1237, 2015.

[2]   S. Seo *et al.*, "An ultra-thin, un-doped NiO hole transporting layer of highly efficient (16.4%) organic–inorganic hybrid perovskite solar cells," *Nanoscale,* vol. 8, no. 22, pp. 11403-11412, 2016.

[3]   A. Garcia, G. C. Welch, E. L. Ratcliff, D. S. Ginley, G. C. Bazan, and D. C. Olson, "Improvement of Interfacial Contacts for New Small‐Molecule Bulk‐Heterojunction Organic Photovoltaics," *Advanced Materials,* vol. 24, no. 39, pp. 5368-5373, 2012.

[4]   A. M. Nardes, M. Kemerink, M. De Kok, E. Vinken, K. Maturova, and R. Janssen, "Conductivity, work function, and environmental stability of PEDOT: PSS thin films treated with sorbitol," *Organic electronics,* vol. 9, no. 5, pp. 727-734, 2008.

[5]   J. Cui *et al.*, "CH3NH3PbI3-based planar solar cells with magnetron-sputtered nickel oxide," *ACS applied materials & interfaces,* vol. 6, no. 24, pp. 22862-22870, 2014.

[6]   X. Yin *et al.*, "Highly efficient flexible perovskite solar cells using solution-derived NiO x hole contacts," *ACS nano,* vol. 10, no. 3, pp. 3630-3636, 2016.

Anomalous Efficiency Scaling with Dark Current in Perovskite Solar Cells

Vikas Nandal and Pradeep R. Nair

Department of Electrical Engineering, Indian Institute of Technology Bombay, Mumbai, Maharashtra, India-400076

Email: nknandal@iitb.ac.in, prnair@ee.iitb.ac.in

Abstract — **Deviation from superposition principle in carrier selective thin film solar cells have attracted immense recent research interest. In this context, we study the effect of dark current (J_o) on efficiency (η) of perovskite solar cells. Interestingly, we find that the efficiency scaling exhibit traditional solar cell trends (i.e. η is inversely proportional to J_o) when the dark current is limited by carrier life time. In contrast, we find that the efficiency scaling with dark current shows anomalous trends when limited by carrier injection and transport in perovskite. These puzzling results are applicable to a broad class of thin film carrier selective solar cells with p-i-n architecture – including perovskites and organic bulk heterojunction solar cells.**

I. INTRODUCTION

Organic and hybrid materials are increasingly explored for photovoltaics due to their low processing cost and excellent absorption properties [1]. Such materials in carrier selective p-i-n device architectures have reported excellent efficiencies – which include the emerging organic-inorganic perovskite, organic based bulk heterojunction solar cells [2]. Various doped high band gap organic and inorganic materials have been utilized to facilitate selective transport of carriers between energy harvesting (active) material and respective contacts [3]. Indeed, the device physics of such P-I-N heterojunction solar cells is quite different and complex as compared to conventional silicon P-N solar cells.

General theory of solar cells states that the current density $J(V)$ under illumination is the sum of dark, light, and recombination current density which can be written as:

$$J(V) = J_o \left(\exp\left(\frac{qV}{nkT}\right) - 1\right) - J_L + J_r(V), \qquad (1)$$

where, J_o is reverse saturation current, q is electronic charge, kT is thermal energy, n is ideality factor, J_L is the voltage independent light current, and $J_r(V)$ is the current due to the voltage dependent recombination of the photo-generated carriers. The principle of superposition holds for systems where - a) J_o is independent of generation rate G and b) $J_r(V) = 0$ [4] . In contrast to conventional silicon solar cells, the principle of superposition might not be valid for P-I-N based thin film carrier selective solar cells. To develop our hypothesis, let us consider two scenarios for recombination current inside P-I-N based cells: (a) $J_r = \alpha V$ and (b) $J_r = J_L(1 - \exp(-\beta V))$. Note that the above are for illustrative purposes, however, case (b) is similar to the results obtained by Crandall [5]. For $\alpha = \beta = 0$ (where

superposition holds), the efficiency indeed decreases with the increase in dark current ($\propto J_o$) of the diode (evident from fig. 1). Interestingly, Fig. 1 also indicates that the efficiency still improves with reduction in J_o even with significant bias dependent recombination (i.e., for non-zero α and β). These findings are in tune with the traditional optimization principle that a decrease in reverse saturation current results in improvement in the efficiency.

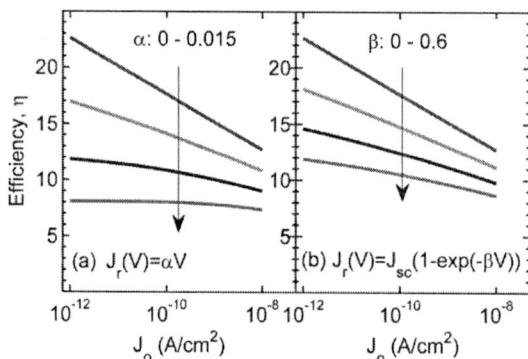

Fig. 1. Effect of dark current (characterized by J_o) on efficiency for devices with different values of - (a) α (0, 0.005, 0.010, 0.015) and (b) β (0, 0.2, 0.4, 0.6). Blue curve represents the device where the principle of superposition holds. Here, we consider that the dark current is limited by recombination of charge carriers (ideality factor, $n = 2$).

In this manuscript, we show that the above long held optimization thumb rule for solar cells is, surprisingly, only partially correct or may even be completely wrong for thin film carrier selective solar cells. For this, we performed detailed numerical simulations on perovskite based solar cells, although the conclusions could be equally valid for all devices with carrier selective P-I-N architecture (like organic bulk heterojunction solar cells) as well. Below, we first describe the model system and numerical simulations after which the device physics behind such anomalous scaling of efficiency is described in detail.

II. MODEL SYSTEM

Fig. 2 shows the schematic and band alignment of the modeled PIN device with perovskite (E$_g$=1.55 eV, 300 nm) as the active material. P ($N_a = 10^{18} cm^{-3}$) and N ($N_d = 10^{17} cm^{-3}$) type doped high bandgap semiconductors acts as

978-1-5090-5606-4/17 $31.00 © 2017 IEEE

hole and electron transport layer (HTL and ETL), respectively. Band alignment of such transport layers (TLs) ensures efficient extraction of photo-generated charge carriers by the corresponding contacts and dark current is limited by recombination of injected charges inside active layer (AL). Recombination flux of charge carriers is modeled by SRH (via mid gap trap states along with carrier life time $\tau = \tau_n = \tau_p$), radiative ($B = 3 \times 10^{-11} cm^3 s^{-1}$), and Auger ($A_n = A_p = 10^{-29} cm^6 s^{-1}$) recombination [6]. Charge carriers are excited by illuminating active material with constant uniform generation rate ($G = 4.75 \times 10^{21} cm^{-3} s^{-1}$, calibrated to achieve $J_{sc} = 22.8\ mA/cm^2$ for 100% collection efficiency). We have assumed equal dielectric constant of different layers and no valence (conduction) band offset at the interface of HTL (ETL)/AL (i.e. $\Delta E_v = 0, \Delta E_c = 0$). Parameters such as μ, τ, and ΔE_c are varied independently from $0.005 - 0.5\ cm^2 V^{-1} s^{-1}$, $10\ ns - 1\mu s$, and $0 - 0.4\ eV$, respectively, to study respective effects on dark and light I-V characteristics. The device characteristics of the modeled device are obtained by solving drift diffusion, continuity and Poisson's equation self-consistently at each voltage step under dark and illumination conditions.

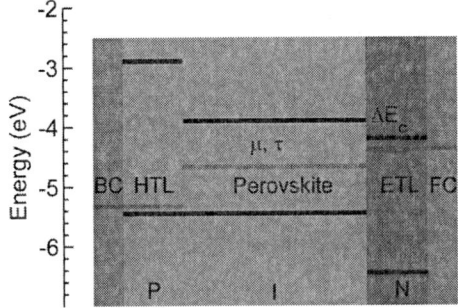

Fig. 2. Band alignment of different layers (BC/HTL/Perovskite/ETL/FC) used in carrier selective perovskite based solar cell. Dotted lines represent fermi energy level in each layer. Effects of parameters such as μ, τ (inside perovskite) and ΔE_c (at the interface of ETL/Perovskite) on dark and light current voltage characteristics are studied.

III. RESULTS AND DISCUSSIONS

Figure 3 shows the device characteristics (both dark and light) due to variation in parameters in active layer parameters like carrier life time τ, carrier mobility μ, and band offset ΔE_c with the ETL. Specifically, part (a) of Fig. 3 shows that the dark current reduces with the increase in τ due to reduction in J_o. Under illumination, we observe that the curvature of light current varies with τ and collection efficiency, below V_{oc}, improves with the increase in τ. Accordingly, the efficiency of the device improves with reduction in τ and hence a reduction in J_o. This observation is in accordance with the traditional analysis of eq. (1) in which a reduction in J_o is always accompanied by an improvement in the efficiency.

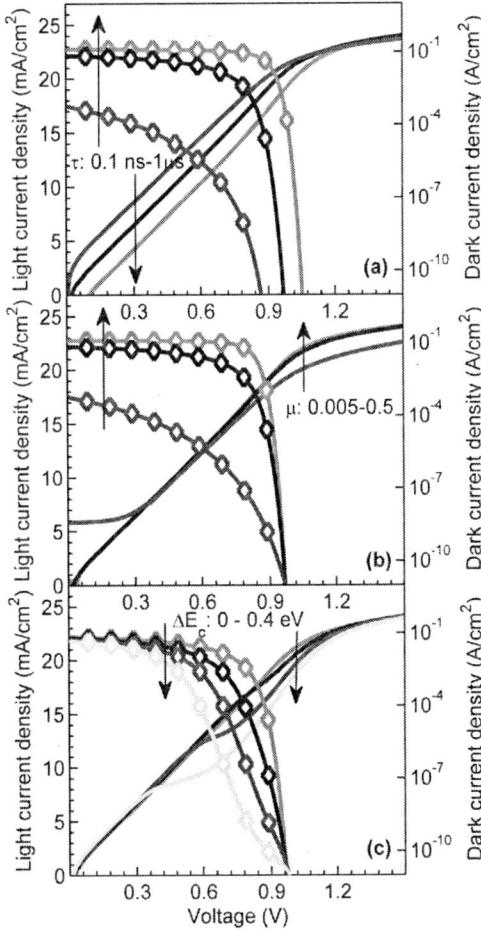

Fig. 3. Current voltage (I-V) characteristics of the modeled device under dark (on right axis with lines) and illumination (on left axis with symbol lines) conditions for different values of (a) carrier mobility μ (0.005,0.05, 0.5 $cm^2 V^{-1} s^{-1}$), (b) carrier life time τ (10ns, 100ns, 1μs), and (c) band offset ΔE_c (0, 0.2,0.3 0.4 eV). Fixed values of μ (for b, c), τ (for a, c), and ΔE_c (for a, b) are 0.05 $cm^2 V^{-1} s^{-1}$, 0.1μs, and 0 eV, respectively. Direction of arrow shows the effect of increased value of associated parameter on the dark and light I-V characteristics. We find that the dark current decreases and light current increases with carrier life time (conventional trend). However, scaling of the dark with light current is non-conventional with carrier mobility (both increases) and band offset (both decreases).

Figure 3 (b) shows the effect of carrier mobility μ on the dark and light I-V characteristics. We find that the dark current is mainly affected by μ in high biasing regime (around built-in voltage, V_{bi}). At very large bias, the dark current increases with μ because of the reduction in the series resistance offered by the active layer. However, note that the reverse saturation current, as extracted in the voltage range near V_{oc}, is rather insensitive to μ. In contrast, effect of μ on light current is more severe in terms of voltage range. Under illumination conditions, the light current depends on efficient extraction of photo-generated charge carriers by the contacts. Such collection efficiency in field driven carrier transport is given by $\eta = l_c(1 - \exp(-l/l_c))/l$ [5]. Here, l is AL

978-1-5090-5606-4/17 $31.00 © 2017 IEEE

thickness and $l_c=\mu\tau E$ (E is electric field inside AL) is carrier drift length. As a result, for $V < V_{oc}$ (open circuit voltage), light current improves with the increase in carrier mobility and resulted in better short circuit density J_{sc} and fill factor. The results indicate that both the dark and light current increases with μ (shown by arrows in fig. 2 a). Interestingly, we observe that a relative invariance in reverse saturation current results in significant improvement of light IV characteristics – an observation not anticipated in traditional analysis of eq. 1 (for $J_r = 0$).

Fig. 3 (c) shows the impact of conduction band offset ΔE_c at the interface of ETL/AL on I-V characteristics under dark and lighted conditions. We find that the dark current reduces with the increase in ΔE_c from 0 to 0.4 eV. This is due to the reduction in carrier injection from the ETL to the active layer. Since the dark current is due to carrier recombination in the active layer, reduction in carrier injection reduces the dark current significantly. However, the collection efficiency of photo-generated charge carriers by the contacts decreases with ΔE_c because electric field E inside active layer reduces with band offset. As a result, fill factor and efficiency degrades with band offset. Surprisingly, we find that the efficiency is directly proportional to dark current – an anomalous behavior which cannot be explained by conventional analysis of eq. 1.

The trends in Fig. 3 are summarized in terms of efficiency and J_o in Figure 4. Note that the J_o has been extracted through a linear fit on a semilog plot over a voltage range (near V_{oc} voltage) in dark I-V characteristics. The results indicate that the variation of efficiency with J_o depends on the limiting factor of dark current. Efficiency scaling with the dark current is more conventional (i.e. efficiency is inversely proportional to J_o) when J_o is determined by carrier life time τ. Further, we find that the effect of mobility on efficiency at a particular J_o can be anticipated by an increase in J_o in eq. 1. However, due to change in ΔE_c, the efficiency shows unique and anomalous (increasing) behavior with dark current. This is due to the simultaneous reduction in J_o and increase in J_r.

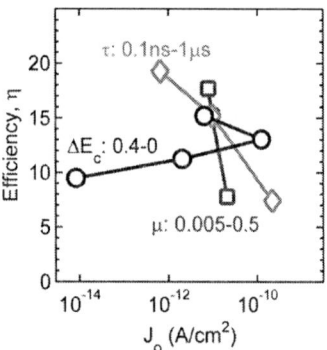

Fig. 4. Effect of the dark current on efficiency of perovskite based solar cell. Each curve (with associated independently varied parameter-τ (red diamond), μ (blue square) and ΔE_c (black circle)) is obtained by extracting efficiency η and reverse saturation current J_o from light and dark I-V characteristics, respectively.

IV. CONCLUSIONS

Through detailed numerical simulations of perovskite based solar cells, we investigated the correlation between dark and lighted I-V characteristics for different values of τ, μ and ΔE_c in carrier selective p-i-n based devices. We find that, due to change in τ, the efficiency scaling with dark current follows traditional solar cell behavior. However, the efficiency trends with dark current are non-traditional with changes in other parameters like band offset ΔE_c and carrier mobility μ. Such anomalous scaling of effciency with dark current illustrates the complex optimization associated with thin film carrier selective solar cells. Indeed, our results are relevant for all p-i-n based thin film solar cells like perovskites and organic bulk heterojunction solar cells.

ACKNOWLEDGEMENT

This article is based upon work supported under the US-India Partnership to Advance Clean Energy-Research (PACE-R) for the Solar Energy Research Institute for India and the United States (SERIIUS), funded jointly by the U.S. Department of Energy (Office of Science, Office of Basic Energy Sciences, and Energy Efficiency and Renewable Energy, Solar Energy Technology Program, under Subcontract DE-AC36-08GO28308 to the National Renewable Energy Laboratory, Golden, Colorado) and the Government of India, through the Department of Science and Technology under Subcontract IUSSTF/JCERDC-SERIIUS/2012 dated 22nd November 2012. The authors also acknowledges Center of Excellence in Nanoelectronics (CEN) and National Center for Photovoltaic Research and Education (NCPRE), IIT Bombay for computational facilities.

REFERENCES

[1] S. De Wolf, J. Holovsky, S.-J. Moon, P. Löper, B. Niesen, M. Ledinsky, F.-J. Haug, J.-H. Yum, and C. Ballif, "Organometallic Halide Perovskites: Sharp Optical Absorption Edge and Its Relation to Photovoltaic Performance.," *J. Phys. Chem. Lett.*, vol. 5, no. 6, pp. 1035–9, Mar. 2014.

[2] National Renewable Energy Laboratory, "Best Research - Cell Efficiencies chart." [Online]. Available: www.nrel.gov/ncpv/images/efficiency_chart.jpg.

[3] G. Yang, H. Tao, P. Qin, W. Ke, and G. Fang, "Recent progress in electron transport layers for efficient perovskite solar cells," *J. Mater. Chem. A*, vol. 4, no. 11, pp. 3970–3990, 2016.

[4] F. A. Lindholm, J. G. Fossum, and E. L. Burgess, "Application of the superposition principle to solar-cell analysis," *IEEE Trans. Electron Devices*, vol. 26, no. 3, pp. 165–171, 1979.

[5] R. S. Crandall, "Modeling of thin film solar cells: Uniform field approximation," *J. Appl. Phys.*, vol. 54, no. 12, 1983.

[6] S. Agarwal and P. R. Nair, "Device engineering of perovskite solar cells to achieve near ideal efficiency," *Appl. Phys. Lett.*, vol. 107, no. 12, p. 123901, Sep. 2015.

978-1-5090-5606-4/17 $31.00 © 2017 IEEE

Numerical simulation and performance optimization of perovskite solar cell

Sai Naga Raghuram Nanduri, Mahbube K. Siddiki, Ghulam M. Chaudhry, Yahya Z. Alharthi

Department of Computer Science and Electrical Engineering, University of Missouri-Kansas City, MO
64110

Abstract — The organic-metal halide perovskite is emerging technology in photo voltaic solar cells. For any solar cell to get the significant efficiency depends on various design parameters such as material thickness, device architecture, doping concentration etc.. There are many solar cell simulation software which can be used to carry out simulation and thus optimization based on those parameters. Here for this research, we are using wxAMPS because it is freely available, simple, efficient and quite popular in solar cell research society. Keeping external parameters constant (one-sun, AM1.5G solar radiation, 1000 w/m² irradiation), numerical simulation and analysis of a perovskite solar cell with configuration ETM/Perovskite/HTM was carried out. In the proposed configuration Zinc oxide (ZnO) was used as Electron Transport Material (ETM) as), mixed halide perovskite ($CH_3NH_3PbI_{3-x}Cl_x$) was used as absorber material, and Copper thiocyanate (CuSCN) was used as Hole transport material (HTM). The performance of the solar cell was optimized using layer by layer optimization techniques. The results show a significant enhancement in conversion efficiency of 14.4% (Jsc = 15.44 mA/cm², Voc = 1.08 V, and FF = 0.85). The solar cell need further optimization which is under progress. However these simulation results can help researchers to reasonably choose materials and optimally design high-performance perovskite solar cells.

I. INTRODUCTION

We have recently witnessed a breakthrough in highly efficient solar cells, where the organolead halide perovskite, $CH_3NH_3PbI_3$, was used as an absorber of sunlight. The perovskite was used to sensitize mesoporous TiO_2 films in a solid-state mesoscopic solar cell to deliver a power conversion efficiency (PCE) of up to 9.7%. A short period later, mesoporous Al_2O_3 was employed as a scaffold to support the formation of continuous thin films of a mixed halide perovskite, $CH_3NH_3PbI_{3-x}Cl_x$, to form non-sensitized solar cells. This so-called "meso-superstructure" perovskite solar cell had a PCE as high as 10.9%. This study showed that the mixed halide perovskite could function as both a light absorber and an electron absorber [1]. Most reported device architectures has adopted spiro-OMeTAD or PEDOT:PSS/ organic hole transporting layer, which is less than ideal owing to their tendency to absorb water, and inability to block electrons. Using P-type inorganic material such as copper thiocyanate (CuSCN) could be a long term solution for constructing stable perovskite-based solar cells [2]. Compared with organic spiro-OMETAD, inorganic p-type CuSCN with similar physical property, good chemical stability, ease of synthesis procedure,

and low cost is an ideal kind of hole transport material. Moreover CuSCN can be easily compatible with flexible substrates, which is deposited with a solution-processed technique at low temperature [3]. In this paper we have proposed a new architecture of a perovskite solar cell using CuSCN as hole transport layer, mixed halide perovskite material $CH_3NH_3PbI_{3-x}Cl_x$ and ZnO as electron transport layer. In order to provide experimental guidance, we performed a theoretical study on band offset properties between perovskite materials and inorganic HTM by using wxAMPS (Analysis of Microelectronic and photonic structure). The wxAMPS program incorporates intra-band tunneling model and trap assisted tunneling model which provides more realistic characteristics for heterojunction solar cells [4].

II. DEVICE SIMULATION PARAMETRS AND PROCEDURE

Many modeling tools support a constant multiplicative reflection factor for the front surface (i.e. RF = 0.1, 10% reflection). Quantum efficiency (QE) is then reduced by this fraction and, if interference effects are neglected, QE will show a fairly flat response at intermediate wavelengths of ~1-RF [5]. The Architecture discussed in this paper is FTO/ZnO/CH₃NH₃PbI₃₋ₓClₓ/CuSCN/Au. Here FTO is fluorine doped tin oxide coated on glass as anti-reflective coating is used on top of the architecture [6]. It also acts as cathode. Also we have used Electron Transport Material (ETM) as Zinc Oxide (ZnO) and Absorber as mixed halide perovskite ($CH_3NH_3PbI_{3-x}Cl_x$) and Hole Transport Material (HTM) as Copper thiocyanate (CuSCN). And at the bottom we have used gold (Au) metal as anode. The light reflection of the top and bottom contacts were set to be 0 and 1 respectively. The material parameters were summarized in table 1, most of which were selected from reported experimental works. The thickness of various layers were varied to find out their optimum values [7].

III. ARCHITECTURE OF PEROVSKITE SOLAR CELLS

Every solar cell have its own architecture contains mainly Hole Transport Material(HTM), Electron Transport Material(ETM) and active layer or absorber layer. The solar cell with a perovskite layer sandwiched between ZnO and CuSCN is shown in fig.1. The perovskite layer is the absorber or active layer used in this solar cell [8]. The presence of HTM not only

978-1-5090-5606-4/17 $31.00 © 2017 IEEE

TABLE I

SIMULATION PARAMETERS OF PEROVSKITE SOLAR CELL IN THIS STUDY

Properties	Symbol	CuSCN [3]	ZnO [3]	$CH_3NH_3PbI_{3-x}Cl_x$ [2]
Dielectric Constant	ε_r	3.9	8.12	30
Band Gap	E_g (eV)	2	3.4	1.5
Electron Affinity	χ(eV)	3.9	4.29	3.93
Electron Mobility	$\mu_n(cm^2V^{-1}S^{-1})$	0.2	2-30	14
Hole Mobility	$\mu_h(cm^2V^{-1}S^{-1})$	0.2	2	14
Acceptor Concentration	$N_A(cm^{-3})$	0	9×10^{16}	6×10^{14}
Donor Concentration	$N_D(cm^{-3})$	2.93×10^{17}	10^{19}	0
Effective Conductive Band Density	$N_c(cm^{-3})$	2.5×10^{21}	3.7×10^{18}	2.5×10^{20}
Effective Valence Band Density	$N_v(cm^{-3})$	2.5×10^{21}	3.7×10^{18}	2.5×10^{20}

favors the hole transport but also blocks the electron transferring from the perovskite layer to the electrode, which can improve the PCE and Voc of the devices.

The instability of perovskite at high relative humidity is one issue that needs to be addressed. One method used to solve this problem is the creation of a mixed halide perovskite. The simple solution mixture of $CH_3NH_3PbI_3$ and $CH_3NH_3PbCl_3$ was reported to result in a solid solution of $CH_3NH_3PbI_{3-x}Cl_x$. Since triiodide and trichloride perovskite have a band gap difference, the solid solution resulted in band gap tuning and color control [9].

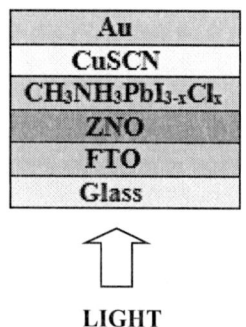

Fig.1 perovskite solar cell architecture

IV. RESULTS AND DISCUSSION

The results are obtained by method of optimization technique. The thickness of the perovskite is varied from 350nm to 750nm. The perovskite thickness optimization curves are shown in fig.2.

A) *Optimization of perovskite thickness:*

Thickness of a material plays a major role in performance of solar cells. The active layer i.e. perovskite layer thickness is varied from 350 nm to 750 nm. As the thickness increases the performance of the solar cell increases which has been shown

in fig.2. At a certain thickness of 600 nm solar cell performance began to decrease.

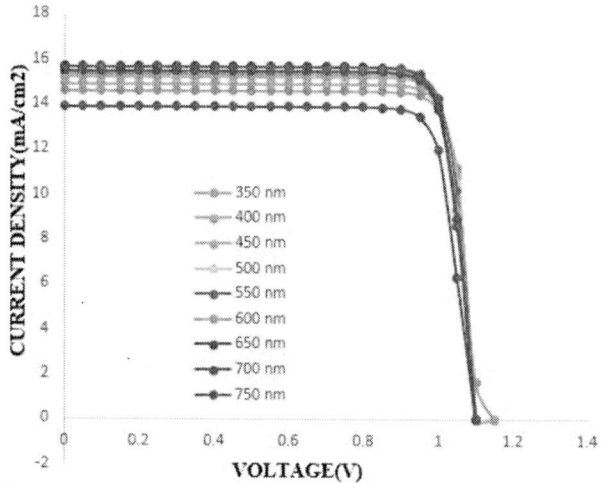

Fig.2 perovskite thickness optimization curves

Therefore the optimum thickness of this perovskite solar cell is 550 nm. The simulated Jsc and PCE of $CH_3NH_3PbI_{3-x}Cl_x$ with varying perovskite thickness are shown in fig.3. The first graph is obtained by giving thickness values on x-axis and current density on y-axis. The current density is highest at 550nm. The second graph is obtained by giving thickness values on x-axis and power conversion efficiency on y-axis. The power conversion efficiency is 14.4% at 550nm.

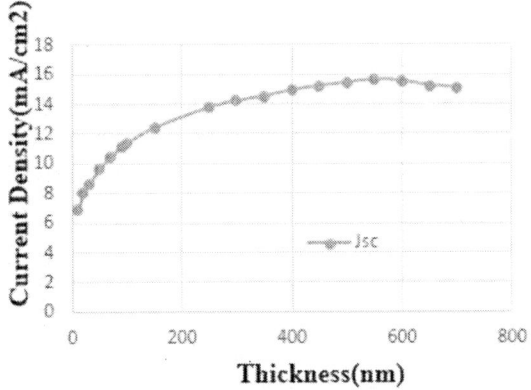

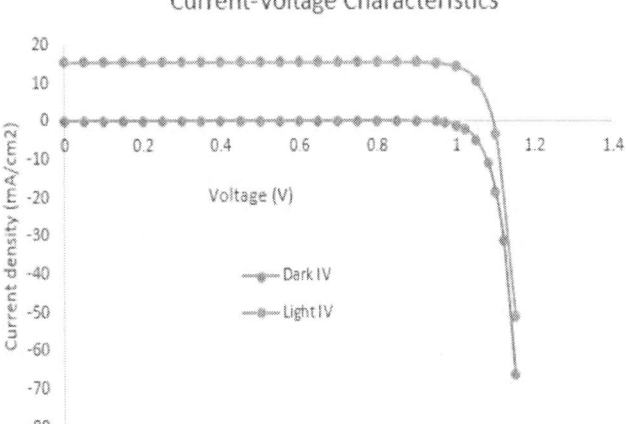

Fig.4. The J-V characteristic curves for solar cells as a function of defect densities in perovskite materials.

By giving the inputs like electrical, optical parameters of each layer such as ETM, Absorber, HTM we have obtained a conversion efficiency of 14.4% (JSC = 15.44 mA/cm^2, VOC = 1.08 V, AND FF = 0.85). The Dark and Illuminated I-V Characteristics of this perovskite solar cell is determined in the graph below.

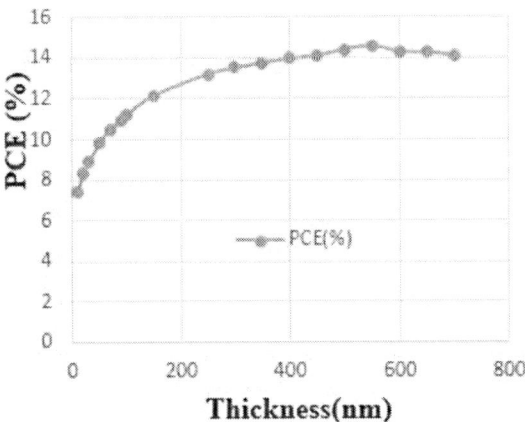

Fig.3. The Jsc and PCE of perovskite solar cell simulated with $CH_3NH_3PbI_{3-x}Cl_x$

B) *Solar cell Performance on defects in perovskite materials*

The density and property of defects in perovskite material play a crucial role due to photoelectrons being mainly generated and recombined in light absorber. The Gaussian distribution is more suitable to describe these defect states because of exiting a lot of defect energy levels in the perovskite[10] . So we assumed that there are both acceptor states and donor states modeled by Gaussian mid-gap states. The simulation results for J-V characteristics with various defect density are shown in fig.4. It shows that PCE , Jsc, Voc attain maximum values when the defect density is 10^{14} cm^{-3}. The increasing of defect density brings about the drop of PCE, Jsc, Voc of perovskite solar cell. The ideal defect density value for perovskite is 10^{14}cm^{-3}. The materials having less defect density are the best materials to use in a solar cell. Here perovskite material have very less defect density when compared to other materials. Hence this is one of the reasons to select perovskite material.

Fig.5 The J-V characteristics curves for perovskite solar cell.

VI. CONCLUSION

Our results indicates that copper thiocyanate was a suitable hole transport material for perovskite solar cells. An efficiency of 14.4% could be achieved, which shows potential of using CuSCN as an alternate hole transport material. This study also provides important theoretical guidance for new mixed halide perovskite has an good absorption properties and researchers can optimize for good results. Coming to future work by

978-1-5090-5606-4/17 $31.00 © 2017 IEEE   1020

employing a narrower bandgap perovskite and adopting a multijunction approach would be likely to push the PCE of perovskite solar cells.

V. REFERENCES

[1] Y. Wang, Z. Xia, Y. Liu, and H. Zhou, "Simulation of perovskite solar cells with inorganic hole transporting materials," in *2015 IEEE 42nd Photovoltaic Specialist Conference, PVSC 2015*, 2015, pp. 3–6.

[2] B. M. Soucase, I. Guaita Pradas, and K. R. Adhikari, "Numerical Simulations on Perovskite Photovoltaic Devices," in *Perovskite Materials - Synthesis, Characterisation, Properties, and Applications*, 1st ed., L. Pan and G. Zhu, Eds. Rijeka, 2016, pp. 445–488.

[3] A. Zhang, Y. Chen, and J. Yan, "Optimal design and simulation of high-performance organic-metal halide perovskite solar cells," *IEEE J. Quantum Electron.*, vol. 52, no. 6, p. 1600106, 2016.

[4] P. P. Boix, K. Nonomura, N. Mathews, and S. G. Mhaisalkar, "Current progress and future perspectives for organic/inorganic perovskite solar cells," *Mater. Today*, vol.

[5] D. K. 5, "Discovery Could Dramatically Boost E ciency of Perovskite Solar Cells," *Berkeley Lab Res.*, 2016.

[6] H. J. Snaith, "Next generation solar cells," no. January, p. 31, 2015.

[7] Y. Ma, S. Wang, L. Zheng, Z. Lu, D. Zhang, Z. Bian, C. Huang, and L. Xiao, "Recent Research Developments of Perovskite Solar Cells," *Chinese J. Chem.*, vol. 32, no. 10, pp. 957–963, 2014.

[8] K. T. Butler, J. M. Frost, and A. Walsh, "Band alignment of the hybrid halide perovskites CH3NH3PbCl3, CH3NH3PbBr3 and CH3NH3PbI3," *Mater. Horizons*, vol. 2, pp. 228–231, 2015.

[9] N. G. Park, "Perovskite solar cells: An emerging photovoltaic technology," *Mater. Today*, vol. 18, no. 2, pp. 65–72, 2015.

[10] P. Loper, B. Niesen, S.-J. Moon, S. Martin de Nicolas, J. Holovsky, Z. Remes, M. Ledinsky, F.-J. Haug, J.-H. Yum, S. De Wolf, and C. Ballif, "Organic-Inorganic Halide Perovskites: Perspectives for Silicon-Based Tandem Solar Cells," *IEEE J. Photovoltaics*, vol. 4, no. 6, pp. 1545–1551, 2014.

Performance Prediction for Large Area Perovskite Solar Cells

Yojak Raote, Hitarth Choubisa and Pradeep R. Nair,
Department of Electrical Engineering
Indian Institute of Technology Bombay, Mumbai, Maharashtra, 400076, India

Abstract— **Perovskite solar cells have attracted great attention due to large efficiencies achieved over the past few years. While both small and large area perovskite solar cells are being explored, majority of the high efficiencies reported are using small area cells (~0.1cm²). Therefore, from a process optimization perspective, it is necessary to predict large area performance of these small area solar cells. In this article, we achieve the same based on compact models for small area cells. The methodology is validated by comparison with experimental data. Our results indicate that transparent conductive oxide (TCO) sheet resistance is major performance limitation for large area perovskite solar cells – which could be improved with the help of metal bus bar.**

Index Terms— **Large area perovskite solar cells, modeling, simulation.**

I. INTRODUCTION

Organic-inorganic perovskite solar cells are considered as a promising candidate towards high efficiency low cost solar cells. Because of nature of materials involved, manufacturing cost of organic-inorganic perovskite solar cells is small as compared to inorganic solar cells. However, to be counted as a mature technology, it is essential for the community to demonstrate high efficiency large area devices with excellent stability. Most of reports with good efficiencies are reported on small area cells (area ~0.1cm²). While this is important as a proof of concept, for commercial applications demonstration of excellent efficiency over large areas is essential. Recently, there has been some reports on large area cells, with efficiency approaching 20% for an aperture area ~1cm² [1]. While this is an exciting development, it is imperative that the community develops modeling tools or schemes to predict the promises and prospects of large area perovskite solar cells from the champion small area solar cells.

In this article, we address the above mentioned technical challenge. Specifically, we identify an appropriate methodology to model the area scaling of perovskite solar cells. To this end, we (a) identify a compact model for small area perovskite solar cells (section II), (b) calibrate the model parameters using experimental results (section III), (c) identify the effect of performance limiting factor like TCO resistance towards area scaling (section IV), and (d) discuss the solutions in terms of metal grid to overcome the above limitation (section V).

II. MODEL SYSTEM

Typical electrical model (series-shunt, SS) used for solar cells is shown in the inset of Figure 1. Here, diode captures the exponential relation between current and voltage. Resistors R_s and R_{sh} represent the series and shunt resistance, respectively. This model assumes that the photo generated current is bias independent and hence superposition holds good between the dark and light IV characteristics. Recently, a physics based analytical model for perovskite solar cells was proposed [2]. This model is valid for devices with heavily doped transport layers – which need not be universally true for perovskite solar cells. Moreover, for perovskite solar cells superposition between dark and light characteristics is not always valid due to the bias dependence of photocurrent [3]. Further, perovskite solar cells show linearity in $J^{1/2}$ (J is the current density) versus applied bias (V) [4], in high bias region of dark current.

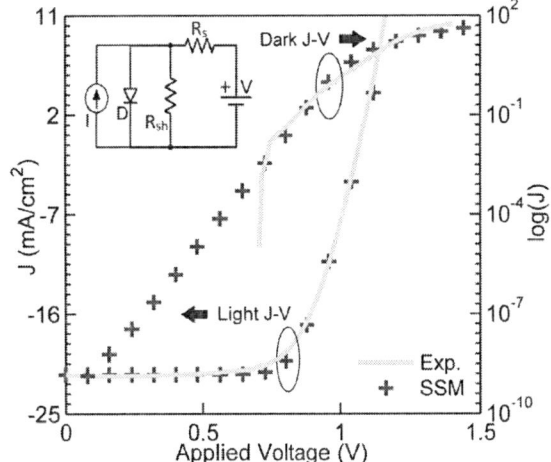

Fig. 1. Comparison between experimental and simulated J-V characteristics for a small area solar cell.

In spite of the above complexities we find the SS model can be used to study area scaling aspects of perovskite solar cells. A comparison of the model with experimental dark and light IV characteristics [5] is provided in Fig. 1. Here the parameters are chosen to match the light IV characteristics. This results in a not so optimal prediction of the dark IV characteristics due to the reasons discussed in the previous paragraph. Since the focus of this manuscript is on the performance prediction of large area solar cells under

978-1-5090-5606-4/17 $31.00 © 2017 IEEE

illumination, it is evident that the SS model is appropriate for the purpose. However, we note that the same model might need significant improvement to study shadow effects where accurate representation of dark IV could significantly influence the results.

A large area solar cell can be considered as a grid of small area cells, as shown in Figure 3 [6]. Each small area cell is connected to its neighbors through the sheet resistance of the Transparent Conducting Oxide (TCO). In fact, non-uniform deposition and sheet resistance of TCO constitute some of major technical challenges associated with large area cells. Non-uniform deposition leads to non-uniform generation and pinholes. Sheet resistance is major performance limiter through resistive power loss [7].

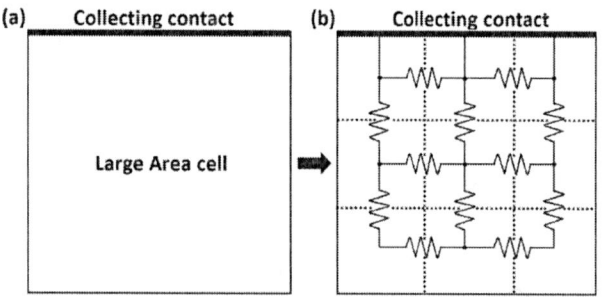

Fig. 2.   Method to simulate large area cell (a) Large area cell (b) Large area cell as matrix of small area cells interconnected through TCO sheet resistance.

For the small area cells, we use the SS model in Figure 1 with the parameters obtained through direct comparison from experimental data. With this calibrated model, the large area cell is simulated as per the scheme in Figure 2 with the current collection through one side as indicated. The entire large area is tessellated into smaller area cells, which are interconnected through the sheet resistance of TCO. Note that the first/top row of small cells is directly connected with collecting contact while all other neighboring connections are completed with TCO sheet resistance. For simulation, it is assumed that all layers of perovskite solar cell are uniform across the active area with uniform illumination so that performance of each small area cell in the grid is identical.

III.   VALIDATION WITH EXPERIMENTAL DATA

Here, we validate the performance predictions by SS model with experimental data. For validation of results, small area cell performance is matched with that of experimental small area cell (0.1cm^2) and then the large area performance is predicted. This prediction is then compared with experimental data of large area cell manufactured using same methodology. The data set is from reference 8 and the device structure is Glass/FTO/c-TiO$_2$/TiO$_2$-perovskite/Spiro-OMeTAD/Au prepared using airflow-assisted PbI$_2$ blade coating deposition process. Sheet resistance of TCO, as mentioned in the same reference, is 8 Ω/□.

Area (cm^2)	Type of data	V_{oc} (mV)	J_{sc} (mA/cm^2)	FF	η (%)
0.1	Exp.[8]	1010	17.4	0.69	12.1
	Sim.	1020	17.36	0.69	12.3
2.52	Exp.[8]	1027	17.17	0.58	10.3
2.5	Sim.	1020	17.33	0.61	10.9

Table 1.   Comparison between simulated and experimentally reported results for both small area and large area solar cells.

Table 1 compares predictions with experimental data for cells of two different area- 0.1cm^2 and 2.5cm^2. As mentioned before, our SS model was calibrated using the data of the small cell and the large area cell was simulated using this compact model and the TCO resistance using the methodology described in Fig. 2. The simulations predict the experimental trends reasonably well which indicate that our modeling methodology and the parameters chosen are appropriate and hence can be used to study the effect of important factors like TCO sheet resistance and metal bus bar.

IV.   EFFECT OF TCO RESISTANCE ON PERFORMANCE OF LARGE AREA CELL

TCO sheet resistance is one of the major performance limiting factor in case of large area cell. Hence, it is important to understand variation in performance with sheet resistance. Indium Tin Oxide (ITO) is generally used as TCO in case of perovskite solar cells. Typical sheet resistance of ITO is 15Ω/□ Figure 3 shows effect of change in TCO resistance for variable area. Here the small area cell parameters are taken to be the same as that in Figure 1 with efficiency close to 15%. Figure 3 indicates that as cell area increases the efficiency degrades. The efficiency degradation increases with TCO resistance. Hence, it is very important to reduce either the TCO resistance or use metal bus bars to overcome this series resistance effect. The latter is discussed in next section.

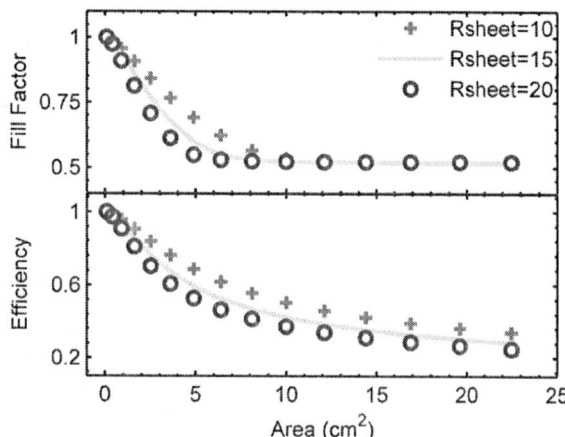

Fig 3.   Effect of TCO sheet resistance on cell performance. Normalized fill factor (top panel) and efficiency (bottom panel) variation is provided as a function of cell area and TCO sheet resistance.

V. PERFORMANCE IMPROVEMENT THROUGH METAL BUS BAR STRUCTURE

The discussion in the previous section indicates that as area increases, the efficiency decreases due to the sheet resistance of TCO. This effective series resistance can be reduced by use of metal grid which offers a low resistance path for current till collection point.

To evaluate the performance, we simulated bus bar type metal grid as shown in Figure 4. Metal grid dimensions and resistances of metal grid are adopted from reference [7]. Distance between two consecutive fingers is around 1mm. But, side length of the unit cell ($0.1cm^2$) used in our previous simulations is ~3mm. So, we divided our cell into 6x6 matrix of extra small cells and adjusted the parameters of those extra small cell model such that the characteristics of that 6x6 matrix of extra small cells match to that of our $0.1cm^2$ cell. With this modified set of parameters, we simulated the performance of a large cell of area $10cm^2$. Sheet resistance was assumed to be $15\Omega/\square$. Loss in efficiency due to the inactive area under the metal grid is accounted in the simulations.

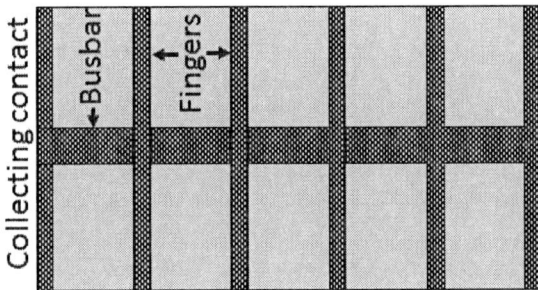

Fig. 4. Metal bus bar schematic to reduce the ill effects of TCO resistance.

Device structure	Area (cm^2)	Normalized			
		V_{oc}	J_{sc}	Fill factor	η
Without metal busbar	10	1	0.81	0.53	0.43
With metal grid	10.85	1	0.92	0.98	0.91

Table 2. Effect of metal grid on cell performance. Data is normalized with respect to small area cell ($0.1cm^2$) with Voc=1.08V, Jsc=21.36mA/cm², Fill Factor=0.68, Efficiency=15.68%.

The results shown in table 2 indicate that metal bus bars are essential for large area perovskite solar cells and can overcome the series resistance issues due to TCO. Specifically, with the metal bus bar, the large area cell is almost as efficient as the small area cell, except for the loss in efficiency due to the area occupied by the metal bus bar. This shows that metal grid helps to overcome TCO sheet resistance losses.

VI. CONCLUSION

To summarize, here we explored the effect of TCO resistance on the area scaling on perovskite solar cells based on well calibrated compact models and simulation methodology. Our results indicate that TCO sheet resistance effect reduces the efficiency by more than 50% as the area is scaled to $10cm^2$. This detrimental effect of TCO sheet resistance can be effectively countered through metal bus bars which results in a moderate reduction of 10% for the same large area solar cell.

ACKOWLEDMENT

This article is based upon work supported under the US-India Partnership to Advance Clean Energy-Research (PACE-R) for the Solar Energy Research Institute for India and the United States (SERIIUS), funded jointly by the U.S. Department of Energy (Office of Science, Office of Basic Energy Sciences, and Energy Efficiency and Renewable Energy, Solar Energy Technology Program, under Subcontract DE-AC36-08GO28308 to the National Renewable Energy Laboratory, Golden, Colorado) and the Government of India, through the Department of Science and Technology under Subcontract IUSSTF/JCERDC-SERIIUS/2012 dated 22nd November 2012. The authors also acknowledge Center of Excellence in Nanoelectronics (CEN) and National Center for Photovoltaic Research and Education (NCPRE), IIT Bombay for computational facilities.

REFERENCES

[1] Xiong Li1, Dongqin Bi, Chenyi Yi, Jean-David Décoppet, Jingshan et al. "A vacuum flash-assisted solution process for high-efficiency large-area perovskite solar cells" *Science* 09 Jun 2016; DOI: 10.1126/science.aaf8060

[2] X. Sun, R. Asadpour, W. Nie, A. Mohite, M. Alam "A Physics-Based Analytical Model for Perovskite Solar Cells," *Photovoltaics, IEEE Journal of* , vol.PP, no.99, pp.1,6; DOI: 10.1109/JPHOTOV.2015.2451000

[3] S. Agarwal and P. Nair "Device engineering of perovskite solar cells to achieve near ideal efficiency" *Applied physics letters* 107, 123901 (2015); DOI: 10.1063/1.4931130

[4] S. Agarwal, M. Seetharaman, et. al. "On the Uniqueness of Ideality Factor and Voltage Exponent of Perovskite-Based Solar Cells", *The Journal of physical chemistry letters,* 2014, 5, 4115–4121.

[5] M. Liu, M. Johnston & H. Snaith "Efficient planar heterojunction perovskite solar cells by vapour deposition" *Nature* 501, 395–398 (19 September 2013). DOI: 10.1038/nature12509

[6] G. T. Koishiyev and J. R. Sites, "Impact of sheet resistance on 2-D modeling of thin -film solar cells," *Sol. Energy Mater. Sol. Cells,* vol. 93, no. 3, pp. 350–354, Mar. 2009. DOI: 10.1016/j.solmat.2008.11.015

[7] S. Choi, W. Potscavage, Jr. and B. Kippelen, "Area-scaling of organic solar cells" *Journal of applied physics,* 106, 054507 (2009) DOI: 10.1063/1.3211850

[8] S. Razza, F. Giacomo, F. Matteocci et.al. "Perovskite solar cells and large area modules (100 cm2) based on an airflow-assisted PbI2 blade coating deposition process" *J. Power Sources*, 277 (2015) 286-291. DOI: 10.1016/j.jpowsour.2014.12.008

Photoconversion Efficiency Modeling in Perovskite Solar Cells

A.V. Sachenko[1], V.P. Kostylyov[1], A.V. Bobyl[2], V.M. Vlasiuk[1], I.O. Sokolovskyi[1, 3], E.I. Terukov[2,4] and M. Evstigneev[3]

[1]V. Lashkaryov Institute of Semiconductor Physics, NAS of Ukraine, 03028 Kiev, Ukraine

[2]Ioffe Institute RAS, 194021 St.-Petersburg, Russia

[3] Department of Physics and Physical Oceanography, Memorial University of Newfoundland, St. John's, NL, A1B 3X7 Canada

[4]TFTC Ioffe R&D Center, 194021 St.-Petersburg, Russia

Abstract — **A theoretical approach to model photoconversion efficiency in perovskite p–i–n structures is developed and used to analyze the experimental results published in the literature. It is shown that the surface of a formamidinium lead iodide (FAPbI$_3$) solar cell (SC) is spontaneously textured, whereas the surface of a CH$_3$NH$_3$PbI$_3$ SC is not. The current-voltage curve ideality factor equals 2 for FAPbI$_3$ and 2.3 for CH$_3$NH$_3$PbI$_3$-based SCs. This value is due not to recombination in the space-charge region. Rather, it can be explained by taking into account the contribution of the rear surface at high injection level. In the limit of negligible Shockley-Read-Hall and surface recombination and in the absence of optical losses, photoconversion efficiency for both materials is 29.2% at i-layer thickness of ca. 1 μm. When these recombination losses are taken into account, the maximal efficiency reduces to about 20%.**

Index Terms — **perovskite solar cells, photoconversion efficiency, surface texturing**

I. INTRODUCTION

In the last few years there has been quite impressive progress in enhancing photoconversion efficiency of perovskite solar cells (SCs) [1]. The record photoconversion efficiency achieved in such SCs exceeds 20% [1], [2]. Yet, there is a lack of publications on the theoretical modeling of these novel energy sources. This prevents one from being able to systematically optimize the parameters of perovskite SCs. In this contribution, we attempt to partially fill in this gap.

In most cases, perovskite SCs are based on the p–i–n structures (see, e.g., [2]). Carrier concentration in the i-region can be as high as 10^{11} cm^{-3}, see [3]. Therefore, this region is designated as i-type purely by convention, because intrinsic carrier concentration in a perovskite semiconductor with the bandgap of 1.6 eV at room temperatures should be about 10^5 cm^{-3}, six orders of magnitude below this figure.

In SCs based on semiconductors with low equilibrium majority carrier concentration n_0 in the base region, the condition $\Delta n \gg n_0$ is fulfilled, where Δn is the excess charge carrier density in the i-region. This happens due to the long Shockley-Read-Hall lifetimes, exceeding 1 ms, and low surface recombination velocity below 10 cm/s, as a result of which Δn in the open-circuit regime may reach the value of the order of 10^{16} cm^{-3}. A similar situation is realized, e.g., in high-efficiency heterojunction solar cells (HJSC) with the base doping level $\leq 10^{15}$ cm^{-3} [4], [5].

Although the estimates indicate that the achievable values of Δn in perovskite SCs are about an order of magnitude smaller than in HJSCs, the concentration of majority carriers in the i-region in perovskite SCs may be one or two orders of magnitude smaller than in the HJSCs as well. As a result, it is possible to fulfil the condition $\Delta n \gg n_0$ in perovskite p–i–n structure.

In the present contribution, an approach to model photoconversion efficiency in these elements is developed. Two cases are analyzed: the limiting case, in which the maximal efficiency is achieved, and the case of realistic SC parameter values. In the latter situation, the theoretical results are compared to the experimental data. The approach proposed here, based on the charge continuity equation and current balance equations, allows one to determine the SC characteristics also for the case of a p$^+$–n–n$^+$ (or n$^+$–p–p$^+$) structure. This case is relevant when the base doping level is high.

II. THEORY

The theory of photoconversion in p–i–n structures presented here is valid if the following criteria are fulfilled: (i) Diffusion length, L_d, is much higher than i-region thickness, d; (ii) Excess electron-hole pair concentration, Δn, due to illumination, notably exceeds the equilibrium carrier concentration in the i-region, n_0. These criteria allow one to use the approach developed for heterojunction solar cells modeling, see [4] and the contribution [5] in the present issue. The first condition (i) implies that excess concentration Δn is uniform in the i-region. The second condition (ii) allows its determination as

$$\Delta n(V) = n_i \left(e^{\frac{q}{mkT}(V - A_{SC}J)R_S} - 1 \right), \quad (1)$$

see equation (3) of [5] in the limit $\Delta n \gg n_0$, p_0 (p_0 is minority carrier concentration). In (1), J is the current density in the presence of illumination, V is the applied voltage, J_{SC} is the short-circuit current density, R_S is the series resistance, q is the elementary charge, kT is thermal energy, A_{SC} is the SC surface area, and $n_i = (N_c N_v)^{1/2} \exp[- E_g/(2kT)]$ is the intrinsic charge carrier concentration in the material with the effective densities of states N_c and N_v in the conduction and valence bands and the bandgap E_g. The ideality factor that follows

TABLE I
PARAMETERS OF PEROVSKITE SCs.

SC material	τ_{SRH}, μs	A, cm^3/s	J_{SC}, mA/cm^2	d, nm	S, cm/s	R_s, Ω	R_{sh}, kΩ	V_{OC}, V	η, %	FF, %	m
FAPbI$_3$	0.2	$9 \cdot 10^{-11}$	24.7	520	15	4.5	1.5	1.104	19.7	72.3	2
CH$_3$NH$_3$PbI$_3$	0.5	$9 \cdot 10^{-11}$	20.1	150	500	1.5	1	1.07	16.3	76.1	2.3

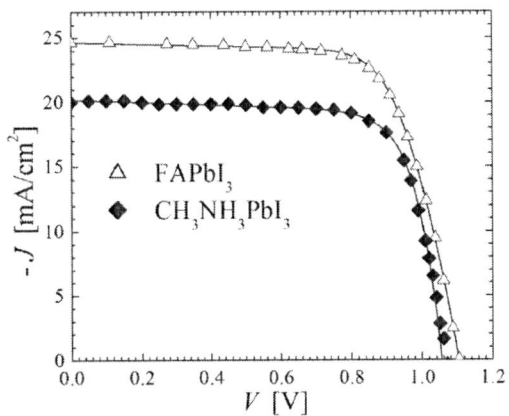

Fig. 1. Illuminated I-V curves for a FAPbI$_3$-based SC from [2] and CH$_3$NH$_3$PbI$_3$-based SC from [6]. Symbols: experiment, solid lines: theory.

from the theory presented in [5] is $m = 2$. This value is due not to the recombination in the space-charge region, but to the high excitation level and the rear surface contribution to the I-V curve. The illuminated *I-V* curve is described by the expression (see equations (7) and (8) from [5])

$$ J(V) = \frac{\left(qd\Delta n / \tau_{eff} - J_{SC} \right) R_{sh} + V A_{SC}^{-1}}{R_s + R_{sh}} , \qquad (2) $$

where R_{sh} is the shunt resistance. We assume that only Shockley-Read-Hall and radiative recombination channels are operative in the bulk. Thus

$$ \frac{1}{\tau_{eff}} = \frac{1}{\tau_{SRH}} + A\Delta n + \frac{S}{d} , \qquad (3) $$

where τ_{SRH} is Shockley-Read-Hall recombination lifetime, A the radiative recombination parameter, and S the surface recombination velocity

The short-circuit current density should be determined experimentally, and its substitution into (2) gives a transcendental equation which should be solved numerically to obtain the current density J for a given voltage V. The open-circuit voltage V_{OC} is obtained from (2) by setting the current density J to zero. By multiplying the current by voltage and

setting the derivative $d(JV)/dV$ to zero, one can determine the photogenerated voltage V_m at maximal power. Substitution of this result into (2) gives the corresponding current J_m. Then, photoconversion efficiency under AM1.5 conditions can be found as $\eta = J_m V_m / P_S$, where P_S is the incident power ($P_S = 0.1$ W/cm^2 for AM1.5g conditions), and I-V curve fill factor is given by $FF = J_m V_m / (J_{SC} V_{OC})$.

III. COMPARISON WITH EXPERIMENT

Here, we compare the results of the theoretical model from the previous section with the experimental results obtained for two perovskite SCs: one based on formamidinium lead iodide (FAPbI$_3$) and the other on CH$_3$NH$_3$PbI$_3$. Experimental data for the former SC material are taken from [2], and for the latter material from [6]. For the Shockley-Read-Hall lifetime and radiative recombination parameter, the values from [7] are used. Table I gives the main parameters used in the modeling. The experimental values of the open-circuit voltage, photoconversion efficiency, and the I-V curve fill factor are reproduced by the model calculations with the accuracy of three significant figures.

A. Current-voltage curves

Shown in Fig. 1 is the experimental photogenerated current density vs. voltage curve, $J(V)$. Focusing on the FAPbI$_3$ curve, the theoretical expression (2) is in excellent agreement with the experimental data if the combination

$$ J_0 = qn_i (S + d / \tau_{SRH}) \qquad (4) $$

related to the dark saturation current is set to $1.15 \cdot 10^{-8}$ mA/cm^2. With the parameters from Table I, we obtain the intrinsic carrier density $n_i \approx 2.6 \cdot 10^5$ cm^{-3}. Given that the bandgap, E_g, in FAPbI$_3$ equals 1.55 eV, this corresponds the densities of states geometric average of $(N_c N_v)^{1/2} \approx 3.7 \cdot 10^{18}$ cm^{-3}. The theory developed here agrees well with the experiment and gives the ideality factor equal to 2, as predicted in [5].

The theoretical $J(V)$ curve for CH$_3$NH$_3$PbI$_3$-based SC agrees with the experiment if one sets $J_0 = 3 \cdot 10^{-7}$ mA/cm^2, $R_S = 1.5$ Ω, $R_{sh} = 1$ kΩ. The ideality factor of the experimental curve equals 2.3, which is slightly higher than the value of 2 predicted in [5]. This increase is presumably related to the effect of the relatively small shunt resistance, because of which the condition $\Delta n \gg n_0$ is fulfilled somewhat worse than in the FAPbI$_3$ case. Then, one should use the full expression (7) from [5] for excess carrier concentration.

978-1-5090-5606-4/17 $31.00 © 2017 IEEE

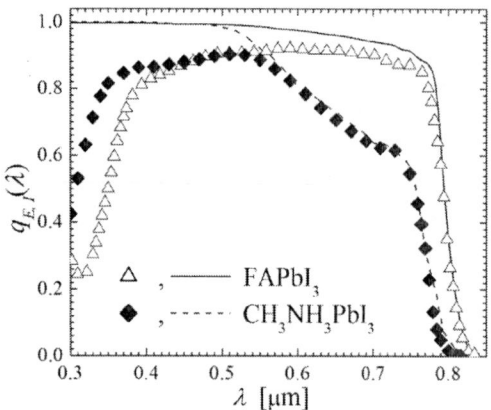

Fig.2. Symbols: experimental external quantum efficiency of a FAPbI$_3$ [2] and CH$_3$NH$_3$PbI$_3$ [6] solar cell. Solid line: theoretical internal quantum efficiency of the FAPbI$_3$ SC according to (6). Dashed line: internal quantum efficiency of the CH$_3$NH$_3$PbI$_3$ SC according to (5).

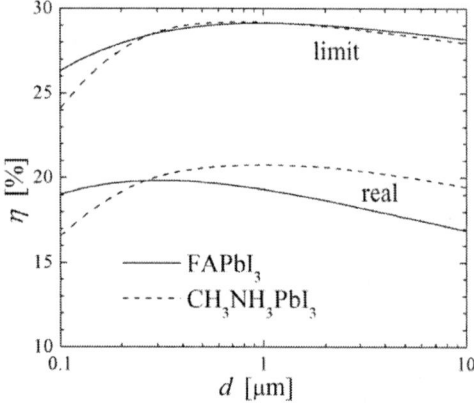

Fig.3. Theoretical photoconversion efficiency vs. i-layer thickness for FAPbI$_3$ (solid curves) and CH$_3$NH$_3$PbI$_3$ (dashed curves) in the limiting and realistic cases.

B. External quantum efficiency and surface texturing

Shown in Fig. 2 are the experimental external quantum efficiency curves in the p–i–n CH$_3$NH$_3$PbI$_3$- and FAPbI$_3$-based structures. In the former case, spectral dependence of the internal quantum efficiency $q_I(\lambda)$ agrees with the external quantum efficiency curves, $q_E(\lambda)$, near the long-wavelength absorption edge, if one assumes for the expression

$$q_I(\lambda) = 1 - e^{-2\alpha(\lambda)d} \qquad (5)$$

valid for a plane-parallel structure. Taking the absorption coefficient $\alpha(\lambda)$ in this material from [8], one can calculate $q_I(\lambda)$, see the dashed line in Fig. 2.

In the case of FAPbI$_3$, the agreement between the two quantum efficiencies is achieved if one assumes the formula

$$q_I(\lambda) = \left(1 + \frac{1}{4\,\alpha(\lambda)\,d\,n_r^2}\right)^{-1}, \qquad (6)$$

which is valid for a textured structure [9]. Here, n_r is the perovskite refractive index.

The theoretical internal quantum efficiency curve for FAPbI$_3$ is built using (6) with the absorption coefficient taken from [10] and refractive index n_r from [11]. This allows one to conclude that due to the non-planar location of grains that compose FAPbI$_3$, its surface is spontaneously textured, which leads to practically complete capture of the incident light due to reabsorption of reflected photons. This conclusion is supported by the value of $q_E(\lambda \approx 0.6\ \mu m)$ is about 92%. If the SCs studied had smooth surfaces, then the light reflection coefficient, given by the Fresnel's expression $(n_r - 1)^2/(n_r + 1)^2$, at $n_r = 2.06$ would have been about 12%, and the corresponding value of $q_E(\lambda)$ would be below 88%, which is smaller than the experimentally measured value.

Thus, there is no spontaneous surface texturing in CH$_3$NH$_3$PbI$_3$-based SCs. Nevertheless, we believe that the net optical losses in CH$_3$NH$_3$PbI$_3$ actually do not exceed those in FAPbI$_3$. Rather, the smaller short-circuit current value in CH$_3$NH$_3$PbI$_3$-based SCs as compared to the FAPbI$_3$-based ones (20.1 vs. 24.7 mA/cm^2) is probably due to the unoptimized thickness of this material. In turn, because the bandgaps in CH$_3$NH$_3$PbI$_3$ and FAPbI$_3$ are practically the same, photoconversion efficiency in CH$_3$NH$_3$PbI$_3$-based SC is smaller than that in the FAPbI$_3$-based SC (19.7 vs. 16.3 %). The arguments in favor of this conjecture are presented in the following section.

C. Optimization of photoconversion efficiency

Fig. 3 shows the theoretical photoconversion efficiency vs. i-layer thickness curves. Calculations were performed for two cases:

(i) The limiting case, where the effects of both parasitic resistances and all recombination channels except for radiative recombination are switched off, i.e. $\tau_{SRH} = \infty$, $S = 0$, $R_S = 0$, $R_{sh} = \infty$ for both materials.

(ii) The realistic case with the parameters given in Table I.

As seen from the figure, the curves are non-monotonic. Efficiency maximum owes to the fact that photogenerated current increases with thickness and eventually saturates, whereas the photogenerated voltage decreases with thickness due to the increase in bulk recombination rate. The maximal efficiency for both materials in the limiting case is 29.2% and is located at about $d = 0.9\ \mu m$.

In the realistic case, $d_{max} = 0.3\ \mu m$ for FAPbI$_3$ and 1 μm for CH$_3$NH$_3$PbI$_3$. For both materials, the maximal efficiency is slightly exceeding 20%.

Although photoconversion efficiency in CH$_3$NH$_3$PbI$_3$-based SC reported in [6] is 16.3%, this value is possibly related to the unoptimized thickness. As seen from Fig. 3, at the optimal thickness $d \approx 1\ \mu m$, the efficiency is 20.5%, which is even slightly higher than that of a FAPbI$_3$-based SC at its optimal thickness of 0.3 μm.

D. Excitonic effects

Because some of the electron-hole pairs may form excitons with binding energy in perovskites E_x between 25 and 50 meV [7], [8], [12], let us discuss their influence on the photoconversion efficiency.

On the one hand, excitons have a positive effect on photoconversion efficiency, because they increase light absorption coefficient. On the other hand, their presence leads to the reduction of the open-circuit voltage and short-circuit current. It is possible to estimate the exciton effect on the former by noticing that, according to [12], $n_{iex} = n_i \exp[E_x /(2kT)]$. This expression together with (2) allows one to estimate the reduction of V_{OC}, which turns out to be about 2-4 mV. Strictly speaking, in the case considered here, saturation current density J_0 is constant, because it is taken from experiment. This means that as n_i increases, bulk recombination rate must decrease, i.e. bulk lifetime should grow. In [3], this lifetime in FAPbI$_3$ is estimated as about $5 \cdot 10^{-7}$ s. If we take this value, then n_i should be equal $5.7 \cdot 10^5$ cm^{-3}, and exciton binding energy is estimated as 43.5 meV.

Let us now estimate the exciton influence on the short-circuit current. Because part of photogenerated electron-hole pairs are bound into excitons, which do not conduct electricity, we should compare the exciton concentration, n_x, with the electron-hole pair concentration Δn. According to [13],

$$ n_x = \frac{N_x \Delta n^2}{N_c N_v} \exp\left(\frac{E_x}{kT}\right), \qquad (7) $$

where $N_{c,v,x} = v_{n,v,x}(2\pi m_{n,p,x}kT/h^3)^{3/2}$, $m_x = m_p + m_n$, and the degeneracy factors are $v_n = v_v = 2$, $v_x = 8$. Estimates show that, depending on the electron and hole effective masses, m_n and m_p, ranging from about 0.2 to 1 electron mass, and on the value of $\Delta n \sim 10^{14} - 10^{15}$ cm^{-3}, the value of n_x turns out to be at least two orders of magnitude smaller than Δn. Thus, in the case considered, excitons reduce the short-circuit current by not more than 1%.

IV. Conclusions

Coming to our conclusions, the analysis of the illuminated I-V curve from [2] yields the ideality factor of 2. The same value is obtained within the theory described here at $\Delta n >> n_0$. Note that this inequality holds quite well, because, according to [3], $n_0 \approx 10^{11}$ cm^{-3} in FAPbI$_3$, whereas the estimated value of Δn exceeds 10^{13} cm^{-3}. In the case of a CH$_3$NH$_3$PbI$_3$-based SC, the ideality factor is 2.3, which is probably related to the relatively low shunt resistance value.

Based on the analysis of the experimentally measured external quantum efficiency of FAPbI$_3$-based SCs, an important conclusion can be made about surface texturing of these SCs, which results in multiple light absorption and strong increase of photoconversion efficiency. In the CH$_3$NH$_3$PbI$_3$-based SC, natural texturing is absent. The reasons for this might be related to the material properties, or to the technological processing details.

It is established that the theoretical efficiency vs. i-layer thickness, $\eta(d)$, has a maximum at $d \sim 0.3 - 0.9$ µm. This maximum is due to the competition between the different dependences of short-circuit current and open-circuit voltage on the i-layer thickness.

Finally, it is established that in perovskite SCs, the surface recombination velocity has a rather low value even in the absence of passivation of the i-region boundaries.

Acknowledgement

M.E. is grateful to the Natural Sciences and Engineering Research Council of Canada (NSERC) and to the Research and Development Corporation of Newfoundland and Labrador (RDC) for financial support.

References

[1] M.A. Green, K. Emery, Y. Hishikawa, W. Warta, and E.D. Dunlop, "Solar cell efficiency tables (version 48)," *Progress in Photovoltaics: Research and Applications*, vol. 24, pp. 905-913, 2016.

[2] W.S. Yang, J.H. Noh, N.J. Jeon, Y.C. Kim, S. Ryu, J. Seo, S.I. Seok, "High-performance photovoltaic perovskite layers fabricated through intramolecular exchange," *Science*, vol. 348, pp. 1234-1237, 2015.

[3] Q. Han , S.-H. Bae , P. Sun , Y.-Ts. Hsieh , Y. Yang , Y.S. Rim, H. Zhao, Q. Chen, W. Shi, G. Li, and Y. Yang, "Single crystal formamidinium lead iodide (FAPbI$_3$): insight into the structural, optical, and electrical properties," *Advanced Materials*, vol. 28, pp. 2253-2258, 2016.

[4] A.V. Sachenko, A.I. Shkrebtii, R.M. Korkishko, V. P. Kostylyov, N. R. Kulish, and I.O. Sokolovskyi, "Features of photoconversion in highly efficient silicon solar cells," *Semiconductors*, vol. 49, pp. 264-269 (2015).

[5] A.V. Sachenko, A.V. Bobyl, V.N. Verbitskiy, V.M. Vlasyuk, D.M. Zhigunov, V.P. Kostylyov, I.O. Sokolovskyi, E.I. Terukov, P.A. Forsh, and M. Evstigneev, "Correlation between Electroluminescence and Photoconversion Efficiency in a-Si:H/c-Si Heterojunction Solar Cells", in *44th IEEE Photovoltaic Specialist Conference*, 2017.

[6] J. Werner, L. Barraud, A. Walter, M. Bräuninger, F. Sahli, D. Sacchetto, N. Tétreault, B. Paviet-Salomon, S.-J. Moon, C. Allebé, M. Despeisse, S. Nicolay, S. De Wolf, B. Niesen, and C. Ballif, "Efficient near-infrared-transparent perovskite solar cells enabling direct comparison of 4‑terminal and monolithic perovskite/silicon tandem cells", *ACS Energy Letters*, vol. 1, pp. 474-480, 2016.

[7] C. Wehrenfenning, G.E. Eperon, M.B. Jonston, H.J. Snaith, and L.M. Herz, "High charge carrier mobilities and lifetimes in organolead trihalide perovskites," *Advanced Materials*, vol. 26, pp. 1584-1589, 2014.

[8] I. Almansouri, A. Ho-Baillie, and M. Green, "Ultimate efficiency limit of single-junction perovskite and dual-junction perovskite/silicon two terminal devices," *Japanese Journal of Applied Physics*, vol. 54, pp. 08KD04:1-6, 2015.

[9] T. Tiedje, E. Yablonovitch, G. D. Cody, and B. G. Brooks, "Limiting efficiency of silicon solar cells," *IEEE Transactions on Electron Devices*, vol. 31, pp. 711-716, 1984.

[10] M. Kato, T. Fujiseki, T. Miyadera, T. Sugita, Sh. Fujimoto, M. Tamakoshi, M. Chikamatsu, and H. Fujiwara, "Universal rules for the visible-light absorption in hybrid perovskite materials," *arXiv*:1605.05124, pp. 1-39, 2016.

[11] P. F. Ndione, Zh. Li, and K. Zhu, "Effects of alloying on the optical properties of organic–inorganic lead halide perovskite thin films," *Journal of Materials Chemistry C*, vol. 4, pp. 7775-7782, 2016.

[12] Y. Chen, H.T. Yi, X. Wu, R. Haroldson, Y.N. Gartstein, Y.I. Rodionov, K.S. Tikhonov, A. Zakhidov, X.-Y. Zhu, and V. Podzorov, "Extended carrier lifetimes and diffusion in hybrid perovskites revealed by Hall effect and photoconductivity measurements", *Nature Communications*, vol. 7, pp. 12253:1-9, 2016.

[13] D.E. Kane and R.M. Swanson, "The effect of excitons on apparent band gap narrowing and transport in semiconductors", *Journal of Applied Physics*, vol. 73, pp. 1193-1197, 1993.

Influence of Mono- and Di-valent Metal Additives on Morphology and Charge Carrier Dynamics of CH₃NH₃PbI₃ Perovskite

Niraj Shrestha, Suneth C. Watthage, Zhaoning Song, Paul J. Roland, Adam B. Phillips, Michael J. Heben, and Randall J. Ellingson

Department of Physics and Astronomy, and Wright Center for Photovoltaics Innovation and Commercialization, The University of Toledo, Toledo, OH 43606, USA.

Abstract — We have applied steady-state and time resolved photoluminescence (PL) to study the influence of monovalent and divalent metal additives on the electronic structure and charge carrier dynamics in $CH_3NH_3PbI_3$ (MAPbI₃) perovskite films fabricated on a layer of compact TiO_2 deposited onto FTO-coated glass substrate. Monovalent metals Li^+, Na^+, K^+ and Cs^+ have a detrimental effect on perovskite crystal grain growth and charge carrier dynamics. In contrast, the divalent additives $Zn+$, $Cd+$ and Hg^+ facilitate larger grain growth in MAPbI₃ perovskite film. Enlarged grain size, reduced trap state density, and the longest photogenerated carrier life time have been observed for the case of Cd^+ incorporated into MAPbI₃ perovskite films.

I. INTRODUCTION

The hybrid organic-inorganic metal halide perovskite class of materials offer great promise in the field of photovoltaics based on their key properties: direct band gap with large absorption coefficient [1, 2], long electron-hole diffusion length [1, 3], low exciton binding energy [4], and low energy inputs required for processing. However, much attention is focused on methylammonium lead triiodide (MAPbI₃) and mixed halide perovskites.

Although organometallic halide perovskite solar cells have already reached an efficiency of 22.1 % [5], instability issues have thus far prevented perovskite PV products from joining the high efficiency commercialized thin film technologies of CdTe and CIGS. The optoelectronic properties of perovskite films depend principally on their electronic structure and properties, and on the crystallinity, grain size, and morphology of the film. It has been reported that metal additives lead to improved surface coverage and crystallinity in polycrystalline perovskite films [6-8]. Such resulting thin films exhibit lower trap state density, which leads to improved power conversion efficiency (PCE). Klug et al. reported that MAPbI₃ perovskite film with lead–cobalt composition 63Pb:1Co yields a highest power conversion efficiency of 15.0 % [8]. Enlarged grain size, strong photoluminescence (PL) intensity, and longer photogenerated charge carrier dynamics have been observed when Cd is incorporated into MAPbI₃ perovskite [7]. In contrast, large densities of voids, aggregates, and smaller grain size have been observed by Wang et al. when NaI was added into the precursor solution during the preparation of perovskite film [6]. Here, we present PL analysis of the effect of monovalent (Li^+, Na^+, K^+ and Cs^+) as well as divalent (Zn^{2+},

Cd^{2+} and Hg^{2+}) metal cations when these are added in small amount to the precursor solution during the fabrication of MAPbI₃ perovskite film by a standard two-step process.

II. EXPERIMENTAL DETAILS

Perovskite films were fabricated on compact TiO_2 coated Fluorine-doped SnO_2 glass (TEC 15) substrates. First, PbI_2 layer was deposited by spin-coating a hot (~ 70 °C) solution of 1 M PbI_2 in anhydrous N,N-dimethlylformamide (DMF) at 100 °C. Then, MAI (methylammonium iodide) solution (with and without metallic salt) spin-coated on the PbI_2 layer at 150 °C. In this study, 250 mM MAI solution and 10 mM metallic salt was used in the second step.

For PL measurements, samples were excited from the film side and measurements were performed on fresh samples at room temperature in ambient air. For steady-state PL measurements, a 532 nm cw laser (beam diameter ~ 90 μm) was used as the source of excitation at 1.3 W/cm². PL signals were detected via Symphony-II CCD detector (integration time = 0.1 s). Time resolved photoluminescence (TRPL) measurements were performed by TCSPC module. For TRPL measurements, samples were excited by a 532 nm pulsed laser (pulse width = 5 ps, beam diameter ~ 150 μm) at ~ 10^{10} photons · pulse⁻¹ · cm⁻². Radiative recombination events were detected by APD/PMT hybrid detector (integration time = 300 s). TCSPC decay curves obtained in TRPL measurement were fitted by iterative re-convolution with the measured system response function. PL decays were biexponential in nature and mean life times were calculated using weighted average method:

$$\text{Mean life time }, \tau = \frac{A_1\tau_1^2 + A_2\tau_2^2}{A_1\tau_1 + A_2\tau_2} \qquad (1)$$

III. RESULTS AND DISCUSSIONS

To investigate the influence of metal additives on the morphology of the perovskite films, we measured the SEM images of the perovskite films prepared by the two-step method incorporating with different metal salts. Fig. 1a-d shows that the addition of monovalent alkali metal ions results in smaller grains compared to the control sample without additives (Fig. 1e), which is consistent with the result of using

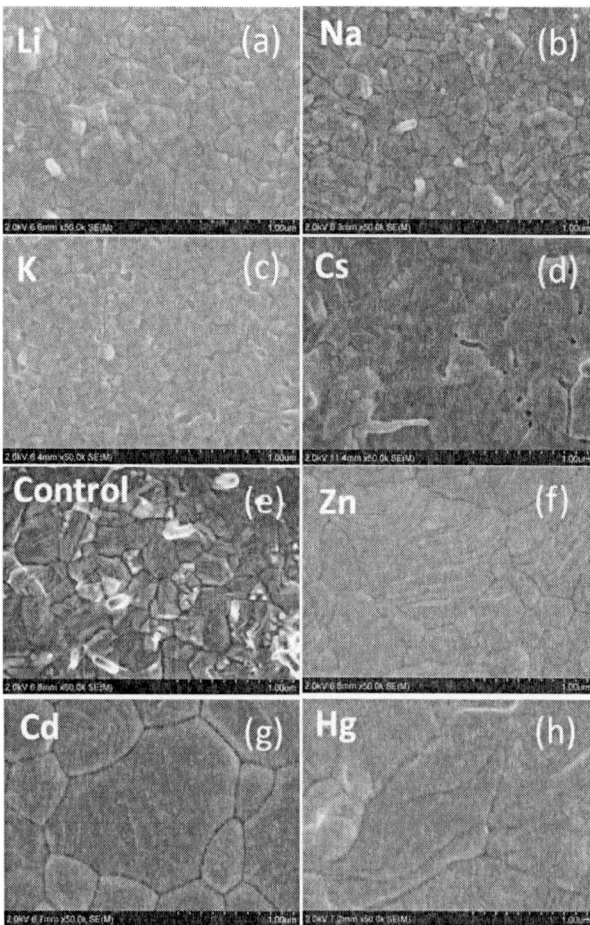

Fig. 1. Scanning electron microscopy (SEM) images of the MAPbI$_3$ perovskite films with different metal additives. (a) Li, (b) Na, (c) K, (d) Cs, (e) Control, (f) Zn, (g) Cd and (h) Hg.

NaI in the precursor solution as reported previously [6]. The small grain size is the result of the rapid perovskite conversion due to the relatively high MAI concentration [9]. In contrast, the incorporation of divalent metal ions (Fig. 1f-h) leads to larger grains due to the formation of an organometallic complex, enabling the solution-liquid-solid type of growth [7]. The larger grain size is understood to typically correlate with longer lifetime and improved device performance.

Fig. 2a-c shows the XRD patterns of MAPbI$_3$ perovskite incorporated with various mono- and divalent metals. The control sample shows characteristic peak positions belonging to the tetragonal perovskite phase [10]. Small amounts of divalent cations Cd^{2+}, Zn^{2+} are likely to substitute for Pb^{2+} in MAPbI$_3$ [11], while the monovalent cation Cs$^+$ favors substitution for the MA$^+$ cation [12]. Since the added divalent and monovalent cations have smaller ionic radii than the corresponding substituted Pb^{2+} and MA$^+$ cations, the perovskite unit cell shrinks [12-15]. This causes XRD peaks at 14.08° and 28.36° to shift towards higher 2θ values as depicted in Fig. 2b-c. The Hg^{2+} added MAPbI$_3$ film shows splitting of

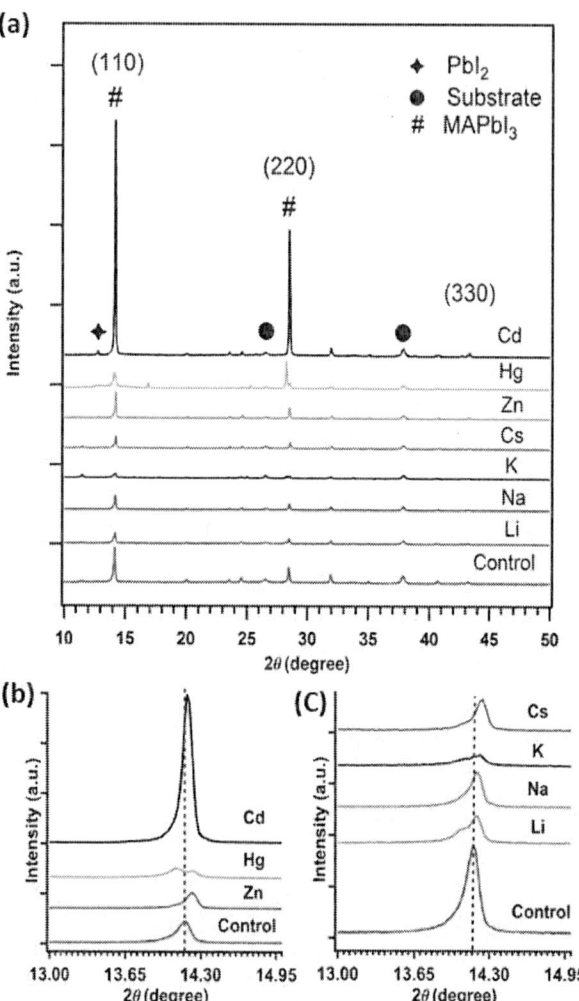

Fig. 2. (a) X-ray diffraction patterns of the MAPbI$_3$ perovskite films with different metal additives. Zoom-in XRD patterns for the (b) divalent and (c) monovalent

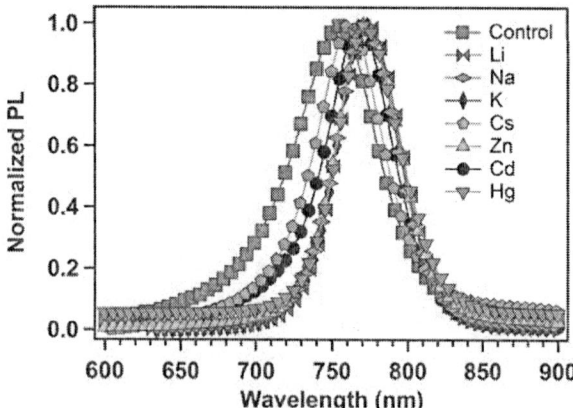

Fig. 3. Normalized photoluminescence (PL) spectra of the MAPbI$_3$ perovskite films with different metal additives.

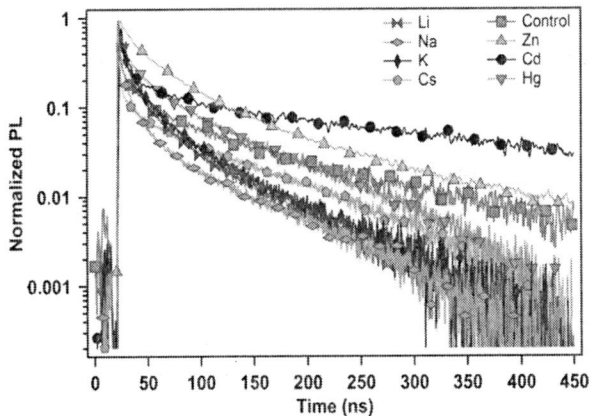

Fig. 4. PL decays of the MAPbI₃ perovskite films with different metal additives.

the XRD peaks which is an indicator of the formation of Hg-based organometal halides such as MAHgI₃ and (MA)₂HgI₄ as shown in Fig. 2b. Unlike divalent metals, Li^+, Na^+, K^+ are smaller in size than MA^+ and reside on interstitial sites [16]. However, these can cause lattice contraction due to ionic attraction between I⁻ ions at octahedral sites and A^+ cation (A^+ = (Li^+, Na^+, K) at the interstitial sites. As a result, the XRD peaks at 14.08° and 28.36° shift towards higher 2θ values as shown in Fig. 2c. More interestingly, we also observe a sharp increase in the intensity of XRD peak in the Cd^{2+} added MAPbI₃ compare to control sample showing better crystallinity in Cd^{2+} incorporated MAPbI₃ than for the other additives.

To probe the radiative recombination processes and dynamics of the perovskite films treated with metal additives, we performed PL and TRPL measurements. Fig. 3 shows the PL spectra of various perovskite films. For the perovskite films incorporated with both monovalent (Li^+, Na^+, K^+ and Cs^+) and the divalent (Zn^{2+}, Cd^{2+} and Hg^{2+}) metals, the perovskite emission peak exhibits a red shift (~10-20 nm) with respect to the control sample (753 nm), indicating the reduced number of surface trap states in the perovskite films.

Fig. 4 shows the TRPL decays of the metal-added perovskite films, and the biexponential fitting results were tabulated in Table I. The small size alkali ions, such as Li^+, Na^+, and K^+, are likely to be present as the interstitial defects in the perovskite crystal. The mobile cations increase the effective hole density, resulting in a higher SRH recombination rate, and consequently, a lower free carrier lifetime (~ 40 ns) with respect to the control sample (96 ns). Unlike the small size alkali ions, the size of Cs^+ (1.67 Å) [15] is comparable to that of MA^+ (1.8 Å) [13, 14]. Therefore, Cs^+ can substitute MA^+ to form the inorganic perovskite CsPbI₃, which possesses optoelectronic properties similar to those of MAPbI₃. Thus, the carrier life time of Cs^+ added MAPbI₃ (72 ns) is similar to that of the control sample. For the divalent metal, Cd^{2+} incorporation leads to the longest photogenerated

carrier life time (256 ns), showing direct correlation with the enhanced grain size. However, Hg^{2+} and Zn^{2+} added films, despite showing larger grains, show shorter carrier life time than the control sample. This is because of metal complexes segregated on the surface of these films, acting as the non-radiative recombination centers [17].

TABLE I
PL AND TRPL RESULTS OF MAPBI₃ PEROVSKITES WITH VARIOUS METAL ADDITIVES

Metal Additives	Detection Wavelength (nm)	τ_1 (ns)	τ_2 (ns)	Mean life time (ns)
Control	753	5	112	96
Li	773	7	54	36
Na	772	4	60	38
K	771	8	55	38
Cs	761	1	83	72
Zn	772	31	155	91
Cd	771	7	285	258
Hg	772	9	78	60

IV. CONCLUSION

The effect of monovalent and divalent metal additives on the morphology and optoelectronic properties of MAPbI₃ perovskite has been studied. Monovalent metal ions have an adverse effect on the grain growth and charge carrier dynamics of the perovskite film. While the divalent Cd^{2+} incorporation yields an enlarged grain size and the longest mean carrier life time.

ACKNOWLEDGEMENT

This work was financially supported by the Air Force Research Laboratory, Space Vehicles Directorate, under Contract #FA9453-11-C-0253, the National Science Foundation under contract #CHE-1230246, and funds from Wright Center Endowment for Photovoltaics Innovation and Commercialization.

REFERENCES

[1] M. I. Saidaminov, A. L. Abdelhady, B. Murali, E. Alarousu, V. M. Burlakov, W. Peng, et al., "High-quality bulk hybrid perovskite single crystals within minutes by inverse temperature crystallization," Nat Commun, vol. 6, 07/06/online 2015.

[2] N.-G. Park, "Perovskite solar cells: an emerging photovoltaic technology," Materials Today, vol. 18, pp. 65-72, 3// 2015.

[3] Q. Dong, Y. Fang, Y. Shao, P. Mulligan, J. Qiu, L. Cao, et al., "Electron-hole diffusion lengths > 175 μm in solution-grown CH₃NH₃PbI₃ single crystals," Science, vol. 347, pp. 967-970, 2015.

[4] A. Miyata, A. Mitioglu, P. Plochocka, O. Portugall, J. T.-W. Wang, S. D. Stranks, et al., "Direct measurement of the exciton binding energy and effective masses for charge carriers in organic-inorganic tri-halide perovskites," Nat Phys, vol. 11, pp. 582-587, 07//print 2015.

[5] *NREL* *chart,* http://www.nrel.gov/ncpv/images/efficiency_chart.jpg, *Accessed 13.03.2016, 2016.*

[6] L. Wang, D. Moghe, S. Hafezian, P. Chen, M. Young, M. Elinski, *et al.*, "Alkali Metal Halide Salts as Interface Additives to Fabricate Hysteresis-Free Hybrid Perovskite-Based Photovoltaic Devices," *ACS Applied Materials & Interfaces,* vol. 8, pp. 23086-23094, 2016/09/07 2016.

[7] S. C. Watthage, Z. Song, N. Shrestha, A. B. Phillips, G. K. Liyanage, P. J. Roland, *et al.*, "Enhanced Grain Size, Photoluminescence, and Photoconversion Efficiency with Cadmium Addition during the Two-Step Growth of $CH_3NH_3PbI_3$," *ACS Applied Materials & Interfaces,* vol. 9, pp. 2334-2341, 2017/01/25 2017.

[8] M. T. Klug, A. Osherov, A. A. Haghighirad, S. D. Stranks, P. R. Brown, S. Bai, *et al.*, "Tailoring metal halide perovskites through metal substitution: influence on photovoltaic and material properties," *Energy & Environmental Science,* vol. 10, pp. 236-246, 2017.

[9] S. C. Watthage, Z. Song, G. K. Liyanage, A. B. Phillips, and M. J. Heben, "Investigation on the nucleation and growth mechanisms of perovskite formation in the two-step solution process," in *2016 IEEE 43rd Photovoltaic Specialists Conference (PVSC)*, 2016, pp. 0831-0834.

[10] Z. Song, S. C. Watthage, A. B. Phillips, B. L. Tompkins, R. J. Ellingson, and M. J. Heben, "Impact of Processing Temperature and Composition on the Formation of Methylammonium Lead Iodide Perovskites," *Chemistry of Materials,* vol. 27, pp. 4612-4619, 2015/07/14 2015.

[11] L. A. Frolova, D. V. Anokhin, K. L. Gerasimov, N. N. Dremova, and P. A. Troshin, "Exploring the Effects of the Pb2+ Substitution in MAPbI3 on the Photovoltaic Performance of the Hybrid Perovskite Solar Cells," *The Journal of Physical Chemistry Letters,* vol. 7, pp. 4353-4357, 2016/11/03 2016.

[12] R. G. Niemann, L. Gouda, J. Hu, S. Tirosh, R. Gottesman, P. J. Cameron, *et al.*, "Cs+ incorporation into $CH_3NH_3PbI_3$ perovskite: substitution limit and stability enhancement," *Journal of Materials Chemistry A,* vol. 4, pp. 17819-17827, 2016.

[13] "Mutations in M2 alter the selectivity of the mouse nicotinic acetylcholine receptor for organic and alkali metal cations," *The Journal of General Physiology,* vol. 100, pp. 373-400, 1992.

[14] C. M. Handley and C. L. Freeman, "A new potential for methylammonium lead iodide," *Physical Chemistry Chemical Physics,* vol. 19, pp. 2313-2321, 2017.

[15] M. Saliba, T. Matsui, K. Domanski, J.-Y. Seo, A. Ummadisingu, S. M. Zakeeruddin, *et al.*, "Incorporation of rubidium cations into perovskite solar cells improves photovoltaic performance," *Science,* vol. 354, pp. 206-209, 2016.

[16] J. Chang, Z. Lin, H. Zhu, F. H. Isikgor, Q.-H. Xu, C. Zhang, *et al.*, "Enhancing the photovoltaic performance of planar heterojunction perovskite solar cells by doping the perovskite layer with alkali metal ions," *Journal of Materials Chemistry A,* vol. 4, pp. 16546-16552, 2016.

[17] S. C. Watthage, Z. Song, N. Shrestha, A. B. Phillips, G. K. Liyanage, P. J. Roland, *et al.*, "Impact of Divalent Metal Additives on the Structural and Optoelectronic Properties of $CH_3NH_3PbI_3$ Perovskite Prepared by the Two-Step Solution Process," *MRS Advances,* pp. 1-6, 2017/001/16 2017.

Effect of Dual Cathode Buffer Layer on Ternary Organic Solar Cell

Ashish Singh[†,*], T. Bhim Raju[#], Anamika Dey[†,#], Ritesh Kant Gupta[†] and Parameswar K. Iyer[#,†,*]

[†]Centre for Nanotechnology, [#]Department of Chemistry, Indian Institute of Technology Guwahati, Guwahati–781039, Assam, India. *E-mail: ashish.iitg17@gmail.com and pki@iitg.ernet.in

ABSTRACT — **Effect of cathode interfacial layer namely, BCP and BPhen on ternary organic solar cell using P3HT, PC$_{71}$BM with CPDT derivative, BH-1, was investigated and compared with the conventional LiF/Al structure. The doping concentration of the BH-1 in P3HT: PC$_{71}$BM blend was systematically standardized from UV-Vis absorption study. The device with ITO/PEDOT: PSS/P3HT: BH-1: PC$_{71}$BM/BCP/LiF/Al configuration showing the best performance with PCE, η=3.23%, J$_{sc}$=12.33 mA/cm^2, V$_{oc}$= 0.54 V and FF=48.10% due to the less recombination at the cathode interface. These results systematically explain the role of dual cathode buffer layer to reduce the S-factor in performance of ternary BHJ solar cell.**

I. INTRODUCTION

The rapidly growing demand on clean and renewable energy resource to prevent environmental pollution, from last few decade, researchers have paid a great attention to the Organic solar cells which is one of the best alternative of conventional mineral materials [1]. However, due to the less absorption capacity of organic polymers, full coverage of the solar spectrum with a single donor polymer blended with fullerene acceptor is highly challenging [2]. A very simple technique to overcome this problem is the use ternary blend consisting of two-donors: single acceptor or single donor: two-acceptors blend as the active material with complementary absorption profiles. Generally in most of the reported ternary solar cell, though the short-circuit current (Jsc) increases significantly due to the additional absorption by the second donor material, the fill factor (FF) is often significantly reduced mainly due to the two reason: (a) different physical and chemical properties of the second donor material with the first one and (b) the charge recombination at the cathode-blend interface.

In this work, reducing the charge carrier recombination with the help dual cathode buffer layer, the performance of ternary solar cell consisting of CPDT based donor polymer, BH-1, blended with P3HT: PC$_{71}$BM were efficiently improved. We incorporated two different organic interfacial layers with the ternary blend, namely, Bathophenanthroline (BPhen) and Bathocuproine (BCP) with LiF/Al as the conventional cathode contact and there corresponding photovoltaic performances were systematically analyzed. Among the all contacts, it was observed that the device with ITO/PEDOT: PSS/ P3HT: BH-1:PC$_{71}$BM/BCP/LiF/Al configuration showing the best performance having PCE, η=3.23%, J$_{sc}$=12.33 mA/cm^2, V$_{oc}$= 0.54 V and FF=48.10% because

of the better hole-blocking capacity and band alignment with the active layer. This results we demonstrated here are successfully explain the role of different organic cathode interfacial layer and their contribution to the overall device performance of ternary solar cell.

II. EXPERIMENTAL PROCEDURE

The schematic of the fabricated device structures are shown in Fig. 1.

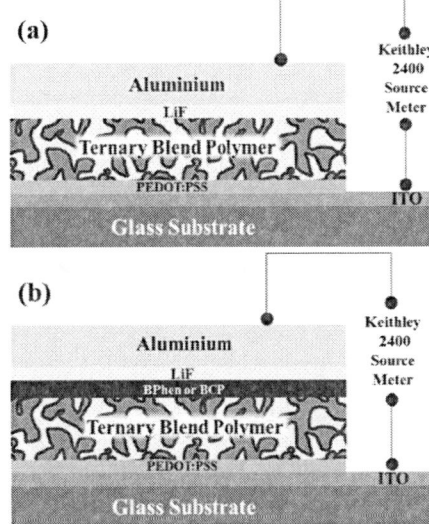

Fig. 1. Schematic of the fabricated devices structures with (a) conventional LiF/Al and (b) BPhen or BCP/LiF/Al cathode contact.

All the organic solar cells were fabricated on commercially available ITO-coated glass substrate. Prior to the device fabrication, the ITO substrate are manually patterned by chemical etching method with the help of scotch tape, diluted HCl (1:1) and Zinc dust, followed by the ultrasonic cleaning in successive solutions of soap, de-ionized water, acetone, and isopropanol. Then the substrates were dried under argon atmosphere and after that they were subjected to UV–ozone treatment for about 20 minutes in order to improve the work function of the substrates. After UV-ozone treatment, PEDOT: PSS, the hole injecting buffer layer, was spin coated at 3000 rpm for 60 seconds and dried at 120°C for 30 minutes in argon atmosphere. The ternary blend of P3HT: BH-1: PC$_{71}$BM solution were prepared in 1, 2-dichlorobenzene (o-DCB) solvent and stirred for 48 h at 60 °C under the glove box. After blending, the solution was filtered with 0.45μm filter and immediately spun on the top of PEDOT: PSS buffer layer. The thickness of active layers was set at 110 ±5 nm, confirmed by Surface Profilometer. The doping concentration of the BH-1 in P3HT:PC$_{71}$BH blend was determined from liquid state UV-absorption study (Fig. 2).

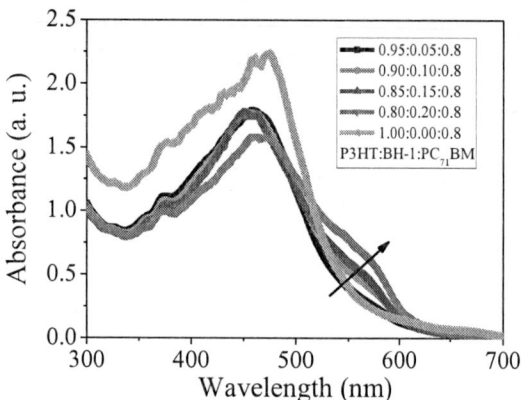

Fig.2.   UV-Vis absorption spectra of P3HT:BH-1:PC$_{71}$BM ternary blend with different weight ratio.

After the spin coating, of the ternary blend active layer, the substrates were annealed at 150 °C for 10 min and, subsequently, the organic interfacial cathode layers, namely BCP or BPhen were thermally evaporated. Finally, ~ 1nm/100 nm thin film of LiF/Al cathode electrode was thermally evaporated under base pressure of ~10^{-6} mbar through a shadow mask determining the active area of the device ~ 6 mm². Schematic energy level diagrams of P3HT: BH-1: PC$_{71}$BM based organic ternary solar with conventional LiF/Al cathode contact and with BPhen/LiF/Al and BCP/LiF/Al as dual cathode buffer layers were shown in Fig. 3 (a) (b) and (c) respectively.

A. Photovoltaic Performance

The current density-voltage (J-V) characteristics of the photovoltaic solar cells were measured both in the dark and under illumination at 100 mW·cm^{-2} by using AM 1.5 solar simulator (Newport, Oriel Sol3A) and using Keithley-2400 digital source meter. The photovoltaic power conversion efficiency of solar cell is determined by following formula

$$\eta = \frac{V_{OC} \times J_{SC} \times FF}{P_{in}} \times 100$$

$$FF = \frac{V_{MAX} \times J_{MAX}}{V_{OC} \times J_{SC}}$$

(1)

V$_{oc}$ is the open circuit voltage, J$_{sc}$ is the short circuit current density and FF is the fill factor. Light intensity is standardized at 100 mW/cm^2 of AM 1.5 spectrum. V$_{max}$ and J$_{max}$ are the voltage and current at the maximum power point. The incident

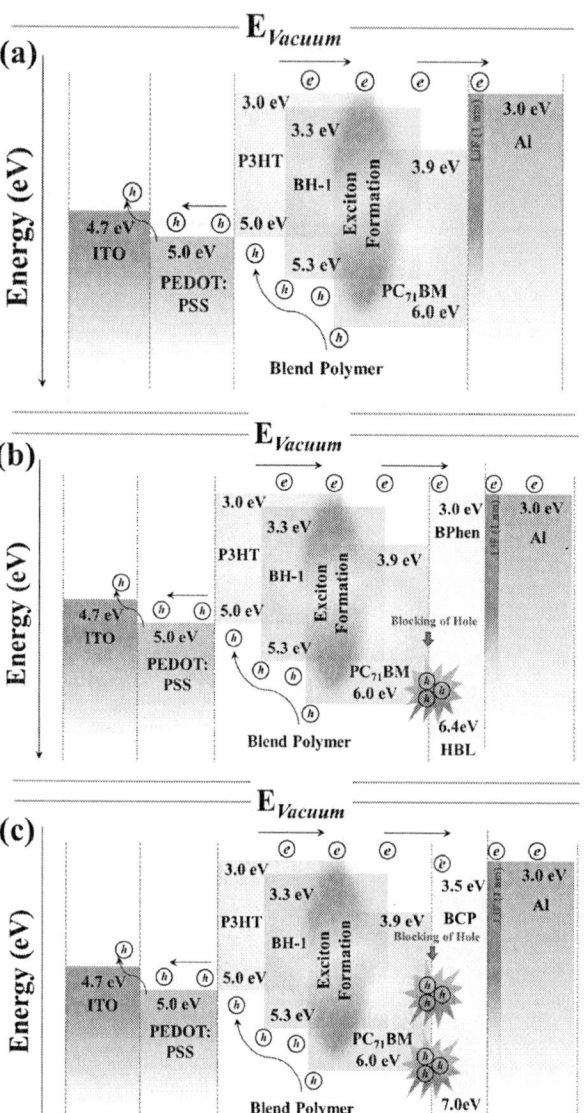

Fig.3.   Band energy diagram of the fabricated ternary solar cell containing (a) LiF/Al and (b) BCP/LiF/Al cathode contacts.

photon-to-current conversion efficiency (IPCE) of the devices was recorded using Newport Oriel IQE-200, where a 250 W quartz tungsten halogen (QTH) lamp was used as the light source.

III. RESULT AND DISCUSSION

Here we have fabricated three different device configuration, viz. (a) ITO/PEDOT: PSS/ P3HT: BH-1:PC$_{71}$BM/ LiF/Al, (b) ITO/PEDOT: PSS/ P3HT: BH-1:PC$_{71}$BM/BPhen/LiF/Al and (c) ITO/PEDOT: PSS/P3HT: BH-1:PC$_{71}$BM/BCP/LiF/Al. The photovoltaic performance parameters of all the devices are summarized in Table 1.

TABLE 1

SUMMARY OF P3HT: BH-1:PC$_{71}$BM SOLAR CELL WITH DIFFERENT DUAL CATHODE BUFFER LAYER

Device configuration	J_{sc} (mA cm^{-2})	V_{oc} (Volt)	FF (%)	Efficiency, η (%)
(a)	13.85	0.50	41.73	2.91
(b)	11.91	0.55	46.28	3.03
(c)	12.33	0.54	48.10	3.23

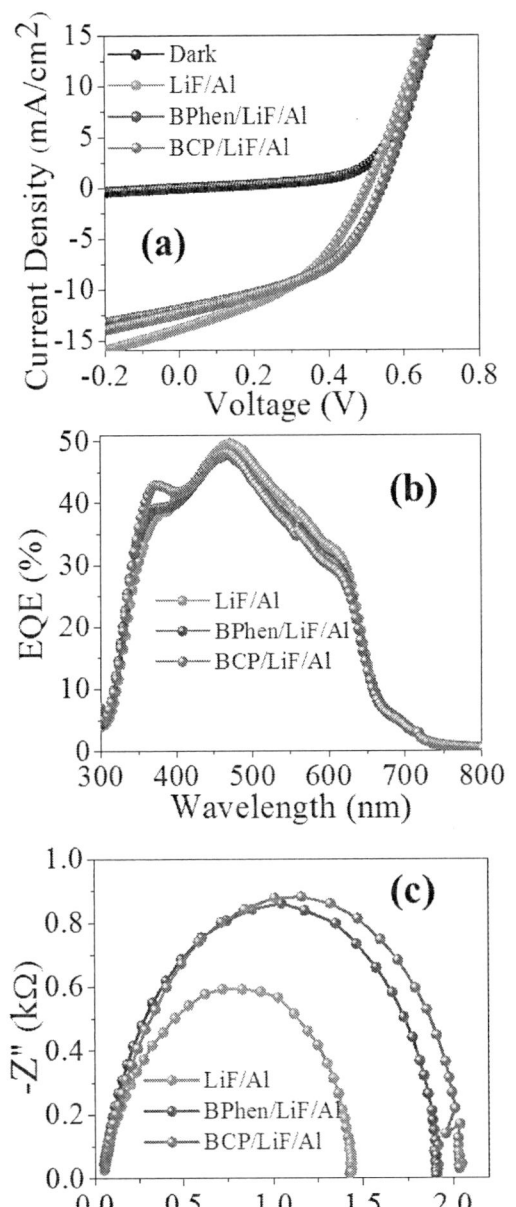

Fig. 4. (a) Current-voltage (J-V) characteristics (b) EQE spectra and (c) Nyquist plots for all the devices

For the device, having conventional configuration (a), fabricated with only LiF/Al electrodes, the power conversion efficiency (PCE) was observed $\eta= 2.91\%$ with J_{sc}=13.85 mA/cm^2, V_{oc}=0.50 V and FF= 41.73%. For the configuration (b) and (c), we introduce BPhen and BCP as hole blocking layer. In both the cases, we get better photovoltaic performance due to the less charge carrier recombination at the blend-cathode interface, which further improve more charge carrier transport to the electrodes. The configuration (c) was showing PCE, η= 3.23% with J_{sc}=12.33 mA/cm^2, Voc=0.54 V and FF= 48.10% whereas PCE, η= 3.03% , J_{sc}=11.91 mA/cm^2, V_{oc}=0.55 V and FF= 46.28% were showing by configuration (b).

BCP, a high band gap material, used as the hole blocking interfacial cathode layer, showing the best performance among the all because of its good band alignment with ternary blend and good selectivity towards blocking the holes to minimizing charge recombination at the interface. Fig. 4 (a), (b) and (c) shows the current-voltage (J-V) characteristics, EQE spectra and the Nyquist plots under the illumination of 100 mW/cm^2 for all the devices. The Nyquist plots, for all the BHJs, show the good agreement for more charge transport at different organic interfacial cathode electrodes.

IV. CONCLUSION

In conclusion, we have demonstrated here the effect of different organic interfacial cathode layer on the performance of P3HT: BH-1:PC$_{71}$BM based ternary organic solar cell. The results indicated that the presence of BCP or BPhen, due to reduce S-factor and less charge recombination at the cathode interface sufficient improvement in term of FF and PCE can be achieved compared to conventional LiF/Al contact. Specially, BCP with Al showing the best performance compared to the other device configuration, due to its better hole blocking capacity compared to BPhen. Finally, this study we demonstrated here, provides important information that organic cathode interfacial layer plays a vital role for enhancing the power convention efficiency ternary organic solar cell.

REFERENCES

[1] A. Singh, A. Dey, D. Das, and P. K. Iyer, "Effect of Dual Cathode Buffer Layer on the Charge Carrier Dynamics of rrP3HT: PCBM Based Bulk Heterojunction Solar Cell," ACS Appl. Mater. Interfaces. vol. 8, pp. 10904, 2016.

[2] G. Li, V. Shrotriya, J. Huang, Y. Yao, T. Moriarty, K. Emery and Y. Yang," High-efficiency solution processable polymer photovoltaic cells by self-organization of polymer blends," Nat. Mater., vol. 4, pp. 864, 2005.

Copper Plated Top Electrode for an Inverted Organic Photovoltaic

Malia Steward, Zhan Shi, Kyoung-Tae Kim, and Seungkeun Choi

University of Washington Bothell, Bothell, WA, 98011, USA

Abstract — **In this work, we have demonstrated improved performances of an inverted OPV by plating Cu on the top Ag electrode. The *FF* is increased by 35% by plating 200 nm of Cu on top of 30 nm of Ag. However, *Jsc* is decreased by 6.5% and this reduction could be due to the interaction between copper plating solution and the photoactive layer of P3HT:PCBM. Despite the decrease of *Jsc*, the PCE increased by 25% thanks to the significant increase of *FF*.**

Index Terms — **Electroplating, copper electrode, organic solar cell, inverted organic photovoltaics.**

I. INTRODUCTION

There have been tremendous interest in new metallization strategies in silicon photovoltaics mainly driven by the increasing silver prices and reduced silicon wafer thicknesses [1]. In silicon photovoltaic technology, industry attractive screen-printing technology has been widely used to coat silver electrode followed by a light induced plating (ILD) of Cu in order to improve the conductance of silver [2]. Therefore, very little Ag is required as a seed layer, hence, decreasing materials cost significantly.

However, in organic photovoltaics (OPV), application of electroplating process has been limited to the bottom electrode of ITO as the ITO significantly limits the performance of large-area OPV due to its lower conductivity, typically two order lower, compared to the opaque top electrodes such as silver or aluminum [3]. However, increased cost of silver also hurdles the large-scale deployment of OPV and reports with this important issue for the OPV are relatively scarce. Particularly, it is difficult to apply an electroplating process to the completed OPV device because of the sensitivity of organic semiconductors to the acids and water.

In this work, we report improved performance of OPV by electroplating Cu on top of Ag electrode as a top electrode. 30 nm of Ag was used a s a top electrode and 200 nm of Cu was electroplated, showing improved *FF* and *PCE* by 35% and 25%, respectively.

II. EXPERIMENT

A. Organic solar cell fabrication

Inverted organic solar cells were fabricated on glass substrates coated with indium-tin-oxide (ITO), which acts as a transparent bottom electrode (cathode). ITO-coated glass was diced into 1" × 1" substrates, and an active area of 0.3 cm² was created by patterning the ITO using photolithography and chemical wet etching. Then, the ITO was cleaned in sequential

ultrasonic baths of acetone and isopropanol for 20 minutes each, followed by drying in the oven for 30 minutes. Polyethylenimine ethoxylated (PEIE) was spin-coated, from a 0.4 wt% solution mixed with 2-methoxyethanol, at 5000 rpm for 60 sec. The active layer of poly(3-hexylthiophene) (P3HT): PCBM (1:0.7, weight ratio) was spin-coated at 3500 rpm for 8 sec in the N_2-filled glove box from a dichlorobenzene solution (60 mg/ml), followed by setting the samples for 40 minutes in an enclosed petri dish to promote solvent annealing process. Then, the samples were annealed on a hotplate at 120 °C for 10 min. The device were transferred to a thermal evaporator and 10 nm of MoO_3 and 30 nm of Ag top-electrodes were deposited through a shadow mask under high vacuum. The completed device had an active area of 0.3 cm² and an inverted OPV structure of Glass/ITO/PEIE/P3HT:PCBM/MoO_3/Ag (Fig. 1).

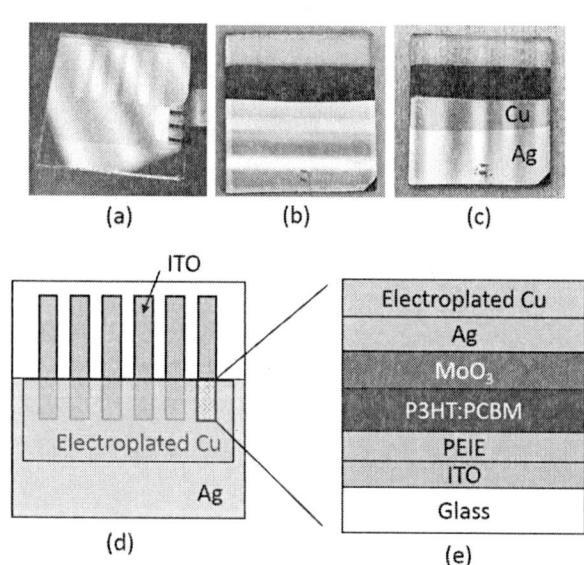

Fig. 1. OPV device structure: (a) ITO patterns, (b) before Cu plating, (c) after Cu plating, (d) layout, (e) cross-sectional structure.

B. Copper electrode electroplating

Copper plating solution was prepared by mixing 250 mg $CuSO_4 \cdot 5H_2O$ and 25 ml H_2SO_4 in 1 liter of deionized water. 200nm of copper was electroplated by applying a dc current density of 10 mA/cm². Photoactive areas were also completely submerged in the solution bath during the copper plating. After plating, the devices were rinsed with DI water and dried with N_2.

TABLE I
SUMMARY OF OPV PARAMETERS

Top electrode configuration	J_{SC} [mA/cm^2]	V_{OC} [V]	FF	R_SA [Ω-cm^2]	η [%]
Ag 30nm	7.7 ± 0.5	0.64 ± 0.00	0.40 ± 0.00	28.2 ± 7.3	2.0 ± 0.1
Ag 30nm + Cu 200nm	7.2 ± 0.4	0.64 ± 0.00	0.54 ± 0.01	10.7 ± 0.0	2.5 ± 0.1
Ag 150nm	9.2 ± 0.1	0.64 ± 0.00	0.49 ± 0.00	16.5 ± 0.3	2.9 ± 0.1
Ag 150nm + Cu 200nm	8.1 ± 0.4	0.64 ± 0.00	0.50 ± 0.01	18.3 ± 2.1	2.6 ± 0.2

C. Characterization

Current density – voltage (J-V) characteristics were measured in ambient air using a source meter (Keithley 2420) in a four-wire connection scheme. A solar simulator (TriSol TSS-156, Class AAA, Newport Oriel) equipped with a 1000W Xenon arc lamp with an air mass (AM) 1.5 filter and providing an irradiance of 100 mW/cm^2 was used as the light source. A Si reference cell (2cm × 2cm OAI verified mono-Si reference cell) was used to calibrate the intensity of the solar simulator.

III. RESULTS AND DISCUSSION

The structure of organic solar cells and the complete device are shown in Fig. 1 (e). The surface of ITO was modified with

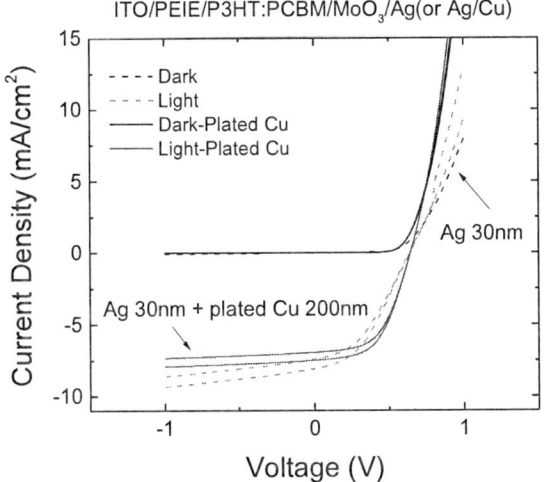

Fig. 2. Current density – Voltage (J-V) characteristics of inverted organic solar cells with- and without electroplated Cu top electrode. Ag thickness is 30 nm.

PEIE in order to reduce its work function; enable it to act as an electron collection electrode [4].

Fig. 2 shows the current density – voltage (J-V) characteristics of solar cells with 30 nm of Ag top electrode tested before- and after-electroplating copper. Under 100 mW/cm^2 of AM 1.5G illumination, the device before the electroplating of copper displays V_{OC} = 0.64 V, J_{SC} = 7.7 ± 0.5 mA/cm^2, and FF = 0.40 ± 0.00, yielding a power conversion efficiency (PCE) = 2.0 ± 0.1 % (Table 1). Averages and standard deviations were based on measurements of 2 devices.

Compared to the device before Cu-plating, the device after Cu-plating exhibits improved overall performance due to a significant decrease and increase of R_SA (28.2 vs. 10.7 Ω-cm^2) and FF (0.40 vs. 0.54), respectively. A reduction of more than half of magnitude was achieved for R_SA, thereby improving FF by 35 %. This improvement attributes to the increased top electrode thickness (Ag 30nm vs. Ag 30nm / Cu 200 nm), thereby increasing conductance of top electrode significantly.

Meanwhile, J_{SC} decreases by 6.5% (7.7 vs. 7.2 mA/cm^2) and this reduction could be due to the interaction between copper plating solution and the photoactive layer of P3HT:PCBM as the entire photoactive regions were submerged during the plating. Although 30nm of Ag protects the photoactive regions from the copper plating solution, the solution can still diffuse into the active layer through the edge of Ag electrode as was shown in Fig. 3.

Fig. 4 shows the current density – voltage (J-V) characteristics of solar cells with 150 nm of Ag top electrode tested before- and after-electroplating copper. The device before the electroplating of copper shows V_{OC} = 0.64 V, J_{SC} = 9.2 ± 0.1 mA/cm^2, and FF = 0.49 ± 0.00, yielding a power conversion efficiency (PCE) = 2.9 ± 0.1 % (Table 1). Because of thicker Ag (150 nm vs. 30 nm), FF improves a lot (0.49 vs. 0.40). Furthermore, J_{SC} is increased as well (9.2 vs. 7.7 mA/cm^2), most likely thanks to the enhanced reflectance from the thick (150 nm) Ag compared to the thin Ag (30 nm).

Compared to the device before Cu-plating (Ag 150 nm only), the device after Cu-plating exhibits decreased J_{SC} by 12% (9.2

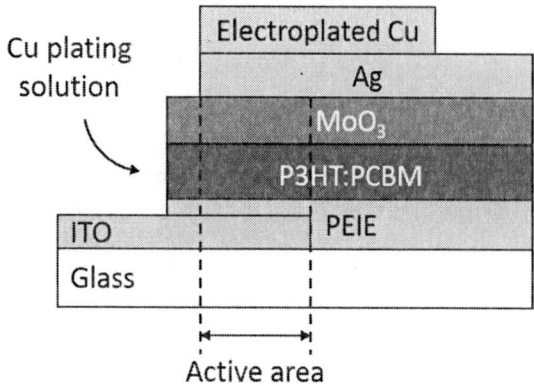

Fig. 3. Device geometry of an inverted OPV with the top electrode comprise evaporated silver and electroplated copper. Although 30nm of Ag protects the photoactive regions from the copper plating solution, the solution can still diffuse into the active layer through the edge of silver electrode.

vs. 8.1 mA/cm^2), again due to the interaction between electroplating solution and the photoactive organic semiconductors. There was almost no improvement in *FF* (0.50 vs. 0.49) since the 150 nm of Ag already provides highly

conductive current path for hole collection. Hence, the PCE decreases from 2.9 % to 2.6 %.

IV. CONCLUSION

In this work, we have demonstrated improved performances of an inverted OPV by plating Cu on the top Ag electrode. The *FF* is increased by 35% by plating 200 nm of Cu on top of 30 nm of Ag. However, J_{SC} is decreased by 6.5% and this reduction could be due to the interaction between copper plating solution and the photoactive layer of P3HT:PCBM. Despite the decrease of J_{SC}, the PCE increased by 25% thanks to the significant increase of *FF*.

The device with 150 nm of Ag electrode shows decreased PCE after plating Cu due to the decreased J_{SC} and nominal increase of *FF*. The additional Cu electrode added to the existing thick Ag electrode (150 nm) was not able to improve the *FF* since the 150 nm of Ag already provides highly conductive current path for hole collection.

ACKNOWLEDGEMENT

The student travel for the 2017 IEEE PVSC presentation was in part supported by the Student Academic Enhancement Fund (SAEF) from the University of Washington at Bothell.

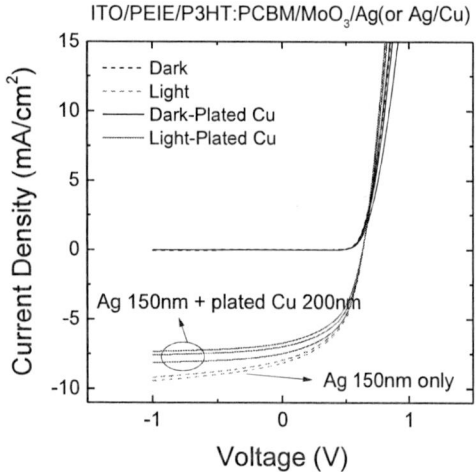

Fig. 4. J-V characteristics of inverted organic solar cells with- and without electroplated Cu top electrode. Ag thickness is 150 nm.

REFERENCES

[1] A. Lennon, Y. Yao, and S. Wenham, "Evolution of metal plating for silicon solar cell metallisation," *Progress in Photovoltaics: Research and Applications*, vol. 21, pp. 1454-1468, 2013.

[2] S. Aksu, S. Pethe, A. Kleiman-Shwarsctein, S. Kundu, and M. Pinarbasi, "Recent advances in electroplating based CIGS solar cell fabrication," 28th *IEEE Photovoltaic Specialists Conference*, pp. 3092-3097, 2012.

[3] S. Choi, W. Potscavage, and B. Kippelen, "Area-scaling of organic solar cells," *Journal of Applied Physics*, Vol. 106, pp. 54507, 2009.

[4] Y Zhou, C. Fuentes-Hernandez, J. Shim, J. Meyer, A. Giordano, H. Li, P. Winget, T. Papadopoulos, H. Cheun, J. Kim, M. Fenoll, A. Dindar, W. Haske, E. Najafabadi, T. Khan, H. Sojoudi, S. Barlow, S. Graham, J-L. Breda, S. Mardar, A. Kahn, and B. Kippelen, "A Universal Method to Produce Low–Work Function Electrodes for Organic Electronics," *Science*, vol. 336, pp. 327-332, 2012.

Interface Band Gap and Charge Trapping in Bulk Heterojunction Solar Cells

Marian Tzolov, and Maxwell McIntyre

Department of Physics, Lock Haven University, Lock Haven, Pennsylvania, 17745, United States)

Abstract — **Bulk heterojunction devices were studied using Fourier Transform Photocurrent Spectroscopy and impedance spectroscopy. The photocurrent below the absorption edge reveals the charge transfer transition and we present experimental evidence confirming that it is due to the presence of the acceptor PCBM. The impedance spectra reveal additional polarization under illumination. We propose a model for photogenerated charges confined in isolated material regions which contribute to this additional polarization.**

Index Terms — **polymer solar cells, bulk heterojunction, photocurrent spectroscopy, impedance spectroscopy, charge transfer transition, recombination.**

I. INTRODUCTION

The bulk heterojunction is one of several successful concepts for organic solar cells. [1] The spontaneous formation of the heterojunction is unique and even more so that the same processing technique of the active layer is used to invert the polarity of the device just by modifying the electrode interfaces. No convincing evidence exists that the bulk of the active layer and hence the ordering of the donor/acceptor interfaces is altered by the modification of the interfaces. This obvious symmetry of the bulk heterojunction is not an obstacle to create rectifying devices and hence to have efficient photocurrent generation. The large built-in electric field leads to short sweep out time [2] for the photogenerated charge carriers thus minimizing the recombination effects. A large value of the built-in electric field is achieved by using thin layers of 100 nm and less. This small thickness together with the organic and disordered nature of the material present challenges for detailed structural studies. Often, the structural models are based on results from studies of the macroscopic properties of the devices. [1]

In this paper, we present the results from two macroscopic experimental methods: photocurrent spectroscopy and impedance spectroscopy. We are showing that they offer unique information about the bulk heterojunction in addition of being nondestructive.

II. EXPERIMENTAL

The device structures were deposited on ITO coated glass substrates. A hole injection layer (HIL), Plexcore® OC CF was deposited onto the substrate. We have used regioregular poly(3-hexyl-thiophene-2,5-diyl) (P3HT) from American Dye Source, Poly[2,6-(4,4-bis-(2-ethylhexyl)-4H-cyclopenta [2,1-b;3,4-b']dithiophene)-alt-4,7(2,1,3-benzothiadiazole)] (PCPDTBT), Poly[N-9'-heptadecanyl-2,7-carbazole-alt-5,5-(4',7'-di-2-thienyl-2',1',3'-benzothiadiazole)] (PCDTBT), and PCBM-C60 from 1-Material. The active layer blend of PCPDTBT/PCBM-C60 and of PCDTBT/PCBM-C60 was dissolved in dichlorobenzene. P3HT/PCBM-C60 was dissolved in chlorobenzene. All films were processed by spin coating in inert atmosphere glovebox, where the water content was kept below 6 ppm. The devices were finalized with thermal evaporation at pressure ~2×10^{-6} Torr of aluminum of about 150 nm. Several test samples were prepared of pure polymer PCPDTBT sandwiched between ITO and aluminum contacts.

The current-voltage characteristics were measured using Keithley 2602 SourceMeter. The admittance spectra were measured using impedance analyzer HP4192A. The photocurrent in the near infrared range was recorded using Fourier Transform Spectroscopy (FTS) which was realized on a Fourier spectrometer Bruker IFS 66 in step scan mode. Long pass filters were used to minimize spurious effects. The photocurrent in the visible and near ultraviolet range was measured on Lambda 650 combined with lock-in amplifier SR830. The solar cells were operated in short circuit mode with the same transimpedance amplifier, Signal Recovery 5182, during all photocurrent measurements. The normalization of the photocurrent spectra was performed using calibrated silicon and germanium photodiodes.

III. RESULTS

All samples with bulk heterojunction polymer/PCBM show good rectification, i.e. more than 100 ratio between the forward and reverse current at voltage bias ±1V. The current-voltage characteristics were recorded before and after the photocurrent and impedance measurements, and the comparison didn't indicate substantial modification as result of these experiments.

A. Photocurrent Spectroscopy

The photocurrent spectrum of PCPDTBT/PCBM in Fig. 1 shows the absorption edge around 1.5 eV with an exponential decrease toward low photon energies. This region has a shoulder around 1.3 eV which has been previously ascribed [1, 3] to charge transfer transition between the HOMO of the polymer to the LUMO of PCBM. In order to check the validity of this interpretation, we have prepared a sample with only polymer (PCPDTBT) sandwiched between ITO and aluminum electrodes. This sample was showing small photocurrent generation, which was enough for recording reliable spectra in the FTS measurements. Unfortunately, the illumination intensity of Lambda 650 was not enough for recording a spectrum in the visible range.

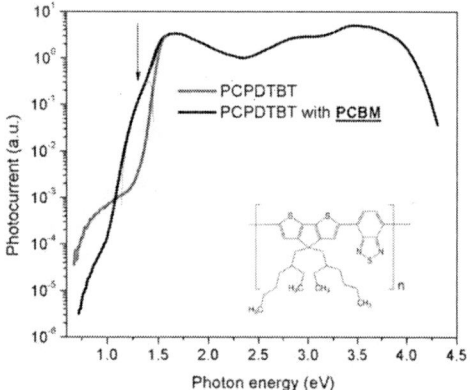

Fig 1. Photocurrent spectra of a device with active layer PCPDTBT/PCBM (black curve) and device with active layer PCPDTBT (red curve).

Fig. 1 shows clear difference between the pure PCPDTBT film and the bulk heterojunction in the range of 1.3 eV. This is a clear evidence that the presence of the bulk heterojunction is causing the shoulder below the absorption edge in support of the interpretation as a charge transfer transition. [1, 3]

It is interesting that the pure PCPDTBT shows much higher subgap absorption, e.g. at 1 eV. We hypothesize that the presence of these subgap defect states promotes the small

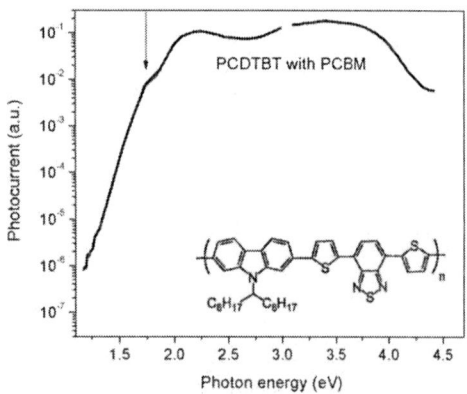

Fig 2. Photocurrent spectra of a device with active layer PCDTBT/PCBM.

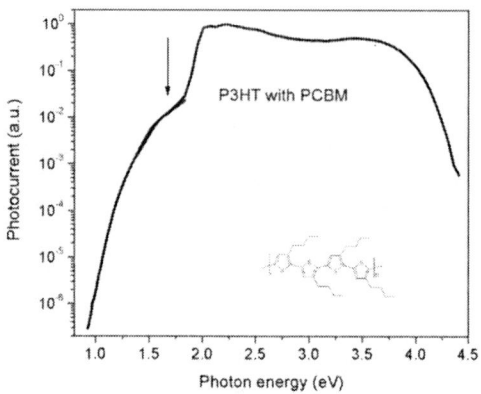

Fig 3. Photocurrent spectra of a device with active layer P3HT/PCBM.

photocurrent generation of this sample. We have studied the FTS in more than twenty PCPDTBT/PCBM devices prepared during the last two years and no subgap absorption of this extent was ever observed in the bulk heterojunction devices. The origin of the relatively large subgap absorption at 1 eV in the pure PCPDTBT devices needs further clarification.

We have studied the photocurrent spectra of devices with different polymers in order to find if similar charge transfer transitions are present. Fig. 2 shows the photocurrent spectrum of PCDTBT/PCBM device. The photocurrent spectrum peaks at 2.2 eV and decreases exponentially toward low photon energy. A shoulder similar to the one in Fig. 1 is observed at 1.75 eV. The width of this feature is not much different than the one observed in Fig. 1 for PCPDTBT.

It is a very different case for the P3HT/PCBM devices. Fig. 3 shows that the maximum of the photocurrent spectrum at 2 eV is followed by an exponential decrease toward low photon energies, followed by a very broad subgap shoulder. We interpret it as due to the charge transfer transition in the P3HT/PCBM bulk heterojunction, which we estimate to be around 1.7 eV. Remarkable is that the charge transfer band is much broader in energy than the corresponding feature in PCPDTBT/PCBM and PCDTBT/PCBM. On the other hand, this feature is about 100 times less intense than the maximum absorption, while this difference for PCPDTBT/PCBM and PCDTBT/PCBM is about 10 times. It is possible that the different structure of the polymer chain (see the insets in Figures 1, 2 and 3) leads to differences in the heterojunction with PCBM. Revealing the nature of these differences would give valuable information about the mechanisms at the bulk heterojunction.

B. Impedance Spectroscopy

Our systematic investigation of more than forty devices prepared during the last two years based on PCPDTBT/PCBM

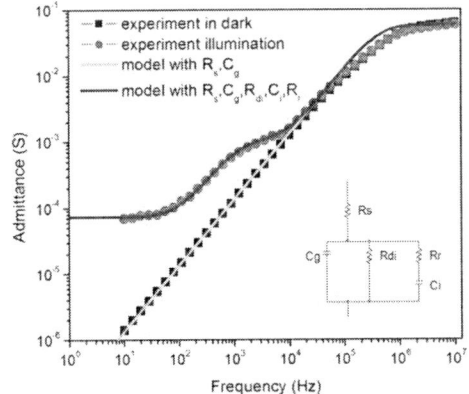

Fig 4. Frequency dependence of the magnitude of the admittance of a device with active layer PCPDTBT/PCBM in dark (black symbols) and under illumination (red symbols). The inset shows the equivalent circuit used to model the admittance spectra.

has indicated a very peculiar feature in the impedance spectroscopy. We are summarizing these finding in a single graph, Fig. 4, and results that are more detailed will be published separately. For convenience, we present the magnitude of the admittance, only. The impedance measurements were performed at short circuit conditions, i.e. not DC offset for the AC stimulus. The admittance spectrum in dark resembles the response of a dielectric film sandwiched between two electrodes with finite resistance. The constant admittance at high frequency is due to the resistance of the connecting ITO and aluminum stripes, R_s. The additional resistance measurements just on the ITO stripes agree with the admittance value at high frequency. At lower frequencies, the admittance varies linearly with frequency as expected for a capacitance and we assign it to the geometrical capacitance (C_g) of the devices, i.e. a dielectric sandwiched between two electrodes. The green line in Fig. 4 shows a very good modeling of the experimental data using a resistance, R_s, and capacitance, C_g, connected in series. It is interesting to note, that modeling the capacitance as parallel plate capacitor and considering the actual electrode area and thickness of the active layer, results in a dielectric constant of about 4, which is consistent with published data. [4]

The light illumination modifies substantially the admittance spectrum at low frequencies. The constant value of the admittance at very low frequency is due to the dynamic resistance of the current-voltage characteristics of the device. The dynamic resistance in dark is very high, which is why it is not detectable within the range of these measurements. The dynamic resistance is equivalent to the shunt resistance, and the inset in Fig. 4 shows it as R_{di}. It carries information about the recombination of photogenerated charge carriers.

The remarkable feature for a device under illumination in Fig. 4 is between 10^2 and 10^4 Hz. In order to simulate this feature, we have introduced an addition series RC group in the equivalent circuit, as shown in the inset in Fig. 4. These new elements, C_i and R_r, reflect the appearance of additional polarization mechanism within the device, C_i, and the associated loss of energy, R_r. Experimentally, this means that if one measures at a fixed frequency, e.g. 10^3 Hz, and applies a simple RC model, the calculation will result in several times higher capacitance, resp. dielectric constant, compared to the measurement without illumination.

We have observed such additional polarization also for the P3HT/PCBM devices, i.e. it appears to be a common feature of the bulk heterojunction devices. Encapsulation of the devices did not modify the additional polarization and the impedance spectrum is generally the same.

IV. DISCUSSION AND CONCLUSIONS

The presented results support the existing interpretation [1, 3] for an absorption below the optical edge as due to charge transfer transitions, i.e. transitions localized in a very narrow space to the physical donor/acceptor interface. In order these transitions to have sizable contribution to the optical properties, the light needs to encounter the donor/acceptor interface multiple times while passing through the device. The random formation of the bulk heterojunction supports such scenario. This randomness implies also that some interfaces are oriented "in reverse" to the photovoltaic operation. The formation of inverse devices just by modifying the interfaces is an indirect support for this.

The random formation of the heterojunction suggests that there may be islands of either donor or acceptor without connectivity to the electrodes. The photogenerated charges in these islands would stay confined in the islands until they recombine. During their lifetime, the confined charges will respond to any external electric field leading to the additional polarization which we have detectred in Fig. 4, and modeled with the equivalent capacitance C_i.

Our admittance measurements with DC bias show that a negative bias, which extracts the photogenerated charges, reduces substantially the additional polarization. This confirms that the polarization is due to charges confined in the active material.

We believe that the measurement of the admittance under illumination is an effective tool to evaluate the presence of isolated regions of material in the bulk heterojunction devices. It is a nondestructive method, and additional rich information can be obtained when combined with biasing of the devices and different levels of illumination. Our detailed results on such studies will be published separately.

The photocurrent spectra below the absorption edge give also access to the interface properties of the bulk heterojunction. We report differences between the charge transfer transition in P3HT/PCBM vs. PCPDTBT/PCBM and PCDTBT/PCBM. It is possible that the bulk heterojunction forms with different properties depending on the polymer. The photocurrent spectroscopy is a convenient tool to evaluate the interface gap, which is the principle upper limit for the open circuit voltage. [1]

ACKNOWLEDGEMENT

The Fourier spectrometer Bruker IFS 66 used in this work was part of shared facilities supported in part by the Penn State MRSEC, Center for Nanoscale Sciences, under the NSF award DMR-1420620.

REFERENCES

[1] M. H. Kang , G. Kim , J. Kim , S. Kwon , H. Kim , and K. Lee "Bulk-Heterojunction Organic Solar Cells: Five Core Technologies for Their Commercialization" *Advanced Materials,* vol 28, pp. 7821-7861, 2016.

[2] S. Cowan, R. Street, S. Cho, and A. Heeger, "Transient photoconductivity in polymer bulk heterojunction solar cells: Competition between sweep-out and recombination." *Physical Review B,* vol. 83, 035205, 2011.

[3] R. Street, K. Song, J. Northrup, and S. Cowan "Photoconductivity measurements of the electronic structure of organic solar cells" *Physical Review B,* vol. 83, 165207, 2011.

[4] M. A. Loi, S. Toffanin, M. Muccini, M. Forster, U. Scherf, and M. Scharber "Charge transfer excitons in bulk heterojunctions of a polyfluorene copolymer and a fullerene derivative" *Advanced Functional Materials*, vol. 17, pp. 2111-2116, 2007.

Fabrication of Efficient CH₃NH₃PbI₃ Solar Cells in Ambient Air

Feng Wang[a,b], Zhongbiao Ye[a,b], Hojjatollah Sarvari[b], Somin Park[c], Kenneth Graham[c], Yuetao Zhao[a,b] and Zhi David Chen[a,b*]

[a]School of Optoelectronic Information, University of Electronic Science and Technology of China, Chengdu, Sichuan, 610054, China

[b]Department of Electrical & Computer Engineering, and Center for Nanoscale Science & Engineering, University of Kentucky, Lexington, Kentucky, 40506, USA

[c]Department of Chemistry, University of Kentucky, Lexington, Kentucky, 40506, USA

*Corresponding author: zhichen@engr.uky.edu

Abstract — **In most research, fabrication of high performance perovskite solar cells (PSCs) has been carried out in a well-controlled environment (glove boxes), which may result in high cost for eventual manufacturing. Therefore, fabrication of efficient perovskite solar cell in ambient air is of great interest. We fabricated CH₃NH₃PbI₃(MAPbI₃) solar cells in ambient air without control of relative humidity (~20-50%RH) using the antisolvent method. The effect of both the antisolvent and the annealing process on the solar cell performance has been investigated. Solar cells with power conversion efficiency of over 15% and negligible hysteresis were obtained, which are close to the performance of solar cells fabricated at nitrogen environment in a glove box. It may pave the way for fabrication of high-performance PSCs in ambient air.**

Index Terms — **perovskite solar cells, CH₃NH₃PbI₃, ambient air, relative humidity, hysteresis.**

I. INTRODUCTION

Solar cells have emerged as an effective alternative energy source for fossil fuels because they harvest solar energy in an environmental-friendly way. Silicon-based solar cells have achieved quite impressive commercial success. However, the cost for silicon solar cell production is still high comparing to fossil fuels. Therefore, new solar cell technology that is able to produce high-performance solar cells in a more economical way is desired. Recently, perovskite solar cells (PSCs) have attracted much attention because of their outstanding photovoltaic performance and low-cost solution processes. Power conversion efficiency (PCE) increased rapidly in the past few years. Nowadays, power conversion efficiency (PCE) of over 21% has been achieved, comparing with PCE of only 3.8% in the first report of PSCs in 2009 [1]-[2]. However, in order to obtain PSCs with high performance, a well-controlled preparation environment is required, gloveboxes filled with N₂ or Ar gas, which may add cost in their commercial manufacturing. Hence, fabrication of PSCs in ambient air is of great interest. Both one-step and two-step methods have been used to fabricate PSCs in ambient air. For two-step, the morphology of PbI₂ was found to be of great importance. Pre-heating the mesoporous TiO₂ layer coated substrates led to

PbI₂ layer of higher crystallization and better infiltration into mesoporous TiO₂ layer, thus better surface coverage of MAPbI₃ layer was achieved [3]. Although the maximum efficiency of 15.76% was obtained, the hysteresis was not negligible. The hysteresis issue also exists when gas flow assisting was adopted to fabricate MAPbI₃ solar cells in the ambient condition of >42%RH [4]. For one step, high-performance MAPbI₃ solar cells were fabricated at<40% RH with a champion efficiency of 16.15% and better hysteresis performance by introducing elaborate temperature treatment [5]. Besides, Pb(SCN)₂ was explored as a new material for PSCs fabrication at high relative humidity of 70%. Average PCE of 13.49% was obtained [6]. It should be noted that fabrication of PSCs in ambient air with high PCE as well as good hysteresis performance remains a challenge. Humidity was considered to be the most important factor in air that plays a key role in uncontrollable influence on crystallization of the perovskite layer, because water molecules could participate in the crystallization process and form hydrate. As a result, making the crystallization process more controllable and thus resistant to water molecule is crucial. It has been demonstrated that by adding dimethyl sulfoxide (DMSO) intermediate of MAI-DMSO-PbI₂ is formed after applying antisolvent solution during spin coating, which remarkably improved the quality of perovskite films [7]. Thus, DMSO is expected to play a positive role in fabrication of high-performance PSCs in ambient air. In this paper, fabrication of MAPbI₃ solar cells at 20-50%RH was conducted using the antisolvent method with equal molar quantity of DMSO added to MAPbI₃ precursor. The effect on the solar cell performance of both antisolvent and the annealing process has been investigated. Solar cells with PCE of over 15% and negligible hysteresis were obtained, which are comparable to the results of solar cells fabricated in the nitrogen environment.

II. EXPERIMENT

A. Materials

Unless mentioned otherwise, all materials and solvents were provided by Sigma-Aldrich and Dyesol, and used without further treatment. Spiro-OMeTAD was purchased from Borun Tech.

B. Device Fabrication

FTO glass(NSG) substrates were cleaned sequentially in detergent, acetone, IPA, and DI water for 15 minutes each, followed by 15 min UVO treatment. A compact layer of TiO_2 was deposited via spray pyrolysis at 450°C from a precursor solution of titanium diisopropoxide bis (acetylacetonate) in ethanol (volume ratio 1:19). The films were annealed at 450°C for 20 minutes and then cooled down to room temperature. A 200-nm mesoporous TiO_2 layer was spin-coated at 4500 rpm for 20s from a diluted 30-nm particle paste (Dyesol) in ethanol; the concentration of TiO_2 solution is 0.14g/ml. After spin coating, the substrates were dried at 100°C for 5mins and then sintered at 500°C for 30 minutes. Li-doping of mesoporous TiO_2 was accomplished as described elsewhere [8].

Equal molar quantity of MAI, DMSO and PbI_2 was dissolved in DMF. Then the precursor solution was spin coated on the meso-TiO_2 layer at 3000 rpm for 25 sec and 0.5 ml of diethyl ether was slowly dripped on the rotating substrate in 3 sec to form a transparent film. For planar PSCs, the $MAPbI_3$ film was deposited on the compact TiO_2 layer after it cools down. The transparent perovskite film was heated at 65°C for 1 min and 100°C for 5 min. After the perovskite ($MAPbI_3$) film was deposited, the spiro-OMeTAD solution was spin coated on the substrates, which consisted of 72.3 mg spiro-MeOTAD, 28.8 ml 4-tert-butylpyridine, 17.5 ml of a solution of lithium bis(trifluoromethylsulphonyl)imide(Li-TFSI) (520 mg Li-TFSI in 1 ml acetonitrile) and 29 ml of a solution of 100mg/ml tris(2-(1H-pyrazol-1-yl)-4-tert-butylpyridine)-cobalt(III) bis(trifluoromethylsulphonyl)imide in acetonitrile in 1 ml of chlorobenzene. Finally, 80nm of Au top electrode was thermally evaporated onto the HTM layer.

C. Characterization

Solar cell performance was measured using a solar simulator (ABET technologies, 11002) at 100 mW/cm² illumination (AM 1.5G). The intensity was adjusted to 100 mW/cm² using a calibrated photodiode with a KG5 filter (ABET technologies). Top-view SEM was measured using Hitachi S-4300. X-ray pattern of perovskite deposited on m-TiO_2 layer was obtained using an X-ray diffractometer (XRD D500 Simens), using Cu-Kα radiation (λ=1.54050Å).

III. RESULTS AND DISCUSSION

The same procedure was adopted for fabrication of PSCs in both the N_2 glovebox and ambient air with relative humidity >40% at 24°C with the device structure of FTO/C-TiO_2/mp-TiO_2/Perovskite/HTM/Au. It was found that the PSCs produced in air have a very competitive performance of

hysteresis and Voc over those obtained in the N_2 glovebox, as shown in Fig.1.

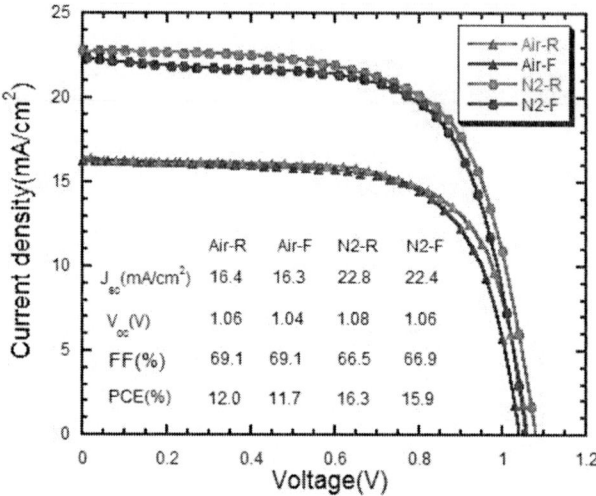

Fig. 1. Mesoporous PSCs with a device structure of FTO/C-TiO_2/mp-TiO_2/Perovskite layer/HTM/Au fabricated in both ambient air(41%RH, 24℃) and a N_2 glovebox. Air stands for fabrication in air; N2 stands for fabrication in the N_2 glovebox; R stands for reverse scan and F stands for forward scan. The perovskite layer made in air is only 400 nm while that made in the N_2 glovebox is 650 nm.

It should be noted that the relatively low current of the perovskite solar cell fabricated in air is due to its thin $MAPbI_3$ film, which is less than 400nm, while the thickness of the perovskite film fabricated in the N_2 glovebox is around 650nm including the mesoporous layer and capping layer, which is thick enough to absorb the light to deliver a decent current. It is believed that the Jsc of PSCs fabricated in air would increase remarkably as the thickness of the perovskite layer increased.

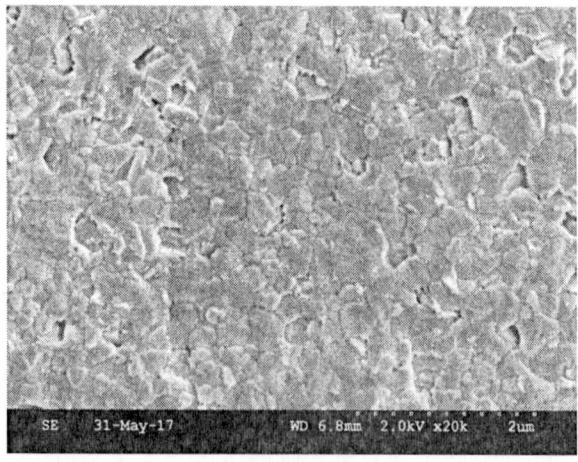

Fig. 2. Top-view SEM image of $MAPbI_3$ films on meso-TiO_2 layer prepared in ambient air.

A top-view scanning electron microscopy(SEM) of MAPbI₃ perovskite films on the mesoporous TiO₂ layer is shown in Fig.2. It is obvious that a compact and pinhole-free perovskite film was formed. The XRD characterization was conducted to analyze the phase of the perovskite film, as shown in Fig.3. It is clearly shown in the XRD pattern that the perovskite film was composed of highly crystallized pure perovskite phase with orthorhombic structure (14.1°, 28.4°, and 32.0°), similar to the MAPbI₃ XRD pattern mentioned elsewhere [9]. PbI_2 trace that was present when fabricating MAPbI₃ in air is disappear here [3]. These complementary characterizations have shown that high-quality MAPbI₃ films was formed.

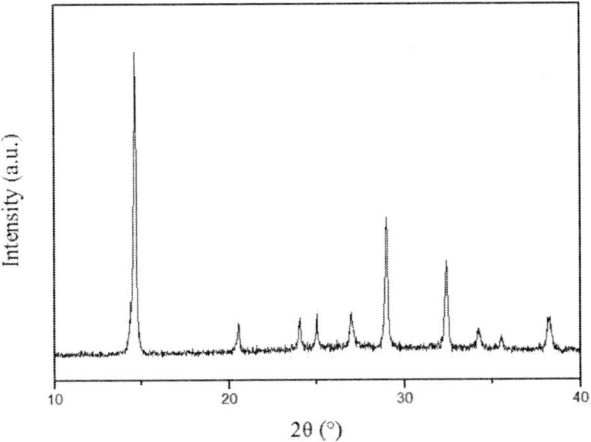

Fig. 3.   XRD pattern of MAPbI₃ films on meso-TiO₂ layer prepared in ambient air.

Planar MAPbI₃ PSCs with a device structure of FTO/C-TiO₂/Perovskite/HTM/Au was also fabricated in ambient air (28%RH, 24°C) with a reverse-scan PCE of 15.2%, as shown in Fig.4. Although the forward scan performance is less satisfied, it should be mentioned that hysteresis is always a problem for planar PSCs because of the unbalance of electron and hole diffusion length. Mesoporous TiO₂ layer and Li-doping have been proved to improve the forward scan performance [9]. As already demonstrated here as shown in Fig.1, MAPbI₃ solar cells fabricated in ambient air with mesoporous TiO₂ layer and Li-doping shows little hysteresis. The solar cell with a thick enough MAPbI₃ perovskite layer was also fabricated in the ambient air with 50%RH at 21°C. The performance of the device under reverse scan(10mv/s) are as follows: short-circuit current density (J_{SC}) = 20.7 mA cm⁻², open-circuit voltage (V_{OC}) = 1.06 V, and fill factor (FF) =0.725, PCE =15.91%. The PCE under forward scan(10mv/s) is 15.44%, which lead to a hysteresis lower than 0.6%(Fig.5). As a result, the feasibility of fabrication efficient MAPbI₃ PSCs with negligible hysteresis in ambient air using antisolvent method under various relative humidity is demonstrated.

It is worth noting that the performance of the perovskite solar cells fabricated in ambient air is comparable to that fabricated in a glovebox where the concentration of both O_2 and humidity is less than 0.5ppm. Interestingly, in most papers those experiments conducted in air, humidity has been proved to be an uncontrollable factor affecting the perovskite crystallization process [5]. There is a big difference between fabrication of PSCs under glovebox and ambient air. In our case, however, the PSCs' performance seems to be insensitive to the relative humidity of fabrication environment.

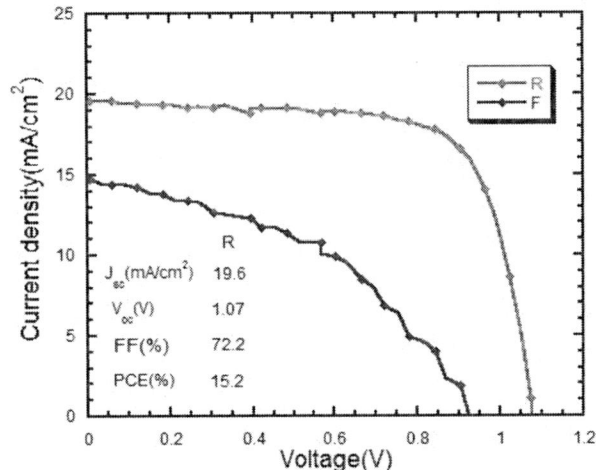

Fig. 4.   Planar PSCs with a device structure of FTO/C-TiO₂/Perovskite/HTM/Au fabricated in ambient air. R stands for reverse scan and F stands for forward scan.

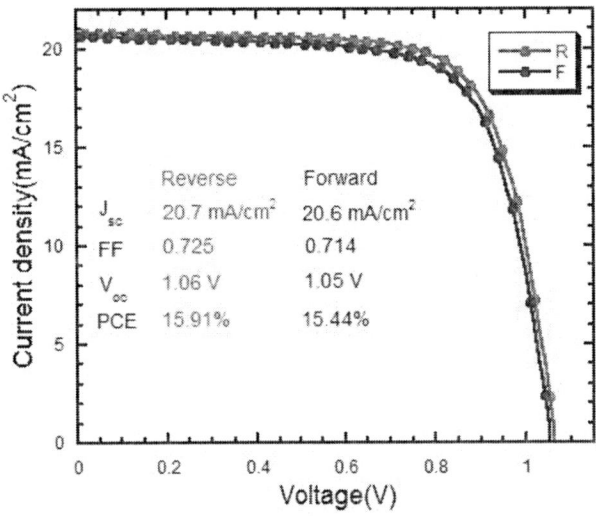

Fig. 5.   Mesoporous PSCs with a device structure of FTO/C-TiO₂/mp-TiO₂/Perovskite layer/HTM/Au fabricated in the ambient air（50%RH, 21°C）.

Experiments were conducted to further investigate the effect of antisolvent and annealing process on the PSCs performance. The results are shown in Fig.6. For the cases using diethyl ether as antisolvent, though various post-annealing processes were adopted, similar performances were obtained and the average PCE is 15.3%. While using chlorobenzene as antisolvent, the performance was severely affected. It can be seen that diethyl ether is more suitable to be adopted as antisolvent than chlorobenzene when fabricating MAPbI₃ in ambient air. The intermediate would be damaged when chlorobenzene was used as antisolvent, because chlorobenzene is miscible with both DMF and DMSO.

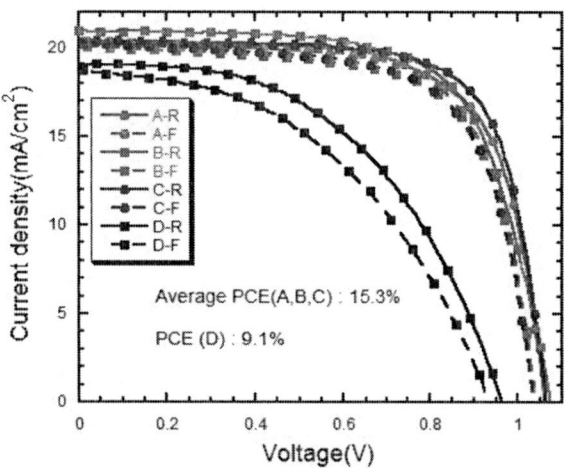

Fig. 6.   Mesoporous PSCs fabricated using different antisolvents and annealing processes. A, B and C used diethyl ether as antisolvent, while D used chlorobenzene as anti-solvent. A: 65°C annealing for 1 min and then 100°C annealing for 5 min; B: 100°C annealing for 6 min; C: natural dry for 1 min, and then 100°C annealing for 5 min; D: 65°C annealing for 1 min and then 100°C annealing for 5 min. Reverse: solid line; Forward: dash line.

From Fig. 6, it can be seen that the MAPbI₃ PSCs treated with various annealing processes after perovskite spin coating show similar performance. It is suggested that a short period of various temperature treatment after spin coating makes negligible influence to the performance of the MAPbI₃ PSCs. It is believed that the MAI-DMSO-PbI₂ intermediate is durable in ambient air and postpones the crystallization process of MAPbI₃. On the other hand, the ignorable influence of annealing treatments on PSCs performance demonstrates that the intermediate helps to prevent the perovskite film from the influence of humidity.

IV. CONCLUSION

In this paper, we fabricated efficient MAPbI₃ solar cells in ambient air with PCE of over 15% and negligible hysteresis. The performance is found to be comparable to that of solar cells prepared in the N₂ glovebox. Antisolvent is found to be

of great importance for fabrication of high-quality perovskite films in ambient air. In our case, diethyl ether is very good to maintain the MAI-DMSO-PbI₂ intermediate. The annealing process was found to have little influence on PSCs' performance. This suggests that MAI-DMSO-PbI₂ intermediate is durable in ambient air. This may provide a strategy for fabrication of high-performance PSCs in ambient air.

ACKNOWLEDGEMENT

Feng Wang is grateful to the China Scholarship Council for the student grant.

REFERENCES

[1] A. Kojima, K. Teshima, Y. Shirai, T. Miyasaka. "Organometal halide perovskites as visible-light sensitizers for photovoltaic cells," *Journal of the American Chemical Society,* vol. 131(17), pp. 6050-6051, 2009.

[2] M. Saliba, T. Matsui, J. Y. Seo, K. Domanski, J. P. C. Baena, M. K. Nazeeruddin, W. Tress, A. Abate, A. Hagfeldtd and M. Grätzel, "Cesium-containing triple cation perovskite solar cells: improved stability, reproducibility and high efficiency," *Energy & Environmental Science,* vol. 9(6), pp. 1989-1997, 2016.

[3] H. S. Ko, J. W. Lee, N. G. Park, "15.76% efficiency perovskite solar cells prepared under high relative humidity: importance of PbI₂ morphology in two-step deposition of CH₃NH₃PbI₃," *Journal of Materials Chemistry A,* vol. 3(6), pp. 8808-8815, 2015.

[4] B. Lei, V. O. Eze, T. Mori, "High-performance CH₃NH₃PbI₃ perovskite solar cells fabricated under ambient conditions with high relative humidity," *Japanese Journal of Applied Physics,* vol. 54(10), pp. 100305, 2015.

[5] M. Lv, X. Dong, X. Fang, B. Lin, S. Zhang, X. Xu, J. Ding, and N. Yuan, "Improved photovoltaic performance in perovskite solar cells based on CH₃NH₃PbI₃ films fabricated under controlled relative humidity," *RSC Advances,* vol.5(114), pp. 93957-93963, 2015.

[6] Q. Tai, P. You, H. Sang, Z. Liu, C. Hu, H. L. Chan, and F. Yan, "Efficient and stable perovskite solar cells prepared in ambient air irrespective of the humidity," *Nature communications,* vol.7, 11105, 2016.

[7] N. Ahn, D. Y. Son, I. H. Jang, S. M. Kang, M. Choi, and N. G. Park, "Highly reproducible perovskite solar cells with average efficiency of 18.3% and best efficiency of 19.7% fabricated via Lewis base adduct of lead (II) iodide," *Journal of the American Chemical Society,* vol. 137(27), pp. 8696-8699, 2015.

[8] F. Giordano, A. Abate, J. P. Correa-Baena, M. Saliba, T. Matsui, S. H. Im, S. M. Zakeeruddin, M. K. Nazeeruddin, A. Hagfeldt, and M. Grätzel "Enhanced electronic properties in mesoporous TiO₂ via lithium doping for high-efficiency perovskite solar cells," *Nature communications,* vol. 7, 10379, 2016.

[9] Q. Chen, H. Zhou, T. Song, S. Luo, Z. Hong, H. Duan, L. Dou, Y. Liu, and Y. Yang, "Controllable self-induced passivation of hybrid lead iodide perovskites toward high performance solar cells," *Nano letters,* vol. 14(7), pp. 4158-4163, 2014.

High Efficiency Perovskite Solar Cells by a Modified Low-Temperature Solution Process Inter-Diffusion Method

Yangyi Yao[co], Wei-Lun Hsu[co], Mario Dagenais

Department of Electrical and Computer Engineering, University of Maryland, College Park,

Maryland, 20742, USA

ABSTRACT — A modified inter-diffusion method to fabricate methylammoium lead iodide (MAPbI3) perovskite solar cell is proposed. A high quality perovskite active layer with thickness over 400 nm and grain sizes over 1 um was fabricated by the controlled solvent-assist annealing technique. (6,6)-Phenyl-C61-butyric acid methyl ester (PCBM) was used as the electron transport layer (ETL) to suppress the hysteresis. The performance dependence of solvent-assisted annealing and PCBM thickness was investigated. The highest power conversion efficiency (PCE) of 16.3% is obtained for a thick perovskite layer with reduced pin-holes and large grain sizes. A large external quantum efficiency (EQE) of 90% is obtained between 400-500 nm and it remains above 70% up to 700 nm.

I. INTRODUCTION

Metal organic halide perovskite solar cells have demonstrated rapid increased energy conversion efficiencies from 3.8% [1] to 22.1% [2] over the last 10 years. Methylammonium (MA) lead halide material is used as the active layer of perovskite solar cell because of its remarkable properties, such as long diffusion length [3], large carrier mobility [4], tunable bandgap [5] and low-temperature solution processable. Two dominant structures have emerged for fabricating perovskite solar cells: the meso-structure and the planar structure. In the meso-structure, the perovskite crystal infiltrates a metal oxide scaffold and forms an absorptive layer on top of the scaffold. The second structure is a planar perovskite film which is sandwiched between n- and p- type layers. The defects in perovskite film will affect the light absorption and carrier extraction; the pin-hole density will determine the current leakage level; the grain size of perovskite crystals dominates the carrier diffusion length [6].

The study of planar heterojunction perovskite solar cell is of significant importance to realize perovskite tandem cells [7] and flexible solar cell [8]. Various growing methods for high quality planar perovskite film have been reported to date. Z. Xiao, et al proposed the inter-diffusion methods [9] where methylammonium lead iodide (MAPbI$_3$) is formed by spin-coating lead iodide (PbI$_2$) and methylammonium iodide (MAI) sequentially. However, the main challenge is to ensure the full reaction of the PbI$_2$ layer. The formation of a perovskite layer on the top surface will prevent the further penetration of MAI, and this implies that the remnant PbI$_2$ will block carrier transport and suppress the light absorption [10]. Thus the thickness of PbI$_2$ is limited and consequently perovskite film cannot be thick enough to fully absorb the sun spectrum.

In our studies, we propose the modified inter-diffusion method with a controlled solvent-assisted annealing technique to grow high quality perovskite film with large grain sizes. The optimized recipe guarantees the successful formation of perovskite film with average thickness greater than 400 nm without PbI$_2$ residue. Large grain sizes and a reduced pin-hole density were realized by the solvent-assisted annealing technique. A PCE of 16.3% was obtained with considerable reproducibility. Hysteresis effect was also suppressed by optimizing the thickness of the electron transport layer (ETL).

II. EXPERIMENTAL SECTION

Fluorine doped Tin Oxide (FTO) glass was cleaned and pre-heated on 120°C hot plate before being spin-coated with Poly(3,4-ethylenedioxythiophene)-Poly(styrenesulfonate) (PEDOT:PSS) at 3750 rpm. The device was annealed for another 15 minutes and then transferred into a glove box. The perovskite film was grown by the modified inter-diffusion method with controlled solvent-assisted annealing technique. Highly concentrated PbI$_2$ solution (500 mg/ml in N,N-Dimethylformamide (DMF)) was spin-coated at 4500 rpm. A 10-minute post annealing at 85°C was required to form a smooth crystalized PbI$_2$ film. Highly concentrated MAI solution (75 mg/ml in 2-propanol) was then spin-coated on the PbI$_2$ layer at 1750 rpm for 2 minutes followed by pure toluene washing. Then the device was moved into the controlled solvent-assisted annealing set-up on a 150°C hot plate shown in Figure 2. The solvent vapor can be DMF or 2-propanal with controlled amount. After 1 hour annealing, a hot PCBM solution (20 mg/ml in chlorobenzene) was spin-coated on the crystallized perovskite film at 3500 rpm, and a 10-minute post- annealing at 100°C was required to dry out the remnant solvent. At the end, 100 nm silver electrode was deposited by electron beam evaporation.

The device area was defined by a shadow mask, and the best performing devices had a sample area of 0.00785 cm^2. I-V characteristics of the perovskite solar cell were measured in ambient with a Keithley 2400 source-meter and a Newport Model 91159 Full Spectrum 150 W Solar Simulator. To study the hysteresis phenomenon, the device was measured alternatively in opposite scanning direction. The X-ray diffraction (XRD) measurement was taken on a Bruker D8/C2 Discover Parallel Beam Diffractometer using Cu Kα radiation. The Scanning Electron Microscope (SEM) images were acquired by an Hitachi SU-70 Schottky Field Emission Gun Scanning Electron Microscope. The External Quantum Efficiency was measured with a self-build photo-spectrometer set up by illuminating the sample area with a modulated monochromatic light beam (Xenon light source), and amplifying the photo-current with a Stanford Research Systems SR570 pre-amplifier and a Stanford Research Systems SR830 lock-in amplifier.

III. RESULTS AND DISCUSSION

Fig. 1 shows the structure and band diagram of the device. FTO is the transparent conducting layer, the opaque Ag contact layer will reflect the incident light and enhance light absorption. MAPbI3 is the active layer, hole/electron transport layers are PEDTO:PSS and PCBM respectively. The sandwich structure of PEDOT:PSS/Perovskite/PCBM forms a p-i-n junction which will allow the build in potential to separate electron/hole pairs and generate photocurrent.

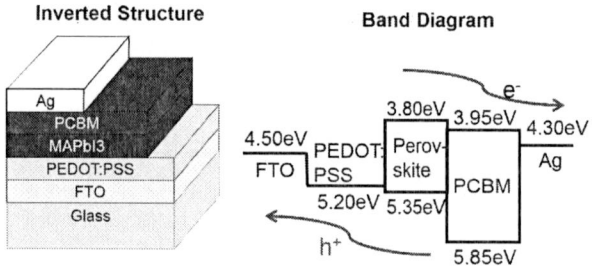

Fig. 1.   Solar cell structure and band diagram.

Much effort was devoted to optimize the recipe of the modified inter-diffusion method. The main drawback of the conventional inter-diffusion method is the difficulty of fully consuming PbI_2 layer. It is hard to get thick perovskite layer (> 300 nm) without leaving a PbI_2 residue by the conventional method [9]. This study proposes the modified inter-diffusion method to achieve high quality thick perovskite film with PbI_2 residue free. A highly concentrated MAI solution was spin-coated on a PbI_2 film using a low spin rate and a relatively long spin time. An obvious color change from light brown to dark brown was observed during spin-coating, which reflects the process of PbI_2 consuming. Interestingly, the remnant MAI on the sample surface was found to deteriorate the morphology of the perovskite during annealing. The sample

was washed by toluene to get rid of MAI remnant and to passivate the perovskite surface before annealing. Fig. 2a shows the XRD pattern of the grown the MAPbI3 film on FTO substrate after annealing. The red line is the XRD pattern of PbI_2 film, which has a strong peak on its (001) plane. The black line indicates the XRD pattern of MAPbI3 film, compared with the red curve, there were no PbI_2 peak present, which reveals the full consumption of PbI_2 layer. There are some common peaks in the XRD patterns of PbI_2 which belong to the FTO substrate.

The controlled solvent-assisted annealing is expected to suppress the trap density by obtaining larger grain size of perovskite crystals with less pin-hole density. As showed in Fig. 2b, the set-up for the solvent-assisted annealing approach could precisely control the solvent vapor. The same amount of solvent (DMF) is dropped on the 4 coverslips and the solvent vapor stays inside the larger beaker. The small beaker is located on the 4 coverslips which leave some space at the bottom to allow vapor exchange. The solvent amount will affect the grain size and the pin-hole density of the perovskite film. Fig. 2c and Fig. 2d compare the SEM images of the perovskite crystal surface without and with solvent assisted annealing. The grain size is less than 1 μm, and pin-holes are observable in Fig. 2c. While with optimized solvent vapor condition, the grain size is enlarged over 1 μm and the pin-hole density is greatly reduced.

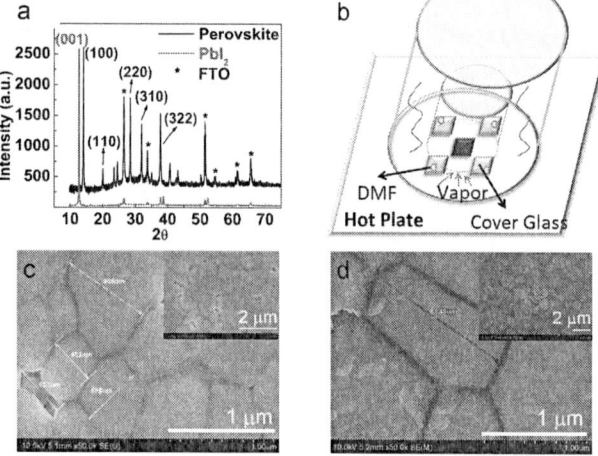

Fig. 2.   (a) XRD spectrum of PbI_2 film and annealed perovskite film. (b) Controlled solvent-assisted annealing set-up. (c) SEM image of perovskite crystals without solvent-assisted annealing. (d) SEM image of perovskite crystals with solvent-assisted annealing

Fig. 3a shows the J-V curve of the champion devices, with a large J_{sc} of 25.11 mA/cm^2, a V_{oc} of 0.92V, a fill factor of 71% and a PCE of 16.3% under 1 sun illumination. Fig. 3b shows the cross-section image of the perovskite solar cell. There is a continuous perovskite layer between the top contact and the

FTO layer with an average thickness of ~450 nm. A large photo-current should result from the thick perovskite layer (> 400 nm) with large grain sizes. Fig. 3c shows the external quantum efficiency (EQE) result of a large area cell with a short circuit current of 19.46 mA/cm^2. The EQE is over 90% between 400 nm and 500 nm, and is dropping slowly to 70% from 500 nm to 700 nm. The integrated J_{sc} from EQE is 19.21 mA/cm^2, which is consistent with the J-V measurement for a large area cell. This is shown in the inset of Fig. 3c. We think that the observed high short circuit current for small area device is over-estimated. We believe that photon recycling will lead to larger carrier transport distances and to an overestimate of the real current density. Fig. 3d demonstrates the performance distribution of 50 devices. The average PCE of 14.1% shows the good reproducibility by the modified inter-diffusion method.

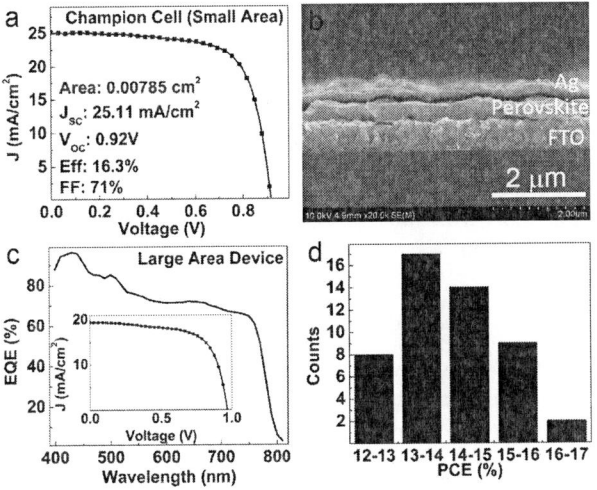

Fig. 3. (a) J-V characteristics of champion cell. (b) Cross-section SEM image. (c) EQE of large area sample and the corresponding J-V as inset. (d) Efficiency distribution over 50 samples

Hysteresis effect has been observed in most perovskite devices using metal oxide as carrier transport layer. Recently, the most convincing explanation for hysteresis effects is the ion migration hypothesis [11]. In our studies, the hysteresis effect was suppressed by PCBM as the top ETL. Although a thinner PCBM layer is preferred to reduce series resistance, it leads to poor coverage and electrode penetration and to more shunt path and current leakage. Thus efforts were taken to optimize the PCBM thickness to get the best performance.

IV. CONCLUSION

In this study, we proposed a modified inter-diffusion method to fabricate MAPbI$_3$ perovskite solar cell. With a controlled solvent-assisted annealing technique, we were able to grow a high quality perovskite active layer. SEM images showed the continuous perovskite layer with a thickness over 400 nm and grain sizes over 1 um. XRD result showed complete conversion from PbI$_2$ to MAPbI$_3$ without a PbI$_2$ residue. PCBM was used as electron transport layer to suppress the hysteresis. The highest PCE of 16.3% was obtained under AM 1.5G irradiation. An EQE measurement shows a high efficiency of 90% in the 400-500 nm spectral region. These results reveal the critical role of interface engineering in the study of perovskite solar cells, and offer a robust method to improve the quality of the perovskite layer.

We acknowledge the support of UMD FabLab, AIMLab and XCC facilities.

REFERENCES

[1] A. Kojima, K. Teshima, Y. Shirai and T. Miyasaka, "Organometal Halide Perovskites as Visible-Light Sensitizers for Photovoltaic Cells", *J. Am. Chem. Soc.*, vol. 131, no. 17, pp. 6050-6051, 2009.

[2] Nrel.gov, 2016. [Online]. Available: http://www.nrel.gov/ncpv/images/efficiency_chart.jpg. [Accessed: 26- Sep- 2016].

[3] G. Xing, N. Mathews, S. Sun, S. Lim, Y. Lam, M. Gratzel, S. Mhaisalkar and T. Sum, "Long-Range Balanced Electron- and Hole-Transport Lengths in Organic-Inorganic CH$_3$NH$_3$PbI$_3$", *Science*, vol. 342, no. 6156, pp. 344-347, 2013.

[4] C. Motta, F. El-Mellouhi and S. Sanvito, "Charge carrier mobility in hybrid halide perovskites", *Sci. Rep.*, vol. 5, p. 12746, 2015.

[5] J. Noh, S. Im, J. Heo, T. Mandal and S. Seok, "Chemical Management for Colorful, Efficient, and Stable Inorganic–Organic Hybrid Nanostructured Solar Cells," *Nano Letters*, vol. 13, no. 4, pp. 1764-1769, 2013.

[6] W. Nie, H. Tsai, R. Asadpour, J. Blancon, A. Neukirch, G. Gupta, J. Crochet, M. Chhowalla, S. Tretiak, M. Alam, H. Wang and A. Mohite, "High-efficiency solution-processed perovskite solar cells with millimeter-scale grains", *Science*, vol. 347, no. 6221, pp. 522-525, 2015.

[7] L. Kranz, A. Abate, T. Feurer, F. Fu, E. Avancini, J. Löckinger, P. Reinhard, S. Zakeeruddin, M. Grätzel, S. Buecheler and A. Tiwari, "High-Efficiency Polycrystalline Thin Film Tandem Solar Cells", *J. Phys. Chem. Lett.*, vol. 6, no. 14, pp. 2676-2681, 2015.

[8] J. You, Z. Hong, Y. Yang, Q. Chen, M. Cai, T. Song, C. Chen, S. Lu, Y. Liu, H. Zhou and Y. Yang, "Low-Temperature Solution-Processed Perovskite Solar Cells with High Efficiency and Flexibility", *ACS Nano*, vol. 8, no. 2, pp. 1674-1680, 2014.

[9] Z. Xiao, C. Bi, Y. Shao, Q. Dong, Q. Wang, Y. Yuan, C. Wang, Y. Gao and J. Huang, "Efficient, high yield perovskite photovoltaic devices grown by interdiffusion of solution-processed precursor stacking layers", *Energy & Environmental Science*, vol. 7, no. 8, p. 2619, 2014.

[10] C. Bi, Y. Shao, Y. Yuan, Z. Xiao, C. Wang, Y. Gao and J. Huang, "Understanding the formation and evolution of interdiffusion grown organolead halide perovskite thin films by thermal annealing", *J. Mater. Chem. A*, vol. 2, no. 43, pp. 18508-18514, 2014.

[11] K. Miyano, M. Yanagida, N. Tripathi and Y. Shirai, "Hysteresis, Stability, and Ion Migration in Lead Halide Perovskite Photovoltaics", J. Phys. Chem. Lett., vol. 7, no. 12, pp. 2240-2245, 2016.

Interfacial Modification of Sol-Gel ZnO/AZO Bilayer as Highly Efficient Electron Transport Layer for Perovskite Solar Cells

Shang-Hsuan Wu[1], Ming-Yi Lin[2], Sheng-Hao Chang[2], Wei-Chen Tu[2], Chi-Wei Chu[1] and Yia-Chung Chang[1*]

[1]*Research Center for Applied Sciences, Academia Sinica, Taipei 115, Taiwan*
[2]*Department of Electronic Engineering, Chung Yuan Christian University, Taoyuan 32023, Taiwan*

Abstract — We report an interfacial modification of sol-gel ZnO/Al-doped ZnO (AZO) bilayer thin film as the electron transport layer (ETL) in the perovskite solar cells. We found that the power conversion efficiency (PCE) of perovskite solar cells was significantly increased from 11.28% to 14.24% with short-circuit current density (J_{sc}) of 19.53 mA cm^{-2}, open-circuit voltage (V_{oc}) of 1.12 V, and fill factor (FF) of 65.1% by employing the AZO thin film as the buffer layer. Moreover, the enhancement in performance was obtained through energy-level offset adjustment of AZO thin film, which allows the better charge separation of perovskite solar cells.

I. INTRODUCTION

Organic-inorganic hybrid perovskite solar cells have attracted much attention in current renewable solar research owing to their excellent semiconductor properties, such as strong light absorption, direct band gap, long carrier diffusion length, and significantly low-cost fabrication process [1-5]. These superior optoelectronic properties enable high power conversion efficiency of planar heterojunction (PHJ) perovskite devices up to 20.1% [6]. In a conventional configuration of PHJ perovskite device, TiO_2 is the most commonly used electron transport layer (ETL). However, several disadvantages of TiO_2 were reported such as low electron mobility and high annealing temperature (above 450°C) in order to crystallize the TiO_2 film into anatase phase [7]. Accordingly, among other n-type metal oxide materials used in inverted solar cells, zinc oxide (ZnO) has been extensively studied to replace titanium dioxide (TiO_2) due to its similar band structure, relatively high electron mobility, high transparency, and versatile nanostructures [8]. Recently, several studies of mesostructured perovskite solar cell based on ZnO nanorod arrays have been used to substitute the mesoporous TiO_2 nanostructures in the conventional perovskite solar cells. For instance, Mahmood et al. have obtained remarkably PCE of 16.1% based on synergistically improved ZnO nanorod arrays [9]. Despite the unstable decomposition process of $CH_3NH_3PbI_3$ to PbI_2 which tends to occur at the ZnO/perovskite interface due to the excessive OH^- groups and chemical residuals on the ZnO surface [10]. The abovementioned problems can be solved via high-temperature annealing, deposition of polymer interlayers between the perovskite and ZnO layer to avoid the direct interaction [11, 12], and the use of aluminum as a dopant to passivate the OH^- group to improve its interface property [13].

In the previous report, Tseng et al. have achieved the highest PCE of 17.6% by using AZO thin films as the ETL which were prepared by sputtering process [14]. Although the physical vapor deposition (PVD) systems can produce better thin film quality including conductivity and crystallinity, they generally require the use vacuum chamber which could increase the production cost and hinder the scale-up technologies used in roll-to-roll processing. In this study, we successfully demonstrated a low temperature sol-gel processed ZnO/AZO bilayer as an ETL for the conventional PHJ perovskite solar cells to achieve the PCE of 14% with V_{oc} of 1.12V. The transport properties of ZnO/AZO bilayer were studied and it was found that the favorable charge transfer between perovskite/AZO contributes to the enhancement in photovoltaic performance of the cell.

II. EXPERIMENTAL

A. Materials, synthesis and characterization

Substrates of the cells are fluorine-doped tin oxide (FTO) conducting glass (Ruilong; thickness 2.2 mm, sheet resistance 14 Ω/square). Before use, FTO glass was first washed with mild detergent, rinsed with distilled water for several times and subsequently with ethanol in an ultrasonic bath, finally dried under air stream. Methylammonium iodide (CH_3NH_3I) preparation Aqueous HI (57 wt% in water), methylamine (CH_3NH_2, 40 wt% in aqueous solution), PbI_2 (99.998%), dimethyl sulfoxide (DMSO), and diethyl ether were purchased from Alfa Aesar and used without further purification. CH_3NH_3I was synthesized by the method described in previous work [15].

B. Synthesis of the ZnO and Al-doped ZnO

The ZnO thin films were synthesized according to a modified procedure reported in literature [12]. Typically, zinc acetate dihydrate, aluminum chloride hexahydrate, 2-methoxethanol, and monoethanolamine (MEA) were used as the starting materials, solvent, and stabilizer, respectively. Zinc acetate dehydrate were first dissolved in a mixture of 2-methoxyethanol. The molar ratio of MEA to zinc acetate dehydrate was maintained at 1.0 and the concentration of zinc acetate was 0.5 M. In order to prepare Al-doped ZnO thin films, the concentration of Al as a dopant were fixed at 1% with respect to Zn. Then, the resulting mixture was stirred at

60°C for two hours. When the mixture was stirred, MEA was added drop by drop. Finally, the clear and transparent homogeneous solutions were formed.

C. Device fabrication and characterization

ZnO/AZO thin films were deposited on the FTO glass substrates using spin coating method, followed by thermal annealing at 200°C for 30 min on a hot plate to evaporate the solvent and remove organic residuals. Next, the substrates were transferred to a glove box for further deposition of the perovskite active layer through the two-step spin-coating method. A hot PbI_2 solution was spin-coated onto ZnO thin film and annealed directly (70°C, 10 min). The hot CH_3NH_3I solution was then spin-coated onto the PbI_2 film; the structure was kept on the hot plate at 100°C for 10 min to form a crystalline perovskite film. The hole-transporting material solution of spiro-OMeTAD in chlorobenzene (CB) and was spin-coated onto the perovskite layer, followed by drying at 70 °C for 10 min. The device was completed through sequential thermal evaporation of MoO_3 and a silver electrode through a shadow mask under vacuum (pressure: 1×10^{-6} torr). The active area of each device was 10 mm². The morphology of ZnO/AZO bilayer ETLs were investigated by field-emission scanning electron microscope (FEI Nova 200, 10 kV). Crytallographic information was obtained using X-ray diffraction (XRD) on a Bruker D8 X-ray diffractometer (2θ range: 10–60°; step size: 0.008°) equipped with a diffracted beam monochromatic set for Cu $K\alpha$ radiation (λ= 1.54056 Å). The photocurrent density-voltage (J-V) characteristics of the cells were illuminated inside a glove box by a Xe lamp as a solar simulator (Thermal Oriel 1000 W), which provided a simulated AM 1.5 spectrum (100 mWcm⁻²). The light intensity was calibrated using a mono-silicon photodiode with a KG-5 color filter (Hamamatsu). All of the measurements were carried out at room temperature (RT).

III. RESULTS AND DISCUSSION

Fig. 1a depicts the device architecture of PHJ perovskite solar cells with ZnO/AZO bilayer used in this work. The device structure is FTO/ZnO/AZO/ $CH_3NH_3PbI_3$/Spiro-oMed/MoO_3/Ag, where the ZnO/AZO bilayer was deposited onto the FTO electrode as ETL. To efficiently extract electrons from perovskite active layer, a 10 nm thick sol-gel processed AZO layer was grown at the ZnO/perovskite interface. In order to verify the energy level diagram of the component materials and evaluate the effect of AZO interfacial modification, UPS measurement was carried out. In the Fig. 1(b), UPS results revealed that the valence band maximum (VBM) and conduction band minimum (CBM) of AZO are slightly higher than those of ZnO (ΔEV = 0.2 eV), thus indicating that this ZnO/AZO bilayer energy band structure could benefit electron extraction from $CH_3NH_3PbI_3$ active layer, suppress charge accumulation in the ZnO layer, and reduce the recombination at ZnO/perovskite interface.

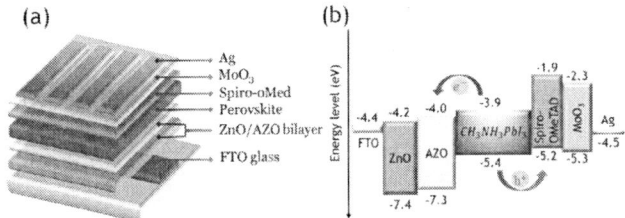

Fig. 1. Schematic diagrams of the (a) device structure and (b) energy levels.

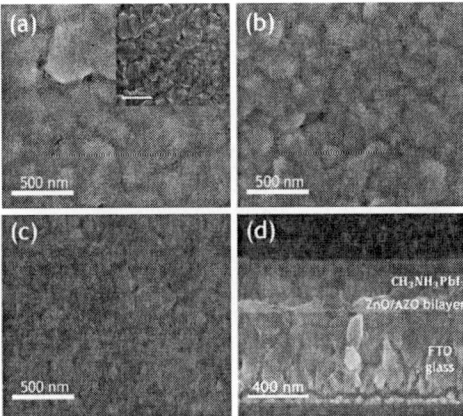

Fig. 2. SEM top-view images of (a) the ZnO monolayer (inset is the bare FTO substrate), (b) the ZnO/AZO bilayer, (c) the perovskite film grown on the ZnO/AZO bilayer and (d) the cross-sectional SEM image of the device using ZnO/AZO bilayer.

The surface morphologies of the ETLs composed of the ZnO monolayer, ZnO/AZO bilayer, and the active layer of $CH_3NH_3PbI_3$, respectively, are shown in Fig. 2 (a-c). As observed in Fig. 2(a), the morphology of ZnO monolayer on the top of FTO substrate shows a clear sol-gel processed nanoparticles feature. However, several cracks can be found in the ZnO layer which correlate with the FTO grain boundaries. These cracks could lead to direct contact between FTO and perovskite, causing serious charge recombination. On the contrary, it is found that AZO thin film modification makes the interface of ETLs/perovskite form into a smooth and homogenous surface. Fig. 2(c-d) show the top view and cross section SEM images of perovskite films deposited on ZnO/AZO bilayer thin films. The perovskite films are uniform and dense-packed with the grain sizes around 150-450 nm. On the other hand, from the cross sectional image of Fig. 2d, it confirms that the uniform and continuous perovskite film was formed on the top of ZnO/AZO bilayer.

The XRD patterns of ZnO/AZO bilayer thin film and perovskite active layer are displayed in Fig. 3. The sol-gel-derived ZnO/AZO bilayer thin film shows no crystallinity (2θ in the range of 10 to 60 degrees) of low-temperature-processed ZnO ETLs. The diffraction pattern of $CH_3NH_3PbI_3$

film reveals peaks, corresponding to the (002), (211), (202), (004), (310), (312), (224), (314) and (404) planes, respectively, confirming the formation of a tetrahedral perovskite crystal structure. The peak positions indicate that a pure perovskite phase was obtained; no secondary phases appeared arising from incomplete formation of perovskite (e.g., PbI_2 or CH_3NH_3I).

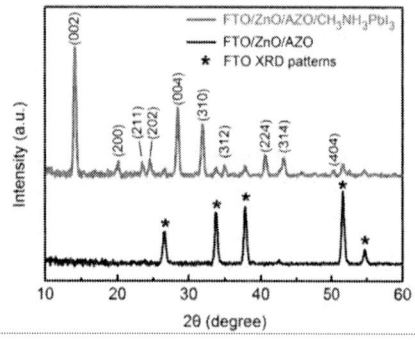

Fig. 3. XRD patterns of ZnO/AZO bilayer and $CH_3NH_3PbI_3$ active layer.

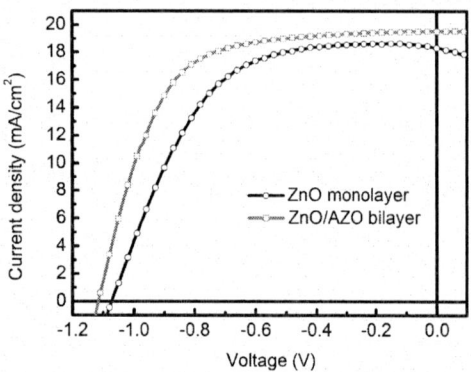

Fig. 4. J-V characteristics of perovskite solar cells incorporating ZnO monolayer and ZnO/AZO bilayer as ETLs measured under AM 1.5G illumination (100mW cm^{-2}).

The PHJ perovskite devices were fabricated using spiro-OMeTAD as a hole transporting layer (HTL) and silver as a top electrode. The J-V characteristics of the cells based on ZnO monolayer and ZnO/AZO bilayer ETLs are displayed in the Fig. 4. The device with ZnO monolayer ETL exhibited a PCE of 11.28% with Jsc of 18.30 mA cm^{-2}, Voc of 1.07 V, and FF of 57.6%. The photovoltaic performances are comparable to those ZnO-based perovskite solar cells in the previous reports. However, the perovskite solar cells containing ZnO/AZO bilayer showed remarkable enhancement in device performance compared with that of the ZnO monolayer devices, resulting in the best PCE of 14.24% with Jsc of 19.53 mA cm^{-2}, Voc of 1.12 V, and FF of 65.1%. Notably, the Jsc and FF values were largely improved by adding AZO as buffer layer. As indicated by the design of

ZnO/AZO bilayer (see Fig. 1b), the enhancement in the device performance can be attributed to the more favorable energy-level alignment between ZnO/AZO bilayer and $CH_3NH_3PbI_3$ for electron extraction. It is noteworthy that the results obtained for our optimized device surpass the state-of-the-art PCE of low temperature solution processed ZnO-based perovskite solar cells.

IV. CONCLUSION

In summary, we have successfully demonstrated the low temperature solution processed ZnO/AZO bilayer ETLs in efficient PHJ perovskite solar cells with PCE above 14%. With introduction of AZO layer, not only forms the uniform surface morphology to fill the defects but also improves the energy level alignment at ZnO/perovskite interface to obtain the better electron extraction, thereby significantly enhanced the device performance. Moreover, the AZO layer is found to passivate the decomposition rate of $CH_3NH_3PbI_3$ to PbI_2. These results offer a new approach for the development of low-cost perovskite solar cell. These results offer a new approach for the development of low-cost perovskite solar cell.

V. REFERENCES

[1] A. Kojima, K. Teshima, Y. Shirai and T. Miyasaka, "Organometal halide perovskites as visible-light sensitizers for photovoltaic cells," J. Am. Chem. Soc., 131, 6050-6051, 2009.

[2] M. M. Lee, J. Teuscher, T. Miyasaka, T. N. Murakami and H. J. Snaith, "Efficient hybrid solar cells based on meso-superstructured organometal halide perovskites," Science, 338, 643, 2012.

[3] J. M. Ball, M. M. Lee, A. Hey and H. J. Snaith, "Low-temperature processed meso-superstructured to thin-film perovskite solar cells," Energy Environ. Sci., 6, 1739, 2013.

[4] J. Burschka, N. Pellet, S. J. Moon, R. Humphry-Baker, P. Gao, M. K. Nazeeruddin and M. Gratzel, "Sequential deposition as a route to high-performance perovskite-sensitized solar cells," Nature, 499, 316, 2013.

[5] M. Z. Liu, M. B. Johnston and H. J. Snaith, "Efficient planar heterojunction perovskite solar cells by vapour deposition," Nature, 501, 395, 2013.

[6] W. S. Yang, J. H. Noh, N. J. Jeon, Y. C. Kim, S. Ryu, J. Seo and S. I. Seok, "High-performance photovoltaic perovskite layers fabricated through intramolecular exchange," Science, 348, 1234, 2015.

[7] J. T. W. Wang, J. M. Ball, E. M. Barea, A. Abate, J. A. Alexander Webber, J. Huang, M. Saliba, I. Mora-Sero, J. Bisquert, H. J Snaith and R. J. Nicholas, "Low-temperature processed electron collection layers of graphene/TiO2 nanocomposites in thin film perovskite solar cells," Nano Lett. 14, 724−730, 2014.

[8] Ü. Özgür, Ya. I. Alivov, C. Liu, A. Teke, M. A. Reshchikov, S. Doğan, V. Avrutin, S.-J. Cho, and H. Morkoç, "A comprehensive review of ZnO materials and devices," J. Appl. Phys. 98, 041301, 2005.

[9] K. Mahmood, B. S. Swain, and A. Amassian, "16.1% efficient hysteresis-free mesostructured perovskite solar cells based on

synergistically improved ZnO nanorod arrays," Adv. Energy. Mater. 5, 17, 1500568, 2015.

[10] Y. H. Cheng, Q. D. Yang, J. Y. Xiao, Q. F. Xue, H. W. Li, Z. Q. Guan, H. L. Yip, and S. W. Tsang, " Decomposition of Organometal Halide Perovskite Films on ZincOxide Nanoparticles," ACS Appl. Mat. Int, 7, 19986−19993, 2015.

[11] J. H. Kim, G. J. Kim, T. K. Kim, S. C. Kwon, H. C. Back, J. H. Lee, S. H. Lee, H. K. Kang, and K. H. Lee, "Efficient planar-heterojunction perovskite solar cells achieved via interfacial modification of a sol−gel ZnO electron collection layer," J. Mater. Chem. A, 2, 17291−17296, 2014.

[12] W. M. Qiu, M. Buffiere, G. Brammertz, U. W. Paetzold, L. Froyen, P. Heremans, D. Cheyns, "High efficiency perovskite solar cells using a PCBM/ZnO double electron transport layer and a short air-aging step," Organic Electronics, 26, 30–35, 2015

[13] X. Y. Zhao, H. P. Shen, Y. Zhang, X. Li, X. C. Zhao, M. Q. Tai, J. F. Li, J. B. Li, X. Li, and H. Lin, "Aluminum-doped zinc oxide as highly stable electron collection layer for perovskite solar cells," ACS Appl. Mat. Int, 8, 7826−7833, 2016.

[14] Z. L. Tseng, C. H. Chiang, S. H. Chang, C. G. Wu, " Surface engineering of ZnO electron transporting layer via Al doping for high efficiency planar perovskite solar cells," Nano Energy, 28, 311-318, 2016.

[15] K. M. Boopathi, R. Mohan, T. Y. Huang, W. Budiawan, M. Y. Lin, C. H. Lee, K. C. Ho, C. W. Chu, " Synergistic improvements in stability and performance of lead iodide perovskite solar cells incorporating salt additives," J. Mater. Chem. A, 4, 1591−1597, 2016.

The Potential of Bifacial Photovoltaics: A Global Perspective

Xingshu Sun,[1] Mohammad R. Khan,[1] Amir Hanna,[2] Muhammad M. Hussain,[2] and Muhammad A. Alam[1]

[1] Network of Photovoltaic Technology, Purdue University, West Lafayette, IN, 47907, USA

[2] Integrated Nanotechnology and IDEA Lab, King Abdullah University of Science and Technology, Thuwal, 23955, Saudi Arabia

Abstract — **With the rapidly growing interest in bifacial photovoltaics (PV), a worldwide perspective of their performance and design is critical to promote further commercialization of this promising technology. The analysis and optimization of bifacial PV panels in previous literature, however, only focused on either few locations or very specific configurations (e.g., vertical panel). It is not clear how to translate these location- and configuration-specific conclusions to the general prospect of the technology across the globe. In this paper, we present a global study and optimization of bifacial PV at both the single-panel and farm level by our rigorous modeling framework. Our simulation indicates that east-west facing vertically mounted bifacial solar panels can outperform conventional south-north facing bifacial panels at zero elevation (ground-mounted) up to the latitude of 30°. For vertical bifacial solar farms, the optimal row-to-row pitch is also found to be ~1 m for 1 m high panel in the region where the vertical installation produces more electricity.**

I. INTRODUCTION

Currently, monofacial solar panels, designed to absorb sunlight only from the front side, dominate the market. Unfortunately, these panels cannot utilize a significant amount of sky-diffused and albedo irradiance at the rear side. Bifacial solar panels, however, can partially recover this loss by making the rear side transparent for light collection. The concept of bifacial has been proposed since the 1960s [1]. Cuevas et al. [2] demonstrated that the extra energy gain from bifacial solar panels could reach up to 50% by fully utilizing the diffuse and ground albedo light, making bifacial solar panels a very attractive candidate to further decreasing the Levelized Cost of Electricity (LCOE) for photovoltaics.

Several studies have analyzed and optimized the performance of bifacial solar panels by simulation and experiment [3], [4]. These studies, however, have been confined to either only few locations or specific bifacial PV configuration (single vertical bifacial panel) [5]. At the farm level, Appelbaum has investigated bifacial solar panel arrays, but again the study only focuses a single location [6]. *A global map for bifacial PV with location- and climate-dependent optimization has not been reported .*

In this paper, we will provide a global investigation and optimization of a variety of bifacial panel configurations (i.e., *conventional south-north facing bifacial solar and east-west facing vertical bifacial with different elevation*) at both the single-panel and farm scales. Our findings will highlight the

importance of elevation in determining the optimal orientation

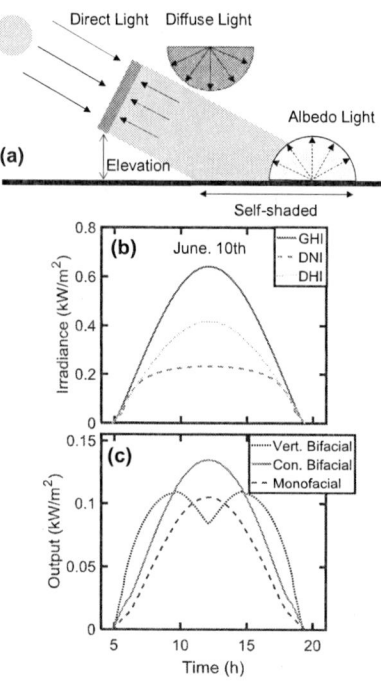

Fig. 1 (a) A schematic showing three different light components (direct, diffuse, and Albedo light) and an elevated bifacial solar panel with self-shading. (b) Hourly intensity of Global Horizontal Irradiance (GHI), Direct Normal Irradiance (DNI), and Diffuse Horizontal irradiance (DHI) on June 10th at Washington DC. (c) Hourly output power of east-west facing vertical bifacial (Vert. bifacial), conventional south facing bifacial (Con. bifacial) tilted at 43°, and monofacial solar panels tilted at 37°. All of them have an elevation of 0.5 m above the ground.

of bifacial solar panels, e.g., *vertical bifacial solar panels can be more desirable than the conventional ones for ground-mounted configuration (zero elevation)* due to self-shading of albedo light. At the farm scale, *we have calculated the optimal pitch between each row for vertical bifacial PV to be around 1 m (panel height of 1 m) in the area of interest (within the latitude of 30°),* to minimize the inter-row shading and maximize the use of land area. The results yield a robust design rule for bifacial PV with full considerations of environmental factors (e.g., geographic location, albedo coefficient, clearness

978-1-5090-5606-4/17 $31.00 © 2017 IEEE

index) to facilitate a broader deployment of such technology around the world.

II. SIMULATION FRAMEWORK

To accurately simulate bifacial solar panels, we start with calculating the temporal sun position and irradiance intensity on *a minute-to-minute basis* using sophisticated algorithms and extensive meteorological database from NASA [7], see Fig. 1(b). Next, the total optical absorption of bifacial solar panels can be determined based on their orientation and elevation, including the losses due to reflection and self-shading. Enabled by our electro-thermal coupled simulator, the power output of bifacial solar panels can be calculated based on the total light absorption.

An example of the calculated performance of bifacial solar panels is summarized in Fig. 1(c) for three different configurations: east-west facing vertical bifacial, south-north facing conventional bifacial, and monofacial solar panels at Washington DC on June 10th. Because of the rear-side absorption, the conventional bifacial and vertical bifacial solar panels outperform the monofacial ones by 30% and 32%, respectively. Interestingly, the result also indicates that vertical bifacial can be more productive than the conventional ones (only for certain period such as June. 10th but not for annually average in Washington DC); the difference is caused by self-shading of albedo light (Fig. 1(a)), which will be discussed in detail in Sec. III.

III. SINGLE BIFACIAL SOLAR PANEL

In Sec. III, we apply our modeling framework to single bifacial solar panels (i.e., not arrays) across the world, while taking the climatic and geographic factors into consideration. As discussed in [3], the effect of self-shading onto the ground, which is critical to determine the energy yield of bifacial solar panels, can be modulated by elevation above the ground. As a result, we divide the optimization of bifacial solar panels into two groups: (1) panels with high elevation (essentially no self-shading effect) and (2) ground-mounted panels (zero elevation). Ground albedo coefficient is set to be 0.5, which is for ground covered by artificial white concrete reflectors.

The optimization of panel orientation involves finding the optimal tilt angle as well as the azimuth angle. For highly elevated bifacial solar panels, simulation indicates that the optimized tilt and azimuth angle follows the same trends as monofacial ones; specifically, tilt angle increases with latitude, see Fig. 2(a). The panel is facing south in the northern hemisphere, and vice versa. The difference of optimal tilt angle between bifacial and monofacial panels (not shown) varies with latitude. In most regions within the latitude of 60°, bifacial solar panels are slightly less tilted (~2°) than monofacial panels to absorb more albedo light. Above latitude of ~60°, it is opposite:

bifacial panels are more tilted than monofacial ones, to absorb more direct light at the rear side.

At zero elevation, self-shading of albedo light from DNI and circumsolar diffuse light can be significant particularly at low tilt angles. Consequently, it is undesirable to place bifacial solar panels horizontally on the ground even at very low latitude. *Accordingly, optimization results of ground-mounted bifacial solar panels in Fig. 2(b) suggests that east-west facing vertical panels are more desirable than the conventional ones in the region within the latitude of 30° due to less self-shading.* For dessert (e.g., Sahara), the actual gain of using vertical bifacial solar panels can be even greater owing to their immunity to soiling, i.e., free of cleaning cost, and high power output in the afternoon when electricity demand peaks [5]. At latitudes beyond 30°, conventional south-north facing bifacial panels produce more energy owing to the reduction of self-shading loss.

Our global optimization of bifacial solar modules suggests that albedo light, which relies on both elevation and albedo coefficient, plays a vital role in optimizing the performance of bifacial solar panel. Indeed, our calculation (not shown) also demonstrates that for albedo of 0.5, the maximum bifacial gain (relative to monofacial panels with optimized orientaiton) with zero elevation can increase by ~30%, while elevated panels may produce up to ~50% more energy.

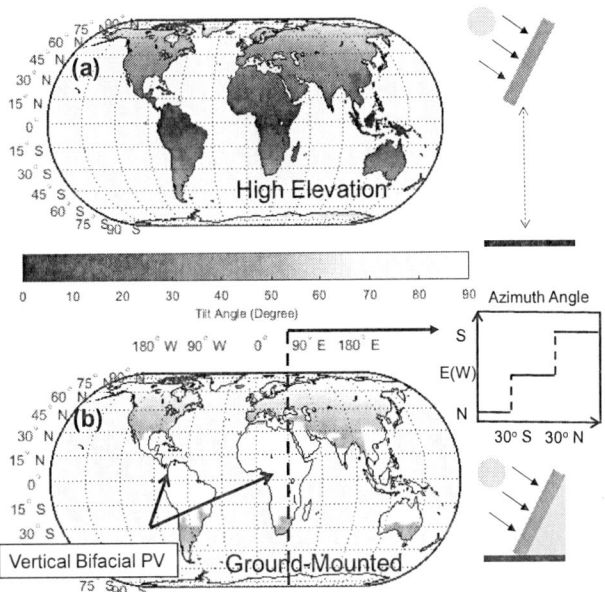

Fig. 2 Global optimization of tilt angle for bifacial solar panels with (a) high elevation (no self-shading) and (b) ground mounted ones (significant self-shading). Ground Albedo reflection is set to be 0.5. The azimuth angle is south facing for panels in the northern hemisphere and vice versa. Vertical bifacial solar modules (Yellow: 90° tilted) are orientated east-west facing.

IV. BIFACIAL SOLAR FARM

When solar panels are arranged in multiple rows to form panel arrays, inter-row shading blocks sunlight as well as induces non-uniform illumination, resulting in significantly reduced power output. This is particularly important for vertical bifacial solar farms, due to the high tilt angle, see Fig. 3(a). Thus, it requires careful optimization of the pitch width between each row. In this section, we extend our study of bifacial solar PV to the farm level, specifically east-west facing vertically mounted panels.

To understand how pitch affects energy yield, we first simulate the average daily electricity yield *per land area* in June as a function of pitch at Dubai (where soiling-resistant vertical bifacial solar panels can be particularly attractive), see Fig. 3(b). At low pitch, electricity output increases with pitch owing to the reduction of inter-row shading, whereas yield shrinks at the high pitch because of overuse of land area. The optimal pitch is found to be ~1.1 m for Dubai in June.

Next, we have calculated the optimal pitch to maximum annual yield for vertical bifacial farm in the area between 30° N and 30° S where vertical bifacial panels can have superior performance, see Fig. 3(c). The optimal pitch appears to vary

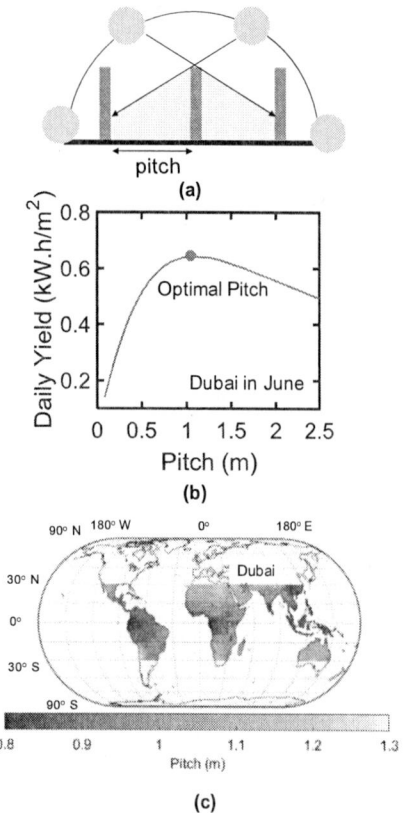

(a)

(b)

(c)

Fig. 3   (a) Vertical bifacial solar farms with inter-row shading. (b) Average daily electricity yield of vertical solar farms in June at Dubai as a function of pitch. (c) Optimized pitch width of vertical solar farms between 30 ° N and 30° S.

slightly around 1 m (simulated panels have a height of 1m) within the region of interest. Only beyond the latitude of 60° (not shown), the pitch starts to increase due to higher solar zenith angle. Notably, in places with higher clearness index (more direct light), e.g., Arabian Peninsula, the pitch become wider (1.2 m) to compensate the loss associated with the more severe inter-row shading. In a cloudy sky with more diffuse light, e.g., China, the pitch decreases to 0.8 m for less use of land area. In practice, optimization of pitch width of solar farms also involves the cost of bifacial modules and land.

V. CONCLUSION

Using our rigorous modeling framework and the extensive meteorological database from NASA, we have performed a worldwide investigation and optimization of bifacial photovoltaics at both the panel and farm scales while fully taking the geographic location, weather condition, as well as ground albedo into account. We found that self-shading plays a critical role in determining the optimal orientation of bifacial solar panels. For ground-mounted bifacial PV, our results suggest that east-west facing vertical bifacial solar panel will outperform the conventional south-north facing bifacial solar panel within the latitude of 30°; at higher elevation (negligible self-shading), the conventional south-north facing bifacial solar panel is more desirable. At the farm level, the optimum pitch for vertical bifacial solar farms has also been calculated to be approximately 1 m (panel height of 1 m) in the region of interest, which minimizes inter-row shading and maximizes the yield per land area. The findings in this paper can provide useful guidance for installing bifacial PV around the world.

REFERENCES

[1]   M. Hiroshi, "Radiation energy transducing device," US3278811A, 1966.

[2]   A. Cuevas, A. Luque, J. Eguren, and J. del Alamo, "50% more output power from an albedo-collecting flat panel using bifacial solar cells," *Sol. Energy*, vol. 29, no. 5, pp. 419–420, 1982.

[3]   U. A. Yusufoglu, T. M. Pletzer, L. J. Koduvelikulathu, C. Comparotto, R. Kopecek, and H. Kurz, "Analysis of the annual performance of bifacial modules and optimization methods," *IEEE J. Photovoltaics*, vol. 5, no. 1, pp. 320–328, 2015.

[4]   G. J. M. M. Janssen, B. B. Van Aken, A. J. Carr, and A. A. Mewe, "Outdoor Performance of Bifacial Modules by Measurements and Modelling," *Energy Procedia*, vol. 77, pp. 364–373, Aug. 2015.

[5]   S. Guo, T. M. Walsh, and M. Peters, "Vertically mounted bifacial photovoltaic modules: A global analysis," *Energy*, vol. 61, pp. 447–454, 2013.

[6]   J. Appelbaum, "Bifacial photovoltaic panels field," *Renew. Energy*, vol. 85, pp. 338–343, 2016.

[7]   "Surface meteorology and Solar Energy: A renewable energy resource web site (release 6.0)," 2017.

Performance Assessment of Stand Alone Bifacial Solar Panel Under Real Time Conditions

Ahmer A.B. Baloch[1], Maher Armoush[2], Basel Hindi[2],
Abdelkader Bousselham[3] and Nouar Tabet[1,3]

1. College of Science and Engineering, Hamad bin Khalifa University, Doha, Qatar
2. Mechanical Engineering Program, Texas A&M University at Qatar, Doha, Qatar
3. Qatar Environment & Energy Research Institute, Hamad bin Khalifa University, Doha, Qatar

Abstract — **Bifacial photovoltaic (PV) is a promising renewable energy technology that can increase the power density by harvesting both incident and albedo radiation. Integration of these resources into the power grid can offer benefits including improved energy efficiency and power continuity. This paper examines the performance of bifacial solar panels in the real-time climate of Qatar under winter and summer. The operating conditions such as ambient temperature, albedo and soiling have been studied for bifacial PV cells. Average percentage improvement for energy yield was found to be 4.3% by changing albedo whereas 29.0% performance enhancement was observed by cleaning the panel daily.**

Index terms — bifacial, solar cell, photovoltaics, experiment.

I. INTRODUCTION

Bifacial photovoltaic(PV) panels have become an active theme of research in the recent past in the field of solar energy. Addition of these systems into the existing electrical network can provide several advantages including improved reliability, better energy productivity, and power steadiness. Bifacial solar panels are a promising renewable energy technology that can increase the power density by harvesting both incident and albedo radiation. They use sunlight to produce electrical energy from the front and rear side of the solar cell by utilizing both the incident and albedo radiation [1]. An overall power conversion efficiency of a Bifacial PV depends primarily on the environmental conditions such as albedo, latitude, wind speed, radiation and dust. Albedo is the solar radiation component reflected from the ground and scattered by diffused radiation. Higher albedo implies more radiation is collected at the rear face of the solar cell resulting in an increase of generation current and overall bifaciality ratio. Bifacial solar cells are albedo collecting devices and are capable of decreasing the LCOE of power plants [2]. Utilizing albedo radiation is preferential for bifacial installations as both beam and diffuse radiation from the atmosphere and earth's surface can be collected. Shoukri et al [3] showed that the cell temperature and albedo significantly affect the performance of PV fields installed. High cell temperature and low albedo were found to reduce the global output of PV.

The operating conditions of bifacial PV cells play a huge role in the system's efficiency [4]. Cell temperature and solar spectrum (front and albedo) affect the generated voltage. However, the photocurrent that flows through the system is proportional to the incident light intensity while the generated voltage decreases as the cell temperature increases. The photocurrent remains unaffected by temperature [5–7]. It is because of the decrease in the output power caused by the increase in temperature that the power generated in summers does not increase even though the irradiance then is high. This loss in efficiency leads to a drop in the system performance

Bifacial solar cells have been predicted to become more dominant in the world energy market. This is due to the recent interest of the international renewable agencies carrying industrial growth workshops, setting up pilot bifacial PV, and standardizing PV systems. By 2024, it has been estimated by the International Technology Roadmap that the global share of bifacial PV systems will be 15% [1]. Abundant solar irradiation in Qatar's region creates a high potential market for such technology and to support the nation's goal to minimize the use of non-renewable energy to power the upcoming FIFA 2022 World Cup. However, the climate and environmental conditions in Qatar pose a threat because of the reduction of the power output of the solar panels. Hence, in order to properly utilize the high potential energy present in the region further research is needed to analyze how the environmental conditions affect the efficiency of solar cells in Qatar. This paper examines the performance of bifacial solar panels in the real-time climate of Qatar under winter and summer seasons. The operating conditions such as cell temperature, albedo and soiling have been investigated.

II. EXPERIMENTAL MATERIAL

Bifacial PV system was analyzed in the climate of Doha Qatar (latitude of 25.31°N and longitude of 51.42°E) to assess the effect of soiling and albedo under two extreme seasons as highlighted by Figure 1. Experimental test rig along with data acquisition system was developed with a mounting rack such that minimum shading was casted on the bifacial PV installed at rooftop. Orientation of the panel was due south with an azimuth angle of 0° with a tilt angle of 26° and height of 1m from the bottom of the PV panel. Data for the ambient conditions and PV parameters were measured and collected over a period of two months, i.e June and December 2016-17 on the site. Climatic measurements such as ambient temperature, wind speed, and global plane-of-array radiation were analyzed along with output parameters of PV panels like I-V characteristics, Maximum Power, Albedo radiation and cell temperature. For soiling, an uncleaned and daily cleaned panel were considered. Albedo and its impact on solar panel output was investigated using materials of concrete and a paper white sheet.

$$Normalized\ power = Bifacial\ Power/Incident\ Power \quad (1)$$
$$Albedo = Albedo\ radiation/Incident\ Power \quad (2)$$

Fig. 1 Experimental Setup and Parameters for Bifacial PV.

Bifacial Standard Testing Conditions (BSTC) from the module data sheet are rated maximum power of 343W with an overall efficiency of 20.5%. Experimental uncertainty for power measured was (1.0% of reading + 6 digits) whereas for radiation it was (1.0% of reading + 5 digits) in the positive and negative space. The main performance parameter of normalized power as defined by Eq. 1 was employed for different comparisons along with albedo (Eq. 2) and total energy yield during day.

III. RESULTS AND DISCUSSION

Performance of bifacial PV installed under different seasons is shown in Fig. 2 with the addition of ambient temperature. It was found that bifacial PV performed better in months of December rather than June due to the effect of ambient conditions. Total energy yield for clean bifacial PV was 1.44 kWhr in June and was increased to 1.72 kWhr in December. This can be attributed mainly to high ambient temperature amongst other variables,

In addition, with an increase in temperature the voltage generated goes down logarithmically while the short circuit current increases slightly thereby reducing the overall output power. It is because of this decrease in output power caused by the increase in temperature that the power produced in summer does not increase even with high irradiance.

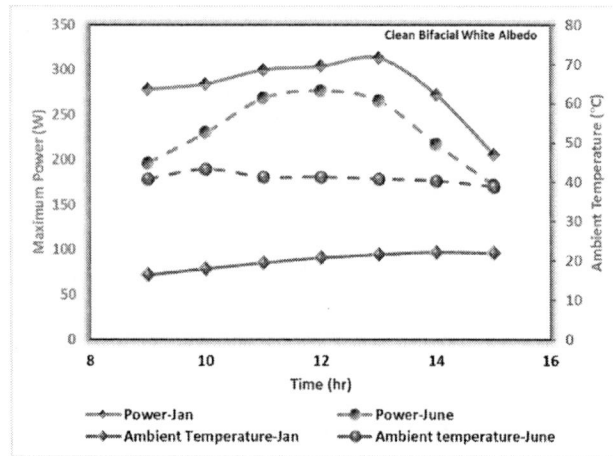

Fig. 2 Key performance parameters for bifacial solar performance.

Figure 3 shows the effect of altering the albedo on the overall performance of bifacial PV. Two materials were chosen for the analysis: concrete and white sheet. Higher albedo implies more radiation is collected at the rear face of the solar cell resulting in an increase of generation current and overall bifaciality ratio. Normalized power was compared for a typical day of June and albedo was measured on site at the rear face of the panel at three different locations. The albedo was found to mainly affect the bifacial PV panel during the hours around solar noon due to enhanced participation from ground reflectance because of high solar altitude angle. Average percentage improvement for energy yield was improved by 4.3% corresponding to an albedo increase of 4.1% from concrete to white sheets.

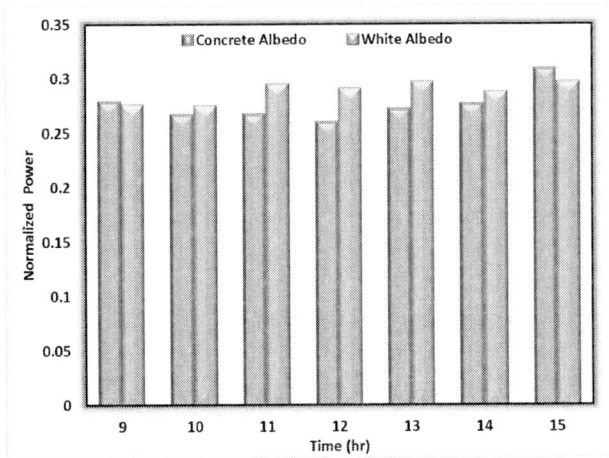

Fig. 3 Normalized power output for clean bifacial PV with concrete and white sheet albedo.

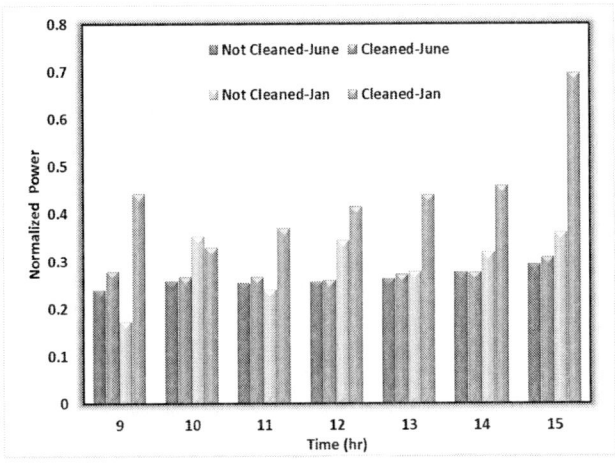

Fig. 4 Effect of soiling on normalized power output for summer and winter.

The effect of soiling is shown in Fig. 4 for a cleaned and uncleaned bifacial PV for the months of summer and winter. Bifacial PV panels cleaned daily, tilted at 26°, showed an enhanced performance than soiled PV because of the accumulation of dust on PV panels and the reduction in the transmittance of light leading to reduced number of photons absorbed. The front side dust deposition was found to mainly affect the bifacial PV panel performance. Average percentage improvement by employing regular cleaning for the normalized power for the month of June and December was 29.0%. These results show that the transmittance of light strongly depends on tilt angle, site location and exposure period.

IV. CONCLUSION

Experimental performance assessment was carried out for standalone rooftop bifacial PV panel for the climate is presented. An Average improvement of the energy yield of 4.3% was achieved by changing the albedo from concrete to white sheets. Furthermore, 29.0% performance enhancement was observed by cleaning the panels daily. These improvements can increase the power yield of the bifacial systems and lead to the reduction of the levelized cost of energy.

REFERENCES

[1] R. Guerrero-Lemus, R. Vega, T. Kim, A. Kimm, L.E. Shephard, Bifacial solar photovoltaics - A technology review, Renew. Sustain. Energy Rev. 60 (2016) 1533–1549. doi:10.1016/j.rser.2016.03.041.

[2] P. Ooshaksaraei, K. Sopian, R. Zulkifli, M.A. Alghoul, S.H. Zaidi, Characterization of a bifacial photovoltaic panel integrated with external diffuse and semimirror type reflectors, Int. J. Photoenergy. 2013 (2013). doi:10.1155/2013/465837.

[3] I. Shoukry, J. Libal, R. Kopecek, E. Wefringhaus, J. Werner, Modelling of bifacial gain for stand-alone and in-field installed bifacial PV modules, 0 (2016) 0–8. doi:10.1016/j.egypro.2016.07.025.

[4] J.E. Castillo-Aguilella, P.S. Hauser, Multi-Variable Bifacial Photovoltaic Module Test Results and Best-Fit Annual Bifacial Energy Yield Model, IEEE Access. 4 (2016) 498–506. doi:10.1109/ACCESS.2016.2518399.

[5] H.M.S. Bahaidarah, A.A.B. Baloch, P. Gandhidasan, Uniform cooling of photovoltaic panels: A review, Renew. Sustain. Energy Rev. 57 (2016) 1520–1544. doi:10.1016/j.rser.2015.12.064.

[6] A.A.B. Baloch, H.M.S. Bahaidarah, P. Gandhidasan, F.A. Al-Sulaiman, Experimental and numerical performance analysis of a converging channel heat exchanger for PV cooling, Energy Convers. Manag. 103 (2015) 14–27. doi:10.1016/j.enconman.2015.06.018.

[7] H.M. Bahaidarah, P. Gandhidasan, A.A.B. Baloch, B. Tanweer, M. Mahmood, A comparative study on the effect of glazing and cooling for compound parabolic concentrator PV systems – Experimental and analytical investigations, Energy Convers. Manag. 129 (2016) 227–239. doi:10.1016/j.enconman.2016.10.028.

Operation and performance assessment of grid-connected PV systems in operation in Maui, Hawaii

Severine Busquet, Jonathan Kobayashi, and Richard E. Rocheleau

Hawaii Natural Energy Institute of the University of Hawaii Manoa, Honolulu, Hawaii, 96822, United States of America

Abstract — Numerous assessments have been conducted around the world to characterize using the performance ratio (PR) the energy production (energy rating) of different PV technologies and manufacturers. Experimental results vary between locations and the single PR criterion is not enough to understand the performance variability. Using present modelling tools, energy rating is possible but there are still uncertainties requiring additional data collection and analysis especially in terms of the effects of spectrum, angle-of-incidence and degradation. The Hawaii Islands have high solar resource, spectrum rich in short wavelength and a hot and humid environment expected to rise degradation rate. Testing in Hawaii provides an ideal environment to support research on PV technologies. A PV test platform, commissioned on the island of Maui in February 2016, tests side by side 15 grid-connected PV systems. A new energy rating analysis is proposed where PR is separated into current (IP) and voltage (VN) performances. These additional criteria help understand the performance differences between PV modules and locations which will lead to better select the PV technology to certain environment. This approach also helps identify the parameters of operation to dissociate the impacts of the environmental and operating conditions (soiling, shading). At the Maui test platform, the yearly average PR is estimated between 87% to 98% depending on the PV systems, with 2 thin films (CIGS, CdTe) over performing the standard and high efficient crystalline modules. IP is evaluated between 95% and 100% for most PV manufacturers except for one low (91%, CIGS) and one high (104%, CdTe) performer. VN ranges between 87% and 98%, the CIGS exhibiting high voltage performance compared to the other technologies.

Index Terms — Outdoor assessment, grid-connected PV systems, PV technologies, energy rating analysis.

I. INTRODUCTION

The State of Hawaii is an isolated island archipelago in the Central Pacific Ocean. It is the most fossil fuel dependent and has the highest electricity rate (0.26 cents/kWh in 2015 [1]) in the United States. The Hawaii Clean Energy Initiative [2] set goals in 2015 to achieve 100% clean energy by 2045. Photovoltaic (PV) generation is an essential renewable energy resource on the islands, accounting for 675 MW in 2016 (21% of total capacity of the Hawaiian Electric Companies) producing 7% of the State's electricity, with a growth projection of 800 MW by 2021 [3]. With the increasing share of PV generation onto the grid, it is crucial to provide research support to the utilities, policy makers, and PV owners. The primary concern of the PV owners is to select the best PV technologies for their installation to optimize the investment.

Energy rating has been assessed in multiple locations around the world [4-7] using the criterion performance ratio (PR) defined in IEC 61724 [8]. Experimental results vary between modules and locations. PR does not provide enough information to understand the performance differences of PV modules operating in different environment. PR is affected by multiple parameters including the temperature, irradiance, spectrum, angle-of-incidence, and degradation. With the present modelling tools, energy rating is possible but there are still uncertainties requiring additional data collection and analysis especially in terms of the effects of spectrum, angle–of-incidence and degradation [9].

Operating conditions and issues need to be identified to extract the performance of the PV modules as a function of the environmental conditions. Operating conditions encompass soiling and shading while "issues" are when the auxiliaries (inverters, optimizers, microinverters) do not operate as expected. This is discussed in the literature related to fault detection which is usually done by comparing the PV system power output with the output of a prediction model [10].

The Hawaiian Islands have high solar resource, a spectrum rich in short-wavelength and a hot and humid environment [11-12] which make the islands an ideal environment to test PV technologies and to support characterization of the spectral effect and degradation. The Hawaii Natural Energy Institute (HNEI) of the University of Hawaii designed a PV test platform to test side-by-side 15 grid-connected PV systems consisting of various PV modules and auxiliaries. The test platform was commissioned in Kihei, Maui in February 2016. A data acquisition system (DAS), designed and built by HNEI, was implemented on the test platform for long term analytical monitoring of grid-connected PV systems. Energy rating is the first focus in analyzing the collected dataset. Our goal is to characterize the operation and performance of the PV systems without the need of complex modelling and understand the relationship between performance and location. To do so, we propose to separate the PR of the PV modules into current (IP) and voltage (VN) performances. Studying IP and VN in addition to PR provides insights to further characterize the PV performance and to diagnose the impact of the environmental and operating conditions.

Section II presents the test platform and the analysis method. Section III shows the results obtained during the first year of operation from April 2016 to March 2017. Results are discussed in section IV prior to the conclusion.

978-1-5090-5606-4/17 $31.00 © 2017 IEEE

II. TESTING AND ANALYSIS

A. PV test platform

The test platform consists of 15 grid-connected PV systems using 10 types of PV modules and 3 system architectures for a total rated power of 22 kW. All PV modules are installed on a carport with an orientation of a 20° tilt and 197°N azimuth. The latitude, longitude, and altitude of the test site are 20.7°N, 156.4°W and 60 meters, respectively. A full description of the PV systems is in Table I including the PV type and technology, the rated power and efficiency at standard test conditions (STC), the number of the PV modules, and the auxiliaries used with each system. The last column in the table indicates the acronym used for each PV system in the subsequent analysis. The tested PV modules comprise of 4 standard (S) p-type mono and polycrystalline, 3 high efficient (H) n-type monocrystalline (including HIT and a bifacial), and 3 thin-films including 1 CIGS (C) and 1 CdTe (D). Comparison of up to 3 system architectures is available for some of the PV modules (S3, H1 and H2). Except for the CdTe modules that were donated by the PV company, all modules and auxiliaries were purchased (in 2015) by the contracted PV installer without informing the PV manufacturers of the assessment. This insures fair testing and unbiased evaluation of commercially available PV technologies.

A new DAS was developed by HNEI and implemented at the test site. It is designed for long term analytical monitoring of grid-connected PV systems using any PV technologies and system architectures. Instrumentation includes a weather station and multiples solar sensors including a secondary standard pyranometer and a spectroradiometer whose dataset is used to calculate the average photon energy (APE) [13].

B. Energy rating analysis

For each day, the average for all measured and computed data is calculated using data with an angle-of-incidence (AOI) lower than 70° to limit error due to the directional response of the solar sensors. The only exceptions are the rainfall and relative humidity that are calculated using the 24h dataset as their impact is not limited to daytime. The irradiance, power and current are integrated over time to calculate the energy collected. PR is calculated (1) by dividing the PV energy production by the solar energy received by the PV modules, and both are normalized using the STC specifications. Here we use a novel approach, where the PR is separated (2) into current performance (IP) and normalized voltage (VN), also referred to as voltage performance. IP is calculated similarly to the PR except that PV current, rather than power, is integrated (3). The average operating voltage is divided by the STC maximum power point voltage of the PV system to obtain VN (4).

$$(1) \; PR = \frac{\int_{\Delta t} P_{PV} \cdot dt}{P_{MP,STC}} \times \frac{G_{STC}}{\int_{\Delta t} G \cdot dt}, \qquad (2) \; PR \approx IP \times VN,$$

$$(3) \; IP = \frac{\int_{\Delta t} I_{PV} \cdot dt}{I_{MP,STC}} \times \frac{G_{STC}}{\int_{\Delta t} G \cdot dt}, \qquad (4) \; VN = \frac{\overline{V_{PV}}}{V_{MP,STC}}$$

where P_{PV} is the PV power of the strings of PV modules, or of a PV module in operation in microinverter or optimizer systems or in test with the IV tracer [W]; G_{STC} is the STC irradiance [=1 kWm^{-2}]; Δt is the period of analysis [h] (using hour to provide energy results directly into Wh); I_{PV} and V_{PV} are the operating current [A] and voltage [V] of the PV system or module; $P_{MP,STC}$, $I_{MP,STC}$ and $V_{MP,STC}$ are the power [W], current [A], and voltage [V] of the maximum power point at STC.

TABLE I

DESCRIPTION OF THE GRID-CONNECTED PV SYSTEMS IN OPERATION AT MEDB, SOUTH MAUI.

PV type (Technology[1]/ brand)	PV technology	Rated power [W]	η [%]	# of PV	Auxiliaries	PV System Acronym[2]
S1	Standard p-type polycrystalline	250	15.4	2	Micro1 (2)	S1M
S2	Standard p-type polycrystalline	250	15.2	2	Micro1 (2)	S2M
S3	Standard p-type polycrystalline	260	15.5	8	Micro1 (8)	S3M
				8	String1 (1), Optimizer1 (8)	S3O
				8	String2 (1)	S3S
S4	Standard p-type monocrystalline	265	16.5	2	Micro1 (2)	S4M
H1	High efficient n-type monocrystalline with heterojunction intrinsic thin layer (HIT)	240	19.0	8	Micro2 (8)	H1M
				8	String2 (1)	H1S
H2	High efficient n-type monocrystalline with rear contact	245	19.7	8	Micro2 (8)	H2M
				8	String2 (1), Optimizer2 (4)	H2O
				8	String2 (1)	H2S
H3	High efficient n-type monocrystalline, Bifacial Hybrid Cell Technology	300	18.2	2	Micro2 (2)	H3M
C1	Copper indium gallium selenide (CIGS)	145	13.3	10	String2 (1)	C1S
C2	CIGS	170	13.8	8	String2 (1)	C2S
D1	Cadmium telluride (CdTe)	77.5	10.8	18	String2 (1)	D1S

[1] PV module technology: S for standard p-type crystalline, H for high efficient n-type crystalline, C for CIGS, D for CdTe

[2] PV system acronyms: the first letter describes the PV technology, the second digit corresponds to the manufacturer, and the last letter indicates the system architecture with M for microinverter, S for string inverter and O for string inverter with optimizers.

The performances PR, IP and VN calculated for the grid-connected PV systems are labelled SYS. The label IVT refers to the PV modules tested individually with the IV tracer. Additional electrical characteristics, including the open-circuit voltage (V_{OC}) and short-circuit current (I_{SC}), are monitored during testing with the IV tracer. I_{SC} is the maximum current that a PV module can produce under specific environmental conditions, and is an indicator of spectral effect [4, 14]. I_{SC} is used to calculate (5) the short-circuit current performance (IP_{SC}) also called the optical performance.

$$(5)\ IP_{SC} = \frac{\int_{\Delta t} I_{SC} \cdot dt}{I_{SC,STC}} \times \frac{G_{STC}}{\int_{\Delta t} G \cdot dt}$$

where I_{SC} and $I_{SC,STC}$ are the operating and STC short-circuit current [A].

The specifications indicated in the manufacturer's datasheet are used for normalization of the PV production. The analysis results therefore include the normalization error related to slight differences between the actual performance of each PV module and the specifications from the datasheet.

Data sets were selected for the following analysis to exclude days when there was unconventional PV operation. This includes days during maintenance of the test platform such as in January 2017. It also encompasses days with grid shortage and reliability issues with the auxiliaries leading to interruption of the PV production.

III. RESULTS

A. One year of operation

Fig. 1 shows the daily average values of the environmental conditions for a year of operation from April 2016 to March 2017. Fig. 1-a shows the irradiation, rain, and AOI while ambient temperature (AT), APE, and airmass (AM) are displayed in Fig. 1-b. Table II presents the statistics on the irradiation, AT and APE. The daily irradiation averaged 5.6±1.1 kWhm^{-2}. Maximum values of the irradiation decrease during solstices in June and December when the AOI is maximal. Daytime temperature averaged 27.5±1.9°C with highs in September and lows in January/February. Daily average APE is estimated at 1.92±0.02 eV with higher values observed opposite to AM. APE is highly affected by cloud cover with the highest APE values recorded in overcast days as seen in early December.

Fig. 2 displays the daily performances of one PV system (S3S) and of the same module (S3) tested with the IV tracer. Fig. 2-a shows the PR of the system and module which vary between 85% and 90% for most days with few peaks up to 102%. Low PR values reach 80% for the module and 70% for the system observed in February. Fig. 2-b shows VN ranging mostly between 85% and 90% with low differences between the system and module except in February where VN is slightly higher for the system than for the module. Fig. 2-c includes IP and IP_{SC} which all vary very similarly except at the end of January where the system IP dropped by 20% while the

current and optical performance of the PV module stayed around 100%. At that time, new solar carports were constructed in the neighboring location leading to shading with varying impact on the systems and modules depending on their location on the carport. Shading impact decreases from February to March 2017, and correlates with increasing sun elevation following the December solstice. During 2016, optical and current performances have low variability except in overcast days exhibiting increased performance values and for periodic incremental decreases reaching up to ~7% at the end of November before jumping back up in December. This latter behavior also visible on PR is related to soiling affecting all modules and systems at the site. Soil accumulation is subsequently washed away by sporadic rainfall (Fig. 1-a), allowing the IP and IP_{SC} to recover.

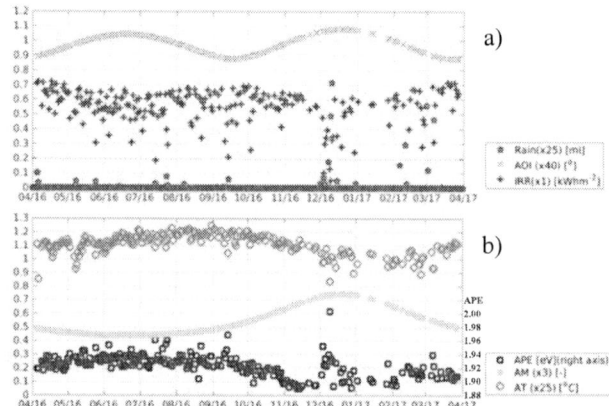

Fig. 1. Variation of the environmental conditions for a year of operation a: irradiation (IRR), rain, and angle-of-incidence (AOI); b: spectral energy (APE), airmass (AM), and ambient temperature (AT).

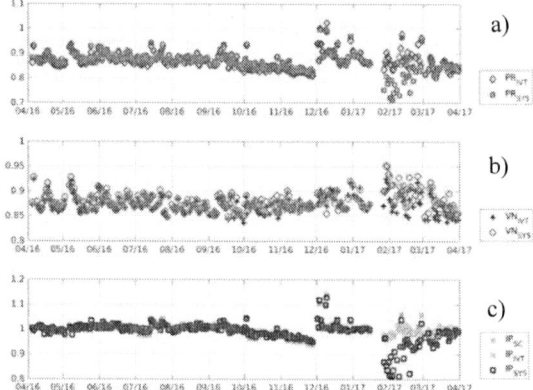

Fig. 2. Variation of the daily performances (c: PR, d: VN, e: IP) for a PV system (S3S) and module (S3) for a year of operation (April 2016 – March 2017).

TABLE II
YEARLY AVERAGE OF THE DAILY ENVIRONMENTAL CONDITIONS (APRIL 2016 – MARCH 2017).

Parameters	Irradiation	AT	APE
Unit	kWhm^{-2}	°C	eV
Average	5.6	27.5	1.92
Standard deviation	1.1	1.9	0.02

B. Correlation between current and optical performances

IP correlated closely with the optical performance for all PV modules and systems. Fig. 3 shows the relationship between the daily average values of the current (IP$_{SYS}$ and IP$_{IVT}$) and optical (IP$_{SC}$) performances for some of the PV systems and modules. For most individually tested PV modules, IP$_{IVT}$ is highly proportional to IP$_{SC}$. Table III details the slope and coefficient of determination (R^2) of the linear least squares regression obtained between IP and IP$_{SC}$ for most PV modules. No IVT data is provided for the CIGS C2 due to overvoltage issues leading to an insufficient dataset. The problem was corrected during the first-year maintenance conducted in January 2017 and datasets for this module will be available for the second year of monitoring. The slopes vary between 1 and 1.1 with R^2 above 95% for all PV except the CIGS C1. For this module, the IP$_{IVT}$ values drop at high IP$_{SC}$ measured in overcast days. The current performance of the systems is mostly similar to IP$_{IVT}$ and differences between these 2 current performances informs on system losses and operating conditions. Most systems exhibit slightly lower performance than the PV module for the highest IP$_{SC}$ values. The PV system H1 exhibits distributed values with IP$_{SYS}$ lower than IP$_{IVT}$ within the full range of performance. This result is related to the shading impact observed in February (Fig. 2-c) also affecting this PV system. Finally, the CIGS PV system C1S exhibits reduced IP at high IP$_{SC}$ values as noted on the individual module. An operating issue identified during the first year of operation is inverter saturation. Saturation decreases IP (such as shading does but with lower effect) and was observed affecting all PV systems using one type of microinverter operating with the standard crystalline modules (S1M, S2M, S3M, S4M). The IV tracer was carefully designed to avoid saturation of the channels at high irradiance.

C. Yearly average performances

Fig. 4 shows the yearly average performances for the PV systems (SYS) and modules tested with the IV tracer (IVT). The difference between PR and the product of IP and VN is estimated below 1% for all PV modules and systems. PR is estimated between 83% and 89% for most PV systems. Higher performance is calculated for 2 thin films, with PR at 95% for the CdTe D1S and 97% for the CIGS C2S. The PV modules tested individually have PR between 87% and 96% exhibiting slightly higher performance than the systems. The difference of performance between the system and module is related to

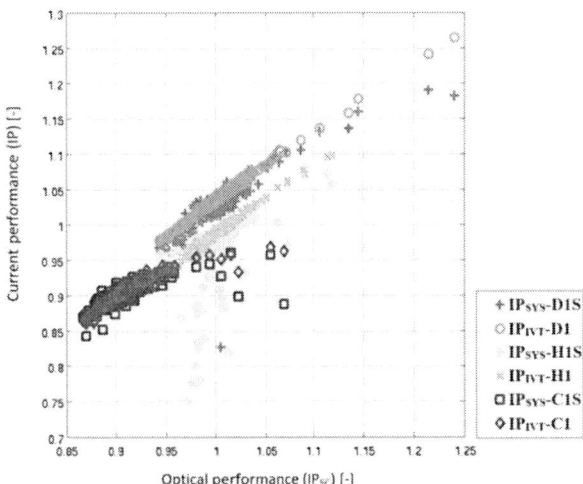

Fig. 3. Current performance (IP$_{SYS}$ and IP$_{IVT}$) as a function of the optical performance (IP$_{SC}$) for some PV systems (D1S, H1S, C1S) and modules (D1, H1, C1).

TABLE III
CORRELATION BETWEEN IP$_{IVT}$ AND IP$_{SC}$ DURING 1 YEAR OF OPERATION FROM APRIL 2016 TO MARCH 2017.

PV types	slope	R^2
S1	1.004	0.954
S2	1.095	0.970
S3	1.057	0.968
S4	1.056	0.970
H1	1.040	0.972
H2	1.078	0.978
H3	1.089	0.960
C1	0.663	0.838
D1	1.002	0.991

normalization error, system losses, and differing operating conditions (shading). Most PV modules and systems (crystalline and CdTe) have VN between 87% to 93%. The 2 CIGS modules and systems have higher voltage performance, above 96%. IP averages between 95% and 100% for most PV technologies except for one low performer with IP around 91% (CIGS C1), and one with high IP at 103-104% (CdTe D1). Optical performance (IP$_{SC}$) of the modules vary from 91% to 102% with most values above 97%. The low IP of C1 is related to a low IP$_{SC}$. However, the high IP of D1 is not associated with a high IP$_{SC}$.

IV. DISCUSSION

The Maui test platform has a mostly sunny environment. The seasonal variation of the irradiation is related to the location and orientation of the carport. With the tilt equal to the latitude of 20.7°N, the variation is small and biannual with peaks in March and September. This is because the sun passes directly above the normal of the PV modules at the equinoxes

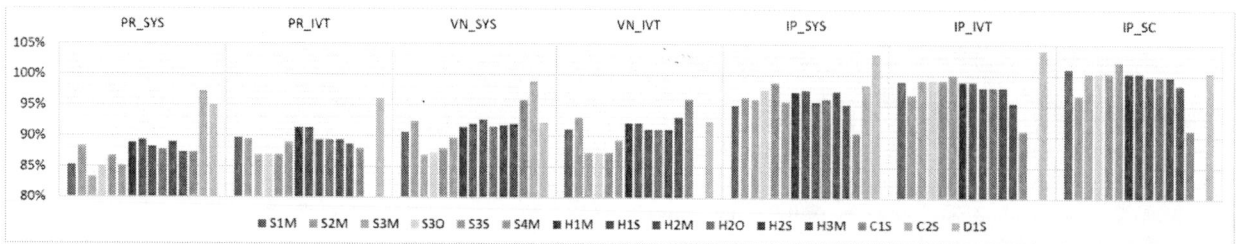

Fig. 4. Yearly average of the daily performances (PR, VN, IP and IP_{SC}) of the grid-connected PV systems (SYS) and PV modules tested with the IV tracer (IVT).

resulting in higher midday irradiance levels in March and September and lower levels in June and December. The test site has a tropical environment with warm temperatures and high spectral energy (1.92 eV) above the STC spectrum (1.88 eV). The solar spectrum is concentrated in short wavelength due to the low latitude and short length of atmosphere crossed by the sunlight (AM). Spectral seasonal variation is, as expected, higher in summer than in winter also in relation with AM. The main parameter of the spectral energy is the irradiation with higher APE in overcast than sunny days. Such environmental conditions are common in Hawaii, similar to 3 of the 5 previous test locations instrumented by HNEI since 2010 [11]. At higher latitude, the seasonal variation of the irradiation is more important [5-7] as the sun never passes above the normal of the PV modules at latitude higher than 23.5°.

Studying IP and VN in addition to PR helps diagnose the impact of the operating conditions. IP is especially informative. It is almost constant yearlong at the location due to mostly sunny conditions. IP increases in overcast days and decreases due to operating conditions and issues such as soiling, shading, and inverter saturation. IP is directly proportional to the optical performance of the PV modules, calculated from the short-circuit current, indicator of spectral effect. For one manufacturer, the relationship between IP and IP_{SC} is not linear which indicates reduced current performance of the modules in overcast days.

PR is simply the product of IP and VN that were found varying between PV technologies and manufacturers. Most PV modules have a yearly PR between 87% and 91% with low difference between standard and high efficient crystalline modules. Two thin films overperform the crystallines with PR above 95%. Thin films were previously suggested to be more suitable in tropical environment [4] and reported with good energy performance in high irradiation conditions [6]. IP averaged between 95% and 100% for most modules except for one module falling low (91%, CIGS C1) and one high performer (104%, CdTe). High IP (105%) was monitored on micromorph modules in Oahu during 2015 [12] but this was supported by high optical performance which related to the spectrally sensitive technology and the high spectral energy in Hawaii. The CdTe modules tested in Maui do not exhibit spectral enhancement of the optical performance therefore we have no explanation at this point for such high IP observed on

both module and system. VN is between 87% and 93% for most technologies with higher values (96%) for the CIGS. Further investigations including the dataset on the second CIGS module will be necessary to confirm and explain such high VN values observed on this technology. The performances of PV systems are affected by system losses and differing operating conditions that will be characterized in further publication.

VI. CONCLUSION

We have introduced a new PV test facility on the island of Maui and a novel energy rating analysis. The selected approach helps diagnose the operating conditions and improves the characterization of the PV performance by determining current and voltage performances.

The Maui test site is a mostly sunny location with warm temperatures and high spectral energy typical of tropical locations. Two thin films outperformed the crystalline technologies due to either high voltage performance (CIGS) or high current performance (CdTe). Standard and high efficient crystalline modules have similar performance at the location. However, the two tested CIGS exhibit high performance differences (10%). The least performing CIGS was diagnosed with low optical performance and reduced current performance in overcast conditions. Next, the sensitivity of the performances to the environmental conditions will be presented describing the differences between PV technologies and manufacturers.

Comparison of the current and voltage performances obtained in different test locations would help better understand the performance variability between PV installations.

ACKNOWLEDGEMENT

HNEI is grateful for the support of the United States Department of Energy (DE-EE0003507) and the Office Naval Research (N00014-10-10310) that funded the development, building and installation of the PV test platforms and the HNEI DAS. Many thanks to Maui Economic and Development Board for hosting our test station in the Maui office parking area. We also recognize the great support of our contractors including the PV installer HNU Energy and the

978-1-5090-5606-4/17 $31.00 © 2017 IEEE

electronic developer Grandalytic. Finally, we are thankful of First Solar for donating a set of their PV modules.

REFERENCES

[1] http://www.eia.gov/electricity/state/hawaii/

[2] http://www.hawaiicleanenergyinitiative.org/

[3] http://www.seia.org/state-solar-policy/hawaii/

[4] K. Akhmad et al., Outdoor performance of amorphous silicon and polycrystalline silicon PV modules, *Solar Energy Materials & Solar Cells* (46), pp. 209-218, 1997.

[5] G. Makrides, B. Zinsser, M. Norton and G. E. Georghiou, Performance of Photovoltaics Under Actual Operating Conditions, *Third Generation Photovoltaics*, InTech, DOI: 10.5772/27386, 2012.

[6] C. Canete, J. Carretero, M. Sidrach-de-Cardone, Energy performance of different photovoltaic module technologies under outdoor conditions, *Energy* (65), pp. 295-302, 2014.

[7] J.A. del Cueto, Comparison of energy production and performance from flat-plate photovoltaic module technologies deployed at fixed tilt, Proceedings of the 29th IEEE PV specialists conference, pp. 1523-1526, 2002.

[8] *IEC 61724*, Photovoltaic system monitoring - Guidelines for measurement, data exchange and analysis, Edition 1.0, 1998-04.

[9] D. Dirnberger at al., PV module energy rating: opportunities and limitations, *Progress in Photovoltaics: Research and Applications* (23), pp. 1754-1770, 2015.

[10] *The Performance of Photovoltaic (PV) Systems – Modelling, Measurement and Assessment*, ISBN: 978-1-78242-354-6, Elsevier, 2017.

[11] *Solar Resource and PV Systems Performance at Selected Test Sites, Subtask 3.1 Photovoltaic Systems: Report 1 - Field Testing and Evaluation of Photovoltaic Technologies* for the United States Department of Energy Hawaii Distributed Energy Resource Technologies for Energy Security, DE-EE0003507, 2014.

[12] *Final technical report* for the Office of Naval Research Hawaii Energy and Environmental Technologies Initiative, N00014-10-10310, 2015.

[13] T. Ishii et al., A simplified methodology for estimating solar spectral influence on photovoltaic energy yield using average photon energy, *Energy Science and Engineering* (1), pp. 18-26, 2013.

[14] M. Alonso-Abella, F. Chenlo, G. Nofuentes, M. Torres-Ramirez, Analysis of spectral effects on the energy yield of different PV technologies: The case of four specific sites, *Energy* (67), pp. 435-443, 2014.

A Novel Multilevel Solar Panel System: Implementation and Verification

Tanmoy Debnath, Syed N.Imtiaz, Syed F. Nawaz, Abdullah Al Mahmud, and Mosaddequr Rahman

Department of Electrical & Electronic Engineering, BRAC University
66 Mohakhali Dhaka 1212, Bangladesh

Abstract — **A prototype multilevel solar panel system with an automated solar tracker has been developed for its use in urban areas. The system consists of three solar panels stacked one above another at some fixed distance to minimize the floor area and shifted horizontally from one another to avoid the shading of the lower panels. Experimental results show that the developed system can harness about 23% more energy while occupying 33% less floor area than that by the conventional fixed panel system of same size. The proposed system is easy to construct and implement, takes less floor space and will be effective for urban areas where not only electricity is in short supply, availability of roof top space to install solar panels is also very limited.**

Index Terms — **electricity, microcontrollers, multilevel systems, solar energy, solar panels.**

I. INTRODUCTION

The large urban cities of the third world countries are faced with acute shortage of energy due to expansion of the cities and ever increasing demand for energy of the urban population. With dwindling fossil fuel and catastrophic environmental conditions resulting from excessive burning of fossil fuels, scientists are forced to look for alternative sources of energy that are clean, pollution free and renewable. Among various forms of renewable energy sources, solar energy has emerged as a key source of renewable energy over the past two or three decades due to its ubiquitous presence and seemingly inexhaustible supply. Generation of electrical energy from solar energy using photovoltaic (PV) system has the promise and potential to solve the electrical energy crisis even for the third world countries as it is technologically simple to manufacture, easy to set up, takes little maintenance, and noise and pollution free.

However, the limited space for setting enough solar photovoltaic (PV) panels to meet the demand of city dwellers is a serious constraint in implementing solar energy system in densely populated urban areas. Therefore, a novel approach was presented in [1] to utilize the available space more efficiently to harness the solar energy using PV panels. In the proposed system, three panels are stacked one above another in a rack to minimize the floor area, while they are shifted horizontally from each other to avoid the shading of the lower panels, and equipped with a sun tracking system to maximize power generation.

In this work, we report on the construction of a prototype of the proposed system with preliminary experimental results. The multilevel system proposed in [1] is slightly modified to make the construction easy and the operation more efficient. The physical structure of the prototype is constructed using stainless steel (SS) bar supporting the three panels equipped with servo motors controlled by a microcontroller to rotate the panels to track the sun. A six channel data logger has been developed to collect the current and voltage data of each of the panels. The experimental results obtained showed significant enhancement in the energy collection by the multilevel system compared to that of the conventional single level fixed panel system and is in good agreement with the theoretical prediction made in [2].

In the forthcoming sections of the paper, section II sheds light on the physical structure, operation and construction of the proposed system along with a description of the automated tracker followed by the description of experimental set-up to collect the outdoor current-voltage data of the system under illumination in section III; section IV provides results and discussions, and section V gives the concluding remarks.

II. MULTILEVEL PANEL SYSTEM: SYSTEM DESCRIPTION

A. Physical Structure

A simple 2-D schematic side view of the multilevel solar panel system proposed in [1] and the modified structure of the proposed system is depicted in Fig. 1. In both the systems, panels are mounted one above another separated by a fixed distance to minimize the floor space, and shifted horizontally by half the panel width to avoid the shading of the lower panels. Each of the panels will be equipped with a micro-controller controlled servo motor to track the sun.

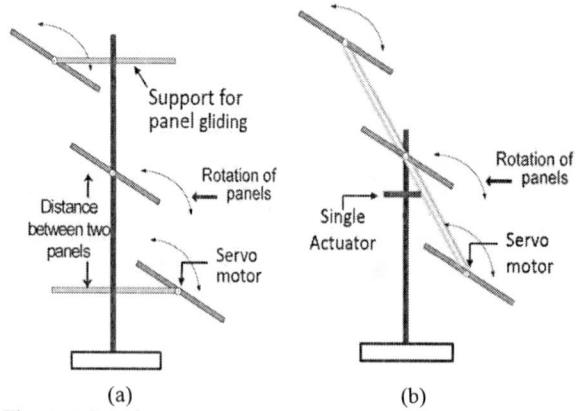

Fig. 1. 2-D schematic side views of the previously proposed (a) and the modified (b) multilevel solar panelsystem.

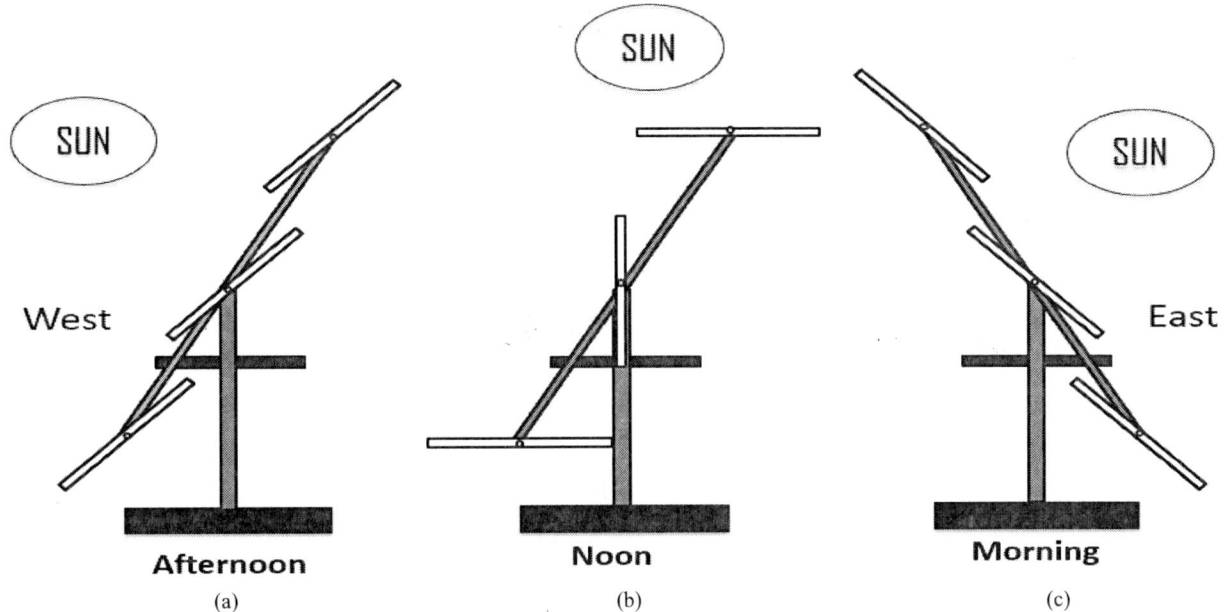

Fig. 2. 2-D schematic side views of the multilevel system showing the panel orientation for three different times in a day (a) afternoon, (b) noon and (c) morning.

In the previously proposed system (Fig. 1, left), top and bottom panels are supported by a frame that is fitted with a gliding mechanism to glide the top panel from left to right and the bottom panel from right to left to realign the panels with the position of the sun as sun moves from east in the morning to west in the afternoon. This orientation will allow the panels to get full exposure to solar radiation. However, in the modified structure, all the panels are supported by a single SS bar. The SS bar with the stack of solar panels is mounted on a vertical stand fixed to a heavy weight iron base in such a way so that the SS bar can easily rotate along a horizontal arc with the operation of the linear actuator as shown in Fig. 1.

For connecting the solar panels with the SS bar, three holes were drilled into the SS bar where iron rods were inserted with small ball bearings so that the iron rods can rotate smoothly around the horizontal axis. Two rectangular iron plates were attached to each of the iron rods to hold the solar panels. For rotating the panels, the end of the iron rods on the other side of the ball bearing is connected to servo motor through servo horns for a better grip.

Three 18 watt panels with dimension 434×277×3 mm^3 are used in the prototype of the proposed multilevel panel system. MG995 servo motors with metal gears are selected to rotate the panels for their extended durability with Mini Maestro 12 channels USB controller as servo motor controller. A DS3231 RTC (Real-Time Clock) which is low-cost, extremely accurate I²C real-time clock with an integrated temperature-compensated crystal oscillator (TCXO) and crystal is used to keep track of the date and time. The device incorporates a battery input, and maintains accurate timekeeping even when main power to the device is interrupted. A 12 V battery powered actuator is used in conjunction with a 2 relay module to toggle the positive and negative power to the actuator in order to determine the direction of piston movement.

B. Operation

Figure 2 shows the 2-D schematic side views of the panel orientation of the multilevel system for three different times in a day: (a) afternoon, (b) noon and (c) morning. In the morning, all the panels are facing east and are gradually shifted right to left from bottom to top by half the panel width (Fig. 2c), so that all the panels can get full sun exposure and shading of the lower panels is avoided.

During noon, when the sun is directly overhead, middle panel will be turned vertically up so that partial shading of the bottom panel is avoided and the upper and lower panels are exposed to the full sunlight (Fig. 2b). While the top and bottom panels rotate throughout the day to track the sun, the middle panel remains in a vertically static position for some time during the noon. The exact time duration for this is determined by the positions of the sun for which the lower panels start getting shaded over by the upper panels. This position of the sun is a function of the inter-panel separation; for large inter-panel separation, the position of the sun will be at a higher altitude in the sky. The higher is the altitude for which lower panels starts getting shaded over, the less is the amount of time the middle panel stays vertically up, and more is the energy harnessed by the system. Following our preliminary analysis, the inter-panel separation is set at 1.5 times the panel width as this enables the middle panel to harness significant amount (69%) of energy with a reasonable system size [1].

All the panels are tilted at latitude angle of Dhaka, Bangladesh, facing south to maximize the energy collection in one whole year. For the sake of clarity this panel tilt along the north-south direction is not shown in the 2-D schematics of

TABLE I
DIMENSIONS OF THE DEVELOPED PROTOTYPE OF THE PROPOSED SYSTEM

18 W Single Crystal Si Solar Panels	
Length	434 mm
Width	277 mm
Thickness	3 mm
Supporting Structure	
Inter-panel separation	415.5 mm
Height of vertical support	554 mm
Supporting bar length	831 mm
Overall height of the system	969.5 mm
Overall lateral dimension	554×434 mm^2

the system. Table I summarizes the detail dimensions of the prototype of the multilevel solar panel system developed.

C. Automated Tracker

A real time digital clock based automated solar tracker has been developed for the multilevel solar panel system. The real time digital clock feeds the data of date and time to the microcontroller. Microcontroller reads the real-time and the day from the digital clock and determines the sunrise (TSR) and sunset (TSS) times using a set of equations [3,4]. The time interval $\Delta T = (TSS - TSR)/12$ at which the panels should be rotated is then computed.

Microcontroller will send signal to the servo-motor controller to rotate the panels by 15° at the fixed time interval ΔT as computed above. When the sun reaches the maximum angular position in the sky at which the lower (middle and bottom) panels start getting shaded over by the upper panels, the microcontroller will send signal to set the middle panel in vertically up position.

Middle panel will be kept static in vertically up position until sun moves past the zenith towards west by the same angle. Just before noon, before the top and bottom panels make their next rotation, microcontroller will send signal to activate the actuator through a relay. The actuator will swing the supporting bar to the other position along east-west direction as shown in Fig. 2 (a & b), to realign the panels with the sun's position in the west in the afternoon so that the panels get the full sun exposer. At sunset, the panels will be reset to their original position for the next day. In this system, 12 positions for the panels have been programmed following the coordinates of the sun throughout the whole day. The detailed algorithm for the microcontroller program is described in [2].

III. EXPERIMENTAL SET-UP

The complete system was installed at the roof top of a 16 storied building of BRAC University. In order to compare its performance with those of conventional fixed panel system,

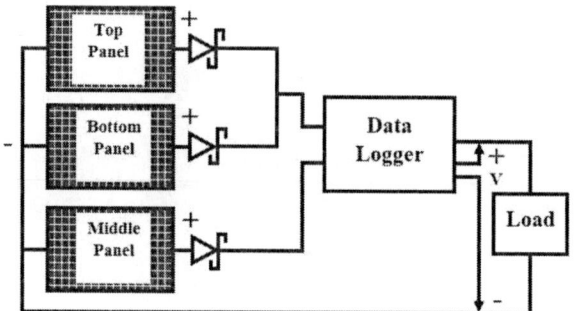

Fig. 3. Experimental setup for the collection of outdoor current-voltage data of the prototype multilevel solar panel system.

another 18W panel with the same specifications as those in the multilevel system was installed next to the multilevel system at the fixed latitude angle facing south.

The current and voltage data form the three panels and that from the fixed panel were collected using a six channel (3 for current and 3 for voltage) data logger, developed in the Thesis Laboratory of EEE Department of BRAC University. Figure 3 shows the connection diagram of the experimental setup to collect the outdoor current-voltage data of the prototype multilevel solar panel system. The output currents from the top and bottom panels are fed into a single current channel as the rotational movement and alignment of these two panels remain same throughout the whole day. The output currents from the middle and the fixed panels are fed into two separate current channels of the data logger. A 40 AH 12V lead-acid battery has been used as the load. Schottky diodes are put in series with each of the panels to prevent the reverse flow of current from battery to panel in case of insufficient sunlight or absence of sunlight.

IV. RESULTS AND DISCUSSIONS

Current and voltage data collected on 7th of August of 2016 have been used to assess the performance of our system. It was a moderate sunny day with average temperature of about 33°C [5].The combined power output from the three panels of the multilevel system (solid line) and that from the three fixed panels (dashed line) of the same size have been plotted in Fig. 4 with respect to time of the day from 8am in the morning to 18pm in the evening. It is obvious from the figure that the multilevel solar panel system equipped with sun tracking system produces considerably higher amount of power than the three fixed panels for most part of the day. During noon from about 11:00 am to 13:00 pm, the power output from the multilevel system drops to a low value as the middle panel is made inactive for that period by keeping it static in a vertically up position.

Figure 5 shows the plot of cumulative energy generated by the multilevel solar panel system (solid line) along with that from the conventional fixed panel system of same size (dashed

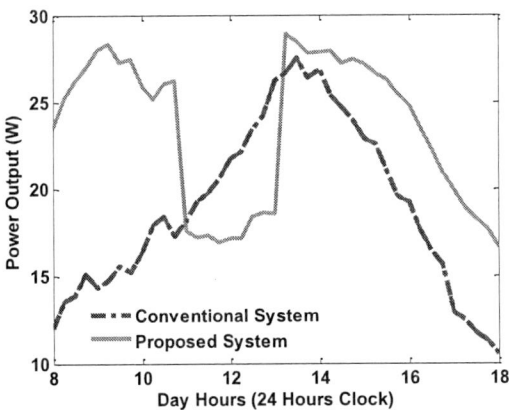

Fig. 4. Plot of combined power output from the three panels of the prototype of the proposed system (solid line) and that from the three fixed panels of conventional system of same size (dashed line) w.r.t. time recorded on 7th August, 2016.

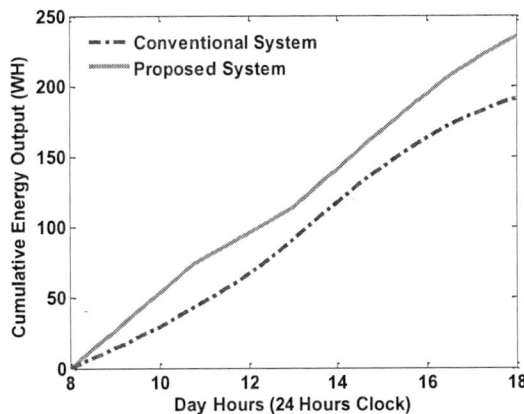

Fig. 5. Plot of cumulative energy output from the three panels of the prototype of the proposed system (solid line) and that from the three fixed panels of conventional system of same size (dashed line) w.r.t. time recorded on 7th August, 2016.

line), obtained by integrating the power output shown in Fig. 4 with respect to time. This figure shows that in spite of a dip in the power output during the time span from 11:00am to 13:00pm, the proposed system can still generate a higher amount of energy than the conventional system, due to its sun tracking system. After 10 hours of experimental time span, the cumulative energy generated by the multilevel solar system is 234.9 watt-hour (WH), and that generated by the conventional system is 191.2 WH. This amounts to about 23% more solar energy harnessed by the developed prototype multilevel solar panel system compared to that of the conventional single level fixed panel system of same size, and is quite close to the theoretically predicted value in [2] for the month of August.

V. CONCLUSION

A prototype multilevel solar panel system has been successfully constructed and tested for assessment of its performance. The system consists of three solar panels stacked one above another at some fixed distance to minimize the floor area and horizontally shifted from one another by half the panel width to avoid the shading of the lower panels.

In the developed prototype, three 18W panels separated by a fixed distance are attached to an SS bar through ball bearings and equipped with servo motors precisely controlled by a RTC based microcontroller controlled control circuit to track the sun. The SS bar is supported by an iron bar erected vertically with a heavy weight iron base in such a way that the former bar can rotate around a certain pivot with the operation of a linear actuator so that panels can be realigned to get full sun exposure as the sun moves from east to west.

The proposed system with three panels can be operated within the space of two panels because of vertical stacking of the panels and can harness more energy than the conventional single level fixed panel system due to the tracking mechanism. Experimental results show that about 23% more energy is harnessed by the prototype of the proposed system than that by the conventional fixed panel system of same size. Energy harnessed can be further enhanced by using specially designed reflectors for the proposed system. Work is under way to develop appropriate reflector for the multi-level panel system that will further enhance the output of the proposed system.

REFERENCES

[1] A. H.Zenan, S. Ahmed, Md. K.Rhaman and M. Rahman, 'A new multilevel solar panel system for urbanareas', In the *Proc. of 39th IEEE Photovoltaic Specialists Conf.* (*39th IEEE PVSC*), Tampa, Florida, USA, pp. 1526-1530, 2013.

[2] H. M.Moniruzzaman, M. Patwary, and M. Rahman. "A three level solar panel system with an automated solar tracker." *3rd Int. Conf. on the Developments in Renewable Energy Tech.,* (*ICDRET 2014*), Dhaka, Bangladesh, pp. 1 − 7, 2014.

[3] Christopher Gronbeck, SunAngle, [Online]. Available: http://www.susdesign.com/sunangle/. (Acessed on Jan 26, 2017)].

[4] Dhaka August Weather 2016 − AccuWeather Forecast for Dhaka Bangladesh [Online]. Available: http://www.accuweather.com/en/bd/dhaka/28143/august-weather/2814. (Acessed on August 10, 2016)

Predicting Power Loss Due to Module Mismatch in Utility-Scale Photovoltaic Systems

Stephen Kaplan and Kendra Passow

First Solar, San Francisco, CA, 94105, United States

Abstract — **A model is proposed for simulating power loss in utility-scale photovoltaic (PV) systems due to mismatch of module electrical characteristics. The main input to the model is a set of current-voltage (I-V) curves of thin-film Cadmium Telluride (CdTe) modules. The impact of the following effects on power loss due to mismatch are observed: the statistical distribution of module electrical characteristics, spatial temperature variation across an array, shading on an array, and module degradation over time. Additionally, two methods for synthetically generating I-V curves to mimic measured data are described and compared.**

Index Terms — **mismatch, photovoltaic systems, degradation**

I. INTRODUCTION

In photovoltaic (PV) systems, modules with the same nameplate ratings at standard test conditions (STC) will vary in electrical characteristics due to manufacturing variability and field degradation [1]. Conditions such as spatial temperature variations and partial shading will also alter the electrical topology of a PV system. Since modules electrically connected in series must operate at the same current and those in parallel must operate at the same voltage, most modules must operate at a point lower than their individual maximum power points (MPP). Thus, the overall power output of an electrically-connected DC array is less than the sum of its individual modules' MPPs. This power loss (P_{loss}) is calculated as a simple error (1):

$$P_{loss} = \frac{\sum P_{max,module} - \sum P_{max,array}}{P_{max,array}} \times 100 \qquad (1)$$

where $P_{max,module}$ is the MPP of an individual module and $P_{max,array}$ is the power output of an array.

Energy prediction tools such as PVsyst [2] and PlantPredict [3] include mismatch as a simple percent derate. Generally, the photovoltaics industry models mismatch loss with values as high as 2% and sometimes less than 1%. Recent analyses show that a lower derate would more accurately estimate mismatch in an array, as explained in the next section.

A model is proposed that allows for specification of each module's electrical characteristics, its configuration in an array, and operating conditions. The effects of varying the statistical distribution of module electrical parameters and other secondary phenomena affecting mismatch are investigated. Both measured and synthetically generated module current-voltage data are used as inputs to the model, and their results are compared. The scope of this work is limited to an investigation of mismatch on the DC side of a single inverter at the utility-scale.

II. MOTIVATION

Bucciarelli [4] uses a statistical approach to propose an equation for estimating mismatch loss which is proportional to the variance of module electrical characteristics within a bin (bin distribution). F. Iannone et al. [5] utilizes Monte Carlo simulation techniques to achieve desired levels of mismatch by randomly generating arrays, validating Bucciarelli's assumptions at low standard deviations. Chamberlin et al. [6] randomly arranges a pool of modules into sets of four and calculates mismatch, achieving results averaging 0.1% loss and never exceeding 0.53%. These results are motivational, but not necessarily applicable to large utility-scale arrays.

MacAlpine et al. suggests that based on recent data, a standard derate factor of 1-2% (or less in some systems) is appropriate for modeling mismatch on the string level [7]. In a similar study based on measured data from groups of 6-24 modules, MacAlpine et al. attributes less than 1% loss annually to mismatch effects [1]. Furthermore, the modules used in these studies were more than ten years old, indicating an even higher degree of conservatism in their conclusions due to trends of decreasing variance in module characteristics in conjunction with increases in efficiency. For instance, First Solar's bin size of ±2.5W has remained constant as module power increased from a 70W Series 2 to a 120W Series 4v3 module over five years. Therefore, the variance as a percentage of the module wattage has dropped significantly with newer module generations. This trend of narrowing bin distributions suggests that a value less than 1.0% might be a more realistic approximation for mismatch losses.

The goal of this work is to examine various factors contributing to mismatch-induced power loss in a DC array. More specifically, this work is motivated by a need to update outdated mismatch assumptions by focusing on utility-scale systems containing I-V curves reflecting the most recent advancements in module technology.

III. MODEL & METHODS

The PV system model is written in MATLAB [8]. It is limited to a single DC array of interconnected modules and does not include inverters or AC-side system components.

978-1-5090-5606-4/17 $31.00 © 2017 IEEE

Each module is assigned a unique I-V curve. The "ideal" power output is equal to the sum of all modules' individual MPP. The overall I-V curve of the system is calculated by graphically adding the modules' I-V curves in series and parallel according to their physical layout. The addition of I-V curves in series is done by adding each voltage value at a common reference of current values. Likewise, the addition of I-V curves in parallel is done by adding each current value at a common reference of voltage values. The MPP of the system I-V curve is calculated. Finally, percent mismatch loss is calculated with (1). Each experiment performed was run forty times with randomized module configurations, using the same pool of I-V curves, and reported as an average result.

In the case of a DC array connected to an inverter with N number of maximum power point trackers (MPPT), the strings are divided into N equal groups. Mismatch loss is calculated for each group, and overall mismatch is assumed to be the average of each calculated loss. All other optional features of the model are detailed in section VI (Impact of Secondary Effects).

A. Using Synthetically Generated I-V Data

One data input option is "synthetic" I-V curves, generated from specified statistical distributions of short-circuit current (I_{sc}), open-circuit voltage (V_{oc}), current at maximum power (I_{maxP}) and voltage at maximum power (V_{maxP}). The benefit of this source of input data is that it allows for full control of the variability in module characteristics. Additionally, it allows for the modeling of future generations of modules before any I-V measurements have been made, or modeling modules for which there is no public data.

A study of nearly 1 million samples of First Solar Series 4v3 module I-V curves determined that the individual module parameters I_{sc}, V_{oc}, I_{maxP}, and V_{maxP} are normally distributed for a single power bin. The data from this study are summarized in Table 1. Figure 1 shows an example of this data.

TABLE I
FS-4117 PARAMETER STATISTICS

Parameter	Mean (μ)	Std. Dev. (σ)	$C_v = \sigma/\mu$
I_{sc}	1.806 A	0.0053 A	0.29%
V_{oc}	92.05 V	0.24 A	0.26%
I_{maxP}	1.652 A	0.0074 A	0.45%
V_{maxP}	72.93 V	0.33 V	0.44%

Therefore, given a standard deviation for each of the four module parameters, and the total number of modules desired for the simulated array, four normal distributions are generated using a MATLAB function. For each module in the array, a value for each module parameter is randomly selected from each of the four generated distributions. It is assumed that there are no underlying correlations in the data between current and voltage. Finally, an I-V curve fitting algorithm is used to generate a unique I-V curve for each module in the array. Section 4 provides a more in-depth discussion of the methods used to simulate I-V curves (including the I-V curve fitting algorithm) and a comparison of their accuracy with respect to measured data.

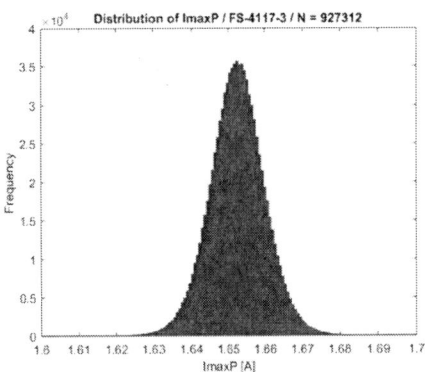

Fig. 1. I_{maxP} data from 927312 FS Series 4v3 117.5W modules.

B. Using Measured I-V Data

The other possible input to the model is measured module I-V curves. In this case, the curves capture subtle differences in module characteristics and manufacturing variability. A fixed module nameplate power bin width is a primary input to the selection of a sample population of module I-V curves. It is calculated by first solving (2) for the desired standard deviation (σ) of module MPPs:

$$C_V = \frac{\sigma}{\mu} \qquad (2)$$

where C_v is the coefficient of variation and μ is the desired mean module MPP.

It is assumed that MPPs of modules in the same bin are distributed uniformly. Production from a single First Solar manufacturing line is normally distributed, but spans two or three bins. While the manufacturing process keeps a relatively tight distribution at any given plant, there is some variation from plant to plant as to where the center of the distribution is. Assuming that a set of modules in the same bin installed together can be from a mix of manufacturing lines and plants, the effective distribution is nearly uniform.

The assumption of a "uniform distribution" conflicts with the data described in the previous subsection, which shows a normal distribution. However, the aforementioned data is from one manufacturing line over a six month period, and is not representative of all First Solar modules of a particular generation manufactured globally. Regardless of this discrepancy, the standard deviations calculated in subsection A can be used to calculate the bin width of a uniform distribution with (3). Furthermore, a brief comparative study showed that calculating mismatch loss with a normal distribution of module parameters distribution provided a more conservative estimate than calculating with a normal distribution.

Given C_v calculated in (2), the equation for the standard deviation of a uniform distribution (3) can be used to solve for the effective fixed bin width (ΔP):

$$\sigma = \frac{\Delta P}{\sqrt{12}} \qquad (3)$$

This is used as a database query parameter to pull a uniformly distributed set of I-V curves, with a well-defined range of MPPs, equal in quantity to the number of modules in the simulated array.

C. Measurement Uncertainty

All I-V curve measurements were performed on modules using Spire SLP 3500, Class AAA, single long pulse solar simulators [9]. First Solar's module I-V curve measurement conditions (STC) were verified by Fraunhofer USA according to IEC 60904 standards in 2014 [10]. First Solar has achieved a companywide mean variation of less than ±1% between its solar simulators [11]. All First Solar modules have a maximum power tolerance of -0/+5W, which excludes a measurement uncertainty of typically ±3%. Table 2 summarizes the measurement uncertainties of the Spire simulator.

TABLE II
SPIRE SLP 3500 UNCERTAINTY

Parameter	Measurement Uncertainty
P_{max}	± 3.1%
I_{sc}	± 3.0%
V_{oc}	± 0.7%

IV. COMPARING METHODS OF GENERATING SYNTHETIC I-V CURVES

Initial efforts to simulate I-V curves relied on the single-diode model. However, plotting measured I-V data against a curve generated from measured single-diode parameters showed a vast discrepancy in curve profile. A more accurate iterative curve-fitting algorithm was employed. The methods are compared in the following sections.

A. Generating I-V Curves with the Single-Diode Model

Given a range of voltage values [0, V_{oc}], the single-diode equivalent circuit model [12] is used to generate a set of current values, as in equation (4):

$$I(V) = I_{ph} - I_0 \left[\exp\left(\frac{qV}{N\gamma kT}\right) - 1 \right] - \frac{V}{R_{sh}} \qquad (4)$$

where $I(V)$ is current (A) as a function of voltage (V), I_{ph} is photocurrent (A), I_0 is diode saturation current (A), q is the charge of one electron (1.602 x 10^{-19} C), N is the number of cells in series within a module, γ is the unitless diode quality factor, k is the Boltzmann constant (1.381 x 10^{-23} J/K), T is temperature (K), and R_{sh} is the shunt (parallel) resistance. In this case, I_0, γ, and R_{sh} all come from the specification sheet of the PV module of interest.

B. Generating I-V Curves by Curve Fitting

The curve-fitting algorithm employed in this model is also based on the single-diode model. It iteratively and incrementally tries values of I_0 such that it solves the nonlinear

equation (5) with constraints on I_{sc}, V_{oc}, I_{maxP}, and V_{maxP}. This ensures that the resulting I-V curve has a high degree of graphical accuracy.

$$V_{maxP} \cdot \log\left(1 + \frac{I_{sc}}{I_0}\right) = V_{oc} \cdot \log\left(\frac{I_{sc} + I_0 - I_{maxP}}{I_0}\right) \qquad (5)$$

A set of current values for the I-V curve is generated using the resulting value of I_0, (6), and a set of voltage values [0, V_{oc}].

$$I(V) = I_{sc} - I_0 \left[\exp\left(\frac{V \cdot \log\left(1 + \frac{I_{sc}}{I_0}\right)}{V_{oc}}\right) - 1 \right] \qquad (6)$$

C. Comparing Both Methods to Measured Data

A short study was performed to test the accuracy of both aforementioned synthetic I-V curve methods with respect to measured I-V data. The measured module parameters for five First Solar Series 4v3 modules were used as inputs to both the curve fitting algorithm and single-diode model to generate two I-V curves. Module parameters were extracted from each of these curves and each set of parameters were compared to those of the original measured data. The results of calculating error for both sets of synthetically generated data with respect to measured data are summarized in Table 3. Figure 2 shows a graphical comparison of the measured, fit-algorithm-generated, and single-diode-generated data.

TABLE III
ERROR ASSOCIATED WITH I-V CURVE METHODS

Parameter	Mean Absolute Error w.r.t Measured Data [%]	
	Curve Fitting	Single-Diode
I_{maxP}	1.0857	2.6097
V_{maxP}	1.1812	0.4187
I_{sc}	0.0316	0.2988
V_{oc}	0.0454	1.4036

The curve fitting algorithm is the preferred method of synthetic I-V curve generation on the basis of a graphical comparison, as well as its ability to match individual module parameters with significantly higher accuracy. Furthermore, the calculated errors between the module parameters extracted from the fitted curve and the measured module parameters are within the measurement uncertainty of the solar simulator used to measure the I-V curve (see Table 2).

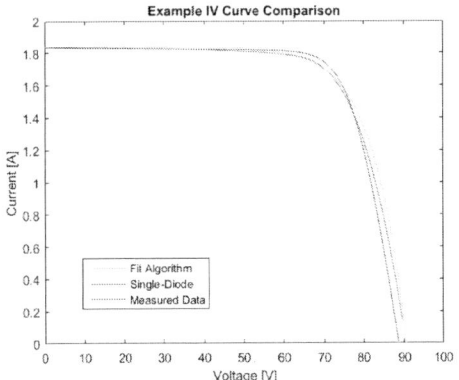

Fig. 2. Comparison of synthetically generated I-V curves from measured module parameters to measured I-V sweep.

V. IMPACT OF MODULE ELECTRICAL PARAMETER VARIANCE

A presentation entitled "Predicting Mismatch Losses in Utility-Scale Photovoltaic Systems", given by S. Kaplan at the 2016 6th PV Performance and Monitoring Workshop in Freiburg, Germany featured a mismatch loss calculation using real I-V data. Sample I-V curves were filtered only by their measured maximum power, and not selected based on a particular set of distributions of module electrical parameters. At 5W (a common tolerance within a given array) for 112.5W modules, mismatch loss is 0.501 ± 0.003% of total power. At bin distributions below 2.5W (the nominal bin distribution), mismatch loss is less than 0.25 ± 0.002% of total power. This trend is illustrated in Fig. 3.

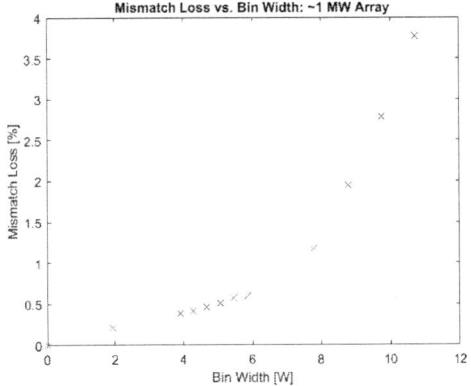

Fig. 3. Mismatch Loss vs. Bin Distribution, First Solar Series 4v3 112.5W modules.

VI. IMPACT OF SECONDARY EFFECTS

Spatial temperature variation, non-uniform shading on an array, and varying rates of degradation are secondary contributors to module mismatch loss. This section features several case studies for each of these categories in an attempt to qualitatively determine the impact of these effects. It should be noted that the I-V curve fitting algorithm was used for this

study, exclusively, as opposed to measured data. This is due to the need for full control of the statistics of the module electrical characteristics. Furthermore, in general the goal of these studies was to simulate *extreme* versions of each effect as to determine an upper limit. Finally, each study assumed that there was *no bin distribution* in order to isolate the effect of interest.

A. Impact of Temperature

For a given array, there will be some degree of variability in module temperature coefficient (T_k). The distribution of T_k for each module electrical parameter will cause some additional mismatch loss even with uniform module surface temperature. In order to perform this analysis, an expected distribution of T_k for V_{oc}, I_{sc}, V_{maxP}, and I_{maxP} was determined from a sample of 20 First Solar Series 4v3 modules that were measured while subjected to a range of temperatures. The slope of a linear fit to the experimental data was taken to be the T_k for a given module parameter. The distributions resulting from this analysis were observed to be uniform, and thus the metric used to describe such distributions was half-width, as summarized in Table 4.

TABLE IV
FS SERIES 4v3 TEMPERATURE COEFFICIENT DISTRIBUTION

Parameter	T_k Mean [%/°C]	T_k Half-Width [%/°C]
I_{sc}	0.041	0.0063
V_{oc}	-0.28	0.012
I_{maxP}	0.022	0.022
V_{maxP}	-0.32	0.014

Mismatch loss due to distribution of temperature coefficient is calculated similarly to the method outlined in Section 3. First, a uniform distribution for each of the four temperature coefficients is generated using a MATLAB function. For each module in the array, a value for each temperature coefficient is randomly selected from each of the four generated distributions. Finally, before calculating mismatch loss, each module parameter is adjusted according to a generic function (7).

$$X' = X_0 + 0.01X_0(T_{mod} - 25^0C) \cdot T_k^i \qquad (7)$$

where X_0 is any nameplate (mean) module parameter for a given module bin, T_{mod} is the module surface temperature (°C), T_k^i is a randomly chosen temperature coefficient from the uniform distribution (%/°C), and X' is the temperature-adjusted module parameter.

The effect of the expected distribution of T_k was calculated for a few different module surface temperatures and reported in Table 5.

TABLE V
MISMATCH LOSS DUE TO EXPECTED TEMPERATURE
COEFFICIENT DISTRIBUTION (FS-4117, 500KW ARRAY)

$T_{module\ surface}$ [^{0}C]	Power Loss [%]
10.0	0.0018
50.0	0.0033
80.0	0.0030

In reality, T_{mod} is not uniform for all modules in an array. Temperature non-uniformity in an array can be modeled using the method mentioned above, with the additional step of varying T_{mod} for each module to reflect an array-wide temperature gradient. For instance, wind flow down an aisle between arrays causes convective cooling for the modules adjacent to that aisle. The resulting temperature gradient used to model this scenario is depicted in Figure 4.

Wind ↑	40 °C	45 °C	50 °C

Fig. 4.   Temperature across one string. The two left squares each represent a module, and the right box represents the rest of the string.

Table 6 summarizes the results of simulating the array edge cooling depicted in Figure 4. The number of strings per row ("string columns") are varied while keeping the total number of strings constant. For example, only half of the strings of a two-column array experience this temperature gradient as compared to a one-column array.

TABLE VI
MISMATCH LOSS DUE TO ARRAY EDGE COOLING
(FS-4117, 500KW ARRAY)

# String Columns	Temperature Mismatch Loss [%]
1	0.0
2	0.0027
4	0.0038

Temperature can also vary with a module's height above the ground. First Solar Series 4v3 modules are commonly installed in groups of four rows ("ranks") on racking structures. While this type of mismatch applies to tracking structures for most of a day, it is assumed that the structure is fixed tilt for this analysis. As a result, the ranks of modules will be progressively farther from the ground, and since the top rank will receive more air flow than the bottom rank, the bottom rank will be the warmest. This scenario is simulated for a range of ΔT (temperature difference between bottom and top rank), and the resulting power losses due to mismatch are summarized in Table 7.

B. Impact of Shading

First Solar modules have a linear shading response due to their lack of bypass diodes. The current values of an I-V curve are scaled by the percent of the module unshaded to model linear shading. For instance, the current values along the I-V curve of a module with 75% of its area shaded (or 75% less light intensity) would be multiplied by 0.25. The visual effect of this operation is the I-V curve "sliding down" the current axis while maintaining its overall shape. Simulation of a partial shading scenario is achieved by programmatically scaling specific curves in an array of I-V curves.

TABLE VII
MISMATCH LOSS DUE TO RANK-TO-RANK TEMPERATURE
GRADIENT (FS-4117, 500KW ARRAY)

ΔT [^{o}C]	Power Loss [%]
1.0	0.0033
2.0	0.0034
3.0	0.0034
4.0	0.0034

Most instances of partial array shading are transient, and cause massive module mismatch and power loss for short periods of time (such as a cloud passing by). These transient scenarios are not factored into this analysis.

The first shading scenario examined is row-to-row shading. Depending on the row spacing in an array, the shaded fraction of the modules on the next tracking structures will vary. Simulations were run for a range of shading geometries and array configurations. For instance, a "single string column" would mean that there is only one string per array row. Every simulation yielded 0% power loss due to shading mismatch. This makes sense since all I-V curves in a string shaded identically will not be mismatched. Furthermore, shading doesn't affect voltage in this case, so there is no row-to-row voltage mismatch.

The second scenario is shading of a fraction of one module in each row of modules in an array. This scenario could possibly result from a building or structure adjacent to an array. A number of simulations were run for a variety of array configurations with the assumption that 25% of one module in each row of strings is shaded. The results are summarized in Table 8.

TABLE VIII
MISMATCH LOSS DUE TO SHADING (FS-4117, 500KW ARRAY)

# String Columns	Shading Mismatch Loss [%]
1	10.37
2	6.614
4	3.702

C. Impact of Degradation

Non-uniform rates of module degradation cause a dispersion in the statistical distribution of electrical characteristics for a group of modules over time. As the standard deviations of I_{sc}, V_{oc}, I_{maxP}, and V_{maxP} increase due to degradation, mismatch loss will increase. It should be noted that while the average values of these parameters degrade over time, the only factor that affects mismatch is the spread in their respective distributions. Therefore, using a constant rate of increase in C_v over time, mismatch loss due to degradation can be calculated at any time interval. Due to limited data on module degradation, it is difficult to calculate accurate rates of degradation for each module electrical parameter. A few years of First Solar Series 4v3 I-V data suggest that 0.1% increase in standard deviation per year is a fair approximation. This approximation is used to calculate mismatch for First Solar Series 4v3 modules at several time intervals Figure 5 shows the results of this study for a 500kW array.

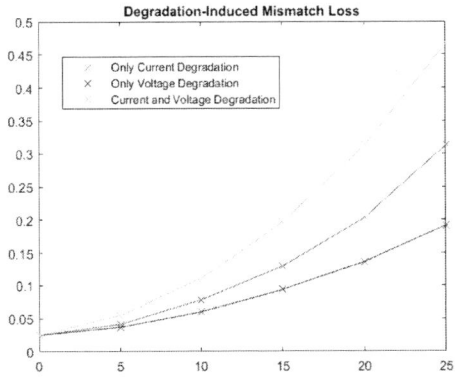

Fig. 5.   Mismatch loss due to degradation, FS Series 4v3 117.5W, 500kW array. RMS dispersion in module parameters is 0.1%/year.

VII. DISCUSSION & CONCLUSIONS

Simulations using real data show that mismatch loss due to bin distribution could be lower than 0.25%. A study of the effect of variance of temperature coefficient along with spatial temperature gradients concludes that temperature effects have a negligible effect on mismatch loss. Both cooling of the outer edges of the array and temperature variation with height above the ground cause minimal mismatch loss. Temperature variation with height on a tracker structure likely causes minimal mismatch loss because all modules in a string are affected similarly. Therefore, current mismatch per string (the dominant factor in mismatch) is relatively unchanged.

Row-to-row shading causes negligible mismatch loss since it affects all modules in a string uniformly. Edge shading of an array can cause drastic mismatch loss. Special consideration and should be made to account for a unique shading geometry.

Module degradation increases mismatch loss over a plant's lifetime. Results suggest that adding approximately 0.1-0.2% to a PV system's overall mismatch loss factor accounts for this trend. This range is chosen because it corresponds with 10-15 years of degradation, which is an approximate average for a module's 25-year warranty-based lifetime.

Although this analysis makes it clear that the mismatch of a power plant array can vary depending on the system configuration, module characteristics, and environmental conditions, the current strategy for modeling mismatch is to use a single number to estimate the overall loss (past work suggested 1% [1]). This current work proposed a model that can be used to model mismatch under various circumstances, but an analysis using typical system design with modern technology can be used to provide an updated estimate for this loss factor. Assuming a 1 MW array size of 117.5 W modules with 2.5 W bin width, results in section V indicate a 0.25% power loss due

to bin distribution. Assuming typical plant configuration with no extreme shading or temperature gradients, secondary effects due to temperature and shading is ignored. Mismatch due to degradation is estimated at year 12.5 in the analysis shown in Fig. 5, indicating 0.16% power loss, and a safety factor of 1.5 is applied to this to account for the fact that simulated I-V data underpredicts mismatch by 50% compared to using measured I-V data. The effect due to bin distribution plus the effect due to mismatch then brings the "typical" mismatch value estimate to 0.5% total loss.

VIII. MODEL VALIDATION

Three First Solar Series 4v3 module electrical responses were individually measured at STC. The modules were then electrically connected in parallel and measured as a composite module. The three individually-measured I-V curves were added in parallel using the model to generate a simulated composite I-V curve. Error was then calculated with respect to the measured I-V curve for a number of I-V curve parameters. The results are summarized in Table 9. Field validation of the model at the array-level is currently under progress in Mesa, Arizona and will be reported upon in future work.

TABLE IX
RESULTS OF SMALL ARRAY VALIDATION

Parameter	Error (Natural Units)	(Error %)
$P_{max\ power}$ (MPP)	0.145 W	0.0421
$I_{max\ power}$	-5.50 mA	-0.278
$V_{max\ power}$	0.564 V	0.322
$I_{short\ circuit}$	-0.867 mA	-0.0389
$V_{open\ circuit}$	0.152 V	0.0684

REFERENCES

[1] S. MacAlpine, et al. "Module Mismatch Loss and Recoverable Power in Unshaded PV Installations," *38th IEEE PVSC,* Austin, TX, 2012.

[2] "PVsyst [software]," PVsyst SA, http://www.pvsyst.com.

[3] "PlantPredict" First Solar, Inc., https://plantpredict.com

[4] L.L. Bucciarelli, "Power loss in photovoltaic arrays due to mismatch in cell characteristics," *Solar Energy*, Vol. 23, pp. 277-288, 1979.

[5] F. Iannone, G. Noviello, A. Sarno, "Monte Carlo techniques to analyse the electrical mismatch losses in large-scale photovoltaic generators," *Solar Energy* Vol. 62, No. 2, pp. 85-92, 1998.

[6] C.E. Chamberlin, et al. "Effects of mismatch losses in photovoltaic arrays," *Solar Energy*, Vol. 54, No. 3, 1995.

[7] S. MacAlpine, et al. "Beyond the Module Model and Into the Array: Mismatch in Series Strings," *38th IEEE PVSC*, Austin, TX, 2012.

[8] "MATLAB software," MathWorks, http://www.mathworks.com

[9] "PD-5-800 FS Series Product Bins and Distribution", First Solar, Inc., Rev. 6.0, 2016.

[10] "PD-5-652 First Solar Module Nameplate Process Review", internal document, First Solar, Inc., 2016.

[11] "PD-5-434 Application Note: Best Practice for Power Characterization", internal document, First Solar, Inc., 2015.

[12] W. De Soto, et al. "Improvement and validation of a model for photovoltaic array performance," *Solar Energy*, 80, 2006

Application of Shaped Reflectors to Increase the Energy Harvest of Bifacial PV Systems - Analyzed with a Miniaturized Test Array

Hartmut Nussbaumer, Markus Klenk, Nico Keller, Dominic Heller, Remo Käslin, Thomas Baumann and Franz Baumgartner

Zurich University of Applied Science, SoE, Institute of Energy Systems and Fluid Engineering
Technikumstrasse 9, 8401 Winterthur, Switzerland *phone: +41 58 934 4799, *e-mail:
hartmut.nussbaumer@zhaw.ch

Abstract — For bifacial module arrays three dimensional, reflective elements in between the rows might be an option in order to achieve a low light concentration resulting in a further increased energy harvest. A simulation of such conditions using the optical behavior of real reflecting foils is hardly possible; measurements at real systems would require considerable effort. In this work we present results obtained by a measuring setup with rotating mini modules. In particular the influence of the ground reflection was studied by using various grounds and reflecting elements with different shapes. Prior to the analysis of the reflector elements the measurement setup is described.

I. INTRODUCTION

The power gain due to the bifacial lay-out is inherently based on the utilization of the radiation which is impinging on the solar cells rear side. A reflecting surface is a prerequisite for a successful application. In almost all applications the reflectors are flat reflecting surfaces. Typically the surrounding ground around the module is used as reflector, but also the backsheet at the laminate rear side can be the reflective surface.

Low concentration was repeatedly proposed for bifacial applications. While low concentration by comparatively simple reflector arrangements around the modules is well known for monofacial solar modules, there are only very limited reports for bifacial applications. This motivates the experimental testing of suchlike simple reflectors below and in between bifacial modules which may be realized in a comparatively simply manner.

The experiments are carried out on a small test rig which is a downscaled model of an existing large test array. This enables a faster and more flexible testing of different reflector types compared to the large test system which is based on standard sized commercial solar modules. Both systems, their set-up and the correspondence of the measurement data had been presented before [1]. The basic set-up of the large as well as of the miniaturized test rig is shown in figure 1. Three rows of modules, or solar cells for the miniaturized version, with manually adjustable distance between the rows, are mounted on vertically adjustable pillars. A unique feature of the measurement set-up is the permanent automatic variation of the tilt angle in steps (12 steps between 0° and 90° in one minute). At each step an I/V- curve of the module in the array`s center is measured. All panels change their tilt angle coordinated with the central row. Height, distance between the rows and reflecting ground can be changed manually.

The benefit of this arrangement, compared to more common test set-ups with fixed stand-alone modules, is the similarity to real world installations with direct shading by other modules and indirect shading of the reflecting ground. This is also the reason for the focus on the center module, because it is best suited to reflect the situation in an extended system.

Due to the tilt angle variation the influence of this factor is systematically investigated. The impact of the tilt angle on the system performance is by far less straightforward for bifacial applications than for standard monofacial ones.

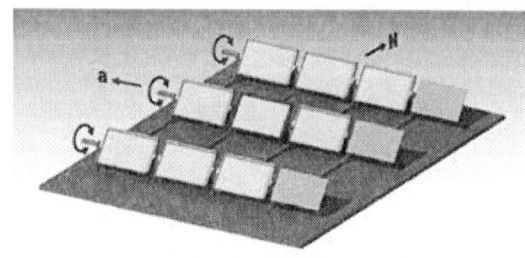

Fig. 1: Measurement setup with permanently moving modules. The most relevant module in the center, which is best suited to represent the actual conditions in real installations, is marked red.

II. APPROACH

The lay-out of the miniaturized test device is a downscaled version of the large test array (scale 1:12). The industrial type 60 cell modules in the large BIFROT array (Megacell) are replaced by cut to size pieces of the same implemented BiSoN solar cells. Thus, the relative dimensions, such as the installation height over the ground, can be maintained. Accordingly it can be assumed that also the influence of these factors will show a comparable dependency. It should thus also be possible to analyze the impact of the installation conditions and their interrelation (height, albedo, tilt angle,..),

978-1-5090-5606-4/17 $31.00 © 2017 IEEE

also for different insolation situations. The general properties of both systems are described elsewhere [1].

Fig. 2: Miniaturized test array with revolving mini-modules, here with 0° tilt angle. The arrangement is the downscaled version of the large test array with commercial modules. Installation height, distance between the rows and reflecting ground can be varied. White reflecting foil is placed below the mini-modules.

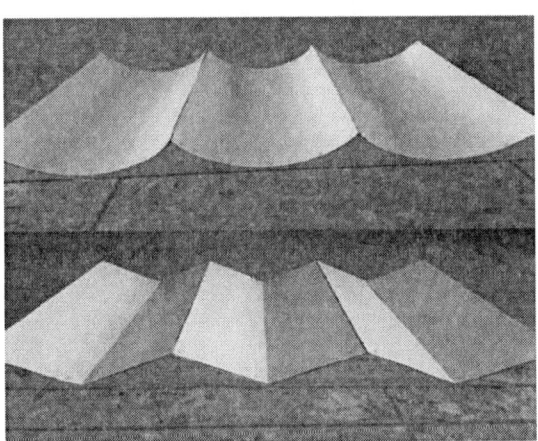

Fig. 3: Differently shaped reflectors made of the same foil material as the flat reflector in figure 2. Versions with different slope were prepared.

In this abstract we focus on a parabolic reflector with two foil materials and compare its properties with the common flat ground lay-out.

III. RESULTS AND SUMMARY

The reflected radiation may not be completely diffuse, but also show direct reflection to a certain extend. This could be detrimental to the homogeneity of the module illumination and cause mismatch effects. Two reflecting foils were analyzed in more detail. Both materials are intended for outdoor use roof sealing and are explicitly dedicated for the use as diffuse

reflector material in bifacial applications. To get a rough appraisal of their reflective properties the angular reflection of an impinging laser was measured, as shown in figures 4 and 5. Without going into detail with regard to the measurement conditions it is obvious that there is a clear difference concerning the reflection properties for the same measurement condition. Foil 2 shows pronounced peaks due to direct reflection, while foil 1 has a more diffuse reflection characteristic.

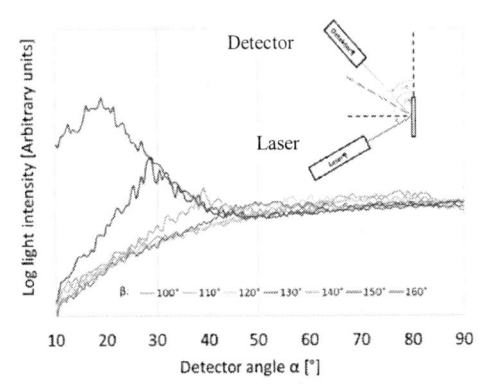

Fig. 4: Foil 1 - Intensity of reflected laser beam versus the angle.

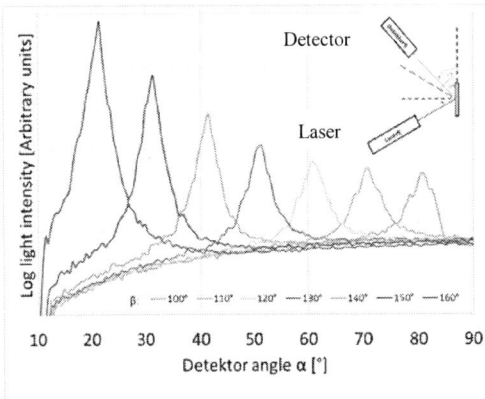

Fig. 5: Foil 2 - Intensity of reflected laser beam versus the angle.

The assumed gain is based on the curved shape of the reflectors. However, the reflectors may also cause an inhomogeneous illumination. In order to investigate the local illumination intensity and homogeneity a small monofacial solar cell of about 1 cm^2 area was placed at different heights to measure the respective light intensity. The focal point of the parabolic reflector corresponds to a mounting height of 6 cm. Also the angle of the cell was varied in order to simulate different tilt angles. With the small monofacial cell this can respectively be done for the bifacial`s module front and rear side. Since the ground reflection mainly affects the modules rear side we present the corresponding results. In the experiments the small, monofacial solar cell was turned from 0° - facing downward – to 90° into vertical position. In order to provide reproducible conditions a large planar LED light source was positioned above of the arrangement. In Figs. 6 and 7 the corresponding results are shown for the flat and a parabolic shaped reflector, both with foil types 1 and 2.

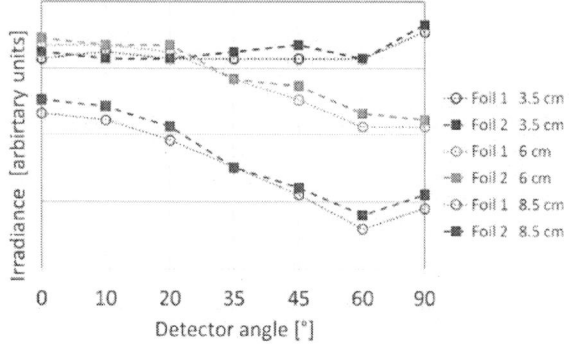

Fig. 6: Intensity vs. tilt angle obtained with a small, downward-facing, monofacial solar cell with varied mounting height in a reflector with parabolic shape. The shape of the reflector has an obvious impact, as revealed by the direct comparison with the flat reflector in figure 7.

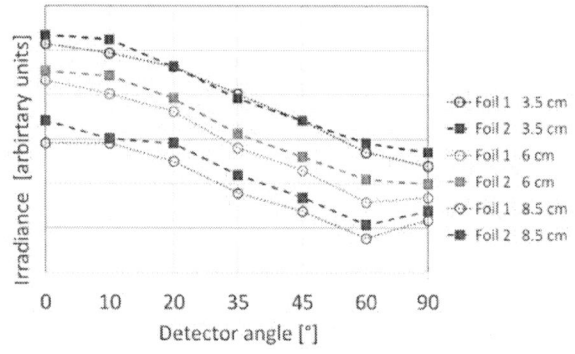

Fig. 7: Intensity vs. tilt angle obtained with a small, downward-facing, monofacial solar cell above of a flat reflector. Results are not fully according to our expectations. We believe, that the limited size of the reflectors may have an influence on the irradiance at higher mounting positions

For the given lighting conditions the difference between the flat and the parabolic reflector is directly visible. Obviously the parabolic shape of the reflector has an effect, which results in an increased intensity for steeply pitched positions. This is however dependent on the mounting height of the small solar cell. The increase is most pronounced for the lowest mounting height, while there is no significant effect for the uppermost position in which the small cell is positioned above the height of the reflectors structure.

The mounting height of the small cell roughly represents the lower central and upper section of a module, if the dimensions are attributed to the real-world situation. This means that there would be an inhomogeneous intensity distribution over the module area. A difference between foil types 1 and 2 is not apparent, in spite of the different reflection properties.

The applied planar LED lighting above the reflector is a reproducible but also a non-standard lighting situation. An outdoor measurement with the bifacial miniaturized test array and parabolic reflector is shown in figure 8. The experiment is carried out on the miniaturized test rig with the bifacial mini-module, as depicted in figure 2. The measurements were carried out in early summer at noon and at clear sky conditions. The reflectors were quickly replaced within minutes to keep the lighting conditions as constant as possible.

Without going into detail concerning the measurement conditions an intensity increase for the parabolic reflector and steep tilt angles is qualitatively visible. Similar results were obtained for differently shaped reflectors and mounting situations [1].

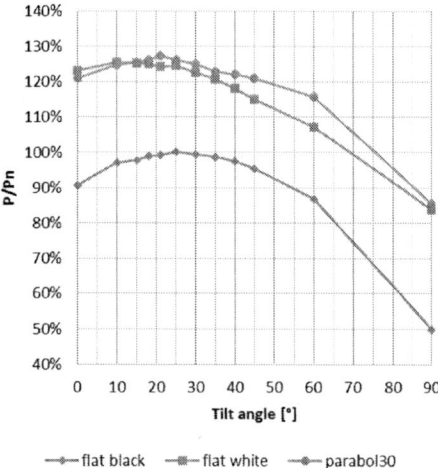

Fig. 8: Pmpp vs. tilt angle as measured with the miniaturized test rig. Flat and parabolic reflector made of foil 2. The highest Pmpp obtained with a flat black surface is chosen as reference value.

Summarized it can be stated that applying shaped reflectors actually has an impact on the intensity distribution and may be beneficial in terms of energy harvest for certain conditions. However, there is also an inhomogeneous illumination distribution over the module area and there is apparently no effect for elevated positions, which is not according to the expectations of simulation programs.

Long-term outdoor measurements still have to be performed in order to reveal more general properties also for other lighting conditions.

REFERENCES

[1] H. Nussbaumer, G. Petrzilek, S. Schartinger, M. Klenk, N. Keller, T. Baumann, F. Carigiet and F. Baumgartner, "INFLUENCE OF LOW CONCENTRATION ON THE ENERGY HARVEST OF PV SYSTEMS USING BIFACIAL MODULES," 32nd European Photovoltaic Solar Energy Conference and Exhibition , pp. 2184-2190, 2016.

Towards new module and system concepts for linear shading response

Kostas Sinapis[1], Tom T.H. Rooijakkers[1], Lenneke H. Slooff[2], Lars A.G. Okel[2], Mark J. Jansen[2], Anna J. Carr[2]

[1]Solar Energy Application Centre, Eindhoven, HTC 21, 5656AE, Eindhoven The Netherlands
[2]Energy research Centre of the Netherlands, P.O. Box 1, 1755 ZG, Petten, The Netherlands

Abstract — **Large scale implementation of PV in urban areas asks for more shadow tolerant modules and more size flexibility. Power optimizers and micro inverters can be used to improve shadow tolerance of standard c-Si modules but the electrical interconnection and the size of cell groups in the standard module design is a limiting factor when it comes to further increase shadow tolerance. In this work another approach is used in which the cells in the module are smaller and series connected in building blocks. The resulting lower currents allow the use of small in laminate diodes and the building blocks give access to more size flexibility and improved shadow robustness. Yearly yield calculations will be shown and compared to a standard 3-sub-string module with either power optimizers or micro inverters. It is shown that for a typical Dutch house the TESSERA concept outperforms the other by up to 6%.**

I. INTRODUCTION

Solar photovoltaic (PV) systems today play a significant role in the energy mix and their penetration is forecasted to keep growing in the future [1]. Particularly the application of building integrated PV (BIPV) and building added PV (BAPV) systems are projected to thrive in the following years as a result of increasing electricity prices for the residential sector and decreasing PV component costs. Residential and small commercial PV systems are typically installed in an urban environment. Roofs and terraces are often affected by shade coming from the close proximity of buildings, poles, antennas, dormers, etc. introducing electrical and thermal mismatch losses between cells and modules. Solar modules are connected in series and thus sharing the same current in a string. This topology is prone to power losses if the solar cells in the module are not operating under the same conditions thereby reducing the current of the module and consequently of the whole string. Partially shaded solar cells may become reverse biased because of the series connection and thus act as a load consuming the power that is generated by the unshaded cells. Two negative effects occur from partially shaded operation of a PV system: power loss and increased temperature of the shaded cells (hot-spot). By-pass diodes have been applied in solar modules to prevent power consumption from shaded cells and to prevent hot-spots by by-passing the shaded substrings of the solar module. Most of the solar modules include one by-pass diode connected anti parallel per 16–24 cells [2]. However, it is known that increased granularity of cell groups can increase performance under partial shading conditions [3].

The TESSERA concept makes use of this knowledge and offers increased granularity. It includes 6 inch Metal Wrap

Through (MWT) c-Si cells which have been cut to produce 16 mini cells. The mini cells are connected in series and protected with an in-laminate low current by-pass diode (Fig. 1a). In total 64 mini cells are connected in series forming an maximum power point (MPP) voltage of around 31V at standard test conditions (STC). Blocks of 64 mini cells are then connected in parallel forming modules with custom size and thus custom current levels. A full size TESSERA module has been built based on ECN's back contact technology and is shown in Fig. 1b. The module consists of a white Tedlar backsheet with Cu foil on top. The cells make contact to the patterned Cu foil via an electrically conductive adhesive. Outside the dots, the cells are isolated from the foil by white

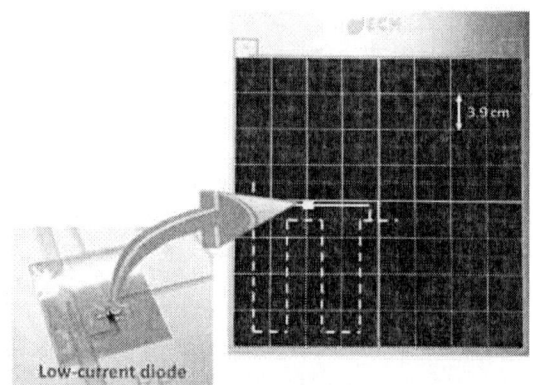

Figure 1a: A building block of the TESSERA concept consisting of 64 mini cells connected in series and protected by 4 low current in laminate diodes; 1b: a full size TESSERA module.

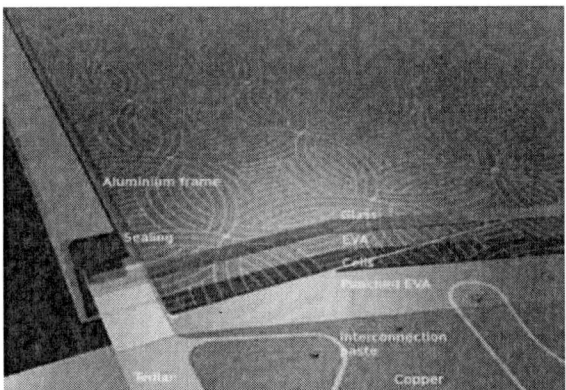

Figure 2: Presentation of the build-up of a MWT back contact module using an interconnecting backsheet.

Figure 3. System layout for the yearly yield calculations. A typical Dutch roof including a dormer, chimney and exhaust poles,

EVA. Transparent EVA is used as the encapsulant between the cells and the front side glass. A schematic representation of the module stack is shown in Fig. 2.

The use of module level power electronic devices (MLPE) has been proposed to mitigate electrical and thermal mismatch losses [4-6] in the field by tracking the maximum power point of individual modules. In general MLPE devices consist of two main categories: micro inverters and power optimizers. In this paper micro inverters (MI) and power optimizers (PO) (boost DC-DC converter) are considered for the TESSERA system.

II. METHODOLOGY

A simulation model is developed [7] to quantify the benefits and drawbacks of different PV system architectures. The simulation model takes into account different module designs (standard cells with 3, 6, 10 by-pass diodes, and TESSERA) and different system architectures (string inverter shadow mode on-off (SI+), micro inverters (MI), power optimizers

(PO)). The characteristics of the inverters and power optimizers used in the calculations are shown in Table I.

In Fig. 3 the 3D design of the roof can be seen including the shading obstacles of the roof (dormer, chimney, exhaust poles). By providing input on the orientation and the inclination of the PV panels as well as the PV module configuration, the determination of the exact position of all the cells in the system can be achieved. This is combined with information on the geographical characteristics and object shading from the location. This determines the shading on all the cells, see Fig 4a, and placed in a lookup table as shading fraction per cell for a specific date and time. The resulting blocked irradiance on an annual level is around 1.7%, see Fig 3b. Next the PV module characteristics and the irradiance (direct and diffuse ratio is taken into account using the Reindl-2-model [8]) and temperature information is added to determine the IV curve of the substrings. The IV curve is generated using a simplified double diode model. To reduce the calculation time it was assumed that the saturation current density of the first and second diode are equal as described in [9]. The resulting diode equation is:

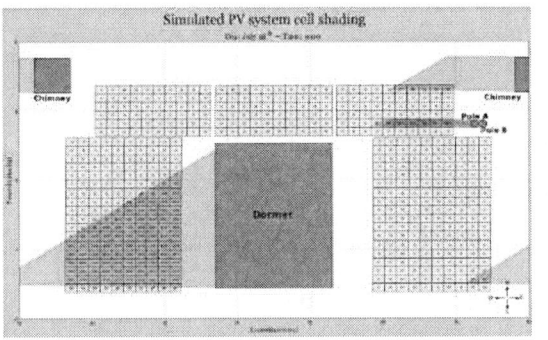

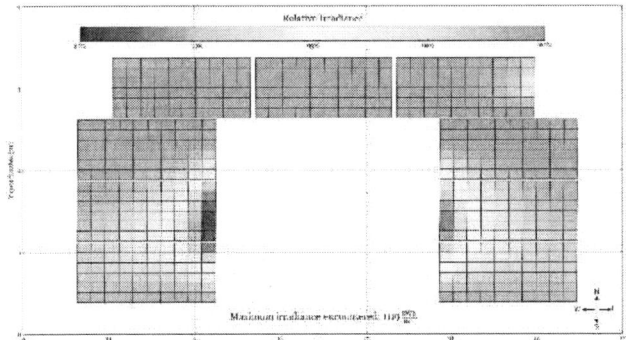

Figure 4. a. Placement of modules on a typical Dutch roof including a dormer, chimney and exhaust poles, b. Effective irradiance reaching the modules in a typical meteorological year for NL.

$$I = I_{ph} - I_o \left(e^{\frac{V_S + IR_S}{V_{th}}} + e^{\frac{V_S + IR_S}{2V_{th}}} - 2 \right) - \left(\frac{V_S + IR_s}{R_{sh}} \right)$$

Subsequently the DC output is determined and after including the conversion losses of the specific AC converter, the AC output is derived. Further settings for the model are given in Table I.

Table I: parameters used in the model.

Parameter	Value
Simulation interval	10 minutes
TMY data interval	60 minutes
TMY data resampling method	Simple interpolation
IAM model	(Snell's and Bougher's laws
Effective cell temperature	Faiman model
Albedo factor	0.15
Irradiance separation model	Perez model (in Meteonorm)
Tilt (in-plane) radiation model	Perez model (in Meteonorm)
Irradiance data period	1991-2010
Ambient temperature data period	2000-2009
POA azimuth tilt angle	180°; 40° (0°; 40° in Meteonorm)
Latitude, longitude, elevation	51.4° N; 5.5° E; 30 m
Other meteorological output	Local wind speed (for temperature)

TMY: Typical Meteorological Year, IAM: Incident Angle Modification, POA: plane of array.

The energy yield of three leading architectures is confirmed (string inverter (SI), power optimizer, micro inverter) for clear and partial shading conditions by means of an outdoor field test [8] and IV curves from a solar simulator [9]. For this work the model has been tuned to accommodate different module and PV system designs in an effort to compare the partial shading response of the TESSERA concept with standard c-Si design of three substrings of 20 cells.

III. RESULTS-DISCUSSION

By using the typical meteorological year's irradiation data of Meteonorm [10], a full year simulation for unshaded and partially shaded scenarios has been performed. Meteonorm provides measured irradiance data for a variety of locations. Moreover, the data can be decomposed and trans-positioned by using known irradiance models. A constant albedo factor of 0.15 has been used for the simulations. The simulation model takes into account different module designs (standard cells with 3, 6, 10 and 60 by-pass diodes, and TESSERA) and different system architectures (string inverter shadow mode on-off, micro inverters, power optimizers). The characteristics

of the inverters and power optimizers used in the calculations are shown in Table II.

	MI	PO	SI
Power electronic devices	Heliox SMI-300	Femtogrid PV300+Kratos Inverter 2400	SMA Sunny Boy SB 2500 TLST-21
Voltage range	16V- 48V	8V – 42V	180V – 500V
European efficiency	93.6%	93.1% (97%*96%)	95.6%
MPP granularity	Module	Module	System (9 modules in series)
Input power/current constraints	> 5.0 W	> 0.03 A; > 6.7 A	> 32 W
Output power/current constraints	< 285.0 W	< 10.5A	< 10.9A
Efficiency determination methodology	Linear-interpolation of V-dependent efficiency curves (MLPM data)	Linear-interpolation of V-dependent efficiency curves (MLPM data)	Double linear interpolation of P, V-dependent efficiency curves (PV Syst)
System elements	9*(4S * 15P)	9*(4S * 15P)	36S * 15P

Table II: characteristics of the inverters and power optimizers used in the calculations.

In Fig. 5 the simulation results can be seen in specific yield kWh/kWp per annum. The total length of the bars indicate the total kWh/kWp that can be generated per year if no shading is present. Under these circumstances, the string inverter system (Fig. 5a). seems to be the best performing system. This is due to the higher efficiency of the string inverter compared to the power optimizer and micro inverter. However, when partial shading is introduced, then the performance drops due to the shading losses. This drop is indicated in Fig. 5 by the black bars. As can be seen, in this case the performance of the power optimizer and the micro inverter are substantially better. The shading losses reduce with about 50% compared to a 3BPD system. The higher granularity of cell groups seems to benefit the string inverter more than the power optimizer or micro inverter due to the wider operational voltage. For the TESSERA concept we can observe an improvement of around 6% when used in combination with a micro inverter or power optimizer in comparison with a typical PV system consisting out of modules connected in series and attached to a string inverter. The performance behavior of the TESSERA concept is similar with that of a standard 60 by pass diode module.

978-1-5090-5606-4/17 $31.00 © 2017 IEEE

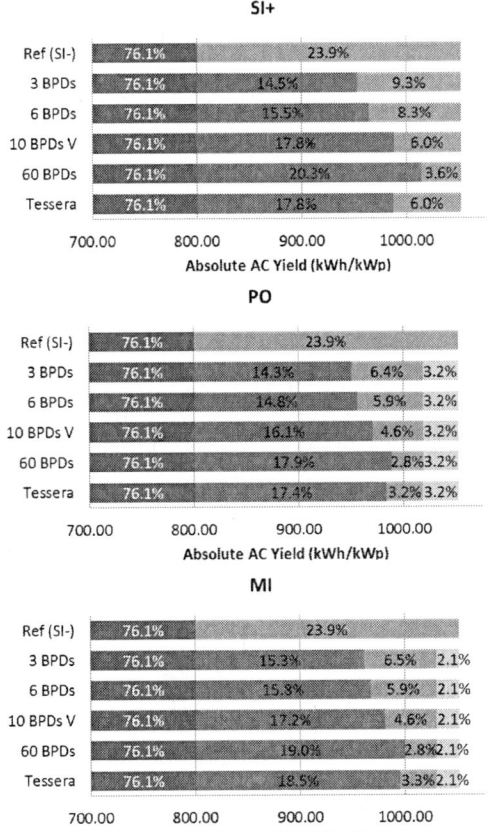

Figure 5: Simulation results for various module and system designs.
The total bar length gives the AC yield of a reference 3 string module
with string inverter without shadow in the module

indicates the annual AC yield of a 60 cell module with
string inverter (without shadow option) with shadow on the
module

is the gain in annual AC yield with respect to the reference
due to improved shadow tolerance

loss due to shadow

conversion loss in the AC converter

Furthermore the TESSERA module manages to keep a steady
MPP voltage throughout the year and independent of the
partial shading conditions. Figure 6 shows the number of
occasions during a year in which a specific Vmpp/Voc is
obtained for the TESSERA module and for a 3 by-pass diode
module (3BPD) for a system with MI and for a system with
PO. The module with 3 bypass-diodes shows 1 or two
additional Voc/Vmpp ranges at which the system is operating
during the year. Although they are relatively low compared to
the main operating condition around 0.8, it still amounts to
2.3% of the year for the MI system and 2.8% for a PO system.

For TESSERA only 0.5% of the time the system operates at a
different Vmpp.

IV. CONCLUSIONS

This works describes a shadow tolerant module concept that
is based on standard back contact foil technology. It consists
of small cells connected in series to form a modular block.
These blocks are connected in parallel to form the module.
This dedicated interconnection makes the module much more
shadow tolerant. Annual yield modeling of the power output
for a typical Dutch house is compared with the output of
standard 3 sub-string modules with string inverters, power
optimizers or micro-inverters. It is shown that the presented
concept outperforms the others by 3-6%.

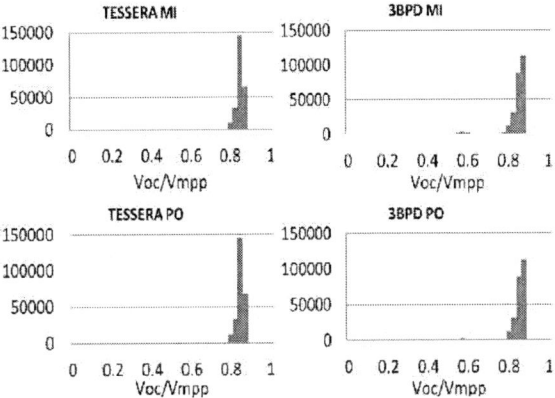

Figure 6. Vmpp/Voc for a 3BPD module with micro-inverter or
power optimizer and a TESSERA module with micro-inverter or
power optimizer

ACKNOWLEDGEMENTS

This work is partly financed through the SSTM project with
financing from the Topsector Energy subsidies from the Dutch
Ministry of Economic Affairs (file nr TEID2150302)

REFERENCES

[1] IEA PVPS ANNUAL REPORT 2015
[2] S.Silvestre, A.Boronat, A.Chouder, "Study of by-pass diodes
     configurations on PV modules," *Appl. Energy*, vol. 86 (9), pp.
     1632–1640, 2009.
[3] K. Sinapis, T.T.H. Rooijakkers, C. Tzikas, G.B.M.A. Litjens,
     M.N. van den Donker, W. Folkerts, W.G.J.H.M. van Sark,
     "Annual Yield Comparison of Module Level Power Electronics
     and String Level PV Systems with Standard and Advanced
     Module Design, " *EUPVSEC proc*, pp. 2011-2015, 2016.
[4] S.Poshtkouhi, V.Palaniappan, M.Fard, O.Trescases, "A general
     approach for quantifying the benefit of distributed power
     electronics for fine grained MPPT in photovoltaic applications

using 3-D modelling", *IEEE Trans. Power Electron.*, vol. 27 (11), pp. 4656–4666, 2012.

[5] A.Woyte, J.Nijs, R.Belmans, "Partial shadowing of photovoltaic arrays with different system configurations: literature review and field test results," *Sol. Energy*, vol. 74, pp. 217–233, 2003.

[6] A.J. Hanson, C. Deline, S.M. MacAlpine, J.T. Staruth, C.R. Sullivan, "Partial shading assessment of photovoltaic installations via module-level monitoring," *IEEE J. Photovolt.*, vol. 4 (6), pp 1618-1625, 2014.

[7] K. Sinapis, C. Tzikas, G. Litjens, M. van den Donker, W. Folkerts, W.G.J.H.M. van Sark, A. Smets, "A comprehensive study on partial shading response of c-Si modules and yield modeling of string inverter and module level power electronics," *Solar Energy*, vol.135, pp. 731–741, 2016.

[8] D. T. Reindl, W. A. Beckman, and J. A. Duffie, *Solar Energy*, "Diffuse fraction correlations", vol 45 (1), pp 1-7, 1990.

[9] K. Ishaque, A. Salam, Syafaruddin, "PV system simulator with partial shading capability based on two-diode model", *Solar Energy*, vol. 85 (9), pp. 2217-2227, 2011

[8] K. Sinapis, G. Litjens, M. van den Donker, W. Folkerts, W.G.J.H.M. van Sark, "Outdoor characterization and comparison of string and module level power electronics under clear and partially shaded conditions," *Energy Sci. Eng.*, vol. 3, pp. 510–519, 2015 http://dx.doi.org/10.1002/ese3.97.

[9] A.J. Carr, K. de Groot, M.J. Jansen, E.E. Bende, J.A.M. van Roosmalen, L.Okels, W.Eerenstein, "Tessera: Scalable, Shade Robust Module" *42th IEEE Photovoltaic Specialists Conference*, 2015.

[10] http://www.meteonorm.com

Partial Shading Abatement Through Cascaded H-Bridge Topology

Steven Tidwell, Joseph Latham, and Dr. Michael McIntyre

University of Louisville, Louisville, KY, 40292, USA

Abstract — **Photovoltaic systems are susceptible to partial shading, which has a disproportional impact on their power production. This paper will examine a system architecture introduced by an earlier paper, and will verify it through a series of simulations. The novel architecture offers a solution to partial shading by utilizing inverters on a cell level. The cascaded H-bridge inverter design provides the maximum power point tracking and power conversion at the cell level. The presented system will be simulated at a basic level, and each simulation after will have an additional portion of the design.**

I. INTRODUCTION

Modern consumer photovoltaic (PV) architectures are susceptible to power loss caused by partial shading. Whether partial shading comes in the form of debris covering a portion of a solar panel, hail damaging a couple individual cells, or tree shading, it can have a significant impact on the power produced by a solar panel. Typically, a solar system will use a string inverter for PV array, or a micro inverter per PV module, to convert the DC voltage from the panel into AC voltage to connect to the grid. The micro inverter method currently provides the best power efficiency on the market [1]. A PV module may have multi-cell strings, utilizing bypass diodes as shown in Fig. 1 below.

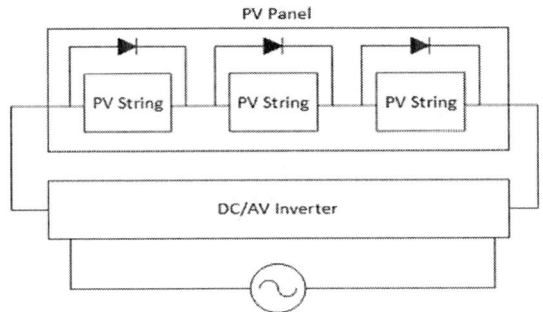

Fig. 1.   Diagram of a solar panel that uses PV strings.

When a cell is shaded, there is a proportional loss in the cell current. An example of this can be seen in Fig. 2. As the incidence of the solar cell is lowered by 20%, the current also drops 20%. If one or two cells in a string are shaded, the rest of the string is only able to operate at the maximum power point (MPP) of the lowest cell. If the current loss is great enough, the bypass diode allows the system to ignore that shaded string, ensuring the greatest amount of power is produced from the remaining non-shaded cells. The micro

inverter then manages the MPP on the panel level. The effects of partial shading are non-linear, as it depends on multiple factors including the amount of shading and where the shading is taking place in relation to the bypass diodes [2]. This makes predicting power loss difficult, though a power loss of as much as a half of the overall output can be expected for a string inverter system [3].

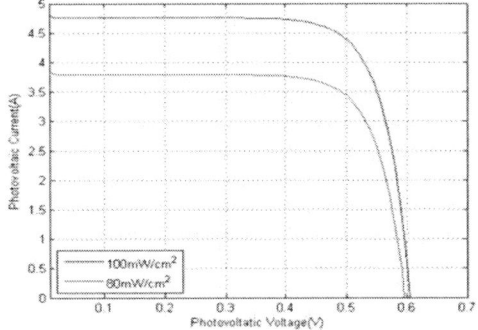

Fig. 2.   An IV curve of a solar cell with different levels of incidence [4].

Current research seeks to improve the efficiency of a solar array via developing a solar panel that directly interfaces with the AC grid without external DC wiring [5]. There was also an examination of a nano level adjustable power supply (nano LAPS) [6]. The use of nano LAPS in a PV system sought to solve the partial shading problem on a cell level by enhancing a cell to optimize the overall system performance, or taking the cell off line if there is a cell failure. In other words, it uses a controllable boost converter with a PV controller to track voltage and current. The controller is responsible for tracking and maintaining the MPP on a cell level, while the converter maintains an AC voltage peak high enough for grid synchronization.

As an alternative solution to the partial shading problem, a cascaded inverter architecture on a cell level was investigated [4]. Inverters would be used on a cell level, similar to the nano LAPS system, producing a cascaded H-bridge architecture. This would allow each cell to operate at its own MPP, and allow the solar panel to produce more energy than it would using a string architecture.

This paper seeks to further examine this cascaded H-bridge design, and provide a simulated demonstration to prove its viability. Section II presents the overall system and experimental design. Section III examines the simulation results.

II. PROJECT OVERVIEW

To prove that the solution is viable, a less complex design will be simulated as the theory may be validated regardless of the scale. The simulation will be completed through the PLECS simulation program. To allow for future real-world experimental replication, the simulation will include three DC current sources to represent the three strings of five cells in series. The five cells allow the effects of partial shading to have a much greater impact on the system, thus making it easier to analyze the results of the experiments. This also helps to analyze partial shading effects, since it is difficult to simulate its effects on a single cell. Each string will have its own inverter.

The design will utilize a proportional integral controller (PI), and a perturb and observe (P&O) controller to determine the gate driver inputs of the inverters. A phase-shifted carrier pulse width modulation (PSCPWM) switching scheme will also be used. PSCPWM allows for sub-level bus voltage control [7] [8]. Finally, the design will use a RL load. A basic system diagram of the real-world experimental design is shown in Fig. 3 below.

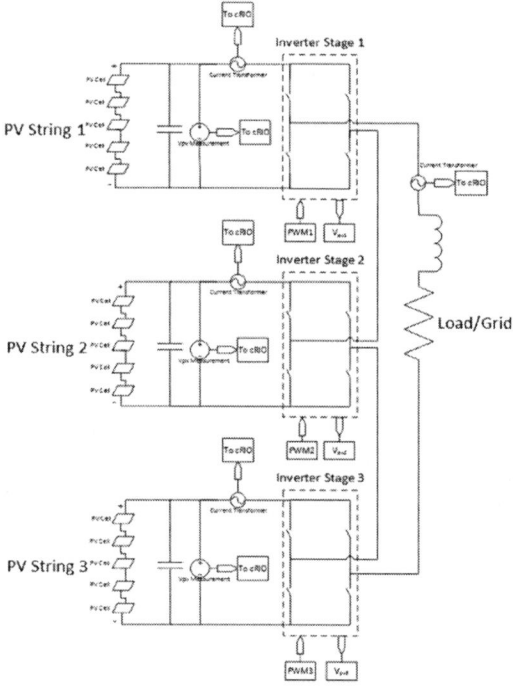

Fig. 3.   System diagram of the proposed design for a real-world test.

A. Control Systems

The control components used for this design are the PI controller, and the P&O algorithm. Fig. 4 below is a block diagram of the system describing the relation between the control systems and the PV system.

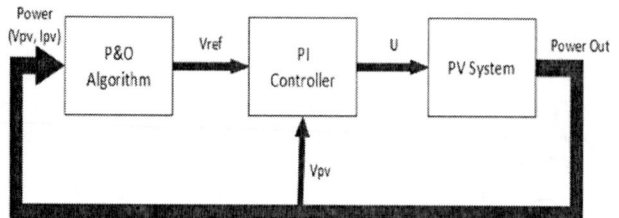

Fig. 4.   Control System Block Diagram

The PI controller determines the gate driver input to either increase or decrease the amplitude (U in the diagram) of the PV bus voltage to match the reference value, given by the P&O. The PI is the fastest of the control loop, as it is responsible for maintaining system control. The P&O increases or decreases the voltage reference to ensure the system always runs at its MPP. This is the slower loop, as it requires the PI to achieve the previous reference value, thus calculating an up-to-date MPP.

III. SIMULATION

Two simulations were conducted to estimate the effectiveness of the proposed design. The first was a baseline test to, and the second was the full cascaded H-bridge system.

A. Simulation 1 - Baseline

Using the PLECS simulation program, a baseline was established using an AC voltage source of 2[V], 60[Hz], to represent a string of five solar cells. The simple circuit included an RL load of 0.25[Ω], 300 [μH], and used probes for voltage, current, and power measurements. The circuit is shown in Fig. 5.

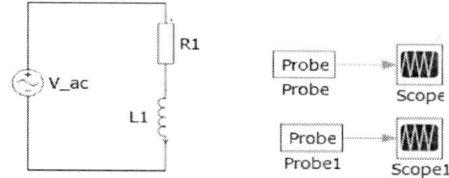

Fig. 5.   Simulation 1 PLECS circuit.

To help verify the simulation results, calculations were made to find the expected results. The load impedance value was calculated using the equation

$$Z = R + j\omega L \qquad (1)$$

R is 0.25[Ω], and L is 300[μH]. Z was found to be 0.274 ∠ 24.34[Ω]. With the impedance found, the expected real current was then calculated to be 7.299[A]. The real power, P, was found to be 13.32[W]. The apparent power, S, was calculated to be 14.498[VA], meaning the power factor of the circuit was 0.912. Therefore, the reactive power, Q, was 6.025[VAR].

The simulation measurements came very close to the calculated values. The output current Iout_max was measured to be 7.655[A], and Vout was 2[V].

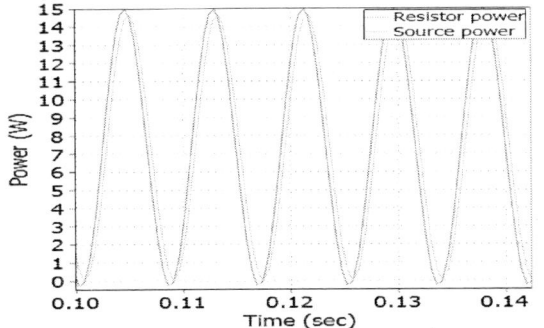

Fig. 6.   Simulation 1 input and output power measurements.

The max input power was 14.99[W], and the real Pout was 14.65[W], which was close to the 13.32[W] expected. The average power was measured to be 7.34[W]. This is shown in Fig. 6 above.

B. Simulation 2 – Cascaded H-Bridge Design

With the basic simulation completed, a final simulation was done for the full multilevel cascaded h-bridge system. This simulation utilized the PSCPWM switching scheme discussed earlier, and a RL load of 1[Ω] and 400[μH].

The current source of the second level was controlled much like the source for the P&O simulation, with an $I_{pv}(t)$ of 3.5[A] from 0 – 2 seconds, and 2.625[A] from 2 – 6 seconds. From 0 – 2 seconds, the system performs as expected, providing a $V_{out}(t)$ of about 6.4[V], and an $I_{out}(t)$ of 6.3[A]. However, from 2 – 3 seconds, V_{pv} of the second level drops drastically to 0.5[V], leaving the remaining two levels to carry a greater load. The system corrects itself around 3 sec, and becomes steady once again. Fig. 7 shows the output waveforms of the system.

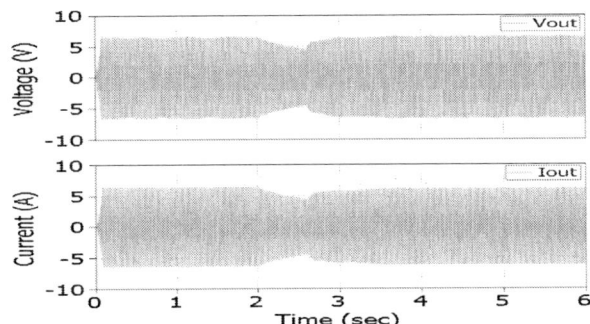

Fig. 7. $V_{out}(t)$ and $I_{out}(t)$ of simulation 2.

In the first 2 seconds, each solar string is providing about 7.5[W]. During the time in which the second level is trying to return to normal, the power it provides drops, while the remaining levels provide about 6[W] each, returning to their previous values around 2.8 sec. This is shown in Fig. 8 below, where the above graph is the V_{ref} provided by the P&O of the altered string, and the power is in the graph below.

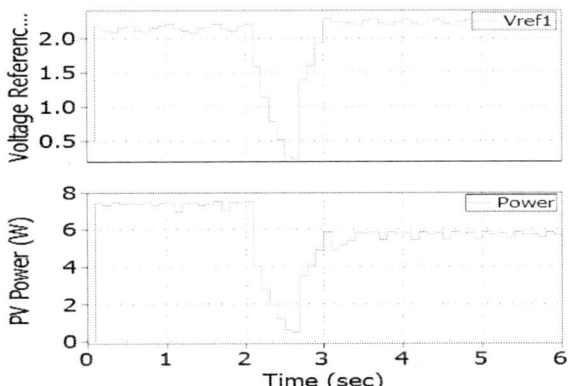

Fig. 8.   Voltage reference and power of the second level.

When the system fully corrects itself around 3.4 sec, the second level provides around 5.7[W], while the other two return to 7.5[W] each. From 0 – 2 seconds, the average real power at the output was about 21.709[W]. After the system becomes stable again around 3.5 seconds to 6 seconds, the average real power at the output was 20.202[W]. Fig. 9 below shows the power at the inputs of each string, and the output.

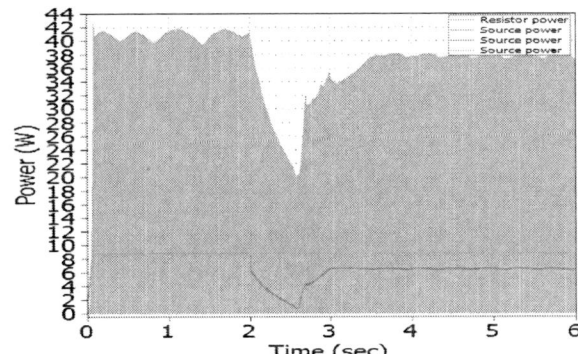

Fig. 9.   Simulation 2 input and output power.

IV. CONCLUSION

Examples of modern PV systems and their limitations regarding partial shading was examined, along with current research into solutions. The plausibility and effectiveness of a cascaded H-bridge inverter topology on a cell level to solve this issue was examined and tested via simulation. These results showed that the system does indeed nearly eliminate the power loss caused by partial shading. The simulation provides a good indicator as to how a physical experimental system may react.

REFERENCES

[1] A. Mukherjee, M. Pahlevaninezhad and G. Moschopoulos, "Flyback Microinverters in Solar Energy Systems," in *Telecommunications Energy Conference 'Smart Power and Efficiency*, Hamburg, 2013.

[2] W. Xiao, O. Nathan and W. G. Dunford, "Topology Study of Photovoltaic Interface for Maximum Power Point Tracking," IEEE TRANSACTIONS ON INDUSTRIAL ELECTRONICS, vol. 54, no. 3, pp. 1696-1704, 2007.

[3] C. Deline, "Partially Shaded Operation of a Grid-Tied PV System," in Photovoltaic Specialist Conference, Philadelphia, 2009.

[4] M. A. Elliot and D. M. McIntyre, "Partial Shading Abatement through Cell Level Inverter System Topology," in IEEE, 2014.

[5] S. Kjaer, P. J. and F. Blaabjerg, "A review of single-phase grid-connected inverters for photovoltaic modules," *IEEE transactions on Industry Applications,* vol. 41, no. 5, pp. 1292-1306, 2005.

[6] E. Watson and D. M. McIntyre, "PV System Architecture Improvement using nano-LAPS Boost Converter to Eliminate Cell Failure Downtime," in IEEE, 2015.

[7] Y. Liang, "A NEW TYPE OF STATCOM BASED ON CASCADING VOLTAGE SOURCE INVERTERS WITH PHASE-SHIFTED UNIPOLAR SPWM," in Industry Applications Conference, ST. Louis, 1998.

[8] B. McGrath, V. Clayton and D. Holmes, "Multicarrier PWM strategies for multilevel inverters," Industrial Electronics, IEEE Transactions on, vol. 49, no. 4, pp. 858-867, 2002

Data analysis for effective monitoring of partially shaded photovoltaic systems.

Odysseas Tsafarakis[1], Kostas Sinapis[2], Wilfried G.J.H.M. van Sark[1]

[1]Utrecht University, Copernicus institute of Sustainable Development, Heidelberglaan 2, 3584 CS, Utrecht, The Netherlands

[2]Solar Energy Application Centre, HTC 21, 5656 AE, Eindhoven, The Netherlands

Abstract — The purpose of this study is the development of an algorithmic tool to automate the process of analyzing monitoring data of partially shaded PV systems. The approach is to compare long-term and high-resolution yield data of a partially shaded Photovoltaic (PV) system (investigated PV) with the yield data of an unshaded PV system or the tilted irradiance (reference PV system) and automatically detect the energy loss due to the expected shadow, caused by any surrounding obstacles and distinguish it from any additional energy loss due to other malfunctions.

I. INTRODUCTION

The majority of the installed PV systems in the Netherlands are relatively small-scale installations, with an average capacity of 3.5 kWp [1]. About 70% of them is placed on residential rooftops [2], where different objects might obstruct the irradiance reaching the PV modules and affects their energy output. Such small-scale, low-cost PV installations are mostly monitored through relatively simple, low-cost, data loggers of AC power, without any further equipment for solar measurements, i.e., pyranometers.

The ultimate aim of our research is to find an effective manner that allows monitoring large numbers of such installations in an automated manner. The proposed method offers the possibility to monitor partly shaded systems by excluding the effects of constant and unavoidable shadow from the performance evaluation and to detect faults that might occur during their operation.

II. SCIENTIFIC INNOVATION

The innovation of the proposed method is its independence on the use of irradiance data. The performance evaluation is not based on the estimation of the energy generation/production of the studied PV systems and simulation of the losses due to shading (with the use of irradiance data and models) and its comparison with the real production [3]–[5] or the comparison of the production with the tilted irradiance, through the calculation of the performance ratio [5], [6]. It is based on the performance comparison and the detection of repeated or non-repeated errors/malfunctions between neighboring PV systems with similar characteristics (tilt and orientation).

III. METHODOLOGY

The basic principle of the proposed method/algorithm, is to study the daily persistence of any occurred error between the investigated PV and the reference PV system and identify these daily repeated errors, characterize them as shadows and create the **"shadow profile"** of the investigated PV, *i.e. a table with the starting and ending dates of the shadow(s) during the year and the expected starting and ending times of the shadow(s) during each day.* Any difference between the yield of the investigated PV and the reference PV system, outside of the limits of these shadow profile time intervals is characterized as a malfunction and within these limits as shadow. An example of the shadow profile is presented in Table 1. In this case, 5 different shadow periods are forming the shadow profile of the studied PV system.

TABLE 1 – SHADOW PROFILE EXAMPLE, WHERE EVERY ROW IS A DIFFERENT SHADOW

N.	Start day	End day	Start time (HH:MM)	End time (HH:MM)	Cause
1	Aug 8	Oct 13	9:05	12:05	Pole
2	Oct 18	Nov 11	9:15	13:55	Pole
3	Jun 1	Jun 29	18:40	19:40	Roof
4	Jul 16	Aug 22	18:50	19:40	Roof
5	Aug 28	Sep 20	19:25	19:40	Roof

During the analysis, the fluctuation of the Mean Absolute Error (MAE) between the production of the systems is studied ($|P_{reference} - P_{shaded}|$). Every timestamp with MAE higher than considered as normal is characterized as error and it is further investigated, in order to be characterized as shadow or malfunction. In order to set a threshold between the normal and the non-normal operation, the iterative method Ran.Sa.C. (Random Sample Consensus) [7] is used.

Due to the nature of earth's orbit around the sun, the time interval of each shadow effect is varying from day to day. Additionally, due to the nature of the diffuse irradiance, the shadow is not present on cloudy days, where 5 days in row without any effect of shadow could easily appear in a country such as the Netherlands. These facts are making the shading detection more difficult than a simple pattern recognition problem since different groups of hours and dates are forming the shadow profile. The clustering method D.B.S.C.A.N. [8] (Density-based Spatial Clustering of Applications with Noise) is used in order to identify shadows during different periods.

Thereafter the continuity of all the shaded periods of the "shadow profile" table is studied and presented in a "Date vs Time" plot, which demonstrates the change of the shadow effect during a long time studied period. We denote this as the **"Shadow story"**. An example of the shadow story is presented in Figure 1.

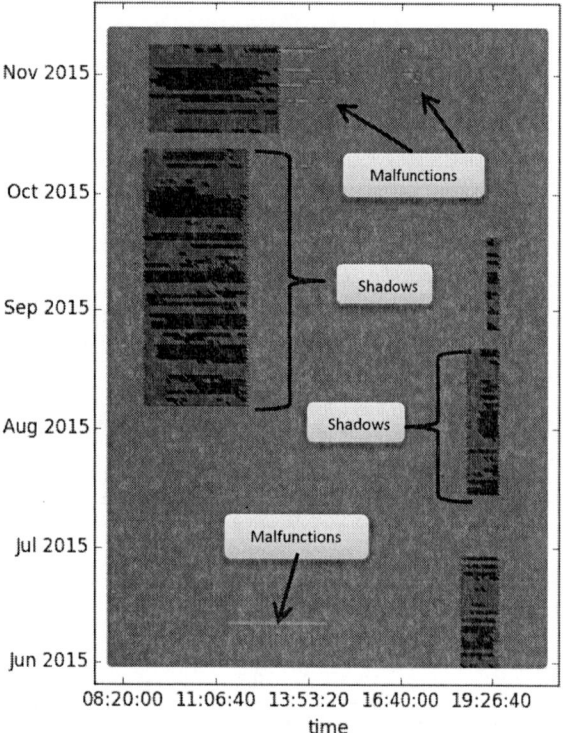

Fig. 1. Example of a "shadow story".

IV. RESULTS

As mentioned above, an example of a partially shaded PV system is compared with a non-shaded PV system and their **"shadow profile"**, i.e. the shadow periods of the system during the year (Table 1) and its **Shadow Story** (Figure 1), i.e. the day vs time plot, are presented. The green color in Figure 1 indicates measurements (date, time) without a presence of error between the studied (partially shaded) and the reference (non-shaded) PV system. A shading obstacle was placed from August 10 to November 11 and impacts the

PV system's performance during the morning hours. Since this error is repeated for long period, it is identified by the algorithm as shadow (black – higher than threshold difference – and blue – lower than threshold difference – marks) and not as malfunction. The shadow from a neighboring rooftop also affects the system in the late afternoon, which is also recognized as shadow.

Major remarks on Figure 1 are:

- At June 12 for external reasons, an obstacle was placed in front of the system, from 11:35 to 14:30. It is clearly within the hours were the pole is affecting the system (when it is placed). However, the algorithm recognized this incident as malfunction, since it is the only performance reduction, which occurred in that period and it is not in the Table 1. A similar incident (roof maintenance) occurred for some days between October 26 and November 6. These dates are in shadow profile (shadow 2 in Table 1), however a larger part of the error was outside of the time intervals of the shadow profile, thus they are marked as red.

- The shadow is shifting from August to November and impacts the system in later hours, due to the fact that the day is becoming shorter. In this case the role of the date clustering through D.B.S.C.A.N. is important, since the shadow period is separated in two smaller periods (shadows 1 & 2, Table 1) and each one has different start and end time.

V. FURTHER DISCUSSION OF THE RESULTS

As depicted in Figure 1, within the time intervals of the shadow profile (blue points) there is a large number of measurements without error. It can be assumed that the reason of this fact is the high defuse irradiance, which, in contrast with the direct, is not obstructed by the shade causing object. This assumption is proved in this paragraph, where the blue points of Figure 1 are further studied.

In Figure 2, Figure 1 is replotted with one major difference. The blue points of figure 1 are further studied and colored depending on the solar irradiance and specifically on the ratio of the diffuse horizontal irradiance (DHI) to the global horizontal irradiance (GHI) or R_{DHI}.

As depicted in Figure 2, in the majority of the points, R_{DHI} is higher than 0.5 (light blue) i.e. the DHI is higher than the direct normal irradiance (DNI). Further analysis of the blue points is presented in the scatterplot of Figure 3. In this scatterplot the three red lines are the 3 quantiles and the green line is the percentile where R_{DHI} is 0.5. As it is clear in Figure 3, 87% of the sample is above 0.5 thus the DHI is higher than the DNI, the irradiance which causes the shadow.

However it is very interesting that the upper right shadow (September, 18:00 to 19:00) of the plots has still only dark blue dots in Figure 2. This fact is explained in Figure 4, where the blue points are studied again and recolored. In this figure the blue dots are colored depending on the global tilted irradiance (GTI) and they are re-colored yellow if GTI < 100 W/m^2. Under this filter, all the blue marks of the specific profile are colored yellow, thus the irradiance is very low, in fact too low to create any shadow effects.

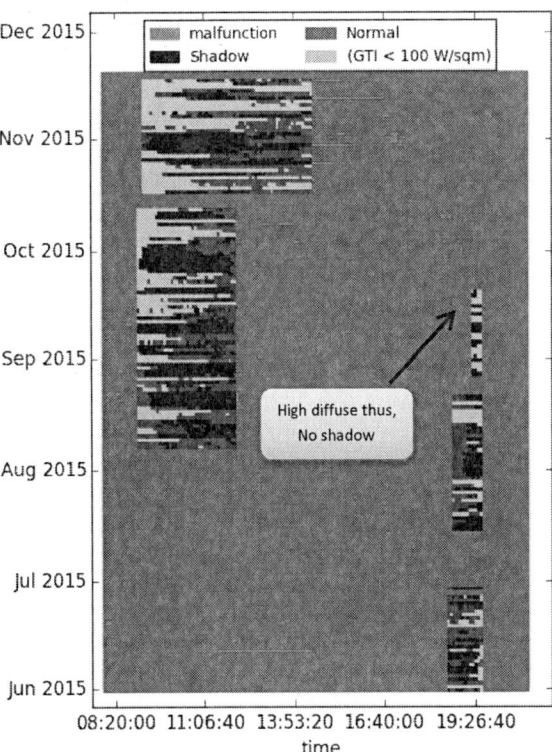

Fig. 4. Shadow story with unshaded measurements inside shadow profile colored depending on the GTI

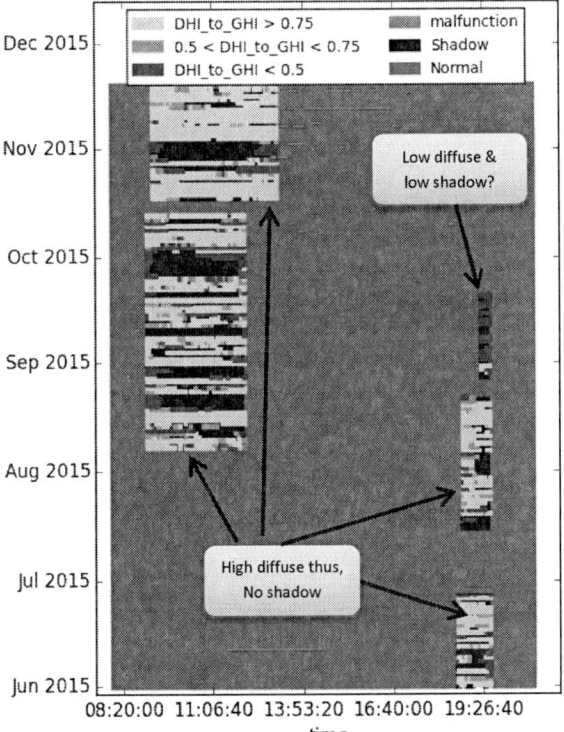

Fig. 2. Shadow story with measurements without fault inside shadow profiled colored depending on the ratio DHI/GHI

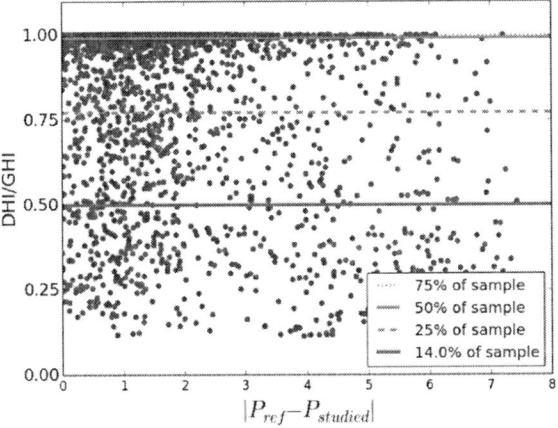

Fig. 3. Further analysis of the non-fault measurements of shadow profiles

VI. SUMMARY OF THE WORK

The proposed method is offering a monitoring solution to small residential PV systems, without the use of high-cost monitoring equipment, which is completely cost ineffective for this kind of installations. It only uses the P_{AC} of two neighboring PV systems, with similar tilt and orientation and not necessarily similar capacity (since the production can be normalized by divided with the rated power (Wp) of the system). The developed algorithm automatically isolates the timestamps with the non-normal operation and studies their concentration during the hours of the day and their continuity during the studied period and provides the owner of the system with the shadow story, a plot which demonstrates the effect of the shadow on his PV system during the year.

REFERENCES

[1] O. Tsafarakis, P. Moraitis, B.B. Kausika, H. van der Velde, S. 't Hart, A. de Vries, P. de Rijk, M.M. de Jong, H.-P. van Leeuwen, W.G.J.H.M. van Sark, "Three years experience in a Dutch public awareness campaign on photovoltaic system performance," *IET-Renewable Power Generation*, vol. 11, 2017 (in press) (doi:10.1049/iet-rpg.2016.1037).

[2] Centraal Bureau voor de Statistiek (CBS), "Bijgeplaatst vermogen zonnestroom bijgesteld." [Online]. Available:

https://www.cbs.nl/nl-nl/achtergrond/2015/51/bijgeplaatst-vermogen-zonnestroom-bijgesteld.

[3] A. Drews, A. C. de Keizer, H. G. Beyer, E. Lorenz, J. Betcke, W. G. J. H. M. van Sark, W. Heydenreich, E. Wiemken, S. Stettler, P. Toggweiler, S. Bofinger, M. Schneider, G. Heilscher, and D. Heinemann, " Monitoring and remote failure detection of grid-connected PV systems based on satellite observations," *Sol. Energy*, vol. 81, no. 4, pp. 548–564, 2007.

[4] S. Silvestre, A. Chouder, and E. Karatepe, "Automatic fault detection in grid connected PV systems," vol. 94, pp. 119–127, 2013.

[5] A. Woyte, M. Richter, D. Moser, M. Green, S. Mau, and H. G. Beyer, "Analytical Monitoring of Grid-connected Photovoltaic Systems," IEA-PVPS, Report IEA-PVPS T13-03:2014, 2014.

[6] R. Eke and A. Senturk, "Monitoring the performance of single and triple junction amorphous silicon modules in two building integrated photovoltaic (BIPV) installations," *Appl. Energy*, vol. 109, pp. 154–162, 2013.

[7] M.A. Fischler and R. C. Bolles, "Random Sample Consensus: A Paradigm for Model Fitting with Applicatlons to Image Analysis and Automated Cartography," *Commun. ACM*, vol. 24, no. 6, pp. 381–395, 1981.

[8] M. Ester, H. P. Kriegel, J. Sander, and X. Xu, "A Density-Based Algorithm for Discovering Clusters in Large Spatial Databases with Noise," *Proc. 2nd Int. Conf. Knowl. Discov. Data Min.*, pp. 226–231, 1996.

Bifacial Photovoltaic Module Energy Yield Calculation and Analysis

Christopher E. Valdivia[1], Chu Tu Li[1], Annie Russell[1], Joan E. Haysom[1], Rui Li[2], David Lekx[2],
Mohsen M. Sepeher[2], Dan Henes[2], Karin Hinzer[1], Henry P. Schriemer[1]

[1] SUNLAB, Centre for Research in Photonics, University of Ottawa, Ottawa, ON, K1N 6N5, Canada
[2] Celestica Inc., 844 Don Mills Rd., Toronto, ON, M3C 1V7, Canada

Abstract — A new computationally-efficient algorithm has been developed for the evaluation of annual energy yields from bifacial photovoltaic panels. The model accounts for detailed anisotropic sky dome and albedo ray tracing with directional reflection, self-shading, and rack shading. The illumination profiles over both front and rear faces of bifacial and mono-facial panels provide realistic solar cell and panel performance calculations over various system configurations. Both landscape panel orientation and panel wiring with multiple parallel horizontal strings provide higher output powers, up to >3% for low ground clearances. Representative conditions for Ottawa, Ontario, Canada, resulted in up to 18% bifacial gain in annual energy yield, while instantaneous power production increased by 13-35% when sunny and 40-70% when cloudy.

Index Terms — atmosphere model, bifaciality, diffuse illumination, annual energy yield, flat panel, mono-facial, optical ray tracing, solar power.

I. INTRODUCTION

Photovoltaic (PV) modules have seen a rapid decrease in cost over the past decade. As a result, the balance of system costs (e.g. racking, cabling), land area, installation and other costs, now account for 63-77% of the total system costs [1], and future PV module cost reductions will have a decreasing impact on the total system cost. Therefore, a reduced levelized cost of electricity (LCOE) can best be achieved by reducing the system size using fewer modules with higher photo-conversion efficiency (which has become more elusive as silicon PV technology matures [2]) or *enhanced energy collection*.

Bifacial PV modules increase energy collection compared to conventional mono-facial panels by absorption of light incident on both their front and rear faces [3], and can be readily integrated into conventional mono-facial PV panel installation schemes. Bifacial panels have demonstrated an energy yield of up to 25% greater than equivalent standard mono-facial panels [4]. While bifacial panels are not new [5], standard testing methodologies are still under development [6].

Bifacial panels provide a next step to solar deployments with higher yields, lower environmental footprints, and higher rate of returns on investment. With expanding use of bifacial panels, accurately quantifying the bifacial gain in energy yield over standard mono-facial panels is crucial for analysis by systems designers choosing the types and configurations of systems to deploy over the hugely varying range of illumination and environmental conditions across the world. To address these needs, this paper presents a comprehensive simulation model [7] that accurately simulates the instantaneous and annual energy yields of both mono- and bi-facial solar panel performance under real-field conditions and deployment configurations.

II. MODEL DESCRIPTION

The model considers the illumination sources, irradiance collection (panel front and rear), and photovoltaic conversion. The tool (built in MATLAB) inputs environmental conditions to determine the output of a single mono- or bi-facial panel, and sums over hourly timestamps to produce the annual energy yield. The *bifacial gain* is defined as an increase in energy yield by a bifacial panel over an equivalent mono-facial panel.

Modeling the performance of a conventional mono-facial photovoltaic panel typically assumes global illumination composed of two components: (1) light in a straight line from the sun (direct), and (2) sunlight scattered by the atmosphere (diffuse). Precise modeling of a bifacial photovoltaic panel requires an additional component: (3) sunlight scattered by the ground or other objects (albedo). This requires coupling between all points in the sky with all points on the ground as well as consideration of ground shading by the photovoltaic panels (self-shading), racking, and other obstructions.

A. Light Sources

<u>Direct beam:</u> The model treats the sun as a point source, and considers shading of both the front and rear of the panel.

<u>Diffuse sky:</u> The diffuse sky dome irradiance is computed using the Perez 1993 sky model [8], providing a continuously-varying anisotropic sky luminance profile (without the direct beam component). Due to unpredictable and continuously changing cloud distributions, this model simplifies the sky as a uniformly cloud-covered dome of varying opacity to represent the sky clearness. Before sunrise and after sunset, a simplified isotropic sky distribution is employed since the Perez model is not defined when the sun is below the horizon.

<u>Albedo:</u> Ground reflections are separated into two further components: (a) reflected direct beam, treated with a directional reflection employing Gueymard's model [9]; and (b) reflected diffuse light, treated with Lambertian reflection.

B. Ray Tracing

The sky, ground, and panel are each divided into computational areas called *patches* (Fig. 1). All rays from a single patch are assumed to be parallel and emanate from its center. Computation of the irradiance on each panel patch must

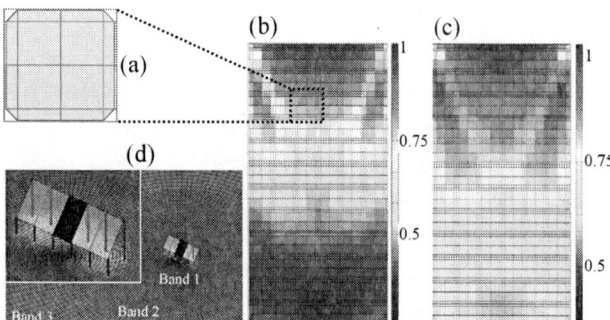

Fig. 1: (a) Solar cell computational zone, with 16 variably-sized patches, is duplicated across the panel in (b-c) showing normalized rear-side irradiance profiles from ground reflections of (b) direct beam and (c) sky diffuse illumination, and includes rack shading and error compensation. The panel is at latitude tilt, on sun, portrait orientation, and evaluated on a sunny day with snow-covered ground (albedo of 0.6). A low ground clearance of 20 cm accentuates the non-uniformity. (d) Schematic of ground patching and racking, showing a functioning panel under test (black) with two non-functioning panels adjacent on each side.

sum the contributions from the direct beam, every sky patch, and every ground patch. Each ground patch is a summation of contributions from the direct beam and from every sky patch. For all pairs of coupled patches, if any portion of the receiving patch is shaded by an obstruction (e.g. panel or racking), that ray is set to zero irradiance. Presently, each ray represents an irradiance value only and therefore neglects spectral effects of scattering and reflection events.

The patching divisions chosen for these simulations are as follows: the sky is divided into 18 azimuth and 10 elevation uniform angular steps; the ground into 18 azimuth and 20 radial divisions, where the sizes of each radial step is divided into 4 bands, with larger steps further away from the panel under test (Fig. 1d); and both the front and rear faces of the cell are divided into 4×4 inhomogeneous rectangular patches (Fig. 1a).

A two-step error compensation scheme is applied to front and rear side panel's ground-reflection irradiance profiles due to the finite size of the ground and panel patches. This compensation allows the reduction of the number of patches in order to increase computational speed while retaining the precision.

C. Photovoltaic behavior

A 1-diode solar cell circuit model was extracted from the current-voltage (*IV*) behavior of a 72-cell bifacial solar panel. With all cells connected in-series in a single string, each cell was fit to an identical *IV* behavior and operating temperature.

The front-side photo-generated current is computed using the front-side irradiance, incidence angle modifier, a typical quantum efficiency response of a mono-facial monocrystalline Si solar cell, and the AM1.5G global solar spectrum. The rear-side current is computed as above, but using the rear-side irradiance multiplied by a bifaciality current de-rating of 95%. The aggregate photo-response (reduced by metallization coverage of ~9% front and ~15% rear) is calculated from a

summation of currents from both sides of each cell, as demonstrated in [10], over all computational patches.

Each cell temperature is also adjusted for both ambient temperature and the irradiance level using the Nominal Operating Cell Temperature (NOCT) [11], which is then fed into equations modifying the short-circuit current, open-circuit voltage, and saturation current.

The total panel *IV* curve is constructed as a voltage sum of the individual temperature-modified curves along each string, and currents summed between parallel strings. The maximum power point is then extracted to determine the energy yield.

D. Panel and Racking Geometry

The panel geometry was modeled after the Sunpreme Inc. 72-cell module, with cells connected in-series in a 12-row by 6-column arrangement. With transparent gaps of 3-4 mm between cells, the total panel area measures 1960×990 mm^2.

The ground-mounted racking used in this model consists of one vertical post at every panel corner, with each post having a cross-section of 6×6 cm^2. The simulated panel is at the center of a single 5-panel row (Fig. 1d). Expanding the field deployment to include additional rows (in both the front and rear) would reduce the amount of albedo and sky diffuse irradiance reaching the panel, and therefore reduce the total output power and bifacial gain.

E. Location

The location for these simulations was chosen to be Ottawa, Ontario, Canada (latitude 45.3° N, longitude 75.67° W, altitude 70 m). The Canadian Weather for Energy Calculations (CWEC) dataset [12], composed of 12 typical meteorological months with 1-hour time-steps, provided environmental (e.g. temperature, pressure) and irradiance (global horizontal, diffuse horizontal, direct normal, and extraterrestrial) data for this location. The albedo was assigned a seasonal-dependence: (a) grass (0.2) in summer (May-Oct); (b) 75% grass and 25% snow in spring (Mar-Apr) and fall (Nov); and (c) half grass and half snow in winter (Dec-Feb). Further, the snow is set as an average of old (0.4) and fresh (0.8) snow. The forward and backward anisotropy coefficients (f_{af}, f_{ab}) for the Gueymard model reflection were set to (1,1), (1.25,0.75), and (1.5,0.5) for the respective seasons.

F. Model Simplifications

In addition to those detailed above, several simplifications were applied to the model, as follows. To reduce computational requirements, several sources of reflection were neglected: (1) primary reflections from panels and racking; (2) secondary reflections; e.g., ground reflections do not feedback to contribute to the sky diffuse irradiance; and (3) internal reflections within the panel module. All rays are assumed to be spectrally-invariant, regardless of source or reflection. The ground is simplified as flat and extending to infinity. The panel ground clearance is assumed constant regardless of snow cover.

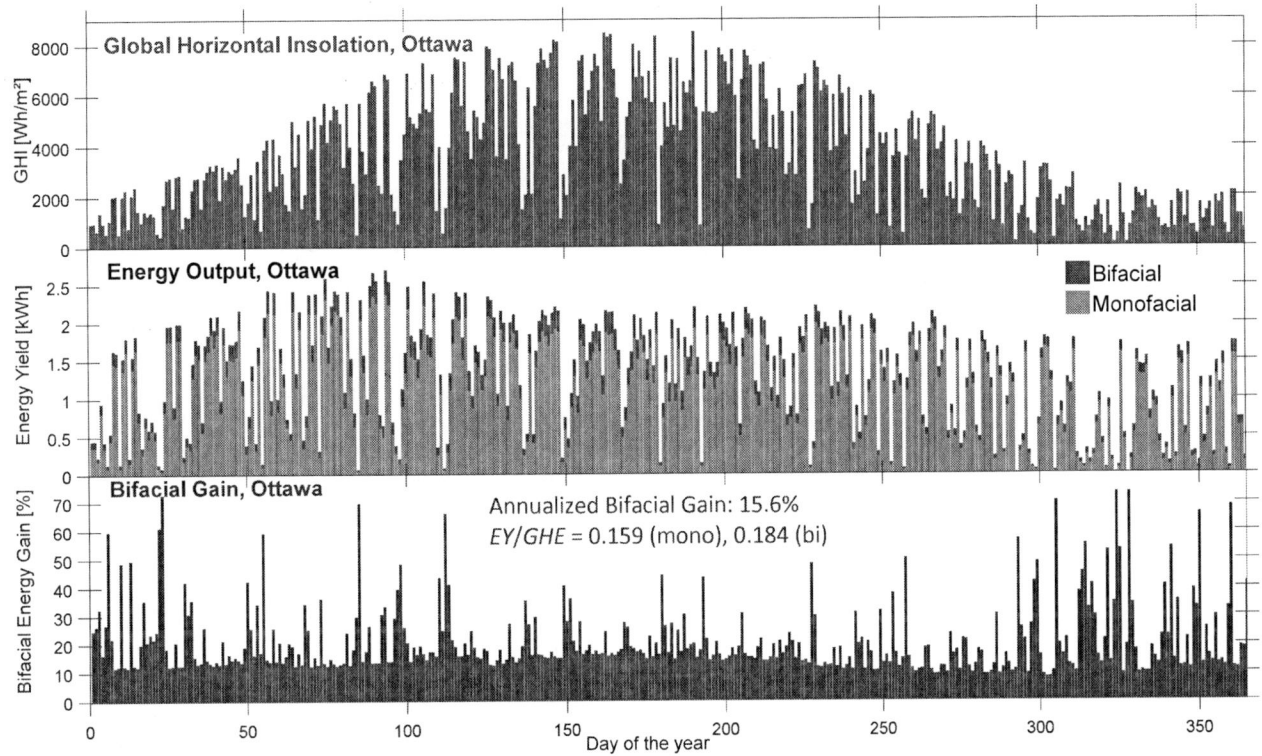

Fig. 2: Daily (a) global horizontal insolation, (b) energy yields of mono- and bifacial panels, and (c) bifacial gain over a year in Ottawa, Ontario, Canada, for a south-facing, latitude-tilt, portrait-oriented panel with 1 m of ground clearance.

Finally, light passing through the gaps between each cell was modeled as a uniform transparency over the entire panel.

III. RESULTS

This bifacial solar panel model was studied for several variations of system design, and below we present performance on representative cloudy and sunny days for comparison. Annual energy yield (EY) calculations over an entire representative year were performed by repeating the procedure by varying the input environmental, irradiance, and geometry parameters for every hour and integrating the energy generation.

A. Baseline Annual Energy Yield

The baseline system consisted of a south-facing, latitude-tilt, portrait-oriented panel with 1-m ground clearance, and yielded a bifacial gain of 15.6% under these conditions, as shown in Fig. 2.

Due to the wide availability of global horizontal irradiance (GHI) datasets, we propose a performance metric for the effective annualized panel efficiency calculated as the summed annual PV energy yield (EY) divided by the summed annual global horizontal irradiance over the single-face panel area, or the global horizontal energy (GHE). It is important to note that

this metric neglects both the area of the rear face and the irradiance incident upon it, but does include the rear face contribution to the energy production. Using this approach and the above conditions, mono- and bi-facial panels studied here have values of 0.159 and 0.184 J/J, respectively. One could instead use the plane-of-array irradiance (translated from GHI or calculated via simulation), but this reduces the ability to compare across variations in the array geometry where the POA irradiance changes due to panel orientation and possibly other factors. The proposed metric is also applicable under solar tracking, where the plane of array is constantly changing and therefore GHI is a simpler concept to apply.

B. Ground Clearance

Previous studies have shown that the height of a bifacial panel above the ground can greatly affect the bifacial energy yield. Our calculations on a snow-covered ground, shown in Fig. 3, have confirmed this trend for panels in both portrait and landscape orientations under (a) sunny and (b) cloudy conditions. Under both conditions, monofacial panel power is nearly flat with increasing ground clearance, but bifacial panel power increases by 23% and 24%, respectively, before flattening off near 5 m. The bifacial power *gain* ranges from 13-35% under sunny conditions and 40-70% under cloudy conditions, depending upon the height of the ground clearance, as shown in Fig. 3(c). This behavior can be understood as a

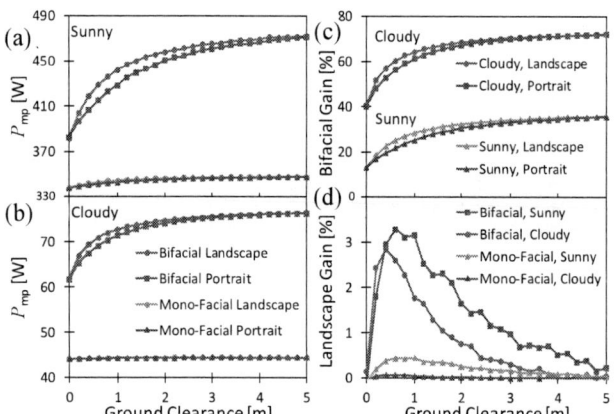

Fig. 3: Impact on maximum power due to ground clearance on a (a) cloudy and (b) sunny hour, for both portrait and landscape panel orientations. The (c) bifacial gain (increase in output power for bifacial as compared to mono-facial panels) and (d) landscape gain (increase in output power for landscape as compared to portrait orientations) due to ground clearance are also shown for cloudy and sunny days. The panel is at latitude tilt, on sun, and evaluated with snow-covered ground (albedo of 0.6).

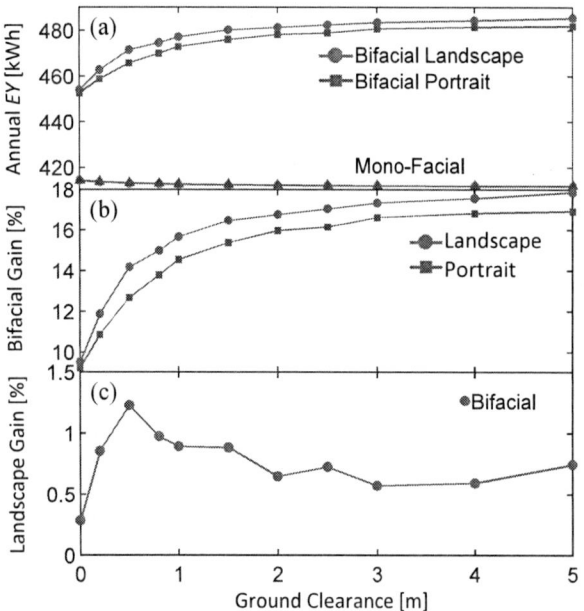

Fig. 4: Impact of ground clearance on annual (a) energy yield (EY), (b) bifacial gain, and (c) landscape gain. The panel is at latitude tilt, wired with a single string of 72 cells.

greater impact from rear-side self-shading for panels low to the ground, and a rear-face vertical irradiance gradient (Fig. 1 b-c).

Positioning the solar panel in a landscape orientation reduces the irradiance difference between the top and bottom cells because the short dimension of the panel is oriented along the irradiance gradient, reducing the current mismatch as compared to the portrait orientation. As a result, the *landscape gain* (defined as the power performance of landscape orientation as

compared to the portrait orientation), shown in Fig. 3(d), reaches >3% for bifacial panels for low ground clearances, and is <0.5% for mono-facial panels under the conditions above.

Similar trends were demonstrated when considering the annual energy yields, shown in Fig. 4, with a ~7% improvement for the highest ground clearances. Bifacial gain also increases by a factor of two, from 9% at ground level to 18% at high clearances. Annual landscape gain reaches >1.2% for low clearances, although it remained consistently above zero even for the greatest clearances tested.

C. Wiring Architecture

The default wiring architecture of the solar panels in the above discussion were all a single-string of 72-cells connected electrically in series. As a result, the panel output is the summation of all cell voltages and a current that is limited by the cell with lowest performance. Due to the vertically non-uniform rear-face irradiance (Fig. 1 b-c), parallel horizontal strings were investigated in order to limit the impact on the overall panel current under snowy conditions (albedo of 0.6). Fig. 5 presents the *parallel gain* (defined as the improvement when using multiple parallel strings in comparison to a single string) for both sunny and cloudy days, using both portrait and landscape panel orientations. Dividing the panel into just two parallel strings realizes a majority of the available gains, with increases in the total panel power of 1.1% in cloudy conditions and 0.7% in sunny conditions, when using a ground clearance of 20 cm. Furthermore, this effect is largest for low ground clearances, and results in almost no additional improvement for clearances of > 1 m due to the reduction in the rear-face non-uniformity.

When considering the annual energy yields, parallel gain shows identical trends (Fig. 6) but with significantly smaller magnitudes (several tenths of a percent), largely impacted by the significantly lower average albedo. Parallel gain is expected to have a more significant impact in annual energy yields for locations with higher annual albedos, including locations with longer periods of snow cover.

IV. CONCLUSION

A new model to calculate the performance of bifacial (and mono-facial) photovoltaic panels has been implemented using a ray-tracing approach to sum the direct, diffuse, and albedo components of the irradiance on both the front and rear faces. The model used a minimum of site-specific data to calculate instantaneous power outputs under specific conditions and annual energy yields, and included the performance impact of the ground clearance, panel orientation, and wiring architecture. In Ottawa under our baseline conditions, an annualized bifacial gain of up to 18% was achievable by employing large ground clearances alone.

The panel's ground clearance impacts both the total irradiance and the non-uniform profile on its rear face. Changing from portrait to landscape panel orientation minimizes the non-uniform illumination, and thus minimizes

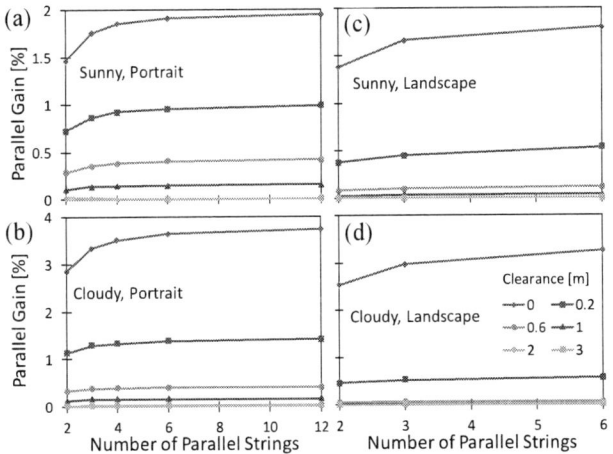

Fig. 5: Impact of the number of parallel horizontal strings on the maximum power of the panel, as compared to a single string configuration, for various ground clearances, both portrait and landscape orientations, and sunny and cloudy conditions. The panel is at latitude tilt, on sun, and evaluated with a snow-covered ground (albedo of 0.6).

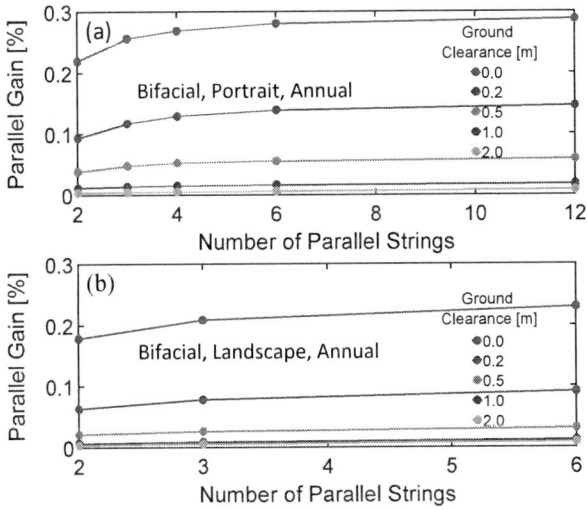

Fig. 6: Impact of wiring architecture (number of parallel strings), on the annualized parallel gain, for (a) portrait and (b) landscape orientation for various ground clearances. The bifacial panel is at latitude tilt.

current mismatched between the series-connected cells. The difference was strongest for low ground clearances (≥3% for both sunny and cloudy conditions, and >1% annually). Adjusting the wiring architecture from a single string (in-series) to several parallel horizontal strings also improved performance by up to ~1% for panels with low ground clearance in both sunny and cloudy conditions. Employing parallel strings made a much lesser impact to the annual energy yield, but this effect may be significantly impacted by the choice of racking and therefore worthwhile for further study. In summary, ground clearances of 1 m or higher, landscape panel orientations, and

parallel-string wiring configurations can each be applied to maximize annual energy yields of bifacial solar panel arrays. Furthermore, these strategies will be most effective in locations with high annual albedo (for instance, with significant snow coverage) which provides strong rear irradiation.

ACKNOWLEDGEMENT

The authors are grateful for funding and support from the Natural Sciences and Engineering Research Council of Canada, CMC Microsystems, Canada Foundation for Innovation, Canada Research Chair, and University of Ottawa.

REFERENCES

[1] D. Chung, C. Davidson, R. Fu, K. Ardani, and R. Margolis, "U.S. Photovoltaic Prices and Cost Breakdowns: Q1 2015 Benchmarks for Residential, Commercial, and Utility-Scale Systems", 2015.

[2] National Renewable Energy Laboratory (USA), "Best Research-Cell Efficiencies Chart". http://www.nrel.gov/pv/assets/images/efficiency-chart.png, accessed on 2017/06/05.

[3] R. Guerrero-Lemus, R. Vega, T. Kim, A. Kimm, and L. E. Shephard, "Bifacial solar photovoltaics - A technology review," *Renew. Sustain. Energy Rev.*, vol. 60, 1533–1549, 2016.

[4] U. A. Yusufoglu, T. M. Pletzer, L. J. Koduvelikulathu, C. Comparotto, R. Kopecek, and H. Kurz, "Analysis of the annual performance of bifacial modules and optimization methods," *IEEE J. Photovoltaics*, vol. 5(1), 320–328, 2015.

[5] A. Cuevas, "The early history of bifacial solar cells," in *20th European Photovoltaic Solar Energy Conference*, pp. 801–805, Munich, Germany, 2005.

[6] C. Deline, S. Macalpine, B. Marion, and J. S. Stein, "Assessment of bifacial photovoltaic module power rating methodologies - Inside and out," *IEEE J. Photovoltaics*, vol. 7(2), 575–580, 2017.

[7] C. T. Li, "Development of field scenario ray tracing software for the analysis of bifacial photovoltaic solar panel performance," M.A.Sc. Thesis, University of Ottawa, 2016.

[8] R. Perez, R. Seals, and J. Michalsky, "All-weather model for sky luminance distribution-Preliminary configuration and validation," *Sol. Energy*, vol. 50(3), 235–245, 1993.

[9] C. Gueymard, "An anisotropic solar irradiance model for tilted surfaces and its comparison with selected engineering algorithms," *Sol. Energy*, vol. 38(5), 367–386, 1987.

[10] J. Johnson, D. Yoon, and Y. Baghzouz, "Modeling and analysis of a bifacial grid-connected photovoltaic system," in *IEEE Power and Energy Society General Meeting*, pp. 1–6, 2012.

[11] M. C. A. García and J. L. Balenzategui, "Estimation of

photovoltaic module yearly temperature and performance based on Nominal Operation Cell Temperature calculations," *Renew. Energy*, vol. 29(2004), 1997–2010, 2010.

[12] Canadian Weather for Energy Calculations (CWEC), Atmosopheric Environment Services (AES), Environment Canada, 2010. http://climate.weather.gc.ca/prods_servs/engineering_e.html, accessed on 2017/02/27.

Design and Development of a Solar Photovoltaic Module Detection Control System based on PLC

Yiwang Wang[1,2,3], Jili Zhang[1], Kanglin Liu[2], Houjun Tang[3], Hui Pan[4], Yan Lin[5], Peter Yang[6], Rui Wang[7]

[1]Suzhou Vocational University,Suzhou,215104,China
[2]The University of Tennessee, Knoxville, TN, 37996,USA
[3]Shanghai Jiaotong University, Shanghai,200240,China
[4]Suzhou Tianye Electric Appliance Co.,Ltd,Suzhou, 215100,China
[5]Luxen Solar Energy Co.,Ltd,Suzhou, 215100,China
[6]Suzhou Smagix Energy Technology Co.,Ltd,Suzhou,215123,China

[7]Suzhou KSTN New Energy Technology Co.,Ltd,Suzhou,215104,China

Abstract — **Solar photovoltaic (PV) modules are the key components of PV systems, in order to enhance the security level of PV modules detection and power generation operation reliability. A novel solar PV modules detection control system based on power line carrier (PLC) is proposed and designed. The system can detect main parameters of single or multiple modules, such as the modules' voltage and temperature etc.. And using a digital control circuit to achieve effective detection and fault protection for PV modules. Then a control method for PV modules protection and control system were designed.An experimental prototype was developed and tested. The designed system has advantages of high precision, low cost and good performance, which can be suitable for applying in future solar power generation system.**

Index Terms — **solar photovoltaic modules, detection control, power line carrier (PLC).**

I. INTRODUCTION

Solar PV modules are an important component converts light energy into electricity, which is directly related to the performance and stability of solar PV power generation system. Working at different application specifications, the modules are installed and operated under various climatic and environmental conditions[1]-[4]. In order to improve the reliability of entire PV power generation system, PV modules are required to be detected in real-time. There are some data acquisition systems which have been developed for applying in PV power generation system[5],[6]. The most conventional PV modules detection system using solar combiner box as shown in Fig.1.

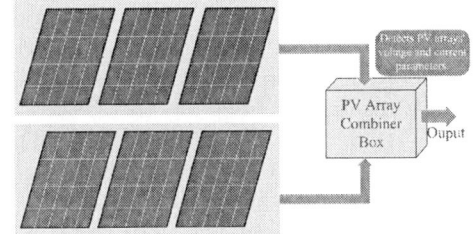

Fig. 1. Conventional PV modules detection system

As the centralized control method, it is difficult to find a single module failure. Especially in certain situations, they even cause failure or loss of the whole PV power generation system. Some losses and failure cases caused by modules or combiner box are shown in Fig.2 [7],[8].

Fig. 2. Some PV modules failure cases

Due to the rapid development of digital control circuits, some novel solar PV modules monitoring system are developed and reported [9]-[20].The ZigBee, GSM and other wireless communication can be designed and applied to the photovoltaic detection or monitoring system. In these systems, the additional communication lines or equipment are required, thus increasing the cost of monitoring and complexity. In addition,because the temperature data of PV modules is very important for photovoltaic power generation,it is necessary to achieve real-time temperature and other operating parameters monitoring.

In this paper, a novel solar PV modules detection control system based on power line communication (PLC) is proposed to achieve the real-time, online, and multi-parameter detection of PV modules.

II. PROPOSED AND DESIGN SYSTEM

In order to improve the operational reliability and enhance the detection level of the entire solar PV power generation system, a novel solar PV module intelligent detection control system based on PLC is proposed and

978-1-5090-5606-4/17 $31.00 © 2017 IEEE

implemented. The system can detect key parameters of single or multiple PV modules, such as the modules' voltage, current and temperature.

The composite architecture diagram of proposed detection control system is illustrated in the Fig.3.The proposed system can be embed within a single or multiple PV modules. Furthermore, an external fault protection circuit is designed for the proposed system. The protection circuit is arranged with the combiner box. when a temperature anomaly is detected, the system can control the switch to protect, and to achieve the galvanic isolation of fault location.

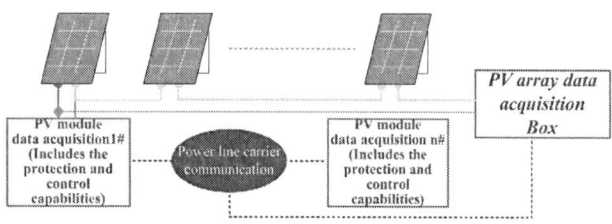

Fig. 3.   The composite architecture diagram of proposed system

The system hardware block diagram is illustrated in Fig.4.The hardware includes data acquisition and receiving & processing. The data receiving part is composed of acquisition circuit, control circuit and communication circuit, and the data processing circuit is composed of communication circuit, control circuit and display circuit. The proposed system uses PLC communication, which doesn't require extra wiring and can reduce system costs.

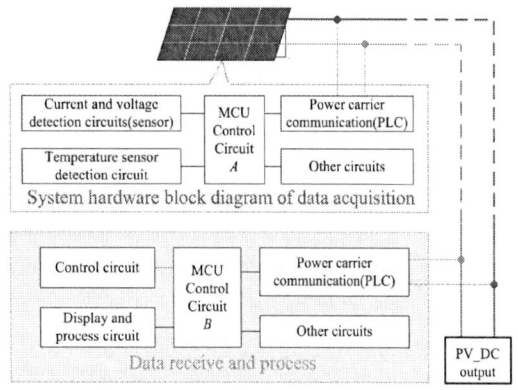

Fig. 4.   The system hardware block diagram

Fig.5 shows the circuit schematics of proposed system, the system uses MCU as core controller, and the power carrier communication circuit using a dedicated digital carrier communication chip module.

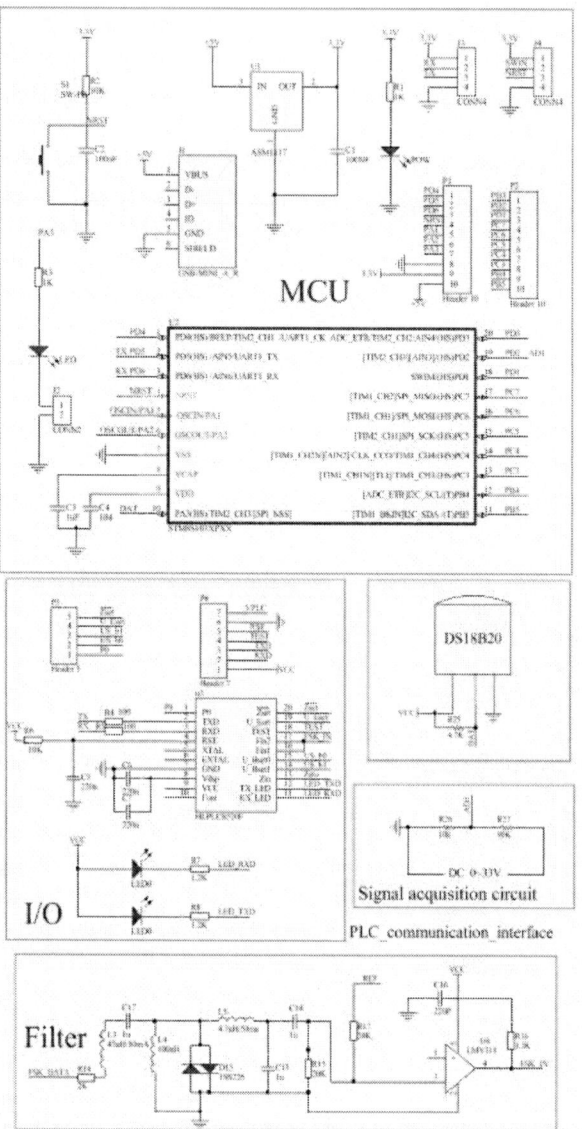

Fig. 5.   The circuit schematic of proposed system

III. CONTROL STRATEGY

The effect of temperature on PV module performance is relatively significant. Therefore, it is necessary to take the influence of temperature into account in system design and control strategy development. A control strategy and method of photovoltaic power generation is proposed based on designed system, which consider the effect of the PV arrays' temperature according to the advance designed detection system. Using the average temperature of PV arrays as a reference factor, a new control strategy for PV system is designed. Additionally, the single PV module will be protectively controlled combined with a set safe temperature value.

The flowchart of control strategy is given in Fig.6.In the beginning,the controller establishes communication connection, and the PV module output voltage, current and temperature are measured. Then, the data is analyzed and processed. if the threshold is exceeded, protection and alarm are performed, otherwise the data outputs and calls control subroutine. Finally the Control implementation and display output.

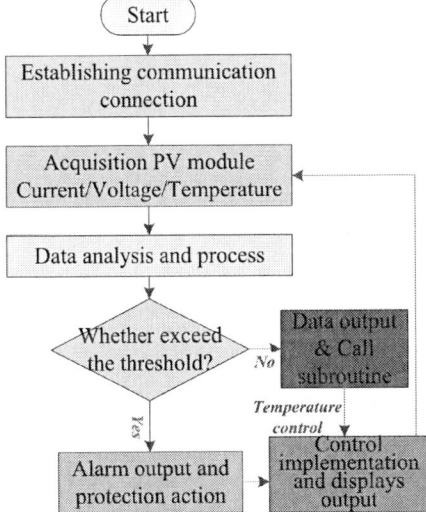

Fig.6. Flowchart of proposed control strategy

IV. EXPERIMENTAL RESULTS

The realization of the proposed system and control algorithm is carried out by developed experimental prototype. The system prototype was constructed accoring to the Fig.4. PV module data acquisition and control algorithm are realized by using MCU controller, and PLC are implemented by HLPLCS520F control chip module. In order to validate the performance of proposed system, the detection experiments are carried out. The results are presented in Fig.7 where the detected voltage, current and temperature are given. It can be seen that the system has realized real-time detection of PV module.

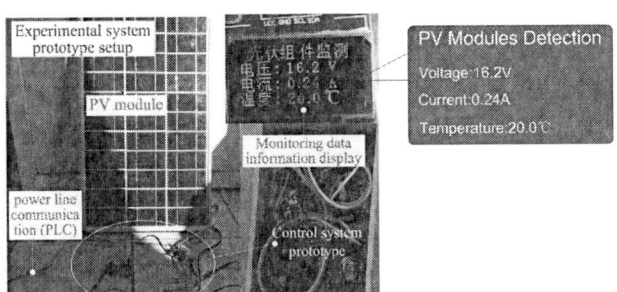

Fig.7. Experimental system prototype and results

V. CONCLUSIONS AND FUTURE WORK

In this paper, a solar PV modules detection control system based on PLC is designed. The protection control circuit hardware and control strategy are depicted. The system using PLC communication can reduce system cost and increase flexibility. The designed system can detect a single and multiple PV module, and monitor the status of each component of PV plant, then alarm timely when a failure occurs, and improve the overall PV system work stability. Additionally, the novel PV module intelligent detection control method using the measurement data from the designed detection control system, can improve control efficiency and accuracy of conventional photovoltaic power generation system.

Product-level integrated system based on proposed scheme will be developed and applied to engineering applications of PV system in the future.

ACKNOWLEDGEMENT

This work was supported by the Jiangsu Provincial University Oversea Education Plan, Science and Technology Planning Project of Suzhou City and Foundation of Suzhou Vocational University.

This work also made use of Engineering Research Center Shared Facilities supported by the Engineering Research Center Program of the National Science Foundation and the Department of Energy under NSF Award Number EEC-1041877 and the CURENT Industry Partnership Program.

REFERENCES

[1] Bruno Andò, Salvatore Baglio, Antonio Pistorio, Giuseppe Marco Tina and Cristina Ventura,"SENTINELLA: A WSN for a smart monitoring of PV systems at module level"in *10th IEEE International Workshop on Measurements & Networking (M&N)*,2013

[2] M. M. Rahman, M. Hasanuzzaman, and N. A. Rahim, "Effects of various parameters on PV-module power and efficiency," *Energy Convers. Manag.*, vol. 103, pp. 348–358, 2015.

[3] Marko Jankovec, Federico Galliano, Eleonora Annigoni, Heng Yu Li, Fanny Sculati-Meillaud, Laure-Emmanuelle Perret-Aebi, Christophe Ballif, and Marko Topic,"In-Situ Monitoring of Moisture Ingress in PV Modules Using Digital Humidity Sensors,"*IEEE JOURNAL OF PHOTOVOLTAICS*, vol. 6, no. 5, pp.1152-1159, 2016

[4] Mohammadreza Aghaei,Francesco Grimaccia, Carlo A. Gonano, and Sonia Leva, "Innovative Automated Control System for PV Fields Inspection and Remote Control,"*IEEE Transactions on Industrial Electronics*, vol.62,no.10, pp.7287-7296, 2015.

[5] N. Forero a, J. Hernandez b, G. Gordillo,"Development of a monitoring system for a PV solar plant,"*Energy Conversion and Management*,no.47,pp.2329–2336,2006

[6] Santiago Silvestre, Mário Aires da Silva, Aissa Chouder, Daniel Guasch, Engin Karatepe,"New procedure for fault detection in grid connected PV systems based on the evaluation of current

and voltage indicators,"*Energy Conversion and Management,*vol. 86,pp. 241-249,2014.

[7] YANG Li-zhong,YANG Hong-yun,ZHANG Tao-lin,ZHOU Xiao-dong."Study on fire risk of photovoltaic panels",*Journal of Thermal Science and Technology , Vol.14,no. 3,pp.173-177,2015*

[8] http://image.baidu.com/

[9] Jinsoo Han, Chang-Sic Choi, Wan-Ki Park, Ilwoo Lee, and Sang-Ha Kim,"PLC-Based Photovoltaic System Management for Smart Home Energy Management System,"*IEEE Transactions on Consumer Electronics*, Vol. 60, No. 2, pp.184-189,2014

[10] Simon Siregar,Duddy Soegiarto."Solar panel and battery street light monitoring system using GSM wireless communication system,"*2nd International Conference on Information and Communication Technology (ICoICT)*,pp.272 - 275,2014

[11] Wen Yen Lin; Kuang Po Hsueh; Wang Hsin Hsu; Liew Gha Yie; Wei Chen Tai."Design and Implementation of Health Monitoring System for Solar Panel in IPv6 Network",*Tenth International Conference on Intelligent Information Hiding and Multimedia Signal Processing*,pp.57-60,2014

[12] Eduardo Roman, Ricardo Alonso, Pedro Ibanez, Sabino Elorduizapatarietxe, and Damian Goitia, "Intelligent PV module for grid-connected PV systems," *IEEE Trans. Industrial Electron.*, vol. 53,no. 4, pp. 1066-1073, Aug. 2006.

[13] C. Ranhotitogamage, S. C. Mukhopadhyay, S. N. Garratt and W. M. Campbell, "Measurement and Monitoring of Performance Parameters of Distributed Solar Panels using Wireless Sensors Network,"*IEEE International Instrumentation and Measurement Technology Conference*,pp1-6,2011

[14] Jinsoo Han,Chang-Sic Choi,Wan-Ki Park,Ilwoo Lee,Sang-Ha Kim, "Smart home energy management system including renewable energy based on ZigBee and PLC,"*IEEE International Conference on Consumer Electronics (ICCE)*,pp.544-545,2014

[15] Y. Rashidi, M. Moallem, and S. Vojdani, "Wireless ZigBee system for performance monitoring of photovoltaic panels," in *37th IEEE Photovoltaic Specialists Conference,* 2011

[16] Cheng Yaling, Zou Xiaoping, Zhuang Guangtao, Teng Gongqing, Qu Weiwei,"Remote Monitoring System for Distributed Photovoltaic Based on ZigBee",in *4th International Conference on Computer Science and Network Technology*,pp.867-871,2015

[17] Michael C. Pacis,Julius T. Sese,Harold A. Blastique,Mariel Dan C. Casibang,Gyver G. Ladisla and Renz Gian D. Villano,"Metering of surplus energy on PV systems using ZigBee Wireless Technology,"in *6th IEEE International Conference on Control System, Computing and Engineering,*2016

[18] Syafii,Muhammad Ilhamdi Rusydi,Roni Putra and Muhammad Hadi Putra,"Real-time measurement of grid connected solar panels based on wireless sensors network,"in *International Conference on Sustainable Energy Engineering and Application,* 2016

[19] Rustu Eke, Ali Senturk,"Monitoring the performance of single and triple junction amorphous silicon modules in two building integrated photovoltaic (BIPV) installations,"*Applied Energy,* vol.109,pp.154-162,2013.

[20] Francisco J. Sánchez-Pacheco, Pedro Juan Sotorrío-Ruiz, Juan R. Heredia-Larrubia,Francisco Pérez-Hidalgo and Mariano Sidrach de Cardona,"PLC-Based PV Plants Smart Monitoring System: Field Measurements and Uncertainty Estimation,"*IEEE Transactions on Instrumentation and Measurement,* vol.63.no.9,pp.2215-2222,2014.

Detecting Calibration Drift at Ground Truth Stations
A Demonstration of Satellite Irradiance Models' Accuracy

Richard Perez[1], James Schlemmer[1], Adam Kankiewicz[2], John Dise[2], Alemu Tadese[2] & Thomas Hoff[2]

[1] Atmospheric Sciences Research Center, SUNY, Albany, New York, 12203, USA
[2] Clean Power Research, Napa, California, 94558, USA

Abstract — **In this article we show how a state-of-the-art satellite irradiance model – SolarAnywhere – is capable of undetected calibration issues at a trusted reference ground truth irradiance measurement station. This evidence suggests that the best satellite models have now achieved a degree of accuracy and versatility that makes them an acceptable, if not a preferred choice, for solar energy engineering applications ranging from long-term site characterization and system monitoring.**

Index Terms —**solar resource, satellite, modeling, irradiance, benchmarking.**

I. INTRODUCTION

Satellite-derived irradiances have several operational advantages over ground measurements. These advantages include a global reach, a geographical resolution ranging from sub-kilometer scale to entire continents, a long-term time reference spanning decades, and a real-time accessibility that can drive operational forecasts. The long-term reference capability, in particular, is widely applied for site characterization and project financing, especially when this is ascertained by short-term measurements tuning.

The known advantage of measurements over satellite data is localized accuracy, <u>but only if instruments are properly calibrated and maintained</u>.

In this article we suggest that state-of-the art satellite models have now achieved such degree of accuracy that they can spot undetected calibration errors in trusted reference stations.

This assertion is based on the comparative analysis of the SolarAnywhere V3 [1] satellite model and the SurfRad station of Fort Peck, Montana. The SurfRad stations are part of WMO's BSRN network [2]. They are considered to be of unquestionable quality, with calibrations traceable to primary standards. These reference stations are frequently used to both validate and tune satellite models.

A SolarAnywhere user triggered the present investigation. The user was comparing SolarAnywhere global irradiance (GHI) data against Fort Peck ground-truth station data. The SolarAnywhere user noticed unexplained discrepancies between the two data sets. The user reported that the year-to-year variability was not always well-matched to Fort Peck ground observations. Figure 1 compares the annual, measured GHI to Solar Anywhere GHI. In particular, SolarAnywhere data are significantly higher than ground measurements in 2015, a difference that slightly exceeds the model's published margin of error.

Our first assumption was that this was an acceptable level of uncertainty. A strength of satellite data, however, is the ability to provide a long-term evaluation resource that accurately accounts for year-to-year variability. This is beneficial to developers and planners. We decided to investigate this issue further.

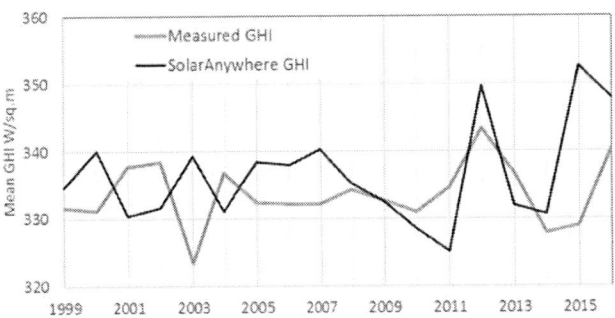

Fig. 1. Annual GHI long-term trends for Fort Peck SURFRAD and SolarAnywhere

II. DATA ANALYSIS

<u>Large 2015 Discrepancy</u>: In Figure 2 we compare SolarAnywhere GHI against Fort Peck GHI measurements in 2013 (when the satellite model slightly underestimated ground measurements) and 2015 (when the model significantly overestimated measurements.)

The 2015 positive bias is visible through clear sky condition events – i.e., the points densely concentrated near the 1-1 line during higher irradiance conditions. There is a close alignment between clear sky points detected by the ground and

978-1-5090-5606-4/17 $31.00 © 2017 IEEE

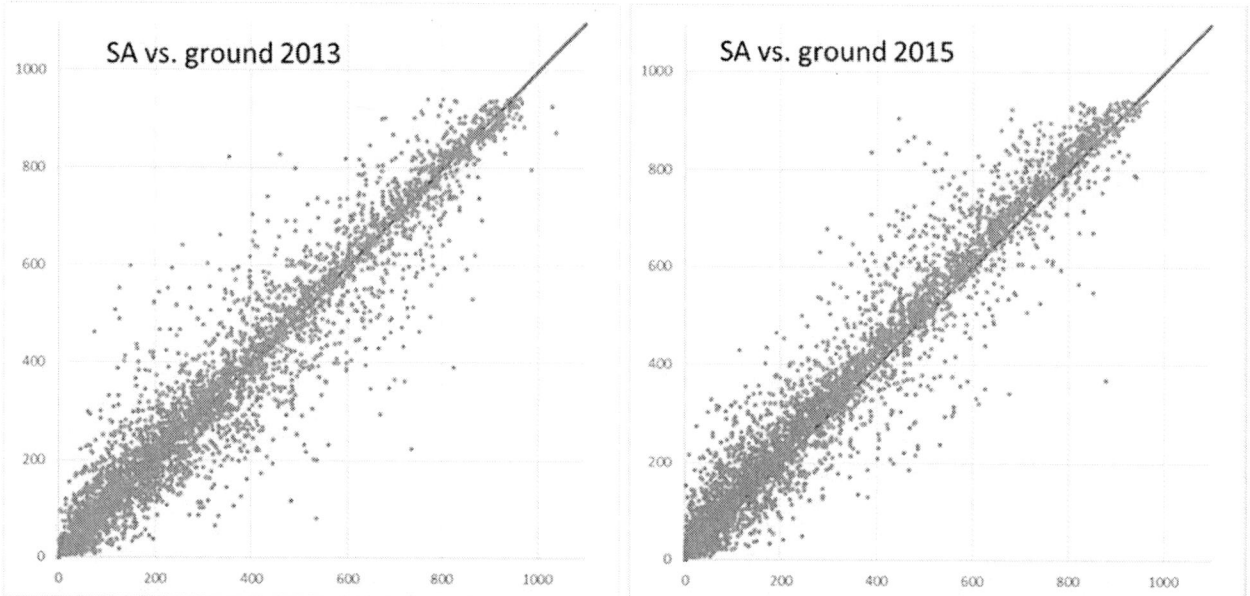

Fig. 2. SolarAnywhere GHI (y axis) vs. SURFRAD GHI (x axis) in 2013 and 2015.

SolarAnywhere in 2013. SolarAnywhere clear sky points are systematically higher than ground measurements in 2015.

There are no significant differences in the satellite model's turbidity inputs (i.e., AOD and precipitable water) underlying the satellite model's clear sky calculations between the two years. Therefore, the difference in Figure 2 must originate from ground observations.

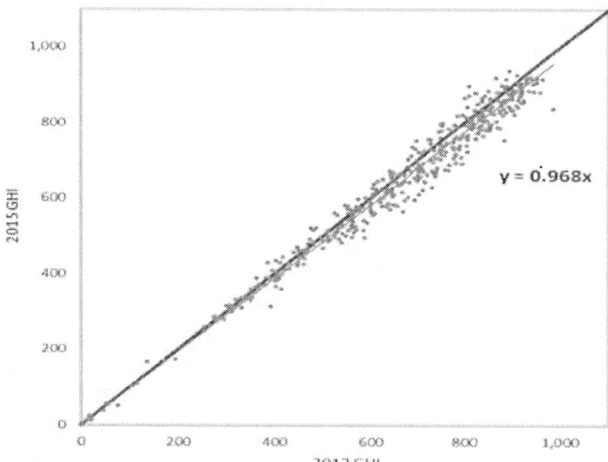

Fig. 3. Date/time coincident clear sky GHI observations in 2015 and 2013.

Focusing on ground measurements alone, Figure 3 compares measured GHI clear sky events in 2013 and 2015. Each point

of the scatter plot corresponds to conditions when GHI was above 92 percent of clear sky conditions (i.e., Kt*> 0.92) in both years for the same calendar day and same time. The sample of coincident points indicates a mean ~ 3.4% decrease of observed clear sky conditions GHI from 2013 to 2015.

What is the reason for this decrease in ground data in 2015? Was it due to higher turbidity in 2015 resulting from an intense fire season in the Western US? Or was it due to instrument calibration change? The former would show that the satellite model has limitations because it is not precise enough to adapt to changing turbidity conditions. The latter would suggest that SolarAnywhere can identify calibration issues.

To answer this question, we examined independently measured DNI. A decrease in clear sky DNI between the two years should be amplified relative to GHI clear sky decrease if the cause is higher turbidity. We repeated the procedure of Figure 3 for DNI, selecting calendar/time of day-coincident points above a clear sky threshold in both years. Figure 4 presents the results.

Figure 4 shows that there is no appreciable change in clear sky DNI events in Fort Peck between 2013 and 2015. It is physically impossible to have a clear sky GHI reduction without a concurrent and larger clear sky DNI reduction. Therefore, we can only conclude that the 2015 calibration of the SurfRad GHI instrument was lower in 2015 than in 2013.

The example shown in Figure 5 both illustrates and ascertains this conclusion. Figure 5 presents measured DNI and GHI and

978-1-5090-5606-4/17 $31.00 © 2017 IEEE

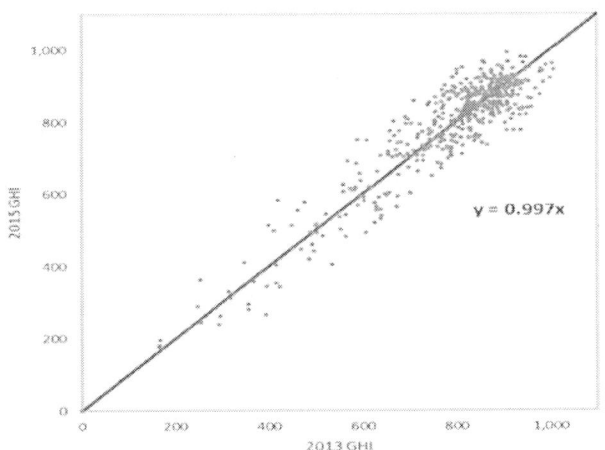

Fig. 4. Date/time coincident clear sky DNI observations in 2015 and 2013

identified it. NOAA has since posted a Data Problem Report following our inquiry [3].

Long Term Reliability of Solaranywhere: Figure 1 shows that, in addition to the large 2015 difference analyzed above there are other noticeable year-to-year differences between the two data streams. In Figure 6, we added indirectly measured GHI to the long term annual trends. Indirectly measured GHI is obtained by summing measured direct horizontal irradiance and measured diffuse irradiance. This indirect GHI measurement is considered to be more accurate than pyranometer measurements because it eliminates issues associated with sensors' cosine response [4]. However it does not eliminate possible calibration issues.

The mean yearly absolute value difference between SolarAnywhere and measured GHI is 2% over the considered 18-year time span. Interestingly, the mean absolute value difference between directly and indirectly measured GHI at

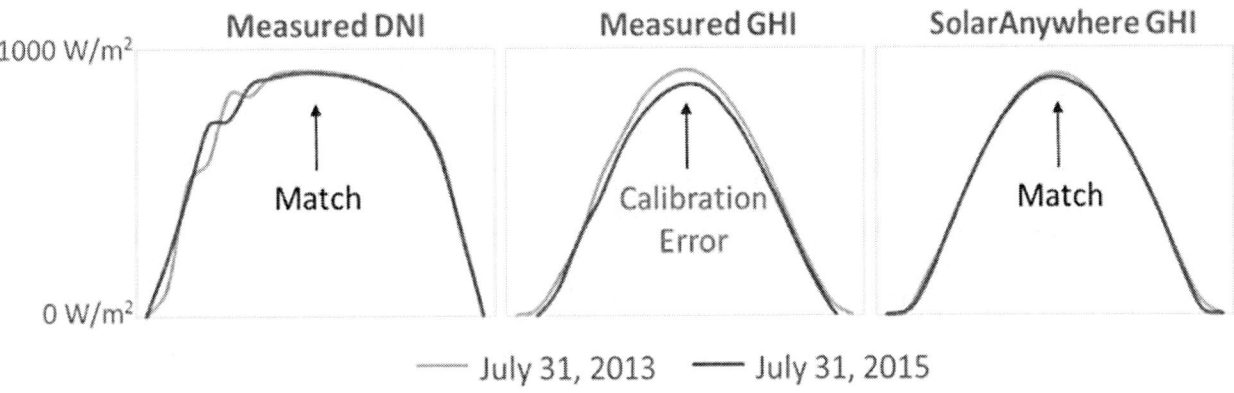

Fig. 5.   Comparing measured DNI, GHI and SolarAnywhere GHI on July 31, 2013 and 2015

satellite-based SolarAnywhere GHI on comparable clear days in 2013 and 2015 (July 31, 2013 and July 31, 2015).

Measured DNI (left graph) is nearly identical for both years. SolarAnywhere GHI (right graph) is nearly identical for both years. Measured GHI (middle graph), however, is not. The measured 2015 GHI is substantially lower than the measured 2013 GHI.

Further, we used independently measured diffuse irradiance (DIF) and DNI to calculate GHI. We found that indirectly measured GHI via diffuse and DNI was nearly identical in both years. This confirms the thesis of a GHI calibration error.

The station operator (NOAA) confirmed that there was indeed a calibration mismatch issue with the GHI sensor that covered much of 2015 and that had not been noted until we

Fort Peck is also 2%. In Figure 7 we plotted the ranked yearly absolute value differences between any two of the three GHI sources. This plot shows that SolarAnywhere's year-to-year uncertainty is equivalent to the uncertainty that exists between two side-by-side GHI measurements obtained with first class instrumentation at a carefully maintained reference station.

Another way to assess SolarAnywhere long-term reliability is presented in Figure 8. This figure respectively plots annual GHI from each of the three considered sources against the two others. SolarAnywhere RMSE vs any of the two measured sources is marginally higher than when comparing one measured source against the other. However, the RMBE is

978-1-5090-5606-4/17 $31.00 © 2017 IEEE

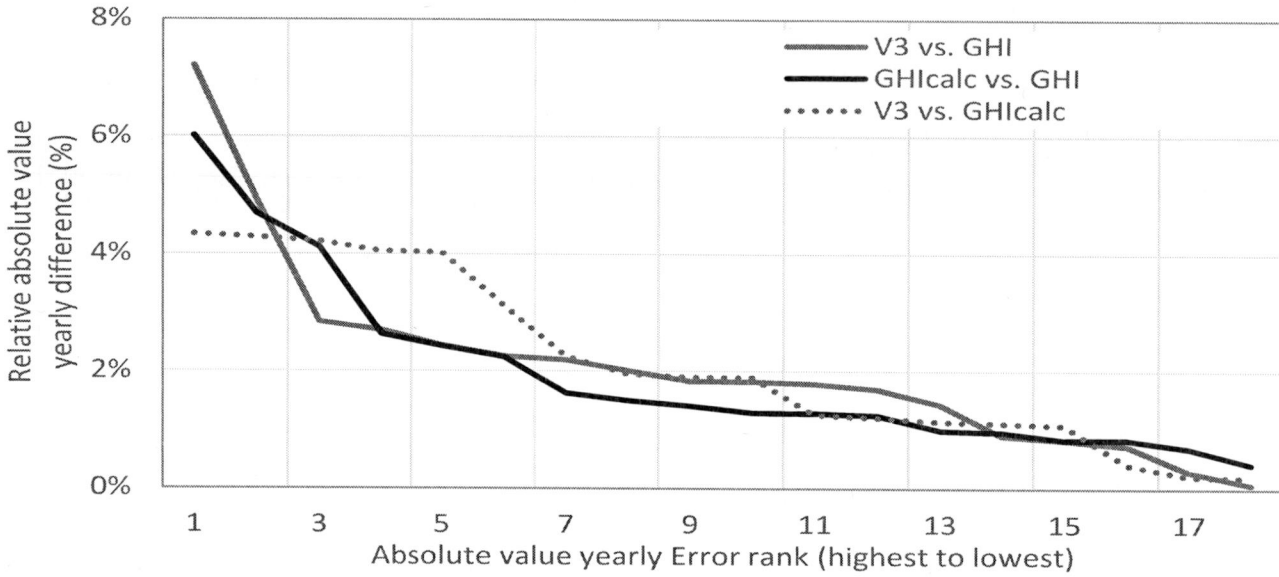

Fig. 7. Ranked absolute value yearly differences between the three sources of GHI

noticeably smaller, positioning this resource as a middle common ground between the two measurement sources.

Taking a detailed look at two one-year periods in addition to the 2015 problem discussed above sheds some light on these differences. Figure 9 contains two sets of scatterplots. The first set (left) includes data spanning October 2007 to September 2008. The second set (right) includes data from October 2008 to September 2009 (Note that we used September as a cutoff date because SURFRAD irradiance sensors are generally switched in September.)

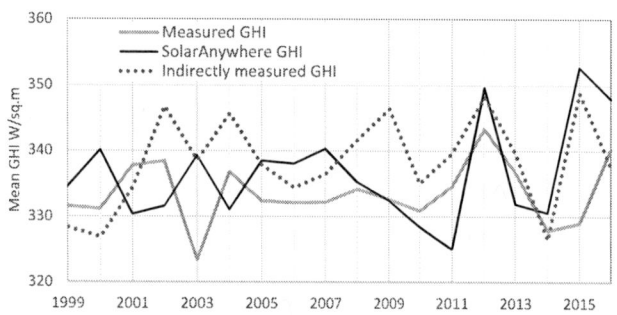

Fig. 6. Comparing annual GHI long-term trends in Fort Peck from SolarAnywhere, direct and indirect ground measurements

The top scatterplots show SolarAnywhere GHI vs. directly measured GHI. The middle plots show SolarAnywhere GHI vs. indirectly measured GHI and the bottom plots compare the direct and indirect GHI measurements.

The bottom plots show the tell-tale sign of pyranometer cosine response. This is apparent through the curvature between the two GHI measurements. In addition, there is substantial calibration difference between two periods for the two sources of GHI amounting to about 2%.

In the middle plots, the behavior of SolarAnywhere is reminiscent of measured GHI in the first period when compared to calculated GHI, while in the second period, the agreement between SolarAnywhere and calculated GHI is excellent.

The top plots show a similar agreement between SolarAnywhere and measured GHI in both periods.

These observations lead to conjecture that, in this case, the source of GHI that is out of pattern is the first period's calculated GHI – the DNI sensor was possibly slightly off calibration that year. This example shows that accurate satellite data such as SolarAnywhere may be a safe common denominator source of quality control between the two measurement methods.

IV. CONCLUSIONS

We draw four conclusions from this investigation.

First, SolarAnywhere is a reliable source of long-term solar resource data. It is precise enough to identify small calibration issues at one of the nation's most trusted irradiance reference data sets. Its year-to-year uncertainty is comparable to the uncertainty between two collocated measurements at a well-maintained station

Second, while it is widely accepted that indirect GHI measurement obtained from DNI and diffuse is preferable to pyranometric GHI measurement for ground truth validations and climate studies [ref] when both sources are available, it does not preclude that, even at the best reference stations, a pyranometric measurement may sometimes be more accurate because of operational calibration uncertainty.

Third, quality remote sensing measurement of irradiance such as SolarAnywhere can be an effective common denominator reference to check the field calibrations and identify/resolve issues between direct and indirect GHI measurements.

Finally, satellite data tuning using ground sensors – a common practice in the industry, e.g., [5], is only as good as the instruments used for tuning. A user must confirm proper calibration and maintenance before proceeding with data tuning.

REFERENCES

[1] Perez, R., J. Schlemmer, K. Hemker, Jr., S. Kivalov, A. Kankiewicz, and C. Gueymard, (2015): Satellite-to-Irradiance Modeling – A New Version of the SUNY Model. in 42nd IEEE PV Specialists Conference, 2014

[2] Ohmura, A., E. Dutton, B. Forgan, C. Fröhlich, H. Gilgen, H. Hegner, A. Heimo, G. König-Langlo, B. McArthur, G. Müller, G. Philipona, R. Pinker, C. Whitlock, K. Dehne, and M. Wild, (1998): Baseline Surface Radiation Network (BSRN/WCRP): New Precision Radiometry for Climate Research. Bulletin of the American Meteorological Society, Vol. 79, No. 10, pp. 2215-22136.

[3] NOAA Earth System Research Laboratory, Global Monitoring Division, (2016): Notice of SURFRAD Network Data Problem Reports. https://www.esrl.noaa.gov/gmd/grad/surfrad/problems.html

[4] Michalsky, J., E. Dutton, M. Rubes, D. Nelson, T. Stoffel, M. Wesley, M. Splitt, and J. Deluisi, (1999): Optimal Measurement of Surface Shortwave Irradiance Using Current Instrumentation. Journal of Atmospheric & Oceanic Technology, 16, pp. 55-69.

[5] Kankiewicz, A., J. Dise, E. Wu and R. Perez, (2014): Reducing Solar Project Uncertainty with an Optimized Resource Assessment Tuning Methodology. Proc. ASES Annual Conference, San Francisco, CA.

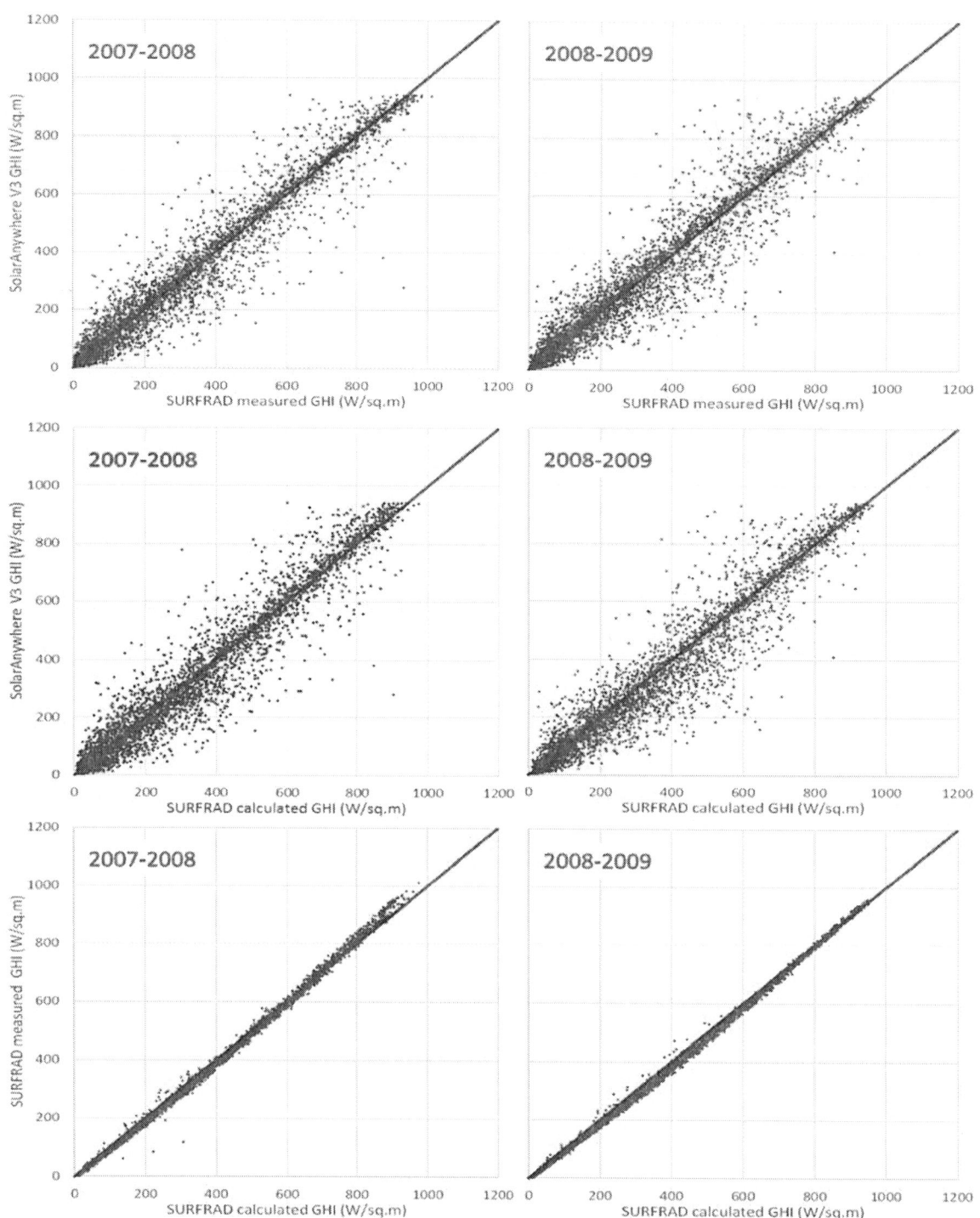

Fig. 9. Hourly SolarAnywhere GHI vs. pyranometer GHI (top), SolarAnywhere GHI vs. indirectly measured GHI (middle), and pyranometer GHI vs. indirectly measured GHI (bottom) for two consecutive one year periods.

978-1-5090-5606-4/17 $31.00 © 2017 IEEE

Performance of Solar Resource Monitoring Stations in Hot Climate Regions

Yahya Z. Alharthi[1], Mahbube K. Siddiki, Ghulam M. Chaudhry, Saad Muaddi, Ahmed Alahmed
Computer Science & Electrical Engineering Department
University of Missouri-Kansas City
Email: Yza6x6@mail.umkc.edu, Siddikim@umkc.edu

Abstract—**This paper presents a study aimed at evaluating and comparing the performance of five selected solar resource monitoring stations in the Kingdom of Saudi Arabia (KSA). Also, present an overview of the current electricity demand since KSA's power sector relies heavily on liquid fuels. The data used in this study have been collected from the new solar atlas which was launched by KACARE for collecting the solar data along with other parameters through the installed stations across the country. The results showed the potential of solar irradiance during the year for different places which would contribute to the implementation of plans for renewable energy (RE) projects. The Selected City in the Northern region of KSA has the highest amount of solar irradiation with an approximate annual of 82450.5 Wh/m²/year, followed by the southern city with 73199.1 Wh/m²/year. Also, GHI and DNI data were compared to the earlier solar atlas data which was published by NREL in 2002 and they were in close agreement with less than 10% difference.**

Index Terms — **Renewable energy, photovoltaic cells, Power System, Solar Power, Saudi Arabia.**

I. INTRODUCTION

Solar data is one of the most important parameters to the implementation of plans for solar power projects. This type of data will play an important role to attract investors and provide companies, researchers, and power project developers with the initial resource data [1]. King Abdullah City for Atomic and Renewable Energy (KACARE) has launched the Renewable Energy Atlas for solar data collection with almost 30 stations across the country[2]. This program collects different weather data through the installed solar resource monitoring stations in KSA. These data can be reached by the Renewable Resource Monitoring and Mapping program (RRMM) at[3]. The parameters are being collected from solar stations are Direct Normal Irradiance (DNI), Global Horizontal Irradiance (GHI), Diffuse Horizontal Irradiance (DHI), Air Temperature, Pressure, Wind Speed, Direction, and Relative Humidity[3].

Various people who can benefit from this data are Power project developers and financers, Researchers Academics, Government entities, Industries, and public. The datasets of this program can be mapped, graphed, and downloaded. By having the chance to access this data, users will have pre-feasibility assessments for their projects[2].

Because of the significant increase in power demand in KSA, renewable energy must get enhanced to reduce dependency on fossil fuels. Over the last 10 years, electricity demand in KSA has recorded a 5-6% increase per year. In 2003, the maximum peak load recorded almost 26,272MW and jumped to

62,260MW in 2015[4]. This increase is because of the rapid population and high economic growth, along with many other factors playing an important role in causing this expanding of power generation [5].

Saudi Electricity Company (SEC) annual report showed the progress in the electricity industry and stated that, the number of customers was 3,622,391 in 2003 and went up to 7,602,279 in 2014 with an average increase of 5.4% per year [5].

Due to the increasing concerns about the future of fossil fuels and climate change, renewable energy technology has become an important project worldwide to save nature and the future generations[6].

In this paper, five cities have been selected to study the performance of their solar resource monitor stations during the year of 2015. Also, different types of solar irradiance and air temperature were compared to NREL old data along with more details and discussion regarding the KACARE stations have been presented.

II. POWER DEMAND OVERVIEW IN KSA

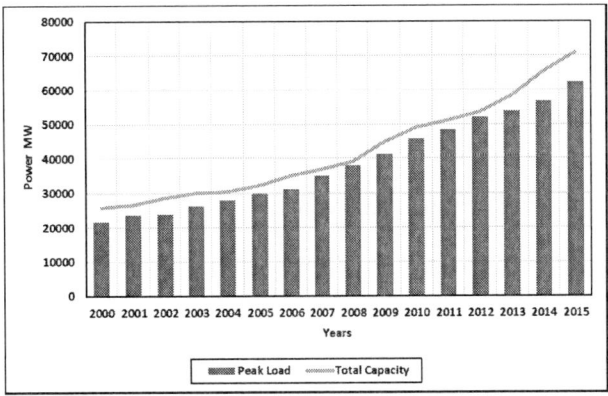

Fig.1. Peak Load Versus Generation Capacity

Due to the significant increase of population and the economic growth every year, KSA power generation have been dramatically increasing over the last decade. The latest report of electricity data showed that the customer's average annual growth rate from 2000 to 2015 is in between 5.4% to 6% [3]. Figure.1 shows the growth of power demand versus total power capacity per year. In 2015, the peak demand recorded almost 62260 MW and the total generation capacity was 69000MW. In KSA, Heavy Fuel Oil (HFO), Natural Gas, Diesel Oil, and Crude Oil are being used as the main source of electricity. The

978-1-5090-5606-4/17 $31.00 © 2017 IEEE

current electricity capacity which provided by different producers in KSA is almost 81603MW generated by almost 81 power plants across the country [4]. At present, RE is getting government attention, particularly solar and wind energy technologies[7]. Further discussion of solar power potential in certain areas is provided next.

III. SOLAR POTENTIAL OF KSA

Solar power in KSA is considered the most effective type of renewable energy among other technologies[7]. Different climate reports showed that such area is one of the most suitable places to implement solar power projects.

TABLE I.
SAUDI ARABIA SOLAR RADIATION

Month	Solar Radiation (KWh / m² / day)
January	5.68
February	6.67
March	6.14
April	6.56
May	7.17
June	7.03
July	6.85
August	7.10
September	7.13
October	6.81
November	5.98
December	5.39
Annual	6.54

Table I illustrates the meteorological data or sometimes referred to peak sun hours (PSH). It can be seen form Table I that the potential of solar power during summer season is in between May and September and that can be considered as the abundance of daily average PSH. Furthermore, results of other solar collectors in KSA showed the yearly recorded data of the maximum solar radiation[8]. Different types of irradiance like DNI , DHI, and GHI, are defined as follows and more details about these components can be found at [1].

A. Direct Normal Irradiance (DNI)
DNI data are important factor for Concentrating Solar Power (CSP) technologies. These type of measurements use a Pyrheliometers to collect the monthly and daily radiation KWh /m²/day [1].

B. Global Horizontal Irradiance (GHI)
This type of measurements represents the solar radiation incident to predict the solar radiation during a certain tilt angle of the collector. An unshaded pyranometer is the instrument used for GHI. This value is important to Photovoltaic (PV) installations and it includes DNI and DHI.

C. Diffuse Horizontal Irradiance (DHI)
In this type of measurements, the data are useful for measuring the amount of radiation that does not reach to the surface directly from the sun.

Users can watch the live irradiance graph of GHI ,DNI, and DHI which updates every minute. By selecting different monitors, users can also see the current solar radiation levels in the selected station in real-time as it can be seen in Figure.2. The data fluctuated at the beginning of the day because it may be affected by clouds, pollution, or dust and in the afternoon tended to be clear until the end of the day. That mean the solar Intensity varies with latitude, climate, and seasons[3].

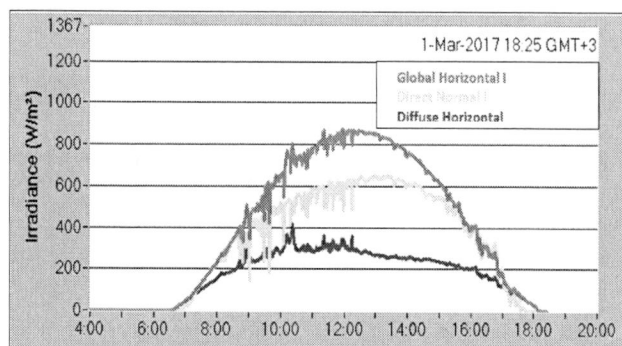

Fig.2. Southern Station Live Data of Solar

IV. METHOD

KACARE's renewable resource monitoring and mapping has been used to do the analysis of the selected cities in this paper. The program collects data from more than 65 different stations installed across the country and then users can map or tabulate different types of data for their study or researches. There are different types of stations which classified as Tier 1 Research station, Tier2 Mid-Range Station, Tier2 Uninstalled Mid-Range Station, and Tier3 Uninstalled simple station as it can be seen in Figure.3[3]. Further details about these type of stations have been discussed at [2].

From RRMM, users can graph single parameter at multiple locations or can select multiple parameters at single location to compare the results. These options have been used as an example to compare the temperature of the selected five cities as it can be seen in Figure.7.

V. RESULTS AND DISCUSSIONS

Figure.4 compares the measured values of the Direct Normal Irradiance of the five selected cities for the year 2015. In this study, different locations that have different topographies have been taken into consideration to test and compare the performance of the solar monitoring stations in each city. The main selected cities are Tabuk-north, AlDhahran-east coast, Jeddah-west coast, Riyadh-middle, and Najran-south. The maximum DNI for the first four locations occurs between June and July with an approximately 9000 Wh/m²/day as the maximum for Tabuk city and 5000-6000 Wh/m²/day for the other cities. From September to November, Najran city station recorded the highest DNI with more than 7000 Wh/m²/day. DNI can be calculated through equation (1) and the incidence angle of the sun rays should be taken into account[9].

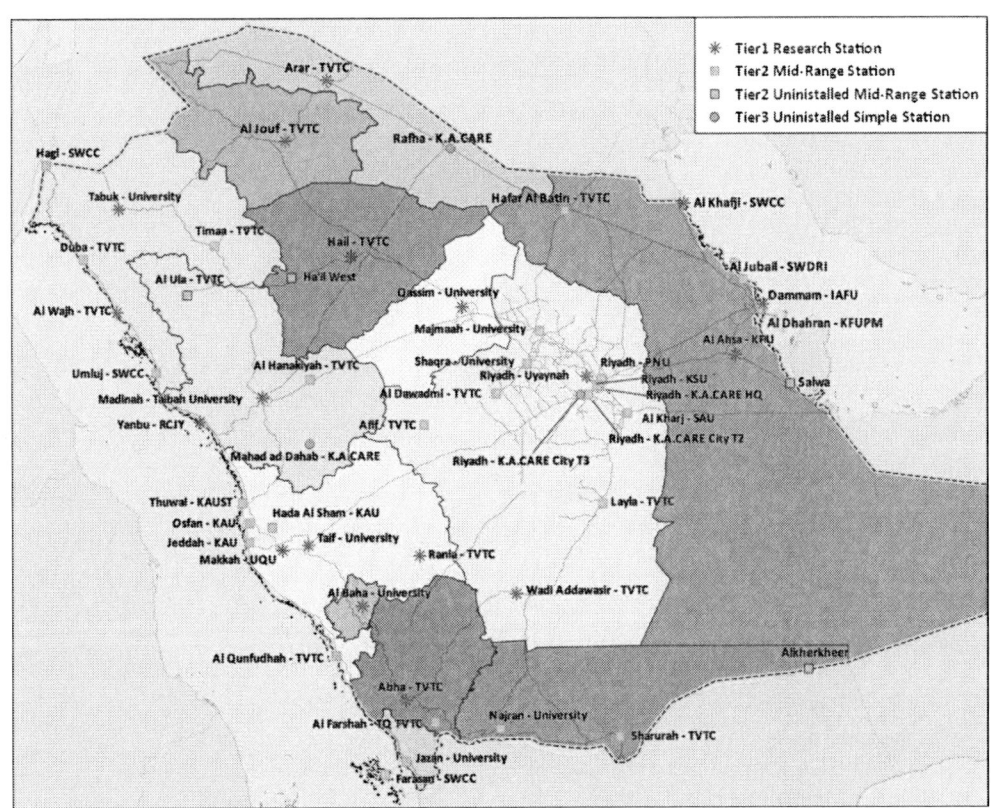

Fig.3. Resource Monitoring Stations in KSA

$$DNI = \frac{GHI - DHI}{cos\theta z} \qquad (1)$$

Where θ_z is the solar zenith angle.

GHI includes the DNI and DHI to get the total amount of radiation by using the Pyranometer (horizontal). It can be calculated by the given formula (2) where (θ) is the solar zenith angle.

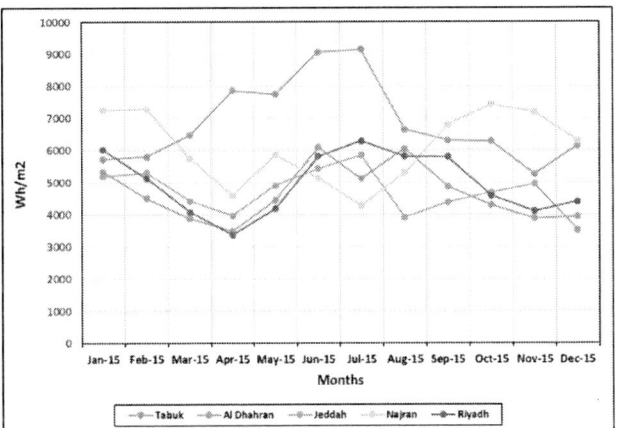

Fig.4. DNI for Different Five Cites in KSA

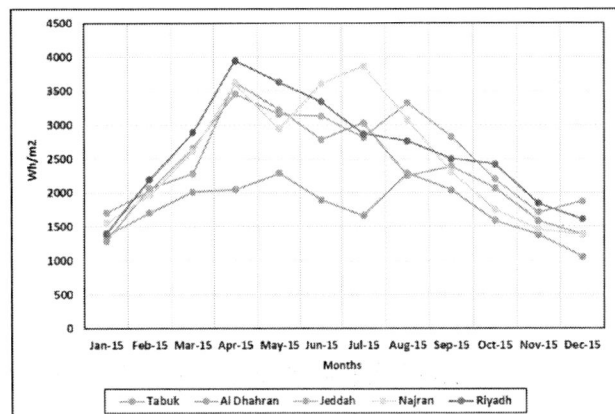

Fig.5. DHI for Different Five Cites in KSA

In Figure.5, the stations recorded the highest amount of DHI radiation at the first quarter of 2015 except Najran city which has the maximum DHI during July. For both, DNI and DHI the minimum reading happened during December as it can be seen from Figure 4 and 5.

$$GHI = DHI + DNI \, cos(\theta) \qquad (2)$$

Figure.6 Showed the GHI for the same five cities. GHI includes both values of DNI and DHI, and it is an important factor to photovoltaic installations. In this part, the GHI of the

978-1-5090-5606-4/17 $31.00 © 2017 IEEE   1112

selected places was compared for the same year of 2015. A uniform pattern for the GHI data can be observed from the graph with maximum values of 6900 Wh/m²/day to 8500 Wh/m²/day during summer season and minimum recorded values during winter in between 3200 Wh/m²/day to 4900 Wh/m²/day. Najran city has the highest annual Global Horizontal Irradiance with 80625 Wh/m²/year and Dhahran city has the lowest amount of GHI with 65892 Wh/m²/year as it can be seen in Figure.9.

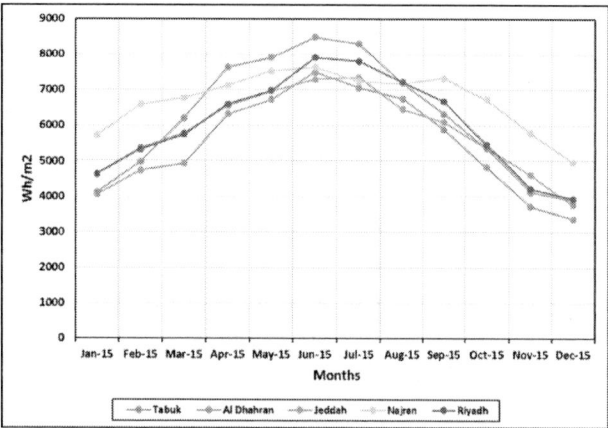

Fig.6. Global Horizontal Irradiance (GHI)

In Figure.7, the results compared the measured average temperature for the year 2015. The line graph was taken from the program after selecting the air temperature option for the same five cities to show how the results looked and compared. Air temperature is an important factor for energy production efficiency[10]. Riyadh and Dhahran recorded the highest average monthly temperature in August with 38-39 °C while Tabuk and Najran has the lowest temperature during the same period. During January, Tabuk has the lowest reading among others with an average of 13 °C. A study has been done in Dhahran city explained how efficiency varies with temperature and proved that PV module efficiency decreases as the temperature increases [11].

Dhahran city's temperature, DNI, and GHI data have been compared in Figure.8 for the year of 2015. Air temperature represented by the dotted line in °C while DNI, and GHI represented by the Orange and Blue lines respectively in Wh/m². It can be seen from the graphs that, GHI got affected directly by air temperature while DNI did not. DNI has a fluctuation between May and August, even though the temperature reached at maximum point during the same period of time. GHI has the maximum amount of irradiance during June with almost 7500 Wh/m² and DNI with 6000 Wh/m² at the same time.

Figure.9 compares the total annual DNI, GHI and DHI for the five selected spots. Tabuk city has the highest amount of DNI Solar Irradiation with an approximate annual of 82450.5 Wh/m²/year followed by Najran 73199.1 Wh/m²/year. Dhahran city and Jeddah have almost the same amount of irradiation per year.

Riyadh city came in third with a total of 59555.9 Wh/m²/year.

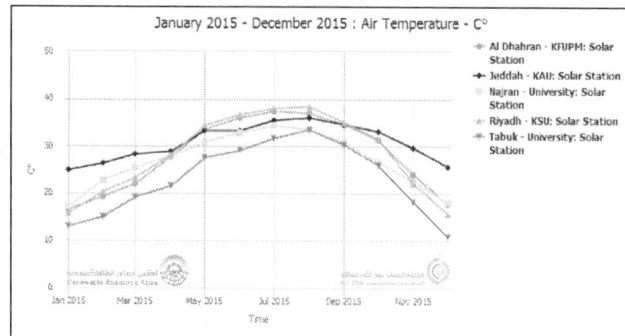

Fig.7. Monthly Average Air Temperature for the Selected Cities

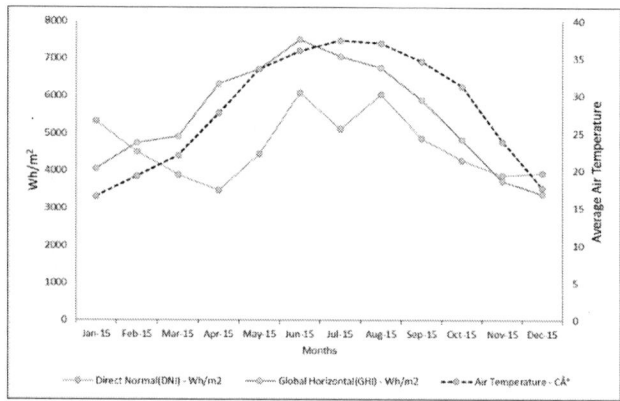

Fig.8. Dhahran City Temperature, GHI, and DNI

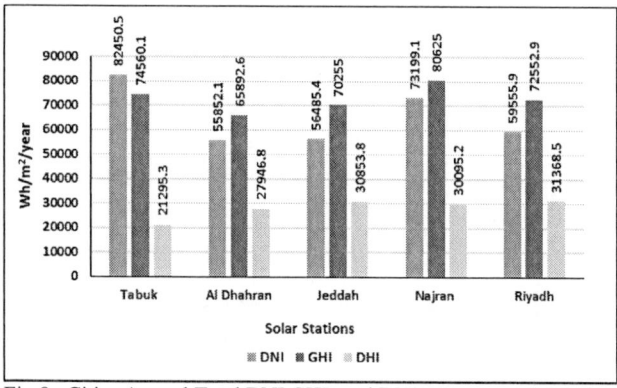

Fig.9. Cities Annual Total DNI,GHI, and DHI

The total annual GHI for the same stations has been compared in the same Figure. As it can be seen, Najran city has the highest annual GHI with 80625 Wh/m²/year followed by Tabuk with 74560.1 Wh/m²/year and Riyadh with a total amount of 72552.9 Wh/m²/year. DHI has the lowest annual variation for the selected locations during the year of 2015 and it considered as an important factor to calculate the GHI.

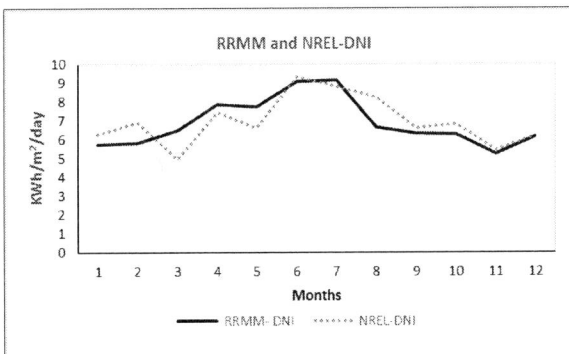

Fig.10. DNI data from RRMM and NREL

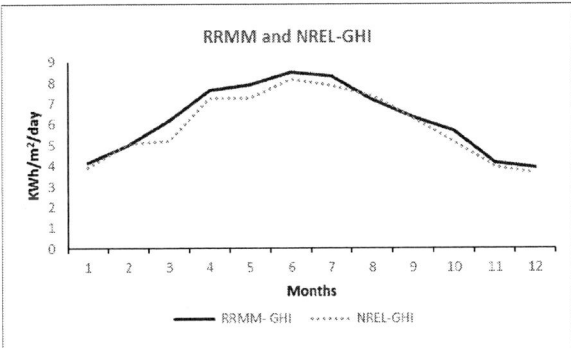

Fig.11. GHI data from RRMM and NREL

Figure.10,11 compare the earlier NREL solar atlas data with the resource monitoring and mapping program data of KACARE. Tabuk city was selected as an example to do this comparison since it has the highest DNI during the year among the other cities with 82.45 KWh/m²/year. The NREL old atlas recorded that the total DNI for same city was 83.67 KWh/m²/year with less than 2% difference between the two stations results. Also, the GHI for both stations are in close agreement with an almost 5% difference between both annual GHI data. The RRMM data used in this comparison is for the year 2015 since 2016 data are not completed yet for all locations to get it downloaded. NREL data was collected from the published report of solar radiation in Saudi Arabia and the data was collected between 1998-2000[12].

VI. STATIONS DETAILS

Table.II Provides more details about the selected stations in this study. These stations are operated and maintained on a regular schedule. This happened under the supervision of KACARE and in partnership with the government agencies, universities, and technical colleges across the country. DHI, GHI, and DHI are not only the data collected by each station, but also other meteorological parameters, such as wind speed and temperature, relative humidity and other parameters are provided for users. Location, station's type, and date of operation are some of the included information in Table II for

the five selected solar resource monitoring stations. Further information can be found at[3].

TABLE II.
STATIONS INFORMATION

Location of the Solar Monitoring Station (Station-1)	Tabuk University
Latitude (N) & Longitude (E)	Latitude (N): 28.38287 Longitude (E): 36.48396
Station Type	Tier 1C
Operating Since:	24 September 2013
Station Elevation	781 meters
Maintenance Schedule	Daily
Location of the Solar Monitoring Station (Station-2)	King Fahad University of Petroleum & Minerals
Latitude (N) & Longitude (E)	Latitude (N): 26.30355 Longitude (E): 50.14412
Station Type	Tier 2
Operating Since:	22 May 2013
Station Elevation	75 meters
Maintenance Schedule	Twice Weekly
Location of the Solar Monitoring Station (Station-3)	King Abdul-Aziz University Main Campus
Latitude (N) & Longitude (E)	Latitude (N) 21.49604 Longitude (E) 39.24492
Station Type	B
Operating Since:	28 May 2013
Station Elevation	75.6 meters
Maintenance Schedule	Daily
Location of the Solar Monitoring Station (Station-4)	Najran University
Latitude (N) & Longitude (E)	Latitude (N): 17.63228 Longitude (E): 44.53735
Station Type	Tier 1C
Operating Since:	12 December 2014
Station Elevation	1187 meters
Maintenance Schedule	Daily
Location of the Solar Monitoring Station (Station-5)	King Saud University
Latitude (N) & Longitude (E)	Latitude (N): 26.30355 Longitude (E): 50.14412
Station Type	2
Operating Since:	22 May 2013
Station Elevation	75 meters
Maintenance Schedule	Twice Weekly

IX. CONCLUSION

This paper has discussed the performance of five solar monitoring stations located in deferent provinces. The data of GHI, DNI, DHI and temperature were used in this study to analyze the stations performance and compare the solar potential of each. Also, an old data from NREL were used to compare them with the new monthly and annual data provided by KACARE atlas. The results showed that the northern city has the highest amount of DNI with an approximate annual of

82450.5 Wh/m^2/year. The maximum GHI was recorded by Najran city with 80625 Wh/m^2/year. Riyadh and Dhahran recorded the highest average monthly air temperature in August with 38-39 °C. Temperature also, compared with GHI and DNI and the results showed that GHI got affected by the raise and drop of temperature while DNI has a fluctuation and do not follow a uniform pattern. The earlier NREL solar atlas GHI and DNI data were in close agreement with the RRMM - KACARE atlas with less than 8% difference.

REFERENCES

[1] S. Alyahya and M. A. Irfan, "Analysis from the new solar radiation Atlas for Saudi Arabia," *Sol. Energy*, vol. 130, pp. 116–127, 2015.

[2] E. Zell, S. Gasim, S. Wilcox, S. Katamoura, T. Stoffel, H. Shibli, J. Engel-cox, and M. Al, "Assessment of solar radiation resources in Saudi Arabia," *Sol. Energy*, vol. 119, pp. 422–438, 2015.

[3] KACARE, "Renewable Resource Atlas." [Online]. Available: https://rratlas.kacare.gov.sa/RRMMPublicPortal/?q=en/Solar.

[4] Y. Z. Alharthi, M. K. Siddiki, and G. M. Chaudhry, "The New Vision and the Contribution of Solar Power in the Kingdom of Saudi Arabia Electricity Production," 2017, pp. 83–88.

[5] SEC, "Electrical Data2000-2014," 2015.

[6] M. M. E. Moula, J. Maula, M. Hamdy, T. Fang, N. Jung, and R. Lahdelma, "Researching social acceptability of renewable energy technologies in Finland," *Int. J. Sustain. Built Environ.*, vol. 2, no. 1, pp. 89–98, 2013.

[7] H. Al, A. Kassem, A. Awasthi, D. Komljenovic, and K. Al-haddad, "A multicriteria decision making approach for evaluating renewable power generation sources in Saudi Arabia," *Sustain. Energy Technol. Assessments*, vol. 16, pp. 137–150, 2016.

[8] I. Tlili, "Renewable energy in Saudi Arabia : current status and future potentials," *Environ. Dev. Sustain.*, pp. 859–886, 2015.

[9] A. Kazantzidis, "Estimation of direct normal irradiance from measured global and corrected diffuse horizontal irradiance," *Energy*, vol. 70, no. June, pp. 382–392, 2014.

[10] H. Bahaidarah, S. Rehman, A. Subhan, P. Gandhidasan, and H. Baig, "Performance evaluation of a PV module under climatic conditions of Dhahran , Saudi Arabia," vol. 33, no. 6, pp. 909–929, 2015.

[11] M. J. Adinoyi and S. A. M. Said, "Effect of dust accumulation on the power outputs of solar photovoltaic modules," *Renew. Energy*, vol. 60, pp. 633–636, 2013.

[12] D. R. Myers, S. M. Wilcox, W. F. Marion, N. M. Al-abbadi, M. Mahfoodh, and Z. Al-otaibi, "Final Report for Annex II — Assessment of Solar Radiation Resources in Saudi Arabia Final Report for Annex II — Assessment of Solar Radiation Resources in Saudi Arabia- 1998–2000," 2002.

First Results of a Low Cost All-Sky Imager for Cloud Tracking and Intra-Hour Irradiance Forecasting serving a PV-based Smart Grid in La Graciosa Island

David Cañadillas[1], Walter Richardson Jr[2], Benjamín González-Díaz[3], Les E. Shephard[4] and Ricardo Guerrero Lemus[1].

[1]Departamento de Física. Universidad de La Laguna. Avenida Astrofísico Francisco Sánchez S/N 38206 S/C de Tenerife. Spain.

[2]Department of Mathematics. University of Texas at San Antonio, One UTSA Circle, San Antonio, TX 78249, USA.

[3]Departamento de Ingeniería Industrial. Universidad de La Laguna. Avenida Astrofísico Francisco Sánchez S/N 38206 S/C de Tenerife. Spain.

[4]Department of Civil Engineering. University of Texas at San Antonio, One UTSA Circle, San Antonio, TX 78249, USA.

Abstract — **A low cost, edge computing, all-sky imager for cloud tracking and intra-hour irradiance forecasting has been deployed on the island of La Graciosa (Canary Islands) (29°13'51.07"N 13°30'11.09"W) in support of a new PV-based smart grid scheduled for deployment before the end of 2017. This work provides a brief description of the current power grid, the unique weather conditions on La Graciosa, the role expected for the all-sky imager in the new smart grid, a brief description of the all-sky imager deployment and operational procedures, including first images and preliminary data analyses. First intra-hour forecasting results were not possible because local authorities have not yet connected the PV plants to the smart grid and supplied the output data required to contrast the predictions with the actual power outputs.**

I. INTRODUCTION

Intra-hour irradiance forecasting is required for large PV integration to predict and manage sudden disturbances in PV production caused by clouds that can significantly impact power supply and power quality on the feeder [1]. These disturbances can be particularly severe when the power grid is small, geographically confined and nearly-isolated. These conditions can be found in insular power systems, defined by the country's incapability, due to size and/or remoteness, to interconnect with other electricity generators and consumers through a wider transmission grid outside its borders [2].

The Canary Islands are composed of 6 insular power grids characterized by different levels of renewable energy integration [3]. For example, El Hierro is expected to reach about a 60% penetration of wind energy in its power grid after the full operation of the Gorona del Viento new hydro-wind power plant [4]. Other power systems, such as the Lanzarote-Fuerteventura only achieve single-digit renewable energy penetration as a result of strong environmental regulations, a weak power grid, and unstable regulations for investing in renewable energies in Spain between 2011 and 2015 [4].

Recently, a new regulatory policy [4] provided a more stable and attractive framework for investment that benefits insular systems like those in the Canary Islands, where conventional generating costs exceed generation costs from PV technologies and where the excess savings can be shared between the owners of the PV systems and the Spanish power systems.

On the other hand, the Canary Islands Government is planning to avoid the use of ground-mounted renewable energies and, instead, install smaller plants in roofs closer to electricity users. However, as the distribution grid is not prepared for a large penetration of PV systems, mainly due to concerns about reverse power flows, voltage instabilities and drops at the end of large lines, the utility (ENDESA) is building a smart grid in a village (Fig. 1) at the far north of the Fuerteventura-Lanzarote insular power system (La Graciosa) as an test facility for large penetrations of PV systems.

The smart-grid is expected to serve as an international test bed for future studies, and it will be a valuable asset for researchers aiming to test their hypothesis (of several areas of study, such as telecommunications, electric engineering, batteries...) in a real environment. This will be particularly appealing, not only for researchers of the University of La Laguna, but also to attract international talent.

Fig. 1. (a) Distribution grid of Lanzarote - La Graciosa, and (b) view of La Graciosa from Lanzarote.

978-1-5090-5606-4/17 $31.00 © 2017 IEEE

This work provides a brief description of the weather conditions in the Caleta de Sebo, where the smart grid is being built, main characteristics of the current power grid, environmental restrictions on the island as a strong limiting factor for large renewable energy deployments, and the role expected for the smart grid and sky imager to enlarge the penetration of distributed PV. In addition, a brief description of the deployment process for the sky imager last December in La Graciosa, the operating procedure, and initial image acquisition and data analysis are included. Finally, first intra-hour forecasting results were not possible because local authorities have not yet connected the PV plants to the smart grid and supplied the output data required to contrast the predictions with the actual power outputs.

II. CHARACTERISTICS OF THE LOCATION

La Graciosa is a 29 km^2 island located in a marine nature reserve to the north of Lanzarote (Fig. 1). About 660 inhabitants are currently living in the capital of the island (Caleta de Sebo). There is another populated area on the island (Casas de Pedro Barba), but inhabited only seasonally, and the area is not connected to the power grid. Average monthly global irradiation reaches 5.157 kWh/kW (1,883 kWh/kW·yr), and the average monthly temperature reaches a high of 20.8° C [5] (Fig. 2).

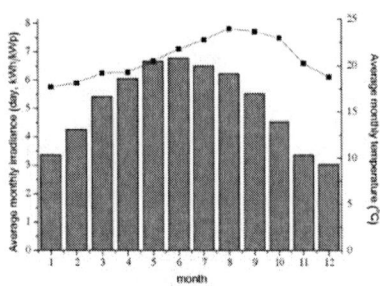

Fig. 2.
Average monthly irradiation (bars) and temperature (symbols) in Caleta de Sebo.

Irradiance is affected by the surrounding orography, as the highest mountain range (Altos de Famara) in Lanzarote (671 meters in height) is located just on the opposite side of the strait between both islands (Fig. 3). This means no direct sun irradiation in the mornings and the regular accumulation of clouds, prevalent for extended periods of the year, and a further analysis is required to understand their impact on PV power production and the interactions with the planned smart grid storage capacity.

Lanzarote's power grid is supplied by 209 MW of conventional power sources (diesel engines and gas turbines) and Fuerteventura is connected to 159 MW of similar power sources. Lanzarote also has 8.8 MW of wind energy (two wind farms) and 13.1 MW in Fuerteventura (three wind farms).

Finally, in Lanzarote there are 7.7 MW of PV connected to the grid and 0.15 MW off-grid. In Fuerteventura there are 13.0 MW of PV connected to the grid and 0.08 MW off-grid [6].

Fig. 3. Solar irradiation map of La Graciosa and north of Lanzarote. The Famara mountain range can be observed in green color as the irradiation diminishes on the western slopes.

The power grid in La Graciosa is supplied by three 20/0.4 kV transformers (630, 400, and 400 kVA) and connected to Lanzarote by a 20 kV seabed cable. Currently, the island has two PV power plants (5 kW and 30 kW), but recently the Lanzarote Government has offered financial support for increasing the PV capacity of La Graciosa, thereby enhancing the attractiveness of a smart grid for managing power supply and demand on the island [4].

III. ATMOSPHERE AND CLOUD DYNAMICS

The geographic location of the Canary archipelago, near the African west coast, with subtropical latitudes and a high influence of the north-east trade winds, determines an atmospheric condition of stability, with a steady vertical structure of the low troposphere. The meteorology of the islands is also highly influenced by the North Atlantic anticyclone (which varies its strength and position throughout the year), the cold oceanic current flowing through the archipelago, the dust transported from the African deserts and the relative pronounced orography of the islands [7][8].

A. Atmosphere structure

The vertical structure found in the troposphere is characterized by the almost constant presence of a thermal inversion, which makes the troposphere over the Canary Islands to be very stratified. Subsidence conditions in the higher layers of the troposphere, along with frequent trade winds flowing in the lowest layer, are responsible for the strong and stable temperature inversion which separates the dry free troposphere from the humid oceanic boundary layer. This thermal inversion varies its characteristics throughout the year, reaching its maximum thickness and temperature gradient during the summer, where the layer is located at an average height of 800 m a.s.l., averaging a thickness of 500 m and a temperature gradient of 5 °C [8][9].

Knowledge of the atmospheric conditions is rather relevant for intra-hour solar forecasting in this case, since these conditions of stability in the lower layers of the atmosphere have a large influence over the cloud dynamics of La Graciosa. In this sense, thermal inversion acts as a stopper, neglecting the ascension of the clouds over the level where the inversion is located, and it results in the formation of stratocumulus clouds [8]. As the trade winds moves towards the Altos the Famara Mountains, stratocumulus clouds tend to stack on the mountain side.

B. Cloud Base Height (CBH)

The situation above explained reveals a clear relation between the height of the base of the thermal inversion and the cloud base height (CBH), which is a necessary input to perform accurate solar forecasting based on ground-based sky-imagery. Existing technologies for CBH measurement includes ceilometers, radiosondes, LIDAR or stereographic cameras. Although one of the final objectives of the project is to obtain the CBH using two cameras in a stereographic approach, during the first stages (and until the second sky-imager is deployed), the determination of the CBH may be made considering the particularities of the atmosphere in the region, using data from previous studies that characterize the height of the thermal inversion and its evolution throughout the year. Although this approach is not common for CBH determination, the special characteristics of the troposphere of the region, as well as the relative easiness to implement it, allow to, at least, try this approach and validate its suitability for future applications.

C. Influence of Saharan dust

Another considerable factor is the influence of the African dust transported from the Sahara and Sahel deserts. The influence of this dust is more pronounced during the winter, because in the summer these masses of air containing high amounts of dust particulates comes at higher altitudes than the thermal inversion, meaning that intrusions of dust are less frequent [9]. Haze dust densities up to $320 \, g/m^3$ and total dust load up to $500 \, mg/m^2$ has been measured in the Canary Islands [10]. Moreover, during the Saharan hazes dust invasions the relative humidity increases (> 90%, while in non-hazes days is around 65%). This increment is interpreted as an adhesion of water molecules to the Saharan haze dust during its trip over the Atlantic Ocean, acting as seed for cloud formation [10]. The dispersion of dust in the atmosphere influences the solar radiation reaching the surface, and has some undesired side effects over the proper function of the sky-imager. On the one hand, the presence of dust in the atmosphere leads to a higher scattering of light that causes a drop in the quality of the images taken by the camera. On the other hand, dust deposition over the enclosure of the camera occurs frequently, introducing some unwanted stains on the pictures that may affect the image processing. Hence, it is necessary to carry on some kind of scheduled cleaning of the enclosure of the camera to avoid this problem.

IV. DEPLOYMENT OF THE ALL-SKY IMAGER

The all-sky imager was installed on the roof of the main municipal building of Caleta de Sebo, where the offices of the local authority and police are located (Fig. 4).

Fig. 4.   Location where the all-sky imager is deployed.

This location was selected mainly because a 5 kW PV plant is located on the same building and will supply power and data to the smart grid (Fig. 5). Also, this location provides a convenient connection to the internet for remote access of the all-sky imager from the University of La Laguna. Due to strict environmental rules, no renewable technologies are permitted on the island outside the urban area of Caleta de Sebo.

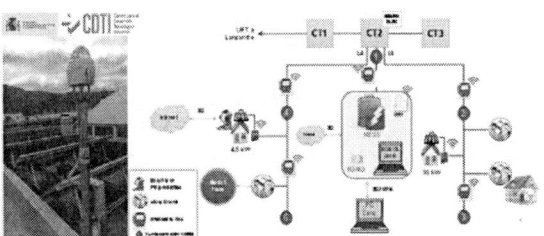

Fig. 5.   (a) View of the all-sky imager located on the roof of the main municipal building in Caleta de Sebo, and (b) scheme of the smart grid.

The enclosure of the camera has been selected to protect sensitive components from harsh climatic conditions that combine high values of humidity and salinity from the ocean, with seasonal dust that originates in the Sahara Desert. The camera is also equipped with a high capacity microprocessor to capture and analyze images, which will provide intra-hour forecasting data to the smart grid (edge computing). Further details of the prediction methodology and hardware can be found in [11].

The first all-sky images show significant obstacles on the horizon due to the Altos de Famara mountains range located in the southeast, and the roofs and antennae of adjacent buildings, as the maximum height of any construction in Caleta de Sebo is limited to one floor (Fig. 6). These obstructions diminish the visual field, but will be properly managed to avoid any loss in the quality of the forecasting.

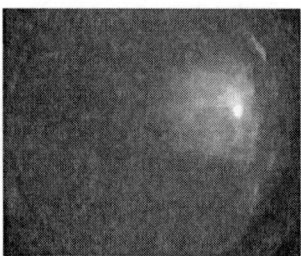

Fig. 6. Sample image of a cloudy period provided by the all-sky imager located in Caleta de Sebo. Obstructions (nearby buildings and mountains) can be observed around the perimeter of the image.

IV. IMAGE CAPTURE AND PROCESSING

The image capture and image processing are two different but not independent issues, since the performance of the latter process will depend on the quality of the images provided by the former.

A. Image capture

The acquisition of the sky images may be an intricate issue, especially when using low-cost camera sensors and lenses. The sky-imagers are focusing directly to the sun, with the associated problems that it has (mainly overexposure of the images that may even damage the camera sensor). Moreover, light conditions are constantly changing all over the day and throughout the year, which makes difficult to find a set of camera properties that performs adequately in every situation. Furthermore, physical camera properties which are usually controlled to avoid these issues (e.g. aperture), cannot be modified in the device. For that reason, photography techniques such as High Dynamic Range (HDR), may be applied to obtain better quality images that will not disrupt the image processing process.

HDR is a technique used in photography consisting in taking several images at different exposures times and merge them together via software to create a new image containing information of the previous ones. HDR increase the computational burden, but it can be found a compromise between the time resolution, the required processing resources to perform the HDR, and the quality of the images needed to continue with the image processing method. The objective is to achieve images that should be consistent with each other (i.e. minimizing the variations in the luminescence of the sequence), which will not disrupt or add complexity to the following steps of the process.

B. Image pre-processing

Prior to proceed with the higher computer-demanding steps, it is worth to evaluate if the image processing is really necessary or not. There are several approaches to decide when the sky imager should initialize the algorithms and perform the forecast and when not. For example, in the persistence method, images are analyzed before making any further processing step to identify if the sky is clear or full covered by clouds. In these two situations, short-term solar forecasting has no sense since no rapid variations in solar irradiance will occur. Once major changes are detected in the images i.e. the clouds are approximating to the region of interest in which the nearby PV power plants will be affected, the image processing process will begin to predict when the variations in irradiance will happen, and ultimately, the magnitude of these changes.

Typical implementations of this method consider the percentage of pixels above a certain threshold (of either luminescence or RGB values). Another approach is to use machine learning classification algorithms to create recognition patterns for the sky-imager, which would be able to evaluate the image and independently decide if it is necessary or not to perform the image processing.

C. Image processing

The image processing consists in a series of computer algorithms/scripts that accomplish the identification of the clouds in the image, detect suitable features to follow between images or frames, track these features in consecutive images providing movement vectors for each feature, and then project the position of those feature points (making use of the movement vectors) to a determined time ahead, to forecast where the clouds will be after a certain time.

The process begins with some previous steps to set up the image: orientating the image to match the true north of the location with the vertical axis of the image and applying a mask to remove unnecessary information from the image matrix (such as buildings or mountains).

Then, the algorithm must identify which parts of the image correspond to clouds, and which ones not. This may result a hard task, since encoding of the images can be quite catchy, and depending on the light conditions present when the picture was taken, the values of the matrix created to store image data may have a large variation in two consecutive frames. Reviewing the literature, a typical approach to overcome this problem is using the "red-to-blue ratio (RBR)" [12], which is the ratio between red and blue channel in the RGB/BGR color spaces. Although it is a contrasted technique, there are certain conditions under which the algorithm tends to fail, especially with clear sky pixels near the sun that may have similar characteristics to those representing the clouds, such as high luminescence, resembling tonality or even equal RBR.

This is when it comes up the idea of using different color spaces (and their respectively channels) to retrieve additional information to include in the original approach. Some authors have used the HSV color space, taking the saturation channel (S) to have a better identification of the borders of the clouds [12][13]. However, the way in which HSV color space is encoded may result in very noisy or blotchy images in the H channel, mainly due to the noise introduced by camera sensors and gray tonalities. This can be seen in Figure 7.

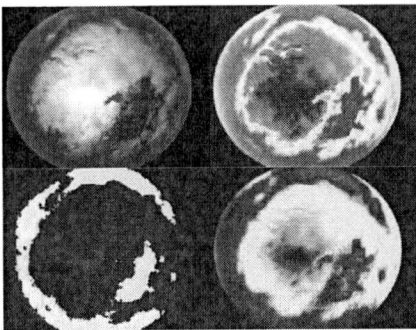

Fig. 7. Sky image in HSV color space. Individual channels are color mapped (a) Original image, (b) V channel, (c) H channel and (d) S channel.

Other color spaces, such as YCbCr or CieLAB may be used for the segmentation of the clouds. Those color spaces use an exclusive channel to store the luminescence of the image (V in HSV, Y in YCrCb and L in CieLAB), using the remaining channels to store color information in different formats. Other image processing techniques, such as adaptive thresholding may also be useful to segment the clouds in the images.

Currently, non-demanding algorithms which make use of different color spaces and various image processing techniques are under development, taking advantage of the best features of each different color space and combining them together to achieve a robust and versatile method with good performance in all different light conditions that may face the sky imager over the different times of the day/year. Figure 8 shows a sample image of one of the cloud-decision algorithms which combines adaptive thresholding with a normalized RBR mask.

Fig. 8. Example of the output of one of the cloud-decision algorithms. (a) Original image, (b) cloud-decision mask and (c) contours of clouds over the original image.

Once clouds are identified, the second objective is to know towards which direction are the clouds moving. To accomplish that, sequences of images taken every 5 seconds are analyzed. Using the segmentation mask created in the former step, pixels of the image are classified as clouds or clear sky (Figure 8.b). The algorithm is then able to find and draw the contour of the clouds (Figure 8.c). Over that curves, potential good features to track are identified (generally pronounced angles), which are used to perform an Optical Flow (OF) analysis.

The OF is a technique used in image processing to estimate the apparent motion of objects in a sequence of images. The analysis can be made by means of different approaches, being more common the differential methods, concretely the Lucas-Kanade (LK) method. The LK method assumes that the pixels identified as features moves around the neighborhood (i.e. displacement of the features is small), which is defined as a square window of n pixels where the LK algorithm will look for possible matches, giving more weight to pixels closer to the center (or feature pixel) [14][15]. This characteristic makes the LK algorithm to perform better with image sequences with higher frame-rates (videos) than with sequence of pictures taken every 5 seconds, since the displacements of pixels are higher. Figure 9 shows the different output in the optical flow using a video and a sequence of pictures as input. The output of the OF are the movement vectors of the pixels which, combined with the frame-rate of the sequence, will result in velocity vectors.

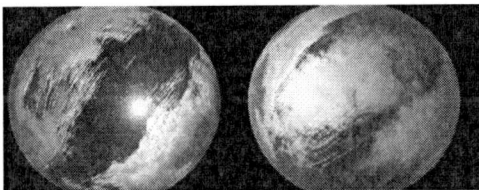

Fig. 9. Lucas-Kanade Optical Flow running (a) over a video and (b) over a sequence of pictures taken every 5 seconds.

The next step involves the projection of the velocity vectors to the required time ahead. The projection of these vectors is based on assumptions that may not be necessarily true, such as clouds will continue moving towards the same direction with the same speed that it did before. This is one of the major drawbacks of this technique which may increase the error in the forecasts. Additionally, when applied to sequence of images taken every few seconds, the error induced by the LK algorithm will be added to that forecast error. In Figure 10, an image of a preliminary forecast where some vectors have no concordance with the others is shown.

Fig. 10. Projection (green) of the movement vectors (red) obtained in the LK optical flow. It can be seen how some vectors does not keep consistency with the rest.

These effects may be mitigated applying some post-processing algorithms which will discard the outlier vectors. Ultimately, the idea is to incorporate advanced computing techniques for machine learning, including neural networks, which will improve the overall performance of the process.

V. RESULTS AND DISCUSSION

The design and operational procedures of a low-cost all-sky imager giving service to a PV-based smart-grid deployed in La Graciosa Island have been described. Descriptions of the location, the power grid and the characteristics of the atmosphere and cloud dynamics have also been presented.

Results of the forecast performed by the sky-imager could not have been provided, since local authorities have not yet supplied the output data from their PV plant located near the device, required to contrast the predictions with the actual power outputs. Hence, the validation of the forecast and further statistical analysis could not be performed yet. Nevertheless, it is expected that this data will be available in the next months, and by end of October of 2017, the sky-imager should be providing forecast data to the smart-grid, once the algorithms are validated.

One of the main goals of the project is to obtain the CBH by means of a pair of sky-imagers, using stereographic techniques. Thus, a second sky-imager is planned to be deployed in the next few months in La Graciosa, which should be paired with the one that is already in the island. This situation will change drastically the approach and the codes being used so far. Despite that, the devices and the codes have been designed to be versatile taking into account this fact, and should not be further problems besides the complexity of the task itself.

For the image capture, it is planned to try some new lenses and sensors, as well as to improve the HDR algorithm. In this sense, some sort of adaptive settings for the camera and the algorithm, which take into account different light conditions, are under development.

Regarding the processing of the images, there is a lot of room for improvement. Future actions will cover the implementation of machine learning techniques in various stages of the process (image pre-processing, analysis of the optical flow output…), as well as exploring different cloud-decision algorithms and segmentation techniques. Also other optical flow algorithms may be implemented to contrast their performance with the LK method in the specific problem of cloud tracking.

A scheduled cleaning of the enclosure of the sky-imager should be done to avoid the adverse effects that the dust deposited on the enclosure has over the quality of the images. Given the influence that dust has over optical devices, PV modules should also be cleaned periodically to fully exploit their capacities.

V. ACKNOWLEDGEMENTS

This work has been funded under the project GRACIOSA (ITC-20151241) by the Centro de Desarrollo Tecnológico Industrial (CDTI), Ministerio de Economía, Industria y Competitividad, and co-financed by the European Union Structural Funds (FEDER).

REFERENCES

[1] D. Nguyen and J. Kleissl, "Research on impacts of distributed versus centralized solar resource on distribution network using power system simulation and solar now-casting with sky imager," in 42nd IEEE Photovoltaic Specialist Conference (PVSC), New Orleans, 14-19 June 2015.

[2] P.A. Fokaides and A. Kulili, "Toward grid parity in insular energy systems: the case of photovoltaics (PV) in Cyprus." Energy Policy 2013;65:223–8.

[3] 3448 ANUNCIO de 13 de septiembre de 2016, por el que se hace pública la aprobación definitiva de las bases reguladoras de subvenciones s Instalaciones de Autoconsumo mediante Sistemas de Energías Renovables en Edificiaciones Conectadas y no Conectadas a la red de distribución. Boletín Oficial de Canarias 187, 26 Sept 2016. Available on Jan 13th 2017 at http://www.gobiernodecanarias.org/boc/2016/187/020.html

[4] R. Guerrero-Lemus, B. González-Díaz, G. Ríos, R. Dib, "Study of the new Spanish legislation applied to an insular system that has achieved grid parity on PV and wind energy," Renewable and Sustainable Energy Reviews 2015;49:226-236.

[5] Sistema de Información Territorial de Canarias. Available on 16 January 2017 at http://visor.grafcan.es/visorweb/default.php?svc=svcStatISTAC&lat=28.3&lng=-15.799999999999954&zoom=8&lang=es

[6] Anuario Energético de Canarias 2014. Available on 16 January 2017 at http://www.gobcan.es/ceic/energia/doc/Publicaciones/AnuarioEnergeticoCanarias/Anuario2014.pdf

[7] C. J. Torres, E. Cuevas, J. C. Guerra, V. Carreño, O. Atmosférico, and D. I. Inm, "Caracterización de las Masas de Aire en la Región Subtropical sobre Canarias," *Proceedings of the V Simposio Nacional de Predicción*, no. 1, pp. 2–7, 1998.

[8] M. V. Marzol, "Temporal characteristics and fog water collection during summer in Tenerife (Canary Islands, Spain)," *Atmos. Res.*, vol. 87, no. 3–4, pp. 352–361, 2008.

[9] M. Viana, X. Querol, A. Alastuey, E. Cuevas, and S. Rodríguez, "Influence of African dust on the levels of atmospheric particulates in the Canary Islands air quality network," *Atmos. Environ.*, vol. 36, no. 38, pp. 5861–5875, Dec. 2002

[10] C. Montes et al., "Effects of the Saharan dust in the performance of multi-MW PV grid connected facilities in the Canary Islands (Spain), 25th European Photovoltaic Solar Energy Conference and Exhibition, pp. 5046-5049, 6-10 September 2010, Valencia, Spain.

[11] W. Richardson Jr, H. Krishnaswami, R. Vega, and M. Cervantes, "A low cost, edge computing , all-sky imager for cloud tracking and intra-hour irradiance forecasting," Sustainability/MDPI, pp. 1–17, 2017.

[12] H. Y. Cheng and C. L. Lin, "Cloud detection in all-sky images via multi-scale neighborhood features and multiple supervised learning techniques," *Atmos. Meas. Tech.*, vol. 10, no. 1, pp. 199–208, 2017.

[13] D. Nguyen and J. Kleissl, "Stereographic methods for cloud base height determination using two sky imagers," *Sol. Energy*, vol. 107, pp. 495–509, 2014.

[14] P. Wood-Bradley, J. Zapata, and J. Pye, "Cloud tracking with optical flow for short-term solar forecasting," *50Th Conf. Aust. Sol. Energy Soc.*, no. November, pp. 2–7, 2012.

[15] F. Barbieri, S. Rajakaruna, and A. Ghosh, "Very short-term photovoltaic power forecasting with cloud modeling: A review," *Renewable and Sustainable Energy Reviews*, vol. 75. pp. 242–263, 2017.

Statistical Analysis of PV Insolation Data

Abdulmunim Guwaeder, and Rama Ramakumar

Oklahoma State University, Stillwater, Oklahoma, 74075, USA

Abstract — In order to develop renewable energy sources technology in power supply plants, it's necessary to provide analysis at any area of suitability in its employment, one of the typical renewable resources considered is photovoltaic (PV). This paper presents the observed radiations at four locations in Libya to analyze the statistical solar radiation data recorded based on analysis of data provided by Centre for Solar Energy Research and Studies of Libya (CSERS). In order to get appropriate probability distribution that best fits the data for a given month of the year. The frequency distributions used include Weibull, Normal, Lognormal and Gamma.

Index Terms — Solar radiation, Probability density function, photovoltaic.

I. INTRODUCTION

Solar energy is highly site specific and readily available, solar technologies are pollution free and therefore act as the leading potential source of alternative energy. Moreover, cost of photovoltaic (PV) modules has been declining significantly at an average rate of 20% with each doubling of sales. The knowledge of available solar radiation of a location is a fundamental requirement before embarking on any solar energy project such as photovoltaics farm, solar thermal system, and passive solar design [1].

Many developing countries do not have the equipment for continuous and exact measurements of solar radiation because of expensive measuring facilities and techniques required. It is therefore necessary to develop alternative methods to estimate the solar radiation of potential locations. Knowing the probability distribution function of any time series data such as solar radiation of a site enables someone to generate data that will have the same characteristic as the actual data for any solar power project.

In this paper, probability distribution functions that best model the solar radiation data of the four cities (Sabha, Bengazi, Alkofra, Abokamash)of Libya are obtained, similar to the application of statistical analysis of solar radiation data in Taiwan [2].

II. AREA OF STUDY

Libya is a crude oil exporting nation and the second largest North African country with an area of 1,759,540 square Km and about 1900Km of costal line. Libya's economy is dominated by the oil sector, around 95% of export revenues is generated by the energy sector. In terms of solar energy, it could be argued that solar energy is the most important renewable energy resource, as Libya enjoys high-level insolation. There are three locations in Libya where data being recorded based on analysis of data provided by Centre for Solar Energy Research and Studies (CSERS). Geographic details of the locations are summarized in Table 1. The inability of the electricity utility company to meet the required electricity demand and the availability of good solar radiation throughout the year makes those cities of Libya good candidate for various solar power projects.

TABLE 1. THE INFORMATION OF THE METEOROGICAL STATIONS

Station	Latitude °N		Longitude °E	
Sabha	27	2	14	25
Bengazi	32	11	20	03
Alkofra	24	17	23	15
Abokamash	33	04	11	44

III. MONTHLY INSOLATION DATA

Modeling monthly solar radiation data by statistical analysis of the distribution functions has been well documented in this paper. Although, several probability distribution functions used and the best four among them reported. The idea was applied to find estimate the best fitted probability distribution. The analysis was conducted on the monthly solar radiation data obtained in the past years to gain much insight of the data to constructively solve the problems as tested. In the coastal regions, the daily average of solar radiation on a horizontal plane could reach $7.1\ kWh/m^2/day$ and $8.1\ kWh/m^2/day$ in the southern region as Libya has a great potential for harnessing solar energy [3]. Table 2 shows the average monthly solar radiation values $kWh/m^2/day$ in four locations.

TABLE 2. AVERAGE SOLAR RADIATION $kWh/m^2/day$ IN FOUR LOCATIONS [4].

Month	Sabha	Bengazi	Alkofra	Abkamash
Jan	3.758388	2.222329	5.714956	3.025136
Feb	4.388402	3.009039	6.409725	3.894885
Mar	5.446902	3.801211	7.0658	4.972774
Apr	6.249389	4.827271	7.468522	5.662379
May	6.598697	5.969165	7.091211	6.620364
Jun	7.200624	6.482689	7.449026	7.017378
Jul	7.563766	7.399022	7.438481	7.544454
Aug	7.340048	6.974948	7.512811	6.902298
Sep	6.24805	5.550657	7.287623	5.54123
Nov	4.924692	3.350155	6.682038	4.108471
Oct	3.886153	2.383708	6.052598	3.275875
Dec	3.263442	1.767222	5.472605	2.767346

IV. PROBABILITY DENSITY FUNCTION

A. Weibull Distribution

Weibull probability density function has been widely used, accepted and recommended to model solar irradiation and to estimate solar energy potential [5]-[6]. The general form of the Weibull density function for solar irradiation is given by the Equation 1:

$$f(x) = \left(\frac{\beta}{\alpha}\right)\left(\frac{x}{\alpha}\right)^{\beta-1}\exp\left[-\left(\frac{x}{\alpha}\right)^{\beta}\right] \tag{1}$$

where $f(x)$ is the probability density function for the random variable solar irradiation (x), β is shape parameter and α is a scale parameter having units of solar radiation $kWh/m^2/day$. The corresponding cumulative distribution function is given by Equation 2:

$$F(x) = 1 - \exp\left[-\left(\frac{x}{\alpha}\right)^{\beta}\right] \tag{2}$$

There are several methods to estimate Weibull parameters as summarized in [7]-[8] values of α and β can be obtained by the method of moments, maximum likelihood, Weibull probability paper (graphic method) and power density. The moment method using the mean solar radiation μ and standard deviation σ is employed in this paper. The applicable Equations are given by 3 and 4:

$$\mu = \alpha\, \Gamma\left(1 + \frac{1}{\beta}\right) \tag{3}$$

$$\left(\frac{\sigma}{\mu}\right)^2 = \left[\frac{\Gamma\left(1+\frac{2}{\beta}\right)}{\Gamma^2\left(1+\frac{1}{\beta}\right)}\right] - 1 \tag{4}$$

where Γ is the Gamma function. Values of α and β can be found by using the approximate relation for equation 4 and the inverse of Equation 3:

$$\beta = \left(\frac{\sigma}{\mu}\right)^{-1.086} \tag{5}$$

$$\alpha = \frac{\mu}{\Gamma\left(1+\frac{1}{\beta}\right)} \tag{6}$$

B. Normal Distribution

The probability density function for normal distribution and the cumulative distribution function is given by [6]:

$$f(x) = \frac{1}{\sigma\sqrt{2\pi}}\exp\left[-\left(\frac{x-\mu}{2\sigma^2}\right)^2\right] \tag{7}$$

$$F(x) = \frac{1}{2}\left[1 + er\, f\left(\frac{x-\mu}{\sigma\sqrt{2}}\right)\right] \tag{8}$$

where the $erf\ (\ )$ is the error function given by:

$$erf(x) = \frac{2}{\sqrt{\pi}}\int_0^x \exp(-t^2)dt \tag{9}$$

C. Lognormal Distribution

The lognormal distribution assumes that the natural logarithm of the random variable is normally distributed with a mean value of μ and standard deviation σ [6]. It should be noted that μ and σ are the mean and standard deviation of the variables natural logarithm, respectively. The Lognormal distribution is given by:

$$f(x) = \frac{1}{x\sigma\sqrt{2\pi}}\exp\left[-\left(\frac{\ln(x)-\mu}{2\sigma^2}\right)^2\right] \tag{10}$$

The cumulative distribution function is given by Equation 11:

$$F(x) = \frac{1}{2}\left[1 + er\, f\left(\frac{\ln(x)-\mu}{\sigma\sqrt{2}}\right)\right] \tag{11}$$

D. Gamma Distribution

The gamma distribution is a two parameter distribution that has properties similar to those of the Weibull distribution. The two parameters, a scale parameter α and shape parameter β can be adjusted to fit observed data [5].the probability density function of gamma distribution is given by Equation 12:

$$f(x) = \frac{x^{\beta-1}}{\alpha^\beta\, \Gamma(\beta)}\exp\left[-\left(\frac{x}{\alpha}\right)\right] \tag{12}$$

where x is the measured solar radiation data. The cumulative distribution function is given by:

$$F(x) = \frac{1}{\alpha^\beta \Gamma(\beta)}\int_0^x t^{\beta-1}\exp\left[-\left(\frac{t}{\alpha}\right)\right]dt \tag{13}$$

V. GOODNESS OF FIT TESTS

In this paper two goodness of fit tests were applied at 5% level of significance to check how accurate distribution model fits the observed data, those goodness of fit tests are Chi-Squared test x^2, and Kolmgorov-Smirnov KS test. Root mean square error and correlation coefficient R, are also used to determine the error and describe the correlation between the data series and each distribution, the distribution which verify highest value of R, and lowest value of RMSE, is the most accurate model fits the observed data.

A. Chi-Squared test

Chi-Squared test is used to determine whether the sample data set fits a specified distribution. It divides the samples into K bins and then calculate a test statistic x^2 gives as [9]

978-1-5090-5606-4/17 $31.00 © 2017 IEEE

$$\chi^2 = \sum_{i=1}^{k} \frac{(O_i - E_i)^2}{E_i} \quad (14)$$

Where O_i denotes the observed frequency, E_i denotes the expected frequency. E_i Can be calculated by [9]

$$E_i = n \int_{xL}^{xU} f(x) = n[F(x_U) - F(x_L)] \quad (15)$$

Where $f(x)$ is the specified probability density function, $F(x)$ is the corresponding cumulative distribution, and x_U, x_L are the lower and upper limits of the ith bin. Value of test statistic x^2 must be less than critical value C.V to be considered that the sample data set follows a specified distribution.

B. Kolmogorov-Smirnov test

This test based on the largest difference between the observed cumulative distribution function. $F_o[x_{(i)}]$ And hypothesized cumulative distribution function, $F_T[x_{(i)}]$. As shown in table 3 values of test statistic d must be less than critical value to be considered that the sample data set follows a specified distribution. The KS test can be estimated by the following expression given as [10]

$$d = max_{i=1}^{n}\{|F_o[x_i] - F_T[x_i]|\} \quad (16)$$

Where

$$F_o[x_i] = \frac{i}{n} \quad (17)$$

C. Root Mean Square Error

Root mean square error (RMSE) is a criterion, which used to know how closely a model fits some data, it's a tool that used to inspect the quality of the fit. The RMSE can be estimated by the following expression [11].

$$RMSE = \sqrt{\frac{1}{k}\sum_{i=1}^{k}(O_i - E_i)^2} \quad (18)$$

The lower the RMSE, the closer model follows data.

D. Correlation Coefficient

The correlation coefficient, R, describes the correlation between the data series, where values between -1, a perfect negative correlation, and 1, a perfect positive correlation. The correlation coefficient is given by [11].

$$R = \frac{1}{k-1}\sum_{i=1}^{k}\frac{(O_i - O_m)(E_i - E_m)}{\sigma_O \, \sigma_E} \quad (19)$$

Where O_m, E_m denote the mean of observed frequency and expected frequency respectively. $\sigma_O \sigma_E$ Indicated the standard deviation of observed frequency and expected frequency respectively.

VI. DATA ANALYSIS AND DISCUSSION

Some standard continuous probability distributions were carefully chosen based on their shape parameters to fit the data, also parameter values for each probability distribution are determined by using the corresponding mean and standard deviation of the solar radiation. The distributions are the Weibull, Normal, Lognormal and Gamma distribution, least square and the maximum likelihood estimation procedures were then applied to estimate the parameter for each distribution from the data. The solar radiation for the individual month has the maximum monthly mean values between 7.20 to 7.51 $KWh/m^2/day$, which arises in June, July and August while a minimum value between 1.76 to 3.7 $KWh/m^2/day$ occurs in December and January. Weibull, Normal, Lognormal and Gamma are fitted to solar radiation data, parameters of these distribution are estimated analytically based on actual data by using formulas described in section 3. Table 3 shows the annual statistical quantities of solar radiation for the four locations in Libya.

TABLE 3. ANNUAL STATISTICAL QUANTITIES OF SOLAR RADIATION IN FOUR LOCATIONS.

Site	Weibull		Normal		Lognormal		Gamma	
	α	β	μ	σ	μ	σ	α	β
Sabha	5.8	4.1	5.57	1.50	5.57	1.50	13.65	0.40
Bngazi	6.0	3.8	4.47	1.97	4.47	1.97	5.13	0.87
Alkofr	6.0	4.4	6.80	0.72	6.80	0.72	87.17	0.07
Abokm	6.3	4.2	5.11	1.68	5.11	1.68	9.19	0.55

Two goodness of fit tests were used to identify the best fit at 5% level of significance, which are Chi-Squared test, and Kolmogorov-Smirnov test, the test statistic values of these tests are presented versus critical values of each test as shown in table 4 and 5. According to x^2 and KS tests it observed that all cities have test statistic less than critical value for all distribution.

TABLE 4. TEST STATISTIC VALUES OF CHI-SQUARED TEST.

Locations	Chi-squared test			
	C.V	Wei	Norm	Lognorm	Gamma
Sabha	9.48	2.69	4.20	3.07	3.31
Bngazi	9.48	4.95	7.12	3..51	3.05
Alkofr	9.48	5.15	3.21	2.98	4.12
Abokm	9.48	3.19	3.96	3.03	2.99

TABLE 5. TEST STATISTIC VALUES OF KOLMOGOROV-SMINROV TEST.

Location	Kolmogorov-Smirnov				
	C.V	Wei	Norm	Lognorm	Gam
Sabha	0.37	0.259	0.256	0.059	0.353
Bngazi	0.37	0.169	0.361	0.429	0.444
Alkofr	0.37	0.189	0.334	0.298	0.246
Abokm	0.37	0.063	0.024	0.168	0.399

As well as, RMSE and R, also used to determine both the error and the correlation, these criterions are also presented in table 6 and 7. It observed that for Weibull distribution, the most values of RMSE are the lowest, and the most values of R are the highest. The standard distribution with the smallest Root Mean Square Error with respect to the data was considered to be the best candidate distribution for the particular city under consideration which is given by the equation 18. In order to determine the most probable statistical that can accurately model the solar radiation for four cities in Libya, various statistical distribution was tested.

TABLE 6. TEST STATISTIC VALUES OF ROOT MEAN SQUARE ERROR.

Location	RMSE			
	Weibull	Norm	Lognorm	Gam
Sabha	0.99	0.98	1.04	1.15
Bngazi	1.00	1.09	1.19	1.13
Alkofr	1.01	1.06	1.10	1.10
Abokm	1.14	1.26	1.21	1.33

TABLE 7. TEST STATISTIC VALUES OF CORRELATION COEFFICIENT

Location	R			
	Wei	Ray	Norm	Gam
Sabha	-0.12	0.06	-0.16	0.0007
Bengazi	-0.58	-0.25	-0.58	0.0010
Alkofra	-0.12	0.05	-0.16	-0.3493
Abokm	-0.25	-0.89	-0.34	-0.3587

Root mean square error of each distribution is plotted in individual city, as shown in Fig 1.

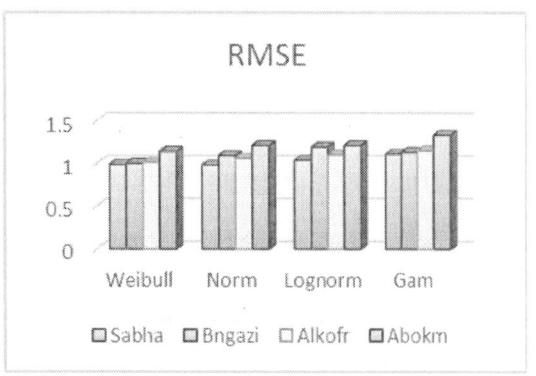

Fig 1. Root mean square error of each distribution

The statistical distribution functions are fitted to the actual solar radiation data as shown in Figures 2-5. It can be observed from the figure that Normal distribution closely matches with the measured data followed by Weibull distribution.

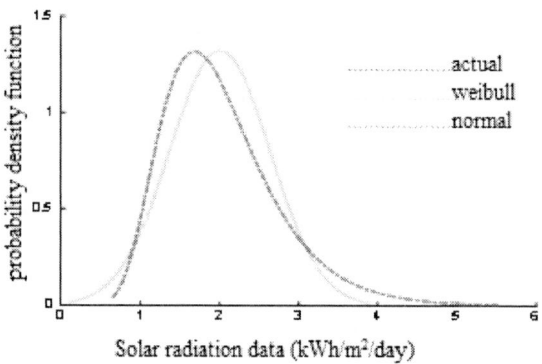

Fig 2. Fitting of pdf to the solar radiation data of Alkofra

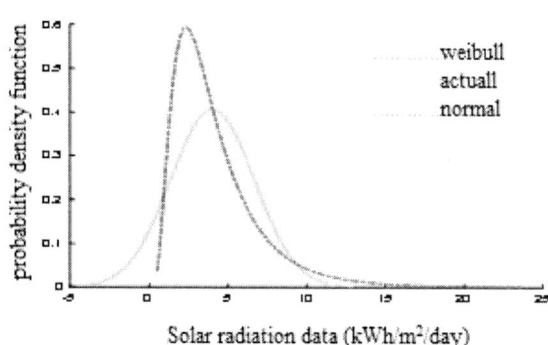

Fig 3. Fitting of pdf to the solar radiation data of Abokmash

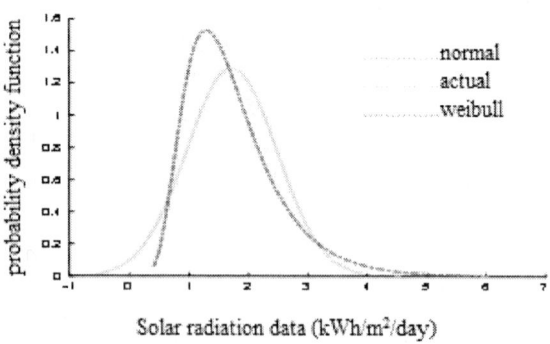

Fig 4. Fitting of pdf to the solar radiation data of Sabha

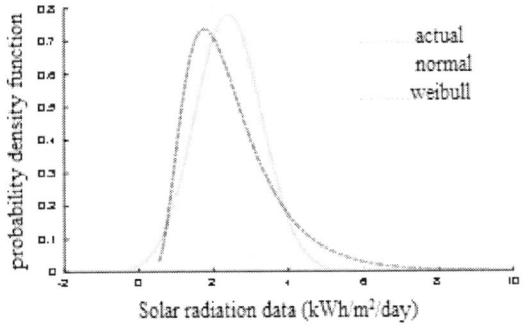

Fig 5. Fitting of pdf to the solar radiation data of Bengazi

VII. CONCLUDING REMARKS

Modeling of solar radiation is important for the evaluation and development of solar renewable energy system. Grid-connected photovoltaics into the Libyan power system will be very useful for those interested in the massive entry of PV. Although energy demand is increasing in Libya and that renewable energy could be the solution to meet some of this demand. It has been found that Photovoltaics could provide an alternative source of energy and an opportunity to generate financial profits as well as decrease the consumption levels of oil and natural gas. The aim of this paper is to compare different probability models namely, Normal, Lognormal, Gamma and Weibull for the monthly solar irradiation data at four locations in Libya based on analysis of data provided by Centre for Solar Energy Research and Studies of Libya (CSERS). Two goodness of fit tests were used to identify best fit at 5% level of significance, which are Chi-Squared test and Kolmogorov-Smirnov test. The standard distribution with the smallest Root Mean Square Error with respect to the data was considered to be the best fit distribution for the particular city and must value of Correlation coefficient verify the highest values. It is concluded that among those models distributions Weibull and Normal distribution gives the best fit for observed solar radiation.

ACKNOWLEDGMENT

This work was supported by the Engineering Energy Laboratory, Oklahoma State University, Stillwater, Oklahoma and the PSO/Albrecht Naeter Professorship in the School of Electrical and Computer Engineering and financially support by the Libyan government.

REFERENCES

[1] H. Bulut and O. Buyukalaca, "Simple model for the generation of daily global solar-radiation data in Turkey". *Applied Energy*, vol. 84, no. 5, pp. 477-491, 2007.

[2] T. Chang, "Investigation on frequency distribution of global radiation using different probability density function,". *Int. J. Appl. Sci. Eng*, vol. 8, no. 2, pp, 99-107 (2010).

[3] S. Ibrahim," Prospects of renewable energy in Libya," *Al-Fateh University,* Tripoli, 2006.

[4] Centre for Solar Energy Research and Studies of Libya (CSERS) (2009). Triopli - Libya

[5] H. Rinne, *The Weibull Distribution a Handbook*. Taylor & Francis Group LLC, 2009.

[6] R. Ramakumar, *Engineering Reliability: Fundamental and Applications.first edition*. Englewood Cliffs, N.J.: Prentice-Hall, 1993.

[7] J.P Hennessy, "A comparison of the Weibull and Rayleigh distribution for estimating wind power potential Wind Eng,". vol. 2, pp. 156-164, 1978.

[8] J.V. Seguro, T.W. Lambert, "Modern estimation of the parameters of the Weibull wind speed distribution for wind energy analysis," *Journal of Wind Engineering and Industrial Aerodynamics,* vol. 85, pp. 75-84, 2000.

[9] Singh, Someshwar; Taylor, and H. James, "Statistical analysis of environment Canada's Wind speed data," *in Electrical & Computer Engineering (CCECE),* 2012 25th IEEE Canadian Conference on IEEE, 2012. P. 1-5.

[10] Soong, T. Tsu, "Fundamentals of Probability and Statistics for Engineers,". *John Wiley & Sons*, 2004.

[11] Papoulis, Athanasios; Pillai, and S. Unnikrishna. "Probability, random variables, and Stochastic Process,". *Tata McGraw-Hill Education,* 2002.

A Comparison of PV Power Forecasts Using PVLib-Python

William F. Holmgren*, Antonio T. Lorenzo*, Clifford Hansen[†]

*Department of Hydrology and Atmospheric Sciences, University of Arizona, Tucson, AZ, 85721, United States
[†]Sandia National Laboratories, Albuquerque, NM, 87185, United States

Abstract—We used the open-source PVLib-Python library to create PV power forecasts for a fleet of utility scale power plants and assessed their accuracies. PVLib-Python allows users to easily retrieve standardized weather forecast data relevant to PV power modeling from NOAA models including the GFS, NAM, RAP, HRRR, and the NDFD. A PV power forecast can then be obtained using the weather data as inputs to the comprehensive modeling capabilities of PVLib-Python. We used these models to benchmark the performance of the University of Arizona's configuration of the Weather Research and Forecasting model. Standardized, open source, reference implementations of forecast methods using publicly available data may help advance the state-of-the-art of solar power forecasting.

Index Terms—forecasting, performance modeling, PV modeling, software

I. INTRODUCTION

PVLib-Python is an open source toolbox for PV modeling [1], [2]. Holmgren *et. al.* developed a forecasting module for PVLib-Python to help the PV modeling community create benchmark solar power forecasts [3]. In this paper, we use the PVLib-Python forecasting tool to create hourly average PV power forecasts for a fleet of utility scale power plants and we compare the forecasts to observed plant generation. We compare the forecasts derived from NOAA weather models, including the GFS, NAM, and RAP with forecasts derived from a high resolution mesoscale model run by the University of Arizona.

II. IRRADIANCE FORECAST DATA

The most critical component of a PV power forecast is the forecast of global horizontal irradiance (GHI). A GHI forecast can be obtained directly from a weather model forecast or it can be inferred from a model's cloud cover forecast [4]. The suitability of each method depends on the parameterizations of the model, the data availability of the model, and the temporal resolution of the desired PV forecast. Parameterization issues include the accuracy of the solar position equation and the impact of aerosols in their radiative transfer algorithms [5]. Model data availability and temporal resolution depends on the data source. The PVLib-Python forecast module described by Holmgren *et. al.* [3] accesses forecast data from the Unidata THREDDS server. However, the Unidata THREDDS server currently only hosts the most recent 2 weeks of forecast model output. To study a longer period of time, we wrote a Python script to download the relevant point forecast data from the NOAA NOMADS THREDDS data service [6]. Table I describes the model variable data availability and temporal

TABLE I
IRRADIANCE AND CLOUD COVER MODEL FIELD DATA AVAILABILITY. FOR NOAA MODELS, THIS DATA ONLY REFLECTS AVAILABILITY ON THE NOMADS THREDDS SERVER. OTHER NOAA DATA SOURCES MAY CONTAIN ADDITIONAL DATA.

Model	GHI	DNI	Cloud cover
GFS, 0.5 deg.	3/6 hr mixed interval average	None	3/6 hr mixed interval average
NAM, 12 km	1 hr for 36 hrs 3 hr for 84 hrs	None	1 hr for 36 hrs 3 hr for 84 hrs
RAP, 13 km	None	None	1 hr instant
UA-WRF, 1.8 km	3 min instant	3 min instant	None

resolutions for the NOAA GFS, NAM, and RAP models on the NOMADS THREDDS service.

For the NOAA models studied in this paper, we use a model proposed by Larson *et. al.* [4] to calculate GHI from cloud cover forecasts:

$$\text{GHI} = (\text{offset} + (1 - \text{offset})(1 - \text{cloud\_cover}))\text{GHI}_{\text{clear}} \quad (1)$$

where offset $= 0.35$, cloud_cover is the total cloud cover, and $\text{GHI}_{\text{clear}}$ is determined by PVLib's clearsky.ineichen function and PVLib's climatological Linke turbidity table. The DISC model is then used to calculate DNI and DHI. Here, we use default values for all functions, however, forecasters may tune the parameterizations to minimize forecast errors.

TABLE II
COMBINATIONS OF WEATHER MODEL AND IRRADIANCE PROCESSING ALGORITHMS STUDIED.

Name	GHI	DNI
GFS-CC	GFS Cloud cover + Larson	GHI + DISC
UA-DISC	WRF	GHI + DISC
UA	WRF + Aeronet	WRF + Aeronet
NAM-CC	NAM Cloud cover + Larson	GHI + DISC
NAM-GHI	NAM GHI	GHI + DISC
RAP-CC	RAP Cloud cover + Larson	GHI + DISC

The University of Arizona Department of Hydrology and Atmospheric Sciences runs a convective-permitting, 1.8 km

978-1-5090-5606-4/17 $31.00 © 2017 IEEE

resolution configuration of the Weather Research and Fore-casting (WRF) model [7] for operational forecasts of weather, solar power, and wind power in Arizona and New Mexico [8]. As an example of the utility of PVLib-Python for creating benchmark forecasts, we will compare PV forecasts derived from the UA-WRF configuration to the PVLib-Python fore-casts. WRF versions 3.7 and 3.8 were used for this study. The model parameterization was adjusted throughout the year, but includes SBU-YLIN and Morrison microphysics schemes, and ACM2 and BouLac planetary boundary layer schemes [9]. UA-WRF namelists are available at [8]. Forecasts from the 0Z, 6Z, and 12Z GFS and NAM runs provide the initial and lateral boundary conditions for a 5.4 km resolution outer domain, which in turn provides initial and boundary conditions for the 1.8 km resolution domain. For this study, we analyzed WRF models initialized with 6Z GFS data.

The UA-WRF model generates 3 minute resolution forecasts of GHI, DNI, 2 m temperature, and 10 m wind speeds, among other variables. This configuration of WRF does not account for the impact of aerosols on irradiance, so we post-processed the model's irradiance forecasts using measurements of the previous day's average aerosol optical depth obtained from the Aeronet site in Tucson, AZ [10]. We calculated daily average broadband AOD, τ_{bb}, from AOD measured at 380 nm and 500 nm [11], and then computed modified DNI and GHI as DNI $= \text{DNI}_{\text{wrf}} \exp(-\tau_{bb}/\cos\theta_z)$ and GHI $= \text{GHI}_{\text{wrf}} \exp(-0.01/\cos^{0.4}\theta_z)$, where θ_z is the solar zenith angle. We also studied UA-WRF derived PV forecasts in which DNI and DHI were inferred from the uncalibrated GHI using the DISC model.

For all models, we linearly interpolate the model forecast data from its native resolution, shown in Table I, to 5 minute resolution. For the GFS, NAM, and RAP models we then apply equation 1 and the DISC model, as discussed above, to determine a forecast GHI, DNI, and DHI. For the NAM model, we also create forecasts directly from its GHI forecasts. Table II summarizes the combinations of weather model data and processing algorithms studied in this paper. Figure 1(top) shows the result of the 3 hourly cloud cover to 5 minute irradiance conversion for the half-degree 2016-01-06-6Z GFS forecast. Figure 1(bottom) shows the UA-WRF GHI forecast and DISC-generated DNI and DHI forecasts. This UA-WRF model was initialized using the same 2016-01-06-6Z GFS forecast as shown in Figure 1(top).

III. FORECASTING PV POWER

We created forecasts for six PV systems in Arizona. The systems included three single axis trackers, totaling 63 MW AC, and three fixed tilt systems, totaling 14 MW AC. Five of the six systems are located near Tucson, Arizona, and the sixth system is located near Kingman, Arizona. Of the five systems in Tucson, three are in the same forecast model grid box. All of the studied PV systems are smaller than a forecast model grid box.

We chose to use a simple PV model (NREL's PVWatts [12]) in which the only required parameters are DC name-

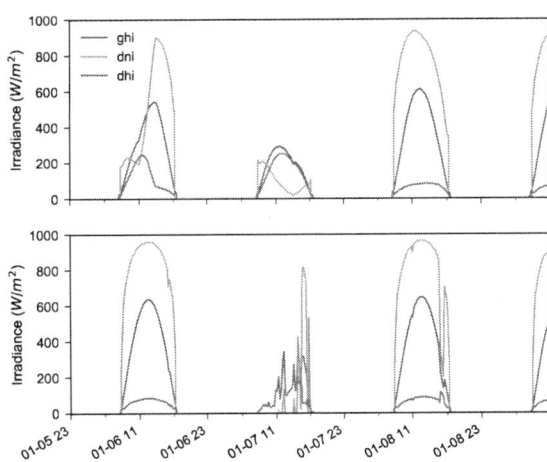

Fig. 1. 3.5 days of GHI (blue), DNI (green), and DHI (red) forecasts derived from a GFS forecast (top) and the UA-WRF initialized with the same GFS forecast (bottom) for Tucson, Arizona. GFS model irradiance is derived using equation 1 to determine GHI and the DISC model to determine its DNI and DHI. The UA-WRF model directly forecasts GHI and, for this figure, the DISC model was used to infer its DNI and DHI.

plate capacity and temperature coefficient. We imposed an additional maximum AC capacity parameter to account for inverter clipping. We determined system parameters by man-ually optimizing forecast model performance for clear days. This process was repeated for each forecast model to minimize the impact of forecast model temperature and wind speed biases. The optimum parameters did not vary by more than 10% from model to model. We resampled and processed the weather forecast data to 5 minutes, applied the PVLib PV modeling tools, and compared 1 hour average forecasts and data. We used the Hay and Davies transposition model [13], PVLib's single axis tracker algorithm, a physical angle of incidence modifier model [14], and the Sandia Array Performance Model's temperature model [15].

The code below demonstrates how the PV forecasts were made using the high-level, object-oriented PVLib-Python API. Readers may consult the PVLib-Python documentation [16] and source code [17] for more information on the function and method calls.

```
location = Location(
    latitude=32.2, longitude=-110.9, altitude=700)
system = SingleAxisTracker(
    module_parameters={
        'pdc0': 10, 'gamma_pdc': -0.0035})
system.peak_ac_power = 9
mc = ModelChain(
    system, location, orientation_strategy=None,
    dc_model='pvwatts', ac_model='pvwatts',
    aoi_model='physical', spectral_model='no_loss',
    temp_model='sapm', losses_model='no_loss')
fx_model = GFS()  # or NAM or RAP
```

```python
for fx_file in nomads_files:
    fx_data = pd.read_csv(fx_file)
    fx_data = fx_data.resample('5min').interpolate()
    fx_data = fx_model.process_data(fx_data)
    mc.run_model(fx_data.index, weather=fx_data)
    ac = mc.ac.clip_upper(system.peak_ac_power)
    ac = ac.resample('1h', label='right').mean()
```

The GFS-CC, NAM-CC, and RAP-CC classes' fx_model methods perform the following steps:

```python
data['temp_air'] = self.kelvin_to_celsius(
    data['temp_air'])
data['wind_speed'] = self.uv_to_speed(data)
irrads = self.cloud_cover_to_irradiance(
    data[cloud_cover], **kwargs)
```

The cloud cover to irradiance method was described above. The NAM-GHI and UA-DISC processing algorithm is:

```python
data['temp_air'] = self.kelvin_to_celsius(
    data['temp_air'])
data['wind_speed'] = self.uv_to_speed(data)
ghi = data['ghi']
solpos = self.location.get_solarposition(
    ghi.index)
dni = disc(
    ghi, solpos['zenith'], ghi.index)['dni']
dhi = ghi - dni * np.cos(
    np.radians(solpos['zenith']))
```

We used the dask.distributed framework to efficiently analyze the data in parallel. Additional processing code is available upon request.

IV. RESULTS

We studied forecast errors for all 2016 6Z GFS, 6Z NAM, and 9Z RAP models that were available on the online NOMADS server. Half-degree GFS data was available for all of 2016, 12 km NAM data was available for August through December 2016, and 13 km RAP data was available for June through December 2016. We downloaded GFS data through 168 hours (out of a possible 384), NAM data through 72 hours (out of a possible 84), and RAP data through 18 hours (out of a possible 18 hours). These initialization times and time ranges ensured that an integer number of local sunrise through sunset periods was available for each model. The initialization times also ensured that data from these models would have been available by local sunrise of the first forecast day. Metered 1 minute resolution PV data was manually filtered for errors, and, where possible, scaled to correct for partial system outages. We focus our analysis on the accuracy of the forecast for all systems added together because this is often a more relevant number for a utility company that manages generation from a fleet of power plants. Hourly average forecasts derived from each day's weather models and the observed power are shown for four days in Figure 2.

We calculated absolute and normalized mean bias error, mean absolute error, and root mean squared error under many conditions, only some of which can be shown here. Additional information is available upon request. Errors were normalized

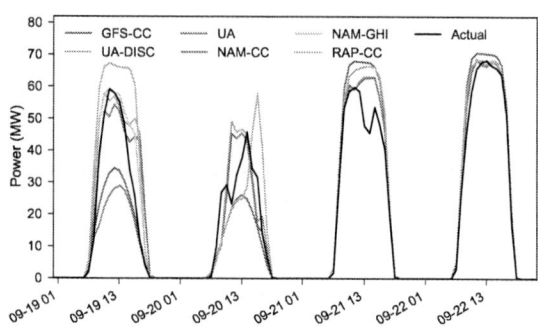

Fig. 2. Four days of PV observed generation (black) and forecasts (colors) derived from the GFS, NAM, RAP, and UA-WRF models using cloud cover (CC) or irradiance forecasts. Generation and forecasts are summed from 6 PV systems in Arizona.

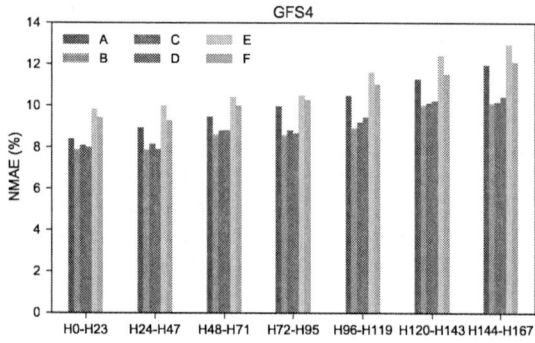

Fig. 3. NMAE forecast errors from the PVLib-Python processed GFS forecasts for six PV systems (labeled A-F) as a function of forecast horizon. The time range for this analysis is Jan. 2016–Dec. 2016.

by the maximum AC generation observed for each plant during 2016.

First, we examined GFS forecast errors as a function of forecast day for each plant in the study period. Figure 3 shows that forecast errors grow as a function of forecast horizon for all systems. Other forecast models demonstrate similar trends for all systems and these trends serve as a sanity check of the algorithms. For different systems, NMAE for hours 0-23 ranges from 8%-10%, while NMAE for hours 144-167 ranges from 9%-12%. The remainder of this paper analyzes the aggregated generation and forecasts for all systems.

We compared the accuracy of all of the NOAA forecast models as a function of the forecast horizon, shown in Figure 4. Times at which any forecast was missing were removed from the comparative analyses. Therefore, the analysis of the NOAA models comprises most dates in August through December 2016. PV forecasts derived from NAM cloud cover and GHI have the lowest errors for hours 0-23, while forecasts derived from RAP cloud cover have the highest errors. The GFS and the two NAM forecasts have similar errors in

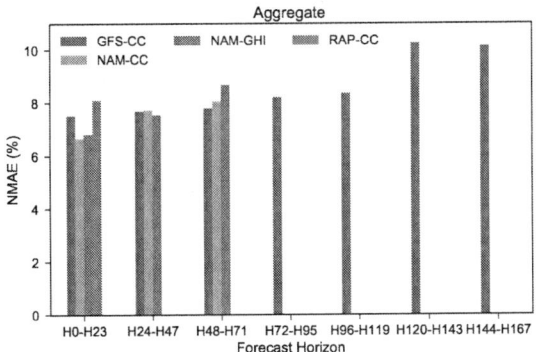

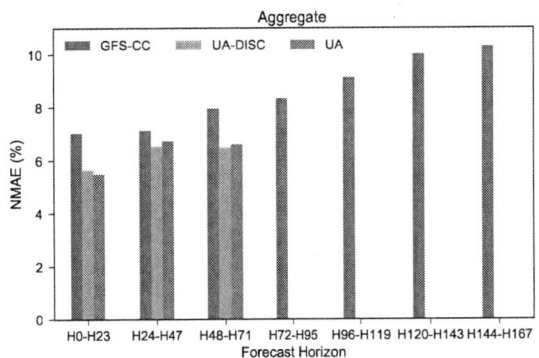

Fig. 4. NMAE forecast errors from the PVLib-Python processed forecasts as a function of forecast horizon for the studied NOAA forecast models. The GFS-CC (blue), NAM-CC (green), and RAP-CC (purple) PV forecasts were derived from cloud cover forecasts, while the NAM-GHI (red) PV forecasts were derived from the NAM's GHI forecast and the DISC model. The time range for this analysis is Aug. 2016–Dec. 2016. Times at which any forecast was missing were removed from the analysis.

Fig. 5. NMAE forecast errors from the PVLib-Python processed forecasts as a function of forecast horizon for the GFS with GHI derived from cloud cover (blue), UA-WRF with DNI derived from DISC (green), and UA-WRF with DNI post-processed with Aeronet data (red). The UA-WRF model was initialized with the GFS forecasts. The time range for this analysis is Jan. 2016–Dec. 2016. Times at which any forecast was missing were removed from the analysis.

hours 24-47, while the GFS slightly outperforms both NAM forecasts in hours 48-71. The cloud cover and GHI-based NAM forecasts perform similarly until forecast hours 48-71, at which the cloud cover based forecast is more accurate than the GHI based forecast. This is likely due to the fact that the NAM's temporal resolution switches from hourly to 3-hourly at 36 hours, and the interpolation of the 3-hourly GHI data has significant errors.

Next, we use the GFS forecasts to benchmark the accuracy of the UA-WRF model initialized on the GFS data. The NMAE in Figure 5 shows that, for the systems studied here, the UA-WRF model's dynamic downscaling of the GFS forecast yields a more accurate PV power forecast than the GFS under most forecast horizons and data subsets. The GFS forecast errors are similar for hours 0-23 and 24-47, and steadily increase beyond that. In contrast, the UA-WRF model's hours 0-23 forecasts are more accurate than its hours 24-47 forecasts, but its hours 48-71 forecasts are no worse than its hours 24-47 forecasts. There is little difference in accuracy between the UA-WRF forecasts based on GHI and the DISC model, and UA-WRF forecasts based on Aeronet-corrected DNI and GHI.

Finally, we examined forecast accuracy as a function of month of year. Figure 6 shows the accuracy of each method for each month. The model errors exhibit similar trends, with some outliers. For most models, forecast accuracy is worse June through September, and best in May and November. These results led us to examine the relationship between forecast accuracy and clear sky condition. We downloaded irradiance observations from the NREL OASIS station located at the University of Arizona [18]. We used PVLib-Python's detect_clear function to determine if a minute is clear or not, summed the number of cloudy minutes in each month, and normalized this by the number of minutes with GHI greater than 1 W/m^2. Figure 7 shows NMAE versus the percentage

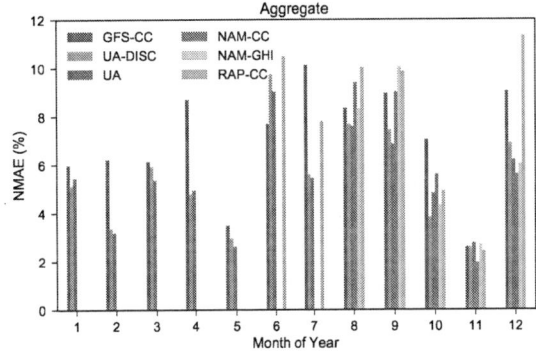

Fig. 6. NMAE intraday forecast errors from the PVLib-Python processed intraday forecasts of each model over 2016.

of cloudy minutes per month. Relatively clear months, such as February, May, October, and November 2016 have lower errors, especially for the UA-WRF model.

V. FUTURE WORK

The emphasis of this work is to illustrate the use of PVLib-Python to facilitate comparison and benchmarking of forecasts. The benchmarks suffer due to the restricted data availability of the NOMADS THREDDS service. NAM and RAP data were only available for half of 2016. The RAP model's GHI variable was not available on the archive. Furthermore, the High Resolution Rapid Refresh model was not available on the NOMADS THREDDS service. A more comprehensive forecast archive would enable more accurate comparisons of forecast accuracy.

It is likely that the benchmark forecasts can be improved with modest effort. The GFS data used here is from the half-degree model. The newer quarter-degree GFS model may yield more accurate predictions, particularly at longer forecast

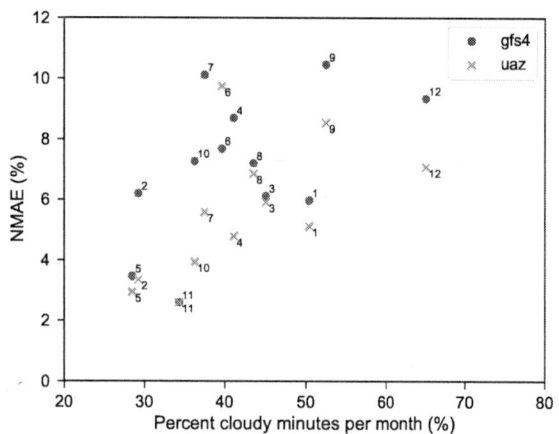

Fig. 7. Aggregate NMAE intraday forecast errors vs. percentage of cloudy minutes per month in Tucson, AZ. Points are labeled by month of year. Forecast errors are correlated with non-clear conditions.

horizons. A bias correction algorithm with a rolling training period could reduce seasonal trends in forecast skill. Scaling clear sky GHI by a different amount for low, mid, and high level cloud cover, rather than one scaling factor for total cloud cover, may yield significant improvements. Forecasts of aerosol optical depth and precipitable water could be used in the clear sky model for GHI.

VI. CONCLUSION

We used the PVLib-Python forecasting module to compare the accuracy of solar power forecasts for a fleet of PV power plants. The PVLib-Python library enables users to easily access weather forecasts and process them into PV power forecasts. The tool creates a standard set of data for PV modeling from the mixed data types provided from weather models. This work supports the standardization of PV power forecast methods, simplification of the use of weather forecast data for PV modeling, and fair and transparent PV power forecast model performance evaluation. As an example, we used PVLib-Python forecasts to benchmark the accuracy of the UA-WRF model.

The PVLib-Python documentation [16] provides examples for how to use the forecasting tool along with general PV system modeling. Readers are encouraged to participate in the PVLib-Python community via its GitHub page [17] and the pvlib tag on stackoverflow.com.

ACKNOWLEDGMENTS

We gratefully acknowledge the Electric Power Research Institute, Tucson Electric Power, Southern Company Services, and Arizona Public Service for funding. We thank Michael Leuthold for running the UA-WRF forecasts. A list of PVLib-Python contributors may be found on the GitHub repository [17] and in the online documentation [16]. We thank the developers and maintainers of the NOAA forecasts, the Unidata

THREDDS service, and the Unidata Siphon project. We thank Brent Holben, Eric Betterton, John Reagon, and Armin Sorooshian for their effort in establishing and maintaining the Tucson Aeronet site. Sandia National Laboratories is a multi-mission laboratory managed and operated by National Technology and Engineering Solutions of Sandia, LLC., a wholly owned subsidiary of Honeywell International, Inc., for the U.S. Department of Energy's National Nuclear Security Administration under contract DE-NA0003525.

REFERENCES

[1] R. W. Andrews, J. S. Stein, C. Hansen, and D. Riley, "Introduction to the open source PV LIB for Python Photovoltaic system modelling package," in *40th IEEE Photovoltaic Specialist Conference*, 2014.

[2] W. F. Holmgren, R. W. Andrews, A. T. Lorenzo, and J. S. Stein, "Pvlib python 2015," in *42nd IEEE Photovoltaic Specialist Conference*, 2015.

[3] W. F. Holmgren and D. G. Groenendyk, "An open source solar power forecasting tool using pvlib python," *43rd IEEE Photovoltaic Specialist Conference*, 2016.

[4] D. P. Larson, L. Nonnenmacher, and C. F. M. Coimbra, "Day-ahead forecasting of solar power output from photovoltaic plants in the american southwest," *Renewable Energy*, vol. 91, pp. 11–20, 6 2016. [Online]. Available: http://www.sciencedirect.com/science/article/pii/S0960148116300398

[5] P. A. Jimenez, J. P. Hacker, J. Dudhia, S. Ellen Haupt, J. A. Ruiz-Arias, C. A. Gueymard, G. Thompson, T. Eidhammer, and A. Deng, "Wrf-solar: An augmented nwp model for solar power prediction. model description and clear sky assessment," *Bulletin of the American Meteorological Society*, 2015/12/14 2015. [Online]. Available: http://dx.doi.org/10.1175/BAMS-D-14-00279.1

[6] W. F. Holmgren. get_nomads. [Online]. Available: https://github.com/wholmgren/get_nomads

[7] . C. Skamarock, J. B. Klemp, J. Dudhia, D. O. Gill, D. M. Barker, M. G. Duda, Xiang-Yu, H. W. Wang, and J. G. Powers, "A description of the advanced research wrf version 3," NCAR, Tech. Rep. NCAR/TN–475+STR, June 2008.

[8] M. Leuthold. Arizona regional wrf model forecasts. [Online]. Available: http://www.atmo.arizona.edu/index.php?section=weather&id=wrf

[9] N. C. for Atmospheric Research, "Wrf arw version 3 modeling system user's guide," National Center for Atmospheric Research, Tech. Rep., July 2015.

[10] B. N. Holben, T. F. Eck, I. Slutsker, D. Tanré, J. P. Buis, A. Setzer, E. Vermote, J. A. Reagan, Y. J. Kaufman, T. Nakajima, F. Lavenu, I. Jankowiak, and A. Smirnov, "Aeronet—a federated instrument network and data archive for aerosol characterization," *Remote Sensing of Environment*, vol. 66, no. 1, pp. 1–16, 1998. [Online]. Available: http://www.sciencedirect.com/science/article/pii/S0034425798000315

[11] R. E. Bird and R. L. Hulstrom, "Direct insolation models," Solar Energy Research Institute, Tech. Rep. SERI/TR-33S-344, 1981.

[12] A. P. Dobos, "Pvwatts version 5 manual," National Renewable Energy Laboratory, Tech. Rep., 2014.

[13] P. Loutzenhiser, H. Manz, C. Felsmann, P. Strachan, T. Frank, and G. Maxwell, "Empirical validation of models to compute solar irradiance on inclined surfaces for building energy simulation," *Solar Energy*, vol. 81, no. 2, pp. 254 – 267, 2007. [Online]. Available: http://www.sciencedirect.com/science/article/pii/S0038092X06000879

[14] W. De Soto, S. A. Klein, and W. A. Beckman, "Improvement and validation of a model for photovoltaic array performance," *Solar Energy*, vol. 80, no. 1, pp. 78–88, 1 2006. [Online]. Available: http://www.sciencedirect.com/science/article/pii/S0038092X05002410

[15] D. King, W. Boyson, and J. Kratochvill, "Photovoltaic array performance model," Sandia National Laboratories, Tech. Rep. SAND2004-3535, 2004.

[16] PVLIB Python contributors. pvlib-python documentation. [Online]. Available: https://pvlib-python.readthedocs.io

[17] ——. pvlib/pvlib-python. [Online]. Available: https://github.com/pvlib/pvlib-python

[18] A. Andreas and S. Wilcox, "Observed atmospheric and solar information system (oasis); tucson, arizona (data)," NREL, Tech. Rep. NREL Report No. DA-5500-56494, 2010.

978-1-5090-5606-4/17 $31.00 © 2017 IEEE

Comparing the Typical GHI Year vs Typical Power Year

Alex Kubiniec, Adam Kankiewicz, and Alemu Tadesse

Clean Power Research, Kirkland, WA, 98003, USA

Abstract — **Siting a solar power plant requires accurate information on the historic solar resource to determine the ideal location and assure that the project will be financially sound. This paper compares the existing and popular method of using a Typical GHI Year (TGY) as a form of historic solar resource assessment, against the method of Typical Power Year.**

I. INTRODUCTION

When siting a solar power plant, an accurate estimate of the long term solar resource is required to insure the financial viability of any PV project. Estimates of historic solar resource can require on site measurements as well as long term measurements. On site measurements are expensive and usually only span a short period of time from the resource assessment perspective (1-2 years). While ground data can provide accurate onsite measurements, these data need to be placed into a longer frame of reference for completeness [1]. Some options to get a longer data record include using Typical Meteorological Year data (TMY2 or TMY3) produced by the National Renewable Energy Laboratory (NREL). This data is publicly available, and derived from data which spans many years. The challenge from a solar resource perspective is that representative TMY months are chosen based on many meteorological variables, not all correlated to solar power, and the stations used to collect the data may be far away from a project site. This can contribute to decreased accuracy and uncertainty in the resource assessment numbers. With the advent of satellite-based irradiance datasets, deriving site-specific Typical GHI Year (TGY) files are feasible based on 19 years of satellite based irradiance observations. [2]

II. LIMITATIONS OF TRADITIONAL CALCULATIONS

The TGY method of assessing historic solar resources is expedient because it focuses solely on the Global Horizontal Irradiance (GHI). A TGY data set is created by calculating the average monthly GHI for span of the data, currently 19 years. Then specific months, closest to the average months, are targeted for inclusion in the TGY dataset (Figure 1). This results in a file with actual data that is as close to the average GHI as possible but preserves the day to day variance expected at a given PV site. The challenge is that by focusing solely on GHI, additional factors which can influence PV energy production, such as temperature and wind speed are not taken into account, or the ratio of direct normal irradiance to diffuse irradiance. All these factors affect the ultimate power generation of any potential solar project and should be

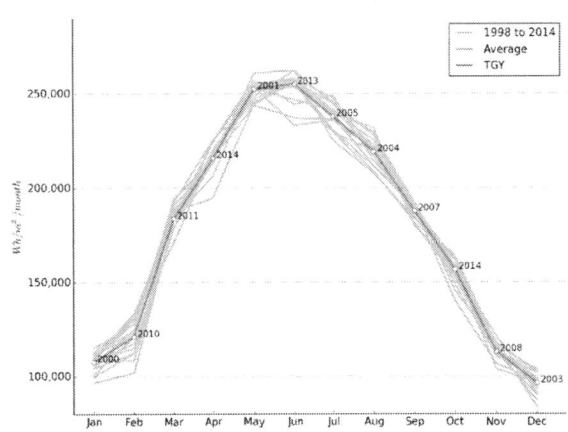

Figure 1: Illustration of Typical GHI Year selection.

considered when attempting to determine the ideal location and financial viability of any project.

Accounting for the specific PV technology and configuration of the proposed project can also impact the initial siting or solar resource assessment. It is important to understand how the ratio of DNI to diffuse irradiance has been historically observed because selected solar power technology for a given project will usually have an ideal DNI and diffuse ratio for power generation. Factors such as tracking arrays vs. fixed arrays are also not accounted for in the TGY file selection process.

III. PROPOSED METHODOLOGY

The end goal of any PV site resource assessment is an accurate estimate of the annual PV energy production. The ideal method for assessing historic resource is to use power as the deciding factor in a typical year process. This would lead to the derivation a Typical Power Year (TPY). The proposed new process would use 19 years of historic satellite-based irradiance, wind speed and temperature to produce a power simulation for all 19 years. Then an average power year would be calculated using the 19 years of power simulations. Finally, specific months would be chosen, that most closely match the arithmetic average month. This would result in a TPY dataset that more accurately represents the historic solar power resource valid at a specific location.

To test if this new TPY method results in increased accuracy over a traditional TGY method the following procedure will be used:

978-1-5090-5606-4/17 $31.00 © 2017 IEEE

1. Select diverse locations where observations of GHI, DNI, diffuse, wind and temperature are present.
2. Calculate the TGY year for those locations
3. Using all available data simulate power for two representative solar power plant configurations. One being a fixed array configuration. The other being a single axis tracker configuration. Site DC-to-AC PV solar ratios will be chosen to minimize inverter clipping.
4. Use 19 years of simulated power data to construct TPY data sets following the standard procedure for calculating a TGY except using power as the variable of interest.
5. Simulate power using the TGY data set for both the fixed and tracking configurations
6. Compare the two datasets representing the TGY power and TPY power. And repeat for a set of climatologically diverse locations and other regions of interest.

IV. RESULTS AND DISCUSSION

When comparing the Bakersfield TGY to the Bakersfield TPY, a very similar annual power profile is observed. The Bakersfield TPY file reported 1.21% more total energy than that returned from the Bakersfield TGY file. Figure 2 shows the annual power profile of both TGY months simulated as

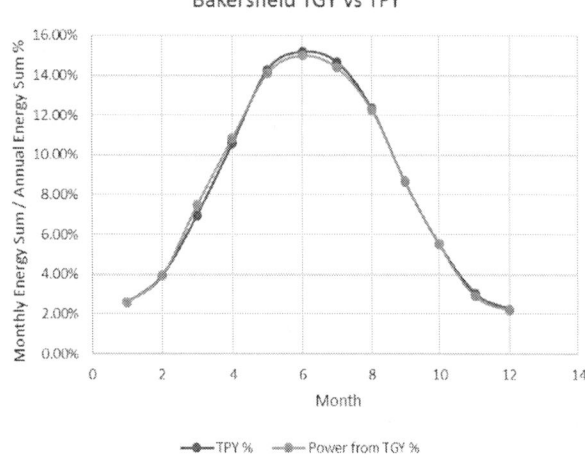

Figure 2: Comparing TGY power to TPY power as annual profiles. TPY power is 1.21% greater than TGY power. These annual power profiles are for the Bakersfield location.

power and TPY months for Bakersfield. When looking at a location in the New Jersey, the difference between the TGY and TPY annual power profiles is greater than in Bakersfield. The TPY method results in a 2.46% higher estimate of power than TGY estimates for the New Jersey location. Figure 3 shows the annual power profile of both TGY months simulated as power and the derived TPY months for the New Jersey location. There appears to be a clear difference between the two methods and the TPY methods tends to result in a different

power estimate. The paper plans to explore additional areas and regions comparing the TGY and TPY approaches. It is likely that different regions and climates within the US will have a different response to a TPY approach. Ideal results would identify locations or conditions where a TPY has significant differences from the TGY approach.

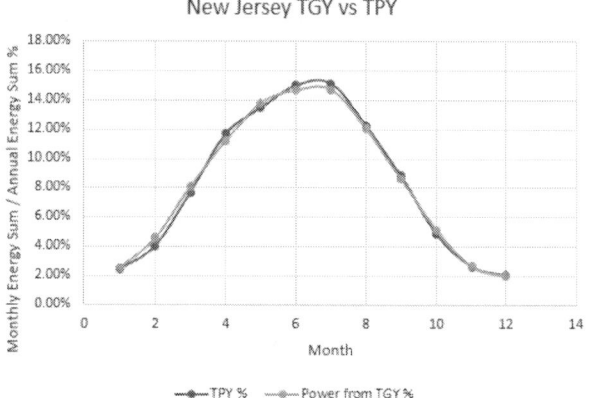

Figure 3: Comparing TGY power to TPY power as annual profiles. TPY power is 2.46% greater than TGY power. These annual power profiles are for the New Jersey location.

V. CONCLUSIONS AND FUTURE RESULTS

There is a difference between TGY and TPY solar resource estimates, illustrated by the two example locations above. TPY-derived power estimations result in a better representation of power for a specific PV system configuration. The paper will attempt to determine and quantify the differences between the TPY and TGY and identify when these differences might have a large impact on solar project development. In many cases the TPY approach results in a more representative estimate of power as it takes temperature, wind speed, and PV system configuration into account. Further research will be focused on different PV configurations and impacts of regional variance and presented in the final paper and presentation.

REFERENCES

[1] R. Perez, A. Kankiewicz, J. Dise, and E. Wu., (2014): Reducing Solar Project Uncertainty with and Optimized Resource Assessment Tuning Methodology. Proc. ASES Annual Conference, San Francisco, CA.

[2] Perez, R., J. Schlemmer, K. Hemker, Jr., S. Kivalov, A. Kankiewicz and C. Gueymard., (2015): Satellite-to-Irradiance Modeling – A new version of the SUNY Model. 42nd IEEE PV Specialists Conference, New Orleans, LA.

The Holy Grail of Resource Assessment: Low Cost Ground-Based Measurements with Good Accuracy

Bill Marion[1], Benjamin Smith[2]

[1]National Renewable Energy Laboratory, Golden, CO, USA

[2]Enphase Energy, Inc., Petaluma, CA, USA

Abstract — **Using performance data from some of the millions of installed photovoltaic (PV) modules with micro-inverters may afford the opportunity to provide ground-based solar resource data critical for developing PV projects. The method used back-solves for the direct normal irradiance (DNI) and the diffuse horizontal irradiance (DHI) from the micro-inverter ac production data. When the derived values of DNI and DHI were then used to model the performance of other PV systems, the annual mean bias deviations were within ±4%, and only 1% greater than when the PV performance was modeled using high quality irradiance measurements. An uncertainty analysis shows the method better suited for modeling PV performance than using satellite-based global horizontal irradiance.**

Index — **direct and diffuse irradiance, performance, model, micro-inverter.**

I. Introduction

Ground-based solar resource measurements are critical for developing PV projects. Unfortunately, accurate measurements at most locations are lacking due to the cost of solar radiation measurement equipment, which can be more than $40,000 for a first class station. In order to provide low or no-cost solar resource data traceable to a ground-based physical measurement at a nearby location, we have been developing a procedure to derive solar resource data from photovoltaic (PV) performance data. Specifically, data such as measured by Enphase Energy Inc. micro-inverters, which have been deployed with millions of PV modules and have been providing reliable data with a 5-minute temporal resolution since 2011, and for some early systems since 2007.

The ac power, P_{ac}, data are used to back-solve for the unknown direct normal irradiance (DNI) and diffuse horizontal irradiance (DHI). The procedure required the development of two key methods: (1) determining the global tilted irradiance (GTI), otherwise known as the plane-of-array (POA) irradiance, from the P_{ac}, and (2) determining the DNI and DHI from the GTI. The DNI and DHI values (or their global horizontal irradiance (GHI) equivalent) may then be used with conventional modeling software, such as PVsyst, Helioscope, and NREL's System Advisor Model (SAM), to estimate the performance of PV systems of any size, or PV array tilt and azimuth orientation, including tracking.

The method to determine the DNI and DHI from the GTI was recently developed and published [1]. It is a modification of the DIRINT model by Perez et al. [2], which separates input values of GHI into their DNI and DHI components. The modification substitutes GTI for GHI, and adds an iterative procedure to adjust the global clearness index to improve the derived values of DNI and DHI. The resulting model is referred to as the GTI-DIRINT model. The GTI-DIRINT model was validated using GTI values measured with Kipp & Zonen CMP11 and CMP22 pyranometers for three climatically diverse locations: Cocoa, Florida; Eugene, Oregon; and Golden, Colorado. For the GTI measured at a small tilt angle from the horizontal (10°), the deviations between the measured DNI and DHI and the GTI-DIRINT modeled DNI and DHI were essentially the same as those for the DIRINT model when using the GHI for model input. For larger tilt angles from horizontal, the deviations between modeled and measured values were larger, but still reasonable.

The method to determine the GTI from the P_{ac} was recently published [3]. It uses inverted PV performance models and solves a quadratic equation for the GTI from the input variables P_{ac}, wind speed, WS, dry bulb temperature, T_a, and the PV module temperature coefficients. A step was added to the GTI-DIRINT model to correct for the presence of angle-of-incidence (AOI) effects due to increased reflection losses from the PV module front cover when the AOI increases. This was accomplished within the GTI-DIRINT model iteration loop by correcting for AOI effects using a method for both direct beam and diffuse radiation [4].

The methods were developed using high quality data: GTIs measured with secondary standard pyranometers, P_{ac} measured with revenue grade meters (±0.2% uncertainty), and WS and T_a measured on-site.

This work evaluated the methods when used with data that is more readily available for universal application of the method. For input values of P_{ac}, we used the micro-inverter data for the PV systems that were measured and archived by Enphase Energy Inc. For input values of T_a and WS, we used Automated Surface Observation Station (ASOS) data.

Results are presented for implementing the overall procedure to derive the DNI and DHI from the P_{ac}, and for then using the derived values of DNI and DHI to model the GTI and P_{ac} for the various tilted orientations.

II. Data

We designed the validation experiment to include five identical PV module/Enphase Energy Inc. micro-inverter systems, each with a different tilt and azimuth orientation. Each of the five systems is instrumented to measure the P_{ac}

Fig. 1. PV modules with micro-inverters installed at NREL. Three PV modules are south-facing, with tilts of 10°, 25°, and 40° from the horizontal. A 4th PV module is tilted 40° and faces 30° west of south. A 5th PV module (not shown) is installed on a nearby two-axis tracker.

and GTI for comparison with the modeled values. The existing DNI and DHI measurements from the NREL's Solar Radiation Research Laboratory are also used for comparison with modeled values. The installed PV systems are shown in Fig. 1.

Non-NREL sources of data were used for input to the methods. For input values of P_{ac}, we used the micro-inverter data for the PV systems that were measured and archived by Enphase Energy Inc. under their Enlighten® program. For input values of T_a and WS, we used the Automated Surface Observation Station (ASOS) network data for the Broomfield, CO station. Broomfield is about 20 km northeast of NREL.

Data are 5-minute averages, except T_a and WS which were interpolated, from hourly data samples, to the midpoint of the 5-minute intervals. The data span the period April 1, 2014 through March 31, 2015.

Although the NREL and Enphase Energy Inc. data acquisition systems were measuring the same P_{ac} produced, differences in values occurred because of the measurement uncertainty, the 5-minute averaging methods, and the synchronization of the data acquisition clocks. The NREL P_{ac} measurements were made with a meter with an accuracy of ±0.2%. The Enphase Energy Inc. P_{ac} measurements, due to economics being more important, have an accuracy of ±2.5%.

Located between the micro-inverters and Enphase's Enlighten web-based monitoring and analysis software, hardware named Envoy is used for monitoring the performance of the micro-inverters and submitting the data to Enlighten via the internet. At nominal 5-minute intervals, the Envoy sequentially polls each micro-inverter in the PV system to obtain the energy produced since last polled. Because the polling takes a second or so per micro-inverter, and the polling doesn't necessarily began exactly at 5-minute increments past the hour, the polling for a micro-inverter could occur up to 2½ minutes from the 5-minute increments past the hour. The Envoy submits the energy produced to Enlighten after

reapportioning the polled energy to evenly spaced 5-minute time stamps. For example, if the micro-inverter is polled at 13:07, two-fifths of the polled energy is included for the energy for the time stamp 13:05 and three-fifths of the polled energy is included for the energy for the time stamp 13:10. This preserves the integrity of the accumulated energy, but the result may differ from the NREL measured 5-minute averages when conditions are changing, such as due to cloud movement. (Enphase 5-minute energy values in joules were converted to average power in watts by dividing by 300 seconds.)

Clock synchronization also contributed to differences between Enphase and NREL measured 5-minute values of ac power. Although the Envoy submitted the micro-inverter data to Enlighten without a problem, NREL's internet security setting did not permit it to update its clock each day as intended, and this was not evident until after the measurement period when NREL had access to Enphase data. (Enlighten keeps track of when the clock is updated.) To agree closer to NREL data, the time stamps for the Enphase data were adjusted by 5 minutes, but a disagreement of up to 2 minutes likely still exists. A statistical comparison of the Enphase and NREL ac power data is provided in the next section.

III. RESULTS

Using the characteristic data for the PV systems; the Enphase Energy Inc. P_{ac} data; the ASOS T_a and WS data; values of DNI and DHI were derived for each of the three south-facing fixed-tilt PV systems. Each set of the derived DNI and DHI values were then used to model the performance (GTI and P_{ac}) of each of the five PV systems for the purpose of comparing the differences between the modeled and measured values.

Modeling results were evaluated using mean bias deviation (MBD) and root-mean-square deviation (RMSD) statistics, with the results expressed as a percent of the mean of the measured value. The deviation is the measured value subtracted from the modeled value. For the MBD, a positive value indicates that the model overestimates on average.

The means of the measured irradiance values are: DNI = 476 W/m²; DHI = 147 W/m²; GHI = 423 W/m²; GTI(10, 180) = 457 W/m²; GTI(25, 180) = 483 W/m²; GTI(40, 180) = 489 W/m²; GTI(40, 210) = 450 W/m²; and GTI(2X Trk) = 652 W/m²; where the GTI(*Tilt, Orientation*) notation of Vignola et al. [5] is used where *Tilt* is in degrees from horizontal and *Orientation* is the azimuth in degrees measured eastward from true north. We designated the two-axis tracking orientation as GTI (2X Trk).

Using the same notation, the means of the ac power values measured by NREL are: $P_{ac}(10, 180)$ = 87.7 W; $P_{ac}(25, 180)$ = 92.2 W; $P_{ac}(40, 180)$ = 91.7 W; $P_{ac}(40, 210)$ = 83.9 W; and $P_{ac}(2X \text{ Trk})$ = 122.9 W.

For comparison, as appropriate, statistics were also determined for model efforts using measured values of DNI and DHI and modeled values of DNI and DHI determined with the DIRINT model.

A. Statistics for Enphase Measured P_{ac}

The P_{ac} values measured by Enphase were compared with the P_{ac} values measured by NREL to understand differences in the mean values reported (indicated by MBD) and the shorter term differences due primarily to averaging methods and time stamp errors from lack of regular clock synchronization (indicated by RMSD). The results are shown in Table 1. The MBDs for P_{ac} were 1.8% or less, which is within the stated uncertainty of 2.5% for the Enphase measurements. RMSDs were less than 10%.

TABLE 1
MEAN BIAS DEVIATION (MBD) AND ROOT-MEAN-SQUARE DEVIATION (RMSD) FOR ENPHASE MEASURED P_{AC} WHEN COMPARED TO NREL MEASURED P_{AC}

P_{ac}	MBD (%)	RMSD (%)
P_{ac} (10, 180)	1.1	7.4
P_{ac} (25, 180)	1.5	8.1
P_{ac} (40, 180)	1.8	7.9
P_{ac} (40, 210)	1.5	8.6
P_{ac} (2X Trk)	0.5	9.8

B. Statistics for Modeling DNI, DHI, and GHI

Using the P_{ac} values measured by Enphase, the GTIs were derived and then values of DNI and DHI were derived from the GTIs using the GTI-DIRINT model. The MBD and RMSD results, when using P_{ac} values for each of three south-facing fixed-tilt PV systems, are shown in Table 2. Also shown in Table 2 are the results for the conventional approach of using the DIRINT model to derive DNI and DHI from a measured GHI. Results for determining the GHI from derived values of DNI and DHI are also provided for the application of the GTI-

DIRINT model (not included when using the DIRINT model because the GHI is a model input).

Compared to DIRINT, the applications using GTI-DIRINT had similar MBDs and RMSDs for DHI and similar MBDs for DNI. The RMSDs for DNI were larger, but still reasonable considering that the RMSD from Table 1 for the Enphase measured P_{ac} likely propagated into the result. Also, because a GHI or GTI may result from many different combinations of DNI and DHI, the RMSDs for derived values of DNI and DHI are relatively large compared to modeling other parameters.

The MBDs for GHI were within about ±1%. The RMSDs were a few percent larger than the RMSDs from Table 1 for the Enphase measured P_{ac}.

C. Statistics for Modeling GTIs Using the DNIs and DHIs

The DNI and DHI values from the preceding section were used with the Perez tilted surface model [6] to model the GTIs for the different PV module orientations. The MBD and RMSD results are shown in Table 3. As expected, using the measured values of DNI and DHI provided the best results, with MBDs within ±1% and RMSDs less than 8%. Next in performance was the DIRINT model when using measured values of GHI for model input, with MBDs within ±1½% and RMSDs less than 11%. When using the DNI and DHI values derived from the P_{ac} values for the various PV module orientations, MBDs were within ±2% and RMSDs were less than 18%. The MBDs of all methods are within the ±3% uncertainty of the GTI measurements.

D. Statistics for Modeling P_{ac}

The GTI values from the preceding section were used to model the P_{ac} for comparison with the NREL measured values of P_{ac}. The MDB and RMSD results are shown in Table 4. When using the GTIs determined from the measured values of DNI and DHI, the MBDs were within ±2½% and RMSDs were less than 10%.

TABLE 2
MEAN BIAS DEVIATION (MBD) AND ROOT-MEAN-SQUARE DEVIATION (RMSD) FOR DIRINT AND GTI-DIRINT MODELED VALUES OF DNI, DHI, AND GHI WHEN USING THE MEASURED GHI WITH DIRINT AND THE ENPHASE MEASURED P_{AC} WITH GTI-DIRINT FOR THE VARIOUS PV MODULE TILT AND AZIMUTH ORIENTATIONS.

Model/Input	DNI (mean = 476 W/m²)		DHI (mean = 147 W/m²)		GHI (mean = 423W/m²)	
	MBD (%)	RMSD (%)	MBD (%)	RMSD (%)	MBD (%)	RMSD (%)
DIRINT/GHI	2.0	19.2	-5.1	35.1	----	----
GTI-DIRINT/						
P_{ac} (10, 180)	2.2	26.8	-3.9	38.0	1.2	10.6
P_{ac} (25, 180)	1.9	27.1	-3.6	38.1	0.9	10.9
P_{ac} (40, 180)	1.7	28.0	-4.3	37.8	0.2	11.7

TABLE 3

MEAN BIAS DEVIATION (MBD) AND ROOT-MEAN-SQUARE DEVIATION (RMSD) FOR MODELING THE GTI FOR THE DIFFERENT PV MODULE ORIENTATIONS WHEN USING THE MEASURED DNI AND DHI; THE DIRINT MODELED DNI AND DHI DERIVED FROM THE MEASURED GHI; AND THE GTI-DIRINT MODELED VALUES OF DNI AND DHI DERIVED FROM THE ENPHASE MEASURED P_{ac} FOR THE VARIOUS PV MODULE TILT AND AZIMUTH ORIENTATIONS.

Source for DNI and DHI	GTI(10,180)		GTI(25,180)		GTI(40,180)		GTI(40,210)		GTI(2X Trk)	
	MBD	RMSD	MBD	RMSD	MBD	RMSD	MBD	RMSD	MBD	RMSD
	(%)	(%)	(%)	(%)	(%)	(%)	(%)	(%)	(%)	(%)
Measured	-0.1	7.4	0.5	7.6	-0.1	7.9	0.1	8.4	0.9	7.5
DIRINT/GHI	0.0	7.4	0.7	8.0	0.2	8.7	0.2	9.5	1.4	10.6
GTI-DIRINT/										
P_{ac} (10, 180)	1.4	9.1	2.2	9.6	1.6	10.0	1.7	11.1	2.0	15.7
P_{ac} (25, 180)	1.0	9.2	1.7	9.2	1.2	9.2	1.2	10.6	1.8	16.4
P_{ac} (40, 180)	0.2	10.0	0.8	9.5	0.2	9.2	0.3	10.8	1.6	17.8

TABLE 4

MEAN BIAS DEVIATION (MBD) AND ROOT-MEAN-SQUARE DEVIATION (RMSD) FOR MODELING THE P_{ac} FOR THE DIFFERENT PV MODULE ORIENTATIONS WHEN USING THE MEASURED DNI AND DHI; THE DIRINT MODELED DNI AND DHI DERIVED FROM THE MEASURED GHI; AND THE GTI-DIRINT MODELED VALUES OF DNI AND DHI DERIVED FROM THE ENPHASE MEASURED P_{ac} FOR THE VARIOUS PV MODULE TILT AND AZIMUTH ORIENTATIONS.

Source for DNI and DHI	P_{ac} (10,180)		P_{ac} (25,180)		P_{ac} (40,180)		P_{ac} (40,210)		P_{ac} (2X Trk)	
	MBD	RMSD	MBD	RMSD	MBD	RMSD	MBD	RMSD	MBD	RMSD
	(%)	(%)	(%)	(%)	(%)	(%)	(%)	(%)	(%)	(%)
Measured	-0.4	7.9	0.3	7.9	1.4	8.5	2.3	9.5	2.3	8.3
DIRINT/GHI	-0.3	7.9	0.5	8.2	1.7	9.3	2.4	10.5	2.8	11.5
GTI-DIRINT/										
P_{ac} (10, 180)	1.1	7.4	1.9	8.3	3.1	9.5	3.9	11.0	3.3	16.5
P_{ac} (25, 180)	0.7	8.1	1.5	7.6	2.7	8.4	3.4	10.2	3.3	17.1
P_{ac} (40, 180)	-0.1	8.8	0.6	7.8	1.8	7.9	2.5	9.9	3.1	18.1

For the GTIs determined from DNI and DHI values from the DIRINT model and using measured values of GHI for model input, MBDs were within ±3% and RMSDs were less than 12%. When using the GTIs determined from DNI and DHI values derived from the Enphase P_{ac} values for the various PV module orientations, MBDs were within ±4% and RMSDs were less than 18%.

E. MBDs for Monthly P_{ac}

Fig. 2 shows the MBD by month for modeling the P_{ac} (40,180) when using: (a) the measured DNI and DHI; and (b) the DNI and DHI values derived from the Enphase measurements for the P_{ac} (25,180) orientation. Based on the work of Lee and Panchula [7], Fig. 2 includes an estimate of the MBD introduced due to spectral irradiance variations when using broadband DNI and DHI measurements to model the performance of a multi-crystalline PV module. The MBD related to spectral effects generally tracks the MBD for when using the measured DNI and DHI, indicating that it contributes to the variation in MBD by month for this method.

MBD variation by month when using the Enphase measurements is less because the spectral characteristics of the PV module used to derive the DNI and DHI values are

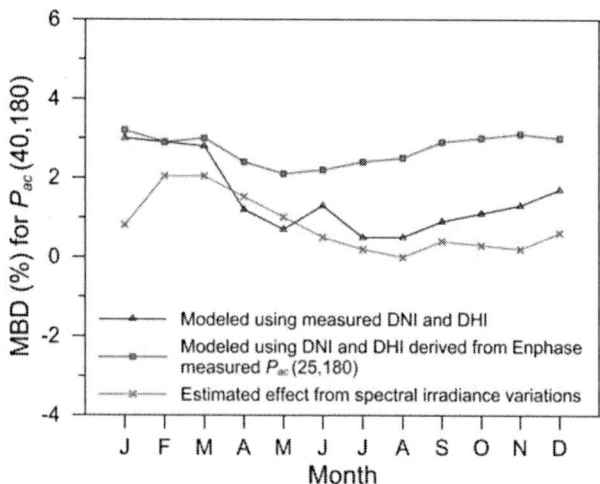

Fig. 2. Mean bias deviation (MBD) by month when modeling the P_{ac} (40,180) using: (a) the measured DNI and DHI; and (b) the DNI and DHI values derived from the Enphase measurements for the P_{ac} (25,180) orientation. The result using the measured DNI and DHI includes the estimated spectral effect. The result using the measured P_{ac} (25,180) shows less seasonal variability because spectral effects are automatically included, but does include the MBD of 1.5% from the Enphase measurement.

essentially the same as for the PV module whose performance is modeled, but shows about a 1% increase in MBD from summer to winter. This may be a consequence of assumptions and estimates related to factors that impact seasonal performance, such as: temperature coefficient, PV module temperature, albedo, and AOI losses. When using the Enphase measurements, the MBD was also increased because of the presence of the MBD of the Enphase measurements (+1.5% from Table 1). Otherwise, the MBDs in Fig. 2 for this method would have been 1.5% less and more similar to the average of the method using the measured DNI and DHI.

IV. UNCERTAINTY ANALYSIS

The uncertainty of the method for predicting the annual P_{ac} was determined and quantified by the source element of the uncertainty. Sources of uncertainty were considered to be the bias type, but not the random type because the large number of data used to determine the annual P_{ac} serves to average out random effects (more than 50,000 five-minute averages). The

uncertainty of the method was also compared with other modeling methods using measured GHI and satellite derived GHI as the sources of the irradiance data. The results are shown in Table 5 and were determined as the root-sum-square of the elemental bias limits [8].

The overall bias limit for modeling the P_{ac} is ±6.3% when using the method using an Enphase measured P_{ac}, ±7.1% when using a satellite-based GHI, and ±5.4% when using a GHI measured with a well maintained secondary standard pyranometer.

Because the elements denoted by asterisks in Table 5 are used twice for the method using the Enphase measured P_{ac}, once when deriving the DNI and DHI from the measured P_{ac} and again when using the derived DNI and DHI to model the P_{ac} of another PV system, their elemental bias limits are greater by a factor of the square root of two than for the other methods. These elements are generally associated with the PV module and inverter specifications. Smaller bias limits were used for elements where the method compensates for their effect. These elements include the spectral effect and soiling.

TABLE 5
UNCERTAINTY ELEMENTS FOR MODELING THE ANNUAL P_{AC} FOR THE VARIOUS METHODS.

Element	Method for Modeling the P_{ac}		
	Measured GHI, DIRINT Model, Perez Tilted Surface Model Bias Limit (±%)	Satellite GHI, DIRINT Model, Perez Tilted Surface Model Bias Limit (±%)	Enphase Measured P_{ac}, GTI-DIRINT Model, Perez Tilted Surface Model Bias Limit (±%)
GHI	2.0	5.0	-
Enphase measured P_{ac}	-	-	2.5
PV module power rating	3.0	3.0	4.2*
PV module binning	1.0	1.0	1.4*
DIRINT model + Perez tilted surface model	1.5	1.5	-
GTI-DIRINT model + Perez tilted surface model	-	-	1.5
Spectral effect	1.8	1.8	0.5
Irradiance effect	1.0	1.0	1.4*
AOI effect	0.5	0.5	0.7
Temperature effect			
γ bias limit ±7.5%	0.8	0.8	1.1*
ASOS location	1.0	1.0	1.4*
Temperature model	1.2	1.2	1.7*
Inverter model	1.0	1.0	1.4*
Inverter clipping	-	-	1.0
Soiling estimate	2.0	2.0	0.5
Root-sum-square of the elemental bias limits	5.4	7.1	6.3

*These bias limits are a factor of the square root of two or 1.4 greater than for the other two methods because they are used twice, once when deriving the DNI and DHI from the measured P_{ac} and again when using the derived DNI and DHI to model the P_{ac} of another PV system.

V. SUMMARY

Using PV performance data from PV modules with micro-inverters affords the opportunity to provide ground-based solar resource values of DNI and DHI critical for developing PV projects. From the measured P_{ac}, values of DNI and DHI were derived, and then used to model the performance of other PV modules with micro-inverters with different azimuth and tilt orientations. The annual MBDs were within ±4%, and only 1% greater than when the PV performance was modeled using high quality irradiance measurements. An uncertainty analysis shows the method's uncertainty for modeling the annual ac energy for a PV system to be ±6.3%, which is less than the ±7.1% uncertainty when modeling the PV performance using satellite-based global horizontal irradiance data

ACKNOWLEDGEMENT

This work was supported by the U.S. Department of Energy under Contract No. DE-AC36-08-GO28308 with the National Renewable Energy Laboratory (NREL). Funding provided by U.S. DOE Office of Energy Efficiency and Renewable Energy Solar Energy Technologies Program. The authors are thankful for the efforts of Bill Sekulic, Jose Rodriguez, and Greg Perrin at NREL, who performed the irradiance and PV performance measurements.

The U.S. Government retains and the publisher, by accepting the article for publication, acknowledges that the U.S. Government retains a nonexclusive, paid-up, irrevocable, worldwide license to publish or reproduce the published form of this work, or allow others to do so, for U.S. Government purposes.

REFERENCES

[1] B. Marion, "A model for deriving the direct normal and diffuse horizontal irradiance from the global tilted irradiance", *Solar Energy* 122: 1037–1046, 2015.

[2] R. Perez, P. Ineichen, E. Maxwell, R. Seals, A. Zelenka, "Dynamic global-to-direct irradiance conversion models", In: ASHRAE Transactions-Research Series, 354–369, 1992.

[3] B. Marion, B. Smith, "Photovoltaic system derived data for determining the solar resource and for modeling the performance of other photovoltaic systems", *Solar Energy* 147: 349–357, 2017.

[4] B. Marion, "Numerical method for angle-of-incidence correction factors for diffuse radiation incident photovoltaic modules", *Solar Energy* 147: 344–348, 2017.

[5] F. Vignola, J. Michalsky, T. Stoffel, Solar and infrared radiation measurements, CRC Press, Boca Raton, 2012.

[6] R. Perez, P. Ineichen, R. Seals, J. Michalsky, "Modeling daylight availability and irradiance components from direct and global irradiance", *Solar Energy* 44: 271–289, 1990.

[7] M. Lee, A. Panchula, "Spectral correction for photovoltaic module performance based on air mass and precipitable water", in 43rd IEEE PVSC, 2016.

[8] R. Dieck, *Measurement Uncertainty—Methods and Application*, Instrument Society of America: Research Triangle Park, 1992.

Global Comparison of the Impact of Temperature and Precipitable Water on CdTe and Silicon Solar Cells

I. M. Peters[1], L. Haohui[2], T. Reindl[2], T. Buonassisi[1]

[1] Massachusetts Institute of Technology, Cambridge, MA 02141, United States of America

[2] Solar Energy Research Institute of Singapore, Singapore 117574, Singapore

Abstract — Solar cells made from different materials are differently affected by the climatic conditions in which they operate. Response to precipitable water and temperature are fundamentally different for base materials with different bandgaps. In this contribution we investigate theoretically how precipitable water and temperature affect the harvesting efficiency of Silicon (Si) and Cadmium Telluride (CdTe) solar cells. Harvesting efficiency calculations capture the differences in local solar cell operation clearly and provide insights beyond those provided by the standard testing condition (STC) efficiency. We use readily-available satellite data to calculate worldwide harvesting efficiencies for a CdTe solar cell with an STC module efficiency of 15.7% and a Si PERC solar cell with an STC efficiency of 20.7%. We find that the different operating conditions found on Earth account for variations of up to ± 23 kWh/m^2 annually in power generation difference when comparing those two solar cell technologies.

I. INTRODUCTION

The standard testing condition (STC) efficiency is widely accepted as the figure of merit for comparing different solar cells. However, when operated in the field, solar cells experience varying operation conditions. In particular, light intensity, temperature and spectrum vary over time. If the sensitivity of two solar cells to these conditions differ, those solar cells will generate different power, even if they have the same STC efficiency. Consequently, the STC efficiency is not a sufficient figure of merit when comparing operation of two solar cell technologies in a given location. The situation is especially relevant when comparing solar cells with different bandgaps, as the bandgap fundamentally impacts the sensitivity of a solar cell to the predominant operating conditions. Understanding specifically how varying operating conditions in different parts of the world impact photovoltaic power generation is, hence, of great relevance for determining the financial benefit of a certain solar cell, but also for the development of new solar cell technologies.

The harvesting efficiency of a solar cell is its energy yield per unit area divided by the total power delivered from the sun onto the same area for a given period of time. Harvesting efficiency calculations capture variation in the power generation due to varying operating conditions graphically, and provide insights beyond those obtained from the standard testing condition (STC) efficiency. The procedure for calculating it is outlined in figure 1. In this paper, we present and compare calculated harvesting efficiencies for a CdTe

solar cell with a STC module efficiency of 15.7% and a Si PERC solar cell with a STC efficiency of 20.7%. CdTe and Si have, partially due to differences in the bandgaps, very different sensitivities towards variations in spectrum and temperature. The different sensitivities are shown in figure 2. We use satellite data to consider insolation and temperature, and we model the incident spectrum globally using the SMARTS [1] spectrum calculator. These inputs are then used to calculate energy yield and harvesting efficiency for the two solar cell technologies. We compare performance ratios to illustrate how much climatic conditions contribute to variations in solar cell performance. Calculations were performed on an area mesh of 1 x 1 deg and for a period of one year (2015) with a time resolution of one data point per day. We find that variations in temperature and in spectrum due to precipitable water are the two leading effects in this comparison. Pearson correlations indicate that temperature has a slightly higher impact than precipitable water. Overall, we find that variations in the operating conditions account for variations of up to ± 23 kWh/m^2 annually in power generation, when comparing the energy output of the two solar cells.

II. HARVESTING EFFICIENCY MODELLING

Figure 1 shows a sketch of the modelling approach applied in this paper. Satellite input data is used to generate the solar spectrum. It is also used as input for irradiance and for temperature calculations. Spectrum, insolation and temperature corrected current voltage characteristics are then used to calculate the power generation for any given location. The procedure is repeated for all locations and for a total of 344 days during 2015 (for the remaining 21 days at least one of the required parameters was not available). Integrating the results over the considered period results in the annual energy yield for a certain location, which in turn is used to calculate the harvesting efficiency, by dividing the obtained energy yield by the annual solar insolation. This harvesting efficiency is then plotted graphically for the entire globe in figure 3. The scale was chosen to represent the same relative variation, showing the much stronger variation in generated power for silicon.

It should be noted that the presented method does not consider the low light performance of the different solar cells, which is expected to result in a performance ratio advantage for thin-film technologies, and hence, shift the graph in favour of CdTe. Effects of degradation exceed the scope of the presented study and are excluded from the presented analysis.

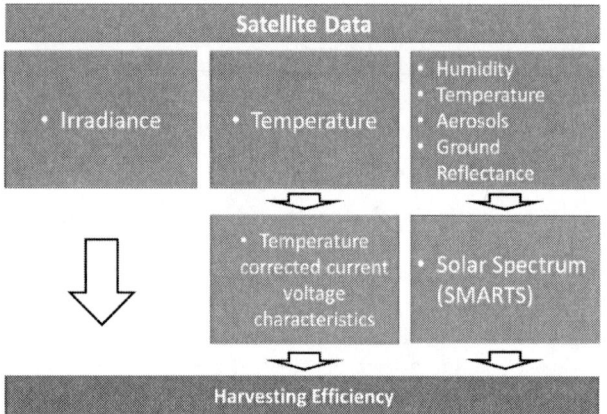

Fig. 1. Sketch of the harvesting efficiency calculation procedure used in this paper. Satellite based meteorological data is used to calculate solar spectra for different dates and locations using the SMARTS calculator [1]. Ambient temperature data is used to predict the module temperature. The module temperature is used to generate temperature corrected IV data, based on published solar cell data. Spectrum, temperature dependent IV, QE and satellite based irradiance data are then used to calculate the harvesting efficiency.

III. INPUT PARAMETERS

All satellite input data used for the calculations presented in this study is openly available from NASA. Insolation data was taken from the Clouds and Earth's Radiant Energy System (CERES) instrument [2], surface air temperature, relative surface humidity and total precipitable water data from the Atmospheric InfraRed Sounder (AIRS) instrument [3], and aerosol and ground reflectance from the Ozone Monitoring Instrument (OMI) [4]. The main focus of this study is on the influence of precipitable water and temperature. Measured temperature coefficients were used for both cells to calculate the impact on performance ratio. The SMARTS model was used to calculate changes in the spectrum due to precipitable water. The sensitivity of the solar cell performance to temperature and precipitable water is shown in figure 2a for Si and in figure 2b for CdTe. Temperature and precipitable water range were chose to include most conditions occurring anywhere on Earth, with the exception of regions in the far North or South, in which module temperature can drop significantly below zero.

We used published data to represent the solar cells. CdTe is represented by a First Solar Series 4™ PV module with a nominal module efficiency of 15.7%. To represent silicon, we used a well-established data set for a PERC type solar cell with a nominal efficiency of 20.7%. Both choices represent efficiencies on the higher end of what is commercially available today for each technology. While the exact results depend on the choice of the specific solar cell, sensitivities to temperature and humidity are to a great extent determined by material parameters, especially the bandgap. It can, therefore, be expected that the trends shown here will generally be found when comparing the two technologies for CdTe and Si. Similar studies using other data sources and solar cells confirm this expectation and find similar trends [5].

An additional aspect that needs to be considered when comparing the efficiencies of CdTe and Si modules is the area coverage with active solar cell area. While CdTe does not reach the efficiencies of the best performing Si solar cells, the area coverage of thin-film solar cells tends to be greater than for crystalline silicon. With ever reducing costs for solar panels, factors like area coverage or low light performance can become decisive when considering different solar cell technologies for a project.

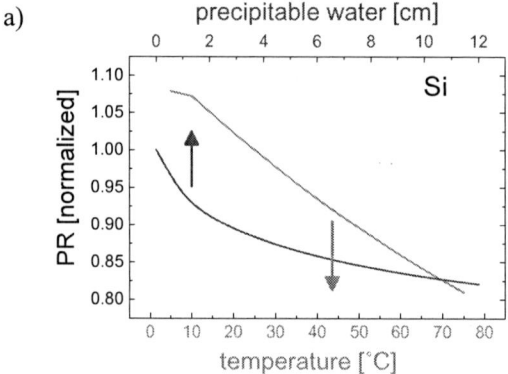

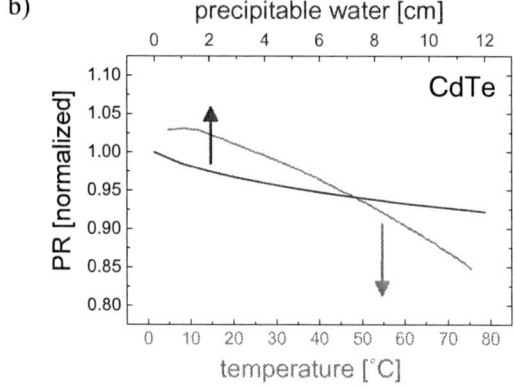

Fig. 2. Performance dependence of Si (a) and CdTe (b) on precipitable water (blue) and temperature (red). Values have been normalized to 25 centigrade and 0 cm precipitable water. Values for CdTe were provided by First Solar, values for silicon were taken from literature [6]. The figure illustrates the stronger dependence of silicon on either parameter. X-axis values were chosen according to values occurring in the calculations.

978-1-5090-5606-4/17 $31.00 © 2017 IEEE   1141

IV   RESULTS AND DISCUSSION

Figure 3 summarizes some of the results of the performed calculations. Figure 3a shows the calculated variation in harvesting efficiency of the CdTe solar cell; figure 3b the corresponding variation of the silicon solar cell. Both figures use the same relative variation in scale to illustrate the more pronounced variation in performance for the silicon solar cell. The regional trends that can be observed for both solar cells are similar, with extremes occurring in very hot/humid and very cold/dry areas. Using the average as reference, the magnitude in variation for the silicon solar cell is approximately a factor three greater than for CdTe. Figure 3c shows the difference in expected annual power generation for the two different solar cells. This difference was calculated using a fixed reference efficiency for both solar cells. The value shown in figure 3c corresponds to how much a comparison based on expected performance would be off, if effects of temperature and precipitable water were neglected. According to our calculations, the error would be up to ± 23 kWh/m^2 annually. Bright colors indicate a performance advantage for CdTe, dark colors for silicon.

High solar insolation often correlates with high temperature and precipitable water levels, which results in a very notable advantage of the higher bandgap CdTe in the hot and humid regions. Silicon shows a notable advantage in high mountain ranges and in areas located in the High North and Deep South. In temperate regions, including Central Europe, Japan and the Central United States neither technology has a strong advantage.

IV   SUMMARY

In this study we investigate the worldwide variation in harvesting efficiency for a representative CdTe- and Si solar cell for field operation in the year 2015 by means of harvesting efficiency modelling. Different harvesting efficiencies in different areas are caused by variations in the local operating conditions. In the present study, we focus on effects caused by precipitable water and by temperature.

We use satellite data to calculate the incident spectrum and module temperature, and use these to calculate harvesting efficiencies. In the presented study, we performed calculations with a temporal resolution of one data point per day on a spatial grid of 1x1 deg. Our results show a difference of up to ± 23 kWh/m^2 in annual power generation, caused by variations in temperature and precipitable water, when comparing the field performance of the two technologies against name plate expectations. Temperature is identified as the leading effect, but precipitable water also contributes. We find that CdTe performs especially well in hot/humid areas, and Si in cold/dry areas.

a) harvesting efficiency CdTe

b) harvesting efficiency Si

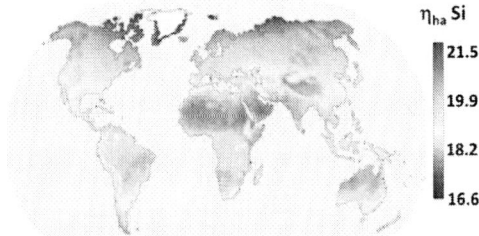

c) estimated performance difference

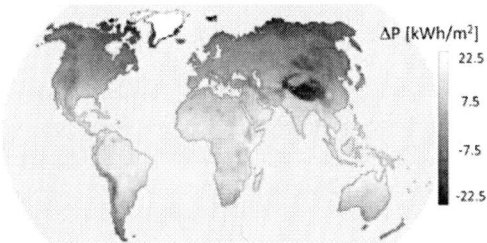

Fig. 3. Calculated global harvesting efficiency of the used CdTe (a) and Si (b) solar cell averaged for the year 2015. The same relative variation is used in both scales. Silicon shows a much stronger variation of performance than CdTe due to the larger sensitivity to precipitable water and temperature. c) Corresponding deviation in power generated annually, when comparing actual yield against nameplate expectation.

REFERENCES

[1] C. A. Gueymard, "SMARTS, A Simple Model of the Atmospheric Radiative Transfer of Sunshine: Algorithms and Performance Assessment". Technical Report No. FSEC-PF-270-95. Cocoa, FL: Florida Solar Energy Center (1995).
[2] https://ceres.larc.nasa.gov/
[3] http://airs.jpl.nasa.gov/
[4] http://www.nasa.gov/mission_pages/aura/spacecraft/omi.html
[5] T. Huld, A. M. Gracia Amillo, "Estimating PV Module Performance over Large Geographical Regions: The Role of Irradiance, Air Temperature, Wind Speed and Solar Spectrum", Energies, 8 (2015), 5159-5181.
[6] S. Ponce-Alcántara. J. P. Connolly. G. Sánchez. J. M. Míguez. V. Hoffmann. R. Ordás, "A Statistical Analysis of the Temperature Coefficients of Industrial Silicon Solar Cells, Energy Procedia, 55 (2014), 578 – 588.

Estimation of mean monthly global solar radiation using model based on sunshine hours for Colombia

Diego J. Rodríguez, Johan Hernández and Adolfo Jaramillo

Universidad Distrital Francisco José de Caldas, Bogotá, djrodriguezp@gmail.com, Colombia

Abstract — For the design of solar systems of flat plate is required to know the average value and extreme values of average global radiation. For the particular Colombian case this information is usually consulted in the NASA database or in the official Atlas of solar radiation. This paper presents a review of the models that relate global radiation to solar brightness and is applied to a particular case study that allows determining which of them presents the greatest effectiveness. Ten models were analyzed in total and greater emphasis was placed on polynomial approximation models of first and second order because it is the simplest and reaches the best approximation. It was observed that due to the dispersion of the data, due to the geographic conditions of the region, the model that best results presents is the order two polynomial approximation.

I. INTRODUCTION

In order to know the solar energy potential it is necessary to determine the average irradiance on the area of implementation. Unfortunately many regions do not have direct measurements of solar radiation; therefore, it is not possible to determine the solar resource accurately, generating errors of design in the solar solutions. For the Colombian case the radiometric observation network has increased its coverage in recent years, but there are no records in certain regions due to the irregular distribution of the pyranometric stations network. This implies that it is necessary to use methods that use indirect variables to model the average daily monthly irradiation [1].

In the literature are registered several types of models for the reconstruction of solar radiation. Due to the large number of factors that influencing solar radiation, it is practically impossible to develop methods to determine the average radiation in periods of less than one month. Several methods have been developed to determine the monthly average daily radiation have based on indirect meteorological variables, such as temperature, pressure and in particular solar brightness.

The number of models that correlate solar radiation with other meteorological variables that have been published is relatively high; this makes it difficult to choose a particular model. The model to be used depends strongly on the characteristics of the site and its purpose. The selection criterion is based on the requirements and the effectiveness of the method. For example, models that calculate the irradiance with cloudy sky are commonly used in solar energy systems, while clear-sky models are used for air conditioning systems, photovoltaic

systems and calibration systems for solar radiation measurement equipment. Regarding the different methods for calculating the solar radiation incident on the Earth's surface, it is based on the physical and statistical models, these being the procedures that are applied to have an estimate close to the magnitude of the reflected solar radiation in a Given surface; However, the use of equipment with direct measurements is also available.

Models for the reconstruction of solar radiation can be divided into three large groups. The first of these are the empirical models, which determine the correlation between solar radiation and measured in-situ variables. Subdivision can be made depend on the group of variables that are studied. The four most relevant subgroups are models based on solar brightness, cloudiness, temperature and other variables such as humidity. The second types of models are based on transfer models that require the knowledge of the transmittals (the amount of energy that goes through a body, or average, per unit of time) of the different atmospheric components. To apply them, spectral measurements of solar radiation are required or to be discriminated by wavelength that are performed with spectro-radiometers. The last group of models is based on the reconstruction of the solar radiation by means of satellite images. The satellites focused on environmental applications allow the evaluation of solar radiation over wide regions, based on estimation models derived from the reflectance observed in its visible radiation detector channel (VIS).

In this work the results of the adjustment of the polynomial models of first and second order based on the data between the year 2000 and 2010 of 79 stations are shown that counts with the instruments to measure global solar radiation and solar brightness of meteorological service of Colombia (IDEAM for its acronym in Spanish). With this information the information of the pyranometric network for the update of the Atlas of solar radiation that has published in the year 2015 is complemented [2].

II. STUDY REGION

The 79 selected meteorological stations are distributed throughout the country. Initially the calibration of radiometric measurement equipment was verified, showing an example of the results and analysis of the calibration performed at a pyranometer by comparison with a reference pyranometer. Subsequently, an application called Analyst stations was developed, with the aim to realizer evaluation and processing of solar radiation data from terrestrial measurement points.

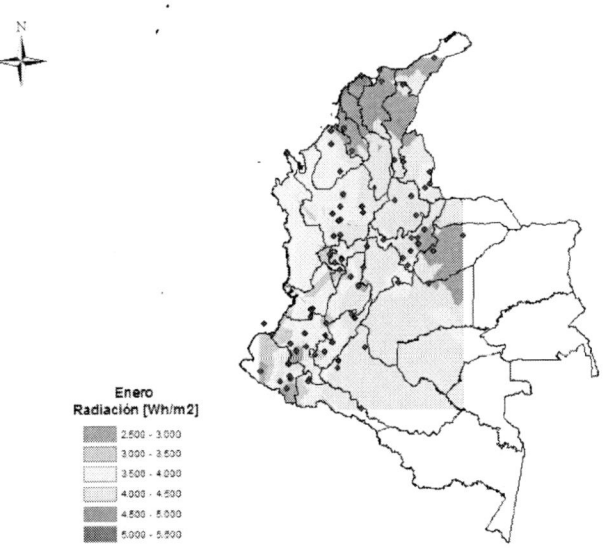

Enero
Radiación [Wh/m2]

2.500 - 3.000
3.000 - 3.500
3.500 - 4.000
4.000 - 4.500
4.500 - 5.000
5.000 - 5.500

Fig. 1: Study region for developed of solar radiation maps, Correspond to average of January

III. METHODS AND MODELING

The solar brightness is constituted as the variable most used in the empirical models to determine the global radiation. The literature shows that these models are based on the ratio of solar brightness (S) to extraterrestrial solar brightness (S_0) to determine the monthly average of global radiation.

A. Model first Order (Angstrom-Prescott model)

This model was published in 1937 [3] to determine the monthly average daily solar radiation by means of a linear correlation of the atmospheric clarity index (H/H_0) con (S/S_0) from the way:

$$\frac{H}{H_0} = a + b\frac{S}{S_0}$$

The value of extraterrestrial solar radiation can be determined by:

$$H_0 = \frac{24}{2\pi} I_{sc} \left[1 + 0.033 \cos\left(\frac{360n}{365}\right)\right] \left[\cos\varphi \cos\delta \sin\omega + \frac{\pi}{180}\omega \sin\varphi \sin\delta\right]$$

Where δ is the solar inclination and ω the main hour angle, where they can be calculated:

$$\delta = 23.45 \sin\left[\frac{360}{365}(284 + n)\right]$$

$$\omega = \cos^{-1}(-tan\varphi tan\delta)$$

For a given month, the maximum solar brightness can be found using (Duffie JA, 1991) the following:

$$S_0 = \frac{2}{15}\omega$$

The relationship of Angstrom-Prescott has been used in several regions of the world. In the (table) registers the values of a and b published for several regions in the world.

TABLE I

VALUES A AND B FOR THE LINEAR MODEL PUBLISHED FOR SEVERAL REGIONS IN THE WORLD.

Region	Autor	Reference	a	b
Italia	Jain	[4]	0,177	0,692
Egipto	El-Metwally	[5]	0,228	0,527
Turquía	Bakirci	[6]	0,2786	0,4160
Amman - Jordania	Alssad	[7]	0,174	0,615
Zambia	Jain	[4]	0,240	0,513
India	Katiyar	[8]	0,2281	0,5093
Tennessee- USA	Lewis	[9]	0,14	0,57
España	Almorox	[10]	0,2170	0,5433
Pakistán	Raja - Twidell	[11]	0,335	0,367
Tíbet-China	Li et al	(Li H, 2011)	0,2223	0,6529

Some articles show how in very close regions but with different climates the values of a and b can differ considerably. Therefore, it implies that the model differs depending on the specific climates of each region; consequently it is necessary to detect regions with similar climates in order to obtain reliable results.

B. Model Second Order (Ogelman Model)

Based on the original idea of the Angström-Prescott model where the solar radiation is linearly related to the solar brightness, Ogelman et al [12] relate the same variables in a quadratic form (Ogelman H., 1984)

$$\frac{H}{H_0} = a + b\frac{S}{S_0} + c\left(\frac{S}{S_0}\right)^2$$

Many authors around the world have applied this model to their regions; in the table 3 we find a review of the authors and parameters of the Ogelman model.

TABLE III

OGELMAN MODEL PARAMETERS PUBLISHED FOR VARIOUS REGIONS IN THE WORLD.

Region	Reference	a	b	c
Turquia	[13]	0,145	0,845	-0,280
España	[14]	0,1840	0,6792	-0,1228
Libia	[15]	0,1	0,874	-0,255
Pakistan	[16]	0,348	0,320	0,070
Oman	[17]	0,9428	-1,2027	0,9336
China	[18]	0,1404	0,6126	0,0351

IV. RESULTS

Two types of adjustments were made according to the models reviewed; these are the Angstrom-Prescott and Ogelman et al that correspond to linear and quadratic models. As an example, the adjustment was made for the data obtained from Granja La Libertad station, located in the eastern plains of the country, with latitude 5.5 and length -70.5. The station has equipment to measure global solar radiation and solar shine at the same time. The results of the adjustment of the polynomial models of first and second order are of the form:

Angstrom-Prescott (Lineal Model):

$$\frac{H}{H_0} = 0.3948\frac{S}{S_0} + 0.2088 \text{ with } R^2 = 0.78322$$

Ogelman et al (Quadratic Model):

$$\frac{H}{H_0} = -0.1827\left(\frac{S}{S_0}\right)^2 + 0.5527\frac{S}{S_0} + 0.1838 \text{ with } R^2 = 0.79214$$

It is observed that the smallest correlation coefficient between solar radiation and solar brightness corresponds to the linear model, although the difference between the two models does not exceed 2%. Therefore both models have a similar validity due to the dispersion of the data itself.

This same procedure was done for each of the stations of the IDEAM that have simultaneous measurements of global solar radiation and solar brightness, in order to determine the correlation between the model and the latitude of the stations. Figure 2 and Figure 3 show the correlation between the parameters of the linear regression and the latitude of the stations.

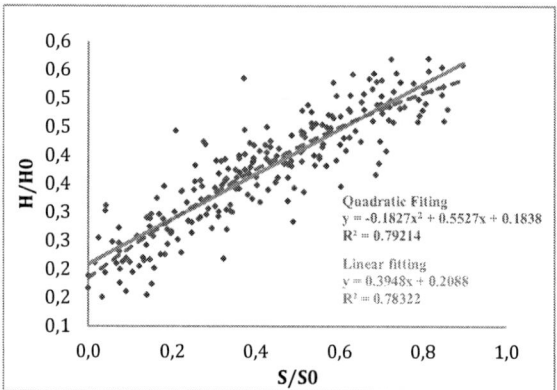

Fig. 2: H/Ho vs S/So for Granja La Libertad Station

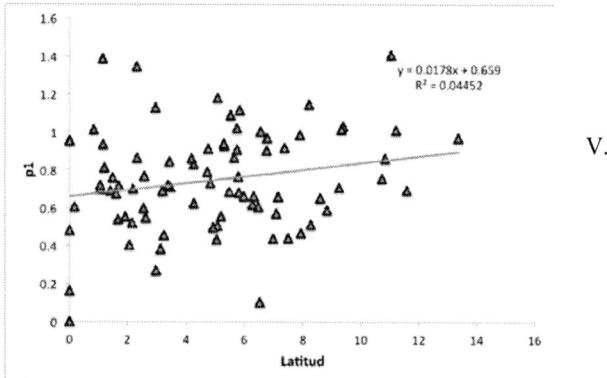

Fig. 3: Correlation between parameter p1 and latitude

CONCLUSIONS

Based on the review of existing models in the literature, it was observed that the most appropriate models to be used, taking into account the IDEAM measurement network, is the first and second order polynomial model. The data were processed from the network stations, which in turn measured global radiation and solar brightness. This processing consists of an analysis on the regularity of the data and statistical coherence. The values of extraterrestrial solar radiation and solar brightness were calculated. With this information the atmospheric clarity index was calculated as the ratio between global terrestrial solar radiation and extraterrestrial solar radiation (H/H_0). The better adjustment is the quadratic model, although the difference with the linear model does not exceed 2%.

REFERENCES

[1] OMM, Guía de prácticas climatológicas, http://www.wmo.int/pages/prog/wcp/ccl /guide/documents/wmo_100_es.pdf, 2011

[2] IDEAM, Atlas de radiación solar interactive, http://atlas.ideam.gov.co/visorAtlasRadiacion.html, 2015

[3] Prescott JA. Evaporation from water surface in relation to solar radiation. Transactions of the Royal Society of Australia 1940;46:114–8.

[4] Jain S, Jain PC. A comparison of the Angstrom-type correlations and the estimation of monthly average daily global irradiation. Solar Energy 1988;40(2):93–8.

[5] El-Metwally M. Sunshine and global solar radiation estimation at different sites in Egypt. Journal of Atmospheric and Solar-Terrestrial Physics 2005;67:1331–42.

[6] Bakirci K. Correlations for estimation of daily global solar radiation with hours of bright sunshine in Turkey. Energy 2009;34:485–501.

[7] Alsaad MA. Characteristic distribution of global radiation for Amman, Jordan. Solar and Wind Technology 1990;7(2/3):261–6.

[8] Lewis G. An empirical relation for estimating global irradiation for Tennes- see. USA Energy Conversion and Management 1992;33(12):1097–9.

[9] Tiris M, Tiris C, Erdalli Y. Water heating systems by solar energy. Marmara Research Centre, Institute of Energy Systems and Environmental Research, NATO TU-COATING, Gebze, Kocaeli, Turkey; 1997 [in Turkish].

[10] Almorox J, Hontoria C. Global solar radiation estimation using sunshine duration in Spain. Energy Conversion and Management 2004;45:1529–35.

[11] Raja IA, Twidell JW. Distribution of global insolation over Pakistan. Solar Energy 1990;44:63–71.

[12] Ogelman H, Ecevit A, Tasdemiroglu E. A new method for estimating solar radiation from bright sunshine data. Solar Energy 1984;33:619–25.

[13] Akinoglu BG, Ecevit A. A further comparison and discussion of sunshine based models to estimate global solar radiation. Energy 1990;15:865–72.

[14] AlmoroxJ, BenitoM, Equation coefficients from measured daily data in Toledo, Spain. Renewable Energy 2005;30:931–6.

[15] Said R, Mansor M, Abuain T. Estimation of global and diffuse radiation at Tripoli. Renewable Energy 1998;14(1–4):221–7.

[16] Ahmad F, Ulfat I. Empirical models for the correlation of monthly average daily global solar radiation with hours of sunshine on a horizontal surface at Karachi, Pakistan. Turkish Journal of Physics 2004;28:301–7.

[17] Ampratwum DB, Dorvlo ASS. Estimation of solar radiation from the number of sunshine hours. Applied Energy 1999;63:161–7.

[18] Jin Z, Yezheng W, Gang Y. General formula for estimation of monthly average daily global solar radiation in China. Energy Conversion and Management 2005;46:257–68.

Implementation of Solar Diffuse CIE Model in Ray Tracing Program for Irradiance Calculations

Liliana Ruiz Diaz, Pierre-Alexandre Blanche, and Robert A. Norwood

College of Optical Sciences, The University of Arizona, Tucson, AZ 85719, USA

Abstract — **In this work, an accurate representation of the sky radiance has been created in MATLAB using the diffuse daylight standard CIE model and exported to ray tracing software for non-sequential optical simulations. This first-order optical model can be applied to most locations at any time of the day and takes into account the weather conditions and cloud distribution. The solar spectrum is incorporated to take into account wavelength-dependent optical effects present in solar energy applications where diffuse sunlight collection is relevant.**

Index Terms — **diffuse radiance, solar model, sunlight, solar irradiance, ray tracing.**

I. INTRODUCTION

The output of solar energy systems that collect solar diffuse irradiance depends on the sky luminance. There exist several sky models to calculate solar diffuse radiance such as the Perez all-weather sky luminance [1], Bird's clear sky model [2], and the International Commission on Illumination (CIE) distribution of daylight model [3]. In particular, the CIE model can be used to compute the angular radiance distribution depending on weather conditions and cloud patterns [3]. However, this model does not provide information about the solar spectrum and is mostly used to calculate total irradiance on a surface. In this work, we incorporate the CIE model into MATLAB to calculate the sky radiance for any location at any time of the day and year. The obtained radiance distribution is exported to non-sequential ray tracing software to simulate the sky irradiance for first order optics applications. We also incorporate the solar spectrum to take into account various wavelength-dependent optical effects. Our application is divided into three tasks:

1) The code calculates the sun position with respect to the coordinates of the selected location. The zenith and the azimuth solar angles are found from standard numerical equations and compared with National Renewable Energy Laboratory (NREL) data.

2) The program generates radiance data implementing the previously calculated sun position into the CIE model. The CIE model depends on two parametric equations that can describe up to 15 different types of solar radiance distributions.

3) The calculated data is used to generate a hemispherical light source in LightTools ray tracing software to simulate the sky

radiance. In addition, the solar spectrum is incorporated to complement the simulations.

This ray tracing model can be used to predict the efficiency/exergy of any solar energy system in which collection of diffuse sunlight is relevant. This becomes particularly important for cases where there is a difference in the efficiency between global and direct normal irradiance collection, as is the case for solar concentration technology.

II. SUN POSITION

The position of the sun in the sky depends on the geolocation of the point of measurement, date, and time of day, due to changes in the earth-sun distance, tilting of the earth, and other effects.

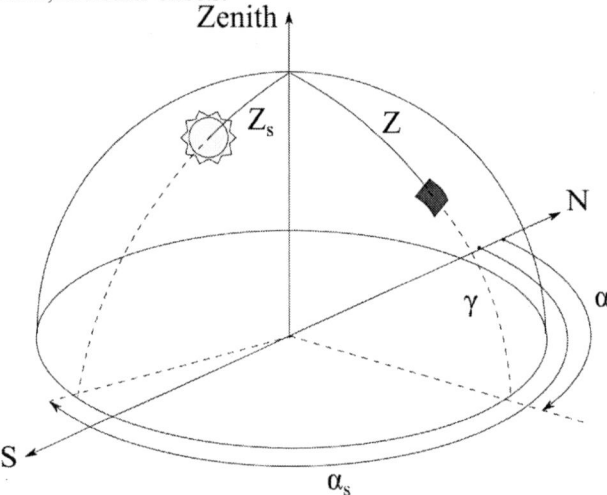

Fig. 1. Sun position and angle definitions. While Z_s and α_s are the solar zenith and solar azimuth angle respectively, Z and α, are the zenith and azimuth angle for an arbitrary point on the hemispherical sky. The angle γ is the elevation angular distance (90°- Z).

The sun position can be defined by the zenith angle Z_s (angle between the sun and the zenith point) and the solar azimuth angle α_s (angular displacement between the projected sun position on the horizontal plane and the north axis) as described in Figure 1. This coordinate system will also be used to locate any point in the hemispherical sky for the diffuse model. These two angles can be found by using the declination of the sun and the solar time equation. For locations with absolute latitudes less than 66.5°, the numerical solar declination equation as defined by Spencer [4] and with Iqbal's notation [5] can be used

$$
\begin{aligned}
\delta = (180/\pi)(&0.006918 - 0.399912\cos(B) \\
&+ 0.070257\sin(B) - 0.006758\cos(2B) \\
&+ 0.000907\sin(2B) - 0.002697\cos(3B) \\
&+ 0.00148\sin(3B))
\end{aligned} \quad (1)
$$

where B depends on the nth day of the year:

$$
B = (n-1)\frac{360}{365}. \quad (2)
$$

The solar zenith angle is related to δ by the relation

$$
\cos Z_s = \cos\alpha\cos\delta\cos\omega + \sin\alpha\sin\delta, \quad (3)
$$

where α is the latitude of the geolocation and ω is the angular displacement due to earth's rotation with respect to the solar noon position and is taken to be 15° per hour (negative before noon and positive afterwards). To increase accuracy [5], the solar time must be used as it differs from the standard time by

$$
\text{Solar time} = \text{standard time} + 4(L_{st} - L) + E, \quad (4)
$$

where L_{st} is the standard meridian of the time zone and L is the exact longitude of the geolocation. The parameter E in minutes as defined by Spencer is

$$
\begin{aligned}
E = 229.18(&0.0000075 + 0.001868\cos(B) \\
&- 0.032077\sin(B) - 0.014615\cos(2B) \\
&- 0.04089\sin(2B))
\end{aligned} \quad (5)
$$

It is noticed that the first term in (5) was corrected by Spencer and some publications do not include this correction [6].

These equations were computed in MATLAB for Phoenix, AZ on May 15th and compared to NREL data measured at the Southwest Solar Research Park [7]. The results of these calculations are shown in Figure 2, where the solar zenith and azimuth angle values change through the day. The root-mean-square errors (RMSE) between the solar zenith angle for the years 2011, 2012, and 2013 and the theoretical calculations were 0.529°, 0.315°, and 0.568°, respectively. The RMSE for the solar azimuth angle values were larger at 0.774°, 1.586°, and 0.803° respectively. Except for 2012, all the RMSE values were within the standard uncertainty of the NREL measurements [7]. This comparison allowed us to modify the variable B to take into account leap years (such as the 2012 case) and other effects not considered in (1) and (5).

III. THE SKY MODEL

The CIE general standard sky model defines 15 types of solar diffuse angular distribution depending on weather conditions, cloud presence, pollution, etc [3]. The model is based on scanned luminance measurements taken at specific places around the world and numerical fitting equations [8].

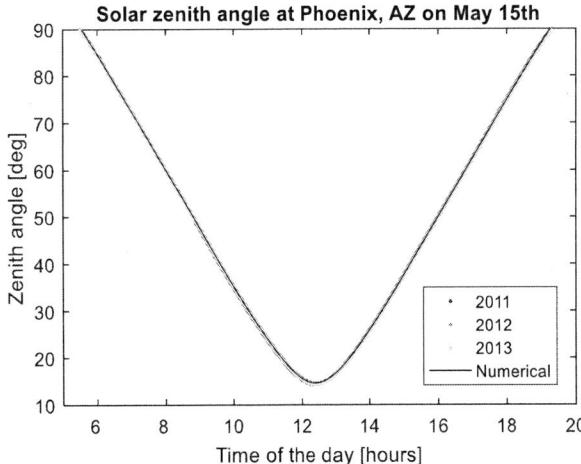

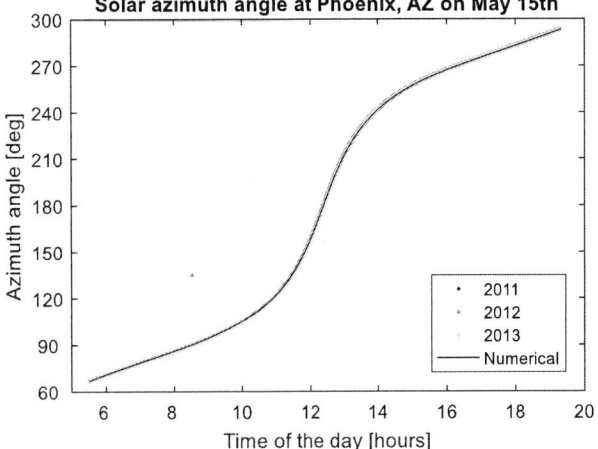

Fig. 2. The calculated Z_s zenith angle and the α_s solar azimuth angle are compared to measurements taken at the NREL Southwest Solar Research Park located at Phoenix, AZ on May 15th for three consecutive years.

The luminance depends on two functions, the luminance gradation function $\varphi(Z)$ and the scattering indicatrix $f(\chi)$, where χ is the absolute angular separation between the sun and a particular position in the hemispherical sky. For any given point in the sky, the luminance ratio with respect to the luminance at the zenith point is

$$
\frac{L_r}{L_z} = \frac{f(\chi)*\varphi(Z)}{f(Z_s)*\varphi(0)}, \quad (6)
$$

where $\varphi(Z)$ and $f(\chi)$ are

$$
\varphi(Z) = 1 + a*\exp\left(\frac{b}{\cos(Z)}\right), \quad (7)
$$

$$
f(\chi) = 1 + c*\left[\exp(d\chi) - \exp\left(d\frac{\pi}{2}\right)\right] + e*\cos^2\chi. \quad (8)
$$

The constants *a - e* are standard parameters corresponding to particular sky conditions. A full description of the different types of sky is found in the CIE standard general sky technical report [3].

By combining the equations for the sun position and the CIE standard model, it is possible to calculate the luminance for a particular location at any time of the day for any given weather conditions. Furthermore, if the luminance at the zenith point is known, the total diffuse luminance can be calculated. For this work, radiometric units are used instead of the photometric units utilized in the CIE report. Moreover, the luminance ratio is normalized and multiplied by 1 W/sr to allow comparison between the different types of sky daylight.

Figure 3 shows three different types of sky radiance angular distributions computed in MATLAB for Tucson, AZ. The graphs are represented in 2D Cartesian projections with identical scale. The first case corresponds to an ideal isotropic sky where the radiant intensity is homogenous for all angles (sky type 5). The second image presents a non-uniform sky with overcast weather, where the sun is mostly obscured by clouds (sky type 9). The third image shows the standard CIE luminance for a cloudless clear day (sky type 12). For most cases, the maximum intensity occurs at the solar zenith angle (13.4° for Tucson at noon on May 15th). These daylight distributions allow us to simulate realistic sky conditions which are vastly different from the isotropic case.

III. RAY TRACING

The sky radiance model was imported into LightTools, a non-sequential optical design software. Combining both codes allows us to generate an accurate representation of the sky for solar energy applications using ray tracing techniques. The pre-computed radiance data is used to generate a hemispherical source by an apodization method in which the surface of a sphere is divided into a finite number of equal angularly spaced areas (polar grid). Each area acts as a small Lambertian source with a relative specific flux magnitude emission. This results in a non-uniform angular radiant source.

In order to create a more realistic simulation of the sky, the Standard Air Mass 1.5 (AM 1.5) solar spectrum was included in the modeled source. The spectral variation through the azimuthal and elevation angle was taken to be homogeneous. The solar spectrum allows for the calculation of various optical effects produced in any type of solar energy collection system, such as reflectivity from photovoltaic cells due to Fresnel reflections, refraction due to concentrating optics, etc.

Figure 4 shows a hemispherical source with a polar grid of LxV dimensions. A small spherical detector is placed inside the source to measure the radiant intensity from all directions. The results of this simulation are shown in the second image, where a non-homogenous radiance distribution (sky type 9) is represented in a 2D Cartesian projection. The third image shows the radiance angular distribution of the sky type 12. In

both cases, the ray-traced radiant intensity shows agreement with the original calculated data from MATLAB.

Diffuse sunlight model at Tucson, AZ on May 15th at noon

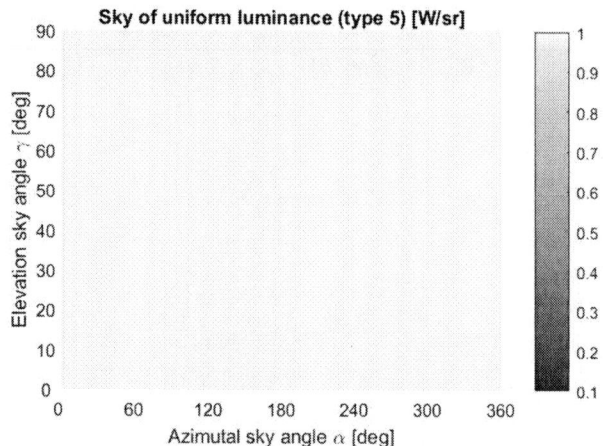

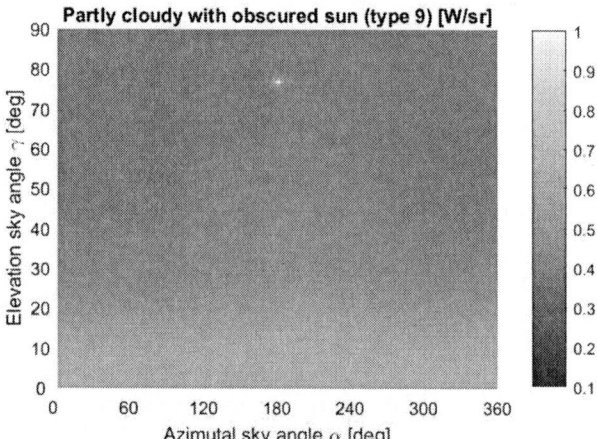

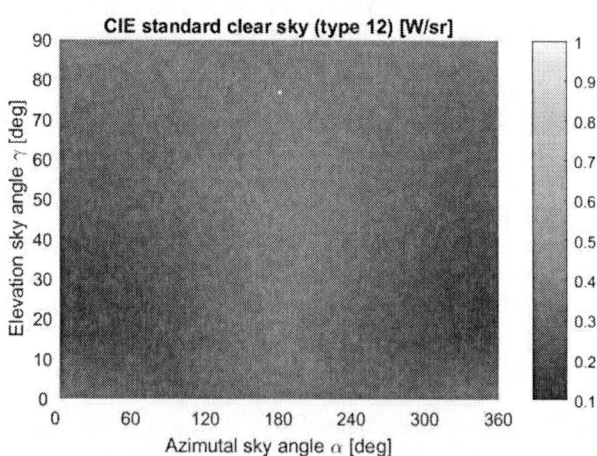

Fig. 3. Cartesian representation of the normalized radiant intensity at Tucson, AZ on May 15th at solar noon for three types of sky condition using the standard CIE daylight model.

Ray tracing simulation of the diffuse sunlight at Tucson, AZ on May 15th at noon

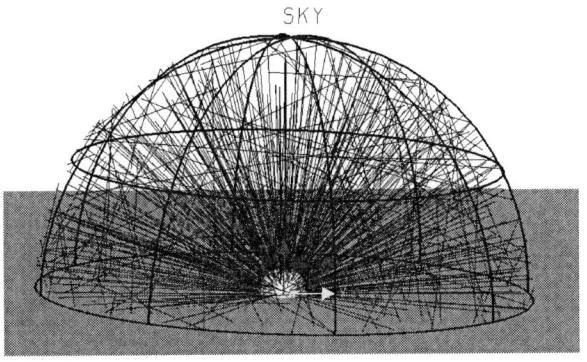

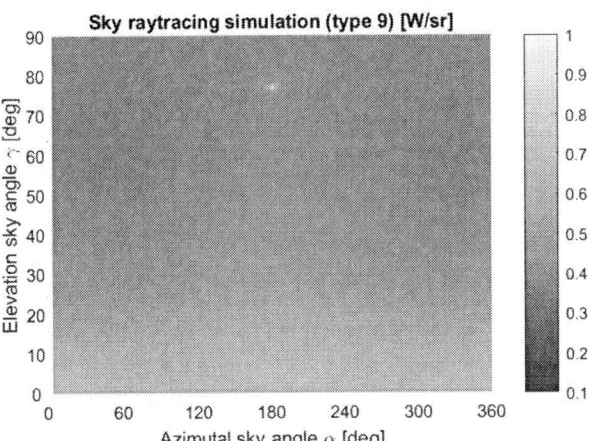

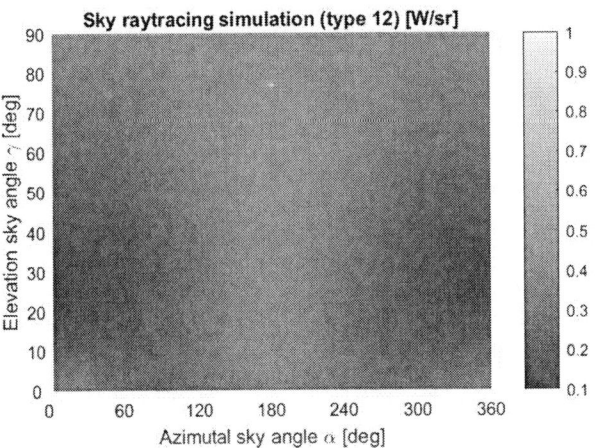

Fig. 4. Ray tracing of hemispherical source with embedded sky radiance information and Cartesian representation of two different normalized radiant intensity distributions. The intensities show agreement with the calculated CIE model for Tucson, AZ on May 15th at solar noon.

Since the ray tracing software is based on a randomization method to generate ray paths, noise is expected in all the simulations. The signal-to-noise ratio depends on several input parameters such as number of rays, size of the detector, resolution, etc. For a ray tracing simulation with 50 million rays, the RMSE compared to the original MATLAB computed sky radiance was 2.5% and 2.6% for sky type 9 and 12, respectively. It was found that the RMSE for non-isotropic sky simulations tends to be larger than for the homogenous case with the same ray tracing parameters.

IV. CONCLUSIONS

An accurate representation of the sky radiance (angular irradiance) was created for several weather conditions (sky types). This model was exported to a ray tracing simulation software using the CIE standard general sky model and numerical equations to locate the sun position. The model can be applied to all locations with absolute latitudes less than 66.5° at any time of the day throughout the year. An AM 1.5 solar spectrum was incorporated in this first-order optical model to reproduce the optical effects that affect the efficiency/exergy of solar energy systems. This ray tracing model can be used to estimate diffuse irradiance for realistic sky daylight conditions.

IV. ACKNOWLEDGEMENT

The authors acknowledge the support of the ARPA-E MOSAIC program through the Department of Energy, Grant DE-AR0000839.

REFERENCES

[1] R. Perez, R. Seals, and J. Michalsky, "All-weather model for sky luminance distribution - preliminary configuration and validation," *Solar Energy*, vol. 50, no. 3, pp. 235-245, 1993.

[2] R.E. Bird and R.L. Hulstrom, "A simplified clear sky model for direct and diffuse insolation on horizontal surfaces," Solar Energy Research Institute. Golden, CO, Tech. Rep. SERI/TR-642-761, Feb. 1981.

[3] Y. Uetani et al. "Spatial distribution of daylight - CIE standard general sky," CIE Commission Internationale de L'eclairage, ISO 15469:2004 (E)/CIE S 011/E:2003, Vienna, Austria, 2003.

[4] J. W. Spencer, "Fourier series representation of the position of the Sun," *Search* 2, no. 5, 1971.

[5] M. Iqbal, *An Introduction to Solar Radiation*. Toronto: Academic Press, 1983, ch.1, pp 6-28.

[6] Oglesby, M. (1998, Feb. 7). *Fourier paper* [Online]. Available: https://www.mail-archive.com/sundial@uni-koeln.de/msg01050.html.

[7] Southwest Solar Research Park, National Renewable Energy Laboratory, http://dx.doi.org/10.5439/1052225.

[8] S. Darula and R. Kittler, "CIE general sky standard defining luminance distributions," *Proceedings Conference eSim*, Montreal, Canada, 2002.

Investigation of City-Level Site-Pair Correlations of Solar Variability using Empirical Satellite Data

Rhythm Singh[a,b], Rangan Banerjee[b]

[a]National Institute of Construction Management and Research, Pune – 411045, Maharashtra, India

[b]Department of Energy Science and Engineering, Indian Institute of Technology Bombay, Mumbai – 400076, Maharashtra, India

Abstract — The paper discusses solar resource variability and geographic smoothing at a city-level by estimating the site-pair correlations of solar variability. The site-pair correlation of solar variability, $\rho_{\Delta Kt}$, is quantified by estimating the correlation between the changes in clear-sky indices, ΔKt, of a pair of sites. The paper studies the variation of $\rho_{\Delta Kt}$ with a) the distance between the sites, and b) both the distance between the sites and the difference between the relative humidity profiles of the sites. The study has been conducted for three Indian cities – Mumbai, Delhi and Bengaluru – on hourly radiation and climatic data spread over a span of five years, 2010 – 2014.

Index Terms — clear-sky index, Pearson's product-moment correlation coefficient, relative humidity, site-pair correlation, solar resource variability.

I. INTRODUCTION

Solar resource variability is a major cause of concern from the point of view of large penetrations of solar PV in a grid. A promising application of solar PV is at the rooftop-scale. The concept of "solar cities", or the large-scale deployment of rooftop solar PV at city-level, has been gaining wide popularity in recent times. Several studies ([1] – [12]), published in the last few years, have focused on estimating the rooftop solar PV potential of urban areas across the globe. Dedicated programs aimed at developing "solar cities" have been launched by governments in different parts of the world. The National Solar Mission of India [13] has an ambitious target of achieving 40,000 MW of installed rooftop solar PV capacity by 2022. At present, though, the application of the technology at city-level is still in the early stages of the learning curve.

However, as the share of rooftop solar PV capacity at urban level will grow, the impacts of solar resource variability would become increasingly important. Solar resource variability, per se, has been the subject of several studies and research papers. Some studies ([14] – [16]) have focused on analyzing the variability of a PV fleet or the site-specific variability. Some studies ([17] – [18]) have analyzed the correlation between solar variability of a pair of sites and its relationship to the distance between them. The impact of cloud speed on solar variability has also been studied [19]. While some studies ([20] – [21]) have discussed the phenomenon of geographical smoothing of solar variability in general, others ([22] – [24])

have studied and quantified the geographical smoothing at a state/province level. There are no papers which analyze the variability metrics at a city-level.

The present paper draws upon the work done in these previous studies ([14], [15], [17], [19], and [21]), and is specifically focused on analyzing and quantifying the correlation between solar variability of a pair of sites at a city-level. The analysis uses empirical satellite data of solar radiation, at a one-hour resolution, obtained from the National Solar Radiation Database (NSRDB) [25], courtesy of the National Renewable Energy Laboratory (NREL), USA. Based on the analysis, the paper investigates the relationship between the site-pair correlations and the distance between the sites.

As stated before, cloud speed has been found to have an impact on the solar variability [19]. However, cloud speed data is not readily available; and determining cloud speed for a certain location is quite difficult, requiring high frequency measurements from a dedicated ground irradiance sensor network. Interestingly, some studies ([26] – [28]) have reported a fairly strong correlation between cloud cover and relative humidity at a location, more so for the tropical regions. The present paper investigates whether the difference in relative humidity of two sites can be used for estimating the site-pair correlation of solar variability.

The study has been conducted on five years of one-hour resolution data for three Indian cities, Mumbai (19°N, 73°E), Delhi (28°N, 77°E) and Bengaluru (13°N, 78°E).

II. METHODOLOGY

Hourly solar radiation and climatic data has been collected for the period 1st January, 2010 – 31st December, 2014, for a) six locations in Mumbai city, b) thirteen locations in Delhi, and c) twelve locations in Bengaluru. The locations of these cities on an Indian map are shown in Fig.1. Figures 2- 4 show the locations of the data points within the respective cities by the intersection points of the lines of latitudes and longitudes shown on the map.

The analysis is based on the clear-sky index, Kt, which is expressed as the ratio of ground-observed GHI to the clear-sky expected GHI, as shown in (1). From this clear-sky index, the change in clear-sky index over a specified time interval Δt, at a given location n, has been estimated; indicated as $\Delta Kt^n_{\Delta t}$, or simply as ΔKt, in brief, as shown in (2).

$$Kt = \frac{GHI_{actual}}{GHI_{clear-sky}} \qquad (1)$$

Fig. 1.   Locations of the cities under the study.

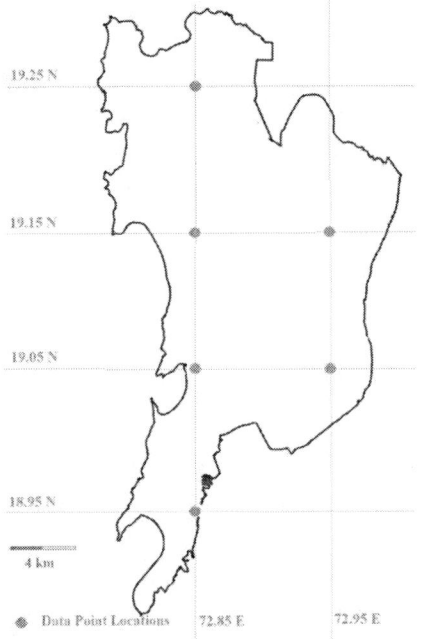

Fig. 2.   Locations of data points in Mumbai city.

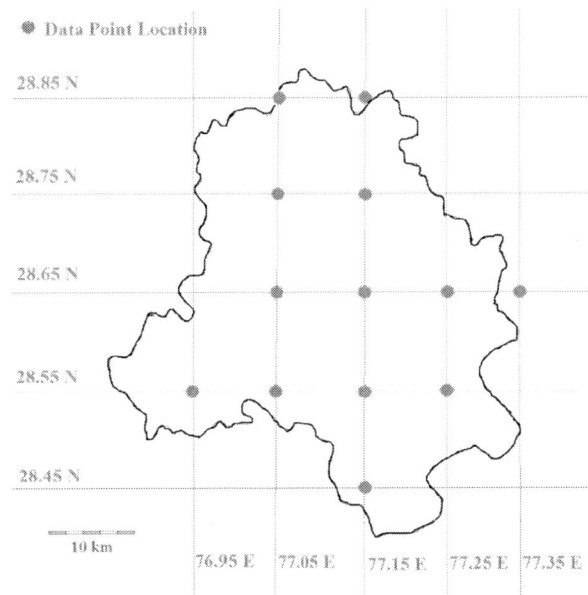

Fig. 3.   Locations of data points in Delhi city.

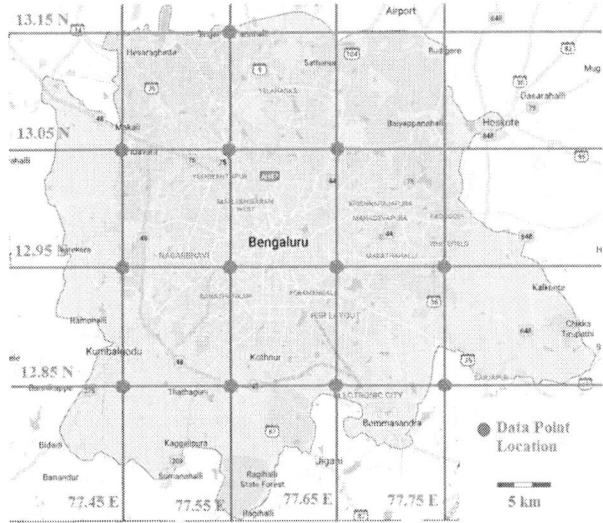

Fig. 4.   Locations of data points in Bengaluru city.

$$\Delta Kt^n_{\Delta t} = Kt^n_{t+\Delta t} - Kt^n_t \qquad (2)$$

For the present analysis Δt is equal to one hour. Thus, ΔKt has been estimated on an hourly resolution for all the sunshine hours (7 am – 5 pm) during the five year period, 1st January, 2010 – 31st December, 2014, for all the data point locations shown in Fig.2 – Fig.4.

The site-pair correlation of solar variability between two sites m and n in the same city, separated by distance $d_{m,n}$, is expressed as $\rho_{\Delta Kt}^{m,n}$. This site-pair correlation has been estimated for each pair of data points in the three cities under the study; it has been estimated in terms of the Pearson's product-moment correlation coefficient between the ΔKt's of the two locations, $\Delta Kt^m_{\Delta t}$ and $\Delta Kt^n_{\Delta t}$, as shown in (3).

$$\rho^{m,n}_{\Delta Kt} = \frac{\text{cov}\left(\Delta Kt^m_{\Delta t}, \Delta Kt^n_{\Delta t}\right)}{\sigma\left(\Delta Kt^m_{\Delta t}\right) \times \sigma\left(\Delta Kt^n_{\Delta t}\right)} \qquad (3)$$

Where cov is the covariance, and σ is the standard deviation.

The variation of the site-pair correlations of solar variability, $\rho_{\Delta Kt}{}^{m,n}$, has been studied with respect to the following two sets of variables: a) the distance between the sites, and b) distance between the sites and the difference between the relative humidity of the sites.

A. Relationship between $\rho_{\Delta Kt}{}^{m,n}$ and $d_{m,n}$

Univariate regression analysis has been used to study the relationship between the variations of the site-pair correlations of solar variability between two locations and the distance between them. Four univariate regression models – linear, log-linear, linear-log and log-log – have been investigated.

B. Relationship between $\rho_{\Delta Kt}{}^{m,n}$, $d_{m,n}$ and Relative Humidity term

In this case, the variations of site-pair correlations of solar variability between two locations have been studied with respect to two independent variables simultaneously: i) the distance between the locations, and ii) the difference between the relative humidity of the sites. This has been done using multiple regression analysis. Four multiple regression models have been investigated: linear, log-linear, linear-log and log-log.

As far as incorporating the difference between the relative humidity, RH, of the two locations m and n is concerned, two different metrics have been investigated. In one approach, the metric used is the Pearson's product-moment correlation coefficient between the hourly RH values of the two locations, expressed as $\rho_{RH}{}^{m,n}$ (shown in (4)). The metric used in the second approach is the standard deviation of the difference between the hourly RH values of locations m and n, $\sigma_{RH^m\text{-}RH^n}$ (shown in (5)).

$$\rho^{m,n}_{RH} = \frac{\text{cov}\left(RH^m, RH^n\right)}{\sigma\left(RH^m\right) \times \sigma\left(RH^n\right)} \qquad (4)$$

$$\sigma_{RH^m-RH^n} =$$

$$\sqrt{\frac{1}{\left(n_{yr} \times 365 \times h_{sh} - 1\right)} \times \sum \left[\left(RH^m - RH^n\right) - \mu_{RH^m-RH^n}\right]^2} \qquad (5)$$

Where RH^m, RH^n are the hourly relative humidity values (fractional) at locations m and n respectively, n_{yr} is the number of years over which data has been taken for analysis, h_{sh} is the number of sunshine hours in a day, and $\mu_{RH}{}^m{}_{-RH}{}^n$ is the mean of the difference between the hourly RH values of locations m and n.

III. RESULTS AND ANALYSIS

ΔKt has been estimated on an hourly resolution for all the sunshine hours (7 am – 5 pm) for the five year period, 1st January, 2010 – 31st December, 2014, for each data point location. Thus, there are a total of 20,075 data points for ΔKt for each location. The variation of ΔKt over the five year period follows a similar profile for the sites in the same city; however, the variation profile is different across different cities; this variation for a sample location each from the cities of Mumbai, Delhi and Bengaluru, has been shown in Fig.5 – Fig.7.

The correlation between the ΔKt's of two locations has been estimated for each pair of data locations in the three cities. Thus, i) $^6C_2$ or 15 site-pair correlations for Mumbai, ii) $^{13}C_2$ or 78 site-pair correlations for Delhi, and iii) $^{12}C_2$ or 66 site-pair correlations for Bengaluru have been estimated.

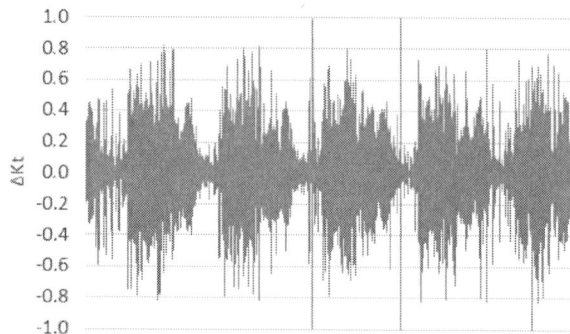

Fig. 5.   Variation of ΔKt for a sample location (19.25° N, 72.85° E) in Mumbai (2010 – 2014).

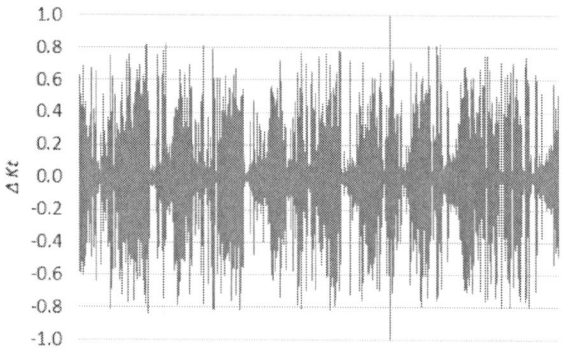

Fig. 6.   Variation of ΔKt for a sample location (28.85° N, 77.15° E) in Delhi (2010 – 2014).

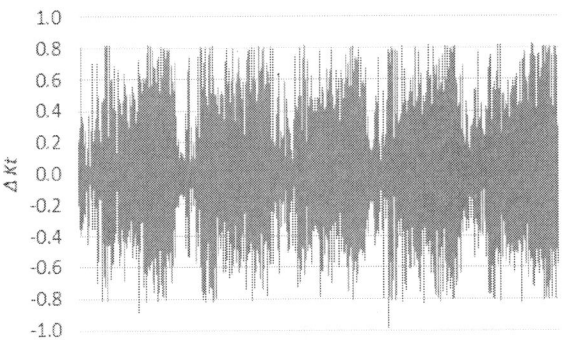

Fig. 7.　Variation of ΔKt for a sample location (12.85° N, 77.55° E) in Bengaluru (2010 – 2014).

A site-pair correlation between the ΔKt's of two locations can be represented graphically as a scatter plot between the two sets of variables. Sample scatter plots for site-pair correlations between sample locations from the three cities are shown in Fig.8 – Fig.10.

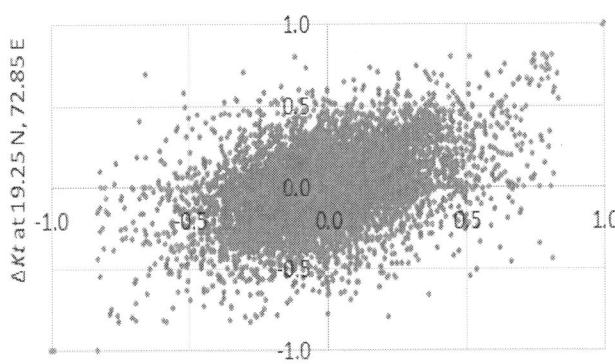

ΔKt at 19.15 N, 72.95 E

Fig. 8.　Scatter plot between ΔKt of two sample locations (19.25° N, 72.85° E and 19.15° N, 72.95° E) in Mumbai (2010 – 2014).

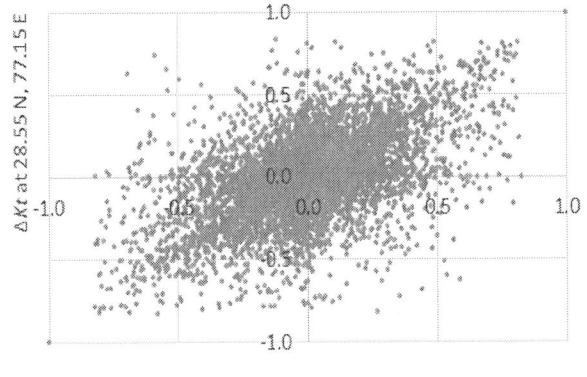

ΔKt at 28.65 N, 77.05 E

Fig. 9.　Scatter plot between ΔKt of two sample locations (28.55° N, 77.15° E and 28.65° N, 77.05° E) in Delhi (2010 – 2014).

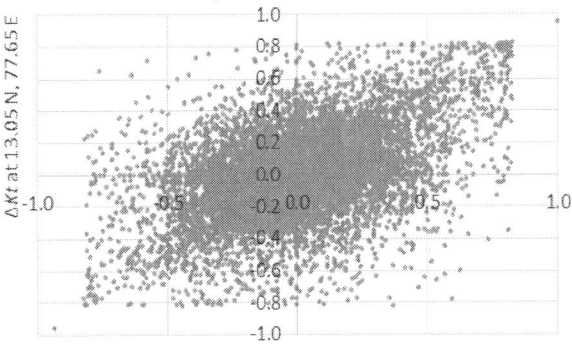

ΔKt at 13.15 N, 77.55 E

Fig. 10.　Scatter plot between ΔKt of two sample locations (13.05° N, 77.65° E and 13.15° N, 77.55° E) in Bengaluru (2010 – 2014).

For estimating the site-pair correlation between two locations m and n, all the 20,075 data points of $\Delta Kt^{m}_{\Delta t}$ and the 20,075 data points of $\Delta Kt^{n}_{\Delta t}$ have been fed to (3). The resultant $\rho_{\Delta Kt}^{m,n}$ value characterizes the particular site-pair correlation. The $\rho_{\Delta Kt}^{m,n}$ values for the sample correlations shown in Fig.8 – Fig.10 are 0.474, 0.539 and 0.478 respectively.

A. Relationship between $\rho_{\Delta Kt}^{m,n}$ and $d_{m,n}$

The variations of the site-pair correlations, $\rho_{\Delta Kt}^{m,n}$, with the distance between the locations, $d_{m,n}$, for the site-pairs in this study (15 pairs for Mumbai, 78 for Delhi and 66 for Bengaluru) are shown in Fig.11 – Fig.13. The values of $\rho_{\Delta Kt}^{m,n}$ and $d_{m,n}$ are found to fit to a log-log regression model, also known as power regression model. Table I gives the adjusted-R^2 values for this model for the three cities. Hence, the relationship between $\rho_{\Delta Kt}^{m,n}$ and $d_{m,n}$ is given by (6) – (8) for Mumbai, Delhi and Bengaluru respectively.

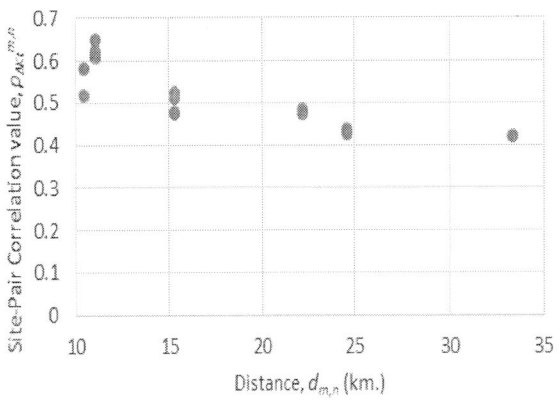

Fig. 11.　Variation of $\rho_{\Delta Kt}^{m,n}$ with distance for the 15 site-pairs of Mumbai city.

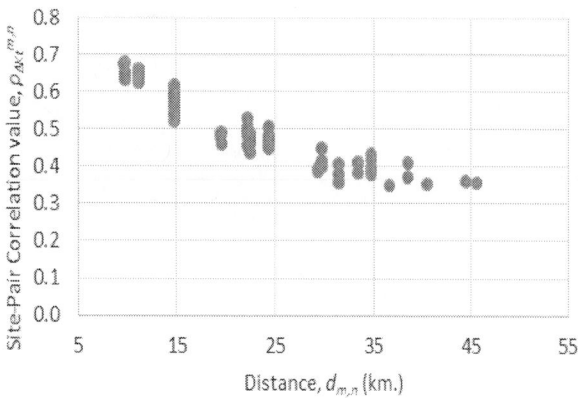

Fig. 12. Variation of $\rho_{\Delta Kt}^{m,n}$ with distance for the 78 site-pairs of Delhi city.

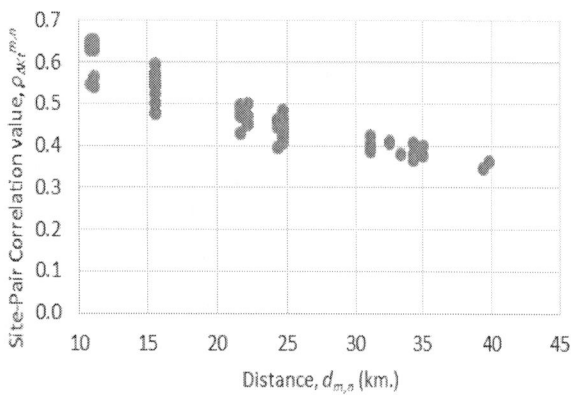

Fig. 13. Variation of $\rho_{\Delta Kt}^{m,n}$ with distance for the 66 site-pairs of Bengaluru city.

TABLE I
UNIVARIATE REGRESSION RESULTS

Dependent Variable		$\rho_{\Delta Kt}^{m,n}$	
Independent Variable		$d_{m,n}$	
Model Type	Adjusted-R^2		
	Mumbai	Delhi	Bengaluru
Log-log	0.76	0.93	0.91

$$\rho_{\Delta Kt}^{m,n} = 1.28 \cdot \left(d_{m,n}\right)^{-0.33} \qquad (6)$$

$$\rho_{\Delta Kt}^{m,n} = 1.78 \cdot \left(d_{m,n}\right)^{-0.43} \qquad (7)$$

$$\rho_{\Delta Kt}^{m,n} = 1.71 \cdot \left(d_{m,n}\right)^{-0.42} \qquad (8)$$

B. Relationship between $\rho_{\Delta Kt}^{m,n}$, $d_{m,n}$ and Relative Humidity term

With $\rho_{RH}^{m,n}$ representing the relative humidity term, $\rho_{\Delta Kt}^{m,n}$, $\rho_{RH}^{m,n}$ and $d_{m,n}$ are found to fit to a log-log regression model; the results are given in Table II, in terms of the adjusted-R^2 values. When $\sigma_{RH^m\text{-}RH^n}$ is used to represent the relative humidity term, the corresponding results for the log-log regression model are given in Table III. Thus, the relationship between $\rho_{\Delta Kt}^{m,n}$, distance and the relative humidity term is given i) by (9) – (11) when $\rho_{RH}^{m,n}$ is used to represent the relative humidity term, and ii) by (12) – (14) when $\sigma_{RH^m\text{-}RH^n}$ is used to represent the relative humidity term for Mumbai, Delhi and Bengaluru respectively.

TABLE II
MULTIPLE REGRESSION RESULTS – I

Dependent Variable		$\rho_{\Delta Kt}^{m,n}$	
Independent Variables		$\rho_{RH}^{m,n}, d_{m,n}$	
Model Type	Adjusted-R^2		
	Mumbai	Delhi	Bengaluru
Log-log	0.74	0.94	0.91

TABLE III
MULTIPLE REGRESSION RESULTS – II

Dependent Variable		$\rho_{\Delta Kt}^{m,n}$	
Independent Variables		$\sigma_{RH^m\text{-}RH^n}, d_{m,n}$	
Model Type	Adjusted-R^2		
	Mumbai	Delhi	Bengaluru
Log-log	0.77	0.94	0.93

$$\rho_{\Delta Kt}^{m,n} = 1.27 \cdot \left(d_{m,n}\right)^{-0.32} \cdot \left(\rho_{RH}^{m,n}\right)^{2.27} \qquad (9)$$

$$\rho_{\Delta Kt}^{m,n} = 1.86 \cdot \left(d_{m,n}\right)^{-0.45} \cdot \left(\rho_{RH}^{m,n}\right)^{-4.21} \qquad (10)$$

$$\rho_{\Delta Kt}^{m,n} = 1.79 \cdot \left(d_{m,n}\right)^{-0.44} \cdot \left(\rho_{RH}^{m,n}\right)^{-2.17} \qquad (11)$$

$$\rho_{\Delta Kt}^{m,n} = 1.20 \cdot \left(d_{m,n}\right)^{-0.30} \cdot \left(\sigma_{RH^m-RH^n}\right)^{-0.03} \qquad (12)$$

$$\rho_{\Delta Kt}^{m,n} = 1.92 \cdot \left(d_{m,n}\right)^{-0.45} \cdot \left(\sigma_{RH^m-RH^n}\right)^{0.02} \qquad (13)$$

$$\rho_{\Delta Kt}^{m,n} = 1.93 \cdot \left(d_{m,n}\right)^{-0.46} \cdot \left(\sigma_{RH^m-RH^n}\right)^{0.03} \qquad (14)$$

The following inferences can be made by comparing the results presented in Table I – Table III and (9) – (14):

(a) For the case of Mumbai city, there is only a slight increase in the adjusted-R^2 value from the univariate regression analysis to the multiple regression analysis when

$\sigma_{RH}{}^{m}{}_{-RH}{}^{n}$ is used to represent the relative humidity term; and there is a decrease in the adjusted-R^2 value when $\rho_{RH}{}^{m,n}$ is used to represent the relative humidity term,

(b) For the case of Delhi and Bengaluru, there is a slight increase in the adjusted-R^2 value from the univariate regression analysis to the multiple regression analysis with either term for relative humidity,

(c) There is an inverse relation between $\rho_{\Delta Kt}{}^{m,n}$ and $\rho_{RH}{}^{m,n}$ in case of Delhi and Bengaluru, whereas for Mumbai the relationship is direct,

(d) There is a direct relation between $\rho_{\Delta Kt}{}^{m,n}$ and $\sigma_{RH}{}^{m}{}_{-RH}{}^{n}$ in case of Delhi and Bengaluru, whereas for Mumbai the relationship is inverse,

(e) The values of adjusted-R^2, coefficients and powers, of both the univariate regression model and the multiple regression models, for Delhi and Bengaluru have been found to be similar; but there is a distinct difference between these and the corresponding values for Mumbai.

It is important to be noted here that i) Mumbai is a coastal city, whereas Delhi and Bengaluru are both inland, and ii) in the present study, the sample set for Mumbai is small in size, consisting of just 15 site-pairs; whereas the sample sets for Delhi and Bengaluru are much larger, consisting of 78 site-pairs and 66 site-pairs respectively.

IV. CONCLUSION

The present paper aims to study the correlation between solar resource variability between pair of locations located in the same city. Such an analysis is important for understanding the extent of solar resource variability reduction and geographical smoothing that can be expected on a city-level. This understanding is vital for designing large-scale rooftop solar photovoltaic scenarios for a city, or popular initiatives like "solar cities" which aim to achieve a significant installation of solar photovoltaic systems in a city.

Variation of clear-sky index has been used as the representative variable for capturing the solar resource variability. The analysis has been done for three Indian cities – Mumbai, Delhi and Bengaluru – on hourly radiation and climatic data, spanning over a five year period from 2010 to 2014. The site-pair correlations of variability have been estimated using the Pearson's product-moment correlation coefficient between the clear-sky index variation profiles of a pair of sites. The variation of these site-pair correlation coefficients with i) the distance between the sites, and ii) both the distance between the sites and the difference in the relative humidity profiles of the sites have been studied.

The results show clearly that the site-pair correlation of variability reduces with distance – hence, sites further distant have more difference in their solar variability profiles, which implies a greater geographic smoothing with increasing spread of the area hosting the rooftop solar photovoltaic systems in a city. The inclusion of the relative humidity terms in the models for explaining the variation of the correlation between the clear-sky indices is not found to have a very significant effect; it only slightly increases the adjusted-R^2 of the model. For Delhi and Bengaluru, the larger the correlation between the relative humidity profiles of the two locations, the lesser the correlation between the solar variability profiles, implying greater geographic smoothing. For Mumbai, this relationship between relative humidity profile and solar variability profile has been found to be inverse of the above.

REFERENCES

[1] Karteris, M., Slini, T., and Papadopoulos, A. M., "Urban solar energy potential in Greece: a statistical calculation model of suitable built roof areas for photovoltaics", *Energy and Buildings*, 62, pp. 459-468, 2013.

[2] Ordóñez, J., Jadraque, E., Alegre, J., and Martínez, G., "Analysis of the photovoltaic solar energy capacity of residential rooftops in Andalusia (Spain)", *Renewable and Sustainable Energy Reviews*, 14(7), pp. 2122-2130, 2010.

[3] Singh, R., and Banerjee, R., "Estimation of roof-top photovoltaic potential using satellite imagery and GIS", In *39th IEEE Photovoltaic Specialists Conference (PVSC)*, 2013, pp. 2343-2347.

[4] Singh, R., and Banerjee, R., "Estimation of rooftop solar photovoltaic potential of a city", *Solar Energy*, 115, pp. 589-602, 2015.

[5] Byrne, J., Taminiau, J., Kurdgelashvili, L., and Kim, K. N., "A review of the solar city concept and methods to assess rooftop solar electric potential, with an illustrative application to the city of Seoul", *Renewable and Sustainable Energy Reviews*, 41, pp. 830-844, 2015.

[6] Freitas, S., C. Catita, P. Redweik, and M. C. Brito, "Modelling solar potential in the urban environment: State-of-the-art review", *Renewable and Sustainable Energy Reviews*, 41, pp. 915-931, 2015.

[7] Jakubiec, J. Alstan, and Christoph F. Reinhart, "A method for predicting city-wide electricity gains from photovoltaic panels based on LiDAR and GIS data combined with hourly Daysim simulations", *Solar Energy*, 93, pp. 127-143, 2013.

[8] Redweik, P., C. Catita, and M. Brito, "Solar energy potential on roofs and facades in an urban landscape", *Solar Energy*, 97, pp. 332-341, 2013.

[9] J. Hofierka and J. Kaňuk, "Assessment of photovoltaic potential in urban areas using open-source solar radiation tools", *Renewable Energy*, 34(10), pp. 2206-2214, 2009.

[10] Redweik, P., C. Catita, and M. Brito, "PV potential estimation using 3D local scale solar radiation model based on urban LIDAR data" In *Proceedings of the 26th European Photovoltaic Solar Energy Conference*, Hamburg, Germany, 2011, pp. 5-9.

[11] Kabir, M.H., Endlicher, W. and Jägermeyr, J., "Calculation of bright roof-tops for solar PV applications in Dhaka Megacity, Bangladesh", *Renewable Energy*, 35(8), pp.1760-1764, 2010.

[12] Strzalka, A., Alam, N., Duminil, E., Coors, V. and Eicker, U., "Large scale integration of photovoltaics in cities", *Applied Energy*, 93, pp.413-421, 2012.

[13] Press Information Bureau, Government of India, "Revision of cumulative targets under National Solar Mission from 20,000 MW by 2021-22 to 100,000 MW", New Delhi, 17 June, 2015.

[14] Hoff, T.E. and Perez, R., "Quantifying PV power output variability", *Solar Energy*, 84(10), pp.1782-1793, 2010.

[15] Hoff, T.E. and Perez, R., "Modeling PV fleet output variability", *Solar Energy*, 86(8), pp.2177-2189, 2012.

[16] Perez, R., Kivalov, S., Schlemmer, J., Hemker, K. and Hoff, T., "Parameterization of site-specific short-term irradiance variability", *Solar Energy*, 85(7), pp.1343-1353, 2011.

[17] Perez, R., Kivalov, S., Schlemmer, J., Hemker, K. and Hoff, T.E., "Short-term irradiance variability: Preliminary estimation of station pair correlation as a function of distance", *Solar Energy*, 86(8), pp.2170-2176, 2012.

[18] Lave, M., Kleissl, J. and Stein, J.S., "A wavelet-based variability model (WVM) for solar PV power plants", *IEEE Transactions on Sustainable Energy*, 4(2), pp.501-509, 2013.

[19] Lave, M. and Kleissl, J., "Cloud speed impact on solar variability scaling–Application to the wavelet variability model", *Solar Energy*, 91, pp.11-21, 2013.

[20] Lave, M., Kleissl, J. and Arias-Castro, E., "High-frequency irradiance fluctuations and geographic smoothing", *Solar Energy*, 86(8), pp.2190-2199, 2012.

[21] Mills, A. and Wiser, R., "Implications of wide-area geographic diversity for short-term variability of solar power." *Lawrence Berkeley National Laboratory*, 2010.

[22] Lave, M. and Kleissl, J., "Solar variability of four sites across the state of Colorado", *Renewable Energy*, 35(12), pp.2867-2873, 2010.

[23] Rowlands, I.H., Kemery, B.P. and Beausoleil-Morrison, I., "Managing solar-PV variability with geographical dispersion: An Ontario (Canada) case-study", *Renewable Energy*, 68, pp.171-180, 2014.

[24] Klima, K. and Apt, J., "Geographic smoothing of solar PV: results from Gujarat", Environmental Research Letters, 10(10), p.104001, 2015.

[25] NSRDB Viewer, National Solar Radiation Database (NSRDB), https://nsrdb.nrel.gov/nsrdb-viewer, visited on 29th December, 2016.

[26] Perlwitz, J. and Miller, R.L., "Cloud cover increase with increasing aerosol absorptivity: A counterexample to the conventional semidirect aerosol effect", *Journal of Geophysical Research: Atmospheres*, 115(D8).

[27] Groisman, P.Y., Bradley, R.S. and Sun, B., "The relationship of cloud cover to near-surface temperature and humidity: Comparison of GCM simulations with empirical data", *Journal of Climate*, 13(11), pp.1858-1878, 2000.

[28] Water Vapor & Cloud Fraction: Global Maps, http://earthobservatory.nasa.gov/GlobalMaps/view.php?d1=MYDAL2_M_SKY_WV&d2=MODAL2_M_CLD_FR, visited on 15th January, 2017.

Ultra-short-term Photovoltaic Generation Forecasting Model Based on Weather Clustering and Markov Chain

Jin Tan, Changhong Deng

School of Electrical Engineering, Wuhan University, Wuhan 430072，China

Abstract — **Random fluctuations of solar radiation result in low precision of ultra-short-term photovoltaic(PV) power prediction. To solve this problem, the modified model based on Adaboost, KNN clustering and Markov Chain is proposed in this paper. First, an improved classifier combining k-Nearest Neighbor (KNN) with Adaboost is adopted to classify the collected historical data. Furthermore, the concept of attenuation coefficient of solar radiation is presented to modify the Hottel model. Then, a weighted Markov chain model is built to predict the solar radiation. Active power output is calculated by PV engineering model with respect to solar radiation. Experimental results indicate that the proposed model can significantly improve the precision of power prediction in cloudy and rainy conditions.**

Index Terms — **photovoltaic(PV) generation, attenuation coefficient, Adaboost, Markov Chain.**

I. INTRODUCTION

Recently, photovoltaic (PV) power generation has developed rapidly because solar energy is clean and renewable. The increasing integration of large-scale distributed PV generation brings great difficulties to the dispatching and operation of power systems. Influenced by a variety of weather factors, PV output power is intermittent and has strong stochastic volatility, which will affect the security and stability operation of power system, especially in a micro grid. A sudden drop of PV power has an effect on the security and stability operation of micro grid with high PV power penetration [1]. As a result, it is essential to forecast PV output power accurately. It can not only reduce the impact of PV output power fluctuations on distribution network, but also enhance the utilization efficiency of PV generation. Besides, it is helpful for dispatchers to make energy storage control strategy, adjust the real-time operation plan and improve the reliability of the grid [2].

Traditional PV power forecasting adopts mathematical statistical methods mostly, including sequential extrapolation method, neural network (ANN) [3]-[5], support vector machine (SVM) [6] and data mining method [7], etc. Considering the significant variation of different weather types, many papers have used a classified approach to classify the collected historical data firstly. [7] used the offline classifier and SVM prediction model to realize ultra-short-term prediction, but the forecast errors in cloudy, overcast and rainy

conditions are normally wide. In [8], the weather index was proposed to reflect the PV power generation fluctuations under different weather types, and neural network prediction model was adopted to predict the PV power generation. Based on the deviation of actual value and forecasting value under different weather types, grey system model was used to correct the prediction results [9]. [10] proposed that the PV output power consisted of trend component and fluctuant component. Trend component was calculated by solar radiation model, and fluctuant component was gained through introducing the factor which described the influence of different weather. In [11]-[13], the weather index and aerosol index were defined to reflect the PV power generation fluctuations under different weather types. They were put into the forecasting model as an additional parameter.

The above references have considered the effect of different weather on PV power output. However, [8-11] did not analyze in depth the correlation of historical PV output power and weather types. Moreover, in these articles, there is just one forecast model being applied in different weather conditions, which have decreased the prediction accuracy. Besides, there was a problem of over-fitting in ANN and it was difficult to determine the number of hidden layer nodes. And it was unable to track the high frequency fluctuations of PV output effectively.

For improving the precision of ultra-short-term power prediction, the modified model based on combination of Adaboost, KNN clustering and Markov Chain is proposed in this paper. And it is applied to a practical PV micro grid in Wuhan.

The rest of the paper is organized as follow. Section II analyzes the characteristics of PV generation. In section III, the modified model based on combination of Adaboost and Markov Chain is proposed. In order to demonstrate the significance of this novel method, the forecast example and simulation results are discussed in section IV. Finally, conclusions are given in section V.

II. CHARACTERISTIC ANALYSIS OF PV GENERATION

A. Classification of PV Generation

According to historical measured data, it is clear that PV output power curves in sunny, cloudy, rainy, overcast and extreme weather conditions are representative and differ from

---

This work is supported by the National Key Research and Development Program of China (2017YFB0903705)

978-1-5090-5606-4/17 $31.00 © 2017 IEEE   1158

each other. As extreme weather condition occurs with low probability and its output power is almost close to zero, this paper establishes four sub-models in sunny, cloudy, rainy and overcast weather conditions respectively. Based on the 2015 ~ 2016 measured data of PV experiment platform which is located in Wuhan China, the PV generation curves of four typical weather types are shown in figure 1.

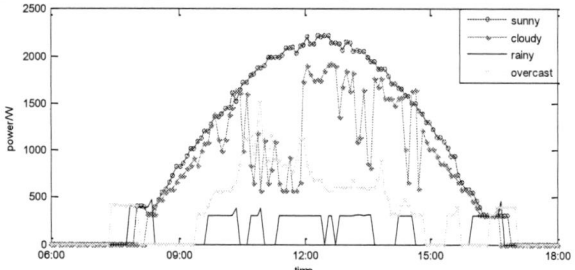

Fig. 1. PV power output curves under typical weathers

From Fig.1 we can see that PV output power changes regularly and smoothly in the weather condition of clear sky. This is because there are little clouds in the sky and the weather condition is relatively steady. Therefore, the surface solar radiation is almost consistent with the standard clear model values. In cloudy condition, the surface solar radiation decreases with increased clouds, which leads to the attenuation and serious fluctuations of PV output power. In rainy condition, the surface solar radiation is greatly weakened and the output is even zero when the sky is full of clouds.

B. Definition of Attenuation Coefficient

Fig.1 shows that in sunny, cloudy, overcast and rain weathers, the attenuation value of solar radiation increases in turn. At the time i, there are different ratios between measured surface solar radiation $I_{mea}(i)$ and reference value $I_{std}(i)$. It is denoted as the attenuation coefficient α.

$$\alpha = \frac{\sum \alpha_i}{N} = \frac{1}{N} \sum \frac{I_{mea}(i)}{I_{std}(i)} \qquad (1)$$

where α_i is the attenuation value of solar radiation at time i; N is the number of sampling point.

This paper chooses the general solar radiation model Hottel [14] to calculate instantaneous solar radiation I_{std}, and take it as the reference value.

$$I_{std} = I_{cb} + I_{cd} \qquad (2)$$

where Icb and Icd are instantaneous direct and diffuse solar radiation.

The modified solar radiation under all kinds of weathers can be calculated as follow.

$$I = \alpha I_{std} \qquad (3)$$

III. Forecasting Model for PV Output Power

k-Nearest Neighbor (KNN) is a kind of supervised learning algorithm. Parameter k represents the closest k clustering centers to the test sample. Based on the calculation formula of distance d, the training sample belongs to the nearest clustering center.

$$d = \sqrt{\sum_{i=1}^{n} (x_i - y_i)^2} \qquad (4)$$

where x_i is the characteristic element of the training sample; y_i is the characteristic element of the clustering center; n is the number of characteristic elements.

Adaboost algorithm [15] is an iterative algorithm of machine learning, which uses basic training sample data to train basic classifiers and then gets a strong classifier through combining all basic classifiers. The improved clustering algorithm based on KNN and Adaboost is as follows.

Step 1, input training data $Z=\{(x_1,y_1), (x_2,y_2),..., (x_n,y_n)\}$. Determine the number of KNN classifier $C(x)$.

Step 2, initialize the weight of each training sample w_{1i} and determine the iteration m of Adaboost algorithm.

$$W_1 = (w_{11}, w_{12,} ..., w_{1n}), w_{1i} = \frac{1}{n}, i = 1, 2, ..., n \qquad (5)$$

Step 3, use weighted data to train KNN classifier, and all samples are classified by $C(x)$. Then calculate their error rate e_m and the weight d_m of classifier in the final classifier.

$$e_m = P(C(x_i) \neq y_i) = \sum_{i=1}^{n} w_{mi} I(C(x_i) \neq y_i) \qquad (6)$$

$$d_m = \frac{1}{2} \log \frac{1 - e_m}{e_m} \qquad (7)$$

It can be seen that d_m decreases with the increase of e_m. Step 4, Sample weight is updated as follow.

$$w_{m+1,i} = \frac{w_{mi}}{Z_t} \exp\left(-d_m y_i C_m(x_i)\right) \qquad (8)$$

$$Z_m = \sum_{i=1}^{n} w_{mi} \exp\left(-d_m y_i C_m(x_i)\right) \qquad (9)$$

Thereafter, return to step 3 until the error rate or iteration reaches the value set in advance. Based on the above, samples are classified into four groups respectively, namely sunny, cloudy, overcast and rainy.

After weather type clustering, four sub-models are established in sunny, cloudy, rainy and overcast weather conditions respectively. According to weather information of the forecasting day, take its attenuation coefficient as input variables into the weighted Markov Chain model [16]-[17] and output the forecast value of solar radiation. Then, PV power is derived from the photovoltaic engineering model [3].

$$P(t) = \eta SI(t)[1 - 0.005(t_0 + 25)] \qquad (10)$$

where η is the efficient of photoelectric conversion (%); S is effective area of PV panel (m^2); $I(t)$ is solar radiance (kW/m^2) at time t; t_0 is the temperature of surroundings (℃).

IV. FORECAST EXAMPLE AND SIMULATION RESULT

The forecast model combining weather clustering with Markov Chain is applied to predicting PV power in Wuhan, China. Based on KNN and Adaboost algorithm, samples are classified into four groups sunny, cloudy, overcast and rainy, respectively. According to the statistical data from 2015 to 2016, the frequency histogram of attenuation coefficient is shown in Fig.2.

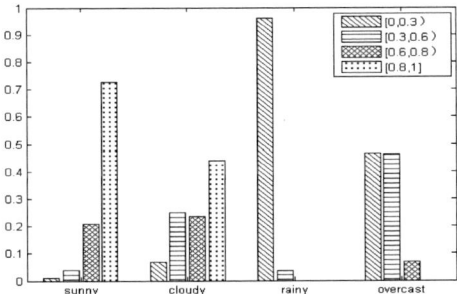

Fig. 2 Interval distribution of attenuation value of solar radiation

In order to verify the effectiveness of improved clustering algorithm, this paper selects valuable data from October 2015 to March 2016 as training and test samples. There are 130 valuable days, and the sampling interval of data is 5minutes. Comparison of results before and after the improvement is shown in table 1 and 2. It shows that improved clustering algorithm is more accurate.

TABLE.1 CLASSIFICATION RESULTS BEFORE IMPROVED

Weather type	Sunny	Cloudy	Overcast	Rainy
Sunny	27	7	4	0
Cloudy	3	22	7	3
Overcast	0	4	21	5
Rainy	0	3	6	18

TABLE.2 CLASSIFICATION RESULTS AFTER IMPROVED

Weather type	Sunny	Cloudy	Overcast	Rainy
Sunny	35	2	1	0
Cloudy	0	32	2	1
Overcast	0	2	27	1
Rainy	0	0	1	26

In this paper, samples are classified into four groups sunny, cloudy, overcast and rainy, respectively. According to weather

of the forecasting day, select corresponding forecast sub-model and take its attenuation coefficient as input variables into the model. The rolling forecast operates every 5 minutes between 9:00-15:00 and the forecast results in four weather conditions are shown in Figs.3-6.

The curves of forecasting value and measured value of sunny day are depicted in Fig.3. It can be seen that the prediction accuracy is very high and forecast results coincide well with the measured value. This is because there are little clouds in the sky and the weather condition is relatively steady. Fig. 4 is the forecasting curve of cloudy day. It is obvious that the proposed model is sensitive to the fall of solar radiation. Therefore, it can track the random fluctuations pretty well. Compared with forecasting result of sunny day, it can be seen that the forecast error is larger. This is because clouds have an effect on PV output power. Its location, as well as thickness, are uncertain and change randomly, which are difficult to forecast accurately.

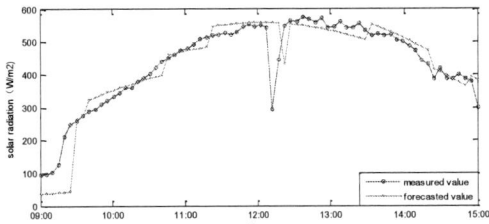

Fig. 3. Forecasted and measured values of sunny day

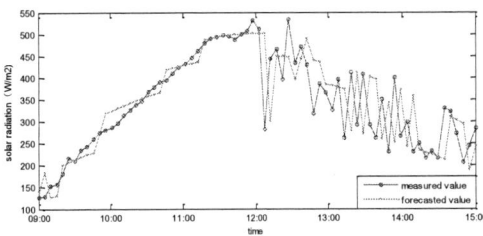

Fig. 4. Forecasted and measured values of cloudy day

The forecasted and actual value curves of rainy day and overcast day are shown in Fig. 5 and Fig.6, respectively. The forecast model proposed in this paper performs pretty well. It is worth to note that actual solar radiations of rainy and overcast day are much lower than sunny day as a whole, especially in the rainy days that solar radiation is less than 300 W/m2. And in this case, small absolute error will result in a large relative error.

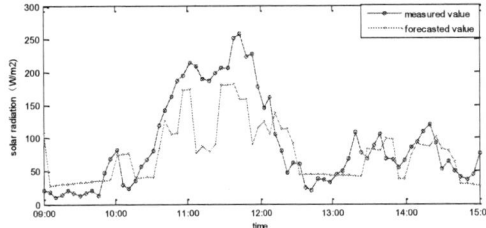

Fig. 5. Forecasted and measured values of rainy day

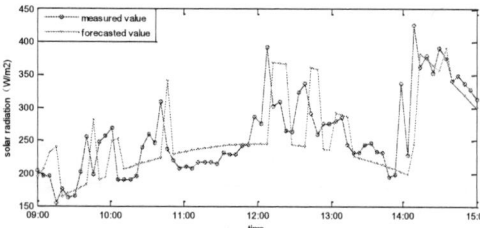

Fig. 6. Forecasted and measured values of overcast day

According to the above analysis, it is necessary to use the mean absolute percentage error (MAPE)[18] and the root mean square error (RMSE) to evaluate the forecast accuracy under different weather types. In this way, the assessment is more significant and can reflect forecasting effect of the model better.

$$E_{MAPE} = \frac{1}{N}\sum_{i=1}^{N}\left|\frac{P^{mear}_{i} - P^{pred}_{i}}{P_{cap}}\times 100\%\right| \qquad (11)$$

$$E_{RMSE} = \sqrt{\frac{1}{N}\sum_{i=1}^{N}(P^{mear} - P^{pred})^2} \qquad (12)$$

where P_i^{mear} is the measured value at time i; P_i^{pred} is the prediction at time i; P_{cap} is the installed capacity of PV generation system; N is the number of samples.

Based on the weather clustering results, the rolling forecast operates every 5 minutes between 9:00-15:00 and the forecast results in four weather conditions are shown in Figs.3-6. MAPE and RMSE results of proposed model are shown in Table.3.

TABLE.3 MAPE AND RMSE RESULTS OF PROPOSED MODEL

Weather type	Sunny	Cloudy	Rainy	Overcast
MAPE（%）	5.8	12	43.5	15.1
RMSE（w/m^2）	44.3	65.9	46.5	51.9

It can be seen in Table.2 that the prediction precision of modified forecasting model is pretty high. The MAPE in sunny day is the minimum and the value is 5.8%. According to Figs.3-5, it is obvious that the proposed model is sensitive to the fall of solar radiation and can track the fluctuation change in cloudy, rainy and overcast day. Although the MAPE in rainy day is up to 43.5%, the RMSE is just about 46.5 W/m^2. The forecasting model operates pretty well in overcast and rainy days. All of the data demonstrate that the modified forecasting model proposed in this paper is practical and can be applied to predict PV output power in PV micro grid system.

V. CONCLUSIONS

A modified model based on combination of Adaboost, KNN clustering and Markov Chain is proposed in this paper. The improved clustering algorithm based on KNN and Adaboost is adopted to cluster weather type. Then, attenuation coefficient of solar radiation is introduced into forecast model, which have improved the precision of prediction effectively. Finally, the ultra-short-term prediction model of PV micro-grid is established based on Markov Chain. The proposed forecast model operates well in Wuhan, China. The experimental results show that it can help dispatchers to adjust the operation plan and make control strategies of energy storage system and ensure the security and stability operation of the power grid. At the same time, it reduces operation costs of PV micro grid because it uses historical data and meteorological information to realize accurate weather classification instead of the numerical weather prediction. Therefore, it is more suitable in practice.

REFERENCES

[1] WANG Chengshan, LI Peng. Development and challenges of distributed generation，the micro-grid and smart distribution system. Automatic of Electric Power System, 2010, 34(2):132-133.

[2] GONG Yingfei, LU Zongxiang, QIAO Ying, et al. An Overview of Photovoltaic Energy System Output Forecasting Technology. Automation of Electric Power Systems. 2016(4):140-151.

[3] Chen Changsong, Duan Shanxu, Yin Jinjun. Design of photovoltaic array power forecasting model based on neutral network. Transactions of China Electrotechnical Society, 2009,24(9):153-158

[4] WANG Shouxiang, ZHANG Na. Short-term output power forecast of photovoltaic based on a grey and neural network hybrid model. Automation of Electric Power Systems. 2012, 36(19):37-41.

[5] Ding Ming, Liu Zhi, Bi Rui, et al. Photovoltaic output prediction based on grey system Correction-Wavelet Neural Network. Power System Technology. 2015, 39(9):2438-2443.

[6] Zhu Yongqiang, Tian Jun. Application of least square suppose vector machine in photovoltaic power forecasting. Power System Technology. 2011(7):54-59.

[7] ZHAO Shuqiang, XIE Yuqi, LIU Dazheng. Short-term solar radiation intensity forecasting based on fuzzy-random theory. Electric Power Automation Equipment, 2015, 35(7):101-105.

[8] Dai Qian, Duan Shanxu, Cai Tao, et al. Short-term PV Generation System Forecasting Model Without Irradiation Based on Weather Type Clustering. Proceeding of the CSEE, 2011, 31(34):28-35.

[9] Wang Fei, Mi Zengqiang, Zhen Zhao, et al. A Classified Forecasting Approach of Power Generation for Photovoltaic Plants Based on Weather Condition Pattern Recognition. Proceeding of the CSEE, 2013(34):75-82.

[10] Zhang Xi, Kang Chongqing, Zhangning, et al. Analysis of mid/long term random characteristics of photovoltaicn power generation. Automatic of Electric Power System, 2014, 38(6):6-13.

[11] Yuan Xiaoling, Shi Junhua, Xu Jieyan. Short-term power forcecasting for photovoltaic generation considering weather type index. Proceedings of the CSEE. 2015, 39(9):2438-2443

[12] Li Le, Liu Tianqi. PV Power Forecasting Based on AP-ESN. Electric Power Automation Equipment. 2016, 36(7).

[13] Liu J, Fang W, Zhang X, et al. An Improved Photovoltaic Power Forecasting Model With the Assistance of Aerosol Index Data[J]. IEEE Transactions on Sustainable Energy, 2015, 6(2):1-9.

[14] Xu Qingshan, Zang Haixiang, Bian Haihong. Establishment and Feasibility Researches of Practical Solar Radiation Model. Acta Energiae Solaris Sinica, 2011, 32(8):1180-1185.

[15] Lu Ting. Comparison of AdaBoost-based Learning Algorithms of Classifiers. South China University of Technology, 2014.

[16] DING Ming, XU Ningzhou. A Method to Forecast Short-Term Output Power of Photovoltaic Generation System Based on Markov Chain. Power System Technology, 2011(1):152-157.

[17] Li Y Z, Niu J C. Forecast of Power Generation for Grid-Connected Photovoltaic System Based on Markov Chain[C]// Industrial Electronics and Applications, 2008. Iciea 2008. IEEE Conference on. IEEE, 2008:1729-1733.

[18] Kobayashi H, Arai J. Short-term forecast for photovoltaic power generation and development of measuring equipment[C]// Information and Automation for Sustainability (ICIAfS), 2014 7th International Conference on. IEEE, 2015:1-6

Daily Solar Irradiance Profile Characterization and Ramp Rate Analysis at Different Time Resolutions

Spyros Theocharides, Venizelos Venizelou, George Makrides and George E. Georghiou

FOSS Research Centre for Sustainable Energy, Photovoltaic Technology Laboratory,
Department of Electrical and Computer Engineering, University of Cyprus
75 Kallipoleos Avenue, P.O. Box 20537, Nicosia, 1678, Cyprus

Abstract — **Characterization of daily solar irradiance profiles and analysis of the fluctuations of the ramp rates exhibited by the power produced by photovoltaic (PV) systems is a crucial topic that adversely affects power quality and reliability especially in weak and island power systems. In this work, a detailed procedure to characterize daily solar irradiance profiles is presented based on the k-means clustering technique and a classification method for daily sky conditions applied to annual global horizontal irradiance (GHI) time series. Both methods were used to analyze the data from a location in Cyprus in order to cluster and characterize the daily GHI profiles. The obtained results of the 3-cluster framework demonstrated that 86 % of the days in a year do not exhibit high ramp rate behavior, while the classification method exhibited that 86.2 % of the year were primarily classified in classes with clear sky and nearly clear sky conditions. Finally, the ramp rate analysis performed demonstrated that time sampling resolution affects the magnitude of the extracted ramp rates.**

Index Terms — **clustering, clearness index, k-means, solar irradiance ramp rate, solar irradiation.**

I. INTRODUCTION

The global installed capacity of grid-connected photovoltaic (PV) systems is rapidly increasing and due to their dependence on weather conditions, the electricity injected to the grid is intermittent in nature with negative impact on the grid operation. The power produced from PV systems mainly depends on the incident global irradiance and for this reason, it is imperative that the location characteristics of the daily solar irradiation patterns and variability are clearly characterized in order to avoid potential power quality issues. In addition, the high variability of solar irradiance is a source of new challenges to the distribution system operators (DSO) and transmission system operators (TSO) concerning planning, production forecasting and unit commitment [1].

The fluctuations arising in power produced from PV systems due to the intermittency of solar irradiation have already been addressed in the past [2, 3] and are currently of prime concern especially to grid operators at locations with high PV penetration rates. Such power fluctuations and rampings may adversely affect power quality and reliability especially in weak and island power systems and for this reason investigations, focusing to characterize solar irradiance and the magnitude of ramping are important [4]. Previous

work concerning the clustering of solar daily patterns was based on daily distributions of the clearness index [5, 6], while others proposed classifying irradiation patterns into three and more classes according to the sampling intervals [7, 8]. In addition, several characterization methods have been proposed based on ground measurements used to construct daily fractal and clearness indices [9] while others focused on cloud ratio and sunshine duration [10]. A recent study further proposed the use of daily clearness index and probability of persistence for robust characterization and classification of the daily sky conditions [11].

In this work, the procedure followed to group daily solar irradiance profiles based on the k-means clustering technique applied to the annual global horizontal irradiance (GHI) time series is presented. Furthermore, the characterization of the solar irradiance conditions of a location by analyzing the daily clearness index and probability of persistence is considered. GHI data-sets acquired over a period of a year from a weather station in Cyprus were used to validate the approach. The obtained clusters and classes present useful information on the daily solar irradiance profiles and their persistence exhibited at a location.

Finally, a Ramp Rate Analysis (RRA) was performed to demonstrate the impact of different time sampling resolutions on the magnitude and frequency of the observed ramp rates.

II. METHODOLOGY

A. Weather Station Global Horizontal Irradiance Data

The annual GHI data was obtained from a weather station located at the southernmost part of Cyprus as shown in Table 1. The GHI data-sets were acquired at a resolution of one second and recorded at one-, fifteen-, thirty- and sixty-minute intervals.

TABLE I
METEOROLOGICAL STATION DETAILS

Description	Value
Latitude1	34.597769°
Longitude	32.987652°
Altitude	17m
Period	2011-2012
Data recording interval	one-, fifteen-, thirty- and sixty-minute

978-1-5090-5606-4/17 $31.00 © 2017 IEEE

B. Daily Solar Irradiance Profile Clustering

The daily solar irradiance clusters were extracted from the annual GHI one-minute time series, which correspond to 1440 samples per day, acquired over the period of a year. The clustering method applied measures the dissimilarity between the time series data-sets in order to extract clusters of common features. The dissimilarity between the data-sets was assessed using an autocorrelation-based dissimilarity function, which utilizes the estimated autocorrelation coefficients (ACFs). The ACF dissimilarity is given as [12]:

$$d_{ACF}\left(X_T, Y_T\right) = \sqrt{(\rho_{X_T} - \rho_{Y_T})^T + \Omega(\rho_{X_T} - \rho_{Y_T})} \quad (1)$$

where ρ_T and ρ_Y are the autocorrelations of X_T and Y_T, respectively and Ω is a matrix of weights. Due to the weight uniformity of Ω, the d_{ACF} becomes the Euclidean distance and the derived uniform autocorrelation coefficient is calculated as follows [12]:

$$d_{AFCU} = \sqrt{\sum_{i=1}^{L} (\rho_{iX_T} - \rho_{iY_T})^2} \quad (2)$$

Once the Euclidean distance was calculated, the k-means clustering technique was applied to the ACF results. K-means clustering is a method of vector quantization that aims to partition n observations into k-clusters, in which each observation belongs to the cluster with the closest mean, attending as an archityple of the cluster [13]. Specifically, the k-means model aims to partition the n observations into k and $S = \{S_1, S_2 ... S_k\}$ so as to minimize the within-cluster sum of squares [14]. The k-means clusters are derived from the following equation:

$$\arg_s \min \sum_{i=1}^{k} \sum_{x \in S_i} \| x - \mu_i \|^2 \quad (3)$$

where, $x_1, x_2, ..., x_n$ is a set of observations and each observation is a d-dimensional vector, μi is the mean of points in Si and k the number of clusters.

C. Daily Solar Irradiance Profile Classification

In order to define classes and characterize solar irradiance at the earth's surface, sky clearness index, k_d, and the probability of persistence, POP_d, were initially calculated from the yearly GHI data-sets [11]. The following equation describes the calculation of k_d:

$$k_d = \frac{I_{GHI}}{I_{EXT}} \quad (4)$$

where I_{GHI} is the instantaneous GHI and I_{EXT} is the intensity of solar irradiance directly outside the earth's atmosphere on a horizontal surface (extraterrestrial irradiance). This index,

captures the instantaneous fluctuations of the solar irradiance and can further be used to calculate the daily solar irradiance the surface of the Earth receives during a day, K_d:

$$K_D = \int_{day} k_d \, dt \quad (5)$$

However, the sky clearness index cannot capture the quality of solar irradiance, which during a day can be calculated using a probabilistic approach. [14]. To record the quality of solar irradiance a probabilistic model is used. At first, the Δk_{day} is calculated as the difference between k_d values, calculated in Eq. 4 within a day:

$$\Delta k_{day} = \left\{ |k_{d1} - k_{d2}|, |k_{d2} - k_{d3}|, ..., |k_{dn-1} - k_{dn}| \right\} \quad (6)$$

The POP_d is described as the probability of Δk_{day} values to be equal to zero. Thus, the higher the value of POP_d of a day the lower vacillations would appear during the day. The POP_d index is given by [15]:

$$POP_d = P(\Delta k_{day} = 0) \quad (7)$$

where $P(\Delta k_{day}=0)$ is the probability of Δk_{day} values equal to zero. The steps followed in the methodology are summarized in the flow-chart of Fig. 1.

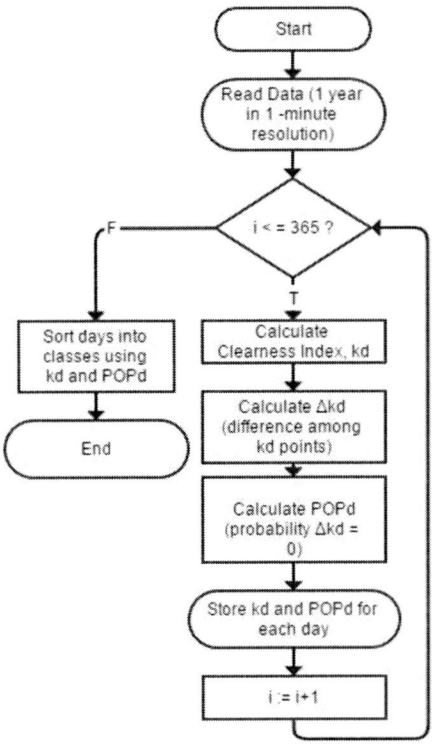

Fig. 1. Flow-chart outline of the methodology followed for the calculation of k_d and POP_d.

Table 2 summarizes the ten classes in which the daily solar radiation conditions can be characterized [11].

TABLE II
CLASSIFICATION PARAMETERS OF SOLAR RADIATION

Class	k_d	POP_d	Description
1	$k_d \geq 0.6$	$POP_d \geq 0.9$	High Quantity High Quality
2	$0.3 \leq k_d < 0.6$	$POP_d \geq 0.9$	Medium Quantity and High Quality
3	$k_d < 0.3$	$POP_d \geq 0.9$	Low Quantity High Quality
4	$k_d \geq 0.6$	$0.7 \leq POP_d < 0.9$	High Quantity Medium Quality
5	$0.3 \leq k_d < 0.6$	$0.7 \leq POP_d < 0.9$	Medium Quantity Medium Quality
6	$k_d < 0.3$	$0.7 \leq POP_d < 0.9$	Low Quantity Medium Quality
7	$k_d \geq 0.6$	$0.5 \leq POP_d < 0.7$	High Quantity Low Quality
8	$0.3 \leq k_d < 0.6$	$0.5 \leq POP_d < 0.7$	Medium Quantity Low Quality
9	$k_d < 0.3$	$0.5 \leq POP_d < 0.7$	Low Quantity Low Quality
10	-	$POP_d < 0.5$	Very Low Quality

D. Solar Irradiance Ramp Rate Analysis

Using the acquired data-sets the GHI ramp rate statistics were derived. The ramp rates are defined as the absolute value of the difference between the instantaneous GHI at the beginning and at the end of the time interval period set. The ramp rate is defined as the value change over a defined time scale (Δt) and can be calculated using the equation:

$$Ramp\ Rate = \left| \frac{I_{GHI_{to}} - I_{GHI_{to+\Delta T}}}{\Delta_t} \right| \qquad (6)$$

where I_{GHI} is the global horizontal irradiance.

The Ramp Rate statistics were subsequently derived for the acquired one-, fifteen-, thirty- and sixty-minute GHI time series in order to extract the ramp rate frequency and magnitude at different time sampling resolutions over a period of a year.

III. RESULTS

The number of k-means clusters of the annual GHI one-minute time series was determined by the sum of squared error (SSE) scree plot, which is a guideline for deciding how many clusters to retain. Fig. 2 shows the scree plot for the annual GHI one-minute time series and the suitable value of clusters to retain was exhibited by the elbow point (3 clusters), which demonstrates how the homogeneity or heterogeneity within the clusters changes for various values of clusters.

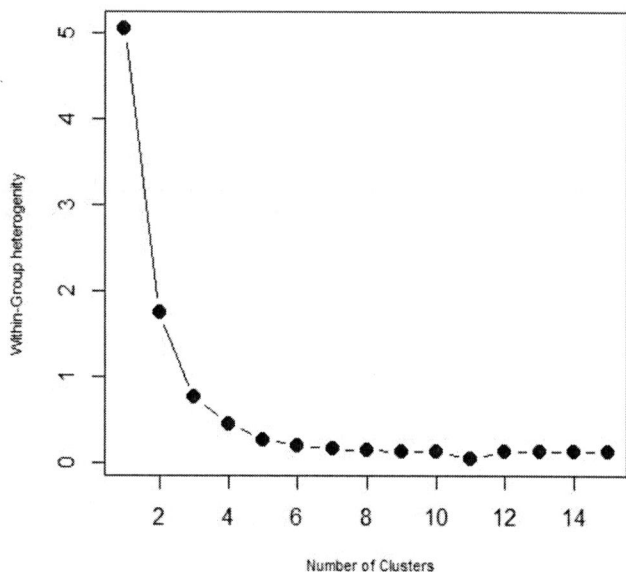

Fig. 2. Scree plot of annual GHI one-minute time series.

The representative daily GHI patterns, within the 3-cluster framework, are depicted in Fig. 3. The resulting clustering analysis showed that 86 % of the days of the year corresponded to cluster 1, which signifies that the majority of days at the examined location do not exhibit high solar irradiance ramp rate behaviours.

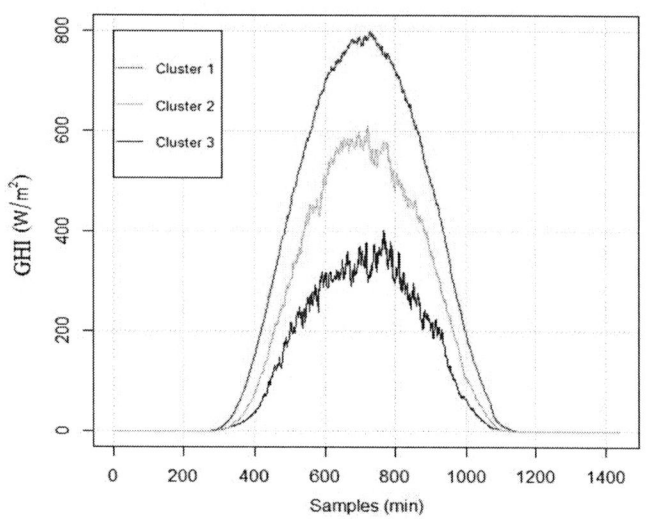

Fig. 3. Clusters of the one-minute daily GHI patterns.

Even though the applied clustering technique demonstrated that the majority of days fall into three clusters with low irradiance ramping, further insight into the behaviour of the GHI daily patterns and the ramping rate persistence was provided from the classification analysis. In this respect, the grouping of days based on the clustering technique, which provided only three generic classes, could be adoptable only for Mediterranean and semi-arid climates. Fig. 4 demonstrates the average daily classes obtained using the K-POP method over a period of a year, which provided more information on the characterization of both the quantity and quality of the daily GHI at the particular location.

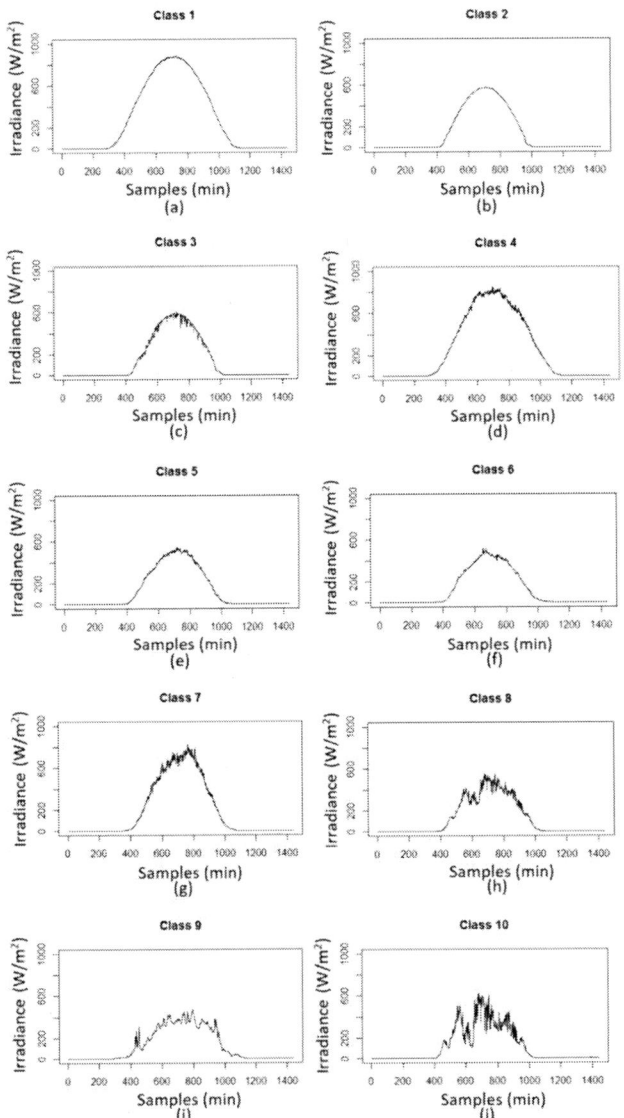

Fig. 4. Classification of daily GHI profiles over the evaluation period into ten classes exhibiting also typical daily GHI plots for each class.

The solar irradiance patterns of the K-POP plots for the data collected at the location in Cyprus over the period 2011 – 2012, are shown in Fig. 5. The black circles represent the K-POP data points for each day of the year. In addition, the highest concentration of data appeared in classes 1, 4 and 5

indicating that most daily solar radiation conditions exhibit a high to medium solar irradiance quantity with rare and infrequent fluctuations.

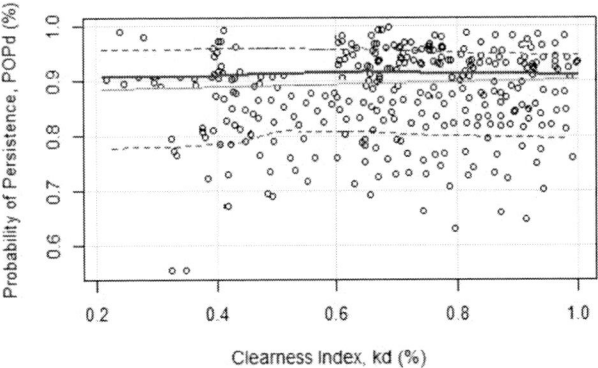

Fig. 5. Classes of annual GHI one-minute time series.

In addition, the distribution of the daily solar irradiance profiles, summarized in Table 3, clearly showed that 86.2 % of the year was predominantly classified in classes C_1, C_4 and C_5, indicating clear sky and nearly clear sky conditions. Moreover, for approximately half of the year (51.4 %) the days were classified as C_1, which represent cloudless days with high solar irradiance.

TABLE III
PERCENTAGE OF OCURENCES FOR CLASSES

Classes	Occurrence (%)
1	51.4
2	3.5
3	1.1
4	22.5
5	12.3
6	1.3
7	2.3
8	4.6
9	0.5
10	0.5

Finally, the histograms of Fig. 6 show the GHI ramp rate frequency of occurrence over a period of a year, for the measurements acquired at different time resolutions (one-, fifteen-, thirty- and sixty-minute). Further analysis of the results showed that the resolved GHI ramp rates mainly occurred in the range of 0 - 100 W/m^2, while for the sixty-minute resolved GHI ramp rates the fluctuations recorded were mainly in the range of 0 – 400 W/m^2. This clearly demonstrated that time sampling resolution affects the magnitude of the extracted ramp rates.

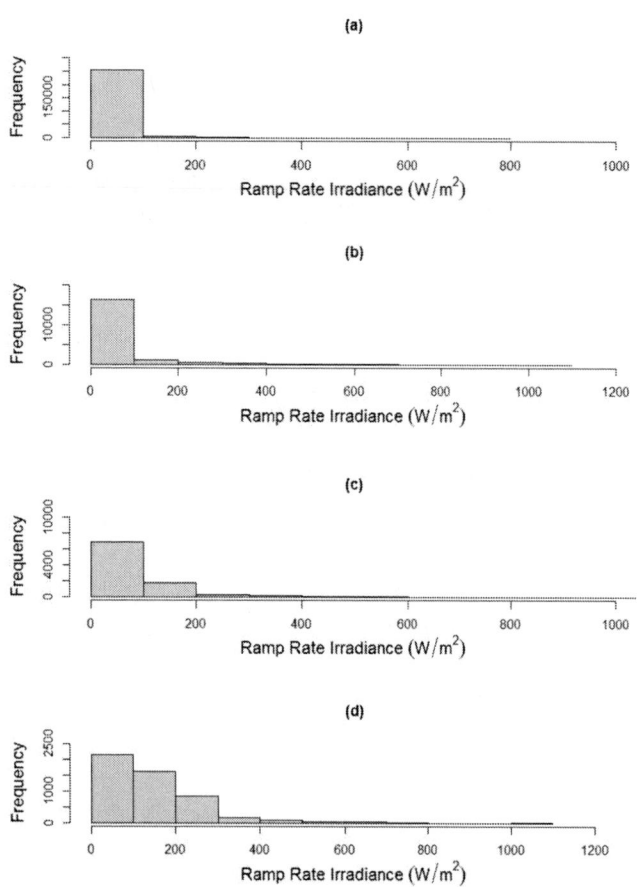

Fig. 6. GHI ramp rate frequency of occurrence for the measurements taken at a (a) one-minute, (b) fifteen-minute, (c) thirty-minutes and (d) sixty-minutes time resolution.

IV. CONCLUSIONS

A novel methodology to characterize daily solar irradiance profiles applied to the annual GHI time series based on the k-means clustering technique and a classification method for daily sky conditions applied to annual GHI time series was described in this work.

The extracted clusters presented useful information on the daily solar irradiance profiles and their persistence exhibited at a certain location. The K-POP method for characterizing the daily GHI provided further insight into the behaviour of the daily patterns and the ramping rate persistence. The distribution of the daily solar irradiance profiles clearly showed that 86.2 % of the year were predominantly classified in classes C_1, C_4 and C_5, indicating clear sky and nearly clear sky conditions.

Finally, the ramp rate analysis performed demonstrated that time sampling resolution affects the magnitude of the extracted ramp rates. Specifically, the results showed that the resolved GHI ramp rates mainly occurred in the range of $0 - 100$ W/m^2

for 1-minute recorded data-sets, while for the sixty-minute resolved GHI ramp rates the fluctuations recorded were mainly in the range of $0 - 400$ W/m^2.

ACKNOWLEDGEMENT

This work was funded through the INFORPV project (KOINA/SOLAR-ERA.NET/1215/02) which was co-financed by the European Regional Development Fund and the Republic of Cyprus through the Cyprus Research Promotion Foundation (DESMI 2009-2010).

REFERENCES

[1] IEA PVPS Task 14, "Characterization of the spatio-temporal variations and ramp rates of solar radiation and PV," 2015.

[2] W. T. Jewell and R. Ramakumar, "The effects of moving clouds on electric utilities with dispersed photovoltaic generation," in *IEEE Transactions on Energy Conversion EC-2*, 1987, pp. 570–576.

[3] W. T. Jewell, T. D. Unruh "Limits on cloud-induced fluctuation in photovoltaic generation," in *IEEE Transactions on Energy Conversion*, 1990, pp. 8–14.

[4] J. Marcos, O. Storkel, L. Marroyo, M. Garcia and E. Lorenzo, "Storage requirements for PV power ramp-rate control," *Solar Energy*, vol. 99, pp. 28-35, 2014.

[5] I. Koumparou, G. Makrides, M. Hadjipanayi, V. Efthymiou, and G. Georghiou, "Characterization and Classification of Daily Sky Conditions in Cyprus and France Based on Ground Measurements of Solar Irradiance," in *EU PVSEC Proc.*, 2016, pp. 1689–1699.

[6] T. Soubdhan, R. Emilion and R. Calif, "Classification of daily solar radiation distributions using a mixture of dirichlet distributions," *Solar Energy*, vol. 83, pp. 1056–1063, 2009.

[7] M. Nijhuis, B. G. Rawn and M. Gibescu, "Classification technique to quantify the significance of partly cloudy conditions for reserve requirements due to photovoltaic plants," in *Proceedings of the 2011 IEEE Trondheim PowerTech Conference*, 2011.

[8] L. Fortuna, G. Nunnari and S. Nunnari, "A new fine-grained classification strategy for solar daily radiation patterns," Pattern Recognition Letters, 2016.

[9] Maafi, S., Harrouni, S., 2003. Preliminary results of the fractal classification of daily solar irradiances. Solar Energy 75, pp. 53–61.

[10] L. Baharuddin, R. S.S.Y. Rahim. Daylight availability in Hong Kong: classification into three sky conditions. Architectural ScienceReview, 2010, 53, pp. 396–407.

[11] B. O Kang, K. Tam, "A new characterization and classification method for daily sky conditions based on ground-based solar irradiance measurement data", Solar Energy, Volume 94, August 2013, pp. 102-118.

[12] P. Montero and J. Vilar, "TSclust: An R Package for Time Series Clustering," *JSS Journal of Statistical Software*, vol. 62, pp. 1–43, 2014.

[13] M. Mohssen, "Machine Learning: Algorithms and Applications" , 2016.

[14] X. Xu, "A New Sub-topics Clustering Method Based on

Semi-supervised Learing," *J. Comput.*, vol. 7, pp. 2471-2478, 2012.

[15] I. Koumparou, G. Makrides, M. Hadjipanayi, V. Efthymiou, G. E. Georghiou, "Characterization and classification of daily sky conditions based on ground measurements of solar irradiance for improved energy forecasting", in *5th International Conference on Renewable Energy Sources and Energy Efficiency*, 2016, pp. 1-6.

A Sky Image Analysis System for Sub-minute PV Prediction

Rodrigo Verschae*, Li Li*, Shohei Nobuhara* and Takekazu Kato*†
*Graduate School of Informatics, Kyoto University, Kyoto, 606-8501, Japan
Email: {rodrigo,li,nob}@vision.kuee.kyoto-u.ac.jp
†Shizouka Institute of Science and Technology, Japan
Email: kato.takekazu@sist.ac.jp

Abstract—**Advanced energy management systems are increasingly gaining importance. These systems will allow the further introduction of Photovoltaic (PV) power generation, but for them to be really effective, sub-minute (1-60 sec.) PV generation prediction is required. In this context we propose a sub-minute PV prediction system based on the analysis of sky images. This is done by analyzing cloud movement, and thus the system does not rely on i) historical PV data, ii) dynamic model of the local weather, nor iii) location dependent information. The proposed system works as follows: from multiple image exposures, high dynamic range images are obtained (one per second), cloud movement is estimated, sky images are predicted, and finally PV generation is estimated using the predicted sky images. The proposed system achieves low error under various weather conditions.**

Index Terms—**photovoltaic, prediction, sky image, sub-minute, high dynamic range, cloud movement.**

I. INTRODUCTION

In recent years, systems that greatly increase the energy management ability of each demand have been proposed. These systems enable the management of generation and consumption, e.g. allowing each end-point of the system to set power consumption targets, to follow fluctuating power sources, and to coordinate consumption and generation with other users (see e.g. [1] [2] [3]).

Another trend has been the rapid introduction of uncontrollable renewables (including PV) due to *i*) the lower cost of renewables and batteries, and *ii*) government policies encouraging their installation (e.g. the Fit-In Tariff (FIT) scheme). As a consequence, more consumers have installed their own generation, storage and energy management systems.

However, restrictions to the further installation of uncontrollable renewables due to the limited controllable generation capacity has been observed in some regions (e.g. in Kyushu, Japan). Moreover, the time-limit of FIT programs [4], may not allow users to continue injecting energy in to the grid, e.g. the FIT program for residential users in Japan lasts 10 years, and it is not clear what these users will do after the FIT program finishes.

To increase the introduction of PV, advanced energy management system are key, and such systems would greatly benefit by having a local sub-minute PV prediction. Thus, we address the problem of such sub-minute prediction PV.

The power generation of a PV system depends mainly on a single environmental factor: the relative position of the sun, the clouds, and the PV station. Thus, by analyzing the movement and position of the clouds with respect to

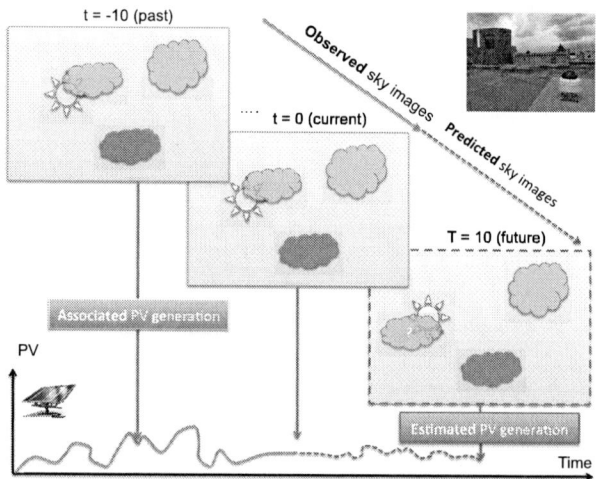

Fig. 1. Proposed Approach. To predict the PV generation, first future sky images are predicted, and later the PV power associated to the predicted sky image is estimated. Sky image prediction is done by analyzing cloud movement, thus his system does not require a historical data, nor dynamical model (of PV generation) nor geo-location information.

TABLE I
OVERVIEW OF APPROACHES FOR PV PREDICTION [5][6].

Approach (Sensor)	Time Horizon	Resolution	RMSE
Weather model [7]	Hour-ahead	1 hour	20-60%
Satellite imaging [8]	Hour-ahead	1 minute	20%
Dynamic PV model [9], [10]	Intra-hour	15 minute	20%
Sky image (cloud mov.)	Intra-minute	1 second	(proposed)

the PV cell and the sun, it should possible to predict the PV generation at any location. In this line, we address the challenge of sub-minute PV prediction (time horizon of 1 - 60 sec.). In particular we investigate how to predict PV power generation using a camera facing up to the sky (see Fig. 1).

II. RELATED WORK

Various approaches have been proposed to predict PV generation [5][6]. These are designed to work at various time scales, time resolutions, and spatial resolutions, with most methods performing a day-ahead prediction for every 30-minute. Most methods can be grouped in 3 categories (see Table 1): weather-based models, satellite imaging systems, and dynamical models (learned from historical PV data).

Numerical weather based models have low time & spatial resolution and are based on physical models. Satellite image methods directly use cloud structure, thus they do not rely on historical data or dynamical models, achieving a higher accuracy and working at smaller time scales (intra-hour PV prediction), but they have low spatial resolution. Satellite imaging has been used for intra-hour PV prediction based on the estimation of cloud structure and motion.

An alternative is to use PV measurements together with models learned from historical data (e.g. dynamical models auto-regressive (AR) models) [10][9]. In general these methods provide intra-hour (a 5 minute-ahead) PV prediction, but they are trained for each location using dynamical models learned using historical data. Given that measuring PV generation at a single location may not be enough, the use of distributed PV measurements has been recently proposed [8] [9]. Distributed PV measurements improve the PV prediction when compared to PV measurements at a single location, achieving intra-hour prediction, however these methods also require a location-specific model learned from historical data.

III. Sub-minute Sky Image and PV Prediction

To prediction PV generation we propose to analyze and predict cloud movement from sky images and to estimate PV generation from these predicted images. An advantage of this approach is that it does not require historical data, dynamical model, nor geo-location. While the analysis of static sky images has been proposed [11], to the best of our knowledge, cloud movement for sub-minute PV prediction has not. The proposed system consists of four main parts:

- PV generation measurement,
- High Dynamic Range (HDR) sky image capture,
- Sky image analysis and prediction, and
- Image based PV power estimation.

A. PV Generation System and Measurement

PV generation data is captured at 1-second resolution at a PV station. The PV cells are parallel to the ground and the station has generation capacity of 2500W. The generated power is sensed using a current transformer (CT) sensor and each sample is time-stamped with a NTP synchronized clock. The station is located 100mt from the image capture system.

B. Sky Image Capture System

1) Capture System and High Dynamic Range Images: The system captures multiple image exposures at a high speed (8 fps) that are combined to obtained the HDR image. The capture program is implemented in C++ and uses multi-thread programming to improve throughput and reduce I/O blocking. The images are cropped in camera (1280x1280 pixels) and captured using a bracketing mode that allows to iterate over 4 shutter speeds. Every half second 4 images are captured (with exposure times in $\{1, 8, 16, 24\} \times 11ms$, see Fig. 3).

In order to reduce cloud movement within a bracket, out of the eight frames captured every second, the first 4 frames are kept. These 4 frames are stored in TIFF format and then

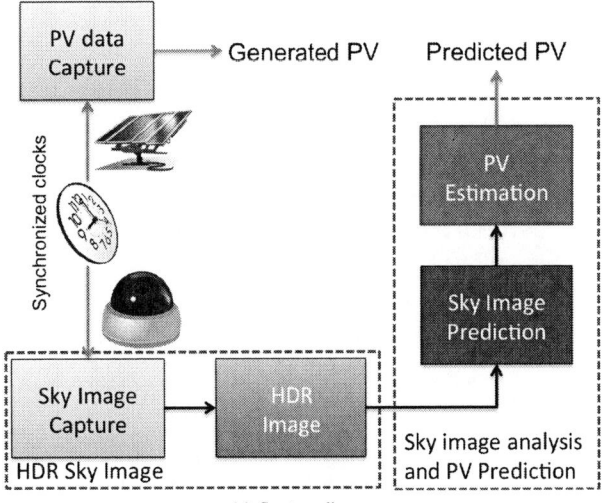

(a) System diagram

(b) Camera system

Fig. 2. System diagram and camera system. The system consists basically of three main parts: PV data capture, Sky image capture and HDR image formation, and Sky image analysis (image prediction and pv estimation). The camera capture system consist of a CMOS Camera, a Fish-eye lens, and a housing to protect the camera (not shown in the image).

encoded as MP4 videos. These videos are later decoded and the exposures and combined to obtain the HDR image [12].

The system's hardware consist of:

- a CM3-U3-31S4C Chameleon3 PointGrey Camera (Sony IMX265, 1/1.8", Color, 2048x1536 (3.2MP), USB3),
- a Spacecom TV1634M, F1.4, 180 Fish-eye lens, and
- an Intel i7, 8 cores, 4 GHz, 32GB RAM PC (Ubuntu).

2) Lens calibration: We use a calibrated model of the camera-lens system that gives us the direction of each light ray being collected by each pixel. The lens is modeled using a 4th degree polynomial [13]:

$$f(\rho) = a_0 + a_1\rho + a_2\rho^2 + a_3\rho^3 + a_4\rho^4, \qquad (1)$$

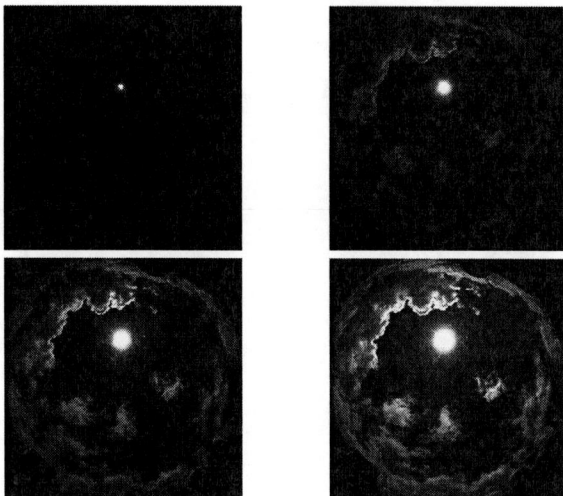

Fig. 3. Multiple exposures ($\in 11 \times \{1, 8, 16, 24\}$msec) are captured to generate a HDR image. Images as obtained from the camera. No color correction was applied, thus the greenish look (the sensor's efficiency varies across color channels, with the green channel having the highest quantum efficiency).

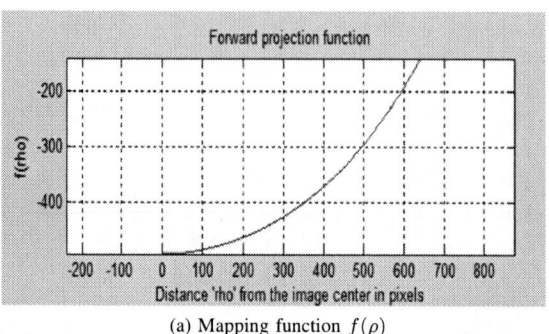

(a) Mapping function $f(\rho)$

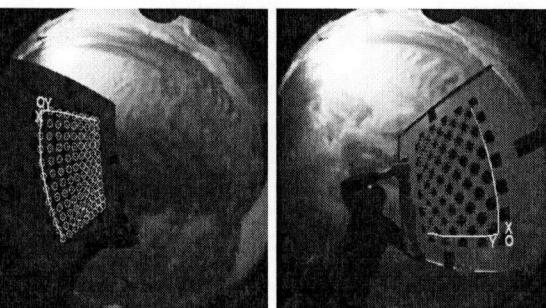

(b) Checkboard and projected lines.

Fig. 4. Lens calibration. The calibrated model closely matches the checkboard patterns (projected lines (green) in (b)).

with ρ the distance from the center of the image ($\rho = \sqrt{u^2 + v^2}$), (u, v) the coordinates in the sensor plane (in pixels). In this model, $f(\rho)$ gives us the coordinate of the 3D point $[u, v, f(\rho)]$ associated to sensor pixel (u, v). Then, given the distance ρ of a pixel from the image center, we estimate the corresponding light direction θ from:

$$\tan \theta = \rho / f(\rho). \qquad (2)$$

This model is calibrated following [13] using a checkboard of known geometry. Fig. (4) presents calibration results.

C. Sky Image Analysis and Prediction

1) Light projection model: We consider a light projection model that maps the light to a plane in the sky (see Fig. 5a) because sub-minute cloud movement can be approximated as linear in this plane (in the original sensor plane, the cloud movement is not linear due to the fish-eye lens). The sun can be assumed to be static (within one minute). Fig. 5 presents an example of this mapping.

2) Movement Analysis: We use optical flow [14] to calculate the movement of each pixel, and we then use linear prediction using the calculate model. This is done in the sky plane projection, where the cloud movement can be assumed to be linear. Empty pixels can be observed (two pixels may move to the same location), and we obtain such empty pixels by interpolating neighbor values. The predicted image is projected back to the original image plane for the later PV Estimation.

D. PV Estimation

To estimate PV generation from a sky image we need to consider the fish-eye lens light projection (in the sensor) and

the cosine law (of the PV cell). Also we need to take into account camera sensor response, and in particular we need to consider that some pixels of the image saturate[1]. Pixel saturation is an important problem, specially during sunny days, as can be observed in Fig. 6. The PV estimation takes into account:

1) Light incidence angle: We integrate the light energy according to the unit projection sphere (Fig. 5a) using the calibrated lens model. For clarity, we use color channel c. Each pixel value $I_c(u, v)$ is weighted considering the light ray direction for the pixel and the PV cell:

$$PV_w = \alpha_c \sum_{(u,v)} I_c(u, v) w(u, v), \qquad (3)$$

where $w(u, v) = w_a(u, v) w_b(u, v)$. Here $w_b(u, v)$ represents the weight of the spherical sector sampled by pixel (u, v) (for our fish-eye lens $w_b(u, v)$ is constant), and $w_a(u, v) = \cos \theta$ is the weight associated the angle of incidence in the PV cell. If we define $\rho = \sqrt{u^2 + v^2}$, the value of θ is estimated from:

$$\tan(\theta) = \rho / f(\rho). \qquad (4)$$

2) Spectral response: The PV cell spectral response and image sensor color channel have a specific spectral response.

---

[1]Note that while the HDR image helps reducing saturation the dynamic range of the light is too large to be handle by a standard camera.

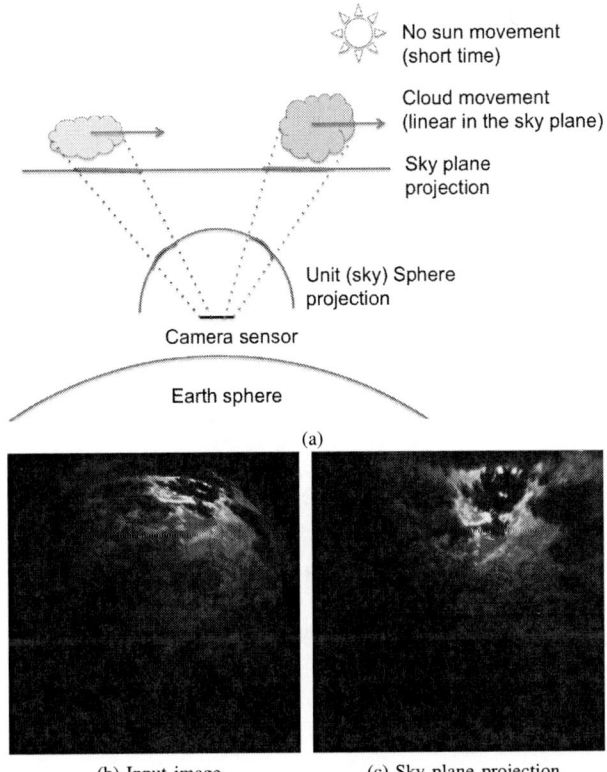

No sun movement
(short time)

Cloud movement
(linear in the sky plane)

Sky plane
projection

Unit (sky) Sphere
projection

Camera sensor

Earth sphere

(a)

(b) Input image   (c) Sky plane projection

Fig. 5. (a) Light projection model: sensor, sphere and sky plane projections. The cloud movement is linear in the sky plane projection, thus it is used for cloud movement prediction. The unit sphere projection is used for PV estimation. (b) Original image (camera sensor) and (c) its sky plane projection.

To take them into account, the intensity of each color channel $c \in \mathbb{C} = \{R, G, B\}$ is included in the PV estimation:

$$PV_{\text{sp}} = \sum_{c \in \mathbb{C}} \alpha_c \sum_{(u,v)} I_c(u,v) w(u,v), \qquad (5)$$

with I_c the intensity of color channel $c \in \mathbb{C} = \{R, G, B\}$ of pixel (u, v). In the following we will refer to the model in Eq. 5 as the **Linear** model for PV estimation (the parameters, α_c, of this model are obtained using linear regression).

3) Saturation: Let us call $\mathbb{S}_c = \{(u,v)|I(u,v) = I_{\max}\}$ the set of saturated pixels, $\tilde{I}_c(u,v)$ the pixel intensity if there were no saturation, and $I_{\max}$ the saturation value. Then we estimate PV generation taking into account saturation using:

$$PV_s = PV_{\text{sp}} + \sum_{c \in \mathbb{C}} \alpha'_c \sum_{(u,v) \in \mathbb{S}_c} (\tilde{I}_c(u,v) - I_{\max}) w(u,v). \quad (6)$$

Given that we do not know the value $\tilde{I}_c(u,v)$ due to pixel saturation, we estimate the PV generation using a **Non-Linear** regression that uses all known terms in Eq. (6):

$$\hat{y} = F\left(\sum_{(u,v) \in \mathbb{N}_c} I_c(u,v) w(u,v), \sum_{(u,v) \in \mathbb{S}_c} w(u,v), |\mathbb{S}_c| \right). \tag{7}$$

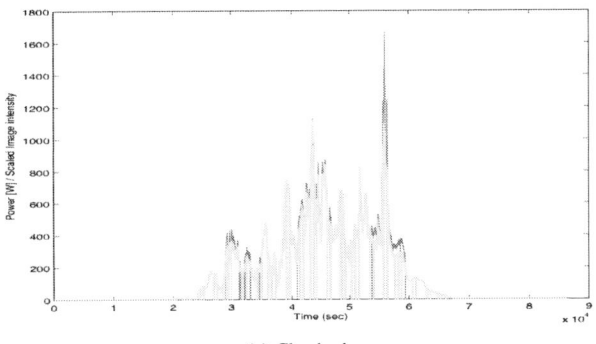

(a) Cloudy day

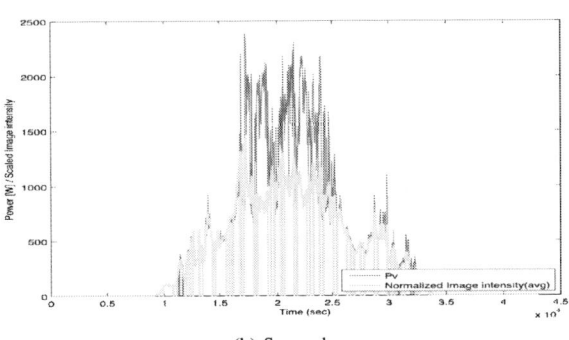

(b) Sunny day

Fig. 6. PV generation (red) vs average image intensity (green). In the case of a cloudy day the average image intensity is a good predictor of the PV generation, but in a sunny day (right) this does not hold due to the camera sensor saturation.

TABLE II
PV ESTIMATION EVALUATION. SEE FIG. 7 FOR PLOTS OF THE RESULTS.

Data (Figure)	RAE E_1 [%]		RMSE E_2 [%]	
	Linear	Non-Linear	Linear	Non-Linear
Cloudy (7a)	11.85	7.78	13.71	9.68
Partly cloudy (7b)	20.11	14.14	30.97	22.58
Partly cloudy (7c)	24.67	14.34	28.64	20.11
Sunny (7d)	25.63	13.36	26.59	14.37
Complete test set	25.52	18.27	32.9	24.16

Namely the weighted sum of non-saturated pixels, the sum of the weights of saturated pixels, and the number of saturated pixels $|\mathbb{S}_c|$), for all colors channels (with $\mathbb{N}_c$ the set of non-saturated pixels for channel c) are used. The rationale is that the number of saturated pixels per color channels can provide information regarding the saturated regions.

4) Experiments in PV estimation: To test the PV estimation we consider the two cases described above: **Linear** and **Non-Linear** regression, i.e. Eq. 5 and Eq. 7 respectively. This is done using Matlab function "fitlm".

We evaluate the accuracy using two measures: the relative absolute error rate (RAE):

$$E_1 = 100 \frac{1}{T} \sum_t |y(t) - \hat{y}(t)|/y(t), \qquad (8)$$

and the root mean squared error rate (RMSE):

$$E_2 = 100\sqrt{\frac{1}{T}\sum_t (y(t) - \hat{y}(t))^2}/\bar{y}, \qquad (9)$$

with $y(t)$ the ground truth, $\hat{y}(t)$ the predicted value, and T the number of samples.

The regression models are estimated using 56 days worth of data, and evaluated using data from 28 days (different from the ones used for building the regression model). Figure 7 and Table II show results under some typical weather conditions. From these results we can see that

- The relative absolute error rate (E_1) for all test videos is 25.52%, and 18.27% for the **Linear** case (Eq. 5) and the **Non-Linear** case (Eq. 7) respectively.
- The root mean squared error rate (E_2) for all test videos is 32.9%, and 24.16% for the **Linear** estimation (Eq. 5) and **Non-Linear** estimation (Eq. 7) respectively.
- For a cloudy day case (Fig. 7a), we observe that E_1 is 11.85%, and 7.78% for the linear estimation (Eq. 5) and the non-linear estimation (Eq. 7) respectively. The good performance is due to the low pixel saturation.
- For a sunny day case (Fig. 7d), i.e. pixel saturation, we observe that E_1 is 25.63%, and 13.36% for the linear estimation (Eq. 5) and the non-linear estimation(Eq. 7) respectively. In this case the error, compared to the cloudy day, is almost doubled.

In summary, the method is accurate when the image is not saturated, and explicitly using information of the saturation further improves when saturation occurs.

IV. Conclusions

We have presented a system for sub-minute (1-60 sec.) PV prediction based on the analysis of sky images and cloud movement. This system is location-independent, and it does not require historical data nor dynamical models.

Target applications of this sub-minute prediction system include: *i*) enhancing energy management systems (at homes, factories, etc.), *ii*) improving the management of distribution & transmission networks, and *iii*) reducing the need of backup generation capacity. Also, sky and cloud analysis can be useful for weather prediction and atmospheric studies.

Future research directions include: layered cloud movement analysis, cloud transparency estimation, prediction of sun occlusion, and formulating a model of the sky image intensity that takes into account Mie scattering and Rayleigh scattering.

Acknowledgments

This work was partially supported by the JSPS Kakenhi grant 17H01922, the i-Energy joint research chair, Kyoto University, by the Research Unit for Smart Energy Managment, Kyoto University, and by Panasonic, Japan.

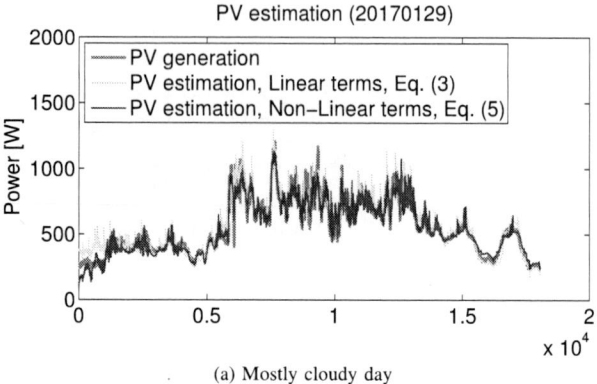

(a) Mostly cloudy day

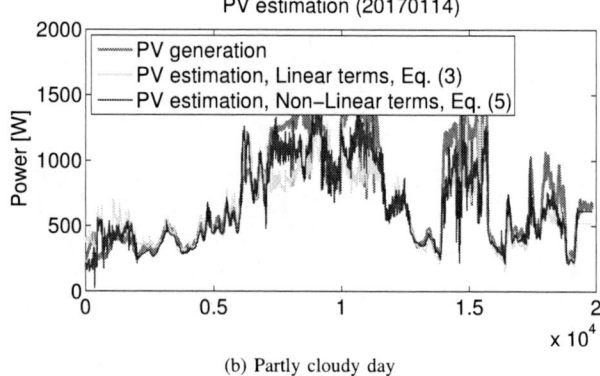

(b) Partly cloudy day

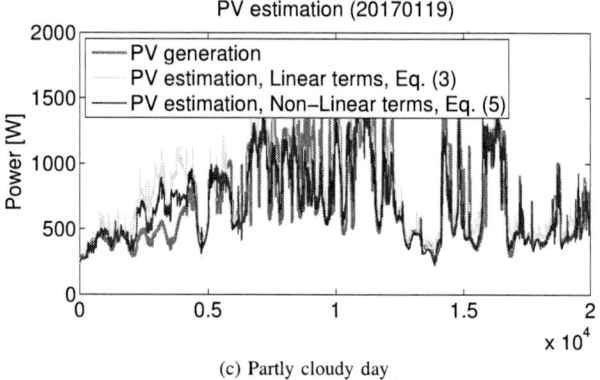

(c) Partly cloudy day

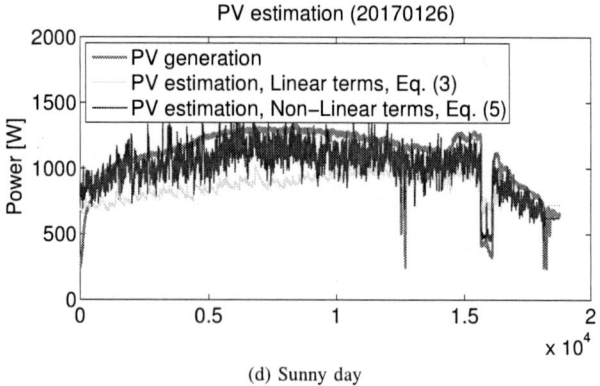

(d) Sunny day

Fig. 7. PV estimation. Linear regression is used for PV estimation. The usage of information regarding pixel saturation (blue) improves the results, in particular during sunny periods.

REFERENCES

[1] T. Kato, K. Tamura, and T. Matsuyama, "Adaptive storage battery management based on the energy on demand protocol," in *2012 IEEE Int. Conf. on Smart Grid Communications (SmartGridComm)*, 2012.

[2] R. Verschae, T. Kato, and T. Matsuyama, "Energy management in prosumer communities: A coordinated approach," *Energies*, vol. 9, no. 7, p. 562, 20 Jul. 2016.

[3] T. Matsuyama, "i-energy: Smart demand-side energy management," in *Smart Grid Applications and Developments*, ser. Green Energy and Technology, D. Mah, P. Hills, V. O. Li, and R. Balme, Eds. Springer London, 2014, pp. 141–163.

[4] K. Ogimoto, I. Kaizuka, Y. Ueda, and T. Oozeki, "A good fit: Japan's solar power program and prospects for the new power system," *Power and Energy Magazine, IEEE*, vol. 11, no. 2, pp. 65–74, March 2013.

[5] V. Kostylev and A. Pavlovski, "Solar power forecasting performance– towards industry standards," in *1st International Workshop on the Integration of Solar Power into Power Systems, Denmark*, 2011.

[6] J. Kleissl, *Solar energy forecasting and resource assessment*. Academic Press, 2013.

[7] S. Pelland, G. Galanis, and G. Kallos, "Solar and photovoltaic forecasting through post-processing of the global environmental multiscale numerical weather prediction model," *Progress in Photovoltaics: Research and Applications*, vol. 21, no. 3, pp. 284–296, 2013.

[8] R. Perez, P. Ineichen, K. Moore, M. Kmiecik, C. Chain, R. George, and F. Vignola, "A new operational model for satellite-derived irradiances: description and validation," *Solar Energy*, vol. 73, no. 5, pp. 307–317, Nov. 2002.

[9] C. Yang, A. A. Thatte, and L. Xie, "Multitime-Scale Data-Driven Spatio-Temporal forecast of photovoltaic generation," *IEEE Transactions on Sustainable Energy*, vol. 6, no. 1, pp. 104–112, Jan. 2015.

[10] T. T. Teo, T. Logenthiran, and W. L. Woo, "Forecasting of photovoltaic power using extreme learning machine," in *2015 IEEE Innovative Smart Grid Technologies - Asia (ISGT ASIA)*, Nov. 2015, pp. 1–6.

[11] B. Urquhart, B. Kurtz, E. Dahlin, M. Ghonima, J. E. Shields, and J. Kleissl, "Development of a sky imaging system for short-term solar power forecasting," *Atmospheric Measurement Techniques*, vol. 8, no. 2, pp. 875–890, 20 Feb. 2015.

[12] P. E. Debevec and J. Malik, "Recovering high dynamic range radiance maps from photographs," *ACM SIGGRAPH 2008 classes*, 2008.

[13] D. Scaramuzza and R. Siegwart, "A Practical Toolbox for Calibrating Omnidirectional Cameras," in *Vision Systems: Applications*, G. Obinata and D. Ashish, Eds. I-Tech, 2007.

[14] B. K. Horn and B. G. Schunck, "Determining optical flow," *Artificial intelligence*, vol. 17, no. 1-3, pp. 185–203, 1981.

Comparison and Analysis of Instruments Measuring Plane of Array Irradiance for One-Axis Tracking PV Systems

Frank Vignola[1], Chun-Yu Chiu[1], Josh Peterson[1], Michael Dooraghi[2], Manajit Sengupta[2]

[1]University of Oregon, Eugene, Oregon, Material Science Institute, 1274-University of Oregon, Eugene, Oregon, 97403-1274,
[2]National Renewable Energy Laboratory, Golden, Colorado, 80301

Abstract – A variety of sensors are studied on a one-axis tracking surface and a horizontal surface in Golden, Colorado and Eugene, Oregon. This is the first year of a long-term study that will look at not only a comparison between the instruments but will also the longer-term degradation in calibration and/or performance. Initially, results from each location will be analyzed, and then the results will be compared between the two locations. A quick comparison at Eugene indicates that reference solar cells yield irradiance values closer to a secondary standard thermopile pyranometer values on a one-axis tracker than photodiode-based pyranometers, especially at large solar zenith angles. More study is needed to characterize and specify this finding.

Index Terms – irradiance, POA, one-axis tracking, reference cell, pyranometer.

I. INTRODUCTION

Evaluations of photovoltaic (PV) system performance are important for many reasons, including analysis of system degradation rates, estimation of long-term system production, and forecasting system performance. The analysis of absolute system performance depends on an accurate measurement of the incident irradiance along with temperature and other meteorological parameters. Some of these questions can be best addressed using knowledge of the relative performance of the system over time. A variety of instruments are used to understand system performance, including thermopile-based pyranometers, photodiode-based pyranometers and reference cells. When analyzing PV system performance and developing models for PV system performance, the myriad devices used to gauge the incident irradiance often yield results that can vary by 10% or more. With such a large discrepancy, the modeling and analysis of PV system performance becomes difficult.

This is especially true with one axis tracking systems wherein a minimum amount of data gathered and compared to the results of various instruments used to monitor the incident irradiance. In 2016, the National Renewable Energy Laboratory (NREL) initiated a program to study irradiance measurements for one-axis tracking and horizontal systems to better understand the measurements from the different instruments. Because instruments are known to perform differently in different locations, similar experiments were set up in Golden, Colorado, and at the University of Oregon in Eugene.

This paper reports preliminary findings after nearly a year's worth of data have been collected. First, the instruments used in the experiment are described along with procedures used to calibrate the instruments. Next, the experimental configuration is illustrated. The findings comparing the measurements are shown along with a brief discussion of the results.

II. INSTRUMENTS USED IN THE EXPERIMENT

At the Eugene location five instruments are compared both on a tracking and a horizontal surface. The instruments mounted on the one-axis tracker are a Kipp & Zonen CMP22 pyranometer, a LI-COR LI-200 SA pyranometer, a Kipp & Zonen SP Lite2 pyranometer, an RCO reference solar cell, and an IMT reference solar cell. The CMP22 is a secondary reference pyranometer; during calibrations, it is reputed to have an absolute accuracy of 1.5% or better at the 95% level of confidence [1].

At the Golden location, a similar set of instruments are used except a CMP 21 pyranometer is used instead of the CMP 22. In addition, other photodiode-based pyranometers are included. For a more comprehensive study, instruments at a fixed tilt are also co-located.

The LI-200 and SP Lite2 pyranometers are photodiode-based pyranometers that are widely used to monitor PV system performance. The responsivity of the photodiode-based pyranometer has been shown to vary with the spectral distribution of the incident irradiance [2–4]. Their irradiance values are expected to vary more during the day in than the CMP22 because the CMP22 is a thermopile pyranometer and has a special dome designed to uniformly transmit irradiance over a wide range of wavelengths. Also, these specially designed double-glass domes produce a more accurate cosine response in the instrument and reduce thermal losses.

The LI-200 SA pyranometer is a photodiode-based pyranometer that essentially monitors the short circuit current of a solar cell under a diffusing lens. The pyranometer body and diffusing lens are designed to minimize deviations from a true cosine response. In addition, some internal circuitry helps minimize the effect of temperature on the pyranometer's performance. The LI-200 SA pyranometer uses blue enhanced solar cells with a higher response in the blue portion of the

978-1-5090-5606-4/17 $31.00 © 2017 IEEE

Fig. 1. One-axis tracker with sensor instruments mounted on plate at the University of Oregon station.

spectrum to more accurately simulate the performance of high-quality thermopile-based pyranometers [5].

The SP Lite2 pyranometer is also a photodiode-based pyranometer, but it has a different body and diffusing lens. Although the LI-200 SA and the SP Lite2 have similar characteristics, their outputs differ measurably.

The reference cells in this study use single crystalline silicon cells that are used to calibrate PV modules on factory production lines. They have also been used in the field to evaluate the performance of PV systems that use the same technology as the reference solar cells. The reference solar cells are expected to have a similar spectral response as the photodiode-based pyranometers because photodiodes and reference cells are both solar cells for which the output is monitored in a short circuit configuration. There are two main differences between the reference cells and photodiode-based pyranometers. First, the photodiode-based pyranometers have a diffusing lens to help provide a true cosine response, whereas the reference cells have a glazing like that used by PV modules (see Fig. 1). Second, the reference cells monitor cell temperature which is used to adjust the temperature dependence of the measured irradiance from the reference cells. Although photodiode-based pyranometers usually do not have an internal temperature measurement that can be used to correct for temperature effects, some photodiode-based pyranometers do have internal circuitry that helps compensate for some temperature effects. In addition, the temperate effects can be modeled on ambient temperature [2].

III. CONFIGURATION AND DATA

The one-axis tracker is a modified LI-200SA automatic tracker (LI-2020) that has been configured to rotate the pyranometers from east to west during the day. The tracker is oriented with the axis aligned along the north-south direction and pointed due east at sunrise and due west at sunset. A plate mounted along the north-south axis holds the pyranometers being evaluated. A similar plate is mounted in a horizontal fixed

position. At solar noon, all the instruments will be horizontal and this can act as a check on the instrument output.

Fig. 1 shows a photo of the experimental setup in Eugene. The instruments at both sites are cleaned five days per week and a maintenance log on the instruments is regularly updated. The instruments are calibrated once per year using an absolute cavity radiometer for the direct normal component and a pyranometer shaded by a sphere for the diffuse component. The reference global irradiance is then calculated from these measurements and the responsivity of the test instruments are obtained by comparison to the reference global measurements.

One-minute average measurements are then gathered using Campbell Scientific CR 1000 data loggers, and the data are downloaded daily. Co-located at the Eugene and Golden sites are a variety of irradiance and meteorological instruments that can be used to supplement the test measurements. Spectral measurements are also available that can be used for future analysis.

IV. CALIBRATIONS

Initially the instruments were calibrated at NREL and shipped to Eugene. Subsequently the Eugene instruments were calibrated in Eugene using an AHF absolute cavity radiometer. Because the atmospheric conditions in Eugene differ from those at NREL and because the AHF cavity has been calibrated against the NREL absolute cavity radiometer, the calibration values determined in Eugene are used for the Eugene data in this study.

Fig. 2 shows the 2016 calibration results at Eugene. The photodiode-based pyranometers are within ±5% throughout the range from 30° to 80°, whereas the CMP22 is within ±2.5%. The reference solar cells are within ±5% to 60°, but they start to deviate significantly from a true cosine response at larger solar zenith angles. These calibrations were performed under clear-sky conditions. Night-time offsets were subtracted from the values before the responsivities were determined. No other adjustments were applied to the data.

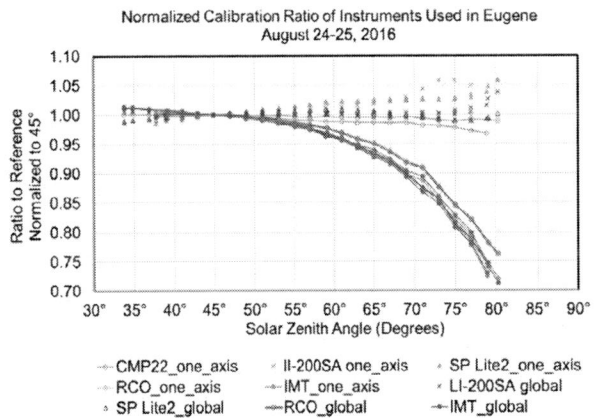

Fig. 2. Calibration comparison of instruments used in this study. The responsivities at 45° are used to normalize the calibrations.

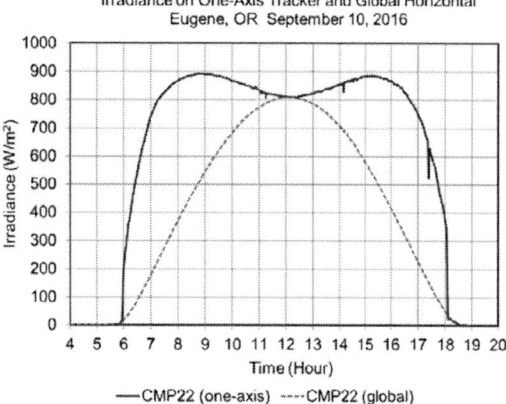

Fig. 3. Comparison of clear-sky GHI and POA irradiance for a one-axis tracking system. This system tracks from east to west. At solar noon, the tracker is horizontal, and the two values should agree when the solar zenith angle is 45°.

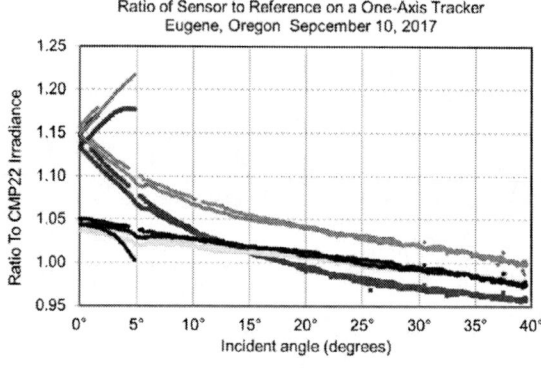

Fig. 4. Ratio of readings of instruments on a one-axis tracker to the reference CMP22 values. The incident angle is the angle between the sun and the normal to the plate on the one-axis tracker. The smallest incident angle in September occurs near sunrise and sunset, and results are likely to be affected by obstructions near the surface.

V. EVALUATING THE DATA

The first comparisons are made under clear-sky conditions. Under clear-sky conditions, the diffuse horizontal irradiance (DfHI) is small compared to the global (or total) tilted irradiance (GTI), and the results are dominated by the characteristics of the direct normal irradiance (DNI). GTI is also called Plane of Array (POA) irradiance and this term is particularly apt for instruments mounted on a one-axis tracker. Fig. 3 shows a typical comparison of global horizontal irradiance (GHI) and the POA irradiance for a one-axis tracking system.

The irradiance as measured by the CMP22 pyranometer has the lowest uncertainty, and thus the CMP22 will be used as the reference instrument. The CMP22 does deviate from a true cosine response and has some thermal offsets, but these are small (Fig. 2); however, if one is using data from this instrument to evaluate other instruments or adjust the readings of other instruments these systematic effects should be kept in mind because these effects can skew the evaluation.

Using the CMP22 as the reference and keeping in mind the uncertainties that are contained in the CMP22 measurements, one can examine the performance of the other sensors during the day by taking the ratio of their measurements to those obtained from the CMP22. Fig. 4 shows this comparison for clear skies on September 10, 2016. The ratio is plotted against the incident angle between the normal to the one-axis tracker and the sun. Because it is September when the sun rises and sets near the equinox and the one axis tracker is pointed due east in the morning and due west in the afternoon, the incident angle between the normal to the one-axis tracker is smallest in the morning and afternoon hours. The spectral distribution also changes the most during the morning and evening hours near sunrise and sunset when the air mass between the sun and the sensors is greatest; consequently, the effects of the changes in spectral distribution on instrument performance are most pronounced during the morning and evening hours.

On September 10, the incident angle is smallest soon after sunrise, when the sun is due east; and soon before sunset, when the sun is due west. Much of the change in ratio between the photodiode-based pyranometers and the reference pyranometer is related to the change in spectral distribution. This effect is shown when the ratio continues to increase even after the incident angle reaches near zero. In contrast, the ratio between the reference cells and the reference pyranometer is linear throughout the day. One might assume that the difference between the two types of instruments is related to temperature effects; however, the values in the morning and evening are nearly identical indicating that the temperature effects are much smaller than the observed changes in ratios throughout the day. Although some temperature effects can be measured, they are small compared to the changes in the responsivities shown in Fig. 4.

To better understand the difference shown in Fig. 4, Fig. 5 shows the ratio of the LI-200SA pyranometer and the reference cell irradiance to that measured by the CMP22 pyranometer. The comparison is made for total horizontal irradiance (GHI).

The results in Fig. 5 are for one day of clear periods per month during one year. As expected, the results are consistent with the calibration results shown in Fig. 2.

For the instruments on the one-axis tracker, the results are expected to be more complicated, and they are. The irradiance incident angle is always smaller for instruments mounted on a one-axis tracker than the incident angle to the horizontal instruments except at solar noon, when all instruments are horizontal. The instruments on the tracker also measure some ground reflected irradiance. Ground-reflected irradiance has its own spectral distribution and as the instruments are tilted more vertically, the ground-reflected irradiance contribution increases. Fig. 6 shows the comparison of a LI-200SA pyranometer and to a RCO reference cell divided by the

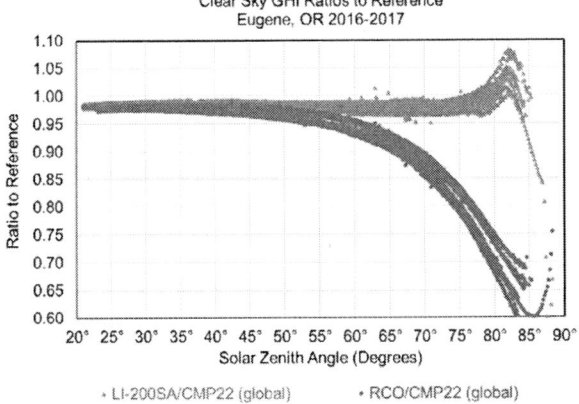

Fig. 5. Ratios of GHI measurements to reference measurements on selected clear periods during the year in Eugene.

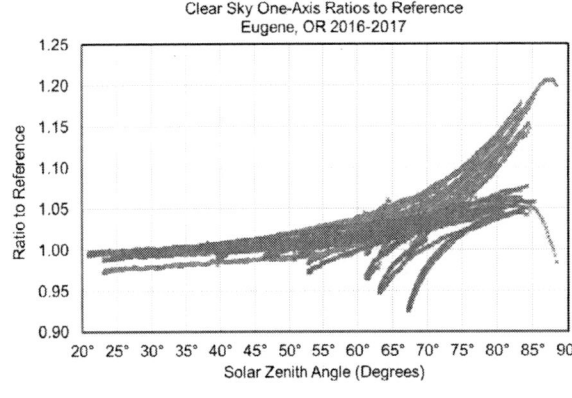

Fig. 6. Comparison of the LI-200SA pyranometer and the RCO reference cell with the output of a CMP 22 pyranometer.

equivalent irradiance from a co-located CMP 22 reference pyranometer mounted on the one-axis tracker.

The low values of the ratio of the RCO reference cell divided by the CMP 22 output on the lower right in Fig. 6 are December 2016 data. The ratio moves up and to the left as the sun gets higher in the sky (during spring and fall). The ratio of the LI-200SA pyranometer increases significantly for larger solar zenith angles. These results are different from those of the GHI measurements.

Another way to plot the data is against the incident angle instead of the solar zenith angle. This is shown in Fig. 7. For the LI-200SA pyranometer there is not a clear relationship between the angle of incident and the ratio to the reference CMP 22; however, there is a clear relationship with the solar zenith angle as shown in Fig. 6, giving credence to the spectral influence. The RCO reference cell ratio might indicate a slight dependence on incident angle, especially at larger incident angles. This is consistent with an angular dependence shown in the calibrations in Fig. 2. The RCO ratio to reference does not show as much dependence.

A. Clear and Cloudy Data

Clearly more than clear-sky data are needed for a comprehensive comparison of measurements. Fig. 8 plots ratio data from the LI-200SA and IMT reference cells for May 2017. There is a fair scatter amount of during the morning and evening hours, resulting from reflections from objects near the horizon and changes in the spectral distribution of incident irradiance. The ratios of the IMT reference cell values typically range between 1.0 to 1.1 during the whole month whereas the ratios of the LI-200SA pyranometer values range between 1.0 and 1.2. The reasons for the different responses need to be further studied. For clarity and to keep the different data sets from being on top of each other, the scale for the LI-200SA, shown on the right, was shifted to show both data sets.

VI. DISCUSSION

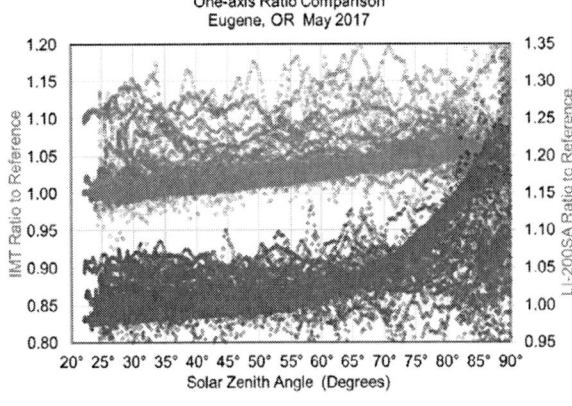

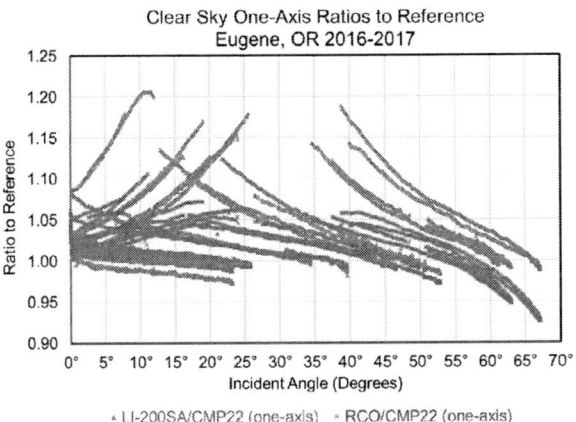

Fig. 7. Comparison of the LI-200SA pyranometer and the RCO reference cell with the output of a CMP 22 pyranometer plotted against incident angle.

Fig. 8. The ratio between the LI-200SA pyranometer and the reference pyranometer measurements is shown by the blue diamonds. Axis on the right. The ratio between the IMT reference cell and reference pyranometer measurements is shown by the red circles. Axis on the left.

The comparisons of the data from the various instruments demonstrate why it is difficult to evaluate and compare various PV systems using different sensors. Even similar irradiance sensors have large uncertainties that make it difficult to develop and validate PV performance models. This is especially true for instruments measuring POA irradiance on a one-axis tracking system. Algorithms to remove systematic effects can be developed, but it is unclear how well these adjustments would work at different locations with different aerosol and ground reflective characteristics.

The deviation of reference cells from a true cosine response is well understood, and the effects of transmission through the glazing can be modeled. Although the reference cells adjust for cell temperature using measured cell temperature, some temperature dependence might remain, and any additional temperature effects need to be evaluated. With knowledge of the spectral responsivity of the reference cells, the spectral effects on the measurements can be modeled [4]; however, the advantage most reported for the reference cells is that they behave much like PV modules made of similar material. Therefore, if one wants to evaluate the performance throughout time using a reference cell, it might not be worth the effort to adjust the reference cell measurements to better simulate the total irradiance incident on the PV module.

On the other hand, if one has a very accurate measurement of the incident irradiance, a model must be used to simulate the performance of the PV module. This is the methodology most used to estimate PV system performance. A problem arises because these models are validated against various types of irradiance sensors and the data gathered so far in Golden and Eugene demonstrate the performance differences among various irradiance sensors.

Because irradiance sensors have a wide variety of characteristics and exhibit systematic effects [1], it might be necessary to specify that the performance model is tested and validated using a given sensor and that data gathered by different sensors should be modified to match the characteristics of the sensor used to develop and validate the model.

Spectral and other meteorological data are also being gathered at the Golden and Eugene stations and this information is important to better identify the causes that result in the diverse measurement results. Untangling the causes and magnitudes of the various effects is the goal of future efforts. In the meantime, it is difficult to recommend the best method to track the performance of a PV system with a high degree of accuracy.

VII. SUMMARY

This study reports the initial results of an analysis of pyranometer measurements on a horizontal and one-axis tracking system. Because PV systems use one-axis tracking technology and it is important to monitor the performance of these systems throughout time, there is concern about the measurements of irradiance on one-axis tracking systems. This is important not only for evaluating the performance of the systems, especially when systems have some benchmark performance standards, but also when evaluating potential solar facility locations. As subsidies for solar facilities are reduced and financing becomes more dependent on revenue from production, accurate evaluation of the incident irradiation becomes increasingly important because the production of electricity is directly tied to the incident irradiance.

This study shows that there is considerable variation between high quality irradiance measurements and other lower cost options. One can choose between either selecting more accurate instruments or developing models to adjust measurements from less accurate instruments. Two things should be considered; the accuracy of measurements under different conditions and whether adjustment algorithms can be reliably developed.

Realizing that with tacking systems a majority of incident angles are less than 45° during the summer months, one might consider using instrument calibrations at an angle other than 45°, a standard for most pyranometer calibrations for instruments that measure GHI irradiance.

This effort provides a trove of high quality data that can determine the uncertainty of irradiance measurements on one-axis tracking systems, an important piece of information for those who are planning, financing, and operating one-axis tracking systems.

ACKNOWLEDGEMENTS

This work was supported by the U.S. Department of Energy under Contract No. DE-AC36-08GO28308 with Alliance for Sustainable Energy, LLC, the Manager and Operator of the National Renewable Energy Laboratory. Funding provided by U.S. Department of Energy Office of Energy Efficiency and Renewable Energy Solar Energy Technologies Office. We would also like to thank the Bonneville Power Administration, the Energy Trust of Oregon, and Portland General Electric for their support of the University of Oregon Solar Radiation Monitoring Laboratory.

REFERENCES

1. Habte, A., Sengupta, M., Andreas, A., Wilcox, S, Stoffel, T., 2016. Intercomparison of 51 radiometers for determining global horizontal irradiance and direct normal irradiance measurements. Sol. Energy 133, 372-393.
2. King, D., Myers, D., 1997, Silicon-photodiode pyranometers operational characteristics, historical experiences, and new calibration procedures. In: 26th IEEE Photovoltaic Specialists Conference, September 29 – October 3, 1997, Anaheim, California
3. King, D., Dratochvil, J.A., Boyson, W. E., 1997. Measuring solar spectral and angle-of-incident effects of photovoltaic modules and solar irradiance sensors. In: 26th IEEE Photovoltaic Specialists Conference, September 29 – October 3, 1997, Anaheim, California
4. Vignola, F., Derocher, Z., Peterson, J., Vuillermier, L., Felix, C., Grobner, J., Kouremeti, N, 2016. Effects of changing spectral

radiation distribution on the performance of photodiode pyranometers. Sol. Energy 129, 224-235.

5. LI-200SA Specifications:
https://www.licor.com/env/products/light/pyranometer.html

Large Area Nanostructure Integration for Broad-Spectrum, Omnidirectional Antireflection Improvements on Polymer Packaged, Mechanically Flexible, Epitaxial Lift-Off III-V Solar Cells

Gabriel Cossio[1], Jihwan Lee[1], Gautham Ragunathan[2], Andre Wibowo[2], Sudersena Rao Tatavarti[2], Kimberly Sablon[3], Edward T. Yu[1]

[1]Microelectronics Research Center, University of Texas, Austin, TX, 78712, USA
[2]Microlink Devices Inc, Niles, IL, 60714, USA
[3]U.S. Army Research Laboratory, 2800 Powder Mill Road, Adelphi, MD, USA

Abstract — We demonstrate the integration of subwavelength moth-eye and Al2O3 nanoisland structures fabricated on, respectively, polymer packaging sheets and conventional Al2O3/TiO2 antireflection coatings on triple junction epitaxial lift-off GaAs solar cells. A scalable, low cost, and high throughput surface patterning process based on nanosphere lithography is presented, and demonstrated experimentally for ELO solar cells and polymer packaging sheets with areas of ~ 20cm². The mechanically flexible cell structures with integrated optoelectronic nanofeatures show substantial improved photovoltaic performance under large angle illumination. An increase in Isc is seen from 5%-72% at normal and 75-degree incident illumination respectively.

I. INTRODUCTION

The epitaxial lift-off (ELO) process [1], which enables the recycling of expensive growth substrates by separation of epitaxially grown thin-film device layers from the substrate surface, can be employed to produce mechanically flexible, low-cost, light-weight, and high-efficiency GaAs thin-film solar cells. High efficiency, light-weight, flexible ELO solar cells are natural candidates to be used in dynamic and mobile solar applications [2]. However, in these photovoltaic applications, Fresnel reflections, particularly at large angles of incidence, become a major factor in limiting the overall energy conversion efficiency.

Various approaches using subwavelength nanostructures have been reported to achieve antireflection properties for photovoltaics that enable broad-spectrum and omnidirectional performance. Tapered moth-eye nanostructures in 2D periodic arrays have been a promising candidate for effective reflection suppression while operating in these difficult conditions. Moth-eye nanostructures use a graded refractive index between air and substrate to reduce Fresnel reflections over a large part of the solar spectrum and wide range of angles of incidence. Conventional patterning techniques such as photolithography, electron beam lithography, or X-ray lithography either provide insufficient resolution or are still too expensive to be used over large areas.

Alternatively, self-assembled 2D monolayer patterns from colloidal templates have been shown to be a very low cost and reproducible method of fabricating antireflection nanostructures. However, current colloidal monolayer formation techniques such as spin coating, Langmuir-Blodgett, layer transferring, and self-assembly at an air/liquid interface have difficulty producing monolayer templates greater than a few cm² in size [5-7]. The inability to scale the fabrication of antireflection nanostructures to larger areas has prevented them from being widely incorporated into industry processes.

In this letter, we demonstrate scaling of such a patterning method to produce very large-area (>100cm²) periodic nanostructure arrays. We also demonstrate the capabilities of this technique to fabricate low-cost, scalable, very-large area, subwavelength antireflection structures to provide broadband, omnidirectional antireflection performance for a polymer packaged triple junction mechanically flexible ELO III-V solar cell. Two key elements are integrated into the final packaged device — a moth-eye textured polyethylene terephthalate (PET) packaging sheet, and Al2O3 nanoislands on a triple-junction GaAs solar cell with conventional Al2O3/TiO2 bilayer antireflection coating.

II. EXPERIMENTAL SECTION

A. Inverted Metamorphic Multijunction Device Layer

All epitaxial structures were grown by metallorganic chemical vapor deposition (MOCVD) at 100 mbar using arsine (AsH₃), phosphine (PH₃), trimethylindium (TMI), and trimethylgallium (TMG) as precursors and using a V/III ratio >50. Inverted metamorphic multijunction (IMM) InGaP/GaAs/InGaAs structures were grown on GaAs substrates. The top InGaP cell has AlInP as the window layer and AlGaInP as the back-surface field (BSF) layer. The growth structure of the GaAs middle cell consisted of an InGaP window and BSF layers, an optically thick GaAs base layer doped with 2E17 cm⁻³ p-type doping, and GaAs emitter with n-type doping in the range of 2E18 cm⁻³. The bottom 1.0eV metamorphic cell was grown on top of an AlGaInAs graded buffer layer. The lattice constant of the buffer layer was graded from the GaAs middle cell value to

In$_{(0.31)}$Ga$_{(0.68)}$As. The graded buffer layer was designed such that the smallest bandgap of the buffer layer was above the GaAs bandgap value of 1.42eV.

B. Epitaxial Lift Off (ELO) Processing

The first layer deposited on the substrate is a thin, AlGaAs release layer. The solar cell epitaxial layers are then deposited, followed by application of a thick (1-2 mil) flexible metal carrier layer. The wafer is then immersed in a concentrated HF-acid chemistry, which selectively dissolves the release layer (the etch selectivity relative to the GaAs epitaxial structure is greater than 1E5). The thin, composite structure consisting of the metal carrier layer and solar cell epitaxial layers is thereby completely separated from the GaAs substrate. The ELO process requires approximately 12 hours to complete, but is amenable to batch processing, enabling scaling up of the process to lift off hundreds of substrates within a 24-hour period.

C. Antireflection Nanostructure Fabrication

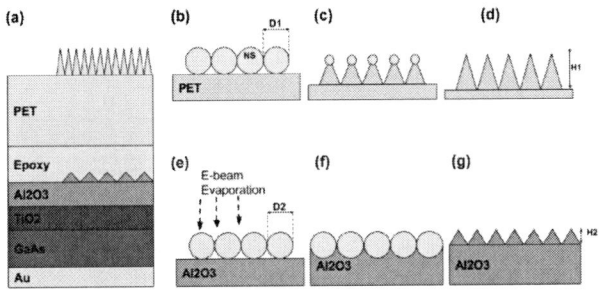

Figure 1. a) Schematic of PET packaged GaAs solar cell coated with conventional Al2O3/TiO2 antireflection coating with Al2O3 nanoislands and moth-eye textured PET packaging sheet. b)-d) Schematic diagrams of process flow for fabricated moth-eye structures on PET substrate. e)-g) Schematic diagram of process flow for fabricating Al2O3 nanoislands.

The final packaged solar cell device is shown schematically in Figure 1 along with key steps of the fabrication process. The moth-eye nanostructures were fabricated on the PET packaging sheet by employing nanosphere lithography (NSL) with 200nm diameter polystyrene nanospheres. The 200nm nanospheres were mixed into a 1:1 solution of nanosphere:ethanol and injected onto an air water interface from a 25 gauge needle by an automated syringe. The ethanol acts as a dispersing agent by lowering the surface tension of the water and allows transport of the nanospheres onto the surface of the water. Continues injection allows the nanospheres to self-assemble into a homogenous, very large area monolayer onto the surface of the water. The nanospheres are transported by the Marangoni force, a mass transfer along

the interface of two fluids because of a surface tension gradient [8]. Fine control of injection rates, the contact angle between the needle and water, depth of injection, number of injection sources and other key parameters allows the growth of nanosphere monolayers to areas >100cm². Our experiments have only used up to two injection sources, therefore we believe that by adding more injection sources will allow the patternable areas to increase to >m². After the formation of the nanosphere monolayer the solution is slowly drained onto the surface of the PET sheet. The nanosphere monolayer is used as a hexagonal array soft mask of close packed spheres. Subsequent O$_2$ reactive ion etches creates the tapered cylindrical structures of height 250nm. The same nanosphere injection method is used to create the monolayer mask for the fabrication of the Al2O3 nanoislands; however, here polystyrene spheres of diameter 1000nm are used instead. A 200nm thick layer of Al2O3 was then deposited by e-beam evaporation, followed by a nanosphere lift-off process by sonication in toluene (Figure1e-f), resulting in a hexagonal array of ~200nm Al2O3 nanoislands (Figure 1g). This patterned cell is then epoxied (Thorlabs NOA68) to the moth-eye patterned PET sheet to create the final packaged solar cel structure shown in Figure 1a.

III. RESULTS AND DISCUSSION

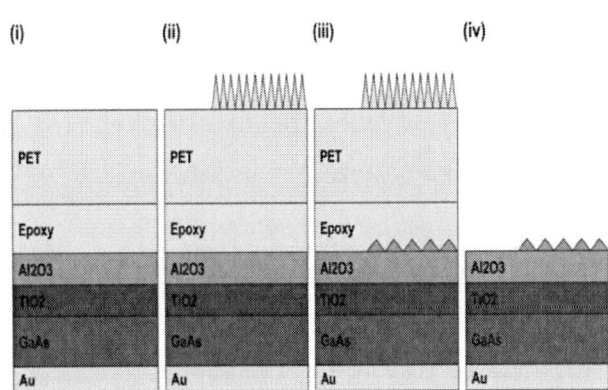

Figure 2. (i)-(iv) Schematic diagram of various integrated structures under comparison. (o) Not shown is a bare ELO cell which contains only the conventional Al2O3/TiO2 antireflection coating which we use for data normalization).

Figure 2 shows the various integrated solar cell structures which we studied including (i) GaAs cell without Al2O3 nanoislands integrated with a nontextured PET packaging sheet; (ii) GaAs cell without Al2O3 nanoislands integrated with moth-eye textured PET packaging sheet; (iii) GaAs cell with Al2O3 nanoislands integrated with moth-eye textured PET packaging sheet; (iv) GaAs cell with Al2O3 nanoislands without packaging or an epoxy layer; GaAs cells with a conventional Al2O3/TiO2 antireflection layer without packaging or epoxy layer which are not shown but were characterized.

978-1-5090-5606-4/17 $31.00 © 2017 IEEE   1182

Short circuit current density (Isc) was measured for each integrated structure with illumination at incident angles varying from 0°-75°. Measurements were obtained with a HP4156A precision semiconductor analyzer, using unpolarized light from a Newport Oriel 9600 solar simulator operating under 100mW/ cm². Figure 3 shows the calculated ratio for Isc of the different integrated structures. Because of variations in current-voltage characteristics from cell to cell prior to any patterning, we normalize the short-circuit current for each structure to the original short-circuit current of the particular cell employed for that measurement to estimate the fractoinal improvement derived, as a function of incident angle, from introduction of motheye patterning on the PET packaging sheet, and from the combination of motheye patterning of the PET sheet and nanoisland fabrication on the original thin-film antireflection coating.

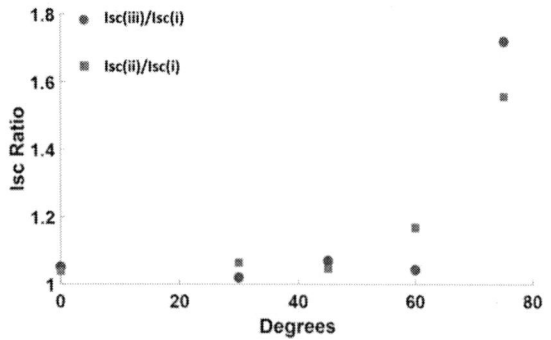

Figure 3. Calculated ratio for Isc of integrated structures (iii) over that of structure (i) and calculated Isc ratio of (ii) over that of (i).

As can be seen from Figure 3, an improvement of Isc for structure (ii) is shown over the entire range of incident angles. This is due to the moth-eye patterned PET sheets, which greatly reduce Fresnel reflections at the air/PET interface. Structure (iii) shows significantly larger improvements to performance at larger angles of incidence, which demonstrates the effectiveness of the integrated optical nanostructures in improving the cells' photovoltaic performance. In particular, we see that at an angle of 75°, a 55% and 72% improvement in the Isc are achieved for moth-eye integrated devices and moth-eye/nanoisland integrated devices, respectively.

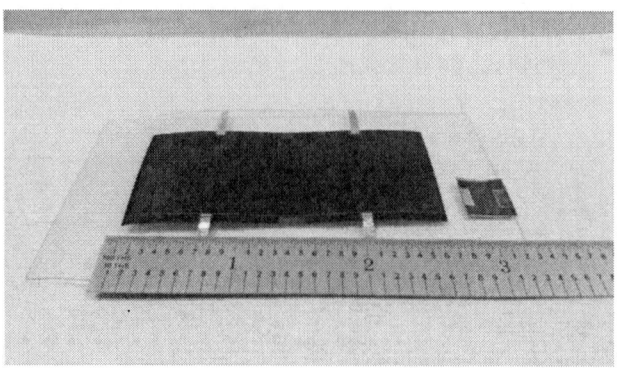

Figure 4. (Left) Fully packaged ~20cm² GaAs triple junction ELO cell used in these experiments. (Right) 1cm² ELO cell for comparison

Figure 4 shows a comparison in size between our moth-eye/nanoisland integrated ELO devices and ELO devices of comparable size from previous reports [7]. Previously, moth-eye/nanoisland patterned solar cells were limited to areas of ~1cm² due to the inability of the Languir-Blodgett deposition methods to create monolayers larger than a few cm². Currently, our method has allowed us to fully pattern a 20 cm² device with great potential for scaling.

In conclusion, we have demonstrated a highly effective antireflection method that can be readily implemented for flexible, ultra-thin ELO GaAs solar cells. ELO GaAs solar cells integrated with both moth-eye textured PET and Al2O3 nanoisland arrays are observed to provide a 5% increase in Isc at normal incidences and a 72% increase at 75° compared to references with only a conventional Al2O3/TiO2 antireflection bilayer. Our fabrication method has also shown to be able to increase deposition area by 20x allowing for higher throughput and scalability. Part of this work was supported by the U.S. Army Research Laboratory, The national Science Foundation (ECCs-1120832, DMR-1311866), and the Judson Swearingen Regents Chair in Engineering at the University of Texas at Austin.

REFERENCES

[1] R. Tavarti, A. Wibowo, G. Martin, F. Tuminello, C. Youtsey, G. Hiller, N. Pan, M.W. Wanlass, M. Romero, InGaP/GaAs/InGaAs inverterted metamorphic (IMM) solar cells on 4'' epitaxial lifted off (ELO) wafers, in: Proceedings of the 35th IEEE PVSC Conference, 2010, pp. 002125-002128

[2] K.M. Trautz, P.P. Jenkins, R.J. Walters, D. Scheiman, R. Hoheisel, R. Tatavarti, R. Chan, H. Miyamoto, J.G.J. Adams, V.C. Elarde, J. Grimsley, Mobile solar power, IEEE J. Photovolt. 3 (1) (2013) 535-541

[3] Shir, D.; Yoon, J.; Chanda, D.; Ryu, J. H.; Rogers, J. A. Nano Lett. 2010, 10 (8) 3041– 3046

[4] Gates, B. D.; Xu, Q. B.; Stewart, M.; Ryan, D.; Willson, C. G.; Whitesides, G. M. Chem. Rev. 2005, 105 (4) 1171– 1196.

[5] Rybczynski J, Ebels U, Giersig M. Large-scale, 2D arrays of magnetic nanoparticles. Colloids and Surfaces A: Physicochemical and Engineering Aspects 2003; 219: 1–6

[6] Li, Xiaohan, et al. "Integration of subwavelength optical nanostructures for improved antireflection performance of mechanically flexible GaAs solar cells fabricated by epitaxial lift-off." *Solar Energy Materials and Solar Cells* 143 (2015): 567-572.

[7] Li, Xiaohan, et al. "Subwavelength nanostructures integrated with polymer-packaged iii–v solar cells for omnidirectional, broad-spectrum improvement of photovoltaic performance." *Progress in Photovoltaics: Research and Applications* 23.10 (2015): 1398-1405.

[8] Scriven, L. E., and C. V. Sternling. "The marangoni effects." *Nature* 187.4733 (1960): 186-188.

Development of Back Surface Texture for Light Management in Epitaxial Lift Off (ELO) Quantum Dot Solar Cells

Brittany L. Smith[1], George T. Nelson[1], Yushuai Dai[1], Michael A. Slocum[1], Andree Wibowo[2], Rao Tatavarti[2], Seth M. Hubbard[1]

[1]Rochester Institute of Technology, Rochester, NY 14623, USA, [2]Microlink Devices, Niles, IL 60714, USA

Abstract— **Quantum dot (QD) solar cells with light management allow for increased QD absorption without the need for excessive strain balancing. Single-junction GaAs cells with 10 QD layers are fabricated with back surface reflectors (BSRs) and removed from the substrate by epitaxial lift off. Devices with different BSRs will be compared: a flat mirror, a periodic texture that varies in one dimension, a periodic texture that varies in two dimensions, and a random texture achieved by crystallographic etch. Initial results show a 40% increase in sub-band current when comparing devices with a 1-D texture to the flat mirror. The final paper will present development and device results from the other two BSR types.**

Index Terms — **epitaxial lift off, GaAs, intermediate band solar cell, light trapping, photovoltaic cell, quantum dot**

I. INTRODUCTION

Quantum dots (QDs) have been applied in photovoltaics to enhance both single junction and multijunction cell performance, since QDs can increase short circuit current (J_{sc}) and improve current-matching [1, 2]. QDs are also an avenue to realizing the intermediate band solar cell, which could exceed conventional single-junction limits with a theoretical maximum efficiency of 44.5% at one sun and over 60% under concentration [3].

InAs QDs in GaAs are well-developed and have been extensively characterized which makes them a useful, known material system for this study. A GaAs cell with InAs QDs has the potential to approach a theoretical maximum of 38% efficiency [3]. The efficiency of a single-junction GaAs cell has been shown to increase by 0.5% with the addition of a 40 period QD superlattice as compared to a control [2]. However, additional QD periods would be needed to further increase sub-band absorption and short circuit current, due to low surface coverage of QDs. Increasing QD periods presents significant challenges regarding growth time as well as strain management, which would negatively impact the open circuit voltage (V_{oc}).

Light management represents a feasible alternative to increase absorption in the QD region of the cell. This would function with a thin device removed from its substrate so that free carriers would not be absorbed and lost in the thick substrate. This could be achieved via an epitaxial lift off (ELO) techniques which permits access to the back of the cell and eliminates the substrate. Historically, ELO technology has been applied to single junction and multijunction cells [4].

Light management in quantum dot solar cells (QDSCs) has been demonstrated using plasmonics as well as gold back surface mirrors [5, 6]. Using a texture to randomize the angles at which the light is reflected can increase the proportion of the light that is totally internally reflected and increase the optical path length. Recent studies on mutiple quantum well solar cells have demonstrated a crystallographic etch to be successful in creating either a periodic or random texture in the rear of the cell which are effective light-trapping structures for this reason [7-9].

This study presents the effects of a textured back surface reflector (BSR) on QDSC performance in comparison to a flat back surface mirror. Experimental and simulated device performance will be evaluated. Growth conditions, layer thicknesses in devices, and geometry of the BSR were optimized.

II. EXPERIMENTAL

Epitaxial layers were grown in a 3x2" Aixtron close-couple showerhead metal organic vapor phase epitaxy (CCS-MOVPE) reactor located at RIT. GaAs cells with a 10 layer QD superlattice were grown inverted on substrates with epitaxial lift off (ELO) layers. ELO layers are thin sacrificial layers grown on the substrate before cell growth that can be selectively etched in order to remove the substrate after the rear of the cell has been metallized and attached to a rigid handle.

The GaAs cells in this study were *nip* configuration with an InGaP window and back surface field. The emitter was 50 nm thick and the base was 2000 nm in order for the cell to be an optically thick absorber. GaAs QD cell growth conditions have been described previously in reference [10]. Strain balancing in the superlattice was achieved by growing a 4 nm GaAs spacer on top of the QDs and wetting layer, a layer of GaP with an optimized thickness as described in [11], and another 4 nm GaAs spacer on which to grow the next QDs.

For back patterning, 4 μm of AlGaAs were grown on the back of the cells. A schematic of cells in this study are shown in Figure 1. One cell was fabricated with a flat back surface to act as a mirror, while the other was textured. Both cells had Cr/Au deposited as the rear reflective material. The back surface texture was created by applying a positive photoresist to the reverse side of the cell and using a dark field

978-1-5090-5606-4/17 $31.00 © 2017 IEEE

photolithography mask with 2 μm thick clear lines separated by 4 μm of chrome. The 2-μm-thick lines of exposed AlGaAs were etched with a 1:1:75 solution of $NH_4OH:H_2O_2:H_2O$ for 14 minutes to achieve a triangular texture or 19 minutes for a rounded texture. ELO and fabrication of the solar cells was then performed. Spectral response data was taken with an Oriel IQE200 monochromator and a Stanford Research SR570 preamplified coupled to a SR830 lock-in amplifier.

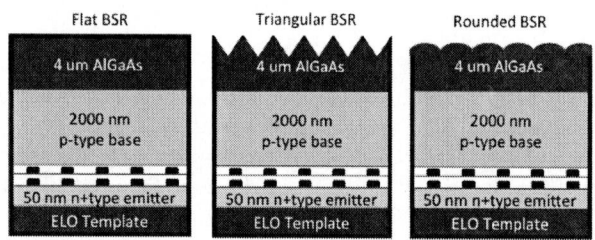

Fig. 1.   Schematic of GaAs QD solar cells with 3 different back surface reflectors: a flat mirror, a periodic triangular texture, and a periodic rounded texture

QD test structures with QDs on the top surface were grown in order to characterize dot size and density. These test structures were grown with conditions identical to the dots embedded in the cells for this study. Atomic force microscopy (AFM) images were taken of the surface layer of dots on these test structures. The density was ~3 x 10^{10} dots/cm^3, and the average height and diameter were 2 and 21 nm, respectively. The dot size and density are consistent with previously successful studies and exhibit low coalescence which would degrade V_{oc} [2, 6].

III. 1-D BACK SURFACE REFLECTORS

A. Back Surface Patterning

The textured BSR etch process described in the experimental section resulted in a uniform texture across the back surface of the cells. Figure 2 shows scanning electron microscope (SEM) images of the textures. The left image is the 14 minutes etch, which displayed a period of 6 um and a pitch of 2.5 um. The right image is of the 19 minutes etch, which displayed a period of 6 um and a pitch of 2 um.

These textures were characterized using a VASE ellipsometer to measure reflectance as a function of detection angle, using an angle of incidence = 10° to normal. Reflectance data for each texture is shown in Figure 3. The triangular texture exhibits more interference fringes as compared to the rounded texture. Furthermore, the angle for total internal reflection (TIR) in GaAs is 16.6°, and the amount of light reflected below this angle is greater for the rounded texture. More specifically, 15.4% is reflected below the TIR angle for the triangular texture, whereas 39.7% of light is reflected below the TIR angle for the rounded texture.

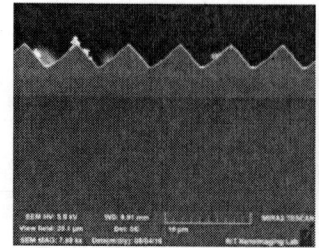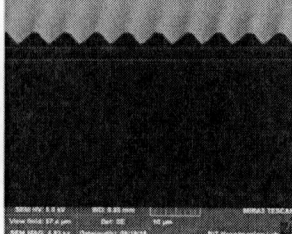

Fig. 2.   SEM images of 1-D period texture BSRs: left is "triangular" with 6 μm period and 2.5 μm pitch, right is "rounded" with 6 μm period and 2 μm pitch.

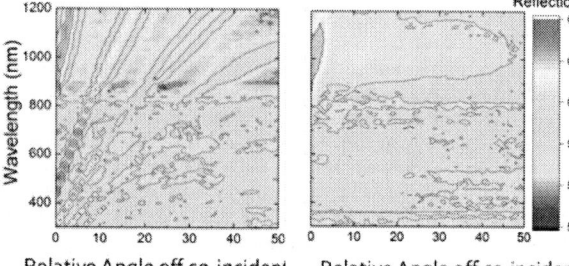

Fig. 3.   Angle-dependent reflectance off the triangular back surface texture (left) and the rounded back surface texture (right), angle of incidence = 15°

B. Devices

The devices shown in Fig. 1 were grown, fabricated and characterized. External quantum efficiency (EQE) data below the GaAs band edge is shown in Figure 4. The sub-band current calculated from integrating the spectral response data is tabulated in Table 1. The current from the rounded texture BSR represents a 40% increase over the current from the flat BSR. Assuming the flat BSR corresponds to an optical path length equal to 2, this indicates the optical path length using the rounded BSR approaches 2.8. However, given that an upright cell with no back reflector from a previous experiment showed a sub-band current of 0.26 mA/cm^2, it appears a flat BSR corresponds to less than 2 passes. This may be due to non-ideal reflectivity of the rear metal contact. It should also be noted that the rounded texture exhibits the greatest EQE near the GaAs band-edge which accounts for its increased sub-band current.

TABLE I

INTEGRATED SUB-BAND SPECTRAL RESPONSE

Device	Current (mA/cm²)
Flat BSR	0.43
Triangular BSR	0.53
Rounded BSR	0.60

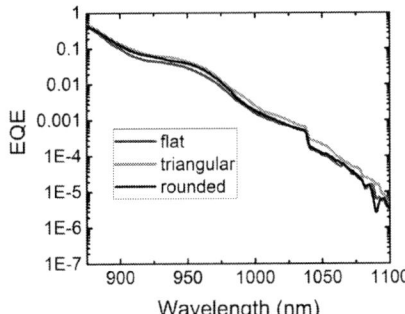

Fig. 4. Sub-band EQE from GaAs QDSCs with different back reflectors: rounded texture shows highest EQE near GaAs band edge

IV. IMPROVED BACK SURFACE REFLECTORS

A. Development of 2-D Textures

In order to approach the Lambertian limit, it is necessary to have a back surface texture that varies in more than one direction. Two processes to create such a texture have been evaluated. The first is a maskless crystallographic etch to achieve a random texture which has previously been demonstrated in GaAs in [8]. Preliminary results display a texture in GaAs with micron-size features, as shown in the scanning electron microscopy (SEM) images in Figure 5, which was etched in 1:4:80 $NH_4OH:H_2O_2:H_2O$ at 5°C for 25 minutes. Roughness of the texture was 60 nm rms, resulting in specular reflectivity less than 5% across the visible and IR range (Figure 6). While the overall reflectivity was similar to the 1D triangular etch, scattered light for the random texture was highly diffuse. Two sets of cells were grown inverted with 4 μm of GaAs as the final layer. One set of cells was textured by maskless etch, and the other serves as a control.

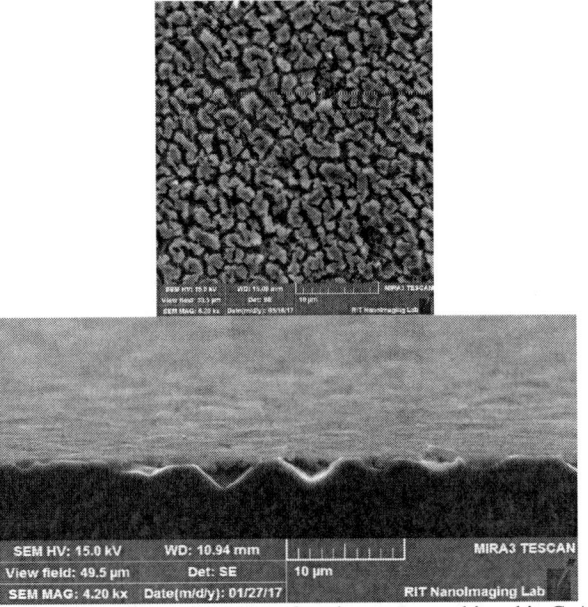

Fig. 5. *(top)* Plan-view SEM of random texture achieved in GaAs by maskless etch 1:4:80 $NH_4OH:H_2O_2:H_2O$ at 5°C. *(bottom)* Cross-

sectional SEM of maskless etch

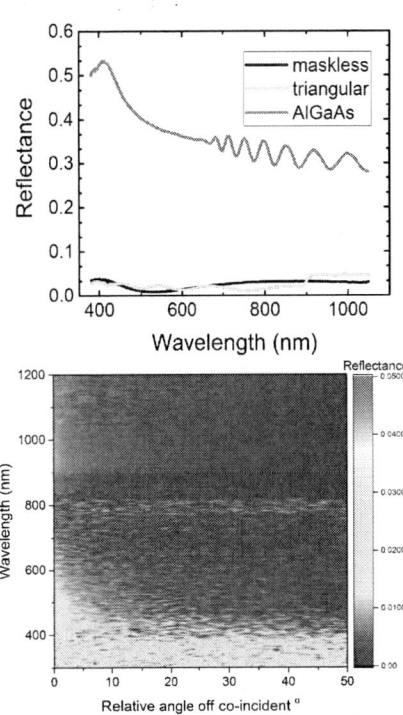

Fig. 6. *(top)* Specular reflectance of the maskless texture as compared to a bare wafer and the 1D triangular etch. *(right)* Reflectance of the maskless texture as a function of wavelength and detector angle, angle of incidence = 15°

The second approach used a Heidelberg DWL66+ direct write laser lithography system to write a mask directly onto AlGaAs to create a 2-D pattern. Two patterns were evaluated: a grid of 4 μm x 4 μm squares separated by 1 μm alleys, similar to the texture reported in [9], and a checkerboard pattern of 4 μm x 4 μm squares.

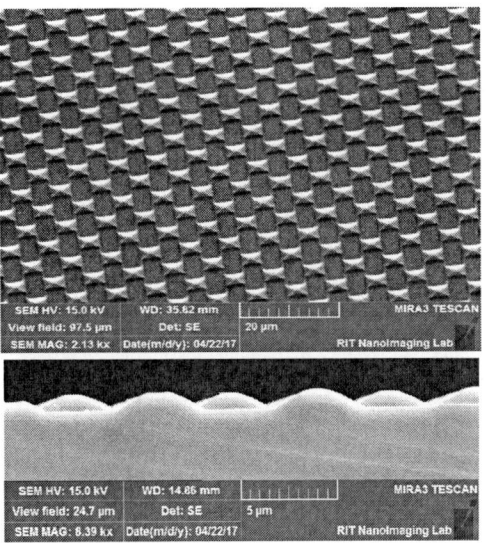

Fig. 7. *(top)* Plan-view SEM of texture achieved in GaAs by direct-write lithography of checkerboard pattern and etch in 1:1:75 $NH_4OH:H_2O_2:H_2O$ *(bottom)* Cross-sectional SEM of checkerboard texture with features 1 µm in height

Each of these patterns were etched for 7 minutes in 1:1:75 $NH_4OH:H_2O_2:H_2O$ at room temperature. Angle-dependent reflectance measurements of both textures were evaluated, and it was determined that the checkerboard texture exhibited lower reflectance across the entire range which indicates that light is being scattered to a greater degree in off-axis directions. For this reason, the checkerboard pattern was used on two sets of cells. Plan-view and cross-sectional SEM images of the checkerboard texture are shown in Figure 7.

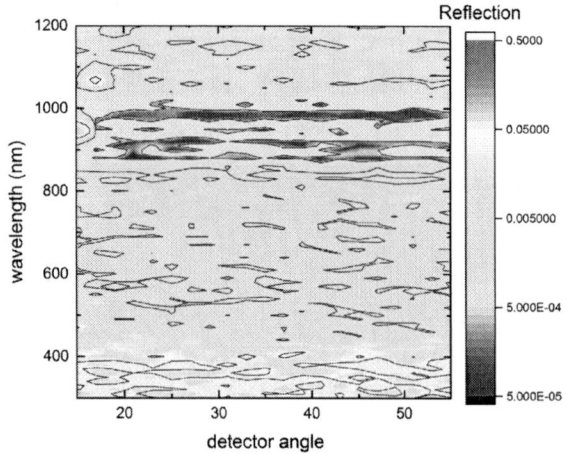

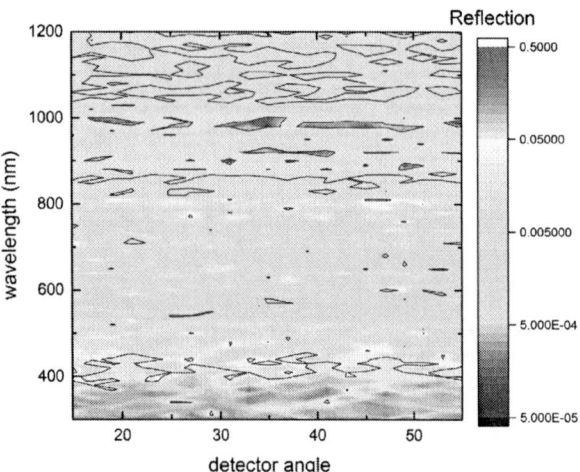

Fig. 8. Reflectance of *(top)* checkerboard and *(bottom)* grid texture as a function of wavelength and detector angle, angle of incidence = 15°

B. Dielectric on BSR Texture

In an effort to enhance the benefits of a BSR, the insertion of a dielectric between the textured back surface of GaAs cell and the reflective metal contact was simulated using the transfer matrix method and TFCalc™ software. The results are shown in Figure 9, where significant benefits are observed for a textured BSR. This has been experimentally demonstrated in [7]. However, due to the ELO process in this study, any oxide or dielectric could be attacked during the lift off etch. One set of cells with the checkerboard texture had 450 nm of SiO_2 deposited on the BSR texture. This oxide was then selectively removed around the perimeter of the wafer and at regular points across the texture by protecting the oxide with photoresist, exposure using a mask with a grid of 2 mm x 2 mm squares spaced 1 cm apart, and then etching the exposed oxide in HF. The photoresist was then removed, and metal was deposited for rear contact to the devices. The metal was intentionally deposited to be greater than 500 nm thick in order to protect oxide sidewalls.

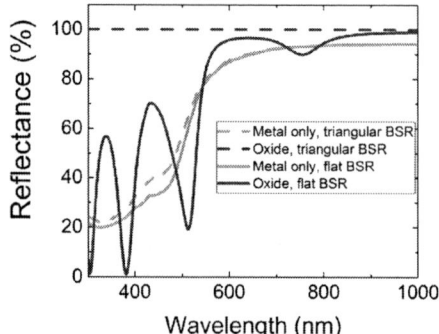

Fig. 9. Simulation of GaAs cell performance as a function of cell thickness, for different back surface configurations. Use of a dielectric enhances cell performance with a textured BSR.

V. CONCLUSIONS

Devices with periodic rear surface texturing showed a 40% EQE enhancement as compared to devices with a flat back surface. Development of rear surface textures that vary in more than one dimension were created by both a chilled maskless etch and direct-write lithography system. Angle-dependent reflectance measurements indicate the maskless etching results in the most diffuse scattering, and the checkerboard pattern created by the direct-write system showed more scattering than a grid pattern. Cells were grown and fabricated with both the maskless texture and the checkerboard texture. Finally, SiO_2 was deposited on the checkerboard texture and selectively removed to enhance reflectivity while enabling rear contact to devices.

ACKNOWLEDGEMENTS

This work was supported by the Air Force Research Laboratory, Space Vehicles Directorate, through STTR FA9453-15-C-0404.

REFERENCES

[1] Kerestes, C., Polly, S., Forbes, D., Bailey, C., Podell, A., Spann, J., Patel, P., Richards, B., Sharps, P. and Hubbard, S., 2014. Fabrication and analysis of multijunction solar cells with a quantum dot (In) GaAs junction. *Progress in Photovoltaics: Research and Applications*, 22(11), pp.1172-1179.

[2] Bailey, C. G.; Forbes, D. V.; Polly, S. J.; Bittner, Z. S.; Dai, Y.; Mackos, C.; Raffaelle, R. P.; Hubbard, S. M., Open-Circuit Voltage Improvement of InAs/GaAs Quantum-Dot Solar Cells Using Reduced InAs Coverage. *Photovoltaics, IEEE Journal of 2012*, 2 (3), 269-275.

[3] Wei, G., Shiu, K.T., Giebink, N.C. and Forrest, S.R., 2007. Thermodynamic limits of quantum photovoltaic cell efficiency. *Applied Physics Letters*, 91(22), p.223507.

[4] Tatavarti, R., Hillier, G., Martin, G., Wibowo, A., Navaratnarajah, R., Tuminello, F., Hertkorn, D., Disabb, M., Youtsey, C., McCallum, D. and Pan, N., 2009, June. Lightweight, low cost InGaP/GaAs dual-junction solar cells on 100 mm epitaxial liftoff (ELO) wafers. In Photovoltaic Specialists Conference, 2009 34th IEEE (pp. 2065-2067).

[5] Lu, H. F.; Mokkapati, S.; Fu, L.; Jolley, G.; Tan, H. H.; Jagadish, C., Plasmonic quantum dot solar cells for enhanced infrared response. Applied Physics Letters 2012, 100 (10), 103505.

[6] Bennett, M. F.; Bittner, Z. S.; Forbes, D. V.; Tatavarti, S. R.; Ahrenkiel, S. P.; Wibowo, A.; Pan, N.; Chern, K.; Hubbard, S.

M., Epitaxial lift-off of quantum dot enhanced GaAs single junction solar cells. Applied Physics Letters 2013, 103, 213902.

[7] Inoue, T.; Watanabe, K.; Toprasertpong, K.; Fujii, H.; Sugiyama, M.; Nakano, Y., Enhanced Light Trapping in Multiple Quantum Wells by Thin-Film Structure and Backside Grooves With Dielectric Interface. Journal of Photvoltaics 2015, 5, 697-703.

[8] Watanabe, K.; Inoue, T.; Sodabanlu, H.; Sugiyama, M.; Nakano, Y. Self-organized texture of GaAs by wet etching for light trapping in MQWS solar cells, In *31st European Photovoltaic Solar Energy Conference*, Hamburg, Germany, 2015; pp 181-184.

[9] Watanabe, K., Kim, B., Inoue, T., Sodabanlu, H., Sugiyama, M., Goto, M., & Nakano, Y. (2014). Thin-film InGaAs/GaAsP MQWs solar cell with backside nanoimprinted pattern for light trapping. *IEEE Journal of Photovoltaics*, 4(4), 1086-1090.

[10] Smith, B.L., Slocum, M.A., Bittner, Z.S., Dai, Y., Nelson, G.T., Hellstroem, S.D., Tatavarti, R. and Hubbard, S.M., 2016, June. Inverted growth evaluation for epitaxial lift off (ELO) quantum dot solar cell and enhanced absorption by back surface texturing. In *Photovoltaic Specialists Conference, 2016 IEEE 43rd* (pp. 1276-1281). IEEE.

[11] Bailey, C.G., Hubbard, S.M., Forbes, D.V. and Raffaelle, R.P., 2009. Evaluation of strain balancing layer thickness for InAs/GaAs quantum dot arrays using high resolution x-ray diffraction and photoluminescence. *Applied Physics Letters*, 95(20), p.203110.

Enabling High-Efficiency InAs/GaAs Quantum Dot Solar Cells by Epitaxial Lift-Off and Light Management

F. Cappelluti*, A. P. Cédola*, A. Khalili*, Farid Elsehrawy*, G. Bauhuis[†], P. Mulder[†], J. Schermer[†],
G. Bissels[‡], T. Aho[§], T. Niemi[§], M. Guina[§], D. Kim[¶], J. Wu[¶], H. Liu[¶]

*Department of Electronics and Telecommunications, Politecnico di Torino, Italy
[†]Institute for Molecules and Materials, Radboud University, Nijmegen, The Netherlands
[‡]tf2 devices B.V., Nijmegen, The Netherlands
[§]Optoelectronics Research Centre, Tampere University of Technology, Tampere, Finland
[¶]Department of Electronic and Electrical Engineering, University College London, London, United Kingdom

Abstract—We report thin-film InAs/GaAs QD solar cells fabricated by epitaxial lift-off of 3-inch wafers containing QD epi-structures with high in-plane QD density. External quantum efficiency measurements demonstrate enhanced QD harvesting in the thin-film configuration. Numerical simulations show that remarkably high increase of the QD photocurrent may be achieved by replacing the planar rear mirror with micro-structured photonic gratings. Measurements of diffraction efficiency of grating prototypes realized on GaAs wafers by nanoimprint lithography are presented.

Index Terms—solar cell, epitaxial lift-off, thin-film, quantum dot, light trapping

I. INTRODUCTION

Quantum dots are very attractive band-engineered materials for the development of high-efficiency single-junction and multi-junction solar cells. Studies of quantum dot solar cells (QDSCs) were motivated by the idea of taking advantage of sub-bandgap transitions to absorb otherwise wasted low energy photons and improve short circuit current [1]. If properly engineered, QDSCs can be exploited to realize the intermediate band operation [2]. Despite their great potential, the improvement in conversion efficiency, if any, with QDs is still marginal. The most challenging issue to demonstrate high-efficiency QDSCs is achieving at the same time a large increase of the short circuit current (J_{sc}) and preservation of high open circuit voltage (V_{oc}). In QDSCs, regardless of their operating regime (thermally-limited or intermediate band), increasing the number of QD layers or the QD density, besides being technologically challenging, causes a larger V_{oc} penalty. Thus, a trade-off exists between maximizing the J_{sc} and conserving the V_{oc}. On the other hand, implementing light-trapping schemes to strongly enhance the QD photon harvesting can yield efficiency comparable to or higher than the one of single-junction cells even for thermally-limited QDSCs [3], and enable the demonstration of QDSCs operating in the intermediate band regime [4].

In this work, we report our recent results in the development of high-efficiency thin-film GaAs cells integrating QDs and photonic nanostructures. These devices can be fabricated using cost-effective and scalable fabrication processes such as Epitaxial Lift-Off [5] for the thin-film processing and NanoImprint Lithography (NIL) to pattern even sub-wavelength period gratings over large areas [6].

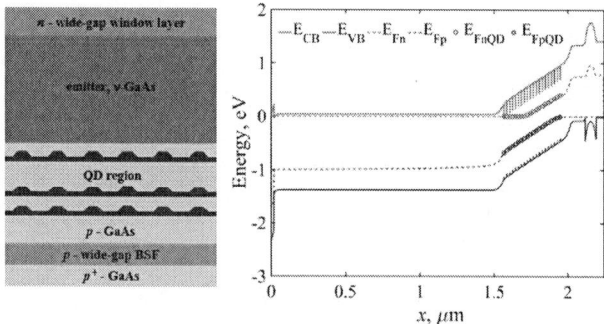

Fig. 1. Cross-section of the analysed QD-based solar cell and calculated energy band diagram at short-circuit condition under AM1.5G illumination.

II. THEORETICAL BACKGROUND AND EXPERIMENT

The photovoltaic performance of InAs/GaAs QDSCs are studied within a drift-diffusion framework accounting for the peculiar carrier transfer dynamics between QD states and transport bands [7]. Such cells work in the thermally limited operation, i.e. the second photon absorption is negligible with respect to the thermal escape of carriers from QD states to transport bands, and the V_{oc} is limited by the radiative recombination through the QD ground-state [7]. Such a penalty, which indicatively ranges between 50-250 mV depending on the QD size and effective Ground State (GS) bandgap [8], [7], [9], can be substantially reduced by partially charging the QD states and inhibiting the carrier capture and subsequent recombination through the GS [10], [11], [12]. A further advantage of QD charging is the passivation of defects induced by the strained epitaxy, thus limiting the V_{oc} degradation due to non radiative recombination in real QD solar cells [12]. By combining QD charging and light-trapping, QDSCs with efficiency higher than their bulk counterpart are theoretically feasible [3], [13].

In this work, we adopt a device design based on a deep junction structure with lightly doped emitter and thin base, as shown in Fig.1. The cell behavior is studied for different photonic configurations: a) wafer-based cell (single-pass); b) thin-film cell with planar metallic reflector at the bottom (double-pass); c) thin-film cell with light-trapping scheme yielding in

978-1-5090-5606-4/17 $31.00 © 2017 IEEE

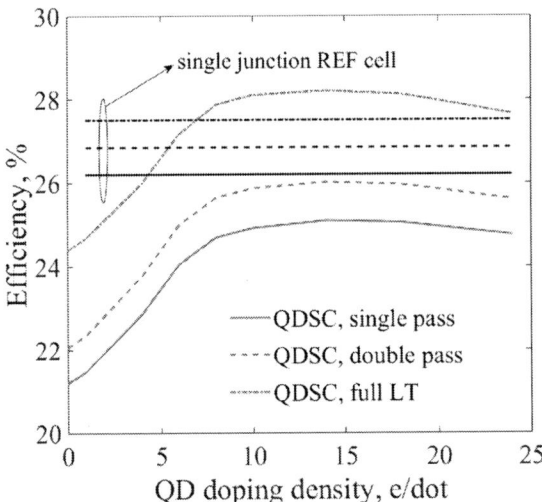

Fig. 2. Efficiency of the QD cell vs. QD doping density for different photonic configurations (see text). The horizontal lines show the efficiency of the reference single-junction cells.

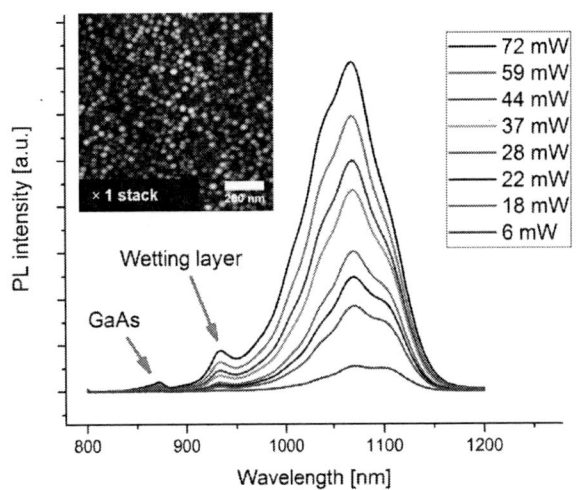

Fig. 3. AFM image of one stack of InAs/GaAs QDs with high in-plane density and photoluminescence spectra at T=300 K and different excitation powers.

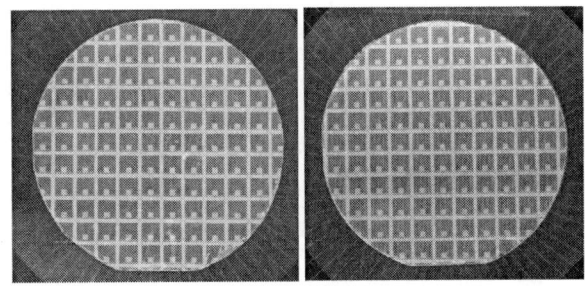

Fig. 4. Single-junction GaAs (left) and QD InAs/GaAs (right) thin-film cells from epitaxial lift-off of 3-inch diameter epi-structures.

the weak absorption limit an optical path enhancement of $2n^2$, n being the GaAs refractive index [3]. QDs are described as a three level system characterized by energy levels identified by the measured photoluminescence spectra (GS bandgap = 1.17 eV) and radiative lifetime of 1 ns [7], [12]. Doping-dependent SRH recombination lifetime of minority carriers is set to 500 ns in the emitter and 20 ns in the base. Fig. 2 shows an overview of the impact of the various photonic configurations and of QD doping (implemented through direct doping) on the achievable efficiency of the QD cells. The efficiency of the corresponding single-junction GaAs cells is also shown. A significant efficiency increase is achieved in both single-junction and QD thin-film cells with respect to their wafer counterpart. Moreover, in the light-trapping configuration, an absolute improvement of about 0.7% is calculated for the thin-film QD cell with optimum doping level (around 14 e/dot) with respect to the thin-film single junction cell.

The fabricated cells include high in-plane density (over 8×10^{10} cm^{-2}, see Fig. 3) QD layers fabricated through the Sb-mediated QD growth technique [14]. Three 3-inch wafers, one for wafer-based processing of the single-junction baseline cell and two for ELO thin-film processing of single-junction and QD solar cells were grown. 3-inch films were released by ELO and processed into cells (area=0.25 cm^2) with planar gold mirror on the backside, as shown in Fig. 4. No ARC was used.

III. RESULTS AND DISCUSSION

Measured current-voltage characteristics are reported in Fig. 5. The thin-film single-junction and QD cells show comparable V_{oc}, with a reduction with respect to the wafer-based cell which is attributed to lattice defects propagating from the substrate to the epilayers. The dark current of all the thin-film cells is in fact dominated by the presence of a diode-like

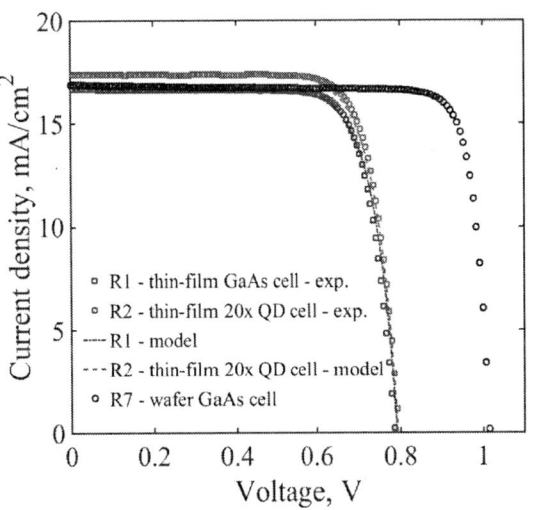

Fig. 5. Current-Voltage characteristics of wafer-based regular GaAs cell and thin-film GaAs and QD cells measured under AM1.5G illumination.

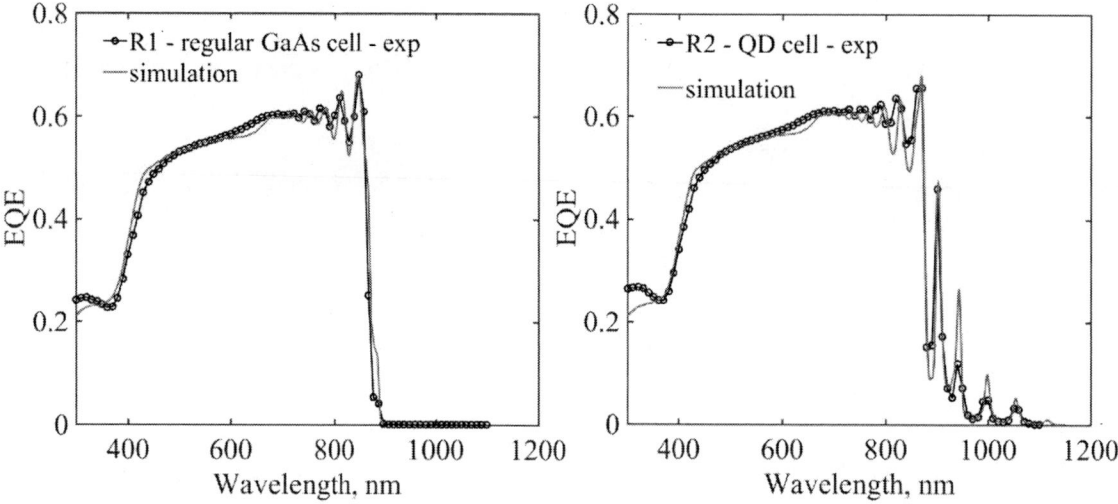

Fig. 6. Measured and simulated EQE spectra of thin-film single-junction GaAs solar cells (left) and thin-film QD solar cells (right).

shunt defect, with reverse saturation current density of about 4 nA/cm² and ideality factor of 2.

External Quantum Efficiency (EQE) spectra of the thin-film cells are reported in Fig. 6, showing the typical extension of the EQE in the long wavelength range due to the insertion of QDs: for $\lambda > 880$ nm, the wafer QDSC - not shown here - provided a short-circuit current of about 0.4 mA/cm², which increases to 0.6 mA/cm² in the thin-film QDSC.

By moving to a micro-structured grating realized on the bottom of the cell, the QD absorbance can be enhanced more significantly. In particular, we are studying diffraction grating configurations where light-trapping is pursued by coupling light into high order diffraction modes propagating outside of the cell escape cone. These gratings require a period length larger than the incident wavelength. Several designs were studied through simulations based on the Rigorous Coupled-Wave Analysis (RCWA) method. The QD photogenerated current was studied as a function of the grating period and height. From the results, the relevant parameter for light-trapping optimization turned to be the aspect ratio (height/period), whose optimum ranges between 0.32 and 0.36 for grating periods between 2-3 μm [15]. By exploiting optimized grating configurations, the simulated absorbance spectra (Fig. 7) show a remarkable increase at the GaAs band edge and in the QD range. With pyramidal gratings, an enhancement of QD photocurrent of about 13 times with respect to the wafer-based configuration is predicted.

Prototype gratings (Fig. 8(a)) were fabricated on a GaAs wafer and diffraction efficiency measured by the scanning angle detector technique (Fig. 8(b)). A comparison between measured and simulated diffraction efficiency of one prototype is reported in Fig. 8(c)-(d) showing good agreement. Significant power coupling to the first two diffraction orders is found. Further optimization of the grating can be carried

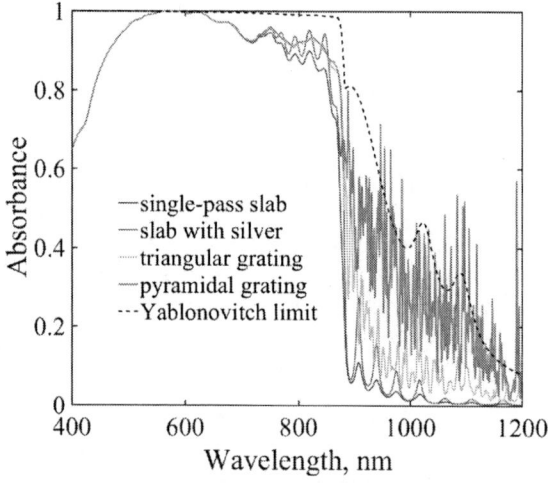

Fig. 7. Calculated absorbance spectra for different photonic configurations. The dashed line indicates the classical Yablonovitch limit [16].

out in order to better match the diffraction spectrum to the QD absorption spectrum. Experimental demonstration of light trapping enhancement by integration of a rear diffraction grating in the thin-film QDSC is ongoing.

IV. ACKNOWLEDGEMENTS

This project has received funding from the European Union's Horizon 2020 research and innovation programme under grant agreement No 687253, www.tfqd.eu.

REFERENCES

[1] V. Aroutiounian, S. Petrosyan, A. Khachatryan, and K. Touryan, "Quantum dot solar cells," *JAP*, vol. 89, no. 4, pp. 2268–2271, 2001.
[2] A. Luque and A. Martí, "Increasing the efficiency of ideal solar cells by photon induced transitions at intermediate levels," *Phys. Rev. Lett.*, vol. 78, no. 26, pp. 5014–5017, 1997.

978-1-5090-5606-4/17 $31.00 © 2017 IEEE

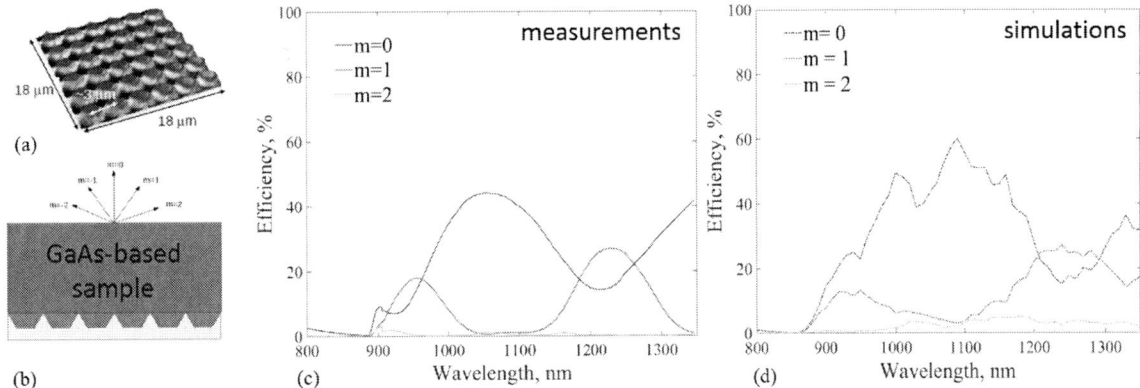

Fig. 8. Profilometer image of back grating of 3 μm period fabricated by NIL (a) and sketch of the prototype samples realized on GaAs wafer (b). Measured (c) and simulated (d) diffraction efficiency.

[3] F. Cappelluti, M. Gioannini, G. Ghione, and A. Khalili, "Numerical study of thin-film quantum-dot solar cells combining selective doping and light-trapping approaches," in *2016 IEEE 43rd Photovoltaic Specialists Conference (PVSC)*, June 2016, pp. 1282–1286.

[4] A. Mellor, A. Luque, I. Tobías, and A. Martí, "The feasibility of high-efficiency inas/gaas quantum dot intermediate band solar cells," *Solar Energy Materials and Solar Cells*, vol. 130, pp. 225–233, 2014.

[5] G. Bauhuis, P. Mulder, E. Haverkamp, J. Huijben, and J. Schermer, "26.1% thin-film gaas solar cell using epitaxial lift-off," *Solar Energy Materials and Solar Cells*, vol. 93, no. 9, pp. 1488–1491, 2009.

[6] J. Tommila, V. Polojärvi, A. Aho, A. Tukiainen, J. Viheriälä, J. Salmi, A. Schramm, J. Kontio, A. Turtiainen, T. Niemi *et al.*, "Nanostructured broadband antireflection coatings on alinp fabricated by nanoimprint lithography," *Solar Energy Materials and Solar Cells*, vol. 94, no. 10, pp. 1845 – 1848, 2010.

[7] M. Gioannini, A. Cedola, N. Di Santo, F. Bertazzi, and F. Cappelluti, "Simulation of quantum dot solar cells including carrier intersubband dynamics and transport," *IEEE J. Photovoltaics*, vol. 3, no. 4, pp. 1271–1278, Oct 2013.

[8] G. Jolley, L. Fu, H. Lu, H. H. Tan, and C. Jagadish, "The role of inter-subband optical transitions on the electrical properties of ingaas/gaas quantum dot solar cells," *Prog. Photovolt: Res. Appl.*, 2012.

[9] C. G. Bailey, D. V. Forbes, R. P. Raffaelle, and S. M. Hubbard, "Near 1 v open circuit voltage inas/gaas quantum dot solar cells," *APL*, vol. 98, no. 16, p. 163105, 2011.

[10] P. Lam, S. Hatch, J. Wu, M. Tang, V. G. Dorogan, Y. I. Mazur, G. J. Salamo, I. Ramiro, A. Seeds, and H. Liu, "Voltage recovery in charged inas/gaas quantum dot solar cells," *Nano Energy*, vol. 6, pp. 159 – 166, 2014.

[11] S. Polly, D. Forbes, K. Driscoll, S. Hellstrom, and S. Hubbard, "Delta-doping effects on quantum-dot solar cells," *IEEE J. Photovoltaics*, vol. 4, no. 4, pp. 1079–10 857, 2014.

[12] F. Cappelluti, M. Gioannini, and A. Khalili, "Impact of doping on inas/gaas quantum-dot solar cells: a numerical study on photovoltaic and photoluminescence behavior," *Solar Energy Materials and Solar Cells*, vol. 157, pp. 209 – 220, 2016.

[13] F. Cappelluti, G. Ghione, M. Gioannini, G. Bauhuis, P. Mulder, J. Schermer, M. Cimino, G. Gervasio, G. Bissels, E. Katsia *et al.*, "Novel concepts for high-efficiency lightweight space solar cells," in *E3S Web of Conferences*, vol. 16. EDP Sciences, 2017, p. 03007.

[14] F. Tutu, J. Wu, P. Lam, M. Tang, N. Miyashita, Y. Okada, J. Wilson, R. Allison, and H. Liu, "Antimony mediated growth of high-density inas quantum dots for photovoltaic cells," *Applied Physics Letters*, vol. 103, no. 4, p. 043901, 2013.

[15] A. Musu, F. Cappelluti, T. Aho, V. Polojärvi, T. K. Niemi, and M. Guina, "Nanostructures for light management in thin-film gaas quantum dot solar cells," in *Light, Energy and the Environment*. Optical Society of America, 2016, p. JW4A.45.

[16] E. Yablonovitch and G. D. Cody, "Intensity enhancement in textured optical sheets for solar cells," *Electron Devices, IEEE Transactions on*,

Characterization of Arsenic Doped CdTe Layers and Solar Cells

Sachit Grover, Xiaoping Li, Wei Zhang, Ming Yu, Gang Xiong, Markus Gloeckler, Roger Malik

First Solar Inc., Santa Clara, CA

Abstract — **Polycrystalline CdTe solar cells with hole carrier-density greater than 10^{16} cm^{-3} have been realized by doping the absorber with Arsenic. The high carrier concentration requires a large excess of Arsenic in the film implying only percent level doping activation. Appearance of a sub-bandgap peak in the photoluminescence (PL) spectrum is linked with the activation of the Arsenic dopant. Low temperature PL measurements confirm an acceptor level at 90 meV from the valence band, which is consistent with reported literature values. Additional features in the PL spectra suggests the presence of sub-band defects. Understanding the origin for these features may enable the high efficiency potential of group-V doping.**

I. INTRODUCTION

First Solar has demonstrated research cell efficiency above 22% for polycrystalline cadmium telluride (px-CdTe) technology. [1] Transfer of research cell improvements to high volume manufacturing has enabled CdTe modules to be at par in efficiency with silicon based modules while sustaining their low-cost advantage. [2] While there is significant room to further improve the CdTe solar-cell-efficiency towards 25%, [3] improvement of open circuit voltage (V_{OC}) for px-CdTe cells has been slow and saturated near 900 mV. [4] Majority-carrier density can have a significant impact on the V_{OC}, [5] however the hole-concentration of copper doped px-CdTe is limited by compensating defect complexes. [6] Doping CdTe with group-V elements is an alternative to copper doping that may permit higher hole-concentration thereby improving V_{OC}.

P-type doping of CdTe can be achieved by substituting the tellurium (Te) site with a group-V atom. Significant progress has been made in the understanding of doping methodology and defect characteristics for phosphorus doped CdTe [7] and Arsenic doped CdTe. [8] In addition $V_{OC} > 1V$ has been demonstrated for sx-CdTe:P based solar cells. [9] Polycrystalline-(px)-CdTe films with Arsenic doping have been grown using MOCVD and made into solar cells with V_{OC} close to 800 mV. [10] Group-V elements can also be introduced in px-CdTe through post-growth high-temperature diffusion [11] or through direct incorporation during CdTe deposition. [12, 13] Here we report on characterization of px-CdTe:As devices that are compatible with First Solar's device architecture and high volume manufacturing. [2]

II. EXPERIMENT

Arsenic doped CdTe films are annealed in a CdCl$_2$ containing atmosphere and finished into solar cells by application of a back-contact. Integrated devices are measured for light-IV characteristics, carrier concentration by capacitance-voltage (CV) and quantum efficiency (QE). PL measurements and low-temperature (LTPL) measurements are carried out using red lasers.

III. RESULTS

Carrier concentration (N_a) vs. depletion width (W) at zero bias is plotted for more than 500 devices in Fig. 1. $N_a > 10^{16}$ cm^{-3} have been consistently achieved with Arsenic doping, exceeding what is possible with Cu doping by at least one order of magnitude. The inset shows representative N_a vs. W curves where the carrier concentration is above 10^{16} cm^{-3}.

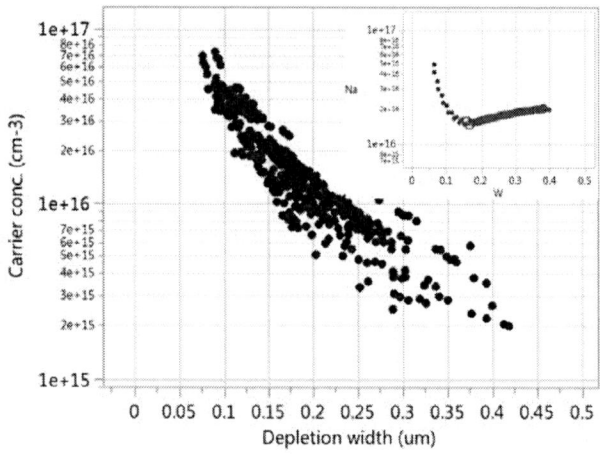

Fig. 1: Carrier concentration (Na) vs. depletion width (W) at zero bias as measured by CV for a group of Arsenic doped devices. Inset shows voltage dependent Na vs. W for two representative solar cells.

In Fig. 2 we compare the external quantum efficiency (EQE) of representative Cu-doped and As-doped (CdTe) devices. The red response of the As-doped cell in the 700 to 850 nm range is lower due to the higher doping of the absorber that leads to a smaller depletion width. This suggests that the current collection is limited by the diffusion length. An extended EQE response from the Arsenic device in the wavelength range of 850 to 900 nm suggests the existence of

978-1-5090-5606-4/17 $31.00 © 2017 IEEE

sub-band-gap defect states. These states are also evidenced in PL measurements shown in Fig. 4, 5.

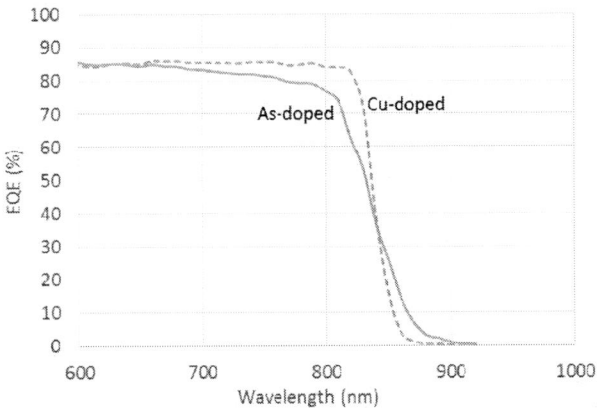

Fig. 2: Quantum efficiency vs. wavelength comparison for Arsenic and copper doped CdTe cells.

Earlier experiments performed on single-crystal CdTe:As, not reported here, showed only percent level Arsenic activation. In Fig. 3 we plot the doping activation for a subset of samples represented in Fig. 1 for which the Arsenic concentration was measured by secondary ion mass spectroscopy. Clearly, an inverse relation is observed between the amount of dopant incorporation and its effectiveness to produce free hole carriers. The high carrier concentration opens a doping regime not accessible with copper doping, however, the large excess of Arsenic present in the films may pose a problem and is therefore further studied utilizing the PL technique.

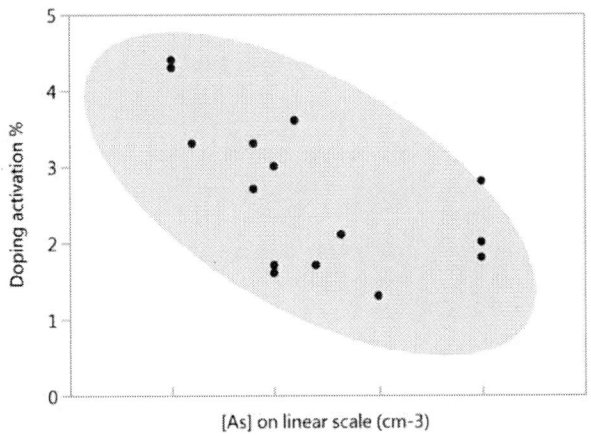

Fig. 3: Doping activation vs. bulk Arsenic concentration measured by secondary ion mass spectroscopy. The inverse relation suggests effective utilization of Arsenic impurities requires lower incorporation.

Figure 4 compares the room temperature PL spectrum measured from the junction side for four devices with varying Arsenic dose. In addition to the CdTe peak near 1.5 eV, we observe a second peak near 1.44 eV. This represents a ~60

mV offset from the band-edge and is in the range of acceptor energy reported for As_{Te}. While it is unusual to observe a dopant energy level in room temperature PL measurements, we observe a strong correlation of this level with group-V doping activation. Reduction in Arsenic dose leads to a decrease in the magnitude of the sub-band peak relative to the CdTe-peak. In conjunction with the sub-band-gap EQE response in Fig. 2, this observation suggests broadening of the CdTe peak with increasing Arsenic.

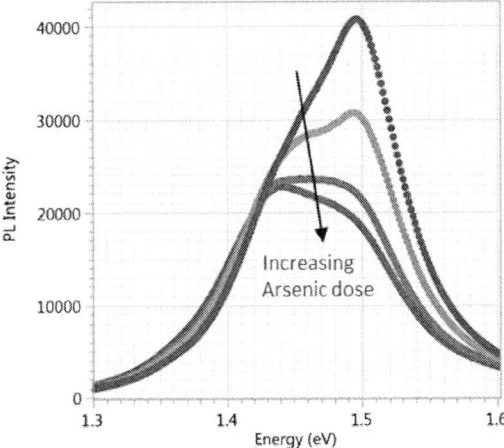

Fig. 4: Room temperature PL intensity vs. energy for CdTe films with varying amount of Arsenic doping.

The presence of As_{Te} acceptor is confirmed by measuring the low-temperature PL as shown in Fig. 5. We observe the acceptor/bound-exciton peak-(3) at 1.51 eV in agreement with literature. [14] The associated donor-acceptor band-(4) is also observed but unlike spectra for sx-CdTe:As, the phonon replica is not trivial to resolve in our measurement.

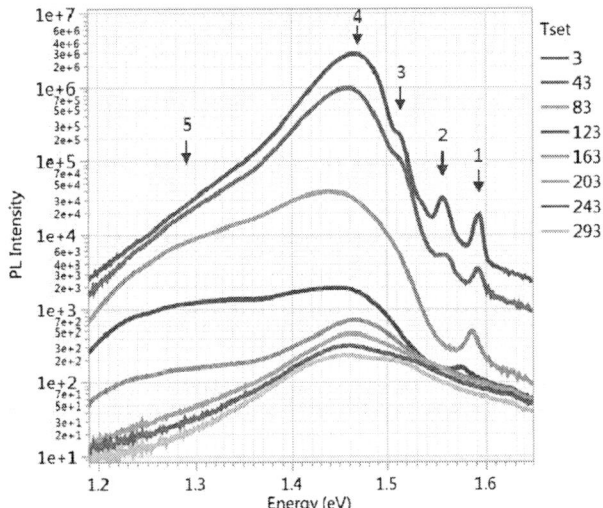

Fig. 5: Photoluminescence spectra measured from liquid-He temperature (3 K) to room temperature (293 K).

Studying the temperature dependence of this band we conclude that the sub-band PL peak observed in Fig. 4 is due to the donor-acceptor transition associate with Arsenic doping. The band-edge exciton is identified as peak-(1) and peak-(2) is likely associated with the V_{Cd}-2Cl_{Te} acceptor. [15] Additionally a defect band-(5) around 1.3 eV is noticed at intermediate measurement temperatures.

IV. DISCUSSION

High p-type carrier concentration in copper-free CdTe is demonstrated and this represents a significant milestone towards making Arsenic doping a potentially viable evolution for CdTe PV. The low doping activation in the presence of high dopant incorporation may suggest defect compensation. The same is true for Copper doping however Arsenic is able to provide a higher hole concentration. Sub-bandgap activity in QE and PL in conjunction with LTPL suggests that the donor-acceptor transition may contribute to device operation at room temperature. Additional sub-band energy states approximately 300 meV from the band-edge are observed. The exact nature and impact of these states is under investigation.

ACKNOWLEDGMENT

We acknowledge the contributions of several colleagues at First Solar for device processing and measurements. We also thank Stuart Irvine and Giray Kartopu from Centre for Solar Energy Research (UK) for informative discussions.

REFERENCES

[1] Best Research-Cell Efficiencies, National Renewable Energy Laboratory (http://nrel.gov/ncpv/images/efficiency_chart.jpg)

[2] M. Gloeckler, "Realization of the potential of CdTe thin-film PV," *43th Photovoltaics Specialist Conference*, 2016

[3] R. Geisthardt, M. Topič, J. Sites, "Status and Potential of CdTe Solar-Cell Efficiency," *IEEE J. Photovoltaics*, vol. 5, pp. 1217-1221, 2015

[4] M. Gloeckler, I. Sankin, Z. Zhao, "CdTe solar cells at the threshold to 20% efficiency", *IEEE J. Photovoltaics*, vol. 3, no. 4, pp. 1389-1393, Oct. 2013.

[5] J. Duenow, et.al., "Relationship of Open-Circuit Voltage to CdTe Hole Concentration and Lifetime," *IEEE J. Photovoltaics*, vol. 6, pp. 1641-1644 , 2016

[6] D. Krasikov, I. Sankin, "Defect interactions and the role of complexes in CdTe solar cell absorber," *J. Mater. Chem. A*, 2017

[7] J. M. Burst et al., "Advances in control of doping and lifetime in single-crystal and polycrystalline CdTe," *40th Photovoltaics Specialist Conference*, 2014

[8] T. Ablekim et.al., "Self-compensation in Arsenic doping of CdTe," *unpublished*

[9] J. M. Burst et al., "CdTe solar cells with open-circuit voltage breaking the 1V barrier," *Nature Energy*, vol. 1, p. 16015, 2016.

[10] S. L. Rugen-Hankey et.al., "Improvement to thin-film CdTe solar cells with controlled back surface oxidation," *Solar Energy Materials & Solar Cells* No. 136 pp. 213–217, 2015

[11] E. Colegrove *et al.*, "Phosphorus doping of polycrystalline CdTe by diffusion," *42nd Photovoltaic Specialist Conference*, 2015

[12] V. Evani, M. Khan, S. Collins, V. Palekis, D. Morel, C. Ferekides, "Phosphorus doping of polycrystalline CdTe", *43rd Photovoltaic Specialists Conference*, pp. 0435-0437, 2016.

[13] E. Colegrove, "Fundamental investigations of CdTe deposited by MBE for applications in thin-film solar photovoltaics," *PhD. Thesis*, 2014

[14] E. Molva et.al., "Photoluminescence studies in N, P, As implanted cadmium telluride", *Solid State Communications*, vol. 48, pp. 955–960, 1983

[15] H. Shin, C. Sun, "Photoluminescence spectra of Cl-doped CdTe crystals," *J. of Crystal Growth*, vol. 186, pp. 354–361, 1998

[16] D. Kuciauskas, et.al., "Recombination Analysis in Cadmium Telluride Photovoltaic Solar Cells With Photoluminescence Spectroscopy", *IEEE J. Photovoltaics*, vol. 6, pp. 313-318, 2016

Enhancing p-type Doping in Polycrystalline CdTe Films

Brian McCandless[1], Wayne Buchanan[1], Gowri Sriramagiri[1], Christopher Thompson[1]

Joel Duenow[2], David Albin[2], Soren Jensen[2], John Moseley[2], M. Al-Jassim, Wyatt K. Metzger[2]

[1]Institute of Energy Conversion, University of Delaware, Newark, DE 19716 U.S.A.

[2]National Renewable Energy Laboratory, 15013 Denver West Parkway, Golden, CO 80401 U.S.A.

Abstract — **An *in-situ* non-equilibrium method to increase acceptor doping in polycrystalline CdTe thin films to 10^{16} cm^{-3} level using group V substitution is presented. Single phase CdTe films doped with P, As, and Sb were deposited at 550°C at 100-200 nm/s onto moving CdS/HRT/TCO/Glass superstrates by vapor transport deposition (VTD) in Cd overpressure from high purity compound sources. Doping levels before and after activation were determined by capacitance-voltage (CV) analysis of devices with no additional treatments. Dopant incorporation levels of 10^{17}-10^{18} atoms/cm^3 were obtained based on SIMS depth profiling. Electronic activation was carried out by post deposition annealing and quenching, increasing acceptor concentrations to $>10^{15}$ cm^{-3} for P and $>10^{16}$ cm^{-3} for As and Sb, compared with mid-10^{14} cm^{-3} acceptor levels for un-doped CdTe films. The acceptor concentration increase by substitutional defect As$_{Te}$ formation was validated for As-doping by cathodoluminsence spectroscopy.**

Index Terms – **CdTe, thin film, doping, photovoltaic**

I. INTRODUCTION

The CdTe thin film solar cell has reached a state of commercial maturity and is competitive with mc-Si and the costs of traditional energy sources, but the technology still has headroom to improve by overcoming fundamental material issues. In polycrystalline thin film devices, V_{OC} remains less than 0.9 V at 25°C. Reaching beyond 1 V will require films having acceptor concentration $>1 \times 10^{16}$ cm^{-3} and aggregate recombination lifetimes of several to tens of ns [1-3]. The challenge with doping CdTe films, n- or p-type, is controlling the point defect response to variations in chemical potential [4] during film growth. Present-generation commercial CdTe solar cells rely to some degree on native point defect control during growth, wherein high deposition temperature favors cadmium vacancies (V_{Cd}) but with maximum as-deposited acceptor levels $\sim 10^{14}$ cm^{-3}. The role of the subsequent and unique cell completion steps, CdCl$_2$ annealing and Cu-diffusion, are in part to improve aggregate lifetime by passivating grain surfaces [2, 5-7]. While these steps may also increase hole concentration, Cu may move and cause stability issues, and stable hole density is typically less than 10^{15} cm^{-3} [8-11].

In intrinsically doped materials, anti-site occupation can compensate free-carrier charge, thereby limiting acceptor concentration. Without charge compensation, a free carrier can be added for each electron added or removed from the system. In extrinsically doped materials, the dopant competes for lattice sites, governed by its electronegativity with respect to the lattice constituents – this determines which lattice sites are preferred and what net charge is delivered to the system.

In bulk and epi-film single crystal CdTe, efforts to substitutional dope with group V elements have shown considerable promise [12]. However, to incorporate the group V atom into the CdTe lattice requires production of an available Te site, V_{Te}. The problem is that high quality polycrystalline CdTe thin films are deposited at high temperatures, greater than 500°C, where the thermodynamic tendency favors production of V_{Cd} over V_{Te} at the growing film surface. Thus, at high temperature, without adjustment of the vapor stoichiometry over the growing CdTe film, thermodynamics favor the production of V_{Cd}, and the equilibrium of the group V dopant species with both Cd and Te lattice sites then governs the system.

In this work, a new non-equilibrium approach for *in-situ* p-type doping CdTe was implemented to substitute the group V elements P, As and Sb onto Te lattice sites by vapor entrainment and delivery to the growing film in a Cd-rich environment, followed by rapid cooling to "freeze-in" the high-temperature defect conditions restricting the degree of re-equilibrium which can occur during cool-down. A similar approach was employed for post-deposition anneals, in Cd vapor, at temperatures near the film growth temperature. CdTe films were deposited at manufacturing rates by vapor transport deposition (VTD) onto translating superstrate plates [13] in a manner analogous to commercial modules [14]. Solid compound doping sources were integrated into the CdTe vapor source within the VTD system, based on modeled thermal vapor decomposition parameters, to achieve desired gas-phase concentrations in the source effluent vapor and impinging onto the superstrate below.

CdTe films with and without extrinsic *in-situ* dopants were deposited on glass coated with a thin film stack consisting of a transparent conductive oxide (TCO), high resistance transparent buffer layer (HRT) and cadmium sulfide (CdS) window layer. Deposited CdTe films were analyzed by glancing incidence x-ray diffraction (GIXRD) to confirm single-phase and secondary ion mass spectrometry (SIMS) to establish dopant incorporation levels. Diagnostic superstrate solar cells were fabricated and assessed using current-voltage (JV), capacitance-voltage (CV) and in the case of As-doping, cathodoluminescence (CL). The present state of this work has yielded acceptor concentrations $>10^{15}$ cm^{-3} for P-doped CdTe films and $>10^{16}$ cm^{-3} for As- and Sb-doped CdTe films. Using As-doped cells before and after activation, CL spectral analysis shows an increase in the substitutional defect As$_{Te}$ and reduction in deep defects after activation in Cd vapor.

II. EXPERIMENTAL DETAILS

A. CdTe Film Deposition and Device Completion

The VTD reactor at the Institute of Energy Conversion (IEC) was modified so that dopant and Cd excess vapor species could be entrained with the Cd and Te vapor generated from high purity CdTe source material. This in-situ entrainment of dopants in the CdTe film during deposition was achieved by addition of a high purity dopant-containing compound, specifically Cd$_3$P$_2$ for phosphorous, Cd$_3$As$_2$ for arsenic, and CdSb for antimony, along with elemental Cd, located within the inlet zone of the CdTe source. The dopant compounds were synthesized in quartz ampoules by the IEC group using 6N purity elements, followed by vapor transport as a final purification step.

The dopant position at the inlet of the deposition source was established to achieve the calculated temperature required to produce a supplemental vapor stream concentration of ~10-100 ppm, which if fully incorporated, would yield ~10^{18} cm^{-3} atomic concentration in the CdTe. In most cases, dopant material remained in the dopant zone of the VTD source after the run, indicating uniform delivery had proceeded throughout the run.

A variant method, which was more successful for As and Sb was also developed in which the VTD source was loaded with a charge of uniformly distributed dopant along its length. This is called the "uniform" source delivery method. For this, high purity CdTe powder was mechanically blended with the dopant compounds Cd$_3$P$_2$, Cd$_3$As$_2$, or CdSb, sealed in quartz ampoules, fired at 840 °C, and then quenched to room temperature. The resulting charge contained CdTe crystals with inclusions of dopant compounds, up to several weight percent.

The CdTe films were deposited at 550°C substrate and 840°C source temperatures in an N$_2$ ambient onto 4"x4" SCHOTT AF45 glass coated with a fluorinated tin oxide/tin oxide bilayer and CdS films. CdS was deposited by chemical surface deposition to a thickness of ~100 nm [15]. The CdS then received a CdCl$_2$ vapor treatment at 415°C for 15 minutes to crystallize and dehydrate the CdS film. Upon completion of the CdTe deposition, the films were rapidly cooled by injection of N$_2$ during a high-speed translation into a separate chamber, resulting in a 200°C temperature drop at a cooling rate of >5°C/s. CdTe film thicknesses of 5-7 microns were achieved. Post-deposition annealing of 1 cm x 1 cm coupons taken from the CdTe-coated plates was performed in a horizontal furnace with measured temperature profile, at 550°C-600°C for 4 minutes. For this, the CdTe coupons were sealed in evacuated quartz ampoules, at less than 1 x 10^{-6} Torr,

together with a 300 mg lump of Cd metal, to provide a Cd vapor excess and a 2 mm^3 chunk of graphite, to getter residual oxygen. The location of the coupons and the Cd lump was varied to control the annealing temperature and the Cd partial pressure.

Solar cells were fabricated as diagnostic devices by performing a light viscous surface etch in ethylenediamine (EDA) to remove residual surface oxides and create an ultra-thin, <10 nm, Te surface. This process was followed by rinsing and then applying a graphite dot contact with Acheson 505SS ink, which was air dried at 30°C for 30 minutes.

B. Analysis

To evaluate film phase composition, x-ray diffraction (XRD) was measured by a Rigaku d/Max diffractometer in both Bragg-Brentano and glancing incidence modes at 1 degree incidence. Grain size was measured by high-resolution optical microscopy and scanning electron microscopy.

The film conductivity type was evaluated by the thermal voltage generated between a heated spring-loaded steel probe and a room temperature probe. A spring-loaded probe heated to 225°C was gently lowered onto the CdTe film in conjunction with a room temperature probe at a distance of 1 cm.

JV measurements were made in the dark and under AM1.5 global simulated light using an Oriel simulator and Keithley 2400 SMU. The voltage was typically swept from -0.5V to +1V in ~7 mV steps.

CV measurements were carried out using an Agilent 4284A LCR meter. DC voltage bias was provided using a Keithley 2400 SMU and the Agilent 16065C external bias adapter. For CV measurements, we use a fixed AC voltage signal and varying DC voltage bias, measuring capacitance as a function of V_{DC} and, since the depletion width is varied with V_{DC}, as a function of W. We calculated the responding charge density, N_{CV} from:

$$N(W)_{CV} = -\frac{C^3}{q\epsilon A^2 \left(\frac{dC}{dV}\right)} = -\frac{2}{q\epsilon A^2}\left[\frac{d(1/C^2)}{dV}\right]^{-1}$$

Where W is the depletion width, A is device area, q is electron charge, and $\varepsilon=13.5\varepsilon_0$. At this time, ascribing the measured charge response to acceptors is based on the N_{CV} versus W behavior compared with an ideal abrupt junction.

Dopant incorporation levels were measured by SIMS depth profiling at the Evans Analytical Group. The elements Sn and Te were used as markers for the extent of the CdTe film and beginning of the tin oxide window layer stack. The detection limit for dopants P, As and Sb were ~ mid 10^{14} atoms/cm^3.

Effective lifetime was measured at NREL using 1 photon excitation time-resolved photoluminescence (TRPL). Laser pulses with a wavelength of 640 nm were fired at a rate of 1.1 MHz through the glass onto the CdTe depletion region.

Photoluminescence was collected and separated from scattered light with a 840-nm bandpass filter and Si avalanche photodiode detector. Decay curves were generated using time-correlated singe photon counting and fit by a biexponential function, where τ_1 and τ_2 describe the exponential decay rates at the initial and final sections of the curve, respectively.

For one dopant, As, cathodoluminescence (CL) was conducted on beveled surfaces, prepared with a focused ion beam-scanning electron microscope (FIB-SEM) [16]. An initial 20°-wedge was milled through the film using a 30-kV ion beam, and then a 5-kV beam prepared the final surface for CL analysis. CL spectrum imaging was performed on a JEOL 7600F Field Emission SEM equipped with a Horiba H-CLUE CL system. Samples were cooled to 6 K on a Gatan CF302 liquid-helium cold stage during CL at electron beam conditions of 5 kV and 4 nA.

III. RESULTS AND DISCUSSION

A. CdTe Film Properties

XRD patterns indicated that all films in this study were single phase CdTe with zincblende structure. Films deposited with P and As exhibited near-random orientation to the substrate, but films deposited with Sb exhibited preferred (111) crystallographic texture. In all cases, the terminating CdTe grains were well-faceted and exhibited width-to-thickness aspect ratio varying from 0.2 to 0.8. Cross-section microscopic measurements showed columnar film growth.

SIMS depth profile analysis of as-deposited films, from the exposed CdTe surface to the CdS interface, showed nearly uniform dopant incorporation at the 10^{17}-10^{18} atoms/cm^3 range throughout the films, which is sufficient to achieve hole densities exceeding 10^{16} cm^{-3}. SIMS analysis was not performed on annealed films, given the short treatment times, 4 minutes compared to the characteristic diffusion length of Group V elements in CdTe at 600°C.

The doped films exhibited positive lateral Seebeck voltage, ranging from +3 to +10 mV, corresponding to a mean Seebeck coefficient of +25 mV/K and p-type conductivity. Comparatively, CdTe films deposited without dopants or oxygen yield Seebeck voltages from 0 to -3 mV, suggesting intrinsic or slightly n-type conductivity type. For these nominally un-doped films, positive voltages were only obtained after post-deposition annealing in CdCl$_2$:O$_2$ ambient at 400-450°C.

B. Dopant Impact on Electronic Behavior

Solar cell devices were analyzed prior to capacitance analysis to verify functioning photodiodes. Cells prepared without treatment in CdCl$_2$ or oxygen exhibited open circuit voltages (V_{OC}) from 300 mV to 650 mV and short-circuit current densities (J_{SC}) from ~1 mA/cm^2 to 17 mA/cm^2.

Detailed analysis of JV characteristics and response to dopant processing will be presented in later work, in which the passivation of surface and interface defects will be considered. The cells were used in the present work to correlate achievable hole density with incorporation levels of the different Group V dopants, prior to adding the complexity of a CdCl$_2$ treatment, Cu diffusion, or similar passivation approaches. A cell made with Sb doping at $N_{CV} = 5 \times 10^{15}$ cm^{-3}, treated in CdCl$_2$:O$_2$ vapor, yielded V_{OC}=714 mV and J_{SC}=23.6 mA/cm^2, suggests a cell optimization path consistent with present module fabrication.

Table I summarizes the mean SIMS dopant concentration, measured free charge response at 0V DC bias, $N_{CV}(0V)$ and TRPL lifetime for VTD films that yielded that highest N_{CV} values, compared to an IEC baseline cell using CdTe deposited in oxygen, and completed with a CdCl$_2$ treatment and Cu-diffusion step prior to contacting. Figure 1 shows N_{CV} versus W for cells with activated CdTe films, for different dopant VTD delivery methods, with $N_{CV}(0V)$ indicated by an open circle on each data set.

In Table I, the cell made with as-deposited un-doped CdTe, corresponding to the first row, exhibited the lowest free charge response $N_{CV}(0V)$. The cell made with un-doped CdTe after annealing at 550°C yielded similar $N_{CV}(0V)$ levels as an IEC 15% efficient "baseline" cell processed with un-doped CdTe and treated with O, Cl and Cu during cell processing, shown in the last row. The Group V doped films yielded cells with elevated N_{CV} prior to activation, compared to un-doped. After activation, N_{CV} increased ~10X, with the highest, As-doped, reaching >10^{16} cm^{-3}. Based on the SIMS incorporation levels and minimum N_{CV} values, doping activation efficiencies of 1.5%, 1.2%, and 10% were obtained for P, As and Sb, respectively.

TABLE I
SUMMARY OF SIMS, CV AND LIFETIME PARAMETERS FOR DIFFERENT DOPANTS

Dopant	Cd vapor Activation	Dopant SIMS (cm^{-3})	$N_{CV}(0V)$ (cm^{-3})	TRPL τ_1/τ_2 (ns/ns)
None	None	-	4.0E13	0.16/0.96
None	550°C	-	3.5E14	-
P	None	2.0E17	1.2E14	0.50/1.2
P	550°C	2.0E17	3.0E15	0.17/1.4
As	None	3.0E18	6.0E15	0.10/1.5
As	600°C	3.0E18	3.6E16	-
Sb	None	1.0E17	5.4E14	0.49/1.6
Sb	550°C	1.0E17	1.0E16	-
O, Cl, Cu	425°C	-	4.5E14	0.55/2.7

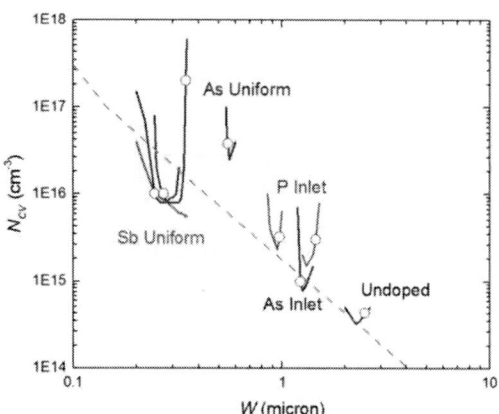

Fig. 1. Free charge response, N_{CV}, vs depletion width, W for cells with dopant-activated CdTe films; the N_{CV} value at zero volts is indicated by the circular mark on each dataset. Dotted line represents trend for abrupt junction and uniform doping. The labels indicate the dopant and the VTD delivery method used.

The plots of N_{CV} versus W in Figure 1, for cells made with films after doping activation for different VTD dopant and delivery methods, show nearly ideal behavior, with data falling near the expected $N_{CV}(W)$ relationship indicated by the dotted line. The zero-volt point indicated by the open circle on each dataset is interpreted as representing doping level N_A when it lies near the ideal line. In most cases the zero-volt points lie near the N_{CV} minimum. One As-doped cell is shown with its zero-volt point at 10^{17} cm^{-3}, located far up the reverse-bias edge of the dataset, anomalous behavior which is attributed to the influence of charges associated with penetration from the back contact. Comparing $N_{CV}(0V)$ or minima shows that all doped films consistently exceeded un-doped film, and highest values were obtained with uniform dopant delivery of As and Sb.

The achievement of 10^{16} cm^{-3} doping levels using As and Sb implies more substitutional modality than P. Within the Group V elements examined, the saturations pressures of As and Sb are sufficiently low at the substrate deposition temperature to minimize re-evaporation, allowing facile control of incorporation into the growing CdTe film. Their larger atomic radii induce less lattice strain when incorporated as a substitutional defect. Phosphorous is more difficult to control, in-part, due to its extremely high saturation pressure at 550˚C, and due to its small atomic radius, which induces more strain than As or Sb. VTD control variables such as source temperature, film growth rate and substrate temperature are thus more critical for incorporation of P compared with As and Sb.

Promising results for P-doping were also obtained using a Zn$_3$P$_2$ doping source during VTD film growth. This yielded a cell with $N_{CV}(0V) = 5.5 \times 10^{15}$ cm^{-3} after activation in Cd vapor. In this case, the VTD process had the additional effect of

forming Cd$_{0.99}$Zn$_{0.01}$Te alloy, verified by SIMS depth profiling and XRD, due to uniform Zn incorporation in CdTe during growth, analogous to results obtained with mixed CdTe:ZnTe sources [17].

The measured TRPL lifetime, shown in the last column of Table I, is lower in doped, un-treated films than in 15% efficient IEC "baseline" VTD CdTe/CdS cells. Such baseline CdTe cells are processed with O$_2$ during CdTe growth, CdCl$_2$ post treatment in air, and Cu diffusion from the back contact and exhibit $\tau_1 \sim 0.6$ ns and τ_2 from 2 to 5 ns. In the present study, cells made with as-deposited doped CdTe grown without CdCl$_2$ have $\tau_1 \sim 0.5$ ns and τ_2 between 1.2 and 1.6 ns. While slightly lower than the baseline, these values are still adequate for reasonable solar cell performance, are very good for cells that have not received CdCl$_2$, and indicate that Sb, As, or P dopants do not severely limit lifetime here [18].

C. Doping Mechanism

Verification of the formation of substitutional defects was obtained using CL spectroscopy for As-doped cells, before and after Cd overpressure anneals intended to place As onto Te sites. Figure 2 shows CL images of the As-doped film before (left) and after activation annealing (right) to compare the spectral defect contributions, thereby helping to ascertain As substitution on Te sites. In these images, CL spectra are taken at every point in the image, and assigned the colors blue, green and red for excitonic, As$_{Te}$, and deep defect peaks, respectively [16]. The uniform green grains in the images illustrate that we have indeed achieved intra-grain As$_{Te}$ substitutional doping throughout the film. Further, and consistent with CV data, the Cd vapor anneal increases the activation of As on Te throughout the film. The measured acceptor concentration, N_{CV} versus depletion width for all the measured samples (Fig. 1) shows a trend close to ideal values for uniform doping, suggesting that the measured N_{CV} values are representative of acceptor levels, N_A.

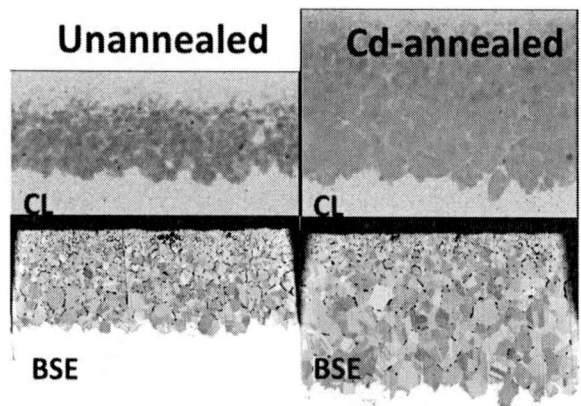

Fig. 2. Backscattered electron images (BSE) and color-coded CL images of beveled CdTe films. Green corresponds to As$_{Te}$ sites, red to deep defects and blue to excitonic peaks [8].

978-1-5090-5606-4/17 $31.00 © 2017 IEEE

IV. CONCLUSIONS

Enhanced p-type doping in polycrystalline CdTe thin films during growth by an *in situ* non-equilibrium process has been demonstrated. Modifying a high growth-rate VTD CdTe deposition process to incorporate group V elements as substitutional dopants, coupled with quenched growth from 550°C and performing a short post-growth anneal+quench in Cd vapor is shown to be a viable approach for raising acceptor levels to 10^{16} cm^{-3} regime without significant loss in minority carrier lifetime. SIMS depth profiles indicate sufficient dopant incorporation, between 10^{17} and 10^{18} atoms/cm^3, attained for P, As, and Sb by vapor mixing from compounds containing the group V elements within the CdTe source. As and Sb enabled acceptor concentrations >10^{16} cm^{-3}. Dopant activation in Cd vapor, with quenching, enhances doping by placing group V elements incorporated during deposition onto Te lattice sites. For example, CL indicates the enhancement of As$_{Te}$ sites throughout the film by Cd overpressure. Without CdCl$_2$ but with group V dopants, TRPL lifetimes were $\tau_1 \sim 0.5$ ns and $\tau_2 \sim 1.5$ ns in polycrystalline CdTe films, while hole densities of greater than 10^{16} cm^{-3} were achieved, indicating a path to overcome longstanding hole density limits in CdTe solar technology.

ACKNOWLEDGEMENTS

The authors at the Institute of Energy Conversion gratefully acknowledge Shannon Fields, at IEC, for VTD implementation, and Alan Fahrenbruch for his generosity in both technical and material support of IEC's research in group II-VI materials. This work was supported by the U.S. Department of Energy under Contract No. DE-AC36-08GO28308 to the National Renewable Energy Laboratory. The U.S. Government retains and the publisher, by accepting the article for publication, acknowledges that the U.S. Government retains a nonexclusive, paid up, irrevocable, worldwide license to publish or reproduce the published form of this work, or allow others to do so, for U.S. Government purposes.

REFERENCES

[1] J. N. Duenow, J. M. Burst, D. S. Albin, M. O. Reese, S. A. Jensen, S.W. Johnston, D. Kuciauskas, S. K. Swain, T. Ablekim, K. G. Lynn, A. L. Fahrenbruch, W. K. Metzger, "Relationship of Open - Circuit Voltage to CdTe Hole Concentration and Lifetime," *IEEE J. Photovolt.*, 6 (6), art. no. 7568988, pp. 1641 - 1644 (2016) DOI: 10.1109/JPHOTOV.2016.2598260.

[2] A. Kanevce, M.O. Reese, T.M. Barnes, S. A. Jensen, W.K. Metzger, "The roles of carrier concentration and interface, bulk, and grain-boundary recombination for 25% efficient CdTe solar cells", *J. Appl. Phys.*, 121, 214506 (2017); doi: http://dx.doi.org/10.1063/1.4984320.

[3] B. E. McCandless, "CdTe Solar Cells: Processing Limits and Defect Chemistry Effects on Open Circuit Voltage," *Conf. Rec. 2013 MRS*, MRS-13-1538-C13-04.R1 (2013).

[4] F. A. Selim and F. A. Kroeger, "Defect Structure of P- Doped CdTe," *J. Electrochem. Soc.*, 124, 401-408, 1977.

[5] J. Moseley et al., "Recombination by grain-boundary type in CdTe," *J. Appl. Phys.* 118 25702 (2015); http://dx.doi.org/10.1063/1.4926726

[6] E. S. Barnard et al., "3D Lifetime Tomography Reveals How CdCl$_2$ Improves Recombination Throughout CdTe Solar Cells," *Adv. Mater.* 29, 1603801 (2017) DOI: 10.1002/adma.201603801

[7] A. Kanevce, J. Moseley, M. Al-Jassim, W.K. Metzger, " Quantitative Determination of Grain Boundary Recombination Velocity in CdTe by Cathodoluminescence Measurements and Numerical Simulations," *IEEE J. Photovolt.*, 5 (6), art. no. 7295555, pp. 1722 - 1726 (2015). DOI: 10.1109/JPHOTOV.2015.2478061

[8] Corwine, C. R., Pudov, A. O., Gloeckler, M., Demtsu, S. H., Sites, J. R., "Copper inclusion and migration from the back contact in CdTe solar cells." *Sol. Energy Mater. Sol. Cells* 82, 481–489 (2004).

[9] Albin, D. S., "Accelerated stress testing and diagnostic analysis of degradation in CdTe solar cells," *Proc. SPIE 7048, Reliability of Photovoltaic Cells, Modules, Components, and Systems*, 70480N (SPIE, 2008).

[10] Burst, J.M. et al, "Carrier density and lifetime for different dopants in single-crystal CdTe," *APL Materials* 4, 116102 (2016). doi: 10.1063/1.4966209

[11] D. Grecu, A. D. Compaan, D. Young, U. Jayamaha, and D. H. Rose, "Photoluminescence of Cu-doped CdTe and related stability issues in CdS/CdTe solar cells," *J. Appl. Phys.* 88, 2490 (2000).

[12] J.M. Burst et al.. "CdTe solar cells with open-circuit breaking the 1 V barrier", *Nature Energy*, Article Number 16015 (2016): DOI: 10:1038/NENERGY.2016.1,.

[13] G. M. Hanket, B. E. McCandless, W. A. Buchanan, S. Fields and R. W. Birkmire, "Design of a vapor transport deposition process for thin film materials," *J. Vac. Sci. Technol. A* 24(5), 1695-1701(2006).

[14] Bülent M. Başol and Brian McCandless, "Brief review of cadmium telluride-based photovoltaic technologies," *J. Photon. Energy.* 4(1), 040996 (Jun 27, 2014). doi:10.1117/1. JPE.4.040996

[15] Brian E. McCandless and William N. Shafarman, "Chemical surface deposition of ultra-thin cadmium sulfide films for high performance and high cadmium utilization," *Conf. Rec. 4th WCPEC*, Osaka, 2LN-C-03, 562-565, (2003).

[16] Guthrey et al., "Spatial luminescence imaging of dopant incorporation in CdTe films," *J. Appl. Phys*, 121, 045304 (2017).

[17] B. McCandless, W. Buchanan, G. Hanket, "Thin Film Cadmium Zinc Telluride Solar Cells," *Conf. Rec. 4th WCPEC*, 483-386 (2006).

[18] W. Metzger, D. Albin, D. Levi, P. Sheldon, X. Li, B. Keyes, R. Ahrenkiel, J., "Time-Resolved Photoluminescence Studies of CdTe Solar Cells," *J. Appl. Phys.* 94, 3549-3555 (2003).

Spectral and Concentration Sensitivity of Multijunction Solar Cells at High Temperature

Daniel J. Friedman[1], Myles A. Steiner[1], Emmett E. Perl[1,2], John Simon[1]

[1]National Renewable Energy Laboratory, Golden, Colorado, 80401, USA

[2]University of California at Santa Barbara, Santa Barbara, CA 93106, USA

Abstract — We model the performance of two-junction solar cells at very high temperatures up to ~400°C for applications such as hybrid PV/solar-thermal power production. We show that high-temperature operation reduces the sensitivity of the cell efficiency to spectral content, but increases the sensitivity to concentration, both of which have implications for energy yield in terrestrial PV applications. For other high-temperature applications such as near-sun space missions, our findings indicate that concentration may be a useful tool to enhance cell efficiency.

I. INTRODUCTION

Although conventional solar cell applications typically call for the cells to operate at temperatures less than ~100°C, there are potentially impactful applications for which the cell operating temperature would be far higher. There has been considerable recent interest in cells operating as a topping cycle in a hybrid solar-thermal concentrator system for terrestrial energy production with storage [1]. These cells would operate at temperatures of ~400°C or above, and as with all terrestrial solar applications would be subject to hourly and seasonal variation of the spectrum and intensity of the solar illumination. Another application of high-temperature solar cell operation would be photovoltaic-powered near-sun missions such as Mercury probes, for which cell operation at ~450°C would be desirable [2]. In this case, the spectrum seen by the cell would be a constant, but the concentration of the light could be treated as a design variable.

In this paper, we study the design of high-efficiency III-V two-junction cells for high-temperature operation, with 400°C chosen as a representative temperature, and identify areas in which the design and performance characteristics behave significantly differently than at more conventional near-room-temperature operating conditions. We show in particular that the sensitivity of the cell efficiency to spectrum and to intensity is notably different at high temperature than at room temperature, and we describe the consequences for solar cell and system design and power production.

II. DEVICE STRUCTURE AND CHARACTERISTICS

The solar cell structure we consider here is an (Al)GaInP/GaAs two-junction tandem cell with a tunnel junction providing a series connection. The top junction thickness, x_1, is adjusted to maximize the efficiency. Here, we consider only idealized junctions with long diffusion lengths and no parasitic absorption from layers such as the tunnel junction or window, but these can readily be accomodated using the same methods. This structure is being studied in the laboratory for high-temperature operation [3] to address the hybrid PV/solar-thermal application.

III. MODEL

To analyze analytically the dependence of the cell performance on temperature, we use a simple model that captures the essential physics. The voltage V of the series-connected tandem cell at a given current density J is the sum of voltages $V_1(J)$ and $V_2(J)$ of the top and bottom junctions,

$$V(J) = V_1(J) + V_2(J). \tag{1}$$

Each individual ith junction is treated with the familiar single-diode model with ideality factor $n=1$ [4-6],

$$J(V_i) = J_{01,i}\left[\exp\left(\frac{eV_i}{k_BT}\right) - 1\right] - J_{SC,i}, \tag{2}$$

$$J_{01,i}(T) = C_{1,i}T^3 e^{-E_{g,i}/k_BT}. \tag{3}$$

Comparison with experiment in the next section shows this model to be an excellent qualitative description for high-quality junctions, especially at high concentration; effects of the n=2 dark current component will be discussed elsewhere. Series and shunt resistance are taken to be negligible.

Equations (2, 3) determine the open-circuit current, V_{OC}:

$$V_{OC} = \frac{k_BT}{e}\ln\left(\frac{J_{SC}}{J_{01}}\right). \tag{4}$$

The value of $C_{1,i}$ is determined empirically by comparison with high-quality experimental junction measurements at room temperature and considered to be temperature-independent. The temperature dependences of the bandgaps $E_{g,i}$ are accounted for via the Varshni coefficients for the respective materials [7]. The QEs are treated with the familiar Hovel equations [4]. The top junction, of thickness x_1 and absorption coefficient $\alpha_1(\lambda)$, filters the bottom junction reducing its QE by a factor $exp[-\alpha_1(\lambda)x_1]$. Other light-absorbing layers such as window layers and tunnel junctions are treated in the same manner.

978-1-5090-5606-4/17 $31.00 © 2017 IEEE

Figure 1 shows the measured V_{OC} of GaAs and AlGaInP junctions under low and high concentrations as a function of temperature from 25–400°C. Details of the fabrication and measurement of the devices are given in [3, 8]. The figure also shows the fit of the model to the data. The one adjustable parameter, the J_{01} coefficient C_1, is determined by self-consistently fitting to the entire V_{OC} vs. temperature and concentration data set. The figure shows good agreement of Eqs. (2,3) with the measured data, including reproducing the effect of concentration on dV_{OC}/dT.

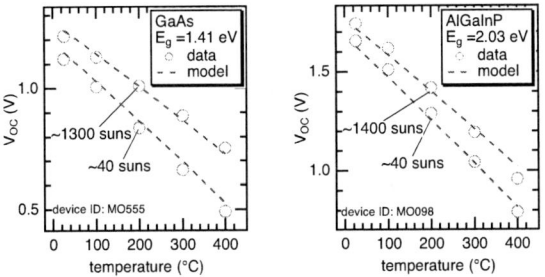

Fig. 1. Measured V_{OC} of GaAs and AlGaInP junctions under low and high concentrations as a function of temperature from 25-400°C from [3], compared with the fit to the model.

IV. CURRENT MATCHING AND SPECTRAL SENSITIVITY

A critical parameter in the design of series-connected multijunction cells is the thickness of each junction, with the thickness chosen to optimize the efficiency of the cell [9]. Figure 2(a) shows the modeled efficiency of an idealized GaInP / GaAs cell with junction bandgaps {1.85eV, 1.41 eV} as a function of the top junction thickness, at 400°C at 1000 suns.

The figure illustrates the well-known behavior that the top-junction photocurrent $J_{SC,1}$ increases and the bottom-junction photocurrent $J_{SC,2}$ decreases with top-junction thickness, leading to an efficiency maximum at roughly the point where $J_{SC,1}$ and $J_{SC,2}$ are matched. This efficiency is re-plotted against

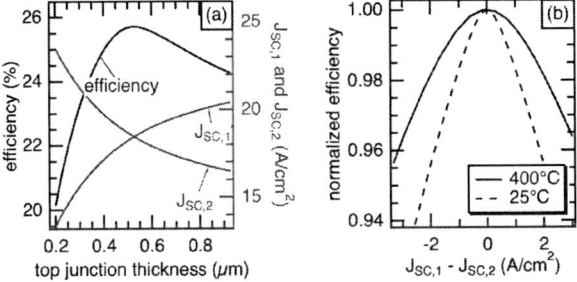

Fig. 2. (a) Cell efficiency and the top- and bottom-junction currents $J_{SC,1}$ and $J_{SC,2}$ as a function of top junction thickness for an idealized 1.85 eV / 1.41 eV two-junction cell, at 400°C, under the ASTM G173 AM1.5 Direct spectrum at 1000 suns. (b) Efficiency vs. current mismatch $J_{SC,1} - J_{SC,2}$ at 400°C as in (a), and at 25°C. The efficiencies are normalized to 1 at their maximum values.

the current mismatch $J_{SC,1} - J_{SC,2}$ in Fig. 2(b), and compared to the efficiency at 25°C. The efficiencies are normalized to 1 at their maximum values to facilitate the comparison of the rate of decrease of efficiency with current mismatch. The comparison shows that the efficiency is much less sensitive to current mismatch at 400°C than at 25°C. This difference arises because junction dark currents increase rapidly with temperature, which decreases fill factors and increases the difference between the short-circuit and maximum-power-point values of the current.

This decreased sensitivity at high operating temperature to current mismatch implies that there should be a corresponding decreased sensitivity to the spectral content of the solar illumination, because the main effect of changes in spectral content is to change the current mismatch. There are many parameters besides air mass that determine a solar spectrum, but focusing on the air mass is a simple and useful proxy for arbitrary spectral fluctuations. To quantify sensitivity of the cell efficiency to spectral variations, we consider how the efficiency of a cell designed for the standard air mass 1.5 direct (AM1.5D) spectrum changes under varying air mass. For the cell under the AM1.5D spectrum at 400°C considered in Fig. 2(a), a top junction thickness of 0.53μm provides the optimal efficiency. Figure 3(a) shows how the efficiency of this AM1.5D-optimized cell varies with air mass. The efficiency peaks at AM=1.5 as expected and falls off at other AM values, but the falloff is quite slow. In comparison, a cell designed for AM1.5D at 25°C and operated at 25°C shows a much faster falloff of efficiency for AM≠1.5, i.e. a much greater spectral sensitivity. Figure 3(b) shows the same efficiencies normalized to their values at AM1.5. For air masses greater than 1.5, the 400°C cell efficiency falls off with increasing air mass at about 2/3 the rate of the 25°C cell

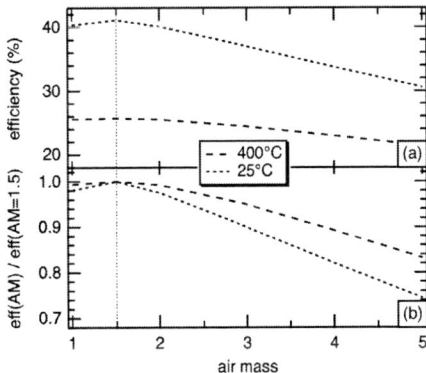

Fig. 3. (a) Modeled efficiencies at 400°C and 25°C at 100 W/cm^2 (1000 suns), and (b) the same efficiencies normalized to their AM=1.5 values, for the direct spectrum over a range of air masses for two-junction cells with top-junction thicknesses chosen for optimal performance at AM1.5D. The spectra were computed using the SMARTS2 code [10]. The spectrum at AM=1.5 is precisely the ASTM standard G173 AM1.5 direct spectrum. The spectra are normalized to 100 W/cm^2.

978-1-5090-5606-4/17 $31.00 © 2017 IEEE

efficiency falloff, confirming a significantly reduced spectral sensitivity at 400°C. Reduced spectral sensitivity results in increased energy harvesting in real-world conditions where the spectrum is not fixed, a promising observation for any application including the high-temperature topping-cycle concept which relies on operation of high-efficiency cells at high temperature under the terrestrial spectrum.

V. CONCENTRATION SENSITIVITY AT HIGH TEMPERATURE

From eq. (4), the rate of increase of V_{OC} with concentration increases with temperature, with the V_{OC} for each junction increasing ~59 mV/decade at 25°C and ~134 mV/decade at 400°C. This effect is accentuated in impact because V_{OC} itself decreases with temperature; both of these phenomena can be seen experimentally in the V_{OC} data of Fig. 1. As a result, the dependence of cell efficiency on concentration is much greater at high temperature than at room temperature. This difference is illustrated in Fig. 4, which shows the modeled relative efficiency of the two-junction cells of Fig. 3 as a function of concentration at 25°C and 400°C. This sensitivity could have a negative impact on energy yield for applications at high temperature under the terrestrial spectrum. Looked at another way, however, concentration may be desirable as a tool to enhance efficiency for high-temperature applications such as near-sun space missions.

VI. SUMMARY

We examined the expected performance of two-junction solar cells for applications at very high temperatures of ~400°C and beyond, under conditions of varying spectral content and concentration. High-temperature operation reduces the sensitivity to spectral content, but increases the sensitivity to concentration, conditions which both must be accounted for to properly assess energy yield under varying

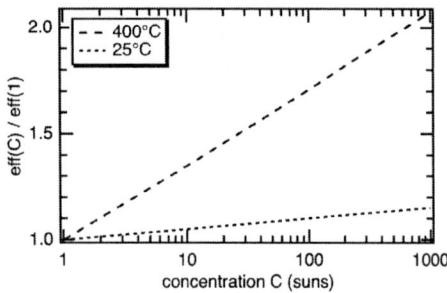

Fig. 4. Relative efficiency of the two-junction cells of Fig. 3 as a function of concentration at 25°C and 400°C.

terrestrial illumination conditions. For other applications such as near-sun space missions, concentration may be a useful tool to enhance cell efficiency.

ACKNOWLEDGEMENT

This work was supported by the ARPA-E FOCUS program under Award DE-AR0000508. The U.S. Government retains and the publisher, by accepting the article for publication, acknowledges that the U.S. Government retains a non-exclusive, paid up, irrevocable, worldwide license to publish or reproduce the published form of this work, or allow others to do so, for U.S. Government purposes.

REFERENCES

[1] H. M. Branz, W. Regan, K. J. Gerst, J. Brian Borak, and E. A. Santoria, "Hybrid solar converters for maximum exergy and inexpensive dispatchable electricity," *Energy Environ. Sci.,* vol. 8, pp. 3083-3091, 2015.

[2] G. A. Landis, D. Merritt, R. P. Raffaelle, and D. Scheiman, "High-temperature solar cell development," in *18th Space Photovoltaic Research and Technology Conference*, 2005, pp. 241-247.

[3] E. E. Perl, J. Simon, J. F. Geisz, M. L. Lee, D. J. Friedman, and M. A. Steiner, "Measurements and Modeling of III-V Solar Cells at High Temperatures up to 400°C," *IEEE Journal of Photovoltaics,* vol. 6, pp. 1345-1352, 2016.

[4] H. Hovel, "Solar Cells," in *Semiconductors and Semimetals.* vol. 11, R. W. a. A. Beers, Ed., ed, pp. 16-19.

[5] M. A. Green, "General temperature dependence of solar cell performance and implications for device modelling," *Progress in Photovoltaics: Research and Applications,* vol. 11, pp. 333-340, 2003.

[6] J. C. Fan, "Theoretical temperature dependence of solar cell parameters," *Solar Cells,* vol. 17, pp. 309-315, 1986.

[7] D. J. Friedman, "Modelling of tandem cell temperature coefficients," in *25th IEEE Photovoltaic Specialists Conference*, Washington D.C., 1996, pp. 89-92.

[8] E. E. Perl, J. Simon, J. F. Geisz, W. Olavarria, M. Young, A. Duda, D. J. Friedman, and M. A. Steiner, "Development of high-bandgap AlGaInP solar cells grown by organometallic vapor-phase epitaxy," *IEEE Journal of Photovoltaics,* vol. 6, pp. 770-776, 2016.

[9] S. R. Kurtz, P. Faine, and J. M. Olson, "Modeling of two-junction, series-connected tandem solar cells using top-cell thickness as an adjustable parameter," *J. Appl. Phys.,* vol. 68, pp. 1890 - 1895, 1990.

[10] C. A. Gueymard, "Parameterized transmittance model for direct beam and circumsolar spectral irradiance," *Solar Energy,* vol. 71, pp. 325-346, 2001.

On the use of transparent conductive oxides in high concentrator III-V multijunction solar cells

Ignacio Rey-Stolle[1,2], Yeonbae Lee[1], Iván García[2], Luis Cifuentes[2],
Kin Man Yu[3], Carlos Algora[2], Wladek Walukiewicz[1]

[1] Lawrence Berkeley National Lab., Materials Science Division, Berkeley, CA (United States)
[2] Universidad Politécnica de Madrid, Instituto de Energía Solar, Madrid (Spain)
[3] City University of Hong Kong, Dept. of Physics and Materials Science, Kowloon (Hong Kong)

Abstract — Transparent conductive oxides are commonplace in many PV technologies but have not been applied to III-V multijunction solar cells so far. In this paper, we analyze the current performance of the latest generation of these materials and study the potential advantages of their use in III-V multijunction solar cells via simulations and experiments. Calculations show that a number of conductive oxides have potential for improving the performance of these devices by boosting the lateral conductance of the emitter without significant optical losses. Initial experimental results reveal that there exist a number of practical challenges to be overcome to reach the improvements predicted by theory.

I. INTRODUCTION

The use of transparent electrodes in electronic devices has experienced a tremendous growth over the last decades. As of today, liquid crystal displays, TFT screens, and a number of different solar cell technologies use transparent electrodes, in particular, those implemented using transparent conductive oxides (TCOs). Focusing on solar cells, and just to name a few examples, Table I presents a summary of the TCOs currently being used or investigated in different PV technologies. As the table shows, Indium Oxides doped with Tin, Gallium or Fluorine are commonplace, and Zinc Oxide and Tin Oxide are also extensively used. Moreover, many other compounds are the subject of intensive research in the lab, in the search for Indium-free alternatives, better conductivities or higher transparencies. Some examples of these include CdO (both intrinsic and doped with In or Ga) and the multiplicity of ternary compounds that can be formed by mixing the binaries in Table I. Detailed reviews on the application of TCOs in photovoltaic and optoeletronic devices can be found in [1-3].

Despite this evident penetration in most PV technologies, historically TCOs have been deemed of little interest for III-V multijunction solar cells. In short, the reason behind this is the lack of sufficient transparency of the films with the thickness needed to attain high enough sheet conductivity. The common belief is that TCOs typically show low parasitic absorption in a limited and well-defined spectral range, normally encompassing the visible part of the spectrum (i.e. from 400

to 700 nm), which is insufficient for devices whose spectral response extends down to ~1850 nm (i.e. Germanium subcell). Moreover, the need of very high conductivities in the top electrode to minimize the ohmic losses associated with the high current densities produced under concentrator operation made the use of TCOs unfeasible. However, these arguments are starting to be questioned as good transparency in wider spectral ranges and resistivity in the range of 10^{-5} $\Omega \cdot cm$ –just an order of magnitude higher than metals– have been reported [3,4]. In this context, the goal of this paper is to present a preliminary assessment about the use of state-of-the-art TCOs in III-V multijunction solar cells, in particular in those designed for high concentrator applications. To do so we first analyze through simulations the effects of including TCOs as lateral conducting layers in the front surface of the cell and then we subsequently present the experimental results of the application of TCO thin films onto GaInP/Ge dual-junction solar cells.

TABLE I
SOME TRANSPARENT CONDUCTIVE OXIDES USED IN SOLAR CELLS

TCO	Dopants	PV Technology
In_2O_3	Sn, Ga, F, H	amorphous Si CdTe, CIGS Silicon HIT
ZnO	Al, In, Ga	amorphous Si CdTe CIGS Silicon HIT
SnO_2	Sb, F	CdTe Dye sensitized Perovskite
TiO_2	Nb	CdTe Dye sensitized Perovskite
CdO	Ga, In	CdTe

978-1-5090-5606-4/17 $31.00 © 2017 IEEE

TABLE II

IMPACT OF DIFFERENT TCOS ON THE SHORT CIRCUIT CURRENT OF TRIPLE JUNCTION SOLAR CELLS

TCO	W_{TCO} [nm]	R_{TCO} [Ω/□]	$\Delta J_{SC,TC}$ [%]	$\Delta J_{SC,MC}$ [%]	$\Delta J_{SC,BC}$ [%]
In$_2$O$_3$:Sn	12.5	120	98.9	100.3	100.7
ITO	25	60	97.2	99.6	99.3
ρ =1.7 10$^{-4}\Omega\cdot cm$	50	30	92.7	95.1	91.5
ZnO:Ga	12.5	240	98.7	100.3	100.6
GZO	25	120	96.7	99.6	99.6
ρ = 4.2·10$^{-4}\Omega\cdot cm$	50	60	91.3	95.4	94.0
	12.5	280	93.8	99.8	100.1
CdO	25	140	87.1	99.6	100.1
ρ = 3.5·10$^{-4}\Omega\cdot cm$	50	70	75.4	99.5	100.4
	12.5	120	98.3	100.2	100.7
CdO:Ga	25	60	96.2	100.1	100.6
ρ = 0.8·10$^{-4}\Omega\cdot cm$	50	30	91.8	99.7	98.8
	12.5	40	99.1	100.3	101.0
CdO:In	25	20	97.9	100.2	100.5
ρ = 0.5·10$^{-4}\Omega\cdot cm$	50	10	95.3	99.1	96.6

The sheet resistances of the TCO layers have been calculated using the representative resistivity values R_{TCO} as included in the third column. The columns labeled ΔJ_{SC} denote the ratio between the short circuit current produced by the subcell with TCO over that produced by the same subcell without TCO. ΔJ_{SC} values over 100% represent situations in which the antireflection performance of the TCO is indeed better than that of the ZnS layer considered in the reference case.

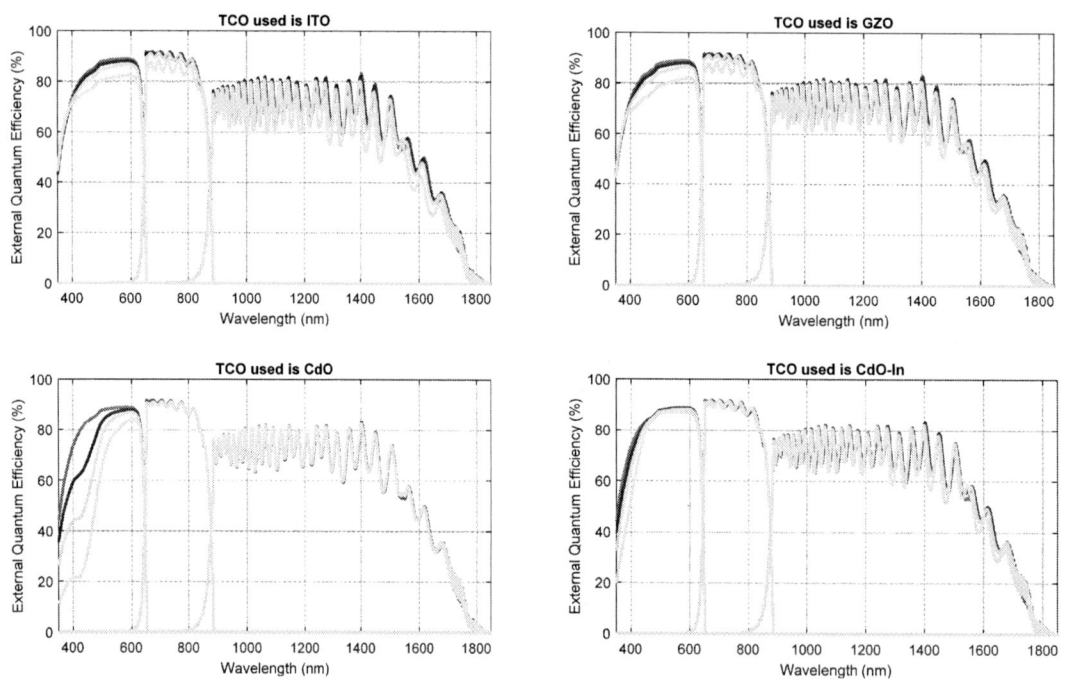

Fig. 1. Simulation of the external quantum efficiency (EQE) of a state-of-the-art triple junction solar cell with different TCOs deposited on the front surfaces. In all graphs the red line corresponds to the solar cell without TCO. The blue, green and cyan lines correspond to solar cells with TCO layers of 12.5 nm, 25nm and 50 nm. The TCO used in each graph is indicated in a label on top of each graph.

II. PERFORMANCE OF MULTIJUNCTION SOLAR CELLS WITH TCOs

Fig. 1 presents the results of the simulation of the external quantum efficiency of state-of-the-art GaInP/Ga(In)As/Ge multijunction solar cells with TCOs, using conventional analytic models [5]. In the calculations, the conductive oxide has been placed between the AlInP window layer of the top cell and the antireflection coating. Oxide thicknesses of 0 nm (i.e. no TCO), 12.5 nm, 25 nm and 50 nm have been considered. For the cell with no TCO, a simple MgF₂/ZnS (100/50 nm) has been used. For the cells with TCO, the thickness of the ZnS layer has been reduced so that ZnS and TCO thickness add up to 50 nm. A shadowing factor of 8% has been considered in all cases.

Table II summarizes some properties of the oxides considered and the results in terms of short circuit current quoted as the ratio of the J_{SC} of the subcell with TCO over the J_{SC} produced by the same subcell without TCO. The analysis of Table II and Fig.1 reveals that the losses for designs using TCO layers of more than 25 nm are significant, for all the oxide compounds considered. These losses mostly affect the GaInP top cell, and to a lower extent to the Ga(In)As middle and Ge bottom subcells (in a similar magnitude in both these subcells). This fact suggest that the most relevant challenge to deal with in order to integrate TCOs in multijunction solar cells is above bandgap absorption and, to a lower extent, free carrier absorption and plasma reflection in the infrared [4, 6]. However, in the other end, almost all TCOs show moderate losses for thin layers of 12.5nm (<2% in top cell J_{SC}, and almost no loss in the middle and bottom subcells). Even more so, ITO and CdO:In show drops of top subcell J_{SC} below 3% for 25nm thick layers. The point worth noting here is that, even with such thin layers, the sheet resistances achievable are attractive. The sheet resistance of the GaInP top cell emitter in triple junction solar cells is typically in the range of 300-500 $\Omega/\square$ (the contribution of the AlInP window is negligible). Hence, values of ~120 $\Omega/\square$ (ITO of 12.5nm) or ~40 $\Omega/\square$ (CdO:In of 12.5nm) would roughly represent a reduction of the "effective" emitter sheet resistance to a third or a factor of ten, respectively, and would thus contribute to the lateral current spreading, provided that the contact resistance between the AlInP window and the TCO is sufficiently low. Moreover, taking advantage of such low sheet resistances, sparser grids with lower shadowing factor could be used, which would in turn compensate for the loss in J_{SC}. In this scenario, lower top cell lateral resistances seem within reach without major drops in photocurrent, paving the way for better concentrator performance.

III. EXPERIMENTAL

In order to provide an initial validation of the simulation results of figure 1, we decided to test the use of CdO layers on double junction GaInP/Ge structures. On the one hand,

according to Table 1, CdO is not the best option but it paves the way to optimize the deposition process for CdO:In or CdO:Ga, which would provide clearly better results. On the other hand, the use of dual-junction GaInP/Ge structures allows monitoring the effects of the TCO on the two most affected subcells, while minimizing the cost and duration of the epi process. Obviously, the next step in this study will be to use CdO:In layers on GaInP/Ga(In)As/Ge structures.

GaInP/Ge solar cell structures were grown in a commercial Aixtron 200/4 MOVPE reactor on p-type Ge(100). Some samples were processed in a conventional way (as controls) whilst in other samples the GaInAs cap layer was chemically etched using NH₄OH:H₂O₂:H₂O (2:1:10) leaving the top AlInP window as the front end of the structure. Subsequently, undoped CdO layers with a nominal thickness of 25nm were deposited in a radio frequency magnetron sputtering system under an Ar background pressure of 5 mTorr. Additionally, soda lime glass slides were also used as substrates to measure the optical and electrical properties of the CdO layers.

On the TCO samples deposited on glass, reflectance and absorptance were measured from 200 to 2500 nm using a Perkin-Elmer Lambda spectrophotometer. Additionally, the mobility and the electron concentration in the oxide were measured at room temperature by Hall Effect in the Van der Pauw configuration using an Ecopia HMS-3000 system with magnetic fields of 0.5 T and 1.0 T.

The TCO/GaInP/Ge samples were processed into solar cells using photolithography and electroplated Au contacts. No anti-reflection coating was deposited on the devices. The solar cells followed a standard characterization process including external quantum efficiency, dark and light I-V. Additionally, the reflectance of the solar cell structures was also measured using the same Perkin-Elmer Lambda spectrophotometer.

IV. RESULTS

A. Optical characterization of the TCOs

Fig. 2 plots the transmittance of a 25nm thick layer of CdO deposited on glass. The agreement between experiment (black dots) and modelling (blue line) is excellent except for the UV region (λ<430nm), where T is overestimated.

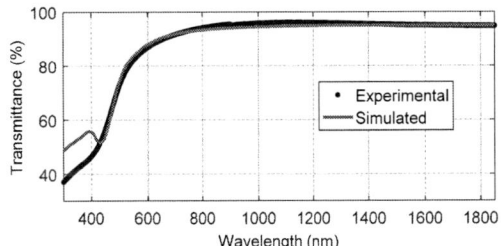

Fig. 2. Transmittance of a 25nm thick layer of CdO deposited on glass. Black dots are experimental data and solid line is the modeling.

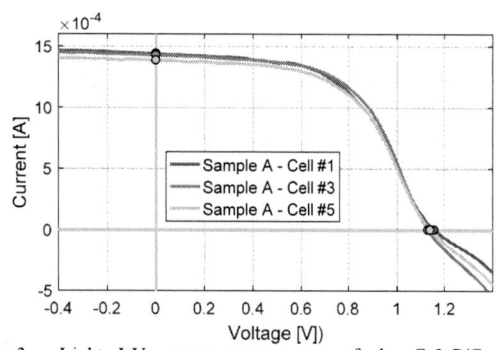 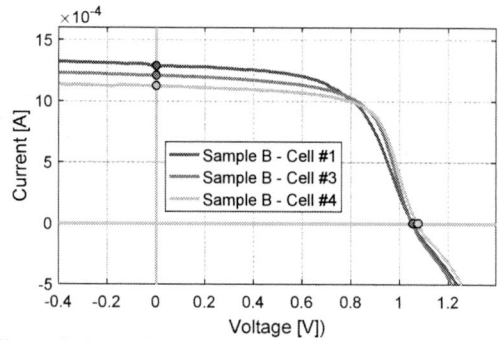

Fig. 3. Light I-V curves at one sun of the GaInP/Ge solar cells studied. (Left) three representative devices of sample A. (Right) Three representative devices of sample B.

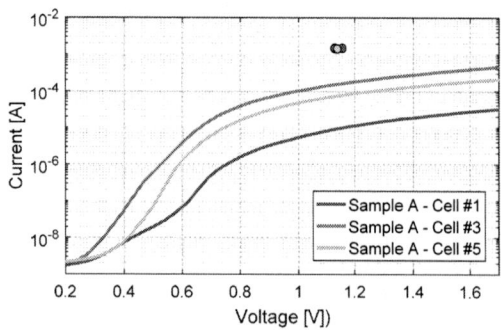 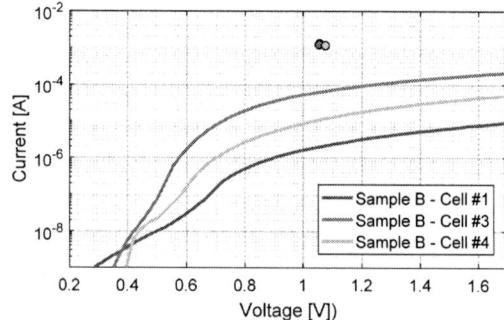

Fig. 4. Dark I-V curves of the GaInP/Ge solar cells studied. (Left) three representative devices of sample A. (Right) Three representative devices of sample B. The devices measured are the same as those included in Fig 3.

B. Electrical characterization of the TCO layers

The results on the Hall Van der Pauw measurements on the CdO samples deposited on glass are summarized on Table III. These results correspond to the average of several devices per sample. On sample A, we obtained a value of carrier concentration lower than expected possibly as a result of O_2 contamination in the sputtering process. The problem was solved in sample B that yielded state of the art values for both mobility and carrier concentration (see Table II) [4].

TABLE III
ELECTRICAL CHARACTERIZATION OF THE CdO LAYERS

Sample	Thickness [nm]	μ [cm²V⁻¹s⁻¹]	n [cm⁻³]	R_{sheet} [Ω/□]
A	25	99	$7 \cdot 10^{19}$	340
B	25	90	$1.8 \cdot 10^{20}$	154

C. Light I-V curves

Fig. 3 shows the I-V curves under one-sun AM1.5d illumination of three illustrative devices in sample A (left graph) and B (right graph). The solar cells plotted were chosen to visually illustrate the device-to-device variability observed in each sample. Table IV summarizes the main solar cell parameters in each sample including the mean I_{SC} and its standard deviation, and the average V_{OC} and its standard deviation

Sample A presents a moderate dispersion in the I_{SC} (~2%); a relative uniformity in the fill factor and a comparatively larger variability in the V_{OC}. In contrast, Sample B presents a significant variability in the I_{SC} (~7%), which is in addition 15% lower than in sample A; a quite variable fill factor; and a significantly lower V_{OC}. However, the most relevant feature in these curves for both samples A and B is the change in slope in the curve beyond V_{OC} that is indicative of a parasitic junction in the structure.

TABLE IV
ONE-SUN I-V CHARACTERIZATION

Sample	I_{SC} [mA]	ΔI_{SC} [%]	V_{OC} [mV]	ΔV_{OC} [mV]
A	1.4	2%	1.142	±15
B	1.2	7%	1.063	±12

D. Dark I-V curves

Fig. 4 shows the dark I-V curves for the same three illustrative devices in sample A (left graph) and B (right graph). The colored circles included in the figures are the I_{SC}-V_{OC} points of each cell obtained from Fig. 3.

Both samples A and B present a very large dispersion in the reverse saturation current (I_0) and in the series resistance (R_S), with sample B having larger values in both parameters. In both cases, I_{SC}-V_{OC} points lie far out of their respective dark I-V curves indicating a lack of superposition between dark an illumination. This fact, together with the high series resistance, is again indicative of the presence of a parasitic junction.

E. External Quantum Efficiency

Fig. 5 depicts the External Quantum Efficiency (EQE) of the GaInP top cells in the three samples studied. The blue line is the *reference sample* with no CdO (blue); the red line is sample A and the green line is sample B, both of them with 25nm of CdO. The shaded region around the red and green curves represents the variability measured in each sample, which is virtually the same as that measured in the light I-V curves (see Table IV).

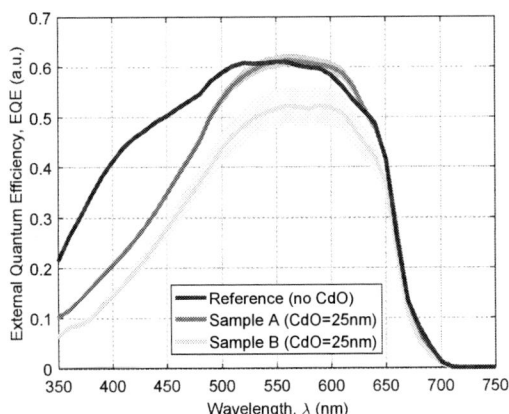

Fig. 5. External Quantum Efficiency of the GaInP top cells studied: Reference device with no CdO (blue); sample A (red) and sample B (green), both with 25nm of CdO. The shadowed area represents the variability measured in each sample.

As compared to the reference cell with no CdO, sample A exhibits the large loss expected from 350 to 500 nm due to absorption in the CdO, as predicted by the simulations in Fig. 1. The slight improvement from 500 to 650 nm was also expected as a result of the antireflection role of the TCO. However, a higher EQE was expected for the 25 nm CdO deposited, according to our simulations. Concerning sample B, the loss in the whole spectral range is evident, as compared to sample A and the Reference.

F. Characterization of the Metal/TCO and TCO/AlInP contacts

In addition to solar cells, TLM patterns were also manufactured on the samples to measure the metal/CdO specific contact resistance. The average specific metal/TCO contact resistance between the gold and the CdO is $2 \cdot 10^{-4}$ $\Omega \cdot cm^2$ in both samples. However, it presents a significant variability with minimum values in the range of $5 \cdot 10^{-5}$ $\Omega \cdot cm^2$ and maximum values around $4 \cdot 10^{-4}$ $\Omega \cdot cm^2$. All contacts were annealed at 200°C for 180s, and no attempt was made to optimize such conditions.

In order to test the current flow between the TCO and the AlInP window, the TLM patterns were etched removing the CdO layer between TLM pads. The CdO layer was still present underneath the pads so current was forced to flow from the CdO into the AlInP layer to circulate from pad to pad. In this experiment rectifying contacts were measured in all cases, confirming the presence of a non-ohmic junction between the CdO and AlInP.

V. DISCUSSION

The CdO layers deposited on the GaInP/Ge structures have affected severely the light and dark I-V curves of the solar cells at two different levels. On the one hand, the heterojunction formed between the CdO and the n-AlInP window produces a non-ohmic contact. This was confirmed by TLM measurements and manifests as a variable R_S in the dark I-V and a leaky diode in reverse bias in the light I-V. The origin of such is still unclear. No cleaning was performed on the samples prior to CdO deposition so a thin layer of oxide is assumed to be present at the interface. In addition, the sputtering temperature of 200°C or the contact annealing (also at 200°C) might favor further oxidation of the AlInP. Currently, the heterojunction formed between these materials is under investigation to clarify the dependence of this barrier on the doping and Al-content of the AlInP window. On the other hand, the increase and dispersion in the reverse saturation current of the cells observable in the dark I-V seems to suggest that structure has been damaged during the sputtering process. The increased recombination resulting from the damage could account for the losses (and variability) observed in both I_{SC} and V_{OC}. However, the big differences between samples A and B, which went through virtually the same sputtering process in terms of power, temperature, duration and distance to the target, seem to indicate that the damage is related to an uncontrolled process variable. At this point, we speculate that the particular position of the sample in the holder plays a role in the final dose of damage received. In this sense, sample A could have received a lower dose of damage and therefore its EQE almost coincides with what simulations predict. Conversely, sample B would have received a higher dose of damage and therefore its EQE is

978-1-5090-5606-4/17 $31.00 © 2017 IEEE

greatly diminished and its reverse saturation current is higher, on average.

VI. SUMMARY AND CONCLUSIONS

The application of transparent conductive oxides on III-V multijunction solar cells has been studied by simulations and experiments. The impact of including layers of several TCOs on the external quantum efficiency of GaInP/Ga(In)As/Ge multijunction solar cells has been analyzed by theoretical simulations using real optical constants of the semiconductor stack and the TCO layers, elucidating the optical losses to be expected. Moderate losses below 2% are found, mainly in the top cell for thin layers of 12.5nm of several TCOs (ITO, GZO, CdO:In, CdO:Ga. However, the electrical properties of such TCO layers offer the possibility of significantly reducing the lateral sheet resistance at the front of the solar cell, which would allow to use sparser grid lines and, hence, reduce the shadowing losses.

Preliminary experiments using undoped CdO to coat GaInP/Ge solar cells have unveiled challenges to overcome in order to exploit the beneficial properties of TCOs. First, the attainment of an ohmic conduction path between the top cell AlInP window layer has not been achieved using this TCO, which obviously spoils the performance of the solar cell. Second, the sputtering process used to deposit the TCO layers appears to damage the solar cells, which exhibit significantly larger reverse saturation currents. However, expectations to overcome these problems by using doped CdOs and redesigned deposition conditions are in place.

In summary, the potential of using TCOs in III-V solar cells to assist in the lateral conduction at the front and improve the performance under concentration has been demonstrated, and the preliminary experiments carried out have shown the challenges to overcome and, therefore, have focused the oncoming efforts to be undertaken.

ACKNOWLEDGEMENTS

This work was supported by the Spanish Ministerio de Economía y Competitividad through projects with code TEC2015-66722-R and projects TEC2014-54260-C3-1-P and by the Comunidad de Madrid through the project MADRID-PV (S2013/MAE-2780). I. García acknowledges the financial support from the Spanish Programa Estatal de Promoción del Talento y su Empleabilidad through the Ramón y Cajal grant with reference RYC-2014-15621. Part of this work was performed at the EMAT, National Center for Electron Microscopy/Molecular Foundry and was supported by the Director, Office of Science, Office of Basic Energy Sciences, Materials Sciences and Engineering Division, of the U.S. Department of Energy under Contract No. DE-AC02-05CH11231. W. Walukiewicz acknowledges funding support from the Singapore National Research Foundation (NRF) through the Singapore Berkeley Research Initiative for Sustainable Energy (SinBeRISE) Programme

REFERENCES

[1] E. Fortunato et. al, '*Transparent Conducting Oxides for Photovoltaics*', MRS Bulletin, 32(3), pp. 242–247 (2007)

[2] A. Klein, et al. "*Transparent Conducting Oxides for Photovoltaics: Manipulation of Fermi Level, Work Function and Energy Band Alignment*", Materials 3, 4892-4914; (2010)

[3] M. Morales-Masis, et al. "*Transparent Electrodes for Efficient Optoelectronics*", Adv. Electron. Mater. 2017, 3, 1600529.

[4] K. M. Yu, et al, "*Ideal transparent conductors for full spectrum photovoltaics*", Journal of Applied Physics 111, 123505 (2012)

[5] I. Rey-Stolle, J. M Olson, C. Algora "*Concentrator Multijunction Solar Cells*", Chapter 2 of Handbook of Concentrator Photovoltaic Technology, John Wiley, 2016

[6] D. S. Ginley and J. D. Perkins, in Handbook of Transparent Conductors, edited by D. S. Ginley, (Springer, 2010), Chapter 1

Component Integration Effects in 4-junction Solar Cells with Dilute Nitride 1eV Subcell

I. García, M. Ochoa, I. Lombardero, L. Cifuentes, P. Caño, M. Hinojosa, I. Rey-Stolle and C. Algora[1]

A. D. Johnson and J. I. Davies[2]

K.H. Tan, W.K. Loke, S. Wicaksono and S. F. Yoon[3]

[1]Instituto de Energía Solar, Universidad Politécnica de Madrid, 28040 Madrid (Spain)
[2]IQE plc, Cardiff (UK)
[3]Nanyang Technological University, 50 Nanyang Avenue, Singapore 639798 (Republic of Singapore)

Abstract — A GaInP/Ga(In)As/GaNAsSb/Ge 4-junction solar cell grown using combined MOVPE+MBE growth is used to analyze the effects during the integration of the subcell components into the full 4J structure. In this preliminary study, the Ge subcell is observed to suffer about 15% J_{sc} drop and ~50 mV V_{oc} loss at 1-sun, while the V_{oc} of the GaNAsSb subcell drops by as much as ~ 140 mV. The degradation of the Ge and GaNAsSb subcells in the current-matched 4J structure can hinder its efficiency potential to a higher extent than in the GaInP/Ga(In)As/Ge 3J. Besides, high quality GaNAsSb and Ge subcells would still limit the current and require redesigning the top subcells to achieve optimum efficiencies.

I. INTRODUCTION

Monolithic and lattice-matched multijunction solar cells are attractive for fabrication simplicity and, hence, lower cost. Using dilute-nitride subcells is promising and has already shown high efficiencies in 3J solar cells grown by MBE [1], and combined MOVPE + MBE [2]. A GaInP/Ga(In)As/ GaNAsSb/Ge 4J solar cell, using the Ge substrate as an active subcell, conceptually consists on just adding the dilute nitride junction to the GaInP/Ga(In)As/Ge 3J solar cell. Currently, attaining a dilute-nitride subcell with longer minority carrier diffusion length so as not to limit the performance of the whole 4J structure is the main limitation to achieving high efficiencies. Therefore, a high fraction of the work to do consists on further developing the semiconductor deposition technique. Ideally, MOVPE would be used to grow the full 4J structure, given its advantages with respect to other growth techniques concerning high volume fabrication cost. However, attaining high quality dilute-nitride subcells by MOVPE has been elusive, although a recent publication has shown positive steps towards achieving this [3]. Conversely, better carrier diffusion lengths have been achieved by MBE, so as to obtain high efficiency 3J solar cells [1]. Thus, the combined MOVPE + MBE method makes sense at this stage of development and should allow the attainment of high efficiencies in the short term.

The 4J structure is complex and involves growing a large number of layers of different materials, growth temperatures and times. It is important to assess the integration of all these components to detect and take care of problems in advance, and facilitate the development of a high efficiency 4J solar cell. The focus has to be put in identifying and quantifying the effects that put in risk the efficiency potential of this 4J solar cell and redesigning, if possible, the growth process to minimize their impact.

In this work we first present our 4J solar cell, based on a prototype structure achieved by the combined MOVPE + MBE method, and using a GaNAsSb junction. This solar cell exhibits a low performance due to the limiting behavior of the dilute nitride subcell, but represents a success in the implementation of the 4J structure. However, this structure is used to analyze the integration of subcells into the full 4J structure as the main point of this work. In this preliminary work, we focus on the GaNAsSb and, mostly, the Ge bottom subcell, which are subjected to long annealing times during the growth of the upper structures. It is found that significant loses are at stake, mainly in the Ge and GaNAsSb subcells, which can limit the potential of this 4J solar cell structure to compete in efficiency with other architectures.

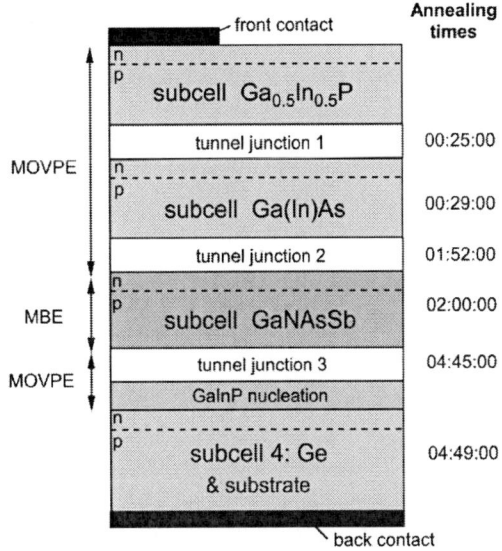

Fig. 1. Structure of the 4J solar cell presented, with indication of the parts grown by MOVPE and MBE, and the annealing times that each component subcell and tunnel junction suffer during the rest of the structure.

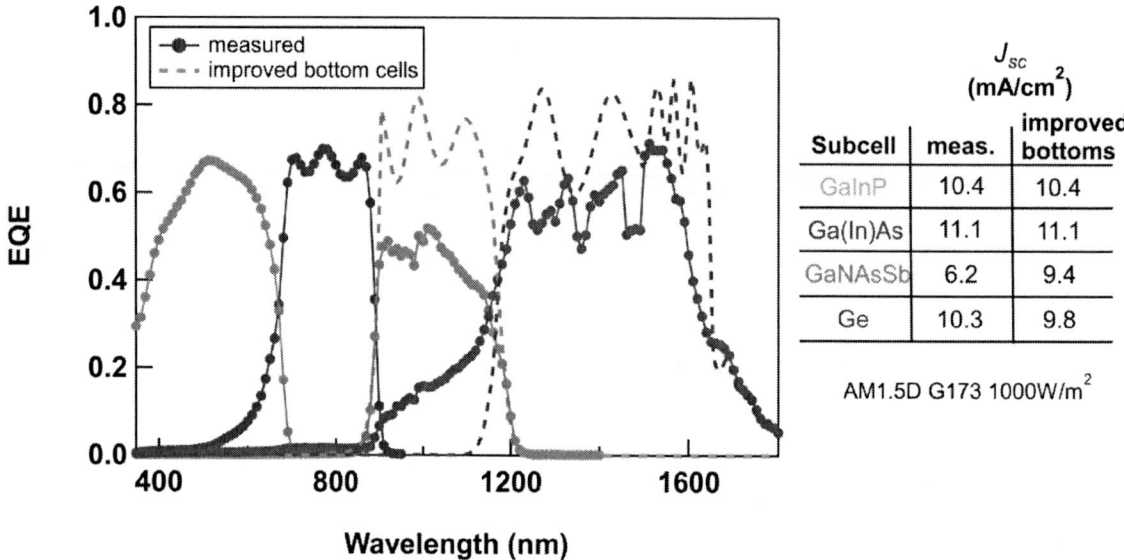

Subcell	meas.	improved bottoms
GaInP	10.4	10.4
Ga(In)As	11.1	11.1
GaNAsSb	6.2	9.4
Ge	10.3	9.8

AM1.5D G173 1000W/m^2

Fig. 2. Symbols: measured EQE of the 4J solar cell implemented; dashed line: modeled GaNAsSb and Ge subcells using minority carrier collection efficiencies of 1 and optically thick GaNAsSb. The table on the right shows the J$_{sc}$ obtained using these EQE and the AM1.5D G173 solar spectrum.

II. MOVPE+MBE 4J SOLAR CELL

The structure consists on a GaInP/Ga(In)As/GaNAsSb/Ge structure grown lattice-matched to a Ge substrate, which is also the 4th junction, as shown in Figure 1. The growth process comprises a sequence of MOVPE (GaInP nucleation on Ge and tunnel junction) + MBE (GaNAsSb subcell) + MOVPE (GaInP/Ga(In)As top junctions, including tunnel junctions). Given the low minority carrier diffusion length in the dilute nitride material, the GaNAsSb subcell absorber was made as thin as 1 μm [4], [5]. These structures were processed into solar cell devices using gold-based metal and standard photolithography techniques. No anti-reflection coating has been applied to these solar cells for this study.

A. External Quantum Efficiency (EQE).

The measured external quantum efficiency (EQE) of this solar cell is shown in Figure 2, with the corresponding J$_{sc}$ calculated using the AM1.5D G173 solar spectrum detailed in the table on the right. The top subcells exhibit the usual EQE with good J$_{sc}$ values, while the low EQE of the nitride subcell limits the J$_{sc}$ of the whole device. The J$_{sc}$ of the Ge bottom junction is also low considering the fact that the light absorption in the GaNAsSb junction is incomplete and the Ge bottom cell has an unrealistic extended response in the short wavelength range. As a first assessment of the potential and loss sources in this EQE, the modeled EQE of improved GaNAsSb and Ge subcell was calculated. The Generalized

Matrix Method was used to compute the absorption in each layer [6] and the EQE was obtained applying a carrier collection efficiency to the absorber layers to fit the experimental EQE. Then, the improved EQE was calculated using an optically thick GaNAsSb subcell and a minority carrier collection efficiency of 1 in both subcells. The resulting EQEs are also shown in Figure 2. A strong effect on the GaNAsSb subcell EQE and J$_{sc}$ is observed, as expected, but, the Ge subcell carrier collection efficiency can also be improved to achieve a higher J$_{sc}$. This can be observed as a higher average value of the EQE. Note that the J$_{sc}$ of the Ge subcell in the "improved bottoms" case is lower than the measured because the GaNAsSb subcell in this case is optically thick. The cause of the low EQE in this Ge subcell is discussed in Section III. Finally, according to the table on the right, the bottom subcells still limit the J$_{sc}$ of the 4J even improving their carrier collection efficiency. This means that realizing the full efficiency potential of this 4J solar cell will involve making the top cells more transparent by changing their thickness or, optimally, their bandgap.

B. Open Circuit Voltage (V$_{oc}$).

Light I-V curves were measured on these 4J solar cells. We have published previous work about the performance and particularities in the shape of the I-V curve of the 4J, predicted using simulations with inputs from the characterization results of GaNAsSb 1J and GaNAsSb/Ge 2J cells [4], [5]. As for this work, we are mostly interested in the open circuit voltage (V$_{oc}$), which we showed to be around 420 mV for the

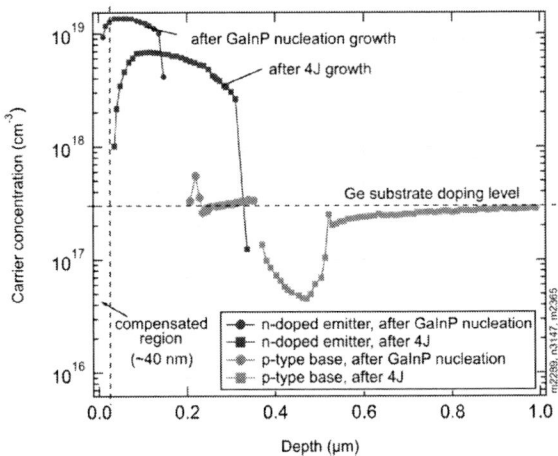

Fig. 3. Carrier concentration profiles of Ge subcells before and after growth of the full 4J structure.

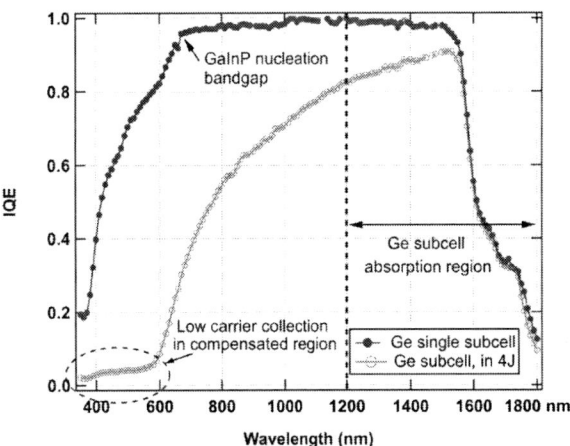

Fig. 4. IQE of the Ge subcell after being grown as a 1-junction, and in the 4-junction solar cell (measured after etching the top layers)

GaNAsSb solar cell as an independent device [4]. The 4J solar cell presented here was measured under a calibrated solar simulator at NREL, and exhibited a V_{oc} at 1-sun of 2.44 V and 2.48 V for the AM1.5d and AM0 solar spectra, respectively. This V_{oc} is rather low, as compared to state-of-the-art GaInP/Ga(In)As/Ge 3J solar cells developed in our lab, which exhibit an V_{oc} of around the same value [7]. This means that the V_{oc} increase in the 4J expected with the insertion of the GaNAsSb subcell does not occur, which eliminates any performance advantage with respect to the 3J solar cell. Part of this voltage loss is due to degradation of the Ge and GaNAsSb subcells, as discussed in next section.

III. COMPONENT INTEGRATION ANALYSIS

In order to elucidate the origin of the voltage loss, and to assess the degradation of other possible solar cell performance parameters such as the EQE, an analysis of the possible sources of subcell degradation when integrating them into a 4J solar cell was carried out. The first suspect in the list of possible causes is the thermal loads applied to the subcells. In Figure 1 we show a list of the annealing times to which each subcell of the 4J structure is subjected to. The growth temperatures range from 550 °C for the tunnel junctions to 675 °C for the subcell growth. The GaNAsSb subcell growth includes a 700 °C annealing step for 5 minutes to improve its properties, similarly as with other dilute-nitride materials [8]–[11]. The insertion of the dilute nitride subcell increases the thermal load to the Ge bottom subcell, as compared to the 3J case. Considering the different growth times and temperatures used for each component in the 4J structure, it is not obvious to predict the real impact of this added thermal load, and we assessed it empirically by examining samples at different stages of the MOVPE + MBE + MOVPE process.

A first test was to measure the dopant diffusion in the Ge junction, which we did by taking electro-chemical capacitance-voltage (ECV) profiles of Ge junctions, shown in Figure 3. The profiles show the n-type and p-type carrier concentration corresponding to the emitter and base, respectively. The base carrier concentration corresponds to the Ge substrate nominal doping, around $3 \cdot 10^{17}$ cm^{-3}. The emitter doping is achieved by diffusion of group-V elements (phosphorus) during the growth of the GaInP nucleation layer. Group-III elements also diffuse and give rise to a lower effective doping at the surface of the emitter. Preliminary SIMS measurements appear to indicate that this is caused mainly by indium diffusion.

The dopant diffusion at the emitter during the growth of the 4J structure is evident in Figure 3: the average carrier concentration level and thickness of the emitter changes substantially. Not less importantly, a strongly compensated region appears at the surface of the emitter, due to further in-diffusion of group-III elements. The carrier concentration level rises sharply to achieve its peak value at a depth of around 100 nm. The electric field created in this region can be expected to act as a sink for minority carriers which are lost by recombination.

IQE measurements were taken on these Ge subcells to assess this, as shown in Figure 4. Firstly, the IQE of the Ge subcell obtained right after the growth of the GaInP nucleation layer shows a value near 1 for all the wavelength range between the Ge and GaInP nucleation layer absorption edges. As for the Ge subcell after the growth of the 4J, an important degradation of the carrier collection efficiency in the lower wavelength region can be observed. The degraded Ge junction can be accurately modeled including the compensated layer (~ 40 nm) with 0 collection efficiency, and a degraded carrier collection in the emitter bulk.

978-1-5090-5606-4/17 $31.00 © 2017 IEEE

It is important to point out that the apparent severity of the IQE degradation shown in Figure 4 is lessened by realizing the fact that, in the 4J, the Ge subcell absorption range falls in the higher photon wavelength region, where the degradation is not so severe. However, for the range of Ge subcell absorption in the 4J, the J_{sc} drop due to the degraded IQE is as high as ~15% calculated for the AM1.5D-G173 solar spectrum.

As expected, the V_{oc} of the Ge subcell exhibits changes too. The study of the degradation of the V_{oc} in the subcells is complicated in these 4J devices: the electroluminescence technique [12] cannot be used due to the undetectable emission of light from the dilute nitride subcell. In this work, the Ge subcell voltages were accessed by etching the upper layers. A drop of ~ 50 mV after the growth of the 4J structure can be observed. We are currently carrying out experiments to elucidate the evolution of the V_{oc} during the growth of the different 4J structure components.

All in all, an important degradation in the Ge junction when integrating it in the 4J structure is observed, which can limit the performance potential of this solar cell architecture. However, measures to minimize this degradation can be taken, such as redesigning the critical steps that determine the phosphorous and indium in-diffusion into the Ge to form the emitter during growth.

As for the GaNAsSb subcell, it is submitted to an intentional annealing in order to improve its material properties, as commented before. Additionally, the growth of the GaInP/Ga(In)As top subcells involves an extra annealing of this subcell, of about 2 hours at ~675°C. However, no significant changes have been observed in its IQE, in the range of photon energies that it absorbs. However, the V_{oc} of GaNAsSb single-junction solar cells drop by as much as ~140 mV at 1-sun. The intentional annealing used to improve the GaNAsSb subcell is normally designed concerning the temperature and time used, so that there exists an optimum combination of these parameters. The additional annealing of our GaNAsSb subcell appears to deviate the total thermal load from the optimum required. This means that the optimization of the intentional annealing will need to take into account the subsequent, unwanted annealing that the GaNAsSb subcell suffers during the growth of the 4J structure.

IV. CONCLUSIONS

We have presented a GaInP/Ga(In)As/GaNAsSb/Ge 4J solar cell grown using MOVPE+MBE combined method. The GaNAsSb subcell quality is not enough to allow a high performance 4J solar cell, yet. However, this 4J structure was used for a preliminary assessment of effects during the integration of the subcell components. The Ge subcell has been found to degrade significantly during the growth of the 4J structure: its emitter carrier concentration profile changes substantially and this is correlated with a significant drop in the IQE (and J_{sc}) and V_{oc}. Moreover, the GaNAsSb subcell

shows a large V_{oc} drop, but no significant change in its IQE. Additional work to complete this analysis and extend it to the other subcells and tunnel junctions is underway. However, we can conclude that the efficiency potential of this 4J architecture can be significantly lowered if the observed degradation of the Ge and GaNAsSb subcells is not minimized by a proper design of the semiconductor structure growth routine. An important remark is that, even in the case of achieving high quality GaNAsSb and Ge subcells, they would still limit the current in the 4J and further optimization would require bandgap/thickness changes in the GaInP and Ga(In)As top junctions. Finally, the conclusions of this paper should be generalizable to other structures using dilute-nitride subcells, since the growth conditions used in our case are not harsher than commonly used.

ACKNOWLEDGEMENTS

This work has been supported by the European Commission by means of the LONGESST project (FP7 grant Agreement Number 607153), by the Spanish MINECO through the projects TEC2014-54260-C3-1-P and PCIN-2015-181-C02-02 and by the Comunidad de Madrid through the project MADRID-PV (S2013/MAE-2780). I. García is funded by the Spanish Programa Estatal de Promoción del Talento y su Empleabilidad through a Ramón y Cajal grant. I. Lombardero and M. Hinojosa are funded by the Spanish MINECO through an FPU14 and FPU15 grants, respectively. The authors want to acknowledge the technical support by J. Bautista. 1-sun I-V measurements with calibrated light spectrum were carried out at NREL, for which the authors are indebted to J. Geisz and D. Friedman.

REFERENCES

[1] R. Jones-Albertus et al., "Using Dilute Nitrides to Achieve Record Solar Cell Efficiencies" MRS Online Proc. Libr. Arch., vol. 1538, pp. 161–166, Jan. 2013.

[2] A. Tukiainen et al., "High-efficiency GaInP/GaAs/GaInNAs solar cells grown by combined MBE-MOCVD technique" Prog. Photovolt. Res. Appl., vol. 24, no. 7, pp. 914–919, Jul. 2016.

[3] Y. Zhang et al., "GaInP/GaInAs/GaInNAs/Ge Four-Junction Solar Cell Grown by Metal Organic Chemical Vapor Deposition with High Efficiency" Chin. Phys. Lett., vol. 33, no. 10, p. 108801, 2016.

[4] M. Ochoa et al., "Advances Towards 4J Lattice-Matched including Dilute Nitride Subcell for Terrestrial and Space Applications" in 2016 IEEE 43rd Photovoltaic Specialist Conference (PVSC), 2016, p. n/a.

[5] M. Ochoa et al., "Modelling of lattice matched dilute nitride 4-junction concentrator solar cells on Ge substrates" AIP Conf. Proc., vol. 1766, no. 1, 2016.

[6] E. Centurioni, "Generalized matrix method for calculation of internal light energy flux in mixed coherent and incoherent multilayers" Appl. Opt., vol. 44, no. 35, pp. 7532–7539, Dec. 2005.

[7] E. Barrigón Montañés, "Development of GaInP/GaInAs/Ge triple-junction solar cells for CPV applications" Phd, E.T.S.I. Telecomunicación (UPM), 2014.

[8] J.-C. Harmand *et al.*, "GaNAsSb: how does it compare with other dilute III–V-nitride alloys?" *Semicond. Sci. Technol.*, vol. 17, no. 8, p. 778, 2002.

[9] R. Kudrawiec *et al.*, "Photoluminescence from as-grown and annealed $GaN_{0.027}As_{0.863}Sb_{0.1}$/GaAs single quantum wells" *J. Appl. Phys.*, vol. 98, no. 6, p. 063527, Sep. 2005.

[10] S. Kurtz, R. King, D. Law, A. Ptak, J. Geisz, and N. Karam, "Effects of in situ annealing on GaInNAs solar cells" in *2013 IEEE 39th Photovoltaic Specialists Conference (PVSC)*, 2013, pp. 2095–2099.

[11] N. Miyashita, N. Ahsan, and Y. Okada, "Improvement of 1.0 eV GaInNAsSb solar cell performance upon optimal annealing" *Phys. Status Solidi A*, vol. 214, no. 3, Mar. 2017.

[12] T. Kirchartz, U. Rau, M. Hermle, A. W. Bett, A. Helbig, and J. H. Werner, "Internal voltages in GaInP/GaInAs/Ge multijunction solar cells determined by electroluminescence measurements" *Appl. Phys. Lett.*, vol. 92, no. 12, p. 123502, Mar. 2008.

Bismuth Surfactant-Mediated Growth of GaNAsSb(Bi) Solar Cells

Aymeric Maros[1], Chaomin Zhang[1], Jongwon Lee[1], Hongfeng Wang[2], Stephen Bremner[2], Nikolai Faleev[1], Christiana B. Honsberg[1], Richard. R. King[1]

[1]Arizona State University, School of Electrical, Computer and Energy Engineering, Tempe, AZ, 85821, USA

[2]University of New South Wales, School of Photovoltaic and Renewable Energy Engineering, Sydney 2052, Australia

Abstract — **In this work, we investigate the effect of using bismuth as a surfactant during the growth of GaNAsSb solar cells lattice-matched to GaAs. First, we present a detailed analysis of the device characteristics grown without Bi. Low current collection, which was attributed to high acceptor background doping and hence, short depletion width, was found to be the main factor limiting the device performance. Bi surfactant was then introduced with relatively high fluxes during the growth of $GaN_xAs_{1-x-y}Sb_y$ films. Preliminary investigation of material quality are presented. Upon tuning of the growth conditions, it is hoped that Bi will reduce the unintentional high p-type background doping and hence, improve the performance of these dilute nitride devices.**

I. INTRODUCTION

GaNAsSb has the potential to be used as the third 1-eV subcell in a lattice-matched four-junction solar cell. It has been proven to be an interesting alternative to the most commonly studied GaInNAs(Sb) dilute nitride material. Due to their difference in band structure, GaNAsSb has the potential to reach lower bandgaps than GaInNAs for the same N content [1]. This is interesting since increasing the N content is known to decrease the minority carrier lifetime [2]. It is also believed that GaNAsSb offers higher nitrogen incorporation in comparison with GaInNAs although our previous work did not necessarily agree with these findings [3, 4]. GaInP/GaAs/GaInNAsSb triple-junction solar cells were the first solar cells to demonstrate an efficiency above 43 % [5], thereby confirming their potential for use in advanced multijunction cell designs. Low minority-carrier diffusion lengths and hence low carrier collection remain one of the main parameters limiting the performance of these dilute nitride solar cells. Even with short diffusion lengths, good current collection can be achieved with wide depletion widths [6], for which low background doping is key. GaNAsSb grown in our MBE chamber revealed a net acceptor concentration close to 1×10^{17} cm^{-3}. In comparison, very low background doping ($< 10^{16}$ cm^{-3}) have been reported in MBE-grown GaInNAs materials [7].

Surfactants are often used during MBE growth and other vapor-phase epitaxy techniques as they tend to reduce surface roughness and improve material quality. Sb has been widely used during the growth of GaInNAs(Sb) and led to record performance for lasers and solar cells [8]. Antimony, however, has been reported to increase the net acceptor concentration of GaInNAs. Additionally, Sb incorporates in the lattice. It was also found that the presence of Sb increased the dark current of GaInNAs(Sb) solar cells [7]. Bi is another surfactant that has been used during the growth of various materials such as AlGaAs, GaInAs, GaNAs and GaInNAs [6,9,10]. Unlike Sb, Bi does not incorporate at typical growth conditions for these materials. While Sb is believed to contribute to the net acceptor concentration, Ptak *et al.* reported that the use of Bi during the growth of GaInNAs(Bi) solar cells increased the net donor concentration and when relatively high Bi fluxes were used, the background doping of GaInNAs switched from p- to n-type [6]. It has also been shown that Bi increases the N incorporation [11] although Ptak *et al.* did not observe this phenomenon. One plausible explanation is that their reported Bi fluxes were 1 to 2 orders of magnitude lower than that reported by other groups and their growth temperature was also slightly higher, which might have had an effect on the growth kinetics [6], [11].

Here, we explore for the first time the use of Bi as a surfactant during the growth of GaNAsSb(Bi) and investigate its effect on device performance. It is hoped that Bi will contribute to reducing the relatively high background acceptor concentration in our GaNAsSb material. The performance of GaNAsSb solar cells grown without Bi are analyzed first. Effect of the post-growth thermal treatment is discussed. The addition of Bi during growth of GaNAs(Bi) and GaNAsSb(Bi) layers is then investigated and preliminary structural and optical properties are reported.

II. EXPERIMENTAL DETAILS

GaNAsSb solar cells were grown in an Applied Epi Mod Gen III molecular beam epitaxy (MBE) system equipped with elemental sources of Ga and Bi while As and Sb were supplied by valved crackers. Nitrogen was supplied by a UNI-bulb radio-frequency (RF) plasma source operated at 300 W with a flow of 0.2 sccm. The growth rate was 1 μm/hr and the growth temperature was 440°C for the GaNAsSb layers, 580°C for the GaAs layers and 600°C for the AlGaAs layers. The V/III ratio was set to 10 for the GaNAsSb layers and to 18 for the (Al)GaAs layers. The solar cell structures were grown on GaAs:Zn wafers and consisted of a 0.45-μm GaAs:Be buffer layer (5×10^{18} cm^{-3}), a 1.0-μm unintentionally doped (UID) GaNAsSb base, a 0.2-μm GaAs:Si emitter (2×10^{18} cm^{-3}) and a 0.03-μm $Al_{0.8}Ga_{0.2}As$ window layer (1×10^{18} cm^{-3}). A reference GaAs solar cell was grown where the GaNAsSb:UID base was

978-1-5090-5606-4/17 $31.00 © 2017 IEEE

replaced by 1.0-µm GaAs:Be (1×10^{17} cm^{-3}) which was grown under the same growth conditions as the GaNAsSb cell (*i.e.*, same substrate temperature and same V/III ratio) to mimic the growth conditions of the dilute nitride layer. Post-growth rapid thermal annealing (RTA) was performed under N$_2$ ambient at 800 °C for 10 min. Standard photolithography and wet-etch processes were used to fabricate the devices. Note that no anti-reflective coating (ARC) layers were applied. Reflection high-energy electron diffraction (RHEED) was used to monitor the surface reconstruction during growth. High-resolution X-ray diffraction (HRXRD) was used to check the lattice-matching (not shown). The external quantum efficiency (EQE) and current-voltage (I-V) characteristics were measured to determine the cell performance. Additionally, Secondary-Ion Mass Spectroscopy (SIMS) was used to measure the concentration of background impurities.

III. RESULTS AND DISCUSSION

A) GaNAsSb solar cells grown without bismuth

The I-V characteristics of the GaNAsSb heterojunction solar cell presented in Fig. 1 were measured under 1-sun conditions, showing a short-circuit current (J$_{sc}$) of 15.5 mA/cm^2, an open-voltage circuit (V$_{oc}$) of 0.39 V and a fill factor (FF) of 66 %. The EQE of this structure is shown in the inset of Fig. 1. The bandgap was determined to be 0.96 eV. These values correspond to a bandgap-voltage offset W$_{oc}$ (W$_{oc}$≡E$_g$/q-V$_{oc}$) of 0.57 V. Although a value closer to 0.4 V is expected for a high quality solar cell [12], this W$_{oc}$ value is comparable with that of other dilute nitride solar cells [13]. For a multijunction solar cell, each subcell should produce close to 14 mA/cm^2 under AM1.5D spectrum. In this case, the corresponding sub-GaAs current density is 4.5 mA/cm^2, indicating that improvement in material quality is still needed. ARC would be expected to boost this current by roughly 30 % but this would still be insufficient to reach the desired current.

Capacitance-Voltage (C-V) and Hall measurements revealed that the acceptor background doping density of the GaNAsSb:UID layer was 8×10^{16} cm^{-3}. SIMS was used to identify the type and concentration of background impurities present in this material. Depth profiling revealed that the concentration of C and O first increased to 8×10^{16} atoms/cc as the substrate temperature decreased from 580 °C to 440 °C and further increased to 8×10^{16} and 2×10^{17} atoms/cc respectively after nitrogen was introduced in the growth. Hydrogen was also measured at 2×10^{17} atoms/cc, which corresponded to the SIMS instrument detection limit and thus indicates that the actual concentration of H might be lower.

Nevertheless, these levels are comparable to those obtained in metalorganic chemical vapor deposition (MOCVD)-grown materials. It has been previously concluded that carbon and hydrogen contamination resulted in high acceptor background doping in MOCVD-grown materials [14]. The MBE materials grown in this work also seem to be subject to high carbon contamination which is most likely the main contributor to the

high p-type background doping. The contribution from O is unclear as its effect on background doping has not been widely reported for dilute nitride materials. The use of higher growth temperatures could also partially reduce the incorporation of background impurities, however our GaNAsSb materials grown at higher temperature usually demonstrate lower photoluminescence intensity.

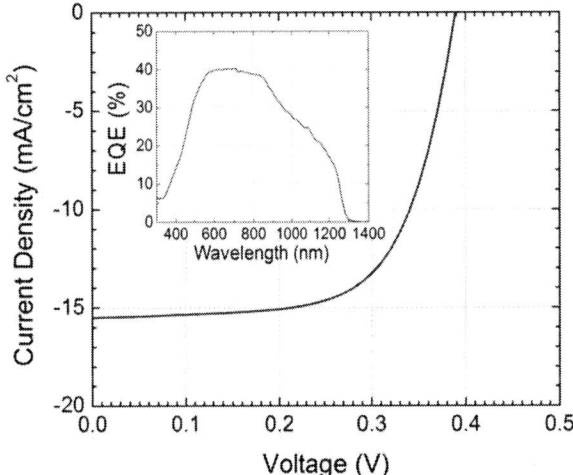

Fig. 1. I-V characteristics of a n-p GaAs/GaNAsSb heterojunction solar cell (structure G16-066). The inset shows the corresponding external quantum efficiency showing a bandgap of 0.96 eV. No ARC was applied.

B) Effect of post-growth annealing on cell performance

Post-growth annealing has been proven to be very effective in reducing the recombination centers in dilute nitride materials and has now become a routine procedure in order to achieve high performance devices [15]. Our previous material study showed that RTA at 800°C for 10 min under N$_2$ ambient led to a drastic increase in PL intensity and reduction of defect density in our GaNAsSb material [4]. One of the requirements for high quality dilute nitride materials is the use of relatively low growth temperature (< 500°C). Under such conditions, a high density of As-related point defects can be expected. We have found that RTA improves the V$_{oc}$ of our GaNAsSb cells but it is unclear if this boost in V$_{oc}$ originates from a reduction of N-related defects or other intrinsic point defects, or both.

To investigate this more thoroughly, GaAs reference cells were grown under the same conditions as the GaNAsSb structures and post-growth RTA was also performed at 800°C for 10 min. As shown in Fig. 2, RTA led to a strong increase of the V$_{oc}$ from 0.84 to 0.92 V. The FF also increased from 71 to 79 %. Owing to the absence of nitrogen in these GaAs cells, we conclude that annealing not only reduces the density of N-related defects but also strongly reduces the density of non-nitrogen related point defects.

However, as shown in Fig. 2, it was found that RTA also led a significant reduction in current density from 19.0 to 15.4 mA/cm^2. As shown in the inset of Fig. 2, this reduction in J$_{sc}$

results from a degradation of the EQE in the short wavelengths region, indicating a degradation of the material quality of the top layers. Given that the GaNAsSb and GaAs solar cells have the same structures, it is apparent that degradation of the blue region must occur also in the GaNAsSb cells after post-growth annealing.

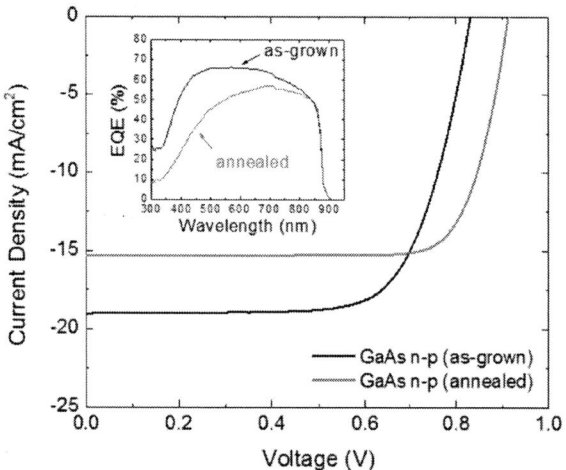

Fig. 2. Effect of post-growth RTA on the performance of a GaAs n-p reference solar cell grown under the same conditions as the dilute nitride cells. The inset shows the quantum efficiency for each case.

In-situ annealing in the MBE chamber under As overpressure, directly following the growth of the GaNAsSb layer, should avoid the need for the post-growth RTA while still improve the cell performance. This represents an area of future work.

C) Modeling of EQE reduction upon annealing

In order to investigate the material parameters limiting the performance of our devices and identify the cause for the EQE degradation after thermal treatment, modeling of the EQE was performed by solving the transport equations in a closed form following work from Hovel and Woodall [16]. This model allows to calculate the current generation in each region of the solar cell, *i.e.*, the base, emitter and depletion regions. For simplicity, only the sum of these contributions is presented in this work.

For this purpose, an additional GaNAsSb solar cell was grown and fabricated. This structure was annealed at 750 °C and 800 °C for 10 min. The best fit to experimental data was obtained by varying the minority carrier diffusion length as well as the front surface recombination velocity while all the other parameters were fixed, similar to what was presented in [17]. As expected from the annealing results of GaAs, increasing the annealing temperature led to a strong decrease of the EQE in the blue region (300 – 700 nm). As shown in Fig. 3, this reduction in EQE was well reproduced by increasing the front surface recombination velocity S_p from 1×10^5 to 3×10^5 cm/s, indicating a reduction in passivation from the front surface

layer. Alternatively, as the annealing temperature increased, the EQE in the long wavelengths region (700 – 1300 nm) increased significantly. This was also well reproduced by simulation by increasing the minority carrier diffusion length L_n in the base region from 300 to 480 nm. It is interesting to note that the V_{oc} for this specific structure increased from 0.29 V to 0.36 V by increasing the annealing temperature from 750 °C to 800 °C, which correlates very well with the increase in diffusion length and hence, confirms the improvement in material quality.

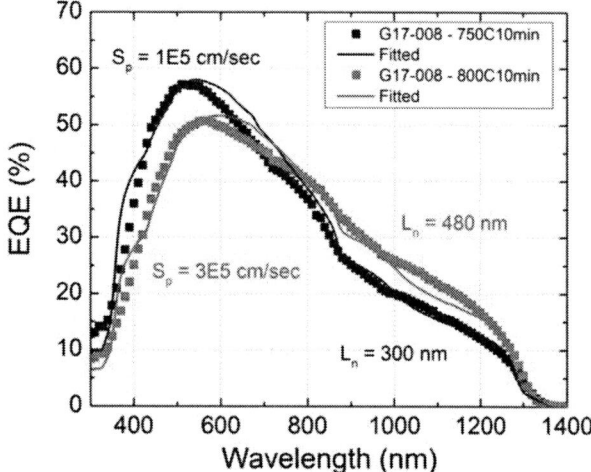

Fig. 3. Simulation of the EQE of a GaNAsSb solar cell (structure G17-008) annealed at two different temperatures. The EQE was fitted using the Hovell's equations [16] and the only varying parameters in the calculations in both cases, S_p and L_n, are denoted for reference.

D) Bi surfactant growth of GaNAs(Bi) and GaNAsSb(Bi)

As shown in Section A, the low current collection is currently the main limiting factor of the dilute nitride solar cells presented in this work. This low current collection is owed to both high unintentional p-type doping and short diffusion lengths. To overcome the low minority carrier diffusion length, dilute nitride solar cells are usually designed based on a p-i-n configuration to take advantage of the field-aided carrier collection. The unintentionally doped GaNAsSb material grown in this work showed high p-type background doping which ultimately led to a short depletion width (in the order of 220 nm) and thus prevented the use of a p-i-n configuration. To improve the current collection of the structures presented in this work it is then crucial that the background doping density be lowered.

Ptak *et al.* showed that the use of a Bi surfactant during the growth of GaInNAs led to an increase in the net donor concentration [6]. Eventually, when a sufficient amount of Bi was provided during growth, the unintentional background doping switched from p- to n-type. We decided to explore the use of Bi as a surfactant during the growth of our GaNAsSb material to investigate the possibility that Bi will also increase donor concentration or reduce the density of p-type defects in this material, which would reduce the net p-type background doping and improve current collection.

Moderately high Bi fluxes (4.9×10^{-8} to 1.5×10^{-7} Torr at beam flux monitor) were used during the growth of 100-nm GaNAs and GaNAsSb films capped with 50-nm of GaAs. Upon the opening of the Bi shutter, the surface reconstruction switched from a (1×2) to a (1×3) pattern, in agreement with previous reports [11]. The surface reconstruction remained (1×3) for all the samples investigated in this work however, as shown by the $\times3$ pattern in Fig. 4, it appears that the quality of the RHEED pattern strongly deteriorated with increasing amount of Bi.

XRD data revealed that Bi led to a strong increase in N incorporation (up to 40 %) until saturation at sufficiently high Bi flux, as shown in Fig. 5a. For the GaNAs film grown without Bi, the N composition was 2.55 % for the growth conditions chosen. Upon introduction of Bi, using the same growth conditions, the composition of N increased up to 3.60 %.

PL measurements, on the other hand, showed that Bi introduced under these growth conditions strongly reduced the PL intensity of GaNAsSb, as shown in Fig. 5b. This is coherent with the observed degradation in the RHEED pattern and might also be correlated with the increased composition of nitrogen. Although still at a preliminary stage, one of the plausible explanations is that the Bi flux used should be lowered by roughly an order of magnitude. Ptak *et al.* demonstrated devices with wide depletion widths (W > 2 μm) in the different dilute nitride material GaInNAs using Bi fluxes lower than 1×10^{-8} Torr [6].

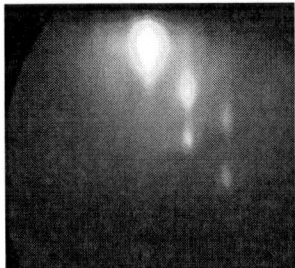 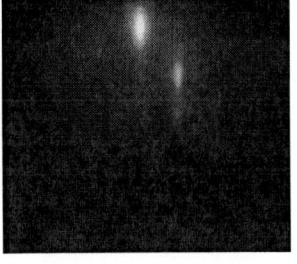

a)   Bi = 4.9×10^{-8} Torr   b)   Bi = 1.5×10^{-7} Torr

Fig. 4. RHEED images showing the $\times3$ pattern during the growth of two GaNAs(Bi) films with different Bi fluxes; a) Bi = 4.9×10^{-8} Torr, and b) Bi = 1.5×10^{-7} Torr. Clear degradation is observed with increased Bi flux.

SIMS analysis of the GaNAsSb(Bi) samples is currently on-going to determine to change in N, As and Sb incorporation as well as identify possible change in impurity incorporation (mainly H, C and O). Bi is not supposed to incorporate at these temperatures but this will also be verified by SIMS.

These results, nevertheless, demonstrate the growth of GaNAsSb(Bi) for the first time. Upon tuning of the growth conditions to achieve material with satisfactory PL, GaAs(Bi) and GaNAsSb(Bi) solar cells will be grown.

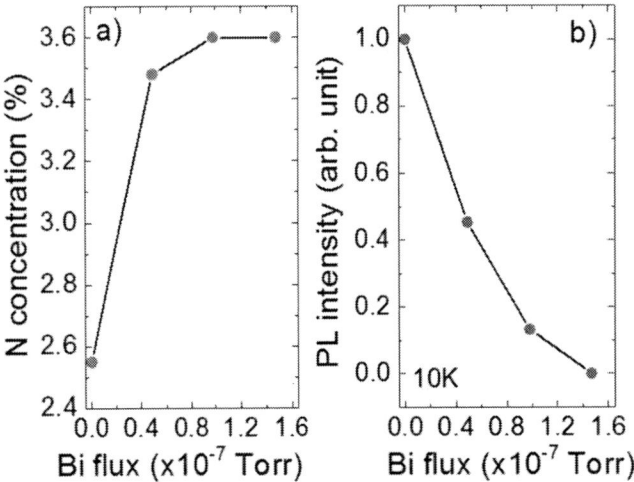

Fig. 5. a) N composition in GaNAs films as a function of Bi flux measured by HRXRD, b) Low-temperature PL intensity of GaNAsSb(Bi) films grown with increasing Bi fluxes.

IV. CONCLUSION

The performance of near 1-eV GaNAsSb solar cells have been investigated. The main factor limiting the performance of these devices was attributed to the unintentionally high acceptor background doping, which was found to limit carrier collection. SIMS analysis suggested that C and O are the main contributors to this high background doping. It was also shown that RTA, although beneficial for improving the V_{oc}, had a detrimental effect on the J_{sc}. Fitting of the EQE before and after annealing revealed that the decrease in J_{sc} was the result of degradation of the top passivation layer while the increase in V_{oc} was found to correlate with an increase in diffusion length in the GaNAsSb layer. Bi surfactant growth was demonstrated in preliminary experiments in indium-free GaNAsSb for the first time. Gradual increase in Bi flux led to a simultaneous increase in N incorporation and deterioration of the material quality.

Further experiments will test whether different growth conditions, in particular lower Bi flux, can reduce the net p-type background doping. This would increase the depletion width and improve current collection.

ACKNOWLEDGMENTS

This material is based upon work primarily supported by the QESST (Quantum Energy and Sustainable Solar Technologies) Engineering Research Center, part of the Engineering Research Center Program of the National Science Foundation and the Office of Energy Efficiency and Renewable Energy of the Department of Energy under NSF Cooperative Agreement No. EEC-1041895.

REFERENCES

[1]   J. C. Harmand *et al.*, "GaNAsSb: how does it compare with other dilute III–V-nitride alloys?," *Semicond. Sci. Technol.*, vol. 17, no. 8, p. 778, 2002.

[2] A. Gubanov, V. Polojärvi, A. Aho, A. Tukiainen, N. V. Tkachenko, and M. Guina, "Dynamics of time-resolved photoluminescence in GaInNAs and GaNAsSb solar cells," *Nanoscale Res. Lett.*, vol. 9, no. 1, pp. 1–4, 2014.

[3] J. C. Harmand, G. Ungaro, L. Largeau, and G. Le Roux, "Comparison of nitrogen incorporation in molecular-beam epitaxy of GaAsN, GaInAsN, and GaAsSbN," *Appl. Phys. Lett.*, vol. 77, no. 16, p. 2482, 2000.

[4] A. Maros, N. Faleev, S. H. Lee, J. S. Kim, C. B. Honsberg, and R. R. King, "1-eV GaNAsSb for multijunction solar cells," *2016 IEEE 43rd Photovolt. Spec. Conf. PVSC Portland OR*, pp. 2306–2309, 2016.

[5] M. A. Green, K. Emery, Y. Hishikawa, W. Warta, and E. D. Dunlop, "Solar cell efficiency tables (version 41): Solar cell efficiency tables," *Prog. Photovolt. Res. Appl.*, vol. 21, no. 1, pp. 1–11, Jan. 2013.

[6] A. J. Ptak, R. France, C.-S. Jiang, and R. C. Reedy, "Effects of bismuth on wide-depletion-width GaInNAs solar cells," *J. Vac. Sci. Technol. B Microelectron. Nanometer Struct.*, vol. 26, no. 3, p. 1053, 2008.

[7] D. B. Jackrel *et al.*, "Dilute nitride GaInNAs and GaInNAsSb solar cells by molecular beam epitaxy," *J. Appl. Phys.*, vol. 101, no. 11, p. 114916, 2007.

[8] J. S. Harris, Kudrawiec, and Yuen, "Development of GaInNAsSb alloys: Growth, band structure, optical properties and applications," *Phys Stat Sol B*, vol. 244, no. 8, pp. 2707–2709, 2007.

[9] S. Tixier, M. Adamcyk, E. C. Young, J. H. Schmid, and T. Tiedje, "Surfactant enhanced growth of GaNAs and InGaNAs using bismuth," *J. Cryst. Growth*, vol. 251, no. 1–4, pp. 449–454, Apr. 2003.

[10] G. Feng, K. Oe, and M. Yoshimoto, "Temperature dependence of Bi behavior in MBE growth of InGaAs/InP," *J. Cryst. Growth*, vol. 301–302, pp. 121–124, Apr. 2007.

[11] E. C. Young, S. Tixier, and T. Tiedje, "Bismuth surfactant growth of the dilute nitride GaNxAs1−x," *J. Cryst. Growth*, vol. 279, no. 3–4, pp. 316–320, Jun. 2005.

[12] R. R. King *et al.*, "Band gap-voltage offset and energy production in next-generation multijunction solar cells," *Prog. Photovolt. Res. Appl.*, vol. 19, no. 7, pp. 797–812, Nov. 2011.

[13] N. Leong *et al.*, "Growth of 1-eV GaNAsSb-based photovoltaic cell on silicon substrate at different As/Ga beam equivalent pressure ratios," *Prog. Photovolt. Res. Appl.*, p. n/a-n/a, 2015.

[14] A. J. Ptak, S. W. Johnston, S. Kurtz, D. J. Friedman, and W. K. Metzger, "A comparison of MBE-and MOCVD-grown GaInNAs," *J. Cryst. Growth*, vol. 251, no. 1, pp. 392–398, 2003.

[15] K. Volz *et al.*, "Optimization of annealing conditions of (GaIn)(NAs) for solar cell applications," *J. Cryst. Growth*, vol. 310, no. 7–9, pp. 2222–2228, Apr. 2008.

[16] H. J. Hovel and J. M. Woodall, "Theoretical and experimental evaluations of GaAlAs-GaAs solar cells," *IEEE Photovolt. Spec. Conf.*, vol. 10, pp. 25–30, 1973.

[17] T. Thomas *et al.*, "GaNAsSb 1-eV solar cells for use in lattice-matched multi-junction architectures," in *Photovoltaic Specialist Conference (PVSC), 2014 IEEE 40th*, 2014, pp. 0550–0553.

Amorphous silicon carbide for silicon surface passivation in carrier-selective contact devices

Mathieu Boccard[1,2], Christophe Ballif[1], Zachary C. Holman[2]

[1] École Polytechnique Fédérale de Lausanne (EPFL), Neuchâtel, Switzerland
[2] School of Electrical, Computer, and Energy Engineering, ASU, Tempe, Arizona, USA

Abstract — **High crystalline silicon solar cell efficiencies are obtained using the excellent passivation property of a-Si:H passivation layers. Yet, this passivation degrades upon heating, limiting post-a-Si:H-deposition steps to low-temperature (< 200 °C) processes; the blue and visible absorption of a-Si:H additionally induces losses in front-contacted devices. We show that a-SiC$_x$:H can be an alternative passivating layer offering a wider bandgap and improved temperature stability, and investigate further here the passivation mechanism of a-SiC$_x$:H when employed as the interfacial layer in passivating stacks. Through modelling, we discuss the evolution of the interface-defect density and of the surface-fixed-charge density in explaining the lifetime trends observed upon annealing for various stacks using the same a-SiC$_x$:H interface layer topped with different a-Si:H-based layers.** *Index Terms* — **amorphous materials, charge carrier lifetime, photovoltaic cells, silicon.**

Index Terms — **amorphous silicon carbide, charge carrier lifetime, defect density, temperature stability.**

I. INTRODUCTION

Hydrogenated amorphous silicon (a-Si:H) is an excellent material to use in very-high-efficiency solar cells because it allows extraordinary surface passivation and charge collection,. Best efficiencies using a-Si:H carrier-selective contacts are reported using a rear-contact architecture due to its relatively strong UV- and visible-light absorption. Also, passivation from a-Si:H degrades when subjecting the wafer to a temperature above 200 °C—especially when used as a passivating layer in a hole-selective contact [1]—restricting post-a-Si:H-deposition steps (such as contact, TCO, metal or ARC deposition) to < 200 °C processes. We recently introduced amorphous silicon carbide (a-SiC$_x$:H) as a passivating layer which is more transparent and more stable towards annealing than a-Si:H [2,3]. In this contribution, we investigate the passivation properties of a-SiC$_x$:H films that were previously successfully implemented in cells [3] and discuss the interface-defect and surface-fixed-charge density evolution upon annealing.

II. RESULTS AND DISCUSSION

Fig. 1 shows the injection-dependence of the lifetime obtained with different stacks of a-SiC$_x$:H / a-Si:H (deposited

by PECVD as described in [3]). Capping the a-SiC$_x$:H layer with an a-Si:H film is beneficial in all cases (intrinsic, p-type and n-type) compared to the use of an a-SiC$_x$:H capping layer. For undoped layers, the lifetime values obtained are an order of magnitude better for the a-Si:H capping layer than for the a-SiC$_x$:H capping layer. We attribute this to hydrogen enrichment of the interface from the deposition of the a-Si:H layer due to the higher affinity to hydrogen incorporation for carbon containing a-SiC$_x$:H alloys.

Concerning the doping of the capping layer, at low annealing temperatures (200 °C, similar to the as-deposited state), a similar lifetime at high injection is observed for i/n and i/p stacks, but i/p stacks perform significantly better at low injection. This suggests a stronger field-effect passivation, potentially due to the presence of a fixed charge in the a-SiC$_x$:H film, as also observed by Martín *et al.* who attributed it to the presence of dangling bonds and highlighted their amphoteric nature, leading to a positive or negative charge depending on the type of the substrate [4]. This result thus suggests that fixed charges also influence the passivation from a-SiC$_x$:H films when used as passivating layers in electron- or hole-selective contacts, with different effects in both cases.

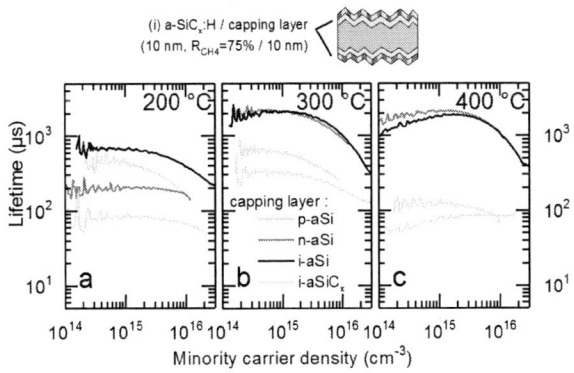

Fig. 1. Lifetimes after successive annealing steps at a) 200 °C, b) 300 °C and c) 400 °C for textured wafers coated with a-SiC$_x$:H / a-Si:H stacks, either p-type, n-type or intrinsic, or a-SiC$_x$:H / a-SiC$_x$:H stack. A very different injection dependence is seen depending on the capping layer.

Fig. 2 shows the calculated surface defect density (N_s) and interface charge density obtained by fitting the lifetime curves shown in Fig. 1 with the model presented in [5]. A high defect density ($\sim 10^{10}$ cm^{-2}) is observed after annealing at 200 °C (as

978-1-5090-5606-4/17 $31.00 © 2017 IEEE

well as initially) for all stacks except for the stack using an intrinsic a-Si:H capping layer (corresponding to the best lifetime). Upon annealing at 300 °C, a decrease of Ns is observed for all stacks, which we attribute to dangling bond passivation through hydrogen redistribution. This decrease of defect density is followed by an increase upon further annealing at 400 °C for the cases of p-type a-Si:H and a-SiC$_x$:H capping layers. These layers impose the largest (negative !) fixed charge density at the surface after annealing at 300 °C as seen in Fig. 2b. Such negative fixed charge density is expected to shift the Fermi-level position close to the valence band, thus favoring the creation of defects through H effusion as discussed in [1]. Contrarily, for the other two cases of n-type a-Si:H and intrinisic a-Si:H capping layers, the defect density remains lower than 2×10^9 cm^{-2} even after annealing at 400 °C. Such defect density is ~10 times lower to the one observed for similar stacks using a standard intrinsic a-Si:H layer in lieu of the a-SiC$_x$:H layer, exhibiting Ns values in the range of 2×10^{10} cm^{-2} for similar structures and annealing temperatures.

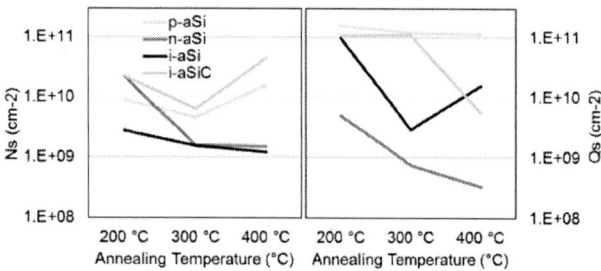

Fig. 2. Surface defect density (Ns) and absolute value of surface charge density (Qs) obtained from fitting the curves shown in Fig. 1 for various a-SiCx:H / a-Si(Cx):H passivation stacks as a function of the annealing temperature. Qs is negative in all cases except for the n-aSi:H capping layer.

IV. Conclusion

Intrinsic amorphous silicon carbide passivating layers can provide good passivation resulting in efficient solar cells when combined to doped a-Si:H contacts. We discussed the passivation mechanism based on calculations of interface defect densities and fixed surface charge densities. The role of a negative fixed charge in a-SiC$_x$:H was highlighted, and leads to improved lifetimes for low-temperature-annealed passivation stacks in hole-selective stacks, but to a detrimental decrease upon annealing at temperaturesabove 300 °C. This improved temperature stability opens possibilities to use a-SiC$_x$:H in carrier-selective contacts combined with high-temperature deposited contacts, allowing processes up to 400 °C as such (lifetime degradaded sharply after a subsequent annealing at 450 °C, not shown). Additionally, a-SiC$_x$:H alloys offer a better stability of the amorphous phase than a-Si:H films (not shown). Though dehydrogenation of a-SiC$_x$:H films occurs after high-temperature annealing (as observed for 500 °C with FTIR), plasma-initiated rehydrogenation was shown to be successful for de-hydrogenated a-Si:H films [6]. The use of a-SiC$_x$:H allows an extension of the temperature range towards higher values even higher than 400 °C.

References

[1] De Wolf, Stefaan, and M. Kondo. "Nature of doped a-Si: H/c-Si interface recombination" *J. of Appl. Phys.* 105. (2009): 103707.

[2] Beyer, W. and Mell, H. in *Disordered Semiconductors*, Institute for Amorphous Studies Series, edited by M. A. Kastner, G. A. Thomas, and S. R. Ovshinsky (Springer US, 1987) pp. 641-658.

[3] Boccard, M. and Z. C. Holman. "Amorphous silicon carbide passivating layers for crystalline-silicon-based heterojunction solar cells" *J. of Appl. Phys.* 118. (2015):065704.

[4] Martín, I., Vetter, M., Garín, M., Orpella, A., Voz, C., Puigdollers, J., & Alcubilla, R. "Crystalline silicon surface passivation with a-SiCx:H films deposited by plasma-enhanced chemical-vapor deposition" *J. Appl. Phys*, **98**. (2005). 114912.

[5] Olibet, S, Vallat-Sauvain, E, and Ballif, C, "Model for a-Si:H/c-Si interface recombination based on the amphoteric nature of silicon dangling bonds." *Phys. Rev. B*, **76** (2007). 035326.

[6] Shi, J., Boccard, M., & Holman, Z. (2016). "Plasma-initiated rehydrogenation of amorphous silicon to increase the temperature processing window of silicon heterojunction solar cells." *Appl. Phys. Lett.*, **109(3)**, 031601.

Surface passivation of boron diffused junctions by borosilicate glass and in situ grown silicon dioxide interface layer

Valentin D. Mihailetchi, Haifeng Chu, Jan Lossen, Radovan Kopecek

ISC Konstanz e.V., Rudolf-Diesel-Str. 15, 78467, Konstanz, Germany

Abstract — An in situ oxidation during the boron tribromide (BBr_3) diffusion process to form p^+ doped junctions on crystalline Si solar cells leads to the formation of stack layer system consisting of borosilicate glass (BSG; a binary B_2O_3-SiO_2 system) and a SiO_2 layer at Si interface. We present a method to passivate the p^+ doped regions by using this in situ grown SiO_2 in combination with a PECVD deposited SiN_x layer. We show that the etching rate of the BSG layer, in a HF acid solution, varies over the wafer, depending on its local B_2O_3 content in BSG, and is markedly higher than that of the SiO_2 layer. This difference in the etching rates can be used to controllably etch back the BSG layer in order to obtain a thin and uniform passivating oxide layer for solar cells application. Using this oxide/SiN_x stack we obtained implied V_{OC} of 705 mV and J_{0e} as low as 14 fA/cm² on symmetrically diffused boron emitters on n-type Cz wafers. These passivation results are comparable, on similar structure and boron emitters, with today's state-of-the-art Al_2O_3 based passivation methods. Moreover, we have successfully implemented this passivation method into mass-production of n-PERT and pilot-production of IBC solar cells, avoiding the need of adopting additional process steps and costs, which are otherwise needed for p^+ boron emitter passivation.

I. INTRODUCTION

Boron tribromide (BBr_3) tube diffusion is today the most commonly used method in large-scale industrial production to form the p^+ diffused regions on wafer based crystalline silicon solar cells [1]. However, high throughput industrial BBr_3 diffusion process and the passivation of the resulting p^+ doped surfaces are often considered to be challenging. The optimization of doping profile, passivation and metallization thereof are essential for further increasing the conversion efficiency of crystalline Si solar cells. The current industrial used methods for passivation of boron doped surfaces involve additional process steps and surface treatments, such as thermal or chemical oxidation, atomic layer deposition (ALD) of Al_2O_3, or plasma enhanced chemical vapor deposition (PECVD) of AlO_x. All these methods, however, increase the production costs and process complexity.

A typical BBr_3 diffusion process consists of a deposition phase where BBr_3 vapor is mixed with oxygen to form B_2O_3 and bromine:

$$4BBr_3 + 3O_2 \rightarrow 2B_2O_3 + 6Br_2 \tag{1}$$

and a subsequent drive-in phase where the deposited B_2O_3 reacts at Si interface to form elemental boron and SiO_2:

$$2B_2O_3 + 3Si \rightarrow 4B + 3SiO_2 \tag{2}$$

Since the vapor pressure of the B_2O_3 at typical deposition temperatures (900-1000 °C) is rather low, liquid B_2O_3 condenses on the silicon wafers as well as on furnace walls [2, 3]. If the diffusion parameters are carefully chosen, the formation of the undesirable boron-rich layer (BRL), which is responsible for lifetime degradation in the Si bulk [4], can be avoided. Then a boron-rich borosilicate glass (BSG), a mixed B_2O_3-SiO_2 system, is formed during the diffusion process. Furthermore, if the drive-in phase occurs partially or totally under oxygen-containing atmosphere, the silicon surface under the BSG is further oxidized by the reaction:

$$Si + O_2 \rightarrow SiO_2 \tag{3}$$

This process results in an additional in situ grown SiO_2 layer at Si/BSG interface. The formation of BSG layer during the deposition phase results inevitable in a relatively thick and non-uniform layer over the silicon wafer and along the diffusion boat. Therefore the utilization of this in situ grown BSG layer in combination with a SiN_x layer, as a passivation and anti-reflection coating (ARC) stack, would lead to solar cells with non uniform optical appearance and lower photocurrent. That is why the BSG layer is typically striped off in a HF acid solution and an additional passivation step is deposited on the p^+ doped surface prior to PECVD deposition of the SiN_x ARC layer.

This work describes the composition and properties of the oxide layers formed in situ during BBr_3 diffusion in crystalline Si. We present a method that yields a thin and uniform in situ grown SiO_2 layer with outstanding surface passivation quality, comparable to that of state-of-the-art Al_2O_3 based passivation method. We show that this method can easily be implemented in mass-production avoiding additional process steps and costs.

II. EXPERIMENTAL PROCEDURE

For our experiments, we used 6-inch n-type Cz wafers with base doping of 10 Ωcm and an as-cut thickness of 200 μm. Prior to the diffusion process, all wafers were etched in alkaline solution and cleaned in a piranha solution. For this study, we choose a typical industrial BBr_3 diffusion recipe, which yields a uniform sheet resistance (R_{sh}) distribution over

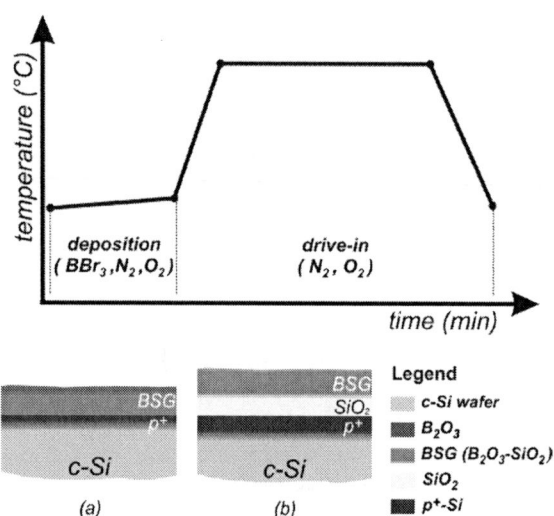

Fig. 1. (Color online) (top) A BBr$_3$ diffusion recipe consisting of a deposition and a drive-in phase. (bottom) The typical layer structure (not to scale) after the deposition (a) and after the drive-in phase (b).

the wafer as well as along the entire diffusion boat. Such a recipe consists of a deposition phase at temperature between 800-900 °C and a subsequent drive-in phase at temperature between 900-1050 °C.

The diffusion parameters, namely the gas flows and duration, were selected according to the best-known-method in our lab, in order to obtain an emitter with a $R_{sh} \approx 135$ Ω/sq. ±5% without formation of the BRL layer. The drive-in phase was done partially in 100% oxygen-containing atmosphere in order to grow an additional in situ SiO$_2$ at the BSG/Si interface and to reduce surface concentration of the diffused boron. In order to estimate the effect of this additional in situ grown SiO$_2$ layer, a thermal SiO$_2$ layer was grown on un-diffused wafers in a separate oxidation tube, using an identical recipe as for the boron diffusion (i.e., temperature profile, gas flows and duration) but without the BBr$_3$ flux. A schematic diagram of the diffusion recipe together with a typical layer structure expected directly after deposition and after the drive-in phase is shown in Fig. 1.

All wafers were separated in three groups. First group was loaded in the diffusion tube, but received only the BBr$_3$ deposition, i.e., the wafers were unloaded under N$_2$ atmosphere directly after deposition phase. The second group received the complete BBr$_3$ diffusion process, and the third group was processed in the separate oxidation furnace as described above. The thickness of the BSG and SiO$_2$ layers was measured on 25 points equally distributed over the wafer using a Sentech SE800-PV ellipsometer assuming a refractive index of $n = 1.46$ for BSG and SiO$_2$ layers [5]. Some of the wafers with boron diffusion were selected for lifetime measurements using QSSPC and received additional PECVD

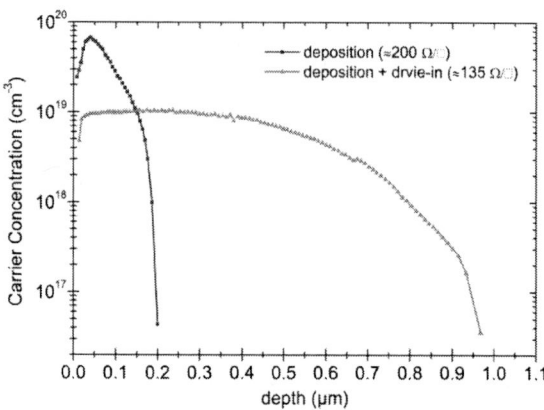

Fig. 2. (Color online) Carrier concentration (ECV) profiles of the diffused boron after the two process phases (see legend).

SiN$_x$ deposition on both sides and a subsequent firing step to activate the passivation. In order to study the etching rate of the BSG and SiO$_2$ layers the wafers were etched by immersion in a 2 vol.% HF acid solution at room temperature for a defined period of time followed by thickness measurements. This process was repeated until the oxide layers were completely etched off.

III. RESULTS AND DISCUSSION

A. Effect of deposition and drive-in on borosilicate glass composition

The carrier concentration profiles, as measured by electrochemical capacitance-voltage (ECV) technique, at the end of deposition phase and after the drive-in phase are shown in Fig. 2. The measurements show a significant boron doping occurring already during the deposition phase, with a $R_{sh} \approx 200$ Ω/sq., despite of the fact that temperature is relatively low (<900 °C) and the deposition time is short (<20 min). During deposition, liquid B$_2$O$_3$ is continuously produced (reaction 1) and condensed on the Si wafers. At the same time B$_2$O$_3$ reacts at Si interface to produce elemental boron and solid SiO$_2$ (reaction 2). At this stage of the diffusion process the B$_2$O$_3$ rapidly dissolves or melts-through the underlying SiO$_2$ and forms the binary BSG; a low viscosity glass layer. Brown and Kennicott estimated that, in this molten phase, the diffusion coefficients of B$_2$O$_3$ in SiO$_2$ are ≈1000 times higher than the corresponding solid-state diffusion coefficients [6], bringing boron dopant to the Si interface with very little time delay. Hence the concentration of boron in Si during the deposition phase is expected to be limited by its solid solubility, which is ≈7×10^{19} cm^{-3} at 900 °C [7]. This seems to be confirmed by the experimental data shown in Fig. 2. A prolong process under this diffusion condition would lead to the formation of BRL at Si interface [3, 4]. Brown et al., showed that attaining boron solid solubility limit in Si would require a B$_2$O$_3$ concentration in the BSG of more than 18 mol % [6, 8]. A BSG layer whose composition exceeds this B$_2$O$_3$ concentration has a glass

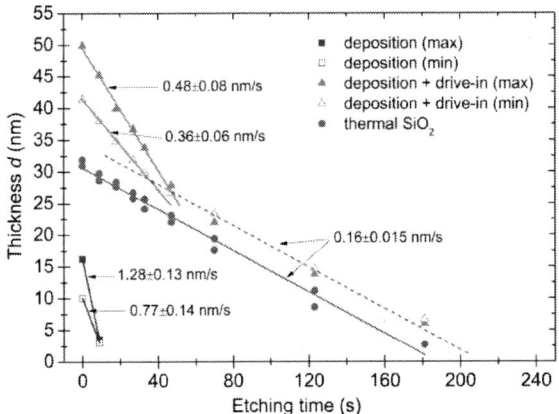

Fig. 3. (Color online) Thickness of different oxide layers studied in this work as a function of etching time in HF solution for different processes (see legend). For the boron diffused samples the area with minimum and maximum BSG thickness is plotted after both: the deposition phase and the drive-in phase. The thermal SiO_2 layer was grown in a separate oxidation tube set with an identical recipe to the boron diffusion (i.e., temperature profile, gas flows and duration) but without the BBr_3 flux.

transition temperature (T_G) below the deposition temperature (850-900 °C) chosen in our diffusion and it is therefore characterized by low viscosity [6]. This seems to be the case for our diffusion recipe at the end of the deposition phase, where the structure of the glass layer is expected to be as shown in Fig. 1a and leading to a surface concentration of no more than $\approx 7 \times 10^{19}$ cm^{-3}.

During the drive-in phase of the diffusion no additional B_2O_3 is produced or deposited on the wafer surface, and the low viscosity BSG layer is being diluted by additions of SiO_2 following the ongoing reaction 2 at Si interface. The dilution of B_2O_3 content in the glass increases its melting point as well as increasing the viscosity. This process is significantly enhanced with the addition of an in situ oxidation (reaction 3). When the concentration of B_2O_3 in BSG drops below that required to maintain a low viscosity phase at the given process temperature, the dissolution of underlying SiO_2 comes to a stop, and a thermal SiO_2 grows beneath BSG layer at Si interface during the remaining oxidation phase (as shown schematically in Fig. 1b). In this region the migration of boron dopant from the BSG layer to the Si interface is governed by the solid-state diffusivities in SiO_2, which are orders of magnitude lower than in a molten glass phase [6, 9]. Hence, this in situ grown SiO_2 layer slows down or could function as a blocking layer for further boron doping of Si providing that a critical thickness is reached. Furthermore, as the solubility of boron is higher in SiO_2 than in Si the oxidation process lowers the surface concentration of boron emitter (as shown in Fig. 2) and the SiO_2 becomes boron doped. This in situ oxidation can efficiently be used to prevent formation of BRL [3].

B. Etching rate of SiO₂/BSG stack layers

In order to be used as antireflective coating the thickness uniformity of the BSG layer must be very good across the wafer as well as along the entire diffusion boat. In practice however, for a diffusion process of relevant throughput with tightly packed wafers in a long boat, this is not the case, the fast decomposition of BBr_3 under the oxidizing atmosphere (reaction 1) results in an uneven condensation of B_2O_3 over the silicon surface and along the boat. The resulting BSG layer, is often much thicker, at the wafer edges and at the boat ends. Therefore, partially etching back of the in situ grown BSG and SiO_2 layers could be a viable and simple process solution for industrial application.

Fig. 3 shows the thickness (d) of the oxide layers grown in this study as a function of etching time in 2 vol.% HF solution (see legend). For the boron diffused samples, the area with minimum and maximum BSG thickness is plotted after both: the deposition ($d_{BSG,min}$, $d_{BSG,max}$) and the drive-in ($d_{total,min}$, $d_{total,max}$) phases. From the sample that received just the thermal oxidation in a separate oxidation furnace, one can see that a SiO_2 layer with a thickness of $d_{SiO2} = 32 \pm 1.16$ nm is grown using the identical process parameters of the boron diffusion, except for BBr_3 flux.

For the boron diffused samples, the in situ grown SiO_2 during the drive-in phase is expected to be similar in thickness and even, despite of the presence of a uneven BSG layer. This is because the diffusion coefficient of oxygen in BSG is higher than in pure SiO_2, and hence the presence of BSG layer presents very little resistance for the diffusion of oxygen to the Si surface [10]. Thus, the growth rate of SiO_2 is not hindered by a BSG layer of varying thicknesses, and therefore the grown SiO_2 layer has a uniform thickness. In fact, this is also confirmed by the experimental data shown in Fig. 3. One can see that, within the experimental error, at $t = 0$ seconds the $d_{total} \approx d_{BSG} + d_{SiO2}$ for both: thickest and thinnest BSG area over the wafer. The wafers that received only the deposition phase were loaded in the same boat position as those that received the complete diffusion process.

The etching rate for each process conditions was determined from the linear fit to the thickness measurements after sequential etching, and is shown by the arrows in Fig. 3. The largest contribution to the etching rate measurement errors comes from our inaccuracy to estimate the actual etching time. The wafers were immersed in HF solution followed by the immersion in DI water, in order to stop the etching process. During the transport between the two baths the surface is hydrophilic and the etching process continues, but probably at slightly lower rate. For the thermally grown SiO_2, the measured etching rate of 0.16 nm/s is in agreement with literature values for comparable etching conditions [11]. For the diffused samples with BSG layer, the etching rate is significant higher. Previous studies showed that the etching rate of BSG layer in HF etchants exponentially increases with increasing B_2O_3 content in the glass [12, 13]. This is because, in HF etchants, the breakage rate of the B-O bond is higher compared to that of the Si-O bond [11]. This is also observed in Fig. 3, where the BSG layer on the samples with only the deposition process show markedly higher etching rate than that of pure SiO_2.

Moreover, the results in Fig. 3 clearly show that the etching rate is higher for areas on the wafer with thicker BSG layer as compared to areas with thinner BSG layer. The areas with thicker BSG layer arose as a result of thicker B_2O_3 film condensed during the BBr_3 deposition and therefore a higher B_2O_3 content in the glass. The same observations are made on the samples that received the complete boron diffusion process (including the drive-in phase); thicker areas are etched faster than thinner areas, but with overall lower etching rates than for as-deposited layer. The etching rate of BSG is a characteristic function of the B_2O_3 content in the glass and therefore it can be used to estimate the glass composition. For instance, by comparing the ratios of etching rates of BSG layers to SiO_2 in Fig. 3 with those reported by Kern and Heim [13], and assuming a linear increase in etching rate with HF etchant concentration [11], one could roughly estimate the B_2O_3 content in the BSG. This gives approximately 17-22 mol % B_2O_3 after deposition phase and about 8-11 mol % after the drive-in phase, with higher values corresponding to the thicker BSG areas.

For boron diffused samples with in situ thermal oxidation during the drive-in phase a second important observation is made: The etching rate decreases significantly after a certain time, and reaches a value equal to thermal SiO_2. Obviously, the high etching rates of the first phase correspond to etching of the BSG film with different B_2O_3 content and different initial thicknesses, while the lower etching rate of the second

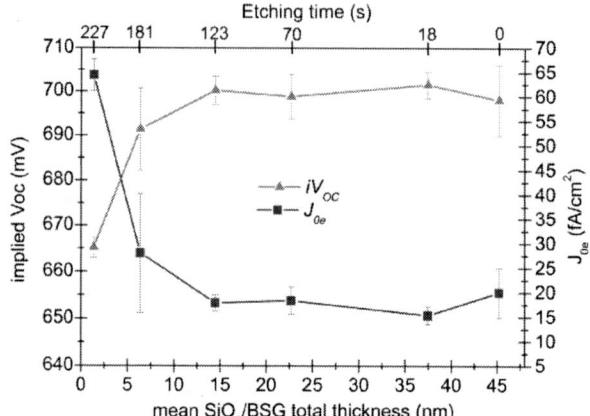

Fig. 5. (Color online) Implied V_{OC} and J_{0e} of the symmetrical diffused boron emitter studied in this work as a function of the total thickness of the in situ grown SiO_2/BSG stack layers. To achieve different thicknesses the diffused wafers were stepwise etched back in 2 vol. % HF acid. The passivation was then completed by PECVD-SiN_x deposition on both sides of the wafers and a subsequent firing step.

phase correspond to etching of the uniform SiO_2 film grown beneath the BSG. The transition to this second phase correspond roughly to the initial thickness of the thermal SiO_2 film, indicating that the BSG layer and the in situ grown SiO_2 layer do not substantially intermix. As shown above, this is the case when B_2O_3 concentration in the BSG drops below that is required to maintain a low viscosity condition.

This increase in etching rate of BSG layers depending of its B_2O_3 content in comparison with pure SiO_2 can be used to advantage. For instance, the in situ grown SiO_2 during the drive-in step of a BBr_3 diffusion process could readily be used as a buffer layer for the etch back of BSG. In this way a very uniform passivation films of a desired thickness can be obtained easily over an entire diffusion boat. This is demonstrated in Fig. 4, which shows the mean and the range thickness values over the wafer as a function of boat position, directly after diffusion process and after ≈130 seconds HF etching respectively. One can see that, before HF thinning, the BSG layer uniformity is obviously not suited in combination with SiN_x as an ARC. The photocurrent generation over the cell and boat would vary ≥ 1.0 mA/cm^2 in this case. In contrast, after HF thinning, the BSG layer is completely etched and a 13.9 ± 1.28 nm of in situ SiO_2 remains on the Si surface, which serves as an interface passivation layer.

A similar diffusion process with in situ oxidation and HF thinning principle described here for boron can also be applied for phosphorous diffusion. The etching rate of binary phosphosilicate glass (PSG) system show a similar behavior as for BSG, i.e., etching rate increase with increasing P_2O_5 content in PSG and is distinctly higher than that of SiO_2 [11, 14]. At high P_2O_5 concentrations in PSG a "melt-through process" of underlying SiO_2 layer, as described for boron, is expected from the P_2O_5–SiO_2 phase diagram [14]. Such an

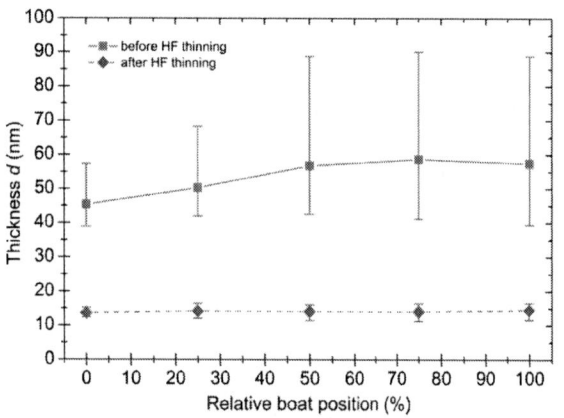

Fig. 4. (Color online) Total BSG/SiO_2 stacks layer thickness directly after diffusion process (deposition + drive-in) and after HF thinning as a function of relative position in diffusion boat (featuring 200 slots). The symbols show the mean thickness values for 25 points/wafer whereas the error bars show the range of thicknesses over a wafer.

978-1-5090-5606-4/17 $31.00 © 2017 IEEE

observation was made recently by Werner et al., who showed that the in situ grown SiO_2 layer thickness is reduced by an additional $POCl_3$ deposition phase, introduced in situ after the drive-in phase [15]. The same glass/SiO_2 stack layer system could also be obtained during annealing of CVD deposited BSG (or PSG) layer in an oxygen containing atmosphere, providing that B_2O_3 (or P_2O_5) content in the glass drops below the melt-through condition.

C. Influence of oxide thickness on surface passivation

In order to investigate the passivation quality of the in situ SiO_2/BSG stack layers we fabricated, in the same batch, lifetime samples on symmetrically diffused n-type Cz wafers. Fig. 5 shows the implied V_{OC} and J_{0e} values as a function of SiO_2/BSG total stack thickness for various etching back levels. The J_{0e} values were extracted from the slope of inverse lifetime at an injection level of 1×10^{16} cm^{-3}. It can be seen from the experimental data that an outstanding surface passivation is obtained, even without HF thinning. With increasing etching time the BSG layer thickness decreases until it is completely removed, after ≈ 50 seconds (as shown in Fig. 3), and ≈ 32 nm of in situ SiO_2 layer remains on the Si surface. Nevertheless, the data in Fig. 5 shows that this SiO_2 layer can be further etched back down to 10-15 nm without sacrificing the passivation quality. This, combined with the results shown in Fig. 4, demonstrates that the in situ grown SiO_2 layer can readily be used as an interface passivation and ARC layer when capped with SiN_x.

Fig. 6 compares the J_{0e} results obtained in this work with several literature reported values by other passivation methods commonly used in industry to passivated boron emitters. Only the experimental data reported for a polished Si surface are plotted in Fig. 6. Although there are certainly differences in sample preparation and emitter doping profiles (e.g., surface

concentration), the in situ grown SiO_2 passivation method shows similar passivation quality with today's state-of-the-art passivation methods, but with the merit of no additional process steps or costs.

This passivation method using the in situ grown SiO_2 during a BBr_3 diffusion process was integrated successfully in the process sequence of n-PERT and IBC Si solar cells [20, 21]. Recently we have demonstrated that this passivation method is viable in mass production of n-PERT bifacial solar cells using a low-pressure BBr_3 furnace with batch size of 1000 wafers per run [21].

IV. CONCLUSION

In this paper we discuss the composition and properties of in situ grown oxide layers during the BBr_3 diffusion process used to form p^+ doped junctions in crystalline Si solar cells. We showed that an in situ thermal oxidation during diffusion process leads to the formation of two-layer stack system consisting of borosilicate glass (BSG; a binary B_2O_3-SiO_2 system) layer and a SiO_2 layer at Si interface. The thickness of this BSG layer grows uneven over the wafer and diffusion boat and it would be unsuitable as ARC for solar cells fabrication.

However, we demonstrated that, due to much higher etching rates of the BSG layer in diluted HF acid solutions, this in situ grown SiO_2 can readily be used as a buffer layer for partially or totally etching back of the non uniform BSG layer. Then the remaining thin and uniformly grown SiO_2 layer at Si interface can be used, in combination with a PECVD SiN_x layer, as a passivation and ARC stack with outstanding passivation quality, comparable to those of state-of-the-art Al_2O_3 based passivation methods.

V. ACKNOWLEDGEMENT

This work was financially supported by the German Federal Ministry for Economic Affairs and Energy within the project PfZ (contract no. 0325840).

REFERENCES

[1] International Technology Roadmap for Photovoltaic, 2017.
[2] R. F. Lever and H. M. Demsky, "Water vapor as an oxidant in BBr_3 open-tube silicon diffusion systems," in *IBM J. Res. Develop.*, pp. 40-46, 1974.
[3] E. Arai et al., "Interface reactions of B2O3-Si system and boron diffusion into silicon," in *J. Electrochem. Soc.*, vol 120, pp. 980–987, 1973.
[4] M. A. Kessler et al., "Charge carrier lifetime degradation in Cz silicon through the formation of a boron-rich layer during BBr_3 diffusion processes," in *Semicond. Sci. Technol.*, vol 25, 055001, 2010.
[5] J. F. Shackelford and W. Alexander, "Materials Science and Engineering Handbook," in *Boca Raton: CRC Press LLC*, 2001.
[6] D. M. Brown and P. R. Kennicott, "Glass source B diffusion in Si and SiO2," in *J. Electrochem. Soc.*, vol 118, pp. 293–300, 1971.

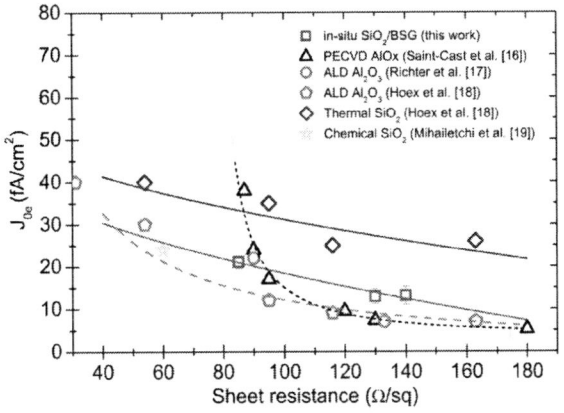

Fig. 6. (Color online) Measured J_{0e} as a function of sheet resistance for boron emitter samples with in situ SiO_2/BSG interface passivation stack and a comparison with several literature reported values for other passivation layers (see legend). The lines are a guide for the eye.

[7] V. E. Borisenko and S. G. Yudin, "Steady-state solubility of substitutional impurities in silicon," in *Phys. Status Solidi, A* 101, pp. 123, 1987

[8] D. M. Brown et al., "Characteristics of doped oxides and their use in silicon device fabrication," in *J Cryst. Growth,* vol 17, pp. 276–287, 1972.

[9] M. Ghezzo and D. M. Brown, "Diffusivity summary of B, Ga, P, As, and Sb in SiO2," in *J. Electrochem. Soc.,* vol 120, pp. 146–148, 1973.

[10] B. E. Deal and M. Sklar "Thermal oxidation of heavily doped silicon," in *J. Electrochem. Soc.*, vol 112, pp. 430–435, 1965.

[11] G. A. C. M. Spierings, "Wet chemical etching of silicate glasses in hydrofluoric acid based solutions," in *Journal of Materials Science,* vol 28, pp. 6261-6273, 1993.

[12] A El-Hoshy, "Measurement of P-etch rates for boron-doped glass films," in *J. Electrochem. Soc.*, vol 117, pp. 1583–1584, 1970.

[13] W. Kern and R. C. Heim, "Chemical vapor deposition of silicate glasses for use with silicon devices," in *J. Electrochem. Soc.*, vol 117, pp. 568–573, 1970.

[14] P. Balk and J. M. Eldridge, "Phosphosilicate glass stabilization of FET devices," in *Proceedings of the IEEE*, vol 57, pp. 1558-1563, 1969.

[15] S. Werner, et. al., "Structure and composition of phosphosilicate glass systems formed by POCl3 diffusion," in *Energy Procedia,* 2017, to be published.

[16] P. Saint-Cast, et. al., "Very low surface recombination velocity of boron doped emitter passivated with plasma-enhanced chemical-vapor-deposited AlOx layers," in *Thin Solid Films,* vol 522, pp. 336-339, 2012.

[17] A. Richter, et al., "Boron emitter passivation with Al2O3 and Al2O3/SiNx stacks using ALD Al2O3," in *IEEE Journal of Photovoltaics*, vol 3, pp. 236-245, 2013.

[18] B. Hoex, et al., "Excellent passivation of highly doped p-type Si surfaces by the negative-charge-dielectric Al2O3," in *Appl. Phys. Lett.*, vol 91, 112107 (2007).

[19] V. D. Mihailetchi, et al., "Nitric acid pretreatment for the passivation of boron emitters for n-type base silicon solar cells," in *Appl. Phys. Lett.*, vol 92, 063510 (2008).

[20] V. D. Mihailetchi, et al., "A comparison study of n-type PERT and IBC cell concepts with screen printed contacts," in *Energy Procedia*, vol 77, pp. 534-539 (2015).

[21] J. Lossen, et al., "From lab to fab: Bifacial n-type cells entering industrial production," in *Proceedings of the 31st European Photovoltaic Solar Energy Conference*, pp. 965-968 (2015).

Improved Light Incoupling in Planar Solar Cells via Improved Texture Morphology of PDMS Scattering Layer

Salman Manzoor[1], Zhenghsan J. Yu[1], Asad Ali[2], Waqar Ali[2] and Zachary C. Holman[1]

Arizona State University, Tempe, Arizona, 85281, USA[1]

U.S.-Pakistan Center for Advanced Studies in Energy, University of Engineering and Technology, Peshawar, 25000, Pakistan[2]

Abstract — **Planar surfaces on the sunward-facing side of solar cells and modules are undesirable because they incur high front-surface reflectance losses and provide poor light trapping. Several emerging and established solar cell technologies, such as perovskite cells, thin-film cells, epitaxial cells, and tandems that include these, have planar surfaces at their front due to restrictions imposed by the fabrication methods employed. Scattering layers made with polydimethylsiloxane (PDMS) are effective in reducing front-surface reflectance when applied to the front of these cells, but show little light trapping enhancement. In this contribution, we use optical simulations to show that tuning the base angle of random pyramids imprinted in the PDMS layer can improve coupling of light into a planar solar cell. A pyramid base angle of 46° gives the highest short-circuit current density.**

Index Terms — **light management, light trapping, ray tracing, silicon heterojunction solar cells, antireflection coating, surface morphology.**

I. INTRODUCTION

The primary goal of a solar cell is to convert all the photons hitting its surface to electricity. However, the front (sunward-facing) surface of a solar cell is a source of reflectance loss because of refractive-index mismatch between the absorber material and air, or the absorber material and the encapsulant material in the case of modules. This problem is remedied by texturing the absorber material for better in-coupling of light. For example, monocrystalline silicon solar cells are textured in an alkaline bath to etch random pyramids in their surfaces [1], multicrystalline silicon solar cells are etched in an acid bath to etch "spherical caps" in their surfaces, and thin-film solar cells are grown on textured substrates such as textured ZnO [2, 3].

These approaches to texture the absorber material directly or to deposit them on textured substrates are either not possible or very challenging for many emerging solar cell technologies. For example, perovskite solar cells are formed by spin-coating, which is incompatible with textured substrates. Similarly, polycrystalline CdTe and CIGS are deposited on planar substrates for better film quality at the cost of higher reflectance and poor light trapping. This problem is also present in epitaxial solar cells such as III-V cells that are grown on polished, lattice-matched substrates. Similarly, promising tandem solar cells such as perovskite/silicon [4, 5] and perovskite/perovskite [6] tandems suffer from large front-surface reflectance loss due to the planar structure of the devices.

Previously, we demonstrated that a polydimethylsiloxane (PDMS) scattering layer carrying random pyramids as texture can successfully reduce the front-surface reflectance of planar silicon and perovskite solar cells if applied to their front surfaces [7]. However, we also observed that these layers did not greatly enhance the infrared response of the planar silicon cells, indicating that they provide little trapping of escaping light. Light trapping—defined here as the path length enhancement in the absence of absorption—is a phenomenon caused by surface textures [8, 9]. Generally, solar absorbers are double-side textured with similar texture at the front and rear, and with no flexibility to tune the front and rear textures separately. PDMS scattering layers applied to planar absorbers, however, open this degree of freedom and their texture can, one would assume, be tuned to better reflect escaping light back into the absorber.

In this contribution we therefore use optical simulations to probe the dependence of the performance of planar silicon solar cells on the morphology of PDMS scattering layers applied to their front surfaces. In particular, we change the base angle of a random pyramid texture imprinted into the PDMS layers and calculate the resulting change in reflectance at each interface and absorbance in each layer. Our results show improved light incoupling of both short- and long-wavelength light for PDMS scattering layers with tuned pyramid base angle, though not primarily from limiting the escape of light from inside the wafer.

II. OPTICAL SIMULATION SETUP

To demonstrate the effect of changing the random pyramid base angle in PDMS scattering layers, we simulated the optics of two planar silicon heterojunction solar cells fabricated in our lab. The first cell has planar front and rear surfaces and is referred to as "Flat/Flat". The second cell has a planar front surface and textured rear surface and is referred to as "Flat/Tex". Both devices are built upon 250-μm-thick *n*-type silicon wafers, with 6- and 5-nm-thick intrinsic and *n*-type amorphous silicon layers, respectively, deposited on the front and with 6- and 11-nm-thick intrinsic and *p*-type amorphous silicon layers, respectively, deposited on the rear. On top of the front amorphous silicon layers is a 115-nm-thick indium tin oxide (ITO) layer that provides lateral conduction of

978-1-5090-5606-4/17 $31.00 © 2017 IEEE

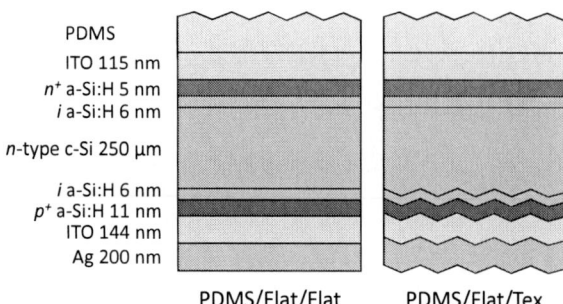

PDMS
ITO 115 nm
n^+ a-Si:H 5 nm
i a-Si:H 6 nm

n-type c-Si 250 μm

i a-Si:H 6 nm
p^+ a-Si:H 11 nm
ITO 144 nm
Ag 200 nm

PDMS/Flat/Flat   PDMS/Flat/Tex

Fig. 1. Schematic diagrams of the two planar devices simulated in this work. "PDMS/Flat/Flat" is a double-side-polished silicon heterojunction solar cell. "PDMS/Flat/Tex" is a front-side-polished and rear-side-textured silicon heterojunction solar cell. Both have PDMS scattering layers at their front surfaces.

generated carriers and also acts as antireflection coating. This ITO thickness is tuned for minimum front-surface reflection when silicon is used as a bottom cell in a four-terminal tandem [10, 11]. On top of the rear amorphous silicon layers is a 144-nm-thick ITO layer, followed by a 200-nm-thick silver layer that acts as a rear reflector.

Optical simulations were performed with the Module Ray Tracer from PV Lighthouse [12]. It combines Monte-Carlo-based ray tracing with thin-film optics. The Module Ray Tracer allows any materials stack to be simulated, provided the complex refractive index and thickness of each layer are specified, and any interface can be textured with random pyramids with arbitrary base angle. The software calculates the front-surface reflectance, escape reflectance, and absorptance in the cell absorber, as well as the absorptance in all other layers of the device.

The complex refractive indices for the doped (n- and p-type) and intrinsic amorphous silicon layers, as well as for the front and rear ITO layers, were characterized using spectroscopic ellipsometry and taken from [11]. The complex refractive indices for the silicon wafer and silver rear reflector were taken from the literature [13, 14].

III. RESULTS AND DISCUSSION

Fig. 1 shows schematics of the Flat/Flat and Flat/Tex solar cells with PDMS on their front surfaces. We simulated these two devices, which were fabricated in the lab, and the results without and with PDMS layers are shown in Fig. 2. The simulation results closely match the measured external quantum efficiency (EQE) and reflectance (R) of the devices, which were reported in Ref. [7]. In these simulations, the base angles of the pyramids in the PDMS scattering layers and the rear-side-textured silicon wafer were taken to be 49.5° and 50.5°, respectively, as experimentally determined with angle-resolved reflectance measurements on the fabricated samples. Fig. 2 shows that after applying a PDMS layer to the front of

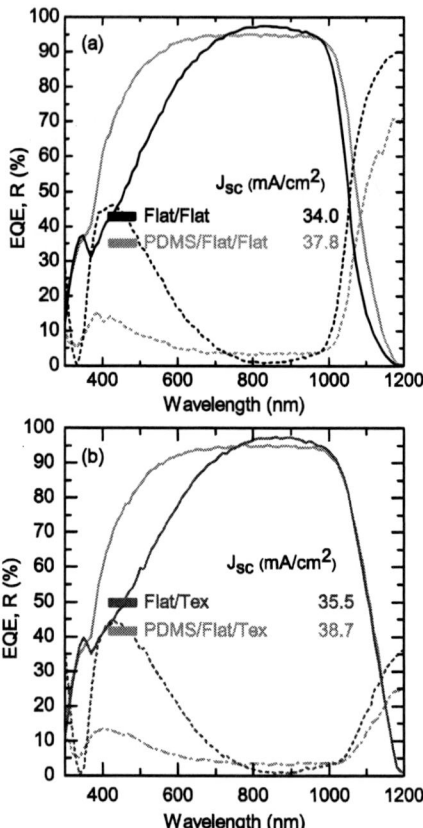

Fig. 2. Simulated reflectance and EQE spectra of (a) Flat/Flat and (b) Flat/Tex devices without and with a PDMS scattering layer.

these devices the reflectance decreases and the EQE increases considerably at visible wavelengths, but the reflectance undesirably increases between 750 and 1000 nm. The reason for this is that the front ITO layer, which acts as a good antireflection coating for the (bare) planar front surfaces, has a refractive index which is no longer the geometric mean of the refractive indices of the layers adjacent to it (PDMS on top and silicon underneath).

To remedy this, we increased the front ITO refractive index in the Module Ray Tracer to its optimum value of $n = 2.3$ and its thickness to 90 nm to account for both this new refractive index and the non-normal angle of incidence of light on the ITO layer after adding PDMS. We then varied the base angle of the pyramids in the PDMS scattering layer from 40 to 56° to investigate its influence on trapping of weakly absorbed light. Fig. 3a plots the absorptance in the wafer (and thus the approximate EQE, given that the collection efficiency of these devices is near unity) for a Flat/Flat device with three values of pyramid base angle: 40, 46, and 56°. The plots show that 46° pyramids have very slightly higher absorptance at long wavelengths (1000–1200 nm), but the effect is minute. This suggests that PDMS pyramid base angle has little influence on light trapping within the wafer—also for the PDMS/Flat/Tex sample (not shown). However, there is an appreciable decrease in absorptance for 40° pyramids between 500 and 950 nm, and

978-1-5090-5606-4/17 $31.00 © 2017 IEEE   1229

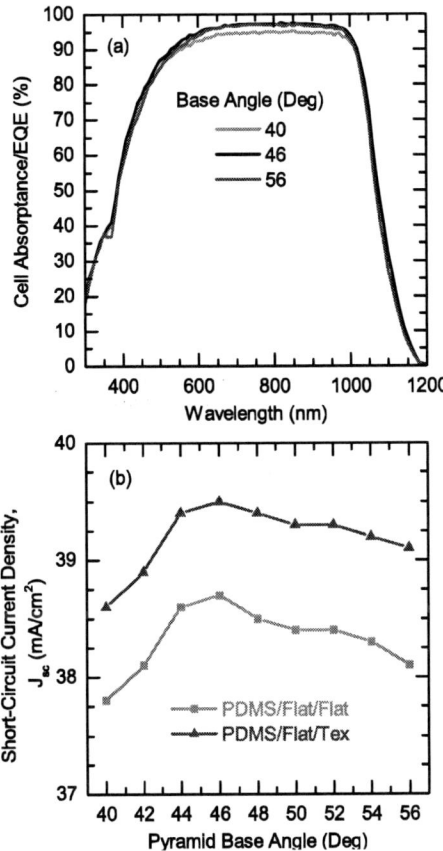

Fig. 3. (a) EQE spectra of the PDMS/Flat/Flat device, and (b) J_{sc} of PDMS/Flat/Flat and PDMS/Flat/Tex devices as the PDMS pyramid base angle is varied.

a smaller decrease for 56° pyramids below 600 nm, both of which warrant further investigation to understand.

Fig. 3b plots the short-circuit current density (J_{sc}) for both the PDMS/Flat/Flat and PDMS/Flat/Tex devices and shows that 0.2 mA/cm^2 or 0.3 mA/cm^2 can be gained, respectively, by reducing the pyramid base angle to 46° from the typical case for silicon for which the random pyramids have approximately 50° base angle (the silicon wafers etched in potassium hydroxide in our experiments have pyramids with 50.5° base angle [1]). In addition, 0.4 mA/cm^2 can be gained for both device structures with 46° compared to pyramids with "ideal" base angle of 54.7°.

To further understand the causes of reduced EQE between 500 and 1000 nm for 40° pyramids, as well as the J_{sc} maxima for 46° pyramids, we parse the contributions from the air/PDMS, PDMS/front-ITO/Si and Si/rear-ITO/Ag interfaces in the device to the total reflectance. The analysis is shown for the PDMS/Flat/Flat device, but is very similar for the Flat/Tex device.

Fig. 4a displays the reflectance from air/PDMS interface as the pyramid base angle is varied. The reflectance has little

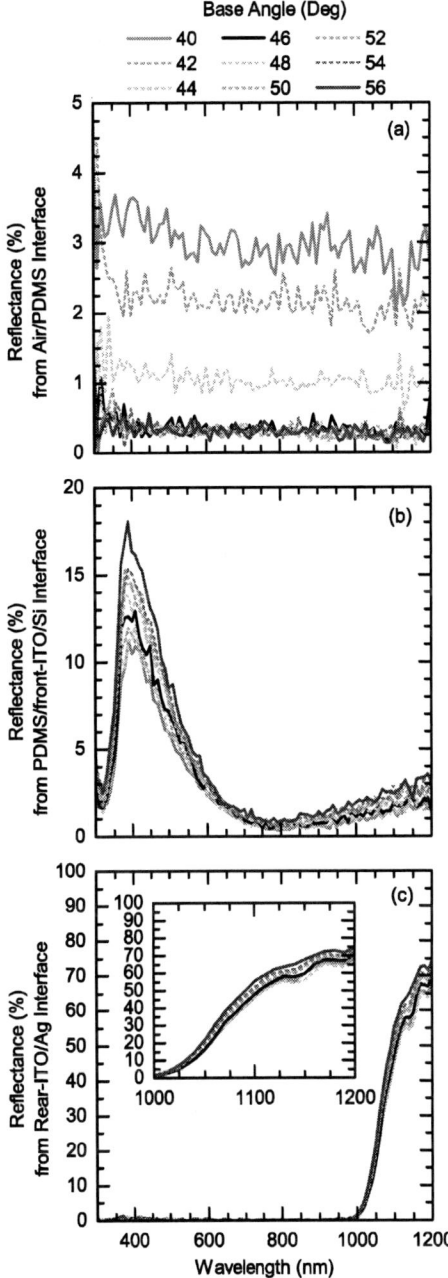

Fig. 4. Simulated reflectance spectra from different interfaces in the PDMS/Flat/Flat device as the PDMS pyramid base angle is varied: (a) air/PDMS interface, (b) PDMS/front-ITO/Si interface, (c) rear-ITO/Ag interface.

wavelength dependence because of the refractive index of PDMS has little dispersion. The reflectance is nearly constant at 0.3% for angles between 46 and 56°, and it increases rapidly for angles below 46° to 3% for 40°. This occurs because incoming light rays intersect only one facet on average, before being reflected away from the cell surface, for base angles below 45° [15]. This explains the lower EQE in Fig. 3a between 500 and 1000 nm for 40° pyramids.

Fig. 4b shows the amount of light that reflects from the PDMS/front-ITO/Si interfaces and escapes out the air/PDMS interface. This was calculated by setting the absorption at the rear of the silicon to unity in the Module Ray Tracer, and subtracting the reflectance from the air/PDMS interface from the resulting reflectance. The trend is opposite that in Fig. 4a: there is least reflectance for 40° and greatest for 56° base angle, especially at shorter wavelengths (300–600 nm) and IR wavelengths (900–1200 nm). This is caused by increasing Fresnal reflection at the PDMS/front-ITO interface as the angle of incidence at that interface increases with base angle. Note, however, that the reflectance for 46° base angle curve is pretty close to that for 40°.

Fig. 4c shows the IR light that reflects from the Si/rear-ITO/Ag interfaces and escapes out the air/PDMS interface. This was calculated as the difference between the total simulated reflectance of the cell and the sum of the reflectances plotted in Figs. 4a and b. The trend is similar to that in Fig. 4b—there is least escape for 40° base angle and greatest for 56°, with 46° being very close to 40°.

Finally, combining all three sets of interfaces and taking a holistic view, it becomes clear that lower base angles are better at coupling in light into the wafer, at all wavelengths, from the PDMS. However, lower angles (40, 42, and 44°) are worse coupling light into the PDMS to begin with—so much so that this front surface reflectance loss overpowers the benefits of improved light in-coupling at other interfaces in the device. Thus, there is a sweet spot where the combined reflectance loss is the lowest: 46° PDMS pyramid base angle.

IV. CONCLUSION

Planar solar cells have large current loss due to front-surface reflectance. PDMS scattering layers can be applied to the front of planar devices such as perovskite, CdTe, GaAs, and other thin-film solar cells to successfully reduce this loss. Application of such scattering layers is especially attractive for tandem solar cells such as two-terminal perovskite/silicon, perovskite/perovskite, and III-V/silicon cells, for the layers can be used to optimize the current matching in sub-cells. The PDMS approach to forming a scattering layer also provides a convenient way to investigate light-management properties of different textures that cannot be produced in the absorber, on a variety of solar cells and non-destructively. However, PDMS scattering layers show little contribution towards light trapping of weakly absorbed wavelengths. This contribution investigated if the light trapping afforded by PDMS layers carrying random inverted pyramids could be improved by varying the base angle of the pyramids. There is in fact not much change in light trapping of long-wavelength light by changing the base angle, but the reflectance at various interfaces within the combine to yield best results for 46° base angle. Highest J_{sc} is expected for solar cells with rear-side texture in conjunction with a PDMS scattering layer with random pyramids having this layer. While it is not trivial to controllably alter the base angle of pyramids etched in silicon and replicated in PDMS, there are demonstrations of tuning both the silicon base angle via alternative etch chemistries [1] and altering the PDMS base angle via the replication process.

ACKNOWLEDGEMENT

This material is based upon work supported primarily by USAID through the U.S.-Pakistan Center for Advanced Studies in Energy Program, under award AID-391-A-15-00001, and through the Engineering Research Center Program of the National Science Foundation and the Office of Energy Efficiency and Renewable Energy of the Department of Energy under NSF Cooperative Agreement No. EEC-1041895.

REFERENCES

high efficiency thin-film solar cells deposited on rough LP-CVD front ZnO," presented at the 22nd European Photovoltaic Solar Energy Conference, Milan, 2007.

[1] S. C. Baker - Finch and K. R. McIntosh, "Reflection distributions of textured monocrystalline silicon: implications for silicon solar cells," *Progress in photovoltaics: Research and Applications,* vol. 21, no. 5, pp. 960-971, 2013.

[2] K. R. McIntosh, T. G. Allen, S. C. Baker-Finch, and M. D. Abbott, "Light trapping in isotextured silicon wafers," *IEEE Journal of Photovoltaics,* vol. 7, no. 1, pp. 110-117, 2017.

[3] P. Buehlmann, A. Billet, J. Bailat, and C. Ballif, "Anti-reflection layer at the TCO/Si interface for

[4] K. A. Bush *et al.*, "23.6%-efficient monolithic perovskite/silicon tandem solar cells with improved stability," *Nature Energy,* vol. 2, p. 17009, 2017.

[5] S. Albrecht *et al.*, "Monolithic perovskite/silicon-heterojunction tandem solar cells processed at low temperature," *Energy & Environmental Science,* vol. 9, no. 1, pp. 81-88, 2016.

[6] G. E. Eperon *et al.*, "Perovskite-perovskite tandem photovoltaics with optimized band gaps," *Science,* vol. 354, no. 6314, pp. 861-865, 2016.

[7] S. Manzoor *et al.*, "Improved light management in planar silicon and perovskite solar cells using PDMS

978-1-5090-5606-4/17 $31.00 © 2017 IEEE

scattering layer (accepted for publication)," *Solar energy materials and solar cells,* 2017.

[8] S. Manzoor, M. Filipič, M. Topič, and Z. Holman, "Revisiting light trapping in silicon solar cells with random pyramids," in *Photovoltaic Specialists Conference (PVSC), 2016 IEEE 43rd,* 2016, pp. 2952-2954: IEEE.

[9] P. Campbell and M. A. Green, "Light trapping properties of pyramidally textured surfaces," *Journal of Applied Physics,* vol. 62, no. 1, pp. 243-249, 1987.

[10] M. Leilaeioun, Z. J. Yu, and Z. Holman, "Optimization of front TCO layer of silicon heterojunction solar cells for tandem applications," in *Photovoltaic Specialists Conference (PVSC), 2016 IEEE 43rd,* 2016, pp. 0681-0684: IEEE.

[11] M. Leilaeioun, Z. J. Yu, S. Manzoor, K. Fisher, J. Shi, and Z. C. Holman, "Design of the Front Transparent Conductive Oxide Layer of Silicon Heterojunction Solar Cells for Four-Terminal Tandem Applications (in preparation).".

[12] (06.19.2017). *PV Lighthouse: Module Ray Tracer.* Available: https://www.pvlighthouse.com.au

[13] M. A. Green, "Self-consistent optical parameters of intrinsic silicon at 300K including temperature coefficients," *Solar Energy Materials and Solar Cells,* vol. 92, no. 11, pp. 1305-1310, 2008.

[14] Y. Jiang, S. Pillai, and M. A. Green, "Realistic Silver Optical Constants for Plasmonics," *Scientific Reports,* vol. 6, 2016.

[15] S. C. Baker - Finch and K. R. McIntosh, "Reflection of normally incident light from silicon solar cells with pyramidal texture," *Progress in Photovoltaics: Research and Applications,* vol. 19, no. 4, pp. 406-416, 2011.

Damage-free laser ablation for emitter patterning of silicon heterojunction interdigitated back-contact solar cells

Menglei Xu[1,2], Twan Bearda[2], Miha Filipič[2], Hariharsudan Sivaramakrishnan Radhakrishnan[2], Maarten Debucquoy[2], Ivan Gordon[2], Jozef Szlufcik[2], Jef Poortmans[1,2,3]

[1]KU Leuven, Kasteelpark Arenberg 10, 3001 Heverlee, Belgium; [2]IMEC, Kapeldreef 75, B-3001 Leuven, Belgium; [3]Universiteit Hasselt, Martelarenlaan 42, 3500 Hasselt, Belgium

Abstract — We present a novel process scheme for a-Si:H patterning using damage-free laser ablation, resulting in a simple, fast, and photolithography-free emitter patterning of silicon heterojunction interdigitated back-contact (SHJ-IBC) solar cells. An a-Si:H laser-absorbing layer and a stack of sacrificial dielectric layers are deposited on top of a-Si:H/c-Si heterocontact to prevent laser damage. Laser ablation only removes the top a-Si:H layer, which limits laser damage to the surface of dielectric layers. These dielectric layers form a distributed Bragg reflector with a high reflectance of 80% at the laser wavelength which results in additional protection of the bottom a-Si:H/c-Si heterocontact. The significant reduction of laser damage is confirmed by atomic-force microscopy and photo-luminescence measurements. Such damage-free laser ablation process was successfully incorporated in a SHJ-IBC process flow and a best efficiency of 21.8% was achieved.

I. INTRODUCTION

Silicon heterojunction interdigitated back-contact (SHJ-IBC) solar cells are the subject of strong interest because of their capability to achieve a very high energy conversion efficiency. Several research groups presented impressive cell efficiencies in the range of 20.2%–25.6% [1]–[7]. Recently, Kaneka has reported the world record efficiency of 26.6% in a SHJ-IBC cell with size of 180 cm^2 [8]. Nevertheless, commercialization of the SHJ-IBC cells is still challenging. One of the key limitations is that the back-contact architecture results in additional fabrication complexity mainly due to the formation of interdigitated n- and p-type amorphous silicon (a-Si:H) strips. Laser ablation is an adequate patterning approach of a-Si:H and has been developed by different groups because of the following advantages: 1) it is a fast, single-side, and contactless process; 2) it has high process precision; 3) it allows a flexible device design [2], [9], [10]. To reduce laser damage at the a-Si:H/c-Si heterocontact during laser ablation, the use of an additional a-Si:H laser-absorbing layer and a SiO$_x$ sacrificial mask layer is reported in literature [9]–[11]. Only the top a-Si:H laser-absorbing layer is ablated, which shifts part of laser damage from the a-Si:H/c-Si heterocontact to the SiO$_x$ surface. However, when scribing line-shaped openings, an overlapping zone (OZ) of adjacent laser pulses has to be considered [11]. After ablation of the a-Si:H laser-absorbing layer, the energy of subsequent laser pulses in such OZs will be transmitted through the SiO$_x$ and absorbed by the

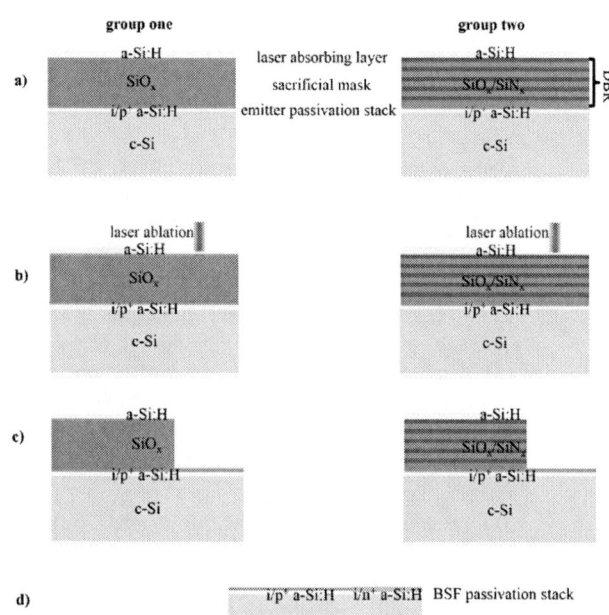

Fig. 1. Patterning process sequence of i/p$^+$ a-Si:H emitter stack and i/n$^+$ a-Si:H BSF stack: a) passivated wafers after deposition of sacrificial mask and laser-absorbing layer; b) laser ablation of absorbing layer; c) wet chemical etching of sacrificial mask; d) etching of i/p$^+$ a-Si:H followed by repassivation of i/n$^+$ a-Si:H and lift-off process.

bottom a-Si:H/c-Si, resulting in significant degradation of a-Si:H/c-Si passivation and issues with re-passivation quality [11].

In this contribution, a novel process scheme for a-Si:H patterning using damage-free laser ablation is presented. A distributed Bragg reflector (DBR) with reflectance of 80% at laser wavelength is used to replace SiO$_x$, substantially reducing laser damage to the underlying a-Si:H/c-Si [12]. The reduction of laser damage has been confirmed by atomic-force microscopy (AFM) and photoluminescence (PL) images. Functional SHJ-IBC cells were fabricated on n-type float zone (FZ) wafers using the developed laser ablation for i/p$^+$ a-Si:H emitter patterning.

978-1-5090-5606-4/17 $31.00 © 2017 IEEE

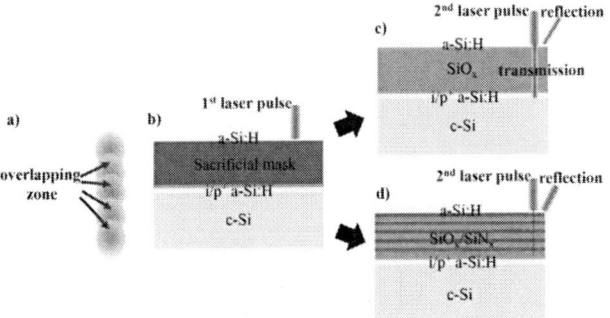

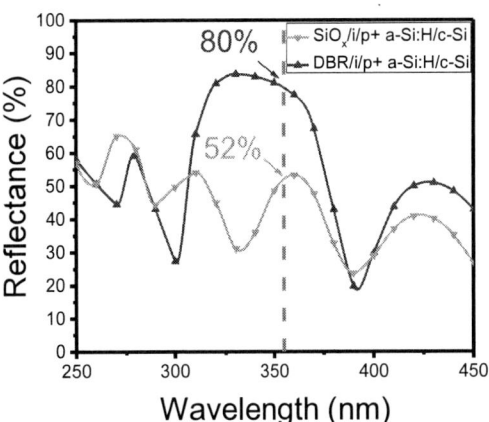

Fig. 2. a) Schematic diagram of laser scribed line. OZ is irradiated by two laser pulses; the cross-sections (not drawn to scale) of samples in OZ after, b) ablation of a-Si:H laser-absorbing layer by 1st laser pulse, c) and d) irradiance by 2nd laser pulse.

Fig. 3. Comparison of reflectance of $SiO_x/i/p^+$ a-Si:H/c-Si and DBR/i/p$^+$ a-Si:H/c-Si. 80% and 52% correspond to the reflectance of SiO_x and DBR samples at the laser wavelength of 355 nm, respectively.

II. RESULTS AND DISCUSSION

A. Process sequence for a-Si:H patterning

The samples were prepared using polished (100) n-type FZ silicon wafers. As shown in Fig. 1, the front side of the wafer was passivated with an intrinsic a-Si:H layer and the rear side was passivated with an i/p$^+$ a-Si:H emitter stack using plasma-enhanced chemical vapor deposition (PECVD). Then the wafers were split into two groups depending on the sacrificial mask layers. Group one had PECVD SiO_x sacrificial mask deposited at the rear side of the wafer, followed by deposition of a-Si:H laser-absorbing layer. Group two had five bi-layers of PECVD SiO_x/SiN_x deposited to form the DBR [13], followed by an identical a-Si:H layer deposition. Laser ablation of line patterns was carried out using a pulsed 355 nm laser (12 ps) with laser energy fluence of 0.2 J/cm^2. Four different laser processing speeds ranging from 0.7 m/s to 1.6 m/s were tested, corresponding to overlap of adjacent laser dots within one laser line from 50% to 0%. The width of the laser scribed lines is around 7 μm. The distance between lines was set to 12 μm to minimize OZ among adjacent lines. A PL image of the wafer was taken after laser ablation. The sacrificial dielectric layers were etched by dipping in HF:HCl:H$_2$O (1:1:20). Then AFM measurements were performed at the laser processed regions to identify laser damage. Finally, the i/p$^+$ a-Si:H stack was etched and the substrate was re-passivated with an i/n$^+$ a-Si:H back surface field (BSF) stack, which was patterned by a self-aligned lift-off process using SiO_x or SiO_x/SiN_x as sacrificial layers.

B. Reflectance measurements

The key factor to reduce laser damage at the i/p$^+$ a-Si:H/c-Si heterocontact is reducing the laser light transmission through the sacrificial dielectric layers, to minimize the laser energy absorption in the bottom i/p$^+$ a-Si:H/c-Si. This is even more important in the OZ where two laser pulses irradiate the same area. As shown in Fig. 3, the reflectance of $SiO_x/i/p^+$ a-Si:H/c-

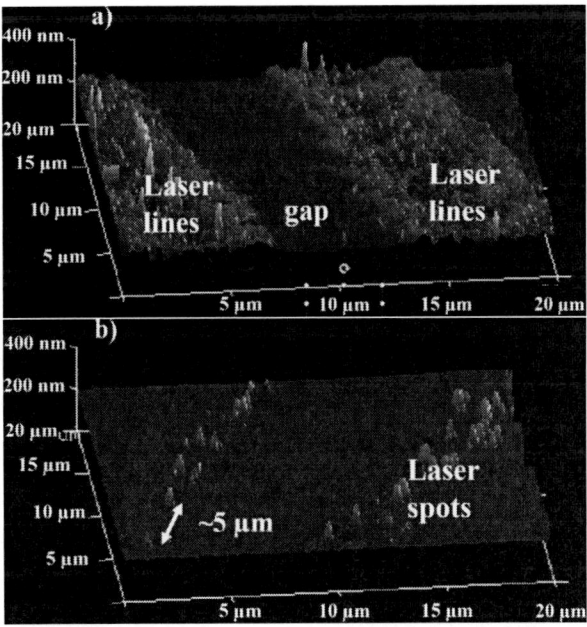

Fig. 4. The 3D AFM images of a) SiO_x, and b) DBR samples after laser ablation and sacrificial mask etching.

Si at laser wavelength of 355 nm is 52%. So approximately half of the laser light that reaches the SiO_x in OZ will be absorbed by the i/p$^+$ a-Si:H/c-Si due to the transparency of SiO_x at 355 nm. However, the reflectance can be increased up to 80% by using the bilayers of SiO_x/SiN_x to form the DBR. Note that our method is not limited to ultraviolet laser (e.g. 355 nm). The process window with high reflectance can be shifted to other wavelengths by simply modifying the thickness of sub-layers in DBR.

C. AFM measurements

A laser processing speed of 0.7 m/s was used for laser ablation of both SiO$_x$ and DBR samples. As shown in Fig. 4 a), for the SiO$_x$ sample, traces of the laser scribed lines are clearly observed in the form of roughness at the i/p$^+$ a-Si:H/c-Si interface. This indicates that the laser light is absorbed by the i/p$^+$ a-Si:H/c-Si and laser damage (e.g. roughness, thermal damage) is induced not only in the OZ but also in the areas processed by one laser pulse. In contrast, on DBR samples only very limited traces of separate laser spots are observed as

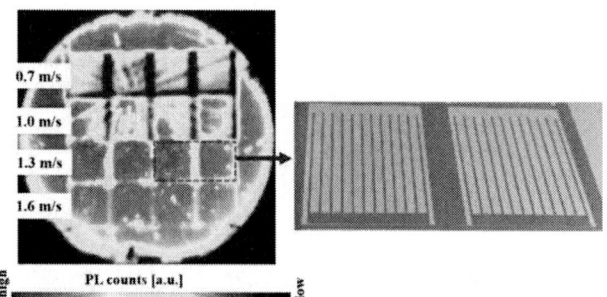

Fig. 5. Uncalibrated PL image of a DBR sample after laser ablation with processing speeds ranging from 0.7 m/s to 1.6 m/s (left); photographic image of two SHJ-IBC cells with active device area of 3.97 cm^2 after emitter patterning using laser ablation and sacrificial mask etching (right).

illustrated in Fig. 4 b). The distance between rough areas is around 5 μm, corresponding to the distance between the OZs. It suggests that laser damage at i/p$^+$ a-Si:H/c-Si interface is prevented at the areas irradiated by one laser pulse. Nevertheless, the presence of limited laser damage in the OZ is still observed. To reduce such damage, the OZ has to be reduced.

D. PL measurements

In order to study this, four different laser processing speeds ranging from 0.7 m/s to 1.6 m/s were tested on a DBR sample. PL images were recorded after laser ablation. As depicted in Fig. 5, dramatic degradation of the surface passivation is observed at the region processed by laser of 0.7 m/s due to laser damage of the bottom i/p$^+$ a-Si:H/c-Si in the OZ (see Fig. 4 b)). The degradation is diminished by increasing the laser speed. For the areas processed by laser of 1.3 m/s and 1.6 m/s, passivation can be maintained. It suggests that laser damage on i/p$^+$ a-Si:H/c-Si has been prevented by using higher laser speed and DBR to replace SiO$_x$. Several methods which are capable of reducing laser damage have been reported [2], [11]. However, in these methods additional etching steps of an a-Si:H laser-absorbing layer and/or SiO$_2$ sacrificial mask layers are always needed after laser ablation. In contrast, our approach is simpler because only the sacrificial DBR layers need to be etched using a short HF dip.

E. Device results

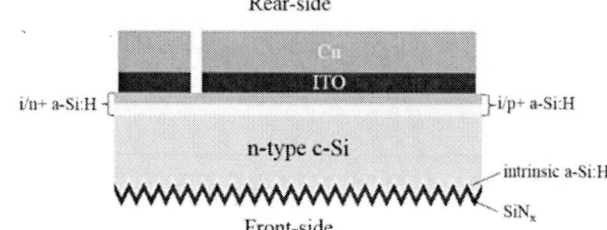

Fig. 6. Schematic of the SHJ-IBC solar cell architecture (not drawn to scale).

TABLE I SUMMARY OF THE CELL RESULTS

Cells	J_{sc} (mA/cm^2)	V_{oc} (mV)	FF (%)	η (%)	pseudo-FF (%)	pseudo-η (%)
Average	41.5	723	71.4	21.4		
Best	41.6	724	72.6	21.8	80.6	24.2

SHJ-IBC solar cells with active area of 3.97 cm^2 were fabricated using DBR mask layers and a process sequence, as shown in Fig. 1. Details of the integration flow including front-side process and rear-side metallization steps have been reported elsewhere [14]. The cross-section of the SHJ-IBC solar cells is depicted in Fig. 6. The cell results are summarized in Table I with the best efficiency of 21.8%. The J-V curve of the best cell is plotted in Fig. 7. As far as we know, this is the highest reported efficiency of SHJ-IBC cell using laser ablation for a-Si:H patterning. The V_{oc} of 724 mV is similar to that of our previous cells, which used photolithography for a-Si:H patterning [3, 5]. It indicates that the impact of laser damage on the c-Si surface passivation quality has been prevented, which is also confirmed by the PL measurements. The efficiencies are limited by moderate FF values (≤72.6%) mainly due to high series resistance (R_s). Suns-V_{oc} measurements predict that efficiencies can potentially reach >24 % by solving the R_s issue.

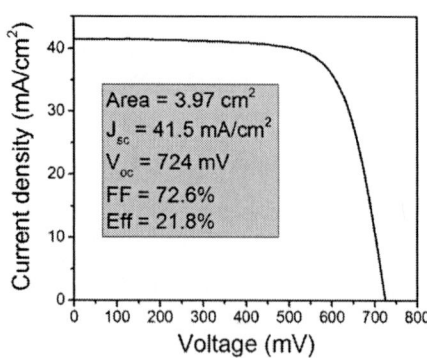

Fig. 7. J-V curve of the best SHJ-IBC cell.

III. CONCLUSION AND OUTLOOK

A damage-free laser ablation process for emitter patterning of SHJ-IBC solar cells has been presented. Laser damage on the i/p$^+$ a-Si:H passivation layer and c-Si substrate is prevented by using optimized laser conditions and a DBR. As such, the i/p$^+$ a-Si:H/c-Si interface passivation can be maintained after laser ablation, as confirmed by PL images. It is worth to notice that our method can be also applied to other laser wavelengths by simply modifying the thickness of sub-layers in the DBR.

Laser ablation and lift-off process result in a simple and photolithography-free a-Si:H patterning approach. To demonstrate the developed process, functional SHJ-IBC cells were fabricated on n-type FZ wafers with the best efficiency of 21.8%. Current developments focus on improvement of FF and further process simplifications, including in situ re-passivation and selective deposition of a-Si:H to replace lift-off process, opening a pathway to efficiencies in the range of 24%

ACKNOWLEDGEMENTS

The authors gratefully acknowledge the financial support of imec's industrial affiliation program for Si-PV. The research leading to these results has received funding from the European Union's Horizon 2020 research and innovation programme under the Marie Sklodowska-Curie grant agreement No 657270. This project has also received funding from the European Union's Horizon 2020 research and innovation programme under grant agreement No 727523.

REFERENCES

[1] J. Haschke, N. Mingirulli, R. Gogolin, R. Ferré, T. Schulze, J. Düsterhöft, N. Harder, and L. Korte, "Interdigitated Back-Contacted Silicon Heterojunction Solar Cells With Improved Fill Factor and Efficiency," IEEE J. Photovoltaics, vol. 1, no. 2, pp. 130-134, 2011.

[2] S. Harrison, O. Nos, G. D'Alonzo, C. Denis, A. Coll and D. Munoz, "Back Contact Heterojunction Solar Cells Patterned by Laser Ablation," Energy Procedia, vol. 92, pp. 730-737, 2016.

[3] M. Xu, T. Bearda, H. Sivaramakrishnan Radhakrishnan, S. K. Jonnak, V. Depauw, K. Van Nieuwenhuysen, M. Filipič, M. Debucquoy, I. Gordon, J. Szlufcik, and J. Poortmans, "Process Development of HJ-IBC cells on Silicon Bonded to Glass," in Proceedings of the 33rd European Photovoltaic Solar Energy Conference, 2016.

[4] A. Tomasi, B. Paviet-Salomon, M.J. Lehmann, D. Lachenal, J. Geissbühler, J. P. Seif, L. Barraud, A. Descoeudres, G. Christmann, N. Badel, H. Watanabe, A. Faes, S. Nicolay, B. Strahm, M. Despeisse, S. De Wolf and C. Ballif, "22.9% Simplified Back-Contacted Silicon Heterojunction Solar Cell," presented at the 33rd European Photovoltaic Solar Energy Conference, 2016.

[5] H. Sivaramakrishnan Radhakrishnan, T. Bearda, M. Xu, S. Jonnak, S. Malik, M. Hasan, V. Depauw, M. Filipič, K. Van Nieuwenhuysen, Y. Abdulraheem, M. Debucquoy, I. Gordon, J. Szlufczik and J. Poortmans, "Module-level Cell Processing of Silicon Heterojunction Interdigitated Back-Contacted (SHJ-IBC) Solar Cells with Efficiencies above 22%: Towards All-Dry Processing," in Proceedings of the 43th IEEE Photovoltaic Specialists Conference, 2016.

[6] J. Nakamura, N. Asano, T. Hieda, C. Okamoto, H. Katayama, and K. Nakamura, "Development of Heterojunction Back Contact Si Solar Cells," IEEE J. Photovoltaics, vol. 4, no. 6, pp. 1491-1495, 2014.

[7] K. Masuko, M. Shigematsu, T. Hashiguchi, D. Fujishima, M. Kai, N. Yoshimura, T. Yamaguchi, Y. Ichihashi, T. Mishima, N. Matsubara, T. Yamanishi, T. Takahama, M. Taguchi, E. Maruyama, and S. Okamoto, "Achievement of More Than 25% Conversion Efficiency With Crystalline Silicon Heterojunction Solar Cell," IEEE J. Photovoltaics, vol. 4, no. 6, pp. 1433–1435, 2014.

[8] Best Research-Cell Efficiencies (NREL, 2016). Available: http://www.nrel.gov/pv/assets/images/e-ciencychart.png. [Accessed: 02-Jun-2017]

[9] T. Desrues, S. De Vecchi, F. Souche, D. Munoz, and P. J. Ribeyron, "SLASH concept: A novel approach for simplified interdigitated back contact solar cells fabrication," in Proceedings of the 38th IEEE Photovoltaic Specialists Conference, 2012.

[10] S. Ring, L. Mazzarella, P. Sonntag, S. Kirner, C. Schultz, U. Schmeißer, J. Haschke, L. Korte, B. Stannowski, B. Stegemann, and R. Schlatmann, "Emitter Patterning for IBC-Silicon Heterojunction Solar Cells Using Laser Hard Mask Writing and Self-aligning," in Proceedings of the 32rd European Photovoltaic Solar Energy Conference, 2015.

[11] S. Ring, S. Kirner, C. Schultz, P. Sonntag, B. Stannowski, L. Korte and R. Schlatmann, "Emitter Patterning for Back-Contacted Si Heterojunction Solar Cells Using Laser Written Mask Layers for Etching and Self-Aligned Passivation (LEAP)," IEEE J. Photovoltaics, vol. 6, no. 4, pp. 894-899, 2016.

[12] M. Xu, T. Bearda, and M. Filipič, European patent application EP 16197123.9, Nov. 3, 2016.

[13] V. Gottschalch, R. Schmidt, B. Rheinländer, D. Pudis, S. Hardt, J. Kvietkova, G. Wagner and R. Franzheld, "Plasma-enhanced chemical vapor deposition of SiOx/SiNx Bragg reflectors," Thin Solid Films, vol. 416, no. 1-2, pp. 224-232, 2002.

[14] M. Xu, T. Bearda, H. Sivaramakrishnan Radhakrishnan, S. K. Jonnak, M. Hasan, S. Malik, M. Filipič, V. Depauw, K. Van Nieuwenhuysen, Y. Abdulraheem, M. Debucquoy, I. Gordon, J. Szlufcik, and J. Poortmans, "Silicon heterojunction interdigitated back-contact solar cells bonded to glass with efficiency > 21%," Sol. Energy Mater. Sol. Cells, vol. 165, pp. 82-87, 2017

Benefits of a thermal drift during atomic layer deposition of Al$_2$O$_3$ for c-Si passivation

Fabien Lebreton[1,2,3], Andy Zauner[4,3], Pavel Bulkin[2,3], François Silva[2,3], Sergej Filonovich[1,3] and Pere Roca i Cabarrocas[2,3]

[1]Total S.A. Renewables, 2 Place Jean Millier, 92078 Paris La Défense cedex, France

[2]LPICM, CNRS, Ecole Polytechnique, Université Paris-Saclay, 91128 Palaiseau, France

[3]Institut Photovoltaïque d'Ile-de-France (IPVF), 8 rue de la renaissance, 92160 Antony, France

[4]Air Liquide Paris-Saclay Research Center, 78350 Les Loges-en-Josas, France

Abstract — **Variation of the substrate temperature within the ALD window during deposition does not significantly impact the growth rate of Al$_2$O$_3$ layers but changes their properties. Reduced initial deposition temperature allows storing hydrogen at the c-Si/Al$_2$O$_3$ interface. This hydrogen is steadily released during the deposition of upper layers due to the increase of substrate temperature. Al$_2$O$_3$ can thus provide SRV <10 cm.s^{-1} in as-deposited state for a thermal budget below 250 °C. Thermal drift also prevents strong blistering caused by post-deposition thermal processes as most of the gases can be released during deposition, in the absence of a thick diffusion barrier.**

I. INTRODUCTION

Passivation of p-type silicon surfaces by atomic layer deposition (ALD) of Aluminum oxide (Al$_2$O$_3$) layers has opened new routes for high efficiency solar cells [1]. Thermal ALD is becoming an industrially viable alternative to current technologies, *e.g.* Plasma-Enhanced Chemical Vapor Deposition (PECVD), owing to the low required layer thickness (~ 10 nm for good chemical and field effect passivation) and the process scalability (batch or spatial ALD) [2]. One of the main advantages of such ALD process for passivation is the deposition temperature, as low as 100 °C [3]. However, Al$_2$O$_3$ provides poor passivation properties in as-deposited state. Annealing above 300 °C is mandatory to achieve surface recombination velocities (SRV) below 10 cm.s^{-1}. This additional thermal step plays against the low thermal budget argument.

Another inconvenience for ALD Al$_2$O$_3$ is that it has to be capped by another material, the mainstream one being hydrogenated silicon nitride (a-SiN$_X$:H) [4]. However, the interaction between Al$_2$O$_3$ and its capping layer can lead to blistering, degrading the optical and electrical properties of such stacks [5]. Gas desorption (hydrogen and water) from Al$_2$O$_3$ is pointed out as one of the blistering origins [6]. Several ways have been proposed to avoid blistering such as: out-gassing prior to capping (annealing> 600 °C) [7], pre-oxidation of c-Si surface to avoid defective island growth during the first ten cycles [8, 9] or even a high deposition temperature for Al$_2$O$_3$ [10].

In this work we show the benefit of a changing substrate temperature during ALD growth on both passivation activation and blistering. This variation of standard ALD is named hereafter Thermal Drift Atomic Layer Deposition (TD-ALD).

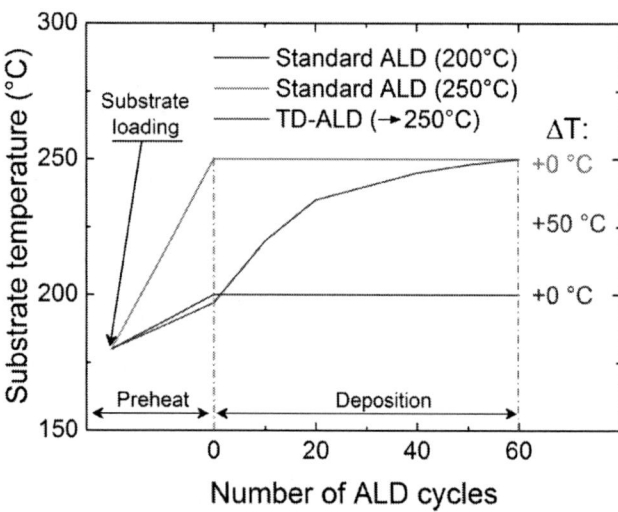

Fig. 1. Conceptual comparison between standard and thermal drift ALD processes. The standard ALD process heats the substrate to the desired temperature and maintains it during deposition. The main point of TD-ALD is to start the deposition at a lower temperature than the one reached at the end of the deposition process.

II. EXPERIMENTAL APPROACH

Symmetrically processed c-Si samples with either a single Al$_2$O$_3$ layer or Al$_2$O$_3$/a-SiN$_X$:H stacks were produced. Al$_2$O$_3$ was deposited using TMA and water. Deposition was done on both sides of the c-Si substrate in the same run owing to our specific substrate holder which allows diffusion of precursors to both surfaces. The c-Si wafers were double-side polished, float-zone, p-type, <100>, with a resistivity of 1-5 Ω.cm and a thickness of 280 μm. After HF dip, Al$_2$O$_3$ deposition was

978-1-5090-5606-4/17 $31.00 © 2017 IEEE

carried out either at a constant substrate temperature (200 °C or 250 °C), later on referred to as standard ALD, or in a thermal drift conditions, i.e. starting around 200 °C and increasing to 250 °C during the 60 first ALD cycles, as schematically represented in Fig. 1. The a-SiN$_X$:H capping has been done by PECVD using silane and ammonia at a substrate temperature of 300°C. As the PECVD is done at a rather low temperature, a passivation activation step has been performed prior to capping for all the layers. Passivation activation was hold for 30 minutes in forming gas atmosphere and its temperature has been varied between 350 °C and 440 °C. After capping, a final thermal step is done at 380 °C for 20 minutes in order to mimic the electroplating metallization step developed at IPVF [11].

The thickness and optical properties of the thin films were measured by spectroscopic ellipsometry. Hydrogen and carbon contents have been assessed by secondary ion mass spectrometry (SIMS) using a calibration sample. SRV was estimated from the measurement of the effective lifetime (τ_{eff}) by quasi steady-state photo-conductance decay using (1) where bulk lifetime (τ_{bulk}) was estimated according to Richter et al. [12], and W is the wafer thickness.

$$S_{eff} = \left(\frac{1}{\tau_{eff}} - \frac{1}{\tau_{bulk}}\right)\frac{W}{2} \qquad (1)$$

III. RESULTS AND DISCUSSION

A. Materials properties

As expected from literature [13], the substrate temperature strongly determines the contamination level (carbon and hydrogen) of the films due to an increased reactivity of the precursors at higher temperature (Fig. 2).

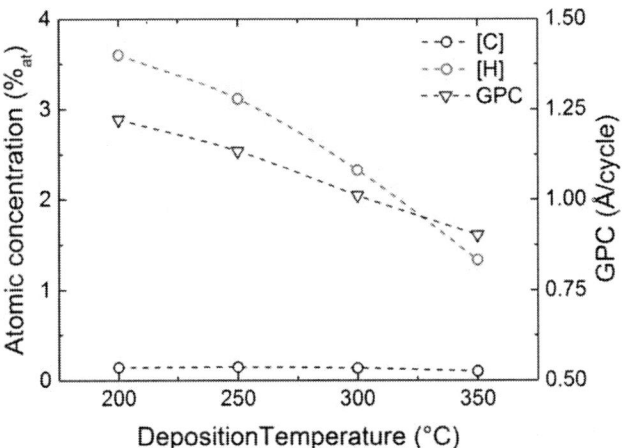

Fig. 2. Atomic concentration of impurities in as-deposited Al$_2$O$_3$ layers as a function of the deposition temperature. Results were obtained by SIMS measurements using a calibration sample itself measured by RBS, ERDA and NRA. Layers were deposited using 1000 ALD cycles. Lines are guides to the eyes.

There is a slight decrease in the growth rate of the film from 1.22 to 0.9 Å/cycle and an increase in the refractive index from 1.646 to 1.656 (at 633 nm). These trends are consistent with lower hydrogen content and denser films at higher deposition temperature. A variation of the substrate temperature during deposition should allow profiling the hydrogen content in the film. Moreover, a temperature drift will somehow anneal the already deposited layers during the deposition of upper layers. Then, hydrogen might be released during the ALD process itself, limiting brutal gas desorption that occurs during post deposition annealing and that often leads to blistering.

B. As-deposited Al$_2$O$_3$

Deposited 200°C, the Al$_2$O$_3$ layer contains higher amount of hydrogen than a layer deposited at 250°C (Fig. 2). However, the variation in hydrogen content does not significantly affect its passivation properties (Fig. 3) in as-deposited state since it remains mainly bonded as -OH groups [14]. As shown on Fig. 3, TD-ALD provides significantly improved as-deposited passivation properties compared to that of standard ALD, leading to SRV < 10 cm.s^{-1} for 200 cycles (~20 nm). Hydroxyl groups left near the c-Si surface during the first cycles react together when the temperature is increased [8, 15]. Released H$_2$ can thus diffuse to c-Si surface or out-diffuse easily from the Al$_2$O$_3$ films as the deposited upper thickness is rather low. Hydrogen content of the film steadily decreases while the temperature is increased.

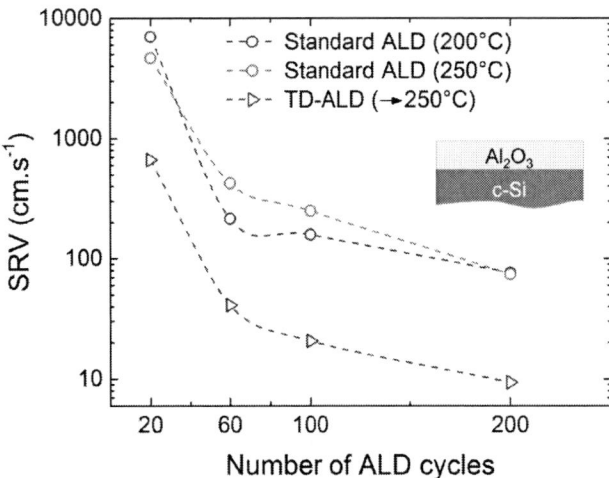

Fig. 3. Surface recombination velocity of passivated samples in as-deposited state for different Al$_2$O$_3$ thicknesses. Lines are guides to the eyes.

C. Stack integration

Even if SRV values lower than 10 cm.s^{-1} can be obtained while never exceeding 250 °C, a thickness below 10 nm is required to satisfy industrial throughput. Layers deposited with 60 cycles perfectly fit this requirement [16].We will focus on the passivation behavior of these layers capped by a-SiN$_X$:H. Stacks with Al$_2$O$_3$ layers deposited at 200 °C led to better passivation levels than those with Al$_2$O$_3$ deposited at 250 °C (Fig. 4). This better passivation level verified over the whole passivation activation temperature range can be attributed to the higher hydrogen content for the deposition temperature of 200 °C. The main drawback of this higher hydrogen content is blistering on all the samples deposited at 200 °C, while samples deposited at 250 °C are blister free. We have discussed the high passivation level in samples presenting blisters in [17].

As previously demonstrated by as-deposited results, thermal drift along the ALD process releases some hydrogen during the deposition. Some of this hydrogen remains at the interface providing passivation after deposition, while the excess of hydrogen may leave the film during deposition. Using TD-ALD for Al$_2$O$_3$ before capping leads to lower SRV (Fig. 4) over the full range of passivation activation temperature than standard ALD. Hydrogen management during TD-ALD provides stacks without any blistering, resulting in a SRV as low as 1.7 cm.s^{-1} for the optimal activation temperature of 380 °C.

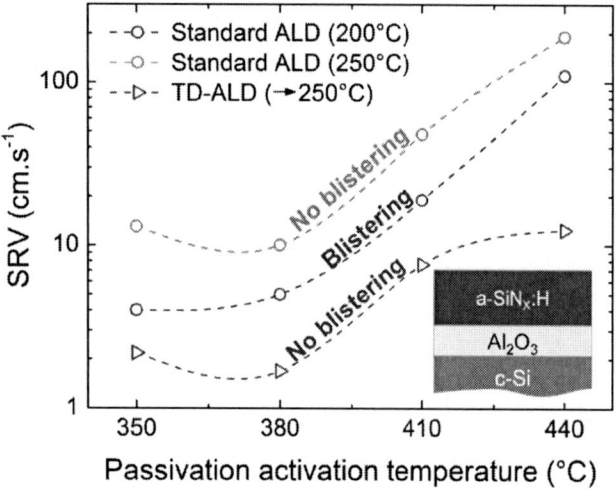

Fig. 4.   Surface recombination velocity for samples passivated with Al$_2$O$_3$/a-SiN$_X$:H stacks. Passivation of Al$_2$O$_3$ is activated for 30 minutes in forming gas atmosphere at different temperatures before a-SiN$_X$:H deposition at 300 °C. Passivation stacks consist of 6 nm of Al$_2$O$_3$ capped by 75 nm of a-SiN$_X$:H. Lines are guides to the eyes.

IV. CONCLUSION

Al$_2$O$_3$ layers deposited by TD-ALD at substrate temperatures below 250 °C provide excellent passivation properties (SRV < 10 cm.s^{-1}) in as-deposited state. The low thermal budget makes TD-ALD of Al$_2$O$_3$ compatible with other materials that degrade under higher temperatures. When these films are capped with a-SiN$_X$:H, the full stacks also

present superior passivation properties compared to the ones deposited by standard ALD. Moreover, blistering can also be avoided using this method, without additional process steps such as out-gassing or pre-oxidation of the substrate. TD-ALD can easily be applied to batch or spatial ALD tools and its temperature rise (ΔT) tuned in order to fit in the solar cells process flow.

ACKNOWLEDGMENTS

This work was carried out in the framework of project A from IPVF (Institut Photovoltaïque d'Ile-de-France). This project has been supported by the French Government in the frame of Programme d'Investissement d'Avenir – ANR-IEED-002-01. Fabien Lebreton thanks ANRT for CIFRE scholarship.

REFERENCES

[1]   J. Benick, B. Hoex, G. Dingemans, W. Kessels, A. Richter, M. Hermle, and S. Glunz, "High-efficiency n-type silicon solar cells with front side boron emitter," in *24th European Photovoltaic Solar Energy Conference*, 2009, pp. 863-870.

[2]   G. Dingemans, and W. M. M. Kessels, "Status and prospects of Al$_2$O$_3$-based surface passivation schemes for silicon solar cells," *Journal of Vacuum Science & Technology A*, vol. 30, no. 4, pp. 040802, 2012.

[3]   J. Frascaroli, G. Seguini, E. Cianci, D. Saynova, J. van Roosmalen, and M. Perego, "Surface passivation for ultrathin Al$_2$O$_3$ layers grown at low temperature by thermal atomic layer deposition," *physica status solidi (a)*, vol. 210, no. 4, pp. 732-736, 2013.

[4]   A. Richter, J. Benick, and M. Hermle, "Boron Emitter Passivation With Al$_2$O$_3$ and Al$_2$O$_3$/SiN$_x$ Stacks Using ALD Al$_2$O$_3$," *IEEE Journal of Photovoltaics*, vol. 3, no. 1, pp. 236-245, 2013.

[5]   A. Richter, F. Souren, D. Schuldis, R. Görtzen, J. Benick, M. Hermle, and S. Glunz, "Thermal stability of spatial ALD deposited Al$_2$O$_3$ capped by PECVD SiN$_x$ for the passivation of lowly- and highly-doped p-type silicon surfaces," in *27th European Photovoltaic Solar Energy Conference*, 2012, pp. 1133-1137.

[6]   B. Vermang, H. Goverde, V. Simons, I. De Wolf, J. Meersschaut, S. Tanaka, J. John, J. Poortmans, and R. Mertens, "A study of blister formation in ALD Al$_2$O$_3$ grown on silicon," in *38th IEEE Photovoltaic Specialists Conference*, 2012, pp. 001135-001138.

[7]   B. Vermang, H. Goverde, A. Lorenz, A. Uruena, G. Vereecke, J. Meersschaut, E. Cornagliotti, A. Rothschild, J. John, J. Poortmans, and R. Mertens, "On the blistering of atomic layer deposited Al$_2$O$_3$ as Si surface passivation," in *37th IEEE Photovoltaic Specialists Conference*, 2011, pp. 003562-003567.

[8]   V. Naumann, M. Otto, R. B. Wehrspohn, and C. Hagendorf, "Chemical and structural study of electrically passivating Al$_2$O$_3$/Si interfaces prepared by atomic layer deposition," *Journal of Vacuum Science & Technology A*, vol. 30, no. 4, pp. 04D106, 2012.

[9]   S. Bordihn, P. Engelhart, V. Mertens, G. Kesser, D. Köhn, G. Dingemans, M. M. Mandoc, J. W. Müller, and W. M.

M. Kessels, "High surface passivation quality and thermal stability of ALD Al_2O_3 on wet chemical grown ultra-thin SiO_2 on silicon," *Energy Procedia,* vol. 8, pp. 654-659, 2011.

[10] T. Lüder, T. Lauermann, A. Zuschlag, G. Hahn, and B. Terheiden, "Al_2O_3/SiN_X-Stacks at Increased Temperatures: Avoiding Blistering During Contact Firing," *Energy Procedia,* vol. 27, no. 0, pp. 426-431, 2012.

[11] J. Couderc, H. e. Belghiti, D. Aureau, J. Dupuis, P.-P. Grand, É. Delbos, A. Etcheberry, and D. Lincot, "Study of one-step annealing for plated nickel-copper contacts on n-pert bifacial monocrystalline silicon solar cells," in *32nd European Photovoltaic Solar Energy Conference*, 2016, pp. 697-702.

[12] A. Richter, S. W. Glunz, F. Werner, J. Schmidt, and A. Cuevas, "Improved quantitative description of Auger recombination in crystalline silicon," *Physical Review B,* vol. 86, no. 16, pp. 165202, 2012.

[13] O. M. Ylivaara, X. Liu, L. Kilpi, J. Lyytinen, D. Schneider, M. Laitinen, J. Julin, S. Ali, S. Sintonen, M. Berdova, E. Haimi, T. Sajavaara, H. Ronkainen, H. Lipsanen, J. Koskinen, S.-P. Hannula, and R. L. Puurunen, "Aluminum oxide from trimethylaluminum and water by atomic layer deposition: the temperature dependence of residual stress, elastic modulus, hardness and adhesion," *Thin Solid Films,* vol. 552, pp. 124-135, 2014.

[14] S.-K. Oh, H.-S. Shin, K.-S. Jeong, M. Li, H. Lee, K. Han, Y. Lee, G.-W. Lee, and H.-D. Lee, "Process Temperature Dependence of Al_2O_3 Film Deposited by Thermal ALD as a Passivation Layer for c-Si Solar Cells," *JSTS:Journal of Semiconductor Technology and Science,* vol. 13, no. 6, pp. 581-588, 2013.

[15] G. Seguini, E. Cianci, C. Wiemer, D. Saynova, J. A. M. van Roosmalen, and M. Perego, "Si surface passivation by Al_2O_3 thin films deposited using a low thermal budget atomic layer deposition process," *Applied Physics Letters,* vol. 102, no. 13, pp. 131603, 2013.

[16] V. I. Kuznetsov, M. A. Ernst, and E. H. A. Granneman, "Al_2O_3 surface passivation of silicon solar cells by low cost ald technology," in *40th IEEE Photovoltaic Specialists Conference*, 2014, pp. 0608-0611.

[17] F. Lebreton, A. Devos, P. Bulkin, F. Silva, S. Filonovich and P. Roca i Cabarrocas, "Blistering of Al_2O_3/a-SiN_X:H stacks: analysis of the submerged part of the iceberg by colored picosecond acoustic microscopy," in *43rd IEEE Photovoltaic Specialists Conference,* 2017

Growth Difference of amorphous silicon between plasma enhanced and catalytic CVD based on silicon heterojunction solar cells

Liping Zhang[1], Renfang Chen[1], Zhuopeng Wu[1], Chenguang Sun[2], Fanying Meng[1], Zhengxin Liu[1]

[1]Shanghai Institute of Microsystem and Information Technology, Chinese Academy of Sciences, Shanghai 200050, China

[2]State Key Laboratory of PV Science and Technology, Trina Solar, Changzhou, Jiangsu 213001, China

Abstract — **The microstructure evolution of hydrogenated amorphous silicon (a-Si:H) films, deposited by plasma enhanced (PE) and catalytic (CAT) chemical vapor deposition, have been investigated with the increasing thickness. It has been found that crystallites prone to present in the initial growth of a-Si:H deposited by CAT method, whose film has a more compact structure and stronger bonding energy than the film deposited by PE method. At a premise of same conversion efficiency, open circuit voltage and fill factor of the silicon heterojunction solar cells fabricated by CAT are higher than that of the cell fabricated by PE.**

Index Terms — **amorphous silicon, passivation, growth difference, silicon heterojunction solar cells.**

I. INTRODUCTION

Silicon heterojunction (SHJ) solar cells have attracted increasing attention owing to their high performance and potential in the research and mass production of photovoltaic devices.[1-4] An bifacial SHJ cell with a conversion efficiency of 25.1% and an open-circuit voltage (V_{oc}) of 738 mV was achieved in 2015.[3] A regular SHJ solar cell comprises crystal silicon (c-Si), i-/n-, i-/p-type of hydrogenated amorphous silicon (a-Si:H) thin film, transparent conducting oxides (TCO), and metal contacts. The bifacial a-Si:H thin film has been proved to be an excellent passivation layer for c-Si owing to its wide bandgap, low fabrication temperature, and low defect density.[5,6] It has been proved that the amorphous-to-nanocrystalline transition phase is an excellent compact network for the surface passivation layer of c-Si.[7,8]

The prevalent technique to deposition a-Si:H is plasma enhanced (PE) chemical vapor deposition (CVD).[3,4] However, ion bombardment frequently occurs in the subsequent deposition, which can lead to damage in the subjacent a-Si:H layer or interface. Therefore, the quality of a-Si:H passivation layer is difficult to be controlled at a thickness scale of less than 20 nm. For obtaining high-quality passivation, H_2-plasma treatment[9-12] used for a-Si:H passivation layers with a highly disordered structure and silicon oxide[13-17] interface layer used for suppressing the epitaxial layer are two effective methods taking responsibility to improve the microstructure of a-Si:H and the passivation quality of the a-Si:H/c-Si heterojunction.[18,19]

Early in 1985, Matsumura et al. have developed a novel low-temperature deposition method, named the catalytic (CAT) CVD method, often called the hot-wire CVD method.[20,21] In this method, deposition gases are decomposed by catalytic cracking reactions on a heated catalyzer located near the substrates. It has reported that the structure of crystalline grains prone to embedded in an a-Si:H matrix closed the interface with the c-Si substrate.[22] The advantage of this CVD technique is rich in Atomic hydrogen generated by cracking H_2 around a heated catalyzer.

In this research, we focus on the growth evolution of a-Si:H deposited by PE and CAT CVD. Intrinsic a-Si:H passivation layers with amorphous-to-nanocrystalline transition phase are chosen to be candidates which passivate the surface of c-Si resulting in two better performance SHJ solar cells. To clarify the growth mechanism of a-Si:H deposited by two CVD methods, the optical constants and the microstructure of a-Si:H with different thicknesses are investigated. The structure difference of a-Si:H and influence on device can be drawn by optical-electrical performance of SHJ solar cells. Finally, the optimization proposals will be given after comparison of two CVD methods.

II. EXPERIMENTAL DETAILS

n-type textured czochralski (CZ) silicon wafers with a resistivity of ~3 Ω·cm, which were used as the substrates of heterojunction passivation and SHJ solar cells, were cleaned by a standard RCA process and then dipped in 2% HF for 2 min. Subsequently, two series of i-a-Si:H thin films were prepared at low substrate temperatures of less than 200 °C by using a radio frequency PE and CAT CVD, respectively. In the case of PE method, power densitiy and deposition pressure are 30 mW/cm^2 and 200 Pa, individually. The electrode distance and $[H_2]/[SiH_4]$ flow ratio are 25 mm and 10, respectively. As for CAT method, $[SiH_4]/[H_2]$ flow ratio is 2 in the few pascal of pressure.

Single-polished float zone (FZ) silicon wafers (100) with resistivity greater than 2000 Ω·cm were used as a substrate so that various optical measurements, such as spectroscopic ellipsometry (SE, J.A. Woollam M-2000XI) and Fourier transformed infrared (FTIR) spectroscopy (PerkinElmer 100), could be carried out. An SHJ solar cell structure of Ag grid/ITO/n-a-Si:H/i-a-Si:H/textured-CZ n-c-Si/i-a-Si:H/p-a-Si:H/ITO/Ag grid was performed to evaluate the a-Si:H passivation layer deposited by two CVD methods.

III. RESULTS AND DISCUSSION

Some a-Si:H thin films of different thicknesses, deposited by PE and CAT CVD, are use to investigate the growth mechanisms on the silicon wafer. Generally, ultrathin a-Si:H film is difficult to be characterized accurately. However, SE as a powerful measurement tool is capable of determining the optical constants of a-Si:H-based film by examining the change in the polarization state after the interaction between the incident light and samples.[10,23-29] Ellipsometric angles Ψ and Δ are related to the magnitude and phase of the complex reflectivity ratio, respectively:[30]

$$\tan\Psi \, e^{i\Delta} = \frac{\tilde{R}_p}{\tilde{R}_s} \tag{1}$$

where $\tilde{R}_p$ and $\tilde{R}_s$ are the complex reflectances of p- and s-light, which are parallel and perpendicular to the plane of incidence, respectively. The microstructure of one i-a-Si:H layer on an FZ c-Si (100) wafer is evaluated by an SE apparatus. Ψ and Δ are acquired in the wavelength range of 300–1000 nm at six incidence angles in the range from 50° to 75° with 5° intervals. The Tauc-Lorentz model is capable of fitting the measured data of amorphous materials. Effective medium approach (EMA) assuming a microscopic mixture of crystalline and amorphous components is used to simulate the initial growth of a-Si:H films. Therefore, some critical optical functions of silicon passivation films, such as bulk layer thickness (d_{bulk}), surface roughness thickness (d_{sr}), crystalline fraction (f_c), refraction index (n), broadening and characteristic energy (E_{01}) can be deduced.

Figure 1 show the variation of critical optical function with the increase of d_{bulk}. The calculated deposition rate (R_d) is normalized to evaluate the precursor concentration contributing to the growth. From figure 1 (a), R_d of a-Si:H grown by PE decreases slowly then saturate at a while that by CAT decreases fast when the thickness is less than 50 Å. High-radiative surface temperature of catalyzer may be the reason of fast decreasing deposition rate with time goes on. In the initial stage of the growth, d_{sr} in figure 1 (b) are all higher owing to the formation of three-dimensional adatom clusters or islands. d_{sr} deposited by CAT rapidly falls from 16 Å to 6 Å which means the growth mode shifts to two-dimensional layer by layer mode. Moreover, f_c of CAT decreases fast but it is still several percent when the thickness is less than 100 Å. However, f_c of PE is almost zero when the thickness is more than 30 Å. Correspondingly, n of CAT, associated the material density, is higher than n of PE due to the high f_c of CAT. The other two evidences of the increasing crystallites in a-Si:H of CAT are broadening and as shown in figure (e) and (f),

respectively. Broadening denotes the disordering degree and E_{01} is resigned to be characteristic bonding energy. Both broadening and E_{01} exhibit opposite tendencies which imply that the disordering degree of films increases and bonding energy decreases with the increasing thickness. The mentioned above illustrate that the structure of a-Si:H deposited by CAT prone to be a structure of crystallites embedded into amorphous silicon network due to the influence of c-Si substrates. An atomic H-rich plasma results in a layer by layer growth of a-Si:H with a part of crystallites, high n and high bonding energy in the case of CAT method. When the thickness is more than 130 Å, the rising d_{sr} implicates that the growth modes of a-Si:H deposited by both two methods transit from layer by layer to layer plus island due to the possible change of surface energies and bonding energy.

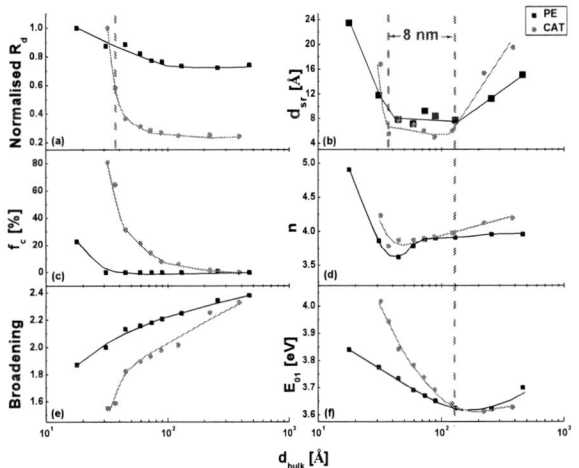

Fig. 1.   Thickness dependences of normalized deposition rate, surface roughness, crystal faction, refractive index, broadening and characteristic Energy.

Figure 2 shows the stretching mode in FTIR spectroscopy of i-a-Si:H thin films on FZ-c-Si (100). Low-stretching mode (LSM) and high-stretching mode (HSM) are located at 1980–2010 cm^{-1} and 2070–2100 cm^{-1}, respectively.[31] Generally, LSM is assigned to the vibration of monohydrides, which represents compact hydrogen incorporation, whereas HSM is attributed to the vibration of clustered hydrogen in monohydrides, dehydrides, or trihydrides at the internal surfaces of voids or the boundaries between the crystalline grains.[31-33] Thicknesses of i-a-Si:H layer deposited by PE and CAT are from ~7.0 nm to ~40.0 nm in figure 2. From figure 2, HSM is rising with the increasing thickness in the case of PE while HSM variation is slight in the case of CAT. When the thickness is thinner than 13.5 nm, C_Hs of the a-Si:H films deposited by two methods are more than 20%. Moreover, same high C_H can be obtained at same a-Si:H thickness by both CVD methods. Combining the optical constants in figure 1 with figure 2, the structure of a-Si:H deposited by CAT is more compact at a same C_H level due to the presence of crystallites. When the thickness is more than 13.5 nm, the reduction of C_H can be found in the both kinds of i-a-Si:H films. The intensity of LSM drops to low levels, which means that the monohydrides decline with the increase of thickness for both methods. In the case of PE, the declination of LSM indicates that a small quantity of weak bonds can be broken owing to the penetration and relaxation of high-energy radicals in the microstructure of a-Si:H layer. Furthermore, the increasing HSM associates the presence of transition boundary towards nanocrystalline. Different from PE, there are no high-energy radicals but highly radiative surface temperature which leads to the dissociation of Si-H bonds in the case of CAT. Thus a decreasing LSM occurs with the increasing thickness of a-Si:H deposited by CAT.

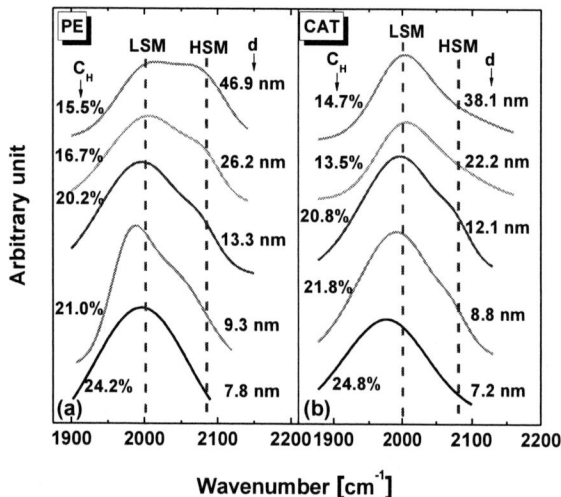

Fig. 2. Absorption coefficient versus wavenumbers of the a-Si:H thin films deposited by PE and CAT CVD technique, respectively.

By applying the two a-Si:H as interface passivation layer mentioned above to the SHJ solar cells, the performance of the SHJ solar cells is listed in table 1. The metal contacts on both sides of SHJ solar cells are silver grids made by screen printing. The total area of the SHJ solar cells is 242 cm². The thickness of the intrinsic layer of two cells is controlled around 5 nm. Optimized field passivation provided by p- and n-type a-Si:H is necessary to give a V_{oc} base of more than 730 mV to support the device research. By using two a-Si:H interface layers, same conversion efficiency of 23.06% can be obtained. The short-circuit current density (J_{sc}) of the SHJ solar cell deposited by PE technique is higher than the cell deposited by CAT technique. However, V_{oc} and fill factor (FF) of the SHJ solar cell deposited by PE technique is lower than the cell deposited by CAT technique. As shown in figure 1, the presence of crystallites in a-Si:H may contribute to low defect and better transportation property of carrier, which are beneficial for the improvement of V_{oc} and FF. Furthermore, the larger f_c of a-Si:H adjacent to c-Si in the CAT method in

figure 1 (c) may be the reason of low J_{sc} of B-CAT due to high absorption coefficient of crystalline silicon. A J_{sc} enhancement of A-PE attributes to the low f_c of a-Si:H adjacent to c-Si in the PE method.

TABLE I
IV CHARACTERISTICS OF SHJ SOLAR CELLS

Cell name	η (%)	V_{oc} (mV)	J_{sc} (mA/cm²)	FF (%)
A-PE	23.06	735.2	38.8	80.9
B-CAT	23.06	738.2	38.4	81.4

IV. CONCLUSIONS

In this work, we have shown the microstructure evolution with the increasing thickness of a-Si:H deposited by PE and CAT method, respectively. It revealed that the growth tends to be a structural epitaxy influence by c-Si in the initial stage of a-Si:H. The a-Si:H film deposited by CAT can rapidly shifts from island mode to layer-by-layer mode, which is implied by the sharply reduction of d_{sr} while the shift thickness of a-Si:H film deposited by PE is obviously thicker. The a-Si:H film deposited by CAT is more compact due to the presence of crystallites. Thus, V_{oc} and FF can be improved by this kinds of compact a-Si:H. But unfortunately, a loss of 0.4 mA/cm² occurs in the cell used by CAT method.

Based on the performance of two SHJ solar cells, a subsequent optimization solution can be given for each CVD methods. The initial f_c of a-Si:H deposited by CAT method should be lower to improve the transmittance while the structure of a-Si:H deposited by PE should be optimized towards a transition phase with small fraction of crystallites to improve the transportation property.

ACKNOWLEDGEMENT

We would like to thank Jinning Liu, Junlin Du, Jianhua Shi, Jian Yu and Yucheng Liu for the process cooperation of solar cells fabrication. This work is supported by International S&T Cooperation Program of China (2015DFA60570) and Key project of Zhangjiang National Innovation Demonstration Zone Special Development Fund (ZJ2015-ZD-001).

REFERENCES

[1] T. Saga, "Advances in crystalline silicon solar cell technology for industrial mass production," *NPG Asia Mater.* vol. 2, no. 3, pp. 96-102, Jul. 2010.
[2] M. Taguchi, A. Yano, S. Tohoda, K. Matsuyama, Y. Nakamura, T. Nishiwaki, K. Fujita, and E. Maruyama,"24.7% Record Efficiency HIT Solar Cell on Thin Silicon Wafer," *IEEE J. Photovolataics*, vol. 4, no. 1, pp. 96-99, Jan. 2014.
[3] D. Adachi, J.L. Hernández and K. Yamamoto, "Impact of carrier recombination on fill factor for large area heterojunction

crystalline silicon solar cell with 25.1% efficiency" Applied Physics Letters, vol. 107 pp.233506, 2015.

[4] E. Kobayashi, Y. Watabe, T. Yamamoto and Y Yamada, "Cerium oxide and hydrogen co-doped indium oxide films for high-efficiency silicon heterojunction solar cells" Solar Energy Materials & Solar Cells, vol. 149, pp. 75-80, 2016.

[5] J.I.Pankove and M.L.Tarng, "Amorphous silicon as a passivant for crystalline silicon," *Appl. Phys. Lett.*, vol. 34, no. 2, pp. 156-157, Jan. 1979.

[6] M. Tanaka, M. Taguchi, T. Matsuyama, T. Sawada, S. Tsuda, S. Nakano, H. Hanafusa, and Y. Kuwano, "Development of New a-Si/c-Si Heterojunction Solar Cells: ACJ-HIT (Artificially Constructed Junction-Heterojunction with Intrinsic Thin-Layer)," *Jpn. J. Appl. Phys.*, vol. 31, no. 11, pp. 3518-3522, Nov. 1992.

[7] A. Descoeudres, L. Barraud, R. Bartlome, G. Choong, S.D. Wolf, F. Zicarelli, and C. Ballif, "Thesilane depletion fraction as an indicator for the amorphous/crystalline silicon interface passivation quality," *Appl. Phys. Lett.*, vol. 97, no. 18, pp. 183505-1-183505-3, Nov. 2010.

[8] J. Ge, Z.P. Ling, J. Wong, R. Stang, A.G. Aberle, and T. Mueller, "Analysis of intrinsic hydrogenated amorphous silicon passivation layer growth for use in heterojunction silicon wafer solar cells by optical emission spectroscopy," *J. Appl. Phys.*, vol. 113, no. 23, pp. 234310-1-234310-7, Jun. 2013.

[9] A. Heya, A. Masuda, and H. Matsumura, "Low-temperature crystallization of amorphous silicon using atomic hydrogen generated by catalytic reaction on heated tungsten," *Appl. Phys. Lett.*, vol. 75, no. 15, pp. 2143-2145, Feb. 1999.

[10] C. Summonte, R. Rizzoli, A. Desalvo, F. Zignani, E. Centurioni, R. Pinghini, and M. Gemmi, "Very high frequency hydrogen plasma treatment of growing surfaces: a study of the p-type amorphous to microcrystalline silicon transition," *J. Non-Cryst. Solids*, vol. 266, pp. 624-629, 2000.

[11] S. Sriraman, S. Agarwal, E.S. Aydil, and D. Maroudas, "Mechanism of hydrogen-induced crystallization of amorphous silicon," *Nature*, vol. 418, no. 6893, pp. 62-65, May 2002.

[12] A. Hadjadj, N. Pham, P. Roca I Cabarrocas, and O. Jbara, "Effect of doping on the amorphous to microcrystalline transition in a hydrogenated amorphous silicon under hydrogen plasma treatment," *Appl. Phys. Lett.*, vol. 94, no. 6, pp. 061909-1-061909-3, Feb. 2009.

[13] H. Fujiwara, T. Kaneko and M. Kondo, "Application of hydrogenated amorphous silicon oxide layers to c-Si heterojunction solar cells," *Appl. Phys. Lett.*, vol. 91, no. 13, pp. 133508-1-133508-3, Sep. 2007.

[14] J. Sritharathikhun, H. Yamamoto, S. Miyajima, A. Yamada and M. Konagai, "Optimization of amorphous silicon oxide buffer layer for high-efficiency p-type hydrogenated microcrystalline silicon oxide/ n-type crystalline silicon heterojunction solar cells," *Jpn. J. Appl. Phys.*, vol. 47, no. 11, pp.8452-8455, Nov. 2008.

[15] F. Einsele, W. Beyer and U. Rau, "Analysis of sub-stoichiometric hydrogenated silicon oxide films for surface passivation of crystalline silicon solar cells," *J. Appl. Phys.*, vol. 112, no. 5, pp. 054905-1-054905-8, Sep. 2012.

[16] T. Mueller, J. Wong and A.G. Aberle, "Heterojunction silicon wafer solar cells using amorphous silicon suboxides for interface passivation," *Energy Procedia*, vol. 15, pp.97-106, 2012.

[17] K. Ding, U. Aeberhard, F. Finger and U. Rau, "Optimized amorphous silicon oxide buffer layers for silicon heterojunction solar cells with microcrystalline silicon oxide contact layers," *J.

Appl. Phys., vol. 113, no. 13, pp. 134501-1-134501-5, Apr. 2013.

[18] A. Descoeudres, L.Barraud, S.D. Wolf, B. Strahm, D. Lachenal, C. Guérin, Z.C. Holman, F. Zicarelli, B. Demaurex, J. Self, J. Holovsky, and C. Ballif, "Improved amorphous/crystalline silicon interface passivation by hydrogen plasma treatment," *Appl. Phys. Lett.*, vol. 99, no. 12, pp. 123506-1-123506-3, Sep. 2011.

[19] M. Mews, T.F. Schulze, N. Mingirulli, and L. Korte, "Hydrogen plasma treatments for passivation of amorphous-crystalline siliconheterojunctions on surfaces promoting epitaxy," *Appl. Phys. Lett.*, vol. 102, no. 12, pp. 122106-1-122106-4, Mar. 2013.

[20] H. Matsumura and H. Tachibana, "Amorphous silicon produced by a new thermal chemical vapor deposition method using intermediate species SiF_2", Appl. Phys. Lett. vol. 47, pp.833 1985.

[21] H. Matsumura, "Silicon nitride produced by catalytic chemical vapor depositon method", J. Appl. Phys. vol. 66, 3612, 1989.

[22] S. Lien and D. Wu, "Simulation and Fabrication of heterojunction silicon solar cells from numerical computer and hot-wire CVD" Progress in Photovoltaics: Research and Applications, vol 17, pp. 489-501, 2009.

[23] D.E. Aspnes, "Optical properties of thin films," Thin Solid Films vol. 89, pp. 249-262, 1982.

[24] M.H. Brodsky, M. Cardona, and J.J. Cuomo, "Infrared and Raman spectra of the silicon-hydrogen bonds in amorphous silicon prepared by glow discharge and sputtering," *Phys. Rev. B*, vol. 16, no. 8, pp. 3556-3571, Oct. 1977.

[25] M. Cardona, "Vibrational spectra of hydrogen in silicon and germanium," *Phys. Status Solidi B*, vol. 118, no. 2, pp. 463-481, 1983.

[26] J. Müllerová, P. Šutta, G. van Elzakker, M. Zeman, and M. Mikula, "Microstructure of hydrogenated silicon thin films prepared from silane diluted with hydrogen," *Appl. Surf. Sci.*, vol. 254, no. 12, pp. 3690-3695, Nov. 2007.

[27] S. Kageyama, M. Akagawa, and H. Fujiwara, "Dielectric function of a-Si:H based on local network structures," *Phy. Rev. B*, vol. 83, no. 19, pp. 195205-1-195205-11, May 2011.

[28] M.Z. Burrows, U.K. Das, R.L. Opila, S. De Wolf, and R.W. Birkmire, "Role of hydrogen bonding environment in a-Si :H films for c-Si surface passivation," *J. Vac. Sci. Technol. A*, vol. 26, no. 4, pp. 683-687, Jun. 2008.

[29] H. Fujiwara and M. Kondo, "Interface structure in a-Si:H/c-Si heterojunction solar cells characterized by optical diagnosis technique," in *Proc. IEEE 4th World Photovoltaic Energy ConversionConf.*, 2006, pp. 1443-1448.

[30] M.F. Saenger, J. Sun, M. Schädel, J. Hilfiker, M. Schubert and J.A. Woollam, "Spectroscopic ellipsometry characterization of SiN_x antireflection films on textured multicrystalline and monocrystalline silicon solar cells," *Thin Solid Films*, vol. 518, no. 7, pp. 1830-1834, Sep. 2009.

[31] A.H.M. Smets, W.M.M. Kessels, and M.C.M. van de Sanden, "Vacancies and voids in hydrogenated amorphous silicon," *Appl. Phys. Lett.*, vol. 82, no.10, pp. 1547-1549, Jan. 2003.

[32] M.F. Saenger, J. Sun, M. Schädel, J. Hilfiker, M. Schubert and J.A. Woollam, "Spectroscopic ellipsometry characterization of SiN_x antireflection films on textured multicrystalline and monocrystalline silicon solar cells," *Thin Solid Films*, vol. 518, no. 7, pp. 1830-1834, Sep. 2009.

[33] G.E. Jellison, Jr. and F.A. Modine, "Parameterization of the optical functions of amorphous materials in the interband region," *Appl. Phys. Lett.*, vol. 69, no. 3, pp. 371-373, May 1996.

Developing an Understanding-Based Selection of Hybrid-Perovskite Compounds and the Cu-In Hybrid-Perovskite (CIHP) Family

Alex Zunger, G. Dalpian, Qihang Liu, L.B Abdalla, and L.L. Kazmerski

University of Colorado Boulder, Renewable and Sustainable Energy Institute (RASEI),
4001 Discovery Drive, Suite N368, Boulder, Colorado 80309 (e-mail: alex.zunger@colorado.edu)

ABSTRACT — The long-term chemical instability and the presence of toxic Pb in otherwise stellar solar absorber $APbX_3$ made of organic molecules on the A site and halides for X have hindered their large-scale commercialization. Previously explored ways to achieve Pb-free halide perovskites involved replacing Pb^{2+} with other similar M^{2+} cations in ns^2 electron configuration, e.g., Sn^{2+} or by Bi^{3+} (plus Ag+), but unfortunately this showed either poor stability (M = Sn) or weakly absorbing oversized indirect gaps (M = Bi), prompting concerns that perhaps stability and good optoelectronic properties might be contraindicated. Herein, we exploit the electronic structure underpinning of classic $Cu[In,Ga]Se_2$ (CIGS) chalcopyrite solar absorbers to design Pb-free halide perovskites by transmuting 2Pb to the pair $[B^{IB} + C^{III}]$ such as [Cu + Ga] or [Ag + In] and combinations thereof. The resulting group of double perovskites with formula A_2BCX_6 (A = K, Cs; B = Cu, Ag; C = Ga, In; X = Cl, Br, I) benefits from the ionic, yet narrow-gap character of halide perovskites, and at the same time borrows the advantage of the strong and rapidly rising Cu(d)/Se(p)→Ga/In(s/p) valence-to-conduction-band absorption spectra known from CIGS. This constitutes a new group of CuIn–based Halide Perovskite (CIHP). Our first-principles calculations guided by such design principles indicate that the CIHPs class has members with clear thermodynamic stability, showing rather strong direct-gap optical transitions, and manifesting a wide-range of tunable gap values (from zero to about 2.5 eV) and combination of light electron and heavy-light hole effective masses. Materials screening of candidate CHIPs then identifies the best-of-class $Cs_2[AgIn]Br_6$ and $K_2[AgIn]Br_6$, having direct band gaps of 1.50 and 1.44 eV, and a theoretical spectroscopic limited maximal efficiency comparable to chalcopyrites and $CH_3NH_3PbI_3$. Our finding offers new routine for designing new-type stable, Pb-free halide perovskite solar absorbers.

Index Terms — hybrid perovskites, theory-guided synthesis, design principles, stability, materials design

I. FAST PROGRESS AND THE QUESTIONS IT RAISED

The family of cubic hybrid organic-inorganic halide AMX_3 perovskites has recently become a rising star in the field of solar energy materials, with power conversion efficiency progressing from the initial demonstration of 3.8% in the year 2009, and exceeding 20% in an unprecedented-short 5-years later. However, following the initial fast progress in achieving very good efficiencies, the quest for equally rapid solutions to some of the outstanding hybrid perovskite emerging problems has been largely dealt with by a shot-gun approach of trying in an Edisonian fashion different combinations of substitutions on the A-site, the M-site, and the X-site, with little guidelines of

where to look and what to look for. Some of the outstanding issues include the need to obtain target band gaps for single and tandem cells; the need to overcome material instability issues of unknown origin, and the need to replace Pb.

II. THE BASIC PROPERTY OF MATERIAL SELECTION USED

We hold the opinion that it is better to select materials for technology based on *microscopic understanding of the limiting factors*, rather than based on a shotgun combinatorial attempt to "make all and test all." Historical examples of poorly understood material instability that came back years later to haunt the technology include the Staebler-Wronski instability in amorphous silicon, or the "dark-spots" instability in CdHgTe night-vision detectors. To select the best AMX_3 materials non-phenomenologically, one needs to establish the critical, materials-specific *'design principles'* (DPs) that render such materials in solar cells superior. DPs refer to microscopic *physical understanding of mechanisms* that control the functionality at hand. *We are not claiming that understanding by itself, or theory by itself, would spontaneously make good devices or invent a cure-all material.* But understanding-based R&D that involves synergism of theory with guided synthesis and characterization would make good and durable solar cells possible, in a much faster time period.

III. SOME OF THE OPEN PUZZLES IN HYBRID PEROVSKITES

Although hybrid halide perovskites have emerged as a promising platform for realizing low-cost thin-film and high-efficiency PV solar cells, there exist several challenges and open questions:

(i) There is a need to improve the efficiency by deliberately tuning the optical band gap of $(CH_3NH_3)PbI_3$ (1.50 eV) and $(CH_3NH_3)PbBr_3$ (2.35 eV) to the optimal value of 1.35 eV from the Shockley–Queisser limit, or 2.2 eV gaps for higher photon-energy absorbing tandem layers. Unfortunately, there is still limited understanding of how the band gaps are controlled by chemical and structural variations in this family of materials. Indeed, $A^+(MX_3)^-$ is an unusual ionic halide that unlike ionic alkali halides has a band gaps that are 3-5 times smaller! Furthermore,

978-1-5090-5606-4/17 $31.00 © 2017 IEEE

although replacing X=I by X=Br would raise the gap, this blessing comes with the curse of light-induced persistent instability of the bromide, preventing thus far its use.

(ii) The devices made by the known materials show poor long-term stability under high temperatures and outdoor illumination, as well as in the presence of moisture or oxygen, requiring various protective shells or treatment whose choice is not based on the understanding of the nature of the instability in the first place. The reason remains totally mysterious. Some say Pb replacement by Sn makes the device poor, but the leading advertised reason—the decomposition of AMX_3 to $AX+MX_2$ is less likely to occur according to first-principles calculations for M = Sn than for M = Pb. So, we must look for the reason elsewhere.

(iii) Another issue pertaining to technology is that the use of Pb-containing materials is an environmental concern, being banned in commercial electronic devices. There is a strong desire to replace toxic Pb with other elements and maintain the high conversion efficiency. But how can one replace the functionality of the S^2 "lone-pair" electrons of Pb by something else? Replacing 2Pb by Ag+Bi in a double unit cell $2APbX_3 = A_2[PbPb]X_6$ leading to $A_2[AgBi]X_6$ created a poorly absorbing *indirect gap* system. Could this have been anticipated? Understood?

(iv) The organic molecule on the A site as a stand-alone molecule does not even absorb light in the solar range but much above it. So, what is really its role, and how do we choose the right A?

IV. WHAT HAS BEEN MISSING ALL ALONG

Given that this group of compounds could be much broader than just the selection A = (CH_3NH_3), B = Pb, Sn, and X =I, one wonders if other members might have advantages. We know that in the area of *chalcogenide* PV materials, successful screening was previously accomplished by applying science-based design metrics, computed via first-principles electronic structure theory. However, the application of such methodologies to the hybrid perovskite group has been historically overlooked (perhaps an important contributing factor to the fact this group was missed by the PV materials community) and is currently at its early stages.

V. TWO LEVELS OF DESIGN

The DP and DM can be divided into categories: (a) design metrics that characterize the ABX_3 material itself ('compound-intrinsic design metrics' or CIDM), as well as (b) PV 'cell-level design metrics' (CLDM), specific to this material configured within a device structure that includes components other than the AMX_3 itself, [such as the hole blocking-layer (HBL) and the electron-blocking-layer (EBL); various interfaces and possible metal contacts]. We believe

that it makes sense to use the compound intrinsic design metrics as the first filter for pre-selecting promising candidates. Clearly, the testing of the PV cell level design metrics should be used as subsequent filter, applied, however, only to a far smaller group of materials than in the first filter. We focus therefore our present search for materials $\{AMX_3\}$ by analyzing the 'compound-intrinsic design metrics of type (a).

VI. THE PURPOSE AND STRATEGY

The idea is to first develop our intuition based on calculating band gaps, effective masses and stability of an array of AMX_3 materials with different A, different M and different X, developing an understanding of the basic chemical trend. **This was accomplished in our recent Ref. [1].** Then use this understanding together with our past experience of what made $CuInSe_2$ a good thin-film absorber, and design novel, never before made double perovskites $A_2[BC]X_6$ where A, B, and C are selected to design optimal properties. **This was accomplished in our recent Ref. [2]. These ideas were leveraged in Ref. [3], in which we studied the stability and properties of the Sn compounds $ASnX_3$ and A_2SnX_6** .*This describes our initial progress, which awaits a close collaboration with the right synthesis chemists and characterization scientist to proceed to the crucial future step of iterative refinement of our materials* [3].

We performed first-principle density functional calculations. The substitution space for screening is based on the AMX_3 stoichiometry, considering 10 inorganic/organic cations for the A site, 3 group-IVA metals (Ge, Sn and Pb) for M site, and 3 halogen elements (Cl, Br and I) for the X site totaling $10\times3\times3 = 90$ compounds. The monovalent (1+) A-site cations chosen have generally comparable ionic sizes, including Cs, ammonium NH_4 (M), hydroxylamine NH_3OH (HA), diamine NH_2NH_3 (DA), methylammonium CH_3NH_3 (MA), formamid NH_3COH (FM), formamidinium $CH(NH_2)_2$ (FA), ethylamine $CH_2CH_2NH_3$ (EA), dimethylamine $NH_2(CH_3)_2$ (DEA) and guanidine amine $C(NH_2)_3$ (GA). Among these 90 compounds, only a few $CsPbI_3$, $(CH_3NH_3)PbI_3$, $(CH_3NH_3)PbBr_3$, $(CH_3NH_3)PbCl_3$, $(CH_3NH_3)PbBr_3$, $(NH_2)_2PbI_3$, are known compounds.

VII. RESULTS SUMMARY [1]-[3]

(i) *Thermodynamic immunity of AMX_3 towards the main decomposition channel into competing phase.*
(ii) *Direct band gap and strong absorption near threshold.*
(iii) *Low electron and hole effective masses at band edges*
(iv) *Screening for major intrinsic defects.*
(v) *A new class of chalcopyrite-inspired hybrid perovskites:*

Previously-explored ways to achieve Pb-free halide perovskites involved replacing Pb^{2+} with other similar M^{2+}

cations in ns^2 electron configuration, *e.g.*, Sn^{2+} or by Bi^{3+} (plus Ag+), but unfortunately this showed either poor stability (M = Sn) or weakly absorbing, oversized indirect gaps (M = Bi), prompting concerns that perhaps stability and good optoelectronic properties might be contraindicated. We exploit the electronic structure underpinning of classic $Cu[In,Ga]Se_2$ (CIGS) chalcopyrite solar absorbers to design Pb-free halide perovskites by transmuting 2Pb to the pair $[B^{IB} + C^{III}]$ such as [Cu + Ga] or [Ag + In] and combinations thereof. The resulting group of double perovskites with formula A_2BCX_6 (A = K, Cs; B = Cu, Ag; C = Ga, In; X = Cl, Br, I) benefits from the ionic, yet narrow-gap character of halide perovskites, and at the same time borrows the advantage of the strong and rapidly rising $Cu(d)/Se(p) \rightarrow Ga/In(s/p)$ valence-to-conduction-band absorption spectra known from CIGS. This constitutes a new group of ***CuIn-based Halide Perovskite*** (CIHP). Our first-principles calculations guided by such design principles indicate that the CIHPs class has members with clear thermodynamic stability, showing rather strong direct-gap optical transitions, and manifesting a wide-range of tunable gap values (from zero to about 2.5 eV) and combination of light electron and heavy-light hole effective masses. Materials screening of candidate CHIPs then identifies the best-of-class $Cs_2[AgIn]Br_6$ and $K_2[AgIn]Br_6$, having direct band gaps of 1.50 and 1.44 eV, and a theoretical spectroscopic limited maximal efficiency comparable to chalcopyrites and $CH_3NH_3PbI_3$. Our finding offers new routine for designing new-type Pb-free, stable halide-perovskite solar absorbers.

Acknowledgements: This work has been performed in close collaboration with L.B Abdalla, L.L. Kazmerski, Qihang Liu and G. Dalpian at CU Boulder and L. Zhang and his students at College of materials science and Engineering, Jilin University, Changchun, China. Work at CU on solar cell aspects was supported by US Department of Energy by EERE PVRD SunShot Initiative, Small Innovative Projects in Solar (SIPS) program (under contract DE-EE0007366), and work at CU on fundamental properties of perovskites was funded by the Office of Science, Basic Energy Science, and MSE Division under Grant No. DE-FG02-13ER46959. Finally, a portion of this research was performed using computational resources sponsored by the Department of Energy's Office of Energy Efficiency and Renewable Energy and located at the National Renewable Energy Laboratory.

REFERENCES

[1] D. Yang, J. Lv, X. Zhao, Q. Xu, Y. Fu, Y. Zhan, Alex Zunger, and Lijun Zhang, "Functionality-directed First-principles Screening of Hybrid Organic-inorganic Perovskites with Target Intrinsic Photovoltaic Functionalities" Chemistry of Materials 29, 524 (2017)

[2] X.G Zhao, D. Yang, Y. Sun, T. Li, Lijun Zhang, Liping Yu, and Alex Zunger, "Cu–In Halide Perovskite Solar Absorbers" J. Am. Chem. Soc. *139* (19), pp 6718–6725 (2017).

[3] G. M. Dalpian, Q. Liu, C. C. Stoumpos, A. P. Douvalis, M. Balasubramanian, Mercouri G. Kanatzidis and Alex Zunger, "Changes in charge density vs changes in formal oxidation states: The case of Sn halide perovskites and their ordered vacancy analogues" Phy. Rev. Materials (Submitted)

Effects of Electron and Proton Radiation on Perovskite Solar Cells for Space Solar Power Application

Jing-Shun Huang[1], Michael D. Kelzenberg[1], Pilar Espinet-González[1], Colin Mann[2], Don Walker[2], Ali Naqavi[1], Nina Vaidya[1], Emily Warmann[1], and Harry A. Atwater[1]

[1]California Institute of Technology, Pasadena, CA 91125, United States

[2]Aerospace Corporation, El Segundo, CA 90245, United States

ABSTRACT — **We report the effects of high energy electron and proton radiation on perovskite solar cells. For irradiation with 1 MeV electrons, at fluence from 10^{12} to 10^{16} cm^{-2}, there are no significant changes in the morphology and the crystal phase in the perovskites, and the perovskite solar cells show only slight degradation in photovoltaic performance and spectral response. The results from Monte Carlo simulations show most of the electrons completely penetrate through all of the layers of solar cells with little scattering. In addition, 50 keV proton radiation with fluence of 10^{12} cm^{-2} has no significant impact on the open-circuit voltage or short-circuit current, and degrades only the fill factor. We have further found that the fill factor can be restored with a vacuum annealing process. The results suggest that perovskites have superior electron and proton radiation tolerance, and thus hold particular promise for space applications.**

I. INTRODUCTION

Although crystalline Si solar modules dominate the photovoltaics market with an affordable price per watt and proven reliability, there are other needs for decentralized or mobile power generation for which other solar technologies are favorable. For example, space solar power, unmanned aerial vehicles (UAVs), and solar blimps require solar cells with high specific power (i.e., power to weight ratio). These applications require a combination of high photovoltaic performance, minimal weight, flexibility, mechanical resilience, working stability, long service life, and, ideally, low cost.

For space solar power applications, photovoltaic cells are subject to high-energy charged particle irradiation from space, which typically causes severe degradation of the cells' photovoltaic performance [1]. To prevent this, a thin layer of cover glass (typically up to hundreds of microns thick) is required to screen the solar cells from radiation, which inevitably increases the module weight and lunch.

Today's mainstream solar technology for space applications, III-V compound solar cells, offers excellent efficiency approaching or exceeding 30% (AM0), and have a long heritage of reliability when adequately protected from radiation in space. However, other types of solar cells may prove superior for space applications if they are capable of achieving even higher specific power, or are intrinsically more radiation resistant, than established technologies, especially if they can be produced at lower cost.

Among various solar cell technologies, organo-lead halide perovskite solar cells have recently emerged as a potentially low-cost material capable of over 22% efficiency (AM1.5G). They have further shown a specific power of 26 W/g [2], a 20-fold increase over that of thin film silicon cells or thin film single-junction GaAs cells (up to 1.3 W/g) [3]. Although their extreme sensitivity to moisture has impeded their large-scale adoption for terrestrial applications, it is interesting to consider their use in space, where they would not be subject to atmospheric moisture after being launched. To our knowledge, their radiation hardness and suitability for space applications have not been thoroughly investigated, but reports to date suggest that they have excellent radiation tolerance [4].

II. EXTERIMENTAL

A. *Photovoltaic Materials and Device Fabrication*

Perovskite solar cells were fabricated on indium tin oxide (ITO) coated quartz superstrates. We used quartz instead of using soda lime glass because the latter would become darkened by the radiation. The ITO had sheet resistance of $10~\Omega$ sq^{-1}. Prior to beginning cell fabrication, the ITO coated quartz substrates were cleaned with detergent, then ultrasonicated in acetone followed by 2-propanol.

In the fabrication process described below, the lead iodide was received from Alfa Aesar. Anhydrous N,N-dimethylformamide (DMF) and 2,2',7,7'-tetrakis (N,N-di-p-methoxyphenylamine)9,9'-spirobifluorene (spiro-OMeTAD) were received from Sigma Aldrich. Methylammonium iodide (MAI) and formamidinium iodide (FAI) were received from Dyesol.

The device architecture adopted in our study of electron irradiation is ITO/TiO2/FAPbI$_3$/Spiro-OMeTAD/Ag. The TiO$_2$ compact layers were deposited on the ITO using electron-beam evaporation at a rate of 0.5 Å/s. The substrates were then transferred into a nitrogen-filled glove box and coated with the perovskite layer via a two-step spin-coating process. A precursor solution containing PbI$_2$ in DMF (600 mg ml^{-1}) was spin-coated first at 6000 rpm for 40 s and dried on a hotplate at 110 °C for 10 min. Then a FAI solution (70 mg ml^{-1} in 2-propanol) was spin-coated onto the PbI$_2$ layer at 6000 rpm for 40 s, following by annealing on hotplate at 150 °C for 30 min. The hole transport layer was deposited by spin-coating a 72 mg ml^{-1} solution of spiro-OMeTAD solution in chlorobenzene, with additives of 18 μl of lithium bis(trifluoromethanesulfonyl)imide (520 mg ml^{-1} 1-butanol solution) and 29 μl of 4-tert-butylpyridine per 1 ml of spiro-OMeTAD solution. Spin-coating was carried out in the glovebox at 2000rpm for 30s. Finally, a 100-nm silver electrode was thermally evaporated under vacuum (~10^{-7} Torr) at a rate of 1 Å/s. A shadow mask was used to pattern the Ag such that each 1x1 cm quartz substrate contained (3) separate cells of 0.1 cm^2 nominal area.

A slightly different device architecture was adopted for our study of proton irradiation: ITO/NiO/ MAPbI$_3$/PCBM/Ag. The NiO layers were spin-cast from 0.1-molar nickel acetate and ethanolamine in ethanol solution and annealed at 300 °C for 1 hour. The substrates were then transferred into a nitrogen-filled glove box and coated with the perovskite layer via a one-step spin-coating process. A precursor solution containing 1-molar PbI$_2$ and MAI in DMF and DMSO (4:1 v/v) was spin-coated at 4000 rpm for 30 s, during which 100 μl of chlorobenzene was dropped onto the spinning sample at the 10th second. Films were dried on a hotplate at 100 °C for 20 min. Then a solution of PCBM solution (20 mg ml^{-1} in chlorobenzene) was spin-coated onto the MAPbI$_3$ layer at 1000 rpm for 60 s. The substrates, ITO, Ag, and cell form factor were otherwise the same as described above.

B. Electron and Proton Radiation Testing

We conducted electron irradiation testing with energy of 1 MeV at room temperature under vacuum of 10^{-5} Torr using a Dynamitron at NASA Jet Propulsion Laboratory. Fluence ranged from 10^{12} to 10^{16} cm^{-2}. The irradiation area was calibrated with a Faraday cup to be uniform (5%) within a 6 inch by 6 inch area to make sure each sample in the chamber receive the same dose. The cells' back contacts were placed to face the electron source to receive direct impacts of the electrons. This was necessary because of our use of relatively thick quartz superstrates, which would shield the cells from the electrons if irradiated from the front. Figure 1 shows the Monte Carlo simulation of the electron trajectories within the cells by using CASINO software package. Most of electrons penetrate the layers of solar cells with little scattering, and stop in the quartz substrate.

The proton irradiation tests were conducted with acceleration energy of 50 keV at the Aerospace Corporation. Fluence was set to 10^{12} cm^{-2}. The cells' back contacts were placed to face the proton source to receive direct impacts of the protons.

It is important to note that, because perovskite solar cells are known to degrade upon exposure to atmospheric moisture, it is possible that the cells' performance changed in the course of their transport to and from testing facilities, or during loading, unloading, and testing, which were performed in ambient air. To minimize this, cells were transported in partially evacuated desiccators, and stored either under

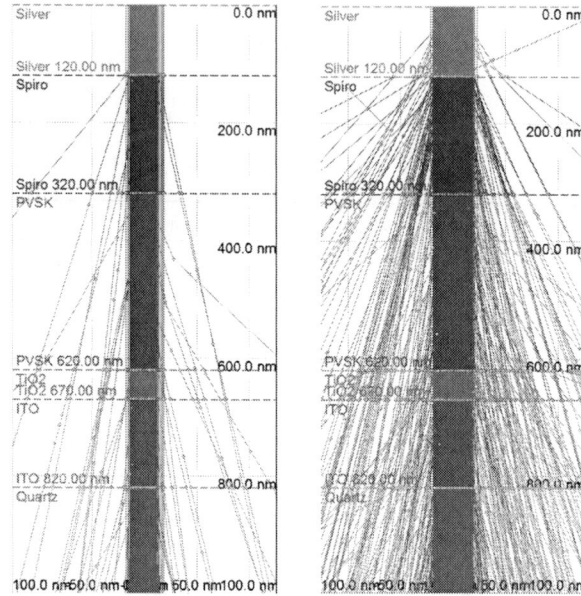

Figure 1. Simulation of 1-MeV electron trajectories with 10^{14} (left) and 10^{15} (right) fluences in the perovskite solar cells, showing most of electrons completely penetrate through all of the layers with little scattering. The electron beam radius is set to 17.845 nm.

vacuum or in nitrogen-purged dry boxes, such that total exposure to atmospheric air prior to final testing was typically on the order of several hours. Rather than testing each cell before and after irradiation, we instead prepared batches of control cells for each experiment. The control cells were fabricated and transported with the test cells, but instead of loading them into the vacuum chambers for irradiation, they were placed in a desiccator during this step. Both the test and the control cells were later characterized during the same measurement session.

C. Characterization

After the irradiation, the perovskite thin films and solar cells were removed from the vacuum chamber and transported to Caltech for characterization. Current density–voltage (J–V) characteristics of the photovoltaic cells were measured in ambient air using a Keithley 238 source meter unit under illumination at ~100 mW cm^{-2} with a simulated AM 1.5G spectrum from an Oriel solar simulator (1 kW Xe arc lamp). The light intensity was calibrated using a hermetically sealed, 2x2 cm Si reference cell (PV Measurements, Inc) with calibration data from NREL. EQE measurements were carried out in ambient air using a Newport Xe arc lamp and monochromator.

III. RESULTS AND DISCUSSION

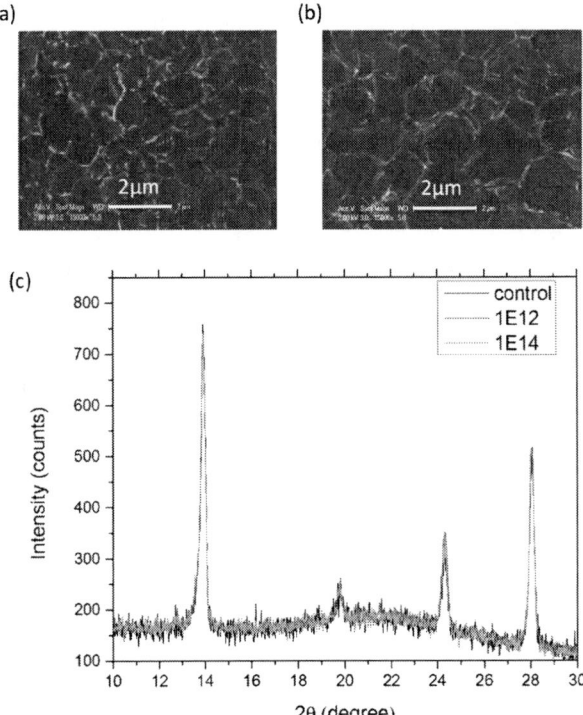

Figure 2. (a) SEM image of perovskite film without electron irradiation. (b) SEM image of perovskite film after 1 MeV electrons at fluence of 10^{14} cm^{-2}. (c) XRD patterns of perovskite films under different fluences of electron irradiation.

Figures 2(a) and (b) show the top-view scanning electron microscope (SEM) images of perovskite films without and with electron irradiation, respectively. The grain size of the perovskite film with electron irradiation is 0.5–2 μm, which is almost identical to that without irradiation. The same is true for their x-ray diffraction (XRD) patterns, as shown in Figure 2(c). The intense diffraction peak at $2\theta = 13.9°$ represents highly oriented crystallinity with a strong preferred orientation of (110). No significant changes were observed from their morphology and crystal phase, which suggests the perovskite is physically resistant to the electron radiation.

Figure 3(a) shows the J–V curves of the perovskite solar cells with different electron radiation fluence. Control devices without electron irradiation exhibit a typical power conversion efficiency (PCE) of 12.2%, with a short-circuit current density (J_{SC}) of 21 mA cm^{-2}, an open-circuit voltage (V_{OC}) of 0.98 V, and a fill factor (FF) of 59%; which comparable with results of planar perovskite solar cells in the literature [5-8]. With electron fluence of 10^{12}, 10^{14}, and 10^{15} cm^{-2}, similar PCE values of 13.3%,

13.4%, and 12.4% were obtained, respectively. Figure 3(b) shows the external quantum efficiency (EQE), which is consistent with the measured J_{SC} in the devices. The spectral response did not change significantly with the fluence of electron radiation. Figure 3(c) summarizes the remaining factors of photovoltaic parameters of the perovskite solar cells with fluence up to 10^{16} cm^{-2}, where the FF and PCE still remain within 90% performance of the control cells. This degradation is substantially less than that reported for Si or III-V solar cells, and is within the range of sample-to-sample variability in our cells. Miyazawa et al. recently reported no detectable degradation of perovskite cell performance at this fluence and energy[4].

For proton testing, we carried out simulations of the irradiation to determine which layers of the cell structure would be affected, using SRIM/TRIM (Stopping and Range of Ions in Matter) software package. Figure 4(a) shows the particle trajectories for 30, 50, and 100 keV proton energy. In the 50-keV case, nearly all protons stop within the cell layers, with few reaching the quartz, but many reaching the ITO and NiO layers. Thus, this is a good energy to probe whether or not any cell layers are particularly sensitive to proton irradiation.

Figure 4(b) shows typical J–V curves of the perovskite solar cells before and after proton irradiation. The photovoltaic device before irradiation exhibits a typical PCE of 12.3%, with a J_{SC} of 17.7 mA cm^{-2}, a V_{OC} of 0.99 V, and a FF of 70%. After the proton irradiation, only the FF changed significantly, dropping to 42%, while the J_{SC} and V_{OC} remained largely unchanged. The apparent series resistance increased dramatically, from 3.8 to 24.5 Ωcm^{-2}. We note at this point the cells were tested under ambient air. We then loaded the cells into a vacuum chamber ($<10^{-6}$ Torr) and annealed them at 90 °C for 3 days. After cooling down to room temperature, we

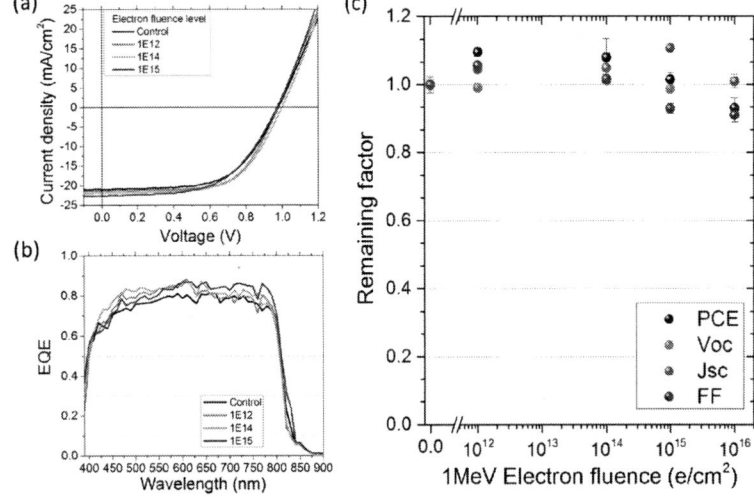

Figure 3. (a) J-V curves of the perovskite solar cells with different electron irradiation doses under one sun illumination. (b) Spectrum responses of these solar cells. (c) Remaining factors of photovoltaic parameters of the perovskite solar cells.

left the cells in the vacuum chamber and tested them under the solar simulator. (The vacuum chamber featured a quartz window and electrical feedthrough to facilitate J–V testing under vacuum.) Surprisingly, the cell performance recovered completely, with a PCE of 12.5% and FF of 67%. We hypothesize that (1) the implanted protons were removed from the cells during the vacuum annealing, and (2) the cations and anions in perovskites were able to migrate to their original locations in the crystal structure under the vacuum annealing. Further investigations are underway to understand the mechanism of damage recovery in perovskite solar cells.

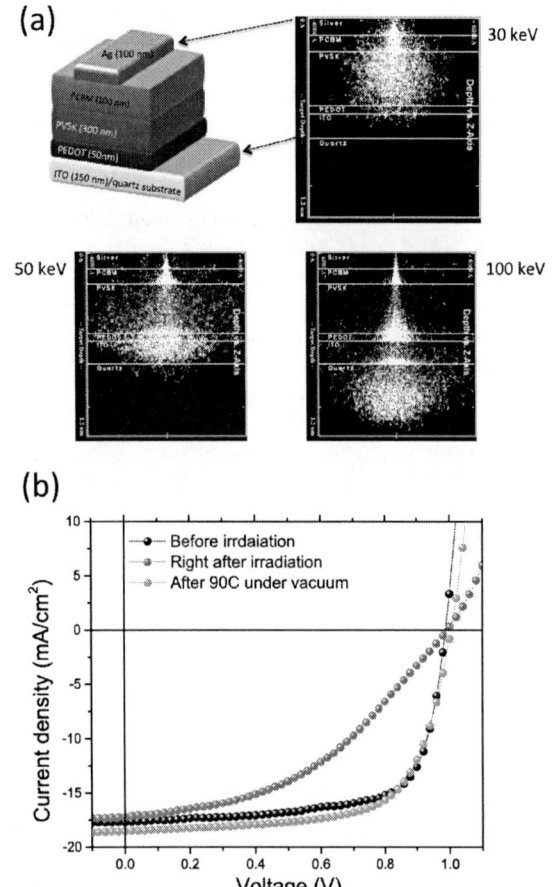

Figure 4. (a) Simulation of proton trajectories at energies from 30 to 100 keV in the perovskite solar cells. (b) J-V curves of the perovskite solar cells with before and after proton irradiation, as well as the recovering after vacuum annealing process.

IV. CONCLUSION

We have shown that perovskite solar cells are remarkably tolerant of 1 MeV electron and 50 keV proton radiation at fluence of up to 10^{16} and 10^{12} cm^{-2}, respectively. Electron irradiation produced no detectable degradation at fluence up to 10^{15} cm^{-2}, and only caused ~10% degradation at 10^{16} cm^{-2}. This exceeds the reported performance of GaAs cells (~40% degradation) and even radiation-hardened InP cells (~20% degradation) at this fluence. [9,10] We further observed no significant changes to the morphology or crystal phase of the perovskite thin films under electron fluence of 10^{14} cm^{-2}.

Proton irradiation initially reduced the cells' efficiency by ~40%, but interestingly, did not significantly impact V_{OC} or J_{SC}. This suggests that neither the cell's radiative efficiency nor its optical properties were damaged by the protons, and indeed, we found that a vacuum annealing process at 90 C completely restored the cell performance. We hypothesize that the cells should tolerate gradual exposure to proton irradiation at this energy, at operating temperatures typical for space solar cells, without requiring an explicit higher temperature annealing step.

The results suggest that perovskites have intrinsically superior radiation tolerance vs. established III-V space solar technologies. Thus, not only can perovskite solar cells offer higher specific power at the cell level, but they might require dramatically less (if any) radiation shielding to operate reliably in space.

ACKNOWLEDGEMENT

This work was funded by Northrop Grumman Corporation in the frame of Space Solar Power Initiative. The authors would like to thank Dennis Thorbourn at JPL for assisting operating Dynamitron.

REFERENCES

[1] M. Yamaguchi, S. J. Taylor, S. Matsuda, O. Kawasaki, "Mechanism for the anomalous degradation of Si solar cells induced by high fluence 1 MeV electron irradiation", *Applied Physics Letters* **68**, 1996, pp. 3141-3143.

[2] M. Kaltenbrunner, G. Adam, E. D. Glowacki, M. Drack, R. Schwodiauer, L. Leonat, D. H. Apaydin, H. Groiss, M. C. Scharber, M. S. White, N. S. Sariciftci, S. Bauer, "Flexible high power-per-weight perovskite solar cells with chromium oxide-metal contacts for improved stability in air", *Nat Mater* **14**, 2015, pp. 1032–1039.

[3] "Alta Devices third generation (Gen3) product," 2016.

[4] Y. Miyazawa, M. Ikegami, T. Miyasaka, T. Ohshima, M. Imaizumi, K. Hirose, "Evaluation of radiation tolerance of perovskite solar cell for use in space," in *Photovoltaic Specialist Conference (PVSC), 2015 IEEE 42nd*, New Orleans, LA, USA, 2015, pp. 1-4.

[5] G. E. Eperon, V. M. Burlakov, P. Docampo, A. Goriely, H. J. Snaith, "Morphological Control for

High Performance, Solution-Processed Planar Heterojunction Perovskite Solar Cells", *Advanced Functional Materials* **24**, 2014, pp. 151-157.

[6] Q. Chen, H. P. Zhou, Z. R. Hong, S. Luo, H. S. Duan, H. H. Wang, Y. S. Liu, G. Li, Y. Yang, "Planar Heterojunction Perovskite Solar Cells via Vapor-Assisted Solution Process", *Journal of the American Chemical Society* **136**, 2014, pp. 622-625.

[7] P.-W. Liang, C.-Y. Liao, C.-C. Chueh, F. Zuo, S. T. Williams, X.-K. Xin, J. Lin, A. K. Y. Jen, "Additive Enhanced Crystallization of Solution-Processed Perovskite for Highly Efficient Planar-Heterojunction Solar Cells", *Advanced Materials* **26**, 2014, pp. 3748-3754.

[8] Y. Wu, A. Islam, X. Yang, C. Qin, J. Liu, K. Zhang, W. Peng, L. Han, "Retarding the crystallization of PbI2 for highly reproducible planar-structured perovskite solar cells via sequential deposition", *Energy & Environmental Science* **7**, 2014, pp. 2934-2938.

[9] M. Yamaguchi, "Radiation-resistant solar cells for space use", *Sol Energ Mat Sol C* **68**, 2001, pp. 31-53.

[10] I. Weinberg, Radiation and temperature effects in gallium arsenide, indium phosphide, and silicon solar cells, National Aeronautics and Space Administration, [Washington, D.C.], 1987.

Towards Perovskite Silicon Tandem Solar Cells with Optimized Optical Properties

Jan Christoph Goldschmidt[1], Alexander J. Bett[1], Patricia S. C. Schulze[1], Nico Tucher[1],
Martin Bivour[1], Markus Kohlstädt[1], Seunghun Lee[1], Simone Mastroianni[1], Laura Mundt[1], Markus Mundus[1],
Paul Ndione[2], Karl Wienands[3], Kristina Winkler[1], Uli Würfel[1], Martin Hermle[1], Stefan W. Glunz[1,3]

[1] Fraunhofer Institute for Solar Energy Systems ISE, Heidenhofstrasse 2, 79110 Freiburg, Germany
[2] National Renewable Energy Laboratory, 15013 Denver West Parkway, Golden, CO 80401, USA
[3] Albert-Ludwigs-University, Faculty of Engineering, Georges-Köhler-Allee 105, 79110 Freiburg, Germany

Abstract — **Perovskite silicon tandem solar cells can overcome the efficiency limit of single junction solar cells. For an optimized optical performance, we propose a 2-terminal device featuring a front-side anti-reflection structure, optimized layer thicknesses, a meso-porous scaffold for the perovskite solar cell and a rear-side light-trapping structure for the silicon solar cell. To maintain the functionality of the underlying layers, we have developed a low-temperature process to realize a meso-porous TiO2 scaffold via ultra-violet (UV) curing. With perovskite solar cell efficiencies >15%, we achieve results comparable to our conventional high-temperature (> 500°C) route. For the optical optimization of the complex tandem device with elements of different feature sizes, we apply a matrix-based formalism and show how layer thickness optimization and the rear-side light trapping can significantly improve the current of the tandem device.**

I. INTRODUCTION

Perovskite silicon tandem solar cells have the potential for efficiencies >30% [1, 2]. After rapid progress, 4-terminal devices have reached efficiencies of 26.4% [3]. This is very close to the record for single-junction silicon solar cells of 26.6%, and 2.5% absolute above the efficiency of the silicon solar cell used in the device. 2-terminal devices have achieved up to 23.6% [4], an increase of 2.2% absolute compared to the used silicon solar cell technology. However, perovskite silicon tandem solar cells are still far from realizing their full potential, which would be necessary for generating a real cost advantage. Optical losses are a major limiting factor: parasitic absorption occurs mostly in the hole transport material in 2-terminal devices, and in the highly doped transparent conductive oxide layers for lateral charge transport in 4-terminal devices [5]. Incomplete absorption close to the bandgap occurs for both perovskite absorber and silicon. This causes unnecessary thermalisation losses for photons that could have been absorbed in the perovskite, while low-energy photons that are not absorbed in the silicon are lost completely.

Optical losses could be mitigated with a tandem device as depicted in Fig. 1. The key optical features are (i) an optical nano-structure, which is capable of reducing reflection losses also in an encapsulated configuration, (ii) adapted layer thicknesses, maximizing useful absorption, while minimizing reflection, (iii) a rear-side light trapping structure.

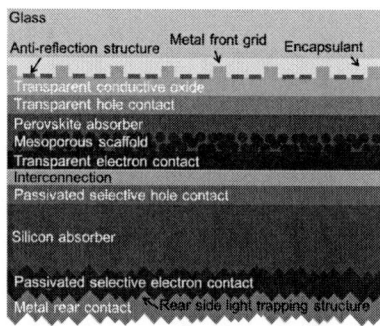

Fig. 1. Sketch of a perovskite silicon tandem solar cell with advanced optics. Reflection is reduced by a front-side nanostructure and adapted layer thicknesses, while absorption of long-wavelength photons is increased by a rear-side light trapping structure.

We aim for 2-terminal tandem solar cells, reducing the number of layers for lateral current transport, and thereby parasitic absorption. Furthermore, the use of a meso-porous scaffold is beneficial. The scaffold reduces the effective distance over which the charge carrier collection occurs, thus allowing for thicker perovskite absorbers [6]. The increased absorption in the thicker perovskite layer mitigates the problem of incomplete absorption close to the perovskite bandgap. Such meso-porous architectures yielded high efficiencies throughout the development of perovskite solar cells [7–9]. Typically, the porous scaffold is sintered at high temperatures (500°C) [10, 11]. However, for monolithic tandem solar cells a low-temperature process helps to preserve the performance of the underlying silicon solar cell. Therefore, we present completely low-temperature processed perovskite solar cells with meso-porous TiO2 scaffold.

Furthermore, the resulting perovskite silicon tandem structure is not trivial to model optically, as it features structures of different length scales. The perovskite solar cell itself can be described as multi-layer thin film, while the silicon solar cell and a pyramidal rear structure are more macroscopic. Therefore, in the second part of the paper we show first optical modeling results for complete tandem devices and indicate the potential for optimization.

978-1-5090-5606-4/17 $31.00 © 2017 IEEE

II. LOW-TEMPERATURE MESOPOROUS PEROVSKITE SOLAR CELLS

A. Solar cell fabrication

Perovskite solar cells were fabricated on fluorine doped tin oxide (FTO) coated glass substrates. The compact TiO_2 electron contact was evaporated using electron beam vapor deposition and a meso-porous TiO_2 scaffold was spin coated. The samples were then exposed to UV irradiation for 200 min. As reference, solar cells were fabricated by spray depositing the compact TiO_2 layer and sintering the meso-porous layer at high temperatures. The process sequences are shown in Fig. 2. Three cells are defined on each substrate by evaporating the gold contact through a shadow mask (area 0.16 cm^2).

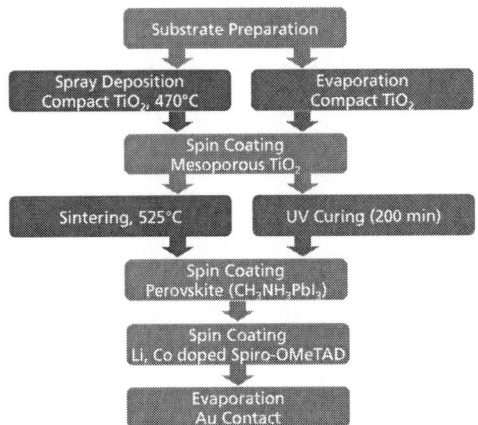

Fig. 2. Processing sequence of high-temperature and low-temperature (low-T) perovskite solar cell fabrication.

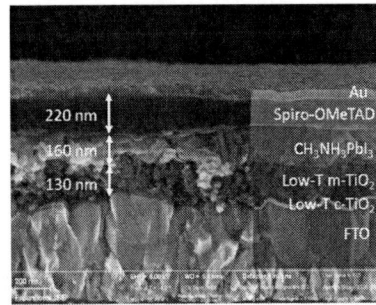

Fig. 3. SEM picture of a low-T processed perovskite solar cell with the compact (c- TiO_2) and the mesoporous TiO_2 (m- TiO_2).

B. Results

Fig. 3 shows a scanning electron microscope (SEM) picture of a perovskite solar cell fabricated by the low-T process. The perovskite absorber nicely infiltrates the meso-porous scaffold. Two batches of 12 solar cells each have been produced for direct comparison of the high- and the low-temperature processing route. Fig. 4 summarizes the results of the electrical characterization. For both routes, efficiencies >15% have been determined in reverse scan direction. For the solar cells

processed at lower temperatures, hysteresis is more pronounced (scan speed 43 mV/s).

After this successful testing of the low-T route, the next step is the production of semi-transparent mesoscopic perovskite solar cells directly on the silicon bottom cell. This requires a contact grid instead of the full-area gold contact and a conductive top layer to laterally transport the current to the grid fingers. We successfully tested molybdenum oxide (MoOx) and indium doped tin oxide (ITO). Details will be available at the conference.

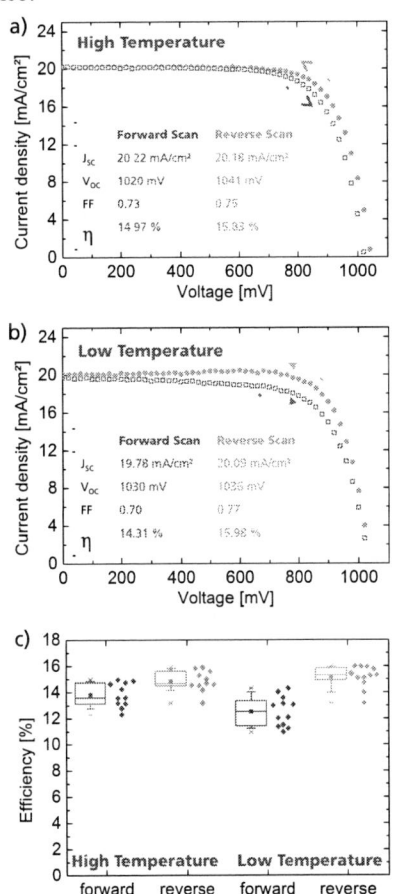

Fig. 4. Electrical characterization of the produced perovskite solar cells. a) Current-voltage curves of the best solar cell fabricated by the high temperature process and b) with the low temperature process. (c) For the whole batch comparable efficiencies have been achieved with the high-T and low-T processing routes, respectively.

IV. OPTICAL MODELING OF PEROVSKITE SILICON TANDEM SOLAR CELLS

To model the optical properties of the tandem structure, we used a modified version of the OPTOS-formalism [12]. This formalism allows for calculating the properties of the individual elements by the most appropriate method and then combines the different features in a matrix-based calculation. The modeled system is displayed in Fig. 5a. Based on experimentally determined typical layer thicknesses, the absorption in the individual sub-cells was calculated and converted into an achievable photo-current (Fig. 5b).

a)

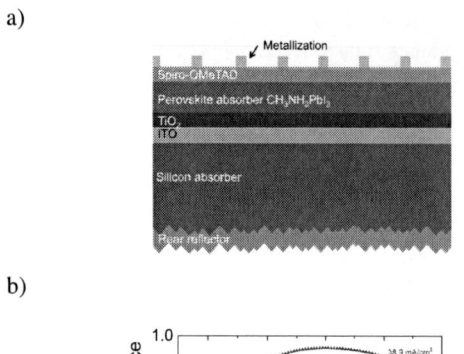

b)

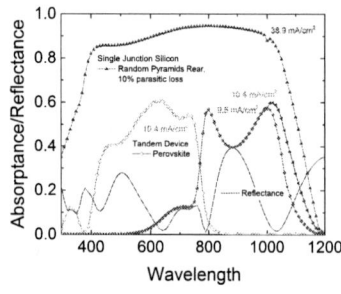

c)

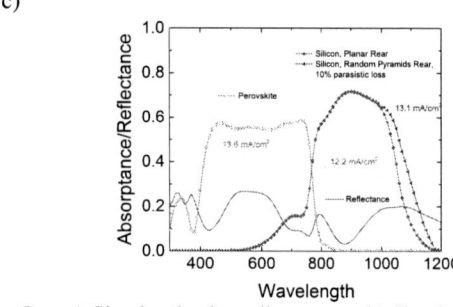

Fig. 5. a) Simulated solar cell structure. b) Simulated absorptance in the individual sub-cells, as well as in the respective single junction silicon solar cell. c) Adapting the layer thicknesses increased the current in the sub-cells. For all configurations, the light trapping structure at the rear increases the current of the silicon solar cell such that current matching is achieved.

The achieved currents of around 10 mA/cm^2 are significantly below that what could have been achieved with the silicon solar cell alone. Reasons are parasitic absorption in the hole-conduction material and high reflection. Reducing the thickness of the hole-conductor reduces parasitic absorption. By adjusting the thickness of the ITO interconnection layer, reflection is reduced (Fig. 5c). The rear-side light trapping structure increases the current of the silicon solar cell by 0.9 mA/cm^2, enabling good current matching. However, optical losses are still significant. For the reduction of reflection losses, we are currently investigating front-side texturing, which promises a very significant current increase. Details will be available at the conference.

SUMMARY

We have developed a low temperature process sequence for the production of mesoscopic perovskite solar cells. Using this process, it is possible to manufacture perovskite silicon tandem solar cells, with high useful absorption in the perovskite absorber. We applied a matrix-based formalism to optimize the layer thicknesses in a perovskite silicon tandem device. Together with a rear-side light trapping structure, a current enhancement of more than 30% is predicted. Work is currently ongoing for further optimization and realization of such an optimized tandem device.

REFERENCES

[1] T. P. White, N. N. Lal, and K. R. Catchpole, "Tandem Solar Cells Based on High-Efficiency c-Si Bottom Cells: Top Cell Requirements for >30% Efficiency," *IEEE Journal of Photovoltaics*, vol. 4, no. 1, pp. 208–214, 2014.

[2] Miha Filipič, Philipp Löper, Bjoern Niesen, Stefaan De Wolf, Janez Krč, Christophe Ballif, Marko Topič, "CH3NH3PbI3 perovskite / silicon tandem solar cells: characterization based optical simulations," *Optical Society of America*, vol. 2015, no. 23, 2015.

[3] T. Duong *et al,* "Rubidium Multication Perovskite with Optimized Bandgap for Perovskite-Silicon Tandem with over 26% Efficiency," *Advanced Energy Materials*, vol. 131, p. 1700228, 2017.

[4] K. A. Bush *et al,* "23.6%-efficient monolithic perovskite/silicon tandem solar cells with improved stability," *Nature Energy*, vol. 2, p. 17009, 2017.

[5] J. Werner *et al,* "Efficient Near-Infrared-Transparent Perovskite Solar Cells Enabling Direct Comparison of 4-Terminal and Monolithic Perovskite/Silicon Tandem Cells," *ACS Energy Lett,* vol. 1, no. 2, pp. 474–480, 2016.

[6] N.-G. Park, M. Grätzel, T. Miyasaka, K. Zhu, and K. Emery, "Towards stable and commercially available perovskite solar cells," *Nature Energy*, vol. 1, p. 16152, 2016.

[7] W. S. Yang *et al,* "High-performance photovoltaic perovskite layers fabricated through intramolecular exchange," (eng), *Science*, vol. 348, no. 6240, pp. 1234–1237, 2015.

[8] J.-P. Correa-Baena *et al,* "Unbroken Perovskite: Interplay of Morphology, Electro-optical Properties, and Ionic Movement," (ENG), *Advanced Materials*, vol. 28, no. 25, pp. 5031–5037, 2016.

[9] A. Kojima, K. Teshima, Y. Shirai, and T. Miyasaka, "Organometal halide perovskites as visible-light sensitizers for photovoltaic cells," (eng), *Journal of the American Chemical Society*, vol. 131, no. 17, pp. 6050–6051, 2009.

[10] F. Hao, C. C. Stoumpos, D. H. Cao, Chang, Robert P. H, and M. G. Kanatzidis, "Lead-free solid-state organic-inorganic halide perovskite solar cells," *Nature Photonics*, vol. 8, no. 6, pp. 489–494, 2014.

[11] J. Werner *et al,* "Sputtered rear electrode with broadband transparency for perovskite solar cells," *Solar Energy Materials & Solar Cells*, vol. 141, pp. 407–413, 2015.

[12] N. Tucher *et al,* "Optical simulation of photovoltaic modules with multiple textured interfaces using the matrix-based formalism OPTOS," *Opt. Express*, vol. 24, no. 14, pp. A1083, 2016.

First-principles density functional theory calculation of metal-substituted lead halide perovskite

Ji-Sang Park,[†] Matthew D. Sampson,[§] Alex B. F. Martinson,[§] and Maria K. Y. Chan[†]

[†]Materials Science Division, Argonne National Laboratory, Argonne, Illinois 60439, United States
[§]Center for Nanoscale Materials, Argonne National Laboratory, Argonne, Illinois 60439, United States

Abstract — We investigated the optoelectronic properties of 3*d* metal-substituted lead halide perovskite by performing first-principles density functional theory calculations to examine whether the materials can be used for intermediate band solar cells. We found that Co substitution for Pb does not change the lattice constant significantly. Co substitution is able to induce intermediate bands that can absorb light, consistent with experiment results. In our calculations, Fe substitution for Pb also can induce intermediate bands, however, it is less energetically favored as compared to Co substitution, and experimentally such extra absorption was not observed.

Index Terms — cobalt, DFT, intermediate band, perovskite halides, solar cells

I. INTRODUCTION

According to Shockley and Queisser, the maximum solar conversion efficiency is strongly correlated with the band gap of absorber material [1]. The band gap should not be too high because that prevents the absorption of significant amount of photons. On the other hand, the band gap should not be too low as that results in low voltage and thus reduced power. This argument is reflected in the fact that many semiconductor materials currently being actively studied such as Si, III-V, CdTe, and $Cu(In,Ga)Se_2$, have band gaps close to the ideal value of 1.34 eV.

There is another type of solar cell called intermediate band (IB) solar cells [2]. Unlike single junction solar cells, an absorber layer with an IB can absorb lights with sub-band-gap photon energies. Therefore, IB solar cells have much higher theoretical maximum efficiency [3], exceeding the Shockley-Queisser limit.

This IB idea has been recently applied to inorganic-organic hybrid perovskite halides by the authors [4],[5]. Hybrid inorganic-organic perovskite halides have attracted much attention [6]-[9], because of their superior material properties and the fact that the record cell efficiency has rapidly increased [10]. We focused on $CH_3NH_3PbBr_3$ because it has a band gap of about 2.3 eV, which is ideal for an IB solar cell [3].

In this work, we performed first-principles density functional theory (DFT) calculations to evaluate the possibility of transition metal-alloyed perovskite halides for IB solar cell applications. We calculated electronic properties of the materials, including the electronic structure and the absorption coefficient, and compared the calculated data to experiment data. We found that Co and Fe can induce IBs in

the band gap, absorbing the light with sub-band-gap photon energies.

II. COMPUTATIONAL METHOD

The electronic properties of metal-substituted lead halide perovskite were obtained using the generalized gradient approximation (GGA) parameterized by Perdew, Burke, and Ernzerhof (PBE) [11] for the exchange correlation potential. Generally speaking, the band gap of semiconductors is largely underestimated with GGA exchange-correlation functionals. The band gap of the hybrid perovskites, however, is calculated with relative accuracy because of error cancellation [9]. We also emphasize that the electronic band gap of the $CH_3NH_3PbBr_3$ is reduced by more than 1 eV by the spin-orbit coupling (SOC), whereas the positions of the IBs with respect to N semi-core level is unchanged. The projector-augmented wave (PAW) potentials [12] were used, as implemented in the VASP code [13]. The cutoff energy for the plane-wave basis was 500 eV. The atomic structures were fully relaxed until the residual forces were less than 0.03 eV/Å. The calculations were performed using the pseudo-cubic unit cell with 12 atoms or 2×2×2 supercells with 96 atoms. For the calculation of a halide alloy, a special quasi-random structure was generated [14]. For the relaxation and dielectric function calculations of the supercell, 3×3×3 and 4×4×4 Monkhorst-Pack meshes were used for Brillouin zone integration, respectively. Absorption coefficients were calculated from the frequency-dependent dielectric function [15].

It is generally considered that the electronic structure of transition metals is not accurately calculated with GGA exchange-correlation. For better description of the metal induced IBs, we also performed GGA+U_d calculations [16]. The appropriate value of U for Co oxides to reproduce reaction energies have been found to be 3.3-3.5 eV [17],[18], and thus it is reasonable to expect that the appropriate U values for chlorides and bromides to be lower than those for oxides. Consistent with this expectation, our calculations for Co fluorides, CoF_2 and CoF_3, show that a U value which reproduces the $CoF_2 + ½ F_2 \leftrightarrow CoF_3$ reaction energy is in the range of 0.3~2 eV. We, on the other hand, found that using U = 7 eV reproduces the optical absorption properties of $CoBr_2$, and in general relatively higher U values are required to reproduce electronic structure properties at the single-particle level [19]. Therefore, we performed the DFT+U calculations with $U = 0$, 3, and 7 eV.

III. RESULTS AND DISCUSSIONS

Stability. It is important to know whether Co and Fe can be incorporated into the hybrid perovskite or not. To examine the stability of Co and Fe in MAPbBr$_3$, we calculated the following formation energy:

$$\Delta E = \frac{E\left(MA_8Pb_7MBr_{24}\right) + E\left(PbBr_2\right) - 8E\left(MAPbBr_3\right) - E\left(MBr_2\right)}{8}$$
$$+ k_B T \left\{ x \ln x + (1-x) \ln(1-x) \right\},$$

where k_B is the Boltzmann constant, x is 1/8, M is either Co or Fe, and T is growth temperature (100 °C). The obtained formation energies are summarized in Table 1. Our calculations show that the value of Δ for Co is either similar or less than the thermal energy at growth temperature (0.032 eV), or negative, indicating that Co can incorporate into MAPbBr$_3$. On the other hand, the value of Δ for Fe incorporation into MAPbBr$_3$ is calculated to be slightly higher as compared to Co incorporation, indicating that Fe is less likely to be incorporated than Co. We similarly calculated the value of Δ for the Co incorporation in a halide alloy, MAPbBr$_{1.5}$Cl$_{1.5}$, by replacing a Pb atom bonded to 3 Br and 3 Cl atoms with a Co atom. The stability of Co in the halide alloy is quantitatively similar to that in MAPbBr$_3$, indicating that Co can incorporate in the inorganic framework even in halide alloys.

Table 1. The formation energy (ΔE) for incorporation of M (M = Co, Fe) from DFT calculations.

Host material	M	U for M $3d$ (eV)	ΔE (eV)
MAPbBr$_3$	Co	0	0.036
		3	0.039
		7	-0.087
	Fe	0	0.047
MAPbBr$_{1.5}$Cl$_{1.5}$	Co	0	0.039
		3	0.040
		7	-0.100

The lattice constant. Table 2 shows the optimized lattice constant of MAPbBr$_3$, MAPbBr$_{1.5}$Cl$_{1.5}$, and those with metal substitution. Since we employed pseudo-cubic supercell, we calculated the cube root of the volume as the lattice constant. In our calculations, the change of the lattice constant upon the metal substitution for 1/8 is less than 0.3%, predicting that Co and Fe substitution for Pb does not change the lattice constant appreciably. This result is also consistent with experimental findings [4]. Our calculations also show that the lattice constant is not sensitive to the U value.

Table 2. The optimized lattice constant of MAPb$_{1-x}$M$_x$Br$_{3-y}$Cl$_y$ (M = Co, Fe).

Material	U for M $3d$ (eV)	Lattice constant (Å)	Change in the lattice constant Δ_{lat} (%)
MAPbBr$_3$	N/A	6.065	
MAPb$_{0.87}$Co$_{0.13}$Br$_3$	0	6.062	-0.05
	3	6.057	-0.13
	7	6.059	-0.10
MAPb$_{0.87}$Fe$_{0.13}$Br$_3$	0	6.057	-0.13
MAPbBr$_{1.5}$Cl$_{1.5}$	N/A	5.933	
MAPb$_{0.87}$Co$_{0.13}$Br$_{1.5}$Cl$_{1.5}$	0	5.916	-0.29
	3	5.922	-0.19
	7	5.915	-0.30

Electronic structure. Our DFT calculations predict gap states induced by metal substitution. Pure MAPbBr$_3$ has band gap of 1.98 eV, whereas MAPb$_{0.87}$Co$_{0.13}$Br$_3$ has three empty IBs and two occupied levels in the host band gap for the majority spin. There is no gap state for the minority spin. The unoccupied IBs are mainly composed of Co d orbitals, with small contributions from the Br p orbitals.

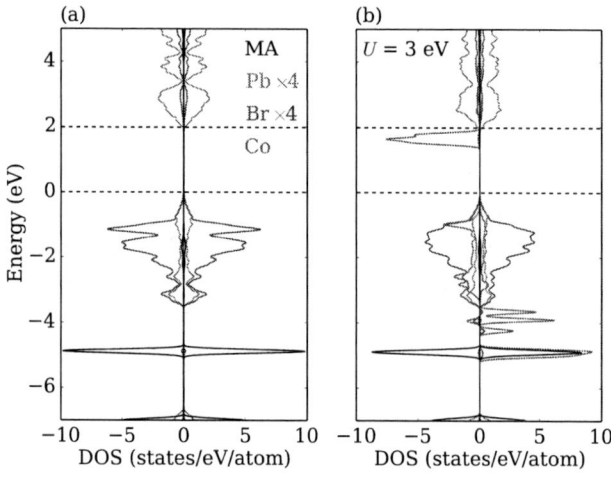

Figure 1. Density of states of (a) pure MAPbBr$_3$ and (b) MAPb$_{0.87}$Co$_{0.13}$Br$_3$. The dashed horizontal lines represent the band edges of pure MAPbBr$_3$ estimated by the N $2s$ semi-core level as reference. The DOS of the perovskite bromide with Co substitution is broader than that of the pure system.

We also found that position of the Co induced states is affected by the Hubbard U value depending on the occupation of the states. Figure 1a,b shows the projected density of states (PDOS) of MAPbBr$_3$ and MAPb$_{0.87}$Co$_{0.13}$Br$_3$ obtained using a Hubbard U of 3 eV. The Hubbard U of 3 eV pushes the occupied levels below the valence band maximum (VBM), while raising the three unoccupied levels. The unoccupied levels, however, remain in the band gap, as shown in Figure 1b. Therefore, PBE+U calculation with a moderate U value also predicts mid-gap states.

When a strong Hubbard U value of 7 eV is applied, the unoccupied levels become higher than the conduction band minimum. In contrast to the experiments, there is no state that can absorb sub-bandgap lights. This indicates that U of 7 eV is too strong to explain the observed the sub-bandgap absorption in experiments.

We point out that the local environment of Co atoms becomes more tetrahedral as the Co atoms are displaced from their ideal position. Four shortened Co–Br bonds are ~2.43 Å in length, whereas two lengthened Co–Br bonds are ~4.07 Å in length. As a result, there are two occupied levels and three unoccupied levels for the majority spin, as shown in Fig. 1.

We find that $MAPb_{0.87}Co_{0.13}Br_{1.5}Cl_{1.5}$ also has IBs in the band gap. Similar to the Co induced IBs in $MAPb_{0.87}Co_{0.13}Br_3$, Co in $MAPb_{0.87}Co_{0.13}Br_{1.5}Cl_{1.5}$ induces three unoccupied bands, and those increase in energy with the U value.

Our DFT calculations also predict that Fe substitution in $MAPbBr_3$ introduces IBs; however, such subgap absorption was not observed experimentally [4]. The discrepancy may be caused by errors in the DFT/DFT+U calculations, and/or lack of incorporation of Fe into the perovskite crystal structure due to the higher formation energy.

We also calculated the energy gap of $MAPbBr_3$ and $MAPb_{0.87}Co_{0.13}Br_3$ by calculating E(N+1) + E(N-1)] − 2E(N), where E is total energy and N is the number of electrons in the supercell. The lowest IB with respect to the VBM of $MAPb_{0.87}Co_{0.13}Br_3$ is calculated as 1.78 and 1.80 eV using $U = 0$ and 3 eV, respectively. Since the band gap of $MAPbBr_3$ is calculated as 2.24 eV, the Co induced states are in the band gap, consistent with the DOS shown above.

IV. CONCLUSION

Stability and the optoelectronic properties of transition metals in hybrid inorganic-organic perovskite halides were investigated by performing the DFT calculations. Our calculations show that Co and Fe substitution for Pb does not change the lattice constant appreciably, and induce IBs in the band gap. Thermodynamic stability of the metal substitutions is examined. It is found that Co substitution is thermodynamically more stable than Fe substitution.

V. ACKNOWLEDGMENTS

This work is supported by Laboratory Directed Research and Development (LDRD) funding from Argonne National Laboratory, provided by the Director, Office of Science, of the U.S. Department of Energy under contract DE-AC02-06CH11357. Use of the Center of Nanoscale Materials, an Office of Science user facility, was supported by the U.S. Department of Energy, Office of Science, Office of Basic Energy Sciences, under contract no. DE-AC02-06CH11357.

REFERENCES

[1] W. Shockley and H. J. Queisser, "Detailed Balance Limit of Efficiency of p-n Junction Solar Cells," *J. Appl. Phys.*, vol. 32, pp. 510-519, 1961.

[2] Y. Okada, N. J. Ekins-Daukes, T. Kita, R. Tamaki, M. Yoshida, A. Pusch, O. Hess, C. C. Phillips, D. J. Farrell, K. Yoshida, N. Ahsan, Y. Shoji, T. Sogabe, and J.-F. Guillemoles, "Intermediate band solar cells: Recent progress and future directions," *Appl. Phys. Rev.,* vol. 2, p. 021302, 2015.

[3] S. P. Bremner, M. Y. Levy, and C. B. Honsberg, "Limiting efficiency of an intermediate band solar cell under a terrestrial spectrum," *Appl. Phys. Lett.*, vol. 92, p. 171110, 2008.

[4] S. Andalibi, A. Rostami, G. Darvish and M. K. Moravvej-Farshi, "Band gap engineering of organo metal lead halide perovskite photovoltaic absorber," Opt. Quant. Electron. vol. 48, p. 258, 2016.

[4] M. D. Sampson, J.-S. Park, R. D. Schaller, M. K. Y. Chan, A. B. F. Martinson, accepted for publication in J. Mater. Chem. A.

[5] J. Burschka, N. Pellet, S. J. Moon, R. Humphry-Baker, P. Gao, M. K. Nazeeruddin, and M. Gratzel, "Sequential Deposition as a Route to High-Performance Perovskite-Sensitized Solar Cells," *Nature*, vol. 499, pp. 316–319, 2013.

[6] N. J. Jeon, J. H. Noh, Y. C. Kim, W. S. Yang, S. Ryu, and S. Il Seol, "Solvent Engineering for High-Performance Inorganic-Organic Hybrid Perovskite Solar Cells," *Nat. Mater.*, vol. 13, pp. 897–903, 2014.

[7] N. J. Jeon, J. H. Noh, W. S. Yang, Y. C. Kim, S. Ryu, J. Seo, and S. I. Seok, "Compositional Engineering of Perovskite Materials for High- Performance Solar Cells," *Nature*, vol. 517, pp. 476–480, 2015.

[8] W.-J. Yin, J.-H. Yang, J. Kang, Y. Yan, and S.-H. Wei, "Halide perovskite materials for solar cells: a theoretical review," *J. Mater. Chem. A*, vol. 3, pp. 8926-8942, 2015.

[9] http://www.nrel.gov/ncpv/

[10] J. P. Perdew, K. Burke, and M. Ernzerhof, "Generalized Gradient Approximation Made Simple," *Phys. Rev. Lett.*, vol. 77, pp. 3865-3868, 1996.

[11] P. E. Blöchl, "Projector augmented-wave method," *Phys. Rev. B*, vol. 50, pp. 17953-17979, 1994.

[12] G. Kresse, and J. Furthmüller, "Efficient iterative schemes for ab initio total-energy calculations using a plane-wave basis set," *Phys. Rev. B*, vol. 54, pp. 11169-11186, 1996.

[13] A. van de Walle, P. Tiwary, M. de Jong, D. L. Olmsted, M. Asta, A. Dick, D. Shin, Y. Wang, L.-Q. Chen, and Z.-K. Liu, "Efficient stochastic generation of special quasirandom structures," *Calphad Journal*, vol. 42, pp. 13-18, 2013.

[14] M. Gajdoš, K. Hummer, G. Kresse, J. Furthmüller, and F. Bechstedt, "Linear optical properties in the PAW methodology," *Phys. Rev. B*, vol. 73, p. 045112, 2006.

[15] S. L. Dudarev, G. A. Botton, S. Y. Savrasov, C. J. Humphreys and A. P. Sutton, "Electron-energy-loss spectra and the structural stability of nickel oxide: An LSDA+U study," *Phys. Rev. B*, vol. 57, pp. 1505-1509, 1998.

[16] Z. Zeng, M. K. Y. Chan, Z.-J. Zhao, J. Kubal, D. Fan, J. Greeley, "Towards First Principles-Based Prediction of Highly Accurate Electrochemical Pourbaix Diagrams," *J. Phys. Chem. C*, vol. 119, pp. 18177-18187, 2015.

[17] L. Wang, T. Maxisch, and G. Ceder, "Oxidation energies of transition metal oxides within the GGA+U framework" *Phys. Rev. B*, vol. 73, p. 195107, 2006.

[18] B. Himmetoglu, A. Floris, S. de Gironcoli, and M. Cococcioni, "Hubbard-corrected DFT energy functionals: The LDA+U description of correlated systems," *Int. J. Quantum Chem.* Vol. 114, pp. 14-49, 2014.

Estimating the Effects of Module Area on Thin-Film Photovoltaic System Costs

Kelsey A. W. Horowitz[1], Ran Fu[1], Xingshu Sun[2], Tim Silverman[1], Michael Woodhouse[1], and Muhammad A. Alam[2]

[1]National Renewable Energy Laboratory, Golden, CO, 80401, USA
[2]Purdue University, West Lafayette, IN, 47907, USA

Abstract — We investigate the potential effects of module area on the cost and performance of photovoltaic systems. Applying a bottom-up methodology, we analyzed the costs associated with thin-film modules and systems as a function of module area. We calculate a potential for savings of up to $0.10/W and $0.13/W in module manufacturing costs for CdTe and CIGS respectively, with large area modules. We also find that an additional $0.04/W savings in balance-of-systems costs may be achieved. Sensitivity of the $/W cost savings to module efficiency, manufacturing yield, and other parameters is presented. Lifetime energy yield must also be maintained to realize reductions in the levelized cost of energy; the effects of module size on energy yield for monolithic thin-film modules are not yet well understood. Finally, we discuss possible non-cost barriers to adoption of large area modules.

Index Terms — cost analysis, large module, photovoltaics, solar economics, thin-film.

I. INTRODUCTION

Area-based economies of scale have been demonstrated in manufacturing of several different technologies, including flat panel displays, coated glass for architectural applications, and wafer-based semiconductor processes. With solar photovoltaic (PV) technology, the vast majority of modules remain relatively small, within the range of 1 to 2 m^2; however, several companies have attempted to leverage area-based economies of scale to reduce PV costs. Perhaps the most-well known and extreme example of such attempts was development of an amorphous silicon (a-Si) SunFab module by Applied Materials. Applied Materials claimed that these modules, which had an area of 5.7 m^2, reduced installed cost of a PV system by more than 20% [1], but then shut down the SunFab line several years later.

It was unclear how much the challenges SunFab faced were attributable to the declining price of incumbent PV, a-Si technology, or the large-area module format. Additionally, despite the fact that very large-area modules have not yet succeeded in the marketplace, interest in the concept has not faded. First Solar, the leading manufacturer of cadmium telluride (CdTe) modules, recently announced plans to move toward much larger area panels, claiming this would reduce capital equipment expenditures (CAPEX) by nearly 40% [2]. REEL Solar Inc. (RSI) has developed a process for electroplating on large-areas to help enable manufacture of large CdTe modules. Siva Power, a start-up in copper indium gallium diselenide (CIGS) module manufacturing, is also developing large area products (2 m^2, which is similar to 72-cell mc-Si modules but larger than other leading thin-film products) [3]. For PV, there are also potential area-based economies for system costs that scale with module count (e.g., installation labor costs). Indeed, this has been observed in comparing labor, electrical, and racking cost per watt for installing 60-cell modules and larger, 72-cell modules [4].

However, the effect of module area on module and system-level costs for larger sizes has not been quantified in the literature. In this paper, we provide an analysis of these costs for the leading commercial thin-film PV technologies: CdTe and CIGS. We focus on the case where rigid glass-glass module architectures are used. Because $/W costs and the levelized cost of energy (LCOE) are strongly influenced by the performance of modules, we also examine the potential effect of module size on efficiency.

II. METHODS

A. Manufacturing Cost Modeling

In this work, we build on NREL's previously developed bottom-up cost models developed for CdTe [5] and CIGS [6] modules of standard size. The CdTe and CIGS models used to generate the results for this paper were last updated in Q4 of 2015 and Q1 2016, respectively. It is important to note that because of this, some costs may have declined. However, there have been no radical technology developments in either of these technologies over the last year, and our cost models are recent enough that they can still be used to illustrate the mechanisms by which module area could drive down cost and estimate the degree of cost reductions that could potentially be achieved.

NREL's bottom-up cost models are based on developing a manufacturing process flow relevant to high volume production and calculating the costs of labor, materials, equipment, facilities, and utilities associated with each step in the process flow. Data is collected from material suppliers, equipment vendors, and PV industry members and is aggregated and anonymized to protect business-sensitive data. NREL is able to obtain high quality data from industry due to existing relationships with members of industry and demonstrated care in handling of sensitive information. Additionally, NREL's analysis center has an established track record of being able to produce cost analyses in-line with publicly available data on costs from PV firms.

In all cases, we assume the large area modules are manufactured at a new manufacturing plant with new equipment, where no subsidies or tax incentives are provided. Large modules can experience increased loading compared to smaller modules, which may cause increased stress. Stress experienced by the cells and module can be controlled through a combination of laminate design, frame design (or frame removal), and mounting. Because very large area modules have not yet been implemented, there is uncertainty around which designs will provide the best performance. For purposes of this analysis, we assume a frameless, glass-glass architecture with the cells located in the neutral axis, which is one possible solution for managing this stress.

For CIGS, we analyzed a series of module sizes based on standard glass sizes used in the flat panel display industry. These include Gen 6, Gen 7, and Gen 8 glass, which measure 1,500 mm x 1,800 mm, 1,870mm x 2200mm, and 2,160 mm x 2,460 mm, respectively, which are currently used to produce a significant fraction of displays. We additionally analyze the cost of CIGS modules on Gen 10 (2,900 mm x 3,100 mm) Gen 10.5 (2,940 mm x 3,370 mm) glass; Gen 10 and Gen 10.5 fabs for display manufacturing are beginning to be built, but are not widely used today. Possible changes in the CIGS manufacturing process and costs that could result from using larger substrates were based on interviews with members of each industry as well as equipment and material suppliers.

For CdTe, due to the limited number of players in the field, we were not able to obtain sufficient data for a similar bottom-up analysis. Instead, we used publically available information from First Solar about their Series 6 product [2] in combination with NREL's existing CdTe cost models. Information in [2] suggests that the Series 6 module is slightly larger than the Series 5 product, which consists of a panel of three 1.2m x 0.6m modules (for a total area of 1.2m x 1.8m). Because of this, we estimate the area of the Series 6 product to be between 1.2m x 1.8m and 1.5m x 1.8m (Gen 6 size) in our analysis.

B. System Cost Modeling

We evaluated the effect of module area on balance-of-system (BOS) costs for utility- scale systems using NREL's established bottom-up cost model [15–16]. In this model, we benchmarked both (1) soft costs (e.g., installation labor costs and engineering, procurement, and construction/developer overhead) and (2) hardware costs (e.g., structural/electrical BOS and inverter costs). We limit the analysis to the case of 100MW (utility-scale) ground-mount installations. Input data for the models is similarly collected from industry members and component suppliers—typically via interviews—as well

as from RSMeans, a standard costing tool used in construction.

A modified system architecture will likely be employed for very large area modules. Several different system architectures for large area modules have been proposed; due to limited field experience, the relative merits of different proposed designs are not well understood. Here, we explore the costs associated with one possible approach, wherein flexible flanges and adhesive material are used in place of traditional fixed clamp connections between modules. This system design was used for our analysis of all large modules (Gen 6 through Gen 10.5). We assume that this configuration allows for the use of one less vertical mounting rail for each module assembly.

Based on our interviews and existing NREL data, we believe that installation of these large modules would require a machine to assist with lifting and placing the modules. In our model, we assume the cost of this machine is equal to that of a standard crane truck used in construction (a truck with a small crane mounted on the bed) plus a 20% premium for the appropriate robots, end effectors, etc. to interface with the module. Our results assume that the machine-assisted module mounting takes approximately the same amount of time as manual module mounting; if this process could be sped up, additional cost savings could be realized. However, as discussed below, the installation labor cost associated with module mounting is a small fraction of the total installation labor costs, so these savings would likely be modest.

Because of the limited experience with large area modules, there is uncertainty around many of the input data for both the manufacturing and installed system cost models. Often, our data sources were making informed estimates about how their costs or processes could be affected by increasing module size. Sensitivity analysis was performed in order to evaluate the potential impact of these uncertainties on our conclusions about the cost of different module areas.

Costs for module shipping are not included, but may be analyzed in greater depth in future work. Preliminary evaluation of the potential of shipping issues associated with very large glass sizes suggests that shipping and handling of glass size Gen 10 or above has been more challenging for the display industry (see [8] for additional discussion).

C. Simulations of Module Efficiency

Rated module efficiency is a key driver of both module and installed system cost per watt. In addition to rated efficiency, LCOE is also strongly influenced by lifetime energy yield. Because of the limited experience with very large area thin-film modules, the effects of module area on these two performance metrics is not well documented. While a comprehensive discussion of these effects requires significantly more research and is well beyond the scope of this paper, we provide simulation of how one of the main causes of non-uniformity in thin-film solar modules—the lognormal distributed shunt [9], which contributes significantly to the cell-to-module efficiency gap [10]—could affect the rated efficiency of CIGS as module area scales. We do this using a physics-based module simulation framework [11-12] and Monte Carlo simulation with different module sizes given a lognormal distribution of shunts. The lognormal distribution of shunts can be inherent in thin-film technologies, which originated from the fact that grain size in poly-crystalline films is lognormal distributed regardless of the choice of materials. Thus, the trends observed in our simulation results may be extensible to other thin-film material systems.

III. RESULTS

Module manufacturing cost reductions for both CIGS and CdTe result from economies-of-scale in the equipment costs, energy, and labor usage, as well as reductions in the cost per watt of components with a unit cost largely independent of module area (e.g. the junction box or j-box) or increase at a slower rate than module power (e.g. busbar costs). For CIGS, we find that a savings of $0.13/W in module manufacturing costs may be possible with Gen 10 or Gen 10.5 (approximately 8m^2) glass is used; 71% of this savings is achieved by moving to Gen 6 sizes, with 91% of the realized with Gen 8 modules. J-box savings are significant at $0.04/W (this assumes the j-box cost per unit is constant, since thin-film j-boxes do not employ bypass diodes like wafer-based designs). Another significant savings is achieved by reduced cost-of-ownership for sputtering processes, which are used for depositing the molybdenum (Mo) back contact layer, the intrinsic and aluminum-doped zinc oxide (i-ZnO/AZO) front contact stack, and the copper (Cu), indium (In), and gallium (Ga) precursors in our modeled process flow. For CdTe, we estimate a potential cost savings of $0.10/W for increasing to Gen 6 module sizes. The modeled difference in installed cost for a 1.2m x 1.8m versus a 1.5m x 1.8m (Gen 6) module was <$0.01/W. Given the level of uncertainty in our other inputs, this difference is negligible.

Figure 1(a) shows the sensitivity of the total manufacturing cost savings for CIGS to ±20% changes in key input parameters. As can be observed from the figure, cost savings are most sensitive to variations in the throughput and equipment cost for the batch selenization step. This step is the single most expensive in the CIGS fabrication process. We assumed in our analysis that the same furnace and process is used for selenization with each module size, but that batch sizes decrease proportional to the module area (i.e. as module area increases, fewer, larger modules loaded into the furnace for each run).

For CdTe, a significant portion of the manufacturing cost savings (34%) achieved by going from current module sizes to Gen6 module sizes results from economies-of-scale in capital equipment (CAPEX) costs. As shown in Figure 1(b), manufacturing cost savings for CdTe are also the most sensitive to uncertainties in CAPEX costs. Unfortunately, because insufficient data could be obtained on the CAPEX of

each tool for the Gen 6 case, we have little insight into the main drivers of this reduction.

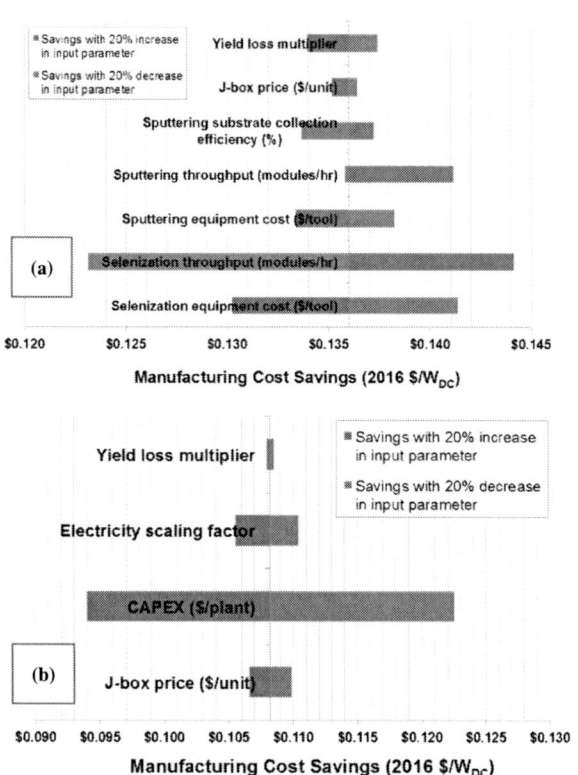

Fig. 1. (a) (top) Sensitivity of CIGS module manufacturing cost savings with Gen 10.5 modules (Gen 10.5 minus reference case cost) to ±20% changes in key input assumptions. (b) (bottom) Sensitivity of CdTe module manufacturing cost savings with Gen 6 modules (Gen 6 minus reference case cost) to ±20% changes in key input assumptions.

Figure 2 shows how module area affects non-module system costs. For Gen 6 modules, we estimate a potential non-module system cost savings of $0.03/W. Increasing module area to Gen 10 and Gen 10.5 could enable an additional BOS savings of $0.005 to $0.01/W. Several factors drive these cost reductions. Material cost savings are observed by replacing typical clamps for glass-glass modules, which cost approximately $2.65/module, with flanges and adhesives, which are estimated to cost $0.80/module. The reduced module count and modified system architecture also result in labor and structural/electrical BOS savings. The reduced module count results in a significant, proportional reduction in module mounting time, but because much of the installation labor is spent on other tasks (e.g. installing the structure, racking, and inverter), the savings as a percent of total installation labor costs are modest. There is also a small additional cost associated with the use of an additional machine to assist with large module placement.

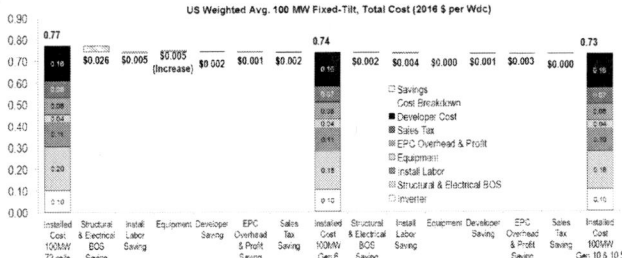

Fig. 2. Modeled impact of module area on non-module installed system costs.

Figure 3 summarizes the total installed cost savings (including both module and non-module savings) for CIGS and CdTe as a function of module area. As can be seen from the figure, CIGS savings begin to asymptote for modules with areas >5m² (beyond Gen 8).

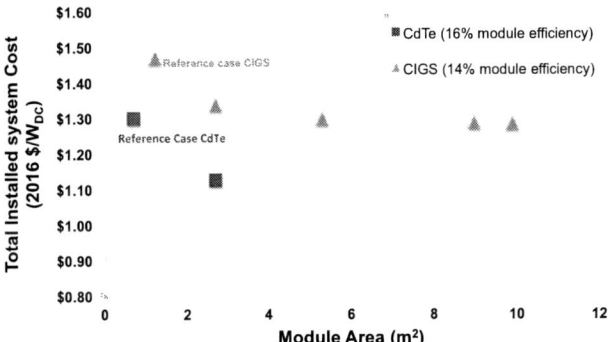

Fig. 3. Effect of module area on U.S.-weighted average total installed system price by technology. Assumes a 100MW utility-scale installation with fixed tilt.

These savings assume that rate module efficiency is constant as area increases. Because large area CdTe or CIGS modules have not yet been publicly demonstrated, this assumption contains significant uncertainty at this point. As discussed above, one key question is how non-uniformities could influence efficiency, and shunts are a primary source of non-uniformities in thin film modules. The results of our efficiency simulations are shown in Figure 4. One observation from the simulation is that modules with a larger size have a narrower distribution of efficiency (i.e., smaller variance). This occurs because the screening effect of poorly shunted cells (at the tail of the log-normal distribution) on the well-performing neighboring cells is reduced as the area of the module increases, which lowers the possibility of producing very inefficient modules. Moreover, as the size of modules expands, it is also less likely to produce a defect-free module with exceptionally high efficiency. Eventually, the efficiency is limited by the mean of shunts for very large modules.

Hence, production of monolithic solar modules with greater areas will affect the binning strategy of the manufacturers (i.e., more uniform power rating and pricing). However, it can reduce the market flexibility of selling solar modules to customers with different needs. For example, at the utility-scale, it is not favorable to deploy expensive solar modules with excessively high efficiency, as it does not offset the BOS cost. In contrast, those highly efficient solar modules are more popular at the residential scale. Prior work suggests that, if a reduction in efficiency due to shunts is observed, novel geometry design and post-process scribing could be employed to isolate defects to improve overall module performance [10].

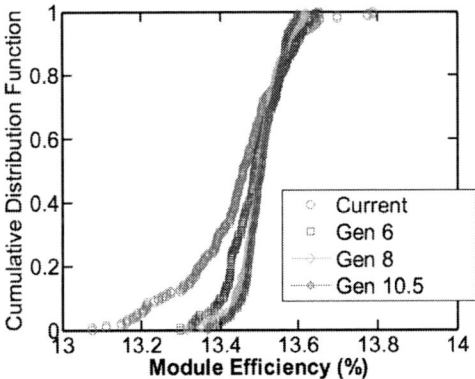

Fig. 4. Effect of module area on the distribution of expected module efficiencies for CIGS modules

Figure 5 shows the sensitivity of the total installed system cost for Gen 10.5 CIGS and Gen 6 CdTe modules to efficiency. We can see from Figure 2 that some cost savings (compared to the reference case size) are still achieved for large area module efficiencies above ~12.5% for CdTe and above ~11% for CIGS, with savings decreasing proportional to the reduction in large module efficiency compared to the reference case. This plot assumes that the module cost in $/m² does not vary with efficiency. However, in reality, higher efficiency modules may be associated with cost and/or price premiums, and vice versa. This figure can inform module cost and efficiency targets that must be achieved in order to realize a savings in total installed system cost with large modules.

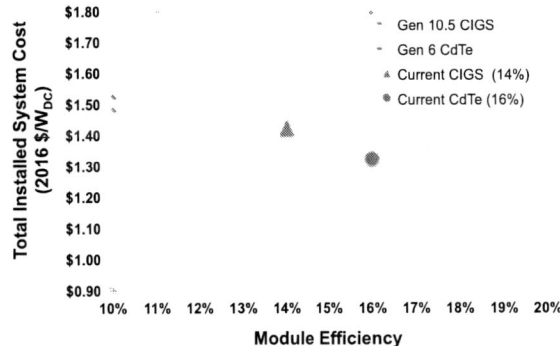

978-1-5090-5606-4/17 $31.00 © 2017 IEEE

Fig. 5. Effect of module efficiency on realized cost savings with large module sizes for thin-film technologies

IV. DISCUSSION AND CONCLUSION

We explored the potential for increased module area to influence module manufacturing and installed system cost per watt of leading thin-film PV technologies. Our analysis indicates that the use of larger area modules has the potential to drive savings at both the module ($0.13/W for CIGS and $0.10/W for CdTe) and system ($0.04/W) levels We observed diminishing returns to scale for sizes above Gen 8 or Gen 10 for module manufacturing cost for CIGS and for sizes above Gen 6 for BOS and installation costs. We did not model costs for manufacturing CdTe modules at sizes larger than Gen6; if efficiencies can be maintained, further cost reductions could be achieved for larger sizes. While there is still uncertainty in many of our assumptions, our overall conclusions are robust over a wide range of potential input values. Further research is required to better understand how module area effects module efficiency and energy yield in order to better understand the LCOE associated with large area CIGS and CdTe modules.

The exact manufacturing cost savings realized will depend on the specific process steps and factory layout involved, whether large modules are manufactured in a new facility or in an upgraded, existing facility, and what equipment is used. Relative savings achieved with larger module size will also depend on the price of materials at any given point in time. Finally, for Gen 10 and Gen 10.5 glass, challenges around logistics and shipping will need to be addressed in order to achieve low-cost.

There may also be some barriers to very large modules that are unrelated to cost per watt. For example, the upfront cost of equipment required to build up the necessary capacity to manufacture large thin film modules at competitive scale, given the large manufacturing capacity that already exists for mc-Si, is substantial. Downstream suppliers may be slow to adopt large area modules if they require additional investment themselves. Increased module size could also require new investments from equipment manufacturers to develop new tools and processes suitable for large-areas. In some cases, knowledge can be borrowed from other industries, as with sputtering equipment currently used in display manufacturing.

REFERENCES

[1] Applied Materials. (2007, Sept. 4). *Applied Materials Revolutionizes Solar Module Manufacturing with Breakthrough SunFab Thin Film Line*. [Press Release]. Retrieved on Aug. 16, 2016 from http://www.appliedmaterials.com/company/news/press-releases/2007/09/applied-materials-revolutionizes-solar-module-manufacturing-with-breakthrough-sunfab-thin-film-line

[2] deJong T. (2016, Apr. 5). 2016 First Solar Analyst Meeting: Manufacturing Update. [Online]. Retrieved on Apr. 6, 2016 from http://files.shareholder.com/downloads/FSLR/1835294976x0x884412/1548B782-59A0-4544-A452-989E1FA42BFE/FS_AnalystDay_ManufacturingUpdate.pdf

[3] Siva Power. (2013–2015). *Technology*. Retrieved on Aug. 23, 2016 from http://www.sivapower.com/#!technology/c23pn.

[4] Woodhouse M, Jones-Albertus R, Feldman D, Fu R, Horowitz K, Chung D, Jordan D, Kurtz S. (2016). *On the Path to SunShot: The Role of Advancements in Solar Photovoltaic Efficiency, Reliability, and Costs.* (NREL/TP-6A20-65872). Retrieved from http://www.nrel.gov/docs/fy16osti/65872.pdf.

[5] Woodhouse M, Goodrich A, Margolis R, James T, Dhere R, Gessert T, Barnes T, Eggert R, Albin D. (2012). Perspectives on the Pathways for Cadmium Telluride Photovoltaic Module Manufacturers to Address Expected Increases in the Price for Tellurium. *Solar Energy Materials and Solar Cells* 2013; **115**: 199–212, DOI: 10.1016/j.solmat.2012.03.023.

[6] Horowitz KAW, Fu R, Woodhouse M. An Analysis of Glass-Glass CIGS Manufacturing Costs. (2016). *Solar Energy Materials and Solar Cells* 2016; **154**: 1–10, DOI: 10.1016/j.solmat.2016.04.029.

[7] Ran Fu, Donald Chung, Travis Lowder, David Feldman, Kristen Ardani, and Robert Margolis. (September 2016) "U.S. Solar Photovoltaic System Cost Benchmark: Q1 2016." NREL Technical Report TP-6A20-66532.

[8] Horowitz, Kelsey A. W., Ran Fu, Xingshu Sun, Tim Silverman, Mike Woodhouse, and Muhammad A. Alam. (April 2017) "An Analysis of the Cost and Perfromance of Photovoltaic Systems as a Function of Module Area." NREL Technical Report NREL/TP-6A20-67006.http://www.nrel.gov/docs/fy17osti/67006.pdf

[9] Dongaonkar S, Loser S, Sheets E, Zaunbrecher K, Argawal R, Marks TJ, Alam MA. Universal Statistics of Parasitic Shunt Formation in Solar Cells, and its Implications for Cell to Module Efficiency Gap. (2013). *Energy & Environmental Science* 2013; 6: 782– 787, DOI: 10.1039/C3EE24167J

[10] Dongaonkar, S, Alam, MA. (2014). In-Line Post-Process Scribing for Reducing Cell to Module Efficiency Gap in Monolithic Thin-Film Photovoltaics. *IEEE Journal of Photovoltaics* 2014; 4: 324–332.

[11] Sun X, Silverman T, Garris R, Deline C, Alam MA. (2016). An Illumination- and Temperature-Dependent Analytical Model for Copper Indium Gallium Diselenide (CIGS) Solar Cells. *IEEE Journal of Photovoltaics* 2016; 6: 1298–1307.

[12] Dongaonkar S, Alam MA. (2012). End to End Modeling for Variability and Reliability Analysis of Thin Film Photovoltaics. Presented at the IEEE International Reliability Physics Symposium, 15–19 April 2012.

Cost Analysis of Tandem Modules

Sarah E. Sofia[1], Jonathan Mailoa[1], Dirk Weiss[2], Tonio Buonassisi[1], Ian Marius Peters[1]

[1]Massachusetts Institute of Technology, Cambridge, MA, 02139, USA; [2]First Solar, Inc., 1035 Walsh Avenue, Santa Clara, California 95050, USA.

Abstract — We develop a bottom up cost model for manufacturing II-VI based tandem PV modules. The architectures we consider are a mechanically stacked CdTe-on-CIGS four-terminal tandem and a monolithic, high-E_g CdTe alloy-on-CIGS two-terminal tandem PV module. Energy yield of these modules is assessed in three locations of distinct climates: dry, temperate, and humid using the method described in [1]. We use our cost model and yield calculations to compare the viability of the two tandem architectures. We find that the four-terminal tandem module has a significantly higher MSP both per area and per watt. Examining the energy yield for each architecture shows that the four-terminal tandem has a better performance ratio than the two-terminal tandem. Due to the high energy yield of the four-terminal tandems, installed systems can cost from 14-19% than the two-terminal for the same LCOE.

Index Terms — multijunction solar cells, tandem solar cells, technoeconomic analysis, CdTe, CIGS

I. INTRODUCTION AND BACKGROUND

There has been significant interest in the development of tandem solar cells due to their potential to reach very high efficiencies. While high efficiency tandems have been demonstrated, a pressing question for industry looking toward the future of PV is whether tandems are cost effective and, if so, which tandem architecture is most preferable.

While efficiency has been shown to be the strongest technical driver to reduce the $/W cost of PV modules [2], increasing efficiency is only worthwhile at certain costs. Manufacturing higher efficiency devices, such as tandems, comes with added cost and complexity. The boost in efficiency must, thus, at least balance out the added cost required to achieve a higher efficiency.

There are two primary tandem architectures: (1) the two-terminal (2T) tandem with monolithically integrated, series-connected sub-cells, and (2) the four-terminal (4T) tandem with mechanically stacked, electrically independent sub-cells. Due to the series-connected sub-cells of the 2T, the tandem is constrained to have equal current flow through all sub-cells. This current matching constraint causes significant loss under a varying spectrum. Since the 4T tandem sub-cells are electrically independent, this architecture is less sensitive to spectral variations than the 2T [3], [4]. The 4T tandem, however, requires more manufacturing steps and added cost due to the need for additional electrical contacts compared to the 2T. Thus, it is unclear which tandem architecture is economically more favorable.

Recent work performed detailed energy yield calculations for II-VI on II-VI tandems, comparing the relative performance of several material combinations and tandem device architectures under realistic, outdoor conditions for several geographic locations [1]. Thin-film on thin-film tandems such as these have the advantage that each sub-cell has a similar manufacturing cost structure. As recent analysis of tandem financial viability has shown, tandems whose sub-cells are a "marriage of equals" with similar device performance and manufacturing cost have the greatest potential to create tandems that have significant financial advantage over the individual sub-cells acting as SJ devices [5].

Thus, in this work we explore the relative merits of the two tandem architectures. We present a bottom up cost model for the two tandem architectures comprised. With this, we examine the relative cost of each architecture. We then examine the impact

TABLE 1. EFFICIENCY AND YIELD DATA FROM [1] FOR EACH CONSIDERED DEVICE ARCHITECTURE AND LOCATION, AND THE CALCULATED MANUFACTURING MINIMUM SUSTAINABLE PRICE (MSP) FOR EACH ARCHITECTURE

Architecture	Efficiency	Yield (kWh/m²/year)		
		Dry	Temperate	Humid
4T	27.1%	704.6	667.9	677.5
2T	25.6%	651.1	610.3	610.6
CdTe SJ	21.2%	554.7	523.9	610.3
CIGS SJ	21.1%	547.3	520.5	524.2

of each device's energy yield in each location on the total PV array cost that enables reaching target LCOE goals.

II. METHODOLOGY

A. Yield Calculations and Architectures

We explore two tandem architectures: mechanically-stacked four-terminal (4T) and monolithic two-terminal (2T), shown in figure 1. The 2T tandem is composed of series-connected sub-cells, integrated monolithically into a single device stack with a tunnel junction interconnecting sub-cells.

The current matching constraint of the 2T tandem causes a 2T CdTe-CIGS tandem to not be viable, as it would be heavily current limited by the CIGS bottom cell resulting in low energy yield. Thus, the 2T architecture we explore employs a high bandgap II-VI CdTe alloy with a bandgap of 1.68 eV as a top cell on a CIGS bottom cell.

The 4T tandem architecture is assumed to have electrically independent top- and bottom-cells, each sub-cell having its own set of front and back contacts and an insulating layer between the two sub-cells. Since the sub-cells are unconstrained, this architecture is much less sensitive to bandgap combinations. Thus, we explore a CdTe-CIGS 4T tandem, utilizing two relatively mature materials as sub-cells.

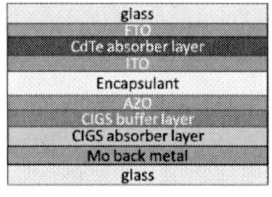 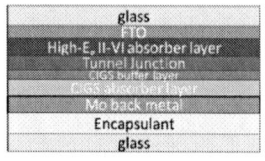

a. 4T tandem   b. 2T tandem

Fig. 1. Schematic of the two tandem device architectures.

We use the methodology for energy yield calculations presented in [1]. While most of the yields used are those directly calculated in [1], in order for these calculations to reflect the newest developments in CIGS and CdTe device performance, we replaced the previously used CdTe quantum efficiency curves with those from the newest record cells by First Solar as given in [6]. This change updates the CdTe cell to a bandgap graded device, as the bandgap graded device results from First Solar were not yet released at the time of the previous publication. We use the EQE data for CIGS from [7]. By using high efficiency results, we explore future module potential.

The energy yield is calculated for three geographic locations that are each characterized by a distinct weather climate: dry (Albuquerque, NM), temperate (Rapid City, SD), and humid (Miami, FL), as the weather conditions affect the spectral variation though the year.

B. Cost Model and LCOE Calculations

i. Module cost model

We have developed a bottom-up cost model for the manufacturing of CdTe and CIGS solar modules. We have derived our costs from supplier quotes, discussions with contacts in industry, and other academic and non-academic sources. From these cost models, we have modularized each step of the manufacturing process and used the components to develop a manufacturing process flow and cost structure for the 2T and 4T tandem comprised of these single-junctions as sub-cells. The high-E_g II-VI CdTe alloy is assumed to be the same cost as CdTe. From these costs, we calculate the minimum sustainable price (MSP) assuming straight-line depreciation and a 14% internal rate of return (IRR) [2].

ii. LCOE

The cost of the module is only a fraction of the total cost of PV system, and STC efficiency is not necessarily indicative of how a device will perform in the field. Thus, LCOE is a more informative metric to consider as it combines total system cost and energy yield. LCOE is given by the equation,

$$LCOE = \frac{total\ lifetime\ cost}{total\ lifetime\ electricity\ production} = \frac{I + \sum_{i=0}^{N} \frac{OM}{(1+r)^i}}{\sum_{i=0}^{N} \frac{E\,(1-d)^i}{(1+r)^i}} \quad (1)$$

where I is the total initial investment to install the PV system (including cost of PV modules, racking, interconnects, labor, permits, etc.), OM is the annual cost for operation and maintenance, E is the (un-degraded) annual energy output by the system as electricity (Table 1), d is the annual module degradation rate, and r is the discount rate (Table 2).

TABLE 1. RELEVANT FINANCIAL PARAMETERS USED FOR LCOE RELATED CALCULATIONS.

Financial Parameters

Nominal Discount rate (installation)	Residential: 8% [10], [11] Utility: 6.5% [11]
Operation and Maintenance (OM)	Utility: $22/kW/yr. [8] Residential: $29/kW/yr. [8]

III. RESULTS AND DISCUSSION

Using the devised module cost models, we compute the $/W module cost for both of the tandem architectures (figure 2).

We see that the 4T tandem has a significantly higher $/W cost than the 2T tandem. This is due to the substantially higher manufacturing cost per area of the 4T tandem. Because the 4T tandem has electrically independent top and bottom cells, each sub cell must have its own set of contacts, thus requiring four contact layers instead of the two required of single-junction devices and the 2T tandem. The 4T tandem has a slightly higher efficiency than the 2T tandem, however this is insufficient to balance the added cost required for the 4T architecture.

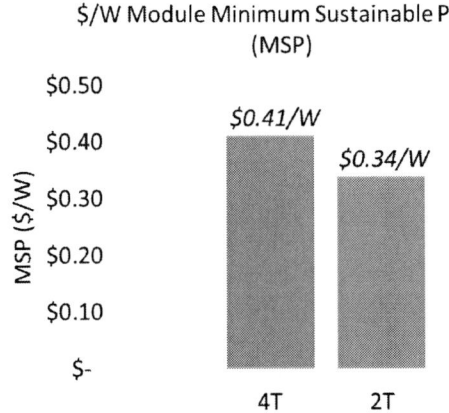

Figure 1. Module minimum sustainable price (MSP) in $/W for each of the two tandem architectures, 2T and 4T.

Impact of Energy Yield

In comparing the two tandem architectures, the 4T appears less cost effective on a module level since the $/W module cost is the highest of all four module architectures – $0.07/W or 21% higher than the 2T tandem. Despite only a 1.5% difference in absolute efficiency (5.8% relative), however, the 4T tandem has a significantly higher energy yield than the 2T for all three locations. The energy yield for the 4T tandem is 8.2%, 9.4%, and 11.0% higher than the 2T tandem energy yield in the dry, temperate, and humid locations, respectively. This shows that the 4T tandem performs at a higher performance ratio than the 2T tandem in each location, as the 4T tandem energy yield is higher relative to the 2T tandem than the STC efficiency. This is due primarily to the current matching constraint of the 2T tandem, causing the efficiency to suffer under a varying input

spectrum. As a result, using the STC efficiency alone does not allow a fair comparison of the two tandem architectures.

To explore the impact of energy yield, we calculate the necessary total initial investment, I, including module, balance-of-systems, and installation costs, for both tandem architectures needed to meet the DOE SunShot 2030 LCOE targets of 5 cents/kWh for residential systems and 3 cents/kWh for utility scale systems (Table 3). To do this, we use equation (1) and the operation and maintenance, discount rate, and degradation given in Table 2 to solve for the total initial investment in each of the three locations for both utility and residential scale systems.

We see that the 4T tandem systems can be significantly more expensive to install and reach the same target LCOE. For the residential scale systems, the 4T tandem installed system can cost 14%, 15%, or 18% more than the 2T tandem system for the dry, temperate, and humid locations, respectively, and 15%, 17%, and 19% for utility scale systems. This shows that, despite the added cost of the 4T tandem module, this tandem architecture may still be cost competitive relative to the 2T tandem due to the high energy yield across climates.

Furthermore, high efficiency reduces the (non-module) system cost of a PV array. Thus, the higher efficiency of the 4T tandem, while small, causes the cost to install a 4T tandem system of a given size to be less than the 2T tandem. This additionally allows for greater 4T tandem module costs.

V. CONCLUSION

We calculate the module cost for two CdTe-CIGS tandem device architectures. We see that the module cost for 4T tandems is significantly higher than the 2T tandem on both a $/m² and $/W level, despite the slightly higher efficiency of the 4T tandem. However, the energy yield calculations for these modules show us that the 4T tandem has a much better performance ratio than 2T tandems, thus using STC efficiency and $/W does not give a fair representation how the two tandems devices perform.

Furthermore, we see that, due to the high energy yield, the installation cost of the 4T tandem can be significantly higher

TABLE 2. CALCULATED INITIAL INVESTMENT REQUIRED TO MEET THE 2030 TARGET LCOE FOR EACH TANDEM ARCHITECTURE FOR RESIDENTIAL AND UTILITY SCALE SYSTEMS BASED ON ENERGY YIELD

	Required Total Installation Cost, I, to meet LCOE target [$/m²]					
Architecture	5 cents/kWh LCOE 39.5 m² Residential System			3 cents/kWh LCOE 0.6 km² Utility System		
	Dry	Temperate	Humid	Dry	Temperate	Humid
4T tandem	256.17	239.29	244.10	157.55	146.34	149.55
2T tandem	225.47	207.38	207.51	137.33	125.34	125.43

than for the 2T tandem while maintaining the same LCOE. This allows for the 4T tandem to be a more expensive module, while still potentially being cost effective. This emphasizes the importance of considering energy yield when comparing the cost competitiveness of 2T and 4T tandem devices. Additionally, the higher efficiency of the 4T tandem will result in reduced system installation costs for an array of a given size. This further gives the 4T tandem room to be a more expensive module and maintain a low LCOE.

REFERENCES

[1]   J. P. Mailoa, M. Lee, I. M. Peters, T. Buonassisi, A. Panchula, and D. Weiss, "Energy-Yield Prediction for II-VI-Based Thin-Film Tandem Solar Cells," *Energy Environ. Sci.*, vol. 9, pp. 2644–2653, 2016.

[2]   D. M. Powell, M. T. Winkler, A. Goodrich, and T. Buonassisi, "Modeling the cost and minimum sustainable price of crystalline silicon photovoltaic manufacturing in the United States," *IEEE J. Photovoltaics*, vol. 3, no. 2, pp. 662–668, Jun. 2013.

[3]   S. R. Kurtz, P. Faine, J. M. Olson, S. R. Kurtz, P. Faine, and J. M. Olson, "Modeling of two -junction , series-connected tandem solar cells using top-cell thickness as an adjustable parameter," *J. Appl. Phys.*, vol. 68, p. 1890, 1990.

[4]   H. Liu, Z. Ren, Z. Liu, A. G. Aberle, T. Buonassisi, and I. M. Peters, "The realistic energy yield potential of GaAs-on-Si tandem solar cells: a theoretical case study.," *Opt. Express*, vol. 23, no. 7, pp. A382-90, Apr. 2015.

[5]   I. M. Peters, S. Sofia, J. Mailoa, and T. Buonassisi, "Techno-economic analysis of tandem photovoltaic systems," *RSC Adv.*, vol. 6, no. 71, pp. 66911–66923, 2016.

[6]   M. A. Green, K. Emery, Y. Hishikawa, W. Warta, and E. D. Dunlop, "Solar cell efficiency tables (version 46)," *Prog. Photovolt Res. Appl.*, vol. 23, pp. 805–812.

[7]   M. Nakamura, N. Yoneyama, K. Horiguchi, Y. Iwata, K. Yamaguchi, H. Sugimoto, and T. Kato, "Recent R&D progress in solar frontier's small-sized Cu(InGa)(SeS)2 solar cells," *2014 IEEE 40th Photovolt. Spec. Conf. PVSC 2014*, pp. 107–110, 2014.

[8]   R. Fu, D. Chung, T. Lowder, D. Feldman, K. Ardani, R. Fu, D. Chung, T. Lowder, D. Feldman, and K. Ardani, "U . S . Solar Photovoltaic System Cost Benchmark : Q1 2016," 2016.

[9]   A. Goodrich, T. James, and M. Woodhouse, "Residential, Commercial, and Utility-Scale Photovoltaic (PV) System Prices in the United States : Current Drivers and Cost-Reduction Opportunities," 2012.

[10]   S. B. Darling, F. You, T. Veselka, and A. Velosa, "Assumptions and the levelized cost of energy for photovoltaics," *Energy Environ. Sci.*, vol. 4, no. 9, p. 3133, 2011.

[11]   D. Feldman, T. Lowder, and P. Schwabe, "PV Project Finance in the United States , 2016," 2016.

978-1-5090-5606-4/17 $31.00 © 2017 IEEE

Cause of Current-Collection Failure Observed in I_{sc}-Reduction Phase of PV Cells and Modules Exposed to Acetic Acid

Tadanori Tanahashi, Norihiko Sakamoto, Hajime Shibata, and Atsushi Masuda

National Institute of Advanced Industrial Science and Technology (AIST), Japan

Abstract — The mechanism of current-collection failure resulting from corrosion of front electrodes is investigated by AC impedance measurement of PV cells and modules degraded by acetic acid. In addition to our previous findings which first incidence of power-loss (with fill factor reduction) is caused by the gap-formation underneath front electrodes, we suggest in this study that the following power-loss with reduction of short-circuit current (I_{sc}) is due to the formation of rectifying contact between silver bulk in front electrodes and emitter in silicon wafer. These suggestions assist further understanding of degradation observed in PV modules deployed in fields.

I. INTRODUCTION

A major degradation-mode in photovoltaic (PV) modules deployed in hot-humid climate has been elucidated as the corrosion of front electrodes, which induces current-collection failure in the pathway from surface of crystalline silicon (Si) PV cell to front grid-lines. And, this corrosion is facilitated by acetic acid (HAc) which is liberated by hydrolysis of ethylene vinyl acetate (EVA), when high-dose of moisture is penetrated into PV modules from ambient environment. Indeed, the high content of HAc is monitored in PV modules exposed in fields for long-term [1], [2], and that is also confirmed in PV modules during damp-heat (DH) stress test [1]–[5]. Furthermore, we have confirmed that, even in the outdoor-exposed PV module without discoloration, the large power-loss with significant decreases in fill factor (FF) and short-circuit current (I_{sc}) is induced. These findings lead a speculation that I_{sc}-reduction contributes to the power-loss, irrespective of irradiance-reduction by discoloration of front glass and/or encapsulant.

We have proposed the two-phase degradation mechanism involved in this corrosion process, which comprises of power-losses induced by FF-reduction (Phase I) and by I_{sc}-reduction (Phase II), by the electrical measurement of bare PV cells exposed to HAc vapor [6]. Although we have already known that the degradation in Phase I is due to the formation of gaps underneath front electrodes, the precise mechanism on degradation in Phase II has not been understood. In this study, we analyzed the characteristics of AC impedance of PV cells and PV modules in Phase II, by DC bias-voltage dependency on them. And, it is suggested that power-loss in Phase II is induced by denaturation or modification of contact (with shift from ohmic contact to rectifying contact) between silver (Ag) bulk and emitter of Si wafer.

II. EXPERIMENTAL PROCEDURES

A. PV cells and PV mini-modules

Commercially available crystalline Si PV cells (156 x 156 mm) were individually tabbed with copper-solder ribbon on each busbar, and these ribbons on each plane of PV cells were connected by broad ribbons to measure current-voltage (I-V) and other electrical characteristics. Initial I-V parameters [maximum power (P_{max}), I_{sc}, open-circuit voltage (V_{oc}), FF, and efficiency (η)] of them were 3.84 ± 0.05 W, 8.44 ± 0.05 A, 0.62 ± 0.00 V, 73.0 ± 0.3%, and 15.8 ± 0.01%, respectively. Using the same PV cell model, PV modules were fabricated by lamination of one PV cell with semi-tempered glass, EVA, and backsheet (polyvinyl fluoride / polyethylene terephthalate / polyvinyl fluoride).

B. Exposure of bare PV cell to HAc vapor and DH test for PV mini-modules

Bare PV cells with ribbons were hung by the clipping of broad ribbon from a wire frame within a glass chamber filled with HAc vapor under high temperature conditions with high humidity [6]. DH test of PV mini-modules was carried out at 85°C / 85% relative humidity (rh) over 9,000 h.

C. Characterization of PV cells and PV mini-modules

I-V characteristics of PV cells and PV mini-modules were periodically assessed by a solar simulator and an EL image analyzer. The characteristics of AC impedance were evaluated by an LCR meter with a superimposing function of variable DC bias-voltage.

III. RESULTS AND DISCUSSION

As reported in our previous proceedings [6], the degradation of PV cells with HAc-vapor is divided into two phases within its process, which comprise of (1) FF-reduction phase (Phase I) and (2) I_{sc}-reduction phase (Phase II). Although the degradation in Phase I is due to gap-formation underneath front electrodes, we have not identified the cause of power-loss in Phase II. Then, we analyzed the effects of DC bias-voltage on all parameters involved in the modeled AC equivalent circuits (Fig. 1), to clarify the origin of current-collection failure observed in this Phase II, because the *p-n* junction of PV cells is hardly damaged even in this phase [7].

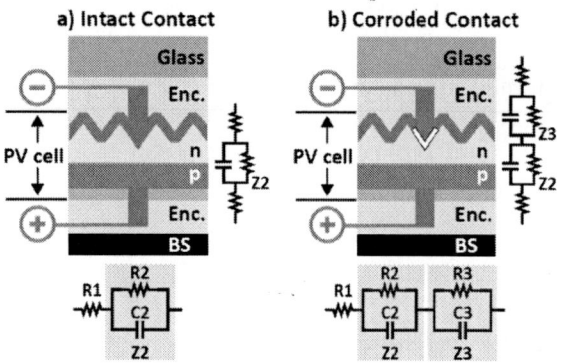

Fig. 1. Formation of gap underneath finger electrode by corrosion, and respective AC equivalent circuits.

The evolutions of *I-V* and AC-impedance parameters in a PV cell exposed to HAc vapor at 85˚C / 80% rh and a PV module under DH stress conditions (85˚C / 85% rh) are summarized in Fig. 2. Since similar transition of the respective parameters were observed in multiple specimens tested, it is revealed in both PV cell/module under respective tests that the gradual power-loss with I_{sc}-reduction is continuously induced even after the extent of FF-reduction has been saturated at about half of its initial value. And, the drastic alterations of AC impedance parameters were developed only in Phase I, especially those were clearly detected when PV cells were exposed to HAc vapor. As shown in Fig. 3, the active area in PV module during DH stress test was anisotropically distributed, although the isotropic degradation in PV cell in HAc vapor was induced. Furthermore, we note that the bright EL spots remaining only on front grid-lines in Phase I of PV cell exposed to HAc vapor had completely disappeared in Phase II, although a part of them were also anisotropically

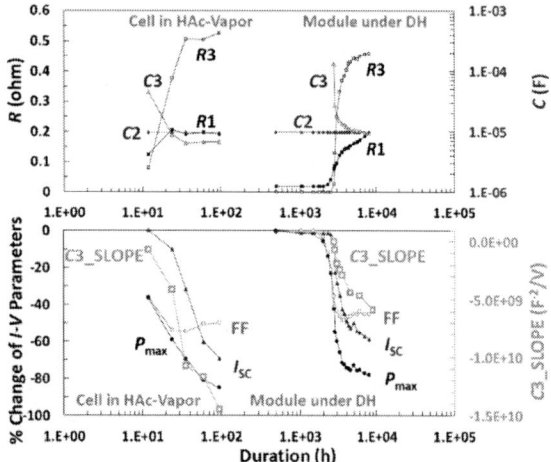

Fig. 2. Evolution of parameters involved in *I-V* and AC impedance.

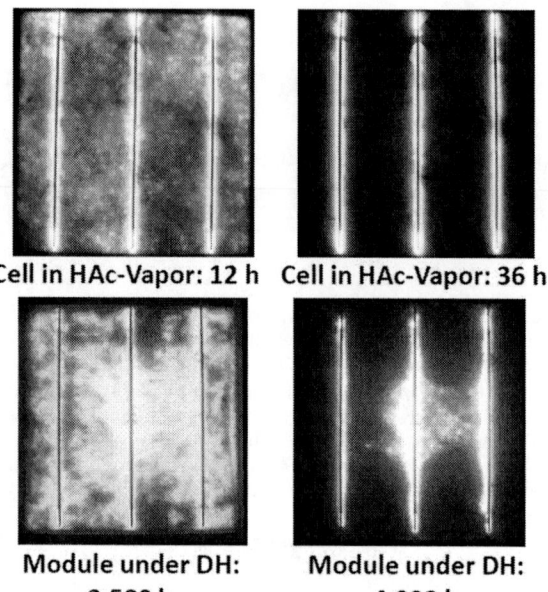

Fig. 3. EL images of PV cell during HAc-vapor exposure and PV module under DH stress test.

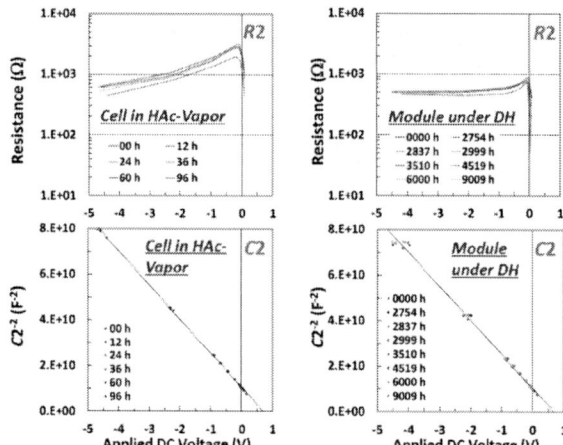

Fig. 4. DC bias-voltage dependency of AC impedance parameters ($R2$ and $C2$) in PV cell during HAc-vapor exposure and PV module under DH stress test.

detected in Phase II of PV module under DH stress test. Therefore, the slow responses of these signals would be attributed in this mosaic distribution of active area.

The intensities of $R2$ and $C2$ completely depended on the applied DC bias-voltage, according to the characteristics of *p-n* diode, i.e., the extent of $R2$ was logarithmically decreased in the low voltage range of forward DC bias, and that of $C2^{-2}$ was linearly decreased in the reverse bias-voltage (Fig. 4). These profiles in PV cell exposed to HAc vapor were not affected by the duration of exposure, as well as in the case of PV module under DH stress conditions. In addition, the extent of DC bias-voltage had no impact on the profiles of $R1$ in PV cell and PV module (data not shown), although the respective intensities were elevated by duration of these tests as shown in Fig. 2. These results indicate that these elements in AC impedance are not involved in the gradual power-loss observed in Phase II.

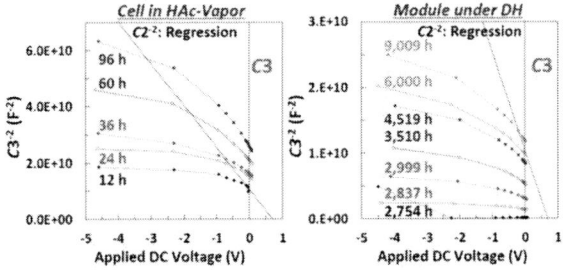

Fig. 5. DC bias-voltage dependency of AC impedance parameter ($C3$) in PV cell during HAc-vapor exposure and PV module under DH stress test.

In contrast, DC bias-voltage dependency of $C3$ in PV cell, which was indicated as $C3^{-2}$ in the left panel of Fig. 5, was drastically changed according to the duration of exposure to HAc vapor, i.e., although DC bias-voltage has little effect on its profile in Phase I (12 h exposure indicated in the left panel in Fig. 5), the dependency on DC bias-voltage was increased according to the duration of HAc-vapor exposure (left panel of Fig. 5). Similar shift of DC bias-dependency was also observed in PV module under DH stress test (right panel of Fig. 5). To quantitatively evaluate the effect of DC bias-voltage on the extent of $C3$, the slopes of these straight lines (-1 to -0.20 V range) were estimated by the linear curve fitting. As shown in the lower panel of Fig. 2, reduction of the estimated slopes ($C3\_SLOPE$) during respective tests was synchronized with each reduction of I_{sc}, but not that of FF. Although the extent of $R3$ slightly depended on DC bias-voltage during both tests, the evolution of DC bias-voltage dependency originating from the duration of both tests was not confirmed, excepting for the elevation of their intensities (data not shown).

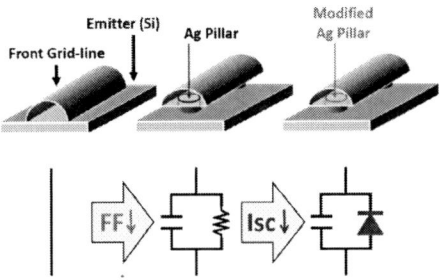

Fig. 6. Putative degradation mechanism of current-collection failure in front electrodes by corrosion.

For these findings, we suggest that the denaturation or modification of direct contact (formation of rectifying contact) between Ag bulk in front electrodes and emitter of Si induces the power-loss in Phase II (I_{sc}-reduction involved degradation), i.e., the current-collection failure in this phase is induced by

the development of a semiconductor-like (diode-like) structure in this interface, under the conditions that this contact (Ag pillar) was directly exposed to HAc by the dissolution of glass layer underneath Ag bulk of front electrodes, as illustrated in Fig. 6.

IV. CONCLUSION

We suggest in this study that the late phase of power-loss (with I_{sc}-reduction), which is induced by corrosion in front electrodes on PV cell encapsulated in PV module under hygrothermal stress conditions, is caused by the formation of rectifying contact between Ag bulk and Si emitter in front electrodes.

ACKNOWLEDGEMENT

This work was supported by the NEDO (New Energy and Industrial Technology Development Organization, Japan) under Contract #15100576-0.

REFERENCES

[1] T. Shioda, "Acetic acid production rate in EVA encapsulant and its influence on performance of PV modules," *presented at 2nd Atlas/NIST Workshop on Photovoltaic Materials Durability*, 2013.

[2] A. Masuda, S. Suzuki, Y. Hara, S. Sakamoto, and T. Doi, "Possible measure of reliability for crystalline-Si photovoltaic modules," in *Proc. 29th Eur. Photovolt. Sol. Energy Conf. Exhib.*, 2014, pp. 2566–2569.

[3] C. Peike, T. Kaltenbach, K.-A. Weiß, and M. Koehl, "Non-destructive degradation analysis of encapsulants in PV modules by Raman spectroscopy," *Sol. Energy Mater. Sol. Cells*, vol. 95, no. 7, pp. 1686–1693, Jul. 2011.

[4] U. Weber, R. Eiden, C. Strubel, T. Soegding, M. Heiss, P. Zachmann, K. Nattermann, H. Engelmann, A. Dethlefsen, and N. Lenck, "Acetic acid production, migration and corrosion effects in ethylene-vinyl-acetate- (EVA-) based PV modules," in *Proc. 27th Eur. Photovolt. Sol. Energy Conf. Exhib.*, 2012, pp. 2992–2995.

[5] A. Masuda, N. Uchiyama, and Y. Hara, "Degradation by acetic acid for crystalline Si photovoltaic modules," *Jpn. J. Appl. Phys.*, vol. 54, no. 4S, p. 04DR04, Apr. 2015.

[6] T. Tanahashi, N. Sakamoto, H. Shibata, and A. Masuda, "Electrical detection of gap formation underneath finger electrodes on c-Si PV cells exposed to acetic acid vapor under hygrothermal conditions," in *Proc. 43rd IEEE Photovoltaic Spec. Conf.*, 2016, pp. 1075–1079.

[7] A. Masuda, C. Yamamoto, T. Tanahashi, H. Sai, and T. Matsui, "Direct evidence for pn junction without degradation in crystalline Si photovoltaic modules under hygrothermal stresses," in *Proc. 43rd IEEE Photovoltaic Spec. Conf.*, 2016, pp. 0904–0906.

Comparison of PV Module Performance Before and After 11, 20, and 25.5 Years of Field Exposure

Jacob Rada, Charles Chamberlin, Peter Lehman, and Arne Jacobson

Humboldt State University (HSU) and the Schatz Energy Research Center (SERC)
Arcata, California, 95521, USA

Abstract — In 1990, 192 ARCO M75 photovoltaic (PV) modules were installed at the HSU Telonicher Marine Lab in Trinidad, California, 150 meters inland from the Pacific Ocean. Current-voltage (IV) tests were performed on each module prior to the array's construction in 1990 [1] and then again in 2001 [2], 2010 [3], and most recently in 2016 after the array was decommissioned. After 25.5 years, 188 of the original 192 modules remained operational. Over their lifetime, the modules' maximum power at the normal operating cell temperature (NOCT) declined by an average 21.6% with a degradation rate of 0.85% per year. The average degradation rate grew from 0.4%/year in the first 11 years to 0.81%/year after 20 years to 0.85%/year after the total 25.5 years.

I. INTRODUCTION

A. Background

The Schatz Solar Hydrogen Project, directed by the Schatz Energy Research Center, was installed in 1990 at the HSU Telonicher Marine Laboratory in the coastal California town of Trinidad. The project's objective was to demonstrate the use of hydrogen to store energy from a renewable energy power system. The system powered the Marine Laboratory aquaria air-compressor either directly from the 9.2 kW-rated solar PV array (with an azimuth of due south and a tilt of 28°) or indirectly using a hydrogen-fueled proton exchange membrane (PEM) fuel cell. The hydrogen for the fuel cell was generated by a Teledyne Energy ALTUS 20 electrolyzer powered by the PV array.

The solar array (see Fig. 1) offered the unique opportunity to measure the performance degradation of the individual modules over 25.5 years of field exposure. Data from this analysis can be used to track the causes of module failures, degradation in power output, and help characterize a timeline of expected performance of such mono-crystalline silicon PV modules. When new, these ARCO M75 modules were rated by the manufacturer at 48 W at standard testing conditions (STC or 1000 W/m^2 at a module temperature of 25°C), which is equivalent to 46.4 W at NOCT (defined here as 1000 W/m^2 at 47°C). Each module is glazed with tempered glass and ethylene vinyl acetate (EVA) encapsulant and contains 33 cells in series with two bypass diodes per module. The lasting effect of the bypass diodes is also analyzed later in this study.

Zoellick [1] set the precedent for the three successive testing cycles. He used a capacitive-based curve tracer connected to a computer interactive data acquisition system to test and record IV curves for each module using methods approved by the American Society of Testing and Materials (ASTM).

Fig. 1. A view of the array during decommissioning after two subarrays had already been removed (Photo Credit: Mark Rocheleau)

B. Previous Findings

Zoellick [1] found in his 1990 testing that the average maximum power (P$_{max}$) was 39.88 W and the average short circuit current (I$_{sc}$) was 3.29 A, about 10% lower than the manufacturer advertised specifications. Thereafter, Zoellick's results were used as the base case data for the following rounds of testing in 2001, 2010, and 2016. The 2001 cycle of tests [2] found that the average P$_{max}$ and I$_{SC}$ had dropped to 38.13 W and 3.15 A, respectively, representing a 4.40% decline over 11 years of operation.

The 2010 tests, after 20 years of field exposure, reported larger degradation for both parameters. The P$_{max}$ experienced a 12.3% decrease from the 2001 testing to 33.43 W, bringing the total loss to 16.1% from the 1990 tests, and the I$_{sc}$ dropped 6.04%, down to 2.96 A, in the second decade for a total lifetime decline of 10.2%. For the maximum power, the average degradation rate over the 20-year interval came to 0.81%/year, but the degradation rate over the second decade (2001-2010) was 1.4%/year, more than three times the degradation rate over the first decade of 0.4%/year. This suggests that the performance of the modules degraded more rapidly as they aged, instead of following the expected pattern of linear degradation.

978-1-5090-5606-4/17 $31.00 © 2017 IEEE

Other IV curve parameters also changed over the first two decades of operation (Fig. 2). The series resistance (R_s), parallel resistance (R_p), and the "ekt" variable (which controls the degree of curvature near the maximum power point) all experienced dramatic variations as evidenced by changes in the slopes and curves of the IV curves. After only 11 years, the average R_p had dropped by 33.75%, the average R_s had increased by 10.66%, and the average ekt rose by 26.38%. The inverse of R_p is the slope of the IV curve as the voltage approaches zero, and the negative inverse of R_s is the slope of the IV curve as the current approaches zero. In contrast, the average open circuit voltage (V_{oc}) hardly changed, falling by only 1.0%, or 0.17 V, over 20 years.

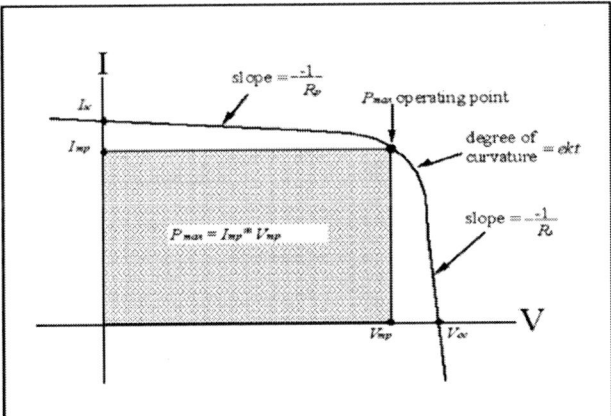

Fig. 2. Five-parameter IV curve showing the location of the maximum power point on the curve [2]

II. PROCEDURE

A. Testing

The conditions in which Zoellick [1] performed his tests were replicated as closely as possible in all subsequent rounds of testing in 2001, 2010, and 2016. Each module was tested outside at a module temperatures centered around 47°C with tests ranging from 25-65°C, a solar insolation of at least 800 W/m² (but closer to 1000 W/m² when achievable), and an air-mass (AM) less than or equal to 1.5. Each module was cleaned prior to testing, eliminating any dust or dirt that would prevent the module from generating at its optimum level for the analysis.

The 1990 and 2016 rounds of testing used a portable and adjustable frame to test each module, but in the 2001 and 2010 testing the array still in operation so each module was tested within the plane of the array. An Eppley PSP pyranometer and a surface mount type-T thermocouple were used to monitor and record the solar insolation in the plane of the module and the module temperature, respectively.

While the different rounds of testing used slightly different methods and technologies to produce IV curves, each one used a capacitive load and was completed in 10 seconds or less.

The 2016 testing used a Mini-KLA PV IV Curve Tracer [4] and its associated MiniLes (R) program to measure and record the pertinent current and voltage data (Fig. 3). This instrument reported the observed P_{max}, V_{oc}, I_{sc}, maximum power voltage (V_{mp}), maximum power current (I_{mp}), the IV curve fill factor (FF), and a plot of the IV curve on its graphical interface. The insolation was measured using an Eppley PSP pyranometer and the module temperature was measured using a type-T surface mount, quick response thermocouple positioned in the center of the back of the module.

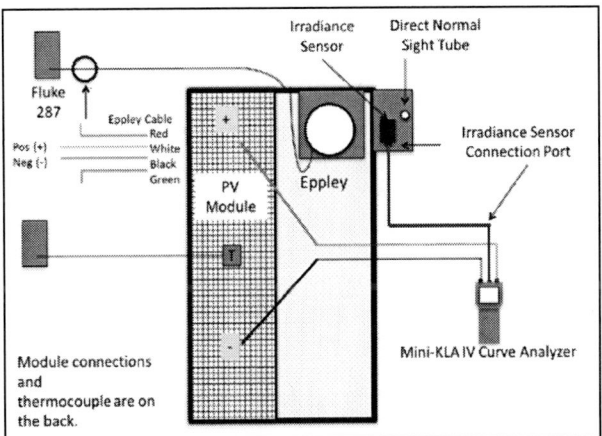

Fig. 3. Schematic showing the 2016 testing apparatus and third party measurement instruments (by Jake Rada)

B. Data collection and standardization

Zoellick [1] adjusted all observations to NOCT conditions based on the observed sensitivity of the module V_{oc} to insolation and module temperature. His and all subsequent corrections adjusted V_{oc} with respect to the measured module temperature (-0.0603 V/°C) and insolation (0.0009296 V*m²/W). Based on a Schottky diode model, these analyses fit the module IV curve to a parameter model, where:

$$I = I_L - \left[\frac{I_L - \frac{V_{oc}}{R_p}}{\exp\left(ektV_{oc}\right) - 1} \right]\left\{ \exp\left[ekt\left(V + R_s I\right)\right] - 1\right\} - \frac{\left(V + R_s I\right)}{R_p} \quad (1)$$

where:

$ekt = q/(nkT)$ [V⁻¹]
I = module current; initial guess [A]
V = module voltage [V]
I_L = light induced module current [A]
V_{oc} = open circuit module voltage [V]
R_s = module series resistance [Ω]
R_p = module parallel resistance [Ω]
q = electronic charge [coulomb]
n = ideality factor per cell [unitless]
k = Boltzmann's constant [Joule/K]
T = temperature [K]

In 2016 testing, the Eppley PSP pyranometer and thermocouple readings were recorded by hand. The MiniLes software takes the data from the Mini-KLA and converts it into a text file. Then software written in Sci-Lab [5], modified from that previously used in the 2010 testing analysis filtered the readings, standardized the data to 1000 W/m² and 47°C using the correction factors from Zoellick and used nonlinear regression to estimate the parameters of Equation 1.

III. RESULTS AND DISCUSSION

After 25.5 years, every single module from this array is discolored and shows signs of delamination. Signs of hot spots with a range of severity have become common in the modules. Fig. 4 shows a comparison between one of the most physically degraded ARCO M75 modules and one of the younger replacement Siemens SM50-H modules. The ARCO module has been exposed to the environment for 26 years old in this image, and the Siemens module has been in the field for 19 years; note the obvious anti-aging improvements in the newer model. This study goes on to investigate whether the increased prevalence of physical degradation led to larger power losses in these modules over the project lifetime.

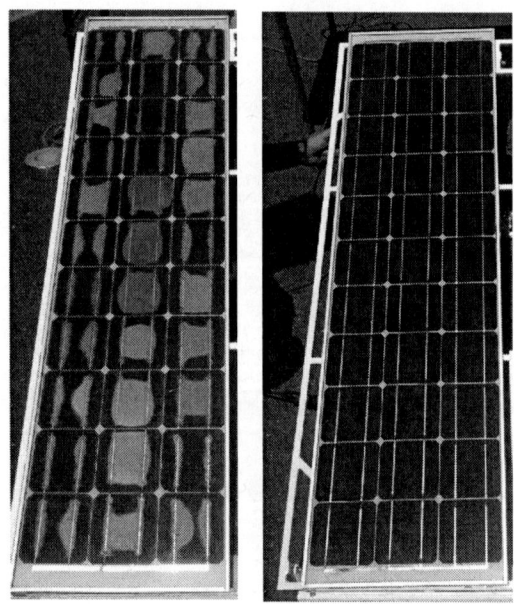

Fig. 4. A physical comparison between a 26-year old ARCO M75 module and a 19-year old Siemens SM50-H (Photos by Jake Rada)

Fig. 5 compares the distributions of the P_{max} of the modules from each testing cycle (i.e., 1990, 2001, 2010, and 2016) in a normal probability plot. The initial 1990 curve remains close to linear throughout the curve, and as the modules aged the standard deviation of the P_{max} among the modules increased. This is exemplified by the steeper slopes and wider P_{max} ranges in the later years that include modules that have lost

significant power. In the 2016 testing cycle, two modules had their P_{mas} drop below 10 W.

Fig. 6 shows the average P_{max} of all the modules for each testing cycle and notates the modules' age during testing. There is clearly less linear degradation from 1990-2001 followed by a steeper degradation from 2001 to the end of the project. The degradation rates (i.e. rates of power loss) for the testing windows include 0.4%/year for the first 11 years (1990-2001), 1.4%/year for the next nine years (2001-2010), and 1.3%/year for the last five and a half years of the project (2010-2016). The average lifetime degradation rate of the 25.5-year project came to 0.85%/year. Fig. 3 has a linear trend line whose equation says that roughly 0.35 W were lost each year for the modules, which, based on the original 1990 average of 39.88 W, is just over 0.85%/year of power loss.

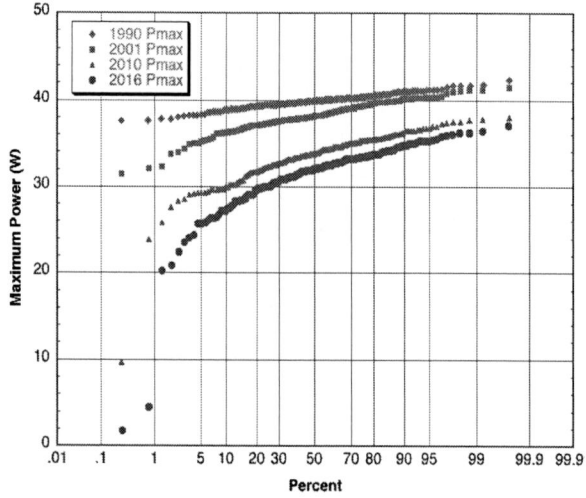

Fig. 5. P_{max} distribution curve for all modules in four test cycles

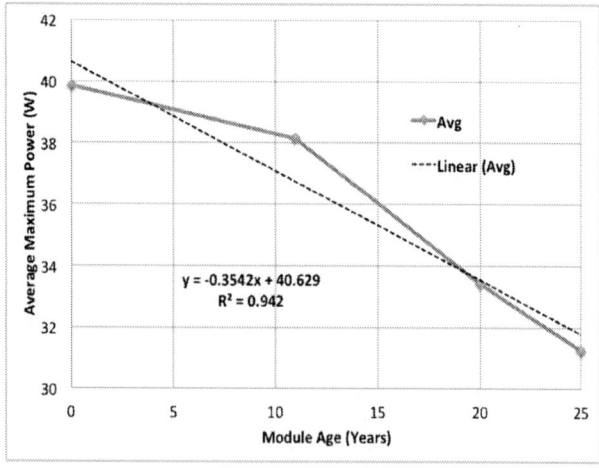

Fig. 6. Average P_{max} based on the age of modules during testing

Each module incorporated two bypass diodes to avoid the creation of localized hot spots, as they direct the current around cells that are either damaged or shaded. The associated effect of the action of the diodes is to reduce the module P_{max}. Fig. 7 shows the IV curves from 1990, 2001, 2010, and 2016 for a single module that lost a small amount of its P_{max} in the first decade of operation but then saw a significant loss in the second decade of field exposure. This second "knee" in the IV curve is a result of the bypass diode directing the current around 22 of the 33 cells in the module. These second knees were uncommon in the 2001 testing, but in the 2016 testing all but 13 of the modules had significant second knees. The fifth curve in this figure shows the effect of removing these bypass diodes from a module with a significant second knee. Results show that this action increases the P_{max}, but the potential for exacerbating hot spots may increase.

Bypass diodes were removed from several modules in 2016 and then retested. Fig. 7 shows the most extreme change caused in the IV curve through this removal of bypass diodes, but this dramatic change to the IV curve only resulted in an increase in the module's P_{max} by 1.3 W, or less than a 5% increase in power generation. This analysis showed that the bypass diodes provided significant over-current protection for the sacrifice of less than 5% of their power generating abilities on average. Modules that did not experience second knees in their IV curves were hardly affected when tested with or without their bypass diodes as the diodes had not be routinely or permanently activated to redirect current around problematic cells, which leads to less cells generating power.

IV. CONCLUSION

After 25.5 years, almost 16 years after the modules' 10-year warranties expired, the modules averaged a power loss of 21.6% from their initially tested power outputs for an average degradation rate of 0.85%/year. The average standard deviation of the power output also increased from 0.92 W to 4.10 W during this time. This calculated degradation rate is only slight higher than the expected range of 0.5-0.8%/year [6]. The 2016 testing supports past analyses conclusions identifying the drop in I_{sc} as the primary source of power loss, which can be attributed to the increased presence of EVA browning, physical delamination, and localized hot spots with the modules. Future studies may have the opportunity to assess to what degree the degradation of the crystalline cells and the EVA encapsulant have on this measured drop in I_{sc} that contributed most heavily to the power loss in these modules.

In 2016 almost half (48%) of the remaining original 188 ARCO M75 modules still generated 80% or more of their initial capabilities tested in 1990, and 90% of them still generate over 70% of their initial power measurements. As one of the oldest, if not the oldest, and best monitored PV arrays of its kind, the results from this 26-year long project can be greatly beneficial to the solar energy industry, as bankability and longevity dictate the successful marketing of this energy generation technology.

REFERENCES

[1] J.I. Zoellick, "Testing and matching photovoltaic modules to maximize solar electric array performance", Senior project presented to the Department of Environmental Resources Engineering Humboldt State University, 1990.
[2] A.M. Reis, N.T. Coleman, M.W. Marshall, P.A. Lehman, and C.E. Chamberlin, "Comparison of PV module performance before and after 11-years of field exposure," in *29th IEEE Photovoltaics Specialists Conference*, 2002.
[3] P. Lehman, C.E. Chamberlin, M.W. Marshall, M. Rocheleau, "Comparison of PV module performance before and after 11 and 20 years of field exposure," in *37th IEEE Photovoltaics Specialists Conference*, 2011.
[4] Ingenieurburo, Menncke, & Tegtmeyer GmBH, "Mini-KLA PV I-V curve analyser", Hameln, 2011.
[5] Scilab Enterprises, "Scilab: free and open source for numerical computation [software], 2012.
[6] D. Jordan and S. Kurtz. "Photovoltaic degradation rates – an analytical review", NREL, in *Progress in Photovoltaics: Research and Applications*, 2011.

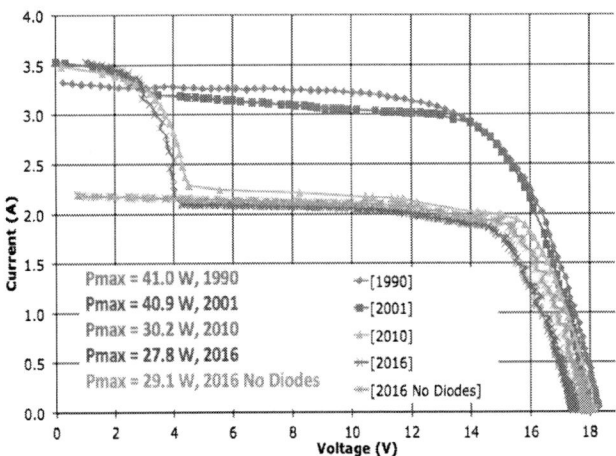

Fig. 7.   IV curve for a module that shows the effect of diodes

Marrying Quality Assurance with Design Engineering – A Winning Partnership! But, a Cultural Divide?

Sarah Kurtz[1], Govind Ramu[2], Robert Cornell[3], Sumanth Lokanath[3], Edward Hsi[4], Tony Sample[5], Masaaki Yamamichi[6], George Kelly[7], Ted Spooner[8], Jonathan Previtali[9], John Wohlgemuth[10]

[1]National Renewable Energy Laboratory, Golden, CO, USA 80401, [2]SunPower, San Jose, CA, USA; [3]First Solar; [4]Swiss RE, Singapore; [5]European Commission, JRC, Ispra, Italy; [6]National Institute of Advanced Industrial Science and Technology (AIST), Tsukuba, Japan; [7]Sunset Technology, Mount Airy, MD, USA; [8]UNSW, Sydney, Australia; [9]Wells Fargo, San Francisco, CA, USA; [10]PowerMark Corporation, USA

Abstract — Investors would like to be confident that a PV plant will work as promised before investing large sums of money. Data showing that PV modules can withstand thousands of hours of accelerated testing are a comfort, but are not sufficient to characterize the economic useful life (EUL) of the whole PV system. An inability to be confident of every aspect of the system can motivate less favorable contract terms. In the end, a winning partnership is obtained when a carefully engineered, durable design is married with a robust quality management system and the marriage is extended across all parts of the value chain. However, this marriage is challenging because of the conventional cultural divide between design engineers and quality management specialists. International standards are being developed to facilitate this partnership and include technology-specific requirements in the quality management systems used for component manufacture and system design, construction and operation.

I. INTRODUCTION

Excellent quality assurance is important to PV module customers because it can 1) improve upfront performance, 2) decrease degradation rates, 3) decrease operating costs, and 4) decrease safety risks. Achieving the desired final result starts with the design, but it doesn't end there.

Success requires that all elements work as promised from day one to the end of the warranty. However, once the design phase is completed, there can be substantial pressure to implement the design with ultra-low-cost materials, components, and mass production techniques, tempting some companies to neglect partnering the design engineers with quality management specialists. Although the design and implementation phases require different skills, both sets of skills are essential to optimize the final design and manufacturing process to provide a functional product at the lowest cost.

Management of quality assurance becomes even more challenging at the PV system level; where both the design and the quality assurance require that elements be selected to be of high quality, to function well together, and to be easily serviceable in the field. Additionally, the installation process must be optimized to give the best outcome at the lowest cost.

This paper discusses the motivation for increased emphasis on quality assurance and describes three new standards that detail requirements for a higher level of quality assurance for 1) PV module manufacturing, 2) power electronics manufac-

turing, and 3) control of the system installation and operation. The paper concludes with a discussion of the challenges and benefits of marrying the quality assurance program with further optimization of the PV system design toward the end goal of maximizing the return on investment (nominally, the electricity production relative to lifecycle costs).

II. WHY QUALITY ASSURANCE SHOULD BE PRIORITIZED

There is substantial evidence that many failures observed in PV systems today result from failures in quality assurance during manufacturing of the components, during installation, or during operation of the system.

Figures 1-3 show examples of failures that may have been caused by poor quality assurance. Figure 1 shows two burn marks (inset) imaged from the backside of the module and corresponding delamination on the front (larger image). Only one cell of one module was affected in this residential system, implying that the problem may have been the result of a manufacturing defect that was not identified during manufacturing. Alternatively, the module may have been damaged during shipping or installation. The observation of burn marks for a small fraction of modules has been reported many times. For example, Degraaff, et al, reported 0.3%, 1.5% and 2.9% failure rates for modules made by three manufacturers, respectively [1]. Although the root causes of these failures were not reported, the observation that only a small fraction of modules were affected suggests that the failures could have been avoided by better quality assurance during manufacturing.

Figure 2 shows a catastrophic failure apparently caused by a ground fault. Although this could have been a design problem, the problem was not observed throughout the array, implying that for the affected modules the cells may have been laminated too close to the frame and were not identified at the time of manufacture. The higher DC voltages being used in today's PV systems can lead to need for tighter quality control to avoid this sort of problem.

978-1-5090-5606-4/17 $31.00 © 2017 IEEE

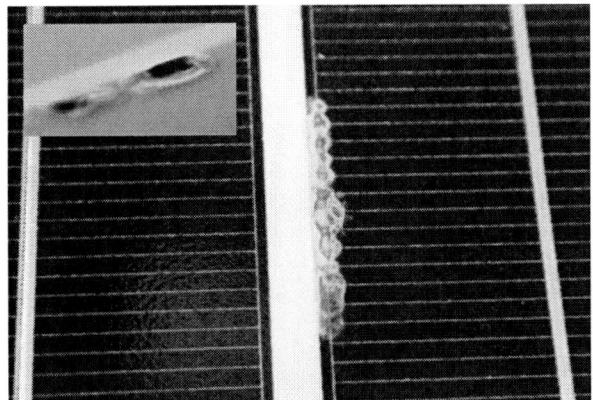

Fig. 1. Quality problem example. The photo shows delamination observed from the front side of the module. The inset shows burn marks on the backside of the module (opposite the delamination).

Fig. 2. Example of problem probably caused apparently by DC arcing between the frame and cells because the cells were too close to the frame to withstand the DC voltage between them.

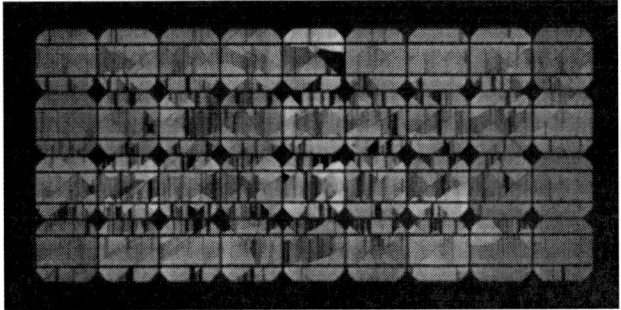

Fig. 3. Third example of apparent problem with quality assurance. The electroluminescent image shows how some parts of cracked cells have become disconnected from the circuit, reducing the power output of the module.

Figure 3 shows an electroluminescent image of a module that was, apparently, damaged (cracking of cells) during manufacture, shipping, or installation, and, after about a year in the field, showed decreased performance, presumably as a result of the pieces of some cracked cells becoming disconnected from the rest of the active circuit [2]. As silicon wafers, module frames, and glass frontsheets have become thinner, reports of cracked cells have increased and this type of power loss (as shown in Fig. 3) has become more common in the field. The extent to which the cracked cells lead to power loss is a subject of research [3]. This issue could be classified as a design problem (e.g. a more robust design might have avoided damage) or a quality-assurance problem (e.g. by controlling the handling process as part of the control plan so that the cells are not broken during manufacture, shipping, or installation). If the cracking is caused by mechanical stress during operation, then the problem is more clearly a design issue. This example underscores the importance of the marriage between the design optimization and quality assurance across all parts of the value chain, as discussed below.

III. KEY ELEMENTS OF NEW SPECIFICATIONS

A. Three specifications

Table I describes the three Technical Specifications that have been or are being developed by the International Electrotechnical Commission (IEC) Technical committee 82 to gain more confidence in quality assurance. Key elements of these are described in the following three sections.

TABLE 1
TARGET PRODUCTS AND ANTICIPATED PUBLICATION DATES
FOR TECHNICAL SPECIFICATIONS BEING DEVELOPED BY IEC

Target	Specification	Publication schedule
PV modules	IEC TS 62941	Released January 2016
PV systems	IEC TS 63049	Anticipated June 2017
Power electronics	IEC TS TBD	Anticipated in 2018

B. PV module quality assurance

The technical specification: IEC/TS 62941 *Terrestrial photovoltaic (PV) modules - Guideline for increased confidence in PV module design qualification and type approval* has been described previously [4,5] and was released in January 2016. This technical specification assumes that the module manufacturer has already satisfied Quality Management System requirements based on ISO 9001 and, hence, describes additional requirements to be included in the IEC/TS 62941 audit including [4,5]:
• Alignment of design lifetime with product warranty,
• Verification of incoming materials to maintain a consistent bill of materials,
• Traceability (so as to notify customers of defective product),
• Ongoing testing program to confirm consistency of design implementation during manufacturing,
• Control of solar simulator calibration for use in determining nameplate power rating,

• Use of statistical methods for monitoring and improving the process, and
• Continual improvement based on field experience.

IECRE is overseeing implementation of IEC/TS 62941 and started accepting applications from Certifying Bodies (for both PV plant inspections and Factory Audits) in 2016. A list of organizations accredited to certify for IEC/TS 62941 may be found on IECRE's website [6].

A set of training materials has been developed to improve the consistency of the interpretation of IEC/TS 62941 internationally. Having been trained with these materials and, later, having participated in peer reviews between the various accredited Factory Auditors, these Auditors can provide consistent implementation of IEC/TS 62941. The validity of IEC/TS 62941 certificates issued by auditors who have not participated on this process is unclear, since the training of these auditors is unknown.

As industry-leading manufacturers demonstrate compliance with IEC/TS 62941, it is anticipated that many more manufacturers will begin to follow, resulting in improvement in PV module reliability worldwide. Careful control of the manufacturing process will become increasingly critical as cost reduction squeezes design margins in coming years.

C. Quality of PV installation and operation

The technical specification: IEC/TS 63049 *Terrestrial photovoltaic (PV) systems – Guideline for effective quality assurance in PV system installation and operations and maintenance* is scheduled for publication in 2017. Similar to IEC/TS 62941, it provides for certification of the organization that is installing or operating the PV system rather than certifying each system separately. However, in contrast to IEC/TS 62941, it does not assume that the organization already has an ISO 9001 certification, so is written to include many ISO 9001 requirements. Key requirements for *installation* include:
• Training of workers,
• Using quality components (switches, overcurrent protection and combiner boxes) that meet relevant standards,
• Ongoing oversight of installation,
• Record keeping and maintaining traceability to enable both learning and correction of mistakes, and
• Tracking of system performance after completion to identify opportunities for continual improvement.

IEC/TS 63049 expects design and installation to be guided by IEC 62548 *Photovoltaic (PV) arrays – Design requirements* and IEC/TS 62738 *Design guidelines and recommendations for ground-mounted photovoltaic power plants*. It requires a final inspection program that is compliant with IEC 62446-1 *Grid connected PV systems – Minimum requirement for system documentation, commissioning tests and inspection* and uses IEC/TS 61724-2 *Photovoltaic system performance – Part 2 Capacity evaluation method* to document initial performance of the completed system.

Key requirements for management of system *operation* are similar, but emphasize continual improvement to reduce the cost of the maintenance and inspection relative to the improved power output while avoiding safety issues. IEC/TS 63049 uses IEC 62446-2 *Grid connected PV systems – Part 2: Maintenance of PV systems* as a guide for defining the maintenance program.

Some PV customers worry that untrained low-cost labor hired to install today's PV systems may be introducing defects. For example, common concerns include (but are not limited to):
• Rough handling of modules (bouncing during transportation, stepping on, dropping, twisting, scratching of back sheets, carrying modules on hard hats, etc.)
• Over/under tightening of assemblies,
• Under tightening of electrical connections, and
• Improper cable management.

Rough handling may crack cells within the module as discussed for the module shown in Fig. 3. Thus, the PV system may perform acceptably at commissioning, but later degrade in performance. Modules that have been dropped or have scratches in the backsheets should be discarded, but some workers may be hesitant to do so. Similarly, over tightening of clamps can lead to broken modules and under tightening of electrical connections can lead to ground-fault problems, Fig. 2. Cables that dangle in such a way that they swing in the breeze may abrade and cables that aren't protected by conduit may be chewed by rodents or otherwise be damaged. These exemplify the many problems that can occur and that are addressed by consistent training and oversight of workers, as described in IEC/TS 63049.

For rapidly evolving PV technology, a very critical element of quality assurance is continual improvement. IEC/TS 63049 requires not only that the installers check to confirm that the PV plant is constructed to meet the design, but that they track the performance of the plant during years of operation. As noted above, failures caused by some installation mistakes may show up after months or years of operation. This requirement may be difficult for installers who usually have no access to the plant after completion. However, while IEC/TS 63049 does not require access to the raw data, it requires confirmation that the plant is performing as designed and without failures over a specified time period. Customers should be reassured by their installer insisting that a mechanism be in place for this information to be communicated.

IEC/TS 63049 describes a different philosophy for quality assurance of the operations and maintenance compared with that for installation. While the goal of the installation process is to approach the 100% metric in assembling the plant correctly, maintenance procedures are designed to optimize the return on investment, which reflects both the electricity generated and the cost of the maintenance. Thus, the goal is to clean the system at a frequency that optimizes the output of the system relative to the cost of the cleaning rather than to strive for cleanliness at all times. Similarly, the frequency of inspection for nests, plant growth, water intrusion, etc., is optimized to balance the maintenance costs with the associated benefits.

The exception to the optimization strategy is when there is a safety risk or preventative maintenance requirement to meet warranty obligations. If delaying maintenance could lead to a fire, an injury (e.g. from a shock hazard), or voiding an equipment warranty, then maintenance is required, even if it increases cost.

D. Quality of inverters and other balance-of-system components

Almost universally, PV system failures are dominated by problems with the inverters. The technical specification: *Guidelines for effective quality assurance of power conversion equipment for photovoltaic systems* is being written to address this problem. It has been submitted as a New Work Item Proposal to the IEC and is expected to be published in 2018. It is largely focused on inverters and is anticipated to emphasize:
• Initial design to provide the desired performance and durability in the expected use environment and to be serviceable (especially important for central inverters)
• Control of the manufacturing process,
• Ongoing testing to ensure that the manufactured product matches the advertised product specifications, and
• Continual improvement of the design and manufacturing process to address problems identified in the field.

It does not explicitly assume that the organization already has a certification to ISO 9001, though that is a common situation. A key challenge for developing a consensus on this document is the specification of design margins. Statistics from field deployment of inverters suggest that failures are related both to inconsistent manufacturing and to poor design. While designs may be greatly improved by IEC 62093 Edition 2, a key element of quality assurance is selecting components that meet the design requirements including a margin to reflect the variability in the applications and construction. This document is still in the early draft stages.

E. Cost pressures on quality systems

Today's oversupplied market is putting painful cost pressures on the industry as a whole. This is manifested in many ways in the designs and in the quality systems. Thinner cells, glass, and frames and less expensive materials are obvious design targets for reducing cost in modules. Less expensive racking, omission of conduit, and less durable materials are example design targets for reducing cost of systems. The design engineer, who plays a primary role in reducing cost, must work with the quality expert to determine whether these design changes will introduce reliability issues. The quality manager has the priority of taking the time to do this assessment correctly, while the design engineer is working to launch the product or complete the PV system as quickly as possible,

demonstrating once again the cultural divide that must be dealt within the partnership.

A second example of the effects of cost pressure is the introduction of warranty exclusions. While the customer may see that the product carries a 25-y warranty, upon closer inspection, the customer may find that some common reasons for replacing the module may not be included or that, although the module may be replaced, the labor and shipping associated with the replacement may not be covered.

Improved quality assurance can help reduce the cost experienced over the life of a product even though the initial cost of the product may be higher, but this benefit can be difficult to quantify at the time of manufacture. The temptation to cut corners because of cost pressure could be reduced if added quality could attract a higher price. Of course, customers prefer higher quality and reliability, but how much more will they pay for it? Large investors and insurance companies develop models to assess anticipated return on investment, risk, necessary reserves, and other financial quantities. Hearing that a component or system has passed a test or an inspection may not give information that is useful. However, if the degradation rate of a system has been demonstrated to be lower than the industry standard, the investor may be willing to estimate a higher return on investment and be willing to negotiate more favorable terms. Similarly, if a warranty reserve can be reduced of because demonstrated low warranty return rates, capital can be freed, providing the value needed to justify the added investment in quality control. The design engineer and quality manager must work together to collect the necessary data to quantify the value of their collective work.

On the other hand, added attention to quality does not always pay for itself. Just as the performance and cost of a product are optimized by the design engineer, the quality assurance program must also be optimized to respond to the cost pressure by streamlining processes without decreasing customer satisfaction. Again, the partnership is key to success.

IV. PARTNERSHIP BETWEEN DESIGN ENGINEERING AND QUALITY ASSURANCE

Successful design engineering requires technical knowledge of the desired function of the product and how to test that the established design will be successful in meeting the warranty in the intended use environment. In contrast, successful quality management during manufacturing sometimes requires motivating workers to execute repetitive tasks consistently, often requiring an in-depth understanding of human psychology. A worker in the field may be taught to discard a module that is dropped, but may be embarrassed by having dropped a module and try to hide it. Successful quality management requires understanding both the technical requirements and having the skills to motivate workers. Experts in quality management may have different training than design engineers and prioritize attendance at different professional meetings than design engineers. Because of the substantial cultural divide, bringing these two professional groups together can be a challenge. The cultural divide is also driven by role-based motivations. While the design engineer is focused on value engineering and time to market, the quality engineer is focused on assuring quality. Herein lies the conflict and tradeoff. The relationship between the two and corporate motives drive outcomes; partnership is essential to success.

IEC and ISO standards writing reflects this cultural difference with IEC taking the mandate to write technical standards while ISO writes the quality management standards. The IEC documents discussed here (Table 1) are intended to form a bridge between these by providing the technical requirements for robust quality assurance.

Design qualification is the first step. For a PV module, IEC 61215 and IEC 61730 are applied to a small number of hand-picked modules to qualify the design from both functional and safety perspectives. If one module fails a test, IEC 61215 allows submission of a second pair of modules. If these two pass, then the test is given a "pass", implying that an IEC 61215 certification may be based on only 75% of the tested modules successfully passing the tests. To improve the consistency, historically, an ISO 9001 certification is obtained as part of the quality assurance program. However, any manufacturing process allows for some variation in the process window. The modules that passed IEC 61215 and IEC 61730 initially are unlikely to represent the full range of products that may result from small variations in the manufacturing process. A recent study by DNVGL [7] found that 6% of more than 300 modules did not pass the IEC 61215 test for thermal cycling (200 cycles). This 6% failure rate could easily explain common statistics of field failures (above, we noted 0.3%, 1.5% and 2.9%) [1]. A marriage between the design engineering and the quality management uses statistical methods for sampling and testing combined with an in-depth technical understanding of the product to facilitate continual improvement of the design and the manufacturing process window. IEC/TS 62941 is designed to facilitate this marriage by identifying important technical details that need to be considered when creating the quality assurance system.

This marriage must extend across the value chain from incoming material control to PV system completion. A change in the bill of materials may not appear to affect a product, but could cause a problem in the field years later. Field failures must be evaluated to identify the root cause and an appropriate improvement plan to be implemented. Putting all of the pieces together to not only connect design engineering with quality assurance but also to connect across the value chain to ensure that the system design fully takes into account variations in the bill of materials and other details is a real challenge. IECRE was created to tackle this bigger challenge [8]. The three documents described here are key elements of IECRE's certifi-

cates for the design of the plant (components must be designed and manufactured under IEC/TS 62941 or the new document for power electronics) and plant commissioning (which must be done by an installer certified to IEC/TS 63049). The IECRE is working to create training materials to ensure consistent implementation of IEC/TS 62941 and has created a process to identify the organizations that are qualified to issue IECRE certificates.

A partnership between the design engineer and the quality management expert can be successful in reducing the cost over the product's lifetime and improving the final outcome. The three Technical Specifications described here provide a first step toward creating that partnership.

ACKNOWLEDGMENT

This work was supported by the U.S. Department of Energy under Contract No. DE-AC36-08GO28308 with the National Renewable Energy Laboratory. Funding provided by U.S. Department of Energy Office of Energy Efficiency and Renewable Energy Solar Energy Technologies Office.

The U.S. Government retains and the publisher, by accepting the article for publication, acknowledges that the U.S. Government retains a nonexclusive, paid-up, irrevocable, worldwide license to publish or reproduce the published form of this work, or allow others to do so, for U.S. Government purposes.

REFERENCES

[1] D. Degraaff, R. Lacerda, Z. Campeau, "Degradation mechanisms in Si module technologies observed in the field; their analysis and statistics," NREL 2011 Photovoltaic Module Reliability Workshop, www.nrel.gov/docs/fy14osti/60170.pdf, p. 20.

[2] A discussion of this problem and several references can be found at http://www.nrel.gov/docs/fy14osti/60950.pdf.

[3] Köntges, M., et al. "The risk of power loss in crystalline silicon based photovoltaic modules due to micro-cracks." *Solar Energy Materials and Solar Cells* 95.4 (2011): 1131-1137.

[4] Yoshihito Eguchi, et al. "Requirements for quality management system for PV module manufacturing." *2014 IEEE 40th Photovoltaic Specialist Conference (PVSC)*. IEEE, 2014.

[5] Govind Ramu, et al. *Updated Proposal for a Guide for Quality Management Systems for PV Manufacturing: Supplemental Requirements to ISO 9001-2008*. No. NREL/TP-5J00-63742. National Renewable Energy Laboratory (NREL), Golden, CO (United States), 2015.

[6] http://www.iecre.org/members/certification

[7] http://forms.greentechmedia.com/af2?LinkID=CH00095679eR0 0000429AD

[8] Kelly, George, et al. "Coordination of international standards with implementation of the IECRE conformity assessment system to provide multiple certification offerings for PV power plants." *in 43rd IEEE Photovoltaic Specialists Conference*, 2016.

Updated evaluation of shock hazards to firefighters working in proximity of PV systems.

Olga Lavrova, Jimmy E. Quiroz, Jack Flicker, Renee Gooding.

Sandia National Laboratories, Albuquerque, NM, 87185, USA

Abstract — Sandia National Laboratories (SNL) is evaluating a variety of Photovoltaic (PV) operating conditions that have raised concern of shock hazard among firefighters and other emergency responders, including a scenario where the array might be illuminated by high power floodlights in a nighttime firefighting event. Theoretical approaches to determining the shock hazards to firefighters from PV arrays under worst-case daytime conditions are described. In order to evaluate the extend of the hazards in a nighttime fire scenario, experimental tests were conducted under full-moon illumination, specifically under a rare Super Moon event. We have monitored available power levels of a PV array power levels under realistic worst-case illumination conditions. The evaluations considered a variety of PV array sizes, proximity to high intensity flood lamps of the type used by firefighter personnel, as well as full-moon illumination. IV traces from individual modules, as well as PV modules connected in parallel and series, were recorded to determine the available power levels. All conditions tested showed that the shock hazard to firefighting personnel under these worst-case conditions is well below the hazard limits defined IEC TS 60479-1.

I. INTRODUCTION

Recently, there has been an increasing concern regarding the risk of shock to firefighters and other emergency response personnel working in the proximity of PV systems. The data made available to support these concerns is often incomplete, and sometimes results in an overly conservative risk assessment. Understanding exact risks is very important to formulate correct mitigation measures. Actual levels of PV power attainable may dictate safety procedures and requirements, such as voltage levels for Rapid System Shutdown (RSS) [1] and other approach boundary voltages.

NEC Article 690 establishes electrical safety requirements for PV installations in the U.S. The 2017 Revision to NEC 690.12 includes new requirements for RSS protection applied to PV arrays [2]. The revision lists the following options to mitigate shock hazard:

A. Listed protection system at the PV array level.
B. 80 V, 30-second limit for controlled conductors internal to the array.
C. PV arrays with no exposed wiring methods, no exposed conductive parts, and installed more than 8ft from exposed grounded conductive parts or ground.

Methods include limiting access to exposed components that might become energized, reducing the voltage difference between energized components, limiting the electrical current that might flow in an electrical circuit involving personnel by increasing circuit resistance, or by a combination of such methods. It should be noted that NEC Article 690 requirements do not apply to ac PV modules since they do not have dc output.

Accidental contact with a damaged array may expose emergency personnel to shock hazards of unknown voltage. Therefore, the new 2017 NEC update made specific requirements for RSS and safe voltages for personnel operations. According to UL 1310 [3], the safe voltage is ≤60V in dry conditions and ≤30V in wet conditions, as listed in 2014 NEC Chapter 9, Table 11(B) [1]. Additionally, UL 62109-1 [4] and 2014 NEC [1] outlined 240 VA as the safe power limit (energy hazard). The 2017 NEC removed the 240 VA reference, but added an 80 V limit.

The 2017 NEC revision prescribes acceptable voltage levels for firefighter personnel for newly installed arrays. However, it does not quantify the hazard to firefighters from arrays commissioned prior to NEC 2017. In order to understand and quantify the hazards of PV arrays (both with 2017 NEC protections and without), Sandia National Laboratories has conducted both theoretical analysis and experimental testing under a variety of environmental conditions. Section II describes the theoretical analysis of worst-case hazards to firefighters under daytime scenarios, Sections III details the results of nighttime lighting tests with worst-case results for ambient illumination (i.e. moonlight) as well as artificial illumination characteristic of a firefighting scenario.

II. THEORETICAL WORST-CASE DAYTIME EVALUATIONS

This section models shock hazard to a firefighter that may come into contact with a PV array under the conditions of the test. Both grounded and ungrounded arrays are considered. The procedure first calculates the effective resistance (R_{Eff}), and then uses the result to estimate the current that a firefighter might experience.

A. Ungrounded Arrays

An ungrounded array is disconnected from ground at the inverter (the array is either ungrounded during normal operation or the ground-tie of the inverter is disconnected during isolation.). This means that the only reference to ground is along the module leakage pathway (R_{module}) and the isolation of the ungrounded array (R_{iso}) is a function of the number of modules in series (S) and parallel (P) and is approximately given by (1):

$$R_{iso}^{ungrounded} \approx R_{module} = \frac{(S-1)P}{R_{leak}} + \frac{2P}{2R_{leak}} = \frac{R_{leak}}{S \cdot P} \qquad (1)$$

For an ungrounded array with the inverter disconnected from the path of current flow, the fault current path through a firefighter (R_{FF}) is in series with R_{iso} (Fig. 1). Therefore, a large

R_{iso} restricts the amount of current flow through the firefighter (I_{FF}).

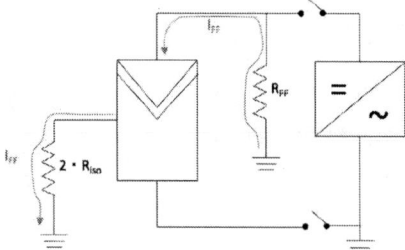

Fig. 1: Simplified diagram of current flow through a firefighter in an ungrounded array. The current flow is in series with the array isolation, R_{iso}.

Since R_{iso} and R_{FF} are in series, the effective impedance of the pathway is equal to (2).

$$R_{Eff} \approx 2 \cdot R_{iso} + R_{FF} \qquad (2)$$

B. Grounded Arrays

A grounded array, on the other hand, is not disconnected from ground at the inverter. This means that the array is referenced to ground both along the module leakage pathway (R_{module}) as well via the ground fault protection device (GFPD). Since the module leakage pathway resistance is typically multiple orders of magnitude greater than R_{GFPD}, the isolation of the grounded array (R_{iso}) can be approximated with $R_{GFPD} \sim 0$ so that $R_{iso} \sim R_{module}$. Since R_{FF} is in parallel with R_{module}, a fraction of the current (F) flows through R_{FF} with the remainder flowing through R_{module}. However, as with R_{iso}, typically $R_{module} \gg R_{FF}$ and $F \sim 1$.

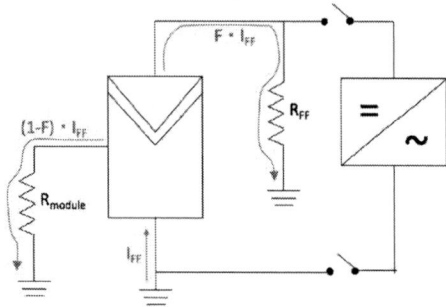

Fig. 2: Simplified diagram of current flow through a firefighter in a grounded array. The current flow is in parallel with the module leakage, R_{module}.

In this grounded case, the effective impedance of the system is equal to (3):

$$R_{Eff} \approx \frac{1}{\frac{1}{R_{iso}} + \frac{1}{R_{FF}}} \qquad (3)$$

As $R_{iso} \gg R_{FF}$, (3) reduces to (4):

$$R_{Eff} \approx R_{FF} \qquad (4)$$

C. Firefighter hazard as a function of array size

The effective resistance of the system (R_{eff}) determines the amount of current flow from the array and hence, the amount of current hazard to a firefighter, as it determines the array load line as shown in Fig. 3.

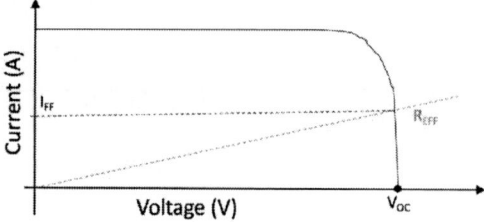

Fig. 3: R_{eff} determines the load line and the amount of current I_{FF} that flows.

If $R_{eff} \gg R_{mp}^{array}$ and the array has a moderate fill factor, then it can be assumed that R_{eff} intersects the array IV curve at array V_{oc}. In this case, the current hazard, I_{FF} can be described by (5):

$$I_{FF} \approx \frac{V_{oc}}{R_{eff}} \qquad (5)$$

This approximation is more appropriate for larger values of R_{eff}. As R_{eff} approaches array R_{mp}, the knee curvature of the array IV curve can no longer be ignored and the intersection of R_{eff} with the IV curve cannot be approximated as V_{oc}.

In general, the current hazard to a firefighter is linearly related to the V_{oc} of the array. For small body impedances (~ 600 Ω [7]) and large array V_{oc} values, the approximation in (5) tends to overestimate the current hazard. This is due to the fact that as modules are added in series, the value of R_{mp}^{array} decreases. For small values for body impedance and large array V_{oc}, the value of R_{eff} begins to approach R_{mp}^{array} and the knee curvature of the array IV curve cannot be ignored. In this case, the intersection of R_{eff} with the IV curve does not occur at V_{oc}, but at some $V < V_{oc}$ and tends to overestimate the current hazard. If $R_{eff} \gg R_{mp}^{array}$, the current hazard to the firefighters is dependent only on array V_{oc} and not array size (I_{sc}).

As we can see, whether the PV system is grounded or ungrounded has a great impact on the level of current that can be expected. For ungrounded systems, a firefighter in the path from exposed DC conductor to ground would essentially be in series with $2^* R_{ISO}$, which can be very large, limiting the current. For grounded systems, the firefighter would instead be in parallel with the module leakage pathway, whose resistance is orders of magnitude larger than body impedance. This means that most of the current would flow through the firefighter.

D. Firefighter danger during daytime operation

To further correctly quantify the electrical hazard to a firefighter, we need to account for additional variables including the following:

1. PV system DC voltage class
2. Isolation from inverter
3. Grounded or ungrounded systems
4. Skin and body impedance
5. Current path through body
6. Duration of contact
7. Surface area of contact

8. Personal protection equipment (PPE) conditions
9. Health and fitness conditions of a firefighter
10. Wet or dry ambient conditions
11. DC physiological effects
12. Fall hazard.

For all our analyses, it is assumed that the inverter is isolated from both the DC as well as the AC sides the array, preventing current flow through the inverter, as is required for "Rapid Shutdown" in [2]. For ungrounded arrays, this would mean the array is disconnected from the inverter ground (GFPD) and the only reference to ground is through the array isolation resistance, R_{ISO} [6]. For grounded arrays, the inverter isolation does not eliminate the connection to ground.

To estimate the current hazard to firefighters, three different PV V_{OC} voltage limits should be considered per [2]:
1. 600 V – small residential PV systems
2. 1000 V – small and mid-size commercial PV systems
3. 1500 V – commercial and utility-scale PV systems

The values for the total body impedance, R_T, as a function of voltage can be found in [7]. R_T can be constructed as a sum of the following components, as shown in Figure 8(b):

R_{C1}, R_{C2} – Contact resistances (skin)

R_I – Internal body resistance

R_{PPE} – PPE resistance

The same R_T calculation can be illustrated in a simplified equivalent circuit diagram shown in Figure 4(a).

The R_T reduces as touch voltage increases. Statistical data exists [7] quantifying R_T for general population, starting from an "average" person, to 5% and 95% percentiles of the population. The 5th and 50th percentiles were considered most appropriate to the average firefighter and are used for further calculations in this work.

To properly apply the effects of direct current passing through the human body per [7], it is assumed that all DC

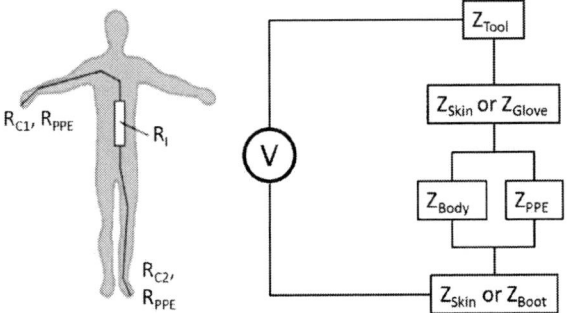

Fig. 4. (a) Illustration of components considered in calculation or the total body impedance, R_T. (b) Simplified equivalent circuit representation of different components of total body impedance and PPE.

currents from PV arrays contains a sinusoidal ripple of no more than 10% r.m.s. Since voltages 600V and above are under consideration, the surface areas of contact defined in [7] become inconsequential. It does however matter when

considering the breakdown of skin, which is dependent on current density (mA/mm²) and duration of current flow. For contact greater than 10 seconds, currents greater than 4800 mA, and small surface areas of contact (100 mm²), the skin can be expected to break down and total body impedance becomes just the internal body impedance, which can be considered the worst case [7].

Physiological effects that can be expected per [6], depending on the contact duration. The thresholds vary up to 2 seconds, after which the physiological effect does not change as a function of time. There are seven different physiological effect zones that can be considered, although some, especially those with no lethal potential, can be evaluated as a group, for simplicity.

The current path through the body can have a large impact on the physiological effects expected [7]. For instance, all possible combinations from hands to feet (left/right, one/both) are considered the baseline path. Other variations, such as left hand to right hand, must consider a "heart-current factor" resulting in greater current tolerance, or chest to left hand resulting in lower current tolerance.

The impedance of the body is reduced if the skin is assumed to be broken. As was previously mentioned, this would require quite an extreme current density, 4800 mA for a contact surface of 100 mm², for greater than 10 seconds. None of the scenarios summarized in Tables I and II came anywhere near being able to provide that level of current, therefore the body impedance with skin intact was assumed.

Finally, the level of PPE was considered, as it also has a tremendous impact on the expected current. Ideally, the probabilities of the following would guide assumptions:
1. Firefighter not wearing boots
2. Firefighter contacting exposed DC conductor with one hand and ground with other body parts other than booted feet
3. Firefighter not wearing gloves
4. Firefighter wearing wet gloves
5. Firefighter not able to cease contact in less than two seconds (different physiological effects) or 10 seconds (skin breakdown)

For this study, the possibility of a firefighter not wearing boots was not considered, as a practical case. Therefore, a left-to-right hand path was investigated initially.

Given all the considerations above, Table I and Table II summarize calculated worst case DC currents (in mA) under different PPE conditions for the different voltage classes, for grounded and ungrounded systems, respectively. The colors associated with Tables I and II correspond to the physiological effects in [7]. The applicable hazard classifications and corresponding thresholds are summarized in Table III.

The currents were calculated with the assumptions listed in Table IV. The impedance percentiles were obtained from [7], which lists impedances not to be exceeded by the given percentage of the population.

Calculation results summarized in Table I and II do indicate potentially high hazard levels for bare hand assumptions at the

TABLE I.
POTENTIAL DC CURRENTS (IN mA) FOR UNGROUNDED SYSTEMS
UNDER DIFFERENT PPE CONDITIONS FOR DIFFERENT PV SYSTEM
DESIGN VOLTAGES

50% Imp	Voltage Class		
PPE	600	1000	1500
Bare Hand	308	513	770
Wet Glove	41	69	103
Dry Glove	2	4	6

TABLE II.
POTENTIAL DC CURRENTS (IN mA) FOR GROUNDED SYSTEMS
UNDER DIFFERENT PPE CONDITIONS FOR DIFFERENT PV SYSTEM
DESIGN VOLTAGES

50% Imp	Voltage Class		
PPE	600	1000	1500
Bare Hand	135	225	338
Wet Glove	35	59	89
Dry Glove	2	4	6

TABLE III.
PERCEIVED SENSITIVITY ZONES BASED ON CURRENT

Current	Effects
< 150 mA	Slight pricking sensation to strong involuntary muscle contractions, no organic damage expected.
150-175 mA	≤ 5% probability of ventricular fibrillation.
> 175 mA	Probability of ventricular fibrillation ≥ 50%

TABLE IV.

Variables	Value
50th percentile impedance	775
Bare hand added impedance (Ω)	0
Wet glove added impedance (Ω)	5000
Dry glove added impedance (Ω)	100000
Heart-current factor for hand-to-hand	0.4
Parallel resistance factor for R_{ISO} of 100 kΩ (grounded systems)	0.994
Series resistance R_{ISO} (Ω, for ungrounded systems)	1000

higher voltage classes. However, it is very important to understand the unlikelihood of this scenario, given both standard operating procedures of firefighters and standard cabling and wiring methods of PV systems. Further work will report on statistical likelihood of such scenarios.

III. NIGHTTIME LIGHTING TESTS

Sandia National Laboratories conducted independent tests to verify maximum (worst case) power generated by a PV system under moonlight conditions. Figure 5 shows a photo of the nighttime testing.

A. Setup

The system used for nighttime testing consisted of a PV array with a single string of 16 modules connected to a grid-tied inverter. The array is installed at a tilt angle of 35 deg. The relevant module and array ratings are listed in Table V.

Fig. 5. Photo of the test setup.

TABLE V
MODULE AND ARRAY PARAMETERS UTILIZED IN TESTING.

Module Rating (W)	245
Module Isc (A)	8.25
Module Voc (A)	37.7
Module V_{MPP} (V)	30.8
Module I_{MPP} (A)	7.96

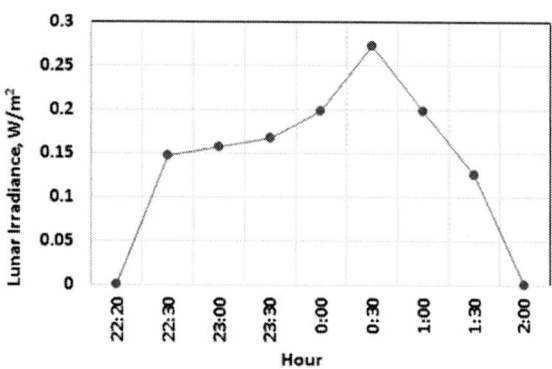

Fig. 6. Measured irradiance during the full ("super") moon in Albuquerque, NM, on November 14th 2016

B. Test conditions

In order to evaluate the worst possible hazards (i.e. the maximum possible power generated by PV panels) presented to firefighters in a nighttime scenario, testing was conducted on a full ,oon night. The date of testing, November 14th 2016 corresponded to a Super Moon" event– a full moon closely coinciding with perigee. The November 14th Supermoon was the closest a full moon has been to Earth since January 26th, 1948 and similar event will not occur until November 25, 2034. As such, the test conditions represent the highest moonlight irradiance level contributing to power produced by a PV array. Additionally, lunar "noon" elevation of 70.6°, and azimuth of

978-1-5090-5606-4/17 $31.00 © 2017 IEEE

180° were optimal angular positions for illumination of our South-facing PV array.

Figure 6 shows measured irradiance during the full moon in Albuquerque, NM, on November 14th 2016. The peak of the irradiance corresponds to the time of lunar "noon" which was at 00:30am local (MST) time. The peak irradiance measured was $0.27W/m^2$. These measurements are consistent with maximum moon power estimates obtained in [5]. Assuming spectral composition of lunar irradiance is the same as solar irradiance, a uniform irradiation of $0.27W/m^2$ would result in less than 10mW of power produced by a typical commercially available Si PV panel. However, spectral composition of lunar irradiance is not identical to that of solar. Further calculations of reduced power produced under moonlight spectra will be reported in a separate publication.

TABLE V
TEST AND CONDITIONS SCHEDULE.

Test	Test Condition	Array Configuration	Floodlight distance
	7 am	3 parallel panels	N/A
	4:30 pm	3 parallel panels	N/A
1	2 x 1000W Halogen	3 parallel panels	56 cm
		single panel	56 cm
2	2500 Lm LED	3 parallel panels	56 cm
		single panel	56 cm
3	2 x 2500 Lm LED	3 parallel panels	56 cm
		single panel	56 cm
4	1000W Halogen 2x 2500Lm LEDs	3 parallel panels	56 cm
		single panel	
5	2x 1000W Halogen 2x 2500 Lm LEDs	3 parallel panels	56 cm
		single panel	
6	Full moon only	Super moon, Elevation 70.6, /AZ 180°	N/A

C. Test procedure

Table V lists the test conditions that were used in nighttime testing. In addition to ambient light measurements, panel and array-level measurements were taken for combinations of two different types of floodlights (halogen and LED) that are typically used for nighttime firefighter operations. The lights were installed at a distance of 56cm from the PV panels. It is important to note that a distance of 56cm is an extreme worst (and probably, impractical) case of a how close a firefighters' light source could possibly be to a PV array; in practical firefighting conditions, lights would be on the firefighters' trucks at least several meters away from a PV array.

For each test, an IV tracer was used to take IV sweeps of individual PV panels as well as PV panels connected in series and in parallel. Current sensitivity of the IV tracer is 30 μA, which is larger than the photocurrent produced by single module in majority of tests. In these cases, PV panels were connected in parallel in order to increase total measurable current.

In addition to panel and array IV curves, an irradiance uniformity map was recorded using standard LI-COR irradiance meter at teach testing condition. As seen from Figure 7, both halogen and LED lights produce localized irradiation

Fig. 7. Photos of the test setup: (a) two 1200W halogen lights; (b) one 2500 Lumens LED light; (c) two 2500Lumens LED lights; (d) top view of illumination by two LED lights, showing significant non-uniformity of the illumination

patterns. Such non-uniform irradiation results in a classical mismatch case of PV cells within the PV module. As a result, PV power produced is even further reduced due to mismatch.

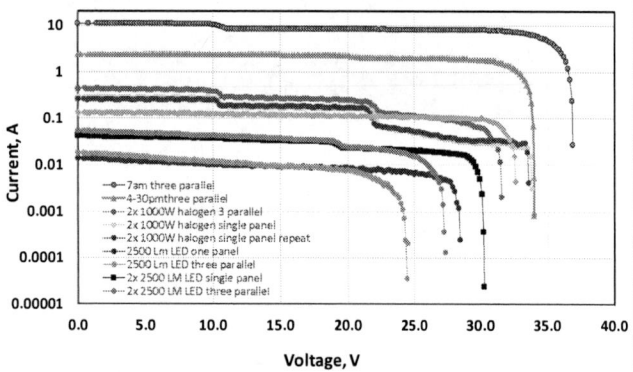

Fig. 8. IV curves recorded under different illumination conditions.

D. Results and Analysis of nighttime testing.

Figure 8 shows a family of IV traces recorded under different illumination conditions as described in Table II. As can be seen from the plots, illumination by halogen floodlights (green, yellow, and navy-blue traces) resulted in higher current response from the PV array compared to LED illumination (pink, purple traces). This is due to better overlap of the Si PV spectral response with the Halogen emission spectra. It should be noted that the majority of firefighter operations vehicles are switching their lighting equipment from halogen to LEDs. Therefore, the LED illumination results are the more relevant result to current firefighter operations and those in the future.

From Figure 8 we can see that, under illumination by two 2500 Lm LED flood lights (pink trace, circle markers), a single PV panel produced approximately 10 mA at 30 V, resulting in a maximum power of $0.3W_{mp}$. A single 2500 LM LED floodlight yields a hazard of less than $0.1W_{mp}$ or $100mW_{mp}$. More importantly, current levels at the maximum power point condition is under 100mA. As described in Section II, current hazard levels to firefighter will be even further reduced due to PPE and R_{iso} of the array.

IV. CONCLUSIONS

We have assessed the worst case hazards to firefighters under power levels which may be present under fire fighting conditions. Detailed measurements were conducted under nighttime conditions. This work has shown that under works-case ambient illumination (full moon) and typical emergency lighting conditions, the current hazards to firefighters is low. All tested conditions showed that worst-case hazard is well below the limits of the hazard by continuous electrical current set by IEC 60479-1.

The theoretical analysis of specific shock hazards for worst-case daytime conditions have been shown. Unlikely scenarios with worst case assumptions (e.g. firefighters with no PPE), considering a dozen potential variables and full array V_{OC}

exposures, show some potential concerns for higher (> 600 V) PV system design voltages. As these results represent worst-case scenarios, it is extremely unlikely that firefighters would be subjected to the current hazards shown in Tables II and IV. The analysis shows that current hazard is directly related to personnel impedance as well as array maximum voltage. PPE resistance is larger than body impedance. Therefore, minimizing current hazards to emergency personnel is dependent on proper PPE (to increase effective impedance) as well as array segmentation strategies to decrease voltage levels.

Further work is being undertaken using simulation as well as experimentation. Complete results will be refined, validated, and published in a future publication. This body of work is intended to further clarify the risks posed by PV installations to firefighter and other emergency personnel. Additionally, developed safety limits and procedures will be relevant to increasing installations combining PV and storage.

ACKNOWLEDGEMENT

This work was partially funded by the U.S. Department of Energy SunShot Initiative under award number DOE-EE-31654. Sandia National Laboratories is a multi-mission laboratory managed and operated by National Technology and Engineering Solutions of Sandia, LLC., a wholly owned subsidiary of Honeywell International, Inc., for the U.S. Department of Energy's National Nuclear Security Administration under contract DE-NA0003525.

REFERENCES

[1] "NFPA 70: National Electrical Code (NEC), 2014 Edition", by National Fire Protection Association.

[2] "NFPA 70: National Electrical Code (NEC), 2017 Edition", by National Fire Protection Association.

[3] Underwriters Laboratories (UL) 1310 standard – "Standard for Class 2 Power Units", 2009.

[4] Underwriters Laboratories (UL) 62109-1 Edition 1 – "Safety of Power Converters for Use in Photovoltaic Power Systems - Part 1: General Requirements", 2014.

[5] Korotkevich, A. O., Galochkina, Z. S., Lavrova, O., & Coutsias, E. A. (2015). "On the comparison of energy sources: Feasibility of radio frequency and ambient light harvesting." Renewable Energy, 81, 804-807.

[6] J. Flicker, J. Johnson, M. Albers, and G. Ball, "Recommendations for Isolation Monitor Ground Fault Detectors on Residential and Utility-Scale PV Systems," Sandia National Laboratories, SAND2015-4667C, 2015.

[7] International Electrotechnical Commission (IEC) TS 60479-1, Edition 4.1 – "Effects of current on human beings and livestock – Part 1: General aspects", 2016.

[8] Schindelholz, E., Yang, B. B., Armijo, K. M., McKenzie, B. B., Taylor, J. M., Sorensen, N. R., & Lavrova, O. (2015, June). "Characterization of fire hazards of aged photovoltaic balance-of-systems connectors". In Photovoltaic Specialist Conference (PVSC), 2015 IEEE 42nd (pp. 1-6). IEEE.

[9] Jones, C. B., Martínez-Ramón, M., Smith, R., Carmignani, C. K., Lavrova, O., Robinson, C., & Stein, J. S. (2016, June). "Automatic fault classification of photovoltaic strings based on an in situ IV characterization system and a Gaussian process algorithm". In Photovoltaic Specialists Conference (PVSC), 2016 IEEE 43rd (pp. 1708-1713). IEEE

[10] Abdollahy, S., Lavrova, O., Heine, N., Mammoli, A., & Poroseva, S. (2013, August). "Integrating heterogeneous distributed energy resources to manage intermittent power at low cost." In Technologies for Sustainability (SusTech), 2013 1st IEEE Conference on (pp. 223-229). IEEE.

Growth and optimization of GaInP/InP nanowire tunnel diode

Xulu Zeng[1], Gaute Otnes[1], Magnus Heurlin[2], Magnus T Borgström[1]

[1]Solid State Physics, NanoLund, Lund University, Lund, Skåne, SE-22100, Sweden

[2]Sol Voltaics AB, Lund, Skåne, SE-22370, Sweden

Abstract — **Nanowire solar cells with single p-n junction have had a rapid increase in energy conversion efficiency. Introducing tandem geometry is the path for higher efficiency, for which nanowire tunnel diode is a critical component. We fabricated, and characterized the GaInP/InP nanowire tunnel diodes with bandgap combinations suitable for a tandem junction solar cell. The electrical contact was improved by using different n type dopant species for the heterojunction and the nanowire contacting area, respectively. A GaInP/InP nanowire tunnel diode with an average measured peak current density of 45 A/cm^2 and maximum peak to valley ratio of 2.4 was demonstrated. The demonstration of GaInP/InP tunnel diodes lays the foundation for the next step of realizing NW tandem junction solar cells.**

Index Terms — **GaInP, InP, nanowires, tandem junction solar cell, tunnel diode.**

I. INTRODUCTION

Recently, the performance of nanowire (NW) solar cells have been gradually catching up with conventional planar cells [1]-[3]. The leading performance is held by axial single p-n junction which has limited maximum theoretical energy conversion efficiency. By combining optimal bandgap combinations, a tandem junction solar cell can harvest a broader range of solar energy, thus overcoming the Shockley-Queisser limit. With the advantage of ease of lattice matching constraints, it is very promising for NWs to combine different materials with high material quality. Such NW tandem junction devices have been reported with either NW top cell on top of planar bottom subcell or tandem homo-junction structure [4]-[6]. However, a tandem junction solar cell within a NW containing materials of different bandgaps has not been reported. For a two-terminal tandem junction solar cell, NW tunnel (Esaki) diodes are critical for providing connections between different subcells with low-electrical resistance and low-optical loss. Due to the challenges of NW growth, especially when involving ternary materials, NW tunnel diodes with material combinations suitable for solar energy harvesting have not been demonstrated.

We fabricated and characterized NW tunnel diodes using the $Ga_{0.3}In_{0.7}P$/InP material system with a bandgap combination of approximately 1.70 eV and 1.35 eV respectively, which is suitable for a tandem geometry design [7]. Electrical measurements demonstrated negative differential resistance (NDR) region in the tunnel diodes with average peak current density of 45 A/cm^2, and a maximum peak to valley current ratio (PVCR) of 2.4 at room temperature. An improvement of contact was achieved by altering the n type dopant species (between S and Sn).

II. EXPERIMENTAL DETAILS

The nanowire samples were grown in a low-pressure (100 mbar) MOVPE system (Aixtron 200/4) with a total flow of 13 l/min. The substrates were Zn doped InP (111)B substrates with a hexagonal pattern of Au disc arrays with a diameter of 200 nm, a pitch of 500 nm and height of 65 nm. For NW growth, trimethylindium (TMIn), trimethylgalium (TMGa), phosphine (PH$_3$), diethylzinc (DEZn), tetraethyltin (TESn) and hydrogen sulfide (H$_2$S) were used as precursors. The NWs were grown to a length of 2 μm with each (GaInP and InP) segment of 1 μm. Dopant flows for degenerate doping were modulated according to a doping evaluation on InP and GaInP NWs, where H$_2$S was chosen as n type dopant precursor at the heterojunction since it was found to reach the highest n doping in our system [8]-[9]. The NW length was monitored by use of *in situ* reflectance spectroscopy [10]-[11]. At the beginning of growth, a pre-anneal nucleation of InP was implemented at 280 °C for 1 min at TMIn and PH$_3$ molar fractions of $\chi_{TMIn} = 8.9 \times 10^{-5}$, and $\chi_{PH3} = 6.92 \times 10^{-3}$ [12]. The samples were then heated to an elevated temperature of 550 °C for 10 min under mixed PH$_3$/H$_2$ atmosphere to desorb surface oxides. Then the temperature was set to 440 °C for NW growth. First, a 30-nm InP stub was grown by adding TMIn to the flow, with $\chi_{TMIn} = 8.9 \times 10^{-5}$, and $\chi_{PH3} = 6.92 \times 10^{-3}$ respectively. After that, hydrogen chloride (HCl) was introduced ($\chi_{HCl} = 4.6 \times 10^{-5}$) as *in situ* etching to avoid radial growth [13]. After another 90 nm of InP growth, the growth of GaInP segment was initialized with: $\chi_{TMIn} = 5.2 \times 10^{-5}$, $\chi_{TMGa} = 3.96 \times 10^{-4}$, $\chi_{PH3} = 6.92 \times 10^{-3}$, and $\chi_{HCl} = 5.4 \times 10^{-5}$ At the end of the GaInP NW growth, DEZn flows were linearly increased during 10 s from low flow ($\chi_{DEZn} = 8.3 \times 10^{-5}$) to high flow ($\chi_{DEZn} = 1.17 \times 10^{-4}$) and kept for 15 s before the heterojunction to achieve degenerate doping. When the growth of GaInP segment was done, the DEZn and TMGa flow was switched off, after which the H$_2$S flow was switched on to grow a n-type InP segment ($\chi_{TMIn} = 4.5 \times 10^{-5}$, $\chi_{PH3} = 6.92 \times 10^{-3}$, and $\chi_{HCl} = 4.6 \times 10^{-5}$) with a high flow ($\chi_{H2S} = 1.55 \times 10^{-5}$) for 15 s. Then the H$_2$S flow was ramped down to a lower flow ($\chi_{H2S} = 1.9 \times 10^{-6}$) during 10 s, after which the flow was kept constant until the NW reached 2 μm in length. The growth was completed by turning TMIn and H$_2$S to the vent flow, and the reactor was cooled down in a PH$_3$/H$_2$ mixture. The as-grown NWs were inspected by a LEO 1560 field-emission scanning electron microscope (SEM). For electrical measurements, electron beam induced current (EBIC) measurement was performed to the as-grown NWs to gain further insight into depletion length and minority carrier diffusion length in the NWs. A tungsten nano-probe in the EBIC setup was put on the Au particle as top contact, and the substrate was used as back contact. In this setup, I-V curves of the as-grown NWs were measured.

III. RESULTS AND DISCUSSIONS

The SEM image (Fig. 1) shows the as-grown NW array with a length of approximately 2 μm and a diameter of 180 ± 5 nm. A contrast in secondary electron signal can be seen between the GaInP and InP segments. A diameter change of NWs can also be observed when the growth of InP segment began, as a result of the change in contact angle of the seed particle

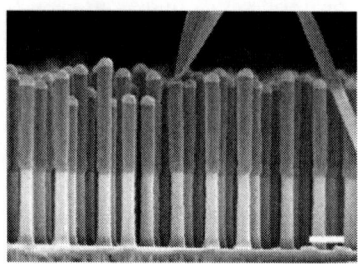

Fig. 1.   Cross-section SEM image of as-grown GaInP:Zn(bottom)/ InP:S(top) junction NW. The nano-probe is contacting one NW. Scale bar is 500 nm.

The peak intensity of the EBIC profile in Fig. 2 (a) depicts the location and extension of the pn-junction depletion region. The depletion region is located approximately in the middle of the NW. The EBIC current along the length of a representative NW is shown in Fig. 2 (b). By fitting a single exponential to the tails of the EBIC profile, effective minority carrier diffusion lengths are estimated to be 52 ± 13 nm in the p-GaInP segment and 58 ± 10 nm in the n-InP segment.

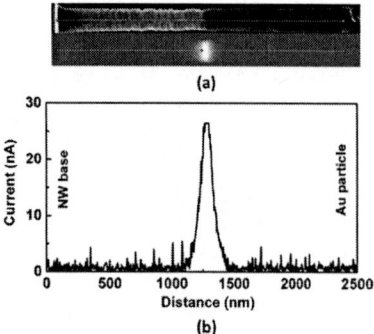

Fig. 2.   (a) EBIC (top) and SEM (bottom) of a representative NW with the configuration of GaInP:Zn/InP:S, The InP substrate is at the

left edge and the Au particle is at the right edge; (b) EBIC line profile along the center of the NW.

The I-V curves measured by contacting NWs with the nano-probe on the GaInP:Zn/InP:S junction show clear NDR region (Fig. 3, sample 1) which is the characteristic of a tunnel current. The peak current ranges from 0.5 to 1.2 nA. Considering the vapor-liquid-solid growth where materials stored in the seed particle are expelled during the cooling-down process, a small extra NW segment is usually grown. Due to the high vapor pressure of H$_2$S, it will evaporate rapidly from the growth front when switched off. This may result in a low n doping level in a small InP segment grown during cool down, known as the NW neck, which can worsen the contact between the NW and the nano-probe.

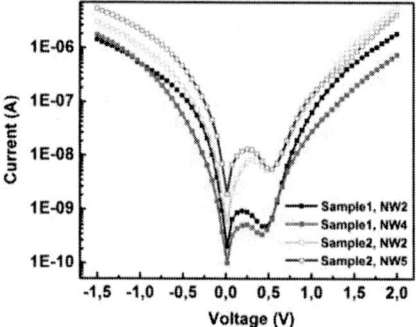

Fig. 3.   I-V curves measured by contacting the as-grown NW with a nano-probe inside the SEM on the GaInP:Zn/InP:S (Sample 1) and GaInP:Zn/InP:S/InP:Sn (Sample 2) configurations, two NWs are shown as representatives for each sample.

To investigate whether such a low doped segment is degrading contact properties, we switched n type dopant species from S to Sn for InP segment after growth of the heterojunction (i.e. a GaInP:Zn/InP:S/InP:Sn configuration). The NW ended up with a smoother particle surface and more uniform diameter through the NW. I-V measurement was done on 4 NWs for statistical comparison of the GaInP:Zn/InP:S and GaInP:Zn/InP:S/InP:Sn configurations. Comparing the I-V curves obtained by using the nanoprobe (Fig. 3), a similar peak voltage and PVCR was obtained, and together with a similar EBIC profile, this reveals a nominally similar tunnel junction as the GaInP:Zn/InP:S configuration. However, a higher peak current was obtained in the GaInP:Zn/InP:S/InP:Sn configuration. Therefore, it might be of practical significance to use different n type dopant species in the tunnel diode heterojunction and NW contact segment to improve the electrical contact and ensure sufficiently high peak current required by a tandem junction solar cell.

As a result, the peak current of the GaInP:Zn/InP:S/InP:Sn junction ranges from 8 to 21 nA, with the average peak current density of 45 A/cm^2. The maximum PVCR measured is 2.4 at room temperature. A more detailed investigation is ongoing on the influence of different dopant species on the material quality and device performance.

IV. CONCLUSIONS

We have synthesized NW tunnel diodes in the GaInP/InP material system on InP substrate. An improvement of tunnel diode peak current was observed when altering S to Sn as n type dopant for the NW contacting area. Electrical measurements showed tunnel diode characteristics with average peak current density of 45 A/cm^2 for the best sample, and maximum PVCR up to 2.4 at room temperature. Our results demonstrate new material combinations for NW tunnel diodes, and suggest a way to improve contact quality.

ACKNOWLEDGEMENT

The research leading to these results was performed within NanoLund at Lund University and supported by Myfab, the Crafoord Foundation, the Swedish Research Council, the Swedish Energy Agency, the European Union's Horizon 2020 research and innovation programme Nano Tandem under grant agreement n°641023, and the People Programme (Marie Curie Actions) of the European Union's Seventh Framework Programme (FP7-People-2013-ITN) under REA grant agreement n°608153, PhD4Energy.

REFERENCES

[1] J. Wallentin, N. Anttu, D. Asoli, M. Huffman, I. Åberg, M. H. Magnusson, G. Siefer, P. Fuss-Kailuweit, F. Dimroth, B. Witzigmann, H. Q. Xu, L. Samuelson, K. Deppert, and M. T. Borgström, "InP nanowire array solar cells achieving 13.8% efficiency by exceeding the ray optics limit," *Science*, vol. 339, pp.1057-1060, 2013.

[2] I. Åberg, G. Vescovi, D. Asoli, U. Naseem, J. P. Gilboy, C. Sundvall, A. Dahlgren, K. E. Svensson, N. Anttu, M. T. Björk, and L. Samuelson, "A GaAs nanowire array solar cell with 15.3% efficiency at 1 sun," *IEEE Journal of Photovoltaics*, vol. 6, pp. 185-190, 2015.

[3] D. van Dam, N. J. J. van Hoof, Y. Cui, P. J. van Veldhoven, E. P. A. M. Bakkers, J. G. Rivas, and J. E. M. Haverkort, "High-

efficiency nanowire solar cells with omnidirectionally enhanced absorption due to self-aligned Indium-Tin-Oxide Mie scatterers," *ACS Nano*, vol. 10, pp. 11414-11419, 2016.

[4] T. J. Kempa, B. Tian, D. R. Kim, J, Hu, X. Zheng, and C. M. Lieber, "Single and tandem axial p-i-n nanowire photovoltaic devices," *Nano Letters*, vol. 8, pp. 3456-3460, 2008.

[5] M. Yao, S. Cong, S. Arab, N. Huang, M. L. Povinelli, S. B. Cronin, P. D. Dapkus, and C. Zhou, "Tandem solar cells using GaAs nanowires on Si: design, fabrication, and observation of voltage addition," *Nano Letters*, vol. 15, pp. 7217-7224, 2015.

[6] M. Heurlin, P. Wickert, S. Fält, M. T. Borgström, K. Deppert, L. Samuelson, and M. Magusson, "Axial InP nanowire tandem junction grown on a silicon substrate," *Nano Letters*, vol. 11, pp. 2028-2031, 2011.

[7] Y. Chen, M.-E. Pistol and N. Anttu, "Design for strong absorption in a nanowire array tandem solar cell," *Scientific Reports*, vol. 6, 32349, 2016.

[8] O. Hultin, G. Otnes, M. T. Borgström, M. Björk, L. Samuelson and K. Storm, "Comparing Hall effect and field effect measurements on the same single nanowire," *Nano Letters*, vol. 16, pp. 205-211, 2016.

[9] F. Lindelöw, M. Heurlin, G. Otnes, V. Dagytė, D. Lindgren, O. Hultin, K. Storm, L. Samuelson, and M. T. Borgström, "Doping evaluation of InP nanowires for tandem junction solar cells," *Nanotechnology*, vol. 27, 065706, 2016.

[10] G. Otnes, M. Heurlin, X. Zeng and M. T. Borgström, "In$_x$Ga$_{1-x}$P Nanowire growth dynamics strongly affected by doping using Diethylzinc," *Nano Letters*, vol. 17, pp. 702-707, 2017.

[11] M. Heurlin, N. Anttu, C. Camus, L. Samuelson, and M. T. Borgström, "In situ characterization of nanowire dimensions and growth dynamics by optical reflectance," *Nano Letters*, vol. 15, pp. 3597-3602, 2015.

[12] G. Otnes, M. Heurlin, M. Graczyk, J. Wallentin, D. Jacobsson, A. Berg, I. Maximov, and M. T. Borgström, "Strategies to obtain pattern fidelity in nanowire growth from large-area surfaces patterned using nanoimprint lithography," *Nano Research*, vol. 9, pp. 2852-2861, 2016.

[13] M. T. Borgström, E. Norberg, P. Wickert, H. A. Nilsson, J. Trägårdh, K. A. Dick, G. Statkute, P. Ramvall, K. Deppert, and L. Samuelson, "Precursor evaluation for in situ InP nanowire doping," *Nanotechnology*, vol. 19, 445602, 2008.

[14] W. Guter, J. Schone, S. P. Philipps, M. Steiner, G. Siefer, A. Wekkli, E. Welser, E. Oliva, A. W. Bett, and F. Dimroth, "Current-matched triple-junction solar cell reaching 41.1% conversion efficiency under concentrated sunlight," *Applied. Physics Letters*, vol. 94, 223504, 2009.

Cathodoluminescence mapping for the determination of n-type doping in single GaAs nanowires

Hung-Ling Chen[1], Chalermchai Himwas[1], Andrea Scaccabarozzi[1,2], Pierre Rale[1], Fabrice Oehler[1], Aristide Lemaître[1], Laurent Lombez[2,3], Jean-François Guillemoles[2,3], Maria Tchernycheva[1], Jean-Christophe Harmand[1], Andrea Cattoni[1], Stéphane Collin[1,2]

[1]Centre of Nanoscience et de Nanotechnology, CNRS, University Paris-Sud/Paris-Saclay, Marcoussis, France

[2]Institut Photovoltaïque d'Ile-de-France (IPVF), Antony, France

[3]Institut de Recherche et Développement sur l'Energie Photovoltaïque (IRDEP) EDF/CNRS/Chimie Paris Tech, Chatou, France

Abstract — **We present a new method to determine the doping level of n-type semiconductors at the nanoscale. Low-temperature and room-temperature cathodoluminescence (CL) measurements are carried out on single Si-doped GaAs nanowires. The spectral shift and the broadening of luminescence spectra are a signature of an increased density of electrons. They are compared to CL spectra of well-calibrated planar Si-doped GaAs layers whose doping levels are determined by Hall measurements and compared to previous experimental studies. We infer a n-type doping of $1 \times 10^{18} \text{cm}^{-3}$ to $2 \times 10^{18} \text{cm}^{-3}$, with a high spatial homogeneity along the nanowire. These results show that cathodoluminescence provides an alternative way to probe carrier concentration in nanostructured and polycrystalline semiconductors, and to map the spatial inhomogeneity of dopants.**

I. INTRODUCTION

Semiconductor nanowires provide a new route toward low cost photovoltaics and multi-junction architectures. They present natural light-trapping properties. Their small lateral dimension allows lattice mismatch growth and enables direct integration of III-V nanowires on silicon cells, or to fabricate flexible devices [1,2]. However, controlling the doping is a major issue in the fabrication of nanowire-based solar cells. It is especially crucial for core-shell structures due to small radial dimension, and numerical calculations have shown that the requirement of doping level is higher for radial junction than axial junction solar cells [3,4].

Hall measurement is the conventional method used for the characterization of doping in planar semiconductor layers. For nanowires, electrical methods can still be applied [5,6]. However, contacting a single nanowire and forming good ohmic contacts requires considerable technical efforts and can hardly be applied to a large number of nanowires. Contactless optical methods based on terahertz spectroscopy and photoluminescence have shown usefulness in measuring the doping level of ensembles of nanowires [7,8], but they cannot be used to assess the homogeneity of semiconductor nanostructures and for the characterization of single nanowires.

In this work, we show that cathodoluminescence (CL) can be used to assess the doping level of n-type GaAs at the nanoscale. CL maps of single GaAs nanowires are measured at 20 K and room temperature. They show a high homogeneity along the wire, and the peaks of CL spectra shifting to higher energy (Burstein-Moss shift) is a signature of increased electron concentration. We compare CL spectra of single nanowires with CL spectra of planar Si-doped GaAs with well-calibrated doping levels, and with data published in previous studies. We infer electron concentrations in GaAs nanowires of $1\text{-}2 \times 10^{18} \text{cm}^{-3}$.

II. -EXPERIMENTAL

GaAs nanowires were grown by molecular beam epitaxy (MBE) on Si(111) substrate. Un-intentionally doped GaAs nanowires were firstly grown by self-catalyzed vapor-liquid-solid method. Ga catalyst droplet was then crystallized by exposing the sample to As flux only. Finally, Si-doped GaAs shell was grown using standard conditions similar to planar films. Nanowire length was measured to be 3-4 μm and diameter about 200 nm. Nanowires are dispersed on a Si substrate for CL measurements.

Cathodoluminescence (CL) measurements were performed with an Attolight "Chronos" quantitative cathodoluminescence microscope. The mean electron current is of the order of 1nA and the acceleration voltage is 6kV. CL spectra were recorded on an Andor Newton CCD camera with a Horiba dispersive spectrometer (grating: 150 grooves/mm). Light is collected through an achromatic reflective objective with a numerical aperture of 0.72. Luminescence spectra are corrected for the optical response of the collection and detection system.

III. RESULTS

A. Cathodoluminescence mapping

978-1-5090-5606-4/17 $31.00 © 2017 IEEE

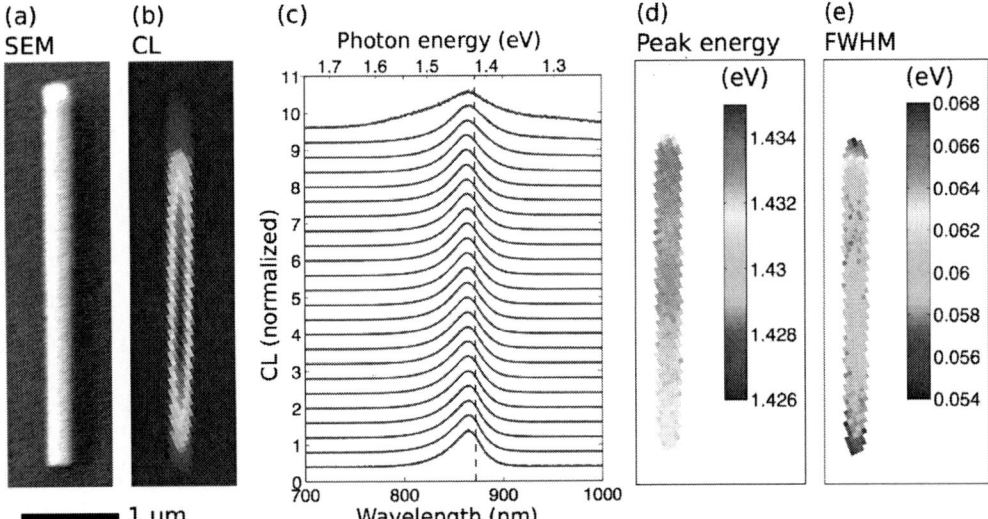

Fig. 1. Cathodoluminescence mapping of a single GaAs nanowire. (a) SEM image of the nanowire. (b) CL map of the total luminescence intensity integrated over the whole emission spectral range. (c) CL emission spectra along the nanowire showing a very homogeneous emission except close to the tip, where the Ga catalyst droplet was crystallized. Doping analysis is done in the homogeneous part of the nanowire. (d) Map of the peak energy of luminescence emission. (e) Map of the full width at half maximum (FWHM) of the CL peaks.

Figure 1 shows cathodoluminescence measurements of a single GaAs nanowire dispersed on a Si substrate. The CL intensity is constant along the major part of the nanowire, except at the nanowire tip (top of the images in Figure 1). This part has been grown by consuming the liquid Ga catalyst droplet, and changing the contact angle of liquid droplet typically produces different crystal phases (zinc-blende or wurtzite) [9]. A broadening of CL spectra is observed in this region (Figure 1c). In the following, we will only analyze CL spectra from the homogeneous region. Figures 1(d) and (e) show the maps of the luminescence peak energy and full width at half mawimum (FWHM).

B. Low-temperature and room-temperature CL spectra

Low-temperature and room-temperature CL spectra were measured on several GaAs nanowires and Si-doped GaAs planar layers with various doping levels, see Figures 2 and 3.

The spectra present a single peak across the band gap of GaAs (1.424 eV at room temperature and 1.519 eV at very low temperature, 1.515 eV corresponds to exciton recombination for low doping samples). We observe a blue shift of the peak energy with increasing electron concentrations (Burstein-Moss shift). It originates from the electron filling in the conduction band. As the electron Fermi level rises above the conduction band minimum, the optical absorption as well as spontaneous emission spectrum shifts to higher energies compared to the nominal band gap values. For many III-V semiconductors, the conduction band has much smaller effective density of states than the valence band. Hence the Burstein-Moss shift is typically observed for n-type

doping. At lower energies, the luminescence tails also grow with increased carrier concentration as a result of Coulombic potential from ionized donor and electron-electron interaction [10]. Both effects contribute to the broadening of the CL peak with increased doping.

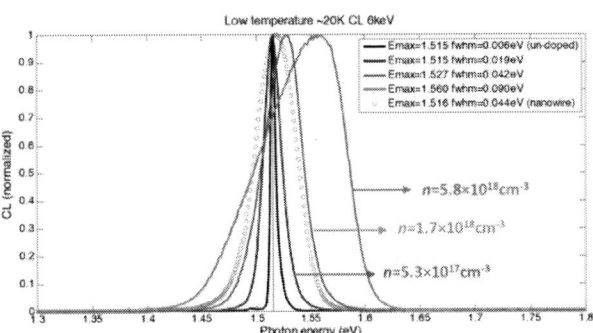

Fig. 2. Low-temperature (20K) cathodoluminescence spectra of planar GaAs layers with various n-type doping concentrations (un-doped and Si-doped), and Si-doped GaAs nanowire.

Luminescence broadening could also be due to inhomogeneity and carrier heating. The homogeneity is ensured by high resolution CL mapping, as shown in Figure 1. The effect of carrier heating was controlled by photoluminescence (PL) measurements performed on planar layers and single nanowires at various excitation intensities (not shown). PL spectra are in very good agreement with CL spectra, except in the long wavelength range were slight carrier heating is visible in CL spectra. Its contribution is negligible with respect to the total peak broadening.

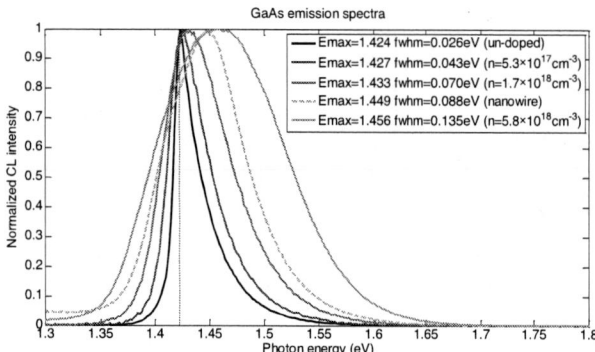

Fig. 3. Room-temperature cathodoluminescence spectra of planar GaAs layers with various n-type doping concentrations (un-doped and Si-doped), and Si-doped GaAs nanowire.

C. Relation between CL spectra characteristics and electron concentration

The two main characteristics of luminescence spectra are the energy position of the maximum peak (peak energy) and the full width at half maximum (FWHM). Figure 4 presents the relation between the peak energy and electron concentration from various experimental data, and Figure 5 shows the relation between FWHM and electron concentration (room temperature).

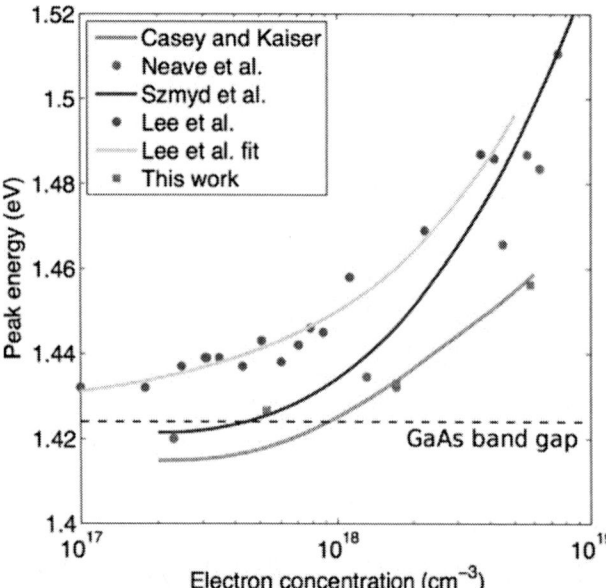

Fig. 4. Peak energy as a function of electron concentration at room temperature. Our CL measurements are compared with experimental data available in the literature: Casey and Kaiser for CL measurements on melt-grown Te-doped GaAs [11], Neave et al. for PL Si-doped GaAs grown by MBE [12], Szmyd et al. for PL Se-doped GaAs grown by MOCVD [13] and Lee et al. for PL Si-doped GaAs grown by MBE [14]. The

horizontal dashed line indicates the band-gap of un-doped GaAs at room temperature.

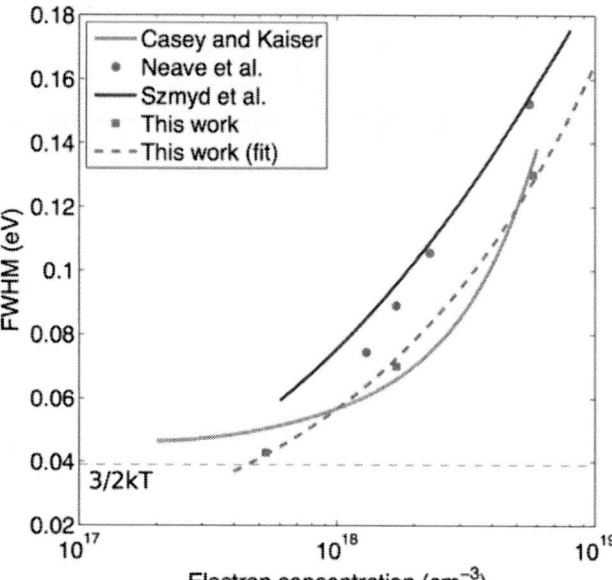

Fig. 5. FWHM as a function of electron concentration at room temperature. Our CL measurements are compared with experimental data available in the literature: Casey and Kaiser [11], Neave et al. [12] and Szmyd et al. [13]. The horizontal line indicates the room temperature thermal energy.

Our measurements are very close to that of Casey and Kaiser [11], while other data show significantly larger blue shift and broadening. Presence of acceptor states related to the growth technique may contribute to dominant radiative recombination, leading to different luminescence features. Casey and Kaiser showed very little dispersion of CL FWHM with electron concentration, so the use of FWHM to determine electron concentration seems simple and accurate [11]. The peak energy showed larger variation, especially at high doping level when the spectra tend to flatten. Therefore, we mainly focus on the use of FWHM for the determination of the n-type doping level, and the peak energy is used as a control.

At low temperature, Neave et al. observed transitions due to acceptor states for MBE-grown Si-doped GaAs samples [12]. In our Si-doped samples, we see no evidence of emission in the lower energy part of the CL spectra (Figure 2). De-Sheng et al. investigated the impact of the low-temperature FWHM with electron concentration n and found a dependence in $n^{2/3}$ [15]. We fit the low-temperature FWHM as a power function of n (Figure 6) and find the power factor very close to 2/3, which can be explained by the electron Fermi level above the parabolic portion of the conduction band [15].

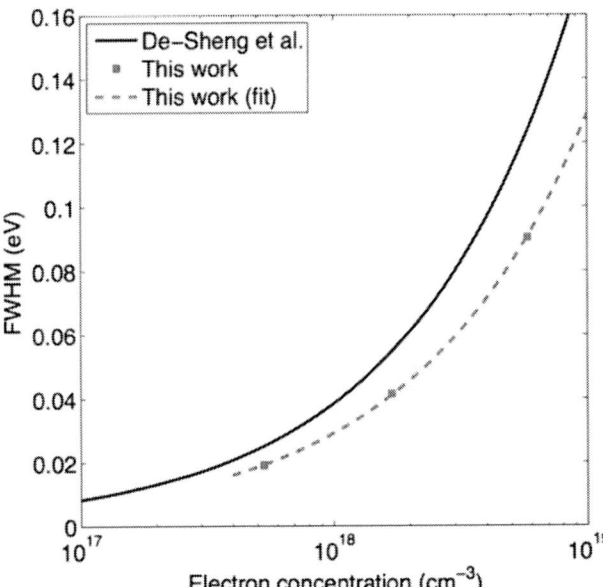

Fig. 6. FWHM as a function of electron concentration at low temperature (20K). Our CL measurements are compared with experimental data from De-Sheng et al. for PL MBE GaAs:Te at 1.8K [15].

D. Determination of n-type doping in single GaAs nanowires

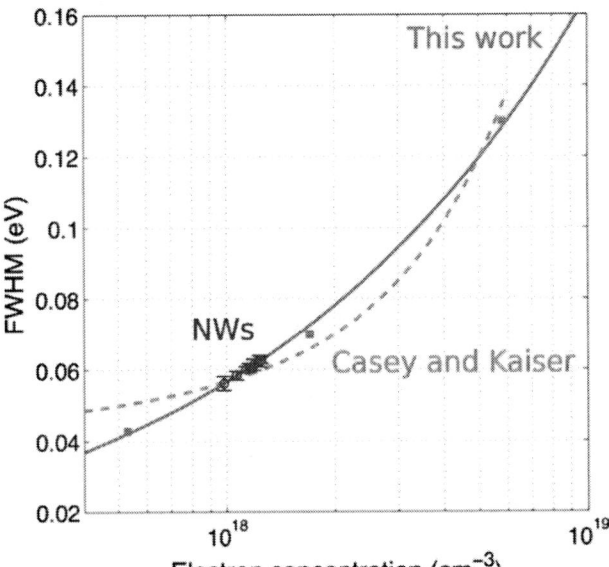

Fig. 7. FWHM of room temperature CL spectra of planar GaAs layers as a function of their electron concentration determined by Hall measurements (red marks). FWHM of CL spectra measured on single GaAs nanowires are compared to planar layers in order to assess the n-type doping level ~1-2×10^{18}cm^{-3} (blue circles and error bars).

The measured FWHM can now be used to determine the electron concentration of nanowires. We checked that there is no clear unexpected luminescence peak and the position of the maximal intensity is blue shifted. Figure 7 reports the room-temperature FWHM of CL spectra measured on planar Si-doped GaAs layers (red). They are used as a reference to determine the electron concentration of different nanowires of the same sample based on the FWHM of the emission peak (blue marks). Doping levels of Si-doped nanowires are of the order of 1×10^{18}cm^{-3} to 2×10^{18}cm^{-3}. Using the spectral shape of luminescence, we then provide a robust way to assess the n-type doping level above degeneration threshold (~5×10^{17}cm^{-3}). More accurate estimation can be achieved by fitting the whole spectrum with physical model [16,17].

V. CONCLUSION

In this work, high-resolution cathodoluminescence mapping of single GaAs nanowires has been performed at low (20K) and room temperature. Cathodoluminescence spectra were compared to planar Si-doped GaAs layers whose doping levels are determined by Hall measurements. The spectral shift and the broadening of luminescence spectra are a signature of an increased density of electrons. They are compared to experimental data available in the literature. We infer n-type doping of 1×10^{18}cm^{-3} to 2×10^{18}cm^{-3}, with a high spatial homogeneity along the nanowire.

This method could be extended p-type nanowires, and polycrystalline thin films developed for photovoltaic applications. It provides an alternative way to probe carrier concentration in nanostructured and polycrystalline semiconductors, and to map the spatial inhomogeneity of dopants.

ACKNOWLEDGMENTS

This work was partly supported by the French ANR projects NANOCELL (ANR-RF-2015-01) and HETONAN. The Attolight cathodoluminescence tool was funded by public grants supported by the Region Ile-de-France in the framework of C'Nano IdF (nanoscience competence center of Paris Region), by the European Union (FEDER 2007-2013), and by the Labex GANEX (ANR-11-LABX-0014) and NanoSaclay (ANR-10-LABX-0035) as part of the "Investissements d'Avenir" program managed by the French National Research Agency (ANR).

REFERENCES

[1] Gaute Otnes, Magnus T. Borgström, "Towards high efficiency nanowire solar cells," *Nano Today*, in press (DOI: 10.1016/j.nantod.2016.10.007), 2017.

[2] M. Yao et al., "Tandem Solar Cells Using GaAs Nanowires on Si: Design, Fabrication, and Observation of Voltage Addition," *NanoLetters*, vol. 15, no. 11, pp. 7217–7224, 2015.

[3] R. R. LaPierre, "Numerical model of current-voltage characteristics and efficiency of GaAs nanowire solar cells," *Journal of Applied Physics*, vol. 109, no. 3, p. 034311, Feb. 2011.

[4] M. Yao *et al.*, "GaAs Nanowire Array Solar Cells with Axial p–i–n Junctions," *Nano Lett.*, vol. 14, no. 6, pp. 3293–3303, juin 2014.

[5] K. Storm *et al.*, "Spatially resolved Hall effect measurement in a single semiconductor nanowire," *Nat Nano*, vol. 7, no. 11, pp. 718–722, Nov. 2012.

[6] O. Hultin, G. Otnes, M. T. Borgström, M. Björk, L. Samuelson, and K. Storm, "Comparing Hall Effect and Field Effect Measurements on the Same Single Nanowire," *Nano Lett.*, vol. 16, no. 1, pp. 205–211, Jan. 2016.

[7] S. Arab, M. Yao, C. Zhou, P. D. Dapkus, and S. B. Cronin, "Doping concentration dependence of the photoluminescence spectra of n-type GaAs nanowires," *Applied Physics Letters*, vol. 108, no. 18, p. 182106, May 2016.

[8] H. J. Joyce, J. L. Boland, C. L. Davies, S. A. Baig, and M. B. Johnston, "A review of the electrical properties of semiconductor nanowires: insights gained from terahertz conductivity spectroscopy," *Semicond. Sci. Technol.*, vol. 31, no. 10, p. 103003, 2016.

[9] F. Glas, J.-C. Harmand, and G. Patriarche, "Why Does Wurtzite Form in Nanowires of III-V Zinc Blende Semiconductors?," *Phys. Rev. Lett.*, vol. 99, no. 14, p. 146101, Oct. 2007.

[10] E. O. Kane, "Band tails in semiconductors," *Solid-State Electronics*, vol. 28, no. 1, pp. 3–10, Jan. 1985.

[11] H. C. Casey and R. H. Kaiser, "Analysis of N☐Type GaAs with Electron☐Beam☐Excited Radiative Recombination," *J. Electrochem. Soc.*, vol. 114, no. 2, pp. 149–153, Feb. 1967.

[12] J. H. Neave, P. J. Dobson, J. J. Harris, P. Dawson, and B. A. Joyce, "Silicon doping of MBE-grown GaAs films," *Appl. Phys. A*, vol. 32, no. 4, pp. 195–200, Dec. 1983.

[13] M. Szmyd, P. Porro, A. Majerfeld, and S. Lagomarsino, "Heavily doped GaAs:Se. I. Photoluminescence determination of the electron effective mass," *Journal of Applied Physics*, vol. 68, no. 5, pp. 2367–2375, Sep. 1990.

[14] N.-Y. Lee *et al.*, "Determination of conduction band tail and Fermi energy of heavily Si-doped GaAs by room-temperature photoluminescence," *Journal of Applied Physics*, vol. 78, no. 5, pp. 3367–3370, Sep. 1995.

[15] De-Sheng, Y. Makita, K. Ploog, and H. J. Queisser, "Electrical properties and photoluminescence of Te-doped GaAs grown by molecular beam epitaxy," *Journal of Applied Physics*, vol. 53, no. 2, pp. 999–1006, Feb. 1982.

[16] Wurfel, "The chemical potential of radiation," *J. Phys. C: Solid State Phys.*, vol. 15, no. 18, p. 3967, Jun. 1982.

[17] J. K. Katahara and H. W. Hillhouse, "Quasi-Fermi level splitting and sub-bandgap absorptivity from semiconductor photoluminescence," *J. Appl. Phys.*, vol. 116, no. 17, p. 173504, 2014.

Optical Optimization of Passivated GaAs Nanowire Solar Cells

Kyle W. Robertson*, Ray R. LaPierre ‡, Jacob J. Krich*†

*Department of Physics, University of Ottawa, Ottawa, ON, K1N 6N5, Canada

†School of Electrical Engineering and Computer Science, University of Ottawa, Ottawa, Ontario, K1N 6N5, Canada

‡Department of Engineering Physics, McMaster University, Hamilton, Ontario, L8S 4L7, Canada

Abstract—We utilize rigorous coupled wave analysis (RCWA) to optimize the optical design of of GaAs nanowire solar cells. RCWA is highly accurate while being less computationally expensive than competing techniques. Using a simplex optimization method, we determine the optimal device geometries that maximize photocurrent density for bare nanowires and contacted nanowires in a dielectric material with a top ITO contact, finding agreement with previous work. We extend these results to include passivating shells, which are required for efficient devices. These optimizations give important guidance for geometric design of nanowire solar cell devices.

I. Introduction

Nanowire solar cells (NWSC) have emerged as promising candidates for highly efficient, inexpensive solar energy harvesters offering many unique benefits. With the correct choice of nanowire array geometry, one can reduce material requirements relative to flat-panel solar cells without sacrificing absorption or device efficiencies.

Nanowire solar cells inherently require less active semiconductor material relative to planar solar cells, making them a potentially less expensive alternative to conventional solar cell designs. The small cross-sectional area of the nanowires allows them to be grown on lattice-mismatched substrates with the resulting strain accommodated without dislocations. This flexibility enables growth on inexpensive substrates and creation of tandem solar cells of lattice-mismatched materials, further reducing cost and improving potential efficiency [1], [2].

The excellent broadband absorption of NWSC has been widely demonstrated [3]–[6] and is due to nanowires having larger absorption cross sections than their physical cross sections. This increased absorption cross section results from a self-concentrating effect [7] that focuses carrier generation into the tip of the nanowire. The optical modes of the nanowire create internal resonances that enhance overall absorption [8]–[10]. The geometric design of the nanowire structure must be properly optimized to maximize the benefit of these performance-enhancing effects.

Despite the promising optical properties of NWSC, their large surface area contributes to high levels of surface recombination in GaAs nanowires. This problem has been solved by introducing a passivating shell of higher bandgap material

around the nanowire core (as illustrated in Figure 1), which has proven to be an effective method for increasing device performance [11]–[13].

While the necessity of a passivating shell for electrical device performance has been well demonstrated [14], little work has been done to investigate the optical consequences of adding such a shell. Because the shell material has a higher bandgap and different optical properties than the core material, it has the potential to change the optical characteristics of nanowire devices and thus the optimal geometric design. Huang and Povinelli consider AlGaAs passivation shells without planarizing dielectric and ITO contact [15]. In this work, we use rigorous coupled wave analysis (RCWA) to perform wave-optics simulations of a square array of cylindrical GaAs nanowires passivated by AlInP shells including conducting ITO and planarizing dielectric, which are important for real devices. We discuss the advantages of the RCWA method and present the shell thickness, core radius, and array periodicity that optimize the maximal photocurrent density J_{ph}. We then give valuable guidance for the design of optical NWSC arrays.

II. Methods

RCWA (also referred to as the Fourier modal method) is a semi-analytic, frequency-domain method for solving Maxwell's equations [16], [17]. Unlike the finite element method (FEM) or finite difference methods, RCWA propagates fields in the z-direction analytically, reducing the number of discretized dimensions from three to two. This analytic propagation requires the device to be composed of discrete layers, where the material parameters must be constant along the z-direction within a layer. Vertical nanowire designs (see Figure 1) are ideal systems for such analysis. Because the axial spatial direction is treated analytically, simulation times are independent of the layer thicknesses and any number of layer thicknesses can be explored during a single simulation essentially for free. We implement our model using the open-source RCWA library S^4 [18]. At each frequency, S^4 expands the electromagnetic fields within each layer using a 2D plane-wave basis with exponential dependence in the axial direction. The fields are made to satisfy continuity conditions at layer interfaces using the S-matrix algorithm [18].

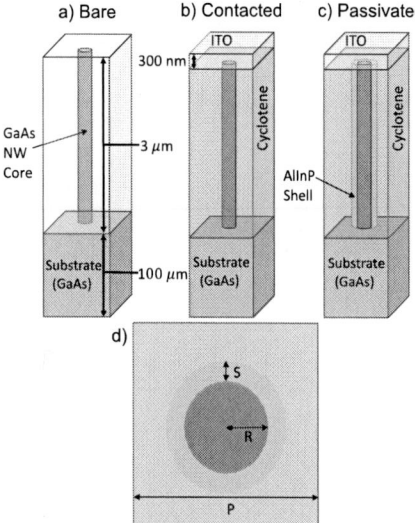

Fig. 1. A single unit cell of the square lattice nanowire array. a) An unpassivated and uncontacted bare nanowire. b) An unpassivated, contacted nanowire planarized with dielectric material. Although there is a non-trivial geometry in the x-y plane, the material parameters remain constant along z within each layer. c) A fully passivated nanowire. d) A top down view of the nanowire layer with core radius R, array period P, and shell thickness S.

We simulate nanowire array absorption by dividing the AM1.5G spectrum into 120 equally spaced frequency bins from 350-900 nm. A circularly polarized plane wave is injected at normal incidence onto the first layer of the device. Due to the circular symmetry of the nanowire, this polarization choice removes the need to simulate two independent polarizations for incoherent illumination [14]. We impose periodic boundary conditions at the x and y limits of the simulation domain to simulate an array of nanowires. The materials are defined through their frequency-dependent refractive index n and extinction coefficient k taken from bulk material [19], [20]. Referring to Fig. 1, we first model a bare nanowire. We then model a nanowire contacted by a top layer of ITO and planarized with a cyclotene dielectric (contacted nanowire). Finally, we model a contacted, planarized, passivated nanowire (passivated nanowire).

We seek to maximize the photocurrent density, where we assume that each absorbed photon produces one excited electron. Thus, we maximize spectral absorptance weighted by photon number

$$\bar{A} = \frac{\int_{\lambda_1}^{\lambda_2} \frac{\lambda}{hc} A(\lambda) I(\lambda) d\lambda}{\int_{\lambda_1}^{\lambda_2} \frac{\lambda}{hc} I(\lambda) d\lambda} \tag{1}$$

where λ is the photon wavelength, h is Planck's constant, c is the speed of light, $A(\lambda)$ is the wavelength dependent absorptance of the NWSC, $I(\lambda)$ is the AM1.5G spectrum in W·m^{-2}·nm^{-1} [21], and hc/λ is the energy of the incident photon. The integration limits are $\lambda_1 = 350$ nm and $\lambda_2 = 900$ nm, which contains all photons with wavelength

shorter than the GaAs band gap of 870.5 nm. Assuming perfect carrier collection, the photocurrent density would be

$$J_{ph} = \frac{q}{hc} \int_{\lambda_1}^{\lambda_2} \lambda A(\lambda) I(\lambda) d\lambda \tag{2}$$

which is directly relatable to Eq. (1) by $J_{ph} = \bar{A} \cdot 33.37$ mA/cm^2. We obtain the wavelength-dependent absorption coefficients from

$$R(\lambda) = \frac{P_{ref}}{P_{in}} \tag{3}$$

$$T(\lambda) = \frac{P_{tran}}{P_{in}} \tag{4}$$

$$A(\lambda) = 1 - R(\lambda) - T(\lambda) \tag{5}$$

where P_{in} is the incident power, and P_{ref} and P_{tran} are the reflected and transmitted powers, respectively.

There exists a subtle but important detail concerning in which regions of the device we choose to calculate the absorptance A. Referring back to Figure 1, there are two reasonable choices. The first choice would be to consider A of the entire device, including the substrate. The second choice would be to instead consider A of only the nanowire, neglecting any absorption in the substrate. Computationally, this comes down to a choice of where we calculate $T(\lambda)$. The first choice places the calculation at the bottom of the substrate, while the second places the calculation at the nanowire-substrate interface.

Collection of carriers generated in the substrate is dependent on the diffusion length of the carrier, and thus also on the material quality of the substrate. Assuming reasonable diffusion lengths in the substrate, most carriers generated in the substrate are not collectable [14]. We therefore want to maximize the number of charge carriers generated in the nanowire near the p-n junction of the device. This choice decreases the distance carriers must travel to reach the junction and thus increases their likelihood of being separated and collected. In axial junction nanowires, the junction is usually placed near the top of the nanowire core. By choosing to optimize A of only the nanowire and ignoring any absorption in the substrate, the optimizer searches for a device geometry that shifts absorption into the nanowire. This geometric optimization achieves the desired effect of concentrating carrier generation near the junction. In this work we choose to calculate $T(\lambda)$ at the nanowire-substrate interface and thus all absorptance values exclude absorption in the substrate. Note these absorptance values include the absorptance of the ITO layer. For the optimized contacted nanowire, ITO absorption only contributes 2.6% to the integrated absorptance and does not significantly affect the resulting optimal geometric parameters.

As a first example of modeling NWSC via the RCWA method, we use Nelder-Mead simplex optimization to find the optimal array period and nanowire radius. This optimization allows us to find the desired geometric parameters to high precision without the unnecessary overhead of blind parameter sweeps. During all optimizations, the ITO thickness is fixed at 300 nm, the nanowire length at 3 μm, and the substrate thickness at 100 μm.

978-1-5090-5606-4/17 $31.00 © 2017 IEEE

TABLE I
Optimizations results for bare and contacted nanowires. J_{ph} includes absorption in the ITO and nanowire layers and excludes the substrate.

System	Array Period [nm]	Core Radius [nm]	J_{ph} [mA cm^{-2}]
Bare Nanowire	338	85	29.37
Contacted nanowire	326	103	28.55

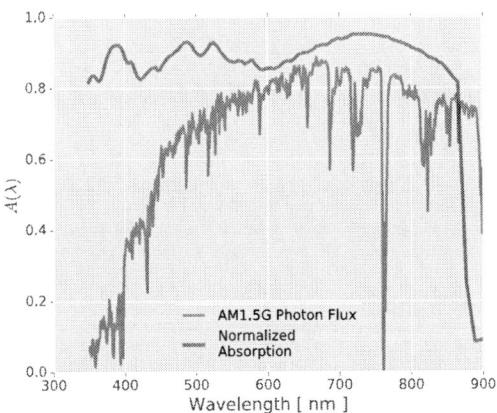

Fig. 2. Absorbance, $A(\lambda)$, in the ITO and nanowire regions of the optimized contacted nanowire (see Table I) plotted with the photon-weighted AM1.5G spectrum, $\frac{\lambda}{hc}I(\lambda)$, in arbitrary units.

We find the array period and core radius that maximize J_{ph} for the bare nanowire (shown in Table I), in good agreement with previous FEM-based results [19]. We use results from this optimization as a starting point to find the optimal parameters for the contacted nanowire. We go on to extend previous work by exploring the effects of adding an AlInP passivating shell of varying thickness to the contacted nanowire and optimizing the array period and core radius for passivated nanowire structures.

III. RESULTS

A. Bare & Contacted Nanowire

When ITO and cyclotene are added to the optimized bare nanowire, J_{ph} drops from 29.37 mA/cm^2 to 28.01 mA/cm^2. Reoptimizing the geometry recovers 40% of the loss to J_{ph}, resulting in a value of 28.55 mA/cm^2. Table I shows the optimal geometric parameters for the bare nanowire and the contacted nanowire, indicating the small change in the optimal array period and 21% change in the optimal core radius when the ITO and dielectric are included. The absorption spectrum of the ITO and nanowire layers for a contacted nanowire using those parameters is shown in Figure 2. As expected, the geometry of the nanowire is tuned to place an absorption peak near the strongest part of the spectrum, between roughly 600-800 nm. Figure 3 shows the generation rate $G(\vec{r})$ along a 2D slice through the contacted nanowire [14],

$$G(\vec{r}) = \sum_i \frac{\epsilon_0}{\hbar}|E(\lambda_i, \vec{r})|^2 n(\lambda_i)k(\lambda_i) \qquad (6)$$

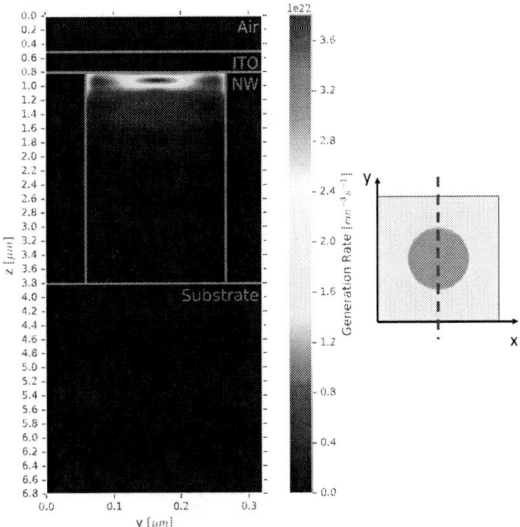

Fig. 3. Generation rate in a contacted nanowire along a fixed y-z plane through the center of the nanowire core, as shown in the schematic on the right. The nanowire is stretched horizontally for clarity. Strong self-concentrating of light at the tip of the nanowire is apparent.

where λ_i are the simulated wavelengths, $E(\lambda_i, \vec{r})$ is the electric field at wavelength λ_i and position vector $\vec{r}$, ϵ_0 is the vacuum permittivity, and $\hbar$ is the reduced Planck's constant. Our optimization has concentrated carrier generation within the desired region at the tip of the nanowire.

B. Passivated Nanowire

Optically, there is no benefit to including a passivating shell. Instead, we look for the best optical designs including the electrically-necessary passivation. First, we sweep through shell thicknesses of 10-80 nm while keeping the total core+shell radius fixed at 103 nm and array period fixed at 326 nm, which are the optimal values for the contacted nanowire (see Table I). Results in Figure 4 show that, for shells thinner than 40 nm, absorption losses are minimal and that dramatic losses occur when including a thicker shell. In terms of overall performance, the passivating shell is a necessity and the small absorption gains accrued from removing it do not outweigh the performance costs due to surface recombination.

Reoptimizing to find the optimal core radius and array period at each shell thickness recovers almost entirely the optical properties of the contacted nanowire (see Figure 4). Here lies the value of device optimization through simulation. We see that by reoptimizing the geometry of the nanowire array, losses to J_{ph} can be almost entirely recovered and it is possible to achieve the electrical benefits of a passivating shell without negatively affecting device absorption.

The parameter values resulting from our optimization are shown in Figure 5. These parameters vary considerably from the contacted nanowire optimal values, and in general one needs to increase the array period and decrease the core radius as the shell thickness increases. A large change in optimal geometric parameters occurs at a 40 nm shell thickness. At a

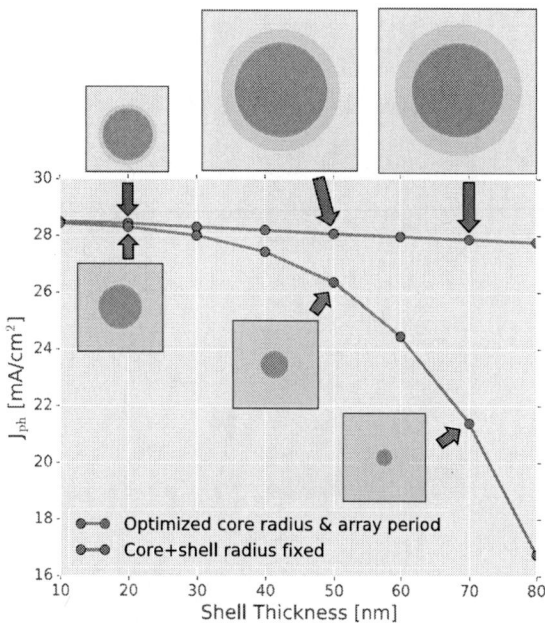

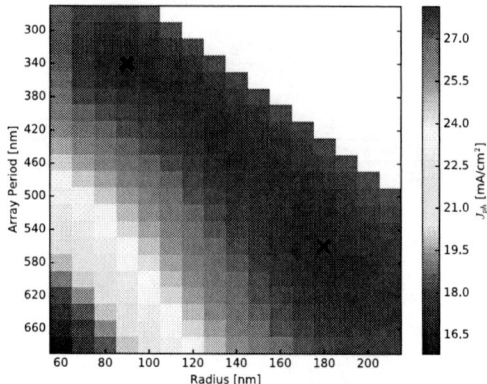

Fig. 6. A heatmap showing the photocurrent density for a shell thickness of 40 nm as a function of the array period and core radius. The white regions correspond to a combined core+shell diameter that would have exceeded the array period at that point. The two local optima are marked with a small x.

Fig. 4. Maximal photocurrent densities for passivated nanowires with fixed shell thickness. The unoptimized data has an array period of 326 nm and the sum of the core radius and shell thickness fixed to 103 nm (see Table I). The optimized data are the maximal photocurrent densities resulting from a reoptimization of the array period and core radius at each shell thickness. Insets are to-scale representations of the top-down view of the nanowire geometries.

IV. CONCLUSION

In this paper, we describe a method for efficient optical simulation of NWSC devices. Our first demonstration of this method is a geometric optimization to maximize photon-weighted optical absorption. This method will be useful for combined optical and electrical simulations [14], with a significant reduction in computational cost. We find that optimal absorption occurs in the absence of a passivating shell, and outline the losses to the photocurrent density as a function of the shell thickness. In the presence of a shell of fixed thickness, we present the array period and core radius that maximize the photocurrent density and show that by reoptimizing the nanowire geometry the losses incurred by adding a shell can be almost completely recovered. This information will be invaluable for use in growing efficient GaAs NWSC.

REFERENCES

[1] K. L. Kavanagh, "Misfit dislocations in nanowire heterostructures," *Semiconductor Science and Technology*, vol. 25, no. 2, p. 024006, 2010.

[2] V. Dhaka, T. Haggren, H. Jussila, H. Jiang, E. Kauppinen, T. Huhtio, M. Sopanen, and H. Lipsanen, "High Quality GaAs Nanowires Grown on Glass Substrates," *Nano Letters*, vol. 12, no. 4, pp. 1912–1918, Apr. 2012.

[3] L. Hu and G. Chen, "Analysis of Optical Absorption in Silicon Nanowire Arrays for Photovoltaic Applications," *Nano Letters*, vol. 7, no. 11, pp. 3249–3252, Nov. 2007.

[4] J. Zhu, Z. Yu, G. F. Burkhard, C.-M. Hsu, S. T. Connor, Y. Xu, Q. Wang, M. McGehee, S. Fan, and Y. Cui, "Optical Absorption Enhancement in Amorphous Silicon Nanowire and Nanocone Arrays," *Nano Letters*, vol. 9, no. 1, pp. 279–282, Jan. 2009.

[5] M. D. Kelzenberg, S. W. Boettcher, J. A. Petykiewicz, D. B. Turner-Evans, M. C. Putnam, E. L. Warren, J. M. Spurgeon, R. M. Briggs, N. S. Lewis, and H. A. Atwater, "Enhanced absorption and carrier collection in Si wire arrays for photovoltaic applications," *Nature Materials*, vol. 9, no. 3, pp. 239–244, Mar. 2010.

[6] M. Yao, N. Huang, S. Cong, C.-Y. Chi, M. A. Seyedi, Y.-T. Lin, Y. Cao, M. L. Povinelli, P. D. Dapkus, and C. Zhou, "GaAs Nanowire Array Solar Cells with Axial pin Junctions," *Nano Letters*, vol. 14, no. 6, pp. 3293–3303, Jun. 2014.

[7] X. Wang, M. R. Khan, M. Lundstrom, and P. Bermel, "Performance-limiting factors for GaAs-based single nanowire photovoltaics," *Optics Express*, vol. 22, no. S2, p. A344, Mar. 2014.

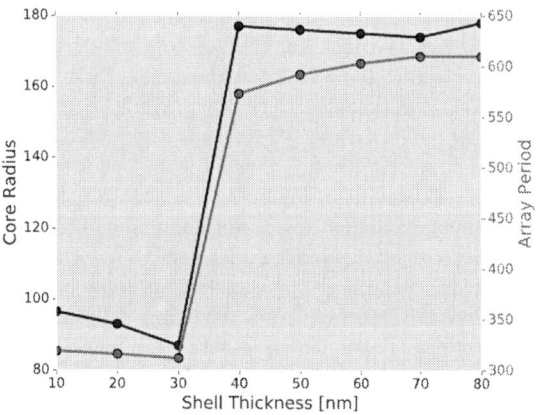

Fig. 5. Reoptimized core radius (left axis) and array period (right axis) that give the maximum possible J_{ph} for fixed shell thicknesses in the passivated nanowires, corresponding to Figure 4. Lines are guides to the eye.

40 nm shell thickness, two local optima exist in the parameter space, which differ in J_{ph} by only 0.004% (see Figure 6). Below a 40 nm shell thickness, the small-period optimum is the global optimum. At and above a 40 nm shell thickness, the large-period optimum is the global optimum. For a shell thickness of 70 nm, the two optima differ by 1.2%.

[8] L. Cao, P. Fan, A. P. Vasudev, J. S. White, Z. Yu, W. Cai, J. A. Schuller, S. Fan, and M. L. Brongersma, "Semiconductor Nanowire Optical Antenna Solar Absorbers," *Nano Letters*, vol. 10, no. 2, pp. 439–445, Feb. 2010.

[9] D. Wu, X. Tang, K. Wang, and X. Li, "An Analytic Approach for Optimal Geometrical Design of GaAs Nanowires for Maximal Light Harvesting in Photovoltaic Cells," *Scientific Reports*, vol. 7, 2017.

[10] K. T. Fountaine, W. S. Whitney, and H. A. Atwater, "Resonant absorption in semiconductor nanowires and nanowire arrays: Relating leaky waveguide modes to Bloch photonic crystal modes," *Journal of Applied Physics*, vol. 116, no. 15, p. 153106, Oct. 2014.

[11] R. R. LaPierre, "Numerical model of current-voltage characteristics and efficiency of GaAs nanowire solar cells," *Journal of Applied Physics*, vol. 109, no. 3, p. 034311, Feb. 2011.

[12] I. Åberg, G. Vescovi, D. Asoli, U. Naseem, J. P. Gilboy, C. Sundvall, A. Dahlgren, K. E. Svensson, N. Anttu, M. T. Björk, and L. Samuelson, "A GaAs Nanowire Array Solar Cell With 15.3%; Efficiency at 1 Sun," *IEEE Journal of Photovoltaics*, vol. 6, no. 1, pp. 185–190, Jan. 2016.

[13] C.-C. Chang, C.-Y. Chi, M. Yao, N. Huang, C.-C. Chen, J. Theiss, A. W. Bushmaker, S. LaLumondiere, T.-W. Yeh, M. L. Povinelli, C. Zhou, P. D. Dapkus, and S. B. Cronin, "Electrical and Optical Characterization of Surface Passivation in GaAs Nanowires," *Nano Letters*, vol. 12, no. 9, pp. 4484–4489, Sep. 2012.

[14] A. H. Trojnar, C. E. Valdivia, R. R. LaPierre, K. Hinzer, and J. J. Krich, "Optimizations of GaAs Nanowire Solar Cells," *IEEE Journal of Photovoltaics*, vol. 6, no. 6, pp. 1494–1501, Nov. 2016.

[15] N. Huang and M. L. Povinelli, "Design of Passivation Layers on Axial Junction GaAs Nanowire Solar Cells," *IEEE Journal of Photovoltaics*, vol. 4, no. 6, pp. 1511–1517, Nov. 2014.

[16] M. G. Moharam and T. K. Gaylord, "Rigorous coupled-wave analysis of planar-grating diffraction," *JOSA*, vol. 71, no. 7, pp. 811–818, Jul. 1981.

[17] M. G. Moharam, T. K. Gaylord, E. B. Grann, and D. A. Pommet, "Formulation for stable and efficient implementation of the rigorous coupled-wave analysis of binary gratings," *JOSA a*, vol. 12, no. 5, pp. 1068–1076, 1995.

[18] V. Liu and S. Fan, "S4 : A free electromagnetic solver for layered periodic structures," *Computer Physics Communications*, vol. 183, no. 10, pp. 2233–2244, Oct. 2012.

[19] Y. Hu, R. R. LaPierre, M. Li, K. Chen, and J.-J. He, "Optical characteristics of GaAs nanowire solar cells," *Journal of Applied Physics*, vol. 112, no. 10, p. 104311, 2012.

[20] S. Adachi, *Optical Constants of Crystalline and Amorphous Semiconductors: Numerical Data and Graphical Information.* New York: Springer Science + Business Media, LLC, 1999.

[21] "Standard Tables for Reference Solar Spectral Irradiances: Direct Normal and Hemispherical on 37 Tilted Surface," ASTM International, Tech. Rep., 2012.

978-1-5090-5606-4/17 $31.00 © 2017 IEEE

High efficiency GaN nanowire/Si photocathode for photoelectrochemical water splitting

Srinivas Vanka[1], Sheng Chu[1], Yichen Wang[1], Ishiang Shih[1], Hong Guo[2], and Zetian Mi[1,3]

[1]Department of Electrical and Computer Engineering, McGill University, Montreal, Quebec, Canada, H3A0E9
[2]Department of Physics, McGill University, Montreal, Quebec, Canada, H3A2T8
[3]Department of Electrical Engineering and Computer Science, Centre for Photonics and Multiscale Nanomaterials, University of Michigan, Ann Arbor, MI, USA, 48105

Abstract —**Photoelectrochemical (PEC) water splitting is one of the most promising technologies to convert solar energy into chemical energy in the form of hydrogen production. III-nitride semiconductors are promising materials to realize high efficiency photoelectrodes: their energy bandgap can be varied across nearly the entire solar spectrum by changing the alloy compositions and the band edge positions straddle water oxidation and reduction potentials under deep visible light irradiation. In this work, we report on the demonstration of a Si-based photocathode, which consists of n^+-GaN nanowire arrays grown on n^+-p Si substrate and enables relatively high efficiency (~9%) and highly stable solar water splitting.**

I. INTRODUCTION

Hydrogen from solar water splitting is an excellent clean fuel which can potentially address the ever-increasing energy problems of the world. PEC cells are the complete version of photovoltaic (PV)–electrolyzer (EL) systems. Both systems require three main steps: light absorption, charge separation and catalysis. In PV-EL systems, light absorption and initial charge transfer occur within the PV and these processes are physically separated from the electrolyte and fuel production. PEC systems are different from PV-EL ones in that the light absorbers are in direct contact with the electrolyte. To date, the highest efficiency reported for PV-EL devices is over 20% [1]. However, there are many limitations for the practical applications of these devices, including high system cost, relatively small size, limited scalability, and difficulty in generating other forms of chemical fuels [2, 3].

PEC water splitting is a promising and environmentally benign method for solar hydrogen generation and has attracted significant attention over past few decades. A self-driven tandem photoelectrode, which has cost advantage over single photoelectrodes working under applied bias, consists of a photocathode and a photoanode, which collect electrons and holes for surface hydrogen and oxygen reactions, respectively. To realize an efficient and stable photocathode, the semiconductor light absorber should have a conduction band minimum (CBM) more negative than that required for

hydrogen evolution reaction (HER). To lower the overpotential for hydrogen production, metal catalysts are usually deposited on the surface of photoelectrodes to catalyze the HER. Fujishima and Honda first demonstrated the concept of PEC water splitting for hydrogen production using TiO_2 photoelectrode [4]. Since then many materials, including Si, GaP, InP and their alloys have been explored to enhance the PEC performance, including the efficiency and stability of photocathodes [5-9]. PEC cells composed of III-V semiconductors have demonstrated solar-to-hydrogen (STH) efficiency as high as 16% [10], while multi-junction silicon PEC cells yielded efficiencies of 4.7% [11] and 7.8% [12]. The use of Si-based photoelectrodes offers the critical advantage of low cost, scalable manufacturing.

Recently, III-nitride semiconductors have been explored for both photocatalytic and PEC water splitting [13-17]. These materials have several distinct advantages compared to other materials: (1) The bandgap can be tuned from 0.65 eV to 3.4 eV which covers nearly the entire solar spectrum; (2) The conduction and valence band edges straddle water redox potentials for indium compositions up to 50% which corresponds to a bandgap of ~ 1.7 eV; (3) High stability in electrolyte due to the recently discovered N-surface passivation [18, 19]. Improved PEC activities have been demonstrated by using GaN nanowire structures, due to the large surface-to-volume ratios, rapid charge carrier separation, and enhanced optical absorption [17]. In this work, we report on the monolithic integration of n^+-GaN nanowire arrays directly on planar n^+-p Si substrate. Such a Si-based photocathode can exhibit an energy conversion efficiency ~ 9%.

II. EXPERIMENTAL

The n^+-p Si substrates were fabricated using thermal diffusion process in a furnace. To test the efficiency of the solar cell, metal contacts were made on p and n sides of the wafer. It was observed that the unpassivated solar cell exhibited an open circuit potential (V_{oc}) ~0.52 V and short circuit current (J_{sc}) ~ 35 mA/cm^2 under AM 1.5G 1 sun illumination. The efficiency of the n^+-p Si solar cell wafer was ~12 % with a fill factor (FF) of ~67%.

978-1-5090-5606-4/17 $31.00 © 2017 IEEE

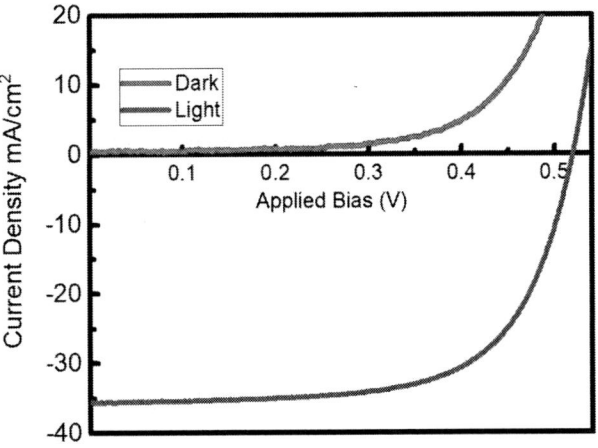

Figure 1. Current voltage characteristics for an n^+-p Si solar cell device under both light (AM 1.5G 1 sun illumination) and dark illumination.

In this study, n^+-GaN nanowire arrays were grown on the n^+-p Si substrates using a Veeco GEN II radio frequency plasma-assisted molecular beam epitaxial growth system. The nanowires were formed spontaneously under nitrogen-rich conditions without using any external metal catalyst. The growth conditions included a substrate temperature of 790 °C, a gallium beam equivalent pressure (BEP) of 6×10^{-8} torr, nitrogen flow rate of 1 standard cubic centimeter per minute (*sccm*) and plasma forward power of 350 W. Before growth initation, oxide desorption was performed at 850 °C. The schematic of n^+-GaN nanowires grown on n^+-p Si solar cells is shown in figure 2(a). The nanowires are vertically aligned and have lengths between 250-300 nm. The SEM image of GaN nanowire arrays grown on Si is shown in figure 2(b). Subsequently, platinum (Pt) co-catalyst nanoparticles were photo-chemically deposited on the nanowires.

(a)

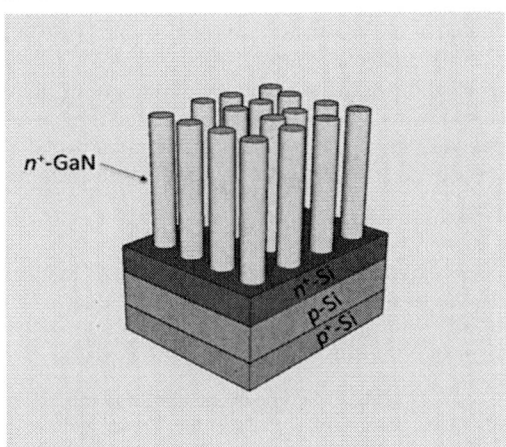

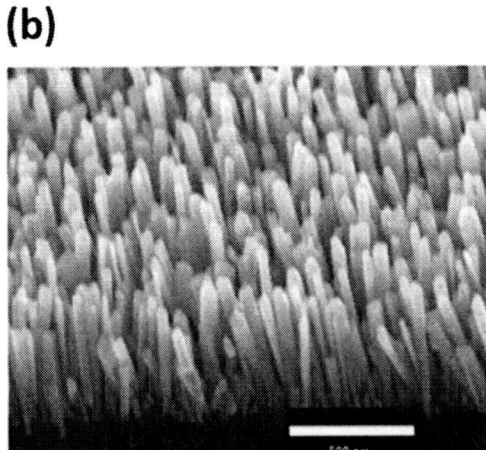

Figure 2. (a)Schematic and (b) SEM image of n^+-GaN nanowires grown on n^+-p Si substrate.

The PEC reaction was first conducted in 1 M HBr solution in a three-electrode configuration. The n^+-GaN/Si nanowires, silver chloride electrode (Ag/AgCl), and Pt wire served as the working, reference, and counter electrodes, respectively. The reaction was conducted in a 450 ml PEC quartz cell. A solar simulator (Newport Oriel) was used as the light source, and the light intensity was calibrated to be 100 mW/cm^2 for all experiments.

III. RESULTS AND DISCUSSION

LSV curves of platinized n^+-GaN nanowires on n^+-p Si substrate and platinized solar cell (n^+-p Si solar cell) are shown in the figure 3(a). The onset potential for the n^+-GaN nanowires on n^+-p Si device is ~ 0.5 V vs NHE. The saturation current is ~ 34 mA/cm^2 at 0.2 V vs NHE under AM 1.5G 1 sun illumination. The maximum applied bias photon-to-current efficiency (ABPE) is ~ 9% at 0.32 V vs NHE (shown in figure 3(b)), which is among the best reported efficiency values Si based photocathodes [5],[9].

(a)

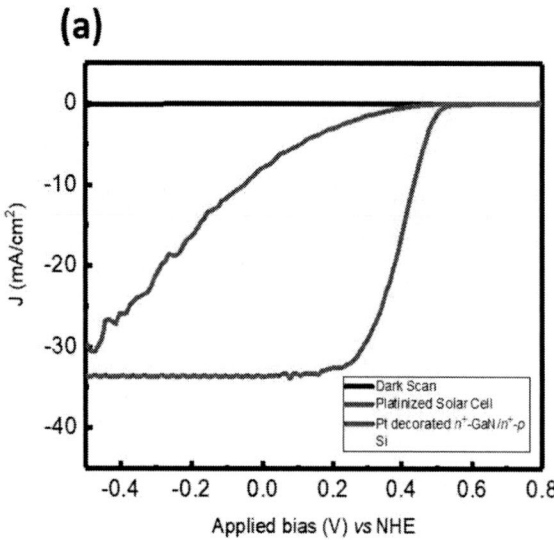

(b)

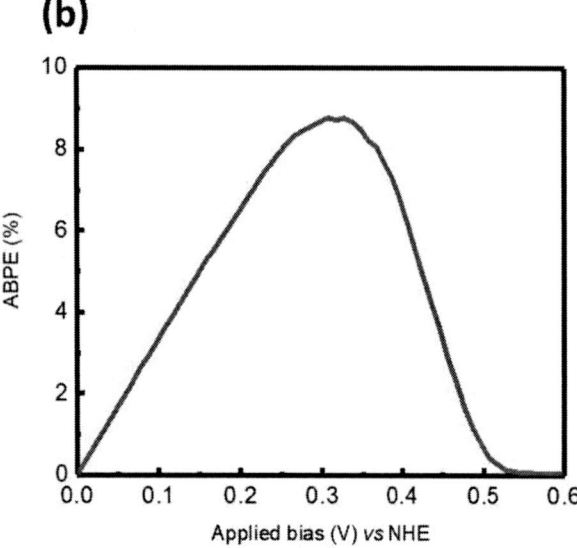

Figure 3. (a) LSV curves for platinized Si solar cell, n^+-GaN nanowires/n^+-p Si under AM 1.5G 1 sun illumination and dark conditions in 1 M HBr. (b) Applied bias photon-to-current efficiency (ABPE) of n^+-GaN nanowires/n^+-p Si device.

It can be seen from figure 3(a) that platinized n^+-p Si without GaN nanowires has a significantly poor onset potential and very small fill factor. In contrast, significantly improved onset potential and fill factor is achieved with the presence of n^+-GaN nanowire arrays. As shown in figure 4, due to the relatively small offset between the Si and GaN conduction band edges and the heavy n-type doping, photoexcited electrons from the underlying Si solar cell are readily injected into the n^+-GaN nanowire segment. Electrons migrate to the surface of nanowires and participate in proton reduction, due to the reduced surface band bending and the highly uniform Pt nanoparticles coverage on the nanowire surfaces. The superior performance of n^+-GaN nanowires on n^+-p Si substrate can be attributed to the light trapping effect of nanowires (anti-reflection effect), enhanced charge carrier extraction of n-GaN nanowires, and improved semiconductor/Pt/electrolyte interface.

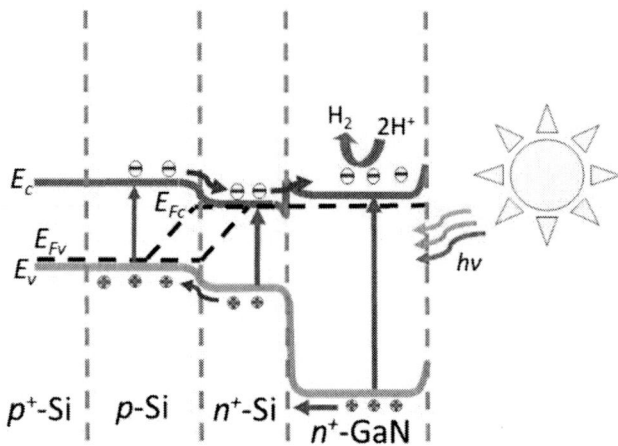

Figure 4. Band-diagram of n^+-GaN nanowires/n^+-p Si.

We further studied the stability of such a Si-based photocathode, which exhibited stability for ~2 hours under AM 1.5G 1 sun illumination in 1 M HBr without any extra surface protection. The high stability of defect free GaN nanowire photocatalyst is attributed to the recently discovered N-rich surface of GaN when grown under N-rich conditions by MBE, which protects against photocorrosion and oxidation [19]. It was also found that such a Si-based photocathode showed similar PEC performance when it was tested in 0.5 M H_2SO_4 under AM 1.5G 1 sun illumination in a three-electrode system. Further studies are required to improve the efficiency of the device.

IV. CONCLUSION

In this work n^+-GaN nanowires were grown on n^+-p Si wafer using PA-MBE. Due to the unique lateral carrier extraction of 1D nanowires proton reduction was significantly enhanced which is complemented by the efficient light absorption of the underlying n^+-p Si substrate. High stability of these nanowires is attributed to the N-rich surfaces of GaN nanowire structures [19]. Further studies are currently being explored for these III-nitrides to improve the efficiency with the use of indium rich InGaN nanowires. It is also important to develop a highly efficient and stable photoanode, which can be paired with the photocathode to achieve unassisted PEC water splitting.

ACKNOWLEDGEMENTS

This work was supported by the Natural Sciences and Engineering Research Council of Canada (NSERC).

REFERENCES

[1] J. Jia, L. C. Seitz, J. D. Benck, Y. Huo, Y. Chen, J. W. Ng, T. Bilir, J. S. Harris and T. F. Jaramillo, "Solar water splitting by photovoltaic-electrolysis with a solar-to-hydrogen efficiency over 30." *Nature Communications*, 2016. 7: p. 13237.

[2] M. Y. Wang, Z. Wang, X. Gong, and Z. Guo, "The intensification technologies to water electrolysis for hydrogen production - A review." *Renewable & Sustainable Energy Reviews*, 2014. 29: p. 573-588.

[3] Y. Kawashima, I. Honda, and T. Hosokawa, "A Study on Hydrogen Generation Using PEM Cell and Solar Battery."

Developments in Chemical Engineering and Mineral Processing, 2003. 11(5-6): p. 521-528.

[4] A. Fujishima and K. Honda, "Electrochemical Photolysis of Water at a Semiconductor Electrode." *Nature*, 1972, 238(5358): 37-38.

[5] S. W. Boettcher, E. L. Warren, M. C. Putnam, E. A. Santori, D. Turner-Evans, M. D. Kelzenberg, M. G. Walter, J. R. McKone, B. S. Brunschwig, H. A. Atwater, and N. S. Lewis, "Photoelectrochemical Hydrogen Evolution Using Si Microwire Arrays." *Journal of the American Chemical Society*, 2011. 133(5): p. 1216-1219.

[6] O. Khaselev and J. A. Turner, "A monolithic photovoltaic-photoelectrochemical device for hydrogen production via water splitting", *Science*, 1998, 280, 425–427.

[7] A. Krawicz, J. H. Yang, E. Anzenberg, J. Yano, I. D. Sharp, and G. F. Moore, "Photofunctional Construct That Interfaces Molecular Cobalt-Based Catalysts for H_2 Production to a Visible-Light-Absorbing Semiconductor." *Journal of the American Chemical Society*, 2013. 135(32): p. 11861-11868.

[8] L. Gao, Y. Cui, J. Wang, A. Cavalli, A. Standing, T. Vu, M. Verheijen, J. Haverkort, E. P. Bakkers and P. H. Notten, "Photoelectrochemical hydrogen production on InP nanowire arrays with molybdenum sulfide electrocatalysts", *Nano Letters*, 2014, 14, 3715−3719.

[9] M.G. Walter, E. L. Warren, J. R. McKone, S. W. Boettcher, Q. X. Mi, E. A. Santori, and N. S. Lewis, "Solar Water Splitting Cells." *Chemical Reviews*, 2010. 110(11): p. 6446-6473.

[10] J. L. Young, M. A. Steiner, H. Döscher, R. M. France, J. A. Turner and T. G. Deutsch "Direct solar-to-hydrogen conversion via inverted metamorphic multi-junction semiconductor architectures." *Nature Energy*, 2017. **2**: p. 17028.

[11] S.Y. Reece, J. A. Hamel, K. Sung, T. D. Jarvi, A. J. Esswein, J. J. H. Pijpers and D. G. Nocera, "Wireless Solar Water Splitting Using Silicon-Based Semiconductors and Earth-Abundant Catalysts." *Science*, 2011. **334**(6056): p. 645-648.

[12] R.E. Rocheleau, E.L. Miller, and A. Misra, "High-efficiency photoelectrochemical hydrogen production using multijunction amorphous silicon photoelectrodes." *Energy & Fuels*, 1998. **12**(1): p. 3-10.

[13] M. G. Kibria, F. A. Chowdhury, S. Zhao, B. AlOtaibi, M. L Trudeau, H. Guo, and Z Mi., "Visible light-driven efficient overall water splitting using *p*-type metal-nitride nanowire arrays." *Nature Communications*, 2015. 6:6797

[14] M. G. Kibria and Z. Mi, "Artificial photosynthesis using metal/nonmetal-nitride semiconductors: current status, prospects, and challenges." *Journal of Materials Chemistry A*, 2016. 4(8): p. 2801-2820.

[15] M.G. Kibria, S. Zhao, F. A. Chowdhury, Q. Wang, H. P. Nguyen, M. L. Trudeau, H. Guo, and Z. Mi, "Tuning the surface Fermi level on *p*-type gallium nitride nanowires for efficient overall water splitting." *Nature Communications*, 2014. 5: p. 3825.

[16] S. Fan, B. AlOtaibi, S. Y. Woo, Y. J. Wang, G. A. Botton, and Z. Mi, "High Efficiency Solar-to-Hydrogen Conversion on a Monolithically Integrated InGaN/GaN/Si Adaptive Tunnel Junction Photocathode." *Nano Letters*, 2015. 15(4): p. 2721-2726.

[17] B. AlOtaibi, H. P. T. Nguyen, S. Zhao, M. G. Kibria, S. Fan and Z. Mi, "Highly stable photoelectrochemical water splitting and hydrogen generation using a double-band InGaN/GaN core/shell nanowire photoanode", *Nano Letters*, 2013, 4356-4361

[18] J. S. Foresi and T. D. Moustakas, "Metal contacts to gallium nitride." *Applied Physics Letters*, 1993, 62(22): 2859.

[19] M. G. Kibria, R. Qiao, W. Yang, I. Boukahil, X. Kong, F. A. Chowdhury, M. L. Trudeau, W. Ji, H. Guo, F. J. Himpsel, L. Vayssieres, and Z. Mi, "Atomic-scale origin of long-term stability and high performance of *p*-GaN nanowire arrays for

photocatalytic overall pure water splitting." *Advanced Materials*, 2016. 28(38): 8388-8397.

Analytic description of the impact of grain boundaries on Voc

Paul Haney[1] and Benoit Gaury[1,2]

[1]Center for Nanoscale Science and Technology National Institute of Standards and Technology
Gaithersburg, MD, 20899, USA
[2]Maryland NanoCenter, University of Maryland College Park, MD 20742

Abstract—The impact of grain boundaries on the performance of polycrsytalline photovoltaics remains an open question. We present a simplified description of dark grain boundary recombination current. The dark current takes the form of a diode equation, and the model provides closed form expressions for the reverse saturation current and ideality factor in terms of grain boundary and system parameters. This model applies under conditions relevant for thin film photovoltaics such as CdTe, namely for p-type absorbers with reasonably high bulk hole mobility, positively charged grain boundaries with high defect density, and grains which are not fully depleted. The dark recombination current can be used to predict the open circuit voltage for a given short circuit density, providing a simple closed form expression which shows how grain boundaries impact V_{oc}.

Index Terms—polycrystalline solar cell, grain boundary recombination, open circuit voltage

I. INTRODUCTION

Thin films photovoltaics like CdTe and Cu(In,Ga)Se$_2$ exhibit high conversion efficiency despite their high defect density [1]. Grain boundaries are a primary source of defects, however so far polycrystalline samples outperform their single crystal counterparts [2]. This apparent dichotomoy between the electrical and structural properties invites the unexpected question: "Can grain boundaries be beneficial for photovoltaic performance?" There is not a clear consensus on this issue. In our view there is not a universal answer to this question, it depends on the details of the grain boundary and bulk properties. Grain boundaries may improve the performance of samples with exceedingly poor bulk properties, but are always detrimental to samples with reasonably good bulk properties. In the best cases, grain boundaries assist carrier collection at zero bias, increasing the short circuit density J_{sc}. This is due to the built-in electric field accompanying charged grain boundaries which separates carriers and inhibits recombination (provided that carriers driven to the grain boundary are majority carriers at the grain boundary core) [3], [4]. However in *all* cases grain boundaries are harmful for the open circuit voltage V_{oc} [5], [6]. This is not surprising since V_{oc} is reduced by recombination, and recombination is enhanced by grain boundaries when the device is under forward bias. However an intuitive, quantitative description of how grain boundaries reduce V_{oc} is still lacking, despite previous numerical and analytical works [5], [6], [7]. This article describes our recently developed models which provide this simple relation between grain boundary properties and V_{oc}.

II. GRAIN BOUNDARY RECOMBINATION

In a recent set of papers [8], [9], we have presented analytical expressions for the grain boundary dark recombination current. These expressions take the form of a general diode equation:

$$J_{GB}(V) = J_0 \exp\left(\frac{qV}{nk_BT}\right), \tag{1}$$

where $q > 0$ is the electron charge, V is the applied voltage, k_B is the Boltzmann constant, and T is the temperature. The result of our work is closed form expressions for the reverse saturation current J_0 and ideality factor n in terms of grain boundary and system parameters. We also demonstrated that the dark recombination current can be used to accurately predict the open circuit voltage for a given short circuit current density. In particular, numerical simulations indicate that the following relation holds:

$$V_{oc} = \frac{nk_BT}{q} \ln\left(\frac{J_{sc}}{J_0}\right). \tag{2}$$

The impact of grain boundary recombination on V_{oc} can therefore be concisely quantified. This may enable rational approaches to mitigating the negative consequences of grain boundary recombination.

Our analysis showed that the system behavior depends on the grain boundary defect type (*e.g.* donor or acceptor), the location of the defect energy level(s) with respect to midgap, the applied voltage, and other factors. The system behavior can be classified into several regimes, where each regime has its own peculiarities and requires its own detailed analysis. The myriad of different cases can obscure the overall picture of grain boundary recombination. Despite the differences between different regimes, a single framework for understanding the system response can be presented, which offers useful perspective when viewing the multiplicity of cases. In this work we aim to provide a more global, qualitative description of our model of grain boundary recombination current.

Before discussing grain boundary recombination, it's useful to rewrite Eq. (1) in a form which helps to frame our analysis. We write the dark recombination current as:

$$J = N\left(\frac{\lambda}{\tau}\right) \exp\left(\frac{qV}{nk_BT}\right). \tag{3}$$

978-1-5090-5606-4/17 $31.00 © 2017 IEEE

The form of Eq. (3) is fully general for thermally activated transport. Current density can always be written as the product of a density N times a velocity: length λ divided by time τ. The exponential factor describes the classic diode voltage dependence with ideality factor n. Although Eq. (3) is generic, when applied to a 1-dimensional p-n junction, the factors in Eq. (3) acquire quite specific physical interpretations. N and τ correspond to the density and effective lifetime of the species controlling the recombination, and λ is the length scale over which recombination occurs. The ideality factor n is determined by the nature of recombination. An ideality factor $n = 1$ corresponds to minority carrier-controlled recombination, while $n = 2$ applies when both species contribute to recombination.

We first demonstrate this interpretation of Eq. (3) by evaluating the well-known 1-dimensional p-n junction dark recombination current. For this system, dark recombination current is the sum of two contributions: the diffusion current and junction recombination current. Diffusion current is associated with minority carrier recombination in the neutral region. For concreteness, we suppose the system is a pn^+ junction and consider recombination in the p-type region. Electrons are minority carriers, so N is given by the equilibrium electron density in the p-type neutral region, which we denote n_p^0. The lifetime is set by the bulk electron lifetime, $\tau = \tau_{\text{bulk}}^e$. Minority carriers undergo simple diffusion in the neutral region, so the length scale of the minority carrier density profile is the diffusion length, $\lambda = \sqrt{D_{\text{bulk}}^e \tau_{\text{bulk}}^e}$, where D_{bulk}^e is the bulk electron diffusivity. Recombination is determined by the minority carrier (electron) density, so the ideality factor $n = 1$. Substituting these factors for Eq. (3) reproduces the well-known result for diffusive dark current. We next apply the same analysis to junction recombination current. In this case the recombination occurs in the depletion region and involves both nonequilibrium electrons and holes. The equilibrium density of electrons and holes in the depletion region is given by $N = n_i$, where n_i is the intrinsic density. The length scale over which recombination occurs is set by the depletion width W of the pn junction, $\lambda \approx W$. The lifetime is determined by bulk recombination, $\tau = \tau_{\text{bulk}}$. Both nonequilibrium electrons and holes participate in recombination, so that $n = 2$. These factors correctly reproduce the junction recombination dark current.

For grain boundary recombination, the same interpretations of the parameters entering Eq. (3) apply. What remains is to identify the recombination species and its lifetime, and to determine the region over which recombination occurs. Doing so requires knowledge of how carriers behave in the 2-dimensional model, which we discuss next.

As in many analytical models, the key aspect of our treatment lies in the approximations we make. The initial problem in its fully general form is not analytically tractable, due to nonlinearities and its 2 (or 3)-dimensional nature. Reducing the problem to a more manageable form requires valid, simplifying assumptions, which in turn require suffi-cient knowledge of system behavior. The rough picture is that positively charged grain boundaries lead to downward band bending in most of the p-type absorber, providing a confinement potential for electrons. We assume one end of the grain boundary is in close proximity with the n-contact, so electrons are efficiently funnelled into the grain boundary and the vast majority of electron current is carried along the grain boundary core. Hole current takes place in the grain bulk, and is directed towards regions of high recombination. We assume grains are not fully depleted. In practice this corresponds to grain sizes which exceed 1 μm in CdTe. Hole current therefore requires very small gradients in the hole quasi-Fermi level. This fact leads to a key assumption: that the hole quasi-Fermi level E^{Fp} is approximately flat everywhere. This assumption is valid only for sufficiently high intragrain hole mobility; for typical material parameter values of CdTe, the intragrain hole mobility μ_p should exceed 30 cm^2/V $\cdot$ s (see ref. [8] for details of this estimate).

We restrict our attention to the recombination at the grain boundary core, reducing the domain of interest to one dimension. However by itself this does not simplify the problem: The solution at the grain boundary depends on the solution in the grain interior, and the problem remains essentially 2-dimensional. The assumption of flat E^{Fp} is crucial here, as it implies $E_{\text{GB}}^{Fp} = E_{\text{bulk}}^{Fp}$. This relation provides a link between the grain boundary and grain bulk and enables the analysis of the two domains to be separated. The dimensionality of the problem is then reduced from 2 to 1.

Our next assumption is that the grain boundary defect density is large. The charge of the grain boundary defect is equal to the defect density multiplied by a statistical factor related to the defect occupation and type (donor or acceptor, see Eq. (7)). If the defect density is very large, the statistical factor must be very small to ensure that the defect charge remains finite. A large defect density also implies that the statistical factor is independent of applied voltage [9]. This leads to a constraint which takes the place of the Poisson equation. The implications of this constraint depend on the details of the grain boundary, such as the position of the defect level with respect to midgap. This is the point at which different cases and their analysis bifurcates. We discuss these cases later. For now the salient point is that this assumption of high defect density provides another simplification to the problem.

Having made these assumptions, the problem can be reduced to a single 1-dimensional effective diffusion equation for the electron quasi-Fermi level $E_{\text{GB}}^{Fn}(x)$, where x is the coordinate along the grain boundary core. It's not surprising that a description of grain boundary recombination would include a continuity equation for E_{GB}^{Fn}: electrons are confined to the grain boundary and carry the recombination current there [5]. The 1-dimensional diffusive motion of electrons along the grain boundary is parameterized by the electron grain boundary diffusivity D_{GB}^e and effective lifetime τ_{GB}^e. These parameters can be expected to differ from their bulk counterparts due to the highly defective grain boundary core.

TABLE I
RECOMBINATION CURRENT PARAMETERS

n-type	p-type	high recombination
$N = p_{\mathrm{GB}}^0$	$N = n_{\mathrm{GB}}^0$	$N = n_i$
$\lambda = L_{\mathrm{GB}}$	$\lambda = x_0 + L_{\mathrm{GB}}^{e\;\mathrm{diff}}$	$\lambda = L_{\mathrm{GB}}^{e\;\mathrm{diff}}$
$S = S_p^{\mathrm{eff}}$	$S = S_n^{\mathrm{eff}}$	$S = \sqrt{S_n^{\mathrm{eff}} S_p^{\mathrm{eff}}}$
$n = 1$	$n = 1$	$n = 2$

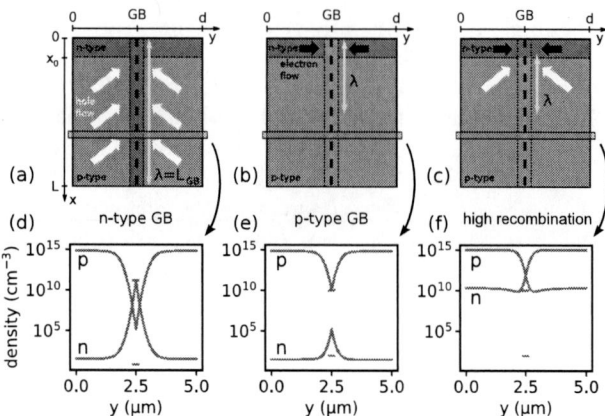

Fig. 1. (a), (b), and (c) show the system geometry and the electron and hole current flow for n-type, p-type, and high recombination grain boundaries, respectively. Also shown is the length λ over which recombination occurs along the grain boundary core. The grain boundary is columnar and positioned at $y = d/2$. (d), (e), and (f) show the electron and hole density under forward bias through a slice of the neutral region perpendicular to the grain boundary for n-type, p-type, and high recombination grain boundaries, respectively. The grain boundary core is located at $y = 2.5\ \mu$m, and the blue (red) tick marks at the grain boundary core position indicate the values of $\overline{n}_{\mathrm{GB}}$ ($\overline{p}_{\mathrm{GB}}$).

The grain boundary-confined electron diffusivity D_{GB}^e is likely reduced from the bulk value due to increased disorder scattering. D_{GB}^e enters as a free parameter in the model. The electron effective lifetime is reduced from its bulk value due to proximity to the grain boundary recombination centers. The recombination strength of the grain boundary is parameterized by an effective surface recombination velocity for electrons (holes) $S_{n(p)}^{\mathrm{eff}}$. The effective lifetime τ_{GB}^e is determined by the length scale of electron confinement L_E according to $\tau_{\mathrm{GB}}^e = L_E/S_n^{\mathrm{eff}}$. The confinement length L_E is determined by the built-in potential around the grain boundary, which is associated with two length scales: the depletion width of the grain boundary and $k_{\mathrm{B}}T/(qE_{\mathrm{GB}})$, where E_{GB} is the magnitude of the electric field at the grain boundary. The appropriate choice for L_E depends on the regime of system behavior, but either choice gives qualitatively similar results. The diffusion length of electrons confined to the grain boundary L_{GB}^e is then given by $L_{\mathrm{GB}}^e = \sqrt{D_{\mathrm{GB}}^e \tau_{\mathrm{GB}}^e} = \sqrt{D_{\mathrm{GB}}^e L_E/S_n^{\mathrm{eff}}}$.

With this picture of the system in mind, we turn again to Eq. (3) and specify the factors entering the formula. The appropriate lifetime τ for Eq. (3) is S^{eff}/d, where d is the grain size (the appropriate S^{eff} depends on the grain boundary type). This amounts to taking all of the recombination centers concentrated on the grain boundary surface and smearing them out uniformly across the entire grain. This is the crudest approximation one can make to account for grain boundary recombination, and has been used rather successfully to describe polycrystalline Si solar cells [11]. For photovoltaics with high grain boundary defect density, simply re-scaling τ and applying 1-dimensional p-n junction theory is inadequate for determining N, λ, and n. For these three parameters, the analysis proceeds differently according to the majority carrier type at the grain boundary core. As always, defect-mediated recombination is controlled by the minority carrier density. The minority carrier type depends on the Fermi level at the grain boundary core, which in turn depends on grain boundary defect properties (mostly the defect energy level).

Beginning with an n-type grain boundary core, holes are minority carriers, so N is the equilibrium grain boundary hole density, which we denote by p_{GB}^0. Since recombination is set by only one species, the ideality factor $n = 1$. The relevant S is that for holes: $S = S_p^{\mathrm{eff}}$. We consider a single donor+acceptor defect, in which case the effective surface recombination velocity is equal to the bare surface recombination velocity: $S_p^{\mathrm{eff}} = S_p$. (For other cases like the single donor defect and the continuum of donor+acceptor defects, the effective surface recombination velocity differs

from the bare surface recombination velocity.) λ is determined by the region over which holes are available to flow into the grain boundary and recombine. The majority of the grain boundary length L_{GB} is embedded in p-type bulk, so holes are available for transport into the grain boundary from the grain bulk over approximately the entire grain boundary. Hence $\lambda = L_{\mathrm{GB}}$ (see Fig. 1(a)).

For a p-type grain boundary, similar reasoning immediately leads to $N = n_{\mathrm{GB}}^0$, $n = 1$, and $S = S_n$. The recombination length λ includes the region over which electrons are available to flow into the grain boundary. We delimit this region with x_0: for $x < x_0$, $n_{\mathrm{bulk}} > p_{\mathrm{bulk}}$. Electrons flow into the grain boundary for $x < x_0$ and propagate via diffusion along the grain boundary core for $x > x_0$, where additional recombination occurs (see Fig. 1(b)). The total length over which recombination occurs is therefore $x_0 + L_{\mathrm{GB}}^{e\;\mathrm{diff}}$, where $L_{\mathrm{GB}}^{e\;\mathrm{diff}}$ is the diffusion length of electrons confined near the grain boundary core. Based on the discussion of electron diffusion along the grain boundary given earlier, we have $\lambda = x_0 + L_{\mathrm{GB}}^{e\;\mathrm{diff}} = x_0 + \sqrt{D_{\mathrm{GB}}^e L_E/S_n^{\mathrm{eff}}}$.

In the case where electron and hole density are of comparable magnitude, we immediately anticipate that $N = n_i$ and $n = 2$. To determine λ, we note that both electrons and holes must be transported to the grain boundary for recombination (see Fig. 1(c)). Holes are available along almost the entire length of the grain boundary, as discussed earlier. However electrons are only available for $x < x_0$. λ is therefore set by the electrons' availability. Recombination is concentrated near the middle of the depletion region, and decays on a length scale of the electron diffusion length, so that $\lambda = L_{\mathrm{GB}}^{e\;\mathrm{diff}}$. These cases are summarized in Table 1, and depicted in Fig. 1 (a)-(c).

Having summarized our understanding of grain boundary recombination in qualitative terms, we next place it in more quantitative context by presenting the relevant governing equations. The key underlying equations provide the grain boundary occupation f^{GB}, grain boundary charge Q^{GB}, and grain boundary recombination R^{GB} in terms of carrier densities and the defect properties. These are all standard equations which can be found in textbooks [12], [13]. We start with the grain boundary occupancy:

$$f^{\mathrm{GB}} = \frac{S_n n_{\mathrm{GB}} + S_p \overline{p}_{\mathrm{GB}}}{S_n \left(n_{\mathrm{GB}} + \overline{n}_{\mathrm{GB}}\right) + S_p \left(p_{\mathrm{GB}} + \overline{p}_{\mathrm{GB}}\right)}. \tag{4}$$

In the above, n_{GB} and p_{GB} are the electron and hole carrier density at the grain boundary, respectively (note that n_{GB}, p_{GB}, and f^{GB} can vary as a function of position along the grain boundary core). $\overline{n}_{\mathrm{GB}}$ and $\overline{p}_{\mathrm{GB}}$ are the electron and hole density one would obtain if the Fermi level is positioned at the defect energy level E_{GB}. If E_{GB} is measured from the valence band, then:

$$\overline{n}_{\mathrm{GB}} = N_c \exp\left[\left(E_{\mathrm{GB}} - E_g\right)/k_{\mathrm{B}}T\right], \tag{5}$$
$$\overline{p}_{\mathrm{GB}} = N_v \exp\left[-E_{\mathrm{GB}}/k_{\mathrm{B}}T\right]. \tag{6}$$

The charge of the grain boundary is given by:

$$Q^{\mathrm{GB}} = q\rho^{\mathrm{GB}} \times \begin{cases} \left(1 - f^{\mathrm{GB}}\right), & \text{(donor)} \\ \left(-f^{\mathrm{GB}}\right), & \text{(acceptor)} \\ \left(1 - 2f^{\mathrm{GB}}\right) & \text{(donor + acceptor)} \end{cases} \tag{7}$$

Here ρ^{GB} is the two-dimensional defect density at the grain boundary core. Note that the charge of the defect state depends on its type (donor or acceptor). Experimental evidence indicates positively charged grain boundaries [10], so we restrict our attention to the donor and donor+acceptor cases (the acceptor case only yields negatively charged grain boundaries). Note that for the donor+acceptor case, both defects are assumed to be present at the same energy E_{GB}. The donor and acceptor defects therefore compensate each other only when the Fermi energy is equal to the defects' common energy level.

The grain boundary charge Q^{GB} is compensated by the surrounding depletion region charge. For a depletion width W, the associated space charge is $2qN_AW$, where N_A is the doping and the factor of 2 arises from having depletion regions on both sides of the grain boundary. The relation $Q^{\mathrm{GB}} = 2qN_AW$ determines the equilibrium band bending between grain boundary and grain bulk.

The assumption we described earlier is that ρ^{GB} is "large". The factors in parentheses in Eq. (7) must therefore be "small" for Q^{GB} to remain finite. For the donor case, this implies $\left(1 - f^{\mathrm{GB}}\right)$ is small, or $f_{\mathrm{GB}} \approx 1$. For the donor+acceptor case, we have $f^{\mathrm{GB}} \approx 1/2$: both donor and acceptor state are approximately half-occupied. As described in Ref. [9], large ρ^{GB} also implies the the factor in parentheses does not change with applied voltage; we can determine f^{GB} in equilibrium and use the same value under forward bias.

Having determined f^{GB}, we can return to Eq. (4) and consider what values of carrier density n_{GB}, p_{GB} are needed to yield the correct value of f^{GB}. We'll consider the donor+acceptor case here as an example, in which $f^{\mathrm{GB}} = 1/2$. In equilibrium, $f^{\mathrm{GB}} = 1/2$ is satisfied by the carrier densities $n_{\mathrm{GB}} = \overline{n}_{\mathrm{GB}}$ and $p_{\mathrm{GB}} = \overline{p}_{\mathrm{GB}}$. This means that the Fermi energy is pinned to E_{GB} (see discussion preceding Eqs. (5)-(6)). In equilibrium, the majority carrier type at the grain boundary core is therefore determined solely by E_{GB}; if E_{GB} is closer to the conduction (valence) band, the grain boundary core is n (p)-type.

Let's consider an n-type grain boundary under forward bias. The terms n_{GB} and $\overline{n}_{\mathrm{GB}}$ are much larger than p_{GB} and $\overline{p}_{\mathrm{GB}}$ in the expression on the right-hand-side of Eq. (4). An applied bias voltage induces nonequilibrium electron and hole densities. Provided p_{GB} is much less than n_{GB}, n_{GB} must remain fixed to the value $\overline{n}_{\mathrm{GB}}$ to maintain $f_{\mathrm{GB}} = 1/2$. This is demonstrated in Fig. 1(d), which shows the electron and hole density under forward bias through a slice of the system perpendicular to the grain boundary core. The small blue (red) tick mark at $y = 2.5$ μm indicates the value of $\overline{n}_{\mathrm{GB}}$ ($\overline{p}_{\mathrm{GB}}$). In this plot, $n_{\mathrm{GB}} = \overline{n}_{\mathrm{GB}}$ while p_{GB} exceeds $\overline{p}_{\mathrm{GB}}$. We can write the grain boundary hole density in terms of the bulk hole density using the assumption we described earlier: $E_{\mathrm{GB}}^{Fp} = E_{\mathrm{bulk}}^{Fp}$. This leads to $p_{\mathrm{GB}} = p_{\mathrm{GB}}^0 \exp\left(qV/k_{\mathrm{B}}T\right)$. The minority carrier density increases exponentially with applied voltage. The same behavior holds for p-type grain boundaries, where majority and minority carrier types are interchanged with respect to the n-type grain boundary (see Fig. 1(e)).

Having determined the majority and minority carrier density, we can compute the grain boundary recombination R_{GB}, given by:

$$R_{\mathrm{GB}} = \frac{S_n S_p \left(n_{\mathrm{GB}} n_{\mathrm{GB}} - n_i^2\right)}{S_n \left(n_{\mathrm{GB}} + \overline{n}_{\mathrm{GB}}\right) + S_p \left(p_{\mathrm{GB}} + \overline{p}_{\mathrm{GB}}\right)} \tag{8}$$

For the n-type grain boundary, Eq. (8) simplifies to $R_{\mathrm{GB}} = S_p p_{\mathrm{GB}}^0 \exp\left(qV/k_{\mathrm{B}}T\right)$. We argued previously that the length scale over which recombination occurs is equal to the entire grain boundary length L_{GB}. Together these factors reproduce the diode equation terms for an n-type grain boundary as given in Table 1.

As the applied voltage increases, the minority carrier density p_{GB} dutifully increases exponentially and will eventually approach n_{GB}. At this point further increases in the applied voltage push both n_{GB} and p_{GB} to exceed $\overline{n}_{\mathrm{GB}}$. Now $f_{\mathrm{GB}} = 1/2$ is satisfied by insisting that $S_n n_{\mathrm{GB}} = S_p p_{\mathrm{GB}}$ (see Eq. (4)). This is the high-recombination regime. Returning to Eq. (8) and utilizing the relation $n_{\mathrm{GB}} p_{\mathrm{GB}} = n_i^2 \exp\left(qV/k_{\mathrm{B}}T\right)$, the maximum value of recombination along the grain boundary core can be found as:

$$R_{\mathrm{GB}}^{\max} = \sqrt{S_n S_p}\, n_i \exp\left(\frac{qV}{2k_{\mathrm{B}}T}\right) \tag{9}$$

In the high recombination regime, both nonequilibrium electrons and holes control the recombination. These carriers are transported to the grain boundary from the bulk. The point of maximum grain boundary recombination occurs in the depletion region. In this regime an electrostatic potential gradient is also developed along the grain boundary which drives the high electron current. We showed that the electron and hole density profiles decay along the grain boundary over a length scale of $L_{\mathrm{GB}}^{e,\mathrm{diff}}$ in this case [8]. This leads to the recombination which decays away from its maximum value with the same length scale, as shown in Fig. 1(c) and the 3rd column of Table 1. This ends the analysis of the donor+acceptor case.

The analysis of the donor cases proceeds along similar lines. One important wrinkle is that $f_{\mathrm{GB}} \approx 1$ in this case (see discussion following Eq. (7)). The defect is nearly fully occupied, which implies that for the same value of E_{GB}, the band bending is larger in the donor case than in the donor+acceptor case. This in turn implies the grain boundary hole density is suppressed in the donor case relative to the donor+acceptor case. For this reason, the effective hole recombination velocity is given by $S_p^{\mathrm{eff}} = (1 - f_{\mathrm{GB}}) S_p$. $(1 - f_{\mathrm{GB}})$ is small by assumption, so the hole-controlled recombination is significantly reduced in the donor defect case [9].

III. Conclusion

We end the overview here, and refer the reader to references for a fuller and more rigorous derivation of the grain boundary dark current, as well as demonstrations that the grain boundary dark current can be used to predict V_{oc}. The type of analysis can be extended to consider a continuum of defect states, grain boundaries with arbitrary orientation, and networks of inhomogeneous grain boundaries. With this degree of model flexibility, we can begin to consider the behavior of realistic, inhomogeneous polycrystalline materials.

Acknowledgment

B. G. acknowledges support under the Cooperative Research Agreement between the University of Maryland and the National Institute of Standards and Technology Center for Nanoscale Science and Technology, Award 70NANB14H209, through the University of Maryland.

References

[1] R. M. Geisthardt, M. Topi Marko J. R. and Sites, "Status and potential of CdTe solar-cell efficiency", *IEEE Journal of photovoltaics* vol. 5, pp. 1217-1221, 2015.

[2] I. Visoly-Fisher, S. R. Cohen, A. Ruzin, D. Cahen, "Polycrystalline Devices Can Outperform Single-Crystal Ones: Thin Film CdTe/CdS Solar Cells", *Advanced Materials* vol. 16, pp. 879883, 2004.

[3] C. Li, Y. Wu, J. Poplawsky, T. J. Pennycook, N. Paudel, W. Yin, S. J. Haigh, M. P. Oxley, A. R. Lupini, M. Al-Jassim, and others, "Grain-boundary-enhanced carrier collection in CdTe solar cells", *Phys. Rev. Lett.*, vol. 112, pp. 156103, 2014.

[4] M. Tuteja, P. Koirala, V. Palekis, S. MacLaren, C. S. Ferekides, R. W. Collins, and A. A. Rockett, "Direct Observation of CdCl2 Treatment Induced Grain Boundary Carrier Depletion in CdTe Solar Cells Using Scanning Probe Microwave Reflectivity Based Capacitance Measurements", *J. Phys. Chem. C*, vol. 120, pp. 7020-7024, 2016.

[5] M. Gloeckler, J. R. Sites, and W. K. Metzger, "Grain-boundary recombination in $\mathrm{Cu(In, Ga)Se_2}$ solar cells", *J. App. Phys.*, vol. 98, pp. 113704, 2005.

[6] K. Taretto and U. Rau, "Numerical simulation of carrier collection and recombination at grain boundaries in $\mathrm{Cu(In, Ga)Se_2}$ solar cells", *J. App. Phys.*, vol. 103, pp. 094523, 2008.

[7] S. A. Edmiston, G. Heiser, A. B. Sproul, and M. A. Green, "Improved modeling of grain boundary recombination in bulk and p-n junction regions of polycrystalline silicon solar cells", *J. App. Phys.*, vol. 80, pp. 6783-6795, 1996.

[8] B. Gaury, and P. M. Haney, "Charged grain boundaries reduce the open-circuit voltage of polycrystalline solar cellsAn analytical description", *J. App. Phys.*, vol. 120, pp. 234503, 2016.

[9] B. Gaury and P. M. Haney, "Anatomy of charged grain boundaries in polycrystalline solar cells", *arXiv preprint arXiv:1704.04234*, 2017.

[10] H. P. Yoon, P. M. Haney, D. Ruzmetov, H. Xu, M. S. Leite, B. Hamadani, A. A. Talin, N. B. Zhitenev, "Local electrical characterization of cadmium telluride solar cells using low-energy electron beam", *Sol. En. Mat. and Sol. Cells*, vol. 117, pp. 499-504, 2013.

[11] H. C. Card, and E. S. Yang, Edward S, "Electronic processes at grain boundaries in polycrystalline semiconductors under optical illumination", *IEEE Transactions on Electron Devices*, vol. 24, pp. 397-402, 1977.

[12] S. J. Fonash, *Solar cell device physics*: Academic Press, 1981.

[13] E. S. Yang, *Fundamentals of semiconductor devices*: McGraw-Hill Companies, 1978.

Role of Tellurium Buffer Layer on CdTe Solar Cells' Absorber/Back-Contact Interface

Tao Song and James R. Sites

Department of Physics, Colorado State University, Fort Collins, CO, 80523, USA

Abstract—The CdTe/back contact interface plays a key role in carrier transport for conventional polycrystalline thin-film CdTe devices. A significant back-contact barrier ϕ_b caused by a metallic contact with a low work function can block hole transport and enhance the forward current, and thus result in a reduced V_{OC}, particularly with fully-depleted CdTe devices. An appropriate buffer layer between the CdTe absorber and metallic contact can mitigate this detrimental impact. The simulation has shown that a thin tellurium (Te) buffer, such as that used successfully at Colorado State, can assume such a role by reducing the downward valence-band bending caused by large ϕ_b. As a result, it can enhance the extraction of charge carriers and produce better cell performance. The simulation has also shown that, by combining a Te buffer with a CdMgTe electron reflector, cell efficiency approaching 24% is predicted to be feasible even with relatively low CdTe doping ($p \sim 10^{14}$ cm^{-3}).

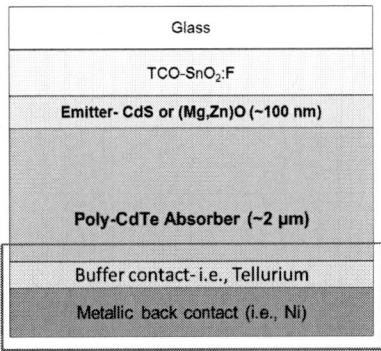

Fig. 1. CdTe device structure used in the numerical simulations.

I. INTRODUCTION

Since CdTe is a large-E_g (~ 1.5 eV) semiconductor with a high electron affinity ($\chi = 4.3$ eV), most metals do not have a sufficiently high work function ϕ_M for an ohmic contact, and hence form a carrier-blocking Schottky barrier ϕ_b at the p-CdTe/metal contact interface. Therefore, instead of directly employing a metal as the back contact, an additional buffer layer adjacent to CdTe absorber usually has been created to mitigate the carrier blockage.

Such a buffer layer is often formed by surface modification (e.g., etching the back CdTe to form a Te-rich surface [1], or copper doping such as a CuCl treatment to highly dope the back CdTe surface [2]), and/or by deposition of an additional layer, such as Te [3] or Cu$_x$Te [2] (an overview of back contact materials can be found in Ref. [4]). However, most previous modeling work only emphasized the metallic contact, simplifying the CdTe/metal as a Schottky diode with opposite polarity to the primary junction [5], [6]. This work is intended to give a systematic study to identify the roles of both the buffer and the metallic back contacts. Meanwhile, it gives insights on how to suppress the detrimental impact of the Schottky barrier with introducing buffer contact layer.

Numerical simulation in this work was performed with SCAPS-1D software [7]. The baseline CdTe device structure in the simulation was from cells fabricated at Colorado State University (CSU), as shown in Fig. 1. The basic structure consist of a 200-nm-thick highly-doped TCO layer, a 100-nm-thick CdS emitter, and a 2-μm-thick CdTe absorber.

II. CDTE/METALLIC CONTACT INTERFACE

The energy-band alignment between the CdTe absorber and the metallic contact is a significant factor for cell performance,

since it can form an unfavorable band alignment that enhances back surface recombination and impedes the photo-generated carrier collection. For CdTe material with a low minority carrier lifetime ($\tau < 0.1$ ns), the presence of a large back barrier ϕ_b can impede hole transport and thus induce the current-limiting effect, known as "rollover", which is reasonably explained with a two-diode model [5], [6]. However, this phenomenon does not take place in good-quality CdTe cells with larger τ (> 1 ns) at room temperature [8]. Since this work is focused on high-efficiency CdTe devices, greater effort was made to study effects of the metallic contact rather than the "rollover" effect.

Fig. 2 shows the energy band diagrams with three values of the back metal work function ϕ_M at 0.8-V cell bias under illumination. Ideally, an ohmic contact is required to achieve an unrestricted carrier transport, as shown in Fig. 2(a). However, it is nearly impossible to find such an appropriate metal with a large ϕ_M (> 5.5 eV) to form an ohmic contact with CdTe absorber. The situations such as Fig. 2(b) and (c) are more realistic. In these cases, a portion of the photo-generated electrons can diffuse to the back and recombine with holes at the back surface. As a result, the forward diode current is enhanced and thus the cell performance is reduced. The corresponding J-V curves with a variety of ϕ_M values are shown in Fig. 3. As expected, the V_{OC} is considerably reduced with a smaller ϕ_M due to increased back barrier ϕ_b enhancing the forward diode current.

III. TE BUFFER LAYER FOR BACK CONTACT

Recently, CSU has fabricated high-efficiency CdTe solar cells with certified $\eta = 18.3\%$ [9] and the application of

978-1-5090-5606-4/17 $31.00 © 2017 IEEE

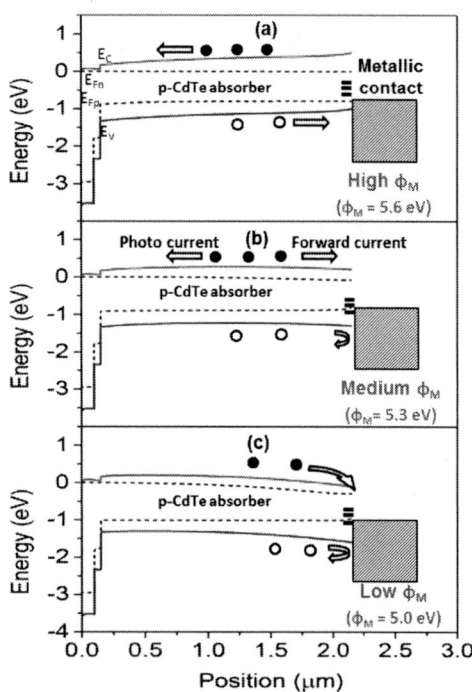

Fig. 2. Energy-band diagrams of CdTe solar cells with three values of back metal work function ϕ_M at 0.8-V bias under illumination.

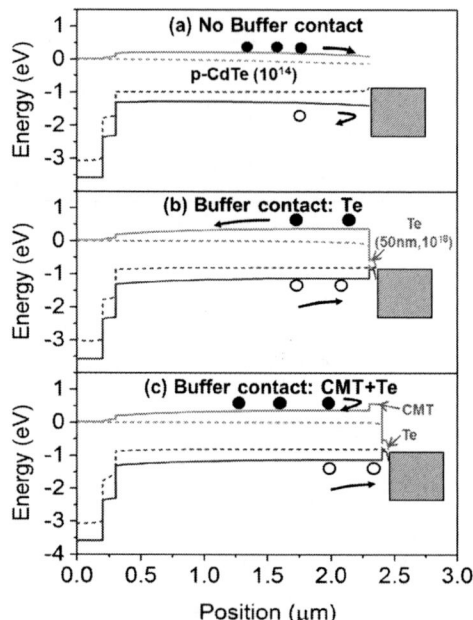

Fig. 4. Energy-band diagrams at 0.8-V bias under illumination for three cases: (a) baseline without Te buffer, (b) with 50-nm Te buffer, and (c) with 100-nm CdMgTe electron reflector ($\Delta E_C = 0.2$ eV) and 50-nm Te buffer. $\phi_M = 5.2$ eV is assumed for all the three cases.

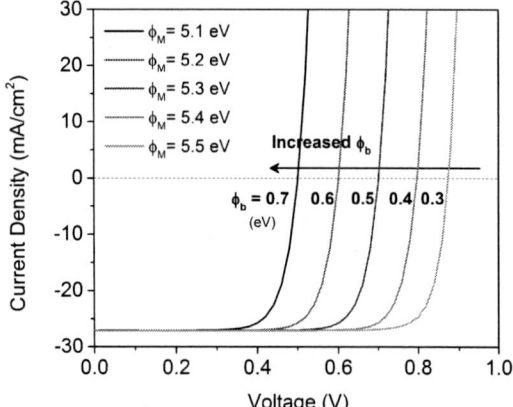

Fig. 3. The corresponding light J-V curves with a variety of metal work function ϕ_M.

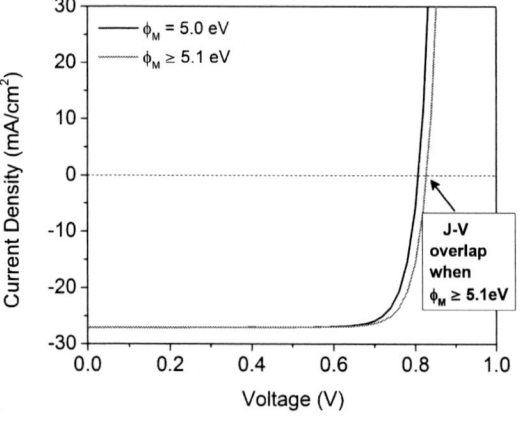

Fig. 5. The corresponding light J-V curves with a 50-nm Te buffer and a variety of metal work function ϕ_M.

the Te buffer layer played an important role in achieving this efficiency. To simulate cells with the Te layer, measured material parameters from films fabricated at CSU were used in the calculations: CdTe absorber doping of 10^{14} cm^{-3} and 50-nm-thick Te with p-type doping of 10^{18} cm^{-3}.

Fig. 4 compares the energy-band diagrams at 0.8-V bias under illumination for three cases: (a) baseline without Te buffer, (b) with 50-nm Te buffer, and (c) with 100-nm Cd-MgTe (CMT) electron reflection (ER) (conduction-band offset

$\Delta E_C = 0.2$ eV) and 50-nm Te buffer. A realistically low $\phi_M = 5.2$ eV is assumed for all the three cases. From the UPS measurements by Niles et al. [10], a valence-band offset $\Delta E_V = 0.27$ eV between CdTe and Te is assumed. In Fig. 4(c), a 100-nm-thick extended-bandgap CdMgTe [11], [12] is inserted before Te buffer to create an additional electron barrier in the conduction band.

The thin Te buffer reduces the voltage drop at the back contact, and suppresses the forward electron current (both

978-1-5090-5606-4/17 $31.00 © 2017 IEEE

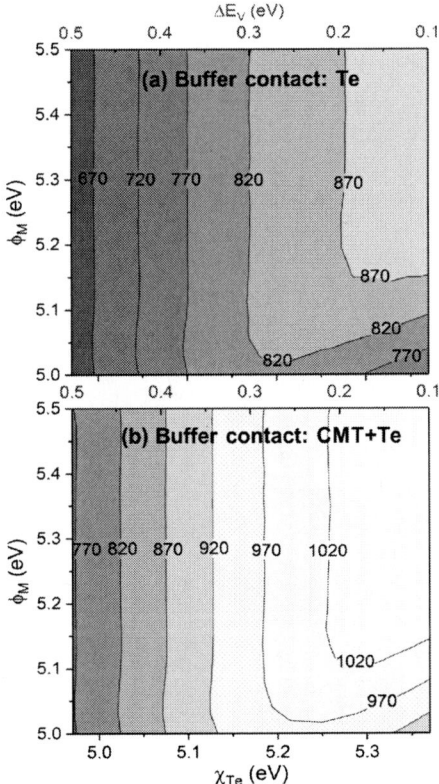

Fig. 6. V_{OC} contour plots of CdTe cells with varying Te electron affinity χ_{Te} and ϕ_M for two types of buffer contact: (a) only 50-nm Te buffer and (b) 100-nm CMT + 50-nm Te.

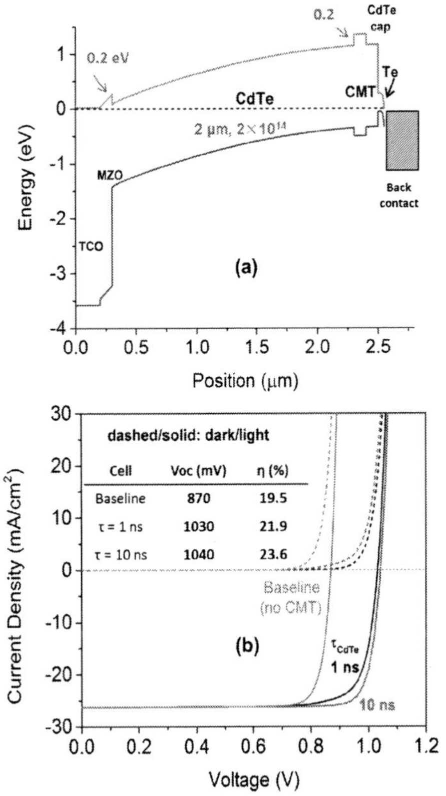

Fig. 7. (a) Energy-band diagrams of the simulated cells at 0-V bias and dark: (a) simulated J-V curves of the cells without CMT ER and with ER for two bulk lifetimes ($\tau = 1$, 10 ns).

Fig. 4b and 4c). Meanwhile, the Te buffer reduces the downward valence-band bending from the low metal work function and thus allows full hole current transport. V_{OC} is therefore nearly independent of ϕ_M, as seen in Fig. 5.

Fig. 6 shows two V_{OC} contour plots calculated over a range of ΔE_V between CdTe (or CdMgTe) and Te (by varying the Te electron affinity χ_{Te}) and a range of metal work function ϕ_M for the two types of buffer back system. For both cases, V_{OC} is nearly independent of the metal work function ϕ_M, except for the lower right region where the beneficial impact of the Te buffer is overshadowed by the low ϕ_M. When $\Delta E_V > 0.3$ eV, V_{OC} is reduced linearly with ΔE_V. This reduction is caused by the downward band bending at the back surface, which enhances forward diode current similar to the situation in Fig. 4a. Additionally, ΔE_V has very slight impact on other performance parameters (not shown here). From Fig. 6b, it is seen that the addition of a CMT electron reflector can further assist with carrier collection by effectively suppressing the electron recombination at the back surface.

IV. Summary and Proposed Structure

A buffer layer between CdTe absorber and metallic contact is highly desirable to mitigate this detrimental impact. The simulations showed that a thin Te buffer layer reduces the downward valence-band bending caused by a large valence-band offset and enhances the extraction of the charge carriers. The metal work function still play a role, but a less significant one. In addition to the Te buffer, the application of a CdMgTe ER can be employed to further improve V_{OC}.

Fig. 7(a) shows the energy band diagram of a proposed device structure with both CMT ER and Te buffer layers. The purpose of the CdTe cap layer is to protect the CMT from oxidation (MgO is easily formed when the film is exposed to atmosphere, causing FF loss [13]). Accordingly, the V_{OC} is enhanced with a CMT ER as shown in Fig. 7(b). Overall, the simulation shows that the combination of an electron reflector and a Te buffer should be an effective approach to increase the cell efficiency to as much as 24% even with a relative low CdTe carrier concentration ($\sim 10^{14}\ cm^{-3}$).

Acknowledgments

The authors would like to thank Dr. W. S. Sampath and Mr. Andy Moore for many helpful discussions. This work was supported by the NSF I/UCRC Program and by DOE SunShot award DE-EE0007543.

REFERENCES

[1] J. Britt and C. Ferekides, "Thin-film cds/cdte solar cell with 15.8% efficiency," *Applied Physics Letters*, vol. 62, no. 22, pp. 2851–2852, 1993.

[2] J. Zhou, X. Wu, A. Duda, G. Teeter, and S. Demtsu, "The formation of different phases of cu x te and their effects on cdte/cds solar cells," *Thin Solid Films*, vol. 515, no. 18, pp. 7364–7369, 2007.

[3] W. Xia, H. Lin, H. N. Wu, C. W. Tang, I. Irfan, C. Wang, and Y. Gao, "Te/cu bi-layer: A low-resistance back contact buffer for thin film cds/cdte solar cells," *Solar Energy Materials and Solar Cells*, vol. 128, pp. 411–420, 2014.

[4] A. Luque and S. Hegedus, *Handbook of photovoltaic science and engineering*. John Wiley & Sons, 2011.

[5] S. Demtsu and J. Sites, "Effect of back-contact barrier on thin-film cdte solar cells," *Thin Solid Films*, vol. 510, no. 1, pp. 320–324, 2006.

[6] A. Niemegeers and M. Burgelman, "Effects of the au/cdte back contact on iv and cv characteristics of au/cdte/cds/tco solar cells," *Journal of Applied Physics*, vol. 81, no. 6, pp. 2881–2886, 1997.

[7] M. Burgelman, P. Nollet, and S. Degrave, "Modelling polycrystalline semiconductor solar cells," *Thin Solid Films*, vol. 361, pp. 527–532, 2000.

[8] J. Pan, M. Gloeckler, and J. R. Sites, "Hole current impedance and electron current enhancement by back-contact barriers in cdte thin film solar cells," *Journal of applied physics*, vol. 100, no. 12, p. 124505, 2006.

[9] A. H. Munshi, "Investigations of process, microstructures, and efficiencies of polycrystalline cdte photovoltaic films and devices," Ph.D. dissertation, Colorado State University, 2017.

[10] D. W. Niles, X. Li, P. Sheldon, and H. Höchst, "A photoemission determination of the band diagram of the te/cdte interface," *Journal of applied physics*, vol. 77, no. 9, pp. 4489–4493, 1995.

[11] T. Song, A. Kanevce, and J. R. Sites, "Design of epitaxial cdte solar cells on insb substrates," *IEEE Journal of Photovoltaics*, vol. 5, no. 6, pp. 1762–1768, 2015.

[12] T. Song, "Design strategies for high-efficiency cdte solar cells," Ph.D. dissertation, Colorado State University, 2017.

[13] D. E. Swanson, "Cdte alloys and their application for increasing solar cell performance," Ph.D. dissertation, Colorado State University, 2016.

Simultaneous Examination of Grain-Boundary Potential, Recombination, and Photocurrent in CdTe Solar Cells Using Diverse Nanometer-Scale Imaging

C.-S. Jiang, H.R. Moutinho, J. Moseley, A. Kanevce, J.N. Duenow, E. Colegrove, C. Xiao, W.K. Metzger, and M.M. Al-Jassim

National Renewable Energy Laboratory (NREL), Golden, CO 80401, USA

Abstract — **We report on the simultaneous measurements of Kelvin probe force microscopy (KPFM), cathodoluminescence (CL), and electron-beam induced current (EBIC) for hundreds of grain boundaries (GBs) after different processing steps. The GB potential steadily increases from tens of mV for as-deposited samples, 80 mV after the Cl treatment, to about 120 mV after Cu diffusion (Cl+Cu). The latter is consistent with the emergence of deep Cu defects observed in CL GB spectra. At the same time, EBIC indicates that GB photocurrent is enhanced by 0% to 50% after Cu is introduced. However, this increase is very modest relative to expectations based on computational modeling and the measured GB potential. The EBIC enhancement coincides with *increased* GB recombination observed by CL. A plausible explanation is that GB defects introduce GB potentials and increase GB recombination; this can reconcile the CL, EBIC, and KPFM data.**

Index Terms — **CdTe solar cell, Grain boundary, Kelvin probe force microscopy (KPFM), Electron-beam induced current (EBIC), Cathodoluminescence (CL).**

I. INTRODUCTION

The effect of grain boundaries (GBs) on polycrystalline solar cells is an important and controversial subject. Electron-beam induced current (EBIC) studies have consistently indicated that GBs enhance photocurrent collection, and some researchers believe this is evidence that GB recombination is reduced or passivated [1,2]. Yet, separate cathodoluminescence (CL) measurements indicate enhanced GB recombination. GB electrostatic potential or band bending can enhance photocurrent near the GBs. However, the potential can similarly increase dark current and negatively impact open-circuit voltage (Voc) and performance. The thin-film community has been left with qualitative conjecture about the role of GBs on thin films for decades because current enhancement and recombination are generally not quantified, or measured together, or subsequently modeled.

In this paper, we report on the simultaneous measurements of nm-scale Kelvin probe force microscopy (KPFM) potential mapping, CL, and EBIC on as-deposited, Cl-treated, and Cu-treated CdTe thin-film devices. Hundreds of GBs in each sample are analyzed. The potential mapping demonstrates positively charged GBs relative to the grain interior (GI). This potential steadily increases from tens of mV for as-deposited (as-dep) samples, 80 mV after the Cl treatment (Cl), to about 120 mV after Cu diffusion (Cl+Cu). The latter is consistent

with the emergence of deep Cu defects observed in CL GB spectra.

At the same time, EBIC indicates that GB photocurrent is enhanced by 0% to 50% after Cu is introduced. However, this increase is very modest relative to expectations based on computational modeling and the measured GB potential. Contrary to some claims in the literature, the EBIC enhancement coincides with *increased* GB recombination observed by CL. A plausible explanation is that GB defects introduce GB potentials and increase GB recombination; this can reconcile the measured GB potentials with the relatively small GB photocurrent enhancement and poor GB luminescence.

II. EXPERIMENTAL

For the device fabrication [3], a bilayer of ~350 nm of

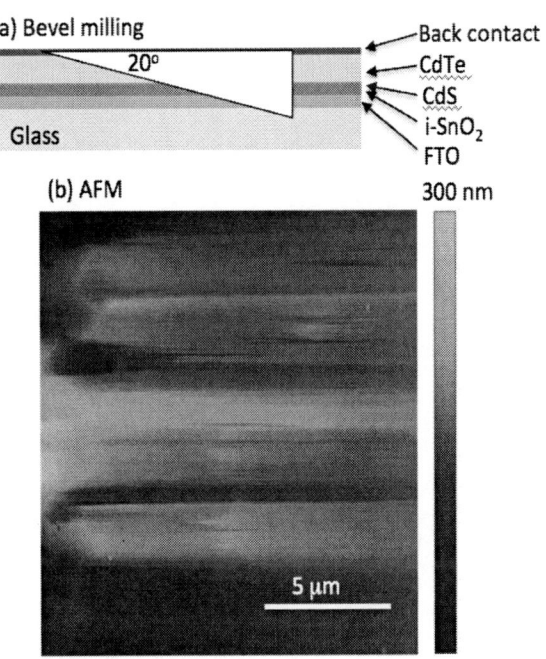

Fig. 1. (a) A schematic showing the FIB-milled bevels for KPFM imaging; (b) An AFM image taken on the beveled area showing the stripes milled by FIB.

conductive (F-doped) and ~150 nm of insulating SnO_2 was deposited by chemical vapor deposition onto Corning 7059 glass substrates maintained at 550°C. CdS:O (~100 nm) was deposited by radio-frequency magnetron sputtering at room temperature in a 6% O_2/Ar ambient. CdTe was deposited by close-spaced sublimation (CSS) at a substrate temperature of 600°C to a thickness of 4–6 µm. A post-deposition vapor $CdCl_2$ heat treatment was performed by CSS at 400°C for 10 min in an ambient of 20% O_2/He. Back contacts were formed by thermally evaporating 5 nm of Cu and 150 nm of Au at room temperature, followed by a contact anneal at 260°C for 30 min.

We milled the device using focused ion beam (FIB). To measure the potential distributions parallel and perpindicular to the film surface, we beveled the sample at an angle of ~20° (Fig. 1a). The AFM image of this beveled area shows stripes along the ion-milling direction (Fig. 1b). Corrugations of the surface (<300 nm) are tolerable for the KPFM measurement; no topographic effect was observed. The sample was annealed at 250°C for 5 min using a vacuum oven to passivate the surface.

KPFM images surface potential in ~30-nm and ~10-mV

spatial and potential resolutions by measuring and nullifying the Coulomb force between the probe and sample using an AC voltage of the second resonant oscillation frequency (300–400 kHz) applied to the probe [4]. Scanning electron microscope-based CL images local near-bandgap emission using an electron beam with voltage of 7.5 kV and current of about 14 nA. EBIC images local current collection of minority carriers generated by electron beam with voltage of 3–10 kV and current of 0.5–1.5 nA.

III. RESULTS AND DISCUSSIONS

Figure 2 shows how the GB potential evolves for as-deposited, Cl-treated, and Cl+Cu samples. The potential map exhibits grain sizes of 1–5 µm. The grain size is smaller in regions close to the junction and larger near the back contact due to grain-growth coalescence in the growth direction. The potential is greater at the GBs; the images show the GB potential as a function of position, and the graphs show a statistical distribution averaged across several images after each processing step. The GB potential illustrates a downward band bending around the GB, which will attract electrons and

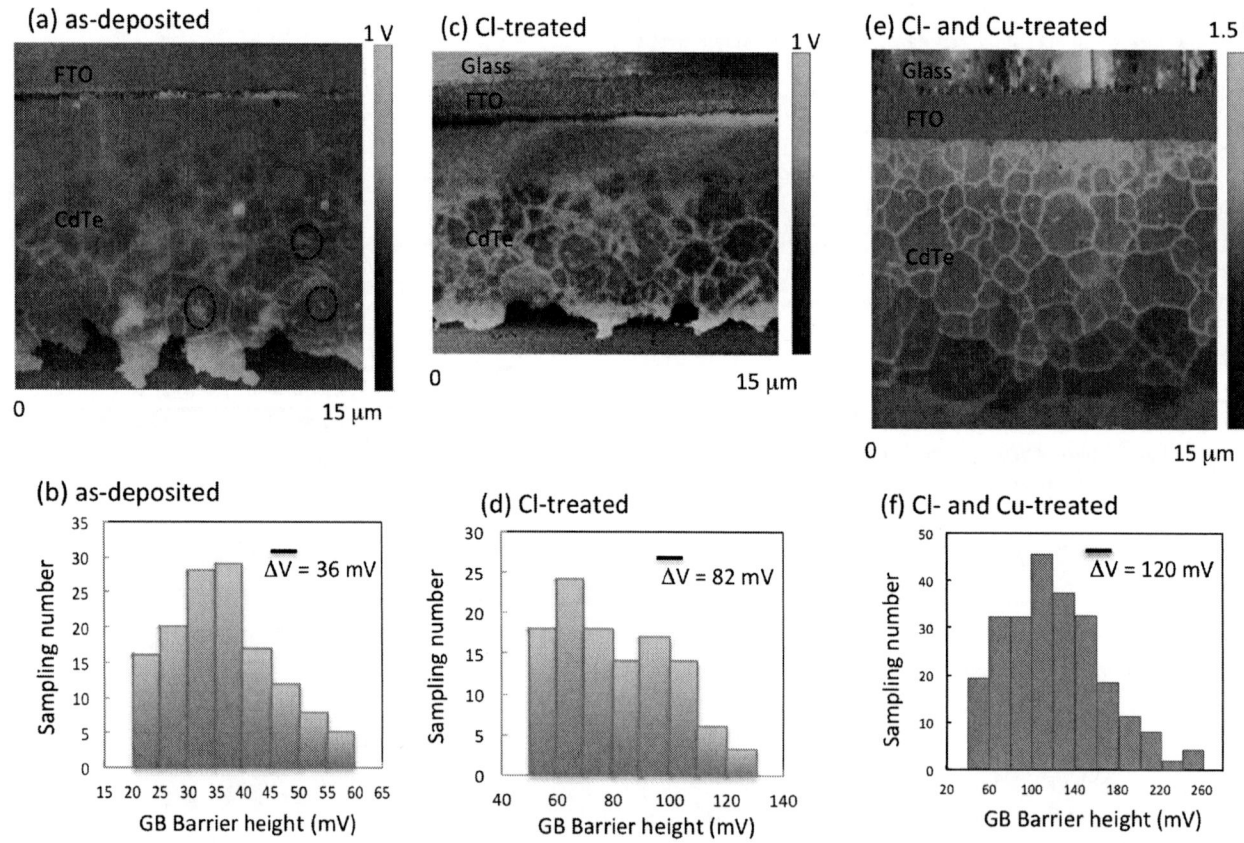

Fig. 2. (a)(c)(e) KPFM potential images and (b)(d)(f) statistical distribution of the GB potential contrast taken on (a)(b) the as-deposited film, (c)(d) films with Cl treatment, and (e)(f) films with both Cl and Cu treatments.

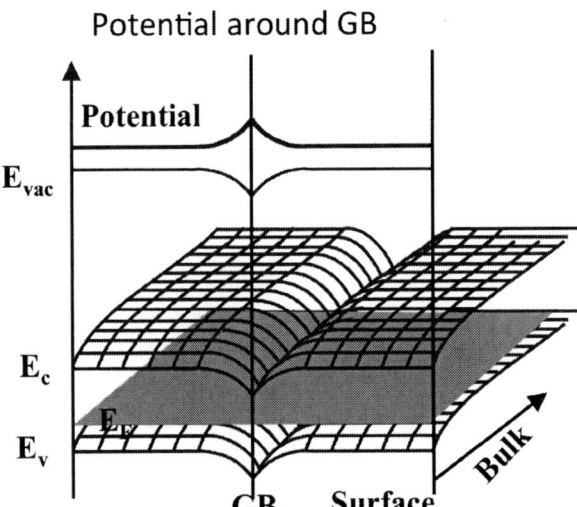

Potential around GB

Fig. 3. A schematic illustrating the effect of surface band bending.

repel holes.

There are clear changes in the potential maps throughout the CdTe films with processing. In the as-deposited samples (Figs. 2a and 2b), the potential appears smeared out near the junction and defective in areas (e.g., dashed circles). The addition of Cl (Figs. 2c and 2d) and Cl+Cu (Figs. 2e and 2f) make the GB potential progressively larger and sharper throughout the CdTe film. The addition of Cl and Cu will place additional impurities (e.g., Cl, S, Cu) in the GBs, which can alter the GB potential. For example, CL spectra clearly indicate an increase of Cu defect peaks along the GBs. The Cl and Cu steps may also adjust the bulk hole density, which can reduce the depletion area around the GBs and sharpen the contrast. Depletion can also reduce the GB contrast near the junction.

Some reports suggest that the GBs in CdTe are inverted and thus form an interpenetrating p-n 3-D network that makes polycrystalline devices superior to single-crystalline devices [2]. However, such strong GB potentials would seriously limit transport across GBs such as those parallel to the p-n junction, and this is generally not observed. Here, the GB contrast after Cl+Cu ($\Delta V=120$ mV) is much smaller than the CdTe bandgap ($\sim$1.45 eV), and it does not indicate inversion within the sensitivity of the technique. Although the sample was annealed, the surface passivation is expected to make the surface charges or surface states uniform across the surface, but cannot eliminate them. The surface band bending

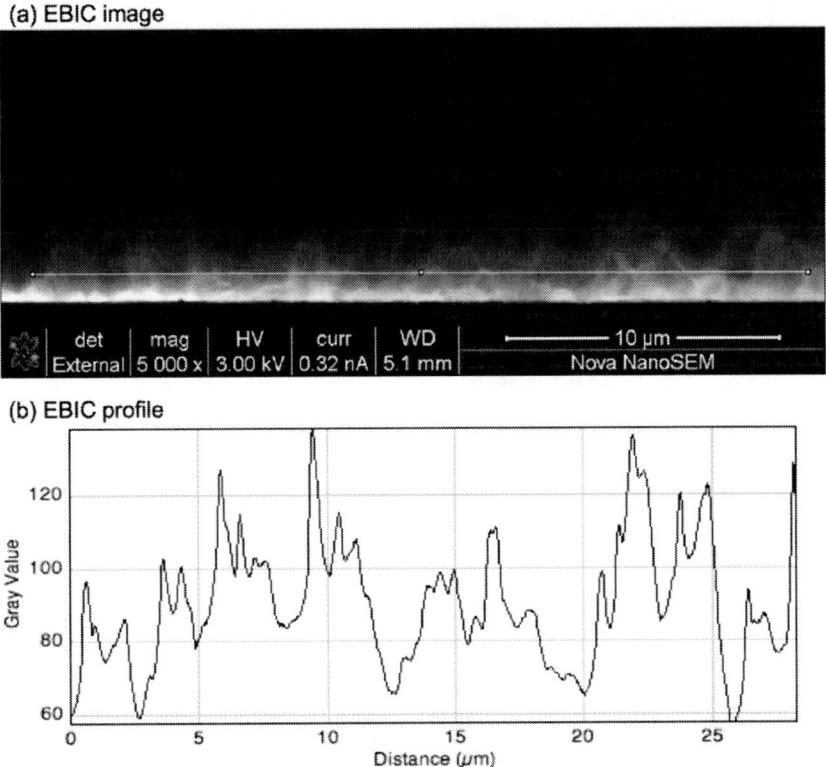

Fig. 4(a) An EBIC image taken on the cross-section of a CdTe device with Cl- and Cu-treatments; (b) shows an EBIC line profile along the yellow line in (a).

originating from the surface charges can make the GB potential contrast on the surface smaller than in the bulk [5], as schematically shown in Fig. 3. Consequently, the GB potential in the bulk cannot be precisely determined by the surface potential measurement. However, this surface-potential imaging qualitatively illustrates that the GB traps positive charges and makes the downward band bending around it, and it is consistent with non-inverted GBs.

The GB potential may enhance short-circuit current density (Jsc) but decrease Voc by also enhancing the flow of electrons from the transparent conducting oxide in forward bias, which increases dark current [6–8]. This can partially contribute to the fact that state-of-the-art polycrystalline CdTe devices have photocurrent near the theoretical limit but a large deficit in Voc. We simulated GBs with a potential value of ~100 mV

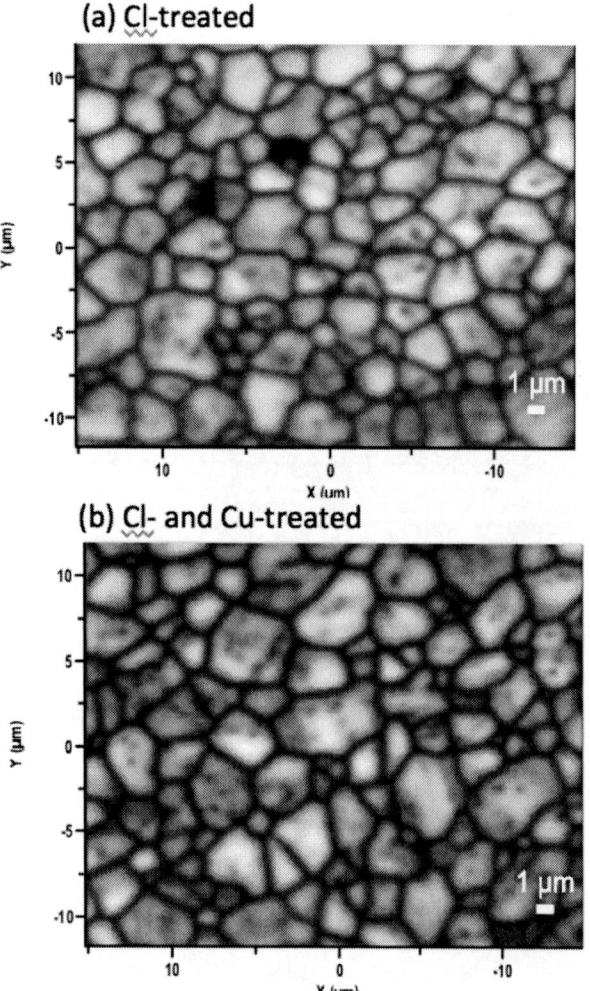

Fig. 5. CL images of overall relative-scale intensity taken on CdTe films (a) with Cl treatment and (b) with both Cl and Cu treatments.

and found enhanced current at the GBs. This GB current qualitatively agrees with our EBIC measurements (e.g., Fig. 4). However, the relative current gain on the GBs is 0%–50%. Modeling indicates that this is consistent with relatively small GB potentials (<200 mV) in the bulk rather than inverted GBs. Future work will present more computational data.

The GB potentials are consistent with the repulsion of holes and attraction of electrons near the GB. Some have stated that this will help reduce recombination; however, Shockley-Reed-Hall kinetics indicates that this will generally increase rather than decrease recombination [6–8]. In CL measurements, the injection levels will screen the GB potentials, so that we can examine recombination independently [7,9]. Figure 5 shows CL images of films with Cl and both Cl+Cu. The images were taken on the film back-surface (not beveled samples). The images of overall CL intensity illustrate darker relative-scale GBs on the (Cl+Cu) film than on the Cl film. Consistent with studies on a wide variety of samples, the CL contrast indicates recombination along the GBs. Thus, GBs appear to be defective, which contributes to GB potentials (more GB charges) and recombination while contributing to some photocurrent around the GBs.

IV. SUMMARY

We report on the simultaneous measurements of KPFM, CL, and EBIC for hundreds of GBs after different processing steps. The GB potential steadily increases from tens of mV for as-deposited samples, 80 mV after the Cl treatment, to about 120 mV after Cu diffusion (Cl+Cu). The latter is consistent with the emergence of deep Cu defects observed in CL GB spectra. At the same time, EBIC indicates that the GB current is enhanced by 0% to 50% after Cu is introduced. However, this increase is very modest relative to expectations based on computational modeling with input from the measured GB potential. The EBIC enhancement coincides with *increased* GB recombination observed by CL. A plausible explanation is that GB defects introduce GB potentials and increase GB recombination; this can reconcile the CL, EBIC, and KPFM data.

ACKNOWLEDGMENT

This work was supported by the U.S. Department of Energy under Contract No. DE-AC36-08GO28308 with the National Renewable Energy Laboratory.

The U.S. Government retains and the publisher, by accepting the article for publication, acknowledges that the U.S. Government retains a nonexclusive, paid-up, irrevocable, worldwide license to publish or reproduce the published form of this work, or allow others to do so, for U.S. Government purposes.

REFERENCES

[1] I. Visoly-Fisher, S.R. Cohen, A. Ruzin, and D. Cahen, "How polycrystalline devices can outperform single-crystal ones: Thin film CdTe/CdS solar cells," *Adv. Mater.* **16**, 879, 2004.

[2] C. Li et al., "Grain-boundary-enhanced carrier collection in CdTe solar cells," *Phys. Rev. Lett.* **112**, 156103, 2014.

[3] D.M. Meysing, C.A. Wolden, M.M. Griffith, H. Mahabaduge, J. Pankow, M.O. Reese, J.M. Burst, W.L. Rance, and T.M. Barness, "Properties of reactively sputtered oxygenated cadmium sulfide (CdS:O) and their impact on CdTe solar cell performance," *J. Vac. Sci. Technol.* **A33**, 021203 (2015).

[4] A. Kikukawa, S. Hosaka, and R. Imura, "Silicon pn junction imaging and characterizations using sensitivity enhanced Kelvin probe force microscopy," *Appl. Phys. Lett.* **66**, 3510 (1995).

[5] C.-S. Jiang, M.A. Contreras, L.M. Mansfield, H.R. Moutinho, B. Eggas, K. Ramanathan, and M.M. Al-Jassim, "Nanometer-scale surface and resistance mapping of wide-bandgap $Cu(In,Ga)Se_2$ thin films," *Appl. Phys. Lett.* **106**, 043901 (2015).

[6] M. Gloeckler, J.R. Sites, and W.K. Metzger, "Grain-boundary recombination in $Cu(In,Ga)Se_2$ solar cells," *J. Appl. Phys.* **98**, 113704 (2005).

[7] W.K. Metzger and M. Gloeckler, "The impact of charged grain boundaries on thin-film solar cells and characterization," *J. Appl. Phys.* **98**, 063701 (2005).

[8] B. Gaury and P.M. Haney, "Charged grain boundaries reduce the open-circuit voltage of polycrystalline solar cells—An analytical description," *J. Appl. Phys.* **120**, 234503 (2016).

[9] W.K. Metzger, R.K. Ahrenkiel, J. Dashdorj, and D.J. Friedman, "Analysis of charge separation dynamics in a semiconductor junction," *Phys Rev.* **B71**, 035301 (2005).

Nanoparticle/metal rear reflectors for low- and high-temperature silicon solar cells

Syeda Qudsia[1,2], Farah Qazi[1,2], Mehwish Azher Javed[1,2], Mathieu Boccard[1], Zhengshan J. Yu[1], Peter Firth[1,3], Jonathan Bryan[1], and Zachary C. Holman[1,3]

[1] Arizona State University, Tempe, Arizona 85281, USA
[2] U.S.-Pakistan Center for Advanced Studies in Energy, National University of Science and Technology, Islamabad, 25000, Pakistan
[3] Swift Coat, Inc., Peoria, Arizona 85383, USA

Abstract — **Metallic rear electrodes in silicon solar cells with textured surfaces absorb infrared light parasitically, stealing current from the cells. The loss can be mitigated, however, with the insertion of a low-refractive-index dielectric layer between the metal and absorber that is at least as thick as the penetration depth of the evanescent wave that results when light arrives outside the escape cone at the rear interface. We investigate silicon nanoparticle coatings with tunable porosity—and thus refractive index—in this role. We show that a nanoparticle/Ag rear reflector achieves an internal reflectance exceeding 99.5%, and, when used in a silicon heterojunction solar cell, boosts the infrared short-circuit current density by 0.4 mA/cm^2 compared to a state-of-the-art ITO/Ag reflector. In addition, the nanoparticle films survive annealing at temperatures up to 800 °C without severe changes in their optical properties, suggesting that they may be used in high-temperature silicon solar cells too.**

I. INTRODUCTION

Present silicon solar cells have infrared (IR) current losses equal to approximately 1 mA/cm^2, or 2–3% of the short-circuit current density [1]. These losses are not due to imperfect light trapping, where light trapping is understood here to mean the path length enhancement of light in the absence of absorption. The random pyramids of textured monocrystalline wafers, for example, have been shown to provide comparable scattering to the Lambertian limit [2-4]. Rather, the losses are caused by parasitic absorption of the weakly absorbed 1000–1200 wavelengths in the thin-film layers adjacent to the wafer absorber as the light interacts with the front and rear surfaces multiple times.

For random pyramid textures, scattered IR light assumes a near-Lambertian distribution within the wafer after only two passes (one bounce off the rear surface), meaning that light subsequently arrives at the front and rear surfaces from within the wafer with a wide range of angles of incidence [2]. Light that arrives at the rear surface at an oblique angle with respect to the local pyramid facet is outside of the escape cone (if a lower-refractive-index dielectric layer is deposited on that surface, as in PERC, TOPCon, and silicon heterojunction solar cells). The light is totally internally reflected if no absorbing layer is present within the penetration depth of the resulting evanescent wave (hundreds of nanometers for typical wavelengths and angles). However, if the evanescent wave encounters an absorbing material, the internal reflectance will be attenuated, stealing current from the solar cell. In particular, the evanescent wave of p-polarized light can be strongly coupled to surface plasmon polaritons in metal rear electrodes, including those of nominally excellent reflectors like silver. This causes the internal rear reflectance to fall precipitously, to, for example, below 90% for no or a very thin dielectric layer and a silver electrode [5].

The solution to this problem is to ensure that high-efficiency solar cells have an IR-transparent dielectric layer at their rear side that is thicker than the decay length of the evanescent wave; simulations and measurements reveal this to be approximately 200 nm [5]. Superior performance is also observed by reducing the refractive index of the dielectric layer, as this reduces the penetration depth and field strength of the evanescent wave. With a MgF$_2$/Ag reflector, internal rear reflectances of over 99.5% are possible (MgF$_2$ has a refractive index of 1.37) [6].

One challenge of this approach, however, is that there are few materials to choose from that have very low refractive index, and those that do are not always easy or inexpensive to deposit as thin films. MgF$_2$, for example, is commonly evaporated but tends to flake off chamber walls and thickness monitors due to internal stress, and evaporation has never successfully been implemented as a process step in silicon solar cell manufacturing despite repeated attempts. In addition, the dielectric layer must be patterned so that the metal electrode can make electrical contact through it to the underlying electron or hole contact.

We recently demonstrated dielectric/Ag rear reflectors in silicon heterojunction solar cells in which the dielectric layer was a porous coating of silicon nanoparticles [7]. As the nanoparticles are approximately 5 nm in diameter, they, and the voids between them in the film, are much smaller than the IR wavelengths that reach the rear of the solar cell, and thus the film behaves as an effective medium with a refractive index that is governed by its porosity. The nanoparticles are deposited by a vacuum spray coating technique in which the film porosity can be controllably tuned, thereby enabling refractive index control. Furthermore, the spraying technique is directional such that a simple shadow mask—in this case a metal mesh—can be used to transfer a pattern for contact formation.

978-1-5090-5606-4/17 $31.00 © 2017 IEEE

Here, we extend this previous work by comparing this nanoparticle-based rear reflector to other dielectric materials in silicon heterojunction solar cells. We also take initial steps to employing these reflectors in silicon solar cells with a screen-printed aluminum rear electrode, such as Al-BSF and PERC cells, since screen-printed aluminum is particularly lossy. In this case, the nanoparticle coating must be able to withstand firing temperatures, and thus we investigate the optical, chemical, and structural properties of the film up to 800 °C.

II. NANOPARTICLE FILM DEPOSITION AND CHARACTERIZATION

Silicon nanoparticle films were deposited via vacuum spray deposition using the tool depicted in Figure 1 [8], [9]. Briefly, silicon nanoparticles were synthesized in a parallel-plate RF plasma reactor turned on its side by running a dusty silane plasma [10]. A constant flow of silane and background helium dragged the nucleating particles out of the plasma zone, arresting their growth: the mean diameter of the nanoparticles is approximately 5 nm and the standard deviation of an ensemble is approximately 15% of the mean diameter. A deposition chamber containing a translated substrate is located downstream of the plasma reactor, and the chamber is continuously evacuated of the helium flowing through the whole system (the silane is consumed in forming the particles). A slit-shaped nozzle placed between the synthesis and deposition chambers creates a critical pressure drop—typical plasma pressures are a few Torr and typical deposition chamber pressures are 10 times lower—such that the gas is accelerated to near-sonic or super-sonic speeds. The gas, in turn, accelerates the nanoparticles in the nozzle through collisions, but the nanoparticles are not able to follow the gas streamlines in the more rarified downstream chamber, and they impact the substrate to form a coating. This process has some similarities to cold gas dynamic spraying ("cold spray") and more similarities to "aerosol deposition", though both of those methods use powder as the particle feedstock (no integrated synthesis) and the particle size is usually considerably larger.

The impact velocity of the impinging nanoparticles can be changed by altering the distance between the nozzle and substrate, the pressure ratio across the nozzle, or the absolute up- or downstream pressures [9]. This, in turn, changes the porosity—and thus the refractive index—of the coating. As the thickness of the coating can be trivially controlled by the speed with which the substrate is rastered under the nozzle and the tool can coat substrates up to 5" on a side, it is possible to integrate low-refractive-index coatings of the desired 200 nm thickness in solar cells. (A typical deposition of this sort takes approximately 1 minute.) Figure 1 shows a photograph of a representative silicon nanoparticle coating, as well as exemplary cross-sectional scanning electron microscopy images of porous ($n = 1.09$) and dense ($n = 1.53$) coatings. Here, and throughout this work, refractive indices were determined via fitting of spectroscopic ellipsometry data with a Bruggeman effective medium approximation combining silicon and void. Only wavelengths longer than 900 nm were fit so as to avoid potential problems induced by the inaccuracy of the (bulk) silicon refractive index at shorter wavelengths related to the (weak) quantum confinement in the small nanoparticles. Spectrophotometry measurements reveal that—unlike bulk crystalline silicon—even thick nanoparticle coatings do not absorb light with wavelengths longer than 700 nm [7]. This, in fact, is why it is possible to use silicon nanoparticles, and not a more conventional transparent dielectric, at the rear of the solar cell without suffering parasitic absorption within the nanoparticles themselves.

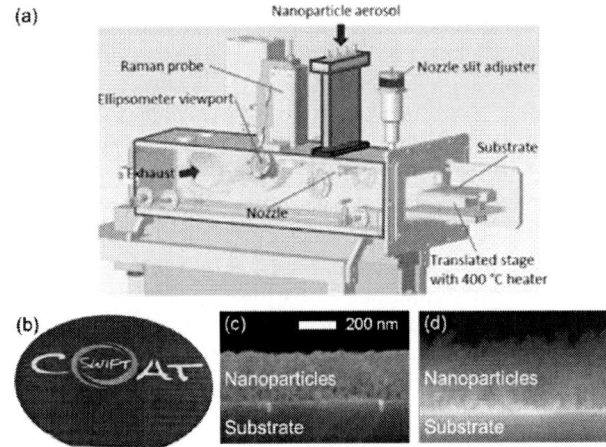

Fig. 1. (a) Schematic of the nanoparticle deposition tool, (b) photograph of a 5" wafer with a blanket (blue) silicon nanoparticle antireflection coating, plus a pattern deposited through a shadow mask, (c) scanning electron microscope cross section of a dense silicon nanoparticle coating on a wafer, and (d) scanning electron microscope cross section of a porous silicon nanoparticle coating on a wafer.

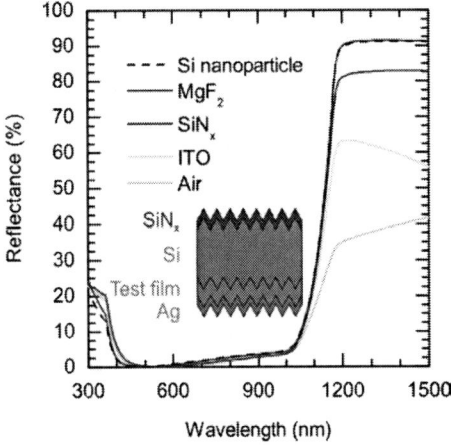

Fig. 2. Measured reflectance spectra of optical test structures with varying rear dielectrics, each approximately 200 nm thick.

978-1-5090-5606-4/17 $31.00 © 2017 IEEE

III. LOW-TEMPERATURE SOLAR CELL RESULTS

Prior to fabricating complete solar cells, we made the optical test structures depicted in Figure 2. These structures are optically similar to silicon heterojunction solar cells (note that the amorphous silicon layers have similar refractive index at long wavelengths to the wafer and can thus be omitted without consequence), but have no IR parasitic absorption except in the rear electrode. With a 150-nm-thick transparent rear indium tin oxide (ITO) layer (electron density of approximately 10^{20} cm^{-1})—our best layer using the typical silicon heterojunction cell structure [11]—only 65% of 1200 nm light, which is just below the bandgap energy of silicon and therefore not absorbed in the wafer, returns out the front as escape reflectance. That is, 35% is absorbed parasitically in the ITO/Ag reflector. When the ITO layer is replaced with SiN$_x$, which has a similar refractive index but is not itself absorptive, the reflectance at 1200 nm climbs to above 80%, and with a lower-refractive-index MgF$_2$ or silicon nanoparticle layer, it exceeds 90%. As the light interacts with the rear surface many times in the textured wafer, this corresponds to an internal rear reflectance exceeding 99.5% according to our calculations. The silicon nanoparticle layer used here had a refractive index of approximately 1.4, as we found that more porous layers—though they have lower refractive index—have surfaces that are sufficiently rough that they induce additional unwanted plasmonic absorption in the no-longer-flat silver, resulting in worse than expected measured total reflectance [7].

To incorporate a nanoparticle/Ag rear reflector into a solar cell, it is necessary to pattern the layer so as to form local openings where the silver may make contact through the insulating nanoparticle coating to the electron or hole contact. We fabricated silicon heterojunction solar cells as described in Ref. [7] and replaced the rear ITO/Ag electrode with an ITO/silicon nanoparticle/Ag stack in which the ITO layer was 20 nm thick and served to transport charge to the nearest opening in the nanoparticle layer, and the nanoparticles were deposited through a metal mesh shadow mask that left 5% uncoated area with a pitch of 500 μm between openings. Indium zinc oxide (IZO) was used as the front transparent conductive oxide (TCO) layer because its comparatively high mobility of approximately 50 cm^2/Vs means that it can meet the target sheet resistance of 70 Ω/sq without incurring as much IR free-carrier absorption as, e.g., lower-mobility ITO.

Figure 3 shows the resulting improvement when such a nanoparticle/Ag rear reflector replaces the best ITO/Ag reflector in otherwise identical silicon heterojunction solar cells. The reflectance at 1200 nm improves by approximately 8% (but falls well short of the 91% obtained in Figure 2 because there are other layers in the complete cell—especially the front IZO—that absorb some IR light) and the IR external quantum efficiency improves slightly, resulting in a 0.4 mA/cm^2 gain in short-circuit current density. This 4 cm^2 cell

had an efficiency of 21.3% despite the batch having very low pseudo fill factors of only about 79%.

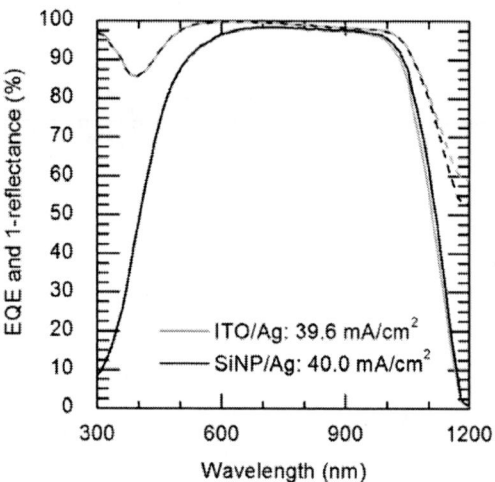

Fig. 3. EQE and total absorbance (1-reflectance) of identical silicon heterojunction solar cells with a baseline ITO/Ag rear reflector or a silicon nanoparticle/Ag rear reflector. The values given are the active-area short-circuit current densities of these cells found by integrating their EQE spectra.

IV. ANNEALED COATINGS FOR HIGH-TEMPERATURE CELLS

While Figure 3 shows a current gain over the present state of the art with a deposition process that is quick, simple, and low CapEx (only rough vacuum is needed), the gain is marginal in large part because silver is already an excellent rear reflector with any dielectric and because parasitic IR absorption in the front TCO layer, which must have high carrier density to successfully transport charge laterally, eats into the savings engineered at the rear side. A different solar cell structure that would take better advantage of the nanoparticle coating would have no TCO and would have a more absorbing rear metal electrode, which is an exact description of an Al-BSF or PERC solar cell, though the latter already has dielectric layers at the rear that work relatively well to enhance IR internal reflectance. The aluminum paste in these cells is particularly absorbing and thus a low-refractive-index nanoparticle layers may provide a significant enhancement.

As Al-BSF and PERC cells are fired at temperatures on the order of 800 °C, we began by investigating the robustness of the silicon nanoparticle layers when annealed up to these temperatures. A rapid thermal annealing furnace with an 80:20 nitrogen:oxygen atmosphere was used for these tests, though lower temperature anneals (< 200 °C) were simply done on a hotplate in air.

Figure 4 displays the Fourier transform infrared (FTIR) spectra of silicon nanoparticle coatings immediately after deposition and following annealing for 20 mins at only 180

°C. The as-deposited nanoparticles have hydrogen-terminated surfaces formed in the silane plasma, as seen by both the stretching and bending/wagging modes observed. After annealing, however, peaks appear corresponding to back-bonded oxygen, and the hydride peaks diminish in intensity. After annealing at 500 or 800 °C (not shown), the oxide peak at approximately 1100 cm^{-1} is all that is observed; no hydrogen remains.

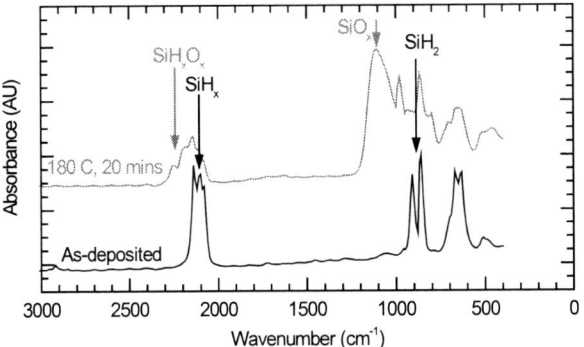

Fig. 4. FTIR spectra of a silicon nanoparticle coating immediately after deposition and after annealing.

One might suspect that the oxidation of the 5-nm-diameter silicon particles would be complete upon annealing at 800 °C and that the coating would become visibly transparent. This, however, is not observed to occur. The absorbance spectra of silicon nanoparticle coatings on quartz before and after annealing for various temperatures and times are shown in Figure 5. Surprisingly, the absorbance *redshifts*, not blueshifts, as the annealing temperature increases (annealing time is less important than temperature). As ellipsometry measurements confirm the FTIR measurements that the nanoparticles have oxidized, the increase in absorbance at visible wavelengths must be caused by a phenomenon other than the compositional change. We hypothesize that the silicon nanoparticles, which are initially small enough to be quantum confined, turn into silicon cores with oxide shells after annealing, and that the silicon cores sinter together, relaxing quantum confinement and narrowing the effective optical bandgap of the material. Scanning electron microscope (SEM) images reveal enlarged pores after annealing at 800 °C that are consistent with this hypothesis (Figure 6). Furthermore, the adhesion of the nanoparticle coating to both the substrate and to itself increases dramatically with annealing temperature, as observed in Taber Abraser tests, supporting particle sintering. As the absorbance still does not extend beyond 900 nm, the annealed coatings should still be suitable for use as transparent IR dielectric layers in rear reflectors, provided that they do not adversely react with the aluminum itself.

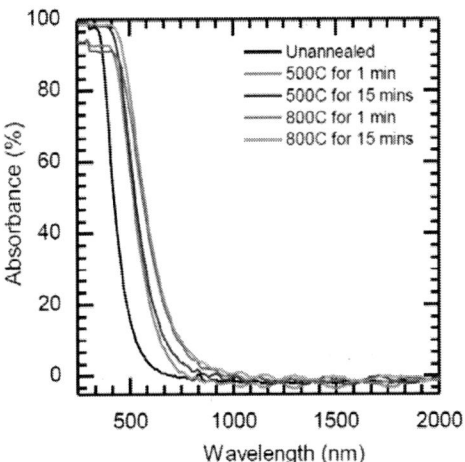

Fig. 5. Absorbance of silicon nanoparticle coatings deposited on quartz as a function of annealing temperature and time.

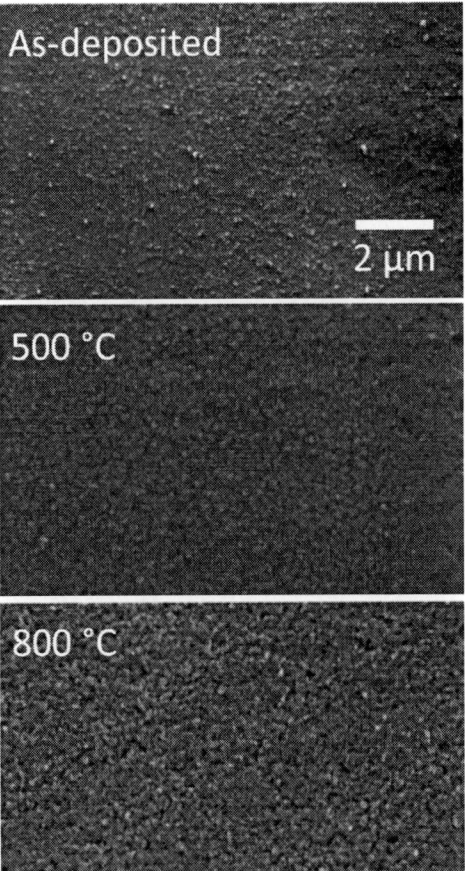

Fig. 6. SEM images of a silicon nanoparticle coating on a silicon wafer before and after annealing.

V. Conclusion

Dielectric layers can reduce infrared plasmonic absorption in lossy metal reflectors at the rear of solar cells if they are approximately 200 nm thick, IR transparent, and composed of a low-refractive-index material. Nanoparticle coatings deposited by vacuum spraying have controllable porosity with particle and pore sizes much smaller than the wavelength of IR light, and thus act as effective media with controllable refractive index. We demonstrated that a sprayed silicon nanoparticle coatings enhances the IR EQE of a silicon heterojunction solar cell with a silver electrode by 0.4 mA/cm^2 compared to the present state of the art. The coatings also have the potential to be used in high-temperature silicon cells with an aluminum rear electrode that is a particularly poor reflector, and our initial studies indicate that they can withstand the firing temperatures of these cells. An important next step will be to investigate whether the aluminum paste diffuses through or alloys with the silicon nanoparticle film when fired, as is expected from the formation of Al-BSF's, and how this changes the optical and electrical properties of the nanoparticle coating. If the aluminum merely dopes the coating without inducing overwhelming free-carrier absorption, it may be possible to conduct current through the nanoparticle dielectric without requiring local patterning.

References

[1] Z. C. Holman, M. Filipič, B. Lipovšek, S. De Wolf, F. Smole, M. Topič, and C. Ballif, "Parasitic absorption in the rear reflector of a silicon solar cell: Simulation and measurement of the sub-bandgap reflectance for common dielectric/metal reflectors," *Sol. Energy Mater. Sol. Cells,* vol. 120, Part A, pp. 426-430, 2014.

[2] S. Manzoor, M. Filipic, M. Topic, and Z. Holman, "Visualizing light trapping inside textured silicon solar cells," (in preparation).

[3] A. Ingenito, O. Isabella, and M. Zeman, "Experimental Demonstration of 4n(2) Classical Absorption Limit in Nanotextured Ultrathin Solar Cells with Dielectric Omnidirectional Back Reflector," *Acs Photonics,* vol. 1, pp. 270-278, 2014.

[4] P. Campbell and M. A. Green, "Light trapping properties of pyramidally textured surfaces," *J. Appl. Phys.,* vol. 62, pp. 243-249, 1987.

[5] Z. C. Holman, S. De Wolf, and C. Ballif, "Improving metal reflectors by suppressing surface plasmon polaritons: A priori calculation of the internal reflectance of a solar cell," *Light: Science & Applications,* vol. 2, p. e106, 2013.

[6] Z. C. Holman, A. Descoeudres, S. De Wolf, and C. Ballif, "Record infrared internal quantum efficiency in silicon heterojunction solar cells with dielectric/metal rear reflectors," *IEEE Journal of Photovoltaics,* vol. 3, pp. 1243-1249, 2013.

[7] M. Boccard, P. Firth, Z. J. Yu, K. C. Fisher, M. Leilaeioun, S. Manzoor, and Z. C. Holman, "Low-refractive-index nanoparticle interlayers to reduce parasitic absorption in metallic rear reflectors of solar cells," *Physica Status Solidi (a),* pp. e201700179.

[8] P. Firth and Z. C. Holman, "Large-area nanoparticle coatings with controllable porosity and 8% non-uniformity formed via aerosol spraying" (in preparation).

[9] Z. C. Holman and U. R. Kortshagen, "A flexible method for depositing dense nanocrystal thin films: impaction of germanium nanocrystals," *Nanotechnology,* vol. 21, p. 335302, 2010.

[10] L. Mangolini, E. Thimsen, and U. Kortshagen, "High-Yield Plasma Synthesis of Luminescent Silicon Nanocrystals," *Nano Letters,* vol. 5, pp. 655-659, 2005.

[11] Z. C. Holman, M. Filipic, A. Descoeudres, S. De Wolf, F. Smole, M. Topic, and C. Ballif, "Infrared light management in high-efficiency silicon heterojunction and rear-passivated solar cells," *J Appl Phys,* vol. 113, p. 013107, 2013.

Absorption in each layer of a silicon heterojunction solar cell

Keith R. McIntosh,[1] Malcolm D. Abbott,[1] Benjamin A. Sudbury,[1]
Salman Manzoor,[2] Zhengshan J. Yu,[2] Mehdi Leilaeioun,[2] Jianwei Shi[2] and Zachary C. Holman[2]

[1]PV Lighthouse, Coledale, NSW 2515, AUSTRALIA
[2]Arizona State University, Tempe, 85281, USA

Abstract—**We study the absorption in a silicon heterojunction solar cell. After determining the texture and film properties of the cell from experimental data, we apply ray tracing to quantify the absorption as a function of wavelength and depth in each of the cell's many layers. By comparing the results to the measured external quantum efficiency, we determine the collection efficiency of the front intrinsic a-Si:H film to be 0.35. We then asses the optimal thickness t_{opt} of the front and rear ITO layers with and without encapsulation under the AM1.5g spectrum, finding that after encapsulation (i) t_{opt} of the front ITO decreases from 70 nm to zero, (ii) t_{opt} of the rear ITO remains at about 100 nm, and (iii) the sensitivity of absorption and reflection to the front ITO thickness decreases. We also conclude that the collection current could be increased by up to 2.7 mA/cm^2 by reducing the absorptance of the ITO and using thinner a-Si films, or by 1.3 mA/cm^2 by increasing the collection efficiency of the front a-Si:H layers. The described approach is applicable to any solar cell containing thin-film stacks.**

I. INTRODUCTION

Amorphous-silicon/crystalline-silicon heterojunction (SHJ) solar cells present an interesting and complicated optimization problem for researchers [1–7]. Both the front and rear surfaces of the cell are typically textured and coated with multiple thin films, where each film absorbs a significant fraction of the incident light, and where some films contribute a fraction of that absorption towards the cell's collection current. Clearly, a sophisticated optical analysis is required to quantify the losses and gains in each film, to assess their many unavoidable trade-offs, and to optimize the film thicknesses.

In this paper, we extend the work published on high-efficiency *n*-type SHJ cells [1–7] by quantifying the absorption in the films as a function of wavelength and depth. It is, to our knowledge, the first demonstration of using ray tracing to determine the absorption profile within thin films on a textured silicon surface. The simulation results assist in (i) the experimental determination of the collection efficiency η_c of the front intrinsic amorphous silicon (a-Si:H) film, (ii) the optimization of film thickness in encapsulated and unencapsulated cells, and (iii) the quantification of the gain that results from increasing the film transparency and η_c. An equivalent approach can be applied to any solar cell that contains one or more thin-film stacks, such as those with double-layer antireflection coatings, a-Si:H passivation layers, and tandem structures—or any thin-film cell architecture.

II. OPTICAL CHARACTERIZATION OF AN SHJ SOLAR CELL

2.1 Sample structure

Fig. 1 presents a schematic of the high-efficiency *n*-type SHJ solar cell examined in this work. It contains three films on the front and four films on the rear. The intrinsic a-Si:H provides surface passivation for the *n*-type c-Si bulk, the *n*- and *p*-type a-Si:H layers form the electron and hole selective contacts, the indium tin oxide (ITO) layer on the front serves as an antireflection coating and supports lateral conduction, the ITO and Ag layers on the rear assist with internal reflectance, and the Ag layer also provides a rear contact to the solar cell. The front contact (not shown in Fig. 1) is comprised of a screen-print Ag paste in the form of a grid.

2.2 Sample fabrication

The SHJ solar cell was fabricated on an *n*-type float-zone Si wafer with a resistivity of 2–3 Ω·cm and a thickness of 135 μm. The wafer was textured in a KOH solution to create random upright pyramids on both sides of the wafer, cleaned in piranha and RCA-B solutions, and dipped in buffered HF to remove any oxide on its surfaces. Intrinsic and *n*-type a-Si:H layers were then deposited onto the front side, and intrinsic and *p*-type a-Si:H layers were deposited onto the rear side by plasma-enhanced chemical vapor deposition using an AMAT P5000 tool. Next, an ITO layer was sputtered onto the front side using an In$_2$O$_3$:SnO$_2$ (9:1) target in an MRC 944 tool, and ITO and Ag layers were sputtered onto the rear side. The device was completed by screen-printing a grid of low-temperature Ag paste onto the front surface and annealing at 200 °C for 20 mins to cure the paste. See [8] for more detail.

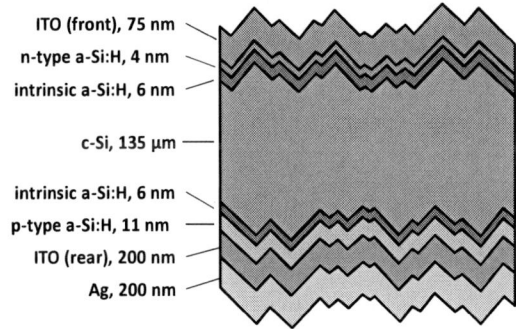

Fig. 1. The layers of the high-efficiency SHJ solar cell.

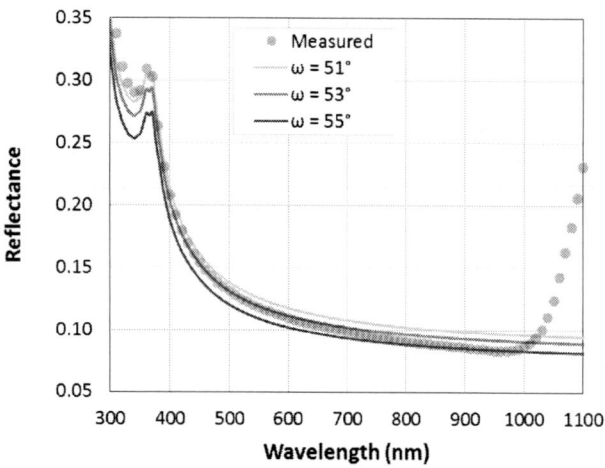

Fig. 2. Reflectance of a bare textured wafer.
Symbols show the measured total reflectance, and
lines show the simulated external reflectance for three ω.

2.3 Characterization of pyramidal texture

We evaluated the effective base angle ω of the pyramids by the procedure described in [9]. This involved measuring the hemispherical reflectance of the sample after texturing and before film deposition, and then determining the value of ω that yielded the best fit between the simulated and measured reflectances.

Fig. 2 plots the experimental reflectance and compares it to the simulated external reflectance $R_{ext}(\lambda)$ calculated by OPAL 2 [10] for $\omega = 51°$, $53°$ and $55°$. The simulations assumed that the incident light was unpolarized and had an incident zenith angle of $\theta = 7°$ (as applied in the experimental setup), that the incident azimuth angle was $\varphi = 0°$, and that the distribution of pyramid heights varied uniformly by a factor of 5 (e.g., from 0.66 to 3.3 μm). Neither of the latter two assumptions had a significant impact on the simulated reflectance.

We find the agreement between simulation and experiment to be poorer than the agreement attained in other publications that apply the same procedure [9, 11]. As evident in Fig. 2, the simulated data fits the experimental data at long wavelengths (e.g., $\lambda = 800–1000$ nm) when ω is near its ideal value of $54.74°$ (purple line); in that case, the disagreement between simulation and experiment is significant at wavelengths below 600 nm. By contrast, the simulated data fits the experimental data at short wavelengths (e.g., 300–500 nm) when $\omega \sim 51°$ (orange line), for which the agreement is poor above 600 nm. It is possible that the observed discrepancy in the 'slope' of $R_{ext}(\lambda)$ arises because the actual pyramids are not adequately represented in the simulation (e.g., the actual pyramids have a variable base angle) [12, 13], but we also do not discount the possibility of there being

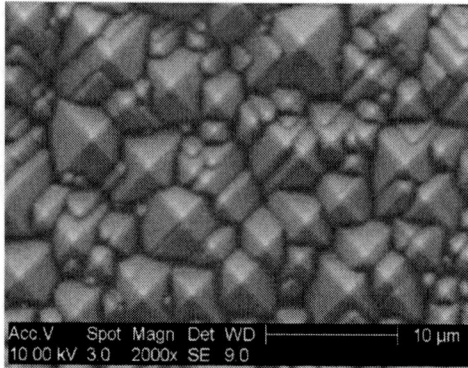

Fig. 3. SEM image of c-Si wafer after texturing.

experimental error related to calibration and to polarization of the incident beam.

Regardless, over the wavelength range where the escape reflectance is negligible ($\lambda < 1000$ nm), we find that $\omega = 53°$ (red line) yields the best agreement between simulation and experiment; the agreement is within $\pm 1\%$ (abs) and $\pm 10\%$ (rel) over the range 370–1000 nm. This value for ω is used for the simulations that follow.

Fig. 3 presents an SEM image of the textured c-Si wafer surface. It shows that there is a variety of pyramid sizes and that the distance across the base of the largest pyramids is about 5 μm. Assuming a base angle of $\omega = 53°$, this gives a maximum pyramid height of 3.3 μm.

2.5 Characterization of the thin films

The thickness and the dispersive complex refractive index ($n(\lambda)$ and $k(\lambda)$) of the films were determined by performing spectroscopic ellipsometry on test samples. The test samples were c-Si polished on one side and coated with 300 nm of thermal SiO_2 and with a single film on the polished surface; the recipe used to deposit each film was the same as that applied when fabricating the solar cell. The purpose of the SiO_2 layer was to enhance optical interference, making the analysis of the experimental data more accurate [14].

The thickness of the films on the textured wafer, which was assumed to be 1.6 times the thickness measured on planar, is listed in Fig. 1 and Table I; their refractive indices are plotted in [7]. Note that the front ITO is more heavily doped than the rear ITO; this makes it more conductive, thereby assisting lateral transport to the fingers, but also more absorptive.

2.6 Characterization of the solar cell

The solar cell's one-sun IV characteristics were measured at ASU to be $V_{oc} = 715.0 \pm 1.5$ mV, $J_{sc} = 38.17 \pm 0.12$ mA/cm^2, $FF = 0.7735 \pm 0.0035$, and $\eta = 21.18\% \pm 0.14\%$ (abs).

Fig. 4 plots (a) the hemispherical reflectance and (b) the external quantum efficiency (EQE) of the solar cell. The symbols plot measured data and the lines plot simulated data.

978-1-5090-5606-4/17 $31.00 © 2017 IEEE

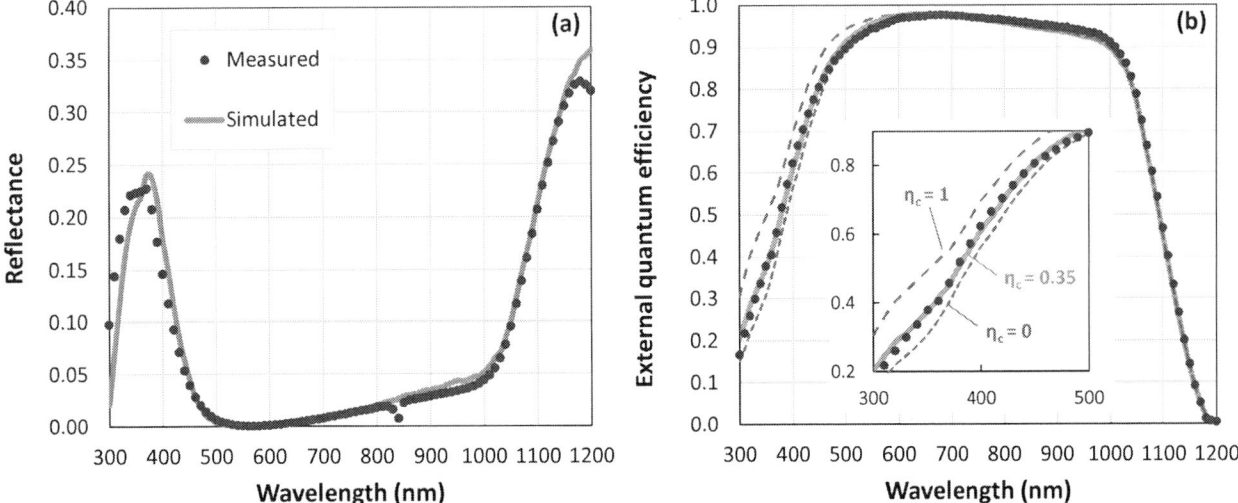

Fig. 4. (a) Hemispherical reflectance and (b) external quantum efficiency (EQE) of the solar cell. Symbols show measurements, lines show simulation. The simulated EQE is calculated with the collection efficiency of the n-type a-Si:H being η_c = 0, 0.35 and 1.

For both measurements the incident beam was located between the fingers so that there was no external reflection from the fingers (although the fingers might still affect the light trapping due to internal reflection). The EQE measurements were conducted at short circuit and did not include a steady-state light bias. The integrated short-circuit current density J_{sc} under the AM1.5g spectrum was calculated from the EQE to be 39.4 mA/cm^2; this is higher than the J_{sc} of the cell because it excludes external reflection from the fingers.

2.7 Ray tracing the solar cell

We simulated the optical behavior of the structure in Fig. 1 using the PV Lighthouse ray tracer [15, 16]. It solves the optical losses with geometric optics and the transfer-matrix method, and distinguishes the absorption loss in each layer of a thin-film stack, providing the results as a function of wavelength and depth.

The simulated incident illumination was composed of a randomly polarized, AM1.5g spectrum incident at θ = 7° and φ = 0° (as used in Section 2.3). Changes in the polarization of each ray were accounted for after each interface interaction. The dispersive refractive index ($n(\lambda)$ and $k(\lambda)$) used in the simulations were taken from the references in Table I. The texture on the front and rear surface of the c-Si wafer was simulated as random upright pyramids with ω = 53° and a height of 3.3 μm. For simplicity, we omitted the effect of the front metallization. The solution was attained with 5 million rays; such a large number of rays was chosen so that there would be negligible noise at each wavelength, as evident in Fig 4.

Thus, in regards to the optical behavior of the cell, there were no free variables with which to 'tune' the ray tracing simulation. It is therefore noteworthy that a close agreement is observed between the simulated and experimental reflectance of Fig. 4(a). The agreement is to within ±1% (abs) over the wavelength range 440–1160 nm. This range accounts for 95% of J_{sc} under the AM1.5g spectrum. Notice that the close agreement includes the escape reflectance (λ > 1000 nm), which suggests that the ray tracing accurately predicted the light trapping within the cell.

Below 440 nm, and representing 4.9% of J_{sc}, the agreement is poorer, whereby the predicted reflectance is up to 8% (abs) different to the experimental reflectance. We refrained from fine-tuning the simulation inputs (such as ω, or the thickness, $n(\lambda)$ or $k(\lambda)$ of the films) to improve the fit.

Above 1160 nm, and representing just 0.1% of J_{sc}, the simulation overestimates the reflectance by at most 4% (abs). The discrepancy at long λ is likely due to the simulation

Table I. Baseline thickness and the reference used for the dispersive refractive index (n and k) of each layer in the simulation.

Layer	Thickness	Reference
Glass	3000 μm	[18]
EVA	450 μm	[19]
ITO (front)	75 nm	[7] (N=4.9 × 10^{20} cm^{-3})
n-type a-Si:H	4 nm	[7]
Intrinsic a-Si:H	6 nm	[7]
c-Si	135 μm	[20], [21]
Intrinsic a-Si:H	6 nm	[7]
p-type a-Si:H	11 nm	[7]
ITO (rear)	200 nm	[7] (N=1.2 × 10^{20} cm^{-3})
Ag	200 nm	[22]

neglecting one or more of the following: free-carrier absorption in the wafer, diffraction, variable ω, and variable pyramid height.

Fig. 4(b) indicates that the simulated and experimental EQE also agree well. We emphasize, however, that the simulated EQE was calculated by (i) assuming some knowledge about the collection of generated carriers in each layer, and (ii) allowing the collection efficiency η_c of the front i-a-Si:H layer to be a free variable and constant with λ (as was applied in [4]). Specifically, we assumed that all carriers generated in the c-Si were collected and therefore contributed to the EQE and J_{sc}, as is consistent with the excellent surface passivation offered by the a-Si:H layers and the high bulk lifetime of the n-type wafer; we also assumed that none of the carriers generated in the ITO or the n-type a-Si:H layers contributed to J_{sc}, as is consistent with the conclusion in [4] deduced from EQE measurements on SHJ cells with a range of a-Si:H thicknesses.

As evident in Fig. 4(b), the simulated EQE only depends on η_c for $\lambda < 600$ nm; this is because at higher wavelengths, absorption in the intrinsic a-Si:H is very small (see Fig. 5 in Section III). The value of η_c that gives the best fit is 0.35. With $\eta_c = 0.35$, the simulated and experimental EQE are within $\pm 2\%$ (abs) over the range 330–600 nm. This η_c is close to the value of 0.3 determined in [4] for a similar SHJ solar cell.

Supporting the accuracy of the ray tracing, and hence this method to determine η_c, Fig. 4(b) indicates that the simulation and experiment agreement to within $\pm 2.5\%$ (abs) over the entire wavelength range. Most pleasing is the agreement over the range $\lambda = 1000–1200$ nm, once again suggesting that the ray-tracing simulation accurately predicts the light trapping within the solar cell.

Before concluding, we remark on several uncertainties in the above approach that have not been evaluated. It is possible that experimental error in the reflectance data arose for the reasons explained in Section 2.4, for which we comment that any disagreement in $R_{ext}(\lambda)$ due to a variable ω would be smaller in Fig. 4(a) than in Fig. 2 due to the suppression of reflectance by the thin films. We also do not discount experimental error that might have arisen in the ellipsometry measurements introducing uncertainty in the thickness and refractive index of the films, or experimental error in the EQE measurements, particularly due to injection dependence, which would have introduced a difference in the EQE determined at the measured intensity compared to the EQE under one-sun conditions. Differences in light trapping between simulation and experiment could also have arisen due to the ray tracing neglecting free-carrier absorption in the wafer, diffraction [17], variability in ω, variability in the height of the pyramids, and the contribution of the fingers to internal reflection. Thus, although the close agreement observed between simulation and experiment is pleasing, we do not discount the possibility that the agreement might have been slightly poorer if the samples were measured or simulated more accurately.

Nevertheless, we conclude that the solar cell is well represented by the ray tracing when it is simulated with a pyramid base angle of 53° and a height of 3.3 μm, with the layers listed in Table I, with 100% carrier collection from the c-Si wafer, $\eta_c = 0.35$ from the front intrinsic a-Si:H, and no collection from the front ITO or the n-type a-Si:H. With these inputs, the simulation predicted $J_{sc} = 39.3$ mA/cm², which is just 0.1 mA/cm² less than the J_{sc} calculated from the experimental EQE; the simulation also predicted the reflectance to within $\pm 1\%$ (abs) in the range 440–1160 nm, and the EQE to within $\pm 2.5\%$ (abs) in the range 320–1200 nm.

III OPTICAL LOSSES AND GENERATION PROFILES

We now use the PV Lighthouse ray tracer to learn more about the optical behavior of the solar cell with and without encapsulation. For simplicity, we omit the effect of the fingers, the busbars, and the spacing between cells in a

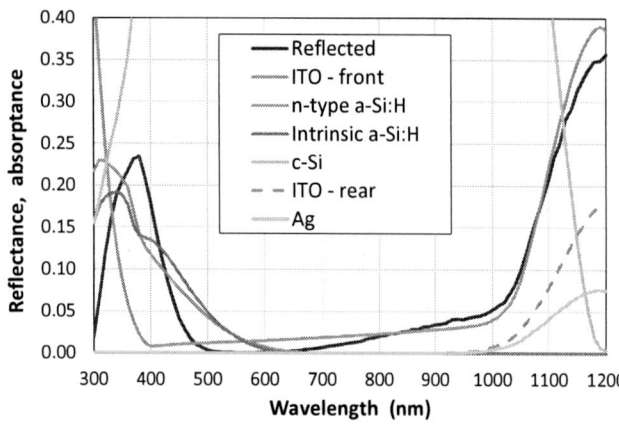

Fig. 5. Reflectance and absorptance in the unencapsulated solar cell.

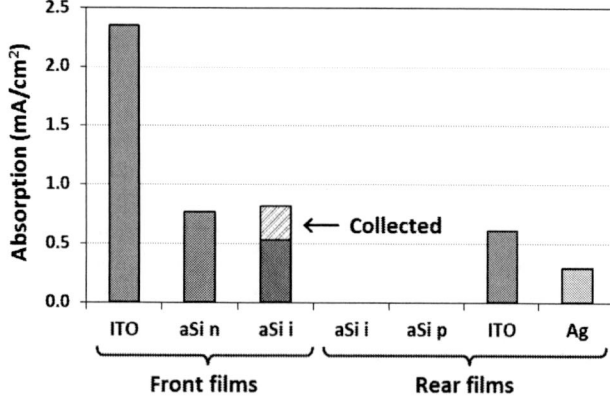

Fig. 6. Absorption in front and rear films under the AM1.5g spectrum integrated from 300–1200 nm. The hashed bar represents the absorbed current that contributes to the cell's collection current.

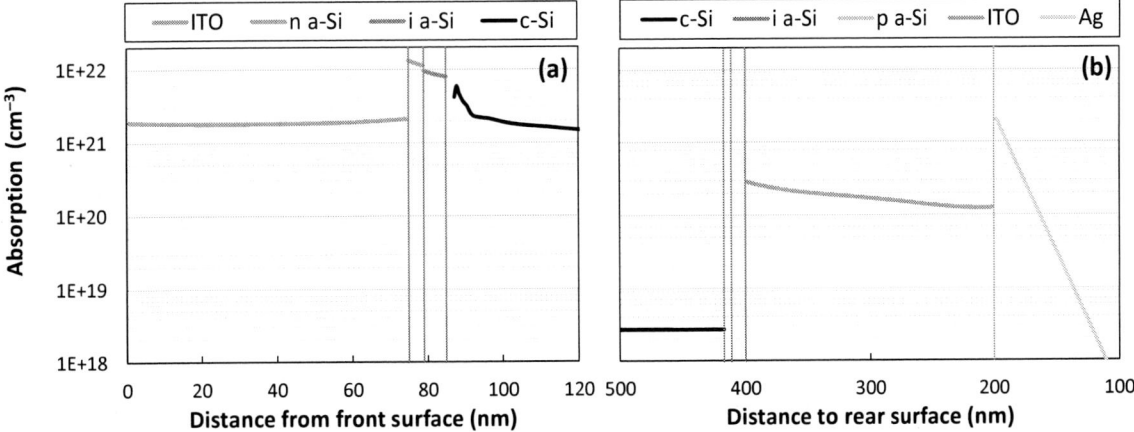

Fig. 7. Simulated photon absorption at the (a) front and (b) rear of the unencapsulated solar cell under a normally incident AM1.5g spectrum.

module, although all of these can all be simulated by the ray tracer. The incident zenith angle was set to $\theta = 0°$ to simulate normally incident light (rather than 7° as used in the previous section to simulate the experimental setup). Solutions were attained with 5,000,000 rays in Section 3.1 and 200,000 rays in Section 3.2.

3.1 Unencapsulated solar cell

The results for the unencapsulated baseline structure are plotted in Figs. 5–7. These simulations included the layers listed in Table I but without the glass and EVA.

Fig. 5 plots the reflectance and absorptance against wavelength, where the absorptance is plotted for all layers except the rear a-Si:H films whose absorptance is negligible. The figure shows that (i) the absorptance in the front ITO layer (solid blue) is significant at all wavelengths, but especially at $\lambda < 400$ nm and $\lambda > 1000$ nm, (ii) the absorptance of the front a-Si:H layers (green and red) is significant at $\lambda < 600$ nm, (iii) the absorptance of the rear ITO (dashed blue) is sufficient to reduce generation in the c-Si at $\lambda > 1000$ nm, and (iv) the absorptance of the Ag (yellow) is about half of the absorptance in the rear ITO.

Fig. 6 puts the absorption in each layer into perspective by plotting it in terms of an equivalent current density (mA/cm^2) integrated from 300 to 1200 nm. It shows that the absorption in the front films amounts to 3.9 mA/cm^2, of which 3.6 mA/cm^2 is lost. This absorption is similar to a previous study [4] where the absorption in each film was determined via a matrix of experimental samples whose film thicknesses were varied. The loss in the rear films is also significant, being equal to 0.9 mA/cm^2.

We emphasize that these absorption losses cannot be entirely regained by increasing the transparency of the films. If the long-wavelength light had not been absorbed in the films, only a fraction of that light would have instead been absorbed in the c-Si and contributed to the cell's J_{sc}; this is due to the low

absorption coefficient of c-Si at $\lambda > 1100$ nm and to imperfect light trapping in the cell. To quantify this aspect, we simulated the same cell but where (i) $k(\lambda)$ in the ITO is zero, (ii) each a-Si:H layer was assigned a thickness of 1 nm, and (iii) the thickness of the front ITO was 85 nm, which is the new optimum for this structure. These modifications led to an increase in J_{sc} of 2.7 mA/cm^2 (from 39.3 to 42.0 mA/cm^2). Thus, we conclude that an upper limit of 2.7 mA/cm^2 could be gained by increasing the transparency of the ITO and reducing the thickness of the a-Si layers. Alternatively, Fig. 6 indicates that an upper limit of 1.3 mA/cm^2 could be gained by increasing the collection efficiency of the a-Si:H layers. (Of course, modifying the film composition would also lead to electronic changes to the cell—such as variation in contact resistance or recombination—but here we limit ourselves to an optical analysis.)

Finally, Fig. 7 plots the absorption profile within each of the thin films in photons/cm^3 (equivalent to a generation profile). Notice the high absorption in the front a-Si:H layers, which is why those layers are made very thin in SHJ solar cells. To our knowledge, this is the first time that the absorption profiles in the individual films of an SHJ cell have been evaluated. With estimates for the carrier mobility, the intrinsic carrier concentration, and the recombination rates within the a-Si:H films, these profiles could be used to determine the collection efficiency of the materials as a function of wavelength.

3.2 Varying ITO thickness, unencapsulated and encapsulated

To demonstrate the value of the optical simulation, Figs. 8 and 9 show how the absorption in the a-Si:H and c-Si depends on the thickness of the front and rear ITO, both with and without encapsulation. In Fig. 8, the results for the front ITO are similar to those published in [6, 7] for the unencapsulated case.

Figs. 8 and 9 show that the thicknesses of the ITO films are close to optimal for this SHJ solar cell when unencapsulated.

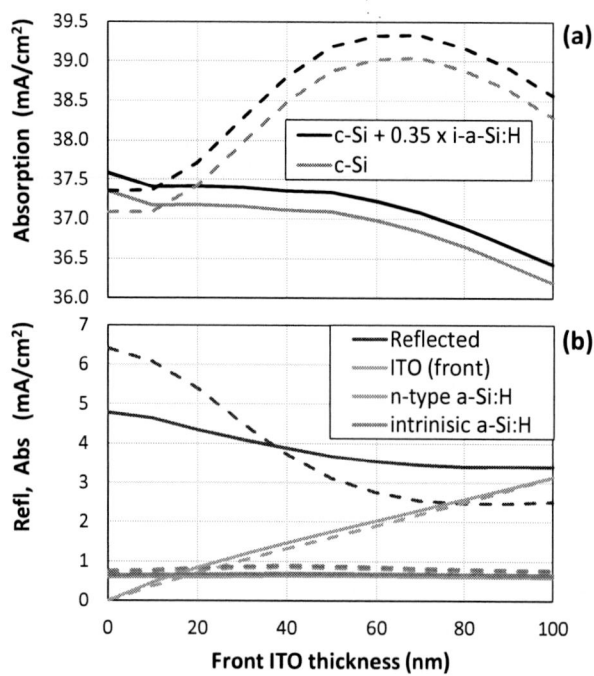

Fig. 8. (a) Absorption that contributes to current collection, with and without collection from the front intrinsic a-Si:H, and (b) reflection and absorption in front films for unencapsulated (dashed) and encapsulated (solid) cases as a function of the front ITO thickness.

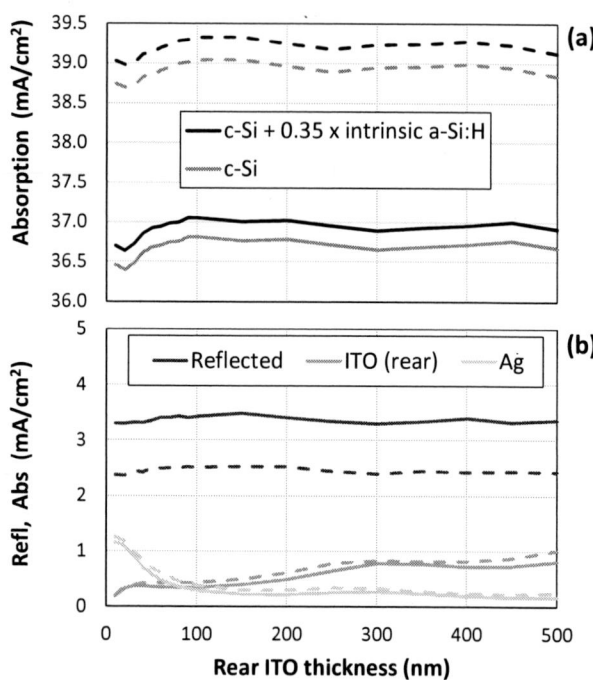

Fig. 9. (a) Absorption that contributes to current collection, with and without collection from the front intrinsic a-Si:H, and (b) reflection and absorption in rear films for unencapsulated (dashed) and encapsulated (solid) cases as a function of the rear ITO thickness.

A slight improvement (~0.1 mA/cm²) might be attained if the front ITO were 70 nm instead of 75 nm, or if the rear ITO were 100 nm instead of 200 nm.

Figs. 8 and 9 also show that—for this cell design and for the ITO deposited on this cell—the thicknesses that maximize photogeneration can change after encapsulation. For the front ITO, the optimal thickness reduces to zero; as evident in Fig. 8, decreasing the ITO thickness increases reflectance but it reduces absorptance even more. (Of course, the ITO is required for lateral conduction but, optically, it would be better to remove it altogether before encapsulation.) For the rear, the optimal thickness remains at about 100 nm. It is also evident that after encapsulation the losses are less sensitive to the thickness of the front ITO but not the rear ITO.

Finally, from a comparison of the black and grey lines, it can be seen that, for all cases tested here, collection from the front intrinsic a-Si:H layer is approximately constant (~0.25 mA/cm² for encapsulated, and ~0.3 mA/cm² for unencapsulated). Consequently, η_c has little influence on the optimal ITO thickness.

We emphasize that these conclusions are specific to the experimental films and cell structure used here. The optimal film thicknesses could be considerably different with different ITO and a-Si:H films, but the approach applied in this work can be used to determine their optimal thicknesses. For example, if the front ITO were less absorbing (like many of

the ITO layers in [4]), then the optimal thickness of the front ITO layer under encapsulation would be finite, although thinner than for the case of no encapsulation.

We note three complications related to the optimization of the thin films: (i) In practice, the refractive index of sputtered ITO depends on its deposition thickness; an assessment of ITO thickness that considers this effect will be published in [7]. (ii) The collection efficiency of an a-Si:H film depends on its thickness. (iii) When operating in the field, the optimal film thicknesses depend on the incident spectra, the angles of incidence, the collection efficiency of the a-Si:H layers, temperature, and more besides. These effects can all be rapidly assessed with the PV Lighthouse ray tracer when the inputs are known.

IV CONCLUSION

We have shown how experimentation and simulation can be applied to determine the absorption within each film of a solar cell as a function of wavelength and depth.

This information assists in (i) the determination of a film's collection efficiency, (ii) the optimization of a cell's film thicknesses, whether the cell is encapsulated or unencapsulated, and (iii) the quantification of losses and gains that result from modifying the thickness and refractive index of a cell's thin films.

For the SHJ solar cell investigated here, we found that (i) the collection efficiency of the front intrinsic c-Si:H was 0.35, (ii) the optimal thickness of the front and rear ITO was 70 and 100 nm before encapsulation and 0 and 100 nm after encapsulation, and (iii) at most, the J_{sc} would increase by 2.7 mA/cm^2 if the ITOs were transparent and the thickness of each a-Si:H film was 1 nm, or by 1.3 mA/cm^2 if the collection efficiency of the front a-Si:H films was 100%.

Although the post-encapsulation optimal thickness of the front ITO was found to be zero, we stress that this conclusion was solely based on the optical analysis. Of course, omitting the front ITO would greatly increase the cell's series resistance. We therefore conclude that after encapsulation, the optimal cell structure would have a thinner (but non-zero) front ITO and a reduced finger spacing.

V REFERENCES

[1] M. Tanaka, M. Taguchi, T. Matsuyama, T. Sawada, S. Tsuda, S. Nakano, H. Hanafusa and Y. Kuwano, Y, "Development of new a-Si/c-Si heterojunction solar cells: ACJ-HIT (artificially constructed junction-heterojunction with intrinsic thin-layer)," *Japanese Journal of Applied Physics* **31**, pp. 3518–3522, 1992.

[2] H. Fujiwara and M. Kondo, "Effects of a-Si:H layer thicknesses on the performance of a-Si:H/c-Si heterojunction solar cells," *Journal of Applied Physics* **101**, 054516, 2007.

[3] D. Zhang, I.A. Digdaya, R. Santbergen, R.A.C.M.M. van Swaaij, P. Bronsveld, M. Zeman, J.A.M. van Roosmalen and A.W. Weeber, "Design and fabrication of a SiO/ITO double-layer anti-reflective coating for heterojunction silicon solar cells, *Solar Energy Materials and Solar Cells* **117**, pp. 132–138, 2013.

[4] Z.C. Holman, A. Descoeudres, L. Barraud, F. Fernandez, J. Seif, S. De Wolf and C. Ballif "Current losses at the front of silicon heterojunction solar cells," *IEEE Journal of Photovoltaics* **2**, pp. 7–15, 2012.

[5] Z.C. Holman, M. Filipic, A. Descoeudres, S. De Wolf, F. Smole, M. Topic and C. Ballif, "Infrared light management in high-efficiency silicon heterojunction and rear-passivated solar cells," *Journal of Applied Physics* **113**, 013107, 2013.

[6] M. Leilaeioun, Z. J. Yu, and Z. Holman, "Optimization of front TCO layer of silicon heterojunction solar cells for tandem applications," *Proc. 43rd IEEE PVSC*, Portland, pp. 681–684, 2016.

[7] M. Leilaeioun, Zhengshan J. Yu, S. Manzoor, K. Fisher, J. Shi and Z. Holman, "Design of the front transparent conductive oxide layer of silicon heterojunction solar for four-terminal tandem applications," in preparation, 2017.

[8] J. Shi, M. Boccard, and Z. Holman, "Plasma-initiated rehydrogenation of amorphous silicon to increase the temperature processing window of silicon heterojunction solar cells." *Applied Physics Letters* **109**, 031601, 2016.

[9] K.R. McIntosh, T.C. Kho, K.C. Fong, S.C. Baker-Finch, Y. Wan, S. Zin, E.T. Franklin, D. Wang, N.E. Grant,

M.D. Abbott, E. Wa ng, M. Stocks and A.W. Blakers, "Quantifying the optical losses in back-contact solar cells," *Proc. 40th IEEE PVSC*, Denver, pp. 115–123, 2014.

[10] K.R. McIntosh and S.C. Baker-Finch, "OPAL 2: Rapid optical simulation of silicon solar cells," *Proc. 38th IEEE PVSC*, Austin, pp. 265–271, 2012.

[11] K.R. McIntosh, T.G. Allen, S.C. Baker-Finch and M.D. Abbott, "Light trapping in isotextured silicon wafers," *IEEE Journal of Photovoltaics* **7**, pp. 110–117, 2017.

[12] H. Mäckel, H. Holst, M. Löhmann, E. Wefringhaus, and P.P. Altermatt. "Detailed Analysis of Random Pyramid Surfaces With Ray Tracing and Image Processing." *IEEE Journal of Photovoltaics* **6**, pp. 1456–1465, 2016.

[13] S. Manzoor, M. Filipič, M. Topič and Z. Holman, "Revisiting light trapping in silicon solar cells with random pyramids," *Proc. 43rd IEEE PVSC*, Portland, pp. 2952–2954, 2017.

[14] J.N. Hilfiker, N. Singh, T. Tiwald, D. Convey, S.M. Smith, J.H. Baker, and H.G. Tompkins. "Survey of methods to characterize thin absorbing films with spectroscopic ellipsometry," *Thin Solid Films* **516**, pp. 7979–7989, 2008.

[15] M.D. Abbott, K.R. McIntosh and B.A. Sudbury, "Optical loss analysis of PV modules," *Proc. 32nd EU PVSEC*, Munich, pp. 976–979, 2016.

[16] www.pvlighthouse.com.au, last accessed 19-Jun-2017.

[17] F.J. Haug, M. Bräuninger, and C. Ballif. "Fourier light scattering model for treating textures deeper than the wavelength," *Optics Express* **25**, A14–A22, 2017.

[18] K.R. McIntosh, G. Lau, J. Cotsell, K. Hanton, D. Batzner, F. Bettiol and B. Richards, "Increase in external quantum efficiency of encapsulated silicon solar cells from a luminescent down-shifting layer," *Progress in Photovoltaics* **17**, pp. 191–197, 2009.

[19] K.R. McIntosh, J. Cotsell, J. Cumpston, A. Norris, N. Powell and B. Ketola, "An optical comparison of silicone and EVA encapsulants for conventional silicon PV modules: A ray-tracing study", *Proc. 34th IEEE PVSC*, Philadelphia, pp. 544–549, 2009.

[20] M.A. Green, "Self-consistent optical parameters of intrinsic silicon at 300K including temperature coefficients," *Solar Energy Materials and Solar Cells*. **92**, pp. 1305–1310, 2008.

[21] H.T. Nguyen, F.E. Rougieux, B. Mitchell and D.H. Macdonald, "Temperature dependence of the band-band absorption coefficient in crystalline silicon from photoluminescence," *Journal of Applid Physics* **115**, 043710, 2014.

[22] Y. Jiang, S. Pillai and M. Green, "Realistic silver optical constants for plasmonics," *Scientific Reports* **6**, p. 30605, 2016.

Investigations on Plasmonic Color Tuning Coating on c-Si Solar Cells

Gerhard Peharz[1], Wolfgang Waldhauser[1], Christine Prietl[1], Bettina Großschädl[1],
Martin C. Schubert[2] and Bernhard Michl[2]

[1]JOANNEUM RESEARCH – MATERIALS, Weiz, 8160, Austria

[2]Fraunhofer ISE, Freiburg, 79110, Germany

Abstract — **Recently an approach was published which describes the color tuning of crystalline Silicon (c-Si) solar cells based on plasmonic coatings. In the current work the influence of the plasmonic coating (Ag nano-particles) on the performance of the solar cells is investigated by mapping the electrical and optical parameters of a half-coated cell. The key findings are that parasitic absorption is the dominating loss mechanism. No significant influences on most of the electrical parameters have been identified.**

Index Terms — c-Si, plasmonics, LBIC, BIPV

I. INTRODUCTION

Building integrated photovoltaics (BIPV) is considered to play an important role in decentralized renewable energy generation [1],[2] and is regarded to be a promising market for PV [3],[4]. Recently an approach was published which describes the color tuning of crystalline Silicon (c-Si) solar cells based on plasmonic coatings [5]. It was found that the color of industrial can be efficiently tuned from blue / black to green, yellow, red and brown when applying a coating of silver nano-particles [5]. The particular color is determined mainly by the size of the nano-particles. In contrast to colors based on dielectric films the plasmonic coating does not depend on the angle of observation, which might be significant advantage for BIPV applications. In another recently published article the application of c-Si solar cells for BIPV modules was investigated [6]. There it was found that plasmonic coloring is feasible to fabricate façade integrated modules having a reasonable power output.

In literature positive impacts of Ag nano-particles in terms of efficiency increase are reported when depositing them on c-Si solar cells [7]. However, in that publication the nano-particle layers are deposited on textured c-Si cells without dielectric anti-reflective coating (ARC). Consequently the reference solar cell device had a rather poor efficiency of 4.5% which was increased to 6.4% by depositing Ag nano-particles and exploiting coupling surface plasmon resonances [7].

In the current work the impact of the plasmonic coating on the performance of industrially produced c-Si solar cells (with ARC) is investigated. The particular scope of the paper is to analyze loss and potential gain mechanisms on individual performance parameters of those cells.

II. EXPERIMENTAL

For the experiments industrially produced monocrystalline Silicon solar cells were used. Onto these cells thin films of silver (Ag) were deposited by pulsed DC sputtering. In particular a sputtering power of 700 W and an Argon flow rate of 30 sccm were applied. The sputtering time was set to 65 seconds which results in an Ag film thickness of about 8 nm on plane substrates. Subsequently the coated solar cells were annealed at 300°C for about 2 minutes in standard atmosphere (air). Due to this thermal annealing, the Ag thin film is converted into a coating of Ag nano-particles. Those nano-particles were characterized by scanning electron microscopy (SEM).

Only one half of the cells have been coated and the other half of the solar cell area was masked during the sputtering process. The half-coated solar cells have been characterized by applying spectrally resolved light-beam induced current measurements (SR-LBIC). In particular the laterally resolved quantum efficiency (QE) and reflection of the half-coated cells was mapped at wavelengths of 405, 532, 670, 780, 940, 1064 nm by using a LOANA measurement set-up from pv-tools. In combination with IV-curve measurements and local spectral response measurements a detailed mapping of the short circuit current (J_{SC}) and a loss-mapping of the half-coated cells was derived [8]. That mapping allowed investigating the impact of the coating on a solar cell without the need of an uncoated reference (self-reference method).

III. RESULTS AND DISCUSSION

The sputtering of a thin layer of Ag followed by subsequent temperature treatment resulted in the formation of Ag nano-particles on the surface of the c-Si solar cells. The shapes of the Ag nano-particles can be described to be spherical and the corresponding diameters are ranging from 50 to 150 nm (see Fig. 1).

978-1-5090-5606-4/17 $31.00 © 2017 IEEE

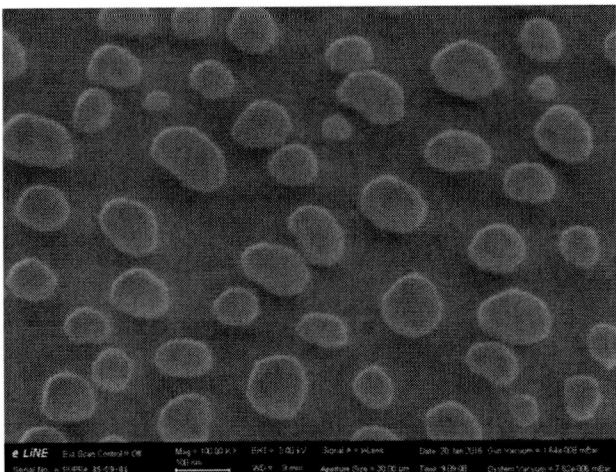

Fig. 1. An SEM image of a c-Si solar cell coated with Ag nano-particles with diameters ranging from 50 to 150 nm is shown.

Impinging light excite surface plasmons on the Ag nano-particles and the corresponding resonance frequency is a function of the dielectric function(s), the particle-diameters and particle-dimensions [9]. Obviously the plasmonic resonance (scattering) frequency of particles applied onto the c-Si solar cells is in the yellow-green spectral range (see Fig. 2).

Fig. 2. A photo of two cells is shown. The lower half of the cells is coated with Ag nano-particles which have a plasmonic scattering resonance frequency in the green spectral range.

The external quantum efficiency (EQE), reflection and internal quantum efficiency (IQE) of a c-Si cell was measured for the coated and the uncoated side (spot-size about 20x20 mm²). Those measurement results are shown in Fig. 3.

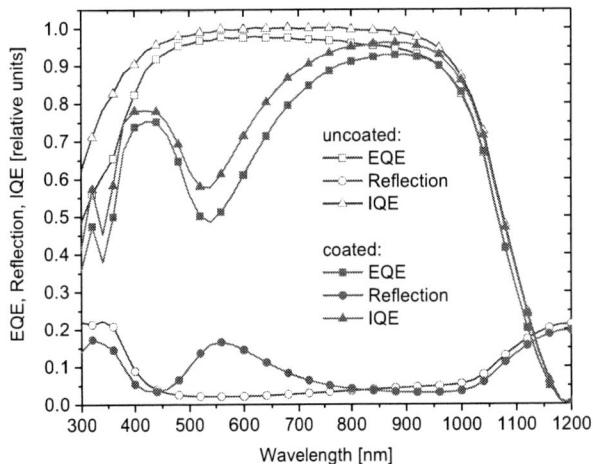

Fig. 3. EQE, reflection and IQE are plotted for a c-Si cell half-coated with Ag nano-particles.

Due to the Ag nano-particle coating the reflection is increased in a wavelength range from 450 to 800 nm (maximum of about 17% at 550 nm). Interestingly the reflection is found to be slightly lower for wavelengths > 800 nm at the coated half of the c-Si cell and in the UV. In contrast the IQE of the uncoated half is found to be higher for most of the measured wavelengths. In particular in the visible spectral range the Ag nano-particles substantially decrease the IQE and the EQE.

The SR-LBIC measurements of a half-coated cell show clearly that that the impact of the coating is highest for the green measurement wavelength (see Fig. 4). In contrast the coated half of the cell cannot be clearly identified at the EQE mapping for the blue wavelength (see Fig. 5) and the NIR wavelengths (see Fig. 6).

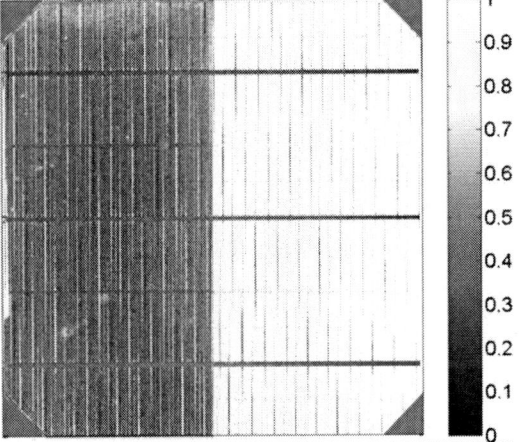

Fig. 4: EQE mapping (**at 532 nm**) of c-Si cell half-coated (left-side) with Ag nano-particles.

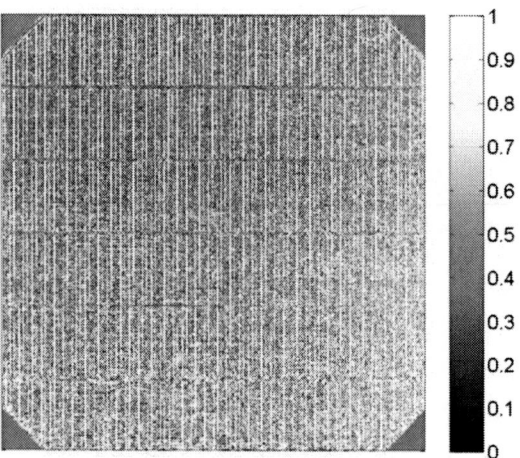

Fig. 5: EQE mapping (**at 405 nm**) of c-Si cell half-coated (left-side) with Ag nano-particles.

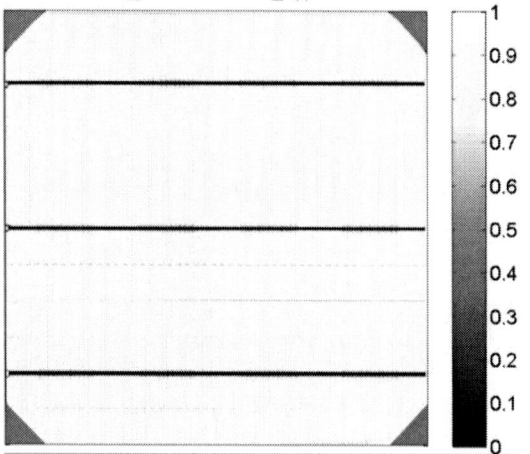

Fig. 6: EQE mapping (**at 940 nm**) of c-Si cell half-coated (left-side) with Ag nano-particles.

A J_{SC} mapping of the half-coated cell shows, that the uncoated area has an average J_{SC} of 38.3 mA/cm² , whereas the J_{SC} in the areas coated with Ag nano-particles is only 32.0 mA/cm² in average. Those losses in J_{SC} are found to mainly due to parasitic absorption losses.

The open circuit voltage (V_{OC}) and the fill factor (FF) of the cell are influenced by that decrease in J_{SC}. A summary of the parameters derived from IV measurements of masked solar cell areas (uncoated and coated half of cell) is shown in table 1. A comparison of the pseudo FFs (pFF) [10], the dark saturation currents (J_{01} and J_{02}) and the shunt resistance (Rsh) do not show a significant impact of the coating in terms of electrically loss mechanisms.

TABLE I
ELECTRICAL PARAMETERS DERIVED FROM HALF-COATED CELL

	uncoated	coated
Efficiency	18.5 %	15.7 %
J_{SC}	38.3 mA/cm²	32.0 mA/cm²
V_{OC}	631 mV	626 mV
FF	76.4 %	78.3 %
pFF	81.0 %	80.7 %
J_{01}	727 fA/cm²	717 fA/cm²
J_{02}	25.5 nA/cm²	26.2 nA/cm²
R_{sh} (bright)	7719 Ohm cm²	8004 Ohm cm²
R_s (FF and FFp)	0.93 Ohm cm²	0.57 Ohm cm²

The series resistance (R_S) of the side of the cell coated with Ag nano-particles is found to be about 40% lower than that of the uncoated half. However, it needs to be noted that due to contacting issues the determination of the R_S was not well reproducible.

IV. DISCUSSION

The results show above indicated that an Ag nano-particle coating mainly influences the optical properties of an industrial produced c-Si cell. On the one hand side the reflection is influenced and in the visible spectral range an increase of reflection is observed, which desired for color tuning of the cells. Moreover, the reflection in the UV and the NIR of the c-Si cells is decreased by the Ag nano-particles. However, that decrease in reflection is compensated by a decrease in IQE and parasitic absorption losses are found to be the main impact on a decrease in J_{SC}.

Absorption losses need to be taken into account in particular for wavelengths around the plasmonic resonance frequency of the Ag nano-particles (about 550 nm in the investigated case). The parasitic absorptions observed in the UV and the NIR might be related to a coupling of the surface plasmon resonances to the substrate. In cells without dielectric ARC these coupling effects resulted in an increase of EQE for off-resonance wavelengths [7]. When depositing the Ag nano-particles on c-Si cells with standard ARC, it seems that such a plasmonic coupling results in an increased absorption within the dielectric ARC and does not contribute to an increased generation of current.

The electrical parameters such as dark saturation currents and the shunt resistance are not influenced by the Ag nano-particles. The series resistance might be decreased by the coating; however, the results are not well reproducible and require further investigation. Potentially the series resistance could be decreased by the additional amount of Ag deposited on the solar cell front side. Since plasmonic scattering is usually only observed at metallic nano-particles which are embedded in a dielectric matrix, a decrease in sheet resistance (i.e. by the introduction of a laterally conductive and semitransparent layer) is not expected. More realistic is that by depositing an additional layer of metal the effective cross-section of the grid fingers is enhanced and the series resistance is decreased.

V. CONCLUSIONS

An optical coating for coloring industrial solar cells based on Ag nano-particles is investigated by applying SR-LBIC and a J_{SC} mapping. It is found that absorption losses are dominating the decrease in J_{SC}. Those absorption losses are highest in the wavelength range around the plasmonic resonance frequency (550 nm), where also the reflection is increased. However, also in the NIR spectral region (> 800 nm) an increase in absorption losses is observed, which outweighs the observed decrease in reflection.

The results shown above indicate, that when standard c-Si solar cells (with ARC) are coated with Ag nano-particles the only positive impact on the cell might be related to a decrease in series resistance. However, the observed decrease in R_s requires additional investigations in order to create a better understanding. From the current perspective the main advantage of coating industrial c-Si cells with plasmonic Ag nano-particles is that their color can be tuned to be green in a simple post-process, which might be of particular interest for façade applications (BIPV).

ACKNOWLEDGEMENTS

This work is financially supported by the Austrian Klima- und Energiefonds within the frame of the "e!Missi0n.at" program (PV@Fassade; project number: 843803).

REFERENCES

[1] D. Knera and D. Heim, "Application of a BIPV to cover net energy use of the adjacent office room," *Manag. Environ. Qual. Int. J.*, vol. 27, no. 6, pp. 649–662, Aug. 2016.

[2] H. P. Ikkurti and S. Saha, "A comprehensive techno-economic review of microinverters for Building Integrated Photovoltaics (BIPV)," *Renew. Sustain. Energy Rev.*, vol. 47, pp. 997–1006, Jul. 2015.

[3] J. Escarré *et al.*, "When PV modules are becoming real building elements: White solar module, a revolution for BIPV," in *2015 IEEE 42nd Photovoltaic Specialist Conference (PVSC)*, 2015, pp. 1–2.

[4] P. Bonomo, A. Chatzipanagi, and F. Frontini, "Overview and analysis of current BIPV products: new criteria for supporting the technological transfer in the building sector," *Vitr. - Int. J. Archit. Technol. Sustain.*, vol. 0, no. 1, pp. 67–85, Dec. 2015.

[5] G. Peharz, B. Grosschädl, C. Prietl, W. Waldhauser, and F. P. Wenzl, "Tuning the colors of c-Si solar cells by exploiting plasmonic effects," 2016, vol. 9937, p. 99370P–99370P–8.

[6] G. Peharz *et al.*, "Application of plasmonic coloring for making building integrated PV modules comprising of green

solar cells," *Renew. Energy*, vol. 109, pp. 542–550, Aug. 2017.

[7] S. K. Sardana *et al.*, "Influence of surface plasmon resonances of silver nanoparticles on optical and electrical properties of textured silicon solar cell," *Appl. Phys. Lett.*, vol. 104, no. 7, p. 073903, Feb. 2014.

[8] M. Padilla, B. Michl, B. Thaidigsmann, W. Warta, and M. C. Schubert, "Short-circuit current density mapping for solar cells," *Sol. Energy Mater. Sol. Cells*, vol. 120, Part A, pp. 282–288, Jan. 2014.

[9] P. Manley, S. Burger, F. Schmidt, and M. Schmid, "Design Principles for Plasmonic Nanoparticle Devices," in *Progress in Nonlinear Nano-Optics*, S. Sakabe, C. Lienau, and R. Grunwald, Eds. Springer International Publishing, 2015, pp. 223–247.

[10] J. Greulich, M. Glatthaar, and S. Rein, "Fill factor analysis of solar cells' current–voltage curves," *Prog. Photovolt. Res. Appl.*, vol. 18, no. 7, pp. 511–515, Nov. 2010.

Investigation of Interface and Bulk Localized States in a-Si:H Solar Cells

Adrien Bidiville, Takuya Matsui, Hitoshi Sai, Koji Matsubara

Research Center for Photovoltaics, National Institute of Advanced Industrial Science and Technology (AIST), Tsukuba, Ibaraki 305-8568, Japan

Abstract — Two series of a-Si:H solar cells with varying absorber thickness (t_i=10-500 nm) and deposition temperature (T_{dep-i}=200-350°C) were investigated by Fourier-transform photocurrent spectroscopy to determine the sub-bandgap absorption originating from the electronic localized states in the a-Si:H absorber i-layer. The thickness series showed that the bulk defect is increased by thickening t_i and by light soaking while no p-i or n-i interface defect prevails. It is demonstrated that the performance of the state-of-the-art a-Si:H p-i-n solar cell is dominated by the total number of (native and light-induced) midgap defects in the bulk i-layer. On the other hand, variation of T_{dep-i} has a different impact on the creation of localized states in the i-layer depending on the layer stack sequence (p-i-n or n-i-p). When depositing a-Si:H i-layer at high temperatures (>200°C), p-i-n cells showed a larger performance decrease than n-i-p cells, along with a broader bandtails and a larger number of defects in the i-layer. The results indicate that the creation of such electronic states is influenced by the local Fermi level position in the i-layer, i.e., the closer the Fermi level to the valence band (p-i interface), the more electronic states are created at high T_{dep-i}. These findings agree with the earlier report that p-doped a-Si:H and c-Si passivated by a p-i stack of a-Si:H layers exhibit a lower thermal stability.

Index Terms — amorphous silicon, solar cells, sub-gap absorption, bandtail, defect, interface, bulk.

I. INTRODUCTION

Undoped (intrinsic) hydrogenated amorphous silicon (a-Si:H) is extensively used for photovoltaic applications, both as a light absorber material in thin-film solar cells and as a passivation layer in c-Si heterojunction (SHJ) cells. In both cases, the electronic quality of intrinsic a-Si:H (i-layer) is of prime importance. In working devices, the a-Si:H i-layer is stacked with p- and n-doped a-Si:H (or its alloys) that act as charge collection layers. The p-i and n-i interfaces are considered the regions where a larger number of electronic localized states are present than in the bulk. The interface where the i-layer deposition starts is particularly crucial because the initial growth of the a-Si:H can differ from its steady-state growth [1]. Moreover, the deposition of the i-layer might interact with the underlying doped layer, and vice versa. For example, inter-diffusion of dopant elements (boron and phosphorous) from doped layer into the i-layer has been considered as one of the possible origins of the poor device performance when using high temperature processes (>200°C) in thin-film silicon solar cells [2]. However, according to the recent detailed SIMS measurements, it has been revealed that

no measurable diffusion takes place [3]. Meanwhile, in SHJ solar cells, it has been reported that the passivation of c-Si surface by a-Si:H is less thermally stable when depositing a p-i stack than when a single i-layer or n-i stack is deposited [4]. This has been attributed to the Fermi level dependence of the Si-H bond rupture energy in the a-Si:H i-layer. To gain comprehensive understanding of these phenomena, a more detailed study is required to assess the interface and bulk electronic localized states and their impact on device performance.

In this contribution, we report on the electronic localized states in a-Si:H i-layers of actual working devices characterized by a sub-bandgap absorption spectroscopy. Based on the measured data and analysis, the relationship between the electronic state distribution in the i-layer and solar cell properties is discussed.

II. EXPERIMENTS

Two series of a-Si:H solar cells were fabricated with respect to the thickness (t_i) and the deposition temperature (T_{dep-i}) of the i-layer: (i) state-of-the-art a-Si:H p-i-n cells whose i-layers were deposited by triode-PECVD with varying absorber thickness (t_i=10-500 nm, T_{dep-i}=200°C) [5,6], (ii) standard p-i-n and n-i-p cells whose i-layers were deposited by diode-PECVD with varying deposition temperature (T_{dep-i}=200-350°C, t_i~250 nm) [7]. In the former t_i series, a record stabilized efficiency of 10.2% has been obtained at t_i~220 nm using the same cell structure but with an improved optical design (i.e., high-haze TCO for enhanced light trapping and antireflection coating) [5,8]. The number of defects in the absorber layer was varied by t_i as well as by light-soaking (3 suns at temperature of 60°C for 6 hours [5]). In the later T_{dep-i} series, the same (optimized) p- and n-doped layers [6,9] were employed whereas the i-layers were deposited in a separate ultra-high vacuum PECVD reactor to minimize the atmospheric impurity incorporation during the deposition. The device structures of these solar cells are shown in Fig. 1, and the details of the experiments are summarized in Table I.

By using Fourier-transform photocurrent spectroscopy (FTPS) [10], we measured the sub-bandgap absorption of the working device, even possible for cells with very thin i-layers (t_i~10 nm). The details of the measurement setup can be found elsewhere [6]. The FTPS curves were calibrated using the

TABLE I
SUMMARY OF THE ABSORBER LAYER DEPOSITIONS FOR SOLAR CELL SAMPLES

series	PECVD electrode configuration	deposition rate (nm/s)	background pressure (Torr)	t_i (nm)	T_{dep-i} (°C)	cell structure	light soaking
absorber thickness: t_i	triode	0.02	3×10^{-8}	10-500	200	p-i-n	Yes
absorber deposition temperature: T_{dep-i}	diode	0.12	8×10^{-10}	250	200-350	p-i-n n-i-p	No

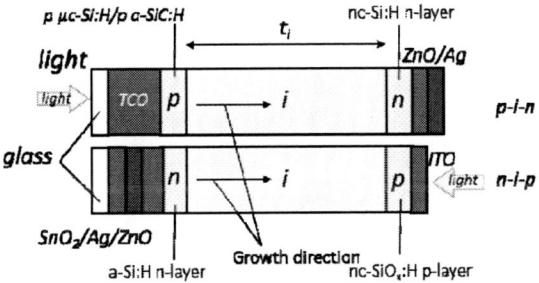

Fig. 1. Schematic solar cell structures studied in this work.

corresponding external quantum efficiency (EQE) spectra measured in the photon energies of >~1.6 eV. Internal quantum efficiency (IQE) was then derived using the reflection spectrum of the cell in order to reduce the optical interferences that appear in the low energy region [6].

III. RESULTS AND DISCUSSION

A. Absorber thickness series

The solar cell parameters of the t_i series are depicted in Fig. 2. In the initial state, J_{sc} expectedly increases with t_i, while it saturates at t_i~300 nm after light-induced degradation (LID). V_{oc} weakly depends on t_i, but does not suffer from LID. This situation is similar to the previously reported results by our group [8] due to the fact that our p- and p-i buffer layers were designed to provide high cell performance after LID, rather than for the initial (annealed) state. The FF is the most decisive parameter, as it shows sizeable differences depending on the thickness as well as the LID. It is seen that FF limits the efficiency particularly for thick solar cells after LID.

Fig. 3 shows the typical sub-bandgap IQE spectra measured by FTPS for an a-Si:H (t_i=220 nm) solar cell in the initial and degraded states. In this study, the bandgap (E_g) of the a-Si:H i-layer is defined as a photon energy that gives IQE=0.1 where the absorption edge from extended to bandtail states

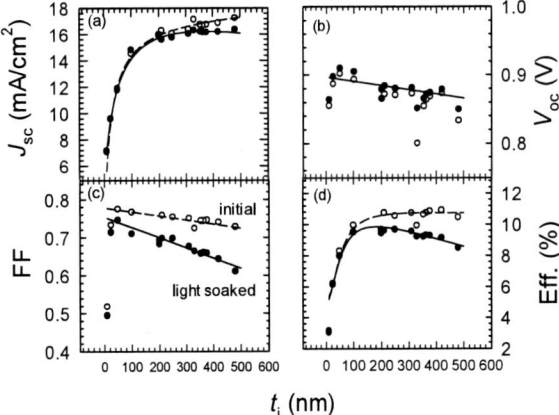

Fig. 2. Solar cell parameters of the p-i-n solar cells as a function of absorber i-layer thickness (t_i) for initial (open) and light-soaked (filled) states. Lines are guides to the eyes.

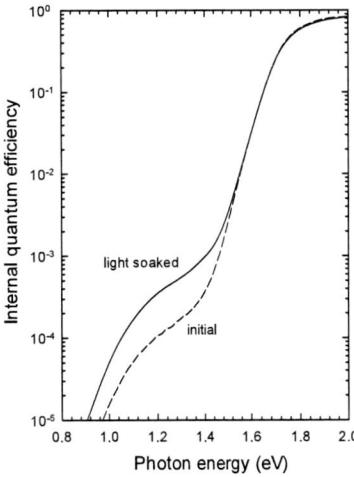

Fig. 3. Measured sub-bandgap IQE spectra of an a-Si:H solar cell (t_i=220 nm) in the initial (dashed line) and light-soaked (solid line) states.

appears. Although the sub-bandgap IQE spectrum depends on the solar cell structure (especially on the light trapping), this estimation of E_g (1.65 eV) roughly matched with the E_g (1.69 eV) of the a-Si:H layer deposited on glass measured by spectroscopic ellipsometry. The band-tail characteristic energy (E_U) is the energy constant of the exponential decrease seen in the region from 1.6 to 1.4 eV, representing the bandtail width. Similarly, the amount of midgap defects in the i-layer can be evaluated as the IQE at an energy 0.5 eV below the band edge, defined as IQE(E_g-0.5 eV). As shown in Fig. 4, the IQE(E_g-0.5 eV) is increased by light soaking due to the light-induced metastable defect creation while the absorption in the bandtail region remains almost unchanged at E_U~40 meV regardless of the light-soaking as well as the t_i [6].

In Fig. 4, the relationship between the t_i and the IQE(E_g-0.5 eV) is shown. It is demonstrated that the p-i and n-i interfaces have a negligible amount of defect, as the intercept of the linear regression for the thin samples (t_i<100 nm) approaches zero. Furthermore, the rate of the midgap defect increase with t_i, represented by the slope ΔIQE(E_g-0.5 eV)/Δt_i, is larger for thin (t_i<100 nm) i-layers than for thick (t_i>200 nm) i-layers, both in the initial and light-soaked states. Thus, it can be considered that the i-layer is composed of three regions: a central region where the defect density is low and two side regions next to the doped layers of a total thickness about 100 nm with a greater defect density. However, according to our device simulation study, the larger slope in the thin solar cells can be attributed to the large occupation of defect states with

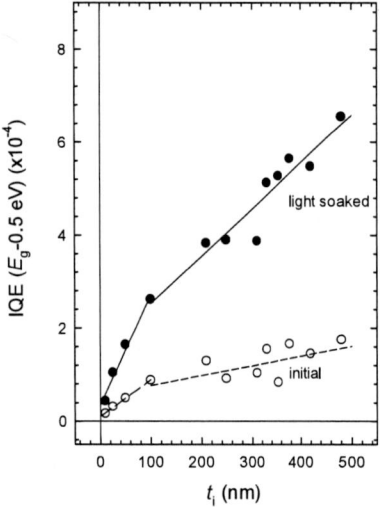

Fig. 4. Sub-bandgap absorption due to midgap defect IQE(E_g-0.5 eV) of the triode deposited a-Si:H p-i-n solar cells as a function of the absorber i-layer thickness in the initial (open symbols) and light-soaked (filled symbols) states. Lines are the linear fits to the data.

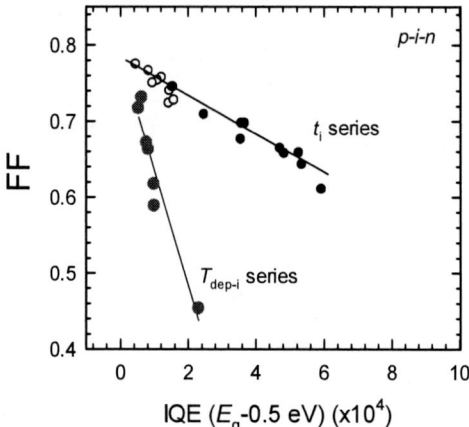

Fig.5.Relationship between the sub-bandgap absorption IQE(E_g-0.5 eV) and FF of the t_i series (black) and the T_{dep-i} series (blue) a-Si:H p-i-n solar cells in their initial (open) and light-soaked (filled) states. The lines are the linear fits to the data.

charge carriers as the Fermi level is mostly away from the midgap in such thin p-i-n solar cells. Thus, the different ΔIQE(E_g-0.5 eV)/Δt_i between the thin and thick i-layers reflects the occupation of defects rather than the defect density itself. In contrast, the change in ΔIQE(E_g-0.5 eV)/Δt_i by the LID reflects the change in the midgap defect density. In the thick region, the slope ΔIQE(E_g-0.5 eV)/Δt_i increases by a factor of 5 by LID. This increase is about 20% smaller compared to the a-Si:H solar cells prepared by the standard diode-PECVD, which qualitatively accounts for the improved light-soaking stability of the a-Si:H by triode-PECVD [5,6].

By plotting FF of solar cells as a function of IQE(E_g-0.5 eV) under both initial and light-soaked conditions, a good correlation is found as shown in Fig. 5 (t_i series), indicating that the cell performance is dominated by the total number of defects present in the whole absorber layer, regardless of thickness and LID. This will be further discussed in the following section together with the results of T_{dep-i} series.

B. Deposition temperature series

In the previous section, we showed that the performance (particularly FF) of the optimized a-Si:H solar cells is dominated by the total amount of (native and light-induced) midgap defects throughout the absorber layer. In contrast, the following T_{dep-i} series provides an alternative example: bandtail state at the p-i interface is predominant. Indeed, the variation of T_{dep-i} has a significant impact on solar cell performance as shown in Fig. 6. For p-i-n cells, the decrease of V_{oc} is roughly proportional to T_{dep-i}, while for n-i-p cells there is almost no V_{oc} change in the range of T_{dep-i} from 200 to

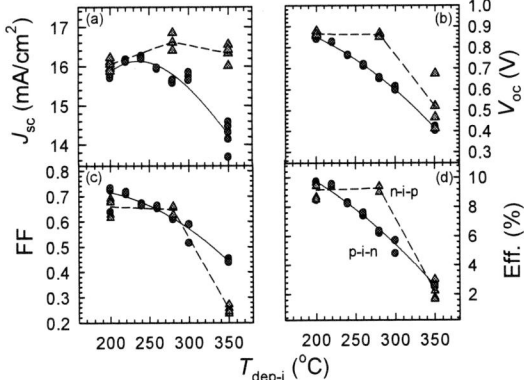

Fig. 6. Solar cell parameters of p-i-n (circles) and n-i-p (triangles) solar cells as a function of i-layer deposition temperature. Lines are guides to the eyes.

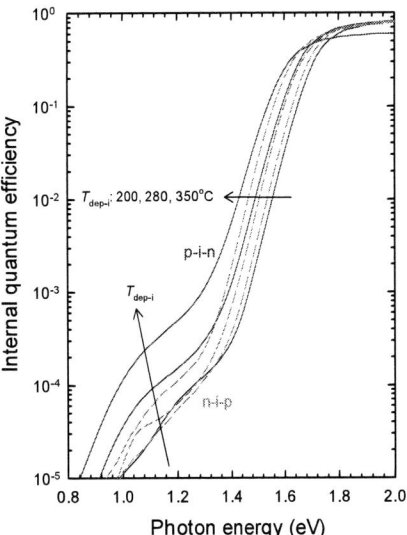

Fig. 7. Sub-bandgap IQE spectra of p-i-n (blue, solid) and n-i-p (red dashed) solar cells for different i-layer deposition temperatures.

280°C. The n-i-p cells prepared at T_{dep-i}=350°C have a much lower V_{oc} because these cells are shunted. Thus, the low V_{oc} of n-i-p cells for T_{dep-i}=350°C does not reflect the electronic quality of the i-layer. (We found that annealing the n-layer alone at 350°C before depositing the i-layer at T_{dep-i}=200°C also results in shunted cells, indicating that some degradation of the n-layer occurs at such high temperature.) In Fig. 6, the J_{sc} first increases with T_{dep-i}, both for p-i-n and for n-i-p cells, but J_{sc} of the p-i-n cells shows a marked decrease when T_{dep-i}>240°C. As shown later, the first J_{sc} increase is due to the bandgap narrowing. For p-i-n cells, the decrease of J_{sc} at T_{dep-i}>240°C is due to a large IQE decrease in the blue region (λ<550 nm, not shown), indicating a pronounced charge carrier recombination near the p-i interface.

In Fig. 7, the sub-bandgap IQE spectra of the p-i-n and n-i-p solar cells with three different deposition temperatures (T_{dep-i}=200, 280 and 350°C) are shown. Both types of solar cells show a redshift of the IQE spectra in the entire energy region, indicating the bandgap decrease with T_{dep-i}. However, it is evident that this effect is greater for p-i-n than for n-i-p cells. In addition, the p-i-n solar cells show a broadening of the bandtail absorption and a larger increase of the midgap defect absorption than n-i-p cells. The three parameters, E_g, E_U and IQE(E_g-0.5 eV), deduced from these sub-bandgap IQE spectra are plotted in Fig. 8 as a function of T_{dep-i}. It is seen that E_g decreases linearly with increasing T_{dep-i}, which is markedly larger for p-i-n cells than for n-i-p cells. Similarly, the changes in E_U and IQE(E_g-0.5 eV) are much greater for p-i-n than for n-i-p cells. Since all the cells have the same thickness, the slope of the lines in Figs. 8(a), (b) and (c) can be reliably compared between the different types of samples.

Above results reveal that the presence of a p-layer during the i-layer deposition induces a significant electronic degradation in its vicinity. This can be linked to the Fermi level position in the absorber layer. When the Fermi level is close to the valence band, broader bandtail and more midgap

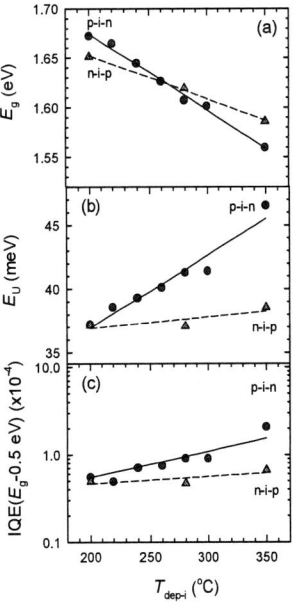

Fig. 8. (a) E_g, (b) E_U and (c) IQE(E_g-0.5 eV) deduced from sub-bandgap IQE spectra as a function of T_{dep-i}.

states are created at high temperature than when the Fermi level is in the midgap or close to the conduction band. This interpretation agrees with the earlier report by De Wolf and Kondo that the passivation of c-Si surface by a-Si:H is less thermally stable for p-i stack than for single i-layer or for n-i stack [4]. Beyer and Wagner have shown that the Si-H rupture energy in a-Si:H is reduced by p-type doping, attributing to the Fermi level shift towards the valence band, not to the dopant element [11]. Indeed, we have measured the hydrogen content in the i-layer of solar cells by Raman scattering spectroscopy and found that when increasing T_{dep-i}, hydrogen content decreases more for p-i-n than for n-i-p solar cells, or for i-layer deposited on glass substrate [7]. This indicates that during the i-layer deposition, the hydrogen desorption from the a-Si:H growth surface is enhanced when the Fermi level in the i-layer is shifted toward valence band by the p-layer underneath. The poor hydrogen incorporation in a-Si:H near the p-i interface leads to the creation of the electronic localized states and thus the enhanced charge carrier trapping and recombination.

By comparing the solar cell properties between the t_i and the T_{dep-i} series, we found different dependencies of FF on IQE(E_g-0.5 eV) as shown in Fig. 5. When varying the T_{dep-i}, the FF decreases linearly with the IQE(E_g-0.5 eV) while its decrease is much sharper than when varying t_i. This is because the increase of E_U affects the FF. Moreover, V_{oc} is known to have a major impact on the FF [12]. In fact, as shown in Fig. 6, V_{oc} decreases from 0.9 to 0.4 V when increasing T_{dep-i} from 200 to 350°C. This massive change in V_{oc} influences the FF greatly. These additional factors account for the enhanced decrease of FF with IQE(E_g-0.5 eV). In our analysis [7], the V_{oc} decrease can be attributed to three major factors: (i) the E_g decrease of the i-layer, (ii) electrical degradation of p-layer by heat (most likely the free-carrier compensation by created defect) and (iii) bandtail broadening in the i-layer. Although it might be rather unusual that the V_{oc} does not depend on the amount of midgap defect, this trend is also seen in the t_i series where the LID does not change V_{oc} at all (Fig. 2). As mentioned earlier, this is because our p- and p-i buffer layers were intentionally tuned to minimize the influence of LID. In fact, the change in IQE(E_g-0.5 eV) for the T_{dep-i} series is only a factor of 3 but that for the t_i series is much more (~a factor of 5). Therefore, in terms of the electronic quality of the a-Si:H i-layer, the degradation of the p-i-n solar cells at high T_{dep-i} mainly originates from the bandtail broadening rather than the increase of the dangling-bond defect.

IV. CONCLUSION

We have changed the amount of the electronic localized states of a-Si:H solar cells by varying the i-layer thickness, light-induced degradation and varying the i-layer deposition temperature. When using an optimum deposition temperature, the amount of the midgap defect increases with thickness and light soaking, while the bandtail characteristic energy stays unchanged. In such case, the bulk defect is predominant in determining the cell performance, especially FF.

When the Fermi level is near the valence band, a-Si:H is more sensitive to increased T_{dep-i}, and more localized states are created. This effect has been explained by the poor hydrogen incorporation that is closely related to the local Fermi level position in the i-layer. Such electronic state creation in the i-layer significantly degrades the all solar cell parameters. These results agree with the earlier report on the behavior of p-doped a-Si:H films and SHJ solar cells, and provide deeper insights into the device physics of a-Si:H solar cells.

ACKNOWLEDGEMENT

The authors are grateful to Y. Miyagi and Y. Sato of AIST for their help in preparing the samples. We also would like to thank J. Melskens and A. Smets of TU Delft, and G. Bugnon of EPFL for helpful discussions about FTPS measurements. We are grateful to the board members of PVTEC for their support. This work was partly supported by the New Energy and Industrial Technology Development Organization (NEDO).

REFERENCES

[1] H. Fujiwara and M. Kondo, "Real-time monitoring and process control in amorphous/crystalline silicon heterojunction solar cells by spectroscopic ellipsometry and infrared spectroscopy," *Appl. Phys. Lett.,* vol. 86, pp. 032112-1-032112-3, 2005.

[2] A. Shah, Thin-Film silicon solar cells, Boca Raton: CRC Press, 2010.

[3] M. Stuckelberger, B.-S. Park, G. Bugnon, M. Despeisse, J.-W. Schüttauf, F.-J. Haug and C. Ballif, "The boron-tailing myth in hydrogenated amorphous silicon solar cells," *Appl. Phys. Lett.,* vol. 107, pp. 201112-1-201112-5, 2015.

[4] S. De Wolf and M. Kondo, "Boron-doped a-Si:H / c-Si interface passivation: Degradation mechanism,"*Appl. Phys. Lett.,* vol. 91, pp. 112109-1-112109-3 (2007).

[5] T. Matsui, A. Bidiville, K. Maejima, H. Sai, T. Koida, T. Suezaki, M. Matsumoto, K. Saito, I. Yoshida and M. Kondo, "High-efficiency amorphous silicon solar cells: Impact of deposition rate on metastability," *Appl. Phys. Lett.,* vol. 106, pp. 053901-1- 053901-5, 2015.

[6] A. Bidiville, T. Matsui and K. Matsubara, "Analysis of bulk and interface defects in hydrogenated amorphous silicon solar cells by Fourier transform photocurrent spectroscopy," *J. Appl. Phys.,* vol. 118, pp. 184506-1-184506-9, 2015.

[7] A. Bidiville, T. Matsui, H. Sai and K. Matsubara, submitted.

[8] T. Matsui, K. Maejima, A. Bidiville, H. Sai, T. Koida, T. Suezaki, M. Matsumoto, K. Saito, I. Yoshida and M.

Kondo, "High-efficiency thin-film silicon solar cells realized by integrating stable a-Si:H absorbers into improved device design," *Jpn. J. Appl. Phys.*, vol. 54, pp. 08KB10-1-08KB10-4, 2015.

[9] H. Sai, T. Matsui and K. Matsubara, "Stabilized 14.0%-efficient triple-junction thin-film silicon solar cell," *Appl. Phys. Lett.*, vol. 109, pp. 183506-1-183506-5, 2016.

[10] M. Vanecek and A. Poruba, "Fourier-transform photocurrent spectroscopy of microcrystalline silicon for solar cells," *Appl. Phys. Lett.*, vol. 80, pp. 719-721, 2002.

[11] W. Beyer and H. Wagner, "Effect of boron-doping on the hydrogen evolution from a-Si:H films," *Solid state communications*, vol. 39, pp. 375-379, 1981.

[12] M. A. Green, "Solar cell fill factors: General graph and empirical expressions,", *Solid-State Electronics*, vol. 24, pp. 788-789, 1981.

Experimental and theoretical study of the infrared emissivity of crystalline silicon solar cells

Alberto Riverola[1], Alexander Mellor[2], Diego Alonso Alvarez[2], Lourdes Ferre Llin[3], Ilaria Guarracino[4], Christos N. Markides[4], Douglas Paul[3], Daniel Chemisana[1], Ned Ekins-Daukes[2]

[1]Applied Physics Section of the Environmental Science Department, University of Lleida, Lleida 25001, Spain
[2]Department of Physics, Imperial College London, London SW7 2AZ, United Kingdom
[3]School of Engineering, University of Glasgow, Glasgow G12 8LT, United Kingdom
[4]Department of Chemical Engineering, Imperial College London, London SW7 2AZ, United Kingdom

Abstract — **Controlling the radiative emissivity of photovoltaic (PV) solar cells and modules in the mid-infrared (MIR) is gaining increasing interest as a means of controlling their operating temperature and performance. However, there is a large uncertainty in our knowledge of the MIR emissivity of existing PV cells. In the present work, we investigate the full radiative spectrum of the crystalline silicon (c-Si) solar cell including both absorption in the UV/VIS/NIR and emission in the MIR. We show that the emissivity of an unencapsulated cell is around 75% in the MIR range, mainly due to free carrier absorption at highly doped layers.**

I. INTRODUCTION

The efficiency of solar cells drops with temperature. Manufacturers or research groups report efficiencies always under standard test conditions (25 °C). Nonetheless, cells hardly operate under these circumstances since the temperature is usually around 50 °C and the operating efficiency is notably reduced.

Operating temperatures have been reduced by either passive or active cooling. More recently, it has been proposed that the temperature can also be significantly reduced by cooling the solar cell radiatively [1,2]. This requires that the radiative emissivity of the solar cell should be engineered to maximize radiative thermal emission into the cold space beyond the atmosphere while preserving the solar absorption properties.

Much to the contrary, for hybrid photovoltaic-thermal (PVT) systems, the thermal efficiency can be improved using spectrally-selective low-emissivity coatings to reduce radiative thermal losses. This will enhance the thermal efficiency with a minimal loss of electrical efficiency in the system, assuming that the cell has a low temperature coefficient [3].

So far, very little has been reported on the emissivity of present commercial silicon solar cells. Sopori et al. [4] modelled the emissivity of polished and textured silicon wafers in the 0.5 – 10 µm range, with and without a thick (1 µm) dielectric coating as a function of doping and temperature. However, their results are based on structures far from real solar cells, considering a constant doping over the cell and with texture on only one side. Santbergen and van Zolingen [5] studied numerically and experimentally the absorptivity, which

is equal to the disperse emissivity via the kirchoff relation, of PV cells up to 1.7 µm, reporting a 90% absorption of the AM1.5 spectrum in this range and the contribution of each layer of the structure. Recently, the emissivity of silicon solar cells up to 25 µm was considered to be equal to the emissivity of a planar p-doped silicon wafer [1]. This wafer is similar to the typical base region on commercial silicon cells. Nonetheless, this assumption does not consider the impact of the texture and the highly doped regions, which are shown in the present work to be crucial to determine the emissivity of an unencapsulated c-Si solar cell.

In this paper, the emissivity of modern silicon solar cells has been measured in the 350 nm – 16 µm range, and a full radiative model of a solar cell considering both absorption in the spectral range of sunlight and thermal emission in the mid-infrared (MIR) has been developed. The model considers the complete cell structure with realistic layer properties and front and back textures. We show that a real c-Si solar cell is highly emissive in the MIR, and that this is mostly due to the highly doped, but very thin, emitter and back-surface field layers, which have been overlooked in previous studies.

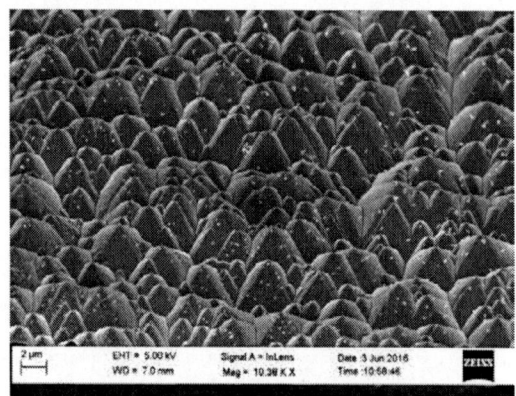

Fig. 1. SEM front surface view tilted 45°.

978-1-5090-5606-4/17 $31.00 © 2017 IEEE

II. METHODS

Absorptivity/emissivity measurements were performed on commercially available monocrystalline c-Si solar cells, purchased from Bolisheng Technology [6], which are considered to be representative of most commercially produced c-Si solar cells. The front-surface texture has been imaged by the scanning electron microscopy (SEM) Leo Gemini 1525 and is shown in Fig. 1. It was textured by a standard etching which is known to produce randomly sized pyramids with an elevation angle of around 55°; this angle has been confirmed by atomic force microscopy.

The dispersive emissivity, E, of a body is equal to its absorptivity, A, via the Kirchoff relation. Emissivity can therefore be measured indirectly by measuring reflectivity, R, and transmissivity, T, and invoking the relation:

$$E(\lambda) = A(\lambda) = 1 - R(\lambda) - T(\lambda) \qquad (1)$$

c-Si solar cells have a surface texture on both faces, and so the reflected and transmitted light will be diffuse. The emissivity (= absorptivity) of a silicon solar cell was therefore measured - in the 350 nm – 16 µm range - via hemispherical reflection and transmission measurements using an integrating sphere, and applying Eq. (1). The measurements were carried out with a Fourier-transform spectrometer Bruker IFS 66 equipped with two integrating spheres (A PTFE coated sphere for the UV/VIS/NIR range (350 – 2400 nm) and a diffuse-gold coated sphere for the MIR (1.5 – 16 µm). Calibrated diffusely reflecting references from the National Institute of Standards and Technology (US) and the National Physical Laboratory (UK) are used for baseline measurements in the UV/VIS/NIR and MIR ranges respectively. The accuracy of the reflectance data is 1% in the solar range and 2% in the IR range.

The absorptivity/emissivity of the c-Si solar cell was modelled over the full experimental range (350 nm – 16 µm) using the OPTOS formalism [7], which is a steady-state scattering-matrix approach that can be formulated to consider all photonic interactions with surface textures, interference coatings, and absorbing/emitting layers. Using this type of approach is of critical importance to accurately model photon absorption and emission from a textured c-Si solar cell. Any photon that is emitted somewhere inside the device can, in general, be internally reflected at any interface and make multiple passes of the solar cell, undergoing numerous reflections and transmission at different interfaces, before either escaping into the surroundings or being reabsorbed. Equivalently, any photon incident from outside can make many passes of the cell before being absorbed or escaping. Therefore, an accurate calculation of the emissivity/absorptivity must take all these interactions into account. The OPTOS formalism performs an angular discretization within the solar cell and considers the irradiance within each angular element as elements of a vector. The emissivity is taken to be equal to the absorptivity via the Kirchoff relation. Properly formulated, this approach treats smaller features wave-optically, and larger features geometric

optically, and is therefore faster and more accurate than so-called full-wave-optical calculations, which make the false assumption of infinite spatial and temporal coherence, and thus predict interference phenomena that are not observed experimentally.

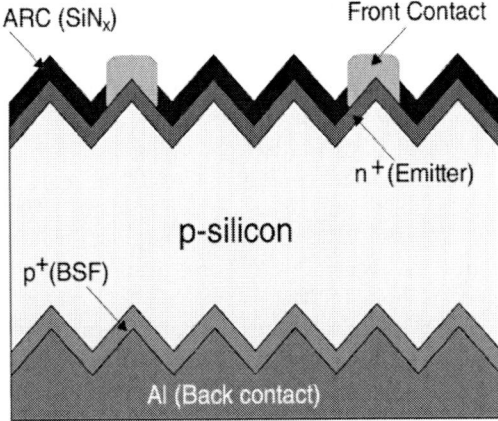

Fig. 2. Structure of our simulated silicon solar cell, which includes an antireflective coating, a highly doped emitter, a silicon wafer, a BSF layer and an aluminum back reflector.

Since commercial cells are considered for this study, the exact cell structure is unknown. However, the type of solar cells that presently dominate manufacture are homojunction cells with diffused front-side emitters, aluminium rear reflectors and back surface field (BSF). This type of technology shares optical features across manufactures that are well known from literature. A cross section of the modelled solar cell structure is shown in Fig. 2. The cells are based on a standard low-doped p-type wafer (the base) with the above described surface texture on both sides. On the front there is a thin heavily-doped n+ c-Si emitter layer and above this a SiN anti-reflection coating. On the rear, there is a heavily-doped p+ c-Si BSF layer, and behind this an Al reflector, which is assumed to be optically thick. The thicknesses and doping concentration of the different layers, taken from various literature sources, are shown in Table 1. The front contact fingers are not considered explicitly in the simulations, but rather assumed to induce a non-dispersive 2% reduction of the emissivity (= absorptivity) due to shading.

TABLE I
THICKNESSES AND DOPING CONCENTRATIONS USED IN THE MODEL

Layer	Thickness (nm)	Doping (cm-3)	Reference
ARC (SiNx)	74	-	[5]
Emitter (n+)	55	3.3×10^{20}	[5]
Bulk (p)	1.5×10^{5}	10^{16}	[5-6]
BSF (p+)	4300	1.5×10^{19}	[8]
Back contact (Al)	1000	-	-

III. RESULTS AND DISCUSSION

The measured emissivity/absorptivity of the unencapsulated c-Si solar cell is shown in Fig. 3. The solar cell is highly absorptive/emissive at all wavelengths in the 350 nm – 16 μm range. The absorption is above 90% between 400 and 1000 nm; indeed, the solar cell is engineered to absorb strongly in this range for photogeneration. The absorptivity drops slightly above 1000 nm - close to the c-Si bandgap, but remains above 80% up to 10 μm, and above 70% out to 16 μm. This range includes the 8 – 13 μm atmospheric transmission window, which also coincides with the thermal emission peak at reasonable solar cell operating temperatures. This is in contrast to the absorptivity/emissivity of a low-doped untextured Si wafer, which drops to below 20% for wavelengths above 1100 nm.

The results of the simulation are shown in Fig. 3, where the dashed blue line represents the total emissivity (= absorptivity). We first observe the excellent agreement between experiment and simulation throughout the 350 nm – 16 μm range, which we view as a validation of the method. Above the bandgap energy (λ < 1100 nm), absorption is dominated by band-to-band absorption in the base as expected. Below the bandgap energy (λ > 1100 nm), band-to-band absorption/emission falls to zero and the absorptivity/emissivity becomes dominated by free carrier absorption/emission in the highly doped emitter and BSF, and in the Al reflector. There is also some free carrier absorption/emission in the low-doped base, but this makes a small contribution. The emissivity contributions at 9 μm are depicted in the pie chart in Fig. 3.

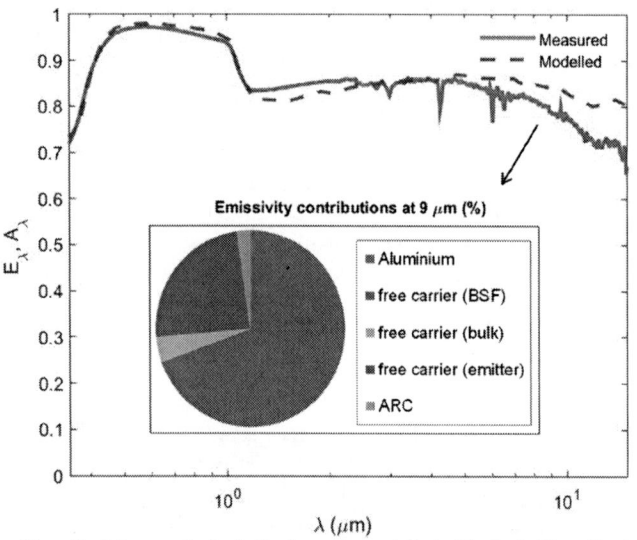

Fig. 3. Measured (red line) and modelled (dashed blue line) emissivity/absorptivity of c-Si solar cells with a pie chart showing the contributions of each layer to the emissivity/absorptivity at 9 μm.

IV. CONCLUSION

The omission of highly doped regions and surface texture have resulted in an understatement of the emissivity of c-Si solar cells in previous work (by around 25% in the MIR range). Our results show that bare silicon solar cells exhibit a larger emissivity (around 75% in the MIR range) and therefore these cells have good properties for radiative cooling purposes, without further processing. However, for PVT applications where such radiative heat loss is undesirable, this emissivity will lead to reduced thermal efficiencies. The design of specific solar cells for hybrid solar systems should address this particular characteristic in order to maximise the thermal efficiency of these systems.

REFERENCES

[1] L. Zhu, A. Raman, K. X. Wang, M. A. Anoma, and S. Fan, "Radiative cooling of solar cells," *Optica*, vol. 1, no. 1, pp. 32–38, 2014.

[2] A. P. Raman, M. A. Anoma, L. Zhu, E. Rephaeli, and S. Fan, "Passive radiative cooling below ambient air temperature under direct sunlight," *Nature*, vol. 515, no. 7528, pp. 540–4, 2014.

[3] A. Mellor, I. Guarracino, L. F. Llin, D. Alonso-Alvarez, A. Riverola, S. Thoms, D. J. Paul, C. N. Markides, D. Chemisana, S. Maier, and N. E.- Daukes, "Specially designed solar cells for hybrid photovoltaic-thermal generators," in *2016 IEEE 43rd Photovoltaic Specialists Conference (PVSC)*, 2016, pp. 2960–2963.

[4] B. L. Sopori, W. Chen, S. Abedrabbo, and N. M. Ravindra, "Modeling emissivity of rough and textured silicon wafers," *J. Electron. Mater.*, vol. 27, no. 12, pp. 1341–1346, 1998.

[5] R. Santbergen and R. C. van Zolingen, "The absorption factor of crystalline silicon PV cells : A numerical and experimental study," *Sol. Energy Mater. Sol. Cells*, vol. 92(4), pp. 432–444, 2008.

[6] B. Technology, "Silicon solar cellsMONO(φ165)S5LV03 125mm."[Online].Available: http://www.bolisheng.com/en/product_show.asp?id=364. [Accessed: 01-Aug-2016].

[7] J. Eisenlohr, N. Tucher, O. Höhn, H. Hauser, M. Peters, P. Kiefel, J. C. Goldschmidt, and B. Bläsi, "Matrix formalism for light propagation and absorption in thick textured optical sheets.," *Opt. Express*, vol. 23, no. 11, pp. A502-18, Jun. 2015.

[8] T. Fellmeth, S. MacK, J. Bartsch, D. Erath, U. Jäger, R. Preu, F. Clement, and D. Biro, "20.1% Efficient Silicon Solar Cell With Aluminum Back Surface Field," *IEEE Electron Device Lett.*, vol. 32, no. 8, pp. 1101–1103, 2011.

High performance molecular donors for organic solar cells, materials design and device optimization.

Paul Geraghty, Haotian Wang, Calvin Lee, Jegadesan Subbiah, David Jones.

School of Chemistry, Bio21 Institute, University of Melbourne, Parkville, Victoria, Australia, 3010

Abstract — We demonstrate the development of high performance molecular donors for the use in organic solar cells. We modified the known BTR small molecule donor to reduce diffusion rates during thermal annealing by chromophore extension to generate BQR. OPV devices containing BQR are thermally stable, with average PCEs of around 10%. Active layers containing BQR can be deposited from industrially relevant solvents using slot die printing. We demonstrate that simple slot die printed 1 cm^2 OPV devices using toluene as a solvent with around 6% PCE.

I. INTRODUCTION

Organic solar cells offer the potential to develop low cost printed organic solar cells (OPV) with rapid recent evolution of the field with solar cells now exceeding 12% power conversion efficiency (PCE) [1]. The rapid advances are due to an improved understanding of materials design and the introduction of new non-fullerene acceptors, allowing better matching of absorption profiles and energy levels. Along with the rapid advance in polymer-non-fullerene acceptor solar cells, and polymer-polymer OPVs [2], there has also been increased interest in small molecule-small molecule (SM/SM) OPVs. This interest is related to improved SM:PCBM OPV devices, now >10% PCE, rivalling those of polymer:PCBM devices [3], and the use of SMs in ternary blend solar cells as morphology or energy cascade modifiers [4]. SMs have distinct advantages over polymers in that, i) their structures are well defined leading to a reproducible synthesis with less batch to batch variation, ii) they often have more organized microstructures leading to higher charge mobilities, and iii) minor structural modifications impacting energy levels, optical absorptions and molecular assembly can be readily modified and the impacts mapped.

II. BACKGROUND

We recently introduced a new class of molecular donors following side-chain engineering of a known A-π-D-π-A structure, where the donor (D) was a substituted benzodithiophene, the π-bridge was a terthiophene, and the acceptor (A) was an N-substituted rhodamine, abbreviated as **BTR** (Figure 1, n = 2) [5]. **BTR** has excellent OFET charge mobility, 0.1 cm^2 V^{-1} s^{-1} (bottom contact, bottom gate), and, when combined with PC$_{71}$BM in the active layer, OPV devices with up to 9.4 % PCE. A key aspect of the devices is the high performance and high fill factors (FF), even for devices with a 350 nm thick active layer, now optimized to 9.4 – 9.5 % PCE. This suggested that, i) these materials may be ideal for

translation to a printed solar cell where film thickness variation would be less of an issue and thicker films are easier to print, and ii) charge recombination must be strongly non-Langevin in **BTR:PC$_{71}$BM** to allow for high FFs in thick films. We have recently demonstrated a Langevin recombination reduction factor β of ≈ 150 in **BTR:PC$_{71}$BM** thin films, this is amongst the highest ever reported [6]. The best **BTR** containing devices are obtained after solvent vapor annealing (SVA) with THF. Transient absorption (TA) spectroscopy has demonstrated excellent charge generation in these devices [7].

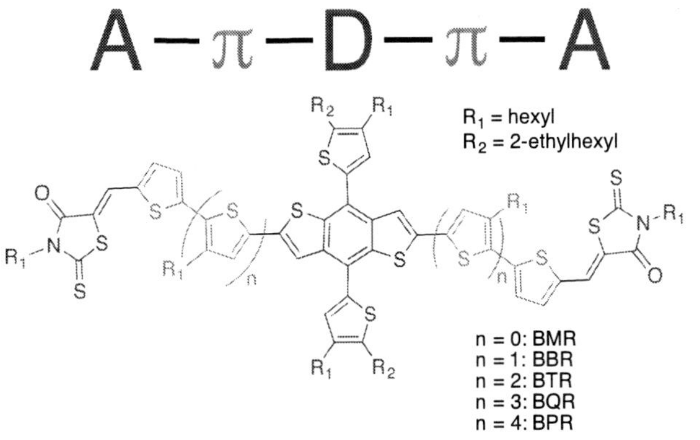

Fig. 1. Chemical structure of the BXR series of p-type molecular donors for OPVs, where A = acceptor unit, π = oligothiophene bridge, and D = donor. BTR (n = 2) and BQR (n = 3).

Although solvent vapor annealed **BTR** containing devices show excellent solar cell performance, recombination and charge generation, the active layer in devices is not thermally stable, with rapid performance loss on thermal annealing. The PCE loss is thought to be due to molecular diffusion and domain size growth resulting in poor charge generation [8]. Therefore, we cannot translate these materials to printed solar cells where temperatures > 80 °C are required for layer drying or printed metal ink curing.

III. RESULTS:

We expected that chain extension may reduce molecular diffusion rates in thin films of **BTR** analogues and thereby overcome thermal instability in **BTR** devices. The extended conjugated π-system would also red-shift the absorption band in the **BTR** analogues, potentially improving performance.

Systematic modification of the chromophore length gave a π-bridge with from 1 to 5 thiophenes, generating the **BXR** series of materials, Figure 1. Initial device screening immediately indicated even a single thiophene extension of the π-bridge significantly improved device thermal stability with the quarterthiophene BQR (Figure 1, n = 3) showing thermal stability at 120 °C with an 8.9 % PCE (Table 1, Entry 6) [9]. We have recently optimized devices containing **BQR** and following a combined SVA (THF 20 sec) and TA (90 °C, 10 minutes) treatment device with an average 9.8 % PCE were obtained with a best of 10.2 % PCE, see Figure 2 (and Table 1, Entry 10). Even with an active layer thicknesses of 230 nm we see FFs of > 70 %, We are currently looking at improving **BQR** containing devices with interlayer treatments to maximize charge extraction.

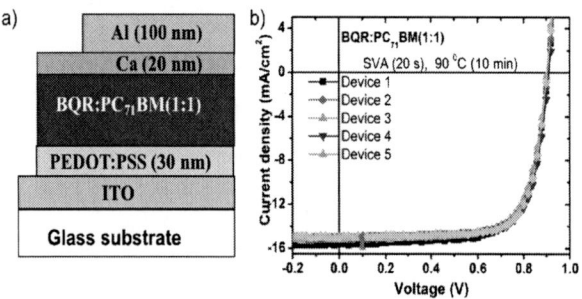

Fig. 2. Optimized OPV deices containing a BQR:PC71BM active layer (230 nm), a) device architecture, and b) J-V curves (9.8 % PCE

Introduction of **BQR** leads to thermally stable devices, and therefore, to the possibility of translation to large scale printed solar cells. We are currently scaling the synthesis of BQR to allow for large scale print trials.

Translation to printed OPV or modules requires and understanding of factors involved in scale up. BQR appeared to be an ideal material to investigate device scaling as it is, i) a high-performance material, ii) available in significant quantities, iii) the active layer thickness dependence for devices is relatively small, and iv) the devices are thermally stable. Meanwhile, OPV device performance is normally reported for areas of around 0.1 cm^2, however the standard for inorganic solar cells is > 1 cm^2, see Figure 3. Our aim was to develop standard protocols for organic solar cells to deliver improved performance and understand scaling issues for OPV. Our preliminary results for 1 cm^2 devices are discussed here.

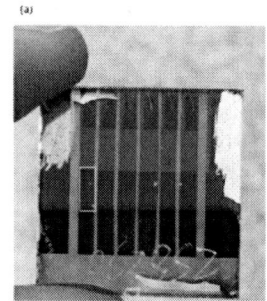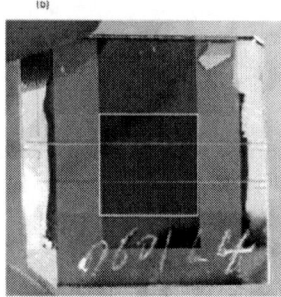

Fig. 3. Photograph of (a) 0.1 cm^2 devices and (b) a 1 cm^2 devices with 2 lines Ag lines for enhanced charge collection. Active area is shown inside of the yellow rectangles. (photo credit: Haotian Wang)

Similarly, translation to an industrially relevant process requires consideration of the solvent and deposition method, as most chlorinated solvents are not acceptable industrially, and spin casting is not a viable option for high throughput large area printing.

A. Solvent optimization

In translating from the laboratory to printed organic solar cells we also look to translate to industrially relevant, non-chlorinated solvents. Comparing chloroform, toluene, and xylene, the last two being industrially acceptable solvents, in laboratory scale spin cast devices (0.1 cm^2) we observed with the use of toluene we can maintain device performance, Table 2, Entries 1-2. Xylene is also effective, however solubility

TABLE I

SUMMARY OF OPV DEVICES CONTAINING BXR MATERIALS UNDER DIFFERENT ANNEALING CONDITIONS

	Active layer	Annealing	J_{sc} (mA cm^{-2})	V_{oc} (V)	FF (%)	PCE (%)
1	BTR:PC$_{71}$BM	As cast	10.8	0.98	43	4.6
2		SVA 20 sec	13.9	0.92	72	9.3
3		TA 120 °C 10 min	11.0	0.88	58	5.7
4	BQR:PC$_{71}$BM	As cast	12.6	0.94	45	5.3
5		SVA 20 sec	15.3	0.88	70	9.4
6		TA 120 °C 10 min	14.9	0.92	65	8.9
7	BPR:PC$_{71}$BM	As cast	7.5	0.9	45	3.0
8		SVA 20 sec	14.3	0.82	74	8.7
9		TA 120 °C 10 min	12.8	0.88	72	8.1
10	BQR:PC$_{71}$BM	SVA 10 sec, TA 90 °C 10 min	15.0	0.9	72	9.8

TABLE II

BQR:PC$_{71}$BM DEVICE OPTIMIZATION USING INDUSTRIALLY RELEVANT SOLVENTS

	Solvent	Area (cm^2)	J_{sc} (mA cm^{-2})	V_{oc} (V)	FF (%)	PCE (%)
1	CHCl$_3$	0.1	0.87	12.42	67.7	7.3 (7.7)
2	Toluene	0.1	0.83	12.05	67.2	6.7 (6.9)
3	Xylene	0.1	0.83	10.89	69.9	6.2 (6.8)
4	CHCl$_3$	1.0	0.87	12.35	58.1	6.2 (6.5)
5	Toluene	1.0	0.84	12.87	57.2	6.2 (6.3)
6	Toluene (slot die)	1.0	0.82	12.31	57.0	5.8 (6.4)

Annealing conditions: SVA 10 sec, TA 90 °C 10 min, PCE average of at least 10 devices (best)

issues meant that this solvent was no longer examined. On scaling to a 1cm^2 device, still using spin coating to deposit the active layer, we observed that no significant difference between deposition from chloroform or toluene, Table 2, Entries 4-5. A slight drop in V_{oc}, 0.87 to 0.84 V, and FF, 58.1 % to 57.2 %, in going from chloroform to toluene is offset by a higher J_{sc}, up from 12.35 to 12.87 mA.cm^{-2}. These are preliminary results and full optimization is currently underway.

B. Translation to slot die printing

Using an industrially relevant deposition process is the next step is the translation. We chose to use slot die printing where the active materials are formulated as an ink and pumped through a slot onto a moving substrate. Film thickness is controlled by varying the pumping speed and the substrate speed past the slot die head, and were adjusted to allow an active layer thickness of ≈200nm to be deposited. Devices deposited using a slot die process are comparable with spin cast films, Table 2, Entries 5 & 6. Device V_{oc} drops slightly for 0.84 to 0.82 V, while the J_{sc} drops from 12.87 to 12.31 mA.cm^{-2}, however the device variability was higher, and the best 1 cm^2 devices with a slot die deposited active layer gave a better performance than the best spin cast device, 6.4 % PCE compared to 6.3 % PCE.

IV. SUMMARY

We have shown that with a detailed understanding of molecular design we can generate high performance p-type donors for organic solar cells, **BTR**. Understanding the underlying physical process in preparing devices, *i.e.* decrease diffusion rates during annealing by increasing the chromophore size, has allowed the development of a high performance material, **BQR**, suitable for translation a large scale printing program. We have demonstrated that we can translate our laboratory process using active materials spin cast from chlorinated solvents to industrially relevant slot die printed active layers using toluene. We have maintained an excellent device performance during this scaling process and we are currently optimizing the printing parameters for these new materials.

REFERENCES

[1] S. Li, L. Ye, W. Zhao, S. Zhang, S. Mukherjee, H. Ade, and J. Hou, "Energy-Level Modulation of Small-Molecule Electron Acceptors to Achieve over 12% Efficiency in Polymer Solar Cells," *Adv. Mater.*, vol. 28, no. 42, pp. 9423–9429, 2016.

[2] L. Gao, Z. Zhang, L. Xue, J. Min, J. Zhang, Z. Wei, and Y. Li, "All-Polymer Solar Cells Based on Absorption-Complementary Polymer Donor and Acceptor with High Power Conversion Efficiency of 8.27%,"*Adv. Mater.*, vol. 28, no. 9, pp. 1884–1890, 2016.

[3] B. Kan, M. Li, Q. Zhang, F. Liu, X. Wan, Y. Wang, W. Ni, G. Long, X. Yang, H. Feng, Y. Zuo, M. Zhang, F. Huang, Y. Cao, T. Russell, and Y. Chen, "A Series of Simple Oligomer-like Small Molecules Based on Oligothiophenes for Solution-Processed Solar Cells with High Efficiency," *J. Am. Chem. Soc.*, vol. 137, no. 11, pp. 3886–3893, 2015.

[4] M. Zhang, F. Zhang, Q. An, Q. Sun, W. Wang, X. Ma, J. Zhang, and W. Tang, "Nematic Liquid Crystal Material as Morphology Regulator for Ternary Small Molecule Solar Cells with Power Conversion Efficiency Exceeding 10%," *J. Mater. Chem.*, 2017. DOI: 10.1039/C7TA00211D

[5] K. Sun, Z. Xiao, S. Lu, W. Zajaczkowski, W. Pisula, E. Hanssen, J. M. White, R. M. Williamson, J. Subbiah, J. Ouyang, A. B. Holmes, W. W. H. Wong, and D. J. Jones, "A molecular nematic liquid crystalline material for high-performance organic photovoltaics," *Nat. Commun.*, vol. 6, 2015. DOI: 10.1038/ncomms7013

[6] A. Armin, J. Subbiah, M. Stolterfoht, S. Shoaee, Z. Xiao, S. Lu, D. J. Jones, and P. Meredith. "Reduced Recombination in High Efficiency Molecular Nematic Liquid Crystalline: Fullerene Solar Cells," *Adv. Energy Mater.*, vol. 6, p. 1600939, 2016.

[7] K. N. Schwarz, P. B. Geraghty, D. J. Jones, T. A. Smith, and K. P. Ghiggino, "Suppressing Subnanosecond Bimolecular Charge Recombination in a High-Performance Organic Photovoltaic Material," *J. Phys. Chem. C*, vol. 120, pp. 24002-24010, 2016.

[8] S. Engmann, H. W. Ro, A. Herzing, C. R. Snyder, L. J. Richter, P. B. Geraghty, and D. J. Jones, "Film morphology evolution during solvent vapor annealing of highly efficient small molecule donor/acceptor blends," *J. Mater. Chem. A*, vol. 4, pp. 15511-15521, 2016.

[9] P. B. Geraghty, C. Lee, J. Subbiah, W. W. H. Wong, J. L. Banal, M. A. Jameel, T. A. Smith, and D. J. Jones, "High performance p-type molecular electron donors for OPV applications via alkylthiophene catenation chromophore extension," *Beilstein J. Org. Chem.*, vol. 12, pp. 2298-2314, 2016.

978-1-5090-5606-4/17 $31.00 © 2017 IEEE

Advanced Optical Modelling of Micro-Textured Solution-Processed Solar Cells with Consideration of Small-Area Effects

Benjamin Lipovšek[1], Marko Jošt[1], Andrej Čampa[1], Fei Guo[2], Christoph J. Brabec[2],
Karen Forberich[2], Janez Krč[1], and Marko Topič[1]

[1]University of Ljubljana, Faculty of Electrical Engineering, Tržaška 25, 1000 Ljubljana
[2]Friedrich-Alexander-Universität Erlangen-Nürnberg (FAU), Martensstr. 7, 91058 Erlangen, Germany

Abstract — We investigate the potential of different types of micro-textured foils applied on top of solution-processed solar cells in superstrate configuration, for the purpose of improved device performance. Both organic as well as perovskite solar cells are included. The results based on numerical simulations show that application of the micro-textured foils can significantly improve the short-circuit current density in both types of solar cells, especially if the corner cube texture is employed. Further on, we test the applicability of the micro-textured foil on an example fabricated organic solar cell. In this case, however, substantial optical losses are discovered due to the limited device area. Using numerical simulations, we evaluate these losses and determine the minimal device area that would be required to avoid them. Finally, we show how to predict the actual performance of the characterized organic cell, as is expected to be achieved in realistic up-scaled PV module application.

Index Terms — organic solar cells, perovskite solar cells, light management, optical simulations.

I. INTRODUCTION

In solution-processed solar cells such as thin-film organic and perovskite cells [1], nano- and micro-scale surface textures can be utilized to enhance light in-coupling and light trapping within these devices [2–4]. As one of such approaches, application of external micro-textured light management (LM) foils directly on top of the fabricated devices in superstrate configuration has been indicated as a promising concept [3–5]. Its main highlights are low cost, ease of fabrication (e.g. UV NIL process), and ease of application that can be done even at some later stage, long after the cell (or PV module) has left the manufacturing line.

To investigate the potential of LM foils in different solar cell technologies, and also to custom design and optimize specific texture profile features, advanced optical models need to be utilized [6,7]. However, due to the relative complexity of the complete solar cell structures enhanced with LM foils, such models typically require a combination of various optical modelling methods to properly analyze light propagation through different parts of the device – e.g. wave optics for thin layers, geometric optics for the thick glass substrate and the micro-textured LM foil, etc. The accuracy of these models depends heavily on the reliability of the input data (optical constants, incident illumination parameters) and an accurate description of the simulation domain.

In this contribution, advanced optical simulations are employed to evaluate the potential of different LM foils employed in example organic and perovskite thin-film solar cells, and to identify the dominant optical mechanisms that enable the enhancement of the device performance. We then test the application of the foils in a realistic fabricated device. However, substantial errors in the measured characteristics are revealed in this case, primarily due to the optical losses related to the limited area of the measured device. Therefore, in the final part of this work, we demonstrate how advanced optical modelling can also be utilized to evaluate and fully quantify the effects caused by geometry constraints of the measured devices and measurement systems, and to indicate the minimum geometry requirements to avoid measurement errors.

II. EXPERIMENTAL AND NUMERICAL PROCEDURES

Two example planar solution-processed solar cells in superstrate (inverted) configuration are investigated in the scope of this contribution – one from the organic and the other from the perovskite solar cell technology. The complete structures of the example solar cells are (top-down):

i) *Organic SC:*	glass (1 mm)
	ITO (360 nm)
	PEDOT:PSS (70 nm)
	pDPP5T-2:PC$_{60}$BM (135 nm)
	Ca/Ag (15/85 nm)

ii) *Perovskite SC:*	glass (1 mm)
	ITO (140 nm)
	PTAA (5 nm)
	Perovskite (270 nm)
	PCBM (50 nm)
	BCP (5 nm)
	Ag (100 nm)

For the purpose of light management, a 0.6 mm thick PET LM foil is laminated on top of the devices. Two different texture profiles are assumed to be realized on the front surface of the LM foil, as shown in Fig. 1: Parabolic, with the period $P = 9$ μm and height $h = 5.5$ μm, and Corner cube, with a cube side length $a = 5.5$ μm. Both are arranged in a hexagonal grid.

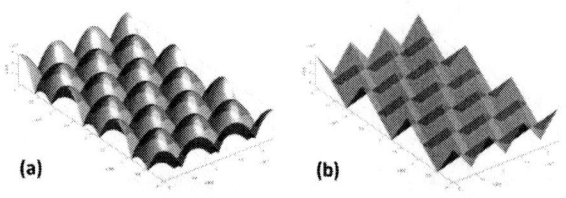

Fig. 1. Studied texture profiles: (a) Parabolic, (b) Corner cube. Both are arranged in a hexagonal grid and have similar feature dimensions.

To perform all optical simulations of complete solution-processed solar cell structures presented in this work, we employed the optical simulator CROWM (Combined Ray Optics / Wave Optics Model) [9–11]. The simulator is based on the combination of two fundamentally different numerical methods for the analysis of light propagation, which are applied separately to different parts of the simulated device.

As test devices, organic solar cells were fabricated as follows [8]: a 1 mm thick glass substrate covered with 360 nm of indium tin oxide (ITO) served as the substrate. On top of it, 70 nm thick poly(3,4-ethylene-dioxythiophene) doped with poly(styrene sulfonate) (PEDOT:PSS) was doctor bladed from its diluted solution (1:3 vol. % in isopropanol) and dried at 140 °C for 5 min. Then, 135 nm thick photoactive blend of diketopyrrolopyrrole (DPP) based polymer (pDPP5T-2) and [6,6]-phenyl-C_{60}-butyric acid methyl ester (PC$_{60}$BM) (1:2 wt. %) was coated from a mixed solvent of dichlorobenzene and chloroform (1:9 vol. %) at a total concentration of 24 mg/mL. The device fabrication was completed by thermal deposition of 15 nm thick calcium (Ca) and 85 nm thick silver (Ag) layers on top of the photoactive absorber layer. The active area (back reflector) of the fabricated solar cell was 4 x 6 mm².

Finally, a commercial LM foil HT-MLA-09 (Holotools GmbH, Germany) was laminated on top of the fabricated solar cells. The top surface of the foil exhibits convex parabolic hexagonal honeycomb micro-texture with the period $P = 9$ μm and the height $h = 5.5$ μm, as shown in Fig. 2. Thus, the texture is almost identical to the parabolic texture shown in Fig. 1(a), which was used in optical simulations.

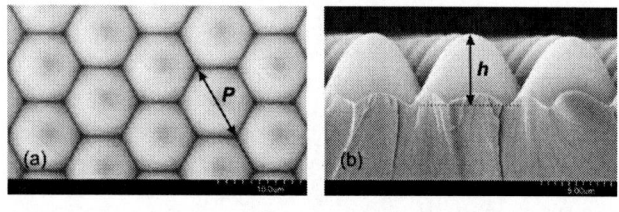

Fig. 2. Top-down (a) and side-view (b) SEM images of the LM foil HT-MLA-09 (images courtesy of Holotools GmbH, Germany).

III. RESULTS AND DISCUSSION

Using the combined optical model, we simulated the spectral absorptance A inside the absorber layer of both solution-processed solar cells under perpendicular illumination conditions. In addition to both studied textures, we also simulated the case of non-textured LM foil as the reference. Finally, by weighting the simulated spectral absorptances by the incident AM1.5 solar spectrum, we calculated the corresponding ideal short-circuit current density (J_{SC}) values for all cases.

The simulated A characteristics are presented in Fig. 3(a) and 3(b) for the organic and the perovskite solar cell, respectively.

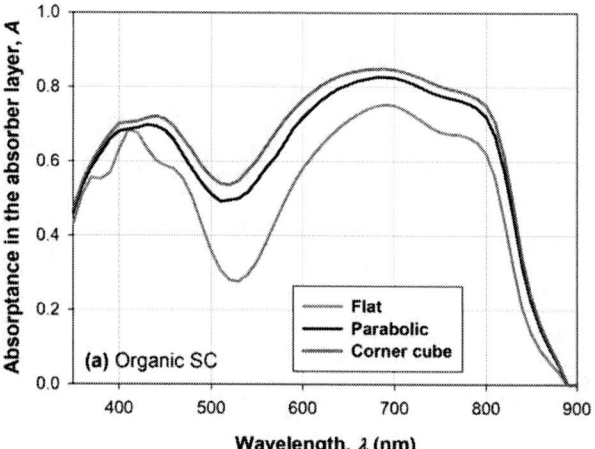

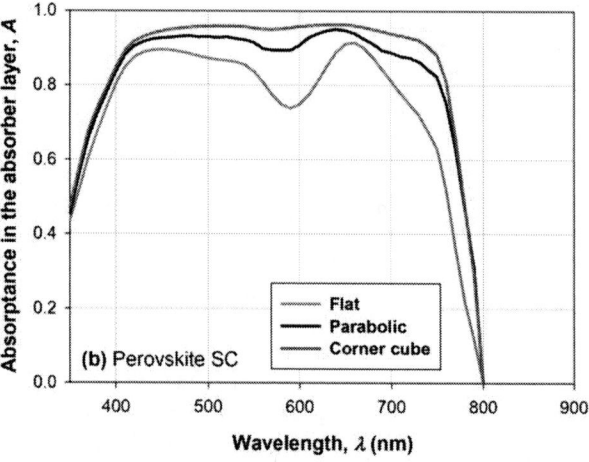

Fig. 3. Simulated absorptance inside the absorber layer of the (a) organic and (b) perovskite solar cell, using both studied textures of the LM foil.

978-1-5090-5606-4/17 $31.00 © 2017 IEEE

The results presented in Fig. 3 show that the application of the LM foil can significantly enhance the performance of both devices. By using the superior corner cube texture, a J_{SC} improvement of 26.6 % is observed for the organic, and 16.9 % for the perovskite solar cell, relative to the non-textured case. The reasons for these improvements are numerous: First, micro-scale textures have the ability to reduce the initial reflection from the front surface of the device (AR effect) and, thus, improve light in-coupling into the absorber layer of the cell. Second, light refraction at the micro-textured surface leads to prolonged optical paths through the absorber layer, resulting in larger photocurrent. And third, the poorly absorbed light reflected from the rear contact of the solar cell can be re-directed back into the device, due to efficient internal reflection at the micro-textured LM foil/air interface.

Further on, we measured the *EQE* curves of the fabricated organic solar cell without and with the LM foil laminated on top of the front glass substrate. The measurement results are presented in Fig. 4 (black and red curve with symbols).

At first glance, only little difference between the measured curves can be observed, which initially leads us to believe that the benefits of the LM foil are modest; the relative J_{SC} increase in this case is only 6.5 %, which is far from the boost predicted by the simulations. However, we found out that the reason for this modest performance lies in the optical losses caused by the limited area of the measured device. Due to the pronounced scattering at the front textured surface of the LM foil, the thick glass substrate, and a limited size of the back contact/reflector, a substantial amount of light can be lost by propagating past the area of the solar cell. Consequently, since this lost light cannot efficiently contribute to the photocurrent, the measured *EQE* values are significantly lower than what they would be in the case of an up-scaled large-area solar cell.

These losses are difficult to evaluate experimentally, but can be tackled using numerical simulations. Therefore, we repeated the simulation of the textured organic solar cell, but this time we also took the exact geometry of the measurement system, including the beam size, the device area size, etc., fully into the account. If we compare it to the previous simulation with no geometry constraints, a substantial 8.6 % J_{SC} difference can be calculated. This represents an estimation of the measurement errors in our system imposed by the limited device area. If we now use this value to correct the measured *EQE*, we can finally predict the actual *EQE* that would be measured in the case of an up-scaled organic solar cell. This predicted result is also shown in Fig. 4 (blue curve). A much larger, almost 17 % J_{SC} increase can now be obtained compared to J_{SC} of the non-textured solar cell, which is a much more relevant result.

As the last step of our studies, beginning with the realistic geometry conditions (device area of 4 x 6 mm²), we gradually increased the device area in the simulations in order to determine the minimum size requirements at which the J_{SC} error caused by the optical side losses would become negligible, relative to the infinite-size device. We studied this dependency for both types of textures while keeping every other parameter of the device unchanged.

The results are presented in Fig. 5. First, the results confirm that a significantly larger device area would be generally required for reducing the measurement J_{SC} error below 1 %. At the same time, however, the results show notable differences between the two textures. While an about 5x larger device would suffice in the case of corner cube texture, a more than 10x larger device would be required in the case of parabolic texture. The reason is that the parabolic texture scatters more light into the very large angles of propagation, and therefore the optical side losses in this case are also larger.

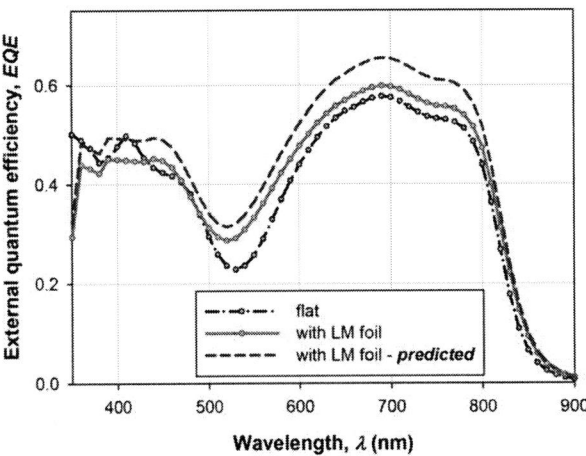

Fig. 4.   Measured *EQE* curves of the organic solar cell (symbols). Also plotted is the curve predicted for a large-area device.

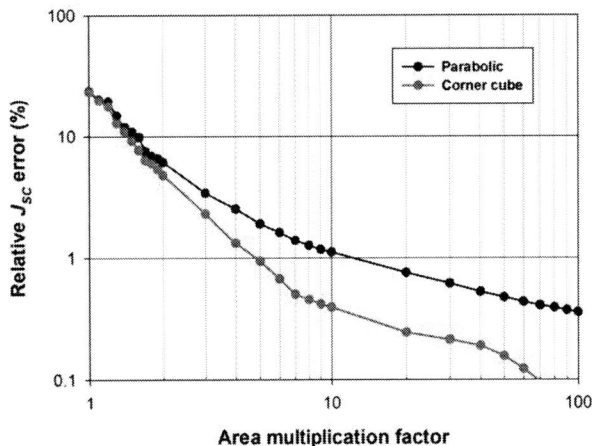

Fig. 5.   Simulated J_{SC} error relative to the infinite-size device as a function of the (proportional) active area scaling factor.

CONCLUSIONS

The application of surface-textured light-management foils on top of solution processed organic and perovskite solar cells in superstrate configuration can significantly improve the performance of these devices. Using the optimal corner cube texture, a 26.6 % J_{SC} gain in the organic, and a 16.9 % J_{SC} gain in the perovskite solar cell was determined based on numerical simulations using an advanced combined optical modelling approach.

However, limited active areas of typical lab-scale solution-processed solar cells together with geometry constraints of the measurement equipment can lead to a substantial amount of errors when measuring the external quantum efficiency of these devices. Consequently, all other characteristics derived from these results, including the short-circuit current density, are also prone to errors. Therefore, the positive effects of the LM foil can become completely obscured by the artifacts of the measuring system.

In our case, by help of realistic simulations, we observed that measurements of a small-size organic solar cell with a parabolically-textured light-management foil applied on top reveal only slightly more than 90 % of the actual photocurrent that would be generated in a large-area device free of optical side losses. Therefore, using numerical simulations, we determined the minimal device area size of the organic solar cell that would be required in order to avoid these errors. Results show that, in order to ensure photocurrent measurement errors below 1 %, at least a 10x and a 5x larger device area would be required if using the LM foil with the parabolic and corner cube texture, respectively.

REFERENCES

[1] B. Azzopardi, *Prog. Photovolt. Res. Appl.*, vol 24, 2016, pp. 261–268.

[2] M. Niggemann, M. Riede, A. Gombert, K. Leo, *Phys. Status Solidi A*, vol. 205, 2008, pp. 2862–2874.

[3] B. Lipovšek, J. Krč, and M. Topič, *IEEE J. Photovolt.*, vol. 4, 2014, pp. 639–646.

[4] R. Santbergen, R. Mishima, T. Meguro, M. Hino, H. Uzu, J. Blanker, K. Yamamoto, M. Zeman, *Opt. Express*, vol. 24, 2016, pp. A1290.

[5] M. Jošt, S. Albrecht, L. Kegelmann, C. M. Wolff, F. Lang, B. Lipovšek, J. Krč, L. Korte, D. Neher, B. Rech, M. Topič, *ACS Photonics*, vol. 4, 2017, pp.1232–1239

[6] M. Topič, M. Sever, B. Lipovšek, A. Čampa, J. Krč, *Sol. Energy Mater. Sol. Cells*, vol. 135, 2015, pp. 57–66.

[7] G. Li, L. Liu, F. Wei, S. Xia, X. Qian, *IEEE J. Photovolt.*, vol. 2, 2012, pp. 320–340.

[8] F. Guo, P. Kubis, T. Stubhan, N. Li, D. Baran, T. Przybilla, E. Spiecker, K. Forberich, C. J. Brabec, *ACS Appl. Mater. Interfaces*, vol. 6, 2014, pp. 18251–18257.

[9] B. Lipovšek, J. Krč, and M. Topič, *Inf. MIDEM - J. Microelectron. Electron. Compon. Mater.*, vol. 41, 2011, pp. 264–271.

[10] M. Kovačič, P.-A. Will, B. Lipovšek, J. Krč, S. Lenk, S. Reineke, M. Topič, *Inf. MIDEM - J. Microelectron. Electron. Compon. Mater.*, vol. 46, 2016, pp. 267–275.

[11] http://lpvo.fe.uni-lj.si/en/software/crowm/

Identification of Degradation Pathways of Organic Solar Cells using Infrared Spectroscopy

S. Shah[a,b], R Biswas[a,b,c], T. Koschny[b], and V L Dalal[a,b]

a) Microelectronics Research Center, and Department of Electrical Engineering, Iowa State University, Ames, IA 50011-3060
b) Ames Laboratory, Ames –Iowa 5001-3060
c) Dept. of Physics, Iowa State University, Ames, IA 50011-3060

Abstract - The degradation of organic solar cells is one of the most urgent problems facing further scientific and commercial development of organic electronics. Degradation can occur in the presence of light exposure together with external oxygen and moisture. We utilize infrared (IR) spectroscopy to identify IR active vibrational modes and the atomistic changes occurring during degradation of organic solar cell films, before and after degradation. We find measurable changes when light exposure is performed in the presence of oxygen or an ambient environment. The low band gap PTB7-PCBM blend and PTB7 films display significant increases of increased absorption at 1727 cm^{-1} attributable to increased C=O modes in conjunction with a broad increase at 3240 cm$^-$ attributed to hydroxyl (OH) groups within polymer. Ab-initio modeling indicates that this can be explained by an oxidation of the PTB7 polymer at the α-C site and a irreversible cleaving of the polymer. Light induced degradation performed in the absence of oxygen/moisture do not lead to large changes in the IR active modes. P3HT-PCBM blends do demonstrate small changes around 2500 cm^{-1} after light soaking, that may be connected to local H-motion induced rearrangements. Films exposed to the ambient atmosphere in the dark do not show IR active changes, identifying photo-excited singlet oxygen to be detrimental. The identification of light induced changes in atomic bonding configurations can open up pathways to stabilizing organic solar cells.

A. INTRODUCTION

Organic photo-voltaics (OPVs) have great promise as a platform for simple, low cost, flexible solar cells. One of the primary roadblocks preventing commercial deployment of OPVs is the instability of the OPVs during light exposure. Stability issues also plague the newly discovered perovskite materials. It is of great interest to assess whether photo-degradation can be reversible or irreversible.

Many studies have reported the photo-degradation of organic solar cells, including the loss of short circuit current (Jsc), open circuit voltage (Voc), fill factor (FF) and efficiency (η) after light soaking in inert ambient environments [1,2]. Measurements of mid-gap states in films found distinct increases in mid-gap densities of states with light and UV exposure, with a hint of stabilization at long time [3]. The oxidation of conjugated vinylene groups has been well studied [1,4] and detailed measurements have identified the diffusion of metal from the cathode, as well as In from the the ITO contact, as potential sources of degradation. Hole-transport

layers such as PEDOT:PSS and the various interfaces in OSC's have been suggested to cause instability [1,4].

To avert the complexity in dealing with the various interfaces, and electron- and hole-transport layers in OSCs, we focus on the stability of the intrinsic absorber layer *alone* in this paper. Changes in device architecture such as use of inverted OScs cannot alter the behavior of the absorber layer. In our approach we deposit organic absorber layers on standard substrates using identical processing conditions used in the fabrication of solar cells. Thus for done-acceptor blends we expect the same bulk hetero-structure in films that were present in devices. There have been very few studies for understanding the atomistic changes occurring after light-degradation. We utilize infrared spectroscopy as the technique for identifying changes in atomic bonding species after light-induced degradation.

B. EXPERIMENTAL DETAILS

We spin coat organic films on salt (NaCl) substrates, which are transparent within the large window from 4000 to 650cm^{-1}. In contrast standard silicon wafers transmit less than ~60% in the mid-IR and suffer from Fabry-Perot fringes that compromise sensitivity. We chose for our studies i) the low band gap polymer PTB7 that is frequently used as a donor in higher efficiency OSCs with efficiency exceeding 8%, ii) P3HT the most common prototypical done used in OSCs with 3-4% efficiency iii) The common acceptor PCBM60 and iv) donor-acceptor blends of PTB7:PCBM60 and P3HT:PCBM60.

Synthesis of PTB7:PCBM60 blend films were prepared with a solution of PTB7 (1-Material, 10 mg) and PCBM60 (Nano-C, 15 mg) in dichlorobenzene solvent (Sigma Aldrich, 0.97 mL) and 1,8-Diiodooctane (DIO) additive (Sigma Aldrich, 0.03 mL (i.e. 3%) in a glove box that was stirred for 12 hours at 75 °C and 300 rpm. The blend was spin-coated on NaCl substrates at 600 rpm for 60s. The sample was then dried at 100 °C for 30 minutes on a hot-plate, to achieve a film thickness of ~100nm nm, typical of absorber layers in OSCs. Spin-coating and drying were performed in a nitrogen filled glove box where H_2O and O_2 levels were a few ppm. For these deposition conditions the film thickness was ~100 nm, typical of the absorber layer in PTB7 solar cells.

Light soaking was performed with an ABET solar simulator capable of 1x to 4x solar-intensity variation. The light soaking was performed inside an environmental chamber with i) an inert

978-1-5090-5606-4/17 $31.00 © 2017 IEEE

N_2 flow ii) dry air containing a mixture of 70:30 N_2/O_2 and iii) ambient air atmosphere containing both O_2 and moisture.

Fourier transform infrared spectroscopy (FTIR) was carried out in vacuum with a Bruker Vertex 70v spectrometer in which the NaCl samples were placed in an evacuated sample chamber. Reference measurements were performed with i) no sample and ii) salt substrate (without any film). The transmission of pristine (as-deposited) films was measured on the same substrate. The films were then light soaked and the IR transmission re-measured utilizing the bare salt as the reference.

C. RESULTS

IR spectroscopy is a technique with a sensitivity of a percent to a fraction of a percent. IR is not expected to measure changes below such a threshold. It should be recognized that in device measurements density of electronic states changes after light soaking are in the range of 10^{17} cm^{-3}, or 1 atom in 10^5 and cannot be observed in IR measurements.

We performed IR spectroscopy of pristine and light-soaked PTB7 films, and PTB7:PCBM blends that were light soaked in N_2, ambient air and dry air in a mixture of 70:30 N_2:O_2. In addition we kept films in the dark in an ambient air atmosphere for one week in the dark and reexamined their IR absorption spectra. The largest measurable changes in the IR absorption were observed in films that were light soaked in dry air or in the ambient atmosphere. Since no measurable changes were observed for films in the dark we concluded that oxygen or moisture alone does not cause degradation. Instead it is necessary to have a combination of oxygen and light. The likely interpretation of this is that light excites triplet oxygen molecules to its excited singlet state (O_2*), which in turn can oxidize or react with organic species. Table 1 qualitatively summarizes these findings

Table 1: General observations with IR spectroscopy

Procedure	Effect
Light soak in N_2	Small IR changes
Light soak in N_2:O_2	IR absorption changes
Light soak in ambient air	IR absorption changes
Exposed to ambient in dark	No significant change

The significant measurements (Fig. 1) show the IR absorption spectrum for PTB7 pristine film compared with the film exposed in 2x solar intensity for 24 hours. Both films display the high frequency asymmetric and symmetric CH$_2$ stretch vibration modes at 2960 and 2933 cm^{-1}, accompanied by the CH stretch from the conjugated back bone at 2867 cm^{-1}. There is a new broad absorption near 3240 cm^{-1} for the light degraded case. There are few species that can contribute to such a high frequency mode. The only viable species is the hydroxyl O-H group, with the OH group bonded within the organic backbone [5], since free water (H_2O) has OH vibrations at higher frequency (3500-3700 cm^{-1}). Although NH vibrations also lie

in this range, N is and amine groups are not present in the organic matrix and can be ruled out.

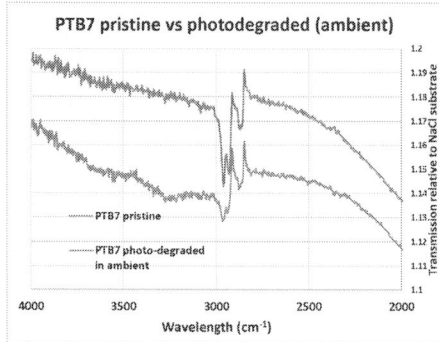

Fig. 1. Measured high-frequency infrared (IR) transmission spectra of PTB7 film showing the pristine as-deposited and photo-degraded film in ambient air atmosphere.

The low frequency IR absorption (Fig. 2) is considerably more complex, having the principal features of C=O stretch (1733 cm^{-1}), CH$_3$ asymmetric bend (1456 cm^{-1}), CH$_2$ scissor bend (cm^{-1}), C-O-C stretch (cm^{-1}). CH$_2$ wag (1153 cm^{-1}). In the film photodegraded in ambient, many of the lower frequency features are bleached away. The noticeable change is the *strengthening* of the 1727 cm^{-1} mode, which we interpret as increased C=O bond density.

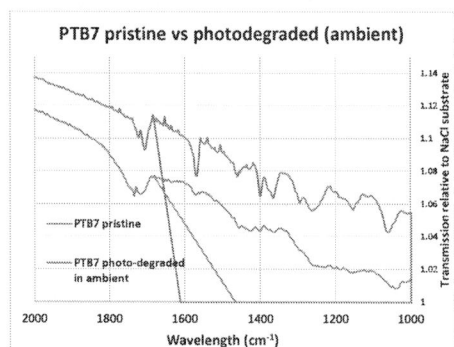

Fig 2. Measured low-frequency infrared (IR) transmission spectra of PTB7 blend films showing the pristine as-deposited and photo-degraded film in ambient air atmosphere.

In contrast we light soaked PTB7 films in an inert N_2 atmosphere (Fig. 3) and found only minor changes in the IR spectrum over all frequency ranges. This argues for the presence of light *and* oxygen to be important for such degradation effects.

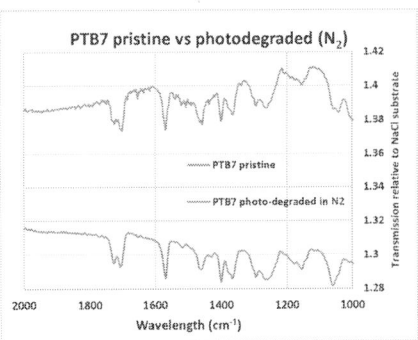

Fig. 3. Measured low-frequency infrared (IR) transmission spectra of PTB7 films showing the pristine as-deposited and photo-degraded film in an inert N_2 atmosphere.

The microscopic interpretation of these changes is based on our ab-initio simulations [6] of the PTB7 polymer. PTB7 consists of an aromatic backbone attached to polymer side-chain through a bridging O atom (Fig. 4). It is well known that the most reactive C site is a-C, the first C in the polymer chain. Our ab-initio simulations identify external O_2 as reacting with a-C forming a O=αC-O-H species. This increases the C=O stretch and creates a broad OH based absorption near 3240 cm^{-1}. The PTB7 polymer cleaves at the αC site – which is likely an irreversible change. Polymers with such a bridging O atom can exhibit such instability.

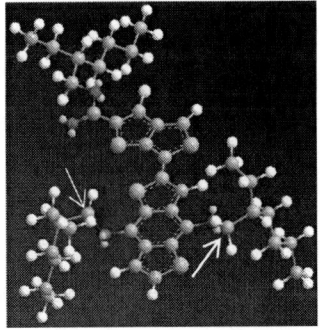

Fig. 4 Atomic structure of the PTB7 polymer with the αC site marked.

We did observe an interesting but small change in the IR absorption of P3HT:PCBM (Fig. 5). Since there is no bridging O there are fewer modes which do include the CH_2 stretch 2800-2900 cm^{-1} and CH_2 bend modes below 1800 cm^{-1}. After light soaking in N_2, there is a small increase in absorption at 2400 cm^{-1}, with negligible changes in other portions of the spectrum.

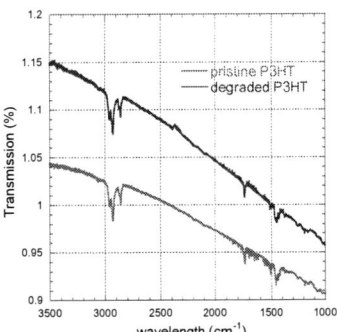

Fig. 5 IR transmission spectra of P3HT:PCBM in the pristine state and light soaked in N_2.

It is likely that the feature at 2400 cm^{-1} may be caused by motion of H from the αC to within the thiophene ring, as suggested in earlier ab-initio simulations by Northrup et al [7] and also examined by us [8].

It is of great interest to characterize the minimum wavelength i.e. the minimum energy of the photons responsible for such degradation effects. Previous theoretical studies [8] place it in the blue wavelengths and future work will quantitatively explore this.

D. CONCLUSION

We have utilized infrared spectroscopy to identify atomic bonding changes occurring after light-degradation of organic absorber films. The largest oxidative changes are observed after light-soaking in ambient air or ldry air and involve new C=O and O-H modes in the material. Subtle changes in H bonding may be observed in light-degradation of P3HT-PCBM.

ACKNOWLEDGEMENT

This work was supported in part (RB, TK) by the U.S. Department of Energy (DOE), Office of Science, basic Energy Sciences, Division of Materials Science and Engineering. Ames Laboratory is operated for the U.S. DOE by Iowa State University under contract No. AC02-07CH1135. This work was also supported in part (SS and VLD) by the National Science Foundation under grant CBET-1336134 and (SS) CMMI-1265844. We acknowledge use of computational resources at the National Energy Research Scientific Computing Center (NERSC) which is supported by the Office of Science of the USDOE under Contract No. DE-AC0205CH11231.

978-1-5090-5606-4/17 $31.00 © 2017 IEEE

REFERENCES

[1] Jorgensen, M.; Norman, K.; F. Krebs, F.; Stability/Degradation of Polymer Solar Cells, *Solar Materials and Solar Cells 2008, 92, 686-714.*

[2] Bhattacharya, J.; Mayer, R. W.; Samiee, M.; and Dalal, V. L., Photo-induced Changes in Fundamental Properties of Polymer Solar Cells, *Appl. Phys. Lett. 2012, 100, 193501.*

[3] Street, R. A.; Krakaris, A.; and Cowan, S. R, Recombination through Different Types of Localized States in Organic Solar Cells, *Advanced Functional Materials, 2012, 22, 4608-4619.*

[4] Jorgensen, M.; Norman, K.; Gevorgyan, S.; Tromholt, T.; Andreasen, B.; Krebs, F.; Stability of Polymer Solar Cells, *Advanced Materials, 2010, 24, 580-612.*

[5] Barbara H. Stuart, Infrared Spectroscopy: Fundamentals and Applications, (John Wiley and Sons, 2004).

[6] S. Shah, R. Biswas, T. Koschny, and V. Dalal, to be published.

[7] Northrup, J. Radiation Induced Hydrogen Rearrangement in Poly(3-alkylthiophene), *Appl Phys Express 6 (2013) 121601.*

[8] S. Shah, and R. Biswas, Fundamental atomic mechanisms underlying intrinsic degradation in organic solar cell materials, *J Physical Chemistry C, 119 (35), 20265–20271 (2015).*

A Device-Independent Screening Technique for Rapidly Identifying Next Generation OPV Materials

Bryon W. Larson, Andrew J. Ferguson, Bertrand J. Tremolet de Villers, Ross E. Larsen

National Renewable Energy Laboratory, Golden, CO, 80401, USA

Abstract — In this paper, we introduce a methodology that enables rapid down selection of promising new OPV polymers, based on a device-independent screening metric that takes into account intrinsic charge generation and transport, as well as charge carrier lifetimes. As a demonstration of our method, we present a case study using a set of well-known high performance polymers based on the benzodithiophene (BDT) moiety: PTB7, PTB7-Th, PBDTTT-CT, and PBDTTPD. Compared to current device-optimization based methods, we are able to identify the champion material in less time, taking only hours rather than months and may even be able to identify promising candidate materials that have been abandoned for device optimization.

Index Terms — charge carrier lifetime, materials screening, microwave conductivity, mobility, organic photovoltaics, polymers, solar cells.

I. INTRODUCTION

Organic semiconducting polymers continue to be the most versatile class of organic semiconductors in the field of organic photovoltaics (OPV), and their fundamental and applied study has helped paved the way for other solution-processed thin film PV technologies. This fact is owed to the incredible diversity in material selection and properties that polymers offer, thanks to the creative efforts of theorists and synthetic chemists alike since the Nobel Prize was awarded in 2000 for the groundbreaking work in semiconducting polymer chemistry and physics. Since then, countless polymers have been synthesized featuring new chromophores, solubilizing side chains, electronics-by-design backbones, structural elements that promote intended bulk ordering, and all combinations thereof.[1]-[3] These efforts clearly exemplify OPV's greatest advantage over any other thin-film solar cell technology: through chemistry, unlimited possibilities of material design and property can be accessed to fulfill specific applied needs. However, this great advantage also poses an enormous challenge for the discovery of new champion PV materials among the myriad options, much like the idiom of finding a needle in a haystack. Despite this materials discovery challenge, it is becoming more and more common to see double-digit lab scale device efficiencies, and maybe more impressively, based on a wide variety of diverse material types and combinations.[4] Arguably, advances in device engineering best practices have played a significant role in the recent breakthroughs in the field. For example, one single active layer composition can be processed in ways such that the device power conversion efficiency (PCE) could range between 1% and 10%, even though the two ingredients (i.e. – polymer donor and fullerene acceptor) remain identical.[5] The begging question then is how can we be sure of whether or not a new material should be considered promising when the most common method by which we evaluate those materials, making devices, is so variable?

Currently, a typical protocol for identifying new OPV donor candidates is to measure UV-vis absorbance spectra, perform electrochemical measurements such as cyclic voltammetry to ensure energetic compatibility with the acceptor, and then build a solar cell. A problem with this approach is the leap from basic characterization of the new material to building a multi-layered, extremely complex bulk-heterojunction composite solar cell just to evaluate the worth of that new material. During the solar cell building phase alone, there is an extensive checklist of tactics used to speed up the "device optimization" phase, such as: solvent choice, solvent additives, processing temperature and blade/spin speed, use of non-innocent ternary additives, post film-deposition treatments, nucleation agents, charge selective blocking layers, and finally choice of electrode composition and architecture. To make matters more complicated, new best practices in solar cell design and processing are being uncovered constantly. In the literature, we see that the vast majority of research time and money are spent on the device optimization phase, with basically no new materials being deemed as interesting (measured by publications) unless they have produced a high-PCE result. The strategy to produce high PCE results has changed recently. Jackson and co-authors pointed out that in the past decade many efforts intended to boost PCE specifically through energy level engineering, however, as they showed, there was no direct correlation between the boosts in efficiency and increases in open circuit voltage (V_{OC}).[6] Instead, the efficiency gains correlated better with short-circuit current (J_{SC}) and fill factor (FF), which are more intimately tied into the generation yield of long-lived charges and their collection efficiency in a device. Clearly there is a benefit to exploring and better understanding the distinction between *extrinsic* factors governing OPV device efficiency, such as device engineering best practices, and the properties *intrinsically* pinned to the active layer composition, such as photon absorption, excited-state management, and local charge mobility.[7] If bulk measurements, such as device performance, could be correlated to the appropriate local properties of an OPV active layer, then the costs in time and money of identifying new technologically important materials could be greatly reduced.

Here, we present a case study of four popular polymers that all contain the commonly employed backbone unit benzodithiophene – BDT).[8] By comparing results from the vast device engineering literature database to our time-resolved microwave conductivity (TRMC) measurements, we gain insight into the impact of intrinsic charge generation and local mobility on bulk solar cell performance. The polymers used in this study (PBDTTPD, PBDTTT-CT, PTB7, and PTB7-Th), whose structures are shown in Figure 1, provide an excellent training set for the screening method proposed herein, since they collectively account for more than 1100 publications in the past seven years, and therefore can reasonably be assigned as more or less optimized, giving a set of benchmark PCEs to compare with our spectroscopy results. We show good agreement between the device results obtained after years of optimization and the rapid TRMC figure of merit obtained for these four structurally similar polymers.

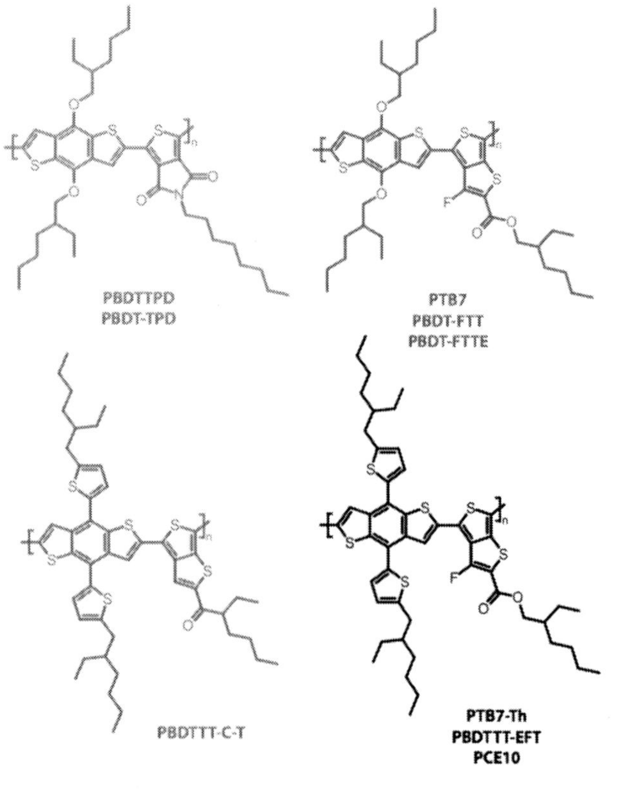

Fig. 1. The structures and common names are shown for the four polymers studied in this work. Each structure is consistently color coded throughout this work to guide the eye. Common alternative names appearing in the literature for the materials are given.

II. EXPERIMENTAL

All OPV materials were purchased from 1-Material, solvents were purchased from Sigma-Aldrich, and used as received.

A. Time-resolved microwave conductivity

TRMC is a pump-probe technique that can be used to measure the photoconductance of a film without the need for charge collection at electrical contacts.[9], [10] The details of

the experimental methodology have been presented elsewhere.[10], [11] In brief, the sample is placed in a microwave cavity at the end of an X-band waveguide operating at ca. 9 GHz, and is photoexcited through a microwave reflector with a 5 ns laser pulse from an OPO pumped by the third harmonic of an Nd:YAG laser. The relative change of the microwave power, P, in the cavity, due to absorption of the microwaves by the photoinduced free electrons and holes, is related to the transient photoconductance, ΔG, by $\Delta P/P = -K\Delta G$, where the calibration factor, K, is experimentally determined individually for each sample. Taking into account that the electrons and holes are generated in pairs, the peak photoconductance during the laser pulse can be expressed as:[10]

$$\Delta G = \beta q_e I_0 F_A (\phi \Sigma \mu) \qquad (1)$$

where q_e is the elementary charge, $\beta = 2.2$ is the geometric factor for the X-band waveguide used, I_0 is the incident photon flux, F_A the fraction of light absorbed at the excitation wavelength, ϕ is the quantum efficiency of free carrier generation per photon absorbed and $\Sigma \mu$ the sum of the mobilities of electrons and holes.[10] Eq. 1 is used to evaluate the Y×M product ($\phi \Sigma \mu$), the quantum efficiency of free carrier generation per photon absorbed multiplied by the local mobility of free carriers, which is related to the photoconductivity of the sample. All films were prepared on quartz according to published optimized or non-optimal processing conditions. Samples containing DIO were pumped on for 1 hour at 10^{-7} torr to remove residual DIO and simulate the processing conditions that the device active layers were exposed to during metal contact evaporation.

B. Device fabrication and testing

ITO patterned glass substrates were cleaned by scrubbing with a Sonicare[TM] toothbrush using a solution of Hellmanex III[TM] in deionized water. After a deionized water rinse and N_2 flush dry, the substrates were subjected to 7 minutes in an ultrasonic bath of acetone, flushed dry with N_2, and then the ITO side was treated with 5 minutes exposure to UV in an ozone chamber. A ZnO precursor solution of 1:3 vol:vol diethyl zinc in toluene:THF was prepared in an inert atmosphere glovebox and then used immediately to prepare the electron transport layer by spinning onto the ITO in air at 5000 rpm for 45 seconds. After baking for 20 minutes at 120 °C, the substrates were transferred to a dry N_2 atmosphere glove box for active layer deposition by spin casting, using the following processing recipes: PBDTTPD = 1:1.5 w:w ratio polymer:PC$_{70}$BM, CHCl$_3$ solvent, 15 mg/mL total concentration, 6000 rpm spin for 20 seconds at r.t., and 3% by volume DIO for additive samples, PBDTTT-CT, PTB7, PTB7-Th = 1:1.5 w:w ratio, 700 rpm from chlorobenzene for 60 seconds, 3% by volume DIO additive, 8 mg/mL polymer concentration. Devices were completed with a 10 nm layer of MoOx and a 120 nm thick layer of Ag metal. All devices were tested using an Oriel Sol3A AM 1.5 solar simulator, calibrated to an NREL-certified Si photodiode, housed within an O_2 and H_2O-free nitrogen atmosphere glovebox. UV-vis absorbance spectra were recorded using a Cary 5000i spectrometer.

III. RESULTS AND DISCUSSION

Figure 2 shows a correlation between the yield mobility products, $\phi\Sigma\mu = Y\times M$, for either the neat polymer (red triangles) or blend films (blue squares) and the current highest reported PCE in the literature for each of the four polymers. The blue data points (corresponding to the left axis) represent the peak signal at the lowest excitation intensity measured for each blend film, and to reiterate, all polymer:fullerene blends were prepared to recreate those reported in the literature by following the published recipe for the highest reported efficiency. For the blends, the best PCE's reported to date are: PBDTTPD = 8.3%[12], PBDTTT-CT = 9.13%[13], PTB7 = 10.02%[14], and PTB7-Th = 10.95%[15]. For the three BDT-TT motif polymers, the measured Y×M also follows the PCE trend observed by device engineers with progressively increasing values of: PBDTTT-CT = 1.3 x 10^{-2} cm²/Vs, PTB7 = 2.6 x 10^{-2} cm²/Vs, and PTB7-Th = 4.6 x 10^{-2} cm²/Vs. For PBDTTPD, the highest reported PCE value is 8.3% yet the Y×M value is 4.6 x 10^{-2} cm²/Vs, the joint-highest of the four blends. One interpretation of this outlier in the trend is that PBDTTPD was discovered and subsequently produced by commercial entities *before* high(er) PCE's were reported for the other three polymers. As such, the number of publications containing device work for PBDTTPD fell concomitantly as the PCE rose for the other three "more promising" polymers.

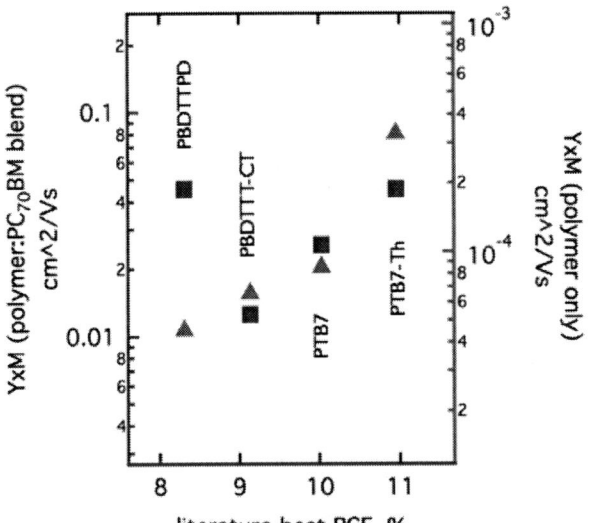

Fig. 2. The yield mobility product measured by TRMC for either pristine polymer (right axis, red) or blend films with PC$_{70}$BM (left axis, blue) is plotted versus the best-reported literature power conversion efficiency.

It would follow that new best-practice device engineering methods responsible for pushing champion PCE's higher were then applied more frequently to the BDT-TT polymers than PBDTTPD. This interpretation is supported by the historical publication record for these polymers; the results of a Sci-Finder® search performed on June 6, 2017 are shown in Figure 3 and show that PTB7 and PTB7-Th have been given substantially more attention than PBDTTPD, again, even though PBDTTPD and PTB7 debuted around the same time. Understandably, the higher PCE polymer at the time simply gained more attention. Another contributing factor to the slightly lower PCE compared to the Y×M value for PBDTTPD could be the related to the frontier molecular orbital energetics. Instead of a modified TT unit linked to the BDT, this polymer features a slightly more electron withdrawing TPD unit, giving rise to a slightly wider bandgap and a lower ionization potential compared to the other three polymers.[16] With that said, our results suggest that the photocurrents (related to Y×M) that could be achievable with PBDTTPD are likely significantly larger than has been achieved thus far, and therefore PBDTTPD may yet have unrealized benchmark potential by PCE standards.

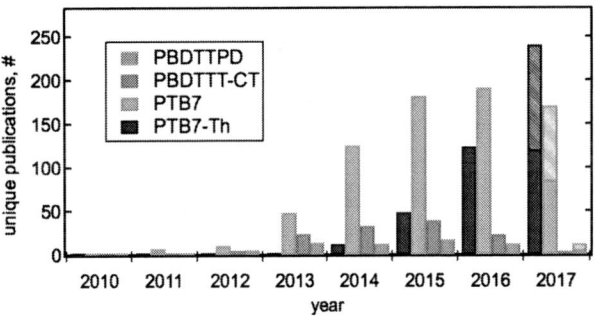

Fig. 3. The results of a literature search performed June 6, 2017 using SciFinder® are shown for each polymer by number of publications per year between 2010 and 2017. The hashed bars for 2017 represent a full year based on doubling the amount of publications in the first six months.

As outlined in the introduction, it is well known that the processing conditions used to form the active layer have a dramatic effect on the performance of the completed device.[17], [18] Processing is often so critical that even "identical" conditions used across different labs can give rise to significantly different device efficiencies. This fact suggests that the early PCEs obtained for devices with new polymers are susceptible to the vagaries of process optimization, meaning that poor performance may have been incorrectly attributed to intrinsically poor properties of the polymer rather than the unsuitability of the chosen device processing conditions. Therefore, we investigated whether our microwave conductivity data is equally as sensitive to processing as the performance of the completed device stack. The data shown in Table 1 summarizes device figures of merit for active layers prepared either with or without DIO (an additive that is known to improve the device performance in each of these active layer blends), compared to the Y×M values obtained for TRMC films that were prepared side by side under identical processing conditions as the device films. In our lab, we were able to reproduce the general differences in solar cell performance of blends with and without DIO (that is, the DIO blends yielded higher PCEs) even though our results did not match exactly the highest reported literature values. Between the DIO and non-DIO devices, the main differences in efficiency come from J_{SC} and FF, also consistent with the literature.[19]

Table 1. Summary of Device and TRMC Results

Sample	Device Performance				TRMC results
	V_{OC} (V)	J_{SC} (mA/cm^2)	FF (%)	PCE (%)	Y×M (cm^2/Vs) x 10^{-2}
PBDTTPD	0.88	11.3	47.9	4.8	5.9
w/DIO	0.86	10.8	60.0	5.5	4.6
PBDTTT-CT	0.64	8.6	46.9	2.7	1.0
w/DIO	0.74	11.2	53.8	4.5	1.3
PTB7	0.73	10.6	46.5	3.6	1.9
w/DIO	0.74	13.1	64.2	6.2	2.5
PTB7-Th	0.80	12.2	56.6	5.5	5.5
w/DIO	0.79	13.3	60.8	6.4	4.6

Unlike the device data, the TRMC results show little to no difference between films prepared with or without DIO; the Y×M values are essentially constant between the two different processing methods. As a reminder, the TRMC signal comes from two contributors: (i) the generation yield of device-relevant charges (on the order of tens to hundreds of nanoseconds), and (ii) the sum of the local mobilities of those charges. We and others have shown that solar cell performance in OPVs can be improved, namely through increases in J_{SC} and FF, by either incorporating molecular design considerations or device processing practices that ideally maximize the yield of highly mobile and long-lived charge carriers. Since TRMC is a local measure (devices are a bulk measure) of charge generation and mobility, it is not surprising that DIO (a bulk processing additive) impacts the device results and not the TRMC results. As such, our TRMC technique represents an excellent methodology to evaluate the potential of novel organic semiconducting polymers for OPV device applications, since it can report on the yield and mobility of carriers, without the need to optimize overall film morphology (i.e., nanostructure and microstructure) or energetics of carrier-selective contacts to maximize carrier collection in a complete device.

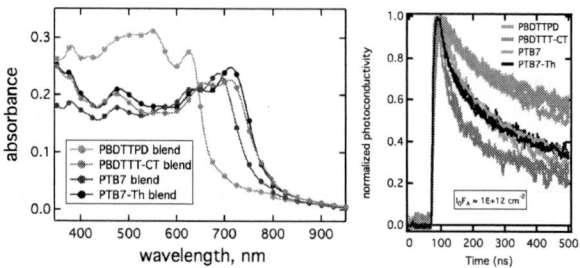

Fig. 4. Absorbance spectra are shown for the active layer blends (with PC$_{70}$BM) for literature-optimized recipes in the left panel. In the right panel, normalized photoconductivity transients over 500 ns are shown for a constant photon flux (based on number of absorbed excitation photons) for each sample. Additional transients are shown for multiple processing conditions.

In addition to ranking each polymer based on the Y×M value, TRMC also provides temporal information that may be used to predict some aspects of the optimized device. The left side of Figure 4 shows the UV-vis absorbance spectrum for each of the literature best recipe active layers; on the right hand side the corresponding photoconductivity transients over 500 ns are plotted. Interestingly, the blend exhibiting the longest-lived photoconductivity, PBDTTPD (orange lines), also has nominally the thickest optimized device active layer, while the PBDTTT-CT blend exhibits the shortest-lived photoconductivity signal and requires a very thin (ca. 70 nm) active layer to optimize device performance. These data suggest that an active layer with both a high Y×M value as well as long lived photoconductivity lifetimes would be a better candidate for extensive device optimization (i.e., process engineering), since it would have an intrinsically high yield of long-lived charges, high local charge mobility, and likely as a result, would be able to generate more current in a device by accommodating a thicker active layer.

IV. CONCLUSIONS AND OUTLOOK

We have used a series of four established OPV polymers in this case study, testing the ability of microwave conductivity measurements to predict which polymer would be the most promising for OPV. In doing so, we find that the optimized PBDTTPD:PC$_{70}$BM blend exhibits the longest lived free carrier signal of the series and a Y×M value as high as the top performing device blend, PTB7-Th (PCE of almost 11%), but comparatively has been studied the least (*vide supra*). For the three BDT-TT polymers that have been most thoroughly studied by groups around the world, there is an excellent agreement between the best reported literature PCE and the TRMC data in terms of ranking them from worst to best (PBDTTT-CT<PTB7<PTB7-Th). The accurate retroactive prediction presented in this case study may translate to predicting the potential of new polymers in the future given at least the TRMC metric baselines established here.

It is evident that the OPV field has taken great steps forward, especially in the past few years, where we now see a great diversity in polymers that produce device efficiencies ≥ 10%, yet still the question looms what is the next big breakthrough for OPV?[6] We believe that a change in the way new materials are screened (through *intrinsic* figures of merit instead of *extrinsic* evaluation) can be the catalyst, and that microwave conductivity measurements appear to quickly identify these key figures of merit. In doing so, the burden of unlimited synthetic tunability that OPV materials offer transforms into a valuable asset. Ultimately, a rapid and device-independent method for screening OPV materials will enable device engineers to prioritize identifying and applying new cutting edge best processing practices to only the most intrinsically promising materials.

ACKNOWLEDGEMENT

The work here was supported by the U.S. Department of Energy under Contract No. DE-AC36-08GO28308 with the National Renewable Energy Laboratory through the DOE Solar Energy Technologies Program.

REFERENCES

[1] A. Facchetti, "π-Conjugated Polymers for Organic Electronics and Photovoltaic Cell Applications †," *Chem. Mater.*, vol. 23, no. 3, pp. 733–758, Feb. 2011.

[2] I. Etxebarria, J. Ajuria, and R. Pacios, "Solution-processable polymeric solar cells: A review on materials, strategies and cell architectures to overcome 10%," *Organic Electronics*, vol. 19, pp. 34–60, Apr. 2015.

[3] H. Spanggaard and F. C. Krebs, "A brief history of the development of organic and polymeric photovoltaics," *Solar Energy Materials and Solar Cells*, vol. 83, no. 2, pp. 125–146, Jun. 2004.

[4] Z. Hu, L. Ying, F. Huang, and Y. Cao, "Towards a bright future: polymer solar cells with power conversion efficiencies over 10%," *Sci. China Chem.*, vol. 60, no. 5, pp. 571–582, Mar. 2017.

[5] G. Luo, X. Ren, S. Zhang, H. Wu, W. C. H. Choy, Z. He, and Y. Cao, "Recent Advances in Organic Photovoltaics: Device Structure and Optical Engineering Optimization on the Nanoscale.," *Small*, vol. 12, no. 12, pp. 1547–1571, Mar. 2016.

[6] N. E. Jackson, B. M. Savoie, T. J. Marks, L. X. Chen, and M. A. Ratner, "The Next Breakthrough for Organic Photovoltaics?," *J. Phys. Chem. Lett.*, vol. 6, no. 1, pp. 77–84, Dec. 2014.

[7] M. P. Griffin, R. Gearba, K. J. Stevenson, D. A. Vanden Bout, and A. Dolocan, "Revealing the Chemistry and Morphology of Buried Donor/Acceptor Interfaces in Organic Photovoltaics.," *J. Phys. Chem. Lett.*, pp. 2764–2773, Jun. 2017.

[8] H. Yao, L. Ye, H. Zhang, S. Li, S. Zhang, and J. Hou, "Molecular Design of Benzodithiophene-Based Organic Photovoltaic Materials," *Chem. Rev.*, vol. 116, no. 12, pp. 7397–7457, Jun. 2016.

[9] Jessica E Kroeze, Tom J Savenije, A. Martien J W Vermeulen, and J. M. Warman, *Contactless Determination of the Photoconductivity Action Spectrum, Exciton Diffusion Length, and Charge Separation Efficiency in Polythiophene-Sensitized TiO2 Bilayers*, vol. 107, no. 31. American Chemical Society, 2003, pp. 7696–7705.

[10] A. J. Ferguson, N. Kopidakis, S. E. Shaheen, and G. Rumbles, "Dark Carriers, Trapping, and Activation Control of Carrier Recombination in Neat P3HT and P3HT:PCBM Blends," *J. Phys. Chem. C*, vol. 115, no. 46, pp. 23134–23148, Nov. 2011.

[11] S. Dayal, N. Kopidakis, and G. Rumbles, "Photoinduced electron transfer in composites of conjugated polymers and dendrimers with branched colloidal nanoparticles.," *Faraday Discuss.*, vol. 155, pp. 323–37– discussion 349–56, 2012.

[12] J. A. Bartelt, J. D. Douglas, W. R. Mateker, A. E. Labban, C. J. Tassone, M. F. Toney, J. M. J. Fréchet, P. M. Beaujuge, and M. D. McGehee, "Controlling Solution-Phase Polymer Aggregation with Molecular Weight and Solvent Additives to Optimize Polymer-Fullerene Bulk Heterojunction Solar Cells," *Adv. Energy Mater.*, vol. 4, no. 9, p. 1301733, Jun. 2014.

[13] X. Guo, M. Zhang, W. Ma, L. Ye, S. Zhang, S. Liu, H. Ade, F. Huang, and J. Hou, "Enhanced photovoltaic performance by modulating surface composition in bulk heterojunction polymer solar cells based on PBDTTT-C-T/PC71 BM.," *Adv. Mater. Weinheim*, vol. 26, no. 24, pp. 4043–4049, Jun. 2014.

[14] X. Ouyang, R. Peng, L. Ai, X. Zhang, and Z. Ge, "Efficient polymer solar cells employing a non-conjugated small-molecule electrolyte," *Nature Photonics*, vol. 9, no. 8, pp. 520–524, Aug. 2015.

[15] J. Huang, J. H. Carpenter, C.-Z. Li, J.-S. Yu, H. Ade, and A. K. Y. Jen, "Highly Efficient Organic Solar Cells with Improved Vertical Donor-Acceptor Compositional Gradient Via an Inverted Off-Center Spinning Method.," *Adv. Mater. Weinheim*, vol. 28, no. 5, pp. 967–974, Feb. 2016.

[16] Y. Zou, A. Najari, P. Berrouard, S. Beaupré, B.-R. Aich, Y. Tao, and M. Leclerc, "A Thieno[3,4-c]pyrrole-4,6-dione-Based Copolymer for Efficient Solar Cells," *J. Am. Chem. Soc.*, vol. 132, no. 15, pp. 5330–5331, Mar. 2010.

[17] K. Wang, C. Liu, T. Meng, C. Yi, and X. Gong, "Inverted organic photovoltaic cells.," *Chem. Soc. Rev.*, vol. 45, no. 10, pp. 2937–2975, May 2016.

[18] A. K. K. Kyaw, D. H. Wang, C. Luo, Y. Cao, T.-Q. Nguyen, G. C. Bazan, and A. J. Heeger, "Effects of Solvent Additives on Morphology, Charge Generation, Transport, and Recombination in Solution-Processed Small-Molecule Solar Cells," *Adv. Energy Mater.*, vol. 4, no. 7, p. 1301469, May 2014.

[19] T. H. Lee, S. Y. Park, B. Walker, S.-J. Ko, J. Heo, H. Y. Woo, H. Choi, and J. Y. Kim, "A universal processing additive for high-performance polymer solar cells," *RSC Adv.*, vol. 7, no. 13, pp. 7476–7482, 2017.

Novel Anthanthrone and Anthanthrene Co-polymers as p-Type Conjugated Semiconductors for Organic Photovoltaics

Suru Vivian John[1,2], Patrick Denk[2], Christoph Ulbricht[2,3], Herwig Heilbrunner[2], Jean-Benoit Giguère[4], Antoine Lafleur-Lambert[4], Jean-Francois Morin[4], Emmanuel Iwuoha[1] and Daniel Ayuk Mbi Egbe[2,3]

[1]SensorLab, Department of Chemistry, University of Western Cape, Robert Sobukwe Road, P. Bag X17, Bellville, 7535, Cape Town, South Africa. [2]Linz Institute for Organic Solar Cells (LIOS), Johannes Kepler University, Altenbergerstr. 69, 4040 Linz, Austria. [3]Institute of Polymeric Materials and Testing (IPMT), Johannes Kepler University, Altenbergerstr. 69, 4040 Linz, Austria. [4]Department of Chemistry, Faculty of Science and Engineering, Pavillon Alexandre-Vachon, Local 1250b, Université Laval, Québec (Québec), G1V 0A6, Canada

Abstract — The design of advanced new generation conjugated materials with improved properties for photovoltaic applications has been the focus of research in recent years. This work reports on novel anthanthrone/anthracene and anthanthrene/anthracene co-polymers as p-type conjugated semiconductors in organic photovoltaics. The presence of anthracene component with the grafted alkyloxy chains enhanced the solubility and process-ability of the co-polymers. The photo-physical and photovoltaic parameters greatly depend on ratio of components of the co-polymers and nature of the grafted chains. Devices using the blends of SV15 and PCBM in 1:2 ratio gave J_{sc} 3.97 mA/cm^2 and high V_{oc} 0.92 V with efficiency 1.7%.

Keywords — Anthanthrones, anthracene, organic photovoltaics, semiconductors, poly(arylene ethynylene), poly(arylene vinylene)

I. INTRODUCTION

At present, polythiophene derivatives such as poly(3-hexylthiophene) (P3HT) and poly (3-octylthiophene) (P3OT) are the commonly used π-conjugated polymers for fabrication of organic photovoltaic (OPV) devices [1-2]. There is a demand for the design of advanced new generation conjugated materials in the form of homo or co-polymers with better absorption and transport properties for photovoltaic applications.

A good understanding of the relationship between chemical structure and morphology of a polymer with its optical response is key to understanding how organic electronics (photovoltaic cells, light emitting diodes and field effect transistors) works [3-4]. Light emission, output yield in light emitting diodes and efficiency of photovoltaic cells [5-6] can be tuned chemically through the linking of monomers in copolymerization or by physical mixing and subsequent co-precipitation [7]. In co-polymer synthesis, it is important to take into consideration intrinsic properties of each monomer in order to get a co-polymer with good properties for electronic devices. In this report, the synthesis, characterization and application in organic electronic devices of two polymeric systems based on polycyclic aromatic compounds (PACs),

anthanthrone/anthracene and anthanthrene/anthracene, respectively, are presented. Using various comonomer compositions eleven polymers with poly(arylene ethynylene)-*alt*-(arylene vinylene) (PAE-PAV) backbones and various side chain configurations were synthesized. Anthracene derivatives have been shown as attractive soluble polymer building blocks with high thermal and device stability [8]. The emission of anthracene-containing polymers can be tuned by the incorporation of other arylene-building blocks during the polymerization [9]. The good photoconductive behavior, high fluorescence quantum yields in thin films and high absorption coefficients around 100 000 M^{-1} cm^{-1} of anthracene-based polymers make them ideal materials for organic electronic devices. They have been investigated for and used in OLEDs [10] as well as in transistors [11] and photovoltaic cells [12]. Anthanthrones on the other hand are polycyclic aromatic compounds with extended conjugation and symmetric structures used as pigments and vat dyes. Their large conjugated planar structure facilitates intermolecular interactions with good optical response and their extended π-conjugation is believed to be an advantage in photovoltaics. However, they have been rarely studied as building blocks for organic semiconductors; except in recent reports by Morin group [13-14]. Anthanthrone and its derivative anthanthrene were incorporated into the backbone of poly(arylene ethynylene)-*alt*-poly(arylene vinylene)s (PAE-PAVs) to give P1 and P2 (Fig. 1b). The polymers show interesting optoelectronic properties with the anthanthrone-based polymer (P1) having the most intriguing absorption properties. However, the polymers show poor photovoltaic responses which is attributed to their poor solubility. To optimize device performance, the advantage offered by the solubility of the anthracene-derived units coupled with the good optical response of anthanthrone and anthanthrene building blocks were exploited to prepare copolymers (SV9-SV16) with different comonomer ratios as shown in Fig. 1b-c from monomers M1-M3, Ma-Mc (Fig. 1a). The photo-physical and photovoltaic performances of the copolymers with respect to (i) the used of constituents, anthanthrone/anthracene or anthanthrene/anthracene, (ii) the ratio of the constituents and (iii) the employed side chains were evaluated and compared to the corresponding homo-polymers P1, P2 and SV7 (Fig. 1b).

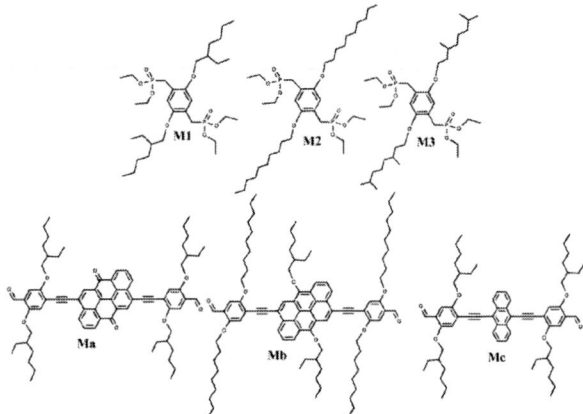

Fig. 1a.   Synthesized monomers.

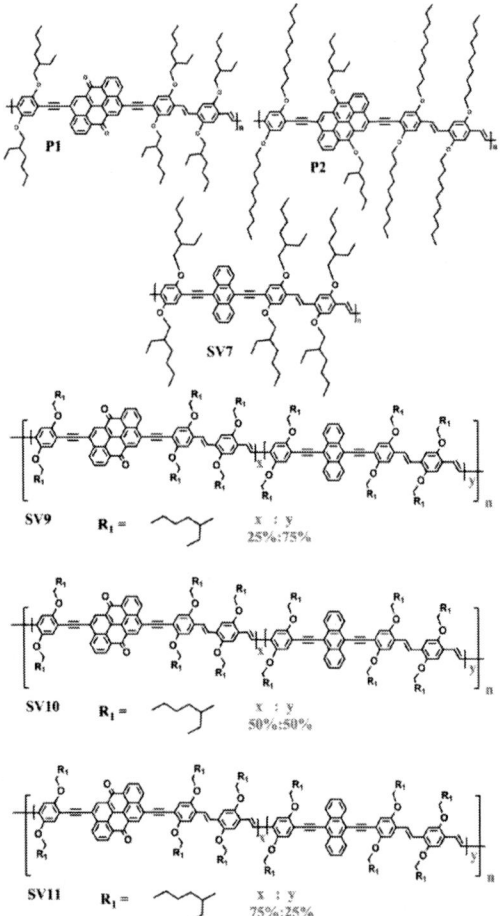

Fig. 1b.   Synthesized homo-polymers P1, P2 and SV7; and co-polymers SV9-SV11.

Fig. 1c.   Synthesized co-polymers SV12-SV16.

II. EXPERIMENTAL

All starting materials were purchased from commercial suppliers (Sigma Aldrich, Merck and Alfa Aesar). The comonomers were prepared in analogy to well-established protocols [15-17]. The polymers were synthesized via the Horner-Wadsworth-Emmons (HWE) olefination reaction of dialdehydes with bisphosphonate esters.

III. RESULTS AND DISCUSSION

A. Photo-physical Properties

The photophysical properties of the "homo" and copolymers were studied by UV-Vis absorption and photoluminescence spectroscopy in chlorobenzene solution as well as in thin film (Fig. 2a and b). All emission data were obtained by exciting at the wavelength of the main absorption band. Copolymers with

25:75, 50:50 and 75:25 anthanthrone/anthracene unit ratios (SV9-SV11) showed absorption spectra combining the absorption characteristics of the anthracene and anthanthrone polymer components in solution as well as in film.

In solution the copolymers show the absorption of the anthracene components spanning from 430 to 615 nm, and, in addition, the absorption expands up to 720 nm, which can be attributed to anthanthrone components. The maximum wavelength of absorption of the anthracene components can be located in the range between 526 and 537 nm and the maximum wavelength of absorption of anthanthrone components was observed in the range between 621 and 623 nm. While for SV10 (50:50) and SV11 (75:25) the anthanthrone absorption characteristics are quite distinct, for SV9 (25:75), the absorption in this region is faint, almost appearing flat.

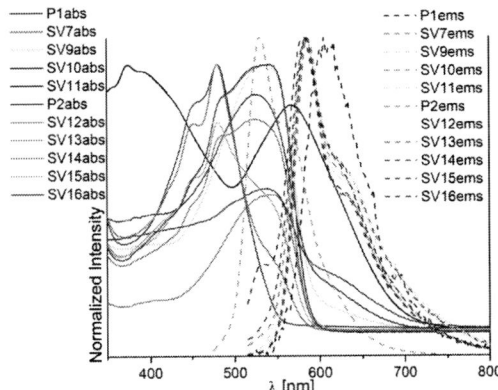

Fig. 2a.   UV-Vis (*solid lines*) and photoluminescence (*dashed lines*) spectra comparing co-polymers SV9-SV16 with homo-polymer P1, P2 and SV7 in solution.

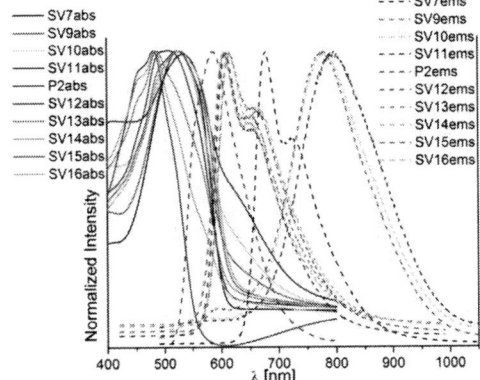

Fig. 2b.   UV-Vis (*solid lines*) and photoluminescence (*dashed lines*) spectra comparing co-polymers SV9-SV16 with homo-polymer P2 and SV7 in thin film.

In thin films however, the absorption of the copolymers including SV9, the contribution of the anthanthrone components appear significantly enhanced exhibiting a slight red shift compared to the solution spectra. The contribution from the anthanthrone components reduced the optical band

gap of the copolymers to 1.7 eV compared to 2.0 eV of SV7 (anthracene-based).

The emission characteristics of SV9-11 in solution are basically identical with the spectra of SV7, and are blue-shifted compared to P1 (anthanthrone-based). The emission peaks around 586 nm and shows a shoulder at about 635 nm. The emission of the copolymers in solution appears dominated by the anthracene components in copolymer. Interestingly, the photoluminescence (PL) spectra of the copolymers in thin film seem to show emission from the anthanthrone components exclusively. Compared to the measurements in solution the PL response of the anthanthrone components shows a strong bathochromic shift. For the SV7 (anthracene-based "homo"-polymer) thin film sample, the emission of the anthracene components appears also bathochromically shifted compared to PL response of the samples in solution showing its maximum at about 608 nm. For the anthanthrone-containing copolymers, no distinct contribution of anthracene components to the emission profile can be spotted regardless of the anthanthrone/anthracene comonomer ratio in the respective polycondensation mixture (SV9, 25:75; SV10, 50:50; SV11, 75:25). However, as the anthanthrone content in the copolymers increases, the emission appears to shift to longer wavelengths, SV9 (λ_{PLmax} = 777 nm) < SV10 (λ_{PLmax} = 781 nm) < SV11 (λ_{PLmax} = 796 nm). While the PL response of the copolymers SV9-11 observed in solution appears highly dominated by strongly emitting anthracene components, in thin film an efficient energy transfer from the anthracene to the anthanthrone components seems to occur.

The absorption and emission spectra of the anthanthrene/anthracene copolymers SV12-16 were evaluated with regard of the comonomer ratio and differences in the side chain configuration. In solution, the absorption spectra show a strongly growing high-energy component with increasing anthanthrene content in the copolymers. Copolymers SV13 and SV14 show structured absorption spectra while highly similar spectra with broad absorption bands were observed for SV12, SV15 and SV16 featuring a anthanthrene/anthracene ratio of 25:75. The main absorption for SV12, SV15 and SV16 appears as a shoulder in SV13 and SV14. This absorption band is dominantly impacted by the anthracene component as the comparison with the SV7 (anthracene-based homo-polymer) spectra reveals, and is absent for the anthanthrene-based homo-polymer P2. The shoulder in SV13 (50% anthracene component) is more pronounced than that in SV14 (25% anthracene component). The higher the anthracene component in the copolymer, the more pronounced the low energy absorption band of the anthracene component becomes All copolymers SV12-16 possess onset absorption similar to SV7 translating into optical bands of 2.1 eV for SV7, SV12, SV13 and SV14, and 2.0 eV for SV15 and SV16.

The thin film absorption follows the same pattern as that of the solution except that they appear red-shifted, which can be assigned to planarization in the solid state. The absorption of SV12-16 varies depending on the grafted side chain as well as the ratio of anthracene to anthanthrene present in the copolymer. Like in solution, the main absorption peak of SV13 and SV14 show rather narrow bands comparable to that of the anthanthrene-based homo-polymer P2 while the absorption of SV12, SV15 and SV16 show broad bands highly similar to the spectra of SV7. The absorption spectra of SV13 and SV14 show two maxima at around 452 nm and 481 nm and a shoulder at about 545 to 548 nm. SV15 and SV16, like in solution, absorb mainly in the same region as SV7. The spectra of SV15 and SV16 are very similar. The spectral response of the copolymer SV12 with the same anthanthrene/anthracene ratio (25:75) as SV15 and SV16, however, appears more affected by the anthanthrene component. This difference in the thin film absorption characteristics of SV12, SV15 and SV16 is likely to originate from the variations in the side chain compositions of these copolymers. Also in thin film the absorption onset of the copolymers is found at resembling values resulting in a band gaps between 1.9 and 2.0 eV.

Despite the variations in absorption of SV12-SV16 in solution and thin film, similar emission spectra were observed for the anthanthrene/anthracene copolymers in solution and in thin film, respectively. For the emissions in solution, a main peak at about 585 nm and a shoulder between 626 and 636 nm is observed for all polymers.

Compared to the solution measurements the PL responses in film appear red-shifted. All the polymers show well-structured vibronic emission similar in shape and in the same spectral region as the PL response of SV7, exhibiting an intense peak in the range of 604 to 615 nm and a less-intense peak at longer wavelengths (651 – 669 nm) showing the dominance of the anthracene components in the PL response. The fine structure (splitting of the emission spectra into two well resolved peaks) observed for the emission spectra both in solution and in thin film and the variation or absence of similarity (mirror image) between absorption and emission spectra, is ascribed to vibronic coupling of the excitons [18]. The emission peaks of SV13 overlap that of SV7 at 608 nm and 653 nm for the first and second peaks respectively.

Although the emission bands of SV12-16 are all sharp and the overlap of the emission and absorption spectra is quite narrow, it is important to draw attention to the slight influence of side chain variation observed for the thin film emission of SV12, SV15 and SV16 (copolymers with 75% anthracene and 25% anthanthrene but vary in the side chain of the attached phenyl bisphosphonates). It is observed that the overlapping area of the film samples slightly increases in the order SV12>SV15>SV16 due to the broadening of the absorption spectra in this order. It is also worthy of note that the emission spectra of SV14 which has 75% anthanthrene is a mirror image of the absorption. This can be ascribed to chain planarization induced by the anthanthrene component and it is an indication that the conformational relaxation is not important for both and their decays take place from lumophores with similar geometry.

The optical bandgap of SV9-16 ranged from 1.7 – 2.1 eV in solution and film. The photophysical data confirm that the structure (linear or branched side chains) and ratio of components have a significant influence on the ground and excited state properties of the copolymers both in solution and film. Very large Stokes shifts were observed for SV9 (250 nm), SV10 (256 nm) and SV11 (259 nm) in film. This indicates excimer formation or aggregation in the solid state [19]. The Stokes shift of anthanthrone or its derivative is so strong that it was impossible to prevent it in the co-polymerization with anthracene. P2, SV12-14 also show large Stokes shift (101 – 120 nm); the higher the ratio of anthanthrene in the copolymer, the higher the Stokes shift. However, those of SV9-11 are the largest for the polymers under study. The ratio of anthanthrone in the copolymer did not show any influence on the Stokes shift in solution but a huge influence was observed in film. We assume that the pronounced tendency to aggregate is due to the carbonyl group present in the anthanthrone component. The photo-physical responses including the maximum absorption (λ_{abs}), the onset absorption (λ_{onset}), optical band gap energy (E_g^{opt}) calculated using $1240/\lambda_{onset}$, wavelength at emission maximum (λ_{PL}) as well as the Stokes' shift both in solution and in film are summarized in Table S2.

B. Photovoltaic Properties

To investigate the photovoltaic performance of the polymers, solar cell devices were fabricated and tested on a regular device configuration of glass/ITO/PEDOT-PSS/polymer:PCBM/LiF/Al. The polymer:PCBM active layers were spin-coated from dichlorobenzene solutions in a 1:2 ratio for all devices. Fig. 3 shows the I-V curves of the fabricated solar cells. The co-polymers SV12-SV16 revealed generally the best photovoltaic parameters in this study.

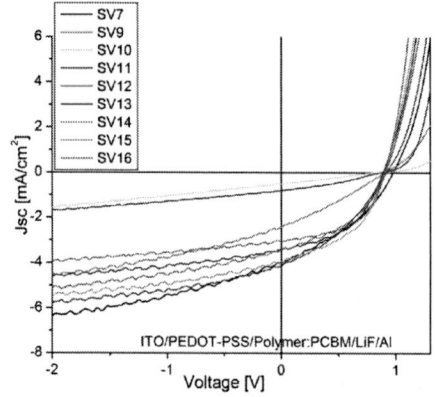

Fig. 3. The current-voltage (I–V) characteristics of (homo-polymer SV7 and co-polymers SV9-SV16):PCBM (1:2) blend devices on PEDOT:PSS configuration under 100 mW/cm² solar simulator irradiation.

For the copolymers SV12-14 with the same phenyl bisphosphonate side chain (2-ethylhexyloxy), the short circuit current (J_{sc}) was observed to decrease from SV12 toSV14 with SV12 (4.53 mA/cm^2)>SV13 (3.50 mA/cm^2)>SV14 (3.01 mA/cm^2). Reverse order was however observed for the fill factor (FF) with SV12 (0.39)<SV13 (0.47)<SV14 (0.54). Going by the photovoltaic parameters of SV7 (anthracene-based homo-polymer with 2-ethylhexyloxy side chains), the J_{sc} of the copolymers SV12-14 can be attributed to contribution of the anthracene components to the copolymer while the relatively good FF is obviously a contribution from the anthanthrene components. This assertion is further buttressed from the fact that SV14 with the highest FF is a combination of 75% anthanthrene and 25% anthracene. In addition to the influence of the ratio of each component in the copolymer on the photovoltaic performance, it can be ascertained that the photovoltaic parameters are also side chain dependent. Comparing the co-polymers SV12, SV15 and SV16 (copolymers with 75% anthracene and 25% anthanthrene) with varying side chain ((SV12; 2-ethylhexyloxy side chain phenyl bisphosphonate), (SV15; decyloxy side chain phenyl bisphosphonate) and (SV16; 3,7-dimethyloctyloxy side chain phenyl bisphosphonate)), the photovoltaic parameters are seen to be slightly influenced by the grafted side chain. For instance, the J_{sc} of SV12 (4.53 mA/cm^2) with bulky 2-ethylhexyl side chain in the phenyl group connected to the copolymer by a vinylene bond is higher than that of SV15 (4.00 mA/cm^2) with a linear decyloxy side chain in the same phenyl position as SV12, while that of SV16 (3.87 mA/cm^2) with 3,7-dimethyloctyloxy is lower than that of SV15. In this group, SV15 exhibits the highest FF (0.47) while SV12 exhibits the highest J_{sc} (4.53 mA/cm^2). The FF of SV16 (0.42) is next to that of SV15 but it gave the least J_{sc} (3.87 mA/cm^2) in this group. SV9-11 show low J_{sc} and relatively low FF. Overall the best performance was observed for SV15 with an efficiency of 1.7%. Table S3 shows the complete set of photovoltaic parameters under AM1.5 illumination for the best solar cell device of each polymer:PCBM blend.

It is obvious that the component ratio and type of side chain has no distinct influence on the open-circuit voltage (V_{oc}) of all investigated polymers. The V_{oc} of the investigated devices were found to be high for all copolymers irrespective of the type and ratio of components or nature of the grafted alkyloxy side chain. In general, high V_{oc} of over 0.9 V were obtained for all investigated polymers including the copolymers SV12-16 (copolymers containing anthanthrene) as against the less than 0.8 V reported for anthanthrene small molecule devices [20].

External quantum efficiency (EQE) measurements of the best devices were carried out to further assess the photoelectrical behavior of the polymers. The EQE reveals the percentage of photons that create charges at a given wavelength that reach the electrodes. Fig. S8a-b shows the EQE of SV7, SV9-SV16. The spectra lie over a wide range of wavelengths (400 – 750 nm) in agreement with the absorption spectra and with two maximum photo-current contributions

around 395 and 525 nm for all polymers investigated. For the maximum around 525 nm, the percentage EQE defined response is in accordance with the photovoltaic performance in terms of the short circuit current (J_{sc}). The intensity or strength of the EQE signal of each polymer clearly correlates with the achieved short-circuits currents. The higher the J_{sc}, the higher the EQE for all the polymers (Fig. S8a and Table S3). SV12 for example, show EQE higher than that of SV7 in the 525 nm range despite having the same efficiency of 1.6%. This is due to the higher J_{sc} generated by SV12 (4.53 mA/cm^2) compared to SV7 (4.15 mA/cm^2). The signals or maxima around 395 nm and the shoulder around 710 nm emanate from PCBM. The EQE response of SV9-SV11 are also in accordance with the output current with SV9>SV11>SV10 (Fig. S8b and Table S3). Between 400 and 500nm, a flat response is observed for the co-polymers (SV9-SV16) while a deep downward peak is observed for SV7. This contribution emanates from the anthanthrone and anthanthrene components in their respective co-polymers.

Going by the ratio of anthracene to anthanthrone and the solubility issues associated with anthanthrone, one would expect to have a better photovoltaic response for SV10 than SV11. Strangely, however, this was not the case. One would also assume that the ratio of the anthanthrone component in the co-polymer may not actually be accurate as a result of the solubility issues which may compromise the actual ratio; resulting in the strange photovoltaic behavior of SV10 and SV11. In order to further confirm the ratio, the NMR spectra were investigated to get an idea of the ratio of the components of the co-polymers. To achieve this, the NMR spectra of P1, SV7 and SV9-SV11 were investigated following the spectra integration (Fig. S9a-f) and equations (equation S1-4).

C. Electroluminescence (EL) Properties

A preliminary investigation of two of the co-polymers (SV12 and SV15) and their corresponding homo-polymers (P2 and SV7) for their electroluminescence behavior was conducted on an ITO/PEI/polymer/MoOx/Ag device configuration. The polymers served as emissive layers in the device. The EL spectra as depicted in Fig. S10a show similarities with those of the photoluminescence of the corresponding polymer films. This similarity can be attributed to emission from identical singlet excited states S_1 in both cases. However, the photoluminescence spectrum of P2 is blue-shifted compared to the electroluminescence spectra. From the EL spectra, we see the devices light up at very low voltages hinting a nicely low turn-on-voltage for the devices. The emissions are stable with EL spectra exhibiting the same shape over a broad voltage range; and the intensity of the emission is varied by the applied voltage. Orange color emission was observed for the co-polymers as shown in the image. Comparing the electroluminescence of co-polymers SV12 and SV15 with their corresponding homo-polymer P1 and SV7, all the polymers emit in the same wavelength region but with different shape (Fig. S10b).

IV. CONCLUSION

This report has shown the successful synthesis and investigation of novel anthanthrone/anthracene and anthanthrene/anthracene copolymers as p-type semiconductors in organic electronics. Improvement in the solubility and processability of anthanthrone- and anthanthrene-based polymers was achieved through the copolymerization with anthracene components yielding copolymers SV9-16. The photophysical behaviors of the copolymers are affected by the grafted side chains and the ratio of the incorporated components. Solar cells fabricated from these polymers were found to show quite different photovoltaic behaviors. The photovoltaic responses of anthanthrene/anthracene copolymers (SV12-16) in blends with PCBM are promising for photovoltaic applications. Although anthracene improved the solubility and processability of anthanthrone polymers, the photovoltaic properties of SV10 and SV11 are not impressive. However, the obtained photovoltaic results are based on non-optimized device configurations, which were fabricated with non-optimized process parameters. An enhancement of the photovoltaic parameters should be achieved through morphological optimization and post-production treatments of the polymer-PCBM blends.

ACKNOWLEDGEMENT

SV John is grateful to the International Centre for Theoretical Physics (ICTP) and the African Network for Solar Energy (ANSOLE), as well as the National Research Foundation (NRF) of South Africa for financial support. SVJ, CU and DAME acknowledge the funding of FWF through Project N°: I 1703-N20.

REFERENCES

[1] T. Erb, S. Raleva, U. Zhokhavets, G. Gobsch, B. Stuhn, M. Spode, and O. Ambacher, "Structural and optical properties of both pure poly(3-octylthiophene)(P3OT) and P3OT/fullerene Films," *Thin Solid Films,* vol. 450, pp. 97-100, 2004.

[2] C. N. Hoth, P. Schilinsky, S. A. Choulis, and C. J. Brabec, "Printing highly efficient organic solar cells," *Nano Lett.,* vol. 8, pp. 2806-2813, 2008.

[3] J. F. de-Deus, G. C. Faria, E. T. Iamazaki, R. M. Faria, T. D. Z. Atvars, and L. Akcelrud, "Polyfluorene based blends for white light emission," *Org. Electron.,* vol. 12, pp. 1493-1504, 2011.

[4] A. M. Assaka, B. Hu, J. Mays, E. T. Iamazaki, T. D. Z. Atvars, and L. Akcelrud, "The effect of complexation with platinum in polyfluorene derivatives: A photo- and electro-luminescence study," *J. Lumin.,* vol. 131, pp. 710-720, 2011.

[5] L. Akcelrud, "Electroluminescent polymers," *Prog. Polym. Sci.,* vol. 28, pp. 875 - 962, 2003.

[6] A. Cirpan, L. Ding, and F. E. Karasz, "Efficient Light Emitting Diodes from Polyfluorene Copolymer Blends," *Synth. Met.,* vol. 150, pp. 195-198, 2005.

[7] B. Nowacki, E. Iamazaki, A. Cirpan, F. Karasz, T. D. Z. Atvars, and L. Akcelrud, "Highly efficient polymer blends from a polyfluorene derivative and PVK for LEDs," *Polymer,* vol. 50, pp. 6057-6064, 2009.

[8] J.-W. Park, P. Kang, H. Park, H.-Y. Oh, J.-H. Yang, Y.-H. Kim, and S.-K. Kwon, "Synthesis and properties of blue-light-emitting anthracene derivative with diphenylamino-fluorene," *Dyes Pigment,* vol. 85, pp. 93-98, 2010.

[9] J. Sun, J. Chen, J. Zou, S. Ren, H. Zhong, D. Zeng, J. Du, E. Xu, and Q. Fang, "π-Conjugated poly(anthracene-alt-fluorene)s with X-shaped repeating units: New blue-light emitting polymers," *Polymer,* vol. 49, pp. 2282-2287, 2008.

[10] P. Raghunath, M. Ananth Reddy, C. Gouri, K. Bhanuprakash, and V. Jayathirtha Rao, "Electronic properties of anthracene derivatives for blue light emitting electroluminescent layers in organic light emitting diodes: a density functional theory study," *Phys. Chem. A,* vol. 110, pp. 1152-1159, 2006.

[11] Y. Li, T.-H. Kim, Q. Zhao, E.-K. Kim, S.-H. Han, Y.-H. Kim, J. Jang, and S.-K. Kwon, "Synthesis and Characterization of a Novel Polymer Based on Anthracene Moiety for Organic Thin Film Transistor," *J. Polym. Sci. Part A Polym. Chem.,* vol. 46, pp. 5115-5122, 2008.

[12] L. Valentini, D. Bagnis, A. Marrocchi, M. Seri, A. Taticchi, and J. M. Kenny, "Novel Anthracene-Core Molecule for the Development of Efficient PCBM-Based Solar Cells," *Chem. Mater.,* vol. 20, pp. 32-34, 2008.

[13] J.-B. Giguère and J.-F. Morin, "Synthesis and Optoelectronic Properties of 6,12-Bis(amino)anthanthrene Derivatives," *J. Org. Chem.,* vol. 78, pp. 12769-12778, 2013.

[14] J.-B. Giguère, J. Boismenu-Lavoie, and J.-F. Morin, "Cruciform Alkynylated Anthanthrene Derivatives: A Structure–Properties Relationship Case Study," *J. Org. Chem.,* vol. 79, pp. 2404-2418, 2014.

[15] D. A. M. Egbe, L. H. Nguyen, K. Schmidtke, A. Wild, C. Sieber, S. Günes, and N. S. Sariciftci, "Combined fffects of conjugation pattern and alkoxy side chains on the photovoltaic properties of thiophene- containing PPE-PPVs," *J. Polym. Sci. Part A Polym. Chem.,* vol. 45, pp. 1619-1631, 2007.

[16] D. A. M. Egbe, S. Sell, C. Ulbricht, E. Birckner, and U.-W. Grummt, "Mixed Alkyl- and Alkoxy-Substituted Poly[(phenylene ethynylene)-alt-(phenylene vinylene)] Hybrid Polymers: Synthesis and Photophysical Properties," *Macromol. Chem. Phys.,* vol. 205, pp. 2105-2115, 2004.

[17] D. A. M. Egbe, S. Turk, S. Rathgeber, F. Kuhnlenz, R. Jadhav, A. Wild, E. Birckner, G. Adam, A. Pivrikas, V. Cimrova, G. Knor, N. S. Sariciftci, and H. Hoppe, "Anthracene Based Conjugated Polymers: Correlation between π-π-Stacking Ability, Photophysical Properties, Charge Carrier Mobility, and Photovoltaic Performance," *Macromolecules,* vol. 43, pp. 1261-1269, 2010.

[18] W. Y. Huang, W. Gao, T. K. Kwei, and Y. Okamoto, "Synthesis and Characterization of Poly(alkyl-substituted p-phenylene ethynylene)s," *Macromolecules,* vol. 34, pp. 1570-1578, 2001.

[19] J. A. Mikroyannidis, V. P. Barberis, D. Vyprachticky, and V. Cimrova´, "Simple Synthesis, Photophysics, and Electroluminescent Properties of Poly[2,7-bis(4-tert-butylstyryl)fluorene-9,9-diyl-alt-alkane-α,ω-diyl]," *J. Polym. Sci. Part A Polym. Chem.,* vol. 45, pp. 809-821, 2006.

[20] J.-B. Giguère, N. S. Sariciftci, and J.-F. Morin, "Polycyclic anthanthrene small molecules: semiconductors for organic field-effect transistors and solar cells applications," *J. Mater. Chem. C,* vol. 3, pp. 601-606, 2015.

Reducing UV induced degradation losses of solar modules with c-Si solar cells featuring dielectric passivation layers

Robert Witteck[1], Henning Schulte-Huxel[1], Boris Veith-Wolf[1], Malte Ruben Vogt[1], Fabian Kiefer[1], Marc Köntges[1], Robby Peibst[1,2], and Rolf Brendel[1,3]

1. Institute for Solar Energy Research Hamelin (ISFH), Am Ohrberg 1, 31860 Emmerthal, Germany

2. Institute of Electronic Materials and Devices (MBE), Leibniz Universität Hannover, Schneiderberg 32, 30167 Hanover, Germany

3. Institute for Solid State Physics, Leibniz Universität Hannover, Appelstraße 2, 30167 Hanover, Germany

Abstract — **We report on the effect of UV irradiation on solar modules featuring crystalline silicon solar cells with various types of passivation layers and encapsulation polymers with varying UV transparency. Our results reveal that solar modules featuring cells with an aluminum oxide/p^+-type silicon passivation interface on the illuminated side are stable within 1500 h UV exposure. Modules featuring bifacial back junction cells with a silicon nitride/n^+-type silicon passivation interface in combination with an ethylene vinyl acetate encapsulation with enhanced UV transparency degrade up to 15% in module power due to UV illumination. We ascribe the UV degradation to an increase in surface recombination. Analytical modeling the degradation of the surface passivation interface indicates that photons with an energy similar to the Si-H bond energy are responsible for UV degradation.**

Index Terms — **UV degradation, solar modules, PERC, surface passivation, PV module reliability.**

I. INTRODUCTION

The crystalline silicon (c-Si) based photovoltaic (PV) industry is currently in a transition replacing the aluminum (Al) back surface field (BSF) by advanced solar cell concepts with dielectric passivation layers on the front and rear side. Such solar cell concepts like PERC (Passivated Emitter and Rear Cell) or PERT (Passivated Emitter Rear Totally-diffused) are predicted to gain market shares over cells with a full area Al BSF [1].

Both advanced cell concepts are suitable for bifacial PV modules, which became of particular interest for the PV industry in the past years [2] [3] [4] [5] [6].

Enhanced surface passivation and the development of metallization pastes improved the spectral response of PERC and PERT cells in the short wavelength range. Hence, encapsulation polymers with an increased transparency in the ultraviolet (UV) wavelength range allow for enhancing the blue response of these cells in a PV module. This is of major importance for PV module manufacturers, since UV transparent encapsulants were shown to have the potential for increasing the module power output by 1.9% [7] [8].

However, UV transparent encapsulants have to be UV stable themselves and also demand for UV stable solar cells. In the past UV degradation in solar modules was primarily observed as a discoloration effect of the ethylene vinyl acetate (EVA)

encapsulation polymer [9] [10]. Today, the stability of encapsulation polymers was improved and no more discoloration is observed with typical modern encapsulation polymers [11].

However, UV induced degradation was observed on non-fired lifetime samples with a silicon nitride SiN_x (abbreviated with SiN in this work) passivation layer on n^+-type Si [12] [13] [14] [15]. This is not representative for industrial PERC since the fast firing step changes the properties of the passivation layer and hence, affects the passivation quality of the SiN [16] [17]. In our previous work [18], we showed that solar modules featuring PERC with various SiN/n^+-type Si passivation interfaces show different degradation characteristics.

We previously reported that the charge carrier lifetime measured on fired c-Si lifetime samples with an aluminum oxide (abbreviated with AlO in this work) passivation layer on p^+-type Si is unaffected by UV light [19].

In this work we investigate the UV degradation of c-Si PV modules with cells featuring SiN and AlO passivation interfaces on the illuminated module side and EVA encapsulation with varying UV transparency.

II. EXPERIMENTAL

A. PERC solar cell processing

For group 1 we process a batch of screen-printed p-type Czochralski (Cz) grown Si PERC according to the process flow in Ref. [20]. We create the phosphorous doped n^+-type front emitter by $POCl_3$ diffusion with a sheet resistance of 90 Ω/sq and a surface donor concentration of 1.8×10^{20} cm^{-3}. For the SiN passivation layer on the n^+-type emitter we vary the $SiH_4/NH_3/H_2$ gas flow rates in the plasma enhanced chemical vapor deposition (PECVD) tool to change the stoichiometry resulting in layers with various refractive indices and hydrogen contents. We adjust the deposition time to obtain layers with similar optical thickness values in order to have the minimum reflectance at similar wavelengths. A typical industrial SiN layer with a refractive index n_{633nm} of 2.07 and a thickness of 100 nm measured on a planar samples is our reference. The other SiN layers have an n_{633nm} of 1.99, 2.29, and 2.53, respectively.

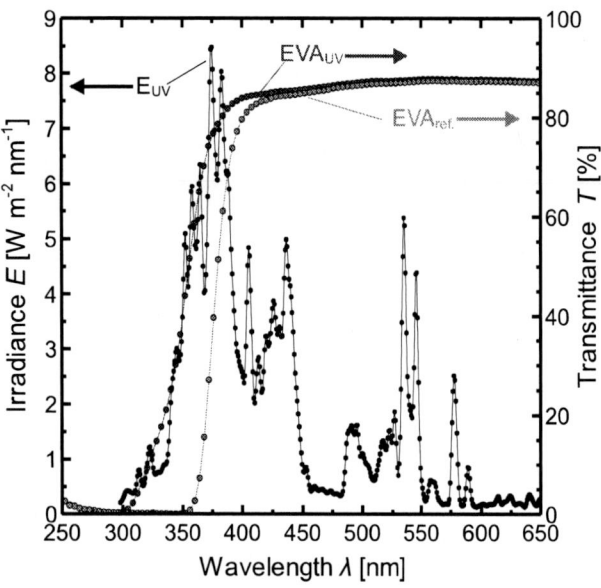

Fig. 1   Spectral irradiance E_{UV} of the employed UV lamps for the UV degradation and measured transmittance data of the EVA_{ref} and EVA_{UV}.

On the rear side of the PERCs we deposit a stack of AlO/SiN. We deposit an AlO layer with a thickness of 5 nm employing spatial atomic layer deposition (SALD) and the reference SiN layer by PECVD. Subsequently, we form the metal contacts employing screen-printing technology and fast firing step in a conveyor belt furnace. After the cell processing we illuminate all cells for 100 h with halogen lamps and an intensity of 1 sun at an temperature of 150 °C to deactivate light induced degradation effects [21].

In the PECVD process we also deposit non-textured float zone (FZ) Si samples with the SiN layers of varying refractive index. On these samples we measure the complex refractive index employing variable angle spectroscopic ellipsometry (VASE).

B. PERT solar cell processing

For group 2 we process a batch of bifacial screen-printed *n*-type Cz-Si PERT solar cells following the process flow in Ref. [22]. We employ ion implantation and subsequent annealing to form the p^+-type emitter and the n^+-type BSF. The implant dose is 2.0×10^{15} cm^{-3} for the boron-doped emitter and 1.25×10^{15} cm^{-3} for the phosphorous-doped BSF layer. We deposit a stack of AlO/SiN on the p^+-type emitter employing SALD and PECVD. The AlO thickness is 10 nm and SiN thickness is 100 nm with $n_{633nm} = 2.07$, measured on a planar device. On the n^+-type BSF we deposit a SiN layer with a thickness of 100 nm and an n_{633nm} of 2.07, measured on a planar device. Finally, we form the metal contacts employing screen-printing technology and a fast firing step.

These cells have rather high bifacial factors of up to 99.4% [22], thus resulting in similar illuminated *IV* characteristics, independently of which side is illuminated.

C. Test module processing

We build eight one-cell test modules from the cells of groups 1 and 2. For all test modules we laminate the cell between two EVA sheets, with a low iron soda lime glass on the illuminated and an opaque backsheet on the rear side. We laminate all test modules for 10 minutes at a peak lamination temperature of 155°C in laboratory laminator.

In the six test modules containing the PERC of group 1 the cell side with the SiN/n^+-type emitter passivation interface always faces the illuminated front side of the module. We denote these test modules with SiN_i in the following, where *i* indicates the corresponding refractive index n_{633nm} of the SiN passivation. In the two test modules with the PERT cells of group 2, either the cell side with the AlO/p^+-type emitter passivation interface or the SiN/n^+-type BSF passivation interface faces the modules' illuminated side. We denote the first module with $PERT_{AlO}$ and the latter with $PERT_{SiN}$.

We apply two different EVAs of varying UV transparency for the encapsulation. Figure 1 shows the spectrally resolved transmittance data of both EVAs. EVA_{ref} is a typical industrial EVA with a UV cut-off wavelength of $\lambda_c = 380$ nm. EVA_{UV} has an improved UV transparency with $\lambda_c = 356$ nm. Note that we choose to define λ_c at the half maximum transmittance. We laminate one PERC of group 1 featuring the reference SiN layer with $n_{633nm} = 2.07$ with the EVA_{ref}. This is our reference module, denoted $SiN_{2.07, ref}$. We laminate all other PERC and PERT cells of groups 1 and 2 with the EVA_{UV}.

D. UV Degradation

Before the UV illumination test we measure the illuminated *IV* characteristics under standard testing conditions (STC) [23] of all test modules employing a HALM module flasher.

Then we illuminate the front side of the test modules with low pressure mercury lamps (denoted as UV lamps) and an illumination intensity of 331±15 W m^{-2} in the spectral range of 290 nm to 400 nm. Figure 1 shows the spectrum of the UV lamps. The metal plate that holds the test modules in the UV chamber has a controlled temperature of 40°C. We measure the illuminated *IV* characteristics under STC after 0 h, 24 h, 150 h, 500 h, 750 h, 1000 h, and 1500 h of UV exposure. In this work a UV exposure time t_{UV} of 1500 h equals a UV dose D of 500 kWh m^{-2}. This corresponds to about 15 years of outdoor exposure in a moderate climate zone (Potsdam, Germany) [23]

III. RESULTS AND DISCUSSION

Fig. 2 (top) shows the relative change of the maximum power P_{mpp} during UV exposure. We observe a power loss that increases with decreasing refractive index n_{633nm} of the SiN layer. For the test module $SiN_{1.99}$ we measure the largest power loss of 4.3% after 1500 h UV exposure of all test modules containing cells of group 1. The power of the test module $SiN_{2.07}$ with the reference SiN and UV transparent EVA degrades by 3.5%. In contrast, module $SiN_{2.07, ref}$ with the same SiN but conventional EVA degrades by 1.3% in module power after a t_{UV} of 1500 h. Test modules $SiN_{2.29}$ and $SiN_{2.53}$ with $n_{633} > 2.29$ show a power degradation of less than 1% within 1500 h of UV exposure.

For the modules with the PERT cells we observe a power loss of 15% for test module $PERT_{SiN}$, where the cell side with the SiN/n^+-type BSF passivation interface faces the UV lamps. In contrast, test module $PERT_{AlO}$ with the AlO/p^+-type emitter passivation interface facing the UV light source shows an increase in P_{mpp} of 3% after a t_{UV} of 1500 h. Results of Veith *et al.* indicate that UV light enhances the negative fixed charge density within the AlO layer [19]. This improves the passivation by AlO, which in turn improves the module power output.

Fig. 2 (bottom) shows the open circuit voltage V_{oc} during UV exposure. The degradation in module power correlates with the degradation in V_{oc}. When the refractive index of the SiN layer increases, the absorption of UV photons in the SiN increases and reduces the number of UV photons reaching the SiN/n^+-type Si passivation interface. The test module with EVA_{UV} shows a higher relative power loss in comparison to the module with EVA_{ref} and equal cells due to its enhanced UV transmittance (see Fig. 1).

The $PERT_{SiN}$ test module shows the highest loss in module power. Here, SiN passivates the n^+-type BSF of the cells' n-type base. The free charge carriers recombine where the splitting of the quasi Fermi levels and thus the carrier concentration is largest. This significantly reduces the modules' voltage and power.

We attribute the degradation in module power to an increase in surface recombination velocity at the SiN/n^+-type Si passivation interface. We determine the optical constants of the various SiN layers from the VASE measurement and take the optical data of the glass and the two EVAs from literature [24]. Our ray tracing software *Daidalos* [25] calculates the number of photons that reach the SiN/n^+-type Si passivation interface after passing the glass, the EVA, and the SiN layer. From the number of photons we determine a photon flux ϕ_B considering only photons over to a certain energy barrier E_B

$$\phi_B = \int_{E_B}^{\infty} \phi(E)\, dE. \tag{1}$$

We presume when a photon of sufficient energy $E > E_B$ is transmitted to the SiN/n^+-type Si passivation interface it affects a passivated bond and turns it into a non-passivated bond. The

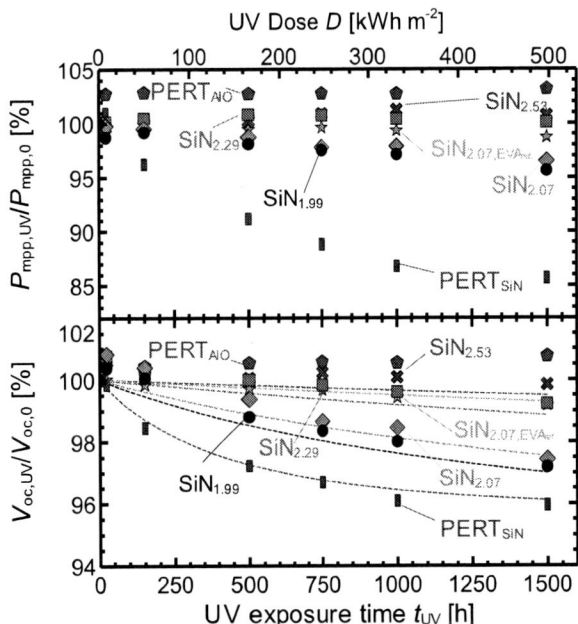

Fig. 2   Relative change in P_{mpp} and V_{oc} for the various test modules during UV exposure. Cells of group 1 are denoted with SiN and the corresponding n_{633nm} as index. Cells of group 2 are labeled with PERT and the passivation layer that faces the illuminated side as index. Dashed lines indicate fits according to Eq. *(6)*

number of passivated bonds N_P changes with UV exposure time t_{uv} according to

$$\frac{dN_P}{dt} = \sigma_{uv}(\lambda) N_P \phi_B(t_{uv}), \tag{2}$$

where σ_{uv} is the capture cross section coefficient of passivated bonds for UV photons. Solving the differential equation we obtain

$$N_P = N_{P0} \exp(-\sigma_{uv}(\lambda)\phi_B t_{uv}), \tag{3}$$

where N_{P0} represents the number of passivated bonds at $t_{uv} = 0$.

Assuming that not all bonds are initially passivated, the number of non-passivated bonds N_{NP} is

$$N_{NP} = N_T - N_{P0} \exp(-\sigma_{uv}(\lambda)\phi_B t_{uv}), \tag{4}$$

where N_T is the total number of bonds. We assume that a dangling bond acts like a single defect state in the mid-gap and that the number of dangling bonds N_{NP} is proportional to the number of interface states N_{it}. From the number of interface states we derive the surface recombination according to the SRH recombination theory [26]. We follow the approach of Del Alamo *et al.* and express the surface recombination velocity in terms of a surface recombination current density J_{0s} [27] [28] [29]. This J_{0s} depends on the surface recombination velocity and thus, the number of interface states by

$$J_{0s} = q \, N_{it} v_{th} \sigma_p \frac{n_i^2}{n_s}, \qquad (5)$$

with the elementary charge q, the thermal velocity v_{th}, the hole capture cross section coefficient σ_p, the intrinsic charge carrier concentration n_i, and the surface charge carrier concentration n_s. Combining Eq. (4) and (5) with the Shockley equation we obtain a relation for the open circuit voltage

$$V_{oc} = \frac{kT}{q} \ln \left(\frac{J_{sc}}{J_{0i} + q \, v_{th} \sigma_p \frac{n_i^2}{n_s} [N_T - N_{P0} \exp(-\sigma_{uv}(\lambda)\phi_B t_{uv})]} \right), \qquad (6)$$

where J_{0i} is the initial diode saturation current prior UV degradation. We fit Eq. (6) to all measured open circuit voltages. For the fit we calculate $\Phi_B(E)$ for all samples and vary the critical photon energy E_B. The dashed lines in Fig. 2 indicate the fits according to Eq. (6). Our model fits best if the critical photon energy is between 3.4 eV to 3.5 eV, which is similar to the bond energy of Si-H according to Yin et al. [28]. They reported a bond energy of 3.34 eV. For further details regarding the model we refer to Ref. [18].

IV. SUMMARY

We investigated the UV stability of test modules with state of the art PERC and PERT solar cells featuring a dielectric passivation layer at the module's illuminated side. We also compared two EVA encapsulations with varying UV cut-off wavelength.

Our results reveal that the UV induced degradation depends on the refractive index of the SiN and the cut-off wavelength of the EVA. The test modules featuring a SiN passivation layer with $n_{633} \geq 2.29$ or an EVA with a cut-off wavelength of $\lambda_c = 380$ nm lose less than 1% in P_{mpp} during 1500 h UV exposure. Combining an EVA with an enhanced UV transparency of $\lambda_c = 356$ nm with a typical industrial SiN layer with $n_{633} = 2.07$ on an n^+-type emitter results in a degradation in module power of 3.5% after 1500 h UV exposure.

Vogt et al. demonstrated that an EVA with an enhanced UV transparency may increase the module performance by 1.9% [7]. Considering this improvement we determine the break-even point, when the initial advantage in power of 1.9% for the module with UV transparent EVA equals the power of the module with the reference EVA due to the UV induced degradation. We calculate a break-even point at 342 kWh m^{-2}, which corresponds to about 10 years outdoor exposure in Potsdam [22].

When SiN passivates the n^+-type BSF of the back junction PERT cell the module power degrades up to 15%.

We find that the UV degradation correlates with the flux of high energy photons ($h\nu \geq 3.4$ eV) reaching the SiN/n^+-type Si passivation interface. These photons de-passivate the SiN/n^+-type Si passivation interface by creating non-passivated bonds. This increases the surface recombination and reduces the module's open circuit voltage and power output.

Modules with PERT front junction cells and AlO/p^+-type emitter passivation interface are stable under UV illumination and even show an increase in module power of 3% after 1500 h UV exposure. This agrees with results reported in Ref. [19] and thus may be explained by an increase in negative fixed charge.

ACKNOWLEDGEMENT

The results were generated in the PERC2Module project funded by German Federal Ministry for Economic Affairs and Energy under contract no. 0325641. We would like to thank Ulrike Sonntag, Sarah Spätlich and the CHIP team for the cell and Susanne Blankemeyer for the module processing.

REFERENCES

[1] A. Metz, M. Fischer, C.-C. Li, A. Hsu, S. Julsrud, T. Chang and B. Tjahjono, "International Technology Roadmap for Photovoltaic Results 2016," SEMI Europe, 2016.

[2] T. Buck, R. Kopecek, J. Libal, A. Herguth, K. Peter, I. Röver, K. Wambach and B. Geerligs, "Industrial screen printed n-type silicon solar cells with front boron emitter and efficiencies exceeding 17%," in 21st EUPVSEC, 2006.

[3] M. R. Vogt, H. Holst, H. Schulte-Huxel, S. Blankemeyer, R. Witteck, P. Bujard, J.-B. Kues, K. Bothe, M. Köntges and R. Brendel, "PV Module Current Gains due to Structured Backsheets," in Silicon PV 2017, 2017.

[4] T. Dullweber, C. Kranz, R. Peibst, U. Baumann, H. Hannebauer, A. Fülle, S. Steckemetz, T. Weber, M. Kutzer, M. Müller, G. Fischer, P. Palinginis and H. Neuhaus, "PERC+: industrial PERC solar cells with rear Al grid enabling bifaciality and reduced Al paste consumption," Progress in Photovoltaics: Research and Applications, vol. 24, no. 12, pp. 1487-1498, oct 2015.

[5] V. D. Mihailetchi, H. Chu, G. Galbiati, C. Comparotto, A. Halm and R. Kopecek, "A Comparison Study of n-type PERT and IBC Cell Concepts with Screen Printed Contacts," Energy Procedia, vol. 77, pp. 534-539, 2015.

[6] H. Schulte-Huxel, F. Kiefer, S. Blankemeyer, R. Witteck, M. R. Vogt, M. Köntges, R. Brendel, J. Krügener and R. Peibst, "Flip-flop Cell Interconnection Enabled By An Extremely High Bifaciality Of Screen-printed Ion Implanted N-pert Si Solar Cells," in 32nd EUPVSEC, 2016.

[7] M. R. Vogt, H. Holst, H. Schulte-Huxel, S. Blankemeyer, R. Witteck, D. Hinken, M. Winter, B. Min, C. Schinke, I. Ahrens, M. Köntges, K. Bothe and R. Brendel, "Optical constants of UV transparent EVA and the impact on the PV module output power under

realistic irradiation," *Energy Procedia*, vol. 92, pp. 523-530, 2016.

[8] C. Schmid, J. Chapon, G. Kinsey, J. Bokria and J. Woods, "Impact Of High Light Transmission Eva-based Encapsulant On The Performance Of Pv Modules," in *27th EUPVSEC*, 2012.

[9] F. J. Pern, "Factors that affect the EVA encapsulant discoloration rate upon accelerated exposure," in *24th IEEE PVSC*, 1994.

[10] A. Badiee, R. Wildman and I. Ashcroft, "Effect of UV aging on degradation of Ethylene-vinyl Acetate (EVA) as encapsulant in photovoltaic (PV) modules," *SPIE Solar Energy+ Technology*, vol. 9179, pp. 91790O-91790O, 2014.

[11] A. Jentsch, K. J. Eichhorn and B. Voit, "Influence of typical stabilizers on the aging behavior of EVA foils for photovoltaic applications during artificial UV-weathering," *Polymer Testing*, vol. 44, pp. 242-247, 2015.

[12] R. Hezel, "Low-Temperature Surface Passivation of Silicon for Solar Cells," *Journal of The Electrochemical Society*, vol. 136, no. 2, p. 518, 1989.

[13] T. Lauinger, J. Moschner, A. G. Aberle and R. Hezel, "UV stability of highest-quality plasma silicon nitride passivation of silicon solar cells," in *25th IEEE PVSC*, 1996.

[14] T. Kamiokal, D. Takail, T. Tachibanal, T. Kojimal and Y. Ohshital, "Plasma Damage Effect on Ultraviolet-Induced Degradation of PECVD SiNx:H Passivation," in *42nd IEEE PVSC*, 2015.

[15] T. Tachibana, D. Takai, Y. Yamashita, N. Ikeno, H. Tokutake, K. Nagata, A. Ogura and Y. Ohshita, "Effects of texture structure on crystalline damage induced by SiNx plasma CVD," in *29th EUPVSEC*, 2016.

[16] J.-F. Lelivre, E. Fourmond, A. Kaminski, O. Palais, D. Ballutaud and M. Lemiti, "Study of the composition of hydrogenated silicon nitride SiNx:H for efficient surface and bulk passivation of silicon," *Solar Energy Materials and Solar Cells*, vol. 93, no. 8, pp. 1281-1289, 2009.

[17] S. Gatz, F. Einsele, T. Dullweber and R. Brendel, "Firing stability of SiNy / SiNx surface passivation stacks for crystalline silicon solar cells," in *26th EUPVSEC*, 2011.

[18] R. Witteck, B. Veith-Wolf, H. Schulte-Huxel, A. Morlier, M. R. Vogt, M. Köntges and R. Brendel, "UV induced degradation of PERC solar modules with UV transparent encapsulation materials," *Progress in Photovoltaics*, vol. 25, no. 6, pp. 409-416, 2016.

[19] B. Veith-Wolf, R. Witteck, A. Morlier, H. Schulte-Huxel and J. Schmidt, "Effect of UV Illumination on the Passivation Quality of AlOx/c-Si Interfaces," in *43rd IEEE PVSC*, 2016.

[20] H. Schulte-Huxel, R. Witteck, H. Holst, M. R. Vogt, S. Blankemeyer, D. Hinken, T. Brendemuhl, T. Dullweber, K. Bothe, M. Kontges and R. Brendel, "High-Efficiency Modules With Passivated Emitter and Rear Solar Cells - An Analysis of Electrical and Optical Losses," *IEEE JPV*, vol. 7, no. 1, pp. 25-31, 2017.

[21] A. Herguth and G. Hahn, "Towards A High Throughput Solution For Boron-oxygen Related Regeneration," in *28th EUPVSEC*, 2013.

[22] F. Kiefer, J. Krügener, F. Heinemeyer, M. Jestremski, H. J. Osten, R. Brendel and R. Peibst, "Bifacial, fully screen-printed n-PERT solar cells with BF2 and B implanted emitters," *Solar Energy Materials and Solar Cells*, vol. 157, pp. 326-330, dec 2016.

[23] I.E.C, "IEC60904-3 - Photovoltaic devices Part 3: Measurement principles for terrestrial photovoltaic (PV) solar devices with reference spectral irradiance data," *VDE Verlag*, 2016.

[24] U. Feister and K. Grasnick, "Solar UV radiation measurements at Potsdam," *Solar Energy*, vol. 49, no. 6, pp. 541-548, 1992.

[25] M. R. Vogt, Development of Physical Models for the Simulation of Optical Properties of Solar Cell Modules, Leibniz Universität Hannover, 2015.

[26] H. Holst, M. Winter, M. R. Vogt, K. Bothe, M. Köntges, R. Brendel and P. P. Altermatt, "Application of a new ray tracing framework to the analysis of extended regions in Si solar cell modules," *Energy Procedia*, vol. 38, pp. 86-93, 2013.

[27] W. Shockley and W. T. Read Jr, "Statistics of the recombinations of holes and electrons," *Physical review*, vol. 87, no. 5, p. 835, 1952.

[28] J. A. Del Alamo and R. Swanson, "The physics and modeling of heavily doped emitters," *IEEE Transactions on Electron Devices*, vol. 31, no. 12, pp. 1878-1888, 1984.

[29] A. G. Aberle, Crystalline silicon solar cells: advanced surface passivation and analysis, Centre for Photovoltaic Engineering. University of New South Wales, 1999.

[30] K. R. McIntosh and L. E. Black, "On effective surface recombination parameters," *Journal of Applied Physics*, vol. 116, p. 014503, 2014.

[31] Z. Yin, "Free energy model for the analysis of bonding in a-SixNyHz alloys," *J. Vac. Sci. Technol. A*, vol. 9, no. 3, p. 972, May 1991.

Large-Area Junction Damage in Potential-Induced Degradation of c-Si Solar Modules

Chuanxiao Xiao,[1,2] Chun-Sheng Jiang,[1] Steve Johnston,[1] Steve P. Harvey,[1] Peter Hacke,[1] Brian Gorman,[2] and Mowafak Al-Jassim[1]

[1] National Renewable Energy Laboratory, Golden, CO 80401 USA

[2] Colorado School of Mines, Golden, CO 80401 USA

Abstract — We report a large area of millimeter-scale p-n junction damage caused by potential-induced degradation (PID) of lab-stressed crystalline-Si modules. Kelvin probe force microscopy results show electrical potential change across the junction, and a recovery was observed after heat treatment. Electron-beam induced current results support the large-area damage instead of local shunts and a much lower collected current for the affected junction area. Furthermore, secondary-ion mass spectrometry indicates that the large-area damage correlates with sodium contamination. The consistent results shed new light on PID mechanisms to investigate that are essentially different than the widely reported local junction shunts.

Index Terms — potential-induced degradation, junction damage, large area, recovery, microscopy.

I. INTRODUCTION

Potential-induced degradation (PID) has been an important degradation component in all types of major solar modules, and it gets more severe as the system voltage of a solar array increases. PID could cause a massive decrease of module performance [1][2][3]. For crystalline-silicon (c-Si)-based modules, PID has been reported to be related to sodium ions diffusing into extended defects that exist in the Si p-n junction, causing local electrical shunting [4][5][6]. However, a nanoscale understanding of PID is still in its infancy. PID is a complicated material and device issue, with many different degradation kinetics and mechanisms among the different solar modules. It is questionable whether local shunting in c-Si modules is the only major degradation mechanism associated with massive degradation or even module failure.

In this paper, we report a large area of p-n junction change by PID that occurred at the millimeter scale. We imaged the electrical potential across the junction of a multicrystalline silicon (mc-Si) cell using Kelvin probe force microscopy (KPFM), current collection using electron-beam induced current (EBIC), and elemental distribution using secondary-ion mass spectrometry (SIMS). We found movement of the apparent electrical field from the p-n junction location to the front surface of the cell. We also observed a recovery of the movement by heat treatment via tracking the potential change in the same location. Characterization of the electric field in heavily PID-shunted regions shows that there is a broad reduction in the junction potential instead of localized perturbations of the electric field at defects. In EBIC measurements of the cross section, a much lower current was collected and the affected area is large instead of just being local shunts. SIMS results indicate a large area of sodium contamination around the damaged junction area. These consistent results suggest a PID behavior, which has not been reported in the literature.

II. EXPERIMENT DETAILS

KPFM measurements require a working cell so that external bias voltage can be applied to the junction (i.e., a sufficiently large shunt resistance). We have performed the KPFM imaging on an in-house mini-module without edge damage, instead of on commercial field-stressed modules that have damage after being cored at selected PID areas in the module. In making the mini-module, a conventional front-junction (n^+/p) mc-Si cell was sandwiched with glass substrate and superstrate and with pre-polymerized ethylene vinyl acetate (EVA) as encapsulant. All the layer materials were held together with clamps without melting the EVA. After laboratory PID stressing, the cell was removed from the mini-module without mechanical damage.

A constant bias voltage of -1,000 V was applied to the shorted mini-module in an environmental chamber at a temperature of 25°C and 10% relative humidity for two weeks. Significant PID was observed by the development of dark current-voltage (I-V) curves captured periodically over the course of the stress. The mini-module was then delaminated to extract the cell. A piece of the cell with heavy PID degradation, as identified by photoluminescence (PL) imaging, was cut out of the cell. The sample contains the most-degraded area in the cell according to the PL image as shown in Fig. 1 (no PL contrast in the same area before the PID). We cleaved the sample across the darkest PL area and used silver epoxy to glue the cell emitter (n^+) face to another piece of silicon. This sandwich sample preparation prevents probe drop-off when scanning the probe close to the front side of the cell or across the junction. The sample was chemical-mechanically polished with deionized water using a set of different diamond pads and finally using silica colloids with a soft cloth. The polishing process must be well controlled to ensure: 1) a flat surface necessary for atomic force microscope (AFM)-based imaging; 2) that the cell has to be alive for the KPFM measurement; and 3) that the heavily PID-stressed area is not polished away.

KPFM is an AFM-based technique that measures electrostatic potential on the sample surface at a spatial resolution of ~30 nm and a potential resolution of ~10 mV [7]. To avoid the effect of surface charges, we applied a bias

978-1-5090-5606-4/17 $31.00 © 2017 IEEE

voltage to a working device and measured the corresponding changes of surface potential. Because the surface-charge configuration should not change with a small bias voltage V_b of 0–2 V, the measured change of the surface potential is about the same as the potential change in the bulk. In this way, we determined the potential change in the bulk by measuring the surface potential change across the p-n junction [8].

EBIC was performed on fresh and heavily degraded samples using identical conditions of an electron-beam voltage of 15 kV and a current of 0.8 nA. SIMS data were acquired from degraded and non-degraded areas on one sample.

III. RESULTS AND DISCUSSION

We investigated the electrical potential across the p-n junction on the cross section around the darkest PL area between the two silver grids over a total length of ~2 mm (Fig. 1). We found three significantly different kinds of areas, labeled as "No PID," "PID-affected area," and "transition area," where the electric-field peak or the potential drop is, respectively, at the normal junction location, at the cell front surface, and between them; details of the KPFM results will be shown later. In contrast to the current knowledge of PID—

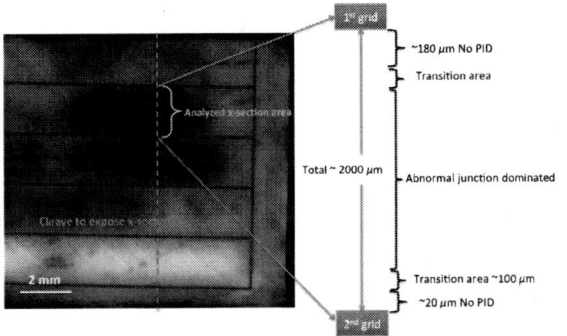

Fig. 1. PL mapping of a PID-shunted area. The sample was cleaved to expose the cross section, indicated by the red dash line. The KPFM analyzed region is between two metal grids with most serious PID-affected area.

that is, junction shunts in highly localized spots within a few micrometers range—we observed a large, continuous PID-affected area (~1.5 mm), with the abnormal electrical potential

in the junction area (indicated in Fig. 1). We also observed normal electrical potential across the junction in the regions under the metal grids and nearby areas. In the transition region, transitional potential behaviors were observed in some local areas, and mixtures of normal and abnormal potentials were observed in some other areas. The large PID-affected area with the abnormal potential is identified by random measurement of the whole area (~1.5 mm) by adequately dense sampling.

We next show the typical KPFM results on the three regions of "No PID," "PID-affected," and "transition area." Figure 2a shows an AFM image on the "No-PID" region, and Fig. 2b shows the corresponding potential image at a reverse bias voltage V_b of 1.5 V. The three black dashed lines from top to bottom in Fig. 2 were drawn following the Si/Ag-epoxy interface, p-n junction location, and depletion edge of the p-type Si, respectively. Figure 2c shows the line profiles of potential averaged from the potential images and aligned with their corresponding AFM line profile. We took the potential images with varying V_b applied to the device to derive the potential change in the bulk by measuring the surface potential changes. The 0-V potential profile was subtracted from the 0.5-V and 1.5-V potential profiles (Fig. 2c) to get the profiles of potential differences (Fig. 2d). The V_b-induced change in the electrical field was obtained by taking the first derivative of the potential difference. The peak of the electric field indicates the p-n junction location (Fig. 2e), which is ~400 nm from the cell surface or epoxy/cell interface, consistent with the junction depth. The junction depletion width is ~500 nm as estimated from the electric-field profile at 0.5 V, which is also consistent with that of a conventional industrial solar cell. Therefore, the p-n junction in this region behaves normally from the perspective of electrical potential.

Fig. 3 shows typical KPFM images across the junction in the "PID-affected" region. The potential image (Fig. 3b) and the electric-field peak (Fig. 3c) all show the potential drop at the cell/Ag-epoxy interface. The V_b applied to the Si cell should drop at the p-n junction if the local cell area is functioning normally. The appearance of the electric-field peak at the cell/epoxy interface suggests a significantly damaged junction; external reverse-bias voltage applied to the cell does not drop at the junction. In this case, because the Ag-epoxy/cell contact resistance should not be very small, the V_b drop at the interface—instead of at the junction—is consistent

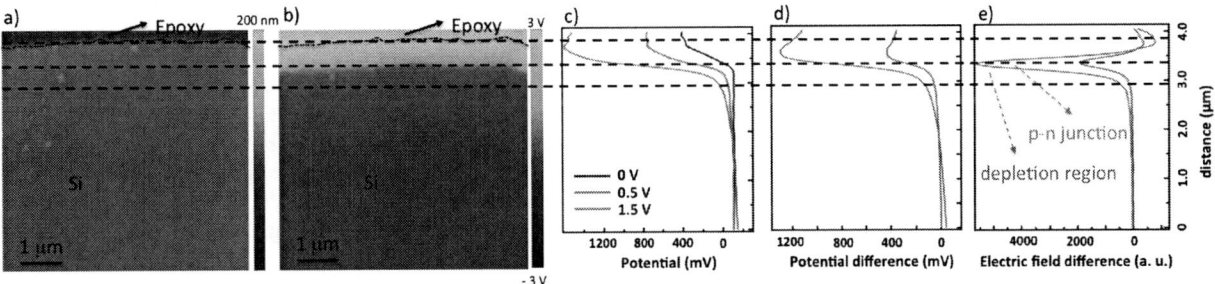

Fig. 2. a) AFM of a good area with normal junction; b) the KPFM image at -1.5 V corresponding to a); c) potential profiles of an area in b); d) potential difference curves; e) electric-field difference curves showig a good p-n junction characteristic.

978-1-5090-5606-4/17 $31.00 © 2017 IEEE

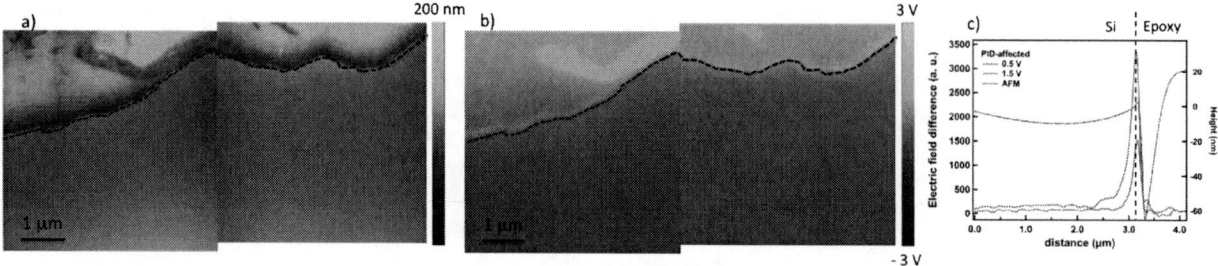

Fig. 3. a) AFM of a PID-affected area; b) the KPFM image potential image at -1.5 V corresponding to a); c) electric-field difference curves showing an abnormal p-n junction characteristic, there is no change in the potential from the cell base to the cell surface at the epoxy.

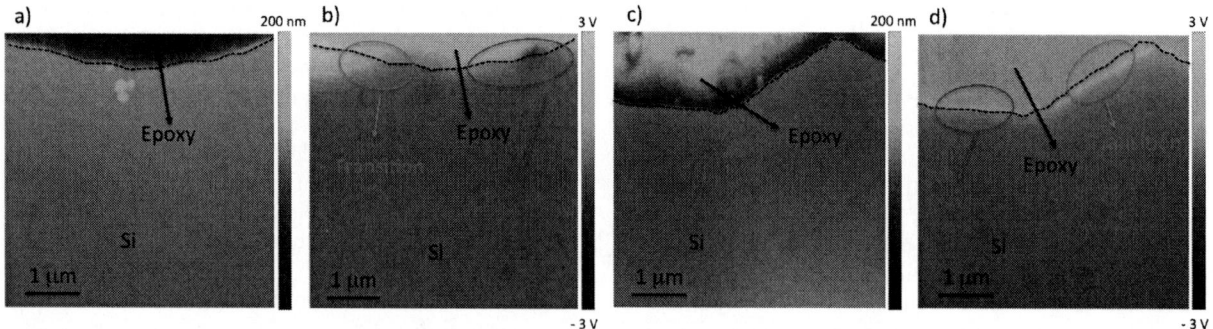

Fig. 4. a) AFM of a transition area; b) the KPFM potential image at 1.5 V corresponding to a); c) AFM of another transition area; d) the potential image at 1.5 V corresponding to c).

with the shunted junction if the equivalent resistance of the shunted junction is much smaller than the contact resistance at the epoxy/cell interface. Another possibility is that the real p-n junction moved due to PID. A junction movement by Ag metallization was reported, where local recrystallization of the emitter at high temperature was responsible for the junction movement and junction damage [3]. However, we do not have a clear picture of how high voltage can cause the junction movement. On the other hand, one may speculate that a highly insulating local epoxy/cell contact can cause the abnormal voltage drop at the interface. This mechanism can be excluded because if the junction were normal, then V_b would drop at the normal junction location through an emitter path, as in the case of an isolated cell itself. Therefore, the abnormal electric potential/field is likely caused by junction damage or shunting.

Figure 4 shows potential imaging in the transition region. Two types of transitions between the normal and abnormal junctions were observed. The first is shown in Figs. 4a and 4b, where the transition occurs in ~1 μm, which is about or slightly larger than the "depletion range" generated by a V_b of -1.5 V applied to the junction. This type should have a relatively abrupt transition between the normal and damaged junctions. The second type is shown in Figs. 4c and 4d, where the location of the electric-field peak is at a location significantly shallower than the normal junction depth. The green circle in Fig. 4d indicates such an area adjacent to two junction-damaged areas. Because we lack a picture of the

kinetics for the real junction movement by high voltage, we prefer an explanation that this apparent movement of electric-field peak is caused by a moderately damaged junction, an equivalent resistance of which is comparable to that at the epoxy/cell interface. The equivalent resistance depends on the quality of the junction, e.g., the gap state density and recombination velocity. In this case, V_b would drop at both the junction and interface. If KPFM measurements cannot resolve the two voltage drops, then the electric-field peak would appear at a location between the junction and interface. Therefore, this type of transition may indicate a moderately damaged junction.

We also investigated the thermal annealing effect on damaged junctions. Interestingly, the PID-affected areas can be recovered by heat treatment, similar to the macroscale I-V recovery reported in the literature [9][10]. The sample was heated on a hot plate at 300°C for 30 minutes. After the heat treatment, most of the PID-affected area was gone and behaves as the normal area, according to the identical KPFM measurement. Figure 5 shows the change of a PID-affected area to a normal area on the same location. Figures 5a and 5b show the AFM images in the same location, with the black dashed lines indicating the Si/epoxy interface. It is clear that before heat treatment, the voltage drops at the epoxy/cell interface (Fig. 5c); but after treatment, the voltage drop moved deeper into the Si to a normal junction depth (Fig. 5d). Further investigation is needed and in-situ stressing of the Si solar cell on the AFM platform is in progress.

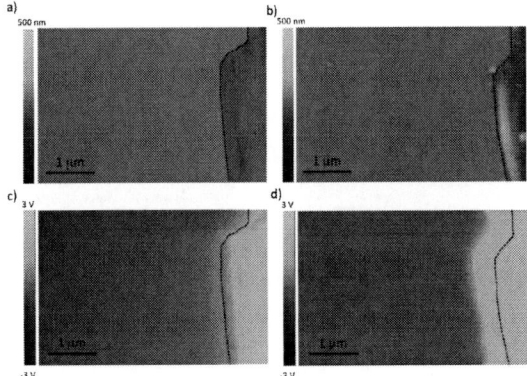

Fig. 5. a) AFM of a PID-affected area; b) potential image at 1.5 V corresponding to a); c) after heat treatment, the AFM image of the same location; d) after heat treatment, the potential at 1.5 V.

EBIC imaging was performed on similar pieces of fresh and heavily degraded areas as the KPFM measurement. Figure 6 shows that the EBIC images are much different between the fresh and degraded samples. Figures 6a and 6b show an SEM and an EBIC image taken on the fresh sample. The wide bright region (~10 μm) around the cell front face is consistent with a normal junction, which can collect current widely and operates as a regular solar cell. However, in a heavily degraded area (Figs. 6c and 6d), the EBIC image shows much smaller current collections with a current intensity two orders of magnitude smaller than that of the normal junction. The vertical stripe-like pattern is likely caused by the cross-sectional sample polishing, which causes the small current-collection contrast. Therefore, the EBIC imaging suggests no significant minority-carrier collection in the heavily degraded area, which is consistent with the large-area junction collapse as observed by the KPFM measurement.

On the same degraded sample as imaged by KPFM, we use SIMS to examine the normal junction area and abnormal area. Figure 7a shows the SIMS image of Si, and Fig. 7b shows the sodium image on the same area. The sodium line profiles

across the junction are shown in Fig. 7c. A large amount of sodium was detected from the abnormal junction areas, especially close to the edge of silicon, where the p-n junction was located; in contrast, the sodium concentration of the normal junction area is just at a background level. These results indicate that the large-area junction damage may correlate with sodium diffusion into the area. Further investigation is required for a deeper understanding of this observed large-area junction damage in PID.

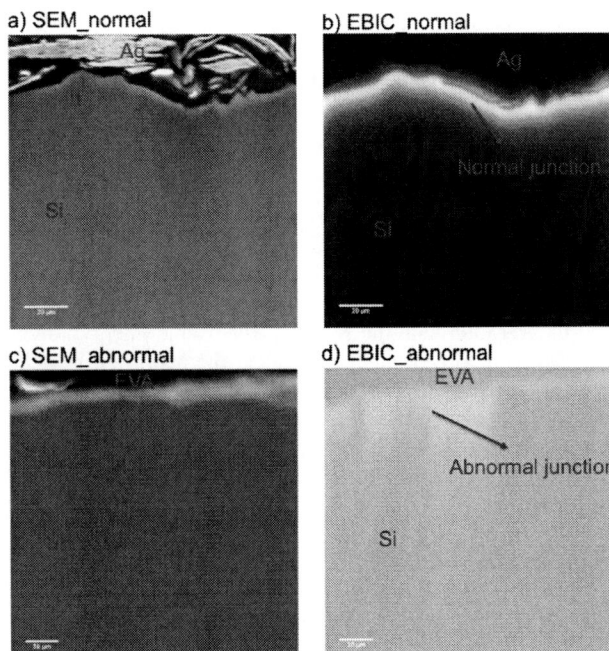

Fig. 6. a) SEM of a fresh sample; b) EBIC image corresponding to a) showing normal junction; c) SEM of a degraded sample; d) EBIC image corresponding to c) showing abnormal junction.

IV. CONCLUSION

We studied the PID of a silicon mini-module by measuring the potential profiling across the p-n junction. We found that the PID-affected region has junction damage in large and

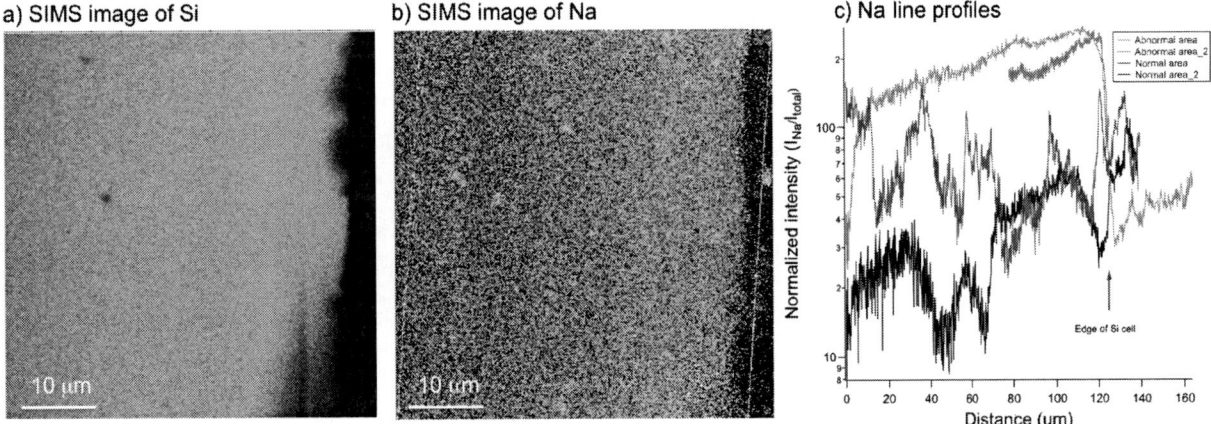

Fig. 7. a) SIMS image of Si on one degraded area; b) SIMS image of Na corresponding to a); c) Na line profiles of four different areas on the cross section, showing high concentration of Na on abnormal areas, but background level of Na on normal areas.

978-1-5090-5606-4/17 $31.00 © 2017 IEEE   1374

continuous areas; this observation is different from the current knowledge that PID shunts are localized in small and separate spots. The KPFM results show that the potential drops at the p-n junction location in a normal junction area, whereas the potential drop is right at the Si/epoxy interface in the PID-affected abnormal area. In the transition region, two types of transition behavior were observed. Furthermore, we observed a recovery of PID after heat treatment, which is consistent with other macroscale recovery studies. EBIC results are consistent with the large-area junction damage by showing a faint current collection in the damaged junction area. SIMS results show high sodium contamination around the abnormal junction area. These findings are essentially different from the widely reported local junction shunts.

ACKNOWLEDGEMENTS

C. X. thanks Dr. Harvey Guthrey (NREL) for EBIC training, Dr. Yuanyue Liu (Caltech) for fruitful discussion on theoretical calculation. This work was supported by the U.S. Department of Energy under Contract No. DE-AC36-08GO28308 with the National Renewable Energy Laboratory. The U.S. Government retains and the publisher, by accepting the article for publication, acknowledges that the U.S. Government retains a nonexclusive, paid up, irrevocable, worldwide license to publish or reproduce the published form of this work, or allow others to do so, for U.S. Government purposes.

REFERENCES

[1] P. Hacke et al., "Characterization of multicrystalline silicon modules with system bias voltage applied in damp heat," 25th EUPVSEC, Spain, 2010.

[2] S. Pingel et al., "Potential induced degradation of solar cells and panels," 35th IEEE PVSC, USA, 2010.

[3] W. Luo et al., "Potential-induced degradation in photovoltaic modules: A critical review," Energy Environ. Sci., 10, 43 (2017).

[4] V. Naumann et al., "Microstructural analysis of crystal defects leading to potential-induced degradation (PID) of Si solar cells," Energy Procedia, 33, 76–83 (2013).

[5] V. Naumann et al., "Explanation of potential-induced degradation of the shunting type by Na decoration of stacking faults in Si solar cells," Solar Energy Materials and Solar Cells, 120, Part A, 383–389 (2014).

[6] S. P. Harvey et al., "Sodium accumulation at potential-induced degradation shunted areas in polycrystalline silicon modules," IEEE Journal of Photovoltaics, 6(6), (Nov. 2016).

[7] C. Xiao et al., "Nanometer-scale electrical potential profiling across perovskite solar cells," 43th IEEE PVSC, USA, 2016.

[8] C.-S. Jiang et al., "Two-dimensional junction identification in multicrystalline silicon solar cells by scanning Kelvin probe force microscopy," J. Appl. Phys. 104, 104501 (2008).

[9] B. Jaeckel et al., "PID effect of c-Si modules: Study of degradation and recovery to more closely mimic field behavior," Israel Journal of Chemistry, 55, 1091 1097 (Oct 2015).

[10] D. Lausch et al., "Sodium outdiffusion from stacking faults as root cause for the recovery process of potential-induced degradation (PID)," Energy Procedia, 55, 486-493 (2014).

Search for Microstructural Defects as Nuclei for PID-Shunts in Silicon Solar Cells

Volker Naumann[1], Otwin Breitenstein[2], Jan Bauer[2], Christian Hagendorf[1]

[1]Fraunhofer Center for Silicon Photovoltaics CSP, D-06120 Halle (Saale), Germany
[2]Max-Planck-Institute for Microstructure Physics, D-06120 Halle (Saale), Germany

Abstract — Up to now, the formation process of stacking faults leading to PID-shunting of silicon solar cells is still unknown. In particular, the type of primordial defects that define the position of PID-shunts is of high interest. In this work it is experimentally proven that stacking faults evolve first under high voltage stress. At the same time, it is shown that stacking faults do not coincide with positions of threading dislocations. Furthermore, a correlation of regions with high PID-shunt densities is discovered at cells made from neighboring multicrystalline silicon wafers. This finding is discussed in terms of surface defect precursors that can act as nuclei for PID-shunts.

I. INTRODUCTION

Potential-induced degradation (PID) of PV modules is a topic of sustained scientific interest [1]. PID of the shunting type (PID-s) exhibits the most detrimental impact on the reliability and yield of installed modules containing crystalline Si solar cells. Shunting of the junction is due to a very high number of scattered, individual shunts. These PID-shunts could be attributed to extended Na decorated, planar crystal defects ("stacking faults") penetrating the p-n junction of solar cells [2]. While the physical nature of the PID-shunts can be assigned to stacking faults with very high probability [3], there is only few knowledge on their origin. After recovery of PID-shunts by thermal annealing, undecorated stacking faults could be identified, leading to the assumption that they could be present before decoration with Na [4]. However, more recent investigations gave strong hints that stacking faults first grow under high voltage stress [5]. The goal to produce potentially inherently PID-s stable cells motivates the quest for extended knowledge on the evolution of PID-s stacking faults.

In this work, a further experimental proof is presented that stacking faults are not existent after the cell production process but first evolve during high voltage stress under PID-s promoting stress conditions (PID-stress). It is also shown that, at least for float zone-monocrystalline cells, the positions of stacking faults do not correlate with primordial threading dislocations in the crystal nor other extended bulk defects. Secondly, a collection of results including new investigations at pairs of cells from originally neighboring multicrystalline wafers are presented, indicating a correlation between defects at the Si wafer front surface (i.e. depth < 10 μm) and intensity/density of PID-shunts. These investigations indicate that PID-shunts may originate preferentially at imperfections of the surface, which might be scratches, texturization-induced surface steps, or dimples.

II. EXPERIMENTAL

A. Set of samples

Defect etch experiments are performed at a ~60 x 12 mm² sized piece of a float zone-monocrystalline solar cell with polished front side, diffused P front side emitter and SiN_x antireflective coating (prone to PID-s). The cell piece contains a busbar and grid fingers.

Correlative electroluminescence on mini module level is done using two standard multicrystalline solar cells prone to PID-s that had been processed from neighboring wafers. The cells are encapsulated into mini modules (single cell modules) without a backsheet foil.

Correlative imaging of PID-s impacted cell regions is done using three PID-s prone multicrystalline solar cells (size 156 x 156 mm², standard industrial screen print layout, acidic front surface texture) that had been processed from neighboring wafers.

B. Degradation of cell samples

A ~1 cm² sized part of the monocrystalline cell piece was degraded using a customized PID cell test setup at 85 °C and at high voltage of +1000 V. The high voltage was applied for 22 hours to the glass surface of the ~10x12 mm² sized glass/EVA stack which is placed on the cell piece. The shunt resistance of the cell sample is monitored by means of in-situ current measurement at a reverse bias of 0.5 V.

The cells inside mini modules were degraded with aluminum foil attached to the front glass over the whole area. The positive high voltage (+600 V) was applied with respect to the grounded cell connectors. The degradation was performed on a hot plate at 85 °C for 24 hours.

Two of the three cells made from neighboring wafers were degraded on an area of 100 x 100 mm² using a glass/EVA stack placed on the front side of the cell. The positive high voltage (+1000 V) was applied to the glass surface with respect to the grounded cell for 24 hours at a temperature of 75 °C using the PIDcon cell test device by Freiberg Instruments.

C. Sample preparation

The cell region of the polished monocrystalline cell piece that was subjected to the PID-stress was cut from the rest of the cell piece and then delaminated from the EVA and glass stack. A reference sample of equal size, but without PID-stressing was cut from the same cell piece.

The glass/EVA stacks were removed from the two degraded multicrystalline cells after the PID test. After initial electroluminescence and dark lock-in thermography characterization, pieces with a size of 32 x 18 mm² were prepared out of the same region of each cell by laser cleaving from the rear side.

D. Analytical instrumentation

Optical microscopy is performed using a Keyence VHX-2000 and a Zeiss Jenatech microscope. Etch marks are evaluated using the ImageJ 1.38x software package. Electroluminescence (EL) images are acquired using an Andor iKon-M PV-Inspector camera equipped with 950 to 1000 nm bandpass filter for reducing photon scattering in the detector. EL images are processed using ImageJ 1.38x software. Dark lock-in thermography (DLIT) images were performed using the PV-LIT system of InfraTec. On the cells embedded in mini modules, the DLIT investigations were made from the rear side. Secondary electron microscopy (SEM) and Electron-beam induced current (EBIC) images were acquired using a JEOL 6400 SEM.

III. RESULTS

A. Absence of stacking fault-like defects before PID-stressing

At first, the important question is addressed whether stacking fault-like extended defects are present at the solar cell surface before PID-stress or if they develop upon PID-stress. A sole microstructural identification of stacking fault-like extended defects was required allowing a stacking fault density analysis before PID testing. For this purpose, defect etching was implemented. In this work, pieces of a monocrystalline solar cell with polished front side have been used. After removal of the SiN$_x$ layer and *Secco* [6] defect etching of the front surface of both monocrystalline cell pieces (one of them with and one of them without PID-shunting), micrographs of 1/4 of the area of each sample were recorded by optical microscopy at 1000x magnification. The images were stitched together and then have been processed by an in-house-developed software tool in order to automatically count the visible etch marks and classify them according to their size and orientation. The histogram in Fig. 1 shows the angular orientation distribution of elongate features (etch marks) for sample pieces without (black) and with PID-s (red). The inset shows a small representative detail area of the sample that exhibits PID-s.

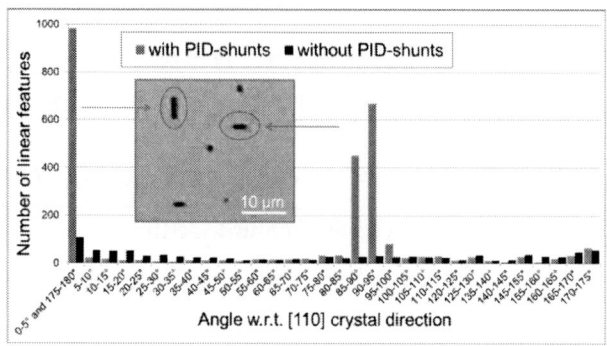

Fig. 1. Frequency of elongate etch marks as a function of their angular distribution for sample pieces without PID-s (black) and with PID-s (red)

The sample with PID-s exhibits a significant surplus amount of etch marks oriented parallel to the <110> crystal directions, represented by the red bars from 175° to 5° and 85° to 95°, which exceed that of the sample without PID-shunts (black bars) by a factor of ~10. The underground counts at odd angles are comparable for both samples. Note that at both samples the same area has been evaluated. This is a strong hint that stacking faults, which are attributed to the etch marks in <110> directions, evolve first after PID-stress. This is further proven by an areal evaluation for the sample with PID-s (Fig. 2a). Obviously, the density of horizontally or vertically oriented etch marks corresponding to stacking faults is significantly increased in the region that was subject to PID-stress before. Therefore, it is concluded that no stacking fault-like extended structural defects are present before application of PID-stress.

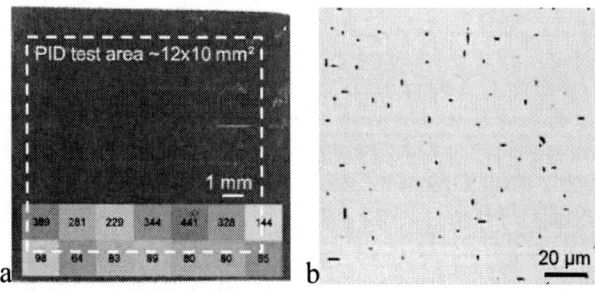

Fig. 2. (a) Laterally resolved counts of etch marks with stacking fault signature. The numbers in the tiles represent the counts of horizontally or vertically oriented etch marks and the color code from green (low density) to red (high density) additionally visualizes their density in correlation with the original PID test area (white dashed border line). (b) Detail view (optical microscopy image) of the region with highest etch mark density.

B. Exclusion of threading dislocations as nuclei of PID-shunts

The next question to be addressed is if there are local structural defects that penetrate the bulk Si material (e.g. dislocations) and thus define the starting point of 2D-extended

PID-shunts. For this, a region with linear etch marks of the latter sample, as shown in Fig. 2b (with PID-shunting), was imaged in detail. Then the surface of the sample was polished down by 5 to 10 µm using an isotropic chemo-mechanical (alkaline colloidal silica) polishing process. After this surface erosion polish the sample was defect etched again. Optical microscopy was repeated at the same position where stacking fault related etch groves have been found before. Fig. 3 shows a comparison of images with etch marks before and after surface erosion. A deep defect structure served as reference point for spatial correlation (blue circle). After surface erosion, there are no etch pits observable at the previous stacking fault positions (red arrows) that may correspond to threading dislocations or other extended bulk defects inside the Si crystal. By this, at least threading dislocations can be excluded to act as nuclei for PID-shunts.

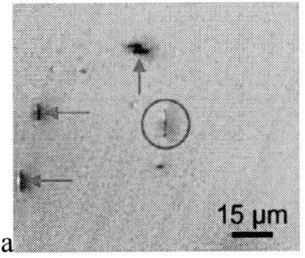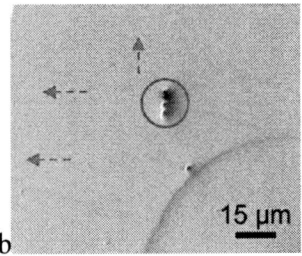

Fig. 3.   (a) Detail view of the position of four etch marks related to stacking faults. (b) No etch marks are visible at the same positions after 5-10 µm surface erosion and subsequent etching. (The blue-marked feature is not a stacking fault.)

C. Correlation of PID-shunt densities with cell properties on macroscopic and microscopic level

With the results shown in section B a direct coherence between threading dislocations and positions of PID-shunts can be excluded. Together with the fact that stacking faults do not exist before application of PID-stress, this implies that the position of PID-shunts must be predefined by surface-near defects. Hints for that were already reported in a recent publication: a significant correlation between scratches at the surface (likely induced before SiN_x deposition) and occurrence of PID-shunts was found on a monocrystalline cell with a polished front surface [5]. This indicates that the probability of PID-shunt formation is highly increased by surface damage or local strain related to scratches, respectively. On the other hand, different surface-near defects that predefine positions of PID-shunts might be caused by bulk related properties (e.g. Si material: crystal orientation, strain, point defect densities; solar cell process damage: sawing, transportation, firing).

In order to follow the possible implications on local strain or locally increased point defect densities, macroscopic patterns have to be found that can be related to such lateral inhomogeneities at the Si surface. A first step into that direction is conducted in the present work. The "PID-shunt map", based on the ratio of EL images after and before PID-stressing, is used for evaluation and correlation of PID-shunt densities on mini module level. The PID-shunt map displays only regions in dark colors that have changed their EL intensity by PID-shunting. Regions with low EL signal, which are not affected by PID-shunting, remain bright in the PID-shunt map. Fig. 4 shows PID-shunt maps of two mini modules, where the PID-tests have been performed on the whole solar cell area, respectively.

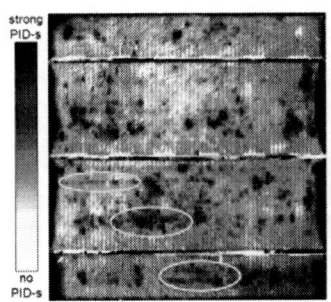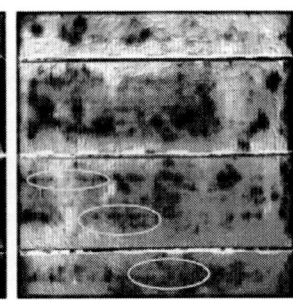

Fig. 4.   PID-shunt maps of two neighboring cells, showing PID-shunts at similar positions and preferentially horizontal shunt signatures (green markings) after PID-testing. The color scale is in a.u.

Confined regions at identical positions on cells made of neighboring wafers seem to be similarly affected, indicating an influence of bulk material properties on the formation of near-surface defect nuclei and thus on the evolution of PID-shunts. This may be related to strain within the crystal or to the crystal orientation of multicrystalline cells. Furthermore, the preferential horizontal orientation of PID-shunted regions (green markings in Fig. 4) indicates process signatures.

The correlative imaging of PID-shunt distribution is repeated with non-encapsulated solar cells that are made from neighboring wafers. In order to have access to the cell surface with PID-shunts, PID tests have been performed using the *PIDcon* cell test setup. This allows microstructural investigations at the PID-s affected cell regions by SEM/EBIC.

Fig. 5 shows EL and DLIT images of the same PID-s affected region (10 cm x 10 cm) of the two neighboring cells after PID testing. The upper left corner of both cells broke during delamination from the EVA/glass layer stack used for the PID cell test. PID-shunt maps could not be evaluated from the EL images of these cells since here shunting was significantly stronger than for the cells shown in Fig. 4. Again, the neighboring cells exhibit a macroscopically similar distribution of PID-s defect density, indicated by the lateral distribution of the shunt intensity, which is high in corresponding dark regions of EL images and in bright regions in DLIT images. Like for the mini modules (Fig. 4), this can be attributed either to bulk material properties, cell process induced surface defects, or – more likely – a mixture of both.

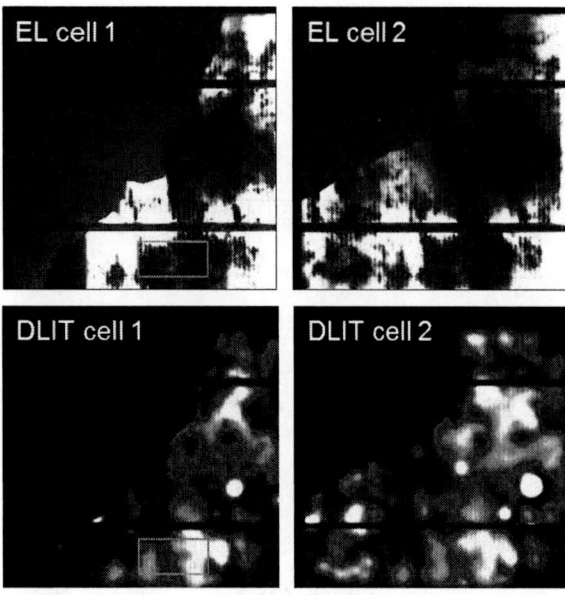

Fig. 5. EL (top) and DLIT images (bottom) of the neighboring cells after similar PID cell tests (image section: 10 cm x 10 cm). The green frame marks the area of cell 1 that is further imaged in Fig. 6.

In order to enlighten possible microstructural features, 32 x 18 mm² sized pieces are cut out of cells a by laser cleaving and investigated with SEM. The main results of electron microscopy investigations done at the cell pieces of the two PID-s affected cells ('1' and '2') and one neighboring reference cell ('R') without PID-s are shown in Fig. 6. The large images on the left hand side of Fig. 6 show the EBIC images of the same region of the cells R and 1, stitched from a number of EBIC images each. Remarkably, there seems to be a weak positive correlation between the density of recombination active grain boundaries (dark lines in EBIC image of cell R) and the PID-shunt density (dark regions in EBIC image of cell 1). This finding fits qualitatively to the results presented in [7], where, however, only a coarse macroscopic correlation between the overall density of defects in the wafer and PID-s strength was reported. Fig. 6 further shows that the PID-shunts do not evolve directly at the grain boundaries, but in the area of grains (higher resolved EBIC images on the right of Fig. 6). Note that the background EBIC intensity and with it the contrast of recombination active grain boundaries depends on the shunt density of respective cell regions. Furthermore, despite the distribution of PID-shunt

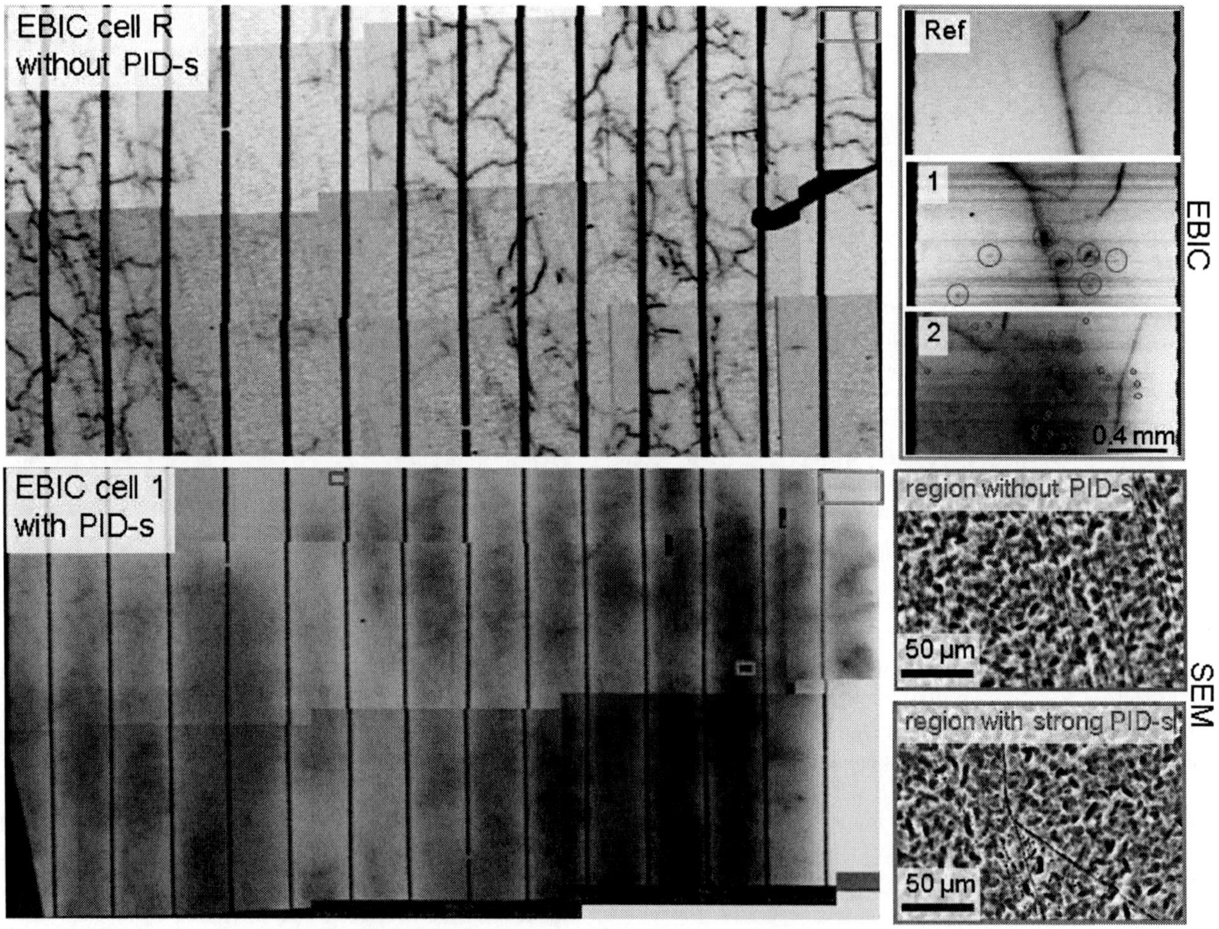

Fig. 6. SEM/EBIC micrographs of cell pieces without previous PID-stress (sample 'R', reference) and of cell pieces after PID-stress (cells 1 and 2) at different magnifications and contrast settings. See text for details.

density correlates between neighboring cells on a macroscopic scale down to some millimeters (Fig. 5), there is no correlation of PID-shunt positions on a microscopic scale, as shown in the magnified EBIC images of cells 1 and 2 in Fig. 6. Here, the density of PID-shunts (marked by red circles) within the same area differs by a factor of ~5 between cell 1 and cell 2. By means of EBIC and SEM imaging on a micrometer scale, at the majority of investigated PID-shunts an etch pit is found, and at others different sharp-edged features are found at the cell surface. The etch pits are a result of the acidic surface texturing process and therefore indicate the positions of surface-near extended crystal defects, that existed already before PID-stressing. Note that for most of previous investigations, cells with alkaline surface texture had been used and therefore such correlations with etch pits were not found before. The two SEM images in the right bottom of Fig. 6 have been acquired in two regions of cell 1 and show typical views of the visible surface structure. The first (blue frame) is of a cell region without PID-shunts (despite it was subject to PID-stress). It exhibits a high number of twin boundaries and relatively few etch pits within grains. The second SEM image is of a cell region with high PID-shunt density (red frame). It exhibits only a few twins, many small and large angle grain boundaries and many etch pits within grains. (Etch pits appear as small black dots, e.g. around the intersection of the three grain boundaries left of the center of the image.)

IV. DISCUSSION AND CONCLUSION

The results shown prove that stacking fault-like extended defects are not present at the Si surface before application of PID stress. It is also found that, at least in monocrystalline cells, common threading dislocations penetrating the Si wafer do not act as potential starting points for the formation of PID-shunts. The present substantial experimental results allow the exclusion of both primordial surface-near (depth < 10 μm) *planar* defects and primordial *extended bulk* defects to be responsible for the evolution of PID-s.

Considering previous work, where PID-shunts preferentially grew at scratches in the Si surface, it is rather assumed that the evolution of PID-shunts is triggered by local strain. The densities of PID-shunts and of original etch pits (formed during acidic surface texturization process) as well as recombination active grain boundaries show a positive correlation. In [8] it was found that etch pits are formed at line defects, which penetrate the surface of the solar cell and are etched during the acidic texturization process. Since we found etch pits in the direct vicinity of PID-shunts in multicrystalline material, such defects may also play a role in the formation of the stacking faults leading to PID-shunts. However, a direct proof of this hypothesis was not achieved in this work and will be part of forthcoming work. At the same time, the density of PID-shunts seems to be anti-correlated to the occurrence of twin boundaries. It is known that twin boundaries relax local strain. This might be a hint that the strain distribution in the grain plays a role for the formation of the stacking faults responsible for PID-shunts. These findings indicate that possible strain induced microstructural surface-near defects represent the points of origin for PID-shunts, i.e. weak points for the incorporation of Na, leading to an evolution of Na filled planar crystal defects.

ACKNOWLEDGEMENTS

Financial support through the German Ministry of Economics within project PID-s (FKZ 0325748A) is gratefully acknowledged. Thanks to Fraunhofer ISE for providing monocrystalline solar cells with polished front surface.

REFERENCES

[1] W. Luo, et al., Potential-induced Degradation in Photovoltaic Modules: A Critical Review, Energy Environ. Sci. 10, 43-68 (2017). DOI: 10.1039/C6EE02271E

[2] V. Naumann et al., Explanation of potential-induced degradation of the shunting type by Na decoration of stacking faults in Si solar cells, Sol. Energ. Mat. Sol. Cells 120, 383 (2014).

[3] S. P. Harvey et al., Sodium Accumulation at Potential-Induced Degradation Shunted Areas in Polycrystalline Silicon Modules, IEEE Journal of Photovoltaics 6 (6), 1440-1445 (2016). DOI: 10.1109/JPHOTOV.2016.2601950

[4] V. Naumann et al., Nanoscopic studies of 2D-extended defects in silicon that cause shunting of Si-solar cells, physica status solidi (c) 12 (8), 1103–1107 (2015).

[5] V. Naumann, et al., Investigations on the formation of stacking fault-like PID-shunts, Energy Procedia 92, 569-575 (2016).

[6] M.H. Jones and S.H. Jones, Wet-chemical etching and cleaning of silicon (2003).

[7] Gou et al., Influence of crystal defect density of silicon wafers on potential-induced degradation (PID) in solar cells and modules, Phys. Status Solidi A, (2017), online version, DOI: 10.1002/pssa.201700006

[8] Breitenstein et al., Defect-Induced Breakdown in Multi-crystalline Silicon Solar Cells, IEEE Transaction on electron devices 57 (9), 2227-2234 (2010). DOI: 10.1109/TED.2010.2053866

Investigating PID Shunting in Polycrystalline Silicon Modules via Multi-Scale, Multi-Technique Characterization

Steven P. Harvey,[1] John Moseley,[1] Adam Stokes,[2] Andrew Norman,[1] Brian Gorman,[2] Peter Hacke,[1] Steve Johnston,[1] Mowafak Al-Jassim[1]

[1] National Renewable Energy Laboratory, Golden, CO, 80401, USA
[2] Colorado School of Mines, Golden, CO 80401, USA

Abstract — **Potential-induced degradation (PID) has been investigated in silicon modules that have degraded during field deployment, as well as in mini-modules stressed in the laboratory. Small cores have been removed from the modules and subjected to analysis. To investigate the root-cause mechanism for PID shunting, we use a combination of photoluminescence and dark lock-in thermography (DLIT) imaging, laser marking, electron-beam induced current measurements, and subsequent focused ion-beam marking to allow analysis of individual defects via electron-beam induced current (EBIC) and time-of-flight secondary-ion mass spectrometry (TOF-SIMS). We see a direct correlation between recombination active shunts and sodium content. The sodium content in shunted areas peaks at the SiN/Si interface, and it is consistently observed at a concentration of 0.1%–1% in shunted areas. PID recovery done *in-situ* in a scanning electron microscope by use of an electron beam was investigated. We can observe the annealing out of previously shunted areas with subsequent DLIT, and TOF-SIMS reveals an order-of-magnitude decrease in sodium content after this *in-situ* annealing/recovery step, in addition to sodium out-diffusion from the defect. We also report for the first time on a potential role of oxygen, and additional species other than sodium, in PID shunting. Scanning transmission electron microscopy energy-dispersive X-ray spectroscopy, atom-probe data, and TOF-SIMS analysis of a PID shunt show a correlation between sodium and also oxygen at concentrations above 1% in a shunted area. TOF-SIMS measurements of grain boundaries where EBIC recombination is noted does not show a correlation between sodium and recombination as observed with PID shunts.**

Index Terms — **PID, TOF-SIMS, EBIC, polysilicon, reliability.**

I. INTRODUCTION

Potential-induced degradation (PID) is a challenging reliability issue for crystalline silicon solar cells. PID of the shunting type can cause significant power loss and even total module failure [1]. PID is believed to be caused by the diffusion of metal ions such as Na^+ from the front glass/module surface that is driven into the silicon by the high voltage present [1].

Several studies have observed an increase of sodium in and around PID shunted areas [1, 2]. We expand upon prior works by continuing to develop methods that enable multi-scale,

multi-technique characterization of PID shunting. We also report for the first time on atom-probe tomography of a PID shunt, which reveals a sodium content consistent with time-of-flight secondary-ion mass spectrometry (TOF-SIMS) data and transmission electron microscopy (TEM) results; but we also report that the atom-probe data revealed a significant amount of oxygen in the shunt region. Subsequent TOF-SIMS three-dimensional (3-D) tomography data taken in negative polarity of similarly shunted areas reveal a high content of oxygen and chlorine in shunted areas, suggesting that these two elements may also be playing a role in PID degradation of polysilicon modules.

II. EXPERIMENTAL

A field-degraded crystalline silicon module was investigated in this work, in addition to a laboratory-stressed mini-module (details of the mini-module stressing were given in a previous publication) [2]. The entire module was imaged with photoluminescence (PL) and dark lock-in thermography (DLIT) imaging to identify shunted areas of interest. Circular cores of the module stack at areas of interest were then cut from the module using a 1"-diamond hole saw, and the silicon was separated from the module glass using a mechanical stack-cleaving technique described in detail in a previous publication [3]. The polysilicon module core was then laser-marked around defects of interest from the photoluminescence (PL) image, which provided a reference point for further characterization at the millimeter length scale. This same concept was completed at the micron length scale by using electron-beam induced current (EBIC) to identify individual shunts within the laser-marked area, and focused ion-beam (FIB) marking around an individual shunt. This allowed cross-technique correlation between PL, DLIT, EBIC, TEM, scanning transmission electron microscopy (STEM), atom probe, and TOF-SIMS for individual shunts.

The shunted areas were analyzed by TOF-SIMS using an ION-TOF TOF SIMS V instrument. Secondary-ion mass spectrometry (SIMS) is a powerful analytical technique for determining elemental and isotopic distributions in solids. TOF-SIMS imaging and 3-D tomography were used to measure the lateral and depth distribution of sodium and other elements potentially responsible for PID. A ccsium or oxygen ion beam with energy varying from 600 eV to 3 keV was used as the sputtering beam (sputtering current 3–60 nA). Secondary ions for analysis were created by a 3-lens 30-keV BiMn ion gun. 3-D tomography was completed using a Bi_3^{++} primary ion-beam cluster (100-ns pulse width, 0.1-pA pulsed beam current); this measurement mode is capable of better than 100-nm lateral resolution. Atom-probe data were completed using laser-assisted, local-electrode atom-probe tomography.

III. RESULTS

An image compilation is shown in Fig 1, which illustrates the methods used to go from characterization at the entire module scale, to chemical information at the nanoscale from one single PID shunt. In this case, the defect that is causing local shunting and recombination (noted in DLIT and EBIC) is correlated to a sodium concentration on the order of 0.1% (1×10^{20} cm^{-3}), which is 1–2 orders of magnitude higher than the sodium content in the surrounding areas. We have examined tens of areas with similar shunts and consistently see similar trends as those observed in Fig. 1. The sodium content peaks at the Si/SiN interface and the concentration in the shunted region is on the order of 0.1%–1% sodium (1×10^{20}–1×10^{21} cm^{-3}). As reviewed in [1], prior TEM work has consistently shown that these shunts are the result of subsurface stacking faults.

The procedure illustrated in Fig. 1 was also used to identify a defect for FIB liftout for TEM lamella and atom-probe tip preparation. Atom-probe data for a PID shunt in a laboratory-stressed mini-module were collected. Similar to what is seen with TOF-SIMS, the shunt area is high in sodium content, and the concentration of ~0.5% correlates well with the TOF-SIMS data. The atom-probe data also showed a similar elevated concentration for oxygen in the shunt area. Similarly, STEM chemical mapping using energy-dispersive X-ray spectroscopy (EDS) showed high sodium and oxygen concentrations of ~1% in the planar defect associated with shunting. Unlike SIMS, STEM EDS and atom probe are not subject to looking at secondary ions of a single polarity. SIMS measurements for PID are typically done in positive polarity so that the sensitivity for sodium is maximized, whereas in this polarity, the sensitivity for oxygen is very poor. TOF-SIMS data were thus collected in negative polarity for a similarly shunted area, and a high concentration of oxygen and chlorine was observed in the presumed shunted areas. This suggests that it may not be sodium alone that is causing PID shunting in degraded polysilicon modules.

During high-resolution EBIC scanning of PID shunts—which we call e-beam annealing—we noticed that the shunt contrast decreased over time. To study this effect further, we conducted a more controlled experiment illustrated in Fig. 4. DLIT imaging (Fig. 4(a)) was first used to find an area with several PID shunts. An EBIC image was taken on the same area (Fig. 4(c)), and a few of the shunts were subsequently subjected to e-beam annealing. As shown in Fig. 4(d), the EBIC contrast at the e-beam annealed shunts is significantly reduced. Another iteration of DLIT imaging indicated that these shunts were no longer visible (Fig. 4(b)), confirming the restorative effects of the e-beam annealing. Finally, the same area was analyzed by TOF-SIMS. The data show an order-of-magnitude decrease in the sodium content at an e-beam annealed shunt compared to a neighboring shunt subjected to less annealing. We postulate that localized heating from the e-beam leads to the observed PID recovery *in-situ*. Indeed, subsequent measurements show evidence for an out-diffusion of sodium from the defects as a result of the electron-beam annealing.

Within the same dataset shown in Fig. 1B there are two grain-boundary areas of interest. This area is shown in finer detail in the EBIC image in Fig. 3. Two areas of the grain boundary are highlighted: one that is an active recombination area, and another that appears to be a boundary where recombination is not occurring. Figure 3 also shows the corresponding TOF-SIMS data from these areas, and it shows a similar sodium content for both the recombination-active and recombination-free grain boundaries. Not only is the sodium content in these areas similar, but it is also present at very low concentration. This is an interesting result and suggests that sodium is not playing the same role in relation to grain-boundary performance as it does with PID, in that it decorates structural defects and leads to the observed shunting.

IV. CONCLUSIONS

We report on further development of multi-scale, multi-technique characterization methods to allow analysis of individual defects to investigate the root-cause mechanisms for PID shunting in field-degraded and laboratory-stressed polysilicon modules. We use a combination of high-resolution DLIT, laser marking, EBIC, and FIB marking to allow TOF-SIMS analysis of specific defects on the micron scale. Using these methods, we see that the shunted areas correspond to planar defects in the silicon and to the sodium content in the shunts peaks at the SiN/Si interface, which is consistently observed at a concentration of 0.1%–1%. We also report on PID recovery done *in-situ* in an SEM by use of an electron beam. TOF-SIMS reveals an order-of-magnitude decrease in sodium content after this *in-situ* annealing/recovery step. We also report on preliminary work that suggests that sodium is not playing a role in recombination observed at grain boundaries.

IV. AKNOWLEDGEMENTS

This work was supported by the U.S. Department of Energy under Contract No. DE-AC36-08GO28308 with the National

Renewable Energy Laboratory. The U.S. Government retains and the publisher, by accepting the article for publication, acknowledges that the U.S. Government retains a nonexclusive, paid up, irrevocable, worldwide license to publish or reproduce the published form of this work, or allow others to do so, for U.S. Government purposes.

V. REFERENCES

[1] W. Luo, Y. S. Khoo, P. Hacke, V. Naumann, D. Lausch, S.P. Harvey, J.P. Singh, J. Chai, Y. Wang, A.G. Aberle, and S. Ramakrishna, "Potential-induced degradation in photovoltaic modules: A critical review," *Energy & Environmental Science,* vol. vol. 10, pp. 43–68, 2017.

[2] S.P. Harvey, J.A. Aguiar, P. Hacke, H. Guthrey, S. Johnston, and M. Al-Jassim, "Sodium accumulation at potential-induced degradation shunted areas in polycrystalline silicon modules," *IEEE Journal of Photovoltaics,* vol. 6, pp. 1440–1445, 2016.

[3] H.R. Moutinho, S. Johnston, B. To, C.-S. Jiang, C. Xiao, P. Hacke, J. Moseley, J. Tynan, N.G. Dhere, and M.M. Al-Jassim, "Development of coring procedures applied to Si, CdTe, and CIGS solar panels," *Solar Energy Materials and Solar Cells,* vol. Submitted, 2017.

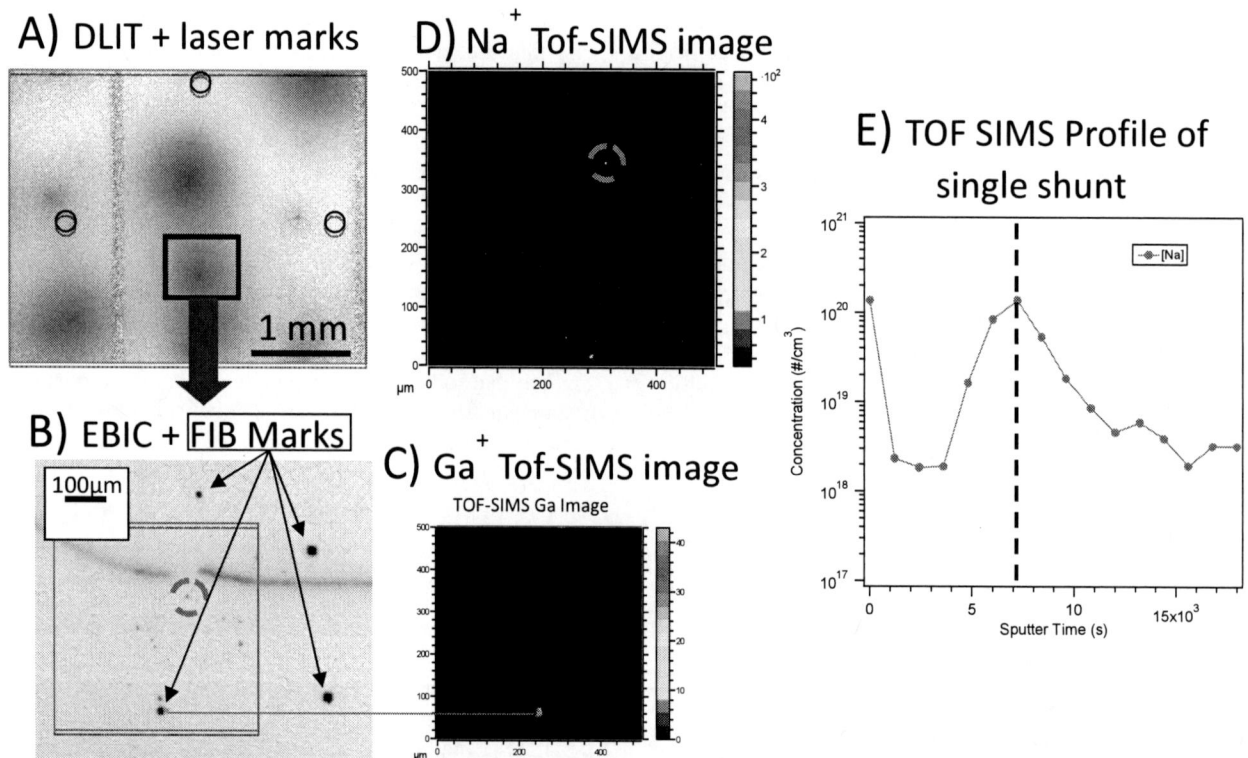

Figure 1. A) DLIT imaging of area with PID shunts and laser markings; B) FIB marks put on and EBIC shows relative to these marks; the red box shows the TOF-SIMS analysis area; C) TOF-SIMS image (500×500 µm), showing the Ga signal (correspond to FIB mark drawn with a line to B); D) TOF-SIMS 2-D image (500×500 µm) of the shunt area. The shunt identified with a red circle in B is also circled here; E) Selected-area depth profile from a shunted region, marked with a circle in D). The sodium concentration peaks at ~0.1% at the SiN/Si interface in the shunted region; this interface is marked with a dashed line in the profile.

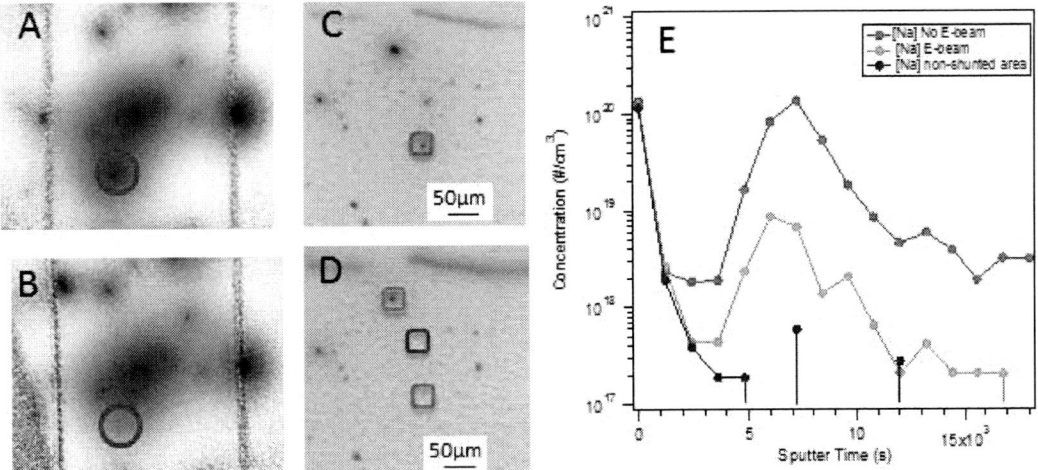

Figure 2. A) DLIT image of shunted areas before e-beam annealing; B) DLIT shows area of shunts that were subjected to e-beam anneal (circled in both images) no longer visible in DLIT; C) EBIC image of grouping of PID shunts; the shunt with the red box was subjected to "e-beam annealing": D) Subsequent EBIC image shows that shunt intensity is reduced. The sample was then FIB-marked for TOF-SIMS analysis; E) TOF-SIMS selected-area depth profiles from areas shown in D); red is the annealed shunt, green is a non-annealed shunt; a nonshunted region is showed for comparison. An order-of-magnitude decrease in sodium content is noted in the high-resolution EBIC scan of a single annealed shunt.

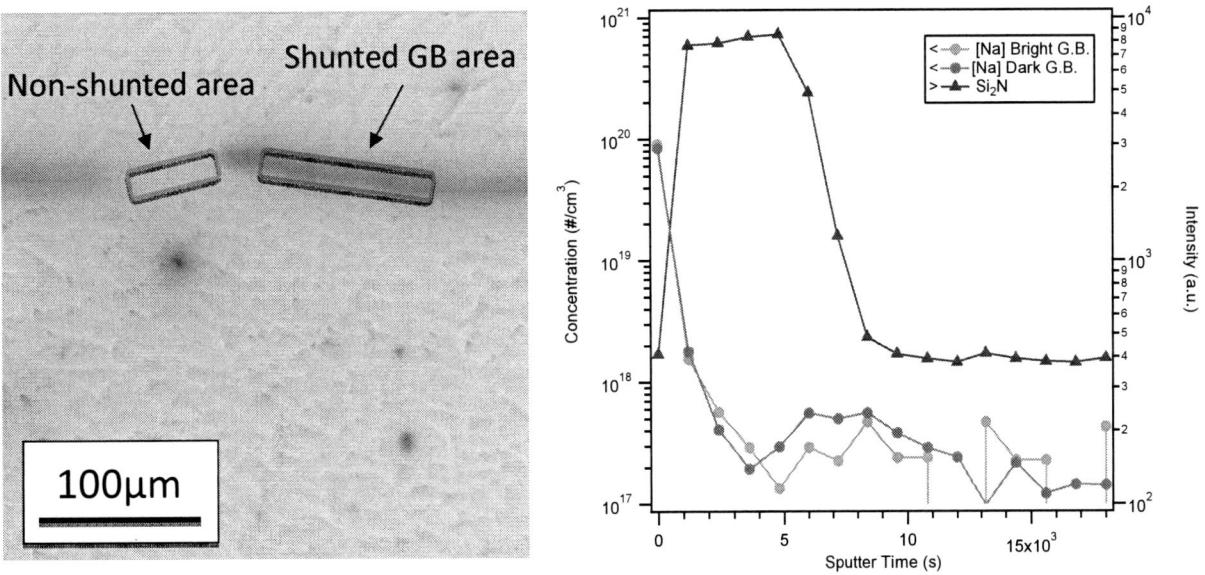

Figure 3. (left) An EBIC image of the same area seen in Fig 1B, focused more on the grain-boundary region. Two areas of the grain boundaries are circled—one section where recombination is active (dark GB – red box), and one where it is not (green box). (right) TOF-SIMS selected-area depth profiles from areas highlighted in the EBIC image; red is the recombination-active shunt, and green is the benign GB area. Despite the clear difference in recombination between the two areas, both grain-boundary regions have similar sodium content.

978-1-5090-5606-4/17 $31.00 © 2017 IEEE

Potential-Induced Degradation of a Si Nitride/Crystalline Si Interface Observed Through Minority Carrier Lifetime Measurement

Naoyuki Nishikawa, Seira Yamaguchi, and Keisuke Ohdaira

Japan Advanced Institute of Science and Technology (JAIST), Nomi, Ishikawa, 923-1292, Japan

Abstract — We directly observed the potential-induced degradation (PID) of a silicon nitride (SiN_x)/crystalline Si (c-Si) interface by measuring the effective minority carrier lifetime (τ_{eff}) of c-Si wafers laminated in a module-like structure. We observed slow τ_{eff} reduction during a negative-bias PID stress and a rapid drop of τ_{eff} by positive-bias application. The former may correspond to the introduction of Na to stacking faults on c-Si, while the latter is probably related to negative charge accumulation in the SiN_x. These behaviors well reproduce the characteristics of PID for n-type rear-emitter (RE) photovoltaic (PV) modules.

Index Terms — potential-induced degradation, photovoltaic module, rear-emitter solar cell, minority carrier lifetime, silicon nitride, surface recombination velocity, crystalline silicon, surface polarization effect, sodium

I. INTRODUCTION

PID, the degradation of PV module performance triggered by a voltage between a module frame and cells, is one of the most significant issues for the long-term operation of PV power plants. The mechanism and behavior of PID strongly depend on cell structures as well as module components. Clarifying the mechanism of PID for PV modules using n-type c-Si wafer-based cells [1–6] is particularly important because of potentially higher efficiency than conventional p-type wafer-based cells and the expectation of increasing market share in the near future. n-type RE PV cells can be fabricated through almost the same process as conventional p-type cells, and rapid popularization is thus expected. We have thus far clarified the behaviors of the PID of n-type RE PV modules [6]. Slight reductions in open-circuit voltage (V_{oc}) and external quantum efficiency (EQE) in a short wavelength region are seen both for negative- and positive-bias applications to cells with respect to a cover glass surface. These are the indications of the enhancement of carrier recombination on the c-Si surface. We have not, however, directly observed the enhancement of surface recombination on and near the surface of c-Si passivated with SiN_x. In this study, we have attempted to observe the change in surface recombination near SiN_x/c-Si interfaces through minority carrier lifetime measurement.

II. EXPERIMENTAL PROCEDURES

Figure 1 shows the schematic diagram of a c-Si sample structure used for the measurement of τ_{eff}. After the standard

RCA cleaning of float-zone (FZ)-grown, 20×20 mm²-sized, mirror-polished c-Si wafers, a highly phosphorus-doped n-type layer was formed on one side of the c-Si wafers by spin coating liquid phosphosilicate glass and successive annealing at 850 °C for 25 min. We then formed ~75-nm-thick SiN_x passivation films with a refractive index of ~2 on both sides of the c-Si wafers by catalytic chemical vapor deposition (Cat-CVD) at SiH_4 and NH_3 flow rates of 8 and 300 sccm, respectively, at a pressure of 15 Pa, and at stage and catalyzer temperatures of 300 and 1800 °C, respectively. Ag electrode was formed on the n^+-layer side by screen printing Ag paste and firing at ~800 °C for Ohmic contact to the c-Si wafers. Interconnector ribbons were then connected to the Ag electrode. The c-Si samples were encapsulated using conventional tempered cover glass, ethylene-vinyl acetate copolymer (EVA), and a poly(vinyl fluoride) (PVF)/poly(ethylene terephthalate) PET/PVF backsheet by putting n^+-layer side down. The lamination of the stacks was performed through a 5-min degassing step at 115 °C and a following 15-min adhesion step at 115 °C. The cross-sectional schematic and surface photograph of the module-like structure are shown in Figs. 2 and 3. PID tests were performed by applying a bias of −1000 V or +1000 V to the laminated c-Si with respect to an Al plate placed on the cover glass surface of the module-like structure. We also performed recovery tests by applying a bias with a polarity opposite to a prior PID test to the laminated c-Si wafers. Temperature was set to 85 °C during the PID and recovery tests, while humidity was not intentionally controlled (<2%RH). Change in the surface recombination by PID and recovery tests was evaluated by measuring the τ_{eff} of the laminated c-Si wafers by microwave photoconductivity decay (μ-PCD). Note that a decrease (increase) in the τ_{eff} of laminated c-Si can be considered as the

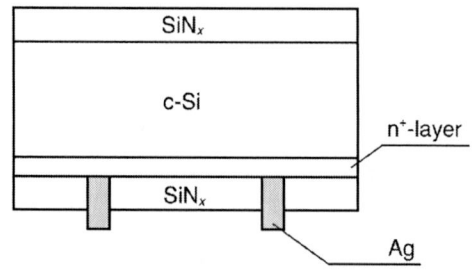

Fig. 1. Schematic diagram of a c-Si wafer with SiN_x passivation films.

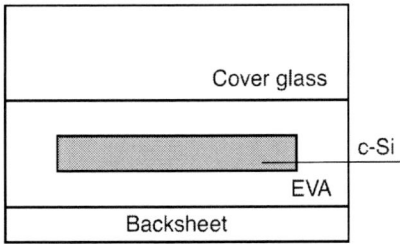

Fig. 2. Cross-sectional schematic diagram of a module-like structure used for the measurement of τ_{eff}.

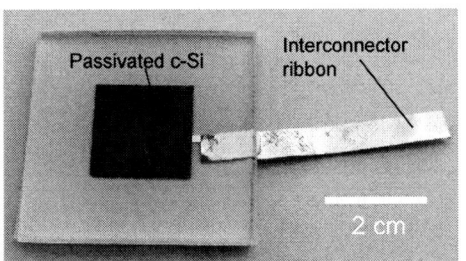

Fig. 3. Surface photograph of a module-like structure used for the measurement of τ_{eff}.

degradation (recovery) of the surface recombination velocity since no degradation of bulk minority carrier lifetime (τ_{bulk}) may have occurred inside c-Si far enough away from the surface during the PID and recovery tests.

III. RESULTS AND DISCUSSION

Figure 4 shows the τ_{eff} of a laminated c-Si wafer as a function of the duration of a PID test applying -1000 V to the

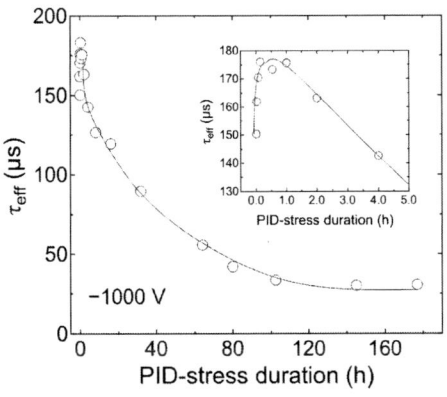

Fig. 4. τ_{eff} of a laminated c-Si wafer as a function of the duration of a PID test applying -1000 V to the c-Si. The solid lines are guides to the eye.

c-Si. The c-Si sample initially showed a τ_{eff} of ~150 µs. The τ_{eff} of the laminated c-Si first rapidly increased to ~180 µs, as shown in the inset of Fig. 4, and then decreased gradually to a saturated value of ~30 µs for ~100 h. The variation in the τ_{eff} of the laminated c-Si is a clear indication of the change of surface recombination velocity. The first rapid improvement in τ_{eff} probably originates from the accumulation of positive charges in the surface SiN_x films and resulting downward band bending near the SiN_x/c-Si interface. The subsequent slow reduction in τ_{eff} may be due to the formation of stacking faults and successive decoration of the stacking faults by Na drifting from cover glass and other materials [7]. The reduction in τ_{eff} may corresponds to the degradation of the performance of n-type RE PV modules by negative bias application [6], although duration needed for the saturation of degradation in the module performance (<24 h) is largely different from that of τ_{eff} observed in this study. The large difference may be due to the difference of surface passivation structure and/or quality of the films.

Figure 5 shows the τ_{eff} of a laminated c-Si wafer as a function of the duration of a PID test applying $+1000$ V to the c-Si. τ_{eff} decreased rapidly from >200 µs to ~100 µs within several minutes and showed no further reduction. The rapid reduction in τ_{eff} is probably due to the accumulation of negative charges in the surface SiN_x, which increases the density of minority carriers, holes, and enhances carrier recombination near the SiN_x/c-Si interface. This phenomenon may be essentially identical to "surface polarization effect", observed in back-contact PV modules [1], and also probably corresponds to a slight reduction in V_{oc} observed in RE PV modules undergoing a positive-bias PID test [6].

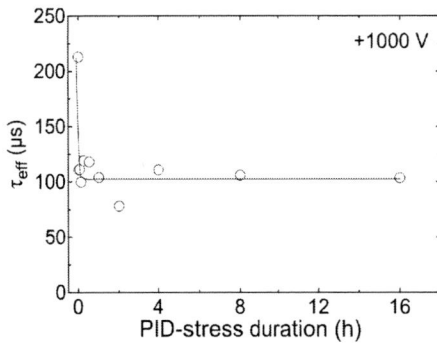

Fig. 5. τ_{eff} of laminated c-Si as a function of the duration of a PID test applying $+1000$ V to the c-Si. The solid line is a guide to the eye.

Figure 6 shows the τ_{eff} of a laminated c-Si wafer receiving a prior 177-hour negative-bias PID test as a function of the duration of positive-bias application. τ_{eff} degraded from the initial value of ~150 µs to ~30 µs during the prior PID test. We can see only a slight increase in τ_{eff} by the application of a positive bias. This is unlike in the case of p-type conventional

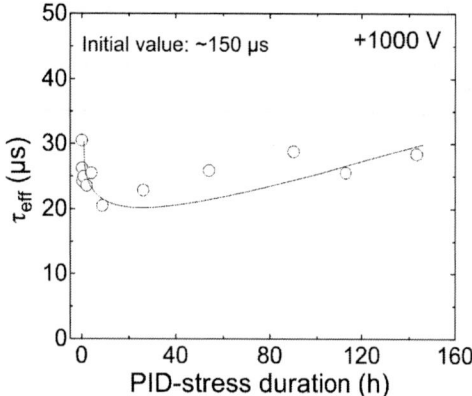

Fig. 6. τ_{eff} of a laminated c-Si wafer receiving a 177-hour negative-bias PID test as a function of the duration of positive-bias application. The solid line is a guide to the eye.

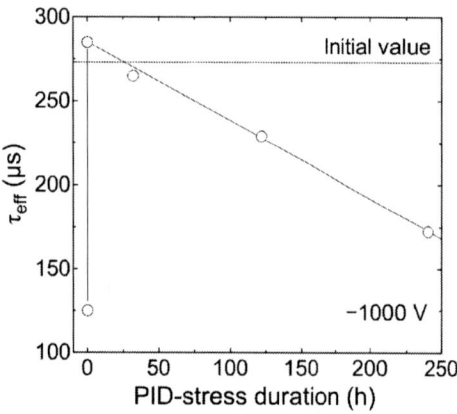

Fig. 7. τ_{eff} of a laminated c-Si wafer undergoing a 30-hour positive-bias PID test as a function of the duration of negative-bias application. The solid line is a guide to the eye.

c-Si PV modules degraded by a negative-bias PID stress showing a clear recovery phenomenon by applying positive bias [8], probably due to the back drifting of Na from the c-Si surface. One possible explanation for the much less significant recovery of τ_{eff} is the existence of stacking faults even after the removal of Na. In the case of p-type c-Si PV modules, Na-decorated stacking faults penetrating p-n junction act as conducting path and induces serious shunting [7]. The shunting behavior can be recovered if Na atoms are removed from the stacking faults and the conducting paths disappear. On the other hand, stacking faults formed on the surface of c-Si during a negative-bias PID stress [9] might act as recombination centers even if they are not decorated by Na. Another possible reason for the insufficient recovery of τ_{eff} is the effect of negative charge accumulation in the SiN$_x$ films. Note that τ_{eff} decreases rapidly in the initial stage of positive

bias application, which is probably an indication of negative charge accumulation in the SiN$_x$. The effect of Na removal on τ_{eff} recovery might thus be compensated by the charge accumulation in the SiN$_x$ films.

Figure 7 shows the τ_{eff} of a laminated c-Si wafer undergoing 30-hour positive-bias PID test as a function of the duration of negative-bias application. The prior PID test deteriorated τ_{eff} to ~125 μs. τ_{eff} rapidly recovered by the application of positive bias. This recovery can be explained by the removal of negative charges accumulated in the surface SiN$_x$ during the prior positive-bias PID-test. Note that τ_{eff} exceeded the initial value immediately after the application of the negative bias. This is probably due to the accumulation of positive charges in the SiN$_x$ similar to the case of the initial behavior observed in the negative-bias PID test. The following reduction in τ_{eff} may be related to Na introduction into stacking faults.

IV. CONCLUSION

We have observed change in the τ_{eff} of laminated c-Si wafers both by negative-bias and positive-bias PID stresses as the direct evidences of the enhancement of surface recombination. Negative-bias application induces gradual τ_{eff} reduction, which may correspond to Na introduction into stacking faults on c-Si surfaces. Rapid reduction in τ_{eff} is seen by applying a positive bias to a laminated c-Si wafer, corresponding to the accumulation of negative charges in SiN$_x$ and resulting attraction of holes near the SiN$_x$/c-Si interface. These behaviors well reproduce the characteristics of PID for n-type RE PV modules.

We have also performed a recovery test by applying a bias with a polarity opposite to a prior PID test. τ_{eff} of laminated c-Si degraded by a positive-bias PID test shows a rapid recovery and exceeds the initial τ_{eff}. This may be due to the removal of negative charges and accumulation of positive charges in the SiN$_x$ film by the recovery test. On the other hand, only a slight τ_{eff} recovery is seen after a positive-bias recovery test for the laminated c-Si undergoing a prior negative-bias PID test. The behavior of insufficient τ_{eff} recovery might originate from stacking faults acting as recombination centers even after the removal of Na and/or compensating effect by the accumulation of negative charges in the SiN$_x$ films.

ACKNOWLEDGMENTS

We would like to thank Dr. Atsushi Masuda of National Institute of Advanced Industrial Science and Technology (AIST) for his fruitful discussion and Mr. Yutaka Komatsu of JAIST for his assistance. This work was supported by the New Energy and Industrial Technology Development Organization (NEDO).

REFERENCES

[1] R. Swanson, M. Cudzinovic, D. De Ceuster, V. Desai, J. Jürgens, N. Kaminar, W. Mulligan, L. R. Barbosa, D. Rose, D.

Smith, A. Terao, and K. Wilson, "The surface polarization effect in high-efficiency silicon solar cells", in *Proceedings of 15th International Photovoltaic Science and Engineering Conference (PVSEC-15)*, 2005.

[2] K. Hara, S. Jonai, and A. Masuda, "Potential-induced degradation in photovoltaic modules based on n-type single crystalline Si solar cells", *Solar Energy Materials and Solar Cells*, vol. 140, pp. 361–365, 2015.

[3] S. Yamaguchi, A. Masuda, and K. Ohdaira, "Progression of rapid potential-induced degradation of n-type single-crystalline silicon photovoltaic modules", *Applied Physics Express*, vol. 9, pp. 112301-1–4, 2016.

[4] S. Bae, W. Oh, K. D. Lee, S. Kim, H. Kim, N. Park, S.-I. Chan, S. Park, Y. Kang, H.-S. Lee, and D. Kim, "Potential induced degradation of n-type crystalline silicon solar cells with p^+ front junction", *Energy Science and Engineering*, vol. 5, pp. 30–37, 2017.

[5] S. Yamaguchi, C. Yamamoto, K. Ohdaira, and A. Masuda, "Reduction in the short-circuit current density of silicon heterojunction photovoltaic modules subjected to potential-induced degradation tests", *Solar Energy Materials and Solar Cells*, vol. 161, pp. 439–443, 2017.

[6] S. Yamaguchi, A. Masuda, and K. Ohdaira, "Changes in the current density–voltage and external quantum efficiency characteristics of n-type single-crystalline silicon photovoltaic modules with a rear-side emitter undergoing potential-induced degradation", *Solar Energy Materials and Solar Cells*, vol. 151, pp. 113–119, 2016.

[7] V. Naumann, D. Lausch, A. Hähnel, J. Bauer, O. Breitenstein, A. Graff, M. Werner, S. Swatek, S. Großer, J. Bagdahn, and C. Hagendorf, "Explanation of potential-induced degradation of the shunting type by Na decoration of stacking faults in Si solar cells", *Solar Energy Materials and Solar Cells*, vol. 120, pp. 383–389, 2014.

[8] A. Masuda, Y. Hara, and S. Jonai, "Consideration on Na diffusion and recovery phenomena in potential-induced degradation for crystalline Si photovoltaic modules", *Japanese Journal of Applied Physics*, vol. 55, pp. 02BF10-1–5, 2016.

[9] V. Naumann, C. Brzuska, M. Werner, S. Großer, and C. Hagendorf, "Investigations on the Formation of Stacking Fault-like PID-shunts", *Energy Procedia*, vol. 55, pp. 569–575, 2016.

Field Inspection of PV Modules:
Quantification of EVA Browning Level using an Image Processing Tool

Sushanth Gudla and GovindaSamy TamizhMani
Arizona State University Photovoltaic Reliability Laboratory (ASU-PRL)
Mesa, Arizona, 85212, USA

Abstract — **PV plant inspections are important to identify and quantitatively determine the impacts of various visual defects on module performance. One of the dominant visual defects is EVA (ethylene vinyl acetate) (encapsulant) browning, which affects the light transmittance and thus output of PV modules. The present study deals with development of an image processing tool to quantify the browning and which is then used to quantify the short circuit current (Isc) of the browned modules without disconnecting the modules from their arrays. This MATLAB tool removes human subjectivity, accurately quantifies browning level, predicts Isc, and reduces time required for field inspection.**

Index Terms — **Browning and PV performance correlation, Browning Index, EVA browning quantification, Image Processing, MATLAB tool.**

I. INTRODUCTION

There are as high as 86 different types of defects that can be found in the PV modules installed in various climates and most of them can be visually observed. However, a quantitative determination of performance impact or financial risk of each of identified defect is a challenging task. Thus, it is utmost important to quantify the risk for each of the visual defects without any human subjectivity. The best way to quantify the risk of each defect is to perform current-voltage (I-V) measurements of the defective modules installed in the plant but it requires disruption of plant operation, use of expensive measuring equipment and intensive human resources.

The NREL visual inspection checklist is one of the best ways to identify visual defects and qualitatively categorize the intensity of each defect [1]. Though this checklist is simple and greatly useful for quick analysis, the defect intensity categorization is very subjective as the perception differs from one person to the other, and may really become a tedious task when considering millions of modules in large power plants. Thus, developing image processing tools which quantify these visual defects is very important for the booming solar PV industry.

One of the most risky and dominant visual defects is EVA (encapsulant) browning, which affects the PV module performance (current and power) due to transmittance loss. The present study deals with the development of an image processing tool to quantify browning level and study the impact of browning on performance without disruptively disconnecting the modules from or at the plant. In this work,

the quantified browning level (browning index [BI]) impact on performance has also been experimentally validated through a correlation study using short-circuit current of browned PV modules retrieved from aged plants/systems installed in the hot-dry climatic condition of Arizona.

II. METHODOLOGY

A. Browning Analysis and Correlation– Process Overview

Photographic images of several field aged modules installed on a manual 2-axis tracker were obtained outdoor during bright sunny times. The MATLAB based image processing tool developed in this work involves two parts. In the first part, as shown in Fig. 2, a six-step image cropping/processing tool was developed and used to obtain the "cells-only-image" of the entire module. In the second part, the browning level (also known as browning index) was determined using the pixel count and pixel weight of the cells-only image of the module. In the last step, the performance parameter (Isc) was correlated with the browning index calculated using the developed tool.

A.1. Image Processing – Part 1

A basic image is made up of small rectangular elements called pixels as shown in Fig. 1. Every pixel carries information about a specific color. Colors are perceived using various color systems (RGB, HSV, HSI, YUV, and others). The other important quality of an image is called resolution. Resolution is the amount of pixel required to make a specific area of an image. Higher resolution implies more information about the image.

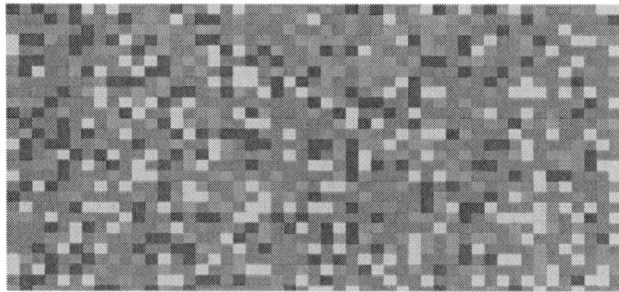

Fig. 1. Example pixel level image [2]

The aim of the first part of this project is to obtain an image that can be processed for browning and thus eliminating other elements in the image.

The six steps of part 1 image processing are as shown in Fig. 2 and are as follows: 1) raw input photograph/image, 2) converting raw photograph image into binary (black and white) raw image using the "thresholding" function, 3) converting the binary raw image of step 2 into another binary raw image using the "Imfill" function which converts all the black patches (in this case, cells) surrounded by white pixels (in this case, inter cell areas) on all sides of the defined area (in this case, module area) into white patches with an exception to the black patches which are surrounded by white pixels and are laying around the edges and away from the defined image area (in this case, all the surroundings except module area), 4) converting the binary raw image of step 3 into another binary raw image using the "filtering" function which is used to convert all the surrounding (of the defined module area) into black pixels, 5) converting the binary raw image of step 4 into a raw module-only photograph image using the "module-only cropping" function which crops out all the area except the module area, and 6) converting the module-only cropped photograph image of step 5 into the cell-only photograph image using the "cell-only cropping" function again. The cropped module image shown in step 5 contains all the components/materials such as, module frame, inter-cell backsheet, fingers, and inter connects which should not be considered for encapsulant browning level analysis. To obtain the step 6 image, a cell isolation factor of 150 is used to get rid of these irrelevant materials. The cell isolation factor is a fixed numerical factor which has been experimentally fixed based on the module samples used. The effect of cell isolation factor is that all the pixels with at least two of the R, G, or B values is greater than or equal to 150 are converted into black pixels. The cell-only image is then used for the part 2 image processing analysis to determine the quantitative browning level/index.

Fig. 2. Cell-only image obtained after the six-step Part 1 image processing of outdoor exposed Siemens M55 modules in Arizona over 18 years (steps 1 through 6 shown in the figure are explained in the text body)

A.2. Image Processing – Part 2

Part 1 of the image processing gives an image that only contains cells and the browning on them. Part 2 deals with calculating the Browning Index (BI). The whole color space is infinite. To quantify and qualify the color space, the color space is divided into finite divisions for understanding and calculating browning index.

The color palette can be developed based on any color model/ system, but works best with HSV or HSI color systems as they are closest to the human perception. In this work, HSV or HSI color systems/ models are used in processing of the final part 1 cell-only image for calculating the Browning Index (BI). The color space is divided in "n" different bins using the color system components and are represented by "n" different colors. Thus, we have a color palette with "n" colors which can be used for processing. The colors ranges were divided based on the H and V component values of the HSV color system as they are very closely replicating the color variation behavior as seen in Fig. 3 and Fig. 4.

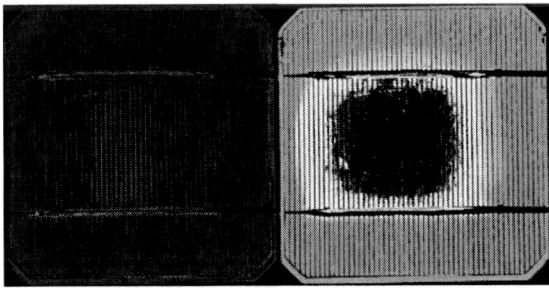

Fig. 3. Comparison of variation of Hue component (on right image) with variation in browning in the original image (on left)

Fig. 4. Comparison of variation of Value component (on bottom image) with variation in browning in the original image (on top)

A.2.1. Pixel Weight

The color palette with "n" colors only describes the number of colors/ bins available but does not give any information about the quality/extent/intensity of browning of a specific bin/ color. Pixel weights are random logical numbers assigned to every color/ bin in the color palette which describes the quality/extent/intensity of browning. The higher the pixel weight, higher the level of browning.

The color palette with different colors, pixel weights is as shown in Fig. 5.

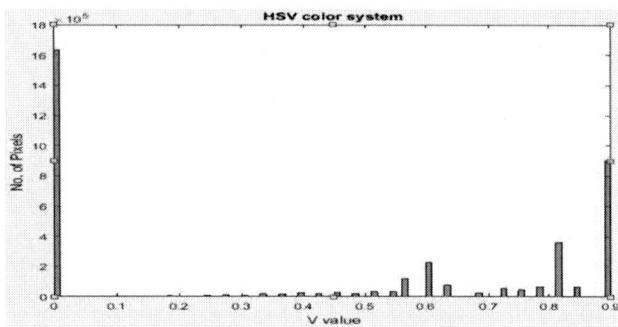

Color code	0	90	84	81	78	75	72	69	66	63
Color palette										
Pixel Weight	0	0.01	0.2	0.4	0.6	0.8	2	4	6	8
Color code	60	57	54	51	48	45	42	39	36	33
Color palette										
Pixel Weight	10	20	30	40	50	60	70	80	90	100
Color code	30	27	24	21	18	15	12	9	6	3
Color palette										
Pixel Weight	200	300	400	500	600	700	800	900	1000	0

Fig. 5. Sample color palette

A.2.2. Pixel Count

The pixel count refers to the number of pixels that fall in a specific bin out of the total number of pixels available in the image. Pixel count are obtained by using the histogram plot of an image component as shown in Fig. 6.

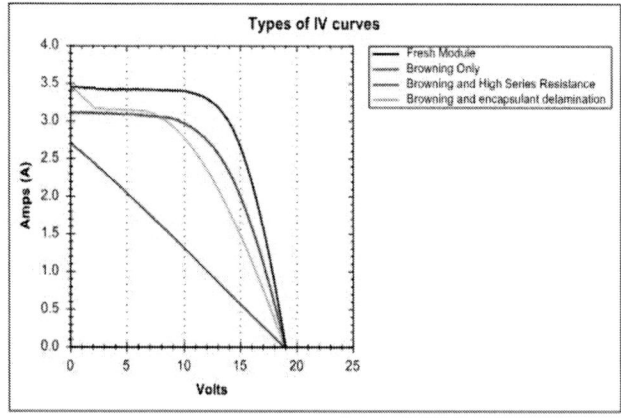

Fig. 6. Histogram based upon V component of a processed HSV image

A.3. Browning Index Calculation

The browning index is calculated is dependent on two parameters, Pixel count and Pixel weight as shown in the below equation.

$$Browning\ index\ (BI) = \frac{\sum Pixel\ count * Pixel\ weight}{\sum Pixel\ count} \quad (1)$$

B. Performance Parameters

I-V curves of all the 21 modules were taken on a day with clear and sunny conditions (800-1000W/m2). Isc of the measurement is used for correlation. Dark I-V have been measured to calculate the series resistance of the modules. Modules with encapsulant delamination have been identified by observing the I-V curves with mismatch as shown in Fig. 7.

Fig. 7. I-V curves for various module defects.

B.1. Iterations Of Tool Development

In order to match and validate the calculated BI with the actual experimental data (Isc loss), the MATLAB tool developed in this work was improved through three developmental iterations. For this tool development and validation, twenty one encapsulant browned Siemens M55 modules exposed in Arizona (hot-dry climate) over 18 years were used. First two iterations were related to module-level images and the third iteration was related to the cell-level images.

B.1.1. First Module-Level Iteration

In the first module-level iteration, the color palette consisted of only 8 colors and colors were segregated only based on hue component values. Analysis encountered three issues: saturation of colors; existence of Moiré pattern; coexistence of two other defects in the module along with browning affecting Isc. The two coexisting defects are encapsulant delamination and excessive series resistence (due most likely to extensive solder bond degradation).

B.1.2. Second Module-Level Iteration

All the three issues found in first module level iteration were addressed by improving the tool for saturation of colors and moire patterns and removing seven modules (from the initial 21 modules) which have coexistence of other defects (delamination and excessive series resistance) interfering the true browning-only Isc loss. The modules with delamination defect can be detected in the field visually or by using a handheld reflectance spectrophotometer and with excessive series resistance issue can be detected using the IR imaging.

Saturation of colors: The saturation as shown in Fig. 8 is due to the limitation of number of colors available in color palette of the first iteration for segregating the brown colors with respect to intensity of browning. To tackle this the number of colors in the second iteration has been increased to 30 colors divided based on H and V values of the HSV color system.

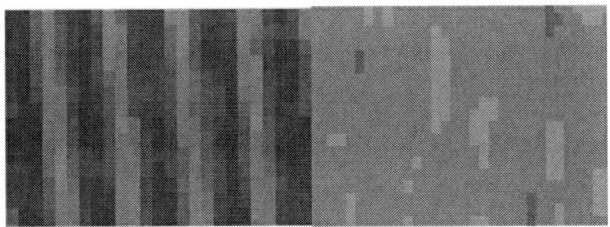

Fig. 8. Saturation of colors clearly seen on processed image (on right) when compared with actual image (on right)

Moire Pattern: The curved white/ shinny lines on the image as shown in Fig. 9 are called moire patterns. The patterns are a result of Aliasing effect. The basic reason for the occurrence of this effect in module image is due to the presence of very closely packed silver finger. MATLAB has inbuilt tools to minimize the effect of moire patterns with Anti-Aliasing filters (AA filters) namely cubic, bicubic, bilinear. Lanczos2, and Lanczos3 filters.

The correlation with browning index was skewed and the reason for this result is with the over refinement of the actual data using various tools and will be discussed in detail in the results and discussion section.

Fig. 9. Moire patterns

B.1.3 Third Iteration- Cell Level

As the module level iterations were having major hindrance with moire patterns, to go around moire patterns cell level

images have been taken, a module with series connected cells, highly browned cells would dominate or dictate the overall Isc of the module. The worst looking cells were picked manually by visual inspection of the module. 2 to 4 different cells were picked from each module for the 3rd iteration. Only Lanczos3 filter was used in the 3rd iteration as it had the best correlation in the second iteration. 24 MP camera with wide and zoom lens were used in this iteration, along wetting of front glass of the module.

The area with maximum BI in a cell would generate the least amount of current. Based on this logic, in the third iteration, only the maximum BI of each cell was considered instead of all the BIs of entire module.

III. RESULTS AND DISCUSSION

The MATLAB tool developed in this work went through a lot of modifications from first iteration to final third iteration. The tool is best understood when the results are studied in sequence for each iteration.

First Iteration: The initial iteration had a color palette in which the whole color range was divided into 8 segments and applied for calculating the BI for the module. The correlation of the calculated BI with measured Isc of the 21 modules came out to be 0.093, by eliminating the modules with high series resistance the correlation was improved to 0.411, by further eliminating the modules with mismatch errors the correlation developed to 0.535, by further manually picking and eliminating the clear outliers the correlation has improved to 0.818 as shown in Fig. 10. The reason for the outliers is believed to be the limitation of the color palette with 8 colors and the moire patterns on the images.

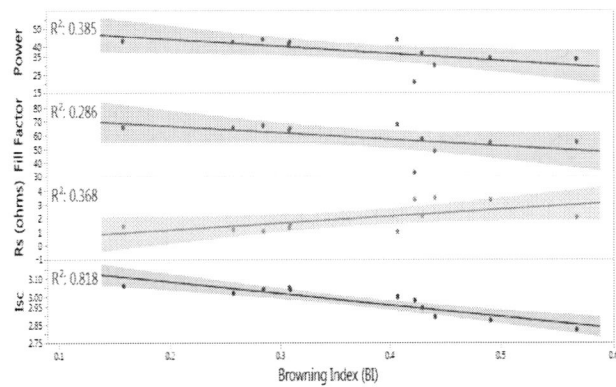

Fig. 10. First iteration - Correlation with no high Rs, mismatch, and outlier modules, 11 AZ Modules, Age – 18 Years

Second Iteration: The second iteration was developed to find a possible solution for moire patterns and color palette limitation of first iteration. The color palette limitation was resolved by developing a new color palette with 30 colors instead of 8 colors and updated pixel weights. The moire patterns were tried to be minimized by using AA filters inbuilt in MATLAB, the procedure visually could minimize the moire patterns on the images as shown in Fig. 11.

Fig. 11. Moire patterns before (top) and after (bottom) using AA filters

The correlation of BI with the performance parameter (Isc) was adversely affected the correlation after the improvements in decreasing the moire patterns. Upon eliminating the data of modules with other defects and manual outliers, the maximum recorded correlation was about 0.503. Lanczos3 filter was found to be the best.

The primary reason for the high drop in correlation was found to be the over purification/ manipulation of the data by using AA filters with low zoom out factor. An AA filter requires 2 specific inputs, zoom out factor and filter type. Zoom out factor is the factor by which the total number of pixels in both X and Y axis of the original image are reduced to as shown in Fig. 12. Filter type (Lanczos3) is the logical algorithm which determines the way in which the complete data from original image is condensed to the zoomed out final image. 0.7 and 0.98 zoom out fractions were experimentally (manual visual inspection of various images) determined to work best for the project. Thus, 0.7 zoom out fraction was used for the second iteration.

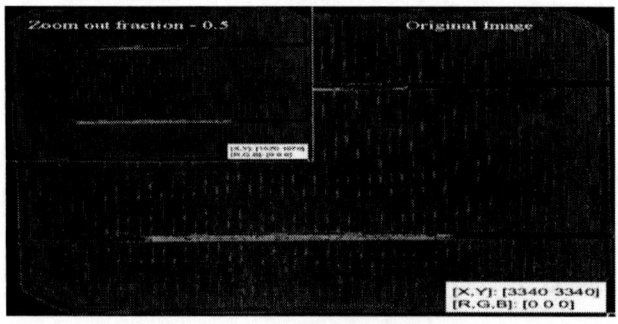

Fig. 12. AA filter – Zoom out fraction

Third Iteration: The primary aim of this iteration was to find a way around moire patterns and improve the correlation between BI and Isc. The Cell level images of the worst browning have been considered for calculating the BI. Lanczos3 filter with 0.98 zoom out factor was used for minimizing any moire patterns. The primary reason for having lesser intensity of moire patterns is due to the use of higher pixel camera (24MP) and cell level images. Cell level images help the camera in clearly differentiating between different fingers on the cell and highly minimizing moire patterns. The images were taken in 3 ways, with zoom lens, wide lens, and with a thin layer of water on the module glass to make it highly transparent (wide lens).

As multiple images were taken for each cell and also multiple cells in a module, maximum BI calculated from each cell is taken as the BI of that cell and maximum of all the maximum BI's calculated for all the cells of the module is taken as the maximum BI for the module (maximum of maximum, referred as "max") and average of all the maximum BI's calculated for all the cells of the module is taken as the average BI for the modules (average of all maxima, referred as "avg"). Upon removal of the outliers and data of modules with other defects, a remarkable Isc correlation coefficient of 0.928 was obtained as shown in the last column of Fig. 13.

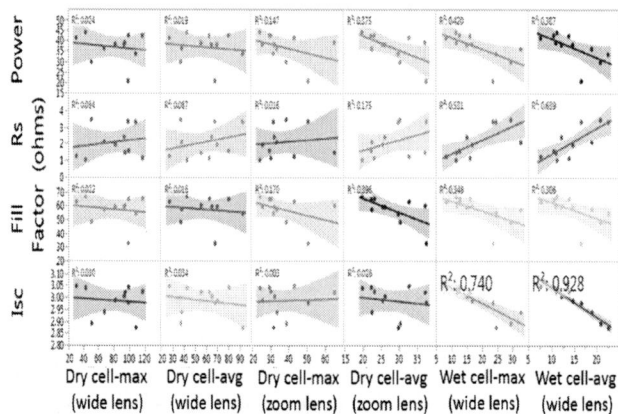

Fig. 13. Third iteration - Correlation with no high Rs, mismatch, and outlier modules, 11 AZ Modules, Age – 18 Years

Thus, it has been experimentally determined that the photographs taken after spraying a thin layer of water on the glass surface (referred as "wet-cell") works the best for the image processing as compared to the dry glass surface (referred as "dry-cell"; Fig. 14). Fig. 15 clearly shows how the processed images obtained using the wet-cell approach have distinct colors and boundaries for two different browned modules. As shown in Fig. 15, the unprocessed dry-cell image (on the left) is mostly covered by both whitish/bluish and brownish colors and is slightly covered by a blackish color at the bottom edge and bottom right corner. The processed dry-cell image (on the right) still shows the blackish color. However, the unprocessed image of the wet-cell does not show any whitish or blackish color at all. Similarly, the processed image does not show any blackish color at all. The exact reason for the appearance of the blackish color of the dry-cell at the bottom edges and bottom right corners is not known but it may be attributed to the surface roughness caused by the wind-born sand blowing on the glass surface over 18 years in the desert climate or to the formation of a thin layer of cemented soil on the glass surface over 18 years.

Fig. 14.   Processed image obtained using the dry-cell approach

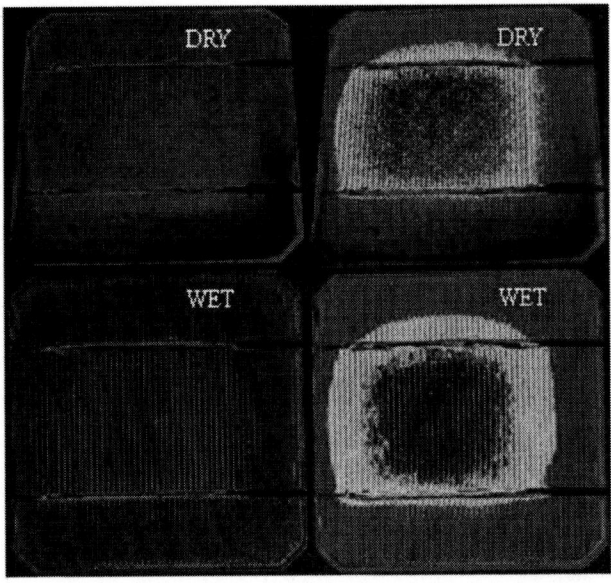

Fig. 15.   Photographic (left) and processed (right) images of a cell in a module obtained using dry-cell and wet-cell approaches

Upon carefully studying all 3 iterations, it is important to note that a combination of iteration 2 with wet-cell approach seems to work the best provided the photograph is taken before the thin water layer gets dried up.

The standard operating procedure (SOP) of the MATLAB tool developed in this work involves only three human input: identify the file location of the photographic image to be processed; input or use default module detection factor (MDF); input or default relative size factor (RSF). The browning index value and processed image are generated in about 30-300 seconds after hitting the "Run" button. Images taken with a cell phone camera (13 MP) takes about 30 seconds to generate the BI. The high resolution DSLR camera (24 MP) images takes about 300 seconds to generate the BI.

IV SUMMARY

An accurate but quick image processing MATLAB tool to quantify the impact of encapsulant browning on the short-circuit current of field aged PV modules is presented through the determination of browning index (BI). This tool is expected to serve as a key inspection tool in the PV power plants. This tool removes the human subjectivity, accurately quantifies the browning level (current/power loss level or annual degradation rate due only to browning) and reduces the workforce time required for field inspection. The photographs taken with the second iteration on wet glass surface is recommended to be used for image processing to improve the accuracy of the results.

REFERENCES

[1] C. Packard, J. Wohlgemuth and S. Kurtz, "Development of a Visual Inspection Data Collection Tool for Evaluation of Fielded PV Module Condition," NREL, Technical Report (NREL/TP-5200-56154), 2012.

[2] webpower, "webpower," [Online]. Available: http://www.webpower-group.com/wp-content/uploads/2014/02/Measuring-conversion-by-conversion-pixels.png.

Preventing Potential-Induced Degradation in Crystalline Silicon PV Modules: Relationship Between Degradation and Bill of Material

Alessandro Virtuani, Eleonora Annigoni, Christophe Ballif

École Polytechnique Fédérale de Lausanne (EPFL), Institute of Microengineering (IMT), Photovoltaics and Thin Film Electronics Laboratory, Neuchâtel, CH-2000, Switzerland

Abstract — In this contribution we investigate the relationship between potential-induced degradation (PID) and the Bill of Material (BOM) used by module manufacturers. The BOM include two types of solar cells (conventional vs PID-free c-Si cells), two types of EVA films with, respectively, low and high resistivity, and two types of backsheets with, respectively, low and high permeability properties. Our results clearly indicate that to have a "PID-free" module the combination of PID-free cells and a high-resistive EVA encapsulants is recommendable. The use of a conventional c-Si cell in combination with a high-resistive EVA encapsulant, is still more effective than the use of PID-free cells in combination with low-quality EVA. These results, however, largely depend on the stress conditions applied during the PID test. Further, our results show that the permeability properties of the backsheet have apparently no influence on PID degradation.

Index Terms — *Potential-Induced-Degradation (PID), Bill-of-Material (BOM), crystalline silicon*

I. INTRODUCTION

Potential Induced Degradation (PID) of crystalline silicon (c-Si) PV modules is a sneaky degradation mechanism, which may seriously affect the performance and return on investment (ROI) of PV installations [1]. The reasons leading to PID degradation, as well as potential mitigating strategies, are multiple and interdependent. These are related to:

(1) the specific *climatic conditions* of the installation site;
(2) the *electrical layout* (i.e. mostly the grounding and polarity) of the PV array/string;
(3) the *inverter choice*;
(4) module design aspects specific to the *mounting solution* (i.e. the presence/absence of a frame or of back rails);
(5) module design aspects specific to the *materials* (encapsulants, cells, glass, backsheet, etc.) used in the fabrication of the sandwich.

In this contribution we focus on the latter point. When coming to module assembly, several strategies related to the choice of the materials are offered to manufacturers, and their customers, to suppress or minimize the negative impact of PID in crystalline silicon (c-Si) framed modules. In particular, the use of:

i. Specific *"PID-free"* cells. One strategy to suppress PID impact is to deposit dielectric anti-reflective (AR) coatings with increased conductivity on the solar cells in a way that the migration of ions from the module glass into the *pn* junction (such as for instance Na+ when applying high

voltages) is avoided or drastically reduced. Other strategies at the cell level focus on increased doping profiles, surface treatments or ion implantation in the cell's emitter [2, 3].

ii. Encapsulating polymers with higher electrical resistivity (ρ) to prevent the diffusion of Na ion from the glass cover into the cell. Several materials such as thermoplastic polyolefin (TPO), ionomer or high-quality ethylene-vinyl acetate (EVA) have been shown to prevent or strongly moderate the insurgence of PID [2, 4].

iii. Special cover glasses with reduced Na content or different processing, such as chemically strengthened glasses [2, 5].

However, the efficacy and the cost-effectiveness of the single strategies (or their combinations) is frequently unknown. In the present contribution, we investigate how PID degradation of encapsulated (and framed) solar cells is affected by the choice of the solar cell and of the encapsulating polymer.

In view of the industrial application of the results and considering the constant market pressure to significantly abate module-manufacturing costs, we focus on the most cost-effective solutions only and do not investigate for example the use of special cover glasses. For the same reason, in this work we focus on EVA as an encapsulant polymer only.

As the properties of the backsheet (BS) may have a strong impact on the physical properties of the encapsulant material (water content and resistivity), in our study we consider as well backsheets with different permeability (i.e. Water Vapor Transmission Rate WVTR) properties. Finally, we briefly discuss cost considerations that might enable to select the best strategy to minimize PID.

II. EXPERIMENTAL WORK

The materials that we use for assembling our test devices are all commercially available. These include two types of crystalline-silicon solar cells (conventional vs PID-free c-Si solar cells, both p-type), two types of EVA films from the same supplier with, respectively, low and high electrical resistivity, and two types of backsheets from the same supplier with, respectively, low and high water vapor permeability properties. We then investigate all possible combinations (i.e. $2^3 = 8$) of this bill of material (BOM) list and we assemble, by means of a conventional lamination process, our one-cell mini-modules, according to the matrix shown in Table I. In order to simulate the presence of a frame, the edges of the mini-modules are covered with a conductive Al tape.

The samples are then subjected to a PID stress test as prescribed by the relevant IEC Technical Specification for c-Si modules (IEC TS 62804-1:2015 [6]) using a climatic chamber. Besides the conventional stress level (i.e. 60°C / 85% RH, -1000 V, or 65/85), to accelerate the degradation in a set of samples, we increase the stress level to 85°C / 85% RH (85/85). The test is run beyond the conventional 96 h up to 192 h. Before and after the degradation, the modules are characterized by means of IV measurements at standard test conditions (STC: 1000 W/m², 25°C, AM1.5) and Electro-Luminescence (EL) imaging.

Additionally, we measure the volume resistivity of our encapsulants as a function of humidity and temperature. Before each measurements, the cured samples were pre-conditioned in a climatic chamber. Measurements are performed according to the IEC 62788-1-2 standard (Method B) [7] using a Keithley electrometer and 8009 resistivity test fixture.

TABLE I

OVERVIEW OF THE DIFFERENT MATERIALS (CELLS, ENCAPSULANTS, BACKSHEETS) USED IN OUR STUDY.

Bill of Material	* from datasheet	
Cells	Standard c-Si **(std-cell)**	PID-free cell **(PID-free)**
Encapsulants	EVA high resistivity **(EVA-high-ρ)** *ρ > 1·10^15 Ω·cm	EVA low resistivity **(EVA-low-ρ)** *ρ > 6·10^14 Ω·cm
Backsheets	BS high permeability **(BS-high-P)** *WVTR ~ 1.8 g/m²/d	BS low permeability **(BS-low-P)** *WVTR ~ 0.7 g/m²/d

III. RESULTS

A. Resistivity measurements

Figure 1 shows the volume resistivity test preformed on the **EVA-high-ρ** and **EVA-low-ρ** encapsulants as a function of relative humidity.

At room temperature (25°C) and RH 40% the **two encapsulants have a measured volume resistivity of 5·10^15 and 8.5·10^14** respectively, slightly higher than datasheet values. The volume resistivity is moderately reduced by increasing humidity. The effect of temperature on ρ (not shown) is apparently more pronounced than that of humidity, as can be seen in Ref. [8] too.

B. PID testing at 60°C, 85% RH

Table II shows the degradation rates (%) of our test samples after PID testing at 60°C, 85% RH. For *conventional cells*, the choice of the EVA polymer has a drastic impact on PID

degradation. Whereas the cells processed with a low resistivity polymer show a huge degradation (~ -80% after 96 and 192 hrs), the ones processed with high-ρ EVA, show nearly no degradation. This is true irrespective of the BS used in sample processing.

Conversely, for the *PID-free cells*, the choice of the EVA, as well as of the BS, at 60°C, 85% RH seems negligible, and for all the possible combinations of materials, the degradation after 96 and 192 hrs is limited to ≤ 1%.

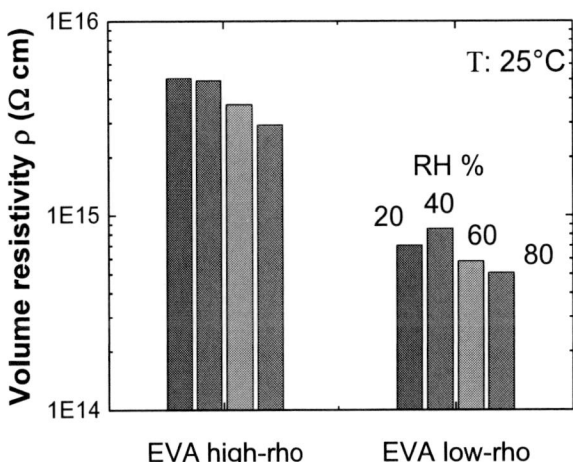

Fig. 1: Volume resistivity for the **EVA-high-ρ** and **EVA-low-ρ** encapsulants as a function of relative humidity (at 25°C room temperature).

TABLE II

POWER DEGRADATION OF THE DIFFERENT TEST SAMPLES AFTER 96 AND 192 HRS OF PID TESTING (60°C, 85% RH, -1kV).

Cells	EVA	BS high P (96 / 192 h) [%]	BS low P (96 / 192 h) [%]
Standard (std) c-Si cells	High ρ	-0.7 / -0.1	-0.8 / -0.4
	Low ρ	-85 / -83.5	-77.9 / -76
PID-free c-Si cells	High ρ	-0.2 / 0	-0.3 / -0.1
	Low ρ	-0.7 / -1	-1.7 / -1

B. PID testing at 85°C, 85% RH

The picture changes when applying more severe testing conditions (85°C, 85% RH) and accelerating the degradation.

In Figures 2 and 3 and Table III, we show the results of extended 85/85 PID testing on our test samples manufactured, respectively, with conventional and PID-free c-Si cells. For *conventional c-Si cells,* the choice of the EVA polymer has a drastic impact on PID degradation. Whereas the cells processed with a low resistivity polymer show a huge degradation (~ -90% after 96 hrs), the ones processed with high-ρ EVA show a moderate degradation of -6.9% and -11.7%, respectively, after 96 and 192 hrs (for the BS-high-P samples). Surprisingly, in the extended testing (85/85) too, the effect of using BS with different permeability properties is nearly negligible.

TABLE III
POWER DEGRADATION OF THE DIFFERENT TEST SAMPLES AFTER 96 AND 192 HRS OF PID TESTING (85°C, 85% RH, -1kV).

Cells	EVA	BS high P (96 / 192 h) [%]	BS low P (96 / 192 h) [%]
Standard (std) c-Si cells	High ρ	-6.9 / -11.7	-9.2 / -10.9
	Low ρ	-89.1 / -88.2	-96.6 / -96
PID-free c-Si cells	High ρ	-1.6 / -1.3	-1.8 /-1.2
	Low ρ	-12.1 / -18.1	-11.7 / -15.4

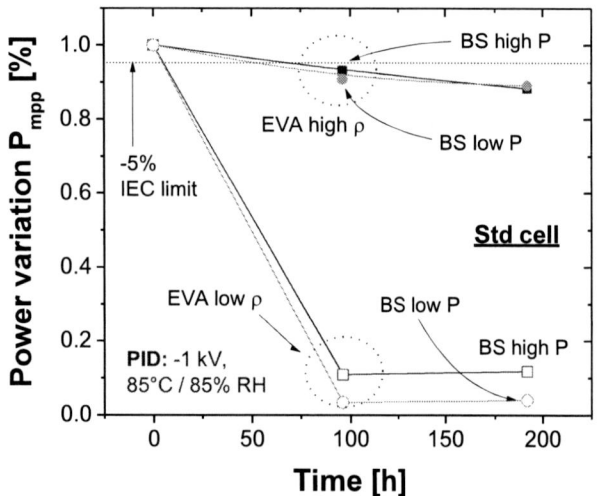

Fig. 2: Degradation after PID test (85°C, 85% RH, -1kV) of encapsulated standard (std) c-Si solar cells, using two different EVA encapsulants with respectively a low and high resistivity ρ and two types of backsheets, with respectively a high and low permeability P.

For the PID-free cells (see Fig. 3), the choice of the EVA polymer has still a significant impact on PID degradation. Whereas the cells processed with a high resistivity polymer show nearly no degradation (~-1.5% after 96 and 192 hrs), the ones processed with low-ρ EVA, show a degradation of -12.1% and -18.1%, respectively, after 96 and 192 hrs (for the BS-high-P samples). This degradation, however, is larger than the one experienced by the samples processed with conventional c-Si cells and high- EVA polymer.

For this set of samples too, the impact of using BS with different permeability properties is nearly negligible.

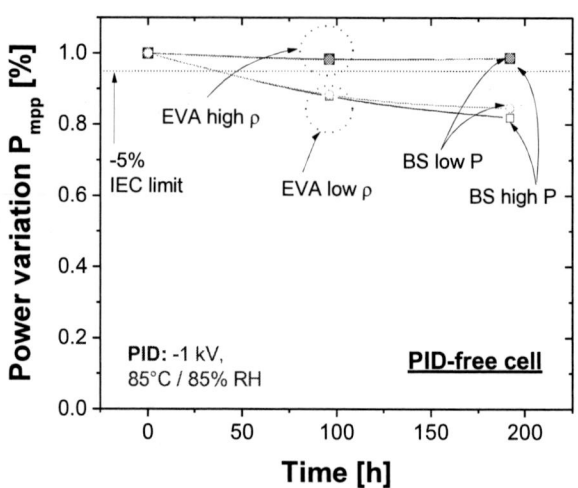

Fig. 3: Degradation after PID test (85°C, 85% RH, -1kV) of commercially available "PID-free" c-Si solar cells, using two different EVA encapsulants and backsheets.

As the effect of the choice of the BS on PID degradation is apparently negligible, in Fig. 4 we merge together, for the BS-low-P samples only, the results obtained for the samples with the conventional cell using high-ρ EVA and for the samples including a PID-free cell (with high-ρ and low-ρ EVA).

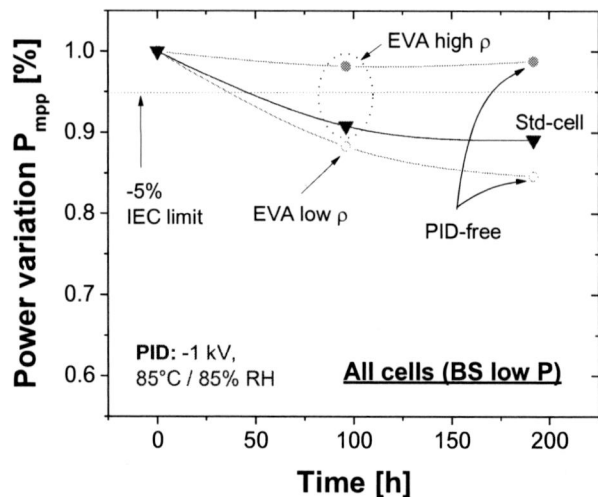

Fig. 4: Degradation after PID test (85°C, 85% RH, -1kV) of encapsulated standards (*std*) cells processed with a high resistivity EVA polymer, and two "PID-free" cell processed with a low and a high ρ EVA encapsulant. Data are taken from Fig.2 and 3, with a modified y-axis scale.

IV. DISCUSSION

The results of extended testing at 85/85 test conditions indicate that the use of PID-free cells is by far the best option to prevent the insurgence of PID, but only if the sandwich is processed in combination with a high resistive EVA polymer.

The use of a conventional c-Si cell in combination with a high-resistive EVA encapsulant, still needs to be favored to the use of PID-free cells in combination with a low-quality EVA encapsulant.

If we stick to the IEC -5% pass/fail criterion, at 85/85 test conditions, however, the only samples experiencing a degradation below the threshold (after 96 and 192 hrs), which will consequently pass the test, are the mini-modules assembled with PID-free cells in combination with the high-ρ EVA.

Conversely, for the 60/85 test conditions, all combinations of samples would pass the test (after 96 and 192 hrs) with the exception of the conventional c-Si cells laminated with low-ρ EVA.

Nevertheless, we want to stress that these results have been obtained on 1-cell mini-modules, so that, by considering the evolution of the PID phenomenon from the module edges (as is frequently reported for the different climatic conditions), they should represent a sort of worst-case situation, exhibiting an accelerated degradation. For this reason, sticking to the -5% IEC pass/fail threshold is possibly not meaningful.

The application of these results to 60- or 72-cell large-area modules is, therefore, not straightforward, nevertheless the results give a clear indication on the possibilities that are offered to mitigate the PID phenomenon at the module level.

Another result worth highlighting is that, apparently, the permeability properties of the backsheet have no significant influence on PID degradation at least in accelerated-aging testing and during the first 192 hrs. As Fig. 1 shows, the volume resistivity of EVA is reduced by increased water content in the film, so that the latter finding is a bit counterintuitive (see Ref. [9] too), but has been systematically confirmed by the experiments.

To investigate the effect of humidity on the evolution of PID further, we are running Finite Elements Method (FEM) simulations to investigate how the resistivity of the encapsulant layer can be influenced by the permeability properties of the backsheets used. These results are not shown in this paper.

Finally, given the strong and constant market dynamics to bring down module manufacturing costs, we highlight some cost considerations that might enable to select the best strategy to minimize PID. These include:

1. The potentially high cost for modified module glass, which makes this option difficult to be adopted;
2. The great commercial advantage for module manufacturer to offer solutions independent on inverter choice and electrical layout and, therefore, suitable for all systems (a sort of *"one-type"* fits all);
3. The relatively low cost for using EVA encapsulants with higher resistivity. Approximately, a +1% cost at module level;
4. The cost at the cell level to arrive to PID-free cells or to buy such cells might vary, even considerably, from one manufacturer to the other.

V. CONCLUSIONS

Our results clearly indicate that to have a "PID-free" module the combination of PID-free cells and a high-resistive EVA encapsulants is the most recommendable The use of a conventional c-Si cell in combination with a high-resistive EVA encapsulant, is still more effective (and may possibly constitute a cheaper solution) than the use of PID-free cells in combination with low-quality EVA. This is true for extended PID testing at 85°C-85% RH.

For the less severe test conditions of 60°C-85% RH, the only combination of materials not passing the test (and showing a considerable degradation in performance, i.e. ~ 80%) is that of conventional c-Si cells laminated with a low-ρ EVA.

Finally, our results show that the permeability properties of the backsheet have apparently no influence on PID degradation (at least in accelerated-aging testing and over the duration of our tests: 192 hrs), even if resistivity measurements on EVA films indicate that the volume resistivity of the polymer is reduced with an increased water content in the film.

VI. ACKNOWLEDGMENTS

We gratefully acknowledge partial financial support from EOS Holding. We are grateful to Xavier Niquille for support in sample manufacturing and to all PV-lab team members. We are grateful to Jörg Horzel, Antonin Faes, and Matthieu Despeisse at CSEM for the fruitful discussions and delivery of testing material.

REFERENCES

[1] S. Pingel, O. Frank, M. Winkler, S. Daryan, T. Geipel and H. B. J. Hoehne, "Potential Induced Degradation of Solar Cells and Panels," 2010.

[2] B. Braisaz, B. et al., "Improved 1500 V PID Resistance: Encapsulants, Cover Glass And Ion Implanted Cells", in Proc. of the *32nd European Photovoltaic Solar Energy Conference and Exhibition*, Munich, 2016.

[3] Han, et al. "Ion implantation - an effective solution to prevent c-Si PV module PID" in Proc. of the *29th European Photovoltaic Solar Energy Conference and Exhibition*, Paris, 2013.

[4] E. Annigoni, M. Jankovec, F. Galliano, H. Li, L. Perret-Aebi, M. Topic, F. Sculati-Meillaud, A. Virtuani and C. Ballif, "Modeling potential-induced degradation (PID) in crystalline silicon solar cells: from accelerated-aging laboratory testing to outdoor prediction," in Proc. of the *32nd European Photovoltaic Solar Energy Conference and Exhibition*, Munich, 2016.

[5] M. Kambe, K. Hara, K. Mitarai, S. Takeda, M. Fukawa, N. Ishimaru, M. Kondo, "Chemically Strengthened Cover Glass for Preventing Potential Induced Degradation of Crystalline Silicon Solar Cells", in Conference Records of the 39th IEEE Photovoltaic Specialists Conference, Tampa, 2013.

[6] IEC TS 62804-1:2015: "Photovoltaic (PV) modules - Test methods for the detection of potential-induced degradation - Part 1: Crystalline silicon", IEC, Geneva.

[7] IEC 62788-1-2:2016: "Measurement procedures for materials used in photovoltaic modules - Part 1-2: Encapsulants -

Measurement of volume resistivity of photovoltaic encapsulants and other polymeric materials", IEC, Geneva.

[8] J. Berghold, S. Koch, B. Frohmann, P. Hacke, and P. Grunow "Properties of Encapsulation Materials and Their Relevance for Recent Field Failures", in Conference Records of the 40th IEEE Photovoltaic Specialists Conference, Denver, 2014.

[9] S. Schulze, A. Apel, R. Meitzner, M. Schak, C. Ehrich, J. Schneider, "Influence of Polymer Properties on Potential Induced Degradation of PV- Modules", in Proc. of the *28th European Photovoltaic Solar Energy Conference and Exhibition*, Paris, 2013.

Identifying Reverse-Bias Breakdown Sites in $CuIn_xGa_{(1-x)}Se_2$

Steve Johnston[1], Elizabeth Palmiotti[2], Andreas Gerber[3], Harvey Guthrey[1], Lorelle Mansfield[1],
Timothy J. Silverman[1], Mowafak Al-Jassim[1], and Angus Rockett[2]

[1]National Renewable Energy Laboratory, Golden, CO, 80401, U.S.A.
[2]Colorado School of Mines, Golden, CO, 80401, U.S.A.
[3]IEK5-Photovoltaics, Forschungszentrum Jülich GmbH, Jülich, 52425, Germany

Abstract — CIGS mini-modules are stressed under reverse bias, resembling partial shading conditions, to predict and characterize where failure occurs. Partial shading can cause permanent damage in the form of "wormlike" defects on thin-film modules due to thermal runaway. We use electroluminescence (EL) imaging and dark lock-in thermography (DLIT) to supplement current-voltage curves when reverse-bias breakdown occurs on various CIGS mini-modules. For better understanding of how these defects originate and propagate, we have implemented a current-limited, reverse-bias stressing technique. This allowed for DLIT-based detection and detailed study of the region where breakdown initiates before thermal runaway leads to permanent damage. A correlation between suspected defect sites and actual breakdown has been determined using this method.

Index Terms — imaging, photovoltaic cells, photoluminescence, electroluminescence, reliability, accelerated aging, II-VI semiconductor materials, thermal analysis, breakdown voltage.

I. INTRODUCTION

When photovoltaic cells are connected in series and packaged into higher voltage modules, partial shading can induce a reverse bias in the shaded cells in order to match current in the series-connected cells. If breakdown occurs with some uniformity throughout the cell, then current density is likely low enough to prevent damage. However, if breakdown occurs in highly localized regions, significant heating due to high current density and power dissipation can permanently damage the device [1, 2]. This leads to long-term power losses of the module. Such degradation is an important concern for thin-film modules [3].

Previous studies have shown that reverse-bias breakdown in CIGS ($CuIn_xGa_{(1-x)}Se_2$) modules due to partial shading has occurred both at or near scribe lines [4, 5] and within the main fields of the cells [6]. Such defects tend to form at a localized point and propagate in seemingly random fashion, forming what are described as "wormlike" defects [4]. These kinds of defects are paths of permanently damaged material where heating due to high current density causes delamination [4]. In this paper, we demonstrate a thermal imaging method that aims to predict where reverse-bias breakdown will occur in CIGS cells.

II. EXPERIMENT

CIGS mini-modules from various sources with cells that are ~4 to 5 mm wide and ~80 to 100 mm long are characterized by probing across two or three cells at a time. Current-voltage (I-V) curves are collected as part of the measurements, while imaging provides spatial information to show locations of breakdown under reverse bias. Photoluminescence (PL) and electroluminescence (EL) images are collected using a Princeton Instruments PIXIS 1024BR camera. Dark lock-in thermography (DLIT) images are collected using a Cedip Silver 660M (FLIR SC5600-M) InSb camera with lock-in data acquisition. A Keithley source-meter has been used to apply a voltage and measure and/or limit the current. I-V curves and EL images could be collected simultaneously.

During DLIT image collection, a voltage is pulsed, and the camera is triggered with the same frequency. In these cases where voltage is applied to the sample, a current limit can be applied using the Keithley source-meter. So, while applied reverse-bias voltage could exceed a sample's breakdown threshold, a current limit can prevent uncontrolled, thermal runaway damage that leads to wormlike defects.

III. RESULTS

An example of applying a reverse bias on a series-connected pair of monolithic CIGS cells within a mini-module is shown in Fig. 1. The image of Fig. 1(a) is a DLIT image with applied reverse bias below the breakdown value. There is some localized heating, or hotspots, observed at points along the scribe lines that separate the cell areas and connect them in series. The image of Fig. 1(b) is a DLIT image with a larger reverse bias exceeding breakdown that has led to thermal runaway and the formation of permanent, wormlike defects. After this image was collected, the cells visually show permanent damage as seen in the room-light photograph of Fig. 1(c).

The effects of the wormlike defects that were generated during the applied reverse bias can be seen in the sample's I-V curve. Figure 2 shows the I-V curves of these cells before and after the applied reverse bias that led to thermal runaway.

978-1-5090-5606-4/17 $31.00 © 2017 IEEE

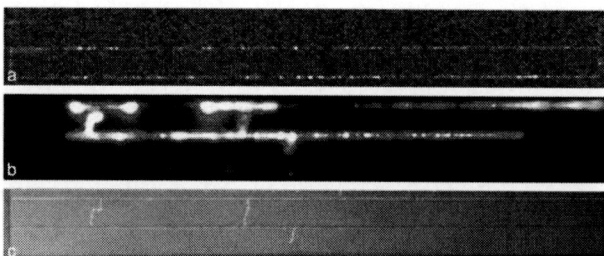

Fig. 1. DLIT images show localized heating in reverse bias for a voltage lower than breakdown (a) and a reverse bias exceeding breakdown (b). A room-light photograph of the cell after wormlike defects are formed due to the bias of (b) is shown in (c).

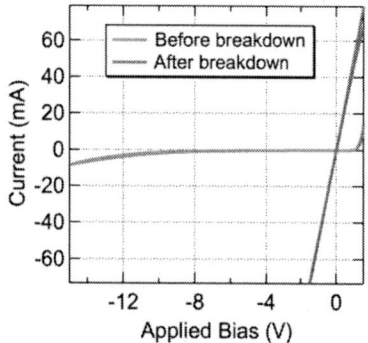

Fig. 2. I-V curves for the sample of Fig. 1 are shown before and after the applied reverse bias that led to thermal runaway and generation of wormlike defects.

The cells have changed from having good diode behavior to having linear characteristics of a resistor.

The heating due to the development of shunts during reverse bias can be captured using thermal imaging, but visible and/or near-infrared light from the energy of carriers at the breakdown defects can also be emitted [1]. In this case, it is possible to identify breakdown locations and image the formation and propagation of wormlike defects using a Si-based camera. The series of images of Fig. 3 illustrate both a room-light image and visible light emission from reverse-bias breakdown sites and how the thermal runaway propagates the defects in wormlike patterns in the cell.

The images of Fig. 3 are collected during a sweep of increasing reverse bias. The first image, Fig. 3(a), is a roomlight view to show the cells and locations of the scribe regions. As breakdown occurs in reverse bias, Figs. 3(b) through 3(h) show emissions of the breakdown, which include light in visible wavelengths. Camera detector saturation is evident when long, bright horizontal lines appear in the images.

The current-voltage (I-V) data collected from the cells shown in Fig. 3 are plotted in Fig. 4, where the voltage begins in forward bias and continues toward increasingly larger reverse bias. Several jumps of increasing reverse current are observed and, presumably, such jitter in the current represents

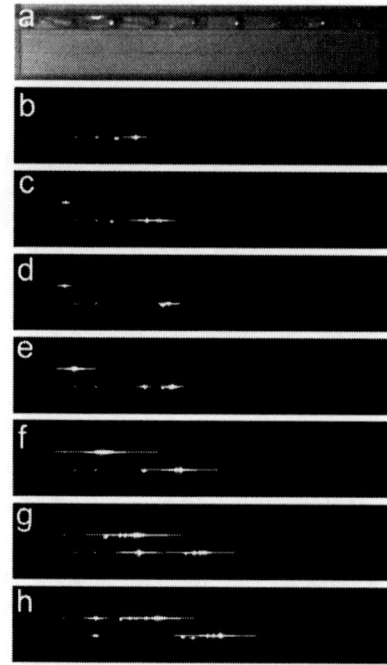

Fig. 3. Images from a Si-based camera capture visible light emitting from breakdown regions of the CIGS cells under reverse bias. A room-light image shows the cells (a) while images (b) through (h) show defect formation and propagation at progressing times with increasing applied reverse bias.

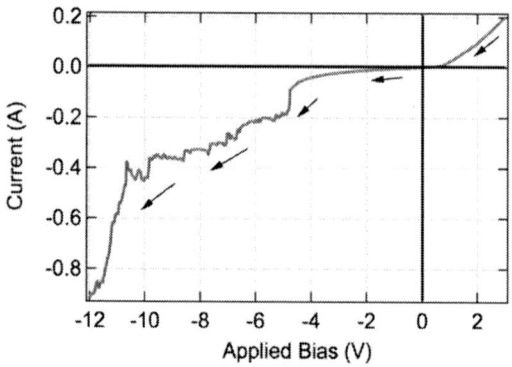

Fig. 4. I-V curve for the sample stressed in Fig. 3. The applied voltage begins in forward bias and progresses to increasing reverse bias, showing jumps in reverse current as wormlike defects are formed and propagate.

various sites of breakdown as thermal runaway and wormlike defects are formed.

We have shown that both a thermal camera sensing heat and a silicon-based camera imaging visible breakdown emissions can spatially track breakdown sites as they occur. Such images can be correlated to current and voltage, which can help characterize the breakdown process. However, these examples lead to permanently damaged material as wormlike defects form. In order to study locations where breakdown initiates, and before thermal runaway occurs and damages the semiconductor device, we use thermal imaging to sense the preliminary heating at a potential breakdown site. We locate suspicious breakdown sites using DLIT imaging under reverse bias with a current limit. As shown in Fig. 5, DLIT is used to collect a thermal image on cells in reverse bias. The reverse bias of 20 V is typically large enough to cause breakdown and lead to propagating wormlike defects. However, in this case, we have set a current limit of ~1 to 2 mA and have prevented the thermal runaway condition leading to wormlike defects. Figure 5(a) shows a thermal image with a sufficient reverse bias to induce breakdown (20 V_{rev}) and a current limit of 1.5 mA. The thermal image of Fig. 5(a) is overlaid on a steady-state thermal image in Fig. 5(b). Potential breakdown sites are circled by marking on the cell with a pen. Within those same areas, no defects are seen on the surface by thermal imaging, as shown in Fig. 5(c), or in a visible-light based microscope. As an example, wormlike defects, if present, are detected by thermal imaging in steady-state. An example is shown in Fig. 6.

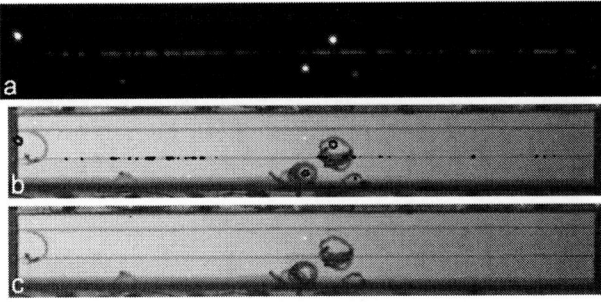

Fig. 5. A reverse-bias DLIT image using a current limit is shown in (a). Image (b) shows the DLIT overlay on a room-light image. Circles are roughly sketched around shunt areas using an ink marker. A room-light image (c) does not show wormlike defects at the shunt locations marked by the drawn circles.

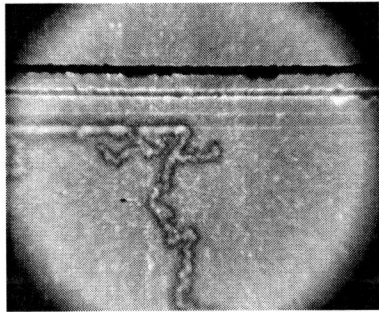

Fig. 6. Example thermal image of a wormlike defect. The field of view is ~2 mm.

This technique of applied breakdown voltage with limited current is used to identify potential breakdown sites. After such locations are documented, the current limit is disabled,

and breakdown can then occur without restriction. Such breakdown typically occurred at values up to 20 V_{rev} and with currents up to ~25 mA. Table 1 shows statistics of several tests on CIGS samples from different sources. The first row labeled "Predicted" lists the number of localized spots that showed a heating signature with current-limited applied breakdown voltage. After unlimited current breakdown, the second row labeled "Correct" lists how many initiated wormlike defects formed at the predicted locations. The third row, "Not predicted" lists additional wormlike defects that were observed when current-limited testing did not show a potential breakdown site. The statistics of Table 1 show that this method is reasonable to predict the wormlike defect origins since ~85% of the breakdown sites were correctly predicted without forming additional damage.

TABLE 1
Statistical Summary of Predicted Breakdown Sites

	Source A			Source B	
	Sample 1	Sample 2	Sample 3	Sample 1	Sample 2
Predicted	4	9	10	12	12
Correct	4	9	10	11	12
Not predicted	0	1	2	4	0
Defect begins on a scribe	2	5	5	9	10
Defect begins within a cell	2	4	3	5	2

Many of the breakdown sites were further characterized by zoomed-in thermal imaging to distinguish if the defect originated at a scribe line or within a cell and perhaps just very near to a scribe line. Table 1 summarizes if the defects are within a scribe or not. For Source A, the defects within scribes and in the cells are roughly equal, while for Source B, a higher percentage of the defects were associated with scribe line origins.

The technique of identifying breakdown sites has been applied to an additional sample in order to further characterize the defects by higher resolution microscopy. A pair of CIGS cells, similar to those shown in Fig. 5, are imaged with 20 V_{rev} and a current limit of ~1.5 mA. Possible breakdown sites are seen where the thermal imaging shows localized heating. Two defects are chosen that do not appear to have any surface features or obvious visible damage when imaging with the zoom lens of the thermal camera or looking through an optical microscope. These defects are marked with small laser burn spots in a cross-hair pattern and sub-mm spacing such that the defect can be accurately located in a scanning electron microscope (SEM). An SEM image of one of the defects is shown in Fig. 7. The overhead view SEM image shows a slight bump in the surface. A focused ion beam (FIB) cut was used to view the defect in cross section. The SEM cross-sectional view of the FIB cut is shown in the inset of Fig. 7. Here, an arc-shaped strip of voids in the CIGS material is seen just above the Mo surface. The length of the voids strip is a few microns. This type of nodule defect has been associated with shunts [7].

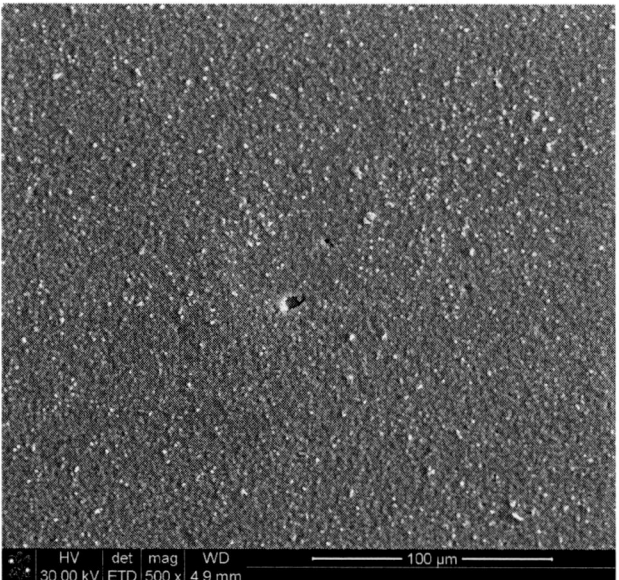

Fig. 8. SEM overhead view of a possible breakdown defect site. This defect is a hole or crater in the CIGS absorber layer.

Fig. 7. SEM overhead view of a possible breakdown defect site. The inset image shows a FIB cross-sectional SEM view of this nodule type of defect.

A second possible breakdown defect site is also characterized, and an SEM image is shown in Fig. 8. This defect is a hole or crater in the CIGS film and is also approximately a few microns in size. This type of crater defect was also previously associated with shunting [7]. Both the nodule and crater defects, as detected here with current-limited thermal breakdown imaging technique, appear to be weak points that are capable of initiating wormlike defects during conditions of reverse bias stress.

IV. SUMMARY

Partial shading on monolithic CIGS modules can lead to degradation when reverse-bias breakdown causes localized heating and permanent, wormlike defects to form. DLIT and EL imaging techniques show the formation and propagation of these defects using their heat and light emission signatures. In order to study how these defects are initiated, we applied reverse bias while limiting the current to prevent thermal runaway. DLIT imaging was used to locate sites where heating shows that breakdown will likely occur. With this technique, a correlation between initial defects and their likelihood to form permanent damage under partial shading

conditions was formed. Both nodule and crater types of CIGS defects, a few microns in size, appear to be weaknesses for applied reverse bias and may be origins for the formation of wormlike defects under conditions of partial shading.

ACKNOWLEDGEMENT

The U.S. Government retains and the publisher, by accepting the article for publication, acknowledges that the U.S. Government retains a nonexclusive, paid-up, irrevocable, worldwide license to publish or reproduce the published form of this work, or allow others to do so, for U.S. Government purposes. This work was supported by the U.S. Department of Energy under Contract No. DE-AC36-08GO28308 with the National Renewable Energy Laboratory, project DE-EE0007141 of PREDICTS2, and by the German ministry for Economic Affairs and Energy (BMWi) under the contract FK0325724A (OptiCIGS) and by the NREL-Helmholtz "Impuls und Vernetzungsprojekt" HNSEI (SO-075).

REFERENCES

[1] J. W. Bishop, "Microplasma breakdown and hot-spots in silicon solar cells," *Solar Cell* vol. 26 pp 335-349, 1989.

[2] T. J. Silverman, M. G. Deceglie, X. Sun, R. L. Garris, M. A. Alam, C. Deline, and S. Kurtz, "Thermal and Electrical Effects of Partial Shade in Monolithic Thin-Film Photovoltaic Modules," *IEEE Journal of Photovoltaics*, vol. 5, no. 6, 2015.

[3] S. Dongaonkar, M. A. Alam, Y. Karthik, S. Mahapatra, D. Wang, and M. Frei, "Identification, Characterization, and Implications of Shadow Degradation in Thin Film Solar Cells," *IEEE International*, 2011.

[4] E. E. van Dyk, C. Radue, and A. R. Gzasheka, "Characterization of Cu(In,Ga)Se$_2$ photovoltaics modules," *Thin Solid Films*, **515**, pp. 6196-6199, 2007.

[5] J. E. Lee, S. Bae, W. Oh, H. Park, S. M. Kim, D. Lee, J. Nam, C. B. Mo, D. Kim, J. Yang, Y. Kang, H. Lee, and D. Kim, "Investigation of damage caused by partial shading of CuIn$_x$Ga$_{(1-x)}$Se$_2$ photovoltaic modules with bypass diodes," *Prog. Photovolt: Res. Appl.*, 2016.

[6] P. O. Westin, U. Zimmermann, L. Stolt, and M. Edoff, "Reverse bias damage in CIGS module," *24th European Photovoltaic Solar Energy Conference and Exhibition*, pp. 2967-2970, 2009.

[7] S. Johnston, T. Unold, I. Repins, A. Kanevce, K. Zaunbrecher, F. Yan, J. V. Li, P. Dippo, R. Sundaramoorthy, K. M. Jones, and B. To, "Correlations of Cu(In,Ga)Se$_2$ Imaging with Device Performance, Defects, and Microstructural Properties," *J. Vac. Sci. Technol. A* **30**, pp. 04D111-1-6, 2012.

Himawari-8 enabled real-time distributed PV simulations for distribution networks

Nicholas A. Engerer*, Jamie M. Bright*, Sven Killinger*

*Fenner School of Environment and Society, The Australian National University, 2601, Canberra, Australia

Abstract—High resolution, next generation satellites such as Himawari-8 show great promise for the provision of accurate estimations of Behind-the-Meter (BtM) solar PV power production. This paper presents a methodology that produces real-time PV power estimates as derived from Himawari-8 satellite imagery, validating them against seven Australian radiation monitoring sites and 78 small-scale BtM solar PV sites in Canberra, Australia. We report an MBE of -7 W m-2 and RMSE of 55 W m-2 for global horizontal radiation values (G_h) and an MBE of 0.04 W/Wp and RMSE of 0.15 W/Wp for estimated actuals at the PV sites. As a capstone, we apply this satellite based radiation modeling tool to a distribution network level distributed PV simulation in a single case-study using 15,500 PV sites.

I. INTRODUCTION AND MOTIVATION

Globally, Solar photovoltaic (PV) installed capacity is growing rapidly. The International Energy Agency (IEA) estimates that by 2050 the global generation of solar electricity will provide 16% of the world's total electricity generation (over 6000 TWh) [ ]. Of the total installed capacity, the IEA estimates that half of the installations will be small-scale, "rooftop" or 'distributed' behind-the-meter (BtM) PV systems.

At high-penetrations, the presence of large amounts of BtM solar poses significant challenges for electrical networks, particularly in distribution networks, which are most often tasked by regulatory bodies with reliability of power quality [ ]. Australia boasts the worlds highest solar PV penetrations [ ] with over 1.66 million installed BtM solar PV systems (as of April 2017 [ ]), resulting in the maximum permissible penetration levels being reached in certain areas of its distribution networks. At high penetration levels, the response is often at detriment to continued penetration of solar energy. For example, Australian distribution network service provider (DNSP) Ergon Energy imposed a 3.5 kW upper limit on new solar installations in 2015 [ ].

Such limiting actions by DNSPs are reflective of the fact that most Australian DNSPs have no active feedback quantifying how much electricity BtM solar installations are currently producing at any given time. In some cases, remotely monitored net metering is available, but this data is both limited and expensive to obtain. This lack of information forces DNSPs to manage their networks in a reactive, rather than proactive manner [ ]. New technologies, however, are making it possible to remove some of the uncertainty facing distribution networks. Real-time irradiance estimates are critical for improving plant and grid operation performance [ ].

The newly operational (mid-2015) Himawari-8 geostationary satellite is of particular interest because it provides 1 to

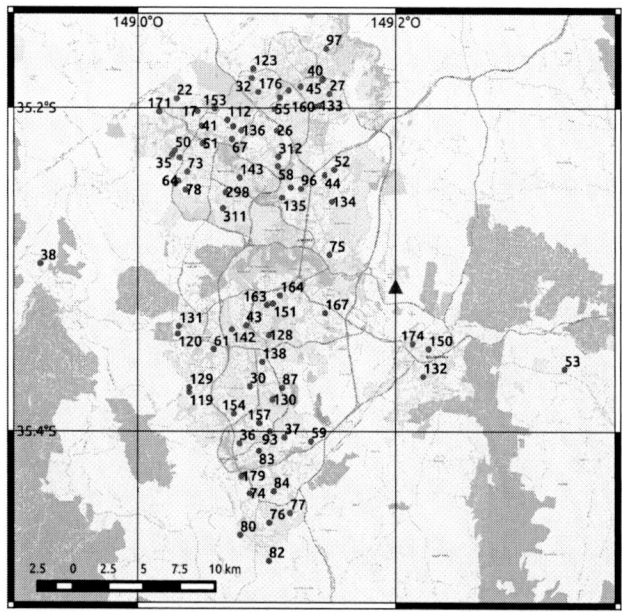

Fig. 1: Map of the 78 PV systems (red circles) in the region of Canberra, Australia. The weather station at Canberra airport is indicated as a blue triangle. Sites include station numbers from the original dataset which were assigned prior to quality control.

2 km imagery data at 10-minute update cycles — the highest resolutions available to-date for solar radiation modelling and forecasting.

With DNSPs resorting to reactive measures to manage PV penetration, and with the possibility of new state-of-the-art satellite imagery, there is room to explore how reliable a satellite based PV simulation system can be for use in estimating the real-time power generation of distributed BtM solar PV systems, with the overarching goal of providing DNSPs with accurate PV simulations in real-time to support their operations. With the development of an accurate real-time PV tool, it is possible to quantify how well real-time estimations of solar irradiance can be produced, which introduces the possibility of advancing PV now-casting for high accuracy solar forecasting.

There are few methods of estimating the solar irradiance in the real-time over large geographical regions. Geographically specific real-time irradiance estimates can be derived from empirical models [ ] or derived from sky imagers [ ], [ ], [ ], however, the only methodologies that offer geographical freedom are those that adopt live satellite data. Many satellite derived irradiance methodologies exist. An extensive comparison of satellite derived irradiance methodologies is performed by Polo et al. (2016) [ ], though the highest temporal resolution identified was 1-hour. The Kosmos-3m model is a satellite enabled methodology that provides real-time irradiance operations functioning at $4km^2$ spatial resolution and 2 to 4-hour temporal resolution [ ]. With the Himawari-8, this resolution is halved and the temporal resolution can be increased to 10-mins. A notable work uses the GOES satellite and WRF numerical prediction model to deliver real-time estimates using feedback from 2-sec ground based power measurements to provide high temporal resolution real-time estimates [ ], however, should a DNSP not have access to these high resolution ground based measurements, the alternative is to rely on lower resolution PV power output estimates. It is therefore the intention of this paper to combine satellite imagery with solar and PV simulation methodologies to produce a high-resolution real-time irradiance and PV power output estimate for the whole city of Canberra.

For the purposes of supporting DNSPs with real-time distributed PV power output estimates, research at The Australian National University has leveraged a unique combination of satellite, solar radiation and PV simulation modelling tools. This paper presents the development of a novel Himawari-8 enabled methodology that provides real-time estimation of distributed BtM PV power output using real PV systems in the distribution network. This approach is subsequently validated.

II. DATA

The methodology requires data from three main inputs. Firstly, satellite imagery is taken from the Himawari-8 satellite, maintained by the Japanese Meteorology Agency and processed by the Australian Bureau of Meteorology (BoM) [ ]. The images have a scan duration of 10-mins and are provided for 16 bands, each with a different scanning wavelength. This model utilises the visible (band 3) and infrared (band 13) images, primarily due to their availability from the BoM — more bands will be utilised in the future. The images extracted from BoM are already projected onto a latitude/longitude grid so that each pixel has attributed coordinates. The images are not parallax corrected, however this will be achieved in future versions. Examples of the images can be seen in Fig. 2.

Secondly, PV metadata of capacity, azimuth and tilt is extracted for the 341 PV installations in Canberra from PVoutput.org [ ], these were cross-checked with data from SolarHub [ ], who provide more accurate metadata with which to check site suitability. Fig. 1 shows the 78 selected PV systems in Canberra that passed the quality checks and are used for validation and demonstration purposes.

Lastly, local temperature observations are provided by BoM Canberra airport weather station (70351) for use in the PV modelling.

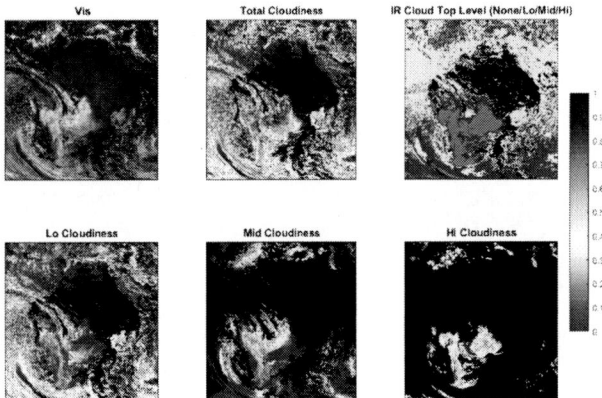

Fig. 2: Himawari-8 satellite imagery based cloud decomposition over the Australian Himawari-8 domain. Top left) Visible image from band 3 of the Himawari-8, exempt from colour scale, black and white image. Top middle) Derived cloud opacity. Top right) cloud top heights detected using IR image from band 13 of the Himawari-8. Bottom row) Clouds present in different banded height regions as low (lo), middle (mid) and high (hi). Color fill indicating cloud opacity is provided at right, with 0 meaning clear skies and a value 1 indicating completely opaque cloud.

For radiation validation of the satellite derived irradiance estimates, solar irradiance observations are taken from seven high-quality sites as maintained by the Australian Bureau of Meteorology (Adelaide, Alice Springs, Cape Grim, Darwin, Melbourne, Rockhampton and Wagga Wagga). These datasets are collected at 1-min resolution, but averaged herein to 10-min periods for validation against the methodology outputs.

III. METHODOLOGY

The methodology described here is simplified and represented as a flow chart as shown in Fig. 3.

The radiation modelling begins with 10-min time step images of visible and IR from the Himawari-8 satellite with pixel size resolution of 2 km². Solar irradiance is derived following the general principles of the Heliosat-2 methodology [ ]. Clouds are decomposed into lower, middle and upper troposphere layers (shown in Fig. 2) as separated by cloud top temperatures that are inferred through utilisation of the IR image identified cloud-top height and numerical weather model isobaric temperatures obtained from the Global Forecast System (GFS) deterministic model [ ]. The use of model isobaric temperatures facilitates improved decomposition of cloud into three height-based layers without contamination from seasonal and daily weather system influence.

Total cloudiness is derived through aggregation of the three cloud layers. Total cloud opacity (represented as an index between 0 and 1) is then derived using albedo differences between the lowest visible return value of albedo recorded for each pixel and a continuous running 50 day sample that

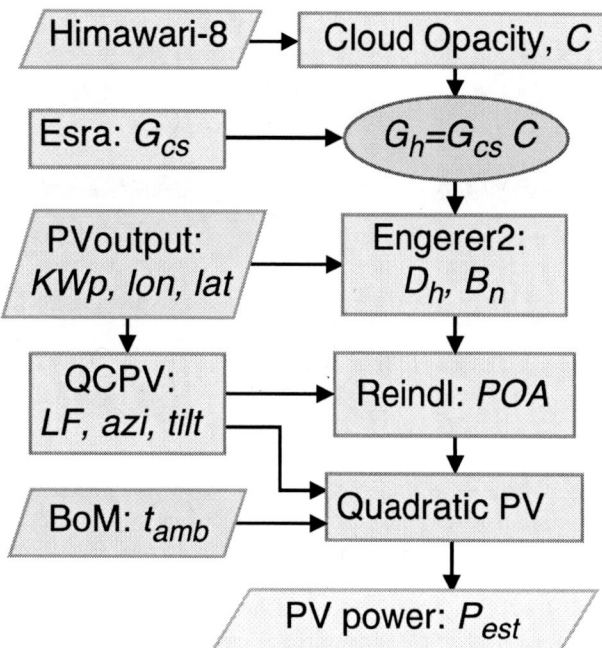

Fig. 3: Flowchart demonstrating the methodology processes. Green parallelograms indicate inputs, blue rectangles indicate modelling steps, yellow parallelograms indicate outputs and the red oval indicates a combining of modelling stage outputs.

is updated daily. Derived cloud opacity is then applied as a linear reduction to the Esra clear sky radiation model [ ] to produce an estimate of G_h at each pixel. The selection of the Esra model was based upon validation against a selection of clear sky radiation models in Australia [ ].

Estimates of G_h are then decomposed into direct normal irradiance (B_n) and diffuse horizontal irradiance (D_h) through the Engerer2 separation model [ ], [ ]. This decomposition then enables plane-of-array irradiance estimates (POA) to be computed for a given solar PV array tilt angle (β) and azimuth angle (α). POA irradiance estimates are generated through the use of the Reindl Transposition model [ ], whose selection is based upon a balance of model performance [ ] and computational simplicity.

To move from POA irradiance estimates to PV system power output estimates, a quadratic PV model is employed [ ], [ ]. The principle importance to modelling accuracy of the POA and PV system modelling steps are PV system installed capacity (kWp), loss factor (LF), α, β. Both α and β dramatically impact the available POA irradiance and location of the daily peak power so their accuracy is important; KWp is a crucial factor in deriving the magnitude of power produced, this is coupled with a LF that incorporates shading, derating, soiling, faults etc. Accurate and reliable kWp values for PV sites are provided by SolarHub [ ], however, LF, α and β are determined empirically via the algorithms established in the QCPV methodology with an accuracy of $\sim 4°$ [ ].

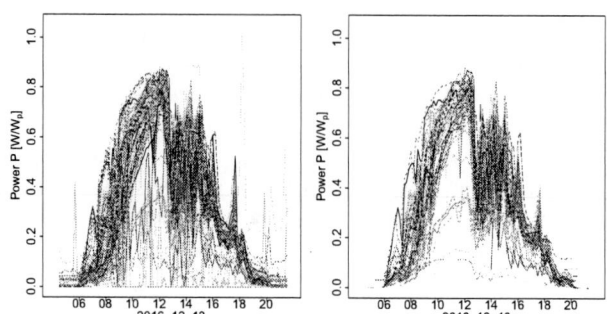

Fig. 4: Plots indicating the raw power observations (P_{meas}; left) and the quality controlled power values (P_{qc}; right) for all 78 PV systems (from Fig. 1) on an exemplary day.

The measured power output data is also quality controlled as a part of this processing step. Two tests of the QCPV routine [ ] are further extended in this paper. Firstly, the "lower limit" can now detect consistent positive values in the morning or afternoon, indicating a shift in the zero line of a PV system. The zero line can be considered as erroneous outputs where the power output during sunlight hours registers $0\ W$, this zero line can migrate by a few watts, though is still an incorrect flat recording. Since some PV systems show such a behavior only for a restricted time period, the test is applied on a daily basis. Secondly, the "persistence" test can now detect periods that periodically fluctuate between zero and non-zero values during the day, indicating potential measurement errors. Since such instances do not necessarily show a high variance, the former "persistence" test or the filter for "spurious data" are not able to find such atypical behavior. The time period between the first and last zero power value for zenith angles $\theta_z < 70°$ during a day is thus marked as issues of power estimation at high solar zenith angles is difficult [ ]. Additionally, all former QCPV checks of the "lower limit" and "persistence" criteria are still applied. The before and after results of the application of the QCPV methodology are displayed in Fig. 4.

The outputs from this methodology are estimated PV power for all PV sites as derived from Himawari-8 satellite imagery. The outputs are validated in the next section for performance in both radiation estimates and PV power output estimates against observation data.

IV. VALIDATION

In order to assess the suitability our proposed methodology for the purposes of simulating distributed BtM PV power output, we perform a standard validation according to radiation modelling standards by calculating and reporting mean bias (MBE) and root mean square errors (RMSE). The validation is completed for both the global horizontal irradiance (G_h) estimates and BtM PV power output estimates against the datasets described in the Data section above.

A. Radiation Validation

For the radiation validation, we report an overall -7 W m-2 and RMSE of 55 W m-2 using data from July and August

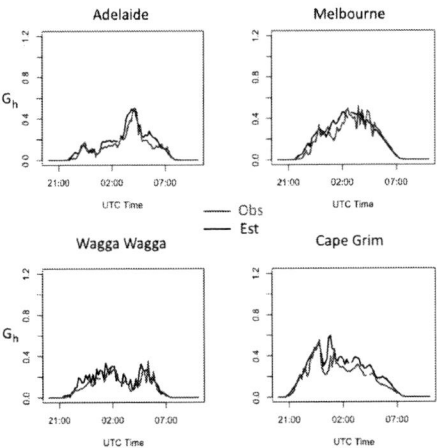

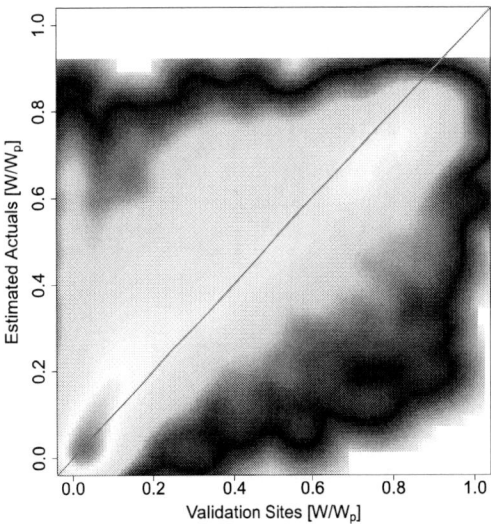

Fig. 5: Example time series validation of G_h estimated actuals at four Bureau of Meteorology sites for 20 July 2016 using 10 minute intervals. Observations are reported in blue, with estimated actuals in black. Across the four sites, a MBE of 6.7 W m-2 and RMSE of 62 W m-2 are reported under complex cloud conditions.

Fig. 7: BtM solar PV validation of estimated actuals across 78 PV sites in Canberra from September 2016 through March 2017. An MBE of 0.04 W/Wp and RMSE of 0.15 W/Wp are reported across the 78 sites.

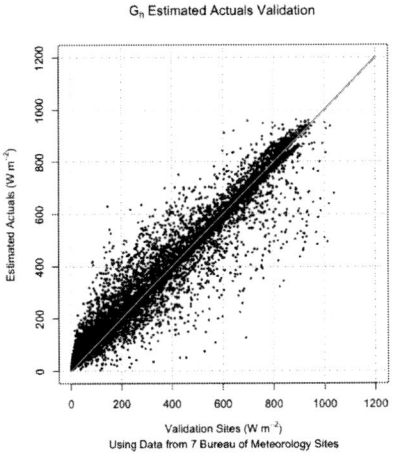

Fig. 6: Estimated actuals versus observations for G_h using two months of data from the seven selected Bureau of Meteorology sites. An MBE of -7 W m-2 and RMSE of 55 W m-2 are reported. Notable overestimation of G_h values at lower observed radiation values is present, but overall fit of the distribution is well-balanced.

2016 (Fig. 6). Despite the low MBE, we note that G_h values are positively biased at lower observed radiation values, with a lesser degree of negative bias at elevated radiation values. This suggests corrections are required to our cloud opacity estimates. However, the overall accuracy of this modelling step meets the requirements of a good radiation model, which

are defined in the literature as having an relative MBE less than 5% (1.8% reported) and relative RMSE less than 15% (10.7% reported) [ ]. Fig. 5 provides an example of time series behaviour from four of the validation sites on 20 July 2016 — a day selected to represent periods of complex cloud cover under active synoptic scale weather patterns.

B. PV Power Validation

Secondly, PV system power estimated actuals are validated based upon the quality controlled PV system metadata and power output data from the 78 Canberra PV sites. Using the period of September 2016 through to March 2017, we report an MBE of 0.04 W/Wp and RMSE of 0.15 W/Wp as demonstrated in Fig. 7. Overall scatter in the validation data shows well behaved results, but with accuracy losses nearly double that of the radiation validation. We note that this validation reports uneven bias across the distribution of PV system measurement values, suggesting further refinement of the quadratic PV model coefficients could be required.

C. DNSP PV Power Simulations for Canberra

The radiation and PV power validation steps provide sufficient confidence to experimentally proceed to Himawari-8 based BtM solar simulations using DNSP network installation data. To execute proof of concept, a simulation from the 23 June 2016 is presented. This was a complex cloud cover day, with northeasterly mean cloud advection featuring broken stratocumulus and altostratus cloud features. In Fig. 8, a single time step from 15,500 distributed BtM solar PVs simulation is provided, with an estimate fleet yield of 18.2 MW at the valid time. In the provided figure, we provide a cross-validation

978-1-5090-5606-4/17 $31.00 © 2017 IEEE   1408

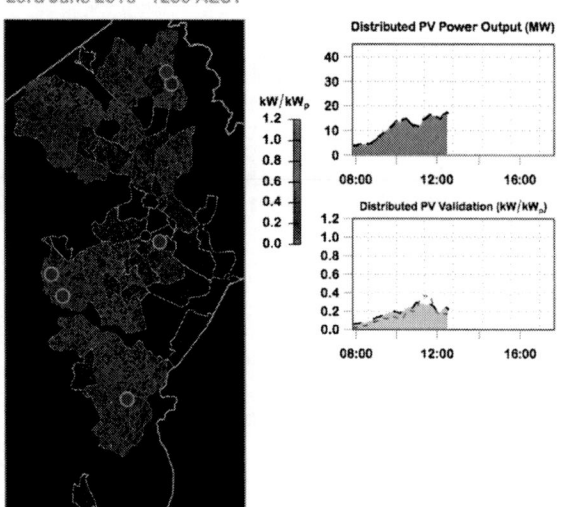

Fig. 8: Distributed BtM PV simulation from 23 June 2016. Each dot is a DNSP distributed PV system, with color coding according to power output. Grey polygons represent local suburbs boundaries. At right, total aggregate power of all PV systems (top) and online cross-validation with PV system observations (bottom) where observed PV power output from the six validation sites is plotted in red.

against 6 of the PV measurement sites. We observe a slightly positive bias in the satellite base PV estimates of 0.038 W/Wp and a relatively low RMSE value of 0.12 W/Wp.

V. Conclusions and Future Work

Real-time modelling of distributed BtM solar PV is challenging, but new tools, such as high-resolution next generation satellites (Himawari-8), show great promise in providing accurate estimates of real-time PV fleet production. We have demonstrated promising initial results from an operational system, which utilises this satellite data, along with several radiation and PV modelling steps. This system, whose purpose is to provide DNSPs with real-time feedback on their PV fleet performance, is in the early stages of its deployment and will undergo much future improvement. At this time, we report an ability to compute estimated actuals of global solar radiation to within 10% accuracy at 10 minute time-steps. For PV system estimated actuals, we report an RMSE of only 0.15 W/Wp.

These results, along with those provided in the radiation and PV system validation above, suggest that similar levels of accuracy can be expected from the DNSP-wide PV simulation. However, significant uncertainties remain that prevent finalising this conclusion. Here, the distribution of actual PV systems orientations are not known, and are instead estimated through a Gaussian modelling approach [20]. The soiling and shading profiles of the unmonitored PV sites are also not known or well-understood. These latter issues of complexity remain outside the scope of the current study, but will be the focus of our future works.

References

[1] IEA, "Technology Roadmap - Solar Photovoltaic Energy," Paris Cedex, France, 2014. [Online]. Available: https://www.iea.org/publications/freepublications/publication/TechnologyRoadmapSolarPhotovoltaicEnergy_2014edition.pdf

[2] F. Vallee, V. Klonari, T. Lisiecki, O. Durieux, F. Moiny, and J. Lobry, "Development of a probabilistic tool using monte carlo simulation and smart meters measurements for the long term analysis of low voltage distribution grids with photovoltaic generation," *International Journal of Electrical Power and Energy Systems*, vol. 53, no. 0, pp. 468 – 477, 2013.

[3] CEC, "Powershift: A blueprint for a 21st century energy system," 2015. [Online]. Available: www.cleanenergycouncil.org.au/powershift

[4] APVI, "Australian Photovoltaic Institute market analysis," 2017. [Online]. Available: pv-map.apvi.org.au/analyses

[5] G. Parkinson, "Ergon imposes new limit on rooftop solar system," Renew Economy, Tech. Rep., June 2015. [Online]. Available: reneweconomy.com.au/ergon-imposes-new-limit-on-rooftop-solar-systems-66926/

[6] ENA, "Electricity network transformation roadmap: final report," Energy Networks Australia, Tech. Rep., 2017. [Online]. Available: http://www.energynetworks.com.au/sites/default/files/entr_final_report_web.pdf

[7] A. Linares-Rodriguez, S. Quesada-Ruiz, D. Pozo-Vazquez, and J. Tovar-Pescador, "An evolutionary artificial neural network ensemble model for estimating hourly direct normal irradiances from meteosat imagery," *Energy*, vol. 91, pp. 264 – 273, 2015.

[8] J. Nou, R. Chauvin, S. Thil, and S. Grieu, "A new approach to the real-time assessment of the clear-sky direct normal irradiance," *Applied Mathematical Modelling*, vol. 40, no. 1516, pp. 7245 – 7264, 2016.

[9] C. W. Chow, B. Urquhart, M. Lave, A. Dominguez, J. Kleissl, J. Shields, and B. Washom, "Intra-hour forecasting with a total sky imager at the {UC} san diego solar energy testbed," *Solar Energy*, vol. 85, no. 11, pp. 2881 – 2893, 2011.

[10] Y. Chu, H. T. Pedro, and C. F. Coimbra, "Hybrid intra-hour {DNI} forecasts with sky image processing enhanced by stochastic learning," *Solar Energy*, vol. 98, Part C, pp. 592 – 603, 2013.

[11] Y. Chu, H. T. Pedro, M. Li, and C. F. Coimbra, "Real-time forecasting of solar irradiance ramps with smart image processing," *Solar Energy*, vol. 114, pp. 91 – 104, 2015.

[12] J. Polo, S. Wilbert, J. Ruiz-Arias, R. Meyer, C. Gueymard, M. Sri, L. Martn, T. Mieslinger, P. Blanc, I. Grant, J. Boland, P. Ineichen, J. Remund, R. Escobar, A. Troccoli, M. Sengupta, K. Nielsen, D. Renne, N. Geuder, and T. Cebecauer, "Preliminary survey on site-adaptation techniques for satellite-derived and reanalysis solar radiation datasets," *Solar Energy*, vol. 132, pp. 25 – 37, 2016.

[13] H. Rezk, I. Tyukhov, and A. Raupov, "Experimental implementation of meteorological data and photovoltaic solar radiation monitoring system," *International Transactions on Electrical Energy Systems*, vol. 25, no. 12, pp. 3573–3585, 2015.

[14] W. F. Holmgren, A. T. Lorenzo, M. Leuthold, C. K. Kim, A. D. Cronin, and E. A. Betterton, "An operational, real-time forecasting system for 250 mw of pv power using nwp, satellite, and dg production data," in *2014 IEEE 40th Photovoltaic Specialist Conference (PVSC)*, June 2014, pp. 0080–0084.

[15] BoM, "Himawari-8 and -9," 2017. [Online]. Available: http://www.bom.gov.au/australia/satellite/himawari.shtml

[16] PVoutput, "PVOutput," 2017. [Online]. Available: https://pvoutput.org/

[17] SolarHub, "SolarHub, we know solar," 2017. [Online]. Available: http://www.solarhub.net.au/

[18] C. Rigollier, M. Lefvre, and L. Wald, "The method heliosat-2 for deriving shortwave solar radiation from satellite images," *Solar Energy*, vol. 77, no. 2, pp. 159 – 169, 2004.

[19] NCEP, "The Global Forecast System (GFS) - Global Spectral Model (GSM)," 2016. [Online]. Available: http://www.emc.ncep.noaa.gov/GFS/doc.php

[20] N. A. Engerer and F. P. Mills, "Validating nine clear sky radiation models in Australia," *Solar Energy*, vol. 120, pp. 9–24, 2015.

[21] N. A. Engerer, "Minute resolution estimates of the diffuse fraction of global irradiance for southeastern Australia," *Solar Energy*, vol. 116, pp. 215–237, 2015.

[22] C. A. Gueymard and J. A. Ruiz-Arias, "Extensive worldwide validation and climate sensitivity analysis of direct irradiance predictions from 1-min global irradiance," *Solar Energy*, 2015.

[23] D. T. Reindl, W. A. Beckman, and J. A. Duffie, "Diffuse fraction correlations," *Solar Energy*, vol. 45, no. 1, pp. 1–7, 1990.

[24] D. Yang, "Solar radiation on inclined surfaces: Corrections and benchmarks," *Solar Energy*, vol. 136, pp. 288 – 302, 2016.

[25] S. Killinger, F. Braam, B. Müller, B. Wille-Haussmann, and R. McKenna, "Projection of power generation between differently-oriented PV systems," *Solar Energy*, vol. 136, pp. 153–165, 2016.

[26] S. Killinger, N. Engerer, and B. Müller, "QCPV: A quality control algorithm for distributed photovoltaic array power output," *Solar Energy*, vol. 143, pp. 120–131, 2017.

[27] C. J. Smith, J. M. Bright, and R. Crook, "Cloud cover effect of clear-sky index distributions and differences between human and automatic cloud observations," *Solar Energy*, vol. 144, pp. 10 – 21, 2017.

[28] C. Gueymard and D. Meyers, "Validation and ranking methodologies for solar radiation models," *Modeling Solar Radiation at the Earth's Surface*, pp. 479–509, 2008.

[29] N. A. Engerer and J. Hansard, "Real-time Simulations of 15,000+ Distributed PV Arrays at Sub-Grid Level using the Regional PV Simulation System (RPSS)," in *ISES Solar World Congress 2015*, Daegu, Korea, 2015.

Reduced measurement uncertainty in PV module batch testing

Blagovest Mihaylov[1,4], Bengt Jaeckel[2], Juergen Arp[3], Ralph Gottschalg[1]

[1]Centre for Renewable Energy Systems Technology, Loughborough University, LE11 3TU, UK

[2]UL International GmbH, Admiral-Rosendahl-Straße 9, 63263 Neu-Isenburg (Zeppelinheim), Germany

[3]PV LAB Germany GmbH, Gartenstrasse 36, 14482 Potsdam, Germany

[4]Research Center for Photovoltaics, AIST, Ibaraki 305-8561, Japan

Abstract — **Measurement uncertainty introduces a risk into the evaluation of PV modules. An approach to nearly half the uncertainty in the measurement of larger batches of modules is presented here. This is achieved via measuring a sample of these modules at multiple laboratories with independent traceability chains. A case study using three laboratories is shown here. The laboratories involved in this trial have measurement uncertainties (k=2) in the range of 2.5-3%. The resulting uncertainty is less than 1.5% following an uncertainty assessment meeting all requirements of the GUM. The presented approach is a cost-effective method to verify large numbers of modules in field tests with the lowest uncertainty currently being on the market.**

Index Terms — **batch testing, measurement uncertainty, photovoltaic modules, silicon.**

I. INTRODUCTION

This paper focuses on the assessment of PV modules from systems in the field, not the highly accurate calibration of a single module. Field measurements typically have uncertainties larger than 2.5% when carried out in a mobile laboratory with thermal control. Measuring large batches with lower uncertainties at state-of-art facilities is often prohibitively expensive.

With PV module prices falling by 30% [1] in 2016 alone, the cost of capital plays an increasingly important role in the overall systems' cost. Financing costs are strongly affected by the perceived risk. While there are numerous risk contributors in the lifecycle of a PV project [2], the most straightforward is associated with the performance estimates based on the DC rating of the modules.

This risk could be borne by the manufacturer, the installer, the investor, the developer or even an insurer depending on the specific project and contractual arrangements. Regardless of which party the risk is allocated to; it increases the overall project cost. To reduce this risk, a batch of modules is typically independently tested to confirm baseline performance, at least for large enough projects. Similarly, a batch of modules is tested in case of having to verify a warranty issue.

Measurement uncertainty can be minimized at a substantial cost, but currently the practical limit of P_{MAX} laboratory measurements seems to be 1.6% [3] with 95% confidence for crystalline PV modules. In fact, typical measurement

uncertainties are from 2.5% to 3%. This number is significant in the context of other associated risks and usual profit margins.

A metrological approach is utilized to minimize the performance risk. This is achieved via multiple laboratory measurements, reducing the measurement uncertainty of the mean value of a batch of PV modules. The proposed methodology results in a lower uncertainty than that of a single laboratory while still being cost-effective. This is especially the case when applied to large PV fields.

II. METHODOLOGY

Respectable PV characterization laboratories have established measurement procedures that ensure high repeatability and reproducibility of their measurements. This is documented as part of an ISO 17025 accreditation. It is typically a small fraction of their overall uncertainties. At the same time, international inter-laboratory comparisons [3]–[7] suggest that the uncertainty estimates are not overly conservative. It can, therefore, be concluded that the contribution of persistent effects, such as reference cell calibration uncertainty, irradiance homogeneity and mismatch factor uncertainty, are dominating the overall uncertainty. A single laboratory cannot decrease its uncertainty further without a very substantial investment aimed at minimizing and/or correcting for persistent effects.

In contrast, it is possible to reduce the uncertainties via measuring modules at different measurement setups with varying persistent effects. In metrology, it is common practice to measure a standard by multiple laboratories and to assign the weighted average as the reference value. The combined information from all measurements results in an uncertainty of the reference value that is lower than any of the individual participant's measurement uncertainties. This was recently demonstrated for primary reference cell calibration [8].

In this paper, the above approach is extended to measuring the mean values of the performance characteristics of a batch of modules by multiple partners. For brevity, the focus is on P_{MAX} measurements, but it is equally applicable to I_{SC} and V_{OC}. Since it is impractical to measure all modules by multiple partners, the method is further modified so that only a small sample of the batch of modules is measured by multiple laboratories.

The methodology is illustrated in Fig. 1. One laboratory measures all modules in a batch. Several partner laboratories measure a few sample modules from this batch. The sample modules' mean, as measured by each laboratory, is then used to calculate a weighted average. This weighted average can have an uncertainty smaller than any of the laboratories' individual uncertainties. The weighted average is used to adjust the mean value of the full batch of modules as measured by a single partner. This final value has an uncertainty that is larger than the weighted average of the sample modules, but smaller than the measurement uncertainty of a single partner. The uncertainty estimation of this approach is outlined in the following section.

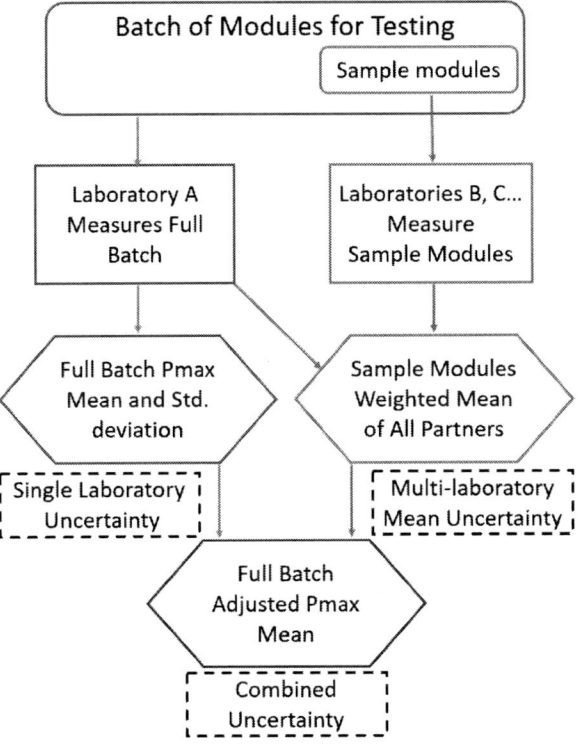

Fig. 1.   Summary of the proposed methodology.

III. UNCERTAINTY ESTIMATION

Following a comprehensive uncertainty analysis of a PV measurement laboratory, it is possible to classify the effects that contribute to P_{MAX} uncertainty as [9]:

- Persistent (i.e. systematic) between consecutive measurements
- Volatile (i.e. random) for each measurement
- A combination of both with an estimate of what part might be persistent.

Henceforth, the terms persistent and systematic are used interchangeably. The same applies to volatile and random.

The relative uncertainty associated with each PV module measurement can therefore be expressed as a sum of uncertainties due to random, $u_r(R)$, and systematic, $u_r(S)$, effects [10],[11]. The uncertainty of the mean of l modules, $\overline{P_l}$, can be calculated as:

$$u_r^2(\overline{P_l}) = u_r^2(S) + \frac{u_r^2(R)\left(\sum_{j=1}^{l} P_j^2\right)}{\left(\sum_{j=1}^{l} P_j\right)^2} \qquad (1)$$

As the number of PV modules increases, the uncertainty of their mean value reduces and approaches the systematic uncertainty of the laboratory, which is the predominant part of the total uncertainty. Repeat measurements cannot reduce that uncertainty beyond this point.

Different laboratories, however, have varying measurement setups thus different spectrum of the solar simulator; operating procedures; reference cells; data acquisition calibration and traceability chains. As such, the mean of PV modules measured on more than one setup can have smaller uncertainties than the systematic uncertainty of a single laboratory. The smallest uncertainty is achieved via using a weighted mean, $\overline{P_w}$, of all laboratories. The weights are chosen so that laboratories with larger uncertainties have smaller weights, i.e. the weights are reciprocal to the variance of each laboratory [11].

It is impractical or at least it would be prohibitively expensive to have a large batch of PV modules tested at multiple laboratories. For this reason, only a small sample of PV modules m, $m < l$, is to be measured by all laboratories and only one laboratory is required to measure the full batch of PV modules. A correction can then be applied to the mean value of the l PV modules, $\overline{P_{i,l}}$. This correction is based on the weighted mean of the m PV modules measured by all laboratories, $\overline{P_w}$, and the measured mean of m PV modules measured by one laboratory, $\overline{P_{i,m}}$:

$$\overline{P_{i,l}cor} = \frac{\overline{P_{i,l}}.\overline{P_w}}{\overline{P_{i,m}}} \qquad (2)$$

Assuming the correlations between mean values cancel out (see Section VI), the relative uncertainty associated with this corrected mean is:

$$u_r^2(\overline{P_{i,l}cor}) = \frac{u_r^2(R_i)}{m} + \frac{u_r^2(R_i)}{l} + u_r^2(\overline{P_w}) \qquad (3)$$

IV. SENSITIVITY ANALYSIS

There are two important questions that need answering: what number of laboratories and what number of PV modules should be measured to optimize the reduction of uncertainty with a minimum increase in cost. A detailed cost function is not yet estimated, however with some approximation these questions can be addressed.

In Fig. 2 the reduction in uncertainty of the mean of a variable number of PV modules are presented. This is done using the uncertainty estimates at CREST. In addition, the reduction in the weighted mean uncertainty with increasing number of laboratories all measuring 20 PV modules is also presented. For simplicity, this is done assuming that all partners have equal uncertainties to those at CREST. These uncertainties are reported in Table I.

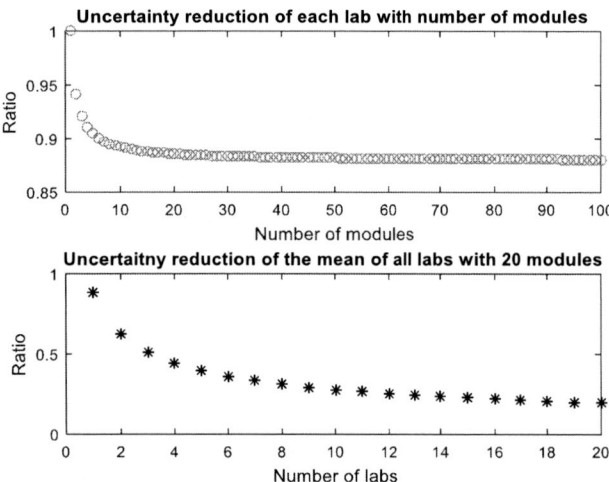

Fig. 2.   Uncertainty reduction with increasing module number and with increasing number of laboratories all measuring 20 modules.

It can be seen in Fig. 2 that measuring any number between 15 and 20 PV modules is sufficient. This number could be further optimized based on the number of particular PV modules that can be shipped together in a single pallet.

Since the full batch of PV modules would not be measured by all laboratories, the total cost would not be proportional to the number of laboratories. Given that the reduction in uncertainty, i.e. the slope, quickly decreases with each additional laboratory and considering potential costs, it is estimated that a cost-effective number of partners would be either two or three.

V. CASE STUDY OF SOLAR PARK DUE DILIGENCE ASSESSMENT

In this section, the approach is exemplified for an assessment of a solar park for warranty cases. The sample batch to be measured by all laboratories was chosen to be 17 PV modules. One laboratory measured the full batch of 120 PV modules, i.e. m =17 and l =120.

The three laboratories that participated were:

- CREST, UK
- UL, DE
- PV LAB, DE

Analyzing the uncertainty of each laboratory following a consistent framework resulted in estimates of the uncertainties due to of random, systematic and mixed effects. This was done employing a similar approach to the one described in [12]. The results of this analysis are summarized in Table I below.

TABLE I
SUMMARY OF LABORATORY UNCERTAINTIES

Laboratory	Relative P_{MAX} Uncertainties at k = 2 in %		
	Total, $U_r(P_i)$,	Systematic, $U_r(S_i)$,	Random, $U_r(R_i)$,
CREST	2.75	2.42	1.31
PV LAB	2.59	2.17	1.40
UL	3.02	2.79	1.17

The mean value and standard deviation of the 17 PV modules as measured by each laboratory alongside the relative uncertainty of the mean are presented in Table II. The calculated weighted mean, $\overline{P_w}$, and relative uncertainty at k=2, $U_r(\overline{P_w})$, of these PV modules are also included.

TABLE II
SUMMARY OF LABORATORY MEASUREMENTS

Laboratory	$\overline{P_{i,m}}$, W	STD, W	$U_r(\overline{P_{i,m}})$,%
CREST	323.56	2.61	2.44
PV LAB	323.51	2.44	2.20
UL	323.32	2.33	2.80
Weighted Mean	$\overline{P_w}$, =323.48	$U_r(\overline{P_w})$ =1.41	

The correction for the mean measurement of the full batch of PV modules, see (2), as measured by PV LAB happened to be negligible at less than 0.01 %. The uncertainty of this corrected mean was 1.44 % at k=2. Therefore, the uncertainty of the batch mean value was reduced by almost half.

VI. ASSUMPTIONS, LIMITATIONS AND OPPORTUNITIES

As described in [13], when using a weighted mean the measurand, in this case i.e. the mean of the sample PV modules, has to be stable. It is a reasonable assumption that if care is taken during shipping of PV modules, the mean of c-Si PV modules measured in a short timespan would be stable. Technologies exhibiting metastability would require an additional validation step of the precondition procedures and potentially an additional uncertainty contribution.

The approach assumes no correlation between measurements at different laboratories. In practice, there could be some level of correlation between partners, if for example employing the same type of simulator and reference cell or having the same traceability chain. If this is the case, a further analysis is required to estimate the size of the correlation. The uncertainty calculation can be modified to include correlations. For the case study considered in this paper, while correlations may not be completely negligible, they are estimated to be insignificant in the context of the claimed uncertainties. When the approach is used for primary reference cell calibration [8] with an order of magnitude lower uncertainties, correlations must be investigated in more detail.

The weighted mean is weakly correlated with both the sample and full batch mean values as measured by the principal laboratory. If assumed equal, due to the difference in the sign of the sensitivity coefficients, these correlations cancel out resulting in (3). In fact, the correlation with the sample mean is stronger and thus this simplification results in a conservative estimate. The uncertainty propagation was simulated with Monte Carlo simulations to validate the formulae presented here. As expected, calculating the uncertainty with the simplified formula is slightly more conservative than the simulation, i.e. by 0.1% at k=1 for the case study.

The opportunities of this approach are clear. Uncertainties of measurements performed in the field of a large sample batch of PV modules can now have significantly lower uncertainty if a sub-sample is measured by other laboratories. This can be useful during commissioning, but also for estimating degradation, which can be smaller than typical measurement uncertainties and thus is hard to prove.

The approach is only valid if robust and comprehensive uncertainty analyses are performed by all partners and if these analyses are consistent with one another. It, therefore, requires disclosure and transparency between partners.

VII. CONCLUSIONS

Typical uncertainties of PV module ratings are still relatively large and represent significant risk associated with the performance of the system. This risk translates into an increased cost of capital. To mitigate this risk, the baseline performance of a batch of PV modules can be independently measured with uncertainties usually ranging between 2.5% and 3%. Similarly, batch PV module measurements are performed to support warranty cases. In this paper, it is demonstrated how this uncertainty can be reduced if a sub-sample of these PV modules is measured by several laboratories that are willing to share their measurement procedures and uncertainty analyses. It is estimated that the most cost-effective uncertainty reduction can be achieved, if up to 20 PV modules are measured by three laboratories. For the presented case study, the approach resulted in a 1.44% (k=2) uncertainty associated with the mean P_{MAX} value of the whole batch. This represents a reduction in uncertainty by a factor of almost 2 and is lower than any uncertainties reported in literature of individual laboratories.

REFERENCES

[1] B. Gallagher, "US PV System Pricing H2 2016 : System Price Breakdowns and Forecasts," GTMResearch Report 2016.

[2] T. Lowder, M. Mendelsohn, B. Speer, R. Hill, "Continuing Developments in PV Risk Management : Strategies, Solutions, and Implications,", NREL Report, 2013.

[3] D. Dirnberger, U. Kräling, H. Müllejans, E. Salis, K. Emery, Y. Hishikawa, and K. Kiefer, "Progress in photovoltaic module calibration: results of a worldwide intercomparison between four reference laboratories," *Meas. Sci. Technol.*, vol. 25, no. 10, p. 105005, 2014.

[4] W. Herrmann, S. Zamini, F. Fabero, T. Betts, N. Van Der Borg, K. Kiefer, G. Friesen, H. Muellejans, and D. Fraile, "PV Module Output Power Characterisation In Test Laboratories And In The PV Industry – Results Of The European Performance Project," in *25th EUPVSEC/WCPEC-5*, 2010, pp. 3879–3883.

[5] Y. Hishikawa, H. Liu, H. Hsieh, T. Inoue, K. Kim, C. Limsakul, S. Kim, R. Ninae, and K. Morita, "Round-robin measurement intercomparison of c-Si PV modules among Asian testing laboratories," *Prog. Photovoltaics Res. Appl.*, vol. 21, no. 5, pp. 1181–1188, 2012.

[6] B. Mihaylov et al., "Results of the Sophia Module Intercomparison Part-1: STC, Low Irradaince Conditions and Temperature Coefficients Measurements of c-Si Technologies," in 29th EUPVSEC, 2014, pp. 2443–2448.

[7] B. Mihaylov et al., "Results of the SOPHIA Module Intercomparison Part-2: STC, Low Irradiance Conditions and Temperature Coefficients Measurements of Thin Film Technologies," in 31th EUPVSEC, 2015, pp. 1871–1876.

[8] H. Müllejans, W. Zaaiman, and E. D. Dunlop, "Reduction of uncertainties for photovoltaic reference cells," Metrologia, vol. 52, pp. 646–653, 2015.

[9] A. Possolo, "Simple Guide for Evaluating and Expressing the Uncertainty of NIST Measurement Results", NIST Technical Note 1900, 2015

[10] J. L. Gardner, "Uncertainties in Photometric Integrals," NMI Technical Report 9, 2005.

[11] E. R. Woolliams, "Determining the uncertainty associated with integrals of spectral quantities," EMPR-ENG05-1.3.1, 2013.

[12] B. Mihaylov, T. R. Betts, A. Pozza, H. Muellejans, and R. Gottschalg, "Uncertainty Estimation of Temperature Coefficient Measurements of PV Modules," IEEE J. Photovoltaics, vol. 6, no. 6, pp. 1554–1563, 2016.

[13] M. G. Cox, "The evaluation of key comparison data," Metrologia, vol. 39, no. 6, pp. 589–595, Dec. 2002.

Cloud Motion Identification Algorithms Based on All-Sky Images to Support Solar Irradiance Forecast

Lydie Magnone, Fabrizio Sossan, Enrica Scolari, Mario Paolone
École Polytechnique Fédérale de Lausanne, Switzerland

Abstract—Cloud motion is a cause of direct irradiance variations at ground level and determines significant fluctuations of PV generation. In this work, we investigate on how integrating information on clouds motion extracted from all-sky images into a time series-based forecasting tool for global horizontal irradiance (GHI) to enhance its prediction performance. We consider three different cloud motion algorithms: heuristic motion detection (HMD), particle image velocimetry (PIV), and a persistent model. The HMD method is originally proposed in this paper. It consists in choosing the cloud motion vector leading to the best cloud map prediction considering the most recent sky images. Results show that integrating the information of the predicted cloud coverage in the circumsolar area leads to a decrease of the width of the GHI prediction intervals up to 2% for prediction horizons in the range 1 to 10 minutes.

I. INTRODUCTION

The trend towards decentralized control and short-term redispatch of conventional generation in power systems has increased the focus on short-term forecasting of stochastic generation at low aggregation level. Examples are in the field of control of microgrids, active distribution networks, and photovoltaic (PV) self-consumption, where the availability of predictions for specific PV installations is required, e.g. [1]. Whereas traditional satellite-based forecasting methods fail to meet spatial and temporal resolution requirements, recent developments in the existing literature have proposed the application of time series based methods to learn patterns of PV production/irradiance from historical observations, see [2]. With respect to the sole use of historical irradiance observations, the integration of information from all-sky images (ASIs) establishes a step further in terms of available knowledge thanks to enabling the identification of clouds position and motion.

The tool-chain to infer irradiance predictions from ASIs generally consists in the following main steps: *i)*, image preprocessing, *ii)*, cloud detection (i.e. deciding whether each pixel of the image correspond to a cloudy or clear sky point), *iii)*, cloud motion identification, and, *iv)*, elaboration of irradiance predictions. In this work, cloud detection is performed by applying Schmidt's algorithm [3].

First, we focus on the performance analysis of cloud motion algorithms. We propose an original method, called heuristic motion detection (HMD), and we compare its performance against those of particle image velocimetry (PIV, from the existing literature) and a persistent benchmark method. Performance is evaluated in terms of misclassification which is achieved when predicting the future position of clouds (so-called cloud map). In order to exclude from the performance assessment the errors due to wrong segmentation, we consider a set of manually segmented images (where the assignment cloud/clear sky pixel is performed by human visual inspection) as a ground truth value.

Second, the predicted cloud map is used to compute the amount of cloudy pixels in a circumsolar area. This information is used as an additional influential variable in a time series based forecasting tool for prediction intervals (PIs) of the global horizontal irradiance (GHI) considering 1, 2, 5 and 10 minutes forecast horizons.

This paper is organized as follows: Section II describes the state-of-the-art, Section III presents the methods, Section IV is for results, and Section V states the conclusions.

II. STATE-OF-THE-ART

The first work focusing on intra-hour irradiance forecast considering information from sky-images is described by Chow et al. in [4]. Cloud shadows projected on the ground are estimated using two-dimensional cloud maps generated from the cloud cover. Cloud motion vectors are generated by processing two consecutive sky images. Marquez and Coimbra, [5], predict the direct normal irradiance (DNI) by considering the clouds in the vicinity of the sun and detect cloud motion with PIV. Their irradiance forecasts outperform the persistence model for forecast horizons in the range 3 to 15 min, with the most accurate forecasts at 5 min. Chen et al. in [6] apply PIV to compute a velocity vector per cloud and develop a multi-scale cloud block matching strategy to account for clouds deformations. Another technique for cloud tracking is *optical flow*, which consists of a collection of apparent velocities of objects in an image. It is applied to cloud motion detection for predicting sun occlusions in [7], using consecutive frames shot at a few seconds distance. Since in this works we consider frames at 1 minute resolution, optical flow is not considered. Quesada-Ruiz et al. in [8] propose a method for cloud tracking applied to intra-hour DNI forecast. A sector-based method is used to detect the direction of motion of potentially sun-blocking clouds, and an adjustable-ladder method focuses on sky regions that potentially affects DNI values. Finally, Bernecker et al. in [9] introduce non-rigid registration for detecting cloud motion.

This research received funding from the Swiss Competence Center for Energy Research (FURIES) and by the SNSF NRP70 "Energy Turnaround".

A sun occlusion probability is filtered by a Kalman filter to obtain continuous GHI forecasts for up to 10 min.

Several works focus on the inclusion of all-sky images features into machine learning methods. In [8], a DNI forecasting model is developed based on an artificial neural network (ANN). It uses cloud cover time series (estimated from ASIs) in combination with DNI historical observations to forecast DNI on the 5 and 10 minutes look-ahead time. A hybrid model to construct DNI prediction intervals is proposed in [10]. It exploits information from ASIs, which are integrated to a support vector machine and ANNs. Authors of [11] propose a model for intra-hour GHI forecasting based on the k-nearest-neighbors (kNN). It exploits local observations and information from ASIs and delivers point predictions and PIs of both GHI and DNI. The inclusion of ASIs improves the definition of the PIs. Authors of [12] incorporate measured irradiance data with features extracted from ASIs to deliver point predictions, achieving to foresee irradiance variations accurately.

III. METHODS

The tool-chain used to compute GHI prediction intervals is shown in Fig. 1 and consists in the following steps (detailed in the rest of this section):

1) Image acquisition.
2) Image pre-processing. The horizon (i.e. topological features and nearby objects) is removed from the image, and the distortion due to the fish-eye lens is corrected with a geometrical transformation. Besides, the sun position is determined as a function of the sun zenith and azimuth values.
3) Cloud detection. Each pixel of the image is labeled as a cloud or clear sky. The result of the segmentation is a binary image called cloud map.
4) Cloud motion identification. It consists in estimating the cloud movement. We consider three methods: (i) HMD, a method proposed in this paper for the first time, (ii) PIV, and (iii) the persistence predictor. Each cloud motion algorithm returns one motion vector (a single one for the whole image), called *global motion vector*.
5) Cloud map forecasting: the cloud map is translated by applying the global motion vector. This leads to the forecasted cloud map.
6) Local cloud cover computation. Given the forecasted cloud map, it consists in computing the percentage of cloudy pixels in a specific area around the sun.
7) GHI prediction intervals computation. We extend a time series-based probabilistic prediction tool from our previous work [2], which was originally fed with GHI measurements only. We add the local cloud cover as an influential variable to explain GHI variations.

A. Image acquisition

Images are from an USB 2 megapixel camera with CMOS sensor and fish-eye lens. It is installed on the roof of our

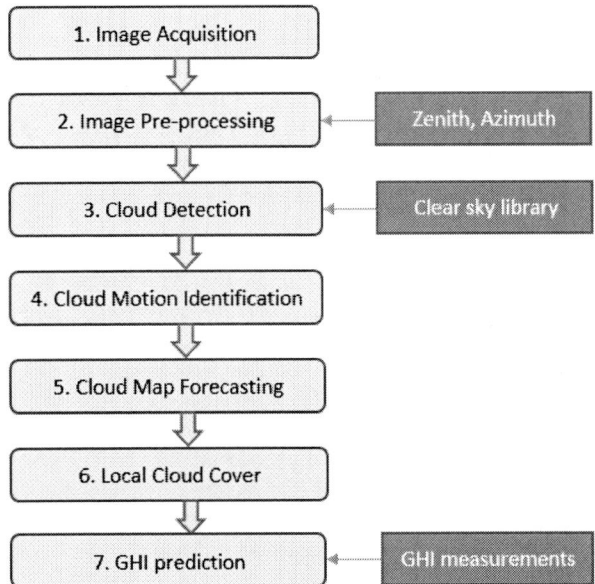

Fig. 1. Process of GHI forecast with all-sky images.

laboratory building, pointing at the zenith. Frames are taken at 1 minute resolution with manual exposure time to limit overexposure of the circumsolar area. The framegrabber is a Raspberry PI computer.

B. Image pre-processing

The sun position is determined by converting the solar zenith and azimuth coordinates (calculated by using the PV Performance Modeling Toolbox by Sandia National Laboratories [13]) into pixels coordinates. We use the Zenithal Equal Area Projection method from [14]. The image distortion is corrected by applying the procedure described in [15], which requires a series of pictures of a chessboard (shot offline). The sun coordinates are then transformed into undistorted coordinates. Finally, the horizon features (trees, buildings, mountain ridges) are removed by using a binary mask, manually determined offline.

C. Cloud detection

It consists in determining if a pixel corresponds to a cloudy or clear sky point. A common procedure for cloud detection consists in computing the red to blue ratio (RBR) of the image color channels, which generally achieves a good degree of separation between clouds (high RBR) and clear sky pixels (low RBR). However, misclassification occurs due to very bright clear sky pixels (high RBR) in the circumsolar area and very dark clouds (low RBR). Authors of [3] proposed an augmented RBR definition. It makes use of a clear sky library to achieve a better segmentation of the circumsolar pixels, and solar saturation and gray intensity level to account for dark clouds. The pixel RBR we use in this work is defined as:

$$R_{mod} = R_{orig} - R_{CSL}(a(2S-1) - b(I-c)), \quad (1)$$

where R_{orig} is the RBR of the considered image, R_{CSL} is the corresponding RBR value on a clear sky picture (selected from a clear sky library such that the sun position is as close as possible to the target picture), S is the saturation and refers to the percentage of saturated pixels in a circumsolar region with radius r, I is the gray intensity level, while a, b and c are tuning parameters. Finally, a pixel is marked as a cloud if its modified RBR value in (1) is larger than a threshold R_{thresh} and vice-versa.

The free parameters a, b, c, r, R_{thresh} involved in the cloud detection procedure are determined in order to give the best performance with our imaging setup. They are chosen with the following heuristic procedure: 40 images referring to periods with different weather conditions and sun position are selected and manually segmented; therefore, the same images are automatically segmented by using different possible combinations of the parameters in the ranges shown in Table I. The best values of the parameters are chosen as those leading to the best cloud map estimation.

TABLE I
SEGMENTATION PARAMETER SETTING

Name	Selected value	Tested range
a	0.6	0.3 to 1, with a step of 0.1
b	0.001	0 to 0.003, with a step of 0.001
c	220	130 to 220, with a step of 10
R_{thresh}	0.2	0.1 to 1, with a step of 0.1
r (pixels)	30	30 to 60, with a step of 10

D. Cloud Motion Identification

In this section, we first describe the formulation of the proposed method for cloud motion identification, the HMD. Later, we summarize the ideas behind PIV and describe the persistence model. As mentioned earlier, we assume in this work that all the visible clouds move uniformly, i.e. with same direction and speed. Under this assumption, there is a unique motion vector, called *global motion vector*.

Let us consider a set of pictures taken at 1 minute resolution. The problem consists in producing a binary image which contains the 1, 2, 5, or 10 minutes ahead predictions of the clouds position.

1) Heuristic Motion Detection (HMD): The HMD procedure is the following:

- we consider two consecutive cloud maps at time t–1 (A) and t (B);
- n binary cloud maps $A_1, \ldots, A_n$ are generated by translating the cloudy map A according to a random motion vector $\mathbf{v} = (u, v)$ with $\mathbf{v} \in \mathbb{R}^2$. We denote by (i, j) the pixel location. Then, the future position of a binary pixel is:

$$[i(t), j(t)] = [i(t-1) + u, j(t-1) + v] \quad (2)$$

In this work we consider 400 random vectors $\mathbf{v}$.

- each cloud map $A_1, \ldots, A_n$ is compared against B in terms of matching error defined as the sum of the misclassified pixels, see Section IV-B.

- the motion vector that generates the lowest matching error is selected.

2) Particle Image Velocimetry (PIV): In brief, PIV consists in comparing two consecutive pictures by evaluating the cross-correlation between portions of the images, called interrogation areas. This allows to infer the most likely particle displacement and to compute the motion vectors. It is also known from [4] as the cross correlation method. In this work, we use the Matlab implementation available in the PIVlab library [16]. Unlike the existing literature, we apply PIV on binary rather than on gray-scale images. The number of interrogation areas in an image, thus the number of vectors, is a parameter of the algorithm. In this case, it is chosen trough a sensitivity analysis as the one which leads to the best performance, as shown in Section IV. The cloud motion vectors of the image are averaged in order to obtain a global motion vector.

3) Persistent method: This method assumes that the clouds are persistent in a short-term horizon, and therefore the global cloud motion vector is zero.

E. Cloud map forecasting

It consists in translating the current cloud map according to the global motion vector, which is scaled in magnitude according to the forecasting horizon to achieve. This leads to the so-called forecasted cloud map. It was observed empirically that averaging the last 5 motion vectors rather than using the last one only leads to better forecasted cloud maps.

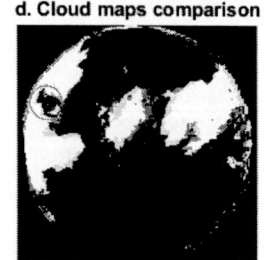

Fig. 2. Example of the forecasted cloud map procedure.

An example of the procedure described until this stage is shown in Fig. 2. Fig. 2a shows the undistorted view of the sky with the sun location (blue circle) and PIV motion vectors (green arrows). The global motion vector, obtained by averaging the PIV vectors, is used to translate the cloud map obtained by segmenting Fig. 2a. The translated cloud map is shown in Fig. 2b, where the white color denotes cloudy pixels, blue clear sky, and yellow circumsolar region. Fig. 2c shows the 1 minute ahead realization. Fig. 2d compares the forecasted cloud map (purple color) against the future ground truth cloud map from Fig. 2c (green color). The white color denotes those pixels which are correctly classified as cloudy.

F. Local cloud cover computation

The forecasted local cloud cover is computed as the percentage of cloudy pixels in the forecasted cloud map in a region around the sun. The region we consider is a disk with 100 pixels radius (at this stage, this is chosen empirically). We consider a circumsolar area rather than the whole picture since this is the region with the largest interest when considering short-term sun occlusions by clouds.

G. GHI prediction intervals computation

The work in [2] describes a method to compute GHI prediction intervals (PIs) with look-ahead time in the range from seconds to minutes starting from historical GHI measurements. We augment this algorithm by including the local cloud cover as an additional influential variable. The PI, which is defined as the interval where the future realization is expected to fall with a designed confidence level α [17], is denoted in the following formulation as $\left(I_{t+1|t}^{\uparrow \alpha}, I_{t+1|t}^{\downarrow \alpha} \right)$. The original method in [2] consists in grouping N historical values of the differentiated GHI time series ΔI into k clusters according to the value of two selected data features:

- the average GHI value on a mobile window of length n, considering the most recent data points:

$$M_i = \frac{1}{n} \sum_{j=i-n}^{i} \Delta I_j, \qquad i = n+1, \ldots, N; \quad (3)$$

- the GHI variability, defined as:

$$V_i = \sqrt{\frac{1}{n} \sum_{j=i-n}^{i} (\Delta I_j - \Delta I_{j-1})^2}, \quad i = n+1, \ldots, N. \quad (4)$$

The clustering process produces k clusters $G_1, \ldots, G_k$ and their centroids $\mathbf{c_1}, \ldots \mathbf{c_k}$; the histograms of the clusters are assumed as the empirical probability distribution function of the variations with respect to the one-step-ahead irradiance realization. During real-time operation, the data features vector at time t, denoted by $\mathbf{p_t} = (M_t, V_t)$, is calculated. The next step is the calculation of the Euclidean distances between $\mathbf{p_t}$ and the centroids $\mathbf{c_l}$

$$d_l = \|\mathbf{c_l} - \mathbf{p_t}\|^2, \quad l = 1, \ldots, k \quad (5)$$

which is used as a similarity criterion to select the cluster representative of the future irradiance value. With respect to the original method, we add the local cloud cover (obtained from the procedure described above) as an additional feature to build the clusters.

Finally, we compare three different combinations of influential variables:

- average irradiance and its variability (henceforth called *GHI measurements* in Section IV-G);
- average irradiance, its variability and the local cloud cover (henceforth called *Images + GHI measurements*);
- local cloud cover only (henceforth called *Images*).

IV. RESULTS

A. Experimental setup

GHI observations are from an Apogee SP-230 all-season pyranometer installed near the camera. Pyranometer GHI measurements are with an error of 2% and 5% at solar zenith angles of 45 and 75 degrees, respectively. ASIs and GHI measurements are synchronized and stored in a database.

B. Image metrics

To evaluate the performance of cloud detection and cloud motion identification, we define the matching error similarly to [4]. Let C be the ground truth cloud map (e.g., from manual segmentation), $\widehat{C}$ the estimated cloud map, and N the total number of pixels in the image. The matching error is:

$$matching\ error = \frac{1}{N} \sum_{i=1}^{I} \sum_{j=1}^{J} \left| C_{ij} - \widehat{C}_{ij} \right| \quad (6)$$

The matching error is used in the work, i), to assign the parameters of the Schmidt's algorithm (as described earlier), ii), cloud detection performance assessment, and, iii), evaluation of the cloud motion algorithms.

C. Forecasting performance metrics

To determine the quality of the prediction intervals of the probabilistic forecasting algorithm, we use the Prediction Interval Coverage Probability (PICP) and the Prediction Interval Normalized Averaged Width (PINAW), [18]. The PICP counts the number of times that the realization falls inside the PI for a given confidence level α:

$$\text{PICP} = \frac{1}{L} \sum_{t=1}^{L} c_t \quad (7)$$

where L is the total number of forecast instances of the testing dataset and

$$c_t = \begin{cases} 1, & \hat{I}_{t+1|t}^{\downarrow \alpha} \leq I_{t+1} \leq \hat{I}_{t+1|t}^{\uparrow \alpha} \\ 0, & \text{otherwise.} \end{cases} \quad (8)$$

where I_{t+1} is the one-step ahead GHI realization. The PINAW measures the width of the prediction interval:

$$\text{PINAW} = \frac{1}{L I_{max}} \sum_{t=1}^{L} (\hat{I}_{t+1|t}^{\uparrow \alpha} - \hat{I}_{t+1|t}^{\downarrow \alpha}), \quad (9)$$

where $I_{max} = 1000\ \text{W/m}^2$.

D. Cloud detection performance assessment

Cloud detection is the first step of the prediction tool-chain. Its accuracy plays a crucial role and it is therefore relevant to quantify it. The average matching error Eq. 6 calculated over the set of the 40 manually segmented pictures is 18%.

E. Selection of the number of motion vectors for PIV

We implement a sensitivity analysis to determine the best number of motion vectors for PIV. In general, the number of vectors depends on the size (in number of pixels) of the first and second interrogation areas. We have tested different combinations for the size of the interrogation areas, as reported in Table II. This allowed to determine the combination leading to the smallest matching error into the forecasted cloud map. The best combinations of interrogation areas are 400×400 and 200×200 pixels for interrogation areas 1 and 2, respectively, for the 1 and 2 minutes forecast horizons, while they are 200×200 and 100×100 pixels for interrogation areas 1 and 2, respectively, for the 5 and 10 minutes forecast horizons. These combinations the sizes used in the following for the PIV interrogation areas.

TABLE II
TESTED PIV VECTORS AND INTERROGATION AREAS

Number of vectors	42×45	29×31	27×29	25×27	20×22
Interrogation area 1	200	280	300	320	400
Interrogation area 2	100	140	150	160	200

F. Performance assessment of cloud motion methods

In this section, the performance of the three considered cloud motion methods (HMD, PIV and persistence) is compared in terms of the matching error of their forecasted cloud maps. Here, the testing image dataset consists in 40 consecutive images captured during partly-cloudy conditions. Results are shown in Fig. 3 and summarized in Table III. We obtain that:

- at 1 minute look-ahead time, the HMD method is the best performing;
- at 2 minutes, the HMD and the PIV have a similar performance and outperform the persistence method;
- at 5 and 10 minutes, neither the HMD nor the PIV outperforms the benchmark persistence model.

The last result is likely due to the fact that the considered cloud motion algorithms do not model phenomena such as cloud generation, dissipation and cloud shape changing, all effects which become more prominent when considering longer forecasting horizon.

G. GHI forecast assessment

To compute the forecasted local cloud cover and test its influence on the GHI forecasting tool, we choose the best performing cloud motion algorithm according to the look-ahead time. From the previous section, these are: HMD at 1 minute, PIV at 2 minutes, and persistence model at 5 and 10 minutes forecast horizon.

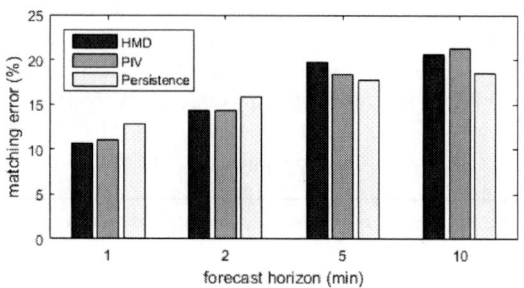

Fig. 3. Cloud motion methods evaluation for different forecast horizons.

TABLE III
MATCHING ERROR (%)

Forecast horizon	1 min	2 min	5 min	10 min
HMD	10.65	14.34	19.72	20.68
PIV	11.08	14.27	18.42	21.23
Persistence	12.83	15.82	17.68	18.44

Table IV shows the Pearson correlation coefficients between the time series of GHI and local cloud cover (considering measurements spanning a 21 days long period) on four distinct forecast horizons (1, 2, 5 and 10 minutes). A high correlation is observed between the two variables, thus suggesting that the local cloud cover is a meaningful variable that can be used to improve the accuracy of GHI forecast methods.

TABLE IV
PEARSON CORRELATION COEFFICIENTS

Forecast horizon	1 min	2 min	5 min	10 min
Cloud motion method	HMD	PIV	Persistence	
Correlation coefficient	-0.7622	-0.7985	-0.7806	-0.7504

We consider the probabilistic forecasting tool described in Section III-G. The training and testing data set consists of measurements for 16 and 5 days, respectively. We select a number of clusters k manually tuned to achieve a coverage probability very close to the target confidence level $\alpha = 95\%$.

Results are summarized in tables V and VI, and shown in Fig. 4. They can be summarized as follows:

- from Table V, all the considered cases have a coverage probability (PICP) close to the target confidence level (95%), denoting that the methods have a good reliability;
- in any case, adding the forecasted local cloud cover to GHI measurements is beneficial as it improves the GHI forecast on all the considered forecasting horizons (or it does not impact negatively, as in the 1 minute case);
- the use of the forecasted local cloud cover as only influential variable outperforms all the other methods at 1, 2, and 5 minutes forecast horizons.

V. CONCLUSION AND FUTURE WORKS

We have carried a first investigation on how to augment a GHI time series-based forecasting tool for solar irradiance

TABLE V
PICP (%)

Forecast horizon	1 min	2 min	5 min	10 min
GHI measurements	96.06	95.6	95.14	97.2
Images + GHI measurements	94.75	95.16	96.45	97.23
Images	94.92	94.56	94.75	95.95

TABLE VI
PINAW (%)

Forecast horizon	1 min	2 min	5 min	10 min
GHI measurements	4.13	7.05	12.69	18.12
Images + GHI measurements	4.13	6.89	12.23	16.69
Images	3.6	5.78	11.87	17.39

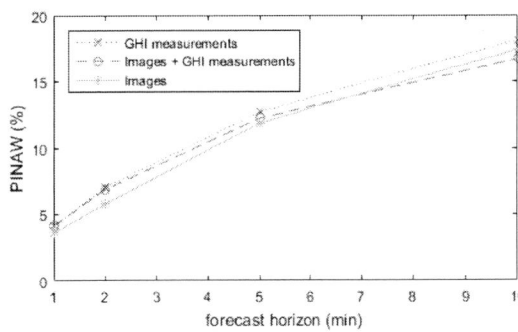

Fig. 4. PINAW of the GHI prediction for different forecast horizons.

by using information on the cloud motion extracted from all-sky images. The cloud motion algorithms considered in the analysis are particle image velocimetry (PIV, from the literature), heuristic motion detection (HMD, an original method described in this work), and a persistent predictor.

It was shown that applying cloud motion identification allows to improve cloud map forecasting performance up to the 2 minutes forecasting horizon. Above the 2 minutes look-ahead time, there is no advantage with respect to using a persistance predictor.

Results on GHI prediction intervals shows that including information on the cloud motion is in general beneficial because it leads to get smaller PIs width.

The future work is in the direction of consolidating the current results by extending the proposed method to an alternative all-sky imager, and comparing the results with those obtained from cloud detection methods based on machine learning strategies.

REFERENCES

[1] R. Palma-Behnke, C. Benavides, F. Lanas, B. Severino, L. Reyes, J. Llanos, and D. Sáez, "A microgrid energy management system based on the rolling horizon strategy," *IEEE Transactions on Smart Grid*, vol. 4, no. 2, pp. 996–1006, 2013.
[2] E. Scolari, F. Sossan, and M. Paolone, "Irradiance prediction intervals for pv stochastic generation in microgrid applications," *Solar Energy*, vol. 139, pp. 116–129, 2016.
[3] T. Schmidt, J. Kalisch, E. Lorenz, and D. Heinemann, "Evaluating the spatio-temporal performance of sky-imager-based solar irradiance analysis and forecasts," *Atmospheric chemistry and physics*, vol. 16, pp. 3399–3412, March 2016.
[4] C. W. Chow, B. Urquhart, M. Lave, A. Dominguez, J. Kleissl, J. Shields, and B. Washom, "Intra-hour forecasting with a total sky imager at the uc san diego solar energy testbed," *Solar Energy*, vol. 85, no. 11, pp. 2881–2893, 2011.
[5] R. Marquez and C. Coimbra, "Intra-hour dni forecasting based on cloud tracking image analysis," *Solar Energy*, vol. 91, no. 99, pp. 327–336, April 2013.
[6] Y. Chen, W. Li, C. Zhang, and C. Hu, "Global velocity constrained cloud motion prediction for short-term solar forecasting," *Application of digital image processing*, vol. 9971, 2016.
[7] S. Sun, E. Ritzhaupt-Kleiss, and T. Chen, "short term cloud coverage prediction using ground based all sky imager," *IEEE international conference on smart grid communications*, vol. 14, 2014.
[8] S. Quesada-Ruiz, Y. Chu, J. Tovar-Pescador, H. Pedro, and C. Coimbra, "Cloud-tracking methodology for intra-hour dni forecasting," *Solar Energy*, vol. 102, pp. 267–275, 2014.

[9] D. Bernecker, C. Riess, A. Elli, and J. Hornegger, "Continuous short-term irradiance forecasts using sky images," *Solar Energy*, vol. 110, pp. 303–315, September 2014.
[10] Y. Chu, M. Li, H. T. Pedro, and C. F. Coimbra, "Real-time prediction intervals for intra-hour dni forecasts," *Renewable Energy*, vol. 83, pp. 234–244, 2015.
[11] H. T. Pedro and C. F. Coimbra, "Nearest-neighbor methodology for prediction of intra-hour global horizontal and direct normal irradiances," *Renewable Energy*, vol. 80, pp. 770–782, 2015.
[12] H.-Y. Cheng, C.-C. Yu, and S.-J. Lin, "Bi-model short-term solar irradiance prediction using support vector regressors," *Energy*, vol. 70, pp. 121–127, 2014.
[13] R. W. Andrews, J. S. Stein, C. Hansen, and D. Riley, "Introduction to the open source pv lib for python photovoltaic system modelling package," in *Photovoltaic Specialist Conference (PVSC), 2014 IEEE 40th*. IEEE, 2014, pp. 0170–0174.
[14] M. R. Calabretta and E. W. Greisen, "Representations of celestial coordinates in fits," *Astronomy & Astrophysics*, vol. 395, no. 3, pp. 1077–1122, 2002.
[15] Z. Zhang, "A flexible new technique for camera calibration," *IEEE Transactions on pattern analysis and machine intelligence*, vol. 22, no. 11, pp. 1330–1334, 2000.
[16] W. Thielicke and E. J. Stamhuis, "Pivlab-toward user-friendly affordable and accurate digital particle image velocimetry in matlab," *Journal of open research software*, vol. 28, no. 99, pp. 1286–1296, April 2014.
[17] P. Pinson, "Estimation of the uncertainty in wind power forecasting," Ph.D. dissertation, École Nationale Supérieure des Mines de Paris, 2006.
[18] A. Khosravi, S. Nahavandi, and D. Creighton, "Prediction intervals for short-term wind farm power generation forecasts," *Sustainable Energy, IEEE Transactions on*, vol. 4, no. 3, pp. 602–610, 2013.

AUTHOR INDEX

Aaditya, Gayathri 604
Abbas, A. 1691, 2457, 3430
Abbas, Ahmed E. 1888
Abbas, Ali 186, 752, 1674
Abbott, Malcolm D 1322, 2576, 2600
Abdalla, L.B 1245
Abdallah, Amir A. 3435
Abdallah, Shaimaa A. 219
Abdellaoui, Imane 900
Abdullah, Ahmad 2128
Aberle, Armin 2318
Aberle, Armin G. 284, 496, 499, 1922
Ablekim, Tursun 3422
Aboubakr, Benazzouz 487
Abouelkhair, Hussain M........................2324
Abtahi, Amir 638
Abudayyeh, Omar K. 88
Acebo, Laura 155
Addamane, S. J. 281
Adewoyin, Adeyinka 2381
Adhikari, Dipendra 2582
Affouda, Chaffra A. 259
Agarwal, Mohit 2330
Agarwal, Sumit................................1777
Agarwal, Vivek 2952, 2981, 2986, 3050
Agbo, Solomon N...............................2114
Ager, Joel 3410
Agrawal, Rakesh 1449
Aguiar, Jeff 2702
Aguiar, Jeffrey 2467
Aguirre, Rodolfo 2419
Ahamioje, Joseph A. 2931
Ahanzhamejhad, Ramez H.......................170
Ahlswede, E. 3260
Ahlswede, Erik 791
Ahmad, Jawad 3096
Ahmed, Benlarabi 487
Ahmed, Nuha..................................658, 2667
Aho, Arto 297, 2520
Aho, T. 1189
Aho, Timo 297
Ahrenkiel, P. 869
Ahrenkiel, Phil 206, 831
Ahrenkiel, Richard K. 3448
Ahrenkiel, S. Phillip 2514
Ahsan, Nazmul 2334
Aierken, Abuduwayiti..........................226
Aindow, Mark 1522
Aïssa, Brahim 3435
Akaki, Yoji..................................2338
Akari, Shunsuke..............................2385
Akarm, Muhammad Nadeem 2776
Ake-Sultan, Bernt 2864
Akiki, Tilda.................................1968
Akimoto, Katsuhiro 33, 160, 900

Akimoto, Naoki................................712
Akiyama, Hidefumi 721, 2781, 3528
Akwari, Chinedum 735, 2446
Al Mahmud, Abdullah 1067
Alahmed, Ahmed 1110
Alam, Giri Wahyu 1498
Alam, Muhammad A. 1055, 1259
Alam, Muhammad Ashraful......................1904
Alberi, Kirstin 2506
Albin, David.................................1196, 3305, 3319
Alcubilla, R. 1781
Alcubilla, Ramón 944
Aleman, Monica...............................2227, 3435
Alexander, Jessica A. 966
Alfadhili, Fadhil K...........................730, 815
Al-Fadhili, Fadhil K. 2462
Al-Ghzaiwat, Mutaz 2593
Algora, C. 1210
Algora, Carlos...............................1204
Alharbi, Fahhad H. 963
Alharthi, Yahya Z. 1018, 1110
Ali, Asad 1228
Ali, Jaffar Moideen Yacob....................2318
Ali, Waqar...................................1228
Alivisatos, A. Paul 1737
Al-Jassim, M. 1196
Al-Jassim, M.M. 1312, 2280, 2785
Al-Jassim, Mowafak 62, 1371, 1381, 1400, 2789, 2887, 3305, 3319
Al-Jassim, Mowafak M. 3147
Aljaziri, Marwa 2011
Alkhayat, Rabee B............................815
Allebé, C. 50, 2073
Allebé, Christophe...........................3254, 3256
Allen, Thomas 2076
Almheiri, Anwar 1946
Almonacid, Florencia 2858
Alrashidi, Hameed............................2858
Altermatt, Pietro P. 1922, 2220, 3304
Alvarez, Diego Alonso 1339
Alvarez, Genesis 2941
Alvarez, José 2453, 2528
Aly, Shahzada P..............................963
Alzahmi, Wadhah..............................1946
Amdemeskel, Mekbib W.........................2672
Anctil, Annick 2124
Anderberg, A. 467
Andler, Joseph 1449
Ando, Daisuke................................931
Ando, Yasutaka 970
Ando, Yuta 192
Andreani, Lucio 290
Angeles-Ordóñez, G. 142
Annigoni, Eleonora...........................1395, 2794
Anoma, Marc Abou 1549

AUTHOR INDEX

Anselmo, Andrew...................74, 2839, 2897
Antony, Aldrin 1755
Anttu, Nicklas...............................2502
Anyadiegwu, Ifeanacho 970
Anyanwu, Uchechi 319
Araki, Hideaki................................2338
Araki, Kenji................359, 412, 1479, 1711,
    1714, 1743, 2498, 2548, 2566
Aranguren, G. 643
Archer, Alexander 771
Arehart, A. R. 30, 2414
Arehart, Aaron R. 215, 2446, 3139
Arinze, Ebuka S. 667
Armour, Eric 827
Armour, Eric A. 210, 2506
Armoush, Maher 1058
Arnold, Daniel B. 3002
Arnou, Panagiota 146, 186
Arora, B. M. 396, 1995, 2716
Arora, Brij M. 3478
Arp, Juergen..................................1411
Arredondo, C. A. 2031
Artegiani, Elisa...............752, 1669, 2372
Aryal, Krishna 182
Asadirad, Mojtaba 866
Asahi, Shigeo 23
Asgharzadeh, Amir...............1537, 1543, 3333
Ashrafee, Tasnuva 735
Aslam, Aasma..................................2355
Asomoza, René 632
Astakhov, Oleksandr 2114
Aswani, U. 1898
Athresh, Eashwer..........................2395, 2399
Atia, Adam A. 3230
Atkins, R. 229
Atlan, Olivier................................626
Atwater, Harry A..............512, 521, 558, 572,
    1248, 1589, 1737, 2236
Augarten, Yael 1651
Augusto, André 1589, 2596
Avasthi, Sushobhan........251, 837, 841, 986, 2395, 2399
Avenet, Julien 1933
Avery, J. E. 1863
Awadallah, Osama 3473
Awasthi, Vishnu 2345
Ayala, Orlando 735
Azkona, N. 2740
Azkona, Nekane 2677
Azzolini, Joseph A. 608
Baba, Masaaki.................................1724
Babbe, Finn 151, 2054
Babcock, Sean J. 2298
Bachman, Benjamin F. 3381
Badel, N. 50
Badosa, Jordi.................................626

Badr, Ikken 487
Bae, Soohyun 935
Baggu, Murali 2991
Baik, Sungsun 2242
Bailey, C. 845
Bailey, Christopher G. 2298
Bailey, J. 2414
Bailey, Jeff.............................1686, 3327
Baines, Tom 742, 1445
Baka, Maro 3343
Baker, Rupesh 3172
Bakhshi, Sara 322
Bakker, Klaas 2875
Balaji, Pradeep 2596
Balakrishnan, G. 281
Balasubramaniam, Kavaipatti R. 1704
Baldus-Jeursen, Christopher...................1908
Ball, Greg 2263
Ballif, C. 50, 2073
Ballif, Christophe 55, 1220, 1395,
    2104, 2794, 3254, 3256, 3435
Baloch, Ahmer A.B.........................963, 1058
Banda, Pedro 1946
Banerje, Rangan 1151
Banerjee, Sanjay K. 363
Barahman, Gil 2285
Barakel, Damien 2255
Barnes, T. M. 138
Barnes, Teresa M.............................3422
Barnett, Allen 315
Barraud, L. 50
Barraud, Loris 3254, 3256
Barrigon, Enrique 2502
Barth, Kurt 424
Bartolo, Robert E. 195
Bartsch, J. 884
Basore, Paul A. 2163
Bastide, Stéphane 3402
Bastola, Ebin 738, 781
Basu, Prabir K...............................396
Battaglia, A. 1747
Baudrit, Mathieu.........................2492, 2562
Bauer, Andreas 791, 2058
Bauer, Jan 1376
Bauhuis, G. 1189
Baumann, Thomas 1077
Baumgartner, Franz...........................1077
Baur, Carsten 541, 2087
Baxter, Jason B. 3143
Bearda, Twan 1233
Beauchemin, Ryan D. 102
Becerril-Romero, Ignacio 155
Becker, Jacob J..........................3366, 3410
Bedair, Salah M. 2195
Belanger, Ted.................................1427

AUTHOR INDEX

Belletête, Marc 1579
Belluardo, Giorgio 3360, 3482
Bemrrr, Andreas............................3500
Benamara, Mourad.........................3370
Benatto, Gisele A. Dos Reis 2672, 2682
Benick, J...................................2064
Benick, Jan.................................2511
Bennett, Dirk..............................2042
Bennett, Mitchell..........................247
Bennett, Mitchell F.............210, 259, 873, 2091
Berardone, Irene...........................402
Berg, Alexander 1773
Berg, Morgann.............................3417
Bermel, Peter.........................1904, 2467
Bernard, Annie........................2870, 2891
Berry, Joseph J............................2176
Bert, J....................................1733
Bertoni, Mariana..............944, 2610, 2854
Bertoni, Mariana I. 2179, 3309
Besanger, Yvon............................3102
Bett, Alexander J..........................1253
Bett, Andreas W...........................2511
Bettenwort, Gerd 1965
Beutel, Paul...............................2511
Beutner, Volker............................1855
Bhaduri, Sonali.......................2799, 3478
Bhan, Mohan Krishan 496
Bhandari, Khagendra P...........738, 748, 781, 815
Bhatia, A..................................1656
Bhatia, Swasti.............................1755
Bhattacharya, Indranil 3083
Bhattacharya, Sitangshu 2376
Bheemreddy, Venkata 2688
Bialek, Tom...............................2991
Bidiville, Adrien 1333
Biedenham, Richard E......................3245
Biegelsen, D. K. 1733
Biiss, M. 2457
Binetti, Simona............................1669
Birch, Max T. 2423
Birkmirc, Robert..........................726
Birkmire, Robert W........................2637
Bishop, Doug 3275
Bishop, Douglas...........................726
Bishop, Douglas M. 1441
Bissels, G.................................1189
Biswas, R.................................1350
Bittau, F..................................3430
Bittau, Francesco 752
Bittner, Zachary..........................677
Bittner, Zachary S............18, 202, 2084
Bivour, M.................................2064
Bivour, Martin............................1253
Beutel, Paul..............................2511
Blakely, Logan............................1573
Blanche, Pierre-Alexandre 1147

Bläsi, Benedikt............................352
Blasi, David 1531
Blum, Adrienne 2692
Blum, Adrienne L..........................2765
Bob, Brion................................2258
Bobela, David C...........................2506
Bobyl, A.V. 1025, 1811
Boca, Andreea.............................2099
Boccard, Mathieu 55, 1220, 1317, 1790, 3366
Boeck, Torta..............................3396
Bohra, Rakesh 1912
Boizot, Bruno 83, 2087
Bolaji, Adewumi 2381
Boley, Allison 2573
Bolke, J. G...............................1656
Bonnassieux, Yvan 626
Bonomo, Pierluigi 2118
Book, Felix...............................1824
Bora, Birinchi 3478
Borgers, Tom 3343
Borgström, Magnus........................2502
Borgström, Magnus T 1286
Borland, John 2947
Borne, Axel...............................2864
Borowik, Lukasz 1516
Bosco, Nick...........................3190, 3200
Bosco, Nick S.............................2864
Bosson, Christopher J.....................2423
Bostock, Peter 2267
Bothe, Karsten 2692
Bourcois, Jérôme 2087
Bourdin, Vincent 626
Bourgoin, Jacques C. 2087
Bourne, Ben C.............................1549
Bousselham, Abdelkader 1058
Bouttcmy, Muriel 2711
Bowden, Stuart.............240, 925, 1797, 2719
Bowden, Stuart G. 1589, 2596
Bowen, Leon...............................1445
Bowers, J.W. 2457, 3430
Bowers, Jake W. 146, 186, 752, 2349
Boyce, Ken................................2000
Boyce, Kenneth 1933
Boyd, Matthew.............................1933
Boyer, Jacob..............................215
Boyer, Jacob T. 2079, 2554
Boyer-Richard, Soline 2192
Brabec, Christoph J. 1346, 3500
Bradshaw, Geoffrey K. 88, 301, 531
Brady, Brendan 3388
Braga, Daniel Sena 2307
Braid, Jennifer L. 1927, 2697, 3456
Brammertz, G..............................3260
Brand, A.A................................884
Brates, Nanu 1728

AUTHOR INDEX

Bräuninger, Matthias 3256
Breitenstein, Otwin 1376
Breitwieser, M. 3135
Bremner, S. P. 953
Bremner, Stephen 858, 1215, 1845, 2186, 2569
Bremner, Stephen P. 948
Brendel, Rolf 1366, 3371
Breus, V. 1752
Bright, Jamie M. 1405
Brinnig, Samuel 2622
Brito, Pedro P. 2307
Britt, Jeffrey 1455
Brittman, Sarah 2245
Broderick, Robert 3008
Broderick, Robert J. 1435, 1555, 1567, 1573, 3025, 3031
Brolo, Alexander G. 3388
Bruckman, Laura 1933
Bruckman, Laura S. 2000
Brückner, Sebastian 2538
Brughera, Céline 2492
Brule, Carlton 1728
Brulo, Gregory S. 1469
Bryan, Jonathan 1317
Buchanan, Wayne 1196
Büchler, A. 884, 3135
Buckner, Jessica 537
Buerhop, Claudia 3500
Bukowsky, Colton R. 1737
Bulkin, P. 1781
Bulkin, Pavel 1237
Bullock, James 59, 2076
Buonassisi, T. 648, 1140, 3295
Buonassisi, Tonio 284, 1264, 1491, 2242, 2532, 2744, 3236, 3290, 3300
Burgers, A.R. 3150
Burgers, Antonius R. 917
Burkhardt, S. 1752
Burnham, Laurie 1435
Burroughs, Scott 272, 1469
Busquet, Severine 1061
Butt, Isaac 182
Cabarrocas, Pere Roca I. 464, 1237, 2528, 2593
Cachet-Vivier, Christine 3402
Caffy, Florent 1516
Calderón-Obaldía, Fausto 626
Calle, Eric 944
Calvo-Barrio, Lorenzo 3285
Campa, Andrej 1346
Campanelli, Mark 437
Campbell, Calli M. 3366, 3410
Campesato, Roberta 76, 541, 545
Campos, Cláudio Dias 2307
Camus, Christian 3500
Cañadillas, David 429, 1116

Canino, A. 1747
Caño, P. 1210
Cao, Huihui 1619
Cao, Wenkai 696
Cao, Xin 2427
Cao, Yunxue 392, 1430, 1873, 2918
Cappelluti, F. 1189
Cardona, Dagoberto 670
Cardwell, D. 3511
Cariou, Romain 2511, 2528
Carlin, John A. 215
Carlson, David E. 3442
Carlson, Emily 701
Carneiro, Lucas M. 417
Carolus, Jome 2875
Carpenter, Bernard 537
Carr, Anna J. 1081
Carriere, Jarrett 2833
Carruthers, Steve 3514
Carter, Catrice M. 3393
Carter, Cedric 2135
Carter, Sam 3514
Casale, Mariacristina 76, 541, 545
Casallas-Moreno, Y. L. 670
Casper, Chadwick 1476
Cassini, Denio A. 1917
Castañeda, Carlos A. Rodríguez 1858
Catthoor, Francky 3343
Cattin, Jean 3435
Cattoni, Andrea 1289
Cavani, Olivier 2087
Cédola, A. P. 1189
Cendagorta-Galarza, Manuel 429
Cepeda, Kyle 876
Cesar, I. 3150
Cesar, Ilkay 917
Chai, Gaoda 976
Chai, Jing 1922
Chakraborty, Sagnik 3300
Chamarthi, Phani Kumar 2952
Chamberlin, Charles 1271
Champliaud, J. 50
Champliaud, Jonathan 3435
Champness, C. H. 2388
Chan, Calvin 3417
Chan, Catherine E. 2576, 2600
Chan, Mandy 2808
Chan, Maria K.Y. 6, 1256, 2759
Chan, R. 3511
Chandralal, Sreeram 1674
Chandran, Deepak 2986
Chang, Jipeng 1873, 2823
Chang, Sheng-Hao 1051
Chang, Via-Chung 1051
Chantana, Jakapan 757, 2385

AUTHOR INDEX

Chaporr, Patrick 2711
Chapuis, Valentin 2104
Chattopadhyay, Kamanio 2811
Chattopadhyay, Shashwata....... 1850, 1858, 2849, 3478
Chaudhry, Ghulam M. 1018, 1110
Chaujar, Rishu 377
Chaurasia, Saloni 837, 841
Chausseau, Matthieu 2711
Chavali, Raghu Vamsi Krishna 1904
Chavez, Jose J. 2419
Chemisana, Daniel 1339
Chen, Benjamin....................................2358
Chen, Chien-Hsun..................................911
Chen, Chun-Chi.....................................1635
Chen, Daniel 2576
Chen, Eric Y. 1598, 3384
Chen, Haiyan 2220
Chen, Hung-Ling 1289
Chen, Junyan............................1835, 2732
Chen, Kaifeng 2185
Chen, Kunji 2656
Chen, Lung-Chien..................................367
Chen, Meixi 326, 999, 2035
Chen, Peng-Wei....................................2660
Chen, Ran..2576
Chen, Renfang......................................1241
Chen, Shi-Wei.......................................1627
Chen, Sung-Yu......................................911
Chen, Tsung-Cheng 329
Chen, Tzu-Yu.......................................1627
Chen, Wanghua....................................2593
Chen, Weijian.......................................2392
Chen, Y. 2785
Chen, Yang...2502
Chen, Yao- Hui.............................893, 2664
Chen, Yifeng...................1922, 2220
Chen, Yunfei..............................761, 2427
Chen, Yusi 1835
Chen, Zhi David 1044
Chen, Zihan...2392
Chendo, Michael 2381
Cheng, Y. 14
Cheng, Yan...667
Cheng, Yuh-Jen....................................1610
Cheng, Zhe...3473
Cheng, Zhongkai...................................3393
Chenna, Shiva Tarun 1674
Chiang, Cho-Chun 893, 2664
Chiang, Fu-Kuo....................................198
Chikhalkar, Abhinav 823, 827
Chin, Ken K. 761, 2427
Chinnusamy, Saravanan 980
Chiu, Chun-Yu.....................................1169
Chiu, P. 2094
Chiu, Philip..2099

Chmielewski, Daniel J. 215, 2079, 2554
Cho, Eunhwan 333, 1824, 1838
Cho, Junsik 810
Cho, Yasuo 3323
Choi, Gyu-Seok 2723
Choi, J. -K 2019
Choi, Rae-Won 2723
Choi, Seungkeun 1037
Choi, Sungjin 1758
Chong, Cheemun 2600
Choubisa, Hitarth 1022
Choudhury, K. R. 2312
Chouhan, Arun Singh 986
Choulat, Patrick 2227, 3435
Chow, E.M. 1733
Chowdhury, Ahrar Ahmed 888
Christians, Jeffrey A. 2176
Christmann, G. 50
Chu, Chi-Wei 1051
Chu, Haifeng 1222
Chu, Sheng 1299
Chua, Soo Jin 284
Chuang, Ta-Wei 343, 367, 893, 2664
Chung, Daniel 2707, 3304
Chung, Haejun 1904
Chung, Simon 696, 2186
Ciesla, Alison M. 2576, 2600
Cifuentes, L. 1210
Cifuentes, Luis 1204
Ciocia, Alessandro 3096
Cirino, Daniel A. Merced 3044
Clayton-Warwick, D. 138
Cleveland, Erin 247, 2091
Clinton, Evan A. 305
Cobo-Yepes, Nicolás 2963
Codd, Daniel S. 3245
Cohen, Bat-El 2170
Cole, Wesley J. 2163
Colegrove, E. 1312
Colegrove, Eric 3147, 3319
Coll, Pablo Guimera 2610
Collin, Stéphane....................................1289, 3147
Collins, Robert W. 807, 2462, 2582, 2646, 3426
Collins, Shamara 802, 1638, 2449, 3413
Comagliotti, Emanuele 3435
Condorelli, G. 1747
Conibeer, Gavin..............696, 2186, 2392
Conlon, Benjamin P. 219
Conrad, Brianna....................................315
Cordeiro, Patricia 2135
Cordova, Adam 1965
Cornagliotti, Emanuele 1804, 2227
Cornaro, Cristina 3482
Cornell, Robert.....................................1275
Correa, J.M. 433

AUTHOR INDEX

Correa-Baena, Juan-Pablo 3300
Cossio, Gabriel 1181
Costa, Sara..1979
Costa, Suellen C. 2307
Côté, Alexandre...1908
Cousar, Larry C. 921
Cravens, R. 2094
Crawford, L. 1733
Crupi, F..2073
Cruz, José Ortega 1959, 1990
Cruz, Leila R. De Oliveira.............................2307
Cruz-Campa, Jose Luis 337
Cuevas, Andres..2076
Cui, Jie..2076
Cui, Min...1765
Cunningham, Daniel W. 1463
Cunningham, Joseph 3161
Cur, Jie..517
Curran, Alan J. 1927, 2697, 3488
Curvat, L. 50
Cushing, Scott K. 417
Da Fonseca, Jérémy 2492, 2562
Dabney, M. S. 138
D'Abrigeon, Laurent.......................................545
Daenen, Michael..2875
Dagenais, Mario 195, 1048
Dagyte, Vilgaile..2502
Dahal, Saroj 309, 3123
Dahal, Som...240
Dai, Yushuai 18, 222, 677, 1184
Dalal, V L...1350
Dalal, Vikram...2247
Dalpian, G...1245
Dam-Hansen, Carsten...........................2672, 2682
Danel, A. 1747
Dang, Hongmei...2432
Dangate, Milind S. 980
Daniil, Andreana...944
Danzl, F.J.K...3150
Darbali-Zamora, Rachid 2957, 2963
Das, Ujjwal 408, 1473, 1761, 1828, 2667
Das, Ujjwal K. 2637
Datas, Alejandro 2562
Dauskardt, Reinhold...............................3190, 3200
Davidsen, Rasmus S. 2672
Davies, J. I. 1210
Davis, Kristopher O. 74, 322, 1804, 3448
Davis, Tracy..537
De Coux, Patricia..464
De Melo, O. 2342
De Nicolas, S. Martin 50
De Oliveira, Michele C. C.................................1917
De Villers, Bertrand J. Tremolet 1354
De Wolf, Stefaan 55, 3256, 3435
De, F. C. Lins Vanessa.....................................1917

Debnath, M. C. 14
Debnath, Tanmoy..1067
Deboever, Jeremiah........................... 1555, 1567
Debrot, F. 50
Debucquoy, Maarten...................................... 1233
Debusschere, Vincent...................................... 3102
Deceglie, Michael........................... 2771, 2789
Deceglie, Michael G.................2488, 2804, 3452
Deckerl, D. 1752
Decobert, Jean... 2528
Deer, Tanya... 1908
Deitz, Julia I.. 3139
Delahoy, Alan E. 761, 2427
Delhotal, J. 3224
Deligiannls, D. 3150
Deline, Chris 116, 1537, 1922, 3184, 3333
Demadrille, Renaud 1516
Demirkan, Korhan 820
Deng, Changhong 1158
Deng, Weiwei 2220
Denk, Patrick 1360
Descoeudres, A... 50
Descoeudres, Antoine 3254
Despeisse, M. 50, 2073
Despeisse, Matthieu...................3254, 3256, 3435
Desrues, Thibaut 2492, 2562
Deutsch, Todd G.. 47
Devos, Arnaud 464
Dewitt, Daniel 1835
Dey, Anamika 1034
Dhakal, Tara P. 989
Dhere, N. 389, 1701
Di Leo, Paolo 3096
Di Mare, Simone 2372
Di Napoli, Simone... 2205
Diaz, Liliana Ruiz... 1147
Diercks, David R... 46
Dimitrievska, Mirjana..................................... 3285
Dimopoulos, Theodoros 178
Dimroth, Frank 2511
Ding, Jie..1937, 2823
Dinger, Justin..2692, 2765
Diniz, Antonia Sônia A. C................... 1917, 2307
Dirriwachter, Antonius B. 3448
Dise, John 132, 1104
Dise, Skip... 1427
Dobrich, Anja 2538
Dobroliubov, Aleksandr................................... 2776
Dobson, Kevin 315
Dobson, Kevin D. 658
Dobson, Weston 2692, 2765
Dogan, Yusuf 229
Doi, T. 441
Dominguez, A... 2342
Dong, Jianfei... 2605

AUTHOR INDEX

Doolittle, William A. 305
Dooraghi, Michael 1169
Döscher, Henning 47
Doty, Matthew F. 1598, 3384
Dougher, Chris 3245
Dougherty, Brian 1933
Drahi, Etienne 464
Drayton, Jennifer A. 164
Drees, M. 3511
Dréon, Julie 55
D'Rozario, Julia 18
Drummy, Lawrence F. 966
Du, Chen-Hsun 911
Du, Xingzhi 2558
Du, Zhongming 198, 767, 1707
Duan, Baosong 392, 2823
Duan, Wenqi 346
Dubey, R. 1995
Dubey, Rajiv 1704, 2849, 3478
Dubois, Anne Migan 626
Duenow, J.N. 1312
Duenow, Joel 1196, 3147
Duerinckx, Filip 2227, 3435
Dugan, Roger C. 3055
Dugdill, Brian 2014
Dumbrell, Robert 420, 3315
Dunham, Scott T. 3119
Dupré, Cécilia 2492, 2562
Durand, Olivier 2192
Durose, Ken 742, 1445
Durstock, Michael F. 966
Durygin, Andriy 3473
Dusane, Rajiv O 2330
Dussarrat, Christian 326
Dutt, A. 370
Dutt, Ateet 2342
Dutta, P. 869
Dutta, Pavel 866, 2368
Duttagupta, S.P. 1898
Eafanti, Joshua 3190
Ebe, Falko 2996
Ebert, Matthieu 1531
Ebong, Abasifreke 888
Ediger, E. 2364
Edinger, Stefan 178
Edoff, Marika 796
Edwards, Daniel J 3514
Eeles, Alexander 146, 186
Efthymiou, Venizelos 3107
Egbe, Daniel Ayuk Mbi 1360
Eggink, Wouter 2109
Ekins-Daukes, Ned 1339
El Assimi, Taha 3402
Elangovan, Hemaprabha 2811
Elanzeery, Hossam 151, 2054

Eldho, T.I. 1898
El-Henawey, Mohamed 2247
Elkhatib, Mohamed 2141, 2969
Elleuch, Omar 359
Ellibee, Donald 1543
Ellingson, Randall 2926
Ellingson, Randall J. 1030
Ellingson, Randy J 738, 748, 781, 815
Ellis, Chase T. 873
Elnosh, Ammar 1946
Elsehrawy, Farid 1189
Emery, K.A. 490
Engerer, Nicholas A. 1405
Eriksen, Ryan 2870
Eriksen, Ryan S. 2891
Ermer, J. 2094
Ermer, James 2099
Ermer, Jim H. 37
Escarra, Matthew D. 37, 3245
Escobar, D. Martínez 1959, 1990
Esfandiari, Parichehr 178
Espinct-Gonzalez, Pilar 558
Espíndola-Rodríguez, Moisés 155, 512, 572, 3265
Espinet-González, Pilar 521, 1248
Essa, Gharibah 2011
Essig, Stephanie 55, 3254, 3371
Etcheberry, Arnaud 2711
Etgar, Lioz 2170
Eugen, Rene 2864
Evani, Vamsi 802, 1638, 2449
Evans, Garrett Z. 921
Evstigneev, M. 663, 1025, 1811
Evstigneev, Mykhaylo A. 690
Eylers, Katharina 3396
Fada, Justin S. 2697, 3456, 3488
Faes, A. 50
Fairbrother, Andrew 1933, 2000, 3204
Faleev, Nikolai 1215, 2573
Fan, S. 3376
Fan, Shanhui 2185, 2732
Fang, Liang 226
Fang, Y. 1603
Fang, Yi 305
Fano, V. 2740
Fano, Vanesa 2677
Faraj, Abdul 2014
Farnung, Boris 2267
Farré, Laia Arqués 3285
Farshchi, Rouin 1459, 1686
Faur, Maria 896
Faur, Orry 896
Favre, W. 1747
Fedina, Maria 2070
Fejfar, A. 2073
Felder, T. 2312

AUTHOR INDEX

Feldmann, F. 2064
Feng, Sheng-Kai 343, 367, 893, 2664
Feng, Shien-Ping 1012
Feng, Zhiqiang 1922, 2220
Fenning, David P. 1494, 2245
Ferekides, Chris 802, 1638, 2449, 2467, 3413
Ferekides, Chris S. 1511
Ferekides, Christos 175
Ferguson, Andrew J. 1354
Ferguson, L. 1863
Fernández, Eduardo F. 2858
Fernandez, R. Mis 1691, 2457
Fetzer, C. 2094
Fiducia, Thomas 424
Fields, Brian J. 2618
Filipic, Miha 1233
Filonovich, Sergej 464, 1237
Firth, Peter 1317
Fischer, A. 1603
Fischer, Alec 823
Fischer, Alec M. 305
Fisher, Brent 210, 272, 1469
Fisher, Dallas 989
Fitzgerald, Eugene A. 213
Fleming, Robert A. 1869
Flicker, J. D. 3224
Flicker, Jack 1280
Florides, Michalis 1941
Foldyna, Martin 2528, 2593
Forberich, Karen 1346
Forbes, David V. 3468
Forchhammer, Soren 2682
Forsh, P.A. 1811
Foster, Robert 2014
Fouchier, Marin 3402
Fournel, Frank 2492, 2562
Fraas, L. M. 1863
Fraas, Lewis 2042
France, Ryan M. 47, 232
Fraser, Ray 337
Frederiksen, Kenn H. B. 2682
Freeman, Janine M. 3494
Freiburger, Brennen M. 1869
French, Roger 1933
French, Roger H. 1927, 2000, 2697, 3456, 3488
Freundlich, Alexandre 236, 673, 1452
Fridman, Lucas 2000
Friedman, Daniel 549, 2543
Friedman, Daniel J. 42, 268, 1201
Friend, Mari Paz 429
Fritzsche, M. 1752
Frontini, Francesco 2118
Fthenakis, V. 2019
Fthenakis, V. M. 3230
Fthenakis, Vasilis 3077

Fu, Ran 1259, 1463
Fuhrich, Alexander 3396
Fuhrmann, Bianca 83
Fujiwara, Koji 1973
Fukuda, Tetsuya 931
Funabiki, Shigeyuki 2906
Fung, Tsun H. 2576
Fuyuki, Takashi 2593
Gabetta, Giuseppe 76, 545
Gabor, Andrew M. 74, 2839, 2897
Gaddy, Edward 585
Gahr, Stefan 178
Gai, Boju 549, 2291
Gaiaschi, Sofia 2711
Gallon, Joshua B. 3448
Galtieri, Jason 2975, 3214
Gambogi, W. 2312
Gao, Hui 226
Gao, Peng 1648
Gao, Wei 226
Gao, Y. 869
Gao, Yijun 2392
Gao, Ying 2368
Gao, Yuan 2048, 2605
Gao, Yujie 2870, 2891
García, I. 1210
Garcia, Iván 1204
Garcia, Juan Lopez 402
Garcia-Linares, Pablo 2562
Garg, Vivek 2345
Garner, Sean 2870
Garner, Sean M. 2891
Garnett, Erik C. 2245
Garreau-Iles, L. 2312
Garrillo, Pablo A. Fernández 1516
Garuz, Richard 2255
Gaury, Benoit 1303, 2438
Gdoutos, Eleftherios E. 558
Geelan-Small, Peter 3304
Gehre, Simon 3500
Geissbühler, J. 50
Geissbuehler, Jonas 3256
Geisz, John 549, 3371
Geisz, John F. 232, 268, 1737, 2195, 3254
Georghiou, George E. 276, 1163, 1941, 1954, 3107
Geraghty, Paul 1342
Gerardi, C. 1747
Gerber, Andreas 1400, 1651
Gerdimenes, Anne 619
Gervasi, Massimo 541
Ghaisas, S.V. 389, 1701
Ghimire, Kiran 993
Ghosh, Kunal 716
Gibbs, Jacob M. 730
Gibelli, François 2192

AUTHOR INDEX

Giebink, Noel C. 1469
Giguère, Jean-Benoit 1360
Gilchrist, James B. 966
Gillispie, Kellen 2762
Giordano, Francesco 3096
Giraldo, Sergio 3265, 3285
Giussani, A. 845
Giussani, Alessandro...............206, 831, 2514
Givot, Bradley L. 2864
Gladden, Christopher 1476
Glasgow, Nate 1427
Glatthaar, M. 884, 3135
Gloeckler, Markus 1193
Glunz, S. 3135
Glunz, S.W. 2064
Glunz, Stefan W. 1253, 2511
Gokkaya, Huseyin Cem 958
Goldschmidt, Jan Christoph 1253
Golembeski, Andrew A. 3143
Goma, Elias Garcia 3462
Gombia, Enos 541
Gona, Michael N. 2349
Gong, Chen 1585
Gong, Jue 2251
Gonzálcz-Díaz, Benjamín 3240
Gonzalez, Maria 259
Gonzalez, S. 3224
Gonzalez, Sigifredo 2147, 3002, 3020
Gonzalez-Díaz, Benjamín 429, 1116
Goodarzi, Mohsen 2707
Gooding, Renee 1280, 1543
Goodnick, S. M. 1603
Goodnick, Stephen 1790
Goodnick, Stephen M. 305, 582, 1797
Gordillo, G. 433, 503
Gordon, Ivan 1233
Gori, Gabriele 76
Gorman, Brian 62, 1371, 1381
Górnez-González, L. A. 2614
Gostein, Michael.........................2808, 2923
Goswamy, Naveen 1908
Gotoh, Kazuhiro 1765, 1794
Gottschalg, Ralph 1411, 2827, 3208
Govaerts, Jonathan 3343
Goverde, Hans 3343
Gowda, Ramesh Rame...............................1912
Graf, Martin 2511
Graham, Kenneth 1044
Grandidier, Jonathan 2099
Grassman, Tyler J........182, 215, 2079, 2554, 3139
Greco, Erminio 76, 541
Grede, Alex J. 1469
Green, Martin............................2213, 2403
Green, Martin A. 858
Green, Michael 2926

Greenhalgh, R.C. 3430
Gregory, Geoffrey 74
Grévin, Benjamin 1516
Grice, Corey R. 771, 1643, 2473, 3426
Grieco, William J................................2618
Griffin, Alecia 2870
Griffin, Alecia C. 2891
Grijalva, Santiago...............1555, 1567
Grini, S. 3269
Große, T. 1752
Großer, Stephan 2232
Grossklaus, Kevin 701
Großschädl, Bettina 1329
Grovenor, Chris 424
Grover, Sachit 1193, 2473
Grübel, B. 884
Gu, Fei 1346
Gu, Tian 1473
Gu, Tingyi 1828
Gu, Xiaohong 1933, 2000, 2844, 3195, 3204
Guarracino, Ilaria 1339
Guay, Nathan 1543
Gudla, Sushanth 1389
Guerrero-Lemus, Ricardo 429
Guillemoles, Jean-François...............1289, 2192
Guillevin, N. 3150
Guillevin, Nicolas 917
Guina, M. 1189
Guina, Mircea 297, 2520
Guischard, Felix 2836
Gunawan, Oki 1441, 3275
Gunnarsson, William B............................2443
Guo, D. 1603, 2816
Guo, Hong 1299
Guo, Q. 3
Guo, Qi 226
Guo, Shuwen..............1430, 1873, 2918
Guo, Yongjie 1719
Gupta, Amit Kumar................2952, 2981, 2986, 3050
Gupta, Mool C. 937
Gupta, Neeti 696
Gupta, Ritesh Kant 1034
Gupta, Shivam 377
Gupta, V. 1733
Gustafsson, Mattias 2025
Guthrey, Harvey..................1400, 2887
Gutiérrez, J. R. 2740
Gutiérrez, R. 643
Gutscher, S. 884
Guwaeder, Abdulmunim 1122
Gwak, Jihye 810
Ha, Dongheon 1585
Habermann, D. 1752
Hack, James 999
Hack, James H. 326

AUTHOR INDEX

Hacke, Peter 1371, 1381, 1421, 1922, 2819, 2854, 3305
Hackl, Wolfeanz 178
Haddad, M. 2094
Haddadian, Rojiar 1927
Hadi, Sabina Abdul............................213, 1741
Hadjipanayi, Maria 276
Hadke, Shreyash 986
Hadley, Wendy......................................2014
Haegel, Nancy M. 62
Hagendorf, Christian 1376, 2232
Hägglund, Carl......................................796
Hahn, Carina E. 175
Hai, Hoang Tri 931
Haight, Richard 1441
Hajimiri, Ali 521, 558
Hajizadeh, Amin 3092
Halbwax, Mathieu 3402
Hall, Allen 1511
Hallam, Brett J. 2576
Halliday, Douglas P...............................2423
Hamadani, Behrang H............263, 437, 508
Hameiri, Ziv 66, 420, 3290, 3315
Hamon, Gwénaëlle 2528
Hamui, L. 2614
Hamzaoui, Saad 900
Hamzavy, Babak T.................................2618
Han, Sang M..88
Han, Xinyue 1719
Han, Youngsik 2242
Hanada, Toru 940
Handwerker, Carol A. 1449
Haney, Paul 1303
Haney, Paul M..2438
Hanley, J. 2094
Hanna, Amir 1055
Hannappel, Thomas..............2524, 2538, 2538
Hanriot, Sergio De Morais 2307
Hansen, Clifford..............1127, 1537, 3184, 3333, 3348
Hansen, Clifford W.................110, 1543, 1549
Hansen, Ole 2672
Hansen, Richard....................................2042
Hansen, Shirley 2042
Hao, Xia 160
Hao, Xiaojing....................858, 2213, 2403
Haohui, L. 1140
Haohui, Liu 2744
Haque, K A S M Ehteshamul 346
Haque, M. D. 552
Hara, Shigeomi....................................1950, 3339
Hara, Tomoya..2548
Harari, Joseph 3402
Hardikar, Kedar 2688
Häring, Adrian 2263
Hariskos, Dimitrios 2058

Harmand, Jean-Christophe 1289
Harris, Christian 319
Harris, James 1835, 2732
Harris, Tom 2991
Harvey, Steven......................................2887
Harvey, Steven P. 1371, 1381, 2702, 3305, 3319
Haschke, Jan 3435
Haslinger, Michael 1804
Hassan, Ibrahim A. I. 2858
Hatch, S. 14
Hatton, Peter D......................................2423
Hauch, Jens 3500
Haug, F.-J. 2073
Hausgen, Paul E. 102
Hausmann, J. 1752
Havu, Ville 2070
Haysom, Joan E. 1094
He, Junwen 1469, 1737
He, Qiuxiang 3304
He, Wenshuang......................................392
Hea, Wenshuang 2823
Heben, Michael 2926
Heben, Michael J. 170, 730, 748, 815, 1030, 2462
Hegedus, Steven................408, 1473, 1761, 1828, 2667
Hegedus, Steven S.................................658
Heidmann, Berit 3396
Heilbrunner, Herwig............................1360
Heilscher, Gerd 2996
Heinz, F. D. 3135
Heinze, Matthias 2263
Heller, Dominic 1077
Henes, Dan 1094
Hentz, Sandrine 966
Hermle, M. 2064
Hermle, Martin 1253, 2511
Hernandez, J. A. 2031
Hernández, Johan 1143
Hernandez, Joseph 3473
Hernandez-Alvidrez, Javier 2153
Hernández-Gutiérrez, C. A. 670
Hernández-Rodríguez, Cecilio 429
Herrera, Daniel J....................................219
Herrmann, W. 107
Herz, Magnus..3360
Heta, Y...2312
Hetterich, Michael 1682, 2216
Hettick, Mark 59, 823, 2076
Heurlin, Magnus 1286
Hickey, Benjamin 1459
Hidaka, Kazuyuki 1973
Higa, M. 441
Hilfiker, M. 2364
Hill, Alex 1893
Himwas, Chalermchai 1289
Hindi, Basel 1058

AUTHOR INDEX

Hinken, David 2692
Hinojosa, M. 1210
Hinzer, Karin 1094
Hirai, Masakazu..................................1769
Hirata, Yoichi.......................................613
Hirose, Kotaro 3323
Hirstl, Louise C. 2091
Hishikawa, Y. 441, 1003
Hishikawa, Yoshihiro 480, 2781
Ho, Jian Wei 496
Ho, Wen-Jeng................343, 367, 893, 2664
Hoang, Bao 96
Ho-Baillie, Anita...............858, 1845, 2569
Hobbs, William B. 2618
Hoerteis, Matthias 914
Hoex, Bram 517
Hoff, Thomas................................132, 1104
Hofmann, Johannes 2407
Hoheisel, Raymond 247, 3514
Höhn, Oliver 352
Holman, Z. C. 3376
Holman, Zachary 1790, 3366
Holman, Zachary C. 1220, 1228, 1317,
    1322, 1820, 3250
Holmgren, William F. 110, 1127
Holzmann, Daniel...............................914
Hong, Chung-Yu 294
Hong, Keunkee 399
Honsberg, Christiana....................827, 3088
Honsberg, Christiana B. 240, 305, 582,
    681, 1215, 1841, 2573
Hopf, Markus 1965
Horenstein, Mark................................2870
Horenstein, N Mark 2891
Horner, Greg S. 3448
Horowitz, Kelsey A.W. 1259, 1463
Horzel, J. 50, 2073
Hoshii, Takuya 2334
Hosokawa, Kazuya...............................613
Hossain, Istiaque 2247
Hossain, Mohammad A. 3456
Hossain, Mohammad I.........................963
Howard, John M. 2443
Hsi, Edward 1275
Hsu, Chia-Jhe 1623
Hsu, Chih An 1638, 2449, 3413
Hsu, Lung-Hsing 1610
Hsu, Shun-Chieh 1606, 1623
Hsu, Shu-Tsung 445, 448, 476
Hsu, Wei-Lun 1048
Hsu, Yu -Chen 888
Hu, Chehao 229
Hu, Cheng-Shun 329
Hu, Hailin 1858
Hu, Juejun 1473

Hu, Lilei...3129
Hu, Long 2392
Hu, Yang 1927
Hu, Yicong 2392
Huang, Jialiang....................................2213
Huang, Jingsheng 1937, 2823
Huang, Jing-Shun............512, 521, 558, 572, 1248
Huang, Shujuan....................................2392
Huang, Vi-Wen 1631
Huang, Weijing 1873, 2918
Huang, Wei-Ming 1627, 1631
Huang, Wen-Hsi 385
Huang, Ying-Yuan 1807
Huang, Yi-Wen 1627
Huang, Yu-Ming 1606
Huang, Yu-Ting 1012
Huang, Z. 3260
Huayamave, Victor..............................2839
Hubbard, S. M. 552, 845, 2755
Hubbard, Seth 677
Hubbard, Seth M............18, 202, 206, 222,
    831, 1184, 2084, 2298, 2514
Huber, Christian 2216
Hudson, A.I. 2755
Huey, Bryan D. 1522
Huffaker, D.L. 2755
Huffaker, Diana 202
Huhn, Vito..1651
Huld, Thomas 2167
Hunault, Philippe 2711
Hung, Yung-Jr 1606
Huo, Yijie 1835
Husein, Sebastian 944
Huss, Alexandra M...............................164
Hussain, Babar 451, 2355
Hussain, Muhammad M. 1055
Hutchings, Douglas..............................1869
Hutchings, Douglas A...........................921
Hutter, Oliver S. 1445
Hwang, James......................................333
Hyvl, M. 2073
Iandolo, Beniamino 2672
Ianno, N.J. 2364
Ichikawa, Yukimi 1769
Idlbi, Basem 2996
Ikki, Osamu 2159
Ilic, Ognjen 1737
Imai, Jun 2906
Imaizumi, Mitsuru 567, 3506
Imtiaz, Syed N. 1067
Ingenhoven, Philip..............................3482
Ingenito, A. 2073
Inns, Daniel 3113
Isabella, Olindo...................................2605
Isbilir, Kenan......................................2827

AUTHOR INDEX

Isherwood, Patrick J. M. 2349
Ishii, Tomoaki 455
Ishino, Yuya...757
Ishizuka, Shogo 33
Islam, Kazi 37
Islam, Muhammad Monirul............................33, 900
Islam, Raisul 1835
Isoaho, Riku..2520
Iwasaki, Kazuya 2338
Iwata, Naotaka 2642
Iwuoha, Emmanuel 1360
Iyer, Abhishek......................................326, 999, 2035
Iyer, Parameswar K. 1034
Izquierdo-Roca, Victor 3265, 3285
Jackson, Christine..215
Jackson, Philip 2205, 2453
Jacob, David 1549
Jacobson, Arne 1271
Jadkar, S.R. 1701
Jaeckel, Bengt..1411
Jae-Yun, Fa-Jun Ma, ...1845
Jagdish, A K 2811
Jäger-Waldau, Arnulf....................................2167
Jahn, Ulrike 3360
Jain, Aditi 333
Jain, Nikhil 42, 46, 232, 578, 2195, 3371
Janoch, Rob 74, 2839, 2897
Jansen, Mark J. 1081
Jany, Christophe 2492, 2562
Janz, Stefan 83, 2407
Jaramillo, Adolfo 1143
Jared, Bradley...1473
Jarmar, T. 30
Jasti, Naga Prathibha 986
Jaswal, Rohit 3172
Javed, Mehwish Azher....................................1317
Javey, Ali 59, 823, 2076
Jeangros, Q...2073
Jenkins, P. P. 845
Jenkins, Phillip P................. 247, 373, 1838, 2091, 3514
Jensen, Brian 2014
Jensen, M. A. 3295
Jensen, Mallory A. 1491, 3290, 3300
Jensen, Soren..1196, 3147
Jeong, Woo-Lim 777, 1665
Jhaveri, Janam 1773
Ji, Liang 1933, 2000
Ji, Yaping...37
Ji, Yaping Vera 3245
Jia, Jieyang 1835, 2732
Jian, Ding-Rung 1627, 1631
Jiang, C. S..1312, 2789
Jiang, C.-S..2280, 2785
Jiang, Chun-Sheng 62, 1371
Jiang, Lian L. 589

Jiang, Lian Lian 120
Jiang, Xuefang..1937
Jiang, Yu 3220
Jimeno, J. C. 643, 2740
Jimeno, Juan Carlos......................................2677
Jin, C. 1781
Jin, Yu 3119
John, Jim J..1946
John, Joachim......................................1804, 2227
John, Suru Vivian...1360
Johnson, A. D..1210
Johnson, E.V...1781
Johnson, Erik V. 2593
Johnson, J. L. 1656, 1661
Johnson, Jay 2135, 2141, 2153, 2969, 3002, 3008
Johnston, S. 2785
Johnston, Steve................................62, 202, 459, 1371,
          1381, 1400, 2213, 2819, 2887, 3305, 3452
Jones, C. Birk....................2618, 3008, 3155, 3488
Jones, David 1342
Joonwichien, Supawan....................................904
Joshi, Madhuwanti S.................2952, 2981, 2986, 3050
Joshi, Pranav 2247
Jošt, Marko 1346
Jovanovic, Raka 963
Juang, B.C..2755
Juang, Bor-Chau 202
Juárez, A. Sánchez 1990
Juárez, Aarón Sánchez....................................1959
Juhl, Mattias K. 420, 3315
Juhl, Mattias Klaus 66
Julien, Scott..1933, 2000
Junci, Wang 496
Junda, Maxwell 771
Junda, Maxwell M.................2462, 2582, 3426, 3468
Jung, Jae Hak 487
Jung, Jiirgen 2864
Jung, Sang Hoon 2723
Jung, Sang Hyun 244
Juso, Hiroyuki 3506
Kabalan, Amal..2358
Kaczynski, Ryan 1455
Kaizu, Toshiyuki 23
Kaizuka, Izumi 2159
Kakosimos, Konstantinos E. 1888
Kalainatharr, Sivaperuman...............................2334
Kalb, J. 1733
Kale, Abhijit...1801
Kale, Abhijit S. 1777
Kallickal, Johnson 1543, 3348
Kalt, Heinz 1682, 2216
Kamata, N. 552
Kamevama, Satoshi......................................2642
Kamino, Brett 3256
Kamins, Ted 1835

AUTHOR INDEX

Kaminski, P.M. 3430
Kaminski, Piotr 1674
Kaminski-Cachopo, Anne 2562
Kamioka, Takefumi 2498, 2548, 2566, 2642
Kanemitsu, Yoshihiko 721, 2781
Kanevce, A. 1312
Kanevce, Ana 3147
Kang, Ho Kwan 244
Kang, Min Gu 356, 1758, 2723
Kang, Yoonmook 935
Kankiewicz, Adam 132, 1104, 1132, 1427
Kannan, C. V. 2716
Kao, Ming-Hsuan 1627
Kaplan, Stephen 600, 1071
Kaplar, R. 3224
Karas, Joseph 925
Karki, Shankar... 182, 735, 807, 2298, 2446, 2646, 3139
Karmarkar, M. 1661
Karpowich, Lindsey 914
Karthik, Shravan 3172
Kashkoush, Ismail 322
Kaslin, Remo 1077
Kasry, Amal 2858
Kasu, Makoto 1950, 3339
Kato, Takekazu 1175
Kato, Takuya 160
Katsube, Ryoji 2361
Kaule, Felix 2622
Kausika, Bala Bhavya 3014, 3167
Kavaipatti, Balasubramaniam 2799
Kawatsu, Tomoyuki 381, 2588
Kazmcrski, Lawrence L. 2799
Kazmerski, L.L. 1245
Kazmerski, Lawrence L. 1917, 2307
Kazumi, Kenji 2361
Keeler, Gordon 1473
Keller, Nico 1077
Kelly, George 1275, 2263
Kelly, Matthew 3514
Kelzenberg, Michael D. 512, 521, 558, 572, 1248
Kempe, M.D. 138
Kempe, Michael 1933, 2000
Kempe, Michael D. 3208
Kephart, Jason 785
Kephart, Jason M. 3417
Kern, Gregory 2147
Kern, Gregory A. 3020
Kesavan, Arul Varman 1614
Kessels, Wilhelmus M.M. 1817
Kessler, Emily 206
Khadimallah, A. 869
Khalili, A. 1189
Khan, Imran 802, 1638, 3413
Khan, Imran S. 2449
Khan, Mohammad R. 1055

Khan, Taj M. 451
Khanna, Raghav 2926
Kharait, Rounak A. 2833
Kharel, Khim 236, 673, 1452
Khatavkar, Sanchit 2716
Khatiwada, D. 869
Khatiwada, Devendra 866
Khatri, Ishwor 192
Khatri, Trijul 377
Khomcnko, Denis V. 690
Khoo, Yong Sheng 1922
Khor, Alan 3172
Khoram, Parisa 2245
Khorenko, Victor 83, 2087
Khoury, R. 1781
Kiefer, Fabian 1366
Killam, Alex 2719
Killinger, Sven 126, 1405
Kim, Boram 2201, 2524, 2538
Kim, Chang Zoo 244
Kim, D. 1189
Kim, Dae Young 626
Kim, Dong Seop 399
Kim, Dong-Ho 2631, 2634
Kim, Donghwan 935, 1758
Kim, Hae-Sun 777, 1665
Kim, Hyo Jin 849
Kim, In-Young 777
Kim, Jae Hyun 363, 2844, 3195, 3204
Kim, Jin-Hyeok 777
Kim, Jisun 399
Kim, Ka-Hyun 1758
Kim, Kangho 244
Kim, Kihwan 810
Kim, Kyoung- Tae 1037
Kim, Min-Soo 487
Kim, Moon 2759
Kim, Sangpyeong..240
Kim, Soo Min 2723
Kim, Woo Kyoung 487
Kim, Yeongho 827
Kim, Yong Bae 2723
Kim, Yong Whan 849
Kim, Youngjo 244
Kimbal, Gregory M. 110
Kimura, Daiki 854
Kindole, Dickson 970
Kindvall, Anna 785
King, Bruce H. 3155, 3488
King, Richard 827
King, Richard R. 301, 823, 1215, 1841
Kingma, Aldo 541
Kini, Roshan 2926
Kinoshita, Kosuke 1504, 2588
Kirk, A. 3511

AUTHOR INDEX

Kita, Takashi 23
Kleider, Jean-Paul 2528
Klein, Talysa R. 2482, 3371, 3439
Kleinschmidt, Peter.......................................2538
Klemm, Hagen W...3396
Klenk, Markus 1077
Klie, Robert F..2759
Klimm, Elisabeth 2836
Klise, Katherine A..3161
Klisel, Geoffrey T...3494
Kluska, S...884, 3135
Knight, Bruce 2014
Knopf, Hannes...1965
Ko, Changhee 326
Kobayashi, Jonathan 1061
Koehl, Michael 3488
Koepgel, Ringo 2622
Kogler, Willi 791
Kohlstädt, Markus 1253
Koike, Junichi 931
Koirala, Prakash 2462, 3426
Kojima, Nobuaki...................359, 2498, 2566
Kojima, Takuto 1504, 2588
Komsa, Hannu-Pekka 2070
Konagai, Makoto..........................1769, 2627
König, M. 1752
Konstantinou, Georgios....................................1941
Kontges, Marc 1366
Kopecek, Radovan 1222
Koschny, T..1350
Kostylyov, V.P..1025, 1811
Kostylyov, Vitaliy P. 690
Kotipalli, Ratan 2209
Kottantharayil, A. 1995
Kottantharayil, Anil............ 396, 716, 1850, 2799, 3478
Kottokkaran, Ranjith 2247
Kotulak, Nicole...............................999, 1838
Kotulak, Nicole A..247
Koyama, Koichi...........................1765, 1787
Kozodoy, Peter 1476
Krabb, Peter 178
Krantz, Patrick W..730
Krasikov, D..2816
Krc, Janez...1346
Krein, Philip T..3214
Krich, Jacob J. 1294
Krishnan, Mani R..1912
Krishnan, Sheeja...76
Krishnaswami, Hariharan.......................2931, 2936
Krogen, J..2094
Krügener, Jan...1494
Krut, Dimitri D. 37
Ku, Chen-Hao...329
Kubiniec, Alex 132, 1132, 1427
Kuciauskas, Darius 1679

Kudriavtsev, Yu.. 670
Kuitche, Joseph 1877, 1883
Kulish, Mykola R.. 690
Kum, Hyun 18, 222, 677, 2084
Kumar, Rajesh.. 3478
Kumar, Shailendra 2345
Kumar, Sukanya Santhosh 980
Kumar, Vijay 2716
Kumari, Khushboo 251
Kuo, Hao-Chung 1610, 1627
Kuo, Po-Tsun 1006
Kuo, Ting-Wei 329
Kurdgelashvili, Lado 2035
Kurihara, Risa 2159
Kurimoto, Yuji 931
Kurokawa, Yasuyoshi 1765
Kurstjens, Rufi 83
Kurtz, Sarah 1275, 1922, 2263, 3190
Kusaki, Kazuki 23
Kuthanazhi, Vivek.. 3478
Kwon, Jung-Dae 2631, 2634
Kwon, Sang Jik 195
Kyureghian, H. 2364
La Centra, Ricci 2870, 2891
Lachaurne, Raphaël 2528, 3402
Lachowicz, A. 50
Lackner, David 2511
Lacroix, Jean-Sébastien 1579
Lafleur-Lambert, Antoine 1360
Lafont, Ombline 2453
Lagumavarapu, Ramesh B. 202
Lai, B. 3295
Lai, Barry............... 1494, 2170, 2179, 2245, 3300, 3309
Lai, Yi 1009
Laine, Hannu S. 1491, 1494, 3236
Lakshmanan, Ramakrishnan 2870, 2891
Landgraf, D.. 1752
Lang, Mario 2216
Lapierre, Ray R. 1294
Larrey, Vincent 2492
Larsen, Ross E. 1354
Larson, Bryon W... 1354
Lasalvia, Vincenzo...............881, 1491, 1801, 2242, 3439
Laschinski, Joachim 1965
Lassise, Maxwell 3410
Latham, Joseph 1086
Lau, Derwin 3220
Lau, Kei May 578
Lave, Matthew 1435, 3008, 3025, 3031, 3184, 3348
Lavrova, Olga 1280, 2618, 3488, 3494
Law, D. 2094
Lazarou, Constantinos 276
Le Corre, Alain 2192
Le Donne, Alessia 1669
Le Gall, Sylvain 3402

AUTHOR INDEX

Le Guen, Vincent 70
Le Rouzo, Judikaël 2255
Lebreton, Fabien 464, 1237
Leclerc, Christophe...........................558, 572
Lecouvey, Christophe 2492, 2562
Ledinek, Dorothea 796
Ledinsky, M. 2073
Lee, Angela 417
Lee, Benjamin G. 881, 1737, 1801, 1832, 2482, 3439
Lee, Calvin 1342
Lee, Dong-Seon.............................777, 1665
Lee, Eunjoo 399
Lee, Eunsang 2124
Lee, Hae-Seok..935
Lee, Hyeonseok....................................1012
Lee, Jaejin 244
Lee, Jeong In 356, 1758
Lee, Ji-Hoon 2631, 2634
Lee, Jihwan...1181
Lee, Jinwoo 1455
Lee, Jongwon................................681, 1215
Lee, Kan-Hua 359, 412, 1479, 1711,
      1714, 1743, 2498, 2566
Lee, Kyumin 1526
Lee, Kyu-Tae 1469
Lee, M. L..3376
Lee, Minjoo 2291
Lee, Minjoo L. 42
Lee, Mitch 600
Lee, Mitchell...595
Lee, Seunghun 1253
Lee, Soonil 363
Lee, Yeonbae 1204
Lee, Yun Seog 1441
Lefebvre, Amy 1933
Lefebvre, Amy L. 2000
Lehman, Peter 1271
Lehr, J..3224
Leilaeioun, M..3376
Leilaeioun, Mehdi 1322, 1790
Leite, Marina S..................1508, 1585, 2443
Lekx, David 1094
Lemaître, Aristide 1289
Lemus, Ricardo Guerrero 1116, 3240
Lennon, Alison 3220
Lennon, Kyle 3384
Lennon, Kyle R. 1598
Leone, Stephen R. 417
Leonhardt, M. 1752
Leow, Shin Woei 3275
Lepkowski, Daniel................................215
Lepkowski, Daniel L. 2079, 2554
Lester, Luke F. 219
Leto, Riccardo 1728
Leu, S. 1752

Levcenco, Sergiu.................................3396
Levi, D.H. 467, 490
Levi, Dean 483
Levrat, J. 50
Levrat, Jacques 3435
Levy, David H. 3442
Li, Chu Tu 1094
Li, Duanhui 1473
Li, Guan-Yi......................343, 367, 893, 2664
Li, Jian 771, 1643
Li, Jian V.......................2473, 2728, 2749
Li, Joel B. 3300
Li, Kexue 424
Li, L. 3295
Li, Lan 1473
Li, Li 1175
Li, Lu 1619
Li, Mengjie 3315
Li, Qiang 578
Li, Rui 1094
Li, Siming 3143
Li, Wenjie 3275
Li, Xiaoping 1193
Li, Xinyi 255
Li, Xueying 2170
Li, Y. 869
Li, Yongkuan 2368
Li, Yunjun 907
Li, Yunpeng 2220
Li, Zhanhang 226
Li, Zhuohui 1598, 3384
Liang, B.L. 2755
Liang, Jianbo 2548, 2551
Liao, Anqi 948
Liao, Yuanxun 696
Liao, Yuaxun 2186
Libby, Cara S. 2618
Licht, Abigail 701
Lichty, Marlene L..................................2298
Lie, Stener 3275
Lim, Bianca 2318
Lin, Albert............................294, 1631
Lin, Albert S. 1627
Lin, Cheng-Shian..................................1006
Lin, Chien-Chung...........1606, 1610, 1623, 1627
Lin, Ching-Fuh 1006, 1009
Lin, Fen 284
Lin, Ming-Yi 1051
Lin, Shang-Pang 1006
Lin, Yan..1100
Lin, Yandan 2048
Lin, Yan-Zhang 1623
Lin, Yida 667
Lin, Yu-Hsuan 911
Lin, Yung-Sheng 329

AUTHOR INDEX

Lin, Zong-Xian 367
Lincoln, Jason 2897
Lincoln, Jason L. 2839
Lincot, Daniel 2453
Linton, John 337
Lipovšek, Benjamin 1346
Lipski, Michael V. 1469
Lisbona, Emilio Fernandez 545
Lisco, F. 1691, 2457
Litjens, Geert 3014
Liu, A. Y. 1485
Liu, Chenxi 3172
Liu, Fang Fang 1648
Liu, Fangyang 2213
Liu, H. 1189
Liu, H.Y. 14
Liu, Haitao 472
Liu, Han-Wen 2660
Liu, Haohui 284, 2532
Liu, Hsiang-Yu 2637
Liu, Huiyun 3370
Liu, Jheng-Jie 343, 893, 2664
Liu, Kanglin 1100
Liu, Mengxia 3129
Liu, Qihang 1245
Liu, Ruimin 2220
Liu, Simon H. 93
Liu, X.Q. 2094
Liu, Xiangxin 198, 767, 1707
Liu, Xinbing 1728
Liu, Xing-Quan 2099
Liu, Zhe 284
Liu, Zhen 2532
Liu, Zhengjun 1494
Liu, Zhengxin 1241
Livera, Andreas 276, 1954
Liyanage, Geethika K. 170, 730, 815, 2462
Llin, Lourdes Ferre 1339
Lloyd, Alexis 2870
Lloyd, Michael A. 726, 3143
Lnr, Yiming 2558
Loach, Andrew J. 2697
Lodha, Saurabh 716
Lokanath, Sumanth 1275
Loke, Samuel P. 512, 521, 558
Loke, W.K. 1210
Lombardero, I. 1210
Lombez, Laurent 70, 1289, 2192
Lonergan, Mark 802
Long, Yean-San 448, 476
Looney, Erin E. 1491, 3236, 3290, 3300
Löper, P. 2073
Löper, Philipp 55
Lopez, Cristina S. Polo 2118
López, G. 1781

Lopez, Roberto 2728, 2749
López-González, J.M. 1781
López-López, M. 670
Lopez-Marino, Simón 155
Lorentzen, Justin 3514
Lorenzo, Antonio T. 1127
Loser, Ulrich 2272
Lossen, Jan 1222
Lotshaw, W.T. 2755
Lou, Chaogang 1619
Loubar, Anais 2711
Loyer, Camille 2000
Lu, Ching-Ying 1835
Lu, Hongbo 255
Lu, J.P. 1733
Lu, Jiawen 2656
Lu, Kyle B. 3448
Lu, Zhou 1728
Lubenow, Tomas 3333
Lujan, R. 1733
Luka, Tabea 2232
Lumb, Matthew P. 210, 247, 259, 272, 873, 2506
Luna, Miguel A. 632
Lunacek, Monte 3008
Lunt, Richard R. 2124
Luo, Shiqiang 976
Luo, Wei 1922
Luo, Yanqi 2170, 2245
Luria, Justin L. 1522
Luther, Joseph M. 2176
Lynn, Kelvin G. 3422
Lyons, Alan 2285
Lyu, Yadong 1933, 2844, 3195, 3204
Lyu, Zheng 1835, 2732
Ma, D. 229
Ma, Fa-Jun 2569
Ma, Xiaokun 1469
Macalpine, Sara 1537
Macco, Bart 1817
Macdonald, D. 1485, 3295
Macdonald, Daniel 2707, 3300
Mack, C. 490
Mack, I. 2073
Mack, Shawn 259, 873
Mackie, Neil 820
Maclaren, Scott 1511
Macmaster, Steven W. 2864
Madani, Keeya 940, 1824
Madsen, C. K. 229
Maeda, P.Y. 1733
Magaña, Ernesto 1494
Magdaleno, R. Santos 1990
Magdaleno, Rocío De La Luz Santos 1959
Magnin, Vincent 3402
Magnone, Lydie 1415

AUTHOR INDEX

Mahadik, N. A. 845
Mahapatra, Chiranjibi 2849
Maia, Cristiana Brasil 2307
Maidaniuk, Yurii 3370
Mailoal, Jonathan 1264
Major, Jonathan D. 742, 1445
Makita, Kikuo 861, 1724
Makoutz, Emily A. 3381
Makrides, George 1163, 1941, 1954, 3107
Malhotra, Raghav 3172
Malik, Roger 1193
Maliya, Heini 226
Malkov, Andrei V. 146, 186
Mallick, Tapas K. 2858
Manda, Surya 761
Mandelis, Andreas 3129
Manganiello, Patrizio 3343
Mangelinck-Noël, Nathalie 1498
Mani, Monto 604
Maniscalco, B. 2457
Maniscalco, Biancamaria 2827
Mann, Colin 1248
Mann, Colin J. 93, 512
Mansfield, Lorelle 1400, 2473
Mansoori, A. 281
Mantel, Claire 2682
Manzoor, Salman 1228, 1322
Marie, Benoit 1498
Marion, Bill 1134, 1537, 1543, 3333, 3348
Markevich, V. P. 1485
Markides, Christos N. 1339
Maros, Aymeric 1215
Marsh, Brett M. 417
Marsillac, Sylvain 182, 735, 807, 2298,
    2446, 2646, 3139
Marsillac, Sylvain X. 2582
Marti, Shilpa 2936
Martín, I. 1781
Martin, Mickaël 2492
Martinez, Aaron D. 2536, 3406
Martinez-Morales, Alfredo A. 2881
Martínez-Pérez, Alejandro 3285
Martín-Martín, D. 3376
Martins, Ana C. 2104
Martinson, Alex B.F. 6, 1256
Masada, Isao 1504
Mascarenhas, Angelo 2506
Maser, Jörg 3309
Maser, Jörg M. 2179
Maskell, Douglas L. 120, 589
Mastroianni, Simone 1253
Masuda, Atsushi 1268
Masuda, Shota 1794
Masutomi, Yasuki 3339
Matei, I. 1733

Mather, Barry 1561
Mathew, Leo 363
Mathew, X. 142
Mathews, N. R. 142
Matsubara, Koji 381, 1333
Matsui, Takuya 381, 1333
Matsumoto, Yasuhiro 632
Matsumura, Hideki 1765, 1787
Matsuo, K. 3
Matthew, Leo 2506
Maximenko, S. I. 845
Maximenko, Sergey 2091
Maximenko, Sergey I. 873
May, Matthias M. 2538
Mayberry, Ryan 914
Mazumder, Malay 2870
Mazumder, Malay K. 2891
Mazur, Yuriy I. 3370
Mccandless, Brian 1196, 3319
Mccandless, Brian E. 726
Mcclary, Scott A. 1449
Mcclung, Larry 2833
Mcclure, E. L. 845
Mcclure, Elisabeth L. 2298
Mcclure, Harumi 2947
Mccndless, Brian E. 3143
Mccomb, David W. 966, 3139
Mcdanal, A.J. 525
Mcdanold, Byron K. 2864
Mcfavilen, Heather 305, 582
Mcintosh, Keith R. 1322
Mcintyre, Maxwell 1040
Mcintyre, Michael 1086
Mckenna, Russell 126
Mcmahon, William E. 268, 3381, 3406
Mcmeans, Philip A. 921
Mcpheeters, Claiborne 42, 525, 2099
Meakin, David 1927
Medic, V. 2364
Medici, Vasco 2118
Meeker, Michael A. 873
Mehlich, H. 1752
Mehta, Hitesh K. 3038
Meier, Florian 2996
Meissner, Dieter 178
Meitl, Matt 272
Meitl, Matthew 873
Melamed, Celeste L. 3406
Melchiorre, Michele 151, 2054
Meleco, A. J. 14
Mellor, Alexander 1339
Melnikov, Alexander 3129
Melvin, Andrew 1
Méndez, Juan A. 3240
Meng, Fanying 1241

AUTHOR INDEX

Meng, Hsin-Fei 1635
Meng, Xiaodong 2854
Menossi, Daniele......................752, 1669, 2372
Menozzi, Roberto 2205
Men-Pérez, E. 370
Meot, Jacques....................................2593
Merdzhanova, Tsvetelina 2114
Merghcim, Julia 3500
Merkle, Agnes....................................3371
Merz, Christopher 1965
Merzlic, Sebastien 1933, 2000
Messer, Alexander 512
Messer, Alexander J............................521, 558
Messerschmidt, Michael 2682
Messmer, C. 2064
Metzger, W.K. 1312
Metzger, Wyatt K. 1196, 3147, 3305, 3319
Meuris, M. 3260
Mewe, Agnes A. 917
Meyers, Bennet..................................3354
Mi, Z. 2388
Mi, Zetian 1299
Mia, Md Dalim 2749
Michaelson, Lynne 925
Micheli, Leonardo...........2301, 2789, 2804, 2858, 2881
Michl, Bernhard 1329
Mihailetchi, Valentin D. 1222
Mihaylov, Blagovest...........................1411
Mikofski, Mark 3354
Mikofski, Mark M. 110
Milakovich, Timothy...........................213
Miller, Bill 1473
Miller, David C.................2789, 2864, 3195, 3208
Miller, Elisa M...................................2536
Milleville, Christopher C......................1598, 3384
Mil'shtein, S. 2411
Min, Jung-Hong 777
Minemoto, Takashi 455, 757, 2385
Minkin, L. 1863
Mints, Paula......................................2039
Miryala, Tejaswini 2646
Mishima, T. D. 14
Mishra, Himani 2376
Misra, Sudhajit..................175, 802, 2467
Mitchell, Bernhard..............................2707, 3304
Mittag, Max 1531
Miyajima, Sakutaro 480
Miyashita, Naoya 854, 2334
Mizuno, Hidenori 1724
Moffett, C.E. 2736
Mohammed, Khaja H. 921
Mohapatra, Soumya Ranjan 3050
Mohr, Christian 83
Monnard, Raphäel...............................3256
Montenegro, Davis 3055

Montes, Carlos...................................429
Montgomery, Kyle H............................531
Montiel-Chicharro, Daniel 3208
Moon, Soo-Jin 3256
Moore, A. 2816
Moore, Andrew...................................1522
Moore, James 1838, 2091
Moore, James E. 259, 272, 373, 2506
Moore, Jay 2947
Moosa, Hassa 2011
Moosa, Maitha 2011
Moradi, Hadis 638, 2941
Moraitis, Panagiotis 3167
Morales, Christophe 2492, 2562
Morales, Cristian 2870, 2891
Morales-Acevedo, A. 670
Morel, Don 802, 1638, 3413
Morgado-Dias, F. 3178
Moriarty, T. 490
Moriarty, Tom 483
Moriki, Akinori 2906
Morin, Jean-Francois 1360
Morishige, Ashley E. 1494, 3236, 3290, 3300
Morita, Hiroshi 1973
Morral, Anna Fontcuberta I....................944
Morris, Jeromie 2996
Morrison, Matthew 229
Mortazavi, Soheyl 2875
Moseley, J. 1312
Moseley, John 62, 1196, 1381, 2887, 3123, 3147
Moser, David 3360, 3482
Moustafa, A. 1747
Moutinho, H.R......................1312, 2280, 2785
Moutinho, Helio 62, 2789, 3305
M'sirdi, Nacer K. 1968
Muaddi, Saad 1110
Mueller, Thomas.........................496, 2318
Mukherjee, Shaibal 2345
Mulder, P. 1189
Muller, Bjorn 126, 2267
Muller, M. 2280
Muller, Matthew.....2294, 2301, 2789, 2804, 2858, 2881
Müller, R. 2064
Munasinghe, Anjali 2124
Munday, Jeremy N. 1585
Mundt, Laura 1253
Mundus, Markus 1253
Munkhammar, Joakim 3067
Muñoz, D. 1747
Munoz, Krystal 925
Munshi, Amit 1674
Munshi, Amita 980
Mur, Pierre 2562
Muralidharan, Pradyumna 1790, 1797
Muramatsu, Kazuo...............................2642

AUTHOR INDEX

Murphy, J. D. 1485
Murugesan, Arumugam 2172
Muskovin, Eric.....................................537
Mutitu, James.......................................315
Mwove, Johnson Kyalo.......................2014
Myers, Matt 525
Nærland, Tine Uberg..........................2610
Nagaoka, Akira 1679
Nagarajan, Adarsh..............................2991
Nage, M..3150
Nagel, H. 3135
Nägelein, Andreas...............................2538
Nair, P. R. 2716
Nair, Pradeep R......................1015, 1022, 1755
Nakada, Tokio 192
Nakamur, Tetsuya 567
Nakamura, Kyotaro.................... 1504, 1794, 2498,
    2566, 2588, 2642
Nakamura, Shigeyuki 2338
Nakamura, Tetsuya........................562, 3506
Nakano, Yoshiaki 854, 2201, 2524, 2538, 3528
Nakata, Tatsuya 854
Nakatsuka, Shigeru............................2385
Nam, Wooseok.....................................2242
Nanda, A. 229
Nandal, Vikas 1015
Nanduri, Sai Naga Raghuram.............1018
Naqavi, Ali.....................512, 521, 558, 572, 1248
Narasimhan, K.L................396, 1850, 1995, 3478
Nardone, Marco..........................309, 3123
Naseem, Hameed A. 921
Natsheh, Ammar 2011
Naumann, Volker 1376
Nawara, Witek 3462
Nawaz, Syed F. 1067
Nayfeh, Ammar 213, 1741
Naylor, Mark.......................................914
Nayshevsky, Illya 2285
Ndione, Paul 1253
Needell, David R.................................1737
Neely, J..3224
Neely, Jason 2141
Neergat, Manoj....................................1704
Nehme, Bechara 1968
Nelson, George T...........202, 206, 222, 1184, 2084
Nemeth, William 1777, 1801, 1817, 1832,
    2242, 2702, 3439
Nespoli, Lorenzo.................................2118
Nett, Zach 1737
Neuschitzer, Markus...........................155
Neuwirth, Markus...............................1682
Newlands, Allan 2042
Ng, Annie 958
Ngan, Lauren.......................................600
Nguven, Dac-Trung.............................2192

Nguyen, H. T. 3295
Nguyen, Tinh 3204
Nickel, Benedikt.................................3388
Nicolay, S... 50
Nicolay, Sylvain 3256
Niemi, T. 1189
Niesen, Bjoern 3256
Nietzold, Tara 944, 2179, 3309
Nii, Kohdai 85
Niki, Shigeru 33
Nilsson, Ulf H. 2864
Nishikawa, Naoyuki 1385
Nishio, M... 3
Nishioka, Kensuke 480, 1479
Noack, Max 2247
Nobre, André M. 3172
Nobuhara, Shohei 1175
Nocerino, John 93
Noda, Naoto 326
Noda, Yoshimasa 970
Nofuentes, Gustavo............................2858
Nogay, G. 2073
Noh, Shinyoung 858
Nonnenmacher, H. J...........................1752
Norman, Andrew...................... 1381, 2887
Norman, Andrew G. 2536, 3406
Norwood, Robert A. 1147
Nose, Yoshitaro....................1679, 2361, 2385
Nowakowski, Marilyn L....................3524
Nsofor, Ugochukwu 1828
Nukala, Tejeswar................................3061
Nunomura, Shota 381
Nurdin, Muhammad 3102
Nussbaumer, Hartmut 1077
Nuzzo, Ralph G...................... 1469, 1737
Nyirjesy, Gabrielle 667
Oberbeck, Lars...................................3370
O'Brien, Greg 1933
O'Brien, Gregory................................2000
Ocaña, Luis 429
O'Carroll, Deirdre M. 3393
Ochoa, M. 1210
Odden, Jan Ove 2651, 2776
Oehler, Fabrice 1289
Ogawa, Tomoki 2548
Ogura, Atsushi.....................1504, 2588, 2642
Ogutman, Kortan 1804, 3448
Oh, Jaewon 1858, 1877, 1883, 2912
Oh, Seung Kyu................................... 866
Oh, Soo-Young....................................487
Ohdaira, Keisuke 1385, 1787
Ohigashi, Takashi...............................2159
Ohshima, H. 441
Ohshima, Takeshi........................ 562, 567
Ohshita, Yoshio 1504, 1794, 2498, 2566, 2588, 2642

AUTHOR INDEX

Ohta, Taisuke 3417
Ok, Young-Woo 333, 1807, 1838
Oka, Naotaka..1973
Okada, Yoshitaka.........................10, 85, 854, 2334
Okafor, Jonathan O. 219
Okano, Y. 3
Okel, Lars A.G. 1081
Oliva, Florian 3265, 3285
Olopade, Muteeu..2381
Olvera, María De La Luz 632
O'Neill, Mark 525
Oney, Michael F. T 2176
Onno, Arthur..3370
Onunkwo, Ifeoma..2135
Oo, W.M. Hlaing..1661
Opila, Robert....................................999, 2035
Opila, Robert L. 315, 326
Oreski, Gemot 178
Orlovskaya, Nina A..2324
Ortega, E. 643
Ortega, Pablo..944
Ortiz, Brenden R. 3406
Ortiz-Rivera, Eduardo I. 2957, 2963
Ory, Daniel 70
Oshima, Ryuji...861
Ososanya, Esther..2432
Osowski, M. 3511
Osterwald, C.R. 467, 490
Ota, Yasuyuki..1479
Otaegi, A. 2740
Otaegi, Aloña..2677
Otnes, Gaute....................................1286, 2502
Ottoson, L. 467, 490
Ouyang, Zi..2403, 3220
Oviedo, Felipe..2744
Ozanne, A. -S...1747
Paap, Scott..1473
Packard, Corinne E. 46
Page, Matthew......................................1777, 2242
Paggi, Marco 402
Palekis, Vasilios.......... 175, 802, 1511, 1638, 2467, 3413
Palekis, Vasilis...2449
Palitzsch, Wolfram..2272
Palmer, Evan 496
Palmiotti, Elizabeth......................................1400
Palmquist, Nathan 667
Pan, Hui 1100
Pan, N. 3511
Pan, Zhen...226
Panchal, A. K. 3061
Panchal, Ashish K. 3038
Pandey, Rahul 377
Paolone, Mario 1415
Parashar, Parag 1627, 1631
Paraskeva, Vasiliki.......................................276

Parenti, Robert C. 3520
Parikh, Anuja V. 3123
Parikh, Harsh 2682
Park, Chinho 487
Park, Ji-Sang 6, 1256
Park, Joo Hyung 810
Park, Kyung Ho 244
Park, S. 2388
Park, Seonyong 2087
Park, Somin 1044
Park, Sungeun 935
Park, Won-Kyu 244
Partain, Larry..2042
Passow, Kendra 595, 600, 1071
Paszuk, Agnieszka 2524, 2538
Patra, Payal...761
Patterson, Robert J 2392
Paudel, Naba R. 2443
Paul, Douglas 1339
Paul, Nicolas..70
Paul, P. K. 30
Paul, Pran K. 2446, 3139
Paul, Sanjoy 2473, 2749
Paulauskas, Tadas 2759
Paull, P. K. 2414
Paull, Sanjoy 2728
Paulsen, Andrew..3514
Pavgi, Ashwini 1877, 1883
Paviet-Salomon, B. 50
Paviet-Salomon, Bertrand 3256
Pavilonis, Michael..1476
Pavlov, Marko 626
Pavlovsky, Igor 907
Pawar, Vaibhav 2986
Payne, David 315
Payne, David N.R. 2576
Peaker, A. R. 1485
Peale, Robert E. 2324
Peharz, Gerhard...................................178, 1329
Peibst, Robby 1366, 3371
Pellegrino, Sergio....................512, 521, 558, 572
Peña, J.L. 1691, 2457
Pena, Juan Luis 1669, 2372
Peña, Ramón 632
Peng, Jun 2076
Peng, Shou 761, 2427
Penning, David P 2170
Peppanen, Jouni 3025
Pera, David 1979
Peraca, Nicolás Márquez 263
Perez, Richard 132, 1104
Pérez-Rodríguez, Alejandro 3265, 3285
Perez-Wurfl, Ivan 315
Perkins, C. 2280
Perkins, Craig...............................2702, 2789

AUTHOR INDEX

Perkins, Craig L. 2294
Perl, E. 3376
Perl, Emmett E. 42, 1201
Perna, Allison 2467
Pesala, Bala 2858
Peschel, Gina 3396
Peshek, Timothy J. 1927, 2697, 3456
Peters, I. M. 648, 1140
Peters, Ian Marius............ 284, 1264, 2532, 2744
Peters, Marius 3236
Petersen, Michael 2682
Peterson, Chris 512
Peterson, Josh 1169
Petoukhoff, Christopher E. 3393
Pfiester, Nicole 701
Phillips, Adam 748
Phillips, Adam B. 170, 730, 815, 1030, 2462
Phillips, Laurie J. 1445
Phillips, Nancy H. 2864
Phinikarides, Alexander 1954
Picard, Sandrine.................................2087
Piccinelli, Fabio 1669, 2372
Pickel, Tobias 3500
Pierro, Marco 3482
Pieters, Bart E. 1651
Pihan, Etienne 1498
Pillai, Supriya 2403
Pistor, Paul 155, 3285
Piszczor, Michael 525
Pitalúa, Nun 632
Platzer-Björkman, C. 3269
Plessing, Lukas 178
Pleus, Albert 1835
Plochowietz, A. 1733
Podraza, Nikolas 2646
Podraza, Nikolas J. 2462, 2582, 2771, 3426, 3468
Poindexter, Jeremy 3300
Poissant, Yves 1908
Pokharel, Nikhil 831, 2514
Polojärvi, Ville 297
Poncho, Corpuz 2947
Poortmans, J. 3260
Poortmans, Jef 1233
Pop, Sergiu C. 921, 1869
Poplavskyy, Dmitry....................1459, 1686
Porter, Ilana J. 417
Potamialis, C. 2457
Pötz, Sandra 178
Pouladi, S. 869
Pouladi, Sara 866
Poulsen, Peter B. 2672, 2682
Powalla, Michael 791
Previtali, Jonathan 1275
Price, Jared S. 1469
Prietl, Christine...................................1329

Printraza, Nikolas J............................. 993
Procel, P. 2073
Ptak, Aaron J. 46, 62, 2275
Puska, Martti J. 2070
Puthanveettil, Suresh E. 76
Qazi, Farah 1317
Qian, Gary 667
Qian, Shen... 958
Qin, Xuefei 1594
Qiu, Botong 667
Qudsia, Syeda 1317
Quinto, Carlos 429
Quiroz, Jimmy E. 1280
Rada, Jacob....................................... 1271
Radhakrishnan, Hariharsudan
Sivaramakrishnan 1233
Raghavan, Srinivasan 837, 841, 986, 2395, 2399
Ragunathan, Gautham........................ 1181
Rahman, Mosaddequr 1067
Rahn, Christopher D........................... 1469
Raiker, Gautam A. 3073
Raj, Samuel 284, 496, 499
Rajan, Grace.............. 182, 735, 807, 2298, 2446, 2646
Rajbhandari, Pravakar P. 989
Rajput, Amit Singh 499
Raju, T. Bhim 1034
Raker, David 2926
Rale, Pierre 1289, 3147
Ramakumar, Rama 1122
Ramamurthy, Praveen C. 1614, 2811
Rambabu, Sugguna 3478
Ramic, Zekija..................................... 2776
Ramírez, A. 503
Ramirez, A.A. 433
Ramírez, E. A. 433, 503
Ramos, Helena Geirinhas 3178
Ramos, Javier 2255
Ramprasad, Sumukh 496
Ramu, Govind................................ 1275, 2263
Rancoita, P.G. 541
Rand, James 925
Ranjan, Rajeev 2395, 2399
Ranjan, Upasna 2811
Ranjbar, S. 3260
Ransome, Steve 652
Rao, Arun D....................................... 1614
Rao, B.V. 1898
Rao, Rajesh 363, 2506
Rao, Roshan R 604
Raorane, Neha 1755
Raote, Yojak 1022
Rashkin, L. 3224
Rastogi, A.C. 3279
Rathi, M. 869
Rathi, Monika 866, 2368

AUTHOR INDEX

Rathore, Sudharm 2902
Rau, Uwe..1651, 2114
Raupp, Christopher 1984
Ravindra, M. 76
Ravindra, Pramod 2395, 2399
Raychaudhuri, S..1733
Razooqi, Mohammed A......................................2462
Recart, Federico 2677
Reddy, Anurag 3528
Reddy, K.S..2858
Reddy, Rekha 3524
Reed, S..869
Reedy, Robert C...881
Reese, M. O...138
Regalado-Pérez, E...142
Reichel, C..2064
Reichert, Andreas 2407
Reinders, Angèle 2109
Reindl, T...1140
Reise, Christian 2267
Rejon, V...1691, 2457
Ren, Zekun 284, 2532, 2744
Ren, Zhiwei..958
Reno, Matthew J. 1555, 1567, 1573, 1579, 2975, 3025, 3031, 3055
Renteria, E. J...281
Repins, Ingrid....................................2728, 3452
Reusser, Jean..2255
Reyes-Banda, M.G. 142
Rey-Stolle, I. 1210
Rey-Stolle, Ignacio 1204
Rhodes, Christopher 1476
Riaz, Hiba 1741
Ribeyron, P. -J..1747
Ricardo, Julian Do Nascimento 3077
Rich, Geoffrey 600
Richards, J...3224
Richardson, Walter...1116
Richter, A..1752, 2064
Richter, Mauricio..3360
Riedel, Nicholas...2672, 2682
Rienacker, Michael 3371
Riesen, Yannick..3435
Rigdon, Terry B. 3448
Riggs, Brian 37
Riggs, Brian C..3245
Riley, Daniel 1537, 3155, 3184, 3348
Riley, Daniel M...1543, 1549
Rimmaudo, I..1691, 2457
Rimmaudo, Ivan...1669, 2372
Rincon-Charris, Amilcar A.................................2963
Ringel, Steven A. 215, 2079, 2446, 2554
Ringleb, Franziska 3396
Rivera, Eduardo I. Ortiz 3044
Riverola, Alberto..1339

Robert, Sofie 1804
Roberts, Jesse..3083
Robertson, John..37
Robertson, Kyle W. 1294
Robinson, Charles D. 3155
Rochat, Raphael 326
Rocheleau, Richard E..1061
Rockett, A. 30
Rockett, Angus.........................182, 1400, 2446
Rockett, Angus A. 1511
Rodrigues, Sandy...3178
Rodriguez, D. J...2031
Rodríguez, Diego J. 1143
Rodríguez, Pedro 2677
Rodríguez-Gallegos, Carlos D. 2318
Roest, Stefan 3462
Roeth, A. J. 14
Rogers, John A. 1469
Rohatgi, Ajeet.................333, 940, 1807, 1824, 1838
Roland, Paul J...1030
Roller, John..508
Romanin, Vince 37, 3245
Romeo, Alessandro........................752, 1669, 2372
Ronoh, Geoffrey Kibiegon 970
Rooijakkers, Tom T.H. 1081
Ropp, Michael..2147, 3020
Rosalcs-Ascnsio, Enrique....................................3240
Rose, Volker 2179, 3300
Ross, N. 3269
Rotoli, P..1747
Rounsaville, Brian 940, 1807, 1824
Routhier, Alexander F. 3088
Rowell, David...3524
Roy, Sam 2358
Roy, Tatiana A.....................................521, 558
Royer, Fabien...558
Rozza, Davide 541
Rubbard, Seth M...3468
Ruffini, Leia 2453
Ruiz, Carmen M..2255
Ruiz, E. O. Ángel 1990
Rummel, S. 467
Rupp, B. 1733
Ruppalt, Laura B...873
Russell, Annie 1094
Russell, Richard...2227, 3435
Russell, Thomas C.R. 2236
Ruth, Daniel 2301
Ryou, J. 869
Ryou, Jae-Hyun 866, 2368
Saavedra, Michael 1473
Sablon, Kimberly 1181
Sabnis, Sanjeev 2849
Sacchetto, Davide 3256
Sachenko, A.V....................................663, 1025, 1811

AUTHOR INDEX

Sachenko, Anatoliy V.......................690
Sáenz, M.J.......................643
Saetre, Tor Oskar 685
Sahayaraj, S.......................3260
Sahli, Florent.......................3256
Sahraei, N. 648
Sai, Hitoshi.......................381, 1333
Saifullah, Muhammad 810
Sainsbury, Cassidy.......................2692, 2765
Saito, K.......................3
Saito, Tomohiro.......................931
Saive, Rebecca.......................1589, 2236
Sakamoto, Katsuyoshi 85
Sakamoto, Norihiko 1268
Sakurai, Takeaki.......................33, 160, 900
Salamo, Gregory J. 3370
Salavei, Andrei.......................2372
Salazar, J.......................370
Salazar-Duque, John E. 2963
Salo, Kristian.......................1494
Salome, Pedro.......................796
Salpakari, Jyri.......................3236
Salvetat, Thierry 2492
Samoilenko, Yegor 1697
Sampath, W.S.......................980, 2736
Sampath, Walajabad 424, 785, 1674
Sampath, Walajabad S. 3417
Sample, Tony.......................1275
Sampson, Matthew D. 6, 1256
Samuelson, Lars 2502
Samundsett, C. 3295
Samundsett, Chris.......................2076
Sánchez, Yudania 155
Sánchez-Pérez, P. A. 1959, 1990
Sanchiz, Joaquín 429
Sandeep, K.......................396
Sang, Baosheng.......................1455
Sang, Shiyu 472, 1430, 2918
Sangjeong, Myeong 356
Sankaran, M.......................76
Sankin, I.......................2816
Santana, G.......................370, 2342, 2614
Santana-Rodríguez, G. 670
Santbergen, Rudi 2605
Santhanam, Parthiban 2185
Santos, M. B.......................14
Santoyo-Salazar, J.......................370
Saraf, Akash 761
Saraswat, Krishna.......................1835
Sargent, Edward H. 3129
Sarmah, Nabin 2858
Sarvari, Hojjatollah 1044, 2432
Sarwar, Jawad 1888
Sasaki, A. 1003
Sastry, O. S.......................3478

Sato, Daisuke 1743
Sato, Shin-Ichiro.......................562
Sato, S-I. 552
Satzinger, Valentin 178
Saucedo, Edgardo.......................155, 3265, 3285
Savin, Hele.......................944, 1494, 3236
Sawallich, S. 3150
Sayed, Islam E.H. 2195
Sayyah, Arash 2891
Scaccabarozzi, Andrea 1289
Scarpulla, M.A. 1656, 1661
Scarpulla, Michael A.......................175, 802, 1679, 2467
Schäfer, Nicolas 2216
Schaller, Richard D. 6
Scheiman, David 1838, 3514
Schelhasl, Laura T 2176
Scheltens, Frank J. 966
Schenller, E.J. 1701
Schermer, J. 1189
Schindler, F. 2064
Schitthelm, F. 1752
Schlemmer, James.......................1104
Schmid, Martina.......................3396
Schmidt, Jan 3371
Schmidt, Thomas.......................3396
Schmieder, Kenneth J. ...210, 259, 272, 873, 2091, 2506
Schnabe, Thomas.......................2216
Schnabel, Erdmut.......................3488
Schnabel, Manuel ... 1817, 2482, 2543, 3254, 3371, 3439
Schnabel, T. 3260
Schnabel, Thomas.......................791
Schneider, Kevin 1476
Schneller, Ej.......................389
Schneller, Eric J.......................2839, 2897, 3448
Schoenfeld, Winston 2839
Schoenfeld, Winston V. 322, 1804
Schoenfelder, Stephan.......................2622
Schoenwald, David 2969
Scholl, Jonathan A.......................1549
Schoop, Urs 1455
Schorch, M. 1752
Schriemer, Henry P.......................1094
Schubert, M. C.......................3135
Schubert, Martin C.......................1329
Schulte, Kevin 62
Schulte, Kevin L. 46, 232, 2275
Schulte-Huxel, Henning 1366, 2543, 3371
Schulz, Gerd.......................914
Schulze, Patricia S.C. 1253
Schwabe, Hartmut.......................2622
Schweiger, M. 107
Sclj, Josefine.......................619
Scofield, A.C. 2755
Scolari, Enrica 1415
Sculati-Meillaud, Fanny.......................2794

AUTHOR INDEX

Seif, Johannes P...................................3435
Seigneur, Hubert 2839, 2897
Sellami, Nazmi...................................2858
Sellers, Andrew 2926
Sellers, Diane G...................................3384
Sellers, I. R...14
Selvamanickam, V. 869
Selvamanickam, Venkat.............866, 2368
Semichaevsky, Andrey.........................319
Sen, Fatih G..2759
Senaud, L.-L...50
Sengar, Brajendra S............................2345
Sengupta, Manajit.......................116, 1169
Senthilarasu, S....................................2858
Sepeher, Mohsen M. 1094
Sera, Dezso..............................1421, 2682
Serra, João M. 1979
Sethia, Saurabh 2902
Seydel, Elisabeth 1682
Shafarman, William N............................26
Shah, S. 1350
Shahirinia, Amir 3092
Shanmugam, Vinodh 2318
Sharma, Ashok K. 396
Sharma, Romika 3300
Sharma, S..2094
Sharps, Paul..................42, 525, 2099
She, Hui...1863
Shen, Chang-Hong 1627
Shen, Zeqing 3393
Shephard, Les E....................................1116
Shervin, Kaveh 1452
Shervin, Shahab 866
Shetty, Nishit 876
Shi, Jianwei..1820
Shi, Jiatiwei 1322
Shi, Xuanyi 3220
Shi, Zhan..1037
Shibata, Hajime............................33, 1268
Shieh, Jia-Min 1627
Shigekawa, Naoteru 2548, 2551
Shih, Cheng-Hao 2035
Shih, I. 2388
Shih, Ishiang 1299
Shima, D. M. 281
Shimura, H. 1003
Shin, Hyun-Beom 244
Shin, Myunghun 2631
Shin, Seunghyun....................................935
Shin, Woo Jung 385
Shinde, O.S. 389, 1701
Shirasawa, Katsuhiko 904, 931
Shkrebtii, Anatoli I. 690
Shoji, Yasushi 10
Shore, Andrew......................................437

Shrestha, Niraj.....................................1030
Shrestha, Santosh 696, 2186
Shu, Chia-Jhe......................................1606
Shu, Jinn-Kong 1606
Shubhrant, Abhishek 2902
Si, Fai Tong 2605
Siddiki, Mahbube K...................1018, 1110
Sidhu, Navjot Kaur 3279
Siebentritt, Susanne 151, 2054, 2205, 2478
Siepchen, Bastian..................................761
Sikchang, Hyo 356
Silva, Francois 464, 1237
Silva, José A. 1979
Silvaggio, Amber C. 2554
Silverman, Timothy....................1259, 1893
Silverman, Timothy J............1400, 2771, 3452
Simon, John 42, 46, 62, 1201, 2275
Simon, Kirby 876
Simpson, L. 2280
Simpson, Lin 2294
Simpson, Lin J. 1893, 2789
Sinapis, Kostas 1081, 1090
Singh, Aparna 2902
Singh, Ashish 1034
Singh, Ashish K. 1704
Singh, Hemant K. 1995, 3478
Singh, Rajeev 2762
Singh, Rhythm 1151
Singh, Rubina 1855
Singh, Sukvhinder 2227
Singlr, Vijay P......................................2432
Sinha, Archana 3478
Sinha, Parikhit 2005
Sinisuka, Ngapuli I 3102
Sink, Joseph 3333
Sinton, Ronald 2707
Sinton, Ronald A. 2692, 2765
Sio, H. C. 3295
Sio, Hang Cheong 3300
Sites, James R. 164, 1308
Slocum, Michael 677
Slocum, Michael A............18, 202, 206, 222, 831, 1184, 2084, 2514, 3468
Slooff, Lenneke H. 1081
Smaglik, Nathan 831, 2514
Smestad, Greg P...................................2858
Smith, Benjamin 1134
Smith, Brittany L. 18, 1184
Smith, David J. 2573
Smith, Mathew 2941
So, Won-Shup 487
Soares, Gabriela De Amorim................2875
Sodabanlu, Hassanet 854
Söderström, T. 1752
Sofia, Sarah E. 1264

AUTHOR INDEX

Sogabe, Tomah 85, 712
Sokolovskyi, I.O. 663, 1025, 1811
Sokolovskyi, Igor O. 690
Solanki, Chetan S. 3478
Solanki, Chetan Singh 1850
Soltanmohammad, Sina..26
Soman, Anishkumar 1828
Søndergaard, Sissel Tind.....................................2651
Song, Dengyuan..1430
Song, Hee-Eun...356, 1758
Song, Myungkwan 2631, 2634
Song, Tao..1308
Song, Zhaoning.................170, 730, 748, 815, 1030
Sonp, Hee-Eun..2723
Sood, Neeru 2858
Sossan, Fabrizio 1415
Soudachanh, A. L...281
Sozzi, Giovanna..2205
Spandana, B..396
Spataru, Sergiu......................1421, 2682, 2819
Spaulding, David...820, 1686
Spertino, Filippo...3096
Spiering, Stefanie 791
Spinelli, P..3150
Spooner, Ted 1275, 2263
Sreekumar, Nimisha 1755
Sridharan, Akirt..999
Sriramagiri, Gowri 658, 1196
Srivatsan, R. 120, 589
Stark, Cameron...1855
Starkl, Hannes...178
Steeman, Rob..337
Steenhoff, Volker..3388
Stefancich, Marco 1946
Steijvers, Henk..2875
Stein, Joshua...........................1537, 3333, 3348
Stein, Joshua S................ 1543, 3155, 3161, 3184, 3488
Steiner, Myles A. 42, 47, 232, 1201, 2195, 3254
Steinfedt, Jeff..525
Stender, C..3511
Stender, Christopher L...3524
Stephan, Jack 2124
Stevens, Margaret 701
Steward, Malia 1037
Stewart, J..3224
Stika, K. 2312
Stiles, Phil 2833
Stoddard, Nathan 2610
Stokes, Adam...1381, 2887
Stolt, L..30
Stone, Kevin H..2176
Stradins, Paul881, 1491, 1777, 1801, 1817, 1832, 2242, 24
Stradins, Pauls...2482, 2702
Strandberg, Rune...706, 2651
Stride, John A...2392

Stuart, Thomas 2926
Stuckelberger, J...2073
Stuckelberger, Michael 2610, 2854, 3309
Stuckelberger, Michael E.....................................2179
Stueve, Bill 2808, 2923
Sturm, James C. 1773
Stutz, Elias Z. 944
Su, Bojie.......................392, 1430, 1873, 2918
Su, Chengfeng 392, 1873
Subbiah, Jegadesan 1342
Subedi, Indra 2771, 3468
Subedi, Kamala Khanal 781
Sudbury, Benjamin A. 1322
Suga, Mitsunobu 567
Sugaya, Takeyoshi..............................562, 861, 1724
Sugimoto, Hiroki 160
Sugiyama, Masakazu 854, 2201, 2524, 2538, 3528
Sugiyama, Mutsumi 192
Sugiyama, Ryo 712
Suhana, Hadi 3102
Sumita, Taishi 3506
Sun, C. 1485
Sun, Ce 2759
Sun, Chang...3300
Sun, Chenguang...1241
Sun, Kaiwen 2213
Sun, Qiang..1648
Sun, Qiming 3129
Sun, S. 869
Sun, Sicong..2368
Sun, Wen-Cheng..2227
Sun, Xiaolin 2656
Sun, Xingshu.....................................1055, 1259, 1904
Sun, Yaojie 2048
Sun, Yubo...2467
Sun, Yukun...2291
Sun, Zeming 937
Supplie, Oliver 2524, 2538
Surya, Charles 958
Sutou, Yuji 931
Sutterlueti, Juergen 652
Suzuki, Ryota 1504, 2588
Swain, Santosh K. 3422
Swartz, Craig H. 2473, 2749
Sweatt, William 1473
Syu, Hong-Jhang 1009
Szabo, Sandor 2167
Szlufcik, Jozef..........................1233, 2227, 3435
Tabet, Nouar...963, 1058, 3435
Tacconi, Mauro 541
Tachibana, Shoji..1504
Tadese, Alemu 1104
Tadesse, Alemu...................................132, 1132, 1427
Tae, Christian 1835
Taekjeong, Kyung 356

AUTHOR INDEX

Takahashi, Akiko 2906
Takahashi, Isao 1765, 1794
Takahashi, Takuji 455
Takahashi, Yasuhito 1973
Takamoto, Tatsuya 3506
Takato, Hidetaka 381, 904, 1724, 3323
Takenouchi, T...441
Tamaki, Ryo 10
Tamboli, Adele 3254, 3371
Tamboli, Adele C. 578, 2482, 2488,
        2536, 2543, 3381, 3406
Tamizhmani, Govindasamy 1389, 1850,
        1858, 1877, 1883, 1959, 1984, 2789, 2912
Tan, Jin 1158
Tan, Joel M. R. 3275
Tan, K.H. 1210
Tan, Xuehai 761
Tanahashi, Katsuto 3323
Tanahashi, Tadanori................................1268
Tanaka, Aki 2642
Tanaka, T...3
Tanaka, Takahiro 2947
Tang, Chiu C. 2423
Tang, Houjun 1100
Tang, Mingchu 3370
Tang, Tao...2558
Tanke-Pedretti, Anna 1473
Tao, Meng...385, 608
Tao, Yuguo 1824
Tappan, Ian A. 2864
Tassone, Christopher J..............................2176
Tatapudi, Sai 1850, 1858, 1877, 1959, 2912
Tatapudi, Sai Ravi Vasista 2789
Tatavarti, Rao 1184, 2084
Tatavarti, Sudersena Rao..........................1181
Tate, John Keith.......................................333
Tayagaki, T...3
Tayyib, Muhammad 2776
Tchemycheva, Maria 1289
Tedeschi, Giampiero 2372
Teena, Percis 3113
Tennyson, Elizabeth M. 1508, 2443
Terheiden, Barbara..................................1824
Terukov, E.I. 1025, 1811
Teubner, Thomas.....................................3396
Teymouri, Arastoo 2403
Thanh, Nguyen Cong...............................1765
Thankalekshmi, Ratheesh R. 3279
Theelen, Mirjam......................................2875
Theigi, San 881
Theingi, San...1832
Theocharides, Spyros....................1163, 3107
Therrien, Francis 1579
Thibeault, Brian......................................315
Thimsen, Elijah 876

Thompson, Christopher 1196
Thompson, Corey S. 1869
Thon, Susanna M......................................667
Thorseth, Anders.........................2672, 2682
Thorsteinsson, Sune 2672, 2682
Thway, Maung..........................284, 2744
Tidwell, Steven.....................................1086
Timò, Gianluca 290
Tirumalai, Tejas 2923
Tischler, Joseph G. 873
Titus, Jochen 820
To, Alexander 517
To, B. 2280
Toberer, Eric S. 2536, 3406
Todorov, Teodor 1441
Togay, M. 2457
Togay, Mustafa 146, 186
Tomasi, A. 50
Tomasi, Andrea 3435
Tomasulo, Stephanie 2091
Tonic, Marko 1346
Toor, Fatima 346, 1537, 1543, 3184, 3333, 3348
Toprasertpong, Kasidit.................2201, 2524
Torralba, Encarnacion.............................3402
Tous, Loïc 2227, 3435
Tracy, Jared.................................3190, 3200
Traverse, Christopher 2124
Trout, T. John 2312
Trupke, Thorsten 66, 420, 2707, 3304, 3315
Tsafarakis, Odysseas 1090
Tsai, Cheng- Ying 3366
Tsai, Jia-Lin 1606
Tsai, Jia-Ling 294
Tseng, Zong-Liang 367
Tsutsumi, S. 3
Tu, Wei-Chen..1051
Tucher, Nico...................352, 1253, 2511
Tukiainen, Antti 297, 2520
Tuminello, F. 3511
Tummala, Abhishiktha 2912
Turek, Marko 2232
Turner, John A. 47
Tuteja, Mohit 1511
Tyagi, Astha 716
Tyler, Kevin 301
Tyson, Tom 925
Tzolov, Marian 1040
Ubukata, Akinori....................................861
Ueda, Kohsuke 3506
Ueda, T...1003
Uematsu, Takumi 1950
Ulbricht, Christoph 1360
Ulicná, Sona 146, 186
Uma, B. R. 76
Umishio, Hiroshi 381

AUTHOR INDEX

Unold, Thomas 3396
Unsur, Veysel..888
Upadhyaya, Ajay D...................................940, 1807
Upadhyaya, Vijay D..333
Upadhyaya, Vijaykumar 940, 1807, 1824
Uprety, Prakash 3468
Urbano, J. Antonio..632
Uruena, Angel...2227, 3435
Usami, Noritaka 1765, 1794
Utsunomiya, Satoshi 904
Vadiee, E...1603
Vadiee, Ehsan 305, 827, 1841
Vagidov, Nizami Z. 531
Vähänissi, Ville 1494
Vaida, Mihai E. 417
Vaidya, Nina 512, 521, 558, 572, 1248
Vaisman, M. 3376
Vaisman, Michelle....................................578, 3381
Vaissiére, Nicolas 2528
Valderrama, Nicolas 1893
Valdivia, Christopher E. 1094
Van Aken, Bas B. 3462
Van Alsburg, Jane..1455
Van De Loo, Bas W.H. 1817
Van Der Heide, Arvid 3343
Van Hest, Maikel F.A.M................2482, 3371, 3439
Van Sark, Wilfried...3014
Van Sark, Wilfried G.J.H.M. 1090, 3167
Vandamme, Nicolas 2453
Vandervelde, Thomas E. 701
Vanka, S...2388
Vanka, Srinivas 1299
Vansant, Kaitlyn 1922, 3452
Vargas, Carlos 3290
Vasi, J...1995
Vasi, Juzer 1850, 3478
Vasileska, D. 1603, 2816
Vasileska, Dragica....................................1790, 1797
Vasilyev, Leonid A. 3448
Vasudevan, Saravanan 2172
Vauche, Laura 2492, 2562
Vedde, Jan 2682
Veettil, Binesh Puthen 2392
Vehse, Martin 3388
Veinberg-Vidal, Elias................................2492, 2562
Veith-Wolf, Boris 1366
Velappan, Krishnakumar....................................761
Venizelou, Venizelos 276, 1163, 1941, 3107
Verbitskiy, V.N..1811
Verlinden, Pierre J...................................1922, 2220
Vermang, B...3260
Vermang, Bart 2209
Verschac, Rodrigo..1175
Vetter, E...1752
Viana, Marcelo Machado 1917, 2307

Vignola, Frank...1169
Vijh, Aarohi 3520
Vilcot, Jean-Pierre...3402
Vincent, Nina 1893
Vines, L. 3269
Vinogradova, Tatiana 512
Vinogradova, Tatiana G. 521, 558, 572
Virtuani, Alessandro 1395, 2104, 2794
Vlasiuk, V.M..1025
Vlasyuk, V.M..1811
Vleugels, J. 3260
Voarino, Philippe 2492, 2562
Vogt, Malte Ruben 1366
Von Gastrow, Guillaume 944
Voroshazi, Eszter 3343
Voss, Henrik 2682
Waddle, John M. 309, 3123
Wade, Andreas 2005
Wagner, Sigurd 1773
Waiis, J.M. 2457
Waldhauser, Wolfgang 1329
Walker, Don 93, 512, 1248
Walls, J.M. 1691, 3430
Walls, John 1674
Walls, John M................146, 186, 752, 2349
Walls, John Michael 2827
Walls, Michael 424
Walter, Arnaud 3256
Walters, Joseph 2839, 2897
Walters, R. J. 845
Walters, Robert 3514
Walters, Robert J. 210, 247, 259, 272,
373, 873, 1838, 2091, 2506
Waltmger, A..1752
Walukiewicz, W. 3
Walukiewicz, Wladek..1204
Wan, Kai-Tak 1933, 2000
Wan, Ronghua 226
Wan, Yimao 59, 2076
Wang, Ao...1937, 2823
Wang, Baomin 1469
Wang, Changlei 993
Wang, Da-Wei 3220
Wang, Deng..2048
Wang, Feng 1044
Wang, Fumei.........................392, 1937, 2823
Wang, Haotian 1342
Wang, He. 226, 392, 1430, 1648, 1873, 1937, 2823, 2918
Wang, Hongfeng 1215
Wang, Laidong 385
Wang, Mu 3370
Wang, Q. 1733
Wang, Rui 1100
Wang, Shenghao 160
Wang, Shizhen 976

AUTHOR INDEX

Wang, Sisi 2600
Wang, Teng-Yu 2660
Wang, Xiaohui 2432
Wang, Y. 1733
Wang, Y. D. 1733
Wang, Yan 1922
Wang, Yichen 1299
Wang, Yiwang 1100
Wang, Yongqian 2220
Wang, Yu 1933, 2000
Wang, Yu-Cian 2498, 2566
Wang, Zigang 1922
Ward, J. Scott...3254
Warmann, Emily..1248
Warmann, Emily C. 512, 521, 558, 572
Warner, Jeffery. H.......................................2091
Warren, Emily...3371
Warren, Emily L. 578, 2482, 2488, 2543, 3381
Washington, Lori 3520
Washio, Hidetoshi 3506
Watanabe, Kentaroh............................854, 3528
Watanabe, Yasuyuki.....................................613
Waters, Martin..2923
Watson, S. 648
Watthage, Suneth C. 170, 730, 748, 815, 1030
Watts, John L.R. 2762
Weeber, Arthur..2875
Weick, Clément 2492, 2562
Weigand, William 1790
Weiss, Charlotte 83, 2407
Weiss, Dirk 1264
Weiss, Karl-Anders......................................2836
Wen, Ching-Chang 329
Wen, Xiaoming..696
Wenham, Stuart R. 2576, 2600
Werner, Florian 2205, 2478
Werner, Jérémie..................................55, 3256
West, Bradley M. 2179, 3309
Western, N. J..953
Western, Ned J...948
Wheeler, Tobias 1476
Whipple, Steven 88
Whiteside, V. R. 14
Wibowo, A. 3511
Wibowo, Andre 1181, 1184, 2084
Wicaksono, S. 1210
Widén, Joakim 3067
Wieghold, Sarah 3300
Wienands, Karl 1253
Wiese, Martin...1531
Wille-Haussmann, Bernhard 126
Williams, J. 1603
Williams, Joshua J. 305, 582
Williams, R. 490
Wilson, Gregory..3236

Wilson, Marshall 322
Wilt, David M. 88, 102, 301, 531
Wilt, Sam 301
Wilterdink, Harrison 2692, 2765
Winkler, Kristina 1253
Winkler, Thilo 3500
Wirsching, Sven 3500
Wirth, Harry.. 1531
Wischmann, Wiltraud 2058
Wissen, A. 1752
Witte, Wolfram 2205
Witteck, Robert 1366
Wohlgemuth, John 1275
Wojtowicz, Anna 164
Wolden, C. A. 138
Wolden, Colin A. 1697
Wolf, Martin 2692
Wolffersdorff, Paul..................................... 595
Wong, Johnson 499, 3113
Wong, Johnson Kai Chi 496
Wong, Lydia H. 3275
Woodhouse, Michael 1259
Woods, Jason 1893
Woods-Robinson, Rachel.............................. 3410
Worrell, Ernst.. 3014
Wright, Lewis D. 146, 186
Wu, Gordon 96
Wu, J. 1189
Wu, Jiang 3370
Wu, Kuen-Yi 911
Wu, Po-Ching 1623
Wu, Ruei-Ying 1635
Wu, Shang-Hsuan 1051
Wu, Teng-Chun 448, 476
Wu, Yonggang 1594
Wu, Yuh-Renn 294
Wu, Zhuopeng 1241
Würfel, Uli 1253
Wyrsch, Nicolas 3435
Wyss, P. 2073
Xia, Hongze... 2392
Xia, Zihuan 1594
Xiao, C. 1312, 2785
Xiao, Chuanxiao 62, 1371
Xiao, T. Patrick 2185
Xiao, Zhi Bin 1648
Xie, Yu 116
Xiong, Gang 1193, 2473
Xiong, Zhen 2220, 3304
Xu, Jun 2656
Xu, Ling... 2656
Xu, Lu 1737
Xu, Menglei 1233
Xu, Qi 37, 3245
Xu, Qianfeng... 2285

AUTHOR INDEX

Xu, Tao 2251
Xu, Xiaojie 3410
Xu, Zhaoran 59
Xue, Muyu 1835, 2732
Yablonovitch, Eli 2185
Yachi, Toshiaki 613
Yadav, Karan Shishir 2902
Yadav, Tarun S. 396
Yakes, Michael K. 873, 2091
Yamada, Noboru 1724, 1743
Yamada, Nobuyuki 2906
Yamagami, Takeru 192
Yamagoe, K. 441
Yamaguchi, Hiroshi 3506
Yamaguchi, Koichi 712
Yamaguchi, Masafumi 359, 412, 1479,
      1711, 1714, 1743, 2498, 2548, 2566
Yamaguchi, Seira 1385
Yamamichi, Masaaki 1275, 2263
Yamaya, Haruki 2159
Yan, Chang 2213
Yan, Di 2076
Yan, Yanfa 771, 993, 1643, 2443, 2473, 3426
Yancey, Billy 2128
Yanchilin, Anton 585
Yang, Fan 2656
Yang, Guangtao 2605
Yang, Hao-Yu 343, 367, 893, 2664
Yang, Hong 392, 1430, 1873, 1937, 2823, 2918
Yang, Jianfeng 2392
Yang, Mohshi 907
Yang, Peter 1100
Yang, Shuying 2697, 3456
Yang, X. 2785
Yang, Yang 2220
Yang, Yi Tong 1648
Yang, Yun-Chie 893, 2664
Yang, Zhihao 74
Yao, Li You 1648
Yao, Y. 869, 1752
Yao, Yangyi 1048
Yao, Yao 866, 2368
Yarnaquchi, Koichi 85
Yates, Peter 1445
Yaung, K. Nay 3376
Yaung, Kevin Nay 284, 2744
Ye, Feng 2220
Ye, J. 2785
Ye, Qilin 948
Yeh, Chun-Ming 911
Yellowhair, Julius 2870
Yellowhair, Julius E. 2891
Yi, Chuqi 2569
Yilmaz, S. 3430
Yoo, Chang Youn 399
Yoon, Howard W. 437
Yoon, Jongseung 549, 2291
Yoon, S. F. 1210
Yoon, Woojun 373, 1838
Yoshiba, Shuhei 1769
Yoshino, Kenji 1679
Yoshita, M. 1003
Yoshita, Masahiro 2781
You, Bang-Jin 367
You, Liang-Chian 1635
Young, David 46
Young, David L 1817, 1832, 2275, 3254
Young, James L. 47
Young, Steven 582
Youssef, Amanda 1491, 2242, 3300
Youtsey, Christopher 3524
Yu, Edward T. 363, 1181
Yu, Jia 2453
Yu, K. M. 3
Yu, Kin Man 1204
Yu, Li-Chieh 3204
Yu, Linwei 2656
Yu, Ming 1193
Yu, Peichen 294, 1610, 1635
Yu, Pei-Chen 1606
Yu, Sun 1522
Yu, Zhengshan J 1228, 1317, 1322, 2039, 3250
Yuan, Bo 315
Yuan, Lin 2392
Yue, Yao 93
Yun, Jae Ho 810
Zachariah, S. 1995
Zachariah, Sachin 2799, 2849, 3478
Zahler, James 1463
Zahler, James M. 3245
Zakaria, Naimi 487
Zamora, Rachid Darbali 3044
Zang, Kai 1835
Zapalac, G. 2414
Zapalac, Geordie 820, 3327
Zauner, Andy 1237
Zech, Tobias 1531
Zelenina, Anastasiya 2054, 2478
Zeman, Miro 2605
Zeng, Guoping 907
Zeng, Xulu 1286
Zeyu, L. 1781
Zhai, Yonghui 472
Zhan, Tien-Chien 294
Zhang, Bao 226
Zhang, C. 1603
Zhang, Chaomin 240, 827, 1215, 1841, 2573
Zhang, Guoqi 2605
Zhang, Hua 3304
Zhang, Huan 2558

AUTHOR INDEX

Zhang, Jili...1100
Zhang, Jing..3384
Zhang, Junjun...............................1937, 2823
Zhang, Lei.................... 408, 1761, 1828, 2667
Zhang, Liang...2247
Zhang, Liping...1241
Zhang, Nian..2432
Zhang, Qiming...226
Zhang, Wei...................................255, 1193
Zhang, Weijie..820
Zhang, X...2094
Zhang, Xiaochen..1567
Zhang, Xue.................... 392, 1430, 1873, 2918
Zhang, Yang...195
Zhang, Yi.......................................696, 2186
Zhang, Yong-Hang......................3366, 3410
Zhang, Yufeng.....................198, 767, 1707
Zhang, Z...2019
Zhang, Zhilong..2392
Zhang, Zongyi...1594
Zhangl, Xiaochen...1555
Zhao, Dewei..993
Zhao, Hui.....................392, 1430, 1873, 2918
Zhao, J...1752
Zhao, Jing...1845
Zhao, Pan....................1430, 1873, 2918
Zhao, Xin-Hao...3366
Zhao, Yuan..3366
Zhao, Yuetao...1044
Zhe, Liu...2744
Zheng, N...869
Zhigunov, D.M. 1811
Zhongbiao, Ye...1044
Zhou, Guomin 472
Zhou, Hang.......................................976, 2558
Zhou, Jian...1594
Zhou, Xiao W. 2419
Zhu, Jiang...3208
Zhu, Lin...721, 3528
Zhu, Yan...66, 3290
Zhu, Ziyao...198
Zide, Joshua M. O.......................................3384
Zielnik, Allen...3208
Zilles, Roberto...1917
Zilouchian, Ali.................................638, 2941
Zimmerman, Jeramy D. 3381
Zin, Ngwe..322
Zinaddinov, M. 2411
Zoppi, Guillaume. 742
Zubia, David..2419
Zunger, Alex..1245

IEEE
445 Hoes Lane
Piscataway, NJ 08854-4141

ISBN 978-1-5090-5606-4